Prepare for Exams with Lehmann's

New Interactive Video Lecture Series

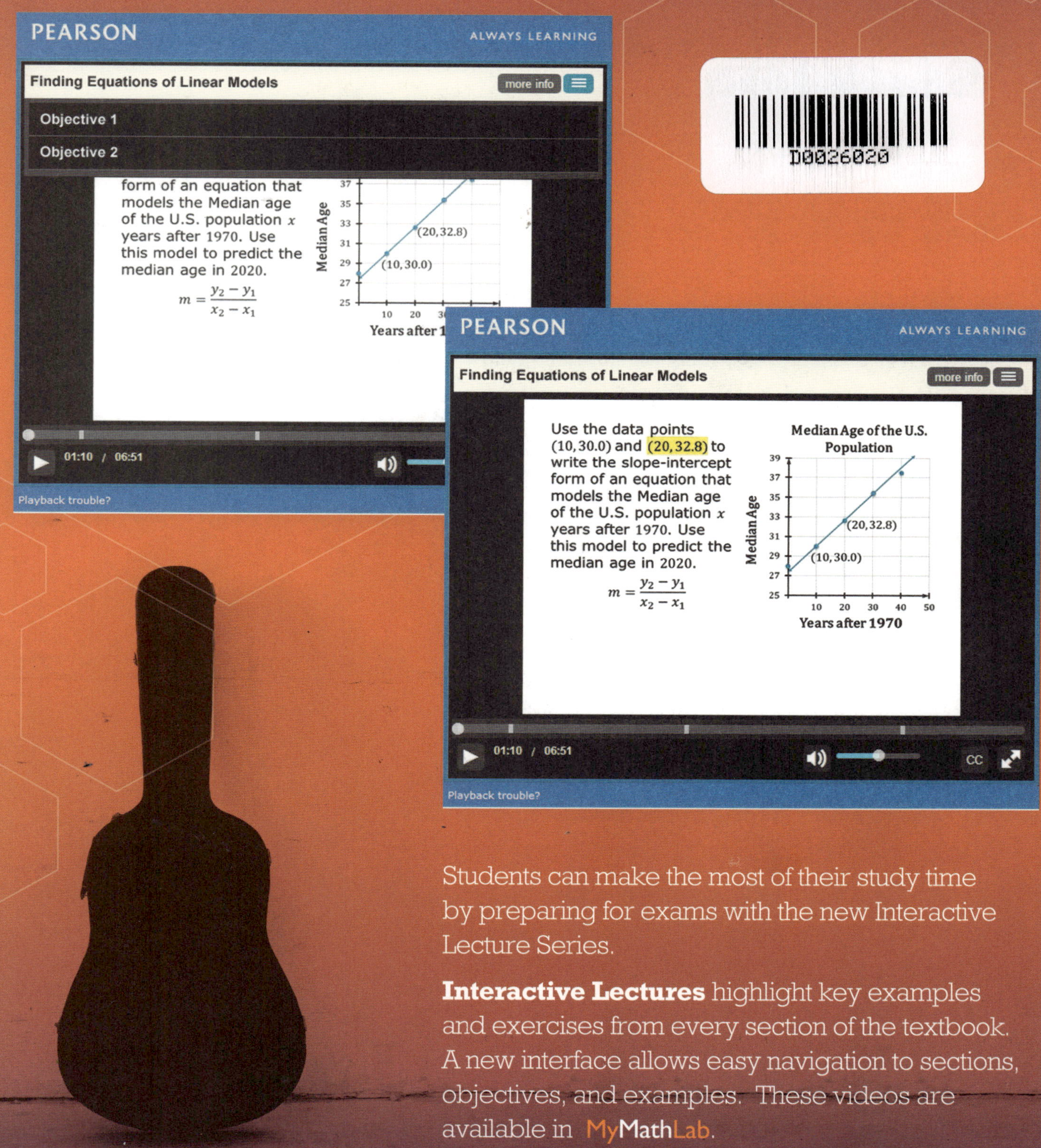

Students can make the most of their study time by preparing for exams with the new Interactive Lecture Series.

Interactive Lectures highlight key examples and exercises from every section of the textbook. A new interface allows easy navigation to sections, objectives, and examples. These videos are available in MyMathLab.

Jay Says...

Before writing my algebra series, it was apparent that my students couldn't relate to the applications in the course; they would repeatedly ask, "What is this good for?" To try to bridge that gap, I wrote some labs, which helped my students collect data, find models via curve fitting, and use the models to make estimates and predictions. My students really loved working with the current, compelling, and authentic data and experiencing how mathematics truly is useful.

My students' response was so strong that I decided to write an algebra series. Little did I know that, to realize this goal, I'd need to embark on a 15-year challenging journey, but the rewards of hearing such excitement from students and faculty across the country have made it all worthwhile! I'm proud to have played even a small role in raising people's respect and enthusiasm for mathematics.

I've tried to honor my inspiration: By working with authentic data, students can experience the power of mathematics. A random-sample study at my college suggests that I'm achieving this goal. The study concludes that students who used my series were more likely to feel that mathematics would be useful in their lives $(p = 0.0061)$ as well as in their careers $(p = 0.024)$.

In addition to curve fitting, my approach includes other types of meaningful modeling, directed-discovery explorations, conceptual questions, and, of course, a large bank of skill problems. The curve-fitting applications serve as a portal for students to see the usefulness of mathematics so that they become fully engaged in the class. Once involved, they're more receptive to all aspects of the course.

Elementary and Intermediate Algebra

Functions & Authentic Applications

Second Edition

Jay Lehmann

College of San Mateo

PEARSON

Boston Columbus Indianapolis New York San Francisco Upper Saddle River
Amsterdam Cape Town Dubai London Madrid Milan Munich Paris Montréal Toronto
Delhi Mexico City São Paulo Sydney Hong Kong Seoul Singapore Taipei Tokyo

Editorial Director: Chris Hoag
Editor in Chief: Michael Hirsch
Senior Acquisitions Editor: Dawn Giovanniello
Senior Content Editor: Lauren Morse
Editorial Assistant: Ashley Yee
Senior Managing Editor: Karen Wernholm
Associate Managing Editor: Tamela Ambush
Digital Assets Manager: Marianne Groth
Media Producer: Shana Siegmund
QA Manager, Assessment Content: Marty Wright
Executive Content Manager: Rebecca Williams
Senior Content Developer: John Flanagan
Executive Marketing Manager: Michelle Renda
Associate Marketing Manager: Alicia Frankel
Marketing Assistant: Emma Sarconi
Liaison Manager, Text Permissions Group: Joseph Croscup
Procurement Specialist: Debbie Rossi
Associate Director of Design, USHE North and West: Andrea Nix
Program Design Lead & Text Design: Beth Paquin
Production Coordination and Composition: Integra
Illustrations: Electronic Publishing Services Inc.
Cover Design: Studio Wink
Cover Image: Front Cover Image: Guitarist Magazine/Future Publishing/Getty Images;
Back Cover Image: 24Novembers/Shutterstock

Library of Congress Cataloging-in-Publication Data
Lehmann, Jay.
Elementary and intermediate algebra: functions & authentic applications/Jay Lehmann, College of San Mateo.—Second edition.
pages cm
ISBN-13: 978-0-321-92272-4 (student edition)
ISBN-10: 0-321-92272-7 (student edition)
1. Algebra—Textbooks. I. Title.
QA152.3.L449 2015
512.9—dc23

2013000873

2 3 4 5 6 7 8 9 10—CRK—17 16 15 14

www.pearsonhighered.com

ISBN-10: 0-321-92272-7
ISBN-13: 978-0-321-92272-4

To Dad, for the WSJ, our laughs about LD, and teaching me that one can accomplish most anything with persistence, thinking outside the box, and a little luck.

Contents

Average Ticket Prices for Top-50-Grossing Concert Tours (pp. 36–37)

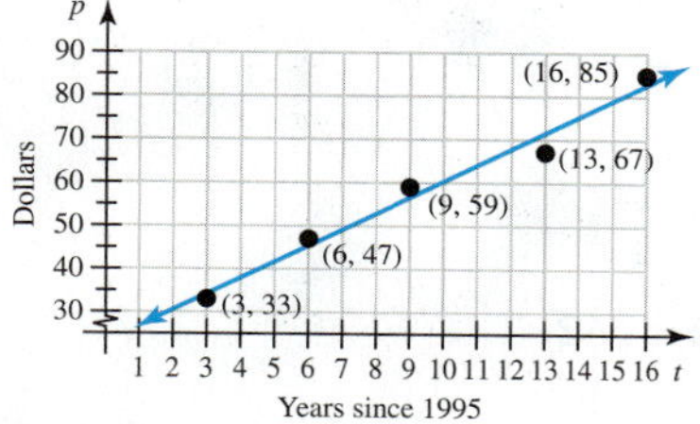

Presidential Election Voter Turnout (p. 89)

Year	Percent of Eligible Voters Who Voted
1980	59.2
1984	59.9
1988	57.4
1992	61.9
1996	54.2
2000	54.7
2004	63.8
2008	63.6
2012	57.5

Percentages of Adult Internet Users Who Use Social Networking Sites (p.169)

Year	Percent
2005	8
2006	16
2008	29
2009	46
2010	61
2011	65

College Freshmen Whose Average Grade in High School Was an A (p. 218)

Year	Percent
1970	19.6
1980	26.6
1985	28.7
1990	29.4
1995	36.1
2000	42.9
2005	46.6
2010	48.4

4 SIMPLIFYING EXPRESSIONS AND SOLVING EQUATIONS 181

Total College Student Discretionary Spending (pp. 279–280)

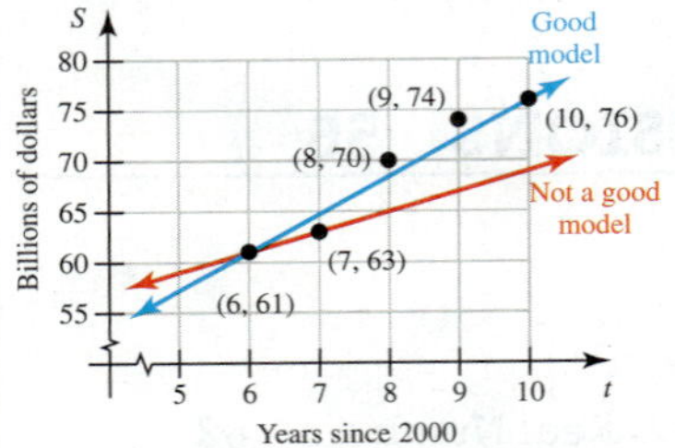

5 LINEAR FUNCTIONS AND LINEAR INEQUALITIES IN ONE VARIABLE 244

U.S. Life Expectancies of Women and Men (pp. 325–326)

Year of Birth	Women (years)	Men (years)
1980	77.4	70.0
1985	78.2	71.1
1990	78.8	71.8
1995	78.9	72.5
2000	79.5	74.1
2005	79.9	74.9
2009	81.3	76.2

6 SYSTEMS OF LINEAR EQUATIONS AND SYSTEMS OF LINEAR INEQUALITIES 321

Average Monthly Cell Phone Bills and Numbers of Subscribers (p. 407)

Year	Average Bill (dollars per month)	Number of Subscribers (millions)
1998	39.43	69.2
2000	45.27	109.5
2002	48.40	140.8
2004	50.64	182.1
2006	50.56	233.0
2008	50.07	270.3
2010	47.21	302.9

Worldwide iPhone Sales (p. 485)

Year	Sales (millions)
2007	1.4
2008	11.6
2009	20.7
2010	40.0
2011	72.3

Bottled-Water Consumption (pp. 552–553)

Year	Bottled-Water Consumption (billions of gallons)
1990	2.2
1995	3.1
2000	4.7
2005	7.5
2009	10.6

Average Ticket Prices to Major League Baseball Games (pp. 623–624)

Year	Average Ticket Price (dollars)
1950	1.54
1960	1.96
1970	2.72
1980	4.45
1991	8.84
2000	16.22
2011	26.91

10 EXPONENTIAL FUNCTIONS 597

Safe Exposure Times to Music at Rock Concerts (p. 699)

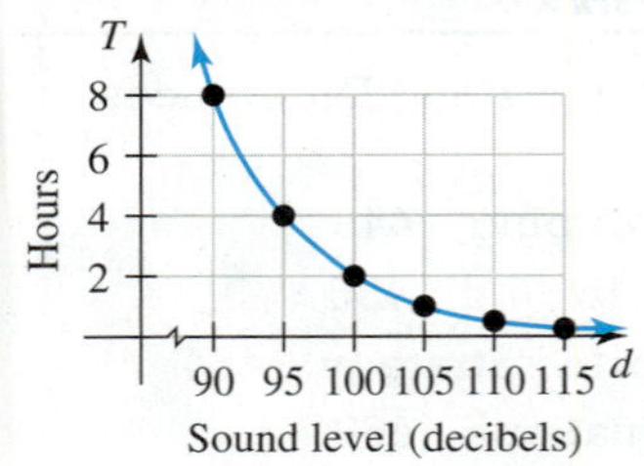

11 LOGARITHMIC FUNCTIONS 656

Numbers of Internet Users in the United States (p. 736)

Year	Number of Internet Users (millions)
2006	204
2007	212
2008	220
2009	228
2010	240

12 RATIONAL FUNCTIONS 728

Percentages of Births "Despite Contraception" (p. 879)

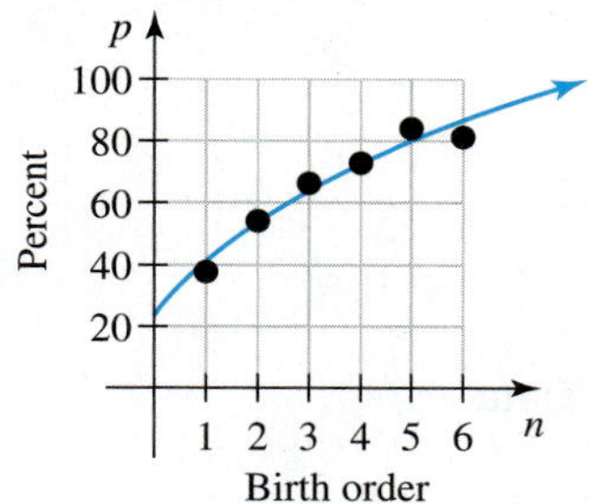

13 RADICAL FUNCTIONS 827

Rates for Math Tutoring (p. 889)

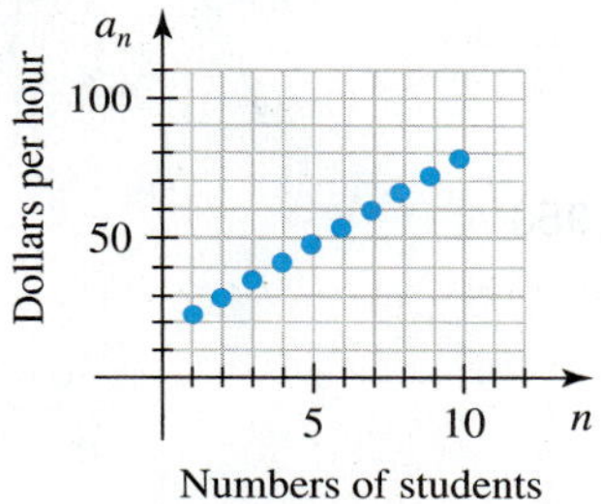

14 SEQUENCES AND SERIES 887

Intercepts of an Ellipse (p. 941)

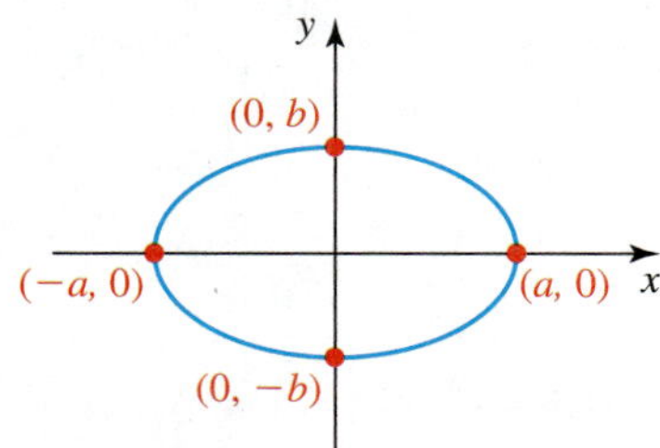

15 ADDITIONAL TOPICS 918

Tracing a Curve

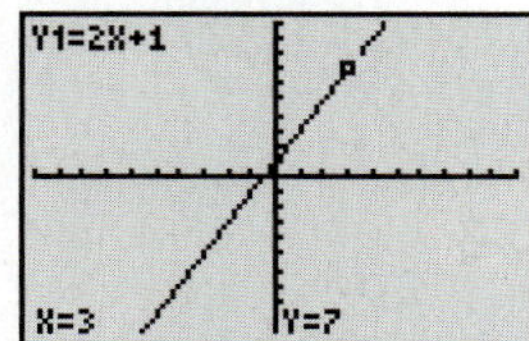

A USING A TI-83 OR TI-84 GRAPHING CALCULATOR *available online only*

Preface

"The question of common sense is always, 'What is it good for?'—a question which would abolish the rose and be answered triumphantly by the cabbage."

—James Russell Lowell

These words seem to suggest that poet and editor James Russell Lowell (1819–1891) took Elementary and Intermediate Algebra. How many times have your students asked, "What is it good for?" After years of responding "You'll find out in the next course," I began an ongoing quest to develop a more satisfying and substantial response to my students' query.

Curve-Fitting Approach Although there are many ways to center Elementary and Intermediate Algebra courses around authentic applications, I chose a curve-fitting approach for several reasons. A curve-fitting approach

- allows great flexibility in choosing interesting, authentic, current situations to model.
- emphasizes concepts related to functions in a natural, substantial way.
- deepens students' understanding of functions, because it requires students to describe functions graphically, numerically, symbolically, and verbally.
- unifies the many diverse topics of typical Elementary and Intermediate Algebra courses.

To fit a curve to data, students learn the following four-step modeling process:

1. Examine the data set to determine which type of model, if any, to use.
2. Find an equation of the model.
3. Verify that the model fits the data.
4. Use the model to make estimates and predictions.

This four-step process weaves together topics that are crucial to the course. Students must notice numerical patterns from data displayed in tables, recognize graphical patterns in scattergrams, find equations of functions, graph and evaluate functions, and solve equations.

Not only does curve fitting foster cohesiveness within chapters, but it also creates a parallel theme for each chapter that introduces and discusses a new function. This structure enhances students' abilities to observe similarities and differences among fundamental functions such as linear functions, quadratic functions, exponential functions, logarithmic functions, rational functions, and radical functions.

Curve fitting serves as a portal for students to see the usefulness of mathematics so they become fully engaged in the class. Once involved, students are more receptive to all aspects of the course.

NEW TO THE SECOND EDITION

Students will benefit from the following changes to the second edition of *Elementary and Intermediate Algebra: Functions and Authentic Applications*:

- Section 11.1 on composite functions has been added.
- The text now has a section on dividing by monomials, dividing polynomials by using long division, and dividing polynomials by using synthetic division (Section 7.5).
- Percent applications have been added to Sections 2.6, 4.3, and 4.4.
- Sections 3.2 and 3.5 now contain graph-related modeling exercises.
- Additional examples and exercises addressing the commutative and associative laws have been added to Section 4.1.

- Modeling exercises that require students to define variables have been added to Sections 4.3, 4.4, 5.5, and 5.6.
- Authentic applications that do not involve curve fitting have been added to Sections 4.4, 5.5, 5.6, and 7.4.
- The number of exercises addressing exponential properties has been increased in Sections 7.1 and 7.2.
- The text now includes an additional 142 conceptual exercises.
- 189 data sets in examples and exercises have been replaced with more compelling and current topics.
- 219 data sets in examples and exercises have been augmented to include values for recent years.
- 55 new data sets have been added to address new concepts.
- Examples on solving general (literal) equations have been added to Sections 5.1, 9.4, 10.4, 11.4, and 13.5.
- All eight Climate Change labs have been updated.
- Grids have been added to most graphs of models so students can better line up inputs and outputs.
- New Interactive Video Lecture Series are available in MyMathLab.
- The number of exercises available in MyMathLab/MathXL has been increased.

CONTINUED FROM THE FIRST EDITION

Unique Organization Many college students who take Elementary and Intermediate Algebra had significant difficulties with the equivalent courses in high school. These students face a greater challenge in the college courses, because they must complete the two courses in two semesters, rather than four. Instead of presenting the material in the "same old way," this textbook provides a unique organization that will better aid students in succeeding.

Removing the Overlap By combining the material from my Elementary Algebra textbook and my Intermediate Algebra textbook and cutting out the overlap, professors have more time to go into depth when discussing concepts and to assign **Taking It to the Lab** experiments and **Group Exploration** activities (both described later in the preface).

Spiraling of Concepts Although removing the overlap offers significant benefits to students, if a concept is never revisited, students may not retain it. Fortunately, curve fitting naturally revisits concepts as students' tool bag of functions grows. In each modeling section, exercises require students to compare the implications of using the various types of functions to model authentic situations. In addition, students' retention of key concepts can be enhanced in Chapters 5–15 by completing two special types of exercises, **Related Review** and **Expressions, Equations, Functions, and Graphs**. These types of exercises are described in greater detail later in the preface.

Micro Combination of Elementary Algebra and Intermediate Algebra Most textbooks that combine Elementary Algebra and Intermediate Algebra intermingle the two courses' content at the chapter or, perhaps, section level. For this textbook, I have performed microsurgery to join the material from the two courses at the Example and exercise level. Each section and Homework set begins at the Elementary Algebra level, yet builds to a solid Intermediate Algebra level.

Making Sure You're Ready for Intermediate Algebra Both Chapters 7 and 8 close with sets of exercises designed to help students who didn't use the textbook for Elementary Algebra smoothly transition into using it for Intermediate Algebra. The exercises also serve as good review for students who did use the textbook for Elementary Algebra.

Modeling Exercises To give this second edition a current and lively feel, the vast majority of the hundreds of modeling exercises in the text have been updated or replaced. Most of the application exercises contain tables of data, but some describe data in paragraph form to give students practice in picking out relevant information and defining variables. Both types of applications are excellent preparation for subsequent courses (especially Statistics).

Early Graphing In Chapter 3, students learn to graph linear equations only in the forms $y = mx + b$ and $x = a$. This way, they can focus on the fundamental concepts of slope and y-intercept. As many professors have reported, students do exceedingly well in Chapter 3. This early-graphing organization postpones simplifying expressions and solving equations, buying students a bit more time to find their "sea legs" before moving on to the more challenging symbolic manipulation work in Chapter 4. By the time that students reach Section 5.1, they are ready to graph equations that are not in slope-intercept form, but can be put into it.

The early-graphing approach also enables students to solve equations graphically as well as symbolically. Most of Chapters 4–15 include exercises that reinforce the connection between graphing and solving equations or systems of equations.

Early Functions Although some textbooks introduce functions early in the course, the concept is rarely included in subsequent sections, and when it is included, the treatment is light. In this textbook, functions are introduced early (Section 5.2) and are emphasized throughout the book in meaningful ways such as by curve fitting, providing students with a solid foundation for subsequent courses such as Trigonometry, College Algebra, and Precalculus.

Early Logarithmic Functions Unlike the organization in most textbooks, exponential functions and logarithmic functions are presented before rational functions and radical functions. Professors who have used the first edition have commented over and over again how much they value an early-logarithm organization. Although rational functions and radical functions present their own challenges, most students have more difficulties with logarithmic functions, and it pays to have them learning about this concept while they still have energy.

Group Explorations Almost all sections of this text contain one or two explorations that support student investigation of a concept. Instructors can use explorations as collaborative activities during class time or as part of homework assignments. Some explorations lead students to think about concepts introduced in the current section. Other, "Looking Ahead" explorations are directed-discovery activities that introduce key concepts to be discussed in the section that follows. The explorations empower students to become active explorers of mathematics and can open the door to the wonder and beauty of the subject.

Taking It to the Lab Sections Laboratory assignments have been included at the end of most chapters, to increase students' understanding of concepts and the scientific method. These labs reinforce the idea that mathematics is useful. They are also an excellent avenue for more in-depth writing assignments.

Some of the labs are about climate change and have been written at a higher reading level than the rest of the text in order to give students a sense of what it is like to perform research. Students will find that by carefully reading (and possibly rereading) the background information, they can comprehend the information and apply concepts they have learned in the course to make estimates and predictions about this compelling, current, and authentic situation.

Balanced Extensive Homework Sections Most exercise sets contain a large number of modeling, skill, and conceptual exercises to allow professors maximum flexibility in setting assignments.

Related Review These exercises (in every section of Chapters 5–11) relate current concepts to previously learned concepts. Such exercises assist students in seeing the "big picture" of the course.

Expressions, Equations, Functions, and Graphs These exercises (in every section of Chapters 5–11) help students gain a solid understanding of those core concepts, including how to distinguish among them.

Technology The text assumes students have access to technology such as the TI-83 or TI-84 graphing calculator. Technology of this sort allows students to create scattergrams and check the fit of a model quickly and accurately. It also empowers students to verify their results from Homework exercises and efficiently explore mathematical concepts in the Group Explorations.

The text supports instructors in holding students accountable for all aspects of the course without the aid of technology, including finding equations of models. (Regression equations are included in the Answers section, because it can be difficult or impossible to anticipate which points a student will choose in trying to find a reasonable equation.)

Appendix A: Using a TI-83 or TI-84 Graphing Calculator Appendix A contains step-by-step instructions for using the TI-83 and TI-84 graphing calculators. A subset of this appendix can serve as a tutorial early in the course. In addition, when the text requires a new calculator skill, students are referred to the appropriate section in Appendix A, available online in MyMathLab and at http://www.pearsonhighered.com/mathstatsresources/.

Exposition If students can't make sense of the prose, it doesn't matter how precise it is. One of my top goals is to write descriptions that are straightforward, accessible, clear, and rigorous.

Tips for Success Many sections close with tips that are intended to help students succeed in the course. A complete listing of these tips is included in the Index.

Additional Topics Chapter Topics typically taught in Elementary and Intermediate Algebra that cannot be connected with a curve-fitting approach at the appropriate level are assembled in Chapter 15. Each section contains a Section Quiz feature. The union of these quizzes can be used as a set of review exercises for Chapter 15. Instructors who wish to "cut and paste" sections from that chapter into earlier chapters can append these quizzes to the appropriate Chapter Review exercises.

RESOURCES FOR INSTRUCTORS

Instructor's Resource Manual This manual contains suggestions for pacing the course and creating homework assignments. It discusses how to incorporate technology and how to structure lab and project assignments. The manual also contains section-by-section suggestions for presenting lectures and for undertaking the explorations in the text.

Instructor's Solutions Manual This manual includes complete solutions to the even-numbered exercises in the homework sections of the text.

MyMathLab® Online Course (access code required) MyMathLab from Pearson is the world's leading online resource in mathematics, integrating interactive homework, assessment, and media in a flexible, easy to use format.

MyMathLab delivers **proven results** in helping individual students succeed.

- MyMathLab has a consistently positive impact on the quality of learning in higher education math instruction. MyMathLab can be successfully implemented in any environment-lab-based, hybrid, fully online, traditional-and demonstrates the quantifiable difference that integrated usage has on student retention, subsequent success, and overall achievement.
- MyMathLab's comprehensive online gradebook automatically tracks your students' results on tests, quizzes, homework, and in the study plan. You can use the gradebook to quickly intervene if your students have trouble, or to provide positive feedback on a job well done. The data within MyMathLab is easily exported to a variety of spreadsheet programs, such as Microsoft Excel. You can determine which points of data you want to export, and then analyze the results to determine success.

MyMathLab provides **engaging experiences** that personalize, stimulate, and measure learning for each student.

- **Exercises:** The homework and practice exercises in MyMathLab are correlated to the exercises in the textbook, and they regenerate algorithmically to give students unlimited opportunity for practice and mastery. The software offers immediate, helpful feedback when students enter incorrect answers.
- **Multimedia Learning Aids:** Exercises include guided solutions, sample problems, animations, videos, and eText access for extra help at point-of-use.
- **Expert Tutoring:** Although many students describe the whole of MyMathLab as "like having your own personal tutor," students using MyMathLab do have access to live tutoring from Pearson, from qualified math and statistics instructors.

And, MyMathLab comes from an **experienced partner** with educational expertise and an eye on the future.

- Knowing that you are using a Pearson product means knowing that you are using quality content. That means that our eTexts are accurate and our assessment tools work. It means we are committed to making MyMathLab as accessible as possible.
- Whether you are just getting started with MyMathLab, or have a question along the way, we're here to help you learn about our technologies and how to incorporate them into your course.

To learn more about how MyMathLab combines proven learning applications with powerful assessment, visit www.mymathlab.com or contact your Pearson representative.

New Ready to Go courses provide students with all the same great MyMathLab features, but make it easier for instructors to get started. Both the Standard and Ready To Go courses include pre-made homework and quizzes, which are pre-assigned in the Ready To Go course to make creating a course even easier.

TestGen TestGen enables instructors to build, edit, print, and administer tests by using a computerized bank of questions developed to cover all the objectives of the text. TestGen is algorithmically based, allowing instructors to create multiple, but equivalent, versions of the same question or test with the click of a button. Instructors can also modify test bank questions or add new questions. Tests can be printed or administered online. The software and testbank are available for download from Pearson Education's online catalogue.

PowerPoint Lecture Slides (download only) Available through www.pearsonhighered.com or inside your MyMathLab course, these fully editable lecture slides include definitions, key concepts, and examples for use in a lecture setting and are available for each section of the text.

RESOURCES FOR STUDENTS

New Interactive Video Lecture Series This series has been completely revised to provide students with extra help for each section of the textbook. The Lecture Series includes:

- Interactive Lectures that highlight key examples and exercises for every section of the textbook.
- A new interface that allows easy navigation to sections, objectives, and examples.

These lectures are available in MyMathLab.

Student's Solutions Manual This manual contains the complete solutions to the odd-numbered exercises in the Homework sections of the text.

GETTING IN TOUCH

I would love to hear from you and would greatly appreciate receiving your comments regarding this text. If you have any questions, please ask them, and I will respond.

Thank you for your interest in preserving the rose.

Jay Lehmann
MathNerdJay@aol.com

Acknowledgments

Writing a modeling text is an endurance run I couldn't have completed without the dedicated assistance of many people. First, I'm greatly indebted to Keri, my wife, who yet again served as an irreplaceable sounding board for the multitude of decisions that went into creating this book. In particular, I credit her internal divining rod in selecting captivating data from a mound of data sets I've collected.

I've received much support from the following professors: Ken Brown, Gary Church, Jon Freedman, Eric Freidenreich, Jenny Freidenreich, Cheryl Gregory, Rick Hough, Evan Innerst, and Tadashi Tsuchida. Over the years, they've given much sound advice in responding to my countless e-mail inquiries.

And thanks to Rick Gilbert for the awesome photograph in the "To the Student" section of the author performing with his band, The Procrastinistas, at the Hotel Utah in San Francisco.

I acknowledge several people at Pearson Education. I'm very grateful to Editor-in-Chief Michael Hirsch, who has shared in my vision for this text and has made significant investments to make that vision happen. The book has been greatly enhanced through the support of Senior Acquisitions Editor Dawn Giovanniello, who has made a multitude of contributions, including assembling an incredible team to develop and produce this text. The team includes Senior Content Editor Lauren Morse, who handled countless tasks to support me in preparing the manuscript for production, leading to a significantly better book.

Heartfelt thanks goes to Integra's Associate Managing Editor Allison Campbell, who orchestrated the many aspects of production.

I thank these reviewers, whose thoughtful, detailed comments helped me sculpt this text into its current form:

Scott Adamson, *Chandler-Gilbert Community College*
Thomas Adamson, *Phoenix College*
Ken Anderson, *Chemeketa Community College*
Gwen Autin, *Southeastern Louisiana University*
Mona Baarson, *Jackson Community College*
Sam Bazzi, *Henry Ford Community College*
Joel Berman, *Valencia Community College—East*
Nancy Brien, *Middle Tennessee State University*
Ronnie Brown, *University of Baltimore*
Barbara Burke, *Hawaii Pacific University*
Laurie Burton, *Western Oregon University*
Paula Castagna, *Fresno City College*
James Cohen, *Los Medanos College*
Jeff Cohen, *El Camino College*
Joseph DeGuzman, *Norco College*
Cynthia Ellis, *Purdue University, Fort Wayne*
Junko Forbes, *El Camino College*
William P. Fox, *Francis Marion University*
Cathy Gardner, *Grand Valley State University*
James Gray, *Tacoma Community College*
Kathryn M. Gundersen, *Three Rivers Community College*
Miriam Harris-Botzum, *Lehigh Carbon Community College*
Stephanie Haynes, *Davis & Elkins College*
Rick Hough, *Skyline College*
Tracey Hoy, *College of Lake County*
Denise Hum, *Cañada College*

Evan Innerst, *Cañada College*
Judy Kasabian, *El Camino College*
Charles Klein, *De Anza College*
Julianne M. Labbiento, *Lehigh Carbon Community College*
Jason Malozzi, *Lehigh Carbon Community College*
Debra Martin, *Purdue University, Fort Wayne*
Diane Mathios, *De Anza College*
Jim Matovina, *Community College of Southern Nevada*
Jane E. Mays, *Grand Valley State University*
Scott McDaniel, *Middle Tennessee State University*
Tim Merzenich, *Chemeketa Community College*
C.R. Messer, *American River College*
Jason L. Miner, *Santa Barbara City College*
Nolan Mitchell, *Chemeketa Community College*
Camille Moreno, *Cosumnes River College*
Ellen Musen, *Brookdale Community College*
Charlie Naffziger, *Central Oregon Community College*
Chris Nord, *Chemeketa Community College*
Donna Marie Norman, *Jefferson Community College*
Denise Nunley, *Glendale Community College*
Karen D. Pain, *Palm Beach State College*
Ernest Palmer, *Grand Valley State University*
Ellen Rebold, *Brookdale Community College*
Jody Rooney, *Jackson Community College*
James Ryan, *State Center Community College District, Clovis*
Barbara Savage, *Roxbury Community College*
Ned Schillow, *Lehigh Carbon Community College*
Ingrid Scott, *Montgomery College*
David Shellabarger, *Lane Community College*
Laura Smallwood, *Chandler-Gilbert Community College*
John Szeto, *Southeastern Louisiana University*
Janet Teeguarden, *Ivy Tech Community College*
Lorna TenEyck, *Chemeketa Community College*
Cindy Vanderlaan, *Purdue University, Fort Wayne*
Lenove Vest, *Lower Columbia College*
Ollie Vignes, *Southeastern Louisiana University*
Linda Wagner, *Purdue University, Fort Wayne*
Karen Wiechelman, *University of Louisiana at Lafayette*
Robin Williams, *Palomar College*
Lisa Winch, *Kalamazoo Valley Community College*

Index of Applications

T

W

Introduction to Modeling

Think about the last concert you attended. What was the ticket price? Was it worth it? The average ticket price for the top-50-grossing concert tours has increased greatly (see Table 1). In Example 2 of Section 1.4, we will predict when the average ticket price will be $105.

Table 1 Average Ticket Prices for Top-50-Grossing Concert Tours

Year	Average Ticket Price (dollars)
1998	33
2001	47
2004	59
2008	67
2011	85

Source: *Pollstar*

In this course, we will discuss how to describe the relationship between two quantities that occur in an authentic situation. For example, we will describe how the percentage of Americans who support same-sex marriage has changed over time. In Chapters 1–6, we will focus on how to use (straight) lines to describe authentic situations. In Chapters 7–15, we will discuss other types of *curves* that can be used to describe authentic situations.

1.1 Variables and Constants

Objectives

- Know the meaning of *variable* and *constant*.
- Know the meaning of *counting numbers, integers, rational numbers, irrational numbers, real numbers, positive numbers,* and *negative numbers*.
- Use a number line to describe numbers.
- Graph data.
- Find the average (or mean) of a group of numbers.
- Know how to describe a concept or procedure.

In this section, we will work with *variables* and *constants,* two extremely important building blocks of algebra. We will also discuss various types of numbers and how to describe numbers visually.

Variables

In arithmetic, we work with numbers. In algebra, we work with *variables* as well as numbers.

Definition Variable

A **variable** is a symbol that represents a quantity that can vary.

For example, we can define h to be the height (in feet) of a specific child. Height is a quantity that varies: As time passes, the child's height will increase. So, h is a variable. When we say $h = 4$, we mean the child's height is 4 feet.

We will discuss other roles of a variable in Sections 2.1 and 4.3.

"The Ramsey account? Should be done any minute..."

Example 1 Using a Variable to Represent a Quantity

1. Let s be a car's speed (in miles per hour). What is the meaning of $s = 60$?
2. Let n be the number of people (in millions) who work from home at least once a week during normal business hours. For the year 2010, $n = 22$ (Source: *World at Work*). What does that mean in this situation?
3. Let t be the number of years since 2010. What is the meaning of $t = 4$?

Solution

1. The speed of the car is 60 miles per hour.
2. In 2010, 22 million people worked from home at least once a week during normal business hours.
3. $2010 + 4 = 2014$; so, $t = 4$ represents the year 2014.

There are many benefits to using variables. For example, in Problem 2 of Example 1, we found that the simple equation "$n = 22$" means the same thing as the wordy sentence "22 million people worked from home at least once a week during normal business hours." Variables can help us describe some situations with a small amount of writing.

In Problem 3 of Example 1, we described the year 2014 by using $t = 4$. So, our definition of t allows us to use smaller numbers to describe various years—an approach that will be helpful throughout the course.

We will see other benefits of variables as we proceed through the course.

Example 2 Using a Variable to Represent a Quantity

Choose a symbol to represent the given quantity. Explain why the symbol is a variable. Give two numbers that the variable can represent and two numbers that it cannot represent.

1. the weight (in pounds) of a baby at birth
2. the number of people who live in a two-bedroom house

Solution

1. Let w be the weight (in pounds) of a baby at birth. The weight of a baby at birth can vary, so w is a variable. For example, w can represent the numbers 6 and 8, because babies can weigh 6 or 8 pounds at birth. The variable w does not represent 0 or 300, because babies cannot weigh 0 or 300 pounds at birth!
2. Let n be the number of people who live in a two-bedroom house. The number of people who live in a two-bedroom house can vary, so n is a variable. For example, n can represent the numbers 2 and 3, because 2 or 3 people can live in a two-bedroom house. The variable n cannot represent the numbers 5000 or $\frac{1}{2}$, because 5000 people cannot live in a two-bedroom house and half of a person doesn't make sense.

In Problem 1 of Example 2, we stated that the units of w are pounds. Without stating the units of w, "$w = 10$" could mean the baby's weight was 10 ounces, 10 pounds, or 10 tons! In defining a variable, it is important to describe the variable's units.

Constants

A variable is a symbol that represents a quantity that can vary. When we use a symbol to represent a quantity that does *not* vary, we call that symbol a *constant*. So, 2, 0, 4.8, and π are constants. The constant π is approximately equal to 3.14.

Definition Constant

A **constant** is a symbol that represents a specific number (a quantity that does *not* vary).

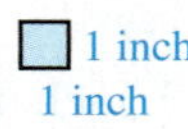

Figure 1 One square inch

In the next example, we will compare the meanings of a variable and a constant while we consider the widths, lengths, and areas of some rectangles. The **area** (in square inches) of a flat surface is the number of square inches that it takes to cover the surface (see Fig. 1). The area of a rectangle is equal to the rectangle's length times its width.

Example 3 Comparing Constants and Variables

A rectangle has an area of 12 square inches. Let W be the width (in inches), L be the length (in inches), and A be the area (in square inches).

1. Sketch three possible rectangles of area 12 square inches.
2. Which of the symbols W, L, and A are variables? Explain.
3. Which of the symbols W, L, and A are constants? Explain.

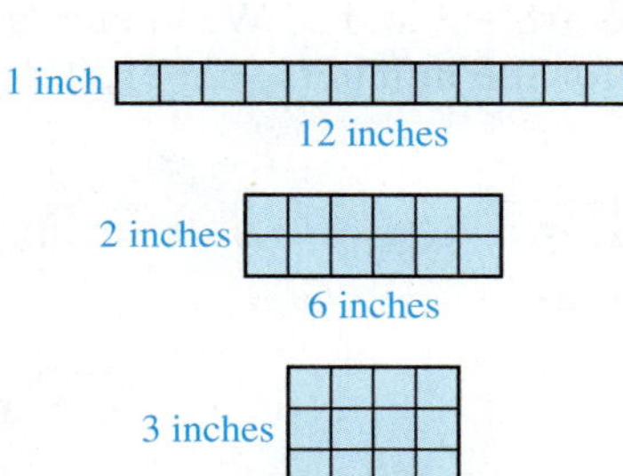

Figure 2 Three possible rectangles of area 12 square inches

Solution

1. We sketch three rectangles for which the width times the length is equal to 12 square inches (see Fig. 2).
2. The symbols W and L are variables, since they represent quantities that vary.
3. The symbol A is a constant, because in this problem the area does not vary—the area is always 12 square inches.

Counting Numbers

When we describe people, it often helps to describe them in terms of certain categories, such as gender, ethnicity, and employment. In mathematics, it helps to describe numbers in terms of categories, too. We begin by describing the *counting numbers,* which are the numbers 1, 2, 3, 4, 5, and so on.

Definition Counting numbers (natural numbers)

The **counting numbers,** or **natural numbers,** are the numbers

$$1, 2, 3, 4, 5, \ldots$$

The three dots mean that the pattern of the numbers shown continues without ending. In this case, the pattern continues with 6, 7, 8, and so on. When a list of numbers goes on forever, we say that there are an *infinite* number of numbers.

Integers

Next, we describe the *integers,* which include the counting numbers and other numbers.

Definition Integers

The **integers** are the numbers

$$\ldots, -3, -2, -1, 0, 1, 2, 3, \ldots$$

The three dots on both sides mean that the pattern of the numbers shown continues without ending in both directions. In this case, the pattern continues with $-4, -5, -6$, and so on, and with 4, 5, 6, and so on.

The **positive integers** are the numbers 1, 2, 3, ... The **negative integers** are the numbers $-1, -2, -3, \ldots$. The integer 0 is neither positive nor negative. So, the integers consist of the counting numbers (which are positive integers), the negative integers, and 0.

The Number Line

We can visualize numbers on a *number line* (see Fig. 3).

Each point (location) on the number line represents a number. The numbers increase from left to right. We refer to the distance between two consecutive integers on the number line as 1 *unit* (see Fig. 3).

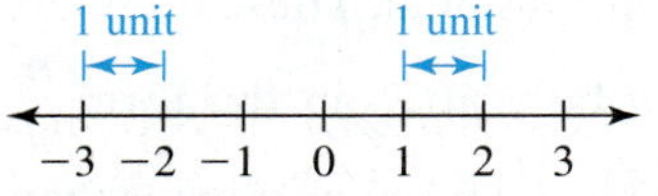

Figure 3 The number line

Example 4 Graphing Integers on a Number Line

Draw dots on a number line to represent the integers between -2 and 3, inclusive.

Solution

The integers between -2 and 3, inclusive, are $-2, -1, 0, 1, 2$, and 3. "Inclusive" means to include the first and last numbers, which in this case are -2 and 3. We sketch a number line and draw dots at the appropriate locations for the numbers $-2, -1, 0, 1, 2$, and 3 (see Fig. 4).

$-5 \quad -4 \quad -3 \quad -2 \quad -1 \quad 0 \quad 1 \quad 2 \quad 3 \quad 4 \quad 5$

Figure 4 Graphing the numbers $-2, -1, 0, 1, 2$, and 3

When we draw dots on a number line, we say that we are "plotting points" or "graphing numbers."

In Example 4, we worked with the integers between -2 and 3, inclusive: $-2, -1, 0, 1, 2$, and 3. Here are the integers between -2 and 3: $-1, 0, 1$, and 2. We did not include -2 or 3, because the word "inclusive" was not used. When working with such problems, it is important to check whether the word "inclusive" is used.

WARNING

Rational Numbers

For a fraction $\frac{n}{d}$, we call n the **numerator** and d the **denominator.** The dash between the numerator and the denominator is the **fraction bar:**

$$\begin{array}{l} \text{Numerator} \longrightarrow n \\ \text{Denominator} \longrightarrow d \end{array} \longleftarrow \text{Fraction bar}$$

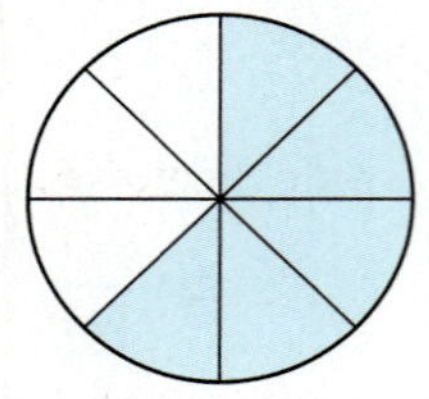

Figure 5 $\frac{5}{8}$ of a pizza

A fraction can be used to describe a part of a whole. For example, consider the meaning of $\frac{5}{8}$ of a pizza. If we divide the pizza into 8 slices of equal area, 5 of the slices make up $\frac{5}{8}$ of the pizza (see Fig. 5).

The number $\frac{5}{8}$ is called a *rational number.*

Definition Rational numbers

The **rational numbers** are the numbers that can be written in the form $\frac{n}{d}$, where n and d are integers and d is nonzero.

We specify that d is nonzero because, as we shall see later, division by zero does not make sense.

Here are some examples of rational numbers:

$$\frac{3}{7} \qquad \frac{-2}{5} \qquad 4 = \frac{4}{1}$$

Rational numbers include all the integers, since any integer n can be written as $\frac{n}{1}$.

Irrational Numbers

There are numbers represented on the number line that are *not* rational. These numbers are called **irrational numbers.** An irrational number *cannot* be written in the form $\frac{n}{d}$, where n and d are integers and d is nonzero. The number $\sqrt{2}$ is the number greater than zero that we multiply by itself to get 2. The number $\sqrt{2}$ is an irrational number. Here are some more examples of irrational numbers:

$$\pi \qquad \sqrt{3} \qquad \sqrt{5}$$

We know that $\sqrt{9} = 3 = \frac{3}{1}$, because $3 \times 3 = 9$. So, $\sqrt{9}$ is rational (not irrational).

Decimals

Any rational number or irrational number can be written as a decimal number.

A rational number can be written as a decimal number that either terminates or repeats:

$$\frac{3}{4} = \underbrace{0.75}_{\text{terminates}} \qquad \frac{3}{11} = \underbrace{0.27272727\ldots}_{\text{repeats}}$$

We can use an overbar to write the repeating decimal $0.272727\ldots = 0.\overline{27}$.

An irrational number can be written as a decimal number that neither terminates nor repeats. It is impossible to write all the digits of an irrational number, but we can approximate the number by rounding. For example, earlier we *approximated* π by rounding to the second decimal place: $\pi \approx 3.14$.

Real Numbers

Recall that each point on the number line represents a number. We call all of the numbers represented by all of the points on the number line the *real numbers*.

Definition Real numbers

The **real numbers** are all of the numbers represented on the number line.

The real numbers are made up of the rational numbers and the irrational numbers. Here are some real numbers:

$$-1.8 \qquad -1 \qquad -\frac{7}{10} \qquad 0 \qquad 0.4 \qquad \frac{6}{5} \qquad \pi$$

We graph these real numbers in Fig. 6.

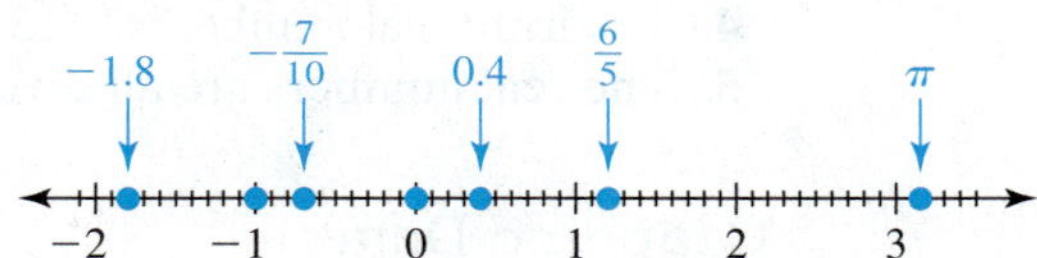

Figure 6 Graphing the real numbers -1.8, -1, $-\frac{7}{10}$, 0, 0.4, $\frac{6}{5}$, and π

We use an arrow in labeling each point that does not fall on a labeled tick mark.

Example 5 Graphing Real Numbers on a Number Line

Graph the number on a number line.

1. $-\frac{7}{4}$

2. 2.3

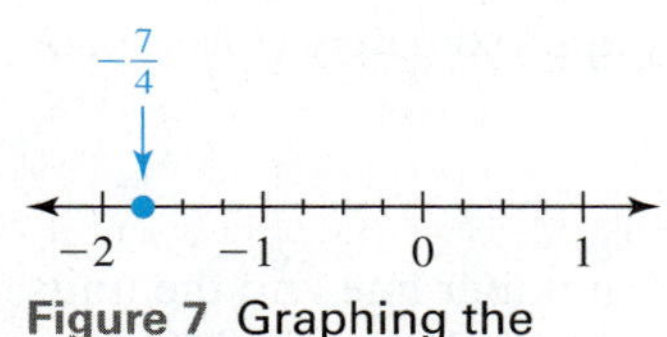

Figure 7 Graphing the number $-\frac{7}{4}$

Solution

1. We draw a number line so that the distance between tick marks is $\frac{1}{4}$ unit (see Fig. 7). To graph $-\frac{7}{4}$, we draw a dot at the seventh tick mark to the left of 0.

Figure 8 Graphing the number 2.3

2. We draw a number line so that the distance between tick marks is $0.1 = \frac{1}{10}$ unit (see Fig. 8). To graph 2.3, we draw a dot at the third tick mark to the right of 2.

Figure 9 illustrates how the various types of numbers we have discussed so far are related. In particular, it shows that every counting number is an integer, every integer is a rational number, and every rational number is a real number. It also shows that irrational numbers are the real numbers that are not rational.

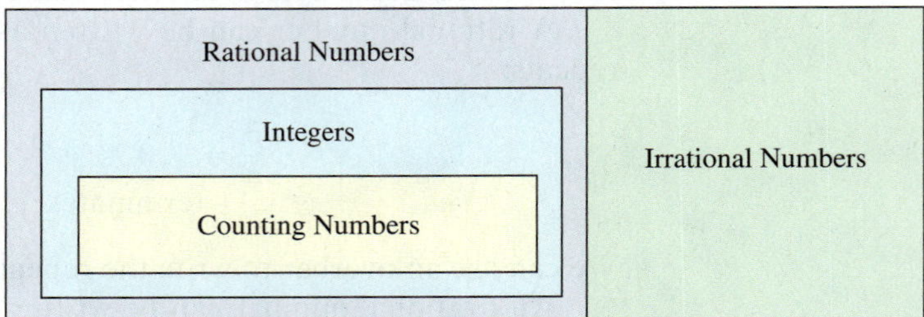

Figure 9 The real numbers

Example 6 Identifying Types of Numbers

Consider the following numbers:

$$-8 \qquad -2.56 \qquad 0 \qquad \frac{3}{5} \qquad 2 \qquad \sqrt{13} \qquad 98$$

Which of these numbers are the given type of number?

1. Counting numbers
2. Integers
3. Rational numbers
4. Irrational numbers
5. Real numbers

Solution

1. The counting numbers are 2 and 98.
2. The integers are -8, 0, 2, and 98.
3. The rational numbers are the numbers -8, -2.56, 0, $\frac{3}{5}$, 2, and 98.
4. The irrational number is $\sqrt{13}$.
5. The real numbers are all seven of the numbers.

Graphing Data

Data are values of quantities that describe authentic situations. We often can get a better sense of data by graphing than by just looking at the data values.

Example 7 Graphing Data

Over the course of a semester, a student took five quizzes. Here are the points he earned on the quizzes, in chronological order: 0, 4, 7, 9, 10. Let q be the number of points earned by the student on a quiz.

1. Graph the student's scores on a number line.
2. Did the quiz scores increase, decrease, stay approximately constant, or none of these?
3. Did the *increases* in the quiz scores increase, decrease, stay approximately constant, or none of these?

Solution

1. We sketch a number line and write "q" to the right of the number line and the units "Points" underneath the number line (see Fig. 10). Then we graph the numbers 0, 4, 7, 9, and 10.

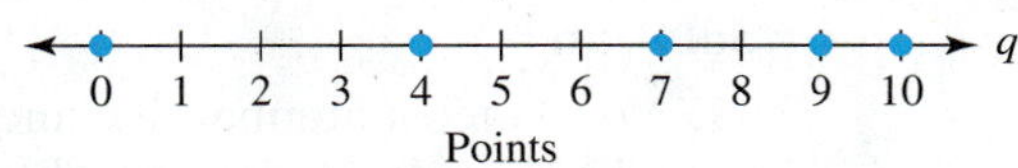

Figure 10 Graphing the quiz scores

2. From the opening paragraph, we know that the quiz scores increased. (From the graph alone, we cannot tell that the quiz scores increased, because the order of the quizzes is not indicated.)
3. As we look from left to right at the points plotted on the graph, we see that the distance between adjacent points decreases. This means that the increases in the quiz scores decreased. That is, the jump from 0 to 4 is greater than the jump from 4 to 7, and so on.

It is often helpful to use a single number to represent a group of numbers. One such number is the *average* (or *mean*).

Definition Average, mean

To find the **average** (or **mean**) of a group of numbers, we divide the sum of the numbers by the number of numbers in the group.

For example, to find the average of the quiz scores included in Example 7, we first add the scores: $0 + 4 + 7 + 9 + 10 = 30$. We then divide the total, 30, by the number of quiz scores, 5:

$$30 \div 5 = 6 \text{ points}$$

So, the average quiz score is 6 points.

In general, the average of a group of numbers estimates the center of the numbers graphed on a number line. Figure 11 illustrates this concept for the quiz data.

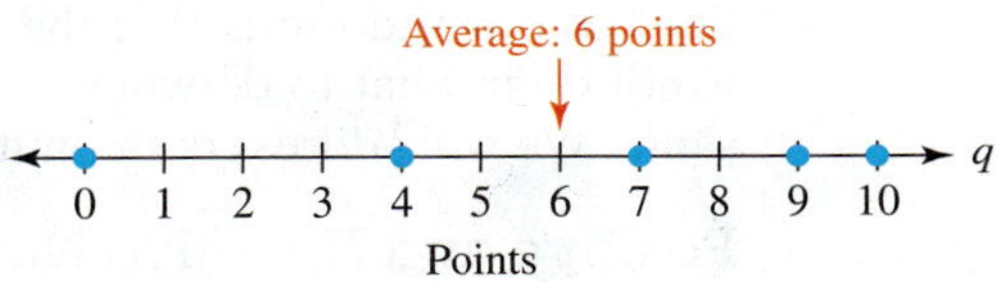

Figure 11 The average quiz score, 6 points, estimates the center of the graphed quiz data

WARNING Always include the units of an average. For example, we say that the average quiz score is *6 points,* not 6.

Averages can be especially helpful in comparing two data sets. For instance, a student whose average quiz score is 9 points would have performed much better, in general, than a student whose average quiz score is 3 points.

Example 8 Graphing Data and the Mean

The numbers (in thousands) of charter schools in the United States for the years 2007, 2008, 2009, 2010, and 2011 are 4.0, 4.3, 4.6, 4.9, and 5.3, respectively (Source: *National Alliance for Public Charter Schools*). Let n be the number (in thousands) of charter schools in a given year.

1. Graph the data. Find the average of the data values and indicate it on the graph.
2. Did the number of charter schools increase, decrease, stay approximately constant, or none of these from 2007 to 2011, inclusive? Explain.
3. Did the *increases* in the number of charter schools increase, decrease, stay approximately constant, or none of these from 2007 to 2011, inclusive? Explain.

Solution

1. We sketch a number line and write "n" to the right of the number line and the units "Thousands of charter schools" underneath the number line (see Fig. 12). Since the data values are between 4.0 and 5.3, inclusive, we write the numbers 4.0, 4.2, 4.4, 4.6, 4.8, 5.0, 5.2, and 5.4 equally spaced on the number line. Then we graph the numbers 4.0, 4.3, 4.6, 4.9, and 5.3.

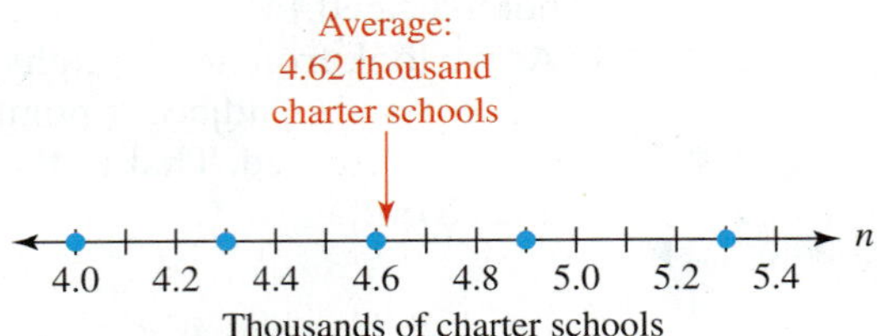

Figure 12 Graphing the data

To find the average, we divide the sum of the data values, $4.0 + 4.3 + 4.6 + 4.9 + 5.3 = 23.1$, by 5: $23.1 \div 5 = 4.62$ thousand charter schools. We indicate the average, 4.62 thousand charter schools, in Fig. 12.

2. From the opening paragraph, we know that the number of charter schools is increasing. (From the graph alone, we cannot tell that the number of charter schools is increasing, because the years are not indicated.)
3. By reading the opening paragraph of this example and viewing the almost equal spacing between dots on the graph, we see that the increases in the number of charter schools were approximately constant.

In Fig. 12, we wrote the numbers 4.0, 4.2, 4.4, 4.6, 4.8, 5.0, 5.2, and 5.4 on the number line. **When we write numbers on a number line, they should increase by a fixed amount and be equally spaced.**

There are two limitations with graphing the data in Example 8 on a *single* number line. One limitation is that the years are not indicated. The other limitation is that it is not clear what to do when the number of charter schools for two of the years is the same. We will address both limitations in Section 1.2.

Positive and Negative Numbers

The **negative numbers** are the real numbers less than 0, and the **positive numbers** are the real numbers greater than 0 (see Fig. 13).

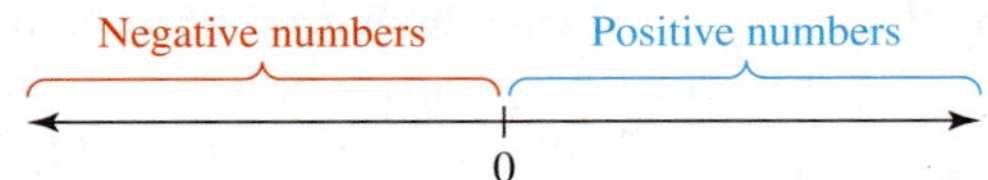

Figure 13 The location of the negative numbers and the positive numbers on the number line

Some examples of negative numbers are -13, -5.2, $-\frac{3}{4}$, and $-\sqrt{2}$. Some examples of positive numbers are 13, 5.2, $\frac{3}{4}$, and π. As we discussed earlier, the number 0 is neither positive nor negative.

The negative numbers include the negative integers, and the positive numbers include the positive integers.

We say that the *sign* of a negative number is negative and that the *sign* of a positive number is positive.

To include zero, we define the *nonnegative* numbers as the positive numbers together with 0. Likewise, we define the *nonpositive* numbers as the negative numbers together with 0.

Example 9 Graphing a Negative Quantity

A person bounces several checks and, as a result, is charged service fees. If b is the balance (in dollars) of the checking account, what value of b means the person owes \$50? Graph the number on a number line.

Solution

Since the person *owes* money, the value of b is negative: $b = -50$. We graph -50 on a number line in Fig. 14.

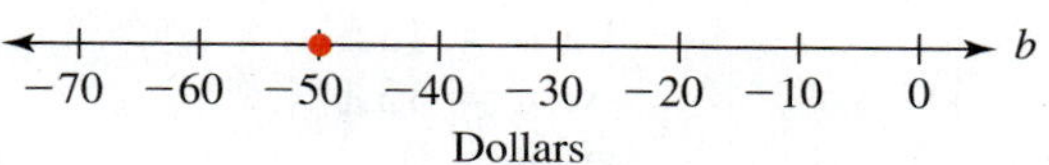

Figure 14 Graphing the number $b = -50$

Describing a Concept or Procedure

In some homework exercises, you will be asked to describe, in general, a concept or procedure.

Guidelines on Writing a Good Response

- Create an example that illustrates the concept or outlines the procedure. Looking at examples or exercises may jump-start you into creating your own example.
- Using complete sentences and correct terminology, describe the key ideas or steps of your example. You can review the text for ideas, but write your description in your own words.
- Describe also the concept or the procedure in general without referring to your example. It may help to reflect on several examples and what they all have in common.
- In some cases, it will be helpful to point out the similarities and the differences between the concept or the procedure and other concepts or procedures.
- Describe the benefits of knowing the concept or the procedure.
- If you have described the steps in a procedure, explain why it's permissible to follow these steps.
- Clarify any common misunderstandings about the concept, or discuss how to avoid making common mistakes when following the procedure.

Example 10 Responding to a General Question about a Concept

Describe the meaning of *variable.*

Solution

Let t be the number of hours that a person works in a department store. The symbol t is an example of a variable, because the value of t can vary. In general, a variable is a symbol that stands for an amount that can vary. A symbol that stands for an amount that does *not* vary is called a constant.

There are many benefits to using variables. We can use a variable to concisely describe a quantity; using the earlier definition of t, we see that the equation $t = 8$ means a person works in a department store for 8 hours. By using a variable, we can also use smaller numbers to describe various years.

In defining a variable, it is important to describe its units.

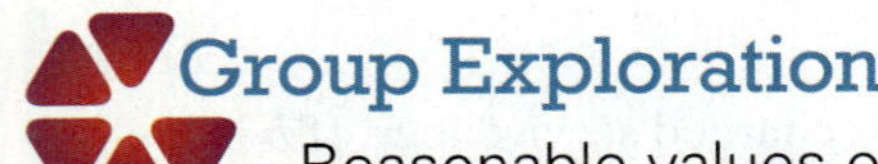

Group Exploration

Reasonable values of a variable

1. Let u be the number of units (credits or hours) a student is currently taking at your college. You may replace the word "units" with "credits" or "hours" if appropriate.
   a. Which of the following values of u are reasonable in this situation? Explain.
      i. $u = 15$
      ii. $u = -5$
      iii. $u = 200$
      iv. $u = 15.5$
      v. $u = 15.1$
      vi. $u = 0$
   b. Describe all of the real numbers that are reasonable values of u. Use a number line, a list of numbers, words, or some other way to describe these numbers.
2. A few months ago, a person bought a Porsche 911 Carrera Turbo for $160,700. It has a 17.7-gallon fuel tank. Let g be the amount of gasoline (in gallons) that is in the tank.
   a. Which of the following values of g are reasonable in this situation? Explain.
      i. $g = 7$
      ii. $g = 19$
      iii. $g = -4$
      iv. $g = 17.7$
      v. $g = 0$
      vi. $g = 10.392$
   b. Describe all of the real numbers that are reasonable values of g.
3. The legal capacity of a club is 180 people. Let n be the number of people who are at the club. You may assume that the number of people in the club never exceeds the legal limit. Describe all of the reasonable values of n.

Tips for Success Take Notes

It is always a good idea to take notes during classroom activities. Not only will you have something to refer to later when doing the homework, but also you will have something to help you prepare for tests. In addition, taking notes makes you become even more involved with the material, which means you will probably increase both your understanding and your retention of it.

Homework 1.1

For extra help ▶ MyMathLab®  Watch the videos in MyMathLab  Download the MyDashboard App

Respond to the questions in Exercises 1–12 by using complete sentences.

1. Let n be the number (in thousands) of fans who attend a Coldplay rock concert. What does $n = 25$ mean in this situation?
2. Let n be the number of home runs hit by Jose Bautista in a season. For 2011, the value of n is 43. What does $n = 43$ mean in this situation?
3. Let n be the number (in millions) of Americans who have a cell phone. For 2011, the value of n is 274 (Source: *Pew Internet*). What does $n = 274$ mean in this situation?
4. Let p be the percentage of children ages 9–13 who participate in organized physical activity. The value of p is about 39 for 2011 (Source: *Infoplease*). What does $p = 39$ mean in this situation?
5. Let s be the annual iPad® sales (in millions). The value of s is 14.8 for 2010 (Source: *Apple*). What does $s = 14.8$ mean in this situation?
6. Let p be the percentage of American workers who are in a union. For 2011, the value of p is 11.8 (Source: *U.S. Census Bureau*). What does $p = 11.8$ mean in this situation?
7. Let p be a company's annual profit (in thousands of dollars). What does $p = -45$ mean in this situation?
8. Let T be the temperature (in degrees Fahrenheit). What does $T = -10$ mean in this situation?
9. Let t be the number of years since 2010. What does $t = 9$ mean in this situation?
10. Let t be the number of years since 2005. What does $t = 13$ mean in this situation?
11. Let t be the number of years since 2005. What does $t = -3$ mean in this situation?
12. Let t be the number of years since 2010. What does $t = -2$ mean in this situation?

For Exercises 13–20, choose a variable name for the given quantity. Give two numbers that the variable can represent and two numbers that it cannot represent.

13. The height (in inches) of a person
14. The amount of time (in hours) that a student prepares for an exam
15. The price (in dollars) of a video game
16. The number of students enrolled in an algebra class

17. The total time (in hours) a person works in a week

18. The temperature (in degrees Fahrenheit) in an oven

19. The annual salary (in thousands of dollars) of a person

20. The value (in thousands of dollars) of a new home

21. A rectangle has an area of 24 square inches. Let W be the width (in inches), L be the length (in inches), and A be the area (in square inches).

a. Sketch three possible rectangles of area 24 square inches.
b. Which of the symbols W, L, and A are variables? Explain.
c. Which of the symbols W, L, and A are constants? Explain.

22. A rectangle has an area of 36 square feet. Let W be the width (in feet), L be the length (in feet), and A be the area (in square feet).

a. Sketch three possible rectangles of area 36 square feet.
b. Which of the symbols W, L, and A are variables? Explain.
c. Which of the symbols W, L, and A are constants? Explain.

23. The *perimeter* of a rectangle is the sum of the lengths of all four sides. A rectangle has a perimeter of 20 feet. Let W be the width, L be the length, and P be the perimeter, all with units in feet.

a. Sketch three possible rectangles of perimeter 20 feet.
b. Which of the symbols W, L, and P are variables? Explain.
c. Which of the symbols W, L, and P are constants? Explain.

24. The *perimeter* of a rectangle is the sum of the lengths of all four sides. A rectangle has a perimeter of 16 inches. Let W be the width, L be the length, and P be the perimeter, all with units in inches.

a. Sketch three possible rectangles of perimeter 16 inches.
b. Which of the symbols W, L, and P are variables? Explain.
c. Which of the symbols W, L, and P are constants? Explain.

25. The length of a rectangle is 3 inches more than the width. Let W be the width (in inches), L be the length (in inches), and A be the area (in square inches).

a. Sketch three possible rectangles of length 3 inches more than the width.
b. Which of the symbols W, L, and A are variables? Explain.
c. Which of the symbols W, L, and A are constants? Explain.

26. The length of a rectangle is twice the width. Let W be the width, L be the length, and P be the perimeter, all with units in inches. [**Hint:** *Twice* means to multiply by 2.]

a. Sketch three possible rectangles in which the length is twice the width.
b. Which of the symbols W, L, and P are variables? Explain.
c. Which of the symbols W, L, and P are constants? Explain.

27. The width of a rectangle is 2 yards. Let W be the width, L be the length, and P be the perimeter, all with units in yards.

a. Sketch three possible rectangles of width 2 yards.
b. Which of the symbols W, L, and P are variables? Explain.
c. Which of the symbols W, L, and P are constants? Explain.

28. The width of a rectangle is 5 centimeters. Let W be the width (in centimeters), L be the length (in centimeters), and A be the area (in square centimeters).

a. Sketch three possible rectangles of width 5 centimeters.
b. Which of the symbols W, L, and A are variables? Explain.
c. Which of the symbols W, L, and A are constants? Explain.

Graph all of the given numbers on one number line.

29. $5, -2, 0, -3, 4, -1$

30. $-4, 1, -6, 2, 7, -3$

31. $-\frac{2}{3}, -1, \frac{7}{3}, 1, -\frac{5}{3}, 2$

32. $\frac{1}{4}, 0, -2, -\frac{5}{4}, \frac{9}{4}, 1$

33. 2.8, 3.6, 0.4, 2.5, 1.1, 1.8

34. 3, 1.5, 2.3, 0.9, 2.7, 3.4

35. $-2, 3.1, 1.2, -1.8, 0.5, 1$

36. $1, 0.2, -2.4, -0.7, 1.9, -1$

Graph the numbers on a number line.

37. Counting numbers between 3 and 8

38. Counting numbers between 1 and 5

39. Integers between -2 and 2, inclusive

40. Integers between -4 and 4, inclusive

41. Integers between -1 and 4, inclusive

42. Integers between -6 and 3, inclusive

43. Negative integers between -4 and 4

44. Positive integers between -4 and 4

For Exercises 45–50, consider the following numbers:

$$-9.7 \quad -4 \quad 0 \quad \frac{3}{5} \quad \sqrt{7} \quad 3 \quad \pi \quad 356$$

Which of these numbers are the given type of number?

45. Counting numbers

46. Integers

47. Negative integers

48. Rational numbers

49. Irrational numbers

50. Real numbers

Give three examples of the following types of numbers.

51. Negative integers

52. Positive integers

53. Negative integers less than -7

54. Negative integers greater than -5

55. Integers that are not counting numbers

56. Rational numbers that are not integers

57. Rational numbers between 1 and 2

58. Irrational numbers between 1 and 10

59. Real numbers between -3 and -2

60. Real numbers that are not rational numbers

Use points on a number line to describe the given values of a variable. Find the average of the values and indicate it on the number line.

61. A student goes to a college for six semesters. Here are the numbers of units (credits or hours) taken per semester: 10, 12, 6, 9, 15, 14. Let u be the number of units taken in one semester.

62. During the summer, a student visits a music website five times. Here are the numbers of songs downloaded: 2, 0, 1, 5, 4. Let n be the number of songs downloaded in one visit to the website.

63. The percentages of disposable personal annual income spent on food for various years are 16%, 14%, 13%, 11%, and 10%. Let p be the percentage of disposable personal annual income spent on food.

64. The percentages of airline flights that are on time for various years are 79%, 82%, 75%, 77%, and 76%. Let p be the percentage of flights in a year that are on time.

For Exercises 65–68, use points on a number line to describe the given values of a variable.

65. The average annual lost time (in hours) due to traffic congestion on highways for various years is 22, 30, 24, 27, and 32. Let L be the average annual lost time (in hours) due to traffic congestion.

66. The U.S. annual per person consumption of sports drinks (in gallons) for various years is 1.9, 2.5, 2.1, 2.3, and 2.2. Let c be the per person consumption (in gallons per year) of sports drinks in a year.

67. The low temperatures (in degrees Fahrenheit) for three days in December in Chicago are 5°F above zero, 4°F below zero, and 6°F below zero. Let F be the low temperature (in degrees Fahrenheit) for one day.

68. Here are a company's annual profits and losses for various years: loss of \$5 million, profit of \$3 million, and loss of \$8 million. Let p be the company's annual profit (in millions of dollars).

69. The revenue (in billions of dollars) of Apple in the years 2007, 2008, 2009, 2010, and 2011 is 38, 43, 65, 108, and 180, respectively (Source: *Apple*). Let r be the annual revenue (in billions of dollars) of Apple.

a. Use points on a number line to describe the given values of r. Find the average of the values and indicate it on the number line.

b. Did the annual revenue increase, decrease, stay approximately constant, or none of these between 2007 and 2011, inclusive? Explain.

c. Did the annual *increases* in the annual revenue increase, decrease, stay approximately constant, or none of these between 2007 and 2011, inclusive? Explain.

70. The prize money (in millions of dollars) for the women's singles Wimbledon champion in the years 2006, 2007, 2008, 2009, 2010, and 2011 is 0.95, 1.01, 1.09, 1.23, 1.45, and 1.77, respectively (Source: *Wimbledon.org*). Let p be the prize money (in millions of dollars).

a. Use points on a number line to describe the given values of p. Find the average of the values and indicate it on the number line.

b. Did the prize money increase, decrease, stay approximately constant, or none of these from 2006 to 2011, inclusive? Explain.

c. Did the *increases* in the prize money increase, decrease, stay approximately constant, or none of these from 2006 to 2011, inclusive? Explain.

71. The percentage of public libraries with wireless Internet access in the years 2006, 2007, 2008, 2009, and 2010 is 37%, 54%, 66%, 76%, and 82%, respectively (Source: *ALA's Public Library Funding & Technology Access Study*). Let p be the percentage of public libraries with wireless Internet access.

a. Use points on a number line to describe the given values of p. Find the average of the values and indicate it on the number line.

b. Did the percentage of public libraries with wireless Internet access increase, decrease, stay approximately constant, or none of these between 2006 and 2010, inclusive? Explain.

c. Did the *increases* in the percentage of public libraries with wireless Internet access increase, decrease, stay approximately constant, or none of these between 2006 and 2010, inclusive? Explain.

72. A *justifiable homicide* is a homicide excused by law for which no criminal punishment is imposed. For example, killing in self defense to avoid getting murdered is a justifiable homicide. The numbers of justifiable homicides in the years 2005, 2006, 2007, 2008, and 2009 are 196, 238, 257, 265, and 266, respectively (Source: *FBI*). Let n be the number of justifiable homicides in a year.

a. Use points on a number line to describe the given values of n. Find the average of the values and indicate it on the number line.

b. Did the number of justifiable homicides increase, decrease, stay approximately constant, or none of these from 2005 to 2009, inclusive? Explain.

c. Did the *increases* in the number of justifiable homicides increase, decrease, stay approximately constant, or none of these from 2005 to 2009, inclusive? Explain.

Concepts

73. Let T be the temperature in degrees Fahrenheit.

a. What value of T represents the temperature that is 5°F below zero?

b. A student says that T represents only positive numbers and zero, because there is no negative sign. Is the student correct? Explain.

74. A student says the integers between 2 and 5 are the numbers 2, 3, 4, and 5. Is the student correct? Explain.

75. a. Find the average of each pair of numbers. Then plot the two given numbers and their average on a number line.

i. 7, 9 **ii.** 1, 5 **iii.** 2, 8

b. What patterns do you notice in your work in part (a)?

c. How many real numbers are there between 0 and 1? Explain. [**Hint:** Use the concept of average to help you show that you can keep finding more numbers.]

76. We can describe how far apart two numbers are on the number line. For example, the numbers 3 and 7 are 4 units apart. How far apart are two consecutive integers on the number line? How far apart are two consecutive even integers? How far apart are two consecutive odd integers?

77. How is a variable different from a constant?

78. The average of a student's four test scores is 90 points. If the student takes another test and the average of all five tests is 90 points, what is the student's score on the fifth test? Explain.

79. List the various types of numbers discussed in this section and describe the meanings of each type. (See page 9 for guidelines on writing a good response.)

80. Describe how to graph a negative quantity. (See page 9 for guidelines on writing a good response.)

1.2 Scattergrams

Objectives

» Know the meaning of *ordered pair, coordinate,* and *coordinate system.*

» Create scattergrams.

» Know the meaning of *independent variable* and *dependent variable.*

» Read bar graphs.

» Plot points on a coordinate system.

In Section 1.1, we used a single number line to describe values of a quantity. In this section, we will use a pair of number lines to describe two quantities that are related. For instance, in Example 3, we will compare the average price of a Super Bowl ticket with the Super Bowl number.

When One Number Line Is Not Enough

A number line is convenient for displaying values of a variable. However, there are limitations to using a *single* number line. For example, suppose a student earns the following points (listed in chronological order) from taking five quizzes: 7, 6, 9, 8, 9. We let q stand for the number of points earned by the student on a quiz. We use a number line to graph the scores in Fig. 15.

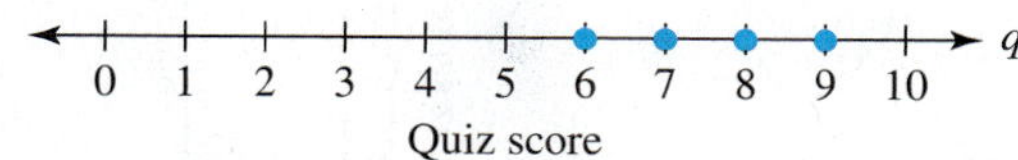

Figure 15 Graphing the scores

There are two limitations with using one number line to graph these scores. First, the graph does not show that there were two scores of 9 points each. Second, the graph does not show which was the first score, the second score, and so on.

The Coordinate System

Table 2 Values of n and q

n	q
1	7
2	6
3	9
4	8
5	9

There is a way to address both limitations. To begin, we let n be the quiz number. For the first quiz, $n = 1$. For the second quiz, $n = 2$, and so on. Next, we organize the values of the variables n and q (the quiz score) in Table 2.

The "1" and "7" in the first row of Table 2 indicate that when $n = 1$, $q = 7$. This means the student's score on the first quiz was 7 points. If we agree to write the quiz number first and the quiz score second, we can use the ordered pair (1, 7) to mean that when $n = 1$, $q = 7$. An **ordered pair** is a pair of numbers (written in parentheses and separated by a comma) for which the order of the numbers is meaningful. We call each of the numbers in an ordered pair a **coordinate.** For (1, 7) in this situation, we call 1 the *n-coordinate* and 7 the *q-coordinate.*

The ordered pair (2, 6) indicates that when $n = 2$, $q = 6$. This means the student's score on the second quiz was 6 points, which agrees with the second row of Table 2.

We call pairs of numbers such as (3, 9) ordered pairs because the order in which the numbers appear matters: The ordered pair (3, 9) means the student's score on the third quiz was 9 points, whereas the ordered pair (9, 3) means the student's score on the ninth quiz was 3 points.

We graph the ordered pairs by using *two* number lines, which are called **axes** (singular: **axis**). To start, we draw a horizontal number line called the n-axis and a vertical number line called the q-axis (see Fig. 16). We refer to such a pair of axes as a

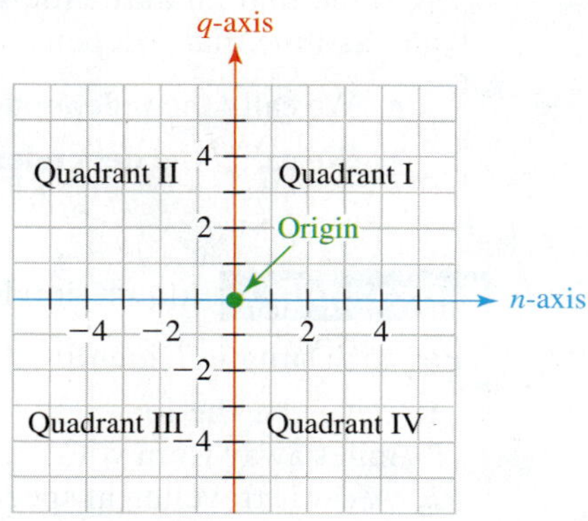

Figure 16 Coordinate system

coordinate system. The **origin** is the intersection point of the axes. The axes divide the coordinate system into four regions called **quadrants,** which we call Quadrants I, II, III, and IV. The quadrants do not include the axes.

Scattergrams

Next, we plot the ordered pair (3, 9) shown in the third row of Table 2. To do so, we start at the origin, look 3 units to the right and 9 units up, and then draw a dot (see Fig. 17). In Fig. 18, we plot all the ordered pairs listed in Table 2.

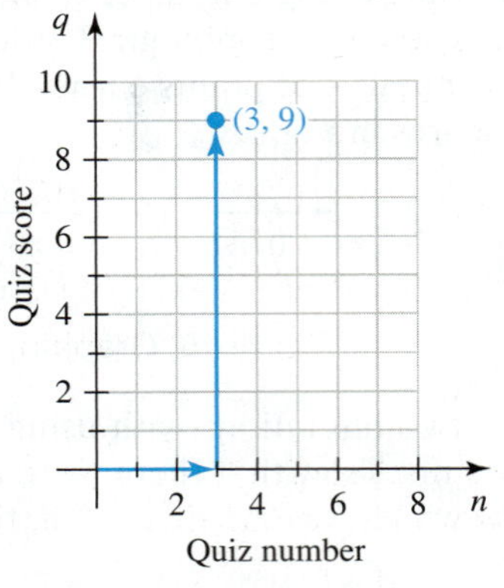

Figure 17 Plot (3, 9)

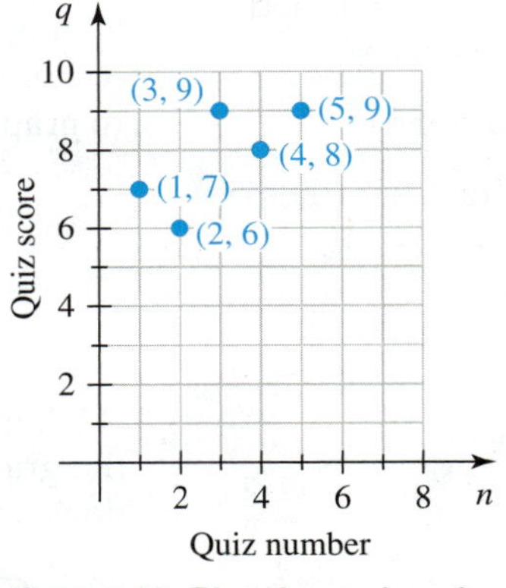

Figure 18 Plot the ordered pairs from Table 2

Note that we have addressed both of the limitations of using just one number line. The coordinate system in Fig. 18 shows that there are two scores of 9, and it also shows which score is from the first quiz, the second quiz, and so on.

As we look at the plotted points in Fig. 18 from left to right, the points, in general, go upward. This means the quiz scores, in general, are increasing.

A graph of plotted ordered pairs, such as the graph in Fig. 18, is called a **scattergram.**

Independent and Dependent Variables

The student's score q on a quiz depends on the quiz number n. For example, if a certain quiz is more difficult than the others, the student may earn a lower score on that quiz than on the other quizzes. Because q depends on n, we call q the *dependent variable*.

The quiz number does not depend on the student's score, however—n is independent of q. We call n the *independent variable.*

Definition Independent and dependent variables

Assume that an authentic situation can be described by using the variables t and p, and assume that p depends on t. Then

- We call t the **independent variable.**
- We call p the **dependent variable.**

Example 1 Identifying Independent and Dependent Variables

For each situation, identify the independent variable and the dependent variable.

1. Let L be the loudness (in decibels) of the sound produced by a foghorn that is d miles away from you.
2. A car is traveling at speed s (in mph) on a dry asphalt road, and the brakes are suddenly applied. Let d be the stopping distance (in feet).

Solution

1. The farther you are from the foghorn, the softer the sound you will hear. So, the loudness L depends on the distance d. Thus, L is the dependent variable and d is the independent variable. (Notice that the distance does *not* depend on the loudness.)
2. The greater the traveling speed, the greater the stopping distance will be. So, the stopping distance d depends on the traveling speed s. Thus, d is the dependent variable and s is the independent variable. (Notice that the traveling speed does *not* depend on the stopping distance.)

For an ordered pair (a, b), we write the value of the independent variable in the first (left) position and the value of the dependent variable in the second (right) position. Note that we did just that for the quiz score application. That is, we listed values of the independent variable n in the first position and values of the dependent variable q in the second position. For example, the ordered pair (4, 8) means that when $n = 4$, $q = 8$. In other words, the student's score on the fourth quiz is 8 points.

Example 2 Determining the Meaning of an Ordered Pair

1. Let n be the total number of Quiznos Sub® restaurants at t years since 2000. What does the ordered pair (11, 2772) mean in this situation?
2. Let p be a runner's pulse rate (in beats per minute) when his speed is s miles per hour. What does the ordered pair (10, 160) mean in this situation?

Solution

1. The total number of restaurants depends on the year. So, n is the dependent variable and t is the independent variable. The ordered pair (11, 2772) means that $t = 11$ and $n = 2772$. There were 2772 restaurants in $2000 + 11 = 2011$.
2. The runner's pulse rate depends on his speed. So, p is the dependent variable and s is the independent variable. The ordered pair (10, 160) means that $s = 10$ and $p = 160$. When the runner's speed is 10 miles per hour, his pulse rate is 160 beats per minute.

Table 3 Values of n and q

n	q
1	7
2	6
3	9
4	8
5	9

For tables of ordered pairs, we list the values of the independent variable in the first (left) column and the values of the dependent variable in the second (right) column. For example, in Table 3 the values of n are in the first column and the values of q are in the second column.

For coordinate systems, we describe the values of the independent variable with the horizontal axis and the values of the dependent variable with the vertical axis. For example, in Fig. 18, the horizontal axis is the n-axis and the vertical axis is the q-axis.

Columns of Tables and Axes of Coordinate Systems

Assume that an authentic situation can be described by using two variables. Then

- For tables, the values of the independent variable are listed in the first column and the values of the dependent variable are listed in the second column (see Table 4).
- For coordinate systems, the values of the independent variable are described by the horizontal axis and the values of the dependent variable are described by the vertical axis (see Fig. 19).

Table 4 Position of the Variables

Independent Variable	Dependent Variable
*	*
*	*
*	*

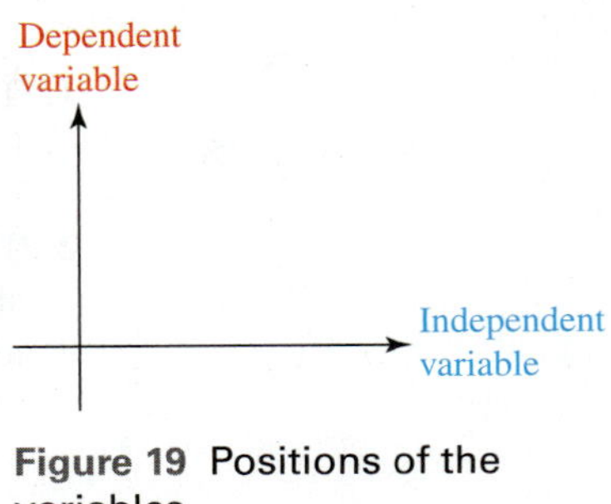

Figure 19 Positions of the variables

More about Scattergrams

Example 3 Creating a Scattergram

The average ticket prices for the Super Bowl are shown in Table 5 for various years. Let p be the average ticket price (in dollars) and n be the Super Bowl number.

1. Draw a scattergram of the data.
2. For the Super Bowls described in Table 5, which Super Bowl had the highest average ticket price? What was that price?
3. Describe any patterns you see in the prices.

Table 5 Average Ticket Prices for the Super Bowl

Super Bowl Number	Average Ticket Price (dollars)
1 (I)	12
5 (V)	15
10 (X)	20
15 (XV)	40
20 (XX)	75
25 (XXV)	150
30 (XXX)	250
35 (XXXV)	325
40 (XL)	550
45 (XLV)	875

Source: *NFL.com*

Solution

1. A scattergram of the data is shown in Fig. 20. It makes sense to think of p as the dependent variable, because the average ticket price depends on the Super Bowl number (and not the other way around). So, we let the vertical axis be the p-axis. Note that we write the variable names "n" and "p" and the units "Super Bowl number" and "Dollars" on the appropriate axes.

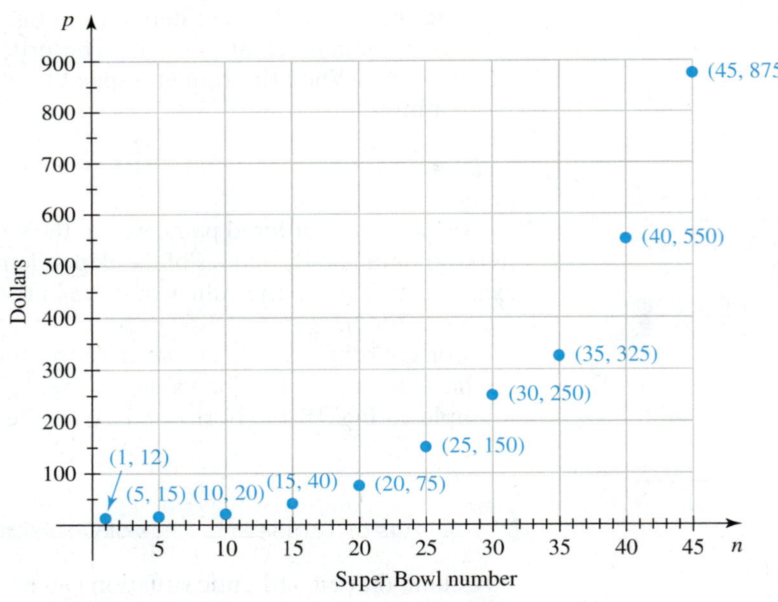

Figure 20 Scattergram of average Super Bowl ticket prices

Recall from Section 1.1 that when we write numbers on an axis, they should increase by a fixed amount and be equally spaced. Since the Super Bowl numbers are between 1 and 45, inclusive, we write the numbers 5, 10, 15, ..., 45 equally spaced on the n-axis. Since the prices shown in Table 5 are between \$12 and \$875, inclusive, we write the numbers 100, 200, 300, ..., 900 on the p-axis.

The ordered pair (1, 12) indicates that the average ticket price was \$12 for Super Bowl I.

2. From Table 5 and the scattergram in Fig. 20, we see that the highest average ticket price was \$875 in Super Bowl XLV.
3. From Table 5 and the scattergram in Fig. 20, we see that average ticket prices have increased. We also see that from Super Bowl I to Super Bowl XXX, inclusive, the *increases* in average ticket prices have increased. That is, as we look from left to right at the first seven points plotted on the coordinate system, we see that the vertical distance between adjacent points increases.

In Example 3, creating a scattergram helped us notice some patterns. Such observations will help us make predictions about authentic situations later in this course.

Example 4 Creating a Scattergram with Age Groups

A householder is the person in whose name a house, condominium, or apartment is owned or rented. The percentages of householders who own a home are listed in Table 6 for various age groups.

Table 6 Percentages of Householders Who Own a Home

Age Group (years)	Age Used to Represent Age Group (years)	Percent
15–24	19.5	18
25–34	29.5	46
35–44	39.5	66
45–54	49.5	75
55–64	59.5	80
65–74	69.5	81
75–84	79.5	77

Source: *U.S. Census Bureau*

Let p be the percentage of householders who own a home when they are at age a years.

1. Draw a scattergram of the data.
2. Describe any patterns in the percentages of householders who own a home.

Solution

1. A look at the first row of Table 6 suggests that we use $a = 19.5$ to represent the age group from 15 years to 24 years. The age 19.5 years is the average of the ages 15 years and 24 years. (Try it.) Likewise, we will use 29.5 to represent the age group from 25 years to 34 years and so on.

   A scattergram of the data is shown in Fig. 21. It makes sense to think of p as the dependent variable, because the percentage of householders depends on the age group (and not the other way around). So, we let the vertical axis be the p-axis. Note that we write the variable names "a" and "p" and the units "Age in years" and "Percent" on the appropriate axes.

   The ordered pair (19.5, 18) indicates that, for the age group from 15 years to 24 years, 18% of householders own a home.

   Since the ages used to represent age groups are between 19.5 years and 79.5 years, inclusive, we write the numbers 10, 20, 30, ..., 80 equally spaced on the a-axis. Since the percents shown in Table 6 are between 18% and 81%, inclusive, we write the numbers 20, 40, 60, 80 and 100 equally spaced on the p-axis.

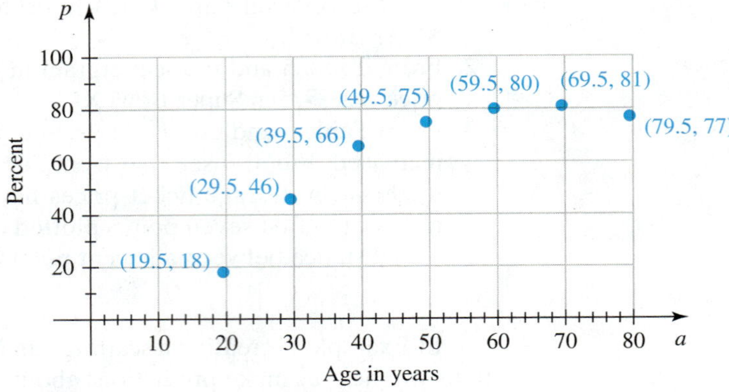

Figure 21 Scattergram of homeowner data

2. From Table 6 and the scattergram in Fig. 21, we see that until about age 70, the percentage of householders who own a home increases as their age increases. After about age 70, the percentage of householders who own a home decreases as their age increases. We also see that until about age 70, the increase in the percentage is decreasing: After every 10 years, the percentage of householders who own a home increases, but by less than it did in the previous 10 years.

In Example 5, we will define a variable to represent time.

Example 5 Defining a Variable for Time

Let t be the number of years since 1990. Find the values of t that represent the years 1990, 1996, 1999, 2000, and 2003.

Solution

We can represent 1990 by $t = 0$, because 1990 is 0 years after 1990. We can represent 1996 by $t = 6$, because 1996 is 6 years after 1990. We list the value of t for each of the years 1990, 1996, 1999, 2000, and 2003 in Table 7.

Table 7 Values of t

Year	Years since 1990 t		
1990	0	because	$1990 - 1990 = 0$
1996	6		$1996 - 1990 = 6$
1999	9		$1999 - 1990 = 9$
2000	10		$2000 - 1990 = 10$
2003	13		$2003 - 1990 = 13$

The values of t in Table 7 are much smaller numbers than the years they represent. When working with authentic situations, we will often perform calculations that involve years. Using definitions similar to the one in Example 5 will enable us to perform those calculations with smaller numbers. It is also easier to label the axes of a coordinate system with smaller numbers.

Table 8 Life Expectancies

Year of Birth	Life Expectancy (years)
1980	73.7
1985	74.7
1990	75.4
1995	75.8
2000	76.9
2005	77.4
2010	78.3

Source: *U.S. Census Bureau*

Example 6 Creating a Scattergram with Zigzag Lines on an Axis

A *life expectancy* is a prediction of how long a person will live. Table 8 shows life expectancies at birth for Americans in various years. Let L be the life expectancy at birth (in years) for an American born t years after 1980.

1. Create a scattergram of the data.
2. Describe any patterns in life expectancies of Americans.

Table 9 Values of t and L

t (years since 1980)	L (years of life)
0	73.7
5	74.7
10	75.4
15	75.8
20	76.9
25	77.4
30	78.3

Solution

1. First, we list the values of t and L in Table 9. For example, $t = 0$ represents 1980, because 1980 is 0 years after 1980. Also, $t = 5$ represents 1985, because 1985 is 5 years after 1980.

   A scattergram of the data is shown in Fig. 22. Since L is the dependent variable, we let the vertical axis be the L-axis. We use zigzag lines on the L-axis to indicate that the part of the axis between 0 and about 71 is not displayed. This is done so we can show a clear view of the data points without having to make the coordinate system exceedingly tall.

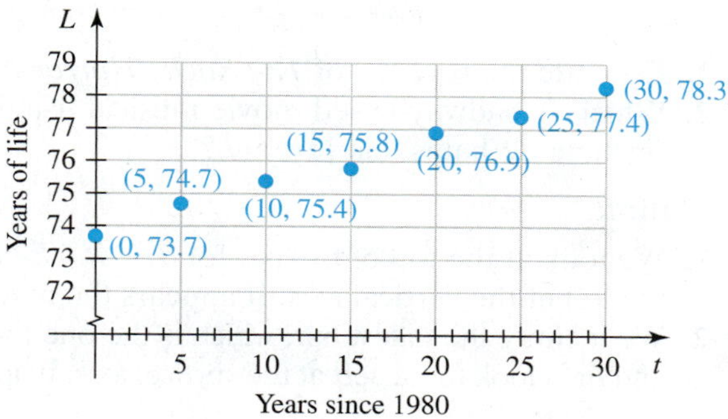

Figure 22 Life expectancy scattergram

2. From Table 9 and Fig. 22, we see that the life expectancy of Americans has been increasing fairly steadily since 1980.

Using zigzag lines on the L-axis in Fig. 22 helped us have a clear view of the data points. To see how poor a view we have without the zigzag lines, see Fig. 23.

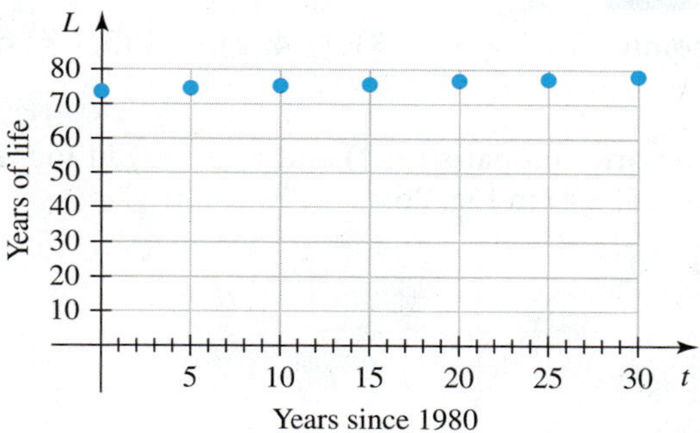

Figure 23 Poor view of life expectancy scattergram (Fig. 22), without zigzag lines

Without the zigzag lines, it is very difficult to plot the data points at the correct height. Also, in Section 1.3, we will begin to estimate coordinates of points by reading graphs, and it would be very difficult to do so accurately by using the graph in Fig. 23. In fact, it is hard to tell that the heights of the points increase from left to right.

Bar Graphs

A **bar graph** is a diagram with two axes that we can use to compare measurements of two or more items (see Fig. 24). Along one axis we list the items, and along the other axis we mark tick marks, write numbers, and write the units of the measurements. We use a bar to indicate the measurement of each item.

Example 7 Reading a Bar Graph

The revenues (in millions of dollars) of some Broadway-based movie musicals are illustrated in the bar graph in Fig. 24 (Source: *Box Office Mojo*).

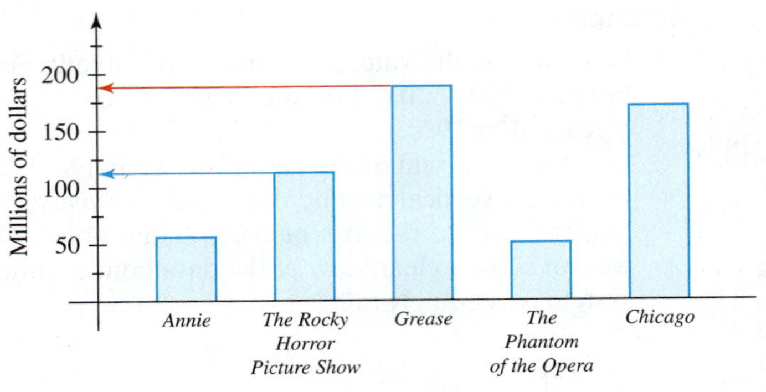

Figure 24 Bar graph of lifetime revenues of some Broadway-based movie musicals

1. Estimate the revenue of *The Rocky Horror Picture Show*.
2. Which Broadway-based movie musical listed in the bar graph has had the largest revenue? What is that revenue?

Solution

1. We look at the top of the bar for *The Rocky Horror Picture Show* and then look to the left at the vertical axis. It appears the revenue is about \$113 million.
2. We identify the tallest bar, which is the one for *Grease*. We look at the top of the bar and then look to the left at the vertical axis. It appears the revenue is about \$188 million.

Plotting Points on a Coordinate System

When we plot points that are not being used to describe authentic situations, we call the horizontal axis the *x-axis* and the vertical axis the *y-axis*. Then x is the independent variable and y is the dependent variable. The ordered pair $(6, 3)$ means $x = 6$ and $y = 3$. So, the x-coordinate is 6 and the y-coordinate is 3.

Example 8 Plotting Points

Plot the points $(3, 4)$, $(-5, -3)$, $(-4, 2)$, and $(5, -4)$ on a coordinate system.

Solution

We plot the ordered pairs $(3, 4)$ and $(-5, -3)$ in Fig. 25, and we plot the ordered pairs $(-4, 2)$ and $(5, -4)$ in Fig. 26.

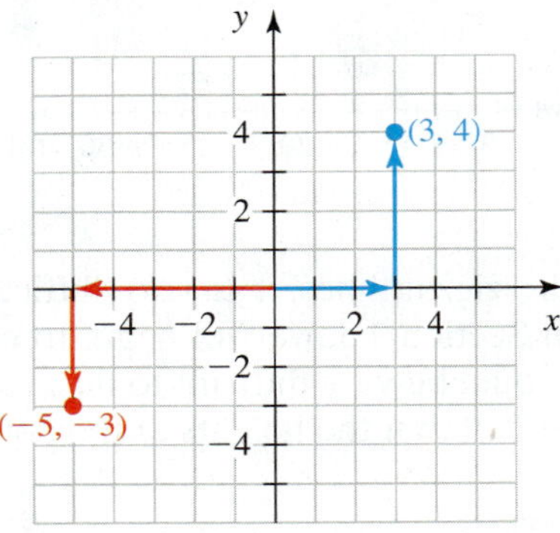

Figure 25 Plotting the ordered pairs $(3, 4)$ and $(-5, -3)$

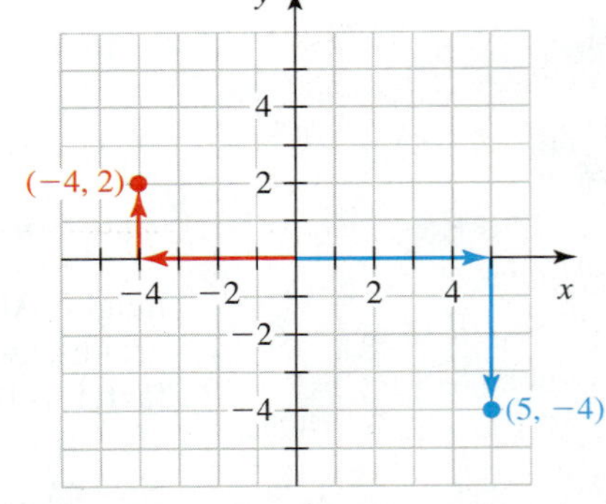

Figure 26 Plotting the ordered pairs $(-4, 2)$ and $(5, -4)$

Group Exploration

Looking ahead: Linear modeling

An airplane is beginning to descend. Let a be the altitude (in thousands of feet) of the airplane at t minutes since it began its descent. Some pairs of values of t and a are shown in Table 10.

1. Create a scattergram of the data.
2. Draw a line that contains the points in your scattergram. We call the line a *linear model*.

Table 10 Altitudes of an Airplane

Time (minutes) t	Altitude (thousands of feet) a
0	24
5	20
10	16
15	12
20	8

3. Use your line to estimate the altitude of the airplane 8 minutes after it began its descent. [**Hint:** On the line, locate the point whose t coordinate is 8.]

4. Use your line to estimate when the airplane reached an altitude of 10 thousand feet.

5. What was the altitude of the airplane when it began its descent?

6. Use your line to estimate when the airplane reached the ground.

Tips for Success Study Time

For each hour of class time, study for at least two hours outside class. If your math background is weak, you may need to spend more time studying.

One way to study is to do what you are doing now: Read the text. Class time is a great opportunity to be introduced to new concepts and to see how they fit together with previously learned ones. However, there is usually not enough time to address details as well as a textbook can. In this way, a textbook can serve as a supplement to what you learn in class.

Homework 1.2

For extra help ▶ MyMathLab® 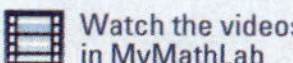 Watch the videos in MyMathLab  Download the MyDashboard App

Plot the given points in a coordinate system.

1. $(5, 1)$ **2.** $(2, 3)$ **3.** $(4, -2)$

4. $(3, -4)$ **5.** $(-5, 4)$ **6.** $(-1, 3)$

7. $(-3, -6)$ **8.** $(-5, -2)$ **9.** $(0, 2)$

10. $(0, -4)$ **11.** $(-3, 0)$ **12.** $(1, 0)$

13. $(2.5, -4.5)$ **14.** $(-3.5, 1.5)$

15. $(-1.3, -3.9)$ **16.** $(-2.4, -4.1)$

17. What is the x-coordinate of the ordered pair $(2, -4)$?

18. What is the y-coordinate of the ordered pair $(2, -4)$?

For Exercises 19–28, identify the independent variable and the dependent variable.

19. Let n be the number of hours a student studies for a quiz, and let s be the student's score (in points) on the quiz.

20. Let t be the number of years a person has worked for a company, and let s be the person's salary (in dollars).

21. Let h be the height (in inches) of a girl, and let a be the age (in years) of the girl.

22. Let p be the percentage of colleges that would accept a student whose grade point average (GPA) is g points.

23. Let T be the tuition (in dollars) for enrolling in c credits (units or hours) of classes.

24. Let p be the percentage of men at age a years who have gray hair.

25. Let A be the floor area (in square feet) of a classroom, and let n be the number of students who can comfortably fit into the classroom.

26. A person cooks a potato in an oven for an hour and then removes the potato and allows it to cool. Let t be the number of minutes since the potato was removed from the oven, and let F be the temperature (in degrees Fahrenheit) of the potato.

27. Let t be the number of seconds after a baseball is hit upward, and let h be the baseball's height (in feet).

28. Let p be the percentage of people at age a years who own a computer.

For Exercises 29–36, describe what the given ordered pair represents.

29. Let n be the average number of magazine subscriptions sold per week by a telemarketer who works t hours per week. What does the ordered pair $(32, 43)$ mean in this situation?

30. Let c be the total cost (in dollars) of buying n pens. What does the ordered pair $(5, 10)$ mean in this situation?

31. Let p be the percentage of Americans at age A years who say they volunteer. What does the ordered pair $(21, 38)$ mean in this situation?

32. Let p be the percentage of Americans who have ever watched a movie by streaming it to their computer at t years since 2005. What does the ordered pair $(6, 42)$ mean in this situation?

33. Let b be the annual defense spending (in billions of dollars) at t years since 2000. What does the ordered pair $(10, 698)$ mean in this situation?

34. Let a be the annual amount of money (in millions of dollars) Google spent to advertise its own products in the United States at t years since 2010. What does the ordered pair $(1, 213)$ mean in this situation?

35. Let p be the percentage of Americans who believe travel websites do a good job of presenting travel choices at t years since 2010. What does the ordered pair $(-1, 33)$ mean in this situation?

36. Let p be the percentage of Americans who are satisfied with the size and power of major corporations at t years since 2010. What does the ordered pair $(-2, 35)$ mean in this situation?

37. Create a scattergram of the ordered pairs listed in Table 11.

Table 11 Some Ordered Pairs

x	y
2	5
7	9
11	10
14	9
16	5

38. Create a scattergram of the ordered pairs listed in Table 12.

Table 12 Some Ordered Pairs

x	y
−5	2
−4	3
−1	5
4	9
11	15

39. Find the coordinates of points A, B, C, D, E, and F shown in Fig. 27.

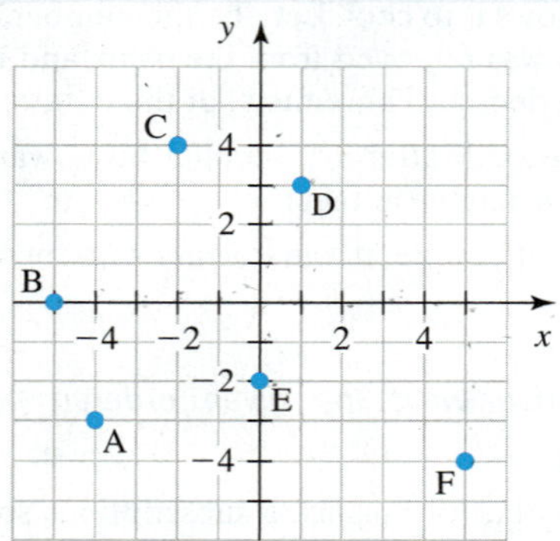

Figure 27 Exercise 39

40. Find the coordinates of points A, B, C, D, E, and F shown in Fig. 28.

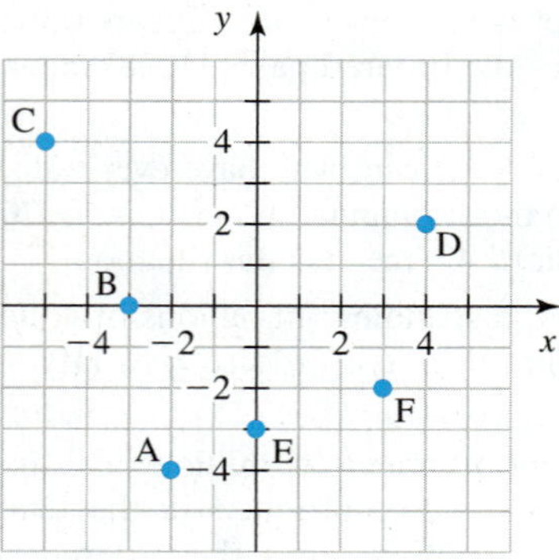

Figure 28 Exercise 40

41. The number of pages in each of the books in the Harry Potter series is listed in Table 13.

Table 13 Numbers of Pages in the Books of the Harry Potter Series

Book Number	Number of Pages
1	309
2	341
3	435
4	734
5	870
6	652
7	784

Let p be the number of pages and b be the corresponding book number.

a. Create a scattergram of the data.

b. Which book has the greatest number of pages?

c. From which book to the next did the number of pages increase the most? Explain how you can tell this by inspecting your scattergram.

42. The average life-spans of various denominations of bills are shown in Table 14. For example, the average life-span of a \$5 bill is 2 years before it is taken out of circulation due to wear and tear.

Table 14 Life-spans of Denominations of Bills

Value of Bill (dollars)	Life-span (years)
1	1.5
5	2
10	3
20	4
50	9

Source: *Federal Reserve System*

Let L be the life-span (in years) of a bill that is worth d dollars.

a. Create a scattergram of the data.

b. Explain why it makes sense that the average life-span of a \$50 bill is greater than the average life-span of a \$1 bill.

c. Each year, many more \$1 bills are printed than \$50 bills. Give at least two reasons why this makes sense.

43. Tiger Woods's golf tournament earnings and numbers of wins are shown in Table 15 for various years.

Table 15 Tiger Woods's Tournament Earnings and Numbers of Wins

Year	Tournament Earnings (millions of dollars)	Wins
2006	9.9	8
2007	10.9	7
2008	5.8	4
2009	10.5	6
2010	1.3	0
2011	0.7	0

Source: *PGA Tour*

a. Let E be Tiger Woods's tournament earnings (in millions of dollars) in the year that is t years since 2000. For example, $t = 0$ represents 2000, and $t = 6$ represents 2006. Create a scattergram of the data.

b. For which of the years shown in Table 15 were Woods's tournament earnings the least? What were those earnings?

c. For which of the years shown in Table 15 were Woods's tournament earnings the greatest? What were those earnings?

d. For the years shown in Table 15, did Woods have his greatest tournament earnings in the same year as his greatest number of wins? Give two reasons why this could be possible.

44. The average price per barrel of crude oil and the total fuel cost for the global airline industry are shown in Table 16 for various years.

Table 16 Average Price per Barrel of Crude Oil and Total Fuel Cost for the Global Airline Industry

Year	Average Price per Barrel of Crude Oil (dollars)	Total Airline Fuel Cost (billions of dollars)
2007	73	135
2008	99	189
2009	62	125
2010	79	139
2011	112	178

Source: *IATA*

Let C be the total airline fuel cost (in billions of dollars) in the year that is t years since 2005. For example, $t = 0$ represents 2005, and $t = 7$ represents 2012.

a. Create a scattergram of the data.

b. For which of the years shown in Table 16 was the total airline fuel cost the least? What was that cost?

c. For which of the years shown in Table 16 was the total airline fuel cost the greatest? What was that cost?

d. For the years shown in Table 16, was the total fuel cost the greatest in the same year that the average price per barrel was the greatest? Explain why this is possible.

45. The total costs to individuals and companies to prepare their taxes are shown in Table 17 for various years.

Table 17 Total Costs to Individuals and Companies to Prepare Taxes

Year	Total Cost (billions of dollars)
1990	80
1995	108
2000	172
2005	260
2010	368

Source: *Tax Foundation*

Let c be the total annual cost (in billions of dollars) to individuals and companies to prepare taxes at t years since 1990.

a. Create a scattergram of the data.

b. Did the total annual cost to prepare taxes increase, decrease, stay approximately constant, or none of these? Explain.

c. Did the five-year *increase* in the total annual cost to prepare taxes increase, decrease, stay approximately constant, or none of these? Explain.

46. The average hourly manufacturing pay is shown in Table 18 for various years.

Table 18 Average Hourly Manufacturing Pay

Year	Average Hourly Pay (dollars)
1970	3.24
1980	7.15
1990	10.78
2000	14.32
2010	18.61

Source: *Bureau of Labor Statistics*

Let p be the average hourly manufacturing pay (in dollars) at t years since 1970.

a. Create a scattergram of the data.

b. Did the average hourly manufacturing pay increase, decrease, stay approximately constant, or none of these? Explain.

c. Did the ten-year *increase* in the average hourly manufacturing pay increase, decrease, stay approximately constant, or none of these? Explain.

47. The numbers of automobile accidents per 1000 licensed drivers per year are shown in Table 19 for various age groups.

Table 19 Automobile Accidents

Age Group (years)	Age Used to Represent Age Group (years)	Accident Rate (number of accidents per 1000 licensed drivers per year)
16	16	190.3
17	17	163.2
18	18	142.9
19	19	127.8
20–29	24.5	91.4
30–39	34.5	54.7
40–49	44.5	43.9
50–59	54.5	36.4
60–69	64.5	31.3
over 69	75	32.1

Source: *National Highway Traffic Safety Administration*

Let r be the automobile accident rate (number of accidents per 1000 licensed drivers per year) for licensed drivers at age a years.

a. Create a scattergram of the data.

b. Which age group shown in Table 19 has the lowest accident rate?

c. Which age group shown in Table 19 has the highest accident rate?

d. Between what two consecutive drivers' ages does there seem to be the greatest change in the accident rate? Explain why we can't be sure this is true, because of the way the data are described in Table 19.

e. Many states put limits on teenage driving. For example, some states do not allow 16-year-old drivers to drive at night. Some states require parental supervision at all times. Why do you think these regulations were adopted?

48. The percentages of Americans of various age groups who are ordering more takeout food than they did two years ago are shown in Table 20. Let p be the percentage of Americans at age a years who are ordering more takeout food than they did two years ago.

Table 20 Percentages of Americans Who Are Ordering More Takeout Food than They Did Two Years Ago

Age Group (years)	Age Used to Represent Age Group (years)	Percent
18–24	21.0	34
25–34	29.5	31
35–44	39.5	27
45–54	49.5	17
55–64	59.5	15
over 64	70.0	7

Source: *National Restaurant Association Survey*

a. Create a scattergram of the data.
b. Which of the points in your scattergram is highest? What does that mean in this situation?
c. Which of the points in your scattergram is lowest? What does that mean in this situation?
d. Do the heights of the points in your scattergram increase or decrease from left to right? What does that mean in this situation?

49. Several inventions are listed in Table 21, along with the years they were invented and how long it took for one-quarter of the U.S. population to use them ("mass use").

Table 21 Number of Years until Inventions Reached Mass Use

Invention	Year Invented	Years until Mass Use
Electricity	1873	46
Telephone	1876	35
Gasoline-Powered Automobile	1886	55
Radio	1897	31
Television	1923	29
Microwave Oven	1953	36
VCR	1965	13
Personal Computer	1975	16
Mobile Phone	1985	11
CD Player	1985	8
World Wide Web	1991	7
DVD Player	1997	5

Source: *Newsweek*

Let M be the number of years elapsed until an invention reached mass use if it was invented at t years since 1870.

a. Create a scattergram of the data.
b. Compare the time it took to reach mass use for recent inventions versus earlier inventions. In your opinion, why did this happen?
c. Does the datum for the microwave oven fit the pattern you described in part (b)? Explain.
d. For a while after the microwave oven was invented, many people feared it would cause radiation poisoning, blindness, or impotence. Discuss the impact of these fears in terms of your response to part (c).
e. Explain why the datum for the gasoline-powered automobile does not fit the pattern you described in part (b). Why do you think this happened?

50. The percentages of adults of various age groups who approve of single men raising children on their own are shown in Table 22. Let p be the percentage of adults at age a years who approve of single men raising children on their own.

Table 22 Percentages of Adults Who Approve of Single Men Raising Children on Their Own

Age Group (years)	Age Used to Represent Age Group (years)	Percentage
18–34	26.0	81
35–44	39.5	73
45–54	49.5	73
55–64	59.5	66
over 64	70	47

Source: *Taylor Nelson Sofres*

a. Create a scattergram of the data.
b. Which age group shown in Table 22 has the most faith in single men raising children on their own?
c. Which age group shown in Table 22 has the least faith in single men raising children on their own?

51. The average starting salaries for employees with a bachelor's degree are illustrated in the bar graph in Fig. 29 for various fields of study (Source: *National Association of Colleges and Employers*).

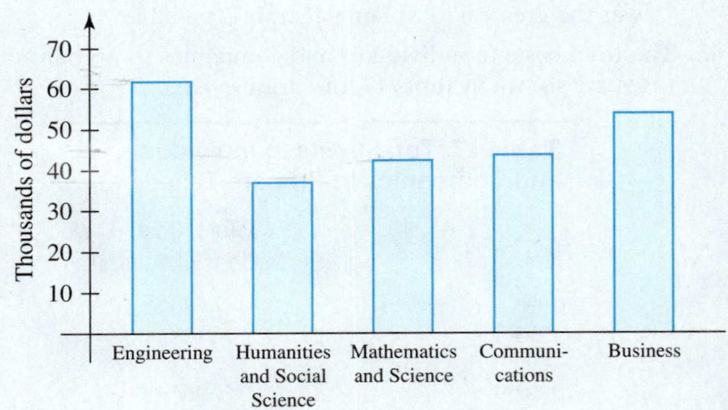

Figure 29 Average starting salaries

a. For which field shown is the average starting salary the highest? What is that salary?
b. For which field shown is the average starting salary the lowest? What is that salary?
c. Estimate the average starting salary for employees with a mathematics or science degree.

52. In baseball, a grand slam is a home run with the bases loaded. The top six numbers of career grand slams for major league baseball players are illustrated in Fig. 30 (Source: *MLB.com*).

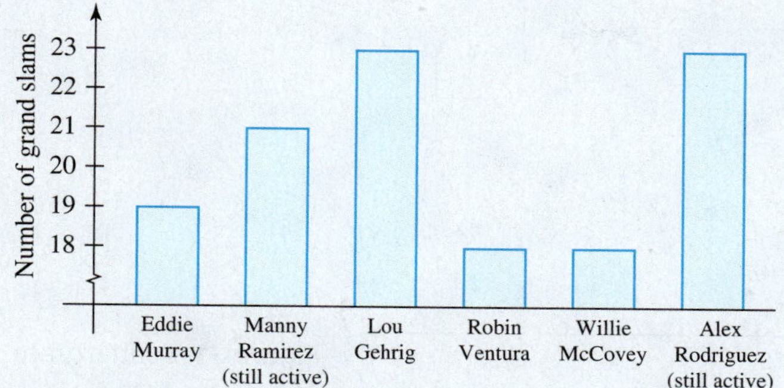

Figure 30 Most career grand slams

a. Estimate Robin Ventura's number of career grand slams.
b. Which two players are tied for the record number of career grand slams? What is that number of grand slams?
c. Of the six baseball players listed in Fig. 30, Manny Ramirez and Alex Rodriguez are the only active major league baseball players. Estimate how many grand slams Ramirez needs to tie the record.

Concepts

53. List five ordered pairs whose x-coordinate is 3. Then create a scattergram of the ordered pairs. What do you notice about the arrangement of the points in your scattergram? Explain why this makes sense.

54. List five ordered pairs whose y-coordinate is 2. Then create a scattergram of the ordered pairs. What do you notice about the arrangement of the points in your scattergram? Explain why this makes sense.

55. The points where the sides of a triangle, rectangle, or any other polygon meet are called *vertices.* A square has vertices at $(2, 1)$ and $(2, 5)$. How many possible positions are there for the other two vertices? Find the coordinates of the verticies for two of these possible positions.

56. The points where the sides of a triangle, rectangle, or any other polygon meet are called *vertices.* A rectangle has vertices at $(2, 4)$ and $(7, 4)$. How many possible positions are there for the other two vertices? Find the coordinates of the verticies for four of these possible positions.

57. Describe the signs of the x-coordinate and the y-coordinate for a point that lies in the given quadrant.
a. Quadrant I
b. Quadrant II
c. Quadrant III
d. Quadrant IV

58. Describe all points on a coordinate system whose x-coordinates are 0.

59. Describe all points on a coordinate system whose y-coordinates are 0.

60. Compare a number line with a coordinate system. When is it useful to describe data by a number line? When is it useful to describe data by a coordinate system?

61. Compare the meaning of *dependent variable* with the meaning of *independent variable.*

1.3 Exact Linear Relationships

Objectives

- Know the meaning of *linearly related, model, linear model, input,* and *output.*
- Use a linear model to make estimates and predictions.
- Use a scattergram to help decide whether to model a situation with a linear model.
- Know the meaning of *intercept.*
- Find intercepts of a line and of a linear model.

In this section, we will use a scattergram to help us sketch a line that can be used to describe an authentic situation, such as the descent of a hot-air balloon. We will then use the line to make estimates and predictions about the situation.

Linear Models

Example 1 Using a Line to Describe an Authentic Situation

A person lowers her hot-air balloon by gradually releasing air from the balloon. Let a be the balloon's altitude (in feet) above the ground after she has released air in the balloon for t minutes. Values of t and a are listed in Table 23.

1. Create a scattergram of the data.
2. Draw the line that contains the points of the scattergram.

Table 23 Altitudes of a Hot-Air Balloon

Time (minutes) t	Altitude (feet) a
0	2200
2	1800
4	1400
6	1000
8	600

Solution

1. We draw a scattergram in Fig. 31.

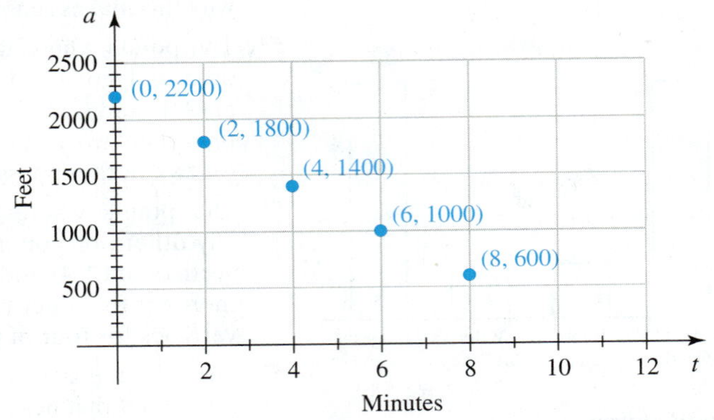

Figure 31 Scattergram of balloon data

2. In mathematics, a "line" means a *straight* line. In Fig. 32, we draw the line that contains the data points shown in Fig. 31.

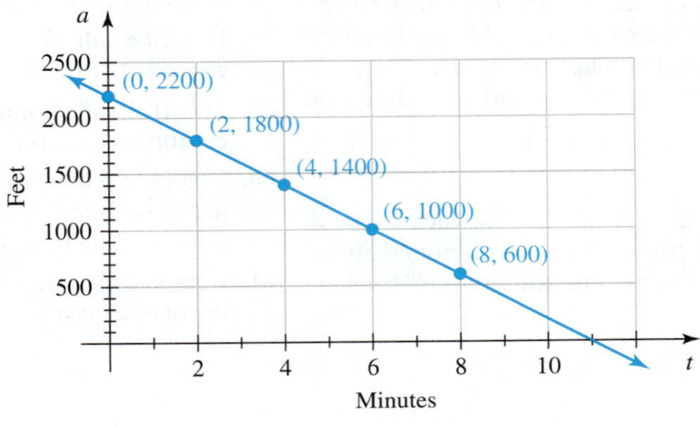

Figure 32 Line that contains the points of the scattergram

The scattergram in Fig. 31 accurately describes the altitude of the balloon at 0, 2, 4, 6, and 8 minutes. But it does not describe the altitude at other times.

If we imagine the line in Fig. 32 to be made up of points, then we can use the line to describe the altitudes at 0, 2, 4, 6, and 8 minutes accurately. We can also use the line to estimate the altitude for other times between 0 and 8 minutes. These results will be good estimates, provided that the altitude of the balloon declined steadily.

In addition, we can use the line to predict altitudes for times a little after 8 minutes. However, these predictions will be accurate only if the altitude of the balloon continued to decline steadily.

If the altitudes of the balloon over a span of time are described accurately by a line, we say that time and altitude (and the variables t and a) are *linearly related* for that span of time.

Definition Linearly related

If two quantities of an authentic situation are described accurately by a line, then the quantities (and the variables representing those quantities) are **linearly related.**

The process of choosing a line to represent the relationship between balloon altitudes and time is an example of *modeling.*

Definition Model

A **model** is a mathematical description of an authentic situation. We say the description *models* the situation.

We call the line in Fig. 32 a *linear model.* In Chapters 7–15, we will discuss other types of models. The term "model" is being used in much the same way as it is used in "airplane model." Just as an airplane designer can use the behavior of an airplane model in a wind tunnel to predict the behavior of an actual airplane, a linear model can be used to predict what might happen in a situation in which two variables are linearly related.

Definition Linear model

A **linear model** is a nonvertical line that describes the relationship between two quantities in an authentic situation.

Using a Linear Model to Make Estimates and Predictions

Example 2 Making Estimates and Predictions

1. Use the linear model shown in Fig. 32 to estimate the balloon's altitude when air has been released for 5 minutes.
2. Use the linear model to predict when the balloon's altitude is 400 feet.

Solution

1. To estimate the altitude of the balloon when air has been released for 5 minutes, we locate the point on the linear model where the t-coordinate is 5 (see Fig. 33). The a-coordinate of that point is 1200. So, *according to the model,* the altitude is 1200 feet when air has been released for 5 minutes.

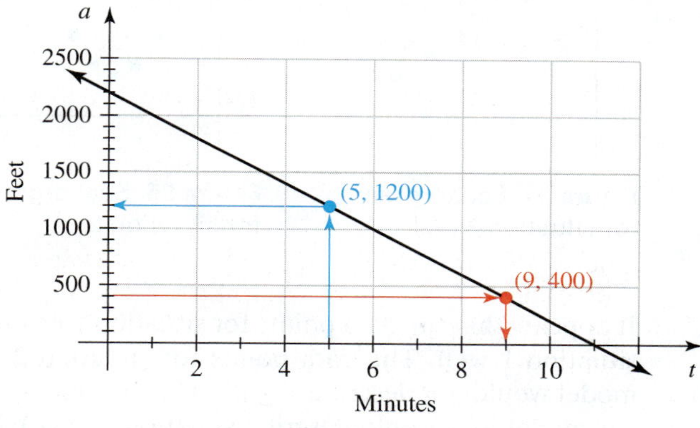

Figure 33 Using the linear model to make an estimate and a prediction

Table 24 Altitudes of a Balloon

Time (minutes) t	Altitude (feet) a
0	2200
2	1800
4	1400
6	1000
8	600

We verify our work by checking that our result is consistent with the values in Table 24. Since the altitude is 1400 feet after 4 minutes and 1000 feet after 6 minutes, it makes sense that the altitude would be between 1000 and 1400 feet when air has been released for 5 minutes, which checks with our result of 1200 feet.

2. To find when the balloon's altitude is 400 feet, we locate the point on the linear model where the a-coordinate is 400 (see Fig. 33). The t-coordinate of that point is 9. So, *according to the linear model,* the altitude is 400 feet when air has been released for 9 minutes.

   Again, we verify our work by checking that our result is consistent with the values in Table 24. Since the altitude is 600 feet when air has been released for 8 minutes, it follows that air would have been released for more than 8 minutes for the altitude to be less than 600 feet, which checks with our result of 9 minutes.

Input and Output

In Example 2, we found that when the value of the independent variable t is 5, the corresponding value of the dependent variable a is 1200. We say the *input* 5 leads to the *output* 1200. The blue arrows in Fig. 33 show the action of the input $t = 5$ leading to the output $a = 1200$.

Definition Input, output

An **input** is a permitted value of the *independent* variable that leads to at least one **output,** which is a permitted value of the *dependent* variable.

For a value to be permitted, it must make physical sense and be defined. For instance, in Example 2, the value -50 is not a permitted value of the variable a, because it does not make sense for the balloon's altitude to be -50 feet. Later in the course we will discuss values that are not permitted for mathematical reasons.

Sometimes we will go "backward," from an output back to an input. For instance, in Example 2, we found that the output $a = 400$ originates from the input $t = 9$. The red arrows in Fig. 33 show the action of going backward from the output $a = 400$ to the input $t = 9$.

When to Use a Line to Model Data

Next, we will discuss how to determine whether an authentic situation can be described well by a linear model.

Example 3 Deciding whether to Use a Line to Model Data

Consider the scattergrams of data for situations 1, 2, and 3 shown in Figs. 34, 35, and 36, respectively. For each situation, determine whether a linear model would describe the situation well.

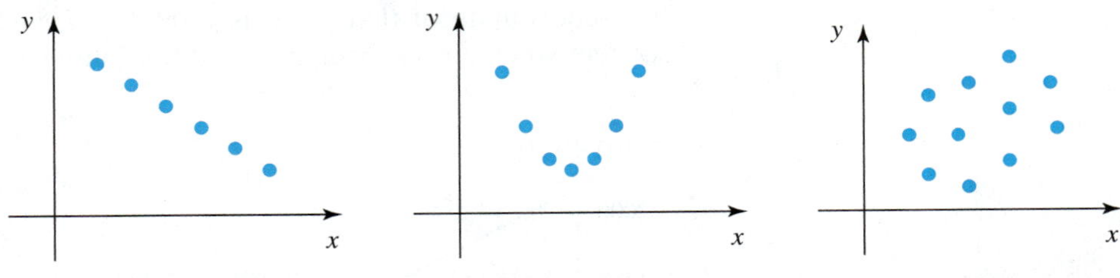

Figure 34 Scattergram for situation 1

Figure 35 Scattergram for situation 2

Figure 36 Scattergram for situation 3

Solution

It appears that the data points for situation 1 lie on a line; a linear model would describe situation 1 well. The data points for situation 2 do not lie close to one line; a linear model would not describe situation 2. (In Chapters 7–9, we will discuss a type of nonlinear model that would describe situation 2 well.) The data points for situation 3 do not lie near a line; a linear model would not describe this situation.

We create a scattergram of data to determine whether the data points lie on a line. If the points lie on a line, then we draw the line and use it to make estimates and predictions.

Intercepts of a Line

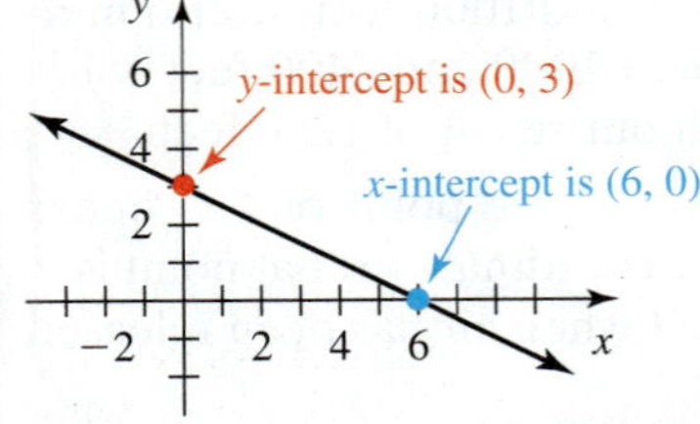

Figure 37 Intercepts of a line

Consider the line sketched in Fig. 37. The line intersects the x-axis at the point (6, 0). The point (6, 0) is called the *x-intercept.* Also, the line intersects the y-axis at the point (0, 3). The point (0, 3) is called the *y-intercept.*

Definition Intercepts of a line

An **intercept** of a line is any point where the line and an axis (or axes) of a coordinate system intersect. There are two types of intercepts of a line sketched on a coordinate system with an x-axis and a y-axis:

- An ***x*-intercept** of a line is a point where the line and the x-axis intersect (see Fig. 38). The y-coordinate of an x-intercept is 0.

- A ***y*-intercept** of a line is a point where the line and the y-axis intersect (see Fig. 38). The x-coordinate of a y-intercept is 0.

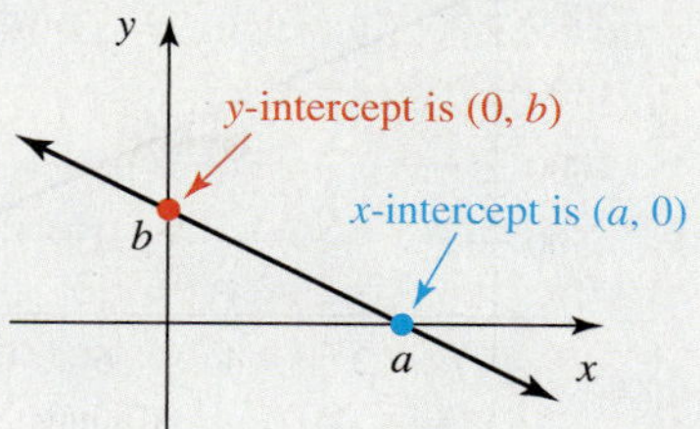

Figure 38 Intercepts of a line

Example 4 Finding Intercepts and Coordinates

Refer to Fig. 39 for the following problems.

1. Find the x-intercept of the line.
2. Find the y-intercept of the line.
3. Find y when $x = -6$.
4. Find x when $y = -3$.

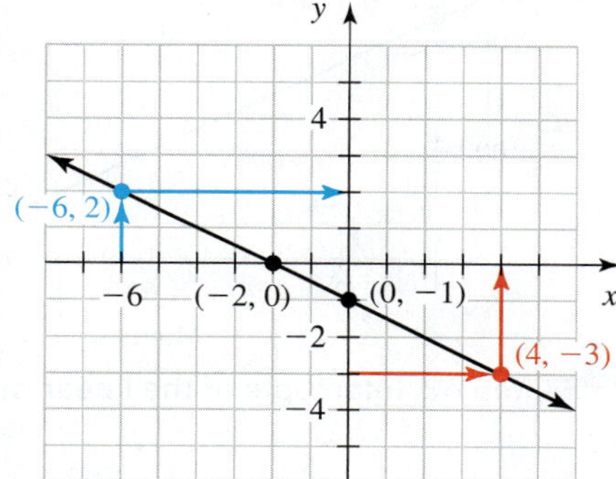

Figure 39 Problems 1–4 of Example 4

Solution

1. The line and the x-axis intersect at $(-2, 0)$. So, the x-intercept is $(-2, 0)$.
2. The line and the y-axis intersect at $(0, -1)$. So, the y-intercept is $(0, -1)$.
3. The blue arrows in Fig. 39 show that the input $x = -6$ leads to the output $y = 2$. So, $y = 2$ when $x = -6$.
4. The red arrows in Fig. 39 show that the output $y = -3$ originates from the input $x = 4$. So, $x = 4$ when $y = -3$.

Intercepts of a Linear Model

Suppose that a linear model describes the relationship between two variables t and p, where p depends on t. Then the t-intercept is a point where the line and the t-axis intersect, and the p-intercept is a point where the line and the p-axis intersect (see Fig. 40).

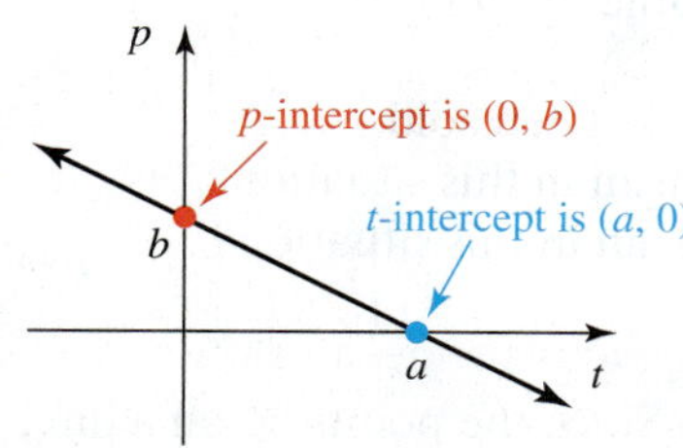

Figure 40 Intercepts of a linear model

Example 5 Finding Intercepts of a Linear Model

1. The linear model from Examples 1 and 2 is shown in Fig. 41. Find the t-intercept of the model. What does it mean in this situation?
2. Find the a-intercept of the model. What does it mean in this situation?

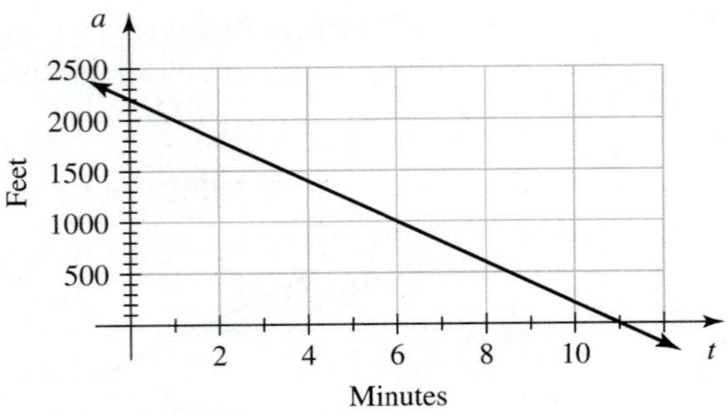

Figure 41 Altitudes of balloon model

Solution

1. The line intersects the t-axis at the point (11, 0). See Fig. 42. So, the t-intercept is (11, 0). This means that when $t = 11$, $a = 0$. The model predicts that the balloon reached the ground 11 minutes after air began to be released from the balloon.

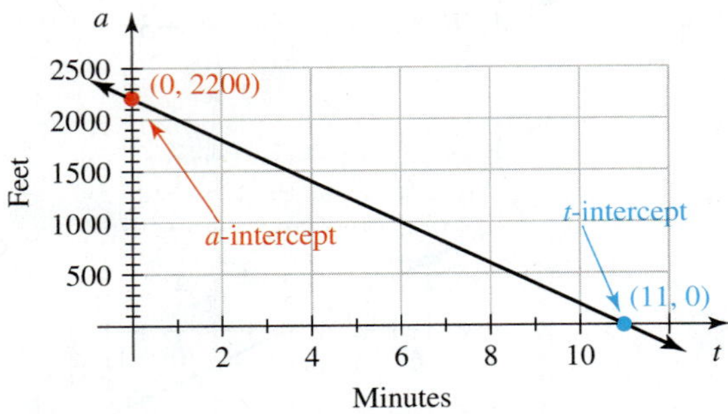

Figure 42 Intercepts of the linear model

2. The line intersects the a-axis at the point (0, 2200). See Fig. 42. So, the a-intercept is (0, 2200). This means that when $t = 0$, $a = 2200$. The model estimates that the balloon's altitude was 2200 feet when air was first released from the balloon. In fact, this estimate is the actual altitude (see Table 24 on p. 27).

Example 6 Using a Linear Model to Make Estimates

An underground rock band manufactures 500 CDs of its original music and tries to sell the CDs at the band's concerts. Let P be the profit (in dollars) from selling a total of n CDs. Some values of n and P are listed in Table 25.

Table 25 Profits from Selling CDs

Sales (number of CDs) n	Profit (dollars) P
50	−1150
100	−800
150	−450
300	600
350	950

1. Sketch a scattergram of the data. Then draw a reasonable model.
2. Estimate the band's profit from selling all 500 CDs.
3. If the band loses \$975, estimate how many CDs will have been sold.
4. Estimate the P-intercept of the model. What does it mean in this situation?
5. Estimate the n-intercept of the model. What does it mean in this situation?

Solution

1. The scattergram is shown in Fig. 43 (the black points). Since the points lie on a line, we use the line as a model.
2. The blue arrows show that the input $n = 500$ leads to the output $P = 2000$. So, if the band sells 500 CDs, the profit will be \$2000.
3. A negative value of P represents a loss. The red arrows show that the output $P = -975$ originates from the input $n = 75$. So, if the band sells 75 CDs, it will lose \$975.

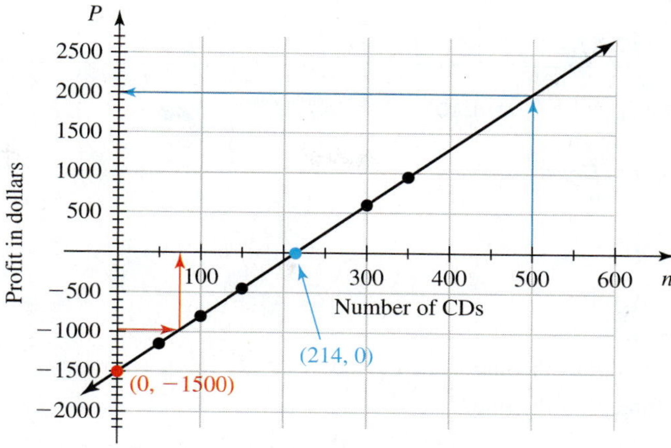

Figure 43 Scattergram and linear model of the CD situation

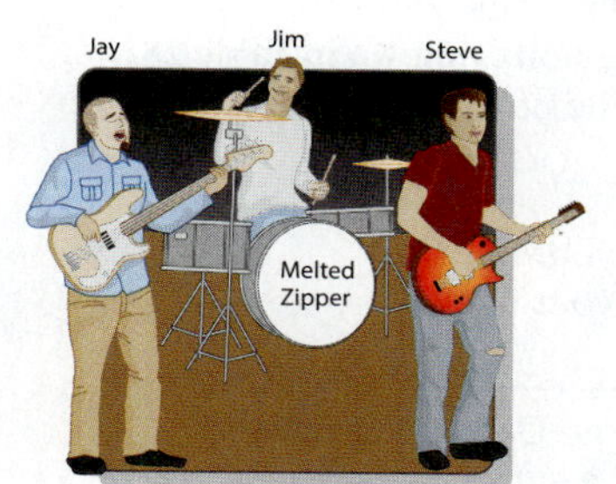

4. The linear model and the P-axis intersect at the point $(0, -1500)$. This means that the band will lose $1500 if no CDs are sold.
5. The linear model and the n-axis intersect at about the point $(214, 0)$. This means that the band will have a profit of about 0 dollars (and hence break even) if 214 CDs are sold.

It is impossible to do a perfect job of sketching models and estimating coordinates of points. Your results for homework exercises will likely be different from the answers provided near the end of this textbook. However, if you do a careful job, your results should be close to those in the book.

Group Exploration

Looking ahead: Approximately linearly related variables

Total amounts of overdraft fees collected by banks are shown in Table 26 for various years. Let A be the total amount (in billions of dollars) of overdraft fees collected in the year that is t years since 2000.

Table 26 Total Amount of Overdraft Fees Collected by Banks

Year	Amount of Overdraft Fees (billions of dollars)
2001	22.2
2003	27.1
2005	29.7
2007	34.1
2009	37.1

Source: *Moebs Services*

1. Create a scattergram of the data.
2. Draw a (straight) line that comes close to all of the data points in your scattergram.
3. Use your line to estimate the total amount of fees collected in 2008.
4. Use your line to estimate when the total amount of fees collected was $32 billion.
5. What is the A-intercept of the linear model? What does it mean in this situation?
6. On July 1, 2010, new overdraft rules went into effect for debit and ATM card users. Now a bank must ask permission to apply its standard overdraft practices, including charging overdraft fees. For customers who do not give permission, their transactions will be declined if there is not enough money in their accounts. Use your line to predict the total amount of fees collected in 2011. Compare your result with $31.6 billion, which is the actual amount. Explain why it is not surprising that your result is an overestimate.

Tips for Success Get in Touch with Classmates

It is wise to exchange phone numbers and e-mail addresses with some classmates. If any of you has to miss class, then you have someone to contact to find out what you missed and what homework was assigned.

Homework 1.3

For extra help ▶ MyMathLab®

For Exercises 1–6, refer to Fig. 44.

1. Find y when $x = -2$.

2. Find y when $x = 4$.

3. Find x when $y = -2$.

4. Find x when $y = 4$.

5. What is the x-intercept of the line?

6. What is the y-intercept of the line?

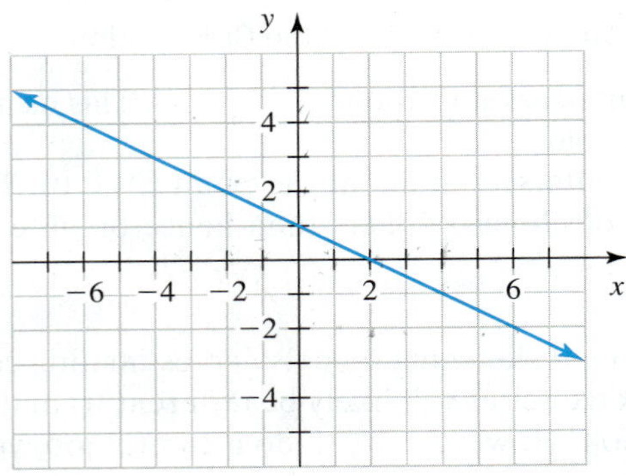

Figure 44 Exercises 1–6

For Exercises 7–12, refer to Fig. 45.

7. Find y when $x = -3$.

8. Find y when $x = 6$.

9. Find x when $y = -3$.

10. Find x when $y = 0$.

11. What is the y-intercept of the line?

12. What is the x-intercept of the line?

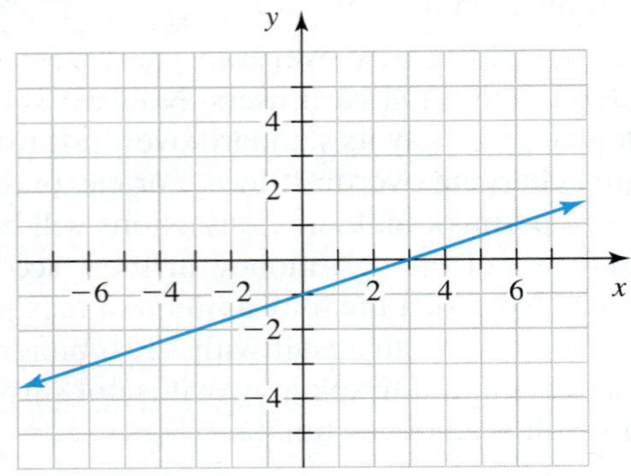

Figure 45 Exercises 7–12

13. Some ordered pairs are listed in Table 27.
  a. Create a scattergram of the points shown in Table 27.
  b. Draw a line that contains the points in your scattergram.
  c. Find y when $x = 3$.
  d. Find x when $y = 6$.
  e. What is the y-intercept of your line?
  f. What is the x-intercept of your line?

Table 27 Some Ordered Pairs

x	y
2	20
4	16
6	12
8	8
10	4

14. Some ordered pairs are listed in Table 28.
  a. Create a scattergram of the points shown in Table 28.
  b. Draw a line that contains the points in your scattergram.
  c. Find y when $x = 4$.
  d. Find x when $y = 17$.
  e. What is the y-intercept of your line?
  f. What is the x-intercept of your line?

Table 28 Some Ordered Pairs

x	y
−6	4
−2	8
2	12
6	16
10	20

15. Water is steadily pumped out of a flooded basement. Let v be the volume of water (in thousands of gallons) that remains in the basement t hours after water began to be pumped. A linear model is shown in Fig. 46.

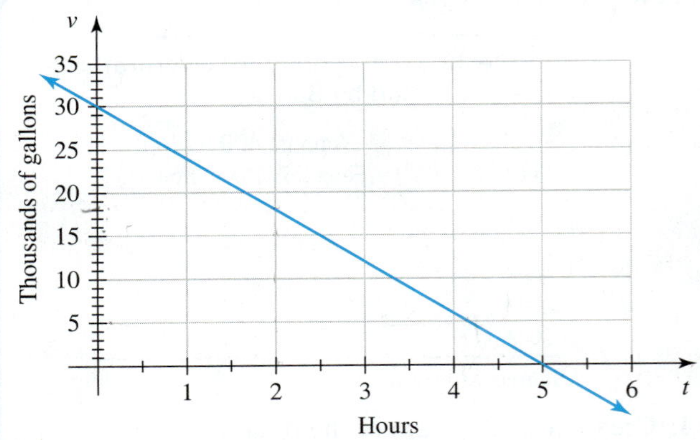

Figure 46 Linear model—Exercise 15

  a. How much water is in the basement after 2 hours of pumping?
  b. After how many hours of pumping will 5 thousand gallons remain in the basement?
  c. How much water was in the basement before any water was pumped out?
  d. After how many hours of pumping will all the water be pumped out of the basement?

16. Let B be the balance (in dollars) of a student's checking account at t months since the student opened the account. A linear model is shown in Fig. 47.

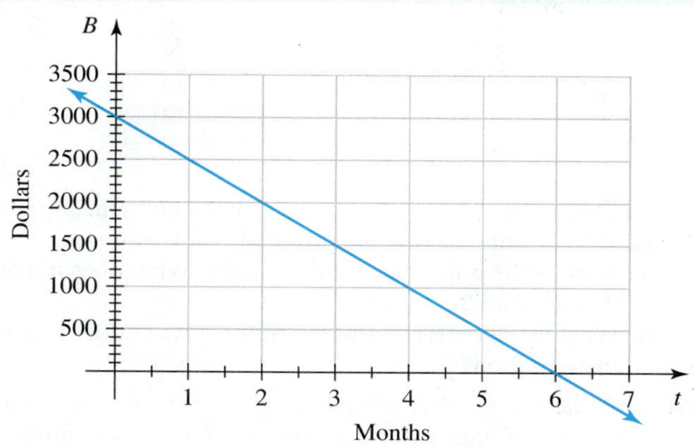

Figure 47 Linear model—Exercise 16

a. What was the balance 3 months after the student opened the account?
b. When was the balance $500?
c. What is the B-intercept of the model? What does it mean in this situation?
d. What is the t-intercept of the model? What does it mean in this situation?

17. A scattergram for a situation is graphed in Fig. 48. Is there a line that is a reasonable model of this situation? If yes, sketch the line. If no, explain why not.

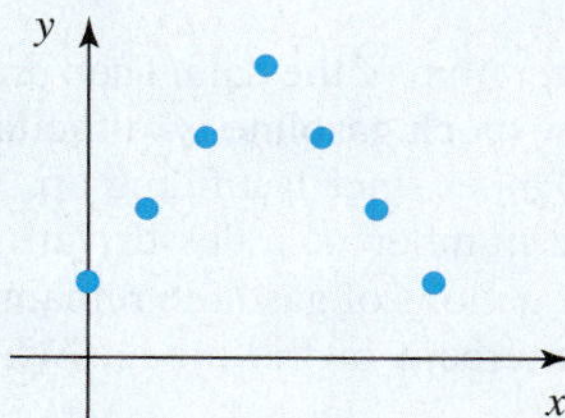

Figure 48 Scattergram—Exercise 17

18. A scattergram for a situation is graphed in Fig. 49. Is there a line that is a reasonable model of this situation? If yes, sketch the line. If no, explain why not.

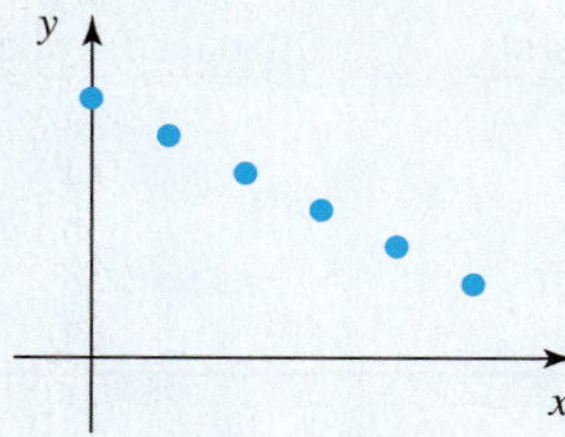

Figure 49 Scattergram—Exercise 18

19. Some ordered pairs are listed in Table 29.
a. Create a scattergram of the data shown in Table 29.
b. Is there a linear relationship between x and y? Explain.

Table 29 Some Ordered Pairs

x	y
0	10
5	5
8	2
10	1
12	2
13	5
14	10

20. Some ordered pairs are listed in Table 30.

Table 30 Some Ordered Pairs

x	y
2	1
3	6
4	9
5	10
6	9
7	6
8	1

a. Create a scattergram of the data shown in Table 30.
b. Is there a linear relationship between x and y? Explain.

21. Let d be the distance traveled (in miles) after a student has driven for t hours (not counting pit stops). Some pairs of values of t and d are shown in Table 31.

Table 31 Times and Distances for a Car

t (hours)	d (miles)
0	0
1	60
2	120
3	180
4	240

a. Create a scattergram of the data. Then draw a linear model.
b. Estimate how far the student has traveled in 2.5 hours.
c. Estimate how long it took the student to travel 210 miles.

22. A student works part time at the college bookstore. Let p be the student's pay (in dollars) for working t hours. Some pairs of values of t and p are shown in Table 32.

Table 32 Pay for Working t Hours

t (hours)	p (dollars)
0	0
5	40
10	80
15	120
20	160

a. Create a scattergram of the data. Then draw a linear model.
b. Estimate the student's pay for working 7 hours.
c. Estimate the number of hours the student must work to earn $96.

23. Let E be a college's enrollment (in thousands of students) at t years since the college began. Some pairs of values of t and E are shown in Table 33.

Table 33 Ages and Enrollments of a College

t (years)	E (thousands of students)
0	5
1	7
2	9
3	11
4	13

a. Create a scattergram of the data. Then draw a linear model.
b. Predict the enrollment when it has been 6 years since the college opened.
c. Predict when the enrollment will reach 19 thousand students.

24. Let s be a person's salary (in thousands of dollars) after he has worked t years at a company. Some pairs of values of t and s are shown in Table 34.

Table 34 Years Worked and Salary

t (years)	s (thousands of dollars)
0	20
2	24
4	28
6	32
8	36

a. Create a scattergram of the data. Then draw a linear model.
b. Estimate the person's salary after he has worked 5 years at the company.
c. Estimate when the person's salary was $34 thousand.
d. What is the s-intercept of the model? What does it mean in this situation?

25. Let v be the value (in dollars) of a company's stock at t years since 2005. Some pairs of values of t and v are shown in Table 35.

Table 35 Values of a Stock

t (years)	v (dollars)
1	28
2	24
4	16
6	8
7	4

a. Create a scattergram of the data. Then draw a linear model.
b. Estimate when the value of the stock was $12.
c. What is the t-intercept of the model? What does it mean in this situation?
d. What is the v-intercept of the model? What does it mean in this situation?

26. Let p be the annual profit (in millions of dollars) of a company at t years since 2005. Some pairs of values of t and p are shown in Table 36.

Table 36 Annual Profits of a Company

t (years)	p (millions of dollars)
1	20
2	18
3	16
5	12
6	10

a. Create a scattergram of the data. Then draw a linear model.
b. Predict when the annual profit will be $2 million.
c. What is the p-intercept of the model? What does it mean in this situation?
d. What is the t-intercept of the model? What does it mean in this situation?

27. Let g be the number of gallons of gasoline that remain in a car's gasoline tank after the car has been driven d miles since the tank was filled. Some pairs of values of d and g are shown in Table 37.

Table 37 Miles Traveled and Gallons of Gasoline

d (miles)	g (gallons)
40	11
80	9
120	7
160	5
200	3
240	1

a. Create a scattergram of the data. Then draw a linear model.
b. Estimate how much gasoline is in the tank after the driver has gone 140 miles since last filling up.
c. Estimate the number of miles driven since the tank was last filled if 2 gallons of gasoline remain in the tank.
d. Find the d-intercept of the model. What does it mean in this situation?
e. Find the g-intercept of the model. What does it mean in this situation?

28. Let v be the value (in thousands of dollars) of a car when it is t years old. Some pairs of values of t and v are listed in Table 38.

Table 38 Ages and Values of a Car

t (years)	v (thousands of dollars)
1	18
3	14
5	10
7	6
9	2

a. Create a scattergram of the data. Then draw a linear model.
b. Estimate the age of the car when it is worth $4 thousand.
c. Estimate the value of the car when it is 6 years old.
d. What is the v-intercept of the model? What does it mean in this situation?
e. What is the t-intercept of the model? What does it mean in this situation?

29. Let r be the annual revenue (in millions of dollars) of a company at t years since 2005. Some pairs of values of t and r are shown in Table 39.

Table 39 Annual Revenues of a Company

t (years)	r (millions of dollars)
0	8
1	11
3	17
5	23
6	26

a. Create a scattergram of the data. Then draw a linear model.
b. Predict the revenue in 2015.
c. Estimate when the annual revenue was \$14 million.
d. What is the r-intercept of the model? What does it mean in this situation?

30. Let v be the value (in dollars) of a company's stock at t years since 2005. Some pairs of values of t and v are shown in Table 40.

Table 40 Values of a Stock

t (years)	v (dollars)
0	15
2	19
4	23
5	25
6	27

a. Create a scattergram of the data. Then draw a linear model.
b. Estimate the value of the stock in 2008.
c. Predict when the value of the stock will be \$35.
d. What is the v-intercept of the model? What does it mean in this situation?

31. Let a be the altitude (in thousands of feet) of an airplane at t minutes since the airplane began its descent. Some pairs of values of t and a are shown in Table 41.

Table 41 Altitudes of an Airplane

t (minutes)	a (thousands of feet)
0	36
5	30
10	24
15	18
20	12

a. Create a scattergram of the data. Then draw a linear model.
b. Use your model to estimate the airplane's altitude 12 minutes after it began its descent.
c. Use your model to estimate when the airplane will reach the ground.
d. Assume that your line does a good job of modeling the airplane's descent up until the last 2 thousand feet, at which point the airplane then descends at a slower rate than before. Is your estimate in part (c) an underestimate or an overestimate? Explain.

32. Let a be the altitude (in feet) of a hot-air balloon after the air in the balloon is released for t minutes. Some pairs of values of t and a are shown in Table 42.

Table 42 Altitudes of a Hot-Air Balloon

t (minutes)	a (feet)
0	1800
1	1600
3	1200
4	1000
6	600

a. Create a scattergram of the data. Then draw a linear model.
b. Estimate the balloon's altitude after air has been released for 5 minutes.
c. Estimate when the balloon will reach the ground.
d. Assume that your line does a good job of modeling the balloon's descent up until the last 400 feet, at which point the balloon then descends at a faster rate than before. Is your estimate in part (c) an underestimate or an overestimate? Explain.

Concepts

33. Let n be the number (in thousands) of cocaine dealers arrested in the year that is t years since 2000. Some pairs of values of t and n are shown in Table 43.

Table 43 Numbers of Cocaine Dealers Arrested

t (years)	n (thousands)
3	11.4
4	12.0
5	13.0
6	13.1
7	12.9
8	12.2
9	11.7
10	10.7

Source: *U.S. Drug Enforcement Administration*

a. Create a scattergram of the data.
b. Are the variables t and n linearly related? Explain.

34. Let p be the percentage of flights that are delayed at t years since 2000. Some pairs of values of t and p are shown in Table 44.

Table 44 Percentages of Flights That Are Delayed

t (years)	p (percent)
3	15
4	22
5	24
6	19
7	24
8	25
9	20
10	19
11	20
12	15

Source: *Bureau of Transportation Statistics*

a. Create a scattergram of the data.
b. Are the variables t and p linearly related? Explain.

35. A student says the y-intercept of the ordered pair $(2, 5)$ is 5. Is the student correct? Explain.
36. A student says the x-intercept of the ordered pair $(-3, 4)$ is -3. Is the student correct? Explain.
37. A student says the x-intercept of a line is $(0, 2)$. Is the student correct? Explain.
38. A student says the y-intercept of a line is $(5, 0)$. Is the student correct? Explain.
39. A student says the x-intercept of a line is 5. Is the student correct? Explain.
40. Are there any lines for which the x-intercept is the same point as the y-intercept? If yes, sketch such a line, and what is that point? If no, explain why not.
41. Sketch three distinct lines that all have the same x-intercept.
42. Sketch three distinct lines that all have the same y-intercept.
43. a. Sketch a nonvertical line in a coordinate system. Find any outputs for the given input. State how many outputs there are for that single input.
    i. the input 2
    ii. the input 4
    iii. the input -3
    b. For your line, a single input leads to how many outputs? Explain.
    c. For *any* nonvertical line, a single input leads to how many outputs? Explain.
44. Explain why the x-coordinate of a y-intercept of a line is 0.
45. Give an example of a linear model other than the ones in the textbook.
46. In your own words, describe the meaning of *linear model*. (See page 9 for guidelines on writing a good response.)
47. Describe how to find a linear model of a situation and how to use the model to make estimates and predictions. (See page 9 for guidelines on writing a good response.)

1.4 Approximate Linear Relationships

Objectives

» Know the meaning of *approximately linearly related.*

» Use a linear model to make estimates and predictions.

» Find errors in estimations.

» Know the meaning of *interpolate, extrapolate,* and *model breakdown.*

» Modify a model.

Table 45 Average Ticket Prices for Top-50-Grossing Concert Tours

Year	Average Ticket Price (dollars)
1998	33
2001	47
2004	59
2008	67
2011	85

Source: *Pollstar*

In this section, we will use a line to model a situation in which data points lie close to the line, but not necessarily on the line.

Modeling when Variables Are Approximately Linearly Related

Example 1 Using a Line to Model Data

The average ticket prices for the top-50-grossing concert tours are shown in Table 45 for various years. Let p be the average ticket price (in dollars) at t years since 1995. Sketch a scattergram of the data, and draw a line that comes close to the points of the scattergram.

Solution

First, we list values of t and p in Table 46 (on the next page). For example, $t = 3$ represents 1998, because 1998 is 3 years after 1995; and $t = 6$ represents 2001, because 2001 is 6 years after 1995.

Next, we sketch a scattergram in Fig. 50. It makes sense to think of p as the dependent variable, so we let the vertical axis be the p-axis. Since t is the independent variable, the horizontal axis is the t-axis.

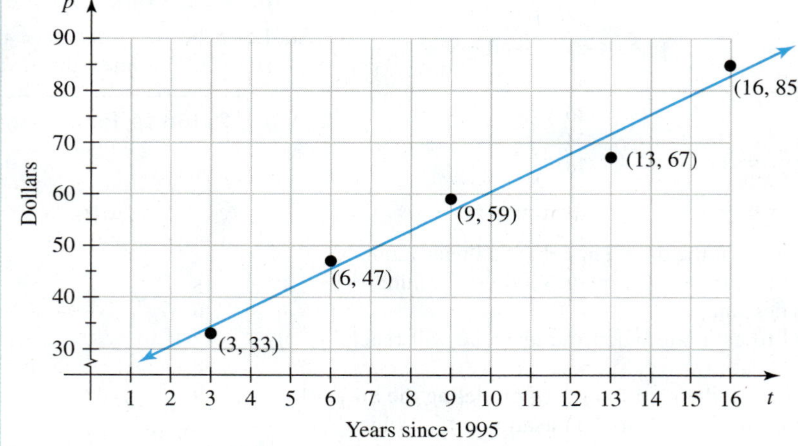

Figure 50 Average ticket price scattergram and model

Then we sketch a line that comes close to points of the scattergram (see Fig. 50).

Table 46 Using Values of t to Stand for the Years

Number of Years since 1995 t	Average Ticket Price (dollars) p
3	33
6	47
9	59
13	67
16	85

The line needn't contain any of the points, but it should come close to all of them. Figure 51 shows that many such lines are possible.

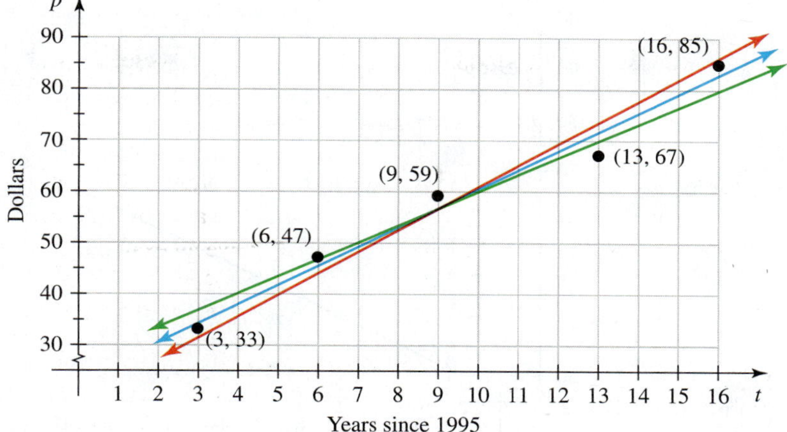

Figure 51 A few of the many reasonable linear models

In Fig. 50, we sketched a line that describes the average-ticket-price situation. However, this description is not exact. For example, the line does not describe exactly what happened in the years 1998, 2001, 2004, 2008, or 2011, because the line does not contain any of the data points. However, the line does come pretty close to these data points, so it suggests pretty good approximations for those years.

If the points in a scattergram of data lie close to (or on) a line, we say the relevant variables are **approximately linearly related.** For the concert tour situation, the variables t and p are approximately linearly related.

Using a Linear Model to Make Estimates and Predictions

Since all of the ticket price data points lie close to our linear model in Fig. 50, it seems reasonable that data points for the years between 1998 and 2011 that are not shown in Table 45 might also lie close to the line. Similarly, it is reasonable that data points for at least a few years before 1998 and for at least a few years after 2011 might also lie near the line.

Example 2 Making Estimates and a Prediction

1. Use the linear model shown in Fig. 50 to estimate the average ticket price in 2000.
2. Use the linear model to estimate the average ticket price in 2010.
3. Use the model to predict in which year the average ticket price will be $105.

Solution

Table 47 Average Ticket Prices

Year	Average Ticket Price (dollars)
1998	33
2001	47
2004	59
2008	67
2011	85

1. The year 2000 corresponds to $t = 5$, because $2000 - 1995 = 5$. The blue arrows in Fig. 52 (on the next page) show that the input $t = 5$ leads to the approximate output $p = 42$. So, according to the model, the average ticket price in 2000 was approximately $42.

   We verify our work by checking that our result is consistent with the values shown in Table 47. Since the average ticket price was $33 in 1998 and $47 in 2001, it follows that the average ticket price in 2000 probably would be between $33 and $47, which checks with our result of $42.
2. The year 2010 corresponds to $t = 15$, because $2010 - 1995 = 15$. The red arrows in Fig. 52 show that the input $t = 15$ leads to the approximate output $p = 79$. So, according to the model, the approximate average ticket price in 2010 was $79. This result is consistent with the values in Table 47.
3. The green arrows in Fig. 52 show that the output $p = 105$ originates from the approximate input $t = 22$. So, according to the linear model, the average ticket price in about $1995 + 22 = 2017$ will be $105.

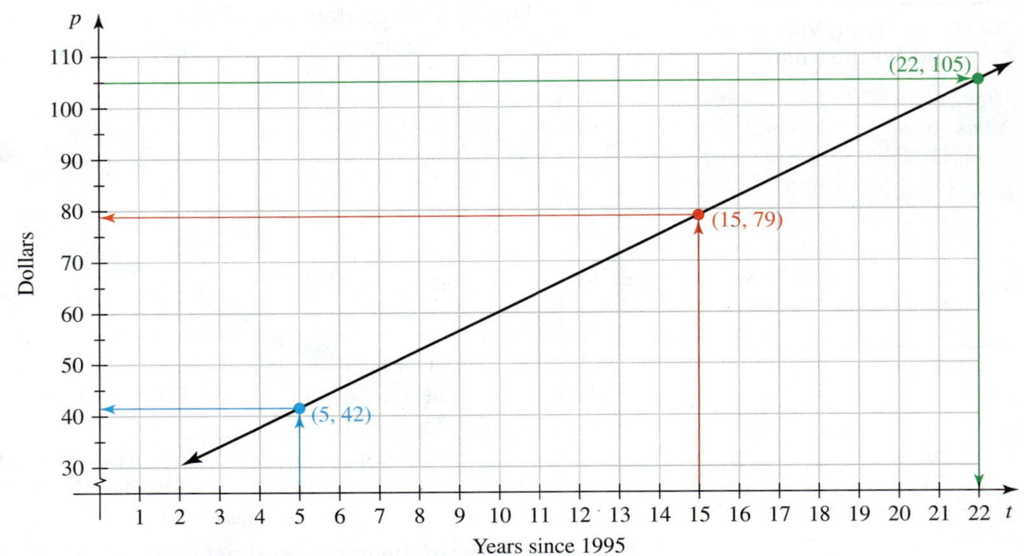

Figure 52 Average-ticket-price model

Since the average ticket price in 2011 was \$85 and average ticket prices have been increasing, it follows that the model would predict that the average ticket price sometime *after* 2011 would be \$105, which checks with our result of 2017.

We create a scattergram of data to determine whether the relevant variables are approximately linearly related. If so, we draw a line that comes close to the data points and use the line to make estimates and predictions.

WARNING

It is a common error to try to find a line that contains the greatest number of points. However, our goal is to find a line that comes close to *all* of the data points. For example, even though model 1 in Fig. 53 does not contain any of the data points shown, it fits the complete set of data points much better than does model 2, which contains three data points.

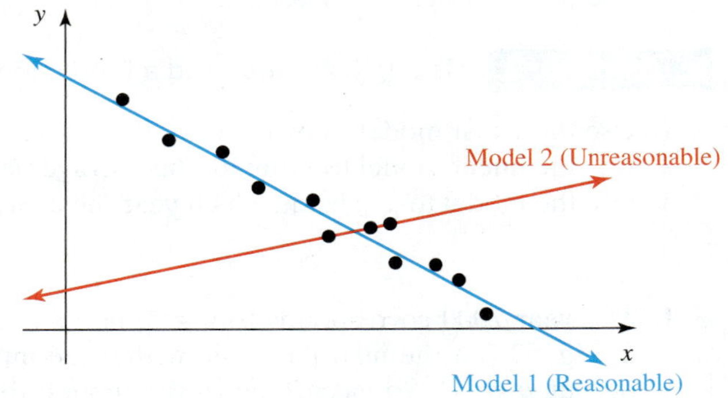

Figure 53 Comparing the fit of two models

Errors in Estimations

WARNING

It is also a common error to confuse the meaning of data points and points that lie on a linear model. Data points are accurate descriptions of an authentic situation. Points on a model may or may not be accurate descriptions.

For example, the data in Table 47 are accurate values for the average ticket prices for the given years. For the linear model in Fig. 50 (page 36), some points on the line describe the situation well, but some points on the line do not. The advantage of using the linear model is that we can estimate the average ticket price for years other than those in Table 47.

The **error** in an estimate is the amount by which the estimate differs from the actual value. For an overestimate, the error is positive. For an underestimate, the error is negative. If the estimate is equal to the actual value, then the error is 0.

Example 3 Calculating Errors

1. In Example 2, we estimated that the average ticket price was \$42 in 2000. The actual average ticket price was \$45. Calculate the error in the estimate.
2. In Example 2, we estimated that the average ticket price was \$79 in 2010. The actual average ticket price was \$75. Calculate the error in the estimate.

Solution

1. Since $45 - 42 = 3$ and we underestimated the actual price, the error is -3 dollars.
2. Since $79 - 75 = 4$ and we overestimated the actual price, the error is 4 dollars.

By viewing the average-ticket-price scattergram and model in the same coordinate system, we can see why our estimate of the average ticket price in 2000 is an underestimate. Since the linear model at (5, 42) is *below* the data point (5, 45), the model *underestimates* the average ticket price in 2000 (see Fig. 54).

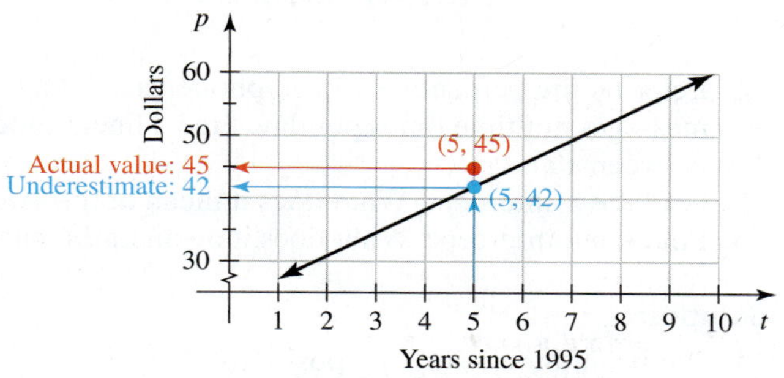

Figure 54 Comparing the data point and the model for 2000

We can also see why our estimate of the average ticket price in 2010 is an overestimate. Since the linear model at (15, 79) is *above* the data point (15, 75), the model *overestimates* the average ticket price in 2010 (see Fig. 55).

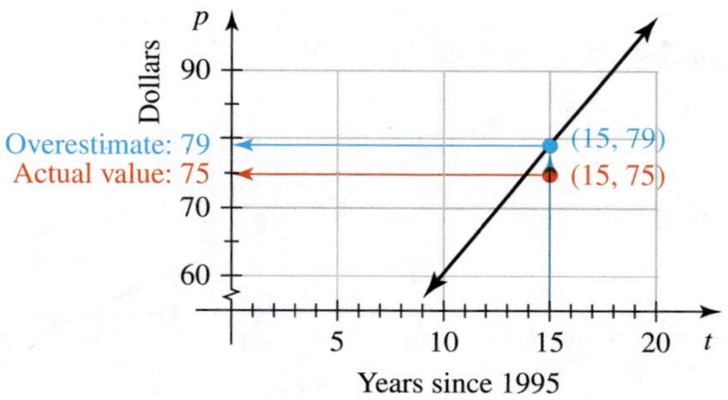

Figure 55 Comparing the data point and the model for 2010

Underestimates and Overestimates

Suppose that an independent variable t and a dependent variable p are approximately linearly related. Then

- If a linear model is below a data point (a, b), the model underestimates the value of p when $t = a$.
- If a linear model is above a data point (a, b), the model overestimates the value of p when $t = a$.

Intercepts of a Model and Model Breakdown

In Example 4, we will find the intercepts of a linear model. By doing so, we will see that even if a model describes *known* data well, the model may not describe the situation well for all values of the independent variable.

Example 4 Intercepts of a Model; Model Breakdown

The percentages of cell phone users who send or receive text messages multiple times per day are shown in Table 48 for various age groups.

Table 48 Percentages of Cell Phone Users Who Send or Receive Text Messages Multiple Times per Day

Age Group (years)	Age Used to Represent Age Group (years)	Percent
18–24	21.0	76
25–34	29.5	63
35–44	39.5	42
45–54	49.5	37
55–64	59.5	17

Source: *Edison Research and Arbitron*

1. Let p be the percentage of cell phone users at age a years who send or receive text messages multiple times per day. Find a linear model that describes the relationship between a and p.
2. Find the p-intercept. What does it mean in this situation?
3. Find the a-intercept. What does it mean in this situation?

Solution

1. We begin by viewing the positions of the points in the scattergram (see Fig. 56). It appears a and p are approximately linearly related, so we sketch a line that comes close to the data points.

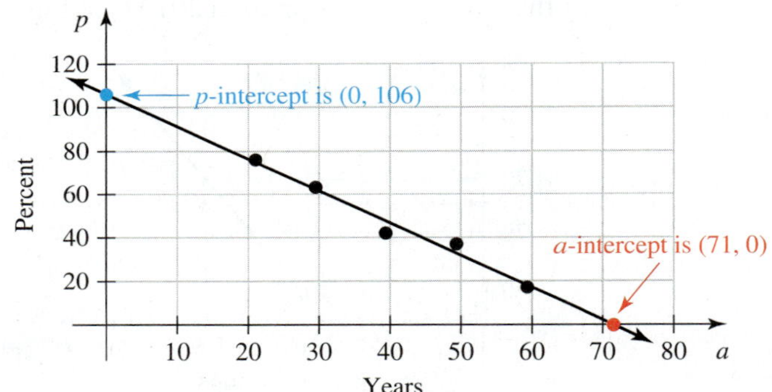

Figure 56 Intercepts of the text message model

2. The p-intercept is $(0, 106)$, or $p = 106$, when $a = 0$. According to the model, 106% of newborns who use cell phones send or receive text messages multiple times per day. Our model gives a false estimate for two reasons: Percentages cannot be larger than 100% in this situation, and newborns cannot send or receive text messages.
3. The a-intercept is $(71, 0)$, or $p = 0$, when $a = 71$. According to the model, no 71-year-old cell phone users send or receive text messages multiple times per day. This is a false estimate. A little research would show some 71-year-old cell phone users send or receive text messages multiple times per day.

To draw the text message model in Fig. 56, we used a scattergram consisting of data points representing various ages from 21.0 years to 59.5 years. In Fig. 57, we draw that portion of the model in blue, and we draw the rest of the model in red. When we use the blue portion of the model to make estimates, we are performing *interpolation*. When we use the red portions of the model, we are performing *extrapolation*.

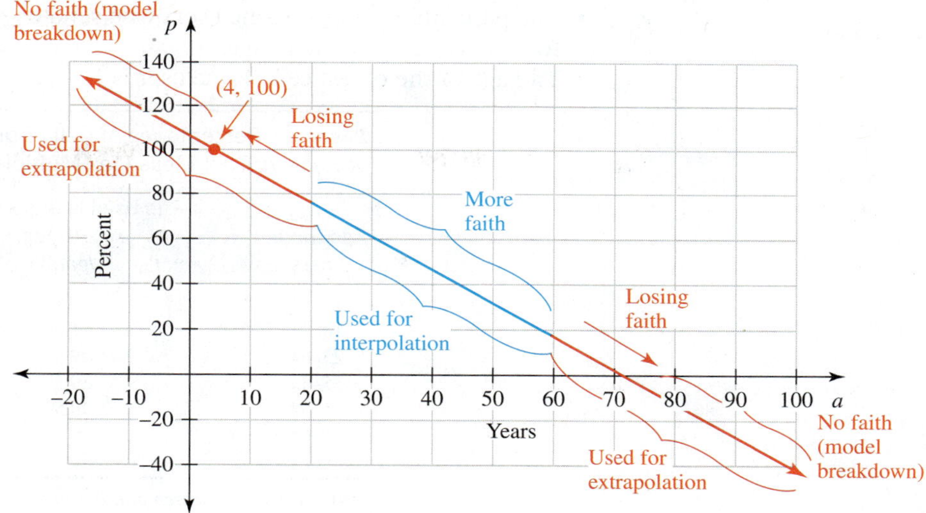

Figure 57 Interpolation versus extrapolation

> **Definition** Interpolation, extrapolation
>
> For a situation that can be described by a model whose independent variable is t,
>
> - We perform **interpolation** when we use a part of the model whose t-coordinates are between the t-coordinates of two data points.
> - We perform **extrapolation** when we use a part of the model whose t-coordinates are not between the t-coordinates of any two data points.

Although we could get large errors from interpolating with the text message model, we have more faith in our results from interpolating than from extrapolating. That's because the blue portion of the model comes close to several known data points, whereas we have no idea whether the red portion of the model comes close to *any* data points.

When we extrapolate, our faith declines more and more as we stray farther and farther from the blue portion of the text message model. In fact, we have no faith in the portion of the model from age 71 years on, because the model estimates nonpositive percentages of adults for these ages. We say *model breakdown* has occurred from age 71 years on. Similarly, model breakdown has occurred for ages less than 4 years, because the model estimates percentages greater than 100% for these ages.

> **Definition** Model breakdown
>
> When a model gives a prediction that does not make sense or an estimate that is not a good approximation, we say **model breakdown** has occurred.

When model breakdown occurs, it is time to modify our model or possibly rethink our modeling process. A different model might give more reasonable predictions. It could be helpful to gather more data to check our choice of model.

Example 5 Modifying a Model

Additional research yields the data shown in the first and last rows of Table 49. Use this data and the following assumptions to modify the model we found in Example 4:

- Children 3 years old and younger do not send or receive text messages multiple times per day.

- The percentage of cell phone users who send or receive text messages levels off at 5% for users over 80 years in age.
- The age of the oldest cell phone user is 116 years.

Table 49 Percentages of Cell Phone Users Who Send or Receive Text Messages Multiple Times per Day

Age Group (years)	Age Used to Represent Age Group (years)	Percent
12–17	14.5	75
18–24	21.0	76
25–34	29.5	63
35–44	39.5	42
45–54	49.5	37
55–64	59.5	17
over 64	70.0	7

Source: *Edison Research and Arbitron*

Solution

Recall that p is the percentage of cell phone users at age a years who send or receive multiple text messages per day. We sketch a scattergram of the data in Table 49, and taking into account the three assumptions, we draw a model that comes close to the data points (see Fig. 58).

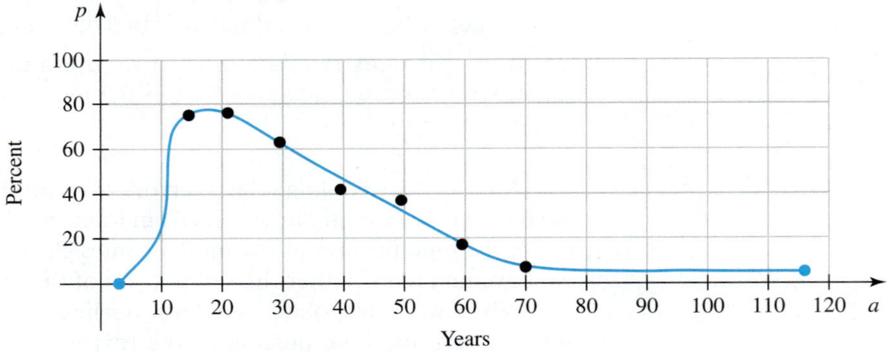

Figure 58 Modified text message model

In Section 1.3, we discussed why your estimates and predictions in the homework exercises will likely be different from the answers given near the end of this textbook. Here we note another reason this will likely happen: If variables are approximately linearly related, there will be many reasonable linear models to choose from. However, if you do a careful job, your results should be close to those in the textbook.

Group Exploration

Looking ahead: Expressions

1. An instructor adds 5 bonus points to each student's test score. For the given original test score, find the original test score plus 5 points.
   a. 74 points
   b. 88 points
   c. r points [**Hint:** Think about how you got your results for parts (a) and (b).]
2. A student gets paid $8 per hour for working at a music store. For the given number of hours worked, find the total amount of money earned (in dollars).
   a. 5 hours **b.** 9 hours **c.** t hours
3. In a lottery, some people won $200, which they will share equally. For the given number of people, find each person's share (in dollars).
   a. 4 people **b.** 10 people **c.** n people

Group Exploration

Identifying types of modeling errors

The interest rates for subsidized student loans are shown in Table 50 for various years.

Table 50 Interest Rates for Subsidized Student Loans

Years	Interest Rate (percent)
2008	6.8
2009	6.0
2010	5.7
2011	4.5
2012	3.4

Source: *New America Foundation*

Here you will explore possible causes of error for estimates and predictions based on a linear model for the interest rate data.

1. Let r be the interest rate (percent) for subsidized student loans at t years since 2005. Create a scattergram of the data.
2. Draw a line that comes close to the points in your scattergram.
3. Use your linear model to estimate the interest rate in 2012. What is the actual interest rate? Calculate the error in your estimate for 2012.
4. Use your linear model to predict the interest rate in 2020. Is this an accurate prediction? Explain.
5. Take another look at your sketch from Problem 2. Is the t-axis perfectly horizontal and the r-axis perfectly vertical? Are the scalings of both axes precise? Is your line straight? How might these considerations relate to the accuracy of an estimation or a prediction? Explain.
6. What are the coordinates of point S plotted in Fig. 59? Do you think you have found the correct first decimal place (tenths place) for these coordinates? How about the second decimal place?

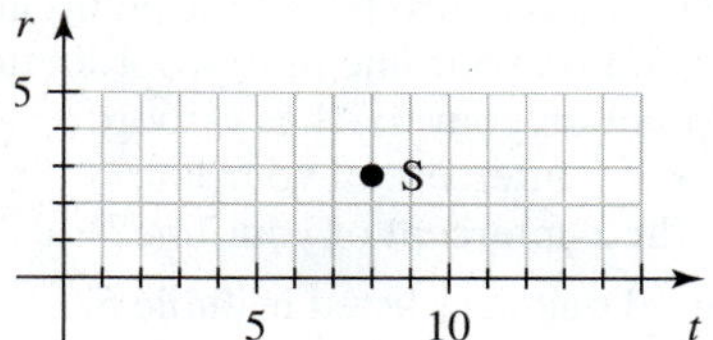

Figure 59 Problem 6

7. Problems 3–6 of this exploration suggest several possible causes of error for estimates and predictions based on a linear model. Describe the possible causes of error.

Tips for Success Practice Exams

When studying for an exam (or quiz), try creating your own exam to take for practice. To create your exam, select several homework exercises from each section on which you will be tested. Choose a variety of exercises that address concepts your instructor has emphasized. Your test should include many exercises that are moderately difficult and some that are challenging. Completing such a practice test will help you reflect on important concepts and pin down what types of problems you need to study more.

It is a good idea to work on the practice exam for a predetermined period. Doing so will help you get used to a timed exam, build your confidence, and lower your anxiety about the real exam.

If you are studying with another student, you can each create a test and then take each other's test. Or you can create a test together and each take it separately.

Homework 1.4

For extra help ▶ MyMathLab®  Watch the videos in MyMathLab  Download the MyDashboard App

1. Some ordered pairs are listed in Table 51.

Table 51 Some Ordered Pairs

x	y
1	13
3	9
5	8
7	4
9	2

a. Create a scattergram of the points shown in Table 51.
b. Are the variables x and y linearly related, approximately linearly related, or neither?
c. Draw a line that comes close to the points in your scattergram.
d. Which point on your line has x-coordinate 8?
e. Which point on your line has y-coordinate 6?
f. What is the y-intercept of your line?
g. What is the x-intercept of your line?

2. Some ordered pairs are listed in Table 52.

Table 52 Some Ordered Pairs

x	y
−8	−5
−5	−2
−2	0
0	5
3	7

a. Create a scattergram of the points shown in Table 52.
b. Are the variables x and y linearly related, approximately linearly related, or neither?
c. Draw a line that comes close to the points in your scattergram.
d. Which point on your line has x-coordinate -1?
e. Which point on your line has y-coordinate -3?
f. What is the y-intercept of your line?
g. What is the x-intercept of your line?

3. The total numbers of animal and plant species in the United States that are listed as endangered or threatened are shown in Table 53 for various years.

Table 53 Numbers of Endangered or Threatened Species

Year	Number of Species Listed
1985	384
1990	596
1995	962
2000	1244
2005	1264
2011	1967

Source: *U.S. Fish and Wildlife Service*

Let n be the number of species that are listed as endangered or threatened at t years since 1980.

a. Create a scattergram of the data.
b. Are the variables t and n linearly related, approximately linearly related, or neither? Explain.
c. Draw a line that comes close to the points in your scattergram.
d. Estimate when 800 species were listed.
e. Predict the number of species that will be listed in 2018.

4. The percentages of Americans who say there should be a ban on the possession of handguns are shown in Table 54 for various years.

Table 54 Percentages of Americans Who Say There Should Be a Ban on the Possession of Handguns

Year	Percent
2004	36
2005	35
2006	32
2007	30
2008	29
2009	28
2010	28
2011	26

Source: *The Gallup Organization*

Let p be the percentage of Americans who say there should be a ban on the possession of handguns at t years since 2000.

a. Create a scattergram of the data.
b. Are the variables linearly related, approximately linearly related, or neither? Explain.
c. Draw a line that comes close to the points in your scattergram.
d. Use your model to predict the percentage of Americans in 2018 who will say there should be a ban on the possession of handguns.
e. Use your model to predict when 20% of Americans will say there should be a ban on the possession of handguns.

5. Due to improved technology and public-service campaigns, the number of collisions at highway–railroad crossings per year has declined since 1992 (see Table 55).

Table 55 Numbers of Collisions at Highway–Railroad Crossings

Year	Number of Collisions (thousands)
1992	4.9
1995	4.6
2000	3.5
2005	3.1
2010	2.0
2011	2.0

Source: *Federal Railroad Administration*

Let n be the number of collisions (in thousands) for the year that is t years since 1990.

a. Create a scattergram of the data. Extend your t-axis and n-axis so you can make estimates about numbers of collisions before 1992 and predictions about future numbers of collisions.
b. Draw a line that comes close to the data points in your scattergram.
c. Use your linear model to estimate the number of collisions in 2007. Did you perform interpolation or extrapolation? Explain.
d. Use your linear model to predict in which year there will be 1.0 thousand collisions. Did you perform interpolation or extrapolation? Explain.
e. Find the n-intercept of your linear model. What does it mean in this situation?
f. Find the t-intercept. What does it mean in this situation? [**Note:** If model breakdown occurs, say so, say where, and explain why.]

6. Repeat Exercise 5, but let n be the number of collisions (in thousands) for the year that is t years *since 1985*. Which of your responses for this exercise are the same as those for Exercise 5? Explain why it makes sense that these responses are the same. Explain why it makes sense that the other responses are different.

7. The numbers of Internet users in the United States are shown in Table 56 for various years.

Table 56 Numbers of Internet Users in the United States

Year	Number of Users (millions)
2006	204
2007	212
2008	220
2009	228
2010	240

Source: *The Nielsen Company*

a. Let n be the number of Internet users (in millions) in the United States at t years since 2005. Create a scattergram of the data.
b. Sketch a line that comes close to the data points.
c. Find the n-intercept of your linear model. What does it mean in this situation? Did you perform interpolation or extrapolation? Explain.
d. Estimate by how much the number of Internet users is increasing per year.
e. Use your linear model to predict when everyone in the United States will be an Internet user. Assume the U.S. population is 315 million. Did you perform interpolation or extrapolation? Explain.

8. The temperature at which water boils (the *boiling point*) depends on elevation: The higher the elevation, the lower is the boiling point. At sea level, water boils at 212°F; at an elevation of 10,000 meters, water boils at about 151°F. Boiling points are listed in Table 57 for various elevations.
a. Let B be the boiling point (in degrees Fahrenheit) at an elevation of E thousand meters. Create a scattergram of the data.
b. Sketch a line that comes close to the data points.

Table 57 Boiling Points of Water

Elevation (in thousands of meters)	Boiling Point (°F)
0	212
1	205
2	200
5	181
10	151
15	123

c. Mount Everest, the highest mountain in the world, reaches 8850 meters at its peak. What is the boiling point of water at the peak? Did you perform interpolation or extrapolation? Explain.
d. We say water is *lukewarm* if its temperature is close to body temperature (about 98.6°F). At what elevation would boiling water feel lukewarm? Did you perform interpolation or extrapolation? Explain.

9. The percentages of Americans who believe marriages between same-sex couples should be recognized by the law as valid are shown in Table 58 for various age groups.

Table 58 Percentages of Americans Who Support Same-Sex Marriage

Age Group (years)	Age Used to Represent Age Group (years)	Percent
18–29	23.5	71
30–39	34.5	56
40–49	44.5	49
50–59	54.5	46
60–69	64.5	42
70–79	74.5	32
over 79	85.0	21

Source: *The Gallup Organization*

Let p be the percentage of Americans at age a years who believe marriages between same-sex couples should be recognized by the law as valid.
a. Create a scattergram of the data.
b. Draw a line that comes close to the data points.
c. Use your linear model to estimate at what age 65% of Americans believe same-sex couples should be recognized by the law as valid.
d. Find the p-intercept of your linear model. What does it mean in this situation?
e. Find the a-intercept of your linear model. What does it mean in this situation?

10. The percentages of Americans who currently have a personal profile page on a social networking website such as Facebook are shown in Table 59 for various age groups. Let p be the percentage of Americans at age a years who have a personal profile page.
a. Create a scattergram of the data.
b. Draw a line that comes close to the data points.
c. Use your linear model to estimate at what age 70% of Americans have a profile page.

Table 59 Percentages of Americans Who Currently Have a Personal Profile Page on a Social Networking Website

Age Group (years)	Age Used to Represent Age Group (years)	Percent
12–17	14.5	78
18–24	21.0	77
25–34	29.5	65
35–44	39.5	51
45–54	49.5	35
55–64	59.5	31
over 64	70.0	13

Source: *Edison Research and Arbitron*

d. Find the p-intercept of your linear model. What does it mean in this situation?

e. Find the a-intercept of your linear model. What does it mean in this situation?

11. The loudness of sound can be measured by using a *decibel scale*. Some examples of sounds at various sound levels are listed in Table 60.

Table 60 Examples of Sound Levels

Sound Level (decibels)	Example
0	Faintest sound heard by humans
20	Whisper
40	Inside a running car
60	Normal conversation
80	Noisy street corner
100	Soft-rock concert
120	Threshold of pain

Source: *From* Math and Music *by Trudi Hammel Garland and Charity Vaughan Kahn, p. 140. Copyright © 1993 Pearson Education, Inc. or its affiliates. Used by permission. All Rights Reserved.*

The sound level of music from a Pioneer MT-2000® stereo-CD-receiver system is controlled by the system's volume number. The sound levels of music for various volume numbers are shown in Table 61.

Table 61 Sound Levels of Music Played by a Stereo

Volume Number	Sound Level (decibels)
6	60
8	66
10	69
12	74
14	78
16	82
18	86
20	90

Source: *J. Lehmann*

Let S be the sound level (in decibels) for a volume number n.

a. Create a scattergram of the data in Table 61.

b. Draw a line that comes close to the points in your scattergram.

c. Use your model to estimate the sound level when the volume number is 19.

d. Use your model to estimate for what volume number the sound level is comparable to that of a noisy street corner (see Table 60).

12. The percentages of teens who have cell phones are shown in Table 62 for various ages.

Table 62 Percentages of Teens Who Have Cell Phones

Age (years)	Percent
13	54
14	68
15	68
16	73
17	79

Source: *JupiterResearch*

Let p be the percentage of teens at age a years who have cell phones.

a. Create a scattergram of the data.

b. Draw a line that comes close to the points in your scattergram.

c. Use your model to estimate the percentage of 18-year-old teens who have cell phones.

d. Use your model to estimate at what age all people have cell phones.

e. Find the a-intercept of the model. What does it mean in this situation?

13. The percentages of Americans who are satisfied with the way things are in the United States are shown in Table 63 for various years.

Table 63 Percentages of Americans Who Are Satisfied

Year	Percent
1992	21
1993	28
1994	33
1995	32
1996	39
1997	49
1998	60
1999	59

Source: *The Gallup Organization*

Let p be the percentage of Americans at t years since 1990 who are satisfied with the way things are.

a. Create a scattergram of the data in Table 63.

b. Draw a line that comes close to the data points.

c. Use your line to estimate the percentage of Americans who were satisfied in 2006.

d. Data for the years 2000–2011 are shown in Table 64. Create a scattergram of the data for the years 1992–2011.

e. Compute the error in the estimation for 2006 that you made in part (c). Explain why the error in your estimate is so large.

Table 64 Percentages of Americans Who Are Satisfied

Year	Percent
2000	60
2002	52
2004	43
2006	31
2008	20
2011	11

Source: *The Gallup Organization*

14. The percentages of Americans living below the poverty level are shown in Table 65 for various years.

Table 65 Percentages of Americans Living Below the Poverty Level

Year	Percent
1993	15.1
1994	14.5
1995	13.8
1996	13.7
1997	13.3
1998	12.7
1999	11.9
2000	11.3

Source: *U.S. Census Bureau*

Let p be the percentage of Americans living below the poverty level at t years since 1990.

a. Create a scattergram of the data in Table 65.

b. Draw a line that comes close to the data points.

c. Use your line to estimate the percentage in 2010.

d. Data for the years 2002–2010 are shown in Table 66. Create a scattergram of the data for the years 1993–2010.

Table 66 Percentages of Americans Living Below the Poverty Level

Year	Percent
2002	12.1
2004	12.7
2006	13.3
2008	13.2
2010	15.3

Source: *U.S. Census Bureau*

e. Compute the error in the estimation for 2010 that you made in part (c). Explain why the estimate is so inaccurate.

15. The annual profits of Alaska Air Group are shown in Table 67 for various years. Let p be the annual profit (in millions of dollars) at t years since 2002.

a. Without graphing, estimate the coordinates of the t-intercept for a line that comes close to the data points. What does that point mean in this situation? If you don't see how to estimate the coordinates, create a scattergram of the data first.

b. Without graphing, estimate the coordinates of the p-intercept for a line that comes close to the data points. What does that point mean in this situation? If you don't see how to estimate the coordinates, create a scattergram of the data first.

Table 67 Annual Profits of Alaska Air Group

Year	Annual Profit (millions of dollars)
2002	−68
2003	−30.8
2005	55.0
2007	91.6
2009	88.7
2010	262.6

Source: *Alaska Air Group*

16. The *windchill* (or *windchill factor*) is a measure of how cold you feel as a result of being exposed to wind. Table 68 provides some data on windchills for various temperatures when the wind speed is 10 mph.

Table 68 Windchills for a 10-mph Wind

Temperature (°F)	Windchill (°F)
−15	−35
−10	−28
−5	−22
5	−10
10	−4
15	3
20	9
25	15

Source: *National Weather Service Forecast Office*

Let w be the windchill (in degrees Fahrenheit) corresponding to a temperature of t degrees Fahrenheit when the wind speed is 10 mph.

a. Without graphing, estimate the coordinates of the t-intercept for a line that comes close to the data points. What does that point mean in this situation? If you don't see how to estimate the coordinates, create a scattergram of the data first.

b. Without graphing, estimate the coordinates of the w-intercept for a line that comes close to the data points. What does that point mean in this situation? If you don't see how to estimate the coordinates, create a scattergram of the data first.

17. The numbers of ride-related injuries at fixed-site amusement parks are shown in Table 69 for various years.

Table 69 Numbers of Ride-Related Injuries at Fixed-Site Amusement Parks

Years	Number of Injuries (thousands)
2002	2.5
2003	2.0
2004	1.6
2005	1.8
2006	1.8
2007	1.7
2008	1.5
2009	1.2
2010	1.3

Source: *National Safety Council*

Let n be the number (in thousands) of ride-related injuries at fixed-site amusement parks in the year that is t years since 2000. A scattergram of the data and a linear model are sketched in Fig. 60.

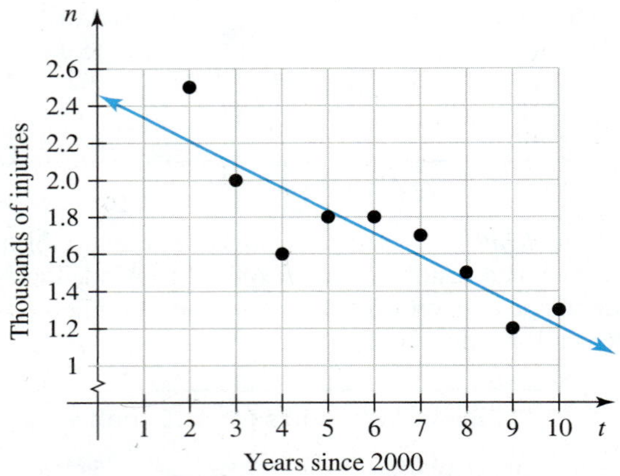

Figure 60 Injury scattergram and model — Exercises 17 and 18

a. Use the linear model to estimate the number of ride-related injuries in 2004.

b. What was the actual number of ride-related injuries in 2004?

c. Is your result in part (a) an underestimate or an overestimate? Explain how you can tell this from the graph of the scattergram and the sketch of the model. Calculate the error in the estimate.

18. Refer to Exercise 17, including Table 69 and Fig. 60.

a. Use the linear model to estimate the number of ride-related injuries in 2002.

b. What was the actual number of ride-related injuries in 2002?

c. Is your result in part (a) an underestimate or an overestimate? Explain how you can tell this from the graph of the scattergram and the sketch of the model. Calculate the error in the estimate.

Concepts

19. Suppose that an independent variable t and a dependent variable p are approximately linearly related. If a data point (c, d) is below a linear model, does the model underestimate or overestimate the value of p when $t = c$? Explain.

20. Suppose that an independent variable t and a dependent variable p are approximately linearly related. If a linear model is below a data point (c, d), does the model underestimate or overestimate the value of p when $t = c$? Explain.

21. When modeling a situation in which the variables are approximately linearly related, different students may all do good work yet not get the same results. Draw a scattergram and at least two reasonable linear models to show how this is possible.

22. Which is more desirable, finding a linear model that contains several, but not all, data points or finding a linear model that does not contain any data points but comes close to all data points? Include in your discussion some sketches of scattergrams and linear models.

23. Compare the meaning of *linearly related* with that of *approximately linearly related.*

24. A student comes up with a shortcut for modeling a situation. Instead of plotting all of the given data points, the student plots only two of the data points and draws a line that contains the two chosen points. Give an example to illustrate what can go wrong with this shortcut.

25. A person collects data by doing research. If the data points lie exactly on a line, will all points on the line describe the situation exactly? Explain.

26. Let t be the number of years since 2010. Some students believe $t = -2$ is an example of model breakdown because time can't be negative. Is this correct? Explain.

27. When using a line to model a situation, do we usually have more faith in a result obtained by interpolation or extrapolation? Explain.

28. Describe how to find a linear model for a situation and how to use the model to make estimates and predictions.

Taking it to the Lab

Climate Change Lab

Many scientists are greatly concerned that the average temperature of the surface of Earth has increased since 1900 (see Table 70).

Although it may not seem that the average temperatures shown in Table 70 have increased much, many scientists believe an increase as small as 3.6°F could be a dangerous climate change.* Global warming would cause the extinction of plants and animals, lead to severe water shortages, create more extreme weather events, increase the number of heat-related illnesses and deaths, and melt glaciers, which would raise ocean levels and, thus, submerge coastlands.

Despite these alarming predictions, not all experts are concerned. Robert Mendelsohn, an environmental economist at Yale University, argues that "global warming will increase agricultural production in the northern half of the United States" and that "the southern half will be able to maintain its current level of production." Even from

*"Meeting the Climate Challenge: Recommendations of the International Climate Change Taskforce," The Institute for Public Policy Research/The Center for American Progress/The Australian Institute, January 2005.

Table 70 Average Surface Temperatures of Earth

Year	Average Temperature (degrees Fahrenheit)	Year	Average Temperature (degrees Fahrenheit)
1900	57.1	1960	57.2
1905	56.8	1965	57.0
1910	56.6	1970	57.3
1915	57.0	1975	57.1
1920	56.9	1980	57.6
1925	56.9	1985	57.3
1930	57.1	1990	57.8
1935	57.0	1995	57.9
1940	57.3	2000	57.8
1945	57.3	2005	58.3
1950	56.9	2010	58.3
1955	57.0	2011	58.1

Source: *NASA–GISS*

a global perspective, he believes, the benefits of global warming will offset the damages.[†]

To test such theories, Peter S. Curtis, an ecologist at Ohio State University, and colleagues ran experiments that showed that increased carbon dioxide levels do in fact increase plant growth. However, the nutritional value of the produce was lower—so much lower that the increase in growth did not make up for the decrease in nutrition.[‡]

Some other theories suggest benefits to global warming. The scientific report *Impacts of a Warming Arctic* points out that it will be easier to extract oil from the Arctic due to less extensive and thinner sea ice. However, the study also says that oil spills are more difficult to clean up in icy seas than in open waters and that many species would suffer from such spills. In addition, potential structural problems such as broken pipelines could mean that the costs outweigh the benefits.[§]

A study done by economist Thomas Gale Moore at Stanford University suggests that a 4.5°F increase in temperature could reduce deaths in the United States by 40,000 per year and that medical costs might be reduced by at least $20 billion annually. Moore also points out that most people prefer warmer climates.[¶]

Most scientists do not share Moore's perspective. On a global scale, they believe warmer climates could increase the spread of diseases such as malaria and, thus, increase the global death rate and medical costs.

Although there may be some relatively small and shortlived benefits to global warming, the vast majority of scientists agree that global warming is already taking its toll and that further warming would bring catastrophic results.

Glaciologists report that, over the past century, glaciers around the globe have been melting.[‖] Biologists note that many species throughout the world have changed their habitats in search of cooler climates.[**] One species, the golden toad, was not able to migrate and as a result has become extinct due to heat stress.[††]

Looking to the future, an international study, the most comprehensive analysis of its kind, predicts that 15 to 37 percent of all species of plants and animals—well over a million species—will become extinct by 2050.[*] Klaus Toepfer, head of the United Nations Environment Programme (UNEP), said, "If one million species become extinct...it is not just the plant and animal kingdoms and the beauty of the planet that will suffer. Billions of people, especially in the developing world, will suffer too as they rely on nature for such essential goods and services as food, shelter and medicines."[†]

Analyzing the Situation

1. Discuss those theories that describe the benefits of global warming and whether they are likely correct.
2. What are some possible costs of global warming? In your opinion, do the possible costs outweigh the possible benefits? Explain.
3. Let A be the average surface temperature of Earth (in degrees Fahrenheit) at t years since 1900. *Carefully* draw a scattergram of the data in Table 70.
4. On the basis of your scattergram, in what year is it first clear that global warming is occurring? Explain. Also, explain why it makes sense that the first World Climate Conference convened in 1979.
5. Are the variables t and A approximately linearly related for the years 1900–2011? Are the variables approximately linearly related for the years 1965–2011? Explain.
6. Draw a line that comes close to the points in your scattergram for the years 1965–2011.
7. Use your linear model to predict the average global temperature in 2020.

Volume Lab

In this lab, you will explore the relationship between the volume of some water in a cylinder and the height of the water. Check with your instructor whether you should collect your own data or use the data listed in Table 71.

[†]From *The Impact of Climate Change on the United States Economy,* R. Mendelsohn and J. Neumann (eds.), Cambridge University Press.

[‡]"Plant Reproduction under Elevated CO_2 Conditions," L. M. Jablonski, X. Wang, and P. S. Curtis, *New Phytologist* (156):9–26, 2002.

[§]Report of the Arctic Climate Impact Assessment, Cambridge University Press, 2004.

[¶]"In sickness or in health: The Kyoto Protocol versus global warming," Hoover Institution, Stanford University, August 2000.

[‖]National Snow and Ice Data Center, 2003.

[**]T. L. Root et al., "Fingerprints of global warming on wild animals and plants," January 2, 2003, *Nature,* 421:57–60.

[††]J. A. Pounds et al., "Biological response to climate change on a tropical mountain," 1999, *Nature (London),* 398(6728):611–615.

[*]C. D. Thomas et al., "Extinction risk from climate change," January 8, 2004, *Nature,* 427:145–148.

[†]Reported by UNEP (United Nations Environment Programme), January 8, 2004.

Table 71 Heights of Water in a Cylinder with Radius 4.45 Centimeters

Height (centimeters)	Volume (ounces)
0	0
0.9	2
1.9	4
2.9	6
3.8	8
4.8	10
5.7	12

Source: *J. Lehmann*

Materials

You will need the following items:

- A "perfect" cylinder (the diameter of the top should equal the diameter of the base) that can hold at least 8 ounces of water
- At least 8 ounces of water
- A $\frac{1}{4}$-cup measuring cup
- A ruler

Recording of Data

Pour $\frac{1}{4}$ cup (2 ounces) of water into the cylinder, and measure the height of the water, using units of centimeters. Then continue adding $\frac{1}{4}$ cup of water and measuring the height after you have added each $\frac{1}{4}$ cup until there is at least 8 ounces of water in the cylinder. Also, measure the height of the cylinder in units of centimeters.

Analyzing the Data

1. Display your data in a table similar to Table 71. If you are using the data in Table 71, the height of the cylinder is 12 centimeters.
2. Let V be the volume of water (in ounces) in the cylinder when the height is h centimeters. Assume that V is the dependent variable. Draw a scattergram of the data.
3. Draw a line that comes close to the points in your scattergram.
4. What is the V-intercept of your model? What does it mean in this situation?
5. Use the model to estimate the volume of water when the height of the water is 3 centimeters.
6. Use the model to estimate the height of 7 ounces of water in the cylinder.
7. What is the height of the cylinder? Use this height and the model to estimate the maximum amount of water that the cylinder can hold.
8. Indicate on your graph of the model where model breakdown occurs. Also, describe in words when model breakdown occurs.

Linear Graphing Lab: Topic of Your Choice

Your objective in this lab is to use a linear model to describe some authentic situation. Choose a situation that has not been discussed in this text. Your first task will be to find some data. Almanacs, newspapers, magazines, and scientific journals are good resources. You may want to try searching on the Internet. Or you can conduct an experiment. Choose something that interests you!

Analyzing the Situation

1. What two quantities did you explore? Define variables for the quantities. Include units in your definitions.
2. Which variable is the dependent variable? Which variable is the independent variable? Explain.
3. Describe how you found your data. If you conducted an experiment, provide a careful description with specific details of how you ran your experiment. If you didn't conduct an experiment, state the source of your data.
4. Include a table of your data.
5. Create a scattergram of your data. (If your data are not approximately linear, find some data that are.)
6. Draw a line that comes close to the points in your scattergram.
7. Choose a value for your independent variable. On the basis of your chosen value for the independent variable, use your model to find a value for your dependent variable. Describe what your result means in the situation you are modeling.
8. Choose a value for your dependent variable. On the basis of your chosen value for the dependent variable, use your model to find a value for your independent variable. Describe what your result means in the situation you are modeling.
9. Comment on your lab experience.
   a. For example, you might address whether this lab was enjoyable, insightful, and so on.
   b. Were you surprised by any of your findings? If so, which ones?
   c. How would you improve your process for this lab if you were to do it again?
   d. How would you improve your process if you had more time and money?

Chapter Summary

Key Points of Chapter 1

Section 1.1 Variables and Constants

Variable	A **variable** is a symbol that represents a quantity that can vary.
Constant	A **constant** is a symbol that represents a specific number (a quantity that does *not* vary).
Counting numbers or natural numbers	The **counting numbers,** or **natural numbers,** are the numbers $1, 2, 3, 4, 5, \ldots$.
Integers	The **integers** are the numbers $\ldots, -3, -2, -1, 0, 1, 2, 3, \ldots$.
Rational numbers	The **rational numbers** are the numbers that can be written in the form $\frac{n}{d}$, where n and d are integers and d is nonzero.
Real numbers	The **real numbers** are all the numbers represented on the number line.
Irrational numbers	The **irrational numbers** are the real numbers that are not rational.
Average or mean	To find the **average** (or **mean**) of a group of numbers, we divide the sum of the numbers by the number of numbers in the group.
Negative numbers and positive numbers	The **negative numbers** are the real numbers less than 0, and the **positive numbers** are the real numbers greater than 0.

Section 1.2 Scattergrams

Scattergram	A **scattergram** is a graph of plotted ordered pairs.
Identifying independent and dependent variables	Assume that an authentic situation can be described by using the variables t and p, and assume that p depends on t. Then • We call t the **independent variable.** • We call p the **dependent variable.** • For an **ordered pair** (a, b), we write the value of the independent variable in the first (left) position and the value of the dependent variable in the second (right) position.
Columns of tables and axes of coordinate systems	Assume that an authentic situation can be described by using two variables. Then • For tables, the values of the independent variable are listed in the first column and the values of the dependent variable are listed in the second column. • For coordinate systems, the values of the independent variable are described by the horizontal axis and the values of the dependent variable are described by the vertical axis.

Section 1.3 Exact Linear Relationships

Linearly related	If two quantities of an authentic situation are described accurately by a line, then the quantities (and the variables representing those quantities) are **linearly related.**
Model	A **model** is a mathematical description of an authentic situation.
Linear model	A **linear model** is a nonvertical line that describes the relationship between two quantities in an authentic situation.
Input and output	An **input** is a permitted value of the *independent* variable that leads to at least one **output,** which is a permitted value of the *dependent* variable.
Determine whether data points lie on a line	We create a scattergram of data to determine whether the data points lie on a line. If the points lie on a line, then we draw the line and use it to make estimates and predictions.

Section 1.3 Exact Linear Relationships (*Continued*)

Intercepts of a line	An **intercept** of a line is any point where the line and an axis (or axes) of a coordinate system intersect. There are two types of intercepts of a line sketched on a coordinate system with an x-axis and a y-axis: • An ***x*-intercept** of a line is a point where the line and the x-axis intersect. The y-coordinate of an x-intercept is 0. • A ***y*-intercept** of a line is a point where the line and the y-axis intersect. The x-coordinate of a y-intercept is 0.

Section 1.4 Approximate Linear Relationships

Approximately linearly related	If the points in a scattergram of data lie close to (or on) a line, we say that the relevant variables are **approximately linearly related.**
Determine whether variables are approximately linearly related	We create a scattergram of data to determine whether the relevant variables are approximately linearly related. If so, we draw a line that comes close to the data points and use the line to make estimates and predictions.
Underestimates and overestimates	Suppose that an independent variable t and a dependent variable p are approximately linearly related. Then • If a linear model is below a data point (a, b), the model underestimates the value of p when $t = a$. • If a linear model is above a data point (a, b), the model overestimates the value of p when $t = a$.
Interpolation and extrapolation	For a situation that can be described by a model whose independent variable is t, • We perform **interpolation** when we use a part of the model whose t-coordinates are between the t-coordinates of two data points. • We perform **extrapolation** when we use a part of the model whose t-coordinates are not between the t-coordinates of any two data points.
Model breakdown	When a model yields a prediction that does not make sense or an estimate that is not a good approximation, we say that **model breakdown** has occurred.

Chapter 1 Review Exercises

1. Let B be the total box office gross (in billions of dollars) from U.S. and Canada movie theaters. For 2010, the value of B is 10.58 (Source: *Rentrak Corporation*). What does that mean in this situation?
2. Let t be the number of years since 1995. What does $t = 21$ represent?
3. Choose a variable name for the percentage of students who are full-time students at a college. Give two numbers that the variable can represent and two numbers that it cannot represent.
4. A rectangle has a perimeter of 40 inches. Let W be the width, L be the length, and P be the perimeter, all with units in inches.
   a. Sketch three possible rectangles with a perimeter of 40 inches.
   b. Which of the symbols W, L, and P are variables? Explain.
   c. Which of the symbols W, L, and P are constants? Explain.
5. Graph the numbers $-2, -\frac{3}{2}, 0, 1, \frac{5}{2}$, and 3 on a number line.
6. Graph the negative integers between -5 and 5 on a number line.
7. Here are a company's profits and losses for various years: profit of \$2 million, loss of \$4 million, loss of \$1 million, profit of \$3 million. Let p be the profit (in millions of dollars). Use points on a number line to describe the profits and losses of the company.
8. Plot the points $(2, 4), (-3, -1), (5, -2)$, and $(-4, 5)$ in a coordinate system.
9. What is the y-coordinate of the ordered pair $(3, -6)$?
10. What is the x-coordinate of the ordered pair $(-4, -7)$?

For Exercises 11 and 12, identify the independent variable and the dependent variable.

11. Let p be the percentage of Americans at age a years who own a home.
12. Let a be the average salary (in dollars) for a person with t years of education.
13. Let n be the total number of U.S. billionaires at t years since 2000. What does the ordered pair (11, 413) mean in this situation?
14. Let r be the annual revenue (in billions of dollars) from ADHD drugs at t years since 2005. What does the ordered pair $(5, 7)$ mean in this situation?
15. Create a scattergram of the ordered pairs listed in Table 72.

Table 72 Some Ordered Pairs

x	y
−5	−2
−3	4
0	1
2	3
4	−5

16. The average gas mileages of cars are shown in Table 73 for various years.

Table 73 Average Gas Mileages of Cars

Year	Average Gas Mileage (miles per gallon)
1970	14
1980	16
1990	20
2000	22
2009	24

Source: *U.S. Federal Highway Administration*

Let g be the average gas mileage (in miles per gallon) of cars at t years since 1970.

a. Create a scattergram of the data.

b. For which of the years shown in Table 73 was the average gas mileage of cars the highest?

c. For which of the years shown in Table 73 was the average gas mileage of cars the lowest?

17. The countries with the top six percentages of electricity generated by nuclear power are shown in the bar graph in Fig. 61 (Source: *International Atomic Energy Agency*).

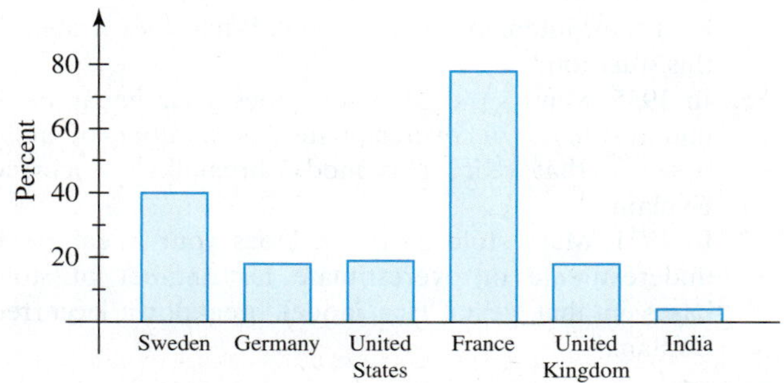

Figure 61 Percentages of electricity generated by nuclear power

a. Which country generates the largest percentage of its electricity by nuclear power? Estimate that percentage.

b. Which of the countries included in Fig. 61 generates the smallest percentage of its electricity by nuclear power? Estimate that percentage.

c. Estimate the percentage of Sweden's electricity that is generated by nuclear power.

For Exercises 18–23, refer to Fig. 62.

18. Find y when $x = -2$.

19. Find y when $x = 6$.

20. Find x when $y = -4$.

21. Find x when $y = 1$.

22. What is the y-intercept of the line?

23. What is the x-intercept of the line?

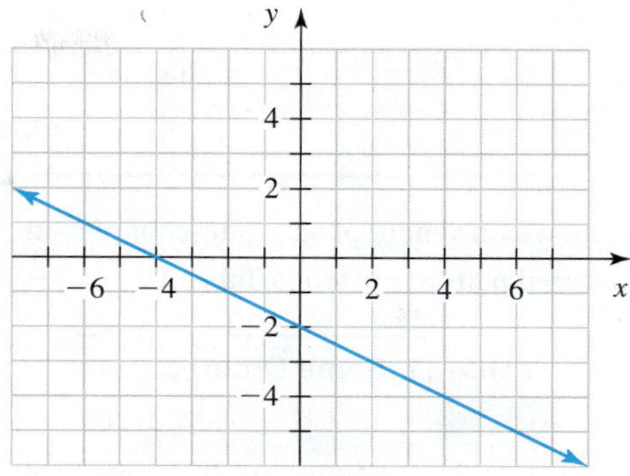

Figure 62 Exercises 18–23

24. Some ordered pairs are listed in Table 74.

Table 74 Some Ordered Pairs

x	y
1	11
2	10
5	7
8	4
9	3

a. Create a scattergram of the points shown in Table 74.

b. Draw a line that contains the points in your scattergram.

c. Find y when $x = 11$.

d. Find x when $y = 5$.

e. What is the x-intercept of your line?

f. What is the y-intercept of your line?

25. Some ordered pairs are listed in Table 75.

Table 75 Some Ordered Pairs

x	y
1	10
2	8
3	6
4	4
5	2
6	0

a. Create a scattergram of the points shown in Table 75.

b. Are the variables x and y linearly related, approximately linearly related, or neither?

26. Let p be the annual profit (in millions of dollars) of a company at t years since 2005. Some pairs of values of t and p are shown in Table 76.

a. Create a scattergram of the data. Then draw a linear model.

b. Predict the profit in 2015.

c. Estimate when the profit was $18 million.

d. What is the p-intercept of the model? What does it mean in this situation?

e. What is the t-intercept of the model? What does it mean in this situation?

Table 76 Profits of a Company

t (years)	p (millions of dollars)
1	20
3	16
4	14
6	10
7	8

27. What is the y-coordinate of an x-intercept of a line?

28. Some ordered pairs are listed in Table 77.

Table 77 Some Ordered Pairs

x	y
1	23
3	17
6	10
8	9
9	4

a. Create a scattergram of the points shown in Table 77.
b. Are the variables x and y linearly related, approximately linearly related, or neither?
c. Draw a line that comes close to the points in your scattergram.
d. Which point on your line has x-coordinate 5?
e. Which point on your line has y-coordinate 20?
f. What is the y-intercept of your line?
g. What is the x-intercept of your line?

29. The percentages of American adults who are obese are shown in Table 78 for various years.

Table 78 Percentages of American Adults Who Are Obese

Year	Percent
2000	31
2002	31
2004	33
2006	34
2008	35
2010	36

Source: *National Center for Chronic Disease Prevention and Health Promotion*

Let p be the percentage of American adults who are obese at t years since 2000.
a. Create a scattergram of the data.
b. Draw a line that comes close to the points in your scattergram.
c. Use your model to predict the percentage of American adults who will be obese in 2016.
d. Use your model to predict when 40% of Americans will be obese.

30. Willie Mays, with all-around talent, was one of the greatest baseball players of all time. Mays's statistics on stolen bases from 1956 to 1963 are shown in Table 79. Let n be Willie Mays's number of stolen bases in the year that is t years since 1955.

Table 79 Willie Mays: Numbers of Stolen Bases

Year	Number of Stolen Bases
1956	40
1957	38
1958	31
1959	27
1960	25
1961	18
1962	18
1963	8

Source: The Sports Encyclopedia: Baseball 2004, *D. S. Neft et al., 2004, St. Martin's Press, NY.*

a. Create a scattergram of the data.
b. Draw a line that comes close to the points in your scattergram.
c. Find the n-intercept of the model. What does it mean in this situation?
d. Find the t-intercept of the model. What does it mean in this situation?
e. In 1955, Mays stole 24 bases. Does your linear model underestimate or overestimate his number of stolen bases in that year? Has model breakdown occurred? Explain.
f. In 1971, Mays stole 23 bases. Does your linear model underestimate or overestimate his number of stolen bases in that year? Has model breakdown occurred? Explain.

Chapter 1 Test

1. A rectangle has an area of 36 square feet. Let W be the width (in feet), L be the length (in feet), and A be the area (in square feet).
a. Sketch three possible rectangles of area 36 square feet.
b. Which of the symbols W, L, and A are variables? Explain.
c. Which of the symbols W, L, and A are constants? Explain.

2. Graph the integers between -4 and 2, inclusive, on a number line.

3. The low temperatures (in degrees Fahrenheit) for four days in January in Indianapolis, Indiana, are 5°F below zero, 7°F above zero, 2°F above zero, and 3°F below zero. Let F be the low temperature (in degrees Fahrenheit) for any one day. Use points on a number line to describe the given values of F.

4. The number of electric cars (in thousands) in use in the United States for various years is 4.5, 5.2, 7.0, 8.7, and 10.4. Let n be the number of electric cars (in thousands) in use. Use points on a number line to describe the given values of n. Find the average of the values and indicate it on the number line.

5. Let c be the total cost (in dollars) of n tickets to a hip-hop concert. What is the dependent variable? Explain.
6. Let s be the salary (in millions of dollars) of Alex Rodriguez of the New York Yankees in the year that is t years since 2010. What does the ordered pair $(2, 30)$ mean in this situation?
7. The percentages of Americans of various age groups who are without health insurance are shown in Table 80.

Table 80 Percentages of Americans Who Are Uninsured

Age Group (years)	Age Used to Represent Age Group (years)	Percent
0–17	8.5	37
18–24	21	50
25–44	34.5	33
45–54	49.5	21
55–64	59.5	17

Source: *The Lewin Group for Families USA*

Let p be the percentage of Americans who are without health insurance at age a years.

a. Create a scattergram of the data.
b. Which point in your scattergram is highest? What does that mean in this situation?
c. Which point in your scattergram is lowest? What does that mean in this situation?

For Exercises 8–11, refer to Fig. 63.

8. Find y when $x = -4$.
9. Find x when $y = 1$.
10. What is the y-intercept of the line?
11. What is the x-intercept of the line?

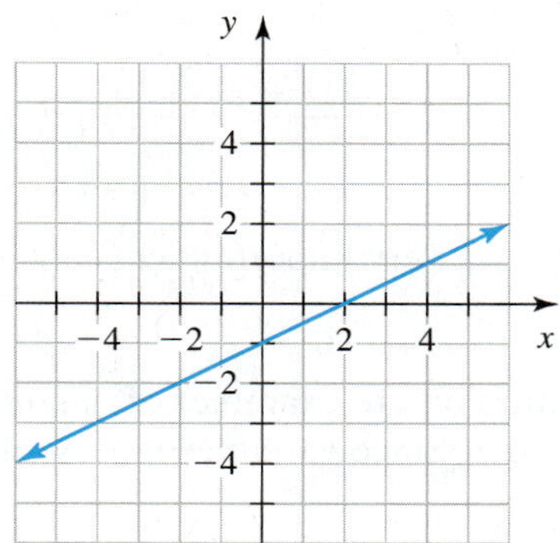

Figure 63 Exercises 8–11

12. Let s be a person's salary (in thousands of dollars) after she has worked t years at a company. Some pairs of values of t and s are shown in Table 81.

Table 81 Years Worked and Salary

t (years)	s (thousands of dollars)
0	21
2	25
3	27
5	31
6	33

a. Create a scattergram of the data. Then draw a linear model.
b. Estimate the person's salary after she has worked 4 years at the company.
c. Estimate when the person's salary will be \$35 thousand.
d. What is the s-intercept of the model? What does it mean in this situation?

13. Describe, in your own words, the meaning of *linear model*.
14. The number of spacecraft, rocket bodies, mission-related debris, and fragmentation debris orbiting the Earth has increased significantly since the beginning of the Space Age in 1957 (see Table 82).

Table 82 Numbers of Space Debris

Year	Number of Space Debris (thousands)
1960	0.02
1970	2.6
1980	5.5
1990	7.5
2000	9.8
2011	16.0

Source: *NASA Orbital Debris Program Office*

Let n be the number of space debris (in thousands) at t years since 1960.

a. Create a scattergram of the data.
b. Draw a line that comes close to the points in your scattergram.
c. Use your model to estimate when there were 9 thousand space debris.
d. Use your model to predict the number of space debris in 2020.

15. Suppose that an independent variable t and a dependent variable p are approximately linearly related. If a data point (c, d) is below a linear model that describes the relationship between t and p, does the model underestimate or overestimate the value of p when $t = c$? Explain.

2 Operations and Expressions

Although the number of AIDS deaths in the United States greatly decreased from 1996 to 2001, reaching a low of 14,175 deaths, it eventually increased to 17,774 deaths in 2009 (see Table 1). Why do you think the decrease and then the increase occurred? In Example 9 in Section 2.4, we will calculate how the number of deaths has changed in various years.

In this chapter, we will describe authentic quantities by using *expressions*. We will discuss how to perform operations and in which order we should perform them, which will help us use expressions to find values of authentic quantities. We will also discuss how to use subtraction and division to compare quantities pertaining to an authentic situation, such as the AIDS data just described.

Table 1 Numbers of AIDS Deaths in the United States

Year	Number of AIDS Deaths
1996	37,787
1997	21,923
1998	17,930
1999	16,273
2000	15,245
2001	14,175
2002	16,948
2003	17,679
2004	17,154
2005	16,823
2006	15,564
2007	14,561
2008	16,084
2009	17,774

Source: *Centers for Disease Control and Prevention*

2.1 Expressions

Objectives

- Know the meaning of *expression* and of *evaluate an expression*.
- Use expressions to describe authentic quantities.
- Evaluate expressions.
- Translate English phrases to and from mathematical expressions.
- Know two roles of a variable.
- Evaluate expressions with more than one variable.

In this section, we will work with expressions—a very important concept in algebra.

Expressions

Addition, subtraction, multiplication, and division are examples of *operations*. In arithmetic, you performed operations with numbers. Since variables represent numbers, we can perform operations with variables, too.

Example 1 Using Operations with Variables and Numbers

Each employee at a small company receives a \$500 bonus at the end of the year. For each employee's annual salary shown, find the employee's annual salary plus bonus.

1. \$28,000 **2.** \$32,000 **3.** s dollars

Solution

1. The employee's annual salary plus bonus is $28{,}000 + 500 = 28{,}500$ dollars.
2. The employee's annual salary plus bonus is $32{,}000 + 500 = 32{,}500$ dollars.
3. In Problems 1 and 2, we added the annual salary and \$500, the bonus, to find the results. So, the employee's annual salary plus bonus (in dollars) is $s + 500$.

In Example 1, we took s to be an employee's annual salary and $s + 500$ to be the employee's annual salary plus bonus. We call s and $s + 500$ *expressions.*

Definition Expression

An **expression** is a constant, a variable, or a combination of constants, variables, operation symbols, and grouping symbols, such as parentheses.

Here are some more examples of expressions:

$$t + 6 \qquad \pi \qquad L + W - 9 \qquad y \qquad 4 \qquad 5 \div (x + 2)$$

In Example 1, we used a variable to represent a quantity from an authentic situation. Sometimes we use variables to represent numbers in a math problem that is not being used to describe an authentic situation. In this case, we often use x for the variable. For example, we could let x represent a number. In this case, x could be *any* number.

To avoid confusing the multiplication symbol $\times$ and the variable name x, we use $\cdot$ or no operation symbol to indicate multiplication. For example, each of the following expressions describes multiplying 2 by 3:

$$2 \cdot 3 \qquad 2(3) \qquad (2)3 \qquad (2)(3)$$

And each of the following expressions describes multiplying 2 by k:

$$2 \cdot k \qquad 2k \qquad 2(k) \qquad (2)k \qquad (2)(k)$$

Using Expressions to Describe Authentic Quantities

We can use expressions to describe authentic quantities. In Example 2, we will find such an expression by noticing a pattern as we calculate values of a quantity.

Example 2 Describing a Quantity

A hot-dog stand sells hot dogs for \$2 apiece. Find the total cost of buying the given number of hot dogs.

1. 3 hot dogs **2.** 5 hot dogs **3.** 8 hot dogs **4.** n hot dogs

Solution

1. Three hot dogs cost $3(2) = 6$ dollars.
2. Five hot dogs cost $5(2) = 10$ dollars.
3. Eight hot dogs cost $8(2) = 16$ dollars.
4. In Problems 1–3, we found the total cost by multiplying the number of hot dogs by 2, the cost (in dollars) per hot dog. So, if there are n hot dogs, the total cost (in dollars) is $n(2)$.

In Example 3, we will use a table to help us find an expression that describes an authentic quantity.

Example 3 Using a Table to Find an Expression

A person drives at a constant speed of 75 miles per hour. Find the distance traveled in 1, 2, 3, and 4 hours of driving at that speed. Show the arithmetic to help you see the pattern. Organize the calculations in a table, and include an expression that stands for the distance traveled in t hours.

Table 2 Driving Times and Distances

Driving Time (hours)	Distance (miles)
1	$75 \cdot 1$
2	$75 \cdot 2$
3	$75 \cdot 3$
4	$75 \cdot 4$
t	$75 \cdot t$

Solution

First, we create Table 2. From the last row of the table, we see that the expression $75t$ represents the distance traveled (in miles) in t hours.

Evaluating Expressions

In Example 3, we used $75t$ to describe the distance traveled (in miles) in t hours. This means if the driving time is 5 hours, the distance traveled is $75(5) = 375$ miles. To find the distance, we substituted 5 for t. We say we have *evaluated* the expression $75t$ for $t = 5$.

Definition Evaluate an expression

We **evaluate an expression** by substituting a number for each variable in the expression and then calculating the result. If a variable appears more than once in the expression, the same number is substituted for that variable each time.

When we evaluate an expression, it is good practice to use parentheses each time a number is substituted for a variable. For example, here we evaluate $5x$ for $x = 3$:

$$5(3) = 15$$

This strategy will be especially helpful when we evaluate an expression for a negative number, which we will begin to do in Section 2.3.

Example 4 Evaluating Expressions

1. In Example 1, we used s to represent an employee's annual salary (in dollars) and $s + 500$ to represent the employee's annual salary plus bonus (in dollars). Evaluate $s + 500$ for $s = 40{,}000$, and describe the meaning of the result.
2. In Example 2, we used n to represent the number of hot dogs bought and $n(2)$ to represent the total cost (in dollars) of n hot dogs. Evaluate $n(2)$ for $n = 4$, and describe the meaning of the result.

Solution

1. We substitute 40,000 for s in $s + 500$:

$$(40{,}000) + 500 = 40{,}500$$

So, the annual salary plus bonus is \$40,500.

2. We substitute 4 for n in $n(2)$:

$$(4)(2) = 8$$

So, the total cost of 4 hot dogs is \$8.

Translating English Phrases to and from Expressions

In Example 2, we used English to describe an authentic situation:

"A hot-dog stand sells hot dogs for \$2 apiece."

Then we translated this information into mathematics:

"So, if there are n hot dogs, the total cost (in dollars) is $n(2)$."

In Problem 2 of Example 4, we performed mathematics by evaluating the expression $n(2)$ for $n = 4$ to find the total cost (in dollars) of 4 hot dogs:

$$(4)(2) = 8$$

Finally, we translated the result into English:

"So, the total cost of 4 hot dogs is \$8."

The following process describes the "big picture" of using mathematics to find results for authentic situations.

Using Mathematics to Find Results for Authentic Situations

1. Describe a situation in English.
2. Translate the English description into mathematics.
3. Perform mathematics to get a desired result.
4. Translate the mathematical result into English.

As you read Example 5, try to identify the four steps just described.

Example 5 Finding and Evaluating an Expression

Some students rent a house together. Each roommate pays an equal share of the $1200 monthly rent.

1. Let n be the number of roommates. Use a table to help find an expression that describes each roommate's share (in dollars) of the rent.
2. Evaluate your expression in Problem 1 for $n = 5$. What does the result mean in this situation?

Table 3 Number of Roommates and Share of Rent

Number of Roommates	Share of Rent (dollars)
1	$1200 \div 1$
2	$1200 \div 2$
3	$1200 \div 3$
4	$1200 \div 4$
n	$1200 \div n$

Solution

1. First, we create Table 3. We show the arithmetic to help us see the pattern. From the last row of the table, we see that the expression $1200 \div n$ represents each roommate's share of the rent (in dollars).
2. We substitute 5 for n in $1200 \div n$:

$$1200 \div (5) = 240$$

So, each roommate pays $240 per month.

When we translate from English to mathematics, or vice versa, the following definitions are helpful:

Definition Product, factor, and quotient

Let a and b be numbers. Then

- The **product** of a and b is ab. We call a and b **factors** of ab.
- The **quotient** of a and b is $a \div b$, where b is not zero.

For example, since $6 \cdot 3 = 18$, the number 18 is the product of 6 and 3 and the numbers 6 and 3 are factors of 18. The quotient of 6 and 3 is $6 \div 3 = 2$.

Here are some examples of English phrases or sentences and mathematical expressions that have the same meaning:

Operation	English Phrase or Sentence	Mathematical Expression
Addition	A number plus 3	$x + 3$
	The sum of a number and 3	$x + 3$
	The total of a number and 3	$x + 3$
	Add a number and 3.	$x + 3$
	3 more than a number	$x + 3$
	A number increased by 3	$x + 3$
Subtraction	A number minus 3	$x - 3$
	The difference of a number and 3	$x - 3$
	Subtract 3 from a number.	$x - 3$
	3 less than a number	$x - 3$
	A number decreased by 3	$x - 3$
Multiplication	Multiply 3 by a number.	$3x$
	3 times a number	$3x$
	The product of 3 and a number	$3x$
	Twice a number	$2x$
	One-third of a number	$\frac{1}{3}x$
Division	Divide a number by 3.	$x \div 3$
	The quotient of a number and 3	$x \div 3$
	The ratio of a number to 3	$x \div 3$

WARNING To subtract 2 from 5, we write $5 - 2$, not $2 - 5$. Suppose you have $5 and you take $2 from the $5. Then you have $5 - 2 = 3$ dollars left. So, subtracting 2 from 5 is $5 - 2$.

Example 6 Translating from English to Mathematics

Let x be a number.

1. Translate the English phrase "The product of 2 and the number" into an expression.
2. Evaluate your result in Problem 1 for $x = 3$.
3. Evaluate your result in Problem 1 for $x = 7$.

Solution

1. The expression is $2x$.
2. $2(3) = 6$
3. $2(7) = 14$

Roles of a Variable

In Chapter 1, we used a variable to represent a quantity that can vary. In Example 6, we used a variable for another reason: In the expression $2x$, the variable x is used as a *placeholder* for a number to be substituted for x. First, we substituted 3 for x. Then we substituted 7 for x.

Roles of a Variable

Here are two roles of a variable:*

1. A variable can represent a quantity that can vary.
2. In an expression, a variable is a placeholder for a number.

Sometimes a variable serves both roles. For example, consider the variable n in Example 5. Recall that n represents the number of roommates, which can vary over time or from house to house. The variable n also serves as a placeholder for a number in the expression $1200 \div n$, which describes each roommate's share of the rent (in dollars).

Example 7 Translating from English to Mathematics

Let x be a number. Translate the English phrase or sentence into an expression. Then evaluate the expression for $x = 6$.

1. The quotient of the number and 3
2. Subtract the number from 8.

Solution

1. The expression is $x \div 3$. Next, we evaluate $x \div 3$ for $x = 6$:

$$(6) \div 3 = 2$$

2. The expression is $8 - x$. Next, we evaluate $8 - x$ for $x = 6$:

$$8 - (6) = 2$$

Example 8 Translating from Mathematics to English

Let x be a number. Translate the expression into an English phrase.

1. $6 - x$
2. $8x$

Solution

1. The difference of 6 and the number
2. The product of 8 and the number

Expressions with More than One Variable

Figure 1 The length L and width W of a rectangle

An expression may contain more than one variable. For example, let W be the width (in feet) and let L be the length (in feet) of a rectangle (see Fig. 1).

*We will discuss one more role of a variable in Section 4.3.

Recall that the area of a rectangle is equal to the length times the width of the rectangle, so the area (in square feet) is equal to the expression LW. We can evaluate the expression LW for $L = 4$ and $W = 3$:

$$(4)(3) = 12$$

So, a 3-foot by 4-foot rectangle has an area of 12 square feet.

Note the power of algebra in that the expression LW *concisely* tells us how to find the area of *any* rectangle, no matter what its dimensions are.

Example 9 Evaluating an Expression in Two Variables

If it takes a student T minutes to complete a test that has n questions, then $T \div n$ is the average time (in minutes) taken to respond to one question. Evaluate $T \div n$ for $T = 48$ and $n = 16$. What does the result mean in this situation?

Solution

We substitute 48 for T and 16 for n in the expression $T \div n$ and then calculate the result:

$$48 \div 16 = 3$$

If it takes a student 48 minutes to respond to 16 questions, the average response time is 3 minutes per question.

Example 10 Translating from English to Mathematics

Write the phrase as a mathematical expression, and then evaluate the result for $x = 8$ and $y = 4$.

1. The sum of x and y

2. The quotient of x and y

Solution

1. The expression is $x + y$. Next, we evaluate $x + y$ for $x = 8$ and $y = 4$:

$$(8) + (4) = 12$$

2. The expression is $x \div y$. Next, we evaluate $x \div y$ for $x = 8$ and $y = 4$:

$$(8) \div (4) = 2$$

Group Exploration

Expressions used to describe a quantity

Consider the expression $x + 2$. Suppose that a child has grown 2 inches within the last year. We could define x to be the child's height (in inches) last year, and then $x + 2$ would be the child's current height (in inches).

Describe a situation in which x represents a meaningful quantity and the expression given describes another meaningful quantity.

1. $x + 3$

2. $x - 4$

3. $3x$

4. $x \div 2$

For each of the four expressions, evaluate it for a reasonable value of x and describe the meaning of the result.

Tips for Success Make Good Use of This Text

You can get more out of this course by making good use of the text. Before class, consider previewing the material for ten minutes. You can do this by reading the objectives and the boxed statements. Even if what you read doesn't make much sense to you, previewing will flag key concepts that you can focus on during class time.

After class, read the relevant section(s). When looking at each example, figure out how it goes from one step to the next.

Then begin working on the homework assignment. If you have difficulty with an exercise, locate a similar example to help guide you. You may need to seek outside help for more challenging exercises. If you needed to look at examples or get outside help for a large number of exercises, then it is important that you keep doing additional exercises until you are self-sufficient. After all, unless your instructor allows open-book tests or collaborative tests, you won't be able to read the text or seek help from others during an exam.

Homework 2.1

For extra help ▶ MyMathLab® 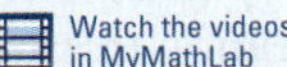 Watch the videos in MyMathLab 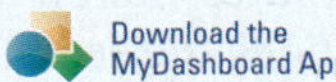 Download the MyDashboard App

For Exercises 1–12, evaluate the expression for $x = 6$.

1. $x + 2$ **2.** $5 + x$ **3.** $9 - x$ **4.** $x - 4$

5. $7x$ **6.** $x(9)$ **7.** $x \div 3$ **8.** $30 \div x$

9. $x + x$ **10.** $x - x$ **11.** $x \cdot x$ **12.** $x \div x$

13. If a person buys n albums, the total cost is $9n$ dollars. Evaluate $9n$ for $n = 4$. What does your result mean in this situation?

14. If a student earns a total of T points on five tests, then $T \div 5$ is the student's average test score (in points). If a student earns a total of 440 points on five tests, what is the student's average test score?

15. For the period 1995–2012, if M is the mean math SAT score for high-school boys, then $M - 34$ is approximately the mean math SAT score for high-school girls (Source: *College Board*). The mean score for boys was 534 points in 2012. Estimate the mean score for girls in 2012.

16. For the period 2005–2009, if M is the number (in millions) of male students in college, then $M + 2.68$ is approximately the number (in millions) of female students in college. There were about 8.77 million male students in college in 2009 (Source: *U.S. National Center for Education Statistics*). Estimate the number of female students in college in 2009.

17. Each share of a certain stock is worth \$5.

a. Complete Table 4 to help find an expression that describes the total value (in dollars) of n shares of the stock. Show the arithmetic to help you see a pattern.

Table 4 Number of Shares and Total Value

Number of Shares	Total Value (dollars)
1	
2	
3	
4	
n	

b. Evaluate the expression you found in part (a) for $n = 7$. What does your result mean in this situation?

18. A pair of socks sells for \$3.

a. Complete Table 5 to help find an expression that describes the total cost (in dollars) of n pairs of socks. Show the arithmetic to help you see a pattern.

b. Evaluate the expression you found in part (a) for $n = 9$. What does your result mean in this situation?

Table 5 Number of Pairs and Total Cost

Number of Pairs	Total Cost (dollars)
1	
2	
3	
4	
n	

19. Each student at a community college pays a student services fee of \$12.

a. Complete Table 6 to help find an expression that describes the total cost (in dollars) of tuition plus the services fee if a student pays t dollars for tuition. Show the arithmetic to help you see a pattern.

Table 6 Tuition and Total Cost

Tuition (dollars)	Total Cost (dollars)
400	
401	
402	
403	
t	

b. Evaluate the expression you found in part (a) for $t = 417$. What does your result mean in this situation?

20. A person is driving 5 miles per hour over the speed limit.

a. Complete Table 7 to help find an expression that describes the driving speed (in miles per hour) if the speed limit is s miles per hour. Show the arithmetic to help you see a pattern.

Table 7 Speed Limit and Driving Speed

Speed Limit (miles per hour)	Driving Speed (miles per hour)
35	
40	
45	
50	
s	

b. Evaluate the expression you found in part (a) for $s = 65$. What does your result mean in this situation?

21. The enrollment fee for a college student is \$87 per hour (unit or credit).

a. Complete Table 8 to help find an expression that describes the total cost (in dollars) of enrolling in n hours of classes. Show the arithmetic to help you see a pattern.

Table 8 Tuition and Total Cost

Number of Hours of Courses	Total Cost (dollars)
1	
2	
3	
4	
n	

b. Evaluate the expression you found in part (a) for $n = 15$. What does your result mean in this situation?

22. The length of a rectangular garden is 20 feet.

a. Complete Table 9 to help find an expression that describes the area (in square feet) of the rectangle if the width is w feet. Show the arithmetic to help you see a pattern.

Table 9 Width and Area

Width (feet)	Area (square feet)
1	
2	
3	
4	
w	

b. Evaluate the expression you found in part (a) for $w = 10$. What does your result mean in this situation?

For Exercises 23–32, let x be a number. Translate the English phrase or sentence into a mathematical expression. Then evaluate the expression for $x = 8$.

23. The number plus 4

24. 8 minus the number

25. The quotient of the number and 2

26. Add 6 and the number.

27. Subtract 5 from the number.

28. 15 more than the number

29. The product of 7 and the number

30. The difference of the number and 7

31. 16 divided by the number

32. Multiply the number by 5.

Let x be a number. Translate the expression into an English phrase.

33. $x \div 2$ **34.** $6 \div x$ **35.** $7 - x$

36. $x - 2$ **37.** $x + 5$ **38.** $4 + x$

39. $9x$ **40.** $x(5)$ **41.** $x - 7$

42. $x + 3$ **43.** $x(2)$ **44.** $x \div 5$

Evaluate the expression for $x = 6$ and $y = 3$.

45. $x + y$ **46.** $y + x$ **47.** $x - y$

48. xy **49.** yx **50.** $x \div y$

For Exercises 51–54, translate the phrase into a mathematical expression. Then evaluate the expression for $x = 9$ and $y = 3$.

51. The product of x and y

52. The sum of x and y

53. The difference of x and y

54. The quotient of x and y

55. If a car travels at a constant speed of r miles per hour for t hours, it will travel rt miles. Evaluate rt for $r = 62$ and $t = 3$. What does your result mean in this situation?

56. Let b be the balance (in dollars) of a checking account. If a check is written for d dollars, then the new balance (in dollars) is $b - d$. Evaluate $b - d$ for $b = 3758$ and $d = 994$. What does your result mean in this situation?

57. If a car can travel m miles on g gallons of gasoline, then the car's gas mileage is $m \div g$ miles per gallon. Evaluate $m \div g$ for $m = 240$ and $g = 12$. What does your result mean in this situation?

58. If T is the total cost (in dollars) for n students to go on a ski trip, then $T \div n$ is the cost (in dollars) per student. Evaluate $T \div n$ for $T = 9000$ and $n = 20$. What does your result mean in this situation?

59. Let C be the total cost (in dollars) of manufacturing some computers and R be the total revenue (in dollars) from selling the computers. Then $R - C$ is the total profit (in dollars). If the total cost of manufacturing some computers is \$315 thousand and the total revenue from selling the computers is \$485 thousand, what is the total profit?

60. For the period 1992–2009, if E is the average verbal SAT score (in points) for a certain year, then the average math SAT score (in points) for that year is approximately $E + t$, where t is the number of years since 1992. The average verbal SAT score was 501 points in 2009. Estimate the average math SAT score in 2009.

Concepts

61. A person gets paid $5t$ dollars for t hours of work.

a. Evaluate $5t$ for $t = 1, t = 2, t = 3$, and $t = 4$. Describe the meaning of these results.

b. Refer to your results from part (a) to determine how much the person gets paid per hour. Explain.

c. Compare your result from part (b) with the expression $5t$. What do you notice?

62. The total price of n loaves of bread is $3n$ dollars.

a. Evaluate $3n$ for $n = 1, n = 2, n = 3$, and $n = 4$. Describe the meaning of these results.

b. Refer to your results from part (a) to determine the cost per loaf of bread. Explain.

c. Compare your result from part (b) with the expression $3n$. What do you notice?

63. A person drives $50t$ miles in t hours.

a. Evaluate $50t$ for $t = 1, t = 2, t = 3$, and $t = 4$. Describe the meaning of your results.

b. Refer to your results from part (a) to determine at what speed the person is traveling. Explain.

c. Compare your result from part (b) with the expression $50t$. What do you notice?

64. An elevator rises $2t$ yards in t seconds.

a. Evaluate $2t$ for $t = 1, t = 2, t = 3$, and $t = 4$. Describe the meaning of your results.

b. Refer to your results from part (a) to determine at what speed the elevator is rising. Explain.

c. Compare your result from part (b) with the expression $2t$. What do you notice?

65. Compare the meaning of *variable* with the meaning of *expression*. (See page 9 for guidelines on writing a good response.)

66. Give an example of an expression containing a variable, and then evaluate it three times to get three different results.

67. Give an example of a variable that is used to represent a quantity that varies.

68. Give an example of a variable that is used as a placeholder for a number in an expression.

69. Describe an authentic situation for the expression $8x$. Include a definition for the variable x in your description.

70. Describe an authentic situation for the expression $200 \div x$. Include a definition for the variable x in your description.

2.2 Operations with Fractions

Objectives

- » Know the meaning of a fraction.
- » Know that division by zero is undefined.
- » Know the rules for $a \cdot 1$, $\frac{a}{1}$, and $\frac{a}{a}$.
- » Perform operations with fractions.
- » Find the prime factorization of a number.
- » Simplify fractions.

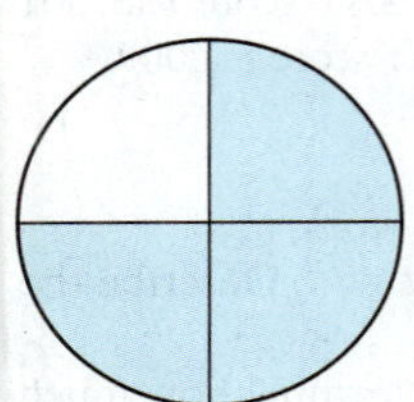

Figure 2 $\frac{3}{4}$ of a pizza

In this section, we perform operations with fractions, which are used in numerous fields, including music, social science, business, computer science, architecture, chemistry, engineering, political science, medicine, and aeronautics.

Meaning of a Fraction

A fraction can be used to describe a part of a whole. For example, consider the meaning of $\frac{3}{4}$ of a pizza. If we divide the pizza into 4 slices of equal area, 3 of the slices make up $\frac{3}{4}$ of the pizza (see Fig. 2).

The fraction $\frac{a}{b}$ means $a \div b$. For example, $\frac{8}{4} = 8 \div 4 = 2$. So 8 quarters of pizza make 2 pizzas with 4 slices each (see Fig. 3).

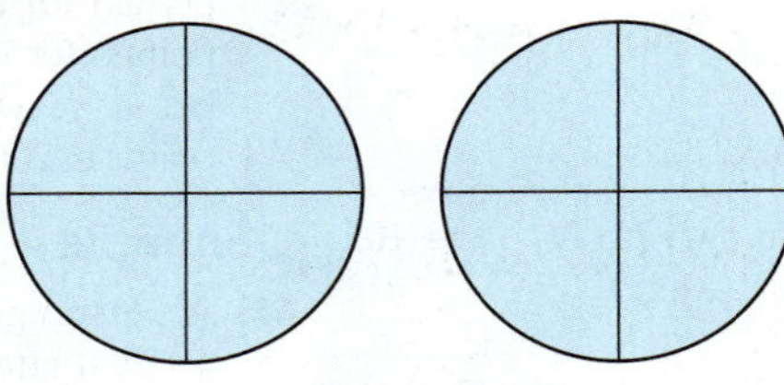

Figure 3 The 8 quarters of pizza make 2 pizzas

Division by Zero

We can think of division in terms of repeated subtraction. For example, $17 \div 5$ is equal to 3 with a remainder of 2 (try it). This means if we subtract 5 from 17 three times, the result is 2 (the remainder):

$$17 - 5 = 12, \qquad 12 - 5 = 7, \qquad 7 - 5 = 2$$

Note that the remainder, 2, is less than the divisor, 5.

As a matter of fact, the remainder must always be less than the divisor. This rule will help us see that division by 0 is undefined. For example, consider $8 \div 0$. No matter how many times we subtract 0 from 8, the result is always 8:

$$8 - 0 = 8, \qquad 8 - 0 = 8, \qquad 8 - 0 = 8, \text{ and so on}$$

If $8 \div 0$ is defined, the remainder would have to be the repeated result 8. Since the remainder must be less than the divisor, it is implied that 8 is less than 0, which is false. So, $8 \div 0$ is undefined. In fact, any number divided by 0 is undefined.

Division by Zero

The fraction $\frac{a}{b}$ is undefined if $b = 0$. Division by 0 is undefined.

For example, $\frac{6}{0}$ is undefined. If you use a calculator to divide by 0, the screen will likely display "Error," "ERR:," "E," or "ERR: Divide by 0" to indicate that division by 0 is undefined.

WARNING

However, the fraction $\frac{0}{6}$ *is* defined. In fact, $\frac{0}{6} = 0$. For example, if a person eats zero sixths of a pizza, this means that the person didn't eat any pizza.

Rules for $a \cdot 1$, $\frac{a}{1}$, and $\frac{a}{a}$

The products $4 \cdot 1 = 4$, $5 \cdot 1 = 5$, and $8 \cdot 1 = 8$ suggest the following property:

Multiplying a Number by 1

$$a \cdot 1 = a$$

In words: A number multiplied by 1 is that same number.

When we write statements such as $a \cdot 1 = a$, we mean if we evaluate $a \cdot 1$ and a for *any* value of a in both expressions, the results will be equal. We say that the expressions $a \cdot 1$ and a are **equivalent expressions.**

The quotients $\frac{4}{1} = 4 \div 1 = 4$, $\frac{5}{1} = 5 \div 1 = 5$, and $\frac{8}{1} = 8 \div 1 = 8$ suggest the following property:

Dividing a Number by 1

$$\frac{a}{1} = a$$

In words: A number divided by 1 is that same number.

Finally, the quotients $\frac{4}{4} = 4 \div 4 = 1$, $\frac{5}{5} = 5 \div 5 = 1$, and $\frac{8}{8} = 8 \div 8 = 1$ suggest the following property:

Dividing a Nonzero Number by Itself

If a is nonzero, then

$$\frac{a}{a} = 1$$

In words: A nonzero number divided by itself is 1.

The properties $a \cdot 1 = a$, $\frac{a}{1} = a$, and $\frac{a}{a} = 1$ (where a is nonzero) will help us when we work with fractions.

Multiplication of Fractions

Figure 4 illustrates that $\frac{1}{2}$ of $\frac{1}{4}$ of a pizza is $\frac{1}{8}$ of a pizza. We can calculate this result by finding the product $\frac{1}{2} \cdot \frac{1}{4}$:

$$\frac{1}{2} \cdot \frac{1}{4} = \frac{1 \cdot 1}{2 \cdot 4} = \frac{1}{8}$$

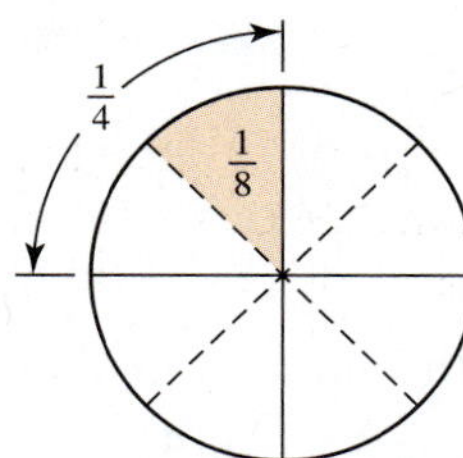

Figure 4 $\frac{1}{2}$ of $\frac{1}{4}$ of a pizza is $\frac{1}{8}$ of a pizza

Multiplying Fractions

If b and d are nonzero, then

$$\frac{a}{b} \cdot \frac{c}{d} = \frac{ac}{bd}$$

In words: To multiply two fractions, write the numerators as a product and write the denominators as a product.

Example 1 Finding the Product of Two Fractions

Find the product $\frac{2}{5} \cdot \frac{3}{7}$.

Solution

$$\frac{2}{5} \cdot \frac{3}{7} = \frac{2 \cdot 3}{5 \cdot 7} \quad \text{Write numerators and denominators as products: } \frac{a}{b} \cdot \frac{c}{d} = \frac{ac}{bd}$$

$$= \frac{6}{35} \quad \text{Find products.}$$

Prime Factorization

When we work with fractions, it can sometimes help to work with prime numbers.

Definition Prime number

A **prime number,** or **prime,** is any counting number larger than 1 whose only positive factors are itself and 1.

Here are the first ten primes:

2, 3, 5, 7, 11, 13, 17, 19, 23, 29

Sometimes when we work with fractions, it is helpful to write a number as a product of primes. We call this product the **prime factorization** of the number.

Example 2 Writing a Number as a Product of Primes

Write 54 as a product of primes.

Solution

$$54 = 6 \cdot 9 \quad \text{Write 54 as a product of two numbers.}$$

$$= 2 \cdot 3 \cdot 3 \cdot 3 \quad \text{Find prime factorizations of 6 and 9.}$$

The prime factorization of 54 is $2 \cdot 3 \cdot 3 \cdot 3$.

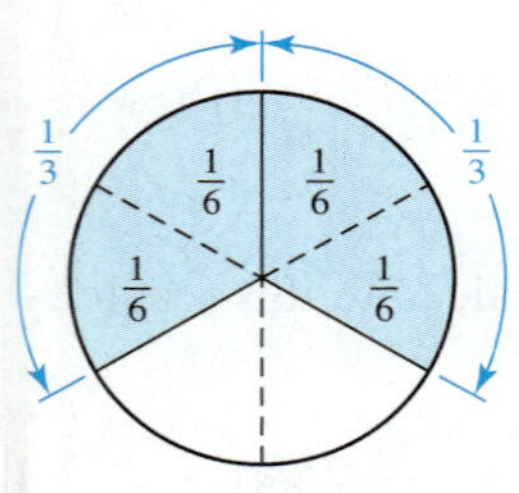

Figure 5 $\frac{4}{6} = \frac{2}{3}$

Simplifying Fractions

Figure 5 illustrates that $\frac{4}{6} = \frac{2}{3}$. We say $\frac{2}{3}$ is *simplified,* because the numerator and denominator do not have positive factors other than 1 in common. The fraction $\frac{4}{6}$ is not simplified, because the numerator and the denominator have a common factor of 2. To **simplify** a fraction, we write it as an equal fraction in which the numerator and the denominator do not have any common positive factors other than 1.

Example 3 Simplifying a Fraction

Simplify $\frac{4}{6}$.

Solution

We begin to simplify $\frac{4}{6}$ by finding the prime factorizations of the numerator 4 and the denominator 6:

$$\begin{aligned}
\frac{4}{6} &= \frac{2 \cdot 2}{2 \cdot 3} && \textit{Find prime factorizations of numerator and denominator.} \\
&= \frac{2}{2} \cdot \frac{2}{3} && \frac{ab}{cd} = \frac{a}{c} \cdot \frac{b}{d} \\
&= 1 \cdot \frac{2}{3} && \textit{Simplify: } \frac{2}{2} = 1 \\
&= \frac{2}{3} && 1 \cdot a = a
\end{aligned}$$

Our result matches with what we found in Fig. 5. By performing long division or using a calculator, we can check that both fractions are equal to the repeating decimal $0.\overline{6}$.

Simplifying fractions can make it easier to work out certain problems. Also, if two fractions are simplified, it is easy to tell whether they are equal. **If the result of an exercise is a fraction, simplify it.**

Example 4 Simplifying a Fraction

Simplify $\frac{30}{42}$.

Solution

We begin to simplify $\frac{30}{42}$ by finding the prime factorizations of the numerator, 30, and the denominator, 42:

$$\begin{aligned}
\frac{30}{42} &= \frac{2 \cdot 3 \cdot 5}{2 \cdot 3 \cdot 7} && \textit{Find prime factorizations of numerator and denominator.} \\
&= \frac{2 \cdot 3}{2 \cdot 3} \cdot \frac{5}{7} && \frac{ac}{bd} = \frac{a}{b} \cdot \frac{c}{d} \\
&= 1 \cdot \frac{5}{7} && \textit{Simplify: } \frac{2 \cdot 3}{2 \cdot 3} = 1 \\
&= \frac{5}{7} && 1 \cdot a = a
\end{aligned}$$

Simplifying a Fraction

To simplify a fraction,

1. Find the prime factorizations of the numerator and denominator.
2. Find an equal fraction in which the numerator and the denominator do not have common positive factors other than 1 by using the property

$$\frac{ab}{ac} = \frac{a}{a} \cdot \frac{b}{c} = 1 \cdot \frac{b}{c} = \frac{b}{c}$$

where a and c are nonzero.

In Example 5, we multiply two fractions and simplify the result.

Example 5 Finding the Product of Two Fractions

Find the product $\frac{8}{9} \cdot \frac{15}{4}$.

Solution

$$\frac{8}{9} \cdot \frac{15}{4} = \frac{8 \cdot 15}{9 \cdot 4}$$ *Write numerators and denominators as products:* $\frac{a}{b} \cdot \frac{c}{d} = \frac{ac}{bd}$

$$= \frac{2 \cdot 2 \cdot 2 \cdot 3 \cdot 5}{3 \cdot 3 \cdot 2 \cdot 2}$$ *Find prime factorizations.*

$$= \frac{2 \cdot 5}{3}$$ *Simplify:* $\frac{2 \cdot 2 \cdot 3}{2 \cdot 2 \cdot 3} = 1$

$$= \frac{10}{3}$$ *Multiply.*

Division of Fractions

The **reciprocal** of $\frac{a}{b}$ is $\frac{b}{a}$. For example, the reciprocal of $\frac{3}{8}$ is $\frac{8}{3}$. We will need to find the reciprocal of a fraction when we divide two fractions.

Dividing Fractions

If b, c, and d are nonzero, then

$$\frac{a}{b} \div \frac{c}{d} = \frac{a}{b} \cdot \frac{d}{c}$$

In words: To divide by a fraction, multiply by its reciprocal.

Example 6 Finding the Quotient of Two Fractions

Find the quotient $\frac{3}{4} \div \frac{1}{8}$.

Solution

$$\frac{3}{4} \div \frac{1}{8} = \frac{3}{4} \cdot \frac{8}{1}$$ *Multiply by reciprocal of* $\frac{1}{8}$*, which is* $\frac{8}{1}$*:* $\frac{a}{b} \div \frac{c}{d} = \frac{a}{b} \cdot \frac{d}{c}$

$$= \frac{3 \cdot 8}{4 \cdot 1}$$ *Write numerators and denominators as products.*

$$= \frac{3 \cdot 2 \cdot 2 \cdot 2}{2 \cdot 2 \cdot 1}$$ *Find prime factorizations.*

$$= \frac{3 \cdot 2}{1}$$ *Simplify:* $\frac{2 \cdot 2}{2 \cdot 2} = 1$

$$= 6$$ $\frac{a}{1} = a$

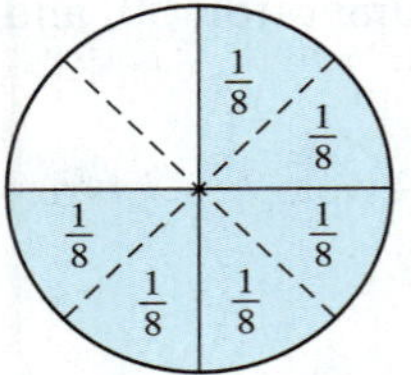

Figure 6 $\frac{3}{4}$ of a pizza divided into slices of size $\frac{1}{8}$ of the pizza gives 6 slices of pizza

Our result makes sense, because $\frac{3}{4}$ of a pizza divided into slices, each of size $\frac{1}{8}$ of the pizza, gives 6 slices (see Fig. 6).

We can use a graphing calculator to check our work in Example 6 (see Fig. 7).

Press (3 ÷ 4) ÷ (1 ÷ 8) ENTER.

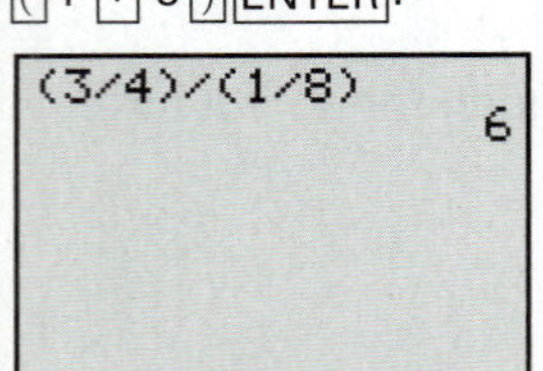

Figure 7 Verify the work

When you use a calculator to check work with fractions, it is good practice to enclose each fraction in parentheses. You will see the importance of using parentheses when we discuss the order of operations in Section 2.6.

To find the reciprocal of 6, we use the fact $6 = \frac{6}{1}$. So, the reciprocal of 6 is $\frac{1}{6}$.

Example 7 Evaluating an Expression

Evaluate $\frac{a}{b} \div c$ for $a = 21$, $b = 2$, and $c = 3$.

Solution

We substitute $a = 21$, $b = 2$, and $c = 3$ into the expression $\frac{a}{b} \div c$:

$$\begin{aligned}
\frac{(21)}{(2)} \div (3) &= \frac{21}{2} \div \frac{3}{1} && \text{Write 3 as a fraction: } 3 = \frac{3}{1} \\
&= \frac{21}{2} \cdot \frac{1}{3} && \text{Multiply by reciprocal of } \frac{3}{1}\text{, which is } \frac{1}{3}\text{: } \frac{a}{b} \div \frac{c}{d} = \frac{a}{b} \cdot \frac{d}{c} \\
&= \frac{21 \cdot 1}{2 \cdot 3} && \text{Write numerators and denominators as products.} \\
&= \frac{3 \cdot 7 \cdot 1}{2 \cdot 3} && \text{Find prime factorizations.} \\
&= \frac{7}{2} && \text{Simplify: } \frac{3}{3} = 1
\end{aligned}$$

In Example 7, the result is $\frac{7}{2}$, which is an improper fraction (that is, the numerator is larger than the denominator). For *nonmodeling* exercises, if a fractional result is in improper form, we will leave it in that form. For *modeling* exercises, if a result is in improper form, we will write it as a mixed number. For example, we say that a car trip takes $3\frac{1}{2}$ hours rather than $\frac{7}{2}$ hours.

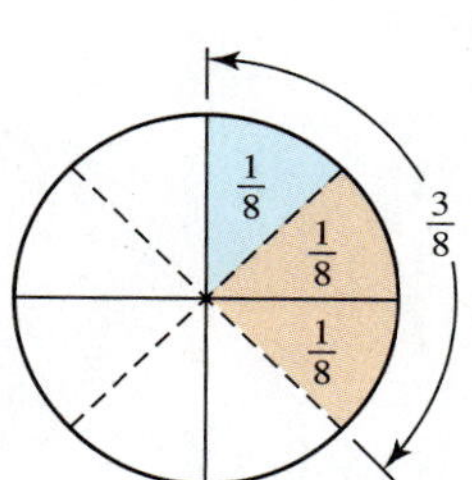

Figure 8 $\frac{1}{8}$ pizza plus $\frac{2}{8}$ pizza is $\frac{3}{8}$ pizza

Addition of Fractions

Figure 8 illustrates that $\frac{1}{8}$ of a pizza plus $\frac{2}{8}$ of a pizza is equal to $\frac{3}{8}$ of a pizza. This illustration suggests that, to find the sum $\frac{1}{8} + \frac{2}{8}$, we add the numerators 1 and 2 and write the result, 3, over the common denominator, 8:

$$\begin{aligned}
\frac{1}{8} + \frac{2}{8} &= \frac{1+2}{8} \\
&= \frac{3}{8}
\end{aligned}$$

Adding Fractions with the Same Denominator

If b is nonzero, then

$$\frac{a}{b} + \frac{c}{b} = \frac{a+c}{b}$$

In words: To add two fractions with the same denominator, add the numerators and write the result above the common denominator.

Example 8 Adding Fractions with the Same Denominator

Find the sum $\frac{4}{15} + \frac{6}{15}$.

Solution

$$\frac{4}{15} + \frac{6}{15} = \frac{4+6}{15} \quad \text{Write numerators as a sum and keep common denominator: } \frac{a}{b} + \frac{c}{b} = \frac{a+c}{b}$$

$$= \frac{10}{15} \quad \text{Find sum.}$$

$$= \frac{2}{3} \quad \text{Simplify.}$$

Least Common Denominators

To find the sum $\frac{1}{4} + \frac{5}{6}$, in which the denominators of the fractions are different, we find an equal sum of fractions in which the denominators are equal. First, we list the multiples of 4 and the multiples of 6:

Multiples of 4: 4, 8, 12, 16, 20, 24, 28, 32, 36, . . .
Multiples of 6: 6, 12, 18, 24, 30, 36, 42, 48, 54, . . .

Common multiples of 4 and 6 are

12, 24, 36, . . .

Note that 12 is the least (lowest) number in the list. We call it the least common multiple of 4 and 6. The **least common multiple** (**LCM**) of a group of numbers is the smallest number that is a multiple of *all* of the numbers in the group.

To find the sum $\frac{1}{4} + \frac{5}{6}$, we use the fact $\frac{a}{a} = 1$, where a is nonzero, to write an equal sum of fractions in which each denominator is equal to the LCM, 12:

$$\frac{1}{4} + \frac{5}{6} = \frac{1}{4} \cdot 1 + \frac{5}{6} \cdot 1 \quad a = a \cdot 1$$

$$= \frac{1}{4} \cdot \frac{3}{3} + \frac{5}{6} \cdot \frac{2}{2} \quad 1 = \frac{a}{a}$$

$$= \frac{3}{12} + \frac{10}{12} \quad \text{Multiply numerators and multiply denominators: } \frac{a}{b} \cdot \frac{c}{d} = \frac{ac}{bd}$$

$$= \frac{13}{12} \quad \text{Add numerators and keep common denominator: } \frac{a}{b} + \frac{c}{b} = \frac{a+c}{b}$$

We also call 12 the least common denominator of $\frac{1}{4}$ and $\frac{5}{6}$. The **least common denominator** (**LCD**) of a group of fractions is the LCM of the denominators of all of the fractions.

Example 9 Adding Fractions with Different Denominators

Find the sum $\frac{5}{8} + \frac{5}{6}$.

Solution

We list multiples of 8 and multiples of 6:

Multiples of 8: 8, 16, 24, 32, 40, 48, . . .
Multiples of 6: 6, 12, 18, 24, 30, 36, . . .

The LCD is 24. We write an equal sum of fractions in which each denominator is 24:

$$\frac{5}{8}+\frac{5}{6}=\frac{5}{8}\cdot\frac{3}{3}+\frac{5}{6}\cdot\frac{4}{4}$$ *LCD is 24.*

$$=\frac{15}{24}+\frac{20}{24}$$ *Multiply numerators and multiply denominators:* $\frac{a}{b}\cdot\frac{c}{d}=\frac{ac}{bd}$

$$=\frac{35}{24}$$ *Add numerators and keep common denominator:* $\frac{a}{b}+\frac{c}{b}=\frac{a+c}{b}$

```
(5/8)+(5/6)
      1.458333333
35/24
      1.458333333
```

Figure 9 Verify the work

We use a graphing calculator to verify the work (see Fig. 9).

Subtraction of Fractions

The rule for subtracting two fractions with the same denominator is similar to the rule for adding such fractions, except we subtract the numerators.

Subtracting Fractions with the Same Denominator

If b is nonzero, then

$$\frac{a}{b}-\frac{c}{b}=\frac{a-c}{b}$$

In words: To subtract two fractions with the same denominator, subtract the numerators and write the result above the common denominator.

Example 10 Subtracting Fractions with the Same Denominator

Find the difference $\frac{5}{8}-\frac{3}{8}$.

Solution

$$\frac{5}{8}-\frac{3}{8}=\frac{5-3}{8}$$ *Write numerators as a difference and keep common denominator:* $\frac{a}{b}-\frac{c}{b}=\frac{a-c}{b}$

$$=\frac{2}{8}$$ *Find difference.*

$$=\frac{1}{4}$$ *Simplify.*

Subtracting fractions with different denominators is similar to adding them. The first step is to rewrite each fraction so that each denominator is the LCD.

Example 11 Subtracting Fractions with Different Denominators

Find the difference $\frac{8}{9}-\frac{3}{5}$.

Solution

We list the multiples of 9 and the multiples of 5:

Multiples of 9: 9, 18, 27, 36, 45, 54, 63, 72, 81, . . .
Multiples of 5: 5, 10, 15, 20, 25, 30, 35, 40, 45, . . .

The LCD is 45. We now rewrite each fraction with the denominator 45:

$$\frac{8}{9}-\frac{3}{5}=\frac{8}{9}\cdot\frac{5}{5}-\frac{3}{5}\cdot\frac{9}{9}$$ *LCD is 45.*

$$=\frac{40}{45}-\frac{27}{45}$$ *Multiply numerators and multiply denominators:* $\frac{a}{b}\cdot\frac{c}{d}=\frac{ac}{bd}$

$$=\frac{13}{45}$$ *Subtract numerators and keep common denominator:* $\frac{a}{b}-\frac{c}{b}=\frac{a-c}{b}$

Adding (or Subtracting) Fractions with Different Denominators

To add (or subtract) two fractions with different denominators, use the fact that $\frac{a}{a} = 1$, where a is nonzero, to write an equal sum (or difference) of fractions for which each denominator is the LCD.

Group Exploration

Illustrations of simplifying fractions and operations with fractions

Draw a picture of a pizza to show that the true statement makes sense. [**Hint:** See Figs. 4, 5, 6, and 8.]

1. $\frac{6}{8} = \frac{3}{4}$
2. $\frac{5}{8} + \frac{2}{8} = \frac{7}{8}$
3. $\frac{5}{6} - \frac{4}{6} = \frac{1}{6}$
4. $\frac{1}{2} \cdot \frac{1}{3} = \frac{1}{6}$
5. $\frac{2}{3} \div \frac{1}{6} = 4$

Tips for Success Review Your Notes as Soon as Possible

How often do you get confused by class notes you wrote earlier the same day, even though the class activities made sense to you? If this happens a lot, review your notes as soon as possible. Even reviewing your notes for just a few minutes between classes will help. This will increase your likelihood of remembering what you learned in class and will give you the opportunity to add new comments to your notes while the class experience is still fresh in your mind.

Homework 2.2

For extra help ▶ MyMathLab®  Watch the videos in MyMathLab  Download the MyDashboard App

1. What is the denominator of $\frac{3}{7}$?

2. What is the numerator of $\frac{2}{5}$?

Write the number as a product of primes.

3. 20 **4.** 18 **5.** 36 **6.** 24
7. 45 **8.** 27 **9.** 78 **10.** 105

Simplify. Then use a calculator to check your work.

11. $\frac{6}{8}$ **12.** $\frac{10}{14}$ **13.** $\frac{3}{12}$ **14.** $\frac{7}{28}$
15. $\frac{18}{30}$ **16.** $\frac{27}{54}$ **17.** $\frac{20}{50}$ **18.** $\frac{49}{63}$
19. $\frac{5}{25}$ **20.** $\frac{9}{81}$ **21.** $\frac{20}{24}$ **22.** $\frac{15}{18}$

Perform the indicated operation. Then use a calculator to check your work.

23. $\frac{1}{3} \cdot \frac{2}{5}$ **24.** $\frac{6}{7} \cdot \frac{4}{9}$ **25.** $\frac{4}{5} \cdot \frac{3}{8}$ **26.** $\frac{2}{3} \cdot \frac{5}{6}$
27. $\frac{5}{21} \cdot 7$ **28.** $\frac{5}{12} \cdot 2$ **29.** $\frac{5}{8} \div \frac{3}{4}$ **30.** $\frac{7}{12} \div \frac{2}{3}$
31. $\frac{8}{9} \div \frac{4}{3}$ **32.** $\frac{4}{7} \div \frac{8}{3}$ **33.** $\frac{2}{3} \div 5$ **34.** $\frac{4}{9} \div 2$
35. $\frac{2}{7} + \frac{3}{7}$ **36.** $\frac{5}{9} + \frac{2}{9}$ **37.** $\frac{5}{8} + \frac{1}{8}$ **38.** $\frac{2}{15} + \frac{8}{15}$
39. $\frac{4}{5} - \frac{3}{5}$ **40.** $\frac{5}{7} - \frac{2}{7}$ **41.** $\frac{11}{12} - \frac{7}{12}$ **42.** $\frac{13}{18} - \frac{9}{18}$
43. $\frac{1}{4} + \frac{1}{2}$ **44.** $\frac{1}{3} + \frac{5}{9}$ **45.** $\frac{5}{6} + \frac{3}{4}$ **46.** $\frac{3}{8} + \frac{1}{6}$
47. $4 + \frac{2}{3}$ **48.** $2 + \frac{3}{7}$ **49.** $\frac{7}{9} - \frac{2}{3}$ **50.** $\frac{3}{4} - \frac{1}{2}$
51. $\frac{5}{9} - \frac{2}{7}$ **52.** $\frac{5}{6} - \frac{4}{7}$ **53.** $3 - \frac{4}{5}$ **54.** $1 - \frac{9}{7}$

Perform the indicated operation. If the fraction is undefined, say so. Then use a calculator to check your work.

55. $\frac{3172}{3172}$ **56.** $\frac{62}{62}$ **57.** $\frac{599}{1}$ **58.** $\frac{215}{1}$
59. $\frac{842}{0}$ **60.** $\frac{713}{0}$ **61.** $\frac{0}{621}$ **62.** $\frac{0}{798}$
63. $\frac{824}{631} \cdot \frac{631}{824}$ **64.** $\frac{173}{190} \cdot \frac{190}{173}$
65. $\frac{544}{293} - \frac{544}{293}$ **66.** $\frac{345}{917} - \frac{345}{917}$

Evaluate the given expression for $w = 4, x = 3, y = 5,$ and $z = 12$.

67. $\frac{w}{z}$ **68.** $\frac{z}{x}$ **69.** $\frac{x}{w} \div \frac{y}{z}$

70. $\frac{y}{z} \cdot \frac{w}{x}$ **71.** $\frac{x}{w} - \frac{y}{z}$ **72.** $\frac{y}{x} + \frac{y}{z}$

Use a calculator to compute. Round the result to two decimal places.

73. $\frac{19}{97} \cdot \frac{65}{74}$ **74.** $\frac{67}{71} \cdot \frac{381}{399}$ **75.** $\frac{684}{795} \div \frac{24}{37}$

76. $\frac{149}{215} \div \frac{31}{52}$ **77.** $\frac{89}{102} - \frac{59}{133}$ **78.** $\frac{614}{701} + \frac{391}{400}$

For Exercises 79 and 80, draw a picture of a pizza to show that the true statement makes sense.

79. $\frac{2}{8} = \frac{1}{4}$ **80.** $\frac{1}{4} + \frac{2}{4} = \frac{3}{4}$

81. A rectangular plot of land has a length of $\frac{2}{5}$ mile and a width of $\frac{1}{4}$ mile. What is the area of this plot?

82. A rectangular picture has a width of $\frac{2}{3}$ foot and a length of $\frac{3}{4}$ foot. What is the perimeter of this picture?

83. For an elementary algebra course, total course points are calculated by adding points earned on homework assignments, quizzes, tests, and the final exam. If the total of scores on tests is worth $\frac{1}{2}$ of the course points and the final exam score is worth $\frac{1}{4}$ of the course points, what fraction of the course points comes from homework assignments and quizzes?

84. A family spends $\frac{1}{3}$ of its income for the mortgage and $\frac{1}{6}$ of its income for food. What fraction of its income remains?

For Exercises 85 and 86, let x be a number. Translate the expression into an English phrase.

85. $\frac{x}{3}$ **86.** $\frac{5}{x}$

87. Some friends pay a total of \$19 for a pizza. Each of the n friends pays an equal share of the cost. Complete Table 10 to help find an expression that describes the cost (in dollars) per person. Show the arithmetic to help you see a pattern.

Table 10 Cost per Person for the Pizza

Number of People	Cost per Person (dollars)
2	
3	
4	
5	
n	

88. A tutor charges \$45 for a tutoring session that lasts for t hours. Complete Table 11 to help find an expression that describes the cost (in dollars) per hour. Show the arithmetic to help you see a pattern.

Table 11 Cost per Hour for the Session

Total Time (hours)	Cost per Hour (dollars per hour)
2	
3	
4	
5	
t	

Concepts

89. a. Perform the indicated operation.

i. $\frac{5}{6} \cdot \frac{2}{3}$ **ii.** $\frac{5}{6} \div \frac{2}{3}$

iii. $\frac{5}{6} + \frac{2}{3}$ **iv.** $\frac{5}{6} - \frac{2}{3}$

b. Compare the methods you used to perform the operations in part (a). Describe how the methods are similar and how they are different.

90. a. Find each product.

i. $\frac{2}{3} \cdot \frac{3}{2}$ **ii.** $\frac{4}{7} \cdot \frac{7}{4}$ **iii.** $\frac{1}{6} \cdot \frac{6}{1}$

b. On the basis of your results from part (a), use words to describe a property of a fraction and its reciprocal. Then describe the property in terms of variables.

91. A student tries to find the product $\frac{1}{2} \cdot \frac{1}{3}$:

$$\begin{aligned} \frac{1}{2} \cdot \frac{1}{3} &= \left(\frac{1}{2} \cdot \frac{3}{3}\right) \cdot \left(\frac{1}{3} \cdot \frac{2}{2}\right) \\ &= \frac{3}{6} \cdot \frac{2}{6} \\ &= \frac{6}{36} \\ &= \frac{1}{6} \end{aligned}$$

What would you tell the student?

92. A student tries to find the sum $\frac{2}{3} + \frac{5}{6}$:

$$\frac{2}{3} + \frac{5}{6} = \frac{2+5}{3+6} = \frac{7}{9}$$

Describe any errors. Then find the sum correctly.

93. A student tries to find the product $2 \cdot \frac{3}{5}$:

$$2 \cdot \frac{3}{5} = \frac{6}{10} = \frac{3}{5}$$

Describe any errors. Then find the product correctly.

94. A student tries to find the product $3 \cdot \frac{7}{2}$:

$$3 \cdot \frac{7}{2} = \frac{7}{3 \cdot 2} = \frac{7}{6}$$

Describe any errors. Then find the product correctly.

95. Greenskeepers have been mowing golf putting surfaces progressively lower over the past half century (see Table 12).

Table 12 Grass Heights on Golf Putting Surfaces

Decade	Year Used to Represent Decade	Grass Height (inches)
1950s	1955	$\frac{1}{4}$
1960s	1965	$\frac{7}{32}$
1970s	1975	$\frac{3}{16}$
1980s	1985	$\frac{5}{32}$
1990s	1995	$\frac{1}{8}$
2000s	2005	$\frac{1}{10}$

Source: *Golf Course Superintendents Association of America*

Let h be the grass height (in inches) at t years since 1950.

a. Create a scattergram of the data.

b. Draw a line that comes close to the data points in your scattergram.

c. Predict the grass height in 2025.

d. Predict when the grass height will be $\frac{1}{16}$ inch.

e. Find the t-intercept. What does it mean in this situation?

96. a. Use a number line to find the distance between 0 and each number.

i. 4 **ii.** -3 **iii.** -6

b. *Without* using a number line, describe in general how you can find the distance between 0 and a number on a number line.

97. Explore how to add two negative numbers:

a. Use a calculator to find each sum of two negative numbers.

i. $-1 + (-5)$ **ii.** $-6 + (-2)$ **iii.** $-3 + (-4)$

b. What pattern do you notice? If you do not see a pattern, continue finding sums of any two negative numbers until you do.

c. Without using a calculator, find the sum $-4 + (-5)$. Then use a calculator to check your work.

d. State a general rule for how to add two negative numbers.

98. Explore how to add two numbers with different signs:

a. First, consider the case in which the positive number is farther from 0 on the number line than the negative number is. Use a calculator to find the following sums:

i. $5 + (-2)$ **ii.** $7 + (-1)$ **iii.** $8 + (-3)$

b. What pattern do you notice in your work from part (a)? If you do not see a pattern, continue finding similar sums until you do.

c. Now consider the case in which the positive number is closer to 0 on the number line than the negative number is. Use a calculator to find the following sums:

i. $2 + (-5)$ **ii.** $1 + (-7)$ **iii.** $3 + (-8)$

d. What pattern do you notice in your work from part (c)?

e. Now consider the case in which the two numbers are the same distance from 0 on the number line, but on opposite sides of 0. Use a calculator to find the following sums:

i. $4 + (-4)$ **ii.** $7 + (-7)$ **iii.** $9 + (-9)$

f. What pattern do you notice in your work from part (e)?

g. Without using a calculator, find each sum. Then use a calculator to check your work.

i. $6 + (-4)$ **ii.** $3 + (-7)$ **iii.** $6 + (-6)$

h. State a general rule for how to add two numbers with different signs.

99. Why is division by 0 undefined?

100. Explain why we do not add the denominators of two fractions when we add the fractions.

2.3 Adding Real Numbers

Objectives

- Find the opposite of a number.
- Find the opposite of the opposite of a number.
- Find the absolute value of a number.
- Add real numbers by thinking in terms of money, the number line, and absolute value.
- Add real numbers pertaining to authentic situations.
- Find an expression to model an authentic quantity.

In this section, our main objective is to add real numbers.

The Opposite of a Number

Note that in Fig. 10 both the numbers -3 and 3 are 3 units from 0 on the number line, but they are on opposite sides of 0. We say that -3 is the *opposite* of 3, that 3 is the *opposite* of -3, and that -3 and 3 are *opposites*.

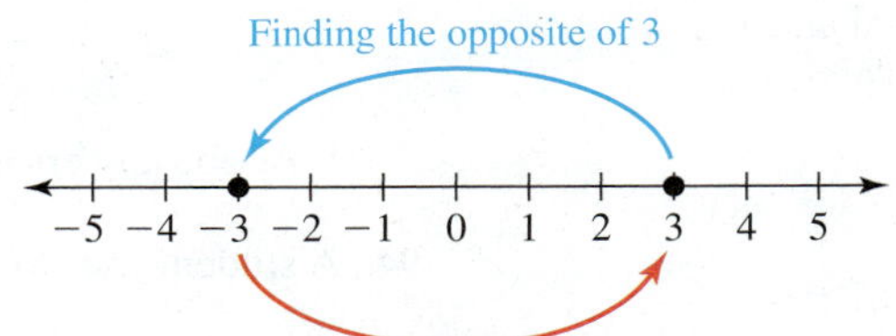

Figure 10 Finding the opposite of 3 and the opposite of -3

Two numbers are called **opposites** of each other if they are the same distance from 0 on the number line, but are on opposite sides of 0. We find the opposite of a number by writing a negative sign in front of the number. For example, the opposite of 3 is -3 (see Fig. 10).

Now consider this true statement:

The opposite of -3 is 3 (see Fig. 10).

In symbols, we write

$$\underbrace{-}_{\text{The opposite of}}\underbrace{(-3)}_{-3}\underbrace{=}_{\text{is equal to}}\underbrace{3}_{3.}$$

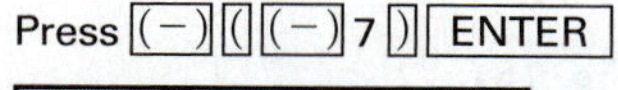

Press [(−)] [(] [(−)] 7 [)] [ENTER].

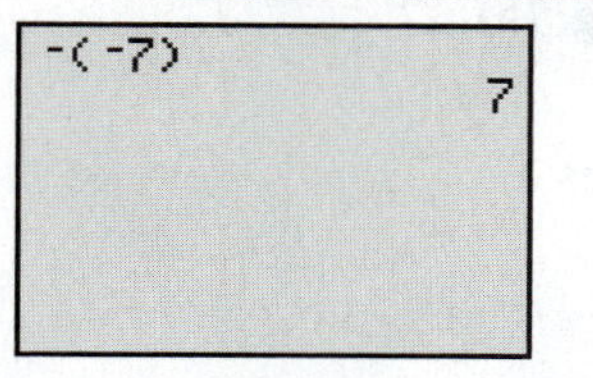

Figure 11 Calculating $-(-7)$

Here are some more examples of finding the opposite of a negative number:

$$-(-2) = 2$$
$$-(-7) = 7$$

We can use a graphing calculator to find $-(-7)$. See Fig. 11. We use the button [−] for subtraction and the button [(−)] for negative numbers and for taking opposites.

We can view $-(-7)$ as finding the opposite of -7 or as finding the opposite of the opposite of 7.

Finding the Opposite of the Opposite of a Number

$$-(-a) = a$$

In words: The opposite of the opposite of a number is equal to that same number.

We use parentheses to separate two opposite symbols or an operation symbol and an opposite symbol.

Example 1 Finding Opposites

Find the opposite.

1. $-(-5)$

2. $-(-(-8))$

Solution

1. $-(-5) = 5$ $\quad -(-a) = a$

2. $-(-(-8)) = -(8)$ $\quad -(-a) = a$

$= -8$ $\quad$ *Write without parentheses.*

Absolute Value

The *absolute value* of a number a, written $|a|$, is the distance the number a is from 0 on the number line.

Definition Absolute value

The **absolute value** of a number is the distance the number is from 0 on the number line.

So $|-3| = 3$, because -3 is a distance of 3 units from 0, and $|3| = 3$, because 3 is a distance of 3 units from 0 (see Fig. 12).

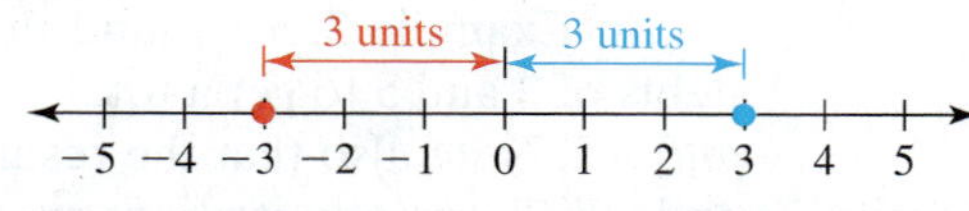

Figure 12 Both -3 and 3 are a distance of 3 units from 0

Press [2nd] 0 [ENTER] [(−)] 3 [)] [ENTER].

abs(-3) 3

Figure 13 Calculating $|-3|$

On a graphing calculator, "abs" stands for absolute value. We find $|-3|$ in Fig. 13.

Example 2 Finding Absolute Values of Numbers

Calculate.

1. $|2|$

2. $|-2|$

3. $-|2|$

4. $-|-2|$

Solution

1. $|2| = 2$, because 2 is a distance of 2 units from 0 (see Fig. 14).

Figure 14 $|2| = 2$

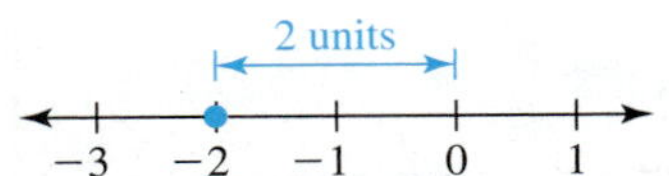

Figure 15 $|-2| = 2$

2. $|-2| = 2$, because -2 is a distance of 2 units from 0 (see Fig. 15).
3. $-|2| = -(2)$ $|2| = 2$
   $= -2$ *Write without parentheses.*
4. $-|-2| = -(2)$ $|-2| = 2$
   $= -2$ *Write without parentheses.*

Press [(−)] [2nd] 0 [ENTER] [(−)] 2 [)] [ENTER].

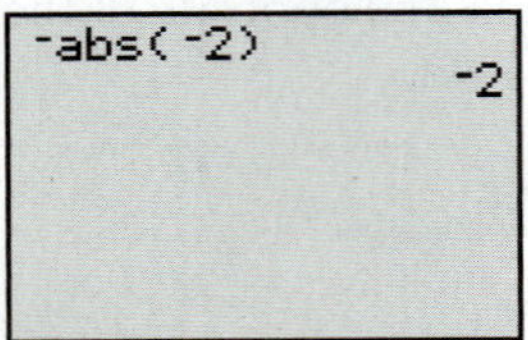

Figure 16 Check that $-|-2| = -2$

We can use a graphing calculator to check that $-|-2| = -2$ (see Fig. 16).

Addition of Two Numbers with the Same Sign

Thinking about credit card balances or the number line can help us see how to add numbers with the same sign.

Example 3 Finding the Sum of Two Numbers with the Same Sign

1. A person has a credit card balance of 0 dollars. If she uses her credit card to make two purchases, one for \$2 and one for \$5, what is the new balance?
2. Write a sum that is related to the computation in Problem 1.
3. Use a number line to illustrate the sum found in Problem 2.

Solution

1. By making purchases for \$2 and \$5, the person now owes \$7. So, the new balance is -7 dollars.
2. Here is the sum:

$$\underbrace{-2}_{\text{Spend \$2}} + \underbrace{(-5)}_{\text{Spend \$5}} = \underbrace{-7}_{\text{Owe \$7}}$$

3. Using the number line, imagine moving 2 units to the left of 0 and then 5 more units to the left. Figure 17 illustrates that $-2 + (-5) = -7$.

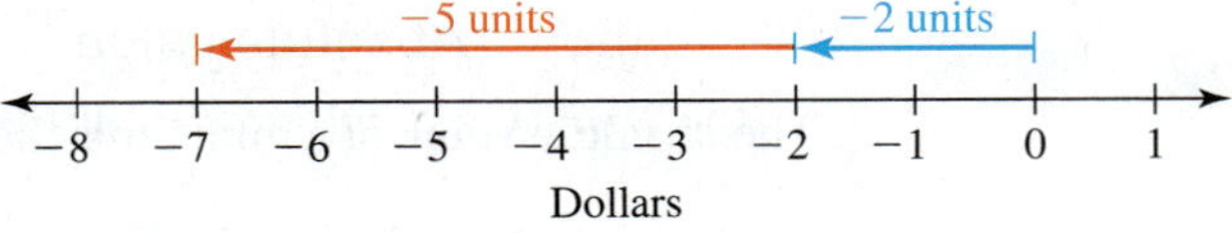

Figure 17 Illustration of $-2 + (-5) = -7$

In Example 3, we found that $-2 + (-5) = -7$. To get this result, we added the debts of 2 and 5 to get a total debt of 7. Note that 2 and 5 are the absolute values of -2 and -5. Note also that the result, -7, of the original sum has the same sign as both -2 and -5. These observations suggest the following procedure:

Adding Two Numbers with the Same Sign

To add two numbers with the same sign,

1. Add the absolute values of the numbers.
2. The sum of the original numbers has the same sign as the sign of the original numbers.

Example 4 Finding the Sum of Two Numbers with the Same Sign

Find the sum.

1. $-3 + (-6)$ **2.** $-\frac{1}{5} + \left(-\frac{3}{5}\right)$

Solution

1. First, we add the absolute values of the numbers -3 and -6: $3 + 6 = 9$. Since both -3 and -6 are negative, their sum is negative. So, $-3 + (-6) = -9$.

2. By adding the absolute values of the fractions, we have $\frac{1}{5} + \frac{3}{5} = \frac{4}{5}$. Since both original fractions are negative, their sum is negative. So,

$$-\frac{1}{5} + \left(-\frac{3}{5}\right) = -\frac{4}{5}$$

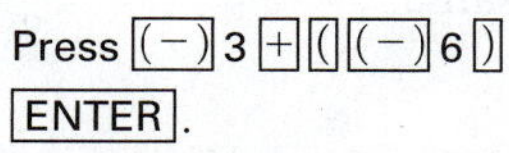

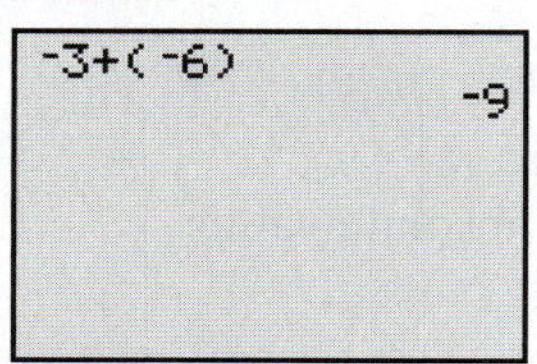

Figure 18 Calculating $-3 + (-6)$

Toward the start of this course, your instructor will likely have you add real numbers without using a calculator. Once you have mastered that skill, your instructor will likely encourage you to use a calculator for longer computations.

After completing an exercise in this section's homework by hand, you can use a graphing calculator to check your work. For example, we can check our work for $-3 + (-6)$, Problem 1 in Example 4 (see Fig. 18).

Addition of Two Numbers with Different Signs

Thinking about the number line or exchanges of money can also help us see how to add numbers with different signs.

Example 5 Finding the Sum of Two Numbers with Different Signs

1. A brother owes his sister \$5. If he then pays her back \$2, how much does he still owe her?

2. Write a sum that is related to your work in Problem 1.

3. Use a number line to illustrate the sum you found in Problem 2.

Solution

1. By owing his sister \$5 and paying her back \$2, the brother now owes his sister \$3.

2. Here's the sum:

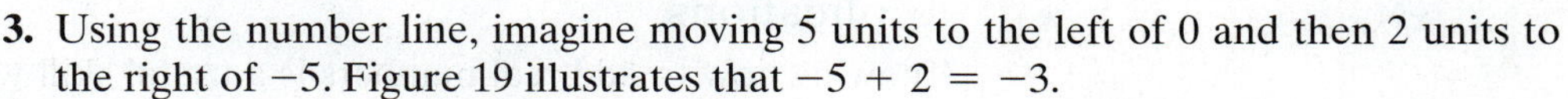

3. Using the number line, imagine moving 5 units to the left of 0 and then 2 units to the right of -5. Figure 19 illustrates that $-5 + 2 = -3$.

2 units
−5 units
−5 −3 0
Dollars

Figure 19 Illustration of $-5 + 2 = -3$

In Problem 2 in Example 5, we found that $-5 + 2 = -3$. We can get this result by first finding the difference of 5 and 2:

$$5 - 2 = 3$$

We can think of this operation as lowering a debt of 5 dollars by 2 dollars to get a debt of 3 dollars, so the result is -3. Note that the result, -3, has the same sign as -5, which has a larger absolute value than 2. These observations suggest the following procedure:

Adding Two Numbers with Different Signs

To add two numbers with different signs,

1. Find the absolute values of the numbers. Then subtract the smaller absolute value from the larger absolute value.

2. The sum of the original numbers has the same sign as the original number with the larger absolute value.

Example 6 Finding the Sum of Two Numbers with Different Signs

Find the sum.

1. $-4 + 7$ **2.** $3 + (-9)$ **3.** $-\frac{5}{6} + \frac{2}{3}$

Solution

1. First, we find that $7 - 4 = 3$. Since 7 has a larger absolute value than -4, and since 7 is positive, the sum is positive: $-4 + 7 = 3$.
2. First, we find that $9 - 3 = 6$. Since -9 has a larger absolute value than 3, and since -9 is negative, the sum is negative: $3 + (-9) = -6$.
3. First, we write the fractions so that the denominators are the same.

$$-\frac{5}{6} + \frac{2}{3} = -\frac{5}{6} + \frac{2}{3} \cdot \frac{2}{2} \quad \text{LCD is 6.}$$

$$= -\frac{5}{6} + \frac{4}{6} \quad \text{Multiply numerators and multiply denominators: } \frac{a}{b} \cdot \frac{c}{d} = \frac{ac}{bd}$$

Next, we subtract the smaller absolute value from the larger absolute value:

$$\frac{5}{6} - \frac{4}{6} = \frac{1}{6}$$

Since $-\frac{5}{6}$ has a larger absolute value than the fraction $\frac{4}{6}$, and since $-\frac{5}{6}$ is negative, the sum is negative:

$$-\frac{5}{6} + \frac{4}{6} = -\frac{1}{6}$$

We have discussed three ways to add real numbers: thinking in terms of money, the number line, and absolute value. **It is good practice to use one method to find a sum and then use another method (or a calculator) as a check.**

Example 7 Translating from English to Mathematics

Let x be a number. Translate the phrase "the sum of -4 and the number" into a mathematical expression. Then evaluate the expression for $x = -2$.

Solution

The expression is $-4 + x$. We substitute -2 for x in the expression $-4 + x$ and then find the sum:

$$-4 + (-2) = -6$$

Applications

Knowing how to add real numbers is a useful skill when you work with quantities that can be negative, such as balances of checking accounts and temperature readings.

Example 8 Applications of Adding Real Numbers

1. A person bounces several checks and is charged service fees such that the balance of the checking account is -90.75 dollars. If the person then deposits 300 dollars, what is the balance?
2. Three hours ago, the temperature was -11°F. If the temperature has increased by 5°F in the last three hours, what is the current temperature?

Solution

1. The balance is $-90.75 + 300$ dollars. To find this sum, we first find the difference $300 - 90.75 = 209.25$. Since 300 has a larger absolute value than -90.75 and since 300 is positive, the sum is positive: $-90.75 + 300 = 209.25$. So, the balance is \$209.25.
2. The temperature is $-11 + 5$ degrees Fahrenheit. To find this sum, we first find the difference $11 - 5 = 6$. Since -11 has a larger absolute value than 5 and since -11 is negative, the sum is negative: $-11 + 5 = -6$. So, the current temperature is -6°F.

Modeling with Expressions

In Example 9, we will use an expression to describe an authentic quantity.

Example 9 Finding and Evaluating an Expression

A person fills up his 15-gallon gasoline tank. Let c be the amount of gasoline (in gallons) consumed after the tank has been filled up.

1. Use a table to help find an expression that describes the amount of gasoline (in gallons) that remains in the tank.
2. Evaluate the expression found in Problem 1 for $c = 7$. What does the result mean in this situation?

Table 13 Gasoline Remaining

Gasoline Consumed (gallons)	Gasoline Remaining (gallons)
1	$15 + (-1)$
2	$15 + (-2)$
3	$15 + (-3)$
4	$15 + (-4)$
c	$15 + (-c)$

Solution

1. First, we create Table 13. We show the arithmetic to help us see the pattern. From the last row of the table, we see that the expression $15 + (-c)$ represents the amount of gasoline (in gallons) that remains.
2. We evaluate $15 + (-c)$ for $c = 7$:

$$15 + (-(7)) = 15 + (-7) = 8$$

If 7 gallons of gasoline are consumed, 8 gallons of gasoline will remain.

Group Exploration

Adding a number and its opposite

1. Evaluate $a + (-a)$ for the given values of a.
   a. $a = 2$ **b.** $a = 3$ **c.** $a = 5$
2. Evaluate $a + (-a)$ for the given values of a. [**Hint:** Use $-(-a) = a$.]
   a. $a = -2$ **b.** $a = -3$ **c.** $a = -5$
3. What do your results in Problems 1 and 2 suggest about $a + (-a)$?

Group Exploration

Looking ahead: Subtracting numbers

1. Find the difference $6 - 4$ and the sum $6 + (-4)$, and compare your results.
2. Find the difference $7 - 3$ and the sum $7 + (-3)$, and compare your results.
3. Find the difference $9 - 2$ and the sum $9 + (-2)$, and compare your results.
4. In Problems 1–3, for each difference, there is a related sum that gives the same result. Write $a - b$ as a sum.
5. In Problem 4, you wrote $a - b$ as a sum. Use this method to find the given difference.
   a. $8 - 3$ **b.** $2 - 5$ **c.** $-4 - 3$

Tips for Success Affirmations

Do you have difficulty with math? If so, do you ever tell yourself (or others) that you are not good at it? This is called *negative self-talk*. The more you say this, the more likely your subconscious will believe it—and you *will* do poorly in math.

You can counteract years of negative self-talk by telling yourself with conviction that you are good at math.

It might seem strange to state that something is true that hasn't happened yet, but it works! Such statements are called *affirmations*.

There are four guiding principles for getting the most out of saying affirmations:

1. Say affirmations that imply the desired event is currently happening. For example, say

"I am good at algebra," not "I will be good at algebra."

2. Say that your desired result is continuing to improve. For example, say

"I am good at algebra, and I continue to improve at it."

3. Say affirmations in the positive rather than in the negative. For example, say

"I attend each class," not "I don't cut classes."

4. Say affirmations with conviction.

If you would like to learn more about affirmations, the book *Creative Visualization* (Bantam Books, 1985), by Shakti Gawain, is an excellent resource.

Homework 2.3

For extra help ▶ MyMathLab®

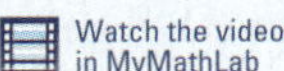

Compute by hand. Then use a calculator to check your work.

1. $-(-4)$ **2.** $-(-9)$ **3.** $-(-(-7))$

4. $-(-(-2))$ **5.** $|3|$ **6.** $|6|$

7. $|-8|$ **8.** $|-1|$ **9.** $-|4|$

10. $-|5|$ **11.** $-|-7|$ **12.** $-|-9|$

Find the sum by hand. Then use a calculator to check your work.

13. $2 + (-7)$ **14.** $5 + (-3)$ **15.** $-1 + (-4)$

16. $-3 + (-2)$ **17.** $7 + (-5)$ **18.** $6 + (-9)$

19. $-8 + 5$ **20.** $-3 + 4$ **21.** $-7 + (-3)$

22. $-9 + (-5)$ **23.** $4 + (-7)$ **24.** $8 + (-2)$

25. $1 + (-1)$ **26.** $8 + (-8)$ **27.** $-4 + 4$

28. $-7 + 7$ **29.** $12 + (-25)$ **30.** $17 + (-14)$

31. $-39 + 17$ **32.** $-89 + 57$

33. $-246 + (-899)$ **34.** $-347 + (-594)$

35. $25{,}371 + (-25{,}371)$ **36.** $127{,}512 + (-127{,}512)$

37. $-4.1 + (-2.6)$ **38.** $-3.7 + (-9.9)$

39. $-5 + 0.2$ **40.** $-0.3 + 7$

41. $2.6 + (-99.9)$ **42.** $37.05 + (-19.26)$

43. $\frac{5}{7} + \left(-\frac{3}{7}\right)$ **44.** $\frac{2}{5} + \left(-\frac{1}{5}\right)$

45. $-\frac{5}{8} + \frac{3}{8}$ **46.** $-\frac{5}{6} + \frac{1}{6}$

47. $-\frac{1}{4} + \left(-\frac{1}{2}\right)$ **48.** $-\frac{2}{3} + \left(-\frac{5}{6}\right)$

49. $\frac{5}{6} + \left(-\frac{1}{4}\right)$ **50.** $\frac{2}{3} + \left(-\frac{3}{4}\right)$

Use a calculator to find the sum. Round the result to two decimal places.

51. $-325.89 + 6547.29$ **52.** $-7498.34 + 6435.28$

53. $-17{,}835.69 + (-79{,}735.45)$

54. $-38{,}487.26 + (-83{,}205.87)$

55. $-\frac{34}{983} + \left(-\frac{19}{251}\right)$ **56.** $-\frac{37}{642} + \left(-\frac{25}{983}\right)$

Evaluate the expression for $a = -4$, $b = 3$, and $c = -2$.

57. $a + b$ **58.** $b + a$ **59.** $a + c$ **60.** $b + c$

For Exercises 61–64, let x be a number. Translate the English phrase into a mathematical expression. Then evaluate the expression for $x = -6$.

61. Two more than the number

62. The number increased by 3

63. The sum of -5 and the number

64. The number plus -8

65. A person bounces several checks and is charged service fees such that the balance of the checking account is -75 dollars. If the person then deposits 250 dollars, what is the balance?

66. A person bounces several checks and is charged service fees such that the balance of the checking account is -112.50 dollars. If the person then deposits 170 dollars, what is the balance?

67. A check register is shown in Table 14. Find the final balance of the checking account.

Table 14 Check Register

Check No.	Date	Description of Transaction	Payment	Deposit	Balance
					−89.00
	7/18	Transfer		300.00	
3021	7/22	State Farm	91.22		
3022	7/22	MCI	44.26		
	7/31	Paycheck		870.00	

68. A check register is shown in Table 15. Find the final balance of the checking account.

Table 15 Check Register

Check No.	Date	Description of Transaction	Payment	Deposit	Balance
					−135.00
	2/31	Paycheck		549.00	
253	3/2	FedEx Kinko's	10.74		
	3/3	ATM	21.50		
254	3/7	Barnes and Noble	17.19		

69. A person has a credit card balance of −5471 dollars. If she sends a check to the credit card company for \$2600, what is the new balance?

70. A person has a credit card balance of −2739 dollars. If he sends a check to the credit card company for \$530, what is the new balance?

71. A student has a credit card balance of −3496 dollars. If he sends a check to the credit card company for \$2500 and then uses his credit card to purchase a bicycle for \$613 and a helmet for \$24, what is the new balance?

72. A student has a credit card balance of −873 dollars. If she sends a check to the credit card company for \$500 and then uses the card to buy a tennis racquet for \$249 and a tennis outfit for \$87, what is the new balance?

73. Three hours ago, it was −5°F. If the temperature has increased by 9°F in the last three hours, what is the current temperature?

74. Four hours ago, it was −12°F. If the temperature has increased by 8°F in the last four hours, what is the current temperature?

75. A person just lost 20 pounds on a diet.

a. Complete Table 16 to help find an expression that describes the person's current weight (in pounds) if the person's weight before the diet was B pounds. Show the arithmetic to help you see a pattern.

Table 16 Weights before and after the Diet

Weight before Diet (pounds)	Weight after Diet (pounds)
160	
165	
170	
175	
B	

b. Evaluate the expression you found in part (a) for $B = 169$. What does your result mean in this situation?

76. An electronics store is offering a weekend sale of \$35 off the retail price of any of its LCD televisions.

a. Complete Table 17 to help find an expression that describes the sale price (in dollars) if the retail price is r dollars. Show the arithmetic to help you see a pattern.

Table 17 Retail and Sale Prices

Retail Price (dollars)	Sale Price (dollars)
350	
400	
450	
500	
r	

b. Evaluate the expression you found in part (a) for $r = 470$. What does your result mean in this situation?

77. The balance in a person's checking account is −80 dollars.

a. Complete Table 18 to help find an expression that describes the new balance (in dollars) if the person deposits d dollars. Show the arithmetic to help you see a pattern.

Table 18 Deposits and New Balances

Deposit (dollars)	New Balance (dollars)
50	
100	
150	
200	
d	

b. Evaluate the expression you found in part (a) for $d = 125$. What does your result mean in this situation?

78. One hour ago, the temperature was −2°F.

a. Complete Table 19 to help find an expression that describes the current temperature (in degrees Fahrenheit) if the temperature decreased by x degrees Fahrenheit in the past hour. Show the arithmetic to help you see a pattern.

Table 19 Decreases in Temperature and Current Temperatures

Decrease in Temperature (degrees Fahrenheit)	Current Temperature (degrees Fahrenheit)
1	
2	
3	
4	
x	

b. Evaluate the expression you found in part (a) for $x = 7$. What does your result mean in this situation?

Concepts

79. If a is negative and b is negative, what can you say about the sign of $a + b$? Use a number line to show this property.

80. If a is positive, b is negative, and b is larger in absolute value than a, what can you say about the sign of $a + b$? Use a number line to show this property.

81. If $a + b = 0$, what can you say about a and b?

82. If $a + b$ is positive, what can you say about a and b?

83. **a.** Evaluate $-a$ for $a = -3$.

b. Evaluate $-a$ for $a = -4$.

c. Evaluate $-a$ for $a = -6$.

d. A student says that $-a$ represents only negative numbers because $-a$ has a negative sign. Is the student correct? Explain.

84. a. Evaluate $a + b$ for $a = -2$ and $b = 5$.
b. Evaluate $b + a$ for $a = -2$ and $b = 5$.
c. Compare your results from parts (a) and (b).
d. Evaluate $a + b$ and $b + a$ for $a = -4$ and $b = -9$, and then compare the results.
e. Evaluate $a + b$ and $b + a$ for values of your choosing for a and b, and then compare the results.
f. Is the statement $a + b = b + a$ true for all numbers a and b? Explain.

2.4 Change in a Quantity and Subtracting Real Numbers

Objectives

- » Find the change in a quantity.
- » Subtract real numbers.
- » Know the sign of the change for an increasing or decreasing quantity.

In this section, we will discuss how to use subtraction to compute how much a quantity has changed. For example, we can compute the increase in a population of wolves or the decrease in the percentage of eligible voters who voted in an election.

Change in a Quantity

If the value of a stock increases from \$5 to \$8, we say the value *changed* by \$3. Finding the change in a quantity is a very important concept in mathematics and has many applications. A company is extremely focused on the change in its profits. During an operation, a surgeon keeps a close eye on the change in a patient's blood pressure. You probably care deeply about a change in your GPA.

Example 1 Finding the Change in a Quantity

1. If a student's CD collection increases from 2 CDs to 7 CDs, find the change in the number of CDs.
2. Write a difference that is related to the computation in Problem 1.

Solution

1. If the number of CDs increases from 2 CDs to 7 CDs, then the change in the number of CDs is 5 CDs (see Fig. 20).

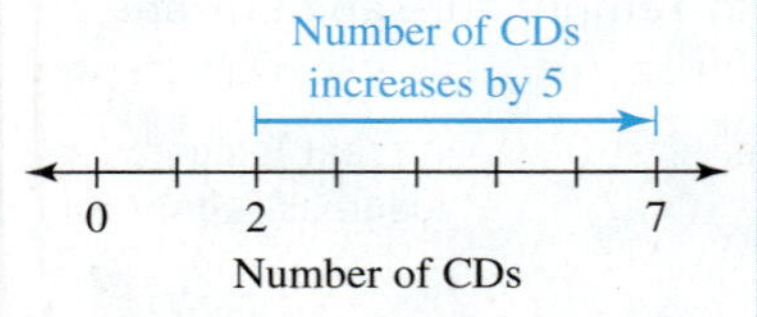

Figure 20 The change in the number of CDs is 5 CDs

2. Here is the difference:

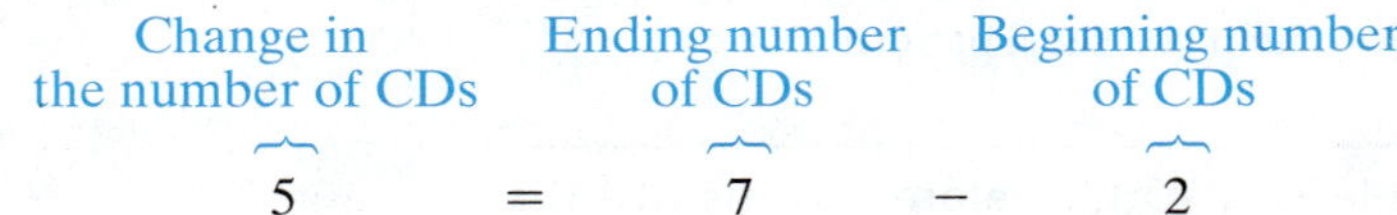

$$\underbrace{5}_{\text{Change in the number of CDs}} = \underbrace{7}_{\text{Ending number of CDs}} - \underbrace{2}_{\text{Beginning number of CDs}}$$

In Example 1, we found the change in the number of CDs by finding the difference of the ending number of CDs and the beginning number of CDs.

Change in a Quantity

The change in a quantity is the ending amount minus the beginning amount:

$$\text{Change in the quantity} = \text{Ending amount} - \text{Beginning amount}$$

In Example 2, we find the changes in a quantity from one year to the next.

Table 20 Average Movie Ticket Prices

Years	Average Price (dollars)
2005	6.41
2006	6.55
2007	6.88
2008	7.18
2009	7.50
2010	7.89

Source: *National Association of Theater Owners*

Example 2 Finding Changes in a Quantity

Average movie ticket prices are shown in Table 20 for various years.

1. Find the change in the average movie ticket price from 2005 to 2006.
2. For the period 2005–2010, find each of the changes in the average movie ticket price from one year to the next.
3. From which year to the next did the average price increase the most?

Solution

1. We find the difference of the average price in 2006 (ending) and the average price in 2005 (beginning):

$$\underbrace{6.55}_{\substack{\text{Ending average price}\\\text{(in dollars)}}} - \underbrace{6.41}_{\substack{\text{Beginning average price}\\\text{(in dollars)}}} = \underbrace{0.14}_{\substack{\text{Change in average price}\\\text{(in dollars)}}}$$

So, the average ticket price from 2005 to 2006 changed by \$0.14.

2. The changes in the average ticket price from one year to the next are listed in Table 21. The changes were found by computing the differences similar to the one found in Problem 1.

Table 21 Changes in Average Movie Ticket Prices from Year to Year

Years	Average Price (dollars)	Change in Average Price (dollars)
2005	6.41	
2006	6.55	$6.55 - 6.41 = 0.14$
2007	6.88	$6.88 - 6.55 = 0.33$
2008	7.18	$7.18 - 6.88 = 0.30$
2009	7.50	$7.50 - 7.18 = 0.32$
2010	7.89	$7.89 - 7.50 = 0.39$

Source: *National Association of Theater Owners*

3. The average ticket price changed by \$0.39 from 2009 to 2010, the greatest change from any year to the next.

Subtraction of Real Numbers

Exploring the change in a quantity can help us see how to subtract real numbers.

Example 3 Finding the Difference of Two Real Numbers

1. A college's enrollment decreases from 7 thousand students to 2 thousand students. What is the change in the enrollment?
2. Write a difference that is related to the computation in Problem 1.

Solution

1. Since the enrollment has decreased from 7 thousand students to 2 thousand students, the change is -5 thousand students. The change is negative because the enrollment is decreasing (see Fig. 21).

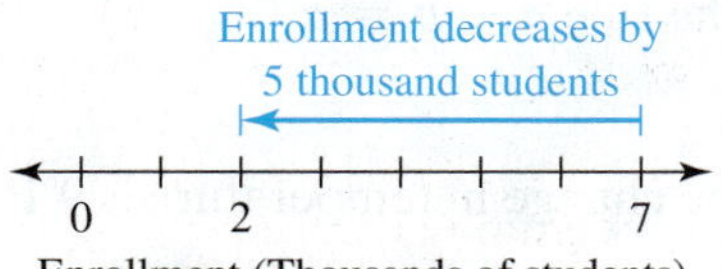

Figure 21 Enrollment decreases from 7 thousand students to 2 thousand students

2. The change in the enrollment is the difference of the ending enrollment and the beginning enrollment:

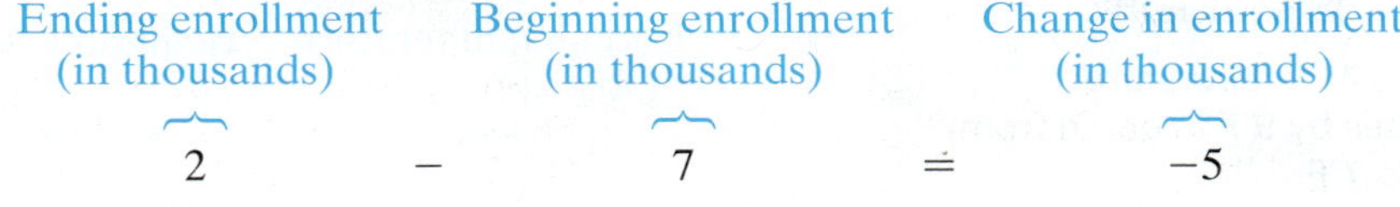

$$\underbrace{2}_{\substack{\text{Ending enrollment}\\\text{(in thousands)}}} - \underbrace{7}_{\substack{\text{Beginning enrollment}\\\text{(in thousands)}}} = \underbrace{-5}_{\substack{\text{Change in enrollment}\\\text{(in thousands)}}}$$

In Example 3, we found that

$$2 - 7 = -5$$

Note that $2 + (-7)$ gives the same result:

$$2 + (-7) = -5$$

This means

$$2 - 7 = 2 + (-7)$$

which suggests that subtracting a number is the same as adding the opposite of that number:

$$\overbrace{2 - 7}^{\text{Subtract 7.}} = \overbrace{2 + (-7)}^{\text{Add the opposite of 7.}}$$

Subtracting a Real Number

$$a - b = a + (-b)$$

In words: To subtract a number, add its opposite.

To subtract real numbers, we first write the difference as a related sum and then find the sum.

Example 4 Finding Differences of Real Numbers

Find the difference.

1. $4 - 6$

2. $\frac{2}{9} - \frac{5}{9}$

Solution

1. $\overbrace{4 - 6}^{\text{Subtract 6.}} = \overbrace{4 + (-6)}^{\text{Add the opposite of 6.}} = -2$ *$a - b = a + (-b)$*

2. $\frac{2}{9} - \frac{5}{9} = \frac{2}{9} + \left(-\frac{5}{9}\right)$ *Add the opposite of $\frac{5}{9}$: $a - b = a + (-b)$*

$= -\frac{3}{9}$

$= -\frac{1}{3}$ *Simplify.*

Considering the change in a quantity can also help us see how to subtract a negative number.

Example 5 Subtracting a Negative Number

1. The temperature increases from −2°F to 7°F. Find the change in temperature.
2. Write a difference that is related to the work in Problem 1.
3. Find the difference obtained in Problem 2 by using the rule $a - b = a + (-b)$.

Solution

1. Since the temperature increased from −2°F to 7°F, the change in temperature is 9°F (see Fig. 22).

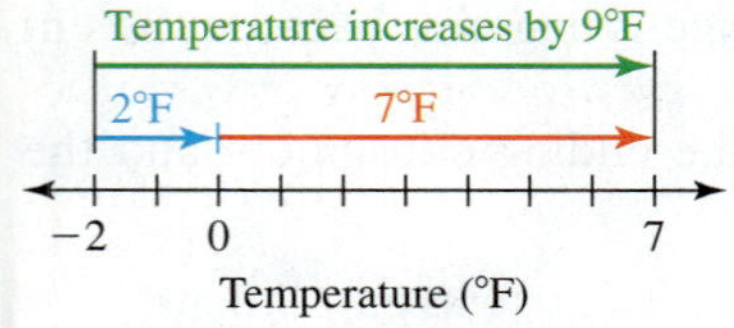

Figure 22 Temperature increases by 9°F in going from −2°F to 7°F

2. The change in temperature is the ending temperature minus the beginning temperature:

$$\overbrace{7}^{\text{Ending temperature (°F)}} - \overbrace{(-2)}^{\text{Beginning temperature (°F)}} = \overbrace{9}^{\text{Change in temperature (°F)}}$$

3.

Subtract -2. Add the opposite of -2. (So, add 2.)

$$7 - (-2) = 7 + 2 \qquad a - b = a + (-b)$$
$$= 9 \qquad \text{Add.}$$

Note that to find the difference $7 - (-2)$, we add 2 and 7. This makes sense, because, in going from -2°F to 0°F, the temperature increases by 2°F, and in continuing from 0°F to 7°F, the temperature increases by another 7°F (see Fig. 22).

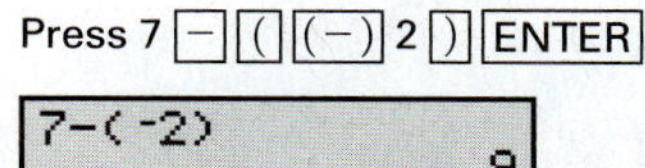

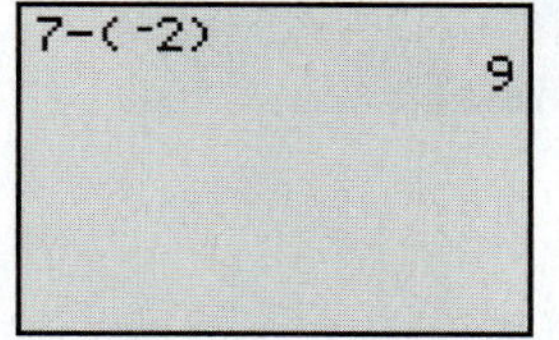

Figure 23 Calculating $7 - (-2)$

We can use a calculator to check our work in Example 5 (see Fig. 23). Recall from Section 2.3 that we use the button [−] for subtraction and the button [(−)] for negative numbers and for taking opposites.

It is good practice to do homework exercises first by hand and then by using a calculator to check your hand results.

Example 6 Subtracting a Negative Number

Find the difference.

1. $4 - (-6)$ **2.** $-9 - (-3)$

Solution

1.

Subtract -6. Add the opposite of -6. (So, add 6.)

$$4 - (-6) = 4 + 6 \qquad a - b = a + (-b)$$
$$= 10 \qquad \text{Add.}$$

2.

Subtract -3. Add the opposite of -3. (So, add 3.)

$$-9 - (-3) = -9 + 3 \qquad a - b = a + (-b)$$
$$= -6 \qquad \text{Add.}$$

Example 7 Translating from English to Mathematics

Translate the phrase "the difference of a and b" into a mathematical expression. Then evaluate the expression for $a = 3$ and $b = -7$.

Solution

The expression is $a - b$. We substitute 3 for a and -7 for b in the expression $a - b$ and then find the difference:

$$(3) - (-7) = 3 + 7 \qquad a - b = a + (-b)$$
$$= 10 \qquad \text{Add.}$$

Elevation

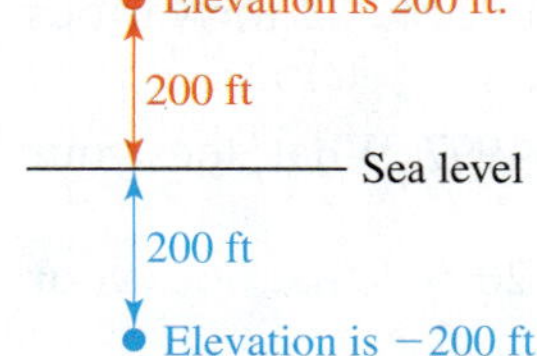

Figure 24 Elevations of 200 ft and -200 ft

In Example 8, you will work with *elevation*. An object that has a *positive* elevation of 200 ft is 200 ft *above* sea level (see Fig. 24). An object that has a *negative* elevation of -200 ft is 200 feet *below* sea level (see Fig. 24).

Example 8 Finding a Change in Elevation

The Golden Gate Bridge has two towers that support the two main cables of the bridge (see Fig. 25). The top of each tower is at an elevation of 746 ft, and the foot of each tower is at an elevation of -136 ft (136 feet below sea level). Find the height of each tower.

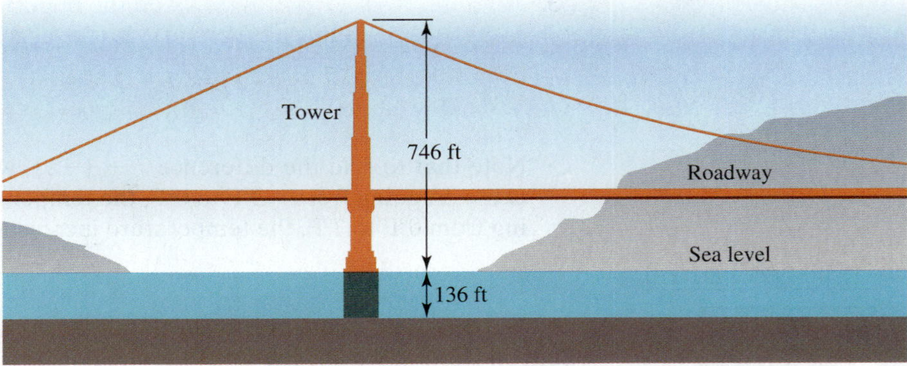

Figure 25 Golden Gate Bridge

Solution

We can find the height of each tower by computing the change in elevation from the bottom of each tower to the top:

$$\underbrace{746}_{\substack{\text{Top elevation} \\ \text{(in feet)}}} - \underbrace{(-136)}_{\substack{\text{Bottom elevation} \\ \text{(in feet)}}} = 746 + 136 \quad a - b = a + (-b)$$

$$= 882 \quad \textit{Add.}$$

So, the height of each tower is 882 ft.

Changes of Increasing and Decreasing Quantities

An increasing quantity has a positive change. For instance, in Example 5, the temperature *increased* from -2°F to 7°F and the change in temperature was *positive* (9°F).

A decreasing quantity has a negative change. For instance, in Example 3, the college's enrollment *decreased* from 7 thousand students to 2 thousand students and the change in enrollment was *negative* (-5 thousand students).

Changes of Increasing and Decreasing Quantities

- An increasing quantity has a positive change.
- A decreasing quantity has a negative change.

In Example 9, we will consider the meaning of a quantity with a positive or negative change.

Table 22 Numbers of AIDS Deaths in the United States

Year	Number of AIDS Deaths
1996	37,787
1997	21,923
1998	17,930
1999	16,273
2000	15,245
2001	14,175
2002	16,948
2003	17,679
2004	17,154
2005	16,823
2006	15,564
2007	14,561
2008	16,084
2009	17,774

Source: *U.S. Centers for Disease Control and Prevention*

Example 9 Finding Changes in Quantities

The numbers of AIDS deaths in the United States are shown in Table 22 for various years.

1. Find the change in the number of AIDS deaths from 1996 to 1997. What does your result mean in terms of the number of AIDS deaths?
2. Find the change in the number of AIDS deaths from 2008 to 2009. What does your result mean in terms of the number of AIDS deaths?
3. Find the change in the number of AIDS deaths from one year to the next, beginning in 1996.
4. Great strides have been made in fighting the AIDS epidemic. However, this progress has brought complacency. Research has shown that some people are now less concerned about getting infected and that risky behaviors, such as unprotected sex and needle sharing, have increased.[†] Do the changes you calculated in Problem 3 support the research findings?

[†]"Combating Complacency in HIV Prevention," *Body Positive*, Feb. 2001, XIV(2).

Solution

1. Since $21{,}923 - 37{,}787 = -15{,}864$, we conclude that the number of AIDS deaths from 1996 to 1997 decreased by 15,864 deaths.
2. Since $17{,}774 - 16{,}084 = 1690$, we conclude that the number of AIDS deaths from 2008 to 2009 increased by 1690 deaths.
3. The changes in number of AIDS deaths from one year to the next are listed in Table 23. The changes were found by computing differences similar to those found in Problems 1 and 2.

Table 23 Changes in AIDS Deaths from Year to Year

Year	Number of AIDS Deaths	Change in the Number of AIDS Deaths from Previous Year to Current Year
1996	37,787	—
1997	21,923	$21{,}923 - 37{,}787 = -15{,}864$
1998	17,930	$17{,}930 - 21{,}923 = -3993$
1999	16,273	$16{,}273 - 17{,}930 = -1657$
2000	15,245	$15{,}245 - 16{,}273 = -1028$
2001	14,175	$14{,}175 - 15{,}245 = -1070$
2002	16,948	$16{,}948 - 14{,}175 = 2773$
2003	17,679	$17{,}679 - 16{,}948 = 731$
2004	17,154	$17{,}154 - 17{,}679 = -525$
2005	16,823	$16{,}823 - 17{,}154 = -331$
2006	15,564	$15{,}564 - 16{,}823 = -1259$
2007	14,561	$14{,}561 - 15{,}564 = -1003$
2008	16,084	$16{,}084 - 14{,}561 = 1523$
2009	17,774	$17{,}774 - 16{,}084 = 1690$

4. Yes, the changes do support the research findings. The annual declines generally decreased from 15,864 for the period 1996–1997 to only 1070 for the period 2000–2001, and the number of deaths increased for both the period 2001–2003 and for the period 2007–2009. In fact, the number of deaths in 2009 was more than the number of deaths in any year of the past ten years. An increase in risky behaviors might explain the reversal.

In Example 10, we will find an expression that describes the change in a quantity.

Example 10 Finding an Expression Describing Change

In 2012, a financial planner had 132 clients. Let n be the number of clients in 2013.

1. Use a table to help find an expression that describes the change in the number of clients from 2012 to 2013.
2. Evaluate the expression found in Problem 1 for $n = 115$. What does the result mean in this situation?

Table 24 Changes in Number of Clients

Number of Clients in 2013	Change in the Number of Clients from 2012 to 2013
133	$133 - 132$
134	$134 - 132$
135	$135 - 132$
136	$136 - 132$
n	$n - 132$

Solution

1. First, we create Table 24. We show the arithmetic to help us see the pattern. From the last row of the table, we see that the expression $n - 132$ represents the change in the number of clients.
2. We evaluate $n - 132$ for $n = 115$:

$$(115) - 132 = 115 + (-132) = -17$$

So, there was a change of -17 clients from 2012 to 2013. In other words, the financial planner's client base decreased by 17 clients.

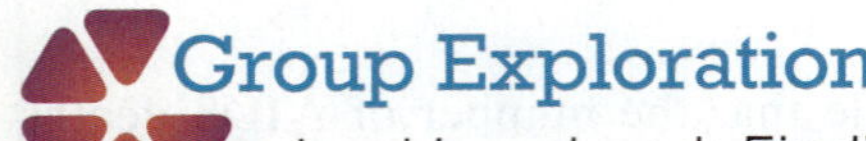

Group Exploration

Looking ahead: Finding the product of a positive number and a negative number

We can think of the multiplication of two counting numbers as a repeated addition. For example, we can think of 4(3) as adding four 3s:

$$4(3) = 3 + 3 + 3 + 3 = 12$$

We can use this idea to help us find the product of a positive number and a negative number.

1. Write each of the products that follow as a repeated sum. Then find the sum.

a. $3(-2)$ **b.** $5(-4)$ **c.** $7(-1)$

2. Are your results in Problem 1 positive or negative? What can you say about the product of a positive number and a negative number? If you are not sure, try some more multiplications.

3. Explain why the observation you made in Problem 2 makes sense. [**Hint:** What can you say about a negative number plus a negative number?]

Group Exploration

Looking ahead: Determining whether a line contains some points

1. a. Plot the points shown in Table 25. Is there one line that contains all the points?

b. Complete the third column of Table 25. Describe any patterns.

Table 25 Some Ordered Pairs

x	y	Change in y (current value minus previous value)
0	1	—
1	3	
2	5	
3	7	
4	9	

2. a. Plot the points shown in Table 26. Is there one line that contains all the points?

b. Complete the third column of Table 26. Describe any patterns.

Table 26 Some Ordered Pairs

x	y	Change in y (current value minus previous value)
0	13	—
1	10	
2	7	
3	4	
4	1	

3. Describe what you have learned so far in this exploration.

4. Let p be the profit (in millions) of a company at t years since 2005. Some pairs of values of t and p are shown in Table 27.

Table 27 Profits of a Company

Years since 2005 t	Profit (millions of dollars) p	Change in Profit (millions of dollars)
0	2	—
1	6	
2	10	
3	14	
4	18	

a. Complete the third column of Table 27.

b. Without plotting the data points, decide whether this situation can be modeled well by a linear model.

c. Check your response to part (b) by creating a scattergram of the data.

Tips for Success Math Journal

Do you tend to make the same mistakes repeatedly throughout a math course? If so, it might help to keep a journal in which you list errors you have made on assignments, quizzes, and tests. For each error you list, include the correct solution, as well as a description of the concept needed to solve the problem correctly. You can review this journal from time to time to help you avoid making these errors.

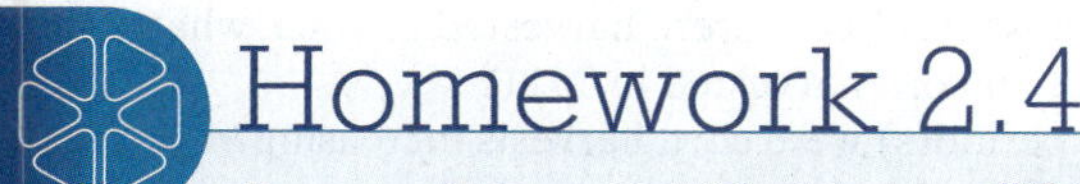

Homework 2.4

For extra help ▶ MyMathLab®  Watch the videos in MyMathLab  Download the MyDashboard App

Find the difference by hand. Then use a calculator to check your work.

1. $6 - 8$

2. $3 - 7$

3. $-1 - 5$

4. $-3 - 9$

5. $2 - (-7)$

6. $5 - (-1)$

7. $-3 - (-2)$

8. $-7 - (-3)$

9. $4 - 7$

10. $-4 - 7$

11. $4 - (-7)$

12. $-4 - (-7)$

13. $-3 - 3$

14. $-7 - 7$

15. $-54 - 25$

16. $-100 - 257$

17. $381 - (-39)$

18. $-1939 - (-352)$

19. $2.5 - 7.9$

20. $5.8 - 3.7$

21. $-6.5 - 4.8$

22. $-1.7 - 7.4$

23. $3.8 - (-1.9)$

24. $3.1 - (-3.1)$

25. $13.6 - (-2.38)$

26. $-159.24 - (-7.8)$

27. $-\frac{1}{3} - \frac{2}{3}$

28. $-\frac{1}{5} - \frac{4}{5}$

29. $-\frac{1}{8} - \left(-\frac{5}{8}\right)$

30. $-\frac{4}{9} - \left(-\frac{7}{9}\right)$

31. $\frac{1}{2} - \left(-\frac{1}{4}\right)$

32. $\frac{5}{12} - \left(-\frac{1}{6}\right)$

33. $-\frac{1}{6} - \frac{3}{8}$

34. $-\frac{2}{3} - \frac{2}{5}$

Perform the indicated operation by hand. Then use a calculator to check your work.

35. $-5 + 7$

36. $-3 + 9$

37. $-6 - (-4)$

38. $-4 - (-3)$

39. $\frac{3}{8} - \frac{5}{8}$

40. $-\frac{5}{6} + \frac{1}{6}$

41. $-4.9 - (-2.2)$

42. $-6.4 + 3.5$

43. $-2 + (-5)$

44. $-5 + (-8)$

45. $10 - 12$

46. $5 - 9$

For Exercises 47–52, use a calculator to perform the indicated operation. Round the result to two decimal places.

47. $-234.913 - 2893.26$

48. $-6178.39 - 52.387$

49. $29{,}643.52 - (-83{,}284.39)$

50. $83{,}451.6 - (-408.549)$

51. $-\frac{17}{89} - \frac{51}{67}$

52. $-\frac{49}{56} - \frac{85}{97}$

53. Three hours ago, the temperature was 7°F. If the temperature has decreased by 19°F in the last three hours, what is the current temperature?

54. Four hours ago, the temperature was −12°F. If the temperature has increased by 18°F in the last four hours, what is the current temperature?

55. Three hours ago, the temperature was −4°F. Now the temperature is 7°F. What is the change in temperature for the past three hours?

56. Four hours ago, the temperature was −2°F. Now the temperature is −13°F. What is the change in temperature for the past four hours?

57. Two hours ago, the temperature was 8°F. The temperature is now −4°F.

a. What is the change in temperature for the past two hours?

b. Estimate the change in temperature for the past hour.

c. Explain why your estimate in part (b) may not be the actual change in temperature for the past hour.

58. Three hours ago, the temperature was −6°F. The temperature is now 9°F.

a. What is the change in temperature for the past three hours?

b. Estimate the change in temperature for the past hour.

c. Explain why your estimate in part (b) may not be the actual change in temperature for the past hour.

59. The lowest elevation in the United States is at Death Valley, California (−282 ft), and the highest elevation is at the top of Mount McKinley, Alaska (20,320 ft). Find the change in elevation from Death Valley to Mount McKinley.

60. The lowest elevation on (dry) land in the world is at the edge of the Dead Sea, along the Israel–Jordan border (−1312 ft), and the highest elevation is at the top of Mount Everest, along the Nepal–Tibet border (29,035 ft). Find the change in the elevation from the Dead Sea to Mount Everest.

61. The U.S. presidential election in 2000 was the closest presidential race in the electoral vote since 1876. Yet, only a little over half of eligible voters chose to cast a vote (see Table 28).

Table 28 Presidential Election Voter Turnout

Year	Percent of Eligible Voters Who Voted
1980	59.2
1984	59.9
1988	57.4
1992	61.9
1996	54.2
2000	54.7
2004	63.8
2008	63.6
2012	57.5

Source: *U.S. Census Bureau, Current Population Study*

a. For the years listed in Table 28, find the changes in percent turnout from one presidential election to the next.

b. What was the greatest increase in percent turnout?

c. In 1993, in an attempt to increase the number of eligible voters, a "motor voter" law was passed that made voter registration a part of the process of applying for a driver's license. As a result, about 11 million new voters were registered. Compare the change in percent turnout between 1992 and 1996 with other changes you found in part (a). On the basis of the information in Table 28 alone, we cannot know for sure, but does it seem that many of these 11 million people voted? Explain.

62. In the 1930s, the gray wolf was hunted to near extinction across the western United States. In 1995, 14 wolves were reintroduced to Yellowstone National Park. In the following year, 17 more wolves were released into the park. By the end of 1996, there had been 20 births and 11 mortalities, leaving 40 wolves. The wolf population in Yellowstone National Park is shown in Table 29 for various years.

Table 29 Wolf Population

Year	Population
2005	118
2006	136
2007	171
2008	124
2009	96
2010	97
2011	98

Source: *National Park Service, Yellowstone National Park*

a. For the years listed in Table 29, find the changes in population from each year to the next.
b. From what year(s) to the next did the population increase the most? What is the change in population?
c. From what year(s) to the next did the population decrease the most? What is the change in population?
d. From 2005 to 2006, the change in the population was 18 wolves. Does that mean that there were 18 births? Explain.

63. The changes in Toyota Prius® car sales (in thousands of cars) from one year to the next are shown in Table 30.

Table 30 Changes in Toyota Prius Sales

Years	Changes in Sales (thousands of cars)
2003–2004	29
2004–2005	54
2005–2006	−1
2006–2007	74
2007–2008	−23
2008–2009	−19
2009–2010	1

Source: *National Renewable Energy Laboratory*

a. If 25 thousand cars were sold in 2003, what were the sales in 2010?
b. During which period(s) were sales increasing?
c. During which period(s) were sales decreasing?

64. The changes in corn harvests (in billions of bushels) in the United States from one year to the next are shown in Table 31.

Table 31 Changes in Corn Harvests

Years	Change in Corn Harvest (billions of bushels)
2005–2006	−0.3
2006–2007	2.4
2007–2008	−0.8
2008–2009	1.1
2009–2010	−0.5
2010–2011	1.1

Source: *Moebs Services*

a. If 10.8 billion bushels of corn were harvested in 2005, what was the amount of corn harvested in 2011?
b. During which period(s) were corn harvests increasing?
c. During which period(s) were corn harvests decreasing?

65. A student scored 87 points on the first exam of the semester.
a. Complete Table 32 to help find an expression that describes the change in score (in points) from the first exam to the second exam if the student scored p points on the second exam. Show the arithmetic to help you see a pattern.

Table 32 Scores on the Second Exam and Changes in Scores

Score on the Second Exam (points)	Change in Score (points)
80	
85	
90	
95	
p	

b. Evaluate the expression you found in part (a) for $p = 81$. What does your result mean in this situation?

66. A year ago, the value of a stock was $35.
a. Complete Table 33 to help find an expression that describes the change in the stock's value (in dollars) if the current value is x dollars. Show the arithmetic to help you see a pattern.

Table 33 Current Values and Changes in Values

Current Value (dollars)	Change in Value (dollars)
30	
35	
40	
45	
x	

b. Evaluate the expression you found in part (a) for $x = 44$. What does your result mean in this situation?

67. Last year the enrollment at a college was 24,500 students.
a. Complete Table 34 to help find an expression that describes the current enrollment if the *change* in enrollment in the past year is c students. Show the arithmetic to help you see a pattern.

Table 34 Changes in Enrollments and Current Enrollments

Change in Enrollment	Current Enrollment
100	
200	
300	
400	
c	

b. Evaluate the expression you found in part (a) for $c = -700$. What does your result mean in this situation?

68. Last year there were 820 deer in a state park.

a. Complete Table 35 to help find an expression that describes the current deer population if the *change* in population in the past year is c deer. Show the arithmetic to help you see a pattern.

Table 35 Changes in Population and Current Populations

Change in Population	Current Population
10	
20	
30	
40	
c	

b. Evaluate the expression you found in part (a) for $c = -25$. What does your result mean in this situation?

Evaluate the expression for $a = -5$, $b = 2$, and $c = -7$.

69. $a + b$ **70.** $a + c$ **71.** $a - b$

72. $c - a$ **73.** $b - c$ **74.** $b - a$

For Exercises 75–80, let x be a number. Translate the English phrase or sentence into a mathematical expression. Then evaluate the expression for $x = -5$.

75. -3 minus the number

76. The number decreased by 4

77. 8 less than the number

78. Subtract 5 from the number.

79. Subtract -2 from the number.

80. The difference of the number and -6

Concepts

81. A student tries to find the difference $7 - (-5)$:

$$7 - (-5) = 7 - 5 = 2$$

Describe any errors. Then find the difference correctly.

82. A student tries to find the difference $2 - 6$:

$$2 - 6 = 6 - 2 = 4$$

Describe any errors. Then find the sum correctly.

83. A quantity increases from amount a to amount b.

a. Find the change in the quantity.

i. $a = 3, b = 5$

ii. $a = 1, b = 9$

iii. $a = 2, b = 7$

b. By referring to your work in part (a), explain why it makes sense that if a quantity increased, then the change will be positive.

84. A quantity decreases from amount a to amount b.

a. Find the change in the quantity.

i. $a = 8, b = 2$

ii. $a = 9, b = 3$

iii. $a = 5, b = 1$

b. By referring to your work in part (a), explain why it makes sense that if a quantity decreased, then the change will be negative.

85. a. Use a calculator to find each product of two numbers with different signs:

i. $-2(5)$

ii. $-4(6)$

iii. $-7(9)$

b. What pattern do you notice? If you do not see a pattern, continue finding products of two numbers with different signs until you do.

c. Without using a calculator, find the product $-3(7)$. Then use a calculator to check your work.

d. State a general rule for how to multiply two numbers with different signs.

86. a. Use a calculator to find each product of two negative numbers.

i. $-2(-5)$

ii. $-4(-6)$

iii. $-7(-9)$

b. What pattern do you notice? If you do not see a pattern, continue finding products of two negative numbers until you do.

c. Without using a calculator, find the product $-3(-7)$. Then use a calculator to check your work.

d. State a general rule for how to multiply two negative numbers.

87. a. Evaluate $a - b$ for $a = 8$ and $b = 5$.

b. Evaluate $b - a$ for $a = 8$ and $b = 5$.

c. Compare your results from parts (a) and (b).

d. Evaluate $a - b$ and $b - a$ for $a = -2$ and $b = 4$, and compare your results.

e. Evaluate $a - b$ and $b - a$ for values of your choosing for a and b, and compare your results.

f. From your work on parts (a) through (e), what connection do you notice between $a - b$ and $b - a$?

88. If the temperature increases from 3°F to 5°F, we find the change in temperature (in degrees Fahrenheit) by performing a subtraction:

$$5 - 3 = 2$$

If the temperature increases from −3°F to 5°F, we find the change in temperature (in degrees Fahrenheit) by eventually performing an addition:

$$5 - (-3) = 5 + 3 = 8$$

Explain why it makes sense that in the first situation we subtract and in the second we eventually add. [**Hint:** How far is -3 from 0 on the number line? How far is 5 from 0 on the number line?]

89. If x is positive and y is negative, find the sign of $x - y$, if possible. If it is impossible to find the sign, explain why.

90. If x and y are both negative, find the sign of $x - y$, if possible. If it is impossible to find the sign, explain why.

2.5 Ratios, Percents, and Multiplying and Dividing Real Numbers

Objectives

- Find the ratio of two quantities.
- Know the meaning of *percent*.
- Convert percentages to and from decimal numbers.
- Find the percentage of a quantity.
- Multiply and divide real numbers.
- Know which fractions with negative signs are equal to each other.

In this section, we will use ratios and percents to describe various quantities. Ratios and percents are important tools used in many fields, including business, cooking, chemistry, political science, aeronautics, fire science technology, humanities, journalism, and psychology. We will also discuss how to multiply and divide real numbers.

The Ratio of Two Quantities

Recall from Section 2.1 that the ratio of a to b is the quotient $a \div b$. Usually, we write the ratio of a to b as the fraction $\frac{a}{b}$ or as $a : b$.

We can use a ratio to compare two quantities. For example, if a person has 6 cats and 2 dogs, then the ratio of cats to dogs is

$$\frac{6 \text{ cats}}{2 \text{ dogs}} = \frac{3 \text{ cats}}{1 \text{ dog}}$$

We say there are "3 cats to 1 dog." This means there are 3 cats per dog. Or we can say there are 3 times as many cats as dogs.

The ratio of 3 cats to 1 dog is an example of a unit ratio. A **unit ratio** is a ratio written as $\frac{a}{b}$ with $b = 1$ or as $a : b$ with $b = 1$.

Example 1 Finding a Unit Ratio

In 2010, the average annual charge for tuition and fees was $5964 at public four-year colleges and $2285 at public two-year colleges (Source: *U.S. National Center for Education Statistics*). Find the unit ratio of the average annual charge at public four-year colleges to the average annual charge at public two-year colleges. What does the result mean?

Solution

We divide the average annual charge at public four-year colleges by the average annual charge at public two-year colleges:

$$\begin{array}{l} \text{public four-year colleges} \longrightarrow \\ \text{public two-year colleges} \longrightarrow \end{array} \frac{\$5964}{\$2285} \approx \frac{2.61}{1}$$

So, the average annual charge for tuition and fees at public four-year colleges is about 2.61 times the average annual charge for tuition and fees at public two-year colleges.

Example 2 Comparing Ratios

The median sales prices of existing homes and the median household incomes in 2010 are shown in Table 36 for four regions of the United States.

Just think, our equity's growing by $100 per mile!
Welcome to California
OVER SIZE LOAD

Table 36 Median Sales Prices of Existing Homes and Median Household Incomes

Region	Median Sales Price of Existing Homes (dollars)	Median Household Income (dollars)
Northeast	243,900	53,283
Midwest	140,800	48,445
South	153,700	45,492
West	220,700	53,142

Sources: *U.S. Census; National Association of Realtors®*

1. Find the unit ratio of the median sales price of existing homes to the median household income in the Northeast. What does the result mean?
2. For each of the four regions, find the unit ratio of the median sales price of existing homes to the median household income. Taking into account the median household income of each region, list the regions in order of affordability of existing homes, from greatest to least.

Solution

1. We divide the median sales price of existing homes in the Northeast by the median household income in the Northeast:

$$\begin{array}{r} \text{Median sales price of existing homes} \longrightarrow \\ \text{Median household income} \longrightarrow \end{array} \frac{\$243{,}900}{\$53{,}283} \approx \frac{4.58}{1}$$

So, the median sales price of an existing home in the Northeast is about 4.58 times the median household income in that region.

2. We find the unit ratios for each region by dividing the region's median sales price of existing homes by the region's median household income (see Table 37).

Table 37 Unit Ratios of Median Sales Prices of Existing Homes to Median Household Incomes

Region	Median Sales Price of Existing Homes (dollars)	Median Household Income (dollars)	Unit Ratio of Median Sales Price of Existing Homes to Median Household Income
Northeast	243,900	53,283	$\frac{243{,}900}{53{,}283} \approx \frac{4.58}{1}$
Midwest	140,800	48,445	$\frac{140{,}800}{48{,}445} \approx \frac{2.91}{1}$
South	153,700	45,492	$\frac{153{,}700}{45{,}492} \approx \frac{3.38}{1}$
West	220,700	53,142	$\frac{220{,}700}{53{,}142} \approx \frac{4.15}{1}$

The lower the unit ratio, the more affordable the existing homes are in the region. So, the regions, in order of affordability of existing homes, from greatest to least, are Midwest, South, West, Northeast.

Meaning of Percent

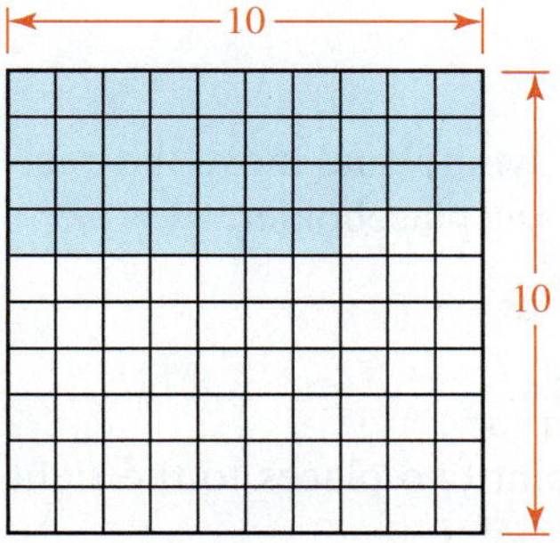

Suppose there are 53 women in a class of 100 students. Then the ratio of the number of women to the total number of students is $\frac{53}{100}$. We say that 53% of the students are women.

Definition Percent

Percent means "for each hundred": $a\% = \frac{a}{100}$

Figure 26 The area of the shaded region is 37% of the area of the large square

For example, 37% means 37 for each 100 (the ratio $\frac{37}{100}$, or the unit ratio $\frac{0.37}{1}$). In Fig. 26, the area of the shaded region is 37% of the area of the large square, because 37 of 100 parts of equal area are shaded.

Converting Percentages to and from Decimal Numbers

Since 37% is the ratio $\frac{37}{100}$, 37% is 37 hundredths, or 0.37:

$$37\% = \frac{37}{100} = 0.\underbrace{\overset{\text{tenths place}}{3}\quad\overset{\text{hundredths place}}{7}}_{\text{37 hundredths}}$$

So, to write 37% as a decimal number, first we remove the percent symbol. Then we divide 37 by 100, which is equivalent to moving the decimal point two places to the left:

$$37\% = 37.0\% = 0.37$$

two places to the left

To write 0.37 as a percentage, first we multiply 0.37 by 100, which is equivalent to moving the decimal point two places to the right. Then we insert a percent symbol:

$$0.37 = 37.0\% = 37\%$$

two places to the right

Converting Percentages to and from Decimal Numbers

- To write a percentage as a decimal number, remove the percent symbol and divide the number by 100 (move the decimal point two places to the left).
- To write a decimal number as a percentage, multiply the number by 100 (move the decimal point two places to the right) and insert a percent symbol.

Example 3 Converting Percentages and Decimal Numbers

Write each percentage as a decimal number, and write each decimal number as a percentage.

1. 86% **2.** 7% **3.** 0.125

Solution

1. To write 86% as a decimal number, we remove the percent symbol and move the decimal point two places to the left:

$$86\% = 86.0\% = 0.86$$

two places to the left

2. To write 7% as a decimal number, we remove the percent symbol and move the decimal point two places to the left, using 0 in the tenths place as a placeholder:

$$7\% = 7.0\% = 0.07$$

two places to the left

3. To write 0.125 as a percentage, we move the decimal point two places to the right and insert a percent symbol:

$$0.125 = 12.5\%$$

two places to the right

WARNING From Problem 2 in Example 3, we see 7% is *not* equal to 0.7. Rather, 7% is equal to 0.07. Remember to move the decimal point *two* places to the left, using 0 in the tenths place as a placeholder.

Percentage of a Quantity

How do we find the percentage of a quantity? For example, consider 75% of 4. That is the same as $\frac{75}{100}$ of 4. To find a fraction of a number, we *multiply* the fraction by that number:

$$\frac{75}{100} \text{ of } 4 = \frac{75}{100} \cdot 4 = \frac{3}{4} \cdot \frac{4}{1} = \frac{3}{1} = 3$$

So, using decimal notation, we find 75% of 4 by *multiplying* 0.75 by 4:

$$75\% \text{ of } 4 = 0.75(4) = 3$$

To see whether our result makes sense, we first form a large square made up of 4 medium-size squares of equal area (see Fig. 27). To find 75% of the four squares, we divide the large square into 100 small squares of equal area and shade 75 of them. The shaded region contains 3 of the 4 medium-size squares, which checks with our earlier computations.

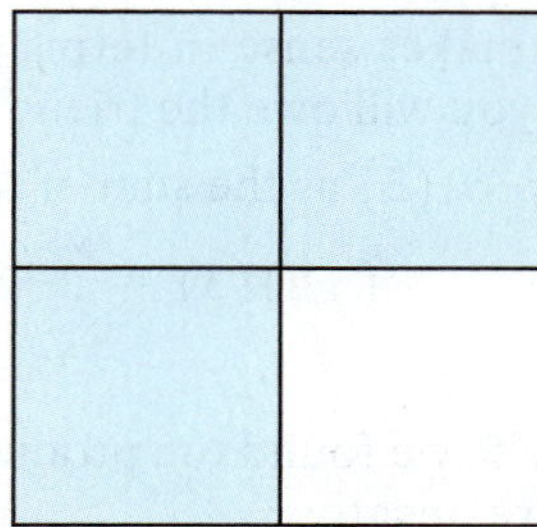

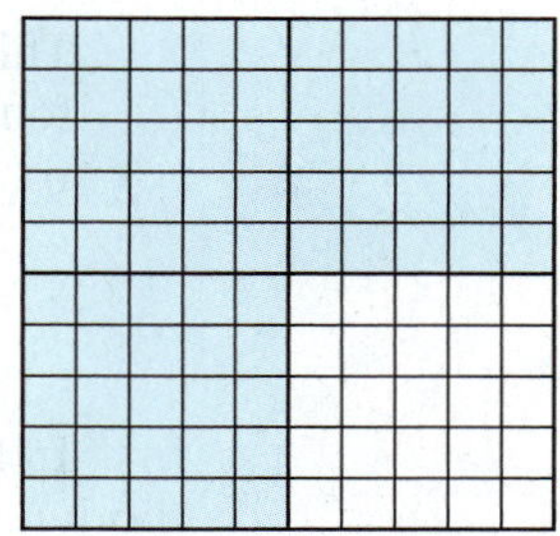

Figure 27 75% of 4 medium-size squares is made up of 75 small squares (in blue), or 3 medium-size squares (in blue)

Finding the Percentage of a Quantity

To find the percentage of a quantity, multiply the decimal form of the percentage and the quantity.

Example 4 Finding the Percentage of a Quantity

1. Find 6% of 3000 students.
2. Find 3.5% of \$6500.

Solution

1. $0.06(3000) = 180$; so, 6% of 3000 students is 180 students.
2. $0.035(6500) = 227.5$; so, 3.5% of \$6500 is \$227.50.

By definition, 100% means 100 for each 100. In other words, 100% of a quantity is *all* of the quantity. For example, 100% of 21 guitars is 21 guitars.

One Hundred Percent of a Quantity

One hundred percent of a quantity is *all* of the quantity.

We will continue to work with ratios and percents as we discuss how to multiply and divide real numbers.

Multiplication of Two Numbers with Different Signs

We can think of multiplication as repeated addition. For example, 3(5) is equal to the sum of three 5s:

$$3(5) = 5 + 5 + 5 = 15$$

Also, 3(5) is equal to the sum of five 3s:

$$3(5) = 3 + 3 + 3 + 3 + 3 = 15$$

We can use the idea of repeated addition to help us find the product of two numbers with different signs.

Example 5 Finding the Product of Two Numbers with Different Signs

Find the product.

1. $4(-2)$
2. $(-6)(3)$

Solution

1. We write $4(-2)$ as the sum of four -2s:

$$4(-2) = (-2) + (-2) + (-2) + (-2) = -8$$

This result makes sense in terms of money. If you borrow 2 dollars from a friend four times, you will owe the friend 8 dollars.

2. We write $(-6)(3)$ as the sum of three -6s:

$$(-6)(3) = (-6) + (-6) + (-6) = -18$$

In Example 5, we found the products of two numbers with different signs. Note that both results were negative:

$$\overbrace{4(-2)}^{\text{Different signs}} = \overbrace{-8}^{\text{Negative}}$$

$$\overbrace{(-6)(3)}^{\text{Different signs}} = \overbrace{-18}^{\text{Negative}}$$

Multiplying Two Numbers with Different Signs

The product of two numbers that have different signs is negative.

Example 6 Finding the Product of Two Numbers with Different Signs

Find the product.

1. $7(-4)$
2. $(-0.2)(0.3)$

Solution

1. Since the signs of 7 and -4 are different, their product is negative: $7(-4) = -28$.
2. Since the signs of -0.2 and 0.3 are different, their product is negative: $(-0.2)(0.3) = -0.06$.

Multiplication of Two Numbers with the Same Sign

We have discussed how to multiply numbers with different signs. What if the signs are the same? To begin this investigation, consider the following pattern:

$$3(-5) = -15$$
$$2(-5) = -10$$
$$1(-5) = -5$$
$$0(-5) = 0$$

This factor decreases by 1. The product increases by 5.

It turns out that this pattern continues. So, we have

$$-1(-5) = 5$$
$$-2(-5) = 10$$
$$-3(-5) = 15$$

This factor decreases by 1. The product increases by 5.

Note that for each of the last three computations, the product of the two negative numbers is positive. This is, in fact, always true. Here we find another product of two negative numbers:

$$\underbrace{(-7)(-9)}_{\text{Same signs}} = \underbrace{63}_{\text{Positive}}$$

Multiplying Two Numbers with the Same Sign

The product of two numbers that have the same sign is positive.

Example 7 Finding the Product of Two Numbers with the Same Sign

Find the product.

1. $-5(-6)$

2. $\left(-\frac{3}{2}\right)\left(-\frac{5}{7}\right)$

Solution

1. Since -5 and -6 have the same sign, their product is positive: $-5(-6) = 30$.

2. Since $-\frac{3}{2}$ and $-\frac{5}{7}$ have the same sign, their product is positive:

$$\left(-\frac{3}{2}\right)\left(-\frac{5}{7}\right) = \frac{15}{14}$$

In Fig. 28, we show a multiplication table for some specific numbers. In Fig. 29, we summarize the multiplication sign rules for all nonzero real numbers.

·	4	−4
2	8	−8
−2	−8	8

Figure 28 Multiplication table for 2, −2, 4, and −4

·	+	−
+	+	−
−	−	+

Figure 29 Multiplication table for all nonzero real numbers

Example 8 Application of Multiplying Real Numbers

A person's credit card balance is −2340 dollars. If the person pays off 30% of the balance, what is the new balance?

Solution

If the person pays off 30% of the balance, then 100% − 30% = 70% of the balance remains. We find 70% of −2340:

$$0.70(-2340) = -1638$$

The new balance is −1638 dollars.

Division of Real Numbers

We can get an idea of how to divide real numbers by writing multiplications as related divisions. For example, consider this statement:

$$2 \cdot 3 = 6 \text{ implies that } 6 \div 3 = 2.$$

We now write similar statements for $(-2)(-3)$ and $-2 \cdot 3$:

$$(-2)(-3) = 6 \text{ implies that } \underbrace{6 \div (-3)}_{\text{Different signs}} = \underbrace{-2.}_{\text{Negative}}$$

$$-2 \cdot 3 = -6 \text{ implies that } \underbrace{-6 \div 3}_{\text{Different signs}} = \underbrace{-2.}_{\text{Negative}}$$

These statements suggest that the quotient of two numbers with different signs is negative.

Now consider the following statement:

$$2(-3) = -6 \text{ implies that } \underbrace{-6 \div (-3)}_{\text{Same signs}} = \underbrace{2.}_{\text{Positive}}$$

This statement suggests that the quotient of two numbers with the same sign is positive.

All three statements suggest that the sign rules for dividing real numbers are similar to those for multiplying real numbers.

Multiplying or Dividing Real Numbers

The product or quotient of two numbers that have different signs is negative. The product or quotient of two numbers that have the same sign is positive.

Example 9 Finding Quotients of Real Numbers

Find the quotient.

1. $-10 \div 2$

2. $-\frac{1}{6} \div \left(-\frac{3}{5}\right)$

Solution

1. Since -10 and 2 have different signs, the quotient is negative: $-10 \div 2 = -5$. This makes sense in terms of money. If we divide a debt of \$10 by 2, the result is a debt of \$5.

2. The quotient of two negative numbers is positive. To find the result, we divide the absolute value of the fractions:

$$\frac{1}{6} \div \frac{3}{5} = \frac{1}{6} \cdot \frac{5}{3} \quad \textit{Multiply by reciprocal of } \frac{3}{5}\textit{, which is } \frac{5}{3}\textit{: } \frac{a}{b} \div \frac{c}{d} = \frac{a}{b} \cdot \frac{d}{c}$$

$$= \frac{5}{18} \quad \textit{Multiply numerators and multiply denominators.}$$

Press (−) (1 ÷ 6) ÷ (−) (3 ÷ 5) ENTER.

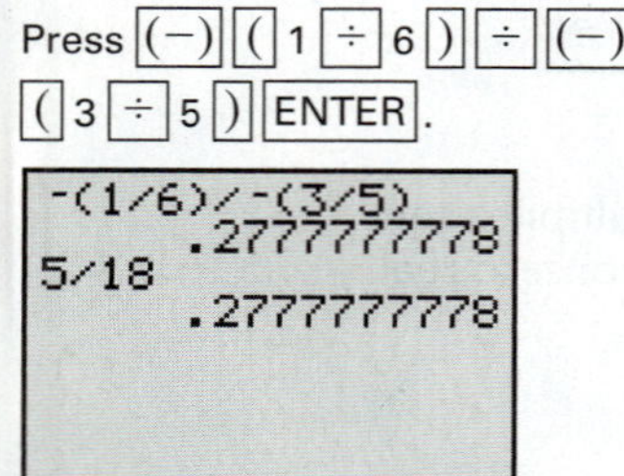

Figure 30 Verify the work

We can use a graphing calculator to check our work for Problem 2 in Example 9 (see Fig. 30).

Example 10 Application of a Ratio of Two Real Numbers

A person has credit card balances of -3950 dollars on a Visa® account and -1225 dollars on a MasterCard® account.

1. Find the unit ratio of the Visa balance to the MasterCard balance.

2. If the person wishes to pay off both accounts gradually in the same amount of time, describe how the result in Problem 1 can help guide the person in making his next payment.

Solution

1. We divide the Visa balance by the MasterCard balance:

$$\frac{-3950}{-1225} \approx \frac{3.22}{1}$$

So, the Visa balance is about 3.22 times the MasterCard balance.

2. For each \$1 the person pays to his MasterCard account, he should pay about \$3.22 to his Visa account. (The ratio will need to be recalculated each month to take into account recent purchases, cash advances, and so on, as well as possible differences in interest rates on the cards.)

Equal Fractions with Negative Signs

We know that $\frac{a}{b}$ means that a is divided by b, for $b \neq 0$:

$$\frac{a}{b} = a \div b$$

For example,

$$\frac{-4}{2} = -4 \div 2 = -2$$

Since $\frac{4}{-2}$ and $-\frac{4}{2}$ are also equal to -2, we can write

$$\frac{-4}{2} = \frac{4}{-2} = -\frac{4}{2}$$

This suggests the following property:

Equal Fractions with Negative Signs

If $b \neq 0$, then

$$\frac{-a}{b} = \frac{a}{-b} = -\frac{a}{b}$$

If the result of a computation is a negative fraction, we write the result in the form $-\frac{a}{b}$ rather than $\frac{-a}{b}$ or $\frac{a}{-b}$.

Example 11 Simplifying Fractions and Adding Fractions

Perform the indicated operation.

1. $\frac{-20}{-5}$

2. $\frac{3}{-5} + \frac{1}{5}$

Solution

1. $\frac{-20}{-5} = -20 \div (-5)$ *$\frac{a}{b} = a \div b$*

$= 4$ *Quotient of two numbers with same sign is positive.*

2. $\frac{3}{-5} + \frac{1}{5} = \frac{-3}{5} + \frac{1}{5}$ *$\frac{a}{-b} = \frac{-a}{b}$*

$= \frac{-3 + 1}{5}$ *Write numerators as a sum and keep common denominator: $\frac{a}{b} + \frac{c}{b} = \frac{a+c}{b}$.*

$= \frac{-2}{5}$ *Find sum.*

$= -\frac{2}{5}$ *$\frac{-a}{b} = -\frac{a}{b}$*

Group Exploration

Looking ahead: Order of operations

If an expression has more than one operation, we can use parentheses to indicate which operation to do first.

1. Perform the indicated operations in $(2 + 3) \cdot 4$ by first doing the addition and then doing the multiplication.
2. Perform the indicated operations in $2 + (3 \cdot 4)$ by first doing the multiplication and then doing the addition.
3. Compare your results for Problems 1 and 2. Does it matter in which order we add and multiply?

For each object, determine whether the area is $(2 + 3) \cdot 4$ or $2 + (3 \cdot 4)$. Explain.

4. 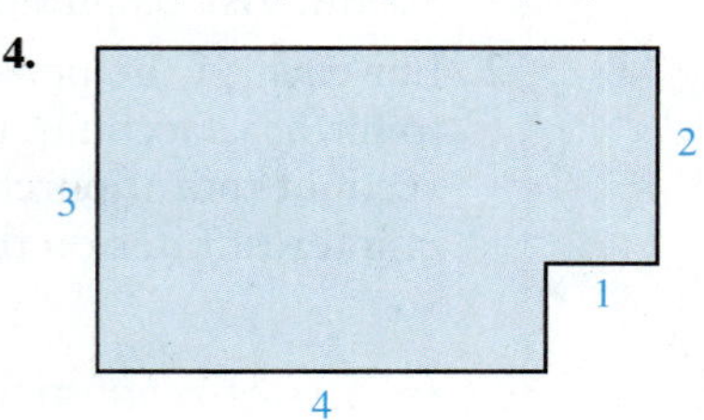

5. 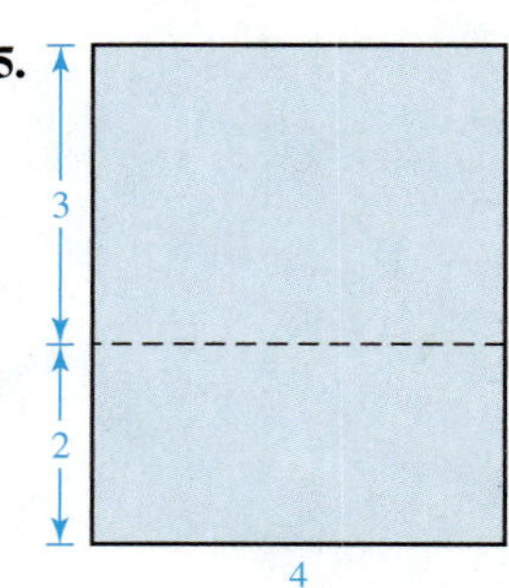

Tips for Success Use Your Instructor's Office Hours

Helping students during office hours is part of an instructor's job. Keep in mind that your instructor wants you to succeed and hopes you take advantage of all opportunities to learn.

It is a good idea to come prepared to office visits. For example, if you are having trouble with a concept, attempt some related exercises and bring your work so your instructor can see where you are having difficulty. If you miss a class, it is helpful to read the material, borrow class notes, and try to complete assigned exercises before visiting your instructor, so you get the most out of the visit.

Homework 2.5

For extra help ▶ MyMathLab®  Watch the videos in MyMathLab  Download the MyDashboard App

For Exercises 1–10, write the percentage as a decimal number or write the decimal number as a percentage, as appropriate.

1. 63% **2.** 91% **3.** 9% **4.** 4% **5.** 0.08
6. 0.01 **7.** 7.3% **8.** 3.8% **9.** 0.052 **10.** 0.089

11. Find 35% of \$8. **12.** Find 67% of \$4.
13. Find 5% of 2500 students. **14.** Find 8% of 4000 students.
15. Find 2.5% of 7000 cars. **16.** Find 6.4% of 3500 cars.

Perform the indicated operation by hand. Then use a calculator to check your work.

17. $-2(6)$ **18.** $-5(4)$ **19.** $-3(-6)$
20. $-8(-9)$ **21.** $1(-1)$ **22.** $5(-2)$
23. $-40 \div 5$ **24.** $-63 \div 7$ **25.** $25 \div (-5)$
26. $24 \div (-3)$ **27.** $-56 \div (-7)$ **28.** $-1 \div (-1)$
29. $-15(-37)$ **30.** $-124(-29)$ **31.** $936 \div (-24)$
32. $1008 \div (-21)$ **33.** $-0.2(-0.4)$ **34.** $-0.3(-0.3)$
35. $2.5(-0.39)$ **36.** $3.7(-5.24)$ **37.** $-0.06 \div 0.2$
38. $-0.12 \div 0.3$ **39.** $\frac{36}{-4}$ **40.** $\frac{9}{-3}$

41. $\frac{-32}{-8}$ **42.** $\frac{-72}{-8}$ **43.** $\frac{1}{2}\left(-\frac{1}{5}\right)$
44. $\frac{1}{3}\left(-\frac{7}{5}\right)$ **45.** $\left(-\frac{4}{9}\right)\left(-\frac{3}{20}\right)$ **46.** $\left(-\frac{7}{25}\right)\left(-\frac{5}{21}\right)$
47. $-\frac{3}{4} \div \frac{7}{6}$ **48.** $-\frac{5}{7} \div \frac{15}{8}$
49. $-\frac{24}{35} \div \left(-\frac{16}{25}\right)$ **50.** $-\frac{3}{8} \div \left(-\frac{9}{20}\right)$

Perform the indicated operation by hand. Then use a calculator to check your work.

51. $6 + (-9)$ **52.** $-9 + (-4)$ **53.** $-39 \div (-3)$
54. $-49 \div 7$ **55.** $4 - (-2)$ **56.** $-2 - 7$
57. $10(-10)$ **58.** $-5(-9)$ **59.** $-\frac{3}{4} + \frac{1}{2}$
60. $-\frac{8}{3} + \left(-\frac{5}{9}\right)$ **61.** $\left(-\frac{10}{7}\right)\left(-\frac{14}{15}\right)$ **62.** $\frac{9}{2}\left(-\frac{4}{21}\right)$
63. $\frac{3}{4} - \frac{5}{3}$ **64.** $-\frac{3}{8} - \left(-\frac{1}{10}\right)$
65. $-\frac{3}{8} \div \frac{5}{6}$ **66.** $-\frac{22}{9} \div \left(-\frac{33}{18}\right)$

Simplify.

67. $\frac{-16}{20}$ **68.** $\frac{-15}{35}$

69. $\frac{-18}{-24}$ **70.** $\frac{-35}{-21}$

Perform the indicated operation by hand. Then use a calculator to check your work.

71. $\frac{3}{-4} + \frac{1}{4}$ **72.** $\frac{5}{-6} + \frac{1}{6}$

73. $\frac{4}{7} - \left(\frac{3}{-7}\right)$ **74.** $\frac{2}{3} - \left(\frac{1}{-3}\right)$

75. $\frac{5}{6} + \frac{7}{-8}$ **76.** $\frac{1}{4} + \frac{5}{-6}$

Use a calculator to perform the indicated operation. Round the result to two decimal places.

77. $-26.87(-381.572)$ **78.** $-489.2(-8.39)$

79. $222.045 \div (-32.76)$ **80.** $64.958 \div (-3.716)$

81. $-\frac{11}{18}\left(-\frac{15}{19}\right)$ **82.** $-\frac{169}{175}\left(-\frac{64}{71}\right)$

83. $-\frac{59}{13} \div \frac{27}{48}$ **84.** $-\frac{75}{22} \div \frac{13}{48}$

Evaluate the expression for $a = -6$, $b = 4$, and $c = -8$.

85. ab **86.** ac **87.** $\frac{a}{b}$ **88.** $\frac{b}{a}$

89. $-ac$ **90.** $-bc$ **91.** $-\frac{b}{c}$ **92.** $-\frac{a}{c}$

Let w be a number. Translate the English phrase into a mathematical expression. Then evaluate the expression for $w = -8$.

93. The quotient of the number and 2

94. The number divided by 4

95. The product of the number and -5

96. -2 times the number

For Exercises 97 and 98, write the ratio as a fraction.

97. the ratio of 6 to 8

98. the ratio of 9 to 15

99. The 1776-foot-tall Freedom Tower at the site where New York's World Trade Center once stood is the world's tallest full-service building (not counting antennas). The John Hancock Tower in Boston is 790 feet tall. Find the unit ratio of the height of the Freedom Tower to the height of the John Hancock Tower. What does your result mean in this situation?

100. About 2.22 million Americans attend Major League Baseball games regularly, and about 1.1 million Americans attend National Football League games regularly (Source: *GfK Mediamark Research & Intelligence, Inc.*). Find the unit ratio of the number of Americans who attend Major League Baseball games regularly to the number of Americans who attend National Football League games regularly. What does your result mean in this situation?

101. There were 412 U.S. billionaires in 2011 and 261 U.S. billionaires in 2001 (Source: Forbes). Find the unit ratio of the number of U.S. billionaires in 2011 to the number of U.S. billionaires in 2001. What does your result mean in this situation?

102. During the week of January 12, 2012, the average number of viewers was 4.7 million viewers per day for the TV show *Good Morning America* and 2.2 million viewers for the competing *The Early Show* (Source: *Nielsen Media Research*). Find the unit ratio of the average number of viewers per day of *Good Morning America* to the average number of viewers per day of *The Early Show*. What does your result mean in this situation?

103. A recipe for roasted red-pepper pasta calls for 4 red bell peppers and 5 black olives. Calculate the given unit ratio. What does your result mean in this situation?

a. The unit ratio of the number of red bell peppers to the number of black olives

b. The unit ratio of the number of black olives to the number of red bell peppers

104. A recipe for beef stroganoff calls for 2 cups of sliced mushrooms and 4 cups of cooked noodles. Calculate the given unit ratio. What does your result mean in this situation?

a. The unit ratio of the number of cups of sliced mushrooms to the number of cups of cooked noodles

b. The unit ratio of the number of cups of cooked noodles to the number of cups of sliced mushrooms

105. The *full-time equivalent enrollment* (FTE enrollment) at a college is the number of full-time students it would take for their total credits (units or hours) to equal the total credits in which both part-time and full-time students combined are enrolled in one semester. The number of *full-time equivalent faculty* (FTE faculty) is the number of full-time faculty it would take to teach all the courses that are taught by both part-time and full-time faculty combined. The FTE enrollments and number of FTE faculty are shown in Table 38 for various colleges.

Table 38 FTE Enrollments and Numbers of FTE Faculty

College	FTE Enrollment	Number of FTE Faculty
Butler University	4168.0	335.1
St. Olaf College	2951.0	248.7
Stonehill College	2351.0	178.0
University of Massachusetts Amherst	17,016.2	982.4
Texas A&M University	37,682.49	1785.9

Sources: *Butler University, St. Olaf College, Stonehill College, University of Massachusetts, Texas A&M University*

a. Find the unit ratio of FTE enrollment at Texas A&M University to the FTE enrollment at St. Olaf College. What does your result mean in this situation?

b. Find the unit ratio of the number of FTE faculty at the University of Massachusetts Amherst to the number of FTE faculty at Butler University. What does your result mean in this situation?

c. Find the unit ratio of FTE enrollment to the number of FTE faculty at each of the colleges listed in Table 38.

d. Which college listed in Table 38 has the largest ratio of FTE enrollment to the number of FTE faculty? Which has the smallest?

e. A person believes that the ratio of FTE enrollment to the number of FTE faculty is lower at Stonehill College than at St. Olaf College, because Stonehill College has the lower FTE enrollment. Is that person correct? Explain.

106. The 2010 populations and land areas are shown in Table 39 for various states.

Table 39 Populations and Land Areas

State	Population	Land Area (square miles)
Alaska	710,231	571,951
California	37,253,956	155,959
Michigan	9,883,640	56,804
New Jersey	8,791,894	7417
New York	19,378,102	47,214

Sources: *U.S. Census Bureau; Infoplease*

a. Find the unit ratio of New York's population to New Jersey's population. What does your result mean in this situation?

b. Find the unit ratio of Alaska's land area to California's land area. What does your result mean in this situation?

c. The unit ratio of population to land area is called the *population density*. Find the population density of each state listed in Table 39.

d. Which state listed in Table 39 has the greatest population density? Which has the least?

e. A person believes that Michigan has a greater population density than New Jersey, because Michigan's population is more than New Jersey's population. Is that person correct? Explain.

107. A person has credit card balances of -4360 dollars on a Discover® account and -1825 dollars on a MasterCard® account.

a. Find the unit ratio of the Discover balance to the MasterCard balance.

b. If the person wishes to pay off both accounts gradually in the same amount of time, describe how the result in part (a) can help guide the person in making her next payment.

108. A person has credit card balances of -6810 dollars on a Visa® account and -2950 dollars on a Sears® account.

a. Find the unit ratio of the Visa balance to the Sears balance.

b. If the person wishes to pay off both accounts gradually in the same amount of time, describe how the result in part (a) can help guide the person in making his next payment.

109. A person's credit card balance is -3720 dollars. If the person pays off 15% of the balance, what is the new balance?

110. A person's credit card balance is -1590 dollars. If the person pays off 35% of the balance, what is the new balance?

111. A student has zero balance on a credit card. The student uses the credit card to buy 12.3 gallons of gasoline at a cost of $2.40 per gallon. What is the new balance?

112. A person has zero balance on a credit card. The person uses the credit card to buy three lamps at a cost of $89.50 per lamp. What is the new balance?

Concepts

113. **a.** Find the sum $-2 + (-4)$.

b. Find the product $-2(-4)$.

c. Consider the following statements:

- Two negative numbers make a positive.
- A negative number times a negative number is equal to a positive number.

Which statement is clearer? Explain.

d. Compare the sign rule for adding two negative numbers with the sign rule for multiplying two negative numbers.

114. **a.** Find the sum $-2 + 3$.

b. Find the product $-2(3)$.

c. Consider the following statements:

- A negative number times a positive number is equal to a negative number.
- A negative and a positive make a negative.

Which statement is clearer? Explain.

d. Compare the sign rule for adding numbers with different signs with the sign rule for multiplying numbers with different signs.

115. Which of the following fractions are equal? (There may be more than one pair of answers.)

$$\frac{a}{b} \quad \frac{-a}{b} \quad \frac{a}{-b} \quad -\frac{a}{b} \quad \frac{-a}{-b} \quad -\frac{-a}{-b}$$

116. **a.** Is $\frac{12}{-4}$ positive or negative? Explain.

b. If a is positive and b is negative, is $\frac{a}{b}$ positive or negative? Explain.

c. A student says that $\frac{a}{b}$ is positive because it has no negative signs. Is the student correct? Explain.

117. Discuss in terms of repeated addition why it makes sense that $3(-6)$ is negative.

118. Discuss in terms of repeated addition why it makes sense that $(-4)(5)$ is negative.

119. If ab is negative, what can you say about a or b?

120. If ab is positive, what can you say about a or b?

121. If $ab = 0$, what can you say about a or b?

122. If $\frac{a}{b}$ is negative, what can you say about a or b?

123. **a.** Perform the indicated operations in $(8 \div 2) \cdot 4$ by first doing the division and then doing the multiplication.

b. Perform the indicated operations in $8 \div (2 \cdot 4)$ by first doing the multiplication and then doing the division.

c. Compare your results for parts (a) and (b). Does it matter in which order we multiply and divide? Explain.

124. **a.** Find each product.

i. $(-1)(-1)$

ii. $(-1)(-1)(-1)$

iii. $(-1)(-1)(-1)(-1)$

iv. $(-1)(-1)(-1)(-1)(-1)$

b. Describe any patterns that you notice in your results in part (a). If you don't see a pattern, find some more products of -1s until you do.

c. Find the product:

$$\underbrace{(-1)(-1)(-1)\cdots(-1)}_{524 \text{ factors of } -1}$$

d. Find the product:

$$\underbrace{(-1)(-1)(-1)\cdots(-1)}_{847 \text{ factors of } -1}$$

125. Some students believe $2x$ is greater than x for *all* real numbers.

a. Find three values of x where $2x$ is greater than x.
b. Find three values of x where $2x$ is less than x.
c. Find one value of x where $2x$ is equal to x.

126. Some students believe $\frac{2}{x}$ is less than 2 for *all* real numbers.

a. Find three values of x where $\frac{2}{x}$ is less than 2.

b. Find three values of x where $\frac{2}{x}$ is greater than 2.

c. Find one value of x where $\frac{2}{x}$ is equal to 2.

2.6 Exponents and Order of Operations

Objectives

» Know the meaning of *exponent.*

» Use the rules for order of operations to perform computations and evaluate expressions.

» Use the rules for order of operations to make predictions.

In this section, we will discuss an operation called *exponentiation.* We will also discuss the order in which we should perform various operations.

Exponents

The notation x^2 stands for $x \cdot x$. So, $7^2 = 7 \cdot 7 = 49$. The notation x^3 stands for $x \cdot x \cdot x$. So, $2^3 = 2 \cdot 2 \cdot 2 = 8$.

Definition Exponent

For any counting number n,

$$x^n = \underbrace{x \cdot x \cdot x \cdot \ldots \cdot x}_{n \text{ factors of } x}$$

We refer to x^n as the **power,** the ***n*th power of *x*,** or ***x* raised to the *n*th power.** We call x the **base** and n the **exponent.**

The expression 2^5 is a power. It is the 5th power of 2, or 2 raised to the 5th power. For 2^5, the base is 2 and the exponent is 5. Here, we label the base and the exponent of 2^5 and compute the power:

$$2^5 = \underbrace{2 \cdot 2 \cdot 2 \cdot 2 \cdot 2}_{5 \text{ factors of } 2} = 32$$

Exponent
Base

When we calculate a power, we say that we are performing **exponentiation.**

Notice that the notation b^1 stands for one factor of b, so $b^1 = b$.

Two powers of x have specific names. We refer to x^2 as the **square of *x*** or ***x* squared.** We refer to x^3 as the **cube of *x*** or ***x* cubed.**

For an expression of the form $-a^n$, we calculate a^n before taking the opposite. For example,

$$-3^4 = -(3^4) = -(3 \cdot 3 \cdot 3 \cdot 3) = -81$$

For -3^4, the base is 3. If we want the base to be -3, we enclose -3 in parentheses:

$$(-3)^4 = (-3)(-3)(-3)(-3) = 81$$

We can use a graphing calculator to check both computations (see Fig. 31).

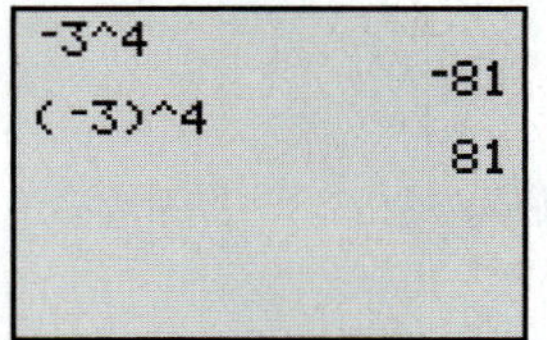

Figure 31 Compute -3^4 and $(-3)^4$

Example 1 Calculating Expressions That Have Exponents

Perform the exponentiation.

1. 5^4 **2.** -3^2 **3.** $(-3)^2$ **4.** $\left(\frac{4}{5}\right)^3$

Solution

1. $5^4 = 5 \cdot 5 \cdot 5 \cdot 5 = 625$ *The base is 5.*

2. $-3^2 = -(3 \cdot 3) = -9$ *The base is 3.*

3. $(-3)^2 = (-3)(-3) = 9$ *The base is −3.*

4. $\left(\frac{4}{5}\right)^3 = \frac{4}{5} \cdot \frac{4}{5} \cdot \frac{4}{5} = \frac{64}{125}$ *The base is* $\frac{4}{5}$.

Order of Operations

We can establish the order of operations by using *grouping symbols,* such as parentheses ( ), absolute-value symbols ‖, and fraction bars. We do operations that lie within grouping symbols before we perform other operations.

But does it matter in which order we perform operations? From the following calculations, it is clear that it *does* matter:

$$(3 + 2) \cdot 4 = 5 \cdot 4 = 20 \quad \text{First add; then multiply.}$$

$$3 + (2 \cdot 4) = 3 + 8 = 11 \quad \text{First multiply; then add.}$$

With a fraction such as $\frac{7+3}{3-1}$, the following use of parentheses is assumed:

$$\frac{7+3}{3-1} = \frac{(7+3)}{(3-1)} = \frac{10}{2} = 5$$

So, we compute both the numerator and the denominator before we divide.

Example 2 Performing Operations

Perform the indicated operations.

1. $(7 + 2)(3 - 8)$ **2.** $\frac{11-3}{1-5}$

Solution

1. $(7 + 2)(3 - 8) = (9)(-5) = -45$

2. $\frac{11-3}{1-5} = \frac{8}{-4} = -2$

We can use a graphing calculator to check our result (see Fig. 32).

```
(11-3)/(1-5)
              -2
```

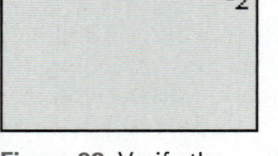

Figure 32 Verify the result

For an expression such as $3 + 2 \cdot 4$, where grouping symbols do not specify the order of operations for all operations in the expression, there is an understood order of operations.

Order of Operations

We perform operations in the following order:

1. First, perform operations within parentheses or other grouping symbols, starting with the innermost group.
2. Then perform exponentiations.
3. Next, perform multiplications and divisions, going from left to right.
4. Last, perform additions and subtractions, going from left to right.

So, for the expression $3 + 2 \cdot 4$, we multiply before adding:

$$3 + 2 \cdot 4 = 3 + 8 = 11$$

Example 3 Performing Operations

Perform the indicated operations.

1. $9 - 8 \div 4$ **2.** $10 \div 5 + (6 - 4) \cdot 5$ **3.** $2 - [7 + 4(3 - 5)]$

Solution

1. $9 - 8 \div 4 = 9 - 2$ *Divide before subtracting.*
$= 7$ *Subtract.*

2. $10 \div 5 + (6 - 4) \cdot 5 = 10 \div 5 + 2 \cdot 5$ *Subtract within parentheses.*
$= 2 + 2 \cdot 5$ *Divide, because the division is to the left of the multiplication.*
$= 2 + 10$ *Multiply before adding.*
$= 12$ *Add.*

```
10/5+(6-4)*5
            12
```

Figure 33 Verify the result

We use a graphing calculator to verify our result (see Fig. 33).

3. $2 - [7 + 4(3 - 5)] = 2 - [7 + 4(-2)]$ *Subtract within innermost parentheses.*
$= 2 - [7 + (-8)]$ *Multiply.*
$= 2 - [-1]$ *Add.*
$= 2 + 1$ *Simplify.*
$= 3$ *Add.*

There is a connection between the order of operations and the strengths of the operations. We explore the strengths of exponentiation, multiplication, and addition by performing these operations on a pair of 10s:

Operation	**Computation with 10s**
Exponentiation	$10^{10} = 10{,}000{,}000{,}000$
Multiplication	$10 \cdot 10 = 100$
Addition	$10 + 10 = 20$

Exponentiation is much more powerful than multiplication, which in turn is more powerful than addition. Since dividing by a number is the same as multiplying by the reciprocal of the number, division is as powerful as multiplication. Since subtracting a number is the same as adding the opposite of the number, subtraction is as powerful as addition. Here is a summary of the strengths of the operations:

Operation	**Strength of Operation**
Exponentiation	Most Powerful
Multiplication and Division	Next Most Powerful
Addition and Subtraction	Weakest

Order of Operations and the Strengths of Operations

After we have performed operations in parentheses, the order of operations goes from the most powerful operation, exponentiation, to the next-most-powerful operations, multiplication and division, to the weakest operations, addition and subtraction.

Knowing the relationship between the order of operations and the strengths of the operations will likely help you remember the order of operations.

Example 4 Performing Operations

Perform the indicated operations.

1. $3 + 2^3 + 4(5)$ **2.** $7 + (2 - 6)^3 - 8 \div (-2)$

Solution

1.

$$\begin{aligned} 3 + 2^3 + 4(5) &= 3 + 8 + 4(5) && \text{Perform exponentiation first: } 2^3 = 2 \cdot 2 \cdot 2 = 8 \\ &= 3 + 8 + 20 && \text{Multiply before adding.} \\ &= 31 && \text{Add.} \end{aligned}$$

2.

$$\begin{aligned} 7 + (2 - 6)^3 - 8 \div (-2) &= 7 + (-4)^3 - 8 \div (-2) && \text{Work within parentheses first.} \\ &= 7 + (-64) - 8 \div (-2) && \text{Perform exponentiation: } (-4)^3 = (-4)\cdot(-4)\cdot(-4) = -64 \\ &= 7 + (-64) - (-4) && \text{Divide.} \\ &= 7 - 64 + 4 && \text{Simplify.} \\ &= -57 + 4 && \text{Subtract.} \\ &= -53 && \text{Add.} \end{aligned}$$

We use a graphing calculator to verify our work (see Fig. 34).

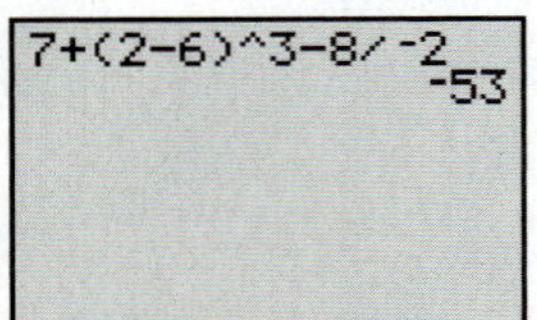

Figure 34 Verify the work

Expressions

Now that we know the order of operations, we can evaluate expressions that involve more than one operation.

Example 5 Evaluating an Expression

Evaluate $\frac{a - b}{c - d}$ for $a = 4$, $b = -2$, $c = -5$, and $d = 4$.

Solution

We begin by substituting 4 for a, -2 for b, -5 for c, and 4 for d:

$$\begin{aligned} \frac{(4) - (-2)}{(-5) - (4)} &= \frac{6}{-9} && \text{Subtract.} \\ &= -\frac{6}{9} && \frac{a}{-b} = -\frac{a}{b} \\ &= -\frac{2}{3} && \text{Simplify.} \end{aligned}$$

Example 6 Evaluating Expressions

1. Evaluate $4x^2$ for $x = -3$.
2. Evaluate $b^2 - 4ac$ for $a = 5$, $b = -3$, and $c = -6$.

Solution

1. We begin by substituting -3 for x in the expression $4x^2$:

$$\begin{aligned} 4(-3)^2 &= 4(9) && \text{Perform exponentiation first.} \\ &= 36 && \text{Multiply.} \end{aligned}$$

2. We begin by substituting 5 for a, -3 for b, and -6 for c in the expression $b^2 - 4ac$:

$$\begin{aligned}(-3)^2 - 4(5)(-6) &= 9 - 4(5)(-6) && \text{Perform exponentiation first.}\\ &= 9 - (-120) && \text{Multiply.}\\ &= 9 + 120 && \text{Simplify.}\\ &= 129 && \text{Add.}\end{aligned}$$

Example 7 Translating from English to Mathematics

Let x be a number. Translate the phrase "6 minus the product of -5 and the number" into a mathematical expression. Then evaluate the expression for $x = -3$.

Solution

First, we translate the given phrase into an expression:

$$\underbrace{6 \;-}_{\text{Six minus}} \; \underbrace{(-5)x}_{\text{the product of } -5 \text{ and the number}}$$

Then we substitute -3 for x in the expression $6 - (-5)x$ and perform the indicated operations:

$$\begin{aligned}6 - (-5)(-3) &= 6 - 15 && \text{Multiply before subtracting.}\\ &= -9 && \text{Subtract.}\end{aligned}$$

Making Predictions

In Example 8, we will find an expression, which we will use to make a prediction.

Example 8 Using a Table to Find an Expression

The value of a stock was $42 in 2005 and has increased by $3 per year since then.

1. Use a table to help find an expression that stands for the value of the stock (in dollars) at t years since 2005.
2. Evaluate the expression you found in Problem 1 for $t = 10$. What does your result mean in this situation?

Solution

1. We create Table 40. We show the arithmetic to help us see a pattern. From the last row of the table, we see that the value of the stock (in dollars) can be represented by $3t + 42$.
2. We substitute 10 for t in the expression $3t + 42$:

$$3(10) + 42 = 72$$

The value of the stock in $2005 + 10 = 2015$ will be $72.

Table 40 Values of a Stock

Years since 2005	Value (dollars)
0	$3 \cdot 0 + 42$
1	$3 \cdot 1 + 42$
2	$3 \cdot 2 + 42$
3	$3 \cdot 3 + 42$
4	$3 \cdot 4 + 42$
t	$3 \cdot t + 42$

In Section 2.5, we found the percentage of a quantity. Now we will find the result of a percentage increase or percentage decrease in a quantity.

Example 9 Solving a Percentage Problem

The attendance at Broadway shows was 12.3 million people in 2009, and it decreased by 2.9% in 2010 (Source: *National Arts Index*). What was the attendance in 2010?

Solution

The decrease in attendance from 2009 to 2010 was 0.029(12.3) million people. To find the attendance (in millions of people) in 2010, we subtract 0.029(12.3) from 12.3:

$$\underbrace{12.3}_{\text{Attendance in 2009 (in millions)}} - \underbrace{0.029(12.3)}_{\text{Decrease in attendance (in millions)}} \approx \underbrace{11.9}_{\text{Attendance in 2010 (in millions)}}$$

The attendance in 2010 was about 11.9 million people.

Group Exploration

Looking ahead: Evaluating an expression and plotting points

1. Evaluate the expression $2x + 1$ at each of the given values of x.

   a. $x = -2$ **b.** $x = -1$ **c.** $x = 0$ **d.** $x = 1$ **e.** $x = 2$

2. Organize the results that you found in Problem 1 in Table 41. The first row has been done for you.

Table 41 Values of x and $2x + 1$

x	$2x + 1$
−2	−3
−1	
0	
1	
2	

3. Treat the values of the expression $2x + 1$ as values of the variable y. So, from the first row of Table 41, we see that $y = -3$ when $x = -2$, which can be represented by the ordered pair $(-2, -3)$. List four ordered pairs that are related to the other four rows of Table 41.

4. Plot the ordered pair $(-2, -3)$ and the four ordered pairs you found in Problem 3 on a coordinate system. Describe any patterns in the position of the points.

5. Repeat the instructions for Problems 1–4, but use the expression $-2x + 3$.

Tips for Success Study in a Test Environment

Do you ever feel that you understand your homework assignments, yet you perform poorly on quizzes and tests? If so, you may not be studying enough to be ready to solve problems *in a test environment*. For example, although it is a good idea to refer to your lecture notes when you are stumped on a homework exercise, you must continue to solve similar exercises until you can solve them *without* referring to your lecture notes (unless your instructor uses open-notebook tests). The same idea applies to getting help from someone, referring to examples in the text, looking up answers in the back of the text, or any other form of support.

Getting support to help you learn math is a great idea. Just make sure you spend the last part of your study time completing exercises without such support. One way to complete your study time would be to make up a practice quiz or test for you to do in a given amount of time.

Homework 2.6

For extra help ▶ MyMathLab® Watch the videos in MyMathLab  Download the MyDashboard App

Perform the exponentiation by hand. Then use a calculator to check your work.

1. 4^3 **2.** 3^4 **3.** 2^5 **4.** 5^3

5. -8^2 **6.** -7^2 **7.** $(-8)^2$ **8.** $(-7)^2$

9. $\left(\frac{6}{7}\right)^2$ **10.** $\left(\frac{3}{5}\right)^3$

Perform the indicated operations by hand. Then use a calculator to check your work.

11. $3 \cdot (5 - 1)$ **12.** $8 \cdot (2 - 6)$

13. $(2 - 5)(9 - 3)$ **14.** $(2 + 8)(3 - 8)$

15. $4 - (3 - 8) + 1$ **16.** $-6 - (4 - 7) + 5$

17. $\dfrac{1-10}{1+2}$

18. $\dfrac{3-9}{2-4}$

19. $\dfrac{2-(-3)}{5-7}$

20. $\dfrac{4-7}{-3-(-1)}$

21. $\dfrac{4-(-6)}{-7-8}$

22. $\dfrac{1-9}{2-(-4)}$

23. $6+8\div 2$

24. $2-3\cdot 5$

25. $-5-4\cdot 3$

26. $1+9\cdot(-4)$

27. $20\div(-2)\cdot 5$

28. $-16\div(-4)\cdot 2$

29. $-9-4+3$

30. $3-7+1$

31. $7-3(5+2)$

32. $2-4(9-6)$

33. $3(5-1)-4(-2)$

34. $2(2-5)+10\div 5$

35. $15\div 3-(2-7)(2)$

36. $6(2+3)-5\cdot 7$

37. $5-[4-7(5-9)]$

38. $-3-[6+2(4-8)]$

39. $3[-2(1+6)-5]$

40. $-2[4(3-8)+1]$

41. $\dfrac{7}{8}-\dfrac{3}{4}\cdot\dfrac{1}{2}$

42. $\dfrac{5}{6}+\dfrac{2}{3}\div\dfrac{2}{5}$

43. $2+5^2$

44. $8-3^2$

45. $-3(4)^2$

46. $8(-2)^3$

47. $\dfrac{2^4}{4^2}$

48. $\dfrac{5^2}{2^5}$

49. 4^3-3^4

50. 5^2+2^5

51. $45\div 3^2$

52. $-20\div 2^2$

53. $(-1)^2-(-1)^3$

54. $4^3-(-4)^3$

55. $-5(3)^2+4$

56. $4(-2)^2-3$

57. $-4(-1)^2-2(-1)+5$

58. $2(-4)^2+3(-4)-7$

59. $\dfrac{9-6^2}{12+3^2}$

60. $\dfrac{10-(-2)^2}{3^3}$

61. $8-(9-5)^2-1$

62. $4+(3-6)^2-2$

63. $8^2+2(4-8)^2\div(-2)$

64. $(9-7)^2\cdot(-3)-2^4$

Use a calculator to perform the indicated operations. Round the result to two decimal places.

65. $13.28-35.2(17.9)+9.43\div 2.75$

66. $8.53\div 5.26+24.91-78.3(45.3)$

67. $5.82-3.16^3\div 4.29$

68. $1.98+8.22^5\cdot 5.29$

69. $\dfrac{(25.36)(-3.42)-17.89}{33.26+45.32}$

70. $\dfrac{(53.25)+99.83}{31.28-(6.31)(89.11)}$

Evaluate the expression for $a=-2$, $b=-4$, and $c=3$.

71. $a+bc$

72. $ac-b$

73. $ac-b\div a$

74. $c\div a+abc$

75. a^2-c^2

76. c^2-b^2

77. b^2-4ac

78. $-cb^2+a^2$

79. $\dfrac{-b-c^2}{2a}$

80. $\dfrac{a^2-b}{c^2-b}$

Evaluate $\dfrac{a-b}{c-d}$ for the given values of a, b, c, and d.

81. $a=3, b=-10, c=-6, d=1$

82. $a=5, b=-1, c=-4, d=7$

83. $a=-3, b=7, c=1, d=-3$

84. $a=-2, b=6, c=5, d=-1$

85. $a=-8, b=-2, c=-15, d=-5$

86. $a=-3, b=-5, c=-8, d=-3$

Evaluate the expression for $x=-3$.

87. $-3x^2$

88. $5x^2$

89. $-x^2+x$

90. $-4x^2+4$

91. $2x^2-3x+5$

92. $4x^2+x-2$

For Exercises 93–96, let x be a number. Translate the English phrase or sentence into a mathematical expression. Then evaluate the expression for $x=-4$.

93. 5 more than the product of -6 and the number

94. -3 minus the quotient of 8 and the number

95. Subtract 3 from the quotient of the number and -2.

96. The number plus the product of -5 and the number

97. Congressional pay in 1975 was \$44.6 thousand and has increased by approximately \$3.8 thousand per year since then (Source: *Bureau of Labor Statistics*).

a. Complete Table 42 to help find an expression that stands for the congressional pay (in thousands of dollars) at t years since 1975. Show the arithmetic to help you see a pattern.

Table 42 Congressional Pay

Years since 1975	Congressional Pay (thousands of dollars)
0	
1	
2	
3	
4	
t	

b. Evaluate the expression you found in part (a) for $t=30$. What does your result mean in this situation?

c. Congress has had a pay freeze since 2009 due to a poor economy. The pay freeze might be lifted in 2014. If the congressional pay is increased by 1% in 2014, what will be the pay in that year?

98. The number of anti-government militia groups was 149 in 2008 and has increased by approximately 375 groups per year since then (Source: *The Southern Poverty Law Center*).

a. Complete Table 43 to help find an expression that stands for the number of militia groups at t years since 2008. Show the arithmetic to help you see a pattern.

Table 43 Numbers of Militia Groups

Years since 2008	Number of Militia Groups
0	
1	
2	
3	
4	
t	

b. Evaluate the expression you found in part (a) for $t = 8$. What does your result mean in this situation?

99. The population of Gary, Indiana, was about 145 thousand in 1980 and has decreased by about 2.2 thousand per year since then (Source: *U.S. Census Bureau*).

a. Complete Table 44 to help find an expression that stands for the population of Gary (in thousands) at t years since 1980. Show the arithmetic to help you see a pattern.

Table 44 Populations of Gary

Years since 1980	Population (thousands)
0	
1	
2	
3	
4	
t	

b. Evaluate the expression you found in part (a) for $t = 37$. What does your result mean in this situation?

100. The percentage of companies offering traditional benefit plans was 85% in 1990 and has decreased by approximately 1.1 percentage points per year since then (Source: *National Compensation Survey*).

a. Complete Table 45 to help find an expression that stands for the percentage of companies offering traditional benefit plans at t years since 1990. Show the arithmetic to help you see a pattern.

Table 45 Percentages of Companies Offering Traditional Benefit Plans

Years since 1990	Percent
0	
1	
2	
3	
4	
t	

b. Evaluate the expression you found in part (a) for $t = 27$. What does your result mean in this situation?

101. The number of text messages sent or received per month was 150 billion messages in 2009, and it increased by 25.3% in 2010 (Source: *CTIA–The Wireless Association*). How many text messages were sent or received per month in 2010?

102. The revenue from electronic gaming software was $16.5 billion in 2010, and it increased by 6.1% in 2011 (Source: *Consumers Electronics Association*). What was the revenue in 2011?

103. The revenue from personal breathalyzers was $130 million in 2009, and it increased by 54% in 2010 (Source: *WinterGreen Research*). What was the revenue in 2010?

104. The number of people participating in the Supplemental Nutrition Assistance Program (formerly known as food stamps) was 40 million people in 2010, and it increased by 10.5% in 2011 (Source: *USDA*). How many people were in the program in 2011?

105. The number of specialty bicycle shops was 4319 shops in 2009, and it decreased by 1.5% in 2010 (Source: *National Bicycle Dealers Association*). What was the number of shops in 2010?

106. The number of U.S. troop deaths in the Afghanistan War in 2010 was 499 deaths, and it decreased by 17% in 2011 (Source: *U.S. Department of Defense*). How many U.S. troop deaths were there in the Afghanistan War in 2011?

107. The revenue from popular music concerts was $4.6 billion in 2009, and it decreased by 7.6% in 2010 (Source: *National Arts Index*). What was the revenue in 2010?

108. The revenue from televisions was $27.5 billion in 2009, and it decreased by 4.7% in 2010 (Source: *Consumer Electronics Association*). What was the revenue from televisions in 2010?

109. If a cube has sides of length s feet, then the volume of the cube is s^3 cubic feet. Find the volume of a cubic box with sides of length 2 feet.

110. If the radius of a sphere is r inches, then the volume of the sphere is $\frac{4}{3}\pi r^3$ cubic inches. Find the volume of a sphere with a radius of 3 inches.

Concepts

111. A student tries to perform the indicated operations in $2(3)^2 + 2(3) + 1$:

$$\begin{aligned} 2(3)^2 + 2(3) + 1 &= 6^2 + 2(3) + 1 \\ &= 36 + 2(3) + 1 \\ &= 36 + 6 + 1 \\ &= 43 \end{aligned}$$

Describe any errors. Then perform the operations correctly.

112. A student tries to evaluate $x^2 + 4x + 5$ for $x = -3$:

$$\begin{aligned} -3^2 + 4(-3) + 5 &= -9 - 12 + 5 \\ &= -21 + 5 \\ &= -16 \end{aligned}$$

Describe any errors. Then evaluate the expression correctly.

113. A student thinks that $-4^2 = 16$, because a negative number times a negative number is equal to a positive number. Is the student correct? Explain.

114. A student tries to perform the indicated operations in $16 \div 2 \cdot 4$:

$$16 \div 2 \cdot 4 = 16 \div 8 = 2$$

Describe any errors. Then perform the operations correctly.

115. In Problem 2 of Example 2, we performed the indicated operations in $\frac{11 - 3}{1 - 5}$ by the following steps:

$$\frac{11 - 3}{1 - 5} = \frac{8}{-4} = -2$$

a. Perform the indicated operations in $(11 - 3) \div (1 - 5)$.

b. Perform the indicated operations in $11 - 3 \div 1 - 5$.

c. In using a calculator to simplify $\frac{11 - 3}{1 - 5}$, a student presses the following buttons:

11 [−] 3 [÷] 1 [−] 5

The result of this calculation is 3 rather than −2. Describe any errors. Then perform the operations correctly.

116. a. Perform the indicated operations in $(12 \div 3) \cdot 2$.
b. Perform the indicated operations in $12 \div (3 \cdot 2)$.
c. For the expression $12 \div 3 \cdot 2$, does it matter whether we multiply and divide from left to right or multiply and divide from right to left? Explain.
d. Perform the indicated operations in $12 \div 3 \cdot 2$.

117. a. Evaluate $(ab)c$ for $a = 2, b = 3$, and $c = 4$.
b. Evaluate $a(bc)$ for $a = 2, b = 3$, and $c = 4$.
c. Compare your results for parts (a) and (b).
d. Evaluate $(ab)c$ and $a(bc)$ for $a = 4, b = -2$, and $c = 5$, and then compare the results.
e. Evaluate $(ab)c$ and $a(bc)$ for a, b, and c values of your choosing, and then compare the results.
f. Do you think that the statement $(ab)c = a(bc)$ is true for all values of a, b, and c? Explain.
g. What does the statement $(ab)c = a(bc)$ imply about the order of operations?

118. a. Evaluate $(a + b) + c$ for $a = 2, b = 3$, and $c = 4$.
b. Evaluate $a + (b + c)$ for $a = 2, b = 3$, and $c = 4$.
c. Compare your results for parts (a) and (b).
d. Evaluate $(a + b) + c$ and $a + (b + c)$ for $a = 4$, $b = -2$, and $c = 5$, and then compare the results.
e. Evaluate $(a + b) + c$ and $a + (b + c)$ for a, b, and c values of your choosing, and then compare the results.
f. Do you think that the statement $(a + b) + c = a + (b + c)$ is true for all values of a, b, and c? Explain.
g. What does the statement $(a + b) + c = a + (b + c)$ imply about the order of operations?

119. a. Evaluate x^2 for x equal to $-2, -1, 0, 1$, and 2.
b. Which of the following describes your results in part (a)?
negative, nonpositive, positive, nonnegative
c. Describe x^2 for any real number x.

120. a. Compute without using a calculator.
i. $(-1)^2$ **ii.** $(-1)^3$
iii. $(-1)^4$ **iv.** $(-1)^5$
v. $(-1)^{87}$ **vi.** $(-1)^{596}$
b. For what counting-number values of n is $(-1)^n$ equal to 1? Explain.
c. For what counting-number values of n is $(-1)^n$ equal to -1? Explain.

121. Some students believe x^2 is greater than x for *all* real numbers.
a. Find three values of x where x^2 is greater than x.
b. Find three values of x where x^2 is less than x.
c. Find two values of x where x^2 is equal to x.

122. Describe the order of operations in your own words.

Taking it to the Lab

Climate Change Lab (continued from Chapter 1)

Given the consensus among most scientists that global warming is occurring and that it could have catastrophic results, scientists have been searching for the cause of the warming (see Table 46).

Table 46 Average Surface Temperatures of Earth

Year	Average Temperature (degrees Fahrenheit)	Year	Average Temperature (degrees Fahrenheit)
1900	57.1	1960	57.2
1905	56.8	1965	57.0
1910	56.6	1970	57.3
1915	57.0	1975	57.1
1920	56.9	1980	57.6
1925	56.9	1985	57.3
1930	57.1	1990	57.8
1935	57.0	1995	57.9
1940	57.3	2000	57.8
1945	57.3	2005	58.3
1950	56.9	2010	58.3
1955	57.0	2011	58.1

Source: *NASA–GISS*

Most scientists believe global warming is largely the result of carbon dioxide emissions from the burning of fossil fuels such as oil, coal, and natural gas. Carbon dioxide emissions in the United States and in the world have increased greatly since 1950 (see Table 47).

Table 47 Carbon Dioxide Emissions from Burning of Fossil Fuels

Year	Carbon Dioxide Emissions (billions of metric tons)	
	United States	World
1950	2.4	5.8
1955	2.7	7.2
1960	2.9	9.4
1965	3.5	11.2
1970	4.3	14.7
1975	4.4	16.5
1980	4.8	19.1
1985	4.6	19.4
1990	5.0	21.6
1995	5.3	22.2
2000	5.9	23.8
2005	6.0	28.4
2010	5.6	30.6

Source: *U.S. Department of Energy*

Not everyone agrees carbon dioxide emissions cause global warming, however. For example, Jonathan Adler, professor of environmental law at Case Western Reserve School of Law, argues that two-thirds of the temperature increase occurred in the first half of the 20th century, yet most of the carbon dioxide emissions occurred in the second half of the century. He concludes that there is no link between carbon dioxide emissions and global warming. Adler believes global warming could be due to slight variations in the Sun's output, combined with fluctuations in Earth's orbit.*

Offering another explanation for global warming, Enric Pallé and his colleagues at the Big Bear Solar Observatory and the California Institute of Technology conducted a 2004 study that suggests the global warming that took place from 1984 to 2000 could have been due to reduced cloud cover.† Other scientists have raised objections to the study's findings.

Although there have been many dissenters along the way, the vast majority of scientists now believe in the warming–emissions connection. Even in 1997 there was strong enough agreement to motivate the United Nations to negotiate a treaty called the Kyoto Protocol. The treaty's goal was to reduce annual greenhouse gas emissions to about 5% to 7% below 1990 levels by 2012. In November 2004, Russia cast the deciding vote to ratify the protocol, which took effect on February 16, 2005.

To create some flexibility in the treaty's requirements, a country that had exceeded its emissions limit could buy emissions credits from a country that was below its emissions limit. Also, a country could receive emissions credits by financing a project to help lower emissions in another country.

The treaty was legally binding for the 128 countries that ratified it. The United States declined to ratify the Kyoto Protocol, saying that reducing emissions to the point called for by the treaty would cripple the U.S. economy. If it had ratified the treaty, the United States would have had to reduce its 2012 greenhouse gas emissions to 7% below its emissions level in 1990.

Instead of ratifying the treaty, in February 2002, the Bush administration adopted a voluntary program, called the Global Climate Change Initiative (GCCI), that includes tax incentives to motivate companies to reduce their emissions. The Bush administration set a goal of lowering the carbon intensity in 2012 to 18% below its level in 2002. *Carbon intensity* is defined as the ratio of annual carbon dioxide emissions to annual economic output.

Critics of the GCCI plan point out that if the U.S. economy is improving, then it is possible for carbon intensity to decrease *even though annual carbon dioxide emissions continue to increase*. In fact, even though annual carbon dioxide emissions have increased during each of the past three decades, carbon intensity has declined by 18%, 23%, and 16% in those decades! However, the projected decrease for 2002–2012 was only 13%, so modest efforts would have had to be implemented to reach the goal of 18%.‡

Critics such as the Earth Policy Institute and the Pew Center on Global Climate Change also say the voluntary plan is unrealistic, because businesses will make modest efforts to reduce emissions when the economy is doing well and will make little or no effort when the economy is doing poorly. These critics argue the United States would have been more likely to respond to the Kyoto Protocol, because it was a legally binding, yet flexible, treaty.

The Obama administration spent $2.5 billion on GCCI from 2010 to 2012, helping developing countries do the following: adapt to climate change, develop clean energy infrastructures, and stop cutting down forests. The Congressional Research Service raised pros and cons of continuing such actions, including questioning whether such monies would be better spent in meeting domestic challenges, especially when the U.S. economy is struggling.§

On December 11, 2011, the United Nations agreed on the Durban Platform, which includes a protocol that will apply to all members, a second commitment period (beyond 2012) for the Kyoto Protocol, and the Green Climate Fund, which is meant to help developing countries in ways similar to the Obama administration's actions.¶

Analyzing the Situation

1. **a.** Find the change in the average global temperature from 1900 to 1950.
   b. Find the change in the average global temperature from 1950 to 2000.
   c. Recall that Adler believes there is no link between carbon dioxide emissions and global warming. What is his argument? Is his reasoning correct? Explain.
2. Let c be U.S. carbon dioxide emissions (in billions of metric tons) in the year that is t years since 1950. Draw a careful scattergram of the data.
3. Draw a line that comes close to the points in your scattergram. Use your model to estimate U.S. carbon dioxide emissions in 2012.
4. Under the Kyoto Protocol, what would have been the largest quantity of U.S. carbon dioxide emissions allowed in 2012? Find the difference between this quantity and your estimate in Problem 3.
5. The numbers in the list 20, 22, 24, 26, 28 are increasing. Decide whether the ratios in the following list are increasing, decreasing, or neither:
$$\frac{20}{1}, \frac{22}{2}, \frac{24}{4}, \frac{26}{8}, \frac{28}{16}$$
6. Explain why your work in Problem 5 illustrates how it is possible for U.S. carbon intensity to decrease while U.S. annual carbon dioxide emissions increase. [**Hint:** Recall that carbon intensity is defined as the *ratio* of annual carbon dioxide emissions to annual economic output.]

*"Global Warming—Hot Problem or Hot Air?" April 1998, *The Freeman*, 48(4), The Foundation for Economic Education, Inc.

†"Changes in Earth's Reflectance over the Past Two Decades," May 28, 2004, *Science* 304: 1299–1301.

‡"Early Release of the Annual Energy Outlook 2003" (November 2002), Energy Information Administration; and "Projected Greenhouse Gas Emissions," May 2002, *U.S. Climate Action Report 2002,* pp. 70–80, U.S. Department of State, Washington, DC.

§Richard K. Lattanzio, "The Global Climate Change Initiative (GCCI): Budget Authority and Request. FY2010-FY2013," Congressional Research Service, *CRS Report for Congress.* March 15, 2012.

¶United Nations, "Establishment of an Ad Hoc Working Group on the Durban Platform for Enhanced Action," December 11, 2011.

7. Which seeks to lower U.S. carbon dioxide emissions more, the GCCI plan or the Kyoto Protocol? Explain. Taking into account the degree to which American companies would have responded to either policy, do you think U.S. carbon dioxide emissions would have been reduced more by the GCCI plan or by the Kyoto Protocol? Explain.

Stocks Lab

Imagine that you have $5000 that you plan to invest in five stocks for one week. In this lab, you will explore some possible outcomes of that investment.**

Collecting the Data

Use a newspaper or the Internet to select five stocks. For each stock, record the company name and the call letters of the stock. Here are some examples:

Gateway Computer has the call letters GTW.
Coca-Cola has the call letters KO.
EMC corporation has the call letters EMC.

Record the value of one share (the beginning share price) of each of the five stocks. Also, record how you distribute your $5000 investment. For example, you may invest all $5000 in one of the five stocks or $1000 in each of the five stocks, or you may opt for some other distribution of the money. The sum of your investments should equal or be close to $5000 by buying whole amounts of stock. Even if you do not invest money in some of the stocks, still record the information about all five stocks.

After one week, record the new value of one share (the ending share price) of each of the five stocks.

Analyzing the Data

1. Complete Table 48. The *profit* from a stock is the money you collect from the stock, minus the money you invested in the stock. What is the total profit from your $5000 investment?
2. For each of the five stocks, create a bar graph displaying the beginning and ending values of one share. You can use the same axes for several bar graphs if the scaling is convenient.
3. Find the change in the share price of each of the five stocks. Which share price had the greatest change? The least? Explain how you can illustrate these changes on your bar graphs.
4. The *percent change* of the value of a stock can be found by dividing the change in value of the stock by the original value of the stock and then converting the decimal result into percent form (by multiplying by 100). For example, suppose that a stock's value increases from $7 to $9. The change in value of this stock is its ending value minus its beginning value: $9 - 7 = 2$ dollars. Here we find the percent change in value:

$$\text{percent change} = \frac{\text{change in value}}{\text{beginning value}} \cdot 100$$

$$= \frac{9-7}{7} \cdot 100 = \frac{2}{7} \cdot 100 \approx 28.57$$

So, the percent change is about 28.57%.

Now find the percent change in value for each of your five stocks. Which stock had the greatest percent change? The least? Explain how you can at least approximately compare the percent changes by viewing your bar graphs.

5. Among your five stocks, is there a pair of stocks for which one stock has the greater change but the other has the greater percent change? If yes, use these stocks to respond to the questions in parts (a)–(c). If no, then use the following values of fictional stocks A and B:

Stock A	Stock B
Increased from $4 to $5	Increased from $20 to $23

a. Find the profit earned from investing all of the $5000 in the stock with the larger change in value.
b. Find the profit earned from investing all of the $5000 in the stock with the larger percent change in value.
c. Which is the better measure of the growth of a stock, change in value or percent change in value? Explain.

6. a. Find the profit earned from each of the following scenarios:
   i. You invest the $5000 in the best-performing stock among the five stocks.
   ii. You invest the $5000 in the worst-performing stock among the five stocks.
   iii. You invest the $5000 by investing $1000 in each of the five stocks.

   b. Describe the benefits and drawbacks to investing your money in a number of stocks rather than in just one stock.

Table 48 Five Stocks' Performances

Call Letters of Stock	Investment in Stock (dollars)	Beginning Share Price (dollars)	Number of Shares	Ending Share Price (dollars)	Money Collected from Stock (dollars)	Profit from Stock (dollars)

**Lab suggested by Jim Ryan, State Center Community College District, Clovis Center, Clovis, CA.

Chapter Summary

Key Points of Chapter 2

Section 2.1 Expressions

Expression	An **expression** is a constant, a variable, or a combination of constants, variables, operation symbols, and grouping symbols, such as parentheses.
Evaluate an expression	We **evaluate an expression** by substituting a number for each variable in the expression and then calculating the result. If a variable appears more than once in the expression, the same number is substituted for that variable each time.

Section 2.2 Operations with Fractions

Division by 0	The fraction $\frac{a}{b}$ is undefined if $b = 0$. Division by 0 is undefined.
Simplify a fraction	To simplify a fraction, **1.** Find the prime factorizations of the numerator and denominator. **2.** Find an equal fraction in which the numerator and the denominator do not have common positive factors other than 1 by using the property $$\frac{ab}{ac} = \frac{a}{a} \cdot \frac{b}{c} = 1 \cdot \frac{b}{c} = \frac{b}{c}$$ where a and c are nonzero.
Simplify results	If the result of an exercise is a fraction, simplify it.
Multiplying fractions	$\frac{a}{b} \cdot \frac{c}{d} = \frac{ac}{bd}$, where b and d are nonzero.
Dividing fractions	$\frac{a}{b} \div \frac{c}{d} = \frac{a}{b} \cdot \frac{d}{c}$, where b, c, and d are nonzero.
Adding fractions	$\frac{a}{b} + \frac{c}{b} = \frac{a + c}{b}$, where b is nonzero.
Subtracting fractions	$\frac{a}{b} - \frac{c}{b} = \frac{a - c}{b}$, where b is nonzero.
How to add or subtract two fractions with different denominators	To add (or subtract) two fractions with different denominators, use the fact that $\frac{a}{a} = 1$, where a is nonzero, to write an equal sum (or difference) of fractions for which each denominator is the LCD.

Section 2.3 Adding Real Numbers

The opposite of the opposite of a number	$-(-a) = a$
Absolute value	The **absolute value** of a number is the distance that the number is from 0 on the number line.
Adding two numbers with the same sign	To add two numbers with the same sign, **1.** Add the absolute values of the numbers. **2.** The sum of the original numbers has the same sign as the sign of the original numbers.
Adding two numbers with different signs	To add two numbers with different signs, **1.** Find the absolute values of the numbers. Then subtract the smaller absolute value from the larger absolute value. **2.** The sum of the original numbers has the same sign as the original number with the larger absolute value.

Section 2.4 Change in a Quantity and Subtracting Real Numbers

Change in a quantity — The change in a quantity is the ending amount minus the beginning amount:

Change in the quantity = Ending amount − Beginning amount

Subtracting a number — To subtract a number, add its opposite:

$$a - b = a + (-b)$$

Increasing quantity — An increasing quantity has a positive change.

Decreasing quantity — A decreasing quantity has a negative change.

Section 2.5 Ratios, Percents, and Multiplying and Dividing Real Numbers

Unit ratio — A **unit ratio** is a ratio written as $\frac{a}{b}$ with $b = 1$ or as $a:b$ with $b = 1$.

Percent — **Percent** means "for each hundred": $a\% = \frac{a}{100}$.

Writing a percentage as a decimal number — To write a percentage as a decimal number, remove the percent symbol and divide the number by 100 (move the decimal point two places to the left).

Writing a decimal number as a percentage — To write a decimal number as a percentage, multiply the number by 100 (move the decimal point two places to the right) and insert a percent symbol.

Finding the percentage of a quantity — To find the percentage of a quantity, multiply the decimal form of the percentage and the quantity.

One hundred percent — One hundred percent of a quantity is *all* of the quantity.

The product or quotient of two numbers — The product or quotient of two numbers that have different signs is negative. The product or quotient of two numbers that have the same sign is positive.

Equal fractions with negative signs — If $b \neq 0$, then $\frac{-a}{b} = \frac{a}{-b} = -\frac{a}{b}$.

Section 2.6 Exponents and Order of Operations

Exponent — For any counting number n,

$$x^n = \underbrace{x \cdot x \cdot x \cdot \ldots \cdot x}_{n \text{ factors of } x}$$

We refer to x^n as the **power,** the ***n*th power of *x*,** or ***x* raised to the *n*th power.** We call x the **base** and n the **exponent.**

Finding $-a^n$ — For an expression of the form $-a^n$, we calculate a^n before taking the opposite.

Order of operations — We perform operations in the following order:

1. First, perform operations within parentheses or other grouping symbols, starting with the innermost group.
2. Then perform exponentiations.
3. Next, perform multiplications and divisions, going from left to right.
4. Last, perform additions and subtractions, going from left to right.

Order of operations and the strengths of operations — After we have performed operations in parentheses, the order of operations goes from the most powerful operation, exponentiation, to the next-most-powerful operations, multiplication and division, to the weakest operations, addition and subtraction.

Chapter 2 Review Exercises

Perform the indicated operations. Then use a calculator to check your work.

1. $8 + (-2)$
2. $(-5) + (-7)$
3. $6 - 9$
4. $8 - (-2)$
5. $8(-2)$
6. $8 \div (-2)$
7. $-24 \div (10 - 2)$
8. $(2 - 6)(5 - 8)$
9. $\dfrac{7 - 2}{2 - 7}$
10. $\dfrac{2 - 8}{3 - (-1)}$
11. $\dfrac{3 - 5(-6)}{-2 - 1}$
12. $3(-5) + 2$
13. $-4 + 2(-6)$
14. $2 - 12 \div 2$
15. $8 \div (-2) \cdot 5$
16. $4 - 6(7 - 2)$
17. $2(4 - 7) - (8 - 2)$
18. $-2(3 - 6) + 18 \div (-9)$
19. $-14 \div (-7) - 3(1 - 5)$
20. $-5 - [3 + 2(1 - 7)]$
21. $4.2 - (-6.7)$
22. $\dfrac{4}{9}\left(-\dfrac{3}{10}\right)$
23. $\left(-\dfrac{8}{15}\right) \div \left(-\dfrac{16}{25}\right)$
24. $\dfrac{5}{9} - \left(-\dfrac{2}{9}\right)$
25. $-\dfrac{5}{6} + \dfrac{7}{8}$
26. $\dfrac{-5}{2} - \dfrac{7}{-3}$
27. $(-8)^2$
28. -8^2
29. 2^4
30. $\left(\dfrac{3}{4}\right)^3$
31. $-6(3)^2$
32. $24 \div 2^3$
33. $(-2)^3 - 4(-2)$
34. $\dfrac{2^3}{3 + 3^2}$
35. $\dfrac{17 - (-3)^2}{5 - 4^2}$
36. $-3(2)^2 - 4(2) + 1$
37. $24 \div (3 - 5)^3$
38. $7^2 - 3(2 - 5)^2 \div (-3)$

Simplify.

39. $\dfrac{-18}{-24}$
40. $\dfrac{-28}{35}$

For Exercises 41 and 42, use a calculator to compute. Round the result to two decimal places.

41. $-5.7 + 2.3^4 \div (-9.4)$
42. $\dfrac{3.5(17.4) - 97.6}{54.2 \div 8.4 - 65.3}$

43. A rectangle has a width of $\frac{1}{4}$ yard and a length of $\frac{5}{6}$ yard. What is the perimeter of the rectangle?

44. A student owes \$4789 to a credit card company. If he sends a check for \$800 and then uses his credit card to purchase a textbook for \$102.99 and a notebook for \$3.50, how much money does he now owe the credit card company?

45. An airplane drops from 32,500 feet to 27,800 feet. Find the change in altitude.

46. Three hours ago the temperature was 4°F. The temperature is now −8°F.

a. What is the change in temperature for the past three hours?

b. Estimate the change in temperature for the past hour.

c. Explain why your estimate in part (b) may not be the actual change in temperature for the past hour.

47. Contributions from individuals to major party nominees of presidential elections are shown in Table 49 for various years.

Table 49 Contributions from Individuals to Major Party Nominees of Presidential Elections

	Contribution (millions of dollars)	
Year	Democrat Nominee	Republican Nominee
1996	28.3	29.6
2000	33.9	91.3
2004	215	259
2008	454	220

Source: *Federal Election Commission*

a. Find the change in the annual individual contributions to the Democratic nominee from 1996 to 2000.

b. Find the change in the annual individual contributions to the Republican nominee from 2004 to 2008.

c. Over which four-year period was there the greatest change in annual individual contributions to the Democratic nominee? What was that change?

d. Over which four-year period was there the greatest change in annual individual contributions to the Republican nominee? What was that change?

48. Text-messaging users sent or received an average of 41.5 messages per day in 2011 and sent or received an average of 29.7 messages per day in 2009 (Source: *The Pew Research Center's Internet & American Life Project*). Find the unit ratio of the number of messages sent or received per day in 2011 to the number of messages sent or received per day in 2009. What does your result mean?

For Exercises 49 and 50, write the percentage as a decimal number.

49. 75%
50. 2.9%

51. Find 87% of \$43.
52. Find 8% of 925 students.

53. A person's credit card balance is −5493 dollars. If the person pays off 20% of the balance, what is the new balance?

Evaluate the expression for $a = 2$, $b = -5$, $c = -4$, and $d = 10$.

54. $ac + c \div a$
55. $b^2 - 4ac$
56. $a(b - c)$
57. $\dfrac{-b - c^2}{2a}$
58. $2c^2 - 5c + 3$
59. $\dfrac{a - b}{c - d}$

For Exercises 60–63, let x be a number. Translate the English phrase into a mathematical expression. Then evaluate the expression for $x = -3$.

60. 5 more than the number

61. The number subtracted from −7

62. 2 minus the product of the number and 4

63. 1 plus the quotient of −24 and the number

64. If T is the total cost (in dollars) for a team to join a softball league and there are n players on the team, then $T \div n$ is the cost (in dollars) per player. Evaluate $T \div n$ for $T = 650$ and $n = 13$. What does your result mean in this situation?

65. A basement is flooded with 400 cubic feet of water. Each hour, 50 cubic feet of water is pumped out of the basement.

a. Complete Table 50 to help find an expression that stands for the volume (in cubic feet) of water in the basement after water has been pumped out for t hours. Show the arithmetic to help you see a pattern.

b. Evaluate the expression that you found in part (a) for $t = 7$. What does your result mean in this situation?

66. The number of Americans who flew to Europe in 2007 was 13.3 million, and it decreased 18.8% by 2011 (Source: *U.S Department of Commerce*). How many Americans flew to Europe in 2011?

Table 50 Volumes of Water

Time (hours)	Volume of Water (cubic feet)
0	
1	
2	
3	
4	
t	

Chapter 2 Test

For Exercises 1–14, perform the indicated operations by hand.

1. $-8 - 5$

2. $-7(-9)$

3. $-3 + 9 \div (-3)$

4. $(4 - 2)(3 - 7)$

5. $\dfrac{4 - 7}{-1 - 5}$

6. $5 - (2 - 10) \div (-4)$

7. $-20 \div 5 - (2 - 9)(-3)$

8. $0.4(-0.2)$

9. $-\dfrac{27}{10} \div \dfrac{18}{75}$

10. $-\dfrac{3}{10} + \dfrac{5}{8}$

11. 3^4

12. -4^2

13. $7 + 2^3 - 3^2$

14. $1 - (3 - 7)^2 + 10 \div (-5)$

15. Simplify $\dfrac{84}{-16}$.

16. Two hours ago the temperature was 5°F. If the temperature has decreased by 9°F in the last two hours, what is the current temperature?

17. The chances of being audited by the Internal Revenue Service (IRS) are shown in Table 51 for various years.

Table 51 Tax Audit Rates

Year	Tax Audit Rate (number of audits per 1000 tax returns)
1999	9.0
2001	5.8
2003	6.5
2005	9.7
2007	10.3
2009	10.3
2011	11.1

Source: *Internal Revenue Service*

a. Find the change in the tax audit rate from 2001 to 2003.

b. Find the change in the tax audit rate from 1999 to 2001.

c. For which two-year period was the change in the tax audit rate the most? What was that change?

18. The average ticket price to major league baseball games was \$9.14 in 1991 and \$26.92 in 2012 (Source: *AP*). Find the unit ratio of the average ticket price in 2012 to the average ticket price in 1991. What does your result mean?

Evaluate the expression for $a = -6$, $b = -2$, $c = 5$, and $d = -1$.

19. $ac - \dfrac{a}{b}$

20. $\dfrac{a - b}{c - d}$

21. $a + b^3 + c^2$

22. $b^2 - 4ac$

For Exercises 23 and 24, let x be a number. Translate the English phrase into a mathematical expression. Then evaluate the expression for $x = -5$.

23. Twice the number minus the product of 3 and the number

24. 6 subtracted from the quotient of -10 and the number

25. U.S. Postal Service first-class mail volume was 78.2 billion pieces in 2010. The first-class mail volume has decreased by about 4.9 billion pieces per year since 2010 (Source: *U.S. Postal Service*).

a. Complete Table 52 to help find an expression that stands for the U.S. Postal Service first-class mail volume (in billions of pieces) in the year that is t years since 2010. Show the arithmetic to help you see a pattern.

Table 52 U.S. Postal Service First-Class Mail Volume

Years since 2010	First-Class Mail Volume (billions of pieces)
0	
1	
2	
3	
4	
t	

b. Evaluate the expression that you found in part (a) for $t = 8$. What does your result mean in this situation?

26. The number of California methamphetamine labs busted in 2010 was 203 labs, and it had decreased 41.9% by 2011 (Source: *California Department of Justice*). How many California meth labs were busted in 2011?

Cumulative Review of Chapters 1 and 2

1. A rectangle has a perimeter of 36 inches. Let W be the width, L be the length, and P be the perimeter, all with units in inches.
   a. Sketch three possible rectangles with a perimeter of 36 inches.
   b. Which of the symbols W, L, and P are variables? Explain.
   c. Which of the symbols W, L, and P are constants? Explain.
2. Graph the integers between -2 and 3, inclusive, on a number line.
3. Here are the changes in a stock's value from one month to the next: increase of 1 dollar, decrease of 3 dollars, increase of 4 dollars, and decrease of 2 dollars. Let C be the change of the stock's value (in dollars) from one month to the next. Use points on a number line to describe the changes in value of the stock.
4. What is the x-coordinate of the ordered pair $(-5, 3)$?
5. A person takes a bath. Let V be the volume (in gallons) of water in the bathtub at t minutes after the person pulls out the plug from the drain. Identify the independent variable and the dependent variable.
6. The revenues of Starbucks are shown in Table 53 for various years.

Table 53 Revenues of Starbucks

Years	Revenue (billions of dollars)
2007	9.4
2008	10.4
2009	9.8
2010	10.7
2011	11.7

Source: *RetailSails*

Let r be the annual revenue (in billions of dollars) at t years since 2005.
   a. Create a scattergram of the data.
   b. For which year shown in Table 53 was the revenue the most?
   c. For which year shown in Table 53 was the revenue the least?
   d. From which year to the next did the annual revenue decrease the most? What was the change in annual revenue?
   e. From which year to the next did the annual revenue increase the most? What was the change in annual revenue?

For Exercises 7–10, refer to Fig. 35.

7. Find y when $x = -4$.
8. Find x when $y = 1$.
9. Find the y-intercept.
10. Find the x-intercept.

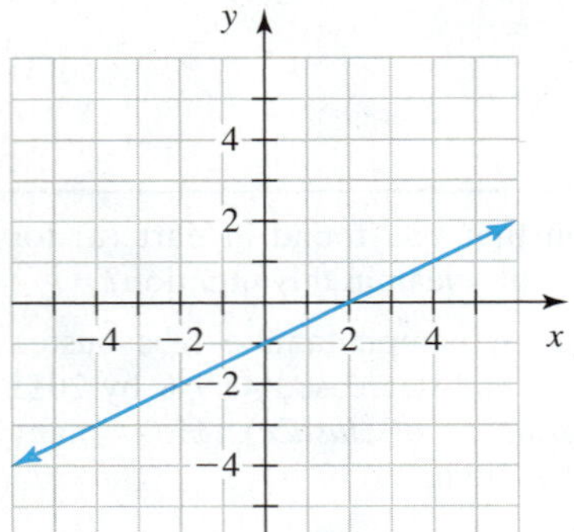

Figure 35 Exercises 7–10

11. A person is laid off from work. Let B be the balance (in thousands of dollars) in her checking account at t months since she was laid off. Some pairs of values of t and B are shown in Table 54.

Table 54 Balances in a Checking Account

t (months)	B (thousands of dollars)
2	16
3	14
5	10
6	8
8	4

   a. Create a scattergram of the data. Then draw a linear model.
   b. What was the balance 4 months after the person was laid off?
   c. When was the balance \$6 thousand?
   d. What is the B-intercept? What does it mean in this situation?
   e. What is the t-intercept? What does it mean in this situation?
12. The average monthly basic rates for cable TV are shown in Table 55 for various years.

Table 55 Cable TV Costs

Year	Average Monthly Basic Rate for Cable TV (dollars)
1990	17
1995	23
2000	30
2005	40
2010	48

Source: *Veronis Suhler Stevenson Communications Industry Forecast*

Let c be the average monthly basic rate (in dollars) for cable TV at t years since 1990.
   a. Create a scattergram of the data.
   b. Draw a line that comes close to the points in your scattergram.
   c. What is the c-intercept? What does it mean in this situation?
   d. Predict when the average monthly basic rate for cable TV will be \$63.
   e. Predict the average monthly basic rate for cable TV in 2018.

For Exercises 13–18, perform the indicated operations by hand. Then use a calculator to check your work.

13. $\dfrac{3(-8) + 15}{2 - 7(2)}$
14. $-4(3) + 6 - 20 \div (-10)$
15. $\left(-\dfrac{14}{15}\right) \div \left(-\dfrac{35}{27}\right)$
16. $\dfrac{3}{8} - \dfrac{5}{6}$
17. $4 - (7 - 9)^4 + 20 \div (-4)$
18. $\dfrac{5 - 3^2}{4^2 + 2}$
19. Two hours ago the temperature was 5°F. The temperature is now −3°F. What is the change in temperature?

20. A student owes \$2692 to a credit card company. If he sends a check to the company for \$850 and then uses his credit card to purchase some gasoline for \$23, how much money does he now owe?

21. Evaluate $\frac{a - b}{c - d}$ for $a = 1$, $b = -4$, $c = -3$, and $d = 7$.

22. Evaluate $b^2 - 4ac$ for $a = 2$, $b = -3$, and $c = -5$.

For Exercises 23 and 24, let x be a number. Translate the English phrase into a mathematical expression. Then evaluate the expression for $x = -4$.

23. The number minus the quotient of -12 and the number

24. 7 more than the product of -2 and the number

25. A stock was worth \$42 last year. Let v be the value (in dollars) of the stock today. The expression

$$\frac{100(v - 42)}{42}$$

represents the percent growth of the investment. Evaluate the expression for $v = 45$. Round your result to the second decimal place. What does your result mean in this situation?

26. A motorcycle company sold 15 thousand motorcycles in 2005, and sales then increased by 4 thousand motorcycles each year thereafter.

a. Complete Table 56 to help find an expression that stands for the annual sales (in thousands of motorcycles) at t years since 2005. Show the arithmetic to help you see a pattern.

Table 56 Numbers of Years and Sales

Years since 2005	Sales (thousands of motorcycles)
0	
1	
2	
3	
4	
t	

b. Evaluate the expression that you found in part (a) for $t = 12$. What does your result mean in this situation?

3 Using Slope to Graph Linear Equations

Do you use social networking sites such as Facebook? The percentage of adult Internet users who use social networking sites has increased greatly (see Table 1). In Exercise 27 of Homework 3.5, you will describe how quickly the percentage has increased over time.

Table 1 Percentages of Adult Internet Users Who Use Social Networking Sites

Year	Percent
2005	8
2006	16
2008	29
2009	46
2010	61
2011	65

Source: *Pew Research Center's Internet & American Life Project surveys*

In Chapter 1, we used a line to describe the relationship between two quantities that are linearly related. In this chapter, we will discuss how to describe such a relationship by a symbolic statement called an *equation*. We will discuss how to use an equation to sketch a line. We will also describe the steepness of a line and how that steepness is related to how quickly one quantity changes in relation to another, such as how quickly yogurt sales have increased in the United States over time.

3.1 Graphing Equations of the Form $y = mx + b$

Objectives

- » For an equation in two variables, know the meaning of *solution, satisfy*, and *solution set*.
- » Know the meaning of the *graph* of an equation.
- » Graph equations of the form $y = mx + b$.
- » Know the meaning of b in equations of the form $y = mx + b$.
- » Know the Rule of Four for equations.

In this section, we will work with equations. An **equation** consists of an equality sign "=" with expressions on both sides. Here are some examples of equations:

$$y = 3x - 5, \qquad 2x - 4y = 8, \qquad x = 5$$

In Sections 1.3 and 1.4, we used lines to model data. It turns out we can describe any line by an equation.

Solutions, Satisfying Equations, and Solution Sets

Consider the equation $y = x + 4$. Let's find y when $x = 3$:

$$\begin{aligned} y &= x + 4 && \text{Original equation} \\ y &= 3 + 4 && \text{Substitute 3 for } x. \\ &= 7 && \text{Add.} \end{aligned}$$

So, $y = 7$ when $x = 3$. Recall from Section 1.2 that the ordered-pair notation $(3, 7)$ is shorthand for saying that when $x = 3$, $y = 7$.

For the equation $y = x + 4$, we found that $y = 7$ when $x = 3$. This means the equation $y = x + 4$ becomes a true statement when we substitute 3 for x and 7 for y:

$$y = x + 4 \quad \text{Original equation}$$
$$7 \stackrel{?}{=} 3 + 4 \quad \text{Substitute 3 for x and 7 for y.}$$
$$7 \stackrel{?}{=} 7 \quad \text{Add.}$$
true

We say that (3, 7) is a *solution* of the equation $y = x + 4$ and that (3, 7) *satisfies* the equation $y = x + 4$.

A *set* is a container. Much as an egg carton contains eggs, a *solution set* contains solutions.

Definition *Solution, satisfy,* and *solution set* of an equation in two variables

An ordered pair (a, b) is a **solution** of an equation in terms of x and y if the equation becomes a true statement when a is substituted for x and b is substituted for y. We say (a, b) **satisfies** the equation. The **solution set** of an equation is the set of all solutions of the equation.

Example 1 Identifying Solutions of an Equation

1. Is (2, 1) a solution of $y = 3x - 5$?
2. Is (4, 9) a solution of $y = 3x - 5$?

Solution

1. We substitute 2 for x and 1 for y in the equation $y = 3x - 5$:

$$y = 3x - 5 \quad \text{Original equation}$$
$$1 \stackrel{?}{=} 3(2) - 5 \quad \text{Substitute 2 for x and 1 for y.}$$
$$1 \stackrel{?}{=} 6 - 5 \quad \text{Multiply before subtracting.}$$
$$1 \stackrel{?}{=} 1 \quad \text{Subtract.}$$
true

So, (2, 1) is a solution of $y = 3x - 5$.

2. We substitute 4 for x and 9 for y in the equation $y = 3x - 5$:

$$y = 3x - 5 \quad \text{Original equation}$$
$$9 \stackrel{?}{=} 3(4) - 5 \quad \text{Substitute 4 for x and 9 for y.}$$
$$9 \stackrel{?}{=} 12 - 5 \quad \text{Multiply before subtracting.}$$
$$9 \stackrel{?}{=} 7 \quad \text{Subtract.}$$
false

So, (4, 9) is *not* a solution of $y = 3x - 5$.

Definition of Graph

Next we will learn how to *graph* an equation. As a first step, we plot some solutions of an equation in the next example.

Example 2 Plotting Some Solutions of an Equation

Find five solutions of $y = 2x - 1$, and plot them in the same coordinate system.

Solution

To find solutions, we are free to choose *any* values we'd like to substitute for x, but it's a good idea to pick integers close to or equal to 0 so the solutions are easy to plot. For example, here we substitute 0, 1, and 2 for x:

$$\begin{aligned} y &= 2(0) - 1 \\ &= 0 - 1 \\ &= -1 \end{aligned}$$
Solution: $(0, -1)$

$$\begin{aligned} y &= 2(1) - 1 \\ &= 2 - 1 \\ &= 1 \end{aligned}$$
Solution: $(1, 1)$

$$\begin{aligned} y &= 2(2) - 1 \\ &= 4 - 1 \\ &= 3 \end{aligned}$$
Solution: $(2, 3)$

Next we substitute -2 and -1 for x:

$$\begin{aligned} y &= 2(-2) - 1 \\ y &= -4 - 1 \\ y &= -5 \end{aligned}$$
Solution: $(-2, -5)$

$$\begin{aligned} y &= 2(-1) - 1 \\ y &= -2 - 1 \\ y &= -3 \end{aligned}$$
Solution: $(-1, -3)$

We organize our findings in Table 2. In Fig. 1, we plot the five solutions.

Table 2 Solutions of $y = 2x - 1$

x	y
-2	$2(-2) - 1 = -5$
-1	$2(-1) - 1 = -3$
0	$2(0) - 1 = -1$
1	$2(1) - 1 = 1$
2	$2(2) - 1 = 3$

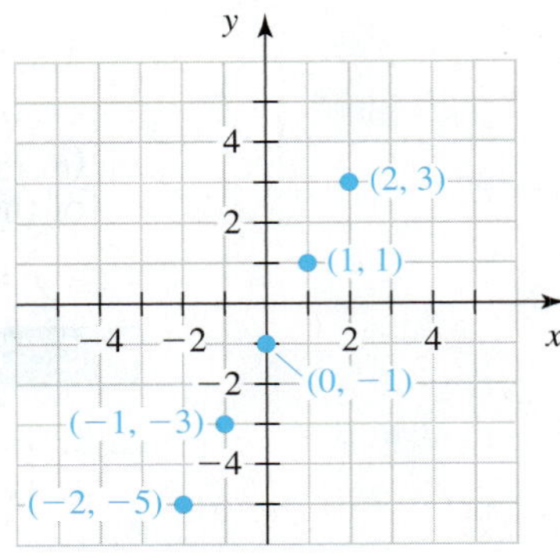

Figure 1 Five solutions of $y = 2x - 1$

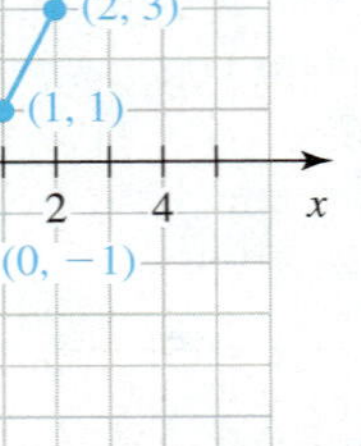

Figure 2 The line contains solutions of $y = 2x - 1$ found in Example 2

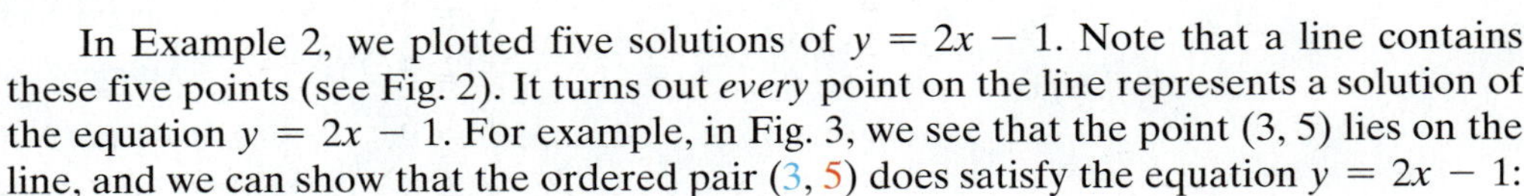

In Example 2, we plotted five solutions of $y = 2x - 1$. Note that a line contains these five points (see Fig. 2). It turns out *every* point on the line represents a solution of the equation $y = 2x - 1$. For example, in Fig. 3, we see that the point $(3, 5)$ lies on the line, and we can show that the ordered pair $(3, 5)$ does satisfy the equation $y = 2x - 1$:

$$\begin{aligned} y &= 2x - 1 && \text{Original equation} \\ 5 &\stackrel{?}{=} 2(3) - 1 && \text{Substitute 3 for } x \text{ and 5 for } y. \\ 5 &\stackrel{?}{=} 6 - 1 && \text{Multiply before subtracting.} \\ 5 &\stackrel{?}{=} 5 && \text{Add.} \end{aligned}$$
true

It also turns out points that do not lie on the line represent ordered pairs that do *not* satisfy the equation. For example, by Fig. 3, we see that the point $(4, 2)$ does not lie on the line, and we can show that the ordered pair $(4, 2)$ does not satisfy the equation $y = 2x - 1$:

$$\begin{aligned} y &= 2x - 1 && \text{Original equation} \\ 2 &\stackrel{?}{=} 2(4) - 1 && \text{Substitute 4 for } x \text{ and 2 for } y. \\ 2 &\stackrel{?}{=} 8 - 1 && \text{Multiply before subtracting.} \\ 2 &\stackrel{?}{=} 7 && \text{Subtract.} \end{aligned}$$
false

We call the line in Fig. 3 the *graph* of the equation $y = 2x - 1$.

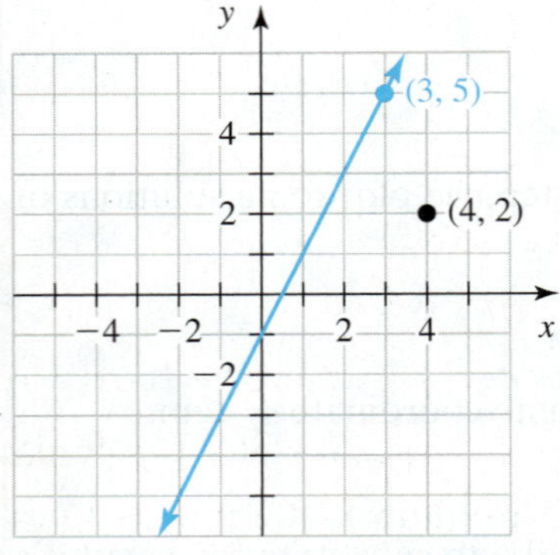

Figure 3 $(3, 5)$ lies on the line, but $(4, 2)$ does not lie on the line

Definition Graph

The **graph** of an equation in two variables is the set of points that correspond to all solutions of the equation.

The graph of an equation in two variables is a visual description of the solutions of the equation. Every point on the graph represents a solution of the equation. Every point *not* on the graph represents an ordered pair that is *not* a solution.

Graphs of Equations of the Form $y = mx + b$

Directly after Example 2, we found that the graph of the equation $y = 2x - 1$ is a line. The equation $y = 2x - 1$, or $y = 2x + (-1)$, is of the form $y = mx + b$, where $m = 2$ and $b = -1$. It turns out for any equation of the form $y = mx + b$, where m and b are constants, the graph is a line.

Graph of $y = mx + b$

The graph of an equation of the form $y = mx + b$, where m and b are constants, is a line.

Here are some equations whose graphs are lines:

$$y = 3x + 7, \quad y = -4x - 5, \quad y = -4x, \quad y = x + 3, \quad y = 2$$

The equation $y = -4x$ is of the form $y = mx + b$ because we can write it as $y = -4x + 0$. The equation $y = 2$ is of the form $y = mx + b$ because we can write it as $y = 0x + 2$.

Table 3 Solutions of $y = -2x + 3$

x	y
0	$-2(0) + 3 = 3$
1	$-2(1) + 3 = 1$
2	$-2(2) + 3 = -1$

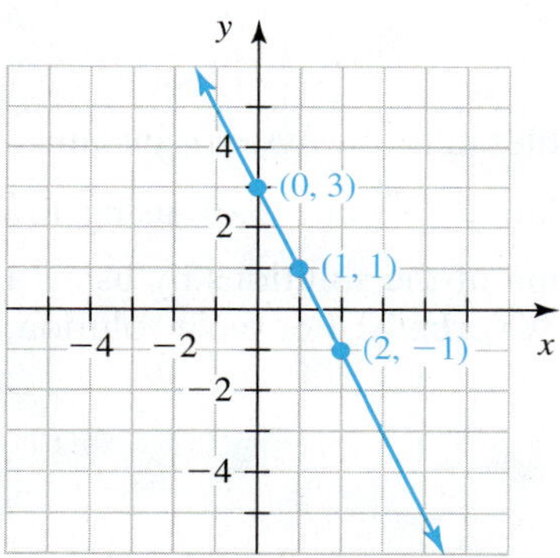

Figure 4 Graph of $y = -2x + 3$

Example 3 Graphing an Equation

Sketch the graph of $y = -2x + 3$. Also, find the y-intercept.

Solution

Since $y = -2x + 3$ is of the form $y = mx + b$, the graph is a line. Although we can sketch a line from as few as two points, we plot a third point as a check. If the third point is not in line with the other two, then we know that we have computed or plotted at least one of the solutions incorrectly.

To begin, we calculate three solutions of $y = -2x + 3$ in Table 3. We use 0, 1, and 2 as values of x because they correspond to points that are easy to plot. Then we plot the three points and sketch the line through them (see Fig. 4).

We use ZStandard followed by ZSquare to verify our graph (see Fig. 5). See Appendix A.3, A.4, and A.6 for graphing calculator instructions.*

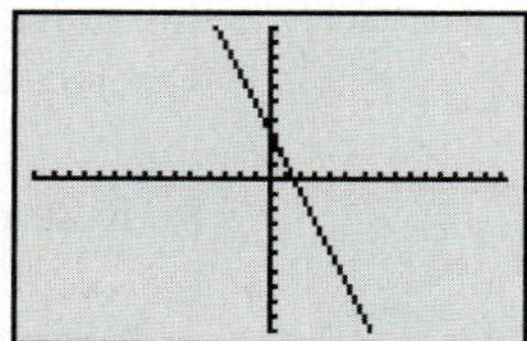

Figure 5 Graph of $y = -2x + 3$

From Table 3 and Fig. 4, we see that the y-intercept is $(0, 3)$.

Table 4 Solutions of $y = \frac{3}{2}x - 2$

x	y
0	$\frac{3}{2}(0) - 2 = -2$
2	$\frac{3}{2}(2) - 2 = 1$
4	$\frac{3}{2}(4) - 2 = 4$

Example 4 Graphing an Equation

Sketch the graph of $y = \frac{3}{2}x - 2$. Also, find the y-intercept.

Solution

In Table 4, we use 0 and multiples of 2 as values of x to avoid fractional values of y. Since $y = \frac{3}{2}x - 2$ is of the form $y = mx + b$, the graph is a line. We plot the points that correspond to the solutions we found and sketch the line that contains them in Fig. 6.

*Appendix A is available online in MyMathLab and at http://www.pearsonhighered.com/mathstatsresources/

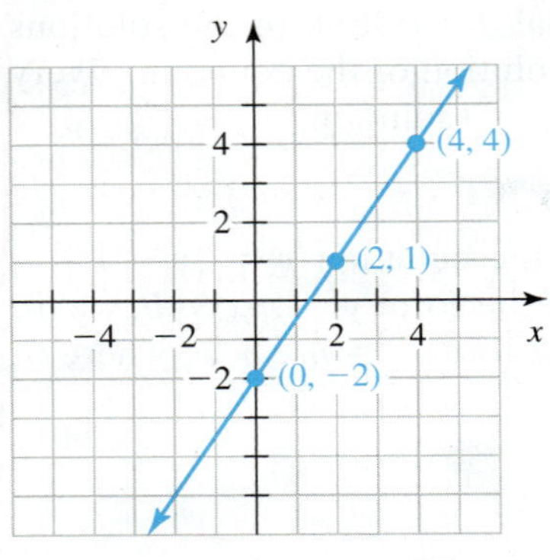

Figure 6 Graph of $y = \frac{3}{2}x - 2$

We use ZDecimal to verify our graph (see Fig. 7). See Appendix A.3, A.4, and A.6 for graphing calculator instructions.

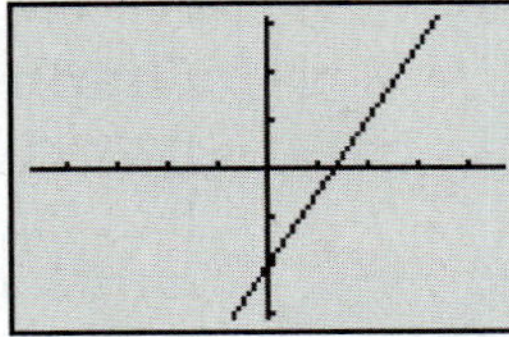

Figure 7 Graph of $y = \frac{3}{2}x - 2$

From Table 4 and Fig. 6, we see that the y-intercept is $(0, -2)$.

The Meaning of b for an Equation of the Form $y = mx + b$

For an equation of the form $y = mx + b$, the y-intercept is $(0, b)$. For instance, in Example 3, we found that the line $y = -2x + 3$ has y-intercept $(0, 3)$. In Example 4, we found that the line $y = \frac{3}{2}x - 2$ has y-intercept $(0, -2)$.

Now consider any equation of the form $y = mx + b$. Substituting 0 for x gives

$$y = m(0) + b = 0 + b = b$$

which shows that the y-intercept is $(0, b)$.

> **y-Intercept of the Graph of $y = mx + b$**
>
> The graph of an equation of the form $y = mx + b$ has y-intercept $(0, b)$.

For $y = -5x + 9$, the y-intercept is $(0, 9)$, and for $y = 8x - 4$, the y-intercept is $(0, -4)$.

Rule of Four for Equations

We can describe the solutions of an equation in two variables in four ways. For instance, in Example 4, we described the solutions of the equation $y = \frac{3}{2}x - 2$ by using the equation and a graph (see Fig. 6). We also described some of the solutions by using a table (see Table 4). Finally, we can describe the solutions verbally: For each solution, the y-coordinate is three-halves of the x-coordinate minus 2.

> **Rule of Four for Solutions of an Equation**
>
> We can describe some or all of the solutions of an equation in two variables with:
>
> **1.** an equation, **2.** a table,
>
> **3.** a graph, or **4.** words.
>
> These four ways to describe solutions are known as the **Rule of Four.**

Example 5 Describing Solutions by Using the Rule of Four

1. List some solutions of $y = 4x$ by using a table.
2. Describe the solutions of $y = 4x$ by using a graph.
3. Describe the solutions of $y = 4x$ by using words.

Solution

1. We list three solutions in Table 5.
2. We plot the solutions listed in Table 5 and sketch the line through them (see Fig. 8).
3. For each solution, the y-coordinate is four times the x-coordinate.

Table 5 Solutions of $y = 4x$

x	y
−1	$4(-1) = -4$
0	$4(0) = 0$
1	$4(1) = 4$

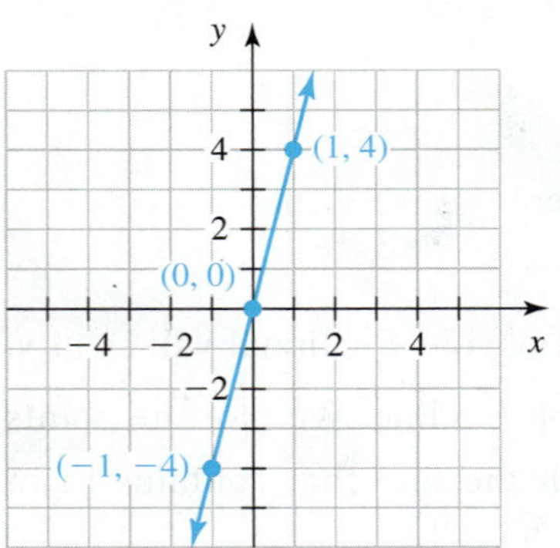

Figure 8 Graph of $y = 4x$

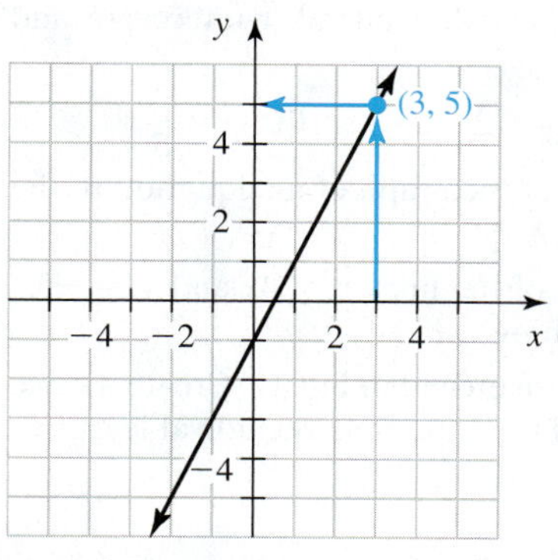

Figure 9 The input $x = 3$ leads to the output $y = 5$

Inputs and Outputs

Recall from Section 1.2 that when we are not describing an authentic situation, we use x as the *independent variable* and y as the *dependent variable*. Recall from Section 1.3 that an *input* is a permitted value of the independent variable that leads to at least one *output*, which is a permitted value of the dependent variable.

In Chapter 1, we used a sketched line to see how an input leads to an output. It will often be easier and more efficient to use an *equation* of a line to perform such a task.

For example, in Fig. 2, we graphed the equation $y = 2x - 1$. From the blue arrows in Fig. 9, we see that the input $x = 3$ leads to the output $y = 5$.

To use the equation, we substitute 3 for x in $y = 2x - 1$:

$$y = 2(3) - 1 = 5$$

This work also shows that the input $x = 3$ leads to the output $y = 5$.

Group Exploration

Solutions of an equation

Consider the equation $y = x - 3$.

1. Use the coordinate system in Fig. 10 to sketch a graph of $y = x - 3$.

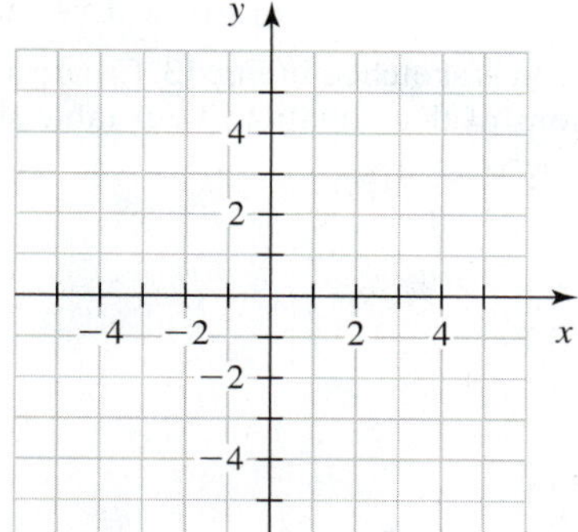

Figure 10 Sketch a graph of $y = x - 3$

2. Pick three points that lie on the graph of $y = x - 3$. Do the coordinates of these points satisfy the equation $y = x - 3$?
3. Pick three points that do not lie on the graph of $y = x - 3$. Do the coordinates of these points satisfy the equation $y = x - 3$?
4. Which ordered pairs satisfy the equation $y = x - 3$? There are too many to list, but describe them in words. [**Hint:** You should say something about the points that do or do not lie on the line.]
5. The graph of an equation is sketched in Fig. 11. Which of the points A, B, C, D, E, and F represent ordered pairs that satisfy the equation?

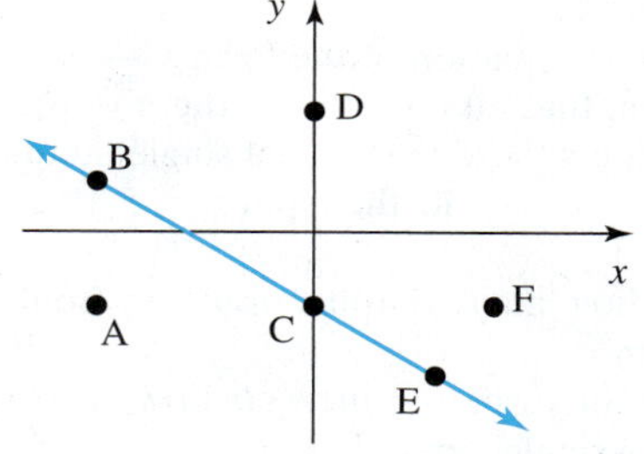

Figure 11 Problem 5

Tips for Success Use the Graphing Calculator Appendix

Remember that Appendix A contains step-by-step graphing calculator instructions for many tools that are helpful for this course. Appendix A includes Section A.26, which describes how to respond to various error messages.

Homework 3.1

For extra help ▶ MyMathLab® 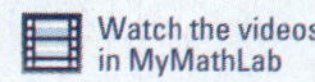 

Which of the given ordered pairs satisfy the given equation?

1. $y = 2x - 4$ $(-3, -10), (1, -3), (2, 0)$
2. $y = 4x - 12$ $(-2, -20), (1, -8), (3, 0)$
3. $y = -3x + 7$ $(-1, 4), (0, 7), (4, -5)$
4. $y = -5x + 8$ $(-2, 3), (0, 8), (3, -7)$

Find the y-intercept. Also, graph the equation by hand. Use a graphing calculator to verify your work.

5. $y = x + 2$
6. $y = x + 4$
7. $y = x - 4$
8. $y = x - 6$
9. $y = 2x$
10. $y = 5x$
11. $y = -3x$
12. $y = -4x$
13. $y = x$

14. $y = -x$

15. $y = \frac{1}{3}x$

16. $y = \frac{1}{2}x$

17. $y = -\frac{5}{3}x$

18. $y = -\frac{3}{2}x$

19. $y = 2x + 1$

20. $y = 3x + 2$

21. $y = 5x - 3$

22. $y = 4x - 1$

23. $y = -3x + 5$

24. $y = -2x + 4$

25. $y = -2x - 3$

26. $y = -4x - 2$

27. $y = \frac{1}{2}x - 3$

28. $y = \frac{1}{3}x - 2$

29. $y = -\frac{2}{3}x + 1$

30. $y = -\frac{4}{3}x + 5$

Concepts

31. Recall that we can describe some or all of the solutions of an equation in two variables with an equation, a table, a graph, or words.

a. Describe three solutions of $y = 2x - 3$ by using a table.
b. Describe the solutions of $y = 2x - 3$ by using a graph.
c. Describe the solutions of $y = 2x - 3$ by using words.

32. Recall that we can describe some or all of the solutions of an equation in two variables with an equation, a table, a graph, or words.

a. Describe three solutions of $y = -4x + 5$ by using a table.
b. Describe the solutions of $y = -4x + 5$ by using a graph.
c. Describe the solutions of $y = -4x + 5$ by using words.

33. a. For the equation $y = 3x + 1$, find all outputs for the given input. State how many outputs there are for that single input.

i. the input $x = 2$ **ii.** the input $x = 4$
iii. the input $x = -2$

b. For $y = 3x + 1$, how many outputs originate from any single input? Explain.

c. Give an example of an equation of the form $y = mx + b$. Using your equation, find all outputs for the given input. State how many outputs there are for that single input.

i. the input $x = 3$ **ii.** the input $x = 5$
iii. the input $x = -3$

d. For your equation, how many outputs originate from any single input? Explain.

e. For *any* equation of the form $y = mx + b$, how many outputs originate from a single input? Explain.

34. a. Graph $y = 2x - 4$ by hand.

b. For the equation $y = 2x - 4$, find all outputs for the given input. Explain by using arrows on your graph in part (a). State how many outputs there are for that single input.

i. the input $x = 3$ **ii.** the input $x = 4$
iii. the input $x = 5$

c. For $y = 2x - 4$, how many outputs originate from any single input? Explain in terms of drawing arrows.

d. Give an example of an equation of the form $y = mx + b$. Graph your equation by hand.

e. Using your equation, find all outputs for the given input. Explain by using arrows on your graph in part (d). State how many outputs there are for that single input.

i. the input $x = 1$ **ii.** the input $x = 3$
iii. the input $x = -2$

f. For your equation, how many outputs originate from any single input? Explain in terms of drawing arrows.

g. For *any* equation of the form $y = mx + b$, how many outputs originate from a single input? Explain in terms of drawing arrows.

35. a. Graph the equation by hand. Find all x-intercepts and y-intercepts.

i. $y = 3x$ **ii.** $y = -2x$ **iii.** $y = \frac{2}{5}x$

b. What are the intercepts of the graph of an equation of the form $y = mx$, where $m \neq 0$?

36. Find the intersection point of the lines $y = 4x$ and $y = -5x$. Try to do this without graphing.

37. The graph of an equation is sketched in Fig. 12. Create a table of ordered-pair solutions of this equation. Include at least five ordered pairs.

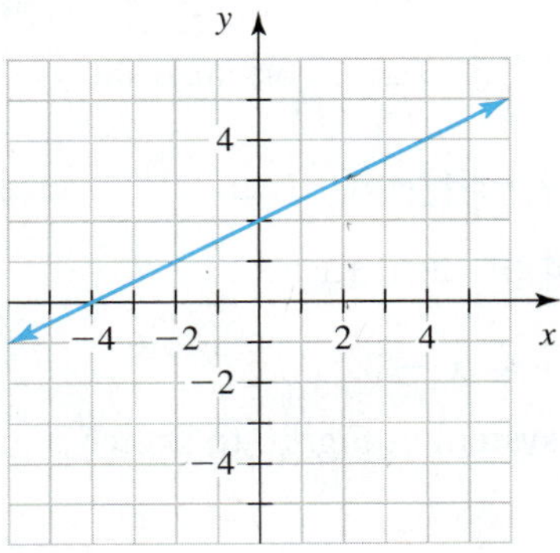

Figure 12 Exercise 37

38. The graph of an equation is sketched in Fig. 13. Create a table of ordered-pair solutions of this equation. Your table should contain at least five ordered pairs.

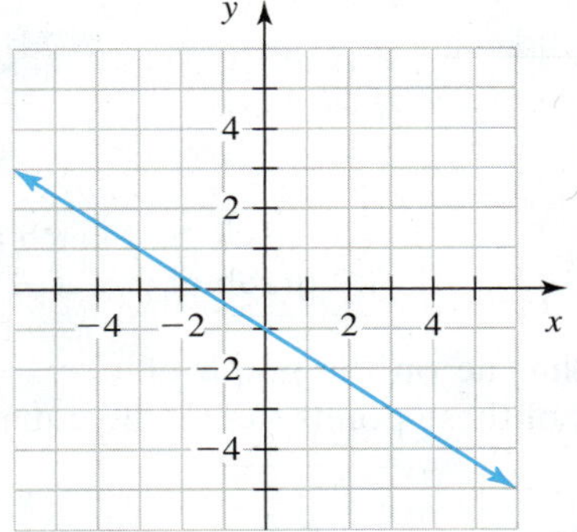

Figure 13 Exercise 38

For Exercises 39–46, refer to the graph sketched in Fig. 14.

39. Find y when $x = -4$.

40. Find y when $x = 0$.

41. Find y when $x = 2$.

42. Find y when $x = -2$.

43. Find x when $y = -1$.

44. Find x when $y = 0$.

45. Find x when $y = 2$.

46. Find x when $y = 3$.

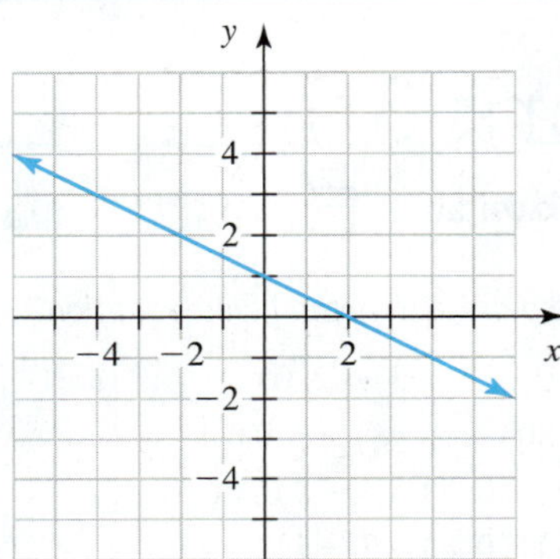

Figure 14 Exercises 39–46

47. The graph of an equation is sketched in Fig. 15. Which of the points A, B, C, D, E, and F represent ordered pairs that satisfy the equation?

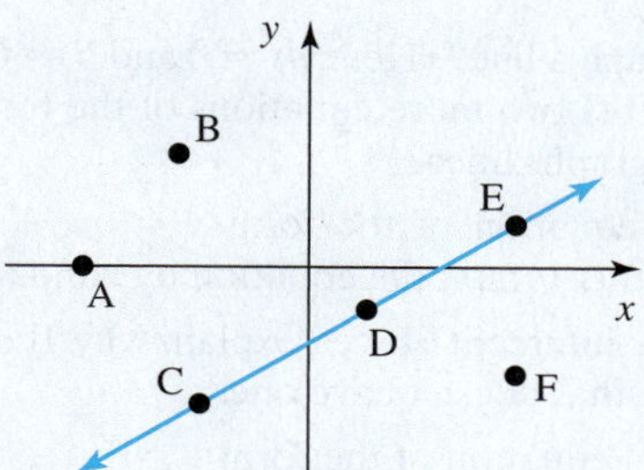

Figure 15 Exercise 47

48. The graphs of equations 1 and 2 are sketched in Fig. 16.

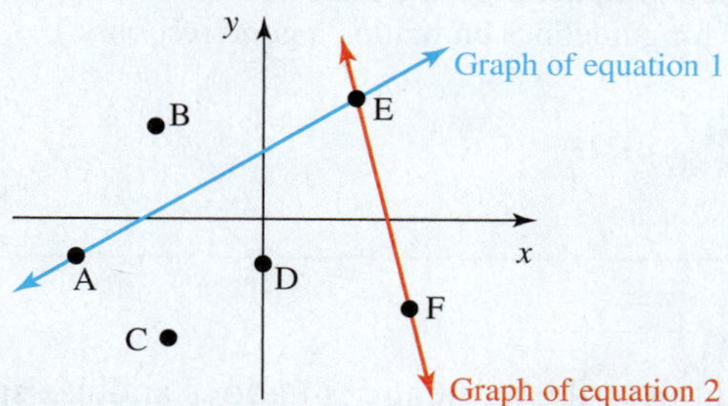

Figure 16 Exercise 48

For each part, decide which one or more of the points A, B, C, D, E, and F represent ordered pairs that

a. satisfy equation 1. **b.** satisfy equation 2.
c. satisfy both equations. **d.** do not satisfy either equation.

49. Find a solution of $y = x + 2$ that lies in Quadrant II. How many solutions of this equation are in Quadrant II?

50. Find a solution of $y = x - 4$ that lies in Quadrant III. How many solutions of this equation are in Quadrant III?

51. Find an equation of a line that contains the points listed in Table 6. [**Hint:** For each point, what number can be added to the x-coordinate to get the y-coordinate?]

Table 6 Points on a Line (Exercise 51)

x	y
0	3
1	4
2	5
3	6
4	7

52. Find an equation of a line that contains the points listed in Table 7. [**Hint:** For each point, what number can be subtracted from the x-coordinate to get the y-coordinate?]

Table 7 Points on a Line (Exercise 52)

x	y
0	−1
1	0
2	1
3	2
4	3

53. Find an equation of a line that contains the points listed in Table 8.

Table 8 Points on a Line (Exercise 53)

x	y
0	0
1	1
2	2
3	3
4	4

54. Find an equation of a line that contains the points listed in Table 9.

Table 9 Points on a Line (Exercise 54)

x	y
0	0
1	−1
2	−2
3	−3
4	−4

55. The graph of an equation is sketched in Fig. 17.

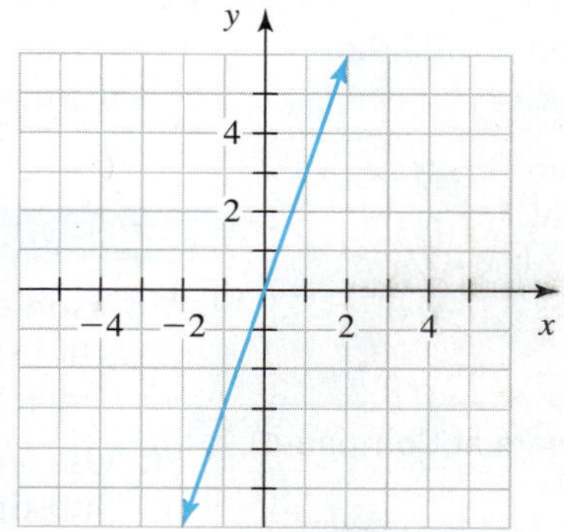

Figure 17 Exercise 55

a. Create a table of ordered-pair solutions of this equation. Your table should contain at least five ordered pairs.

b. Find an equation of the line. [**Hint:** Recognize a pattern from the table you created in part (a).]

56. The graph of an equation is sketched in Fig. 18.

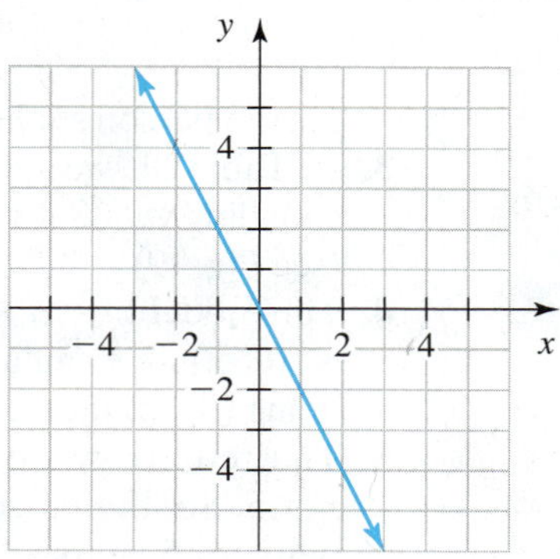

Figure 18 Exercise 56

a. Create a table of ordered-pair solutions of this equation. Your table should contain at least five ordered pairs.

b. Find an equation of the line. [**Hint:** Recognize a pattern from the table you created in part (a).]

57. Graph $x + y = 5$ by hand. [**Hint:** Assume that the graph is a line. Think of pairs of numbers whose sum is 5.]

58. Graph $x - y = 5$ by hand. [**Hint:** Assume that the graph is a line. Think of pairs of numbers whose difference is 5.]

59. Why does the graph of the equation $y = mx + b$ have y-intercept $(0, b)$?

60. Find an ordered pair that satisfies both of the equations $y = x + 1$ and $y = -2x + 4$. [**Hint:** Graph the equations.]

61. **a.** Use a graphing calculator to graph each equation.

i. $y = 2$ **ii.** $y = -2$ **iii.** $y = 5.4896$

b. Describe the graph of $y = b$, where b is a constant.

62. Use a graphing calculator to sketch the graphs of equations of the form $y = mx + b$.

a. Graph $y = -4.1x + 8.7$. Is the graph a line? (Here, $m = -4.1$ and $b = 8.7$.)

b. Graph $y = 6$. Is the graph a line? (Here, $m = 0$ and $b = 6$.)

c. Create and graph at least two more equations of the form $y = mx + b$. Are the graphs lines?

63. Give an example of an equation of the form $y = mx + b$, where m and b are constants. Graph the equation by hand.

64. Does every line have an x-intercept? If yes, explain why. If no, give an equation of a line that doesn't have one.

65. Describe how to graph an equation of the form $y = mx + b$. (See page 9 for guidelines on writing a good response.)

66. In your own words, describe the Rule of Four for equations. (See page 9 for guidelines on writing a good response.)

3.2 Graphing Linear Models; Unit Analysis

Objectives

- » Find an equation of a linear model and make predictions.
- » Know the Rule of Four for linear models.
- » Perform a unit analysis of a linear model.
- » Graph equations of the form $y = b$ and $x = a$.

In this section, we will graph equations that describe authentic situations and use such graphs to make predictions. We will find the units of both sides of such equations. We will also graph equations of the form $y = b$ and $x = a$.

Linear Model

In Section 1.3, we defined a linear model as a nonvertical line that describes two quantities in an authentic situation. An equation of such a line is also called a *linear model.*

Example 1 Using a Linear Model to Make Predictions

A person earns a starting salary of \$32 thousand at a company. Each year, he receives a \$2 thousand raise. Let s be the person's salary (in thousands of dollars) after he has worked at the company for t years.

1. Use a table to help find an equation for t and s.
2. Substitute 8 for t in the equation. What does the result mean in this situation?
3. Graph the equation.
4. What is the s-intercept? What does it mean in this situation?
5. When will the salary be \$42 thousand?

Solution

1. We create Table 10. From the last row of the table, we see that the salary s (in thousands of dollars) can be represented by $2t + 32$. So, $s = 2t + 32$.
2. We substitute 8 for t in the equation $s = 2t + 32$:

$$s = 2(8) + 32 = 48$$

So, the person's salary is \$48 thousand after he has worked at the company for 8 years.

3. In Table 11, we substitute values for t in the equation $s = 2t + 32$ to find the corresponding values for s. Then, we plot the points and sketch a line that contains the points (see Fig. 19).
4. The model $s = 2t + 32$ is of the form $s = mt + b$, where $b = 32$. So, the s-intercept is $(0, 32)$. We can also find the s-intercept from Table 11 and Fig. 19. The s-intercept being $(0, 32)$ means that the starting salary is \$32 thousand.
5. The red arrows in Fig. 20 show that the output $s = 42$ originates from the input $t = 5$. So, the person's salary will be \$42 thousand after he has worked at the company for 5 years.

Table 10 Years at Company and Salaries

Years at Company t	Salary (thousands of dollars) s
0	$2 \cdot 0 + 32$
1	$2 \cdot 1 + 32$
2	$2 \cdot 2 + 32$
3	$2 \cdot 3 + 32$
4	$2 \cdot 4 + 32$
t	$2 \cdot t + 32$

Table 11 Years at Company and Salaries

t	s
0	$2(0) + 32 = 32$
1	$2(1) + 32 = 34$
2	$2(2) + 32 = 36$
3	$2(3) + 32 = 38$
4	$2(4) + 32 = 40$

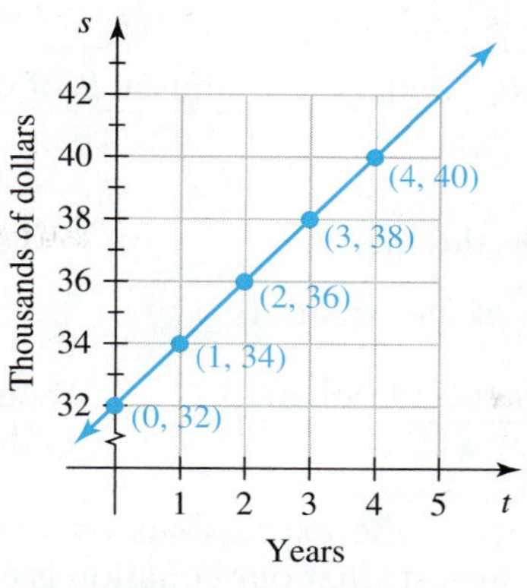

Figure 19 Salary model

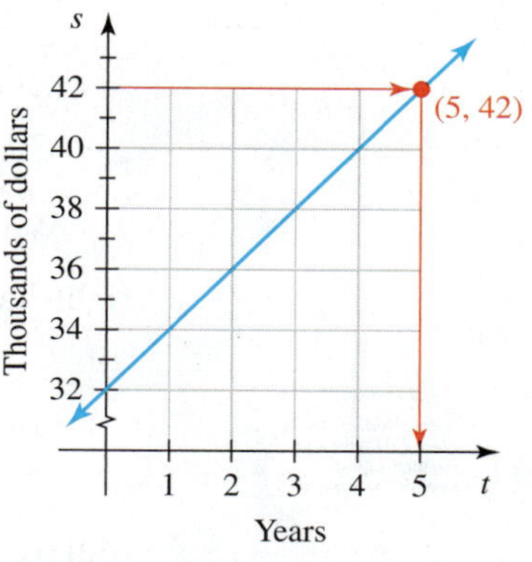

Figure 20 Using the income model to make a prediction

We use a graphing calculator to verify our work (see Fig. 21). See Appendix A.4, A.5, and A.7 for graphing calculator instructions.

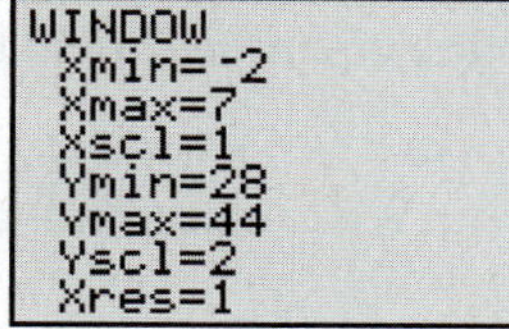

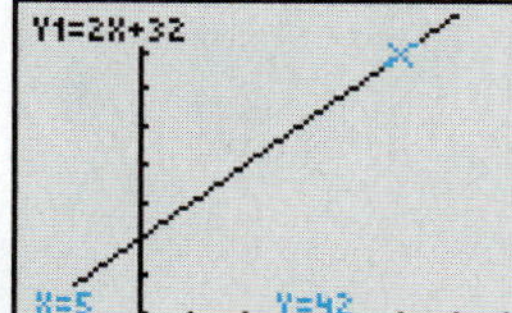

Figure 21 Verify the work

We can add two numbers in either order and get the same result. For example, $2 + 5 = 7$ and $5 + 2 = 7$. We can also multiply two numbers in either order and get the same result. For example, $3 \cdot 8 = 24$ and $8 \cdot 3 = 24$. Depending on how we had arranged the arithmetic in Table 10, we could have found any one of the following equations:

$$s = 2t + 32 \qquad s = 32 + 2t \qquad s = t(2) + 32 \qquad s = 32 + t(2)$$

All of these equations describe the same relationship between t and s and have the same graph.

Rule of Four for Linear Models

Just as we can use equations, tables, graphs, and words to describe solutions of equations in two variables, we can use these four ways to describe linear models. For instance, in Example 1, we first described a linear model by using words:

> A person earns a starting salary of $32 thousand at a company. Each year, he receives a $2 thousand raise.

We also described the model by using the equation $s = 2t + 32$, a table (see Table 11), and a graph (see Fig. 19).

Rule of Four for Linear Models

We can describe a linear model with

1. an equation,
2. a table,
3. a graph, or
4. words.

Unit Analysis of a Linear Model

In Example 1, we found the linear model $s = 2t + 32$. We can perform a *unit analysis* of a model by determining the units of the expressions on both sides of the equation:

$$\underbrace{s}_{\text{thousands of dollars}} = \underbrace{2}_{\frac{\text{thousands of dollars}}{\text{year}}} \cdot \underbrace{t}_{\text{years}} + \underbrace{32}_{\text{thousands of dollars}}$$

We can use the fact that $\frac{\text{years}}{\text{year}} = 1$ to simplify the units of the expression on the right-hand side of the equation:

$$\frac{\text{thousands of dollars}}{\text{year}} \cdot \text{years} + \text{thousands of dollars} = \text{thousands of dollars} + \text{thousands of dollars}$$

So, the units of the expressions on both sides of the equation are thousands of dollars, which suggests that our equation is correct.

"How'd you lose the weight?"
"I exercised, drank plenty of fluids, and avoided eating anything that even remotely had flavor."

Definition Unit analysis

We perform a **unit analysis** of a model's equation by determining the units of the expressions on both sides of the equation. The units of the expressions on both sides of the equation should be the same.

We can perform a unit analysis of a model's equation to help verify the equation.

Example 2 Using a Linear Model to Make a Prediction

Before starting a diet, a person weighs 160 pounds. While on the diet, she loses 3 pounds per month. Let w be the person's weight (in pounds) after she has been on the diet for t months.

1. Use a table to help find an equation for t and w.
2. Perform a unit analysis of the equation you found in Problem 1.
3. Predict the person's weight after she has been on the diet for 6 months.

Table 12 Numbers of Months and Weights

Time on Diet (months) t	Weight (pounds) w
0	$160 - 3 \cdot 0$
1	$160 - 3 \cdot 1$
2	$160 - 3 \cdot 2$
3	$160 - 3 \cdot 3$
4	$160 - 3 \cdot 4$
t	$160 - 3 \cdot t$

Solution

1. We create Table 12. From the bottom row of Table 12, we see that we can model the situation by using the equation $w = 160 - 3t$.
2. Here is a unit analysis of the equation $w = 160 - 3t$:

$$\underbrace{w}_{\text{pounds}} = \underbrace{160}_{\text{pounds}} - \underbrace{3}_{\frac{\text{pounds}}{\text{month}}} \cdot \underbrace{t}_{\text{months}}$$

We can use the fact that $\frac{\text{months}}{\text{month}} = 1$ to simplify the units of the expression on the right-hand side of the equation:

$$\text{pounds} - \frac{\text{pounds}}{\text{month}} \cdot \text{months} = \text{pounds} - \text{pounds}$$

So, the units on both sides of the equation are pounds, which suggests that our equation is correct.

3. We substitute 6 for t in the equation $w = 160 - 3t$:

$$w = 160 - 3(6) = 142$$

The person will weigh 142 pounds after she has been on the diet for 6 months.

Horizontal Linear Models

We will now model an authentic situation in which one of two quantities does not change.

Example 3 Finding and Graphing a Linear Model

In 2009, the minimum wage was increased to \$7.25. For the period 2009–2013, the minimum wage was constant. At the time of this writing, an increase in the minimum wage was under debate for 2014. Let w be the minimum wage (in dollars) at t years since 2009. Find an equation that models the wages for the period 2009–2013. Also, graph the equation.

Solution

For the period 2009–2013, the minimum wage was a constant \$7.25. So, the equation of the model is

$$w = 7.25$$

To graph the equation, we first list some corresponding values of t and w in Table 13. Then we plot the points and sketch a line that contains the points (see Fig. 22).

Table 13 Values of t and w

t (years since 2009)	w (dollars)
0	7.25
1	7.25
2	7.25
3	7.25
4	7.25

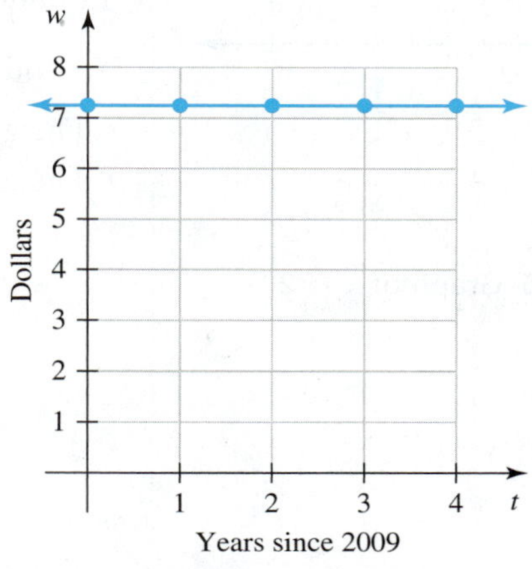

Figure 22 Minimum-wage model

Note that the model is a horizontal line. Recall from Section 1.4 that when a model yields an estimate that is not a good approximation or when it gives a prediction that does not make sense, we say *model breakdown* has occurred. Model breakdown has occurred in the years before 2009. If the minimum wage does increase in 2014, then model breakdown will have also occurred in the years after 2013.

Horizontal and Vertical Lines

Graphing the horizontal linear model in Example 3 will suggest how to graph another horizontal line in Example 4.

Table 14 Solutions of $y = 3$

x	y
−2	3
−1	3
0	3
1	3
2	3

Example 4 Graphing an Equation of the Form $y = b$

Sketch the graph of $y = 3$.

Solution

Note that y must be 3, but x can have any value. Some solutions of $y = 3$ are listed in Table 14. We plot the corresponding points and sketch the line through them (see Fig. 23). The graph of $y = 3$ is a horizontal line.

We can use a graphing calculator to verify our graph (see Fig. 24).

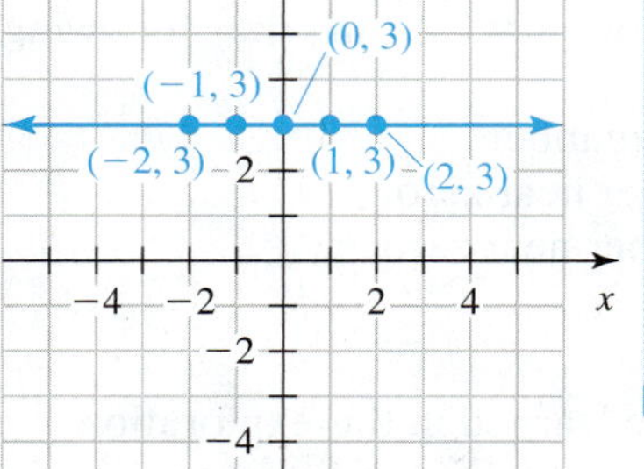

Figure 23 Graph of $y = 3$

Plot1 Plot2 Plot3
\Y1=3
\Y2=
\Y3=
\Y4=
\Y5=
\Y6=
\Y7=

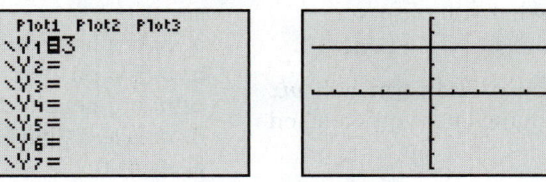

Figure 24 The graph of $y = 3$

In Example 4, we saw that the graph of the equation $y = 3$ is a horizontal line. Any equation that can be written in the form $y = b$, where b is a constant, has a horizontal line as its graph.

Table 15 Solutions of $x = 2$

x	y
2	−2
2	−1
2	0
2	1
2	2

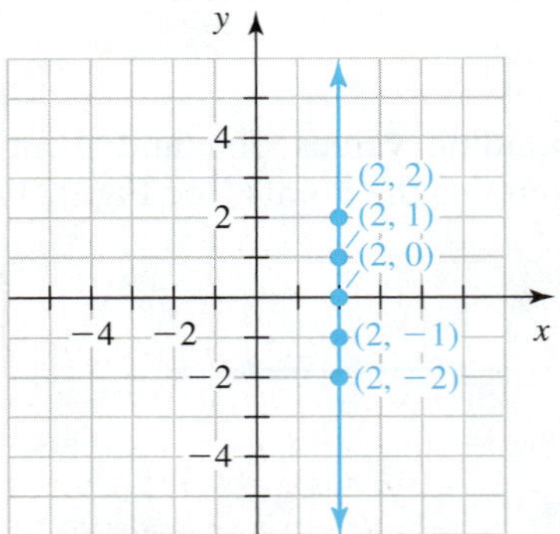

Figure 25 Graph of $x = 2$

Example 5 Graphing an Equation of the Form $x = a$

Sketch the graph of $x = 2$.

Solution

Note that x must be 2, but y can have any value. Some solutions of $x = 2$ are listed in Table 15. We plot the corresponding points and sketch the line through them (see Fig. 25). The graph of $x = 2$ is a vertical line.

In Example 5, we saw that the graph of the equation $x = 2$ is a vertical line. Any equation that can be written in the form $x = a$, where a is a constant, has a vertical line as its graph.

Equations for Horizontal and Vertical Lines

If a and b are constants, then

- The graph of $y = b$ is a horizontal line (see Fig. 26).
- The graph of $x = a$ is a vertical line (see Fig. 27).

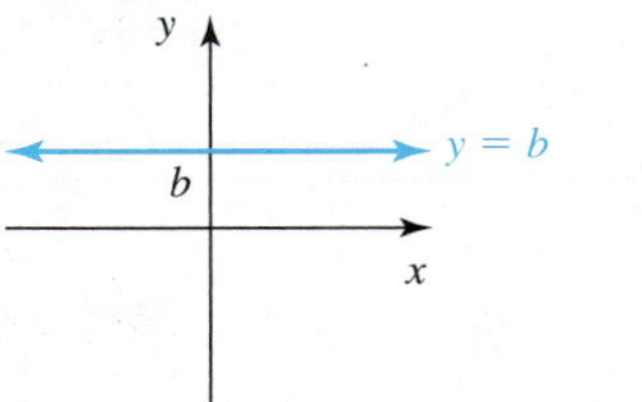

Figure 26 Graph of $y = b$

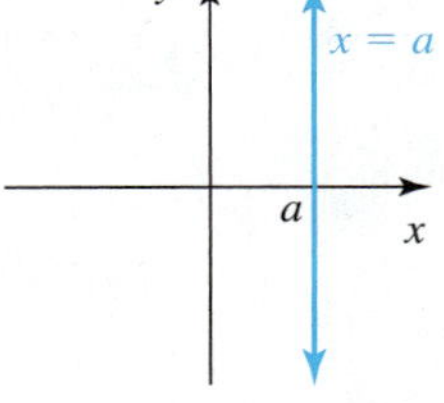

Figure 27 Graph of $x = a$

For example, the graphs of the equations $y = 5$ and $y = -2$ are horizontal lines. The graphs of the equations $x = 6$ and $x = -4$ are vertical lines.

Equations Whose Graphs Are Lines

If an equation can be put into the form

$$y = mx + b \quad \text{or} \quad x = a$$

where m, a, and b are constants, then the graph of the equation is a line. We call such an equation a **linear equation in two variables.**

Any equation that can be put into the form $x = a$ has a vertical line as its graph. Any equation that can be put into the form $y = mx + b$ has a nonvertical line as its graph.

Group Exploration

Looking ahead: Graphical significance of m for $y = mx$

1. Use ZDecimal to graph these equations of the form $y = mx$ in order, and describe what you observe:

$$y = x, \quad y = 2x, \quad y = 3x, \quad y = 4x$$

2. Give an example of an equation of the form $y = mx$ whose graph is a line steeper than the lines you sketched in Problem 1.
3. Use ZDecimal to graph these equations in order, and describe what you observe:

$$y = -x, \quad y = -2x, \quad y = -3x, \quad y = -4x$$

4. Describe the graph of $y = mx$ in the following situations:
   a. m is a large positive number
   b. m is a positive number near zero
   c. m is a negative number near zero
   d. m is less than -10
   e. $m = 0$
5. Describe what you have learned in this exploration.

Tips for Success Study with a Classmate

It can be helpful to meet with a classmate and discuss what happened in class that day. Not only can you ask questions *of* each other, but you will learn just as much by explaining concepts *to* each other. Explaining a concept to someone else forces you to clarify your own understanding of the concept.

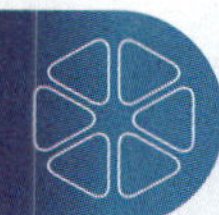

Homework 3.2

For extra help ▶ MyMathLab®  Watch the videos in MyMathLab  Download the MyDashboard App

1. A person pays a \$3 cover charge to hear a hip-hop band. Let T be the total cost (in dollars) of the cover charge plus d dollars spent on drinks.

a. Complete Table 16 to help find an equation that describes the relationship between d and T. Show the arithmetic to help you see a pattern.

Table 16 Drink Cost and Total Cost

Drink Cost (dollars) d	Total Cost (dollars) T
2	
3	
4	
5	
d	

b. Perform a unit analysis of the equation you found in part (a).

c. Graph the equation by hand.

d. What is the T-intercept of the linear model? What does it mean in this situation?

e. If \$10 is spent on drinks, find the total cost of the cover charge and drinks. Explain by using arrows on your graph in part (c).

2. A company offers a \$2 mail-in rebate on a shaver. The *retail price* of a shaver is the price paid at the store (not including the \$2 rebate). The *net price* is the price of the shaver, taking into account the money saved by the rebate. Let n be the net price (in dollars) of a shaver whose retail price is r dollars.

a. Complete Table 17 to help find an equation that describes the relationship between r and n. Show the arithmetic to help you see a pattern.

Table 17 Retail Price and Net Price

Retail Price (dollars) r	Net Price (dollars) n
4	
5	
6	
7	
r	

b. Perform a unit analysis of the equation you found in part (a).

c. Graph the equation by hand.

d. If the net price is \$6, what is the retail price? Explain by using arrows on your graph in part (c).

3. Chemeketa Community College charged \$87 per credit (unit or hour) for tuition in spring semester 2012. Let T be the total cost (in dollars) of tuition for enrolling in c credits of classes.

a. Complete Table 18 to help find an equation that describes the relationship between c and T. Show the arithmetic to help you see a pattern.

Table 18 Numbers of Credits and Total Costs

Number of Credits c	Total Cost (dollars) T
3	
6	
9	
12	
c	

b. Perform a unit analysis of the equation you found in part (a).

c. Graph the equation by hand.

d. What is the total cost of tuition for 15 credits of classes? Explain by using arrows on your graph in part (c).

4. The average gas mileage of a Toyota Camry Hybrid XLE® is 40 miles per gallon. Let g be the number of gallons used to drive d miles.

a. Complete Table 19 to help find an equation that describes the relationship between d and g. Show the arithmetic to help you see a pattern.

Table 19 Distances and Gasoline Consumed

Distance (miles) d	Gasoline Consumed (gallons) g
40	
80	
120	
160	
d	

b. Perform a unit analysis of the equation you found in part (a).

c. Graph the equation by hand.

d. How many gallons are used to travel 100 miles? Explain by using arrows on your graph in part (c).

5. A person earns a starting salary of \$24 thousand at a company. Each year, she receives a \$3 thousand raise. Let s be the salary (in thousands of dollars) after she has worked at the company for t years.

a. Complete Table 20 to help find an equation for t and s. Show the arithmetic to help you see a pattern.

Table 20 Years at Company and Salaries

Time at Company (years) t	Salary (thousands of dollars) s
0	
1	
2	
3	
4	
t	

b. Perform a unit analysis of the equation you found in part (a).

c. Graph the equation by hand.

d. What is the s-intercept? What does it mean in this situation?

e. When will the person's salary be \$42 thousand? Explain by using arrows on your graph in part (c).

6. A kitchen appliances company sold 27 thousand ovens in 2010, and sales then increased by 4 thousand ovens each year. Let s be the annual sales (in thousands of ovens) at t years since 2010.

a. Complete Table 21 to help find an equation for t and s. Show the arithmetic to help you see a pattern.

Table 21 Years and Annual Sales

Years Since 2010 t	Annual Sales (thousands of ovens) s
0	
1	
2	
3	
4	
t	

b. Perform a unit analysis of the equation you found in part (a).

c. Graph the equation by hand.

d. Find the year when the sales were 51 thousand ovens. Explain by using arrows on your graph in part (c).

7. The number of U.S. traffic deaths was 33 thousand in 2010 and has declined by about 2 thousand deaths per year since then (Source: *National Highway Traffic Safety Administration*). Let n be the number (in thousands) of traffic deaths in the year that is t years since 2010.

a. Complete Table 22 to help find an equation for t and n. Show the arithmetic to help you see a pattern.

b. Perform a unit analysis of the equation you found in part (a).

c. Graph the equation by hand.

d. What is the n-intercept? What does it mean in this situation?

e. Find the year when there will be 19 thousand traffic deaths. Explain by using arrows on your graph in part (c).

Table 22 Years and Numbers of U.S. Traffic Deaths

Years Since 2010 t	Number of Traffic Deaths (thousands) n
0	
1	
2	
3	
4	
t	

8. The percentage of Americans who are satisfied with the size and power of the federal government was 30% in 2012 and has declined by about 3 percentage points per year since then (Source: *Gallup Organization*). Let p be the percentage of Americans who are satisfied with the size and power of the federal government at t years since 2012.

a. Complete Table 23 to help find an equation for t and p. Show the arithmetic to help you see a pattern.

Table 23 Years and Percentages of Americans Who Are Satisfied with the Size and Power of the Federal Government

Years Since 2012 t	Percent p
0	
1	
2	
3	
4	
t	

b. Perform a unit analysis of the equation you found in part (a).

c. Graph the equation by hand.

d. What is the p-intercept? What does it mean in this situation?

e. Find the year when 15% of Americans will be satisfied with the size and power of the federal government. Explain by using arrows on your graph in part (c).

9. To make fudgelike brownies, a person bakes a brownie mix for 5 minutes less than the baking time suggested on the box. Let r be the suggested baking time (in minutes) and a be the actual baking time (in minutes).

a. Find an equation that describes the relationship between r and a. Assume that a is the dependent variable. [**Hint:** If you have trouble finding the equation, create a table of values for r and a.]

b. Perform a unit analysis of the equation you found in part (a).

c. Graph the equation by hand.

d. If the actual baking time is 23 minutes, what is the baking time suggested on the box? Explain by using arrows on your graph in part (c).

10. A person pays \$4 for parking at an arts-and-crafts fair. Let T be the total cost (in dollars) of parking plus v dollars spent on a vase.

a. Find an equation that describes the relationship between v and T. [**Hint:** If you have trouble finding the equation, create a table of values for v and T.]

b. Perform a unit analysis of the equation you found in part (a).

c. Graph the equation by hand.

d. If the person spends \$25 on the vase, find the total cost of parking and the vase. Explain by using arrows on your graph in part (c).

11. A person drives at a constant speed of 60 miles per hour. Let d be the distance traveled (in miles) after the person has driven for t hours.
   a. Find an equation that describes the relationship between t and d. [**Hint:** If you have trouble finding the equation, create a table of values for t and d.]
   b. Perform a unit analysis of the equation you found in part (a).
   c. Graph the equation by hand.
   d. What is the d-intercept? What does it mean in this situation?
   e. After how many hours will the person have traveled 150 miles? Explain by using arrows on your graph in part (c).

12. A bicycle shop charges \$12 per hour to rent a bicycle. Let C be the total cost (in dollars) of renting a bicycle for t hours.
   a. Find an equation that describes the relationship between t and C. [**Hint:** If you have trouble finding the equation, create a table of values for t and C.]
   b. Perform a unit analysis of the equation you found in part (a).
   c. Graph the equation by hand.
   d. If a person paid \$18 for renting a bicycle, how long did the person ride the bicycle? Explain by using arrows on your graph in part (c).

13. For spring semester 2012, Cosumnes River College charged \$36 per unit (credit or hour) for tuition. All students paid a \$1 student representation fee each semester. Students who drove to school paid a \$30 parking fee each semester. Let T be the total one-semester cost (in dollars) of tuition and fees for a student who drove to school and took u units of classes.
   a. Find an equation for u and T. [**Hint:** If you have trouble finding the equation, try creating a table of values for u and T.]
   b. Perform a unit analysis of the equation you found in part (a).
   c. What is the total one-semester cost of tuition and fees for a student who drove to school and took 15 units of classes?

14. For spring semester 2012, undergraduates taking 15 hours (units or credits) of courses at Southeastern Louisiana University paid \$2393.28 for tuition. Students could rent textbooks for \$35 per course. Let T be the total one-semester cost (in dollars) of tuition and renting textbooks for an undergraduate who took n courses for a total of 15 hours.
   a. Find an equation for n and T. [**Hint:** If you have trouble finding the equation, try creating a table of values for n and T.]
   b. Perform a unit analysis of the equation you found in part (a).
   c. What was the total one-semester cost of tuition and renting textbooks for an undergraduate who took five 3-hour courses?

15. The pressure in a bike tire is 62 pounds per square inch (psi). A person uses a bike pump to add air to the tire. The pressure increases by 3 psi with each pump.
   a. Find an equation of a linear model to describe the situation. Explain what your variables represent.
   b. Perform a unit analysis of the equation you found in part (a).
   c. What is the pressure after 17 pumps?

16. Just before some bad publicity was released about a company, the company's stock was worth \$95. Now that the bad publicity has been released, the value of the stock has been declining by \$4 per week.
   a. Find an equation of a linear model to describe the situation. Explain what your variables represent.
   b. Perform a unit analysis of the equation you found in part (a).
   c. What is the value of the stock 14 weeks after the bad publicity was released?

17. A person fills up his car's 11-gallon gasoline tank and then drives at a constant 60 mph. For each hour of driving, the car uses 2 gallons of gasoline. Let g be the amount of gasoline (in gallons) in the tank after the person has driven for t hours.
   a. Find an equation for t and g.
   b. Perform a unit analysis of the equation you found in part (a).
   c. Graph the equation by hand.
   d. For how long can the person drive before refueling if he wants to refuel when 1 gallon of gasoline is left in the tank? Explain by using arrows on your graph in part (c).
   e. Estimate the amount of gasoline in the tank after 8 hours of driving. Explain by using arrows on your graph in part (c). [**Hint:** Remember, if you think that model breakdown occurs, say so, say where, and explain why.]

18. When a small airplane begins to descend, its altitude is 4 thousand feet. The airplane descends $\frac{1}{2}$ thousand feet each minute. Let h be the altitude (in thousands of feet) of the airplane t minutes after it has begun to descend.
   a. Find an equation for t and h.
   b. Perform a unit analysis of the equation you found in part (a).
   c. Graph the equation by hand.
   d. When will the airplane reach an altitude of 1 thousand feet? Explain by using arrows on your graph in part (c).
   e. Predict the airplane's altitude 11 minutes after the airplane has begun its descent. Explain by using arrows on your graph in part (c). [**Hint:** Remember, if you think that model breakdown occurs, say so, say where, and explain why.]

19. *Fair-trade coffee* guarantees farmers a minimum fair price for their crops. Although most segments of the U.S. coffee business are stagnant, the sales of fair-trade coffee are widely viewed as the fastest-growing niche. Sales of fair-trade coffee were 138.6 million pounds in 2011 and have increased by about 15.7 million pounds per year since then (Source: *Fair Trade USA*). Let s be the annual sales (in millions of pounds) of fair-trade coffee at t years since 2011. Recall that we can describe a linear model using an equation, a table, a graph, or words.
   a. Use a table of values of t and s to describe the situation.
   b. Use an equation to describe the situation.
   c. Use a graph to describe the situation.

20. The number of albums downloaded was 83.1 million in 2010 and has increased by 9.8 million albums per year since then (Source: *Recording Industry Association of America*). Let n be the number of albums (in millions) downloaded in the year that is t years since 2010. Recall that we can describe a linear model using an equation, a table, a graph, or words.
   a. Use a table of values of t and n to describe the situation.
   b. Use an equation to describe the situation.
   c. Use a graph to describe the situation.

21. A person has had a constant 45 novels since 2012. Let n be the number of novels that the person owns at t years since 2012. Find an equation that models the number of novels owned by the person for 2012 and thereafter. Also, graph the equation by hand.

22. A small company has had a constant 35 employees since 2010. Let n be the number of employees at t years since 2010. Find an equation that models the number of employees for 2010 and thereafter. Also, graph the equation by hand.

23. Let n be the number (in thousands) of independent CD and record stores at t years since 2005. A reasonable linear model is shown in Fig. 28 (Source: *National Arts Index*).

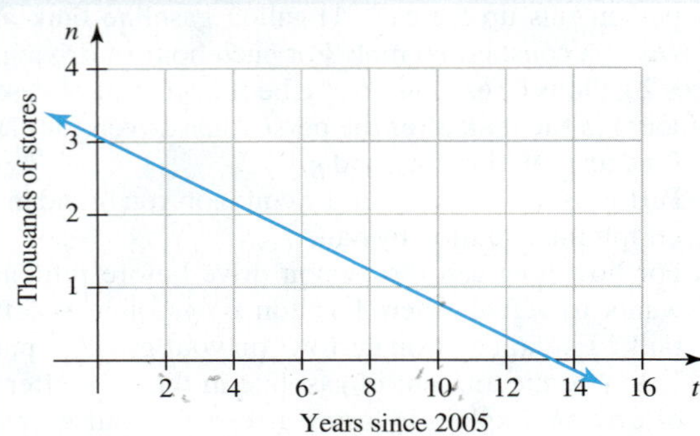

Figure 28 Exercise 23

a. Estimate when there were 2.6 thousand independent CD and record stores.

b. Predict the number of independent CD and record stores in 2015.

c. What is the n-intercept? What does it mean in this situation?

d. What is the t-intercept? What does it mean in this situation?

24. Let v be the total value (in billions of dollars) of unused gift cards at t years since 2005. A reasonable linear model is shown in Fig. 29 (Source: *CEB TowerGroup*).

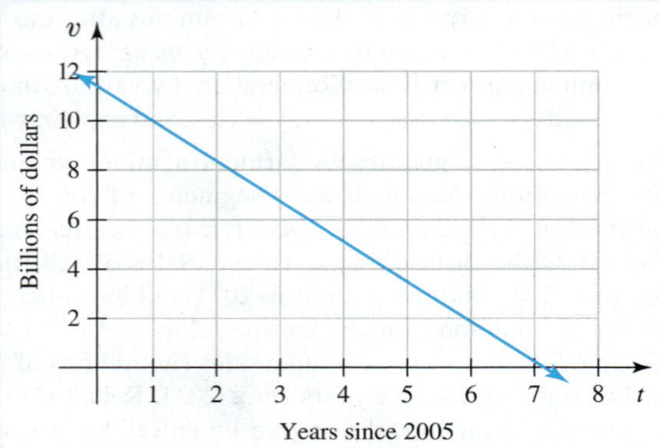

Figure 29 Exercise 24

a. Estimate in which year the total value of unused gift cards was \$6 billion.

b. Estimate the total value of unused gift cards in 2011.

c. What is the t-intercept? What does it mean in this situation?

d. What is the v-intercept? What does it mean in this situation?

Graph the equation by hand.

25. $x = 3$ **26.** $x = 6$ **27.** $y = 1$ **28.** $y = 5$

29. $y = -2$ **30.** $y = -4$ **31.** $x = -1$ **32.** $x = -3$

33. $x = 0$ **34.** $y = 0$

Graph the equation by hand. Use a graphing calculator to verify your work when possible.

35. $y = x - 2$ **36.** $y = x - 4$

37. $y = 2$ **38.** $y = -3$

39. $y = -3x + 1$ **40.** $y = -2x - 2$

41. $y = \frac{3}{5}x$ **42.** $y = \frac{4}{3}x$

43. $y = -\frac{5}{3}x + 1$ **44.** $y = -\frac{3}{2}x + 4$

45. $y = 4x - 3$ **46.** $y = 3x - 5$

47. $x = -4$ **48.** $x = 1$

Concepts

49. Find an equation of the line sketched in Fig. 30.

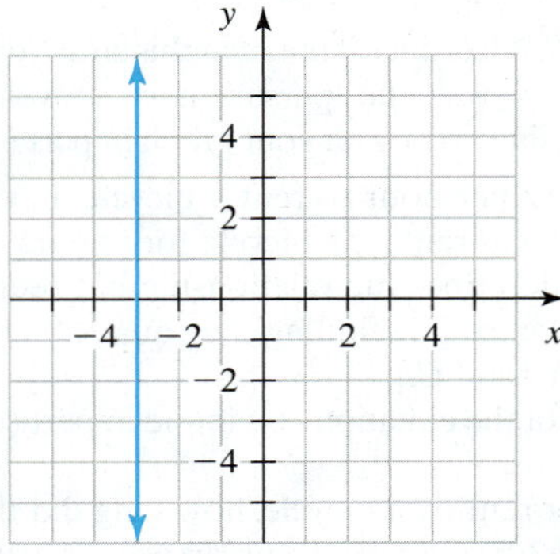

Figure 30 Exercise 49

50. Find an equation of the line sketched in Fig. 31.

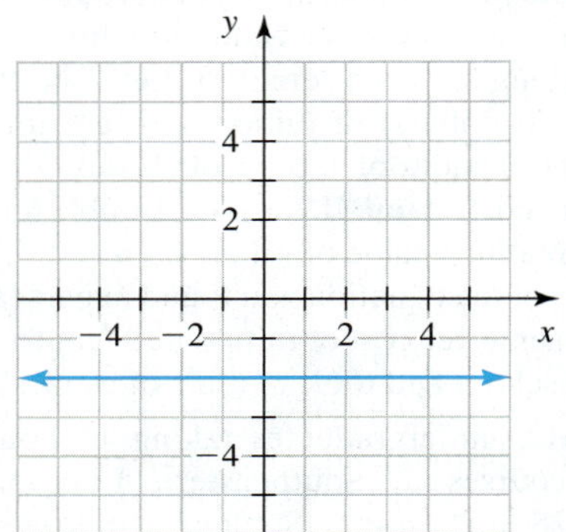

Figure 31 Exercise 50

51. The points $(1, 2)$ and $(5, 2)$ are plotted in Fig. 32.

a. In going from point $(1, 2)$ to point $(5, 2)$, find the change

i. in the x-coordinate. **ii.** in the y-coordinate.

b. In going from point $(5, 2)$ to point $(1, 2)$, find the change

i. in the x-coordinate. **ii.** in the y-coordinate.

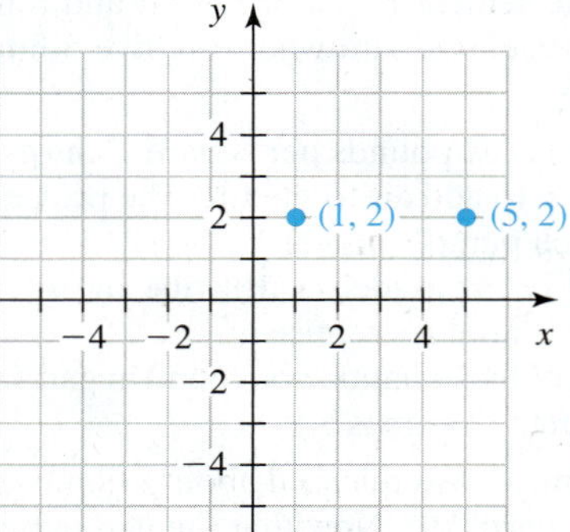

Figure 32 Exercise 51

52. The points $(3, -4)$ and $(3, 2)$ are plotted in Fig. 33.

a. In going from point $(3, -4)$ to point $(3, 2)$, find the change

i. in the x-coordinate. **ii.** in the y-coordinate.

b. In going from point $(3, 2)$ to point $(3, -4)$, find the change

i. in the x-coordinate. **ii.** in the y-coordinate.

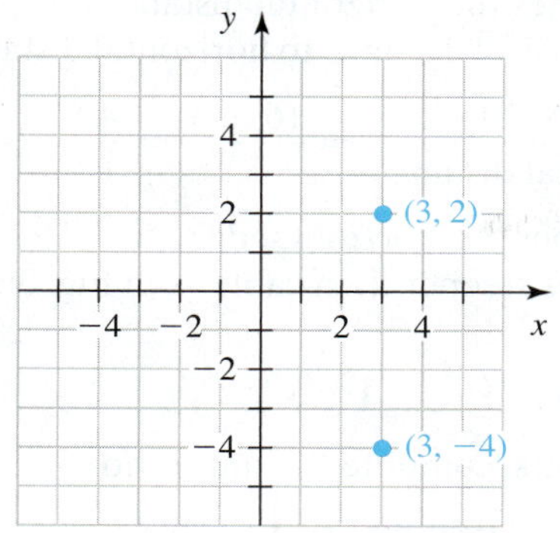

Figure 33 Exercise 52

53. The points $(-3, 1)$ and $(2, 4)$ are plotted in Fig. 34.

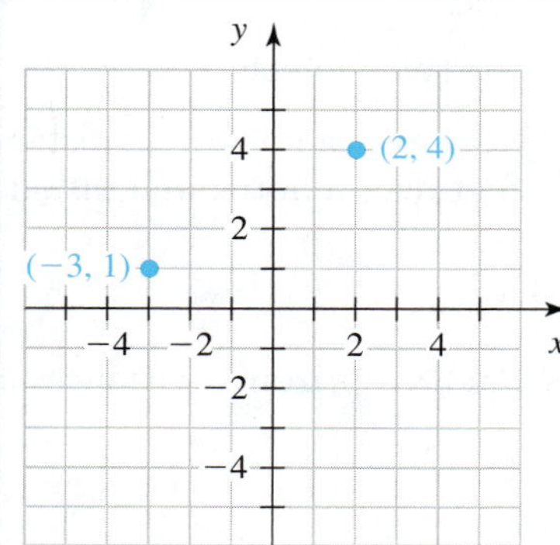

Figure 34 Exercise 53

a. In going from point $(-3, 1)$ to point $(2, 4)$, find the change
i. in the x-coordinate. **ii.** in the y-coordinate.

b. In going from point $(2, 4)$ to point $(-3, 1)$, find the change
i. in the x-coordinate. **ii.** in the y-coordinate.

54. The points $(1, 4)$ and $(5, -3)$ are plotted in Fig. 35.

a. In going from point $(1, 4)$ to point $(5, -3)$, find the change
i. in the x-coordinate. **ii.** in the y-coordinate.

b. In going from point $(5, -3)$ to point $(1, 4)$, find the change
i. in the x-coordinate. **ii.** in the y-coordinate.

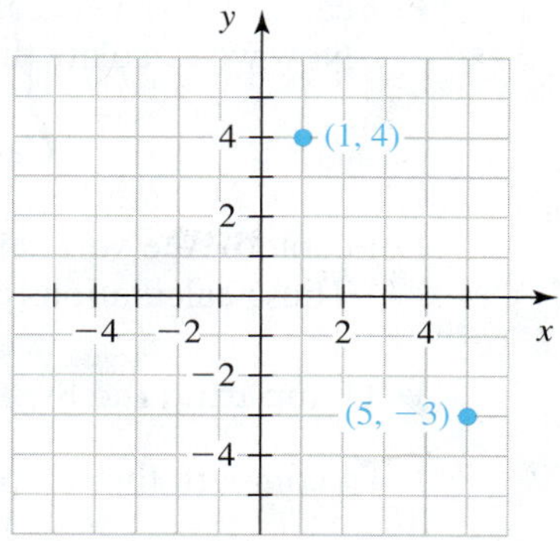

Figure 35 Exercise 54

55. a. Use a graphing calculator to graph the equations $y = x$, $y = 2x$, and $y = 3x$ in the same viewing window.
b. List the lines $y = x$, $y = 2x$, and $y = 3x$ in order of steepness, from least to greatest.
c. What pattern does your list in part (b) suggest?
d. Give an equation of a line that is steeper than the lines $y = x$, $y = 2x$, and $y = 3x$.

56. a. Use a graphing calculator to graph the pair of equations in the same viewing window.
i. $y = 2x + 1, y = 2x + 3$ **ii.** $y = 3x - 2, y = 3x - 5$
iii. $y = -2x + 4, y = -2x + 3$
b. Describe what pattern your work in part (a) suggests.
c. Find an equation whose graph is parallel to the line $y = 4x - 5$.

57. Assume a is a constant. Is the graph of $x = a$ a horizontal line, a vertical line, or neither? Explain.

58. Assume b is a constant. Is the graph of $y = b$ a horizontal line, a vertical line, or neither? Explain.

59. Does every line have an x-intercept? If yes, explain why. If no, give an equation of a line that doesn't have one.

60. Does every line have a y-intercept? If yes, explain why. If no, give an equation of a line that doesn't have one.

61. In your own words, describe the Rule of Four for linear models.

62. Describe how to perform a unit analysis of a model's equation. Explain why a unit analysis can be used to help verify the equation.

3.3 Slope of a Line

Objectives

- Use a ratio to compare the steepness of two objects.
- Know the meaning of, and how to calculate, the *slope* of a nonvertical line.
- Know the sign of the slope of an increasing line and of a decreasing line.
- Know that the slope of a horizontal line is zero and that the slope of a vertical line is undefined.

How do we measure the steepness of an object such as a ladder? In this section, we will discuss the *slope* of a line. This key concept has many applications in business, engineering, nursing, surveying, physics, social science, mathematics, and lots of other fields.

Comparing the Steepness of Two Objects

Consider the sketch of two cables (guy wires) running from the ground to a telephone pole in Fig. 36. Which cable is steeper?

Cable B is steeper than cable A, even though both cables reach a point at the same height on the building. To measure the steepness of each cable, we compare the *vertical* distance from the base of the pole to the top end of the cable with the *horizontal* distance from the bottom end of the cable to the building. In Section 2.5, we used a unit ratio to compare two quantities. Here, we calculate the unit ratio of vertical distance to horizontal distance for cable A:

$$\text{Cable A:}\quad \frac{\text{vertical distance}}{\text{horizontal distance}} = \frac{16\text{ feet}}{8\text{ feet}} = \frac{2}{1}$$

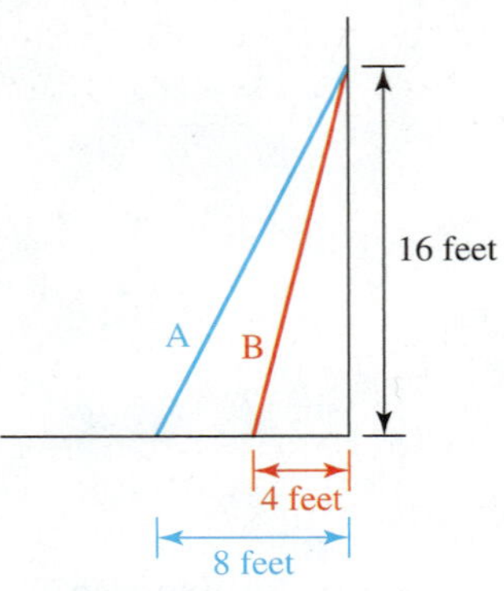

Figure 36 Cable A and cable B running from the ground to a telephone pole

For cable A, the vertical distance is 2 times the horizontal distance.

Now we calculate the unit ratio of vertical distance to horizontal distance for cable B:

$$\text{Cable B:} \quad \frac{\text{vertical distance}}{\text{horizontal distance}} = \frac{16 \text{ feet}}{4 \text{ feet}} = \frac{4}{1}$$

For cable B, the vertical distance is 4 times the horizontal distance.

These calculations confirm cable B is steeper than cable A in Fig. 36.

Comparing the Steepness of Two Objects

To compare the steepness of two objects, compute the unit ratio

$$\frac{\text{vertical distance}}{\text{horizontal distance}}$$

for each object. The object with the larger ratio is the steeper object.

Example 1 Comparing the Steepness of Two Objects

A portion of road A climbs steadily for 120 feet over a horizontal distance of 4250 feet. A portion of road B climbs steadily for 95 feet over a horizontal distance of 2875 feet. Which road is steeper? Explain.

Solution

Figure 37 shows sketches of the two roads, but the horizontal distances and vertical distances are not drawn to scale.

Figure 37 Roads A and B

Here, we calculate the unit ratio of the vertical distance to the horizontal distance for each road:

$$\text{Road A:} \quad \frac{\text{vertical distance}}{\text{horizontal distance}} = \frac{120 \text{ feet}}{4250 \text{ feet}} \approx \frac{0.028}{1}$$

$$\text{Road B:} \quad \frac{\text{vertical distance}}{\text{horizontal distance}} = \frac{95 \text{ feet}}{2875 \text{ feet}} \approx \frac{0.033}{1}$$

Road B is a little steeper than road A, because road B's ratio of vertical distance to horizontal distance is greater than road A's.

The **grade** of a road is the ratio of the vertical distance to the horizontal distance, written as a percentage. Recall from Section 2.5 that to write a decimal number as a percentage, we move the decimal point two places to the right and insert the percent symbol. In Example 1, the grade of road A is about 2.8% and the grade of road B is about 3.3%.

Finding a Line's Slope

To measure the steepness (also called *slope*) of a nonvertical line, we will also use a ratio, but we will work with *changes* in a quantity, which can be negative, rather than with distances, which are always positive.

Consider the line that contains the points (4, 2) and (6, 5) sketched in Fig. 38. To go from point (4, 2) to point (6, 5), we look 2 units to the right and then look 3 units up. So, the horizontal change, called the *run,* is 2 and the vertical change, called the *rise,* is 3. The slope of the line is the ratio of the rise to the run:

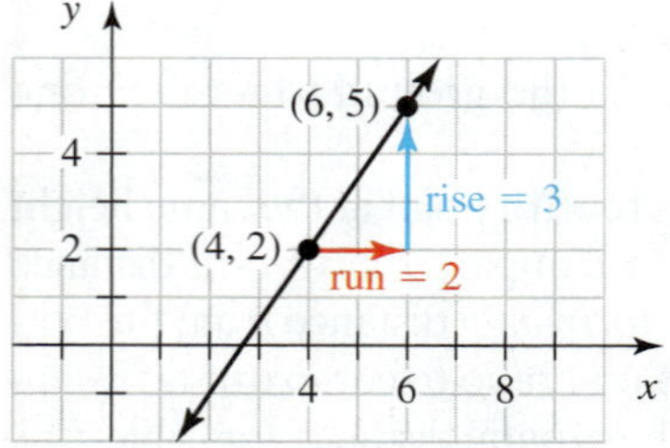

Figure 38 The slope is $\frac{3}{2}$

$$\text{slope} = \frac{\text{vertical change}}{\text{horizontal change}} = \frac{\text{rise}}{\text{run}} = \frac{3}{2}$$

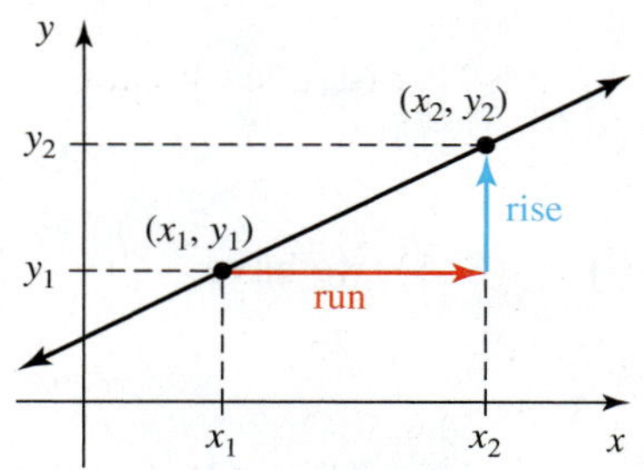

Figure 39 The rise and run between any two points

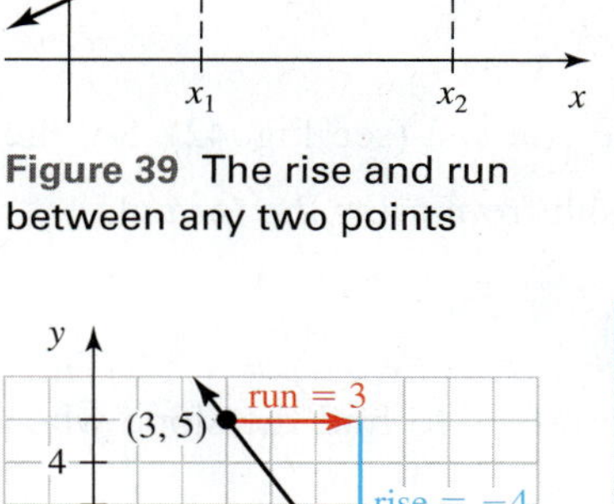

In general, for any line, the **rise** is the vertical change and the **run** is the horizontal change in going from one point on the line to another point on the line. We use the letter m to represent the slope.

Definition Slope of a nonvertical line

$$m = \text{slope} = \frac{\text{vertical change}}{\text{horizontal change}} = \frac{\text{rise}}{\text{run}}$$

In words: The **slope** of a nonvertical line is equal to the ratio of the rise to the run (in going from one point on the line to another point on the line). See Fig. 39.

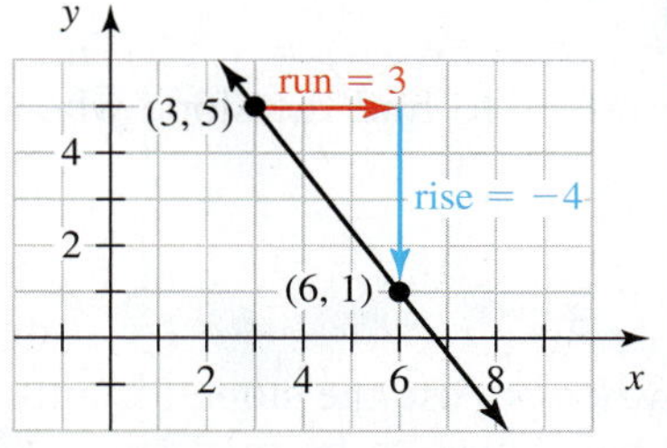

Figure 40 The slope is $\frac{-4}{3} = -\frac{4}{3}$

Example 2 Finding the Slope of a Line

Find the slope of the line that contains the points (3, 5) and (6, 1).

Solution

We begin by plotting the points (3, 5) and (6, 1) and sketching the line that contains them (see Fig. 40). To go from point (3, 5) to point (6, 1), we must look 3 units to the right and then 4 units down. So, the run is 3 and the rise is −4. The slope of the line is the ratio of the rise to the run:

$$\text{slope} = \frac{\text{vertical change}}{\text{horizontal change}} = \frac{\text{rise}}{\text{run}} = \frac{-4}{3} = -\frac{4}{3}$$

In Example 2, we found that, from point (3, 5) to point (6, 1), the run is 3. We can calculate this run by computing the change in the x-coordinates. Recall from Section 2.4 that we can find the change in a quantity by computing the ending amount minus the beginning amount:

$$\text{Change in } x\text{-coordinates} = \begin{matrix} x\text{-coordinate} \\ \text{of ending point} \end{matrix} - \begin{matrix} x\text{-coordinate} \\ \text{of beginning point} \end{matrix} = 6 - 3 = 3$$

In Example 2, we also found that, from point (3, 5) to point (6, 1), the rise is −4. We can calculate this rise by computing the change in the y-coordinates:

$$\text{Change in } y\text{-coordinates} = \begin{matrix} y\text{-coordinate} \\ \text{of ending point} \end{matrix} - \begin{matrix} y\text{-coordinate} \\ \text{of beginning point} \end{matrix} = 1 - 5 = -4$$

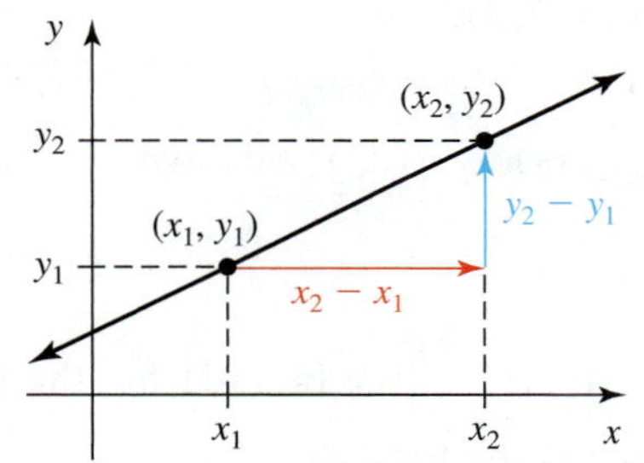

Figure 41 The slope of the line is $m = \frac{y_2 - y_1}{x_2 - x_1}$

How do we calculate the slope of *any* nonvertical line if we are given two points on the line? First, we use the subscript 1 to identify x_1 and y_1 as the coordinates of the first point, (x_1, y_1). Likewise, we identify x_2 and y_2 as the coordinates of the second point, (x_2, y_2). When we look from point (x_1, y_1) to point (x_2, y_2), the run is the difference $x_2 - x_1$ and the rise is the difference $y_2 - y_1$ (see Fig. 41).

Calculating Slope

Let (x_1, y_1) and (x_2, y_2) be two distinct points of a nonvertical line. The slope of the line is

$$m = \frac{\text{vertical change}}{\text{horizontal change}} = \frac{y_2 - y_1}{x_2 - x_1}$$

(see Fig. 41).

A *formula* is an equation that contains two or more variables. We will refer to the equation $m = \frac{y_2 - y_1}{x_2 - x_1}$ as the **slope formula.**

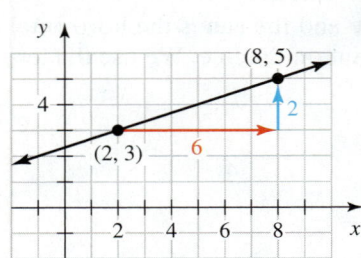

Figure 42 The slope is $\frac{2}{6} = \frac{1}{3}$

▶ Example 3 Finding the Slope of a Line

Find the slope of the line that contains the points (2, 3) and (8, 5).

Solution

Using the slope formula with $(x_1, y_1) = (2, 3)$ and $(x_2, y_2) = (8, 5)$, we have

$$m = \frac{y_2 - y_1}{x_2 - x_1} = \frac{5 - 3}{8 - 2} = \frac{2}{6} = \frac{1}{3}$$

By plotting points, we find that if the run is 6, then the rise is 2 (see Fig. 42). So, the slope is $m = \frac{\text{rise}}{\text{run}} = \frac{2}{6} = \frac{1}{3}$, which is the same as our result from using the formula.

In Example 3, we calculated the slope of a line for $(x_1, y_1) = (2, 3)$ and $(x_2, y_2) = (8, 5)$. Here, we switch the roles of the two points to find the slope when $(x_1, y_1) = (8, 5)$ and $(x_2, y_2) = (2, 3)$:

$$m = \frac{y_2 - y_1}{x_2 - x_1} = \frac{3 - 5}{2 - 8} = \frac{-2}{-6} = \frac{1}{3}$$

The result is the same as that in Example 3. In general, when we use the slope formula with two points on a line, it doesn't matter which point we choose to be first, (x_1, y_1), and which point we choose to be second, (x_2, y_2).

WARNING

It is a common error to make incorrect substitutions into the slope formula. Carefully consider why the middle and right-hand formulas are incorrect:

Correct	Incorrect	Incorrect
$m = \frac{y_2 - y_1}{x_2 - x_1}$	$m = \frac{y_2 - y_1}{x_1 - x_2}$	$m = \frac{x_2 - x_1}{y_2 - y_1}$

▶ Example 4 Finding the Slope of a Line

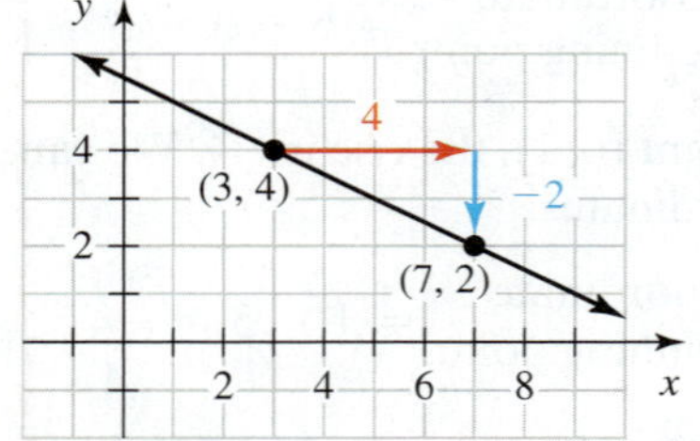

Figure 43 The slope is $\frac{-2}{4} = -\frac{1}{2}$

Find the slope of the line that contains the points (3, 4) and (7, 2).

Solution

Using the formula for slope with $(x_1, y_1) = (3, 4)$ and $(x_2, y_2) = (7, 2)$, we have

$$m = \frac{y_2 - y_1}{x_2 - x_1} = \frac{2 - 4}{7 - 3} = \frac{-2}{4} = -\frac{1}{2}$$

By plotting points, we find that when the run is 4, the rise is −2 (see Fig. 43). So, the slope is $\frac{-2}{4} = -\frac{1}{2}$, which is the same as our result from using the formula.

Increasing and Decreasing Lines

Since the line in Fig. 44 goes upward from left to right, we say the line (and the graph) is **increasing.** A sign analysis of the rise and run in Fig. 44 shows that the slope of the line is positive.

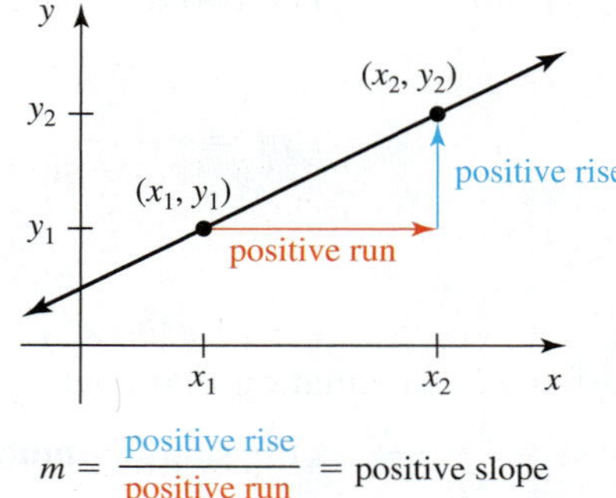

$$m = \frac{\text{positive rise}}{\text{positive run}} = \text{positive slope}$$

Figure 44 Increasing lines have positive slope

Since the line in Fig. 45 goes downward from left to right, we say the line (and the graph) is **decreasing.** A sign analysis of the rise and run in Fig. 45 shows that the slope of the line is negative.

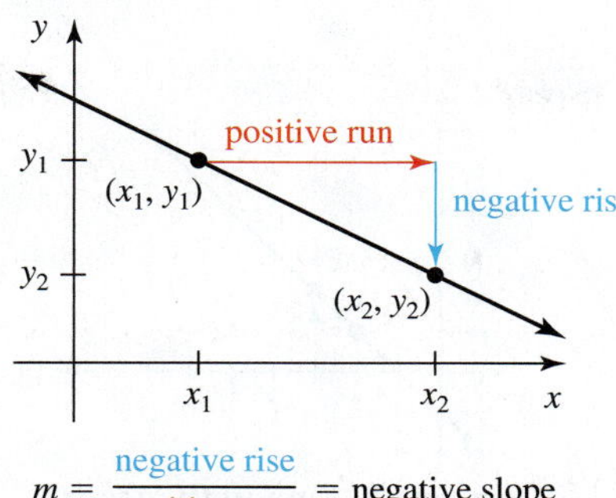

$$m = \frac{\text{negative rise}}{\text{positive run}} = \text{negative slope}$$

Figure 45 Decreasing lines have negative slope

Slopes of Increasing or Decreasing Lines

- An increasing line has positive slope (see Fig. 44).
- A decreasing line has negative slope (see Fig. 45).

When we compute the slope of two points that have negative coordinates, it can help to write first

$$\frac{(\ \ ) - (\ \ )}{(\ \ ) - (\ \ )}$$

and then insert the coordinates of the two points into the appropriate parentheses.

Example 5 Finding the Slope of a Line

Find the slope of the line that contains the points $(-5, 2)$ and $(3, -4)$.

Solution

$$m = \frac{(-4) - (2)}{(3) - (-5)} = \frac{-4 - 2}{3 + 5} = \frac{-6}{8} = -\frac{6}{8} = -\frac{3}{4}$$

Since the slope is negative, the line is decreasing.

Example 6 Finding the Slope of a Line

Find the approximate slope of the line that contains the points $(-4.9, -3.5)$ and $(-2.3, 5.8)$. Round the result to the second decimal place.

Solution

$$m = \frac{(5.8) - (-3.5)}{(-2.3) - (-4.9)} = \frac{9.3}{2.6} \approx 3.58$$

So, the slope is approximately 3.58. Since the slope is positive, the line is increasing.

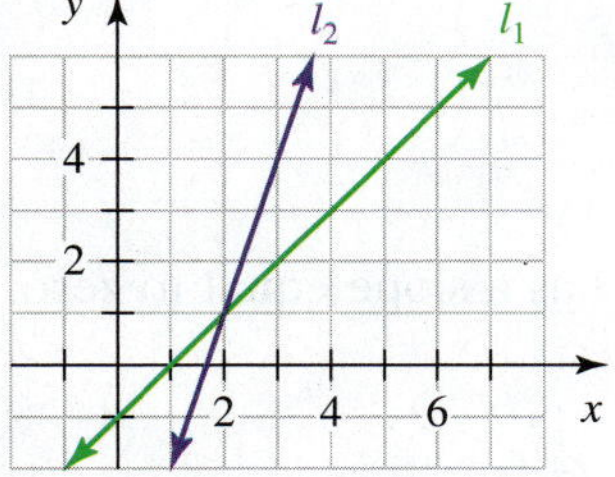

Figure 46 Find the slopes of the two lines

Example 7 Comparing the Slopes of Two Lines

Find the slopes of the two lines sketched in Fig. 46. Which line has the greater slope? Explain why this makes sense in terms of the steepness of a line.

Solution

In Fig. 47, we see that for line l_1, if the run is 1, the rise is 1. We calculate the slope of line l_1:

$$\text{Slope of line } l_1 = \frac{\text{rise}}{\text{run}} = \frac{1}{1} = 1$$

In Fig. 48, we see that for line l_2, if the run is 1, the rise is 3. We calculate the slope of line l_2:

$$\text{Slope of line } l_2 = \frac{\text{rise}}{\text{run}} = \frac{3}{1} = 3$$

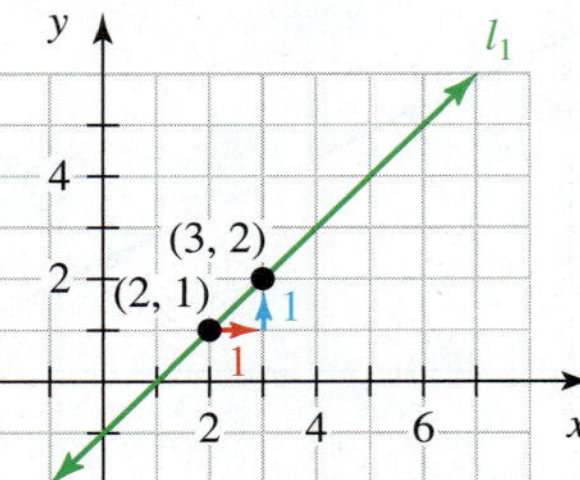

Figure 47 Line with lesser slope

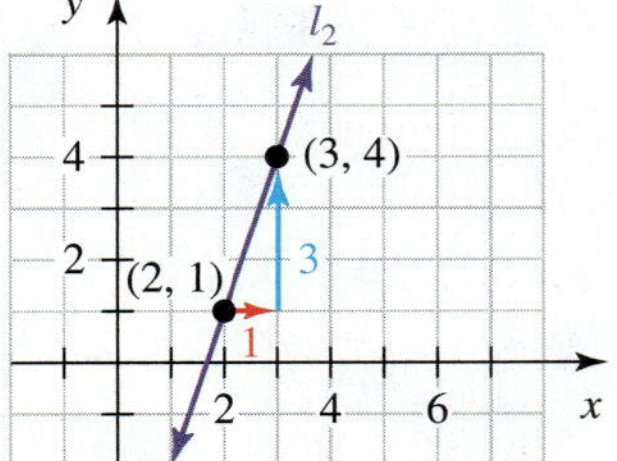

Figure 48 Line with greater slope

Note that the slope of line l_2 is greater than the slope of line l_1, which is what we would expect because line l_2 is steeper than line l_1.

In general, **for two nonparallel increasing lines, the steeper line has the greater slope.**

We show three decreasing lines and three increasing lines and their slopes in Fig. 49.

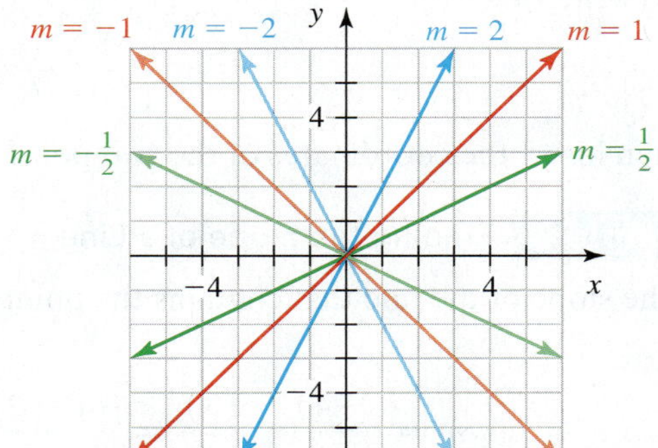

Figure 49 Slopes of some lines

Horizontal and Vertical Lines

So far, we have discussed slopes of lines that are increasing or decreasing. What about the slope of a horizontal line or the slope of a vertical line?

Example 8 Finding the Slope of a Horizontal Line

Find the slope of the line that contains the points (3, 2) and (7, 2).

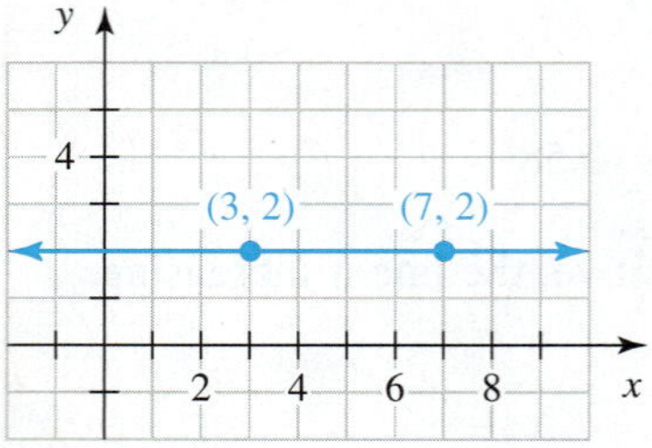

Solution

We plot the points (3, 2) and (7, 2) and sketch the line that contains the points (see Fig. 50). The formula for slope gives

$$m = \frac{2 - 2}{7 - 3} = \frac{0}{4} = 0$$

So, the slope of this horizontal line is zero.

Figure 50 The horizontal line has zero slope

In Example 8, we saw that the horizontal line in Fig. 50 has a slope equal to zero. Note that *any* horizontal line has no rise, so

$$\text{Slope of a horizontal line} = \frac{\text{rise}}{\text{run}} = \frac{0}{\text{run}} = 0$$

The slope of any horizontal line is zero.

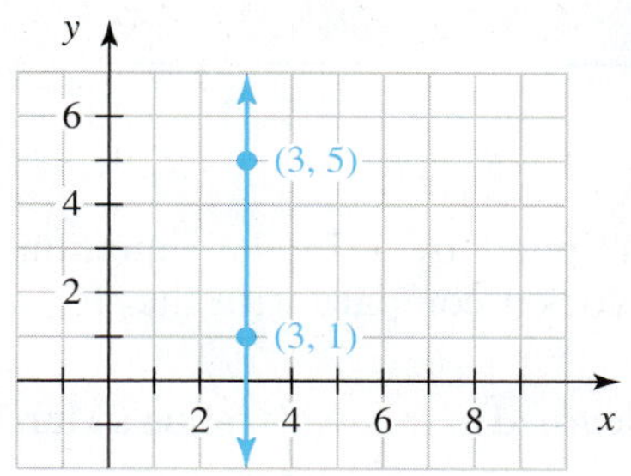

Figure 51 The vertical line has undefined slope

Example 9 Finding the Slope of a Vertical Line

Find the slope of the line that contains the points (3, 1) and (3, 5).

Solution

We plot the points (3, 1) and (3, 5) and sketch the line that contains the points (see Fig. 51). The formula for slope gives

$$m = \frac{5-1}{3-3} = \frac{4}{0}$$

Since division by zero is undefined, the slope of the vertical line is *undefined.*

In Example 9, we saw that the vertical line in Fig. 51 has undefined slope. Note that *any* vertical line has zero run, so

$$\text{Slope of a vertical line} = \frac{\text{rise}}{\text{run}} = \left.\frac{\text{rise}}{0}\right\} \text{undefined}$$

The slope of any vertical line is undefined.

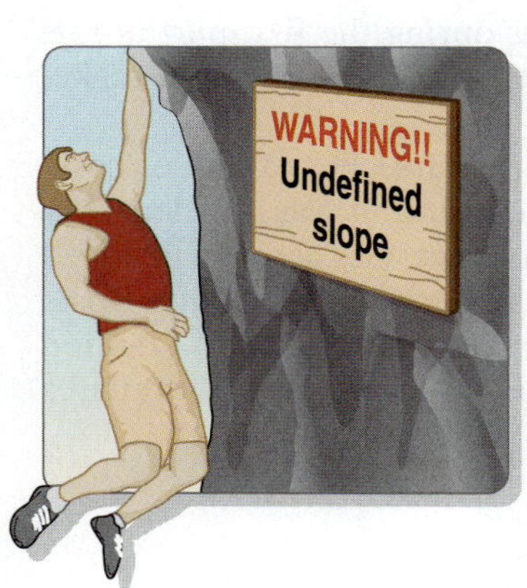

Slopes of Horizontal and Vertical Lines

- A horizontal line has a slope equal to zero (see Fig. 52).
- A vertical line has undefined slope (see Fig. 53).

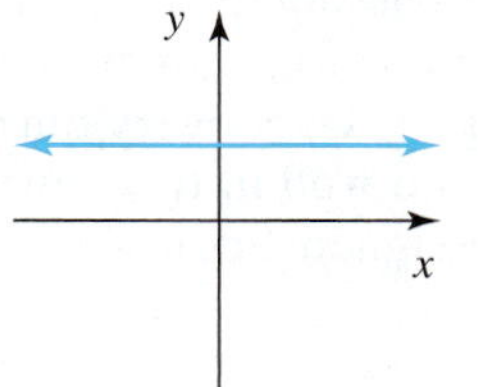

Figure 52 Horizontal lines have a slope equal to zero

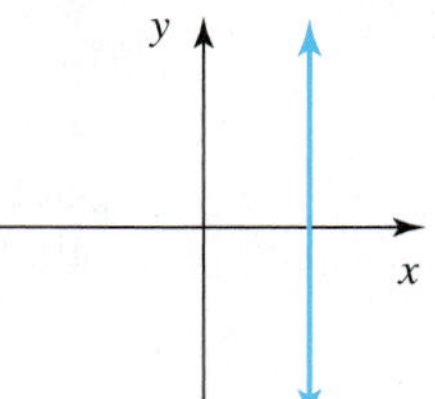

Figure 53 Vertical lines have undefined slope

Group Exploration

For a line, rise over run is constant

1. A line is sketched in Fig. 54. Plot the points $(-2, -5)$, $(1, 1)$, and $(3, 5)$. (Plotted correctly, these points will lie on the line.)
2. Using the points $(-2, -5)$ and $(1, 1)$, find the slope of the line.
3. Using the points $(1, 1)$ and $(3, 5)$, find the slope of the line.
4. Using the points $(-2, -5)$ and $(3, 5)$, find the slope of the line.
5. Using two other points of your choice, find the slope of the line.
6. What do you notice about the slopes you have calculated? Does it matter which two points on a line are used to find the slope of the line?

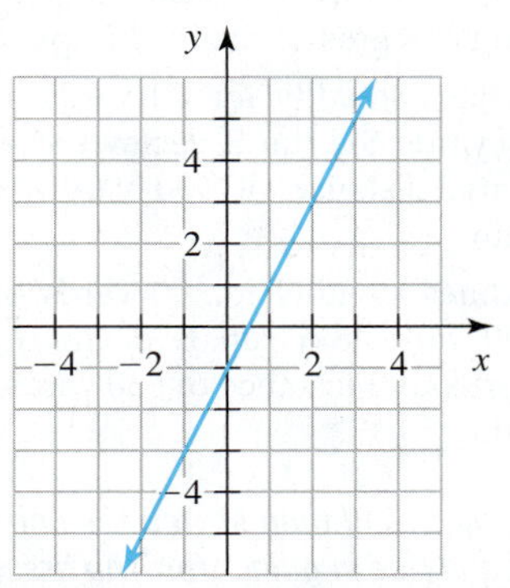

Figure 54 Use different pairs of points to calculate the slope

Group Exploration

Looking ahead: The meaning of m in the equation $y = mx + b$

1. **a.** Carefully sketch a graph of the line $y = 2x - 1$.
   b. Using the formula $m = \frac{\text{rise}}{\text{run}}$, find the slope of the line you sketched.
   c. What number is multiplied by x in the equation $y = 2x - 1$? How does it compare with the slope you found in part (b)?
2. **a.** Carefully sketch a graph of the line $y = -3x + 5$.
   b. Using the formula $m = \frac{\text{rise}}{\text{run}}$, find the slope of the line you sketched.
   c. What number is multiplied by x in the equation $y = -3x + 5$? How does it compare with the slope you found in part (b)?
3. Describe what you have learned in this exploration so far.
4. Without graphing, determine the slope of each line.
   a. $y = 4x - 7$ **b.** $y = -2x + 4$
   c. $y = \frac{2}{5}x - 3$ **d.** $y = x - 2$
   e. $y = 3$

Tips for Success Make Changes

If you have not had passing scores on tests and quizzes during the first part of this course, it is time to evaluate what the problem is, what changes you should make, and whether you can commit to making those changes.

Sometimes students must change how they study for a course. For example, Rosie did poorly on exams and quizzes for the first third of the course. It was not clear why she was not passing the course, since she had good attendance, was actively involved in classroom work, and was doing the homework assignments. Suddenly, Rosie started getting As on every quiz and test. What had happened? Rosie said, "I figured out that, to do well in this course, it was not enough for me to do just the exercises that were assigned. So now I do a lot of extra exercises from each section."

Homework 3.3

For extra help ▶ MyMathLab®  Watch the videos in MyMathLab  Download the MyDashboard App

1. A portion of road A climbs steadily for 210 feet over a horizontal distance of 3500 feet. A portion of road B climbs steadily for 275 feet over a horizontal distance of 5000 feet. Which road is steeper? Explain.
2. While taking off, airplane A climbs steadily for 4800 feet over a horizontal distance of 9000 feet. Airplane B climbs steadily for 5800 feet over a horizontal distance of 10,500 feet. Which plane is climbing at a greater incline? Explain.
3. Ski run A declines steadily for 80 yards over a horizontal distance of 400 yards. Ski run B declines steadily for 90 yards over a horizontal distance of 600 yards. Which ski run is steeper? Explain.
4. Ski run A declines steadily for 90 yards over a horizontal distance of 500 yards. Ski run B declines steadily for 130 yards over a horizontal distance of 650 yards. Which ski run is steeper? Explain.

Plot the two given points and then sketch the line that contains the points. Find the run and rise in going from the first point listed to the second point listed. Find the slope of the line.

5. $(2, 3)$ and $(4, 6)$
6. $(1, 4)$ and $(6, 5)$
7. $(3, 6)$ and $(5, 2)$
8. $(2, 5)$ and $(6, 3)$
9. $(-4, 1)$ and $(2, 5)$
10. $(3, -4)$ and $(5, 2)$
11. $(-4, -2)$ and $(-2, -6)$
12. $(-3, -2)$ and $(-1, -4)$

Use the slope formula to find the slope of the line that passes through the two given points. State whether the line is increasing, decreasing, horizontal, or vertical.

13. $(1, 5)$ and $(3, 9)$
14. $(2, 3)$ and $(5, 12)$
15. $(3, 10)$ and $(5, 2)$
16. $(5, 8)$ and $(7, 2)$
17. $(2, 1)$ and $(8, 4)$
18. $(3, 2)$ and $(7, 4)$
19. $(2, 5)$ and $(8, 3)$
20. $(1, 7)$ and $(9, 5)$
21. $(-2, 4)$ and $(3, -1)$
22. $(-3, 4)$ and $(1, -2)$
23. $(5, -2)$ and $(9, -4)$
24. $(2, -3)$ and $(8, -6)$
25. $(-7, -1)$ and $(-2, 9)$
26. $(-6, -8)$ and $(-4, 2)$
27. $(-6, -9)$ and $(-2, -3)$
28. $(-5, -2)$ and $(-1, -3)$
29. $(6, -1)$ and $(-4, 7)$
30. $(4, -5)$ and $(-2, 10)$
31. $(-2, -11)$ and $(7, -5)$
32. $(-3, -1)$ and $(9, -3)$
33. $(0, 0)$ and $(4, -2)$
34. $(-6, -9)$ and $(0, 0)$
35. $(3, 5)$ and $(7, 5)$
36. $(-4, -6)$ and $(3, -6)$
37. $(-3, -1)$ and $(-3, -2)$
38. $(4, 2)$ and $(4, 7)$

For Exercises 39–46, find the approximate slope of the line that contains the two given points. Round your result to the second decimal place. State whether the line is increasing, decreasing, horizontal, or vertical.

39. $(-3.2, 5.1)$ and $(-2.8, 1.4)$

40. $(-1.9, 4.8)$ and $(-3.1, 5.5)$

41. $(4.9, -2.7)$ and $(6.3, -1.1)$

42. $(9.7, -6.8)$ and $(4.5, -2.7)$

43. $(-4.97, -3.25)$ and $(-9.64, -2.27)$

44. $(-3.22, -8.54)$ and $(-7.29, -6.13)$

45. $(-2.45, -6.71)$ and $(4.88, -1.53)$

46. $(-3.99, -2.49)$ and $(1.06, -3.76)$

47. Find the slope of the line sketched in Fig. 55.

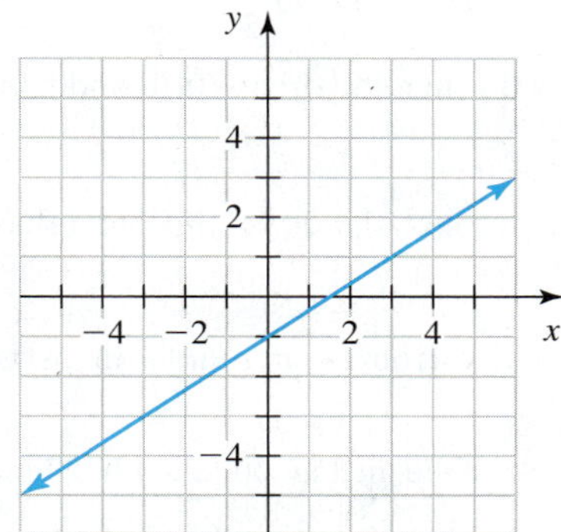

Figure 55 Exercise 47

48. Find the slope of the line sketched in Fig. 56.

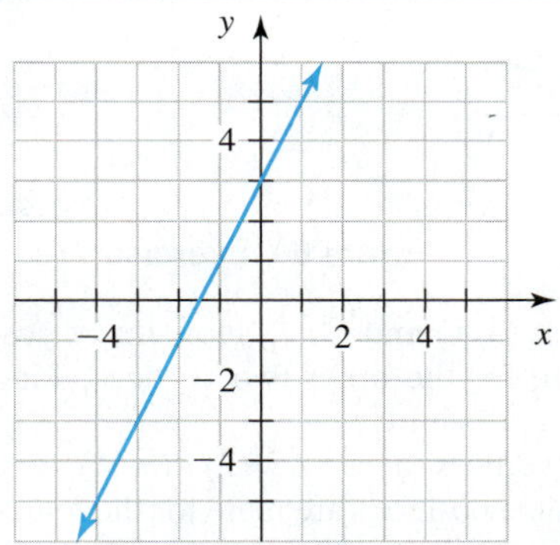

Figure 56 Exercise 48

49. Find the slope of the line sketched in Fig. 57.

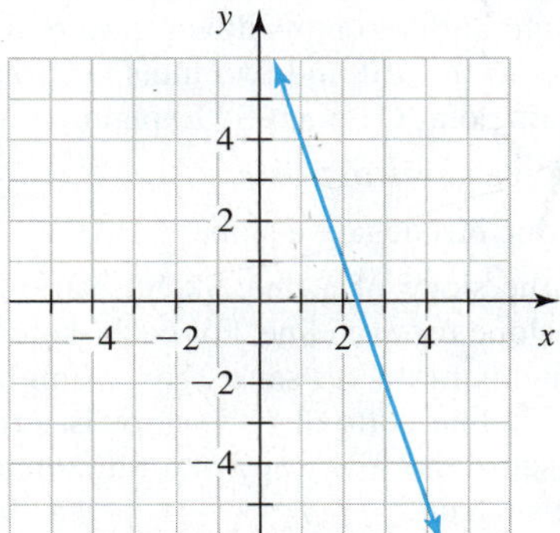

Figure 57 Exercise 49

50. Find the slope of the line sketched in Fig. 58.

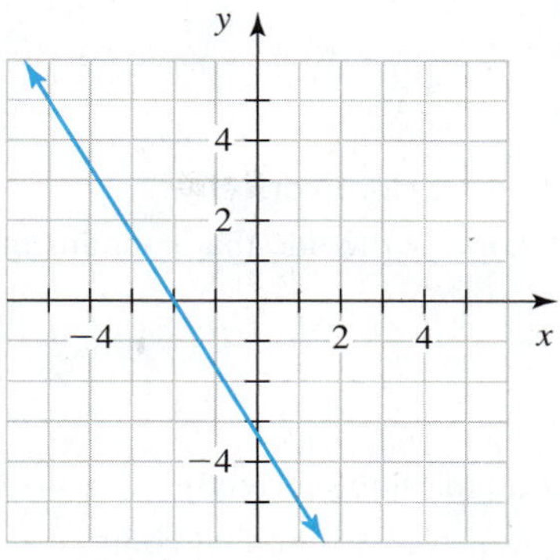

Figure 58 Exercise 50

Concepts

51. For each line sketched in Fig. 59, determine whether the line's slope is positive, negative, zero, or undefined.

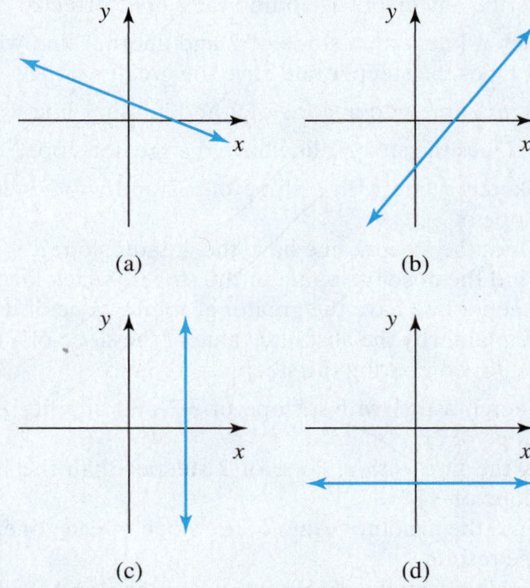

Figure 59 Exercise 51

52. Find the slope of the line with x-intercept $(-3, 0)$ and y-intercept $(0, 4)$.

Sketch a line that meets the given description. Find the slope of the line.

53. An increasing line that is nearly horizontal

54. A decreasing line that is nearly horizontal

55. A decreasing line that is nearly vertical

56. An increasing line that is nearly vertical

For Exercises 57–60, sketch a line that meets the given description.

57. The slope is a large positive number.

58. The slope is a positive number near zero.

59. The slope is a negative number near zero.

60. The slope is less than -5.

61. A student tries to find the slope of the line that contains the points $(1, 3)$ and $(4, 7)$:

$$\frac{4-1}{7-3} = \frac{3}{4}$$

Describe any errors. Then find the slope correctly.

62. A student tries to find the slope of the line that contains the points (6, 4) and (3, 9):

$$\frac{9-4}{6-3} = \frac{5}{3}$$

Describe any errors. Then find the slope correctly.

63. A student tries to find the slope of the line that contains the points $(-1, -5)$ and $(3, 8)$:

$$\frac{8-5}{3-1} = \frac{3}{2}$$

Describe any errors. Then find the slope correctly.

64. A student tries to find the slope of the line that contains the points (2, 5) and (6, 7):

$$\frac{7-5}{6-2} = \frac{2}{4}$$

Describe any errors. Then find the slope correctly.

65. Sketch a line with a slope of 2 and another line with a slope of 3. Does the steeper line have the greater slope?

66. Sketch a line with a slope of 1 and another line with a slope of $\frac{1}{2}$. Does the steeper line have the greater slope?

67. a. Sketch a line with a slope of -2 and another line with a slope of -3.

b. Does the steeper line have the greater slope?

c. Find the absolute value of the slope of each line. Does the steeper line have the greater absolute value of the slope?

d. Explain why the absolute value of the slope of a line is useful for comparing the steepness of lines.

68. a. Sketch a line with a slope of -2 and another line with a slope of 2.

b. Is the line with a slope of 2 steeper than the line with a slope of -2?

c. Find the absolute value of the slope of each line. Compare the results.

d. Explain why the absolute value of the slope of a line is useful for comparing the steepness of lines.

69. A line contains the points (2, 1) and (3, 4). Find three more points that lie on the line. [**Hint:** Find the slope of the line. Then use the slope and a point on the line to help you find other points on the line.]

70. A line contains the points (1, 5) and (4, 3). Find three more points that lie on the line. [**Hint:** See Exercise 69.]

71. Explain why the slope of a vertical line is undefined.

72. Explain why the slope of a horizontal line is 0.

73. Explain why the slope of an increasing line is positive.

74. Explain why the slope of a decreasing line is negative.

75. a. Carefully graph the given equation by hand. Then find the slope of the line by using the ratio $\frac{\text{rise}}{\text{run}}$.

i. $y = 2x + 1$ **ii.** $y = 3x - 5$ **iii.** $y = -2x + 6$

b. Compare the slope of each line with the coefficient of x in the corresponding equation.

76. a. Use the expression $\frac{y_2 - y_1}{x_2 - x_1}$ to find the slope of the line that contains the points $(x_1, y_1) = (2, 3)$ and $(x_2, y_2) = (7, 5)$.

b. Use the expression $\frac{y_1 - y_2}{x_1 - x_2}$ to find the slope of the line that contains the points $(x_1, y_1) = (2, 3)$ and $(x_2, y_2) = (7, 5)$.

c. Compare your results from parts (a) and (b).

d. Show that $\frac{y_2 - y_1}{x_2 - x_1} = \frac{y_1 - y_2}{x_1 - x_2}$, where (x_1, y_1) and (x_2, y_2) are two distinct points of a nonvertical line. [**Hint:** $a - b = -(b - a)$]

e. When using two given points on a line to calculate the slope of the line, does it matter which point we choose to be first, (x_1, y_1), and which second, (x_2, y_2)? Explain.

77. Explore the relationship among three lines that pass through the origin (0, 0), where the slope of one of the lines is the reciprocal of the slope of one of the other lines.

a. By hand, carefully sketch the lines that pass through the origin (0, 0) and that have slopes 5, 1, and $\frac{1}{5}$.

b. Sketch the lines that pass through the origin (0, 0) and that have slopes $\frac{2}{5}$, 1, and $\frac{5}{2}$.

c. Sketch the lines that pass through the origin (0, 0) and that have slopes $\frac{3}{4}$, 1, and $\frac{4}{3}$.

d. What pattern do you notice from your graphs in parts (a)–(c)?

e. A line with slope m is sketched in Fig. 60. Sketch a line with slope $\frac{1}{m}$ that passes through the origin (0, 0). Assume both axes are scaled the same.

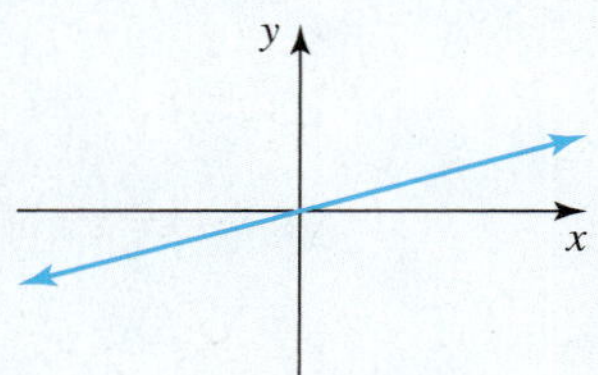

Figure 60 Exercise 77e

78. a. A square has vertices at (3, 1) and (3, 7). How many possible positions are there for the other two vertices? Find the coordinates for each possibility.

b. A parallelogram has vertices at $(-7, -2)$, (3, 1), and $(-4, 2)$. How many possible positions are there for the fourth vertex? Find the coordinates for each possibility. [**Hint:** Try drawing different line segments between the given vertices.]

79. Suppose that a line with slope $-\frac{2}{3}$ contains a point P. A point Q lies three units to the right and two units down from point P. A point S lies three units to the left and two units up from point P. Does the line contain point Q? point S? Explain.

80. Draw a sketch of a road with a grade of 5%.

81. Explain why a decreasing line has negative slope.

82. Describe the meaning of the slope of a line. Sketch various types of lines and give the slope for each line. For each sketch, explain why the slope assignment makes sense. For example, you could sketch a horizontal line, state that the slope is zero, and explain why it makes sense that the slope of a horizontal line is zero in terms of rise and run.

3.4 Using Slope to Graph Linear Equations

Objectives

- » Use the slope and the y-intercept of a line to sketch the line.
- » Know the meaning of m for an equation of the form $y = mx + b$.
- » Graph an equation of the form $y = mx + b$ by using the line's slope and y-intercept.
- » Graph an equation of a linear model by using the model's slope and y-intercept.
- » Find an equation of a line from its graph.
- » Know the relationship between slopes of parallel lines.
- » Know the relationship between slopes of perpendicular lines.

In Sections 3.1 and 3.2, we graphed a linear equation in two variables by first finding solutions of the equation. In this section, we will discuss a more efficient way to graph such equations.

Using the Slope and the y-Intercept to Sketch a Line

We can use the slope and the y-intercept of a line to sketch the line.

Example 1 Sketching a Line

Sketch the line that has slope $m = -\frac{2}{5}$ and y-intercept $(0, 3)$.

Solution

We first plot the y-intercept, $(0, 3)$. The slope is $-\frac{2}{5} = \frac{-2}{5} = \frac{\text{rise}}{\text{run}}$. From $(0, 3)$, we count 5 units to the right and 2 units down, where we plot the point $(5, 1)$. See Fig. 61. We then sketch the line that contains these two points (see Fig. 62).

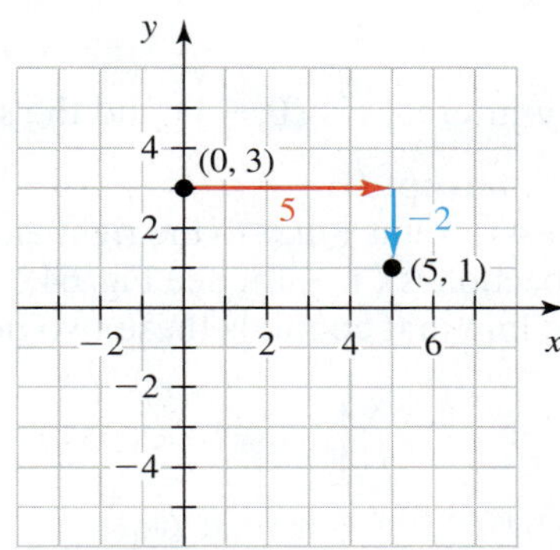

Figure 61 Plot (0, 3). Then count 5 units to the right and 2 units down to plot (5, 1)

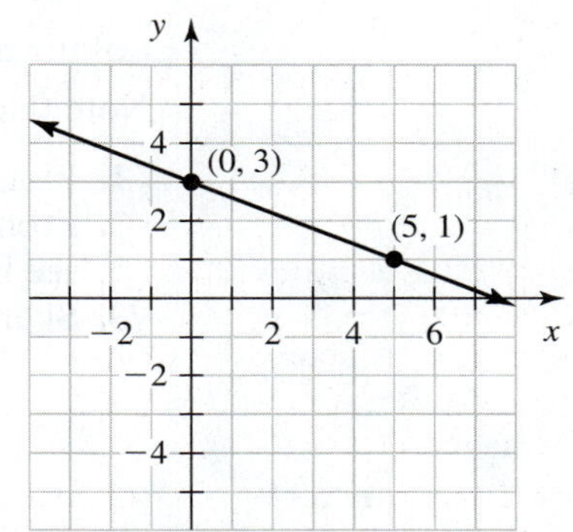

Figure 62 Sketch the line containing (0, 3) and (5, 1)

Because the slope is negative, we can verify our work by checking that the line is decreasing.

Table 24 Solutions of $y = 2x + 1$

x	y
0	$2(0) + 1 = 1$
1	$2(1) + 1 = 3$
2	$2(2) + 1 = 5$

The Meaning of m for $y = mx + b$

In Example 1, we saw that if we know the slope and the y-intercept of a line, we can sketch that line. Next, we will discuss how to determine the slope and y-intercept of a line from the line's equation.

Example 2 Finding the Slope of a Line

Find the slope of the line $y = 2x + 1$.

Solution

We list some solutions in Table 24 and sketch the graph of the equation in Fig. 63.

If the run is 1, the rise is 2 (see Fig. 63). So, the slope is

$$m = \frac{\text{rise}}{\text{run}} = \frac{2}{1} = 2$$

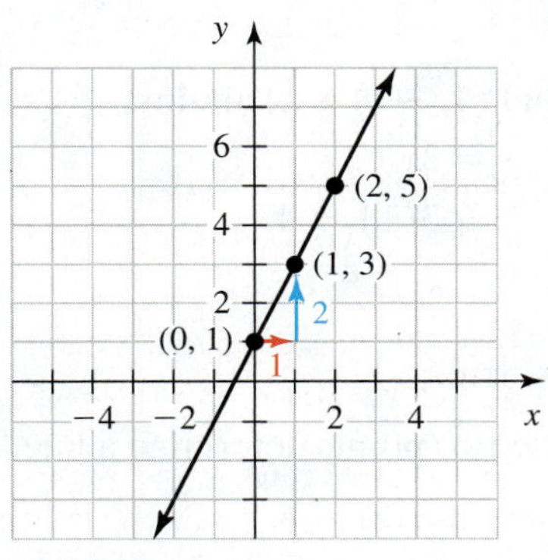

Figure 63 Graph of $y = 2x + 1$

In Example 2, we found that the line $y = 2x + 1$ has slope 2. Note that 2 is also the number multiplied by x in the equation $y = 2x + 1$. This observation suggests a general property about a linear equation of the form $y = mx + b$.

Finding the Slope and y-Intercept from a Linear Equation

For a linear equation of the form $y = mx + b$,

- the slope of the line is m and
- the y-intercept of the line is $(0, b)$.

We say that this equation is in **slope–intercept form.**

For example, the equation $y = -4x + 5$ is in slope–intercept form with $m = -4$ and $b = 5$. The graph of this equation is a line with slope -4 and y-intercept $(0, 5)$. The line $y = 8x - 2$ has slope 8 and y-intercept $(0, -2)$.

Graphing Equations of the Form $y = mx + b$

In Example 3, we will graph an equation in slope–intercept form.

Example 3 Graphing an Equation

Sketch the graph of $y = 3x - 4$.

Solution

Note that the y-intercept is $(0, -4)$ and the slope is $3 = \frac{3}{1} = \frac{\text{rise}}{\text{run}}$. To graph the line:

1. Plot the y-intercept $(0, -4)$.
2. From $(0, -4)$, count 1 unit to the right and 3 units up to plot a second point, which we see by inspection is $(1, -1)$. See Fig. 64.
3. Sketch the line that contains these two points (see Fig. 65).

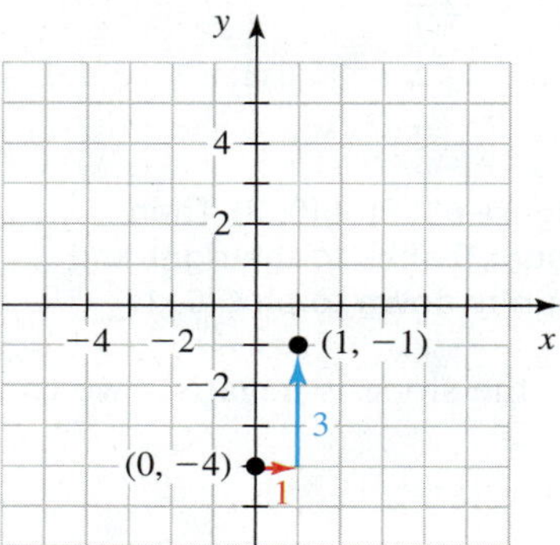

Figure 64 Plot $(0, -4)$. Then count 1 unit to the right and 3 units up to plot $(1, -1)$

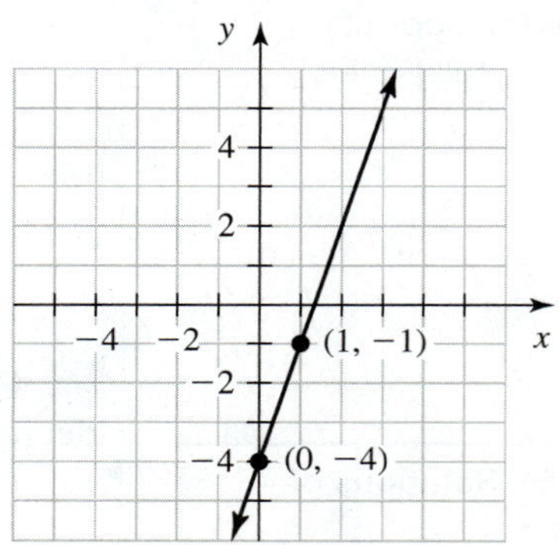

Figure 65 Sketch the line containing $(0, -4)$ and $(1, -1)$

Recall from Section 3.1 that every point on the graph of an equation represents a solution of the equation. We can verify our result by checking that both $(0, -4)$ and $(1, -1)$ are solutions of $y = 3x - 4$:

Check that $(0, -4)$ is a solution

$$y = 3x - 4$$
$$-4 \stackrel{?}{=} 3(0) - 4$$
$$-4 \stackrel{?}{=} 0 - 4$$
$$-4 \stackrel{?}{=} -4$$
true

Check that $(1, -1)$ is a solution

$$y = 3x - 4$$
$$-1 \stackrel{?}{=} 3(1) - 4$$
$$-1 \stackrel{?}{=} 3 - 4$$
$$-1 \stackrel{?}{=} -1$$
true

We check two ordered pairs (rather than just one) because two points determine a line.

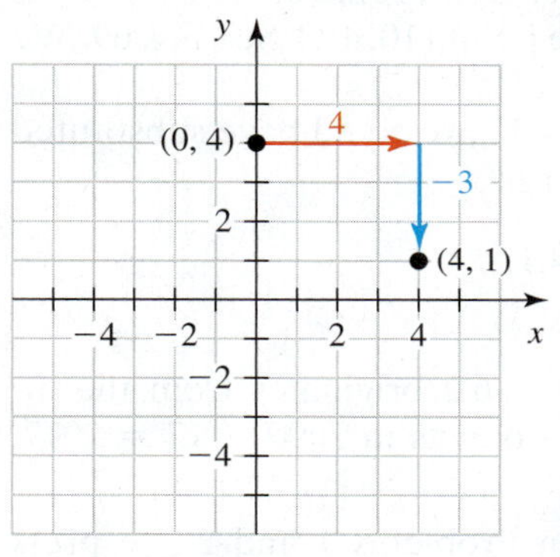

Figure 66 Plot (0, 4). Then count 4 units to the right and 3 units down to plot (4, 1)

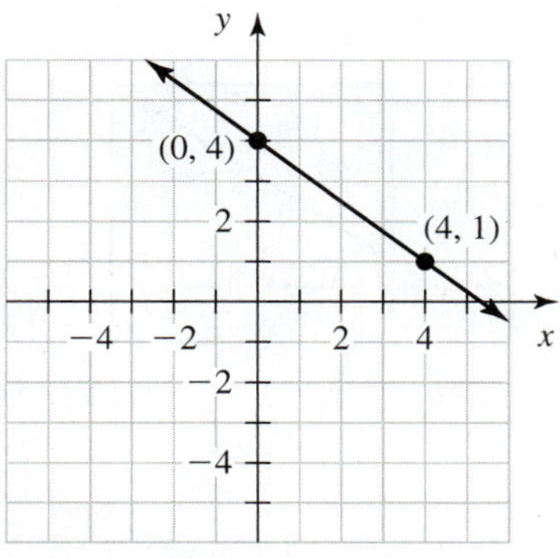

Figure 67 Sketch the line containing (0, 4) and (4, 1)

Graphing an Equation in Slope–Intercept Form

To graph an equation of the form $y = mx + b$,

1. Plot the y-intercept $(0, b)$.
2. Use $m = \frac{\text{rise}}{\text{run}}$ to plot a second point. For example, if $m = \frac{2}{3}$, then count 3 units to the right (from the y-intercept) and 2 units up to plot another point.
3. Sketch the line that passes through the two plotted points.

Example 4 Graphing an Equation

Sketch the graph of $y = -\frac{3}{4}x + 4$.

Solution

The y-intercept is $(0, 4)$, and the slope is $-\frac{3}{4} = \frac{-3}{4} = \frac{\text{rise}}{\text{run}}$. To graph the line:

1. Plot the y-intercept $(0, 4)$.
2. From $(0, 4)$, count 4 units to the right and 3 units down to plot a second point, which we see by inspection is $(4, 1)$. See Fig. 66.
3. Sketch the line that contains these two points (see Fig. 67).

We use a graphing calculator to verify our work (see Fig. 68).

Plot1 Plot2 Plot3
\Y1■-(3/4)X+4
\Y2=
\Y3=
\Y4=
\Y5=
\Y6=
\Y7=

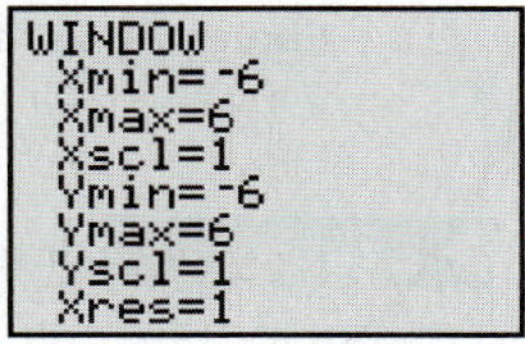

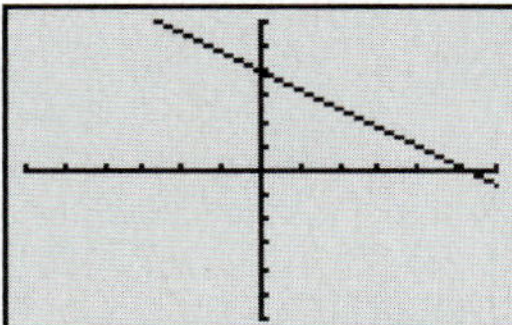

Figure 68 Verify the work

Table 25 Percentages of U.S. Food Sales That Are Organic Products

Year	Percent
2000	1.2
2002	1.6
2004	2.2
2006	2.9
2008	3.6
2010	4.0

Source: *Organic Trade Association*

We now know two methods for graphing a linear equation: We can first find solutions of the equation (as discussed in Sections 3.1 and 3.2), or we can first find the slope and y-intercept (as discussed in this section).

Graphing an Equation of a Linear Model

We can use either method to graph an equation of a model.

Example 5 Graphing a Model's Equation

The percentages of U.S. food sales that are organic products are shown in Table 25 for various years. Let p be the percentage of U.S. food sales that are organic products at t years since 2000. A reasonable model is

$$p = 0.3t + 1.1$$

1. Graph the model.
2. Predict when 6.2% of food sales will be organic products.

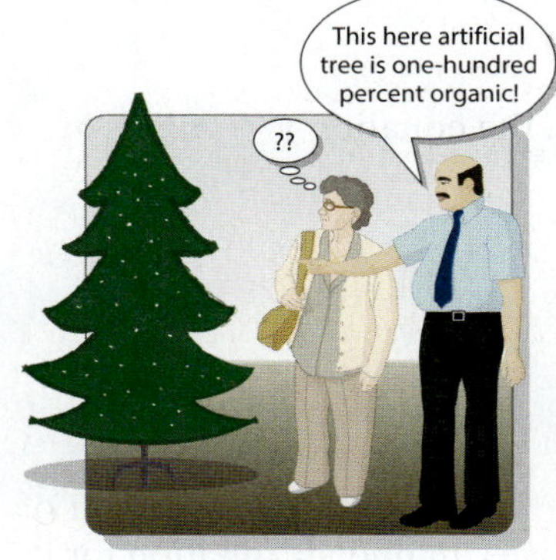

Solution

1. The p-intercept is $(0, 1.1)$, and the slope is $0.3 = \frac{0.3}{1} = \frac{\text{rise}}{\text{run}}$. It will be easier to graph the model if we multiply the slope by $\frac{10}{10} = 1$ so the rise and run are larger and both are integers:

$$0.3 = \frac{0.3}{1} \cdot \frac{10}{10} = \frac{3}{10} = \frac{\text{rise}}{\text{run}}$$

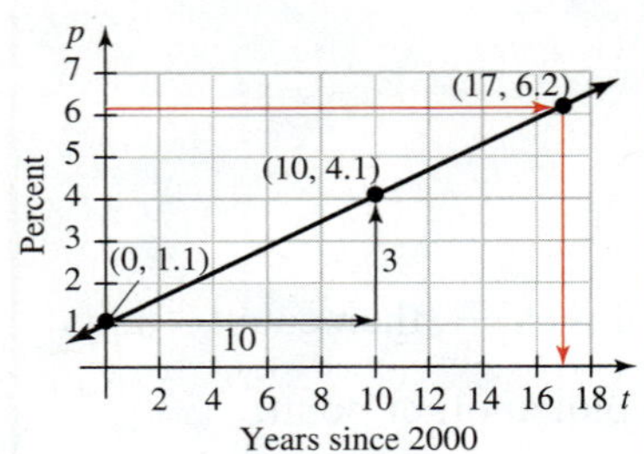

Figure 69 Organic products model

To graph the model, we first plot the p-intercept (0, 1.1). From (0, 1.1), we count 10 units to the right and 3 units up, where we plot the point (10, 4.1). See Fig. 69. We then sketch the line that contains the two points.

Instead of using the slope to find the point (10, 4.1), we could have substituted 10 for t in the equation $p = 0.3t + 1.1$ and solved for p:

$$p = 0.3(10) + 1.1 = 4.1$$

So, the point (10, 4.1) is a point on the linear model.

2. The red arrows in Fig. 69 show that the output $p = 6.2$ originates from the input $t = 17$. So, 6.2% of food sales will be organic products in $2000 + 17 = 2017$, according to the model.

To use a graphing calculator to verify our work in Problems 1 and 2, we press WINDOW and set Xmin to be -10 (for 1990), set Xmax to be 25 (for 2025), and use ZoomFit to set the values for Ymin and Ymax automatically (see Fig. 70). Then we use TRACE to check that (17, 6.2) is a point on the linear model. See Appendix A.5, A.6, and A.7 for graphing calculator instructions.

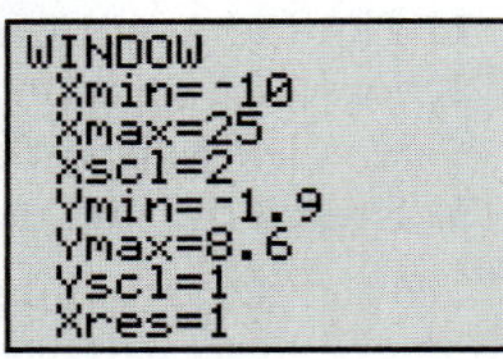

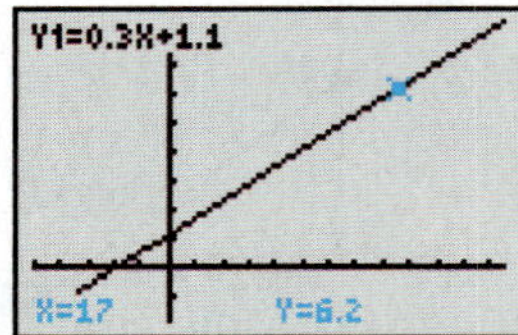

Figure 70 Verify the work

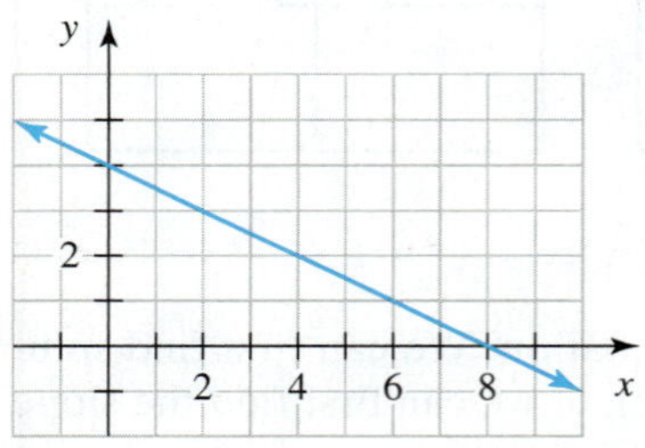

Figure 71 A decreasing line with y-intercept above the origin

Example 6 Interpreting the Signs of m and b

Graph an equation of the form $y = mx + b$ where m is negative and b is positive.

Solution

We should sketch a decreasing line, because the slope m is negative. We should draw a line whose y-intercept, $(0, b)$, is above the origin, because b is positive. We sketch such a line in Fig. 71.

There is nothing special about the line in Fig. 71—any line that is decreasing and has its y-intercept above the origin will do.

Finding an Equation of a Line from a Graph

Given the slope and y-intercept of a nonvertical line, we can find an equation of the line.

Example 7 Finding an Equation of a Line from Its Slope and y-Intercept

Find an equation of the line that has slope $\frac{2}{3}$ and y-intercept $(0, -5)$.

Solution

To find an equation, we substitute $\frac{2}{3}$ for m and -5 for b in the equation $y = mx + b$:

$$y = \frac{2}{3}x + (-5)$$

$$y = \frac{2}{3}x - 5$$

Given an equation in slope–intercept form, $y = mx + b$, we can find the graph of the equation. We can also go backward: Given the graph, we can find an equation for it.

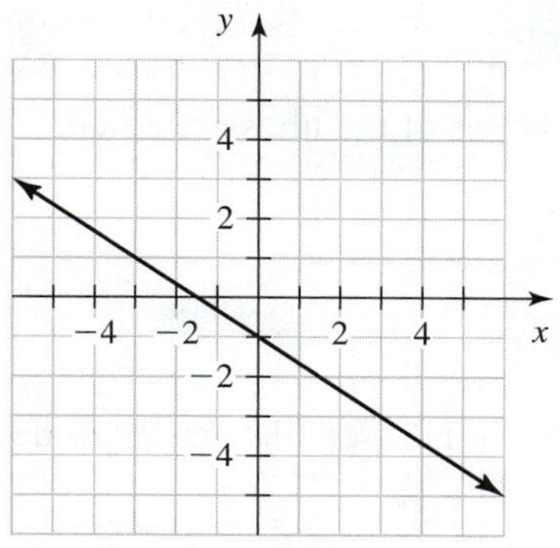

Figure 72 Graph of a line

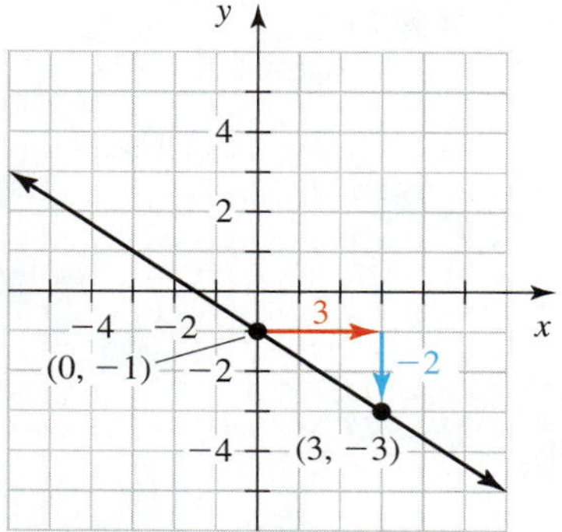

Figure 73 Finding the slope of the line

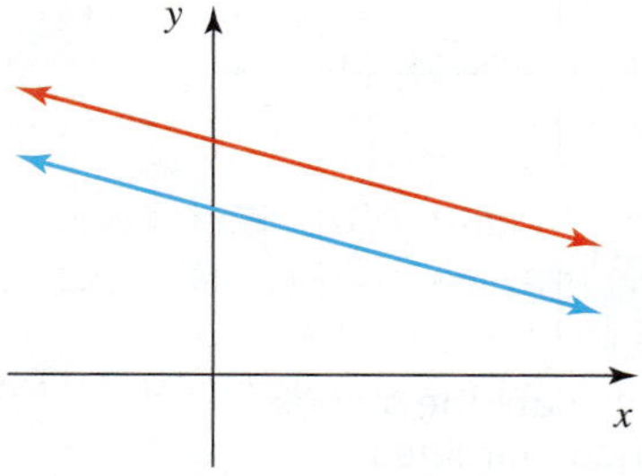

Figure 75 Two parallel lines

Example 8 Finding an Equation of a Line from Its Graph

Find an equation of the line sketched in Fig. 72.

Solution

From Fig. 73, we see that the y-intercept of the line is $(0, -1)$. We also see that if the run is 3, then the rise is -2. So, the slope is $\frac{-2}{3} = -\frac{2}{3}$. By substituting $-\frac{2}{3}$ for m and -1 for b in the equation $y = mx + b$, we have $y = -\frac{2}{3}x - 1$.

We can verify our equation by checking that both $(0, -1)$ and $(3, -3)$ satisfy the equation. Or we can use a graphing calculator to verify our work (see Fig. 74).

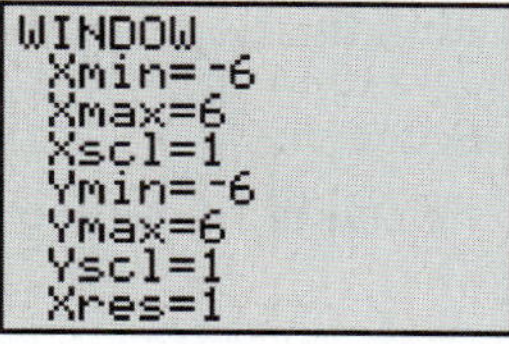

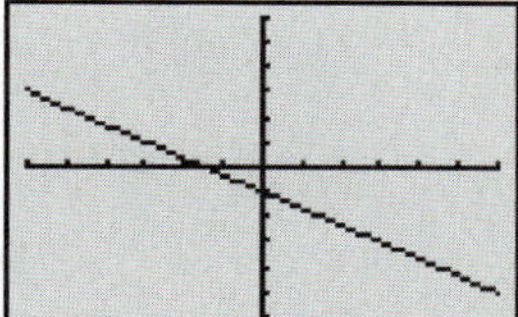

Figure 74 Verify the work

Finding an Equation of a Line from a Graph

To find an equation of a line from a graph,

1. Determine the slope m and the y-intercept $(0, b)$ from the graph.
2. Substitute your values for m and b into the equation $y = mx + b$.

Parallel Lines

Two lines are called **parallel** if they do not intersect (see Fig. 75). How do the slopes of two parallel lines compare?

Example 9 Find the Slopes of Two Parallel Lines

Find the slopes of the parallel lines l_1 and l_2 sketched in Fig. 76.

Solution

For both lines, if the run is 2, the rise is 1 (see Fig. 77). So, the slope of both parallel lines is

$$m = \frac{\text{rise}}{\text{run}} = \frac{1}{2}$$

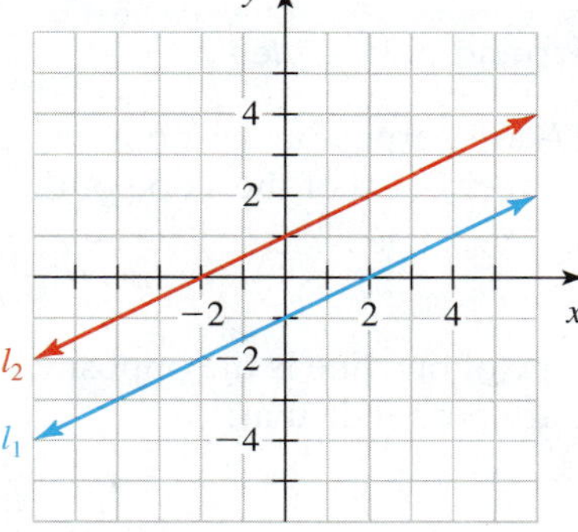

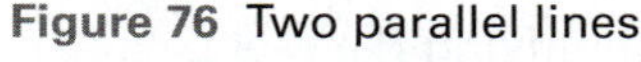
Figure 76 Two parallel lines

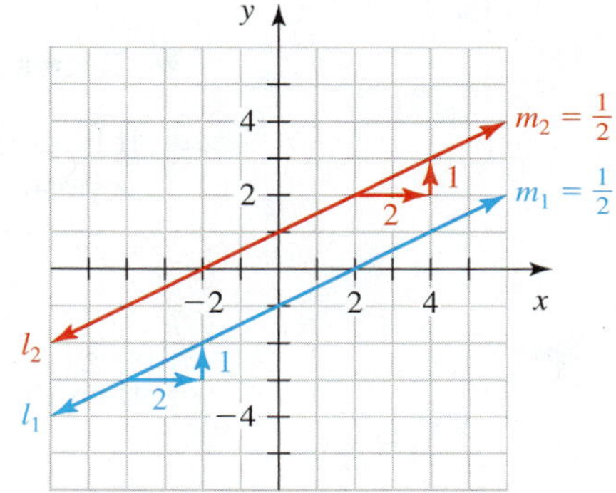

Figure 77 Calculate the slopes of the parallel lines

It makes sense that parallel nonvertical lines have equal slope, since parallel lines have the same steepness.

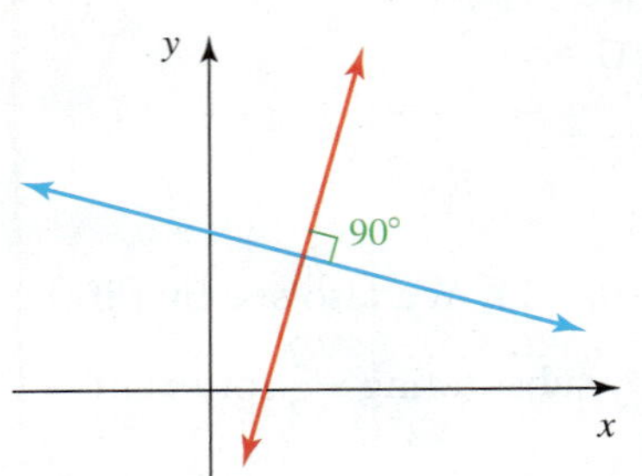

Figure 78 Two perpendicular lines

Slopes of Parallel Lines

If lines l_1 and l_2 are parallel nonvertical lines, then the slopes of the lines are equal:

$$m_1 = m_2$$

Also, if two distinct lines have equal slope, then the lines are parallel.

Perpendicular Lines

Two lines are called **perpendicular** if they intersect at a 90° angle (see Fig. 78). How do the slopes of two perpendicular lines compare?

Example 10 Find the Slopes of Two Perpendicular Lines

Find the slopes of the perpendicular lines l_1 and l_2 sketched in Fig. 79.

Solution

From Fig. 80 we see that the slope of line l_1 is

$$m_1 = \frac{2}{3}$$

and the slope of line l_2 is

$$m_2 = \frac{-3}{2} = -\frac{3}{2}$$

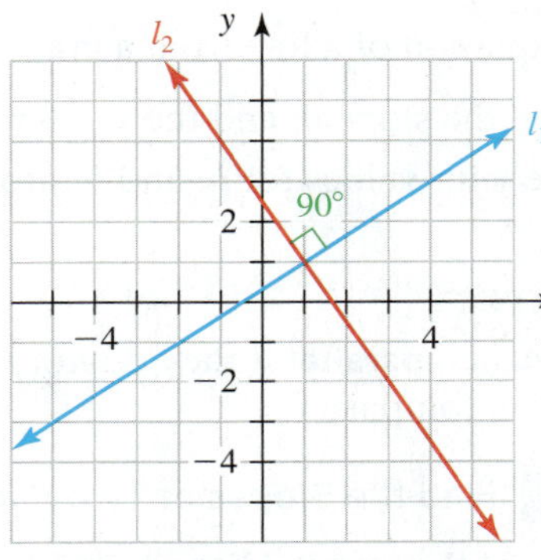

Figure 79 Two perpendicular lines

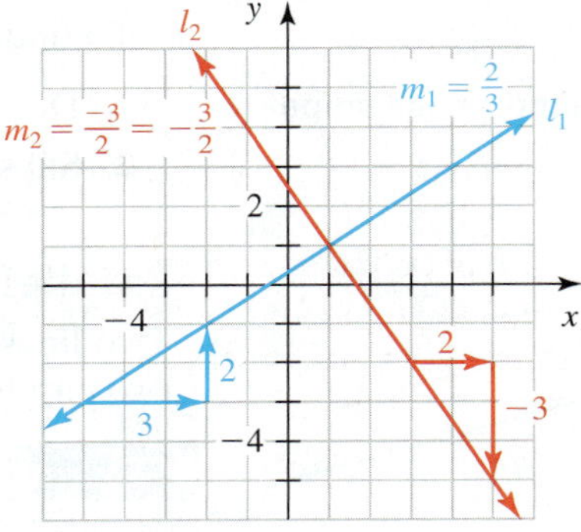

Figure 80 Calculate the slopes of the perpendicular lines

In Example 10, the slope $-\frac{3}{2}$ is the opposite of the reciprocal of the slope $\frac{2}{3}$. A similar relationship applies to any pair of perpendicular nonvertical lines.

Slopes of Perpendicular Lines

If lines l_1 and l_2 are perpendicular nonvertical lines, then the slope of one line is the opposite of the reciprocal of the slope of the other line:

$$m_2 = -\frac{1}{m_1}$$

Also, if the slope of one line is the opposite of the reciprocal of another line's slope, then the lines are perpendicular.

Example 11 Identifying Parallel Lines and Perpendicular Lines

Determine whether the pair of lines is parallel, perpendicular, or neither.

1. $y = 3x - 5$ and $y = 3x + 2$

2. $y = \frac{8}{5}x - 3$ and $y = -\frac{5}{8}x + 6$

Solution

1. The slope of each line is 3. Since the slopes of the lines are equal, the lines are parallel. We use ZStandard followed by ZSquare to verify this in Fig. 81.

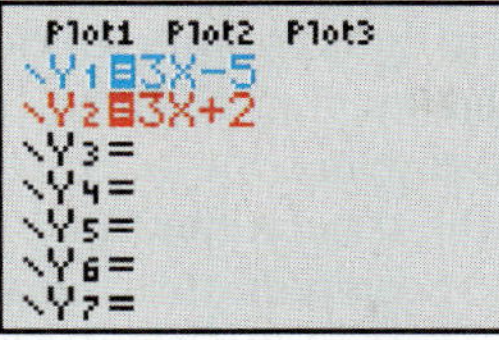

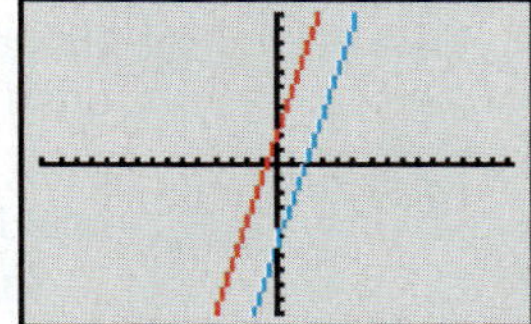

Figure 81 The lines appear to be parallel

2. The slopes of the lines are $\frac{8}{5}$ and $-\frac{5}{8}$. Since the slope $-\frac{5}{8}$ is the opposite of the reciprocal of the slope $\frac{8}{5}$, the lines are perpendicular. We use ZStandard followed by ZSquare to verify this in Fig. 82.

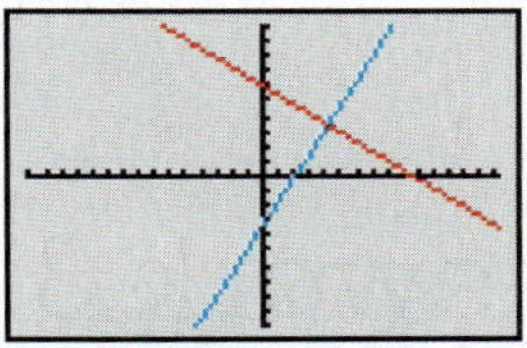

Figure 82 The lines appear to be perpendicular

Group Exploration

Drawing lines with various slopes

1. On a graphing calculator, graph a group of lines (a *family of lines*) to make a starburst like the one in Fig. 83. List the equations of your lines.

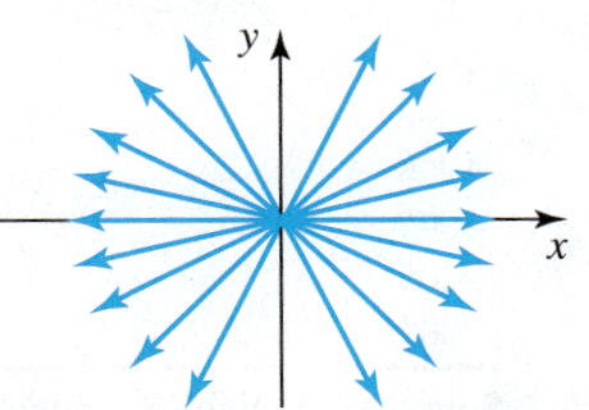

Figure 83 A starburst

2. On a graphing calculator, graph a family of lines to make a starburst like the one in Fig. 84. List the equations of your lines.

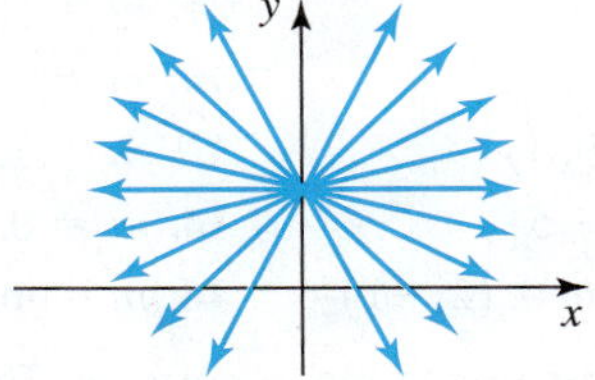

Figure 84 Another starburst

3. Summarize what you have learned about slope from this exploration, this section, and Section 3.3.

Group Exploration

Finding an equation of a line from its graph

In this exploration, you will need to work with one other student.

1. First, both you and your partner should create a linear equation of the form $y = mx + b$, but do not show each other your equations. For m, use a rational number between -3 and 3. For b, use an integer between -3 and 3.
2. Use ZDecimal to graph the equation.
3. Exchange graphing calculators.
4. By inspecting the graph, determine the equation that your partner created. You can use TRACE to find coordinates of a point on your line. Or you can use a grid by pressing [2nd] [FORMAT]. Press [▽] twice, then press [▷], and then press [ENTER]. The "GridOn" choice should now be highlighted. Next, press [GRAPH]. A grid should now be visible.
5. Press [Y=] to check the equation you found.

Tips for Success Visualize

To help prepare themselves mentally and physically for competition, many exceptional athletes visualize themselves performing well at their event many times throughout their training period. For example, a runner training for the 100-meter dash might imagine getting set in the starting blocks, taking off right after the gun goes off, being in front of the other runners, and so on, right up until the moment of breaking the tape at the finish line.

In an experiment, three groups of basketball players were used to test the effectiveness of visualization. The first group warmed up by shooting baskets before a game. The second group visualized shooting baskets, but did not shoot any baskets during the pregame warm-up. The third group did not warm up or visualize before the game. The visualization group not only outperformed the group that did not warm up, but also did better than the group that warmed up by shooting baskets!

You can do visualizations, too. Visualize doing all the things you feel you need to do to succeed in the course. If you do this regularly, you will have better follow-through with what you intend to do. You will also feel more confident about succeeding.

Homework 3.4

For extra help ▶ MyMathLab®

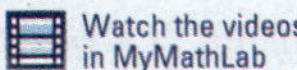

Sketch the line that has the given slope and contains the given point.

1. $m = \frac{2}{3}, (0, 1)$ **2.** $m = \frac{3}{5}, (0, 2)$

3. $m = -\frac{5}{2}, (0, 4)$ **4.** $m = -\frac{3}{4}, (0, -2)$

5. $m = -\frac{3}{2}, (0, 0)$ **6.** $m = -\frac{1}{2}, (0, 0)$

7. $m = 2, (0, 1)$ **8.** $m = 4, (0, -3)$

9. $m = -3, (0, -2)$ **10.** $m = -5, (0, 4)$

11. $m = -1, (0, 3)$ **12.** $m = 1, (0 - 2)$

13. $m = 0, (4, -5)$ **14.** $m = 0, (6, 3)$

15. m is undefined, $(2, -1)$ **16.** m is undefined, $(-1, -3)$

Determine the slope and the y-intercept. Then use the slope and the y-intercept to graph the equation by hand. Verify your work with a graphing calculator.

17. $y = \frac{2}{3}x - 1$ **18.** $y = \frac{1}{5}x + 2$ **19.** $y = -\frac{1}{3}x + 4$

20. $y = -\frac{3}{2}x - 1$ **21.** $y = \frac{4}{3}x + 2$ **22.** $y = \frac{5}{2}x - 3$

23. $y = -\frac{4}{5}x - 1$ **24.** $y = -\frac{1}{4}x + 2$ **25.** $y = \frac{1}{2}x$

26. $y = \frac{1}{4}x$ **27.** $y = -\frac{5}{3}x$ **28.** $y = -\frac{4}{5}x$

For Exercises 29–46, determine the slope and the y-intercept. Then use the slope and the y-intercept to graph the equation by hand. Verify your work by checking that two points on your line satisfy the given equation.

29. $y = 4x - 2$ **30.** $y = 2x - 4$ **31.** $y = -2x + 4$

32. $y = -3x + 5$ **33.** $y = -4x - 1$ **34.** $y = -2x - 2$

35. $y = x + 1$ **36.** $y = x - 4$ **37.** $y = -x + 3$

38. $y = -x + 2$ **39.** $y = -3x$ **40.** $y = 4x$

41. $y = x$ **42.** $y = -x$ **43.** $y = -3$

44. $y = -2$ **45.** $y = 0$ **46.** $y = 1$

47. The percentages of degrees in medicine that are earned by women are shown in Table 26 for various years.

Table 26 Percentage of Degrees in Medicine Earned by Women

Year	Percent
1985	30.4
1990	34.2
1995	38.8
2000	42.7
2005	47.3
2009	48.9

Source: *U.S. National Center for Education Statistics*

Let p be the percentage of degrees in medicine that are earned by women at t years since 1985. A reasonable model of this situation is

$$p = 0.8t + 30.5$$

a. Graph the model by hand.

b. Predict when 56% of the degrees in medicine will be earned by women. Explain by using arrows on your graph in part (a).

48. The average monthly employee contribution required to cover a family in an employer-sponsored health plan has increased greatly since 2000 (see Table 27).

Let a be the average monthly employee contribution (in dollars) required to cover a family in an employer-sponsored health plan at t years since 2000. A reasonable model of the situation is

$$a = 19t + 139$$

Table 27 Costs to Cover a Family in an Employer-Sponsored Health Plan

Year	Average Monthly Employee Contribution (dollars)
2000	135
2002	178
2004	222
2006	248
2008	280
2010	333
2011	344

Source: *Kaiser Family Foundation*

a. Graph the model by hand.
b. Use your graph to predict when the average monthly employee contribution will be $480. Explain by using arrows on your graph in part (a).

49. J. D. Power's Initial Quality Study surveys about 40,000 car buyers 3 years after they purchase a new car. Participants grade their car or truck on 135 attributes, such as trim quality, wind noise, and fuel economy. The average number of problems incurred per 100 vehicles during the past 12 months is shown in Table 28 for various years.

Table 28 Average Number of Vehicle Problems per 100 Vehicles during the Past 12 Months

Year	Average Number of Vehicle Problems
2008	206
2009	170
2010	155
2011	151
2012	132

Source: *J. D. Power & Associates*

Let n be the average number of problems incurred per 100 vehicles during the past 12 months at t years since 2005. A reasonable model of the situation is

$$n = -17t + 246$$

a. Graph the model by hand.
b. In 2012, the brand Lexus® had the fewest problems, 86 per 100 cars during the past year. Predict when the average number of problems for all vehicles (of any brand) will reach that level. Explain by using arrows on your graph in part (a).

50. The number of injuries on farms to people under age 20 has dropped significantly since 1998 (see Table 29).

Let n be the number (in thousands) of injuries on farms to people under age 20 in the year that is t years since 1995. A reasonable model of this situation is

$$n = -1.9t + 43$$

a. Graph the model by hand.
b. Predict in which year 3 thousand people under age 20 will be injured on farms. Explain by using arrows on your graph in part (a).

Table 29 Numbers of Injuries on Farms to People under Age 20

Year	Number of Injuries (in thousands)
1998	38
2001	29
2004	27
2006	23
2009	16

Source: *National Institute for Occupational Heath and Safety*

51. Graphs of four linear equations are shown in Fig. 85. State whether m and b are positive, negative, zero, or undefined for the $y = mx + b$ form of each equation.

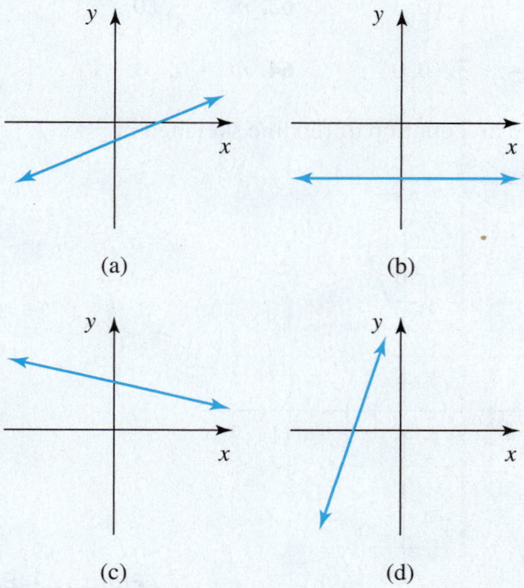

Figure 85 Exercise 51

52. Graphs of four linear equations are shown in Fig. 86. State the signs of the constants m and b for the $y = mx + b$ form of each equation.

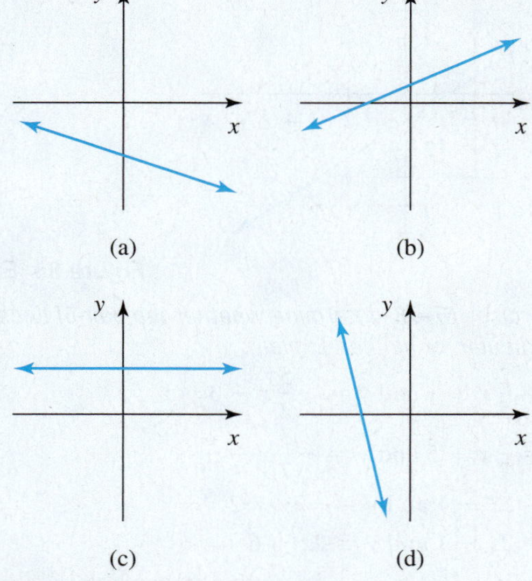

Figure 86 Exercise 52

Graph by hand an equation of the form $y = mx + b$ that meets the given criteria for m and b. Also, find an equation of each graph.

53. m is positive and b is positive

54. m is positive and b is negative

55. m is negative and b is negative

56. m is negative and $b = 0$

57. $m = 0$ and b is negative

58. $m = 0$ and $b = 0$

For Exercises 59–64, find an equation of a line that has the given slope and contains the given point.

59. $m = 3, (0, -4)$ **60.** $m = -2, (0, 5)$

61. $m = -\frac{6}{5}, (0, 3)$ **62.** $m = \frac{3}{4}, (0, -2)$

63. $m = -\frac{2}{7}, (0, 0)$ **64.** $m = 0, (0, -1)$

65. Find an equation of the line sketched in Fig. 87.

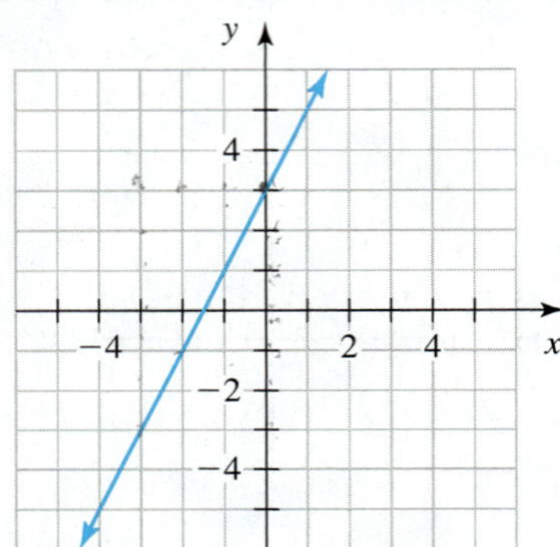

Figure 87 Exercise 65

66. Find an equation of the line sketched in Fig. 88.

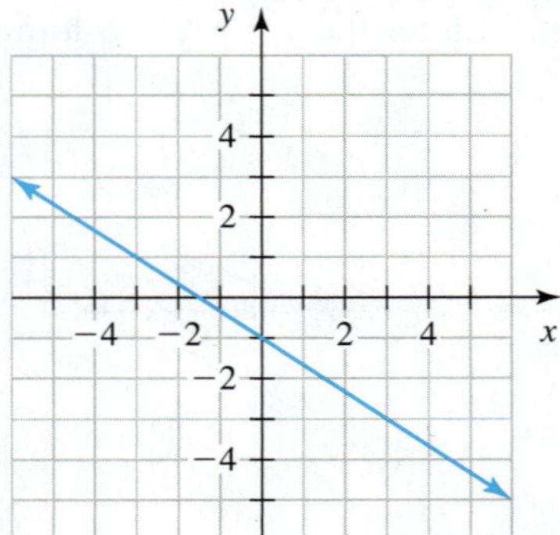

Figure 88 Exercise 66

For Exercises 67–78, determine whether the pair of lines is parallel, perpendicular, or neither. Explain.

67. $y = \frac{2}{5}x + 1$ and $y = -\frac{5}{2}x - 3$

68. $y = \frac{3}{7}x + 5$ and $y = -\frac{3}{7}x - 6$

69. $y = 3x - 1$ and $y = -3x + 2$

70. $y = 2x - 1$ and $y = 2x + 6$

71. $y = -4x + 2$ and $y = -4x + 3$

72. $y = -7x + 5$ and $y = \frac{1}{7}x - 8$

73. $y = \frac{2}{3}x - 1$ and $y = \frac{3}{2}x + 3$

74. $y = -\frac{4}{11}x + 2$ and $y = \frac{11}{4}x + 5$

75. $y = 2$ and $y = -4$ **76.** $x = -2$ and $y = 1$

77. $x = 0$ and $y = 0$ **78.** $x = -2$ and $x = 4$

79. Are the lines sketched in Fig. 89 parallel? Explain.

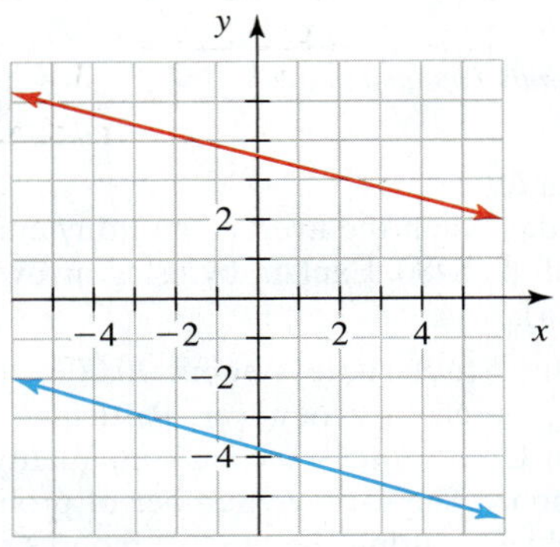

Figure 89 Exercise 79

80. Are the lines sketched in Fig. 90 perpendicular? Explain.

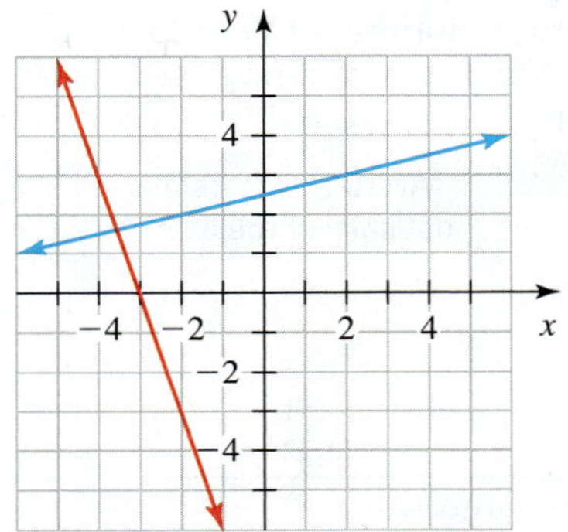

Figure 90 Exercise 80

For Exercises 81–86, refer to Fig. 91.

81. Find y when $x = -3$. **82.** Find x when $y = -3$.

83. Find x when $y = 0$. **84.** Find y when $x = 0$.

85. Find the slope of the line. **86.** Find an equation of the line.

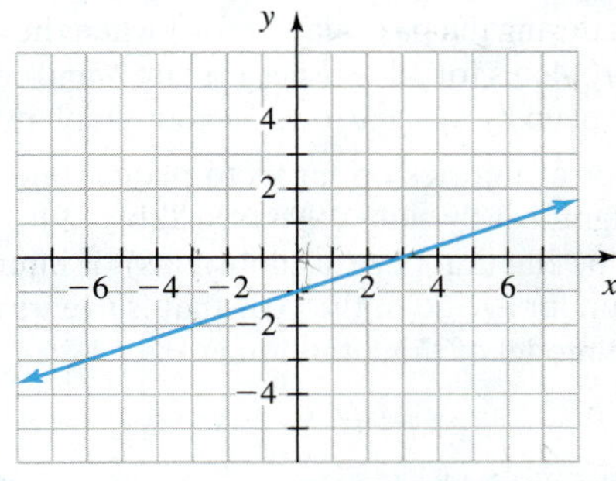

Figure 91 Exercises 81–86

Concepts

87. A student says that the slope of the line $y = 2x + 1$ is $2x$. Is the student correct? Explain.

88. A student says that the y-intercept of the line $y = 3x - 2$ is $(0, 2)$. Is the student correct? Explain.

89. Use the slope and y-intercept of the line $y = 2x - 1$ to graph the equation $y = 2x - 1$ by hand. Then choose two points that lie on the graph, and show that both of the corresponding ordered pairs are solutions of the equation $y = 2x - 1$.

90. Use the slope and y-intercept of the line $y = -2x + 3$ to graph the equation $y = -2x + 3$ by hand. Then choose two points that lie on the graph, and show that both of the corresponding ordered pairs are solutions of the equation $y = -2x + 3$.

91. Recall that we can describe some or all of the solutions of an equation in two variables with an equation, a table, a graph, or words.

a. Describe the solutions of $y = \frac{1}{2}x + 2$ by using a graph.

b. Describe three solutions of $y = \frac{1}{2}x + 2$ by using a table.

c. Describe the solutions of $y = \frac{1}{2}x + 2$ by using words.

92. Recall that we can describe some or all of the solutions of an equation in two variables with an equation, a table, a graph, or words.

a. Describe the solutions of $y = \frac{1}{3}x - 1$ by using a graph.

b. Describe three solutions of $y = \frac{1}{3}x - 1$ by using a table.

c. Describe the solutions of $y = \frac{1}{3}x - 1$ by using words.

93. A student tries to graph the equation $y = -3x + 1$ (see Fig. 92).

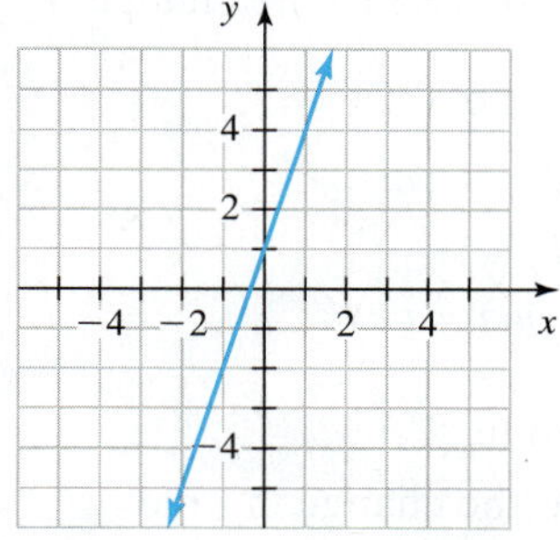

Figure 92 Exercise 93

Choose two points that lie on the line, and show that at least one of these points is not a solution of $y = -3x + 1$. Explain why your work shows that Fig. 92 is incorrect. Then sketch the correct graph.

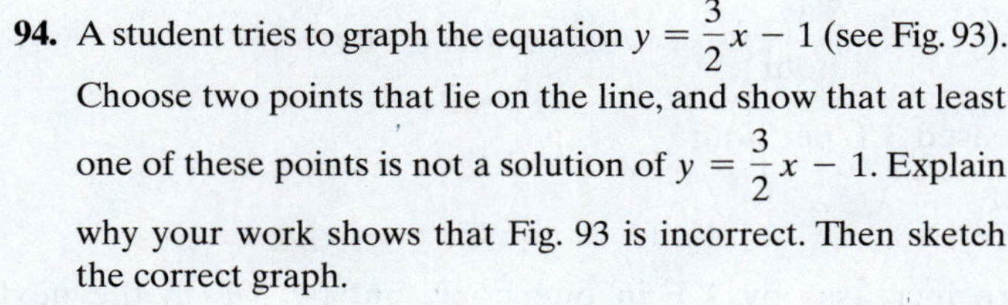

94. A student tries to graph the equation $y = \frac{3}{2}x - 1$ (see Fig. 93). Choose two points that lie on the line, and show that at least one of these points is not a solution of $y = \frac{3}{2}x - 1$. Explain why your work shows that Fig. 93 is incorrect. Then sketch the correct graph.

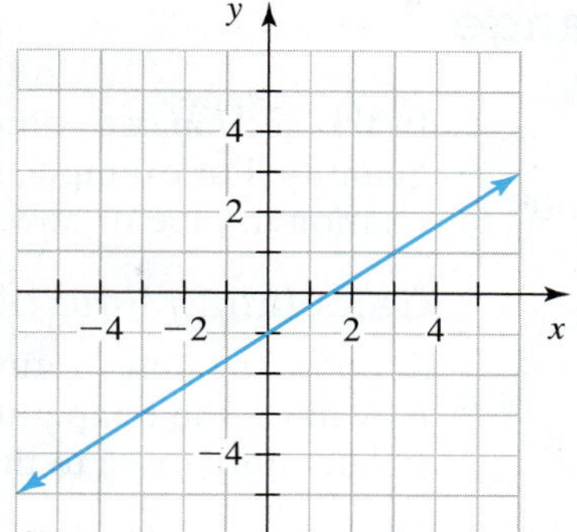

Figure 93 Exercise 94

95. A line passes through the point $(-1, 1)$ and has a slope of 2.

a. Sketch the line.

b. Find an equation of the line. [**Hint:** Use part (a) to find b in $y = mx + b$.]

96. A line passes through the point $(1, 2)$ and has a slope of -3.

a. Sketch the line.

b. Find an equation of the line. [**Hint:** Use part (a) to find b in $y = mx + b$.]

97. a. Find the slope of each line: $y = 3$, $y = -5$, and $y = 0$.

b. Find the slope of the graph of any linear equation of the form $y = k$, where k is a constant.

98. a. Find the slope of each line: $x = 2$, $x = -4$, and $x = 0$.

b. Find the slope of the graph of any equation of the form $x = k$, where k is a constant.

99. Graphs of the equations $y = mx + b$ and $y = kx + c$ (where m, b, k, and c are constants) are sketched in Fig. 94.

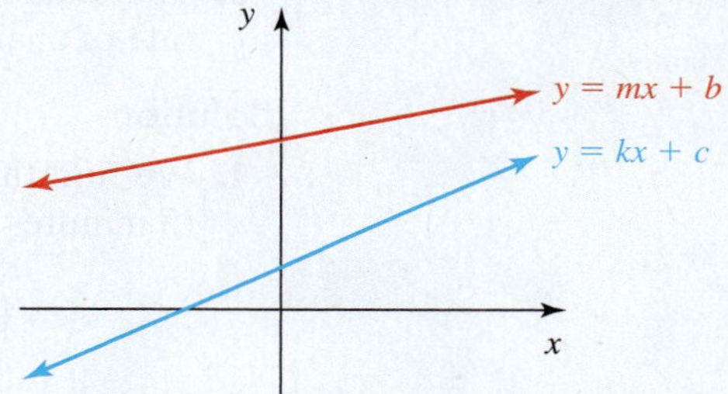

Figure 94 Exercise 99

a. Which is greater, m or k? Explain.

b. Which is greater, b or c? Explain.

100. Explain why distinct lines with equal slopes are parallel.

101. A line passes through the point $(0, -3)$ and has slope 2. What is an equation of the line?

102. Create an equation of the form $y = mx + b$. Find two points on the graph of your equation by substituting two values for x. Use the two points to calculate the slope of the line and compare the result with your chosen value of m.

103. Create an equation of the form $y = mx + b$. Find two points on the graph of your equation by substituting two values for x. Use the two points to calculate the slope of the line, and compare the result with your chosen value of m.

104. Describe how to use the method discussed in this section to graph an equation of the form $y = mx + b$.

3.5 Rate of Change

Objectives

- Calculate the rate of change of a quantity.
- Understand why slope is a rate of change.
- Use rate of change to help find a linear model.
- Know the slope addition property.

In this section, we will describe how quickly a quantity changes in relation to another quantity. For example, we can describe how quickly the tuition of a college increases in relation to time or how quickly the altitude of an airplane declines in relation to time.

Calculating Rate of Change

Suppose the temperature increased *steadily* by 6°F in the past 3 hours. We can compute how much the temperature changed *per hour* by finding the unit ratio of the change in temperature (6°F) to the change in time (3 hours):

$$\frac{6°\text{F}}{3\text{ hours}} = \frac{2°\text{F}}{1\text{ hour}}$$

So, the temperature increased by 2°F per hour. This is an example of a *rate of change*. We say the rate of change of temperature with respect to time is 2°F per hour. The rate of change is a *constant* because the temperature increased *steadily*.

Here are some other examples of rates of change:

- The number of friends on a student's Facebook account increases by 10 friends per month.
- The revenue of a company decreases by $3 million per year.
- A college charges $70 per unit (hour or credit).

Example 1 Finding Rates of Change

1. An airplane's altitude increases steadily by 10,000 feet over a 5-minute period. Find the rate of change of altitude with respect to time.
2. The temperature increased steadily from 60°F at 6 A.M. to 72°F at 10 A.M. Find the rate of change of temperature with respect to time.

Solution

1. We find the unit ratio of the change in altitude (10,000 feet) to the change in time (5 minutes):

$$\frac{\text{change in altitude}}{\text{change in time}} = \frac{10{,}000\text{ feet}}{5\text{ minutes}} \quad \textit{Find ratio.}$$

$$= \frac{2000\text{ feet}}{1\text{ minute}} \quad \textit{Find unit ratio.}$$

So, the airplane climbed at a rate of 2000 feet per minute.

2. We find the unit ratio of the change in temperature to the change in time:

$$\frac{\text{change in temperature}}{\text{change in time}} = \frac{72°\text{F} - 60°\text{F}}{10{:}00 - 6{:}00} \quad \textit{Change in a quantity is ending amount minus beginning amount.}$$

$$= \frac{12°\text{F}}{4\text{ hours}} \quad \textit{Subtract.}$$

$$= \frac{3°\text{F}}{1\text{ hour}} \quad \textit{Find unit ratio.}$$

So, the temperature increased 3°F per hour.

Suppose the temperature increases by 3°F in one hour, but by 5°F in the next hour. We can find the *average rate of change* of temperature with respect to time by finding the unit ratio of the *total* change in temperature (8°F) to the *total* change in time (2 hours):

$$\frac{8°\text{F}}{2\text{ hours}} = \frac{4°\text{F}}{1\text{ hour}}$$

So, the average rate of change is 4°F per hour.

Formula for Rate of Change and Average Rate of Change

Suppose that a quantity y changes steadily from y_1 to y_2 as a quantity x changes steadily from x_1 to x_2. Then the **rate of change** of y with respect to x is the ratio of the change in y to the change in x:

$$\frac{\text{change in } y}{\text{change in } x} = \frac{y_2 - y_1}{x_2 - x_1}$$

If either quantity does not change steadily, then the preceding formula is the **average rate of change** of y with respect to x.

We often refer to rate of change *with respect to time* simply as "rate of change."

Example 2 Finding Rates of Change

1. The number of pedestrian fatalities in the United States declined approximately steadily from 4795 fatalities in 2006 to 4092 fatalities in 2009 (Source: *U.S. National Highway Traffic Safety Administration*). Find the average rate of change of the number of pedestrian fatalities.
2. The total cost of 12 karate classes and an enrollment fee is \$158. The total cost of 20 karate classes and the same enrollment fee is \$230. The charge per class is the same, regardless of the number of classes for which you pay. Find the rate of change of the total cost with respect to the number of classes.

Solution

1.
$$\frac{\text{change in number of fatalities}}{\text{change in time}}$$
$$= \frac{4092\text{ fatalities} - 4795\text{ fatalities}}{\text{year }2009 - \text{year }2006}$$ *Change in a quantity is ending amount minus beginning amount.*
$$= \frac{-703\text{ fatalities}}{3\text{ years}}$$ *Subtract.*
$$\approx \frac{-234\text{ fatalities}}{1\text{ year}}$$ *Find unit ratio.*

The average rate of change of the number of pedestrian fatalities was about -234 fatalities per year. The number of fatalities had an average yearly decline of about 234 fatalities.

2. To be consistent in finding the signs of the changes, we assume that the number of classes increases from 12 to 20 and that the total cost increases from \$158 to \$230:

$$\frac{\text{change in total cost}}{\text{change in number of classes}} = \frac{230\text{ dollars} - 158\text{ dollars}}{20\text{ classes} - 12\text{ classes}}$$ *Change in a quantity is ending amount minus beginning amount.*
$$= \frac{72\text{ dollars}}{8\text{ classes}}$$ *Subtract.*
$$= \frac{9\text{ dollars}}{1\text{ class}}$$ *Find unit ratio.*

The rate of change of the total cost with respect to the number of classes is \$9 per class. So, the cost of each class is \$9.

From our work in Example 2, we can see a connection between the sign of a rate of change and whether a quantity is increasing or decreasing. In Problem 2, the rate of change was *positive* because the total cost *increases* (as the number of classes increases). In Problem 1, the average rate of change was *negative* because the number of pedestrian fatalities was *decreasing* (as time increased).

Increasing and Decreasing Quantities

Suppose that a quantity p depends on a quantity t. Then

- If p increases steadily as t increases steadily, then the rate of change of p with respect to t is positive.
- If p decreases steadily as t increases steadily, then the rate of change of p with respect to t is negative.

Slope Is a Rate of Change

You may have noticed that the expression

$$\frac{y_2 - y_1}{x_2 - x_1}$$

we have been using to calculate rate of change is the same expression we use to calculate the slope of a line. In other words, the slope of a linear model is a rate of change. We will explore this important concept in Example 3.

Example 3 Comparing Slope with a Rate of Change

Table 30 Times and Distances

Time (hours) t	Distance (miles) d
0	0
1	50
2	100
3	150
4	200
5	250

Suppose that a student travels at a constant rate on a road trip. Let d be the distance (in miles) that the student can drive in t hours. Some values of t and d are shown in Table 30.

1. Create a scattergram. Then draw a linear model.
2. Find the slope of the linear model.
3. Find the rate of change of the distance traveled in each given period. Compare each result with the slope of the linear model.
   a. From $t = 3$ to $t = 4$ **b.** From $t = 0$ to $t = 5$

Solution

1. We draw a scattergram and then draw a line that contains the data points (see Fig. 95).

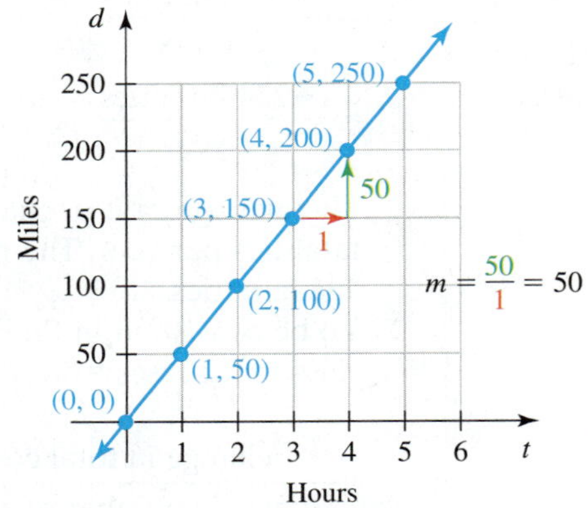

Figure 95 Car model and scattergram

2. Recall from Section 3.3 that the formula for slope is $m = \frac{y_2 - y_1}{x_2 - x_1}$. So, with the variables t and d, we have

$$\frac{d_2 - d_1}{t_2 - t_1}$$

We arbitrarily use the points (3, 150) and (4, 200) to calculate the slope of the linear model:

$$\frac{200 - 150}{4 - 3} = \frac{50}{1} = 50$$

So, the slope is 50. This checks with the calculation shown in Fig. 95.

3. a. First we calculate the rate of change of distance traveled from $t = 3$ to $t = 4$:

$$\frac{\text{change in distance}}{\text{change in time}} = \frac{200 \text{ miles} - 150 \text{ miles}}{4 \text{ hours} - 3 \text{ hours}}$$ *Change in a quantity is ending amount minus beginning amount.*

$$= \frac{50 \text{ miles}}{1 \text{ hour}}$$ *Subtract.*

$$= 50 \text{ miles per hour}$$ *Divide.*

The rate of change (50 miles per hour) is equal to the slope (50).

b. Now we calculate the rate of change of distance traveled from $t = 0$ to $t = 5$:

$$\frac{\text{change in distance}}{\text{change in time}} = \frac{250 \text{ miles} - 0 \text{ miles}}{5 \text{ hours} - 0 \text{ hours}}$$ *Change in a quantity is ending amount minus beginning amount.*

$$= \frac{250 \text{ miles}}{5 \text{ hours}}$$ *Subtract.*

$$= \frac{50 \text{ miles}}{1 \text{ hour}}$$ *Find unit ratio.*

$$= 50 \text{ miles per hour}$$ *Divide.*

The rate of change (50 miles per hour) is equal to the slope (50).

In Example 3, we found that the time t and the distance traveled d are linearly related. We also found that the slope of the linear model is equal to the rate of change of distance traveled with respect to time.

Slope Is a Rate of Change

If there is a linear relationship between the quantities t and p, and if p depends on t, then the slope of the linear model is equal to the rate of change of p with respect to t.

In Problem 3 of Example 3, we calculated the same rate of change (50 miles per hour) for two different periods. In fact, the rate of change is 50 miles per hour for *any* period within the first five hours. This makes sense because the rate of change is equal to the slope of the line (50), which is a constant.

Constant Rate of Change

Suppose that a quantity p depends on a quantity t. Then

- If there is a linear relationship between t and p, then the rate of change of p with respect to t is constant.
- If the rate of change of p with respect to t is constant, then there is a linear relationship between t and p.

Finding an Equation of a Linear Model

We can use what we have learned about rate of change to help us find an equation of a linear model.

Example 4 Finding a Model

In 2000, a college's enrollment was 20 thousand students. Each year, the enrollment increases by 2 thousand students. Let E be the enrollment (in thousands of students) at t years since 2000.

1. Is there a linear relationship between t and E? Explain.
2. Find the E-intercept of a linear model. What does it mean in this situation?
3. Find the slope of the linear model. What does it mean in this situation?
4. Find an equation of the linear model.

Solution

Table 31 College Enrollments

Years since 2000 t	Enrollment (thousands of students) E
0	20
1	22
2	24
3	26
4	28
5	30

1. Since the rate of change of enrollment per year is a *constant* 2 thousand students per year, the variables t and E are linearly related.
2. We list some values of t and E in Table 31. We plot the corresponding points and sketch the line that contains the points in Fig. 96.

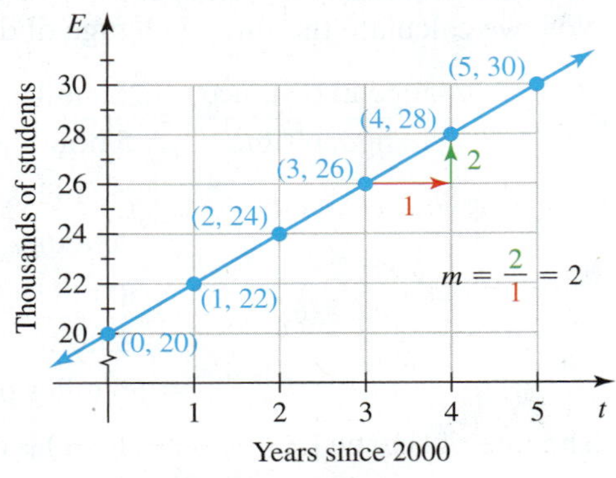

Figure 96 Enrollment scattergram and model

From the table and the graph, we see that the E-intercept is (0, 20). This means that the enrollment was 20 thousand students in the year 2000.

3. The rate of change of enrollment is 2 thousand students per year. So, the slope of the linear model is 2. This checks with the calculation shown in Fig. 96.
4. An equation of a line can be written in slope–intercept form, $y = mx + b$. Using t and E, we have $E = mt + b$. Since the slope is 2 and the E-intercept is (0, 20), we have $E = 2t + 20$.

To check our work with a graphing calculator, we begin by entering our model (see Fig. 97). Then we check that the entries in the graphing calculator table in Fig. 98 equal the entries in Table 31. We also check that the graph of our equation contains the points of the scattergram of the data (see Fig. 99).

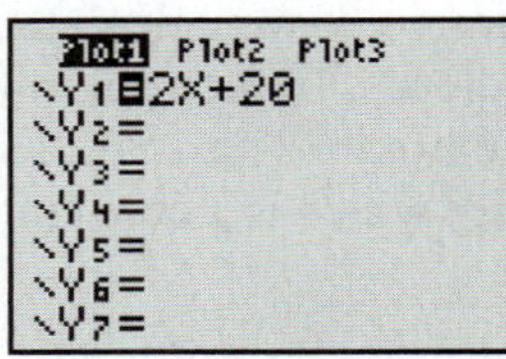

Figure 97 Enter the model

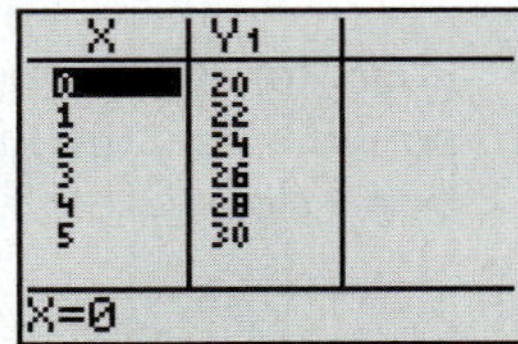

Figure 98 Use a table to verify the model

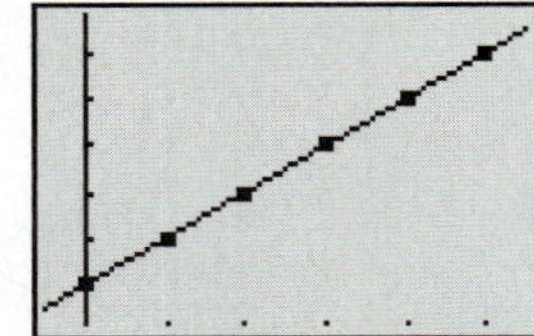

Figure 99 Use a graph to verify the model

For graphing calculator instructions on creating tables, see Appendix A.13. For instructions on creating scattergrams, see Appendix A.8 and A.10.

In Example 4, we found the model $E = 2t + 20$. Here is the connection between parts of the equation and the situation:

$$E = \underbrace{2}_{\substack{\text{rate of}\\ \text{change of}\\ \text{enrollment}}} \cdot t + \underbrace{20}_{\substack{\text{enrollment at}\\ t = 0}}$$

Example 5 Finding a Model

The temperature at which water boils (the *boiling point*) depends on elevation: the higher you are, the lower the boiling point is. At sea level (elevation 0), water boils at 212°F. The boiling point declines by 5.9°F for each thousand-meter increase in elevation (Source: *Thermodynamics, an Engineering Approach* by Cengal & Boles). Let B be the boiling point (in degrees Fahrenheit) at an elevation of E thousand meters.

1. Is there a linear relationship between E and B? If so, find the slope.
2. Find an equation of the model.
3. Perform a unit analysis of the equation.
4. Mount McKinley, the highest mountain in the United States, reaches 6.194 thousand meters. What is the boiling point of water at the peak?

Solution

1. Since the boiling point depends on the elevation, we will consider the rate of change of the boiling point with respect to elevation. Because this rate of change is a *constant* −5.9°F per thousand meters, the variables E and B are linearly related and the slope is −5.9.
2. We want an equation of the form $B = mE + b$. Since $B = 212$ when $E = 0$, the B-intercept is (0, 212). Recall that $m = -5.9$, so an equation is $B = -5.9E + 212$.
3. Here is a unit analysis of the equation $B = -5.9E + 212$:

$$\underbrace{B}_{\text{degrees Fahrenheit}} = \underbrace{-5.9}_{\frac{\text{degrees Fahrenheit}}{\text{thousand meters}}} \cdot \underbrace{E}_{\text{thousand meters}} + \underbrace{212}_{\text{degrees Fahrenheit}}$$

We can use the fact that $\frac{\text{thousand meters}}{\text{thousand meters}} = 1$ to simplify the units of the expression on the right-hand side of the equation:

$$\frac{\text{degrees Fahrenheit}}{\text{thousand meters}} \cdot \text{thousand meters} + \begin{matrix}\text{degrees}\\ \text{Fahrenheit}\end{matrix} = \begin{matrix}\text{degrees}\\ \text{Fahrenheit}\end{matrix} + \begin{matrix}\text{degrees}\\ \text{Fahrenheit}\end{matrix}$$

So, the units of the expressions on both sides of the equation are degrees Fahrenheit, which suggests that our equation is correct.

4. To find the boiling point, we substitute the input 6.194 for E in the equation $B = -5.9E + 212$:

$$B = -5.9(6.194) + 212 \quad \textit{Substitute 6.194 for E.}$$
$$\approx 175.46 \quad \textit{Perform indicated operations.}$$

So, the boiling point is about 175°F at the peak of Mount McKinley.

Table 32 Yogurt Sales

Year	Sales (billions of dollars)
2001	2.3
2003	2.7
2005	3.0
2008	3.5
2010	4.1

Source: *ACNielsen*

Example 6 Analyzing a Model

Yogurt sales in the United States are shown in Table 32 for various years. Let s be annual yogurt sales (in billions of dollars) at t years since 2000. A model of the situation is

$$s = 0.19t + 2.09$$

1. Use a graphing calculator to draw a scattergram and the model in the same viewing window. Check whether the line comes close to the data points.
2. What is the slope of the model? What does it mean in this situation?
3. Find the rates of change in sales from one year in Table 32 to the next one listed. Compare the rates of change with the result in Problem 2.
4. What is the s-intercept? What does it mean in this situation?
5. Predict the sales in 2018.

Solution

1. We draw the scattergram and the model in the same viewing window (see Fig. 100). For graphing calculator instructions on drawing scattergrams and models, see Appendix A.8 and A.10.

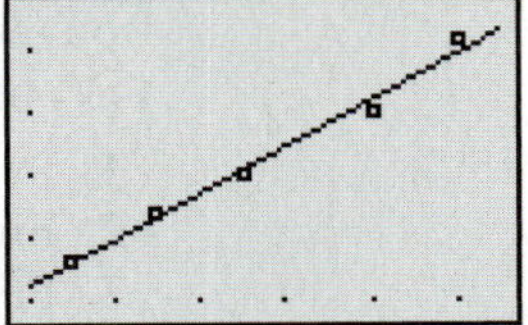

Figure 100 Yogurt scattergram and model

The line comes close to the data points, so the model is a reasonable one.

2. The slope is 0.19, because $s = 0.19t + 2.09$ is of the form $y = mx + b$ and $m = 0.19$. According to the model, sales are increasing by 0.19 billion dollars per year.
3. The rates of change of sales are shown in Table 33. All of the rates of change are fairly close to 0.19 billion dollars per year.

Table 33 Rates of Change of Yogurt Sales

Year	Sales (billions of dollars)	Rate of Change of Sales from Previous Year (billions of dollars per year)
2001	2.3	—
2003	2.7	$(2.7 - 2.3) \div (2003 - 2001) = 0.20$
2005	3.0	$(3.0 - 2.7) \div (2005 - 2003) = 0.15$
2008	3.5	$(3.5 - 3.0) \div (2008 - 2005) \approx 0.17$
2010	4.1	$(4.1 - 3.5) \div (2010 - 2008) = 0.30$

4. The s-intercept is $(0, 2.09)$, because $s = 0.19t + 2.09$ is of the form $y = mx + b$ and $b = 2.09$. According to the model, yogurt sales were \$2.09 billion in 2000.
5. We substitute the input 18 for t in the equation $s = 0.19t + 2.09$:

$$s = 0.19(18) + 2.09 = 5.51$$

According to the model, yogurt sales will be about \$5.5 billion in 2018.

In Example 6, we found that the slope of the yogurt model is 0.19, which means that, according to the model, yogurt sales have increased by 0.19 billion dollars per year. In reality, yogurt sales did not necessarily increase by 0.19 billion dollars in any of the years between 2001 and 2010, inclusive. However, 0.19 billion dollars per year *is* a reasonable estimate of the *average* yearly increase.

Slope Is an Average Rate of Change

If two quantities t and p are approximately linearly related and p depends on t, then the slope of a reasonable linear model is approximately equal to the average rate of change of p with respect to t.

WARNING A common error in describing the meaning of the slope of a model is vagueness. For example, a description such as

The slope means that it is increasing.

neither specifies the quantity that is increasing nor the rate of increase. The following statement includes the missing information:

The slope of 0.19 means sales are increasing by 0.19 billion dollars per year.

Slope Addition Property

So far, we have discussed rate of change for linear models. Now we will explore rate of change for linear equations that are not used as models.

Table 34 Some Solutions of $y = 2x + 1$

x	y
0	1
1	3
2	5
3	7
4	9

Table 35 Some Solutions of $y = -3x + 8$

x	y
0	8
1	5
2	2
3	−1
4	−4

Example 7 Interpreting Slope as a Rate of Change

What is the slope of the line $y = 2x + 1$? Interpret the slope as a rate of change.

Solution

The slope is 2. So, the rate of change of y with respect to x is 2. This means that if the value of x increases by 1, the value of y increases by 2 (see Table 34).

Example 8 Interpreting Slope as a Rate of Change

What is the slope of the line $y = -3x + 8$? Interpret the slope as a rate of change.

Solution

The slope is -3. So, the rate of change of y with respect to x is -3. This means that if the value of x increases by 1, the value of y decreases by 3 (see Table 35).

Our observations made in Examples 7 and 8 suggest yet another way to think about slope: the **slope addition property.**

Slope Addition Property

For a linear equation of the form $y = mx + b$, if the value of x increases by 1, then the value of y changes by the slope m. In other words, if an input increases by 1, the output changes by the slope.

For the line $y = 4x - 9$, the slope is 4. If the value of x increases by 1, then the value of y changes by 4. For the line $y = -5x - 2$, the slope is -5. If the value of x increases by 1, then the value of y changes by -5.

Example 9 Identifying Possible Linear Equations

Some solutions of four equations are listed in Table 36. Which of the equations could be linear?

Table 36 Some Solutions of Four Equations

Equation 1		Equation 2		Equation 3		Equation 4	
x	y	x	y	x	y	x	y
1	23	4	12	0	3	50	8
2	20	5	17	1	6	51	8
3	17	6	22	2	12	52	8
4	14	7	27	3	24	53	8
5	11	8	32	4	48	54	8

Solution

1. Equation 1 could be linear: Each time the value of x increases by 1, the value of y changes by -3.
2. Equation 2 could be linear: Each time the value of x increases by 1, the value of y changes by 5.
3. Equation 3 is not linear: Each time the value of x increases by 1, the value of y does not change by the same value.
4. Equation 4 could be $y = 8$, which is a linear equation: Each time the value of x increases by 1, the value of y changes by 0.

Table 37 Points on a Line

x	y
0	15
1	9
2	3
3	−3
4	−9

Example 10 Finding an Equation of a Line

A line contains the points listed in Table 37. Find an equation of the line.

Solution

For $y = mx + b$, the graph has y-intercept $(0, b)$. From Table 37, we see that the y-intercept is $(0, 15)$, so $b = 15$. As the value of x increases by 1, the value of y changes by -6. By the slope addition property, we know that $m = -6$. Therefore, an equation of the line is $y = -6x + 15$.

We use a graphing calculator table to verify our work in Fig. 101.

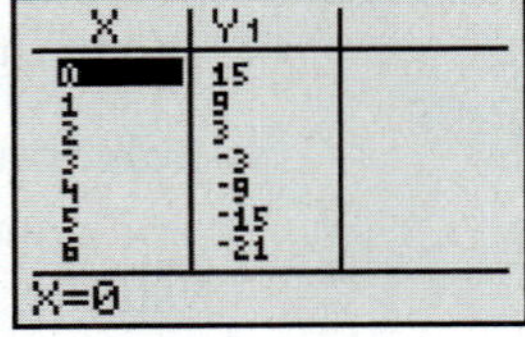

Figure 101 Verify the equation

Group Exploration

Averaging rates of change

The numbers of service members in the armed forces diagnosed as overweight are shown in Table 38 for various years.

Table 38 Numbers of Service Members Diagnosed as Overweight

Years	Number of Overweight Service Members (thousands)
2006	56
2007	62
2008	72
2009	79
2010	86

Source: *Defense Department*

1. Find the average rate of change of the number of overweight service members from 2006 to 2010. [**Hint:** Use the 2006 and 2010 dates in Table 38.]
2. Find the average rate of change in the number of overweight service members from each year to the next, beginning in 2006. [**Hint:** You should find four results.]
3. Find the average of the four rates of change you found in Problem 2. [**Hint:** Divide the sum of the four rates of change by 4.]
4. Compare your result in Problem 3 with your result in Problem 1.
5. The following expression is an example of a *telescoping sum*:

   $(62 - 56) + (72 - 62) + (79 - 72) + (86 - 79)$

   Explain why the above sum is equal to $86 - 56$.
6. Use your result in Problem 5 to help explain why the following statement is true:

   $$\frac{\dfrac{62 - 56}{2007 - 2006} + \dfrac{72 - 62}{2008 - 2007} + \dfrac{79 - 72}{2009 - 2008} + \dfrac{86 - 79}{2010 - 2009}}{4} = \frac{86 - 56}{2010 - 2006}$$

   Also, explain how this statement is related to your comparison in Problem 4.
7. Describe what you have learned from doing this exploration.

Group Exploration

Looking ahead: Laws of operations

1. **a.** Evaluate $a(b + c)$ for $a = 2, b = 3$, and $c = 5$.

   b. Evaluate $ab + ac$ for $a = 2, b = 3$, and $c = 5$.

   c. Compare your results for Problems 1 and 2.

   d. Evaluate both $a(b + c)$ and $ab + ac$ for $a = 4$, $b = 2$, and $c = 6$, and then compare the results.

   e. Evaluate both $a(b + c)$ and $ab + ac$ for values of your choosing for a, b, and c, and then compare the results.

   f. Make an educated guess as to whether the statement $a(b + c) = ab + ac$ is true for all numbers a, b, and c.

2. Evaluate both $a(bc)$ and $(ab)(ac)$ for values of your choosing for a, b, and c. Is the statement $a(bc) = (ab)(ac)$ true for all numbers a, b, and c?
3. Evaluate both $a(bc)$ and $(ab)c$ for values of your choosing for a, b, and c. Make an educated guess as to whether the statement $a(bc) = (ab)c$ is true for all numbers a, b, and c.
4. Evaluate both $a + b$ and $b + a$ for values of your choosing for a and b. Make an educated guess as to whether the statement $a + b = b + a$ is true for all numbers a and b.
5. What are the main points of this exploration?

Tips for Success **Get the Most Out of Working Exercises**

If you work an exercise by referring to a similar example in your notebook or in the text, it is a good idea to try the exercise again without referring to your source of help. If you need to refer to your source of help to solve the exercise a second time, consider trying the exercise a third time without help. When you finally complete the exercise without help, reflect on which concepts you used to work the exercise, where you had difficulty, and what the key idea was that opened the door of understanding for you.

A similar strategy can be used in getting help from a student, an instructor, or a tutor.

If this sounds like a lot of work, it is! But it is well worth it. Although it is important to complete each assignment, it is also important to learn as much as possible while progressing through it.

Homework 3.5

For extra help ▶ MyMathLab®

Watch the videos in MyMathLab

Download the MyDashboard App

1. A person's annual salary increases by \$12,400 over an 8-year period. Find the average rate of change of the salary.
2. A person downloaded 185 songs over a 5-year period. Find the average rate of change of the number of songs downloaded.
3. An airplane's altitude declines steadily by 24,750 feet over a 15-minute period. Find the rate of change of the airplane's altitude.
4. The temperature decreases steadily by 12°F over a 4-hour period. Find the rate of change of temperature.
5. The percentage of all e-mail that is spam increased approximately linearly from 30% in 2004 to 90% in 2010 (Source: *Symantec*). Find the average rate of change of this percentage.
6. The number of skier/snowboarder visits to U.S. slopes has increased approximately linearly from 57 million visits in 2005 to 61 million visits in 2011 (Source: *National Ski Areas Association*). Find the average rate of change of the number of visits.
7. The Steller sea lion population decreased approximately linearly from about 300 thousand in 1980 to about 40 thousand in 2012 (Source: *National Marine Mammal Laboratory*). Find the average rate of change of the Steller sea lion population.
8. Median household income (adjusted for inflation) decreased approximately linearly from \$52,823 in 2007 to \$49,445 in 2010 (Source: *U.S. Census Bureau*). Find the average rate of change of median household income.
9. The average number of fast-food meals consumed by a young adult (ages 18–24) decreased approximately linearly from 245 meals in 2006 to 192 meals in 2011 (Source: *The NPD Group*). Find the average rate of change of the annual number of fast-food meals a young adult eats.
10. The percentage of Americans who say they or their spouses have saved money for retirement decreased approximately linearly from 75% in 2009 to 66% in 2012 (Source: *Employee Benefit Research Institute*). Find the average rate of change of this percentage.
11. In-district students at Triton College pay \$882 for 9 credit hours (units) of classes and \$1176 for 12 credit hours of classes (Source: *Triton College*). Find the average rate of change of the total cost of classes with respect to the number of credit hours of classes.
12. In Manhattan, the average price of a two-bedroom house is \$1,694,547, and that of a four-bedroom house is \$6,315,496 (Source: *Trulia*). Find the average rate of change of price with respect to the number of bedrooms.
13. In order for a family living in New York to qualify for the health insurance program Family Health Plus, the family's annual income must be less than a maximum level. In 2012, the maximum income level of a three-person family was \$28,635 and the maximum income level of a seven-person family was \$52,395 (Source: *New York State Department of Health*). Find the average rate of change of maximum income level with respect to family size.

14. A person stacks some cups of uniform shape and size (one placed inside the next). The height of 3 stacked cups is 17.5 centimeters and of 5 stacked cups is 23.0 centimeters. Find the average rate of change of the height of the stacked cups with respect to the number of cups.

15. A company's revenue for 2010 was $3 million. Each year its revenue increases by $4 million. Let *r* be the annual revenue (in millions of dollars) at *t* years since 2010.

a. Is there a linear relationship between *t* and *r*? Explain. If the relationship is linear, find the slope and describe what it means in this situation.

b. Recall that we can describe a linear model using an equation, a table, a graph, or words.

i. Use an equation to describe the situation.

ii. Use a table of values of *t* and *r* to describe the situation.

iii. Use a graph to describe the situation.

16. As of February 1, a garage band knows 5 songs. Each week, the band members learn 2 more songs. Let *n* be the number of songs that the band knows at *t* weeks since February 1.

a. Is there a linear relationship between *t* and *n*? Explain. If the relationship is linear, find the slope and describe what it means in this situation.

b. Recall that we can describe a linear model using an equation, a table, a graph, or words.

i. Use an equation to describe the situation.

ii. Use a table of values of *t* and *n* to describe the situation.

iii. Use a graph to describe the situation.

17. In 2009, the number of drive-in movie sites in the United States was 381. The number has been decreasing by about 4.9 sites per year since then (Source: *National Association of Theatre Owners*). Let *n* be the number of drive-in movie sites at *t* years since 2009.

a. Is there an approximate linear relationship between *t* and *n*? Explain. If the relationship is approximately linear, find the slope and describe what it means in this situation.

b. What is the *n*-intercept of the model? What does it mean in this situation?

c. Find an equation of the model.

d. Perform a unit analysis of the equation you found in part (c).

e. Predict the number of sites in 2018.

18. Total spending for Father's Day gifts was $9.8 billion in 2010 and has increased by about $0.4 billion per year since then (Source: *National Retail Federation*). Let *s* be the annual total spending (in billions of dollars) for Father's Day gifts at *t* years since 2010.

a. Is there an approximate linear relationship between *t* and *s*? Explain. If the relationship is approximately linear, find the slope and describe what it means in this situation.

b. What is the *s*-intercept of the model? What does it mean in this situation?

c. Find an equation of the model.

d. Perform a unit analysis of the equation you found in part (c).

e. Predict the total spending in 2017.

19. A student's savings account has a balance of $4700 on September 1. Each month, the balance declines by $650. Let *B* be the balance (in dollars) at *t* months since September 1.

a. Find the slope of the linear model that describes this situation. What does it mean in this situation?

b. What is the *B*-intercept of the model? What does it mean in this situation?

c. Find an equation of the model.

d. Perform a unit analysis of the equation you found in part (c).

e. Find the balance on March 1 (6 months after September 1).

20. A person owns a propane-gas barbecue grill with a tank that holds 5 gallons of propane. The person always sets the temperature to 350°F, which uses 0.125 gallon of propane per hour. Let *g* be the number of gallons of propane that remain in the tank after *t* hours of cooking since the tank was last filled.

a. Find the slope of the linear model that describes this situation. What does it mean in this situation?

b. What is the *g*-intercept of the model? What does it mean in this situation?

c. Find an equation of the model.

d. The person fills the propane tank and then uses the grill for 3 hours. How much propane remains in the tank?

e. Estimate the amount of propane that will remain in the tank after 50 hours of cooking since the tank was last filled.

21. For the spring semester 2013, part-time students at Centenary College paid $550 per credit (unit or hour) for tuition and paid a mandatory part-time student fee of $15 per semester.

a. Find an equation of a linear model to describe the situation. Explain what your variables represent.

b. Perform a unit analysis of the equation you found in part (a).

c. What is the slope of your model? What does it mean in this situation?

d. What was the total one-semester cost of tuition plus part-time student fee for 9 credits of classes?

22. For the spring semester 2012, California residents paid an enrollment fee of $36 per unit (credit or hour) at Santa Barbara City College. Students were also required to pay a $17 health fee each semester.

a. Find an equation of a linear model to describe the situation. Explain what your variables represent.

b. Perform a unit analysis of the equation you found in part (a).

c. What is the slope of your model? What does it mean in this situation?

d. What was the total one-semester cost of tuition plus health fee for 15 units of classes?

23. A person drives her Toyota Prius® on a road trip. At the start of the trip, she fills up the 11.9-gallon tank with gasoline. During the trip, the car uses about 0.02 gallon of gasoline per mile. Let *G* be the number of gallons of gasoline remaining in the tank after driving *d* miles.

a. What is the slope of the linear model that describes this situation? What does it mean in this situation?

b. What is the *G*-intercept of the model? What does it mean in this situation?

c. Find an equation of the model.

d. Perform a unit analysis of the equation you found in part (c).

e. If the person drives 525 miles before she refuels the car, how much gasoline is required to fill up the tank?

24. Although the United States and Great Britain use the Fahrenheit (°F) temperature scale, most countries use the Celsius (°C) scale. The temperature reading 0°C is equivalent to the Fahrenheit reading 32°F. An increase of 1°C is equivalent to an increase of 1.8°F. Let F be the Fahrenheit reading that is equivalent to a Celsius reading of C degrees. Assume that F is the dependent variable.

a. Find the slope of a linear model that describes this situation. What does it mean in this situation?

b. What is the F-intercept of the model? What does it mean in this situation?

c. Find an equation of the model.

d. Perform a unit analysis of the equation you found in part (c).

e. If the temperature is 30°C, what is the Fahrenheit reading?

25. Let n be the number of new U.S. offshore wells being drilled in the year that is t years since 2005. A reasonable model of the number of new U.S. offshore wells is $n = -18.59t + 146.86$ (Source: *Energy Information Administration*).

a. What is the slope? What does it mean in this situation?

b. What is the n-intercept? What does it mean in this situation?

c. Estimate the number of new U.S. offshore wells drilled in 2012.

26. Let n be the number of firefighters who died on duty in the year that is t years since 2010. A reasonable model of the number of firefighters who have died is $n = -1.75t + 82.72$ (Source: *U.S. Fire Administration*).

a. What is the slope? What does it mean in this situation?

b. What is the n-intercept? What does it mean in this situation?

27. The percentages of adult Internet users who use social networking sites are shown in Table 39 for various years.

Table 39 Percentages of Adult Internet Users Who Use Social Networking Sites

Year	Percent
2005	8
2006	16
2008	29
2009	46
2010	61
2011	65

Source: *Pew Research Center's Internet & American Life Project surveys*

Let p be the percentage of adult Internet users who use social networking sites at t years since 2005. A model of the situation is $p = 10t + 5.7$.

a. Use a graphing calculator to draw a scattergram and the model in the same viewing window. Does the line come close to the data points?

b. What is the slope? What does it mean in this situation?

c. Find the rate of change of the percentage of adult Internet users who use social networking sites for each of the periods 2005–2006, 2006–2009, and 2009–2011. Compare the rate of the change of each of the three periods with your result in part (b).

d. What is the p-intercept? What does it mean in this situation?

e. Use the model to predict the percentage of adult Internet users who will use social networking sites in 2015.

28. Traffic congestion on highways has continued to increase over time. The average annual person-hours (the number of hours each person loses due to traffic delays per year) are listed in Table 40 for various years. Let L be the average annual person-hours lost to traffic delays for the year that is t years since 1980. A model of the situation is $L = 1.01t + 11.18$.

Table 40 Average Annual Person-Hours of Lost Time

Year	Average Annual Lost Time (hours)
1982	14
1992	22
1996	27
1998	30
2001	32
2005	38
2007	38

Source: *Federal Highway Administration*

a. Use a graphing calculator to draw a scattergram and the model in the same viewing window. Does the line come close to the data points?

b. What is the slope? What does it mean in this situation?

c. Find the rate of change of the average person-hours lost per year for each of the periods 1982–1992, 1996–2001, and 2001–2007. Compare the rate of change of each of the three periods with your result in part (b).

d. What is the L-intercept? What does it mean in this situation?

e. Predict the average person-hours lost for 2011. The actual average person-hours lost was 34. Is your result an underestimate or an overestimate? Explain why this might be due to the poor economy during the period 2008–2011.

29. In an attempt to raise the low graduation rates of college football and basketball players, Division I institutions require that prospective athletes must meet new grade point average (GPA) standards (based on a maximum of 4.0) to play as freshmen, and enrolled athletes must meet these new standards to keep playing (see Table 41).

Table 41 Core GPAs Needed to Qualify to Play, for Given SAT Scores

SAT Score	Core GPA
620	3.0
700	2.8
780	2.6
860	2.4
940	2.2
1010	2.0

Source: *NCAA*

Let G be the qualifying core GPA for an SAT score of s points. A model of the situation is $G = -0.00254s + 4.58$.

a. Use a graphing calculator to draw a scattergram and the model in the same viewing window. Does the line come close to the data points?

b. If an athlete's SAT score is 400 points, the lowest possible score, estimate the student's qualifying core GPA. The actual qualifying core GPA is 3.55. Compute the error in the estimate.

c. What is the slope? What does it mean in this situation?

d. What is the G-intercept? What does it mean in this situation?

30. The average selling prices for a Subaru Outback® of various ages are shown in Table 42. Let p be the average price (in dollars) of a Subaru Outback of age a years. A model of the situation is $p = -3284.68a + 30{,}700.57$.

Table 42 Average Selling Prices of Subaru Outbacks

Age (years)	Average Price (dollars)
1	27,080
2	25,240
3	21,112
4	16,950
5	13,400
6	9992
7	9159

Source: *AutoTrader®*

a. Use a graphing calculator to draw a scattergram and the model in the same viewing window. Does the line come close to the data points?
b. What is the slope? What does it mean in this situation?
c. Estimate the average price of an 8-year-old Subaru Outback.

31. A person is on a car trip. Let d be the distance (in miles) traveled in t hours of driving. A reasonable model is shown in Fig. 102.

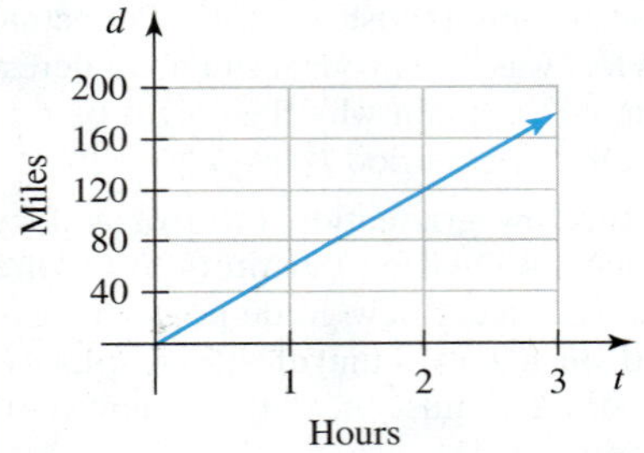

Figure 102 Car model

a. Is the car traveling at a constant speed? Explain.
b. What is the speed of the car?

32. A person is on an airplane. Let d be the distance (in miles) traveled in t hours. A reasonable model is shown in Fig. 103.

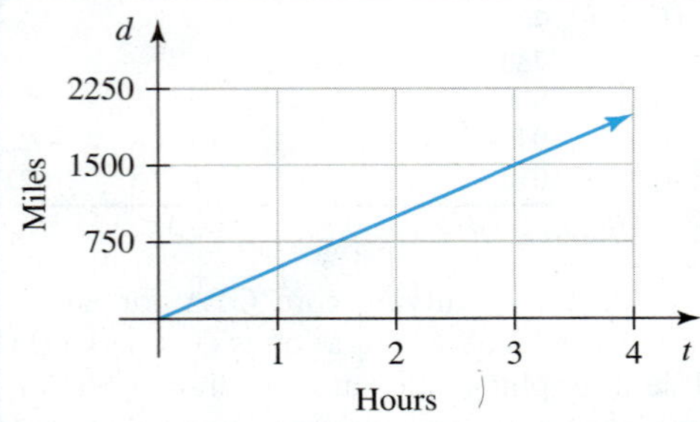

Figure 103 Airplane model

a. Is the airplane traveling at a constant speed? Explain.
b. What is the speed of the airplane?

33. A person is on a car trip. Let V be the volume (in gallons) of gasoline that remains in the gasoline tank after t hours of driving. A reasonable model is shown in Fig. 104.

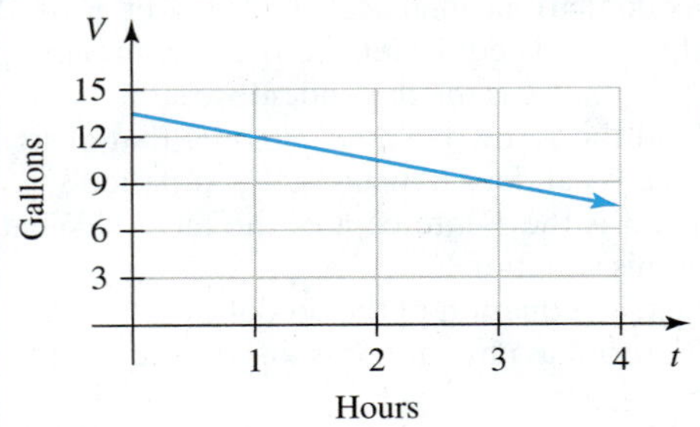

Figure 104 Gasoline model

a. Is the rate of change of gasoline remaining in the tank constant? Explain.
b. What is the rate of change of gasoline remaining in the tank?

34. Let F be the temperature (in degrees Fahrenheit) at t hours after noon. A reasonable model is shown in Fig. 105.

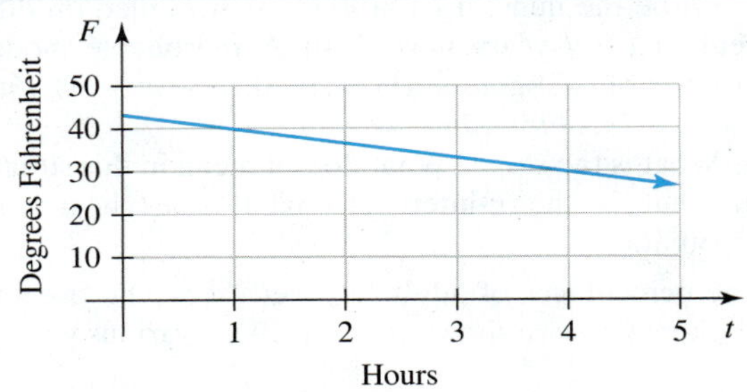

Figure 105 Temperature model

a. Is the rate of change of temperature constant? Explain.
b. What is the rate of change of temperature?

35. Let V be the value (in dollars) of a stock at t weeks since October 1. A reasonable linear model is shown in Fig. 106. What is the slope of the line? What does it mean in this situation?

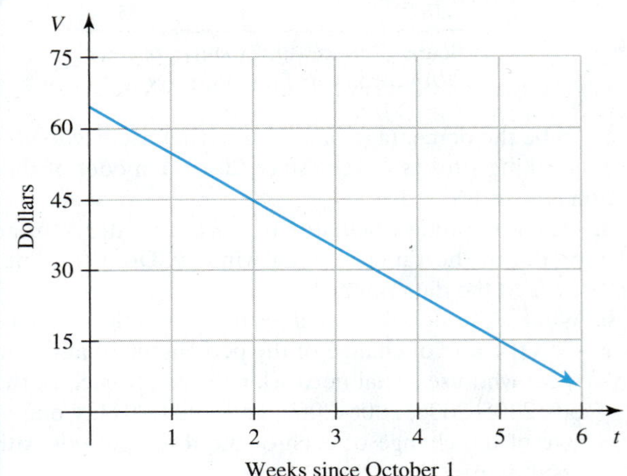

Figure 106 Stock model

36. Let p be the population (in thousands) of a city at t years since 1950. A reasonable linear model is shown in Fig. 107. What is the slope of the line? What does it mean in this situation?

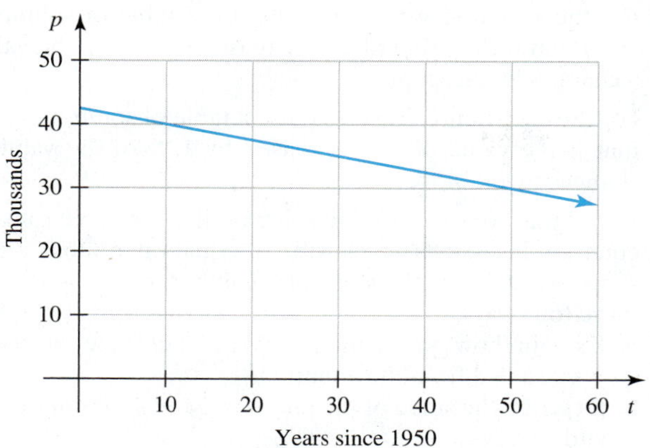

Figure 107 Population model

37. Let p be the median sales price (in thousands of dollars) of a home with n bedrooms in San Diego, California, from May 12 to July 12 in 2012. A reasonable model is shown in Fig. 108. (Source: *Trulia*). What is the slope of the line? What does it mean in this situation?

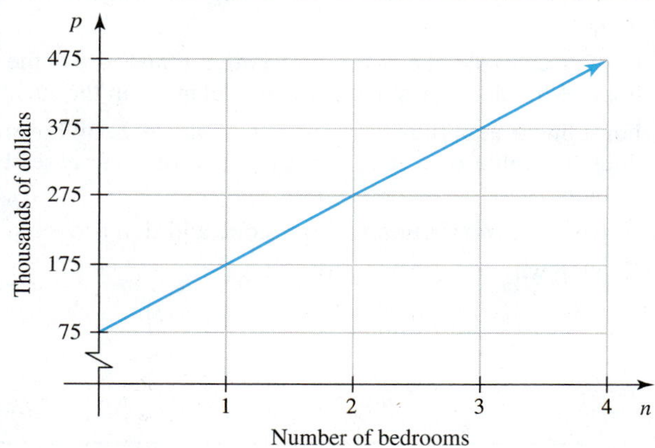

Figure 108 Home model

38. Let c be the total cost (in dollars) of a party if n guests attend. A reasonable model is shown in Fig. 109. What is the slope of the line? What does it mean in this situation?

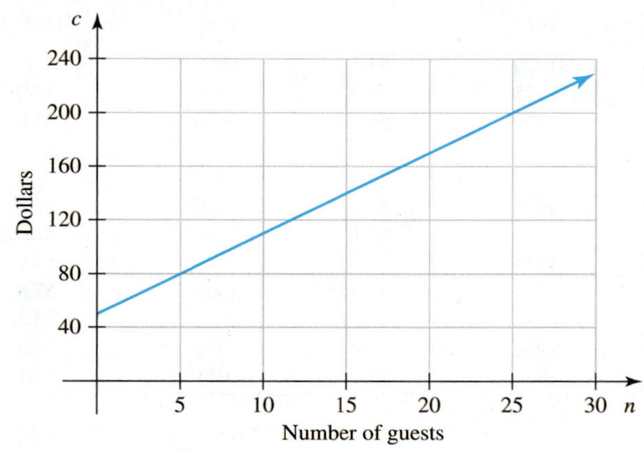

Figure 109 Party model

39. A math tutor charges \$30 per hour. Let c be the total charge (in dollars) for t hours of tutoring. Are t and c linearly related? If so, find the slope and describe what it means in this situation.

40. Gasoline sells for \$3 per gallon at a certain gas station. Let T be the total charge (in dollars) for g gallons of gas. Are g and T linearly related? If so, find the slope and describe what it means in this situation.

41. Some solutions of four equations are listed in Table 43. Which of the equations could be linear?

Table 43 Solutions of Four Equations

Equation 1		Equation 2		Equation 3		Equation 4	
x	y	x	y	x	y	x	y
0	2	0	5	2	5	5	20
1	5	1	6	3	3	6	19
2	8	2	8	4	1	7	17
3	11	3	9	5	−1	8	14
4	14	4	11	6	−3	9	10

42. Some solutions of four equations are listed in Table 44. Which of the equations could be linear?

Table 44 Solutions of Four Equations

Equation 1		Equation 2		Equation 3		Equation 4	
x	y	x	y	x	y	x	y
0	2	0	35	4	−7	23	25
1	4	1	31	5	−3	24	23
2	8	2	27	6	1	25	20
3	16	3	23	7	5	26	18
4	32	4	19	8	9	27	15

43. Some solutions of four linear equations are listed in Table 45. Complete the table.

Table 45 Solutions of Four Linear Equations

Equation 1		Equation 2		Equation 3		Equation 4	
x	y	x	y	x	y	x	y
0	3	0	99	21	16	43	
1	8	1	92	22		44	
2		2		23	12	45	23
3		3		24		46	
4		4		25		47	29

44. Some solutions of four linear equations are listed in Table 46. Complete the table.

Table 46 Solutions of Four Linear Equations

Equation 1		Equation 2		Equation 3		Equation 4	
x	y	x	y	x	y	x	y
0	100	0	2	13	32	75	4
1	94	1	9	14		76	
2		2		15		77	
3		3		16	26	78	
4		4		17		79	16

45. Four sets of points are described in Table 47. For each set, find an equation of a line that contains the points.

Table 47 Solutions of Four Equations

Set 1		Set 2		Set 3		Set 4	
x	y	x	y	x	y	x	y
0	5	0	20	0	21	0	9
1	7	1	17	1	29	1	4
2	9	2	14	2	37	2	−1
3	11	3	11	3	45	3	−6
4	13	4	8	4	53	4	−11

46. Four sets of points are described in Table 48. For each set, find an equation of a line that contains the points.

Table 48 Solutions of Four Equations

Set 1		Set 2		Set 3		Set 4	
x	y	x	y	x	y	x	y
0	22	0	3	0	15	0	−12
1	19	1	7	1	6	1	−10
2	16	2	11	2	−3	2	−8
3	13	3	15	3	−12	3	−6
4	10	4	19	4	−21	4	−4

Concepts

47. For the equation $y = 7x + 2$, describe the change in the value of y as the value of x is increased by 1.

48. For the equation $y = -9x - 5$, describe the change in the value of y as the value of x is increased by 1.

49. For the equation $y = -6x + 40$, use a table of solutions to show that if the value of x is increased by 1, then the value of y changes by the slope.

50. For the equation $y = 8x + 1$, use a table of solutions to show that if the value of x is increased by 1, then the value of y changes by the slope.

51. Recall that we can describe some or all of the solutions of an equation in two variables with an equation, a table, a graph, or words. Describe the slope of the line $y = 3x - 4$ in each of these four ways:

a. Explain how you can determine the slope of the line $y = 3x - 4$ from the equation.

b. Describe the slope of the line $y = 3x - 4$ in terms of "run" and "rise."

c. Describe the slope of the line $y = 3x - 4$ in terms of the slope addition property.

d. Describe the meaning of the slope of the line $y = 3x - 4$ in your own words.

52. If p increases steadily as t increases steadily, then the rate of change of p with respect to t is positive. Explain why this makes sense.

53. If p decreases steadily as t increases steadily, then the rate of change of p with respect to t is negative. Explain why this makes sense.

54. Give an example of a linear model other than those in the textbook. What does the slope of the model mean in the situation?

55. For a linear equation of the form $y = mx + b$, by how much does the value of y change if the value of x increases by 2? Explain.

56. In your own words, describe the slope addition property.

57. Explain why slope is a rate of change.

Taking it to the Lab

Climate Change Lab (continued from Chapter 2)

Scientists estimate that the average global temperature has increased by about 1°F in the past century (see Table 49). Recall from the Climate Change Lab in Chapter 1 that many scientists believe an increase as small as 3.6°F could be dangerous. Because the planet's average temperature increased by 1°F in the past century, it might seem it would take about another two-and-a-half centuries for Earth to reach a dangerous temperature level.

However, the Intergovernmental Panel on Climate Change (IPCC), composed of hundreds of scientists around the world, predicted that Earth's average temperature will rise 2.5°F to 10.4°F in the coming century.* Scientists from the United States and Europe have predicted that there is a 9-out-of-10 chance the global average temperature will increase 3°F to 9°F, with a range of 4°F to 7°F most likely.†

Table 49 Average Surface Temperatures of Earth

Year	Average Temperature (degrees Fahrenheit)	Year	Average Temperature (degrees Fahrenheit)
1900	57.1	1960	57.2
1905	56.8	1965	57.0
1910	56.6	1970	57.3
1915	57.0	1975	57.1
1920	56.9	1980	57.6
1925	56.9	1985	57.3
1930	57.1	1990	57.8
1935	57.0	1995	57.9
1940	57.3	2000	57.8
1945	57.3	2005	58.3
1950	56.9	2010	58.3
1955	57.0	2011	58.1

Source: *NASA–GISS*

One result of the last century of global warming is that glaciers are receding and ocean levels are rising. For example, NASA scientist Bill Krabill estimates that Greenland's

*IPCC, 2001: Summary for Policymakers.

†T. M. L. Wigley and S. C. B. Raper, "Interpretation of high projections for global-mean warming," *Science* 293 (5529):451–454, July 20, 2001.

ice cap, the world's second largest, may be losing ice at a rate of 50 cubic kilometers per year.[‡] The ice cap contained 2.85 million cubic kilometers of ice in 2000. Climatologist Jonathan Gregory of the University of Reading, in the United Kingdom, believes that by 2050 the ice cap may start an irreversible runaway melting. A total meltdown could take 1000 years.[§]

Jonathan Overpeck, director of the Institute for the Study of Planet Earth at the University of Arizona, believes that Greenland's ice cap could melt completely in as little as 150 years and that a partial melting of Greenland's ice cap by 2100 could raise ocean levels by more than 1 meter.[¶] Coastal engineers estimate that a 1-meter rise would translate into a loss of about 100 meters of land.[‖] In some areas of the world, land loss could be much greater.[**] For example, most of Florida is less than 1 meter above sea level.

Most scientists predict that global warming will also cause the extinction of plants and animals, bring severe water shortages, create extreme weather events, and increase the number of heat-related illnesses and deaths.

Analyzing the Situation

1. Let A be the average global temperature (in degrees Fahrenheit) at t years since 1900. Refer to the scattergram that you drew in Problem 3 of the Climate Change Lab of Chapter 1. (If you did not draw this scattergram earlier, create a *careful* one now, using the data found in Table 49.) Does it appear that Earth's average temperature increased much from 1900 to 1965? Explain.
2. **a.** Estimate the average global temperature from 1900 to 1965. [**Hint:** Divide the sum of the temperatures by the number of temperature readings.]
   b. Estimate the average global temperature from 1990 to 2000. [**Hint:** Divide the sum of the temperatures by the number of temperature readings.]
   c. Use your results in parts (a) and (b) to estimate the change in Earth's average temperature in the past century. Compare your result with the scientists' estimate of 1°F.
3. In Problem 5 of the Climate Change Lab of Chapter 1, you showed that the planet's average temperature increased approximately linearly from 1965 to 2011. (If you didn't do this earlier, do so now.) Find the rate of change of the average global temperature from 1965 to 2011.
4. Use the rate of change found in Problem 3 to predict the change in average temperature for the coming century. Does your result fall within the 2.5–10.4°F IPCC range? Does it fall within either the 3–9°F or 4–7°F range predicted by the team of scientists from the United States and Europe? If yes, which one(s)?
5. Explain why, by using terminology such as "9 out of 10" and "most likely," scientists have been able to predict narrower ranges of increase in global average temperatures in the coming century. Explain how this terminology, coupled with narrower ranges, will help policymakers better understand the risks involved in various courses of action or inaction.
6. Let I be the amount of ice (in cubic kilometers) in Greenland's ice cap at t years since 2000. Assuming that Greenland's ice cap continues to melt at the current rate, find an equation that models the situation.
7. **a.** Use your model to predict the amount of ice in Greenland's ice cap in 1000 years. If a total meltdown occurs in 1000 years, as predicted by Gregory, what must happen to the rate of melting?
   b. If a total meltdown occurs in 150 years, as predicted by Overpeck, what must happen to the rate of melting?

[‡]W. Krabill et al., "Greenland ice sheet: High-elevation balance and peripheral thinning," *Science* 289:428–430, 2000.

[§]J. M. Gregory et al., "Climatology: Threatened loss of the Greenland ice sheet," *Nature* (April 8, 2004):426–616.

[¶]American Geophysical Union meeting, October 2002.

[‖]National Oceanic and Atmospheric Administration.

[**]R. J. Nicholls and F. M. J. Hoozemans, "Vulnerability to sea-level rise with reference to the Mediterranean region," *Medcoast 95*, Vol. II, October 1995.

Workout Lab

In this lab, you will explore your walking or running speed. Check with your instructor about whether you should collect your own data or use the data listed in Table 50.

Table 50 Times for Walking 440-Yard Laps

Lap Number	Distance (yards)	Time (seconds)
0	0	0
1	440	217
2	880	436
3	1320	656
4	1760	878
5	2200	1095
6	2640	1308

Source: *J. Lehmann*

Materials

You will need the following items:

- A timing device
- A pencil or pen
- A small pad of paper

Preparation

Locate a running track on which you can walk or run. For most tracks, one lap is 440 yards; so, four laps are 1760 yards, or 1 mile. You may select some other type of route, provided that you know the distance of one lap and you can easily complete six laps. Or map out a route in your neighborhood and estimate the distance by measuring it with the odometer in a car.

Recording of Data

Start your timing device and begin walking or running. Complete six laps of your course. Each time you complete a lap, record the *total* elapsed time. It will be easier to have a friend record the times for you. You may go slowly or quickly, but try to move at a constant speed throughout this experiment.

Analyzing the Data

1. Describe your route and the distance of one lap.
2. Use a table to describe the six total elapsed times for the six laps.
3. Let d be the distance (in yards) after you have walked or run for t seconds. Throughout this lab, treat d as the dependent variable and t as the independent variable. Show the six pairs of values of t and d in a table.
4. Create a scattergram for the variables t and d.
5. For each of the six laps, calculate your average speed. Did you move at a steady rate, slow down, speed up, or engage in a combination of these?
6. Explain how your six average speeds are related to the position of the points in your scattergram.
7. Find the average of the six average speeds you found in Problem 5. Compare this result with the average speed for the entire workout.
8. What is your average speed for the entire workout, in units of miles per hour?
9. Use your result from Problem 8 to predict how long it would take you to walk or run 2 miles.
10. Use your result from Problem 8 to predict how long it would take you to walk or run a marathon, which is 26.2 miles long. Has model breakdown occurred? Explain.

Balloon Lab

In this lab, you will explore how long it takes for air to be released from a balloon when it is inflated with various amounts of air. Check with your instructor about whether you should collect your own data or use the data listed in Table 51.

Table 51 Average Release Times for a Balloon Inflated with a Single Breath Having a Volume of 20 Ounces

Number of Breaths	Volume (ounces)	Average Release Time (seconds)
0	0	0
5	100	1.9
10	200	3.5
15	300	6.1
20	400	6.0
25	500	7.7
30	600	10.6

Source: *J. Lehmann*

Materials

You will need one helper and the following items:

- A balloon
- A timing device
- A bucket, sink, or bathtub
- Water to fill the bucket, sink, or bathtub
- A transparent 1-cup or larger measuring cup (a larger cup is more convenient)

Preparation

Inflate and deflate the balloon fully several times to stretch it out. Each time you inflate the balloon, practice blowing into it with uniform-size breaths. There is no need to breathe deeply into the balloon. Medium-size breaths will work well for this lab.

Recording of Data

Perform the following tasks:

1. Count how many medium-size breaths it takes to fill the balloon.
2. For each trial of this experiment, you will time how long it takes for the balloon to release all the air inside it. Run three trials for each of six different volumes. Decide for which volumes you will have trials. For example, if it takes 30 breaths to fill the balloon, you could run three trials for each of the volumes of 5, 10, 15, 20, 25, and 30 breaths. To run a trial, first fill the balloon to the desired volume. Then begin timing as you release the balloon. Stop timing when the balloon is deflated completely. Record the time and the corresponding volume (in number of breaths).
3. To find the volume of a medium-size breath, fill a bucket, sink, or bathtub with water. Have one person fully submerge the measuring cup, so that there is no air in it, and then turn the cup upside down while it is still under water. The cup should not rest on the bottom of the container. Next, have a second person blow once into the balloon and then carefully release the air into the submerged cup. The air from the balloon will displace the water in the cup. The volume of air in the balloon will likely be more than what can fit in the cup, so the second person will have to stop releasing air from the balloon, and then the first person can empty out the air by resubmerging the cup. The second person should continue releasing air from the balloon into the cup until the balloon is empty. Depending on the size of the breath and the cup, the first person may have to resubmerge the cup several times. Then compute the volume of a medium-size breath by measuring the amount of air in the cup and taking into account how many times you filled the cup with air.

Analyzing the Data

1. For each of the volumes of 5, 10, 15, 20, 25, and 30 breaths, compute the average of the three release times. Then record the numbers of breaths and the average release times in columns 1 and 3 of a table similar to Table 51. Use the volume of one breath to help you find the entries for the second column of the table.
2. Let T be the time (in seconds) it takes for the balloon to deflate completely when the initial volume is n ounces. Create a scattergram of the data.
3. Draw a line that comes close to the data.
4. What is the T-intercept of the linear model? What does it mean in this situation? If model breakdown occurs, draw a better model.

5. Find the slope of the linear model. What does it mean in this situation?
6. Find an equation of the model.
7. Use the model to estimate the release time for a volume other than any of the volumes you used for the trials.
8. As you inflated the balloon, did it take about the same amount of effort to breathe in each time, or did it get progressively easier or harder? Thinking about how that effort is related to release times, does it suggest that the data points should lie close to a line or to a curve that bends? Explain. Do the (actual) data points support your theory? Explain.

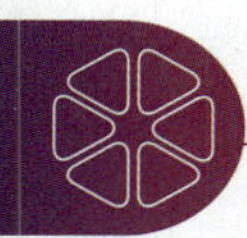

Chapter Summary

Key Points of Chapter 3

Section 3.1 Graphing Equations of the Form $y = mx + b$

Solution, satisfy, and solution set of an equation	An ordered pair (a, b) is a **solution** of an equation in terms of x and y if the equation becomes a true statement when a is substituted for x and b is substituted for y. We say (a, b) **satisfies** the equation. The **solution set** of an equation is the set of all solutions of the equation.
Graph	The **graph** of an equation in two variables is the set of points that correspond to all solutions of the equation.
Graph of $y = mx + b$	The graph of an equation of the form $y = mx + b$, where m and b are constants, is a line.
Rule of Four	We can describe some or all of the solutions of an equation with an equation, a table, a graph, or words. These four ways to describe solutions are known as the **Rule of Four.**

Section 3.2 Graphing Linear Models; Unit Analysis

Rule of Four	We can describe an authentic situation with an equation, a table, a graph, or words. These four ways to describe authentic situations are known as the **Rule of Four.**
Unit analysis	We perform a **unit analysis** of a model's equation by determining the units of the expressions on both sides of the equation. The units of the expressions on both sides of the equation should be the same.
Equations of horizontal and vertical lines	If a and b are constants, then • The graph of $y = b$ is a horizontal line. • The graph of $x = a$ is a vertical line.
Equations whose graphs are lines	If an equation can be put into the form $y = mx + b$ or $x = a$, where m, a, and b are constants, then the graph of the equation is a line. We call such an equation a **linear equation in two variables.**

Section 3.3 Slope of a Line

Comparing the steepness of two objects	To compare the steepness of two objects, compute the unit ratio $\frac{\text{vertical distance}}{\text{horizontal distance}}$ for each object. The object with the larger ratio is the steeper object.

Section 3.3 Slope of a Line (*Continued*)

Slope of a nonvertical line	Let (x_1, y_1) and (x_2, y_2) be two distinct points of a nonvertical line. Then the **slope** of the line is $$m = \frac{\text{vertical change}}{\text{horizontal change}} = \frac{\text{rise}}{\text{run}} = \frac{y_2 - y_1}{x_2 - x_1}$$
Slopes of increasing and decreasing lines	An increasing line has positive slope. A decreasing line has negative slope.
Slopes of horizontal and vertical lines	A horizontal line has a slope equal to zero. A vertical line has undefined slope.

Section 3.4 Using Slope to Graph Linear Equations

Slope and *y*-intercept of a linear equation of the form $y = mx + b$; slope–intercept form	For a linear equation of the form $y = mx + b$, • the slope of the line is m and • the *y*-intercept of the line is $(0, b)$. We say that this equation is in **slope–intercept form.**
Using slope to graph an equation of the form $y = mx + b$	To graph an equation of the form $y = mx + b$, **1.** Plot the *y*-intercept $(0, b)$. **2.** Use $m = \frac{\text{rise}}{\text{run}}$ to plot a second point. **3.** Sketch the line that passes through the two plotted points.
Find an equation of a line from a graph	To find an equation of a line from a graph, **1.** Determine the slope m and the *y*-intercept $(0, b)$ from the graph. **2.** Substitute your values for m and b into the equation $y = mx + b$.
Slopes of parallel lines	If lines l_1 and l_2 are parallel nonvertical lines, then the slopes of the lines are equal: $$m_1 = m_2$$ Also, if two distinct lines have equal slope, then the lines are parallel.
Slopes of perpendicular lines	If lines l_1 and l_2 are perpendicular nonvertical lines, then the slope of one line is the opposite of the reciprocal of the slope of the other line: $$m_2 = -\frac{1}{m_1}$$ Also, if the slope of one line is the opposite of the reciprocal of another line's slope, then the lines are perpendicular.

Section 3.5 Rate of Change

Rate of change and average rate of change	Suppose that a quantity y changes steadily from y_1 to y_2 as a quantity x changes steadily from x_1 to x_2. Then the **rate of change** of y with respect to x is the ratio of the change in y to the change in x: $$\frac{\text{change in } y}{\text{change in } x} = \frac{y_2 - y_1}{x_2 - x_1}$$ If either quantity does not change steadily, then the preceding formula is the **average rate of change** of y with respect to x.
Increasing and decreasing quantities	Suppose that a quantity p depends on a quantity t. Then • If p increases steadily as t increases steadily, then the rate of change of p with respect to t is positive. • If p decreases steadily as t increases steadily, then the rate of change of p with respect to t is negative.

Section 3.5 Rate of Change (*Continued*)

Slope is a rate of change	If there is a linear relationship between the quantities t and p, and if p depends on t, then the slope of the linear model is equal to the rate of change of p with respect to t.
Constant rate of change	Suppose that a quantity p depends on a quantity t. Then • If there is a linear relationship between t and p, then the rate of change of p with respect to t is constant. • If the rate of change of p with respect to t is constant, then there is a linear relationship between t and p.
Slope is an average rate of change	If two quantities t and p are approximately linearly related and p depends on t, then the slope of a reasonable linear model is approximately equal to the average rate of change of p with respect to t.
Slope addition property	For a linear equation of the form $y = mx + b$, if the value of x increases by 1, then the value of y changes by the slope m. In other words, if an input increases by 1, the output changes by the slope.

Chapter 3 Review Exercises

1. Which of the ordered pairs $(-3, 9)$, $(1, 2)$, and $(4, -5)$ satisfy the equation $y = -2x + 3$?

For Exercises 2–7, refer to the graph sketched in Fig. 110.

2. Find y when $x = 2$.
3. Find y when $x = -2$.
4. Find y when $x = 0$.
5. Find x when $y = -3$.
6. Find x when $y = -4$.
7. Find x when $y = 0$.

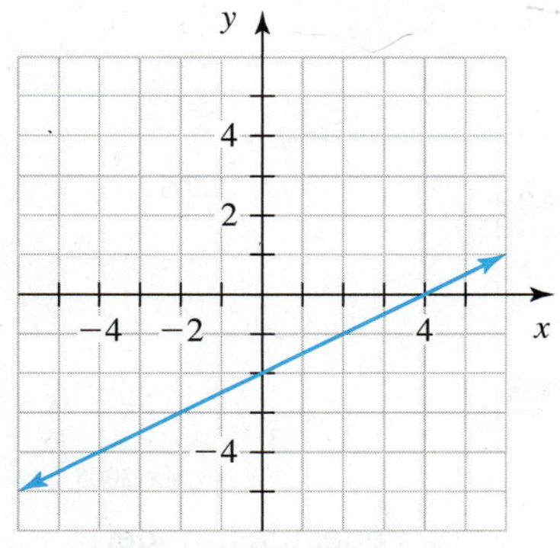

Figure 110 Exercises 2–7

8. While taking off, airplane A climbs steadily for 6500 feet over a horizontal distance of 12,700 feet. Airplane B climbs steadily for 7400 feet over a horizontal distance of 15,600 feet. Which plane is climbing at a greater incline? Explain.

Find the slope of the line passing through the two given points. State whether the line is increasing, decreasing, horizontal, or vertical.

9. $(-3, 1)$ and $(2, 11)$
10. $(-2, -4)$ and $(1, -7)$
11. $(4, -3)$ and $(8, -1)$
12. $(-6, 0)$ and $(0, -3)$
13. $(-5, 5)$ and $(2, -2)$
14. $(-10, -3)$ and $(-4, -5)$
15. $(-5, 2)$ and $(3, -7)$
16. $(-4, -1)$ and $(2, -5)$
17. $(-4, 7)$ and $(-4, -3)$
18. $(-5, 2)$ and $(-1, 2)$

Find the approximate slope of the line that contains the two given points. Round your result to the second decimal place. State whether the line is increasing, decreasing, horizontal, or vertical.

19. $(5.4, 7.9)$ and $(8.3, -2.6)$

20. $(-8.74, -2.38)$ and $(-1.16, 4.77)$

21. Sketch a line whose slope is a negative number near zero.

Sketch the line that has the given slope and contains the given point.

22. $m = 3, (0, -4)$
23. $m = \frac{4}{3}, (0, 1)$
24. $m = 0, (2, -3)$

Determine the slope and the y-intercept. Use them to graph the equation by hand. Use a graphing calculator to verify your work.

25. $y = \frac{3}{4}x - 1$
26. $y = -\frac{1}{2}x + 3$
27. $y = -\frac{2}{5}x - 1$
28. $y = \frac{2}{3}x$
29. $y = -4x$
30. $y = 2x - 4$
31. $y = -3x + 1$
32. $y = x + 2$
33. $y = -5$

For Exercises 34 and 35, graph the equation by hand.

34. $x = -3$
35. $y = 2$

36. Recall that we can describe some or all of the solutions of an equation in two variables with an equation, a table, a graph, or words.
a. Describe three solutions of $y = -2x + 1$ by using a table.
b. Describe the solutions of $y = -2x + 1$ by using a graph.
c. Describe the solutions of $y = -2x + 1$ by using words.

37. When a certain person loses his job, the balance in his checking account is \$19 thousand. While the person is unemployed, the balance declines by \$3 thousand per month.

a. Let B be the balance (in thousands of dollars) in the checking account after t months of unemployment. Find an equation for t and B.

b. Graph by hand the equation you found in part (a).

c. What is the B-intercept? What does it mean in this situation?

d. How long has the person been unemployed if the balance is \$4 thousand? Explain by using arrows on your graph in part (b).

38. Find an equation of the line sketched in Fig. 111.

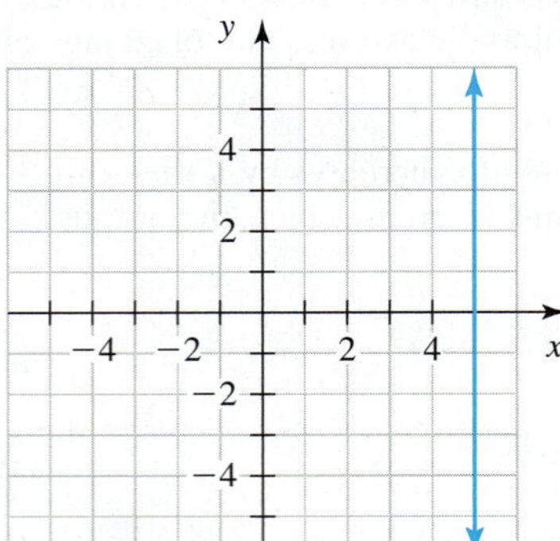

Figure 111 Exercise 38

39. Find an equation of the line sketched in Fig. 112.

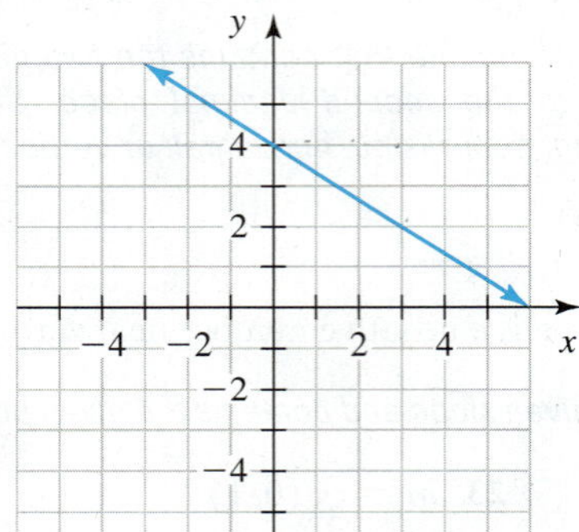

Figure 112 Exercise 39

For Exercises 40–43, determine whether the pair of lines is parallel, perpendicular, or neither. Explain.

40. $y = 3x - 2$ and $y = \frac{1}{3}x + 6$

41. $y = \frac{4}{7}x + 1$ and $y = -\frac{7}{4}x - 5$

42. $x = -2$ and $y = 5$

43. $x = -4$ and $x = 1$

44. The temperature declines by 6°F over a 4-hour period. Find the average rate of change of temperature.

45. U.S. revenue of Viagra decreased approximately linearly from \$1.7 billion in 2004 to \$1.0 billion in 2010 (Source: *IMS Health*). Find the average rate of change of revenue.

46. In New York City, taxis charge a flat fee of \$2.50 plus \$2 per mile (40 cents per $\frac{1}{5}$ mile). Let c be the total charge (in dollars) for a trip of d miles.

a. Find an equation for d and c.

b. Perform a unit analysis of the equation you found in part (a).

c. What is the cost of a 17-mile trip?

47. A person weighs 195 pounds when he begins a weight-loss program. After beginning the program, he loses 4 pounds per month. Let w be his weight (in pounds) t months after he has begun the program.

a. Find the slope of the linear model of the situation. What does it mean in this situation?

b. Find an equation of the model.

c. The person reaches his goal after he has dieted for 6 months. What was his goal?

48. The average monthly day-care cost in 2011 was \$972. The cost has increased by \$45.60 per year (Source: *National Association of Child Care Resource & Referral Agencies*). Costs are based on a three-year-old attending a day-care center eight hours a day, five days a week. Let c be the average monthly cost (in dollars) at t years since 2011.

a. Find the slope of the linear model of the situation. What does it mean in this situation?

b. Find an equation of the model.

c. Predict the average monthly cost in 2018.

49. The suggested retail price for a TI-84 Plus® graphing calculator is \$139. Let C be the total cost (in dollars) of purchasing n of these calculators at the retail price. Are n and C linearly related? If so, what is the slope? What does the slope mean in this situation?

50. Let n be the number (in thousands) of pedestrian deaths in the year that is t years since 2005. A reasonable linear model is shown in Fig. 113 (Source: *NHTSA*).

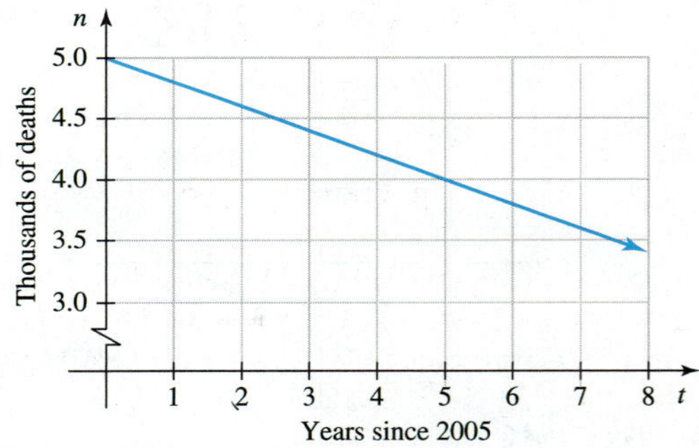

Figure 113 Pedestrian model

a. Estimate the number of pedestrian deaths in 2007.

b. Estimate in which year there were 3.6 thousand pedestrian deaths.

c. What is the n-intercept? What does it mean in this situation?

d. What is the slope? What does it mean in this situation?

51. Some solutions of four equations are listed in Table 52. Which of the equations could be linear?

Table 52 Solutions of Four Equations

Equation 1		Equation 2		Equation 3		Equation 4	
x	y	x	y	x	y	x	y
0	17	0	2	3	4	1	−8
1	14	1	6	4	4	2	−3
2	11	2	9	5	4	3	2
3	8	3	11	6	4	4	7
4	5	4	12	7	4	5	12

52. Some solutions of four linear equations are listed in Table 53. Complete the table.

Table 53 Solutions of Four Linear Equations

Equation 1		Equation 2		Equation 3		Equation 4	
x	y	x	y	x	y	x	y
0	50	0	12	61	25	26	−4
1	41	1	16	62		27	
2		2		63		28	
3		3		64	19	29	
4		4		65		30	8

53. For the equation $y = -6x + 39$, describe the change in the value of y as the value of x is increased by 1.

Chapter 3 Test

For Exercises 1–4, refer to the graph sketched in Fig. 114.

1. Find y when $x = -3$.

2. Find x when $y = -1$.

3. Find the y-intercept.

4. Estimate the x-intercept.

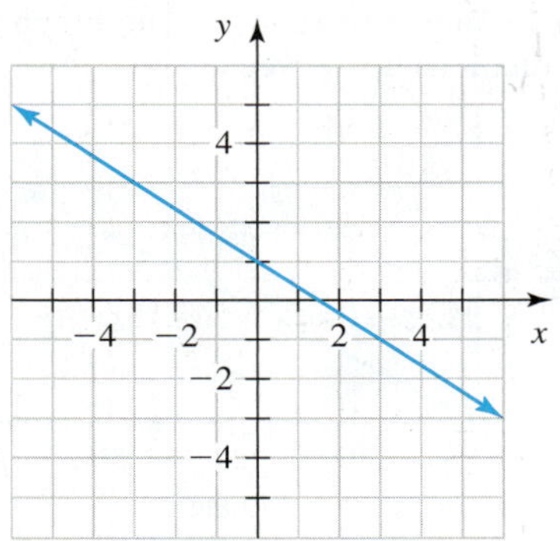

Figure 114 Exercises 1–4

5. Ski run A declines steadily for 115 yards over a horizontal distance of 580 yards. Ski run B declines steadily for 150 yards over a horizontal distance of 675 yards. Which ski run is steeper? Explain.

For Exercises 6–9, find the slope of the line passing through the two given points. State whether the line is increasing, decreasing, horizontal, or vertical.

6. $(3, -8)$ and $(5, -2)$

7. $(-4, -1)$ and $(2, -4)$

8. $(-5, 4)$ and $(1, 4)$

9. $(-2, -7)$ and $(-2, 3)$

10. Find the approximate slope of the line that contains the points $(-5.99, -3.27)$ and $(2.83, 8.12)$. Round your result to the second decimal place. State whether the line is increasing, decreasing, horizontal, or vertical.

11. Sketch the line that has a slope of $\frac{2}{5}$ and contains the point $(0, -3)$.

For Exercises 12–16, determine the slope and the y-intercept. Use them to graph the equation by hand.

12. $y = -\frac{3}{2}x + 2$

13. $y = \frac{5}{6}x$

14. $y = 3x - 4$

15. $y = 2$

16. $y = -2x + 3$

17. Find an equation of the line sketched in Fig. 115.

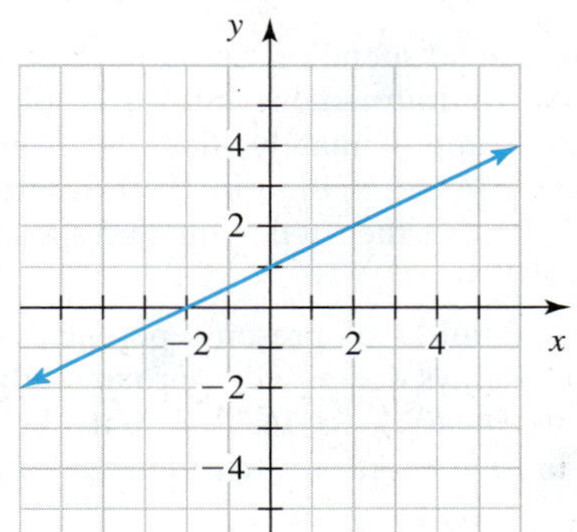

Figure 115 Exercise 17

18. A used car is worth \$17 thousand. Each year, the car's value declines by \$2 thousand. Let v be the car's value (in thousands of dollars) after t years.

a. Find an equation for t and v.

b. Graph by hand the equation you found in part (a).

c. What is the v-intercept? What does it mean in this situation?

d. When will the value of the car be \$5 thousand dollars? Explain by using arrows on your graph in part (b).

19. Graphs of four equations are shown in Fig. 116. State whether m and b are positive, negative, zero, or undefined for the $y = mx + b$ form of each equation.

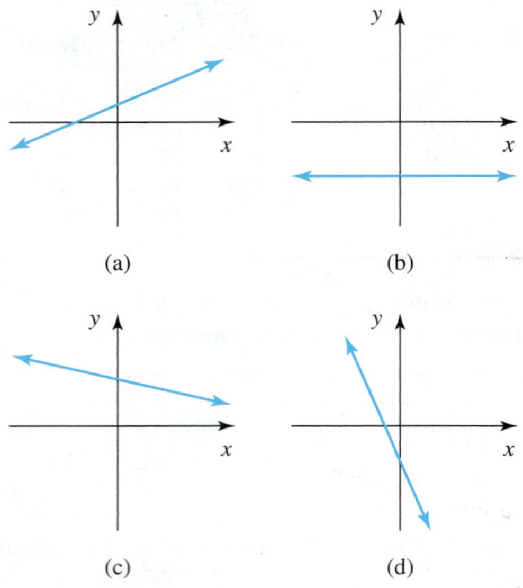

Figure 116 Exercise 19

For Exercises 20 and 21, determine whether the pair of lines is parallel, perpendicular, or neither. Explain.

20. $y = \frac{2}{5}x + 3$ and $y = \frac{5}{2}x - 7$

21. $y = -3x + 8$ and $y = -3x - 1$

22. The total annual ad spending (in millions of dollars) for the NCAA basketball tournament (known as "March Madness") has increased approximately linearly from \$239.1 million in 1998 to \$613.8 million in 2010 (Source: *TNS Media Intelligence*). Predict the total annual ad spending for March Madness in 2017.

23. The average number of presents people said they would buy for the holidays decreased approximately linearly from 23.1 presents in 2007 to 16.8 presents in 2010 (Source: *Deloitte*). Find the average rate of change of the number of presents.

24. The median compensation (including salary and bonus) of college presidents was \$870 thousand in 2010 and has increased by about \$37 thousand per year since then (Source: *IRS*). Let C be the median annual compensation (in thousands of dollars) of college presidents at t years since 2010.

a. Is there an approximate linear relationship between t and C? Explain. If the relationship is approximately linear, find the slope and describe what it means in this situation.
b. What is the C-intercept? What does it mean in this situation?
c. Find an equation of the model.
d. Predict the median compensation in 2018.

25. The cooking times of a turkey in a 325°F oven are shown in Table 54 for various weights. Let T be the cooking time (in hours) of a turkey that weighs w pounds. A model of the situation is $T = 0.24w + 1.64$.

Table 54 Cooking Times of a Turkey in a 325°F Oven

Weight Group (pounds)	Weight Used to Represent Weight Group (pounds)	Cooking-Time Group (hours)	Time Used to Represent Cooking-Time Group (hours)
6–8	7	3.0–3.5	3.25
8–12	10	3.5–4.5	4
12–16	14	4.5–5.5	5
16–20	18	5.5–6.5	6
20–24	22	6.5–7.0	6.75

Source: *About, Inc.*

a. Use a graphing calculator to draw a scattergram and the model in the same viewing window. Does the line come close to the data points?
b. What is the slope? What does it mean in this situation?
c. What is the T-intercept? What does it mean in this situation?
d. Estimate the cooking time of a 19-pound turkey.

26. A person goes out for a run. Let d be the distance (in miles) traveled in t minutes of running. The graph in Fig. 117 describes the relationship between t and d.

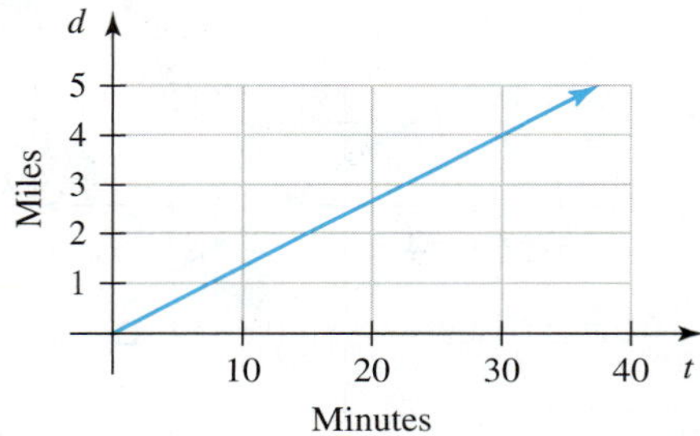

Figure 117 Runner model

a. Is the person running at a constant speed? Explain.
b. How fast is the person running? Use units of miles per hour.

27. Four sets of points are described in Table 55. For each set, find an equation of a line that contains the points.

Table 55 Solutions of Four Equations

Set 1		Set 2		Set 3		Set 4	
x	y	x	y	x	y	x	y
0	25	0	2	0	12	0	47
1	22	1	6	1	7	1	53
2	19	2	10	2	2	2	59
3	16	3	14	3	−3	3	65
4	13	4	18	4	−8	4	71

28. For the equation $y = 3x - 8$, describe the change in the value of y as the value of x is increased by 1.

Simplifying Expressions and Solving Equations

What was your average grade in high school? The percentage of college freshmen whose average grade in high school was an A has increased greatly since 1970 (see Table 1). Do you think the increase is due to students learning more, teachers lowering their standards, or some other reason? In Exercise 51 of Homework 4.4, you will estimate when half of all college freshmen earned an average grade of A in high school.

In this chapter, we will discuss how to write expressions more simply. We will also discuss how to find solutions of equations in *one* variable. Once we have learned these skills, we will be able to make predictions efficiently about the independent variable (or the dependent variable) in authentic situations, such as the grade data just described.

Table 1 College Freshmen Whose Average Grade in High School Was an A

Year	Percent
1970	19.6
1980	26.6
1985	28.7
1990	29.4
1995	36.1
2000	42.9
2005	46.6
2010	48.4

Source: *Higher Education Research Institute*

4.1 Commutative, Associative, and Distributive Laws

Objectives

- » Know the commutative, associative, and distributive laws.
- » Know the meaning of *equivalent expressions.*
- » Know how to *simplify expressions.*
- » Subtract two expressions.
- » Know how to show that a statement is false.
- » Compare the distributive and associative laws for multiplication.

In this section, we will discuss three laws of operations that will help us write expressions more simply.

Commutative Laws

Consider the equations $2 + 6 = 6 + 2$ and $2 \cdot 6 = 6 \cdot 2$. These true statements suggest the *commutative laws.*

Commutative Laws for Addition and Multiplication

Commutative law for addition: $a + b = b + a$
Commutative law for multiplication: $ab = ba$

In words: We can add two numbers in either order and get the same result, and we can multiply two numbers in either order and get the same result.

Example 1 Using the Commutative Law for Addition

Use the commutative law for addition to write the expression in another form.

1. $3 + x$ **2.** $5 + 3w$

Solution

1. $3 + x = x + 3$ *Commutative law for addition:* $a + b = b + a$
2. $5 + 3w = 3w + 5$ *Commutative law for addition:* $a + b = b + a$

Example 2 Using the Commutative Law for Multiplication

Use the commutative law for multiplication to write the expression in another form.

1. cd 2. $-7x$ 3. $k \cdot 8 + 2$

Solution

1. $cd = dc$ *Commutative law for multiplication: ab = ba*
2. $-7x = x(-7)$ *Commutative law for multiplication: ab = ba*
3. $k \cdot 8 + 2 = 8k + 2$ *Commutative law for multiplication: ab = ba*

A **term** is a constant, a variable, or a product of a constant and one or more variables raised to powers. Here are some terms:

$$4x \qquad -7 \qquad y \qquad -5xy^2 \qquad 97x^6y^3$$

The expression $7xy + 4x - 5y + 3$ has four terms: $7xy$, $4x$, $5y$, and 3. Note that, in the expression, the terms are separated by addition and subtraction symbols.

Variable terms are terms that contain variables. **Constant terms** are terms that do not contain variables. For example, $-5x$ is a variable term, and 8 is a constant term. We usually write a sum of both variable and constant terms with the variable terms to the left of the constant terms. So, for $5 - 8x$, we write

$$5 - 8x = 5 + (-8x) \qquad a - b = a + (-b)$$
$$= -8x + 5 \qquad \text{Commutative law for addition: } a + b = b + a$$

Associative Laws

As we discussed in Section 2.6, we work from left to right when we perform additions and subtractions. However, if a sum is of the form $a + b + c$, we can get the same result by performing the addition on the left first or the one on the right first. For example,

$$(4 + 2) + 3 = 6 + 3 = 9$$
$$4 + (2 + 3) = 4 + 5 = 9$$

A similar law is true for an expression of the form *abc*:

$$(4 \cdot 2) \cdot 3 = 8 \cdot 3 = 24$$
$$4 \cdot (2 \cdot 3) = 4 \cdot 6 = 24$$

These examples suggest the *associative laws*.

Associative Laws for Addition and Multiplication

Associative law for addition: $a + (b + c) = (a + b) + c$
Associative law for multiplication: $a(bc) = (ab)c$

In words: For an expression $a + b + c$, we get the same result by performing the addition on the right first or the one on the left first. For an expression *abc*, we get the same result by performing the multiplication on the right first or the one on the left first.

It is important to know the difference between the associative laws and the commutative laws. The associative laws change the order of *operations,* whereas the commutative laws change the order of *terms* or even *expressions*.

For example, consider the expression $(3 + x) + 1$. By the associative law of addition, we can change the order of the additions:

$$(3 + x) + 1 = 3 + (x + 1)$$

By the commutative law of addition, we can change the order of the terms 3 and x:

$$(3 + x) + 1 = (x + 3) + 1$$

We can also use the commutative law to change the order of the expressions $x + 3$ and 1:

$$(x + 3) + 1 = 1 + (x + 3)$$

Example 3 Using the Associative Laws

Use an associative law to write the expression in another form.

1. $(x + 2) + y$ **2.** $w + (9 + k)$ **3.** $3(mp)$ **4.** $(wx)y$

Solution

1. $(x + 2) + y = x + (2 + y)$ *Associative law for addition:* $(a + b) + c = a + (b + c)$

2. $w + (9 + k) = (w + 9) + k$ *Associative law for addition:* $a + (b + c) = (a + b) + c$

3. $3(mp) = (3m)p$ *Associative law for multiplication:* $a(bc) = (ab)c$

4. $(wx)y = w(xy)$ *Associative law for multiplication:* $(ab)c = a(bc)$

We can use a combination of the commutative law and the associative law, both for addition, to write the terms of $a + b + c$ in different orders:

$$a + c + b \qquad b + a + c \qquad b + c + a \qquad c + a + b \qquad c + b + a$$

We **rearrange the terms** of an expression by writing the terms in a different order.

Likewise, we **rearrange the factors** of an expression by writing the factors in a different order. Here we rearrange the factors of abc to get the following products:

$$acb \qquad bac \qquad bca \qquad cab \qquad cba$$

Example 4 Rearranging Terms

Rearrange the terms of $9 - 2x - 5 + 3$ so that the numbers can be added.

Solution

$$\begin{aligned} 9 - 2x - 5 + 3 &= 9 + (-2x) + (-5) + 3 && a - b = a + (-b) \\ &= -2x + 9 + (-5) + 3 && \text{Rearrange terms.} \\ &= -2x + 7 && \text{Add constant terms.} \end{aligned}$$

It is helpful to practice problems such as the one in Example 4 until you can do each problem in one step.

Example 5 Using the Commutative and Associative Laws

Use the commutative and associative laws to remove the parentheses. Also, multiply and add numbers when possible.

1. $-3(8x)$ **2.** $(5 + 7p) + 8$

Solution

1. $$\begin{aligned} -3(8x) &= (-3 \cdot 8)x && \text{Associative law for multiplication: } a(bc) = (ab)c \\ &= -24x && \text{Multiply.} \end{aligned}$$

2. $$\begin{aligned} (5 + 7p) + 8 &= (7p + 5) + 8 && \text{Commutative law for addition: } a + b = b + a \\ &= 7p + (5 + 8) && \text{Associative law for addition: } (a + b) + c = a + (b + c) \\ &= 7p + 13 && \text{Add.} \end{aligned}$$

Distributive Law

For the expression $2(3 + 4)$, we perform the addition before the multiplication. However, compare the result with that for computing $2 \cdot 3 + 2 \cdot 4$:

$$2(3 + 4) = 2(7) = 14$$
$$2 \cdot 3 + 2 \cdot 4 = 6 + 8 = 14$$

Since both results are equal to 14, we can write

$$2(3 + 4) = 2 \cdot 3 + 2 \cdot 4$$

The blue curves indicate what numbers we multiply by 2. We say that we *distribute* the 2 to both the 3 and the 4.

Finally, we compare $2(7 - 3)$ with $2 \cdot 7 - 2 \cdot 3$:

$$2(7 - 3) = 2(4) = 8$$
$$2 \cdot 7 - 2 \cdot 3 = 14 - 6 = 8$$

Since both results are equal to 8, we can write

$$2(7 - 3) = 2 \cdot 7 - 2 \cdot 3$$

Here we have distributed the 2 to both the 7 and the 3. These examples suggest the **distributive law.**

Distributive Law

$$a(b + c) = ab + ac$$

In words: To find $a(b + c)$, distribute a to both b and c.

Example 6 Using the Distributive Law

Find the product.

1. $3(x + 5)$ **2.** $5(2x + 4y)$

Solution

1. $3(x + 5) = 3x + 3 \cdot 5$ *Distributive law: $a(b + c) = ab + ac$*
$= 3x + 15$ *Multiply.*

2. $5(2x + 4y) = 5 \cdot 2x + 5 \cdot 4y$ *Distributive law: $a(b + c) = ab + ac$*
$= 10x + 20y$ *Multiply.*

WARNING In Problem 1 of Example 6, we found that $3(x + 5) = 3x + 15$. In applying the distributive law to an expression such as $3(x + 5)$, remember to distribute the 3 to *every* term in the parentheses. For example, the expression $3x + 5$ is an incorrect result.

Since subtracting a number is the same as adding the opposite of the number, it is also true that

$$a(b - c) = ab - ac$$

We can use the distributive law and the commutative law for multiplication to show that we can "distribute from the right" as well as from the left (see Exercises 115 and 116):

$$(a + b)c = ac + bc$$
$$(a - b)c = ac - bc$$

Example 7 Using the Distributive Law

Find the product.

1. $4(3x - 2y)$ **2.** $(x - 6)(3)$ **3.** $-2(6 + t)$ **4.** $-5(w - 3)$

Solution

1. $4(3x - 2y) = 4 \cdot 3x - 4 \cdot 2y$ *Distributive law: $a(b - c) = ab - ac$*

$= 12x - 8y$ *Multiply.*

2. $(x - 6)(3) = x \cdot 3 - 6 \cdot 3$ *Distributive law: $(a - b)c = ac - bc$*

$= 3x - 18$ *Commutative law for multiplication: $ab = ba$; multiply.*

3. $-2(6 + t) = -2(6) + (-2t)$ *Distributive law: $a(b + c) = ab + ac$*

$= -12 + (-2t)$ *Multiply.*

$= -2t + (-12)$ *Commutative law for addition: $a + b = b + a$*

$= -2t - 12$ *$a + (-b) = a - b$*

4. $-5(w - 3) = -5w - (-5)(3)$ *Distributive law: $a(b - c) = ab - ac$*

$= -5w + 15$ *Multiply.*

It is extremely helpful to practice problems such as those in Example 7 until you can do them in one step.

We can also use the distributive law when there are more than two terms inside the parentheses.

Example 8 Distributive Law

Find the product $3(2t - 5w + 4)$.

Solution

$3(2t - 5w + 4) = 3 \cdot 2t - 3 \cdot 5w + 3 \cdot 4$ *Distributive law*

$= 6t - 15w + 12$ *Multiply.*

Equivalent Expressions

In Problem 1 of Example 6, we found that $3(x + 5) = 3x + 15$. In Example 9, we will explore the meaning of this statement.

Example 9 Evaluating Expressions

Evaluate both of the expressions $3(x + 5)$ and $3x + 15$ for the given values of x.

1. $x = 2$ **2.** $x = 4$

3. $x = 0, x = 1, x = 2, x = 3, x = 4, x = 5$, and $x = 6$

Solution

1. First, we evaluate $3(x + 5)$ for $x = 2$:

$$3(2 + 5) = 3(7) = 21$$

Then we evaluate $3x + 15$ for $x = 2$:

$$3(2) + 15 = 6 + 15 = 21$$

Both results are equal to 21.

2. First, we evaluate $3(x + 5)$ for $x = 4$:

$$3(4 + 5) = 3(9) = 27$$

Then we evaluate $3x + 15$ for $x = 4$:

$$3(4) + 15 = 12 + 15 = 27$$

Both results are equal to 27.

3. We use a graphing calculator to create a table for the equations $y = 3(x + 5)$ and $y = 3x + 15$ (see Fig. 1). See Appendix A.14 for graphing calculator instructions.

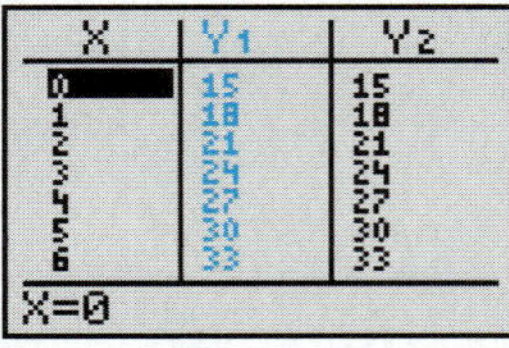

Figure 1 Evaluate the expressions $3(x + 5)$ and $3x + 15$ for several values of x

For each value of x in the table, the values of y for the two equations are equal. So for each of these values of x, the two results of evaluating the two expressions are equal.

In Example 9, we found that for each value that we used to evaluate the expressions $3(x + 5)$ and $3x + 15$, the two results were equal.

> **Definition** Equivalent expressions
>
> Two or more expressions are **equivalent expressions** if, when each variable is evaluated for *any* real number (for which all the expressions are defined), the expressions all give equal results.

Simplifying an Expression

We **simplify** an expression by using the laws of operations to remove any parentheses and to rearrange terms so that we can add any constant terms. We will discuss other ways to simplify expressions throughout this text. The result of simplifying an expression is a **simplified expression,** which is equivalent to the original expression.

Simplifying an expression often makes it easier to evaluate the expression and to graph an equation that contains the expression. We will see other benefits of simplifying expressions throughout the text.

Example 10 Simplifying Expressions

Simplify.

1. $4(2t - 3) - 7$ **2.** $5 - 6(x - 2)$ **3.** $1 - 2(5w + 3k - 6)$

Solution

1.
$$\begin{aligned} 4(2t - 3) - 7 &= 8t - 12 - 7 && \textit{Distributive law: } a(b - c) = ab - ac \\ &= 8t - 19 && \textit{Subtract constant terms.} \end{aligned}$$

2.
$$\begin{aligned} 5 - 6(x - 2) &= 5 - 6x + 12 && \textit{Distributive law: } a(b - c) = ab - ac \\ &= -6x + 5 + 12 && \textit{Rearrange terms.} \\ &= -6x + 17 && \textit{Add constant terms.} \end{aligned}$$

We use a graphing calculator table to verify our work (see Fig. 2). For each value of x in the table, the two results of evaluating the two expressions are equal; this is strong evidence that the expressions are equivalent.

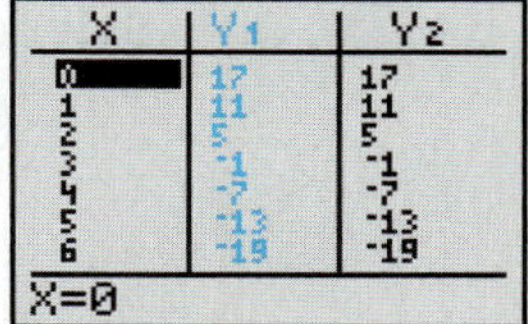

Figure 2 Verify the work

3. $1 - 2(5w + 3k - 6) = 1 - 10w - 6k + 12$ *Distributive law: $a(b + c) = ab + ac$*

$= -10w - 6k + 1 + 12$ *Rearrange terms.*

$= -10w - 6k + 13$ *Add.*

After we simplify an expression, it is wise to use a graphing calculator table to verify that the result and the original expression are indeed equivalent.

Example 11 Translating from English to Mathematics

Let x be a number. Translate the phrase "1, plus 3 times the difference of twice the number and 4" to an expression. Then simplify the expression.

Solution

The expression is

1, plus	3 times	the difference of twice the number and 4
$1 \ +$	$3 \cdot$	$(2x - 4)$
		($2x$: twice the number)

Next, we simplify the expression:

$1 + 3 \cdot (2x - 4) = 1 + 6x - 12$ *Distributive law: $a(b - c) = ab - ac$*

$= 6x + 1 - 12$ *Rearrange terms.*

$= 6x - 11$ *Subtract constant terms.*

The products $-1 \cdot 2 = -2$, $-1 \cdot 5 = -5$, and $-1 \cdot 6 = -6$ suggest the following property:

Multiplying a Number by -1

$$-1a = -a$$

In words: -1 times a number is equal to the opposite of the number.

To remove parentheses in $-(x + 3)$, we use the fact that $-a = -1a$ to write

$-(x + 3) = -1(x + 3)$ *$-a = -1a$*

$= -1x + (-1)3$ *Distributive law: $a(b + c) = ab + ac$*

$= -x - 3$ *$-1a = -a$*

Example 12 Simplifying an Expression

Simplify $-(x - 4y + 7)$.

Solution

$$
\begin{aligned}
-(x - 4y + 7) &= -1(x - 4y + 7) && -a = -1a \\
&= -1x - (-1)4y + (-1)7 && \text{Distributive law: } a(b - c) = ab - ac \\
&= -x + 4y - 7 && -1a = -a
\end{aligned}
$$

Subtracting Two Expressions

Note that $a - 1b = a - b$. We can use this fact, written $a - b = a - 1b$, to help us subtract two expressions.

Example 13 Simplifying an Expression

Simplify $7 - (x + 9)$.

Solution

$$
\begin{aligned}
7 - (x + 9) &= 7 - 1(x + 9) && a - b = a - 1b \\
&= 7 - 1x + (-1)9 && \text{Distributive law: } a(b + c) = ab + ac \\
&= 7 - x - 9 && -1a = -a \\
&= -x - 2 && \text{Rearrange terms; subtract constant terms.}
\end{aligned}
$$

We use a graphing calculator table to verify our work (see Fig. 3).

Figure 3 Verify the work

False Statements

We have discussed the commutative laws for addition and multiplication. Are there similar laws for subtraction and division? Example 14 will show that there are no such laws. **To show that an equation is false, we need find only *one* set of numbers that, when substituted for any variables, gives a result of the form $a = b$, where a and b are different real numbers.**

Example 14 Showing That a Statement Is False

Show that the statement is false.

1. $a - b = b - a$ **2.** $a \div b = b \div a$

Solution

1. We substitute the arbitrary number 5 for a and the arbitrary number 2 for b in the statement $a - b = b - a$:

$$
\begin{aligned}
a - b &= b - a \\
5 - 2 &\stackrel{?}{=} 2 - 5 && \text{Substitute 5 for } a \text{ and 2 for } b. \\
3 &\stackrel{?}{=} -3 && \text{Subtract.}
\end{aligned}
$$

false

So, the statement $a - b = b - a$ is false.

2. We substitute the arbitrary number 6 for a and the arbitrary number 3 for b in the statement $a \div b = b \div a$:

$$a \div b = b \div a$$

$$6 \div 3 \stackrel{?}{=} 3 \div 6 \quad \text{Substitute 6 for } a \text{ and 3 for } b.$$

$$2 \stackrel{?}{=} \frac{3}{6} \quad \text{Divide; } a \div b = \frac{a}{b}.$$

$$2 \stackrel{?}{=} \frac{1}{2} \quad \text{Simplify.}$$

false

So, the statement $a \div b = b \div a$ is false.

In Example 14, we showed that there is no commutative law for either subtraction or division. In Exercises 111 and 112, you will show that there is no associative law for either subtraction or division as well.

Comparing the Distributive Law with the Associative Law for Multiplication

It is helpful to compare the distributive law with the associative law for multiplication to avoid confusing the two laws. For the expression $a(b + c)$, we distribute a to both of the terms b and c:

$$a(b + c) = ab + ac \quad \text{Distributive law}$$

For the expression $a(bc)$, we do *not* distribute the a to both of the factors b and c. Rather, we distribute a to just one of the factors:

$$a(bc) = (ab)c \quad \text{Associative law for multiplication}$$

Group Exploration

Equivalent expressions

1. Evaluate both $3(x + 2)$ and $x + 6$ for $x = 0$.
2. Your results in Problem 1 should be equal. If not, check your work. If the results are equal, does that mean that $3(x + 2)$ and $x + 6$ are equivalent? Explain.
3. Evaluate both $3(x + 2)$ and $x + 6$ for $x = 5$. Compare your results and explain what your comparison tells you.
4. Explain why it is good practice to evaluate expressions several times, each time for a different value of x, to be more confident that the expressions are equivalent.
5. Simplify $3(x + 2)$. Then evaluate both $3(x + 2)$ and your simplified result for several values of x to check that your work is correct.

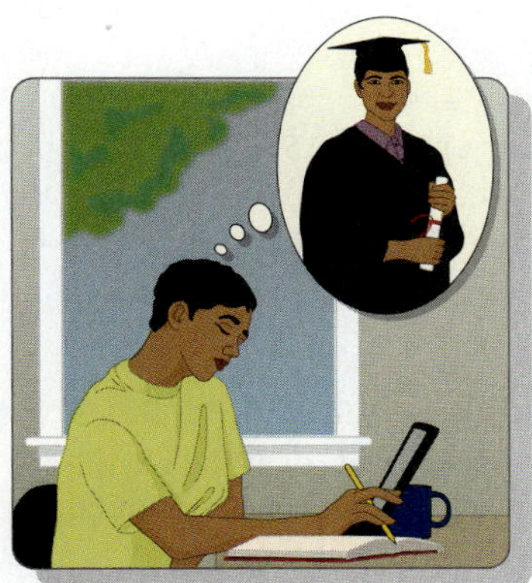

Tips for Success — Desire and Faith

To accomplish anything worthwhile, including succeeding in this course, requires substantial effort *and* faith that you will succeed. The following quote describes what it takes to achieve a goal:

> *"The secret of making something work in your lives is, first of all, the deep desire to make it work. Then the faith and belief that it can work. Then to hold that clear definite vision in your consciousness and see it working out step by step without one thought of doubt or disbelief."*
>
> —From *Footprints on the Path* by Eileen Caddy, © 1976, 1991. Used by permission of Findhorn Press, Scotland, UK.

Your deep desire to succeed in this course might be to earn a degree so you can earn more money. Or perhaps you will be motivated to learn algebra for the love of learning or to experience setting a goal and reaching it. Your faith and belief can come from knowing you, your instructor, and your college will do everything possible to ensure your success. To hold your vision of success "without one thought of doubt or disbelief" is a tall order, but the more you look for ways to succeed rather than feel discouraged, the better are your chances of success.

Homework 4.1

For extra help ▶ MyMathLab® 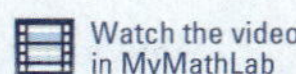 Watch the videos in MyMathLab  Download the MyDashboard App

Use the commutative law for addition to write the expression in another form.

1. $5 + x$ **2.** $x + 8$ **3.** $2p + 7$ **4.** $6 + 8w$

Use the commutative law for multiplication to write the expression in another form.

5. xy **6.** pw **7.** $-2x$
8. $-9x$ **9.** $15 + 4m$ **10.** $bn + 12$

Use an associative law to write the expression in another form.

11. $(x + 4) + y$ **12.** $(8 + x) + y$
13. $(4b)c$ **14.** $(7p)w$
15. $x + (y + 3)$ **16.** $9 + (k + d)$
17. $a(bc)$ **18.** $k(pw)$

Use the commutative and associative laws to simplify the expression. Use a graphing calculator table to verify your work.

19. $2(5x)$ **20.** $3(6x)$
21. $(p + 4) + 3$ **22.** $(k + 8) + 1$
23. $-4(-9x)$ **24.** $-5(-7x)$
25. $2 + (3b + 9)$ **26.** $1 + (8m + 4)$
27. $\frac{1}{2}(-8x)$ **28.** $\frac{2}{3}(-12x)$
29. $7\left(\frac{x}{4}\right)$ **30.** $6\left(\frac{x}{5}\right)$

Simplify.

31. $5 - 3x + 4 - 7$ **32.** $2 - 5x - 6 + 1$
33. $-4 + 2k + 8p - 3$ **34.** $5 - 3k + 2p - 9$
35. $x - \frac{9}{7} + y - \frac{5}{7}$ **36.** $x - \frac{5}{3} - y + \frac{2}{3}$

Use the distributive law to simplify the expression. Use a graphing calculator to verify your work when possible.

37. $3(x + 9)$ **38.** $6(x + 7)$
39. $(x - 5)2$ **40.** $(x - 8)4$
41. $-2(t + 5)$ **42.** $-4(w + 3)$
43. $-5(6 - 2x)$ **44.** $-8(5 - 3x)$
45. $(4x + 7)(-6)$ **46.** $(8x + 2)(-3)$
47. $2(3x - 5y)$ **48.** $4(2x - 8y)$
49. $-5(5x + 3y - 8)$ **50.** $-6(2x - 4y + 5)$
51. $-0.3(x + 0.2)$ **52.** $-0.5(x + 0.4)$
53. $\frac{1}{2}(4x + 8)$ **54.** $\frac{1}{3}(6x - 9)$
55. $-\frac{3}{5}(45 - 35x)$ **56.** $-\frac{5}{6}(30 + 12x)$

Simplify. Use a graphing calculator to verify your work when possible.

57. $3 + 2(x + 1)$ **58.** $1 + 5(x + 3)$
59. $4(a + 3) + 7$ **60.** $2(t + 4) + 1$
61. $-3(4x - 2) + 3$ **62.** $-5(2x - 3) + 7$
63. $4 - 3(3a - 5)$ **64.** $-2 - 6(2p - 3)$
65. $-3.7 + 4.2(2.5x - 8.3)$ **66.** $6.4 + 3.9(6.1x - 4.4)$
67. $-(t + 2)$ **68.** $-(a + 5)$
69. $-(8x - 9y)$ **70.** $-(4x - 3y)$
71. $-(5x + 8y - 1)$ **72.** $-(3x - 7y + 4)$
73. $-(3x + 2) + 5$ **74.** $-(4x + 1) + 3$
75. $8 - (x + 3)$ **76.** $4 - (x + 1)$
77. $-2 - (2t - 5)$ **78.** $-3 - (4a - 2)$
79. $7 - 2(4x - 7y + 5)$ **80.** $3 - 6(2x + 3y - 9)$

Simplify the expression. Then evaluate both the expression and your simplified result for x = 2 to check your work. Finally, evaluate both expressions for a value of x of your choosing (other than 2) to be even more confident that your work is correct.

81. $2(4x)$ **82.** $3(5x)$
83. $5(x - 7)$ **84.** $4(x - 3)$
85. $5 - 3(x + 4)$ **86.** $1 - 6(x + 3)$

Let x be a number. Translate the English phrase to a mathematical expression. Then simplify the expression.

87. 3 times the sum of the number and 2
88. -5 times the difference of the number and 7
89. -4 times the difference of twice the number and 5
90. 2 times the sum of twice the number and 6
91. 5, minus 2 times the difference of the number and 4
92. 6, minus 3 times the sum of the number and 1
93. -2, plus 7 times the sum of twice the number and 1
94. 3, minus 2 times the difference of twice the number and 5

Concepts

95. **a.** Simplify $2(x - 3) + 4$.
b. Evaluate $2(x - 3) + 4$ for $x = 5$.
c. Evaluate your result in part (a) for $x = 5$.
d. Compare your results in parts (b) and (c). If the results are equal, what does this tell you? If the results are not equal, what does that tell you?

96. **a.** Simplify $7 - 4(x + 2)$.
b. Evaluate $7 - 4(x + 2)$ for $x = 3$.
c. Evaluate your result in part (a) for $x = 3$.
d. Compare your results in parts (b) and (c). If the results are equal, what does this tell you? If the results are not equal, what does that tell you?

97. A student tries to simplify $3(x + 4)$:

$$3(x + 4) = 3x + 4$$

Evaluate both $3(x + 4)$ and $3x + 4$ for $x = 2$ and compare your results. What does this comparison tell you?

98. A student tries to simplify $-2(x-5)$:

$$-2(x-5) = -2x - 10$$

Evaluate both $-2(x-5)$ and $-2x-10$ for $x=3$ and compare your results. What does this comparison tell you?

99. It is a common error to write $a(bc) = (ab)(ac)$.

a. Evaluate both $a(bc)$ and $(ab)(ac)$ for $a=2$, $b=3$, and $c=4$.

b. Explain why your results show that the statement $a(bc) = (ab)(ac)$ is false.

c. Write a true statement that involves $a(bc)$.

100. It is a common error to write $a(b+c) = ab + c$.

a. Evaluate both $a(b+c)$ and $ab+c$ for $a=2, b=3$, and $c=4$.

b. Explain why your results show that the statement $a(b+c) = ab+c$ is false.

c. Write a true statement that involves $a(b+c)$.

Simplify the right-hand side of the equation. Then graph the equation by hand. Finally, use graphing calculator graphs to verify your work.

101. $y = 2(x-3)$ **102.** $y = -4(x+2)$

103. $y = -3(x+1)$ **104.** $y = -2(x-2)$

105. $y = -2(3x)$ **106.** $y = 4(2x)$

107. A student works 7 hours on Monday, 3 hours on Tuesday, and 6 hours on Wednesday. The student earns \$10 per hour. Show two ways to compute the student's total earnings. Use the distributive law to explain why both methods give the same result.

108. When we write $-(x+5)$ as $-x-5$, what *number* have we distributed? Show the missing steps.

109. Give three examples of expressions that are equivalent to the expression $2x-6$.

110. Give an example of two expressions that are not equivalent, but that happen to give the same result when they are evaluated for $x=0$.

111. Show that the statement $a-(b-c) = (a-b)-c$ is false by substituting 7 for a, 5 for b, and 1 for c. Explain why your work shows that the statement is false and that there is no associative law for subtraction.

112. Show that the statement $a \div (b \div c) = (a \div b) \div c$ is false by substituting 8 for a, 4 for b, and 2 for c. Explain why your work shows that the statement is false and why there is no associative law for division.

113. Explain what law was used in each step to rearrange terms of $a+b+c$ as follows:

$$\begin{aligned} a+b+c &= (a+b)+c && \text{Add from left to right.} \\ &= (b+a)+c \\ &= b+(a+c) \\ &= b+(c+a) \\ &= (b+c)+a \\ &= b+c+a && \text{Add from left to right.} \end{aligned}$$

114. Explain what law was used in each step to rearrange factors of abc as follows:

$$\begin{aligned} abc &= (ab)c && \text{Multiply from left to right.} \\ &= a(bc) \\ &= a(cb) \\ &= (ac)b \\ &= (ca)b \\ &= cab && \text{Multiply from left to right.} \end{aligned}$$

115. Explain what law was used in each step to show that $(a+b)c = ac+bc$:

$$\begin{aligned} (a+b)c &= c(a+b) \\ &= ca+cb \\ &= ac+bc \end{aligned}$$

116. Use the distributive law $a(b-c) = ab-ac$ and other laws to show that $(a-b)c = ac-bc$. [**Hint:** See Exercise 115.]

117. In Section 2.5, we learned that the product of two numbers with the same sign is positive. Explain what property or law was used in each step to show why $(-2)(-3) = 6$:

$$\begin{aligned} (-2)(-3) &= (-1\cdot 2)(-3) \\ &= -1(2\cdot(-3)) \\ &= -1(-6) \\ &= -(-6) \\ &= 6 \end{aligned}$$

118. Describe the meaning of *equivalent expressions*. (See page 9 for guidelines on writing a good response.)

119. Give an example of two equivalent expressions.

4.2 Simplifying Expressions

Objectives

- Learn to combine like terms.
- Simplify expressions.
- Locate errors in simplifying expressions by evaluating expressions.

In Section 4.1, we discussed several ways to simplify an expression. Now we will discuss yet another technique used to simplify an expression: *combining like terms.*

Combining Like Terms

The **coefficient** of a term is the constant factor of the term. For example, the coefficient of the term $-3x$ is -3. Since $x = 1\cdot x$, the coefficient of the term x is 1. Since $-x = -1x$, the coefficient of the term $-x$ is -1. For the term 4, the coefficient is 4.

Like terms are either constant terms or variable terms that contain the same variable(s) raised to exactly the same power(s). For example, 5 and 9 are like terms; so are $3x$ and $8x$. Also, $2y^3$, $5y^3$, and $-6y^3$ are like terms.

If terms are not like terms, we say that they are **unlike terms.** For example, $3x$ and $8y$ are unlike terms, because $3x$ contains an x but $8y$ does not and because $8y$ contains a y but $3x$ does not. The terms $4x^2$ and $7x^3$ are unlike terms, because the exponents of x are different.

For a sum of like terms, such as $2x + 4x$, we can use the distributive law to write the sum as one term:

$$\begin{aligned} 2x + 4x &= (2 + 4)x && \text{Distributive law: } ac + bc = (a + b)c \\ &= 6x && \text{Add.} \end{aligned}$$

When we write a sum or difference of like terms as one term, we say we have **combined like terms.**

X	Y1	Y2
0	0	0
1	11	11
2	22	22
3	33	33
4	44	44
5	55	55
6	66	66

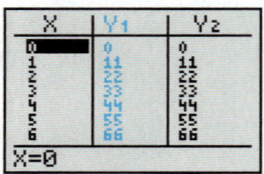

X=0

Figure 4 Verify the work

Example 1 Combining Two Like Terms

Combine like terms.

1. $4x + 7x$ **2.** $8x - 5x$ **3.** $6a + a$

Solution

1. $$\begin{aligned} 4x + 7x &= (4 + 7)x && \text{Distributive law: } ac + bc = (a + b)c \\ &= 11x && \text{Add.} \end{aligned}$$

We use a graphing calculator table to verify our work (see Fig. 4).

2. $$\begin{aligned} 8x - 5x &= (8 - 5)x && \text{Distributive law: } ac - bc = (a - b)c \\ &= 3x && \text{Subtract.} \end{aligned}$$

3. $$\begin{aligned} 6a + a &= 6a + 1a && a = 1a \\ &= (6 + 1)a && \text{Distributive law: } ac + bc = (a + b)c \\ &= 7a && \text{Add.} \end{aligned}$$

In Problem 1 of Example 1, we found that $4x + 7x = 11x$. We can find the sum $4x + 7x$ in one step by adding the coefficients of $4x$ and $7x$ (that is, 4 and 7) to get the coefficient of the result, 11:

$$4x + 7x = 11x$$

$$4 + 7 = 11$$

In Problem 2 of Example 1, we found that $8x - 5x = 3x$. Note that if we write $8x - 5x = 8x + (-5x)$, then we can add the coefficients of $8x$ and $-5x$ (that is, 8 and -5) to get the coefficient of the result, 3:

$$8x - 5x = 3x$$

$$8 + (-5) = 3$$

Combining Like Terms

To combine like terms, add the coefficients of the terms and keep the same variable factors.

Simplifying an Expression

We simplify an expression by removing parentheses and combining like terms.

Example 2 Simplifying Expressions

Simplify.

1. $2x + 5y - 6x + 2 + 3y + 7$ **2.** $-2(x + 5y) - 3(2x - 4y)$

3. $4(x - 2) - (x + 3)$

Solution

1. $$2x + 5y - 6x + 2 + 3y + 7 = 2x - 6x + 5y + 3y + 2 + 7 \quad \textit{Rearrange terms.}$$
$$= -4x + 8y + 9 \quad \textit{Combine like terms.}$$

2. $$-2(x + 5y) - 3(2x - 4y) = -2x - 10y - 6x + 12y \quad \textit{Distributive law}$$
$$= -2x - 6x - 10y + 12y \quad \textit{Rearrange terms.}$$
$$= -8x + 2y \quad \textit{Combine like terms.}$$

3. We write $4(x - 2) - (x + 3)$ as $4(x - 2) - 1(x + 3)$ so that we can distribute the -1:

$$4(x - 2) - (x + 3) = 4(x - 2) - 1(x + 3) \quad a - b = a - 1b$$
$$= 4x - 8 - 1x - 3 \quad \textit{Distributive law}$$
$$= 4x - 1x - 8 - 3 \quad \textit{Rearrange terms.}$$
$$= 3x - 11 \quad \textit{Combine like terms.}$$

We use a graphing calculator table to verify the work (see Fig. 5).

X	Y1	Y2
0	-11	-11
1	-8	-8
2	-5	-5
3	-2	-2
4	1	1
5	4	4
6	7	7

X=0

Figure 5 Verify the work

Example 3 Translating from English to Mathematics

Let x be a number. Translate the phrase "3 times the number, minus 4, plus twice the number" to an expression. Then simplify the expression.

Solution

The expression is

$$\underbrace{3x}_{\text{3 times the number,}} \quad \underbrace{-4}_{\text{minus 4,}} \quad \underbrace{+}_{\text{plus}} \quad \underbrace{2x}_{\text{twice the number}}$$

Next, we simplify the expression:

$$3x - 4 + 2x = 3x + 2x - 4 \quad \textit{Rearrange terms.}$$
$$= 5x - 4 \quad \textit{Combine like terms.}$$

In Example 3, we translated an English phrase into a mathematical expression. In Example 4, we will translate a mathematical expression into an English phrase.

Example 4 Translating from Mathematics to English

Let x be a number. Translate the expression $x - 5 \cdot (x - 2)$ into an English phrase. Then simplify the expression.

Solution

Here is the translation:

$$\underbrace{x \; -}_{\text{the number, minus}} \quad \underbrace{5 \; \cdot}_{\text{5 times}} \quad \underbrace{(x \; - \; 2)}_{\text{the difference of the number and 2}}$$

One of many possible correct translations is "the number, minus 5 times the difference of the number and 2." Next, we simplify the expression:

$$x - 5 \cdot (x - 2) = x - 5x + 10 \quad \textit{Distributive law}$$
$$= -4x + 10 \quad \textit{Combine like terms.}$$

Locate Errors in Simplifying Expressions

So far, we have used a graphing calculator table to verify our work in simplifying an expression. If we determine that we have made an error, we still have to find it to correct it. In Example 5, we will discuss how to pinpoint the step(s) in which any errors have been made.

Example 5 Locating an Error

Consider the following incorrect work:

$$\begin{aligned} 5x + 6 + 3(x + 4) &= 5x + 6 + 3x + 4 \quad \textit{Incorrect} \\ &= 5x + 3x + 6 + 4 \\ &= 8x + 10 \end{aligned}$$

1. Use a graphing calculator table to show that the work is incorrect.
2. Pinpoint where the error was made by evaluating the expression in each step for $x = 2$.

Solution

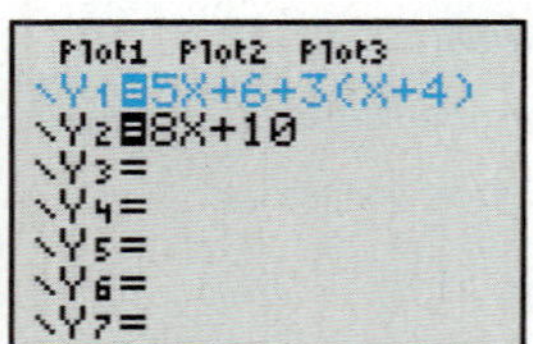

X	Y1	Y2
0	18	10
1	26	18
2	34	26
3	42	34
4	50	42
5	58	50
6	66	58

X=0

Figure 6 The work is incorrect

1. From Fig. 6, we can tell that the expressions $5x + 6 + 3(x + 4)$ and $8x + 10$ are *not* equivalent. So, an error has been made.
2. We evaluate each expression for $x = 2$:

Expression	Evaluate the Expression for $x = 2$
$5x + 6 + 3(x + 4)$	$5(2) + 6 + 3(2 + 4) = 34$
$= 5x + 6 + 3x + 4$	$5(2) + 6 + 3(2) + 4 = 26$
$= 5x + 3x + 6 + 4$	$5(2) + 3(2) + 6 + 4 = 26$
$= 8x + 10$	$8(2) + 10 = 26$

Since the result of evaluating $5x + 6 + 3(x + 4)$ for $x = 2$ is different from the result of evaluating $5x + 6 + 3x + 4$ for $x = 2$, we conclude that an error was made in writing $5x + 6 + 3(x + 4) = 5x + 6 + 3x + 4$. (Note that the right-hand side of the equation *should* be $5x + 6 + 3x + 12$.)

Locating an Error in Simplifying an Expression

After simplifying an expression, use a graphing calculator table to check whether the result is equivalent to the original expression. If the expressions are not equivalent, an error was made. To pinpoint the error, evaluate all of the expressions by using the same number(s) and find the pair of consecutive expressions for which the results are different.

Group Exploration

Laws of operations

For the work shown, carefully describe any errors.

1. $\begin{aligned} 3(x + 5) + 4x &= 3x + 5 + 4x \\ &= 3x + 4x + 5 \\ &= 7x + 5 \end{aligned}$

2. $\begin{aligned} 2(xy) &= (2x)(2y) \\ &= 2(2)xy \\ &= 4xy \end{aligned}$

3. $\begin{aligned} 5 + x - 3 &= 5 + 3 - x \\ &= 8 - x \\ &= x - 8 \end{aligned}$

Tips for Success Read Your Response

After writing a response to a question, read your response to make sure that it says what you intended it to say. Reading your response aloud, if you are in a place where you feel comfortable doing so, can help.

Homework 4.2

For extra help ▶ MyMathLab® Watch the videos in MyMathLab  Download the MyDashboard App

Simplify. Use a graphing calculator table to verify your work when possible.

1. $2x + 5x$ **2.** $3x + 6x$

3. $9x - 4x$ **4.** $6x - 5x$

5. $-8w - 5w$ **6.** $-4p + 7p$

7. $-t + 5t$ **8.** $a + 3a$

9. $6.6x - 7.1x$ **10.** $-4.5x - 2.9x$

11. $\frac{2}{3}x + \frac{5}{3}x$ **12.** $\frac{9}{5}x - \frac{2}{5}x$

13. $2 + 4x - 5 - 7x$ **14.** $-8x - 1 + 3x - 4$

15. $-3p + 2 + p - 9$ **16.** $7 - w - 10 - 2w$

17. $3y + 5x - 2y - 2x + 1$ **18.** $5 - 3x + 7y + 6x$

19. $-4.6x + 3.9y + 2.1 - 5.3x - 2.8y$

20. $4.7 - 3.5y + 8.8x - 6.2y + 1.9x$

21. $-3(a - 5) + 2a$ **22.** $5(t - 8) - 6t$

23. $5.2(8.3x + 4.9) - 2.4$ **24.** $-3.8(2.7x - 5.5) - 8.4$

25. $4(3a - 2b) - 5a$ **26.** $-2(8m + 4n) - 3n$

27. $8 - 2(x + 3) + x$ **28.** $3 - 4(x + 1) - x$

29. $6x - (4x - 3y) - 5y$ **30.** $8x - (3x + 7y) - 2y$

31. $2t - 3(5t + 2) + 1$ **32.** $3a - 2(3a + 4) + 5$

33. $6 - 2(x + 3y) + 2y$ **34.** $4 - 5(x + 2y) + 7y$

35. $-3(x - 2) - 5(x + 4)$ **36.** $-2(x - 3) - 7(x + 2)$

37. $6(2x - 3y) - 4(9x + 5y)$

38. $2(7x - 5y) - 5(3x + y)$

39. $-(x - 1) - (1 - x)$

40. $-(6x - 7) - (7 - 6x)$

41. $2x - 5y - 3(2x - 4y + 7)$

42. $4x + 3y - 2(5x + 2y + 8)$

43. $5(2x - 4y) - (3x - 7y + 2)$

44. $3(4x - 3y) - (9x + 2y - 5)$

45. $\frac{2}{7}(a + 1) - \frac{4}{7}(a - 1)$ **46.** $\frac{3}{5}(t - 1) + \frac{1}{5}(t + 1)$

47. $5x - \frac{1}{2}(4x + 6)$ **48.** $7x - \frac{1}{3}(6x + 9)$

Let x be a number. Translate the English phrase into a mathematical expression. Then simplify the expression.

49. The number plus the product of 5 and the number

50. The number minus the product of the number and 3

51. 4 times the difference of the number and 2

52. −6 times the sum of the number and 4

53. The number, plus 3 times the difference of the number and 7

54. The number, minus 5 times the sum of the number and 2

55. Twice the number, minus 4 times the sum of the number and 6

56. Twice the number, plus 9 times the difference of the number and 4

Let x be a number. Translate the expression into an English phrase. Then simplify the expression.

57. $2x + 6x$ **58.** $3x - 8x$

59. $7(x - 5)$ **60.** $-2(x + 4)$

61. $x + 5(x + 1)$ **62.** $x - 8(x - 3)$

63. $2x - 3(x - 9)$ **64.** $2x + 6(x - 4)$

65. Find the sum of $3x - 7$ and $5x + 2$.

66. Find the sum of $6x - 3$ and $8x - 9$.

67. Find the difference of $4x + 8$ and $7x - 1$.

68. Find the difference of $5x - 9$ and $2x + 6$.

Simplify. Then evaluate both the expression and your result for $x = 4$ to check your work. Finally, evaluate both expressions for an x-value of your choosing (other than 4) to be even more confident that your work is correct.

69. $-2x + 5 - 3 + 7x$ **70.** $4x - 8 - 2 + x$

71. $4(x + 2) - (x - 3)$ **72.** $-(x - 7) + 4(x + 2)$

Concepts

73. a. Simplify $3(x + 4) + 5x$.
b. Evaluate $3(x + 4) + 5x$ for $x = 2$.
c. Evaluate your result in part (a) for $x = 2$.
d. Compare your results in parts (b) and (c). If the results are equal, what does this tell you? If the results are not equal, what does that tell you?

74. a. Simplify $4(x - 2) - 3(x + 1)$.
b. Evaluate $4(x - 2) - 3(x + 1)$ for $x = 3$.
c. Evaluate your result in part (a) for $x = 3$.
d. Compare your results in parts (b) and (c). If the results are equal, what does this tell you? If the results are not equal, what does that tell you?

75. Which of the following expressions are equivalent?

$-2x - 3$ $\quad -2(x - 3)$ $\quad 2(3 - x)$ $\quad -2x - 6$ $\quad -3(x - 2) + x$ $\quad -2x + 6$

76. Which of the following expressions are equivalent?

$-(2x-3)$ $\quad 3-2x$ $\quad -(3-2x)$ $\quad -2x+3$

$-(3x-2)+x+1$ $\quad -(x-1)-(x-2)$

77. A student incorrectly simplifies $4(x-3)+5x-1$:

Expression	Evaluate the Expression for $x = 2$
$4(x-3)+5x-1$	
$= 4x-3+5x-1$	
$= 4x+5x-3-1$	
$= 9x-4$	

Evaluate each of the four expressions for $x = 2$ to pinpoint where the error was made. Then simplify $4(x-3)+5x-1$ correctly.

78. A student incorrectly simplifies $2(x+1)+x-4$:

Expression	Evaluate the Expression for $x = 3$
$2(x+1)+x-4$	
$= 2x+2+x-4$	
$= 2x+x+2-4$	
$= 2x-2$	

Evaluate each of the four expressions for $x = 3$ to pinpoint where the error was made. Then simplify $2(x+1)+x-4$ correctly.

79. Give three examples of expressions equivalent to the expression $2(x-5)+3(x+1)$.

80. Give three examples of expressions equivalent to x.

For Exercises 81–86, simplify the right-hand side of the equation. Then graph the equation by hand. Finally, use graphing calculator graphs to verify your work.

81. $y = 3x - 5x$

82. $y = x + 2x$

83. $y = 9x - 4 - 7x$

84. $y = 4x + 3 - 6x$

85. $y = 4(2x-1) - 5x$

86. $y = -2(3x-2) + 5x$

87. a. Pick a number. Next, subtract 3. Then double the result. Then subtract the original number. Finally, add 6. Compare your result with the original number.

b. Pick another number and follow the instructions in part (a), including comparing your result with the original number.

c. Let x be a number and follow the instructions in part (a). Your result should be an expression. Simplify the expression to explain your observations in parts (a) and (b).

88. Give an example of combining like terms. Then use the distributive law to show why we can do this.

89. Describe how to simplify an expression.

4.3 Solving Linear Equations in One Variable

Objectives

- » Know the meaning of *linear equation in one variable.*
- » For an equation in one variable, know the meaning of *satisfy, solution, solution set,* and *solve.*
- » Know three roles of variables.
- » Know the meaning of *equivalent equations.*
- » Know the addition property of equality and the multiplication property of equality.
- » Solve a linear equation in one variable.
- » Solve a percentage problem.
- » Use a graph or a table to solve a linear equation in one variable.
- » Think of solving an equation in terms of "undoing" operations.

So far, we have discussed how to graph linear equations in *two* variables, such as

$$y = x + 2 \qquad y = -5x \qquad y = 4x - 7 \qquad y = 2x - 6$$

In this section and Section 4.4, we will work with *linear equations in one variable,* such as:

$$0 = x + 2 \qquad 9 = -5x \qquad 4x - 7 = 3 \qquad x + 1 = 2x - 6$$

We will see in this section and in Section 4.4 that we can put each of these four equations in the form $mx + b = 0$, where m and b are constants and $m \neq 0$.

Definition Linear equation in one variable

A **linear equation in one variable** is an equation that can be put into the form

$$mx + b = 0$$

where m and b are constants and $m \neq 0$.

Working with linear equations in one variable will enable us to make efficient estimates and predictions about authentic situations.

Meaning of Satisfy, Solution, Solution Set, and Solve

Consider the linear equation

$$x + 1 = 6$$

This equation becomes a false statement if we substitute 2 for x:

$$x + 1 = 6 \quad \textit{Original equation}$$
$$(2) + 1 \stackrel{?}{=} 6 \quad \textit{Substitute 2 for x.}$$
$$3 \stackrel{?}{=} 6 \quad \textit{Add.}$$
$$\text{false}$$

However, the equation $x + 1 = 6$ becomes a true statement if we substitute 5 for x:

$$x + 1 = 6 \quad \textit{Original equation}$$
$$(5) + 1 \stackrel{?}{=} 6 \quad \textit{Substitute 5 for x.}$$
$$6 \stackrel{?}{=} 6 \quad \textit{Add.}$$
$$\text{true}$$

We say 5 *satisfies* the equation $x + 1 = 6$ and 5 is a *solution* of the equation. In fact, 5 is the only solution of $x + 1 = 6$, because 5 is the only number that, when increased by 1, is equal to 6. We call the set containing only this number the *solution set* of the equation.

Definition *Solution, satisfy, solution set,* and *solve* for an equation in one variable

A number is a **solution** of an equation in one variable if the equation becomes a true statement when the number is substituted for the variable. We say the number **satisfies** the equation. The set of all solutions of the equation is called the **solution set** of the equation. We **solve** the equation by finding its solution set.

Example 1 Identifying Solutions of an Equation

1. Is 3 a solution of the equation $5(x - 1) = 10 + 2x$?
2. Is 5 a solution of the equation $5(x - 1) = 10 + 2x$?

Solution

1. We begin by substituting 3 for x in $5(x - 1) = 10 + 2x$:

$$5(x - 1) = 10 + 2x \quad \textit{Original equation}$$
$$5(3 - 1) \stackrel{?}{=} 10 + 2(3) \quad \textit{Substitute 3 for x.}$$
$$5(2) \stackrel{?}{=} 10 + 6 \quad \textit{Simplify.}$$
$$10 \stackrel{?}{=} 16 \quad \textit{Simplify.}$$
$$\text{false}$$

So, 3 is not a solution of the equation $5(x - 1) = 10 + 2x$.

2. We begin by substituting 5 for x in $5(x - 1) = 10 + 2x$:

$$5(x - 1) = 10 + 2x \quad \textit{Original equation}$$
$$5(5 - 1) \stackrel{?}{=} 10 + 2(5) \quad \textit{Substitute 5 for x.}$$
$$5(4) \stackrel{?}{=} 10 + 10 \quad \textit{Simplify.}$$
$$20 \stackrel{?}{=} 20 \quad \textit{Simplify.}$$
$$\text{true}$$

So, 5 is a solution of the equation $5(x - 1) = 10 + 2x$.

Roles of a Variable

In Section 2.1, we discussed two roles of a variable. In one role, a variable represents a quantity that can vary. For example, if we let s represent the speed of an airplane, then the variable s can vary between 0 and 2200 miles per hour. In another role, a variable can serve as a placeholder in an expression. For example, the variable x is a placeholder for a number in the expression $2x + 5$.

A variable can be used in yet another way: In an equation, a variable is used to represent any number that is a solution of the equation. For instance, in Example 1, the variable x is used to represent any number that satisfies the equation $5(x - 1) = 10 + 2x$. In Problem 2 of Example 1, we found that 5 is such a number.

Roles of a Variable

Here are three roles of a variable:

1. A variable represents a quantity that can vary.
2. In an expression, a variable is a placeholder for a number.
3. In an equation, a variable represents any number that is a solution of the equation.

Here we emphasize the third role of a variable.

Equivalent Equations

Consider the equation $x = 2$. We add 5 to both sides of the equation:

$$x = 2 \quad \textit{Original equation}$$
$$x + 5 = 2 + 5 \quad \textit{Add 5 to both sides.}$$
$$x + 5 = 7 \quad \textit{Add.}$$

Note that 2 satisfies all three equations:

Equation	Does 2 satisfy the equation?	
$x = 2$	$(2) \stackrel{?}{=} 2$	*true*
$x + 5 = 2 + 5$	$(2) + 5 \stackrel{?}{=} 2 + 5$	*true*
$x + 5 = 7$	$(2) + 5 \stackrel{?}{=} 7$	*true*

In fact, 2 is the *only* number that satisfies any of these equations. So, the equations $x = 2$, $x + 5 = 2 + 5$, and $x + 5 = 7$ have the same solution set. We say that the three equations are *equivalent*.

Definition Equivalent equations

Equivalent equations are equations that have the same solution set.

Addition Property of Equality

The fact that the equations $x = 2$ and $x + 5 = 2 + 5$ have the same solution set suggests that adding a number to both sides of an equation does not change an equation's solution set. This property is called the **addition property of equality.**

Addition Property of Equality

If A and B are expressions and c is a number, then the equations $A = B$ and $A + c = B + c$ are equivalent.

To solve an equation in one variable, x, we can sometimes use the addition property of equality to get x alone on one side of the equation. Then we can identify solutions of the equation. For example, for the equation $x = 3$, we can see the solution is 3.

Example 2 Solving an Equation by Adding a Number to Both Sides

Solve $x - 2 = 3$.

Solution

To get x alone on the left side, we add 2 to *both* sides:

$$
\begin{aligned}
x - 2 &= 3 && \text{Original equation} \\
x - 2 + 2 &= 3 + 2 && \text{Addition property of equality: Add 2 to both sides.} \\
x + 0 &= 5 && \text{Simplify: } -a + a = 0 \\
x &= 5 && \text{Simplify: } a + 0 = a
\end{aligned}
$$

Next, we check that 5 satisfies the original equation, $x - 2 = 3$:

$$
\begin{aligned}
x - 2 &= 3 && \text{Original equation} \\
(5) - 2 &\stackrel{?}{=} 3 && \text{Substitute 5 for } x. \\
3 &\stackrel{?}{=} 3 && \text{Subtract.} \\
&\text{true}
\end{aligned}
$$

So, the solution is 5.

After solving an equation, check that all of your results satisfy the equation.

In Example 2, we worked with the equations $x - 2 = 3$, $x - 2 + 2 = 3 + 2$, $x + 0 = 5$, and $x = 5$. Although we could find the solution set with any of these equivalent equations, it is easiest to determine the solution set from the equation $x = 5$, which has the variable alone on the left side. **Our strategy in solving linear equations in one variable will be to use properties to get the variable alone on one side of the equation.**

Assume that A and B are expressions and that c is a number. Since subtracting a number is the same as adding the opposite of the number, the addition property of equality implies that the equations $A = B$ and $A - c = B - c$ are equivalent.

Example 3 Solving an Equation by Subtracting a Number from Both Sides

Solve $x + 4 = 6$.

Solution

To get x alone on the left side, we subtract 4 from *both* sides:

$$
\begin{aligned}
x + 4 &= 6 && \text{Original equation} \\
x + 4 - 4 &= 6 - 4 && \text{Subtract 4 from both sides.} \\
x &= 2 && a + 0 = a
\end{aligned}
$$

Next, we check that 2 satisfies the original equation, $x + 4 = 6$:

$$
\begin{aligned}
x + 4 &= 6 && \text{Original equation} \\
(2) + 4 &\stackrel{?}{=} 6 && \text{Substitute 2 for } x. \\
6 &\stackrel{?}{=} 6 && \text{Add.} \\
&\text{true}
\end{aligned}
$$

The solution is 2.

Multiplication Property of Equality

Consider the equation $x = 3$. We multiply both sides of $x = 3$ by 7:

$$
\begin{aligned}
x &= 3 && \text{Original equation} \\
7 \cdot x &= 7 \cdot 3 && \text{Multiply both sides by 7.} \\
7x &= 21 && \text{Multiply.}
\end{aligned}
$$

Note that 3 satisfies all three equations:

Equation	Does 3 satisfy the equation?	
$x = 3$	$(3) \stackrel{?}{=} 3$	true
$7 \cdot x = 7 \cdot 3$	$7 \cdot (3) \stackrel{?}{=} 7 \cdot 3$	true
$7x = 21$	$7(3) \stackrel{?}{=} 21$	true

In fact, 3 is the *only* number that satisfies any of these equations. So, the equations $x = 3$, $7 \cdot x = 7 \cdot 3$, and $7x = 21$ are equivalent. The fact that the equations $x = 3$ and $7 \cdot x = 7 \cdot 3$ are equivalent suggests the **multiplication property of equality.**

Multiplication Property of Equality

If A and B are expressions and c is a nonzero number, then the equations $A = B$ and $Ac = Bc$ are equivalent.*

The multiplication property of equality is often helpful in solving equations that contain fractions. It is useful to know that a fraction times its reciprocal is equal to 1. For example,

$$\frac{2}{7} \cdot \frac{7}{2} = \frac{14}{14} = 1$$

Example 4 Solving an Equation by Multiplying Both Sides by a Number

Solve $\frac{4}{5}x = 8$.

Solution

To get x alone on the left side, we multiply *both* sides by the reciprocal of $\frac{4}{5}$, which is $\frac{5}{4}$:

$$\begin{aligned} \frac{4}{5}x &= 8 && \textit{Original equation} \\ \frac{5}{4} \cdot \frac{4}{5}x &= \frac{5}{4} \cdot 8 && \textit{Multiplication property of equality: Multiply both sides by } \frac{5}{4}. \\ 1x &= 10 && \textit{Simplify.} \\ x &= 10 && 1a = a \end{aligned}$$

The solution is 10. We use a graphing calculator table to check that, when 10 is substituted for x in the expression $\frac{4}{5}x$, the result is 8 (see Fig. 7). For graphing calculator instructions on using "Ask" in a table, see Appendix A.15.

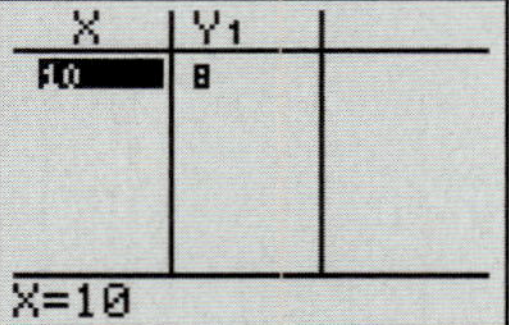

Figure 7 Verify the work

Assume that A and B are expressions and c is a nonzero number. Since dividing by a number is the same as multiplying by its reciprocal, the multiplication property of equality implies the equations $A = B$ and $\frac{A}{c} = \frac{B}{c}$ are equivalent.

*If $c = 0$, the equations may not be equivalent. For example, $x = 5$ and $x \cdot 0 = 5 \cdot 0$ (or $0 = 0$) are not equivalent. The only solution of $x = 5$ is 5, but the material in Section 4.5 will help us see that the solution set of $0 = 0$ is the set of all real numbers.

Example 5 Solving an Equation by Dividing Both Sides by a Number

Solve $-12 = -3t$.

Solution

The variable term $-3t$ is on the right-hand side this time. To get t alone on this side, we divide *both* sides by -3:

$$-12 = -3t \quad \text{Original equation}$$

$$\frac{-12}{-3} = \frac{-3t}{-3} \quad \text{Divide both sides by } -3.$$

$$4 = t \quad \text{Simplify.}$$

The solution is 4. We can check that 4 satisfies the original equation (try it).

The last step of Example 5 is $4 = t$. Note that the equation $t = 4$ is equivalent to $4 = t$ and that we can see from either equation that the solution is 4.

Example 6 Solving an Equation by Multiplying Both Sides by -1

Solve $-w = 5$.

Solution

To get w alone on the left side, we multiply both sides by -1:

$$-w = 5 \quad \text{Original equation}$$

$$-1(-w) = -1(5) \quad \text{Multiply both sides by } -1.$$

$$w = -5 \quad \text{Multiply.}$$

Next, we check that -5 satisfies the original equation, $-w = 5$:

$$-w = 5 \quad \text{Original equation}$$

$$-(-5) \stackrel{?}{=} 5 \quad \text{Substitute } -5 \text{ for } w.$$

$$5 \stackrel{?}{=} 5 \quad -(-a) = a$$

$$\text{true}$$

The solution is -5.

Example 7 Solving an Equation by Multiplying Both Sides by a Number

Solve $\dfrac{2x}{3} = \dfrac{5}{6}$.

Solution

Since $\dfrac{2}{3}x = \dfrac{2}{3} \cdot \dfrac{x}{1} = \dfrac{2x}{3}$, we first write $\dfrac{2x}{3}$ as $\dfrac{2}{3}x$. Then we multiply both sides by the reciprocal of $\dfrac{2}{3}$, which is $\dfrac{3}{2}$, to get x alone on the left side:

$$\frac{2x}{3} = \frac{5}{6} \quad \text{Original equation}$$

$$\frac{2}{3}x = \frac{5}{6} \quad \text{Write } \frac{2x}{3} \text{ as } \frac{2}{3}x.$$

$$\frac{3}{2} \cdot \frac{2}{3}x = \frac{3}{2} \cdot \frac{5}{6} \quad \text{Multiply both sides by } \frac{3}{2}.$$

$$1x = \frac{3 \cdot 5}{2 \cdot 2 \cdot 3} \quad \text{Simplify: } \frac{a}{b} \cdot \frac{c}{d} = \frac{ac}{bd}$$

$$x = \frac{5}{4} \quad \text{Simplify: } \frac{3}{3} = 1$$

X	Y1	Y2
1.25	.83333	.83333

X=

Figure 8 Verify the work.

The solution is $\frac{5}{4}$. We use a graphing calculator table to check that, when $\frac{5}{4} = 1.25$ is substituted for x in the expression $\frac{2x}{3}$, the result is $\frac{5}{6} \approx 0.83333$ (see Fig. 8).

Solving Percentage Problems

Recall that to find the percentage of a quantity, we multiply the decimal form of the percentage and the quantity (Section 2.5).

Example 8 Solving a Percentage Problem

A person deposits 28% of her weekly earnings into her savings account. If she deposits \$448 into her savings account each week, how much is her weekly pay?

Solution

We let p be the weekly pay (in dollars). Since 28% of the weekly pay is equal to \$448, we write

$$\overbrace{0.28p}^{\text{28\% of weekly pay}} \quad \overbrace{=}^{\text{is equal to}} \quad \overbrace{448}^{448}$$

Now we solve the equation:

$$\begin{aligned} 0.28p &= 448 \\ \frac{0.28p}{0.28} &= \frac{448}{0.28} && \text{Divide both sides by 0.28.} \\ p &= 1600 && \text{Simplify; divide.} \end{aligned}$$

The weekly pay is \$1600.

In most application problems in this text so far, variable names and their definitions have been provided. In Example 8, a key step was to create the variable name p and define it.

Using Graphing to Solve an Equation in One Variable

In Example 2, we showed that the solution of the equation $x - 2 = 3$ is 5. How can we use graphing to solve this equation? Here are three steps:

Step 1. We set y equal to the left side, $x - 2$, to form the equation $y = x - 2$, and we set y equal to the right side, 3, to form the equation $y = 3$. Then we graph the two equations $y = x - 2$ and $y = 3$ in the same coordinate system (see Fig. 9).

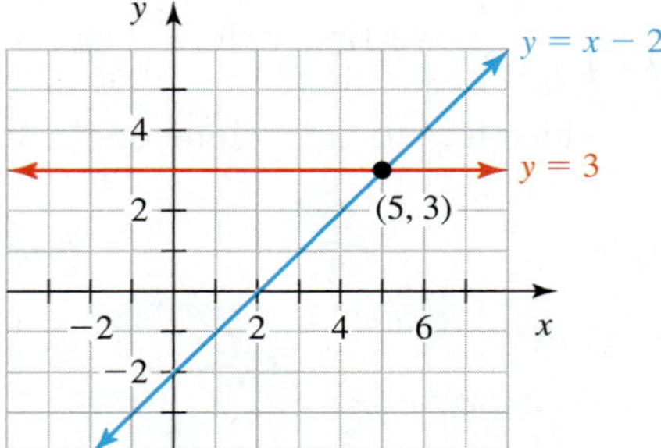

Figure 9 The intersection point is (5, 3)

Step 2. We find the intersection point of the two graphs, which is the point (5, 3).

Step 3. The solution of the original equation $x - 2 = 3$ is the x-coordinate of the intersection point (5, 3). So, the solution is 5.

In general, the solutions are all of the x-coordinates of any intersection points.

Using Graphing to Solve an Equation in One Variable

To use graphing to solve an equation $A = B$ in one variable, x, where A and B are expressions,

1. Graph the equations $y = A$ and $y = B$ in the same coordinate system. (For example, if the original equation is $5x - 9 = 3x + 7$, then we would graph the equations $y = 5x - 9$ and $y = 3x + 7$.)
2. Find all intersection points.
3. The x-coordinates of those intersection points are the solutions of the equation $A = B$.

Example 9 Solving an Equation in One Variable by Graphing

The graphs of $y = \frac{3}{2}x + 1$, $y = 4$, and $y = -5$ are shown in Fig. 10. Use these graphs to solve the given equations.

1. $\frac{3}{2}x + 1 = 4$
2. $\frac{3}{2}x + 1 = -5$

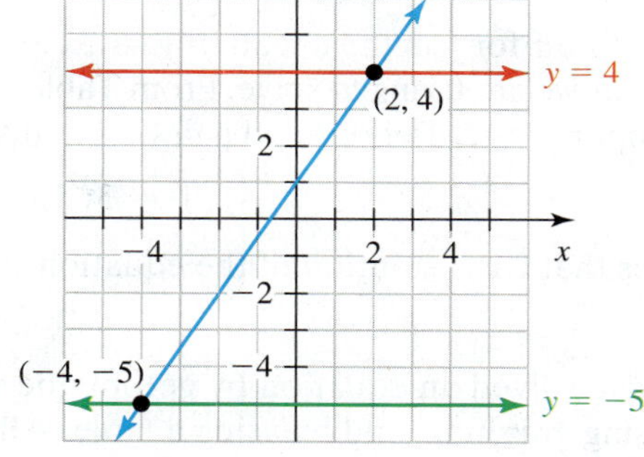

Figure 10 Solving $\frac{3}{2}x + 1 = 4$ and $\frac{3}{2}x + 1 = -5$

Solution

1. The graphs of $y = \frac{3}{2}x + 1$ and $y = 4$ intersect only at the point (2, 4). The intersection point, (2, 4), has x-coordinate 2. So, 2 is the solution of $\frac{3}{2}x + 1 = 4$. We can verify our work by checking that 2 satisfies the equation (try it).
2. The graphs of $y = \frac{3}{2}x + 1$ and $y = -5$ intersect only at the point $(-4, -5)$. The intersection point, $(-4, -5)$, has x-coordinate -4. So, -4 is the solution of $\frac{3}{2}x + 1 = -5$. We can verify our work by checking that -4 satisfies the equation (try it).

Example 10 Solving an Equation in One Variable by Graphing

Use "intersect" on a graphing calculator to solve the equation $-2x + 4 = x - 5$.

Solution

We use a graphing calculator to graph the equations $y = -2x + 4$ and $y = x - 5$ in the same coordinate system and then use "intersect" to find the intersection point, which turns out to be $(3, -2)$. See Fig. 11. See Appendix A.18 for graphing calculator instructions.

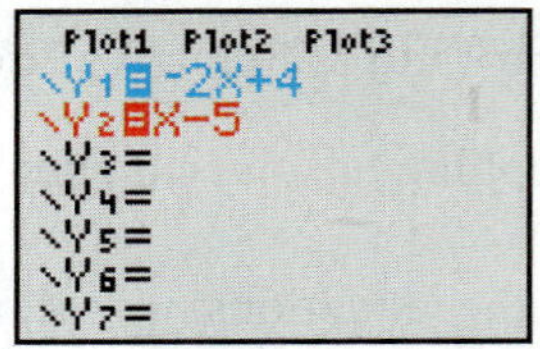

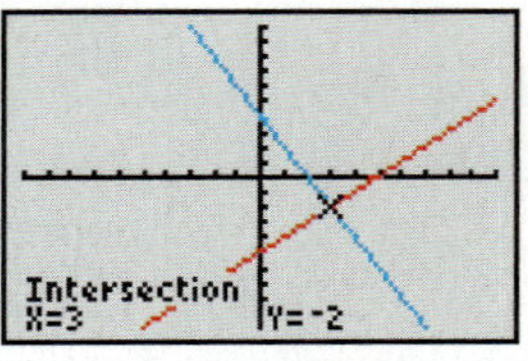

Figure 11 Using "intersect" to solve an equation in one variable

The intersection point, $(3, -2)$, has x-coordinate 3. So, the solution of the original equation is 3. We can verify our work by checking that 3 satisfies the equation $-2x + 4 = x - 5$ (try it).

Using a Table to Solve an Equation

We can use a table of solutions of an equation in two variables to help us solve an equation in one variable.

Table 2 Solutions of $y = 3x - 5$

x	y
−1	−8
0	−5
1	−2
2	1
3	4

Example 11 Solving an Equation in One Variable by Using a Table

Some solutions of $y = 3x - 5$ are shown in Table 2. Use the table to solve the equation $1 = 3x - 5$.

Solution

If we substitute 1 for y in the equation $y = 3x - 5$, the result is the equation $1 = 3x - 5$, which is what we are trying to solve. From Table 2, we see that the output $y = 1$ originates from the input $x = 2$. The ordered pair $(2, 1)$ satisfies the equation $y = 3x - 5$:

$$1 = 3(2) - 5$$

This means that 2 is a solution of the equation $1 = 3x - 5$.

We have solved an equation by getting the variable alone on one side of the equation, by using graphing, and by using a table. All three methods give the same results.

Undoing Operations

Consider the following four equations and the first step to get x alone on the left side of the equation:

Equation	First Step in Solving Equation	How to Get x Alone on the Left Side
$x + 5 = 2$	Subtract 5 from both sides.	Undo addition by subtracting.
$x - 3 = 6$	Add 3 to both sides.	Undo subtraction by adding.
$\frac{x}{2} = 3$	Multiply both sides by 2.	Undo division by multiplying.
$4x = 8$	Divide both sides by 4.	Undo multiplication by dividing.

For the left side of each equation, we get x alone by "undoing" the operation that involves x and the number. For example, we "undo" the addition of 5 to x by subtracting 5 from that sum:

$$(x + 5) - 5 = x$$

This logic also works for equations such as $\frac{4}{5}x = 8$. We undo the multiplication of x by $\frac{4}{5}$ by dividing that product by $\frac{4}{5}$. But dividing by $\frac{4}{5}$ is the same as multiplying by the reciprocal of $\frac{4}{5}$, which is $\frac{5}{4}$. In fact, we did just that in Example 4.

Remember that **whether adding, subtracting, multiplying, or dividing, what is done to one side of the equation must also be done to the other side of the equation.**

Group Exploration

Locating an error in solving an equation

A student tries to solve the equation $x - 3 = 5$:

$$x - 3 = 5$$
$$x - 3 - 3 = 5 - 3$$
$$x - 0 = 2$$
$$x = 2$$

Substitute 2 for x in each of the four equations to determine which equations have 2 as a solution. Explain how your work shows that the student made an error and how your work helps you pinpoint the step in which the error was made.

Group Exploration

Looking ahead: Solving linear equations

1. Solve $x + 3 = 11$.
2. Solve $2x = 8$.
3. Solve $2x + 3 = 11$. [**Hint:** First, subtract 3 from both sides.]
4. Solve $6x - 4x + 3 = 11$. [**Hint:** First, combine like terms on the left side of the equation.]
5. Solve the equation $2x + 4 + 3x = 14$.

Tips for Success Review Material

At various times throughout this course, you can improve your understanding of algebra by reviewing material that you have learned so far. Your review should include solving problems, redoing explorations, and reexamining concepts and techniques from previous sections.

Homework 4.3

For extra help ▶ MyMathLab® 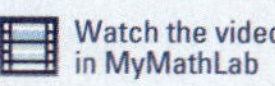 Watch the videos in MyMathLab 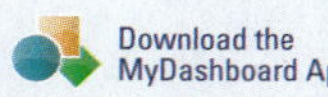 Download the MyDashboard App

Determine whether 2 is a solution of the equation.

1. $3x + 1 = 7$

2. $-2x + 5 = 4$

3. $5(2x - 1) = 0$

4. $-3(x - 2) = 0$

5. $12 - x = 2(4x - 3)$

6. $5 - x = 3(2x - 1)$

For Exercises 7–50, solve. Verify that your result satisfies the equation.

7. $x - 3 = 2$

8. $x - 4 = 9$

9. $x + 6 = -8$

10. $x + 1 = -5$

11. $t - 9 = 15$

12. $a - 5 = 12$

13. $x + 11 = -17$

14. $x + 18 = -13$

15. $-5 = x - 2$

16. $-3 = x - 4$

17. $x - 3 = 0$

18. $x - 5 = 0$

19. $6r = 18$

20. $4w = 24$

21. $-3x = 12$

22. $-5x = 20$

23. $15 = 3x$

24. $24 = 2x$

25. $6x = 8$

26. $4x = 6$

27. $-10x = -12$

28. $-14x = -6$

29. $-2x = 0$

30. $-5x = 0$

31. $\frac{1}{3}t = 5$

32. $\frac{1}{2}w = 4$

33. $-\frac{2}{7}x = -3$

34. $-\frac{4}{9}x = -5$

35. $-9 = \frac{3x}{4}$

36. $-8 = \frac{2x}{5}$

37. $\frac{2}{5}p = -\frac{4}{3}$

38. $\frac{4}{7}b = -\frac{5}{21}$

39. $-\frac{3x}{8} = -\frac{9}{4}$

40. $-\frac{5x}{4} = -\frac{15}{8}$

41. $-x = 3$

42. $-x = 2$

43. $-\frac{1}{2} = -x$

44. $-\frac{3}{4} = -x$

45. $x + 4.3 = -6.8$

46. $x + 7.5 = -2.8$

47. $25.17 = x - 16.59$

48. $5.27 = x - 28.85$

49. $-3.7r = -8.51$

50. $-2.9w = 13.34$

51. In Chicago, a person must pay 9.5% of the price of a couch in sales tax. If the sales tax is \$77.90, what is the price of the couch?

52. A person tips a waiter 18% of the bill. If the tip is \$4.50, how much is the bill?

53. A student earns 85% of the points on a test. If the student earns 34 points, how many points is the test worth?

54. A book agent charges 15% to represent an author. If the agent earns \$4.5 thousand from a book deal, how much does the author earn?

55. In 2009, there were 186 thousand students who earned bachelor's degrees in business, which was 20% of the students who earned bachelor's degrees in any field in that year (Source: *U.S. Center for Education Statistics*). How many students earned bachelor's degrees in 2009?

56. In 2011, there were 34.1 million satellite television subscribers, which was 33% of all television subscribers in that year (Source: *Bernstein Research*). What was the number of television subscribers in 2011?

57. In 2011, Americans consumed an average of 506 servings of carbonated sweetened drinks per person, which was 69.7% of the average per-person consumption of all sweetened drinks (both carbonated and noncarbonated) in that year (Source: *Centers for Disease Control and Prevention*). What was the average number of servings of sweetened drinks consumed per person in 2011?

58. In 2010, there were 223 thousand new cases of lung cancer, which was 14.6% of the new cases of all types of cancer in that year (Source: *U.S. National Institutes of Health*). How many new cases of cancer were there in 2010?

For Exercises 59–62, use the graph of $y = -\frac{1}{2}x + 1$, shown in Fig. 12, to solve the given equation.

59. $-\frac{1}{2}x + 1 = 3$ **60.** $-\frac{1}{2}x + 1 = 2$

61. $-\frac{1}{2}x + 1 = -1$ **62.** $-\frac{1}{2}x + 1 = -2$

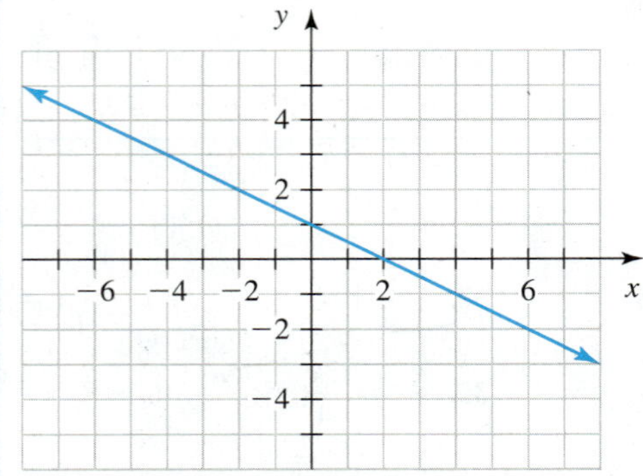

Figure 12 Exercises 59–62

Use "intersect" on a graphing calculator to solve the equation.

63. $x + 2 = 7$ **64.** $x - 3 = 4$

65. $2x - 3 = 5$ **66.** $-3x + 5 = -7$

67. $-4(x - 1) = -8$ **68.** $2(x + 5) = 6$

69. $\frac{2}{3}t - \frac{3}{2} = -\frac{7}{2}$ **70.** $\frac{5}{2}w - \frac{5}{3} = \frac{10}{3}$

For Exercises 71–74, solve the given equation by referring to the solutions of $y = 5x - 3$ shown in Table 3.

71. $5x - 3 = 12$ **72.** $5x - 3 = 7$

73. $5x - 3 = -13$ **74.** $5x - 3 = -8$

Table 3 Exercises 71–74

x	y
−3	−18
−2	−13
−1	−8
0	−3
1	2
2	7
3	12

*Solve. [**Hint:** Combine like terms on the left side.]*

75. $2x + 5x = 14$ **76.** $3x + x = 20$

77. $4x - 5x = -2$ **78.** $2x - 8x = -4$

Concepts

79. A student tries to solve the equation $2x = 10$:

$$\begin{aligned} 2x &= 10 \\ 2x - 2 &= 10 - 2 \\ x &= 8 \end{aligned}$$

Describe any errors. Then solve the equation correctly.

80. A student tries to solve the equation $x + 6 = 9$:

$$\begin{aligned} x + 6 &= 9 \\ x + 6 - 6 &= 9 \\ x + 0 &= 9 \\ x &= 9 \end{aligned}$$

Describe any errors. Then solve the equation correctly.

81. A student solves the equation $4x = 12$:

$$\begin{aligned} 4x &= 12 \\ \frac{4x}{4} &= \frac{12}{4} \\ x &= 3 \end{aligned}$$

Show that 3 satisfies each of the three equations.

82. A student solves the equation $x - 4 = 7$:

$$\begin{aligned} x - 4 &= 7 \\ x - 4 + 4 &= 7 + 4 \\ x + 0 &= 11 \\ x &= 11 \end{aligned}$$

Show that 11 satisfies each of the four equations.

83. A student tries to solve the equation $x + 2 = 7$:

$$\begin{aligned} x + 2 &= 7 \\ x + 2 - 7 &= 7 - 7 \\ x - 5 &= 0 \\ x - 5 + 5 &= 0 + 5 \\ x &= 5 \end{aligned}$$

Is the work correct? If yes, show a way to solve the equation in fewer steps. If no, describe the student's error(s) and then solve the equation correctly.

84. A student tries to solve the equation $\frac{3}{7}x = 2$:

$$\frac{3}{7}x = 2$$
$$7 \cdot \frac{3}{7}x = 7 \cdot 2$$
$$\frac{7}{1} \cdot \frac{3}{7}x = 14$$
$$3x = 14$$
$$\frac{3x}{3} = \frac{14}{3}$$
$$x = \frac{14}{3}$$

Is the work correct? If yes, show a way to solve the equation in fewer steps. If no, describe the student's error(s) and then solve the equation correctly.

85. Give an example of an equation for which it is helpful to add 7 to both sides. Then solve the equation.

86. Give an example of an equation for which it is helpful to subtract 7 from both sides. Then solve the equation.

87. Give an example of an equation for which it is helpful to multiply both sides by 7. Then solve the equation.

88. Give an example of an equation for which it is helpful to divide both sides by 7. Then solve the equation.

89. Are the equations $\frac{x}{3} = 2$ and $x - 1 = 4$ equivalent?

90. Are the equations $2x = 10$ and $x + 2 = 7$ equivalent?

91. Give three examples of equations that have 4 as their solution.

92. Give three examples of equations that have -2 as their solution.

93. Give an example of three equations that are equivalent. Explain.

94. Give an example of two equations that are not equivalent. Explain.

95. Are the equations $\frac{x-5}{2} = \frac{3}{x+1}$ and $\frac{x-5}{2} + 6 = \frac{3}{x+1} + 6$ equivalent? Explain.

96. Are the equations $x(x + 1) = 6$ and $2x(x + 1) = 12$ equivalent? Explain.

97. **a.** Solve $x + 2 = 7$.
b. Solve $x + 5 = 9$.
c. Solve $x + b = k$, where b and k are constants. [**Hint:** Refer to parts (a) and (b). The solution will be in terms of b and k.]

98. **a.** Solve $2x = 7$.
b. Solve $5x = 9$.
c. Solve $mx = p$, where m and p are constants and m is nonzero. [**Hint:** Refer to parts (a) and (b). The solution will be in terms of m and p.]

99. Describe in your own words what the addition property of equality means.

100. Describe in your own words what the multiplication property of equality means.

101. Explain why it is not necessary to state a subtraction property of equality: If A and B are expressions and c is a number, then the equations $A = B$ and $A - c = B - c$ are equivalent.

102. Explain why it is not necessary to state a division property of equality: If A and B are expressions and c is a nonzero number, then the equations $A = B$ and $\frac{A}{c} = \frac{B}{c}$ are equivalent.

4.4 Solving More Linear Equations in One Variable

Objectives

- Solve a linear equation in one variable.
- Make estimates and predictions by solving equations.
- Translate English sentences to and from mathematical equations.
- Use a graph or a table to solve a linear equation in one variable.
- Know the meaning of *conditional equation, inconsistent equation,* and *identity* and how to solve these types of equations.
- Locate errors in solving linear equations.

In this section, we will solve more complicated linear equations in one variable than those in Section 4.3. We will solve such equations to help us make predictions about authentic situations.

Solving Linear Equations in One Variable

In Section 4.3, we used either the addition property of equality or the multiplication property of equality to solve a linear equation. In this section, we will use a combination of these properties to solve linear equations.

Example 1 Using the Addition and Multiplication Properties of Equality

Solve $2x - 3 = 5$.

Solution

We begin by adding 3 to both sides to get $2x$ alone on the left side:

$$2x - 3 = 5 \quad \textit{Original equation}$$
$$2x - 3 + 3 = 5 + 3 \quad \textit{Add 3 to both sides.}$$
$$2x = 8 \quad \textit{Combine like terms.}$$
$$\frac{2x}{2} = \frac{8}{2} \quad \textit{Divide both sides by 2.}$$
$$x = 4 \quad \textit{Simplify.}$$

Next, we check that 4 satisfies the equation $2x - 3 = 5$;

$$\begin{aligned} 2x - 3 &= 5 && \text{Original equation} \\ 2(4) - 3 &\stackrel{?}{=} 5 && \text{Substitute 4 for } x. \\ 8 - 3 &\stackrel{?}{=} 5 && \text{Multiply.} \\ 5 &\stackrel{?}{=} 5 && \text{Subtract.} \\ &\text{true} \end{aligned}$$

So, the solution is 4.

It is wise to check that a result (or the results) of solving an equation does indeed satisfy the equation.

In Example 1, we used the addition property of equality so that a variable term was alone on one side of the equation and a constant term was on the other side. Then we used the multiplication property of equality to get the variable alone on one side of the equation.

We will do the same thing in Example 2, but first we will simplify one side of the equation.

Example 2 Combining Like Terms to Help Solve an Equation

Solve $4x - 7x + 2 = 17$.

Solution

First, we combine like terms on the left side:

$$\begin{aligned} 4x - 7x + 2 &= 17 && \text{Original equation} \\ -3x + 2 &= 17 && \text{Combine like terms.} \\ -3x + 2 - 2 &= 17 - 2 && \text{Subtract 2 from both sides to get } -3x \text{ alone on left side.} \\ -3x &= 15 && \text{Combine like terms.} \\ \frac{-3x}{-3} &= \frac{15}{-3} && \text{Divide both sides by } -3 \text{ to get } x \text{ alone on left side.} \\ x &= -5 && \text{Simplify.} \end{aligned}$$

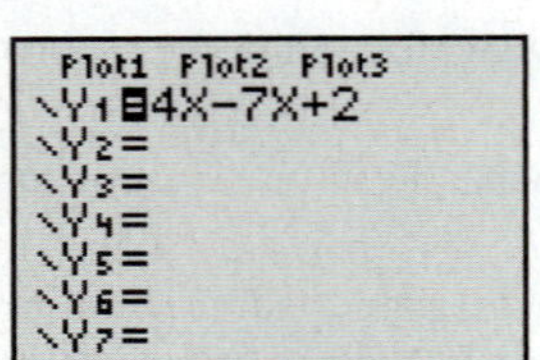

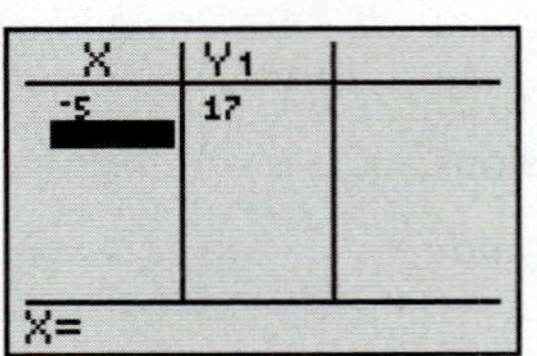

Figure 13 Verify the work

The solution is -5. We use a graphing calculator table to check that if -5 is substituted for x in the expression $4x - 7x + 2$, the result is 17 (see Fig. 13).

In Example 3, we will solve an equation that contains both variable terms and constant terms on each side of the equation. To solve such an equation, we first use the addition property of equality to write the variable terms on one side of the equation and constant terms on the other side.

Example 3 Solving an Equation with Variable Terms on Both Sides

Solve $-x + 7 = 2x - 2$.

Solution

First, we subtract $2x$ from both sides to get all of the variable terms on the left side:*

$$\begin{aligned} -x + 7 &= 2x - 2 && \text{Original equation} \\ -x + 7 - 2x &= 2x - 2 - 2x && \text{Subtract } 2x \text{ from both sides.} \\ -3x + 7 &= -2 && \text{Combine like terms.} \\ -3x + 7 - 7 &= -2 - 7 && \text{Subtract 7 from both sides.} \\ -3x &= -9 && \text{Combine like terms.} \\ \frac{-3x}{-3} &= \frac{-9}{-3} && \text{Divide both sides by } -3. \\ x &= 3 && \text{Simplify.} \end{aligned}$$

*The addition property of equality implies that we can add a number to both sides of an equation without changing the equation's solution set. We can also add (or subtract) a variable term of the form mx, where m is a constant, to both sides of an equation without changing the equation's solution set.

The solution is 3. We use a graphing calculator table to check that if 3 is substituted for x in the expressions $-x + 7$ and $2x - 2$, the two results are equal (see Fig. 14).

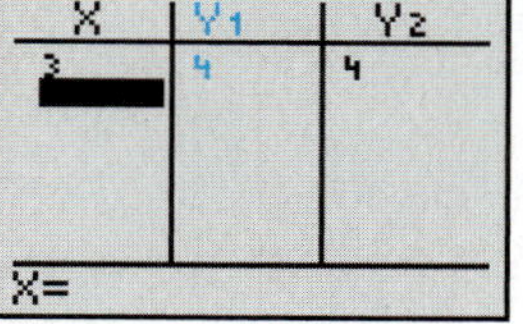

Figure 14 Verify the work

When possible, we simplify each side of an equation before using properties such as the addition property of equality or the multiplication property of equality.

Example 4 Using the Distributive Law to Help Solve an Equation

Solve $-2(a - 3) + 4 = 4(a - 2)$.

Solution

First, we use the distributive law on each side:

$$\begin{aligned}
-2(a - 3) + 4 &= 4(a - 2) && \textit{Original equation} \\
-2a + 6 + 4 &= 4a - 8 && \textit{Distributive law} \\
-2a + 10 &= 4a - 8 && \textit{Combine like terms.} \\
-2a + 10 - 4a &= 4a - 8 - 4a && \textit{Subtract 4a from both sides.} \\
-6a + 10 &= -8 && \textit{Combine like terms.} \\
-6a + 10 - 10 &= -8 - 10 && \textit{Subtract 10 from both sides.} \\
-6a &= -18 && \textit{Combine like terms.} \\
\frac{-6a}{-6} &= \frac{-18}{-6} && \textit{Divide both sides by } -6. \\
a &= 3 && \textit{Simplify; divide.}
\end{aligned}$$

Next, we check that 3 satisfies $-2(a - 3) + 4 = 4(a - 2)$:

$$\begin{aligned}
-2(a - 3) + 4 &= 4(a - 2) && \textit{Original equation} \\
-2(3 - 3) + 4 &\stackrel{?}{=} 4(3 - 2) && \textit{Substitute 3 for a.} \\
-2(0) + 4 &\stackrel{?}{=} 4(1) && \textit{Simplify.} \\
4 &\stackrel{?}{=} 4 && \textit{Simplify.} \\
&\text{true}
\end{aligned}$$

So, 3 is the solution.

A key step in solving an equation that contains fractions is to multiply both sides of the equation by the LCD so there are no fractions on either side of the equation. An equation that is "cleared of fractions" will be easier to solve than the original equation.

Example 5 Solving an Equation That Contains Fractions

Solve $\frac{2}{3}x + \frac{1}{6} = \frac{3}{4}$.

Solution

To find the LCD of the three fractions in the equation, we list the multiples of 3, the multiples of 6, and the multiples of 4:

Multiples of 3: 3, 6, 9, 12, 15, 18, 21, . . .
Multiples of 6: 6, 12, 18, 24, 30, 36, 42, . . .
Multiples of 4: 4, 8, 12, 16, 20, 24, 28, . . .

The LCD is 12. Next, we multiply both sides of $\frac{2}{3}x + \frac{1}{6} = \frac{3}{4}$ by 12 to clear the equation of fractions:

$$\frac{2}{3}x + \frac{1}{6} = \frac{3}{4} \quad \text{Original equation}$$

$$12 \cdot \left(\frac{2}{3}x + \frac{1}{6}\right) = 12 \cdot \frac{3}{4} \quad \text{Multiply both sides by the LCD, 12.}$$

$$12 \cdot \frac{2}{3}x + 12 \cdot \frac{1}{6} = 12 \cdot \frac{3}{4} \quad \text{Distributive law}$$

$$8x + 2 = 9 \quad \text{Simplify.}$$

$$8x + 2 - 2 = 9 - 2 \quad \text{Subtract 2 from both sides.}$$

$$8x = 7 \quad \text{Combine like terms.}$$

$$\frac{8x}{8} = \frac{7}{8} \quad \text{Divide both sides by 8.}$$

$$x = \frac{7}{8} \quad \text{Simplify.}$$

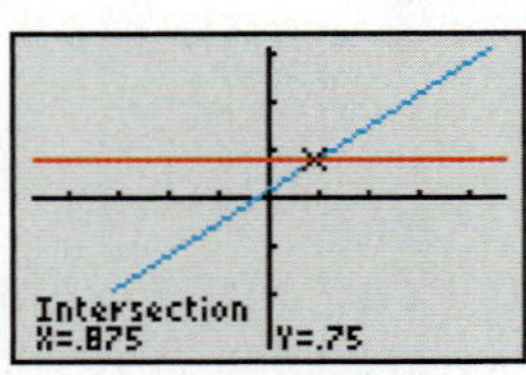

Figure 15 Verify the work

The solution is $\frac{7}{8}$. We check our work by using ZDecimal to graph the equations $y = \frac{2}{3}x + \frac{1}{6}$ and $y = \frac{3}{4}$ in the same coordinate system and by using "intersect" to find the intersection point, (0.875, 0.75). See Fig. 15. So, the solution of the original equation is $\frac{7}{8} = 0.875$, which checks.

Example 6 Solving an Equation That Contains Fractions

Solve $\frac{3x - 1}{2} = \frac{4x + 2}{3}$.

Solution

We multiply both sides of $\frac{3x - 1}{2} = \frac{4x + 2}{3}$ by the LCD, 6, to clear the equation of fractions:

$$\frac{3x - 1}{2} = \frac{4x + 2}{3} \quad \text{Original equation}$$

$$6 \cdot \frac{3x - 1}{2} = 6 \cdot \frac{4x + 2}{3} \quad \text{Multiply both sides by the LCD, 6.}$$

$$3(3x - 1) = 2(4x + 2) \quad \text{Simplify.}$$

$$9x - 3 = 8x + 4 \quad \text{Distributive law}$$

$$9x - 3 - 8x = 8x + 4 - 8x \quad \text{Subtract 8x from both sides.}$$

$$x - 3 = 4 \quad \text{Combine like terms.}$$

$$x - 3 + 3 = 4 + 3 \quad \text{Add 3 to both sides.}$$

$$x = 7 \quad \text{Combine like terms.}$$

We can use "intersect" on a graphing calculator to check our work.

To solve some true-to-life problems, we will need to solve equations that contain decimal numbers. When solving such problems, we round results to two decimal places.

Example 7 Solving an Equation That Contains Decimal Numbers

Solve $2.71t = -3.4(5.9t - 4.8)$. Round any solutions to two decimal places.

Solution

$$
\begin{aligned}
2.71t &= -3.4(5.9t - 4.8) && \text{Original equation} \\
2.71t &= -20.06t + 16.32 && \text{Distributive law} \\
2.71t + 20.06t &= -20.06t + 16.32 + 20.06t && \text{Add 20.06t to both sides.} \\
22.77t &= 16.32 && \text{Combine like terms.} \\
\frac{22.77t}{22.77} &= \frac{16.32}{22.77} && \text{Divide both sides by 22.77.} \\
t &\approx 0.72 && \text{Round right-hand side to two decimal places.}
\end{aligned}
$$

The solution is approximately 0.72. We can check that 0.72 approximately satisfies the original equation (try it).

Making Predictions by Solving Equations

Now that we know how to solve linear equations in one variable, we can make predictions about the dependent variable of a linear model.

Example 8 Using a Model to Make Predictions

The percentage of Americans who believe gay/lesbian relations are morally acceptable was 52% in 2010 and has increased by about 1.8 percentage points per year since then (Source: *Gallup Organization*). Let p be the percentage of Americans who believe gay/lesbian relations are morally acceptable at t years since 2010.

1. Find a model of the situation.
2. Predict what percentage of Americans will believe gay/lesbian relations are morally acceptable in 2017.
3. Predict when 60% of Americans will believe gay/lesbian relations are morally acceptable.

Solution

1. Since the percentage is increasing by about a *constant* 1.8 percentage points per year, the variables t and p are approximately linearly related. We want an equation of the form $p = mt + b$. Because the slope is 1.8 and the p-intercept is $(0, 52)$, a reasonable model is

$$p = 1.8t + 52$$

2. We substitute the input 7 for t in the equation $p = 1.8t + 52$ and solve for p:

$$p = 1.8(7) + 52 = 64.6$$

About 65% of Americans will believe gay/lesbian relations are morally acceptable in 2017, according to the model.

3. We substitute 60 for p in the equation $p = 1.8t + 52$ and solve for t:

$$
\begin{aligned}
60 &= 1.8t + 52 && \text{Substitute 60 for } p. \\
60 - 52 &= 1.8t + 52 - 52 && \text{Subtract 52 from both sides.} \\
8 &= 1.8t && \text{Combine like terms.} \\
\frac{8}{1.8} &= \frac{1.8t}{1.8} && \text{Divide both sides by 1.8.} \\
4.44 &\approx t
\end{aligned}
$$

In $2010 + 4.44 \approx 2014$, 60% of Americans will believe gay/lesbian relations are morally acceptable, according to the model. We verify our work in Problems 2 and 3 by using a graphing calculator table (see Fig. 16).

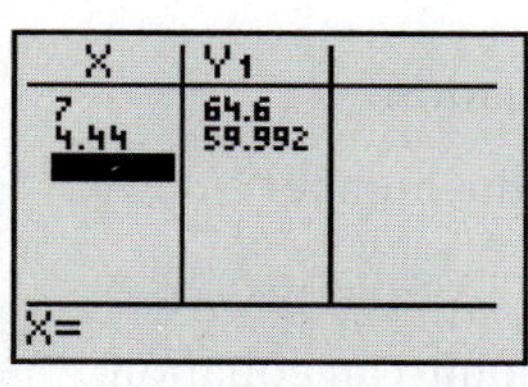

Figure 16 Verify the work

Example 9 Making a Prediction

The sales of musical instruments were \$6.39 million in 2010 and are decreasing by about \$0.28 million per year (Source: *National Arts Index*). Predict in which year sales will be \$4.4 million.

Solution

We let t be the number of years since 2010. An expression that describes the sales (in millions of dollars) is $-0.28t + 6.39$. To find when the sales will be \$4.4 million, we set the expression $-0.28t + 6.39$ equal to 4.4 and solve:

$$
\begin{aligned}
-0.28t + 6.39 &= 4.4 && \text{Set } -0.28t + 6.39 \text{ equal to 4.4.} \\
-0.28t + 6.39 - 6.39 &= 4.4 - 6.39 && \text{Subtract 6.39 from both sides.} \\
-0.28t &= -1.99 && \text{Combine like terms.} \\
\frac{-0.28t}{-0.28} &= \frac{-1.99}{-0.28} && \text{Divide both sides by } -0.28. \\
t &\approx 7.11 && \text{Simplify.}
\end{aligned}
$$

In $2010 + 7.11 \approx 2017$, sales of musical instruments will be \$4.4 million.

Example 10 Solving a Percentage Problem

In 2010, the number of digital cameras sold was 23.6 million cameras, down 6% from 2009 (Source: *Consumer Electronics Association*). What was the number of digital cameras sold in 2009?

Solution

We let n be the number (in millions) of digital cameras sold in 2009. Since the sales decreased by 6%, we write

$$\underbrace{n}_{\text{Millions of cameras sold in 2009}} - \underbrace{0.06n}_{\text{Decrease in sales (in millions of cameras)}} = \underbrace{23.6}_{\text{Millions of cameras sold in 2010}}$$

Now we solve the equation:

$$
\begin{aligned}
n - 0.06n &= 23.6 \\
0.94n &= 23.6 && \text{Combine like terms.} \\
\frac{0.94n}{0.94} &= \frac{23.6}{0.94} && \text{Divide both sides by 0.94.} \\
n &\approx 25.11
\end{aligned}
$$

The sales in 2009 were about 25.1 million cameras.

Translating Sentences to and from Equations

Let x represent a number. Here are some sentences that have the same meaning as $x = 4$:

- The number is 4.
- The number is equal to 4.
- The number is the same as 4.

Example 11 Translating an English Sentence to an Equation

Five, minus 2 times the sum of a number and 3 is 11. What is the number?

Solution

Let x be the number. Next, we translate the given information into an equation:

$$\underbrace{5\ -}_{\text{Five, minus}}\quad \underbrace{2\ \cdot}_{\text{2 times}}\quad \underbrace{(x\ +\ 3)}_{\text{the sum of a number and 3}}\quad \underbrace{=\ 11}_{\text{is 11.}}$$

Then we solve the equation.

$$\begin{aligned} 5 - 2(x + 3) &= 11 && \textit{Original equation} \\ 5 - 2x - 6 &= 11 && \textit{Distributive law} \\ -2x - 1 &= 11 && \textit{Combine like terms.} \\ -2x - 1 + 1 &= 11 + 1 && \textit{Add 1 to both sides.} \\ -2x &= 12 && \textit{Combine like terms.} \\ \frac{-2x}{-2} &= \frac{12}{-2} && \textit{Divide both sides by } -2. \\ x &= -6 && \textit{Simplify.} \end{aligned}$$

So, the number is −6.

In Example 11, we translated an English sentence into a mathematical equation, solved the equation, and then translated the result into an English sentence.

Example 12 Translating an Equation to an English Sentence

Let x be a number. Translate the equation $7x - 1 = 4x + 5$ into an English sentence.

Solution

$$\underbrace{7x}_{\text{Seven times the number,}}\quad \underbrace{-\ 1}_{\text{minus 1,}}\quad \underbrace{=}_{\text{is equal to}}\quad \underbrace{4x}_{\text{4 times the number,}}\quad \underbrace{+\ 5}_{\text{plus 5.}}$$

One of many correct translations is "Seven times the number, minus 1, is equal to 4 times the number, plus 5." The solution of the equation is 2 (try it).

Using Graphing to Solve an Equation in One Variable

In Section 4.3, we used graphing to solve some linear equations in one variable. We can use graphing to solve more complicated linear equations in one variable.

Example 13 Solving an Equation in One Variable by Graphing

The graphs of $y = -\frac{1}{2}x - 1$ and $y = -\frac{5}{4}x + 2$ are shown in Fig. 17. Use these graphs to solve the equation $-\frac{1}{2}x - 1 = -\frac{5}{4}x + 2$.

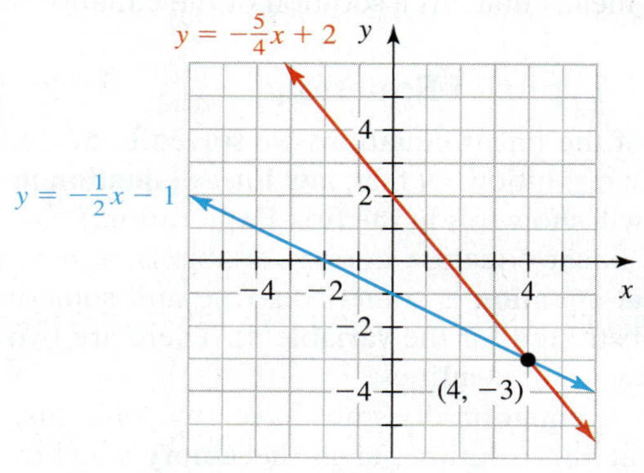

Figure 17 The intersection point is (4, −3)

Solution

The two lines intersect only at $(4, -3)$, whose x-coordinate is 4. So, 4 is the solution of the equation $-\frac{1}{2}x - 1 = -\frac{5}{4}x + 2$.

Example 14 Solving an Equation in One Variable by Graphing

Use "intersect" on a graphing calculator to solve the equation $-\frac{3}{5}x + \frac{5}{2} = \frac{2}{3}x + \frac{16}{3}$, with the solution rounded to the second decimal place.

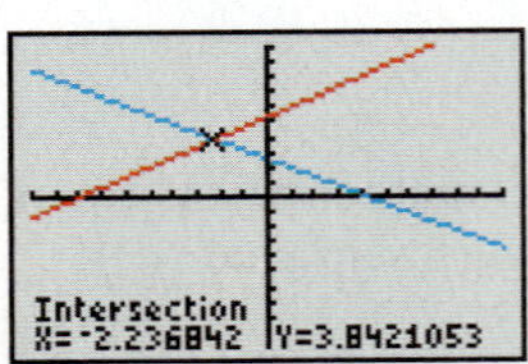

Figure 18 Using "intersect" to solve an equation in one variable

Solution

We use a graphing calculator to graph the equations $y = -\frac{3}{5}x + \frac{5}{2}$ and $y = \frac{2}{3}x + \frac{16}{3}$ on the same coordinate system and then use "intersect" to find the approximate intersection point, $(-2.24, 3.84)$. See Fig. 18. See Appendix A.18 for graphing calculator instructions.

The approximate intersection point $(-2.24, 3.84)$ has x-coordinate -2.24. So, the approximate solution of the original equation is -2.24.

Using Tables to Solve an Equation in One Variable

Recall from Section 4.3 that we can also solve equations by using tables.

Example 15 Solving an Equation in One Variable by Using Tables

The solutions of $y = 2x - 5$ and $y = -4x + 13$ are shown in Tables 4 and 5, respectively. Use the tables to solve the equation $2x - 5 = -4x + 13$.

Table 4 Solutions of $y = 2x - 5$

x	y
0	−5
1	−3
2	−1
3	1
4	3

Table 5 Solutions of $y = -4x + 13$

x	y
0	13
1	9
2	5
3	1
4	−3

Solution

From Tables 4 and 5, we see that for both of the equations $y = 2x - 5$ and $y = -4x + 13$, the input 3 leads to the output 1:

$$2(3) - 5 = 1 \quad \text{and} \quad -4(3) + 13 = 1$$

It follows that

$$2(3) - 5 = -4(3) + 13$$

which means that 3 is a solution of the equation $2x - 5 = -4x + 13$.

Three Types of Equations

Each of the linear equations we solved in Section 4.3 and in this section so far has exactly one solution. In fact, **any linear equation in one variable has exactly one solution.** (You will show this in the first Exploration.)

A linear equation in one variable is an example of a conditional equation. A **conditional equation** is sometimes true and sometimes false, depending on which values are substituted for the variable(s). There are two other types of equations: inconsistent equations and identities.

If an equation does not have any solutions, we call the equation **inconsistent** and say that the solution set is the **empty set.** For example, the equation $x = x + 1$ is inconsistent, because no number is 1 more than itself. Here we subtract x from both sides of the equation:

$$x = x + 1 \quad \text{Original equation}$$
$$x - x = x + 1 - x \quad \text{Subtract } x \text{ from both sides.}$$
$$0 = 1 \quad \text{Combine like terms.}$$
$$\text{false}$$

When we apply the usual steps to solve an inconsistent equation, the result is always a false statement.

An equation that is true for all permissible values of the variable(s) it contains is called an **identity.** For example, $x + 5 = x + 2 + 3$ is an identity, because it is true for all real numbers. Here we add the numbers 2 and 3 and then subtract x from both sides:

$$x + 5 = x + 2 + 3 \quad \text{Original equation}$$
$$x + 5 = x + 5 \quad \text{Add.}$$
$$x + 5 - x = x + 5 - x \quad \text{Subtract } x \text{ from both sides.}$$
$$5 = 5 \quad \text{Combine like terms.}$$
$$\text{true}$$

When we apply the usual steps to solve an identity, the result is always a true statement of the form $a = a$, where a is a constant.

Example 16 Solving Three Types of Equations

Solve the equation. State whether the equation is a conditional equation, an inconsistent equation, or an identity.

1. $2x + 1 = 9$ **2.** $3x + 2 = 3x + 7$ **3.** $2(x + 4) + 1 = 2x + 9$

Solution

1.

$$2x + 1 = 9 \quad \text{Original equation}$$
$$2x + 1 - 1 = 9 - 1 \quad \text{Subtract 1 from both sides.}$$
$$2x = 8 \quad \text{Combine like terms.}$$
$$\frac{2x}{2} = \frac{8}{2} \quad \text{Divide both sides by 2.}$$
$$x = 4 \quad \text{Simplify.}$$

The number 4 is the only solution, so the equation is conditional.

2.

$$3x + 2 = 3x + 7 \quad \text{Original equation}$$
$$3x + 2 - 3x = 3x + 7 - 3x \quad \text{Subtract } 3x \text{ from both sides.}$$
$$2 = 7 \quad \text{Combine like terms.}$$
$$\text{false}$$

Since $2 = 7$ is a false statement, we conclude that the original equation is inconsistent and its solution set is the empty set.

3.

$$2(x + 4) + 1 = 2x + 9 \quad \text{Original equation}$$
$$2x + 8 + 1 = 2x + 9 \quad \text{Distributive law}$$
$$2x + 9 = 2x + 9 \quad \text{Combine like terms.}$$
$$2x + 9 - 2x = 2x + 9 - 2x \quad \text{Subtract } 2x \text{ from both sides.}$$
$$9 = 9 \quad \text{Combine like terms.}$$
$$\text{true}$$

Since $9 = 9$ is a true statement (of the form $a = a$), we conclude that the original equation is an identity and its solution set is the set of all real numbers.

Locating Errors in Solving Linear Equations in One Variable

In Example 17, we will check whether a result satisfies equations. Doing this will help us locate an error.

Example 17 Locating an Error

Consider the following incorrect work in an attempt to solve the equation $2x + 1 = 5$:

$$\begin{aligned} 2x + 1 &= 5 \\ 2x + 1 - 1 &= 5 - 1 \\ 2x &= 4 \\ \frac{2x}{2} &= 4 \qquad \textit{Incorrect} \\ x &= 4 \end{aligned}$$

1. Show that the work is incorrect by substituting the result 4 for x in the original equation.
2. Pinpoint the error by substituting 4 for x in all five equations.

Solution

1. We substitute 4 for x in the equation $2x + 1 = 5$:

$$\begin{aligned} 2x + 1 &= 5 \qquad \textit{Original equation} \\ 2(4) + 1 &\overset{?}{=} 5 \qquad \textit{Substitute 4 for x.} \\ 9 &\overset{?}{=} 5 \qquad \textit{Simplify.} \\ &\text{false} \end{aligned}$$

Since 4 does not satisfy the original equation $2x + 1 = 5$, we know that an error was made in solving that equation.

2. We substitute 4 for x in all five equations and note which pair of consecutive equations gives a false statement followed by a true statement.

Equation	Does 4 satisfy the equation?	
$2x + 1 = 5$	$2(4) + 1 \overset{?}{=} 5$	*false*
$2x + 1 - 1 = 5 - 1$	$2(4) + 1 - 1 \overset{?}{=} 5 - 1$	*false*
$2x = 4$	$2(4) \overset{?}{=} 4$	*false*
$\frac{2x}{2} = 4$	$\frac{2(4)}{2} \overset{?}{=} 4$	*true*
$x = 4$	$(4) \overset{?}{=} 4$	*true*

Since substituting 4 for x in $2x = 4$ gives a false statement, but substituting 4 for x in the next equation, $\frac{2x}{2} = 4$, gives a true statement, we know an error was made when only the left side of the equation $2x = 4$ was divided by 2:

$$\begin{aligned} 2x &= 4 \qquad \textit{Third equation of the work} \\ \frac{2x}{2} &= 4 \qquad \textit{Incorrect} \end{aligned}$$

Instead, *both* sides of the equation $2x = 4$ should be divided by 2:

$$\begin{aligned} 2x &= 4 \qquad \textit{Third equation of the work} \\ \frac{2x}{2} &= \frac{4}{2} \qquad \textit{Divide both sides by 2.} \\ x &= 2 \qquad \textit{Simplify.} \end{aligned}$$

So, the solution is 2, not 4. We can check that 2 satisfies the original equation $2x + 1 = 5$ (try it).

Locating an Error in Solving a Linear Equation

If the result of an attempt to solve a linear equation in one variable does not satisfy the equation, an error was made. To pinpoint the error, substitute the result into all of the equations of the work and find which pair of consecutive equations gives a false statement followed by a true statement.

Group Exploration

Any linear equation in one variable has exactly one solution.

1. Solve $7x + 5 = 0$.
2. Solve $4x + 3 = 0$.
3. Solve $5x + 2 = 0$.
4. Solve $mx + b = 0$, where m and b are constants and $m \neq 0$. [**Hint:** Perform steps similar to those you followed for Problems 1–3.]
5. Describe a linear equation in one variable.
6. Does using the addition property of equality or the multiplication property of equality change an equation's solution set? Explain.
7. Explain why your responses to Problems 4–6 show that any linear equation in one variable has exactly one solution.
8. Use your result from Problem 4 to solve $8x + 3 = 0$ in one step.

Group Exploration

Looking ahead: Comparing expressions and equations

1. Simplify the expression $2(x + 3)$.
2. Solve the equation $2(x + 3) = 0$.
3. Compare your results from Problems 1 and 2. Also, explain how to substitute values for x to check your work in the problems.
4. Simplify the expression $3x + 4x$.
5. Solve the equation $3x + 4x = 15 + 2x$.
6. Compare your results from Problems 4 and 5. Also, explain how to substitute values for x to check your work in the problems.
7. In general, compare the results of simplifying an expression with the results of solving an equation.

Tips for Success Choose a Good Time and Place to Study

To improve your effectiveness at studying, consider taking stock of when and where you are best able to study. Tracy, a student who lives in a sometimes distracting household, completes her assignments at the campus library just after she attends her classes. Gerome, a morning person, gets up early so that he can study before classes. Being consistent in the time and location for studying can help, too. Studies have shown that, after repeating a daily activity for about 21 days, the activity becomes a habit. Even if it takes some willpower to shuffle your schedule so that you can study at your prime time and location, things will start to feel comfortable and familiar within three weeks.

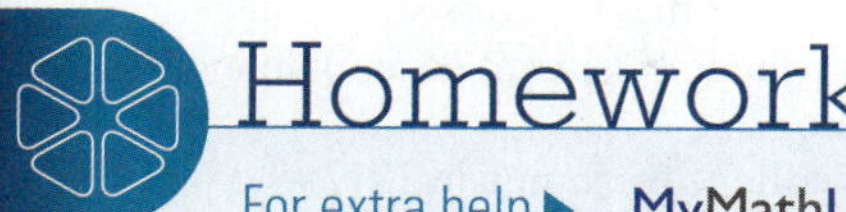

Homework 4.4

For extra help ▶ MyMathLab®  Watch the videos in MyMathLab  Download the MyDashboard App

Solve. Use a graphing calculator to verify your result.

1. $3x - 2 = 13$ **2.** $5x - 1 = 9$ **3.** $-4x + 6 = 26$

4. $-2x + 7 = 23$ **5.** $-5 = 6x + 3$ **6.** $-7 = 4x - 1$

7. $8 - x = -4$ **8.** $2 - x = -9$ **9.** $2x + 6 - 7x = -4$

10. $4x + 3 - 9x = -22$ **11.** $5x + 4 = 3x + 16$

12. $7x + 5 = 4x + 17$ **13.** $-3r - 1 = 2r + 24$

14. $-6w - 3 = 4w + 17$ **15.** $9 - x - 5 = 2x - x$

16. $8 - 2x - 2 = 3x + x$ **17.** $2(x + 3) = 5x - 3$

18. $-3(x - 4) = 2x + 2$ **19.** $1 - 3(5b - 2) = 4 - (7b + 3)$

20. $3 - 4(3p + 2) = 7 - (9p - 1)$

21. $4x = 3(2x - 1) + 5$

22. $2x = 5(2x + 9) + 3$

23. $3(4x - 5) - (2x + 3) = 2(x - 4)$

24. $-2(5x + 3) - (4x - 1) = 5(x + 2)$

25. $\frac{x}{2} - \frac{3}{4} = \frac{1}{2}$ **26.** $\frac{x}{9} - \frac{1}{3} = \frac{2}{9}$

27. $\frac{5x}{6} + \frac{2}{3} = 2$ **28.** $\frac{3x}{8} - \frac{1}{2} = 1$

29. $\frac{5}{6}k = \frac{3}{4}k + \frac{1}{2}$ **30.** $\frac{3}{8}t = \frac{5}{6}t - \frac{1}{4}$

31. $\frac{7}{12}x - \frac{5}{3} = \frac{7}{4} + \frac{5}{6}x$ **32.** $\frac{5}{4} + \frac{9}{2}x = \frac{3}{8}x - \frac{1}{4}$

33. $\frac{4}{3}x - 2 = 3x + \frac{5}{2}$ **34.** $\frac{2}{5}x - 4 = 2x - \frac{3}{4}$

35. $\frac{3(x - 4)}{5} = -2x$ **36.** $\frac{5(x - 2)}{3} = -4x$

37. $\frac{4x + 3}{5} = \frac{2x - 1}{3}$ **38.** $\frac{3x + 2}{2} = \frac{6x - 3}{5}$

39. $\frac{4m - 5}{2} - \frac{3m + 1}{3} = \frac{5}{6}$

40. $\frac{2p + 4}{3} - \frac{5p - 7}{6} = \frac{11}{12}$

For Exercises 41–46, solve. Round the result to the second decimal place. Use a graphing calculator to verify your work.

41. $0.3x + 0.2 = 0.7$

42. $0.6x - 0.1 = 0.4$

43. $5.27x - 6.35 = 2.71x + 9.89$

44. $8.25x - 17.56 + 4.38x = 25.86$

45. $0.4x - 1.6(2.5 - x) = 3.1(x - 5.4) - 11.3$

46. $3.2x + 0.5(7.3 - x) = 4.7 - 6.4(x - 2.1)$

47. The number of Americans without health insurance was 46.7 million in 2010 and has increased by 1.04 million per year since then (Source: *Centers for Disease Control and Prevention*). Let n be the number (in millions) of Americans without health insurance at t years since 2010.

a. Find an equation of a linear model to describe the situation.

b. The population of the Northeast was 55.3 million in 2010. Use the model to predict when there will be that many people without health insurance.

c. A new law goes into effect in 2014 that requires people to get health insurance or pay a penalty if they don't. Use the model to predict how many Americans would've been uninsured if the law hadn't been passed.

48. The average salary for a hockey player in the National Hockey League was \$2.4 million in 2011 and has increased by about \$0.1 million per year since then (Source: *National Hockey League Players Association*). Let s be the average salary (in millions of dollars) at t years since 2011.

a. Find an equation of a linear model to describe the situation.

b. Predict the average salary in 2017.

c. Predict when the average salary will be \$3.2 million.

49. Martha Stewart went to prison for selling nearly 4000 shares of ImClone Systems stock because she had inside information that the company was about to announce bad news that would result in the stock's value plummeting. Surprisingly, the value of a share of Martha Stewart Living Omnimedia stock increased by about \$3.36 per month after she was sentenced. The value of a share of her stock was \$8.16 just before she was sentenced (Source: *Bloomberg Financial Markets*). Let v be the value (in dollars) of a share of the stock at t months since Stewart was sentenced.

a. Find an equation of a model to describe the situation.

b. Estimate the value of a share of the stock 3 months after the sentencing, which is about when Stewart reported to prison.

c. Estimate when the value of a share of the stock reached \$28.32.

50. General Motors (GM) has been struggling in recent years. In fact, GM's share of new-vehicle sales in the United States is decreasing by about 1.39 percentage points per year. In 2010, GM's share was 19.1% (Source: *GM*). Let s be GM's share of new-vehicle sales in the United States at t years since 2010.

a. Find an equation of a model to describe the situation.

b. Predict when GM's share of new-vehicle sales will be 11%.

c. Predict GM's share of new-vehicle sales in 2018.

51. The percentages of college freshmen whose average grade in high school was an A are shown in Table 6 for various years.

Table 6 College Freshmen Whose Average Grade in High School Was an A

Year	Percent
1970	19.6
1980	26.6
1985	28.7
1990	29.4
1995	36.1
2000	42.9
2005	46.6
2010	48.4

Source: *Higher Education Research Institute*

Let p be the percentage of college freshmen whose average grade in high school was an A at t years since 1970. A model of the situation is $p = 0.76t + 18.06$.

a. Use a graphing calculator to draw a scattergram and the model in the same viewing window. Does the line come close to the data points?

b. What is the slope? What does it mean in this situation?

c. Estimate when half of all college freshmen earned an average grade of A in high school.

d. Predict the percentage of college freshmen in 2017 who will have earned an average grade of A in high school.

e. Give at least two possible explanations of why the percentage of college freshmen who have earned an average grade of A is increasing.

Well, according to my calculating, if we wait 20 years, we'll be guaranteed that our child will get straight A's in high school!
Honey, don't you think it's time for us to have a child?

52. The average selling prices of a home sold in San Bruno, California, are shown in Table 7 for various square footages.

Table 7 Average Selling Prices of Homes

Number of Square Feet	Number Used to Represent Square Feet	Average Selling Price (thousands of dollars)
500–1000	750	632
1001–1500	1250	733
1501–2000	1750	814
2001–2500	2250	894
2501–3000	2750	1025

Source: *Green Banker*

Let p be the average selling price (in thousands of dollars) of a home measuring s square feet. A model of the situation is $p = 0.19s + 488.15$.

a. Use a graphing calculator to draw a scattergram and the model in the same viewing window. Does the line come close to the data points?

b. What is the slope? What does it mean in this situation?

c. Estimate the square footage for which the average selling price is $950 thousand.

53. Chicago taxis charge $2.25 plus $1.80 for each mile traveled.

a. Find an equation of a linear model to describe the situation. Explain what your variables represent.

b. If a person paid $25.65 for a cab fare in Chicago, how far was the ride?

54. Houston taxis charge $2.55 plus $2.20 for each mile traveled.

a. Find an equation of a linear model to describe the situation. Explain what your variables represent.

b. If a person paid $37.75 for a cab fare in Houston, how far was the ride?

55. In 2012, a one-year-old Subaru Outback was worth about $25,772, with a depreciation of about $1976 per year (Source: *AutoTrader.com*).

a. Find an equation of a linear model to describe the situation. Explain what your variables represent.

b. Estimate when the car will be worth $6000.

56. In 2012, a one-year-old Nissan Pathfinder was worth about $23,748, with a depreciation of about $1302 per year (Source: *AutoTrader.com*).

a. Find an equation of a linear model to describe the situation. Explain what your variables represent.

b. Estimate when the car will be worth $8000.

57. A student has $37 set aside for an Xbox 360 4GB console, which sells for $199.99 (Source: *Microsoft*). If the student saves $15 each week, how long will it take until he can afford the console?

58. An employee's starting salary is $29,200. If the employee receives a raise of $1700 per year, after how many years will the employee's salary be $43,000?

59. In 2011, the percentage of private-sector workers who were in a union was 6.7% and has since decreased by 0.25 percentage point per year (Source: *Bloomberg BNA*). Predict when the percentage will be 5%.

60. The total number of overnight visits to national parks was 13.7 million visits in 2011 and has since decreased by 0.21 million visits per year (Source: *National Park Service*). Predict in which year there will be 12 million overnight visits.

61. In 2010, total U.S. debit card purchases were $1.6 trillion, up 33% from 2009 (Source: *The Nilson Report*). What were the total U.S. debit card purchases in 2009?

62. In 2011, the number of farmers markets was 7175 markets, up 17% from 2010 (Source: *U.S. Department of Agriculture*). What was the number of farmers markets in 2010?

63. In 2010, the average per-person annual consumption of cheese was 33.3 pounds, up 11.7% from 2000 (Source: *USDA*). What was the average per-person annual consumption of cheese in 2000?

64. In 2011, the number of trips taken by Americans on public transportation was 10.4 billion trips, up 10.6% from 2010 (Source: *American Public Transportation Association*). What was the number of trips taken by Americans in 2010?

65. In 2012, the American Red Cross received 5.9 million pints of donated blood, down 10.6% from 2009 (Source: *American Red Cross*). How much blood was donated in 2009?

66. In 2010, Bank of America revenue was $134 billion, down 10.7% from 2009 (Source: *Bank of America*). What was the revenue in 2009?

67. In 2012, the average office space per worker was 176 square feet, down 21.8% from 2010 (Source: *CoreNet Global*). What was the average office space per worker in 2010?

68. In 2012, the number of cattle in the United States was 90.8 million cattle, down 2% from 2011 (Source: *U.S. Department of Agriculture*). How many cattle were there in 2011?

For Exercises 69–76, find the number.

69. Three more than the product of 5 and a number is 18.

70. Four more than the product of 3 and a number is 11.

71. Three times the difference of a number and 2 is −18.

72. Six times the sum of a number and 2 is −12.

73. Three subtracted from a number is equal to twice the number plus 1.

74. Four, plus 7 times a number is equal to 5 times the number, minus 3.

75. One, minus 4 times the sum of a number and 5 is 9.

76. Three, plus 5 times the difference of a number and 2 is 18.

Let x be a number. Translate the given equation into an English sentence. Then solve the equation.

77. $2x - 3 = 7$

78. $3x + 8 = 2$

79. $6x - 3 = 8x - 4$

80. $2x + 9 = 7x + 12$

81. $2(x - 4) = 10$

82. $3(x + 6) = 12$

83. $4 - 7(x + 1) = 2$

84. $9 + 5(x - 3) = 4$

For Exercises 85–88, solve the given equation by referring to the solutions of $y = -3x + 7$ and $y = 5x + 15$ shown in Tables 8 and 9, respectively.

85. $-3x + 7 = 5x + 15$

86. $-3x + 7 = 4$

87. $5x + 15 = 5$

88. $5x + 15 = 25$

Table 8 Solutions of $y = -3x + 7$

x	y
-2	13
-1	10
0	7
1	4
2	1

Table 9 Solutions of $y = 5x + 15$

x	y
-2	5
-1	10
0	15
1	20
2	25

Use "intersect" on a graphing calculator to solve the equation. Round the solution to the second decimal place.

89. $-4x + 8 = 2x - 9$

90. $-2x - 7 = x + 1$

91. $2.5x - 6.4 = -1.7x + 8.1$

92. $-1.5x - 9.3 = 3.1x + 2.1$

93. $\frac{23}{75}x - \frac{99}{38} = -\frac{52}{89}x - \frac{67}{9}$

94. $-\frac{54}{35}x + \frac{26}{21} = \frac{67}{95}x - \frac{72}{31}$

For Exercises 95–100, solve the given equation by referring to the graphs of $y = -\frac{3}{2}x + 2$ and $y = \frac{1}{2}x - 2$ shown in Fig. 19.

95. $-\frac{3}{2}x + 2 = \frac{1}{2}x - 2$

96. $-\frac{3}{2}x + 2 = -4$

97. $-\frac{3}{2}x + 2 = 5$

98. $\frac{1}{2}x - 2 = -3$

99. $\frac{1}{2}x - 2 = 0$

100. $-\frac{3}{2}x + 2 = 2$

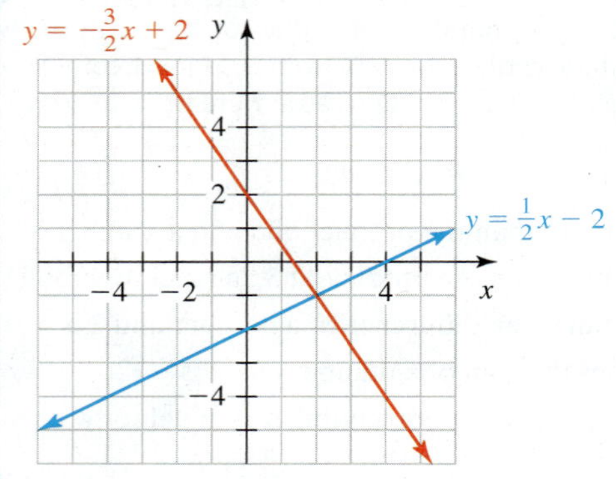

Figure 19 Exercises 95–100

For Exercises 101–106, solve the given equation by referring to the graphs of $y = -\frac{1}{3}x + \frac{5}{3}$ and $y = \frac{3}{2}x + \frac{7}{2}$ shown in Fig. 20.

101. $-\frac{1}{3}x + \frac{5}{3} = \frac{3}{2}x + \frac{7}{2}$

102. $-\frac{1}{3}x + \frac{5}{3} = 3$

103. $-\frac{1}{3}x + \frac{5}{3} = 1$

104. $\frac{3}{2}x + \frac{7}{2} = -4$

105. $\frac{3}{2}x + \frac{7}{2} = -1$

106. $-\frac{1}{3}x + \frac{5}{3} = 0$

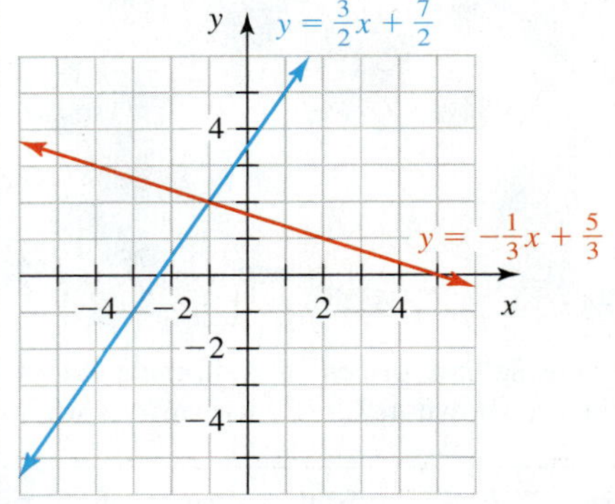

Figure 20 Exercises 101–106

Solve the equation. State whether the equation is a conditional equation, an inconsistent equation, or an identity.

107. $3x + 4x = 7x$

108. $9x - 3x = 6x$

109. $4x - 5 - 2x = 2x - 1$

110. $3x + 8x - 7 = 15x - 3 - 4x$

111. $3k + 10 - 5k = 4k - 2$

112. $5 - r + 8r = 3r + 1$

113. $2(x + 3) - 2 = 2x + 4$

114. $5(x - 2) + 1 = 5x - 9$

115. $5(2x - 3) - 4x = 3(2x - 1) + 6$

116. $4(3x + 2) - 6x = 2(3x - 4) + 1$

Concepts

117. A student incorrectly solves $3(x - 3) = 12$:

$$3(x - 3) = 12$$
$$3x - 9 = 12$$
$$3x - 9 - 9 = 12 - 9$$
$$3x = 3$$
$$\frac{3x}{3} = \frac{3}{3}$$
$$x = 1$$

Substitute 1 for x in each of the six equations, and explain how the results help you pinpoint the error. Describe the error. Then solve the equation correctly.

118. A student incorrectly solves $-2x + 1 = 3$:

$$-2x + 1 = 3$$
$$-2x + 1 - 1 = 3 - 1$$
$$-2x = 2$$
$$\frac{-2x}{-2} = 2$$
$$x = 2$$

Substitute 2 for x in each of the five equations, and explain how the results help you pinpoint the error. Describe the error. Then solve the equation correctly.

119. A student tries to solve $2(x - 5) = x - 3$:

$$2(x - 5) = x - 3$$
$$2x - 10 = x - 3$$
$$2x - 10 + 10 = x - 3 + 10$$
$$2x = x + 7$$
$$\frac{2x}{2} = \frac{x + 7}{2}$$
$$x = \frac{x + 7}{2}$$

Describe any errors. Then solve the equation correctly.

120. A student tries to solve the equation $\frac{1}{2}x + 3 = \frac{5}{2}$:

$$\frac{1}{2}x + 3 = \frac{5}{2}$$
$$2 \cdot \frac{1}{2}x + 3 = 2 \cdot \frac{5}{2}$$
$$x + 3 = 5$$
$$x + 3 - 3 = 5 - 3$$
$$x = 2$$

Describe any errors. Then solve the equation correctly.

121. Two students try to solve the equation $5x + 3 = 3x + 7$:

Student A	**Student B**
$5x + 3 = 3x + 7$	$5x + 3 = 3x + 7$
$5x + 3 - 3x = 3x + 7 - 3x$	$5x + 3 - 3 = 3x + 7 - 3$
$2x + 3 = 7$	$5x = 3x + 4$
$2x + 3 - 3 = 7 - 3$	$5x - 3x = 3x + 4 - 3x$
$2x = 4$	$2x = 4$
$\frac{2x}{2} = \frac{4}{2}$	$\frac{2x}{2} = \frac{4}{2}$
$x = 2$	$x = 2$

Compare the methods of solving the equation. Is one method better than the other? Explain.

122. Two students try to solve the equation $2x + 4 = 0$:

Student 1	**Student 2**
$2x + 4 = 0$	$2x + 4 = 0$
$2x + 4 + 6 = 0 + 6$	$2x + 4 - 4 = 0 - 4$
$2x + 10 = 6$	$2x = -4$
$2x + 10 - 10 = 6 - 10$	$\frac{2x}{2} = \frac{-4}{2}$
$2x = -4$	$x = -2$
$2x \cdot 3 = -4 \cdot 3$	
$6x = -12$	
$\frac{6x}{6} = \frac{-12}{6}$	
$x = -2$	

Did either student, both students, or neither student solve the equation correctly? Explain.

123. Solve $5(x - 2) = 2x - 1$. Show that your result satisfies each of the equations in your work.

124. Solve $3(x + 1) + 2 = x + 7$. Show that your result satisfies each of the equations in your work.

125. Consider the following equations:

$$x = 3$$
$$x + 2 = 3 + 2$$
$$x + 2 = 5$$
$$3(x + 2) = 3 \cdot 5$$
$$3(x + 2) = 15$$
$$3(x + 2) + 1 = 15 + 1$$
$$3(x + 2) + 1 = 16$$

What is the solution of the equation $3(x + 2) + 1 = 16$? What is the solution of the equation $3(x + 2) = 15$? Explain how you can respond to these questions without doing any work besides the work already included in this problem.

126. If an equation contains fractions, why is it helpful to multiply both sides of the equation by the LCD?

127. Consider the equation $7x + 4 = mx + b$. For what values of m and b is the following true?

a. There is exactly one solution.
b. The solution set is the set of all real numbers.
c. The solution set is the empty set.

4.5 Comparing Expressions and Equations

Objectives

- Compare the meanings of *simplifying expressions* and *solving equations.*
- Translate English phrases and sentences to mathematical expressions and equations.

In this section, we compare the meanings of mathematical expressions and equations.

Comparing Expressions and Equations

First, we define a *linear expression in one variable*.

Definition Linear expression in one variable

A **linear expression in one variable** is an expression that can be put into the form

$$mx + b$$

where m and b are constants and $m \neq 0$.

Next, we recall the definition of a linear equation from Section 4.3: A *linear equation in one variable* is an equation that can be put into the form $mx + b = 0$, where m and b are constants and $m \neq 0$.

For example, $4x + 7$ is a linear expression in one variable and $4x + 7 = 0$ is a linear equation in one variable. Also, the expression $2(x - 5) + 3x$ is linear, because it can be put into the form $mx + b$, where $m \neq 0$. (Try it.) The equation $2(x - 5) + 3x = 8x - 4$ is linear, because it can be put into the form $mx + b = 0$, where $m \neq 0$. (Try it.)

Example 1 Simplifying an Expression and Solving an Equation

1. Simplify $5(x + 2) - 2x$.
2. Solve $5(x + 2) - 2x = 2x + 14$.
3. Use a graphing calculator table to check the work in Problem 1.
4. Use a graphing calculator table to check the work in Problem 2.

Solution

1. $5(x + 2) - 2x = 5x + 10 - 2x$ *Distributive law*

$= 3x + 10$ *Combine like terms.*

The simplified expression is $3x + 10$.

2. $5(x + 2) - 2x = 2x + 14$ *Original equation*

$5x + 10 - 2x = 2x + 14$ *Distributive law*

$3x + 10 = 2x + 14$ *Combine like terms.*

$3x + 10 - 2x = 2x + 14 - 2x$ *Subtract 2x from both sides.*

$x + 10 = 14$ *Combine like terms.*

$x + 10 - 10 = 14 - 10$ *Subtract 10 from both sides.*

$x = 4$ *Simplify.*

The solution is 4.

3. We use a graphing calculator to find a table for $y = 5(x + 2) - 2x$ and $y = 3x + 10$, using the original expression and the simplified result, respectively (see Fig. 21). For each value of x in the table, the values of y for both expressions are equal. So, we are confident that the two expressions are equivalent.

Figure 21 Verify the work

4. We use a graphing calculator table to check that if 4 is substituted for x in the expressions $5(x + 2) - 2x$ and $2x + 14$ (both sides of the original equation), the two results are equal (see Fig. 22).

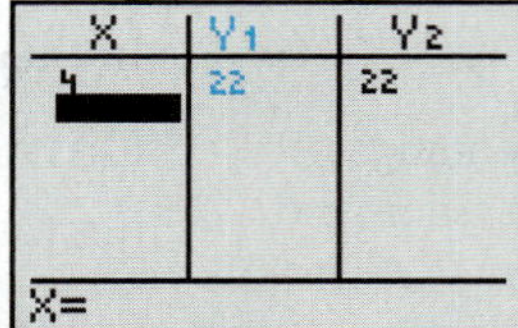

Figure 22 Verify the work

Note that if any other number besides 4 is substituted for x in the expressions $5(x + 2) - 2x$ and $2x + 14$, the two results will *not* be equal. That is because 4 is the *only* solution of the equation $5(x + 2) - 2x = 2x + 14$.

From our work in Example 1, we can make two important observations: First, the result of simplifying the *expression* $5(x + 2) - 2x$ is $3x + 10$, an *expression;* second, the result of solving the *equation* $5(x + 2) - 2x = 2x + 14$ is 4, a *number*.

Results of Simplifying an Expression and Solving an Equation

The result of simplifying an expression is an expression. The result of solving a linear equation in one variable is a number.

Example 2 Identifying an Error in Solving an Equation

A student tries to solve the equation $2(x - 3) = x$:

$$\begin{aligned} 2(x - 3) &= x \\ 2x - 6 &= x \\ 2x - 6 + 6 &= x + 6 \\ 2x &= x + 6 \\ x &= \frac{x + 6}{2} \end{aligned}$$

What would you tell the student?

Solution

The student found an equation that is equivalent to the original equation, but the student did not find the *solution* of the original equation. The solution of a linear equation is a number, not an expression that contains a variable. So, the solution cannot be $\frac{x + 6}{2}$. Notice that x still appears on both sides of the equation. By getting x alone on one side of the equation, we can show that the solution is 6. (Try it.)

In Example 1, we simplified $5(x + 2) - 2x$ to $3x + 10$. So, if we use the same number to evaluate both expressions, the two results will be equal. This implies that *any* number satisfies the equation $5(x + 2) - 2x = 3x + 10$. For example, in Fig. 21, we see that the numbers 0, 1, 2, 3, 4, 5, and 6 are solutions of this equation. Since any number satisfies the equation, the equation is an identity.

We also found that 4 is the solution of the linear equation $5(x + 2) - 2x = 2x + 14$. This means 4 is the *only* number that satisfies that equation. In general, a linear equation in one variable has exactly one solution.

Comparing Simplifying an Expression with Solving an Equation

- If an expression A in one variable is simplified to an expression B, then every real number (for which both expressions are defined) is a solution of the equation $A = B$. In other words, the equation $A = B$ is an identity.
- Exactly one number is a solution of a linear equation in one variable.

Example 3 Comparing Expressions and Equations

1. The simplified form of $4(2a - 5) - 2(3a + 4)$ is $2a - 28$. What does this mean?
2. The solution of $4(2a - 5) - 2(3a + 4) = 2$ is 15. What does this mean?

Solution

1. This means the linear expressions $4(2a - 5) - 2(3a + 4)$ and $2a - 28$ are equivalent. It also means every real number is a solution of the equation $4(2a - 5) - 2(3a + 4) = 2a - 28$. So, the equation is an identity.
2. This means 15 is the only number that satisfies the linear equation $4(2a - 5) - 2(3a + 4) = 2$. (Try it.)

Example 4 Comparing Expressions and Equations

1. Simplify $\frac{2}{5}x + \frac{1}{3}x$.
2. Solve $\frac{2}{5}x + \frac{1}{3}x = \frac{4}{5}$.
3. Compare the first steps of your work in Problems 1 and 2.

Solution

1.
$$\begin{aligned} \frac{2}{5}x + \frac{1}{3}x &= \frac{3}{3}\cdot\frac{2}{5}x + \frac{5}{5}\cdot\frac{1}{3}x && \textit{The LCD of } \tfrac{2}{5} \textit{ and } \tfrac{1}{3} \textit{ is 15.} \\ &= \frac{6}{15}x + \frac{5}{15}x && \textit{Multiply numerators and multiply denominators: } \frac{a}{b}\cdot\frac{c}{d} = \frac{ac}{bd} \\ &= \frac{11}{15}x && \textit{Combine like terms.} \end{aligned}$$

2.
$$\begin{aligned} \frac{2}{5}x + \frac{1}{3}x &= \frac{4}{5} && \textit{Original equation} \\ 15\cdot\left(\frac{2}{5}x + \frac{1}{3}x\right) &= 15\cdot\frac{4}{5} && \textit{Multiply both sides by 15.} \\ 15\cdot\frac{2}{5}x + 15\cdot\frac{1}{3}x &= 15\cdot\frac{4}{5} && \textit{Distributive law} \\ 6x + 5x &= 12 && \textit{Simplify.} \\ 11x &= 12 && \textit{Combine like terms.} \\ \frac{11x}{11} &= \frac{12}{11} && \textit{Divide both sides by 11.} \\ x &= \frac{12}{11} && \textit{Simplify.} \end{aligned}$$

3. To start simplifying the expression in Problem 1, we multiplied the term $\frac{2}{5}x$ by $\frac{3}{3} = 1$ and multiplied the term $\frac{1}{3}x$ by $\frac{5}{5} = 1$. To start solving the equation in Problem 2, we multiplied both sides of the equation by 15, which is *not* equal to 1.

Problem 3 in Example 4 suggests the following comparison of multiplying an expression by a number and multiplying both sides of an equation by a number:

Multiplying Expressions and Both Sides of Equations by Numbers

In simplifying an expression, the only number that we can multiply the expression or part of it by is 1. In solving an equation, we can multiply both sides of the equation by *any* number except 0.

By the property $a \cdot 1 = a$, we know multiplying an expression by 1 gives an equivalent expression. By the multiplication property of equality, we know multiplying both sides of an equation by *any* nonzero number gives an equivalent equation.

Translating from English to Mathematics

In Section 2.1, we discussed how to translate an English phrase or sentence into a mathematical expression, and in Section 4.4, we discussed how to translate an English sentence into an equation. We use an expression to represent a quantity, and we use an equation to state that one quantity is equal to another.

Example 5 Translating from English to Mathematics

Let x be a number. Translate the following into an expression or an equation, as appropriate, and then simplify the expression or solve the equation:

1. Five, minus 3 times the sum of the number and 4
2. The product of 6 and the number is equal to the number subtracted from 2.

Solution

1. The phrase describes a single quantity. So, we translate the phrase to an expression:

$$\underbrace{5\ -}_{\text{Five, minus}}\quad \underbrace{3\ \cdot}_{\text{3 times}}\quad \underbrace{(x+4)}_{\text{the sum of the number and 4}}$$

Next, we simplify the expression:

$$\begin{aligned} 5 - 3\cdot(x+4) &= 5 - 3x - 12 && \text{Distributive law} \\ &= -3x - 7 && \text{Combine like terms.} \end{aligned}$$

2. The sentence states that one quantity is equal to another. So, we translate the sentence to an equation:

$$\underbrace{6x}_{\text{The product of 6 and the number}}\quad \underbrace{=}_{\text{is equal to}}\quad \underbrace{2 - x}_{\text{the number subtracted from 2.}}$$

Next, we solve the equation:

$$\begin{aligned} 6x &= 2 - x && \text{Original equation} \\ 6x + x &= 2 - x + x && \text{Add } x \text{ to both sides.} \\ 7x &= 2 && \text{Combine like terms.} \\ \frac{7x}{7} &= \frac{2}{7} && \text{Divide both sides by 7.} \\ x &= \frac{2}{7} && \text{Simplify.} \end{aligned}$$

So, the solution is $\frac{2}{7}$.

Group Exploration

Simplifying versus solving

1. Two students try to solve the equation $2 = \frac{x}{2} + \frac{x}{3}$:

Student A

$$\begin{aligned} 2 &= \frac{x}{2} + \frac{x}{3} \\ 6\cdot 2 &= 6\left(\frac{x}{2} + \frac{x}{3}\right) \\ 12 &= 6\cdot\frac{x}{2} + 6\cdot\frac{x}{3} \\ 12 &= 3x + 2x \\ 12 &= 5x \\ \frac{12}{5} &= \frac{5x}{5} \\ \frac{12}{5} &= x \end{aligned}$$

Student B

$$\begin{aligned} 2 &= \frac{x}{2} + \frac{x}{3} \\ &= \frac{x}{2} + \frac{x}{3} - 2 \\ &= \frac{x}{2}\cdot\frac{3}{3} + \frac{x}{3}\cdot\frac{2}{2} - 2 \\ &= \frac{3x}{6} + \frac{2x}{6} - 2 \\ &= \frac{3x + 2x}{6} - 2 \\ &= \frac{5x}{6} - 2 \end{aligned}$$

Did either student, both students, or neither student solve the equation correctly? Explain.

2. Three students try to simplify the expression $\frac{x}{2} + \frac{x}{3}$:

Student C

$$\begin{aligned} \frac{x}{2} + \frac{x}{3} &= 6\left(\frac{x}{2} + \frac{x}{3}\right) \\ &= 6\cdot\frac{x}{2} + 6\cdot\frac{x}{3} \\ &= 3x + 2x \\ &= 5x \end{aligned}$$

Student D

$$\begin{aligned} \frac{x}{2} + \frac{x}{3} &= \frac{3}{3}\cdot\frac{x}{2} + \frac{2}{2}\cdot\frac{x}{3} \\ &= \frac{3x}{6} + \frac{2x}{6} \\ &= \frac{3x + 2x}{6} \\ &= \frac{5x}{6} \end{aligned}$$

Student E

$$\frac{x}{2} + \frac{x}{3} = 0$$
$$6\left(\frac{x}{2} + \frac{x}{3}\right) = 6 \cdot 0$$
$$6 \cdot \frac{x}{2} + 6 \cdot \frac{x}{3} = 0$$
$$3x + 2x = 0$$
$$5x = 0$$
$$\frac{5x}{5} = \frac{0}{5}$$
$$x = 0$$

Which students, if any, simplified the expression correctly? Explain.

Tips for Success **Ask Questions**

When you have a question during class time, do you ask it? Many students are reluctant to ask questions for fear of slowing the class down or feeling embarrassed. If you tend to shy away from asking questions, keep in mind that the main idea of school is for you to learn through open communication with your instructor and other students.

If you are confused about some concept, it's highly likely that other students in your class are confused, too. If you ask your question, everyone else who is confused will be grateful that you did so. Most instructors want students to ask questions. It helps an instructor know when students understand the material and when they are having trouble.

Homework 4.5

For extra help ▶ MyMathLab®

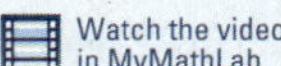

Watch the videos in MyMathLab

Download the MyDashboard App

Is the following a linear expression or a linear equation?

1. $3x + 7x = 8$

2. $2x + 4x - 7$

3. $3x + 7x$

4. $2x + 4x - 7 = 5$

5. $4 - 2(x - 9)$

6. $3(x - 1) = 7$

7. $4 - 2(x - 9) = 5$

8. $3(x - 1)$

Simplify the expression or solve the equation, as appropriate. Use a graphing calculator to verify your work.

9. $3x + 4x = 14$

10. $2x - 7x = 15$

11. $3x + 4x$

12. $2x - 7x$

13. $b - 5(b - 1)$

14. $-6k + 2(k + 4)$

15. $b - 5(b - 1) = 0$

16. $-6k + 2(k + 4) = 0$

17. $3(3x - 5) + 2(5x + 4) = 0$

18. $-4(2x + 1) - 2(3x - 6) = 0$

19. $3(3x - 5) + 2(5x + 4)$

20. $-4(2x + 1) - 2(3x - 6)$

21. $3(x - 2) - (7x + 2) = 4(3x + 1)$

22. $5(2x + 3) - (3x - 8) - 2(4x - 5)$

23. $3(x - 2) - (7x + 2) - 4(3x + 1)$

24. $5(2x + 3) - (3x - 8) = 2(4x - 5)$

25. $7.2p - 4.5 - 1.3p$

26. $8.3t + 9.2 - 7.7t$

27. $7.2p - 4.5 - 1.3p = 20.5 - 6.6p$

28. $8.3t + 9.2 - 7.7t = 3.5 - 1.2t$

29. $-3.5(x - 8) - 2.6(x - 2.8) = 13.93$

30. $-4.8(x + 3) + 6.5(x - 1.2)$

31. $-3.5(x - 8) - 2.6(x - 2.8)$

32. $-4.8(x + 3) + 6.5(x - 1.2) = -25.09$

33. $-\frac{6w}{8}$

34. $-\frac{15t}{10}$

35. $-\frac{6w}{8} = \frac{3}{2}$

36. $-\frac{15t}{10} = -\frac{5}{4}$

37. $\frac{5x}{6} + \frac{1}{2} - \frac{3x}{4}$

38. $\frac{2x}{3} - \frac{5x}{2} - 1 = 0$

39. $\frac{5x}{6} + \frac{1}{2} - \frac{3x}{4} = 0$

40. $\frac{2x}{3} - \frac{5x}{2} - 1$

41. $\frac{7}{2}x - \frac{5}{6} = \frac{1}{3} + \frac{3}{4}x$

42. $\frac{3}{5}x + 2x - \frac{5}{2} - \frac{7}{10}x$

43. $\frac{7}{2}x - \frac{5}{6} - \frac{1}{3} + \frac{3}{4}x$

44. $\frac{3}{5}x + 2x = \frac{5}{2} - \frac{7}{10}x$

Let x be a number. Translate each of the following into an expression or an equation, as appropriate, and then simplify the expression or solve the equation.

45. The sum of 3 and twice the number is -10.

46. The difference of 3 times the number and 5 is -12.

47. Four, minus 6 times the difference of the number and 2

48. Six, plus 3 times the sum of the number and 8

49. The product of -9 and the number is equal to the difference of the number and 5.

50. The sum of the number and 4 is equal to the number subtracted from 1.

51. The quotient of the number and 2 is equal to 3 times the difference of the number and 5.

52. The quotient of the number and 3 is equal to 2 plus the quotient of the number and 6.

53. The number plus the product of the number and 6

54. The number minus the product of the number and -4

55. The number plus the quotient of the number and 2

56. Four minus the quotient of the number and 5

Concepts

For Exercises 57 and 58, give an example of each. Then simplify or solve, as appropriate.

57. linear expression

58. linear equation in one variable

59. Two students try to solve the equation $\frac{3}{4}x - \frac{5}{6} = \frac{1}{3}$:

Student A

$$\begin{aligned}
\frac{3}{4}x - \frac{5}{6} &= \frac{1}{3}\\
\frac{3}{4}x &= \frac{1}{3} + \frac{5}{6}\\
\frac{3}{3}\cdot\frac{3}{4}x &= \frac{4}{4}\cdot\frac{1}{3} + \frac{2}{2}\cdot\frac{5}{6}\\
\frac{9}{12}x &= \frac{4}{12} + \frac{10}{12}\\
\frac{9}{12}x &= \frac{14}{12}\\
\frac{3}{4}x &= \frac{7}{6}\\
\frac{4}{3}\cdot\frac{3}{4}x &= \frac{4}{3}\cdot\frac{7}{6}\\
1x &= \frac{2\cdot 2\cdot 7}{3\cdot 2\cdot 3}\\
x &= \frac{14}{9}
\end{aligned}$$

Student B

$$\begin{aligned}
\frac{3}{4}x - \frac{5}{6} &= \frac{1}{3}\\
12\left(\frac{3}{4}x - \frac{5}{6}\right) &= 12\cdot\frac{1}{3}\\
12\cdot\frac{3}{4}x - 12\cdot\frac{5}{6} &= 4\\
9x - 10 &= 4\\
9x &= 14\\
x &= \frac{14}{9}
\end{aligned}$$

Compare the methods of solving the equation. Is one method better than the other? Explain.

60. A student believes that 5 is the solution (and the only solution) of the equation $-2(x + 4) = -2x - 8$, because 5 satisfies that equation:

$$\begin{aligned}
-2(5 + 4) &\stackrel{?}{=} -2(5) - 8\\
-18 &\stackrel{?}{=} -18 \qquad \text{true}
\end{aligned}$$

Is the student correct? Explain.

61. A student believes that $x + 6$ is the solution of an equation. Is the student correct? Explain.

62. A student tries to simplify an expression. The student writes, "The solution of the expression is 5." Is the student correct? Explain.

63. A student tries to simplify the expression $\frac{1}{4}x + \frac{1}{3}x$:

$$\begin{aligned}
\frac{1}{4}x + \frac{1}{3}x &= 12\left(\frac{1}{4}x + \frac{1}{3}x\right)\\
&= 12\cdot\frac{1}{4}x + 12\cdot\frac{1}{3}x\\
&= 3x + 4x\\
&= 7x
\end{aligned}$$

Describe any errors. Then simplify the expression correctly.

64. A student tries to simplify the expression $3x + 5x - 16$:

$$\begin{aligned}
3x + 5x - 16 &= 0\\
8x - 16 &= 0\\
8x &= 16\\
x &= 2
\end{aligned}$$

Describe any errors. Then simplify the expression correctly.

65. The simplified form of $7 + 2(x + 3)$ is $2x + 13$. What does this mean?

66. The solution of $7 + 2(x + 3) = 19$ is 3. What does this mean?

67. A student tries to simplify $2(5 + x)$:

$$2(5 + x) = 10 + x$$

To check the work, the student evaluates $2(5 + x)$ and $10 + x$, using 0 for x:

Evaluate $2(5 + x)$ for $x = 0$: $2(5 + 0) = 10$

Evaluate $10 + x$ for $x = 0$: $10 + 0 = 10$

Since the results of evaluating the expressions are equal, the student decides that the work is correct. What would you tell the student about checking the work?

68. A student evaluates both of the expressions $3x + 1$ and $21 - 2x$, using 4 for x:

Evaluate $3x + 1$ for $x = 4$: $3(4) + 1 = 13$

Evaluate $21 - 2x$ for $x = 4$: $21 - 2(4) = 13$

Since the results of evaluating the expressions are equal, the student concludes that the expressions are equivalent. Is the student correct? Explain.

69. Give three examples of an equation whose solution is 5.

70. Give three examples of an expression equivalent to the expression 5.

71. When simplifying an expression, what is the only number we can multiply it by? Why?

72. When solving an equation, why can we multiply both sides by *any* nonzero number?

73. Explain the meaning of simplifying an expression. (See page 9 for guidelines on writing a good response.)

74. Explain the meaning of solving an equation. (See page 9 for guidelines on writing a good response.)

4.6 Formulas

Objectives

- Know area and perimeter formulas of a rectangle and a total-value formula.
- Use formulas to solve various types of problems.
- Translate an English sentence into a formula.
- Solve a formula for a variable.

Recall from Section 3.3 that a **formula** is an equation that contains two or more variables. We can use formulas to find quantities such as the area and perimeter of rectangular objects, the total value of a group of objects, and the average test score of a group of tests.

Area Formula of a Rectangle

Recall from Section 1.1 that the area of a rectangle is equal to the length times the width. We can write $A = LW$, where A is the area, L is the length, and W is the width (see Fig. 23). The equation $A = LW$ is a formula.

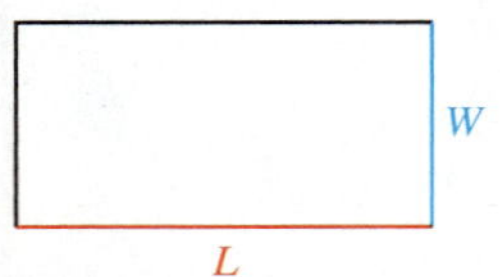

Figure 23 The length L and width W of a rectangle

Example 1 Using the Area Formula of a Rectangle

An architect is designing a school building so that each classroom contains 35 student desks. Fire codes require that there be 18 square feet per student desk. If there is enough room for classrooms of length 28 feet, what is the smallest width permitted by the fire codes?

Solution

To find the area of the floor of one room, we multiply the area needed per student desk by the number of student desks: $18(35) = 630$ square feet. Next, we must find the width of a rectangle that has an area of 630 square feet and a length of 28 feet (see Fig. 24). To do so, we substitute 630 for A and 28 for L in the formula $A = LW$, where the units of L and W are feet and the units of A are square feet. Then we solve for W:

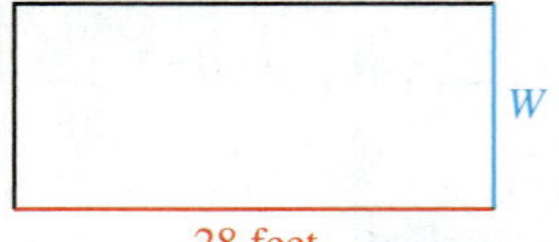

Figure 24 The length is 28 feet, and the width is W feet

$$A = LW \quad \text{Area formula of a rectangle}$$

$$630 = (28)W \quad \text{Substitute 630 for A and 28 for L.}$$

$$\frac{630}{28} = \frac{28W}{28} \quad \text{Divide both sides by 28.}$$

$$22.5 = W \quad \text{Simplify.}$$

The smallest width permitted is 22.5 feet.

To check, we find the product of the length and width: $28(22.5) = 630$, which is equal to the area (in square feet).

In Example 1, we substituted values for A and L in the formula $A = LW$ and then solved for the variable W. In general, **to find a single value of a variable in a formula, we often substitute numbers for all of the other variables and then solve for that one variable.**

Perimeter Formula of a Rectangle

The **perimeter** of a polygon, which is a geometrical object such as a triangle, rectangle, or trapezoid, is the total distance around the object. For example, consider a rectangle with length L and width W (see Fig. 25).

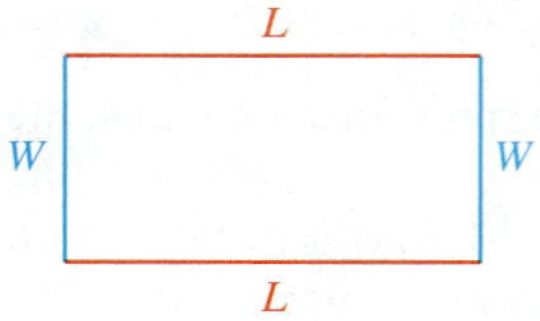

Figure 25 The length L and width W of a rectangle

The perimeter P of a rectangle is equal to the sum of the lengths of the four sides:

$$P = L + W + L + W$$

Combining like terms on the right-hand side of the formula gives

$$P = 2L + 2W$$

Area and Perimeter of a Rectangle

For a rectangle with length L, width W, area A, and perimeter P,

- $A = LW$
- $P = 2L + 2W$

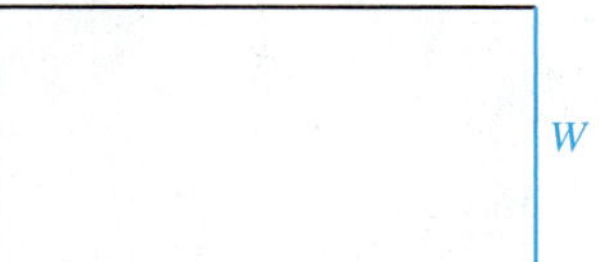

Figure 26 The length is 37.5 feet, and the width is W feet

Example 2 Using the Perimeter Formula of a Rectangle

A landscaper has budgeted enough money for 100 feet of fencing to enclose a rectangular garden. If the length of the garden is to run the full length of the property, which is 37.5 feet, what will be the width of the garden (see Fig. 26)?

Solution

Since the 100-foot-long fencing is on the border of the garden, the perimeter of the garden is 100 feet. So, we substitute 100 for P and 37.5 for L in the formula $P = 2L + 2W$, where the units of P, L, and W are feet. Then we solve for W:

$$
\begin{aligned}
P &= 2L + 2W && \textit{Perimeter formula of a rectangle} \\
100 &= 2(37.5) + 2W && \textit{Substitute 100 for P and 37.5 for L.} \\
100 &= 75 + 2W && \textit{Simplify.} \\
100 - 75 &= 75 + 2W - 75 && \textit{Subtract 75 from both sides.} \\
25 &= 2W && \textit{Combine like terms.} \\
\frac{25}{2} &= \frac{2W}{2} && \textit{Divide both sides by 2.} \\
12.5 &= W && \textit{Write } \frac{25}{2} \textit{ as a decimal number.}
\end{aligned}
$$

The width of the garden will be 12.5 feet.

To check, we find twice the length of 37.5 feet plus twice the width of 12.5 feet: $2(37.5) + 2(12.5) = 100$ feet, which is equal to the perimeter.

When describing an authentic quantity, we use a decimal form rather than a fractional form. For example, in Example 2 we say that the width is 12.5 feet rather than $\frac{25}{2}$ feet.

Total-Value Formula

Three quarters are worth $25 \cdot 3 = 75$ cents. Note that we found the total value of the quarters by multiplying the value of one quarter (25 cents) by the number of quarters (3).

> **Total-Value Formula**
>
> If n objects each have value v, then their total value T is given by
>
> $$T = vn$$
>
> In words: The total value is equal to the value of one object times the number of objects.

Example 3 Total-Value Formula

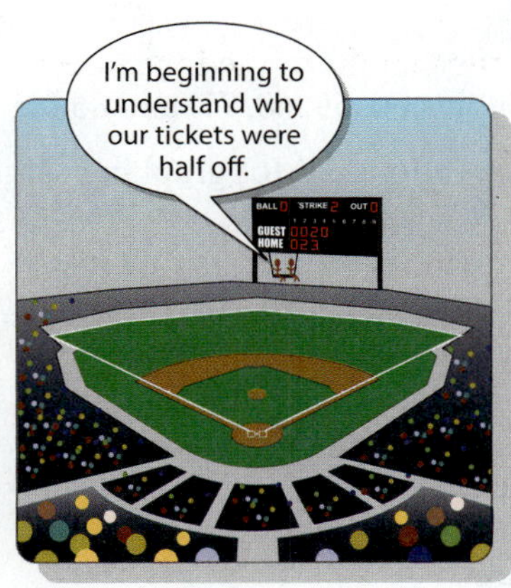

I'm beginning to understand why our tickets were half off.

In 2012, Chicago Cubs baseball tickets for upper-deck reserved outfield seating at Wrigley Field cost \$8 each.

1. Find a formula of the total cost T (in dollars) of n of these tickets.
2. Perform a unit analysis of the formula found in Problem 1.
3. Substitute 136 for T in the formula from Problem 1 and solve for n. What does the result mean in this situation?

Solution

1. We substitute 8 for v in the formula $T = vn$:

$$T = 8n$$

2. Here is a unit analysis of the formula $T = 8n$:

$$\underbrace{T}_{\text{dollars}} = \underbrace{8}_{\frac{\text{dollars}}{\text{ticket}}} \underbrace{n}_{\text{number of tickets}}$$

The units of the expressions on both sides of the equation are dollars, which checks.

3. We substitute 136 for T in the formula $T = 8n$ and solve for n:

$$T = 8n \quad \text{Total-value formula}$$
$$136 = 8n \quad \text{Substitute 136 for } T.$$
$$17 = n \quad \text{Divide both sides by 8.}$$

The total cost of 17 upper-deck reserved tickets at Wrigley Field is \$136.

Translating from English to a Mathematics Formula

In the next example, we will translate some information into a formula and then use the formula to find a quantity.

Example 4 Translating from English to Mathematics

A Celsius temperature reading is equal to $\frac{5}{9}$ times the difference of the Fahrenheit temperature reading and 32.

1. Write a formula of the Celsius temperature in terms of the Fahrenheit temperature.
2. Convert 50°F to the equivalent Celsius temperature.
3. Use a graphing calculator to convert 50°F, 59°F, 68°F, 77°F, and 86°F to the equivalent Celsius temperatures.

Solution

1. We let C be the Celsius reading and F be the equivalent Fahrenheit reading. Next, we translate the given information into a formula:

$$\underbrace{C}_{\text{A Celsius reading}} \underbrace{=}_{\text{is equal to}} \underbrace{\frac{5}{9}}_{\frac{5}{9}} \underbrace{\cdot}_{\text{times}} \underbrace{(F - 32)}_{\text{the difference of the Fahrenheit reading and 32.}}$$

So, the formula is $C = \frac{5}{9}(F - 32)$.

2. We substitute 50 for F in the formula $C = \frac{5}{9}(F - 32)$ and solve for C:

$$C = \frac{5}{9}(50 - 32) = 10$$

So, 10°C is equivalent to 50°F.

3. We use a graphing calculator table to substitute the inputs 50, 59, 68, 77, and 86 for F in the formula $C = \frac{5}{9}(F - 32)$. See Fig. 27. See Appendix A.13 for graphing calculator instructions.

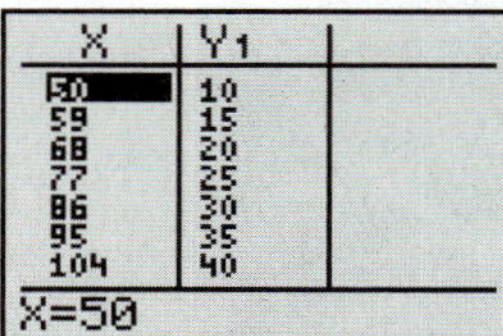

Figure 27 Finding Celsius temperatures

By viewing the second column of the graphing calculator table, we see that the values of C are 10, 15, 20, 25, and 30. So, 50°F, 59°F, 68°F, 77°F, and 86°F are equivalent to 10°C, 15°C, 20°C, 25°C, and 30°C, respectively.

Solving a Formula

In Problem 3 of Example 4, we converted five Fahrenheit temperature readings to equivalent Celsius temperature readings. How can we convert several Celsius temperature readings to equivalent Fahrenheit temperature readings? We must first solve $C = \frac{5}{9}(F - 32)$ for F and then use a graphing calculator to do the conversions. We will do just that in Example 8, but first we work with some simpler formulas in Examples 5, 6, and 7.

Example 5 Solving a Formula for One of Its Variables

Solve $A = LW$ for W.

Solution

We can solve the formula $A = LW$ for W in much the same way that we solved the equation $630 = (28)W$ in Example 1:

$630 = (28)W$	$A = LW$	*Area formula*
$\frac{630}{28} = \frac{28W}{28}$	$\frac{A}{L} = \frac{LW}{L}$	*Divide both sides by L.*
$22.5 = W$	$\frac{A}{L} = W$	*Simplify.*

The result is $W = \frac{A}{L}$. We are done, because W is alone on one side of the formula and does not appear on the other side.

Example 6 Solving a Formula for One of Its Variables

Solve $P = 2L + 2W$ for W.

Solution

We can solve the formula $P = 2L + 2W$ for W in much the same way that we solved the equation $100 = 75 + 2W$ in Example 2:

$100 = 75 + 2W$	$P = 2L + 2W$	*Perimeter formula*
$100 - 75 = 75 + 2W - 75$	$P - 2L = 2L + 2W - 2L$	*Subtract 2L from both sides.*
$25 = 2W$	$P - 2L = 2W$	*Combine like terms.*
$\frac{25}{2} = \frac{2W}{2}$	$\frac{P - 2L}{2} = \frac{2W}{2}$	*Divide both sides by 2.*
$\frac{25}{2} = W$	$\frac{P - 2L}{2} = W$	*Simplify.*

The result is $W = \frac{P - 2L}{2}$.

The equation $2x - 5y = 10$ is a formula, because it is an equation that contains two (or more) variables. In Example 7, we will solve this equation for y.

Example 7 Writing a Linear Equation in Slope–Intercept Form

Write the equation $2x - 5y = 10$ in slope–intercept form $(y = mx + b)$.

Solution

We solve $2x - 5y = 10$ for y:

$$2x - 5y = 10 \qquad \textit{Original equation}$$

$$2x - 5y - 2x = 10 - 2x \qquad \textit{Subtract } 2x \textit{ from both sides to get } -5y \textit{ alone on left side.}$$

$$-5y = -2x + 10 \qquad \textit{Combine like terms; rearrange terms.}$$

$$\frac{-5y}{-5} = \frac{-2x}{-5} + \frac{10}{-5} \qquad \textit{Divide both sides by } -5 \textit{ to get } y \textit{ alone on left side.}$$

$$y = \frac{2}{5}x - 2 \qquad \textit{Simplify.}$$

In Section 5.1, we will perform work similar to our work in Example 7 to help us graph linear equations.

Example 8 Solving a Formula for One of Its Variables

1. Solve the Fahrenheit–Celsius model $C = \frac{5}{9}(F - 32)$ for F.
2. Convert 10°C to the equivalent Fahrenheit temperature.
3. Use a graphing calculator to convert 10°C, 15°C, 20°C, 25°C, and 30°C to the equivalent Fahrenheit temperatures.

Solution

1.
$$C = \frac{5}{9}(F - 32) \qquad \textit{Celsius formula}$$

$$\frac{9}{5} \cdot C = \frac{9}{5} \cdot \frac{5}{9}(F - 32) \qquad \textit{Multiply both sides by } \frac{9}{5}.$$

$$\frac{9}{5}C = F - 32 \qquad \textit{Simplify.}$$

$$\frac{9}{5}C + 32 = F - 32 + 32 \qquad \textit{Add 32 to both sides.}$$

$$\frac{9}{5}C + 32 = F \qquad \textit{Combine like terms.}$$

The result is $F = \frac{9}{5}C + 32$.

2. We substitute 10 for C in the formula $F = \frac{9}{5}C + 32$ and solve for F:

$$F = \frac{9}{5}(10) + 32 = 50$$

So, 10°C is equivalent to 50°F.

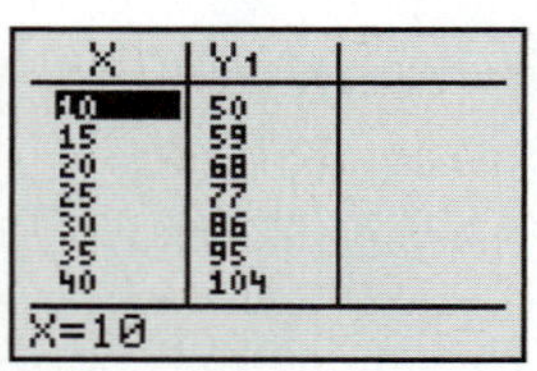

Figure 28 Finding Fahrenheit temperatures

3. We use a graphing calculator table to substitute 10, 15, 20, 25, and 30 for C in the formula $F = \frac{9}{5}C + 32$ (see Fig. 28). By viewing the second column of the graphing calculator table, we see that the values of F are 50, 59, 68, 77, and 86. So, 10°C, 15°C, 20°C, 25°C, and 30°C are equivalent to 50°F, 59°F, 68°F, 77°F, and 86°F, respectively.

In Example 4, we used the formula $C = \frac{5}{9}(F - 32)$ to convert five Fahrenheit temperature readings to five equivalent Celsius temperature readings. In Example 8, we used the formula $F = \frac{9}{5}C + 32$ to convert the same five Celsius temperature readings to the five equivalent Fahrenheit temperature readings. We can make two observations:

1. Both formulas describe the same relationship between F and C. In general, **solving for a variable in a formula will not change the relationship between the variables in the formula.**
2. To find Celsius readings, it is convenient to use the formula $C = \frac{5}{9}(F - 32)$, because C is alone on one side of the equation (and does not appear on the other side). Similarly, to find Fahrenheit readings, it is convenient to use the formula $F = \frac{9}{5}C + 32$. In general, **to find several values of a variable in a formula, we usually solve the formula for that variable before we make any substitutions.**

Example 9 Solving a Linear Model for One of Its Variables

The average fee banks charged noncustomers to use their ATMs was \$2.40 in 2011 and has increased by approximately \$0.14 per year since then (Source: *Bankrate.com*). Let F be the average ATM fee (in dollars) at t years since 2011.

1. Find an equation of a linear model to describe the situation.
2. Solve the equation that we found in Problem 1 for t.
3. Predict when the average ATM fee will be \$3.
4. Use a graphing calculator table to predict in which years the average ATM fee will be \$3.20, \$3.40, \$3.60, \$3.80, and \$4.00.

Solution

1. Since the average ATM fee increases by about \$0.14 each year, we can model the situation by using a linear model with slope 0.14. Since the average ATM fee was \$2.40 in 2011, the F-intercept is (0, 2.4). So, a reasonable model is

$$F = 0.14t + 2.4$$

2.

$$\begin{aligned} F &= 0.14t + 2.4 && \text{Original equation} \\ F - 2.4 &= 0.14t + 2.4 - 2.4 && \text{Subtract 2.4 from both sides.} \\ F - 2.4 &= 0.14t && \text{Combine like terms.} \\ \frac{F - 2.4}{0.14} &= \frac{0.14t}{0.14} && \text{Divide both sides by 0.14.} \\ \frac{F - 2.4}{0.14} &= t && \text{Simplify.} \end{aligned}$$

The result is the formula $t = \frac{F - 2.4}{0.14}$.

3. We substitute 3 for F in $t = \frac{F - 2.4}{0.14}$:

$$t = \frac{3 - 2.4}{0.14} \approx 4.29$$

The average ATM fee will be \$3 in 2011 + 4 = 2015, according to the model.

4. We use a graphing calculator table to substitute 3.20, 3.40, 3.60, 3.80, and 4.00 for F in the equation $t = \frac{F - 2.4}{0.14}$ (see Fig. 29). By viewing the second column of the graphing calculator table, we see that the approximate values of t (rounded to the ones place) are 6, 7, 9, 10, and 11. The model predicts that the average ATM fee will be \$3.20, \$3.40, \$3.60, \$3.80, and \$4.00 in the years 2017, 2018, 2020, 2021, and 2022, respectively.

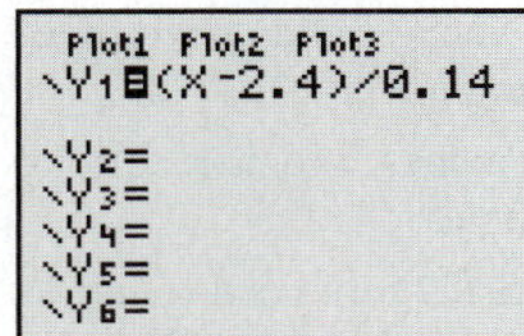

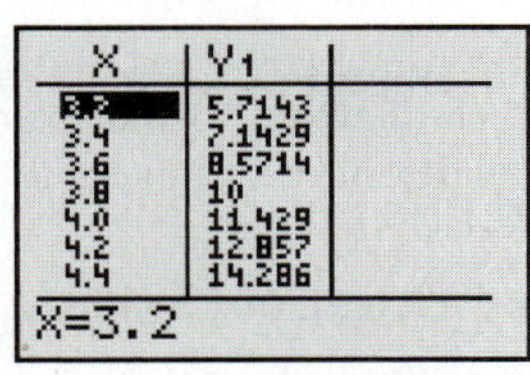

Figure 29 Finding the years

Group Exploration

Looking ahead: Graphing linear equations

1. a. Solve the equation $3x + 4y = 8$ for y.
   b. Use the result you found in part (a) to help you graph $3x + 4y = 8$ by hand.
2. a. Solve the equation $5x - 3y - 9 = 0$ for y.
   b. Use the result you found in part (a) to help you graph $5x - 3y - 9 = 0$ by hand.

Tips for Success Relax during a Test

If you get flustered during a test, close your eyes, take a couple of deep breaths, and think about something pleasant or nothing at all for a moment. This short break from the test might give you some perspective and help you relax.

Homework 4.6

For extra help ▶ MyMathLab®
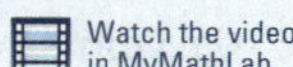
Watch the videos in MyMathLab

Download the MyDashboard App

Find a formula of the perimeter P of the polygon.

1. See Fig. 30.

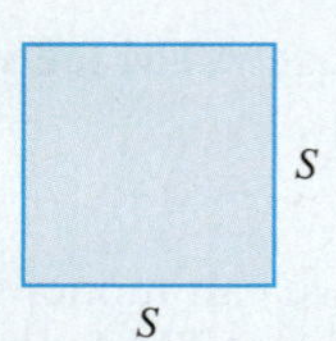

Figure 30 Exercise 1

2. See Fig. 31.

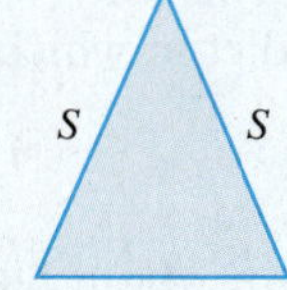

Figure 31 Exercise 2

3. See Fig. 32.

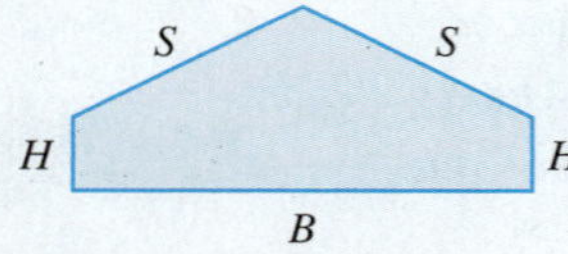

Figure 32 Exercise 3

4. See Fig. 33.

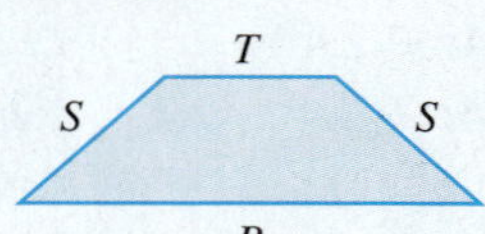

Figure 33 Exercise 4

5. See Fig. 34.

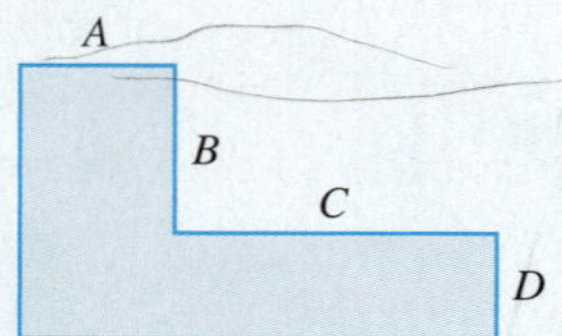

Figure 34 Exercise 5

6. See Fig. 35.

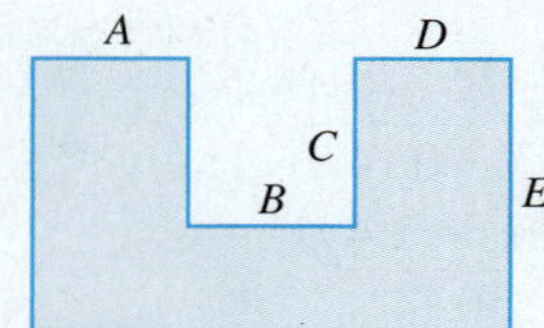

Figure 35 Exercise 6

For Exercises 7–16, substitute the given values for the variables and then solve the equation for the remaining variable. Round approximate results to the second decimal place.

7. $P = VI$; $P = 20$, $I = 4$ (power in an electrical circuit)

8. $PV = nRT$; $P = 0.97$, $V = 2.85$, $R = 0.082$, $T = 295$ (pressure, volume, and temperature of a gas)

9. $A = \frac{1}{2}BH$; $A = 6$, $H = 3$ (area of a triangle)

10. $U = -\frac{GmM}{r}$; $U = -50$, $G = 9.8$, $M = 7$, $r = 2$ (gravitational potential energy)

11. $v = gt + v_0$; $v = 80$, $g = 32.2$, $v_0 = 20$ (speed of a projectile)

12. $A = P + Prt$; $A = 850$, $P = 500$, $r = 0.1$ (simple interest)

13. $S = 2WL + 2WH + 2LH$; $S = 52$, $W = 2$, $H = 4$ (surface area of a rectangular box)

14. $S = 2WL + 2WH + 2LH$; $S = 76$, $L = 2$, $H = 4$ (surface area of a rectangular box)

15. $A = \frac{a + b + c}{3}$; $A = 5$, $a = 2$, $c = 6$ (average length of a side of a triangle)

16. $\frac{x}{a} + \frac{y}{a} = 1$; $a = 2$, $y = 6$ (equation of a line)

17. A rectangular carpet has an area of 116 square feet and a width of 8 feet. Find the length of the carpet.

18. A rectangular floor has an area of 207 square feet and a length of 18 feet. Find the width of the room.

19. A rectangle has a perimeter of 52 inches and a width of 10 inches. Find the length of the rectangle.

20. A rectangle has a perimeter of 88 inches and a length of 25 inches. Find the width of the rectangle.

21. A portable volleyball court includes a 177-foot cord to use as the rectangular boundary of the official-size court. The official width of a volleyball court is 29.5 feet. What is the official length of the court?

22. A photographer plans to use 55 inches of wood to make a rectangular frame for a photo. If the width of the frame is to be 12.5 inches, what will its length be?

23. A rectangle has a width of 3 inches and a length of x inches.

a. Find a formula of the area A (in square inches) of the rectangle.

b. Find a formula of the perimeter P (in inches) of the rectangle.

c. Perform a unit analysis of the formula you found in part (b).

24. A rectangle has a width of x feet and a length of 7 feet.
a. Find a formula of the area A (in square feet) of the rectangle.
b. Find a formula of the perimeter P (in feet) of the rectangle.
c. Perform a unit analysis of the formula you found in part (b).

25. A landscaper is going to dig a rectangular garden and enclose the garden with 40 feet of fencing. He can plant one flower per square foot of ground.
a. Give three examples of rectangular gardens that could be enclosed with 40 feet of fencing. State the length and width of each garden.
b. Find the area of each of the three gardens you described in part (a).
c. Which of the gardens you described in part (a) would hold the greatest number of flowers? Explain.

26. A landscaper is going to dig a small rectangular garden that has an area of 36 square feet.
a. Give three examples of rectangular gardens that have an area of 36 square feet. State the length and width of each garden.
b. Find the perimeter of each of the three gardens you described in part (a).
c. The landscaper plans to enclose the garden with fencing. Which of the gardens you described in part (a) would require the least amount of fencing? How much fencing is that?

27. **a.** What is the value (in cents) of 3 dimes?
b. What is the value (in cents) of 4 dimes?
c. Find a formula of the total value T (in cents) of d dimes.
d. Perform a unit analysis of the formula you found in part (c).

28. **a.** What is the value (in dollars) of 3 nickels?
b. What is the value (in dollars) of 4 nickels?
c. Find a formula of the total value T (in dollars) of n nickels.
d. Perform a unit analysis of the formula you found in part (c).

29. Find a formula of the total sales T (in dollars) of selling n guitars at \$725 per guitar.

30. Find a formula of the total sales T (in dollars) of selling n gallons of gasoline at \$3.95 per gallon.

31. Wye Oak played at the Crystal Ballroom in Portland, Oregon, on July 25, 2012. Tickets purchased in advance sold for \$20 each.
a. Assuming all the tickets were purchased in advance, find a formula of the total sales T (in dollars) if x people bought tickets.
b. Substitute 30,000 for T in the formula you found in part (a), and solve for x. What does your result mean in this situation?

32. T-Mobile charges \$69.99 per month for Classic Unlimited™ Plus, which includes unlimited texting, talking, and data (up to 2 GB high speed).
a. Find a formula of the total charge T (in dollars) for m months of service.
b. Substitute 559.92 for T in the formula you found in part (a), and solve for m. What does your result mean in this situation?

33. Baseball tickets for general seating at Yankee Stadium in the Bronx ranged from \$15 to \$575 in 2012.
a. Find a formula of the total cost C (in dollars) of k tickets that sold for \$15 per ticket.
b. Find a formula of the total cost E (in dollars) of n tickets that sold for \$575 per ticket.
c. Find a formula of the total cost T (in dollars) of k tickets that sold for \$15 per ticket and n tickets that sold for \$575 per ticket.
d. Use the formula you found in part (c) to find the value of n when $k = 8000$ and $T = 177{,}500$. What does your result mean in this situation?

34. Football season tickets at University of California, Berkeley, ranged from \$98 to \$300 in 2012.
a. Find a formula of the total cost C (in dollars) of k season tickets that sold for \$98 per ticket.
b. Find a formula of the total cost E (in dollars) of n season tickets that sold for \$300 per ticket.
c. Find a formula of the total cost T (in dollars) of k season tickets that sold for \$98 per ticket and n season tickets that sold for \$300 per ticket.
d. Use the formula you found in part (c) to find the value of n when $k = 825$ and $T = 305{,}850$. What does your result mean in this situation?

35. One cubic foot is shown in Fig. 36. The *volume,* in cubic feet, of an object is the number of cubic feet that it takes to fill that object. Let L be the length, W be the width, and H be the height, all in feet, of a rectangular box (see Fig. 37). The volume V (in cubic feet) of the rectangular box is equal to the length times the width times the height of the box.

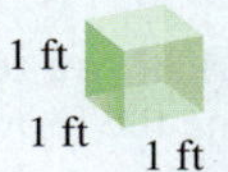

Figure 36 One cubic foot

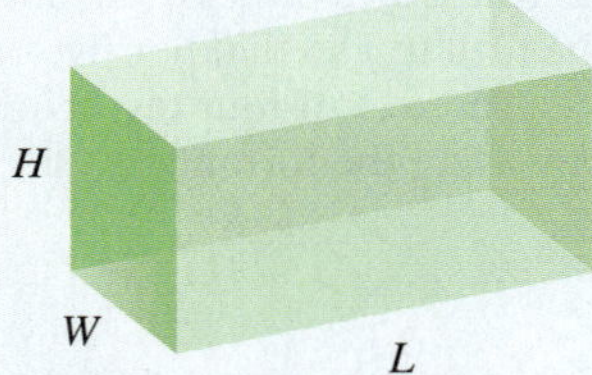

Figure 37 The length L, width W, and height H of a rectangular box.

a. Write a formula of the volume of the rectangular box.
b. Find the height of the rectangular box if the volume is 48 cubic feet, the length is 3 feet, and the width is 2 feet.

36. Let B be the length of the base, T be the length of the top, and H be the height of a trapezoid (see Fig. 38). The area of a trapezoid is equal to $\frac{1}{2}$ of the height times the sum of the lengths of the base and top.

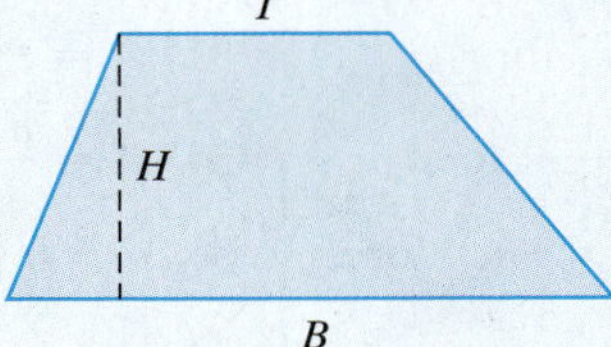

Figure 38 The base B, top T, and height H of a trapezoid

a. Write a formula of the area A of a trapezoid.
b. Find the base of a trapezoid with an area of 20 square inches, a height of 4 inches, and a top length of 3 inches.

37. The interest I (in dollars) earned from a simple-interest bank account is equal to the money P (in dollars) invested, times the decimal annual interest rate r, times the number of years t of the investment.

a. Write a formula of I.

b. Find the interest earned from investing \$5000 in a 4% simple-interest account for 3 years. [**Hint:** If the interest rate is 4%, then the decimal interest rate is 0.04.]

c. The balance B (in dollars) of a bank account is equal to the original money invested plus the interest. Find a formula of B in terms of P, r, and t. [**Hint:** Build on your formula from part (a).]

d. Find the balance after 4 years in a 5% simple-interest account in which \$2000 was originally invested.

38. The force F (in newtons) needed to lift an object is equal to the product of the object's mass m (in kilograms) and the constant a (in meters per second squared), which is the object's acceleration due to gravity.

a. Write a formula of the force needed to lift an object.

b. Describe the unit newton (N) in terms of kilograms (kg), meters (m), and seconds (s). [**Hint:** Perform a unit analysis of the right-hand side of the formula you found in part (a).]

c. If a force of 25 newtons is required to lift an object and the constant a is 9.8 meters per second squared, find the mass of the object.

39. The average test score A of five test scores t_1, t_2, t_3, t_4, and t_5 can be computed by using the formula

$$A = \frac{t_1 + t_2 + t_3 + t_4 + t_5}{5}$$

If a student has test scores of 74, 81, 79, and 84, what score does she need on the fifth test so that her five-test average is 80?

40. The average test score A of four test scores t_1, t_2, t_3, and t_4 can be computed by using the formula

$$A = \frac{t_1 + t_2 + t_3 + t_4}{4}$$

If a student has test scores of 87, 92, and 86, what score does he need on the fourth test so that his four-test average is 90?

Solve the equation for the specified variable.

41. $A = LW$, for W

42. $P = VI$, for I

43. $PV = nRT$, for T

44. $PV = nRT$, for n

45. $U = -\frac{GmM}{r}$, for M

46. $U = -\frac{GmM}{r}$, for m

47. $A = \frac{1}{2}BH$, for B

48. $A = \frac{1}{2}BH$, for H

49. $v = gt + v_0$, for t

50. $y = mx + b$, for x

51. $A = P + Prt$, for r

52. $A = P + Prt$, for t

53. $A = \frac{a + b + c}{3}$, for b

54. $A = \frac{a + b + c}{3}$, for c

55. $y - k = m(x - h)$, for x

56. $y - k = m(x - h)$, for h

57. $\frac{x}{a} + \frac{y}{a} = 1$, for y

58. $\frac{x}{a} + \frac{y}{a} = 1$, for x

Write the linear equation in slope–intercept form.

59. $3x + 4y = 16$

60. $2x + 3y = 24$

61. $2x + 4y - 8 = 0$

62. $7x + 3y - 21 = 0$

63. $5x - 2y = 6$

64. $4x - 5y = 30$

65. $-3x - 7y = 5$

66. $-5x - 9y = 2$

67. The percentages of new-vehicle buyers who use the Internet during the shopping process are shown in Table 10 for various years.

Table 10 Percentages of New-Vehicle Buyers Who Use the Internet during the Shopping Process

Year	Percent
2000	54
2002	60
2004	64
2006	68
2008	75
2010	79

Source: *J. D. Power and Associates*

Let p be the percentage of new-vehicle buyers who use the Internet during the shopping process at t years since 2000. A model of the situation is $p = 2.5t + 54.2$.

a. Use a graphing calculator to draw a scattergram and the model in the same viewing window. Does the line come close to the data points?

b. Solve the equation $p = 2.5t + 54.2$ for t.

c. Use the equation you found in part (b) to estimate when 77% of new-vehicle buyers used the Internet during the shopping process.

d. Use a graphing calculator table to predict in which years the percentage of new-vehicle buyers who will use the Internet during the shopping process will be 89%, 92%, 94%, 97%, and 99%.

68. Computer and video game sales (in billions of dollars) in the United States are shown in Table 11 for various years.

Table 11 Computer and Video Game Sales

Year	Computer and Video Game Sales (billions of dollars)
2005	6.9
2006	7.5
2007	9.5
2008	11.7
2009	16.0
2010	16.9
2011	16.6

Source: *NPD Group*

Let s be computer and video game sales (in billions of dollars) in the year that is t years since 2000. A model of the situation is $s = 1.94t - 3.4$.

a. Use a graphing calculator to draw a scattergram and the model in the same viewing window. Does the line come close to the data points?

b. Solve the equation $s = 1.94t - 3.4$ for t.

c. Use the equation you found in part (b) to predict in which year computer and video game sales will be \$28 billion.

d. Use a graphing calculator table to predict in which years computer and video game sales will be 30, 32, 34, 36, and 38, all in billions of dollars.

69. Kia® sold 356 thousand automobiles in 2010, and sales are increasing by about 20.1 thousand automobiles per year (Source: *Auto Pacific*). Let s be the annual sales (in thousands of automobiles) at t years since 2010.

a. Find an equation of a linear model to describe the situation.
b. Solve the equation you found in part (a) for t.
c. Use the equation you found in part (b) to predict in which year Kia's automobile sales will be 475 thousand automobiles.
d. Use a graphing calculator table to predict in which years Kia's automobile sales will be 500, 525, 550, 575, and 600, all in thousands of automobiles.

70. The percentage of consumers whose purchasing decisions are influenced by products made in the United States was 60% in 2011 and has increased by 4.3 percentage points per year (Source: *Schneider Associates*). Let p be the percentage of consumers whose purchasing decisions are influenced by products made in the United States at t years since 2011.

a. Find an equation of a linear model to describe the situation.
b. Solve the equation you found in part (a) for t.
c. Use the equation you found in part (b) to predict when the percentage of consumers whose purchasing decisions will be influenced by products made in the United States will be 75%.
d. Use a graphing calculator table to predict the years when the percentage of consumers whose purchasing decisions will be influenced by products made in the United States will be 80%, 85%, 90%, 95%, and 100%.

71. Due to the airline industry charging more fees, the percentage of revenue from fares was only 71% in 2010 and has decreased by 0.8 percentage point per year (Source: *Bureau of Transportation Statistics*). Let p be the percentage of revenue from fares at t years since 2010.

a. Find an equation of a linear model to describe the situation.
b. Use the equation you found in part (a) to predict in which year 66% of revenue will come from fares.
c. Solve the equation you found in part (a) for t.
d. Use the equation you found in part (c) to predict in which year 66% of revenue will come from fares.
e. Compare your results from parts (b) and (d). Which equation was easier to use? Explain.
f. Which equation is easier to use to predict the percentage of revenue from fares in 2018? Explain. Then find that percentage.

72. Some 213.9 million pounds of fireworks were used in the United States in 2009, and that amount is decreasing by about 20.1 million pounds per year (Source: *American Pyrotechnics Association*). Let F be the amount of fireworks used (in millions of pounds) in the year that is t years since 2009.

a. Find an equation of a linear model to describe the situation.
b. Use the equation you found in part (a) to estimate in which year 130 million pounds of fireworks were used.
c. Solve the equation you found in part (a) for t.
d. Use the equation you found in part (c) to estimate in which year 130 million pounds of fireworks were used.
e. Compare your results from parts (b) and (d). Which equation was easier to use? Explain.
f. Which equation is easier to use to predict the amount of fireworks used in 2014? Explain. Then find that amount.

73. A person travels d miles at a constant speed s (in miles per hour) for t hours.

a. Complete Table 12 to help find a formula that describes the relationship among s, t, and d. Show the arithmetic to help you see a pattern.

Table 12 Speed, Time, and Distance

Speed (miles per hour) s	Time (hours) t	Distance (miles) d
50	4	
70	3	
65	2	
55	5	
s	t	

b. Perform a unit analysis of the formula you found in part (a).
c. Solve the formula you found in part (a) for t.
d. Use the formula you found in part (c) to find the value of t when $d = 315$ and $s = 70$. What does your result mean in this situation?
e. A student plans to drive from Albuquerque Technical Vocational Institute in Albuquerque, New Mexico, to Denver, Colorado. The speed limit is 75 mph in New Mexico and 65 mph in Colorado. The trip involves 229.9 miles of travel in New Mexico followed by 219.2 miles in Colorado. If the student drives at the posted speed limits, estimate the driving time.

Concepts

74. a. Solve the equation $mx + b = 0$, where $m \neq 0$, for x.
b. Explain why a linear equation in one variable has exactly one solution.

75. An object travels d miles at a constant speed s (in miles per hour) for t hours. A student believes that the formula of s is

$$s = dt \quad \text{Incorrect}$$

Perform a unit analysis to show that the formula is incorrect.

76. A student believes that a formula of the total value T (in dollars) of n objects, each worth v dollars, is

$$n = Tv \quad \text{Incorrect}$$

Perform a unit analysis to show that the formula is incorrect.

77. a. How does doubling the width and doubling the length of a rectangle affect the perimeter of the rectangle?
b. How does doubling the width and doubling the length of a rectangle affect the area of the rectangle?

78. a. How does increasing the length of a rectangle by 1 inch affect the perimeter (in inches) of the rectangle?
b. How does increasing the length of a rectangle by 1 inch affect the area (in square inches) of the rectangle?

79. For a rectangle of length L, width W, and perimeter P, explain why the perimeter is given by $P = 2L + 2W$.

80. If n objects each have value v, then their total value T is given by $T = vn$. Give an example of this.

81. In what cases would you first solve a formula for a variable and then make substitutions for any other variables, and in what cases would you first make substitutions for all but one variable and then solve for the remaining variable?

82. Give an example of an equation in one variable and an example of a formula. How are these equations different?

Chapter Summary

Key Points of Chapter 4

Section 4.1 Commutative, Associative, and Distributive Laws

Commutative law for addition	$a + b = b + a$
Commutative law for multiplication	$ab = ba$
Term	A **term** is a constant, a variable, or a product of a constant and one or more variables raised to powers.
Associative law for addition	$a + (b + c) = (a + b) + c$
Associative law for multiplication	$a(bc) = (ab)c$
Distributive law	$a(b + c) = ab + ac$
Equivalent expressions	Two or more expressions are **equivalent expressions** if, when each variable is evaluated for *any* real number (for which all the expressions are defined), the expressions all give equal results.
Multiplying a number by -1	$-1a = -a$
Showing that an equation is false	To show that an equation is false, we need find only *one* set of numbers that, when substituted for any variables, gives a result of the form $a = b$, where a and b are different real numbers.

Section 4.2 Simplifying Expressions

Like terms	**Like terms** are either constant terms or variable terms that contain the same variable(s) raised to exactly the same power(s).
Combining like terms	To combine like terms, add the coefficients of the terms and keep the same variable factors.
Simplifying an expression	We simplify an expression by removing parentheses and combining like terms.
Locating an error in simplifying an expression	After simplifying an expression, use a graphing calculator table to check whether the result is equivalent to the original expression. If the expressions are not equivalent, an error was made. To pinpoint the error, evaluate all of the expressions by using the same number(s) and find the pair of consecutive expressions for which the results are different.

Section 4.3 Solving Linear Equations in One Variable

Linear equation in one variable	A **linear equation in one variable** is an equation that can be put into the form $mx + b = 0$, where m and b are constants and $m \neq 0$.
Solution, satisfy, solution set, and solve for an equation in one variable	A number is a **solution** of an equation in one variable if the equation becomes a true statement when the number is substituted for the variable. We say that the number **satisfies** the equation. The set of all solutions of the equation is called the **solution set** of the equation. We **solve** the equation by finding its solution set.
Roles of a variable	Here are three roles of a variable: **1.** A variable represents a quantity that can vary. **2.** In an expression, a variable is a placeholder for a number. **3.** In an equation, a variable represents any number that is a solution of the equation.

Section 4.3 Solving Linear Equations in One Variable (*Continued*)

Equivalent equations	**Equivalent equations** are equations that have the same solution set.
Addition property of equality	If A and B are expressions and c is a number, then the equations $A = B$ and $A + c = B + c$ are equivalent.
Checking results after solving an equation	After solving an equation, check that all of your results satisfy the equation.
Multiplication property of equality	If A and B are expressions and c is a nonzero number, then the equations $A = B$ and $Ac = Bc$ are equivalent.
Using graphing to solve an equation in one variable	To use graphing to solve an equation $A = B$ in one variable, x, where A and B are expressions, **1.** Graph the equations $y = A$ and $y = B$ in the same coordinate system. **2.** Find all intersection points. **3.** The x-coordinates of those intersection points are the solutions of the equation $A = B$.
Getting the variable alone on one side of the equation	Our strategy in solving linear equations in one variable will be to use properties to get the variable alone on one side of the equation. For example, to solve the equation $x + 3 = 7$, $x - 3 = 7$, $3x = 7$, or $\frac{x}{3} = 7$, we get x alone by "undoing" the operation that involves x and 3.

Section 4.4 Solving More Linear Equations in One Variable

Solving an equation that contains fractions	A key step in solving an equation that contains fractions is to multiply both sides of the equation by the LCD so that there are no fractions on either side of the equation.
Number of solutions	Any linear equation in one variable has exactly one solution.
Conditional equation, inconsistent equation, and identity	There are three types of equations: **1.** A **conditional equation** is sometimes true and sometimes false, depending on which values are substituted for the variable(s). **2.** If an equation does not have any solutions, we call the equation **inconsistent** and say that the solution is the **empty set.** When we apply the usual steps to solve an inconsistent equation, the result is always a false statement. **3.** An equation that is true for all permissible values of the variable(s) it contains is called an **identity.** When we apply the usual steps to solve an identity, the result is always a true statement of the form $a = a$, where a is a constant.
Locating an error in solving a linear equation	If the result of an attempt to solve a linear equation in one variable does not satisfy the equation, an error was made. To pinpoint the error, substitute the result into all of the equations in the work and find which pair of consecutive equations gives a false statement followed by a true statement.

Section 4.5 Comparing Expressions and Equations

Linear expression in one variable	A **linear expression in one variable** is an expression that can be put into the form $mx + b$, where m and b are constants and $m \neq 0$.
Results of simplifying an expression and solving an equation	The result of simplifying an expression is an expression. The result of solving a linear equation in one variable is a number.
Connection between simplifying an expression and an identity	If an expression A in one variable is simplified to an expression B, then every real number (for which both expressions are defined) is a solution of the equation $A = B$. In other words, the equation $A = B$ is an identity.
Multiplying expressions and both sides of equations by numbers	In simplifying an expression, the only number that we can multiply the expression or part of it by is 1. In solving an equation, we can multiply both sides of the equation by *any* number except 0.

Section 4.6 Formulas

Formula	A **formula** is an equation that contains two or more variables.
Finding a single value of a variable in a formula	If we want to find a single value of a variable in a formula, we often substitute numbers for all of the other variables and then solve for that one variable.
Perimeter	The **perimeter** of a polygon, such as a triangle, rectangle, or trapezoid, is the total distance around the object.
Area and perimeter of a rectangle	For a rectangle with length L, width W, area A, and perimeter P, • $A = LW$ **Area formula** • $P = 2L + 2W$ **Perimeter formula**
Total-value formula	If n objects each have value v, then their total value T is given by $T = vn$.
Finding several values of a variable in a formula	If we want to find several values of a variable in a formula, we usually solve the formula for that variable before making any substitutions.

Chapter 4 Review Exercises

1. Use the commutative law for addition to write the expression $5w + 9$ in another form.
2. Use the commutative law for multiplication to write the expression $8 + wp$ in another form.

Use an associative law to write the expression in another form.

3. $2 + (k + y)$
4. $(bx)w$

Simplify. Use a graphing calculator table to verify your work when possible.

5. $-5(4x)$
6. $-3(8x + 4)$
7. $\frac{4}{5}(15y - 35)$
8. $-(3x - 6y - 8)$
9. $\frac{2}{9}x + \frac{5}{9}x$
10. $5a + 2 - 13b - a + 4b - 9$
11. $-5y - 3(4x + y) - 6x$
12. $-2.6(3.1x + 4.5) - 8.5$
13. $-(2m - 4) - (3m + 8)$
14. $4(3a - 7b) - 3(5a + 4b)$

For Exercises 15 and 16, let x be a number. Translate the English phrase to a mathematical expression. Then simplify the expression.

15. -4 times the difference of the number and 7
16. -7, plus 2 times the sum of the number and 8
17. Use values for a, b, and c to show that $a(b + c) = ab + c$ is false. Then write a true statement that involves $a(b + c)$.
18. Give three examples of expressions that are equivalent to $3x - 9$.
19. Which of the following expressions are equivalent?

$$-5x - 20 \quad -5(x - 4) \quad 5(4 - x) \quad -5x - 4 \quad -2(x - 10) - 3x \quad -5x + 20$$

For Exercises 20 and 21, graph the equation by hand.

20. $y = 2x + 3 - 4x$
21. $y = -3(x - 2)$
22. Determine whether 3 is a solution of the linear equation $2 - 5x = -3(4x - 7)$.

For Exercises 23–33, solve. Use a graphing calculator to verify your result.

23. $a + 5 = 12$
24. $-4x = 20$
25. $-p = -3$
26. $-\frac{7}{3}a = 14$
27. $4.5x - 17.2 = -5.05$
28. $5x - 9x + 3 = 17$
29. $8m - 3 - m = 2 - 4m$
30. $8x = -7(2x - 3) + x$
31. $6(4x - 1) - 3(2x + 5) = 2(5x - 3)$
32. $\frac{w}{8} - \frac{3}{4} = \frac{5}{6}$
33. $\frac{3p - 4}{2} = \frac{5p + 2}{4} + \frac{7}{6}$
34. A student tries to solve the equation $x - 5 = 2$:

$$x - 5 = 2$$
$$x - 5 + 5 = 2$$
$$x + 0 = 2$$
$$x = 2$$

Describe any errors. Then solve the equation correctly.

35. Give three examples of equations that have the solution -6.
36. Solve $-2.5(3.8x - 1.9) = 83.7$. Round the result to two decimal places.
37. The number of prisoners under sentence of death was 3189 prisoners in 2012 and has decreased by about 38 prisoners per year since then (Source: *Death Penalty Information Center*). Let n be the number of death-row prisoners at t years since 2012.
    a. Find an equation of a model.
    b. Predict when there will be 3000 death-row prisoners.
    c. Predict the number of death-row prisoners in 2019.

For Exercises 38 and 39, find the number.

38. Four times the difference of a number and 6 is 15.
39. Two, minus 3 times the sum of a number and 8 is 95.
40. Use "intersect" on a graphing calculator to solve the equation $\frac{1}{2}x + \frac{5}{3} = -\frac{2}{3}x - \frac{1}{4}$. Round the solution to the second decimal place.

For Exercises 41–44, solve the given equation by referring to the solutions of $y = -2x + 17$ and $y = 5x - 4$ shown in Tables 13 and 14.

41. $-2x + 17 = 5x - 4$

42. $-2x + 17 = 15$

43. $5x - 4 = 6$

44. $5x - 4 = -4$

Table 13 Solutions of $y = -2x + 17$

x	y
0	17
1	15
2	13
3	11
4	9

Table 14 Solutions of $y = 5x - 4$

x	y
0	−4
1	1
2	6
3	11
4	16

For Exercises 45–47, solve the equation. State whether the equation is a conditional equation, an inconsistent equation, or an identity.

45. $7x - 4 + 3x = 2 + 10x - 6$

46. $6(2x - 3) - (5x + 2) = -2(4x - 1)$

47. $2(x - 5) + 3 = 2x - 4$

48. A student incorrectly solves $4(x - 5) = 28$:

$$\begin{aligned} 4(x - 5) &= 28 \\ 4x - 20 &= 28 \\ 4x - 20 - 20 &= 28 - 20 \\ 4x &= 8 \\ \frac{4x}{4} &= \frac{8}{4} \\ x &= 2 \end{aligned}$$

Substitute 2 for x in each of the six equations, and explain how the results help you pinpoint the error. Describe the error. Then solve the equation correctly.

Is the following a linear expression or a linear equation?

49. $8 - 3(x + 5)$

50. $8 - 3(x + 5) = 4x$

For Exercises 51–58, simplify the expression or solve the equation, as appropriate.

51. $6t - 8t$

52. $0.1 + 0.5a - 0.3a = 0.7$

53. $6t - 8t = 10$

54. $0.1 + 0.5a - 0.3a$

55. $9(2p - 5) - 3(7p + 3)$

56. $\frac{5}{6}r - \frac{3}{4} = \frac{1}{6} + \frac{7}{2}r$

57. $9(2p - 5) - 3(7p + 3) = 0$

58. $\frac{5}{6}r - \frac{3}{4} - \frac{1}{6} + \frac{7}{2}r$

59. A student is trying to simplify an expression. The student writes, "The solution of the expression is 2." Is the student correct? Explain.

60. A student tries to simplify the expression $\frac{2}{3}x + \frac{7}{5}$:

$$\begin{aligned} \frac{2}{3}x + \frac{7}{5} &= 15\left(\frac{2}{3}x + \frac{7}{5}\right) \\ &= 15 \cdot \frac{2}{3}x + 15 \cdot \frac{7}{5} \\ &= 10x + 21 \end{aligned}$$

Describe any errors. Then simplify the expression correctly.

For Exercises 61 and 62, let x be a number. Translate the English phrase or sentence into an expression or an equation. Then simplify the expression or solve the equation, as appropriate.

61. Four times the difference of 6 and the number is 17.

62. The number minus the quotient of the number and 2

63. Find a formula of the perimeter P of the polygon shown in Fig. 39.

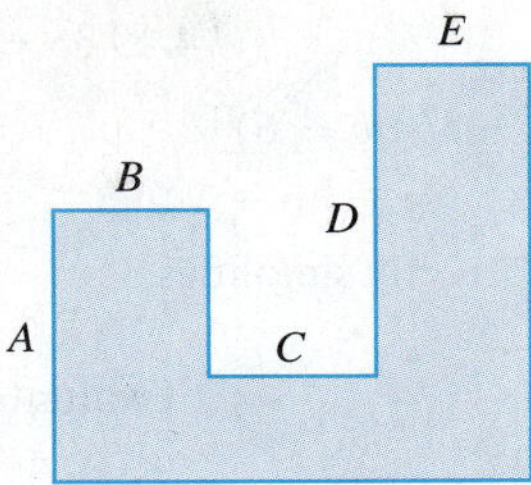

Figure 39 Exercise 63

64. a. Find a formula of the total cost T (in dollars) of n tickets that sell for \$15 per ticket and w tickets that sell for \$25 per ticket.

b. Use the formula you found in part (a) to find the value of w when $n = 370$ and $T = 11{,}050$. What does your result mean in this situation?

For Exercises 65–68, solve the equation for the specified variable.

65. $C = 2\pi r$, for r

66. $P = a + b + c$, for c

67. $3x - 6y = 18$, for y

68. $A = \frac{1}{2}H(B + T)$, for T

69. The number of visits to U.S. libraries was 1.59 billion in 2009 and increases by 0.05 billion visits per year (Source: *Institute of Museum and Library Services*). Let v be the number of visits (in billions) to U.S. libraries in the year that is t years since 2009.

a. Find an equation of a linear model to describe the situation.

b. Solve the equation you found in part (a) for t.

c. Use the equation you found in part (a) to predict in which year there will be 2 billion visits.

d. Use the equation you found in part (b) to predict in which year there will be 2 billion visits.

e. Compare your results from parts (c) and (d). Which equation was easier to use? Explain.

f. Which equation is easier to use to predict the number of U.S. library visits in 2019? Explain. Then find that number of visits.

70. In 2011, the average salary of public school teachers was \$56,643 and has since increased by \$1199 per year (Source: *National Education Association*). Predict when the average salary will be \$64,000.

71. In 2011, the number of Little League baseball participants was 2.1 million participants, down 8.7% from 2004 (Source: *Little League International*). What was the number of participants in 2004?

Chapter 4 Test

1. Use the commutative law for addition to write the expression $4 + 3p$ in another form.
2. Use an associative law to write the expression $3(xy)$ in another form.

For Exercises 3–6, simplify.

3. $-\frac{2}{3}(6x - 9)$
4. $9.36 - 2.4(1.7x + 3.5)$
5. $-5(2w - 7) - 3(4w - 6)$
6. $-(3a + 7b) - (8a - 4b + 2)$
7. A student incorrectly simplifies $3(x - 2) - (5x + 4)$:

Expression	Evaluate the Expression for $x = 3$
$3(x - 2) - (5x + 4)$	
$= 3x - 6 - 5x + 4$	
$= 3x - 5x - 6 + 4$	
$= -2x - 2$	

Evaluate each of the four expressions for $x = 3$ to pinpoint where the error was made. Then simplify $3(x - 2) - (5x + 4)$ correctly.

8. Graph $y = -2(x + 1)$ by hand.

For Exercises 9–14, solve.

9. $6x - 3 = 19$
10. $\frac{3}{5}x = 6$
11. $9a - 5 = 8a + 2$
12. $8 - 2(3t - 1) = 7t$
13. $3(2x - 5) - 2(7x + 9) = 49$
14. $\frac{7}{8}x + \frac{3}{10} = \frac{1}{4}x - \frac{1}{2}$
15. Solve $8.21x = 3.9(4.4x - 2.7)$. Round your result to two decimal places.
16. Four, minus 2 times the sum of a number and 7 is 54. Find the number.

For Exercises 17 and 18, simplify the expression or solve the equation, as appropriate.

17. $9(3x + 2) - (4x - 6)$
18. $9(3x + 2) - (4x - 6) = x$
19. A student believes that $x - 3$ is the solution of an equation. Is the student correct? Explain.
20. Give three examples of an expression equivalent to the expression 4.

For Exercises 21 and 22, let x be a number. Translate the following into an expression or an equation, as appropriate. Then simplify the expression or solve the equation.

21. Five times the difference of the number and 2 is 29.
22. Two, plus 4 times the sum of 3 and the number

For Exercises 23–26, solve the given equation by referring to the graphs of $y = \frac{3}{2}x - 4$ and $y = \frac{1}{2}x - 2$ shown in Fig. 40.

23. $\frac{3}{2}x - 4 = \frac{1}{2}x - 2$
24. $\frac{3}{2}x - 4 = 2$
25. $\frac{1}{2}x - 2 = -3$
26. $\frac{1}{2}x - 2 = 0$

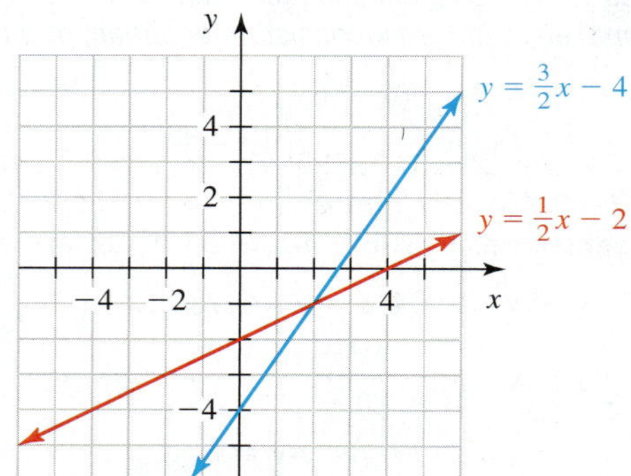

Figure 40 Exercises 23–26

27. The number of U.S. patent applications was 520 thousand in 2010 and increases by about 21.7 thousand per year (Source: *U.S. Patent and Trademark Office*). Let n be the number of patent applications (in thousands) during the year that is t years since 2010.
    a. Find an equation of a model.
    b. Predict the number of patent applications in 2018.
    c. Predict when there will be 650 thousand patent applications.
28. A person has \$8350 in a savings account. If the person withdraws \$125 per month, after how many months will there be \$6975 in the account?
29. In 2011, the number of complaints of online crime was 314 thousand complaints, up 3.3% from 2010 (Source: *Internet Crime Complaint Center*). What was the number of complaints in 2010?
30. A person plans to enclose a rectangular garden with fencing. If the width of the garden is 8 feet, and if 52 feet of fencing is required to enclose the garden, what is the length of the garden?
31. Solve the equation $A = \frac{a + b}{2}$ for the variable a.

Cumulative Review of Chapters 1–4

1. Choose a variable name for the number of pages in a book. Give two numbers that the variable represents and two numbers that it does not represent.
2. Graph the numbers $-3, -\frac{5}{2}, 1,$ and $\frac{3}{2}$ on one number line.
3. Let n be the average number (in millions) of unique monthly visitors to the website YouTube in the year that is t years since 2000. What does the ordered pair $(11, 790)$ represent?

For Exercises 4–7, perform the indicated operations by hand. Then use a calculator to check your work.

4. $4 + 3(-2)$

5. $-8 \div 4 - 2(7 - 10)$

6. $\frac{15}{8} \cdot \left(-\frac{4}{25}\right)$

7. $\left(-\frac{3}{10}\right) + \left(-\frac{7}{8}\right)$

8. Simplify $\frac{27}{-45}$.

9. A rectangle has a width of $\frac{3}{4}$ foot and a length of $\frac{5}{6}$ foot. What is the perimeter of the rectangle?

10. A student scores 92 points on one test and 85 points on the next test. What is the change in the scores?

11. Evaluate the expression $a(b - c)$ for $a = -3$, $b = 5$, and $c = -4$.

For Exercises 12 and 13, find the slope of the line passing through the two given points. State whether the line is increasing, decreasing, horizontal, or vertical.

12. $(-5, -2)$ and $(-1, -4)$

13. $(-4, -5)$ and $(-4, 3)$

14. Road A climbs steadily for 150 feet over a horizontal distance of 5000 feet. Road B climbs steadily for 95 feet over a horizontal distance of 3500 feet. Which road is steeper? Explain.

For Exercises 15–18, graph the equation by hand.

15. $y = -\frac{2}{3}x + 4$

16. $y = 2x - 3$

17. $x = -5$

18. $y = 3$

19. Find an equation of the line sketched in Fig. 41.

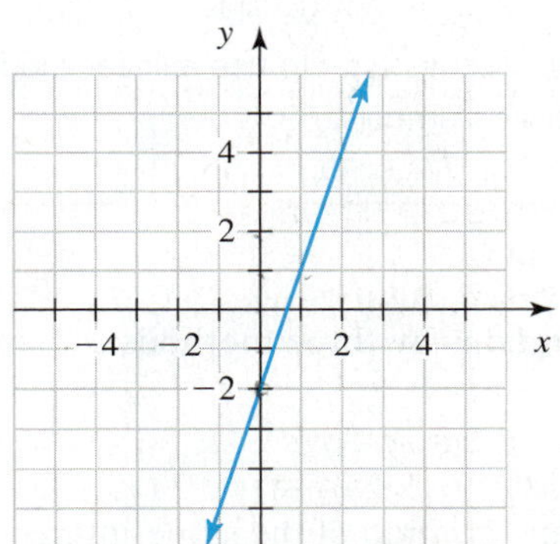

Figure 41 Exercise 19

20. A person's savings account balance decreased from \$7500 to \$2840 in 5 months. Find the average rate of change of the balance.

21. The number of reports filed with the U.S. Food and Drug Administration (FDA) about patients' adverse reactions to drugs increased approximately linearly from 422 thousand reports in 2004 to 759 thousand reports in 2010 (Source: *FDA*). Find the rate of change of the number of reports.

22. Sales for *Call of Duty* video games were \$1.06 billion in 2011 and have increased approximately linearly by \$0.14 billion per year (Source: *The NPD Group*). Let s be the annual sales (in billions of dollars) at t years since 2011.

a. Find the slope of the linear model that describes the situation. What does it mean in this situation?

b. Find an equation of the model.

c. Predict the sales in 2017.

d. Predict in which year the sales will be \$2.1 billion.

23. Some solutions of four equations are listed in Table 15. Which of the equations could be linear?

Table 15 Solutions of Four Equations

Equation 1		Equation 2		Equation 3		Equation 4	
x	y	x	y	x	y	x	y
0	11	0	56	3	35	1	1
1	13	1	53	4	44	2	1
2	16	2	50	5	53	3	1
3	20	3	47	6	62	4	1
4	25	4	44	7	71	5	1

24. The numbers of firearms discovered at TSA airport checkpoints are shown in Table 16 for various years.

Table 16 Numbers of Firearms Discovered at TSA Airport Checkpoints

Year	Number of Firearms
2007	803
2008	928
2009	976
2010	1123
2011	1310

Source: *Transportation Security Administration*

Let n be the number of firearms discovered at TSA airport checkpoints in the year that is t years since 2005. A model of the situation is $n = 120.9t + 544.4$.

a. Use a graphing calculator to draw a scattergram and the model in the same viewing window. Does the line come close to the data points?

b. What is the slope? What does it mean in this situation?

c. What is the n-intercept? What does it mean in this situation?

d. Predict in what year 2000 firearms will be discovered at TSA airport checkpoints.

e. Predict the number of firearms that will be discovered at TSA airport checkpoints in 2018. How many firearms is that per day, on average?

Simplify the expression or solve the equation, as appropriate. Use a graphing calculator to verify your work.

25. $3r + 4 = 7r - 8$

26. $2(3x - 2) = 4(3x + 5) - 3x$

27. $4a - 5b + 6 - 2b - 7a$

28. $7 - 2(3p - 5) + 5(4p - 2)$

29. $\frac{2}{3}r - \frac{5}{6} = \frac{1}{2}$

30. $-(2a + 5) - (4a - 1)$

31. Solve $25.93 - 7.6(2.1x + 8.7) = 53.26$. Round the result to the second decimal place.

For Exercises 32 and 33, let x be a number. Translate the following into an expression or an equation, as appropriate. Then simplify the expression or solve the equation.

32. The number, plus 9 times the quotient of the number and 3

33. Two times the difference of 7 and twice the number is 87.

Solve the equation for the specified variable.

34. $A = 2\pi rh$, for h

35. $4x - 6y = 12$, for y

5 Linear Functions and Linear Inequalities in One Variable

Are many of the married couples you know happily married? The percentages of married persons who say they are "very happy" with their marriages are shown in Table 1 for various years. In Exercise 7 of Homework 5.6, you will predict when 60% of married persons will say they are very happy with their marriages.

Table 1 Percentages of Married Persons "Very Happy" with Their Marriages

Year	Percent
1970	67.0
1980	67.5
1990	65.0
2000	62.0
2008	62.0

Source: *Institute for American Values*

In Chapter 4, we discussed how to simplify expressions and solve equations. In this chapter, we will use these skills to help us increase our effectiveness in graphing equations, finding equations, and modeling authentic situations such as the data on total student loan amounts. We will also discuss how to describe certain relationships between two variables by using an extremely important concept called a *function*. Finally, we will discuss how to use a linear model to predict when a quantity will be more than (or less than) a certain amount.

5.1 Graphing Linear Equations

Objectives

- » Graph a linear equation by solving for y.
- » Graph a linear equation by finding the intercepts of its graph.
- » Find coordinates of solutions of a linear equation.
- » Know the difference between equations in one variable and equations in two variables.

In this section, we will discuss how to graph linear equations by three methods: solving for y, finding intercepts, and finding three points.

Graphing Equations by Solving for y

In Section 3.4, we discussed how to use slope to graph an equation of the slope-intercept form $y = mx + b$. In Example 1, we will discuss how to graph equations that are not in this form, but that can be put into it.

Example 1 Graphing an Equation by Solving for y

Sketch the graph of $2x + 3y = 9$.

Solution

We will put the equation in $y = mx + b$ form so that we can use the y-intercept and the slope to help us graph the equation. To begin, we get y alone on one side of the equation:

$$2x + 3y = 9 \qquad \text{Original equation}$$

$$2x + 3y - 2x = 9 - 2x \qquad \text{Subtract } 2x \text{ from both sides to get } 3y \text{ alone on left-hand side.}$$

$$3y = -2x + 9 \qquad \text{Combine like terms; rearrange right-hand side.}$$

$$\frac{3y}{3} = \frac{-2x}{3} + \frac{9}{3} \qquad \text{Divide both sides by 3 to get } y \text{ alone on left-hand side.}$$

$$y = -\frac{2}{3}x + 3 \qquad \text{Simplify.}$$

Since $2x + 3y = 9$ can be put into the form $y = mx + b$ (as $y = -\frac{2}{3}x + 3$), we know that the graph of $2x + 3y = 9$ is a line. The y-intercept is $(0, 3)$ and the slope is $-\frac{2}{3}$. The graph is shown in Fig. 1.

We can verify our result by checking that both $(0, 3)$ and $(3, 1)$ are solutions of $2x + 3y = 9$.

Before we can use the y-intercept and the slope to graph a linear equation, we must solve for y to put the equation into the form $y = mx + b$.

WARNING

It is a common error to think the slope of an equation such as $2x + 3y = 9$ is 2, because 2 is the coefficient of x. However, we must solve for y (and get $y = -\frac{2}{3}x + 3$) to determine that the slope is $-\frac{2}{3}$ (see Example 1).

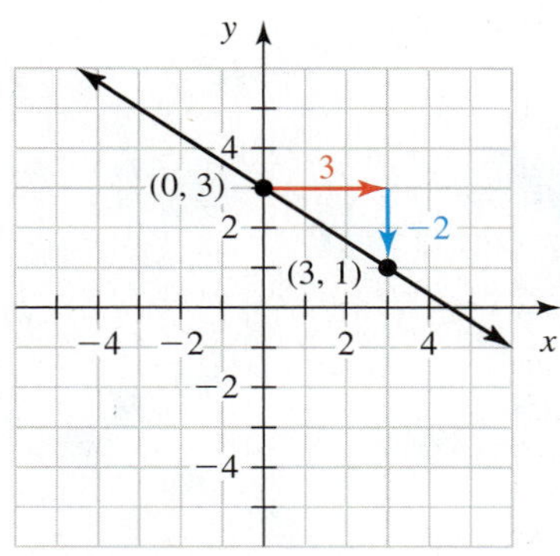

Figure 1 Graph of $2x + 3y = 9$

Example 2 Graphing an Equation by Solving for y

Sketch the graph of $2(3x - 2y) = 9x - 5y - 1$.

Solution

First, we use the distributive law on the left side of the equation:

$$\begin{aligned} 2(3x - 2y) &= 9x - 5y - 1 && \text{Original equation} \\ 6x - 4y &= 9x - 5y - 1 && \text{Distributive law} \\ 6x - 4y - 6x &= 9x - 5y - 1 - 6x && \text{Subtract } 6x \text{ from both sides.} \\ -4y &= 3x - 5y - 1 && \text{Combine like terms.} \\ -4y + 5y &= 3x - 5y - 1 + 5y && \text{Add } 5y \text{ to both sides.} \\ y &= 3x - 1 && \text{Combine like terms.} \end{aligned}$$

Since $2(3x - 2y) = 9x - 5y - 1$ can be put into the slope–intercept form $y = mx + b$ (as $y = 3x - 1$), we know that the graph of $2(3x - 2y) = 9x - 5y - 1$ is a line. The y-intercept is $(0, -1)$ and the slope is 3. The graph is shown in Fig. 2.

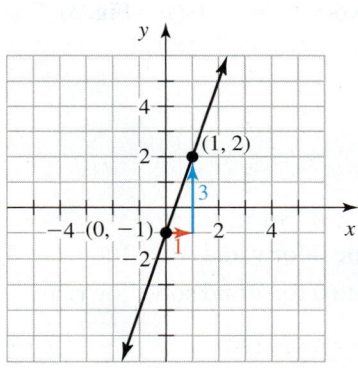

Figure 2 Graph of $2(3x - 2y) = 9x - 5y - 1$

Example 3 Working with a General Linear Equation

1. Determine the slope and the y-intercept of the graph of $ax + by = c$, where a, b, and c are constants and b is nonzero.
2. Find the slope and the y-intercept of the graph of $3x + 7y = 5$.

Solution

1.

$$\begin{aligned} ax + by &= c && \text{Original equation} \\ ax + by - ax &= c - ax && \text{Subtract } ax \text{ from both sides.} \\ by &= -ax + c && \text{Combine like terms; rearrange terms.} \\ \frac{by}{b} &= \frac{-ax}{b} + \frac{c}{b} && \text{Divide both sides by } b. \\ y &= -\frac{a}{b}x + \frac{c}{b} && \text{Simplify; } \frac{-a}{b} = -\frac{a}{b} \end{aligned}$$

The slope is $-\frac{a}{b}$ and the y-intercept is $\left(0, \frac{c}{b}\right)$.

2. We substitute 3 for a, 7 for b, and 5 for c in our results from Problem 1 to find that the slope is $-\frac{3}{7}$ and the y-intercept is $\left(0, \frac{5}{7}\right)$.

Example 4 Working with a General Linear Equation

Determine the slope and y-intercept of the graph of $a(y + c) = x$, where a and c are constants and a is nonzero.

Solution

To begin, we get y alone on one side of the equation:

$$a(y + c) = x \qquad \textit{Original equation}$$

$$\frac{a(y + c)}{a} = \frac{x}{a} \qquad \textit{Divide both sides by } a.$$

$$y + c = \frac{x}{a} \qquad \textit{Simplify.}$$

$$y + c - c = \frac{x}{a} - c \qquad \textit{Subtract } c \textit{ from both sides.}$$

$$y = \frac{1}{a}x - c \qquad \textit{Combine like terms: } \frac{x}{a} = \frac{1}{a}\cdot\frac{x}{1} = \frac{1}{a}x$$

The slope is $\frac{1}{a}$, and the y-intercept is $(0, -c)$.

Finding Intercepts

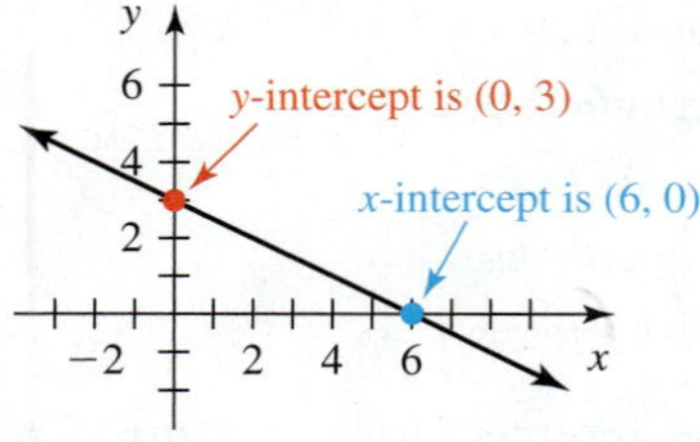

Figure 3 Intercepts of a line

Another way to graph a linear equation is to use the intercepts of its graph. In Fig. 3, we show the x-intercept and y-intercept of a line.

Recall from Section 1.3 that for an x-intercept, the y-coordinate is 0 (see Fig. 3). For a y-intercept, the x-coordinate is 0 (see Fig. 3).

Intercepts of the Graph of an Equation

For an equation containing the variables x and y,

- To find the x-coordinate of each x-intercept, substitute 0 for y and solve for x.
- To find the y-coordinate of each y-intercept, substitute 0 for x and solve for y.

Example 5 Graphing an Equation by Using Intercepts

Consider the equation $5y - 2x = 10$.

1. Find the x-intercept.
2. Find the y-intercept.
3. Sketch the graph of the equation.

Solution

1. To find the x-intercept, we substitute 0 for y and solve for x:

$$5y - 2x = 10 \qquad \textit{Original equation}$$

$$5(0) - 2x = 10 \qquad \textit{Substitute 0 for } y.$$

$$-2x = 10 \qquad \textit{Simplify.}$$

$$\frac{-2x}{-2} = \frac{10}{-2} \qquad \textit{Divide both sides by } -2.$$

$$x = -5 \qquad \textit{Simplify.}$$

The x-intercept is $(-5, 0)$.

2. To find the y-intercept, we substitute 0 for x and solve for y:

$$\begin{aligned} 5y - 2x &= 10 && \text{Original equation} \\ 5y - 2(0) &= 10 && \text{Substitute 0 for } x. \\ 5y &= 10 && \text{Simplify.} \\ \frac{5y}{5} &= \frac{10}{5} && \text{Divide both sides by 5.} \\ y &= 2 && \text{Simplify.} \end{aligned}$$

The y-intercept is $(0, 2)$.

3. The equation $5y - 2x = 10$ can be put into the form $y = mx + b$ (as $y = \frac{2}{5}x + 2$; try it). So, the graph of the equation $5y - 2x = 10$ is a line.

Before sketching the line, we find another solution to check our work. Here we substitute 3 for x and solve for y:

$$\begin{aligned} 5y - 2x &= 10 && \text{Original equation} \\ 5y - 2(3) &= 10 && \text{Substitute 3 for } x. \\ 5y - 6 &= 10 && \text{Multiply.} \\ 5y &= 16 && \text{Add 6 to both sides.} \\ y &= \frac{16}{5} && \text{Divide both sides by 5.} \end{aligned}$$

So, $\left(3, \frac{16}{5}\right)$ is a solution.

Finally, in Table 2, we list the three solutions of $5y - 2x = 10$ that we found, and in Fig. 4 we sketch the graph.

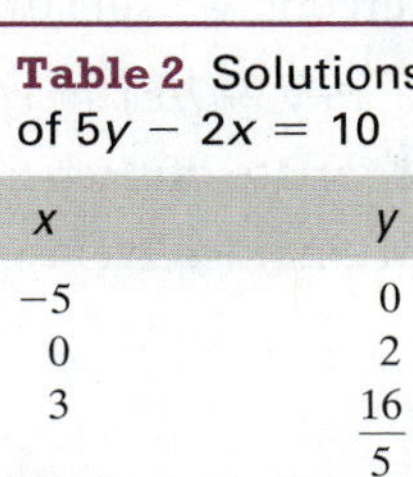

Table 2 Solutions of $5y - 2x = 10$

x	y
-5	0
0	2
3	$\frac{16}{5}$

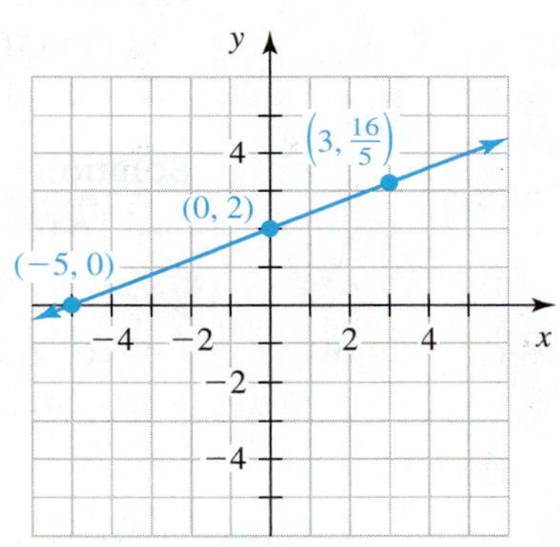

Figure 4 Graph of $5y - 2x = 10$

We can verify our work by checking that the ordered pairs $(-5, 0)$ and $(0, 2)$ satisfy the equation $5y - 2x = 10$.

In Example 5, we used the intercepts of the line $5y - 2x = 10$ and another point on the line to graph the equation $5y - 2x = 10$.

Using Intercepts to Graph an Equation

To graph a linear equation whose graph has exactly two intercepts,

1. Find the intercepts.
2. Plot the intercepts and a third point on the line, and graph the line that contains the three points.

When working with models, we will sometimes want to find intercepts that have decimal coordinates.

Example 6 Finding Approximate Intercepts

Find the indicated intercept of the graph of $y = -2.67x + 8.95$. Round the coordinates to the second decimal place.

1. x-intercept
2. y-intercept

Solution

1. To find the x-intercept, we substitute 0 for y and solve for x:

$$\begin{aligned} y &= -2.67x + 8.95 && \text{Original equation} \\ 0 &= -2.67x + 8.95 && \text{Substitute 0 for } y. \\ 0 + 2.67x &= -2.67x + 8.95 + 2.67x && \text{Add } 2.67x \text{ to both sides.} \\ 2.67x &= 8.95 && \text{Combine like terms.} \\ \frac{2.67x}{2.67} &= \frac{8.95}{2.67} && \text{Divide both sides by 2.67.} \\ x &\approx 3.35 && \text{Round right-hand side to second decimal place.} \end{aligned}$$

Figure 5 Verify the work

The approximate x-intercept is $(3.35, 0)$. We use "zero" on a graphing calculator to verify our work (see Fig. 5). See Appendix A.21 for graphing calculator instructions.

2. For an equation of the form $y = mx + b$, as is $y = -2.67x + 8.95$, the graph has y-intercept $(0, b)$. So, the y-intercept is $(0, 8.95)$. We use TRACE on a graphing calculator to verify our work (see Fig. 6). See Appendix A.5 for graphing calculator instructions.

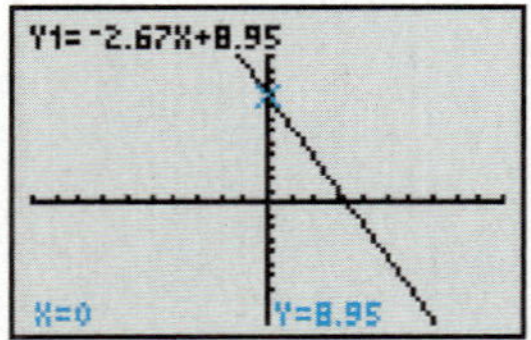

Figure 6 Verify the work

Example 7 Finding Intercepts of a General Linear Equation

Assume the graph of $a(y - bx) = c$ has an x-intercept and a y-intercept.

1. Find the x-intercept.
2. Find the y-intercept.

Solution

1. To find the x-intercept, we substitute 0 for y and solve for x:

$$\begin{aligned} a(y - bx) &= c && \text{Original equation} \\ a(0 - bx) &= c && \text{Substitute 0 for } y. \\ -abx &= c && \text{Simplify.} \\ \frac{-abx}{-ab} &= \frac{c}{-ab} && \text{Divide both sides by } -ab. \\ x &= -\frac{c}{ab} && \text{Simplify.} \end{aligned}$$

The x-intercept is $\left(-\frac{c}{ab}, 0\right)$.

2. To find the y-intercept, we substitute 0 for x and solve for y:

$$\begin{aligned} a(y - bx) &= c && \text{Original equation} \\ a(y - b \cdot 0) &= c && \text{Substitute 0 for } x. \\ ay &= c && \text{Simplify.} \\ \frac{ay}{a} &= \frac{c}{a} && \text{Divide both sides by } a. \\ y &= \frac{c}{a} && \text{Simplify.} \end{aligned}$$

The y-intercept is $\left(0, \frac{c}{a}\right)$.

Finding Coordinates of Solutions

In Section 3.1, we graphed equations of the form $y = mx + b$ by plotting points. After learning how to solve equations in one variable in Chapter 4, we can now graph a linear equation in any form by plotting points.

Example 8 Graphing an Equation by Plotting Points

Complete the following steps to graph the equation $3x + 2y = 8$:

1. Find y when $x = 4$.
2. Find x when $y = 1$.
3. Sketch the graph of $3x + 2y = 8$.

Solution

1. We substitute 4 for x in the equation $3x + 2y = 8$ and solve for y:

$$\begin{aligned} 3x + 2y &= 8 && \text{Original equation} \\ 3(4) + 2y &= 8 && \text{Substitute 4 for } x. \\ 12 + 2y &= 8 && \text{Multiply.} \\ 12 + 2y - 12 &= 8 - 12 && \text{Subtract 12 from both sides.} \\ 2y &= -4 && \text{Combine like terms.} \\ \frac{2y}{2} &= \frac{-4}{2} && \text{Divide both sides by 2.} \\ y &= -2 && \text{Simplify.} \end{aligned}$$

So, $y = -2$ when $x = 4$. The ordered pair $(4, -2)$ is a solution.

2. We substitute 1 for y in the equation $3x + 2y = 8$ and solve for x:

$$\begin{aligned} 3x + 2y &= 8 && \text{Original equation} \\ 3x + 2(1) &= 8 && \text{Substitute 1 for } y. \\ 3x + 2 &= 8 && \text{Multiply.} \\ 3x + 2 - 2 &= 8 - 2 && \text{Subtract 2 from both sides.} \\ 3x &= 6 && \text{Combine like terms.} \\ \frac{3x}{3} &= \frac{6}{3} && \text{Divide both sides by 3.} \\ x &= 2 && \text{Simplify.} \end{aligned}$$

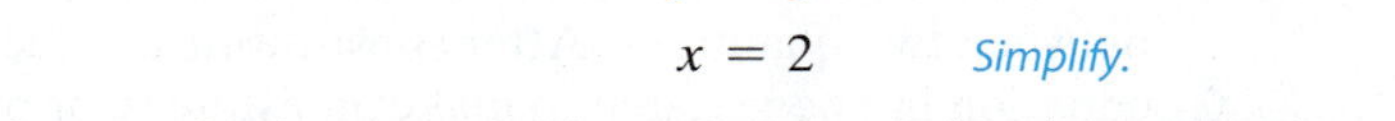

So, $x = 2$ when $y = 1$. The ordered pair $(2, 1)$ is a solution.

3. First, we find that the y-intercept is $(0, 4)$. (Try it.) Then we plot the points $(4, -2)$, $(2, 1)$, and $(0, 4)$ and sketch the line that contains them (see Fig. 7).

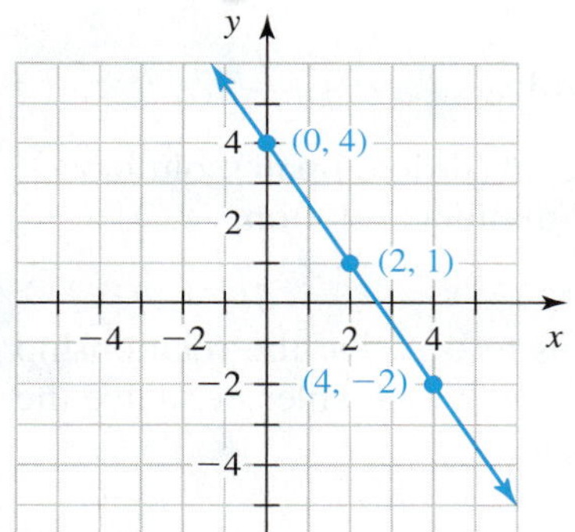

Figure 7 Graph of $3x + 2y = 8$

Using Three Solutions to Graph a Linear Equation

To graph any linear equation,

1. Find three solutions of the equation.
2. Plot the three solutions and graph the line that contains them.

Examples 1, 5, and 8 show three ways to graph an equation in which y is not alone on one side of the equation. For an equation in which y is alone on one side, the y-intercept and the slope are usually best to use to graph the equation.

Comparing Equations in One and Two Variables

In Example 9, we will compare solutions of an equation in one variable with solutions of an equation in two variables.

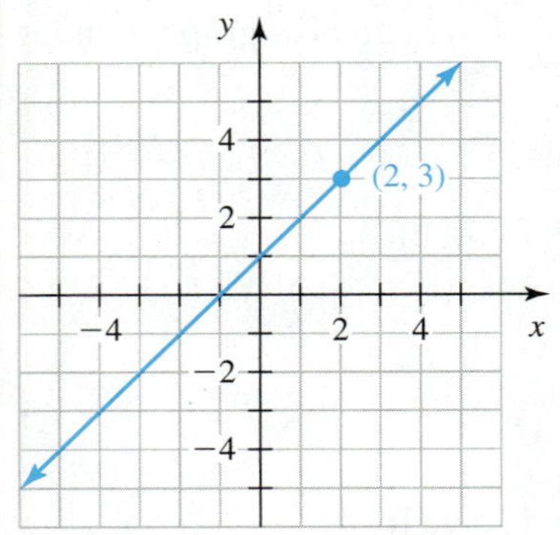

Figure 8 Graph of $y = x + 1$

Example 9 Comparing Equations in One and Two Variables

1. Describe the solution(s) of the equation $y = x + 1$, an equation in two variables.
2. Describe the solution(s) of the equation $3 = x + 1$, an equation in one variable.

Solution

1. The equation $y = x + 1$ is a linear equation in *two* variables, x and y. We can use the graph of the equation (a line) to describe the solutions (see Fig. 8). Since there is an infinite number of points on the line, there is an infinite number of solutions of the equation $y = x + 1$.
2. The equation $3 = x + 1$ is a linear equation in *one* variable, x. So, exactly one number is a solution:

$$3 = x + 1 \quad \textit{Original equation}$$
$$3 - 1 = x + 1 - 1 \quad \textit{Subtract 1 from both sides.}$$
$$2 = x \quad \textit{Combine like terms.}$$

The solution is the number 2.

There is a connection between the equations $y = x + 1$ and $3 = x + 1$. If we substitute 3 for y in the equation $y = x + 1$, the result is the equation $3 = x + 1$, whose solution is $x = 2$. So, $(2, 3)$ is one of an infinite number of solutions of $y = x + 1$ (see Fig. 8).

The graph of any linear equation in two variables is a line, which contains an infinite number of points. So, a linear equation in two variables has an infinite number of (ordered-pair) solutions. Recall from Section 4.4 that a linear equation in one variable has exactly one (real-number) solution.

Comparing Linear Equations in One and Two Variables

A linear equation in two variables has an infinite number of (ordered-pair) solutions. A linear equation in one variable has exactly one (real-number) solution.

When modeling, we use an equation in two variables to describe the relationship between two quantities. After substituting a value for one of the variables, we solve the equation in one variable to make an estimate or prediction.

Group Exploration

Comparing two graphing techniques

In this exploration, you will compare two methods of graphing an equation.

1. Consider the equation $4x + 5y = 20$.
   a. Graph the equation by finding the slope and the y-intercept.
   b. Graph the equation by finding both intercepts.
   c. Compare your graphs from parts (a) and (b). Decide which method you prefer for graphing the equation $4x + 5y = 20$.
2. Consider the equation $y = \frac{2}{5}x - 1$.
   a. Graph the equation by finding the slope and the y-intercept.
   b. Graph the equation by finding both intercepts.
   c. Compare your graphs from parts (a) and (b). Decide which method you prefer for graphing the equation $y = \frac{2}{5}x - 1$.
3. In general, which types of linear equations do you prefer to graph by using the slope and the y-intercept? Which types do you prefer to graph by finding the intercepts? Explain.

Tips for Success Keep a Positive Attitude

Sometimes when students have difficulty in doing mathematics, they take this as evidence that they cannot learn it. However, getting stumped when trying to solve a math problem is something everyone has experienced—even your math instructor! Students who are successful at mathematics realize this is part of the process of learning the subject. When they have trouble learning something, they keep a positive attitude and continue working hard, knowing they will eventually learn the material.

Homework 5.1

For extra help ▶ MyMathLab®

Watch the videos in MyMathLab

Download the MyDashboard App

For Exercises 1–40, determine the slope and the y-intercept. Use the slope and the y-intercept to graph the equation by hand. Use a graphing calculator to verify your work.

1. $y = 2x - 3$ **2.** $y = -4x + 1$ **3.** $y = -\frac{3}{5}x - 2$

4. $y = \frac{1}{3}x + 4$ **5.** $y = -4$ **6.** $y = 0$

7. $y + x = 3$ **8.** $y + x = 1$ **9.** $y + 2x = 4$

10. $y + 3x = -2$ **11.** $y - 2x = -1$ **12.** $y - 3x = 2$

13. $3y = 2x$ **14.** $2y = -5x$ **15.** $2y = -3x - 4$

16. $3y = -4x + 6$ **17.** $5y = 4x - 15$ **18.** $4y = 7x - 20$

19. $3x - 4y = 8$ **20.** $4x + 3y = 9$ **21.** $6x - 15y = 30$

22. $6x - 8y = 16$ **23.** $x + 4y = 4$ **24.** $-x + 5y = -20$

25. $-4 = x + 2y$ **26.** $9 = x + 3y$

27. $4x + y + 2 = 0$ **28.** $2x - y - 3 = 0$

29. $6x - 4y + 8 = 0$ **30.** $15x + 12y - 36 = 0$

31. $0 = 5x + 3y$ **32.** $0 = 4x - 6y$

33. $y - 3 = 0$ **34.** $y + 5 = 0$

35. $3(x - 2y) = 9$ **36.** $2(y - 3x) = 8$

37. $4x - 5y + 3 = 2x - 2y - 3$

38. $3y - 6x + 2 = 7y - x - 6$

39. $1 - 3(y - 2x) = 7 + 3(x - 3y)$

40. $8 - 2(y - 3x) = 2 + 4(x - 2y)$

Determine the slope and y-intercept of the graph of the given equation, where a, b, c, and d are nonzero constants.

41. $ax - by = c$ **42.** $ax + by + c = 0$ **43.** $ay = b(x - d)$

44. $ay = b(x + d)$ **45.** $a(y + b) = x$ **46.** $a(y - b) = x$

47. $a(x - y) = d$ **48.** $a(x + y) = d$

49. $\frac{x}{a} + \frac{y}{a} = 1$ **50.** $\frac{y + b}{a} = x$

Find the x-intercept and y-intercept. Then graph the equation by hand.

51. $x - 3y = 6$ **52.** $2x - y = 6$ **53.** $15 = 3x + 5y$

54. $20 = 4x - 5y$ **55.** $2x - 3y + 12 = 0$ **56.** $2x - 4y - 8 = 0$

57. $y = -3x + 6$ **58.** $y = x + 2$ **59.** $\frac{1}{2}x - \frac{1}{3}y = 2$

60. $\frac{1}{4}x - \frac{1}{2}y = -1$ **61.** $\frac{x}{3} + \frac{y}{5} = 1$ **62.** $\frac{x}{4} + \frac{y}{3} = 1$

Find the approximate x-intercept and the approximate y-intercept. Round the coordinates to the second decimal place.

63. $6.2x + 2.8y = 7.5$ **64.** $8.1x + 3.9y = 14.2$

65. $6.62x - 3.91y = -13.55$ **66.** $-7.29x + 4.72y = 26.36$

67. $y = -4.5x + 9.32$ **68.** $y = 3.5x - 4.8$

69. $y = -2.49x - 37.21$ **70.** $y = -8.79x - 92.58$

Assuming the graph of the equation has an x-intercept and a y-intercept, find both intercepts.

71. $y = mx + b$ **72.** $ax + by = c$

73. $a(bx + y) = c$ **74.** $a(x - by) = c$

75. $ax = b(cy - d)$ **76.** $ay = b(c + dx)$

77. $\frac{x}{a} + \frac{y}{b} = 1$ **78.** $\frac{y - b}{m} = x$

For Exercises 79–86, graph the equation by hand. Use any method.

79. $2x - y = 5$ **80.** $3x + y = 1$

81. $3y = 4x - 3$ **82.** $5y = 3x + 10$

83. $4y - 3x = 0$ **84.** $2y + 5x = 0$

85. $2x - 3y - 12 = 0$ **86.** $3x + 4y + 24 = 0$

87. Consider the equation $6x + 5y = -13$:
 a. Find y when $x = -3$.
 b. Find x when $y = -5$.
 c. Use your results from parts (a) and (b) to help you graph the equation $6x + 5y = -13$ by hand.

88. Consider the equation $y = -2x + 4$:
 a. Find y when $x = 3$.
 b. Find x when $y = 6$.
 c. Use your results from parts (a) and (b) to help you graph the equation $y = -2x + 4$ by hand.

Concepts

89. A student thinks the graph of $3y + 2x = 6$ has slope 2 because the coefficient of the $2x$ term is 2. Is the student correct? If yes, explain why. If no, find the slope.

90. A student says the y-intercept of the line $2y = 3x + 4$ is $(0, 4)$ because the constant term is 4. Is the student correct? Explain.

91. Recall that we can describe some or all of the solutions of an equation in two variables with an equation, a table, a graph, or words.
 a. Describe the solutions of the equation $3x - 5y = 10$ by using a graph.

b. Describe three solutions of the equation $3x - 5y = 10$ by using a table.

c. Describe the solutions of the equation $3x - 5y = 10$ by using words.

92. Recall that we can describe some or all of the solutions of an equation in two variables with an equation, a table, a graph, or words.

a. Describe the solutions of the equation $3x + 4y = 8$ by using a graph.

b. Describe three solutions of the equation $3x + 4y = 8$ by using a table.

c. Describe the solutions of the equation $3x + 4y = 8$ by using words.

93. **a.** Find the intercepts of the line $\frac{x}{5} + \frac{y}{7} = 1$.

b. Find the intercepts of the line $\frac{x}{4} + \frac{y}{6} = 1$.

c. Find the intercepts of the line $\frac{x}{a} + \frac{y}{b} = 1$, where a and b are nonzero constants. [**Hint:** Do the same steps as in parts (a) and (b).]

d. Find an equation of a line whose x-intercept is $(2, 0)$ and whose y-intercept is $(0, 5)$. [**Hint:** See your work in part (c).]

94. **a.** Find the slope of the line $3x + 5y = 7$.

b. Find the slope of the line $2x + 7y = 3$.

c. Find the slope of the line $ax + by = c$, where a, b, and c are constants and b is nonzero. [**Hint:** Do the same steps as in parts (a) and (b).]

95. For an equation containing the variables x and y, why do we substitute 0 for y to find the x-intercept?

96. **a.** Solve the equation $2y - 6 = 4x$ for y.

b. Find two solutions of the equation $y = 2x + 3$.

c. Show that the two ordered pairs you found in part (b) satisfy both of the equations $2y - 6 = 4x$ and $y = 2x + 3$.

d. Explain why it makes sense that the graphs of $2y - 6 = 4x$ and $y = 2x + 3$ are the same.

97. If the graph of a linear equation has a defined slope, describe how to find the slope and the y-intercept, and use this information to graph the equation. (See page 9 for guidelines on writing a good response.)

98. If the graph of a linear equation has exactly two intercepts, describe how to find the intercepts, and use the intercepts to graph the equation. (See page 9 for guidelines on writing a good response.)

Related Review

For Exercises 99–106, refer to the graph sketched in Fig. 9.

99. Find y when $x = 6$.

100. Find y when $x = -2$.

101. Find x when $y = -1$.

102. Find x when $y = -5$.

103. Find the x-intercept.

104. Find the y-intercept.

105. Find the slope of the line.

106. Find the equation of the line.

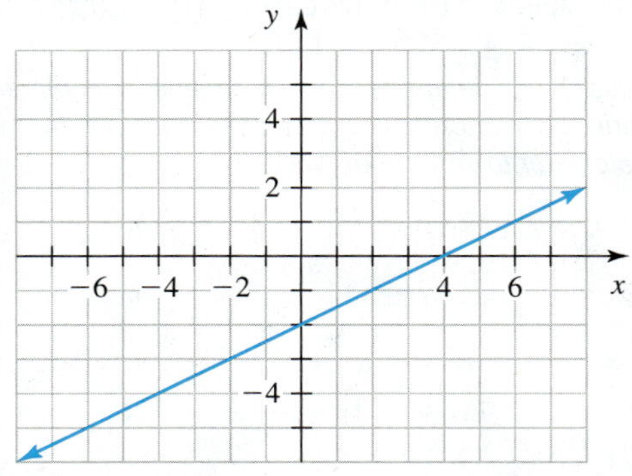

Figure 9 Exercises 99–106

Expressions, Equations, and Graphs

Perform the indicated instruction. Then use words such as linear, one variable, *and* two variables *to describe the expression or equation. For instance, to describe $2x = 8$, you could say "$2x = 8$ is a linear equation in one variable."*

107. Simplify $5 - 3(2x - 7)$.

108. Graph $y = -2x + 6$ by hand.

109. Solve $5 - 3(2x - 7) = 8$.

110. Solve $10 = -2x + 6$.

5.2 Functions

Objectives

- » Know the meanings of *relation, domain, range,* and *function.*
- » Identify functions by using the *vertical line test.*
- » Know the definition of a *linear function.*
- » Know the Rule of Four for functions.
- » Use the graph of a function to find the function's domain and range.

In Chapters 1–4 and Section 5.1, we described relationships between two variables. In this section, we will discuss how to describe some of these relationships by using an extremely important concept called a *function.*

Relation, Domain, Range, and Function

In Chapters 1–4 and Section 5.1, we have used graphs, tables, and equations to describe the relationship between two variables. To illustrate, Table 3 describes a relationship between the variables x and y. This relationship is also described graphically in Fig. 10.

We call the set of ordered pairs listed in Table 3 a *relation.* This relation consists of the ordered pairs $(3, 2)$, $(4, 1)$, $(5, 3)$, and $(5, 4)$. The *domain* of the relation is the set of all values of x (the independent variable)—in this case, 3, 4, and 5. The *range* of the relation is the set of all values of y (the dependent variable)—here, 1, 2, 3, and 4.

Table 3 A Relationship Described by a Table

x	y
3	2
4	1
5	3
5	4

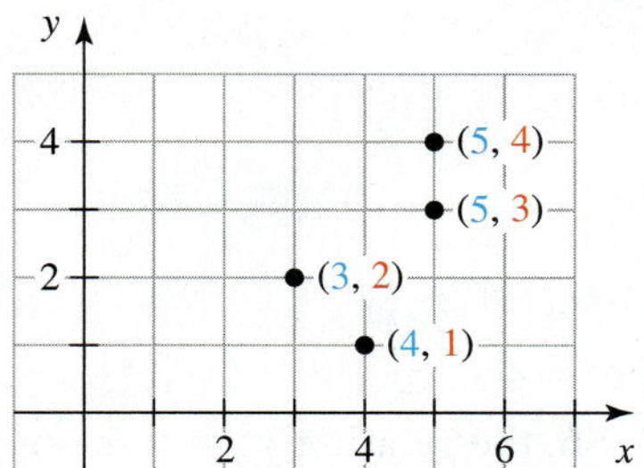

Figure 10 The relationship of Table 3 described by a graph

Definition Relation, domain, and range

A **relation** is a set of ordered pairs. The **domain** of a relation is the set of all values of the independent variable, and the **range** of the relation is the set of all values of the dependent variable.

We can think of a relation as a machine in which values of x are "inputs" and values of y are "outputs." In general, each member of the domain is an **input,** and each member of the range is an **output.**

For the relation described in Table 3, we can think of the values of x as being sent to the values of y (see Fig. 11).

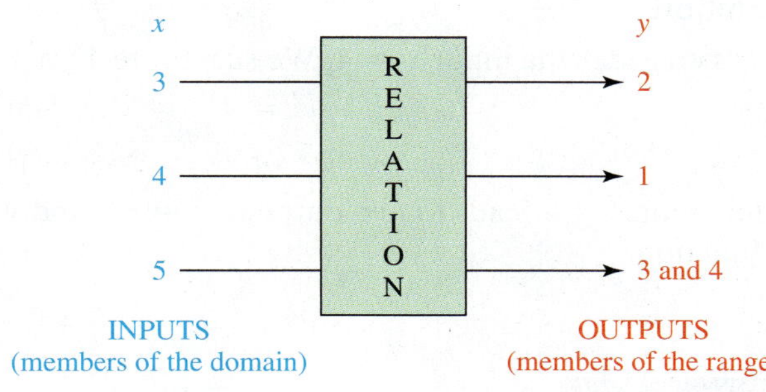

Figure 11 Think of a relation as an input–output machine

Note that the input $x = 5$ is sent to *two* outputs: $y = 3$ and $y = 4$. In a special type of relation called a *function,* each input is sent to exactly *one* output. The relation described in Table 3 is not a function.

Definition Function

A **function** is a relation in which each input leads to exactly one output.

The equation $y = x + 2$ describes a relation consisting of an infinite number of ordered pairs. We will determine whether the relation is a function in Example 1.

Example 1 Deciding whether an Equation Describes a Function

Is the relation $y = x + 2$ a function? Find the domain and range of the relation.

Solution

Let's consider some input–output pairs (in Fig. 12).

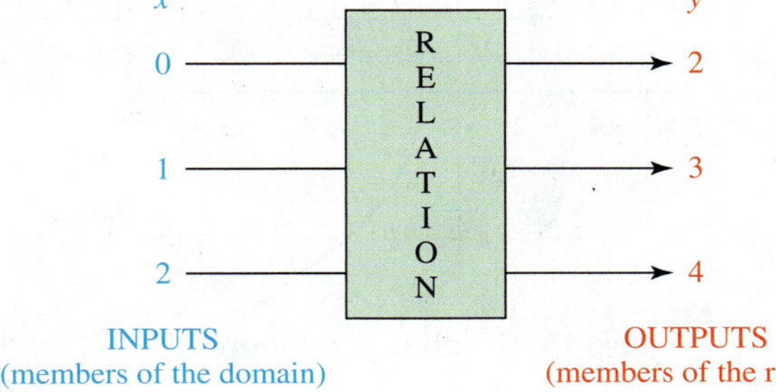

Figure 12 The "increasing by 2" relation: $y = x + 2$

Each input leads to just *one* output—namely, the input increased by 2—so the relation $y = x + 2$ is a function.

The domain of the relation $y = x + 2$ is the set of all real numbers, since we can add 2 to *any* real number. The range of $y = x + 2$ is also the set of real numbers, since any real number is the output of the number that is 2 units less than it.

Example 2 Deciding whether an Equation Describes a Function

Is the relation $y = \pm x$ a function?

Solution

If $x = 1$, then $y = \pm 1$. So, the input $x = 1$ leads to *two* outputs: $y = -1$ and $y = 1$. Therefore, the relation $y = \pm x$ is not a function.

Example 3 Deciding whether an Equation Describes a Function

Is the relation $y^2 = x$ a function?

Solution

Let's consider the input $x = 4$. We substitute 4 for x and solve for y:

$$y^2 = 4 \quad \text{Substitute 4 for } x.$$

$$y = -2 \quad \text{or} \quad y = 2 \quad (-2)^2 = 4, 2^2 = 4$$

The input $x = 4$ leads to *two* outputs: $y = -2$ and $y = 2$. So, the relation $y^2 = x$ is not a function.

Table 4 Input–Output Pairs of a Relation

x (input)	y (output)
0	2
1	3
1	5
2	7
3	10

Example 4 Deciding whether a Table Describes a Function

Is the relation described by Table 4 a function?

Solution

The input $x = 1$ leads to *two* outputs: $y = 3$ and $y = 5$. So, the relation is not a function.

Example 5 Deciding whether a Graph Describes a Function

Is the relation described by the graph in Fig. 13 a function?

Solution

The input $x = 3$ leads to *two* outputs: $y = -4$ and $y = 4$ (see Fig. 14). So, the relation is not a function.

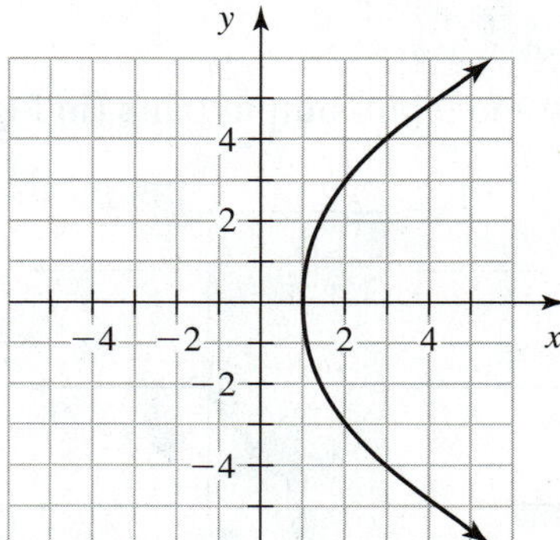

Figure 13 Graph of a relation

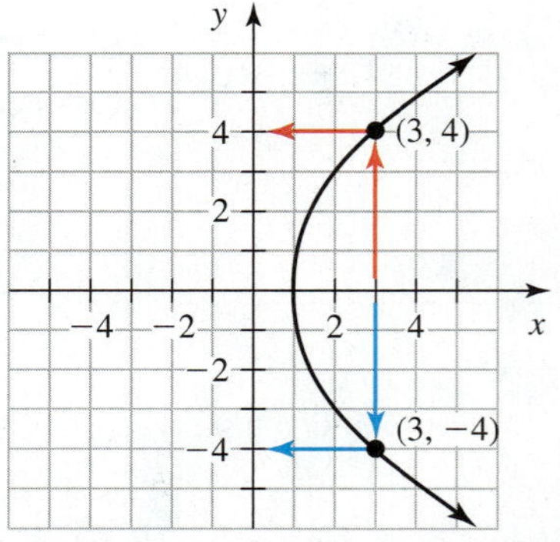

Figure 14 The input $x = 3$ gives two outputs: $y = -4$ and $y = 4$

Vertical Line Test

Notice that the relation described in Example 5 is not a function because some vertical lines would intersect the graph more than once.

> **Vertical Line Test**
>
> A relation is a function if and only if each vertical line intersects the graph of the relation at no more than one point. We call this requirement the **vertical line test.**

Example 6 Deciding whether a Graph Describes a Function

Determine whether the graph represents a function.

1.

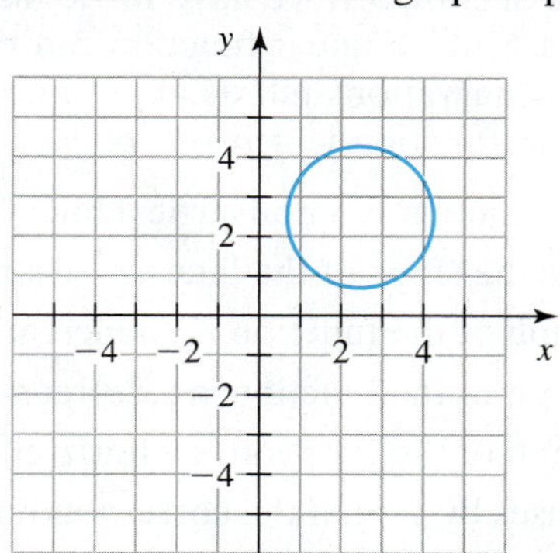

2.

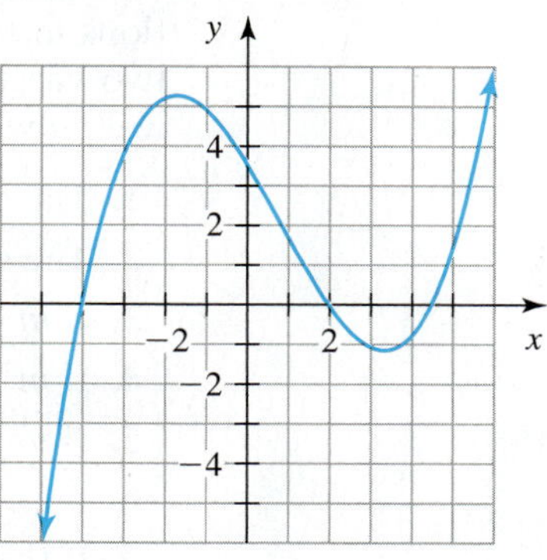

Solution

1. Since the vertical line sketched in Fig. 15 intersects the circle more than once, the relation is not a function.
2. Each vertical line sketched in Fig. 16 intersects the curve at one point. In fact, *any* vertical line would intersect this curve at just one point. So, the relation is a function.

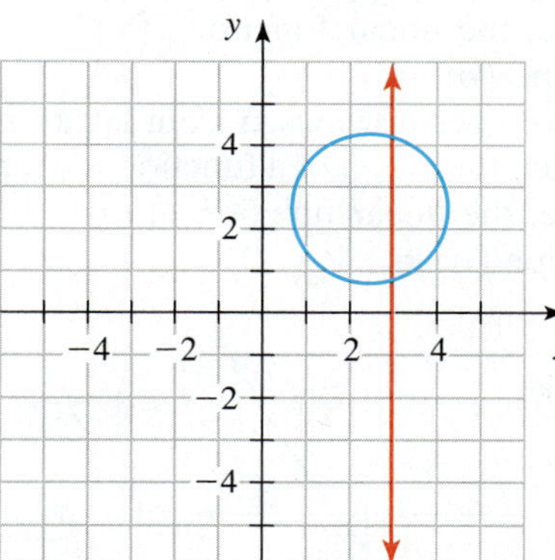

Figure 15 The circle does not describe a function

Figure 16 The curve describes a function

Example 7 Deciding whether an Equation Describes a Function

Is the relation $y = 2x + 1$ a function?

Solution

We begin by sketching the graph of $y = 2x + 1$ in Fig. 17. Note that each vertical line would intersect the line $y = 2x + 1$ at just one point. So, the relation $y = 2x + 1$ is a function.

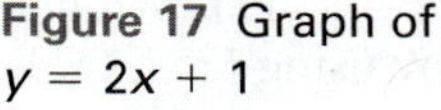

Figure 17 Graph of $y = 2x + 1$

Linear Functions

In Example 7, we saw that the line $y = 2x + 1$ is a function. In fact, any nonvertical line is a function, since it passes the vertical line test.

Definition Linear function

A **linear function** is a relation whose equation can be put into the form

$$y = mx + b$$

where m and b are constants.

In Chapter 3 and Section 5.1, we have made many observations about linear equations in two variables. Since a linear function can be described by a linear equation in two variables, these observations tell us about linear functions. Let's summarize what we know about a linear function $y = mx + b$:

1. The graph of the function is a nonvertical line.
2. The constant m is the slope of the line, a measure of the line's steepness.
3. If $m > 0$, the graph of the function is an increasing line.
4. If $m < 0$, the graph of the function is a decreasing line.
5. If $m = 0$, the graph of the function is a horizontal line.
6. If an input increases by 1, then the corresponding output changes by the slope m.
7. If the run is 1, the rise is the slope m.
8. The y-intercept of the line is $(0, b)$.

Finally, since a linear equation of the form $y = mx + b$ is a *function,* we know each input leads to exactly one output.

If a curve goes upward from left to right, we say that the curve is **increasing** (see Fig. 18). When the graph of a function is increasing, we say that the function is **increasing.** For example, the linear function $f(x) = 3x - 2$ is increasing, because the graph is an increasing line $(m = 3 > 0)$.

If a curve goes downward from left to right, we say that the curve is **decreasing** (see Fig. 19). When the graph of a function is decreasing, we say that the function is **decreasing.** For example, the linear function $g(x) = -2x + 5$ is decreasing, because the graph is a decreasing line $(m = -2 < 0)$.

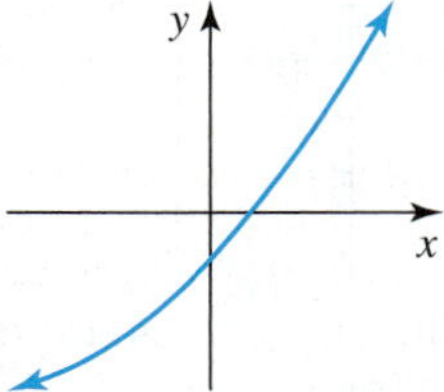

Figure 18 An increasing curve

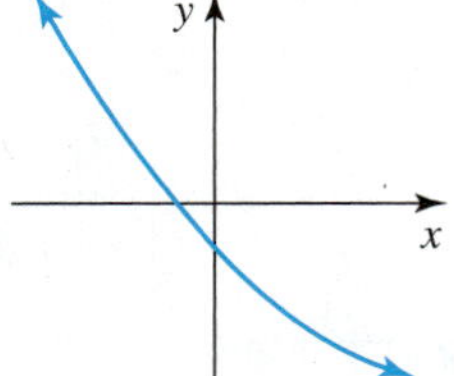

Figure 19 A decreasing curve

For example, the linear function $y = 4x - 7$ is increasing, because the graph is an increasing line $(m = 4 > 0)$. The linear function $y = -3x + 6$ is decreasing, because the graph is a decreasing line $(m = -3 < 0)$.

Table 5 Input–Output Pairs for $y = 2x + 1$

x	y
0	1
1	3
2	5
3	7
4	9

Rule of Four for Functions

We can describe functions in four ways. For instance, in Example 7, we described the function $y = 2x + 1$ by using (1) the equation and (2) a graph (see Fig. 17). We can also describe some of the input–output pairs for the same function by using (3) a table (see Table 5). Finally, we can describe the function (4) verbally: In this case, for each input–output pair, the output is 1 more than twice the input.

Rule of Four for Functions

We can describe some or all of the input–output pairs of a function by means of

1. an equation,
2. a graph,
3. a table, or
4. words.

These four ways to describe input–output pairs of a function are known as the **Rule of Four** for functions.

Example 8 Describing a Function by Using the Rule of Four

1. Is the relation $y = -2x - 1$ a function?
2. List some input–output pairs of $y = -2x - 1$ by using a table.
3. Describe the input–output pairs of $y = -2x - 1$ by using a graph.
4. Describe the input–output pairs of $y = -2x - 1$ by using words.

Solution

1. Since $y = -2x - 1$ is of the form $y = mx + b$, it is a (linear) function.
2. We list five input–output pairs in Table 6.
3. We graph $y = -2x - 1$ in Fig. 20.
4. For each input–output pair, the output is 1 less than -2 times the input.

Table 6 Input–Output Pairs of $y = -2x - 1$

x	y
-2	$-2(-2) - 1 = 3$
-1	$-2(-1) - 1 = 1$
0	$-2(0) - 1 = -1$
1	$-2(1) - 1 = -3$
2	$-2(2) - 1 = -5$

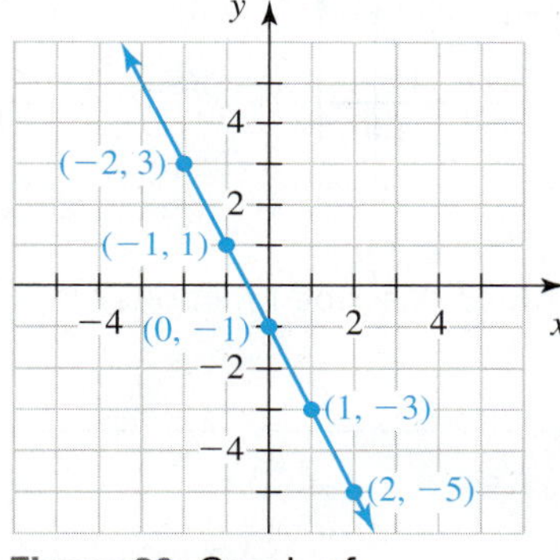

Figure 20 Graph of $y = -2x - 1$

Using a Graph to Find the Domain and Range of a Function

To describe the domain or range of a function, it is sometimes helpful to use the *inequality symbols* $\leq$ and $\geq$. The symbol $\leq$ means "is less than or equal to"; the symbol $\geq$ means "is greater than or equal to." For example, the inequality $x \leq 4$ means all values of x are less than or equal to 4. And the inequality $y \geq 7$ means all values of y are greater than or equal to 7.

The inequality $5 \leq x$ means 5 is *less* than or equal to all values of x. Notice that it is more natural to say all values of x are *greater* than or equal to 5, which is true.

The inequality $2 \leq x \leq 6$ means $2 \leq x$ *and* $x \leq 6$: All values of x are *both* greater than or equal to 2 *and* less than or equal to 6. In other words, all values of x are between 2 and 6, inclusive.

Example 9 Finding the Domain and Range

Use the graph of the function to determine the function's domain and range.

1.

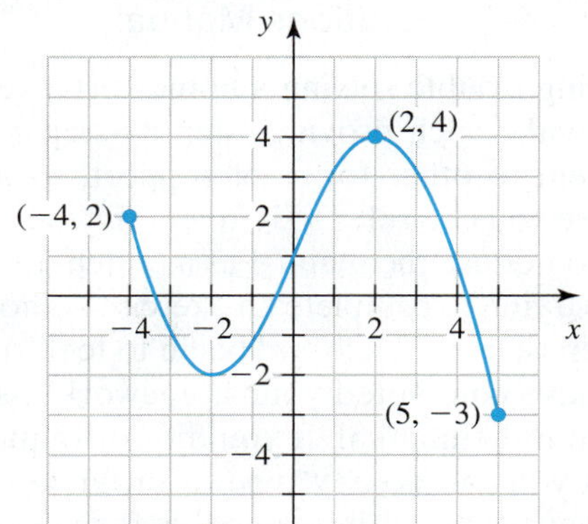

2.

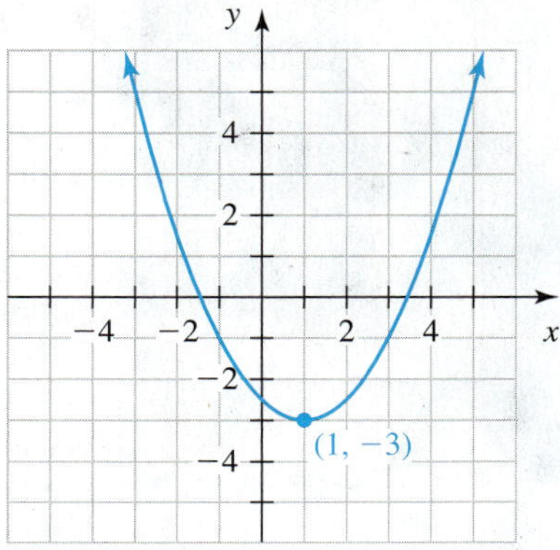

Solution

1. The domain is the set of all x-coordinates of points in the graph. Since there are no breaks in the graph, and since the leftmost point is $(-4, 2)$ and the rightmost point is $(5, -3)$, the domain is $-4 \leq x \leq 5$.

The range is the set of all y-coordinates of points in the graph. Since the lowest point is $(5, -3)$ and the highest point is $(2, 4)$, the range is $-3 \le y \le 4$.

2. The graph extends to the left and right indefinitely without breaks, so every real number is an x-coordinate of some point in the graph. The domain is the set of all real numbers.

The output -3 is the smallest number in the range, because $(1, -3)$ is the lowest point in the graph. The graph also extends upward indefinitely without breaks, so every number larger than -3 is also in the range. The range is $y \ge -3$.

Group Exploration

Vertical line test

1. Consider the relation described by Table 7. Is the relation a function? Explain. Now plot the points on a coordinate system. What do you notice about them?

Table 7 A Relation Described by a Table

x	y
2	1
2	5
2	7

2. Consider the relation described by Table 8. Is the relation a function? Explain. Now plot the points on a coordinate system. What do you notice about them?

Table 8 A Relation Described by a Table

x	y
4	2
4	3
4	6

3. Describe the graph of a relation that is not a function.

4. Determine whether each graph in Fig. 21 is the graph of a function. Explain.

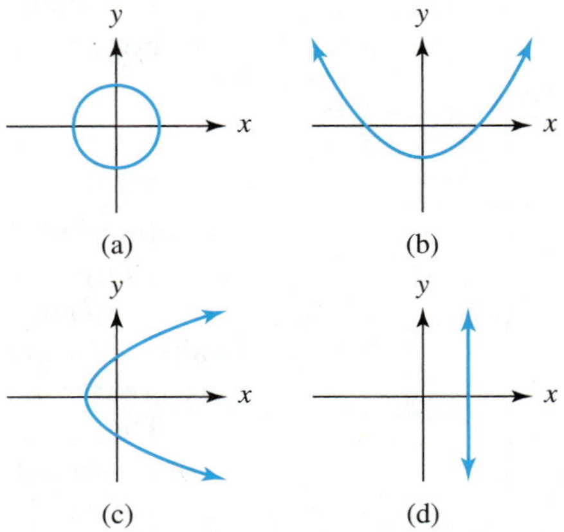

Figure 21 Which graphs describe functions?

Tips for Success Solutions Manual

If you are having trouble solving a homework exercise, it can help to refer to the Solutions Manual, which provides step-by-step solutions to the exercises. It is a nice complement to other forms of support, such as your instructor, a tutor, or friends, because you can refer to it at any time, day or night.

However, do not use the manual as a crutch. It is a common error to consult the manual before trying to complete an exercise without it. But if you have been trying to solve an exercise for a while and begin to feel frustrated, reach for the manual.

Once you have completed your homework assignment, assess how often you sought help from the manual. If you did so frequently, then continue to do more exercises until you can do any type of problem in the homework without referring to the manual. The ultimate goal is for you to understand the material, not just complete the assignment.

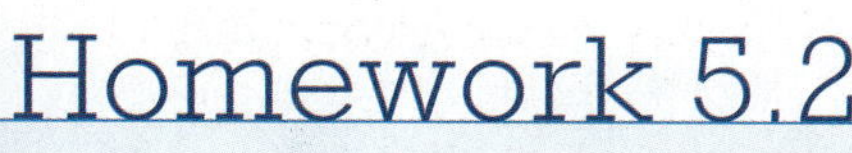

Homework 5.2

For extra help ▶ MyMathLab®  Watch the videos in MyMathLab  Download the MyDashboard App

1. Some ordered pairs of four relations are listed in Table 9. Which of these relations could be functions? Explain.

Table 9 Which Relations Might Be Functions? (Exercise 1)

Relation 1		Relation 2		Relation 3		Relation 4	
x	y	x	y	x	y	x	y
1	1	3	27	0	4	5	10
2	3	4	24	1	4	6	20
3	5	5	21	2	4	7	30
3	7	6	18	3	4	8	40
4	9	7	15	4	4	8	50

2. Some ordered pairs of four relations are listed in Table 10.

Table 10 Which Relations Might Be Functions? (Exercise 2)

Relation 1		Relation 2		Relation 3		Relation 4	
x	y	x	y	x	y	x	y
1	3	5	27	0	50	3	11
2	4	5	24	1	45	4	13
3	5	5	21	2	40	5	17
3	6	5	18	3	35	6	25
4	7	5	15	4	30	7	40

a. Which of the relations could be functions? Explain.

b. Which could be linear functions? Explain.

3. For a certain relation, an input leads to two different outputs. Could the relation be a function? Explain.

4. For a certain relation, two different inputs lead to the same output. Could the relation be a function? Explain.

5. A relation's graph contains the points (2, 3) and (5, 3). Could the relation be a function? Explain.

6. A relation's graph contains the points (4, 5) and (4, 9). Could the relation be a function? Explain.

Determine whether the graph represents a function. Explain.

7.

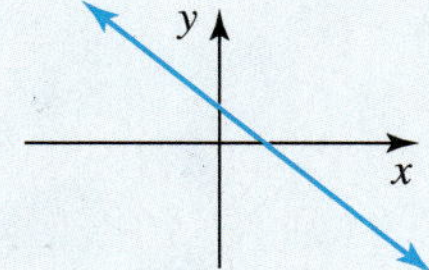

8.

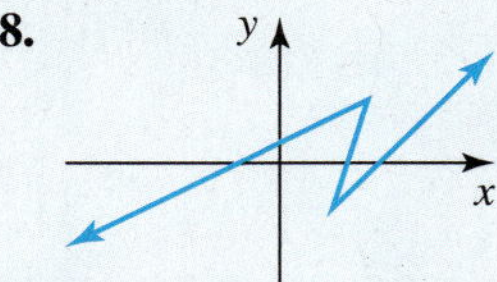

9.

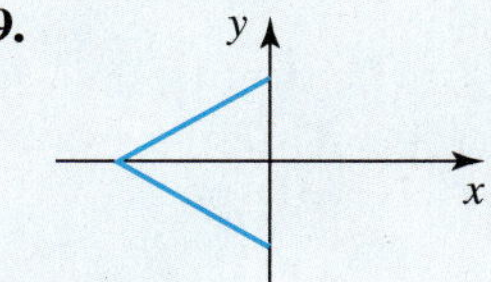

10.

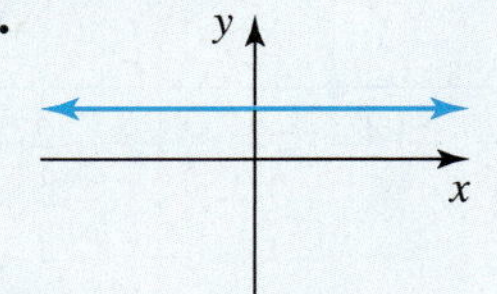

11.

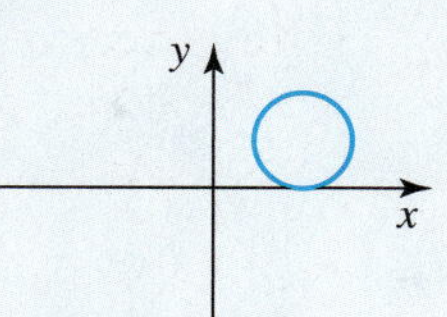

12.

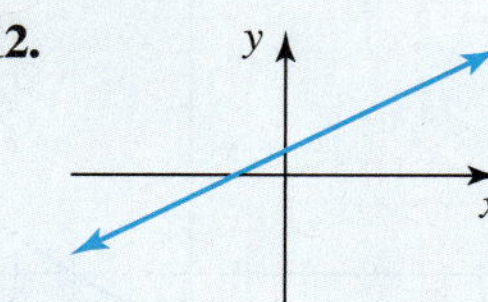

13.

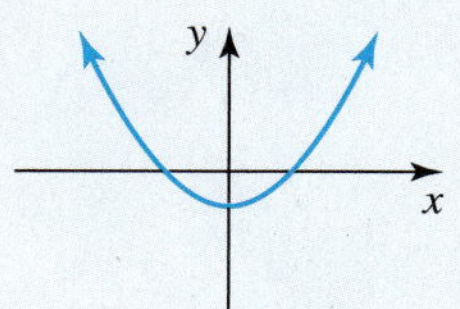

14.

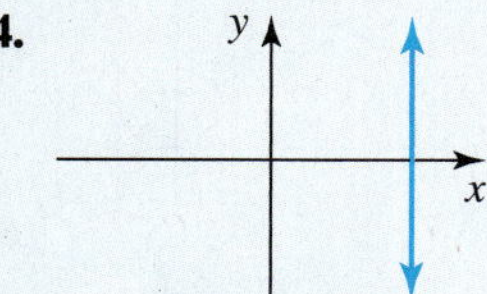

For Exercises 15–24, determine whether the relation is a function. Explain.

15. $y = 5x - 1$

16. $y = -3x + 8$

17. $2x - 5y = 10$

18. $4x + 3y = 24$

19. $y = 4$

20. $y = -1$

21. $x = -3$

22. $x = 0$

23. $7x - 2y = 21 + 3(y - 5x)$

24. $2x + 5y = 9 - 4(x + 2y)$

25. Is a nonvertical line the graph of a function? Explain.

26. Is a vertical line the graph of a function? Explain.

27. Is a circle the graph of a function? Explain.

28. Is a semicircle that is the "upper half" of a circle the graph of a function? Explain.

29. Recall that we can describe some or all of the input–output pairs of a function by means of an equation, a graph, a table, or words.

a. Describe five input–output pairs of $y = 3x - 2$ by using a table.

b. Describe the input–output pairs of $y = 3x - 2$ by using a graph.

c. Describe the input–output pairs of $y = 3x - 2$ by using words.

30. Recall that we can describe some or all of the input–output pairs of a function by means of an equation, a graph, a table, or words.

a. Describe five input–output pairs of $y = \frac{1}{2}x + 2$ by using a table.

b. Describe the input–output pairs of $y = \frac{1}{2}x + 2$ by using a graph.

c. Describe the input–output pairs of $y = \frac{1}{2}x + 2$ by using words.

For Exercises 31–42, use the graph of the function to determine the function's domain and range.

31.

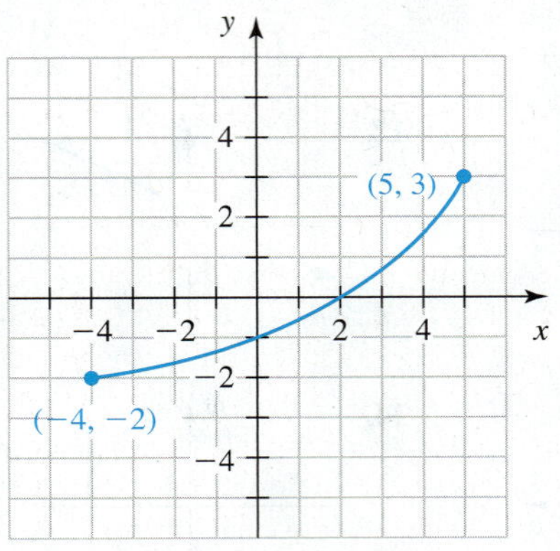

32.

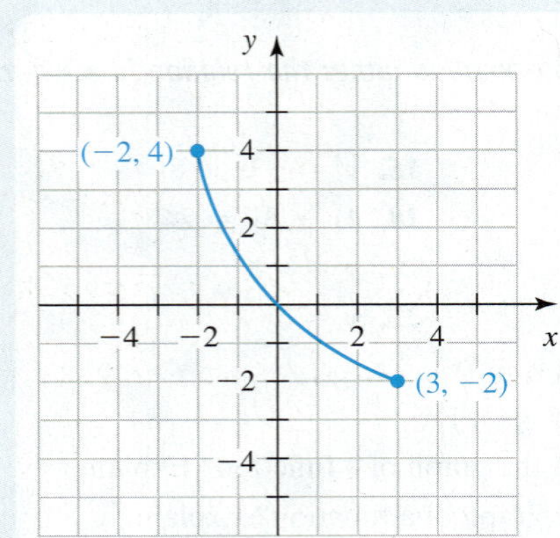

33.

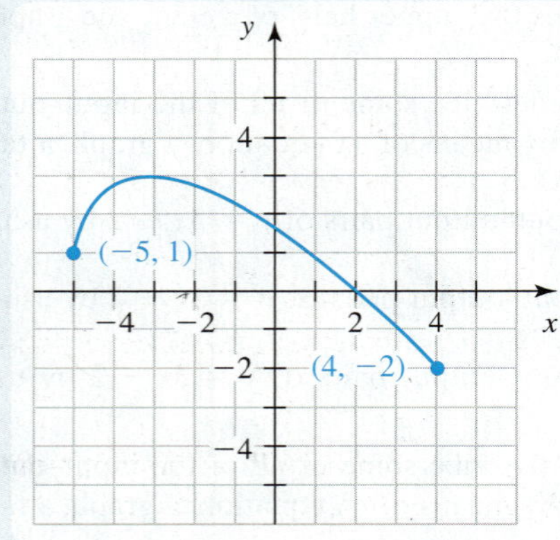

34.

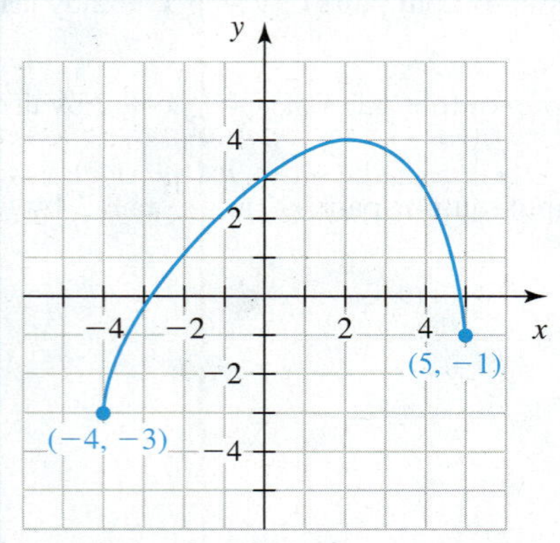

35.

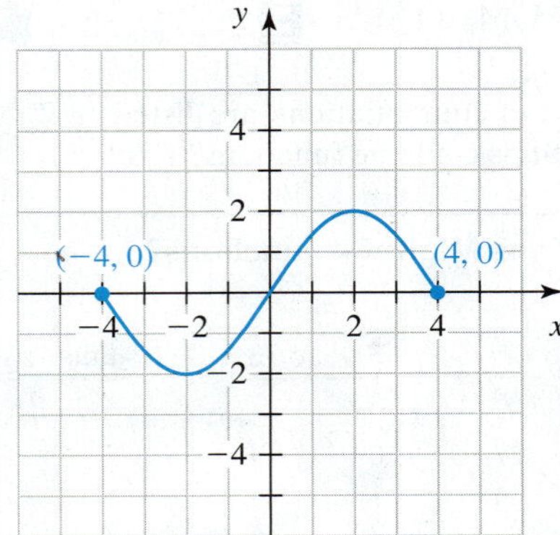

36.

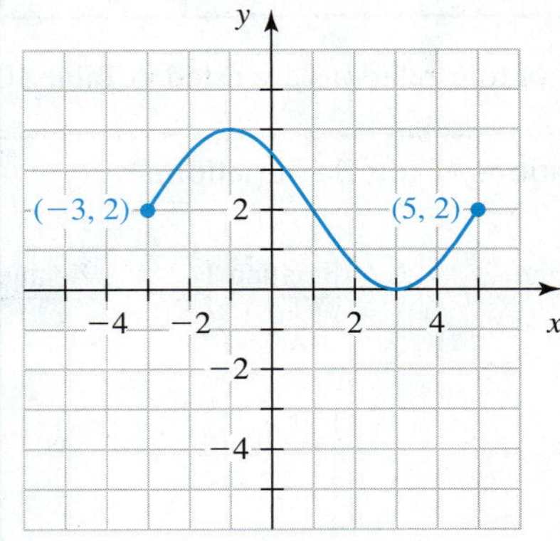

37.

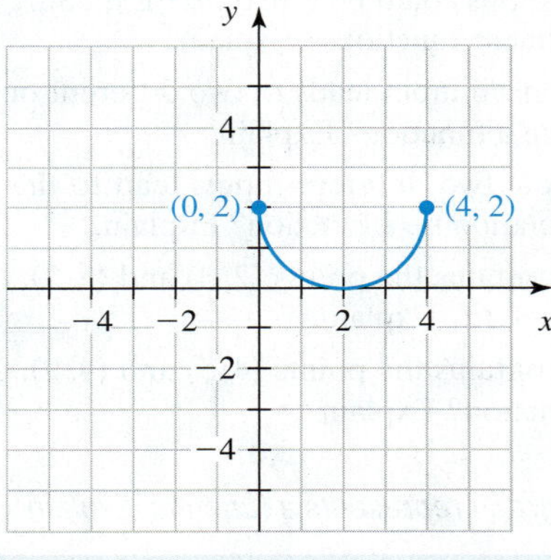

38.

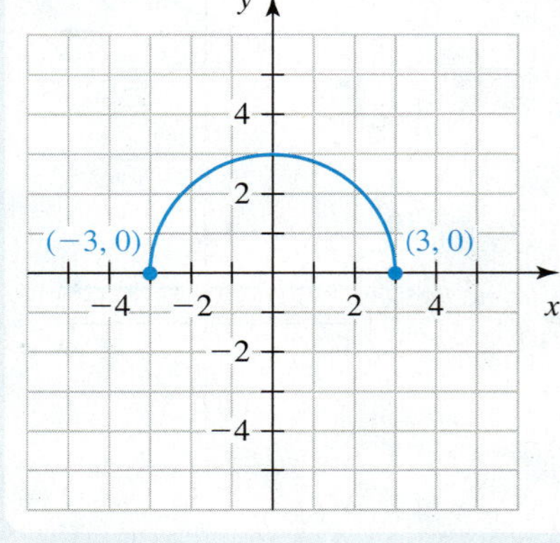

39.

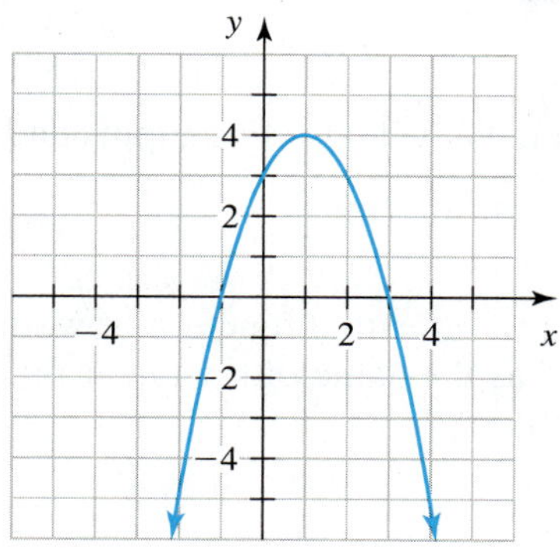

40.

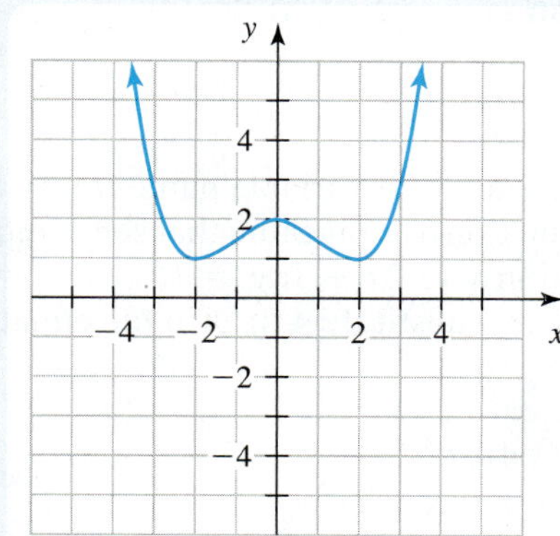

41.

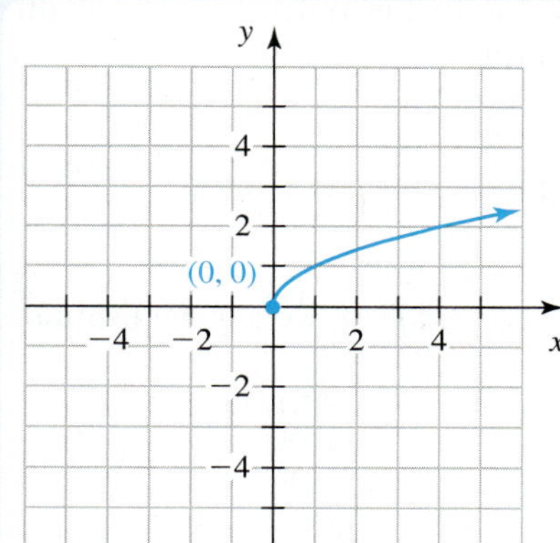

42.

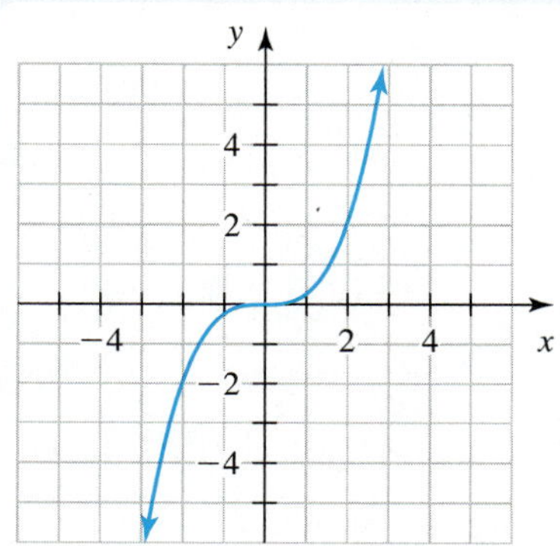

Decide whether the relation is a function. Explain.

43. $y = \sqrt{x}$ **[Hint:** Sketch a graph.]

44. $y = x^4$ **[Hint:** Sketch a graph.]

45. $y^4 = x$ **[Hint:** Substitute 16 for x; then solve for y.]

46. $y^3 = x$ **[Hint:** Substitute 0, 1, and 8 for x, and solve for y after each substitution.]

Concepts

47. Describe the input–output pairs of a function (different from those in this section) by using an equation, a graph, and words. Describe also five input–output pairs of the function by using a table. Explain why your relation is a function.

48. Sketch the graph of a relation (different from those in this section) that is not a function. Next, create a table that lists five ordered pairs of the relation. Explain why your relation is not a function.

49. Sketch the graph of a relation for which the input $x = 2$ gives exactly two outputs and the input $x = 6$ gives exactly one output. Is the relation a function? Explain.

50. Sketch the graph of a relation for which the input $x = -4$ gives exactly three outputs and the input $x = 5$ gives exactly one output. Is the relation a function? Explain.

51. Sketch the graph of a function whose domain is $-3 \le x \le 5$ and whose range is $-2 \le y \le 4$.

52. Sketch the graph of a function whose domain is $-2 \le x \le 4$ and whose range is $-1 \le y \le 5$.

53. Sketch the graph of a function whose domain is the set of all real numbers and whose range is $y \ge 2$.

54. Sketch the graph of a function whose domain is the set of all real numbers and whose range is $y \le 3$.

55. Which types of lines are linear functions? Explain why other lines are not functions.

56. Explain why the vertical line test works.

57. A student tries to determine whether the relation $y = x^2$ is a function. She finds that both inputs $x = -3$ and $x = 3$ give the same output, $y = 9$. The student concludes the relation is not a function. Is her conclusion correct? Explain.

58. Explain how you can determine whether a relation is a function.

Related Review

For Exercises 59–62, the slope of a line is given. Determine whether the line represents a function.

59. positive **60.** negative

61. undefined **62.** zero

Expressions, Equations, Functions, and Graphs

Perform the indicated instruction. Then use words such as linear, function, one variable, *and* two variables *to describe the expression or equation. For instance, to describe $y = 4x - 7$, you could say "$y = 4x - 7$ is a linear function."*

63. Simplify $2(4x - 5) - 4(3x + 2)$.

64. Evaluate $3x - 5y - 20$ for $x = 2$ and $y = -4$.

65. Solve $2(4x - 5) - 4(3x + 2) = 0$.

66. Graph $3x - 5y - 20 = 0$.

5.3 Function Notation

Objectives

- » Use *function notation.*
- » Find inputs and outputs of a function.
- » Use function notation with models.

In this section, we will discuss how to name a function. We will also find inputs and outputs of a function.

Function Notation

Rather than use an equation, table, graph, or words to refer to a function, it would be easier to name the function. For example, to use "f" as the name of the linear function $y = 2x + 1$, we use "$f(x)$" (read "f of x") to represent y:

$$y = f(x)$$

We refer to "$f(x)$" as *function notation.* To use function notation to write the equation of this function, we substitute $f(x)$ for y in the equation $y = 2x + 1$:

$$f(x) = 2x + 1$$

WARNING The notation "$f(x)$" does *not* mean f times x. It is another variable name for y.

Recall from Section 5.2 that we can think of a function as a machine that sends inputs to outputs. Here we substitute 4 for x in the equation $y = 2x + 1$: $y = 2(4) + 1 = 9$. So, the input $x = 4$ leads to the output $y = 9$. Now we substitute 4 for x in the equation $f(x) = 2x + 1$:

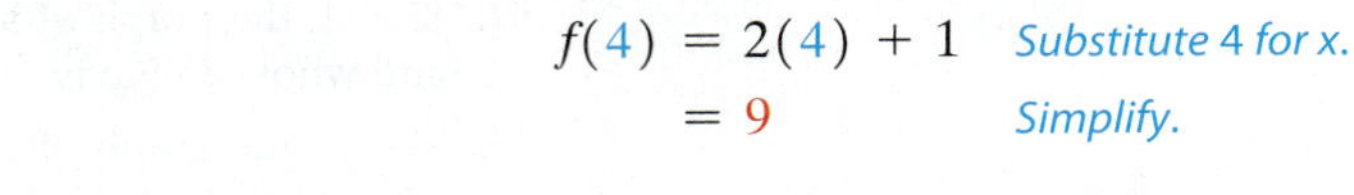

$$\begin{aligned} f(x) &= 2x + 1 && \textit{Equation of f} \\ f(4) &= 2(4) + 1 && \textit{Substitute 4 for x.} \\ &= 9 && \textit{Simplify.} \end{aligned}$$

Figure 22 A function "machine"

The equation $f(4) = 9$ means the input $x = 4$ leads to the output $y = 9$. Figure 22 shows the "machine" f sending the input 4 to the output 9.

Notice that $f(4) = 9$ is of the form

$$f(\text{input}) = \text{output}$$

This is true for any function f.

The number $f(4)$ is the value of y when x is 4. To find $f(4)$, we say we **evaluate** the function f at $x = 4$.

Example 1 Evaluating a Function

Evaluate $f(x) = -4x + 2$ at 5.

Solution

$$\begin{aligned} f(x) &= -4x + 2 && \textit{Equation of f} \\ f(5) &= -4(5) + 2 && \textit{Substitute 5 for x.} \\ &= -18 && \textit{Simplify.} \end{aligned}$$

We can also use "g" to name the function $y = -4x + 2$:

$$g(x) = -4x + 2$$

The most commonly used symbols to name functions are f, g, and h.

Example 2 Evaluating Functions

For $f(x) = 2x^2 - 3x$, $g(x) = \dfrac{4x - 2}{5x - 1}$, and $h(x) = 3x - 5$, find the following:

1. $f(-2)$ **2.** $g(3)$ **3.** $h(a)$ **4.** $h(a - 2)$

Solution

1.

$$
\begin{aligned}
f(-2) &= 2(-2)^2 - 3(-2) && \text{Evaluate } f \text{ at } -2.\\
&= 2(4) - 3(-2) && \text{Perform exponentiation first: } (-2)^2 = (-2)(-2) = 4\\
&= 8 + 6 && \text{Multiply.}\\
&= 14 && \text{Add.}
\end{aligned}
$$

2.

$$
\begin{aligned}
g(3) &= \frac{4 \cdot 3 - 2}{5 \cdot 3 - 1} && \text{Evaluate } g \text{ at } 3.\\
&= \frac{12 - 2}{15 - 1} && \text{Multiply first.}\\
&= \frac{10}{14} && \text{Subtract.}\\
&= \frac{5}{7} && \text{Simplify.}
\end{aligned}
$$

3. To find $h(a)$, we substitute a for x in the equation $h(x) = 3x - 5$:

$$h(a) = 3a - 5 \quad \text{Evaluate } h \text{ at } a.$$

4.

$$
\begin{aligned}
h(a - 2) &= 3(a - 2) - 5 && \text{Evaluate } h \text{ at } a - 2.\\
&= 3a - 6 - 5 && \text{Distributive law}\\
&= 3a - 11 && \text{Combine like terms.}
\end{aligned}
$$

So far, we have used equations to evaluate functions. Next, we will use a table to find an output and an input of a function.

Table 11 Input–Output Pairs of g

x	$g(x)$
3	12
4	9
5	8
6	9
7	12

Example 3 Using a Table to Find an Output and an Input

Some input–output pairs of a function g are shown in Table 11.

1. Find $g(7)$.
2. Find x when $g(x) = 9$.

Solution

1. From Table 11, we see the input $x = 7$ leads to the output $y = 12$. So, $g(7) = 12$.
2. To find x when $g(x) = 9$, we need to find all inputs in the table that lead to the output $y = 9$. From Table 11, we see both inputs $x = 4$ and $x = 6$ lead to the output $y = 9$. So, the values of x are 4 and 6.

In Example 4, we will use an equation to find an output and an input of a function.

Example 4 Using an Equation to Find an Output and an Input

Let $f(x) = \dfrac{3}{2}x - 1$.

1. Find $f(4)$.
2. Find x when $f(x) = -4$.

Figure 23 Putting table into "Ask" mode

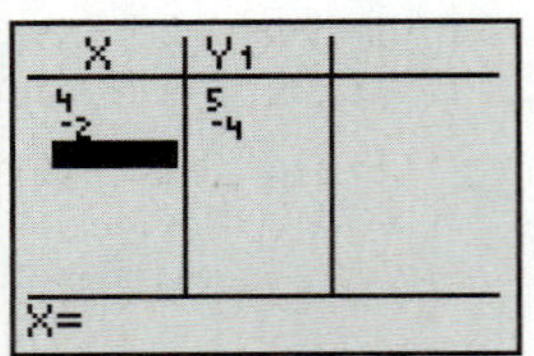

Figure 24 Verify the work

Solution

1.
$$\begin{aligned} f(4) &= \frac{3}{2}(4) - 1 && \text{Substitute 4 for } x. \\ &= 6 - 1 && \frac{3}{2}(4) = \frac{3}{2}\cdot\frac{4}{1} = 6 \\ &= 5 && \text{Subtract.} \end{aligned}$$

2. We substitute -4 for $f(x)$ in $f(x) = \frac{3}{2}x - 1$ and solve for x:

$$\begin{aligned} -4 &= \frac{3}{2}x - 1 && \text{Substitute } -4 \text{ for } f(x). \\ 2(-4) &= 2\cdot\frac{3}{2}x - 2\cdot 1 && \text{Multiply both sides by LCD, 2.} \\ -8 &= 3x - 2 && \text{Multiply; simplify.} \\ -6 &= 3x && \text{Add 2 to both sides.} \\ -2 &= x && \text{Divide both sides by 3.} \end{aligned}$$

We can verify our work in Problems 1 and 2 by putting a graphing calculator table into Ask mode (see Figs. 23 and 24). For graphing calculator instructions, see Appendix A.15.

WARNING

Example 4 asks for both a value of y (in Problem 1) and a value of x (in Problem 2). Be sure you know which value you need to find. When you are asked for $f(x)$, what you are looking for is a value of y, not a value of x.

For $f(x) = \frac{3}{2}x - 1$, we found in Problem 1 of Example 4 that $f(4) = 5$. Since $y = 5$ when $x = 4$, we know that the ordered pair $(4, 5)$ is a solution of $f(x) = \frac{3}{2}x - 1$ and that the point $(4, 5)$ is on the graph of f (see Fig. 25). We use blue arrows to show that the input $x = 4$ leads to the output $y = 5$.

In Problem 2 of Example 4, we found that $x = -2$ when $f(x) = -4$. So, the point $(-2, -4)$ is on the graph of f. We use red arrows in Fig. 25 to show that the output $y = -4$ originates from the input $x = -2$.

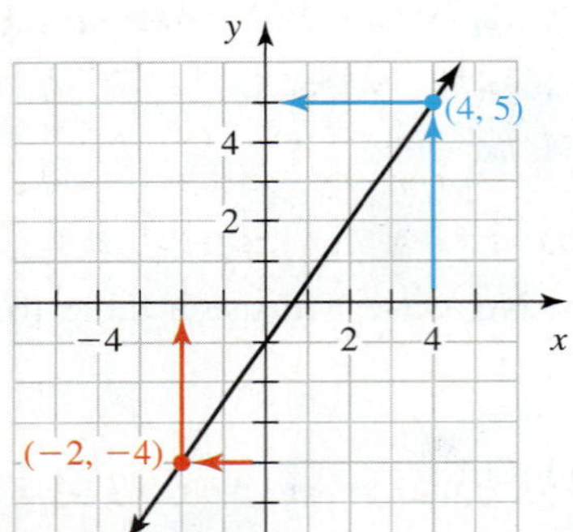

Figure 25 Graph of $f(x) = \frac{3}{2}x - 1$

Example 5 Using a Graph to Find Values of x or $f(x)$

A graph of a function f is sketched in Fig. 26.

1. Find $f(4)$.
2. Find $f(0)$.
3. Find x when $f(x) = -2$.
4. Find x when $f(x) = 0$.

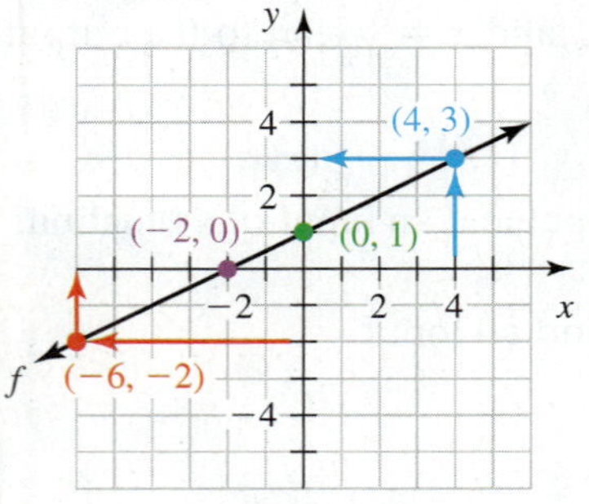

Figure 26 Problems 1–4 of Example 5

Solution

1. Recall that $y = f(x)$. The notation $f(4)$ refers to $f(x)$ when $x = 4$. So, we want the value of y when $x = 4$. The blue arrows in Fig. 26 show the input $x = 4$ leads to the output $y = 3$. Hence, $f(4) = 3$.
2. To find $f(0)$, we want the value of y when $x = 0$. The line contains the point $(0, 1)$, so $f(0) = 1$.
3. We have $y = f(x) = -2$. Thus, $y = -2$. So, we want the value of x when $y = -2$. The red arrows in Fig. 26 show the output $y = -2$ originates from the input $x = -6$. Hence, $x = -6$.
4. We have $y = f(x) = 0$. Thus, $y = 0$. The line contains the point $(-2, 0)$, so $x = -2$.

Example 6 Finding Intercepts

Find the indicated intercept of the graph of $f(x) = 2x - 6$.

1. x-intercept
2. y-intercept

Solution

1. To find the x-intercept, we substitute 0 for $f(x)$, which is y, and solve for x:

$$
\begin{aligned}
f(x) &= 2x - 6 && \textit{Original equation} \\
0 &= 2x - 6 && \textit{Substitute 0 for } f(x). \\
0 + 6 &= 2x - 6 + 6 && \textit{Add 6 to both sides.} \\
6 &= 2x && \textit{Combine like terms.} \\
\frac{6}{2} &= \frac{2x}{2} && \textit{Divide both sides by 2.} \\
3 &= x && \textit{Simplify.}
\end{aligned}
$$

So, the x-intercept is $(3, 0)$. We use "zero" on a graphing calculator to verify our work (see Fig. 27).

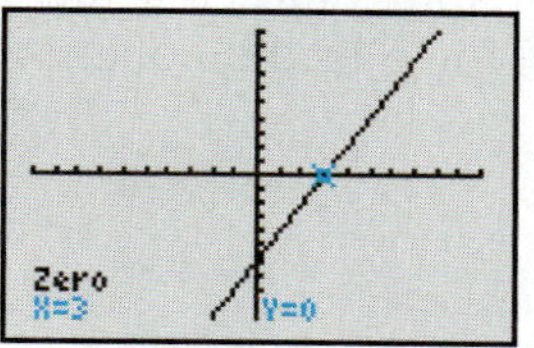

Figure 27 Verify the work

2. For an equation of the form $f(x) = mx + b$, as is $f(x) = 2x - 6$, the graph has y-intercept $(0, b)$. So, the y-intercept is $(0, -6)$. We use TRACE on a graphing calculator to verify our work (see Fig. 28).

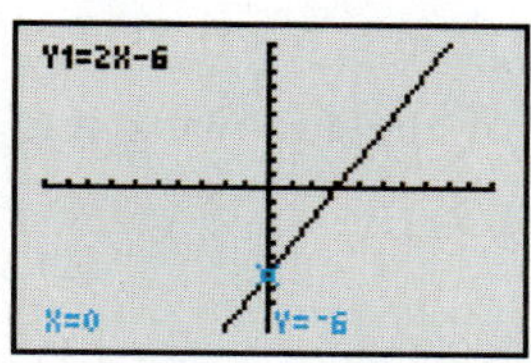

Figure 28 Verify the work

Using Function Notation with Models

Recall from Section 1.2 that when we are *not* describing an authentic situation, we treat x as the independent variable and y as the dependent variable. Here we label the independent variable, dependent variable, and function name of the equation $y = f(x)$:

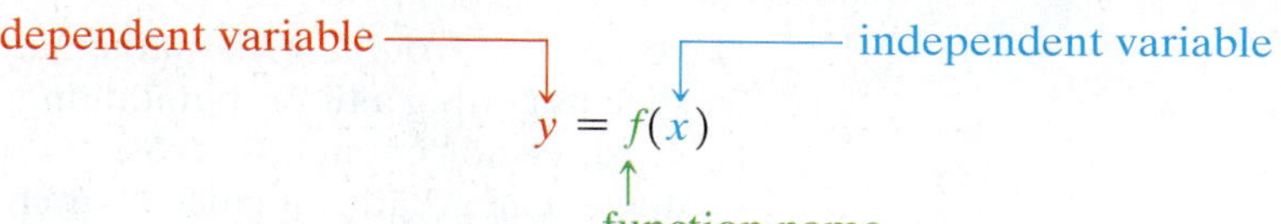

We follow this same format for a function f that *is* a model.

> **Definition** Function notation
>
> The dependent variable of a function f can be represented by the expression formed by writing the independent variable name within the parentheses of $f(\,)$:
>
> $$\text{dependent variable} = f(\text{independent variable})$$
>
> We call this representation **function notation.**

In Section 1.3, we defined a linear model as a nonvertical line that describes two quantities in an authentic situation. Because any nonvertical line is a linear function, it follows that every linear model is a linear function.

However, not every linear function is a linear model. Models are used only to describe situations. Functions are used both to describe situations *and* to describe certain *mathematical* relationships between two variables. For example, if the equation $y = 2x$ is not being used to describe a situation, then it is a function, not a model.

Example 7 Using Function Notation with a Linear Model

In Example 8 of Section 4.4, we found the model $p = 1.8t + 52$, where p is the percentage of Americans who believe gay/lesbian relations are morally acceptable at t years since 2010. Rewrite the equation $p = 1.8t + 52$ with the function name f.

Solution

Here, t is the independent variable and p is the dependent variable. Since the function name is f, we can write $p = f(t)$. Then we substitute $f(t)$ for p in the equation $p = 1.8t + 52$:

$$f(t) = 1.8t + 52$$

Group Exploration

Formula for slope

1. Let $f(x) = 2x + 1$. Find each of the following and compare all three results. [**Hint for part (b):** First, find $f(5)$ and $f(3)$. Then subtract. Finally, divide.]
   a. the slope of the graph of f
   b. $\dfrac{f(5) - f(3)}{5 - 3}$
   c. $\dfrac{f(7) - f(4)}{7 - 4}$
2. Let $g(x) = 3x + 5$. Find each of the following and compare all three results:
   a. the slope of the graph of g
   b. $\dfrac{g(3) - g(1)}{3 - 1}$
   c. $\dfrac{g(4) - g(0)}{4 - 0}$
3. Let f be a function of the form $f(x) = mx + b$. Describe $\dfrac{f(c) - f(d)}{c - d}$, where $c \neq d$. Explain.

Tips for Success Do Five New Study Activities

Have you had trouble with mathematics before? Did you try out a new study activity such as getting a tutor, but it didn't seem to help? Sometimes employing one new behavior is not enough to cross the threshold to success. Incorporating *five* such strategies can greatly enhance your chances.

Imagine taking part in five new activities such as asking more questions in class, doing extra problems, forming a study group, visiting your professor's office hours, and taking practice exams. When you get involved with so many activities, the benefits of one activity tend to increase the benefits of all of the others. It also sends a message to your psyche that you mean business about passing your algebra course, which can do wonders.

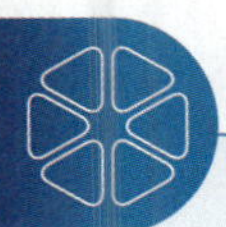

Homework 5.3

For extra help ▶ MyMathLab® 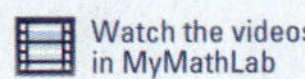 Watch the videos in MyMathLab  Download the MyDashboard App

Evaluate $f(x) = 6x - 4$ at the given value of x.

1. $f(5)$ **2.** $f(-2)$ **3.** $f\left(\frac{2}{3}\right)$

4. $f\left(\frac{5}{2}\right)$ **5.** $f(a + 2)$ **6.** $f(a - 3)$

Evaluate $g(x) = 2x^2 - 5x$ at the given value of x.

7. $g(2)$ **8.** $g(3)$ **9.** $g(-3)$ **10.** $g(-2)$

Evaluate $h(x) = \dfrac{3x - 4}{5x + 2}$ at the given value of x.

11. $h(2)$ **12.** $h(-4)$ **13.** $h(a - 3)$ **14.** $h(3a)$

For $f(x) = -2x + 7$, $g(x) = -3x^2 + 2x$, and $h(x) = -4$, find the following.

15. $g(-2)$ **16.** $g(-1)$ **17.** $f(5)$

18. $f(-4)$ **19.** $h(7)$ **20.** $h(-9)$

Evaluate $f(x) = -4x - 7$ at the given value of x.

21. $f(5a)$
22. $f(-3a)$
23. $f\left(\frac{a}{2}\right)$
24. $f\left(\frac{3a}{2}\right)$
25. $f(a + 4)$
26. $f(a - 4)$
27. $f(a - h)$
28. $f(a + h)$

For $f(x) = -3x + 7$, find the value of x that leads to the given value of $f(x)$.

29. $f(x) = 6$
30. $f(x) = 0$
31. $f(x) = \frac{5}{2}$
32. $f(x) = -\frac{4}{3}$
33. $f(x) = a$
34. $f(x) = a + 2$

For Exercises 35–38, let $f(x) = -5.95x + 183.22$. Round any results to the second decimal place.

35. Find $f(10.91)$.
36. Find $f(17.28)$.
37. Find x when $f(x) = 99.34$.
38. Find x when $f(x) = 72.06$.

For Exercises 39–42, refer to Table 12.

39. Find $f(2)$.
40. Find $f(4)$.
41. Find x when $f(x) = 2$.
42. Find x when $f(x) = 4$.

Table 12 Values of f (Exercises 39–42)

x	f(x)
0	0
1	2
2	4
3	2
4	0

For Exercises 43–54, refer to Fig. 29.

43. Estimate $f(-6)$.
44. Estimate $f(0)$.
45. Estimate $f(2.5)$.
46. Estimate $f\left(-\frac{11}{2}\right)$.
47. Estimate x when $f(x) = 0$.
48. Estimate x when $f(x) = 1$.
49. Estimate x when $f(x) = 3$.
50. Estimate x when $f(x) = 3.5$.
51. Estimate x when $f(x) = \frac{1}{2}$.
52. Estimate x when $f(x) = \frac{5}{2}$.
53. Find the domain of f.
54. Find the range of f.

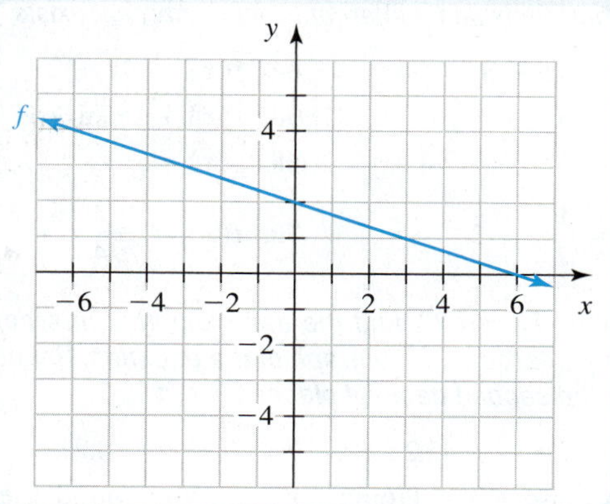

Figure 29 Exercises 43–54

For Exercises 55–58, refer to Fig. 30.

55. Find $g(-2)$.
56. Find x when $g(x) = 3$.
57. Find the domain of g.
58. Find the range of g.

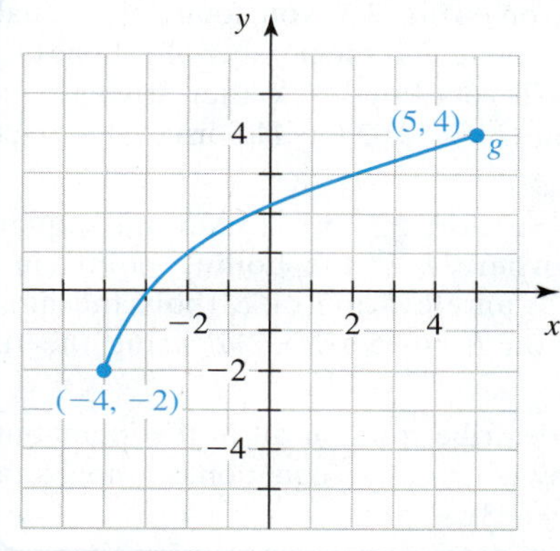

Figure 30 Exercises 55–58

For Exercises 59–62, refer to Fig. 31.

59. Find $h(1)$.
60. Find x when $h(x) = -1$.
61. Find the domain of h.
62. Find the range of h.

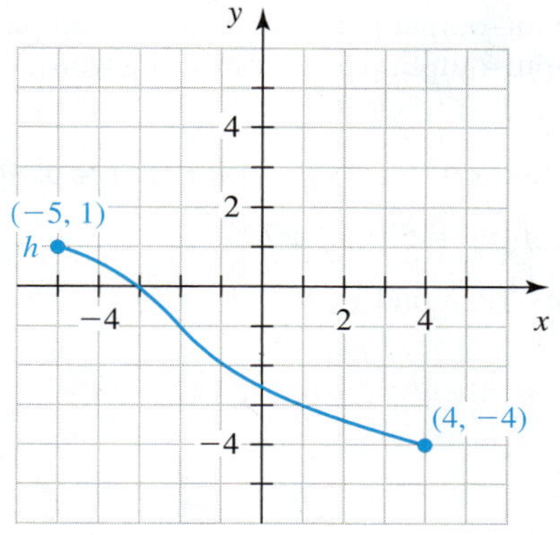

Figure 31 Exercises 59–62

Find all x-intercepts and y-intercepts. If no intercept exists, say so.

63. $f(x) = 5x - 8$

64. $f(x) = 4x + 2$

65. $f(x) = 3x$

66. $f(x) = -7x$

67. $f(x) = 5$

68. $f(x) = -2$

69. $f(x) = \frac{1}{2}x - 3$

70. $f(x) = -\frac{3}{4}x + \frac{1}{2}$

For Exercises 71 and 72, find the approximate x-intercept and approximate y-intercept of the graph of the equation. Round the coordinates to the second decimal place.

71. $f(x) = 2.58x - 45.21$

72. $f(x) = -4.29x + 37.58$

73. In Exercise 47 of Homework 4.4, you found the equation $n = 1.04t + 46.7$, where n is the number (in millions) of Americans without health insurance at t years since 2010. Rewrite the equation $n = 1.04t + 46.7$ using the function name f.

74. In Exercise 50 of Homework 4.4, you found the equation $s = -1.39t + 19.1$, where s is GM's share of new-vehicle sales in the United States at t years since 2010. Rewrite the equation $s = -1.39t + 19.1$ using the function name f.

75. In Exercise 23 of Homework 3.5, you found the equation $G = -0.02d + 11.9$, where G is the number of gallons of gasoline remaining in a Toyota Prius tank after driving d miles. Rewrite the equation $G = -0.02d + 11.9$ using the function name h.

76. In Example 5 of Section 3.5, we found the equation $B = -5.9E + 212$, where B is the boiling point (in degrees Fahrenheit) at an elevation of E thousand meters. Rewrite the equation $B = -5.9E + 212$, using the function name g.

77. Recall that we can describe some or all of the input–output pairs of a function by means of an equation, a graph, a table, or words. Let $g(x) = -3x + 4$.

a. Describe five input–output pairs of g by using a table.
b. Describe the input–output pairs of g by using a graph.
c. Describe the input–output pairs of g by using words.

78. Recall that we can describe some or all of the input–output pairs of a function by means of an equation, a graph, a table, or words. Let $f(x) = -\frac{3}{5}x - 1$.

a. Describe five input–output pairs of f by using a table.
b. Describe the input–output pairs of f by using a graph.
c. Describe the input–output pairs of f by using words.

Concepts

79. A student tries to find x when $f(x) = 5$ for $f(x) = x + 2$:

$$f(5) = 5 + 2 = 7$$

Describe any errors. Then find x correctly.

80. A student tries to find $g(-5)$, where $g(x) = x^2$:

$$g(-5) = -5^2 = -25$$

Describe any errors. Then find $g(-5)$ correctly.

81. a. For $f(x) = 4x$, find $f(3)$, $f(5)$, and $f(8)$. Is the equation $f(3 + 5) = f(3) + f(5)$ a true statement?
b. For $f(x) = x^2$, find $f(2)$, $f(3)$, and $f(5)$. Is the equation $f(2 + 3) = f(2) + f(3)$ a true statement?
c. For $f(x) = \sqrt{x}$, find $f(9)$, $f(16)$, and $f(25)$. Is $f(9 + 16) = f(9) + f(16)$ a true statement?
d. Is $f(a + b) = f(a) + f(b)$ a true statement for every function f?

82. a. For $f(x) = 3x + 2$, find $f(5) - f(4)$. Compare your result with the slope of the graph of f. [**Hint:** Find $f(5)$ and $f(4)$. Then subtract.]
b. For $f(x) = 2x + 5$, find $f(7) - f(6)$. Compare your result with the slope of the graph of f.
c. For $f(x) = 4x + 1$, find $f(3) - f(2)$. Compare your result with the slope of the graph of f.
d. For $f(x) = mx + b$, find $f(a + 1) - f(a)$. Compare your result with the slope of the graph of f, and discuss what this result means. [**Hint:** Consider the slope addition property.]

83. A student says $f(4)$ means that $y = 4$, because $y = f(x)$. Is the student correct? Explain.

84. For a function f, assume $f(3) = 5$. Name an input and an output of f. Also, find three possible equations of f.

Related Review

Find the slope of the graph of the linear function f that meets the following conditions:

85. $f(-5) = 2$ and $f(3) = -4$

86. $f(-6) = 1$ and $f(-2) = -5$

Find the intercepts of the graph of the linear function f that meets the following conditions:

87. $f(0) = -3$ and $f(4) = 0$

88. $f(-5) = 0$ and $f(0) = -1$

Expressions, Equations, Functions, and Graphs

Perform the indicated instruction. Then use words such as linear, function, one variable, *and* two variables *to describe the expression or equation. For instance, to describe $2x + 7x$, you could say "$2x + 7x$ is a linear expression in one variable."*

89. For $f(x) = 2x - 4$, find x when $f(x) = 2$.

90. For $f(x) = 2x - 4$, find $f(2)$.

91. Solve $6 = 2x - 4$.

92. Graph $f(x) = 2x - 4$.

5.4 Finding Linear Equations

Objectives

- Find an equation of a line by using the slope and a point.
- Find an equation of a line by using two points.
- Know the point–slope form of a linear equation.

In this section, we will discuss two methods of finding an equation of a line. We will first use the slope–intercept form. Then we will use another form of an equation of a line called the *point–slope form.*

Method 1: Using the Slope–Intercept Form

In Example 1, we will use the concept that a point that lies on a line satisfies an equation of the line.

Example 1 Using the Slope and a Point to Find a Linear Equation

Find an equation of the line that has slope $m = 2$ and contains the point $(4, 3)$.

Solution

An equation of a nonvertical line can be put into the form $y = mx + b$. Since $m = 2$, we have

$$y = 2x + b$$

To find b, recall from Section 3.1 that any point on the graph of an equation represents a solution of that equation. In particular, the point $(4, 3)$ should satisfy the equation $y = 2x + b$:

$$\begin{aligned} y &= 2x + b && \text{Slope is } m = 2. \\ 3 &= 2(4) + b && \text{Substitute 4 for } x \text{ and 3 for } y. \\ 3 &= 8 + b && \text{Multiply.} \\ 3 - 8 &= 8 + b - 8 && \text{Subtract 8 from both sides.} \\ -5 &= b && \text{Combine like terms.} \end{aligned}$$

Now we substitute -5 for b in $y = 2x + b$:

$$y = 2x - 5$$

To verify our work, we check that the coefficient of the variable term $2x$ is 2 (the given slope) and we see whether $(4, 3)$ satisfies the equation $y = 2x - 5$:

$$\begin{aligned} y &= 2x - 5 && \text{The equation we found} \\ 3 &\overset{?}{=} 2(4) - 5 && \text{Substitute 4 for } x \text{ and 3 for } y. \\ 3 &\overset{?}{=} 8 - 5 && \text{Multiply.} \\ 3 &\overset{?}{=} 3 && \text{Subtract.} \end{aligned}$$

true

Finding an Equation of a Line by Using the Slope, a Point, and the Slope–Intercept Form

To find an equation of a line by using the slope and a point,

1. Substitute the given value of the slope m into the equation $y = mx + b$.
2. Substitute the coordinates of the given point into your equation from step 1 and solve for b.
3. Substitute the value of b from step 2 into your equation from step 1.
4. Check that the graph of your equation contains the given point.

Example 2 Using the Slope and a Point to Find a Linear Equation

Find an equation of the line that has slope $m = -\frac{2}{5}$ and contains the point $(-4, 1)$.

Solution

Since the slope is $-\frac{2}{5}$, the equation has the form $y = -\frac{2}{5}x + b$. To find b, we substitute the coordinates of the point $(-4, 1)$ into the equation $y = -\frac{2}{5}x + b$ and solve for b:

$$\begin{aligned} y &= -\frac{2}{5}x + b && \textit{Slope is } -\tfrac{2}{5}. \\ 1 &= -\frac{2}{5}(-4) + b && \textit{Substitute } -4 \textit{ for x and 1 for y.} \\ 1 &= \frac{8}{5} + b && \textit{Simplify.} \\ 5 &= 8 + 5b && \textit{Multiply both sides by LCD, 5, to clear equation of fractions.} \\ 5 - 8 &= 8 + 5b - 8 && \textit{Subtract 8 from both sides.} \\ -3 &= 5b && \textit{Combine like terms.} \\ -\frac{3}{5} &= b && \textit{Divide both sides by 5.} \end{aligned}$$

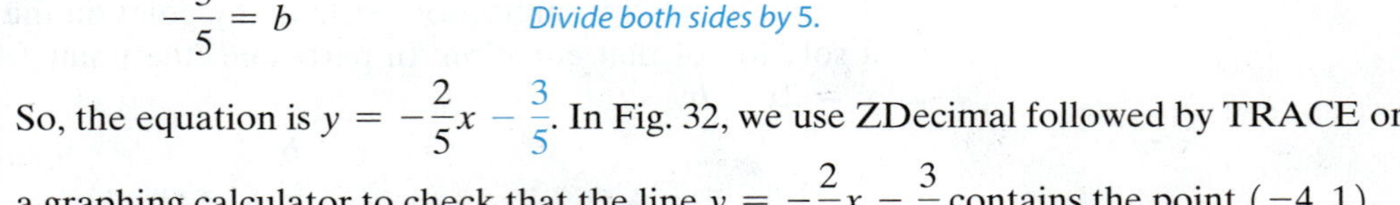

Figure 32 Check that the line contains $(-4, 1)$

So, the equation is $y = -\frac{2}{5}x - \frac{3}{5}$. In Fig. 32, we use ZDecimal followed by TRACE on a graphing calculator to check that the line $y = -\frac{2}{5}x - \frac{3}{5}$ contains the point $(-4, 1)$.

In Examples 1 and 2, we found an equation of a line by using the slope of the line and a point. We can also find an equation of a line by using two points.

Example 3 Using Two Points to Find a Linear Equation

Find an equation of the line that passes through $(-1, 4)$ and $(2, -5)$.

Solution

First, we find the slope of the line:

$$m = \frac{y_2 - y_1}{x_2 - x_1} = \frac{-5 - 4}{2 - (-1)} = \frac{-5 - 4}{2 + 1} = \frac{-9}{3} = -3$$

So, we have $y = -3x + b$. Next, we will find the value of b. Since the line contains the point $(-1, 4)$, we substitute -1 for x and 4 for y:

$$\begin{aligned} y &= -3x + b && \textit{Slope is } m = -3. \\ 4 &= -3(-1) + b && \textit{Substitute } -1 \textit{ for x and 4 for y.} \\ 4 &= 3 + b && \textit{Multiply.} \\ 4 - 3 &= 3 + b - 3 && \textit{Subtract 3 from both sides.} \\ 1 &= b && \textit{Combine like terms.} \end{aligned}$$

The equation is $y = -3x + 1$. To verify our equation, we check that both $(-1, 4)$ and $(2, -5)$ satisfy the equation $y = -3x + 1$:

Check that $(-1, 4)$ is a solution

$$\begin{aligned} y &= -3x + 1 \\ 4 &\stackrel{?}{=} -3(-1) + 1 \\ 4 &\stackrel{?}{=} 3 + 1 \\ 4 &\stackrel{?}{=} 4 \end{aligned}$$

true

Check that $(2, -5)$ is a solution

$$\begin{aligned} y &= -3x + 1 \\ -5 &\stackrel{?}{=} -3(2) + 1 \\ -5 &\stackrel{?}{=} -6 + 1 \\ -5 &\stackrel{?}{=} -5 \end{aligned}$$

true

In Example 3, we substituted the coordinates of the given point $(-1, 4)$ into the equation $y = -3x + b$ to help us find the constant b. If we had used the other given point, $(2, -5)$, we would have found the same value of b. (Try it.)

Finding an Equation of a Line by Using Two Points and the Slope–Intercept Form

To find an equation of the line that passes through two given points whose x-coordinates are different,

1. Use the formula $m = \frac{y_2 - y_1}{x_2 - x_1}$, or use a graph to determine $\frac{\text{rise}}{\text{run}}$, to find the slope of the line containing the two points.
2. Substitute the m value you found in step 1 into the equation $y = mx + b$.
3. Substitute the coordinates of one of the given points into the equation from step 2 and solve for b.
4. Substitute the b value you found in step 3 into your equation from step 2.
5. Check that the graph of your equation contains the two given points.

Example 4 Using Two Points to Find a Linear Equation

Find an equation of the line that passes through the points $(-9, -2)$ and $(-3, 7)$.

Solution

We begin by finding the slope of the line:

$$m = \frac{y_2 - y_1}{x_2 - x_1} = \frac{7 - (-2)}{-3 - (-9)} = \frac{9}{6} = \frac{3}{2}$$

So, we have $y = \frac{3}{2}x + b$. To find b, we substitute the coordinates of $(-3, 7)$ into the equation $y = \frac{3}{2}x + b$ and solve for b:

$$y = \frac{3}{2}x + b \qquad \textit{Slope is } \frac{3}{2}.$$

$$7 = \frac{3}{2}(-3) + b \qquad \textit{Substitute } -3 \textit{ for } x \textit{ and } 7 \textit{ for } y.$$

$$7 = -\frac{9}{2} + b \qquad \textit{Simplify.}$$

$$14 = -9 + 2b \qquad \textit{Multiply both sides by LCD, 2, to clear equation of fractions.}$$

$$14 + 9 = -9 + 2b + 9 \qquad \textit{Add 9 to both sides.}$$

$$23 = 2b \qquad \textit{Combine like terms.}$$

$$\frac{23}{2} = b \qquad \textit{Divide both sides by 2.}$$

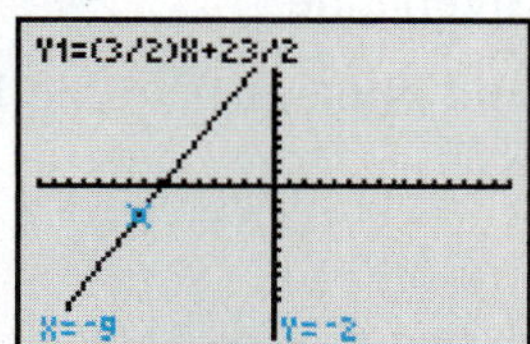

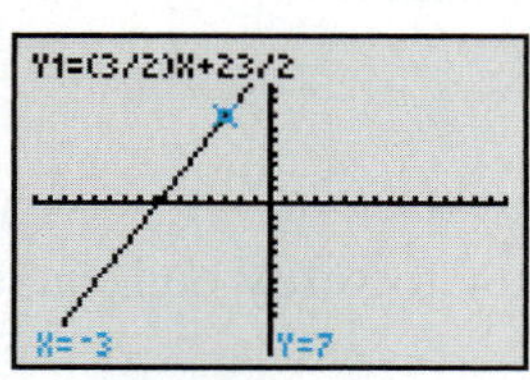

Figure 33 Check that the line contains both $(-9, -2)$ and $(-3, 7)$

The equation is $y = \frac{3}{2}x + \frac{23}{2}$. We use ZStandard followed by ZSquare to check that the line $y = \frac{3}{2}x + \frac{23}{2}$ contains the points $(-9, -2)$ and $(-3, 7)$. See Fig. 33. For graphing calculator instructions, see Appendix A.3–A.6.

In Example 5, we will work with decimal numbers to prepare us to find equations of models in Section 5.5.

Example 5 Finding an Approximate Equation of a Line

Find an approximate equation of the line that contains the points $(-3.1, 5.7)$ and $(1.6, -4.8)$.

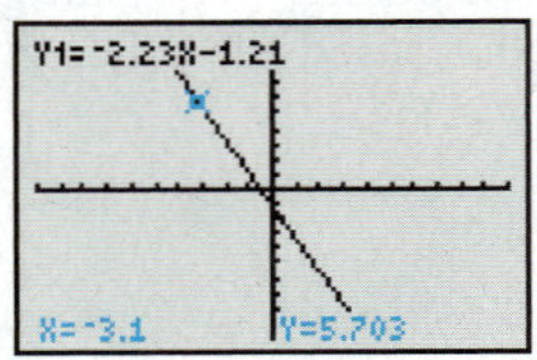

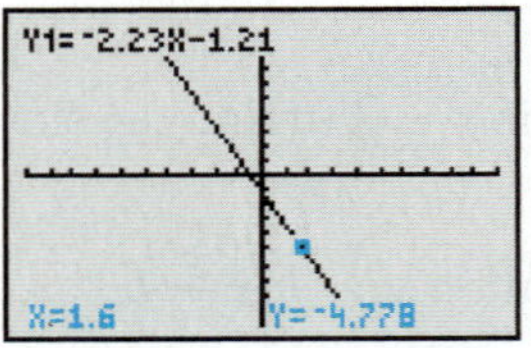

Figure 34 Check that the line comes very close to $(-3.1, 5.7)$ and $(1.6, -4.8)$

Solution

First, we find the slope of the line:

$$m = \frac{y_2 - y_1}{x_2 - x_1} = \frac{-4.8 - 5.7}{1.6 - (-3.1)} = \frac{-10.5}{4.7} \approx -2.23$$

So, the equation has the form $y = -2.23x + b$. To find b, we substitute the coordinates of the point $(-3.1, 5.7)$ into the equation $y = -2.23x + b$ and solve for b:

$$\begin{aligned} y &= -2.23x + b && \textit{Slope is approximately } -2.23. \\ 5.7 &= -2.23(-3.1) + b && \textit{Substitute } -3.1 \textit{ for } x \textit{ and } 5.7 \textit{ for } y. \\ 5.7 &= 6.913 + b && \textit{Multiply.} \\ 5.7 - 6.913 &= 6.913 + b - 6.913 && \textit{Subtract 6.913 from both sides.} \\ -1.21 &\approx b && \textit{Combine like terms.} \end{aligned}$$

The approximate equation is $y = -2.23x - 1.21$.

We use a graphing calculator to check that the line $y = -2.23x - 1.21$ comes very close to the points $(-3.1, 5.7)$ and $(1.6, -4.8)$. See Fig. 34.

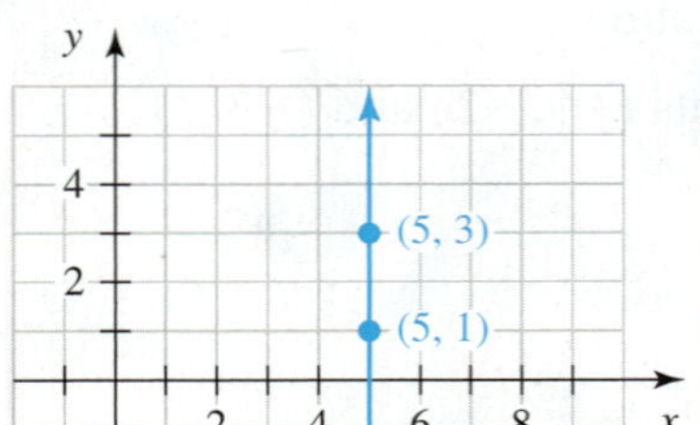

Figure 35 The line that contains $(5, 1)$ and $(5, 3)$

Example 6 Finding an Equation of a Vertical Line

Find an equation of the line that contains the points $(5, 1)$ and $(5, 3)$.

Solution

Since the x-coordinates of the given points are equal (both 5), the line that contains the points is vertical (see Fig. 35). An equation of the line is $x = 5$.

In Example 7, we will find an equation of a line that contains a given point and is parallel to a given line. Recall from Section 3.4 that nonvertical parallel lines have equal slopes.

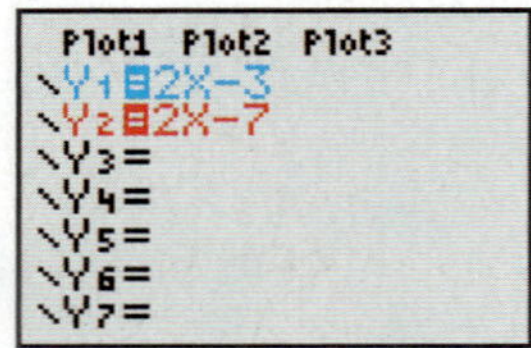

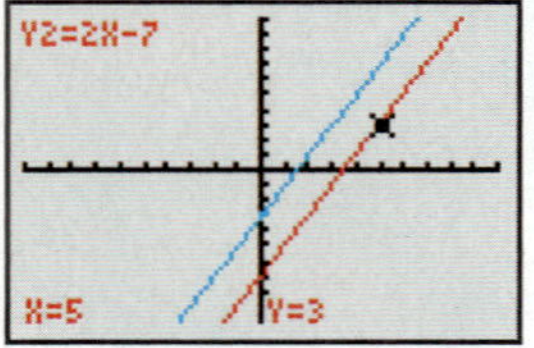

Figure 36 Check that the line contains (5, 3) and is parallel to $y = 2x - 3$

Example 7 Finding an Equation of a Line Parallel to a Given Line

Find an equation of a line l that contains the point $(5, 3)$ and is parallel to the line $y = 2x - 3$.

Solution

For the line $y = 2x - 3$, the slope is 2. So, the slope of parallel line l is also 2. An equation of line l is $y = 2x + b$. To find b, we substitute the coordinates of $(5, 3)$ into the equation $y = 2x + b$:

$$\begin{aligned} 3 &= 2(5) + b && \textit{Substitute 5 for } x \textit{ and 3 for } y. \\ -7 &= b && \textit{Multiply; subtract 10 from both sides.} \end{aligned}$$

An equation of l is $y = 2x - 7$. We use a graphing calculator to verify our equation (see Fig. 36).

Recall from Section 3.4 that if two nonvertical lines are perpendicular, then the slope of one line is the opposite of the reciprocal of the slope of the other line.

Example 8 Finding an Equation of a Line Perpendicular to a Given Line

Find an equation of the line l that contains the point $(6, -7)$ and is perpendicular to the line $-2x + 5y = 10$.

Solution

First, we isolate y in the equation $-2x + 5y = 10$:

$$\begin{aligned} -2x + 5y &= 10 && \text{Original equation: line perpendicular to line } l \\ -2x + 5y + 2x &= 10 + 2x && \text{Add } 2x \text{ to both sides.} \\ 5y &= 2x + 10 && \text{Simplify.} \\ \frac{5y}{5} &= \frac{2x}{5} + \frac{10}{5} && \text{Divide both sides by 5.} \\ y &= \frac{2}{5}x + 2 && \text{Simplify.} \end{aligned}$$

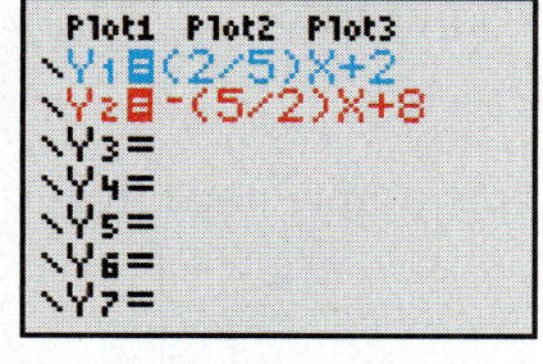

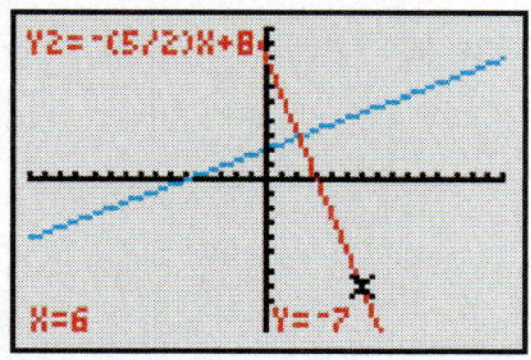

Figure 37 Check that the line contains $(6, -7)$ and is perpendicular to $-2x + 5y = 10$

For the line $y = \frac{2}{5}x + 2$, the slope is $m = \frac{2}{5}$. The slope of the line l must be the opposite of the reciprocal of $\frac{2}{5}$, or $-\frac{5}{2}$. An equation of l is $y = -\frac{5}{2}x + b$. To find b, we substitute the coordinates of the given point $(6, -7)$ into $y = -\frac{5}{2}x + b$:

$$\begin{aligned} -7 &= -\frac{5}{2}(6) + b && \text{Substitute 6 for } x \text{ and } -7 \text{ for } y. \\ -7 &= -15 + b && -\frac{5}{2}(6) = -\frac{5}{2}\left(\frac{6}{1}\right) = -\frac{5 \cdot 2 \cdot 3}{2} = -\frac{15}{1} = -15 \\ 8 &= b && \text{Add 15 to both sides.} \end{aligned}$$

An equation of l is $y = -\frac{5}{2}x + 8$. We use ZStandard followed by ZSquare to verify our work (see Fig. 37).

In Example 9, we will use a graphical approach to find the equation of a line.

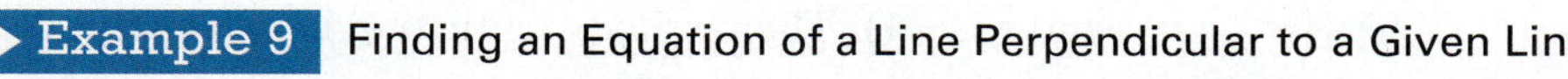

Example 9 Finding an Equation of a Line Perpendicular to a Given Line

Find an equation of the line l that contains $(4, 3)$ and is perpendicular to the line $x = 2$.

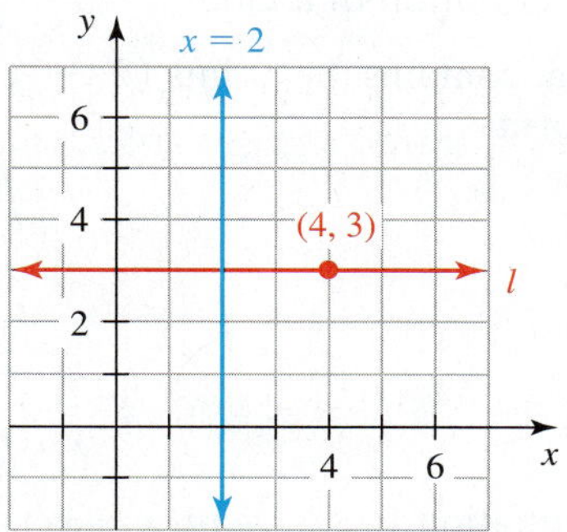

Figure 38 Find an equation of the line l

Solution

The graph of $x = 2$ is a vertical line (see Fig. 38). A line perpendicular to it must be horizontal, so there is an equation of l of the form $y = b$. To find b, we substitute the y-coordinate of the given point $(4, 3)$ into $y = b$ and get $3 = b$. So, an equation of l is $y = 3$.

Method 2: Using Point–Slope Form

We can find an equation of a line by another method. Suppose a nonvertical line has slope m and contains the point (x_1, y_1). Then, if (x, y) represents a different point on the line, the slope of the line is

$$\frac{y - y_1}{x - x_1} = m$$

Multiplying both sides of the equation by $x - x_1$ gives

$$\begin{aligned} \frac{y - y_1}{x - x_1} \cdot (x - x_1) &= m(x - x_1) \\ y - y_1 &= m(x - x_1) \end{aligned}$$

We say this linear equation is in **point–slope form.***

*Although we assumed that (x, y) is different from (x_1, y_1), note that (x_1, y_1) is a solution of the equation $y - y_1 = m(x - x_1)$: $y_1 - y_1 = m(x_1 - x_1)$, or $0 = 0$, a true statement.

Point–Slope Form

If a nonvertical line has slope m and contains the point (x_1, y_1), then an equation of the line is

$$y - y_1 = m(x - x_1)$$

Example 10 Using the Point–Slope Form to Find an Equation of a Line

Use the point–slope form to find an equation of the line that has slope $m = -3$ and contains the point $(-4, 2)$. Then write the equation in slope–intercept form.

Solution

We begin by substituting $x_1 = -4$, $y_1 = 2$, and $m = -3$ into the point–slope form $y - y_1 = m(x - x_1)$:

$$\begin{aligned}
y - y_1 &= m(x - x_1) && \textit{Point–slope form}\\
y - 2 &= -3(x - (-4)) && \textit{Substitute } x_1 = -4, y_1 = 2, \textit{ and } m = -3.\\
y - 2 &= -3(x + 4) && \textit{Simplify.}\\
y - 2 &= -3x - 12 && \textit{Distributive law}\\
y - 2 + 2 &= -3x - 12 + 2 && \textit{Add 2 to both sides.}\\
y &= -3x - 10 && \textit{Combine like terms.}
\end{aligned}$$

So, the equation is $y = -3x - 10$.

Example 11 Using the Point–Slope Form to Find an Equation of a Line

Use the point–slope form to find an equation of the line that contains the points $(2, -6)$ and $(5, -4)$. Then write the equation in slope–intercept form.

Solution

We begin by finding the slope of the line:

$$m = \frac{-4 - (-6)}{5 - 2} = \frac{2}{3}$$

Then we substitute $x_1 = 2$, $y_1 = -6$, and $m = \frac{2}{3}$ into the equation $y - y_1 = m(x - x_1)$:

$$\begin{aligned}
y - y_1 &= m(x - x_1) && \textit{Point–slope form}\\
y - (-6) &= \frac{2}{3}(x - 2) && \textit{Substitute, } x_1 = 2, y_1 = -6, \textit{ and } m = \frac{2}{3}.\\
y + 6 &= \frac{2}{3}x - \frac{4}{3} && \textit{Simplify; distributive law}\\
y + 6 - 6 &= \frac{2}{3}x - \frac{4}{3} - 6 && \textit{Subtract 6 from both sides.}\\
y &= \frac{2}{3}x - \frac{22}{3} && \textit{Combine like terms; } -\frac{4}{3} - 6 = -\frac{4}{3} - \frac{18}{3} = -\frac{22}{3}
\end{aligned}$$

So, the equation is $y = \frac{2}{3}x - \frac{22}{3}$.

Group Exploration

Deciding which points to use to find an equation of a line

1. **a.** Use the method shown in Example 3 to find an equation of the line that contains the points $(1, 2)$ and $(3, 8)$.
   b. In part (a), you used either point $(1, 2)$ or point $(3, 8)$ to find the constant b for the equation $y = 3x + b$. Now use the other point to find b.
   c. Does it matter which point is used to find the constant b? Explain.

2. Imagine any line that is not parallel to either axis. Choose four points on the line. Name the points A, B, C, and D.
   a. Use points A and B to find an equation of the line. Write your equation in slope–intercept form.
   b. Use points C and D to find an equation of the line. Write your equation in slope–intercept form.
   c. Are the equations you found in parts (a) and (b) the same? Explain.

Group Exploration

Finding equations of lines

The objective of this game is to earn 19 credits. You earn credits by finding equations of lines that pass through one or more of the following points:

$$(-3, 2), (-3, 0), (-2, -7), (-2, -1), (-1, 4), (0, 2),$$
$$(1, -1), (2, -2), (3, 1), (3, 3)$$

If a line passes through exactly one point, then you earn one credit. If a line passes through exactly two points, then you earn three credits. If a line passes through exactly three points, then you earn five credits. You may use five equations. You may use points more than once. [**Hints:** First, plot the points. After finding your equations, use your graphing calculator to check that they are correct.]

Tips for Success Show What You Know

Even if you don't know how to do one step of a problem, you can still show the instructor that you understand the other steps of the problem. Depending on how your instructor grades tests, you may earn partial credit even though you pick an incorrect number to be the result for a particular step, as long as you then show what you would do with that number in the remaining steps of the solution. Check with your instructor first.

For example, suppose you want to find an equation of the line that passes through the points $(1, 5)$ and $(2, 8)$, but you have forgotten how to find the slope. You could still write,

I've drawn a blank on finding slope. However, assuming that the slope is 2, then

$$\begin{aligned} y &= 2x + b \\ 5 &= 2(1) + b \\ 5 &= 2 + b \\ 5 - 2 &= 2 + b - 2 \\ 3 &= b \end{aligned}$$

Therefore, $y = 2x + 3$.

You could point out that you know your result is incorrect, because the graph of $y = 2x + 3$ does *not* pass through the point $(2, 8)$. Also, seeing your result (with the graph) may jog your memory about finding the slope and allow you to go back and do the problem correctly.

Homework 5.4

For extra help ▶ MyMathLab® 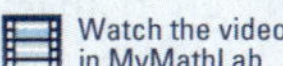 Watch the videos in MyMathLab  Download the MyDashboard App

Find an equation of the line that has the given slope and contains the given point. If possible, write your equation in slope–intercept form. Check that the ordered pair that represents the given point satisfies your equation.

1. $m = 2, (3, 5)$ **2.** $m = 3, (2, 4)$

3. $m = -3, (1, -2)$ **4.** $m = -5, (3, -8)$

5. $m = -6, (-2, -3)$ **6.** $m = -1, (-7, -4)$

7. $m = \frac{2}{5}, (3, 1)$ **8.** $m = \frac{1}{2}, (5, 3)$

9. $m = -\frac{3}{4}, (-2, -5)$ **10.** $m = -\frac{5}{3}, (-4, -2)$

11. $m = 0, (5, 3)$ **12.** $m = 0, (-1, -3)$

13. m is undefined, $(-2, 4)$ **14.** m is undefined, $(3, -2)$

Find an approximate equation of the line that has the given slope and contains the given point. Write your equation in slope–intercept form. Round the constant term to two decimal places. Use a graphing calculator to verify your equation.

15. $m = 2.1, (3.7, -5.9)$ **16.** $m = -1.3, (-6.6, 3.8)$

17. $m = -6.59, (-2.48, -1.61)$ **18.** $m = -2.07, (-4.73, -9.60)$

Find an equation of the line that passes through the two given points. If possible, write your equation in slope–intercept form. Check that the graph of your equation contains the given points.

19. $(3, 2)$ and $(5, 6)$ **20.** $(1, 4)$ and $(2, 1)$

21. $(-1, -7)$ and $(2, 8)$ **22.** $(-2, -10)$ and $(3, 5)$

23. $(-5, -4)$ and $(-2, -10)$ **24.** $(-3, -2)$ and $(-1, -8)$

25. $(0, 9)$ and $(2, 1)$ **26.** $(-3, 1)$ and $(0, -5)$

27. $(3, 2)$ and $(5, 2)$ **28.** $(-5, -3)$ and $(-1, -3)$

29. $(-4, -1)$ and $(-4, 3)$ **30.** $(7, 1)$ and $(7, 6)$

31. $(4, 3)$ and $(8, 5)$ **32.** $(2, 3)$ and $(6, 1)$

33. $(-3, 2)$ and $(3, 1)$ **34.** $(2, -2)$ and $(6, -5)$

35. $(-2, 1)$ and $(5, -1)$ **36.** $(-6, 5)$ and $(-2, 2)$

37. $(-4, -2)$ and $(6, 4)$ **38.** $(-1, -2)$ and $(5, 6)$

39. $(-4, -8)$ and $(-2, -5)$ **40.** $(-6, -9)$ and $(-2, -4)$

Find an approximate equation of the line that passes through the two given points. Write your equation in slope–intercept form. Round the slope and the constant term to two decimal places. Use a graphing calculator to verify your equation.

41. $(-4.5, 2.2)$ and $(1.2, -7.5)$

42. $(-8.1, -5.3)$ and $(3.3, 2.7)$

43. $(-4.57, -8.29)$ and $(7.17, -2.69)$

44. $(-8.99, 4.82)$ and $(-5.85, -3.92)$

Find an equation of the line that contains the given point and is parallel to the given line. Use a graphing calculator to verify your result.

45. $(4, 5), y = 3x + 1$ **46.** $(1, 4), y = 4x - 6$

47. $(-3, 8), y = -2x + 7$ **48.** $(2, -3), y = -x + 2$

49. $(3, 4), 3x - 4y = 12$ **50.** $(4, -1), 5x + 2y = 10$

51. $(-3, -2), 6y - x = -7$

52. $(-1, -4), 3y + 5x = -11$

53. $(2, 3), y = 6$ **54.** $(3, -1), y = -4$

55. $(-5, 4), x = 2$ **56.** $(-2, -5), x = 1$

Find an equation of the line that contains the given point and is perpendicular to the given line. Use ZStandard followed by ZSquare with a graphing calculator to verify your result.

57. $(3, 8), y = 2x + 5$ **58.** $(2, 1), y = 5x - 4$

59. $(-1, 7), y = -3x + 7$ **60.** $(-3, -2), y = -6x - 13$

61. $(10, 3), 4x - 5y = 7$ **62.** $(6, -1), 5x + 2y = -9$

63. $(-3, -1), -2x + 3y = 5$ **64.** $(-1, 2), -3x - 4y = 12$

65. $(2, 3), x = 5$ **66.** $(-4, -2), x = -1$

67. $(2, 8), y = -3$ **68.** $(1, -1), y = 7$

69. Find an equation of the line sketched in Fig. 39. [**Hint:** Choose two points whose coordinates appear to be integers.]

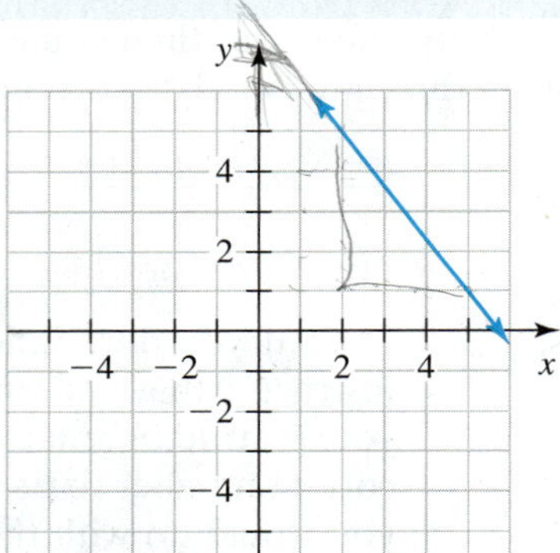

Figure 39 Exercise 69

70. Find an equation of the line sketched in Fig. 40. [**Hint:** Choose two points whose coordinates appear to be integers.]

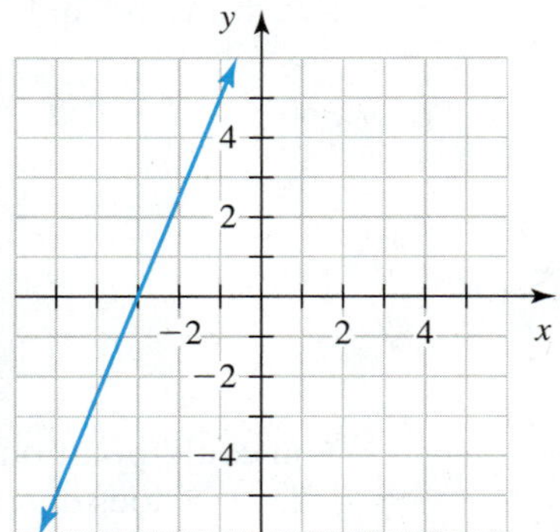

Figure 40 Exercise 70

Concepts

71. A student thinks if a line has slope 2 and contains the point $(3, 5)$, then the equation of the line is $y = 2x + 5$, because the slope is 2 (the coefficient of x) and the y-coordinate of $(3, 5)$ is 5 (the constant term). What would you tell the student?

72. A student tries to find an equation of the line that contains the points $(1,5)$ and $(3,9)$. The student believes an equation of the line is $y = 4x + 1$. The student then checks whether $(1,5)$ satisfies $y = 4x + 1$:

$$y = 4x + 1$$
$$5 \stackrel{?}{=} 4(1) + 1$$
$$5 \stackrel{?}{=} 5$$
$$\text{true}$$

The student concludes that $y = 4x + 1$ is an equation of the line. Find any errors. Then find an equation correctly.

73. Consider the line that contains the points $(-2,-6)$ and $(4,3)$.
a. Find an equation of the line.
b. Sketch the line on a coordinate system.
c. Use a table to list five ordered pairs that correspond to points on the line.

74. Consider the line that contains the points $(-3,4)$ and $(6,-2)$.
a. Find an equation of the line.
b. Sketch the line on a coordinate system.
c. Use a table to list five ordered pairs that correspond to points on the line.

75. Decide whether it is possible for a line to have the indicated number of x-intercepts. If it is possible, find an equation of such a line. If it is not possible, explain why not.
a. no x-intercepts
b. exactly one x-intercept
c. exactly two x-intercepts
d. an infinite number of x-intercepts

76. Decide whether it is possible for a line to have the indicated number of y-intercepts. If it is possible, find an equation of such a line. If it is not possible, explain why not.
a. no y-intercepts
b. exactly one y-intercept
c. exactly two y-intercepts
d. an infinite number of y-intercepts

77. Let E be the enrollment (in thousands of students) at a college t years after the college opens. Some pairs of values of t and E are listed in Table 13.

Table 13 Enrollments

Age of College (years) t	Enrollment (thousands of students) E
2	9
4	13
7	19
9	23
12	29

Find an equation that describes the relationship between t and E.

78. Let s be a person's savings (in thousands of dollars) at t years since 2000. Some pairs of values of t and s are listed in Table 14.

Table 14 A Person's Savings

Years t	Savings (thousands of dollars) s
1	5
4	14
5	17
7	23
10	32

Find an equation that describes the relationship between t and s.

79. a. Use each of the following forms to find an equation of the line that contains the points $(2, 1)$ and $(4, 7)$. Write each result in slope–intercept form.
i. slope–intercept form
ii. point–slope form
b. Compare your results from parts (ai) and (aii).

80. a. Use each of the following forms to find an equation of the line that contains the points $(-4, 3)$ and $(2, -5)$. Write each result in slope–intercept form.
i. slope–intercept form
ii. point–slope form
b. Compare your results from parts (ai) and (aii).

81. a. Find an equation of a line with slope -2. [**Hint:** There are *many* correct answers.]
b. Find an equation of a line with y-intercept $(0,4)$.
c. Find an equation of a line that contains the point $(3,5)$.
d. Determine whether there is a line that has slope -2 and y-intercept $(0,4)$ and contains the point $(3,5)$. Explain.

82. Suppose you are trying to find the equation of a line that contains two given points, and you find the slope is undefined. What type of line is it? Explain.

83. Create a table of seven pairs of values of x and y for which
a. each point lies on the line $y = 3x - 6$.
b. each point lies close to, but not on, the line $y = 3x - 6$.
c. the points do not lie close to the line $y = 3x - 6$, but all of them lie close to another line. In addition to creating the table, provide an equation of the other line.

84. Suppose a set of points all lie 0.5 unit above the line $y = -4x + 3$. Find an equation of the line that passes through the points of the set.

85. Find an equation of the line that contains the points $(3,7)$ and $(4, 9)$. After finding the slope, you used one of the points to complete the process. Now use the other point to complete the process. Did you get the same result? Why does this makes sense?

86. Describe how to find an equation of a line that contains two given points. Also, explain how you can check that the graph of your equation contains the two points.

Related Review

87. Four sets of points are described in Table 15. For each set, find an equation of a line that contains the points.

Table 15 Four Sets of Points

Set 1		Set 2		Set 3		Set 4	
x	y	x	y	x	y	x	y
0	25	0	12	0	77	0	3
1	23	1	16	1	72	1	3
2	21	2	20	2	67	2	3
3	19	3	24	3	62	3	3
4	17	4	28	4	57	4	3

88. Sketch a vertical line on a coordinate system. Find an equation of the line. What is the slope of the line?

89. Sketch a decreasing line that is nearly horizontal on a coordinate system. Find an equation of the line.

90. Find an equation of a line that has no solutions in Quadrant I.

91. a. Graph $y = 2x - 3$ by hand.

b. Choose two points that lie on the graph. Then use the two points to find an equation of the line that contains them. Compare your equation with the equation $y = 2x - 3$.

92. A line contains the points $(2, 4)$ and $(4, 1)$. Find three more points that lie on the line.

Expressions, Equations, Functions, and Graphs

Perform the indicated instruction. Then use words such as linear, function, one variable, *and* two variables *to describe the expression or equation. For instance, to describe* $2x + 7x$, *you could say "*$2x + 7x$ *is a linear expression in one variable."*

93. Graph $3x + 2y = 6$ by hand.

94. Simplify $\frac{5x}{8} + \frac{3}{4} - \frac{7x}{2}$.

95. Evaluate $3x + 2y$ for $x = 4$ and $y = -5$.

96. Solve $\frac{5x}{8} + \frac{3}{4} = \frac{7x}{2}$.

5.5 Finding Equations of Linear Models

Objectives

- » Find an equation of a linear model by using data described in words.
- » Find an equation of a linear model by using data displayed in a table.

In Section 5.4, we used two given points to find an equation of a line. In this section, we use this skill to find an equation of a linear model.

Finding a Model by Using Data Described in Words

In Example 1, we will use data described in words to find an equation of a linear model.

Example 1 Finding an Equation of a Linear Model

The number of Chrysler employees has decreased approximately linearly from 63 thousand employees in 2002 to 26 thousand employees in 2011 (Source: *Chrysler*). Let n be the number of Chrysler employees (in thousands) at t years since 2000.

1. Find an equation of a linear model.
2. What is the slope? What does it mean in this situation?
3. What is the n-intercept? What does it mean in this situation?

Solution

Table 16 Known Values of t and n

Years since 2000 t	Number of Employees (thousands) n
2	63
11	26

1. Known values of t and n are shown in Table 16.

A linear function can be put into the form $y = mx + b$, where y depends on x. Since the variables t and n are approximately linearly related and n depends on t, we will find an equation of the form $n = mt + b$. We can use the data points $(2, 63)$ and $(11, 26)$ to find the values of m and b.

First, we use $(2, 63)$ and $(11, 26)$ to find the slope:

$$m = \frac{26 - 63}{11 - 2} \approx -4.11$$

So, we can substitute -4.11 for m in the equation $n = mt + b$:

$$n = -4.11t + b$$

To find the constant b, we substitute the coordinates of the point $(2, 63)$ into the equation $n = -4.11t + b$ and then solve for b:

$$63 = -4.11(2) + b \quad \text{Substitute 2 for } t \text{ and 63 for } n.$$
$$63 = -8.22 + b \quad \text{Multiply.}$$
$$63 + 8.22 = -8.22 + b + 8.22 \quad \text{Add 8.22 to both sides.}$$
$$71.22 = b \quad \text{Combine like terms.}$$

Now we can substitute 71.22 for b in the equation $n = -4.11t + b$:

$$n = -4.11t + 71.22$$

We verify our equation by using a graphing calculator to check that our line comes very close to the points $(2, 63)$ and $(11, 26)$. See Fig. 41.

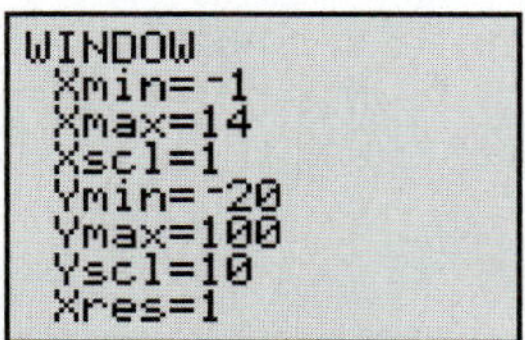

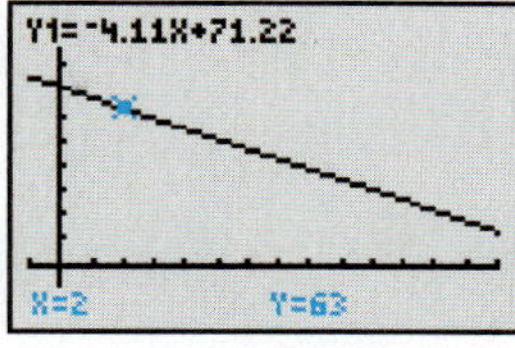

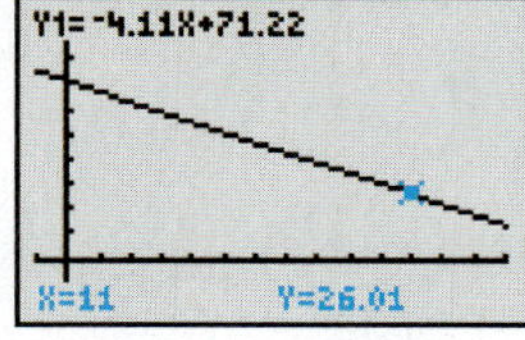

Figure 41 Checking that the model contains both $(2, 63)$ and $(11, 26)$

2. The slope is -4.11. According to the model, the number of Chrysler employees decreases by 4.11 thousand employees per year.
3. Since the model $n = -4.11t + 71.22$ is in slope–intercept form, the n-intercept is $(0, 71.22)$. So, the model estimates there were 71.22 thousand Chrysler employees in 2000.

Finding a Linear Model by Using Data Displayed in a Table

We begin by finding an equation of a model by using data shown in a table.

Example 2 Finding an Equation of a Linear Model

College student *discretionary spending* consists of expenditures other than necessary ones such as tuition, boarding, and textbooks. Total college student discretionary spending is shown in Table 17 for various years. Let S be total college student discretionary spending (in billions of dollars) in the year that is t years since 2000. Find an equation of a line that comes close to the points in the scattergram of the data.

Table 17 Total College Student Discretionary Spending

Year	Spending (billions of dollars)
2006	61
2007	63
2008	70
2009	74
2010	76

Source: *Alloy Media*

Solution

We begin by viewing the positions of the points in the scattergram (see Fig. 42).

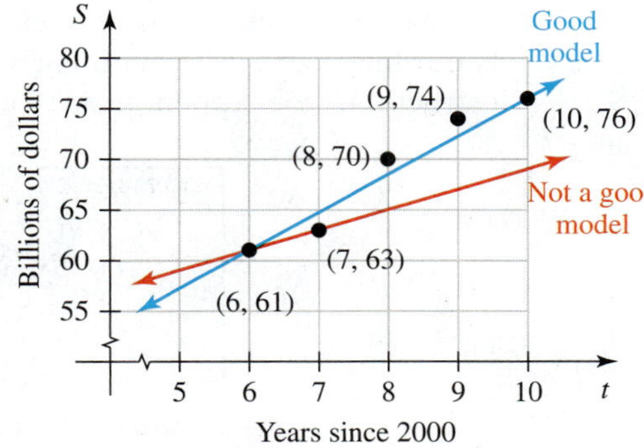

Figure 42 Discretionary spending scattergram

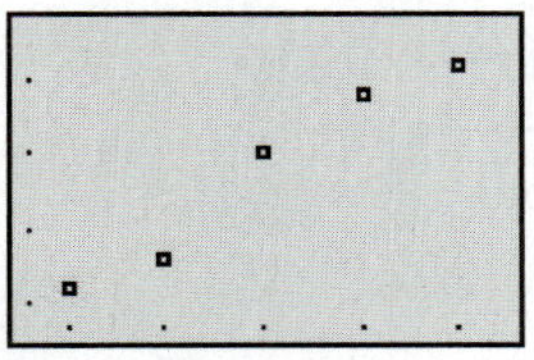

Figure 43 Graphing calculator scattergram

To save time and improve the accuracy of plotting points, we can use a graphing calculator to view the scattergram (see Fig. 43). For graphing calculator instructions on drawing scattergrams, see Appendix A.8, A.9, and A.12.

Our task is to find an equation of a line that comes close to the data points. It is not necessary to use two *data points* to find an equation, although it will often be convenient and satisfactory to do so.

The red line that contains the points $(6, 61)$ and $(7, 63)$ does *not* come close to the other data points (see Fig. 42). However, the blue line that passes through the points $(6, 61)$ and $(10, 76)$ appears to come close to the rest of the points. We will find the equation of this line.

To obtain an equation of the form $S = mt + b$, we use the points $(6, 61)$ and $(10, 76)$ to find the values of m and b. First, we calculate the slope of this line (see Fig. 44):

$$m = \frac{76 - 61}{10 - 6} = \frac{15}{4} = 3.75$$

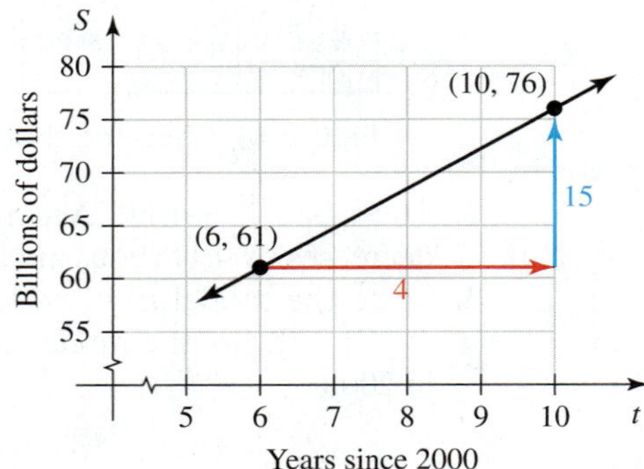

Figure 44 Points $(6, 61)$ and $(10, 76)$ of the discretionary spending scattergram

So, 3.75 can be substituted for m in the equation $S = mt + b$:

$$S = 3.75t + b$$

To find the constant b, we substitute the coordinates of the point $(6, 61)$ into the equation $S = 3.75t + b$ and then solve for b:

$$\begin{aligned} 61 &= 3.75(6) + b && \text{Substitute 6 for } t \text{ and 61 for } S. \\ 61 &= 22.5 + b && \text{Multiply.} \\ 61 - 22.5 &= 22.5 + b - 22.5 && \text{Subtract 22.5 from both sides.} \\ 38.5 &= b && \text{Combine like terms.} \end{aligned}$$

Now we substitute 38.5 for b in the equation $S = 3.75t + b$:

$$S = 3.75t + 38.5$$

We can check the correctness of our equation by using a graphing calculator to verify our line contains the points $(6, 61)$ and $(10, 76)$. See Fig. 45. For graphing calculator instructions, see Appendix A.10 and A.11.

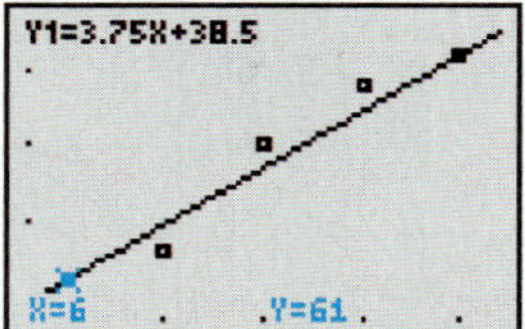

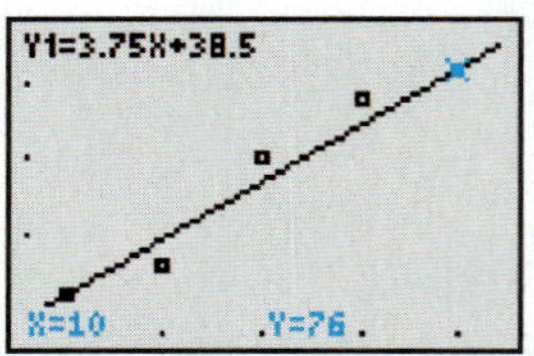

Figure 45 Checking that the model contains both $(6, 61)$ and $(10, 76)$

We also see our model fits the data well, which is our objective.

We benefit in many ways by viewing a scattergram of data. First, we can determine whether the data are approximately linearly related. Second, if the data are approximately linearly related, viewing a scattergram helps us choose two good points with which to find an equation of a linear model. Third, by graphing the model with the scattergram, we can assess whether the model fits the data reasonably well.

Example 2 outlines four steps to take to find an equation of a linear model.

Finding an Equation of a Linear Model

To find an equation of a linear model, given some data,

1. Create a scattergram of the data.
2. Determine whether there is a line that comes close to the data points. If so, choose two points (not necessarily data points) that you can use to find the equation of a linear model.
3. Find an equation of the line.
4. Use a graphing calculator to verify that the graph of your equation contains the two chosen points and comes close to all of the points of the scattergram.

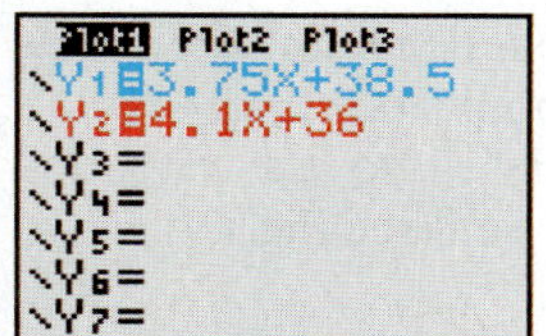

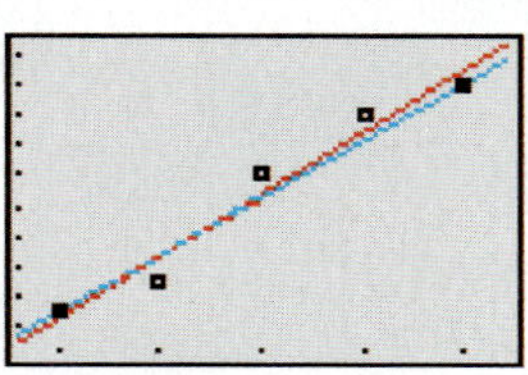

Figure 46 Comparing two discretionary spending models

What should you do if you discover that a model does not fit a data set well? A good first step is to check for any graphing or calculation errors. If your work appears to be correct, then one option is to try using different points to derive your equation. Another option is to increase or decrease the slope m and/or the constant term b until the fit is good. You can practice this "trial-and-error" process by completing the exploration in this section.

In Example 2, we used two points to find the model $S = 3.75t + 38.5$. There is another way to find a model. Most graphing calculators have a built-in **linear regression** feature for finding an equation of a linear model. Linear regression gives the equation $S = 4.1t + 36.0$. In Fig. 46, we see both models fit the data well. For graphing calculator instructions, see Appendix A.16.

A linear equation found by linear regression is called a **linear regression equation,** and the function described by the equation is called a **linear regression function.** The graph is called a **regression line.** (You can learn more about linear regression in a statistics course.)

Now you have two ways to find the equation of a linear model. No matter which method you use, the objective is the same: Find an equation of a line that comes close to the data points.

Table 18 Percentages of American Adults Who Smoke

Year	Percent Who Smoke
1970	37.4
1980	33.2
1990	25.3
2000	23.1
2010	19.4

Source: *National Center for Health Statistics*

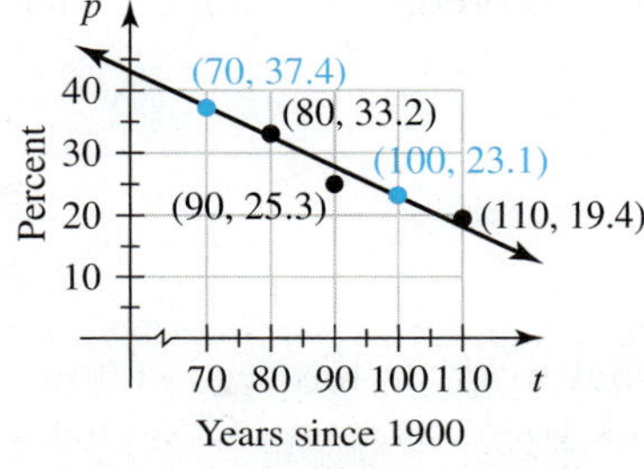

Figure 47 Smoking scattergram and linear model

Example 3 Finding the Equation of a Linear Model

Cigarette smoking has been on the decline for the past several decades (see Table 18). Let p be the percentage of American adults who smoke at t years since 1900.

1. Use two well-chosen points to find an equation of a model that describes the relationship between t and p.
2. Find the linear regression equation and line by using a graphing calculator. Compare this model with the one you found in Problem 1.

Solution

1. We see from the scattergram in Fig. 47 that a line containing the points (70, 37.4) and (100, 23.1) comes close to the rest of the data points. The zigzag lines on the t-axis near the origin indicate part of the scale is missing. This is done to show a clearer view of the data points.

   For an equation of the form $p = mt + b$, we first use the points (70, 37.4) and (100, 23.1) to find m:

$$m = \frac{23.1 - 37.4}{100 - 70} = -0.48$$

So, the equation has the form

$$p = -0.48t + b$$

To find b, we use the point (70, 37.4) and substitute 70 for t and 37.4 for p in the equation $p = -0.48t + b$:

$$37.4 = -0.48(70) + b \quad \text{Substitute 70 for } t \text{ and 37.4 for } p.$$
$$37.4 = -33.6 + b \quad \text{Multiply.}$$
$$37.4 + 33.6 = -33.6 + b + 33.6 \quad \text{Add 33.6 to both sides.}$$
$$71 = b \quad \text{Simplify.}$$

So, the equation is $p = -0.48t + 71$.

We can use a graphing calculator to verify the linear model contains the points (70, 37.4) and (100, 23.1) and comes close to the other data points (see Fig. 48).

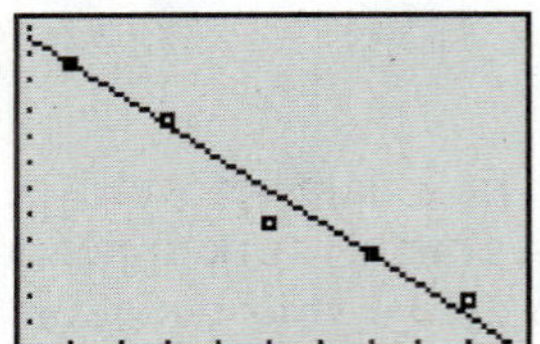

Figure 48 Verifying the smoking model

2. The regression equation is $p = -0.46t + 69.17$. It is "close" to the equation $p = -0.48t + 71$. Further, both models appear to fit the data well (see Fig. 49).

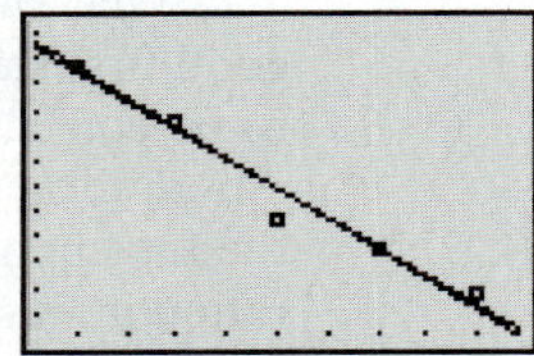

Figure 49 Comparing two smoking models' graphs that are so close together that there appears to be only one line

Group Exploration

Adjusting the fit of a model

The winning times for the men's Olympic 100-meter freestyle swimming event are shown in Table 19 for various years.

Table 19 Winning Times for the Men's Olympic 100-Meter Freestyle

Year	Swimmer	Country	Winning Time (seconds)
1980	Jörg Woithe	E. Germany	50.40
1984	Rowdy Gaines	USA	49.80
1988	Matt Biondi	USA	48.63
1992	Aleksandr Popov	Unified Team	49.02
1996	Aleksandr Popov	Russia	48.74
2000	Pieter van den Hoogenband	Netherlands	48.30
2004	Pieter van den Hoogenband	Netherlands	48.17
2008	Alain Bernard	France	47.21
2012	Nathan Adrian	USA	47.52

Source: *The New York Times Almanac*

Let w be the winning time (in seconds) at t years since 1980.

1. Use a graphing calculator to draw a scattergram of the data. Do the variables t and w appear to be approximately linearly related?
2. The linear model $w = -0.0325t + 48.95$ can be found by using the data points (20, 48.30) and (24, 48.17). Draw the line and the scattergram in the same viewing window. Check that the line contains these two points.
3. The model $w = -0.0325t + 48.95$ does not fit the seven data points very well. Adjust the equation by increasing or decreasing the slope -0.0325 and/or the constant term 48.95 so that your new model will fit the data better. Keep adjusting the model until it fits the data points reasonably well.
4. Use your improved model to predict the winning time in the 2016 Olympics.

Tips for Success Verify Your Work

Remember to use your graphing calculator to verify your work. In this section, for example, you can use your graphing calculator to check your equations. Checking your work increases your chances of catching errors and thus will likely improve your performance on homework assignments, quizzes, and tests.

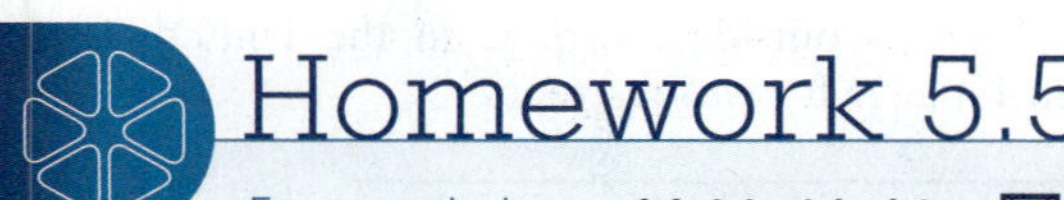

Homework 5.5

For extra help ▶ MyMathLab®  Watch the videos in MyMathLab  Download the MyDashboard App

1. The number of African-American federal and state legislators has increased approximately linearly from 1136 positions in 2001 to 1256 positions in 2009 (Source: *National Black Caucus of State Legislatures*). Let n be the number of African-American legislators at t years since 2000. Find an equation of a linear model to describe the data.

2. The percentage of "media time" Americans spent shopping online increased approximately linearly from 11% in 2006 to 19% in 2011 (Source: *GroupM*). Let p be the percentage of "media time" Americans spend shopping online at t years since 2000. Find an equation of a linear model to describe the data.

3. The number of inmates younger than 18 held in state prisons has decreased approximately linearly from 3147 inmates in 2001 to 2295 inmates in 2011 (Source: *U.S. Department of Justice*). Let n be the number of inmates younger than 18 held in state prisons at t years since 2000. Find an equation of a linear model to describe the data.

4. The average time per day an American spends reading a daily newspaper has decreased approximately linearly from 38 minutes per day in 2008 to 26 minutes per day in 2011 (Source: *Census Bureau*). Let n be the average number of minutes an American spends reading a daily newspaper per day in the year that is t years since 2000. Find an equation of a linear model to describe the data.

5. The percentage of sexual harassment charges filed by men has increased approximately linearly from 9.1% in 1992 to 16.4% in 2010 (Source: *Equal Employment Opportunity Commission*). Let p be the percentage of sexual harassment charges filed by men at t years since 1990.

a. Find an equation of a linear model to describe the data.
b. What is the slope? What does it mean in this situation?
c. What is the p-intercept? What does it mean in this situation?

6. The number of city and county ordinances that restrict outdoor smoking has increased approximately linearly from 30 ordinances in 1999 to 1532 ordinances in 2012 (Source: *American Nonsmokers' Rights Foundation*). Let n be the number of ordinances at t years since 1990.

a. Find an equation of a linear model to describe the data.
b. What is the slope? What does it mean in this situation?
c. What is the n-intercept? What does it mean in this situation?

7. The number of suicides among soldiers has increased approximately linearly from 67 suicides in 2004 to 155 suicides in 2010 (Source: *U.S. Army*). Let n be the number of suicides among soldiers in the year that is t years since 2000.

a. Find an equation of a linear model to describe the data.
b. What is the slope? What does it mean in this situation?
c. What is the n-intercept? What does it mean in this situation?

8. The market share of American automakers has decreased approximately linearly from 66% in 2000 to 47% in 2011 (Source: *Ward's AutoInfoBank*). Let p be American automakers' market share at t years since 2000.

a. Find an equation of a linear model to describe the data.
b. What is the slope? What does it mean in this situation?
c. What is the p-intercept? What does it mean in this situation?

9. The average length of a 6-year-old striped bass is 27 inches, and that of a 20-year-old striped bass is 54.5 inches (Source: *daybreakfishing.com*). A striped bass's age and average length are approximately linearly related. Let L be the average length (in inches) of a striped bass at age a years. Find an equation of a linear model to describe the data.

10. For Pacific albacore tuna, the weight and mercury concentration are approximately linearly related. A 4-kilogram tuna has an average mercury concentration of 0.10 part per million. A 10-kilogram tuna has an average mercury concentration of 0.19 part per million. Let c be the average mercury concentration (in parts per million) of a Pacific albacore tuna that weighs w kilograms. Find an equation of a linear model to describe the data.

11. Find an equation of a line that comes close to the points listed in Table 20. Then use a graphing calculator to check that your line comes close to the points.

Table 20 Find an Equation

x	y
4	8
5	10
6	13
7	15
8	18

12. Find an equation of a line that comes close to the points listed in Table 21. Then use a graphing calculator to check that your line comes close to the points.

Table 21 Find an Equation

x	y
1	6
5	13
7	14
10	17
12	22

13. Find an equation of a line that comes close to the points listed in Table 22. Then use a graphing calculator to check that your line comes close to the points.

Table 22 Find an Equation

x	y
3	18
7	14
9	9
12	7
16	4

14. Find an equation of a line that comes close to the points listed in Table 23. Then use a graphing calculator to check that your line comes close to the points.

Table 23 Find an Equation

x	y
2	14
5	10
6	8
8	4
11	1

15. Table 24 shows how old a 20- to 50-pound dog is in relation to human years.

Table 24 Dog Years Compared with Human Years

Dog Years	Human Years
1	15
2	24
3	29
5	38
7	47
9	56
11	65
13	74
15	83

Source: *Fred Metzger, State College, PA*

Let H be the number of human years that is equivalent to d dog years.

a. Use a graphing calculator to draw a scattergram of the data.
b. Find an equation of a linear model to describe the data.
c. Draw your line and the scattergram in the same viewing window. Verify that the line passes through the two points that you chose in finding the equation in part (b) and that it comes close to all of the data points.

16. The enrollments in an elementary algebra course at the College of San Mateo are shown in Table 25 for various Tuesdays leading up to the first day of class on Tuesday, January 17.

Table 25 Enrollments in an Elementary Algebra Course

Date	Number of Weeks since November 22	Enrollment
November 22	0	8
November 29	1	9
December 6	2	12
December 13	3	13
December 20	4	16
December 27	5	18
January 3	6	19

Source: *J. Lehmann*

Let E be the enrollment in the elementary algebra course at t weeks since November 22.

a. Use a graphing calculator to draw a scattergram of the data.
b. Find an equation of a linear model to describe the data.
c. Draw your line and the scattergram in the same viewing window. Verify that the line passes through the two points that you chose in finding the equation in part (b) and that it comes close to all of the data points.

17. The percentages of births outside marriage in the United States are shown in Table 26 for various years.

Table 26 Births Outside Marriage

Year	Percent of Births outside Marriage
1970	10.7
1975	14.3
1980	18.4
1985	22.0
1990	28.0
1995	32.2
2000	33.2
2005	36.8
2010	40.8

Source: *National Center for Health Statistics*

Let p be the percentage of births outside marriage in the United States at t years since 1900.

a. Use a graphing calculator to draw a scattergram of the data.
b. Find an equation of a linear model to describe the data.
c. Draw your line and the scattergram in the same viewing window. Verify that the line passes through your two chosen points and that it comes close to all of the data points.

18. Repeat Exercise 17, but let p be the percentage of births outside marriage in the United States at t years *since 1970*. Compare the slope of your model with the slope of the model you found in Exercise 17. Compare the p-intercepts. Explain why your comparisons make sense.

19. The prices of ski rental packages from Gold Medal Sports® are shown in Table 27 for various numbers of days.

Table 27 Prices of Ski Rental Packages

Number of Days	Price of Package (dollars)
1	15.00
2	30.00
3	45.00
4	56.00
5	70.00
6	78.00

Source: *Gold Medal Sports*

Let p be the price (in dollars) of a ski rental package for n days.

a. Use a graphing calculator to draw a scattergram of the data.
b. Find an equation of a linear model to describe the data.
c. Draw your line and the scattergram in the same viewing window. Verify that the line passes through the two points you chose in finding the equation in part (b) and that it comes close to all of the data points.

20. The life expectancies at birth of Americans are shown in Table 28 for various years.

a. Let L be the life expectancy at birth (in years) of an American born t years after 1980. Use a graphing calculator to draw a scattergram of the data.

Table 28 Life Expectancies at Birth

Year of Birth	Life Expectancy (years)
1980	73.7
1985	74.7
1990	75.4
1995	75.8
2000	77.0
2005	77.9
2010	78.7

Source: *U.S. Census Bureau*

b. Find an equation of a linear model to describe the data.
c. Draw your line and the scattergram in the same viewing window. Verify that the line passes through the two points you chose in finding the equation in part (b) and that it comes close to all of the data points.

21. Table 29 shows, for various years, the percentage of Americans who are baseball fans.

Table 29 Percentages of Americans Who Are Baseball Fans

Year	Percent
1999	54
2001	51
2003	50
2005	48
2007	44
2008	43

Source: *The Gallup Organization*

Let p be the percentage of Americans who are baseball fans at t years since 1990.
a. Use a graphing calculator to draw a scattergram of the data.
b. Find an equation of a linear model to describe the data.
c. Draw your line and the scattergram in the same viewing window. Verify that the line passes through the two points you chose in finding the equation in part (b) and that it comes close to all of the data points.

22. The percentages of Americans who have been diagnosed with diabetes are shown in Table 30 for various age groups.

Table 30 Percentages of Americans Diagnosed with Diabetes, by Age Group

Age Group (years)	Age Used to Represent Age Group (years)	Percent
35–39	37	2
40–44	42	4
45–49	47	5
50–54	52	8
55–59	57	10
60–64	62	13
65–69	67	14

Source: *National Health Interview Survey*

Let p be the percentage of Americans at age a years who have been diagnosed with diabetes at some point in their lives.

a. Use a graphing calculator to draw a scattergram of the data.
b. Find an equation of a linear model to describe the data.
c. Draw your line and the scattergram in the same viewing window. Verify that the line passes through the two points you chose in finding the equation in part (b) and that it comes close to all of the data points.
d. Find the regression equation to describe the data.
e. Use a graphing calculator to graph the equations you found in parts (b) and (d) in the same viewing window. Compare the graphs.

23. Table 31 lists world record times for the women's 400-meter run. Let r be the record time (in seconds) at t years since 1900.

Table 31 Women's 400-Meter Run Record Times

Year	Runner	Country	Record Time (seconds)
1957	Marlene Mathews	Australia	57.0
1959	Maria Itkina	USSR	53.4
1962	Shin Geum Dan	North Korea	51.9
1969	Nicole Duclos	France	51.7
1972	Monika Zehrt	E. Germany	51.0
1976	Irena Szewinska	Poland	49.29
1979	Marita Koch	E. Germany	48.60
1983	Jarmila Kratochvílová	Czechoslovakia	47.99
1985	Marita Koch	E. Germany	47.60

Source: *International Association of Athletics Federations*

a. Use a graphing calculator to draw a scattergram of the data.
b. Find an equation of a linear model to describe the data.
c. Draw your line and the scattergram in the same viewing window. Verify that the line passes through the two points you chose in finding the equation in part (b) and that it comes close to all of the data points.

24. Table 32 lists world record times for the men's 400-meter run. Let r be the record time (in seconds) at t years since 1900.

Table 32 Men's 400-Meter Run Record Times

Year	Runner	Country	Record Time (seconds)
1900	Maxie Long	USA	47.8
1916	Ted Meredith	USA	47.4
1928	Emerson Spencer	USA	47.0
1932	Bill Carr	USA	46.2
1941	Graver Klemmer	USA	46.0
1950	George Rhoden	Jamaica	45.8
1960	Carl Kaufmann	Germany	44.9
1968	Lee Evans	USA	43.86
1988	Harry Reynolds	USA	43.29
1999	Michael Johnson	USA	43.18

Source: *International Association of Athletics Federations*

a. Use a graphing calculator to draw a scattergram of the data.
b. Find an equation of a linear model to describe the data.
c. Draw your line and the scattergram in the same viewing window. Verify that the line passes through the two points you chose in finding the equation in part (b) and that it comes close to all of the data points.

25. In Exercises 23 and 24, you found equations for the women's and men's 400-meter run record times. Equations that model the data well are

$$r = -0.27t + 70.45 \quad \text{(women's model)}$$
$$r = -0.053t + 48.08 \quad \text{(men's model)}$$

where r represents the record time (in seconds) at t years since 1900.
a. Graph both models by hand for the years from 1900 to 2050. (If you are able to use a graphing calculator to do this exercise, you may do so.)
b. Do the models predict that the women's record time will ever equal the men's record time? If so, what is that record time, and when will the record be set?
c. Do the models predict that the women's record time will ever be less than the men's record time? If so, in what years?

26. A runner's *stride rate* is the number of steps per second. The average stride rates of the top female and male runners are shown in Table 33 for various speeds.

Table 33 Top Female and Male Runners' Speeds and Average Stride Rates

Speed (feet per second)	Average Stride Rate (number of steps per second)	
	Women	Men
15.86	3.05	2.92
16.88	3.12	2.98
17.50	3.17	3.03
18.62	3.25	3.11
19.97	3.36	3.22
21.06	3.46	3.31
22.11	3.55	3.41

Source: *Biomechanical comparison of male and female runners, R. C. Nelson et al., 1977*

a. Let r be the average stride rate of a woman running at s feet per second. Find the regression equation to describe the data.
b. Let r be the average stride rate of a man running at s feet per second. Find the regression equation to describe the data.
c. Which of your two models has the r-intercept with the larger r-coordinate? What does this tell you about the graphs of the models?
d. Which model has the larger slope? What does this tell you about the graphs of the models?
e. Explain why your work in parts (c) and (d) suggests the graphs of your two models do not intersect in quadrant I (where s and r are positive). What does that mean in this situation?

27. To enroll in Intermediate Algebra, a student at the College of San Mateo (CSM) must score at least 21 points (out of 50) on a placement test. Using four semesters of data, the CSM Mathematics Department computed the percentages of students who succeeded in Intermediate Algebra (grade of A, B, or C) for various groups of scores on the placement test (see Table 34).

Table 34 Percentages of Intermediate Algebra Students Who Succeeded

Placement Score Group	Score Used to Represent Score Group	Percentage Who Succeeded in Intermediate Algebra
21–25	23	34
26–30	28	47
31–35	33	55
36–40	38	71
41–45	43	84
46–50	48	*

Source: *College of San Mateo Mathematics Department*
*There were not enough students in this group to give useful data.

a. Let p be the percentage of Intermediate Algebra students succeeding in the course who scored x points on the placement test. Use a graphing calculator to draw a scattergram of the data.
b. Find an equation of a line that you think comes close to the points in the scattergram.
c. Draw your line and the scattergram in the same viewing window. Verify that the line contains the two points you chose in finding the equation in part (b) and that it comes close to all of the data points.

28. The "crime index" refers to the number of incidents of crime. The numbers of burglaries, aggravated assaults, and all types of crime per 100,000 Americans are shown in Table 35 for various years.

Table 35 Crime Indexes

Year	Crime Index (number of incidents per 100,000 people)		
	Burglary	Aggravated Assault	All Types of Crime
1993	1099	440	5484
1995	987	418	5276
1997	919	382	4930
1999	770	334	4267
2001	742	319	4163
2003	741	295	4067
2005	727	291	3899
2007	726	287	3749
2009	716	263	3466

Source: *FBI*

a. Let A be the crime index of aggravated assaults for the year that is t years since 1990. Can the data be modeled well by a linear model? If yes, find such a model. If no, explain why not.
b. Let B be the crime index of burglaries for the year that is t years since 1990. Can the data be modeled well by a linear model? If yes, find such a model. If no, explain why not.
c. Let C be the crime index of all types of crime for the year that is t years since 1990. Can the data be modeled well by a linear model? If yes, find such a model. If no, explain why not.
d. Economist Steven D. Levitt believes crime and the economy are not related. The U.S. economy performed well from 1993 to 2000 and from 2003 to 2008, and did poorly from 2000 to 2003 and from 2008 to 2012.
   i. Explain why the aggravated-assault data support Levitt's theory.

ii. Explain why the burglary data are less supportive (than the aggravated-assault data) of Levitt's theory.

iii. Explain why the data for all types of crime support Levitt's theory.

Concepts

29. Three students are to find a linear model of the data in Table 36. Student A uses points (1, 5.9) and (2, 6.4), student B uses points (3, 9.0) and (4, 11.0), and student C uses points (5, 12.1) and (6, 15.5). Which student seems to have made the best choice of points? Explain.

Table 36 Three Students Model Data

x	y
1	5.9
2	6.4
3	9.0
4	11.0
5	12.1
6	15.5
7	16.5

30. Three students are to find a linear model of the data in Table 37. Student A uses points (3, 13.8) and (4, 10.1), student B uses points (6, 7.8) and (7, 4.3), and student C uses points (5, 9.1) and (8, 3.1). Which student seems to have made the best choice of points? Explain.

Table 37 Three Students Model Data

x	y
3	13.8
4	10.1
5	9.1
6	7.8
7	4.3
8	3.1
9	1.1

For Exercises 31 and 32, consider the scattergram of data and the graph of the model $y = mx + b$ in the indicated figure. Sketch the graph of a linear model that describes the data better. Then explain how you would adjust the values of m and b of the original model so it would describe the data better.

31. See Fig. 50.

32. See Fig. 51.

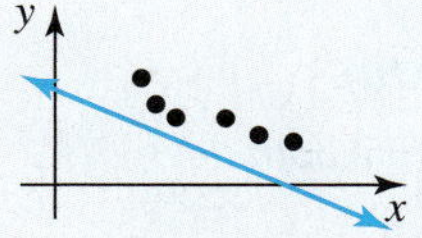

Figure 50 Exercise 31

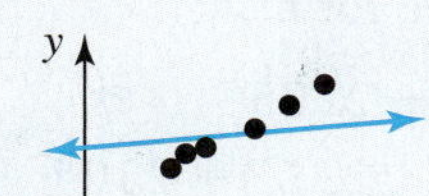

Figure 51 Exercise 32

33. Explain how to find an equation of a linear model for a given situation. Also, explain how to verify that the linear function models the situation reasonably well.

34. A student comes up with a shortcut for modeling a situation described by a table that contains several rows of data. Instead of creating a scattergram of the data, the student chooses two data points at random and uses them to find an equation of a line. Give at least two examples to illustrate what can go wrong with this shortcut.

Related Review

35. In 2005, a stock is worth \$10. Each year thereafter, the value of the stock increases by \$2. Let V be the value (in dollars) of the stock at t years since 2005. Recall that we can describe a linear model using an equation, a table, a graph, or words.

a. Find an equation of a linear model to describe the situation. Then perform a unit analysis of the equation.

b. Use a graph to describe the linear model.

c. Use a table of values of t and V to describe the linear model.

36. A person is on a car trip. At the start of the trip, there are 12 gallons of gasoline in the gasoline tank. The car consumes 3 gallons of gasoline per hour of driving. Let V be the volume of gasoline (in gallons) in the gasoline tank after t hours of driving. Recall that we can describe a linear model using an equation, a table, a graph, or words.

a. Find an equation of a linear model to describe the situation. Then perform a unit analysis of the equation.

b. Use a graph to describe the linear model.

c. Use a table of values of t and V to describe the linear model.

37. The average attendance at college bowl games was 50,435 in 2012 and is decreasing by about 731 spectators per year (Source: *NCAA*). Let n be the average attendance in the year that is t years since 2012.

a. Find an equation of a linear model to describe the situation.

b. What is the slope? What does it mean in this situation?

c. Perform a unit analysis of the equation you found in part (a).

38. In 2010, 10.1% of the U.S. workforce had manufacturing jobs. The percentage of the workforce having manufacturing jobs is decreasing by about 0.43 percentage point per year (Source: *U.S. Bureau of Labor Statistics*). Let p be the percentage of the workforce who has manufacturing jobs at t years since 2010.

a. Find an equation of a linear model to describe the data.

b. What is the slope? What does it mean in this situation?

c. Perform a unit analysis of the equation that you found in part (a).

39. To rent a 20-foot truck from Metro Truck Rental for one day, it costs \$49.95 plus \$0.69 per mile (Source: *Metro Truck Rental*). If the total cost to rent such a truck for one day was \$109.29, how far was the truck driven?

40. To rent a 20-foot truck from U-Haul for one day, it costs \$39.95 plus \$1.19 per mile (Source: *U-Haul*). If the total cost to rent such a truck for one day was \$129.20, how far was the truck driven?

Expressions, Equations, Functions, and Graphs

Perform the indicated instruction. Then use words such as linear, function, one variable, *and* two variables *to describe the expression or equation. For instance, to describe $y = 4x$, you could say "$y = 4x$ is a linear equation in two variables."*

41. Solve $3 = -\frac{2}{5}x + 4$.

42. Simplify $4x - 3(2x - 5) + 1$.

43. Graph $y = -\frac{2}{5}x + 4$ by hand.

44. Solve $4x - 3(2x - 5) + 1 = 0$.

5.6 Using Function Notation with Linear Models to Make Estimates and Predictions

Objectives

- Use function notation to make estimates and predictions.
- Find intercepts of a model.
- Use data described in words to make estimates and predictions.
- Know the meaning of *domain* and *range* of a model.

In this section, we will use function notation together with models to make estimates and predictions. We will also use an equation of a model to find the intercepts of the model. Finally, we will find the *domain* and the *range* of a model.

Use Function Notation to Make Estimates and Predictions

Recall from Section 5.3 that, for a function f, we can write

$$\text{dependent variable} = f(\text{independent variable})$$

In Example 1, we will use function notation in making predictions.

Example 1 Using Function Notation to Make Estimates and Predictions

Table 38 shows the average salaries of faculty members at public colleges and universities. Let s be the faculty members' average salary (in thousands of dollars) at t years since 1980. A possible model is

$$s = 1.82t + 22.25$$

Table 38 Average Salaries of Faculty Members at Public Colleges and Universities

Year	Average Salary (thousands of dollars)
1980	22.1
1985	31.2
1990	41.9
1995	49.1
2000	57.7
2005	66.9
2010	78.0

Source: *American Association of University Professors*

1. Verify that the function $s = 1.82t + 22.25$ models the data well.
2. Rewrite the equation $s = 1.82t + 22.25$ with the function name f.
3. Predict the average salary in 2018.
4. Predict when the average salary will be $95,000.

Solution

1. We draw the graph of the model and the scattergram of the data in the same viewing window (see Fig. 52). The function appears to model the data quite well.

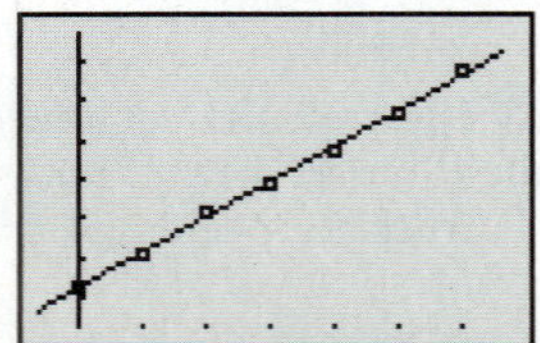

Figure 52 Check how well $s = 1.82t + 22.25$ models the data

2. Here, t is the independent variable and s is the dependent variable. Since the function name is f, we can write $s = f(t)$. Then we substitute $f(t)$ for s in the equation $s = 1.82t + 22.25$:

$$f(t) = 1.82t + 22.25$$

3. We represent the year 2018 by $t = 38$. To find the average salary, we substitute 38 for t in the equation $f(t) = 1.82t + 22.25$:

$$\begin{aligned} f(38) &= 1.82(38) + 22.25 && \text{Substitute 38 for } t. \\ &= 91.41 && \text{Simplify.} \end{aligned}$$

The model predicts the average salary will be about $91.4 thousand in 2018.

4. We can represent the salary $95,000 by $s = 95$. Since $s = f(t)$, we can write $f(t) = 95$. To find the year, we substitute 95 for $f(t)$ in the equation $f(t) = 1.82t + 22.25$ and solve for t:

$$\begin{aligned} 95 &= 1.82t + 22.25 && \text{Substitute 95 for } f(t). \\ 95 - 22.25 &= 1.82t + 22.25 - 22.25 && \text{Subtract 22.25 from both sides.} \\ 72.75 &= 1.82t && \text{Combine like terms.} \\ \frac{72.75}{1.82} &= \frac{1.82t}{1.82} && \text{Divide both sides by 1.82.} \\ 39.97 &\approx t && \text{Simplify.} \end{aligned}$$

According to the model, the average salary will be $95,000 in $1980 + 40 = 2020$. We can verify our work in both Problems 3 and 4 by using a graphing calculator table (see Fig. 53).

Or we can graphically verify our work by using TRACE (see Fig. 54).

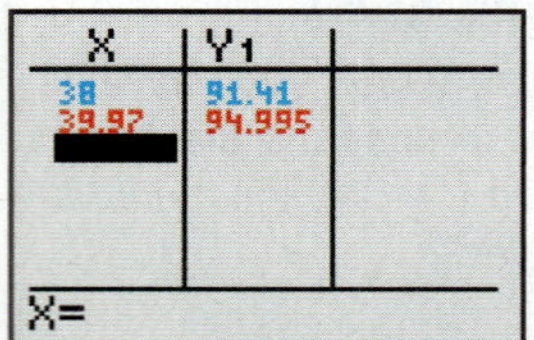

Figure 53 Verify the predictions

Here we summarize how to use an equation of a model to make predictions (or estimates).

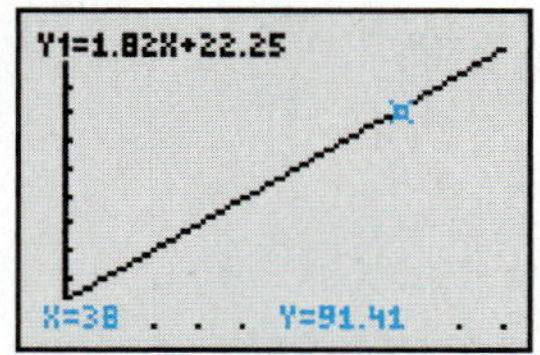

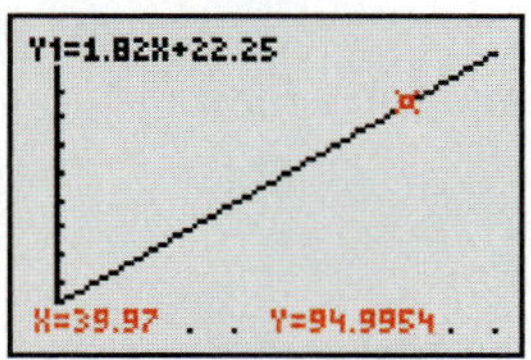

Figure 54 Verify predictions by using TRACE

Using an Equation of a Linear Model to Make Predictions

- When making a prediction about the dependent variable of a linear model, substitute a chosen value for the independent variable in the model. Then solve for the dependent variable.
- When making a prediction about the independent variable of a linear model, substitute a chosen value for the dependent variable in the model. Then solve for the independent variable.

In Section 5.5 and this section, we discussed how to find linear models and how to use these models to make estimates and predictions. Here is a summary of this process.

Four-Step Modeling Process

To find a linear model and make estimates and predictions,

1. Create a scattergram of the data to determine whether there is a nonvertical line that comes close to the data points. If so, choose two points (not necessarily data points) that you can use to find the equation of a linear model.
2. Find an equation of your model.
3. Verify your equation by checking that the graph of your model contains the two chosen points and comes close to all of the data points.
4. Use the equation of your model to make estimates, make predictions, and draw conclusions.

In Example 1, we used f to name the function $f(t) = 1.82t + 22.25$, where $f(t)$ represents the average salary (in thousands of dollars) of faculty members at *public* colleges and universities at t years since 1980. When we use more than one function to model situations, naming the functions helps us distinguish among them. For example, we can also use a linear function to model the average salaries of faculty members at *private* colleges and universities. A good model is $s = 2.56t + 15.76$, where s is the faculty members' average salary (in thousands of dollars) at t years since 1980. We can distinguish this function from f by using g as its name:

$$g(t) = 2.56t + 15.76$$

Finding Intercepts of a Model

In Example 2, we will use a model to make a prediction and an estimate, find the intercepts of the model, and interpret the meaning of the intercepts.

Example 2 Using Function Notation; Finding Intercepts

In Example 3 of Section 5.5, we found the equation $p = -0.48t + 71$, where p is the percentage of American adults who smoke at t years since 1900 (see Table 39).

Table 39 Percentages of American Adults Who Smoke

Year	Percent
1970	37.4
1980	33.2
1990	25.3
2000	23.1
2010	19.4

Source: *National Center for Health Statistics*

1. Rewrite the equation $p = -0.48t + 71$ with the function name g.
2. Find $g(117)$. What does the result mean in this situation?
3. Find the value of t when $g(t) = 30$. What does it mean in this situation?
4. Find the p-intercept of the model. What does it mean in this situation?
5. Find the t-intercept of the model. What does it mean in this situation?

Solution

1. To use the name g, we substitute $g(t)$ for p in the equation $p = -0.48t + 71$:

$$g(t) = -0.48t + 71$$

2. To find $g(117)$, we substitute 117 for t in the equation $g(t) = -0.48t + 71$:

$$\begin{aligned} g(t) &= -0.48t + 71 && \text{Equation of } g \\ g(117) &= -0.48(117) + 71 && \text{Substitute 117 for } t. \\ &= 14.84 && \text{Simplify.} \end{aligned}$$

So, $p = 14.84$ when $t = 117$. According to the model, about 14.8% of American adults will smoke in 2017.

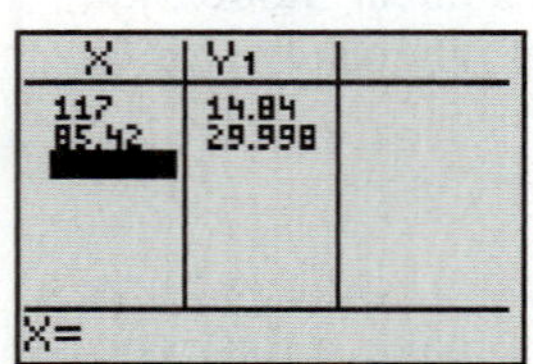

Figure 55 Verify the work

3. We substitute 30 for $g(t)$ in the equation $g(t) = -0.48t + 71$ and solve for t:

$$\begin{aligned} g(t) &= -0.48t + 71 && \text{Equation of } g \\ 30 &= -0.48t + 71 && \text{Substitute 30 for } g(t). \\ 30 - 71 &= -0.48t + 71 - 71 && \text{Subtract 71 from both sides.} \\ -41 &= -0.48t && \text{Combine like terms.} \\ \frac{-41}{-0.48} &= \frac{-0.48t}{-0.48} && \text{Divide both sides by } -0.48. \\ 85.42 &\approx t && \text{Simplify.} \end{aligned}$$

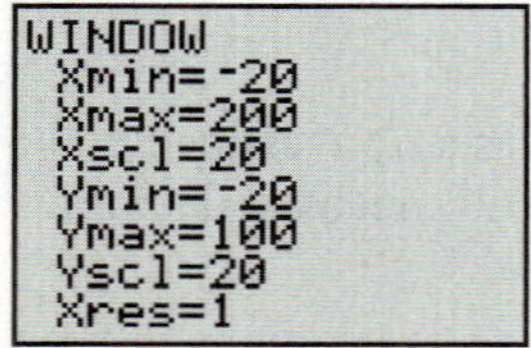

The model estimates 30% of Americans smoked in $1900 + 85.42 \approx 1985$. We can verify our work in Problems 2 and 3 by using a graphing calculator table (see Fig. 55).

4. Since the model $g(t) = -0.48t + 71$ is in slope–intercept form, the p-intercept is $(0, 71)$. So, the model estimates 71% of American adults smoked in 1900. Research would show this estimate is too high; model breakdown has occurred.

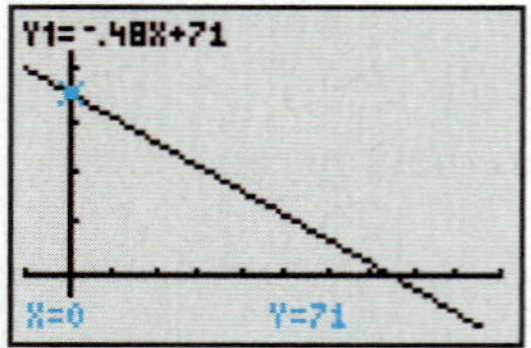

5. To find the t-intercept, we substitute 0 for $g(t)$ and solve for t:

$$\begin{aligned} 0 &= -0.48t + 71 && \text{Substitute 0 for } g(t). \\ 0 + 0.48t &= -0.48t + 71 + 0.48t && \text{Add } 0.48t \text{ to both sides.} \\ 0.48t &= 71 && \text{Combine like terms.} \\ \frac{0.48t}{0.48} &= \frac{71}{0.48} && \text{Divide both sides by 0.48.} \\ t &\approx 147.92 && \text{Simplify.} \end{aligned}$$

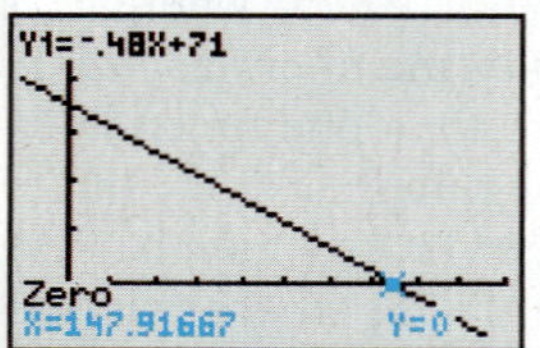

Figure 56 Verify the intercepts

The t-intercept is $(147.92, 0)$. So, the model predicts no American adults will smoke in $1900 + 147.92 \approx 2048$. However, common sense suggests this event probably won't occur.

We can use TRACE to verify the p-intercept and the "zero" option to verify the t-intercept (see Fig. 56).

Using an Equation of a Linear Model to Find Intercepts

If a function of the form $p = mt + b$, where $m \neq 0$, is used to model a situation, then

- The p-intercept is $(0, b)$.
- To find the t-coordinate of the t-intercept, substitute 0 for p in the model's equation and solve for t.

Using Data Described in Words to Make Predictions

In most application problems in this text so far, we have been provided variable names and their definitions. In Example 3, a key step will be to create variable names and define the variables.

Example 3 Making a Prediction

The percentage of Americans with student loans who defaulted on their loans within two years of when the loans came due has increased approximately linearly from 5.0% in 2005 to 9.1% in 2010 (Source: *Department of Education*). Predict when 15% of Americans with student loans will default on their loans within two years of when the loans came due.

Table 40 Known Values of t and p

Years since 2000 t	Percent p
5	5.0
10	9.1

Solution

Let p be the percentage of Americans with student loans who defaulted on their loans within two years of when the loans came due at t years since 2000. Known values of t and p are shown in Table 40.

Since the variables t and p are approximately linearly related, we want an equation of the form $p = mt + b$. First, we use the values in Table 40 to find the slope of the model:

$$\frac{9.1 - 5.0}{10 - 5} = 0.82$$

Then we substitute 0.82 for m in the equation $p = mt + b$:

$$p = 0.82t + b$$

To find b, we use the point (10, 9.1) and substitute 10 for t and 9.1 for p and then solve for b:

$$\begin{aligned} 9.1 &= 0.82(10) + b && \textit{Substitute 10 for t and 9.1 for p.} \\ 9.1 &= 8.2 + b && \textit{Multiply.} \\ 9.1 - 8.2 &= 8.2 + b - 8.2 && \textit{Subtract 8.2 from both sides.} \\ 0.9 &= b && \textit{Combine like terms.} \end{aligned}$$

Plot1 Plot2 Plot3
\Y1=.82X+.9
\Y2=
\Y3=
\Y4=
\Y5=
\Y6=
\Y7=

Then we substitute 0.9 for b in the equation $p = 0.82t + b$:

$$p = 0.82t + 0.9$$

Finally, to predict when the percent will be 15%, we substitute 15 for p in the equation $p = 0.82t + 0.9$ and solve for t:

$$\begin{aligned} 15 &= 0.82t + 0.9 && \textit{Substitute 15 for p.} \\ 15 - 0.9 &= 0.82t + 0.9 - 0.9 && \textit{Subtract 0.9 from both sides.} \\ 14.1 &= 0.82t && \textit{Combine like terms.} \\ 17.20 &\approx t && \textit{Divide both sides by 0.82.} \end{aligned}$$

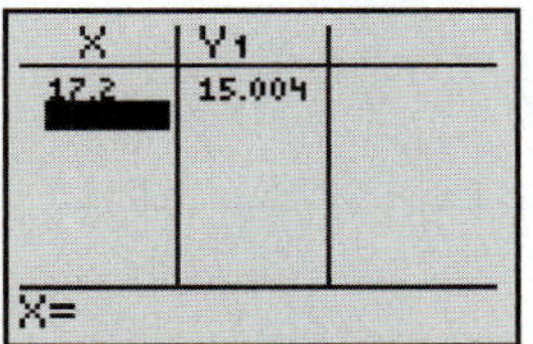

Figure 57 Verify the work

The model predicts that in 2000 + 17 = 2017, 15% of Americans with student loans will default on their loans within two years of when the loans came due. We can verify our work by using a graphing calculator table (see Fig. 57).

In Example 3, we defined t to be the number of years since 2000. If we had defined t to be the number of years since *1950* (or any other year), we would have obtained the same prediction of 2017, although the equation of our model would have been different. **If an exercise does not state a year from which to begin counting, choose any year.**

Domain and Range of a Model

Recall from Section 5.2 that the domain of a function is the set of all inputs and that the range of a function is the set of all outputs. For the **domain** and **range** of a model, we consider input–output pairs only when both the input and the output make sense in the situation. The domain of the model is the set of all such inputs, and the range of the model is the set of all such outputs.

Example 4 Finding the Domain and Range of a Model

A store is open from 9 A.M. to 5 P.M., Mondays through Saturdays. Let $I = f(t)$ be an employee's weekly income (in dollars) from working t hours each week at $10 per hour.

1. Find an equation of the model f.
2. Find the domain and range of the model f.

Solution

1. The employee's weekly income (in dollars) is equal to the pay per hour times the number of hours worked per week:

$$f(t) = 10t$$

2. To find the domain and range of the model f, we consider input–output pairs only when both the input and the output make sense in this situation. Time is the input. Since the store is open 8 hours a day, 6 days a week, the employee can work up to 48 hours each week. So, the domain is the set of numbers between 0 and 48, inclusive: $0 \le t \le 48$.

Income is the output. Since the number of hours worked is between 0 and 48 hours, inclusive, and the pay is \$10 per hour, the range is the set of numbers between 0 and $10(48) = 480$, inclusive: $0 \le f(t) \le 480$.

In Fig. 58, we illustrate the inputs 22, 35, and 48 being sent to the outputs 220, 350, and 480, respectively. We also label the part of the t-axis that represents the domain and the part of the I-axis that represents the range.

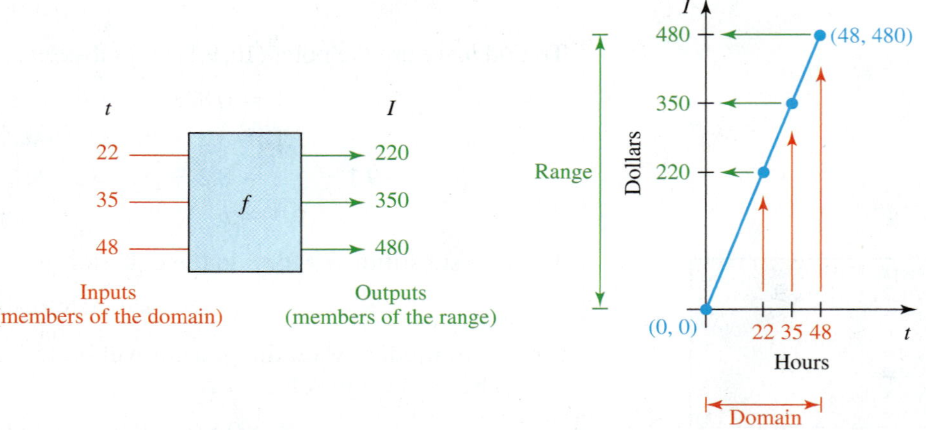

Figure 58 Domain and range of the employee income model

Group Exploration

Looking ahead: Properties of inequalities

The symbol "<" means *is less than*. For example, the statement $2 < 5$ means that 2 is less than 5, which is true. The statement $-5 < -3$ means that -5 is less than -3, which is also true.

The symbol ">" means *is greater than*. For example, the statement $6 > 1$ means that 6 is greater than 1, which is true.

Statements of the form $a < b$ or $a > b$ are examples of *inequalities*.

1. Decide whether each inequality is true.
   a. $4 < 9$ **b.** $3 > 8$ **c.** $-1 < -7$ **d.** $-2 > -6$
2. Decide whether performing the given operation on both sides of the true inequality $4 < 6$ will give a true statement.
   a. Add 2 to both sides.
   b. Add -2 to both sides.
   c. Multiply both sides by 2.
   d. Multiply both sides by -2.
3. For any true inequality (such as $3 < 7$), can you add the given type of number to both sides of the inequality and then still have a true inequality?
   a. positive number **b.** negative number **c.** zero
4. For any true inequality, can you multiply both sides of the inequality by the given type of number and then still have a true inequality?
   a. positive number **b.** negative number **c.** zero
5. Recall from Section 2.4 that subtracting a number is the same as adding the opposite of that number. What does this tell you about subtracting a number from both sides of an inequality?
6. Recall from Section 2.2 that dividing by a nonzero number is the same as multiplying by the reciprocal of that number. What does this tell you about dividing both sides of an inequality by a nonzero number?

Tips for Success Stick with It

If you are having difficulty doing an exercise, don't panic! Reread the exercise and reflect on what you have already sorted out about the problem—what you know and where you need to go. Your solution to the problem may be just around the corner.

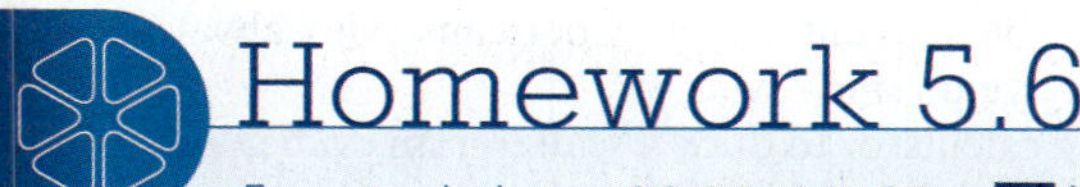

Homework 5.6

For extra help ▶ MyMathLab® 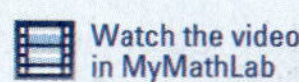 Watch the videos in MyMathLab  Download the MyDashboard App

1. In Exercise 17 of Homework 5.5, you found an equation close to $p = 0.76t - 42.04$ that models the percentage p of births outside marriage in the United States at t years since 1900 (see Table 41).

Table 41 Births Outside Marriage

Year	Percent of Births outside Marriage
1970	10.7
1975	14.3
1980	18.4
1985	22.0
1990	28.0
1995	32.2
2000	33.2
2005	36.8
2010	40.8

Source: *National Center for Health Statistics*

a. Rewrite the equation $p = 0.76t - 42.04$ with the function name f.

b. Find $f(118)$. What does your result mean in this situation?

c. Find the value of t so that $f(t) = 47$. What does your result mean in this situation?

d. According to the model, in what year will all births be outside marriage?

e. Estimate the percentage of births outside marriage in 1997. The actual percentage was 32.4%. What is the error in your estimate? (The error is the difference between the estimated value and the actual value.)

2. In Exercise 18 of Homework 5.5, you found an equation close to $p = 0.76t + 11.09$ that models the percentage p of births outside marriage in the United States at t years *since 1970*. Repeat Exercise 1, but use the model $p = 0.76t + 11.09$. In what instances are your answers the same? When are they different?

3. In Exercise 21 of Homework 5.5, you found an equation close to $p = -1.19t + 64.86$, where p is the percentage of Americans who are baseball fans at t years since 1990 (see Table 42).

Table 42 Percentages of Americans Who Are Baseball Fans

Year	Percent
1999	54
2001	51
2003	50
2005	48
2007	44
2008	43

Source: *The Gallup Organization*

a. Rewrite the equation $p = -1.19t + 64.86$ with the function name f.

b. Find $f(28)$. What does it mean in this situation?

c. Find t when $f(t) = 28$. What does it mean in this situation?

d. Find the p-intercept of the model. What does it mean in this situation?

e. Find the t-intercept. What does it mean in this situation?

4. First-class mail volumes are shown in Table 43 for various years.

Table 43 First-Class Mail Volumes

Year	First-Class Mail Volume (billions of pieces)
2007	95.9
2008	91.7
2009	83.8
2010	78.2
2011	73.5

Source: *U.S. Postal Service*

a. Let $v = f(t)$ be the first-class mail volume (in billions of pieces) in the year that is t years since 2000. Find an equation of f. Does the model fit the data well?

b. Find $f(18)$. What does it mean in this situation?

c. Find t when $f(t) = 18$. What does it mean in this situation?

d. Find the v-intercept. What does it mean in this situation?

e. Find the t-intercept. What does it mean in this situation?

f. Give some possible reasons why first-class mail volume has decreased so much from 2007 to 2011.

5. In Exercise 19 of Homework 5.5, you found an equation close to $p = 12.74n + 4.40$, where p is the price (in dollars) of a Gold Medal Sports ski rental package for n days (see Table 44).

Table 44 Prices of Ski Rental Packages

Number of Days	Price of Package (dollars)
1	15.00
2	30.00
3	45.00
4	56.00
5	70.00
6	78.00

Source: *Gold Medal Sports*

a. Rewrite the equation $p = 12.74n + 4.40$ with the function name f.

b. Find the p-intercept of the model. What does it mean in this situation?

c. Use f to estimate the price of renting skis for 7 days.

d. Use a graphing calculator table to find $f(1), f(2), f(3), \ldots,$ and $f(6)$. Then estimate by how much p increases each time n is increased by 1. Compare these results with the slope of the graph of f. What does that mean in this situation?

e. Let $g(n) = \dfrac{f(n)}{n}$. Find $g(1)$, $g(2)$, $g(3), \ldots,$ and $g(6)$. Which of these results is the least? What does that mean in this situation? [**Hint:** To find $g(2)$, first find $f(2)$. Then divide the result by 2 to get $g(2) = \dfrac{f(2)}{2}$.]

f. For ski rental packages for over 6 days, Gold Medal Sports charges a *daily* fee of $13.00 *per day*. Compare this result with your results in parts (d) and (e).

6. In Exercise 20 of Homework 5.5, you found an equation close to $L = 0.16t + 73.71$, where L is the life expectancy at birth (in years) of an American born t years after 1980 (see Table 45).

Table 45 Life Expectancies at Birth

Year of Birth	Life Expectancy (years)
1980	73.7
1985	74.7
1990	75.4
1995	75.8
2000	77.0
2005	77.9
2010	78.7

Source: *U.S. Census Bureau*

a. Use the linear model to predict the life expectancy of an American born in 2017.

b. Use the linear model to predict the birth year in which the life expectancy of an American will be 80 years.

c. Use the linear model to estimate what your (or someone else's) life expectancy was at birth. Given that you have made it to your current age, do your think your life expectancy is now less than, the same as, or more than it was at your year of birth?

d. Find the t-intercept. What does it mean in this situation?

7. The percentages of married persons who say they are "very happy" with their marriages are shown in Table 46 for various years. Let $p = f(t)$ be the percentage of married persons who say they are "very happy" with their marriages at t years since 1970.

Table 46 Percentages of Married Persons "Very Happy" with Their Marriages

Year	Percent
1970	67.0
1980	67.5
1990	65.0
2000	62.0
2008	62.0

Source: *Institute for American Values*

a. Use a graphing calculator to draw a scattergram of the data.

b. Find an equation of f.

c. Use your model to predict when 60% of married persons will say they are "very happy" with their marriages.

d. Find the p-intercept. What does it mean in this situation?

e. Find the t-intercept. What does it mean in this situation?

8. The percentages of Americans who always wear seat belts are shown in Table 47 for various years.

Table 47 Percentages of Americans Who Always Wear Seat Belts

Year	Percent
1992	70
1996	75
2000	79
2004	83
2008	87
2011	91

Source: *Harris Interactive*

Let $p = f(t)$ be the percentage of Americans who always wear seat belts at t years since 1990.

a. Use a graphing calculator to draw a scattergram of the data.

b. Find an equation of f.

c. What is the p-intercept? What does it mean in this situation?

d. Predict the percentage of Americans in 2017 who will always wear seat belts.

e. Use your model to predict when every American will always wear seat belts.

9. The number of commercial airline boardings on domestic flights increased steadily during the 1990s (see Table 48).

Table 48 Numbers of Commercial Airline Boardings on Domestic Flights

Year	Number of Boardings (millions)
1991	452
1993	487
1995	547
1997	599
1999	635
2000	666

Source: *Bureau of Transportation Statistics*

a. Let $f(t)$ be the number of commercial airline boardings on domestic flights (in millions) for the year that is t years since 1990. Use a graphing calculator to draw a scattergram of the data.

b. Find an equation of f. Does your model fit the data well?

c. Use your model f to estimate the number of boardings in 2001. The actual number was 622 million. What is the error in your estimate? (The error is the difference between the estimated value and the actual value.)

d. The number of boardings in 2001 was low due to the terrorist attacks on September 11, 2001. By making the following assumptions, estimate the amount of money airlines lost in 2001.

- All trips were round trips.
- The average number of boardings for a round trip was four (two flights out, two back).
- The average round-trip fare was $340.

10. The average gasoline taxes (in 2010 dollars) per 1000 miles driven are shown in Table 49.

Table 49 Average Gasoline Taxes per 1000 Miles Driven

Year	Average Gasoline Tax Per 1000 Miles Driven (2010 dollars)
1995	28
1998	27
2001	24
2004	23
2007	20
2010	19

Source: *Bureau of Economic Analysis; Bureau of Transportation Statistics, Bureau of Labor Statistics*

a. Let $A = f(t)$ be the average gasoline tax (in 2010 dollars) per 1000 miles driven at t years since 1990. Use a graphing calculator to draw a scattergram of the data.

b. Find an equation of f. Does your model fit the data well?

c. Find the t-intercept. What does it mean in this situation?

d. Since 1995, inflation has occurred and the average gas mileage of cars has increased. Explain how these factors have affected the gasoline tax (in 2010 dollars) per 1000 miles driven.

e. i. Teenagers drive an average of 7624 miles per year (Source: *Federal Highway Administration*). Use your model to predict the total money (in 2010 dollars) a teenager will pay in gasoline taxes during the period 2016–2018.

ii. How much *additional* money (in 2010 dollars) would a teenager pay in gasoline taxes during the period 2016–2018 if the gasoline tax were still at the 1995 level of $28 (in 2010 dollars) per 1000 miles driven?

11. The United States and Great Britain use the Fahrenheit temperature scale, but most countries use the Celsius temperature scale. The Celsius reading 0°C is the temperature at which water freezes, and 100°C is the temperature at which water boils (at sea level). Table 50 shows equivalent Celsius and Fahrenheit temperatures.

Table 50 Equivalent Temperature Readings

Celsius Reading (°C)	Fahrenheit Reading (°F)
0	32
20	68
40	104
60	140
80	176
100	212

a. Let $F = f(C)$ be the Fahrenheit reading corresponding to a Celsius reading of C degrees. Find an equation of f.

b. The average high temperature during July in Paris is 24°C (Source: *The Weather Channel*). What is the Fahrenheit reading?

c. The average high temperature during September in Quebec is 64°F (Source: *The Weather Channel*). What is the Celsius reading?

d. The theoretical lowest possible temperature is −273.15°C. What is the Fahrenheit temperature?

12. The rate at which a cricket chirps depends on the temperature of the surrounding air. You can estimate the air temperature by counting chirps! Some data are provided in Table 51.

Table 51 Rates of Cricket Chirping

Temperature (°F)	Rate (number of chirps per minute)
50	43
60	86
70	129
80	172
90	215

Source: Eric Sloane's Weather Book, *Eric Sloane, 2005*

a. Let $g(F)$ be the number of chirps per minute a cricket makes when the temperature is F degrees Fahrenheit. Find an equation of g. Verify that the graph of your equation comes close to the points in the scattergram of the data.

b. Find $g(73)$. What does it mean in this situation?

c. Find the value of F, where $g(F) = 100$. What does your result mean in this situation?

d. What are the possible air temperatures at a field where the crickets are not chirping?

13. In Exercise 15 of Homework 5.5, you found an equation close to $H = 4.5d + 15.3$, where H is the number of "human years" that is equivalent to d "dog years" for dogs that weigh between 20 pounds and 50 pounds, inclusive (see Table 52).

Table 52 Dog Years Compared with Human Years

Dog Years	Human Years
1	15
2	24
3	29
5	38
7	47
9	56
11	65
13	74
15	83

Source: *Fred Metzger, State College, PA*

a. Rewrite the equation $H = 4.5d + 15.3$ with the function name f.

b. The life expectancy of an American is about 78 years. Use the model to estimate the life expectancy of dogs.

c. The oldest dog ever was Bluey, an Australian cattle dog, who died at age 29 years. How long did Bluey live in human years?

d. The oldest American ever was Sarah Knauss, who died at age 119 years. How long did Knauss live in dog years?

e. What is the slope? What does it mean in this situation?

f. Although the information in Table 52 is much more accurate, it has long been believed that each dog year is equivalent to 7 human years. Use this old rule of thumb to find another model that describes the relationship between d and H.

g. What is the slope of the graph of your equation in part (f)? What does it mean in this situation? Compare this slope with the slope you found in part (e).

14. If you could stop time and live forever in good health at a particular age, what age would you choose? The average ideal ages chosen by various age groups are shown in Table 53.

Table 53 Ideal Ages

Age Group (years)	Age Used to Represent Age Group (years)	Average Ideal Age (years)
18–24	21	27
25–29	27	31
30–39	34.5	37
40–49	44.5	40
50–64	57	44
over 64	75	59

Source: *Harris Poll*

Let $f(a)$ be the average ideal age (in years) chosen by people whose actual age is a years.

a. Use a graphing calculator to draw a scattergram of the data.
b. Find an equation of f.
c. Use your model to estimate the average ideal age chosen by 18-year-olds.
d. What is the slope? What does it mean in this situation?
e. What is the age of people whose ideal age is equal to their actual age?

15. In Exercise 27 of Homework 5.5, you found an equation close to $p = 2.48x - 23.64$, where p is the percentage of Intermediate Algebra students at the College of San Mateo (CSM) succeeding in the course (grade of A, B, or C) who scored x points on the placement test (see Table 54).

Table 54 Percentages of Intermediate Algebra Students Who Succeeded

Placement Score Group	Score Used to Represent Score Group	Percent Who Succeeded in Intermediate Algebra
21–25	23	34
26–30	28	47
31–35	33	55
36–40	38	71
41–45	43	84
46–50	48	*

Source: *College of San Mateo Mathematics Department*
*There were not enough students in this group to give useful data.

a. Rewrite the equation $p = 2.48x - 23.64$ with the function name f.
b. Students who score below 21 points (out of 50) on the placement test cannot enroll in Intermediate Algebra. Use the model f to estimate how high the cutoff score would have to be to ensure that all students succeed in the course.
c. Use the model f to estimate for which scores no students would succeed in the course.
d. If, in one semester, 145 students scored in the 16–20-point range on the placement test, predict how many of these students would have succeeded in the course if they had been allowed to enroll in it. Would you advise CSM to lower the placement score cutoff to 16? Explain.
e. Table 55 shows the numbers of students in various placement score groups for one semester. For students who scored at least 21 points on the placement test that semester, estimate how many succeeded in the course.

Table 55 Placement Test Scores for One Semester

Placement Score Group	Number of Students
21–25	94
26–30	44
31–35	19
36–40	12
41–45	9
46–50	4

16. In Exercise 22 of Homework 5.5, you found an equation close to $p = 0.42a - 13.91$, where p is the percentage of Americans at age a years who have been diagnosed with diabetes at some point in their lives (see Table 56).

Table 56 Percentages of Americans Diagnosed with Diabetes, by Age Group

Age Group (years)	Age Used to Represent Age Group (years)	Percent
35–39	37	2
40–44	42	4
45–49	47	5
50–54	52	8
55–59	57	10
60–64	62	13
65–69	67	14

Source: *National Health Interview Survey*

a. Rewrite the equation $p = 0.42a - 13.91$ with the function name f.
b. Estimate the percentage of 40-year-old Americans who have been diagnosed with diabetes.
c. Estimate at what age 7% of Americans have been diagnosed with diabetes.
d. Find the a-intercept of the model. What does it mean in this situation?
e. The chance of any one person being diagnosed increases as the person grows older. However, 13% of all Americans over the age of 70 have been diagnosed at some point in their lives—less than the percentage for ages 65–69 years. How is this possible?

17. Public school per-student expenditures increased approximately linearly from \$2.2 thousand in 1980 to \$10.8 thousand in 2008 (Source: *National Education Association*). Predict the per-student expenditure in 2017.

18. The number of words in the federal tax code increased approximately linearly from 3.4 million words in 2005 to 3.8 million words in 2012 (Source: *The Tax Foundation*). Predict the number of words in the federal tax code in 2018.

19. Blood donations to American Red Cross decreased approximately linearly from 6.6 million pints in 2009 to 5.9 million pints in 2012 (Source: *American Red Cross*). Predict in which year the blood donations will be 4.5 million pints.

20. The percentage of female workers who prefer a female boss over a male boss increased approximately linearly from 10% in 1975 to 27% in 2011 (Source: *The Gallup Organization*). Predict when 30% of female workers will prefer a female boss.

21. The revenue of Kodak decreased approximately linearly from \$19.0 billion in 1990 to \$6.0 billion in 2011 (Source: *FactSet*).

a. The company is trying to emerge from bankruptcy protection in 2013. Predict the revenue in that year.

b. Predict when the revenue will be 0 dollars (and the company will go out of business).

22. Despite the No Child Left Behind Act of 2001, students' average math score on the SAT decreased approximately linearly from 518 points in 2006 to 514 points in 2012 (Source: *College Board*).

a. Predict the average math score on the SAT in 2017.

b. Predict when the average math score on the SAT will be 509 points.

23. The average score on the National Assessment of Educational Progress test in U.S. history was 195 points for fourth-graders who studied history about 45 minutes per week. The average score was 211 points for fourth-graders who studied history about 150 minutes per week. There is an approximate linear relationship between the number of hours fourth-graders study history per week and the average score on the test (Source: *U.S. Department of Education*). Estimate the average score for fourth-graders who study history about 200 minutes per week.

24. In Mississippi, a child is eligible for the Children's Health Insurance Program (CHIP) if the child's family meets an income limit. For a family of four, family monthly income must be no more than \$3067. For a family of six, family monthly income must be no more than \$4114. There is a linear relationship between family size and the income limit (Source: *CHIP*). What is the income limit of a family of seven?

25. A basement is flooded with 640 cubic feet of water. It takes 4 hours to pump out the water. Let $f(t)$ be the number of cubic feet of water that remains in the basement after t hours of pumping.

a. Find a linear equation of f. [**Hint:** You are given information about two points that can be used to find an equation.]

b. Graph f by hand. Use a graphing calculator to verify your graph.

c. What are the domain and range of the model? Explain.

26. It takes a person 5 minutes to eat all 12 ounces of ice cream in a cup. Let $f(t)$ be the number of ounces of ice cream remaining in the cup t minutes after the person began eating the ice cream.

a. Find a linear equation of f. [**Hint:** You are given information about two points that can be used to find an equation.]

b. Graph f by hand. Use a graphing calculator to verify your graph.

c. What are the domain and range of the model? Explain.

Concepts

27. Describe in your own words the four-step modeling process.

28. Compare the meanings of *domain* and *range* of a model with the meanings of *domain* and *range* of a function.

Related Review

29. The number of Americans who live alone was 31.4 million in 2010 and has increased by about 0.42 million each year since then (Source: *U.S. Census Bureau*). Let $n = f(t)$ be the number (in millions) of Americans who live alone at t years since 2010.

a. Find an equation of f.

b. What is the n-intercept? What does it mean in this situation?

c. Perform a unit analysis of the equation you found in part (a).

d. The population of California, the most populous state, is 37.3 million. Predict when 37.3 million people will live alone.

30. The median age of buyers of Harley-Davidson motorcycles was 49 years in 2011 and has increased by about 0.31 year each year since then (Source: *Harley-Davidson Motor Co.*). Let $f(t)$ be the median age (in years) of buyers at t years since 2011.

a. Find an equation of f.

b. Perform a unit analysis of the equation you found in part (a).

c. Predict when the median age of buyers will be 51 years.

d. Estimate the median age of buyers in 2019.

31. Annual sales of echinacea were \$132 million in 2009 and have decreased by about \$6.9 million per year since then (Source: *Nutrition Business Journal*). Let $s = f(t)$ be the annual echinacea sales (in millions of dollars) at t years since 2009.

a. Find an equation of f.

b. What is the s-intercept? What does it mean in this situation?

c. Perform a unit analysis of the equation you found in part (a).

d. Estimate when annual echinacea sales will be \$75 million.

32. The number of households that watched the *Miss Universe Pageant* was 4.49 million in 2011 and has decreased by about 0.13 million per year since then (Source: *Nielsen Media Research*). Let n be the number of households (in millions) that watched the *Miss Universe Pageant* at t years since 2011.

a. Find an equation of f.

b. Perform a unit analysis of the equation you found in part (a).

c. Predict the number of households that will watch the *Miss Universe Pageant* in 2018.

d. What is the t-intercept? What does it mean in this situation?

33. The percentage of U.S. births from minorities was 50.4% in 2011 and is increasing by 0.73 percentage point per year (Source: *Census Bureau*). Predict when 56% of U.S. births will be from minorities.

34. The number of special operations soldiers was 24,145 soldiers in 2010 and is increasing by about 1017 soldiers per year (Source: *U.S. Army*). Predict when there will be 31,300 special operations soldiers.

Expressions, Equations, Functions, and Graphs

Perform the indicated instruction. Then use words such as linear, function, one variable, *and* two variables *to describe the expression or equation.*

35. Solve $-4(3x - 5) = 3(2x + 1)$.

36. Graph $4x - 3y + 3 = 0$ by hand.

37. Simplify $-4(3x - 5) - 3(2x + 1)$.

38. Evaluate $4x - 3y + 3$ for $x = -2$ and $y = -3$.

5.7 Solving Linear Inequalities in One Variable

Objectives

- » Know the meaning of *inequality symbols* and *inequality*.
- » Know how to graph an inequality.
- » Know the properties of inequalities.
- » Know the meaning of *satisfy*, *solution*, and *solution set* for a *linear inequality in one variable*.
- » Solve a linear inequality in one variable, and graph the solution set.
- » Solve a *three-part inequality in one variable*, and graph the solution set.
- » Use linear inequalities to make predictions about authentic situations.

In Section 5.6, we predicted when a quantity would reach a certain amount. In this section, we will discuss how to use a model to predict when a quantity will be more than (or less than) a certain amount. For instance, in Example 13, we will predict the years when the annual worldwide music industry revenue will be more than $77 billion.

Inequality Symbols

We use the **inequality symbols** $<$, $\leq$, $>$, and $\geq$ to compare the sizes of two quantities. Here are the meanings of these symbols and some examples of *inequalities*:

Symbol	Meaning	Examples of Inequalities
$<$	is less than	$2 < 5, 0 < 5, -6 < -1$
$\leq$	is less than or equal to	$4 \leq 7, 2 \leq 2, -3 \leq 0$
$>$	is greater than	$9 > 2, -4 > -6, 2 > 0$
$\geq$	is greater than or equal to	$8 \geq 3, 5 \geq 5, -2 \geq -8$

An **inequality** contains one of the symbols $<$, $\leq$, $>$, and $\geq$ with expressions on both sides. Here are some more examples of inequalities:

$$2x + 5 < 3x - 9 \qquad x \leq 5 \qquad 7 > 2 \qquad 4x - 1 \geq 6$$

Example 1 Inequalities

Decide whether the inequality statement is true or false.

1. $3 \leq 6$ **2.** $-5 > -2$ **3.** $8 \geq 8$ **4.** $9 < 9$

Solution

1. Since 3 is less than 6, the statement $3 \leq 6$ is true.
2. Since -5 lies to the left of -2 on the number line, -5 is less than -2. So, -5 is *not* greater than -2, and the statement $-5 > -2$ is false.
3. Since 8 is equal to itself, the statement $8 \geq 8$ is true.
4. Since 9 is not less than itself, the statement $9 < 9$ is false.

Graphing Inequalities

Figure 59 Graph of $x \leq 2$

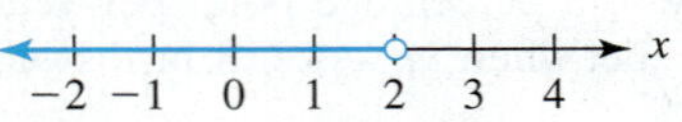

Figure 60 Graph of $x < 2$

Consider the inequality $x \leq 2$. This inequality says the values of x are less than or equal to 2. We can represent these values graphically on a number line by shading the part of the number line that lies to the left of 2 (see Fig. 59). We draw a *filled-in* circle at 2 to indicate that 2 is a value of x, too.

To graph the inequality $x < 2$, we shade the part of the number line that lies to the left of 2, but draw an *open* circle at 2 to indicate that 2 is *not* a value of x (see Fig. 60).

We use **interval notation** to describe a set of numbers. For example, we describe the numbers greater than 3 by $(3, \infty)$. We describe the numbers greater than or equal to 3 by $[3, \infty)$. We describe the set of real numbers by $(-\infty, \infty)$. More examples of inequalities and interval notation are shown in Fig. 61.

In Words	Inequality	Graph	Interval Notation
numbers less than 3	$x < 3$		$(-\infty, 3)$
numbers less than or equal to 3	$x \leq 3$		$(-\infty, 3]$
numbers greater than 3	$x > 3$		$(3, \infty)$
numbers greater than or equal to 3	$x \geq 3$		$[3, \infty)$

Figure 61 Words, inequalities, graphs, and interval notation

Example 2 Graphing an Inequality

Write the inequality $x > -2$ in interval notation, and graph the values of x.

Solution

The inequality $x > -2$ means that the values of x are greater than -2. We describe these numbers in interval notation by $(-2, \infty)$. To graph the values of x, we shade the part of the number line that lies to the right of -2 and draw an open circle at -2 (see Fig. 62).

Figure 62 Graph of $x > -2$

Addition Property of Inequalities

What happens if we add 3 to both sides of the inequality $4 < 6$?

$$4 < 6 \quad \text{Original inequality}$$
$$4 + 3 \stackrel{?}{<} 6 + 3 \quad \text{Add 3 to both sides.}$$
$$7 \stackrel{?}{<} 9 \quad \text{Simplify.}$$
true

What happens if we add -3 to both sides of the inequality $4 < 6$?

$$4 < 6 \quad \text{Original inequality}$$
$$4 + (-3) \stackrel{?}{<} 6 + (-3) \quad \text{Add } -3 \text{ to both sides.}$$
$$1 \stackrel{?}{<} 3 \quad \text{Simplify.}$$
true

These examples suggest the following property:

Addition Property of Inequalities

If $a < b$, then $a + c < b + c$.

Similar properties hold for $\leq$, $>$, and $\geq$.

Similar rules hold for subtraction, since subtracting a number is the same as adding the opposite of the number.

We can use a number line to illustrate that if $a < b$, then $a + c < b + c$. From Figs. 63 and 64, we see that if a lies to the left of b $(a < b)$, then $a + c$ lies to the left of $b + c$ $(a + c < b + c)$. In Fig. 63, c is negative; in Fig. 64, c is positive.

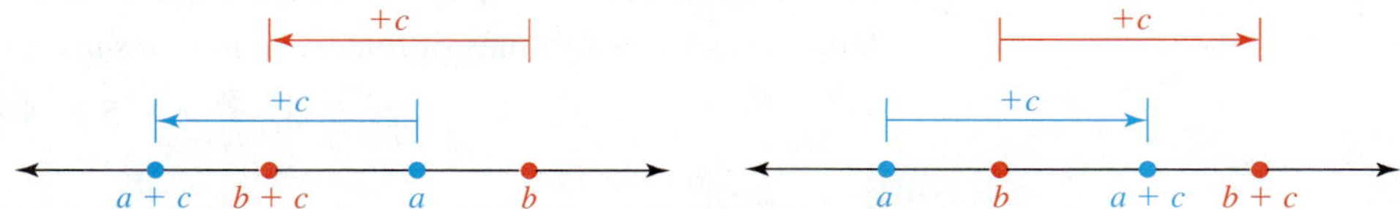

Figure 63 Adding c where c is negative **Figure 64** Adding c where c is positive

Multiplication Property of Inequalities

What if we multiply both sides of the inequality $4 < 6$ by 3?

$$4 < 6 \quad \text{Original inequality}$$
$$4(3) \stackrel{?}{<} 6(3) \quad \text{Multiply both sides by 3.}$$
$$12 \stackrel{?}{<} 18 \quad \text{Simplify.}$$
true

Finally, what happens if we multiply both sides of $4 < 6$ by -3?

$$4 < 6 \quad \text{Original inequality}$$
$$4(-3) \overset{?}{<} 6(-3) \quad \text{Multiply both sides by } -3.$$
$$-12 \overset{?}{<} -18 \quad \text{Simplify.}$$
$$\text{false}$$

The result is the false statement $-12 < -18$. We can get a *true* statement if we *reverse the inequality symbol* when we multiply both sides of $4 < 6$ by -3:

$$4 < 6 \quad \text{Original inequality}$$
$$4(-3) \overset{?}{>} 6(-3) \quad \text{Reverse inequality symbol.}$$
$$-12 \overset{?}{>} -18 \quad \text{Simplify.}$$
$$\text{true}$$

So, when we multiply both sides of an inequality by a *negative* number, we *reverse* the inequality symbol.

Multiplication Property of Inequalities

- For a *positive* number c, if $a < b$, then $ac < bc$.
- For a *negative* number c, if $a < b$, then $ac > bc$.

Similar properties hold for $\leq$, $>$, and $\geq$. In words: When we multiply both sides of an inequality by a positive number, we keep the inequality symbol. When we multiply both sides by a negative number, we reverse the inequality symbol.

Similar rules apply for division, since dividing by a nonzero number is the same as multiplying by its reciprocal. Therefore, **when we multiply or divide both sides of an inequality by a negative number, we reverse the inequality symbol.**

Consider multiplying both sides of $a < b$ by -1:

$$a < b \quad \text{Original inequality}$$
$$-1a > -1b \quad \text{Reverse inequality symbol.}$$
$$-a > -b \quad -1a = -a$$

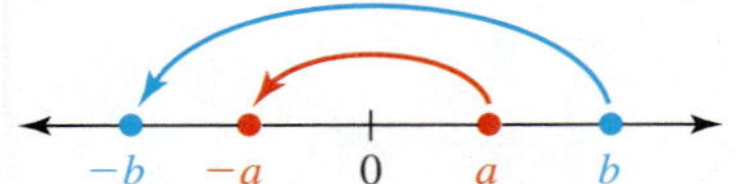

Figure 65 The points for a, b, $-a$, and $-b$

So, if $a < b$, then $-a > -b$. We can use a number line to illustrate this fact. To plot the point for $-a$, we move the point for a to the other side of the origin so that the points for $-a$ and a are the same distance from the origin (see Fig. 65).

From Fig. 65 we see that if the point for a lies to the *left* of the point for b $(a < b)$, then the point for $-a$ lies to the *right* of the point for $-b$ $(-a > -b)$.

Solving Linear Inequalities in One Variable

Here are some examples of *linear inequalities in one variable:*

$$3x + 5 < 8, \qquad 2x \leq 5, \qquad x - 5 > 4 - 2x, \qquad 5(x - 3) \geq 1$$

Definition Linear inequality in one variable

A **linear inequality in one variable** is an inequality that can be put into one of the forms

$$mx + b < 0, \qquad mx + b \leq 0, \qquad mx + b > 0, \qquad mx + b \geq 0$$

where m and b are constants and $m \neq 0$.

We say a number **satisfies** an inequality in one variable if the inequality becomes a true statement after we have substituted the number for the variable.

Example 3 Identifying Solutions of an Inequality

1. Does the number 4 satisfy the inequality $2x - 3 < 7$?
2. Does the number 6 satisfy the inequality $2x - 3 < 7$?

Solution

1. We substitute 4 for x in the inequality $2x - 3 < 7$:

$$2(4) - 3 \stackrel{?}{<} 7 \quad \text{Substitute 4 for } x.$$
$$8 - 3 \stackrel{?}{<} 7 \quad \text{Multiply.}$$
$$5 \stackrel{?}{<} 7 \quad \text{Subtract.}$$
$$\text{true}$$

So, 4 satisfies the inequality $2x - 3 < 7$.

2. We substitute 6 for x in the inequality $2x - 3 < 7$:

$$2(6) - 3 \stackrel{?}{<} 7 \quad \text{Substitute 6 for } x.$$
$$12 - 3 \stackrel{?}{<} 7 \quad \text{Multiply.}$$
$$9 \stackrel{?}{<} 7 \quad \text{Subtract.}$$
$$\text{false}$$

So, 6 does not satisfy the inequality $2x - 3 < 7$.

Definition *Solution, solution set,* and *solve* for an inequality in one variable

We say that a number is a **solution** of an inequality in one variable if it satisfies the inequality. The **solution set** of an inequality is the set of all solutions of the inequality. We **solve** an inequality by finding its solution set.

To solve a linear inequality in one variable, we apply properties of inequalities to get the variable alone on one side of the inequality.

Example 4 Solving a Linear Inequality

Solve $2x - 3 < 7$. Describe the solution set as an inequality, in interval notation, and in a graph.

Solution

We get x alone on one side of the inequality:

$$2x - 3 < 7 \quad \text{Original inequality}$$
$$2x - 3 + 3 < 7 + 3 \quad \text{Add 3 to both sides to get } 2x \text{ alone on left side.}$$
$$2x < 10 \quad \text{Combine like terms.}$$
$$\frac{2x}{2} < \frac{10}{2} \quad \text{Divide both sides by 2 to get } x \text{ alone on left side.}$$
$$x < 5 \quad \text{Simplify.}$$

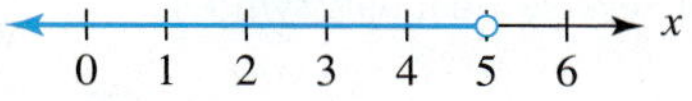

The solution set is the set of all numbers less than 5, which we describe in interval notation as $(-\infty, 5)$. We graph the solution set on a number line in Fig. 66.

0 1 2 3 4 5 6 x

Figure 66 Graph of $x < 5$

In Example 3, we found that 4 is a solution of the inequality $2x - 3 < 7$ but 6 is not. This checks with our work in Example 4, because 4 is on the graph in Fig. 66 and 6 is not.

Example 5 Solving a Linear Inequality

Solve the inequality $-3x \geq -12$. Describe the solution set as an inequality, in interval notation, and in a graph.

Solution

We divide both sides of the inequality by -3, a negative number:

$$-3x \geq -12 \quad \text{Original inequality}$$

$$\frac{-3x}{-3} \leq \frac{-12}{-3} \quad \text{Divide both sides by } -3\text{; reverse inequality symbol.}$$

$$x \leq 4 \quad \text{Simplify.}$$

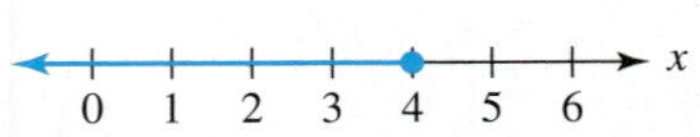

Figure 67 Graph of $x \leq 4$

Since we divided both sides of the inequality by a negative number, we reversed the inequality symbol. The solution set is $(-\infty, 4]$. We graph the solution set in Fig. 67.

WARNING It is a common error to forget to reverse an inequality symbol when multiplying or dividing both sides of an inequality by a negative number. For instance, in Example 5, it is important that we reversed the inequality symbol $\geq$ when we divided both sides of the inequality $-3x \geq -12$ by -3.

Example 6 Solving a Linear Inequality

Solve the inequality $3x \geq -12$. Describe the solution set as an inequality, in interval notation, and in a graph.

Solution

We divide both sides of the inequality by 3, a positive number:

$$3x \geq -12 \quad \text{Original inequality}$$

$$\frac{3x}{3} \geq \frac{-12}{3} \quad \text{Divide both sides by 3.}$$

$$x \geq -4 \quad \text{Simplify.}$$

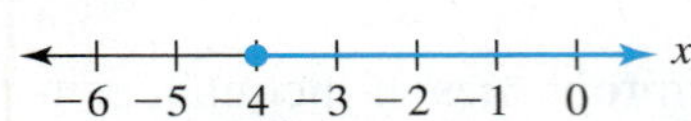

Figure 68 Graph of $x \geq -4$

Since we divided both sides of the inequality by a positive number, we did *not* reverse the inequality symbol. We graph the solution set, $[-4, \infty)$, in Fig. 68.

Example 7 Solving a Linear Inequality

Solve $2x - 5 > 6x + 3$. Describe the solution set as an inequality, in interval notation, and in a graph.

Solution

$$2x - 5 > 6x + 3 \quad \text{Original inequality}$$

$$2x - 5 - 6x > 6x + 3 - 6x \quad \text{Subtract } 6x \text{ from both sides.}$$

$$-4x - 5 > 3 \quad \text{Combine like terms.}$$

$$-4x - 5 + 5 > 3 + 5 \quad \text{Add 5 to both sides.}$$

$$-4x > 8 \quad \text{Combine like terms.}$$

$$\frac{-4x}{-4} < \frac{8}{-4} \quad \text{Divide both sides by } -4\text{; reverse inequality symbol.}$$

$$x < -2 \quad \text{Simplify.}$$

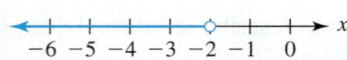

Figure 69 Graph of $x < -2$

We graph the solution set, $(-\infty, -2)$, in Fig. 69.

To verify our result, we check that for some values of x less than -2, the value of $2x - 5$ is greater than the value of $6x + 3$ (see Fig. 70). We do this by setting up the table so x begins at -2 and increases by 1. Then we scroll up 3 rows so we can view values of x that are less than -2 and values of x that are greater than -2.

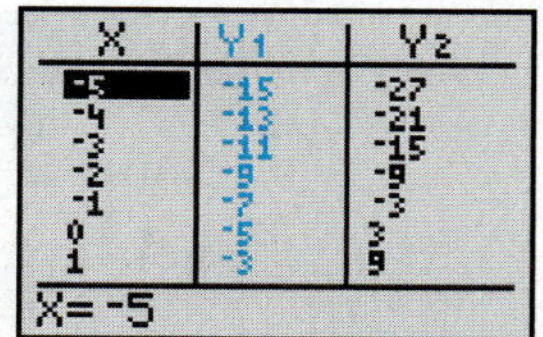

Figure 70 Verify the work

Example 8 Solving a Linear Inequality

Solve $3(x-2) < 5x-3$. Describe the solution set as an inequality, in interval notation, and in a graph.

Solution

$3(x-2) < 5x-3$	*Original inequality*
$3x-6 < 5x-3$	*Distributive law*
$3x-6-5x < 5x-3-5x$	*Subtract 5x from both sides.*
$-2x-6 < -3$	*Combine like terms.*
$-2x-6+6 < -3+6$	*Add 6 to both sides.*
$-2x < 3$	*Combine like terms.*
$\dfrac{-2x}{-2} > \dfrac{3}{-2}$	*Divide both sides by −2; reverse inequality symbol.*
$x > -\dfrac{3}{2}$	*Simplify.*

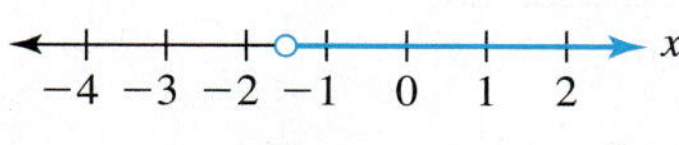

Figure 71 Graph of $x > -\frac{3}{2}$

We graph the solution set, $\left(-\frac{3}{2}, \infty\right)$, in Fig. 71.

To verify our result, we check that for some values of x greater than $-\frac{3}{2}$, the value of $3(x-2)$ is less than the value of $5x-3$ (see Fig. 72). We do this by setting up the table so x begins at -1.5 and increases by 1. Then we scroll up 3 rows so we can view values of x that are less than -1.5 and values of x that are greater than -1.5.

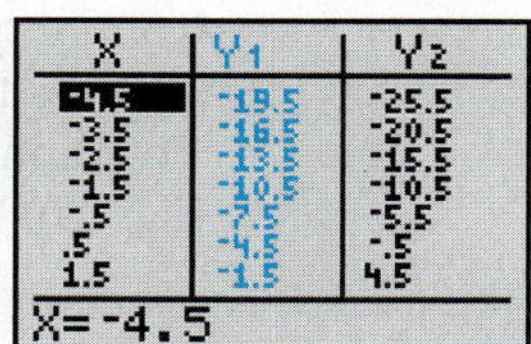

Figure 72 Verify the work

Example 9 Solving a Linear Inequality That Contains Decimals

Solve $-2.9x + 4.1 \le -6.05$. Describe the solution set as an inequality, in interval notation, and in a graph.

Solution

$-2.9x+4.1 \le -6.05$	*Original inequality*
$-2.9x+4.1-4.1 \le -6.05-4.1$	*Subtract 4.1 from both sides.*
$-2.9x \le -10.15$	*Combine like terms.*
$\dfrac{-2.9x}{-2.9} \ge \dfrac{-10.15}{-2.9}$	*Divide both sides by −2.9; reverse inequality symbol.*
$x \ge 3.5$	*Simplify.*

0 1 2 3 4 5 6 x

Figure 73 Graph of $x \ge 3.5$

We graph the solution set, $[3.5, \infty)$, in Fig. 73.

We can use graphing calculator tables to check our work.

Example 10 Solving a Linear Inequality

Solve $\frac{3x+5}{4} - \frac{2x-7}{6} < \frac{11}{3}$. Describe the solution set as an inequality, in interval notation, and in a graph.

Solution

First, we multiply both sides of the inequality by the LCD, 12:

$$12\left(\frac{3x+5}{4} - \frac{2x-7}{6}\right) < 12 \cdot \frac{11}{3} \quad \text{Multiply both sides by LCD, 12.}$$

$$12 \cdot \frac{3x+5}{4} - 12 \cdot \frac{2x-7}{6} < 12 \cdot \frac{11}{3} \quad \text{Distributive law}$$

$$3(3x+5) - 2(2x-7) < 44 \quad \text{Simplify.}$$

$$9x + 15 - 4x + 14 < 44 \quad \text{Distributive law}$$

$$5x + 29 < 44 \quad \text{Combine like terms.}$$

$$5x + 29 - 29 < 44 - 29 \quad \text{Subtract 29 from both sides.}$$

$$5x < 15 \quad \text{Combine like terms.}$$

$$\frac{5x}{5} < \frac{15}{5} \quad \text{Divide both sides by 5.}$$

$$x < 3 \quad \text{Simplify.}$$

We graph the solution set, $(-\infty, 3)$, in Fig. 74.

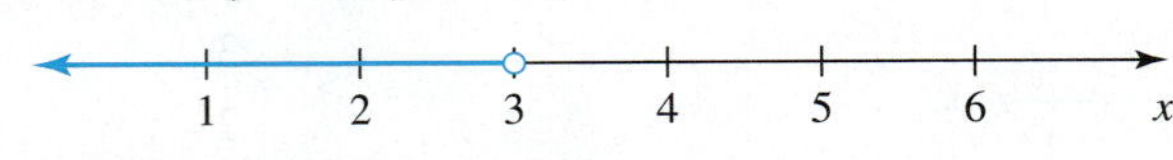

Figure 74 Graph of $x < 3$

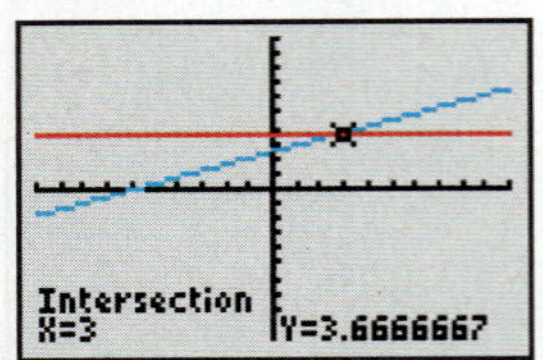

Figure 75 Verify the work

To verify our result, we check that, for values of x less than 3, the graph of the equation $y = \frac{3x+5}{4} - \frac{2x-7}{6}$ is below the horizontal line $y = \frac{11}{3}$. See Fig. 75.

Three-Part Inequalities

Now we will work with *three-part inequalities in one variable,* such as $3 \le x \le 7$. Recall from Section 5.2 that $3 \le x \le 7$ means the values of x are *both* greater than or equal to 3 *and* less than or equal to 7. In other words, all values of x are between 3 and 7, inclusive. To graph the solutions, we shade the part of the number line that lies between 3 and 7 (see Fig. 76). We draw filled-in circles at 3 and 7 to indicate that 3 and 7 are solutions, too.

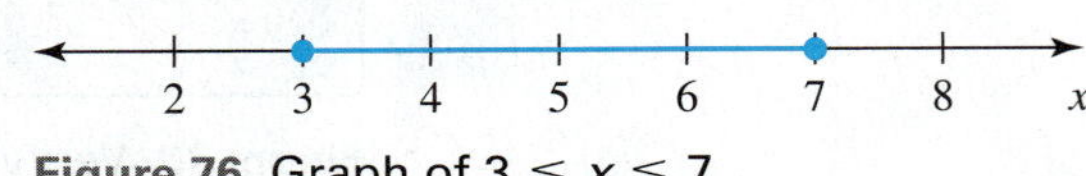

Figure 76 Graph of $3 \le x \le 7$

We describe the numbers between 3 and 7, inclusive, in interval notation by [3, 7]. More examples of three-part inequalities, with matching graphs and interval notation, are shown in Fig. 77.

In Words	Inequality	Graph	Interval Notation
Numbers between 1 and 3	$1 < x < 3$	1 3 x	(1, 3)
Numbers between 1 and 3, inclusive	$1 \le x \le 3$	1 3 x	[1, 3]
Numbers between 1 and 3, as well as 1	$1 \le x < 3$	1 3 x	[1, 3)
Numbers between 1 and 3, as well as 3	$1 < x \le 3$	1 3 x	(1, 3]

Figure 77 Words, inequalities, graphs, and interval notations

WARNING We use notation such as $(3, 7)$ in two ways: When we work with one variable, the *interval* $(3, 7)$ is the set of numbers between 3 and 7; when we work with two variables, such as x and y, the *ordered pair* $(3, 7)$ means $x = 3$ and $y = 7$.

Example 11 Solving a Three-Part Inequality

Solve $-5 < 2x - 1 < 7$.

Solution

We can get x alone in the "middle part" of the inequality by applying the same operations to all three parts of the inequality:

$$-5 < 2x - 1 < 7 \quad \text{Original inequality}$$
$$-5 + 1 < 2x - 1 + 1 < 7 + 1 \quad \text{Add 1 to all three parts.}$$
$$-4 < 2x < 8 \quad \text{Combine like terms.}$$
$$\frac{-4}{2} < \frac{2x}{2} < \frac{8}{2} \quad \text{Divide all three parts by 2.}$$
$$-2 < x < 4 \quad \text{Simplify.}$$

So, the solution set is the set of numbers between -2 and 4. We can graph the solution set on a number line (see Fig. 78), or we can describe the solution set in interval notation as $(-2, 4)$.

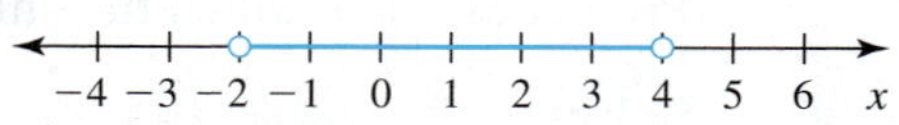

Figure 78 Graph of $-2 < x < 4$

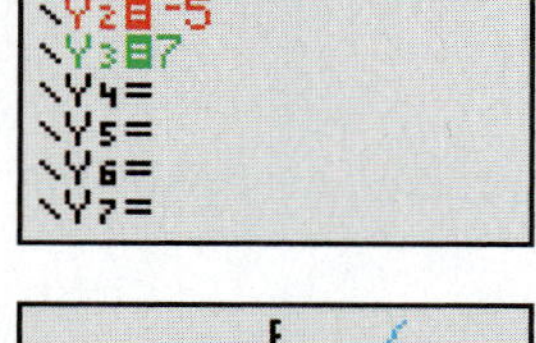

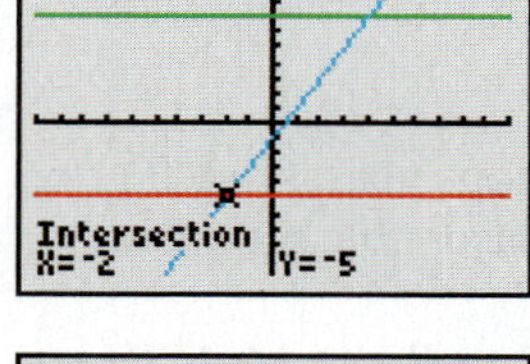

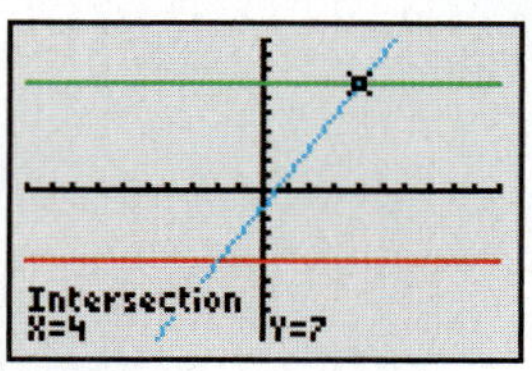

Figure 79 Verify the work

To verify our result, we check that, for values of x between -2 and 4, the graph of $y = 2x - 1$ is between the horizontal lines $y = -5$ and $y = 7$ (see Fig. 79).

Example 12 Solving a Three-Part Inequality

Solve $\frac{1}{2} \le 5 - \frac{3}{2}w \le 4$.

Solution

$$\frac{1}{2} \le 5 - \frac{3}{2}w \le 4 \quad \text{Original inequality}$$
$$2 \cdot \frac{1}{2} \le 2 \cdot 5 - 2 \cdot \frac{3}{2}w \le 2 \cdot 4 \quad \text{Multiply all three parts by LCD, 2.}$$
$$1 \le 10 - 3w \le 8 \quad \text{Simplify.}$$
$$1 - 10 \le 10 - 3w - 10 \le 8 - 10 \quad \text{Subtract 10 from all three parts.}$$
$$-9 \le -3w \le -2 \quad \text{Combine like terms.}$$
$$\frac{-9}{-3} \ge \frac{-3w}{-3} \ge \frac{-2}{-3} \quad \text{Divide all three parts by } -3\text{; reverse inequality symbols.}$$
$$3 \ge w \ge \frac{2}{3} \quad \text{Simplify.}$$
$$\frac{2}{3} \le w \le 3 \quad \text{Write in form } a \le w \le b.$$

So, the solution set is the set of numbers between $\frac{2}{3}$ and 3, inclusive. We can graph the solution set on a number line (see Fig. 80), or we can describe the solution set in interval notation as $\left[\frac{2}{3}, 3\right]$.

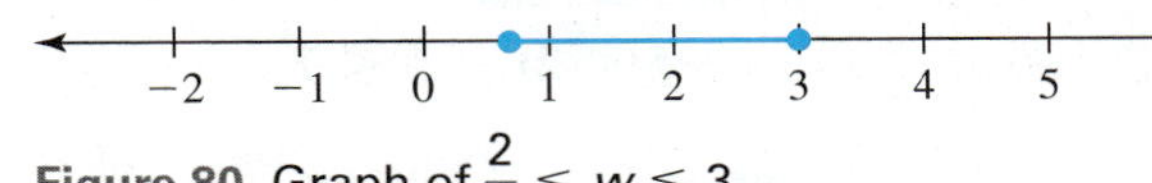

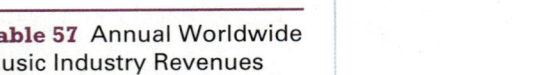

Figure 80 Graph of $\frac{2}{3} \le w \le 3$

Table 57 Annual Worldwide Music Industry Revenues

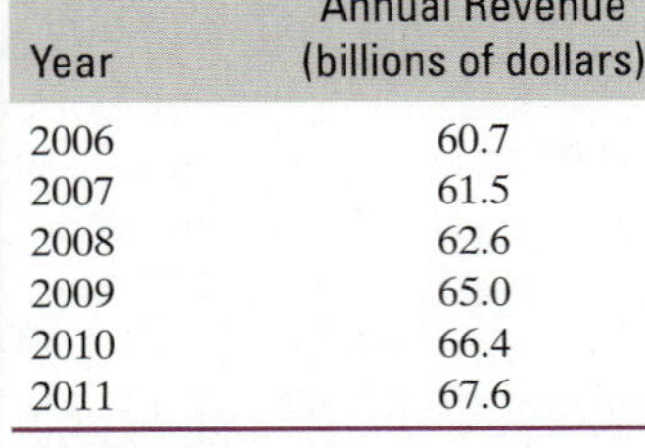

Year	Annual Revenue (billions of dollars)
2006	60.7
2007	61.5
2008	62.6
2009	65.0
2010	66.4
2011	67.6

Source: *eMarketer*

Using Linear Inequalities to Make Predictions

When working with a linear model, we can make certain types of predictions by solving a linear inequality that is related to the linear model.

Example 13 Making a Prediction

Worldwide music industry revenues are shown in Table 57 for various years. Let $f(t)$ be the annual revenue (in billions of dollars) at t years since 2000. A reasonable model is

$$r = 1.47t + 51.44$$

Predict the years when the annual revenue will be more than \$77 billion.

Solution

To predict when the annual revenue will be more than \$77 billion, we find the values of t such that the expression $1.47t + 51.44$ is more than 77:

$$1.47t + 51.44 > 77 \quad \textit{Original inequality}$$
$$1.47t + 51.44 - 51.44 > 77 - 51.44 \quad \textit{Subtract 51.44 from both sides.}$$
$$1.47t > 25.56 \quad \textit{Combine like terms.}$$
$$\frac{1.47t}{1.47} > \frac{25.56}{1.47} \quad \textit{Divide both sides by 1.47.}$$
$$t > 17.39 \quad \textit{Simplify; divide.}$$

So, the annual revenue will be more than \$77 billion after 2017 ($t > 17$).

Group Exploration

Meaning of the solution set of an inequality

We solve the inequality $-3x + 7 < 1$:

$$-3x + 7 < 1$$
$$-3x + 7 - 7 < 1 - 7$$
$$-3x < -6$$
$$\frac{-3x}{-3} > \frac{-6}{-3}$$
$$x > 2$$

1. Choose a number greater than 2. Check that your number satisfies the inequality $-3x + 7 < 1$.
2. Choose two more numbers greater than 2. Check that both of these numbers satisfy the inequality $-3x + 7 < 1$.
3. Choose three numbers that are *not* greater than 2. Show that each of these numbers does *not* satisfy the inequality $-3x + 7 < 1$.
4. Explain what it means when we write $x > 2$ as the last step in solving the inequality $-3x + 7 < 1$.

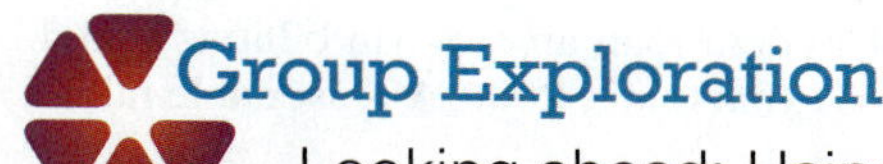

Group Exploration

Looking ahead: Using a system of equations to model a situation

The U.S. market shares of reduced-sugar beverages from Diet Pepsi and from Coca-Cola Zero are shown in Table 58 for various years. The market shares D from Diet Pepsi and C from Coca-Cola Zero are modeled by the system

$$D = -0.78t + 19.78$$
$$C = 1.27t + 1.12$$

where t is the number of years since 2005.

Table 58 Market Shares of Diet Pepsi and Coca-Cola Zero

	Market Share (percent)	
Year	Diet Pepsi	Coca-Cola Zero
2005	19.7	1.3
2006	19.2	2.2
2007	18.2	3.4
2008	17.2	5.3
2009	16.8	6.1

Source: *Euromonitor*

1. Find the D-intercept of the Diet Pepsi model and the C-intercept of the Coca-Cola Zero model. What do these intercepts mean in this situation?
2. Find the slopes of both models. What do these slopes mean in this situation?
3. Your responses to Problems 1 and 2 should suggest an event that will happen in the future. Describe that event.
4. By using Zoom Out and "intersect" on a graphing calculator, estimate the coordinates of the point where the graphs of the models intersect. (See Appendix A.6 and A.18 for graphing calculator instructions.) In terms of market share, what does it mean that the graphs intersect at this point? State the year and market share for this event.

Tips for Success Scan Test Problems

When you take a test, scan the test problems quickly, pick the ones with which you feel most comfortable, and complete those problems first. By doing so, you will warm up and gain confidence, and you may do better on the rest of the test. Also, you will probably have a better idea of how to allot your time on the remaining problems.

Homework 5.7

For extra help ▶ MyMathLab®

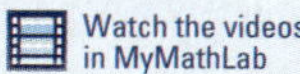

Watch the videos in MyMathLab

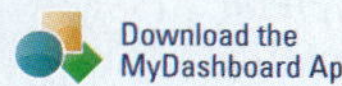

Download the MyDashboard App

Decide whether the given inequality is true or false.

1. $-3 > -5$ **2.** $-6 \le -2$ **3.** $4 \ge 4$ **4.** $-5 < -5$

Sketch the graph of the given inequality.

5. $x < 4$ **6.** $x < -5$ **7.** $x \ge -1$ **8.** $x \ge 1$

9. $x \le -2$ **10.** $x \le 4$ **11.** $x > 6$ **12.** $x > -3$

13. Use words, inequalities, graphs, and interval notation to complete Fig. 81.

In Words	Inequality	Graph	Interval Notation
		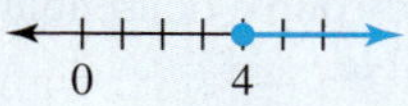 0 4 x	
numbers less than or equal to -2			
			$(-\infty, 1)$
	$x > -5$		

Figure 81 Exercise 13

14. Use words, inequalities, graphs, and interval notation to complete Fig. 82.

In Words	Inequality	Graph	Interval Notation
	$x \le -6$		
numbers greater than 1			
			$[-4, \infty)$
		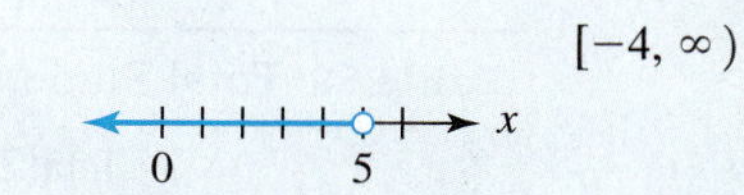	

Figure 82 Exercise 14

Which of the given numbers satisfy the given inequality?

15. $3x + 5 \ge 14;\ 2, 3, 6$

16. $-4x - 7 > 1;\ -3, -2, 0$

17. $2x < x + 2;\ -4, 2, 3$

18. $x - 9 \le 4x;\ -3, 2, 5$

Solve the inequality. Describe the solution set as an inequality, in interval notation, and in a graph. Then use a graphing calculator table to verify your result.

19. $x + 2 > 3$ **20.** $x + 5 \le 9$ **21.** $x - 1 < -4$

22. $x - 3 \ge -1$ **23.** $2x \le 6$ **24.** $3x > 9$

25. $4x \ge -8$ **26.** $2x < -10$ **27.** $-3t \ge 6$

28. $-2w \le 2$ **29.** $-2x > 1$ **30.** $-4x < -2$

31. $5x \le 0$ **32.** $-3x > 0$ **33.** $-x < 2$

34. $-x \le -1$ **35.** $-\frac{2}{3}x \ge 2$ **36.** $-\frac{5}{2}x \le 10$

37. $3x - 1 \ge 2$ **38.** $4x + 7 < 15$ **39.** $5 - 3x < -7$

40. $8 - 2x \le 6$ **41.** $3c - 6 \le 5c$ **42.** $7w + 4 > 3w$

43. $5x \ge x - 12$ **44.** $4x < 6 - 2x$

45. $-3.8x + 1.9 > -7.6$ **46.** $-2.4x + 5.8 \le 8.92$

47. $3b + 2 > 7b - 6$ **48.** $5k - 1 \le 2k + 8$

49. $4 - 3x < 9 - 2x$ **50.** $8 - x \ge 2 - 3x$

51. $2(x + 3) \le 8$ **52.** $5(x - 2) \ge 15$

53. $-(a - 3) > 4$ **54.** $-(t + 5) < -2$

55. $3(2x - 1) \le 2(2x + 1)$ **56.** $4(3x - 5) > 5(2x - 3)$

57. $4(2x - 3) + 1 \ge 3(4x - 5) - x$

58. $-2(5x + 3) - 2x < -3(2x - 4) + 2$

59. $4.3(1.5 - x) \ge 13.76$ **60.** $3.1(2.7 - x) > -1.55$

61. $\frac{1}{2}y + \frac{2}{3} \ge \frac{3}{2}$ **62.** $\frac{3}{4}t - \frac{1}{2} \le \frac{1}{4}$

63. $\frac{5}{3} - \frac{1}{6}x < \frac{1}{2}$ **64.** $\frac{1}{4} - \frac{2}{3}x > \frac{7}{12}$

65. $-\frac{1}{2}x - \frac{5}{6} \ge \frac{1}{3} + \frac{3}{2}x$ **66.** $-\frac{3}{4}x + \frac{5}{2} < \frac{7}{8} - \frac{5}{2}x$

67. $\frac{4c - 5}{6} \le \frac{3c + 7}{4}$ **68.** $\frac{6p + 2}{8} \ge \frac{4p - 1}{6}$

69. $\frac{3x + 1}{6} - \frac{5x - 2}{9} > \frac{2}{3}$ **70.** $\frac{4x - 7}{15} + \frac{2x + 3}{10} < \frac{2}{5}$

For Exercises 71–78, use a symbolic method to solve the inequality. Describe the solution set as an inequality, in interval notation, and in a graph. Then, use graphing calculator tables or graphs to verify your result.

71. $4 < x + 3 < 8$ **72.** $-2 < x - 4 < 3$

73. $-15 \le 2x - 5 \le 7$ **74.** $-5 \le 3x + 1 \le 13$

75. $-17 < 3 - 4x \le 15$ **76.** $7 < 5 - 2x \le 13$

77. $\frac{1}{3} \le 4 - \frac{2}{3}x < 2$ **78.** $\frac{3}{4} \le 1 - \frac{1}{4}x < 3$

79. Total student loan amounts are shown in Table 59 for various years.

Table 59 Total Student Loan Amounts

Year	Total Student Loan Amount (billions of dollars)
2006	83
2007	87
2008	96
2009	98
2010	107

Source: *College Board; Bureau of Labor Statistics*

Let $f(t)$ be the total student loan amount (in billions of dollars) in the year that is t years since 2000. A reasonable model is $f(t) = 5.9t + 47$.

a. What is the slope? What does it mean in this situation?

b. In which years will the total student loan amount be more than $153 billion?

80. The percentages of obstetricians and gynecologists who are women are shown in Table 60 for various years.

Table 60 Percentages of Obstetricians and Gynecologists Who Are Women

Year	Percent
1980	12
1985	19
1990	23
1995	30
2000	34
2002	38
2009	47

Source: *American Medical Association*

Let $f(t)$ be the percentage of obstetricians and gynecologists who are women at t years since 1980. A reasonable model is $f(t) = 1.17t + 12.07$.

a. What is the slope? What does it mean in this situation?

b. In which years will more than 54% of physicians specializing in obstetrics and gynecology be women?

81. The numbers of married-couple households (in millions) and the percentages of households that are married-couple households are shown in Table 61 for various years.

Table 61 Married-Couple Households

Year	Number of Married-Couple Households (millions)	Percent of Households That Are Married-Couple Households
2000	55.3	52.8
2003	57.3	51.5
2005	58.0	51.2
2008	58.3	50.0
2010	58.4	49.7

Source: *U.S. Census Bureau*

Let $f(t)$ be the percentage of households that are married-couple households at t years since 2000.

a. Find an equation of f.

b. In which years were more than 49% of households married-couple households?

c. Although the number of married-couple households is *increasing*, the percentage of households that are married-couple households is *decreasing*. Explain how this is possible.

82. The percentages of Americans who think they'll be able to live comfortably when they retire are shown in Table 62 for various years.

Table 62 Percentages of Americans Who Think They'll Be Able to Live Comfortably when They Retire

Year	Percent
2002	59
2004	59
2006	50
2008	46
2010	46
2012	38

Source: *Gallup Organization*

Let $p = f(t)$ be the percentage of Americans at t years since 2000 who think they'll be able to live comfortably when they retire.

a. Find an equation of f.

b. Predict when one-third of Americans will think they'll be able to live comfortably when they retire.

c. What is the p-intercept? What does it mean in this situation?

d. What is the t-intercept? What does it mean in this situation?

e. Predict when less than 25% of Americans will think they'll be able to live comfortably when they retire.

83. The teenage birthrate in the United States has declined since 1990 (see Table 63).

Table 63 American Teenage Birthrate

Year	Birthrate (number of births per 1000 women ages 15–19)
1990	59.9
1995	56.0
2000	47.7
2005	40.5
2010	34.3

Source: *U.S. National Center for Health Statistics*

Let $f(t)$ be the American teenage birthrate (number of births per 1000 women ages 15–19) at t years since 1990.

a. Find an equation of f.

b. Predict the teenage birthrate in 2020. Then predict the *number* of births to women ages 15–19 in 2020. Use the U.S. Census Bureau's prediction that there will be 11,007,000 women ages 15–19 in that year.

c. The American teenage birthrate is 2 to 10 times larger than that in most other Western countries. For example, the birthrate in France is 10.2 births per 1000 women ages 15–19. Predict in which years the American birthrate will be less than 10.2 births per 1000 women ages 15–19.

84. A dollar store is a no-frills retailer that sells many goods for one dollar. In 2002, there were 13,000 dollar stores in the United States—three times as many as in 1993. The percentages of households in various income groups that shop at dollar stores are shown in Table 64.

Table 64 Households That Shop at Dollar Stores

Household Income Group (thousands of dollars)	Income Used to Represent Income Group (thousands of dollars)	Percent of Households That Shop at Dollar Stores
0–19.999	10	74
20–29.999	25	71
30–39.999	35	67
40–49.999	45	64
50–69.999	60	58
70+	100	45

Source: *ACNielsen Homescan Panel*

Let $f(t)$ be the percentage of households with an income of d thousand dollars that shop at dollar stores.

a. Find an equation of f.

b. What percentage of households with an income of \$27 thousand shop at dollar stores?

c. At what incomes do more than half of households shop at dollar stores?

Concepts

85. A student tries to solve $-3x < 15$:

$$-3x < 15$$
$$\frac{-3x}{-3} < \frac{15}{-3}$$
$$x < -5$$

Describe any errors. Then solve the inequality correctly.

86. A student tries to solve $4x < -24$:

$$4x < -24$$
$$\frac{4x}{4} > \frac{-24}{4}$$
$$x > -6$$

Describe any errors. Then solve the inequality correctly.

87. a. List three numbers that satisfy $3x - 7 < 5$.

b. List three numbers that do not satisfy $3x - 7 < 5$.

88. a. List three numbers that satisfy $2(x + 3) > 17$.

b. List three numbers that do not satisfy $2(x + 3) > 17$.

89. a. Solve $x + 1 = -2x + 10$.

b. Solve $x + 1 < -2x + 10$.

c. Solve $x + 1 > -2x + 10$.

d. Graph the solutions in parts (a), (b), and (c) on the same number line. Use three colors to identify the different solutions. Make observations about the solutions.

90. a. Is the following statement true?

$$\text{If } a < b, \text{ then } a - c < b - c.$$

Explain.

b. Is the following statement true?

$$\text{If } a < b \text{ and } c \neq 0, \text{ then } \frac{a}{c} < \frac{b}{c}.$$

Explain.

91. Use the number line to show that if $a < b$, then $2a < 2b$.

92. Use the number line to show that if $a < b$, then $-2a > -2b$.

93. Give an example of a linear inequality of the form $mx + b \le c$, where m, b, and c are constants. Then solve the inequality. Describe the solution set as an inequality, in interval notation, and in a graph.

94. Describe how to solve a linear inequality in one variable. Describe when you need to reverse an inequality symbol. Explain why it is necessary to reverse the symbol in this case. Finally, explain what you have accomplished by solving an inequality.

Related Review

Solve the equation or inequality. If the statement is an inequality, then describe the solution set as an inequality, in interval notation, and in a graph.

95. $-2x + 6 = 3x - 14$

96. $-3(2x - 5) = 21$

97. $-2x + 6 > 3x - 14$

98. $-3(2x - 5) \le 21$

Let x be a number. Translate the English sentence into an inequality. Then solve the inequality. Describe the solution set as an inequality, in interval notation, and in a graph.

99. The sum of the number and 5 is greater than 2.

100. The difference of 7 and the number is less than 3.

101. Twice the number is less than or equal to 5 times the number, minus 6.

102. Three times the number is greater than or equal to the difference of the number and 8.

Expressions, Equations, Functions, and Graphs

Give an example of the following. Then solve, simplify, or graph, as appropriate.

103. linear equation in one variable

104. linear expression in one variable with four terms

105. linear inequality in one variable

106. linear function

Taking it to the Lab

Climate Change Lab (continued from Chapter 3)

Recall from the Climate Change Lab in Chapter 1 that many scientists believe that an increase in average global temperature as small as 3.6°F could be a dangerous climate change (see Table 65). Recall from the Climate Change Lab in Chapter 2 that most scientists believe global warming is largely the result of carbon dioxide emissions from the burning of fossil fuels (see Table 66).

Table 65 Average Surface Temperatures of Earth

Year	Average Temperature (degrees Fahrenheit)	Year	Average Temperature (degrees Fahrenheit)
1900	57.1	1960	57.2
1905	56.8	1965	57.0
1910	56.6	1970	57.3
1915	57.0	1975	57.1
1920	56.9	1980	57.6
1925	56.9	1985	57.3
1930	57.1	1990	57.8
1935	57.0	1995	57.9
1940	57.3	2000	57.8
1945	57.3	2005	58.3
1950	56.9	2010	58.3
1955	57.0	2011	58.1

Source: *NASA–GISS*

Table 66 Carbon Dioxide Emissions from Burning of Fossil Fuels

	Carbon Dioxide Emissions (billions of metric tons)	
Year	United States	World
1950	2.4	5.8
1955	2.7	7.2
1960	2.9	9.4
1965	3.5	11.2
1970	4.3	14.7
1975	4.4	16.5
1980	4.8	19.1
1985	4.6	19.4
1990	5.0	21.6
1995	5.3	22.2
2000	5.9	23.8
2005	6.0	28.4
2010	5.6	30.6

Source: *U.S. Department of Energy*

Finally, recall from the Climate Change Lab in Chapter 2 that the goal of the Kyoto Protocol was to cut developed countries' carbon dioxide emissions to about 5% to 7% below 1990 levels by 2012. This was a crucial first goal, but the Intergovernmental Panel on Climate Change (IPCC) is calling for carbon dioxide emissions in 2050 to be 60% less than carbon dioxide emissions in 1990.

Many countries are condemning the United States because the Bush and Obama administrations have refused to ratify the Kyoto Protocol. Although the U.S. population in 2010 was only 5% of the world population, 18% of world carbon dioxide emissions that year was produced by the United States (see Table 67).

Critics of the Kyoto Protocol say a fairer pact would be for all countries to commit to the same level of carbon

Table 67 United States and World Populations

Year	Population (billions)	
	United States	World
1960	0.18	3.04
1970	0.21	3.71
1980	0.23	4.45
1990	0.25	5.29
2000	0.28	6.09
2010	0.31	6.85

Source: *U.S. Census Bureau*

dioxide emissions *per person*. Using the IPCC's recommendation for 2050 carbon dioxide emissions, coupled with the United Nations' prediction of 9.3 billion people in 2050, carbon dioxide emissions should be about 0.9 metric ton per person that year.*

In 2010, annual carbon dioxide emissions were about 4.5 metric tons of carbon dioxide per person. Average annual carbon dioxide emissions for developing countries is 2.7 metric tons per person, which is three times IPCC's recommendation. Even worse, average annual carbon dioxide emissions for developed countries is 10.2 metric tons per person, more than ten times IPCC's recommendation (see Table 68).

Table 68 GDP Ranks, per-Person GDPs, and per-Person Carbon Dioxide Emissions

Country	2010 GDP Rank	2010 per-Person GDP (thousands of international dollars)	2010 per-Person Carbon Emissions (metric tons)
Sweden	22	39.0	5.3
Switzerland	19	46.4	5.4
France	5	34.1	5.5
Italy	8	32.0	6.7
United Kingdom	6	35.7	7.9
Austria	25	40.0	8.1
Denmark	31	40.2	8.4
Japan	3	33.7	8.9
Germany	4	37.4	9.3
Belgium	21	37.6	9.9
Norway	23	57.2	10.5
Australia	15	38.2	16.0
United States	1	47.2	18.1
Netherlands	16	42.2	31.9

Source: *World Bank; Carbon Dioxide Information Analysis Center*

GDP, the gross domestic product, is a measure of a country's economic strength.†

With annual carbon dioxide emissions of 18.1 metric tons per person, the United States would have to reduce emissions by 95% to meet the standard of 0.9 metric ton per person. This means Americans could emit only 5% of the carbon dioxide they currently emit. Imagine driving your car, heating and cooling your home, using your home appliances (including your refrigerator), using your computer, using your lights, and watching your television only 5% (one-twentieth) of the time you currently do.

Now that the Kyoto Protocol has expired, it is time for a new climate deal. Recall from the Climate Change Lab in Chapter 2 that the United Nations has agreed to the Durban Platform for Enhanced Action. The details of the plan are still being worked out, and the agreement won't come into effect until 2020. Unlike the Kyoto Protocol, the Durban Platform will be legally binding to all countries.‡

Progress on working out the details is slow, mostly because many countries fear that reducing greenhouse emissions will hurt their economies.

Some experts, however, such as the engineer Alan Pears, codirector of the environmental consultancy Sustainable Solutions, believe that it is possible for emissions to be significantly reduced without harming a country's economy. Norway, for instance, has a larger per-person GDP than the United States has, as well as significantly lower per-person carbon dioxide emissions. In fact, with the exception of the Netherlands and Australia, all of the countries listed in Table 68 have strong economies and significantly lower per-person carbon dioxide emissions than the United States has. A scattergram of the data would show that countries with higher per-person GDP do not necessarily have higher carbon dioxide emissions.

Many states have taken the matter into their own hands by adopting policies to reduce carbon dioxide emissions. And by using alternative sources of energy, many countries have slowed or reversed the growth of carbon dioxide emissions in recent years.§

In addition to national, state, and even corporate actions, individuals can help lower carbon dioxide emissions by purchasing hybrid automobiles, major appliances with the Energy Star logo, solar thermal systems to help provide hot water, and compact fluorescent light bulbs. Individuals can also car pool or use public transportation.

Analyzing the Situation

1. a. Let $f(t)$ be the average global temperature (in degrees Fahrenheit) at t years since 1900. Use a graphing calculator to draw a scattergram of the data in Table 65 and then find an equation of f *for the years 1965 to 2000.* Then verify that your model fits the data well for those years.

b. In Problem 2a of the Climate Change Lab in Chapter 3, you found the average global temperature from 1900 to 1965. If you didn't do this, do so now.

*United Nations Population Division.

†International Energy Agency, "CO_2 Emissions from Fuel Combustion Highlights," 2011.

‡United Nations, "Establishment of an Ad Hoc Working Group on the Durban Platform for Enhanced Action," December, 11, 2011.

§Pamela Person, "Reducing Greenhouse Gas Emissions," Maine Center for Economic Policy, *Choices*, VII(9), Oct. 11, 2001.

[**Hint:** Divide the sum of the temperatures by the number of temperature readings.] Use your model to predict when the planet's average temperature will have increased by 3.6°F—a potentially dangerous climate change.

2. Use Tables 66 and 67 to verify the claims that although the U.S. population in 2010 was only 5% of world population, 18% of annual world carbon dioxide emissions was produced by the United States in that year.
3. Use the United Nations' prediction that the world population will be 9.3 billion in 2050 to verify the claim that per-person carbon dioxide emissions that year should be about 0.9 metric ton per person for the IPCC recommendation of a 60% reduction by then.
4. Let $g(t)$ be the U.S. population (in billions) at t years since 1950. Create a scattergram of the data and then find an equation of g. Finally, verify that your model fits the data well.
5. Let $h(t)$ be U.S. carbon dioxide emissions (in billions of metric tons) in the year that is t years since 1950. Use a graphing calculator to draw a scattergram of the data and then find an equation of h. Finally, verify that your model fits the data well.
6. **a.** Use your model g of Problem 4 to estimate U.S. population in 2011.
   b. Use your model h of Problem 5 to estimate U.S. carbon dioxide emissions in 2011.
   c. Use your results from parts (a) and (b) to estimate U.S. *per-person* carbon dioxide emissions in 2011.
   d. The actual U.S. per-person carbon dioxide emissions in 2011 were 18.0 metric tons. Is your result in part (c) an underestimate or an overestimate? Explain why this can be explained at least in part by the poor economy during the period 2008–2011.
7. Let G be the per-person GDP (in international dollars) of a country with per-person carbon emissions c (in metric tons). Use a graphing calculator to draw a scattergram of the data. Do countries with higher per-person GDPs always have higher per-person carbon emissions? Explain.

Golf Ball Lab

In this lab, you will explore the relationship between the height of a golf ball before dropping it and its height after one bounce.*

Materials

You will need at least three people and the following items:

1. a tape measure
2. a golf ball

*The Golf Ball Lab adapted from a lab written by Jim Ryan, State Center Community College District, Willow International College Center, Clovis, CA. Used by permission of James Ryan.

Recording the Data

The same person should drop the golf ball each time. A second person should measure the height of the golf ball (from the bottom of the ball) before the first person drops it. The ball should be dropped from an initial height of 12 inches. A spotter should estimate the bounce height of the golf ball. Repeat this process three times. Then compute the average of the three bounce heights. Next, find average bounce heights of the golf ball for initial heights of 24 inches, 36 inches, 48 inches, 60 inches, and 72 inches. If your instructor prefers, use the data listed in Table 69.

Table 69 Drop and Bounce Heights of a Golf Ball

Drop Height (inches)	Bounce Height (inches)
12	10.0
24	20.3
36	31.0
48	44.5
60	52.0
72	64.0

Source: *J. Lehmann*

Analyzing the Data

1. Display your golf ball data in a table.
2. Let B be the bounce height (in inches) after the ball was dropped from an initial height of H inches. Use a graphing calculator to draw a scattergram of the golf ball data.
3. Find an equation of a linear model to describe the situation. Write your equation with the function name f.
4. Find the B-intercept of your model. What does it mean in this situation? If you can find a linear model with a better B-intercept, do so.
5. Use a graphing calculator to draw a graph of your model and the scattergram in the same viewing window. Also, graph the model and scattergram by hand. How well does f model the data?
6. Use your model to estimate the bounce height for a drop height of 80 inches.
7. On a golf course, a golf ball is hit to a maximum height of 50 feet. What does your model estimate the bounce height to be after one bounce? Do you think this estimate is accurate? If not, will it be an underestimate or an overestimate? Explain.
8. Find the slope of your model. What does the slope mean in this situation? Explain.
9. Estimate the bounce height after three bounces for a drop height of 90 inches.

Taking It One Step Further

10. Redo the experiment with a rubber ball and then with a tennis ball. Then repeat Parts 1–5. Finally, compare the slopes of your three linear models and explain why the comparison makes sense.

Rope Lab

In this lab, you will explore the relationship between the number of knots tied in a rope and the rope's length.

Check with your instructor whether you should collect your own data or use the data listed in Table 70.

Table 70 Lengths of a Rope with Diameter about 7 Millimeters

Number of Knots	Length of Rope (centimeters)
0	60.0
1	53.2
2	45.8
3	38.3
4	30.6

Source: *J. Lehmann*

Materials

1. A 60-centimeter-long piece of rope with diameter about 7 millimeters
2. A meterstick or other measuring device with units of millimeters

Recording of Data

Pull the rope taut and measure its length (in centimeters). Then tie a knot close to one end of the rope and measure the length of the rope again. Next, tie another knot next to the first one and measure the length of the rope. Continue tying knots, working your way along the rope and measuring the rope's length after you have tied each knot. Tie a total of four knots.

Analyzing the Situation

1. Display your data in a table or use the data in Table 70.
2. Let $L = f(n)$ be the length (in centimeters) of the rope with n knots. Use a graphing calculator to draw a scattergram of the data. Copy the scattergram by hand.
3. Find an equation of f.
4. Use a graphing calculator to draw a graph of f and the scattergram in the same viewing window. Copy the graph of f and the scattergram by hand.
5. Is f increasing or decreasing? What does that mean in this situation?
6. Find the L-intercept of your model. What does it mean in this situation?
7. Find the slope of your model. What does it mean in this situation?
8. Use f to estimate the length of the rope with five knots.
9. Check whether your result in Problem 8 is an underestimate or an overestimate by tying a fifth knot in the rope and then measuring the rope's length.
10. Continue tying knots in the rope. When does model breakdown first occur? Explain.
11. Find the n-intercept of your model. What does it mean in this situation?

Shadow Lab

In this lab, you will compare the relationship between an object's height and the length of its shadow.

Check with your instructor whether you should collect your own data or use the data listed in Table 71.

Table 71 Heights of Objects and the Lengths of Their Shadows

Object	Height (inches)	Length of Shadow (inches)
Nothing	0	0
Wine bottle	15.5	10.3
Toy putter	20.8	13.0
Box	36.5	23.6
Mop	47.8	31.0
Person's shoulder	55.0	36.5
Person	63.0	41.5

Source: *J. Lehmann*

Materials

1. Six objects of various heights up to 7 feet
2. A building, pole, tree, or other tall object with height greater than 15 feet
3. A tape measure or other measuring device

Recording of Data

Run the experiment when the objects (including the tall object) have noticeable and measurable shadows. For each object, measure its height and the length of its shadow. Record the beginning and ending time of the experiment. Also, record the length of the shadow of the tall object. It is important that you record all the data quickly.

Analyzing the Situation

1. Display your data in a table similar to Table 71. Those data were collected from 2:10 P.M. to 2:20 P.M., and the tall object is a tree whose shadow has a length of 49.5 feet.
2. Let h be the height (in inches) of an object and $L = f(h)$ be the length (in inches) of the object's shadow. Use a graphing calculator to draw a scattergram of the data. Copy the scattergram by hand.
3. Find an equation of f.
4. Use a graphing calculator to draw the graph of f and the scattergram in the same viewing window. Copy the graph of f and the scattergram by hand.
5. Is f increasing or decreasing? What does that mean in this situation?
6. Find the L-intercept of your model. What does it mean in this situation?
7. Find the slope of your model. What does it mean in this situation?
8. Use the length of the shadow of the tall object to estimate the object's height.
9. Explain why it was important that you run the experiment quickly.

10. Suppose you had run the experiment half an hour later. How would that have affected the slope of your model? Explain. (If the Sun would have set by then, describe the impact on the slope if the experiment had been performed half an hour earlier.) Would this change in time have resulted in a different estimate of the height of the tall object? Explain.

Linear Lab: Topic of Your Choice

Your objective in this lab is to use a linear model to describe some authentic situation. Find some data on two quantities that describe a situation that has not been discussed in this text. Almanacs, newspapers, magazines, scientific journals, and the Internet are good resources. Or you can conduct an experiment. Choose something that interests you!

Analyzing the Situation

1. What two quantities did you explore? Define variables for the quantities. Include units in your definitions.
2. Which variable is the dependent variable? Which variable is the independent variable? Explain.
3. Describe how you found your data. If you conducted an experiment, provide a careful description with specific details of how you ran your experiment. If you didn't conduct an experiment, state the source of your data.
4. Include a table of your data.
5. Use a graphing calculator to draw a scattergram of your data. (If your data are not approximately linear, find some data that are.)
6. Find an equation of a linear model to describe the data.
7. What is the slope of your linear model? What does it mean in this situation?
8. Does it make sense that your variables are approximately linearly related in terms of the situation you chose to model? Explain.
9. Choose a value for your independent variable. On the basis of that chosen value, use your model to find a value for your dependent variable. Describe what your result means in the situation you are modeling.
10. Choose a value for your dependent variable. On the basis of that chosen value, use your model to find a value for your independent variable. Describe what your result means in the situation you are modeling.
11. Find the intercepts of your linear model. What do they mean in the situation you are modeling? Has model breakdown occurred at the intercepts?
12. Comment on your lab experience.
    a. For example, you might address whether the lab was enjoyable, insightful, and so on.
    b. Were you surprised by any of your findings? If so, which ones?
    c. How would you improve your process for this lab if you were to do it again?
    d. How would you improve your process if you had more time and money?

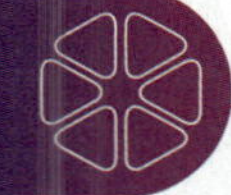

Chapter Summary

Key Points of Chapter 5

Section 5.1 Graphing Linear Equations

Solving for y to graph	Before we can use the y-intercept and the slope to graph a linear equation, we must solve for y to put the equation into the form $y = mx + b$.
Intercepts of the graph of an equation	For an equation containing the variables x and y, • To find the x-coordinate of each x-intercept, substitute 0 for y and solve for x. • To find the y-coordinate of each y-intercept, substitute 0 for x and solve for y.
Using intercepts to graph an equation	To graph a linear equation whose graph has exactly two intercepts, 1. Find the intercepts. 2. Plot the intercepts and a third point on the line, and graph the line that contains the three points.
Comparing linear equations in one and two variables	A linear equation in two variables has an infinite number of (ordered-pair) solutions. A linear equation in one variable has exactly one (real-number) solution.

Section 5.2 Functions

Relation, domain, and range	A **relation** is a set of ordered pairs. The **domain** of a relation is the set of all values of the independent variable, and the **range** of the relation is the set of all values of the dependent variable.
Input and output	Each member of the domain is an **input,** and each member of the range is an **output.**
Function	A **function** is a relation in which each input leads to exactly one output.
Vertical line test	A relation is a function if and only if each vertical line intersects the graph of the relation at no more than one point.
Linear function	A **linear function** is a relation whose equation can be put into the form $y = mx + b$, where m and b are constants.
Rule of Four for functions	We can describe some or all of the input–output pairs of a function by means of (1) an equation, (2) a graph, (3) a table, or (4) words. These four ways to describe input–output pairs of a function are known as the **Rule of Four** for functions.

Section 5.3 Function Notation

Function notation	The dependent variable of a function f can be represented by the expression formed by writing the independent variable name within the parentheses of $f(\ )$: dependent variable $= f($independent variable$)$ We call this representation **function notation.**

Section 5.4 Finding Linear Equations

Finding an equation of a line by using the slope, a point, and the slope–intercept form	To find an equation of a line by using the slope and a point, **1.** Substitute the given value of the slope m into the equation $y = mx + b$. **2.** Substitute the coordinates of the given point into your equation from step 1 and solve for b. **3.** Substitute the value of b from step 2 into your equation from step 1. **4.** Check that the graph of your equation contains the given point.
Finding an equation of a line by using two points and the slope–intercept form	To find an equation of the line that passes through two given points whose x-coordinates are different, **1.** Use the formula $m = \frac{y_2 - y_1}{x_2 - x_1}$, or use a graph to determine $\frac{\text{rise}}{\text{run}}$, to find the slope of the line containing the two points. **2.** Substitute the m value you found in step 1 into the equation $y = mx + b$. **3.** Substitute the coordinates of one of the given points into the equation from step 2 and solve for b. **4.** Substitute the b value you found in step 3 into your equation from step 2. **5.** Check that the graph of your equation contains the two given points.
Point–slope form	If a nonvertical line has slope m and contains the point (x_1, y_1), then an equation of the line is $y - y_1 = m(x - x_1)$.

Section 5.5 Finding Equations of Linear Models

Finding an equation of a linear model	To find an equation of a linear model, given some data, **1.** Create a scattergram of the data. **2.** Determine whether there is a line that comes close to the data points. If so, choose two points (not necessarily data points) that you can use to find the equation of a linear model. **3.** Find an equation of the line. **4.** Use a graphing calculator to verify that the graph of your equation contains the two chosen points and comes close to all of the points of the scattergram.

Section 5.6 Using Function Notation with Linear Models to Make Estimates and Predictions

Using an equation of a linear model to make predictions	When making a prediction about the dependent variable of a linear model, substitute a chosen value for the independent variable in the model and then solve for the dependent variable. When making a prediction about the independent variable of a linear model, substitute a chosen value for the dependent variable in the model and then solve for the independent variable.
Four-step modeling process	To find a linear model and then make estimates and predictions, **1.** Create a scattergram of the data to determine whether there is a nonvertical line that comes close to the points. If so, choose two points (not necessarily data points) that you can use to find an equation of a linear model. **2.** Find an equation of your model. **3.** Verify your equation by checking that the graph of your model contains the two chosen points and comes close to all of the data points. **4.** Use the equation of your model to make estimates, make predictions, and draw conclusions.
Using an equation of a linear model to find intercepts	If a function of the form $p = mt + b$, where $m \neq 0$, is used to model a situation, then • The p-intercept is $(0, b)$. • To find the t-coordinate of the t-intercept, substitute 0 for p in the model's equation and then solve for t.
Domain and range of a model	For the **domain** and **range** of a model, we consider input–output pairs only when both the input and the output make sense in the situation. The domain of the model is the set of all such inputs, and the range of the model is the set of all such outputs.

Section 5.7 Solving Linear Inequalities in One Variable

Addition property of inequalities	If $a < b$, then $a + c < b + c$. Similar properties hold for $\leq$, $>$, and $\geq$.
Multiplication property of inequalities	• For a *positive* number c, if $a < b$, then $ac < bc$. • For a *negative* number c, if $a < b$, then $ac > bc$. Similar properties hold for $\leq$, $>$, and $\geq$.
Linear inequality in one variable	A **linear inequality in one variable** is an inequality that can be put into one of the forms $mx + b < 0, \quad mx + b \leq 0, \quad mx + b > 0, \quad mx + b \geq 0$ where m and b are constants and $m \neq 0$.
Solution, solution set, and solve for an inequality in one variable	We say that a number is a **solution** of an inequality in one variable if it satisfies the inequality. The **solution set** of an inequality is the set of all solutions of the inequality. We **solve** an inequality by finding its solution set.

Chapter 5 Review Exercises

For Exercises 1–7, determine the slope and the y-intercept. Use the slope and the y-intercept to graph the equation by hand. Use a graphing calculator to verify your work.

1. $3y = 5x$

2. $3x - 2y = -6$

3. $x + 3y = 6$

4. $2x + 5y - 20 = 0$

5. $y - 4 = 0$

6. $-3(y + 2) = 2x + 9$

7. $3x - 2(2y - 1) = 8x - 3(x + 2)$

8. Determine the slope and y-intercept of the graph of the equation $a(x - y) = c$.

For Exercises 9–12, find the x-intercept and y-intercept. Then graph the equation by hand.

9. $4x - 5y = 20$

10. $3x + 4y + 12 = 0$

11. $y = 2x - 4$

12. $\frac{1}{3}x - \frac{1}{2}y = 1$

13. Find the x-intercept and y-intercept of the graph of $ax + b = cy$.

For Exercises 14 and 15, find the approximate x-intercept and approximate y-intercept. Round the coordinates to the second decimal place.

14. $9.2x - 3.8y = 87.2$

15. $y = 2.56x + 97.25$

16. **a.** Find the intercepts of the graph of $y = 3x + 7$.
b. Find the intercepts of the graph of $y = 2x + 9$.
c. Find the intercepts of the graph of $y = mx + b$, where b is a constant and m is a nonzero constant.

17. Consider $2x - 4y = 8$.

a. Graph the equation by finding the slope and the y-intercept.

b. Graph the equation by finding the intercepts.

c. Compare your graphs from parts (a) and (b). Decide which method you prefer for graphing the equation $2x - 4y = 8$.

18. Some ordered pairs for four relations are listed in Table 72. Which of these relations could be functions? Explain.

Table 72 Which Relations Might Be Functions? (Exercise 18)

Relation 1		Relation 2		Relation 3		Relation 4	
x	y	x	y	x	y	x	y
1	12	3	27	0	7	2	1
2	15	4	24	1	7	2	2
3	18	4	21	2	7	2	3
4	21	5	18	3	7	2	4
5	24	6	15	4	7	2	5

19. Determine whether the graph in Fig. 83 represents a function. Explain.

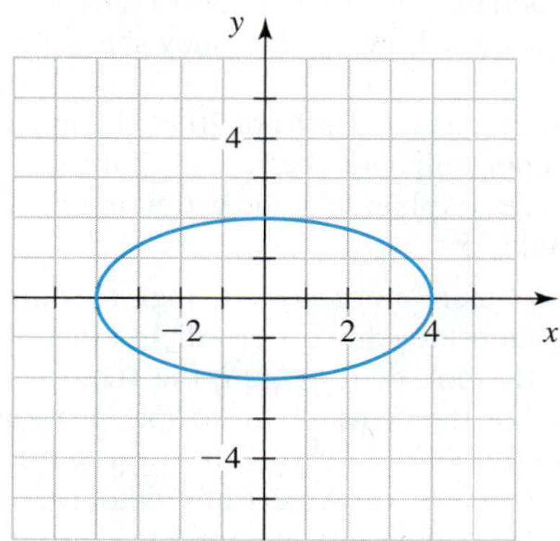

Figure 83 Exercise 19

For Exercises 20–22, determine whether each relation is a function. Explain.

20. $5x - 6y = 3$ **21.** $x = 9$ **22.** $y^2 = x$

23. Use the graph of the function in Fig. 84 to determine the function's domain and range.

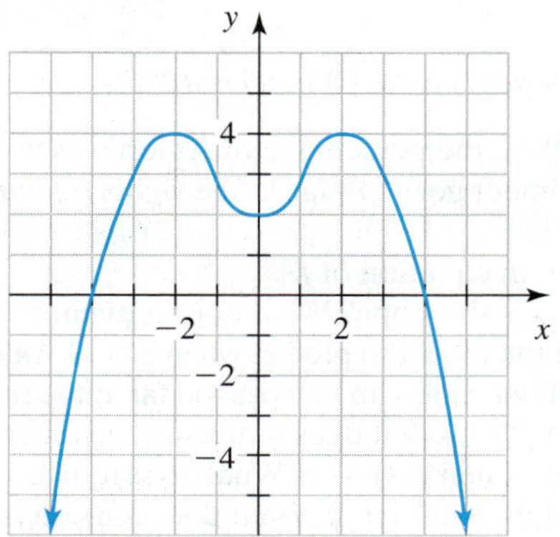

Figure 84 Exercise 23

For $f(x) = 3x^2 - 7$, $g(x) = \frac{2x + 5}{3x + 6}$, *and* $h(x) = -10x - 3$, *find the following.*

24. $f(-3)$ **25.** $g(2)$ **26.** $h(a + 3)$

For $f(x) = 2x + 3$, *find the value of x that corresponds to the given value of* $f(x)$.

27. $f(x) = \frac{2}{3}$ **28.** $f(x) = a + 7$

For Exercises 29–34, refer to Fig. 85.

29. Estimate $f(0)$.

30. Estimate $f(-3)$.

31. Estimate x when $f(x) = 0$.

32. Estimate x when $f(x) = -1$.

33. Find the domain of f.

34. Find the range of f.

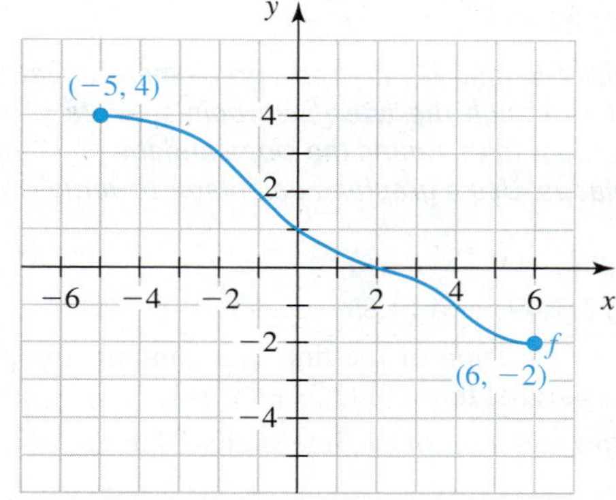

Figure 85 Exercises 29–34

For Exercises 35 and 36, refer to Table 73.

35. Find $f(2)$.

36. Find x when $f(x) = 2$.

Table 73 Values of f (Exercises 35 and 36)

x	f(x)
0	1
1	2
2	4
3	3
4	0

Find all x-intercepts and y-intercepts of the graph of the function.

37. $f(x) = -7x + 3$ **38.** $f(x) = -\frac{4}{7}x + 2$

Find an equation of the line that has the given slope and contains the given point. If possible, write your equation in slope–intercept form. Check that the graph of your equation contains the given point.

39. $m = -4$, $(2, -1)$

40. $m = -\frac{2}{3}$, $(-6, -4)$

41. m is undefined, $(2, 5)$

42. $m = 0, (-1, -4)$

Find an approximate equation of the line that has the given slope and contains the given point. Write your equation in slope–intercept form. Round the constant term to the second decimal place. Use a graphing calculator to verify your equation.

43. $m = -5.29, (-4.93, 8.82)$

44. $m = 1.45, (-2.79, -7.13)$

Find an equation of the line that passes through the two given points. If possible, write your equation in slope–intercept form. Check that the graph of your equation contains the given points.

45. $(-2, -7)$ and $(1, 2)$

46. $(2, -5)$ and $(4, 5)$

47. $(-3, 9)$ and $(6, -6)$

48. $(-4, -10)$ and $(-2, -7)$

49. $(5, -3)$ and $(5, 2)$

50. $(-4, -3)$ and $(-1, -3)$

For Exercises 51 and 52, find an approximate equation of the line that passes through the two given points. Write your equation in slope–intercept form. Round the slope and the constant term to two decimal places. Use a graphing calculator to verify your equation.

51. $(3.5, 9.2)$ and $(8.7, 4.8)$

52. $(-5.22, 2.49)$ and $(1.83, -3.99)$

53. Find an equation of the line that contains the point $(-2, 5)$ and is parallel to the line $3x - y = 6$.

54. Find an equation of the line sketched in Fig. 86.

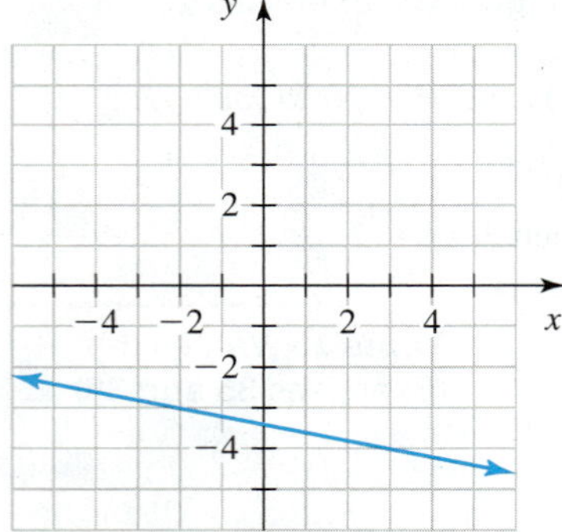

Figure 86 Exercise 54

55. Find an equation of a line that comes close to the points listed in Table 74. Then use a graphing calculator to check that your line comes close to those points.

Table 74 Find an Equation

x	y
1	28
4	23
6	16
9	13
10	8

56. The Internal Revenue Service (IRS) standard mileage rate is a way of computing an automobile expense deduction on a tax return. Mileage rates for businesses are provided in Table 75 for various years.

Table 75 IRS Standard Mileage Rates for Businesses

Year	Standard Mileage Rate (cents per mile)
2000	32.5
2002	36.5
2004	37.5
2006	44.5
2008	54.5
2010	50.0
2012	55.5

Source: *IRS*

Let $M = f(t)$ be the standard mileage rate (in cents per mile) at t years since 2000.

a. Find an equation of f.

b. What is the slope? What does it mean in this situation?

c. Find the M-intercept. What does it mean in this situation?

d. Predict when the standard mileage rate will be 69 cents per mile.

e. If a person will drive 12,500 miles on business trips in 2017, predict how much money she will be able to deduct for driving expenses.

f. For the first six months of 2008, the mileage rate was 50.5 cents per mile. For the rest of that year, it was 58.5 cents per mile. Explain why the estimate of 54.5 cents was used in Table 75.

57. The First Amendment says: "Congress shall make no law respecting an establishment of religion or prohibiting the free exercise thereof, or abridging the freedom of speech or of the press, or the right of the people peaceably to assemble, and to petition the government for a redress of grievances." The percentages of Americans who feel the amendment goes too far in the rights it guarantees are shown in Table 76.

Table 76 Percentages of Americans Who Feel the First Amendment Goes Too Far in the Rights It Guarantees

Year	Percent
2001	39
2003	34
2005	23
2007	25
2009	19
2011	18

Source: *State of the First Amendment 2011*

Let $f(t)$ be the percentage of Americans who think the First Amendment goes too far in the rights it guarantees at t years since 2000.

a. Find an equation of f.

b. What is the slope? What does it mean in this situation?

c. Use the model to predict when 3% of Americans will think the First Amendment goes too far in the rights it guarantees.

d. Find $f(8)$. What does it mean in this situation?

e. Find t when $f(t) = 8$. What does it mean in this situation?

f. Find the t-intercept. What does it mean in this situation?

58. Average U.S. personal income increased approximately linearly from \$14,420 in 1990 to \$27,589 in 2008 (Source: *U.S. Census Bureau*). Estimate the average personal income in 2010. The actual personal income in that year was \$26,059. Is your estimate an underestimate or an overestimate? Explain why this might be due to the poor economy from 2008 to 2010.

59. The average cost per 30-second ad slot during the Academy Awards has increased approximately linearly from \$1.3 million in 2009 to \$1.7 million in 2012 (Source: *Kantar Media*). Predict when the average cost will be \$2.4 million.

For Exercises 60–65, solve the inequality. Describe the solution set as an inequality, in interval notation, and in a graph. Then, use a graphing calculator table to verify your result.

60. $x - 3 \geq -4$

61. $-4x < 8$

62. $5w - 3 > 3w - 9$

63. $-3(2a + 5) + 5a \geq 2(a - 3)$

64. $\dfrac{2b - 4}{3} \leq \dfrac{3b - 4}{4}$

65. $1 \leq 2x + 5 < 11$

66. Violent-crime rates in the United States are shown in Table 77 for various years.

Table 77 Violent-Crime Rates

Year	Number of Violent Crimes per 100,000 People
1990	730
1995	685
2000	507
2005	469
2009	429

Source: *U.S. Department of Justice*

Let $f(t)$ be the violent-crime rate (number of violent crimes per 100,000 people) at t years since 1990.

a. Find an equation of f.

b. What is the slope? What does it mean in this situation?

c. Predict the years when the violent-crime rate will be less than 270 violent crimes per 100,000 people.

Chapter 5 Test

For Exercises 1–3, determine the slope and the y-intercept. Use the slope and the y-intercept to graph the equation by hand.

1. $2x - 5y = 10$

2. $y - 5 = 0$

3. $2(2x - y) = 2x + 9 + y$

4. Find the x-intercept and y-intercept of the graph of the equation $6x - 3y = 18$. Then graph the equation by hand.

5. Find the x-intercept and y-intercept of the graph of the equation $\dfrac{x}{2} + \dfrac{y}{7} = 1$.

6. Determine whether the relation described by $y = \pm x$ is a function. Explain.

7. Determine whether the relation described by $y = -2x + 5$ is a function. Explain.

8. Use the graph of the relation in Fig. 87 to determine the relation's domain and range. Determine whether the relation is a function.

Figure 87 Exercise 8

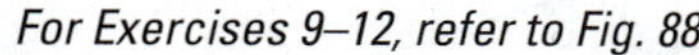

For Exercises 9–12, refer to Fig. 88.

9. Estimate $f(-3)$.

10. Estimate x when $f(x) = 0$.

11. Find the domain of f.

12. Find the range of f.

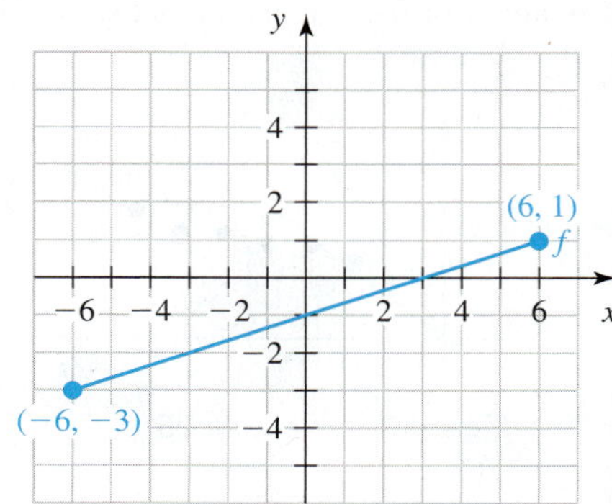

Figure 88 Exercises 9–12

For Exercises 13–16, let $f(x) = -4x + 7$.

13. Find $f(-3)$.

14. Find $f(a - 5)$.

15. Find x when $f(x) = 2$.

16. Find x when $f(x) = a$.

Find all x-intercepts and y-intercepts.

17. $f(x) = 3x - 7$

18. $k(x) = \dfrac{1}{3}x - 8$

For Exercises 19 and 20, find an equation of the line that has the given slope and contains the given point. If possible, write your equation in slope–intercept form.

19. $m = 7, (-2, -4)$ **20.** $m = -\frac{2}{3}, (6, -1)$

21. Find an equation of the line that passes through the points $(-4, 6)$ and $(2, 3)$.

22. Find an approximate equation of the line that passes through the points $(-3.4, 2.9)$ and $(1.8, -7.1)$. Write your equation in slope–intercept form. Round the slope and the constant term to two decimal places.

23. Find the equation of a line that contains the point $(4, -1)$ and is perpendicular to the line $3x - 5y = 20$.

24. Find an equation of the line sketched in Fig. 89.

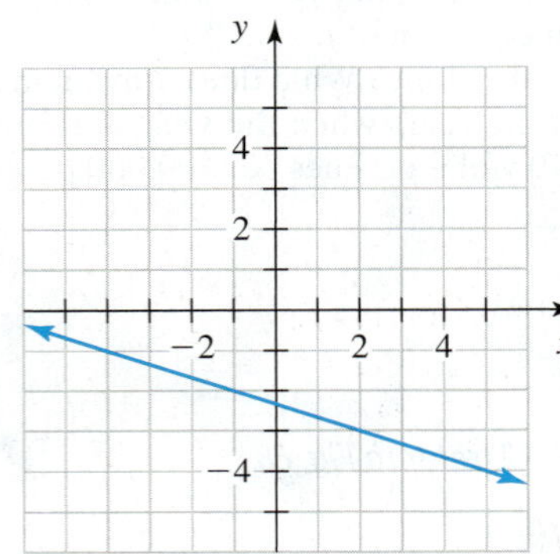

Figure 89 Exercise 24

25. Consider the scattergram of data and the graph of the model $y = mt + b$ in Fig. 90. Sketch a linear model that describes the data better, and then explain how you would adjust the values of m and b of the original model so that it would describe the data better.

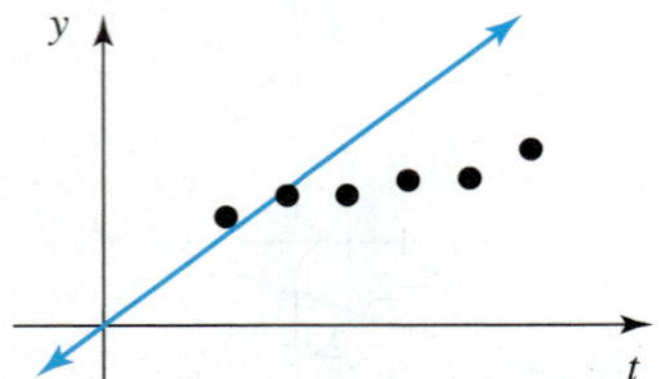

Figure 90 Exercise 25

26. Find an equation of a line that comes close to the points listed in Table 78. Then use a graphing calculator to check that your line comes close to those points.

Table 78 Find an Equation

x	y
1	7
3	11
7	16
12	20
14	23

27. The percentages of Fortune 100 companies that offer a pension are shown in Table 79 for various years.

Table 79 Percentages of Fortune 100 Companies That Offer a Pension

Year	Percent
1998	68
2002	49
2004	40
2007	28
2010	17

Source: *Hewitt Associates*

Let $p = f(t)$ be the percentage of Fortune 100 companies that offer a pension at t years since 1995.

a. Use a graphing calculator to draw a scattergram of the data.

b. Find an equation of f.

c. What is the slope? What does it mean in this situation?

d. What is the p-intercept? What does it mean in this situation?

e. What is the t-intercept? What does it mean in this situation?

f. When did all Fortune 100 companies offer a pension, according to the model?

g. Estimate in which years more than half of all Fortune 100 companies offered pensions.

Solve the inequality. Describe the solution set as an inequality, in interval notation, and in a graph.

28. $3(2x + 1) \leq 4(x + 2) - 1$

29. $\frac{1}{2} < 3 - \frac{5}{2}x \leq 13$

Systems of Linear Equations and Systems of Linear Inequalities

Even though the women's current record time for the 400-meter run is on par with the men's record times set far back in the early 1900s, women have been improving much faster than men (see Table 1). In Example 2 of Section 6.4, we will predict when the women's record time and the men's record time will be equal.

In Chapters 1–5, we used a single linear model to describe an authentic situation. In this chapter, we will use *two* linear models to describe authentic situations. In particular, we will predict when two quantities will be equal, such as the average annual U.S. per-person consumption of milk and soft drinks.

Table 1 400-Meter Run Record Times

Women		Men	
Year	Record Time (seconds)	Year	Record Time (seconds)
1957	57.0	1900	47.8
1959	53.4	1916	47.4
1962	51.9	1928	47.0
1969	51.7	1932	46.2
1972	51.0	1941	46.0
1976	49.29	1950	45.8
1979	48.60	1960	44.9
1983	47.99	1968	43.86
1985	47.60	1988	43.29
		1999	43.18

Source: *International Association of Athletics Federations*

6.1 Using Graphs and Tables to Solve Systems

Objectives

- Know the meaning of *solution* and *solution set* of a *system of linear equations in two variables.*
- Use a graphical approach to solve systems of linear equations.
- Use graphing to make predictions about situations that can be modeled by two linear models.
- Know the three types of linear systems of two equations.
- Find the solution of a system of linear equations from tables of solutions of such equations.

In this section, we will work with graphs and tables that describe solutions of two linear equations in two variables. We will also work with situations that can be modeled by using two linear models.

Systems of Two Linear Equations

A **system of linear equations in two variables,** or a **linear system,** for short, consists of two or more linear equations in two variables. Here is an example of a system of two linear equations in two variables:

$$y = 3x - 1$$
$$y = -2x + 4$$

We will work with such systems throughout this chapter.

Recall from Section 3.1 that every point on the graph of an equation represents a solution of the equation and that every point *not* on the graph represents an ordered pair that is *not* a solution. Knowing the meaning of a graph will help us greatly in this section.

Example 1 Finding Ordered Pairs That Satisfy Both of Two Given Equations

Find all ordered pairs that satisfy both of the equations in the system

$$y = 3x - 1$$
$$y = -2x + 4$$

Solution

To begin, we graph each equation on the same coordinate system (see Fig. 1).

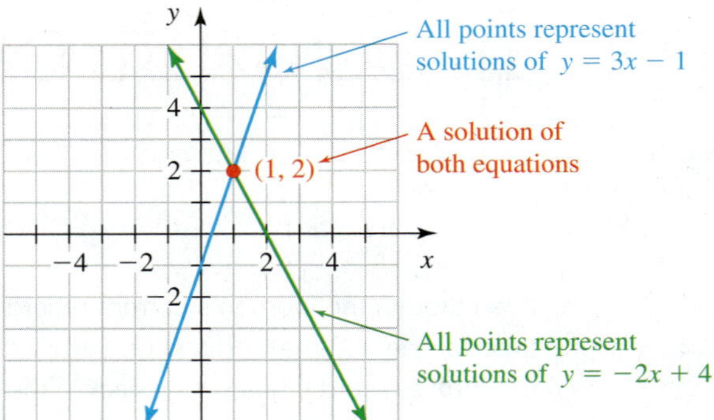

Figure 1 The intersection point is (1, 2)

For an ordered pair to be a solution of *both* equations, it must represent a point that lies on *both* lines. The intersection point (1, 2) is the only point that lies on both lines. So, the ordered pair (1, 2) is the only ordered pair that satisfies both equations.

We can verify that (1, 2) satisfies both equations:

$y = 3x - 1$	$y = -2x + 4$
$2 \stackrel{?}{=} 3(1) - 1$	$2 \stackrel{?}{=} -2(1) + 4$
$2 \stackrel{?}{=} 2$	$2 \stackrel{?}{=} 2$
true	true

Solution Set of a System

In Example 1, we worked with the system

$$y = 3x - 1$$
$$y = -2x + 4$$

We found that the only point whose coordinates satisfy both equations is the intersection point (1, 2). We call the set containing only (1, 2) the *solution set of the system.*

Definition *Solution, solution set,* and *solve* for a system

We say that an ordered pair (a, b) is a **solution** of a system of two equations in two variables if it satisfies both equations. The **solution set** of a system is the set of all solutions of the system. We **solve** a system by finding its solution set.

In general, **the solution set of a system of two linear equations can be found by locating any intersection point(s) of the graphs of the two equations.**

Example 2 Solving a System of Two Linear Equations by Graphing

Solve the system

$$y = 2x + 1$$
$$y = \frac{1}{2}x - 2$$

Solution

The graphs of the equations are sketched in Fig. 2.

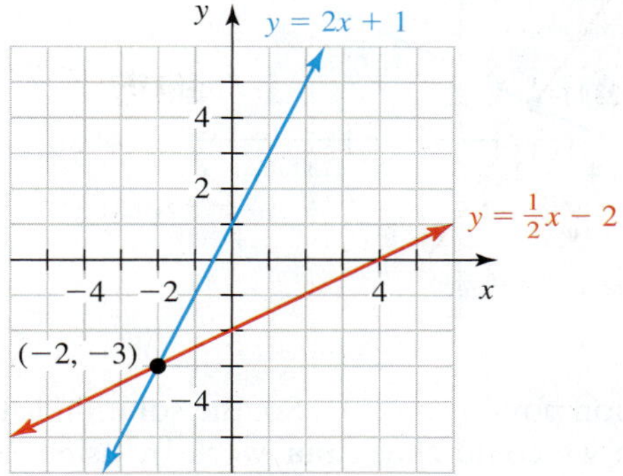

Figure 2 The intersection point is $(-2, -3)$

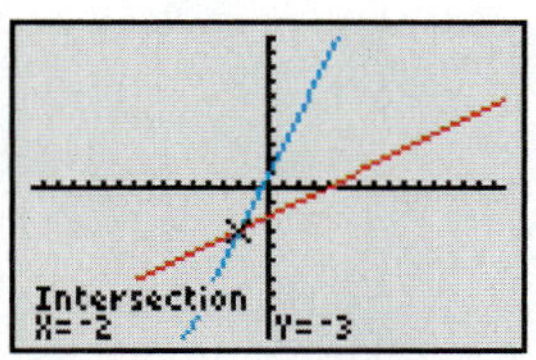

Figure 3 Verify that the intersection point is $(-2, -3)$

The intersection point is $(-2, -3)$. So, the solution is the ordered pair $(-2, -3)$.

We can check that $(-2, -3)$ satisfies both equations:

$$y = 2x + 1 \qquad\qquad y = \frac{1}{2}x - 2$$

$$-3 \stackrel{?}{=} 2(-2) + 1 \qquad\qquad -3 \stackrel{?}{=} \frac{1}{2}(-2) - 2$$

$$-3 \stackrel{?}{=} -3 \qquad\qquad -3 \stackrel{?}{=} -1 - 2$$

$$\text{true} \qquad\qquad -3 \stackrel{?}{=} -3$$

$$\text{true}$$

We can also check our work by using "intersect" on a graphing calculator (see Fig. 3). See Appendix A.18 for graphing calculator instructions.

WARNING After solving a system of two linear equations, you commit a common error if you check that a result satisfies only one of the two equations. It is important to check that your result satisfies *both* equations.

In Example 2, we solved a system in which both equations are in $y = mx + b$ form. If any equations in a system are not in $y = mx + b$ form, we usually begin by writing them in $y = mx + b$ form.

Example 3 Solving a System of Two Linear Equations by Graphing

Solve the system

$$3x + 2y = -4 \qquad \textit{Equation (1)}$$

$$y = \frac{1}{2}x + 2 \qquad \textit{Equation (2)}$$

Solution

First, we write equation (1) in slope–intercept form by solving the equation for y:

$$3x + 2y = -4 \qquad \textit{Equation (1)}$$

$$2y = -3x - 4 \qquad \textit{Subtract 3x from both sides.}$$

$$\frac{2y}{2} = \frac{-3x}{2} - \frac{4}{2} \qquad \textit{Divide both sides by 2.}$$

$$y = -\frac{3}{2}x - 2 \qquad \textit{Simplify.}$$

Next, we sketch a graph of the equations $y = -\frac{3}{2}x - 2$ and $y = \frac{1}{2}x + 2$ (see Fig. 4).

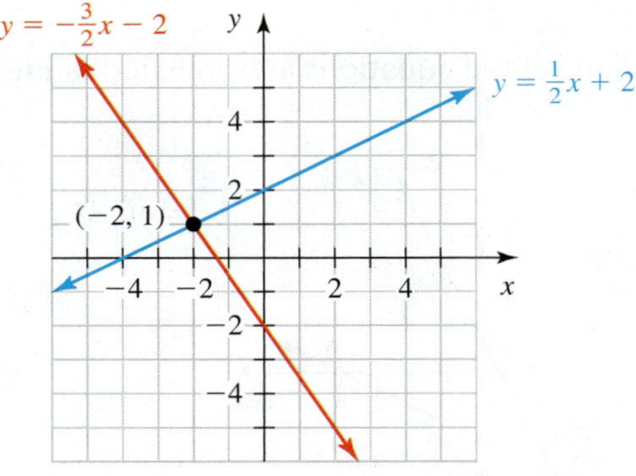

Figure 4 The solution is (−2, 1)

The intersection point is (−2, 1). So, the solution is the ordered pair (−2, 1).

Although we could check our work by using "intersect" on a graphing calculator with the equations $y = -\frac{3}{2}x - 2$ and $y = \frac{1}{2}x + 2$, this check would not verify that we solved $3x + 2y = -4$ for y correctly. To check all of our work, we check that (−2, 1) satisfies both of the *original* equations:

$$\begin{aligned} 3x + 2y &= -4 \\ 3(-2) + 2(1) &\stackrel{?}{=} -4 \\ -6 + 2 &\stackrel{?}{=} -4 \\ -4 &\stackrel{?}{=} -4 \\ &\text{true} \end{aligned} \qquad \begin{aligned} y &= \frac{1}{2}x + 2 \\ 1 &\stackrel{?}{=} \frac{1}{2}(-2) + 2 \\ 1 &\stackrel{?}{=} -1 + 2 \\ 1 &\stackrel{?}{=} 1 \\ &\text{true} \end{aligned}$$

Using a Graphing Calculator to Solve a System

So far, we have solved systems with simple equations that can be accurately graphed by hand in a reasonable amount of time. In Example 4, we will use a graphing calculator to help us solve a system with equations that are more difficult to graph.

Example 4 Using a Graphing Calculator to Solve a System

Use a graphing calculator to find any approximate solutions of the system

$$\begin{aligned} y &= 0.65x - 5.29 \\ y &= -1.82x + 4.47 \end{aligned}$$

Solution

We use "intersect" to find the approximate solution (3.95, −2.72). See Fig. 5.

We check that (3.95, −2.72) approximately satisfies both of the original equations:

$$\begin{aligned} y &= 0.65x - 5.29 \\ -2.72 &= 0.65(3.95) - 5.29 \\ -2.72 &\approx -2.7225 \end{aligned} \qquad \begin{aligned} y &= -1.82x + 4.47 \\ -2.72 &= -1.82(3.95) + 4.47 \\ -2.72 &\approx -2.719 \end{aligned}$$

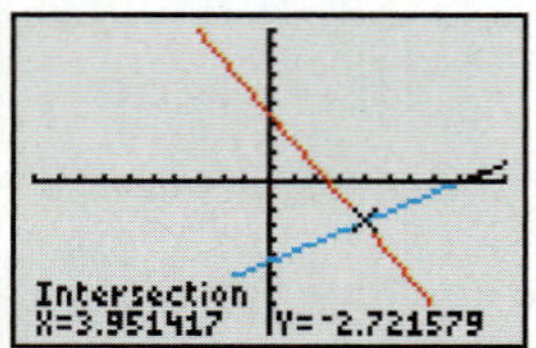

Figure 5 The approximate solution is (3.95, −2.72)

Example 5 Using a Graphing Calculator to Solve a System

Use a graphing calculator to find any approximate solutions of the system

$$\begin{aligned} \frac{1}{2}x + \frac{5}{4}y &= \frac{5}{2} \qquad \text{Equation (1)} \\ y &= 3x - 4 \qquad \text{Equation (2)} \end{aligned}$$

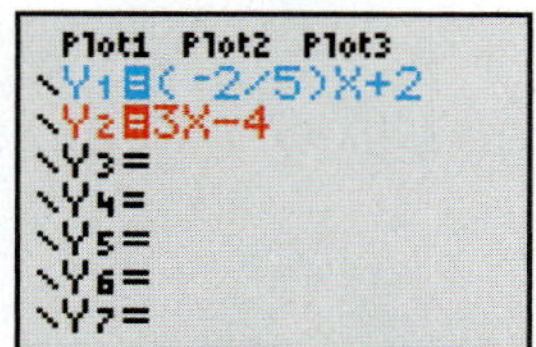

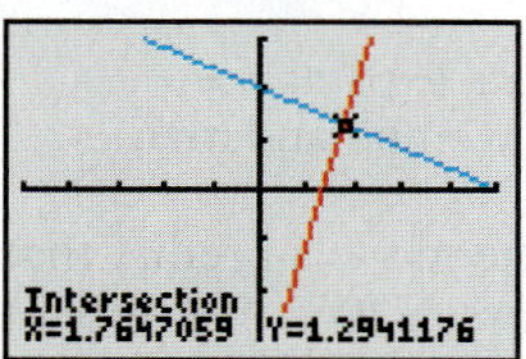

Figure 6 Using "intersect" to find the intersection point

Solution

First, we multiply both sides of equation (1) by the LCD, 4, to clear the equation of fractions, and then we solve for y:

$$\frac{1}{2}x + \frac{5}{4}y = \frac{5}{2} \qquad \text{Equation (1)}$$

$$4\left(\frac{1}{2}x + \frac{5}{4}y\right) = 4 \cdot \frac{5}{2} \qquad \text{Multiply both sides by LCD, 4.}$$

$$4 \cdot \frac{1}{2}x + 4 \cdot \frac{5}{4}y = 4 \cdot \frac{5}{2} \qquad \text{Distributive law}$$

$$2x + 5y = 10 \qquad \text{Simplify.}$$

$$5y = -2x + 10 \qquad \text{Subtract 2x from both sides.}$$

$$y = -\frac{2}{5}x + 2 \qquad \text{Divide both sides by 5.}$$

Next, we enter the equations $y = -\frac{2}{5}x + 2$ and $y = 3x - 4$ in a graphing calculator and use "intersect" to find the approximate solution (1.76, 1.29). See Fig. 6.

We check that (1.76, 1.29) approximately satisfies both of the original equations:

$$\frac{1}{2}x + \frac{5}{4}y = \frac{5}{2} \qquad\qquad y = 3x - 4$$

$$\frac{1}{2}(1.76) + \frac{5}{4}(1.29) \overset{?}{=} \frac{5}{2} \qquad\qquad 1.29 \overset{?}{=} 3(1.76) - 4$$

$$2.4925 \approx 2.5 \qquad\qquad 1.29 \approx 1.28$$

Because (1.76, 1.29) satisfies both equations approximately, we know (1.76, 1.29) is a good approximation of the *exact* solution, which we will learn to find in Section 6.2.

Using Two Linear Models to Make a Prediction

We can use graphing to make predictions about some authentic situations that can be modeled by two linear functions.

Example 6 Using Two Models to Make a Prediction

In the United States, life expectancies of women have been longer than life expectancies of men for many years (see Table 2). The life expectancies (in years) $W(t)$ and $M(t)$ of women and men, respectively, are modeled by the system

$$L = W(t) = 0.115t + 77.44$$
$$L = M(t) = 0.208t + 69.86$$

where t is the number of years since 1980. Use graphs of W and M to predict when life expectancies of women and men will be equal.

Table 2 U.S. Life Expectancies of Women and Men

Year of Birth	Women (years)	Men (years)
1980	77.4	70.0
1985	78.2	71.1
1990	78.8	71.8
1995	78.9	72.5
2000	79.5	74.1
2005	79.9	74.9
2009	81.3	76.2

Source: *U.S. Census Bureau*

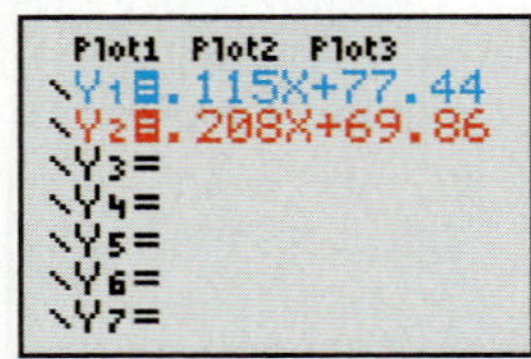

Figure 8 Enter the functions

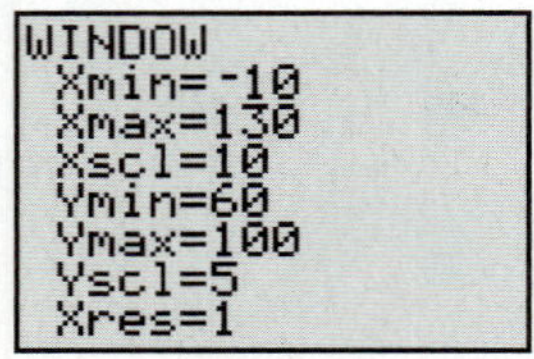

Figure 9 Set up the window

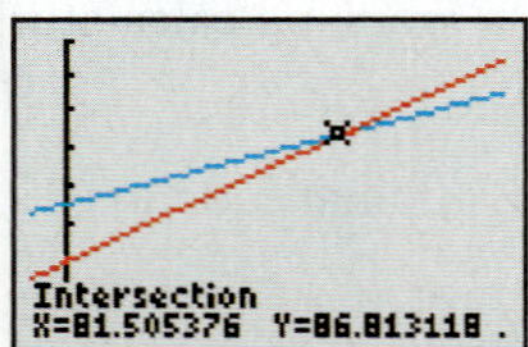

Figure 10 Find the intersection point

Solution

We begin by sketching graphs of W and M on the same coordinate system (see Fig. 7).

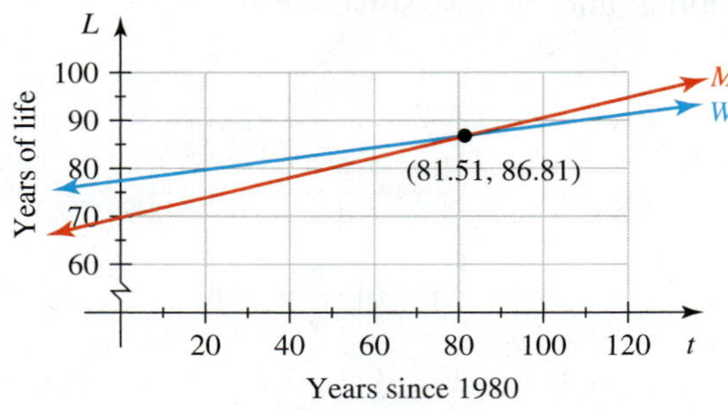

Figure 7 Life expectancy models of women and men

The intersection point is approximately (81.51, 86.81). So, the models predict that the life expectancy of both women and men will be about 86.8 years in 2062. We are not very confident about this prediction, however, because it is so far into the future.

We verify our work by using "intersect" on a graphing calculator (see Figs. 8–10). For graphing calculator instructions, see Appendix A.7 and A.18.

> **Intersection Point of the Graphs of Two Models**
>
> If the independent variable of two models represents time, then an intersection point of the graphs of the two models indicates when the quantities represented by the dependent variables were or will be equal.

Three Types of Linear Systems of Two Equations

Each of the systems in Examples 1–6 has one solution. Some systems do not have exactly one solution, however.

Example 7 A System Whose Solution Is the Empty Set

Solve the system

$$y = 2x - 1$$
$$y = 2x - 3$$

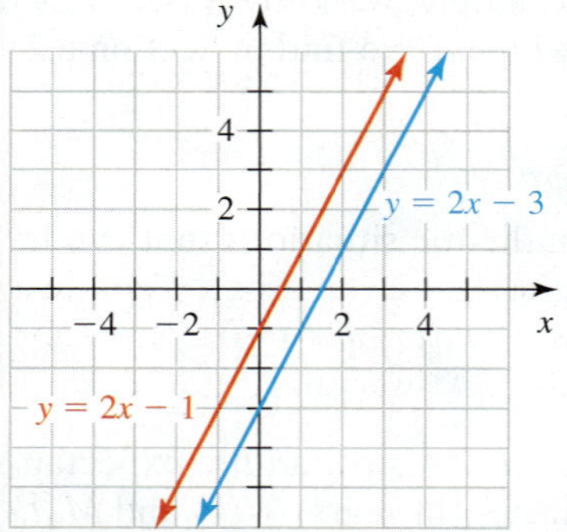

Figure 11 The solution set is the empty set

Solution

Since the distinct lines have equal slopes, the lines are parallel (see Fig. 11). Parallel lines do not intersect, so there is no ordered pair that satisfies both equations. The solution set is the empty set.

A linear system whose solution is the empty set is called an **inconsistent system.**

Example 8 A System That Has an Infinite Number of Solutions

Solve the system

$$y = 3x - 5 \quad \textit{Equation (1)}$$
$$6x - 2y = 10 \quad \textit{Equation (2)}$$

Solution

We write equation (2) in slope–intercept form:

$$\begin{aligned} 6x - 2y &= 10 && \textit{Equation (2)} \\ -2y &= -6x + 10 && \textit{Subtract 6x from both sides.} \\ y &= 3x - 5 && \textit{Divide both sides by } -2. \end{aligned}$$

So, the graph of $6x - 2y = 10$ and the graph of $y = 3x - 5$ are the same line. The solution set of the system is the set of the infinite number of ordered pairs represented by points on the line $y = 3x - 5$ and the (same) line $6x - 2y = 10$.

A linear system that has an infinite number of solutions is called a **dependent system.** In Examples 1, 7, and 8, we have seen three types of linear systems. We now describe them.

Types of Linear Systems

There are three types of linear systems of two equations:

1. One-solution system: The lines intersect in one point. The solution of the system is the ordered pair that corresponds to that point. See Fig. 12.
2. Inconsistent system: The lines are parallel. The solution set of the system is the empty set. See Fig. 13.
3. Dependent system: The lines are identical. The solution set of the system is the set of the infinite number of solutions represented by points on the same line. See Fig. 14.

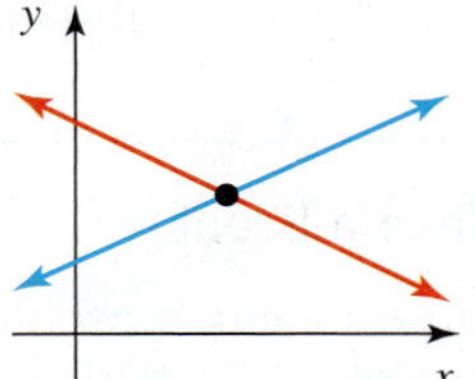

Figure 12 Graphs of the equations in a system with one solution

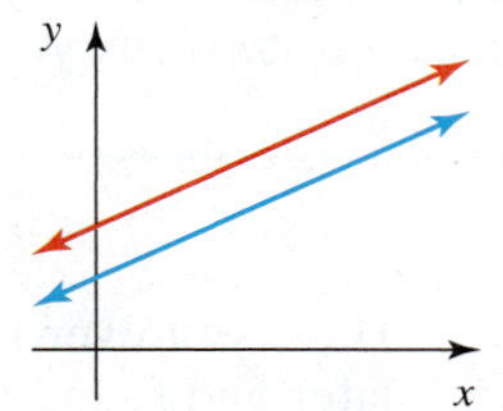

Figure 13 Graphs of the equations in an inconsistent system

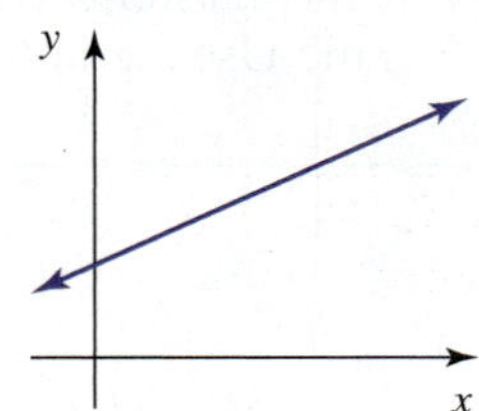

Figure 14 Graph of the equations in a dependent system

Finding the Solution of a System from a Table of Solutions

By the Rule of Four, we know we can use tables to describe some solutions of a linear equation in two variables. In Example 9, we will use such tables to help us find the solution of a system.

Example 9 Using a Table to Solve a System

Use a table of solutions to solve the following system of two equations:

$$y = 2x - 3$$
$$y = -3x + 7$$

Solution

Some solutions of the two equations are shown in Table 3.

Table 3 Some Solutions of $y = 2x - 3$ and $y = -3x + 7$

x	0	1	2	3	4
$y = 2x - 3$	−3	−1	1	3	5
$y = -3x + 7$	7	4	1	−2	−5

Since the ordered pair (2, 1) is a solution of both equations, it is a solution of the system of equations.

The lines $y = 2x - 3$ and $y = -3x + 7$ have different slopes, so there is only one intersection point. Thus, the ordered pair (2, 1) is the *only* solution of the system.

In Example 9, we used a table of solutions of two linear equations to help us find the solution of the linear system. **If an ordered pair is listed in a table as a solution of both of two linear equations, then that ordered pair is a solution of the system.**

Group Exploration

Comparing the three types of systems

1. Is the system

$$y = -2x + 3$$
$$y = -2x + 5$$

a dependent system, an inconsistent system, or a one-solution system? Explain.

2. Consider the system

$$y = 3x - 5$$
$$y = mx + b$$

where m and b are constants.

a. Find values of m and b such that the given system is inconsistent. What is the solution set of your system? Use a graphing calculator to verify your work.

b. Find values of m and b such that the given system is dependent. What is the solution set of your system?

c. Find values of m and b such that the given system is a one-solution system. Use "intersect" on a graphing calculator to find the solution.

3. Now consider this general system of two linear equations:

$$y = m_1x + b_1$$
$$y = m_2x + b_2$$

Discuss dependent systems, inconsistent systems, and one-solution systems in terms of $m_1, m_2, b_1,$ and b_2.

Tips for Success Take a Break

Have you ever had trouble solving a problem but returned to the problem hours later and found it easy to solve? By taking a break, you can return to the exercise with a different perspective and renewed energy. You've also given your unconscious mind a chance to reflect on the problem while you take your break. You can strategically take advantage of this phenomenon by allocating time to complete your homework assignment at two different points in your day.

Homework 6.1

For extra help ▶ MyMathLab®

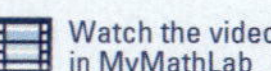
Watch the videos in MyMathLab

Download the MyDashboard App

*Determine which of the given ordered pairs is a solution of the given system. [**Hint:** There is no need to graph.]*

1. $(2, 3), (1, -1), (-4, 6)$
$y = 4x - 5$
$y = -2x + 1$

2. $(-3, 5), (-7, 3), (3, -7)$
$y = -x - 4$
$y = -3x + 2$

3. $(-1, 8), (3, -2), (7, 1)$
$5x + 2y = 11$
$3x - 4y = 17$

4. $(3, -1), (-4, -2), (-2, -5)$
$4x - 5y = 17$
$3x + 2y = -16$

For Exercises 5–26, find the solution set of the system by graphing the equations by hand. If the system is inconsistent or dependent, say so. If your result is one ordered pair, check that it satisfies both equations.

5. $y = 2x + 2$
$y = -3x + 7$

6. $y = x - 5$
$y = -2x + 4$

7. $y = -\frac{1}{2}x + 3$
$y = \frac{1}{3}x + 8$

8. $y = -\frac{1}{4}x + 3$
$y = \frac{1}{2}x$

9. $y = -3x$
$y = 4x$

10. $y = x$
$y = -x$

11. $y = 3(x - 1)$
$y = -2x + 7$

12. $y = 2(x - 3)$
$y = -3x + 4$

13. $y = -2x + 3$
$4y - 12 = -8x$

14. $y = 3x - 7$
$9x - 3y = 21$

15. $4x - 6y = 24$
$6x - 9y = 18$

16. $20x - 8y = 16$
$-15x + 6y = 18$

17. $x + 4y = 20$
$2x - 4y = -8$

18. $2x - 3y = -6$
$x + 3y = -3$

19. $x = 3$
$y = -2$

20. $x = 0$
$y = 0$

21. $y = 2x - 1$
$2(2x - y) = 2$

22. $y = 2x + 1$
$3y - 1 = 2(3x + 1)$

23. $5(y - 2) = 21 - 2(x + 3)$
$y = 3(x - 1) + 8$

24. $3(y - 2) = 2(3x + 1) + 7$
$y = 7 - 4(x + 2)$

25. $\frac{1}{2}x - \frac{1}{2}y = 1$
$\frac{1}{4}x + \frac{1}{2}y = 2$

26. $\frac{1}{3}x + \frac{1}{2}y = -3$
$\frac{1}{2}x - \frac{1}{3}y = -\frac{7}{3}$

Use "intersect" on a graphing calculator to solve the system, with the coordinates of the solution rounded to the second decimal place. Check that your result approximately satisfies both equations.

27. $y = 2.18x - 5.34$
$y = -3.53x + 1.29$

28. $y = 4.95x + 7.51$
$y = -0.84x - 5.38$

29. $y = \frac{5}{4}x + 2$
$y = -\frac{1}{4}x - 5$

30. $y = -\frac{7}{3}x + 3$
$y = -\frac{2}{3}x - 2$

31. $y = \frac{3}{4}x - 8$
$5x + 3y = 6$

32. $y = -\frac{2}{3}x - 5$
$7y - 3x = 7$

33. $-2x + 5y = 15$
$6x + 14y = -14$

34. $12x + 9y = 18$
$-x - 4y = 20$

35. $\frac{1}{2}x - \frac{1}{2}y = 1$
$\frac{1}{3}x + \frac{2}{3}y = 2$

36. $\frac{4}{3}x - \frac{7}{3}y = -5$
$\frac{1}{4}x + \frac{3}{4}y = 6$

37. The winning times for the Olympic 500-meter speed-skating event have generally decreased since 1972 (see Table 4).

Table 4 Olympic 500-Meter Speed-Skating Times

	Winning Time (seconds)	
Year	**Women**	**Men**
1972	43.33	39.44
1976	42.76	39.17
1980	41.78	38.03
1984	41.02	38.19
1988	39.10	36.45
1992	40.33	37.14
1994	39.25	36.33
1998	38.21	35.59
2002	37.375	34.615
2006	38.285	34.880
2010	38.050	34.910

Source: *The Universal Almanac*

The winning times (in seconds) $W(t)$ and $M(t)$ for women and men, respectively, are modeled by the system

$$w = W(t) = -0.153t + 43.19$$
$$w = M(t) = -0.135t + 39.64$$

where t is the number of years since 1970.

a. Use the equations of the models to estimate the winning time for women and the winning time for men in 2010. Find the errors in your estimates.

b. Compare the slopes of the two models. What does your comparison tell you about this situation?

c. Explain why your work in parts (a) and (b) suggests that there may be a time when the women's winning time will be equal to the men's winning time.

d. Use "intersect" on a graphing calculator with the window shown in Fig. 15 to predict when the women's winning time will be equal to the men's winning time. What will be that winning time?

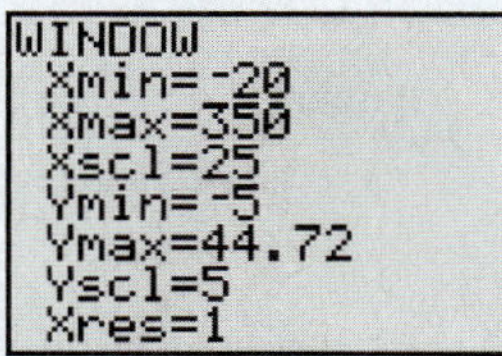

Figure 15 Window for skating models

38. Average annual U.S. per-person consumption of chicken and red meat (in pounds per person) is described for various years in Table 5.

Table 5 Average Annual U.S. Per-Person Consumption of Chicken and Red Meat

	Average Annual Consumption (pounds per person)	
Year	**Chicken**	**Red Meat**
1970	40.3	145.8
1980	48.0	136.8
1990	61.5	120.0
2000	78.0	120.7
2010	83.7	108.7
2011	84.2	104.3

Source: *U.S. Department of Agriculture*

Let $C(t)$ be the average annual per-person consumption of chicken and $R(t)$ be the average annual per-person consumption of red meat, both in pounds per person, in the year that is t years since 1970. The consumption of each can be modeled by the system

$$C(t) = 1.13t + 39.29$$
$$R(t) = -0.94t + 144.88$$

a. Use the equations of the models to estimate the consumptions of chicken and red meat in 2011. Find the errors in your estimates.

b. Compare the slopes of the two models. What does your comparison tell you about the situation?

c. Explain why your work in parts (a) and (b) suggests that there may be a time when the consumption of chicken will be equal to the consumption of red meat.

d. Use "intersect" on a graphing calculator with the window shown in Fig. 16 to predict when the consumption of chicken will equal that of red meat. What will be that consumption? How confident are you?

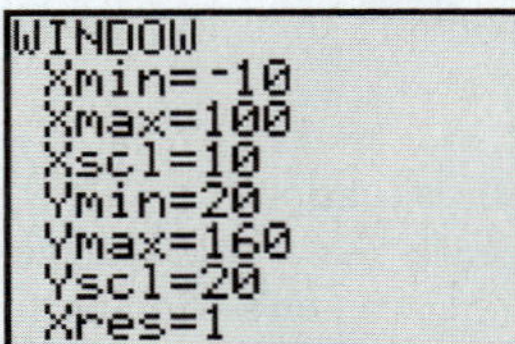

Figure 16 Window for consumption models

39. The percentages of bills paid by checks and online payments are shown in Table 6 for various years.

Table 6 Percentages of Bills Paid by Checks and Online Payments

	Percent	
Year	Checks	Online Payments
2007	34	39
2008	31	42
2009	30	44
2010	26	45
2011	23	50

Source: *Fiserv, Inc.*

a. Let $C(t)$ be the percentage of bills paid by checks and $W(t)$ be the percentage of bills paid by online payments, both at t years since 2000. Find equations of C and W.

b. Use "intersect" on a graphing calculator with the window shown in Fig. 17 to estimate when the percentage of bills paid by checks was equal to the percentage of bills paid by online payments. What was that percentage?

```
WINDOW
 Xmin=-2
 Xmax=10
 Xscl=2
 Ymin=15
 Ymax=60
 Yscl=5
 Xres=1
```

Figure 17 Window for bill models

c. Predict the total percentage of bills that will be paid by checks and online payments in 2018.

40. The worldwide numbers of fixed telephone lines and cell-phone subscriptions per 100 people are shown in Table 7 for various years.

Table 7 Worldwide Numbers of Fixed Telephone Lines and Cell-Phone Subscriptions per 100 People

	Number per 100 People	
Year	Fixed Telephone Lines	Cell-Phone Subscriptions
2006	19.2	40.6
2007	19.0	50.1
2008	18.3	59.3
2009	17.7	67.9
2010	17.3	76.2

Source: *ITU World Telecommunication*

a. Let $F(t)$ be the number of fixed telephone lines per 100 people and $C(t)$ be the number of cell-phone subscriptions per 100 people, both at t years since 2000. Find equations of F and C.

b. Use "intersect" on a graphing calculator with the window shown in Fig. 18 to estimate when the number of fixed telephone lines per 100 people was equal to the number of cell-phone subscriptions per 100 people.

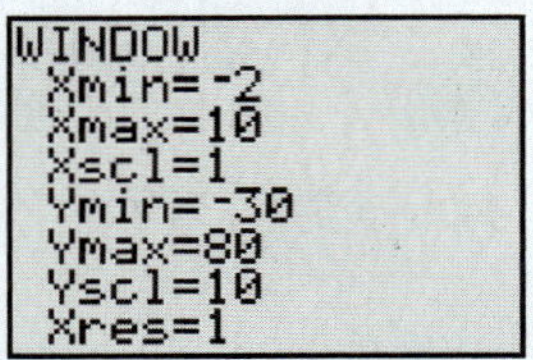

Figure 18 Window for phone models

c. Predict the total number of fixed telephone lines and cell-phone subscriptions per 100 people in 2017. Is your result greater than 100? Explain why this might be possible.

41. The percentages of households that have Internet access and broadband Internet access are shown in Table 8 for various years.

Table 8 Percentages of Households That Have Internet Access and Broadband Internet Access

	Percent	
Year	Internet Access	Broadband Internet Access
2001	50	9
2003	55	20
2007	62	51
2009	69	64
2010	71	68

Source: *U.S. Census Bureau*

a. Let $I(t)$ be the percentage of households that have Internet access and $B(t)$ be the percentage of households that have broadband Internet access, both at t years since 2000. Find equations of I and B.

b. Use "intersect" on a graphing calculator with the window shown in Fig. 19 to estimate when all households with Internet access had broadband Internet access.

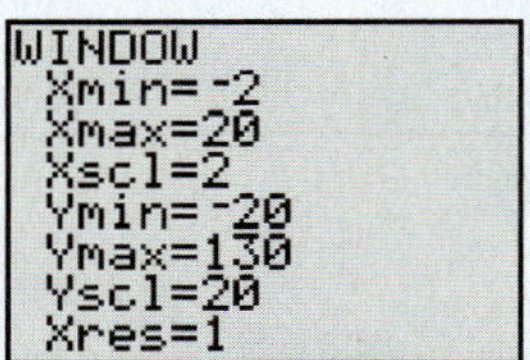

Figure 19 Window for Internet models

c. Explain why there is model breakdown for at least one of the models I and B after the year you found in part (b).

42. The percentages of women and men in the United States who are married are shown in Table 9 for various years.

a. Let $W(t)$ be the percentage of women who are married and $M(t)$ be the percentage of men who are married, both at t years since 1990. Find equations of W and M.

b. Use "intersect" on a graphing calculator with the window shown in Fig. 20 to predict when the percentage of women who are married will equal the percentage of men who are married. What will be that percentage? How confident are you?

Table 9 Percentages of Women and Men Who Are Married

Year	Percent of Women Who Are Married	Percent of Men Who Are Married
1990	59.7	64.3
1995	59.2	62.7
1997	57.9	61.5
2000	57.6	61.5
2004	57.1	60.3
2006	56.5	59.9
2010	55.2	58.0

Source: *U.S. Census Bureau*

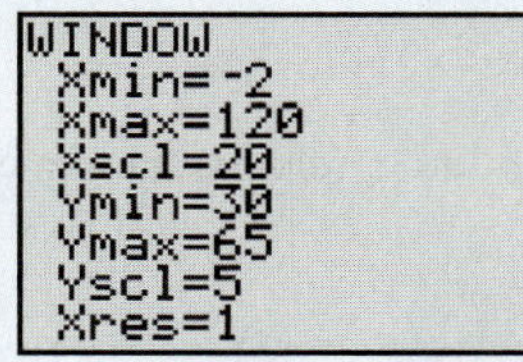

Figure 20 Window for married models

c. In 2015, there will be about 165.1 million women and 160.4 million men in the United States. Predict the total *number* of people who will be married in 2015.

43. Figure 21 shows the graphs of two linear equations. To the first decimal place, estimate the coordinates of the solution of the system.

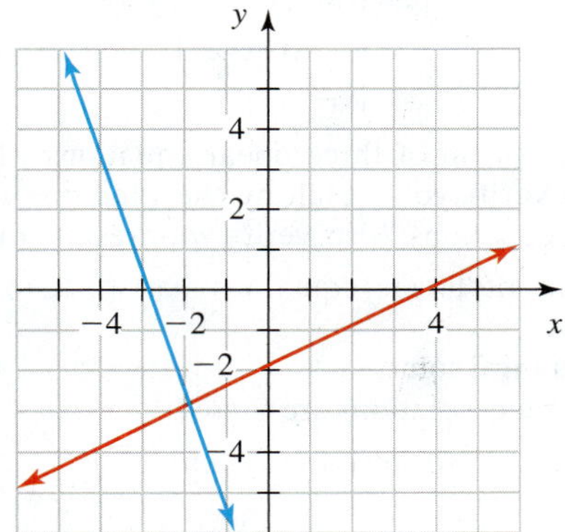

Figure 21 Exercise 43

44. Figure 22 shows the graphs of two linear equations. To the first decimal place, estimate the coordinates of the solution of the system.

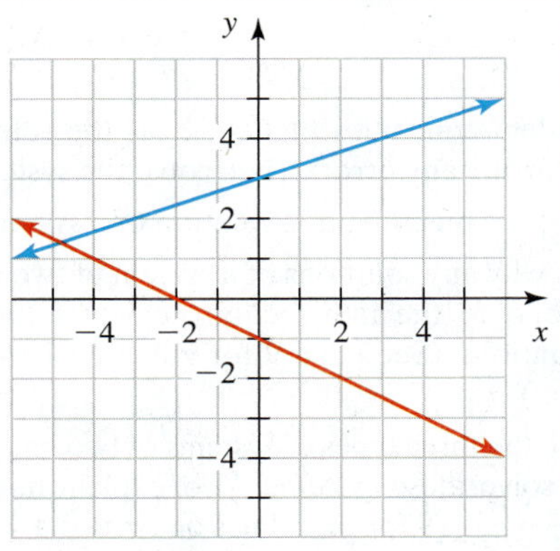

Figure 22 Exercise 44

45. Figure 23 shows the graphs of two linear equations. To the nearest integer, estimate the coordinates of the solution of the system. Explain. [**Hint:** Use the slope of each line.]

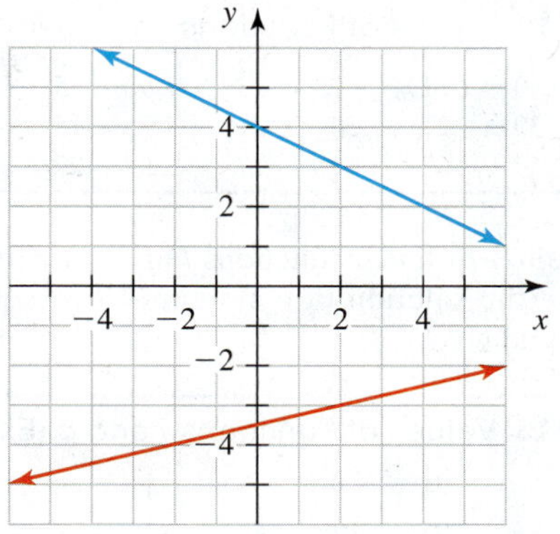

Figure 23 Exercise 45

46. Figure 24 shows the graphs of two linear equations. To the nearest integer, estimate the coordinates of the solution of the system. Explain. [**Hint:** Use the slope of each line.]

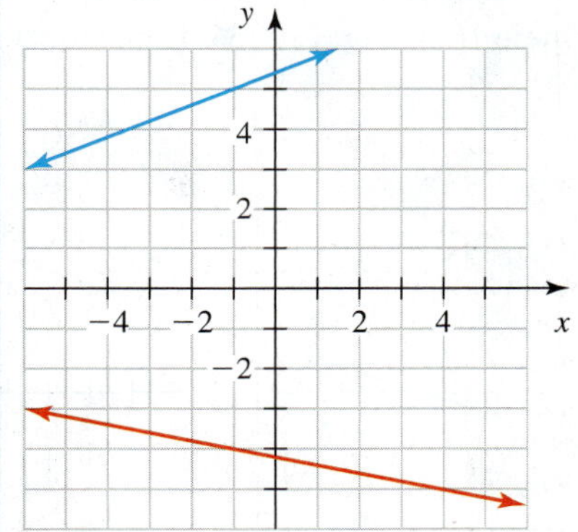

Figure 24 Exercise 46

47. Use Table 10 to solve the system

$$y = -5x + 11$$
$$y = 7x - 25$$

Table 10 Some Solutions of $y = -5x + 11$ and $y = 7x - 25$ (Exercise 47)

x	0	1	2	3	4
$y = -5x + 11$	11	6	1	−4	−9
$y = 7x - 25$	−25	−18	−11	−4	3

48. Use Table 11 to solve the system

$$y = 4x - 1$$
$$y = -3x + 6$$

Table 11 Some Solutions of $y = 4x - 1$ and $y = -3x + 6$ (Exercise 48)

x	0	1	2	3	4
$y = 4x - 1$	−1	3	7	11	15
$y = -3x + 6$	6	3	0	−3	−6

49. Some values of linear functions f and g are listed in Table 12. Estimate the solution of a system of two equations that describe f and g.

Table 12 Values of Functions f and g (Exercise 49)

x	0	1	2	3	4	5	6	7	8
$f(x)$	30	27	24	21	18	15	12	9	6
$g(x)$	2	7	12	17	22	27	32	37	42

50. Some values of linear functions f and g are listed in Table 13. Estimate the solution of a system of two equations that describe f and g.

Table 13 Values of Functions f and g (Exercise 50)

x	0	1	2	3	4	5	6	7	8
$f(x)$	99	95	91	87	83	79	75	71	67
$g(x)$	3	5	7	9	11	13	15	17	19

For Exercises 51–56, refer to Fig. 25.

51. Find $f(-4)$.

52. Find $g(-4)$.

53. Find x, where $f(x) = 3$.

54. Find x, where $g(x) = -1$.

55. Find x, where $f(x) = g(x)$.

56. Estimate x, where $g(x) = 0$.

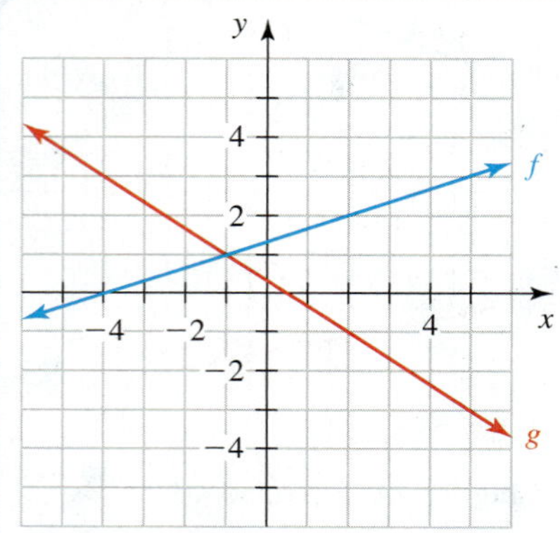

Figure 25 Exercises 51–56

Concepts

57. The graphs of $y = ax + b$ and $y = cx + d$ are sketched in Fig. 26.

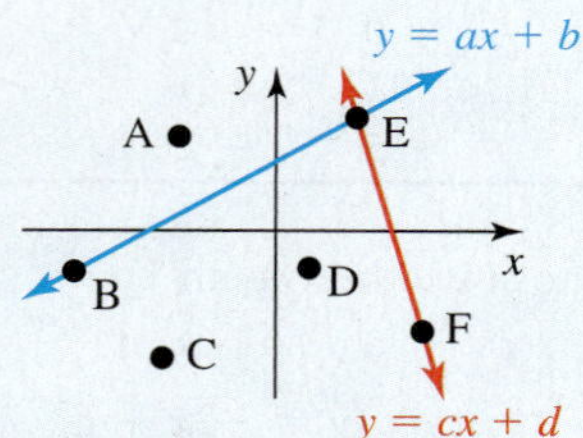

Figure 26 Exercise 57

For each part that follows, decide which one or more of the points A, B, C, D, E, or F represent ordered pairs that

a. satisfy the equation $y = ax + b$.
b. satisfy the equation $y = cx + d$.
c. satisfy both equations.
d. do not satisfy either equation.

58. Consider the system

$$y = 3x - 7$$
$$y = -2x + 3$$

Find an ordered pair that

a. satisfies $y = 3x - 7$ but does not satisfy $y = -2x + 3$.
b. satisfies $y = -2x + 3$ but does not satisfy $y = 3x - 7$.
c. satisfies both equations.
d. does not satisfy either equation.

59. Create a system of two linear equations as indicated. Verify your system graphically.

a. The solution of the system is $(2, 1)$.
b. The system is inconsistent.
c. The system is dependent.

60. Solve the system. [**Hint:** Sketch the graphs on the same coordinate system.]

$$y = 2$$
$$x = -3$$

61. Find all ordered pairs that satisfy all three equations:

$$y = x + 1$$
$$y = -2x + 7$$
$$y = 3x - 3$$

62. Find all ordered pairs that satisfy all three equations:

$$y = 2x - 5$$
$$y = 0.6x + 1$$
$$y = -1.2x + 5$$

63. Create a system of three linear equations whose solution is $(-4, 3)$. Verify your result by checking that $(-4, 3)$ satisfies all three equations. Also, verify your result graphically.

64. A system of linear equations has $(-2, 3)$ and $(4, 1)$ as solutions.

a. Find a third solution.
b. How many solutions are there?

65. A student tries to solve the system

$$y = 3x - 1$$
$$y = -2x + 9$$

After graphing the equations, she believes the solution is $(1, 2)$. She then checks whether $(1, 2)$ satisfies $y = 3x - 1$:

$$y = 3x - 1$$
$$2 \stackrel{?}{=} 3(1) - 1$$
$$2 \stackrel{?}{=} 2$$
true

The student concludes that $(1, 2)$ is the solution of the system. Describe any errors. Then solve the system correctly.

66. Describe the meaning of a solution of a system.

67. Explain why any solutions of a system of two linear equations correspond to the intersection points of the graphs of the two equations. (See page 9 for guidelines on writing a good response.)

68. Describe the three types of systems of two linear equations and how to solve these systems. Also, explain how to verify your work. (See page 9 for guidelines on writing a good response.)

Related Review

69. **a.** Determine whether the lines $y = 3x - 1$ and $y = 3x + 2$ are parallel.

b. Do parallel lines that are not the same line have an intersection point?

c. Solve the system

$$y = 3x - 1$$
$$y = 3x + 2$$

70. **a.** Determine whether the lines $3x - 7y = 14$ and $y = \frac{3}{7}x - 4$ are parallel.

b. Do parallel lines that are not the same line have an intersection point? Explain.

c. Solve the system

$$3x - 7y = 14$$
$$y = \frac{3}{7}x - 4$$

For Exercises 71–76, solve the given equation or system by referring to the graphs shown in Fig. 27.

71. $\frac{1}{3}x + \frac{5}{3} = x - 1$

72. $\frac{1}{3}x + \frac{5}{3} = -3x - 5$

73. $\frac{1}{3}x + \frac{5}{3} = 2$

74. $\frac{1}{3}x + \frac{5}{3} = 0$

75. $y = \frac{1}{3}x + \frac{5}{3}$
$y = -3x - 5$

76. $y = \frac{1}{3}x + \frac{5}{3}$
$y = x - 1$

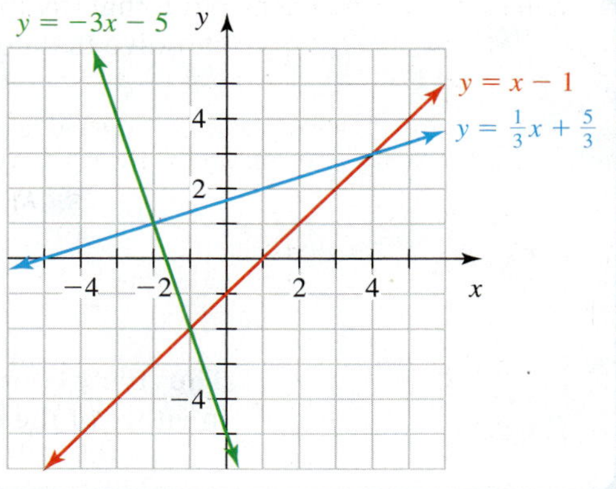

Figure 27 Exercises 71–76

Expressions, Equations, Functions, and Graphs

Perform the indicated instruction. Then use words such as linear, function, one variable, *and* two variables *to describe the expression, equation, or system.*

77. Solve the following:

$$y = 2x + 8$$
$$y = -3x - 2$$

78. Simplify $(2x + 8) - (-3x - 2)$.

79. Solve $2x + 8 = -3x - 2$.

80. Graph $y = -3x - 2$ by hand.

6.2 Using Substitution to Solve Systems

Objectives

- Use substitution to solve a system of two linear equations.
- Isolate a variable in an equation to help solve a system by substitution.
- Solve inconsistent and dependent systems by substitution.

In Section 6.1, we used graphs and tables to solve systems. In this section, we will use *equations* to solve systems.

The Substitution Method for Solving Systems

In Example 1, we will discuss how to solve a system by using a technique called *substitution.*

Example 1 Making a Substitution for y

Solve the system

$$y = x + 3 \quad \text{Equation (1)}$$
$$2x + 3y = 14 \quad \text{Equation (2)}$$

Solution

From equation (1), we know that for any solution of the system, the value of y is equal to the value of $x + 3$. So, we substitute $x + 3$ for y in equation (2):

$$2x + 3y = 14 \quad \text{Equation (2)}$$
$$2x + 3(x + 3) = 14 \quad \text{Substitute } x + 3 \text{ for } y.$$

Note that, by making this substitution, we now have an equation in *one* variable. Next, we solve that equation for x:

$$
\begin{aligned}
2x + 3(x + 3) &= 14 \\
2x + 3x + 9 &= 14 && \text{Distributive law} \\
5x + 9 &= 14 && \text{Combine like terms.} \\
5x &= 5 && \text{Subtract 9 from both sides.} \\
x &= 1 && \text{Divide both sides by 5.}
\end{aligned}
$$

Thus the x-coordinate of the solution is 1. To find the y-coordinate, we substitute 1 for x in either of the original equations and solve for y:

$$
\begin{aligned}
y &= x + 3 && \text{Equation (1)} \\
y &= 1 + 3 && \text{Substitute 1 for } x. \\
y &= 4 && \text{Add.}
\end{aligned}
$$

So, the solution is (1, 4). We can check that (1, 4) satisfies both of the system's equations:

$$
\begin{array}{ll}
y = x + 3 & 2x + 3y = 14 \\
4 \stackrel{?}{=} 1 + 3 & 2(1) + 3(4) \stackrel{?}{=} 14 \\
4 \stackrel{?}{=} 4 & 2 + 12 \stackrel{?}{=} 14 \\
\text{true} & \text{true}
\end{array}
$$

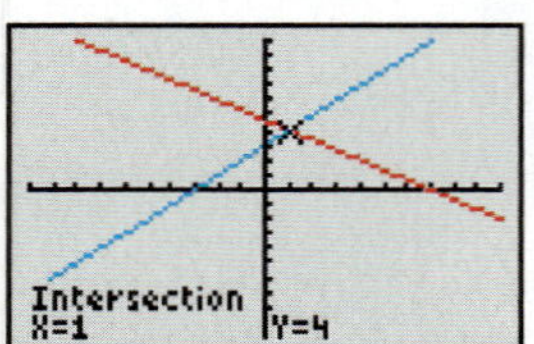

Figure 28 Verify that the solution is (1, 4)

Or we can verify that (1, 4) is the solution by graphing equations (1) and (2) and checking that (1, 4) is the intersection point of the two lines (see Fig. 28). To do so on a graphing calculator, we must first solve equation (2) for y:

$$
\begin{aligned}
2x + 3y &= 14 && \text{Equation (2)} \\
3y &= -2x + 14 && \text{Subtract } 2x \text{ from both sides.} \\
\frac{3y}{3} &= \frac{-2x}{3} + \frac{14}{3} && \text{Divide both sides by 3.} \\
y &= -\frac{2}{3}x + \frac{14}{3} && \text{Simplify.}
\end{aligned}
$$

Using Substitution to Solve a Linear System

To use **substitution** to solve a system of two linear equations,

1. Isolate a variable to one side of either equation.
2. Substitute the expression for the variable found in step 1 into the other equation.
3. Solve the equation in one variable found in step 2.
4. Substitute the solution found in step 3 into one of the original equations, and solve for the other variable.

A system of equations can be solved by substitution as well as by graphing. The methods give the same result.

In Example 1, we solved a system by making a substitution for y. In Example 2, we will discuss how to solve a system by making a substitution for x.

Example 2 Making a Substitution for x

Solve the system

$$x = 2y - 8 \quad \text{Equation (1)}$$
$$3x + 4y = 6 \quad \text{Equation (2)}$$

Solution

Since x is alone on one side of the equation $x = 2y - 8$, we begin by substituting $2y - 8$ for x in equation (2):

$$\begin{aligned} 3x + 4y &= 6 && \text{Equation (2)} \\ 3(2y - 8) + 4y &= 6 && \text{Substitute } 2y - 8 \text{ for } x. \\ 6y - 24 + 4y &= 6 && \text{Distributive law} \\ 10y - 24 &= 6 && \text{Combine like terms.} \\ 10y &= 30 && \text{Add 24 to both sides.} \\ y &= 3 && \text{Divide both sides by 10.} \end{aligned}$$

The y-coordinate of the solution is 3. To find the x-coordinate, we substitute 3 for y in equation (1) and solve for x:

$$\begin{aligned} x &= 2y - 8 && \text{Equation (1)} \\ x &= 2(3) - 8 && \text{Substitute 3 for } y. \\ x &= -2 && \text{Simplify.} \end{aligned}$$

The solution is $(-2, 3)$. We could then check that $(-2, 3)$ satisfies *both* of the original equations.

In Example 1, we made a substitution for y because y was alone on one side of one of the given equations. In Example 2, we made a substitution for x because x was alone on one side of one of the given equations.

In Example 3, we will solve a system of equations that have decimal coefficients. This will help prepare us for Section 6.4, where we will use systems to model situations.

Example 3 Solving a System of Equations

Solve the system

$$y = 1.72x - 4.38 \quad \text{Equation (1)}$$
$$y = -0.53x + 6.94 \quad \text{Equation (2)}$$

Solution

We start by substituting $1.72x - 4.38$ for y in equation (2):

$$\begin{aligned} y &= -0.53x + 6.94 && \text{Equation (2)} \\ 1.72x - 4.38 &= -0.53x + 6.94 && \text{Substitute } 1.72x - 4.38 \text{ for } y. \\ 2.25x - 4.38 &= 6.94 && \text{Add } 0.53x \text{ to both sides.} \\ 2.25x &= 11.32 && \text{Add 4.38 to both sides.} \\ x &\approx 5.03 && \text{Divide both sides by 2.25.} \end{aligned}$$

Next, we substitute 5.03 for x in equation (1):

$$\begin{aligned} y &= 1.72x - 4.38 && \text{Equation (1)} \\ y &= 1.72(5.03) - 4.38 && \text{Substitute 5.03 for } x. \\ y &\approx 4.27 && \text{Simplify.} \end{aligned}$$

So, the approximate solution is $(5.03, 4.27)$. We use "intersect" on a graphing calculator to check our work (see Fig. 29).

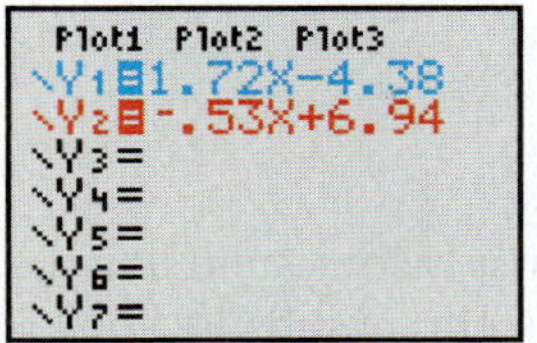

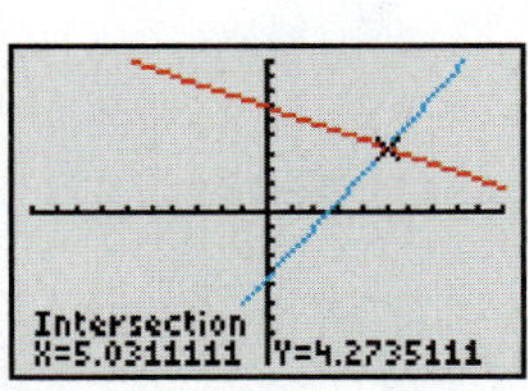

Figure 29 Verify that the approximate solution is (5.03, 4.27)

Isolating a Variable

If no variable is alone on one side of either equation in a system of two linear equations, then we must first solve for a variable in one of the equations before we can make a substitution.

Example 4 Isolating a Variable and Then Using Substitution

Solve the system

$$3x + y = -7 \quad \text{Equation (1)}$$
$$2x - 5y = 1 \quad \text{Equation (2)}$$

Solution

We begin by solving for one of the variables in one of the equations. We can avoid fractions by choosing to solve equation (1) for y:

$$\begin{aligned} 3x + y &= -7 && \text{Equation (1)} \\ y &= -3x - 7 && \text{Subtract } 3x \text{ from both sides.} \end{aligned}$$

Next, we substitute $-3x - 7$ for y in equation (2) and solve for x:

$$\begin{aligned} 2x - 5y &= 1 && \text{Equation (2)} \\ 2x - 5(-3x - 7) &= 1 && \text{Substitute } -3x - 7 \text{ for } y. \\ 2x + 15x + 35 &= 1 && \text{Distributive law} \\ 17x + 35 &= 1 && \text{Combine like terms.} \\ 17x &= -34 && \text{Subtract 35 from both sides.} \\ x &= -2 && \text{Divide both sides by 17.} \end{aligned}$$

Finally, we substitute -2 for x in the equation $y = -3x - 7$ and solve for y:

$$\begin{aligned} y &= -3(-2) - 7 && \text{Substitute } -2 \text{ for } x. \\ y &= 6 - 7 && \text{Multiply.} \\ y &= -1 && \text{Subtract.} \end{aligned}$$

The solution is $(-2, -1)$. We could verify our work by checking that $(-2, -1)$ satisfies both of the original equations.

Solving Inconsistent and Dependent Systems by Substitution

Each system in Examples 1–4 has one solution. What happens when we solve an inconsistent system (empty solution set) or a dependent system (infinitely many solutions) by substitution?

Example 5 Applying Substitution to an Inconsistent System

Consider the linear system

$$y = 3x + 2 \quad \text{Equation (1)}$$
$$y = 3x + 4 \quad \text{Equation (2)}$$

The graphs of the equations are parallel lines (why?), so the system is inconsistent and the solution set is the empty set. What happens when we solve this system by substitution?

Solution

We substitute $3x + 2$ for y in equation (2) and solve for x:

$$\begin{aligned} y &= 3x + 4 && \text{Equation (2)} \\ 3x + 2 &= 3x + 4 && \text{Substitute } 3x + 2 \text{ for } y. \\ 2 &= 4 && \text{Subtract } 3x \text{ from both sides.} \end{aligned}$$

false

We get the *false* statement $2 = 4$.

Inconsistent System of Equations

If the result of applying substitution to a system of equations is a false statement, then the system is inconsistent; that is, the solution set is the empty set.

Example 6 Applying Substitution to a Dependent System

In Example 8 of Section 6.1, we found that the system

$$y = 3x - 5 \quad \text{Equation (1)}$$
$$6x - 2y = 10 \quad \text{Equation (2)}$$

is dependent and that the solution set is the infinite set of solutions of the equation $y = 3x - 5$. What happens when we solve this system by substitution?

Solution

We substitute $3x - 5$ for y in equation (2) and solve for x:

$$6x - 2y = 10 \quad \text{Equation (2)}$$
$$6x - 2(3x - 5) = 10 \quad \text{Substitute } 3x - 5 \text{ for } y.$$
$$6x - 6x + 10 = 10 \quad \text{Distributive law}$$
$$10 = 10 \quad \text{Combine like terms.}$$
$$\text{true}$$

We get the *true* statement $10 = 10$.

Dependent System of Two Linear Equations

If the result of applying substitution to a linear system of two equations is a true statement that can be put into the form $a = a$, then the system is dependent; that is, the solution set is the set of ordered pairs represented by every point on the (same) line.

Group Exploration

Comparing techniques for solving systems

1. Consider the system

$$y = -2x + 3$$
$$y = 4x + 3$$

   a. Solve the system by graphing the equations by hand.
   b. Now solve the system by substitution.
   c. Compare your results from parts (a) and (b).
   d. Decide which method you prefer for this system. Explain.

2. Consider the system

$$3x - 2y = -4$$
$$y = -2x + 9$$

   a. Solve the system by graphing the equations by hand.
   b. Now solve the system by substitution.
   c. Compare your results from parts (a) and (b).
   d. Decide which method you prefer for this system. Explain.

3. In general, are there any systems that you prefer to solve by graphing by hand? If yes, describe them and give an example. If no, explain why not.

4. In general, are there any systems that you prefer to solve by substitution? If yes, describe them and give an example. If no, explain why not.

Tips for Success Write a Summary

Consider writing, after each class meeting, a summary of what you have learned. Your summaries will increase your understanding, as well as your memory, of concepts and procedures and will also serve as a good reference for quizzes and exams.

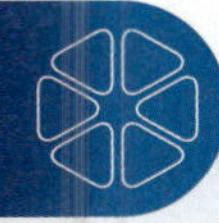

Homework 6.2

For extra help ▶ MyMathLab®  Watch the videos in MyMathLab · Download the MyDashboard App

Solve the system by substitution. Verify your solution by checking that it satisfies both equations of the system.

1. $y = 2x$
$3x + y = 10$

2. $y = 4x$
$2x + y = 18$

3. $x - 4y = -3$
$x = 2y - 1$

4. $x + 2y = 4$
$x = 3y - 1$

5. $2x + 3y = 5$
$y = x + 5$

6. $4x - 3y = 13$
$y = x - 4$

7. $-5x - 2y = 17$
$x = 4y + 1$

8. $3x + 2y = -9$
$x = 1 - 2y$

9. $2x - 5y - 3 = 0$
$y = 2x - 7$

10. $4x + 3y + 2 = 0$
$y = 1 - 3x$

11. $x = 2y + 6$
$-4x + 5y + 12 = 0$

12. $x = -2y + 5$
$7x + 2y + 13 = 0$

13. $y = \frac{1}{2}x - 5$
$2x + 3y = -1$

14. $4x - 7y = 3$
$x = \frac{2}{3}y + 4$

Solve the system by substitution. Verify your solution by using "intersect" on a graphing calculator.

15. $y = -2x - 1$
$y = 3x + 9$

16. $y = 4x + 2$
$y = -3x - 5$

17. $y = 2x$
$y = 3x$

18. $y = x$
$y = -x$

19. $y = 2(x - 4)$
$y = -3(x + 1)$

20. $y = 5(x - 1)$
$y = 2(x + 2)$

Use substitution to solve the system, with coordinates of solutions rounded to the second decimal place. Verify your work by using "intersect" on a graphing calculator.

21. $y = 2.57x + 7.09$
$y = -3.61x - 5.72$

22. $y = -1.45x - 6.18$
$y = 2.63x - 2.73$

23. $y = -3.17x + 8.92$
$y = 1.65x - 7.24$

24. $y = -0.51x - 2.64$
$y = -2.79x + 5.94$

Solve the system by substitution. Verify your solution by checking that it satisfies both equations of the system.

25. $2x + y = -9$
$5x - 3y = 5$

26. $-3x + y = -14$
$2x - 7y = 22$

27. $4x - 7y = 15$
$x - 3y = 5$

28. $2x + 3y = -1$
$x + 3y = -5$

29. $4x + 3y = 5$
$x - 2y = -7$

30. $6x - 5y = -8$
$x + 3y = 14$

31. $2x - y = 1$
$5x - 3y = 5$

32. $4x - 3y = -7$
$3x - y = -9$

33. $3x + 2y = -3$
$2x = y + 5$

34. $4x - 5y = -2$
$3x = y - 7$

For Exercises 35–44, solve the system by substitution. If the system is inconsistent or dependent, say so.

35. $x = 4 - 3y$
$2x + 6y = 8$

36. $x = 2y - 1$
$4x - 8y = -4$

37. $5x - 2y = 18$
$y = -3x + 2$

38. $-3x - 2y = -10$
$y = 4 - 2x$

39. $y = -5x + 3$
$15x + 3y = 6$

40. $y = 2x - 5$
$-4x + 2y = -6$

41. $-4x + 12y = 4$
$x = 3y - 1$

42. $3x + 12y = 6$
$x = -4y + 2$

43. $y = 3x + 2$
$12x - 4y = 9$

44. $y = 4x - 1$
$-8x + 2y = -5$

45. Some values of linear functions f and g are listed in Table 14. Find the solution of a system of two equations that describes the functions f and g. [**Hint:** Find equations of f and g.]

Table 14 Values of Functions f and g (Exercise 45)

x	0	1	2	3	4	5	6	7	8
$f(x)$	3	7	11	15	19	23	27	31	35
$g(x)$	50	44	38	32	26	20	14	8	2

46. Some values of linear functions f and g are listed in Table 15. Find the solution of a system of two equations that describes the functions f and g. [**Hint:** Find equations of f and g.]

Table 15 Values of Functions f and g (Exercise 46)

x	0	1	2	3	4	5	6	7	8
$f(x)$	201	204	207	210	213	216	219	222	225
$g(x)$	6	11	16	21	26	31	36	41	46

Concepts

47. Find the coordinates of the points A, B, C, and D as shown in Fig. 30. The equations of lines l_1 and l_2 are provided, but no attempt has been made to sketch the lines accurately except for showing the intersection points. Use TRACE and "intersect" on a graphing calculator to verify your work.

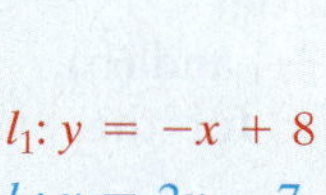

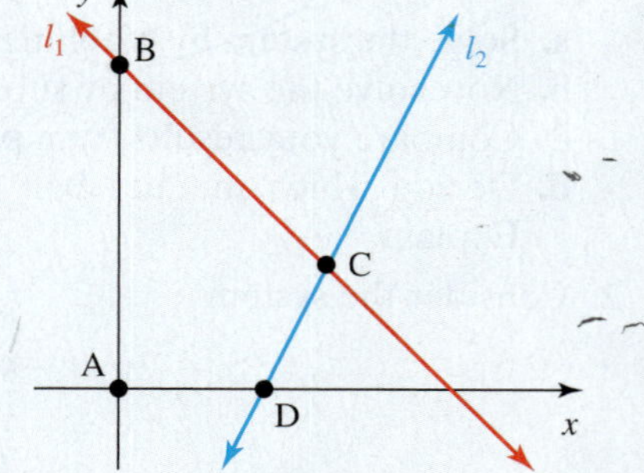

Figure 30 Exercise 47

48. Find the coordinates of the points A, B, C, and D as shown in Fig. 31. The equations of lines l_1 and l_2 are provided, but no attempt has been made to sketch the lines accurately except for showing the intersection points. Use TRACE and "intersect" on a graphing calculator to verify your work.

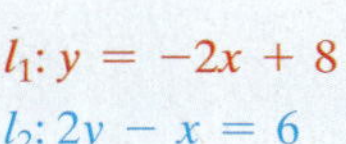

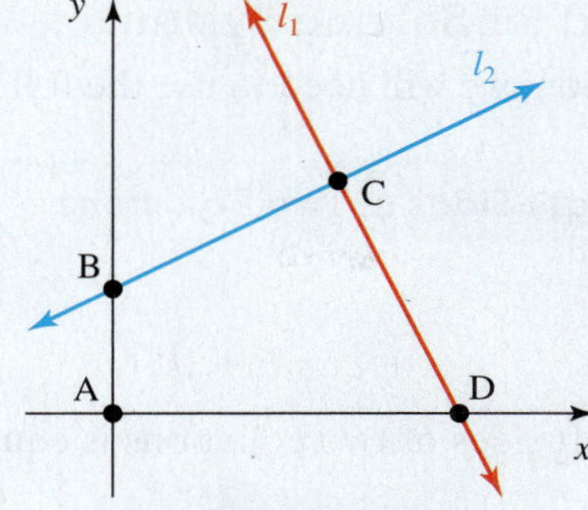

Figure 31 Exercise 48

49. Consider the system

$$x + y = 3$$
$$3x + 2y = 7$$

a. Isolate x to one side of one of the equations, and then solve the system by substitution.
b. Isolate y to one side of one of the equations, and then solve the system by substitution.
c. Compare your results in parts (a) and (b).

50. Consider the system

$$2x + y = 8$$
$$x + 3y = 9$$

a. Isolate x to one side of one of the equations, and then solve the system by substitution.
b. Isolate y to one side of one of the equations, and then solve the system by substitution.
c. Compare your results in parts (a) and (b).

51. Consider the system

$$y = 3x$$
$$2x + y = 5$$

A student tries to solve the system:

$$2x + 3x = 5$$
$$5x = 5$$
$$x = 1$$

Describe any errors. Then solve the system correctly.

52. a. When using substitution to solve a system, how can you tell if the solution set is the empty set?
b. When using substitution to solve a system, how can you tell if there are an infinite number of solutions?

53. Describe how to solve a system by substitution.

54. Is it more reliable to use substitution or graphing by hand to find exact solutions of a system? Explain.

Related Review

55. a. Use "intersect" on a graphing calculator to solve the system

$$y = 2x - 3$$
$$y = -3x + 7$$

b. Solve the same system by substitution.
c. Solve the equation $2x - 3 = -3x + 7$.
d. Compare your result in part (c) with the x-coordinate of the result in part (b).
e. Use "intersect" to help you solve $x - 1 = -2x + 8$. Then solve this equation without using graphing. Compare your results.
f. Summarize the main concepts addressed in this exercise. Explain how you can use "intersect" to verify your work.

56. a. Find an equation of the line that has slope 3 and contains the point (3, 4).
b. Find an equation of the line that has slope -2 and contains the point (3, 4).
c. Solve the system of the two linear equations you found in parts (a) and (b).
d. Explain how your work in part (c) is related to your work in parts (a) and (b).

Use "intersect" on a graphing calculator to solve the equation. Round the solution to the second decimal place.

57. $5x - 4 = -2x + 7$ **58.** $-3x - 8 = 4x + 1$

59. $-0.39x - 4.98 = 1.04x - 1.52$

60. $0.48x - 5.37 = 1.32x - 8.14$

61. $\frac{57}{94}x - \frac{71}{14} = \frac{95}{32}x + \frac{19}{43}$

62. $\frac{25}{73}x + \frac{99}{23} = -\frac{47}{68}x - \frac{34}{13}$

Expressions, Equations, Functions, and Graphs

Perform the indicated instruction. Then use words such as linear, function, one variable, *and* two variables *to describe the expression, equation, or system.*

63. Graph $y = -3x + 5$ by hand.

64. Solve $7 = -3x + 5$.

65. Solve the following:

$$9x + 2y = 1$$
$$y = -3x + 5$$

66. Simplify $9x + 2y - 3x + 5$.

6.3 Using Elimination to Solve Systems

Objectives

» Use elimination to solve a system of two linear equations.

» Compare three ways to solve a system.

So far, we have solved systems by graphing and by substitution. In this section, we will use a third technique, called *elimination* or the *addition method*.

The Elimination Method for Solving Systems

To solve systems by elimination, we will need to use the following property:

Adding Left Sides and Right Sides of Two Equations

If $a = b$ and $c = d$, then

$$a + c = b + d$$

In words, the sum of the left sides of two equations is equal to the sum of the right sides.

For example, if we add the left sides and add the right sides of the equations $3 = 3$ and $5 = 5$, we obtain the true statement $3 + 5 = 3 + 5$.

Example 1 Solving a System by Elimination

Solve the system

$$\begin{aligned} 3x + 2y &= 9 && \text{Equation (1)} \\ -3x + 5y &= 12 && \text{Equation (2)} \end{aligned}$$

Solution

We begin by adding the left sides and adding the right sides of the two equations:

$$\begin{aligned} 3x + 2y &= 9 && \text{Equation (1)} \\ -3x + 5y &= 12 && \text{Equation (2)} \\ \hline 0 + 7y &= 21 && \text{Add left sides and add right sides; combine like terms.} \end{aligned}$$

By "eliminating" the variable x, we now have an equation in *one* variable. Next, we solve that equation for y:

$$\begin{aligned} 0 + 7y &= 21 \\ 7y &= 21 && 0 + a = a \\ y &= 3 && \text{Divide both sides by 7.} \end{aligned}$$

Then, we substitute 3 for y in either of the original equations and solve for x:

$$\begin{aligned} 3x + 2y &= 9 && \text{Equation (1)} \\ 3x + 2(3) &= 9 && \text{Substitute 3 for } y. \\ 3x + 6 &= 9 && \text{Multiply.} \\ 3x &= 3 && \text{Subtract 6 from both sides.} \\ x &= 1 && \text{Divide both sides by 3.} \end{aligned}$$

The solution is (1, 3). We check that (1, 3) satisfies *both* equations (1) and (2):

$$\begin{aligned} 3x + 2y &= 9 \\ 3(1) + 2(3) &\stackrel{?}{=} 9 \\ 3 + 6 &\stackrel{?}{=} 9 \\ &\text{true} \end{aligned} \qquad \begin{aligned} -3x + 5y &= 12 \\ -3(1) + 5(3) &\stackrel{?}{=} 12 \\ -3 + 15 &\stackrel{?}{=} 12 \\ &\text{true} \end{aligned}$$

Example 2 Solving a System by Elimination

Solve the system

$$\begin{aligned} -5x + 3y &= -9 && \text{Equation (1)} \\ 3x + 6y &= 21 && \text{Equation (2)} \end{aligned}$$

Solution

If we add the left sides and add the right sides of the equations as given to us, neither variable would be eliminated. Therefore, we first multiply both sides of equation (1) by -2, yielding the system

$$\begin{aligned} 10x - 6y &= 18 \quad \text{Multiply both sides of equation (1) by } -2. \\ 3x + 6y &= 21 \quad \text{Equation (2)} \end{aligned}$$

Now that the coefficients of the y terms are equal in absolute value and opposite in sign, we add the left sides and add the right sides of the equations and solve for x:

$$\begin{aligned} 10x - 6y &= 18 \\ 3x + 6y &= 21 \\ \hline 13x + 0 &= 39 \quad \text{Add left sides and add right sides; combine like terms.} \\ 13x &= 39 \quad a + 0 = a \\ x &= 3 \quad \text{Divide both sides by 13.} \end{aligned}$$

We substitute 3 for x in equation (1) and solve for y:

$$\begin{aligned} -5x + 3y &= -9 \quad \text{Equation (1)} \\ -5(3) + 3y &= -9 \quad \text{Substitute 3 for } x. \\ -15 + 3y &= -9 \quad \text{Simplify.} \\ 3y &= 6 \quad \text{Add 15 to both sides.} \\ y &= 2 \quad \text{Divide both sides by 3.} \end{aligned}$$

The solution is (3, 2). We could check that (3, 2) satisfies both of the original equations.

Using Elimination to Solve a Linear System

To use **elimination** to solve a system of two linear equations,

1. Use the multiplication property of equality (Section 4.3) to get the coefficients of one variable to be equal in absolute value and opposite in sign.
2. Add the left sides and add the right sides of the equations to eliminate one of the variables.
3. Solve the equation in one variable found in step 2.
4. Substitute the solution found in step 3 into one of the original equations, and solve for the other variable.

Example 3 Solving a System by Elimination

Solve the system

$$\begin{aligned} 5x - 4y &= 13 \quad \text{Equation (1)} \\ 9x + 3y &= 3 \quad \text{Equation (2)} \end{aligned}$$

Solution

To eliminate the y terms, we multiply both sides of equation (1) by 3 and multiply both sides of equation (2) by 4, yielding the system

$$\begin{aligned} 15x - 12y &= 39 \quad \text{Multiply both sides of equation (1) by 3.} \\ 36x + 12y &= 12 \quad \text{Multiply both sides of equation (2) by 4.} \end{aligned}$$

The coefficients of the y terms are now equal in absolute value and opposite in sign. Next, we add the left sides and add the right sides of the equations and solve for x:

$$\begin{aligned} 15x - 12y &= 39 \\ 36x + 12y &= 12 \\ \hline 51x + 0 &= 51 \quad \text{Add left sides and add right sides; combine like terms.} \end{aligned}$$

$$51x = 51 \quad a + 0 = a$$
$$x = 1 \quad \text{Divide both sides by 51.}$$

Substituting 1 for x in equation (1) gives

$$5x - 4y = 13 \quad \text{Equation (1)}$$
$$5(1) - 4y = 13 \quad \text{Substitute 1 for } x.$$
$$5 - 4y = 13 \quad \text{Simplify.}$$
$$-4y = 8 \quad \text{Subtract 5 from both sides.}$$
$$y = -2 \quad \text{Divide both sides by } -4.$$

The solution is $(1, -2)$.

In Example 3, we eliminated y by getting the coefficients of the y terms to be equal in absolute value and opposite in sign. Note that this process is similar to finding a least common multiple.

Example 4 Using Elimination to Solve a System with Fractions

Solve the system

$$\frac{1}{3}x - \frac{1}{2}y = \frac{1}{6} \quad \text{Equation (1)}$$
$$\frac{1}{5}x + \frac{1}{4}y = \frac{13}{20} \quad \text{Equation (2)}$$

Solution

First, we clear the fractions in equation (1) by multiplying both sides by the LCD, 6, and clear the fractions in equation (2) by multiplying both sides by the LCD, 20:

Equation (1)	**Equation (2)**
$\frac{1}{3}x - \frac{1}{2}y = \frac{1}{6}$	$\frac{1}{5}x + \frac{1}{4}y = \frac{13}{20}$
$6 \cdot \left(\frac{1}{3}x - \frac{1}{2}y\right) = 6 \cdot \frac{1}{6}$	$20 \cdot \left(\frac{1}{5}x + \frac{1}{4}y\right) = 20 \cdot \frac{13}{20}$
$6 \cdot \frac{1}{3}x - 6 \cdot \frac{1}{2}y = 6 \cdot \frac{1}{6}$	$20 \cdot \frac{1}{5}x + 20 \cdot \frac{1}{4}y = 20 \cdot \frac{13}{20}$
$2x - 3y = 1$	$4x + 5y = 13$

Next, we use elimination to solve the system

$$2x - 3y = 1 \quad \text{Equation (3)}$$
$$4x + 5y = 13 \quad \text{Equation (4)}$$

To eliminate the variable x, we multiply both sides of equation (3) by -2:

$$-4x + 6y = -2 \quad \text{Multiply both sides of equation (3) by } -2.$$
$$4x + 5y = 13$$
$$0 + 11y = 11 \quad \text{Add left sides and add right sides; combine like terms.}$$
$$y = 1 \quad \text{Divide both sides by 11.}$$

To find the value of x, we can substitute 1 for y in any of the equations that have both x and y. We substitute 1 for y in equation (3) and solve for x:

$$2x - 3(1) = 1 \quad \text{Substitute 1 for } y \text{ in equation (3).}$$
$$2x = 4 \quad \text{Add 3 to both sides.}$$
$$x = 2 \quad \text{Divide both sides by 2.}$$

The solution is $(2, 1)$.

Solving Inconsistent Systems and Dependent Systems by Elimination

When we apply elimination to an inconsistent system or a dependent system, the results are similar to the results obtained from applying substitution to such systems.

Solving Inconsistent Systems and Dependent Systems by Elimination

If the result of applying elimination to a linear system of two equations is

- a false statement, then the system is inconsistent; that is, the solution set is the empty set.
- a true statement that can be put into the form $a = a$, then the system is dependent; that is, the solution set is the set of ordered pairs represented by every point on the (same) line.

Example 5 Using Elimination to Solve a Dependent System

Solve

$$2x - 7y = 5 \quad \text{Equation (1)}$$
$$6x - 21y = 15 \quad \text{Equation (2)}$$

Solution

To eliminate the x terms (and the y terms), we multiply both sides of equation (1) by -3, yielding the system

$$-6x + 21y = -15 \quad \text{Multiply both sides of equation (1) by } -3.$$
$$6x - 21y = 15 \quad \text{Equation (2)}$$

Now that the coefficients of the x terms (and those of the y terms) are equal in absolute value and opposite in sign, we add the left sides and add the right sides of the equations:

$$\begin{aligned} -6x + 21y &= -15 \\ 6x - 21y &= 15 \\ \hline 0 + 0 &= 0 && \text{Add left sides and add right sides; combine like terms.} \\ 0 &= 0 && \text{Simplify.} \end{aligned}$$

Since $0 = 0$ is a true statement of the form $a = a$, we conclude that the system is dependent and that the solution set of the system is the set of ordered pairs represented by the points on the line $2x - 7y = 5$ and the (same) line $6x - 21y = 15$.

Comparing Three Ways to Solve a System

In the next example, we will solve the same system that we solved by substitution in Example 3 of Section 6.2, but now we will solve it by elimination.

Example 6 Solving a System by Elimination

Solve the system

$$y = 1.72x - 4.38 \quad \text{Equation (1)}$$
$$y = -0.53x + 6.94 \quad \text{Equation (2)}$$

Solution

First we multiply both sides of equation (1) by -1, yielding the system

$$-y = -1.72x + 4.38 \quad \text{Multiply equation (1) by } -1.$$
$$y = -0.53x + 6.94 \quad \text{Equation (2)}$$

Now that the coefficients of y are equal in absolute value and opposite in sign, we add the left sides and add the right sides of the equations and solve for x:

$$\begin{aligned} -y &= -1.72x + 4.38 \\ y &= -0.53x + 6.94 \\ \hline 0 &= -2.25x + 11.32 && \text{Add left sides and add right sides; combine like terms.} \\ 2.25x &= 11.32 && \text{Add } 2.25x \text{ to both sides.} \\ x &\approx 5.03 && \text{Divide both sides by 2.25.} \end{aligned}$$

Then we substitute 5.03 for x in equation (1):

$$\begin{aligned} y &= 1.72x - 4.38 && \text{Equation (1)} \\ y &= 1.72(5.03) - 4.38 && \text{Substitute 5.03 for } x. \\ y &\approx 4.27 && \text{Simplify.} \end{aligned}$$

So, the approximate solution is (5.03, 4.27), which is the same result that we obtained in Example 3 of Section 6.2.

In both Sections 6.2 and 6.3, we solved the system

$$\begin{aligned} y &= 1.72x - 4.38 \\ y &= -0.53x + 6.94 \end{aligned}$$

by graphing, substitution, and elimination. We obtained the same approximate solution by all three methods. In fact, **any linear system of two equations can be solved by graphing, substitution, or elimination. All three methods give the same result.**

How do we determine which method to use? When we solve a system of two equations, if a variable is alone or is easy to isolate to one side of either equation, substitution is often convenient. Otherwise, elimination is usually easiest. With most systems, we avoid solving by graphing or by using tables, except as a check, because these methods often lead only to approximate solutions.

Group Exploration

Comparing techniques of solving systems

Consider the system

$$\begin{aligned} x + 2y &= 4 \\ y &= 3x - 5 \end{aligned}$$

1. Use substitution to solve the system.
2. Use elimination to solve the system.
3. Solve the system by graphing the equations by hand.
4. Compare your results from Problems 1, 2, and 3.
5. Give an example of a system that is easiest to solve by substitution. Also, give an example of a system that is easiest to solve by elimination. Finally, give an example of a system that is easiest to solve by graphing by hand. Explain. Solve your three systems.

Tips for Success Cross-Checks

If you finish a quiz or an exam early, it pays to verify your answers with cross-checks. For example, suppose you determine by elimination that the solution of the system

$$\begin{aligned} 5x + 2y &= 9 \\ 3x + 4y &= 11 \end{aligned}$$

is (1, 2). There are several ways to verify your answer. You could check that (1, 2) satisfies both equations. You could graph each equation and check that the intersection point is (1, 2). Or you could solve the system by substitution.

Homework 6.3

For extra help ▶ MyMathLab®  Watch the videos in MyMathLab  Download the MyDashboard App

Solve the system by elimination. Verify your solution by checking that it satisfies both equations in the system.

1. $2x + 3y = 7$
$-2x + 5y = 1$
2. $-4x + 5y = 11$
$4x + 3y = 13$
3. $5x - 2y = 2$
$-3x + 2y = 2$
4. $3x + 5y = 19$
$2x - 5y = -4$
5. $x + 2y = -4$
$3x - 4y = 18$
6. $x - 4y = 5$
$5x - 2y = -11$
7. $2x - 3y = 8$
$5x + 6y = -7$
8. $3x + 4y = -5$
$-5x + 2y = 17$
9. $6x - 5y = 4$
$2x + 3y = -8$
10. $4x - 7y = 25$
$8x + 3y = -1$
11. $5x + 7y = -16$
$2x - 5y = 17$
12. $6x + 5y = -14$
$-4x - 7y = 2$
13. $-8x + 3y = 1$
$3x - 4y = 14$
14. $5x - 2y = 8$
$2x - 3y = 1$
15. $y = 3x - 6$
$y = -4x + 1$
16. $y = 2x + 7$
$y = -x - 8$
17. $3x + 6y - 18 = 0$
$17 = 7x + 9y$
18. $3 = -2x + 9y$
$5x - 5y - 10 = 0$
19. $0.9x + 0.4y = 1.9$
$0.3x - 0.2y = 1.3$
20. $0.2x - 0.5y = 0.2$
$0.8x + 1.5y = -6.2$
21. $3(2x - 1) + 4(y - 3) = 1$
$4(x + 5) - 2(4y + 1) = 18$
22. $2(x - 3) - 3(y + 1) = -5$
$-4(x - 2) + 5(y + 3) = 13$
23. $\frac{2}{3}x + \frac{1}{2}y = \frac{1}{6}$
$\frac{1}{2}x + \frac{5}{4}y = \frac{11}{4}$
24. $\frac{1}{4}x + \frac{5}{2}y = 2$
$\frac{5}{6}x - \frac{1}{3}y = -2$
25. $\frac{2}{9}x + \frac{1}{3}y = 4$
$\frac{1}{2}x - \frac{2}{5}y = -\frac{5}{2}$
26. $\frac{1}{2}x - \frac{5}{4}y = \frac{9}{2}$
$\frac{3}{8}x + \frac{1}{2}y = \frac{1}{2}$

Solve the system by elimination. If the system is inconsistent or dependent, say so.

27. $4x - 7y = 3$
$8x - 14y = 6$
28. $4x + 6y = 10$
$-6x - 9y = -15$
29. $8x - 6y = 4$
$12x - 9y = 5$
30. $3x + 2y = 5$
$-12x - 8y = -17$
31. $3x - 2y = -14$
$6x + 5y = -19$
32. $8x - 3y = -2$
$5x + 12y = -29$
33. $6x - 15y = 7$
$-4x + 10y = -5$
34. $10x - 12y = 5$
$-15x + 18y = -8$
35. $3x - 9y = 12$
$-4x + 12y = -16$
36. $9x - 6y = 15$
$12x - 8y = 20$

Use elimination to solve the system, with coordinates of solutions rounded to the second decimal place. Verify your work by using "intersect" on a graphing calculator.

37. $y = 4.29x - 8.91$
$y = -1.26x + 9.75$
38. $y = 1.28x + 2.05$
$y = 3.94x - 8.83$
39. $y = -2.15x + 8.38$
$y = 1.67x + 2.57$
40. $y = 3.28x + 1.43$
$y = 0.56x + 6.72$

Solve the system by either elimination or substitution. Verify your work by using "intersect" on your graphing calculator or by checking that your result satisfies both equations of the system.

41. $4x - y = -12$
$3x + 5y = 14$
42. $6x + y = -13$
$3x - 2y = -19$
43. $-2x + 7y = -3$
$x = 3y + 2$
44. $4x - 5y = 23$
$y = 1 - 2x$
45. $2x + 7y = 13$
$3x - 4y = -24$
46. $5x - 2y = 7$
$4x - 3y = 7$
47. $2x - 7y = -1$
$-x - 3y = 7$
48. $3x - 8y = -4$
$x - 5y = -6$
49. $y = -2x - 3$
$y = 3x + 7$
50. $y = -4x - 9$
$y = 2x + 3$
51. $8x + 5y = 7$
$7y = -6x + 15$
52. $-3x + 2y = 19$
$5y = -4x - 10$
53. $3(2x - 5) + 4y = 11$
$5x - 2(3y + 1) = 1$
54. $4(2x - 7) + 3y = 7$
$7x - 5(2y + 3) = 3$
55. $y = \frac{1}{2}x + 3$
$2y - x = 6$
56. $y = -\frac{1}{3}x + 4$
$x + 3y = 12$
57. $\frac{5}{6}x + \frac{1}{4}y = 3$
$-\frac{1}{3}x + \frac{5}{2}y = 4$
58. $\frac{1}{2}x - \frac{3}{4}y = -4$
$\frac{2}{3}x - \frac{1}{3}y = -4$
59. $\frac{x + 2y}{3} - \frac{x - y}{2} = \frac{13}{6}$
$\frac{x + 3y}{2} + \frac{x + y}{4} = \frac{17}{4}$
60. $\frac{2x + y}{3} - \frac{3x - y}{6} = 1$
$\frac{x + y}{3} + \frac{2x - y}{4} = \frac{31}{12}$

Concepts

For Exercises 61 and 62, solve the system of equations three times, once by each of the three methods: graphing by hand, substitution, and elimination. Decide which method you prefer for this system. Explain.

61. $3x + 2y = 8$
$2x - y = 3$
62. $y = -2x + 7$
$5x - 2y = 4$
63. Consider the system

$$5x + 3y = 11$$
$$2x - 4y = -6$$

a. Solve the system by eliminating the x terms.
b. Solve the system by eliminating the y terms.
c. Compare your results in parts (a) and (b).

64. Consider the system

$$4x - 7y = 15$$
$$5x + 3y = 7$$

a. Solve the system by eliminating the x terms.
b. Solve the system by eliminating the y terms.
c. Compare your results in parts (a) and (b).

65. Some values of linear functions f and g are listed in Table 16. Find the solution of a system of two equations that describes the functions f and g. [**Hint:** Find equations of f and g.]

Table 16 Values of Functions f and g (Exercise 65)

x	0	1	2	3	4	5	6
$f(x)$	93	90	87	84	81	78	75
$g(x)$	−22	−20	−18	−16	−14	−12	−10

66. Some values of linear functions f and g are listed in Table 17. Find the solution of a system of two equations that describes the functions f and g. [**Hint:** Find equations of f and g.]

Table 17 Values of Functions f and g (Exercise 66)

x	0	1	2	3	4	5	6
$f(x)$	−76	−72	−68	−64	−60	−56	−52
$g(x)$	85	82	79	76	73	70	67

67. Find the coordinates of points A, B, C, D, E, and F as shown in Fig. 32. The equations of lines l_1–l_4 are provided, but no attempt has been made to sketch the lines accurately, except for showing the intersection points. Verify your results graphically.

l_1: $y = 2x + 3$
l_2: $3y + x = 30$
l_3: $y + 3x = 26$
l_4: $y = 2x - 10$

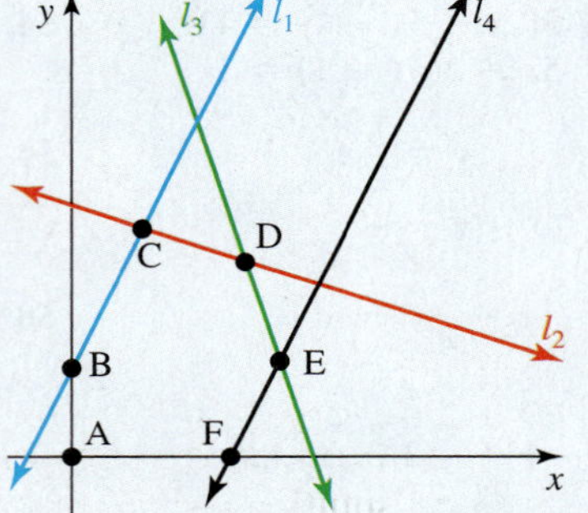

Figure 32 Exercise 67

68. Find the coordinates of points A, B, C, D, E, and F as shown in Fig. 33. The equations of lines l_1–l_4 are provided, but no attempt has been made to sketch the lines accurately, except for showing the intersection points. Verify your results graphically.

l_1: $y = 3x + 2$
l_2: $2y + x = 18$
l_3: $y + 4x = 23$
l_4: $y = 3x - 12$

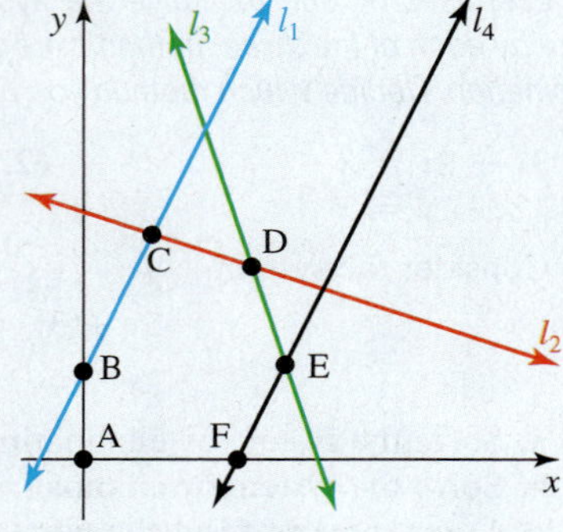

Figure 33 Exercise 68

69. a. Let a, b, c, d, k, and p be constants such that the two equations form a one-solution system. Solve the system.

$$ax + by = c$$
$$kx + py = d$$

b. Use your result from part (a) to solve this system:

$$3x + 5y = 2$$
$$4x + 3y = 4$$

70. To solve the system

$$3x - y = 3$$
$$-2x + y = -1$$

a student adds the left sides and the right sides to get $x = 2$. He thinks that 2 is the solution. Is the student correct? Explain.

71. Describe how to solve a system by elimination. (See page 9 for guidelines on writing a good response.)

72. Describe how to determine whether to solve a system by graphing, substitution, or elimination. (See page 9 for guidelines on writing a good response.)

Related Review

73. In this exercise, you will solve a system of equations to help find an equation of the line that contains the points (2, 5) and (4, 9).

a. Substitute 2 for x and 5 for y in the equation $y = mx + b$.
b. Substitute 4 for x and 9 for y in the equation $y = mx + b$.
c. Consider the system of equations formed by your results from parts (a) and (b). Solve this system to find values of m and b.
d. Substitute the values you found for m and b in the equation $y = mx + b$.
e. Verify your equation by using a graphing calculator.

74. In this exercise, you will solve a system of equations to help find an equation of the line that contains the points (−1, 5) and (2, −4).

a. Substitute −1 for x and 5 for y in the equation $y = mx + b$.
b. Substitute 2 for x and −4 for y in the equation $y = mx + b$.
c. Consider the system of equations formed by your results from parts (a) and (b). Solve this system to find values of m and b.
d. Substitute the values you found for m and b in the equation $y = mx + b$.
e. Verify your equation by using a graphing calculator.

For Exercises 75–80, solve the given equation or system by referring to the solutions of the functions shown in Table 18.

75. $\frac{1}{2}x + \frac{7}{2} = \frac{4}{5}x + 2$

76. $\frac{1}{2}x + \frac{7}{2} = \frac{11}{10}x + \frac{17}{10}$

77. $\frac{11}{10}x + \frac{17}{10} = 5$

78. $\frac{1}{2}x + \frac{7}{2} = 4$

79. $y = \frac{4}{5}x + 2$
$y = \frac{11}{10}x + \frac{17}{10}$

80. $y = \frac{1}{2}x + \frac{7}{2}$
$y = \frac{4}{5}x + 2$

Table 18 Some Solutions of Three Functions (Exercises 75–80)

x	0	1	2	3	4	5	6
$y = \frac{1}{2}x + \frac{7}{2}$	3.5	4	4.5	5	5.5	6	6.5
$y = \frac{4}{5}x + 2$	2	2.8	3.6	4.4	5.2	6	6.8
$y = \frac{11}{10}x + \frac{17}{10}$	1.7	2.8	3.9	5	6.1	7.2	8.3

Expressions, Equations, Functions, and Graphs

Perform the indicated instruction. Then use words such as linear, function, one variable, *and* two variables *to describe the expression, equation, or system.*

81. Simplify $2(5x + 4) - 6(3x + 2)$.

82. Graph $5x + 4y = 8$ by hand.

83. Solve $2(5x + 4) - 6(3x + 2) = 0$.

84. Solve the following:

$$5x + 4y = 8$$
$$3x + 2y = 2$$

6.4 Using Systems to Model Data

Objectives

» Use substitution and elimination to make predictions about situations described by a table of data.

» Use substitution and elimination to make predictions about situations described by rates of change.

In Section 6.1, we used a graphical approach to find the intersection point of the graphs of two linear models. In this section, we discuss how to use substitution and elimination to find such an intersection point.

Using a Table of Data to Find a System for Modeling

In Example 1, we will use two linear models from Section 6.1 and work with a table of data.

Table 19 U.S. Life Expectancies of Women and Men

Year of Birth	Women (years)	Men (years)
1980	77.4	70.0
1985	78.2	71.1
1990	78.8	71.8
1995	78.9	72.5
2000	79.5	74.1
2005	79.9	74.9
2009	81.3	76.2

Source: *U.S. Census Bureau*

Example 1 Solving a System to Make a Prediction

In Example 6 of Section 6.1, we modeled the life expectancies (in years) $W(t)$ and $M(t)$ of U.S. women and men, respectively, by the system

$$L = W(t) = 0.115t + 77.44$$
$$L = M(t) = 0.208t + 69.86$$

where t is the number of years since 1980 (see Table 19). Use a symbolic method to predict when the life expectancies of women and men will be equal.

Solution

In Example 6 of Section 6.1, we found that the intersection point of the graphs of the two models represents the event when the life expectancies of women and men will be equal. We can find this intersection point by substitution (or elimination). To solve by substitution, we substitute $0.115t + 77.44$ for L in the equation $L = 0.208t + 69.86$ and solve for t:

$$0.115t + 77.44 = 0.208t + 69.86 \quad \text{Substitute } 0.115t + 77.44 \text{ for } L.$$
$$0.115t - 0.208t = 69.86 - 77.44 \quad \text{Subtract } 0.208t \text{ and } 77.44 \text{ from both sides.}$$
$$-0.093t = -7.58 \quad \text{Combine like terms.}$$
$$t \approx 81.51 \quad \text{Divide both sides by } -0.093.$$

Next, we substitute 81.51 for t in the equation $L = 0.115t + 77.44$:

$$L = 0.115(81.51) + 77.44 \approx 86.81$$

So, the approximate solution of the system is (81.51, 86.81), the same result we found in Example 6 of Section 6.1. According to the models, the life expectancy of both women and men will be about 86.8 years in 2062.

In Example 1, we used substitution to predict when the life expectancy of women will equal the life expectancy of men. Graphically, this event is described by the intersection point of the graphs of the two life-expectancy models. In general, **we can find an intersection point of two linear models by graphing, substitution, or elimination.**

Example 2 Solving a System to Make a Prediction

In Exercises 23 and 24 of Homework 5.5, you modeled world record times for the 400-meter run (see Table 20).

Table 20 400-Meter Run Record Times

Women		Men	
Year	Record Time (seconds)	Year	Record Time (seconds)
1957	57.0	1900	47.8
1959	53.4	1916	47.4
1962	51.9	1928	47.0
1969	51.7	1932	46.2
1972	51.0	1941	46.0
1976	49.29	1950	45.8
1979	48.60	1960	44.9
1983	47.99	1968	43.86
1985	47.60	1988	43.29
		1999	43.18

Source: *International Association of Athletics Federations*

The record times (in seconds) $W(t)$ and $M(t)$ for women and men, respectively, are modeled by the system

$$\begin{aligned} r &= W(t) = -0.27t + 70.45 \\ r &= M(t) = -0.053t + 48.08 \end{aligned}$$

where t is the number of years since 1900. Predict when the women's record time and the men's record time will be equal.

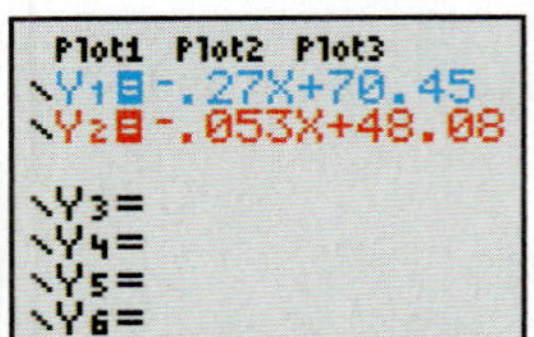

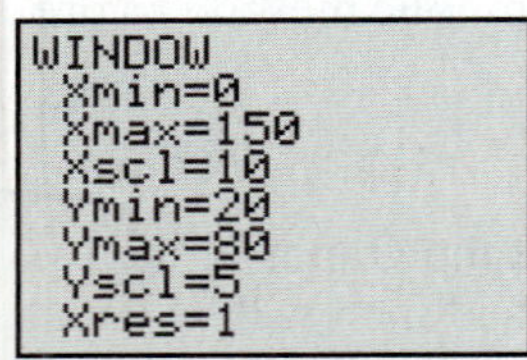

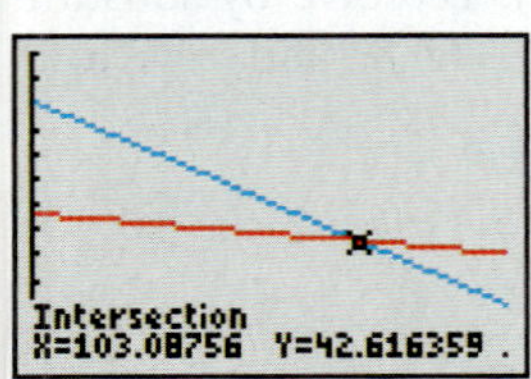

Figure 34 Verify that the intersection point is approximately (103.09, 42.62)

Solution

We solve the system

$$\begin{aligned} r &= -0.27t + 70.45 \\ r &= -0.053t + 48.08 \end{aligned}$$

by substitution. We do so by substituting $-0.27t + 70.45$ for r in the equation $r = -0.053t + 48.08$:

$$\begin{aligned} -0.27t + 70.45 &= -0.053t + 48.08 && \text{Substitute } -0.27t + 70.45 \text{ for } r. \\ -0.27t + 0.053t &= 48.08 - 70.45 && \text{Add } 0.053t \text{ to both sides; subtract } 70.45 \text{ from both sides.} \\ -0.217t &= -22.37 && \text{Combine like terms.} \\ t &\approx 103.09 && \text{Divide both sides by } -0.217. \end{aligned}$$

Next, we substitute 103.09 for t in the equation $r = -0.27t + 70.45$ and solve for r:

$$r = -0.27(103.09) + 70.45 \approx 42.62$$

So, according to the models, the world record times for both women and men were 42.62 seconds in 2003. Model breakdown has occurred, because in 2012 the women's record time was still 47.60 seconds and the men's record time was still 43.18 seconds.

We can verify our result by using "intersect" on a graphing calculator (see Fig. 34).

Using Rate of Change to Find a System for Modeling

Recall from Section 3.5 that if the rate of change of the dependent variable with respect to the independent variable is constant, then there is a linear relationship between the variables. And for the linear model that describes the relationship, the slope is equal to the constant rate of change. We use these ideas in Example 3.

Example 3 Solving a System to Make a Prediction

In 2012, a 2011 Acura RL cost about \$37,120 and a 2011 Chevrolet Impala LS sedan cost about \$14,050. The RL depreciates by \$4245 per year, and the Impala depreciates by \$1200 per year (Source: *MotorTrend*). When will these 2011 cars have the same value?

Solution

Let $V = R(t)$ be the value (in dollars) of a 2011 RL and $V = C(t)$ be the value (in dollars) of a 2011 Impala, both at t years since 2012.

Because a 2011 RL's value decreases by a *constant* \$4245 each year, the function R is linear and its slope is -4245. The V-intercept is $(0, 37{,}120)$, since the car is worth \$37,120 at year $t = 0$. So, an equation of R is

$$V = R(t) = -4245t + 37{,}120$$

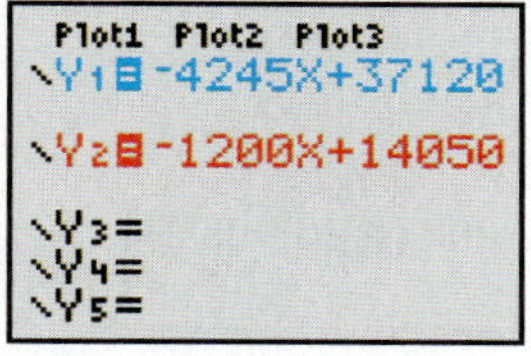

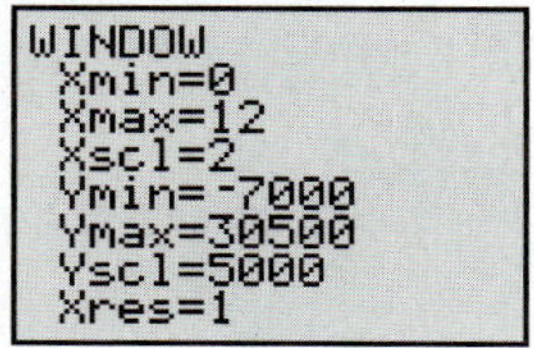

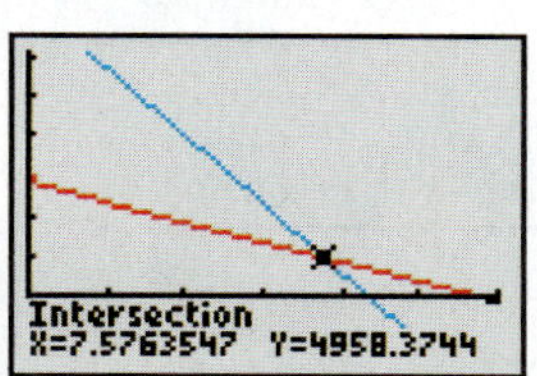

Figure 35 Verify the work

Similar work in finding the equation of the function C gives

$$V = C(t) = -1200t + 14{,}050$$

Next, we substitute $-4245t + 37{,}120$ for V in the equation $V = -1200t + 14{,}050$:

$$-4245t + 37{,}120 = -1200t + 14{,}050$$

Then we solve for t:

$$-4245t + 1200t = 14{,}050 - 37{,}120 \quad \text{Add } 1200t \text{ to both sides; subtract } 37{,}120 \text{ from both sides.}$$
$$-3045t = -23{,}070 \quad \text{Combine like terms.}$$
$$t \approx 7.58 \quad \text{Divide both sides by } -3045.$$

We conclude that the cars will have the same value in approximately 8 years (in 2020).

Next, we find the common value of the cars:

$$C(7.58) = -1200(7.58) + 14{,}050 = 4954$$

So, both cars will be worth about \$4954 in 2020, according to the models. Using "intersect" on a graphing calculator gives the result that both cars will be worth about \$4958 in 2020 (see Fig. 35). Our result of \$4954 is \$4 less than the calculator result because we evaluated C at a rounded value of t.

Group Exploration

Using a difference to make a prediction

1. Suppose that A and B are real numbers.
   a. If $A - B = 2$, which is larger, A or B? How much larger?
   b. If $A - B = -3$, which is larger, A or B? How much larger?
   c. If $A - B = 0$, what is true about A and B?
2. In Example 3, we worked with the models

   $$R(t) = -4245t + 37{,}120 \quad \text{2011 Acura RL}$$
   $$C(t) = -1200t + 14{,}050 \quad \text{2011 Chevrolet Impala}$$

   Find the difference of the expressions $-4245t + 37{,}120$ and $-1200t + 14{,}050$. [**Hint:** Use parentheses and then simplify.]
3. Evaluate the result in Problem 2 for $t = 5$. What does your result mean in this situation?
4. Evaluate the result in Problem 2 for $t = 10$. What does your result mean in this situation?
5. For what values of t is the result in Problem 2 equal to 0? What does your result mean in this situation? [**Hint:** Solve an equation.]
6. Compare your result in Problem 5 with the result in Example 3.

Tips for Success Reread a Problem

After you think you have solved a problem, reread it to make sure you have answered its question(s). Also, reread the problem with the solution in mind. If what you read makes sense, then you've provided another check of your result(s).

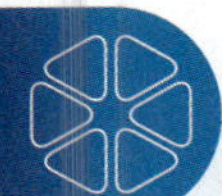

Homework 6.4

For extra help ▶ MyMathLab®  Watch the videos in MyMathLab  Download the MyDashboard App

1. In Exercise 37 of Homework 6.1, the winning times (in seconds) $W(t)$ and $M(t)$ for women and men, respectively, in Olympic 500-meter speed skating are modeled by the system

$$y = W(t) = -0.153t + 43.19$$
$$y = M(t) = -0.135t + 39.64$$

where t is the number of years since 1970 (see Table 21). Use substitution or elimination to predict when the women's winning time will be equal to the men's winning time. What will be that winning time?

Table 21 Olympic 500-Meter Speed-Skating Times

Year	Winning Time (seconds)	
	Women	Men
1972	43.33	39.44
1976	42.76	39.17
1980	41.78	38.03
1984	41.02	38.19
1988	39.10	36.45
1992	40.33	37.14
1994	39.25	36.33
1998	38.21	35.59
2002	37.375	34.615
2006	38.285	34.880
2010	38.050	34.910

Source: *The Universal Almanac*

2. In Exercise 38 of Homework 6.1, the average annual U.S. per-person consumption (in pounds per person) $C(t)$ and $R(t)$ of chicken and red meat, respectively, are modeled by the system

$$y = C(t) = 1.13t + 39.29$$
$$y = R(t) = -0.94t + 144.88$$

where t is the number of years since 1970 (see Table 22).

Table 22 Average Annual U.S. Per-Person Consumption of Chicken and Red Meat

Year	Average Annual Consumption (pounds per person)	
	Chicken	Red Meat
1970	40.3	145.8
1980	48.0	136.8
1990	61.5	120.0
2000	78.0	120.7
2010	83.7	108.7
2011	84.2	104.3

Source: *U.S. Department of Agriculture*

Use substitution or elimination to predict when the annual consumption of chicken will equal that of red meat. What will be that consumption? How confident are you in this prediction?

3. In Exercise 41 of Homework 6.1, the percentages $I(t)$ and $B(t)$ of households that have Internet access and broadband Internet access, respectively, are modeled by the system

$$p = I(t) = 2.30t + 47.6$$
$$p = B(t) = 6.83t + 1.4$$

where t is the number of years since 2000 (see Table 23).

Table 23 Percentages of Households That Have Internet Access and Broadband Internet Access

Year	Percent	
	Internet Access	Broadband Internet Access
2001	50	9
2003	55	20
2007	62	51
2009	69	64
2010	71	68

Source: *U.S. Census Bureau*

a. Find the slopes of the graphs of I and B. What do they mean in this situation?

b. Since the percentage of U.S. households that have Internet access is increasing, it is not surprising that the percentage of households that have broadband Internet access is also increasing. Explain why your results in part (a) show that this is not the only reason the percentage of households with broadband Internet access is increasing.

c. Use substitution or elimination to estimate when all households with Internet access had broadband Internet access.

4. In Exercise 40 of Homework 6.1, the worldwide numbers $F(t)$ and $C(t)$ of fixed telephone lines and cell-phone subscriptions per 100 people, respectively, are modeled by the system

$$n = F(t) = -0.51t + 22.38$$
$$n = C(t) = 8.9t - 12.38$$

where t is the number of years since 2000 (see Table 24). Use substitution or elimination to estimate when the number of fixed telephone lines per 100 people was equal to the number of cell-phone subscriptions per 100 people.

Table 24 Worldwide Numbers of Fixed Telephone Lines and Cell-Phone Subscriptions per 100 People

	Number per 100 People	
Year	Fixed Telephone Lines	Cell-Phone Subscriptions
2006	19.2	40.6
2007	19.0	50.1
2008	18.3	59.3
2009	17.7	67.9
2010	17.3	76.2

Source: *ITU World Telecommunication*

5. The average annual U.S. per-person consumption of milk and soft drinks is shown in Table 25 for various years.

Table 25 Average Annual U.S. Per-Person Consumption of Milk and Soft Drinks

Year	Average Annual Consumption of Milk (gallons per person)	Year	Average Annual Consumption of Soft Drinks (gallons per person)
1970	31.3	1950	10.8
1980	27.5	1960	13.4
1985	26.7	1970	24.3
1990	25.7	1980	35.1
1995	23.9	1990	46.2
2000	22.5	2000	49.3
2005	21.0	2005	51.5

Source: *USDA/Economic Research Service*

a. Let $M(t)$ be the average annual per-person consumption of milk and $S(t)$ be the average annual per-person consumption of soft drinks, both in gallons per person, in the year that is t years since 1900. Find equations of M and S.

b. Use substitution or elimination to estimate when the annual consumption of milk was equal to the annual consumption of soft drinks. What was that consumption?

c. Use "intersect" on a graphing calculator to verify your work.

d. Because the annual per-person consumption of milk decreased from 1970 to 2005, it's not surprising that the annual per-person soft drink consumption increased during that period. But people turning from milk to soft drinks can't be the only reason annual per-person consumption of soft drinks increased. Explain by referring to the slopes of the two models.

e. In 2010, per-person consumption of milk was 20.4 gallons and per-person consumption of soft drinks was 44.7 gallons. Compare these data to those for 2005 in Table 25 and explain how the comparison relates to your response in part (d).

6. The average fuel efficiencies of domestic and imported light-duty passenger cars are shown in Table 26 for various years.

a. Let $D(t)$ be the average fuel efficiency of domestic cars and $I(t)$ be the average fuel efficiency of imported cars, both in miles per gallon, at t years since 2000. Find equations of D and I.

b. Use substitution or elimination to estimate when the fuel efficiency of domestic cars was equal to the fuel efficiency of imported cars. What was that fuel efficiency?

Table 26 Average Fuel Efficiencies of Domestic and Imported Cars

	Average Fuel Efficiency (miles per gallon)	
Year	Domestic	Imported
2006	30.3	29.7
2007	30.6	32.2
2008	31.2	31.8
2009	32.1	33.8
2010	33.1	35.2

Source: *U.S. Department of Transportation*

c. According to CAFE emissions standards, average fuel efficiency of passenger cars must be 37.8 miles per gallon in 2016. According to the models, will domestic passenger cars meet this standard? How about imported passenger cars?

d. Find the average of $D(16)$ and $I(16)$. If there will be more domestic passenger cars sold in the United States than imported passenger cars in 2016, is your result more likely an underestimate or an overestimate of the actual average fuel efficiency of all passenger cars (both domestic and imported)? Explain.

7. Scores on tests that evaluate general knowledge and vocabulary and scores on tests that evaluate memory and information-processing speed are shown in Table 27 for various ages. (The higher the score, the better is the mental function. A score of 0 is average.)

Table 27 Age versus Mental Functioning

Age Group	Age Used to Represent Age Group	General Knowledge and Vocabulary Score	Memory and Information-Processing Speed Score
20–30	25	−0.4	1.0
30–40	35	−0.3	0.7
40–50	45	0	0.3
60–70	65	0.2	−0.2
70–80	75	0.3	−0.4

Source: *Denise Park, University of Illinois, Champaign–Urbana*

a. Let $K(a)$ be the score on general knowledge and vocabulary tests and $M(a)$ be the score on memory and information-processing speed tests, both at age a years. Find equations of K and M.

b. Use substitution or elimination to estimate at what age a person's score on general knowledge and vocabulary tests is equal to the person's score on memory and information-processing speed tests. What is that score?

c. Use "intersect" on a graphing calculator to verify your work.

8. The percentages of Americans who gamble online and at traditional casinos are shown in Table 28 for various age groups.

a. Let $I(t)$ be the percentage of Americans who gamble online and let $C(t)$ be the percentage of Americans who gamble at traditional casinos, both at age a years. Find equations of I and C.

b. Use substitution or elimination to estimate at what age the percentage of Americans who gamble online is equal to the percentage of Americans who gamble at traditional casinos. What is that percentage?

c. Use "intersect" on a graphing calculator to verify your work.

Table 28 Percentages of Americans Who Gamble Online and at Traditional Casinos

Age Group (years)	Age Used to Represent Age Group (years)	Percent	
		Online	Traditional Casinos
21–29	25	9	43
30–39	34.5	14	26
40–49	44.5	18	14
50–59	54.5	20	15
over 59	70	37	2

Source: *American Gaming Association*

9. The *Denver Post* and the *Rocky Mountain News* were competing newspapers in Denver, Colorado. Their circulations are listed in Table 29 for various years.

Table 29 Newspaper Circulations

Year	*Denver Post* Circulation (thousands)	*Rocky Mountain News* Circulation (thousands)
1992	255	355
1993	270	350
1994	285	345
1995	295	340

Source: *Audit Bureau of Circulations*

Let $C = D(t)$ be the circulation (in thousands) of the *Denver Post* and $C = R(t)$ be the circulation (in thousands) of the *Rocky Mountain News,* where t is the number of years since 1990. Reasonable equations of D and R are

$$C = D(t) = 13.5t + 229$$
$$C = R(t) = -5t + 365$$

a. Use substitution or elimination to estimate when the two newspapers had equal circulation.

b. The newspaper with greater circulation can usually charge more for advertisements, thus increasing its revenue. On March 1–2, 1997, both newspapers included articles claiming the other newspaper had made false reports of its circulation to the audit bureau. Use your result in part (a) to explain why it is not surprising that the competition between the two newspapers heated up in 1997.

c. Throughout the years, both newspapers increased their circulation by giving away "bonus issues" free to those who subscribed to less than a full week of newspapers. During the circulation wars from 1997 to 2000, both newspapers increased their distribution of bonus issues dramatically. In 2000, the combined circulation of the two newspapers was 826 thousand. Estimate the combined increase in the number of bonus issues from the two newspapers due to the circulation wars. [**Hint:** Begin by finding $D(10)$ and $R(10)$.]

d. Use the models D and R to estimate the combined circulation of the two newspapers in 2001.

e. In January 2001, the two newspapers joined their revenue streams under a single entity, the Denver Newspaper Agency (DNA). In 2001, the combined circulation was 638 thousand. Is your estimate in part (d) an underestimate or an overestimate? Give a possible explanation for this in terms of the circulation wars, DNA, and bonus issues.

10. World record times for the 1500-meter run are listed in Table 30.

Table 30 1500-Meter Run Record Times

Women		Men	
Year	Record Time (seconds)	Year	Record Time (seconds)
1927	318	1926	231
1936	287	1941	227
1946	277	1955	220
1957	269	1980	211
1962	259	1995	207
1993	230	1998	206

Source: *International Association of Athletics Federations*

a. Let $W(t)$ and $M(t)$ be the record times (in seconds) of women and men, respectively, at t years since 1900. Find the regression equations of W and of M.

b. Use substitution or elimination to predict when the women's record time will equal the men's record time. Verify your result by using "intersect" on a graphing calculator.

c. Now find the regression equation of the women's record times, excluding the record set in 1927. Also, find the regression equation of the men's times, excluding the record set in 1926. Use these equations to predict when the women's record time will equal the men's record time.

d. Explain why your result in part (b) is so different from that in part (c).

11. In 2012, a 2011 Ford Fusion S Sedan® cost about $14,290 and a 2011 Cadillac DTS sedan cost about $30,450. The Fusion depreciates by about $1414 per year, and the DTS depreciates by about $3740 per year (Source: *MotorTrend*).

a. Let $V = F(t)$ be the value (in dollars) of a 2011 Ford Fusion and $V = D(t)$ be the value (in dollars) of a 2011 Cadillac DTS, both at t years since 2012. Find equations of F and D.

b. Use substitution or elimination to predict when the cars will have the same value. What is that value?

c. Use "intersect" on a graphing calculator to verify your work in part (b).

12. In 2012, the price of a 2011 Honda Accord® was about $15,905, with a depreciation of about $1231 per year (Source: *MotorTrend*). A student had $500 in 2012 and saves $1700 each year. Assume the student does not earn interest on her savings.

a. Let $H(t)$ be the value (in dollars) of a 2011 Honda Accord and $S(t)$ be the student's total savings (in dollars), both at t years since 2012. Find equations of H and S.

b. Use substitution or elimination to predict when the student will be able to buy a 2011 Honda Accord for cash.

c. Use "intersect" on a graphing calculator to verify your work in part (b).

d. Now assume the student earns interest on her savings. Is your answer in part (b) an underestimate or an overestimate? Explain.

13. Nutrisystem® offers a weight-loss program that charges $91 per week. The program includes food and phone access to weight-loss counselors (Source: *Nutrisystem*). Weight Watchers® offers a program that charges $12 per week for meetings plus a one-time membership fee of $20 (Source: *Weight Watchers*). Members purchase their own food, which usually costs about $75 per week.

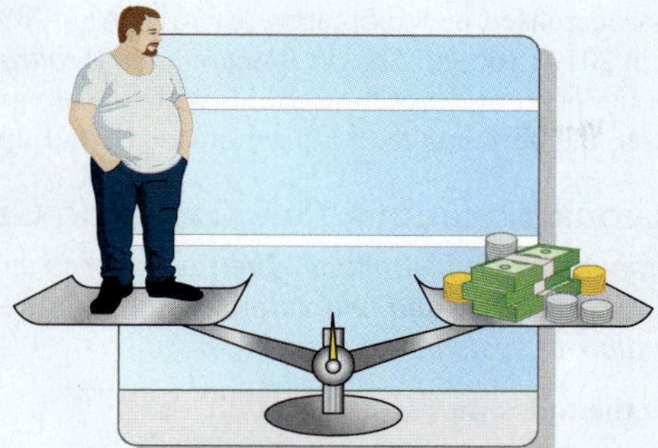

a. Let $N(t)$ be the per-person cost (in dollars) of the Nutrisystem program (including food) for t weeks. Let $W(t)$ be the per-person cost (in dollars) of the Weight Watchers program *plus food* for t weeks. Find equations of N and W.

b. Perform a unit analysis of the equations of N and W.

c. Use substitution or elimination to estimate how many weeks it will take for the total cost at Nutrisystem to equal the total cost at Weight Watchers (plus the cost of food). What is that total cost?

d. Use "intersect" on a graphing calculator to verify your work in part (c).

e. If a person joins Nutrisystem or Weight Watchers but quits 7 weeks later, which of the two weight-loss programs should he have joined? Explain.

14. Gold's Gym® offers two payment options. Option 1 requires an initial investment of \$149.99 plus a monthly fee of \$29.99. Option 2 requires an initial investment of \$49.99 plus a monthly fee of \$39.99 (Source: *Gold's Gym*). Let $f(t)$ be the total cost (in dollars) of Option 1 and $g(t)$ be the total cost (in dollars) of Option 2, both for t months.

a. Find equations of f and g.

b. Perform a unit analysis of the equations of f and g.

c. Use substitution or elimination to estimate when the total cost of each payment option will be equal. What is that total cost?

d. Use "intersect" on a graphing calculator to verify your work in part (c).

e. If a person joins Gold's Gym but quits 10 months later, which payment option should she have chosen? Explain.

15. The number of students who earned a bachelor's degree in communications was 83.1 thousand in 2009 and has since increased by about 2.0 thousand degrees each year. The number of students who earned a bachelor's degree in education was 101.7 thousand students in 2009 and has since decreased by 0.7 thousand degrees each year (Source: *U.S. National Center for Education Statistics*).

a. Let $C(t)$ be the number (in thousands) of bachelor's degrees earned in communications and let $E(t)$ be the number (in thousands) of bachelor's degrees earned in education, both at t years since 2009. Find equations of C and E.

b. Use substitution or elimination to predict when the same number of bachelor's degrees will be earned in communications and education. What is that number of degrees?

c. Use "intersect" on a graphing calculator to verify your work in part (b).

16. The number of students who earned bachelor's degrees in computer science was 38.0 thousand in 2009 and has since decreased by about 4.3 thousand degrees each year. The total number of students who earned bachelor's degrees in mathematics and statistics was 15.5 thousand in 2009 and has since increased by about 0.5 thousand degrees each year (Source: *U.S. National Center for Education Statistics*).

a. Let $C(t)$ be the number (in thousands) of bachelor's degrees earned in computer science and let $M(t)$ be the total number (in thousands) of bachelor's degrees earned in mathematics and statistics, both at t years since 2009. Find equations of C and M.

b. Use substitution or elimination to predict when the number of bachelor's degrees in computer science will equal the total number of bachelor's degrees in mathematics and statistics. What is that number of degrees?

c. Use "intersect" on a graphing calculator to verify your work in part (b).

17. The percentage of U.S. electricity generated from natural gas was 24% in 2010 and has increased by about 0.9 percentage point per year (Source: *U.S. Energy Information Administration*). The percentage of U.S. electricity generated from coal was 44% in 2010 and has decreased by about 0.7 percentage point per year. Predict when the percentage of U.S. electricity generated from natural gas will be equal to that from coal. What is that percentage?

18. Total spending by advertisers on Internet ads was \$21 million in 2007 and has increased by about \$2 million per year (Source: *Interactive Advertising Bureau*). Total spending by advertisers on print newspaper ads was \$42 million in 2007 and has decreased by about \$6 million per year (Source: *Newspaper Association of America*). Estimate in which year total spending on Internet ads was equal to total spending on newspaper ads. What was that total spending?

19. The number of unique monthly visitors to Facebook was 165 million visitors in 2012 and has since increased by about 36 million visitors each year. The number of unique monthly visitors to Myspace was 30 million visitors in 2012 and has since decreased by about 16 million visitors each year (Source: *comScore Media Metrix*). Use substitution or elimination to estimate when there were the same number of unique monthly visitors to Facebook and Myspace. What was that number of visitors?

20. The percentage of 30–34-year-olds who are married was 60.5% in 2010 and has since decreased by about 0.9 percentage point each year. The percentage of 30–34-year-olds who have never married was 31.8% in 2010 and has since increased by about 0.8 percentage point each year (Source: *U.S. Census Bureau*). Use substitution or elimination to predict when the percentage of 30–34-year-olds who are married will equal the percentage of 30–34-year-olds who have never married. What is that percentage?

Concepts

21. To solve the system

$$y = f(t) = m_1t + b_1$$
$$y = g(t) = m_2t + b_2$$

where m_1, b_1, m_2, and b_2 are constants, a student eliminates y and finds a noninteger value of t. He rounds the value of t *up* to an integer I.

a. He is confused, because $f(I)$ is not equal to $g(I)$. Draw a graph to illustrate what happened.

b. If m_1 is larger than m_2, which will be larger, $f(I)$ or $g(I)$? Draw a graph to illustrate this.

22. Describe how you can find a system of linear equations to model a situation. Also, explain how you can use the system to make an estimate or prediction about the situation.

Related Review

23. The percentage of Americans who own an iPod increased approximately linearly from 4% in 2005 to 21% in 2010. The percentage of Americans who own a non-iPod portable MP3 player increased approximately linearly from 8% in 2005 to 14% in 2010 (Source: *Edison Research and Arbitron*). Estimate when the percentage of Americans who own an iPod was equal to the percentage of Americans who own a non-iPod portable MP3 player. What was that percentage?

24. The percentage of households with broadband Internet access increased approximately linearly from 21% in 2002 to 84% in 2010. The percentage of households with dial-up Internet access decreased approximately linearly from 78% in 2002 to 13% in 2010 (Source: *Edison Research and Arbitron*). Estimate when the percentage of households with broadband access was equal to the percentage of households with dial-up access.

Expressions, Equations, Functions, and Graphs

Perform the indicated instruction. Then use words such as linear, function, one variable, *and* two variables *to describe the expression, equation, or system.*

25. Solve the following:

$$y = -2x + 6$$
$$y = 3x + 1$$

26. Evaluate $-2x + 6$ for $x = 1$.

27. Solve $-2x + 6 = 3x + 1$.

28. Simplify $(-2x + 6) + (3x + 1)$.

6.5 Perimeter, Value, Interest, and Mixture Problems

Objectives

» Know a five-step problem-solving method.

» Use a system of two linear equations or a linear function to solve perimeter, value, interest, and mixture problems.

In this section, we will use a five-step method to solve problems that involve the perimeter of a rectangle, the (dollar) value of an object, the interest from an investment, and the percentage of a substance in a mixture.

Five-Step Problem-Solving Method

To solve some problems in which we want to find two quantities, it is useful to perform the following five steps:

- ***Step 1: Define each variable.*** For each quantity we are trying to find, we usually define a variable to represent that unknown quantity.
- ***Step 2: Write a system of two equations.*** We find a system of two equations by using the variables from step 1. We can usually write each equation either by translating the information stated in the problem into mathematics or by making a substitution into a formula.
- ***Step 3: Solve the system.*** We solve the system of equations from step 2.
- ***Step 4: Describe each result.*** We use a complete sentence to describe the quantities we found.
- ***Step 5: Check.*** We reread the problem and check that the quantities we found agree with the given information.

Perimeter Problems

Recall from Section 4.6 that the formula of the perimeter P of a rectangle with length L and width W is $P = 2L + 2W$.

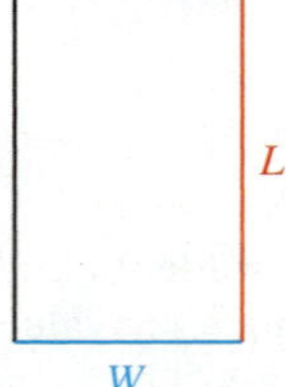

Figure 36 A golden rectangle drawn to scale

Example 1 Solving a Perimeter Problem

Throughout history, rectangles whose length is about 1.62 times their width, called *golden rectangles,* have been viewed as the most pleasing to the eye (see Fig. 36). Many famous structures, including the Parthenon in Athens and the Great Pyramid of Giza in Cairo, incorporate golden rectangles into their design. For Leonardo Da Vinci's *Mona Lisa,* the edges of the painting form a golden rectangle.

Suppose an artist wants a piece of canvas in the shape of a golden rectangle with a perimeter of 9 feet. Find the dimensions of the canvas.

Solution

Step 1: Define each variable. Let W be the width (in feet) and L be the length (in feet). See Fig. 36.

Step 2: Write a system of two equations. Since the length must be 1.62 times the width, our first equation is

$$L = 1.62W$$

Because the perimeter is 9 feet, we find our second equation by substituting 9 for P in the perimeter formula $P = 2L + 2W$:

$$9 = 2L + 2W$$

The system is

$$\begin{aligned} L &= 1.62W && \text{Equation (1)} \\ 2L + 2W &= 9 && \text{Equation (2)} \end{aligned}$$

Step 3: Solve the system. We can solve the system by substitution. We substitute $1.62W$ for L in equation (2) and then solve for W:

$$\begin{aligned} 2L + 2W &= 9 && \text{Equation (2)} \\ 2(1.62W) + 2W &= 9 && \text{Substitute } 1.62W \text{ for } L. \\ 3.24W + 2W &= 9 && \text{Simplify.} \\ 5.24W &= 9 && \text{Combine like terms.} \\ \frac{5.24W}{5.24} &= \frac{9}{5.24} && \text{Divide both sides by 5.24.} \\ W &\approx 1.72 && \text{Round } \frac{9}{5.24} \text{ to second decimal place.} \end{aligned}$$

To find the approximate length, we substitute $\frac{9}{5.24}$ for W in equation (1):

$$L = 1.62\left(\frac{9}{5.24}\right) \approx 2.78$$

Step 4: Describe each result. The approximate width is 1.72 feet, and the approximate length is 2.78 feet.

Step 5: Check. We add the lengths of the four sides: $1.72 + 2.78 + 1.72 + 2.78 = 9$, which checks for a perimeter of 9 feet. We can also check that 2.78 is about 1.62 times 1.72 (try it).

Value Problems

Recall from Section 4.6 that if n objects each have value v, then their total value T is given by

$$T = vn$$

When some objects are sold, we refer to the total money collected as the **revenue** from selling the objects.

Example 2 Solving Value Problems

A vendor charges \$5 per hamburger and \$3 per hot dog.

1. What is the vendor's total revenue from selling 40 hamburgers and 85 hot dogs?
2. If the vendor sells a total of 135 hamburgers and hot dogs for a total revenue of \$495, how many hamburgers and hot dogs did he sell?

Solution

1. The revenue from hamburgers is equal to the price per hamburger times the number of hamburgers sold: $5 \cdot 40 = 200$ dollars. The revenue from hot dogs is equal to the price per hot dog times the number of hot dogs sold: $3 \cdot 85 = 255$ dollars. We add the revenue of the hamburgers and that of the hot dogs to find the total revenue:

$$\underbrace{5}_{\frac{\text{dollars}}{\text{hamburger}}} \cdot \underbrace{40}_{\text{hamburgers}} + \underbrace{3}_{\frac{\text{dollars}}{\text{hot dog}}} \cdot \underbrace{85}_{\text{hot dogs}} = \underbrace{455}_{\text{total revenue in dollars}}$$

So, the total revenue from hamburgers and hot dogs is \$455. The units of the expressions on both sides of the equation are dollars, which suggests that our work is correct.

2. ***Step 1: Define each variable.*** Let x be the number of hamburgers sold and y be the number of hot dogs sold.

Step 2: Write a system of two equations. Our work in Problem 1 suggests that the formula of the total revenue T (in dollars) is

$$T = \underbrace{5}_{\frac{\text{dollars}}{\text{hamburger}}} \cdot \underbrace{x}_{\text{hamburgers}} + \underbrace{3}_{\frac{\text{dollars}}{\text{hot dog}}} \cdot \underbrace{y}_{\text{hot dogs}}$$

To obtain our first equation, we substitute 495 for T:

$$495 = 5x + 3y$$

Since the vendor sells a total of 135 hamburgers and hot dogs, our second equation is

$$x + y = 135$$

The system is

$$\begin{aligned} 5x + 3y &= 495 && \text{Equation (1)} \\ x + y &= 135 && \text{Equation (2)} \end{aligned}$$

Step 3: Solve the system. We can use elimination to solve the system. To eliminate the y terms, we multiply both sides of equation (2) by -3, yielding the system

$$\begin{aligned} 5x + 3y &= 495 && \text{Equation (1)} \\ -3x - 3y &= -405 && \text{Multiply both sides of equation (2) by } -3. \end{aligned}$$

Then we add the left sides and add the right sides of the equations and solve for x:

$$\begin{aligned} 5x + 3y &= 495 \\ -3x - 3y &= -405 \\ \hline 2x + 0 &= 90 && \text{Add left sides and right sides; combine like terms.} \\ 2x &= 90 && a + 0 = a \\ x &= 45 && \text{Divide both sides by 2.} \end{aligned}$$

Next, we substitute 45 for x in equation (2) and solve for y:

$$\begin{aligned} x + y &= 135 && \text{Equation (2)} \\ 45 + y &= 135 && \text{Substitute 45 for } x. \\ y &= 90 && \text{Subtract 45 from both sides.} \end{aligned}$$

Step 4: Describe each result. The vendor sold 45 hamburgers and 90 hot dogs.

Step 5: Check. First, we find the sum $45 + 90 = 135$, which is equal to the total number of hamburgers and hot dogs sold. Next, we find the total revenue from selling 45 hamburgers and 90 hot dogs: $5 \cdot 45 + 3 \cdot 90 = 495$, which checks.

Notice that the arithmetic $5 \cdot 40 + 3 \cdot 85 = 455$ in Problem 1 of Example 2 suggests the equation $5x + 3y = 495$ that we formed in Problem 2. **If you have difficulty forming an equation, try making up numbers for unknown quantities and performing arithmetic to compute a related quantity. Your computation may suggest how to form the desired equation.**

Example 3 Solving a Value Problem

An auditorium has 500 balcony seats and 2100 main-level seats. If tickets for balcony seats cost \$18 less than tickets for main-level seats, what should the prices be for each type of ticket so that the total revenue from a sellout performance will be \$128,800?

Solution

Step 1: Define each variable. Let b be the price (in dollars) for balcony seats and m be the price (in dollars) for main-level seats.

Step 2: Write a system of two equations. Since tickets for balcony seats will cost \$18 less than tickets for main-level seats, our first equation is

$$\underbrace{b}_{\text{balcony ticket price}} \quad \underbrace{=}_{\text{is}} \quad \underbrace{m - 18}_{\text{\$18 less than main-level ticket price}}$$

Since the total revenue is \$128,800, our second equation is

$$\underbrace{b}_{\frac{\text{dollars}}{\text{balcony ticket}}} \cdot \underbrace{500}_{\text{balcony tickets}} + \underbrace{m}_{\frac{\text{dollars}}{\text{main-level ticket}}} \cdot \underbrace{2100}_{\text{main-level tickets}} = \underbrace{128{,}800}_{\text{total revenue in dollars}}$$

The units of the expressions on both sides of the equation are dollars, which suggests that our work is correct. The system is

$$\begin{aligned} b &= m - 18 && \text{Equation (1)} \\ 500b + 2100m &= 128{,}800 && \text{Equation (2)} \end{aligned}$$

Step 3: Solve the system. We can use substitution to solve the system. We substitute $m - 18$ for b in equation (2) and solve for m:

$$\begin{aligned} 500b + 2100m &= 128{,}800 && \text{Equation (2)} \\ 500(m - 18) + 2100m &= 128{,}800 && \text{Substitute } m - 18 \text{ for } b. \\ 500m - 9000 + 2100m &= 128{,}800 && \text{Distributive law} \\ 2600m - 9000 &= 128{,}800 && \text{Combine like terms.} \\ 2600m &= 137{,}800 && \text{Add 9000 to both sides.} \\ m &= 53 && \text{Divide both sides by 2600.} \end{aligned}$$

Then we substitute 53 for m in equation (1) and solve for b:

$$b = 53 - 18 = 35$$

Step 4: Describe each result. Tickets for balcony seats should be priced at \$35 per ticket, and tickets for main-level seats should be priced at \$53 per ticket.

Step 5: Check. First, we find the difference in the ticket prices: $53 - 35 = 18$ dollars, which checks. Then we compute the total revenue from selling 500 of the \$35 tickets and 2100 of the \$53 tickets: $35 \cdot 500 + 53 \cdot 2100 = 128{,}800$ dollars, which checks.

In Examples 1–3, we analyzed *one* aspect of a situation by working with a linear system. **If we want to analyze many aspects of a certain situation, it can help to use a system of equations to find a linear function. We can then use the function to analyze the situation in various ways.**

Example 4 Using a Function to Model a Value Situation

A 10,000-seat amphitheater will sell general-seat tickets at \$45 and reserved-seat tickets at \$65 for a Foo Fighters concert. Let x and y be the number of tickets that will sell for \$45 and \$65, respectively. Assume the show will sell out.

1. Let $T = f(x)$ be the total revenue (in dollars) from selling the \$45 and \$65 tickets. Find an equation of f.
2. Use a graphing calculator to sketch a graph of f for $0 \le x \le 10{,}000$. What is the slope? What does it mean in this situation?
3. Find $f(8500)$. What does it mean in this situation?
4. Find $f(11{,}000)$. What does it mean in this situation?
5. The total cost of the production is \$350,000. How many of each type of ticket must be sold to make a profit of \$150,000?

Solution

1. We add the revenues from the general tickets and the reserved tickets to find an equation of the total revenue T:

$$T = \overbrace{45}^{\frac{\text{dollars}}{\text{general ticket}}} \cdot \overbrace{x}^{\text{general tickets}} + \overbrace{65}^{\frac{\text{dollars}}{\text{reserved ticket}}} \cdot \overbrace{y}^{\text{reserved tickets}}$$

So far, we have described T in terms of x and y. Next, we describe T in terms of just x. The total number of tickets sold for a sellout performance is 10,000:

$$x + y = 10{,}000$$

Now we get y alone on one side of the equation:

$$y = 10{,}000 - x$$

Next, we substitute $10{,}000 - x$ for y in the equation $T = 45x + 65y$:

$$\begin{aligned} T &= 45x + 65(10{,}000 - x) && \text{Substitute } 10{,}000 - x \text{ for } y. \\ &= 45x + 650{,}000 - 65x && \text{Distributive law} \\ &= -20x + 650{,}000 && \text{Combine like terms.} \end{aligned}$$

So, an equation of f is

$$f(x) = -20x + 650{,}000$$

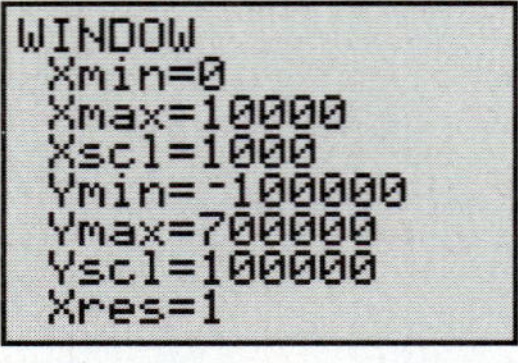

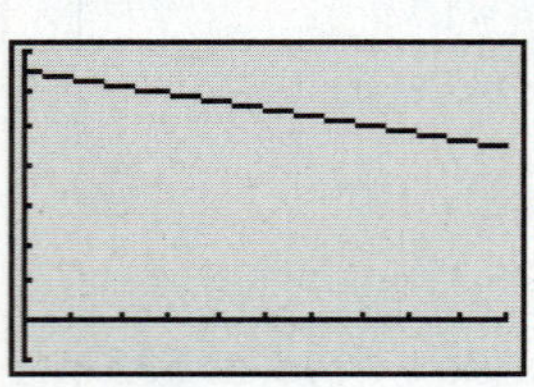

Figure 37 Graphing the revenue model

2. We draw a sketch of f in Fig. 37. The graph of f is a decreasing line with slope -20. This means that if one more ticket is sold for \$45 (and one less ticket is sold for \$65), the revenue will decrease by \$20.
3. Here, $f(8500) = -20(8500) + 650{,}000 = 480{,}000$. This means that if 8500 tickets sell for \$45 (and 1500 tickets sell for \$65), the total revenue will be \$480,000.
4. Here, $f(11{,}000) = -20(11{,}000) + 650{,}000 = 430{,}000$. This means that if 11,000 tickets sell for \$45, the total revenue will be \$430,000. Since there are only 10,000 seats, model breakdown has occurred.
5. To make a profit of \$150,000, the revenue would need to be $350{,}000 + 150{,}000 = 500{,}000$ dollars. We substitute 500,000 for T in the equation $T = -20x + 650{,}000$ and solve for x:

$$\begin{aligned} 500{,}000 &= -20x + 650{,}000 && \text{Substitute 500,000 for } T. \\ -150{,}000 &= -20x && \text{Subtract 650,000 from both sides.} \\ 7500 &= x && \text{Divide both sides by } -20. \end{aligned}$$

So, 7500 \$45 tickets and $10{,}000 - 7500 = 2500$ \$65 tickets would need to be sold for the profit to be \$150,000.

Interest Problems

Money deposited in an account, such as a savings account, certificate of deposit (CD), or mutual fund, is called **principal.** A person invests money in hopes of later getting back the principal plus additional money, called **interest,** which is a percentage of the principal (see Fig. 38). The **annual simple-interest rate** is the percentage of the principal that equals the interest earned per year. So, if we invest \$100 and earn \$5 per year, then the annual simple-interest rate is 5%.

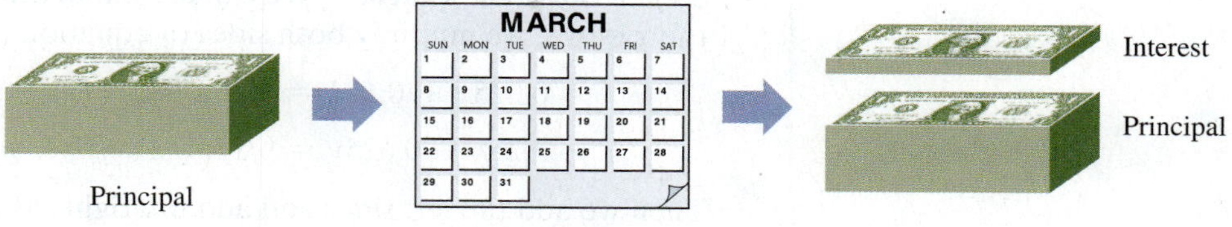

Figure 38 Invest the principal; time passes; get back the principal plus interest

Example 5 Finding the Interest from an Investment

How much interest will a person earn by investing \$2500 in an account at 6% annual interest for one year?

Solution

We find 6% of 2500:

$$0.06(2500) = 150$$

The person will earn \$150 in interest.

Some people invest in a variety of accounts, including lower-risk accounts and some higher-risk accounts. These investors might earn large amounts of interest from the higher-risk accounts, but will have a safety net of principal and interest from the lower-risk accounts.

Example 6 Solving Interest Problems

1. How much interest will a person earn from investing \$6000 in a Presidential Bank Internet CD account at 2.5% annual interest and \$1000 in a John Hancock Health Sciences A mutual fund account at 12% annual interest for one year?
2. A person plans to invest a total of \$7000. She will invest in the two accounts described in Problem 1. How much money should she invest in each account to earn a total interest of \$400 in one year?

Solution

1. We add the interest from both accounts to compute the total interest:

$$\overbrace{0.025(6000)}^{\text{interest from 2.5\% account}} + \overbrace{0.12(1000)}^{\text{interest from 12\% account}} = \overbrace{270}^{\text{total interest}}$$

2. ***Step 1: Define each variable.*** Let x be the money (in dollars) invested at 2.5% annual interest, and let y be the money (in dollars) invested at 12% annual interest.

   Step 2: Write a system of two equations. Our work in Problem 1 suggests the first equation:

$$\overbrace{0.025x}^{\text{interest from 2.5\% account}} + \overbrace{0.12y}^{\text{interest from 12\% account}} = \overbrace{400}^{\text{total interest}}$$

Since the total investment is \$7000, our second equation is

$$x + y = 7000$$

The system is

$$\begin{aligned} 0.025x + 0.12y &= 400 && \text{Equation (1)} \\ x + y &= 7000 && \text{Equation (2)} \end{aligned}$$

Step 3: Solve the system. We can use elimination to solve the system. To eliminate the x terms, we multiply both sides of equation (2) by -0.025, yielding the system

$$\begin{aligned} 0.025x + 0.12y &= 400 && \text{Equation (1)} \\ -0.025x - 0.025y &= -175 && \text{Multiply both sides of equation (2) by } -0.025. \end{aligned}$$

Then we add the left sides and add the right sides of the equations and solve for y:

$$\begin{aligned} 0.025x + 0.12y &= 400 \\ -0.025x - 0.025y &= -175 \\ \hline 0 + 0.095y &= 225 && \text{Add left sides and right sides; combine like terms.} \\ 0.095y &= 225 && 0 + a = a \\ y &= 2368.42 && \text{Divide both sides by 0.095.} \end{aligned}$$

Next, we substitute 2368.42 for y in equation (2) and solve for x:

$$\begin{aligned} x + y &= 7000 && \text{Equation (2)} \\ x + 2368.42 &= 7000 && \text{Substitute 2368.42 for } y. \\ x &= 4631.58 && \text{Subtract 2368.42 from both sides.} \end{aligned}$$

Step 4: Describe each result. The person should invest \$4631.58 at 2.5% annual interest and \$2368.42 at 12% annual interest.

Step 5: Check. First, we find the sum $4631.58 + 2368.42 = 7000$, which checks with the total money invested. Next, we calculate the total interest earned from investing \$4631.58 at 2.5% and \$2368.42 at 12%:

$$0.025(4631.58) + 0.12(2368.42) \approx 400$$

which checks.

Example 7 Using a Function to Model a Situation Involving Interest

A person plans to invest a total of \$6000 in a Gabelli ABC mutual fund that has a three-year average annual interest of 6% and in a Presidential Bank Internet CD account at 2.25% annual interest. Let x and y be the money (in dollars) invested in the mutual fund and the CD, respectively.

1. Let $I = f(x)$ be the total interest (in dollars) earned from investing the \$6000 for one year. Find an equation of f.
2. Use a graphing calculator to draw a graph of f for $0 \le x \le 6000$. What is the slope of f? What does it mean in this situation?
3. Use a graphing calculator to create a table of values of f. Explain how such a table could help the person decide how much money to invest in each account.
4. How much money should be invested in each account to earn \$300 in one year?

Solution

1. The interest earned from investing x dollars in an account earning 6% annual interest is $0.06x$, and the interest earned from investing y dollars in an account earning 2.25%

annual interest is $0.0225y$ dollars. We add the two interest earnings to find the total interest earned:

$$\underbrace{I}_{\text{total interest}} = \underbrace{0.06x}_{\substack{\text{interest from}\\ 6\%\text{ account}}} + \underbrace{0.0225y}_{\substack{\text{interest from}\\ 2.25\%\text{ account}}}$$

Next, we describe I in terms of just x. The person plans to invest \$6000:

$$x + y = 6000$$

Then, we isolate y:

$$y = 6000 - x$$

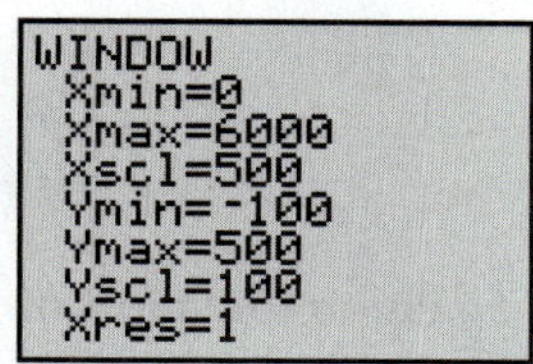

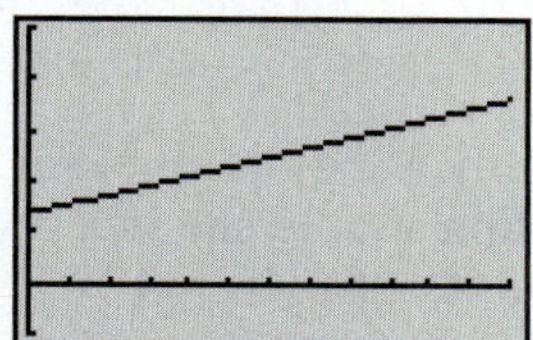

Figure 39 The revenue model

X	Y1
0	135
1000	172.5
2000	210
3000	247.5
4000	285
5000	322.5
6000	360

X=0

Figure 40 Table of values of f

Now we substitute $6000 - x$ for y in the equation $I = 0.06x + 0.0225y$:

$$\begin{aligned} I &= 0.06x + 0.0225(6000 - x) && \text{Substitute } 6000 - x \text{ for } y. \\ &= 0.06x + 135 - 0.0225x && \text{Distributive law} \\ &= 0.0375x + 135 && \text{Combine like terms.} \end{aligned}$$

So, our equation of f is

$$f(x) = 0.0375x + 135$$

2. We draw a graph of f in Fig. 39. The graph of f is an increasing line with slope 0.0375. This means that if one more dollar is invested at 6% (and one less dollar is invested at 2.25%), the total interest will increase by 3.75 cents.
3. We create a table in Fig. 40. The person may know she wants to invest some of the \$6000 in the safe Presidential CD account and the rest in the riskier Gabelli mutual fund, but she may not be clear as to exactly how much risk she is willing to take. By seeing some possible interest earnings, she may get a clearer idea of how much money to invest in each account.
4. We substitute 300 for I in the equation $I = 0.0375x + 135$ and solve for x:

$$\begin{aligned} 300 &= 0.0375x + 135 && \text{Substitute 300 for } I. \\ 165 &= 0.0375x && \text{Subtract 135 from both sides.} \\ 4400 &= x && \text{Divide both sides by 0.0375.} \end{aligned}$$

The person should invest \$4400 in the Gabelli mutual fund account and $6000 - 4400 = 1600$ dollars in the Presidential CD account.

Mixture Problems

Many fields, including chemistry, cooking, and medicine, involve mixing different substances. Suppose that 2 ounces of lime juice is mixed with 6 ounces of water to make 8 ounces of unsweetened limeade. Note that $\frac{2}{8} = 0.25 = 25\%$ of the limeade is lime juice. We call the limeade a *25% lime-juice solution.*

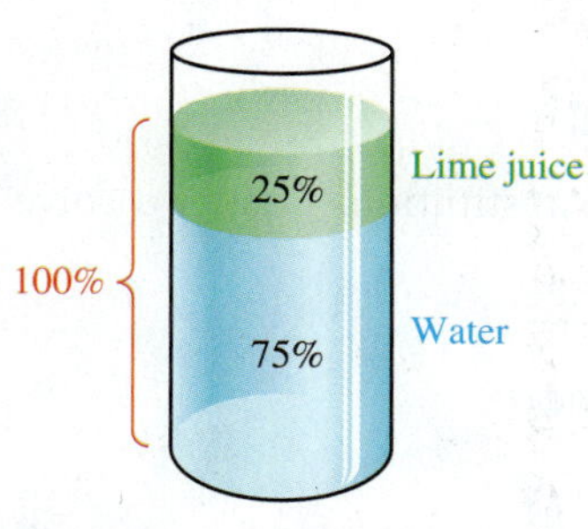

Figure 41 A 25% lime-juice solution

Then, the remaining $\frac{6}{8} = 0.75 = 75\%$ of the limeade is water. The percentage of the solution that is lime juice plus the percentage of the solution that is water is equal to 100% (all) of the solution: $25\% + 75\% = 100\%$. See Fig. 41.

Therefore, in a 20% lime-juice solution, 20% of the solution is lime juice and $100\% - 20\% = 80\%$ of the solution is water. In general, **for an x% solution of two substances that are mixed, x% of the solution is one substance and $(100 - x)$% is the other substance.**

Example 8 Solving Mixture Problems

1. How much lime juice is in 6 ounces of a 15% lime-juice solution?
2. If 5 gallons of a 40% antifreeze solution is mixed with 3 gallons of a 75% antifreeze solution, how much pure antifreeze is in the mixture?

Solution

1. We find 15% of 6:

$$0.15(6) = 0.9$$

There is 0.9 ounce of lime juice in the 6 ounces of lime-juice solution.

2. The total amount of pure antifreeze doesn't change, regardless of how it is distributed in the two solutions. To find the amount of pure antifreeze in the mixture, we add the pure amounts of antifreeze in the 40% solution and the 75% solution:

$$\underbrace{0.40(5)}_{\substack{\text{number of gallons of}\\ \text{pure antifreeze in}\\ \text{40\% solution}}} + \underbrace{0.75(3)}_{\substack{\text{number of gallons of}\\ \text{pure antifreeze in}\\ \text{75\% solution}}} = \underbrace{4.25}_{\substack{\text{number of gallons of}\\ \text{pure antifreeze}\\ \text{in mixture}}}$$

There are 4.25 gallons of pure antifreeze in the mixture.

Our work in Problem 2 of Example 8 suggests how to find an equation for step 2 of our work in Example 9.

Example 9 Solving a Mixture Problem

A chemist needs 8 liters of a 20% alcohol solution, but she has only a 15% alcohol solution and a 35% alcohol solution. How many liters of each solution should she mix to make the desired 8 liters of 20% alcohol solution?

Solution

Step 1: Define each variable. Let x be the number of liters of 15% alcohol solution, and let y be the number of liters of 35% alcohol solution.

Step 2: Write a system of two equations. Since she wants 8 liters of the total mixture, our first equation is

$$x + y = 8$$

We find our second equation from the fact that the sum of the amounts of pure alcohol in the 15% alcohol solution and the 35% alcohol solution is equal to the amount of pure alcohol in the mixture:

$$\underbrace{0.15x}_{\substack{\text{pure alcohol in}\\ \text{15\% solution}}} + \underbrace{0.35y}_{\substack{\text{pure alcohol in}\\ \text{35\% solution}}} = \underbrace{0.20(8)}_{\substack{\text{pure alcohol}\\ \text{in mixture}}}$$

The system is

$$\begin{aligned} x + y &= 8 && \textit{Equation (1)} \\ 0.15x + 0.35y &= 1.6 && \textit{Equation (2)} \end{aligned}$$

Step 3: Solve the system. We can solve the system by substitution. First, we solve equation (1) for y:

$$\begin{aligned} x + y &= 8 && \textit{Equation (1)} \\ y &= 8 - x && \textit{Subtract x from both sides.} \end{aligned}$$

Then, we substitute $8 - x$ for y in equation (2) and solve for x:

$$\begin{aligned} 0.15x + 0.35y &= 1.6 && \textit{Equation (2)} \\ 0.15x + 0.35(8 - x) &= 1.6 && \textit{Substitute } 8 - x \textit{ for } y. \\ 0.15x + 2.8 - 0.35x &= 1.6 && \textit{Distributive law} \\ -0.20x + 2.8 &= 1.6 && \textit{Combine like terms.} \\ -0.20x &= -1.2 && \textit{Subtract 2.8 from both sides.} \\ x &= 6 && \textit{Divide both sides by } -0.20. \end{aligned}$$

Next, we substitute 6 for x in the equation $y = 8 - x$ and solve for y:

$$y = 8 - 6 = 2$$

Step 4: Describe each result. Six liters of the 15% alcohol solution and 2 liters of the 35% alcohol solution are required.

Step 5: Check. First, we compute the total amount (in liters) of pure alcohol in 6 liters of the 15% alcohol solution and 2 liters of the 35% alcohol solution:

$$0.15(6) + 0.35(2) = 1.6$$

Next, we compute the amount (in liters) of pure alcohol in 8 liters of the 20% alcohol solution:

$$0.20(8) = 1.6$$

Since the two results are equal, they check. Also, $6 + 2 = 8$, which checks with the chemist wanting 8 liters of the 20% solution.

Example 10 Solving a Mixture Problem

A chemist needs 6 ounces of a 5% acid solution, but has only a 40% acid solution. How much of the 40% solution and water should he mix to form the desired 6 ounces of the 5% solution?

Solution

Step 1: Define each variable. Let x be the number of ounces of the 40% solution, and let y be the number of ounces of water.

Step 2: Write a system of two equations. Since he wants 6 ounces of the total mixture, our first equation is

$$x + y = 6$$

There is no acid in pure water, so we find our second equation from the fact that the amount of pure acid in the 40% acid solution is equal to the amount of pure acid in the mixture:

$$\overbrace{0.40x}^{\text{amount of pure acid in 40\% solution}} = \overbrace{0.05(6)}^{\text{amount of pure acid in mixture}}$$

The system is

$$\begin{aligned} x + y &= 6 && \text{Equation (1)} \\ 0.40x &= 0.3 && \text{Equation (2)} \end{aligned}$$

Step 3: Solve the system. We begin by solving equation (2) for x:

$$\begin{aligned} 0.40x &= 0.3 && \text{Equation (2)} \\ x &= 0.75 && \text{Divide both sides by 0.40.} \end{aligned}$$

Next, we substitute 0.75 for x in equation (1):

$$\begin{aligned} x + y &= 6 && \text{Equation (1)} \\ 0.75 + y &= 6 && \text{Substitute 0.75 for } x. \\ y &= 5.25 && \text{Subtract 0.75 from both sides.} \end{aligned}$$

Step 4: Describe each result. The chemist needs to mix 0.75 ounce of the 40% acid solution with 5.25 ounces of water.

Step 5: Check. First, we find $0.75 + 5.25 = 6$, which checks with the chemist wanting 6 ounces of the 5% solution. Next, we compute the amount of pure acid in 0.75 ounce of the 40% solution: $0.40(0.75) = 0.3$ ounce. Finally, we compute the amount of pure acid in the 5% solution: $0.05(6) = 0.3$ ounce. The computed amounts of pure acid in the 40% solution and the 5% solution are equal, so they check.

Group Exploration

Looking ahead: Graphing an inequality in two variables

1. Graph $y = x + 1$ carefully by hand.
2. Find 4 points that lie above the line, 2 points that lie on the line, and 4 points that lie below the line. Choose your 10 points so that there are at least 2 points in each of the four quadrants. List the ordered pairs for these points so that it is clear which points are above, below, or on the line.
3. Consider the inequality $y > x + 1$. We say that the ordered pair (2, 5) *satisfies* the inequality, because the inequality becomes a true statement when we substitute 2 for x and 5 for y:

$$y > x + 1$$
$$5 \stackrel{?}{>} 2 + 1$$
$$5 \stackrel{?}{>} 3$$
true

Which of the ordered pairs found in Problem 2 satisfy the given inequality or equation?

a. $y > x + 1$ **b.** $y < x + 1$ **c.** $y = x + 1$

4. Describe all of the ordered pairs (not just those in Problem 2) that satisfy the given inequality or equation. Include in your description any of the three categories (above, below, or on the line) that are relevant.

a. $y > x + 1$ **b.** $y < x + 1$ **c.** $y = x + 1$

5. Draw a dashed line where the graph of $y = -\frac{1}{2}x - 1$ would be in a coordinate system. Then use shading to indicate points whose ordered pairs satisfy the inequality $y < -\frac{1}{2}x - 1$.

Homework 6.5

For extra help ▶ MyMathLab®  Watch the videos in MyMathLab  Download the MyDashboard App

1. The length of a golden rectangle is equal to about 1.62 times the width. If an architect wants to design the base of a building to be a golden rectangle, what are the dimensions of the base if the perimeter is to be 600 feet?
2. The length of a golden rectangle is equal to about 1.62 times the width. An architect wants to design a room so that the floor is a golden rectangle with a perimeter of 50 feet. What are the dimensions of the floor?
3. A landscaper plans to dig a rectangular garden whose length is to be 5 feet more than the width. If the landscaper has 42 feet of fencing to enclose the garden, what should be the dimensions of the garden?
4. An official football field is a rectangle whose length (including end zones) is 2.25 times the width. If a little-league football field is to be constructed with a perimeter of 200 yards, what should be the width and length of the field?
5. An official tennis court is a rectangle whose length is 6 feet more than twice the width. The perimeter of the court is 228 feet. Find the dimensions of the court.
6. An official table tennis (ping pong) table is a rectangle whose length is 1 foot less than twice the width. The perimeter of the table is 28 feet. Find the dimensions of the table.
7. The length of a rectangle is 3 inches less than twice the width. If the perimeter is 108 inches, find the dimensions of the rectangle.
8. The length of a rectangle is 5 inches more than twice the width. If the perimeter is 112 inches, find the dimensions of the rectangle.
9. A 2000-seat theater has tickets for sale at \$15 and \$22. How many tickets should be sold at each price for a sellout performance to generate a total revenue of \$33,500?
10. A 10,000-seat amphitheater has tickets for sale at \$45 and \$65. How many tickets should be sold at each price for a sellout performance to generate a total revenue of \$560,000?
11. The album *The Best of Radiohead,* by Radiohead, has a list price of \$14.99, and the album *The King of Limbs,* by Radiohead, has a list price of \$9.99. If iTunes' total sales of both albums in one week is 253 albums and the total revenue of both albums in that week is \$2722.47, how many of each album were sold?

12. The album *Blood Pressures,* by The Kills, has a list price of \$9.99, and the EP *Live Session,* by The Kills, has a list price of \$4.95. If iTunes' total sales of the album and the EP in one month are 373 recordings, and the total revenue of both recordings in that month is \$2647.71, how many albums and EPs were sold?

13. An auditorium has 300 balcony seats and 1400 main-level seats. If tickets for balcony seats are priced at \$12 less than the price of main-level seats, what should the prices be for each type of ticket so that the total revenue from a sellout performance will be \$40,600?

14. An auditorium has 450 balcony seats and 1700 main-level seats. If tickets for balcony seats are priced at \$15 less than the price of main-level seats, what should the prices be for each type of ticket so that the total revenue from a sellout performance will be \$100,750?

15. An amphitheater has 8000 general seats and 4000 reserved seats. If tickets for general seats are priced at \$25 less than the price of reserved seats, what should the prices be for each type of ticket so that the total revenue from a sellout performance will be \$544,000?

16. An amphitheater has 9500 general seats and 3200 reserved seats. If tickets for general seats are priced at \$17 less than the price of reserved seats, what should the prices be for each type of ticket so that the total revenue from a sellout performance will be \$435,400?

17. A 20,000-seat amphitheater will sell tickets at \$50 and \$75 for a Green Day concert. Let x and y be the number of tickets that will sell for \$50 and \$75, respectively. Assume the show will sell out.

a. Let $R = f(x)$ be the total revenue (in dollars) from selling the \$50 and \$75 tickets. Find an equation of f.

b. Use a graphing calculator to draw a graph of the function f for $0 \le x \le 20{,}000$. What is the slope? What does it mean in this situation?

c. Find $f(16{,}000)$. What does it mean in this situation?

d. The total cost of the production is \$475,000. How many of each ticket must be sold to make a profit of \$600,000?

18. An 8000-seat amphitheater will sell tickets at \$30 and \$55 for a Garbage concert. Let x and y be the number of tickets that will sell for \$30 and \$55, respectively. Assume the show will sell out.

a. Let $R = f(x)$ be the total revenue (in dollars) from selling the \$30 and \$55 tickets. Find an equation of f.

b. Use a graphing calculator to draw a graph of f for $0 \le x \le 8000$. What is the slope? What does it mean in this situation?

c. Find $f(6500)$. What does it mean in this situation?

d. How many of each ticket must be sold for the revenue to be \$250,000?

19. A 12,000-seat amphitheater will sell tickets at \$45 and \$70 for a Cake concert. Let x and y be the number of tickets that will sell for \$45 and \$70, respectively. Assume the show will sell out.

a. Let $R = f(x)$ be the total revenue (in dollars) from selling the \$45 and \$70 tickets. Find an equation of f.

b. Use a graphing calculator table to find the values $f(0)$, $f(2000)$, $f(4000)$, ..., $f(12{,}000)$. What do they mean in this situation?

c. Describe the various possible total revenues from selling the tickets.

d. How many of each ticket must be sold for the revenue to be \$602,500?

20. A 25,000-seat amphitheater will sell tickets at \$55 and \$85 for a U2 concert. Let x and y be the number of tickets that will sell for \$55 and \$85, respectively. Assume the show will sell out.

a. Let $R = f(x)$ be the total revenue (in dollars) from selling the \$55 and \$85 tickets. Find an equation of f.

b. Use a graphing calculator table to find the values $f(0)$, $f(5000)$, $f(10{,}000)$, ..., $f(25{,}000)$. What do they mean in this situation?

c. Describe the various possible total revenues from selling the tickets.

d. How many of each ticket must be sold for the revenue to be \$1,510,000?

21. For flights between Los Angeles and San Francisco, United Airlines® usually uses a 737 airplane that has 8 first-class seats and 126 coach seats. The average price of round-trip first-class tickets is \$242 more than the average price of round-trip coach tickets. Assume a round-trip flight is sold out. Let x and y be the prices (in dollars) of coach and first-class tickets, respectively.

a. Let $R = f(x)$ be the total revenue (in dollars) from selling the tickets. Find an equation of f.

b. Find the slope of the graph of f. What does it mean in this situation?

c. If United Airlines wants the total revenue from the tickets for the round trip to be \$14,130, what should be the average selling prices of both types of tickets?

Yes sir, the price yesterday was \$350. But now it's \$800, and it will go up to \$1450 tomorrow. Wait that's a black-out date for that price. It will be \$1725, nonrefundable, with a \$500 fee for any changes.
Ahhh... will there be a movie?

22. An orchestra will perform in an auditorium with 1400 main-level seats and 300 balcony seats. Tickets for the balcony seats will cost \$12 less than tickets for the main-level seats. Assume there is a sellout performance. Let x and y be the prices (in dollars) for the main-level seats and the balcony seats, respectively.

a. Let $R = f(x)$ be the total revenue (in dollars) from ticket sales. Find an equation of f.

b. Find $f(55)$. What does it mean in this situation?

c. Find the value of x when $f(x) = 101{,}800$. What does it mean in this situation?

23. A person invests the given amount of money in an account at 8% annual interest. Find the interest in one year.
a. 2500 dollars **b.** 3500 dollars **c.** d dollars

24. A person invests the given amount of money in an account at 5% annual interest. Find the interest in one year.
a. 4000 dollars **b.** 5000 dollars **c.** d dollars

25. Find the total interest earned in one year from the given investments and annual interest rates.
a. A person invests \$2000 in a 3% account and \$7000 in a 6% account.
b. A person invests \$4000 in a 3% account and \$5000 in a 6% account.
c. A person invests x dollars in a 3% account and y dollars in a 6% account.

26. Find the total interest earned in one year from the given investments and annual interest rates.
a. A person invests \$3000 in a 4% account and \$6000 in a 12% account.
b. A person invests \$4000 in a 4% account and \$8000 in a 12% account.
c. A person invests x dollars in a 4% account and $2x$ dollars in a 12% account.

For Exercises 27–42, all interest rates are five-year averages.

27. A person plans to invest a total of \$20,000 in a First Funds TN Tax-Free I account at 6% annual interest and a W & R International Growth C account at 11% annual interest. How much should she invest in each account so that the total interest in one year will be \$1500?

28. A person plans to invest a total of \$3500 in a cdbank.com account at 3% annual interest and a Turner Small Cap Value account at 16% annual interest. How much should he invest in each account so that the total interest in one year will be \$200?

29. A person plans to invest a total of \$8500 in a Middlesex Savings Bank CD account at 3.6% annual interest and a First Funds Growth & Income I account at 15% annual interest. How much should he invest in each account so that the total interest in one year will be \$990?

30. A person plans to invest a total of \$6500 in a Connecticut Bank CD account at 3% annual interest and an Artisan International account at 14% annual interest. How much should she invest in each account so that the total interest in one year will be \$400?

31. A person plans to invest three times as much in a Limited Term NY Municipal X account at 5% interest as in a Calvert Income A account at 10% annual interest. How much will the person have to invest in each account to earn a total of \$625 in one year?

32. A person plans to invest twice as much money in a BankUSA CD account at 3.5% annual interest as in a Fidelity Select Health Care account at 17% annual interest. How much will the person have to invest in each account to earn a total of \$700 in one year?

33. A person plans to invest an equal amount of money in a USAA Tax Exempt Short-Term account at 5% annual interest and a Putnam Global Equity B account at 9% annual interest. How much should the person invest in each account so that the total interest in one year will be \$700?

34. A person plans to invest twice as much money in a LaSalle Bank CD account at 2.5% annual interest as in a Franklin Strategic Mortgage account at 8% annual interest. How much should the person invest in each account so that the total interest for one year will be \$650?

35. A person plans to invest a total of \$10,000 in a Charter One Bank CD at 2.87% annual interest and a Dodge & Cox Balanced mutual fund that has a three-year average annual interest rate of 8.10%. Let x and y be the money (in dollars) invested in the CD and mutual fund, respectively.
a. Let $I = f(x)$ be the total interest (in dollars) earned from investing the \$10,000 for one year. Find an equation of f.
b. Use a graphing calculator to draw a graph of the function f for $0 \le x \le 10{,}000$. What is the slope? What does it mean in this situation?
c. How much money should be invested in each account to earn a total of \$400 in one year?

36. A person plans to invest a total of \$5000 in a Citizens Bank CD at 2% annual interest and a Calamos Market Neutral A mutual fund that has a three-year average annual interest rate of 9.45%. Let x and y be the money (in dollars) invested in the CD and the mutual fund, respectively.
a. Let $I = f(x)$ be the total interest (in dollars) earned from investing the \$5000 for one year. Find an equation of f.
b. Use a graphing calculator to draw a graph of f for $0 \le x \le 5000$. What is the slope? What does it mean in this situation?
c. How much money should be invested in each account to earn a total of \$350 in one year?

37. A person plans to invest a total of \$9000 in an ING Direct CD at 2.5% annual interest and a Thompson Plumb Balanced mutual fund that has a three-year average annual interest rate of 9.45%. Let x and y be the money (in dollars) invested in the CD and the mutual fund, respectively.
a. Let $I = f(x)$ be the total interest (in dollars) earned from investing the \$9000 for one year. Find an equation of f.
b. Find $f(500)$. What does it mean in this situation?
c. Find the value of x when $f(x) = 500$. What does it mean in this situation?
d. Find $f(10{,}000)$. What does it mean in this situation?

38. A person plans to invest a total of \$12,000 in a Savings Bank of Manchester CD at 1.92% annual interest and a Vanguard Wellesley Income mutual fund that has a three-year average annual interest rate of 8%. Let x and y be the money (in dollars) invested in the CD and the mutual fund, respectively.
a. Let $I = f(x)$ be the total interest (in dollars) earned from investing the \$12,000 for one year. Find an equation of f.
b. Find $f(800)$. What does it mean in this situation?
c. Find the value of x when $f(x) = 800$. What does it mean in this situation?
d. Find $f(15{,}000)$. What does it mean in this situation?

39. A person plans to invest a total of \$8000 in a LaSalle Bank CD at 1.5% annual interest and a Bridgeway Aggressive Investors 1 mutual fund that has a three-year average annual interest rate of 11.6%. Let x and y be the money (in dollars) invested in the CD and the mutual fund, respectively.
a. Let $I = f(x)$ be the total interest (in dollars) earned from investing the \$8000 for one year. Find an equation of f.
b. A minimum principal of \$2500 is required for the LaSalle Bank CD account. Describe the various possible total interest earnings from investing the \$8000 in the accounts for one year.

c. How much of the $8000 should be invested in each account so the total interest earned in one year is $400?

40. A person plans to invest a total of $7000 in a Security Bank USA CD at 2.55% annual interest and an Artisan Mid Cap mutual fund that has a three-year average annual interest rate of 8%. Let x and y be the money (in dollars) invested in the CD and the mutual fund, respectively.

a. Let $I = f(x)$ be the total interest (in dollars) earned from investing the $7000 for one year. Find an equation of f.

b. A minimum principal of $500 is required for the Security Bank USA CD account. Describe the various possible total interest earnings from investing the $7000 in the accounts for one year.

c. How much of the $7000 should be invested in each account so the total interest earned in one year is $300?

41. A person plans to invest a total of $6000 in a Nexity Bank CD at 2.85% annual interest and an FMI Focus mutual fund that has a three-year average annual interest rate of 9%. Let x and y be the money (in dollars) invested in the CD and the mutual fund, respectively.

a. Let $I = f(x)$ be the total interest (in dollars) earned from investing the $6000 for one year. Find an equation of f.

b. Find the I-intercept of the model. What does it mean in this situation?

c. Find the x-intercept. What does it mean in this situation?

d. What is the slope? What does it mean in this situation?

42. A person plans to invest a total of $14,000 in a North Middlesex Savings Bank CD at 2.27% annual interest and a Calamos Growth A mutual fund that has a three-year average annual interest rate of 15%. Let x and y be the money (in dollars) invested in the CD and the mutual fund, respectively.

a. Let $I = f(x)$ be the total interest (in dollars) earned from investing the $14,000 for one year. Find an equation of f.

b. Find the I-intercept of the model. What does it mean in this situation?

c. Find the x-intercept. What does it mean in this situation?

d. What is the slope? What does it mean in this situation?

43. A salad dressing is a 65% oil solution. Determine the amount of oil in the given amount of oil solution.

a. 2 ounces of oil solution

b. 3 ounces of oil solution

c. x ounces of oil solution

44. Some milk is a 2% butterfat solution. Determine the amount of butterfat in the given amount of 2% milk.

a. 4 cups of milk

b. 8 cups of milk

c. M cups of milk

45. a. If 4 ounces of a 35% alcohol solution is mixed with 3 ounces of a 10% alcohol solution, how much pure alcohol is in the mixture?

b. How many ounces of a 35% acid solution and a 10% acid solution need to be mixed to make 15 ounces of a 20% solution?

46. a. If 2 liters of a 24% alcohol solution is mixed with 5 liters of an 8% alcohol solution, how much pure alcohol is in the mixture?

b. How many liters of a 24% acid solution and an 8% acid solution need to be mixed to make 8 liters of a 20% solution?

47. A chemist wants to mix a 20% acid solution and a 30% acid solution to make a 22% acid solution. How many quarts of each solution must be mixed to make 5 quarts of the 22% solution?

48. A chemist wants to mix a 30% acid solution and a 50% acid solution to make a 42% acid solution. How many quarts of each solution must be mixed to make 10 quarts of the 42% solution?

49. How many gallons of a 10% antifreeze solution and a 25% antifreeze solution need to be mixed to make 3 gallons of a 20% antifreeze solution?

50. How many liters of a 25% antifreeze solution and a 40% antifreeze solution need to be mixed to make 6 liters of a 30% antifreeze solution?

51. A chemist wants to mix a 15% acid solution and a 35% acid solution to make a 30% acid solution. How many quarts of each solution must be mixed to make 4 quarts of the 30% solution?

52. A chemist wants to mix a 5% acid solution and a 20% acid solution to make a 10% acid solution. How many quarts of each solution must be mixed to make 6 quarts of the 10% solution?

53. A chemist needs 5 ounces of a 12% alcohol solution, but has only a 20% alcohol solution. How much 20% solution and water should she mix to make the desired 5 ounces of 12% solution?

54. A chemist needs 6 liters of a 25% alcohol solution, but has only a 30% alcohol solution. How much 30% solution and water should he mix to make the desired 6 liters of 25% solution?

Concepts

55. For a sellout performance at a 300-seat theater,

a. Find the total revenue if all tickets sell for $10.

b. Find the total revenue if all tickets sell for $15.

c. If some, none, or all tickets sell for $10 and the remaining tickets, if any, sell for $15, is it possible for the total revenue to be the amount that follows? If yes, how many of each ticket must be sold? If no, explain.

i. $2500

ii. $4875

iii. $3250

56. A person plans to invest a total of $3000, part of it in an account at 5% annual interest and the rest in an account at 10% annual interest.

a. Find the interest earned in one year from the following investments:

i. The person invests all $3000 in the 5% account.

ii. The person invests all $3000 in the 10% account.

b. Is it possible for her to earn the given amount of interest in one year? If yes, how much should she invest in each account? If no, explain why not.

i. $120

ii. $330

iii. $180

57. If equal amounts of a 10% alcohol solution and a 20% alcohol solution are mixed, what percentage of the mixture is alcohol? Explain.

58. If a person invests equal amounts of money in an account at 5% annual interest and an account at 9% annual interest for one year, what percentage of the total money invested will be the total interest earned? Explain.

In many examples, we wrote a system of two equations that helped us find quantities for an authentic situation. For the following exercises, you will work backward. For example, given the system

$$xy = 48$$
$$y = x - 2$$

we can relate the following to the system:

A person wants to build a rectangular garden with an area of 48 square feet in which the width is 2 feet less than the length. Find the dimensions of the garden.

Describe an authentic situation for the given system. Find the unknown quantities for your situation.

59. $y = x + 3$
$2x + 2y = 50$

60. $x + y = 500$
$15x + 25y = 8300$

61. $y = 2x$
$0.03x + 0.07y = 340$

62. $x + y = 8$
$0.35x + 0.55y = 0.40(8)$

Related Review

Solve the equation for the specified variable.

63. $P = 2(L + W)$, for L

64. $P = 2(L + W)$, for W

65. $A = P(1 + rt)$, for r

66. $A = P(1 + rt)$, for P

Expressions, Equations, Functions, and Graphs

Perform the indicated instruction. Then use words such as linear, function, one variable, *and* two variables *to describe the expression, equation, or system.*

67. Simplify $-4(7x - 3)$.

68. Solve the following:

$$3x + 2y = 6$$
$$y = 2x - 11$$

69. Solve $-4(7x - 3) = 5x + 2$.

70. Graph $3x + 2y = 6$ by hand.

6.6 Linear Inequalities in Two Variables; Systems of Linear Inequalities in Two Variables

Objectives

- » Graph a linear inequality in two variables.
- » Solve a system of linear inequalities in two variables.
- » Use a system of linear inequalities in two variables to make estimates.

Recall from Section 3.2 that a linear equation in two variables is an equation that can be put into the form $y = mx + b$ or $x = a$, where m, a, and b are constants. A **linear inequality in two variables** is an inequality that can be put into the form

$$y < mx + b \quad \text{or} \quad x < a$$

(or with $<$ replaced with $\le$, $>$, or $\ge$), where m, a, and b are constants. Some examples are $y < 3x - 5$, $y \le -2x + 4$, $5x + 2y > 10$, and $x \ge 2$. In the first part of this section, we will work with one linear inequality in two variables. In the second part, we will work with two or more such inequalities.

Linear Inequality in Two Variables

We use the terms *solution, satisfy,* and *solution set* for an inequality in two variables in much the same way that we have used them for equations in one or two variables and inequalities in one variable.

Definition *Satisfy, solution, solution set,* and *solve* for an inequality in two variables

If an inequality in the two variables x and y becomes a true statement when a is substituted for x and b is substituted for y, we say the ordered pair (a, b) **satisfies** the inequality and call (a, b) a **solution** of the inequality. The **solution set** of an inequality is the set of all solutions of the inequality. We **solve** the inequality by finding its solution set.

We describe the solution set of an inequality by graphing all of the solutions.

Example 1 Sketching the Graph of an Inequality

Graph $y > x - 1$.

Solution

We begin by sketching the graph of $y = x - 1$ (see Fig. 42).

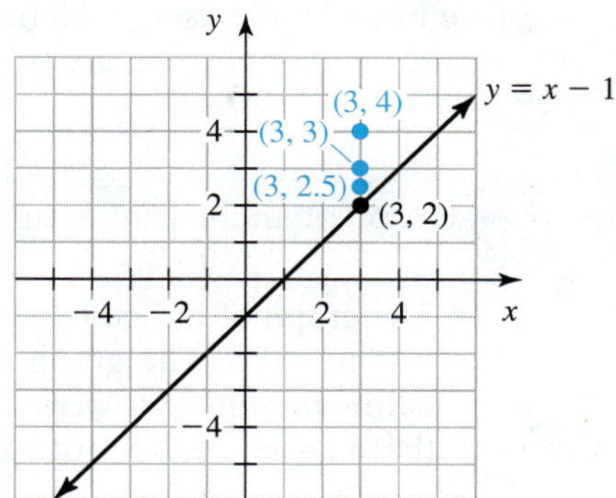

Figure 42 Some solutions of $y > x - 1$ (in blue)

To investigate how to solve $y > x - 1$, we choose a value of x (say, 3) and find several solutions with an x-coordinate of 3. For $y = x - 1$, if $x = 3$, then $y = (3) - 1 = 2$. So, the point (3, 2) is on the line $y = x - 1$.

For $y > x - 1$, if $x = 3$, then

$$y > (3) - 1$$
$$y > 2$$

So, if $x = 3$, some possible values of y are $y = 2.5$, $y = 3$, and $y = 4$. Note that the points (3, 2.5), (3, 3), and (3, 4) lie *above* the point (3, 2), which is on the line $y = x - 1$ (see Fig. 42).

We could choose other values of x besides 3 and go through a similar argument. These investigations would suggest that the solutions of $y > x - 1$ lie *above* the graph of $y = x - 1$. This is, in fact, true. In Fig. 43, we shade the region that contains all of the points that represent solutions of $y > x - 1$.

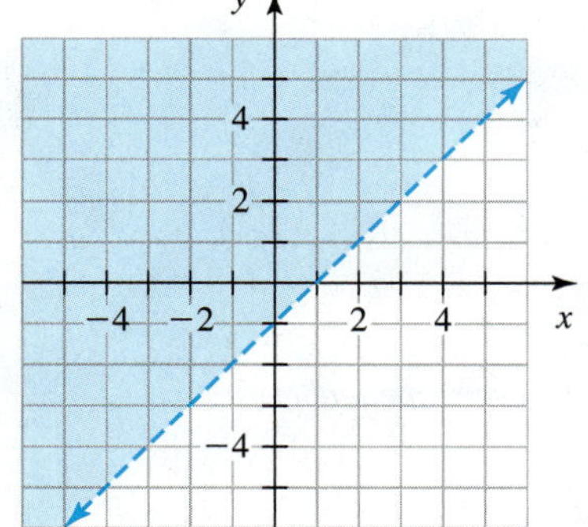

Figure 43 Graph of $y > x - 1$

We make the line $y = x - 1$ dashed in Fig. 43 to indicate that its points are *not* solutions of $y > x - 1$. For example, the point (3, 2) on the line $y = x - 1$ does *not* satisfy the inequality $y > x - 1$:

$y > x - 1$	*Original inequality*
$2 \overset{?}{>} (3) - 1$	*Substitute 3 for x and 2 for y.*
$2 \overset{?}{>} 2$	*Subtract.*
false	

We can draw a graph of $y > x - 1$ by using a graphing calculator (see Fig. 44), but we have to imagine that the border $y = x - 1$ is drawn with a dashed line. To shade above a line, press [Y=] and then press [◁] twice. Next press [ENTER] as many times as necessary for the triangle shown in Fig. 44 to appear.

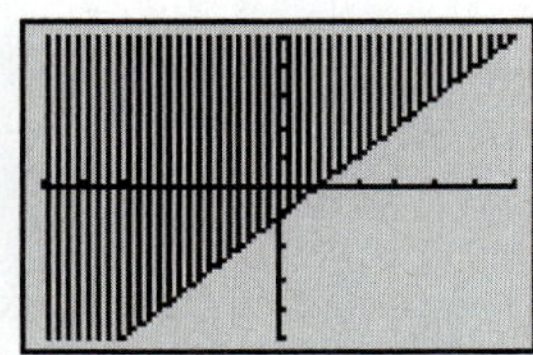

Figure 44 Graph of $y > x - 1$ (where we imagine that the border line is dashed)

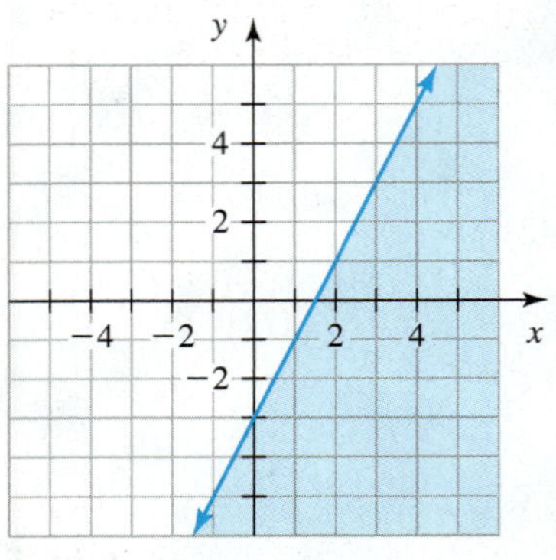

Figure 45 Graph of $y \le 2x - 3$

Example 2 Sketching the Graph of an Inequality

Sketch a graph of $y \le 2x - 3$.

Solution

The graph of $y \le 2x - 3$ is the line $y = 2x - 3$, as well as the region below that line (see Fig. 45). We use a *solid* line along the border $y = 2x - 3$ to indicate that the points on the line $y = 2x - 3$ are solutions of $y \le 2x - 3$.

Graph of an Inequality in Two Variables

- The graph of an inequality of the form $y > mx + b$ is the region above the line $y = mx + b$. The graph of an inequality of the form $y < mx + b$ is the region below the line $y = mx + b$. For either inequality, we use a dashed line to show that $y = mx + b$ is not part of the graph.
- The graph of an inequality of the form $y \ge mx + b$ is the line $y = mx + b$, as well as the region above that line. The graph of an inequality of the form $y \le mx + b$ is the line $y = mx + b$, as well as the region below that line.

To sketch a graph of an inequality in two variables, if the variable y is not alone on one side of the inequality, we begin by isolating it. Recall from Section 5.7 that when we multiply or divide both sides of an inequality by a negative number, we must reverse the inequality symbol.

Example 3 Sketching the Graph of an Inequality

Sketch a graph of $-2x - 5y > 10$.

Solution

First, we get y alone on one side of the inequality:

$$-2x - 5y > 10 \qquad \text{Original inequality}$$

$$-2x - 5y + 2x > 10 + 2x \qquad \text{Add } 2x \text{ to both sides.}$$

$$-5y > 2x + 10 \qquad \text{Combine like terms.}$$

$$\frac{-5y}{-5} < \frac{2x}{-5} + \frac{10}{-5} \qquad \text{Divide both sides by } -5\text{; reverse inequality symbol.}$$

$$y < -\frac{2}{5}x - 2 \qquad \text{Simplify.}$$

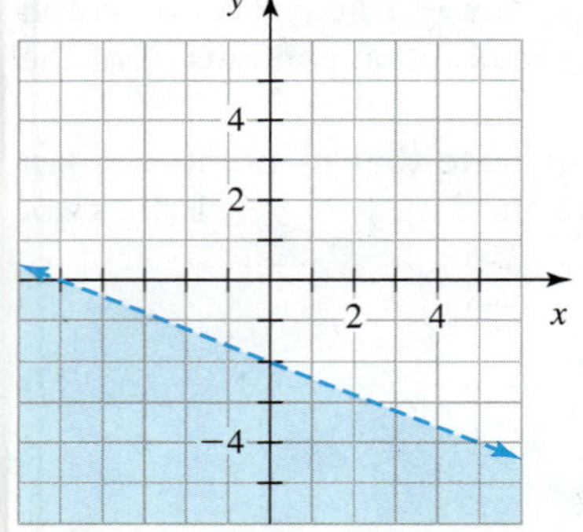

Figure 46 Graph of $-2x - 5y > 10$

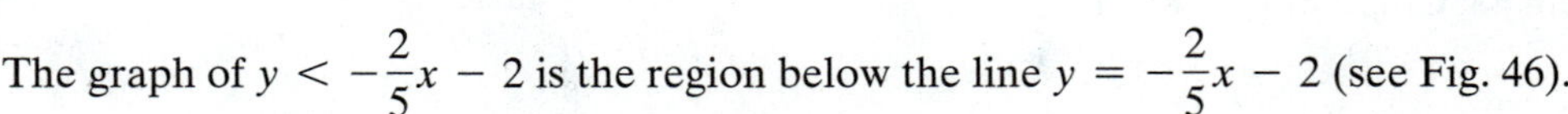

The graph of $y < -\frac{2}{5}x - 2$ is the region below the line $y = -\frac{2}{5}x - 2$ (see Fig. 46).

To verify our work, we choose a point on our graph—say, $(-4, -2)$—and check that it satisfies the original inequality:

$$-2x - 5y > 10 \qquad \text{Original inequality}$$

$$-2(-4) - 5(-2) \overset{?}{>} 10 \qquad \text{Substitute } -4 \text{ for } x \text{ and } -2 \text{ for } y.$$

$$18 \overset{?}{>} 10 \qquad \text{Simplify.}$$

$$\text{true}$$

To check further, we could choose several other points on the graph and check that each point satisfies the inequality.

We could also choose several points that are *not* on the graph and check that each point does *not* satisfy the inequality. For example, note that $(0, 0)$ is not on the graph and that this point does not satisfy the original inequality $-2x - 5y > 10$:

$$-2x - 5y > 10 \quad \textit{Original inequality}$$

$$-2(0) - 5(0) \stackrel{?}{>} 10 \quad \textit{Substitute 0 for x and 0 for y.}$$

$$0 \stackrel{?}{>} 10 \quad \textit{Simplify.}$$

false

WARNING It is a common error to think the graph of an inequality such as $-2x - 5y > 10$ is the region *above* the line $-2x - 5y = 10$, because the symbol ">" means "is greater than." However, we must first isolate y on the left side of a linear inequality before we can determine whether the graph includes the region that is above or below a line. In fact, in Example 3 we wrote the inequality $-2x - 5y > 10$ as $y < -\frac{2}{5}x - 2$ and concluded that the graph is the region *below* the line $y = -\frac{2}{5}x - 2$.

Example 4 Sketching the Graph of an Inequality

Sketch the graph of the inequality.

1. $y \geq -2$

2. $x < 3$

Solution

1. The graph of $y \geq -2$ is the horizontal line $y = -2$, as well as the region above that line (see Fig. 47).
2. Ordered pairs with x-coordinates *less* than 3 are represented by points that lie to the *left* of the vertical line $x = 3$. So, the graph of $x < 3$ is the region to the left of $x = 3$ (see Fig. 48).

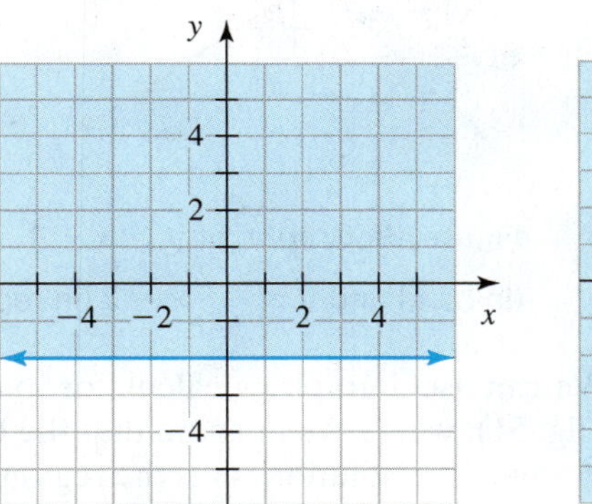

Figure 47 Graph of $y \geq -2$

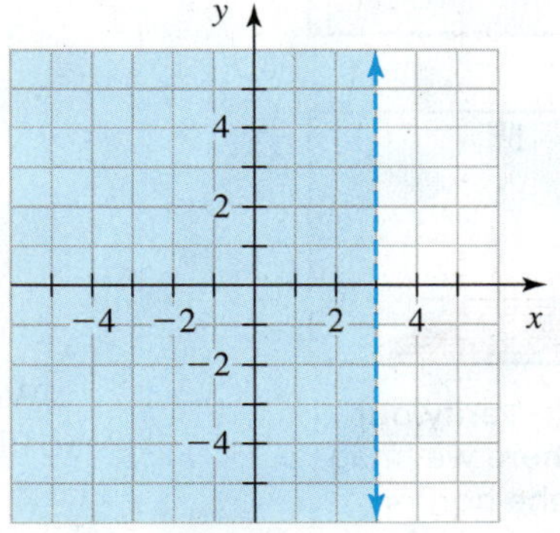

Figure 48 Graph of $x < 3$

System of Inequalities

A **system of inequalities in two variables** consists of two or more linear inequalities in two variables. Here is an example:

$$y > x - 1$$
$$y \leq -\frac{2}{3}x - 2$$

Definition *Solution, solution set,* and *solve* for a system of inequalities in two variables

We say an ordered pair (a, b) is a **solution** of a system of inequalities in two variables if it satisfies all of the inequalities of the system. The **solution set** of a system is the set of all solutions of the system. We **solve** a system by finding its solution set.

Recall from Section 6.1 that we can find the solution set of a system of two linear *equations* in two variables by locating any intersection point(s) of the graphs of the two equations. Similarly, **the solution set of a system of inequalities in two variables can be found by locating the intersection of the graphs of all of the inequalities.** This makes sense, because a solution is an ordered pair that satisfies all of the inequalities, which means the point that represents the ordered pair lies on the graphs of all of the inequalities.

In this text, we use a graph to describe the solution set of a system of inequalities.

Example 5 Graphing the Solution Set of a System of Inequalities

Solve the system

$$y > x - 1$$
$$y \le -\frac{2}{3}x - 2$$

Solution

First, we sketch the graph of $y > x - 1$ (see Fig. 49, blue region) and the graph of $y \le -\frac{2}{3}x - 2$ (see Fig. 49, red region). The graph of the solution set of the system is the intersection of the graphs of the inequalities, which is shown in Fig. 50.

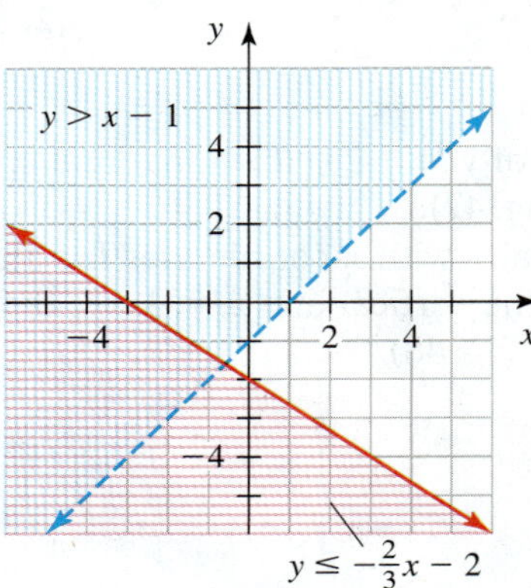

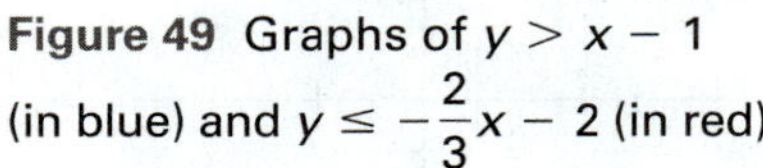

Figure 49 Graphs of $y > x - 1$ (in blue) and $y \le -\frac{2}{3}x - 2$ (in red)

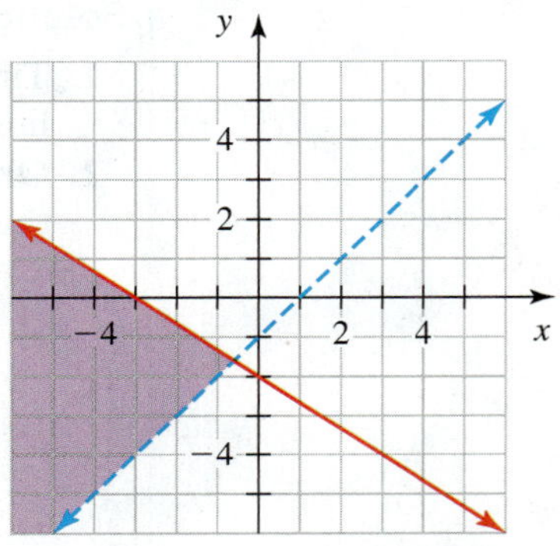

Figure 50 Graph of the solution set of the system

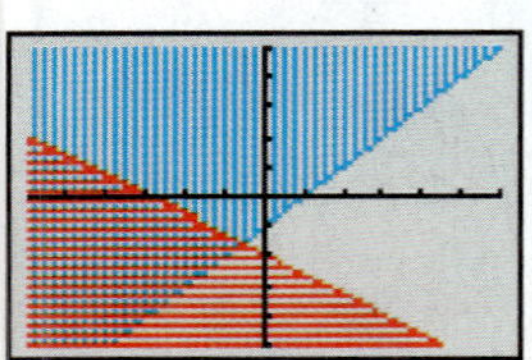

Figure 51 Verify our work (where we imagine that the border $y = x - 1$ is dashed)

We can use a graphing calculator to draw a graph of the solution set of the system (see Fig. 51), where we imagine that the border $y = x - 1$ is drawn with a dashed line. The graph of the solution set is the region shaded by vertical lines (in blue) and horizontal lines (in red).

Example 6 Graphing the Solution Set of a System of Inequalities

Graph the solution set of the system

$$y > -\frac{3}{2}x$$
$$-2x + 5y < 5$$

Solution

First, we sketch the graph of $y > -\frac{3}{2}x$ (see Fig. 52, blue region) and the graph of $-2x + 5y < 5$ (see Fig. 52, red region). The graph of the solution set of the system is the intersection of the graphs of the inequalities, which is shown in Fig. 53.

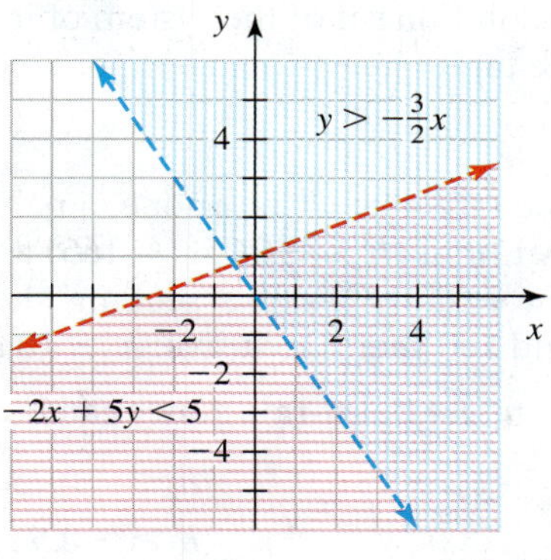

Figure 52 Graphs of $y > -\frac{3}{2}x$ (in blue) and $-2x + 5y < 5$ (in red)

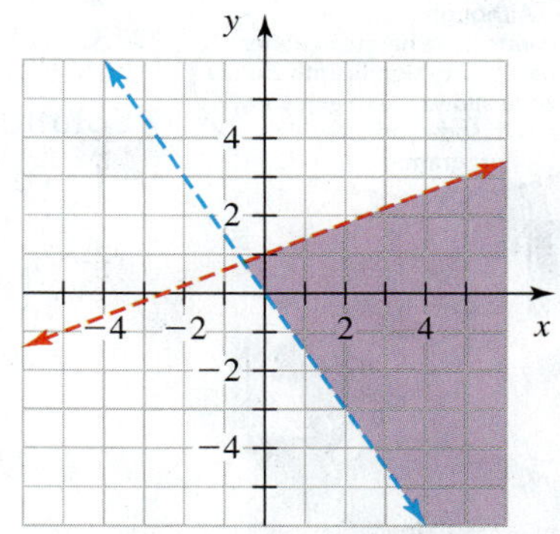

Figure 53 Graph of the solution set of the system

Example 7 Graphing the Solution Set of a System of Inequalities

Graph the solution set of the system

$$\begin{aligned} -x + 3y &\le 9 \\ -x + 3y &\ge 3 \\ x &\ge 1 \\ x &\le 4 \end{aligned}$$

Solution

In Fig. 54, we use arrows to indicate the graphs of each of the four inequalities. The solution set of the system is the intersection of the four graphs.

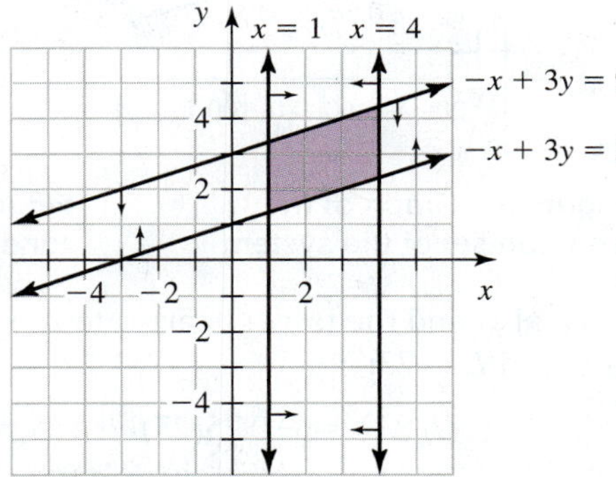

Figure 54 Graph of the solution set of the system

Table 31 Target Heart-Rate Zone

Age	Target Heart-Rate Zone (beats per minute) Lower Limit	Upper Limit
20	100	150.0
30	95	142.5
40	90	135.0
50	85	127.5
60	80	120.0
70	75	112.5
80	70	105.0
90	65	97.5

Source: *American Heart Association*

Modeling with Systems of Inequalities

We can use a system of inequalities to help us make estimates about authentic situations.

Example 8 Using a System of Inequalities to Make Estimates

For you to get the most benefits, yet be safe, while exercising, your heart rate should be within a certain range, called the *target heart-rate zone*. Table 31 shows the lower and upper limits of the target heart-rate zones for various ages.

Let $h = U(a)$ be the upper limit of the target heart-rate zone and $h = L(a)$ be the lower limit of the target heart-rate zone, both in beats per minute, for a person who is a years of age. The linear regression equations of U and L are, respectively,

$$U(a) = -0.75a + 165$$
$$L(a) = -0.5a + 110$$

1. Find a system of inequalities that describes the target heart-rate zones for people between the ages of 10 years and 100 years.

Although the target heart-rate zone has its benefits, I find that staying significantly below the zone allows me to enjoy my ice-cold soda and television program more.
???!

2. Graph the solution set of the system of inequalities that you found in Problem 1.
3. What is the target heart-rate zone for a person who is 23 years old?

Solution

1. To be in the target heart-rate zone, a person's heart rate must be less than or equal to the upper limit ($h \le -0.75a + 165$) and greater than or equal to the lower limit ($h \ge -0.5a + 110$). We are seeking the zones for people between the ages of 10 years and 100 years, inclusive: $a \ge 10$ and $a \le 100$.

   So, the target heart-rate zones can be described by the system

$$\begin{aligned} h &\le -0.75a + 165 \\ h &\ge -0.5a + 110 \\ a &\ge 10 \\ a &\le 100 \end{aligned}$$

2. In Fig. 55, we use arrows to indicate the graphs of each of the four inequalities. The solution set of the system is the intersection of the four graphs.
3. The target heart-rate zone for a 23-year-old person is represented by the blue vertical line segment above $a = 23$ on the a-axis in Fig. 56. From this line segment, we see that the target heart-rate zone for a 23-year-old person is between 98.5 beats per minute and 147.75 beats per minute, inclusive.

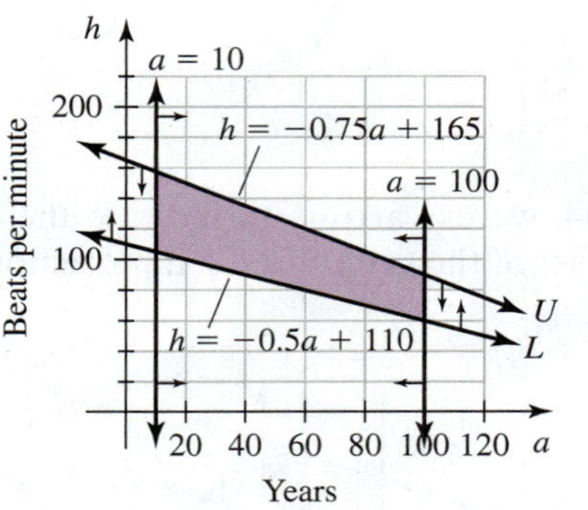

Figure 55 Graph of the solution set of the system

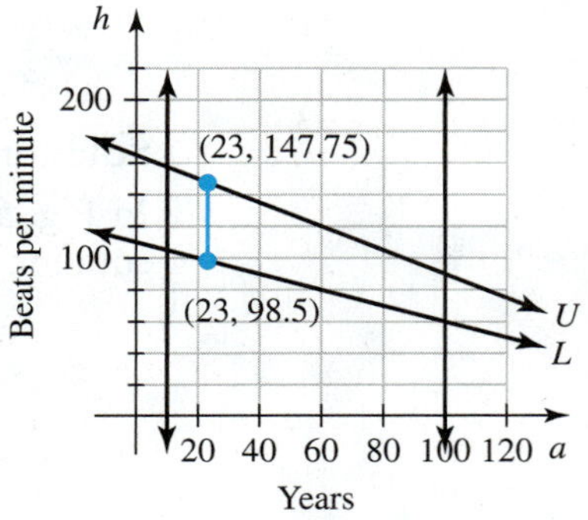

Figure 56 The target heart-rate zone for a 23-year-old person

We can also find the target heart-rate zone for a 23-year-old person by evaluating both U and L at 23:

$$\begin{aligned} U(23) &= -0.75(23) + 165 = 147.75 && \text{Upper limit} \\ L(23) &= -0.5(23) + 110 = 98.5 && \text{Lower limit} \end{aligned}$$

This means the target heart-rate zone is between 98.5 beats per minute and 147.75 beats per minute, inclusive, which checks.

Group Exploration

Looking ahead: Graphing quadratic equations in two variables

Graph the equation by hand. To begin, substitute the values −3, −2, −1, 0, 1, 2, and 3 for x. Make other substitutions as necessary. Then, use a graphing calculator to verify your work.

1. $y = x^2$
2. $y = x^2 + 1$
3. $y = x^2 - 3$
4. $y = 2x^2$
5. $y = -x^2$

Tips for Success The Value of Learning Mathematics

Imagine someone who is a math expert. Are you imagining a person wearing broken glasses that are taped together, a pocket protector, and wrinkled, unfashionable clothing—in other words, a math nerd? Of course, this stereotype does not accurately describe most mathematicians. But perhaps it is because of the stereotype of a math nerd that many students don't want to become "too good" at mathematics.

Learning mathematics will not transform you into a nerd. Rather, it will transform you into a more educated, well-rounded person. Learning mathematics will equip you to be more effective in many lines of work, as well as when you do math-intensive activities such as investing or filling out a tax return. Most people have a high respect for mathematicians due to their commitment to a useful and challenging subject.

Homework 6.6

For extra help ▶ MyMathLab® 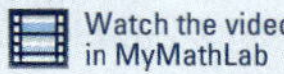 Watch the videos in MyMathLab 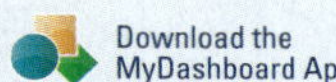  Download the MyDashboard App

Which of the given ordered pairs satisfy the given inequality?

1. $y < 2x - 3$ $(4, 1), (-2, 3), (-3, -1)$

2. $y > -3x + 2$ $(4, -1), (-4, 3), (-2, -5)$

3. $2x - 5y \geq 10$ $(-3, -1), (-1, -4), (5, 0)$

4. $3y \leq -4x$ $(2, 1), (-3, 4), (-2, -1)$

Graph the inequality by hand.

5. $y > x - 3$

6. $y < 2x - 4$

7. $y \leq -2x + 3$

8. $y \geq -3x + 4$

9. $y < \frac{1}{3}x + 1$

10. $y > \frac{2}{3}x - 5$

11. $y \geq -\frac{3}{5}x - 1$

12. $y \leq -\frac{1}{2}x + 3$

13. $y > x$

14. $y < -x$

15. $y - 2x < 0$

16. $y + 3x \leq 0$

17. $3x + y \leq 2$

18. $2x + y \geq 3$

19. $4x - y > 1$

20. $x - y < 3$

21. $4x - 3y \geq 0$

22. $5x - 2y \leq 0$

23. $4x + 5y \geq 10$

24. $2x + 4y > 4$

25. $2x - 5y < 5$

26. $3x - 4y > 12$

27. $3(x - 2) + y \leq -2$

28. $y > -2(x - 1) + 3$

29. $y > 3$

30. $y < -2$

31. $x \geq -3$

32. $x \leq 4$

33. $x < -2$

34. $y > 1$

35. $y \leq 0$

36. $x \geq 0$

For Exercises 37–64, graph the solution set of the system of inequalities by hand.

37. $y \geq \frac{1}{3}x - 3$, $y \leq -\frac{3}{2}x + 2$

38. $y \leq \frac{2}{3}x + 1$, $y \geq -\frac{4}{3}x - 2$

39. $y > 2x - 3$, $y > -\frac{3}{4}x + 1$

40. $y < -3x + 5$, $y < \frac{2}{5}x - 3$

41. $y \leq -\frac{2}{3}x - 3$, $y > 2x + 1$

42. $y \leq \frac{4}{3}x - 2$, $y > -x + 4$

43. $y \geq -2x - 1$, $y > \frac{1}{3}x + 2$

44. $y \leq 3x - 3$, $y < -\frac{1}{4}x + 1$

45. $y > -3x + 4$, $y \leq 2x + 3$

46. $y > -4x + 5$, $y \leq -x + 1$

47. $y \leq \frac{2}{3}x$, $y < -\frac{2}{5}x$

48. $y \leq -\frac{1}{2}x$, $y < \frac{4}{3}x$

49. $5x - 3y \leq 12$, $-2y < x$

50. $x - 2y < 6$, $3y \geq 6x$

51. $x - y \leq 2$, $2x + y < 1$

52. $x + y < 1$, $3x - y \geq 2$

53. $2x - 3y > 3$, $3x + 5y \geq 10$

54. $-3x + 4y \leq 4$, $5x - 2y < 2$

55. $1 + y \geq \frac{1}{2}(x - 4)$, $3 - y > 2(x - 1)$

56. $5 - y > \frac{1}{3}(x + 6)$, $y - 7 \leq 2(x - 5)$

57. $y < 3$, $x \geq -2$

58. $y \geq 2$, $x < -1$

59. $y \leq 1$, $y \geq -2$, $x \geq -3$, $x \leq 4$

60. $y \leq -1$, $y \geq -4$, $x \leq 2$, $x \geq -5$

61. $2x - 5y \geq -5$, $2x - 5y \leq 15$, $x \geq -1$, $x \leq 3$

62. $3x + 4y \leq 16$, $3x + 4y \geq 8$, $x \geq 1$, $x \leq 4$

63. $y < -x + 5$
$y \le x + 5$
$y > \frac{1}{2}x + 1$

64. $y > -x - 4$
$y \ge 2x + 6$
$y \le \frac{1}{2}x + 6$

65. A person's life expectancy predicts how many *remaining* years the person will live. The life expectancies of U.S. males at birth and at age 20 years are shown in Table 32 for various calendar years.

Table 32 Life Expectancies of U.S. Males at Birth and at Age 20 Years

Year	Life Expectancy (years) At Birth	At Age 20 Years
1980	70.1	51.9
1985	71.1	52.6
1990	71.8	53.3
1995	72.5	53.8
2000	74.3	55.3
2005	74.9	55.9
2008	75.5	56.5

Source: *U.S. National Center for Health Statistics*

Let $L = B(t)$ be the life expectancy (in years) at birth and $L = T(t)$ be the life expectancy (in years) at age 20 years, both of U.S. males at t years since 1980.

a. Find equations of B and T.

b. Find a system of inequalities that describes the life expectancies of U.S. males from 0 years through 20 years old from 1980 to 2020.

c. Graph the solution set of the system of inequalities you found in part (b).

d. Predict the life expectancies of U.S. males from 0 years through 20 years old in 2015.

66. Recall from Exercise 16 of Homework 1.4 that the windchill is a measure of how cold you feel when exposed to wind. Table 33 gives the windchills for various temperatures when the wind speed is 10 mph or 20 mph. Let $w = f(t)$ be the windchill when the wind speed is 10 mph and $w = g(t)$ be the windchill when the wind speed is 20 mph, both in degrees Fahrenheit at a temperature t in degrees Fahrenheit.

Table 33 Windchills for 10-mph and 20-mph Winds

Temperature (degrees Fahrenheit)	Windchill (degrees Fahrenheit) 10-mph Wind	20-mph Wind
−15	−35	−42
−10	−28	−35
−5	−22	−29
0	−16	−22
5	−10	−15
10	−4	−9
15	3	−2
20	9	4

Source: *National Weather Service*

a. Find equations of f and g.

b. Find a system of inequalities that describes the windchills for temperatures between −20°F and 30°F, inclusive, with wind speeds between 10 mph and 20 mph, inclusive.

c. Graph the solution set of the system of inequalities you found in part (b).

d. If the temperature is 7°F, what are the windchills for wind speeds between 10 mph and 20 mph, inclusive?

67. Recommended ski lengths are shown in Table 34 for skiers of various weights and abilities.

Table 34 Recommended Ski Lengths

Weight (pounds)	Ski Length (centimeters) Beginner	Intermediate	Advanced	Expert
100.0	130	135	140	145
112.5	135	140	145	150
130.5	140	145	150	155
143.0	145	150	155	160
158.0	150	155	160	165
172.5	155	160	165	170
185.0	165	170	175	180
195.0	175	180	185	190

Source: *backcountry.com*

Let $L = B(w)$ be the recommended ski length for a beginner skier and $L = I(w)$ be the recommended ski length for an intermediate skier, both in centimeters, for a skier who weighs w pounds.

a. Find linear equations of B and I.

b. Find a system of inequalities that describes the recommended ski lengths for beginning to intermediate skiers who weigh between 130 and 150 pounds, inclusive.

c. Graph the solution set of the system of inequalities you found in part (b).

d. Estimate all recommended ski lengths for beginning to intermediate 140-pound skiers.

68. Let $L = A(w)$ be the recommended ski length for an advanced skier and $L = E(w)$ be the recommended ski length for an expert skier, both in centimeters, for a skier who weighs w pounds (see Table 34).

a. Find linear equations of A and E.

b. Find a system of inequalities that describes the recommended ski lengths for advanced to expert skiers who weigh between 140 and 160 pounds, inclusive.

c. Graph the solution set of the system of inequalities you found in part (b).

d. Estimate all recommended ski lengths for advanced to expert 145-pound skiers.

Concepts

69. Find an inequality in two variables whose graph is shown in Fig. 57.

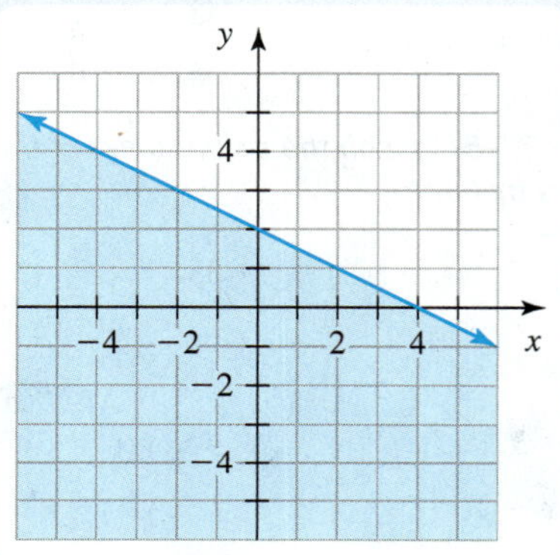

Figure 57 Exercise 69

70. Find a system of two linear inequalities whose solution set is shown in Fig. 58.

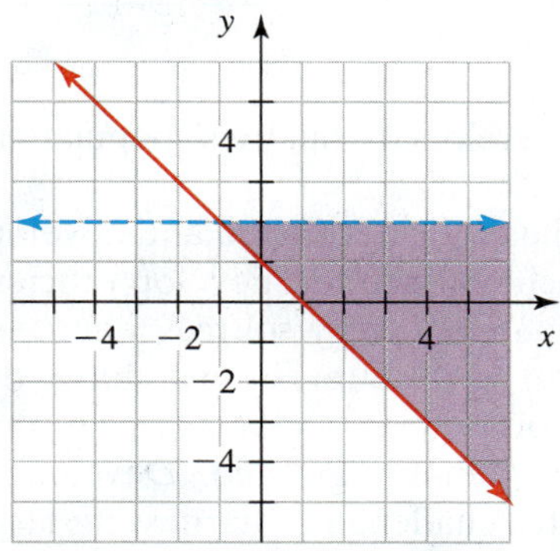

Figure 58 Exercise 70

71. A student believes that the graph of $5x - 2y < 6$ is the region below the line $5x - 2y = 6$. Is the student correct? Explain.

72. A student believes that the graph of $y \leq 3x - 1$ is the region below the line $y = 3x - 1$. Is the student correct? Explain.

73. The graphs of $y = ax + b$ and $y = cx + d$ are sketched in Fig. 59.

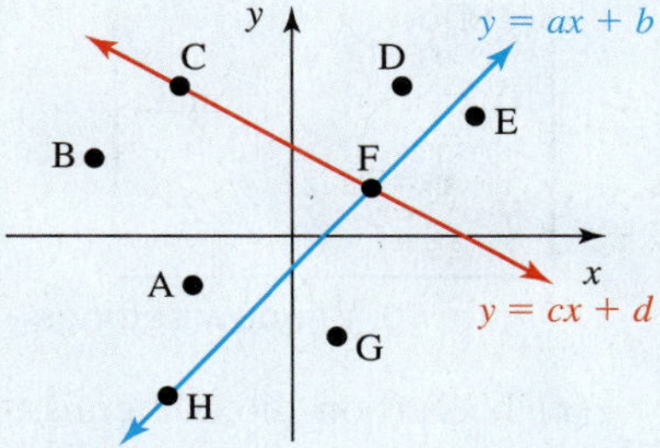

Figure 59 Exercise 73

Which of the points A, B, C, D, E, F, G, and H represent an ordered pair that

a. satisfies the inequality $y > ax + b$?
b. satisfies the inequality $y \leq cx + d$?
c. satisfies both $y > ax + b$ and $y \leq cx + d$?
d. satisfies neither $y > ax + b$ nor $y \leq cx + d$?

74. Consider the system of inequalities

$$y \geq 2x + 1$$
$$y < -x - 2$$

Find an ordered pair that

a. satisfies $y \geq 2x + 1$, but does not satisfy $y < -x - 2$.
b. satisfies $y < -x - 2$, but does not satisfy $y \geq 2x + 1$.
c. satisfies both inequalities.
d. does not satisfy either inequality.

75. Find three ordered pairs that are solutions of the inequality $y \geq 2x - 3$. Also, find three ordered pairs that are not solutions.

76. Find three ordered pairs that are solutions of the inequality $2x - y < 4$. Also, find three ordered pairs that are not solutions.

77. Give an example of an inequality in terms of x and y for which $(3, 4)$ is a solution and $(4, 3)$ is not.

78. Give an example of an inequality in terms of x and y for which $(3, 3)$, $(-3, 3)$, and $(-3, -3)$ are solutions and $(3, -3)$ is not.

79. Graph by hand the solution set of the system

$$y \geq 2x + 1$$
$$y \leq 2x + 1$$

80. Describe the solution set of the system

$$y > 2x + 1$$
$$y < 2x + 1$$

81. Why is the graph of an inequality of the form $y > mx + b$ the region above the line $y = mx + b$?

82. Why does the graph of an inequality of the form $y \geq mx + b$ include the line $y = mx + b$?

83. Explain how to graph an inequality in two variables by hand.

84. Explain how to solve a system of inequalities in two variables.

Related Review

85. Graph $y = 2x - 1$ by hand.

86. Graph $3x - 4y = 4$ by hand.

87. Graph the inequality $y \leq 2x - 1$ by hand.

88. Graph the inequality $3x - 4y > 4$ by hand.

89. Solve the inequality $3 \leq 2x - 1$, an inequality in *one* variable. Describe the solution set as an inequality, in interval notation, and in a graph.

90. Solve the inequality $3x - 4 > 4$, an inequality in *one* variable. Describe the solution set as an inequality, in interval notation, and in a graph.

Graph the solution set of the system of linear inequalities by hand.

91. $y \leq 2x - 1$
$y \geq -x + 5$

92. $3x - 4y > 4$
$x + 2y < 8$

Solve the system of linear equations.

93. $y = 2x - 1$
$y = -x + 5$

94. $3x - 4y = 4$
$x + 2y = 8$

Expressions, Equations, Functions, and Graphs

Give an example of the following, and then solve, simplify, or graph, as appropriate:

95. linear equation in one variable

96. linear equation in two variables

97. linear inequality in one variable

98. linear inequality in two variables

99. system of two linear equations in two variables

100. system of two linear inequalities in two variables

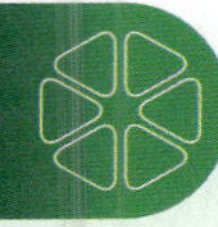

Taking it to the Lab

Climate Change Lab (continued from Chapter 5)

Recall from the Climate Change Lab in Chapter 5 that, to avoid a climate catastrophe, the IPCC has recommended that carbon dioxide emissions be lowered to 0.9 metric ton per person per year by 2050. Developed countries' carbon dioxide emissions are about 10.2 metric tons per person per year. So, these countries will have to reduce their per-person emissions significantly while trying to sustain their relatively strong economies. The United States, with carbon dioxide emissions of 18.1 metric tons per person per year, will have to reduce its emissions drastically.*

Developing countries' carbon dioxide emissions are about 2.7 metric tons per person per year. Due to their already low per-person emissions, it seems these countries will have an easier time meeting the IPCC's goal than developed countries. However, by 2050 many such countries will have become *developed* countries, and their economies will have become much stronger. As a result of this growth, their carbon dioxide emissions will greatly increase without intervention.†

For example, China's carbon dioxide emissions were only 1.15 metric tons per person in 1990. Due to China's booming economy, however, the country's carbon dioxide emissions in 2050 may reach 41.5 metric tons per person per year—more than 36 times the 1990 per-person level. This large increase in per-person carbon dioxide emissions will be amplified by a population increase of 480 million people in those 60 years (from 0.98 billion to 1.46 billion).‡

An analysis of the data in Table 35 shows that developing countries' carbon dioxide emissions are growing at a greater rate than developed countries' carbon dioxide emissions. This is occurring not only because developing countries are becoming more industrialized, but also because their populations are increasing significantly. Developed countries' economies and populations are growing at a much slower rate; in fact, most developed countries' populations will begin to decline slowly after 2020.§

Many challenges lie ahead for developed and developing countries, both of which will need to develop efficient systems that rely on alternative energy sources whenever possible. Citizens of developed countries will have the extra challenges of foregoing certain conveniences that up until now have been taken for granted. Developing countries will have the extra challenge of harnessing large population growths.

Table 35 Carbon Dioxide Emissions

Year	Carbon Dioxide Emissions (billion metric tons carbon)	
	Developed Countries	Developing Countries
1980	15.0	4.1
1985	14.5	4.9
1990	15.2	6.4
1995	14.1	8.1
2000	14.6	9.2
2005	15.0	13.3
2009	14.1	16.2

Source: *Carbon Dioxide Information Analysis Center (CDIAC)*

Analyzing the Situation

1. Let $c = f(t)$ be carbon dioxide emissions (in billions of metric tons) of developed countries in the year that is t years since 1980. Draw a scattergram by using a graphing calculator with the window shown in Fig. 60. Then find a model of the situation.

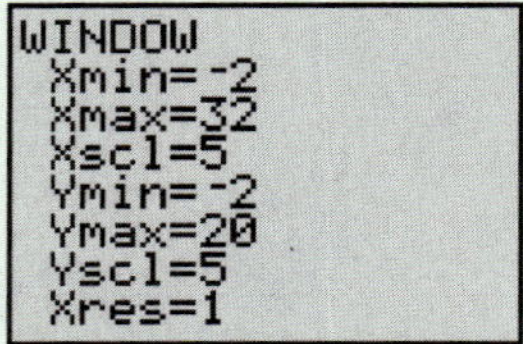

Figure 60 Window settings

2. Let $c = g(t)$ be carbon dioxide emissions (in billions of metric tons) of developing countries in the year that is t years since 1980. Use a graphing calculator to draw a scattergram, and then find a linear equation of g.
3. Compare the slopes of both models. What does the comparison tell you about the situation?
4. Explain why developing countries' carbon dioxide emissions are increasing at a greater rate than developed countries' carbon dioxide emissions.
5. Use substitution or elimination to estimate when developing countries' carbon dioxide emissions were equal to developed countries' carbon dioxide emissions. What is that carbon dioxide emission value?
6. Use your result from Problem 5 to predict the per-person carbon dioxide emissions in the year when developing countries' carbon dioxide emissions were equal to developed countries' carbon dioxide emissions. Assume world population was 6.7 billion in that year.¶ Given that the carbon dioxide emissions in 2000 were 3.9 metric tons per person, did per-person carbon dioxide emissions increase or decrease from 2000 to the year you found in Problem 5? Explain.
7. What are the challenges that lie ahead in trying to reduce carbon dioxide emissions? What challenges are unique to developed countries? To developing countries?

*International Energy Agency, "CO_2 Emissions from Fuel Combustion Highlights," 2011.
†Carbon Dioxide Information Analysis Center (CDIAC).
‡United Nations Population Division.
§U.S. Census Bureau.
¶U.S. Census Bureau.

Sports Lab

In Example 2 of Section 6.4, we predicted when the women's record time would equal the men's record time in the 400-meter run. In this lab, you will research a sporting event of your choice and use linear functions to model your data for the women's and men's versions of the event. The following websites may help you perform your research:

Track and Field Statistics
http://www.saunalahti.fi/~sut/eng/

Index to the Olympics
http://www.hickoksports.com/history/olympix.shtml

USA Swimming
www.usaswimming.org/DesktopDefault.aspx

Respond to the following instructions and questions:

1. Include tables of the women's data and the men's data you collected. State the source of your data.
2. Define any variables you used to model your data.
3. Draw sketches of scattergrams for the women's and men's data by hand in the same coordinate system. Make sure that it is clear which data points are for the women and which are for the men.
4. Find a linear function that models the women's data and a linear function that models the men's data. If you can't model your data well with linear functions, choose another sporting event.
5. Draw graphs of your functions in the same coordinate system as your scattergrams.
6. Compare the intercepts on the vertical axis for both of your linear models. What does your comparison mean in this situation?
7. Compare the slopes of your two linear models. What does your comparison mean in this situation?
8. Is there a point in time when your two models estimate (predict) that the women's performance was (will be) equal to the men's performance? If so, when did or will this happen?
9. Sketch a reasonable curve on paper that describes the relationship between the performance of the women and years *for all time*. Sketch a similar curve for the men. Sketch both curves in the same coordinate system.

Truck Lab

You are the owner of a large appliance store and are considering whether to continue renting delivery trucks or purchase a truck.

On average, you rent a truck twice a week and drive 100 miles on each delivery day. The rental truck costs \$19.99 per day, plus \$0.49 per mile. You must also pay 8% sales tax.

You can buy a delivery truck for \$36,450, plus 8% sales tax. The truck will require \$450 per year in maintenance, including sales tax. You estimate that repair costs will add about \$600 in costs each year, including sales tax.

Because you must pay for gas and insurance regardless of whether you rent or buy, you decide that these costs will not influence your decision.

1. Let $R(t)$ be the total money (in dollars) you will pay for costs related to renting a truck for t years. Let $O(t)$ be the total money (in dollars) you will pay for costs related to owning a truck for t years. Find equations of both R and O.
2. If you choose to buy a truck, how long would you have to own the truck so that owning would be as costly as renting?
3. If you buy a truck, you can sell it later. Use the rule of thumb that a new truck loses 20% of its value as soon as you drive it off the lot. Assume that the value of the truck will decrease by \$1200 each year. How long would you have to own the truck before selling it so that owning would be as costly as renting?
4. Compare your result in Part 2 with your result in Part 3. Explain why it makes sense that the results differ in this way.
5. Inflation will affect some of the costs, and if you choose to buy a truck, you may have to pay interest on a loan. Without performing calculations, discuss how these two factors would affect your answer to Part 3. Explain.

Chapter Summary

Key Points of Chapter 6

Section 6.1 Using Graphs and Tables to Solve Systems

Solution, solution set, and solve for a system	We say that an ordered pair (a, b) is a **solution** of a system of two equations in two variables if it satisfies both equations. The **solution set** of a system is the set of all solutions of the system. We **solve** a system by finding its solution set.
Using graphing to solve a system	The solution set of a system of two linear equations can be found by locating any intersection point(s) of the graphs of the two equations.

Section 6.1 Using Graphs and Tables to Solve Systems (*Continued*)

Intersection point of the graphs of two models	If the independent variable of two models represents time, then an intersection point of the graphs of the two models indicates when the quantities represented by the dependent variables were or will be equal.
Types of linear systems	There are three types of linear systems of two equations: **1. One-solution system:** The lines intersect in one point. The solution of the system is the ordered pair that corresponds to that point. **2. Inconsistent system:** The lines are parallel. The solution set of the system is the empty set. **3. Dependent system:** The lines are identical. The solution set of the system is the set of the infinite number of solutions represented by points on the same line.
Using tables to solve a linear system	If an ordered pair is listed in a table as a solution of both of two linear equations, then that ordered pair is a solution of the system.

Section 6.2 Using Substitution to Solve Systems

Using substitution to solve a linear system	To use **substitution** to solve a system of two linear equations, **1.** Isolate a variable to one side of either equation. **2.** Substitute the expression for the variable found in step 1 into the other equation. **3.** Solve the equation in one variable found in step 2. **4.** Substitute the solution found in step 3 into one of the original equations, and solve for the other variable.
Solving inconsistent systems and dependent systems by substitution	If the result of applying substitution to a linear system of two equations is • a false statement, then the system is inconsistent; that is, the solution set is the empty set. • a true statement that can be put into the form $a = a$, then the system is dependent; that is, the solution set is the set of ordered pairs represented by every point on the (same) line.

Section 6.3 Using Elimination to Solve Systems

Using elimination to solve a linear system	To use **elimination** to solve a system of two linear equations, **1.** Use the multiplication property of equality to get the coefficients of one variable to be equal in absolute value and opposite in sign. **2.** Add the left sides and add the right sides of the equations to eliminate one of the variables. **3.** Solve the equation in one variable found in step 2. **4.** Substitute the solution found in step 3 into one of the original equations and solve for the other variable.
Solving inconsistent systems and dependent systems by elimination	If the result of applying elimination to a linear system of two equations is • a false statement, then the system is inconsistent; that is, the solution set is the empty set. • a true statement that can be put into the form $a = a$, then the system is dependent; that is, the solution set is the set of ordered pairs represented by every point on the (same) line.
Methods of solving a linear system	Any linear system of two equations can be solved by graphing, substitution, or elimination. All three methods give the same result.

Section 6.4 Using Systems to Model Data

Methods of finding an intersection point of two linear models	We can find an intersection point of two linear models by graphing, substitution, or elimination.

Section 6.5 Perimeter, Value, Interest, and Mixture Problems

Five-step problem-solving method	To solve some problems in which we want to find two quantities, it is useful to perform the following five steps: • ***Step 1: Define each variable.*** • ***Step 2: Write a system of two equations.*** • ***Step 3: Solve the system.*** • ***Step 4: Describe each result.*** • ***Step 5: Check.***

Section 6.5 Perimeter, Value, Interest, and Mixture Problems (*Continued*)	
Using arithmetic to suggest how to form an equation	If you have difficulty forming an equation, try making up numbers for unknown quantities and performing arithmetic to compute a related quantity. Your computation may suggest how to form the desired equation.
Using a function to analyze a situation	If you want to analyze many aspects of a certain situation, it can help to use a system of equations to find a linear function. We can then use the function to analyze the situation in various ways.
Annual simple interest rate	The **annual simple interest rate** is the percentage of the **principal** that equals the **interest** earned per year.

Section 6.6 Linear Inequalities in Two Variables; Systems of Linear Inequalities in Two Variables	
Satisfy, solution, solution set, and solve for an inequality in two variables	If an inequality in the two variables x and y becomes a true statement when a is substituted for x and b is substituted for y, we say that the ordered pair (a, b) **satisfies** the inequality and call (a, b) a **solution** of the inequality. The **solution set** of an inequality is the set of all solutions of the inequality. We **solve** the inequality by finding its solution set.
Graph of an inequality in two variables	The graph of an inequality of the form $y > mx + b$ is the region above the line $y = mx + b$. The graph of an inequality of the form $y < mx + b$ is the region below the line $y = mx + b$. For either inequality, we use a dashed line to show that $y = mx + b$ is not part of the graph. The graph of an inequality of the form $y \geq mx + b$ is the line $y = mx + b$, as well as the region above that line. The graph of an inequality of the form $y \leq mx + b$ is the line $y = mx + b$, as well as the region below that line.
Solution, solution set, and solve for a system of inequalities in two variables	We say that an ordered pair (a, b) is a **solution** of a system of inequalities in two variables if it satisfies all of the inequalities of the system. The **solution set** of a system is the set of all solutions of the system. We **solve** a system by finding its solution set.
Finding the solution set of a system of inequalities	The solution set of a system of inequalities in two variables can be found by locating the intersection of the graphs of all of the inequalities.

Chapter 6 Review Exercises

Find the solution of the system by graphing the equations by hand.

1. $y = 2x - 3$
   $y = -3x + 7$
2. $y = \frac{3}{2}x + 4$
   $y = -\frac{1}{2}x - 4$
3. $y = \frac{2}{5}x$
   $y = -2x$
4. $4x - 2y = 5$
   $-6x + 3y = -9$
5. $4x + 2y = 16$
   $y = -2(x - 4)$
6. $x - 3y = 3$
   $2x + 3y = -12$

Solve the system by substitution. Verify your solution by checking that it satisfies both equations of the system.

7. $3x - 2y = 11$
   $y = 5x - 16$
8. $4x - 3y - 5 = 0$
   $x = 4 - 2y$
9. $y = -5x$
   $y = 2x$
10. $y = -3(x + 2)$
    $y = 4(x - 5)$
11. $x + y = -1$
    $2x - y = 4$
12. $x + 2y = 5$
    $4x + 2y = -4$

Use substitution to solve the system, with coordinates of solutions rounded to the second decimal place. Verify your work by using "intersect" on a graphing calculator.

13. $y = -2.19x + 3.51$
    $y = 1.54x - 6.22$
14. $y = -4.98x - 1.18$
    $y = -0.57x + 4.08$

Solve the system by elimination. Verify your work by using "intersect" on a graphing calculator or by checking that your result satisfies both equations of the system.

15. $x - 2y = -1$
    $3x + 5y = 19$
16. $2x - 5y = -3$
    $4x + 3y = -19$
17. $3x + 8y = 2$
    $5x - 2y = -12$
18. $4x - 3y = -6$
    $-7x + 5y = 11$
19. $-2x - 5y = 2$
    $3x + 6y = 0$
20. $y = 3x - 5$
    $y = -2x + 5$
21. $2(x + 3) + y = 6$
    $x - 3(y - 2) = -1$
22. $\frac{1}{2}x - \frac{2}{3}y = -\frac{5}{3}$
    $\frac{1}{3}x - \frac{3}{2}y = -\frac{13}{6}$

Use elimination to solve the system, with coordinates of solutions rounded to the second decimal place. Verify your work by using "intersect" on a graphing calculator.

23. $y = 4.59x + 1.25$
$y = 0.52x + 4.39$

24. $y = 0.91x - 3.57$
$y = -3.58x + 6.05$

For Exercises 25–36, solve the system by either substitution or elimination. If the system is inconsistent or dependent, say so.

25. $2x - 7y = -13$
$5x + 3y = -12$

26. $4x + 7y = 8$
$x = 3 - 2y$

27. $-3x + 7y = 6$
$6x + 2y = -12$

28. $y = -x + 7$
$y = 2x - 5$

29. $4x + 5y = -6$
$2y = -3x - 8$

30. $y = x - 2$
$3x + 5y - 30 = 0$

31. $2x - 3y = 0$
$5x - 7y = -1$

32. $y = -4x + 3$
$8x + 2y = 6$

33. $2x - 6y = 4$
$-3x + 9y = -3$

34. $2(4x - 3) - 5y = 12$
$5(3x - 1) + 2y = 6$

35. $2(3x - 4) + 3(2y - 1) = -5$
$-3(2x + 1) + 4(y + 3) = -7$

36. $\frac{3}{5}x - \frac{2}{3}y = 4$
$-\frac{6}{5}x + \frac{8}{3}y = -4$

37. Consider the system

$$y = -2x + 7$$
$$y = 3x - 3$$

Find an ordered pair that

a. satisfies $y = -2x + 7$, but does not satisfy $y = 3x - 3$.
b. satisfies $y = 3x - 3$, but does not satisfy $y = -2x + 7$.
c. does not satisfy either equation.
d. satisfies both equations.

38. Figure 61 shows the graphs of two linear equations. To the first decimal place, estimate the coordinates of the solution of the system.

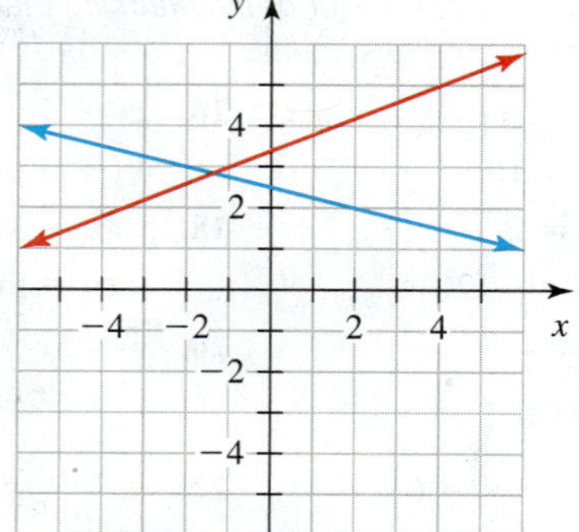

Figure 61 Exercise 38

39. Some values of functions f and g are given in Table 36. Determine two approximate solutions of the system of equations that describe the functions f and g.

Table 36 Some Values of Functions f and g (Exercise 39)

x	−4	−3	−2	−1	0	1	2	3	4
$f(x)$	2	5	8	11	14	17	20	23	26
$g(x)$	31	24	19	16	15	16	19	24	31

40. Determine the constants a and b such that $(5, 3)$ is the solution of the system.

$$2x + 3y = a$$
$$6x - 4y = b$$

Verify your result by checking that $(5, 3)$ satisfies each of your equations. Also, verify your result graphically.

41. Solve the system of equations three times, once by each of the three methods: graphing by hand, substitution, and elimination. Decide which method you prefer for this system.

$$3x + 4y = 15$$
$$2y = -5x + 11$$

42. The percentages of households with landlines (with or without wireless telephones) and those with only wireless phones are shown in Table 37 for various years.

Table 37 Percentages of Households with Phone Landlines and Those with Only Wireless Phones

	Percent	
Year	Landlines	Wireless-Only
2006	86.8	10.5
2007	84.4	13.6
2008	80.0	17.5
2009	75.3	22.7
2010	71.0	26.6
2011	66.2	31.6

Source: *U.S. Centers for Disease Control and Prevention*

a. Let $L(t)$ be the percentage of households with landlines and $W(t)$ be the percentage of households with only wireless phones, both at t years since 2000. Find equations of L and W.
b. Find the slopes of the two models. What do the slopes mean in this situation?
c. Predict when the percentage of households with phone landlines will be equal to the percentage of households with only wireless phones. What is that percentage?
d. Give a reason why the percentage you found in part (c) is not 50%.
e. Find values of t where $L(t) < W(t)$. What do they mean in this situation?

43. The television ratings of the baseball World Series and the average television ratings of prime-time shows on the major networks are shown in Table 38 for various years.

a. Let $w(t)$ be the rating of the World Series and $p(t)$ be the average rating of prime-time shows on the major networks, both at t years since 1970. Find equations of w and p.
b. Use substitution or elimination to predict when the rating of the World Series will equal the average rating of prime-time shows on the major networks. What will that rating be?

Table 38 Television Ratings of World Series and of Prime-Time Shows

Year	Rating of World Series	Average Rating of Prime-Time Shows on Major Networks
1973	30.7	19.2
1983	23.3	16.7
1993	17.3	12.5
2003	12.8	7.3
2007	10.6	7.9

Sources: *Fox Sports; Nielsen Media Research*

44. One Rent-A-Wreck® office rents 16-foot trucks for a one-day fee of \$75 plus \$0.22 per mile. One U-Haul office charges a one-day fee of \$29.95 plus \$0.69 per mile.

- **a.** Let $R(d)$ be Rent-A-Wreck's charge (in dollars) and $U(d)$ be U-Haul's charge (in dollars), both for driving d miles in one day. Find equations of R and U.
- **b.** For how many miles driven is the one-day charge at Rent-A-Wreck equal to the charge at U-Haul? What is that charge?
- **c.** For how many miles driven is the one-day charge at Rent-A-Wreck less than the charge at U-Haul?

45. In 2012, the average price of a home in a community was \$250,000, and it has increased by about \$9000 each year. A family had \$12,000 on January 1, 2012, and planned to save \$230 each month. Predict how long it should take the family to pay a 10% down payment on an average-priced house in the community.

46. The length of a rectangle is 2 feet more than three times the width. If the perimeter is 44 feet, find the dimensions of the rectangle.

47. An 8000-seat amphitheater has tickets for sale at \$22 and \$39. How many tickets should be sold at each price for a sell-out performance to generate a total revenue of \$201,500?

48. How many gallons of a 10% antifreeze solution and a 20% antifreeze solution must be mixed to make 10 gallons of a 16% antifreeze solution?

49. A person plans to invest \$8000. She will invest in both a Hartford Global Leaders Y account at 6.8% annual interest and a Mutual Discovery Z account at 13.0% annual interest. Both interest rates are five-year averages. Let x and y be the money (in dollars) invested in the 6.8% account and the 13.0% account, respectively.

- **a.** Let $I = f(x)$ be the total interest (in dollars) earned from investing the \$8000 for one year. Find an equation of f.
- **b.** Find $f(575)$. What does it mean in this situation?
- **c.** Find the value of x when $f(x) = 575$. What does it mean in this situation?

Graph the inequality in two variables by hand.

50. $y \le 3x - 5$ **51.** $y \ge -2x + 4$ **52.** $3x - 2y > 4$

53. $2y - 5x < 0$ **54.** $x \ge 3$ **55.** $y < -2$

56. $-2(y + 3) + 4x \ge -8$

For Exercises 57–61, graph the solution set of the system of inequalities by hand.

57. $y > x + 1$, $y \le -2x + 5$

58. $y \ge \frac{3}{5}x + 1$, $x < -1$

59. $3x - 4y \ge 12$, $5y \le -3x$

60. $x > 2$, $y \le -1$

61. $x - y < 3$, $x + y < 5$, $x > 0$, $y > 0$

62. The lower and upper limits of ideal weights of men with a medium frame are listed in Table 39 for various heights. Assume that the men are wearing 5 pounds of clothing, including shoes with 1-inch heels.

Table 39 Ideal Weights of Men with a Medium Frame

Height (inches)	Ideal Weight (pounds) Lower Limit	Ideal Weight (pounds) Upper Limit
65	137	148
67	142	154
69	148	160
71	154	166
73	160	174
76	171	187

Source: *Metropolitan Life Insurance Company*

Let $w = L(h)$ be the lower limit (in pounds) and $w = U(h)$ be the upper limit (in pounds), both of the ideal weight (in pounds, including clothing) of a man who has a medium frame and a height of h inches (including shoes).

- **a.** Find equations of L and U.
- **b.** Find a system of inequalities that describes the ideal weights of men who have a medium frame and are between the heights of 63 inches and 78 inches, inclusive.
- **c.** Graph by hand the solution set of the system of inequalities that you found in part (a).
- **d.** What is the ideal weight range of men who have a medium frame and a height of 68 inches?

63. Find three ordered pairs that are solutions of the inequality $4x - 3y > 9$. Also, find three ordered pairs that are not solutions.

Chapter 6 Test

1. Find the solution of this system by graphing the equations by hand:

$$y = -\frac{2}{5}x - 1$$
$$y = -2x + 7$$

2. Use "intersect" on a graphing calculator to solve this system, with the coordinates of the solution rounded to the second decimal place:

$$y = \frac{2}{3}x + 4$$
$$3x + 4y = -2$$

Solve the system by substitution.

3. $5x - 2y = 4$
$y = 3x - 1$

4. $3x + 4y = 9$
$x - 2y = -7$

Solve the system by elimination.

5. $-7x - 2y = -8$
$5x + 4y = -2$

6. $2x - 5y = -18$
$3x + 4y = -4$

For Exercises 7–12, solve the system by substitution or elimination. If the system is inconsistent or dependent, say so.

7. $3x - 5y = -21$
$x = 2(2 - y)$

8. $2x - 3y = 4$
$-4x + 6y = -8$

9. $x = 2y - 3$
$3x - 6y = 12$

10. $4x - 7y = 6$
$-5x + 2y = -21$

11. $-4(x + 2) + 3(2y - 1) = 21$
$5(3x - 2) - (4y + 3) = -59$

12. $\frac{2}{5}x - \frac{3}{4}y = 8$
$\frac{3}{5}x + \frac{1}{4}y = 1$

13. Use substitution or elimination to solve this system, with coordinates of the solution rounded to the second decimal place:

$$y = -1.94x + 8.62$$
$$y = 1.25x - 2.38$$

14. Solve the system of equations three times, once by each of the three methods of elimination, substitution, and graphing:

$$4x - 3y = 9$$
$$y = 2x - 5$$

15. Consider the solution set of the system

$$y = 5x - 13$$
$$y = mx + b$$

where m and b are constants. If the system's solution set is the empty set, what can you say about m? About b?

16. The graphs of $y = ax + b$ and $y = cx + d$ are sketched in Fig. 62.

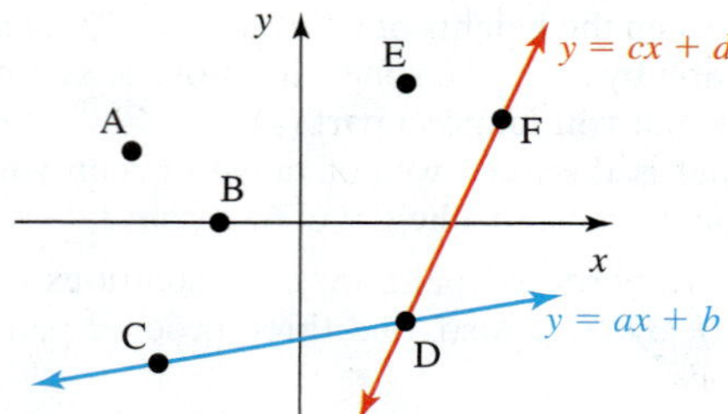

Figure 62 Exercise 16

Which of the points A, B, C, D, E, and F

a. satisfy the equation $y = ax + b$?
b. satisfy the equation $y = cx + d$?
c. satisfy both equations?
d. do not satisfy either equation?

17. Figure 63 shows the graphs of two linear equations. Find the solution of the system.

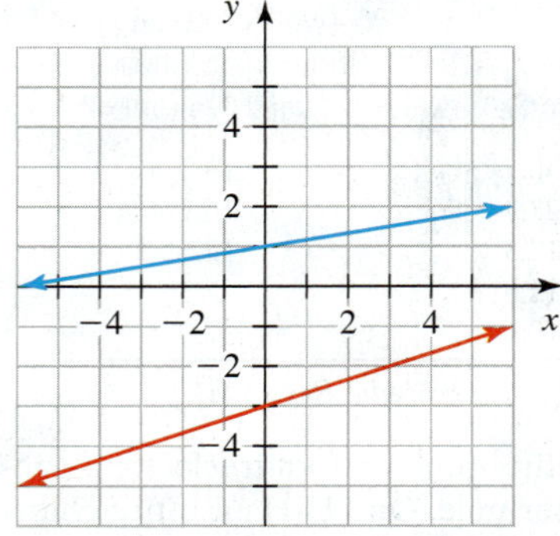

Figure 63 Exercise 17

18. Find the coordinates of the points A, B, C, D, E, and F as shown in Fig. 64. The equations of the sketched lines are provided, but no attempt has been made to sketch the lines accurately except for showing the intersection points.

l_1: $y = 3x + 4$
l_2: $3y + 2x = 34$
l_3: $y + 4x = 28$
l_4: $y = 3x - 14$

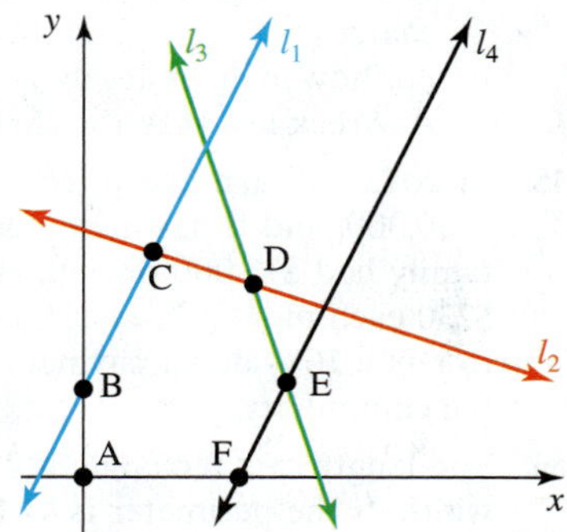

Figure 64 Exercise 18

19. The clarity of Lake Tahoe in California is measured by the depth at which a white disk with diameter 10 inches remains visible when lowered beneath the water's surface. The clarities of Lake Tahoe in the winter and summer are shown in Table 40 for various years.

Table 40 Clarity of Lake Tahoe

Year	Clarity (feet) Winter	Summer
1970	99.4	93.5
1980	90.8	74.9
1990	84.6	75.5
2000	70.6	64.1
2010	73.0	51.9

Source: *UC Davis Tahoe Environmental Research Center*

a. Let $w(t)$ be the clarity of Lake Tahoe in the winter and $s(t)$ be the clarity of Lake Tahoe in the summer, both in feet, at t years since 1970. Find equations of w and s.
b. Estimate in what year the clarity in the winter was equal to the clarity in the summer. What was that clarity?
c. Predict the average clarity throughout all of 2018.

20. The revenue of Amazon from books, CDs, and DVDs increased approximately linearly from \$4.6 billion in 2007 to \$7.0 billion in 2010. The revenue of Borders decreased approximately linearly from \$3.7 billion in 2007 to \$2.3 billion in 2010 (Source: *Amazon, Borders*).

a. Let $A(t)$ be the annual revenue of Amazon from books, CDs, and DVDs, and let $B(t)$ be the annual revenue of Borders, both in billions of dollars, at t years since 2000. Find equations of A and B.

b. Estimate when the annual revenue of Amazon from books, CDs, and DVDs was equal to the annual revenue of Borders. What was that revenue?

c. For which years was the revenue of Amazon from books, CDs, and DVDs less than the revenue of Borders?

d. What is the slope of the graph of B? What does it mean in this situation? Explain why it is not surprising that Borders went out of business in 2011.

21. A 20,000-seat theater has tickets for sale at \$55 and \$70. How many tickets should be sold at each price for a sellout performance to generate a total revenue of \$1,197,500?

22. A person plans to invest a total of \$7000, part of it in an account at 3% annual interest and the rest in an account at 7% annual interest. How much should the person invest in each account so that the total interest in one year will be \$410?

23. A 10,000-seat amphitheater will sell tickets at \$35 and \$50 for a Blonde Redhead concert. Let x and y be the number of tickets that will sell for \$35 and \$50, respectively. Assume the show will sell out.

a. Let $R = f(x)$ be the total revenue (in dollars) from selling the \$35 and \$50 tickets. Find an equation of f.

b. Use a graphing calculator to draw a graph of f for $0 \le x \le 10{,}000$. What is the slope? What does it mean in this situation?

c. How many of each ticket must be sold for the revenue to be \$390,500?

Graph the inequality in two variables by hand.

24. $5x - 2y \le 6$ **25.** $y < -3$

Graph the solution set of the system of inequalities by hand.

26. $y \le -3x + 4$, $x - 3y > 6$ **27.** $y > 2$, $x \ge -3$

Cumulative Review of Chapters 1–6

1. The low temperatures in New York City for the first four days of March are 4°F, −2°F, −1°F, and 3°F. Let F be the temperature in Fahrenheit degrees. Use points on a number line to describe these low temperatures.

2. The rate at which a cricket chirps depends on the temperature. Let n be the number of times a cricket chirps per minute when the temperature is t degrees Fahrenheit. What does the ordered pair (70, 129) represent?

For Exercises 3 and 4, perform the indicated operations. Use a calculator to verify your work.

3. $-\frac{26}{27} \cdot \frac{12}{13}$ **4.** $\frac{5}{7} - \left(-\frac{3}{5}\right)$

5. Evaluate $a^2 - bc + b^2$, where $a = -3$, $b = -2$, and $c = 4$.

6. Find the slope of the line that contains the points $(-3, -2)$ and $(5, -8)$. State whether the line is increasing, decreasing, horizontal, or vertical.

7. Some solutions of four linear equations are provided in Table 41. Find an equation of each of the four lines.

Table 41 Solutions of Four Linear Equations

Equation 1		Equation 2		Equation 3		Equation 4	
x	y	x	y	x	y	x	y
0	49	0	11	3	39	2	14
1	41	1	15	4	37	4	20
2	33	2	19	5	35	5	23
3	25	3	23	6	33	8	32
4	17	4	27	7	31	9	35

8. Graphs of four equations are shown in Fig. 65. State whether m and b are positive, negative, zero, or undefined for the $y = mx + b$ form of each equation.

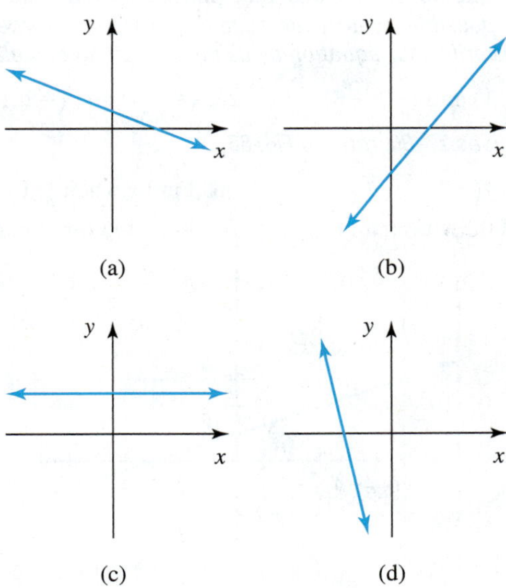

Figure 65 Exercise 8

For Exercises 9–12, simplify the expression or solve the equation, as appropriate. Use a graphing calculator to verify your work.

9. $2 - 5(4x + 8) = 3(2x - 7) + 3$

10. $-3(2p - w) - (7p + 2w) + 5$

11. $\frac{2}{3}(6w + 9y - 15)$

12. $\frac{3m}{5} - \frac{2}{3} = \frac{4}{5}$

13. Solve $ax + by = c$ for y.

Let x be a number. Translate the following to an expression or an equation, as appropriate, and then simplify the expression or solve the equation.

14. Six, plus 3 times the sum of 4 and the number

15. Four minus the quotient of the number and 3 is 2.

Determine the slope and the y-intercept, and use them to graph the equation by hand. Use a graphing calculator to verify your work.

16. $y = 2x - 4$

17. $x - 2y = 6$

18. $5x + 2y - 12 = 0$

19. $y = -3$

For Exercises 20 and 21, find the x-intercept and y-intercept. Then graph the equation by hand.

20. $2x - 5y = 10$

21. $y = -2x + 4$

22. Graph the equation $x = 4$.

23. Determine whether the lines $2x + 5y = 7$ and $y = \frac{2}{5}x - 3$ are parallel.

24. Find an equation of the line that has slope $-\frac{2}{5}$ and contains the point $(3, -2)$. Write your equation in slope–intercept form. Check that the point $(3, -2)$ satisfies your equation.

Find an equation of the line that passes through the two given points. If possible, write your equation in slope–intercept form. If possible, verify your equation by using a graphing calculator.

25. $(-5, 1)$ and $(-2, -3)$

26. $(-2, 8)$ and $(-2, 1)$

For Exercises 27–30, refer to Fig. 66.

27. Find $f(-3)$.

28. Find x when $f(x) = -1$.

29. Find the x-intercept.

30. Find an equation of the line.

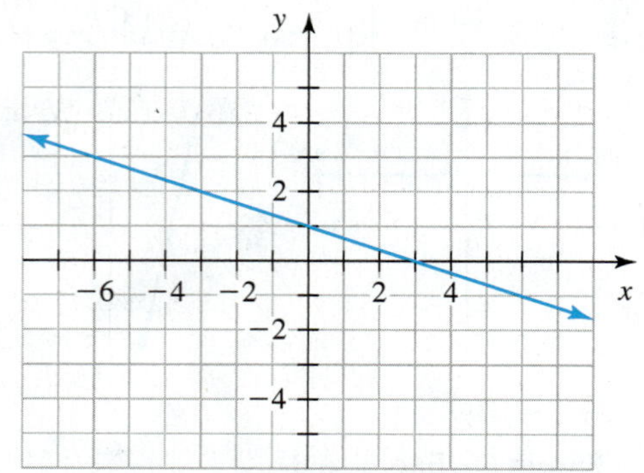

Figure 66 Exercises 27–30

For Exercises 31–33, refer to the relation graphed in Fig. 67.

31. Find the domain of the relation.

32. Find the range of the relation.

33. Is the relation a function? Explain.

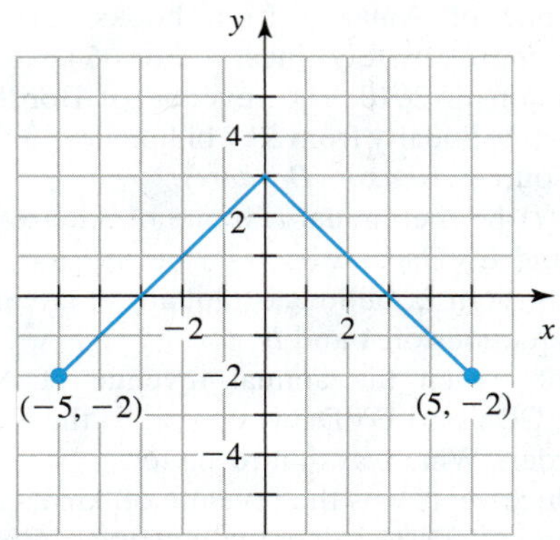

Figure 67 Exercises 31–33

34. Solve the inequality $-3(x - 5) > 18$, an inequality in one variable. Describe the solution set as an inequality, in interval notation, and in a graph.

35. Find the solution of this system by graphing the equations by hand:

$$y = 2x - 3$$
$$x + y = 3$$

36. Use "intersect" on a graphing calculator to solve this system, with the coordinates of the solution rounded to the second decimal place:

$$y = -2.9x + 7.8$$
$$y = 1.3x - 6.1$$

For Exercises 37 and 38, solve the system by substitution or elimination. Use "intersect" on a graphing calculator to verify your work.

37. $3x + 5y = -1$
$2x - 3y = 12$

38. $4x - y - 9 = 0$
$y = 5 - 3x$

39. Use elimination or substitution to solve the system, with coordinates of solutions rounded to the second decimal place. Verify your work by using "intersect" on a graphing calculator.

$$y = -2.9x + 97.8$$
$$y = 3.1x - 45.6$$

40. Solve the system three times, once by each of the three methods: graphing by hand, substitution, and elimination. Decide which method you prefer for this system.

$$2x - 3y = 19$$
$$y = 4x - 13$$

41. Some solutions of two linear equations are shown in Tables 42 and 43. Find the solution of the system of the two equations.

Table 42 Solutions of One Equation

x	y
0	97
1	93
2	89
3	85
4	81

Table 43 Solutions of the Other Equation

x	y
0	7
1	9
2	11
3	13
4	15

42. Graph the inequality $y < -\frac{2}{3}x + 3$.

43. Graph the solution set of the system

$$3x - 5y \leq 10$$
$$x > -4$$

44. Traffic safety is usually measured by the fatality rate, which is the number of traffic-related deaths per 100 million miles traveled. Fatality rates are shown in Table 44 for various years.

Table 44 Traffic Fatality Rates

Year	Fatalities (thousands)	Fatality Rate (number of deaths per 100 million miles)
1998	41.5	1.58
1999	41.7	1.55
2000	41.9	1.53
2001	42.2	1.51
2002	43.0	1.51
2003	42.6	1.48
2004	42.8	1.46
2005	43.4	1.46
2006	42.6	1.42
2007	41.1	1.36
2008	37.3	1.25
2009	33.8	1.15
2010	32.9	1.10

Source: *National Center for Statistics & Analysis*

Let $r = f(t)$ be the traffic fatality rate (in number of deaths per 100 million miles traveled) at t years since 1990.

a. Explain how it is possible for the fatality rate to be generally decreasing from 1998 to 2002 when the number of fatalities is increasing in that period.

b. Use a graphing calculator to draw a scattergram of the data that describes the values of t and r.

c. Find an equation of f that describes the relationship between t and r for the years 2006–2010.

d. Find the r-intercept. What does it mean in this situation? Is it likely that this is an accurate estimate? Explain by referring to the scattergram.

e. In 2003, Transportation Secretary Norman Mineta proposed a plan to lower the traffic fatality rate to 1.0 death per 100 million vehicle miles traveled by 2008. Estimate when the fatality rate reached 1.0 death per 100 million miles traveled.

f. Predict the fatality rate in 2017.

g. Find the t-intercept. What does it mean in this situation?

45. The number of bicyclists younger than 16 who were hit and killed by motor vehicles was 82 children in 2009 and has declined by about 9 children each year (Source: *U.S. Department of Transportation*).

a. Let $n = f(t)$ be the number of bicyclists younger than 16 who were hit and killed by motor vehicles in the year that is t years since 2009. Find an equation of f.

b. What is the slope of the model? What does it mean in this situation?

c. Find the n-intercept. What does it mean in this situation?

d. Find the t-intercept. What does it mean in this situation?

e. For which years in the future is there model breakdown for certain? Explain.

46. The *days sales of inventory* (DSI) measures how long it takes a company to turn its inventory into sales. The DSI of Home Depot was 86.8 days in 2010 and has increased by about 2.2 days per year. The DSI of Lowes was 95.9 days in 2010 and has increased by about 3.2 days per year (Source: *Morningstar.com*).

a. Let $H(t)$ be the DSI of Home Depot and $L(t)$ be the DSI of Lowes, both in days, at t years since 2010. Find equations of H and L.

b. What are the slopes of your two models? What do they mean in this situation? Which would be better for the companies, a positive or negative slope? Explain.

c. Use substitution or elimination to estimate when the DSI of Home Depot was equal to the DSI of Lowes. What was that DSI?

d. Solve the inequality $H(t) < L(t)$. What does your result mean in this situation?

47. World record times for the 200-meter run are listed in Table 45.

Table 45 200-Meter Run Record Times

Women		Men	
Year	Record Time (seconds)	Year	Record Time (seconds)
1973	22.38	1951	20.6
1974	22.21	1963	20.3
1978	22.06	1967	20.14
1984	21.71	1979	19.72
1988	21.34	1996	19.32
		2009	19.19

Source: *International Association of Athletics Federations*

a. Let $r = W(t)$ and $r = M(t)$ be the record times (in seconds) for women and men, respectively, at t years since 1900. Find equations of W and M.

b. Find $W(118)$ and $M(118)$. What do they mean in this situation?

c. Compare the slopes of the two models. What does your comparison tell you about the situation?

d. Explain why your results from parts (b) and (c) suggest there may be a time when the women's record time will be equal to the men's record time.

e. Predict when the women's record time will equal the men's record time. What will that time be?

f. Solve $W(t) > M(t)$. What does your result mean in this situation?

g. Find the t-intercepts of the two models. What do your results mean in this situation?

48. How many quarts of a 16% acid solution and a 28% acid solution need to be mixed to form 12 quarts of a 20% acid solution?

Polynomial Functions and Properties of Exponents

Do you have a cell phone? If so, how much do you usually pay each month to use it? The average monthly cell phone bill has actually decreased since 2004 (see Table 1). In Exercise 95 of Homework 7.2, you will predict the total amount of money paid by all cell phone subscribers in a certain year.

So far, we have studied linear expressions and linear functions. In this chapter, we will work with *polynomial expressions, polynomial functions,* and *power expressions.* We will perform operations with polynomial expressions and *polynomial functions.* We will use polynomial functions to model authentic situations, such as the total cost of state corrections (prisons and related costs). We will also simplify power expressions.

Table 1 Average Monthly Cell Phone Bills and Numbers of Subscribers

Year	Average Bill (dollars per month)	Number of Subscribers (millions)
1998	39.43	69.2
2000	45.27	109.5
2002	48.40	140.8
2004	50.64	182.1
2006	50.56	233.0
2008	50.07	270.3
2010	47.21	302.9

Source: *Cellular Telecommunications & Internet Association*

7.1 Adding and Subtracting Polynomial Expressions and Functions

Objectives

- Know the meaning of *term, monomial, polynomial, degree, coefficient, like terms, polynomial function, quadratic function, parabola, vertex,* and *cubic function.*
- Add and subtract polynomials.
- Evaluate polynomial functions.
- Recognize typical graphs of quadratic functions and cubic functions.
- Find *sum functions* and *difference functions.*
- Use sum functions and difference functions to describe authentic situations.

In this section, we will discuss how to add and subtract expressions called *polynomials.* We will discuss how to combine two *polynomial functions* to form a "new" function and how to use some of these new functions to model authentic situations.

Polynomials

Recall from Section 4.1 that a **term** is a constant, a variable, or a product of a constant and one or more variables raised to powers. Here are some examples of terms: $5x^4$, x, 17, $-6x^8y^2$, and x^{-3}.

A **monomial** is a constant, a variable, or a product of a constant and one or more variables raised to *counting-number* powers. Here are some examples of monomials:

$$-2x^7,\ y,\ -3,\ \frac{2}{7}x^4y^9,\text{ and } x^3$$

A **polynomial,** or **polynomial expression,** is a monomial or a sum of monomials. Here are some examples of polynomials:

$$3x^2 + 5x + 2 \qquad 9x^5y^3 - 6x^3y^2 - xy + 5 \qquad -2x + 5 \qquad 7 \qquad 3x^4$$

The polynomial $7x^3 - 5x^2 + 8x - 1$ is a *polynomial in one variable.* It has four terms: $7x^3$, $-5x^2$, $8x$, and -1. We usually write polynomials in one variable so that the exponents of the terms decrease from left to right, an arrangement called **decreasing order.** If a polynomial contains more than one variable, we usually write the polynomial so that the exponents of one of the variables decrease from left to right.

The **degree of a term** in one variable is the exponent on the variable. For example, the degree of the term $2x^7$ is 7. The degree of a term in two or more variables is the sum of the exponents on the variables. For example, the term $5x^4y^2$ has degree $4 + 2 = 6$. The **degree of a polynomial** is the largest degree of any nonzero term of the polynomial. For example, the polynomial $7x^4 - 2x^2 + 5$ has degree 4. A constant polynomial such as 5 has degree 0.

Polynomials with degrees 1, 2, or 3 have special names:

Degree	Name	Examples
1	linear (first-degree) polynomial	$-3x + 7, 2x$
2	quadratic (second-degree) polynomial	$8x^2 - 5x + 2, x^2 - 4$
3	cubic (third-degree) polynomial	$4x^3 + 2x^2 - 6x + 5, -6x^3 + x$

Example 1 Describing Polynomials

Use words such as *linear, quadratic, cubic, polynomial, degree, one variable,* and *two variables* to describe the expression.

1. $5x^2 - 3x + 7$ **2.** $-7x^3 + 8x^2 - 1$ **3.** $4a^3b^2 - 2a^2b^2 + 9ab^2$

Solution

1. The term $5x^2$ has degree 2, which is larger than the degrees of the other terms. So, $5x^2 - 3x + 7$ is a quadratic (or second-degree) polynomial in one variable.
2. The term $-7x^3$ has degree 3, which is larger than the degrees of the other terms. So, $-7x^3 + 8x^2 - 1$ is a cubic (or third-degree) polynomial in one variable.
3. The term $4a^3b^2$ has degree $3 + 2 = 5$, which is larger than the degrees of the other terms. So, $4a^3b^2 - 2a^2b^2 + 9ab^2$ is a fifth-degree polynomial in two variables.

Combining Like Terms

Recall from Section 4.2 that the **coefficient** of a term is the constant factor of the term. For the term $-3x^2$, the coefficient is -3. For the term x^3, the coefficient is 1, because $x^3 = 1x^3$. For the term 6, the coefficient is 6. The coefficients of a polynomial are the coefficients of the terms. For example, the coefficients of the polynomial $7x^3 - 5x^2 + 8x - 1$ are 7, -5, 8, and -1. The **leading coefficient** of a polynomial is the coefficient of the term with the largest degree. For $7x^3 - 5x^2 + 8x - 1$, the leading coefficient is 7.

Recall from Section 4.2 that **like terms** are either constant terms or variable terms that contain the same variable(s) raised to exactly the same power(s). For example, the terms $9x^2y^7$ and $4x^2y^7$ are like terms, because both terms have an x with the exponent 2 and a y with the exponent 7. Recall from Section 4.2 that if terms are not like terms, we say that they are **unlike terms.** For example, $3x^3$ and $5x^2$ are unlike terms, because the exponents of x are different.

Recall also from Section 4.2 that we can combine like terms of linear polynomials by using the distributive law. For example,

$$3x + 5x = (3 + 5)x = 8x$$

Likewise, we can *combine like terms* of polynomials of higher degree, such as $4x^2 + 6x^2$, by using the distributive law:

$$4x^2 + 6x^2 = (4 + 6)x^2 = 10x^2$$

Note that we can find $4x^2 + 6x^2$ in one step by adding the coefficients:

$$4x^2 + 6x^2 = 10x^2$$

$$4 + 6 = 10$$

We can also find $6x^3 - 2x^3$ by adding the coefficients:

$$6x^3 - 2x^3 = 4x^3$$

$$6 + (-2) = 4$$

However, we can't add the coefficients in $8x^4 + 3x^2$, because $8x^4$ and $3x^2$ are not like terms. There is no helpful way to use the distributive law for unlike terms.

Combining Like Terms

To combine like terms, add the coefficients of the terms.

Example 2 Combining Like Terms

Combine like terms when possible.

1. $7x^2 + 5x^2$
2. $-4x^3y^2 - 6x^3y^2$
3. $5a^2b^4 + 3a^4b^2$
4. $-6x^2 + x^2$

Solution

1. $7x^2 + 5x^2 = 12x^2$ *Add coefficients:* $7 + 5 = 12$
2. $-4x^3y^2 - 6x^3y^2 = -10x^3y^2$ *Add coefficients:* $-4 + (-6) = -10$
3. Although $5a^2b^4$ and $3a^4b^2$ have the same variables, the variables do not have the same exponents. They are not like terms, so we cannot combine them.
4. $-6x^2 + x^2 = -6x^2 + 1x^2 = -5x^2$ *Add coefficients:* $-6 + 1 = -5$

Example 3 Combining Like Terms

Combine like terms when possible.

1. $5x^3 - 4x^2 + 2x^3 - x^2$
2. $3p^3t^2 + p^2t - 8p^3t^2 + 7p^2t$

Solution

1. We rearrange the terms so the terms with x^3 are adjacent and the terms with x^2 are adjacent:

$$\begin{aligned} 5x^3 - 4x^2 + 2x^3 - x^2 &= 5x^3 + 2x^3 - 4x^2 - x^2 && \text{Rearrange terms.} \\ &= 7x^3 - 5x^2 && \text{Combine like terms.} \end{aligned}$$

2.

$$\begin{aligned} &3p^3t^2 + p^2t - 8p^3t^2 + 7p^2t && \text{Rearrange terms.} \\ &= 3p^3t^2 - 8p^3t^2 + p^2t + 7p^2t \\ &= -5p^3t^2 + 8p^2t && \text{Combine like terms.} \end{aligned}$$

Addition of Polynomials

Now that we know how to combine like terms, we can add polynomials.

Adding Polynomials

To add polynomials, combine like terms.

Example 4 Adding Polynomials

Find the sum.

1. $(4x^2 + 3x) + (6x^2 - 2x)$
2. $(5x^3 - 3x^2 + 8x - 1) + (7x^3 - x^2 - 3x + 4)$
3. $(2x^3 - 5x) + (7x^2 - 4x)$
4. $(4x^2 + 3xy - 5y^2) + (3x^2 - 7xy + 9y^2)$

Solution

1. $(4x^2 + 3x) + (6x^2 - 2x) = 4x^2 + 6x^2 + 3x - 2x$ *Rearrange terms.*
$= 10x^2 + x$ *Combine like terms.*

2. $(5x^3 - 3x^2 + 8x - 1) + (7x^3 - x^2 - 3x + 4)$
$= 5x^3 + 7x^3 - 3x^2 - x^2 + 8x - 3x - 1 + 4$ *Rearrange terms.*
$= 12x^3 - 4x^2 + 5x + 3$ *Combine like terms.*

3. $(2x^3 - 5x) + (7x^2 - 4x) = 2x^3 + 7x^2 - 5x - 4x$ *Rearrange terms.*
$= 2x^3 + 7x^2 - 9x$ *Combine like terms.*

We use a graphing calculator table to verify the work (see Fig. 1).

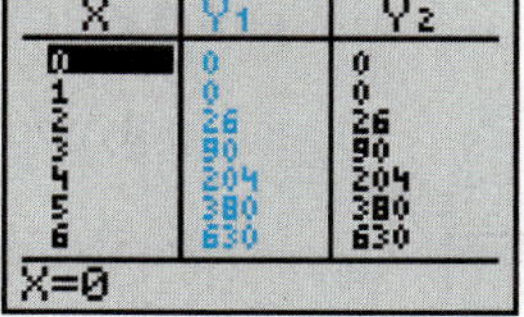

Figure 1 Verify the work

4. $(4x^2 + 3xy - 5y^2) + (3x^2 - 7xy + 9y^2)$
$= 4x^2 + 3x^2 + 3xy - 7xy - 5y^2 + 9y^2$ *Rearrange terms.*
$= 7x^2 - 4xy + 4y^2$ *Combine like terms.*

Subtraction of Polynomials

In Section 4.1, we discussed how to subtract expressions; those expressions were actually linear polynomials. In this section, we take similar steps to subtract polynomials of higher degree.

Example 5 Subtracting Polynomials

Find the difference.

1. $(8x^3 + 5x^2) - (6x^3 + 2x^2)$
2. $(2a^2 - 5ab + 7b^2) - (6a^2 - 4ab + 3b^2)$

Solution

1. To begin, we write $(8x^3 + 5x^2) - (6x^3 + 2x^2)$ as $(8x^3 + 5x^2) - 1(6x^3 + 2x^2)$ and then distribute the -1:

$(8x^3 + 5x^2) - (6x^3 + 2x^2) = (8x^3 + 5x^2) - 1(6x^3 + 2x^2)$ *$a - b = a - 1b$*
$= 8x^3 + 5x^2 - 6x^3 - 2x^2$ *Distributive law*
$= 8x^3 - 6x^3 + 5x^2 - 2x^2$ *Rearrange terms.*
$= 2x^3 + 3x^2$ *Combine like terms.*

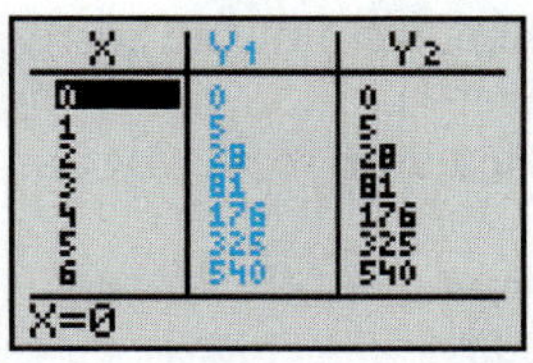

Figure 2 Verify the work

We use a graphing calculator table to verify the work (see Fig. 2).

2. $(2a^2 - 5ab + 7b^2) - (6a^2 - 4ab + 3b^2)$
$= (2a^2 - 5ab + 7b^2) - 1(6a^2 - 4ab + 3b^2)$ *$a - b = a - 1b$*
$= 2a^2 - 5ab + 7b^2 - 6a^2 + 4ab - 3b^2$ *Distributive law*
$= 2a^2 - 6a^2 - 5ab + 4ab + 7b^2 - 3b^2$ *Rearrange terms.*
$= -4a^2 - ab + 4b^2$ *Combine like terms.*

In Problem 1 of Example 5, a key step in finding $(8x^3 + 5x^2) - (6x^3 + 2x^2)$ is to write this difference as $(8x^3 + 5x^2) - 1(6x^3 + 2x^2)$, so that we can distribute the -1.

Subtracting Polynomials

To subtract polynomials, first distribute -1 and then combine like terms.

Now that we are familiar with polynomial expressions, we turn our attention to *polynomial functions*.

Quadratic Functions

The following are examples of *polynomial functions:*

$$f(x) = 3x^5 - 9x^4 + 8x^2 + 1 \qquad g(x) = 4x^2 - 3x + 5 \qquad h(x) = \frac{2}{3}x$$

A **polynomial function** is a function whose equation can be put into the form $f(x) = P$, where P is a polynomial in terms of the variable x. If P is a quadratic (second-degree) polynomial, we call the function a *quadratic function*.

Definition Quadratic function

A **quadratic function** is a function whose equation can be put into the form

$$f(x) = ax^2 + bx + c$$

where $a \neq 0$. This form is called the **standard form.**

For example, the function $f(x) = 8x^2 - 2x + 7$ is quadratic, because it is in the form $f(x) = ax^2 + bx + c$, with $a = 8$, $b = -2$, and $c = 7$.

Example 6 Evaluating a Quadratic Function

For $f(x) = -2x^2 + 5x - 1$, find the following.

1. $f(4)$ **2.** $f(-3)$ **3.** $f(0)$

Solution

1. $f(4) = -2(4)^2 + 5(4) - 1 = -2(16) + 5(4) - 1 = -13$

2. $f(-3) = -2(-3)^2 + 5(-3) - 1 = -2(9) + 5(-3) - 1 = -34$

3. $f(0) = -2(0)^2 + 5(0) - 1 = -2(0) + 5(0) - 1 = -1$

We use a graphing calculator table to verify our work (see Fig. 3).

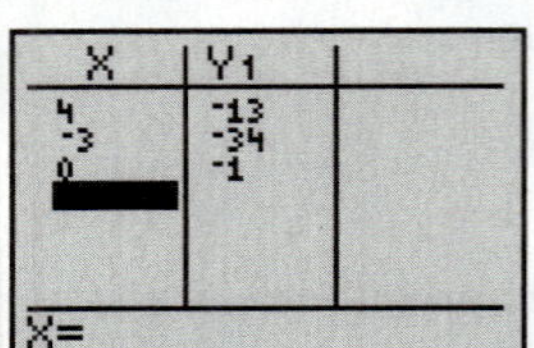

Figure 3 Verify the work

The function $f(x) = x^2$ is quadratic, because it is in the form $f(x) = ax^2 + bx + c$, with $a = 1$, $b = 0$, and $c = 0$.

Example 7 Graphing a Quadratic Function

Sketch the graph of $f(x) = x^2$.

Solution

First, we list some input–output pairs of $f(x) = x^2$ in Table 2. Then, we plot the corresponding points and sketch a curve that contains the points (see Fig. 4).

We use ZStandard followed by ZSquare on a graphing calculator to verify our graph (see Fig. 5).

Table 2 Input–Output Pairs of $f(x) = x^2$

x	$f(x)$
-3	$(-3)^2 = 9$
-2	$(-2)^2 = 4$
-1	$(-1)^2 = 1$
0	$0^2 = 0$
1	$1^2 = 1$
2	$2^2 = 4$
3	$3^2 = 9$

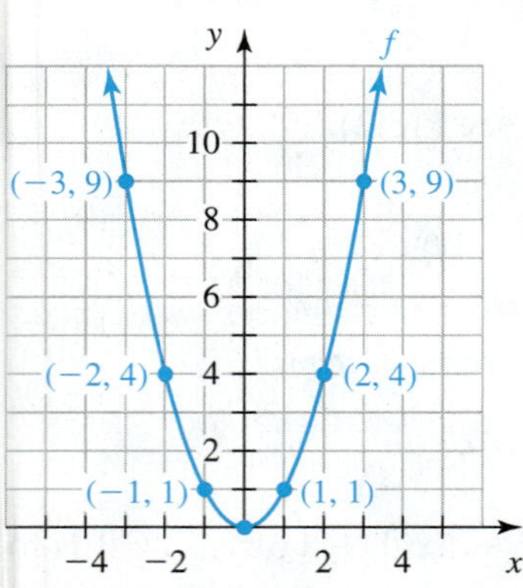

Figure 4 Graph of $f(x) = x^2$

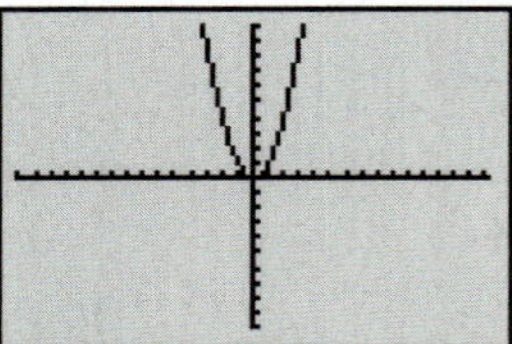

Figure 5 Verify the work

The graph of a quadratic function is called a **parabola.** The curve sketched in Fig. 4 is a parabola. Two more examples of parabolas are sketched in Figs. 6 and 7.

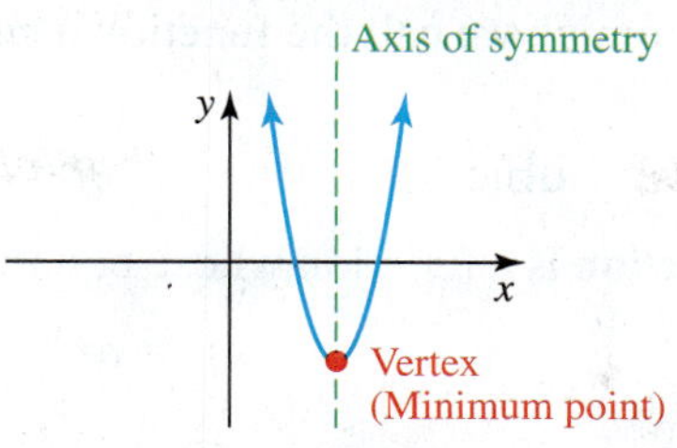

Figure 6 A parabola that opens upward

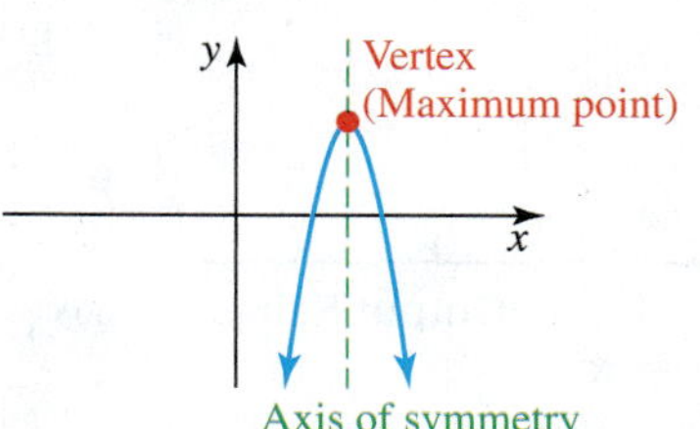

Figure 7 A parabola that opens downward

There is a difference between parabolas and lines that is worth noting. The lowest point of a parabola that *opens upward* (see Fig. 6) is called the **minimum point.** The highest point of a parabola that *opens downward* (see Fig. 7) is called the **maximum point.** The minimum point or maximum point of a parabola is called the **vertex** of the parabola. In contrast, lines do not have a lowest or highest point.

The vertical line that passes through a parabola's vertex is called the **axis of symmetry** (see Figs. 6 and 7). The part of the parabola that lies to the left of the axis of symmetry is the mirror reflection of the part that lies to the right.

Example 8 Using a Graph to Find Values of a Function

A graph of a quadratic function f is sketched in Fig. 8.

1. Find $f(4)$.
2. Find x when $f(x) = -3$.
3. Find x when $f(x) = 5$.
4. Find x when $f(x) = 6$.

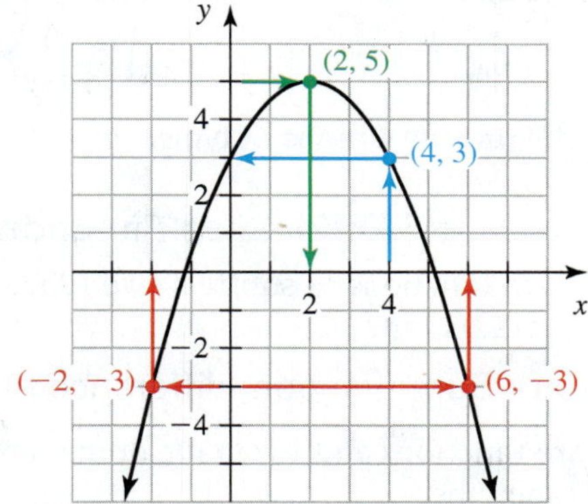

Figure 8 Graph of a quadratic function

Solution

1. The blue arrows in Fig. 8 show that the input $x = 4$ leads to the output $y = 3$. So, $f(4) = 3$.
2. The red arrows in Fig. 8 show that the output $y = -3$ originates from the two inputs $x = -2$ and $x = 6$. That is, $f(-2) = -3$ and $f(6) = -3$. So, the values of x are -2 and 6 when $f(x) = -3$.
3. The green arrows in Fig. 8 show that the output $y = 5$ originates from the single input $x = 2$. That is, $f(2) = 5$. So, the value of x is 2 when $f(x) = 5$. (There is a single input because the vertex $(2, 5)$ is the only point on the parabola that has a y-coordinate equal to 5.)
4. No point of the downward-opening parabola is above the vertex, which has a y-coordinate of 5. So, there is no point on the parabola with $f(x) = 6$.

By considering the shape of any parabola that opens upward or downward, we see that each (output) value of y in the range originates from either one or two (input) values of x in the domain.

Cubic Function

If a polynomial function can be put into the form $f(x) = P$, where P is a cubic (third-degree) polynomial, we call the function a *cubic function.*

Definition Cubic function

A **cubic function** is a function whose equation can be put into the form

$$f(x) = ax^3 + bx^2 + cx + d$$

where $a \neq 0$.

Here are some examples of cubic functions:

$$f(x) = -5x^3 + 9x^2 - 7x + 2 \qquad g(x) = 8x^3 - 5x + 1 \qquad h(x) = x^3 - 4x^2$$

Example 9 Graphing a Cubic Function

Sketch the graph of $f(x) = x^3$.

Solution

First, we list some input–output pairs of $f(x) = x^3$ in Table 3. Then, we plot the corresponding points and sketch a curve that contains the points (see Fig. 9).

Table 3 Input–Output Pairs of $f(x) = x^3$

x	$f(x)$
-3	$(-3)^3 = -27$
-2	$(-2)^3 = -8$
-1	$(-1)^3 = -1$
0	$0^3 = 0$
1	$1^3 = 1$
2	$2^3 = 8$
3	$3^3 = 27$

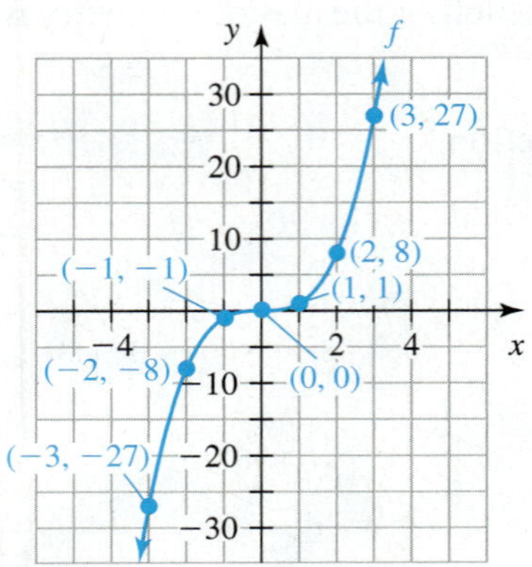

Figure 9 Graph of $f(x) = x^3$

Four graphs of typical cubic functions are shown in Fig. 10.

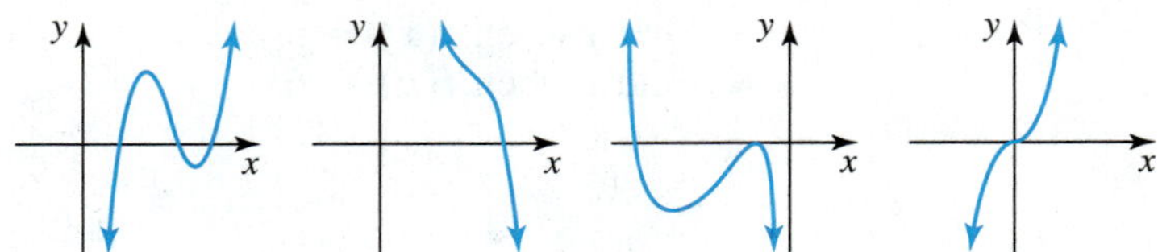

Figure 10 Graphs of typical cubic functions

Sum Function and Difference Function

We can add two functions or subtract two functions to form another function.

Definition Sum function, difference function

If f and g are functions and x is in the domain of both functions, then we can form the following functions:

- **Sum function $f + g$**, where $(f + g)(x) = f(x) + g(x)$
- **Difference function $f - g$**, where $(f - g)(x) = f(x) - g(x)$

For $f(x) = 5x$ and $g(x) = 3x$, we have $(f + g)(x) = f(x) + g(x) = 5x + 3x = 8x$. So, $(f + g)(x) = 8x$. And $(f - g)(x) = f(x) - g(x) = 5x - 3x = 2x$. So, $(f - g)(x) = 2x$.

Example 10 Finding a Sum Function and a Difference Function

Let $f(x) = 5x^2 - x + 3$ and $g(x) = -2x^2 + 9x - 7$.

1. Find an equation of the sum function $f + g$.
2. Find $(f + g)(2)$.
3. Find an equation of the difference function $f - g$.
4. Find $(f - g)(2)$.

Solution

1. $(f+g)(x) = f(x) + g(x)$ *Definition of sum function*

$= (5x^2 - x + 3) + (-2x^2 + 9x - 7)$ *Substitute $5x^2 - x + 3$ for $f(x)$ and $-2x^2 + 9x - 7$ for $g(x)$.*

$= 5x^2 - 2x^2 - x + 9x + 3 - 7$ *Rearrange terms.*

$= 3x^2 + 8x - 4$ *Combine like terms.*

2. $(f+g)(2) = 3(2)^2 + 8(2) - 4 = 24$

3. $(f-g)(x) = f(x) - g(x)$ *Definition of difference function*

$= (5x^2 - x + 3) - (-2x^2 + 9x - 7)$ *Substitute $5x^2 - x + 3$ for $f(x)$ and $-2x^2 + 9x - 7$ for $g(x)$.*

$= (5x^2 - x + 3) - 1(-2x^2 + 9x - 7)$ *$a - b = a - 1b$*

$= 5x^2 - x + 3 + 2x^2 - 9x + 7$ *Distributive law*

$= 5x^2 + 2x^2 - x - 9x + 3 + 7$ *Rearrange terms.*

$= 7x^2 - 10x + 10$ *Combine like terms.*

4. $(f-g)(2) = 7(2)^2 - 10(2) + 10 = 18$

Modeling with Sum Functions and Difference Functions

Suppose A and B represent quantities. Then $A + B$ represents the sum of the two quantities.

The difference $A - B$ tells us how much more there is of one quantity than the other quantity. For example, suppose $A = 8$ and $B = 5$. Then $A - B = 8 - 5 = 3$ tells us A is 3 more than B. Now suppose $A = 5$ and $B = 8$. Then $A - B = 5 - 8 = -3$ tells us A is 3 less than B.

The Meaning of the Sign of a Difference

If a difference $A - B$ is positive, then A is more than B. If a difference $A - B$ is negative, then A is less than B.

In Example 11, we use a sum function and a difference function to describe an authentic situation.

Example 11 Using a Sum Function and a Difference Function to Model a Situation

Table 4 College Enrollments

Year	Enrollment (millions) Women	Men
1995	7.9	6.3
2000	8.6	6.7
2005	10.0	7.5
2008	10.9	8.2
2010	12.1	9.1

Source: *National Center for Education Statistics*

Women's and men's enrollments at U.S. colleges and universities are shown in Table 4 for various years. The enrollments (in millions) $W(t)$ and $M(t)$ for women and men, respectively, are modeled by the system

$$W(t) = 0.014t^2 - 0.08t + 7.96$$
$$M(t) = 0.012t^2 - 0.12t + 6.65$$

where t is the number of years since 1990.

1. Find an equation of the sum function $W + M$.
2. Perform a unit analysis of the expression $W(t) + M(t)$.
3. Find $(W + M)(27)$. What does it mean in this situation?
4. Find an equation of the difference function $W - M$.
5. Find $(W - M)(27)$. What does it mean in this situation?

Solution

1. $(W + M)(t) = W(t) + M(t)$ *Definition of sum function*

$= (0.014t^2 - 0.08t + 7.96) + (0.012t^2 - 0.12t + 6.65)$ *Substitute $0.014t^2 - 0.08t + 7.96$ for $W(t)$ and $0.012t^2 - 0.12t + 6.65$ for $M(t)$.*

$= 0.026t^2 - 0.20t + 14.61$ *Combine like terms.*

2. For the expression $W(t) + M(t)$, we have

$$\underbrace{W(t)}_{\text{millions of female students}} + \underbrace{M(t)}_{\text{millions of male students}}$$

The units of the expression are millions of students.

3. $(W + M)(27) = 0.026(27)^2 - 0.20(27) + 14.61 \approx 28.16$

This means the total enrollment for women and men in $1990 + 27 = 2017$ will be about 28.2 million students, according to the model.

4.
$$\begin{aligned}(W - M)(t) &= W(t) - M(t) && \text{Definition of difference function} \\ &= (0.014t^2 - 0.08t + 7.96) - 1(0.012t^2 - 0.12t + 6.65) && \text{Substitute } 0.014t^2 - 0.08t + 7.96 \text{ for } W(t) \text{ and } 0.012t^2 - 0.12t + 6.65 \text{ for } M(t);\ a - b = a - 1b \\ &= 0.014t^2 - 0.08t + 7.96 - 0.012t^2 + 0.12t - 6.65 && \text{Distributive law} \\ &= 0.002t^2 + 0.04t + 1.31 && \text{Combine like terms.}\end{aligned}$$

5. $(W - M)(27) = 0.002(27)^2 + 0.04(27) + 1.31 \approx 3.85$

This means in 2017 women's enrollment will exceed men's enrollment by about 3.9 million students, according to the model.

Group Exploration

Using a difference function to solve a system

1. Let $f(x) = -3x + 7$ and $g(x) = 2x - 3$.
   a. Graph by hand the functions f and g on the same coordinate system.
   b. Find an equation of the difference function $f - g$.
   c. Find $(f - g)(1)$. Refer to your graphs in part (a) to explain why your result is positive.
   d. Find $(f - g)(3)$. Refer to your graphs in part (a) to explain why your result is negative.
   e. Find $(f - g)(2)$. Refer to your graphs in part (a) to explain why your result is 0.
   f. Refer to your graphs in part (a) to solve the system

$$\begin{aligned} y &= -3x + 7 \\ y &= 2x - 3 \end{aligned}$$

   Explain why your work in part (e) shows that the x-coordinate of the solution of the system is 2.

2. Let $f(x) = -x + 5$ and $g(x) = 2x - 4$.
   a. Find an equation of the difference function $f - g$.
   b. Find x when $(f - g)(x) = 0$.
   c. Without using graphing, substitution, or elimination, state the x-coordinate of the solution of the system that follows. Then use any method to find the y-coordinate of the solution.

$$\begin{aligned} y &= -x + 5 \\ y &= 2x - 4 \end{aligned}$$

Group Exploration

The degree of a sum of polynomials

In this exploration, you will explore the degree of the sum of two polynomials.

1. Find the degrees of the given expressions. Next, add the expressions. Then determine the degree of the sum.
   a. $2x + 6$ and $4x^2 + 5x + 4$
   b. $4x + 2$ and $3x^3 + 5x^2 + x + 6$
   c. $3x^2 + 6x + 1$ and $4x^3 + 2x^2 + 5x + 7$
2. What is the degree of the sum of a polynomial of degree 1 and a polynomial of degree 2?
3. What is the degree of the sum of a polynomial of degree 1 and a polynomial of degree 3?
4. What is the degree of the sum of a polynomial of degree 2 and a polynomial of degree 3?
5. Find the degrees of the given expressions. Next, add the expressions. Then determine the degree of the sum.
   a. $3x^2 + 2x + 6$ and $5x^2 + 7x + 4$
   b. $2x^2 + 3x + 5$ and $-2x^2 + 4x + 1$
6. If two polynomials have the same degree, what can you say about the degree of the sum of the two polynomials?

Homework 7.1

For extra help ▶ MyMathLab® 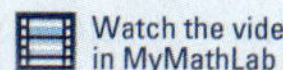 Watch the videos in MyMathLab  Download the MyDashboard App

Use words such as linear, quadratic, cubic, polynomial, degree, one variable, *and* two variables *to describe the expression.*

1. $3x^2 - 4x + 2$
2. $9x^2 + 5x^3 - 1$
3. $-7x^3 - 9x - 4$
4. $-4x^3 + 6x^2 - 9x + 8$
5. $3p^5q^2 - 5p^3q^3 + 7pq^4$
6. $8m^7n + 2m^4n^3 + 7m^2n^4$

Combine like terms when possible. Use a graphing calculator table to verify your work when possible.

7. $3t^2 + 5t^2$
8. $-4a^2 + 9a^2$
9. $-8a^4b^3 - 3a^4b^3$
10. $9x^2y^5 - 2x^2y^5$
11. $4x^2 + x^2$
12. $-8x^2 - x^2$
13. $7x^2 - 3x$
14. $6x^2 + 2x$
15. $5b^3 - 8b^3$
16. $-4m^5 + 2m^5$
17. $-x^6 + 7x^6$
18. $-x^4 + 6x^4$
19. $2t^3w^5 + 4t^5w^3$
20. $4a^7b^2 - 7a^2b^7$
21. $6x^2 - 3x - 2x^2 + 4x$
22. $4x^2 - 7x + 3x^2 - 9x$
23. $-5x^3 - 4x + 2x^2 - 7x^3 + 5 - x$
24. $-7x^2 + 6x - 4x^3 - 1 + 2x^3 - x^2$
25. $4a^4b^2 - 7ab^3 - 9a^4b^2 + 2ab^3$
26. $2m^3n + 4mn^2 - 8m^3n - 7mn^2$
27. $2x^4 - 4x^3y + 2x^2y^2 + x^3y - 2x^2y^2 + xy^3$
28. $3r^4 - r^3t - 4r^2t^2 + 6rt^2 + 4r^2t^2 - 4rt^3$

Perform the addition. Use a graphing calculator table to verify your work when possible.

29. $(3x^2 - 5x - 2) + (6x^2 + 2x - 7)$
30. $(5x^2 + 3x - 6) + (-7x^2 - 5x + 1)$
31. $(-2x^3 + 4x - 3) + (5x^3 - 6x^2 + 2)$
32. $(-5x^3 - 8x^2 + 4) + (-4x^3 + x - 9)$
33. $(8a^2 - 7ab + 2b^2) + (3a^2 + 4ab - 7b^2)$
34. $(6t^2 + 2tw - w^2) + (-4t^2 - 9tw - 3w^2)$
35. $(2m^4p + m^3p^2 - 7m^2p^3) + (m^3p^2 + 7m^2p^3 - 8mp^3)$
36. $(x^3y - 5x^2y^2 + 2xy^3) + (5x^2y^2 - 4xy^3 + 7y^4)$

Perform the subtraction. Use a graphing calculator table to verify your work when possible.

37. $(2x^2 + 4x - 7) - (9x^2 - 5x + 4)$
38. $(6x^2 + 3x - 1) - (4x^2 - 8x + 3)$
39. $(6x^3 - 3x^2 + 4) - (-7x^3 + x - 1)$
40. $(5x^3 - x + 6) - (-9x^3 + 4x^2 - 6)$
41. $(8m^2 + 3mp - 5p^2) - (-2m^2 - 7mp - 4p^2)$
42. $(b^2 - 6bc + 4c^2) - (5b^2 + 3bc - 2c^2)$
43. $(a^3b - 5a^2b^2 + ab^3) - (5a^2b^2 - 7ab^3 + b^3)$
44. $(6x^4y + x^3y^2 - 3x^2y^3) - (4x^3y^2 - 3x^2y^3 - 5xy^4)$

For the functions $f(x) = -2x^2 - 5x + 3$, $g(x) = 3x^2 - 8x - 1$, *and* $h(x) = 2x^3 - 4x$, *find the following.*

45. $f(3)$
46. $f(-4)$
47. $g(-4)$
48. $g(2)$
49. $f(0)$
50. $g(0)$
51. $h(3)$
52. $h(2)$
53. $h(-2)$
54. $h(-1)$

For Exercises 55–60, refer to Fig. 11.

55. Estimate $f(-1)$.
56. Estimate $f(1)$.
57. Estimate a when $f(a) = 3$.
58. Estimate a when $f(a) = 0$.
59. Estimate a when $f(a) = -1$.
60. Estimate a when $f(a) = -2$.

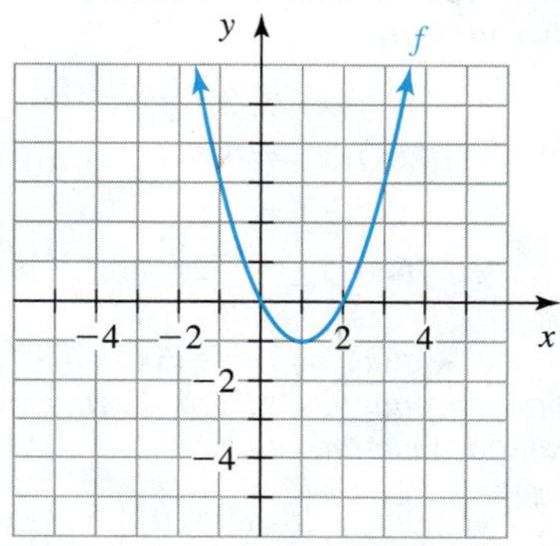

Figure 11 Exercises 55–60

For Exercises 61–66, refer to Table 5, which lists values of a quadratic function f.

61. Find $f(4)$.
62. Find $f(6)$.
63. Find x when $f(x) = 19$.
64. Find x when $f(x) = 3$.
65. Find x when $f(x) = 1$.
66. Find x when $f(x) = 0$.

Table 5 Some Values of a Quadratic Function f (Exercises 61–67)

x	$f(x)$
0	19
1	9
2	3
3	1
4	3
5	9
6	19

67. Consider the quadratic function f described in Table 5.
    a. Find x when $f(x) = 9$.
    b. Explain why f does not have an inverse function.

68. The values of a quadratic function f are listed in Table 6. Estimate the x-intercept(s) and the y-intercept(s).

Table 6 Values of a Function f (Exercise 68)

x	$f(x)$
−3	5.35
−2	0.17
−1	−3.61
0	−5.99
1	−6.97
2	−6.55
3	−4.73
4	−1.51
5	3.11
6	9.13

Graph the function by hand. To begin, substitute the values −2, −1, 0, 1, and 2 for x. Make other substitutions as necessary. Use a graphing calculator to verify your work.

69. $f(x) = 2x^2$

70. $f(x) = 3x^2$

71. $f(x) = -2x^2$

72. $f(x) = -3x^2$

73. $f(x) = 3x^3$

74. $f(x) = 2x^3$

75. $f(x) = -3x^3$

76. $f(x) = -2x^3$

For $f(x) = 4x^2 - 2x + 8$, $g(x) = 7x^2 + 5x - 1$, *and* $h(x) = -3x^2 - 4x - 9$, *find an equation of the given function; then evaluate the function at the indicated value.*

77. $f + g, (f + g)(3)$

78. $f + h, (f + h)(3)$

79. $f - h, (f - h)(4)$

80. $g - h, (g - h)(4)$

For $f(x) = 2x^3 - 4x + 1$, $g(x) = -3x^2 + 5x - 3$, *and* $h(x) = x^3 - 3x^2 + 2x$, *find an equation of the given function; then evaluate the function at the indicated value.*

81. $f + g, (f + g)(2)$

82. $f + h, (f + h)(2)$

83. $f - h, (f - h)(-1)$

84. $g - h, (g - h)(-1)$

85. In Exercise 5 of Homework 6.4, the annual U.S. consumption (in gallons per person) $M(t)$ and $S(t)$ of milk and soft drinks, respectively, is modeled by the system

$$M(t) = -0.28t + 36.67$$
$$S(t) = 0.82t + 8.96$$

where t is the number of years since 1950 (see Table 7).

Table 7 Annual U.S. Consumption of Milk and Soft Drinks

Year	Average Annual Consumption of Milk (gallons per person)	Year	Average Annual Consumption of Soft Drinks (gallons per person)
1970	31.3	1950	10.8
1980	27.5	1960	13.4
1985	26.7	1970	24.3
1990	25.7	1980	35.1
1995	23.9	1990	46.2
2000	22.5	2000	49.3
2005	21.0	2005	51.5

Source: *USDA/Economic Research Service*

a. Find an equation of the sum function $M + S$.

b. Perform a unit analysis of the expression $M(t) + S(t)$.

c. Find $(M + S)(53)$. What does it mean in this situation?

d. Find an equation of the difference function $M - S$.

e. Find $(M - S)(53)$. What does it mean in this situation?

86. In Exercise 38 of Homework 6.1, annual U.S. consumption (in pounds per person) $C(t)$ and $R(t)$ of chicken and red meat, respectively, is modeled by the system

$$C(t) = 1.13t + 39.29$$
$$R(t) = -0.94t + 144.88$$

where t is the number of years since 1970 (see Table 8).

Table 8 Annual U.S. Consumption of Chicken and Red Meat

	Annual Consumption (pounds per person)	
Year	Chicken	Red Meat
1970	40.3	145.8
1980	48.0	136.8
1990	61.5	120.0
2000	78.0	120.7
2010	83.7	108.7
2011	84.2	104.3

Source: *U.S. Department of Agriculture*

a. Find an equation of the sum function $C + R$.

b. Perform a unit analysis of the expression $C(t) + R(t)$.

c. Find $(C + R)(48)$. What does it mean in this situation?

d. Find an equation of the difference function $C - R$.

e. Find $(C - R)(48)$. What does it mean in this situation?

87. If you are driving and spot an object in the road, the distance it will take you to stop is equal to the sum of the following:

- The **reaction distance** is the distance you will continue to travel before you hit the brakes.
- The **braking distance** is the distance you will travel as you are braking.

The reaction and braking distances are shown in Table 9 for various driving speeds.

a. Let $R(s)$ be the reaction distance (in feet) when driving at s miles per hour. Find an equation of R.

b. The braking distance (in feet) $B(s)$ can be modeled by the function $B(s) = 0.063s^2$, where s is the driving speed (in mph) just before braking. Find an equation of the sum function $R + B$.

Table 9 Reaction and Braking Distances

Driving Speed (miles per hour)	Reaction Distance (feet)	Braking Distance (feet)
20	44	25
30	66	57
40	88	101
50	110	158
60	132	227
70	154	310
80	176	404

Source: *National Highway Traffic Safety Administration*

c. Perform a unit analysis of the expression $R(s) + B(s)$.

d. Find $(R + B)(26)$. What does it mean in this situation?

e. Low-beam headlights can illuminate objects up to 160 feet away. Suppose a person is driving 38 mph at night and then slams on the brakes when she sees a tree lying across the road. Will she hit the tree? Explain.

88. The United States' market share $U(t)$ of world manufacturing output can be modeled by the function $U(t) = -0.47t + 24.68$, where t is the number of years since 2000 (see Table 10). China's market share $C(t)$ can be modeled by the function $C(t) = 0.13t^2 - 0.54t + 11.24$, where t is the number of years since 2000.

Table 10 United States' and China's Market Shares of Manufacturing Output

	Market Share (percent)	
Year	United States	China
2004	22.8	11.1
2005	22.5	11.9
2006	21.7	12.8
2007	21.4	13.9
2008	20.8	15.0
2009	20.6	17.2
2010	20.0	18.9

Source: *United Nations*

a. Find an equation of the sum function $U + C$.

b. Perform a unit analysis of the expression $U(t) + C(t)$.

c. Find $(U + C)(17)$. What does it mean in this situation?

d. Find an equation of the difference function $U - C$.

e. Find $(U - C)(17)$. What does it mean in this situation?

Concepts

89. A student tries to find the difference of the polynomials $6x^2 + 8x + 5$ and $2x^2 + 4x + 3$:

$$\begin{aligned}(6x^2 + 8x + 5) - (2x^2 + 4x + 3) \\ = 6x^2 + 8x + 5 - 2x^2 + 4x + 3 \\ = 4x^2 + 12x + 8\end{aligned}$$

Describe any errors. Then find the difference correctly.

90. Let f and g be functions, both with independent variable x. A student says that $(f + g)(x) = f(x) + g(x)$ as a result of the distributive law. Explain why the student is incorrect. Then create two functions f and g so you can show the meaning of $(f + g)(x) = f(x) + g(x)$.

91. Let $f(x) = 3x + 7$ and $g(x) = 5x + 2$.

a. Find equations of the difference function $f - g$ and the difference function $g - f$.

b. Find $(f - g)(2)$ and $(g - f)(2)$. Compare the results.

c. Find $(f - g)(4)$ and $(g - f)(4)$. Compare the results.

d. Find $(f - g)(7)$ and $(g - f)(7)$. Compare the results.

e. Summarize your findings from parts (b)–(d). Explain why this makes sense.

92. a. Is it possible for the sum of two quadratic polynomials to be the given type of polynomial? If yes, give an example. If no, explain.

i. cubic

ii. quadratic

iii. linear

iv. constant

b. Summarize the possible results for the sum of two quadratic polynomials.

93. If $A - B = -10$, explain why this means A is 10 less than B.

94. Use the distributive law to explain why $3x^4 + 2x^4 = 5x^4$.

95. Describe how to add two polynomials. Describe how to subtract two polynomials. (See page 9 for guidelines on writing a good response.)

96. Explain how it is sometimes useful to use a sum function or difference function to model an authentic situation. (See page 9 for guidelines on writing a good response.)

Related Review

For Exercises 97–100, match the given type of function to the appropriate graph in Fig. 12.

97. $f(x) = mx + b, m > 0$ and $b < 0$

98. $f(x) = mx + b, m < 0$ and $b > 0$

99. $f(x) = ax^2 + bx + c$

100. $f(x) = ax^3 + bx^2 + cx + d$

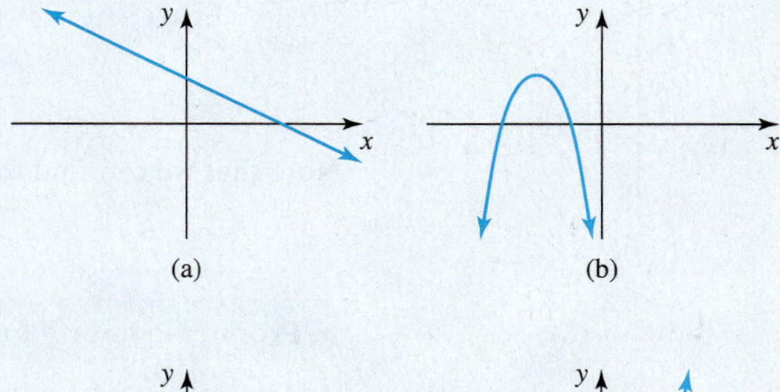

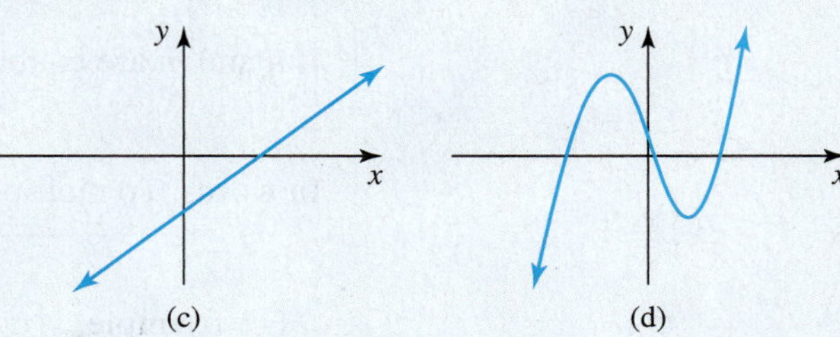

Figure 12 Exercises 97–100

For Exercises 101–104, let x be a number. Translate the English phrase to a mathematical expression. Then simplify the expression.

101. The number squared, minus 3 times the number, plus 5 times the number squared

102. The number squared plus twice the number squared minus 3

103. Four times the number squared, minus the sum of the number squared and the number

104. Six times the number squared, plus the difference of the number squared and 5

Expressions, Equations, Functions, and Graphs

Perform the indicated instruction. Then use words such as linear, quadratic, cubic, polynomial, degree, function, one variable, *and* two variables *to describe the expression, equation, or system.*

105. Solve the following:

$$3x - 2y = -4$$
$$4x + 5y = 33$$

106. Find $f(-9)$ where $f(x) = -5x + 2$.

107. Graph $3x - 2y = -4$ by hand.

108. For $f(x) = -5x + 2$, find x when $f(x) = -9$.

7.2 Multiplying Polynomial Expressions and Functions

Objectives

- » Know the meaning of *binomial* and *trinomial.*
- » Multiply polynomials.
- » Find *product functions.*
- » Use product functions to describe authentic situations.

In Section 7.1, we added and subtracted polynomials and worked with sum functions and difference functions. In this section, we will multiply polynomials and work with *product functions*.

Monomials, Binomials, and Trinomials

We refer to a polynomial as a **monomial,** a **binomial,** or a **trinomial,** depending on whether it has one, two, or three nonzero terms, respectively:

Name	Examples	Meaning
monomial	$4x^7$, x^3, $-9x^2y^3$, -3	one nonzero term
binomial	$4x^2 - 9$, $5x^3 + x^2$, $-2y + 5$	two nonzero terms
trinomial	$5x^3 - 3x + 6$, $3x^2 + 7x + 2$	three nonzero terms

Product Property for Exponents

Consider the product $x^2 \cdot x^3$. We can write this product as a single power:

$$x^2 \cdot x^3 = (x \cdot x)(x \cdot x \cdot x) \quad \text{Write factors without exponents.}$$
$$= x \cdot x \cdot x \cdot x \cdot x \quad \text{Remove parentheses.}$$
$$= x^5 \quad \text{Simplify.}$$

Note that we can find the same product by adding the exponents:

$$x^2 \cdot x^3 = x^{2+3} = x^5$$

Product Property for Exponents

If n and m are counting numbers, then

$$x^m x^n = x^{m+n}$$

In words: To multiply two powers of x, keep the base and add the exponents.

For example, $x^4x^8 = x^{4+8} = x^{12}$. Also, $x^5x = x^5x^1 = x^{5+1} = x^6$.

Multiplication of Monomials

We can use the product property to help us multiply monomials.

Example 1 Finding Products of Monomials

Find the product.

1. $2x(7x^3)$

2. $3x^4(-5x^2)$

Solution

1. We rearrange factors so that the coefficients are adjacent and the powers of x are adjacent:

$$\begin{aligned} 2x(7x^3) &= (2 \cdot 7)(x^1x^3) && \text{Rearrange factors; } x = x^1 \\ &= 14x^4 && \text{Add exponents: } x^mx^n = x^{m+n} \end{aligned}$$

2.
$$\begin{aligned} 3x^4(-5x^2) &= 3(-5)(x^4x^2) && \text{Rearrange factors.} \\ &= -15x^6 && \text{Add exponents: } x^mx^n = x^{m+n} \end{aligned}$$

WARNING The expression $3x^4(-5x^2)$ is *not* the same as the expression $3x^4 - 5x^2$. The expression $3x^4(-5x^2)$ is a product, whereas $3x^4 - 5x^2$ is a difference. We can tell $3x^4(-5x^2)$ is a product because there is no operation symbol between the expressions $3x^4$ and $(-5x^2)$.

Example 2 Finding the Product of Two Monomials

Find the product $3a^2b(2a^3b^2)$.

Solution

We rearrange factors so the coefficients are adjacent, the powers of a are adjacent, and the powers of b are adjacent:

$$\begin{aligned} 3a^2b(2a^3b^2) &= (3 \cdot 2)(a^2a^3)(b^1b^2) && \text{Rearrange factors.} \\ &= 6a^5b^3 && \text{Add exponents: } x^mx^n = x^{m+n} \end{aligned}$$

Product of a Monomial and a Polynomial

We can use the distributive law to help us find the product of a monomial and a polynomial.

Example 3 Finding the Product of a Monomial and a Polynomial

Find the product.

1. $2x(3x + 5)$

2. $-4t(8t^2 + 3t)$

3. $7p^2t(3p^2t - 2pt^2 + 5t^3)$

Solution

1.
$$\begin{aligned} 2x(3x + 5) &= 2x \cdot 3x + 2x \cdot 5 && \text{Distributive law} \\ &= 6x^2 + 10x && x \cdot x = x^2 \end{aligned}$$

2.
$$\begin{aligned} -4t(8t^2 + 3t) &= -4t \cdot 8t^2 - 4t \cdot 3t && \text{Distributive law} \\ &= -32t^3 - 12t^2 && \text{Add exponents: } x^mx^n = x^{m+n} \end{aligned}$$

3.
$$\begin{aligned} &7p^2t(3p^2t - 2pt^2 + 5t^3) \\ &= 7p^2t \cdot 3p^2t - 7p^2t \cdot 2pt^2 + 7p^2t \cdot 5t^3 && \text{Distributive law} \\ &= 21p^4t^2 - 14p^3t^3 + 35p^2t^4 && \text{Add exponents: } b^mb^n = b^{m+n} \end{aligned}$$

Multiplication of Two Polynomials

We can also use the distributive law to help us find the product of two binomials. For instance, we can find the product $(a + b)(c + d)$ by using the distributive law three times:

$$\begin{aligned} (a + b)(c + d) &= a(c + d) + b(c + d) && \text{Distribute } (c + d). \\ &= ac + ad + bc + bd && \text{Distribute } a \text{ and distribute } b. \end{aligned}$$

By examining the expression $ac + ad + bc + bd$, we can find the product $(a + b)(c + d)$ more directly by adding the four products formed by multiplying each term in the first sum by each term in the second sum:

$$(a + b)(c + d) = ac + ad + bc + bd$$

After using this technique, we combine like terms if possible.

Multiplying Two Polynomials

To multiply two polynomials, multiply each term in the first polynomial by each term in the second polynomial. Then combine like terms if possible.

Example 4 Finding Products of Binomials

Find the product.

1. $(x + 3)(x + 5)$
2. $(3x + 4y)(2x - 5y)$
3. $(-2.8x + 3.5)(4.1x + 9.7)$
4. $(x - 4)(x^2 - 3)$

Solution

1.
$$\begin{aligned}(x + 3)(x + 5) &= x \cdot x + x \cdot 5 + 3 \cdot x + 3 \cdot 5 && \text{Multiply pairs of terms.}\\ &= x^2 + 5x + 3x + 15 && x \cdot x = x^2\\ &= x^2 + 8x + 15 && \text{Combine like terms.}\end{aligned}$$

2.
$$\begin{aligned}(3x + 4y)(2x - 5y) &= 3x \cdot 2x - 3x \cdot 5y + 4y \cdot 2x - 4y \cdot 5y && \text{Multiply pairs of terms.}\\ &= 6x^2 - 15xy + 8xy - 20y^2 && x \cdot x = x^2\\ &= 6x^2 - 7xy - 20y^2 && \text{Combine like terms.}\end{aligned}$$

3.
$$\begin{aligned}&(-2.8x + 3.5)(4.1x + 9.7) && \text{Multiply pairs of terms.}\\ &= -2.8x(4.1x) - 2.8x(9.7) + 3.5(4.1x) + 3.5(9.7)\\ &= -11.48x^2 - 27.16x + 14.35x + 33.95 && x \cdot x = x^2\\ &= -11.48x^2 - 12.81x + 33.95 && \text{Combine like terms.}\end{aligned}$$

4.
$$\begin{aligned}(x - 4)(x^2 - 3) &= x \cdot x^2 - x \cdot 3 - 4 \cdot x^2 + 4 \cdot 3 && \text{Multiply pairs of terms.}\\ &= x^3 - 3x - 4x^2 + 12 && \text{Add exponents: } x^m x^n = x^{m+n}\\ &= x^3 - 4x^2 - 3x + 12 && \text{Rearrange terms.}\end{aligned}$$

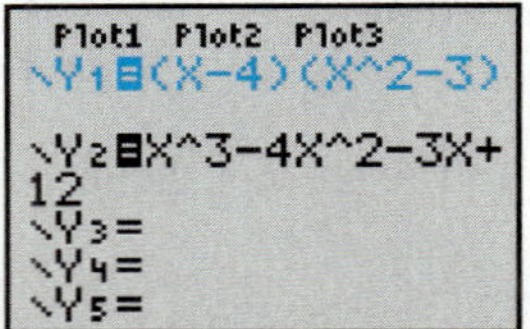

X	Y1	Y2
0	12	12
1	6	6
2	-2	-2
3	-6	-6
4	0	0
5	22	22
6	66	66

X=0

Figure 13 Verify the work

We use a graphing calculator table to verify our work (see Fig. 13).

Example 5 Finding Products

Find the product.

1. $(2a^2 - 5b^2)(4a^2 - 3b^2)$
2. $4x(x^2 + 2)(x - 3)$

Solution

1.
$$\begin{aligned}&(2a^2 - 5b^2)(4a^2 - 3b^2)\\ &= 2a^2 \cdot 4a^2 - 2a^2 \cdot 3b^2 - 5b^2 \cdot 4a^2 + 5b^2 \cdot 3b^2 && \text{Multiply pairs of terms.}\\ &= 8a^4 - 6a^2b^2 - 20a^2b^2 + 15b^4 && \text{Add exponents: } b^m b^n = b^{m+n}\\ &= 8a^4 - 26a^2b^2 + 15b^4 && \text{Combine like terms.}\end{aligned}$$

2. We begin by multiplying $4x$ by both terms in the binomial $x^2 + 2$:

$$\begin{aligned} 4x(x^2+2)(x-3) &= (4x\cdot x^2 + 4x\cdot 2)(x-3) && \textit{Distributive law} \\ &= (4x^3 + 8x)(x-3) && \textit{Add exponents: } b^m b^n = b^{m+n} \\ &= 4x^3\cdot x - 4x^3\cdot 3 + 8x\cdot x - 8x\cdot 3 && \textit{Multiply pairs of terms.} \\ &= 4x^4 - 12x^3 + 8x^2 - 24x && \textit{Add exponents: } b^m b^n = b^{m+n} \end{aligned}$$

We use a graphing calculator table to verify our work (see Fig. 14).

X	Y1	Y2
0	0	0
1	-24	-24
2	-48	-48
3	0	0
4	288	288
5	1080	1080
6	2736	2736

X=0

Figure 14 Verify the work

We can find the product of two polynomials of any degree in a similar fashion. The key is to multiply each term in the first polynomial by each term in the second polynomial. Then combine like terms.

Example 6 Finding the Products of Polynomials

Find the product.

1. $(3a - 2)(a^2 + 4a + 5)$
2. $(2x + y)(5x^2 - 3xy + 4y^2)$

Solution

1. To begin, we multiply each term in the first polynomial by each term in the second polynomial:

$$\begin{aligned} &(3a-2)(a^2+4a+5) && \textit{Multiply pairs of terms.} \\ &= 3a\cdot a^2 + 3a\cdot 4a + 3a\cdot 5 - 2\cdot a^2 - 2\cdot 4a - 2\cdot 5 \\ &= 3a^3 + 12a^2 + 15a - 2a^2 - 8a - 10 && \textit{Add exponents: } x^m x^n = x^{m+n} \\ &= 3a^3 + 12a^2 - 2a^2 + 15a - 8a - 10 && \textit{Rearrange terms.} \\ &= 3a^3 + 10a^2 + 7a - 10 && \textit{Combine like terms.} \end{aligned}$$

2. $(2x + y)(5x^2 - 3xy + 4y^2)$

$$\begin{aligned} &= 2x\cdot 5x^2 - 2x\cdot 3xy + 2x\cdot 4y^2 + y\cdot 5x^2 - y\cdot 3xy + y\cdot 4y^2 && \textit{Multiply pairs of terms.} \\ &= 10x^3 - 6x^2y + 8xy^2 + 5x^2y - 3xy^2 + 4y^3 && \textit{Add exponents: } b^m b^n = b^{m+n} \\ &= 10x^3 - 6x^2y + 5x^2y + 8xy^2 - 3xy^2 + 4y^3 && \textit{Rearrange terms.} \\ &= 10x^3 - x^2y + 5xy^2 + 4y^3 && \textit{Combine like terms.} \end{aligned}$$

Example 7 Finding the Products of Two Polynomials

Find the product $(x^2 - 3x + 2)(x^2 + x - 5)$.

Solution

$$\begin{aligned} &(x^2 - 3x + 2)(x^2 + x - 5) \\ &= x^2\cdot x^2 + x^2\cdot x - x^2\cdot 5 - 3x\cdot x^2 - 3x\cdot x + 3x\cdot 5 + 2\cdot x^2 + 2\cdot x - 2\cdot 5 \\ &= x^4 + x^3 - 5x^2 - 3x^3 - 3x^2 + 15x + 2x^2 + 2x - 10 \\ &= x^4 + x^3 - 3x^3 - 5x^2 - 3x^2 + 2x^2 + 15x + 2x - 10 \\ &= x^4 - 2x^3 - 6x^2 + 17x - 10 \end{aligned}$$

We use a graphing calculator table to verify our work (see Fig. 15).

X	Y1	Y2
0	-10	-10
1	0	0
2	0	0
3	14	14
4	90	90
5	300	300
6	740	740

X=0

Figure 15 Verify the work

Product Function

In Section 7.1, we worked with sum functions and difference functions. Now we will work with *product functions*. For example, if $f(x) = 2x$ and $g(x) = x + 3$, then it follows that $f(x) \cdot g(x) = 2x(x + 3) = 2x^2 + 6x$. The function $f(x) \cdot g(x) = 2x^2 + 6x$ is called the product function of f and g.

Definition Product function

If f and g are functions and x is in the domain of both functions, then we can form the **product function $f \cdot g$:**

$$(f \cdot g)(x) = f(x) \cdot g(x)$$

Example 8 Finding the Product Function

Let $f(x) = 3x + 7$ and $g(x) = 5x - 2$.

1. Find an equation of the product function $f \cdot g$.
2. Find $(f \cdot g)(2)$.

Solution

1.
$$\begin{aligned}(f \cdot g)(x) &= f(x) \cdot g(x) && \textit{Definition of product function}\\ &= (3x + 7)(5x - 2) && \textit{Substitute } 3x + 7 \textit{ for } f(x) \textit{ and } 5x - 2 \textit{ for } g(x).\\ &= 15x^2 - 6x + 35x - 14 && \textit{Multiply pairs of terms.}\\ &= 15x^2 + 29x - 14 && \textit{Combine like terms.}\end{aligned}$$

2. $(f \cdot g)(2) = 15(2)^2 + 29(2) - 14 = 104$

Sometimes we can find a meaningful model by finding the product of two models.

Example 9 Using a Product Function to Model a Situation

The annual cost of state corrections (prisons and related costs) per person in the United States can be modeled by the function

$$C(t) = 3.3t + 89$$

where $C(t)$ is the annual cost (in dollars per person) at t years since 1990 (see Table 11). The U.S. population can be modeled by the function

$$P(t) = 2.8t + 254$$

where $P(t)$ is the population (in millions) at t years since 1990.

Table 11 Costs of State Corrections; U.S. Population

Year	Cost (dollars per person)	U.S. Population (millions)
1998	113	275.9
2000	125	282.2
2002	135	287.8
2004	134	293.0
2006	135	298.6
2008	154	304.4

Source: *U.S. Census Bureau*

1. Check that the models fit the data well.
2. Find an equation of the product function $C \cdot P$.
3. Perform a unit analysis of the expression $C(t) \cdot P(t)$.
4. Find $(C \cdot P)(28)$. What does it mean in this situation?
5. Use a graphing calculator graph to determine whether the function $C \cdot P$ is increasing, decreasing, or neither for values of t between 0 and 30. What does your result mean in this situation?

Solution

1. We check the fit of the cost model in Fig. 16 and the fit of the population model in Fig. 17. The models appear to fit the data fairly well.

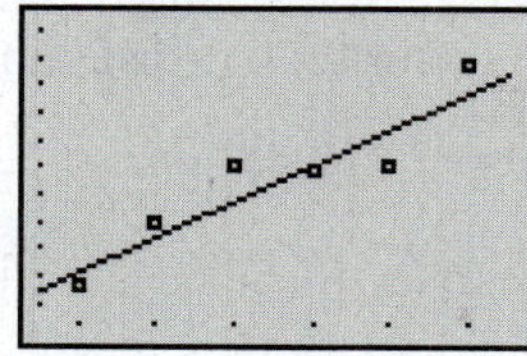

Figure 16 Cost scattergram and model

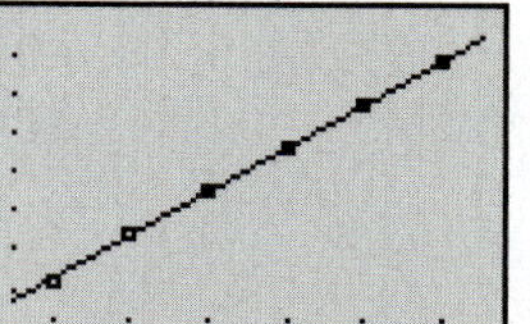

Figure 17 Population scattergram and model

2. $(C \cdot P)(t) = C(t) \cdot P(t)$ — *Definition of product function*

$= (3.3t + 89)(2.8t + 254)$ — *Substitute $3.3t + 89$ for $C(t)$ and $2.8t + 254$ for $P(t)$.*

$= 9.24t^2 + 838.2t + 249.2t + 22{,}606$ — *Multiply pairs of terms.*

$= 9.24t^2 + 1087.4t + 22{,}606$ — *Combine like terms.*

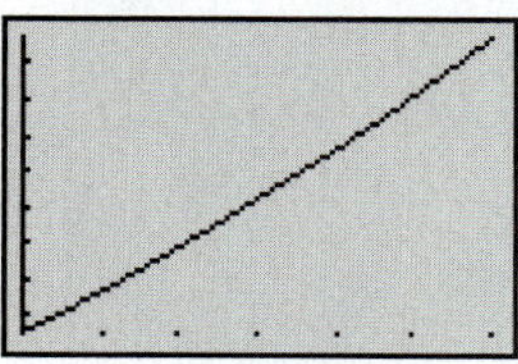

3. For the expression $C(t) \cdot P(t)$, we have

$$\underbrace{C(t)}_{\frac{\text{dollars}}{\text{person}}} \cdot \underbrace{P(t)}_{\text{millions of people}}$$

The units of the expression are millions of dollars.

4. $(C \cdot P)(28) = 9.24(28)^2 + 1087.4(28) + 22{,}606 \approx 60{,}297$. This means the total cost of state corrections will be about \$60,297 million (\$60.297 billion) in 2018, according to the model.

5. To graph the model, we enter the function, then press WINDOW, and set Xmin to be 0 and Xmax to be 30. Then we use ZoomFit to draw the graph (see Fig. 18). For values of t between 0 and 30, the model is increasing. This means the total cost of state corrections has been increasing since 1990 and will continue to increase until 2020.

Figure 18 Graph of the function $C \cdot P$

Group Exploration

Looking ahead: Product of binomial conjugates

1. Find the products.
   a. $(x + 2)(x - 2)$ b. $(x + 3)(x - 3)$
   c. $(x + 4)(x - 4)$
2. What pattern do you notice in your work from Problem 1?
3. Use the pattern you described in Problem 2 to help you find each of the following products:
   a. $(x + 5)(x - 5)$ b. $(x + 9)(x - 9)$
   c. $(x + k)(x - k)$
4. Compare the results of finding the product $(x + k)(x - k)$ with the results of finding the product $(x - k)(x + k)$. Explain in terms of one of the laws of operations why the results are the same.

Tips for Success "Work Out" by Solving Problems

Although there are many things you can do to enhance your learning, there is no substitute for solving problems. Your mathematical ability will respond to solving problems in much the same way that your muscles respond to lifting weights. Muscles increase greatly in strength when you work out intensely, frequently, and consistently.

Just as with building muscles, to learn math, you must be an *active* participant. No amount of watching weight lifters lift, reading about weight-lifting techniques, or conditioning yourself psychologically can replace working out by lifting weights. The same is true of learning math: No amount of watching your instructor do problems, reading your text, or listening to a tutor can replace "working out" by solving problems.

Homework 7.2

For extra help ▶ MyMathLab®  Watch the videos in MyMathLab  Download the MyDashboard App

Find the product. Use a graphing calculator table to verify your work when possible.

1. x^4x^3 **2.** x^3x^5 **3.** w^8w **4.** pp^8

5. $9x^5(4x^7)$ **6.** $3x^8(5x^6)$

7. $-5x^4(-6x^3)$ **8.** $-6x^9(-2x^2)$

9. $x^6y^3(x^9y^4)$ **10.** $x^7y^5(x^6y^2)$

11. $3c^4d^6(8c^{10}d^2)$ **12.** $6a^8b^3(4ab^5)$

13. $4p^2t(-9p^3t^2)$ **14.** $2a^3b^2(-7ab^2)$

15. $\frac{4}{5}x^3\left(-\frac{7}{2}x^2\right)$ **16.** $-\frac{3}{8}x^5\left(-\frac{4}{5}x\right)$

17. $3w(w-2)$ **18.** $6y(y+1)$

19. $-4x(2x^2+3)$ **20.** $-5x(3x^2-7)$

21. $2mn^2(3m^2+5n)$ **22.** $4pq(3p+2q^2)$

23. $2x(3x^2-2x+7)$ **24.** $4x(x^2+3x-1)$

25. $-3t^2(2t^2+4t-2)$ **26.** $-2a^2(3a^2-a-4)$

27. $2xy^2(3x^2-4xy+5y^2)$ **28.** $3p^2q(4p^2-pq+2q^2)$

Find the product. Use a graphing calculator table to verify your work when possible.

29. $(x+2)(x+4)$

30. $(x+3)(x+7)$

31. $(x-2)(x+5)$

32. $(x+4)(x-6)$

33. $(a-3)(a-2)$

34. $(p-6)(p-5)$

35. $(x+6)(x-6)$

36. $(x-8)(x+8)$

37. $(x-5.3)(x-9.2)$

38. $(x-1.7)(x+4.3)$

39. $(5y-2)(3y+4)$

40. $(6a+3)(2a-4)$

41. $(2x+4)(2x+4)$

42. $(5x+2)(5x+2)$

43. $(3x-1)(3x-1)$

44. $(4x-3)(4x-3)$

45. $(3x-5y)(4x+y)$

46. $(2x+7y)(3x-2y)$

47. $(2a-8b)(3a-4b)$

48. $(3m-7n)(2m-4n)$

49. $(3x-4)(3x+4)$

50. $(2x+7)(2x-7)$

51. $(9x+4y)(9x-4y)$

52. $(6a-5b)(6a+5b)$

53. $(2.5x+9.1)(4.6x-7.7)$

54. $(4.8x-2.5)(3.1x-8.9)$

55. $(x+6)(x^2-3)$

56. $(x-4)(x^2+2)$

57. $(2t^2-5)(3t-2)$

58. $(3a^2-3)(4a-5)$

59. $(3a^2+5b^2)(2a^2-3b^2)$

60. $(4m^2-7n^2)(3m^2-2n^2)$

61. $3x^2(2x-5)(4x+1)$

62. $2x^2(3x+2)(2x-5)$

63. $5x(x^2+3)(x-4)$

64. $2x(x^2-7)(x+3)$

65. $(x+2)(x^2+3x+5)$

66. $(x+3)(x^2+2x+4)$

67. $(x+2)(x^2-2x+4)$

68. $(x-4)(x^2+4x+16)$

69. $(2b^2-3b+2)(b-4)$

70. $(3y^2-y+5)(y-2)$

71. $(a+b)(a^2-ab+b^2)$

72. $(a-b)(a^2+ab+b^2)$

73. $(4x-3y)(2x^2-xy+5y^2)$

74. $(3x-5y)(4x^2+2xy-y^2)$

75. $(2x^2+3)(3x^2-x+4)$

76. $(5x^2+2)(2x^2-2x-3)$

77. $(2x^2+4x-1)(3x^2-x+2)$

78. $(4x^2-x+3)(3x^2+2x+1)$

79. $(2x^2 + xy - 3y^2)(x^2 - 2xy + y^2)$

80. $(x^2 + 2xy - y^2)(3x^2 - xy - 2y^2)$

For $f(x) = 2x - 3$, $g(x) = 3x + 2$, $h(x) = 2x^2 - 4x + 3$, and $k(x) = 3x^2 + x - 5$, find an equation of the given product function; then evaluate the product function at the indicated value.

81. $f \cdot g, (f \cdot g)(3)$

82. $g \cdot h, (g \cdot h)(3)$

83. $f \cdot h, (f \cdot h)(2)$

84. $h \cdot k, (h \cdot k)(2)$

85. $f \cdot f, (f \cdot f)(4)$

86. $g \cdot g, (g \cdot g)(4)$

For $f(x) = 4x + 1$, $g(x) = 5x + 3$, $h(x) = 3x^2 - x - 2$, and $k(x) = 2x^2 - 4x + 3$, find an equation of the given product function; then evaluate the product function at the indicated value.

87. $f \cdot g, (f \cdot g)(-1)$

88. $g \cdot f, (g \cdot f)(-1)$

89. $f \cdot h, (f \cdot h)(-2)$

90. $g \cdot k, (g \cdot k)(-2)$

91. $h \cdot h, (h \cdot h)(1)$

92. $k \cdot k, (k \cdot k)(1)$

93. The average value per acre of U.S. farmland (in dollars per acre) $V(t)$ is modeled by the function $V(t) = 62t + 508$, where t is the number of years since 1990 (see Table 12). The amount of U.S. farmland (in millions of acres) $A(t)$ is modeled by the function $A(t) = -3.3t + 978$, where t is the number of years since 1990.

Table 12 Values and Acres of U.S. Farmland

Year	Average Value (dollars per acre)	Amount of Farmland (millions of acres)
1993	740	968
1995	840	961
1997	930	956
1999	1030	948
2001	1150	941
2003	1270	939
2005	1407	928
2007	1634	921
2008	1646	920

Sources: *National Agriculture Statistics Service; Agricultural Statistics Board*

a. Check that the models fit the data well.
b. Find an equation of the product function $V \cdot A$.
c. Perform a unit analysis of the expression $V(t) \cdot A(t)$.
d. Find $(V \cdot A)(27)$. What does it mean in this situation?
e. Use a graphing calculator graph to determine whether the function $V \cdot A$ is increasing, decreasing, or neither for values of t between 0 and 27. What does your result mean in this situation? Explain how that is possible, given that the amount of farmland has been decreasing.

94. World Internet population (in billions) is modeled by $W(t) = 0.084t + 0.32$, where t is the number of years since 2000 (see Table 13). The number of web pages viewed per person per day is modeled by $N(t) = t + 31$, where t is the number of years since 2000.

Table 13 World Internet Populations and Numbers of Web Pages Viewed per Person per Day

Year	World Internet Population (in billions)	Pages Viewed per Person per Day
2006	0.817	37
2007	0.903	38
2008	0.988	39
2009	1.072	40
2010	1.152	41

Source: *J.P. Morgan*

a. Check that the models fit the data well.
b. Find an equation of the product function $W \cdot N$.
c. Perform a unit analysis of the expression $(W \cdot N)(t)$.
d. Find $(W \cdot N)(17)$. What does it mean in this situation?
e. Predict the total number of pages that will be viewed by everyone in the world *throughout* 2018.
f. Use a graphing calculator graph to determine whether the function $W \cdot N$ is increasing, decreasing, or neither for values of t between 0 and 20. What does your result mean in this situation?

95. The average monthly cell phone bill (in dollars per month) $B(t)$ is modeled by the function $B(t) = -0.197t^2 + 6.15t + 3$, where t is the number of years since 1990 (see Table 14). The number of cell phone subscribers (in millions) $N(t)$ is modeled by the function $N(t) = 20t - 92$, where t is the number of years since 1990.

Table 14 Average Monthly Cell Phone Bills and Numbers of Subscribers

Year	Average Bill (dollars per month)	Number of Subscribers (millions)
1998	39.43	69.2
2000	45.27	109.5
2002	48.40	140.8
2004	50.64	182.1
2006	50.56	233.0
2008	50.07	270.3
2010	47.21	302.9

Source: *Cellular Telecommunications & Internet Association*

a. Check that the models fit the data well.
b. Find an equation of the product function $B \cdot N$.
c. Perform a unit analysis of the expression $B(t) \cdot N(t)$.
d. Find $(B \cdot N)(24)$. What does it mean in this situation?
e. Use a graphing calculator graph to determine whether the function $B \cdot N$ is increasing, decreasing, or neither for values of t between 14 and 20. What does your result mean in this situation? Explain how this is possible, given that the average bill decreased from 2004 to 2010.

96. In Example 1 of Section 5.6, we modeled the average salaries of faculty members at public colleges and universities. A reasonable model is $S(t) = 1.8t + 22$, where $S(t)$ is the average salary (in thousands of dollars) at t years since 1980 (see Table 15). The number of such faculty members (in thousands) $N(t)$ is modeled by the function $N(t) = 0.6t^2 + 9t + 670$, where t is the number of years since 1980.

Table 15 Average Salaries and Numbers of Faculty Members

Year	Average Salary (thousands of dollars per faculty member)	Year	Number of Faculty Members (thousands)
1980	22.1	1980	686
1985	31.2	1985	715
1990	41.9	1991	826
1995	49.1	1995	932
2000	57.7	2001	1174
2005	66.9	2005	1290
2010	78.0	2009	1439

Sources: *American Association of University Professors; U.S. National Center for Education Statistics*

a. Check that the models fit the data well.
b. Find an equation of the product function $S \cdot N$.
c. Perform a unit analysis of the expression $S(t) \cdot N(t)$.
d. Find $(S \cdot N)(38)$. What does it mean in this situation?
e. Use a graphing calculator graph to determine whether the function $S \cdot N$ is increasing, decreasing, or neither for values of t between 0 and 40. What does your result mean in this situation?

Concepts

97. A student tries to find the product $6x(-4x)$:

$$6x(-4x) = 6x - 4x = 2x$$

Describe any errors. Then find the product correctly.

98. A student tries to find the product $(x + 3)(x + 5)$:

$$\begin{aligned}(x + 3)(x + 5) &= 2x + 5x + 3x + 8 \\ &= 10x + 8\end{aligned}$$

Describe any errors. Then find the product correctly.

99. a. Find the given product. Then decide whether the result is a linear, quadratic, or cubic polynomial.
i. $(2x + 3)(4x + 5)$
ii. $(3x - 7)(5x + 2)$
b. Create two linear polynomials and find their product. Is the result a linear, quadratic, or cubic polynomial?
c. In general, is the product of two linear polynomials a linear, quadratic, or cubic polynomial? Explain.

100. a. Find the given product. Then decide whether the result is a linear, quadratic, or cubic polynomial.
i. $(x + 2)(x^2 + 3x + 5)$
ii. $(2x - 3)(3x^2 + 4x - 1)$
b. Create a linear polynomial and a quadratic polynomial, and find their product. Is the result a linear, quadratic, or cubic polynomial?
c. In general, is the product of a linear polynomial and a quadratic polynomial a linear, quadratic, or cubic polynomial? Explain.

101. a. Find the product $(x + 4)(x + 7)$.
b. Find the product $(x + 7)(x + 4)$.
c. Explain in terms of the laws of operations why it makes sense that $(x + 4)(x + 7)$ and $(x + 7)(x + 4)$ are equivalent expressions.

102. Use the distributive law three times to help you show that $(x + 2)(x + 7)$ and $x^2 + 9x + 14$ are equivalent expressions.

103. Which expressions are equivalent?

$(2x - 5)(3x + 4)$ $\quad 6x^2 + 7x - 20$ $\quad 3x(2x - 2) - x - 20$
$(3x - 4)(2x + 5)$ $\quad 6x^2 - 7x - 20$ $\quad (3x + 4)(2x - 5)$

104. Which expressions are equivalent?

$(8x + 1)(x - 4)$ $\quad (4x - 2)(2x + 2)$ $\quad 8x^2 - 31x - 4$
$8x^2 + 31x - 4$ $\quad (x - 4)(8x + 1)$ $\quad 4(2x^2 - 1) - 31x$

Related Review

Perform the indicated operation. Use a graphing calculator table to verify your work.

105. $(3x - 5)(2x^2 - 4x + 2)$
106. $(2x^2 - x - 8)(4x - 1)$
107. $(3x - 5) - (2x^2 - 4x + 2)$
108. $(2x^2 - x - 8) - (4x - 1)$

Simplify the right-hand side of the equation to help you decide whether the equation is linear or quadratic. State whether the graph of the equation is a line or a parabola. Use a graphing calculator graph to verify your decision.

109. $y = 3x(x - 2)$
110. $y = -4x(2x - 1)$
111. $y = (x + 2) - (3x + 5)$
112. $y = (4x - 3) + (-9x + 7)$
113. $y = (2x + 1)(5x - 2)$
114. $y = (3x - 5)(6x - 4)$

Expressions, Equations, Functions, and Graphs

Perform the indicated instruction. Then use words such as linear, quadratic, cubic, power, polynomial, degree, function, one variable, *and* two variables *to describe the expression, equation, or system.*

115. Solve $2x - 5 = 7x + 5$.
116. Graph $y = 2x - 5$ by hand.
117. Find the product $(2x - 5)(7x + 5)$.
118. Solve the following:

$$\begin{aligned} y &= 2x - 5 \\ y &= 7x + 5 \end{aligned}$$

7.3 Powers of Polynomials; Product of Binomial Conjugates

Objectives

- » Know how to raise a product to a power.
- » Find the power of a monomial.
- » Simplify the square of a binomial.
- » Evaluate quadratic functions, and write them in standard form.
- » Find the product of two binomial conjugates.

Recall from Section 2.6 that we refer to x^n as "the nth power of x." For most of this section, we will discuss how to find powers of polynomials.

Raising a Product to a Power

Consider the power $(xy)^3$. We can write this power as a product of monomials:

$$\begin{aligned}(xy)^3 &= (xy)(xy)(xy) && \textit{Write power without exponents.}\\ &= (x \cdot x \cdot x)(y \cdot y \cdot y) && \textit{Rearrange factors.}\\ &= x^3y^3 && \textit{Simplify.}\end{aligned}$$

Note that we can get this same result by distributing the exponent 3 to both factors of the base xy:

$$(xy)^3 = x^3y^3$$

Raising a Product to a Power

If n is a counting number, then

$$(xy)^n = x^ny^n$$

In words: To raise a product to a power, raise each factor to the power.

For example, $(xy)^6 = x^6y^6$.

Power of a Monomial

We can use the property of raising a product to a power to help us find the power of a monomial.

Example 1 Finding Powers of Monomials

Perform the indicated operation.

1. $(xy)^7$ **2.** $(7p)^2$ **3.** $(-2xy)^4$

Solution

1. $(xy)^7 = x^7y^7$

2. $(7p)^2 = 7^2p^2 = 49p^2$

3. $(-2xy)^4 = (-2)^4x^4y^4 = 16x^4y^4$

WARNING

The expressions $5x^2$ and $(5x)^2$ are *not* equivalent expressions:

$$5x^2 = 5x \cdot x$$
$$(5x)^2 = (5x)(5x) = 25x \cdot x$$

For $5x^2$, the base is the variable x. For $(5x)^2$, the base is the product $5x$.

Here we show a typical error and the correct work performed to find the power $(5x)^2$:

$$(5x)^2 = 5x^2 \quad \textit{Incorrect}$$
$$(5x)^2 = 5^2x^2 = 25x^2 \quad \textit{Correct}$$

Since the base $5x$ is a product, we need to distribute the exponent 2 to *both* factors, 5 and x.

In general, when finding a power of the form $(AB)^n$, don't forget to distribute the exponent n to both of the factors A and B.

Square of a Binomial

Next, we will discuss how to square a sum. The square of the binomial $x + 4$ is $(x + 4)^2$:

$$
\begin{aligned}
(x+4)^2 &= (x+4)(x+4) && y^2 = yy \\
&= x^2 + 4x + 4x + 16 && \text{Multiply pairs of terms.} \\
&= x^2 + 8x + 16 && \text{Combine like terms.}
\end{aligned}
$$

We have simplified $(x + 4)^2$ by writing it as $x^2 + 8x + 16$. We **simplify the square of a binomial** by writing it as an expression that does not have parentheses.

Now we generalize and simplify $(A + B)^2$:

$$
\begin{aligned}
(A+B)^2 &= (A+B)(A+B) && y^2 = yy \\
&= A^2 + AB + BA + B^2 && \text{Multiply pairs of terms.} \\
&= A^2 + 2AB + B^2 && \text{Combine like terms.}
\end{aligned}
$$

So, $(A + B)^2 = A^2 + 2AB + B^2$. We can use similar steps to find the property $(A - B)^2 = A^2 - 2AB + B^2$.

Squaring a Binomial

$$
\begin{aligned}
(A+B)^2 &= A^2 + 2AB + B^2 && \text{Square of a sum} \\
(A-B)^2 &= A^2 - 2AB + B^2 && \text{Square of a difference}
\end{aligned}
$$

In words: The square of a binomial equals the first term squared, plus (or minus) twice the product of the two terms, plus the second term squared.

We can instead simplify $(x + 4)^2$ by substituting x for A and 4 for B in the formula for the square of a sum:

$$
\begin{aligned}
(A+B)^2 &= A^2 + 2 \cdot A \cdot B + B^2 \\
&\downarrow \\
(x+4)^2 &= x^2 + 2 \cdot x \cdot 4 + 4^2 && \text{Substitute.} \\
&= x^2 + 8x + 16 && \text{Simplify.}
\end{aligned}
$$

The result is the same as our result from writing $(x + 4)^2$ as $(x + 4)(x + 4)$ and then multiplying. So, there are two ways to simplify $(x + 4)^2$. Similarly, there are two ways to simplify the square of any binomial. If you experiment with both methods on several exercises in the homework, you will be able to make an informed choice of methods for solving future problems.

Example 2 Simplifying Squares of Binomials

Simplify.

1. $(x + 3)^2$

2. $(x - 7)^2$

Solution

1. We substitute x for A and 3 for B:

$$
\begin{aligned}
(A+B)^2 &= A^2 + 2 \cdot A \cdot B + B^2 \\
&\downarrow \\
(x+3)^2 &= x^2 + 2 \cdot x \cdot 3 + 3^2 && \text{Substitute.} \\
&= x^2 + 6x + 9 && \text{Simplify.}
\end{aligned}
$$

Another way to simplify $(x + 3)^2$ is to write it as $(x + 3)(x + 3)$ and then multiply each term in the first binomial by each term in the second binomial:

$$
\begin{aligned}
(x+3)^2 &= (x+3)(x+3) && b^2 = bb \\
&= x^2 + 3x + 3x + 9 && \text{Multiply pairs of terms.} \\
&= x^2 + 6x + 9 && \text{Combine like terms.}
\end{aligned}
$$

2. We substitute x for A and 7 for B:

$$\begin{aligned}(A - B)^2 &= A^2 - 2 \cdot A \cdot B + B^2 \\ (x - 7)^2 &= x^2 - 2 \cdot x \cdot 7 + 7^2 && \textit{Substitute.} \\ &= x^2 - 14x + 49 && \textit{Simplify.}\end{aligned}$$

Another way to simplify $(x - 7)^2$ is to write it as $(x - 7)(x - 7)$ and then multiply each term in the first binomial by each term in the second binomial:

$$\begin{aligned}(x - 7)^2 &= (x - 7)(x - 7) && b^2 = bb \\ &= x^2 - 7x - 7x + 49 && \textit{Multiply pairs of terms.} \\ &= x^2 - 14x + 49 && \textit{Combine like terms.}\end{aligned}$$

In Example 2, we found that $x^2 + 6x + 9$ and $x^2 - 14x + 49$ are *squares* of binomials. Both of these *trinomials* are called perfect-square trinomials. A **perfect-square trinomial** is a trinomial that is equivalent to the square of a binomial.

X	Y1	Y2
0	16	16
1	1	1
2	36	36
3	121	121
4	256	256
5	441	441
6	676	676

X=0

Figure 19 Verify the work

Example 3 Simplifying Squares of Binomials

Simplify.

1. $(5x - 4)^2$ **2.** $(3x + 5y)^2$ **3.** $(3r^2 + 2y^2)^2$

Solution

1. We substitute $5x$ for A and 4 for B:

$$\begin{aligned}(A - B)^2 &= A^2 - 2 \cdot A \cdot B + B^2 \\ (5x - 4)^2 &= (5x)^2 - 2 \cdot 5x \cdot 4 + 4^2 && \textit{Substitute.} \\ &= 25x^2 - 40x + 16 && \textit{Simplify.}\end{aligned}$$

We use a graphing calculator table to verify our work (see Fig. 19).

2. We substitute $3x$ for A and $5y$ for B:

$$\begin{aligned}(A + B)^2 &= A^2 + 2 \cdot A \cdot B + B^2 \\ (3x + 5y)^2 &= (3x)^2 + 2 \cdot 3x \cdot 5y + (5y)^2 && \textit{Substitute.} \\ &= 9x^2 + 30xy + 25y^2 && \textit{Simplify.}\end{aligned}$$

3. $(3r^2 + 2y^2)^2 = (3r^2)^2 + 2(3r^2)(2y^2) + (2y^2)^2$ — *$(A + B)^2 = A^2 + 2AB + B^2$; substitute $3r^2$ for A and $2y^2$ for B.*

$= 9r^4 + 12r^2y^2 + 4y^4$ — *Simplify.*

Here we compare squaring a product with squaring a sum:

$$\begin{aligned}(AB)^2 &= A^2B^2 && \textit{Square of a product} \\ (A + B)^2 &= A^2 + 2AB + B^2 && \textit{Square of a sum}\end{aligned}$$

WARNING It is a common error to omit the middle term in squaring a binomial. Here we show not only a typical error made in simplifying $(x + 6)^2$, but also the correct result:

$$\begin{aligned}(x + 6)^2 &= x^2 + 36 && \textit{Incorrect} \\ (x + 6)^2 &= x^2 + 12x + 36 && \textit{Correct}\end{aligned}$$

When simplifying $(A + B)^2$, don't omit the middle term, $2AB$, of $A^2 + 2AB + B^2$. Likewise, when simplifying $(A - B)^2$, don't omit the middle term, $-2AB$, of $A^2 - 2AB + B^2$.

Function Notation

So far in this section, we have been working with polynomial expressions. Now we will apply the skills we have learned to polynomial functions.

Example 4 Evaluating a Quadratic Function

For $f(x) = x^2 - 5x$, find the following.

1. $f(a - 3)$
2. $f(a + 2) - f(a)$

Solution

1. $f(a - 3) = (a - 3)^2 - 5(a - 3)$ *Substitute $a - 3$ for x.*

$= a^2 - 6a + 9 - 5a + 15$ *$(A - B)^2 = A^2 - 2AB + B^2$; distributive law*

$= a^2 - 11a + 24$ *Combine like terms.*

2. $f(a + 2) - f(a)$

$= \left[(a + 2)^2 - 5(a + 2)\right] - (a^2 - 5a)$ *Substitute $a + 2$ for x; substitute a for x.*

$= a^2 + 4a + 4 - 5a - 10 - a^2 + 5a$ *$(A + B)^2 = A^2 + 2AB + B^2$; distributive law; subtract.*

$= 4a - 6$ *Combine like terms.*

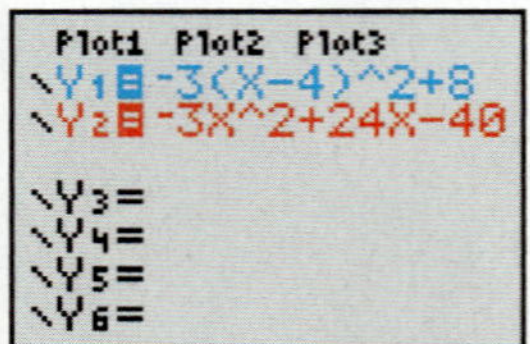

Recall from Section 7.1 that a quadratic function in the form $f(x) = ax^2 + bx + c$ is in standard form.

Example 5 Writing a Quadratic Function in Standard Form

Write $f(x) = -3(x - 4)^2 + 8$ in standard form.

Solution

We begin by simplifying $(x - 4)^2$, because we work with exponents before we multiply or add:

$$f(x) = -3(x - 4)^2 + 8 \quad \textit{Original equation}$$
$$= -3(x^2 - 8x + 16) + 8 \quad (A - B)^2 = A^2 - 2AB + B^2$$
$$= -3x^2 + 24x - 48 + 8 \quad \textit{Distributive law}$$
$$= -3x^2 + 24x - 40 \quad \textit{Combine like terms.}$$

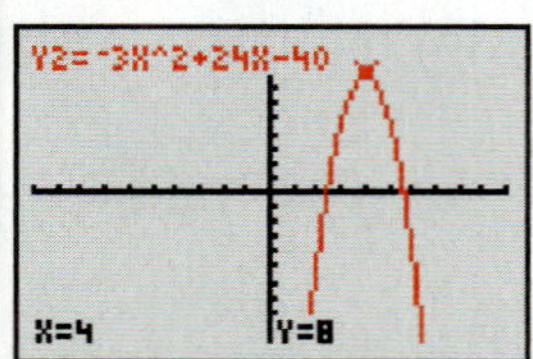

Figure 20 Verify the work

We use graphing calculator graphs to verify our work (see Fig. 20).

Multiplication of Binomial Conjugates

We say that the sum of two terms and the difference of the same two terms are **binomial conjugates** of each other. For instance, $5x + 7$ and $5x - 7$ are binomial conjugates. To find the binomial conjugate of a binomial, we change the plus symbol to a minus symbol or vice versa.

How do we find the product of two binomial conjugates? To begin, we find the product of the binomial conjugates $x + 3$ and $x - 3$:

$$(x + 3)(x - 3) = x^2 - 3x + 3x - 9 \quad \textit{Multiply pairs of terms.}$$
$$= x^2 - 9 \quad \textit{Combine like terms.}$$

Now we generalize and find the product of the binomial conjugates $A + B$ and $A - B$:

$$(A + B)(A - B) = A^2 - AB + AB - B^2 \quad \textit{Multiply pairs of terms.}$$
$$= A^2 - B^2 \quad \textit{Combine like terms.}$$

We see that the product of $A + B$ and $A - B$ is the binomial $A^2 - B^2$, which is called a **difference of two squares.**

Product of Binomial Conjugates

$$(A + B)(A - B) = A^2 - B^2$$

In words: The product of two binomial conjugates is the difference of the square of the first term and the square of the second term.

By the commutative law of multiplication, we have

$$(A - B)(A + B) = (A + B)(A - B) = A^2 - B^2.$$

So, it is also true that

$$(A - B)(A + B) = A^2 - B^2$$

Example 6 Finding the Product of Binomial Conjugates

Find the product.

1. $(x + 5)(x - 5)$
2. $(3x - 8)(3x + 8)$
3. $(2a - 7b)(2a + 7b)$
4. $(4m^2 - 7rt)(4m^2 + 7rt)$
5. $(x + 3)(x - 3)(x^2 + 9)$

Solution

1. We substitute x for A and 5 for B:

$$\begin{aligned} (A + B)(A - B) &= A^2 - B^2 \\ (x + 5)(x - 5) &= x^2 - 5^2 && \textit{Substitute.} \\ &= x^2 - 25 && \textit{Simplify.} \end{aligned}$$

2. We substitute $3x$ for A and 8 for B:

$$\begin{aligned} (A - B)(A + B) &= A^2 - B^2 \\ (3x - 8)(3x + 8) &= (3x)^2 - 8^2 && \textit{Substitute.} \\ &= 9x^2 - 64 && \textit{Simplify.} \end{aligned}$$

X	Y1	Y2
0	-64	-64
1	-55	-55
2	-28	-28
3	17	17
4	80	80
5	161	161
6	260	260

X=0

Figure 21 Verify the work

We use a graphing calculator table to verify our work (see Fig. 21).

3. We substitute $2a$ for A and $7b$ for B:

$$\begin{aligned} (A - B)(A + B) &= A^2 - B^2 \\ (2a - 7b)(2a + 7b) &= (2a)^2 - (7b)^2 && \textit{Substitute.} \\ &= 4a^2 - 49b^2 && \textit{Simplify.} \end{aligned}$$

4. $$\begin{aligned} (4m^2 - 7rt)(4m^2 + 7rt) &= (4m^2)^2 - (7rt)^2 && (A - B)(A + B) = A^2 - B^2 \\ &= 16m^4 - 49r^2t^2 && \textit{Simplify.} \end{aligned}$$

5. $$\begin{aligned} (x + 3)(x - 3)(x^2 + 9) &= (x^2 - 9)(x^2 + 9) && (A + B)(A - B) = A^2 - B^2 \\ &= (x^2)^2 - 9^2 && (A - B)(A + B) = A^2 - B^2 \\ &= x^4 - 81 && \textit{Simplify.} \end{aligned}$$

Group Exploration

Looking ahead: Properties of exponents

1. a. Write each expression as a product of fractions. Then simplify so that the result is a single fraction.

 i. $\left(\frac{x}{y}\right)^2$ ii. $\left(\frac{x}{y}\right)^3$ iii. $\left(\frac{x}{y}\right)^4$

 b. Write $\left(\frac{x}{y}\right)^n$ in another form. [**Hint:** Refer to your results in part (a).]

 c. Use your result in part (b) to write $\left(\frac{x}{y}\right)^{28}$ in another form in one step.

2. a. Write each expression as a product of powers. Then simplify so that the result is a single power.

 i. $(x^2)^3$ [**Hint:** Write as a product of three x^2 factors.]

 ii. $(x^5)^2$

 iii. $(x^3)^4$

 b. Write $(x^m)^n$ in another form. [**Hint:** Refer to part (a) and determine whether we add, subtract, multiply, or divide the exponents m and n.]

 c. Use your result in part (b) to write $(x^6)^9$ in another form in one step.

3. a. We can write the power x^3 as $x \cdot x \cdot x$ without using exponents. Write each expression without using exponents. Then simplify the result.

 i. $\frac{x^5}{x^3}$ ii. $\frac{x^6}{x^2}$ iii. $\frac{x^8}{x^5}$

 b. Write $\frac{x^m}{x^n}$ in another form. [**Hint:** Refer to part (a) and determine whether we add, subtract, multiply, or divide the exponents m and n.]

 c. Use your result in part (b) to write $\frac{x^{25}}{x^{21}}$ in another form in one step.

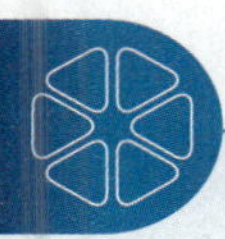

Homework 7.3

For extra help ▶ MyMathLab®  Watch the videos in MyMathLab  Download the MyDashboard App

Perform the indicated operation. Use a graphing calculator table to verify your work when possible.

1. $(xy)^8$ 2. $(xy)^5$ 3. $(6x)^2$ 4. $(9x)^2$
5. $(4p)^3$ 6. $(2w)^2$ 7. $(-8x)^2$ 8. $(-5x)^2$
9. $(-3x)^3$ 10. $(-2x)^5$ 11. $(-a)^5$ 12. $(-w)^4$
13. $(2mp)^4$ 14. $(5bw)^2$ 15. $(-3xy)^4$ 16. $(-2xy)^3$

Simplify. Use a graphing calculator table to verify your work when possible.

17. $(x+5)^2$ 18. $(x+2)^2$ 19. $(x-4)^2$
20. $(x-3)^2$ 21. $(2x+3)^2$ 22. $(3x+7)^2$
23. $(5y-2)^2$ 24. $(4a-9)^2$ 25. $(2a+5b)^2$
26. $(3t+7w)^2$ 27. $(8x-3y)^2$ 28. $(4m-3n)^2$
29. $(2x^2-6y^2)^2$ 30. $(5m^2-4p^2)^2$
31. $-2x(2x+5)^2$ 32. $-5x(3x+2)^2$

For $f(x) = x^2 - 3x$, find the following.

33. $f(5b)$ 34. $f(6b)$ 35. $f(c+4)$
36. $f(c+1)$ 37. $f(b-3)$ 38. $f(b-2)$
39. $f(a+2) - f(a)$
40. $f(a+3) - f(a)$
41. $f(a+h) - f(a)$
42. $f(a) - f(a-h)$

Write the quadratic function in standard form. Use graphing calculator graphs to verify your work.

43. $f(x) = (x+6)^2$ 44. $f(x) = (x-8)^2$
45. $f(x) = (x-3)^2 + 1$ 46. $f(x) = (x+4)^2 - 5$
47. $f(x) = 2(x+4)^2 - 3$ 48. $f(x) = 3(x-1)^2 + 2$
49. $f(x) = -3(x-1)^2 - 2$ 50. $f(x) = -2(x-2)^2 - 1$

Find the product. Use a graphing calculator table to verify your work when possible.

51. $(x+4)(x-4)$ 52. $(x+6)(x-6)$
53. $(t-7)(t+7)$ 54. $(a-8)(a+8)$
55. $(7a+9)(7a-9)$ 56. $(5y+4)(5y-4)$
57. $(2x-3y)(2x+3y)$ 58. $(4b-7c)(4b+7c)$
59. $(3rt-9w)(3rt+9w)$ 60. $(7ab-5c)(7ab+5c)$
61. $(8a^2+3b^2)(8a^2-3b^2)$ 62. $(5p^2+7q^2)(5p^2-7q^2)$
63. $(x-2)(x+2)(x^2+4)$ 64. $(x-1)(x+1)(x^2+1)$
65. $(3a+2b)(3a-2b)(9a^2+4b^2)$
66. $(2m-5n)(2m+5n)(4m^2+25n^2)$

Perform the indicated operation. Use a graphing calculator table to verify your work when possible.

67. $(4x^2-5x)-(2x^3-8x)$ 68. $(7x^3-2x)-(3x^2-6x)$
69. $5t(-2t^2)$ 70. $-3w^2(-4w)$

71. $(3x + 4)(x^2 - x + 2)$
72. $(2x - 5)(x^2 + 4x - 1)$
73. $(2tw - 3p)(2tw + 3p)$
74. $(5ab - 2c)(5ab + 2c)$
75. $2xy^2(4x^2 - 8x - 5)$
76. $-3xy(8x^2 - 2xy + 9y^2)$
77. $(-6x^2 - 4x + 5) + (-2x^2 + 3x - 8)$
78. $(5x^3 - 2x + 1) + (-9x^3 - 5x^2 - 8x)$
79. $(4w - 8)^2$
80. $(6c - 4)^2$
81. $(3x - 7y)(2x + 3y)$
82. $(2a + 3b)(4a - 5b)$
83. $(6x - 7)(6x + 7)$
84. $(5x - 2)(5x + 2)$
85. $(2t^2 + 5w^2)^2$
86. $(4k^2 + 3p^2)^2$

Concepts

87. A student tries to find the power $(4x)^2$:

$$(4x)^2 = 4x^2$$

Describe any errors. Then find the power correctly.

88. A student tries to find the power $(-3x)^2$:

$$(-3x)^2 = -9x^2$$

Describe any errors. Then find the power correctly.

89. A student tries to simplify $(x + 7)^2$:

$$(x + 7)^2 = x^2 + 7^2$$
$$= x^2 + 49$$

Describe any errors. Then simplify the expression correctly.

90. A student tries to simplify $(x - 9)^2$:

$$(x - 9)^2 = x^2 - 9^2$$
$$= x^2 - 81$$

Describe any errors. Then simplify the expression correctly.

91. Use graphing calculator tables to show that $(x - 5)^2$ is not equivalent to $x^2 - 5^2$ and to show that $(x - 5)^2$ is not equivalent to $x^2 + 5^2$. Then simplify $(x - 5)^2$, and verify your work by using a graphing calculator table.

92. Use a graphing calculator table to show that $(x + 5)^2$ is not equivalent to $x^2 + 5^2$. Then simplify $(x + 5)^2$, and verify your work by using a graphing calculator table.

93. **a.** Use a graphing calculator table to show that $(x + 4)^2$ and $x^2 + 4^2$ are not equivalent.
b. Simplify $(x + 4)^2$.
c. Use a graphing calculator table to show that $(x + 4)^2$ and your result in part (b) are equivalent.

94. **a.** Use a graphing calculator table to show that $(x - 3)^2$ and $x^2 - 3^2$ are not equivalent.
b. Simplify $(x - 3)^2$.
c. Use a graphing calculator table to show that $(x - 3)^2$ and your result in part (b) are equivalent.

95. Show that the statement $(A + B)^2 = A^2 + B^2$ is false by substituting 2 for A and 3 for B. Explain why your work shows that the statement is false. Then write an important true statement that has the expression $(A + B)^2$ on the left side.

96. Show that the statement $(A - B)^2 = A^2 - B^2$ is false by substituting 5 for A and 3 for B. Explain why your work shows that the statement is false. Then write an important true statement that has the expression $(A - B)^2$ on the left side.

97. Which of the following expressions are equivalent?

$(x - 2)^2$ $\quad (x + 2)^2 - 4x$ $\quad x(x - 4) + 4$ $\quad (x - 2)(x + 2)$
$x^2 - 4x + 4$ $\quad (x - 1)(x - 4)$

98. Which of the following expressions are equivalent?

$(x + 3)^2$ $\quad x^2 + 6x + 9$ $\quad (x + 3)^2 + 6x$ $\quad (x + 3)(x + 3)$
$(x - 3)^2 + 12x$ $\quad (x + 3)(x - 3) + 6x$

99. Suppose that a student asks you, "Where did the term $14x$ come from?" in the equation $(x + 7)^2 = x^2 + 14x + 49$. How would you respond?

100. A student tries to simplify the expression $2(x + 4)^2$:

$$2(x + 4)^2 = (2x + 8)^2$$
$$= (2x)^2 + 2(2x)(8) + 8^2$$
$$= 4x^2 + 32x + 64$$

Describe any errors. Then simplify the expression correctly.

101. **a.** Simplify $(A + B + C)^2$. [**Hint:** $D^2 = DD$.]
b. Compare the property $(A + B)^2 = A^2 + 2AB + B^2$ with your result from part (a).

102. **a.** Is it possible for the product of two binomials to be equivalent to a trinomial? If yes, give an example. If no, why not?
b. Is it possible for the product of two binomials to be equivalent to a binomial? If yes, give an example. If no, why not?

103. In this exercise, you will explore a reasonable definition of b^0, where b is a nonzero real number.
a. **i.** Perform the exponentiation for $2^5, 2^4, 2^3, 2^2$, and 2^1.
ii. What pattern do you notice in your results in part (i)?
iii. Based on your results in part (i), what would be a reasonable value for 2^0?
b. Perform the exponentiations for $3^4, 3^3, 3^2$, and 3^1. Based on these values, what would be a reasonable value for 3^0?
c. If b is a nonzero real number, what would be a reasonable value for b^0?

104. **a.** Perform the exponentiations for $(-2)^2, (-2)^4, (-2)^6$, and $(-2)^8$. State whether each result is positive, negative, or 0.
b. Perform the exponentiations for $(-2)^3, (-2)^5, (-2)^7$, and $(-2)^9$. State whether each result is positive, negative, or 0.
c. For which counting numbers n is $(-2)^n$ positive? negative? Explain.
d. Let k be a negative number. For which counting numbers n is k^n positive? negative? Explain.

Related Review

Simplify the right-hand side of the equation to help you decide whether the equation is linear or quadratic. State whether the graph of the equation is a line or a parabola. Use a graphing calculator graph to verify your work.

105. $f(x) = 2x(5x - 3)$
106. $f(x) = -3(7x + 2)$
107. $f(x) = (4x - 3)(6x - 5)$
108. $f(x) = 2x(x + 5) - 2x^2$
109. $f(x) = x^2 - (x - 3)^2$
110. $f(x) = x - (x + 2)^2$

Expressions, Equations, Functions, and Graphs

Perform the indicated instruction. Then use words such as linear, quadratic, cubic, power, polynomial, degree, function, one variable, *and* two variables *to describe the expression, equation, or system.*

111. Solve the following:

$$2x - 5y = 15$$
$$y = 3x - 16$$

112. Simplify $(3a - 8)^2$.
113. Graph $2x - 5y = 15$ by hand.
114. Solve $2(3a - 8) = 2a + 7$.

7.4 Properties of Exponents

Objectives

» Know properties of exponents.

» Recognize whether a power expression is simplified.

» Use the properties of exponents to simplify power expressions.

» Know the meaning of the exponent zero.

In Sections 7.2 and 7.3, we discussed two properties of exponents. In this section, we will discuss several more properties of exponents.

Product Property and Raising a Product to a Power

We begin by reviewing the product property for exponents, from Section 7.2, and how to raise a product to a power, from Section 7.3.

If n and m are counting numbers, then

$$x^m x^n = x^{m+n} \quad \text{Product property for exponents}$$

$$(xy)^n = x^n y^n \quad \text{Raising a product to a power}$$

To multiply two powers of x, we keep the base and add the exponents. For example, $x^2x^5 = x^{2+5} = x^7$. To raise a product to a power, we raise each factor to the power. For example, $(xy)^5 = x^5y^5$.

Example 1 Performing Operations

Perform the indicated operation.

1. $(2x^3y^6)(3x^8y^7)$ **2.** $(-2xy)^3$

Solution

1. We rearrange the factors so that the coefficients are adjacent, the powers of x are adjacent, and the powers of y are adjacent:

$$\begin{aligned}(2x^3y^6)(3x^8y^7) &= (2\cdot 3)(x^3x^8)(y^6y^7) && \text{Rearrange factors.}\\ &= 6x^{11}y^{13} && \text{Add exponents: } x^mx^n = x^{m+n}\end{aligned}$$

2.
$$\begin{aligned}(-2xy)^3 &= (-2)^3x^3y^3 && \text{Raise each factor to third power: } (xy)^n = x^ny^n\\ &= -8x^3y^3 && \text{Simplify.}\end{aligned}$$

Recognizing whether a Power Expression Is Simplified

In Example 1, we worked with the power expressions $(2x^3y^6)(3x^8y^7)$ and $(-2xy)^3$. A **power expression** is an expression that contains one or more powers. Here are some more power expressions:

$$(9x^7)(5x^8) \qquad 3x^5(2xy)^3 \qquad \frac{8x^6y^9}{6x^2y^4} \qquad \left(\frac{7x^8}{x^3}\right)^2 \qquad x^3y + \frac{x}{y^2}$$

We can use the two properties about exponents that we have discussed, as well as three more properties that we will discuss in this section, to *simplify a power expression*.

Simplifying a Power Expression

A power expression is simplified if

1. It includes no parentheses.
2. In any monomial, each variable or constant appears as a base at most once. For example, for nonzero x, we write x^3x^5 as x^8.
3. Each numerical expression (such as 5^2) has been calculated, and each numerical fraction has been simplified.

Quotient Property for Exponents

Consider the quotient $\frac{x^5}{x^2}$. We can simplify this quotient by first writing the expression without exponents:

$$\begin{aligned}\frac{x^5}{x^2} &= \frac{x \cdot x \cdot x \cdot x \cdot x}{x \cdot x} && \textit{Write quotient without exponents.}\\ &= \frac{x \cdot x}{x \cdot x} \cdot \frac{x \cdot x \cdot x}{1} && \frac{ac}{bd} = \frac{a}{b} \cdot \frac{c}{d}\\ &= 1 \cdot x \cdot x \cdot x && \textit{Simplify: } \frac{x \cdot x}{x \cdot x} = 1\\ &= x^3 && \textit{Simplify.}\end{aligned}$$

Note that we can get the same result by subtracting the exponents 5 and 2:

$$\frac{x^5}{x^2} = x^{5-2} = x^3$$

Quotient Property for Exponents

If m and n are counting numbers and x is nonzero, then

$$\frac{x^m}{x^n} = x^{m-n}$$

In words: To divide two powers of x, keep the base and subtract the exponents.

For example, $\frac{x^9}{x^4} = x^{9-4} = x^5$.

Example 2 Quotient Property

Simplify.

1. $\frac{6x^9}{4x^5}$

2. $\frac{12x^6y^8}{4xy^7}$

Solution

1. $$\begin{aligned}\frac{6x^9}{4x^5} &= \frac{6}{4} \cdot \frac{x^9}{x^5} && \frac{ac}{bd} = \frac{a}{b} \cdot \frac{c}{d}\\ &= \frac{3}{2} \cdot x^{9-5} && \textit{Simplify; subtract exponents: } \frac{x^m}{x^n} = x^{m-n}\\ &= \frac{3}{2} \cdot \frac{x^4}{1} && \textit{Simplify.}\\ &= \frac{3x^4}{2} && \textit{Multiply numerators; multiply denominators: } \frac{A}{B} \cdot \frac{C}{D} = \frac{AC}{BD}\end{aligned}$$

2. $$\begin{aligned}\frac{12x^6y^8}{4xy^7} &= \frac{12}{4} \cdot \frac{x^6}{x^1} \cdot \frac{y^8}{y^7} && \frac{ac}{bd} = \frac{a}{b} \cdot \frac{c}{d}\\ &= 3 \cdot x^{6-1} \cdot y^{8-7} && \textit{Simplify; subtract exponents: } \frac{x^m}{x^n} = x^{m-n}\\ &= 3x^5y^1 && \textit{Simplify.}\\ &= 3x^5y && y^1 = y\end{aligned}$$

Zero as an Exponent

What is the meaning of 2^0? Computing powers of 2 can suggest the meaning:

$$2^4 = 16$$
$$2^3 = 8$$
$$2^2 = 4$$
$$2^1 = 2$$
$$2^0 = 1$$

The exponent decreases by 1. The value is divided by 2.

Each time we decrease the exponent by 1, the value is divided by 2. This pattern suggests that $2^0 = 1$.

Similar work with any other nonzero base x would suggest that $x^0 = 1$.

Definition Zero exponent

For nonzero x,

$$x^0 = 1$$

So, for nonzero x and y, $4^0 = 1$, $(xy)^0 = 1$, and $\left(\frac{5x^2}{y^4}\right)^0 = 1$.

Raising a Quotient to a Power

Consider the power expression $\left(\frac{x}{y}\right)^4$. We begin to write this expression in another form by writing it without exponents:

$$\left(\frac{x}{y}\right)^4 = \frac{x}{y} \cdot \frac{x}{y} \cdot \frac{x}{y} \cdot \frac{x}{y} \quad \textit{Write power expression without exponents.}$$
$$= \frac{x \cdot x \cdot x \cdot x}{y \cdot y \cdot y \cdot y} \quad \textit{Multiply numerators; multiply denominators: } \frac{A}{B} \cdot \frac{C}{D} = \frac{AC}{BD}$$
$$= \frac{x^4}{y^4} \quad \textit{Simplify.}$$

Note that we can get the same result by distributing the exponent 4 to both the numerator and the denominator of the base $\frac{x}{y}$:

$$\left(\frac{x}{y}\right)^4 = \frac{x^4}{y^4}$$

This method is called *raising a quotient to a power.*

Raising a Quotient to a Power

If n is a counting number and y is nonzero, then

$$\left(\frac{x}{y}\right)^n = \frac{x^n}{y^n}$$

In words: To raise a quotient to a power, raise both the numerator and the denominator to the power.

Example 3 Raising a Quotient to a Power

Simplify.

1. $\left(\frac{x}{y}\right)^8$ **2.** $\left(\frac{x}{2}\right)^3$

Solution

1. $\left(\frac{x}{y}\right)^8 = \frac{x^8}{y^8}$ *Raise numerator and denominator to eighth power:* $\left(\frac{x}{y}\right)^n = \frac{x^n}{y^n}$

2. $\left(\frac{x}{2}\right)^3 = \frac{x^3}{2^3}$ *Raise numerator and denominator to third power:* $\left(\frac{x}{y}\right)^n = \frac{x^n}{y^n}$

$= \frac{x^3}{8}$ *Simplify.*

Raising a Power to a Power

Consider the power expression $(x^2)^3$. We can write this expression as the power x^6:

$$(x^2)^3 = x^2 \cdot x^2 \cdot x^2 \quad \text{Write expression without exponent 3.}$$
$$= x^{2+2+2} \quad \text{Add exponents: } x^m x^n = x^{m+n}$$
$$= x^6 \quad \text{Simplify.}$$

Note that we can get the same result by multiplying the exponents 2 and 3:

$$(x^2)^3 = x^{2\cdot 3} = x^6$$

This method is called *raising a power to a power.*

Raising a Power to a Power

If m and n are counting numbers, then

$$(x^m)^n = x^{mn}$$

In words: To raise a power to a power, keep the base and multiply the exponents.

Example 4 Raising a Power to a Power

Simplify.

1. $(x^2)^7$ **2.** $(x^5)^8$

Solution

1. $(x^2)^7 = x^{2\cdot 7}$ *Multiply exponents:* $(x^m)^n = x^{mn}$

$= x^{14}$ *Simplify.*

2. $(x^5)^8 = x^{5\cdot 8}$ *Multiply exponents:* $(x^m)^n = x^{mn}$

$= x^{40}$ *Simplify.*

Using Combinations of Properties of Exponents

Next, we use more than one property of exponents to simplify an expression.

Example 5 Simplifying Power Expressions

Simplify.

1. $(2x^8y^5)^3$

2. $2x^7(4x^4)^2$

3. $\dfrac{8x^3x^4}{2x^7}$

Solution

1. $(2x^8y^5)^3 = 2^3(x^8)^3(y^5)^3$ Raise each factor to third power: $(xy)^n = x^ny^n$

$= 8x^{24}y^{15}$ Multiply exponents: $(x^m)^n = x^{mn}$

2. $2x^7(4x^4)^2 = 2x^7[4^2(x^4)^2]$ Raise each factor to second power: $(xy)^n = x^ny^n$

$= 2x^7[16x^8]$ Multiply exponents: $(x^m)^n = x^{mn}$

$= (2\cdot 16)(x^7x^8)$ Rearrange factors.

$= 32x^{15}$ Add exponents: $x^mx^n = x^{m+n}$

3. $\dfrac{8x^3x^4}{2x^7} = \dfrac{4x^7}{x^7}$ Simplify; add exponents: $x^mx^n = x^{m+n}$

$= 4x^{7-7}$ Subtract exponents: $\dfrac{x^m}{x^n} = x^{m-n}$

$= 4x^0$ Simplify.

$= 4$ $x^0 = 1$.

Example 6 Simplifying Power Expressions

Simplify.

1. $\left(\dfrac{2x^4}{3y^7}\right)^3$

2. $\dfrac{(2x^3y)^5}{x^8y}$

Solution

1. $\left(\dfrac{2x^4}{3y^7}\right)^3 = \dfrac{(2x^4)^3}{(3y^7)^3}$ Raise numerator and denominator to third power: $\left(\dfrac{x}{y}\right)^n = \dfrac{x^n}{y^n}$

$= \dfrac{2^3(x^4)^3}{3^3(y^7)^3}$ Raise factors to third power: $(xy)^n = x^ny^n$

$= \dfrac{8x^{12}}{27y^{21}}$ Multiply exponents: $(x^m)^n = x^{mn}$

2. $\dfrac{(2x^3y)^5}{x^8y} = \dfrac{2^5(x^3)^5y^5}{x^8y}$ Raise factors to fifth power: $(xy)^n = x^ny^n$

$= \dfrac{32x^{15}y^5}{x^8y^1}$ Multiply exponents: $(x^m)^n = x^{mn}; y = y^1$

$= 32x^{15-8}y^{5-1}$ Subtract exponents: $\dfrac{x^m}{x^n} = x^{m-n}$

$= 32x^7y^4$ Simplify.

Group Exploration

Properties of exponents

1. For the statement $x^2x^3 = x^5$, a student wants to know why there are two x's on the left-hand side of the equation and only one x on the right-hand side. What would you tell the student?

2. A student tries to simplify the expression $(3x^3)^2$:
$$(3x^3)^2 = (3\cdot 2)x^{3\cdot 2} = 6x^6$$
Describe any errors. Then simplify the expression correctly.

3. In simplifying power expressions, a student is confused about when to add exponents and when to multiply exponents. What would you tell the student?

4. A student tries to simplify $\dfrac{(2x^5)^4}{x^2}$:
$$\frac{(2x^5)^4}{x^2} = (2x^{5-2})^4 = (2x^3)^4 = 2^4(x^3)^4 = 16x^{12}$$
Describe any errors. Then simplify the expression correctly.

5. Simplify $x^3 + x^3$. Then simplify x^3x^3. Explain why your two results have different exponents.

Tips for Success **Form a Study Group to Prepare for the Final Exam**

To prepare for the final exam, it may be helpful to form a study group. The group could list important concepts of the course and discuss the meanings of these concepts and how to apply them. You could also list important techniques learned in the course and practice those techniques. Set aside some solo study time after the group study session to make sure you can do the mathematics without the help of other members of the study group.

Homework 7.4

For extra help ▶ MyMathLab®

Simplify.

1. x^3x^5
2. x^2x^7
3. r^5r
4. yy^8
5. $(5x^4)(3x^5)$
6. $(-2x^7)(6x^8)$
7. $(-4b^3)(-8b^5)$
8. $(3t^5)(-7t^4)$
9. $(6a^2b^5)(9a^4b^3)$
10. $(10c^7d^2)(3c^5d)$
11. $(rt)^7$
12. $(ab)^4$
13. $(8x)^2$
14. $(2x)^4$
15. $(2xy)^5$
16. $(3xy)^3$
17. $(-2a)^4$
18. $(-3t)^3$
19. $(9xy)^0$
20. $(-3x)^0$
21. $\dfrac{a^5}{a^2}$
22. $\dfrac{w^8}{w^6}$
23. $\dfrac{6x^7}{3x^3}$
24. $\dfrac{8x^6}{4x^5}$
25. $\dfrac{15x^6y^8}{12x^3y}$
26. $\dfrac{14x^5y^9}{21x^3y^6}$
27. $\left(\dfrac{t}{w}\right)^7$
28. $\left(\dfrac{a}{b}\right)^4$
29. $\left(\dfrac{3}{t}\right)^3$
30. $\left(\dfrac{b}{7}\right)^2$
31. $\left(\dfrac{x}{3}\right)^0$
32. $\left(\dfrac{2}{x}\right)^0$
33. $(r^2)^4$
34. $(a^3)^8$
35. $(x^4)^9$
36. $(x^6)^9$

For Exercises 37–72, simplify.

37. $(6x^3)^2$
38. $(4x^5)^3$
39. $(-t^3)^4$
40. $(-a^2)^3$
41. $(2a^2a^7)^3$
42. $(2w^3w^4)^5$
43. $(x^2y^3)^4x^5y^8$
44. $(xy^6)^3x^2y^4$
45. $5x^4(3x^6)^2$
46. $4x^2(2x^5)^3$
47. $-3c^6(c^4)^5$
48. $-7w^4(w^3)^6$
49. $(xy^3)^5(xy)^4$
50. $(x^2y)^3(xy^3)^2$
51. $\dfrac{10t^5t^7}{8t^4}$
52. $\dfrac{15y^9y^2}{20y^8}$
53. $\dfrac{18x^{10}}{24x^4x^6}$
54. $\dfrac{12x^{15}}{21x^9x^6}$
55. $\left(\dfrac{y}{2x}\right)^3$
56. $\left(\dfrac{8x}{y}\right)^2$
57. $\left(\dfrac{x^2}{y^5}\right)^4$
58. $\left(\dfrac{x^8}{y^2}\right)^5$
59. $\left(\dfrac{r^6}{6}\right)^2$
60. $\left(\dfrac{2}{a^4}\right)^3$
61. $\left(\dfrac{2a^4}{3b^2}\right)^3$
62. $\left(\dfrac{4w^6}{7t^3}\right)^2$
63. $\left(\dfrac{3x^4}{5y^7}\right)^0$
64. $\left(\dfrac{-5y^2}{9x^4}\right)^0$
65. $\left(\dfrac{2a^6b}{3c^5}\right)^3$
66. $\left(\dfrac{5xy^4}{7w^3}\right)^2$

67. $\dfrac{(x^4y)^4}{x^5}$

68. $\dfrac{(x^3y^7)^5}{y^{10}}$

69. $\dfrac{(w^3)^4}{(2w)^5}$

70. $\dfrac{(3t)^3}{(t^2)^5}$

71. $\dfrac{(4x^5y^8)^2}{8x^8y^9}$

72. $\dfrac{(2x^4y^3)^3}{12x^5y^4}$

73. A person invests \$5000 in an account at 3% interest compounded annually. The value V (in dollars) of the account after t years is described by $V = 5000(1.03)^t$. Find the value of the account after 10 years.

74. A person invests \$8000 in an account at 7% interest compounded annually. The value V (in dollars) of the account after t years is described by $V = 8000(1.07)^t$. Find the value of the account after 6 years.

75. If an object falls from rest for t seconds, the distance d (in feet) the object will fall is described by the equation $d = 16t^2$, assuming no air resistance. Estimate how far a sky diver will have fallen 5 seconds after jumping out of an airplane.

76. The weight w (in ounces) of a thick-crust pizza with three toppings at Papa Del's in Champaign, Illinois, is described by the equation $w = 0.41d^2$, where d is the diameter (in inches) of the pizza. Estimate the weight of such a pizza with diameter 12 inches.

77. The power P (in watts) generated by a windmill is described by the equation $P = 0.8r^2v^3$, where r is the radius (in meters) of the windmill and v is the wind speed (in meters per second). Estimate the power a windmill with radius 0.57 meter can generate if the wind speed is 12.5 meters per second.

78. The intensity I (in watts per square meter) of a television signal at a distance of d kilometers is described by the equation $I = \dfrac{180}{d^2}$. Find the intensity of the signal at a distance of 5 kilometers.

79. The volume V (in cubic centimeters) of a cylinder is described by the equation $V = \pi r^2h$, where h is the height (in centimeters) of the cylinder and r is the radius (in centimeters) of the bottom (see Fig. 22). Find the volume of a can of chili with height 11 centimeters and radius 3.5 centimeters.

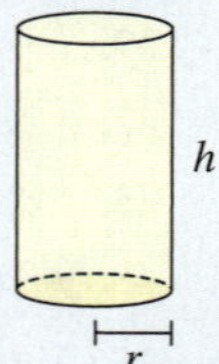

Figure 22 The height h and radius r of a cylinder

80. The volume V (in cubic feet) of a rectangular box with a square base is described by the equation $V = L^2H$, where H is the height (in feet) of the box and L is the length (in feet) of the base (see Fig. 23). Find the volume of a 4-foot-tall cardboard box whose square base has length 2 feet.

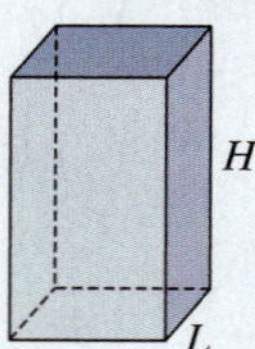

Figure 23 The height H and length L of a rectangular box with a square base

Concepts

81. A student tries to simplify x^3x^5:

$$x^3x^5 = x^{3 \cdot 5} = x^{15}$$

Describe any errors. Then simplify the expression correctly.

82. A student tries to simplify $(x^4)^6$:

$$(x^4)^6 = x^{4+6} = x^{10}$$

Describe any errors. Then simplify the expression correctly.

83. A student tries to simplify $(5x^3)^2$:

$$(5x^3)^2 = 5(x^3)^2 = 5x^6$$

Describe any errors. Then simplify the expression correctly.

84. A student tries to simplify $2(3x)^2$:

$$2(3x)^2 = (6x)^2 = 36x^2$$

Describe any errors. Then simplify the expression correctly.

We can explain why $(xy)^3 = x^3y^3$ by writing

$$(xy)^3 = (xy)(xy)(xy) = (xxx)(yyy) = x^3y^3$$

For Exercises 85 and 86, use a similar approach to show that the given statement is true.

85. $\left(\dfrac{x}{y}\right)^3 = \dfrac{x^3}{y^3}$

86. $\dfrac{x^6}{x^4} = x^2$

87. A student tries to simplify $(2x^2)^4$:

$$(2x^2)^4 = (2 \cdot 4)x^{2 \cdot 4} = 8x^8$$

Describe any errors. Then simplify the expression correctly.

88. Describe what it means to use properties of exponents to simplify a power expression. Include several examples.

Related Review

Simplify.

89. x^3x^2

90. x^5x^3

91. $x^3 + x^2$

92. $x^5 + x^3$

93. $2x^4 + 3x^4$

94. $5x^3 - 2x^3$

95. $(2x^4)(3x^4)$

96. $(5x^3)(-2x^3)$

97. $(3x)^2$

98. $(6x)^2$

99. $(3 + x)^2$

100. $(6 + x)^2$

Expressions, Equations, Functions, and Graphs

Perform the indicated instruction. Then use words such as linear, quadratic, cubic, power, polynomial, degree, function, one variable, *and* two variables *to describe the expression, equation, or system.*

101. Solve $\dfrac{2}{3}x - \dfrac{5}{6} = \dfrac{1}{2}x$.

102. Find the product $(2p - 1)(3p^2 + p - 2)$.

103. Combine like terms: $\dfrac{2}{3}x - \dfrac{5}{6} - \dfrac{1}{2}x$.

104. Graph $y = x^2 - 4x + 3$ by hand.

7.5 Dividing Polynomials: Long Division and Synthetic Division

Objectives

- » Divide a polynomial by a monomial.
- » Use long division to divide a polynomial by a binomial.
- » Perform synthetic division.

In this chapter, we have discussed how to add, subtract, and multiply polynomials. In this section, we will divide polynomials.

Dividing by a Monomial

First, we will discuss how to divide a polynomial by a monomial. Recall the following rule about adding fractions with a common denominator:

$$\frac{A}{B} + \frac{C}{B} = \frac{A + C}{B}, \quad \text{where } B \text{ is nonzero}$$

To divide by a monomial, we will go backward.

Dividing by a Monomial

If A, B, and C are monomials and B is nonzero, then

$$\frac{A + C}{B} = \frac{A}{B} + \frac{C}{B}$$

In words: To divide a polynomial by a monomial, divide each term of the polynomial by the monomial.

WARNING It is a common error to divide only some of the terms of the polynomial by the monomial. Remember to divide *every* term of the polynomial by the monomial.

Example 1 Dividing by a Monomial

Find the quotient $\frac{15w^2 + 35w}{5w}$.

Solution

$$\frac{15w^2 + 35w}{5w} = \frac{15w^2}{5w^1} + \frac{35w^1}{5w^1} \quad \text{Divide each term by } 5w\text{: } \frac{A + C}{B} = \frac{A}{B} + \frac{C}{B}$$

$$= 3w^{2-1} + 7w^{1-1} \quad \text{Simplify; subtract exponents: } \frac{x^m}{x^n} = x^{m-n}$$

$$= 3w + 7 \quad \text{Simplify; } x^{1-1} = x^0 = 1$$

Since $\frac{6}{2} = 3$, it follows that $2 \cdot 3 = 6$. We can use this concept to verify our work in Example 1. To check that

$$\frac{15w^2 + 35w}{5w} = 3w + 7$$

we multiply $5w$ and $3w + 7$:

$$5w(3w + 7) = 15w^2 + 35w$$

which checks.

Example 2 Dividing by a Monomial

Find the quotient $\frac{6x^4 - 7x^3 + 8}{2x^3}$.

Solution

$$\frac{6x^4 - 7x^3 + 8}{2x^3} = \frac{6x^4}{2x^3} - \frac{7x^3}{2x^3} + \frac{8}{2x^3}$$ *Divide each term by $2x^3$: $\frac{A - C + D}{B} = \frac{A}{B} - \frac{C}{B} + \frac{D}{B}$*

$$= 3x^{4-3} - \frac{7x^{3-3}}{2} + \frac{4}{x^3}$$ *Simplify; subtract exponents: $\frac{x^m}{x^n} = x^{m-n}$*

$$= 3x - \frac{7}{2} + \frac{4}{x^3}$$ *Simplify.*

We use a graphing calculator table to verify our work (see Fig. 24).

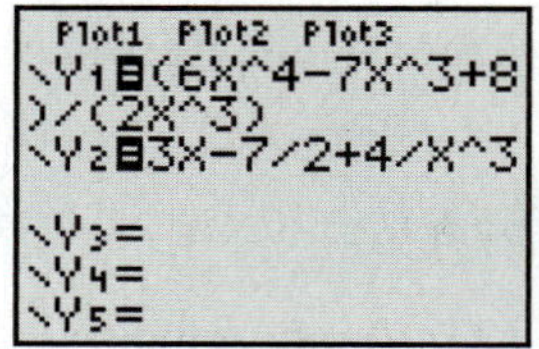

X	Y1	Y2
1	3.5	3.5
2	3	3
3	5.6481	5.6481
4	8.5625	8.5625
5	11.532	11.532
6	14.519	14.519
7	17.512	17.512

X=1

Figure 24 Verify the work

Example 3 Dividing by a Monomial

Find the quotient $\frac{9x^4y + 15x^3y^2 - 6x^2y^3}{-3x^2y}$.

Solution

$$\frac{9x^4y + 15x^3y^2 - 6x^2y^3}{-3x^2y}$$

$$= \frac{9x^4y}{-3x^2y} + \frac{15x^3y^2}{-3x^2y} - \frac{6x^2y^3}{-3x^2y}$$ *Divide each term by $-3x^2y$: $\frac{A + C - D}{B} = \frac{A}{B} + \frac{C}{B} - \frac{D}{B}$*

$$= -3x^{4-2}y^{1-1} - 5x^{3-2}y^{2-1} + 2x^{2-2}y^{3-1}$$ *Simplify; subtract exponents: $\frac{x^m}{x^n} = x^{m-n}$*

$$= -3x^2 - 5xy + 2y^2$$ *Simplify.*

We verify our work by finding the product $-3x^2y(-3x^2 - 5xy + 2y^2)$:

$$-3x^2y(-3x^2 - 5xy + 2y^2) = 9x^4y + 15x^3y^2 - 6x^2y^3$$

Using Long Division to Divide by a Binomial

We can use long division to divide a polynomial by a binomial. The steps are similar to performing long division with numbers. Here we review how to use long division to divide 3981 by 17:

```
                   234   ← Quotient
Divisor →  17 ) 3981     ← Dividend
               −34↓      2·17 = 34
                 58      Subtract and bring down next digit of dividend.
                −51↓     3·17 = 51
                  71     Subtract and bring down next digit of dividend.
                 −68     4·17 = 68
Remainder ——→      3     Subtract.
```

We conclude

$$\frac{3981}{17} = 234 + \frac{3}{17}$$

Recall that we can verify our work by checking that

$$\underset{\text{Divisor}}{17} \cdot \underset{\text{Quotient}}{234} + \underset{\text{Remainder}}{3} \stackrel{?}{=} \underset{\text{Dividend}}{3981}$$

which is true.

Example 4 Dividing by a Binomial

Divide: $\dfrac{3x^2 + 10x + 8}{x + 2}$.

Solution

The steps are similar to performing long division with numbers. To begin, we divide $3x^2$ (the first term of $3x^2 + 10x + 8$) by x (the first term of $x + 2$): $\dfrac{3x^2}{x} = 3x$.

$$\begin{array}{r} 3x \\ x + 2 \overline{)\, 3x^2 + 10x + 8} \\ 3x^2 + 6x \end{array} \qquad \frac{3x^2}{x} = 3x \qquad 3x(x + 2) = 3x^2 + 6x$$

To subtract $3x^2 + 6x$, we change the signs of $3x^2$ and $6x$ and add:

$$\begin{array}{r} 3x \\ x + 2 \overline{)\, 3x^2 + 10x + 8} \\ {}^{-}3x^2 \mathbin{\not+} 6x \\ \hline 4x \end{array} \qquad \text{Add.}$$

Change signs.

Next, we bring down the 8:

$$\begin{array}{r} 3x \\ x + 2 \overline{)\, 3x^2 + 10x + 8} \\ -3x^2 - 6x \\ \hline 4x + 8 \end{array} \qquad \text{Bring down the 8.}$$

Then we repeat the process:

$$\begin{array}{r} 3x + 4 \\ x + 2 \overline{)\, 3x^2 + 10x + 8} \\ -3x^2 - 6x \\ \hline 4x + 8 \\ {}^{-}4x \mathbin{\not+} 8 \\ \hline 0 \end{array} \qquad \frac{4x}{x} = 4 \qquad 4(x + 2) = 4x + 8 \qquad \text{Add.}$$

Change signs.

We conclude

$$\frac{3x^2 + 10x + 8}{x + 2} = 3x + 4$$

We verify our work by checking that

$$\underset{\text{Divisor}}{(x + 2)} \cdot \underset{\text{Quotient}}{(3x + 4)} + \underset{\text{Remainder}}{0} \stackrel{?}{=} \underset{\text{Dividend}}{3x^2 + 10x + 8}$$

which is true.

Example 5 Dividing by a Binomial

Divide: $\dfrac{6x^2 - 23x + 27}{3x - 4}$.

Solution

To begin, we divide $6x^2$ (the first term of $6x^2 - 23x + 27$) by $3x$ (the first term of $3x - 4$): $\dfrac{6x^2}{3x} = 2x$.

$$\begin{array}{r|l} 3x - 4\,\overline{)\,6x^2 - 23x + 27}\quad \text{quotient: } 2x & \dfrac{6x^2}{3x} = 2x \\ \text{Change signs.}\quad -6x^2 + 8x \quad\downarrow & 2x(3x-4) = 6x^2 - 8x \\ \hline -15x + 27 & \text{Add. Then bring down the 27.} \end{array}$$

Then we repeat the process:

$$\begin{array}{r|l} 3x - 4\,\overline{)\,6x^2 - 23x + 27}\quad \text{quotient: } 2x - 5 & \dfrac{-15x}{3x} = -5 \\ -6x^2 + 8x & \\ \hline -15x + 27 & \\ \text{Change signs.}\quad +15x - 20 & -5(3x-4) = -15x + 20 \\ \hline 7 & \text{Add.} \end{array}$$

We conclude

$$\frac{6x^2 - 23x + 27}{3x - 4} = 2x - 5 + \frac{7}{3x - 4}$$

To verify our work, we simplify $(3x - 4)(2x - 5) + 7$:

Divisor ↓ Quotient ↓ Remainder ↓

$$\begin{aligned} (3x - 4) \cdot (2x - 5) + 7 &= 6x^2 - 15x - 8x + 20 + 7 \\ &= 6x^2 - 23x + 27 \leftarrow \text{Dividend} \end{aligned}$$

Since the result is the dividend, this checks.

Example 6 Dividing a Polynomial with Missing Terms

Divide: $\dfrac{8x^3 - 27}{2x - 3}$.

Solution

Since the dividend, $8x^3 - 27$, doesn't have x^2 or x terms, we use $0x^2$ and $0x$ as placeholders:

$$8x^3 + 0x^2 + 0x - 27$$

This way, like terms will line up when we perform long division:

$$\begin{array}{r|l} 2x - 3\,\overline{)\,8x^3 + 0x^2 + 0x - 27}\quad \text{quotient: } 4x^2 + 6x + 9 & \text{Use } 0x^2 \text{ and } 0x \text{ as placeholders.} \\ \text{Change signs.}\quad -8x^3 + 12x^2 \quad\downarrow & 4x^2(2x-3) = 8x^3 - 12x^2 \\ \hline 12x^2 + 0x & \text{Add and bring down } 0x. \\ \text{Change signs.}\quad -12x^2 + 18x \quad\downarrow & 6x(2x-3) = 12x^2 - 18x \\ \hline 18x - 27 & \text{Add and bring down } -27. \\ \text{Change signs.}\quad -18x + 27 & 9(2x-3) = 18x - 27 \\ \hline 0 & \text{Add.} \end{array}$$

We conclude

$$\frac{8x^3 - 27}{2x - 3} = 4x^2 + 6x + 9$$

We use a graphing calculator table to verify our work (see Fig. 25).

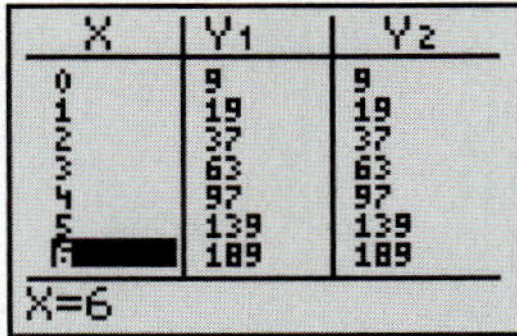

Figure 25 Verify the work

Example 7 Dividing a Polynomial with Missing Terms

Divide: $\dfrac{1 + 5x^3 - 4x + 2x^2}{x^2 - 3}$.

Solution

First, we write the dividend in decreasing order: $5x^3 + 2x^2 - 4x + 1$. Also, since the divisor, $x^2 - 3$, doesn't have an x term, we use $0x$ as a placeholder:

$$x^2 + 0x - 3$$

Just like in Example 6, this will allow like terms to line up:

$$\begin{array}{rl}
 & 5x + 2 \\
x^2 + 0x - 3 \,\big) & 5x^3 + 2x^2 - 4x + 1 \\
 & \underline{-5x^3 \mp 0x^2 \pm 15x} \quad\downarrow \\
 & 2x^2 + 11x + 1 \\
 & \underline{-2x^2 \mp 0x \pm 6} \\
 & 11x + 7
\end{array}$$

Change signs. — $5x(x^2 + 0x - 3) = 5x^3 + 0x^2 - 15x$

Add; then bring down the 1.

Change signs. — $2(x^2 + 0x - 3) = 2x^2 + 0x - 6$

Add.

We conclude

$$\frac{5x^3 + 2x^2 - 4x + 1}{x^2 - 3} = 5x + 2 + \frac{11x + 7}{x^2 - 3}$$

To verify our work, we simplify $(x^2 - 3)(5x + 2) + (11x + 7)$:

$$\begin{array}{ccccccl}
\text{Divisor} & & \text{Quotient} & & \text{Remainder} & & \\
\downarrow & & \downarrow & & \downarrow & & \\
(x^2 - 3) & \cdot & (5x + 2) & + & (11x + 7) & = & 5x^3 + 2x^2 - 15x - 6 + 11x + 7 \\
 & & & & & = & 5x^3 + 2x^2 - 4x + 1 \leftarrow \text{Dividend}
\end{array}$$

Since the result is the dividend, this checks.

Synthetic Division

When we divide a polynomial by a binomial of the form $x - a$, we can use a method called *synthetic division*, which uses the ideas of long division but is more efficient. To take our first step toward synthetic division, consider dividing $3x^3 - 7x^2 + 10x - 12$ by

$x - 2$. Here we perform long division twice, the first time showing the variables and the second time not showing them:

$$
\begin{array}{r|l}
 & 3x^2 - x + 8 \\ \hline
x - 2 & 3x^3 - 7x^2 + 10x - 12 \\
 & \underline{{}^{-}3x^3 \overset{+}{\not-} 6x^2} \\
 & \quad -x^2 + 10x \\
 & \quad \underline{\overset{+}{\not-}x^2 \overset{-}{\not+} 2x} \\
 & \qquad\quad 8x - 12 \\
 & \qquad\quad \underline{{}^{-}8x \overset{+}{\not-} 16} \\
 & \qquad\qquad\quad 4
\end{array}
\qquad\qquad
\begin{array}{r|l}
 & 3 \quad -1 \quad\ \ 8 \\ \hline
1 - 2 & 3 - 7 \quad 10 - 12 \\
 & \underline{{}^{-}3 \overset{+}{\not-} 6} \\
 & \ \ -1 \quad 10 \\
 & \ \ \underline{\overset{+}{\not-}1 \quad {}^{-}2} \\
 & \qquad\quad 8 - 12 \\
 & \qquad\quad \underline{{}^{-}8 \overset{+}{\not-} 16} \\
 & \qquad\qquad\quad 4
\end{array}
$$

We conclude

$$\frac{3x^3 - 7x^2 + 10x - 12}{x - 2} = 3x^2 - x + 8 + \frac{4}{x - 2}$$

In Example 8, we will perform the division yet one more time, now using synthetic division.

Example 8 Performing Synthetic Division

Use synthetic division to divide: $\dfrac{3x^3 - 7x^2 + 10x - 12}{x - 2}$.

Solution

When performing synthetic division, there is no need to write the coefficient of x of the divisor $x - 2$. The method already takes the divisions by x into account. Also, instead of dividing by the constant term -2, we divide by its opposite: 2. That way, we can add without having to change signs first.

$$
\begin{array}{r|rrrr}
2 & 3 & -7 & 10 & -12 \\
 & \downarrow & & & \\ \hline
 & 3 & & &
\end{array}
$$

Write 2 of $x - 2$ and the coefficients of $3x^3 - 7x^2 + 10x - 12$.

Bring down the 3.

$$
\begin{array}{r|rrrr}
2 & 3 & -7 & 10 & -12 \\
 & & 6 & & \\ \hline
 & 3 & & &
\end{array}
$$

Multiply 3 by 2 to get 6.

$$
\begin{array}{r|rrrr}
2 & 3 & -7 & 10 & -12 \\
 & & 6 & & \\ \hline
 & 3 & -1 & &
\end{array}
$$

Add -7 and 6 to get -1.

$$
\begin{array}{r|rrrr}
2 & 3 & -7 & 10 & -12 \\
 & & 6 & -2 & \\ \hline
 & 3 & -1 & 8 &
\end{array}
$$

Multiply -1 by 2 to get -2.

Add 10 and -2 to get 8.

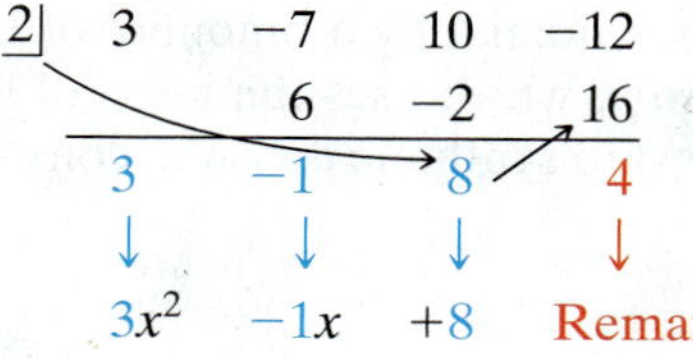

$$
\begin{array}{r|rrrr}
2 & 3 & -7 & 10 & -12 \\
 & & 6 & -2 & 16 \\ \hline
 & 3 & -1 & 8 & 4 \\
 & \downarrow & \downarrow & \downarrow & \downarrow \\
 & 3x^2 & -1x & +8 & \text{Remainder}
\end{array}
$$

Multiply 8 by 2 to get 16.

Add -12 and 16 to get 4.

So, the result is $3x^2 - x + 8$ with remainder 4. We conclude

$$\frac{3x^3 - 7x^2 + 10x - 12}{x - 2} = 3x^2 - x + 8 + \frac{4}{x - 2}$$

This checks with our work just before this example.

Recall that **to perform synthetic division, the divisor must be of the form $x - a$.** The first step is to write the dividend in descending order. Then we write a inside the $\rfloor$ and write the coefficients of the dividend outside it.

Example 9 Performing Synthetic Division

Use synthetic division to divide: $\dfrac{x^2 - 1 - 39x + 3x^3}{x + 4}$.

Solution

First, we write the dividend in descending order: $3x^3 + x^2 - 39x - 1$. Next we use synthetic division. Note that $x + 4 = x - (-4)$. So, we write -4 inside the $\rfloor$:

$$\begin{array}{r|rrrr} -4 & 3 & 1 & -39 & -1 \\ & & -12 & 44 & -20 \\ \hline & 3 & -11 & 5 & -21 \\ & \downarrow & \downarrow & \downarrow & \downarrow \\ & 3x^2 & -11x & +5 & \text{Remainder} \end{array}$$

The quotient is $3x^2 - 11x + 5$ with remainder -21. We conclude

$$\frac{x^2 - 1 - 39x + 3x^3}{x + 4} = 3x^2 - 11x + 5 - \frac{21}{x + 4}$$

Example 10 Performing Synthetic Division when There Are Missing Terms

Use synthetic division to divide: $\dfrac{-2p^3 + 22p - 12}{p - 3}$.

Solution

Because the dividend $-2p^3 + 22p - 12$ does not have a p^2 term, we use 0 as a placeholder:

$$\begin{array}{r|rrrr} 3 & -2 & 0 & 22 & -12 \\ & & -6 & -18 & 12 \\ \hline & -2 & -6 & 4 & 0 \end{array}$$

We conclude

$$\frac{-2p^3 + 22p - 12}{p - 3} = -2p^2 - 6p + 4$$

Group Exploration

Looking ahead: Factoring the difference of two squares

Note that $(x + 5)(x - 5) = x^2 - 25$. When we work backward and write $x^2 - 25 = (x + 5)(x - 5)$, we say we have *factored* $x^2 - 25$. So we factor a polynomial by writing it as a product.

1. Multiply: $(x + 4)(x - 4)$.
2. Factor $x^2 - 16$.
3. Factor $x^2 - 9$.
4. Factor $w^2 - 4$.
5. Use long division to find the quotient $\dfrac{w^2 - 4}{w - 2}$. To verify your work, check that the product of the divisor and the quotient is the dividend. Explain how your work is related to your work in Problem 4.
6. Factor $25x^2 - 49$.
7. Use long division to find the quotient $\dfrac{25x^2 - 49}{5x + 7}$. To verify your work, check that the product of the divisor and the quotient is the dividend. Explain how your work is related to your work in Problem 6.
8. Factor $9w^2 - 4y^2$.

Homework 7.5

For extra help ▶ MyMathLab®

Watch the videos in MyMathLab

Download the MyDashboard App

Divide. Use a graphing calculator table to verify your work.

1. $\dfrac{3x^5 + 7x^3}{x}$
2. $\dfrac{2x^6 + 9x^2}{x^2}$
3. $\dfrac{6x^3 + 12x^2}{3x}$
4. $\dfrac{8x^3 + 20x}{4x}$
5. $\dfrac{20p^9 - 5p^6 + 15p^3}{5p^3}$
6. $\dfrac{10c^5 + 8c^3 - 4c^2}{2c^2}$
7. $\dfrac{16x^5 - 3x^4 - 24x^3}{-8x^2}$
8. $\dfrac{18x^5 + 5x^4 - 6x^3}{-6x^3}$

Divide. Verify your work by multiplying.

9. $\dfrac{4x^5 - 8x^4 + 7}{2x^4}$
10. $\dfrac{20x^6 - 35x^4 - 10}{5x^3}$
11. $\dfrac{-8k^4 - 6k^3 + 12}{-4k^2}$
12. $\dfrac{-2b^5 + 9b^4 - 15}{-3b^3}$
13. $\dfrac{x^4y + x^3y^2 - xy^4}{xy}$
14. $\dfrac{x^5y - x^4y^2 - x^2y^4}{xy}$
15. $\dfrac{10m^4r^2 + 3m^2r^3 + 4mr^4}{2mr^2}$
16. $\dfrac{6w^5p + 12w^4p^2 + 2w^3p^3}{6w^2p}$
17. $\dfrac{12x^6y^5 + 18x^5y^4 - 8x^4y^2}{-4x^3y^2}$
18. $\dfrac{18x^5y^2 - 9x^2y^3 + 7xy^4}{-3xy^2}$

Perform long division. To verify your work, check that the product of the divisor and the quotient plus the remainder is the dividend.

19. $\dfrac{3x^2 + 14x + 8}{x + 4}$
20. $\dfrac{2x^2 + 11x + 15}{x + 3}$
21. $\dfrac{4x^2 - 22x + 10}{x - 5}$
22. $\dfrac{5x^2 - 7x - 6}{x - 2}$
23. $\dfrac{6p^2 + 19p + 12}{3p + 2}$
24. $\dfrac{8w^2 + 26w + 7}{2w + 6}$
25. $\dfrac{4x^2 + 2x - 23}{2x + 5}$
26. $\dfrac{15x^2 + 14x - 6}{3x + 4}$
27. $\dfrac{10x^2 - 7x + 3}{5x - 1}$
28. $\dfrac{12x^2 - 23x + 2}{4x - 1}$
29. $\dfrac{2m - 9 + 20m^2}{4m - 2}$
30. $\dfrac{10b^2 - 35 + 27b}{5b - 4}$

Perform long division. Use a graphing calculator table to verify your work.

31. $\dfrac{4x^3 + 12x^2 + 15x + 13}{2x + 3}$
32. $\dfrac{15x^3 - 4x^2 - 24x - 11}{5x + 2}$
33. $\dfrac{17p + 12p^3 - 14 - 22p^2}{3p - 4}$
34. $\dfrac{7 - 13w^2 + 6w^3 - 9w}{2w - 5}$
35. $\dfrac{x^2 + 7}{x + 3}$
36. $\dfrac{x^2 + 5}{x + 2}$
37. $\dfrac{8x^3 - 125}{2x - 5}$
38. $\dfrac{27x^3 - 64}{3x - 4}$
39. $\dfrac{3x^3 - 10x^2 + 7}{3x - 1}$
40. $\dfrac{6x^3 - 11x^2 + 5}{2x - 1}$
41. $\dfrac{3x^3 - 4x^2 + 9x - 16}{x^2 + 3}$
42. $\dfrac{2x^3 - 5x^2 + 4x - 7}{x^2 + 2}$
43. $\dfrac{6y^3 + 9y^2 - 4y - 7}{3y^2 - 2}$
44. $\dfrac{10b^3 - 6b^2 - 20b + 17}{2b^2 - 4}$

Perform synthetic division. Use a graphing calculator table to verify your work.

45. $\dfrac{3x^3 - 10x^2 + 10x - 4}{x - 2}$
46. $\dfrac{2x^3 - 11x^2 + 18x - 9}{x - 3}$
47. $\dfrac{-x^3 - 4x^2 + 5x + 23}{x + 4}$
48. $\dfrac{4x^3 + 18x^2 - 7x + 15}{x + 5}$

49. $\dfrac{3k^2 - 7k + 1}{k - 1}$

50. $\dfrac{2w^2 - 5w - 10}{w + 1}$

51. $\dfrac{-11x^2 - 17 + 8x + 2x^3}{x - 5}$

52. $\dfrac{7 + 3x^3 - 9x - 2x^2}{x - 2}$

53. $\dfrac{3x^3 + 11x^2 + 14}{x + 4}$

54. $\dfrac{2x^3 - 7x^2 + 5}{x + 3}$

Concepts

55. A student tries to find a quotient:

$$\frac{6x^4 + 7x^3}{2x^2} = \frac{6x^4}{2x^2} + 7x^3$$
$$= 3x^{4-2} + 7x^3$$
$$= 3x^2 + 7x^3$$

Describe any errors. Then find the quotient correctly.

56. A student tries to find a quotient:

$$\frac{6x^6 - 5x^3}{2x^2} = \frac{6x^6}{2x^2} - \frac{5x^3}{2x^2}$$
$$= 3x^3 - \frac{5x}{2}$$

Describe any errors. Then find the quotient correctly.

57. A student tries to find the quotient $\dfrac{2x^2 + 10x + 12}{x + 3}$:

$$\begin{array}{r} -2x - 4 \\ x + 3 \overline{)\,2x^2 + 10x + 12} \\ \underline{-2x^2 - 6x} \\ 4x + 12 \\ \underline{-4x - 12} \\ 0 \end{array}$$

$$\frac{2x^2 + 10x + 12}{x + 3} = -2x - 4$$

Describe any errors. Then find the quotient correctly.

58. A student tries to find the quotient $\dfrac{3x^3 - 7x^2 + 3x - 2}{x + 2}$:

$$\begin{array}{r|rrrr} 2 & 3 & -7 & 3 & -2 \\ & & 6 & -2 & 2 \\ \hline & 3 & -1 & 1 & 0 \end{array}$$

$$\frac{3x^3 - 7x^2 + 3x - 2}{x + 2} = 3x^2 - x + 1$$

Describe any errors. Then find the quotient correctly.

59. A student tries to find the quotient $\dfrac{4x^3 - 10x^2 - 9x + 15}{x - 3}$:

$$\begin{array}{r|rrrr} 3 & 4 & -10 & -9 & 15 \\ & & 12 & 6 & -9 \\ \hline & 4 & 2 & -3 & 6 \end{array}$$

$$\frac{4x^3 - 10x^2 - 9x + 15}{x - 3} = 4x^3 + 2x^2 - 3x + 6$$

Describe any errors. Then find the quotient correctly.

60. **a.** Find the product $(x + 2)(x + 5)$.

b. Use long division to find the quotient $\dfrac{x^2 + 7x + 10}{x + 5}$.

c. Explain how your work in parts (a) and (b) is related.

61. Give an example of a polynomial and a monomial such that the quotient of the polynomial and the monomial is $2x^2 - 3x + 7$.

62. Give an example of a polynomial and a binomial such that the quotient of the polynomial and the binomial is $5x - 2$.

63. When a certain polynomial is divided by the binomial $x - 3$, the result is $2x - 5 + \dfrac{4}{x - 3}$. What is the polynomial?

64. If the division of two polynomials has remainder 0, what does the product of the divisor and quotient equal? Explain.

65. When we divide a polynomial by a monomial, we divide each term of the polynomial by the monomial. Explain why this make sense.

66. Give several reasons why long division of polynomials is similar to long division of numbers.

67. Describe how to divide a polynomial by a monomial.

68. Describe how to divide a polynomial by a binomial.

69. Describe when synthetic division can be used and how to perform it.

70. Describe how to verify your work after you divide a polynomial by a binomial.

Related Review

Perform the indicated operation. Use a graphing calculator table to verify your work.

71. $(6x^2 - x - 2)(3x - 2)$

72. $(8x^2 - 10x - 3)(4x + 1)$

73. $(6x^2 - x - 2) - (3x - 2)$

74. $(8x^2 - 10x - 3) - (4x + 1)$

75. $\dfrac{6x^2 - x - 2}{3x - 2}$

76. $\dfrac{8x^2 - 10x - 3}{4x + 1}$

77. $(3x - 2)^2$

78. $(4x + 1)^2$

Expressions, Equations, Functions, and Graphs

Give an example of the following. Then solve, simplify, or graph, as appropriate.

79. A quadratic function

80. A linear function

81. A quadratic polynomial in one variable with four terms

82. A linear equation in one variable

83. A cubic polynomial in one variable with five terms

84. A system of two linear equations in two variables

Taking it to the Lab

Climate Change Lab (continued from Chapter 6)

Throughout our study of climate change, we have used the IPCC's yardstick that, by 2050, carbon dioxide emissions should be 0.9 metric ton per person per year. Yet it is difficult to imagine developed countries, with average annual carbon dioxide emissions of 10.2 metric tons per person today, reducing their emissions by 91% to meet the IPCC's recommendation. In particular, it is difficult to imagine that the United States will reduce its annual carbon dioxide emissions of 18.1 metric tons per person by 95% to reach that desired level.

Recall from the Climate Change Lab in Chapter 5 that the IPCC's yardstick is equivalent to recommending that 2050 carbon dioxide emissions be 60% less than the carbon dioxide emissions in 1990. We can reach this goal without reducing per-person carbon dioxide emissions at all—if we reduce world population.

Reducing world population seems impossible, because it is currently growing by leaps and bounds. The United Nations (U.N.) predicts that, from 2000 to 2050, the world population will increase from 6.09 billion to 9.3 billion—a 53% increase (see Table 16).

Table 16 World Population

Year	Population (billions)	Year	Population (billions)
A.D. 1	0.17	1600	0.55
200	0.19	1800	0.81
400	0.19	2000	6.09
600	0.20	2025	7.9
800	0.22	2050	9.3
1000	0.25	2075	9.9
1200	0.36	2100	10.1
1400	0.35		

Source: *The Cambridge Factfinder*

Despite the large population growth in the first half of the century, the population will probably not grow nearly as quickly in the second half. The U.N. has predicted the population will increase from 9.3 billion in 2050 to 10.1 billion in 2100—a mere 9% increase. In fact, the U.N. has predicted that the rate of change in 2100 will be very close to zero and might even be negative.

The U.N. uses a complicated model to make these predictions. A simpler model that matches well with the U.N. model for the years 2000 to 2100 is the quadratic equation $p = -0.00046t^2 + 0.086t + 6.08$, where p is world population (in billions) at t years since 2000.

Using the yardstick of 2050 carbon dioxide emissions at 60% less than the 21.6 billion metric tons of carbon dioxide emissions in 1990, we can calculate the largest carbon dioxide emissions that Earth can handle each year:

$$0.40(21.6) = 8.64 \text{ billion metric tons}$$

Next, we will consider two scenarios that stay within this limit.

For our first scenario, let's assume that in the future each person in the world emits 18.1 metric tons per year, the current per-person annual emissions rate for Americans (see Table 18). Then we could still be at the yearly limit of 8.64 billion metric tons of carbon dioxide if the world population were a mere 477 million. By modeling the data in Table 17, we could show that U.S. population alone will reach 477 million before 2080. Common sense dictates that countries would never voluntarily lower their populations enough to help reach this level.

Table 17 United States Population

Year	U.S. Population (billions)
1960	0.18
1970	0.21
1980	0.23
1990	0.25
2000	0.28
2010	0.31

Source: *U.S. Census Bureau; The Cambridge Factfinder*

For our second scenario, let's make an assumption about how low carbon dioxide emissions could be. Although 0.9 metric ton per person per year seems unreachable,

Table 18 Gross Domestic Product (GDP) Ranks, per-Person GDPs, and per-Person Annual Carbon Dioxide Emissions

Country	2010 GDP Rank	2010 per-Person GDP (thousands of dollars)	2010 per-Person Carbon Dioxide Emissions (metric tons)
Sweden	22	39.0	5.3
Switzerland	19	46.4	5.4
France	5	34.1	5.5
Italy	8	32.0	6.7
United Kingdom	6	35.7	7.9
Austria	25	40.0	8.1
Denmark	31	40.2	8.4
Japan	3	33.7	8.9
Germany	4	37.4	9.3
Belgium	21	37.6	9.9
Norway	23	57.2	10.5
Australia	15	38.2	16.0
United States	1	47.2	18.1
Netherlands	16	42.2	31.9

Source: *World Bank; Carbon Dioxide Information Analysis Center*

perhaps 5.4 metric tons per person per year is attainable. After all, Sweden, Switzerland, and France already have per-person annual carbon dioxide emissions at or near 5.4 metric tons (see Table 18). If all countries annually produced 5.4 metric tons of carbon dioxide per person, Earth could handle 1.6 billion people.

The first scenario reveals that we will not be able to get carbon dioxide emissions under control merely by reducing world population. Earlier, we observed that simply reducing per-person emissions will not be acceptable. So, as the second scenario suggests, it is only by reducing both world population and per-person emissions that we can reach the IPCC's goal.

Analyzing the Situation

1. Let w be world population (in billions of people) in the year t. For example, $t = 2005$ represents the year 2005.
   a. Use a graphing calculator to draw a scattergram of world population data.
   b. Calculate the changes in world population for every 200 (or 199) years up until 2000.
   c. When has world population grown most? Explain why it is surprising world population will increase by only 0.2 billion from 2075 to 2100.
2. Now let w be world population (in billions of people) at t years *since 2000*.
   a. Create a scattergram for the years 2000–2100 that are listed in Table 16. In the same viewing window, draw a graph of the scattergram and the quadratic world population model provided earlier.
   b. Use the quadratic model to predict world population for the years 2000, 2025, 2050, 2075, and 2100.
   c. Does the quadratic model make predictions that are similar to the U.N. predictions listed in Table 16? Explain.
3. In Problem 4 of the Climate Change Lab in Chapter 5, you found a model of the U.S. population (in billions) at t years since 1950. If you didn't find such an equation, find it now. Then use it to predict when the U.S. population will reach 477 million. Finally, compare your result with the claim that this will happen before 2080.
4. Perform a unit analysis for the estimate of the amount of carbon dioxide emissions (in billions of tons of carbon dioxide) that Earth can withstand each year:

   $$0.40(21.6) = 8.64 \text{ billion metric tons}$$

   Explain why this computation gives the amount of carbon dioxide emissions that Earth can withstand each year.
5. Show the work for the claim that if everyone in the world emitted carbon dioxide as Americans do, then Earth could support only 477 million people.
6. Verify the claim that if worldwide annual carbon dioxide emissions were 5.4 tons of carbon dioxide per person, then Earth could support 1.6 billion people.

Projectile Lab

In this lab, you will estimate your vertical throwing speed.

Materials

You will need at least three people and the following items:

1. A baseball or other ball (a solid, heavy ball works best)
2. A digital stopwatch or some other timing device

Recording of Data

The first person should throw the ball straight up and say "Mark" at the moment the ball is released. The second person should hold his or her hand at the position of the ball's release and should say "Mark" when the ball returns to its initial height. The third person should begin timing when the first person says "Mark" and stop timing when the second person says "Mark."

Theory

Throughout this lab, we will use "height" to mean distance above the ground. Let h_0 be the ball's height (in feet) at the moment of its release. Let v_0 be the ball's speed (in feet per second) at that moment. We call h_0 the *initial height* and v_0 the *initial velocity*. The ball's height h (in feet) is given by

$$h = -16t^2 + v_0t + h_0$$

where t is the time (in seconds) since the ball was released. So, t and h are variables and v_0 and h_0 are constants.

Analyzing the Situation

1. Substitute 0 for t in the equation $h = -16t^2 + v_0t + h_0$ and solve for h. Explain why your work shows that h_0 represents the initial height.
2. What was the actual initial height of the ball in your experiment? Substitute this value for h_0 in the equation $h = -16t^2 + v_0t + h_0$.
3. How long did it take for the ball to return to its initial height? Substitute that time for t and the initial height for h in the equation you found in Problem 2. Then solve for v_0. What does your value for v_0 mean in this situation?
4. Substitute the value for v_0 that you found in Problem 3 into the equation you found in Problem 2. [**Hint:** Your equation should still have the variables t and h in it.]
5. Use a graphing calculator to draw a graph of your model. Use a pencil and paper to copy the graph.
6. What is the h-intercept of the model? What does it mean in this situation?
7. Use the model to estimate the height of the ball at 1 second.

Chapter Summary

Key Points of Chapter 7

Section 7.1 Adding and Subtracting Polynomial Expressions and Functions

Term	A **term** is a constant, a variable, or a product of a constant and one or more variables raised to powers.
Monomial	A **monomial** is a constant, a variable, or a product of a constant and one or more variables raised to counting-number powers.
Polynomial or polynomial expression	A **polynomial,** or **polynomial expression,** is a monomial or a sum of monomials.
Like terms	**Like terms** are either constant terms or variable terms that contain the same variable(s) raised to exactly the same power(s).
Combining like terms	To combine like terms, add the coefficients of the terms.
Adding polynomials	To add polynomials, combine like terms.
Subtracting polynomials	To subtract polynomials, first distribute -1 and then combine like terms.
The meaning of the sign of a difference	If a difference $A - B$ is positive, then A is more than B. If a difference $A - B$ is negative, then A is less than B.

Section 7.2 Multiplying Polynomial Expressions and Functions

Binomial and trinomial	We refer to a polynomial as a **binomial** or a **trinomial,** depending on whether it has two or three nonzero terms, respectively.
Product property for exponents	If n and m are counting numbers, then $x^m x^n = x^{m+n}$.
Multiplying two polynomials	To multiply two polynomials, multiply each term in the first polynomial by each term in the second polynomial. Then combine like terms if possible.

Section 7.3 Powers of Polynomials; Product of Binomial Conjugates

Raising a product to a power	If n is a counting number, then $(xy)^n = x^n y^n$.
Square of a sum	$(A + B)^2 = A^2 + 2AB + B^2$
Square of a difference	$(A - B)^2 = A^2 - 2AB + B^2$
Product of binomial conjugates	$(A + B)(A - B) = A^2 - B^2$

Section 7.4 Properties of Exponents

Simplifying power expressions	A power expression is simplified if **1.** It includes no parentheses. **2.** In any monomial, each variable or constant appears as a base at most once. **3.** Each numerical expression (such as 7^2) has been calculated, and each numerical fraction has been simplified.

Section 7.4 Properties of Exponents (*Continued*)

Quotient property for exponents	If n and m are counting numbers and x is nonzero, then $\frac{x^m}{x^n} = x^{m-n}$.
Zero exponent	For nonzero x, $x^0 = 1$.
Raising a quotient to a power	If n is a counting number and y is nonzero, then $\left(\frac{x}{y}\right)^n = \frac{x^n}{y^n}$.
Raising a power to a power	If m and n are counting numbers, then $(x^m)^n = x^{mn}$.

Section 7.5 Dividing Polynomials: Long Division and Synthetic Division

Dividing a polynomial by a monomial	To divide a polynomial by a monomial, use the property $\frac{A + C}{B} = \frac{A}{B} + \frac{C}{B}$, where B is nonzero, to divide each term of the polynomial by the monomial.
Dividing a polynomial by a binomial	To divide a polynomial by a binomial, use long division.
Synthetic division	To perform synthetic division, the divisor must be of the form $x - a$. The first step is to write the dividend in descending order. Then we write a inside the $\lrcorner$ and write the coefficients of the dividend outside it.

Chapter 7 Review Exercises

Combine like terms when possible. Use a graphing calculator to verify your work when possible.

1. $-4x^3 + 5x - 2x^2 - 8x + x^3$

2. $3a^4b - 2a^3b^2 + 5a^2b^3 - 7a^3b^2 - 9a^2b^3$

For Exercises 3 and 4, perform the operation.

3. $(-7x^3 + 5x^2 - 9) + (2x^3 - 8x^2 + 3x)$

4. $(5a^3b - 2a^2b^2 + 9ab^3) - (8a^3b + 4a^2b^2 - ab^3)$

5. For $f(x) = 3x^2 - 5x + 2$, find $f(-2)$.

For Exercises 6–9, refer to Fig. 26.

6. Find $f(-3)$.

7. Find x when $f(x) = 3$.

8. Find x when $f(x) = 1$.

9. Find x when $f(x) = -3$.

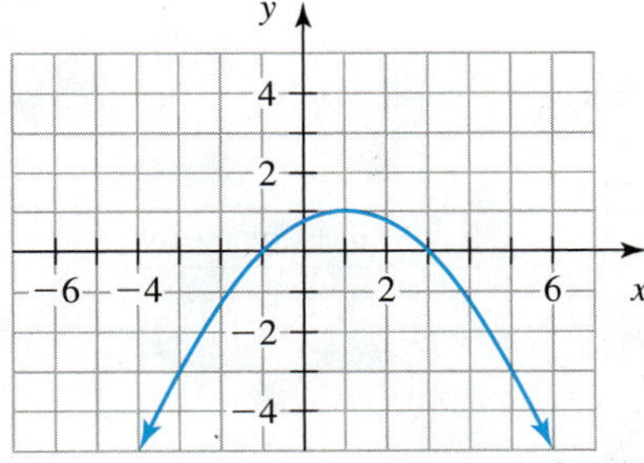

Figure 26 Exercises 6–9

Values of a quadratic function f are listed in Table 19. For Exercises 10–13, refer to this table.

10. Find $f(2)$.

11. Find x when $f(x) = 9$.

12. Find x when $f(x) = 10$.

13. Find x when $f(x) = 11$.

Table 19 Some Values of a Quadratic Function f (Exercises 10–13)

x	y
0	1
1	6
2	9
3	10
4	9
5	6
6	1

For $f(x) = 3x^3 - 7x^2 - 4x + 2$ *and* $g(x) = -2x^3 + 5x^2 - 3x + 1$, *find an equation of the given function; then evaluate the function at the indicated value.*

14. $(f + g)(x), (f + g)(2)$

15. $(f - g)(x), (f - g)(-3)$

16. The number (in thousands) of bank tellers is modeled by the function $B(t) = -2.42t^2 + 80.8t - 76$, where t is the number of years since 1990 (see Table 20). The number (in thousands) of ATMs is modeled by the function $A(t) = -2.29t^2 + 76.8t - 227$, where t is the number of years since 1990.

Table 20 Numbers of Bank Tellers and ATMs

Year	Number of Bank Tellers (thousands)	Number of ATMs (thousands)
1999	453	273
2001	533	352
2003	539	383
2005	599	395
2007	608	425
2009	577	402

Source: *American Bankers Association*

a. Find an equation of the sum function $B + A$.
b. Perform a unit analysis of the expression $B(t) + A(t)$.
c. Find $(B + A)(22)$. What does it mean in this situation?
d. Find an equation of the difference function $B - A$.
e. Find $(B - A)(22)$. What does it mean in this situation?

Find the product or simplify, as appropriate.

17. $(x - 7)(x + 7)$
18. $8a^2b(-5a^3b^5)$
19. $(4p + 9t)(2p - 5t)$
20. $(4x - 3)(5x^2 - 2x + 4)$
21. $(3x + 7y)^2$
22. $(6p^2 - 9t^3)(6p^2 + 9t^3)$
23. $-3rt^3(2r^2 - 5rt + 3t^2)$
24. $-4x(3x - 2)^2$
25. $(3m^2 - mp + 2p^2)(2m^2 + 3mp - 4p^2)$
26. $(2a^2 + 3b^2)(5a^2 - 2b^2)$
27. $(3p - 4t)(3p + 4t)(9p^2 + 16t^2)$
28. $(x - 4)^2$
29. $(4b^2 - b + 3)(2b - 7)$
30. $(w - 3)(w - 9)$
31. $-5(c + 2)^2$
32. $(4m - 7n)(4m + 7n)$

For Exercises 33 and 34, if $f(x) = x^2 - 2x$, find the following.

33. $f(a - 4)$
34. $f(a + 3) - f(a)$
35. Write the function $f(x) = -2(x - 4)^2 + 3$ in standard form.
36. For $f(x) = 3x - 7$ and $g(x) = 2x^2 - 4x + 3$, find an equation of the product function $f \cdot g$. Then find $(f \cdot g)(3)$.
37. The annual consumption (in gallons consumed per person) of sports drinks $G(t)$ can be modeled by the equation $G(t) = 0.47t - 3$, where t is the number of years since 1990 (see Table 21). In Example 9 of Section 7.2, we found the equation $P(t) = 2.8t + 254$, where $P(t)$ is the U.S. population (in millions of people) at t years since 1990.

a. Find an equation of the product function $G \cdot P$.
b. Perform a unit analysis of the expression $G(t) \cdot P(t)$.

Table 21 U.S. Per-Person Annual Consumption of Sports Drinks; U.S. Population

Year	Annual Consumption (gallons per person)	Year	Population (millions)
2002	2.7	1998	275.9
2003	3.0	2000	282.2
2004	3.4	2002	287.8
2005	4.1	2004	293.0
2006	4.5	2006	298.6
		2008	304.4

Sources: *Beverage Digest; U.S. Census Bureau*

c. Find $(G \cdot P)(27)$. What does it mean in this situation?
d. Use a graphing calculator graph to determine whether the function $G \cdot P$ is increasing, decreasing, or neither for values of t between 10 and 30. What does your result mean in this situation?

38. Is the product of a linear polynomial and a quadratic polynomial a linear, quadratic, or cubic polynomial? Give an example.

Simplify.

39. $(-x^5)^2$
40. $(2x^3)(6x^4)$
41. $(8a^2b^3)(-5a^4b^9)$
42. $\dfrac{8x^4y^8}{16x^3y^5}$
43. $\left(\dfrac{x}{2}\right)^3$
44. $(2x^9y^3)^5$
45. $3x^6(5x^4)^2$
46. $\dfrac{15c^2c^7}{10c^4}$
47. $\left(\dfrac{a^4}{9}\right)^2$
48. $\left(\dfrac{-9x^5}{5y^7}\right)^0$
49. $\dfrac{(3x^5y^4)^2}{6x^7y^3}$
50. $\left(\dfrac{3x^5}{4x^2}\right)^3$

Divide.

51. $\dfrac{6x^5 - 2x^3 + 9}{-3x^2}$
52. $\dfrac{20w^4p + 15w^3p^2 - 35w^2p^3}{5w^2p}$

Perform long division.

53. $\dfrac{8x^2 - 6x - 8}{2x + 1}$
54. $\dfrac{6x^3 - 16x^2 + 17x - 2}{3x - 2}$
55. $\dfrac{64b^3 - 27}{4b - 3}$

Perform synthetic division.

56. $\dfrac{2x^3 - 10x^2 + 15x - 4}{x - 3}$
57. $\dfrac{5y^2 - 8 + 3y^3 - 6y}{y + 2}$
58. $\dfrac{x^3 + x^2 + 16}{x + 3}$

Chapter 7 Test

Perform the operation.

1. $(4a^3b - 9a^2b^2 - 2ab^3) + (-5a^3b + 4a^2b^2 + 3ab^3)$
2. $(2x^3 - 4x^2 + 7x) - (6x^3 - 3x^2 + 9x)$

For Exercises 3–6, refer to the graph sketched in Fig. 27.

3. Find $f(-5)$.
4. Find a when $f(a) = -3$.
5. Find a when $f(a) = -4$.
6. Find a when $f(a) = -5$.

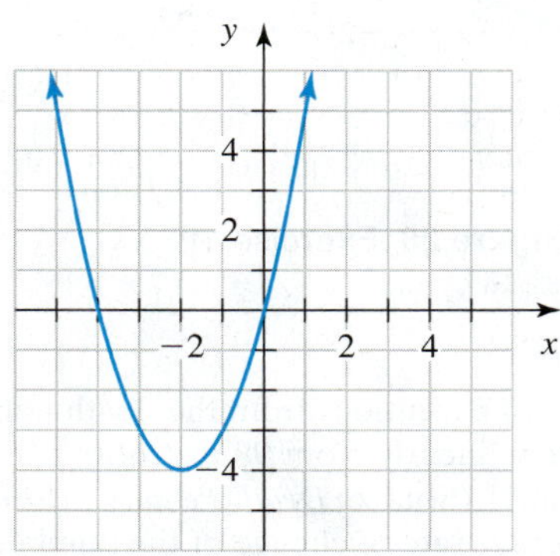

Figure 27 Exercise 3–6

7. For $f(x) = 4x^2 + 5x - 9$ and $g(x) = 6x^2 - 3x + 7$, find an equation of the difference function $f - g$; then find $(f - g)(-2)$.
8. The number (in millions) of women who live alone $W(t)$ is modeled by the equation $W(t) = 0.21t + 11.58$, where t is the number of years since 1980 (see Table 22). The number (in millions) of men who live alone $M(t)$ is modeled by the equation $M(t) = 0.24t + 6.75$, where t is the number of years since 1980.

Table 22 Numbers of Women and Men Who Live Alone

Year	Number Living Alone (millions) Women	Men
1980	11.3	7.0
1985	12.7	7.9
1990	14.0	9.0
1995	14.6	10.1
2000	15.6	11.2
2005	17.3	12.8
2010	17.4	14.0

Source: *U.S. Census Bureau*

   a. Find an equation of the sum function $W + M$.
   b. Perform a unit analysis of the expression $W(t) + M(t)$.
   c. Find $(W + M)(38)$. What does it mean in this situation?
   d. Find an equation of the difference function $W - M$.
   e. Find $(W - M)(38)$. What does it mean in this situation?

For Exercises 9–15, find the product or simplify, as appropriate.

9. $-2xy^2(7x^2 - 3xy + 6y^2)$
10. $(4x - 7y)(3x + 5y)$
11. $(2w - 5t)(3w^2 - wt + 4t^2)$
12. $(x - 7)^2$
13. $3x(2x + 3)^2$
14. $(3x^2 + x - 5)(2x^2 + 4x - 1)$
15. $(4x^2 + 9y^2)(4x^2 - 9y^2)$
16. For $f(x) = x^2 - 3x$, find $f(a - 5)$.
17. For $f(x) = 2x^2 - 5x + 4$ and $g(x) = 3x - 2$, find an equation of the product function $f \cdot g$; then find $(f \cdot g)(3)$.
18. The death rate from heart disease (number of deaths per 100,000 people) $R(t)$ in the United States is modeled by the function $R(t) = -7.8t + 564$, where t is the number of years since 1960 (see Table 23). The U.S. population (in 100,000s of people) $P(t)$ is modeled by the equation $P(t) = 26t + 1784$, where t is the number of years since 1960.

Table 23 Death Rates from Heart Disease; U.S. Population

Year	Death Rate (number of deaths per 100,000 people)	Population (in 100,000s)
1960	559	1807
1970	493	2051
1980	412	2277
1990	322	2501
2000	258	2824
2008	187	3048

Sources: *U.S. Center for Health Statistics; U.S. Census Bureau*

   a. Check that the models fit the data well.
   b. Find an equation of the product function $R \cdot P$.
   c. Perform a unit analysis of the expression $R(t) \cdot P(t)$.
   d. Find $(R \cdot P)(58)$. What does it mean in this situation?
   e. Use a graphing calculator graph to determine whether the function $R \cdot P$ is increasing, decreasing, or neither for values of t between 5 and 60. What does your result mean in this situation? Explain how that is possible, given that U.S. population has been increasing.

19. Write $f(x) = -3(x - 1)^2 + 5$ in standard form.

For Exercises 21–26, simplify.

20. $\dfrac{6x^7y^4}{8x^3y^9}$
21. $(4a^3b^5)^3a^6b$
22. $\left(\dfrac{x^3}{y^4}\right)^6$
23. $(7x^3)^2$
24. $\left(\dfrac{x^3y^6}{2w^3}\right)^4$
25. $\dfrac{(2p^5t^2)^3}{4p^2t^3}$
26. A person plays an electric guitar outside. The sound level (in decibels) $f(d)$ is described by the equation $f(d) = \dfrac{58{,}000}{d^2}$, where d is the distance (in feet) from the amplifier. What is the sound level at 30 feet?
27. Find the quotient $\dfrac{9w^5y - 12w^4y^3 + 6w^3y^4}{3w^2y}$.
28. Divide: $\dfrac{12x^3 - 7x^2 + 9x + 2}{4x - 1}$.
29. Perform synthetic division: $\dfrac{x^3 - 7x^2 + 13x - 8}{x - 2}$.

Making Sure You're Ready for Intermediate Algebra: A Review of Chapters 1–7

The exercises that follow can be used in both Elementary Algebra and Intermediate Algebra. The exercises can be used

- at the end of Elementary Algebra to help you prepare for the final exam.
- at the start of Intermediate Algebra to make sure that you're fully prepared.

The exercises can steer you to material that you may need to study. After completing an exercise, look up the answer in the Answer section. If your answer is correct and *you feel confident about that material,* go on to the next exercise. If your answer is incorrect, find and read the example that is referenced in parentheses. If the example does not provide enough help, read the text surrounding the example or read the entire referenced section, depending on how much help you need. Once you feel that you have the idea, don't make any assumptions! It is essential that you practice similar exercises in the section's homework until you consistently can get correct results.

1. Let p be the percentage of Americans at age a years who work.
   a. What is the independent variable? What is the dependent variable? (Section 1.2, Example 1)
   b. What does the ordered pair $(45, 85)$ mean in this situation? (Section 1.2, Example 2)
2. An hour ago the temperature was 1°F. The temperature is now −5°F. What is the change in temperature? (Section 2.4, Examples 1, 2, 3, and 5)

Perform the indicated operations.

3. $5(-2) - (-6)^2 + 4$ (Section 2.6, Examples 2–4)
4. $9 - (6 - 8)^3 + 4 \div (-2)$ (Section 2.6, Examples 2–4)

For Exercises 5 and 6, evaluate the given expression for $a = -4$, $b = -2$, and $c = 5$.

5. $b^2 - 4ac$ (Section 2.6, Examples 5 and 6)
6. $\dfrac{c + a^2}{a - b^2}$ (Section 2.6, Examples 5 and 6)
7. Graph $x = -3$ by hand. (Section 3.2, Example 5)
8. Find the slope of the line that passes through the points $(-3, -1)$ and $(5, 3)$. State whether the line is increasing, decreasing, horizontal, or vertical. (Section 3.3, Examples 2–9)

For Exercises 9 and 10, graph the equation.

9. $y = 4x - 5$ (Section 3.4, Example 3)
10. $y = -\frac{2}{5}x + 1$ (Section 3.4, Example 4)
11. Find an equation of the line sketched in Fig. 28. (Section 3.4, Example 8)

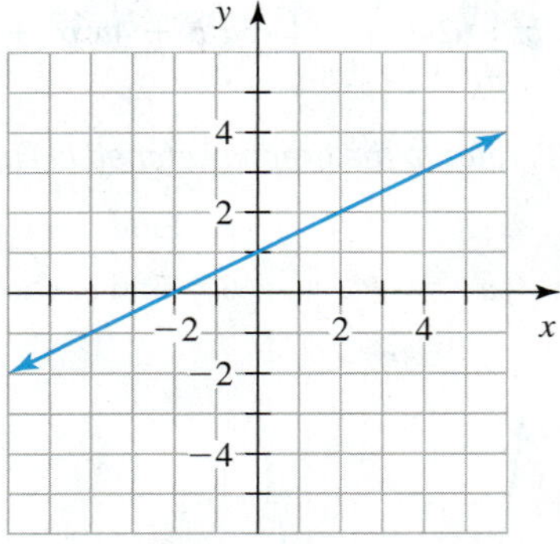

Figure 28 Exercise 11

12. The number of executions from the death penalty decreased approximately linearly from 98 executions in 1999 to 43 executions in 2011 (Source: *Death Penalty Information Center*). Find the average rate of change of the number of executions. (Section 3.5, Examples 1 and 2)
13. The number of U.S. households that paid bills online was 66 million in 2010 and has increased by about 5.3 million per year (Source: *Forrester Research*). Let n be the number of households (in millions) that paid bills online at t years since 2010. (Section 3.5, Examples 4 and 5)
    a. Is there an approximate linear relationship between t and n? Explain. If the relationship is approximately linear, find the slope and describe what it means in this situation.
    b. What is the n-intercept of the model? What does it mean in this situation?
    c. Find an equation of the model.
    d. Perform a unit analysis of the equation you found in part (c).
    e. Predict the number of households that will pay bills online in 2016.
14. Simplify $7(2x - 5y) - 4(6x + 3y)$. (Section 4.2, Example 2)
15. Let x be a number. Translate the phrase "Seven minus 4 times the quotient of the number and 2" to an expression. Then simplify the expression. (Section 4.2, Example 3)

For Exercises 16–18, solve.

16. $5t - 10 = 4 - 7t$ (Section 4.4, Example 3)
17. $3(4x - 1) - (7x + 2) = 8(x - 3)$ (Section 4.4, Example 4)
18. $\frac{3}{4}x - \frac{7}{2} = \frac{3}{8}x$ (Section 4.4, Example 5)
19. Three times the sum of a number and 5 is 9. Find the number. (Section 4.4, Example 11)

For Exercises 20 and 21, solve the given equation or system by referring to the graphs shown in Fig. 29.

20. $-\frac{1}{4}x - \frac{3}{2} = -3$ (Section 4.3, Example 9)

21. $-\frac{1}{4}x - \frac{3}{2} = 2x + 3$ (Section 4.4, Example 13)

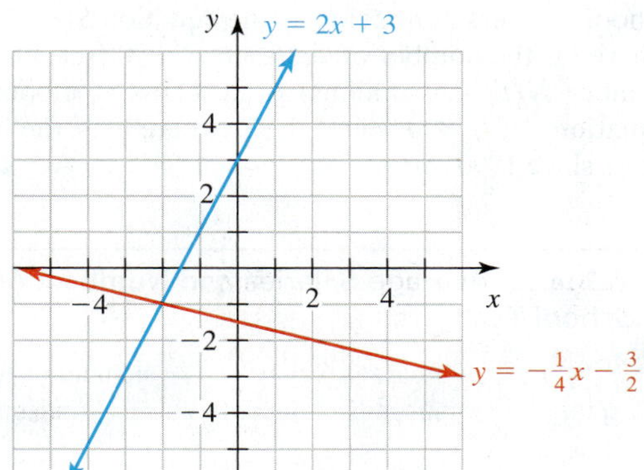

Figure 29 Exercises 20 and 21

22. Solve the formula $S = 2\pi r^2 + rh$ for h. (Section 4.6, Examples 5–8)

23. Graph $2x - 4y = 8$ by hand. (Section 5.1, Example 1)

24. Find the x-intercept and the y-intercept of the graph of $2x - 5y = 20$. (Section 5.1, Example 5)

For Exercises 25–27, refer to the relation graphed in Fig. 30.

25. Find the domain of the relation. (Section 5.2, Example 9)

26. Find the range of the relation. (Section 5.2, Example 9)

27. Is the relation a function? Explain. (Section 5.2, Examples 5–7)

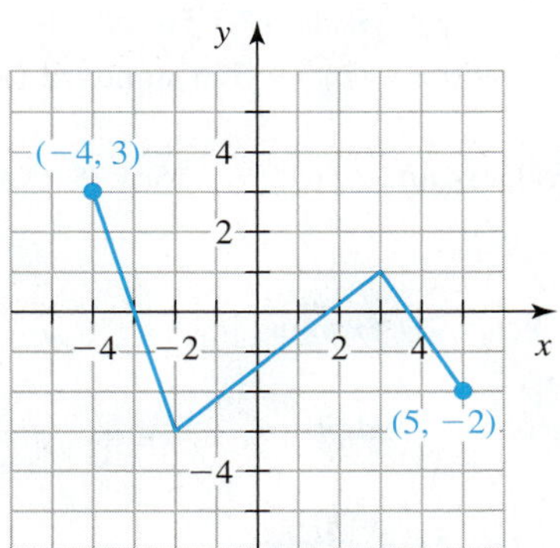

Figure 30 Exercises 25–27

28. For $f(x) = -3x^2 + 7x$, find $f(-2)$. (Section 5.3, Examples 1 and 2)

29. For $g(x) = 6x - 2$, find x when $g(x) = -5$. (Section 5.3, Example 4)

30. Find an equation of the line that has slope $\frac{2}{3}$ and contains the point $(-2, -6)$. Write your equation in slope–intercept form. (Section 5.4, Examples 1 and 2)

31. Find an equation of the line that contains the points $(-4, 7)$ and $(2, -3)$. Write your equation in slope–intercept form. (Section 5.4, Examples 3 and 4)

32. The percentage of high school students who dropped out of school are shown in Table 24 for various years.

Table 24 Percentages of High School Students Who Dropped Out of School

Year	Percent
2004	12.1
2005	11.3
2006	11.0
2007	10.2
2008	9.3
2009	9.4

Source: *U.S. Census Bureau*

Let $p = f(t)$ be the percentage of high school students who have dropped out of school at t years since 2000.

a. Use a graphing calculator to draw a scattergram of the data. (Section 5.5, Example 2)

b. Find an equation of a linear model to describe the data. (Section 5.5, Examples 2 and 3)

c. Predict when 4% of high school students will drop out of school. (Section 5.6, Example 1)

d. What is the p-intercept? What does it mean in this situation? (Section 5.6, Example 2)

e. What is the t-intercept? What does it mean in this situation? (Section 5.6, Example 2)

33. The percentage of Americans who say religion is not very important in their lives has decreased approximately linearly from 61% in 2003 to 55% in 2011 (Source: *Gallup Organization*). Predict when half of Americans will say religion is not very important in their lives. (Section 5.6, Example 3)

34. Solve the inequality $2(x - 1) + 4 < 5(x + 3) - 1$. Describe the solution set as an inequality, in interval notation, and in a graph. (Section 5.7, Example 8)

35. Solve the inequality $-10 \leq 3x - 7 \leq 2$. Describe the solution set as an inequality, in interval notation, and in a graph. (Section 5.7, Example 11)

36. Figure 31 shows the graphs of two linear equations. Find the solution of the system of the two equations. (Section 6.1, Examples 1–3)

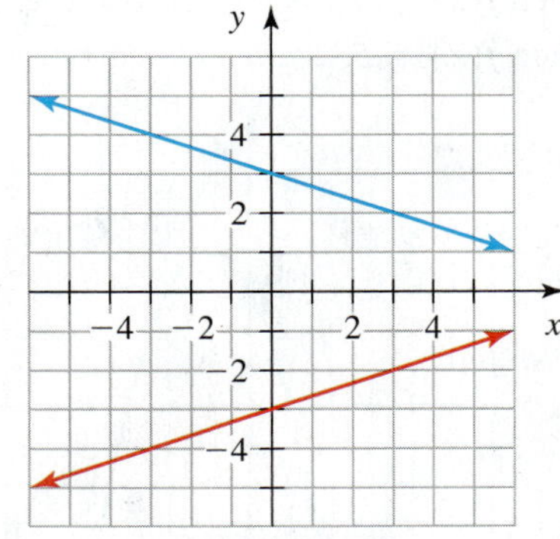

Figure 31 Exercise 36

37. Solve the system: (Section 6.2, Example 1)

$$2x - 3y = 7$$
$$y = 4x - 9$$

38. Solve the system: (Section 6.3, Example 3)

$$3x + 4y = 4$$
$$7x - 5y = 38$$

39. Sales of digital downloads of music were 1247 million units in 2010 and have since increased by 184 million units per year. Sales of physical media (CDs, vinyl, and cassettes) of music were 231 million units in 2010 and have since decreased by 91 million units per year (Source: *Recording Industry Association of America*). (Section 6.4, Example 3)

a. Let $D(t)$ be the annual sales (in millions of units) of digital downloads of music at t years since 2010. Find an equation of D.

b. Let $P(t)$ be the annual sales (in millions of units) of physical media of music at t years since 2010. Find an equation of P.

c. Estimate in which year the sales of digital downloads of music were equal to the sales of physical media of music. What are those sales?

40. A 6000-seat theater has tickets for sale at \$20 and \$35. How many tickets should be sold at each price for a sell-out performance to generate a total revenue of \$147,000? (Section 6.5, Examples 2 and 3)

41. Graph the inequality $4x - 5y \geq 20$ by hand. (Section 6.6, Example 3)

42. Graph the solution set of the system. (Section 6.6, Examples 5 and 6)

$$y > -\frac{1}{2}x + 2$$
$$y < 3$$

43. Find the sum $(3x^2 - 5x) + (-7x^2 - 2x + 9)$. (Section 7.1, Example 4)

44. Find the difference $(7a^2 - ab - 4b^2) - (2a^2 - 5ab + 3b^2)$. (Section 7.1, Example 5)

45. For $f(x) = 4x^2 - 8x - 3$ and $g(x) = -2x^2 - x + 5$, find an equation of the difference function $f - g$; then find $(f - g)(-3)$. (Section 7.1, Example 10)

For Exercises 46–49, refer to the graph sketched in Fig. 32.

46. Find $f(4)$. (Section 7.1, Example 8)

47. Find x when $f(x) = 4$. (Section 7.1, Example 8)

48. Find x when $f(x) = 3$. (Section 7.1, Example 8)

49. Find x when $f(x) = 5$. (Section 7.1, Example 8)

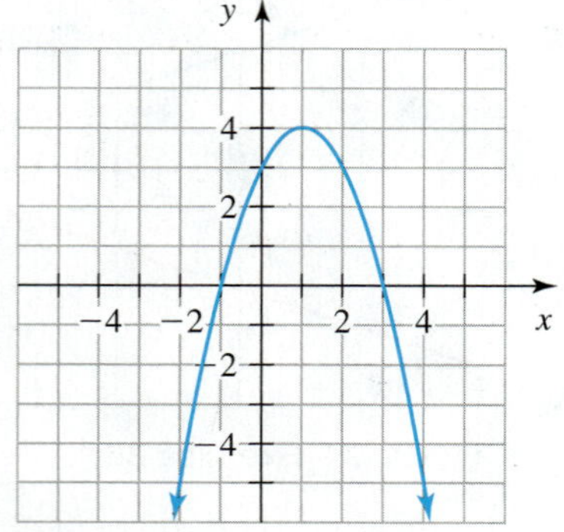

Figure 32 Exercises 46–49

50. Find the product $(8p - 3t)(2p + 5t)$. (Section 7.2, Example 4)

51. Find the product $(x^2 - 2x + 4)(2x^2 + x - 3)$. (Section 7.2, Example 7)

52. For $f(x) = 3x - 5$ and $g(x) = 2x^2 - 4x + 3$, find an equation of the product function $f \cdot g$; then find $(f \cdot g)(2)$. (Section 7.2, Example 8)

53. The average salary (in thousands of dollars) $S(t)$ of public school teachers is modeled by the equation $S(t) = 1.2t + 31$, where t is the number of years since 1990 (see Table 25). The number $N(t)$ (in millions) of teachers is modeled by the equation $N(t) = 0.046t + 2.83$, where t is the number of years since 1990. (Section 7.2, Example 9)

Table 25 Average Salaries and Numbers of Public School Teachers

Year	Average Salary (thousands of dollars)	Number of Teachers (millions)
1990	31.4	2.8
1995	37.3	3.0
2000	41.8	3.4
2005	47.5	3.6
2009	54.3	3.6

Source: *National Center for Education Statistics; National Education Association*

a. Find an equation of the product function $S \cdot N$. Round the coefficients of your result to the second decimal place.

b. Perform a unit analysis of the expression $S(t) \cdot N(t)$. [**Hint:** What is one thousand times one million?]

c. Find $(S \cdot N)(28)$. What does your result mean in this situation?

54. Simplify $(x + 6)^2$. (Section 7.3, Example 2)

55. Simplify $(3p - 7q)^2$. (Section 7.3, Example 3)

56. Write $f(x) = -3(x - 2)^2 + 5$ in standard form. (Section 7.3, Example 5)

57. Find the product $(8b - 5c)(8b + 5c)$. (Section 7.3, Example 6)

Simplify.

58. $3x^5(2x^2)^3$ (Section 7.4, Example 5)

59. $\left(\dfrac{x^4w}{2y^8}\right)^5$ (Section 7.4, Example 6)

60. $\dfrac{(3x^2y^3)^4}{x^7y^2}$ (Section 7.4, Example 6)

61. Find the quotient $\dfrac{8x^5 - 10x^3 + 7}{-6x^2}$. (Section 7.5, Example 2)

62. Divide: $\dfrac{15x^3 - 7x^2 - 8x + 1}{3x - 2}$. (Section 7.5, Example 5)

Factoring Polynomials and Solving Polynomial Equations

Do you or one of your friends own an iPhone? Sales of iPhones have increased greatly in the past few years (see Table 1). In Exercise 7 of Homework 8.6, you will predict when iPhone sales will reach 212 million phones.

In Chapter 7, we discussed how to multiply polynomials. In this chapter, we will discuss how to do the reverse process, *factoring*, which will help us solve new types of equations. In turn, solving these equations will help us use quadratic functions to make predictions about authentic situations, such as when the annual revenue from gambling in the United States will be $24 billion.

Table 1 iPhone Sales

Year	Sales (millions)
2007	1.4
2008	11.6
2009	20.7
2010	40.0
2011	72.3

Source: *Apple, Inc.*

8.1 Factoring Trinomials of the Form $x^2 + bx + c$ and Differences of Two Squares

Objectives

- » Know that multiplying and *factoring* are reverse processes.
- » Factor a trinomial of the form $x^2 + bx + c$.
- » Know the meaning of a *prime* polynomial.
- » Factor a difference of two squares.

We multiply 3 and 7 as follows: $3 \cdot 7 = 21$. We factor the number 21 as follows: $21 = 3 \cdot 7$. Thus, factoring 21 is the reverse process of multiplying 3 and 7. Next, we will make similar observations about polynomials.

Multiplying Polynomials versus Factoring Polynomials

In Chapter 7, we found products of polynomials—for example,

$$\begin{aligned}(x + 3)(x + 5) &= x^2 + 5x + 3x + 15 && \text{Multiply pairs of terms.}\\ &= x^2 + 8x + 15 && \text{Combine like terms.}\end{aligned}$$

In this section, we will learn how to go backward. That is, we will learn how to write $x^2 + 8x + 15$ as a product. This process is called *factoring*.

We **factor** a polynomial by writing it as a product. We say that $(x + 3)(x + 5)$ is a **factored polynomial** and that both $(x + 3)$ and $(x + 5)$ are factors of the polynomial.

Multiplying versus Factoring

Multiplying and factoring are reverse processes. For example,

Multiplying →

$$(x + 3)(x + 5) = x^2 + 8x + 15$$

← Factoring

Factoring a Trinomial of the Form $x^2 + bx + c$

To see how to factor $x^2 + 8x + 15$, let's take another look at how we find the product $(x + 3)(x + 5)$:

last terms

$$\begin{aligned}(x + 3)(x + 5) &= x^2 + 5x + 3x + 3 \cdot 5 \\ &= x^2 + 8x + 15\end{aligned}$$

sum of last terms $3 + 5 = 8$ — product of last terms $3 \cdot 5 = 15$

For $x^2 + 8x + 15$, notice that the coefficient of x is 8, which is the sum of the *last terms* of $(x + 3)(x + 5)$, 3 and 5. Also, the constant term of $x^2 + 8x + 15$ is 15, which is the product of 3 and 5. Now we find the product $(x + p)(x + q)$:

$$\begin{aligned}(x + p)(x + q) &= x^2 + qx + px + pq && \text{Multiply pairs of terms.} \\ &= x^2 + px + qx + pq && \text{Rearrange terms.} \\ &= x^2 + (p + q)x + pq && \text{Distributive law}\end{aligned}$$

In the result, we see that the coefficient of x is the sum of the last terms p and q and that the constant term is the product of the last terms p and q. This observation can help us factor some quadratic trinomials.

Trinomials with Positive Constant Terms

For the trinomial $x^2 + 8x + 15$, the constant term, 15, is positive. In Examples 1–3, we will factor three more trinomials whose constant term is positive.

Example 1 Factoring Trinomials of the Form $x^2 + bx + c$

Factor $x^2 + 6x + 8$.

Solution

To factor $x^2 + 6x + 8$, we need two integers whose product is 8 and whose sum is 6. We try only positive integers, since both their product and their sum have to be positive. The only pairs of factors of 8 whose product is 8 are 1 and 8 or 2 and 4:

Product = 8	**Sum = 6?**
$1(8) = 8$	$1 + 8 = 9$
$2(4) = 8$	$2 + 4 = 6$ ⟵ Success!

Since $2(4) = 8$ and $2 + 4 = 6$, we conclude that the last terms of the factors are 2 and 4:

$$x^2 + 6x + 8 = (x + 2)(x + 4)$$

We check the result by finding the product $(x + 2)(x + 4)$:

$$(x + 2)(x + 4) = x^2 + 4x + 2x + 8 = x^2 + 6x + 8$$

By the commutative law, $(x + 2)(x + 4) = (x + 4)(x + 2)$, so we can write the factors $x + 2$ and $x + 4$ in either order.

We now summarize how to factor a trinomial of the form $x^2 + bx + c$.

Factoring $x^2 + bx + c$

To factor $x^2 + bx + c$, look for two integers p and q whose product is c and whose sum is b. That is, $pq = c$ and $p + q = b$. If such integers exist, the factored polynomial is

$$(x + p)(x + q)$$

Example 2 Factoring Trinomials of the Form $x^2 + bx + c$

Factor $x^2 + 8x + 12$.

Solution

To factor $x^2 + 8x + 12$, we need two integers whose product is 12 and whose sum is 8. We try only positive integers, since both their product and their sum have to be positive. Here are the possibilities:

Product = 12	**Sum = 8?**	
$1(12) = 12$	$1 + 12 = 13$	
$2(6) = 12$	$2 + 6 = 8$	⟵ Success!
$3(4) = 12$	$3 + 4 = 7$	

Since $2(6) = 12$ and $2 + 6 = 8$, we conclude that the last terms of the factors are 2 and 6:

$$x^2 + 8x + 12 = (x + 2)(x + 6)$$

We check the result by finding the product $(x + 2)(x + 6)$:

$$(x + 2)(x + 6) = x^2 + 6x + 2x + 12 = x^2 + 8x + 12$$

Example 3 Factoring a Trinomial of the Form $x^2 + bx + c$

Factor $x^2 - 12x + 36$.

Solution

To factor $x^2 - 12x + 36$, we need two integers whose product is 36 and whose sum is -12. Since the product 36 is positive, the two integers must have the same sign. Therefore, both integers must be negative, because the sum -12 is negative. Here are the possibilities:

Product = 36	**Sum = −12?**	
$-1(-36) = 36$	$-1 + (-36) = -37$	
$-2(-18) = 36$	$-2 + (-18) = -20$	
$-3(-12) = 36$	$-3 + (-12) = -15$	
$-4(-9) = 36$	$-4 + (-9) = -13$	
$-6(-6) = 36$	$-6 + (-6) = -12$	⟵ Success!

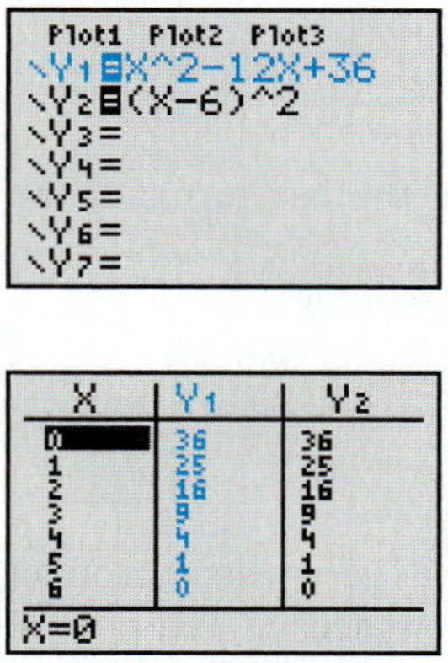

Since $-6(-6) = 36$ and $-6 + (-6) = -12$, we conclude that the last terms of the factors are -6 and -6:

$$x^2 - 12x + 36 = (x - 6)(x - 6) = (x - 6)^2$$

Note that our result, $(x - 6)^2$, is the square of a binomial. So, the original expression $x^2 - 12x + 36$ is a perfect-square trinomial (Section 7.3).

We use a graphing calculator table to verify our work (see Fig. 1).

Figure 1 Verify the work

When the constant term of a trinomial is positive, we need to consider only certain possibilities for the factors of that constant term. In Examples 1 and 2, we worked with only positive factors of the positive constant term, because the coefficient of the middle term was positive. In Example 3, we worked with only negative factors of the positive constant term, because the coefficient of the middle term was negative.

Factoring $x^2 + bx + c$ with c Positive

To factor a trinomial of the form $x^2 + bx + c$ with a positive constant term c,

- If b is positive, look for two *positive* integers whose product is c and whose sum is b. For example,

$$x^2 + 10x + 21 = (x + 3)(x + 7)$$

Positive b ↑ Positive c ↑ Both last terms are positive. ↑↑

- If b is negative, look for two *negative* integers whose product is c and whose sum is b. For example,

$$x^2 - 11x + 28 = (x - 7)(x - 4)$$

Negative b — Positive c — Both last terms are negative.

Trinomials with Negative Constant Terms

How do we factor a quadratic trinomial whose constant term is negative?

Example 4 Factoring Trinomials of the Form $x^2 + bx + c$

Factor $w^2 - w - 20$.

Solution

To factor $w^2 - 1w - 20$, we need two integers whose product is -20 and whose sum is -1. Since the product -20 is negative, the two integers must have different signs. Here are the possibilities:

Product = −20	**Sum = −1?**	
$1(-20) = -20$	$1 + (-20) = -19$	
$2(-10) = -20$	$2 + (-10) = -8$	
$4(-5) = -20$	$4 + (-5) = -1$	⟵ Success!
$5(-4) = -20$	$5 + (-4) = 1$	
$10(-2) = -20$	$10 + (-2) = 8$	
$20(-1) = -20$	$20 + (-1) = 19$	

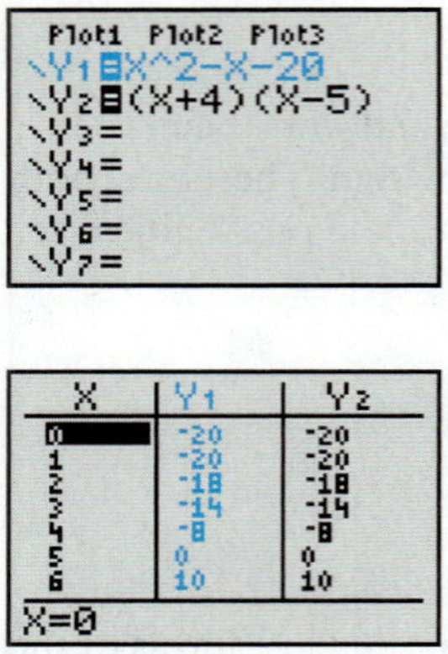

Since $4(-5) = -20$ and $4 + (-5) = -1$, we conclude that the last terms of the factors are 4 and -5:

$$w^2 - w - 20 = (w + 4)(w - 5)$$

We use a graphing calculator table to verify our work (see Fig. 2).

```
Plot1 Plot2 Plot3
\Y1=X^2-X-20
\Y2=(X+4)(X-5)
\Y3=
\Y4=
\Y5=
\Y6=
\Y7=
```

X	Y1	Y2
0	-20	-20
1	-20	-20
2	-18	-18
3	-14	-14
4	-8	-8
5	0	0
6	10	10

X=0

Figure 2 Verify the work

When the constant term of a trinomial is negative, any two integers whose product equals that negative constant term have different signs. For instance, in Example 4 we factored the trinomial $w^2 - w - 20$ by working with integers with different signs whose product is -20.

Factoring $x^2 + bx + c$ with c Negative

To factor a trinomial of the form $x^2 + bx + c$ with a negative constant term c, look for two integers with *different* signs whose product is c and whose sum is b. For example,

$$x^2 + 2x - 24 = (x - 4)(x + 6)$$

Negative c — The last terms have different signs.

We can use a similar method to factor trinomials that have two variables.

Example 5 Factoring Trinomials with Two Variables

Factor $x^2 + 5xy + 6y^2$.

Solution

To help us find the last terms, we write the trinomial in the form $x^2 + (5y)x + 6y^2$. We need two monomials whose product is $6y^2$ and whose sum is $5y$. So, the last terms are $2y$ and $3y$:

$$x^2 + 5xy + 6y^2 = (x + 2y)(x + 3y)$$

We check by finding the product $(x + 2y)(x + 3y)$:

$$(x + 2y)(x + 3y) = x^2 + 3xy + 2xy + 6y^2 = x^2 + 5xy + 6y^2$$

Prime Polynomials

Just as a prime number has no positive factors other than itself and 1, a polynomial that cannot be factored is called **prime**.

For example, consider the polynomial $x^2 + 4x + 6$. To factor this polynomial, we need two integers whose product is 6 and whose sum is 4. We try only positive integers, since both their product and sum have to be positive. Here are the possibilities:

Product = 6	**Sum = 4?**
$1(6) = 6$	$1 + 6 = 7$
$2(3) = 6$	$2 + 3 = 5$

None of the possible sums equal 4, so we conclude that the trinomial $x^2 + 4x + 6$ is prime.

Example 6 Identifying a Prime Polynomial

Factor $3x - 15 + x^2$.

Solution

First, we write $3x - 15 + x^2$ in descending order to avoid confusion about the coefficients:

$$x^2 + 3x - 15$$

To factor $x^2 + 3x - 15$, we need two integers whose product is -15 and whose sum is 3. Since the product -15 is negative, the two integers must have different signs. Here are the possibilities:

Product = −15	**Sum = 3?**
$1(-15) = -15$	$1 + (-15) = -14$
$3(-5) = -15$	$3 + (-5) = -2$
$5(-3) = -15$	$5 + (-3) = 2$
$15(-1) = -15$	$15 + (-1) = 14$

Since none of the sums equal 3, we conclude that the trinomial $x^2 + 3x - 15$ is prime. So, the original trinomial, $3x - 15 + x^2$, is prime.

Factoring the Difference of Two Squares

In Section 7.3, we found the product of two binomial conjugates by using the property $(A + B)(A - B) = A^2 - B^2$. The expression $A^2 - B^2$ is the difference of two squares. To factor a difference of two squares, we can use that property in reverse.

Difference of Two Squares

$$A^2 - B^2 = (A + B)(A - B)$$

In words: The difference of the squares of two terms is the product of the sum of the terms and the difference of the terms.

Example 7 Factoring Differences of Two Squares

Factor.

1. $x^2 - 16$ **2.** $9m^2 - 4$ **3.** $25p^2 - 49q^2$

Solution

1. Since $x^2 - 16 = (x)^2 - (4)^2$, we substitute x for A and 4 for B:

$$\begin{array}{ccccc} A^2 & - & B^2 & = & (A + B)(A - B) \\ \downarrow & & \downarrow & & \downarrow \quad \downarrow \quad \downarrow \quad \downarrow \\ x^2 - 16 = x^2 & - & 4^2 & = & (x + 4)\,(x - 4) \end{array}$$

2. Since $9m^2 - 4 = (3m)^2 - (2)^2$, we substitute $3m$ for A and 2 for B:

$$\begin{array}{ccccc} A^2 & - & B^2 & = & (A + B)(A - B) \\ \downarrow & & \downarrow & & \downarrow \quad \downarrow \quad \downarrow \quad \downarrow \\ 9m^2 - 4 = (3m)^2 & - & 2^2 & = & (3m + 2)(3m - 2) \end{array}$$

3. Since $25p^2 - 49q^2 = (5p)^2 - (7q)^2$, we substitute $5p$ for A and $7q$ for B:

$$\begin{array}{ccccc} A^2 & - & B^2 & = & (A + B)(A - B) \\ \downarrow & & \downarrow & & \downarrow \quad \downarrow \quad \downarrow \quad \downarrow \\ 25p^2 - 49q^2 = (5p)^2 & - & (7q)^2 & = & (5p + 7q)(5p - 7q) \end{array}$$

WARNING

The binomial $x^2 + 16$ is prime. Some students think this polynomial can be factored as $(x + 4)^2$, but simplify $(x + 4)^2$; you'll see that the result is $x^2 + 8x + 16$, not $x^2 + 16$. In general, **a polynomial of the form $x^2 + k^2$, where $k \neq 0$, is prime.**

Example 8 Factoring Differences of Two Squares

Factor $16p^4 - 1$.

Solution

The binomial $16p^4 - 1$ is a difference of squares, since $(4p^2)^2 = 16p^4$ and $1^2 = 1$. So, we proceed as follows:

$$\begin{aligned} 16p^4 - 1 &= (4p^2)^2 - 1^2 && \textit{Write as difference of squares.} \\ &= (4p^2 + 1)(4p^2 - 1) && A^2 - B^2 = (A + B)(A - B) \\ &= (4p^2 + 1)(2p + 1)(2p - 1) && 4p^2 + 1 \textit{ is prime: } A^2 - B^2 = (A + B)(A - B) \end{aligned}$$

We use a graphing calculator table to verify our work (see Fig. 3).

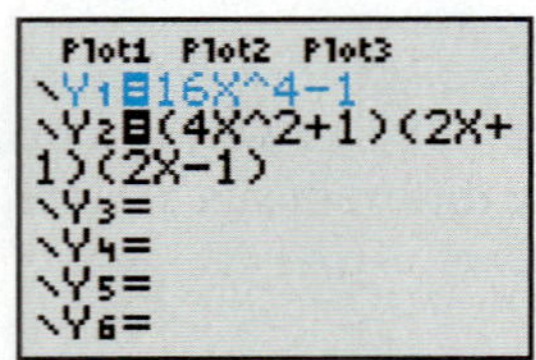

X	Y1	Y2
0	-1	-1
1	15	15
2	255	255
3	1295	1295
4	4095	4095
5	9999	9999
6	20735	20735

X=0

Figure 3 Verify the work

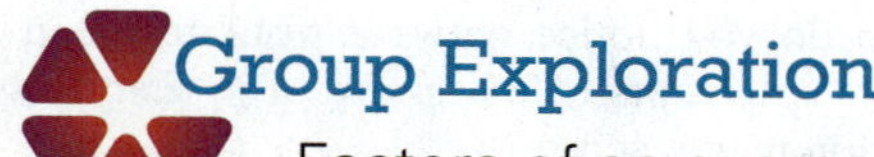

Group Exploration

Factors of an expression and *x*-intercepts of the graph of a function

In this exploration, you will explore a connection between factors of a polynomial of the form $x^2 + bx + c$ and x-intercepts of the graph of the polynomial function $f(x) = x^2 + bx + c$.

1. Factor $x^2 + x - 2$.
2. Use ZDecimal to draw a graph of the function $f(x) = x^2 + x - 2$. What are the x-intercepts?
3. What connection do you notice between your result in Problem 1 and the x-intercepts of the graph of f? Explain why this connection makes sense.
4. The graph of a function $g(x) = x^2 + bx + c$ is sketched in the indicated figures. Use the graph to help you factor $x^2 + bx + c$. Then find the values of b and c.

a. Fig. 4

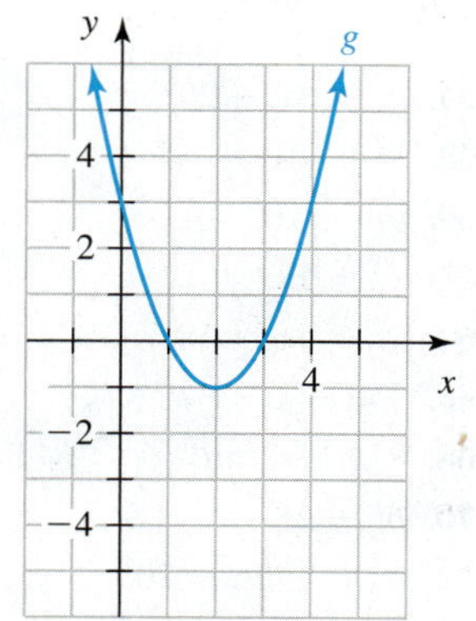

Figure 4 Graph of $g(x) = x^2 + bx + c$

b. Fig. 5

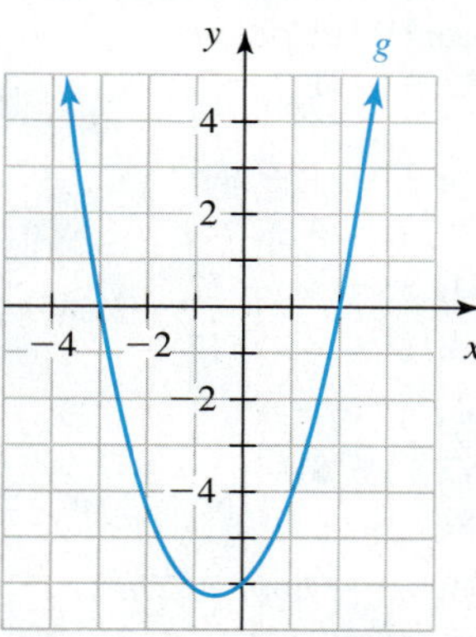

Figure 5 Graph of $g(x) = x^2 + bx + c$

Group Exploration

Looking ahead: Factoring out the greatest common factor

1. Simplify the following.
   a. $2(x + 3)$ **b.** $5(x - 4)$ **c.** $3(x^2 - 5x + 2)$
2. Factor the following. [**Hint:** Refer to your work in Problem 1.]
   a. $2x + 6$ **b.** $5x - 20$ **c.** $3x^2 - 15x + 6$
3. Factor the following.
   a. $3x + 12$ **b.** $2x - 10$ **c.** $5x^2 - 20x + 15$
4. Find the product of the following.
   a. $x(3x + 9)$ **b.** $2x^2(5x - 3)$ **c.** $7x(2x^2 + 5x + 4)$
5. Factor the following. [**Hint:** Refer to your work in Problem 4.]
   a. $3x^2 + 9x$ **b.** $10x^3 - 6x^2$ **c.** $14x^3 + 35x^2 + 28x$
6. Factor the following.
   a. $x^2 + 5x$ **b.** $6x^3 - 15x^2$ **c.** $2x^3 - 8x^2 - 10x$

Homework 8.1

For extra help ▶ MyMathLab®
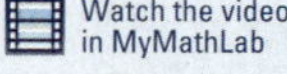
Watch the videos in MyMathLab

Download the MyDashboard App

Factor when possible. If a polynomial is prime, say so. Verify that you have factored correctly by finding the product of your factored polynomial.

1. $x^2 + 5x + 6$ **2.** $x^2 + 9x + 8$ **3.** $t^2 + 9t + 20$

4. $w^2 + 10w + 16$ **5.** $x^2 + 8x + 16$ **6.** $x^2 + 14x + 49$

7. $x^2 - 2x - 8$ **8.** $x^2 - 3x - 10$ **9.** $a^2 - 6a - 16$

10. $w^2 - 5w - 24$ **11.** $x^2 + 5x - 24$ **12.** $x^2 + 7x - 30$

13. $x^2 + 8x - 12$ **14.** $x^2 + 9x - 20$ **15.** $3t - 28 + t^2$

16. $y^2 - 14 + 5y$ **17.** $x^2 - 10x + 16$ **18.** $x^2 - 11x + 28$

19. $24 - 11x + x^2$ **20.** $36 + x^2 - 15x$ **21.** $x^2 - 3x + 10$

22. $x^2 - 2x + 8$ **23.** $r^2 - 10r + 25$ **24.** $a^2 - 6a + 9$

25. $x^2 + 10xy + 9y^2$ **26.** $a^2 + 8ab + 12b^2$

27. $m^2 - mn - 6n^2$ **28.** $r^2 + rt - 20t^2$

29. $a^2 - 7ab + 6b^2$ **30.** $m^2 - 7mn + 10n^2$

31. $p^2 + 3pq - 28q^2$ **32.** $r^2 + 4rt - 12t^2$

33. $p^2 - 6pq - 16q^2$ **34.** $m^2 - mr - 42r^2$

Factor when possible. Use a graphing calculator table to verify your work when possible.

35. $x^2 - 25$ **36.** $x^2 - 49$ **37.** $x^2 - 81$ **38.** $x^2 - 36$

39. $x^2 + 36$ **40.** $x^2 + 4$ **41.** $4x^2 - 25$ **42.** $9x^2 - 49$

43. $81r^2 - 1$ **44.** $25p^2 - 64$ **45.** $36x^2 + 49$ **46.** $25x^2 + 9$

47. $49p^2 - 100q^2$ **48.** $4a^2 - 9b^2$

49. $64m^2 - 9n^2$ **50.** $25b^2 - 9c^2$

51. $16x^4 - 81$ **52.** $81x^4 - 1$ **53.** $t^4 - w^4$ **54.** $a^4 - 81b^4$

Factor when possible. Use a graphing calculator table to verify your work when possible.

55. $x^2 - 3x - 18$ **56.** $x^2 - 8x - 20$

57. $x^2 + 14x + 49$ **58.** $x^2 + 10x + 25$

59. $a^2 - 4$ **60.** $m^2 - 100$

61. $x^2 + 4x + 12$ **62.** $x^2 + 12x + 24$

63. $x^2 - 8x + 12$ **64.** $x^2 - 9x + 18$

65. $-2w - 48 + w^2$ **66.** $-4t - 60 + t^2$

67. $t^4 - 81k^4$ **68.** $81m^4 - 16n^4$

69. $w^2 + 49$ **70.** $b^2 + 25$

71. $m^2 - 6mn - 27n^2$ **72.** $a^2 - 19ab - 20b^2$

73. $32 - 18x + x^2$ **74.** $30 - 13x + x^2$

75. $100p^2 - 9t^2$ **76.** $64a^2 - 49b^2$

77. $a^2 + 12ab + 36b^2$ **78.** $m^2 + 18mn + 81n^2$

Concepts

79. A student tries to factor the polynomial $x^2 + 9$:

$$x^2 + 9 = (x + 3)(x + 3) = (x + 3)^2$$

Describe any errors. Then factor the polynomial correctly.

80. Two students try to factor the polynomial $x^2 + 14x + 48$:

Student A

$$x^2 + 14x + 48 = (x + 6)(x + 8)$$

Student B

$$x^2 + 14x + 48 = (x + 8)(x + 6)$$

Are both students, one student, or neither student correct? Explain.

81. Which of the following expressions are equivalent?

$(x - 3)(x + 7)$ $x^2 - 21$ $x^2 - 4x - 21$

$x^2 + 4x - 21$ $(x + 7)(x - 3)$

82. Which of the following expressions are equivalent?

$x^2 - 2x - 24$ $(x - 6)(x + 4)$ $x^2 - 10x - 24$

$(x + 4)(x - 6)$ $x^2 + 2x - 24$

83. Factor the expression $x^2 - 5x - 24$. Then find the product of the result. What do you observe?

84. Find the product $(x - 2)(x + 8)$. Then factor the result. What do you observe?

85. a. Factor $x^2 - 5x + 4$.

b. Use ZStandard on a graphing calculator to graph the function $f(x) = x^2 - 5x + 4$. What are the x-intercepts?

c. What connection do you notice between your result in part (a) and the x-intercepts of the graph of f? Explain why this connection makes sense.

86. a. Factor $x^2 - 6x + 8$.

b. Use ZStandard on a graphing calculator to graph the function $f(x) = x^2 - 6x + 8$. What are the x-intercepts?

c. What connection do you notice between your result in part (a) and the x-intercepts of the graph of f? Explain why this connection makes sense.

87. Give three examples of quadratic polynomials in which $x + 5$ is a factor.

88. Give three examples of quadratic polynomials in which $x - 3$ is a factor.

89. Find all possible values of k so that $x^2 + kx + 12$ can be factored.

90. Find all possible values of k so that $x^2 + kx - 20$ can be factored.

91. Evaluate $x^2 - 6x + 8$ for $x = 2$ and for $x = 4$. Why does it make sense that both of your results are 0? [**Hint:** Factor $x^2 - 6x + 8$.]

92. Evaluate $x^2 + 5x + 6$ for $x = -3$ and for $x = -2$. Why does it make sense that both of your results are 0? [**Hint:** Factor $x^2 + 5x + 6$.]

93. Compare the process of factoring an expression with that of finding the product of an expression. (See page 9 for guidelines on writing a good response.)

94. Describe how to factor a difference of two squares. (See page 9 for guidelines on writing a good response.)

Related Review

If the expression is not factored, then factor it. If the expression is factored, then find the product.

95. $(x - 9)(x + 2)$ **96.** $(x - 1)(x - 8)$

97. $x^2 - 15x + 50$ **98.** $x^2 + 19x - 20$

99. $(3x - 7)(3x + 7)$ **100.** $(4x + 1)(4x - 1)$

101. $25x^2 - 36$ **102.** $49x^2 - 81$

Expressions, Equations, Functions, and Graphs

Perform the indicated instruction. Then use words such as linear, quadratic, cubic, polynomial, degree, function, one variable, *and* two variables *to describe the expression, equation, or system.*

103. Find the difference $(5p + 7w) - (2p - 4w)$.

104. Solve $6 = 3(x - 5)$.

105. Find the product $(5p + 7w)(2p - 4w)$.

106. Graph $y = 3(x - 5)$ by hand.

107. Factor $p^2 - 11pw + 18w^2$.

108. Solve the following:

$$y = 3(x - 5)$$
$$7x + 4y = -3$$

8.2 Factoring Out the GCF; Factoring by Grouping

Objectives

- Factor out the *greatest common factor (GCF)* of a polynomial.
- Factor polynomials completely.
- Factor out the opposite of the GCF of a polynomial.
- Factor a polynomial by grouping.

In Section 8.1, we discussed some ways to factor polynomials. In this section, we will discuss more factoring techniques.

Factoring Out the GCF

Consider the polynomial $2x + 10$. Note that 2 is a common factor of both $2x = 2 \cdot x$ and $10 = 2 \cdot 5$, so we have:

$$2x + 10 = 2 \cdot x + 2 \cdot 5$$

We use the distributive law to factor out the common factor 2:

$$2x + 10 = 2 \cdot x + 2 \cdot 5 = 2(x + 5)$$

So, we have factored $2x + 10$ as $2(x + 5)$. We check the result by finding the product $2(x + 5)$:

$$2(x + 5) = 2x + 10$$

Example 1 Factoring Out a Common Factor

Factor.

1. $3x + 21$ **2.** $8x^2 - 6x$ **3.** $6x^3 + 12x^2$

Solution

1. The number 3 is a common factor of $3x = 3 \cdot x$ and $21 = 3 \cdot 7$. So, we use the distributive law to factor out 3:

$$3x + 21 = 3 \cdot x + 3 \cdot 7 \quad \text{3 is a common factor.}$$
$$= 3(x + 7) \quad \text{Factor out 3.}$$

We can verify our work by finding the product $3(x + 7)$:

$$3(x + 7) = 3x + 21$$

2. The expression $2x$ is a common factor of $8x^2 = 2x \cdot 4x$ and $6x = 2x \cdot 3$. So, we use the distributive law to factor out $2x$:

$$8x^2 - 6x = 2x \cdot 4x - 2x \cdot 3 \quad 2x \text{ is a common factor.}$$
$$= 2x(4x - 3) \quad \text{Factor out } 2x.$$

3. The expression $6x^2$ is a common factor of $6x^3 = 6x^2 \cdot x$ and $12x^2 = 6x^2 \cdot 2$. So, we use the distributive law to factor out $6x^2$:

$$6x^3 + 12x^2 = 6x^2 \cdot x + 6x^2 \cdot 2 \quad 6x^2 \text{ is a common factor.}$$
$$= 6x^2(x + 2) \quad \text{Factor out } 6x^2.$$

In Problem 3 of Example 1, notice that $2x$ is also a common factor of $6x^3 = 2x \cdot 3x^2$ and $12x^2 = 2x \cdot 6x$. So, we could have factored $6x^3 + 12x^2$ by factoring out $2x$ rather than $6x^2$:

$$6x^3 + 12x^2 = 2x(3x^2 + 6x)$$

However, the resulting expression is not factored completely: We can still factor $3x^2 + 6x$ by factoring out $3x$:

$$6x^3 + 12x^2 = 2x(3x^2 + 6x) = 2x \cdot 3x(x + 2) = 6x^2(x + 2)$$

Although we have found the same final result, it was more efficient to factor out $6x^2$, which has a larger coefficient and a higher degree than $2x$. We call $6x^2$ the *greatest common factor* of $6x^3$ and $12x^2$.

Definition Greatest common factor

The **greatest common factor (GCF)** of two or more terms is the monomial with the largest coefficient and the largest degree that is a factor of all the terms.

For each polynomial in Example 1, the common factor that we factored out of the polynomial was the GCF. In Example 2, we factor some more polynomials by factoring out the GCF.

Example 2 Factoring Out the GCF

Factor.

1. $20x^2 + 35x$

2. $14p^3 - 21p^2$

Solution

1. We begin by factoring $20x^2$ and $35x$:

$$20x^2 = 2 \cdot 2 \cdot 5 \cdot x \cdot x$$
$$35x = 5 \cdot 7 \cdot x$$

Both 5 and x are common factors. So, the GCF is $5x$:

$$20x^2 + 35x = 5x \cdot 4x + 5x \cdot 7 \quad \text{5x is the GCF.}$$
$$= 5x(4x + 7) \quad \text{Factor out 5x.}$$

2. We begin by factoring $14p^3$ and $21p^2$:

$$14p^3 = 2 \cdot 7 \cdot p \cdot p \cdot p$$
$$21p^2 = 3 \cdot 7 \cdot p \cdot p$$

There are three common factors: 7, p, and p. So, the GCF is $7p^2$:

$$14p^3 - 21p^2 = 7p^2 \cdot 2p - 7p^2 \cdot 3 \quad 7p^2 \text{ is the GCF.}$$
$$= 7p^2(2p - 3) \quad \text{Factor out } 7p^2.$$

We use a graphing calculator table to verify our work (see Fig. 6).

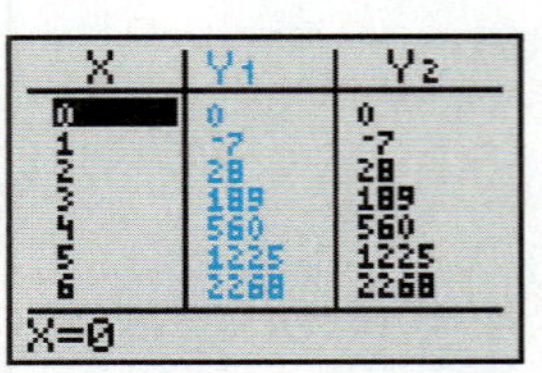

Figure 6 Verify the work

After you factor a polynomial, verify your work by finding the product of your result or by using a graphing calculator table.

So far, we have factored out the GCF for some binomials with one variable. We can also factor out the GCF for some polynomials with more than two terms and more than one variable.

Example 3 Factoring Out the GCF

Factor $12x^4y^2 - 6x^2y^3 + 15xy^2$.

Solution

We begin by factoring $12x^4y^2$, $6x^2y^3$, and $15xy^2$:

$$12x^4y^2 = 2 \cdot 2 \cdot 3 \cdot x \cdot x \cdot x \cdot x \cdot y \cdot y$$
$$6x^2y^3 = 2 \cdot 3 \cdot x \cdot x \cdot y \cdot y \cdot y$$
$$15xy^2 = 3 \cdot 5 \cdot x \cdot y \cdot y$$

There are four common factors: 3, x, y, and y. So, the GCF is $3xy^2$:

$$12x^4y^2 - 6x^2y^3 + 15xy^2 = 3xy^2 \cdot 4x^3 - 3xy^2 \cdot 2xy + 3xy^2 \cdot 5 \quad 3xy^2 \text{ is the GCF.}$$
$$= 3xy^2(4x^3 - 2xy + 5) \quad \text{Factor out } 3xy^2.$$

Factoring Polynomials Completely

After we factor the GCF out of a polynomial, we must check whether the result can be further factored by using factoring techniques discussed in Section 8.1. If a result cannot be further factored, it is said to be **factored completely.**

Example 4 Factoring Polynomials Completely

Factor $4x^2 - 36$.

Solution

The GCF of $4x^2$ and 36 is 4, so:

$$4x^2 - 36 = 4(x^2 - 9) \quad \text{Factor out GCF, 4.}$$
$$= 4(x + 3)(x - 3) \quad A^2 - B^2 = (A + B)(A - B)$$

To factor $4x^2 - 36$ in Example 4, we first factored out the GCF, 4, and then factored the resulting difference of two squares, $x^2 - 9$. These steps require less work than first using the property for the difference of two squares:

$$4x^2 - 36 = (2x + 6)(2x - 6) \quad A^2 - B^2 = (A + B)(A - B)$$
$$= 2(x + 3)(2)(x - 3) \quad \text{Factor out GCF, 2.}$$
$$= 4(x + 3)(x - 3) \quad \text{Simplify.}$$

Not only are there fewer steps in Example 4, but it is easier to use the property for the difference of squares to factor $x^2 - 9$ than $4x^2 - 36$. In general, **when the leading coefficient of a polynomial is positive and the GCF is not 1, we first factor out the GCF.** (We will soon discuss what to do when the leading coefficient of a polynomial is negative.)

Example 5 Factoring Polynomials Completely

Factor $2x^2 + 14x + 24$.

Solution

The GCF of $2x^2$, $14x$, and 24 is 2, so

$$2x^2 + 14x + 24 = 2(x^2 + 7x + 12) \quad \text{Factor out the GCF, 2.}$$
$$= 2(x + 3)(x + 4) \quad \text{Find two integers whose product is 12 and whose sum is 7.}$$

WARNING It is a common error when factoring a polynomial to forget to factor it *completely*. In Example 5, we factored the GCF, 2, out of $2x^2 + 14x + 24$:

$$2x^2 + 14x + 24 = 2(x^2 + 7x + 12) \quad \text{Not factored completely}$$

However, we were not done factoring, because we could still factor $x^2 + 7x + 12$:

$$2(x^2 + 7x + 12) = 2(x + 3)(x + 4) \quad \text{Factored completely}$$

When factoring a polynomial, always factor it *completely*.

Example 6 Factoring Polynomials Completely

Factor.

1. $12x^3 - 75x$ **2.** $36x + 4x^3 - 24x^2$

Solution

1. The GCF of $12x^3$ and $75x$ is $3x$, so:

$$12x^3 - 75x = 3x(4x^2 - 25) \quad \text{Factor out GCF, 3x.}$$
$$= 3x(2x + 5)(2x - 5) \quad A^2 - B^2 = (A + B)(A - B)$$

2. We begin by writing $36x + 4x^3 - 24x^2$ in descending order:

$$\begin{aligned} 36x + 4x^3 - 24x^2 &= 4x^3 - 24x^2 + 36x && \text{Write in descending order.} \\ &= 4x(x^2 - 6x + 9) && \text{Factor out GCF, } 4x. \\ &= 4x(x - 3)(x - 3) && \text{Find two integers whose product is 9 and whose sum is } -6. \\ &= 4x(x - 3)^2 && bb = b^2 \end{aligned}$$

Example 7 Factoring Polynomials Completely

Factor $2x^4y - 6x^3y - 20x^2y$.

Solution

The GCF of $2x^4y$, $6x^3y$, and $20x^2y$ is $2x^2y$, so:

$$\begin{aligned} 2x^4y - 6x^3y - 20x^2y &= 2x^2y(x^2 - 3x - 10) && \text{Factor out GCF, } 2x^2y. \\ &= 2x^2y(x - 5)(x + 2) && \text{Find two integers whose product is } -10 \text{ and whose sum is } -3. \end{aligned}$$

Factoring Out the Opposite of the GCF of a Polynomial

How do we factor a polynomial in which the leading coefficient is negative? For example, consider the polynomial $-5x^3 + 30x^2 - 40x$, which has a negative leading coefficient, -5.

How to Factor when the Leading Coefficient Is Negative

When the leading coefficient of a polynomial is negative, we first factor out the opposite of the GCF.

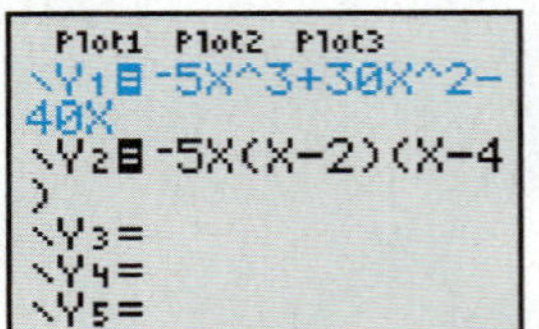

X	Y1	Y2
0	0	0
1	-15	-15
2	0	0
3	15	15
4	0	0
5	-75	-75
6	-240	-240

X=0

Figure 7 Verify the work

Example 8 Factoring Out the Opposite of the GCF

Factor $-5x^3 + 30x^2 - 40x$.

Solution

For $-5x^3 + 30x^2 - 40x$, the GCF is $5x$. Since the leading coefficient, -5, is negative, we factor out the opposite of the GCF:

$$\begin{aligned} -5x^3 + 30x^2 - 40x &= -5x(x^2 - 6x + 8) && \text{Factor out } -5x\text{, opposite of GCF.} \\ &= -5x(x - 2)(x - 4) && \text{Find two integers whose product is 8 and whose sum is } -6. \end{aligned}$$

We use a graphing calculator table to verify our work (see Fig. 7).

Example 9 Factoring Out the Opposite of the GCF

Factor $49 - w^2$.

Solution

First we write $49 - w^2$ in descending order: $-w^2 + 49$. The GCF of $-w^2 + 49$ is 1. Since the leading coefficient, -1, is negative, we factor out the opposite of the GCF:

$$\begin{aligned} -w^2 + 49 &= -1(w^2 - 49) && \text{Factor out } -1\text{, opposite of GCF.} \\ &= -1(w + 7)(w - 7) && A^2 - B^2 = (A + B)(A - B) \\ &= -(w + 7)(w - 7) && -1a = -a \end{aligned}$$

Factoring by Grouping

So far, we have factored out a GCF when the GCF is a monomial. For example, here we factor out the monomial p from the polynomial $x^2(p) + 2(p)$:

$$x^2(p) + 2(p) = (x^2 + 2)(p)$$

We can also factor out a GCF that is a binomial. For example, here we factor out the binomial $x + 3$ from the polynomial $x^2(x + 3) + 2(x + 3)$:

$$x^2(x + 3) + 2(x + 3) = (x^2 + 2)(x + 3)$$

We can factor some polynomials that contain four terms by first factoring the first two terms and the last two terms—for example,

$$\underbrace{x^3 + 3x^2}_{\text{factor}} + \underbrace{2x + 6}_{\text{factor}} = x^2(x + 3) + 2(x + 3) \quad \text{Factor both pairs of terms.}$$

$$= (x^2 + 2)(x + 3) \quad \text{Factor out GCF, } x + 3.$$

We call this method *factoring by grouping.*

Example 10 Factoring by Grouping

Factor $2x^3 - 8x^2 - 3x + 12$.

Solution

$$2x^3 - 8x^2 - 3x + 12 = 2x^2(x - 4) - 3(x - 4) \quad \text{Factor both pairs of terms.}$$

$$= (2x^2 - 3)(x - 4) \quad \text{Factor out GCF, } x - 4.$$

WARNING

It is a common error to think a polynomial such as $2x^2(x - 4) - 3(x - 4)$ in Example 10 is factored. Even though both of the terms $2x^2(x - 4)$ and $3(x - 4)$ are factored, the entire expression $2x^2(x - 4) - 3(x - 4)$ is a difference, not a product. The polynomial $(2x^2 - 3)(x - 4)$ in Example 10 *is* factored, because it is a product.

We now describe in general how to factor a polynomial by grouping.

Factoring by Grouping

For a polynomial with four terms, we **factor by grouping** (if it can be done) by

1. Factoring the first two terms and the last two terms.
2. Factoring out the binomial GCF.

When trying to factor a polynomial with four terms, consider trying to factor it by grouping.

Example 11 Factoring by Grouping

Factor $10x^3 - 6x^2 + 5x - 3$.

Solution

$$10x^3 - 6x^2 + 5x - 3 = 2x^2(5x - 3) + 1(5x - 3) \quad \text{Factor both pairs of terms.}$$

$$= (2x^2 + 1)(5x - 3) \quad \text{Factor out GCF, } 5x - 3.$$

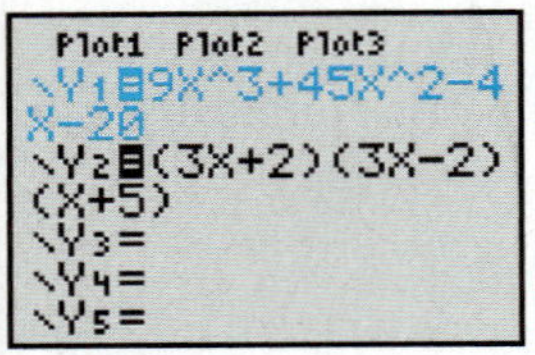

X	Y1	Y2
0	-20	-20
1	30	30
2	224	224
3	616	616
4	1260	1260
5	2210	2210
6	3520	3520

X=0

Figure 8 Verify the work

Example 12 Factoring by Grouping

Factor $9x^3 + 45x^2 - 4x - 20$.

Solution

$$\begin{aligned} 9x^3 + 45x^2 - 4x - 20 &= 9x^2(x + 5) - 4(x + 5) && \text{Factor both pairs of terms.} \\ &= (9x^2 - 4)(x + 5) && \text{Factor out GCF, } x + 5. \\ &= (3x + 2)(3x - 2)(x + 5) && A^2 - B^2 = (A + B)(A - B) \end{aligned}$$

We use a graphing calculator table to verify our work (see Fig. 8).

Example 13 Factoring by Grouping

Factor $2a - 2b + xa - xb$.

Solution

$$\begin{aligned} 2a - 2b + xa - xb &= 2(a - b) + x(a - b) && \text{Factor both pairs of terms.} \\ &= (2 + x)(a - b) && \text{Factor out GCF, } a - b. \end{aligned}$$

In summary, there are two aspects to factoring that are important to remember:

1. If the leading coefficient of a polynomial is positive and the GCF is not 1, first factor out the GCF. If the leading coefficient is negative, first factor out the opposite of the GCF.
2. Always factor a polynomial *completely*.

Group Exploration

Equivalent expressions and locating errors

1. Consider the following correct work for factoring $3x^3 + 2x^2 - 27x - 18$:

$$\begin{aligned} 3x^3 + 2x^2 - 27x - 18 &= x^2(3x + 2) - 9(3x + 2) \\ &= (x^2 - 9)(3x + 2) \\ &= (x + 3)(x - 3)(3x + 2) \end{aligned}$$

Use graphing calculator tables to show that all of the preceding four expressions are equivalent.

2. Consider the following incorrect work for factoring $4x^3 - 3x^2 - 16x + 12$:

$$\begin{aligned} 4x^3 - 3x^2 - 16x + 12 &= x^2(4x - 3) - 4(4x + 3) \\ &= (x^2 - 4)(4x - 3)(4x + 3) \\ &= (x + 2)(x - 2)(4x - 3)(4x + 3) \end{aligned}$$

Use graphing calculator tables to find the step(s) in which an error was made. Then do the factoring correctly.

Tips for Success Complete the Rest of the Assignment

If you have spent a good amount of time trying to solve an exercise but can't, consider going on to the next exercise in the assignment. You may find that the next exercise involves a different concept or involves a more familiar situation. You may even find that, after completing the rest of the assignment, you are able to complete the exercise(s) you skipped. One explanation of this phenomenon is that you may have learned or remembered some concept in a later exercise that relates to the exercise with which you were struggling.

Homework 8.2

For extra help ▶ MyMathLab®  Watch the videos in MyMathLab  Download the MyDashboard App

Factor. Use a graphing calculator table to verify your work when possible.

1. $6x + 8$
2. $3x + 15$
3. $20w^2 + 35w$
4. $28p^2 + 21p$
5. $12x^3 - 30x^2$
6. $18x^4 - 12x^2$
7. $6a^2b - 9ab$
8. $8pq^3 - 6p^2q$
9. $8x^3y^2 + 12x^2y^3$
10. $27x^3y - 45x^2y^3$
11. $15x^3 - 10x - 30$
12. $28x^3 + 12x - 20$
13. $12t^4 + 8t^3 - 16t$
14. $6r^4 - 9r^2 - 12r$
15. $10a^4b - 15a^3b + 25ab$
16. $8p^4q + 6p^2q - 4pq$

Factor. Verify that you have factored correctly by finding the product of your factored polynomial.

17. $2x^2 - 18$
18. $3x^2 - 75$
19. $3m^2 + 21m + 30$
20. $5p^2 - 5p - 60$
21. $2x^2 - 18x + 36$
22. $4x^2 - 44x + 72$
23. $4r^3 - 16r^2 - 20r$
24. $2t^3 - 12t^2 - 32t$
25. $6x^4 - 24x^2$
26. $2x^4 - 50x^2$
27. $8m^4n - 18m^2n$
28. $75p^4y - 27p^2y$
29. $5x^4 + 10x^3 - 120x^2$
30. $8x^4 - 24x^3 + 16x^2$
31. $36t^2 + 32t + 4t^3$
32. $60y - 40y^2 + 5y^3$
33. $-12x^3 + 27x$
34. $-4x^4 + 4x^2$
35. $-3x^3 - 18x^2 + 48x$
36. $-4x^3 + 20x^2 - 24x$
37. $-x^2 + 11x - 10$
38. $-x^2 - 2x + 35$
39. $6a^4b + 36a^3b + 54a^2b$
40. $4m^4n - 40m^3n + 100m^2n$
41. $4x^4y - 12x^3y^2 - 40x^2y^3$
42. $3a^3b - 3a^2b^2 - 36ab^3$
43. $-2x^3y^2 + 16x^2y^3 - 32xy^4$
44. $-5x^4y + 20x^3y^2 - 20x^2y^3$

Factor. Use a graphing calculator table to verify your work when possible.

45. $5x^2(x - 3) + 2(x - 3)$
46. $8x^2(x + 4) + 3(x + 4)$
47. $6x^2(2x + 5) - 7(2x + 5)$
48. $5x^2(4x - 1) - 3(4x - 1)$
49. $2p^3 + 6p^2 + 5p + 15$
50. $12r^3 + 3r^2 + 8r + 2$
51. $6x^3 - 2x^2 + 21x - 7$
52. $4x^3 - 16x^2 + 3x - 12$
53. $15w^3 + 5w^2 - 6w - 2$
54. $6b^3 - 10b^2 - 21b + 35$
55. $16x^3 - 12x^2 - 36x + 27$
56. $50x^3 + 125x^2 - 8x - 20$
57. $2b^3 - 5b^2 - 18b + 45$
58. $4t^3 - 7t^2 - 16t + 28$
59. $x^3 - x^2 - x + 1$
60. $x^3 + x^2 - x - 1$
61. $ax - 3ay - 2bx + 6by$
62. $ax - 5ay - 3bx + 15by$
63. $5a^2x + 2a^2y - 5bx - 2by$
64. $3ax + 7ay - 3b^2x - 7b^2y$
65. $81x^2 - 25$
66. $9x^2 - 100$
67. $w^2 - 10w + 16$
68. $p^2 + 3p - 40$
69. $24 - 10x + x^2$
70. $-14 - 5x + x^2$
71. $20a^2b - 15ab^3$
72. $14xy^2 + 21x^2y$
73. $x^2 + xy - 30y^2$
74. $x^2 - 3xy - 28y^2$
75. $-6r^3 + 24r^2 - 24r$
76. $-5k^3 + 40k^2 - 80k$
77. $64x^3 - 49x$
78. $2x^3 - 162x$
79. $-m^2 + 6m - 9$
80. $-w^2 + 16w - 64$
81. $x^3 + 9x^2 - 4x - 36$
82. $28x^3 - 35x^2 + 8x - 10$
83. $2m^3n - 10m^2n^2 + 12mn^3$
84. $4p^3y + 8p^2y^2 - 12py^3$

Concepts

85. A student tries to factor $6x^3 + 8x^2 + 15x + 20$:

$$6x^3 + 8x^2 + 15x + 20 = 2x^2(3x + 4) + 5(3x + 4)$$

Describe any errors. Then factor the polynomial correctly.

86. Two students try to factor $15x^3 + 3x^2 - 35x - 7$:

Student A

$$15x^3 + 3x^2 - 35x - 7 = 3x^2(5x + 1) - 7(5x + 1) = (3x^2 - 7)(5x + 1)$$

Student B

$$15x^3 + 3x^2 - 35x - 7 = 15x^3 - 35x + 3x^2 - 7 = 5x(3x^2 - 7) + 1(3x^2 - 7) = (5x + 1)(3x^2 - 7)$$

Are both students, one student, or neither student correct?

87. A student tries to factor $4x^3 + 28x^2 + 40x$:

$$4x^3 + 28x^2 + 40x = 4x(x^2 + 7x + 10)$$

Describe any errors. Then factor the polynomial correctly.

88. A student tries to factor $5x^3 - 45x$:

$$5x^3 - 45x = 5x(x^2 - 9)$$

Describe any errors. Then factor the polynomial correctly.

89. A student tries to factor $4x^2 - 100$:

$$4x^2 - 100 = (2x + 10)(2x - 10) = 2(x + 5)(2)(x - 5) = 4(x + 5)(x - 5)$$

What would you tell the student?

90. A student tries to factor $64x^2 - 36$:

$$64x^2 - 36 = (8x + 6)(8x - 6) = 2(4x + 3)(2)(4x - 3) = 4(4x + 3)(4x - 3)$$

What would you tell the student?

91. Give three examples of cubic polynomials in which $2x$ is a factor.

92. Give three examples of quadratic polynomials in which $-5x$ is a factor.

93. A student tries to factor $2x^2 + 10x + 12$:

$$2x^2 + 10x + 12 = 2(x^2 + 5x + 6)$$

The student then checks that tables for $y = 2x^2 + 10x + 12$ and $y = 2(x^2 + 5x + 6)$ are the same. Explain why the student's work is incorrect even though the tables for the two equations are the same.

94. Explain why, when factoring a polynomial, it is a good idea to factor out the GCF first if it is not 1 and the leading coefficient of the polynomial is positive.

Related Review

If the expression is not factored, then factor it. If the expression is factored, then find the product.

95. $2x(x - 3)(x + 4)$

96. $-5x(x - 2)(x - 7)$

97. $5x^3 - 40x^2 + 80x$

98. $3x^3 + 12x^2 + 12x$

99. $6x^3 - 9x^2 - 4x + 6$

100. $8x^3 + 6x^2 - 20x - 15$

101. $(x - 3)(x^2 + 5)$

102. $(x^2 - 2)(x - 8)$

Expressions, Equations, Functions, and Graphs

Perform the indicated instruction. Then use words such as linear, quadratic, cubic, polynomial, degree, function, one variable, *and* two variables *to describe the expression, equation, or system.*

103. Graph $y = -4x + 1$ by hand.

104. Find the product $(3x - 2)(4x^2 - x + 5)$.

105. Solve $-4x + 1 = 2x - 5$.

106. Find the sum $(3x - 2) + (4x^2 - x + 5)$.

107. Solve the following:

$$y = -4x + 1$$
$$y = 2x - 5$$

108. Factor $36x^2 - 81y^4$.

8.3 Factoring Trinomials of the Form $ax^2 + bx + c$

Objectives

» Factor a trinomial by trial and error.

» Know how to rule out possibilities when factoring by trial and error.

» Factor a trinomial by grouping.

In Section 8.1, we factored trinomials of the form $ax^2 + bx + c$, where $a = 1$. In this section, we will factor trinomials of the form $ax^2 + bx + c$, where $a \neq 1$. We will discuss two methods: factoring by trial and error and factoring by grouping. These two methods give equivalent results.

Method 1: Factoring Trinomials by Trial and Error

One way to factor trinomials of the form $ax^2 + bx + c$, where $a \neq 1$, is to make educated guesses at the factorization and then find the product of these guesses to see if any of them work. This method is called **factoring by trial and error.**

Example 1 Factoring by Trial and Error

Factor $3x^2 + 14x + 8$.

Solution

If we can factor $3x^2 + 14x + 8$, the result will be of the form

$$(3x + ?)(x + ?)$$

The product of the last terms must be 8, so the last terms must be 1 and 8 or 2 and 4, where we can write each pair in either order. We can rule out negative last terms in the factors, because the middle term of $3x^2 + 14x + 8$ has the positive coefficient 14. We decide between the two pairs of possible last terms by multiplying:

$$(3x + 1)(x + 8) = 3x^2 + 24x + x + 8 = 3x^2 + 25x + 8$$
$$(3x + 8)(x + 1) = 3x^2 + 3x + 8x + 8 = 3x^2 + 11x + 8$$
$$(3x + 2)(x + 4) = 3x^2 + 12x + 2x + 8 = 3x^2 + 14x + 8 \leftarrow \text{Success!}$$
$$(3x + 4)(x + 2) = 3x^2 + 6x + 4x + 8 = 3x^2 + 10x + 8$$

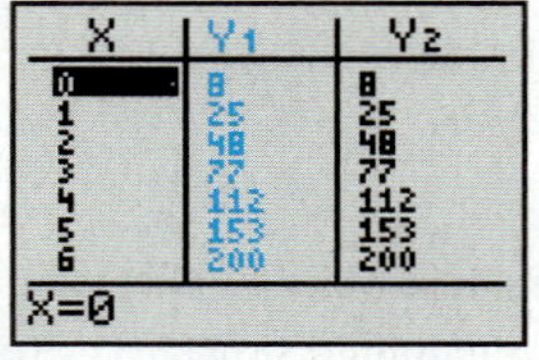

Figure 9 Verify the work

So, $3x^2 + 14x + 8 = (3x + 2)(x + 4)$. We use a graphing calculator table to verify our work (see Fig. 9).

In trying to factor a polynomial, once we find the factored polynomial, there is no need to multiply the other possibilities. In Example 1, we multiplied all possible factorizations of $3x^2 + 14x + 8$ only to show how to organize the work in case the last possibility is the correct one.

In order to use the method shown in Example 1, it is helpful to be able to multiply two binomials in one step. Consider the product of $2x + 3$ and $4x + 5$:

$$\begin{aligned}(2x + 3)(4x + 5) &= 8x^2 + 10x + 12x + 15 \\ &= 8x^2 + 22x + 15\end{aligned}$$

To find the product in one step, we must combine the like terms $10x$ and $12x$ mentally. Note that these like terms come from the product of the two *outer terms* and the product of the two *inner terms* of $(2x + 3)(4x + 5)$:

outer terms — Add these terms mentally.

$$\begin{aligned}(2x + 3)(4x + 5) &= 8x^2 + 10x + 12x + 15 \\ &= 8x^2 + 22x + 15\end{aligned}$$

inner terms

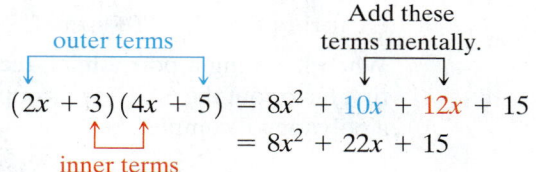

Example 2 Factoring by Trial and Error

Factor $2x^2 - 3x - 9$.

Solution

If we can factor $2x^2 - 3x - 9$, the result will be of the form

$$(2x + ?)(x + ?)$$

The product of the last terms is -9, so the last terms must be 1 and -9, 3 and -3, or -1 and 9, where each pair can be written in either order. We decide among the three pairs of possible last terms by multiplying:

$$\begin{aligned}(2x + 1)(x - 9) &= 2x^2 - 17x - 9 \\ (2x - 9)(x + 1) &= 2x^2 - 7x - 9 \\ (2x + 3)(x - 3) &= 2x^2 - 3x - 9 \longleftarrow \text{Success!} \\ (2x - 3)(x + 3) &= 2x^2 + 3x - 9 \\ (2x - 1)(x + 9) &= 2x^2 + 17x - 9 \\ (2x + 9)(x - 1) &= 2x^2 + 7x - 9\end{aligned}$$

Therefore, $2x^2 - 3x - 9 = (2x + 3)(x - 3)$.

Factoring $ax^2 + bx + c$ by Trial and Error

To **factor a trinomial** of the form $ax^2 + bx + c$ **by trial and error,** if the trinomial can be factored as a product of two binomials, then the product of the coefficients of the first terms of the binomials is equal to a and the product of the last terms of the binomials is equal to c. For example,

Coefficients of first terms: $3 \cdot 2 = 6 = a$

$$6x^2 + 23x + 20 = (3x + 4)(2x + 5)$$

$a = 6$ $b = 23$ $c = 20$

Last terms: $4 \cdot 5 = 20 = c$

To find the correct factored expression, multiply the possible products and identify the one for which the coefficient of x is b.

For example,

outer terms

$$(3x+4)(2x+5) = 6x^2 + 15x + 8x + 20$$
$$= 6x^2 + 23x + 20$$

inner terms

$b = 23$ is correct.

Factoring Out the GCF and Then Factoring by Trial and Error

When factoring a polynomial, recall from Section 8.2 that if the GCF is not 1, then we first factor out the GCF (or its opposite) and continue factoring if possible. Always factor a polynomial completely.

Example 3 Factoring a Polynomial Completely

Factor $15x^4 - 39x^3 + 18x^2$.

Solution

To factor $15x^4 - 39x^3 + 18x^2$, we first factor out the GCF, $3x^2$:

$$3x^2(5x^2 - 13x + 6)$$

If we can factor further, the desired result is of the form

$$3x^2(5x + ?)(x + ?)$$

The product of the last terms has to be 6, so the last terms must be -1 and -6 or -2 and -3, where each pair can be written in either order. We decide by multiplying:

(We have temporarily put aside the GCF, $3x^2$.)

$$(5x-1)(x-6) = 5x^2 - 31x + 6$$
$$(5x-6)(x-1) = 5x^2 - 11x + 6$$
$$(5x-2)(x-3) = 5x^2 - 17x + 6$$
$$(5x-3)(x-2) = 5x^2 - 13x + 6 \longleftarrow \text{Success!}$$

So, $15x^4 - 39x^3 + 18x^2 = 3x^2(5x^2 - 13x + 6) = 3x^2(5x-3)(x-2)$.

WARNING When we factor a polynomial by trial and error, we can easily forget about the GCF by the time we have found the other factors. **If there is more factoring to be done after factoring out the GCF, write a note several lines down that reminds you to include the GCF in your result.**

Ruling Out Possibilities while Factoring by Trial and Error

Example 4 shows how to rule out possible factorizations to help speed up the process of factoring.

Example 4 Ruling Out Possibilities

Factor $6x^2 - 19x + 8$.

Solution

If we can factor $6x^2 - 19x + 8$, the result is of one of the following two forms:

$$(6x + ?)(x + ?) \qquad (3x + ?)(2x + ?)$$

The product of the last terms has to be 8, so the last terms must be -1 and -8, or -2 and -4, where each pair can be written in either order. We can rule out positive last terms, because the middle term, $-19x$, has a negative coefficient, -19.

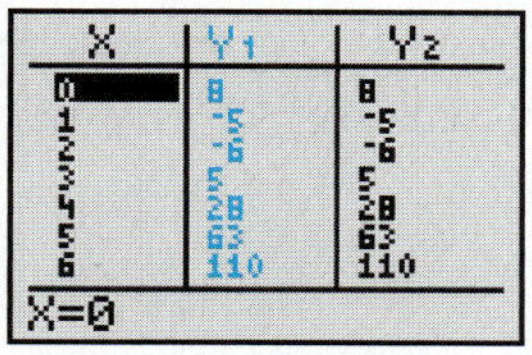

Since the terms of $6x^2 - 19x + 8$ do not have a common factor of 2, we can also rule out products that have a factor of 2. For example, we can rule out $(6x - 8)(x - 1)$, because it has a factor of 2:

$$(6x - 8)(x - 1) = 2(3x - 4)(x - 1)$$

We decide among the remaining possible last terms by multiplying:

$$(6x - 1)(x - 8) = 6x^2 - 49x + 8$$

Contains a factor of 2, rule out: $(6x - 2)(x - 4)$

Contains a factor of 2, rule out: $(6x - 4)(x - 2)$

Contains a factor of 2, rule out: $(3x - 1)(2x - 8)$

$$(3x - 8)(2x - 1) = 6x^2 - 19x + 8 \longleftarrow \text{Success!}$$

Contains a factor of 2, rule out: $(3x - 2)(2x - 4)$

Contains a factor of 2, rule out: $(3x - 4)(2x - 2)$

So, $6x^2 - 19x + 8 = (3x - 8)(2x - 1)$. We use a graphing calculator table to verify our work (see Fig. 10).

Figure 10 Verify the work

We can use a similar method to factor trinomials that have two variables.

Example 5 Factoring a Trinomial That Has Two Variables

Factor $10p^2 + 19pw + 6w^2$.

Solution

If we can factor $10p^2 + 19pw + 6w^2$, the result is of one of the following two forms:

$$(10p + ?)(p + ?) \qquad (5p + ?)(2p + ?)$$

The product of the last terms has to be $6w^2$, so the last terms must be $6w$ and w or $3w$ and $2w$, where each pair can be written in either order.

Since the terms of $10p^2 + 19pw + 6w^2$ do not have a common factor of 2, we can rule out products that have a factor of 2. We decide among the remaining possible last terms by multiplying:

Contains a factor of 2, rule out: $(10p + 6w)(p + w)$

$$(10p + w)(p + 6w) = 10p^2 + 61pw + 6w^2$$

$$(10p + 3w)(p + 2w) = 10p^2 + 23pw + 6w^2$$

Contains a factor of 2, rule out: $(10p + 2w)(p + 3w)$

$$(5p + 6w)(2p + w) = 10p^2 + 17pw + 6w^2$$

Contains a factor of 2, rule out: $(5p + w)(2p + 6w)$

Contains a factor of 2, rule out: $(5p + 3w)(2p + 2w)$

$$(5p + 2w)(2p + 3w) = 10p^2 + 19pw + 6w^2 \longleftarrow \text{Success!}$$

So, $10p^2 + 19pw + 6w^2 = (5p + 2w)(2p + 3w)$.

Example 6 Factoring a Polynomial Completely

Factor $6x^3y^2 + 26x^2y^3 + 24xy^4$.

Solution

To factor $6x^3y^2 + 26x^2y^3 + 24xy^4$, we first factor out the GCF, $2xy^2$:

$$2xy^2(3x^2 + 13xy + 12y^2)$$

If we can factor further, the result will be in the form

$$2xy^2(3x + ?)(x + ?)$$

The product of the last terms must be $12y^2$, so the last terms must be y and $12y$, $2y$ and $6y$, or $3y$ and $4y$, where we can write each pair in either order. Since we have factored out the GCF, we rule out any possibility in which one of the binomials has a monomial factor. We decide among the remaining possibilities by multiplying:

(We have temporarily put aside the GCF, $2xy^2$.)

$$(3x + y)(x + 12y) = 3x^2 + 37xy + 12y^2$$

Contains a factor of 3, rule out: $(3x + 12y)(x + y)$

$$(3x + 2y)(x + 6y) = 3x^2 + 20xy + 12y^2$$

Contains a factor of 3, rule out: $(3x + 6y)(x + 2y)$

Contains a factor of 3, rule out: $(3x + 3y)(x + 4y)$

$$(3x + 4y)(x + 3y) = 3x^2 + 13xy + 12y^2 \leftarrow \text{Success!}$$

So, $6x^3y^2 + 26x^2y^3 + 24xy^4 = 2xy^2(3x^2 + 13xy + 12y^2) = 2xy^2(3x + 4y)(x + 3y)$.

Method 2: Factoring Trinomials by Grouping

Instead of using trial and error to factor a trinomial, we can factor by grouping.

To factor a trinomial of the form $x^2 + bx + c$, recall from Section 8.1 that we look for two integers whose product is c and whose sum is b. To factor a trinomial of the form $ax^2 + bx + c$, we must look for two integers whose product is ac and whose sum is b.

Factoring $ax^2 + bx + c$ by Grouping

To **factor a trinomial** of the form $ax^2 + bx + c$ **by grouping** (if it can be done),

1. Find pairs of numbers whose product is ac.
2. Determine which of the pairs of numbers from step 1 has the sum b. Call this pair of numbers m and n.
3. Write the bx term as $mx + nx$:

$$ax^2 + bx + c = ax^2 + mx + nx + c$$

4. Factor $ax^2 + mx + nx + c$ by grouping.

Another name for this technique is the ***ac* method.**

Example 7 Factoring a Trinomial by Grouping

Factor $3x^2 + 14x + 8$ by grouping.

Solution

Here, $a = 3$, $b = 14$, and $c = 8$.

Step 1: Find the product ac: $ac = 3(8) = 24$.

Step 2: We want to find two numbers m and n that have the product $ac = 24$ and the sum $b = 14$:

Product = 24	Sum = 14?
$1(24) = 24$	$1 + 24 = 25$
$2(12) = 24$	$2 + 12 = 14 \leftarrow$ Success!
$3(8) = 24$	$3 + 8 = 11$
$4(6) = 24$	$4 + 6 = 10$

Since $2(12) = 24$ and $2 + 12 = 14$, we conclude that the two numbers m and n are 2 and 12.

Step 3: We write the bx term, $14x$, as the sum $mx + nx$:

$$3x^2 + 14x + 8 = 3x^2 + 2x + 12x + 8$$

Step 4: We factor $3x^2 + 2x + 12x + 8$ by grouping:

$$\begin{aligned} 3x^2 + 2x + 12x + 8 &= x(3x + 2) + 4(3x + 2) && \textit{Factor both pairs of terms.} \\ &= (x + 4)(3x + 2) && \textit{Factor out GCF, } (3x + 2). \end{aligned}$$

In step 3 of Example 7, we could switch the mx and nx terms to get $3x^2 + 12x + 2x + 8$ and still be able to factor by grouping in step 4. (Try it.)

In Example 1, we used trial and error to factor $3x^2 + 14x + 8$ as $(3x + 2)(x + 4)$. In Example 7, we factored it as $(x + 4)(3x + 2)$ by using grouping. The two results are equivalent. In general, the results from factoring a trinomial by trial and error and factoring a trinomial by grouping are equivalent.

Example 8 Factoring a Trinomial by Grouping

Factor $6x^2 - 7x + 2$ by grouping.

Solution

Here, $a = 6$, $b = -7$, and $c = 2$.

Step 1: Find the product ac: $ac = 6(2) = 12$.

Step 2: We want to find two numbers m and n that have the product $ac = 12$ and the sum $b = -7$:

Product = 12	**Sum = −7?**
$-1(-12) = 12$	$-1 + (-12) = -13$
$-2(-6) = 12$	$-2 + (-6) = -8$
$-3(-4) = 12$	$-3 + (-4) = -7$ ← Success!

Since $-3(-4) = 12$ and $-3 + (-4) = -7$, we conclude that the two numbers m and n are -3 and -4.

Step 3: We write $6x^2 - 7x + 2 = 6x^2 - 3x - 4x + 2$.

Step 4: We factor $6x^2 - 3x - 4x + 2$ by grouping:

$$\begin{aligned} 6x^2 - 3x - 4x + 2 &= 3x(2x - 1) - 2(2x - 1) && \textit{Factor both pairs of terms.} \\ &= (3x - 2)(2x - 1) && \textit{Factor out GCF, } (2x - 1). \end{aligned}$$

Example 9 Factoring Out the GCF, Then Factoring by Grouping

Factor $20x^4 - 40x^3 - 25x^2$.

Solution

First, we factor out the GCF, $5x^2$:

$$20x^4 - 40x^3 - 25x^2 = 5x^2(4x^2 - 8x - 5)$$

Next, we use grouping to try to factor $4x^2 - 8x - 5$, where $a = 4$, $b = -8$, and $c = -5$.

Step 1: Find the product ac: $ac = 4(-5) = -20$.

Step 2: We want to find two numbers m and n that have the product $ac = -20$ and the sum $b = -8$:

(We have temporarily put aside the GCF, $5x^2$.)

Product = −20	**Sum = −8?**
$1(-20) = -20$	$1 + (-20) = -19$
$2(-10) = -20$	$2 + (-10) = -8$ ← Success!
$4(-5) = -20$	$4 + (-5) = -1$
$5(-4) = -20$	$5 + (-4) = 1$
$10(-2) = -20$	$10 + (-2) = 8$
$20(-1) = -20$	$20 + (-1) = 19$

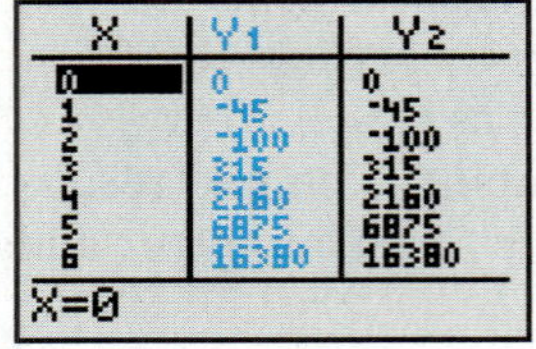

Figure 11 Verify the work

Since $2(-10) = -20$ and $2 + (-10) = -8$, we conclude that the two numbers m and n are 2 and -10.

Step 3: We write $4x^2 - 8x - 5 = 4x^2 + 2x - 10x - 5$.

Step 4: We factor $4x^2 + 2x - 10x - 5$ by grouping:

$$4x^2 + 2x - 10x - 5 = 2x(2x + 1) - 5(2x + 1) \quad \text{Factor both pairs of terms.}$$
$$= (2x - 5)(2x + 1) \quad \text{Factor out GCF, } (2x + 1).$$

So, $20x^4 - 40x^3 - 25x^2 = 5x^2(4x^2 - 8x - 5) = 5x^2(2x - 5)(2x + 1)$. We use a graphing calculator table to verify our work (see Fig. 11).

Group Exploration

Factoring polynomials

1. A student tries to factor $2x^2 - 17x - 30$:
$$2x^2 - 17x - 30 = (2x - 5)(x - 6)$$
Multiply $(2x - 5)(x - 6)$ to show the work is incorrect. Then factor $2x^2 - 17x - 30$ correctly.

2. A student tries to factor $2x^2 + 10x + 12$:
$$2x^2 + 10x + 12 = (2x + 4)(x + 3)$$
Explain why the work is not correct. Then factor the polynomial correctly.

3. A student tries to factor $2x^2 - x - 6$. Since the product of -3 and 2 is -6 and the sum of -3 and 2 is -1, the student does the following work:
$$2x^2 - x - 6 = (2x - 3)(x + 2)$$
Find the product $(2x - 3)(x + 2)$ to show the work is incorrect. Explain what is wrong with the student's reasoning. Then factor the polynomial correctly.

Group Exploration

Looking ahead: Developing a factoring strategy

In this exploration, you will summarize what you have learned about factoring.

1. When factoring a polynomial, what should you try to do first? Give an example.
2. Describe various techniques that you can use to factor a polynomial with the given number of terms. For each technique, give an example of factoring a polynomial.
   a. two terms **b.** three terms **c.** four terms
3. Explain how you know when you are done factoring a polynomial.

Homework 8.3

For extra help ▶ MyMathLab® 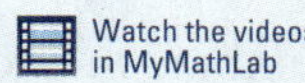 Watch the videos in MyMathLab  Download the MyDashboard App

Factor if possible. Verify that you have factored correctly by finding the product of your factored polynomial.

1. $2x^2 + 7x + 3$
2. $3x^2 + 16x + 5$
3. $5x^2 + 11x + 2$
4. $4x^2 + 13x + 3$
5. $3x^2 + 8x + 4$
6. $5x^2 + 13x + 6$
7. $2t^2 + t - 6$
8. $3b^2 - 10b + 8$
9. $6x^2 - 13x + 6$
10. $8x^2 - 14x - 15$
11. $4x^2 + 20x + 25$
12. $9x^2 + 12x + 4$
13. $2r^2 + 5r + 4$
14. $3a^2 - 2a - 6$
15. $18x^2 + 21x - 4$
16. $12x^2 - 17x + 6$
17. $3m^2 - 22m + 24$
18. $10t^2 - 27t + 18$
19. $2x^2 - 21x + 40$
20. $4x^2 + 81x + 20$
21. $1 + 9w^2 - 6w$
22. $-10m + 1 + 25m^2$
23. $2a^2 + 5ab + 3b^2$
24. $3m^2 + 8mn + 4n^2$

25. $5x^2 + 18xy - 8y^2$
26. $2p^2 - 5pt - 12t^2$
27. $6b^2 - 15bc + 6c^2$
28. $8r^2 - 33rw + 4w^2$
29. $4r^2 - 20ry + 25y^2$
30. $9b^2 - 24bc + 16c^2$

Factor. Use a graphing calculator table to verify your work when possible.

31. $4x^2 + 26x + 30$
32. $6x^2 + 27x + 30$
33. $20a^2 - 40a + 15$
34. $12b^2 - 56b + 32$
35. $24x^2 + 15x - 9$
36. $12x^2 - 27x + 15$
37. $-20x^2 + 22x + 12$
38. $-24x^2 + 54x - 27$
39. $-12x^2 + 3x + 9$
40. $-10x^2 - 8x + 24$
41. $12x - 32x^2 + 16x^3$
42. $-65x^2 + 25x + 30x^3$
43. $4w^4 - 6w^3 - 12w^2$
44. $9t^4 - 21t^3 - 12t^2$
45. $10x^4 - 5x^3 - 50x^2$
46. $36x^4 - 30x^3 + 6x^2$
47. $6a^2 - 34ab - 12b^2$
48. $6m^2 + 38mn + 12n^2$
49. $12r^3 + 40r^2w + 32rw^2$
50. $15x^3 + 36x^2y + 12xy^2$
51. $20a^3b^2 + 30a^2b^3 - 140ab^4$
52. $15a^3b - 36a^2b^2 + 12ab^3$

Factor when possible. Use a graphing calculator table to verify your work when possible.

53. $x^2 - 6x - 27$
54. $x^2 - 4x - 45$
55. $-48x^2 + 40x$
56. $-35x^2 - 42x$
57. $x^2 + 9$
58. $4x^2 + 25$
59. $4x^2 - 12x + 9$
60. $25x^2 - 10x + 1$
61. $-17p^2 + 17$
62. $-5t^2 + 20$
63. $24 + 10x + x^2$
64. $-12 + x + x^2$
65. $b^2 - 3bc - 28c^2$
66. $p^2 + 4pt - 32t^2$
67. $8t^2 - 10t + 3$
68. $6a^2 - 19a + 15$
69. $7x^4 - 28x^2$
70. $2x^4 - 50x^2$
71. $3x^4 - 21x^3y - 54x^2y^2$
72. $6a^2b^3 - 36ab^4 + 48b^5$
73. $12p^3 - 4p^2 - 27p + 9$
74. $50m^3 - 75m^2 - 2m + 3$
75. $x^2 - 6x + 12$
76. $x^2 - 3x - 8$
77. $3x^4 - 21x^3 + 30x^2$
78. $4x^4 - 12x^3 + 8x^2$
79. $20x^2 + 16x^4 - 42x^3$
80. $-14x^3 + 6x^4 - 40x^2$
81. $36a^2 - 49b^2$
82. $64m^2 - 25n^2$
83. $-2x^2y + 8xy + 24y$
84. $-6x^2y + 24xy - 24y$
85. $10p^3t^2 + 22p^2t^3 - 24pt^4$
86. $24w^4y - 44w^3y^2 - 40w^2y^3$

Concepts

87. A student tries to factor the polynomial $2x^2 + 7x + 10$. Since the integers 2 and 5 have the product $2(5) = 10$ and the sum $2 + 5 = 7$, the student writes

$$2x^2 + 7x + 10 = (2x + 5)(x + 2)$$

Is the student correct? Explain.

88. A student tries to factor the polynomial $3x^2 + 12$:

$$3x^2 + 12 = 3(x^2 + 4) = 3(x + 2)(x + 2) = 3(x + 2)^2$$

Describe any errors. Then factor the polynomial correctly.

89. A student tries to factor the polynomial $8x^2 + 28x + 12$:

$$8x^2 + 28x + 12 = (4x + 2)(2x + 6)$$

The student then uses a graphing calculator table to verify that the expressions $8x^2 + 28x + 12$ and $(4x + 2)(2x + 6)$ are equivalent. Explain why the student's work is incorrect even though the expressions are equivalent.

90. A student tries to factor the polynomial $6x^2 + 28x - 10$. First, the student divides the polynomial by 2: $3x^2 + 14x - 5$. Then the student factors $3x^2 + 14x - 5$:

$$3x^2 + 14x - 5 = (3x - 1)(x + 5)$$

The student thinks that the factorization of $6x^2 + 28x - 10$ is $(3x - 1)(x + 5)$. What would you tell the student? Also, explain how using a graphing calculator table to check the work can help the student identify the error.

91. To factor $4x^2 + 28x + 48$, why is it a good idea to first factor out the GCF before using any other factoring technique?

92. Factor $x^2 + 9x + 20$ using grouping. Then factor $x^2 + 9x + 20$ by the method discussed in Section 8.1. Which method is easier?

93. Give three examples of quadratic polynomials in which $2x - 3$ is a factor.

94. Give three examples of cubic polynomials in which $3x + 1$ is a factor.

95. Which of the following expressions are equivalent?

$$2(x - 2)(x - 6) \quad 2(x^2 - 8x + 12) \quad 2x^2 - 16x + 24$$
$$(x - 2)(2x - 12) \quad (x - 2)(2x - 6) \quad (2x - 2)(x - 6)$$
$$2(x - 4)^2 - 8 \quad (2x - 4)(x - 6)$$

96. Describe the various factoring techniques addressed in this section and in Sections 8.1 and 8.2. Give an example to illustrate each technique. Explain how to recognize polynomials to which each technique applies.

Related Review

If the expression is not factored, then factor it. If the expression is factored, then find the product.

97. $3x^2 + 16x - 12$
98. $5x^2 - 29x + 20$
99. $(4x - 7)(3x - 1)$
100. $(2x - 9)(5x + 3)$
101. $(x - 3)(2x^2 + 3x - 5)$
102. $(3x^2 - x - 2)(2x + 3)$
103. $6x^3 + 10x^2 - 4x$
104. $6x^4 - 33x^3 + 36x^2$

Expressions, Equations, Functions, and Graphs

Perform the indicated instruction. Then use words such as linear, quadratic, cubic, polynomial, degree, function, one variable, *and* two variables *to describe the expression, equation, or system.*

105. Graph $y = 2x^2$ by hand.

106. Solve the following:

$$7x + 2y = -6$$
$$3x - 4y = -22$$

107. Factor $x^2 - 2x - 3$.

108. Graph $7x + 2y = -6$ by hand.

109. Evaluate $x^2 - 2x - 3$ for $x = -5$.

110. Find the product $(7x + 2y)(3x - 4y)$.

8.4 Sums and Differences of Cubes; A Factoring Strategy

Objectives

- » Factor a sum or difference of two cubes.
- » Know a factoring strategy.

In Section 8.1, we discussed how to factor a difference of two squares. Here, first we will discuss how to factor a sum or difference of two cubes. Then we will discuss how to sift through the many factoring techniques we have discussed in this chapter and select the best ones to factor a given polynomial completely.

The Sum or Difference of Two Cubes

To see how to factor the sum of two cubes, we begin by multiplying the expressions $A + B$ and $A^2 - AB + B^2$:

$$\begin{aligned}(A + B)(A^2 - AB + B^2) &= A \cdot A^2 - A \cdot AB + A \cdot B^2 + B \cdot A^2 - B \cdot AB + B \cdot B^2 \\ &= A^3 - A^2B + AB^2 + A^2B - AB^2 + B^3 \\ &= A^3 + B^3\end{aligned}$$

So, $(A + B)(A^2 - AB + B^2) = A^3 + B^3$. Note that the right-hand side of the equation, $A^3 + B^3$, is a sum of two cubes. By similar work, we can also find a property for the difference of two cubes. You will do this in Exercise 93.

> **Sum or Difference of Two Cubes**
>
> $A^3 + B^3 = (A + B)(A^2 - AB + B^2)$ *Sum of two cubes*
>
> $A^3 - B^3 = (A - B)(A^2 + AB + B^2)$ *Difference of two cubes*

We can use these two properties to factor any polynomial that is a sum or difference of two cubes. In order to use the properties, it will help to memorize the following cubes:

$2^3 = 8 \qquad 3^3 = 27 \qquad 4^3 = 64 \qquad 5^3 = 125 \qquad 10^3 = 1000$

Example 1 Factoring a Sum and a Difference of Two Cubes

Factor.

1. $x^3 + 8$ **2.** $x^3 - 64$

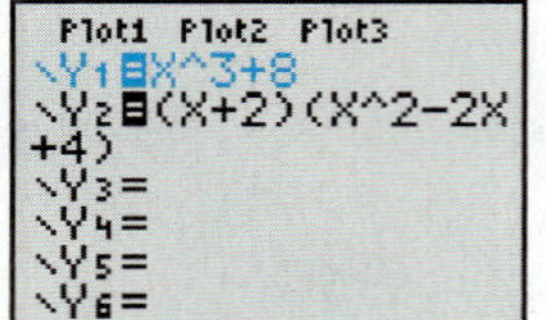

X	Y1	Y2
0	8	8
1	9	9
2	16	16
3	35	35
4	72	72
5	133	133
6	224	224

X=0

Figure 12 Verify the work

Solution

1.

$$\begin{aligned}A^3 + B^3 &= (A + B)(A^2 - A\,B + B^2) \\ x^3 + 8 = x^3 + 2^3 &= (x + 2)(x^2 - x \cdot 2 + 2^2) \quad \textit{Factor.} \\ &= (x + 2)(x^2 - 2x + 4) \quad \textit{Simplify.}\end{aligned}$$

The trinomial $x^2 - 2x + 4$ is prime, so we have factored $x^3 + 8$ completely. We use a graphing calculator table to verify our work (see Fig. 12).

2.

$$\begin{aligned}A^3 - B^3 &= (A - B)(A^2 + A\,B + B^2) \\ x^3 - 64 = x^3 - 4^3 &= (x - 4)(x^2 + x \cdot 4 + 4^2) \quad \textit{Factor.} \\ &= (x - 4)(x^2 + 4x + 16) \quad \textit{Simplify.}\end{aligned}$$

The trinomial $x^2 + 4x + 16$ is prime, so we have factored $x^3 - 64$ completely.

Example 2 Factoring a Sum and a Difference of Two Cubes

Factor.

1. $64t^3 + 27w^3$ **2.** $3x^5 - 24x^2y^3$

Solution

1. $64t^3 + 27w^3 = (4t)^3 + (3w)^3$ — *Write as a sum of cubes.*

$= (4t + 3w)((4t)^2 - 4t \cdot 3w + (3w)^2)$ — *$A^3 + B^3 = (A + B)(A^2 - AB + B^2)$*

$= (4t + 3w)(16t^2 - 12tw + 9w^2)$ — *Simplify.*

The trinomial $16t^2 - 12tw + 9w^2$ is prime, so we have factored $64t^3 + 27w^3$ completely.

2. For $3x^5 - 24x^2y^3$, first we factor out the GCF, $3x^2$:

$3x^5 - 24x^2y^3 = 3x^2(x^3 - 8y^3)$ — *Factor out GCF, $3x^2$.*

$= 3x^2(x^3 - (2y)^3)$ — *Write as a difference of cubes.*

$= 3x^2(x - 2y)(x^2 + x \cdot 2y + (2y)^2)$ — *$A^3 - B^3 = (A - B)(A^2 + AB + B^2)$*

$= 3x^2(x - 2y)(x^2 + 2xy + 4y^2)$ — *Simplify.*

The trinomial $x^2 + 2xy + 4y^2$ is prime, so we have factored $3x^5 - 24x^2y^3$ completely.

Consider the properties for the sum of cubes and the difference of cubes:

$$A^3 + B^3 = (A + B)(A^2 - AB + B^2) \qquad A^3 - B^3 = (A - B)(A^2 + AB + B^2)$$

Provided we have first factored out the GCF (or its opposite), and A and B are first-degree monomials, we can assume the trinomials $A^2 - AB + B^2$ and $A^2 + AB + B^2$ are prime.

Example 3 Factoring the Difference of Two Sixth Powers

Factor $n^6 - p^6$.

Solution

Since $n^6 = (n^3)^2$ and $p^6 = (p^3)^2$, we can begin to factor the binomial by using the property for a difference of two squares:

$n^6 - p^6$

$= (n^3)^2 - (p^3)^2$ — *Write as a difference of two squares.*

$= (n^3 + p^3)(n^3 - p^3)$ — *$A^2 - B^2 = (A + B)(A - B)$*

$= (n + p)(n^2 - np + p^2)(n - p)(n^2 + np + p^2)$ — *$A^3 + B^3 = (A + B)(A^2 - AB + B^2)$; $A^3 - B^3 = (A - B)(A^2 + AB + B^2)$*

Although we could have started to factor $n^6 - p^6$ by writing it as a difference of two cubes, $(n^2)^3 - (p^2)^3$, this would have led to very challenging factoring after we had used the property for a difference of two cubes.

A Factoring Strategy

We will now discuss a five-step factoring strategy that will help us determine the best factoring techniques to use to factor a given polynomial completely.

Five-Step Factoring Strategy

These five steps can be used to factor many polynomials (steps 2–4 can be applied to the entire polynomial or to a factor of the polynomial):

1. If the leading coefficient is positive and the GCF is not 1, factor out the GCF. If the leading coefficient is negative, factor out the opposite of the GCF.
2. For a binomial, try using one of the properties for the difference of two squares, the sum of two cubes, or the difference of two cubes.

3. For a trinomial of the form $ax^2 + bx + c$,
   a. If $a = 1$, try to find two integers whose product is c and whose sum is b.
   b. If $a \neq 1$, try to factor by using trial and error or by grouping.
4. For an expression with four terms, try factoring by grouping.
5. Continue applying steps 2–4 until the polynomial is factored completely.

Example 4 Factoring a Polynomial

Factor $x^4 - 2x^3 + 1000x - 2000$.

Solution

Since $x^4 - 2x^3 + 1000x - 2000$ has four terms, we try to factor it by grouping:

$$\begin{aligned} & x^4 - 2x^3 + 1000x - 2000 \\ &= x^3(x-2) + 1000(x-2) \\ &= (x^3 + 1000)(x-2) && \text{Factor by grouping.} \\ &= (x+10)(x^2 - 10x + 100)(x-2) && A^3 + B^3 = (A+B)(A^2 - AB + B^2) \end{aligned}$$

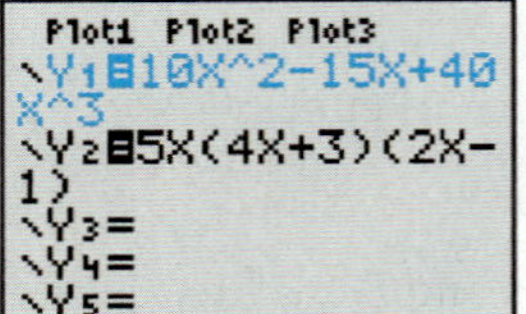

X	Y1	Y2
0	0	0
1	35	35
2	330	330
3	1125	1125
4	2660	2660
5	5175	5175
6	8910	8910

X=0

Figure 13 Verify the work

Example 5 Factoring a Polynomial

Factor $10x^2 - 15x + 40x^3$.

Solution

First, we write $10x^2 - 15x + 40x^3$ in descending order:

$$\begin{aligned} 10x^2 - 15x + 40x^3 &= 40x^3 + 10x^2 - 15x && \text{Rearrange terms.} \\ &= 5x(8x^2 + 2x - 3) && \text{Factor out GCF, } 5x. \\ &= 5x(4x+3)(2x-1) && \text{Factor by trial and error.} \end{aligned}$$

We use a graphing calculator table to verify our work (see Fig. 13).

Example 6 Factoring a Polynomial

Factor $50t^2w^2 - 8w^4$.

Solution

For $50t^2w^2 - 8w^4$, the GCF is $2w^2$. First, we factor out $2w^2$:

$$50t^2w^2 - 8w^4 = 2w^2(25t^2 - 4w^2)$$

Since the factor $25t^2 - 4w^2$ has two terms, we check to see whether it is the difference of two squares, which it is. So, we have

$$50t^2w^2 - 8w^4 = 2w^2(25t^2 - 4w^2) = 2w^2(5t + 2w)(5t - 2w)$$

Example 7 Factoring a Polynomial

Factor $3a^3b - 21a^2b^2 + 18ab^3$.

Solution

For $3a^3b - 21a^2b^2 + 18ab^3$, the GCF is $3ab$. First, we factor out $3ab$:

$$3a^3b - 21a^2b^2 + 18ab^3 = 3ab(a^2 - 7ab + 6b^2)$$

Since the factor $a^2 - 7ab + 6b^2$ is a trinomial with leading coefficient 1, we try to find two monomials whose product is $6b^2$ and whose sum is $-7b$. The monomials are $-b$ and $-6b$, so we have

$$3a^3b - 21a^2b^2 + 18ab^3 = 3ab(a^2 - 7ab + 6b^2) = 3ab(a - b)(a - 6b)$$

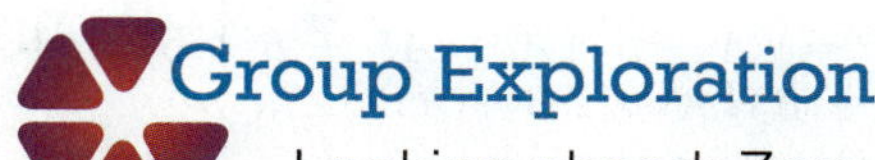

Group Exploration

Looking ahead: Zero factor property

1. What can you say about A or B if $AB = 0$?
2. What can you say about A or B if $A(B - 1) = 0$?
3. What can you say about x if $x(x - 1) = 0$?
4. Solve $x^2 - x = 0$. [**Hint:** Does this have something to do with Problem 3?]
5. Solve $2x^2 - 6x = 0$.
6. Solve $x^2 - 8x + 15 = 0$.

Tips for Success Create an Example

When learning a definition or property, try to create an example. For example, while studying the material in Sections 8.1–8.4, you could create polynomials that might be factored by the following methods:

- Factor out the GCF.
- Factor a cubic polynomial of the form $ax^3 + bx^2 + cx + d$ by grouping.
- Factor a difference of two squares.
- Factor a sum of two cubes.
- Factor a difference of two cubes.
- Factor a quadratic trinomial of the form $x^2 + bx + c$.
- Factor a quadratic trinomial of the form $ax^2 + bx + c$, where $a \neq 1$, by using trial and error or by grouping.
- Factor a polynomial completely by using a combination of methods.

You could create these examples by finding products of polynomials. For instance, create a trinomial of the form $ax^2 + bx + c$ by finding the product $(2x + 5)(3x - 4)$. After creating many polynomials, let time pass so that you won't remember their factored forms. Then practice factoring them. While factoring each polynomial, reflect on the polynomial's type and on why you chose to use a particular method to factor that type.

Creating examples will shed light on many details of a concept and will personalize the information. It will also help you understand the similarities and differences between related concepts.

Homework 8.4

For extra help ▶ MyMathLab®

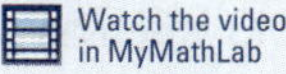
Watch the videos in MyMathLab

Download the MyDashboard App

Factor. Use a graphing calculator table to verify your work when possible.

1. $x^3 + 27$ **2.** $x^3 + 64$ **3.** $x^3 + 125$ **4.** $x^3 + 1000$

5. $x^3 - 8$ **6.** $x^3 - 27$ **7.** $x^3 - 1$ **8.** $x^3 - 125$

9. $8t^3 + 27$ **10.** $27p^3 + 64$ **11.** $27x^3 - 8$ **12.** $8x^3 - 125$

13. $5x^3 + 40$ **14.** $2x^3 + 2$ **15.** $2x^3 - 54$ **16.** $3x^3 - 24$

17. $8x^3 + 27y^3$ **18.** $27m^3 + 64n^3$ **19.** $64a^3 - 27b^3$

20. $125p^3 - 8t^3$ **21.** $2x^4 - 54xy^3$ **22.** $4x^5 - 32x^2y^3$

23. $k^6 - 1$ **24.** $t^6 - 64$ **25.** $64x^6 - y^6$

26. $x^6 - y^6$

Factor when possible. Use a graphing calculator table to verify your work when possible.

27. $x^2 - 64$ **28.** $x^2 - 1$ **29.** $2t^2 + 2t - 24$

30. $3a^2 - 21a + 30$ **31.** $x^2 + 49$ **32.** $x^2 + 4$

33. $-3x^2 + 24x - 45$ **34.** $-2x^2 + 22x - 60$ **35.** $1 + 15p^2 - 8p$

36. $-6t - 8 + 5t^2$ **37.** $x^2 - 2x + 1$ **38.** $x^2 - 8x + 16$

39. $24r^2 + 4r - 4$ **40.** $20y^3 - 12y^2 - 8y$ **41.** $-4ab^3 + 6a^2b^2$

42. $-6a^2b + 9ab^3$ **43.** $a^2 - ab - 20b^2$ **44.** $m^2 - 12mn + 32n^2$

45. $64a^3 + 27$ **46.** $27r^3 + 8$ **47.** $8x^3 - 20x^2 - 2x + 5$

48. $5x^3 - 2x^2 - 5x + 2$ **49.** $-12x^4 - 4x^3$

50. $-25x^4 + 35x^3$ **51.** $15a^4 + 25a^3 + 10a^2$

52. $6t^4 - 3t^3 - 30t^2$ **53.** $24 - 14x + x^2$

54. $9 - 6x + x^2$ **55.** $2w^4 + 4w^3 - 8w^2$

56. $3p^4 - 9p^3 + 24p^2$ **57.** $12x^4 - 27x^2$

58. $98x^4 - 18x^2$ **59.** $x^2 + 10x + 25$

60. $x^2 + 4x + 4$ **61.** $2x^2 + x - 21$

62. $4x^2 + 20x + 21$ **63.** $81p^4 - 16q^4$

64. $16m^4 - n^4$ **65.** $3x^3 + 24$

66. $7x^3 - 7$ **67.** $m^3 - 13m^2n + 36mn^2$

68. $p^2q + 5pq^2 - 12q^3$ **69.** $4x - 5 + 2x^2$

70. $7x - 4 + 3x^2$ **71.** $100x^2 - 9y^2$

72. $49b^2 - 36c^2$ **73.** $4x^2 + 12x + 9$

74. $16x^2 - 8x + 1$

75. $18x^3 + 27x^2 - 8x - 12$

76. $12x^3 + 9x^2 + 8x + 6$

77. $3a^3 - 10a^2b + 8ab^2$

78. $5p^2t + 18pt^2 - 8t^3$

79. $x^2 - 9x - 20$

80. $x^2 - 13x - 30$

81. $x^3 - 1000$

82. $x^3 + 1$ **83.** $12x^2y - 26xy^2 - 10y^3$ **84.** $36x^4 - 21x^3y + 3x^2y^2$

Concepts

For Exercises 85–88, discuss which technique(s) you should consider using when factoring the given type of polynomial. Also, refer to an exercise in Homework 8.4 that illustrates this technique.

85. binomial

86. trinomial of the form $x^2 + bx + c$

87. trinomial of the form $ax^2 + bx + c$, where $a \neq 1$

88. polynomial with four terms

89. A student tries to factor the polynomial $x^3 - 27$:

$$x^3 - 27 = (x - 3)(x^2 + 6x + 9)$$

Describe any errors. Then factor the polynomial correctly.

90. A student tries to factor the polynomial $x^3 + 8$:

$$x^3 + 8 = (x + 2)(x^2 + 2x + 4)$$

Describe any errors. Then factor the polynomial correctly.

91. Is the following statement true? Explain.

$$A^3 + B^3 = (A + B)(A^2 - 2AB + B^2)$$

92. When factoring a polynomial with a negative leading coefficient, what should you do? Give an example.

93. Show that $A^3 - B^3 = (A - B)(A^2 + AB + B^2)$ by finding the product $(A - B)(A^2 + AB + B^2)$.

94. Show that $A^2 + AB + B^2$ is prime when $A = x$ and $B = y$.

95. When factoring a polynomial, does it matter whether you factor out the GCF in the first step or in the last step? Explain.

96. Describe a factoring strategy in your own words.

Related Review

If the expression is not factored, then factor it. If the expression is factored, then find the product.

97. $(5x - 7)(5x + 7)$

98. $(2x + 5)(x^2 - x + 4)$

99. $81x^2 - 16$

100. $x^2 - 5x - 36$

101. $3x^3 + 9x^2 - 12x$

102. $30x^2 - 24$

103. $-2x(7x^2 - 5x + 1)$

104. $(4x - 3)(2x + 9)$

Expressions, Equations, Functions, and Graphs

Perform the indicated instruction. Then use words such as linear, quadratic, cubic, polynomial, degree, function, one variable, *and* two variables *to describe the expression, equation, or system.*

105. Graph $y = 4x - 5$ by hand.

106. Factor $x^3 - 3x^2 - 4x + 12$.

107. Solve $4x - 5 = 3x + 2$.

108. Simplify $(2x - 7)^2$.

109. Find the product $(4x - 5)(3x + 2)$.

110. Graph $y = 3x^2$ by hand.

8.5 Using Factoring to Solve Polynomial Equations

Objectives

- Know the *zero factor property*.
- Use factoring to solve *quadratic equations in one variable*.
- Find the x-intercept(s) of the graph of a polynomial function.
- Know a connection between the x-intercepts of the graph of a function and the solutions of a related equation in one variable.
- Use factoring to solve *cubic equations in one variable*.
- Use graphing to solve a polynomial equation in one variable.
- Compare solving polynomial equations with factoring polynomials.

In this section, we will discuss how to use factoring to help solve equations. In Section 8.6, we will use this skill to make efficient predictions with some quadratic models.

Zero Factor Property

In Section 7.1, we graphed quadratic equations in *two* variables. In this section, we will solve quadratic equations in *one* variable, such as

$$x^2 - 3x - 28 = 0 \qquad 25x^2 - 49 = 0 \qquad 12x = 9x^2 + 4$$

A **quadratic equation in one variable** is an equation that can be put into the form

$$ax^2 + bx + c = 0$$

where a, b, and c are constants and $a \neq 0$. The connection between solving a quadratic equation and factoring an expression lies in the *zero factor property*.

Zero Factor Property

Let A and B be real numbers.

If $AB = 0$, then $A = 0$ or $B = 0$.

In words, if the product of two numbers is zero, then at least one of the numbers must be zero.

Solving Quadratic Equations in One Variable

We can use the zero factor property to help us solve some quadratic equations in one variable.

Example 1 Solving a Quadratic Equation

Solve $(x-5)(x+2)=0$.

Solution

$$(x-5)(x+2)=0 \quad \text{Original equation}$$
$$x-5=0 \quad \text{or} \quad x+2=0 \quad \text{Zero factor property}$$
$$x=5 \quad \text{or} \quad x=-2 \quad \text{Add 5 to both sides./Subtract 2 from both sides.}$$

We check that both 5 and -2 satisfy the original equation:

Check $x = 5$

$$(x-5)(x+2)=0$$
$$(5-5)(5+2)\overset{?}{=}0$$
$$0(7)\overset{?}{=}0$$
$$0\overset{?}{=}0$$
true

Check $x = -2$

$$(x-5)(x+2)=0$$
$$(-2-5)(-2+2)\overset{?}{=}0$$
$$-7(0)\overset{?}{=}0$$
$$0\overset{?}{=}0$$
true

So, the solutions are 5 and -2.

Example 2 Solving a Quadratic Equation

Solve $w^2-2w-8=0$.

Solution

$$w^2-2w-8=0 \quad \text{Original equation}$$
$$(w+2)(w-4)=0 \quad \text{Factor left side.}$$
$$w+2=0 \quad \text{or} \quad w-4=0 \quad \text{Zero factor property}$$
$$w=-2 \quad \text{or} \quad w=4 \quad \text{Subtract 2 from both sides./Add 4 to both sides.}$$

We check that both -2 and 4 satisfy the original equation:

Check $w = -2$

$$w^2-2w-8=0$$
$$(-2)^2-2(-2)-8\overset{?}{=}0$$
$$4+4-8\overset{?}{=}0$$
$$0\overset{?}{=}0$$
true

Check $w = 4$

$$w^2-2w-8=0$$
$$4^2-2(4)-8\overset{?}{=}0$$
$$16-8-8\overset{?}{=}0$$
$$0\overset{?}{=}0$$
true

So, the solutions are -2 and 4.

The key step in solving a quadratic equation of the form $ax^2+bx+c=0$ is to factor the left side of the equation so we can apply the zero factor property.

Connection between x-Intercepts and Solutions

There is an important connection between x-intercepts of the graph of a function and the solutions of a related equation in one variable. We begin to investigate this connection in Example 3.

Example 3 Finding x-Intercepts of the Graph of a Quadratic Function

Find the x-intercepts of the graph of $f(x)=x^2-7x+10$.

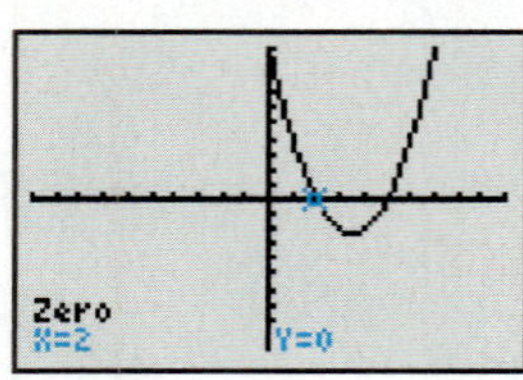

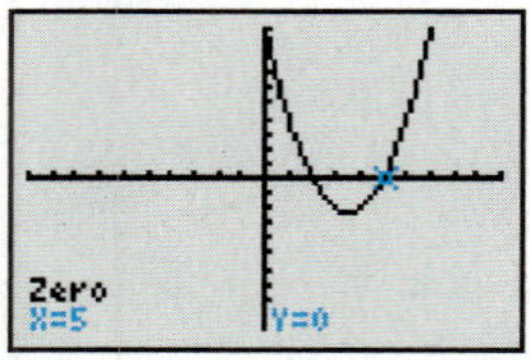

Figure 14 Verify the work

Solution

To find the x-intercepts, we substitute 0 for $f(x)$ and solve for x:

$$
\begin{aligned}
0 &= x^2 - 7x + 10 && \textit{Substitute 0 for } f(x). \\
0 &= (x-5)(x-2) && \textit{Factor right-hand side.} \\
x - 5 = 0 \quad &\text{or} \quad x - 2 = 0 && \textit{Zero factor property} \\
x = 5 \quad &\text{or} \quad x = 2
\end{aligned}
$$

So, the x-intercepts are $(2, 0)$ and $(5, 0)$. We use "zero" on a graphing calculator to verify our work (see Fig. 14).

In general, we find the x-intercepts of the graph of $f(x) = ax^2 + bx + c$ by solving the equation $ax^2 + bx + c = 0$.

Connection Between x-Intercepts and Solutions

Let f be a function. If k is a real-number solution of the equation $f(x) = 0$, then $(k, 0)$ is an x-intercept of the graph of the function f. Also, if $(k, 0)$ is an x-intercept of the graph of f, then k is a solution of $f(x) = 0$.

This property suggests that we can verify our solutions of a quadratic equation $ax^2 + bx + c = 0$ by using a graphing calculator to find the x-intercepts of the graph of $f(x) = ax^2 + bx + c$.

Solving More Quadratic Equations

In Example 4, we will solve quadratic equations and use a graphing calculator to verify our work.

Example 4 Solving Quadratic Equations

Solve.

1. $4x^2 + 26x + 30 = 0$ **2.** $x^2 - 10x + 25 = 0$

Solution

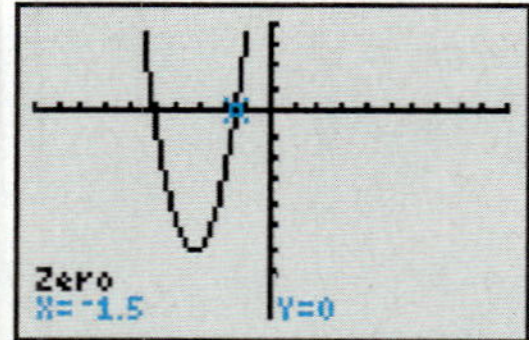

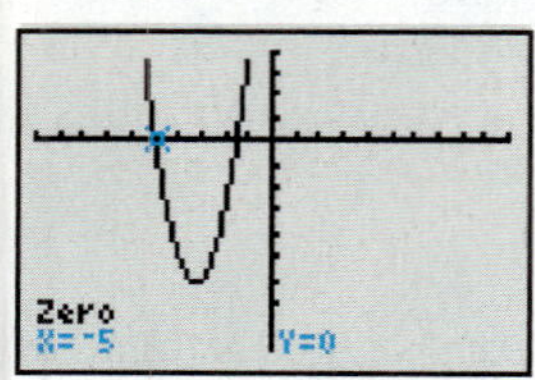

Figure 15 Verify the work

1.

$$
\begin{aligned}
4x^2 + 26x + 30 &= 0 && \textit{Original equation} \\
2(2x^2 + 13x + 15) &= 0 && \textit{Factor out GCF, 2.} \\
2(2x + 3)(x + 5) &= 0 && \textit{Factor left side completely.}
\end{aligned}
$$

Now we can apply a variation of the zero factor property: If $2AB = 0$, then $A = 0$ or $B = 0$. Here, we take $2x + 3$ to be A and $x + 5$ to be B:

$$
\begin{aligned}
2x + 3 = 0 \quad &\text{or} \quad x + 5 = 0 && \textit{Zero factor property} \\
2x = -3 \quad &\text{or} \quad x = -5 \\
x = -\frac{3}{2} \quad &\text{or} \quad x = -5
\end{aligned}
$$

We use "zero" on a graphing calculator to check that the x-intercepts of the graph of $f(x) = 4x^2 + 26x + 30$ are $\left(-\frac{3}{2}, 0\right)$ and $(-5, 0)$. See Fig. 15.

2.

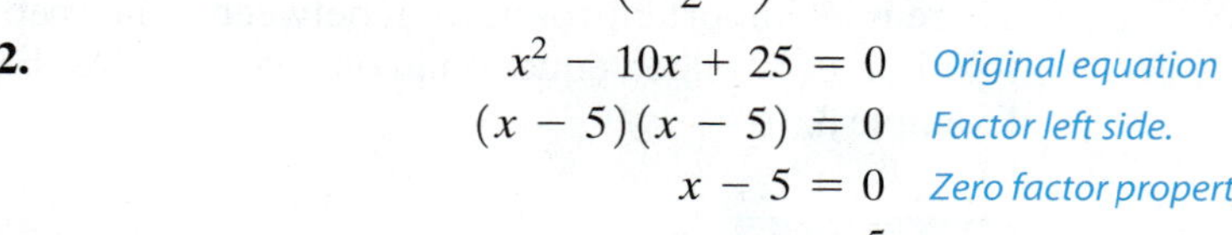

$$
\begin{aligned}
x^2 - 10x + 25 &= 0 && \textit{Original equation} \\
(x-5)(x-5) &= 0 && \textit{Factor left side.} \\
x - 5 &= 0 && \textit{Zero factor property} \\
x &= 5
\end{aligned}
$$

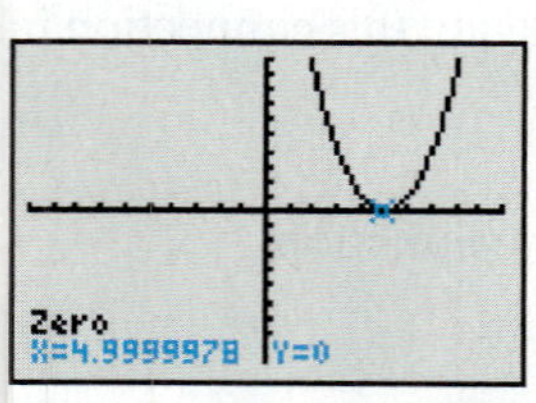

Figure 16 Verify the work

Since using the zero factor property yields the *one* equation $x - 5 = 0$, there is one solution: 5. We use "zero" on a graphing calculator to check that the x-intercept of the graph of $f(x) = x^2 - 10x + 25$ is $(5, 0)$. See Fig. 16.

In Example 4, we found that the equation $4x^2 + 26x + 30 = 0$ has two solutions, whereas the equation $x^2 - 10x + 25 = 0$ has one solution. What are the possible numbers of solutions of a quadratic equation? To decide, note that the graph of a quadratic function can have two x-intercepts, one x-intercept, or no x-intercepts (see Fig. 17).

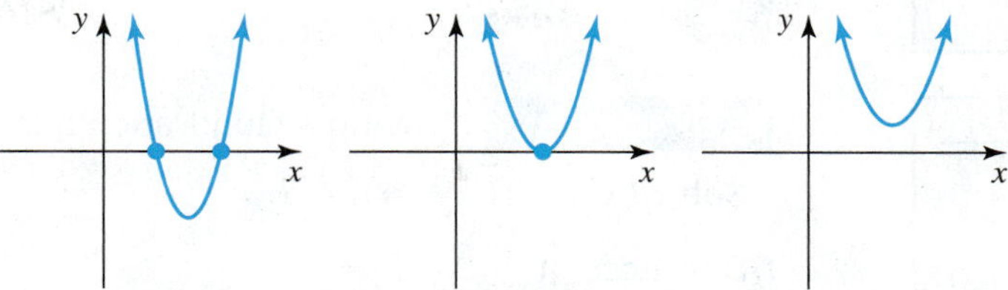

Figure 17 A quadratic function can have two, one, or no x-intercepts

Since the number of real-number solutions of an equation $ax^2 + bx + c = 0$ is equal to the number of x-intercepts of the graph of the function $f(x) = ax^2 + bx + c$, we conclude that **the solution set of a quadratic equation in one variable may contain two real numbers, one real number, or no real numbers.** We will solve quadratic equations that have no real-number solutions in Chapter 9.

In Example 4, both original equations had one side that was 0. If neither side of a quadratic equation in one variable is 0, we must first use properties of equality to get one side to be 0 so we can then apply the zero factor property.

Example 5 Solving a Quadratic Equation by Factoring

Solve $6x^2 = 18x$.

Solution

$$
\begin{aligned}
6x^2 &= 18x && \textit{Original equation} \\
6x^2 - 18x &= 0 && \textit{Write in } ax^2 + bx = 0 \textit{ form.} \\
6x(x - 3) &= 0 && \textit{Factor left side.} \\
6x = 0 \quad &\text{or} \quad x - 3 = 0 && \textit{Zero factor property} \\
x = 0 \quad &\text{or} \quad x = 3
\end{aligned}
$$

So, the solutions are 0 and 3. We can check that the solutions satisfy the original equation. (Try it.)

WARNING In Example 5, we solved the equation $6x^2 = 18x$ by first writing it in the form $ax^2 + bx = 0$ (there is no c term) and then factoring the left side of the equation. It is a common error to try to solve the equation $6x^2 = 18x$ by dividing both sides by $6x$ to get the result $x = 3$. But note that this incorrect method gives only one of the two solutions 0 and 3.

In general, when solving quadratic equations, we can't divide by x because we don't know if the value of x is zero. Recall from Section 2.2 that division by zero is undefined.

Example 6 Solving a Quadratic Equation

Solve $2x^2 - 8x = 5x - 20$.

Solution

$$
\begin{aligned}
2x^2 - 8x &= 5x - 20 && \textit{Original equation} \\
2x^2 - 13x + 20 &= 0 && \textit{Write in } ax^2 + bx + c = 0 \textit{ form.} \\
(2x - 5)(x - 4) &= 0 && \textit{Factor left side.} \\
2x - 5 = 0 \quad &\text{or} \quad x - 4 = 0 && \textit{Zero factor property} \\
2x = 5 \quad &\text{or} \quad x = 4 \\
x = \frac{5}{2} \quad &\text{or} \quad x = 4
\end{aligned}
$$

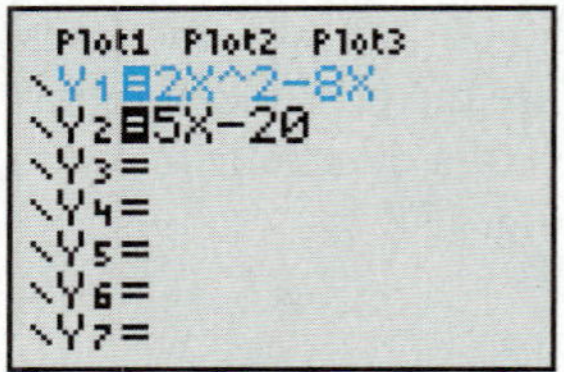

X	Y1	Y2
2.5	-7.5	-7.5
4	0	0

X=2.5

Figure 18 Verify the work

To verify that $\frac{5}{2}$ is a solution of $2x^2 - 8x = 5x - 20$, we can use a graphing calculator table to check that, for input $\frac{5}{2}$, the output for $y = 2x^2 - 8x$ is equal to the output for $y = 5x - 20$. We do similarly for input $x = 4$ (see Fig. 18).

Example 7 Solving a Quadratic Equation

Solve $(x + 2)(x - 4) = 7$.

Solution

Although the left-hand side of $(x + 2)(x - 4) = 7$ is factored, the right-hand side is not zero. First, we find the product on the left-hand side of the equation:

$$(x + 2)(x - 4) = 7 \qquad \textit{Original equation}$$
$$x^2 - 2x - 8 = 7 \qquad \textit{Find product on left side.}$$
$$x^2 - 2x - 15 = 0 \qquad \textit{Write in } ax^2 + bx + c = 0 \textit{ form.}$$
$$(x - 5)(x + 3) = 0 \qquad \textit{Factor left-hand side.}$$
$$x - 5 = 0 \quad \text{or} \quad x + 3 = 0 \qquad \textit{Zero factor property}$$
$$x = 5 \quad \text{or} \quad x = -3$$

Therefore, the solutions are -3 and 5. To verify the work, we can enter the equations $y = (x + 2)(x - 4)$ and $y = 7$ and use "intersect" to find the intersection points $(-3, 7)$ and $(5, 7)$. See Fig. 19. The x-coordinates of these points, -3 and 5, are the solutions of the original equation, which checks.

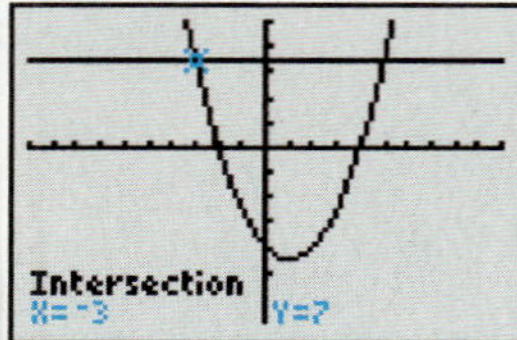

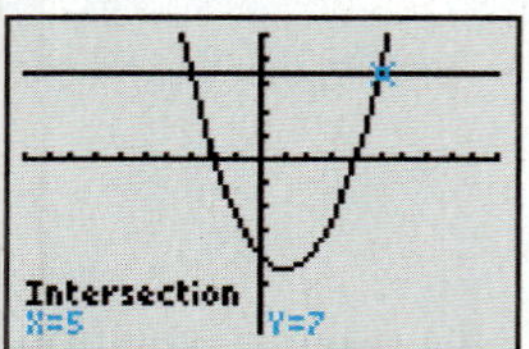

Figure 19 Verify the work

Example 8 Solving a Quadratic Equation That Contains Fractions

Solve $\frac{1}{2}t^2 + \frac{1}{3} = \frac{5}{6}t$.

Solution

To clear the equation of fractions, we multiply both sides by the least common denominator (LCD), 6:

$$\frac{1}{2}t^2 + \frac{1}{3} = \frac{5}{6}t \qquad \textit{Original equation}$$
$$6 \cdot \frac{1}{2}t^2 + 6 \cdot \frac{1}{3} = 6 \cdot \frac{5}{6}t \qquad \textit{Multiply both sides by LCD, 6.}$$
$$3t^2 + 2 = 5t \qquad \textit{Simplify.}$$
$$3t^2 - 5t + 2 = 0 \qquad \textit{Write in } ax^2 + bx + c = 0 \textit{ form.}$$
$$(3t - 2)(t - 1) = 0 \qquad \textit{Factor left side.}$$
$$3t - 2 = 0 \quad \text{or} \quad t - 1 = 0 \qquad \textit{Zero factor property}$$
$$3t = 2 \quad \text{or} \quad t = 1$$
$$t = \frac{2}{3} \quad \text{or} \quad t = 1$$

Now that we can solve some quadratic equations in one variable, we can find inputs of a quadratic function for some outputs.

Example 9 Finding an Input and an Output of a Quadratic Function

Let $f(x) = x^2 - 3x - 23$.

1. Find $f(5)$. **2.** Find x when $f(x) = 5$.

Solution

1. $f(5) = 5^2 - 3(5) - 23 = 25 - 15 - 23 = -13$
2. We substitute 5 for $f(x)$ in the equation $f(x) = x^2 - 3x - 23$:

$$
\begin{aligned}
5 &= x^2 - 3x - 23 && \textit{Substitute 5 for } f(x). \\
0 &= x^2 - 3x - 28 && \textit{Write in } 0 = ax^2 + bx + c \textit{ form.} \\
0 &= (x - 7)(x + 4) && \textit{Factor right-hand side.} \\
x - 7 = 0 \quad &\text{or} \quad x + 4 = 0 && \textit{Zero factor property} \\
x = 7 \quad &\text{or} \quad x = -4 &&
\end{aligned}
$$

Next, we verify that $f(7) = 5$ and $f(-4) = 5$:

Check $f(7) = 5$

$$
\begin{aligned}
f(x) &= x^2 - 3x - 23 \\
f(7) &= 7^2 - 3(7) - 23 \\
&= 5
\end{aligned}
$$

Check $f(-4) = 5$

$$
\begin{aligned}
f(x) &= x^2 - 3x - 23 \\
f(-4) &= (-4)^2 - 3(-4) - 23 \\
&= 5
\end{aligned}
$$

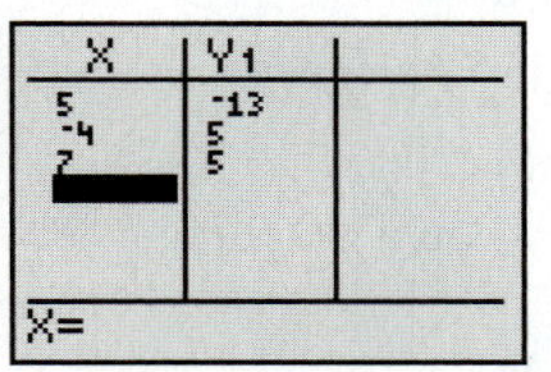

Figure 20 Verify the work

Or we can verify our work in Problems 1 and 2 by using a graphing calculator table (see Fig. 20).

Cubic Equations in One Variable

So far in this section, we have solved quadratic equations. We will now solve some cubic equations, such as

$$4x^3 - 2x^2 - 36x + 18 = 0 \qquad 2x^3 = 42x + 8x^2$$

A **cubic equation in one variable** is an equation that can be put into the form

$$ax^3 + bx^2 + cx + d = 0$$

where a, b, c, and d are constants and $a \neq 0$. We can solve some cubic equations by applying the zero factor property to three factors:

$$\text{If } ABC = 0, \text{ then } A = 0, B = 0, \text{ or } C = 0.$$

Example 10 Solving a Cubic Equation

Solve $2x^3 = 42x + 8x^2$.

Solution

$$
\begin{aligned}
2x^3 &= 42x + 8x^2 && \textit{Original equation} \\
2x^3 - 8x^2 - 42x &= 0 && \textit{Write in } ax^3 + bx^2 + cx + d = 0 \textit{ form.} \\
2x(x^2 - 4x - 21) &= 0 && \textit{Factor out GCF, } 2x. \\
2x(x - 7)(x + 3) &= 0 && \textit{Factor left side completely.} \\
2x = 0 \quad \text{or} \quad x - 7 = 0 \quad &\text{or} \quad x + 3 = 0 && \textit{Zero factor property} \\
x = 0 \quad \text{or} \quad x = 7 \quad &\text{or} \quad x = -3 &&
\end{aligned}
$$

So, the solutions are -3, 0, and 7.

Note that solving a cubic equation in one variable is similar to solving a quadratic equation in one variable. We now summarize the steps used to solve either type of equation.

Solving Quadratic or Cubic Equations by Factoring

If an equation can be solved by factoring, we solve it by the following steps:

1. Write the equation so one side of the equation is equal to zero.
2. Factor the nonzero side of the equation.
3. Apply the zero factor property.
4. Solve each equation that results from applying the zero factor property.

Example 11 Finding x-Intercepts of the Graph of a Cubic Function

Find the x-intercepts of the graph of $f(x) = x^3 - 5x^2 - 4x + 20$.

Solution

We substitute 0 for $f(x)$ and solve for x:

$$x^3 - 5x^2 - 4x + 20 = 0 \qquad \text{Substitute 0 for } f(x).$$

$$\left.\begin{aligned} x^2(x-5) - 4(x-5) &= 0 \\ (x^2-4)(x-5) &= 0 \end{aligned}\right\} \qquad \text{Factor by grouping.}$$

$$(x+2)(x-2)(x-5) = 0 \qquad A^2 - B^2 = (A+B)(A-B)$$

$$x + 2 = 0 \quad \text{or} \quad x - 2 = 0 \quad \text{or} \quad x - 5 = 0 \qquad \text{Zero factor property}$$

$$x = -2 \quad \text{or} \quad x = 2 \quad \text{or} \quad x = 5$$

So, the x-intercepts are $(-2, 0)$, $(2, 0)$, and $(5, 0)$. We use "zero" on a graphing calculator to verify our work (see Fig. 21).

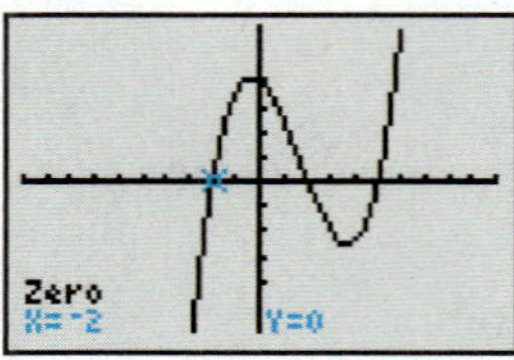

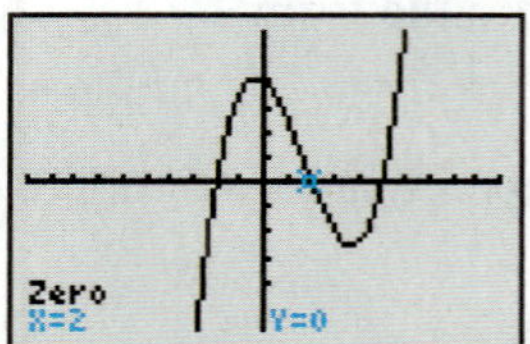

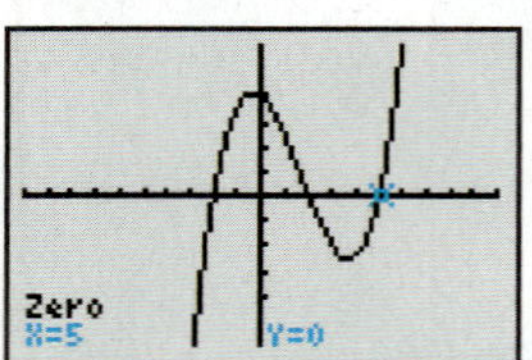

Figure 21 Verify the work

WARNING It is a common error to try to apply the zero factor property to an equation such as $x^2(x-5) - 4(x-5) = 0$ and incorrectly conclude that the solutions are 0 and 5. The expression $x^2(x-5) - 4(x-5)$ is *not* factored, because it is a difference, not a product. Only after we factor the left side of the equation $x^2(x-5) - 4(x-5) = 0$ can we apply the zero factor property.

In Example 11, we worked with the cubic function $f(x) = x^3 - 5x^2 - 4x + 20$. Recall from Section 7.1 that a cubic function is a function whose equation can be put into the form $f(x) = ax^3 + bx^2 + cx + d$, where $a \neq 0$.

The cubic function sketched in graph (a) of Fig. 22 has exactly three x-intercepts, the cubic function sketched in graph (b) has exactly two x-intercepts, and cubic functions sketched in graphs (c) and (d) have exactly one x-intercept. It turns out the graph of any cubic function has exactly one, two, or three x-intercepts.

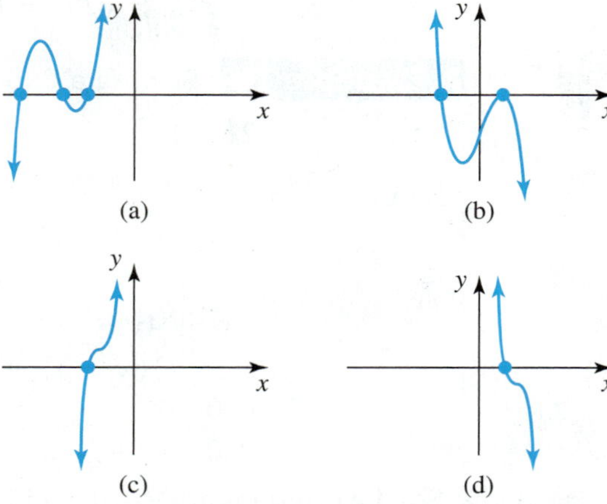

Figure 22 Graphs of typical cubic functions

The number of real-number solutions of a cubic equation $ax^3 + bx^2 + cx + d = 0$ is equal to the number of x-intercepts of the graph of the function $f(x) = ax^3 + bx^2 + cx + d$, so we conclude that **the solution set of a cubic equation in one variable may contain one, two, or three real numbers.**

Solving Polynomial Equations in One Variable by Graphing

We can solve quadratic equations by graphing in much the same way we solved linear equations by graphing in Sections 4.3 and 4.4.

Example 12 Solving a Quadratic Equation in One Variable by Graphing

The graphs of $y = x^2 - 10x + 28$, $y = 7$, $y = 3$, and $y = 1$ are shown in Fig. 24. Use these graphs to solve the given equation.

1. $x^2 - 10x + 28 = 7$ **2.** $x^2 - 10x + 28 = 3$ **3.** $x^2 - 10x + 28 = 1$

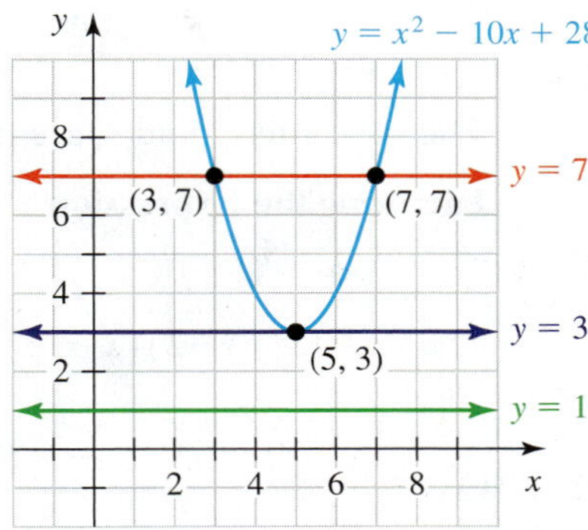

Figure 24 Solving three quadratic equations in one variable

Solution

1. The graphs of $y = x^2 - 10x + 28$ and $y = 7$ intersect only at the points (3, 7) and (7, 7), which have x-coordinates 3 and 7. So, 3 and 7 are the solutions of $x^2 - 10x + 28 = 7$.
2. The graphs of $y = x^2 - 10x + 28$ and $y = 3$ intersect only at the point (5, 3), which has x-coordinate 5. So, 5 is the solution of $x^2 - 10x + 28 = 3$.
3. The graphs of $y = x^2 - 10x + 28$ and $y = 1$ do not intersect. So, no real number is a solution of $x^2 - 10x + 28 = 1$.

Consider the equation $x^2 - x - 7 = -x^2$. If we add x^2 to both sides, the result is $2x^2 - x - 7 = 0$. Since the left side of the equation, $2x^2 - x - 7$, is prime (try it), we cannot solve the equation by factoring. However, we will show in Example 13 that we can solve it by graphing.

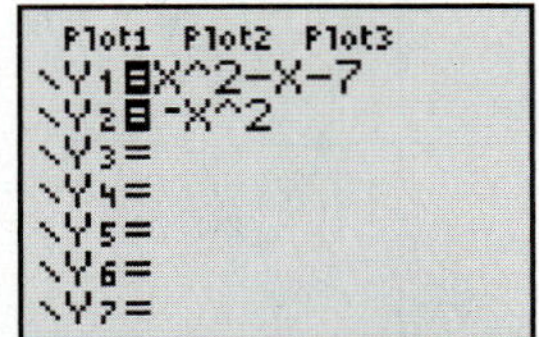

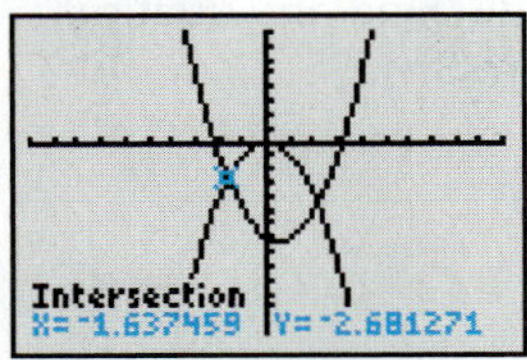

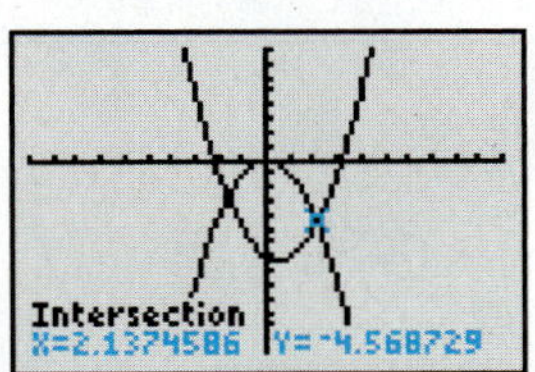

Figure 23 Solve the system

Example 13 Using Graphing to Solve an Equation in One Variable

Use graphing to solve $x^2 - x - 7 = -x^2$.

Solution

We use "intersect" on a graphing calculator to find the solutions of the system

$$y = x^2 - x - 7$$
$$y = -x^2$$

See Fig. 23.

The approximate solutions of the system are $(-1.64, -2.68)$ and $(2.14, -4.57)$. The x-coordinates of these ordered pairs, -1.64 and 2.14, are the approximate solutions of the equation $x^2 - x - 7 = -x^2$.

In Sections 9.5 and 9.6, we will discuss two symbolic methods that can be used to solve $x^2 - x - 7 = -x^2$.

Solving Polynomial Equations versus Factoring Polynomials

In Section 4.5, we compared solving a linear equation with simplifying a linear expression. **When we solve any equation, our objective is to find all *numbers* that satisfy the equation. When we factor a polynomial, our objective is to write the polynomial as a product of polynomials, which is an *expression*.**

Here we compare solving a polynomial equation with factoring a polynomial:

Solving the Equation $x^2 - 5x + 6 = 0$

$$x^2 - 5x + 6 = 0 \quad \text{Original equation}$$
$$(x - 2)(x - 3) = 0 \quad \text{Factor left side.}$$
$$x - 2 = 0 \quad \text{or} \quad x - 3 = 0 \quad \text{Zero factor property}$$
$$x = 2 \quad \text{or} \quad x = 3$$

The solutions, 2 and 3, are numbers.

Factoring the Expression $x^2 - 5x + 6$

$$x^2 - 5x + 6 = (x - 2)(x - 3)$$

The result, $(x - 2)(x - 3)$, is an expression.

Group Exploration

Finding equations of quadratic functions

1. Use ZStandard to sketch graphs of the functions
$$f(x) = (x - 2)(x + 3),\ g(x) = 2(x - 2)(x + 3),$$
$$h(x) = \frac{1}{2}(x - 2)(x + 3), \text{ and } k(x) = -(x - 2)(x + 3).$$
   a. What do you notice about the x-intercepts of the graphs of f, g, h, and k?
   b. For a function of the form $y = a(x - 2)(x + 3)$, describe the effect the value of a has on the graph of the function. Sketch more graphs with varying values of a if you are unsure.
2. Find a possible equation of the function sketched in Fig. 25. Use a graphing calculator to verify your work.
3. Find an equation of the function sketched in Fig. 26. Use a graphing calculator to verify your work. [**Hint:** A point on the graph of an equation satisfies the equation.]

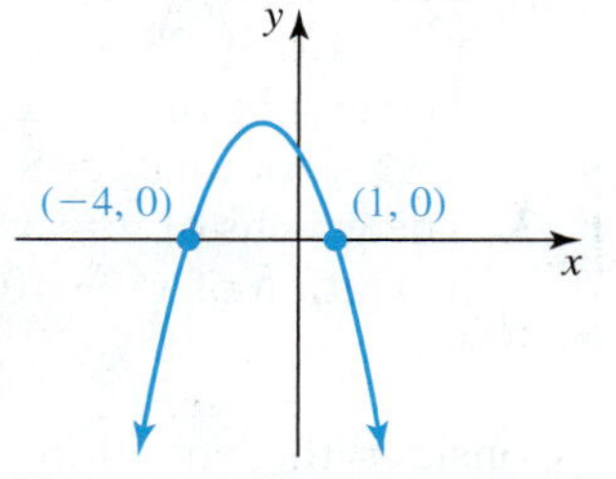

Figure 25 Find a possible equation (Problem 2)

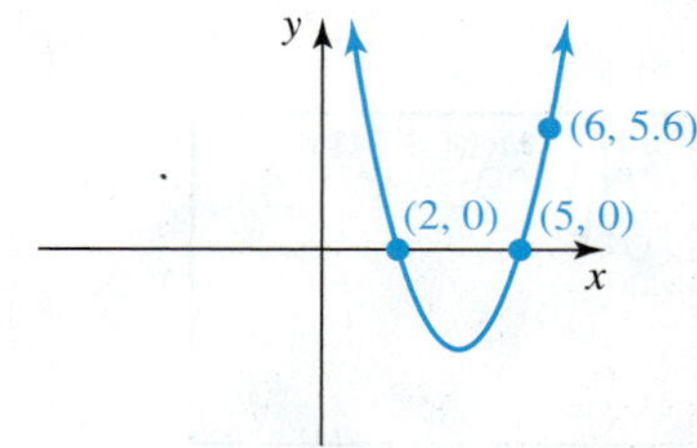

Figure 26 Find an equation (Problem 3)

Homework 8.5

For extra help ▶ MyMathLab®  Watch the videos in MyMathLab  Download the MyDashboard App

Solve. Verify any results by checking that they satisfy the equation.

1. $(x + 4)(x - 7) = 0$
2. $(x + 3)(x - 9) = 0$
3. $x^2 + 7x + 10 = 0$
4. $x^2 + 5x + 6 = 0$
5. $w^2 + w - 12 = 0$
6. $r^2 + 3r - 28 = 0$
7. $x^2 - 8x + 15 = 0$
8. $x^2 - 6x + 5 = 0$
9. $14x + 49 + x^2 = 0$
10. $16 + x^2 + 8x = 0$
11. $-24 - 2t + t^2 = 0$
12. $-36 + y^2 - 5y = 0$
13. $25x^2 - 49 = 0$
14. $64x^2 - 9 = 0$
15. $6m^2 - 11m + 3 = 0$
16. $4p^2 - 3p - 10 = 0$
17. $3x^2 + 3x - 90 = 0$
18. $2x^2 + 6x - 80 = 0$
19. $8x^3 - 12x^2 - 20x = 0$
20. $12x^3 - 2x^2 - 2x = 0$

Solve. Use a graphing calculator table or graph to verify your work.

21. $x^2 = 5x + 14$
22. $x^2 = 11x + 12$
23. $4x^2 - 8x = 32$
24. $10x^2 - 30x = -20$
25. $12t - 36 = t^2$
26. $4w - 4 = w^2$
27. $x^2 = 49$
28. $x^2 = 4$
29. $16x^2 = 25$
30. $49x^2 = 4$
31. $6x^3 - 24x = 0$
32. $4x^3 - 100x = 0$
33. $3r^2 = 6r$
34. $5p^2 = 35p$
35. $9x = -2x^2 + 5$
36. $10x = -8x^2 - 3$
37. $2x^3 = 6x^2 + 36x$
38. $36x = 24x^2 - 4x^3$
39. $18y^3 + 3y^2 = 6y$
40. $8x^3 - 14x^2 = 4x$
41. $\dfrac{a^2}{2} - \dfrac{a}{6} = \dfrac{1}{3}$
42. $\dfrac{t^2}{5} - \dfrac{t}{2} = -\dfrac{1}{5}$
43. $-\dfrac{1}{3}x^2 + \dfrac{1}{3}x + 10 = 6$
44. $-\dfrac{1}{2}x^2 - \dfrac{5}{2}x + 12 = 5$
45. $x^2 - \dfrac{1}{25} = 0$
46. $x^2 - \dfrac{1}{49} = 0$
47. $(x + 2)(x + 5) = 40$
48. $(x + 3)(x - 2) = 24$

49. $4r^3 - 2r^2 - 36r + 18 = 0$

50. $y^3 + 3y^2 - 4y - 12 = 0$

51. $9x^3 - 12 = 4x - 27x^2$

52. $3x^2 - 4x = 12 - x^3$

53. $2x(x + 1) = 5x(x - 7)$

54. $x(x - 3) = 3x(x - 4)$

55. $4p(p - 1) - 24 = 3p(p - 2)$

56. $2w^2 + 2(w - 1) = w(1 - w)$

57. $(x^2 + 5x + 6)(x^2 - 5x - 24) = 0$

58. $(x^2 - 7x + 6)(x^2 + 3x - 4) = 0$

Find all x-intercepts. Verify your intercept(s) graphically by using a graphing calculator.

59. $f(x) = x^2 - 9x + 20$

60. $f(x) = x^2 - 4x - 21$

61. $f(x) = 36x^2 - 25$

62. $f(x) = 16x^2 - 81$

63. $f(x) = 24x^3 - 14x^2 - 20x$

64. $f(x) = 6x^3 + 15x^2 + 6x$

65. $f(x) = x^3 + 2x^2 - x - 2$

66. $f(x) = 4x^3 - 12x^2 - 9x + 27$

For Exercises 67–70, let $f(x) = x^2 - x - 6$.

67. Find $f(3)$.

68. Find $f(-4)$.

69. Find x when $f(x) = 14$.

70. Find x when $f(x) = 6$.

For Exercises 71–74, use the graph of $y = \frac{1}{2}x^2 + x - \frac{7}{2}$ shown in Fig. 27 to solve the given equation.

71. $\frac{1}{2}x^2 + x - \frac{7}{2} = 4$

72. $\frac{1}{2}x^2 + x - \frac{7}{2} = -2$

73. $\frac{1}{2}x^2 + x - \frac{7}{2} = -4$

74. $\frac{1}{2}x^2 + x - \frac{7}{2} = -5$

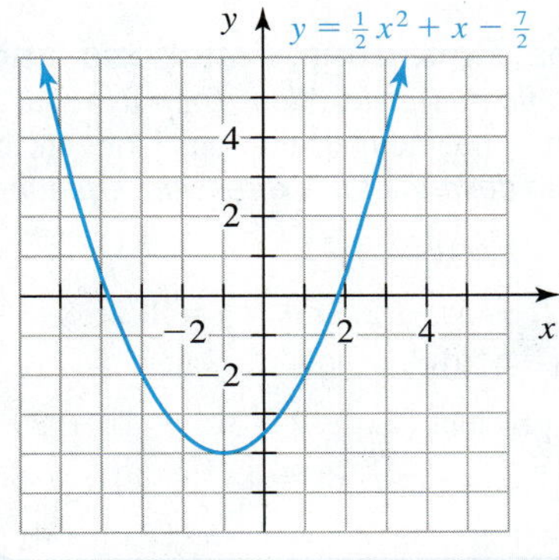

Figure 27 Exercises 71–74

For Exercises 75–78, use the graphs of $y = x^3 - 3x^2 + 1$ and $y = x - 2$ shown in Fig. 28 to solve the given equation or system.

75. $x^3 - 3x^2 + 1 = -3$

76. $x^3 - 3x^2 + 1 = 1$

77. $x^3 - 3x^2 + 1 = x - 2$

78. $y = x^3 - 3x^2 + 1$
$y = x - 2$

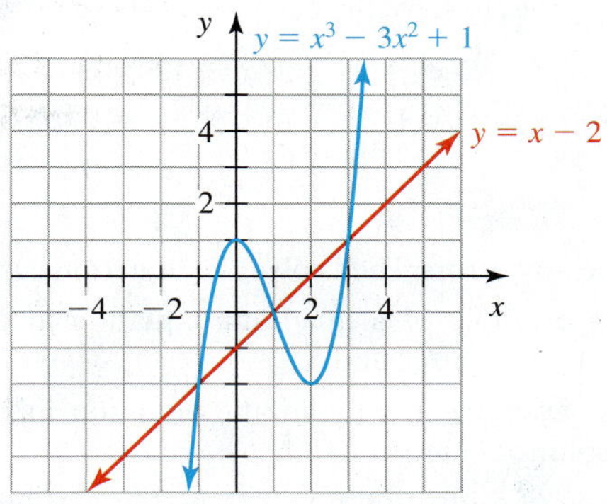

Figure 28 Exercises 75–78

Solve by using "intersect" on a graphing calculator. Round the solution(s) to the second decimal place.

79. $2x^2 - 5x - 3 = -x + 5$

80. $-x^2 + 2x + 4 = x - 1$

81. $-x^3 + 4x^2 - 2x = 4 - x$

82. $x^3 - 4x^2 + 5 = x - 5$

For Exercises 83–86, solve by referring to the values of the function $y = 3x^2 - 6x + 1$ shown in Table 2.

83. $3x^2 - 6x + 1 = 25$

84. $3x^2 - 6x + 1 = 1$

85. $3x^2 - 6x + 1 = -3$

86. $3x^2 - 6x + 1 = -2$

Table 2 Exercises 83–86

x	y
−2	25
−1	10
0	1
1	−2
2	1
3	10
4	25

Concepts

87. A student tries to solve the equation $x^2 = x$:

$$x^2 = x$$
$$\frac{x^2}{x} = \frac{x}{x}$$
$$x = 1$$

Describe any errors. Then solve the equation correctly.

88. A student tries to solve the equation $x^2 = 25$:

$$x^2 = 25$$
$$x = 5$$

Describe any errors. Then solve the equation correctly.

89. A student tries to solve the equation $2x^2 - 26x + 80 = 0$:

$$2x^2 - 26x + 80 = 0$$
$$2(x^2 - 13x + 40) = 0$$
$$2(x - 5)(x - 8) = 0$$
$$x = 2, x = 5, \text{ or } x = 8$$

Describe any errors. Then solve the equation correctly.

90. A student tries to solve the equation $x^3 + 3x^2 - 4x - 12 = 0$:

$$x^3 + 3x^2 - 4x - 12 = 0$$
$$x^2(x + 3) - 4(x + 3) = 0$$
$$x + 3 = 0$$
$$x = -3$$

Describe any errors. Then solve the equation correctly.

91. Give an example of a quadratic equation in one variable whose solutions are 2 and 5.

92. Give an example of a quadratic equation in one variable whose solutions are -4 and 1.

93. The graph of a function h is sketched in Fig. 29. Find a possible equation of h. Verify your equation by using a graphing calculator.

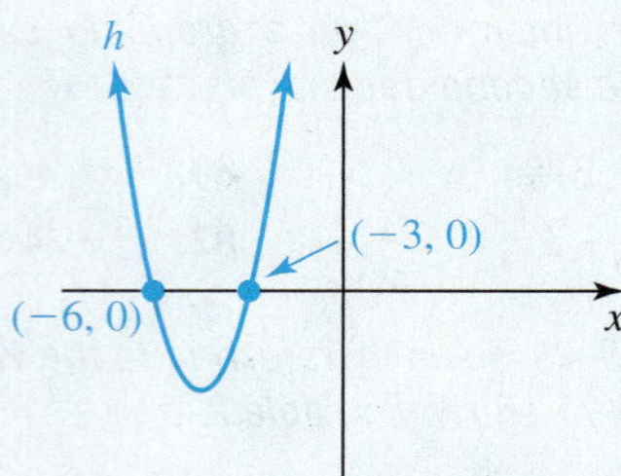

Figure 29 Exercise 93

94. The graph of a function g is sketched in Fig. 30. Find a possible equation of g. Verify your equation by using a graphing calculator.

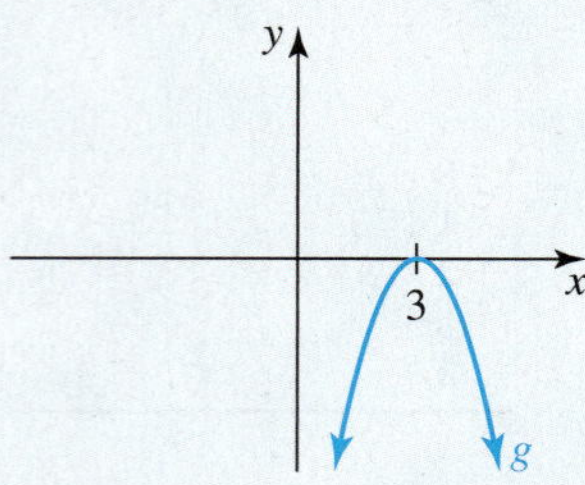

Figure 30 Exercise 94

95. If $(x + 5)(x + 6) = 0$, then $x + 5 = 0$ or $x + 6 = 0$. Explain why.

96. If an equation in one variable has exactly one solution, must the equation be linear? If yes, explain why. If no, give an example to show why not.

97. Consider the equation $y = x^2 - 6x + 8$.

a. Find x when $y = 3$. **b.** Find x when $y = -1$.

c. Use your results from parts (a) and (b) to help you graph the equation $y = x^2 - 6x + 8$ by hand. [**Hint:** Your work in part (b) should help you determine the vertex of the graph. (Why?)]

98. Here is the work done in solving the equation $x^2 = 5x - 6$:

$$x^2 = 5x - 6$$
$$x^2 - 5x + 6 = 0$$
$$(x - 2)(x - 3) = 0$$
$$x - 2 = 0 \quad \text{or} \quad x - 3 = 0$$
$$x = 2 \quad \text{or} \quad x = 3$$

Show that *each line* of the work becomes a true statement when 2 or 3 is substituted for x.

99. Compare factoring a quadratic polynomial with solving a quadratic equation. (See page 9 for guidelines on writing a good response.)

100. Explain in your own words how to solve a quadratic equation in one variable. (See page 9 for guidelines on writing a good response.)

Related Review

For Exercises 101–108, factor or solve, as appropriate.

101. $x^2 + 5x + 6$

102. $25x^2 - 64 = 0$

103. $x^2 + 5x + 6 = 0$

104. $25x^2 - 64$

105. $3p^3 + 8p^2 + 4p = 0$

106. $4w^3 - 20w^2 - 9w + 45$

107. $3p^3 + 8p^2 + 4p$

108. $4w^3 - 20w^2 - 9w + 45 = 0$

109. Solve the formula $A = P + PRT$ for P. [**Hint:** Begin by factoring the right-hand side of the formula.]

110. Solve the formula $S = 2WL + 2WH + 2LH$ for L. [**Hint:** You will need to use factoring eventually.]

Expressions, Equations, Functions, and Graphs

Perform the indicated instruction. Then use words such as linear, quadratic, cubic, polynomial, degree, function, one variable, *and* two variables *to describe the expression, equation, or system.*

111. Simplify $-2x(3x - 5)^2$.

112. Solve $3(2t - 5) - 2(2t + 3) = 12$.

113. Factor $8x^3 - 40x^2 + 50x$.

114. Find the product $(5x - 2)(3x^2 - 4x + 2)$.

115. Find $f(-2)$, where $f(x) = 8x^3 - 40x^2 + 50x$.

116. Find the difference $(5x - 2) - (3x^2 - 4x + 2)$.

8.6 Using Factoring to Make Predictions with Quadratic Models

Objectives

- Use an equation of a quadratic model to make estimates and predictions.
- Model projectile motion and the area of rectangular objects.

Using a Quadratic Function to Model a Situation

In Sections 7.1 and 7.2, we used linear and quadratic functions to model authentic situations. Now we focus on using quadratic functions to perform modeling.

Definition Quadratic model

A **quadratic model** is a quadratic function, or its graph, that describes the relationship between two quantities in an authentic situation.

Table 3 Annual Revenue of Restaurants

Year	Annual Revenue (billions of dollars)
1970	42.8
1980	119.6
1990	239.3
2000	379.0
2012	631.8

Source: *National Restaurant Association*

Example 1 Modeling with a Quadratic Function

Annual revenues of restaurants are shown in Table 3 for various years. Let $f(t)$ be the annual revenue (in billions of dollars) of restaurants at t years since 1970. A model of the situation is $f(t) = \frac{1}{5}t^2 + 5t + 54$.

1. Use a graphing calculator to draw the graph of the model and, in the same viewing window, the scattergram of the data. Does the model fit the data well?
2. Predict the revenue in 2018.
3. Predict when the annual revenue was $84 billion.

Solution

1. The graph of the model and the scattergram of the data are shown in Fig. 31. The model appears to fit the data quite well.
2. To predict the revenue in 2018, we find $f(48)$:

$$f(48) = \frac{1}{5}(48)^2 + 5(48) + 54 = 754.8$$

The revenue in 2018 will be $754.8 billion, according to the model.

3. To estimate when the annual revenue was $84 billion, we substitute 84 for $f(t)$ in $f(t) = \frac{1}{5}t^2 + 5t + 54$ and solve for t:

$$84 = \frac{1}{5}t^2 + 5t + 54 \quad \text{Substitute 84 for } f(t).$$

$$0 = \frac{1}{5}t^2 + 5t - 30 \quad \text{Write in } 0 = at^2 + bt + c \text{ form.}$$

$$0 = t^2 + 25t - 150 \quad \text{Multiply both sides by the LCD, 5.}$$

$$0 = (t - 5)(t + 30) \quad \text{Factor right-hand side.}$$

$$t - 5 = 0 \quad \text{or} \quad t + 30 = 0 \quad \text{Zero Factor Property}$$

$$t = 5 \quad \text{or} \quad t = -30$$

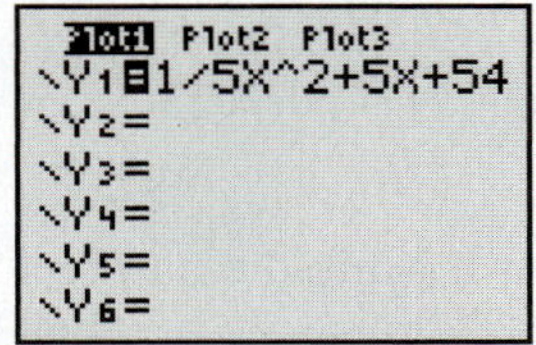

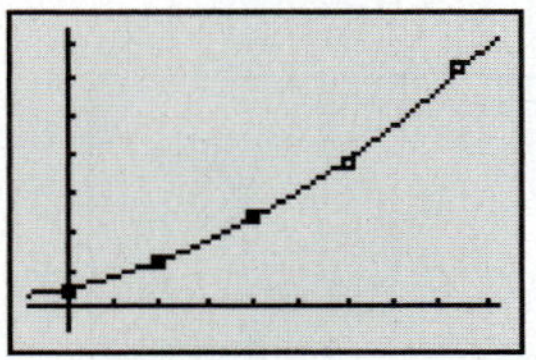

Figure 31 Check the fit

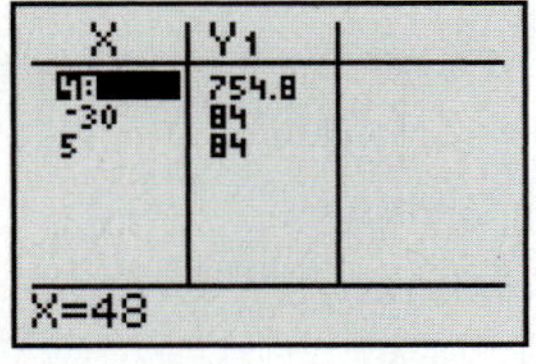

Figure 32 Verify the work

The inputs −30 and 5 represent the years 1940 and 1975, respectively. The estimate of 1940 is an example of model breakdown, as a little research would show the revenue in 1940 was much less than $84 billion. Therefore, we estimate that it was 1975 when the revenue was $84 billion. We use a graphing calculator table to verify our work in Problems 2 and 3 (see Fig. 32).

To make a prediction about the dependent variable of a quadratic model, we substitute the chosen value of the independent variable into the equation and solve for the dependent variable.

To make a prediction about the independent variable of a quadratic model, we substitute the chosen value of the dependent variable into the equation and solve for the independent variable. That involves writing the equation so one side is zero and factoring the nonzero side.

Projectile Motion

Any thrown object is called a *projectile*. If you launch a projectile such as a baseball or stone into the air, the relationship between time and the height of the projectile can be described well by a quadratic model. Such a model will not work well for an object that encounters a lot of air resistance—a feather, for example.

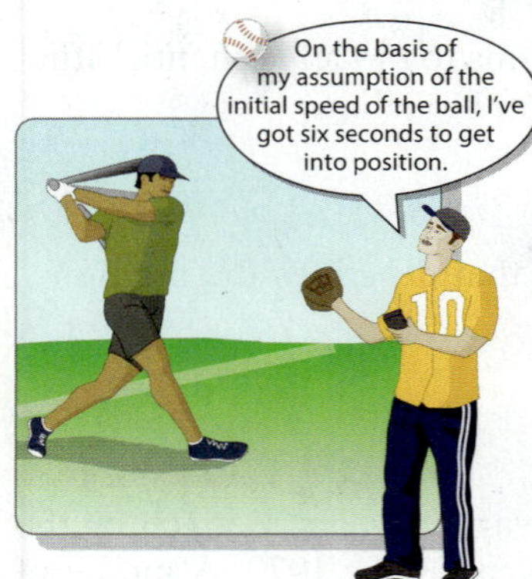

On the basis of my assumption of the initial speed of the ball, I've got six seconds to get into position.
10

Example 2 Making Estimates

A batter hits a baseball into the air. The height $f(t)$ (in feet) of the baseball at t seconds after the ball is hit is given by

$$f(t) = -16t^2 + 96t + 4$$

1. When is the baseball at a height of 84 feet? Explain why there are two such times.
2. When is the baseball at a height of 148 feet? Explain why there is one such time.
3. When is the baseball at a height of 150 feet?

Solution

1. We substitute the output 84 for $f(t)$ in the equation $f(t) = -16t^2 + 96t + 4$ and solve for t:

$$
\begin{aligned}
84 &= -16t^2 + 96t + 4 && \textit{Substitute 84 for } f(t). \\
0 &= -16t^2 + 96t - 80 && \textit{Write in } 0 = ax^2 + bx + c \textit{ form.} \\
0 &= -16(t^2 - 6t + 5) && \textit{Factor out } -16 \textit{ (on right-hand side).} \\
0 &= -16(t - 1)(t - 5) && \textit{Factor right-hand side completely.} \\
t - 1 = 0 \quad &\text{or} \quad t - 5 = 0 && \textit{Zero factor property} \\
t = 1 \quad &\text{or} \quad t = 5
\end{aligned}
$$

So, 1 second after the baseball is hit, and again 5 seconds after it is hit, the baseball is at a height of 84 feet. It makes sense that there are two times, because the baseball reaches 84 feet both on its way up and on its way down (see Fig. 33).

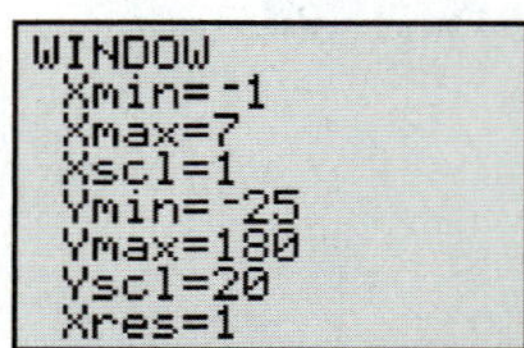

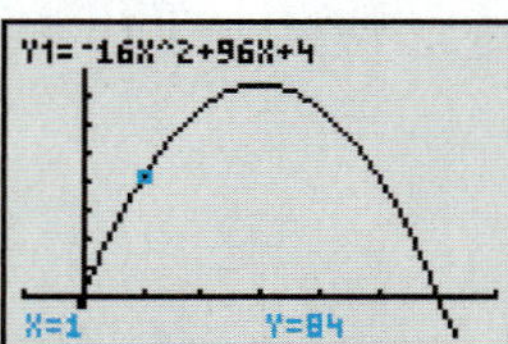

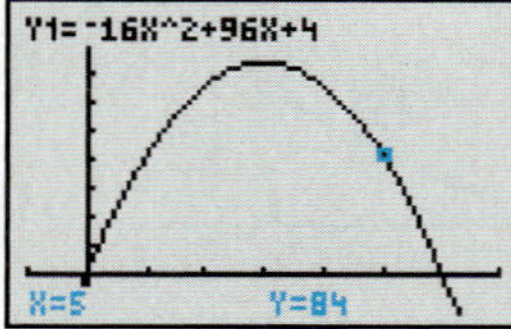

Figure 33 Finding when the baseball is at a height of 84 feet

2. We substitute the output 148 for $f(t)$ in the equation $f(t) = -16t^2 + 96t + 4$ and solve for t:

$$
\begin{aligned}
148 &= -16t^2 + 96t + 4 && \textit{Substitute 148 for } f(t). \\
0 &= -16t^2 + 96t - 144 && \textit{Write in } 0 = ax^2 + bx + c \textit{ form.} \\
0 &= -16(t^2 - 6t + 9) && \textit{Factor out } -16 \textit{ (on right-hand side).} \\
0 &= -16(t - 3)(t - 3) && \textit{Factor right-hand side completely.} \\
t - 3 &= 0 && \textit{Zero factor property} \\
t &= 3
\end{aligned}
$$

So, 3 seconds after the baseball is hit, it is at a height of 148 feet. It makes sense that there is only one time *if* the baseball reaches a height of 148 feet only at the top of its climb—in other words, if the vertex of the parabola is (3, 148). Figure 34 confirms the situation.

Figure 34 Finding when the baseball is at a height of 148 feet

3. Since the maximum height of the baseball is 148 feet, the baseball never reaches a height of 150 feet (see Fig. 34).

Consider the following observations about a quadratic model, a parabola, and a quadratic equation in one variable:

- Our work in Example 2 suggests that the baseball reaches each height either two times, one time, or never.
- In Section 7.1, we found that, for a parabola opening upward or downward, each value of y corresponds to two, one, or no values of x.
- In Section 8.5, we learned that a quadratic equation in one variable has either two, one, or no real-number solutions.

These observations suggest an important common thread between quadratic models, parabolas, and quadratic equations in one variable. For the purposes of this section, that common thread means that when we use a quadratic model to predict the value of an independent variable for a specific value of the dependent variable, there will be either two, one, or no real-number values.

Area of Rectangular Objects

Recall from Section 4.6 that the area A of a rectangle is given by the formula $A = LW$, where L is the length and W is the width.

Example 3 Using the Area of a Rectangular Object

A rectangular garden has an area of 40 square feet. If the length is 3 feet more than the width, find the dimensions of the garden.

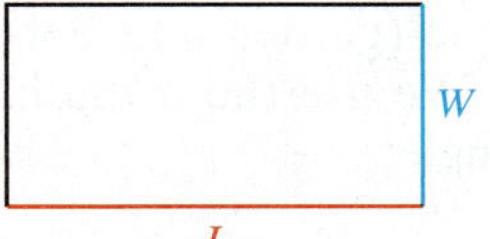

Figure 35 The length L and width W of a rectangular garden

Solution

We use the five-step problem-solving method of Section 6.5.

Step 1: Define each variable. Let L be the length (in feet) and W be the width (in feet). See Fig. 35.

Step 2: Write a system of two equations. Since the length is 3 feet more than the width, our first equation is $L = W + 3$. Because the area is 40 square feet, we find our second equation by substituting 40 for A in the area formula $A = LW$: $40 = LW$. The system is

$$L = W + 3 \quad \text{Equation (1)}$$
$$LW = 40 \quad \text{Equation (2)}$$

Step 3: Solve the system. We can solve the system by substitution. We substitute $W + 3$ for L in equation (2) and then solve for W:

$$\begin{aligned} LW &= 40 && \text{Equation (2)} \\ (W+3)W &= 40 && \text{Substitute } W+3 \text{ for } L. \\ W^2 + 3W &= 40 && \text{Distributive law} \\ W^2 + 3W - 40 &= 0 && \text{Subtract 40 from both sides.} \\ (W+8)(W-5) &= 0 && \text{Factor left side.} \\ W + 8 = 0 \quad \text{or} \quad W - 5 &= 0 && \text{Zero factor property} \\ W = -8 \quad \text{or} \quad W &= 5 \end{aligned}$$

The result $W = -8$ means that the width is negative; model breakdown has occurred, because a width must be positive. So, the width is 5 feet (our other result). To find the length, we substitute 5 for W in equation (1):

$$L = W + 3 = 5 + 3 = 8$$

Step 4: Describe each result. The width is 5 feet and the length is 8 feet.

Step 5: Check. The product of 5 and 8 is 40, which checks for an area of 40 square feet. Also, 8 is indeed 3 more than 5.

Example 4 Solving an Area Problem

A person has a rectangular garden with a width of 9 feet and a length of 12 feet. She plans to place mulch outside of the garden to form a border of uniform width. If she has just enough mulch to cover 100 square feet of land, determine the width of the border.

Solution

We use the five-step problem-solving method of Section 6.5, but for step 2 we find an equation in one variable rather than a system of two equations in two variables (and solve the equation in step 3).

Step 1: Define each variable. Let x be the width (in feet) of the mulch border (see Fig. 36).

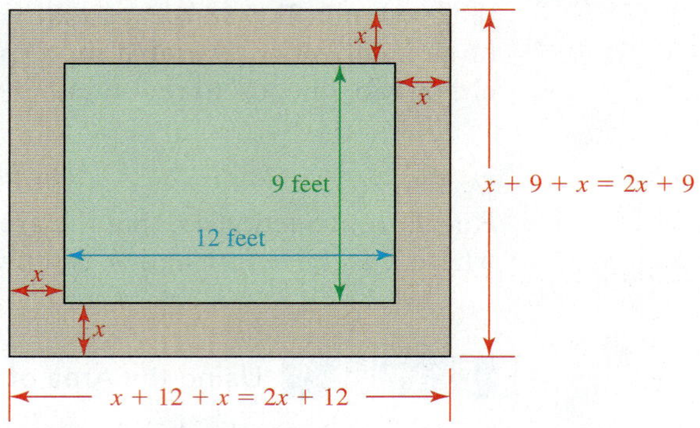

Figure 36 Garden with border

Step 2: Write an equation in one variable. For the outer rectangle (garden and border), the length is $2x + 12$ and the width is $2x + 9$ (see Fig. 36). We use the formula $A = LW$ to describe the area A (in square feet) of the outer rectangle:

$$\underbrace{A}_{\text{area of outer rectangle}} = \underbrace{(2x + 12)}_{\text{length}} \cdot \underbrace{(2x + 9)}_{\text{width}}$$

The area of the garden is $12 \cdot 9 = 108$ square feet, and the area of the border is given as 100 square feet. The area of the garden plus the area of the border is equal to the area of the outer rectangle:

$$\underbrace{108}_{\text{area of garden}} + \underbrace{100}_{\text{area of border}} = \underbrace{(2x + 12)(2x + 9)}_{\text{area of outer rectangle}}$$

$$208 = (2x + 12)(2x + 9)$$

Step 3: Solve the equation.

$$\begin{aligned}
208 &= (2x + 12)(2x + 9) && \textit{Equation from step 2}\\
208 &= 4x^2 + 18x + 24x + 108 && \textit{Find product on right side.}\\
208 &= 4x^2 + 42x + 108 && \textit{Combine like terms.}\\
0 &= 4x^2 + 42x - 100 && \textit{Write in } 0 = ax^2 + bx + c \textit{ form.}\\
0 &= 2(2x^2 + 21x - 50) && \textit{Factor out GCF, 2.}\\
0 &= 2(2x + 25)(x - 2) && \textit{Factor by trial and error.}
\end{aligned}$$

$$\begin{aligned}
2x + 25 &= 0 \quad \text{or} \quad x - 2 = 0 && \textit{Zero factor property}\\
2x &= -25 \quad \text{or} \quad x = 2\\
x &= -\frac{25}{2} \quad \text{or} \quad x = 2
\end{aligned}$$

Step 4: Describe each result. For the result $x = -\frac{25}{2}$, the border width is negative. Model breakdown has occurred, because a width must be positive. The border width is 2 feet.

Step 5: Check. If the border width is 2 feet, then the outer rectangle has a width of 13 feet and a length of 16 feet. Therefore, the total area of the outer rectangle is $13 \cdot 16 = 208$ square feet. This checks with our calculation near the beginning of our work.

Group Exploration

Looking ahead: Significance of a, h, and k for $y = a(x - h)^2 + k$

1. Use ZStandard followed by ZSquare to draw a graph of $y = x^2$.

2. Graph these equations of the form $y = x^2 + k$ in order, and describe, in terms of k, how you could "move" the graph of $y = x^2$ to get each graph:

$y = x^2 + 1,\ y = x^2 + 2,\ y = x^2 + 3,$ and $y = x^2 + 4$

Do the same with these equations:

$y = x^2 - 1,\ y = x^2 - 2,\ y = x^2 - 3,$ and $y = x^2 - 4$

3. Graph these equations of the form $y = (x - h)^2$ in order, and describe, in terms of h, how you could "move" the graph of $y = x^2$ to get each graph:

$y = (x - 1)^2,\ y = (x - 2)^2,\ y = (x - 3)^2,$ and $y = (x - 4)^2$

Do the same with these equations:

$y = (x + 1)^2,\ y = (x + 2)^2,\ y = (x + 3)^2,$ and $y = (x + 4)^2$

4. In this problem, you will explore the graphical significance of the constant a in functions of the form $y = ax^2$. From Problems 2 and 3, you should have an idea of how to go about it. Do this now and describe what you find. Don't forget to try negative values of a as well as values of a between 0 and 1.

5. a. Graph these equations in order, and explain how the graphs relate to your observations in Problems 2, 3, and 4:

$y = x^2, y = 0.5x^2, y = -0.5x^2, y = -0.5(x + 1)^2,$ and $y = -0.5(x + 1)^2 - 6$

b. Using your graph of $y = -0.5(x + 1)^2 - 6$, find the coordinates of the vertex. Compare these coordinates with the equation $y = -0.5(x + 1)^2 - 6$. What do you notice?

6. Summarize your findings about a, h, and k in terms of how you could move or adjust the graph of $y = x^2$ to get the graph of $y = a(x - h)^2 + k$. Also, discuss how the coordinates of the vertex are related to a, h, and k. If you are unsure, continue exploring.

Tips for Success Retake Quizzes and Exams to Prepare for the Final Exam

To study for your final exam, consider retaking your quizzes and other exams. These quizzes and exams can reveal your weak areas. If you have difficulty with a certain concept, you can refer to Homework exercises that address that concept. Reflect on *why* you are having such difficulty, rather than just doing more Homework exercises that address the concept.

Homework 8.6

For extra help ▶ MyMathLab®  Watch the videos in MyMathLab  Download the MyDashboard App

1. The numbers of participants in the Ford Ironman World Championship in Hawaii are shown in Table 4 for various years. Participants attempt to swim 2.4 miles, bike 112 miles, and run 26.2 miles (a marathon). Let n be the number of participants at t years since 1990. A linear model of the situation is $L(t) = 31t + 1301$. A quadratic model is $Q(t) = t^2 + 8t + 1376$.

a. Use a graphing calculator to draw the graphs of both models and, in the same viewing window, the scattergram of the data. Which model describes the situation better?

b. Use ZOOM OUT on a graphing calculator to help you determine which model predicts the larger participation for years between 2010 and 2020.

Table 4 Numbers of Participants in the Ironman World Championship

Year	Number of Participants
1990	1387
1994	1405
1998	1487
2002	1607
2006	1786
2010	1927
2012	2039

Source: *Ford Ironman World Championship*

c. Find the n-intercept of each model. Which n-intercept describes the situation better?
d. Use Q to estimate when there were 1396 participants.

2. The numbers of U.S. households with VCRs are shown in Table 5 for various years.

Table 5 Numbers of U.S. Households with VCRs

Year	Number of Households with VCRs (millions)
2005	98.9
2006	97.7
2007	95.2
2008	88.8
2009	82.5
2010	74.3

Source: *The Nielsen Company*

Let $n = f(t)$ be the number (in millions) of U.S. households with VCRs at t years since 2000.
a. Use a graphing calculator to draw a scattergram to describe the data. Is it better to use a linear or quadratic function to model the data? Explain.
b. Use a graphing calculator to draw the graph of the model $f(t) = -t^2 + 10t + 74$ and, in the same viewing window, the scattergram of the data. Does the model fit the data well?
c. What is the n-intercept? What does it mean in this situation?
d. Predict when only 18 million households will have VCRs.

3. Moody's Investors Service® evaluates the investment quality of companies. Although a B2 rating is six steps above Moody's worst rating, many companies rated B2 eventually default on their bonds (see Table 6).

Table 6 Percentage of Companies with a B2 Rating That Defaulted on Bonds

Years after Being Rated B2	Percent
2	12
4	24
6	32
8	36
10	41

Source: *Moody's Investors Service*

Let $f(t)$ be the percentage of companies with a B2 rating that defaulted on their bonds at t years since being rated B2. A model of the situation is $f(t) = -\frac{1}{3}t^2 + \frac{22}{3}t$.
a. Use a graphing calculator to draw the graph of the model and, in the same viewing window, the scattergram of the data. Does the model fit the data well?
b. Find $f(7)$. What does it mean in this situation?
c. Find t when $f(t) = 7$. What does it mean in this situation?
d. Find the t-intercepts. What do they mean in this situation?

4. The United States' annual coal exports are shown in Table 7 for various years. Let $c = f(t)$ be U.S. annual coal exports (in million metric tons) at t years since 2000. A model of the situation is $f(t) = t^2 - 9t + 43$.

Table 7 U.S. Coal Exports

Year	Coal Exports (million metric tons)
2004	23
2006	26
2008	39
2010	52
2011	63

Source: *Stifle, Nicolaus, and Co.*

a. Use a graphing calculator to draw the graph of the model and, in the same viewing window, the scattergram of the data. Does the model fit the data well?
b. Find the c-intercept. What does it mean in this situation?
c. Find $f(17)$. What does it mean in this situation?
d. Find t when $f(t) = 79$. What does it mean in this situation?
e. In 2011, the United States imported 14 million metric tons of coal (Source: *U.S. Energy Information Administration*). Find the ratio of U.S. coal imports to U.S. coal exports in 2011. Assuming this ratio will be the same in 2018, predict the number of metric tons of coal that will be *imported* in 2018.

5. The percentages of Americans who think labor unions in the United States will become stronger are shown in Table 8 for various years.

Table 8 Percentages of Americans Who Think Labor Unions Will Become Stronger

Year	Percent
2004	21
2005	19
2007	19
2009	22
2011	25

Source: *The Gallup Organization*

Let $p = f(t)$ be the percentage of Americans at t years since 2000 who think labor unions in the United States will become stronger. A model of the situation is $f(t) = \frac{1}{4}t^2 - 3t + 28$.
a. Use a graphing calculator to draw the graph of the model and, in the same viewing window, the scattergram of the data. Does the model fit the data fairly well?
b. Find the p-intercept. What does it mean in this situation?
c. Predict the percentage of Americans in 2017 who will think labor unions will become stronger.
d. Predict when 44% of Americans will think labor unions will become stronger.
e. Let $f(t)$ be the percentage of private-sector workers who are in a union at t years since 2000. A reasonable model is $f(t) = -0.25t + 9.45$ (Source: *Bloomberg BNA*). Explain why it is surprising that the percentage of Americans who thought unions would become stronger increased from 2007 to 2011.

6. First-quarter (January through March) Internet advertising revenues are shown in Table 9 for various years. Let $r = f(t)$ be first-quarter Internet advertising revenues (in billions of dollars) at t years since 2000. A model of the situation is $f(t) = \frac{1}{25}t^2 + \frac{2}{25}t + \frac{8}{5}$.

Table 9 First-Quarter Internet Advertising Revenues

Year	First-Quarter Revenues (billions of dollars)
2000	2.0
2002	1.5
2004	2.2
2006	3.6
2008	5.7
2010	5.9
2012	8.4

Source: *Interactive Advertising Bureau*

a. Use a graphing calculator to draw the graph of the model and, in the same viewing window, the scattergram of the data. Does the model fit the data well?
b. Find the *r*-intercept. What does it mean in this situation?
c. Predict the first-quarter Internet advertising revenues in 2018.
d. Estimate in which year first-quarter Internet advertising revenues were $3 billion.

7. Table 10 shows worldwide iPhone sales for various years.

Table 10 Worldwide iPhone Sales

Year	Sales (millions)
2007	1.4
2008	11.6
2009	20.7
2010	40.0
2011	72.3

Source: *Apple, Inc.*

Let S be worldwide sales (in millions) of iPhones in the year that is t years since 2000. A linear model of the situation is $S = 17.02t - 123.98$. A quadratic model of the situation is $S = 4t^2 - 54t + 184$.
a. Use a graphing calculator to draw a scattergram of the data and both models in the same viewing window. Which model comes closer to the points in the scattergram?
b. Use first the linear model and then the quadratic model to estimate iPhone sales in 2009. Which is the better estimate? Explain.
c. Use the quadratic model to predict iPhone sales in 2017.
d. Use the quadratic model to predict when iPhone annual sales will be 212 million phones.

8. The revenues from gambling in the United States are shown in Table 11 for various years.

Table 11 U.S. Annual Gambling Revenues

Year	Annual Revenue (billions of dollars)
2006	38.0
2007	39.4
2008	38.4
2009	37.6
2010	36.1

Source: *Bureau of Economic Analysis*

Let $f(t)$ be the annual revenue (in billions of dollars) from gambling at t years since 2000.
a. Use a graphing calculator to draw a scattergram of the data. Can the data be modeled better by a linear equation or a quadratic equation? Explain.
b. Use a graphing calculator to draw the graph of the model $f(t) = -\frac{1}{2}t^2 + \frac{15}{2}t + 11$ and, in the same viewing window, the scattergram of the data. Does the model fit the data well?
c. Predict the revenue in 2014.
d. Estimate when the annual revenue was or will be $24 billion.

9. The total attendance at Broadway shows in New York is shown in Table 12 for various years.

Table 12 Total Attendance at Broadway Shows

Year	Total Attendance (thousands)
2005	11,527
2006	12,003
2007	12,312
2008	12,267
2009	12,250
2010	11,890

Source: *National Arts Index 2012*

Let $f(t)$ be the total attendance (in thousands) at Broadway shows in the year that is t years since 2005. A model of the situation is $f(t) = -98t^2 + 561t + 11{,}536$.
a. Use a graphing calculator to draw the graph of the model and, in the same viewing window, the scattergram of the data. Does the model fit the data well?
b. Estimate the total attendance in 2012.
c. Estimate in which year(s) the total attendance was 11,536 thousand (11.536 million).
d. One possible reason for the decrease in attendance from 2008 to 2010 is that the economy performed poorly during that period. Assuming the economy improves and attendance increases in the future, will f model the situation well? Explain.

10. New-home sales rates (in thousands of homes per year) are shown in Table 13 for July of various years.

Table 13 New-Home Sales Rates in July

Year	Sales Rate (thousands of homes per year)
2005	1348
2006	1030
2007	800
2008	479
2009	402
2010	289
2011	300
2012	381

Source: *Commerce Department*

Let $f(t)$ be the new-home sales rate (in thousands of homes per year) in July at t years since 2005. A model of the situation is $f(t) = 34t^2 - 383t + 1370$.

a. Use a graphing calculator to draw the graph of the model and, in the same viewing window, the scattergram of the data. Does the model fit the data well?

b. Predict the sales rate in July 2017.

c. Predict in which year the sales rate in July will be 1370 thousand homes (1.37 million homes) per year.

d. The decrease in new-home sales rates in July from 2005 to 2010 is likely due to the high-risk home loans and the poor economy during that period. Assuming the economy improves and new-home sales rates increase in the future, is it possible f will model the situation well? Explain.

11. A company's profit can be modeled by $p = t^2 - 3t + 5$, where p is the profit (in thousands of dollars) for the year that is t years since 2010. Estimate when the annual profit was or will be \$23 thousand.

12. A company's profit can be modeled by $p = t^2 - 6t + 17$, where p is the profit (in thousands of dollars) for the year that is t years since 2010. Estimate when the annual profit was or will be \$24 thousand.

13. A company's revenue can be modeled by $r = 2t^2 - 13t + 25$, where r is the revenue (in millions of dollars) for the year that is t years since 2010. Estimate when the annual revenue was or will be \$10 million.

14. A company's revenue can be modeled by $r = 3t^2 - 19t + 35$, where r is the revenue (in millions of dollars) for the year that is t years since 2010. Estimate when the annual revenue was or will be \$29 million.

15. A batter hits a baseball into the air. The height h (in feet) of the baseball after t seconds is given by the quadratic equation $h = -16t^2 + 64t + 3$.

a. When is the baseball at a height of 3 feet? Explain why there are two such times. [**Hint:** Use a graphing calculator to graph the model.]

b. When is the baseball at a height of 51 feet? Explain why there are two such times.

c. When is the baseball at a height of 67 feet? Explain why there is just one such time.

16. A person throws a ball into the air. The height h (in feet) of the ball after t seconds is given by $h = -16t^2 + 80t + 4$.

a. When is the baseball at a height of 68 feet? Explain why there are two such times. [**Hint:** Use a graphing calculator to graph the model.]

b. When is the baseball at a height of 100 feet? Explain why there are two such times.

c. When is the baseball at a height of 104 feet? Explain why there is just one such time.

17. A rectangular boardroom table has an area of 60 square feet. If the length is 7 feet more than the width, find the dimensions of the table.

18. A rectangular rug has an area of 80 square feet. If the length is 2 feet more than the width, find the dimensions of the rug.

19. A rectangular rug has an area of 60 square feet. If its length is 2 feet more than twice its width, find the dimensions of the rug.

20. A rectangular garden has an area of 65 square feet. If its length is 3 feet less than twice its width, find the dimensions of the garden.

21. The length of a rectangle is 4 centimeters more than the width. If both the width and length were doubled, the area would be 48 square centimeters. Find the dimensions of the original rectangle.

22. The length of a rectangle is 6 meters more than the width. If both the width and length were doubled, the area would be 108 square meters. Find the dimensions of the original rectangle.

23. A person has a rectangular garden with a width of 6 feet and a length of 10 feet. To form a border of uniform width, he plans to place mulch around the outside of the garden. If he has just enough mulch to cover 80 square feet of land, determine the width of the border.

24. A person has a rectangular garden with a width of 8 feet and a length of 12 feet. To form a border of uniform width, she plans to put sod around the outside of the garden. If she has just enough sod to cover 44 square feet of land, determine the width of the border.

25. A rectangular painting (not including the frame) has a width of 10 inches and a length of 14 inches. If the area of the frame (of uniform width) is 52 square inches, what is the width of the frame?

26. A rectangular painting (not including the frame) has a width of 9 inches and a length of 15 inches. If the area of the frame (of uniform width) is 112 square inches, what is the width of the frame?

Concepts

27. For a given set of data, describe how you can determine whether to model the situation by a linear model, a quadratic model, or neither.

28. Explain how to make a prediction for the dependent variable of a quadratic model. Explain how to make a prediction for the independent variable.

Related Review

29. The average tax refund has increased since 1997 (see Table 14).

Table 14 Average Tax Refunds

Year	Average Tax Refund (dollars)
1997	1295
2000	1624
2003	1973
2007	2699
2011	2913

Source: *Internal Revenue Service*

Let $f(t)$ be the average tax refund (in dollars) at t years since 1990.

a. Use a graphing calculator to draw a scattergram of the data. Can the data be modeled better by a linear equation or a quadratic equation? Explain.

b. Find an equation of f.

c. Predict when the average tax refund will be \$3750.

30. The revenues of Wal-Mart®, the largest private employer in the United States, are shown in Table 15 for various years.

Table 15 Wal-Mart's Annual Revenues

Year	Annual Revenues (billions of dollars)
2001	220
2002	247
2003	259
2004	288
2005	316
2006	351
2007	379
2008	406
2009	408

Sources: *Wal-Mart; Redbook*

Let r be the annual revenue (in billions of dollars) at t years since 2000.

a. Use a graphing calculator to draw a scattergram of the data. Can the data be modeled better by a linear equation or a quadratic equation? Explain.

b. Find an equation of a model to describe the data.

c. Predict when the annual revenue will be \$650 billion.

31. The revenue from U.S. sugar-free/reduced-sugar products was \$3.34 billion in 2010 and has increased by about \$0.27 billion per year since then (Source: *Euromonitor*). Let $g(t)$ be the annual revenue (in billions of dollars) from U.S. sugar-free/reduced-sugar products at t years since 2010.

a. Find an equation of g.

b. Predict when the annual revenue will be \$5.5 billion.

32. The number of health care fraud prosecutions was 731 in 2010 and has increased by about 28 prosecutions each year since then (Source: *Justice Department*). Let $f(t)$ be the number of health care fraud prosecutions in the year that is t years since 2010.

a. Find an equation of f.

b. Predict in which year there will be 930 prosecutions.

33. The average cost of room and board at public universities has increased approximately linearly from \$4900 for the school year ending in 2001 to \$8260 for the school year ending in 2010 (Source: *The College Board*). Predict the average cost of room and board for the school year ending in 2018.

34. The percentage of boys ages 3 to 17 who have been diagnosed as having attention deficit hyperactivity disorder (ADHD) has increased approximately linearly from 10.3% in 2002 to 11.2% in 2010 (Source: *Centers for Disease Control and Prevention*). Predict the percentage of boys who will be diagnosed as having ADHD in 2017.

Expressions, Equations, Functions, and Graphs

Give an example of each. Then solve, simplify, or graph, as appropriate.

35. cubic expression in one variable with five terms

36. quadratic expression in one variable with four terms

37. linear expression in one variable with four terms

38. quadratic equation in one variable

39. system of two linear equations in two variables

40. linear equation in one variable

41. quadratic function

42. linear function

Chapter Summary

Key Points of Chapter 8

Section 8.1 Factoring Trinomials of the Form $x^2 + bx + c$ and Differences of Two Squares

Factor	We **factor** a polynomial by writing it as a product.
Multiplying versus factoring	Multiplying and factoring are reverse processes.
Factoring $x^2 + bx + c$	To factor $x^2 + bx + c$, look for two integers p and q whose product is c and whose sum is b. That is, $pq = c$ and $p + q = b$. If such integers exist, the factored polynomial is $(x + p)(x + q)$.
Factoring $x^2 + bx + c$ with c positive	To factor a trinomial of the form $x^2 + bx + c$ with a positive constant term c, • If b is positive, look for two *positive* integers whose product is c and whose sum is b. • If b is negative, look for two *negative* integers whose product is c and whose sum is b.

Section 8.1 Factoring Trinomials of the Form $x^2 + bx + c$ and Differences of Two Squares (*Continued*)

Factoring $x^2 + bx + c$ with c negative	To factor a trinomial of the form $x^2 + bx + c$ with a negative constant term c, look for two integers with *different* signs whose product is c and whose sum is b.
Prime polynomial	A polynomial that cannot be factored is called **prime.**
Difference of two squares	$A^2 - B^2 = (A + B)(A - B)$
Polynomial $x^2 + k^2$, where $k \neq 0$	A polynomial of the form $x^2 + k^2$, where $k \neq 0$, is prime.

Section 8.2 Factoring Out the GCF; Factoring by Grouping

GCF	The **greatest common factor (GCF)** of two or more terms is the monomial with the largest coefficient and the largest degree that is a factor of all the terms.
Factor out GCF	When the leading coefficient of a polynomial is positive and the GCF is not 1, we first factor out the GCF.
Factor completely	When factoring a polynomial, always factor it *completely*.
Factoring when the leading coefficient is negative	When the leading coefficient of a polynomial is negative, we first factor out the opposite of the GCF.
Factor by grouping	For a polynomial with four terms, **factor by grouping** (if it can be done) by **1.** Factoring the first two terms and the last two terms. **2.** Factoring out the binomial GCF.

Section 8.3 Factoring Trinomials of the Form $ax^2 + bx + c$

Factor $ax^2 + bx + c$ by trial and error	To **factor a trinomial** of the form $ax^2 + bx + c$ **by trial and error,** if the trinomial can be factored as a product of two binomials, then the product of the coefficients of the first terms of the binomials is equal to a and the product of the last terms of the binomials is equal to c. To find the correct factored expression, multiply the possible products and identify the one for which the coefficient of x is b.
Factoring $ax^2 + bx + c$ by grouping (ac method)	To **factor a trinomial** of the form $ax^2 + bx + c$ **by grouping** (if it can be done), **1.** Find pairs of numbers whose product is ac. **2.** Determine which of the pairs of numbers from step 1 has the sum b. Call this pair of numbers m and n. **3.** Write the bx term as $mx + nx$: $ax^2 + bx + c = ax^2 + mx + nx + c$ **4.** Factor $ax^2 + mx + nx + c$ by grouping. Another name for this technique is the ***ac* method.**

Section 8.4 Sums and Differences of Cubes; A Factoring Strategy

Sum of two cubes	$A^3 + B^3 = (A + B)(A^2 - AB + B^2)$
Difference of two cubes	$A^3 - B^3 = (A - B)(A^2 + AB + B^2)$
Factoring strategy	The five steps that follow can be used to factor many polynomials. Steps 2–4 can be applied to the entire polynomial or to a factor of the polynomial. **1.** If the leading coefficient is positive and the GCF is not 1, we factor out the GCF. If the leading coefficient is negative, we factor out the opposite of the GCF. **2.** For a binomial, try using one of the properties for the difference of two squares, the sum of two cubes, or the difference of two cubes. **3.** For a trinomial of the form $ax^2 + bx + c$, **a.** If $a = 1$, try to find two integers whose product is c and whose sum is b. **b.** If $a \neq 1$, try to factor by trial and error or by grouping. **4.** For an expression with four terms, try factoring by grouping. **5.** Continue applying steps 2–4 until the polynomial is factored completely.

Section 8.5 Using Factoring to Solve Polynomial Equations	
Quadratic equation in one variable	A **quadratic equation in one variable** is an equation that can be put into the form $ax^2 + bx + c$, where a, b, and c are constants and $a \neq 0$.
Zero factor property	Let A and B be real numbers: If $AB = 0$, then $A = 0$ or $B = 0$.
Connection between x-intercepts and solutions	Let f be a function. If k is a real-number solution of the equation $f(x) = 0$, then $(k, 0)$ is an x-intercept of the graph of the function f. Also, if $(k, 0)$ is an x-intercept of the graph of f, then k is a solution of $f(x) = 0$.
Number of solutions: quadratic equation	The solution set of a quadratic equation in one variable may contain two real numbers, one real number, or no real numbers.
Cubic equation	A **cubic equation in one variable** is an equation that can be put into the form $$ax^3 + bx^2 + cx + d = 0$$ where a, b, c, and d are constants and $a \neq 0$.
Solving quadratic or cubic equations by factoring	If an equation can be solved by factoring, we solve it by the following steps: **1.** Write the equation so one side of the equation is equal to zero. **2.** Factor the nonzero side of the equation. **3.** Apply the zero factor property. **4.** Solve each equation that results from applying the zero factor property.
Number of solutions: cubic equation	The solution set of a cubic equation in one variable may contain one, two, or three real numbers.

Section 8.6 Using Factoring to Make Predictions with Quadratic Models	
Quadratic model	A **quadratic model** is a quadratic function, or its graph, that describes the relationship between two quantities in an authentic situation.
Making a prediction about the dependent variable	In making a prediction about the dependent variable of a quadratic model, substitute a chosen value for the independent variable in the model. Then solve for the dependent variable.
Making a prediction about the independent variable	In making a prediction about the independent variable of a quadratic model, substitute a chosen value for the dependent variable in the model. Then solve for the independent variable, which involves writing the equation so that one side of it is zero and then factoring the nonzero side.

Chapter 8 Review Exercises

For Exercises 1–30, factor when possible. Use a graphing calculator table to verify your work when possible.

1. $x^2 + 9x + 20$
2. $6x^2 - 2x - 8$
3. $x^2 + 14x + 49$
4. $-18t^4 - 33t^3 + 30t^2$
5. $p^2 - 3pq - 54q^2$
6. $32 - 12x + x^2$
7. $-9x^2 + 4$
8. $4w^2 + 25$
9. $20m^2n - 45mn^3$
10. $16x^2 + 14x + 2x^3$
11. $16x^3 - 32x^2 + 16x$
12. $24x^3 - 32x^2$
13. $5x^4y - 35x^3y + 60x^2y$
14. $-m^2 - 2m + 35$
15. $4r^3 - 10r^2 + 6r - 15$
16. $81t^4 - 16w^4$
17. $2y^3 - 54$
18. $x^2 - 9x + 20$
19. $6t^2 + 11ty - 10y^2$
20. $2x^3 - 50x$
21. $x^2 - 10x + 25$
22. $p^2 - 81$
23. $8w^2 - 12w + 3$
24. $4x^2 + 20x + 25$
25. $12w^3 - 50w^2 + 8w$
26. $49a^2 - 9b^2$
27. $x^2 - 7x - 12$
28. $x^3 + 3x^2 - 4x - 12$
29. $r^3 + 8$
30. $2ax - 10ay - 3bx + 15by$

31. A student tries to factor $x^2 + 25$:

$$x^2 + 25 = (x + 5)(x + 5) = (x + 5)^2$$

Describe any errors. Then factor the polynomial correctly.

32. A student tries to factor $5x^3 + 35x^2 + 60x$:

$$5x^3 + 35x^2 + 60x = 5x(x^2 + 7x + 12)$$

Describe any errors. Then solve the equation correctly.

33. Find all possible values of k so that $x^2 + kx + 24$ can be factored.

For Exercises 34–49, solve the given equation. Verify your results by checking that they satisfy the equation.

34. $2x^3 + 16x^2 = -24x$

35. $(m - 3)(m + 2) = -4$

36. $t^2 - 6t + 9 = 0$

37. $x^2 - 3x = 5x - 15$

38. $25x^2 - 81 = 0$

39. $2x^3 - 7x^2 - 2x + 7 = 0$

40. $6x^2 + x - 2 = 0$

41. $3x^2 = 15x$

42. $8r^2 - 18r + 9 = 0$

43. $a^2 = 2a + 35$

44. $\frac{1}{3}x^2 - \frac{1}{3}x - 4 = 6$

45. $3x^3 - 2x^2 = 27x - 18$

46. $a^2 = 4$

47. $5p^2 + 20p - 60 = 0$

48. $\frac{m^2}{2} - \frac{7m}{6} + \frac{1}{3} = 0$

49. $4p(5p - 6) = (2p + 3)(2p - 3)$

50. Find all x-intercepts of the graph of $f(x) = 3x^3 + 3x^2 - 18x$.

51. Use "intersect" on a graphing calculator to solve the equation $2x^2 + 4x - 5 = 2x + 3$. Round the solution(s) to the second decimal place.

For Exercises 52–55, solve the given equation by referring to the solutions of $y = -3x^2 + 6x + 20$ shown in Table 16.

52. $-3x^2 + 6x + 20 = 23$

53. $-3x^2 + 6x + 20 = -4$

54. $-3x^2 + 6x + 20 = 27$

55. $-3x^2 + 6x + 20 = 20$

Table 16 Exercises 52–55

x	y
−2	−4
−1	11
0	20
1	23
2	20
3	11
4	−4

For Exercises 56 and 57, find all x-intercepts. Verify your work by using your graphing calculator.

56. $f(x) = x^2 - 49$

57. $f(x) = 8x^2 - 14x - 15$

58. Give an example of a quadratic equation whose solutions are 3 and −6.

59. Give an example of a cubic equation whose solutions are −2, 0, and 1.

60. The numbers of worldwide malaria deaths are shown in Table 17 for various years.

Table 17 Numbers of Worldwide Malaria Deaths

Year	Number of Deaths (thousands)
2000	755
2002	789
2004	810
2006	782
2008	711
2010	655

Source: *World Health Organization (WHO)*

Let $f(t)$ be the number (in thousands) of worldwide malaria deaths in the year that is t years since 2000.

a. Use a graphing calculator to draw a scattergram of the data. Can the data be modeled better by a linear equation or a quadratic equation? Explain.

b. Use a graphing calculator to draw the graph of the model $f(t) = -4t^2 + 28t + 758$ and, in the same viewing window, the scattergram of the data. Does the model fit the data well?

c. Predict the number of malaria deaths in 2017.

d. Estimate in which year(s) the number of malaria deaths was or will be 798 thousand deaths.

e. Use "maximum" on a graphing calculator to find the vertex of the model. What does the vertex mean in this situation?

f. An article in the British health journal *The Lancet* reports 1.24 million (1240 thousand) people died of malaria in 2010. Is this estimate close to WHO's? (See Table 17.) *The Lancet* collected its data by interviewing people about the symptoms of relatives and friends who had died possibly from malaria. WHO collected its data by inspecting medical records. Explain why the two estimates might be so different.

61. A batter hits a baseball into the air. The height (in feet) $f(t)$ of the baseball after t seconds is given by $f(t) = -16t^2 + 80t + 4$. When is the baseball at a height of 68 feet?

62. A rectangular banner has area 30 square feet. If the length is 4 feet more than twice the width, find the dimensions of the banner.

Chapter 8 Test

For Exercises 1–11, factor.

1. $x^2 - 3x - 40$

2. $24 + x^2 - 10x$

3. $8m^2n^3 - 10m^3n$

4. $p^2 - 14pq + 40q^2$

5. $25p^2 - 36y^2$

6. $3x^4y - 21x^3y + 36x^2y$

7. $8x^3 + 20x^2 - 18x - 45$

8. $8x^2 - 26x + 15$

9. $-16x^2 - 26x + 12$

10. $16a^4b - 36a^3b^2 + 18a^2b^3$

11. $54m^3 + 128p^3$

12. Which of the following polynomials are equivalent?

$x^2 - 7x - 10 \qquad (x - 5)(x + 2) \qquad x^2 - 3x - 10$

$x^2 + 3x - 10 \qquad (x + 2)(x - 5)$

13. A student tries to factor $5x^3 + 3x^2 - 20x - 12$:

$$5x^3 + 3x^2 - 20x - 12 = x^2(5x + 3) - 4(5x + 3)$$

Describe any errors. Then factor the polynomial correctly.

For Exercises 14–20, solve.

14. $x^2 - 13x + 36 = 0$

15. $49x^2 - 9 = 0$

16. $(2x - 7)(x - 3) = 10$

17. $\frac{1}{4}p^2 - \frac{1}{2}p - 6 = 0$

18. $3x^3 - 12x = 8 - 2x^2$

19. $2x^3 = 8x^2 + 10x$

20. $3x(2x - 5) + 4x = 2(x - 3)$

21. Find all x-intercepts of the graph of $f(x) = 10x^2 - 19x + 6$.

For Exercises 22–25, use the graph of $y = x^2 + 6x + 7$ shown in Fig. 37 to solve the given equation.

22. $x^2 + 6x + 7 = 2$

23. $x^2 + 6x + 7 = -1$

24. $x^2 + 6x + 7 = -2$

25. $x^2 + 6x + 7 = -4$

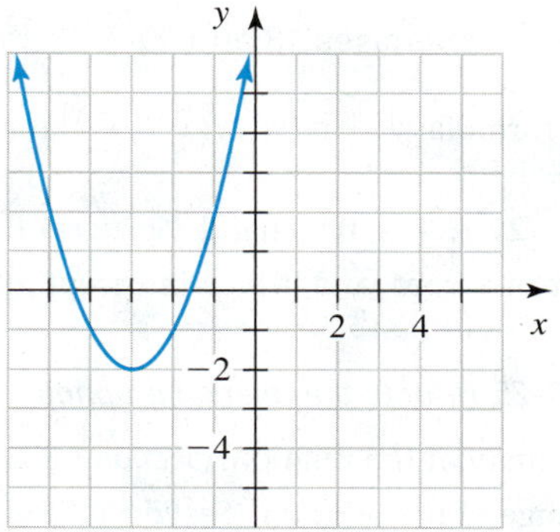

Figure 37 Exercises 22–25

26. Find the x-intercepts of the graph of $f(x) = 10x^2 - 11x - 6$.

27. Give an example of a quadratic equation whose solutions are -3 and 8.

28. The numbers of troops in the Afghanistan national army are shown in Table 18 for various years.

Table 18 Numbers of Troops in the Afghanistan National Army

Year	Number of Troops (thousands)
2003	6
2005	26
2007	50
2009	100
2010	138

Source: *Department of Defense*

Let $f(t)$ be the number (in thousands) of troops in the Afghanistan national army at t years since 2000.

a. Use a graphing calculator to draw a scattergram of the data. Can the data be modeled better by a linear equation or a quadratic equation? Explain.

b. Use a graphing calculator to draw the graph of the model $f(t) = 2t^2 - 8t + 16$ and, in the same viewing window, the scattergram of the data. Does the model fit the data well?

c. Estimate when there were or will be 80 thousand troops in the current form of the Afghanistan national army, which began in 2002.

d. Estimate the number of troops in the Afghanistan national army in 2012.

29. A rectangular painting (not including the frame) has a width of 11 inches and a length of 15 inches. If the area of the frame (of uniform width) is 120 square inches, what is the width of the frame?

Making Sure You're Ready for Intermediate Algebra: A Review of Chapters 1–8

The exercises that follow can be used in both Elementary Algebra and Intermediate Algebra. The exercises can be used

- at the end of Elementary Algebra to help you prepare for the final exam.
- at the start of Intermediate Algebra to make sure that you're fully prepared.

The exercises can steer you to material that you may need to study. After completing an exercise, look up the answer in the Answer section. If your answer is correct and *you feel confident about that material,* go on to the next exercise. If your answer is incorrect, find and read the example that is referenced in parentheses. If the example does not provide enough help, read the text surrounding the example or read the entire referenced section, depending on how much help you need. Once you feel that you have the idea, don't make any assumptions! It is essential that you practice similar exercises in the section's homework until you consistently can get correct results.

1. Let p be the percentage of Americans who own a home at age a years.

a. What is the independent variable? What is the dependent variable? *(Section 1.2, Example 1)*

b. What does the ordered pair (50, 75) mean in this situation? *(Section 1.2, Example 2)*

2. Last week, the balance of a checking account was -65 dollars. The balance is now 145 dollars. What is the change in the balance? *(Section 2.4, Examples 1 and 2)*

For Exercises 3 and 4, evaluate the given expression for $a = 3$, $b = -4$, and $c = -2$.

3. $\frac{a - b}{c - a}$ *(Section 2.6, Example 5)*

4. $b^2 - 4ac$ *(Section 2.6, Example 6)*

5. Graph $y = -4$ by hand. *(Section 3.2, Example 4)*

6. Find the slope of the line that passes through the points $(-2, 5)$ and $(4, -3)$. State whether the line is increasing, decreasing, horizontal, or vertical. (Section 3.3, Examples 2–9)

For Exercises 7 and 8, graph the equation by hand.

7. $y = -2x + 4$ (Section 3.4, Example 3)

8. $y = \frac{3}{2}x - 3$ (Section 3.4, Example 4)

9. Find an equation of the line sketched in Fig. 38. (Section 3.4, Example 8)

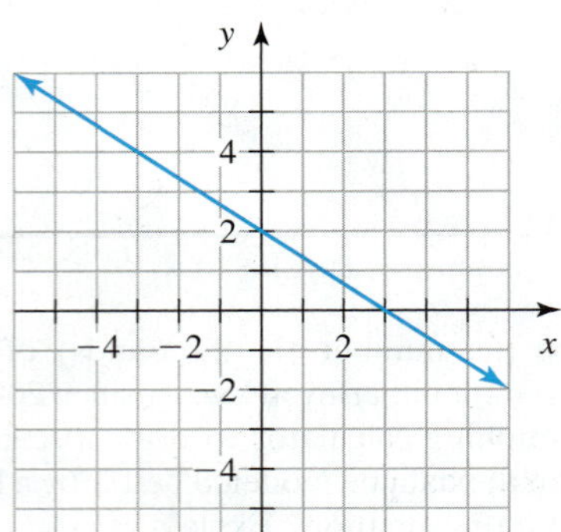

Figure 38 Exercise 9

10. In response to rising concerns about identity theft, Fellowes, Inc.®, increased approximately steadily the number of models of paper shredders it manufactures from 2 models in 1990 to 41 models in 2012 (Source: *Fellowes*). Find the average rate of change of the number of shredder models. (Section 3.5, Examples 1 and 2)

11. A person drives her Honda CR-V® on a road trip. At the start of the trip, she fills up the 15.3-gallon tank with gasoline. During the trip, the car uses about 0.05 gallon of gas per mile. Let $g(x)$ be the number of gallons of gasoline remaining in the tank (which has not been refilled) after she has driven x miles. (Section 3.5, Examples 4 and 5)
   a. What is the slope of the graph of g? What does it mean in this situation?
   b. Find an equation of g.
   c. Find the x-intercept of the model. What does it mean in this situation?
   d. What are the domain and range of g?
   e. If the driver will refuel the car when 1 gallon of gasoline remains in the tank, how far can she drive the car before refueling?

12. Simplify $2(6x + 3y) - (4x - 5y + 9)$. (Section 4.2, Example 2)

13. Let x be a number. Translate the phrase "Five, minus 3 times the difference of the number and 4" to an expression. Then simplify the expression. (Section 4.2, Example 3)

For Exercises 14–16, solve.

14. $3t - 14 = 6 - 8t$ (Section 4.4, Example 3)

15. $2 - 5(4t - 3) = 9 - (6t - 2)$ (Section 4.4, Example 4)

16. $\frac{2x - 1}{5} = \frac{3x + 4}{2}$ (Section 4.4, Example 6)

17. Four times the difference of a number and 7 is 25. Find the number. (Section 4.4, Example 11)

For Exercises 18 and 19, solve the given equation by referring to the graphs of $y = \frac{1}{2}x - \frac{5}{2}$ and $y = -\frac{2}{3}x - \frac{4}{3}$ shown in Fig. 39.

18. $-\frac{2}{3}x - \frac{4}{3} = 2$ (Section 4.3, Example 9)

19. $\frac{1}{2}x - \frac{5}{2} = -\frac{2}{3}x - \frac{4}{3}$ (Section 4.4, Example 13)

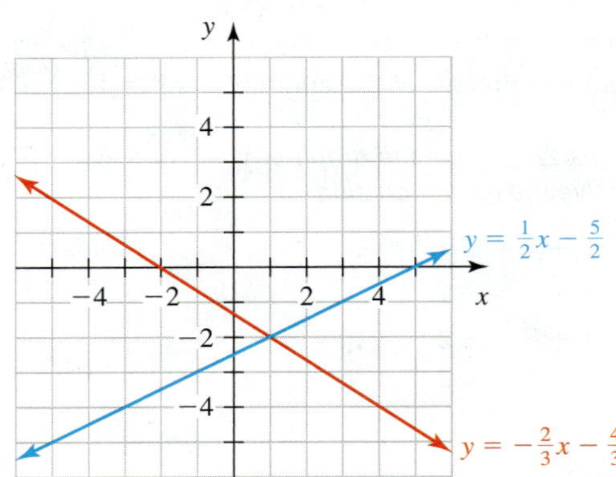

Figure 39 Exercises 18 and 19

20. Solve the formula $y - y_1 = m(x - x_1)$ for x. (Section 4.6, Examples 5–8)

21. Graph $3x - 2y + 4 = 0$ by hand. (Section 5.1, Example 1)

22. Find the x-intercept and the y-intercept of the graph of $3x - 7y = 28$. (Section 5.1, Example 5)

For Exercises 23–25, refer to the relation graphed in Fig. 40.

23. Find the domain of the relation. (Section 5.2, Example 9)

24. Find the range of the relation. (Section 5.2, Example 9)

25. Is the relation a function? Explain. (Section 5.2, Examples 5–7)

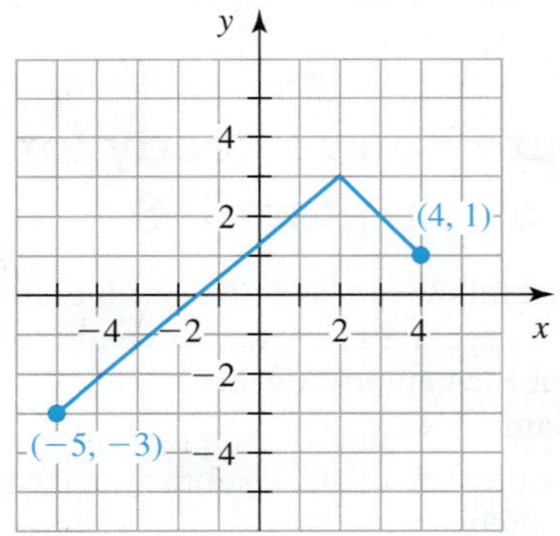

Figure 40 Exercises 23–25

26. For $f(x) = 2x^2 - 5x$, find $f(-3)$. (Section 5.3, Examples 1 and 2)

27. For $g(x) = \frac{2}{3}x + 5$, find x when $g(x) = -4$. (Section 5.3, Example 4)

28. Find an equation of the line that has slope $-\frac{3}{5}$ and contains the point $(4, -2)$. Write your equation in slope–intercept form. (Section 5.4, Examples 1 and 2)

29. Find an equation of the line that contains the points $(-5, -3)$ and $(1, -6)$. Write your equation in slope–intercept form. (Section 5.4, Examples 3 and 4)

30. The percentages of Americans who say they have "a great deal of confidence" in Congress are shown in Table 19 for various years.

Table 19 Percentages of Americans Who Say They Have "A Great Deal of Confidence" in Congress

Years	Percent
2001	18
2003	20
2005	16
2007	10
2009	9
2011	6

Source: *Harris Interactive*

Let $p = f(t)$ be the percentage of Americans who say they have "a great deal of confidence" in Congress at t years since 2000.

a. Use a graphing calculator to draw a scattergram of the data. (Section 5.5, Example 2)

b. Find an equation of f. (Section 5.5, Examples 2 and 3)

c. What is the slope? What does it mean in this situation? (Section 3.5, Example 6)

d. Predict what percentage of Americans will say they have "a great deal of confidence" in Congress in 2014. (Section 5.6, Example 1)

e. Estimate when 13% of Americans said they have "a great deal of confidence" in Congress. (Section 5.6, Example 1)

f. Find the p-intercept. What does it mean in this situation? (Section 5.6, Example 2)

g. Find the t-intercept. What does it mean in this situation? (Section 5.6, Example 2)

31. The number of prescriptions for the sleep aid Ambien® increased from 39 million prescriptions in 2008 to 44 million prescriptions in 2012 (Source: *IMS Health*). Predict when there will be 50 million Ambien prescriptions. (Section 5.6, Example 3)

32. Solve the inequality $5 - 4(3x - 3) \le -19$. Describe the solution set as an inequality, in interval notation, and in a graph. (Section 5.7, Example 8)

33. Solve the inequality $-1 < 2x + 7 < 11$. Describe the solution set as an inequality, in interval notation, and in a graph. (Section 5.7, Example 11)

34. Use graphing to solve the following system. (Section 6.1, Example 3)

$$x - 2y = -8$$
$$y = -\frac{5}{2}x - 2$$

35. Solve the following system. (Section 6.2, Example 2)

$$3x + 7y = 11$$
$$x = 2y - 5$$

36. Solve the following system. (Section 6.3, Example 2)

$$2x - 5y = 16$$
$$4x + 3y = 6$$

37. The *participation rate* is the percentage of people over the age of 15 who are currently employed or actively seeking employment. The participation rates of women and men are shown in Table 20 for various years. (Section 6.4, Examples 1 and 2)

Table 20 Participation Rates

	Participation Rate (percent)	
Year	**Women**	**Men**
1948	32.7	86.6
1958	37.1	84.2
1968	41.6	80.1
1978	50.0	77.9
1988	56.6	76.2
1998	59.8	74.9
2008	59.5	73.0
2011	58.1	70.5

Source: *Bureau of Labor Statistics*

a. Let $W(t)$ be the participation rate of women and $M(t)$ be the participation rate of men, both at t years since 1900. Find equations of W and M.

b. Use substitution or elimination to predict when the participation rates for women and men will be equal. What is that participation rate?

c. Due to the poor economy from 2008 to 2011, many unemployed workers stopped actively seeking employment during that period. Explain how you can tell this from the scattergrams of the women's data and the men's data.

38. In 2012, a one-year-old Acura® CL coupe was worth about \$18,249, with a depreciation of about \$1903 per year. A one-year-old Subaru® Legacy sedan was worth about \$14,564, with a depreciation of about \$1225 per year (Source: Edmund's Automobile Buyer's Guide). (Section 6.4, Example 3)

a. Let $f(t)$ be the value (in dollars) of the Acura CL and $g(t)$ be the value (in dollars) of the Subaru Legacy, both at t years since 2012. Find equations of f and g.

b. Use substitution or elimination to predict when the two cars will have the same value. What is that value?

39. A person plans to invest twice as much in a UBS Global Equity Y account at 7.2% annual interest as in a Fidelity Worldwide account at 9.4% interest. Both interest rates are five-year averages. How much will the person have to invest in each account to earn a total of \$595 in one year? (Section 6.5, Example 6)

40. Graph the inequality $x - 2y > 4$ by hand. (Section 6.6, Example 3)

41. Graph the solution set of the following system. (Section 6.6, Examples 5 and 6)

$$y \ge \frac{1}{3}x - 2$$
$$y \le -\frac{3}{4}x + 1$$

42. Perform the operations indicated in $3x - 7x^3 + 5x^2 - 9x + x^3$. (Section 7.1, Example 3)

43. Find the difference $(2a^2 + 3ab - 7b^2) - (5a^2 - 6ab + 3b^2)$. (Section 7.1, Example 5)

44. For $f(x) = 3x^2 + 5x - 2$ and $g(x) = 5x^2 - 2x - 3$, find an equation of the difference function $f - g$; then find $(f - g)(-2)$. (Section 7.1, Example 10)

For Exercises 45–48, refer to the graph sketched in Fig. 41.

45. Find $f(-3)$. (Section 7.1, Example 8)

46. Find x when $f(x) = -2$. (Section 7.1, Example 8)

47. Find x when $f(x) = -3$. (Section 7.1, Example 8)

48. Find x when $f(x) = -4$. (Section 7.1, Example 8)

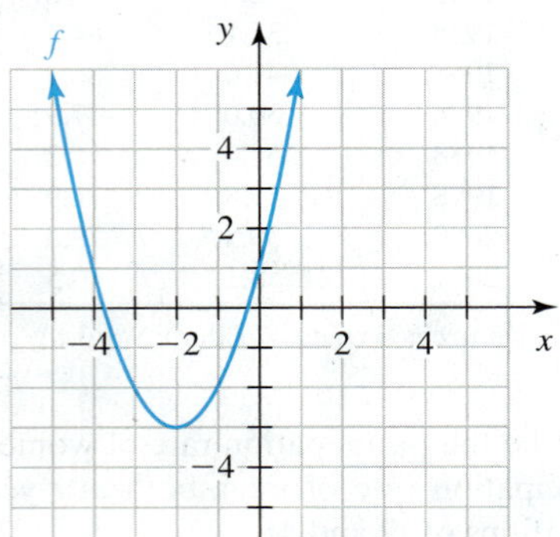

Figure 41 Exercises 45-48

49. Graph $y = -3x^2$ by hand. (Section 7.1, Example 7)

50. Find the product $(5r - 6t)(3r - 2t)$. (Section 7.2, Example 4)

51. Find the product $(2x^2 - x + 3)(4x^2 + 3x - 2)$. (Section 7.2, Example 7)

52. For $f(x) = 4x + 2$ and $g(x) = 3x^2 - 5x - 1$, find an equation of the product function $f \cdot g$; then find $(f \cdot g)(-1)$. (Section 7.2, Example 8)

53. Simplify $(2x - 5)^2$. (Section 7.3, Example 3)

54. Simplify $(4a + 3b)^2$. (Section 7.3, Example 3)

55. Write $f(x) = 2(x - 3)^2 - 2$ in standard form. (Section 7.3, Example 5)

56. Find the product $(3x^2 - 8)(3x^2 + 8)$. (Section 7.3, Example 6)

Simplify.

57. $2x^4(3x^3)^4$ (Section 7.4, Example 5)

58. $\left(\dfrac{3x^3}{y^7w^4}\right)^3$ (Section 7.4, Example 6)

59. $\dfrac{(2x^4y^2)^4}{x^9y^5}$ (Section 7.4, Example 6)

60. Find the quotient $\dfrac{12p^5 - p^3 - 6}{9p^3}$. (Section 7.5, Example 2)

61. Divide: $\dfrac{8x^3 + 18x^2 + 17x + 4}{4x + 3}$. (Section 7.5, Example 5)

Factor.

62. $w^2 + 5w - 14$ (Section 8.1, Example 4)

63. $4m^2 - 49n^2$ (Section 8.1, Example 7)

64. $a^2 - 3ab - 40b^2$ (Section 8.1, Example 5)

65. $3x^4 - 33x^3y + 54x^2y^2$ (Section 8.2, Example 7)

66. $8x^3 - 27$ (Section 8.4, Examples 1 and 2)

For Exercises 67–69, solve.

67. $x^2 = 2x + 35$ (Section 8.5, Example 6)

68. $5x^2 + 18x - 8 = 0$ (Section 8.5, Example 4)

69. $4r^3 - 9r = 18 - 8r^2$ (Section 8.5, Examples 10 and 11)

70. Find all x-intercepts of the graph of $y = x^2 + 4x - 21$. (Section 8.5, Example 3)

71. A company's profit can be modeled by $p = t^2 - 2t + 8$, where p is the profit (in millions of dollars) for the year that is t years since 2010. Estimate when the profit was \$32 million and predict when the profit will be \$32 million. (Section 8.6, Example 1)

72. A rectangular rug has an area of 84 square feet. If the length is 8 feet more than the width, find the dimensions of the rug. (Section 8.6, Example 3)

73. The percentages of teenagers and adults who have watched online video in the past week are shown in Table 21 for various years. (Section 8.6, Example 1)

Table 21 Percentages of Teenagers and Adults Who Have Watched Online Video in the Past Week

Year	Percent
2004	7
2005	8
2006	12
2007	15
2008	18
2009	27
2010	29

Source: *Edison Research and Arbitron*

Let $f(t)$ be the percentage of teenagers and adults who have watched online video in the past week at t years since 2000. A quadratic model of the situation is $f(t) = \frac{1}{3}t^2 - t + 6$.

a. Use a graphing calculator to draw the graph of the model and, in the same viewing window, the scattergram of the data. Does the model fit the data well?

b. Predict the percentage of teenagers and adults who will have watched online video in the past week in 2017.

c. Estimate when 42% of teenagers and adults watched online video in the past week.

9

Quadratic Functions

Would you say you are very happy, in general? The percentages of Americans who say they are "very happy" are shown in Table 1 for various years. Interestingly, the percentage decreased from 1990 to 2000, even though the economy performed quite well during that period. In Exercise 65 of Homework 9.4, we will predict the percentage of people who will say they are "very happy" in 2019.

Table 1 Percentages of Americans Who Say They Are "Very Happy"

Year	Percent
1972	30
1980	35
1990	36
2000	34
2010	29

Source: *National Opinion Research Center*

In Chapters 7 and 8, we worked with polynomial expressions and equations. In this chapter, we will narrow our focus to quadratic equations. In particular, in Sections 9.1 and 9.2, we will graph quadratic functions. In Sections 9.4–9.6, we will discuss three nonfactoring methods of solving quadratic equations in one variable. In Sections 9.7 and 9.8, we will solve *systems of linear equations in three variables* to help us find equations of parabolas and quadratic models. Finally, in Section 9.9, we will use many of the skills learned earlier in this chapter to help us make estimates and predictions about authentic situations, such as the age at which Americans are most likely to forget to do something special for their significant other on Valentine's Day.

9.1 Graphing Quadratic Functions in Vertex Form

Objectives

- Know the *vertex form* of a quadratic function.
- Graph a quadratic function in vertex form.
- Find the domain and range of a quadratic function.
- Find a quadratic model in vertex form.

In this section, we will work with parabolas. As we saw in Section 7.1, each parabola has a vertex and an axis of symmetry (see Figs. 1 and 2). Recall that the part of the parabola that lies to the left of the axis of symmetry is the mirror reflection of the part that lies to the right.

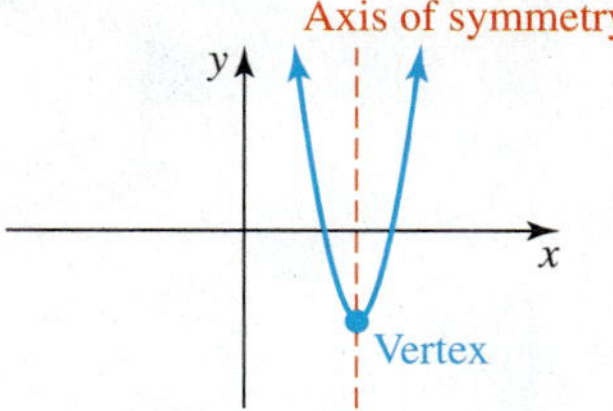

Figure 1 A parabola that opens upward

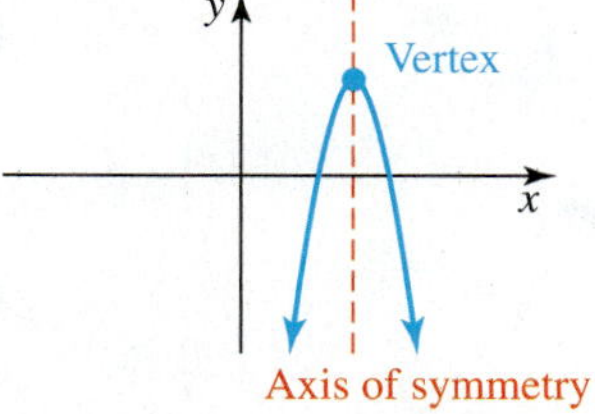

Figure 2 A parabola that opens downward

Vertex Form

Recall from Section 7.1 that a quadratic function is a function whose equation can be put into the standard form $f(x) = ax^2 + bx + c$, where $a \neq 0$, and that the graph of the function is a parabola. Here we will work with equations in another form, called *vertex form:* $f(x) = a(x-h)^2 + k$, where $a \neq 0$. Any equation in either of these two forms can be written in the other form. So, a function in vertex form is quadratic, and its graph is a parabola.

Vertex Form of a Quadratic Function

Let $f(x) = a(x-h)^2 + k$, where $a \neq 0$. Then f is a quadratic function, and its graph is a parabola. We say the equation is in **vertex form.**

Here are some quadratic functions in vertex form:

$$f(x) = 3(x-5)^2 + 4 \qquad g(x) = -4(x+9)^2 - 1 \qquad h(x) = 6x^2 - 8 \qquad k(x) = x^2$$

We will see later in this section that, for a quadratic function in vertex form, we can find the vertex quickly.

Graphs of Quadratic Functions of the Form $f(x) = ax^2$

To graph quadratic functions in vertex form $f(x) = a(x-h)^2 + k$, we begin by exploring graphs of equations of the form $f(x) = ax^2$ (where $h = 0$ and $k = 0$).

Table 2 Input–Output Pairs of $f(x) = x^2$ and $g(x) = 2x^2$

x	$f(x) = x^2$	$g(x) = 2x^2$
−3	9	18
−2	4	8
−1	1	2
0	0	0
1	1	2
2	4	8
3	9	18

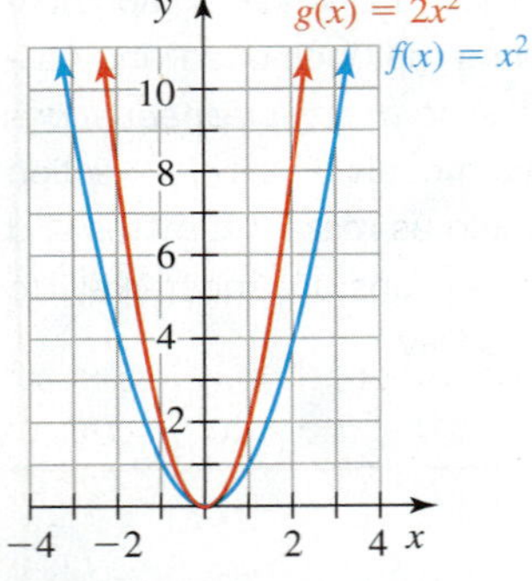

Figure 3 Graphs of $f(x) = x^2$ and $g(x) = 2x^2$

Example 1 Stretching a Graph Vertically

Compare the graph of $g(x) = 2x^2$ with the graph of $f(x) = x^2$.

Solution

We list input–output pairs of f and g in Table 2 and sketch the graphs of f and g in Fig. 3. For example, $g(-2) = 2(-2)^2 = 2(4) = 8$. Therefore, $(-2, 8)$ is a point on the graph of g.

For each value of x, the value of y is twice as large for $g(x) = 2x^2$ as it is for $f(x) = x^2$ (see Table 2). Therefore, the graph of g appears steeper (narrower) than the graph of f. Also, notice that the vertex for both functions is the point $(0, 0)$.

Example 2 Reflecting a Graph across the x-Axis

Compare the graph of $f(x) = \frac{1}{2}x^2$ with the graph of $g(x) = -\frac{1}{2}x^2$.

Solution

We list input–output pairs of f and g in Table 3 and sketch the graphs of f and g in Fig. 4.

Table 3 Input–Output Pairs of $f(x) = \frac{1}{2}x^2$ and $g(x) = -\frac{1}{2}x^2$

x	$f(x) = \frac{1}{2}x^2$	$g(x) = -\frac{1}{2}x^2$
−3	4.5	−4.5
−2	2	−2
−1	0.5	−0.5
0	0	0
1	0.5	−0.5
2	2	−2
3	4.5	−4.5

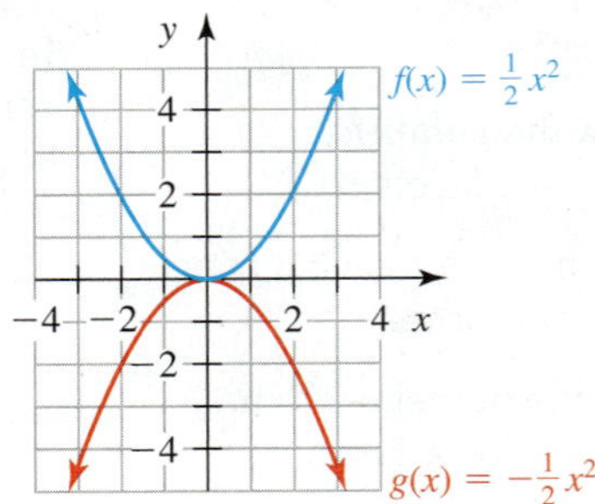

Figure 4 Graphs of $f(x) = \frac{1}{2}x^2$ and $g(x) = -\frac{1}{2}x^2$

In Table 3, we see that for each value of x, the outputs of g are the opposites of the outputs of f. Because of this, the graph of g is the **reflection,** or mirror image, of the graph of f, with the mirror along the x-axis. We can find the graph of g by *reflecting the graph of f across the x-axis.*

The observations we made in Examples 1 and 2 suggest the following properties.

Graphs of Quadratic Functions of the Form $f(x) = ax^2$

For a function of the form $f(x) = ax^2$,

- The graph is a parabola with vertex $(0, 0)$.
- If $|a|$ is a large number, then the parabola is steep. (It is narrow.)
- If a is near zero, then the parabola is not steep. (It is wide.)
- If $a > 0$, then the parabola opens upward.
- If $a < 0$, then the parabola opens downward.
- The graph of $y = -ax^2$ is the reflection of the graph of $f(x) = ax^2$ across the x-axis.

Translating Graphs

Next, we investigate graphs of functions of the form $f(x) = x^2 + k$.

Example 3 Translating a Graph Vertically (Up–Down Shifts)

Compare the graphs of $f(x) = x^2 - 3$, $g(x) = x^2$, and $h(x) = x^2 + 3$.

Solution

We list input–output pairs of f, g, and h in Table 4 and sketch the graphs of the functions in Fig. 5.

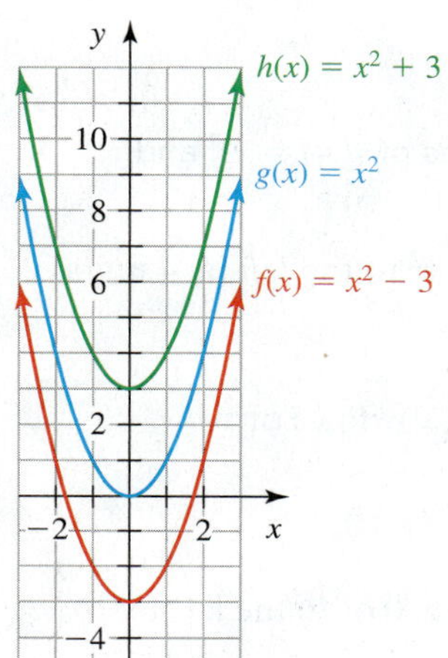

Figure 5 Graphs of $f(x) = x^2 - 3$, $g(x) = x^2$, and $h(x) = x^2 + 3$

Table 4 Input–Output Pairs of $f(x) = x^2 - 3$, $g(x) = x^2$, and $h(x) = x^2 + 3$

x	$f(x) = x^2 - 3$	$g(x) = x^2$	$h(x) = x^2 + 3$
−3	6	9	12
−2	1	4	7
−1	−2	1	4
0	−3	0	3
1	−2	1	4
2	1	4	7
3	6	9	12

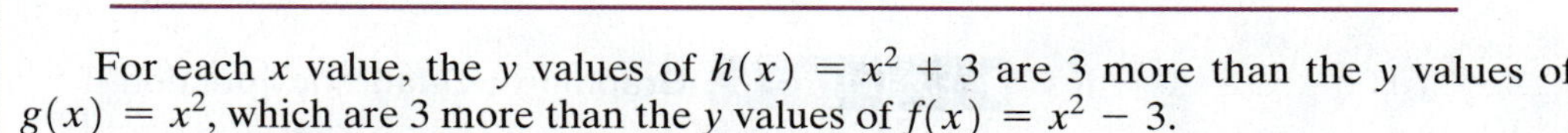

For each x value, the y values of $h(x) = x^2 + 3$ are 3 more than the y values of $g(x) = x^2$, which are 3 more than the y values of $f(x) = x^2 - 3$.

To sketch the graph of $h(x) = x^2 + 3$, we *translate* (move) the graph of $g(x) = x^2$ up by 3 units. To sketch the graph of $f(x) = x^2 - 3$, we translate the graph of $g(x) = x^2$ down by 3 units.

Example 4 Translating a Graph Horizontally (Left–Right Shifts)

Compare the graph of $g(x) = (x - 5)^2$ with the graph of $f(x) = x^2$.

Solution

We list input–output pairs of g in Table 5 and sketch the graphs of f and g in Fig. 6.

Table 5 Input–Output Pairs of $g(x) = (x - 5)^2$

x	$g(x)$
−1	36
0	25
1	16
2	9
3	4
4	1
5	0 ← Vertex
6	1
7	4
8	9

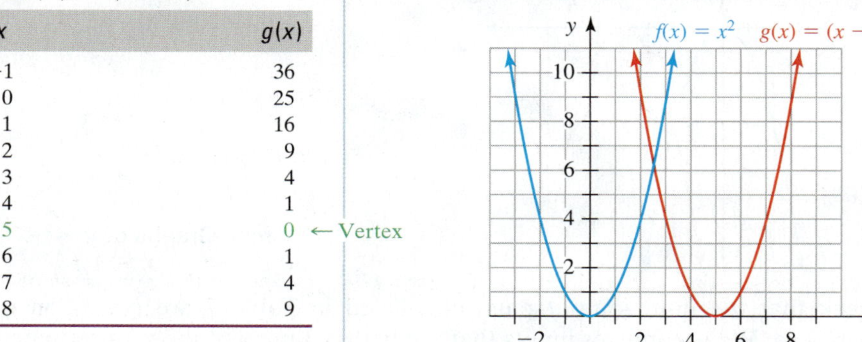

Figure 6 Graphs of $f(x) = x^2$ and $g(x) = (x - 5)^2$

To graph $g(x) = (x - 5)^2$, we translate the graph of $f(x) = x^2$ to the *right* by 5 units.

Example 5 Translating a Graph Horizontally (Left–Right Shifts)

Compare the graph of $g(x) = (x + 4)^2$ with the graph of $f(x) = x^2$.

Solution

We list input–output pairs of g in Table 6 and sketch the graphs of f and g in Fig. 7.

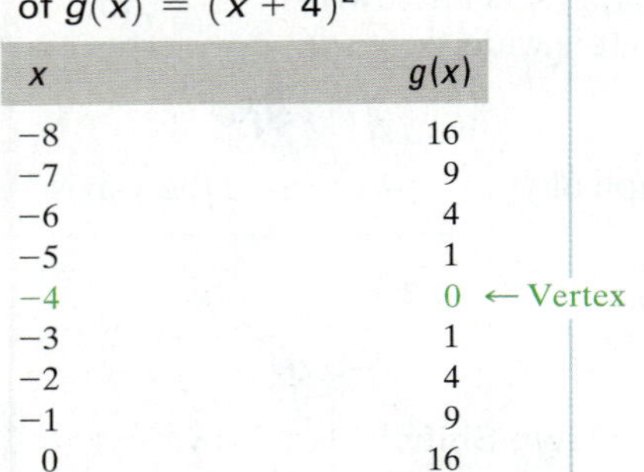

Table 6 Input–Output Pairs of $g(x) = (x + 4)^2$

x	$g(x)$
−8	16
−7	9
−6	4
−5	1
−4	0 ← Vertex
−3	1
−2	4
−1	9
0	16
1	25

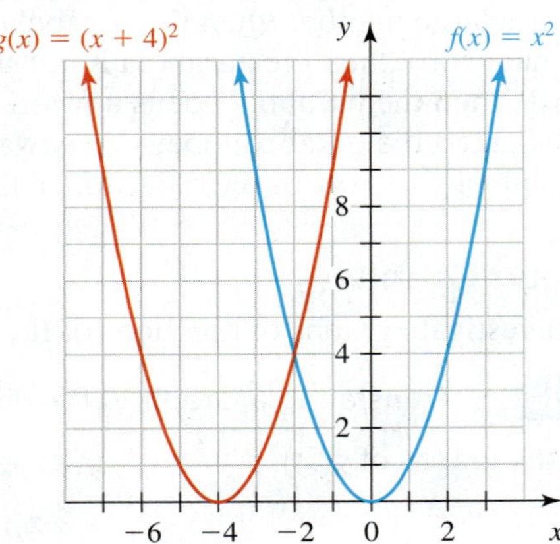

Figure 7 Graphs of $f(x) = x^2$ and $g(x) = (x + 4)^2$

To graph $g(x) = (x + 4)^2$, we translate the graph of $f(x) = x^2$ to the *left* by 4 units.

Three-Step Method of Graphing a Quadratic Function in Vertex Form

To sketch the graph of $f(x) = a(x - h)^2 + k$, where $a \neq 0$,

1. Sketch the graph of $y = ax^2$.
2. Translate the graph from step 1 to the right by h units if $h > 0$ or to the left by $|h|$ units if $h < 0$.
3. Translate the graph from step 2 up by k units if $k > 0$ or down by $|k|$ units if $k < 0$.

Example 6 Graphing a Quadratic Function

Sketch the graph of $f(x) = -(x - 4)^2$.

Solution

The equation of f is already in $f(x) = a(x - h)^2 + k$ form, with $a = -1$, $h = 4$, and $k = 0$. We follow the three-step graphing method:

Step 1. We sketch the graph of $y = -x^2$ in Fig. 8.

Step 2. Since $h = 4$, we translate the graph from step 1 to the right by 4 units.

Step 3. Since $k = 0$, there is no vertical translation.

Table 7 Input–Output Pairs of $f(x) = -(x - 4)^2$

x	$f(x)$
1	−9
2	−4
3	−1
4	0 ← Vertex
5	−1
6	−4
7	−9

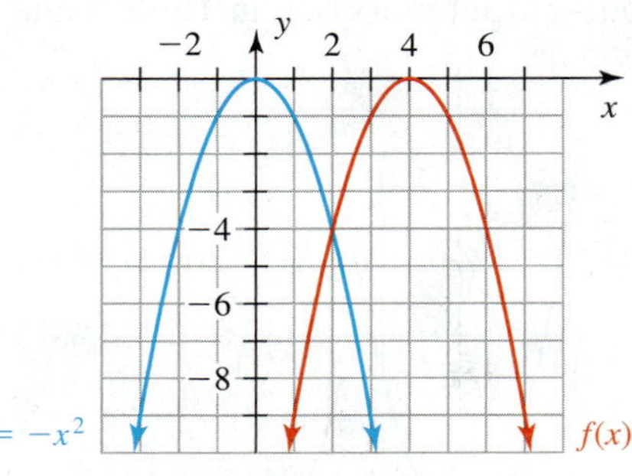

Figure 8 Graphs of $y = -x^2$ and $f(x) = -(x - 4)^2$

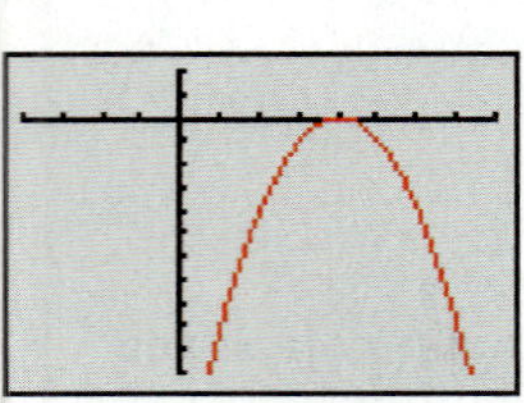

Figure 9 Graph of $f(x) = -(x - 4)^2$

We check that the input–output pairs of f listed in Table 7 are points on our sketched parabola. We use integer inputs that are within 3 units of the x-coordinate of the vertex, 4. Also, we use a graphing calculator to verify our sketch (see Fig. 9).

Example 7 Graphing a Quadratic Function

Sketch the graph of $f(x) = (x + 3)^2 + 1$.

Solution

First, we write the equation of f in $f(x) = a(x - h)^2 + k$ form:

$$f(x) = [x - (-3)]^2 + 1$$

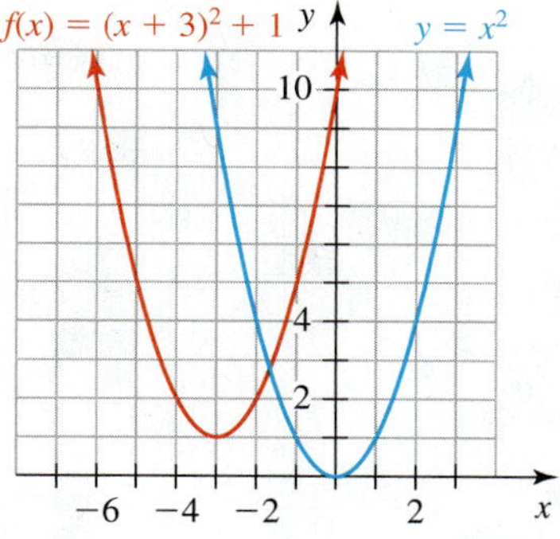

Figure 10 Graphs of $y = x^2$ and $f(x) = (x + 3)^2 + 1$

Then we follow the three-step graphing method:

Step 1. We sketch the graph of $y = x^2$ in Fig. 10.

Step 2. Since $h = -3$, we translate the graph from step 1 to the left by 3 units.

Step 3. Since $k = 1$, we translate the graph from step 2 up by 1 unit.

We check that the input–output pairs of f listed in Table 8 are points on our sketched parabola. We use integer inputs that are within 3 units of the x-coordinate of the vertex, −3. Also, we can use a graphing calculator to verify our work (see Fig. 11).

Table 8 Input–Output Pairs of $f(x) = (x + 3)^2 + 1$

x	$f(x)$	
−6	10	
−5	5	
−4	2	
−3	1	← Vertex
−2	2	
−1	5	
0	10	

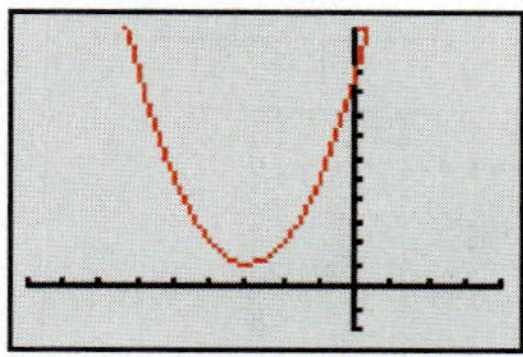

Figure 11 Graph of $f(x) = (x + 3)^2 + 1$

In Fig. 10, we see that the vertex of $f(x) = (x + 3)^2 + 1$ is the point $(-3, 1)$. This makes sense, since the vertex $(0, 0)$ of the graph of $y = x^2$ has been translated to the left by 3 units and up by 1 unit.

Vertex of the Graph of a Quadratic Function

The vertex of the graph of a quadratic function in vertex form

$$f(x) = a(x - h)^2 + k$$

is the point (h, k).

For example, the vertex of the graph of the function $f(x) = -3(x - 1)^2 + 5$ is $(1, 5)$. To find the vertex of the graph of the function $g(x) = 6(x + 2)^2 - 7$, we write the equation as $g(x) = 6[x - (-2)]^2 + (-7)$. So, the vertex is $(-2, -7)$.

Domain and Range of a Quadratic Function

Recall from Section 5.2 that the domain of a function is the set of values of x (the independent variable) and that the range of a function is the set of values of y (the dependent variable).

Example 8 Graphing a Quadratic Function

Sketch the graph of $f(x) = -2(x + 6)^2 - 2$, and find the domain and range of f.

Solution

First, we write the equation of f in $f(x) = a(x - h)^2 + k$ form:

$$f(x) = -2[x - (-6)]^2 + (-2)$$

Then we follow the three-step graphing method:

Step 1. We sketch the graph of $y = -2x^2$ in Fig. 12.

Step 2. Since $h = -6$, we translate the graph from step 1 to the left by 6 units.

Step 3. Since $k = -2$, we translate the graph from step 2 down by 2 units.

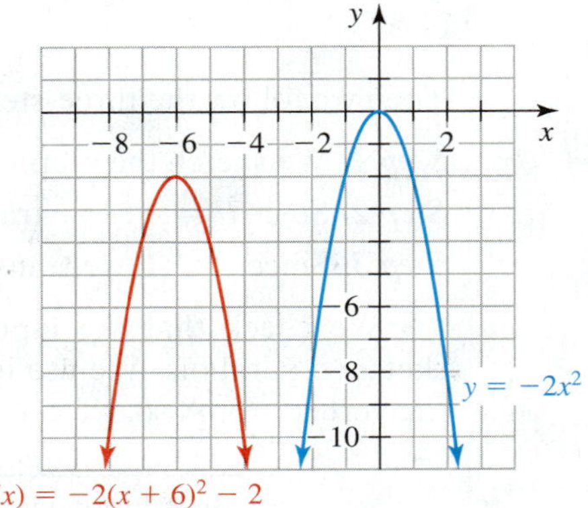

Figure 12 Graphs of $y = -2x^2$ and $f(x) = -2(x + 6)^2 - 2$

Table 9 Input–Output Pairs of $f(x) = -2(x + 6)^2 - 2$

x	$f(x)$	
−8	−10	
−7	−4	
−6	−2	← Vertex
−5	−4	
−4	−10	

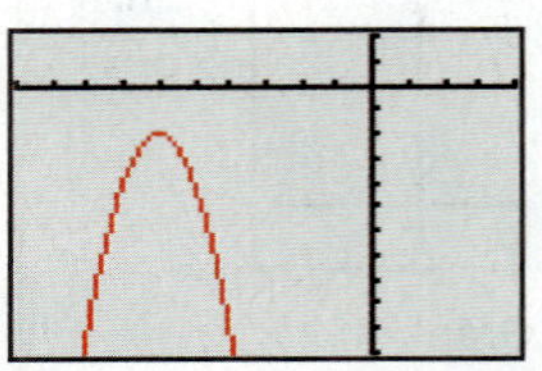

Figure 13 Graph of $f(x) = -2(x + 6)^2 - 2$

We check that the input–output pairs of f listed in Table 9 are points on our sketched parabola. Also, from the equation $f(x) = -2[x - (-6)]^2 + (-2)$, we see that the vertex is $(-6, -2)$, which matches our sketched parabola's vertex. Finally, we use a graphing calculator to verify our sketch (see Fig. 13).

Since we can compute a value of $-2(x + 6)^2 - 2$ for any real number x, the domain of $f(x) = -2(x + 6)^2 - 2$ is the set of all real numbers. From the graph of f, we see that the vertex $(-6, -2)$ is the maximum point and that values of y are less than or equal to -2. So, the range of f is the set of numbers y where $y \leq -2$.

Example 9 Graphing a Quadratic Function

Sketch the graph of $f(x) = \frac{1}{3}(x - 5)^2 + 3$, and find the domain and range of f.

Solution

We follow the three-step graphing method:

Step 1. We sketch the graph of $y = \frac{1}{3}x^2$ in Fig. 14.

Step 2. Since $h = 5$, we translate the graph from step 1 to the right by 5 units.

Step 3. Since $k = 3$, we translate the graph from step 2 up by 3 units.

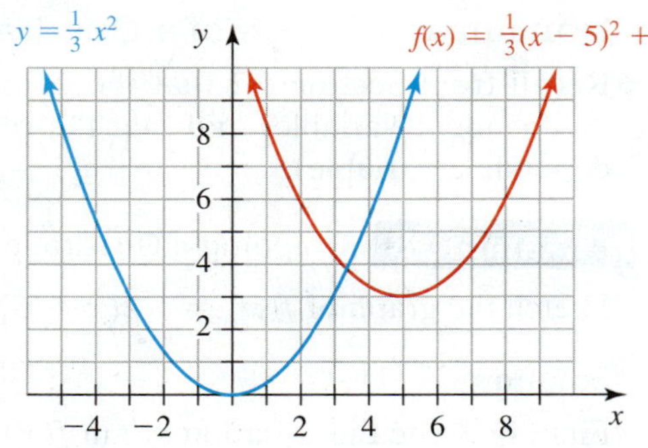

Figure 14 Graphs of $y = \frac{1}{3}x^2$ and $f(x) = \frac{1}{3}(x - 5)^2 + 3$

Table 10 Input–Output Pairs of $f(x) = \frac{1}{3}(x - 5)^2 + 3$

x	$f(x)$
2	6
5	3 ← Vertex
8	6

We check that the input–output pairs of f listed in Table 10 are points on our sketched parabola. Also, from the equation $f(x) = \frac{1}{3}(x - 5)^2 + 3$, we see that the vertex is $(5, 3)$, which matches our sketched parabola's vertex. Finally, we use a graphing calculator to verify our work (see Fig. 15).

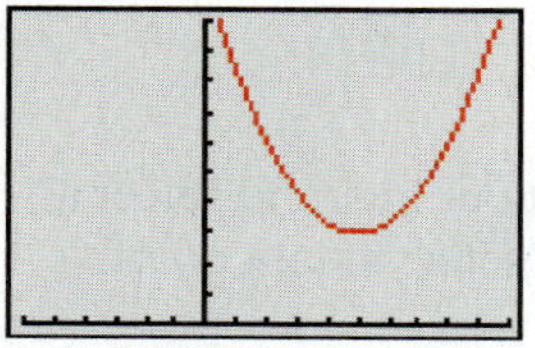

Figure 15 Graph of $f(x) = \frac{1}{3}(x - 5)^2 + 3$

Since we can compute a value of $\frac{1}{3}(x - 5)^2 + 3$ for any real number x, the domain of $f(x) = \frac{1}{3}(x - 5)^2 + 3$ is the set of all real numbers. From the graph of f, we see that the vertex $(5, 3)$ is the minimum point and that values of y are greater than or equal to 3. So, the range of f is the set of numbers y where $y \geq 3$.

Find a Quadratic Model in Vertex Form

We can use our knowledge of graphing quadratic functions in vertex form to help us find an equation of a quadratic model.

Example 10 Finding an Equation of a Quadratic Model

The percentages of women between the ages of 25 years and 54 years who work are shown in Table 11 for various calendar years. Let $f(t)$ be the percentage of women between the ages of 25 years and 54 years who work at t years since 1980.

1. Find an equation of f.
2. What is the vertex of the graph of f? What does it mean in this situation?
3. Use f to estimate the percentage of women between the ages of 25 years and 54 years who worked in 2008.

Table 11 Percentages of Women Between the Ages of 25 Years and 54 Years Who Work

Year	Percent
1980	64.0
1985	69.6
1990	74
1995	75.6
2000	76.7
2005	75.3
2010	75.1

Source: *Bureau of Labor Statistics*

Solution

1. We use a graphing calculator to draw a scattergram of the data (see Fig. 16). It appears that a quadratic function would describe the situation much better than a linear function. We will find a quadratic function in vertex form, $f(t) = a(t - h)^2 + k$. Although it is not necessary to select the highest data point, $(20, 76.7)$, to be the vertex (see Fig. 16), it is convenient and satisfactory to do so. This means $h = 20$ and $k = 76.7$:

$$f(t) = a(t - 20)^2 + 76.7$$

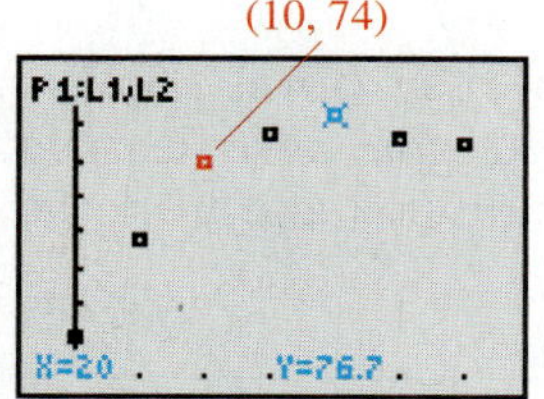

Figure 16 Working women scattergram

Next, we imagine a parabola with vertex $(20, 76.7)$ that comes close to (or contains) the data points. Such a parabola might be the one that contains the data point $(10, 74)$. See Fig. 16. To find a, we substitute 10 for t and 74 for $f(t)$ in the equation $f(t) = a(t - 20)^2 + 76.7$ and solve for a:

$$\begin{aligned} 74 &= a(10 - 20)^2 + 76.7 && \text{Substitute 10 for } t \text{ and 74 for } f(t). \\ 74 &= a(-10)^2 + 76.7 && \text{Subtract.} \\ 74 &= 100a + 76.7 && \text{Simplify.} \\ -2.7 &= 100a && \text{Subtract 76.7 from both sides.} \\ -0.027 &= a && \text{Divide both sides by 100.} \end{aligned}$$

The approximate equation is $f(t) = -0.027(t - 20)^2 + 76.7$. We check how well the model fits the data in Fig. 17. Since the graph appears to come close to (or contain) the data points, we conclude that f is a reasonable model of the situation.

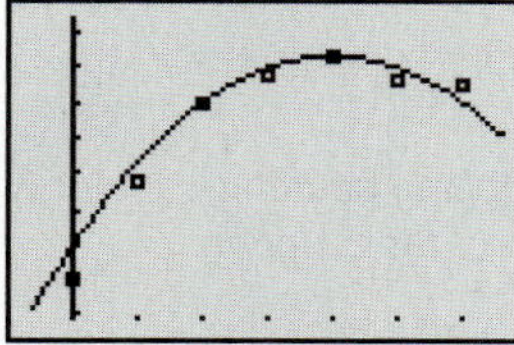

Figure 17 Check the fit

2. The vertex of the graph of $f(t) = -0.027(t - 20)^2 + 76.7$ is $(20, 76.7)$, which means 76.7% of women between the ages of 25 years and 54 years worked in 2000. That is the highest percentage in any year, according to the model.

3. We evaluate f at $t = 28$:

$$f(28) = -0.027(28 - 20)^2 + 76.7 = 74.972$$

In 2008, about 75.0% of women between the ages of 25 years and 54 years worked, according to the model.

Finding a Quadratic Model in Vertex Form

To find a quadratic model in vertex form, given some data,

1. Create a scattergram of the data.
2. Imagine a parabola that comes close to (or contains) the data points, and select a point (h, k) to be the vertex. Although it is not necessary to select a data point, it is often convenient and satisfactory to do so.
3. Select a nonvertex point (not necessarily a data point) of the parabola, substitute the point's coordinates into the equation $f(t) = a(t - h)^2 + k$, and solve for a.
4. Substitute the result you found for a in step 3 into $f(t) = a(t - h)^2 + k$.

Group Exploration

Drawing families of parabolas

1. On a graphing calculator, graph eight quadratic functions to make a design like the one in Fig. 18. List the equations of your parabolas.

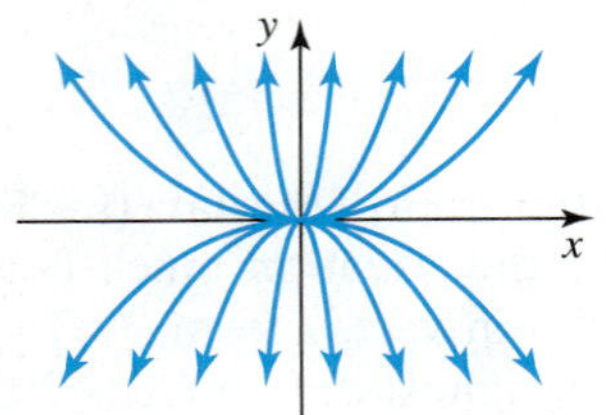

Figure 18 A family of parabolas

2. Now make a design like the one in Fig. 19. List the equations of your parabolas.

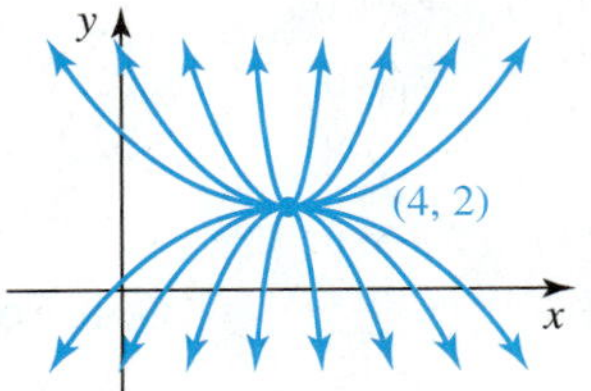

Figure 19 A family of parabolas with vertex (4, 2)

3. For a quadratic function of the form $f(x) = a(x - h)^2 + k$, summarize what you have learned about a, h, and k from this exploration and from the rest of this section.

Homework 9.1

For extra help ▶ MyMathLab®  Watch the videos in MyMathLab  Download the MyDashboard App

Graph the function by hand. Give the coordinates of the vertex. Then use a graphing calculator to verify your work.

1. $f(x) = 3x^2$

2. $f(x) = 2x^2$

3. $f(x) = -\frac{3}{2}x^2$

4. $f(x) = -\frac{2}{3}x^2$

5. $f(x) = -2x^2 + 5$

6. $f(x) = 2x^2 - 4$

7. $f(x) = (x - 1)^2$

8. $f(x) = (x + 3)^2$

9. $f(x) = -(x + 2)^2$

10. $f(x) = -3(x - 3)^2$

11. $f(x) = (x + 2)^2 - 6$

12. $f(x) = (x + 5)^2 - 7$

13. $f(x) = (x - 1)^2 + 3$

14. $f(x) = (x - 4)^2 + 2$

15. $f(x) = 2(x + 6)^2 - 6$

16. $f(x) = 3(x - 4)^2 + 1$

17. $f(x) = -3(x - 6)^2 - 2$

18. $f(x) = -(x + 1)^2 - 2$

19. $f(x) = \frac{1}{3}(x - 2)^2 + 3$

20. $f(x) = -\frac{1}{2}(x + 4)^2 - 2$

Graph the function by hand. Then use a graphing calculator to verify your work. Also, find the domain and range of the function.

21. $f(x) = 2x^2 - 4$

22. $f(x) = 3x^2 + 1$

23. $f(x) = -x^2 - 3$

24. $f(x) = -2x^2 + 5$

25. $f(x) = (x + 4)^2$ **26.** $f(x) = (x - 2)^2$

27. $f(x) = (x + 6)^2 + 2$ **28.** $f(x) = (x - 3)^2 + 1$

29. $f(x) = 2(x - 1)^2 - 4$ **30.** $f(x) = 3(x + 4)^2 - 7$

31. $f(x) = -(x - 5)^2 + 2$ **32.** $f(x) = -2(x + 2)^2 + 5$

33. U.S. Department of Defense spendings are shown in Table 12 for various years.

Table 12 U.S. Department of Defense Spendings

Year	Defense Spending (billions of dollars)
1992	298
1995	272
1998	268
2001	305
2004	456
2007	551
2010	720

Sources: *Department of Defense; Government Accountability Office*

Let $f(t)$ be the U.S. Department of Defense spending (in billions of dollars) in the year that is t years since 1990.

a. Find a quadratic equation of f in vertex form.

b. What is the vertex of the model? What does it mean in this situation?

c. Estimate defense spending in 2008.

d. In 2011, the federal government spent $68.3 billion on education. Estimate the ratio of federal spending on defense to federal spending on education in 2011.

34. The percentages of people living in the United States who are immigrants are shown in Table 13 for various years.

Table 13 Percentages of People Living in the United States Who Are Immigrants

Year	Percent
1930	12
1940	9
1950	7
1960	6
1970	5
1980	6
1990	8
2000	11
2010	13

Source: *U.S. Census Bureau*

Let $p = f(t)$ be the percentage of people living in the United States who are immigrants at t years since 1900.

a. Find a quadratic equation of f in vertex form.

b. What is the vertex of the model? What does it mean in this situation?

c. Predict the percentage of people living in the United States who will be immigrants in 2018.

d. Predict the *number* of immigrants in the United States in 2020. Assume the U.S. population will be 341.4 million in that year. Compare your result to 37.3 million, which is the current population of California, the most populous state in the United States.

35. The percentages of Americans who are obese are shown in Table 14 for various ages.

Table 14 Percentages of Americans Who Are Obese

Age Group (years)	Age Used to Represent Age Group (years)	Percent
under 25	18	11
25–34	29.5	25
35–44	39.5	28
45–54	49.5	30
55–64	59.5	30
65–74	69.5	27
75–84	79.5	19
over 84	90	8

Source: *The Gallup Organization*

Let $f(t)$ be the percentage of Americans who are obese at age t years.

a. Find a quadratic equation of f in vertex form.

b. What is the vertex of the model? What does it mean in this situation?

c. Estimate the percentage of Americans who are obese at age 73.

d. Use a graphing calculator to find the t-intercepts of the model. What do they mean in this situation? [**Hint:** First Zoom Out. Then see Appendix A.21 for graphing calculator instructions on how to use "zero."]

36. The pregnancy rates for American women are listed in Table 15 for various age groups.

Table 15 Pregnancy Rates for American Women

Age Group (years)	Age Used to Represent Age Group (years)	Pregnancy Rate (number of pregnancies per 1000 women)
15–17	16	42
18–19	18.5	119
20–24	22	164
25–29	27	169
30–34	32	135
35–39	37	76
40–44	42	17

Source: *National Center for Health Statistics*

Let $r = f(t)$ be the pregnancy rate (number of pregnancies per 1000 women) for women at age t years.

a. Find a quadratic equation of f in vertex form.

b. What is the vertex of the model? What does it mean in this situation?

c. Estimate the pregnancy rate for 20-year-old women.

d. Use a graphing calculator to find the t-intercepts of the model. What do they mean in this situation? [**Hint:** Zoom Out. Then see Appendix A.21 for graphing calculator instructions on how to use "zero."]

37. Recall that we can describe some or all of the input–output pairs of a function by means of an equation, a graph, a table, or words. Let $f(x) = 2(x - 1)^2 - 3$.

a. Describe the input–output pairs of f by using a graph.

b. Describe five input–output pairs of f by using a table.
c. Describe the input–output pairs of f by using words.

38. Recall that we can describe some or all of the input–output pairs of a function by means of an equation, a graph, a table, or words. Let $g(x) = -3(x + 4)^2 + 7$.
a. Describe the input–output pairs of g by using a graph.
b. Describe five input–output pairs of g by using a table.
c. Describe the input–output pairs of g by using words.

39. Let $f(x) = (x - 3)^2 + 2$.
a. Graph f by hand.
b. Find x when $f(x) = 3$.
c. Find x when $f(x) = 2$.
d. Find x when $f(x) = 1$.

40. Let $f(x) = -(x + 2)^2 + 5$.
a. Graph f by hand.
b. Find x when $f(x) = 6$.
c. Find x when $f(x) = 5$.
d. Find x when $f(x) = 4$.

Concepts

Write an equation of a parabola that meets the given criteria.

41. opens downward and has vertex $(-3, 4)$

42. opens upward and has vertex $(2, -5)$

43. Four functions of the form $y = a(x - h)^2 + k$ are graphed in Fig. 20. For each function, determine whether the constants a, h, and k are positive, negative, or zero.

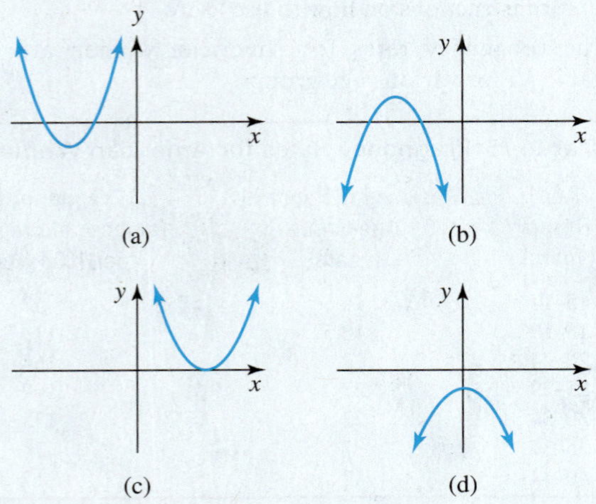

Figure 20 Exercise 43

44. For a quadratic function $f(x) = a(x - h)^2 + k$, for what values of a, h, and k does the graph of f have a maximum point? a minimum point? Describe the maximum or minimum point in terms of a, h, and k.

45. Use a graphing calculator to draw a family of parabolas similar to the family shown in Fig. 21. List the equations of your parabolas.

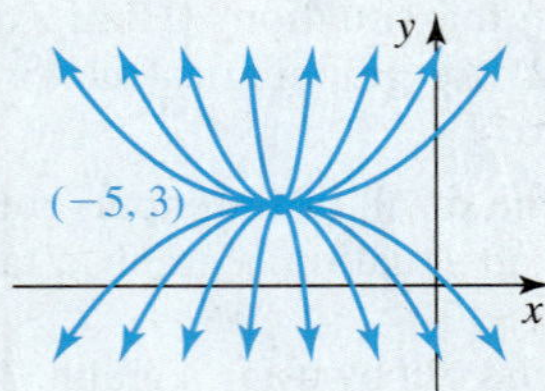

Figure 21 A family of parabolas with vertex $(-5, 3)$

46. Use a graphing calculator to draw a family of parabolas similar to the family shown in Fig. 22. List the equations of your parabolas.

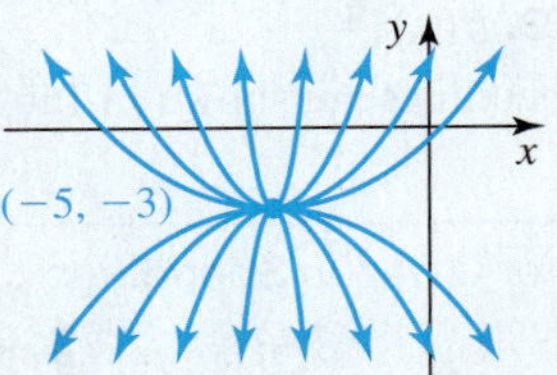

Figure 22 A family of parabolas with vertex $(-5, -3)$

47. Find an equation of the function f graphed in Fig. 23.

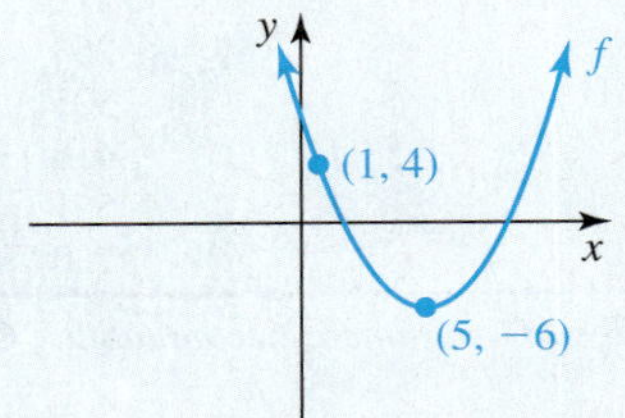

Figure 23 Exercise 47

48. The graph of the function $f(x) = 2.1(x - 2.73)^2 - 3.71$ is shown in Fig. 24. Find an equation of the function g, also graphed in Fig. 24.

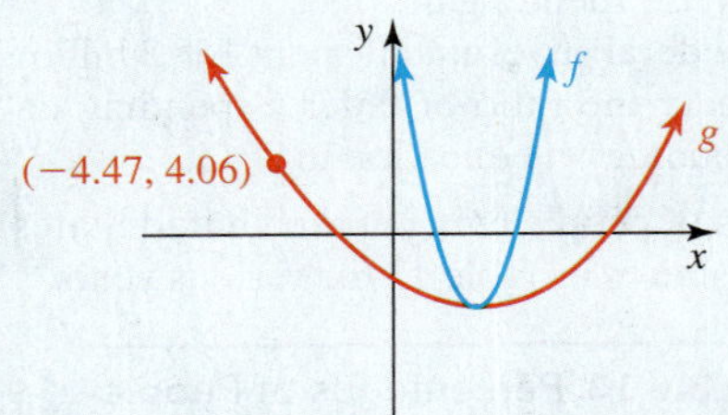

Figure 24 Exercise 48

49. The graph of the function $f(x) = 2.1(x - 2.73)^2 - 3.71$ is shown in Fig. 25. Find an equation of the function g, also graphed in Fig. 25. Assume the graphs of f and g have the same shape.

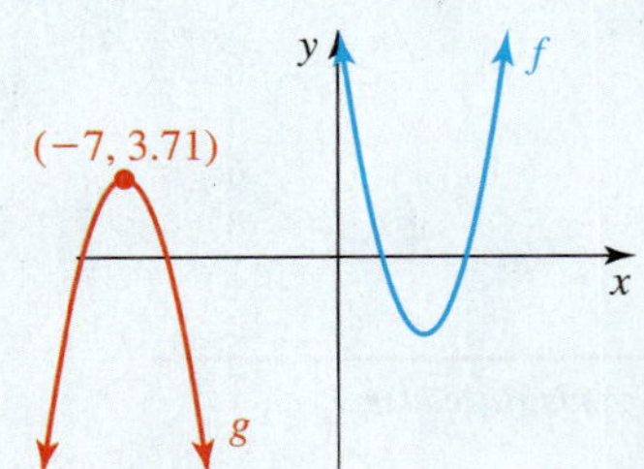

Figure 25 Exercise 49

50. Find equations of the four functions whose graphs produce the design shown in Fig. 26.

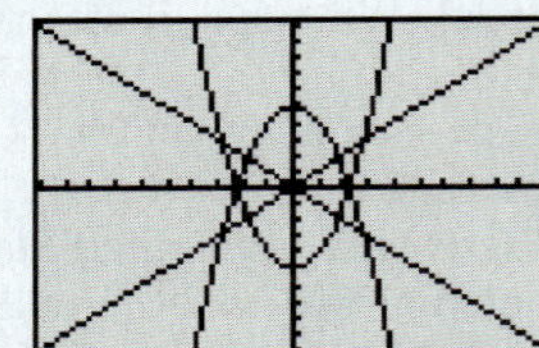

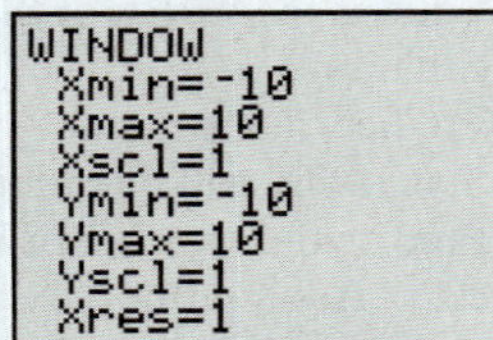

Figure 26 A design created by graphs of functions

For Exercises 51–54, decide whether it is possible for a parabola to have the indicated number of x-intercepts. If it is possible, find an equation of such a parabola. If it is not possible, explain why.

51. no x-intercepts

52. one x-intercept

53. two x-intercepts

54. three x-intercepts

55. Solve the system by finding ordered pairs that satisfy both equations. [**Hint:** Graph both equations on the same coordinate system.]

$$y = 2(x - 2)^2 + 5$$
$$y = -3(x - 2)^2 + 5$$

56. Use ZDecimal to draw the graph of $f(x) = 0.7x^2 + 2x - 1$. Then use TRACE to complete Table 16. Verify your table entries by using a graphing calculator table.

Table 16 Values of $f(x) = 0.7x^2 + 2x - 1$

x	$f(x)$
-3	
-2	
-1	
0	
1	

57. Use a graphing calculator to graph $y = x^2$.

a. Use the window settings displayed in Fig. 27. (You can get these settings by using ZDecimal.)

b. Use the window settings displayed in Fig. 28. Compare what you see with what you saw in part (a).

c. Use the window settings displayed in Fig. 29. Compare what you see with what you saw in part (a).

d. Explain your results in parts (b) and (c).

```
WINDOW
 Xmin=-4.7
 Xmax=4.7
 Xscl=1
 Ymin=-3.1
 Ymax=3.1
 Yscl=1
 Xres=1
```

Figure 27 Window for Exercise 57a

```
WINDOW
 Xmin=-4.7
 Xmax=4.7
 Xscl=1
 Ymin=-1
 Ymax=1
 Yscl=1
 Xres=1
```

Figure 28 Window for Exercise 57b

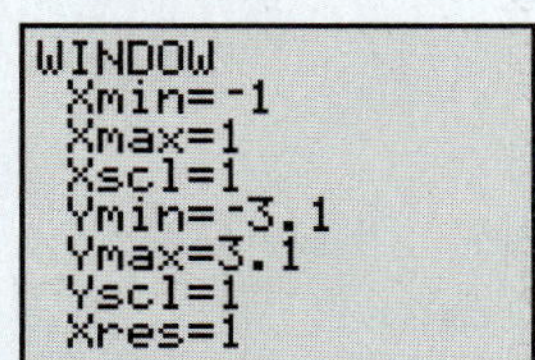

Figure 29 Window for Exercise 57c

58. Use a graphing calculator to graph the function. Record window settings that allow you to see the graph, including the vertex, on the calculator screen.

a. $f(x) = 1000x^2$

b. $g(x) = 1000(x - 1000)^2$

c. $h(x) = (x + 10{,}000)^2 + 10{,}000$

59. A student says that to graph $y = (x - 4)^2$, we translate the graph of $y = x^2$ to the left by 4 units. Is the student correct? Explain.

60. A student says the slope of the graph of $y = 2x^2$ is 2. Is the student correct? Explain.

61. A student uses ZStandard on a graphing calculator to graph $f(x) = 0.0001x^2 + 5$. He thought the graph should be a parabola, but the calculator displays a horizontal line. What would you tell him?

62. The vertex of a parabola is $(5, 3)$, and the parabola passes through the point $(8, 10)$. Find a third point that lies on the parabola.

63. Sketch, on the same coordinate system, the graph of the given function and the graph of $f(x) = a(x - h)^2 + k$ as shown in Fig. 30. Be sure to label which graph is which.

a. $g(x) = a(x - 2h)^2 + k$ **b.** $p(x) = a(x - h)^2 + 2k$
c. $q(x) = 2a(x - h)^2 + k$ **d.** $l(x) = 2a(x - 2h)^2 + 2k$

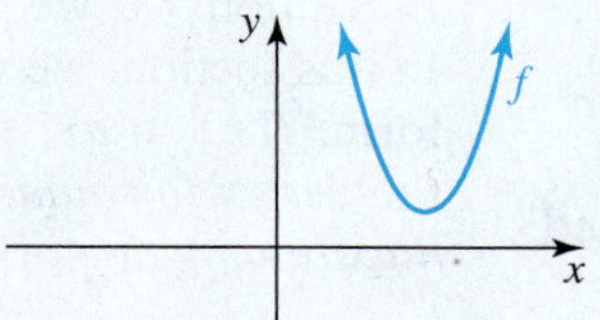

Figure 30 Exercise 63

64. Find two other quadratic functions that have the same domain and range as the function $f(x) = -2(x + 4)^2 + 7$.

65. Describe how to sketch the graph of a function of the form $f(x) = a(x - h)^2 + k, a \neq 0$. Include the effects that the values of a, h, and k have on the graph. (See page 9 for guidelines on writing a good response.)

66. To graph the quadratic function $y = a(x - h)^2 + k$, we translate the graph of $y = a(x - h)^2$ up by k units if $k > 0$ and down by $|k|$ units if $k < 0$. Explain. (See page 9 for guidelines on writing a good response.)

Related Review

For Exercises 67–70, graph the equation by hand. Then use a graphing calculator to verify your work.

67. $f(x) = -2x - 1$

68. $f(x) = 3x - 2$

69. $f(x) = -2x^2 - 1$

70. $f(x) = 3x^2 - 2$

71. a. Graph $y = 2x$.

b. Graph $y = x^2$.

c. Compare the y-intercepts of the graphs of $y = 2x$ and $y = x^2$.

d. Which of the graphs of $y = 2x$ and $y = x^2$ appears to be "steeper" for large values of x? Explain.

72. a. Create a table of values of x and y for the equation $y = 2x$. Use the values 1, 100, 200, 300, 400, and 500 for x.

b. Create a table of values of x and y for the equation $y = x^2$. Use the values 1, 100, 200, 300, 400, and 500 for x.

c. As the values of x increase beyond 1, do the values of y increase at a greater rate for the equation $y = 2x$ or for the equation $y = x^2$? Explain.

Expressions, Equations, Functions, and Graphs

Perform the indicated instruction. Then use words such as linear, quadratic, cubic, polynomial, degree, function, one variable, *and* two variables *to describe the expression, equation, or system.*

73. Solve $8(3x - 2) = 4(x - 5)$.

74. Graph $f(x) = 2x^2 - 4$ by hand.

75. Simplify $8(3x - 2) - 4(x - 5)$.

76. Graph $g(x) = 2x - 4$ by hand.

77. Find the product $8(3x - 2)(x - 5)$.

78. Find $f(-3)$, where $f(x) = 2x^2 - 4$.

9.2 Graphing Quadratic Functions in Standard Form

Objectives

- Graph a quadratic function in standard form by using two symmetric points to find the vertex.
- Graph a quadratic function in standard form by using the *vertex formula* to find the vertex.
- Find the *minimum value* or *maximum value* of a quantity.

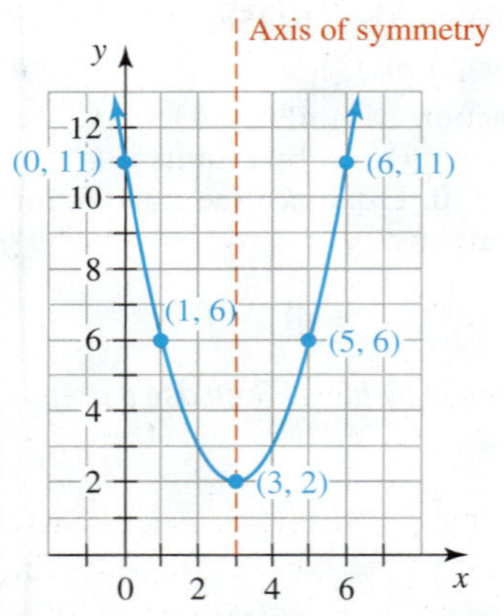

Figure 31 Locating symmetric points on a parabola

In Section 9.1, we graphed quadratic functions in vertex form, $f(x) = a(x - h)^2 + k$. In this section, we will discuss two methods of graphing quadratic functions in standard form, $f(x) = ax^2 + bx + c$. First, we will use two *symmetric points*. Then we will use the *vertex formula*. After discussing these methods, we will find the *minimum value* or *maximum value* of a quantity.

Method 1: Graphing by Using Two Symmetric Points to Find the Vertex

Recall from Section 7.1 that the part of the parabola that lies to the left of the axis of symmetry is the mirror reflection of the part that lies to the right (see Fig. 31). For any point on one side of the axis of symmetry, there is a point on the other side that is the "mirror reflection" of the first point. Such a pair of **symmetric points** corresponds to ordered pairs with equal y-coordinates.

For example, in Fig. 31, the point $(1, 6)$ and the point $(5, 6)$ are symmetric points. Since the vertex lies on the axis of symmetry, the x-coordinate of the vertex, 3, is equal to the average of the x-coordinates of the points $(1, 6)$ and $(5, 6)$:

$$\frac{1 + 5}{2} = 3$$

The average of the x-coordinates of *any* two symmetric points is equal to the x-coordinate of the vertex. As another example, we find the average of the x-coordinates of the y-intercept $(0, 11)$ and its symmetric point $(6, 11)$:

$$\frac{0 + 6}{2} = 3$$

The only point that does not have a symmetric point is the vertex, which lies on the axis of symmetry.

Using Two Symmetric Points to Find the Vertex

To find the vertex of a parabola in which (p, s) and (q, s) are symmetric points,

1. Find the x-coordinate by using the formula
$$x = \frac{p + q}{2}$$
In words, the x-coordinate is the average of the x-coordinates of any two symmetric points.
2. Find the y-coordinate by evaluating f at the value found in step 1. That is, find
$$f\left(\frac{p + q}{2}\right)$$

In short, the vertex is $\left(\frac{p+q}{2}, f\left(\frac{p+q}{2}\right)\right)$.

Example 1 Finding the x-Coordinate of a Vertex

Find the x-coordinate of the vertex of the parabola sketched in the indicated figure. Write the result in decimal form.

1. Fig. 32

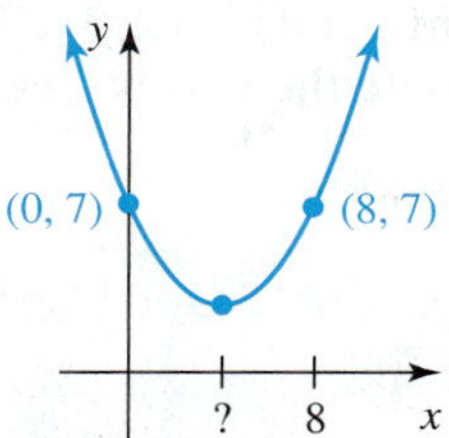

Figure 32 Problem 1

2. Fig. 33

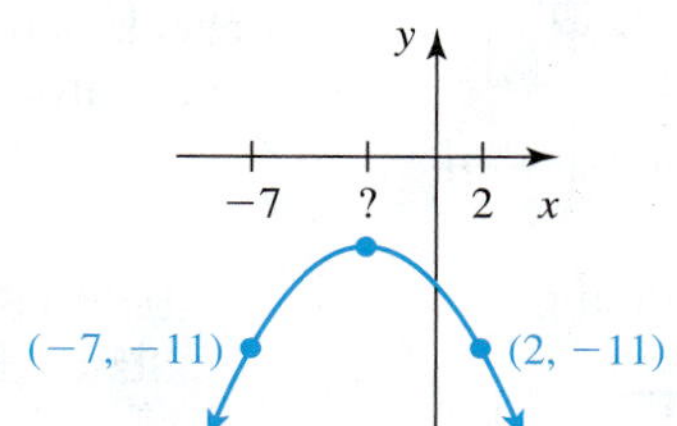

Figure 33 Problem 2

Solution

1. The x-coordinates of the symmetric points $(0, 7)$ and $(8, 7)$ are 0 and 8, respectively. Since $\frac{0+8}{2} = 4$, the x-coordinate of the vertex is 4.
2. The x-coordinates of the symmetric points $(-7, -11)$ and $(2, -11)$ are -7 and 2, respectively. Since $\frac{-7+2}{2} = -2.5$, the x-coordinate of the vertex is -2.5.

Averaging the x-coordinates of the y-intercept and its symmetric point to find the x-coordinate of the vertex is a key step in sketching the graph of a quadratic function in standard form.

Example 2 Graphing a Quadratic Function

Sketch the graph of $g(x) = x^2 - 4x + 7$.

Solution

First, we find the y-intercept of the graph of $g(x) = x^2 - 4x + 7$:

$$g(0) = 0^2 - 4(0) + 7 = 7$$

The y-intercept is $(0, 7)$. See Fig. 34.

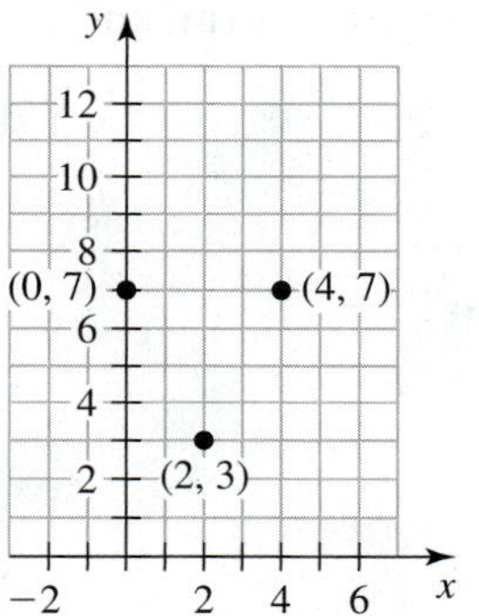

Figure 34 The vertex and two symmetric points

Next, we find the symmetric point of the y-intercept. Since symmetric points have the same height, we know the y-coordinate of the symmetric point is 7. We find the x-coordinate by substituting 7 for y in the equation $g(x) = x^2 - 4x + 7$ and solving for x:

$$\begin{aligned} 7 &= x^2 - 4x + 7 && \text{Substitute 7 for } y. \\ 0 &= x^2 - 4x && \text{Write in } 0 = ax^2 + bx \text{ form.} \\ 0 &= x(x-4) && \text{Factor right-hand side.} \\ x = 0 \quad &\text{or} \quad x - 4 = 0 && \text{Zero factor property} \\ x = 0 \quad &\text{or} \quad x = 4 \end{aligned}$$

Therefore, the points that have height 7 are $(0, 7)$ and $(4, 7)$; these are symmetric points. To find the x-coordinate of the vertex, we average the x-coordinates of the symmetric points:

$$x\text{-coordinate of vertex} = \frac{0+4}{2} = 2$$

To find the y-coordinate of the vertex, we find $g(2)$:

$$g(2) = 2^2 - 4(2) + 7 = 4 - 8 + 7 = 3$$

So, the vertex is $(2, 3)$.

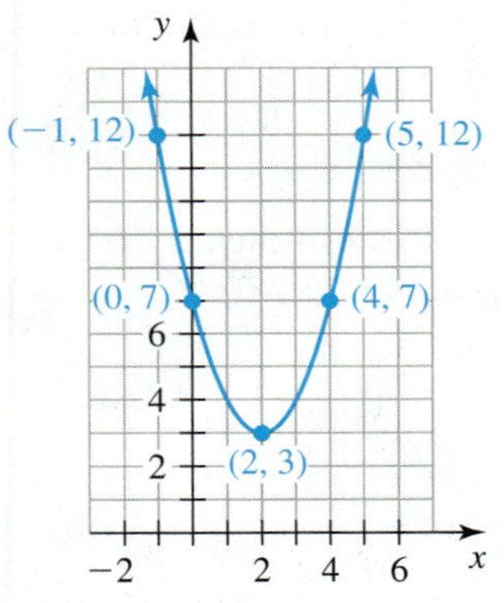

Figure 35 Graph of g

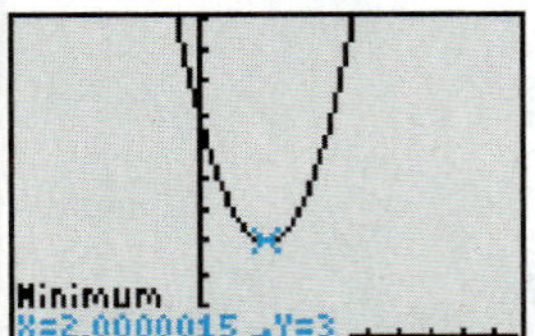

Figure 36 Verify the graph of g

We can find another pair of symmetric points on the graph by evaluating g at the values of x that are 3 units from $x = 2$ on either side—namely, at $x = -1$ and $x = 5$. At $x = 5$,

$$g(5) = 5^2 - 4(5) + 7 = 25 - 20 + 7 = 12$$

Thus, the graph passes through $(5, 12)$ and its symmetric point, $(-1, 12)$, We plot these points to assist us in sketching a graph of g (see Fig. 35).

We use a graphing calculator to verify our sketch (see Fig. 36). In particular, we check that the vertex (minimum point) is $(2, 3)$. The "minimum" choice on a graphing calculator will closely approximate the vertex (see Appendix A.19).

Using Symmetric Points to Graph a Quadratic Function in Standard Form (Method 1)

To sketch a graph of a quadratic function $f(x) = ax^2 + bx + c$, where $b \neq 0$,

1. Find the y-intercept.
2. Find the y-intercept's symmetric point.
3. Average the x-coordinates of the two symmetric points to find the x-coordinate of the vertex.
4. Find the y-coordinate of the vertex.
5. Depending on how accurate your sketch is to be, find additional points on the parabola, as needed.
6. Sketch a parabola that contains the points found.

We use these steps to graph a quadratic function $f(x) = ax^2 + bx + c$, where $b \neq 0$. If $b = 0$, the y-intercept is the vertex and, therefore, does not have a symmetric point. In this case, however, $f(x) = ax^2 + 0x + c = ax^2 + c$ is in vertex form, and we can readily use the methods of Section 9.1 to sketch the graph.

Example 3 Graphing a Quadratic Function

Sketch the graph of $f(x) = -0.9x^2 - 5.8x - 5.7$.

Solution

We find the y-intercept by finding $f(0)$:

$$f(0) = -0.9(0)^2 - 5.8(0) - 5.7 = -5.7$$

So, the y-intercept is $(0, -5.7)$.

Next, we find the symmetric point of the y-intercept. We substitute -5.7 for $f(x)$ and solve for x:

$$\begin{aligned}
-5.7 &= -0.9x^2 - 5.8x - 5.7 && \text{Substitute } -5.7 \text{ for } f(x). \\
0 &= -0.9x^2 - 5.8x && \text{Write in } 0 = ax^2 + bx \text{ form.} \\
0 &= -x(0.9x + 5.8) && \text{Factor right-hand side.} \\
-x = 0 \quad &\text{or} \quad 0.9x + 5.8 = 0 && \text{Zero factor property} \\
x = 0 \quad &\text{or} \quad 0.9x = -5.8 \\
x = 0 \quad &\text{or} \quad x = \frac{-5.8}{0.9} \\
& \qquad\quad\; x \approx -6.44
\end{aligned}$$

So, $(0, -5.7)$ and $(-6.44, -5.7)$ are approximate symmetric points. The approximate symmetric point of the y-intercept $(0, -5.7)$ is $(-6.44, -5.7)$.

We find the approximate x-coordinate of the vertex by averaging the x-coordinates of the points $(0, -5.7)$ and $(-6.44, -5.7)$:

$$\frac{0 + (-6.44)}{2} = -3.22$$

We find the approximate y-coordinate of the vertex by computing $f(-3.22)$:

$$f(-3.22) = -0.9(-3.22)^2 - 5.8(-3.22) - 5.7 \approx 3.64$$

So, the vertex is approximately $(-3.22, 3.64)$. See Fig. 37.

Although we could find and plot additional points, we can sketch a fairly accurate graph of f from the three points already found (see Fig. 38). We use a graphing calculator to verify our sketch (see Fig. 39). In particular, we check that the approximate vertex (maximum point) is $(-3.22, 3.64)$.

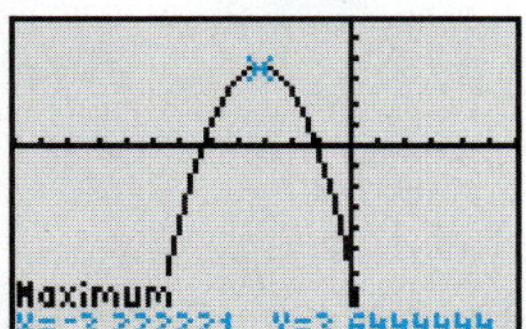

Figure 39 Verify the graph of f

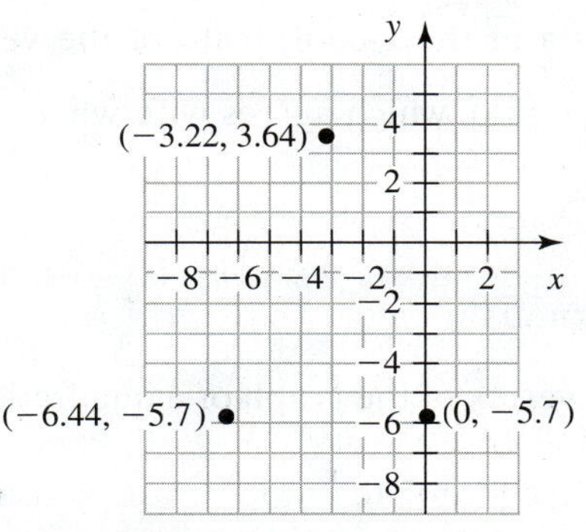

Figure 37 The vertex and two symmetric points

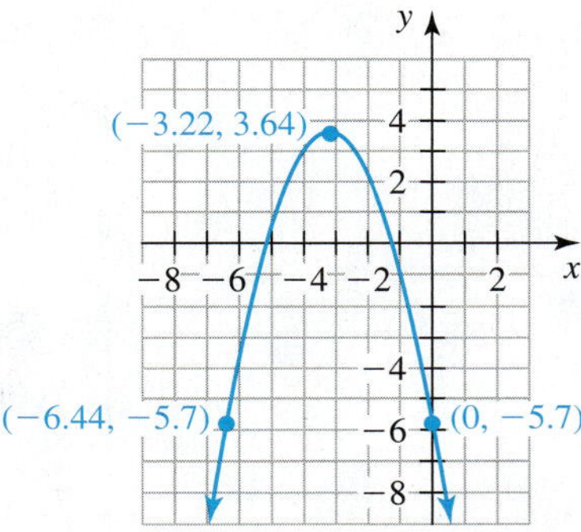

Figure 38 Graph of f from three points

Method 2: Graphing by Using the Vertex Formula to Find the Vertex

We can find a formula of the x-coordinate of the vertex of a parabola $f(x) = ax^2 + bx + c$, where $a \neq 0$. If $b = 0$, the vertex is the y-intercept, so the x-coordinate of the vertex is 0.

If $b \neq 0$, then we can find the x-coordinate of the vertex from the first three steps of the graphing method (method 1) we have been using in this section. To begin, we find the y-intercept by finding $f(0)$:

$$f(0) = a(0)^2 + b(0) + c = c$$

The y-intercept is $(0, c)$.

Next, we find the symmetric point of the y-intercept by substituting c for $f(x)$ and solving for x:

$$\begin{aligned}
c &= ax^2 + bx + c && \textit{Substitute c for f(x).}\\
0 &= ax^2 + bx && \textit{Subtract c from both sides.}\\
0 &= x(ax + b) && \textit{Factor right-hand side.}\\
x = 0 \quad &\text{or} \quad ax + b = 0 && \textit{Zero factor property}\\
x = 0 \quad &\text{or} \quad ax = -b && \textit{Subtract b from both sides.}\\
x = 0 \quad &\text{or} \quad x = -\frac{b}{a} && \textit{Divide both sides by a.}
\end{aligned}$$

The x-coordinate of the symmetric point is $-\frac{b}{a}$.

We find the x-coordinate of the vertex by averaging the x-coordinates of the y-intercept and symmetric point:

$$x = \frac{0 + \left(-\frac{b}{a}\right)}{2} = \frac{-\frac{b}{a}}{2} \qquad 0 + d = d$$

$$= -\frac{b}{a} \div 2 \qquad \frac{Q}{R} = Q \div R$$

$$= -\frac{b}{a} \cdot \frac{1}{2} \qquad \text{Multiply by reciprocal of 2, which is } \frac{1}{2}.$$

$$= -\frac{b}{2a} \qquad \text{Multiply numerators and multiply denominators.}$$

So, the formula of the x-coordinate of the vertex is $x = -\frac{b}{2a}$. If $b = 0$, this formula gives $x = -\frac{0}{2a} = 0$, which agrees with what we said earlier. So, the formula works for any value of b.

Vertex Formula

To find the vertex of the graph of a quadratic function $f(x) = ax^2 + bx + c$,

1. Find the x-coordinate of the vertex by using the **vertex formula** $x = -\frac{b}{2a}$.
2. Find the y-coordinate of the vertex by evaluating f at the value found in step 1. That is, find $f\left(-\frac{b}{2a}\right)$.

In short, the vertex is $\left(-\frac{b}{2a}, f\left(-\frac{b}{2a}\right)\right)$.

In Example 4, we will use the vertex formula to find the vertex of the parabola we sketched in Example 2.

Example 4 Using the Vertex Formula to Find the Vertex

Find the vertex of the graph of $g(x) = x^2 - 4x + 7$.

Solution

Comparing $g(x) = x^2 - 4x + 7$ with $f(x) = ax^2 + bx + c$, we see that $a = 1$ and $b = -4$. We substitute these values of a and b into the formula $x = -\frac{b}{2a}$:

$$x = -\frac{-4}{2(1)} = 2$$

The x-coordinate of the vertex is 2. To find the y-coordinate, we find $f(2)$:

$$f(2) = 2^2 - 4(2) + 7 = 3$$

So, the vertex is $(2, 3)$, as we saw in Example 2.

In general, for any quadratic function of the form $f(x) = ax^2 + bx + c$, where $b \neq 0$, both methods we have discussed for finding the vertex give the same result.

Table 17 Input–Output Pairs of $f(x) = 2x^2 + 10x + 7$

x	$f(x)$
−4	−1
−3	−5
−2	−5
−1	−1

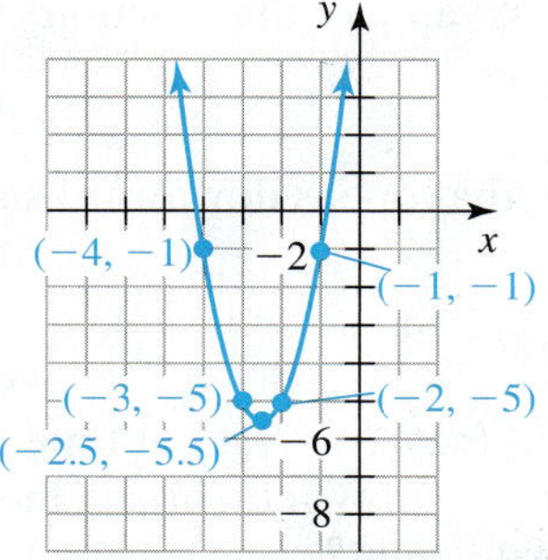

Figure 40 Graph of f from five points

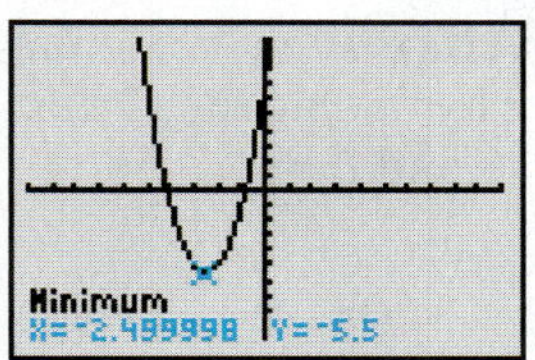

Figure 41 Verify the graph of f

Example 5 Using the Vertex Formula to Graph a Quadratic Function

Sketch the graph of $f(x) = 2x^2 + 10x + 7$.

Solution

To find the x-coordinate of the vertex, we substitute 2 for a and 10 for b in the vertex formula $x = -\dfrac{b}{2a}$:

$$x = -\frac{10}{2(2)} = -2.5$$

The x-coordinate of the vertex is -2.5. To find the y-coordinate, we find $f(-2.5)$:

$$f(-2.5) = 2(-2.5)^2 + 10(-2.5) + 7 = -5.5$$

The vertex is $(-2.5, -5.5)$.

We find additional input–output pairs of f in Table 17. Notice that we select integer input values that are close to the vertex's x-coordinate, -2.5. We plot these found points and then sketch a parabola that contains them (see Fig. 40).

We use a graphing calculator to verify our sketch (see Fig. 41). In particular, we check that the approximate vertex (minimum point) is $(-2.5, -5.5)$.

Example 6 Using the Vertex Formula to Graph a Quadratic Function

Sketch the graph of $f(x) = -2.2x^2 + 6.1x + 1.4$.

Solution

To find the x-coordinate of the vertex, we substitute -2.2 for a and 6.1 for b in the vertex formula $x = -\dfrac{b}{2a}$:

$$x = -\frac{6.1}{2(-2.2)} \approx 1.39$$

The x-coordinate of the vertex is about 1.39. To find the y-coordinate, we find $f(1.39)$:

$$f(1.39) = -2.2(1.39)^2 + 6.1(1.39) + 1.4 \approx 5.63$$

The approximate vertex is $(1.39, 5.63)$.

We find additional input–output pairs of f in Table 18. We plot these found points and then sketch a parabola that contains them (see Fig. 42).

Table 18 Input–Output Pairs of $f(x) = -2.2x^2 + 6.1x + 1.4$

x	$f(x)$
0	1.4
1	5.3
2	4.8
3	−0.1

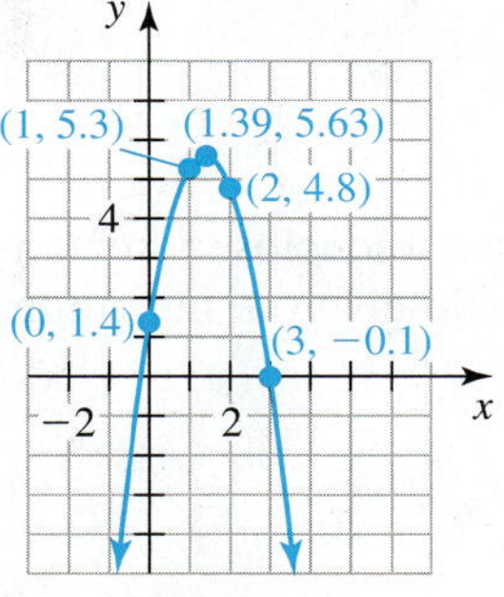

Figure 42 Graph of f using five points

Minimum or Maximum Value

Earlier in this section, we located the vertex to help ourselves graph quadratic functions in standard form. Recall from Section 7.1 that the vertex is also the minimum or maximum point of the parabola. If the vertex is the maximum point, we call the y-coordinate of the vertex the **maximum value** of the function (see Fig. 43). If the vertex is the minimum point, we call the y-coordinate of the vertex the **minimum value** of the function (see Fig. 44).

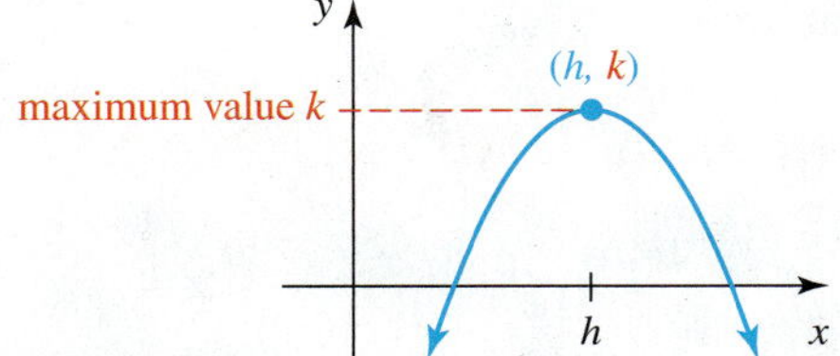

Figure 43 Quadratic function with a maximum value k but no minimum value

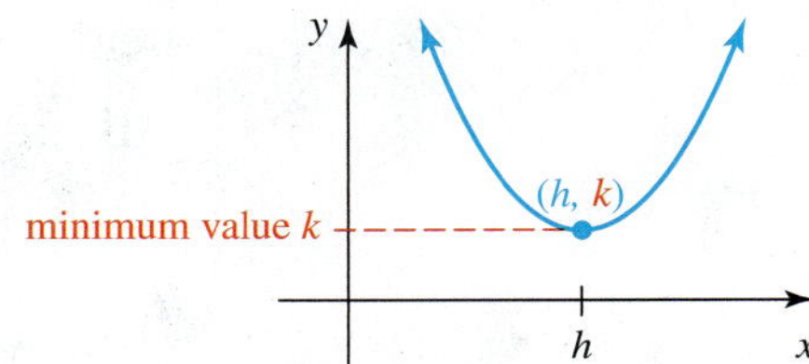

Figure 44 Quadratic function with a minimum value k but no maximum value

If we convert a quadratic function in standard form, $f(x) = ax^2 + bx + c$, into vertex form, $f(x) = a(x - h)^2 + k$, the constant a has the same value for both forms.

So, what we have determined about the graphical significance of a for quadratic functions in vertex form also applies to quadratic functions in standard form.

Maximum or Minimum Value of a Function

For a quadratic function $f(x) = ax^2 + bx + c$ whose graph has vertex (h, k),

- If $a < 0$, then the parabola opens downward and the maximum value of the function is k (see Fig. 43).
- If $a > 0$, then the parabola opens upward and the minimum value of the function is k (see Fig. 44).

In Example 7, we will find the vertex of a quadratic model that opens downward so that we can then estimate the maximum value of a quantity.

Example 7 Finding the Maximum Value of a Quantity

A fireworks shell is launched into the air. The shell's height (in feet) $h = f(t)$ at time t seconds is modeled well by the function $f(t) = -16t^2 + 200t + 1$. When should the shell explode so it goes off at the maximum height? What is that height?

Solution

The function f is of the form $f(t) = at^2 + bt + c$, with $a = -16$ (a negative number), so the graph of f is a parabola that opens downward. Therefore, the vertex is the maximum point. We can find the time the shell will reach the maximum height by finding the vertex.

To find the t-coordinate of the vertex, we substitute $a = -16$ and $b = 200$ in the vertex formula $t = -\dfrac{b}{2a}$:

$$t = -\frac{200}{2(-16)} = 6.25$$

The t-coordinate is 6.25. To find the h-coordinate, we evaluate f at 6.25:

$$f(6.25) = -16(6.25)^2 + 200(6.25) + 1 = 626$$

So, the vertex is $(6.25, 626)$. We verify our work by using "maximum" on a graphing calculator (see Fig. 45). For graphing calculator instructions, see Appendix A.19.

Figure 45 Verify that the vertex is (6.25, 626)

A vertex of $(6.25, 626)$ means the shell will reach a maximum height of 626 feet at 6.25 seconds and should explode then.

The function $f(t) = -16t^2 + 200t + 1$ describes the shell's height in terms of time, *not* in terms of horizontal distance. So, the graph of f does *not* describe the path of the shell. Its path can be modeled by a quadratic function that describes the height in terms of horizontal distance.

Example 8 Finding the Maximum Value of a Quantity

A person plans to use 60 feet of fencing and the side of his house to enclose a rectangular garden (see Fig. 46). What dimensions of the rectangle would give the maximum area? What is that area?

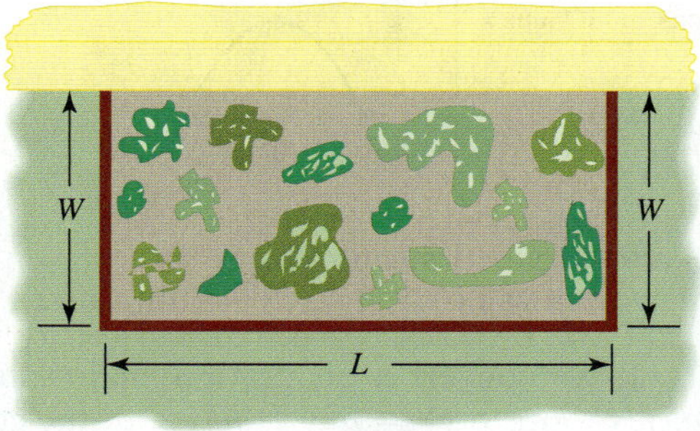

Figure 46 A garden

Solution

We use the five-step problem-solving method of Section 6.5, but in step 2 we use a system of equations to build a function and in step 3 we find the maximum point of the graph of the function.

Step 1: Define each variable. We let W be the width (in feet), L be the length (in feet), and A be the area (in square feet) of the rectangle (see Fig. 46).

Step 2: Use a system of two equations to build a function. No fencing is needed along the house, so the 60 feet of fencing can be used for the three sides of length W, W, and L:

$$W + W + L = 60$$

To obtain our first equation, we combine like terms on the left side:

$$2W + L = 60$$

We use the area formula, $A = LW$, as our second equation. The system is

$$2W + L = 60 \quad \textit{Equation (1)}$$
$$A = LW \quad \textit{Equation (2)}$$

Next, we will build a function that describes the area of the rectangle as a function of the width. To begin, we solve equation (1) for L:

$$2W + L = 60 \quad \textit{Equation (1)}$$
$$L = -2W + 60 \quad \textit{Subtract 2W from both sides.}$$

Then we substitute $-2W + 60$ for L in equation (2):

$$A = LW \quad \textit{Equation (2)}$$
$$A = (-2W + 60)W \quad \textit{Substitute } -2W + 60 \textit{ for } L.$$
$$A = -2W^2 + 60W \quad \textit{Distributive law}$$

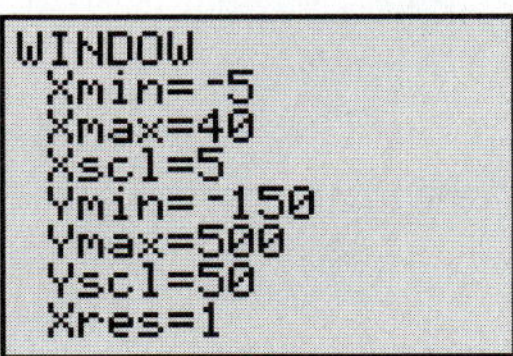

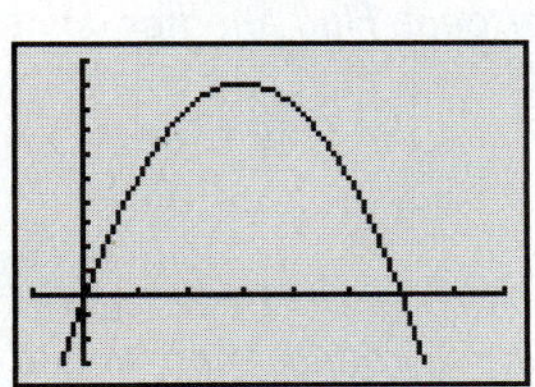

Figure 47 Graph of $A = -2W^2 + 60W$

Step 3: Find the maximum point of the graph of the function. The function $A = -2W^2 + 60W$ is of the form $A = aW^2 + bW + c$, where $a = -2$ (a negative number), so the graph is a parabola that opens downward (see Fig. 47). Therefore, the vertex is the maximum point. We can find the width of the rectangle with maximum area by finding the vertex.

To find the W-coordinate of the vertex, we substitute $a = -2$ and $b = 60$ in the vertex formula $W = -\frac{b}{2a}$:

$$W = -\frac{60}{2(-2)} = 15$$

To find the A-coordinate of the vertex, we substitute 15 for W in the equation $A = -2W^2 + 60W$:

$$A = -2(15)^2 + 60(15) = 450$$

To find the length of the rectangle, we substitute 15 for W in the equation $L = -2W + 60$:

$$L = -2(15) + 60 = 30$$

Step 4: Describe each result. The width is 15 feet, the length is 30 feet, and the area is 450 square feet.

Step 5: Check. The amount of fencing is $15 + 15 + 30 = 60$ feet, which checks. To check that the largest area is 450 square feet, we find the area of a couple of other rectangles that would involve 60 feet of fencing (see Fig. 48 on the next page).

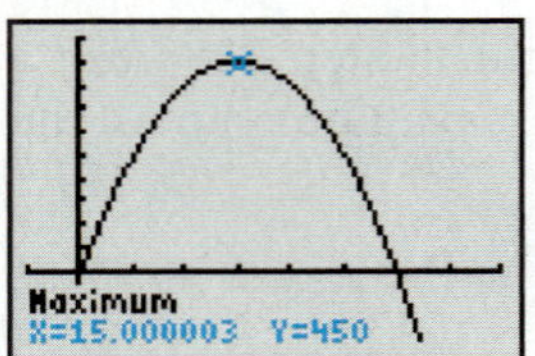

Figure 49 Verify that the maximum point is (15, 450)

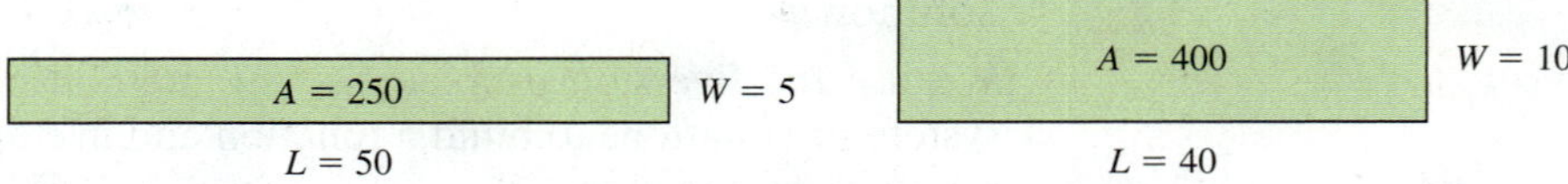

Figure 48 Some other rectangles involving 60 feet of fencing

Since 450 square feet is larger than the area of either of the two rectangles in Fig. 48, it seems reasonable that our work is correct. Finally, we use "maximum" on a graphing calculator to check that the maximum point of the graph of $A = -2W^2 + 60W$ is $(15, 450)$. See Fig. 49.

Group Exploration

Comparing methods of graphing quadratic functions

1. a. Graph the function $f(x) = 2(x - 3)^2 - 5$ by using the method discussed in Section 9.1.
   b. Simplify the right-hand side of the equation $f(x) = 2(x - 3)^2 - 5$.
   c. Graph the equation you found in part (b) by using two symmetric points to find the vertex (method 1).
   d. Graph the equation you found in part (b) by using the vertex formula to find the vertex (method 2).
   e. Compare your graphs from parts (a), (c), and (d).
2. a. Graph the function $g(x) = x^2 + 10x + 25$ by using two symmetric points to find the vertex.
   b. Graph g by using the vertex formula to find the vertex.
   c. Factor the right-hand side of the equation $g(x) = x^2 + 10x + 25$.
   d. Use the method discussed in Section 9.1 to graph the equation you found in part (c).
   e. Compare your graphs from parts (a), (b), and (d).

Homework 9.2

For extra help ▶ MyMathLab® Watch the videos in MyMathLab  Download the MyDashboard App

Find the x-coordinate of the vertex of the parabola sketched in the figure.

1. Fig. 50

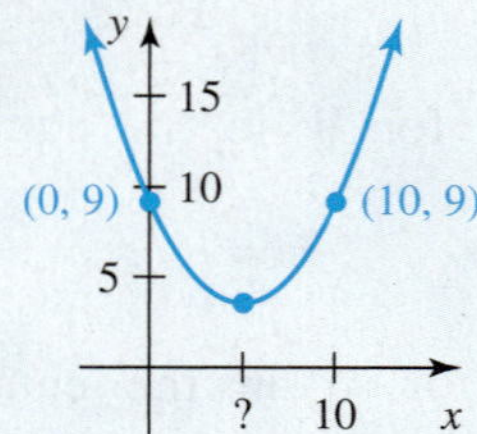

Figure 50 Exercise 1

2. Fig. 51

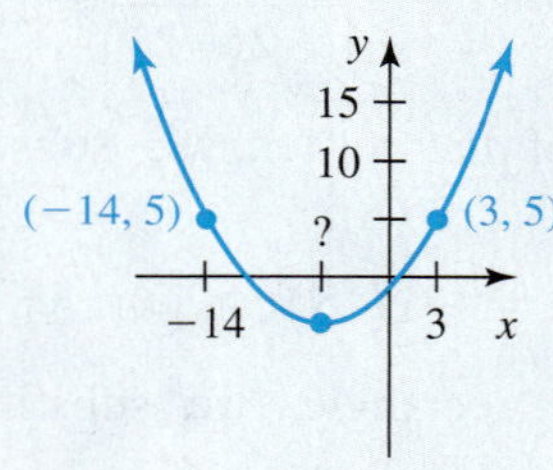

Figure 51 Exercise 2

Find the x-coordinate of the vertex of a parabola that passes through the given points. Round approximate results to the second decimal place.

3. $(0, 8)$ and $(6, 8)$
4. $(0, 5)$ and $(8, 5)$
5. $(0, -3)$ and $(-7, -3)$
6. $(0, -6)$ and $(-9, -6)$
7. $(0, 2)$ and $(7.29, 2)$
8. $(0, -7)$ and $(15.37, -7)$

A parabola has the given vertex and y-intercept. Find another point on the parabola.

9. vertex $(2, 5)$ and y-intercept $(0, 9)$
10. vertex $(-3, -8)$ and y-intercept $(0, 4)$

Graph the function by hand. Find the vertex; round approximate coordinates to the second decimal place. Verify your sketch by using a graphing calculator.

11. $y = x^2 - 6x + 7$
12. $y = x^2 - 4x + 5$
13. $y = x^2 + 8x + 9$
14. $y = x^2 + 2x - 7$
15. $y = -x^2 + 8x - 10$
16. $y = -x^2 + 6x - 9$
17. $y = 3x^2 + 6x - 4$
18. $y = 2x^2 - 4x + 1$
19. $y = -3x^2 + 12x - 5$
20. $y = -2x^2 - 8x - 3$
21. $y = -4x^2 - 9x - 5$
22. $y = -2x^2 + 5x + 3$
23. $y = 2x^2 - 7x + 7$
24. $y = 3x^2 - 8x - 1$
25. $4x^2 - y + 6 = 8x$
26. $6x^2 = 3y - 24x - 15$
27. $y = 2.8x^2 - 8.7x + 4$
28. $y = -1.6x^2 - 4.8x + 3$
29. $y = 3.9x^2 + 6.9x - 3.4$
30. $y = -2.4x^2 + 6.1x - 7.8$

31. $3.6y - 26.3x = 8.3x^2 - 7.1$

32. $5.3 - 2.1y = 9.8x^2 - 3.4x + 8.3$

A parabola has the given x-intercepts. What is the x-coordinate of the vertex? Write your result in decimal form.

33. $(2, 0)$ and $(6, 0)$ **34.** $(-4, 0)$ and $(3, 0)$

35. $(-9, 0)$ and $(4, 0)$ **36.** $(-7, 0)$ and $(-3, 0)$

Find the x-intercepts and y-intercept. Next, find the vertex. Write the coordinates in decimal form. Then graph the function by hand. Verify your result by using a graphing calculator.

37. $y = 5x^2 - 10x$ **38.** $y = 2x^2 - 8x$

39. $y = -2x^2 + 6x$ **40.** $y = -3x^2 - 6x$

41. $y = x^2 - 10x + 24$ **42.** $y = x^2 - 4x - 5$

43. $y = x^2 - 8x + 7$ **44.** $y = x^2 - 10x + 16$

45. $y = x^2 - 9$ **46.** $y = x^2 - 1$

47. A batter hits a baseball. The ball's height (in feet) $h(t)$ after t seconds is given by $h(t) = -16t^2 + 140t + 3$.

a. What is the ball's height when the batter makes contact with it?

b. What is the maximum height of the ball? When does it reach that height?

c. Graph h by hand.

48. A person on the edge of a cliff throws a stone so it hits the ground near the base of the cliff. The stone's height (in feet above the base) $h(t)$ after t seconds is given by $h(t) = -16t^2 + 30t + 200$.

a. Find the vertex. What does it mean in this situation?

b. Estimate the height of the cliff. State any assumptions you make.

c. Did the person throw the stone upward or downward? Explain.

49. The percentages of households with outstanding student debt are shown in Table 19 for various years.

Table 19 Percentages of Households with Outstanding Student Debt

Year	Percent
1995	12
1998	11
2001	12
2004	13
2007	15
2010	19

Source: *Pew Research Center*

Let $f(t)$ be the percentage of households with outstanding student debt at t years since 1990.

a. Use a graphing calculator to draw a scattergram to describe the data. Is it better to use a linear function or a quadratic function to model the data? Explain.

b. Use a graphing calculator to draw the graph of the model $f(t) = 0.058t^2 - 0.98t + 15.43$ and, in the same viewing window, the scattergram of the data. Does the model fit the data well?

c. Predict the percentage of households that will have outstanding student debt in 2017.

d. Estimate when the percentage of households with outstanding student debt was the least. What was that minimum percentage?

e. In 2009, there were about 117 million households. Estimate the *number* of households that had outstanding student debt in 2009.

50. The numbers of people who died as a result of police action or while in police custody in South Africa are shown in Table 20 for various years.

Table 20 Numbers of People Who Died as a Result of Police Action or While in Police Custody in South Africa

Year	Number of Deaths
1998	552
2000	429
2002	328
2004	370
2006	408
2008	603
2010	799

Source: *Pew Research Center*

Let $f(t)$ be the number of people who died as a result of police action or while in police custody in South Africa in the year that is t years since 1990.

a. Use a graphing calculator to draw a scattergram to describe the data. Is it better to use a linear function or a quadratic function to model the data? Explain.

b. Use a graphing calculator to draw the graph of the model $f(t) = 9.13t^2 - 235t + 1849$ and, in the same viewing window, the scattergram of the data. Does the model fit the data well?

c. Estimate the number of people who died as a result of police action or while in police custody in 2012.

d. Estimate when the number of people who died as a result of police action or while in police custody was the least. What was that number of people?

51. Americans' average annual expenditures are shown in Table 21 for various age groups.

Table 21 Americans' Average Annual Expenditures

Age Group (years)	Age Used to Represent Age Group (years)	Average Annual Expenditure (thousands of dollars)
under 25	20.0	24.3
25–34	29.5	40.3
35–44	39.5	48.3
45–54	49.5	48.7
55–64	59.5	44.3
65–74	69.5	32.2

Source: *Bureau of Labor Statistics, Consumer Expenditure Survey*

Let $f(t)$ be the average annual expenditure (in thousands of dollars) of Americans at age t years.

a. Use a graphing calculator to draw a scattergram to describe the data. Is it better to use a linear funtion or a quadratic function to model the data? Explain.

b. Use a graphing calculator to draw the graph of the model $f(t) = -0.035t^2 + 3.25t - 26.34$ and, in the same viewing window, the scattergram of the data. Does the model fit the data well?

c. Estimate the average annual expenditure of 18-year-old Americans.

d. Estimate the age of Americans with the highest average annual expenditure. What is that expenditure?

52. The percentages of Americans who buy newspapers (print or online) are shown in Table 22 for various age groups.

Table 22 Percentages of Americans Who Buy Newspapers

Age Group (years)	Age Used to Represent Age Group (years)	Percent
25–34	29.5	33
35–44	39.5	43
45–54	49.5	48
55–64	59.5	53
over 65	70	51

Source: *Bureau of Labor Statistics*

Let $f(t)$ be the percentage of Americans at age t years who buy newspapers.

a. Use a graphing calculator to draw a scattergram to describe the data. Is it better to use a linear function or a quadratic function to model the data? Explain.

b. Use a graphing calculator to draw the graph of the model $f(t) = -0.017t^2 + 2.15t - 15.57$ and, in the same viewing window, the scattergram of the data. Does the model fit the data well?

c. Estimate the percentage of 21-year-old Americans who buy newspapers.

d. Estimate the age of Americans who are most likely to buy newspapers. What percentage of these Americans buy newspapers?

e. In 2010, there were about 3.5 million 30-year-old Americans. Estimate the *number* of those Americans who bought newspapers.

53. Student-to-faculty ratios at Bates College are shown in Table 23 for various years.

Table 23 Student-to-Faculty Ratios at Bates College

Year	Student-to-Faculty Ratio
1985	13.8
1990	11.9
1995	11.0
2000	10.9
2004	10.7
2007	11.0

Source: *Bates College*

Let r be the student-to-faculty ratio at Bates College at t years since 1980. A linear model of the situation is $L(t) = -0.12t + 13.53$. A quadratic model is $Q(t) = 0.0119t^2 - 0.5t + 15.9$.

a. Use a graphing calculator to draw the graphs of both models and, in the same viewing window, the scattergram of the data. Which model better describes the situation?

b. Use Zoom Out on a graphing calculator to help you determine which model predicts the lower student-to-faculty ratios for future years.

c. Use Q to predict when the student-to-faculty ratio will be 15.9.

d. Use Q to estimate when the student-to-faculty ratio was the least. What is that ratio?

e. The enrollment in 2011 was 1769 students. Use Q to estimate the *number of faculty* in that year.

54. The numbers of recreational boating fatalities are shown in Table 24 for various years.

Table 24 Numbers of Recreational Boating Fatalities

Year	Number of Fatalities
1980	1392
1985	1104
1990	880
1995	829
2000	701
2005	697
2011	758

Source: *U.S. Coast Guard*

Let n be the number of recreational boating fatalities in the year that is t years since 1980. A linear model of the situation is $L(t) = -20t + 1212$. A quadratic model is $Q(t) = 1.24t^2 - 58.18t + 1377$.

a. Use a graphing calculator to draw the graphs of both models and, in the same viewing window, the scattergram of the data. Which model better describes the situation?

b. Use each of the two models to find two estimates of the number of fatalities in 2000. Which is the better estimate? Explain.

c. Use Zoom Out on a graphing calculator to help you determine which model predicts the lower number of fatalities for any year beyond 2006.

d. Use Q to estimate in which year the number of recreational boating fatalities was the least. What is that number of fatalities?

55. A person plans to use 80 feet of fencing to enclose a rectangular garden. What dimensions of the rectangle would give the maximum area? What is that area?

56. A person plans to use 60 feet of fencing to enclose a rectangular patio. What dimensions of the rectangle would give the maximum area? What is that area?

57. A rancher plans to use 400 feet of fencing and a side of his barn to form a rectangular boundary for cattle (see Fig. 52). What dimensions of the rectangle would give the maximum area? What is that area?

Figure 52 Exercise 57

58. A farmer plans to use 200 feet of fencing and a side of her barn to enclose a rectangular garden (see Fig. 53). What dimensions of the rectangle would give the maximum area? What is that area?

Figure 53 Exercise 58

59. Recall that we can describe some or all of the input–output pairs of a function by means of an equation, a graph, a table, or words. Let $f(x) = x^2 - 10x + 18$.

a. Describe the input–output pairs of f by using a graph.
b. Describe five input–output pairs of f by using a table.
c. Describe the input–output pairs of f by using words.

60. Recall that we can describe some or all of the input–output pairs of a function by means of an equation, a graph, a table, or words. Let $g(x) = -2x^2 + 4x + 1$.

a. Describe the input–output pairs of g by using a graph.
b. Describe five input–output pairs of g by using a table.
c. Describe the input–output pairs of g by using words.

For Exercises 61–68, refer to the graph sketched in Fig. 54.

61. Find $f(-5)$.

62. Find $f(-3)$.

63. Find x when $f(x) = 3$.

64. Find x when $f(x) = 4$.

65. Find x when $f(x) = 2$.

66. Find x when $f(x) = -1$.

67. Find the maximum value of f.

68. Find the vertex of the graph of f.

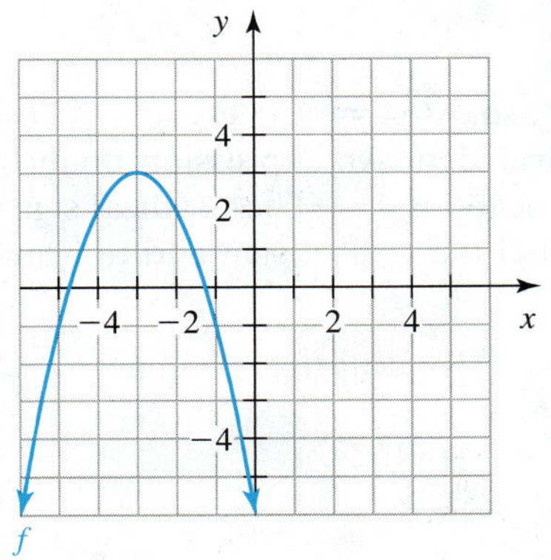

Figure 54 Exercises 61–68

Concepts

69. a. Find the x-coordinate of the vertex of the parabola $f(x) = x^2 + 4x - 12$ by averaging the x-coordinates of the y-intercept and its symmetric point.

b. Find the x-coordinate of the vertex of the parabola $f(x) = x^2 + 4x - 12$ by averaging the x-coordinates of the x-intercepts.

c. Compare the methods you used in parts (a) and (b). Are your results the same?

d. Which method from parts (a) and (b) is easier to use to find the x-coordinate of the vertex of the parabola $g(x) = 54x^2 - 195x - 216$? Explain.

e. Which method(s) can be used to find the x-coordinate of the vertex of the parabola $h(x) = x^2 + 4x + 6$? Explain.

f. Summarize your findings in this exercise.

70. In this exercise, you will discover how to convert the standard form of a quadratic function to its vertex form.

a. Find the vertex of the parabola $f(x) = 3x^2 - 6x + 7$.

b. Recall that if (h, k) is the vertex of the graph of a quadratic function, then $f(x) = a(x - h)^2 + k$. Use your result from part (a) to determine h and k for $f(x) = 3x^2 - 6x + 7$.

c. Substitute the values of h and k you found from part (b) in $f(x) = a(x - h)^2 + k$.

d. Compare $f(x) = 3x^2 - 6x + 7$ with your result in part (c). Determine the value of a. [**Hint:** You may see this immediately. If not, write your result in part (c) in standard form and compare again with $f(x) = 3x^2 - 6x + 7$.]

e. Substitute your value of a into your result from part (c).

f. Use a graphing calculator to verify your result by comparing the graph of $f(x) = 3x^2 - 6x + 7$ with the graph of $f(x) = a(x - h)^2 + k$ for the values of a, h, and k you found.

71. Input–output pairs for four quadratic functions f, g, h, and k are listed in Table 25. For each function, decide whether $(3, 2)$ is the vertex. If $(3, 2)$ is not the vertex, estimate the coordinates of the vertex.

Table 25 Values of Four Quadratic Functions

x	$f(x)$	$g(x)$	$h(x)$	$k(x)$
1	10	8.2	19.3	11.6
2	4	2.9	7.3	4.4
3	2	2	2	2
4	4	5.5	3.4	4.4
5	10	13.4	11.5	11.6

72. Suppose that $f(x) = ax^2 + bx + c$, where $a > 0$ and g is a linear function. If $f(2) = g(2)$ and $f(5) = g(5)$, which is larger, $f(4)$ or $g(4)$? Explain. [**Hint:** Think graphically.]

73. Assume the graph of a quadratic function $f(x) = ax^2 + bx + c$ has vertex (h, k). If $a < 0$, then the maximum value of f is k. Explain.

74. Assume the graph of a quadratic function $f(x) = ax^2 + bx + c$ has vertex (h, k). If $a > 0$, then the minimum value of f is k. Explain.

75. Explain how to sketch the graph of a quadratic function $f(x) = ax^2 + bx + c$, where $b \neq 0$.

76. The x-coordinate of the vertex of the graph of a quadratic function is equal to the average of the x-coordinates of two symmetric points. Explain. Include a sketch of a graph of a quadratic function.

Related Review

Graph the equation by hand. Use a graphing calculator to verify your work.

77. $y = -3(x + 3)^2 + 5$

78. $y = -3x^2 - 6x + 5$

79. $y = -3x + 5$

80. $2x - 3y - 12 = 0$

81. We saw in Section 9.1 that the vertex of the graph of $f(x) = a(x - h)^2 + k$, where $a \neq 0$, is (h, k). In this exercise, we will prove that this is true.

a. Simplify the right-hand side of $f(x) = a(x - h)^2 + k$ to show that $f(x) = ax^2 - 2ahx + c$, where $c = ah^2 + k$.

b. Use the vertex formula to show that the vertex of the graph of f is (h, k).

82. Consider the system

$$y = x^2 - x - 7$$
$$y = x + 1$$

a. Use "intersect" on a graphing calculator to solve the system. [**Hint:** There are two solutions.]

b. Use substitution to solve the system. [**Hint:** After finding the x-coordinates of the two solutions, don't forget to find the y-coordinates.]

Expressions, Equations, Functions, and Graphs

Perform the indicated instruction. Then use words such as linear, quadratic, cubic, polynomial, degree, function, one variable, *and* two variables *to describe the expression, equation, or system.*

83. Simplify $\dfrac{(3x^4y^5)^3}{18x^7y^2}$.

84. Solve $x(3x - 5) + 2 = x^2 + 5$.

85. Simplify $(3x^2 - 2y^2)^2$.

86. Find the product $-2x(3x - 4)(5x - 2)$.

87. Solve $\dfrac{at - b}{c} = d$ for t.

88. Factor $6x^3 + 3x^2 - 18x$.

9.3 Simplifying Radical Expressions

Objectives

- Know the meaning of *square root* and *principal square root.*
- Approximate a principal square root.
- Know the *product property for square roots.*
- Use the product property for square roots to simplify radicals.
- Know the *quotient property for square roots.*
- Simplify a radical expression.
- *Rationalize the denominator* of a fraction.
- Know how to simplify a radical expression completely.

So far, we have worked with polynomial expressions. In this section, we will work with a new type of expression called a *radical expression.* This work will prepare us to solve more quadratic equations in Sections 9.4–9.6.

Principal Square Roots

What number squared is equal to 9? The numbers that "work" are -3 and 3:

$$(-3)^2 = 9 \qquad 3^2 = 9$$

We say that -3 and 3 are *square roots* of 9. A **square root** of a number a is the number we square to get a.

Of the square roots of 9, only the nonnegative square root, 3, is the *principal square root* of 9. Using symbols, we write $\sqrt{9} = 3$.

Definition Principal square root

If a is a nonnegative number, then $\sqrt{a}$ is the nonnegative number we square to get a. We call $\sqrt{a}$ the **principal square root** of a.

For example, $\sqrt{25} = 5$, since $5^2 = 25$. Also, $\sqrt{64} = 8$, because $8^2 = 64$.

The symbol "$\sqrt{\ }$" is called a **radical sign.** An expression under a radical sign is called a **radicand.** For $\sqrt{4x - 7}$, the radicand is $4x - 7$. A radical sign together with a radicand is called a **radical.** Here we label the radical sign and radicand of the radical $\sqrt{2x + 5}$:

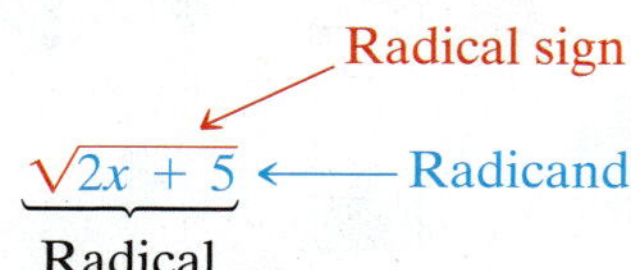

Here are some more radicals: $\sqrt{5}, \sqrt{x}, \sqrt{9x + 4}$.

An expression that contains a radical is called a **radical expression.** Here are some radical expressions:

$$\sqrt{5}, \quad \sqrt{x}, \quad \sqrt{2x-5}, \quad 4\sqrt{x+8}-8x, \quad \left(3\sqrt{x}-2\right)\left(\sqrt{x}+7\right)$$

WARNING

Is a square root of a negative number a real number? Consider $\sqrt{-9}$. Note that $\sqrt{-9} \neq -3$, because $(-3)^2$ is equal to 9, not -9. Since any number squared is nonnegative, we see that $\sqrt{-9}$ is not a real number. In general, a square root of a negative number is not a real number.

Example 1 Finding Square Roots

Find the square root.

1. $\sqrt{49}$ **2.** $\sqrt{-49}$ **3.** $-\sqrt{49}$ **4.** $-\sqrt{-49}$

Solution

1. $\sqrt{49} = 7$, because $7^2 = 49$.
2. $\sqrt{-49}$ is not a real number, because the radicand -49 is negative.
3. $-\sqrt{49} = -7$.
4. $-\sqrt{-49}$ is not a real number, because the radicand -49 is negative.

Rational and Irrational Square Roots

Table 26 Perfect Squares

x	Perfect Square x^2
0	0
1	1
2	4
3	9
4	16
5	25
6	36
7	49
8	64
9	81
10	100
11	121
12	144
13	169
14	196
15	225

Recall from Section 1.1 that a rational number is a number that can be written in the form $\frac{n}{d}$, where n and d are integers and d is nonzero. A **perfect square** is a number whose principal square root is rational. For example, 25 is a perfect square, because $\sqrt{25} = 5 = \frac{5}{1}$ is rational. By squaring the integers from 0 to 15, we can find the integer perfect squares between 0 and 225, inclusive (see Table 26). You should memorize the perfect squares shown in this table, because you will work with them again and again.

For a number that is not a perfect square, any square root of the number is not rational. Recall from Section 1.1 that we call such a number *irrational.* For example, $\sqrt{7}$ is irrational. We know that $\sqrt{7}$ is a number between 2 and 3, because $2^2 = 4$ and $3^2 = 9$. To use a calculator to get the estimate $\sqrt{7} \approx 2.645751311$, press [2nd] [√‾] 7 [)] [ENTER] (see Fig. 55).

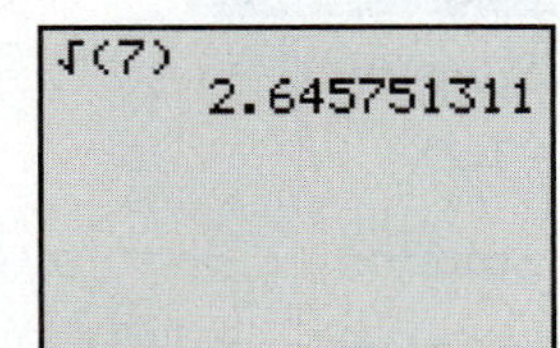

Figure 55 Estimating $\sqrt{7}$

Example 2 Approximating Square Roots

State whether the square root is rational or irrational. If the square root is rational, find the (exact) value. If the square root is irrational, estimate its value by rounding to the second decimal place.

1. $\sqrt{19}$ **2.** $\sqrt{169}$

Solution

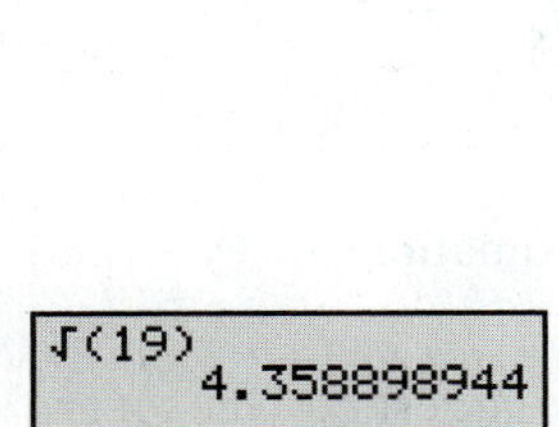

Figure 56 Estimating $\sqrt{19}$

1. The number 19 is not a perfect square, so $\sqrt{19}$ is irrational. We use a calculator to compute $\sqrt{19} \approx 4.36$ (see Fig. 56).
2. The number 169 is a perfect square, so $\sqrt{169}$ is rational. In fact, $\sqrt{169} = 13$, because $13^2 = 169$.

Product Property for Square Roots

Consider the following computations with radicals:

$$\sqrt{4 \cdot 9} = \sqrt{36} = 6$$
$$\sqrt{4}\sqrt{9} = 2 \cdot 3 = 6$$

Since both results are equal, we can write

$$\sqrt{4 \cdot 9} = \sqrt{4}\sqrt{9}$$

This equation suggests the following **product property for square roots,** which can help us work with the principal square root of a product.

> **Product Property for Square Roots**
>
> For nonnegative numbers a and b,
>
> $$\sqrt{ab} = \sqrt{a}\sqrt{b}$$
>
> In words: The square root of a product is the product of the square roots.

For example, $\sqrt{16 \cdot 49} = \sqrt{16}\sqrt{49} = 4 \cdot 7 = 28$.

Using the Product Property for Square Roots to Simplify Radicals

We can sometimes write a radical as an expression with a smaller radicand. For example, consider $\sqrt{24}$. Note that the perfect square 4 is a factor of 24: $4 \cdot 6 = 24$. We can use the product property for square roots to write

$$\sqrt{24} = \sqrt{4 \cdot 6} = \sqrt{4}\sqrt{6} = 2\sqrt{6}$$

When writing an expression such as $2\sqrt{6}$, we write the "2" to the left of "$\sqrt{6}$." If we were to write the expression as $\sqrt{6}2$, the radicand might appear to be 62, not 6.

A square root is **simplified** when the radicand does not have any perfect-square factors other than 1.

Example 3 Simplifying Radicals

Simplify.

1. $\sqrt{12}$ **2.** $\sqrt{32}$ **3.** $7\sqrt{45}$

Solution

```
√(12)
        3.464101615
2√(3)
        3.464101615
```

Figure 57 Verify the work

1. Note that 4 is the largest perfect-square factor of 12. We write 12 as a product of 4 and 3 and then apply the product property for square roots:

$$\begin{aligned} \sqrt{12} &= \sqrt{4 \cdot 3} && \text{4 is the largest perfect-square factor.} \\ &= \sqrt{4}\sqrt{3} && \text{Product property: } \sqrt{ab} = \sqrt{a}\sqrt{b} \\ &= 2\sqrt{3} && \sqrt{4} = 2 \end{aligned}$$

We verify our work by using a calculator to compare approximations for $\sqrt{12}$ and $2\sqrt{3}$ (see Fig. 57).

2. Both 4 and 16 are perfect-square factors of 32. Since 16 is the largest perfect-square factor, we write 32 as the product of 16 and 2 and then apply the product property for square roots:

$$\begin{aligned} \sqrt{32} &= \sqrt{16 \cdot 2} && \text{16 is the largest perfect-square factor.} \\ &= \sqrt{16}\sqrt{2} && \text{Product property: } \sqrt{ab} = \sqrt{a}\sqrt{b} \\ &= 4\sqrt{2} && \sqrt{16} = 4 \end{aligned}$$

3. $7\sqrt{45} = 7\sqrt{9 \cdot 5}$ *9 is the largest perfect-square factor.*

$= 7\sqrt{9}\sqrt{5}$ *Product property:* $\sqrt{ab} = \sqrt{a}\sqrt{b}$

$= 7 \cdot 3 \cdot \sqrt{5}$ $\sqrt{9} = 3$

$= 21\sqrt{5}$ *Multiply.*

In Problem 2 of Example 3, we used the largest perfect-square factor, 16, of the radicand 32 to simplify $\sqrt{32}$. What would happen if we used the smaller perfect-square factor, 4?

$$\sqrt{32} = \sqrt{4 \cdot 8} = \sqrt{4}\sqrt{8} = 2\sqrt{8}$$

The radicand, 8, has a perfect-square factor of 4, so we continue to simplify:

$$\sqrt{32} = 2\sqrt{8} = 2\sqrt{4 \cdot 2} = 2\sqrt{4}\sqrt{2} = 2 \cdot 2\sqrt{2} = 4\sqrt{2}$$

The result is the same as our result in Example 3. This exploration suggests two things:

- The most efficient way to simplify a square root is to use the *largest* perfect-square factor of the radicand.
- If we don't use the largest perfect-square factor of the radicand, we can simplify the square root by continuing to use perfect-square factors until the radicand has no perfect-square factors other than 1.

Simplifying a Square Root with a Small Radicand

To simplify a square root in which it is easy to determine the largest perfect-square factor of the radicand,

1. Write the radicand as the product of the *largest* perfect-square factor and another number.
2. Apply the product property for square roots.

In Example 4, we will discuss how to simplify a radical that has a large radicand.

Example 4 Simplifying a Radical with a Large Radicand

Simplify $\sqrt{72}$.

Solution

We begin by finding the prime factorization of the radicand, 72:

$\sqrt{72} = \sqrt{2 \cdot 2 \cdot 2 \cdot 3 \cdot 3}$ *Find prime factorization of 72.*

$= \sqrt{2 \cdot 2 \cdot 3 \cdot 3 \cdot 2}$ *Rearrange factors to highlight pairs of identical factors.*

$= \sqrt{2 \cdot 2} \cdot \sqrt{3 \cdot 3} \cdot \sqrt{2}$ *Product property:* $\sqrt{ab} = \sqrt{a}\sqrt{b}$

$= 2 \cdot 3 \cdot \sqrt{2}$ $\sqrt{2 \cdot 2} = \sqrt{4} = 2;\ \sqrt{3 \cdot 3} = \sqrt{9} = 3$

$= 6\sqrt{2}$ *Multiply.*

Simplifying a Square Root with a Large Radicand

To simplify a square root in which it is difficult to determine the largest perfect-square factor of the radicand,

1. Find the prime factorization of the radicand.
2. Look for pairs of identical factors of the radicand and rearrange the factors to highlight them.
3. Use the product property for square roots.

Quotient Property for Square Roots

Here we compute the square root of a quotient and a quotient of square roots:

Square root of a quotient: $\sqrt{\frac{16}{49}} = \frac{4}{7}$ because $\left(\frac{4}{7}\right)^2 = \frac{16}{49}$

Quotient of square roots: $\frac{\sqrt{16}}{\sqrt{49}} = \frac{4}{7}$

Since the two results are equal, we can write

$$\sqrt{\frac{16}{49}} = \frac{\sqrt{16}}{\sqrt{49}}$$

This equation suggests the **quotient property for square roots.**

Quotient Property for Square Roots

For a nonnegative number a and a positive number b,

$$\sqrt{\frac{a}{b}} = \frac{\sqrt{a}}{\sqrt{b}}$$

In words: The square root of a quotient is the quotient of the square roots.

Using the Quotient Property for Square Roots to Simplify Radicals

If a radical has a fractional radicand, we **simplify the radical** by writing it as an expression whose radicand is not a fraction.

Example 5 Simplifying Radicals

Simplify.

1. $\sqrt{\frac{5}{9}}$ 2. $\sqrt{\frac{50}{81}}$

Solution

1. $\sqrt{\frac{5}{9}} = \frac{\sqrt{5}}{\sqrt{9}}$ *Quotient property:* $\sqrt{\frac{a}{b}} = \frac{\sqrt{a}}{\sqrt{b}}$

$= \frac{\sqrt{5}}{3}$ $\sqrt{9} = 3$

We verify our work by using a graphing calculator to compare approximations for $\sqrt{\frac{5}{9}}$ and $\frac{\sqrt{5}}{3}$ (see Fig. 58).

```
√(5/9)
          .7453559925
√(5)/3
          .7453559925
```

Figure 58 Verify the work

2. $\sqrt{\frac{50}{81}} = \frac{\sqrt{50}}{\sqrt{81}}$ *Quotient property:* $\sqrt{\frac{a}{b}} = \frac{\sqrt{a}}{\sqrt{b}}$

$= \frac{\sqrt{25 \cdot 2}}{9}$ *25 is a perfect square;* $\sqrt{81} = 9$

$= \frac{\sqrt{25}\sqrt{2}}{9}$ *Product property:* $\sqrt{ab} = \sqrt{a}\sqrt{b}$

$= \frac{5\sqrt{2}}{9}$ $\sqrt{25} = 5$

Rationalizing the Denominator of a Radical Expression

We simplify an expression of the form $\frac{p}{\sqrt{q}}$ by leaving no denominator as a radical expression. We call this process **rationalizing the denominator.**

Example 6 Rationalizing the Denominator

Simplify.

1. $\frac{5}{\sqrt{7}}$ 2. $\frac{7}{\sqrt{18}}$

Solution

1. $\frac{5}{\sqrt{7}} = \frac{5}{\sqrt{7}} \cdot 1$ — $a = a \cdot 1$

$= \frac{5}{\sqrt{7}} \cdot \frac{\sqrt{7}}{\sqrt{7}}$ — *Rationalize denominator.*

$= \frac{5\sqrt{7}}{\sqrt{49}}$ — *Product property:* $\sqrt{a}\sqrt{b} = \sqrt{ab}$

$= \frac{5\sqrt{7}}{7}$ — $\sqrt{49} = 7$

2. $\frac{7}{\sqrt{18}} = \frac{7}{3\sqrt{2}}$ — $\sqrt{18} = \sqrt{9 \cdot 2} = \sqrt{9}\sqrt{2} = 3\sqrt{2}$

$= \frac{7}{3\sqrt{2}} \cdot \frac{\sqrt{2}}{\sqrt{2}}$ — *Rationalize denominator.*

$= \frac{7\sqrt{2}}{3\sqrt{4}}$ — *Product property:* $\sqrt{a}\sqrt{b} = \sqrt{ab}$

$= \frac{7\sqrt{2}}{6}$ — $3\sqrt{4} = 3 \cdot 2 = 6$

As shown in Example 6, **to rationalize the denominator of a fraction of the form $\frac{p}{\sqrt{q}}$, where q is positive, we multiply the fraction by 1 in the form $\frac{\sqrt{q}}{\sqrt{q}}$.**

Example 7 Rationalizing the Denominator

Simplify.

1. $\sqrt{\frac{7}{3}}$ 2. $\sqrt{\frac{3}{20}}$

Solution

1. We first use the quotient property for square roots and then rationalize the denominator.

$\sqrt{\frac{7}{3}} = \frac{\sqrt{7}}{\sqrt{3}}$ — *Quotient property:* $\sqrt{\frac{a}{b}} = \frac{\sqrt{a}}{\sqrt{b}}$

$= \frac{\sqrt{7}}{\sqrt{3}} \cdot \frac{\sqrt{3}}{\sqrt{3}}$ — *Rationalize denominator.*

$= \frac{\sqrt{21}}{\sqrt{9}}$ — *Product property:* $\sqrt{a}\sqrt{b} = \sqrt{ab}$

$= \frac{\sqrt{21}}{3}$ — $\sqrt{9} = 3$

2. $\sqrt{\frac{3}{20}} = \frac{\sqrt{3}}{\sqrt{20}}$ *Quotient property:* $\sqrt{\frac{a}{b}} = \frac{\sqrt{a}}{\sqrt{b}}$

$= \frac{\sqrt{3}}{2\sqrt{5}}$ $\sqrt{20} = \sqrt{4 \cdot 5} = \sqrt{4}\sqrt{5} = 2\sqrt{5}$

$= \frac{\sqrt{3}}{2\sqrt{5}} \cdot \frac{\sqrt{5}}{\sqrt{5}}$ *Rationalize denominator.*

$= \frac{\sqrt{15}}{2\sqrt{25}}$ *Product property:* $\sqrt{a}\sqrt{b} = \sqrt{ab}$

$= \frac{\sqrt{15}}{10}$ $2\sqrt{25} = 2 \cdot 5 = 10$

Simplifying a Radical Quotient

Next we will **simplify a radical quotient.** We will use this skill when solving quadratic equations in Section 9.6.

Example 8 Simplifying a Radical Quotient

Simplify $\frac{6 + 3\sqrt{2}}{12}$.

Solution

We first factor the numerator and the denominator, and then simplify the expression:

$\frac{6 + 3\sqrt{2}}{12} = \frac{3(2 + \sqrt{2})}{3(4)}$ *Factor out 3.*

$= \frac{3}{3} \cdot \frac{2 + \sqrt{2}}{4}$ $\frac{ac}{bd} = \frac{a}{b} \cdot \frac{c}{d}$

$= \frac{2 + \sqrt{2}}{4}$ $\frac{3}{3} = 1$; *simplify.*

Example 9 Simplifying a Radical Quotient

Simplify $\frac{8 - \sqrt{28}}{10}$.

Solution

We begin by simplifying the radical $\sqrt{28}$:

$\frac{8 - \sqrt{28}}{10} = \frac{8 - 2\sqrt{7}}{10}$ $\sqrt{28} = \sqrt{4 \cdot 7} = \sqrt{4}\sqrt{7} = 2\sqrt{7}$

$= \frac{2(4 - \sqrt{7})}{2(5)}$ *Factor out 2.*

$= \frac{2}{2} \cdot \frac{4 - \sqrt{7}}{5}$ $\frac{ac}{bd} = \frac{a}{b} \cdot \frac{c}{d}$

$= \frac{4 - \sqrt{7}}{5}$ $\frac{2}{2} = 1$; *simplify.*

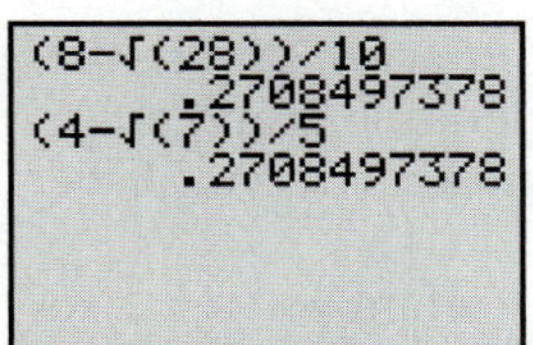

Figure 59 Verify the work

We verify our work by using a calculator to compare approximations for $\frac{8-\sqrt{28}}{10}$ and $\frac{4-\sqrt{7}}{5}$ (see Fig. 59). When we perform such a check, it is important that all parentheses be entered as shown.

In Example 9, we obtained the result $\frac{4-\sqrt{7}}{5}$. Keep in mind this radical expression is a number. From our check in Fig. 59, we see that $\frac{4-\sqrt{7}}{5} \approx 0.27$.

Summary of Simplifying a Radical Expression

We have discussed various ways to simplify a radical expression. What follows is a summary of these methods.

Simplifying a Radical Expression

To simplify a radical expression,

1. Use the quotient property for square roots so that no radicand is a fraction.
2. Use the product property for square roots so that no radicands have perfect-square factors other than 1.
3. Rationalize denominators so that no denominator is a radical expression.
4. Use the property $\frac{a}{a} = 1$, where a is nonzero, to simplify.
5. Continue applying steps 1–4 until the radical expression is simplified completely.

Group Exploration

Estimating the value of a radical expression

1. For parts (a) through (c), do not use a calculator.
   a. Between what two consecutive counting numbers is $\sqrt{19}$? Explain.
   b. Between what two consecutive counting numbers is $3+\sqrt{19}$? Explain. [**Hint:** Build on your result from part (a).]
   c. Between what two rational numbers is $\frac{3+\sqrt{19}}{2}$? Explain. Your results should be no more than 0.5 apart. [**Hint:** Build on your result from part (b).]
   d. Use a calculator to estimate $\frac{3+\sqrt{19}}{2}$. Is your work in part (c) correct? Explain.
2. a. Without using a calculator, determine between what two rational numbers is $\frac{9-\sqrt{29}}{2}$. Your results should be no more than 0.5 apart.
   b. Use a calculator to estimate $\frac{9-\sqrt{29}}{2}$. Is your work in part (a) correct? Explain.

Tips for Success Use 3-by-5 Cards

Do you have trouble memorizing definitions and properties? If so, try writing a word or phrase on one side of a 3-by-5 card, and, on the other side, put its definition or state its property and how it can be applied. For example, you could write "product property for square roots" on one side of a card and "For nonnegative numbers a and b, $\sqrt{ab} = \sqrt{a}\sqrt{b}$" on the other side. You could also describe, in your own words, the meaning of the property and how you can apply it.

Once you have completed a card for each definition and property, shuffle the cards and quiz yourself until you are confident you know the definitions and properties and how to apply them. Quiz yourself again later to make sure you have retained the information.

In addition to memorizing definitions and properties, it is important you strive to understand their meanings and how to apply them.

Homework 9.3

For extra help ▶ MyMathLab® 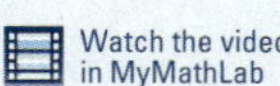 Watch the videos in MyMathLab  Download the MyDashboard App

Find the square root.

1. $\sqrt{81}$ 2. $\sqrt{36}$ 3. $\sqrt{121}$ 4. $\sqrt{100}$
5. $\sqrt{144}$ 6. $\sqrt{225}$ 7. $-\sqrt{81}$ 8. $-\sqrt{169}$
9. $\sqrt{-9}$ 10. $\sqrt{-4}$ 11. $-\sqrt{-25}$ 12. $-\sqrt{-16}$

State whether the square root is rational or irrational. If the square root is rational, find the (exact) value. If the square root is irrational, estimate its value by rounding to the second decimal place.

13. $\sqrt{30}$ 14. $\sqrt{62}$ 15. $\sqrt{78}$
16. $\sqrt{256}$ 17. $\sqrt{196}$ 18. $\sqrt{95}$

Simplify. Use a calculator to verify your work.

19. $\sqrt{20}$ 20. $\sqrt{28}$ 21. $\sqrt{45}$ 22. $\sqrt{18}$
23. $\sqrt{27}$ 24. $\sqrt{8}$ 25. $\sqrt{50}$ 26. $\sqrt{32}$
27. $\sqrt{300}$ 28. $\sqrt{75}$ 29. $-\sqrt{98}$ 30. $-\sqrt{80}$
31. $4\sqrt{72}$ 32. $5\sqrt{52}$ 33. $3\sqrt{120}$ 34. $2\sqrt{135}$

Simplify. Use a graphing calculator table to verify your work when possible.

35. $\sqrt{\frac{25}{36}}$ 36. $\sqrt{\frac{9}{49}}$ 37. $\sqrt{\frac{121}{4}}$ 38. $\sqrt{\frac{196}{9}}$
39. $\sqrt{\frac{7}{25}}$ 40. $\sqrt{\frac{3}{16}}$ 41. $\sqrt{\frac{19}{64}}$ 42. $\sqrt{\frac{23}{81}}$
43. $-\sqrt{\frac{8}{49}}$ 44. $-\sqrt{\frac{27}{16}}$ 45. $\sqrt{\frac{20}{81}}$ 46. $\sqrt{\frac{45}{4}}$
47. $\sqrt{\frac{75}{36}}$ 48. $\sqrt{\frac{32}{9}}$ 49. $\sqrt{\frac{80}{49}}$ 50. $\sqrt{\frac{200}{81}}$
51. $\frac{2}{\sqrt{3}}$ 52. $\frac{5}{\sqrt{2}}$ 53. $\frac{3}{\sqrt{24}}$ 54. $\frac{7}{\sqrt{45}}$
55. $\sqrt{\frac{2}{7}}$ 56. $\sqrt{\frac{3}{5}}$ 57. $\sqrt{\frac{11}{2}}$ 58. $\sqrt{\frac{5}{6}}$
59. $\sqrt{\frac{3}{8}}$ 60. $\sqrt{\frac{5}{18}}$ 61. $\sqrt{\frac{3}{50}}$ 62. $\sqrt{\frac{7}{32}}$

Simplify. Use a calculator to verify your work.

63. $\frac{9 + 3\sqrt{2}}{6}$ 64. $\frac{10 + 6\sqrt{5}}{8}$ 65. $\frac{8 - 4\sqrt{7}}{4}$
66. $\frac{15 - 5\sqrt{2}}{5}$ 67. $\frac{8 + 12\sqrt{13}}{6}$ 68. $\frac{21 + 6\sqrt{17}}{9}$
69. $\frac{4 + \sqrt{12}}{8}$ 70. $\frac{6 + \sqrt{20}}{4}$ 71. $\frac{10 - \sqrt{50}}{20}$
72. $\frac{6 - \sqrt{27}}{12}$ 73. $\frac{9 - \sqrt{45}}{6}$ 74. $\frac{25 - \sqrt{75}}{10}$

75. Suppose that, while traveling on a dry road, a driver slams on the brakes. Let D be the distance (in feet) that the car will skid and S be the speed (in miles per hour) before braking. The relationship between D and S is described by the model $S = \sqrt{30FD}$, where F is a drag factor, which is a measure of the roughness of the surface of the road. The drag factor on new concrete is 0.95, and the drag factor on polished concrete or asphalt is 0.75.

    A motorist who was involved in an accident claims he was driving at the posted speed limit of 60 miles per hour. A police officer measures the car's skid marks on the asphalt road to be 210 feet long. Assuming the motorist applied the brakes suddenly, estimate the speed at which the motorist was traveling before braking.

76. The size of a rectangular television screen is usually described by the length of its diagonal, given (in inches) by $d = \sqrt{w^2 + h^2}$, where w is the screen's width and h is the screen's height, both in inches. Find the length of the diagonal of a television screen whose width is 17 inches and height is 11 inches.

77. Let $f(h)$ be the amount of time (in seconds) it takes a baseball to fall to the ground when the ball is dropped from h feet above the ground. A good model is $f(h) = \sqrt{\frac{h}{16}}$.
    a. Simplify the right-hand side of the equation.
    b. How long would it take for a baseball to reach the ground if it were dropped from the top of Chicago's Willis Tower, which is 1450 feet tall?

78. The distance (in miles) $f(h)$ to the horizon at an altitude of h feet above sea level is given by the equation $f(h) = \sqrt{\frac{3h}{2}}$.
    a. Simplify the right-hand side of the equation.
    b. New York City's Empire State Building is 1250 feet tall. What would be the distance from the top of the building to the horizon? Assume that the base of the building is at sea level.
    c. If an airplane flies at an altitude of 34,000 feet over the skyscraper, what is the distance from the airplane to the horizon?

Concepts

79. A student thinks that $\sqrt{-25} = -5$. Is the student correct? Explain.

80. A student thinks that $-\sqrt{-9} = 3$. Is the student correct? Explain.

Without using a calculator, find the two consecutive integers nearest to the given radical. Verify your result by using a calculator.

81. $\sqrt{22}$ **82.** $\sqrt{15}$ **83.** $\sqrt{71}$ **84.** $\sqrt{43}$

85. a. Which of the two numbers is larger?
i. $2, \sqrt{2}$ **ii.** $5, \sqrt{5}$ **iii.** $8, \sqrt{8}$
b. Describe any patterns in your results for part (a).
c. Which of the two numbers is larger? [**Hint:** Use a calculator to be sure.]
i. $0.2, \sqrt{0.2}$ **ii.** $0.5, \sqrt{0.5}$ **iii.** $0.8, \sqrt{0.8}$
d. Describe any patterns in your results for part (c).
e. For which type of number does finding the principal square root give a smaller result? A larger result?

86. a. Square the given number. Then find the principal square root of the result.
i. 3 **ii.** 5 **iii.** 6
b. Find the square root.
i. $\sqrt{3^2}$ **ii.** $\sqrt{5^2}$ **iii.** $\sqrt{6^2}$
c. Describe any patterns from your work in parts (a) and (b).
d. Explain why $\sqrt{x^2} = x$, where x is nonnegative.

Sketch the graph of the equation. Verify your result by using a graphing calculator.

87. $y = \sqrt{x}$ **88.** $y = -\sqrt{x}$

89. Explain why the square root of a negative number is not a real number.

90. Choose nonnegative numbers for a and b, and show $\sqrt{ab} = \sqrt{a}\sqrt{b}$.

91. A student tries to simplify $\dfrac{5}{\sqrt{3}}$:

$$\frac{5}{\sqrt{3}} = \left(\frac{5}{\sqrt{3}}\right)^2 = \frac{5^2}{(\sqrt{3})^2} = \frac{25}{3}$$

Describe any errors. Then simplify the expression correctly.

92. A student thinks that $\dfrac{\sqrt{14}}{7} = 2$. Is the student correct? Explain.

93. A student tries to simplify $\dfrac{3}{\sqrt{20}}$:

$$\frac{3}{\sqrt{20}} = \frac{3}{\sqrt{20}} \cdot \frac{\sqrt{20}}{\sqrt{20}} = \frac{3\sqrt{20}}{20} = \frac{3\sqrt{4 \cdot 5}}{20} = \frac{3(2)\sqrt{5}}{2(10)} = \frac{3\sqrt{5}}{10}$$

Is the work correct? Is there an easier approach? Explain.

94. Choose a nonnegative number for a and a positive number for b, and show $\sqrt{\dfrac{a}{b}} = \dfrac{\sqrt{a}}{\sqrt{b}}$.

95. To simplify $\dfrac{1}{\sqrt{7}}$, why don't we square the fraction?

96. Describe how to simplify a radical expression.

Related Review

Factor.

97. $12 + 8x$ **98.** $10 - 35x$ **99.** $12 + 8\sqrt{3}$ **100.** $10 - 35\sqrt{2}$

Expressions, Equations, Functions, and Graphs

Perform the indicated instruction. Then use words such as linear, quadratic, cubic, power, radical, polynomial, degree, function, one variable, *and* two variables *to describe the expression, equation, or system.*

101. Graph $f(x) = -2x^2 + 4x + 3$ by hand.

102. Factor $4r^3 - 12r^2 - r + 3$.

103. Solve $x^2 = 6x - 8$.

104. Simplify $(7m + 2n)^2$.

105. Simplify $\sqrt{68}$.

106. Solve the following:

$$6x + 5y = 8$$
$$2x - 3y = -16$$

9.4 Using the Square Root Property to Solve Quadratic Equations

Objectives

- Use the *square root property* to solve quadratic equations.
- Make predictions with a quadratic function in vertex form.
- Know the meaning of *pure imaginary number, complex number,* and *imaginary number.*

In this section, we will discuss how to solve some equations that we cannot solve by factoring. We will then be able to make predictions about the independent variable of a quadratic model in vertex form. Finally, we will define numbers that are not real numbers but can be solutions of quadratic equations.

Solving Quadratic Equations of the Form $x^2 = k$

Consider the equation $x^2 = 9$. We first use factoring to solve this equation:

$x^2 = 9$	*Original equation*
$x^2 - 9 = 0$	*Subtract 9 from both sides.*
$(x + 3)(x - 3) = 0$	*Factor left side.*
$x + 3 = 0$ or $x - 3 = 0$	*Zero factor property*
$x = -3$ or $x = 3$	

So, the solutions are -3 and 3. We can use the notation ± 3 to stand for the numbers -3 and 3.

It is not necessary to use factoring to solve the equation $x^2 = 9$, however. Notice that a solution of this equation is any number that, when squared, is equal to 9. So, the solutions are the square roots of 9; in symbols, we write $x = \pm\sqrt{9} = \pm 3$.

Here we solve $x^2 = 16$:

$$\begin{aligned} x^2 &= 16 \\ x &= \pm\sqrt{16} \\ x &= \pm 4 \end{aligned}$$

This work suggests the *square root property.*

Square Root Property

Let k be a nonnegative constant. Then $x^2 = k$ is equivalent to

$$x = \pm\sqrt{k}$$

Example 1 Using the Square Root Property to Solve Equations

Solve.

1. $x^2 = 25$ **2.** $x^2 = 3$ **3.** $x^2 = -9$

Solution

1. $x^2 = 25$ *Original equation*

$x = \pm\sqrt{25}$ *Square root property*

$x = \pm 5$ *Simplify.*

2. $x^2 = 3$ *Original equation*

$x = \pm\sqrt{3}$ *Square root property*

3. Since the square of a number is nonnegative, we conclude that $x^2 = -9$ has no real-number solutions.

WARNING It is a common error to confuse solving an equation such as $x^2 = 25$ with computing a principal square root such as $\sqrt{25}$. In Problem 1 of Example 1, we found that the solutions of $x^2 = 25$ are the *two* numbers -5 and 5. Recall from Section 9.3 that $\sqrt{25}$ is the *one* number 5.

Example 2 Using the Square Root Property to Solve Equations

Solve.

1. $x^2 - 24 = 0$ **2.** $3x^2 - 4 = 3$

Solution

1. First, we isolate x^2 to (get x^2 alone on) one side of the equation:

$x^2 - 24 = 0$ *Original equation*

$x^2 = 24$ *Add 24 to both sides.*

$x = \pm\sqrt{24}$ *Square root property*

$x = \pm 2\sqrt{6}$ *Simplify.*

2. First, we isolate x^2 to one side of the equation:

$$3x^2 - 4 = 3 \quad \text{Original equation}$$
$$3x^2 = 7 \quad \text{Add 4 to both sides.}$$
$$x^2 = \frac{7}{3} \quad \text{Divide both sides by 3.}$$
$$x = \pm\sqrt{\frac{7}{3}} \quad \text{Square root property}$$
$$x = \pm\frac{\sqrt{7}}{\sqrt{3}} \quad \text{Quotient property: } \sqrt{\frac{a}{b}} = \frac{\sqrt{a}}{\sqrt{b}}$$
$$x = \pm\frac{\sqrt{7}}{\sqrt{3}} \cdot \frac{\sqrt{3}}{\sqrt{3}} \quad \text{Rationalize denominator.}$$
$$x = \pm\frac{\sqrt{21}}{3} \quad \text{Simplify.}$$

To verify our work, we use a graphing calculator to graph $y = 3x^2 - 4$ and $y = 3$ in the same coordinate system and then use "intersect" to find the approximate intersection points $(-1.53, 3)$ and $(1.53, 3)$, which have respective x-coordinates -1.53 and 1.53 (see Fig. 60). So, the approximate solutions of the original equation are -1.53 and 1.53, which check.

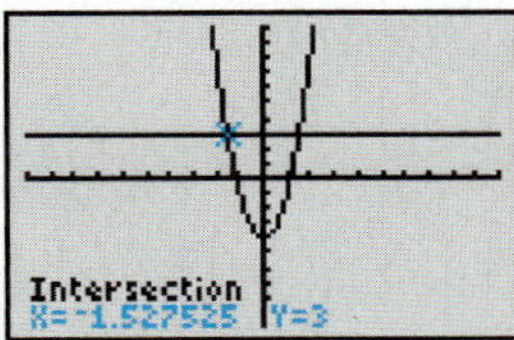

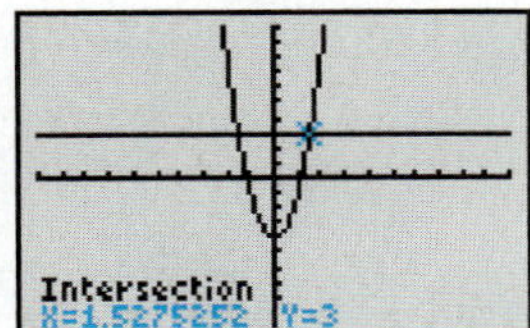

Figure 60 Verify the work

From Example 2, we see that **to solve an equation of the form $ax^2 + c = k$, we isolate x^2 to one side of the equation before using the square root property.**

Solving Quadratic Equations of the Form $(x + p)^2 = k$

We can also use the square root property to solve equations of the form $(x + p)^2 = k$.

Example 3 Using the Square Root Property to Solve Equations

Solve.

1. $(x + 4)^2 = 36$ **2.** $(x - 7)^2 = 50$

Solution

1. Note that the base of $(x + 4)^2$ is $x + 4$. We can still use the square root property to solve the equation $(x + 4)^2 = 36$.

$$(x + 4)^2 = 36 \quad \text{Original equation}$$
$$x + 4 = \pm\sqrt{36} \quad \text{Square root property}$$
$$x + 4 = \pm 6 \quad \text{Simplify.}$$
$$x + 4 = -6 \quad \text{or} \quad x + 4 = 6 \quad \text{Write as two equations.}$$
$$x = -10 \quad \text{or} \quad x = 2$$

2.
$$(x - 7)^2 = 50 \quad \text{Original equation}$$
$$x - 7 = \pm\sqrt{50} \quad \text{Square root property}$$
$$x - 7 = \pm 5\sqrt{2} \quad \text{Simplify.}$$
$$x = 7 \pm 5\sqrt{2} \quad \text{Add 7 to both sides.}$$

So, the solutions are $7 - 5\sqrt{2}$ and $7 + 5\sqrt{2}$. We use graphing calculator tables to check that if $7 - 5\sqrt{2}$ or $7 + 5\sqrt{2}$ is substituted for x in the expression $(x - 7)^2$, the result is 50 (see Fig. 61).

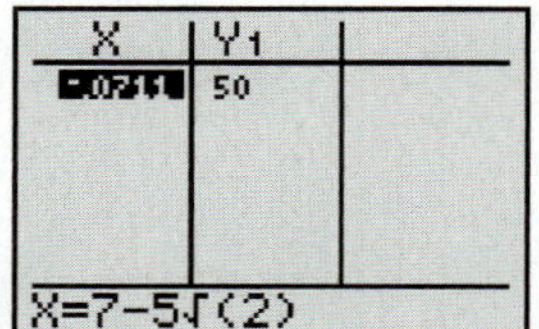

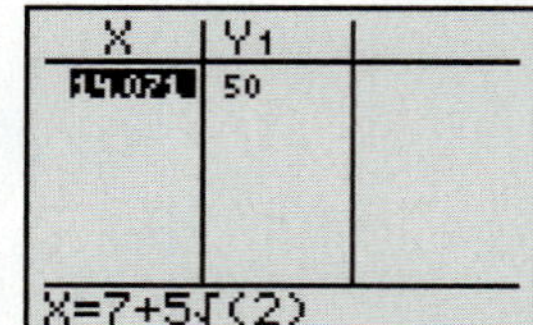

Figure 61 Verify the work

Example 4 Using the Square Root Property to Solve an Equation

Solve $(3w - 4)^2 = 36$.

Solution

The base of $(3w - 4)^2$ is $3w - 4$. We can use the square root property to solve the equation $(3w - 4)^2 = 36$:

$$(3w - 4)^2 = 36 \quad \text{Original equation}$$

$$3w - 4 = \pm\sqrt{36} \quad \text{Square root property}$$

$$3w - 4 = \pm 6 \quad \sqrt{36} = 6$$

$$3w - 4 = -6 \quad \text{or} \quad 3w - 4 = 6 \quad \text{Write as two equations.}$$

$$3w = -2 \quad \text{or} \quad 3w = 10$$

$$w = -\frac{2}{3} \quad \text{or} \quad w = \frac{10}{3}$$

Example 5 Using the Square Root Property to Solve an Equation

Solve $\left(x + \frac{5}{2}\right)^2 = \frac{31}{4}$.

Solution

$$\left(x + \frac{5}{2}\right)^2 = \frac{31}{4} \quad \text{Original equation}$$

$$x + \frac{5}{2} = \pm\sqrt{\frac{31}{4}} \quad \text{Square root property}$$

$$x + \frac{5}{2} = \pm\frac{\sqrt{31}}{\sqrt{4}} \quad \text{Quotient property: } \sqrt{\frac{a}{b}} = \frac{\sqrt{a}}{\sqrt{b}}$$

$$x + \frac{5}{2} = \pm\frac{\sqrt{31}}{2} \quad \sqrt{4} = 2$$

$$x = -\frac{5}{2} \pm \frac{\sqrt{31}}{2} \quad \text{Subtract } \frac{5}{2} \text{ from both sides.}$$

$$x = \frac{-5 \pm \sqrt{31}}{2} \quad \text{Add/subtract numerators and keep common denominator.}$$

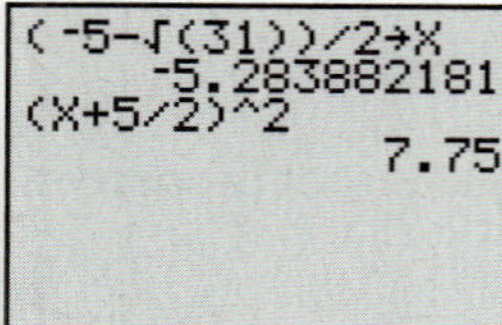


```
(-5+√(31))/2→X
        .2838821814
(X+5/2)^2
               7.75
```

Figure 62 Verify the work

To verify our work, we store each result as x and check that $\left(x + \frac{5}{2}\right)^2$ is equal to $\frac{31}{4} = 7.75$ (see Fig. 62). See Appendix A.20 for graphing calculator instructions.

Example 6 Solving a General Quadratic Equation

Solve $(px + q)^2 = k$ for x. Assume $k \geq 0$ and $p \neq 0$.

Solution

$$
\begin{aligned}
(px + q)^2 &= k && \text{Original equation} \\
px + q &= \pm\sqrt{k} && \text{Square root property} \\
px &= -q \pm \sqrt{k} && \text{Subtract } q \text{ from both sides.} \\
x &= \frac{-q \pm \sqrt{k}}{p} && \text{Divide both sides by } p.
\end{aligned}
$$

Example 7 Finding x-Intercepts

Find the x-intercepts of the parabola $f(x) = -2(x - 7)^2 + 20$.

Solution

To find the x-intercepts, we substitute 0 for $f(x)$:

$$
\begin{aligned}
-2(x - 7)^2 + 20 &= 0 && \text{Substitute 0 for } f(x). \\
-2(x - 7)^2 &= -20 && \text{Subtract 20 from both sides.} \\
(x - 7)^2 &= 10 && \text{Divide both sides by } -2. \\
x - 7 &= \pm\sqrt{10} && \text{Square root property} \\
x &= 7 \pm \sqrt{10} && \text{Add 7 to both sides.}
\end{aligned}
$$

The x-intercepts are $(7 - \sqrt{10}, 0)$ and $(7 + \sqrt{10}, 0)$. In Fig. 63, we use "zero" on a graphing calculator to check that the approximate x-intercepts are $(3.84, 0)$ and $(10.16, 0)$. See Appendix A.21 for graphing calculator instructions.

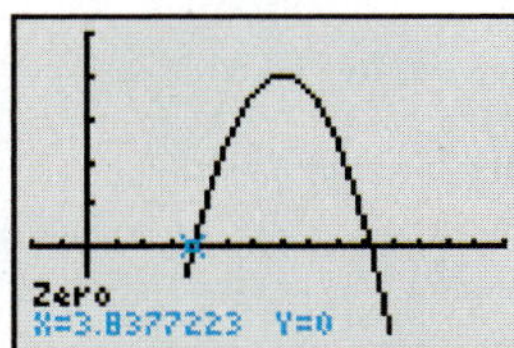

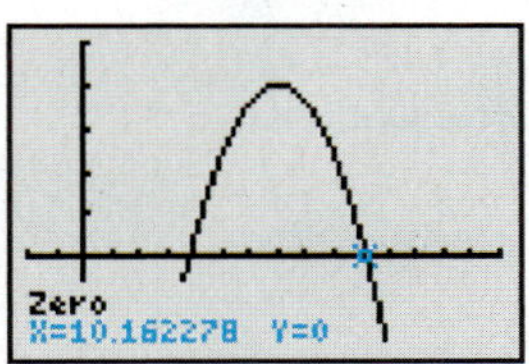

Figure 63 Verify the work

Making Predictions with a Quadratic Model in Vertex Form

Our work in this section enables us to use a quadratic model in vertex form $f(t) = a(t - h)^2 + k$ to make predictions for the independent variable, t.

Table 27 Numbers of International Adoptions in the United States

Year	Number of Adoptions (thousands)
2000	18.9
2002	21.5
2004	23.0
2006	20.7
2008	17.5
2010	11.1
2011	9.3

Source: *U.S. Department of State*

Example 8 Using a Quadratic Model to Make Predictions

The numbers of international adoptions in the United States are shown in Table 27 for various years. Let $f(t)$ be the number (in thousands) of international adoptions in the United States in the year that is t years since 2000.

1. Find an equation of f.
2. Estimate in which years there were 15 thousand international adoptions.

Solution

1. We use a graphing calculator to draw a scattergram of the data (see Fig. 64). It appears that a quadratic function will fit the data much better than a linear model. Recall from Section 9.1 that we can find a quadratic function in vertex form $f(t) = a(t - h)^2 + k$. To begin, we select the highest data point, $(4, 23)$, to be the vertex (see Fig. 64). Here, $h = 4$ and $k = 23$:

$$f(t) = a(t - 4)^2 + 23$$

Next, we imagine a parabola with vertex $(4, 23)$ that comes close to (or contains) the data points. Such a parabola might be the one that contains the data point

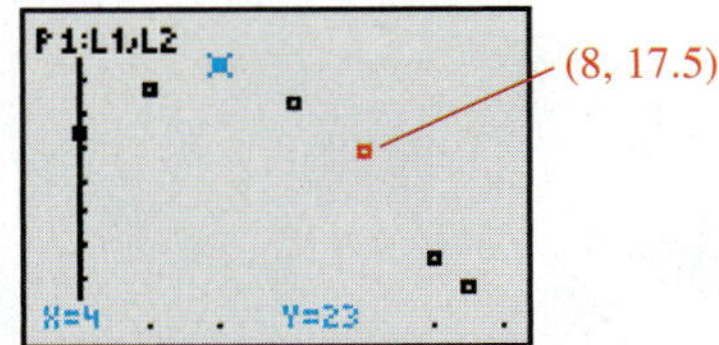

Figure 64 Scattergram of adoption data

$(8, 17.5)$. See Fig. 64. To find a, we substitute 8 for t and 17.5 for $f(t)$ in the equation $f(t) = a(t - 4)^2 + 23$:

$$
\begin{aligned}
17.5 &= a(8 - 4)^2 + 23 && \text{Substitute 8 for } t \text{ and 17.5 for } f(t).\\
17.5 &= a(4)^2 + 23 && \text{Subtract.}\\
17.5 &= 16a + 23 && \text{Simplify.}\\
-5.5 &= 16a && \text{Subtract 23 from both sides.}\\
-0.34 &\approx a && \text{Divide both sides by 16.}
\end{aligned}
$$

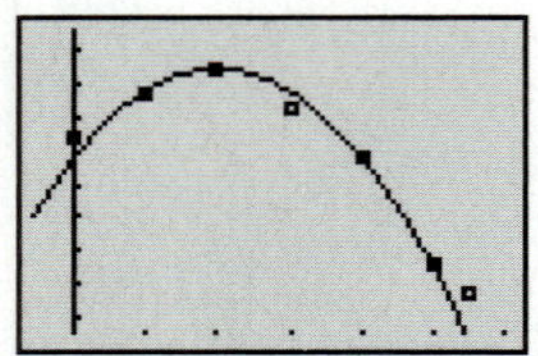

Figure 65 Check the fit

The approximate equation is $f(t) = -0.34(t - 4)^2 + 23$. The model fits the data well (see Fig. 65).

2. We substitute 15 for $f(t)$ in the equation $f(t) = -0.34(t - 4)^2 + 23$ and solve for t:

$$
\begin{aligned}
-0.34(t - 4)^2 + 23 &= 15 && \text{Substitute 15 for } f(t).\\
-0.34(t - 4)^2 &= -8 && \text{Subtract 23 from both sides.}\\
(t - 4)^2 &= \frac{-8}{-0.34} && \text{Divide both sides by } -0.34.\\
t - 4 &= \pm\sqrt{\frac{8}{0.34}} && \text{Square root property; simplify fraction.}\\
t &= 4 \pm \sqrt{\frac{8}{0.34}} && \text{Add 4 to both sides.}
\end{aligned}
$$

$$
\begin{aligned}
t = 4 - \sqrt{\frac{8}{0.34}} \quad &\text{or} \quad t = 4 + \sqrt{\frac{8}{0.34}} && \text{Write as two equations.}\\
t \approx 4 - 4.85 \quad &\text{or} \quad t \approx 4 + 4.85 && \text{Approximate square root.}\\
t \approx -0.85 \quad &\text{or} \quad t \approx 8.85
\end{aligned}
$$

The model estimates that there were 15 thousand international adoptions in the United States in both 1999 and 2009.

Complex Numbers

So far in this section, we have discussed equations of the form $x^2 = k$, where k is nonnegative. What happens when k is a negative number, such as -1?

Since the square of *any* real number is nonnegative, we conclude that the equation $x^2 = -1$ has no real-number solutions. Now we define a number that is *not* a real number but *is* a solution of $x^2 = -1$.

We define $\sqrt{-1}$ to be a number whose square is -1. We represent this number as i.

Definition Imaginary unit i

The **imaginary unit,** written i, is the number whose square is -1. That is,

$$i^2 = -1 \quad \text{and} \quad i = \sqrt{-1}$$

Next, we define the square root of any negative number.

Definition Square root of a negative number

If p is a positive real number, then

$$\sqrt{-p} = i\sqrt{p}$$

If b is a nonzero real number, then we call an expression of the form bi a **pure imaginary number.**

Example 9 Writing Numbers in *bi* Form

Write the number in *bi* form, where *b* is a real number. Simplify the result.

1. $\sqrt{-49}$ **2.** $-\sqrt{-24}$

Solution

1. $\sqrt{-49} = i\sqrt{49} = 7i$ **2.** $-\sqrt{-24} = -i\sqrt{24} = -i\sqrt{4 \cdot 6} = -2i\sqrt{6}$

We can combine real numbers *a* and *b* with the imaginary unit *i* to form a *complex number* of the form

$$a + bi$$

Definition Complex number

A **complex number** is a number of the form

$$a + bi$$

where *a* and *b* are real numbers.

Here are some examples of complex numbers:

$$2 + 5i, \quad 3 - 4i, \quad 5 + 0i = 5, \quad 0 - 7i = -7i$$

Since $a = a + 0i$ and $bi = 0 + bi$, we see that all real numbers and all pure imaginary numbers are complex numbers.

A complex number that is not a real number is called an *imaginary number*.

Definition Imaginary number

An **imaginary number** is a number $a + bi$, where *a* and *b* are real numbers and $b \neq 0$.

The complex number $3 + 7i$ is an imaginary number, since $b = 7 \neq 0$. The complex number $4 = 4 + 0i$ is not an imaginary number, since $b = 0$.

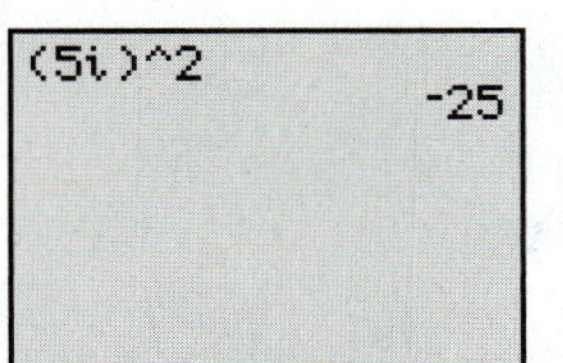

Figure 66 Check the work

We perform operations with complex numbers in much the same way that we perform operations with polynomials. Here we compare squaring the monomial $5x$ with squaring the pure imaginary number $5i$:

$$(5x)^2 = 5^2x^2 = 25x^2 \qquad (5i)^2 = 5^2i^2 = 25(-1) = -25$$

We use a graphing calculator to verify our work of squaring $5i$ (see Fig. 66). To compute $(5i)^2$, press (5 2nd .) ^ 2 ENTER.

Solving Quadratic Equations of the Form $x^2 = k$, where $k < 0$

We can extend the square root property to the case in which *k* is negative.

Square Root Property

Let *k* be a real-number constant. Then $x^2 = k$ is equivalent to

$$x = \pm\sqrt{k}$$

Example 10 Using the Square Root Property to Solve Equations

Solve.

1. $y^2 = -9$ **2.** $x^2 = -12$

Solution

1.

$$\begin{aligned} y^2 &= -9 && \textit{Original equation} \\ y &= \pm\sqrt{-9} && \textit{Square root property} \\ y &= \pm i\sqrt{9} && \sqrt{-p} = i\sqrt{p}, \textit{ where } p > 0 \\ y &= \pm 3i && \sqrt{9} = 3 \end{aligned}$$

We check that both $-3i$ and $3i$ satisfy the original equation:

Check $y = -3i$	**Check $y = 3i$**
$y^2 = -9$	$y^2 = -9$
$(-3i)^2 \stackrel{?}{=} -9$	$(3i)^2 \stackrel{?}{=} -9$
$(-3)^2 i^2 \stackrel{?}{=} -9$	$3^2 i^2 \stackrel{?}{=} -9$
$9(-1) \stackrel{?}{=} -9$	$9(-1) \stackrel{?}{=} -9$
$-9 \stackrel{?}{=} -9$	$-9 \stackrel{?}{=} -9$
true	true

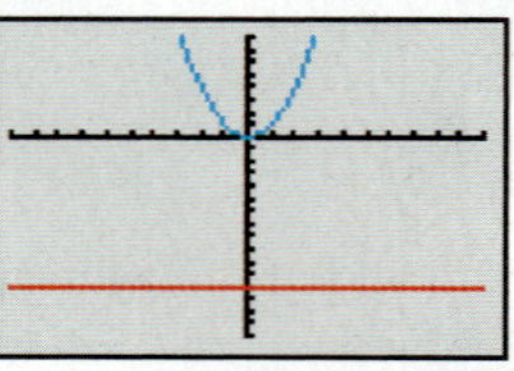

Figure 67 Check the work

2.

$$\begin{aligned} x^2 &= -12 && \textit{Original equation} \\ x &= \pm\sqrt{-12} && \textit{Square root property} \\ x &= \pm i\sqrt{12} && \sqrt{-p} = i\sqrt{p}, \textit{ where } p > 0 \\ x &= \pm i(2\sqrt{3}) && \sqrt{12} = \sqrt{4 \cdot 3} = 2\sqrt{3} \\ x &= \pm 2i\sqrt{3} && \textit{Rearrange factors.} \end{aligned}$$

The graphical check in Fig. 67 shows that the graphs of $y = x^2$ and $y = -12$ do not intersect. This means that the equation $x^2 = -12$ has no real-number solutions. It *does* have the two imaginary-number solutions $\pm 2i\sqrt{3}$.

Example 11 Using the Square Root Property to Solve an Equation

Solve $(x + 3)^2 = -28$.

Solution

$$\begin{aligned} (x+3)^2 &= -28 && \textit{Original equation} \\ x + 3 &= \pm\sqrt{-28} && \textit{Square root property} \\ x + 3 &= \pm i\sqrt{28} && \sqrt{-p} = i\sqrt{p}, \textit{ where } p > 0 \\ x + 3 &= \pm 2i\sqrt{7} && \sqrt{28} = \sqrt{4 \cdot 7} = 2\sqrt{7} \\ x &= -3 \pm 2i\sqrt{7} && \textit{Subtract 3 from both sides.} \end{aligned}$$

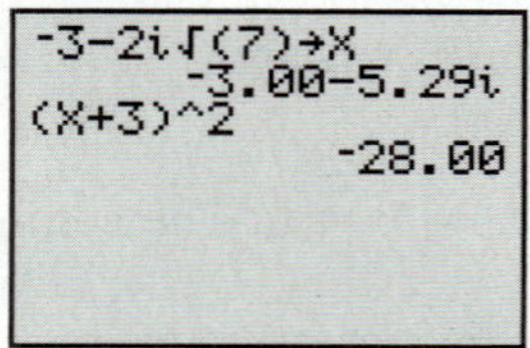

Figure 68 Verify the work

To verify our work, we store each result as x and check that $(x + 3)^2$ is equal to -28 (see Fig. 68). We press MODE and set FLOAT to 2, so the numbers in the result are rounded to the second decimal place.

Group Exploration

Deriving a formula for solving quadratic equations in $a(x - h)^2 + k = p$ form

1. Solve $2(x - 5)^2 + 7 = 10$.
2. Solve the equation $a(x - h)^2 + k = p$ for x. Assume $a \neq 0$ and $\dfrac{p - k}{a} \geq 0$. [**Hint:** Follow the same steps as in Problem 1.]
3. Use your result from Problem 2 to solve the equation $3(x - 4)^2 + 2 = 7$. [**Hint:** Substitute the appropriate values for a, h, k, and p into your formula from Problem 2.]

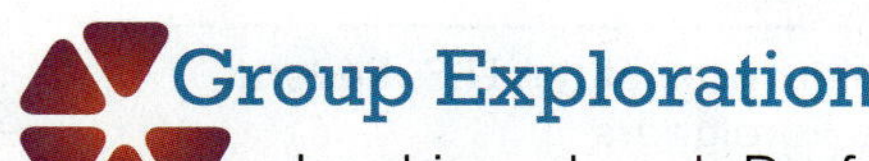

Group Exploration

Looking ahead: Perfect-square trinomials

1. Consider the true statement
$$(x + k)^2 = x^2 + (2k)x + k^2$$
For the trinomial on the right-hand side, the coefficient of x is $2k$ and the constant term is k^2. Describe how to use operations such as adding, subtracting, multiplying, dividing, and/or squaring to change $2k$ into k^2.
2. Show that your description in Problem 1 works for the true statement
$$(x + 3)^2 = x^2 + 6x + 9$$
3. Simplify $(x - k)^2$. Give a description similar to the one you made in Problem 1. Show that your description works for the true statement
$$(x - 4)^2 = x^2 - 8x + 16$$
4. Find a value of c such that the trinomial can be factored as the square of a binomial $(x + k)^2$ or $(x - k)^2$. Then factor the trinomial.

a. $x^2 + 8x + c$ **b.** $x^2 + 10x + c$
c. $x^2 - 14x + c$ **d.** $x^2 - 18x + c$

Homework 9.4

For extra help ▶ 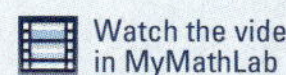 MyMathLab®  Watch the videos in MyMathLab Download the MyDashboard App

Solve. Use a graphing calculator to verify your work.

1. $x^2 = 4$ **2.** $x^2 = 16$ **3.** $x^2 = 0$
4. $x^2 = 1$ **5.** $w^2 = 15$ **6.** $r^2 = 17$
7. $x^2 = 20$ **8.** $x^2 = 18$ **9.** $b^2 - 28 = 0$
10. $c^2 - 40 = 0$ **11.** $4x^2 = 5$ **12.** $9x^2 = 11$
13. $5x^2 = 7$ **14.** $7x^2 = 3$ **15.** $8m^2 = 5$
16. $12y^2 = 11$ **17.** $2x^2 + 4 = 7$ **18.** $3x^2 + 1 = 8$
19. $5x^2 - 3 = 11$ **20.** $7x^2 - 8 = -3$

Solve. Use a graphing calculator to verify your work.

21. $(x + 2)^2 = 16$ **22.** $(x + 4)^2 = 25$ **23.** $(x - 7)^2 = 13$
24. $(x - 4)^2 = 35$ **25.** $(x + 2)^2 = 18$ **26.** $(x + 7)^2 = 45$
27. $(8y + 3)^2 = 36$ **28.** $(4m + 7)^2 = 81$ **29.** $(9x - 5)^2 = 0$
30. $(5x - 4)^2 = 0$ **31.** $\left(x + \frac{3}{4}\right)^2 = \frac{41}{16}$
32. $\left(x + \frac{5}{6}\right)^2 = \frac{51}{36}$ **33.** $\left(w - \frac{7}{3}\right)^2 = \frac{5}{9}$
34. $\left(p - \frac{2}{5}\right)^2 = \frac{17}{25}$ **35.** $5(x - 6)^2 + 3 = 33$
36. $2(x - 4)^2 + 5 = 27$
37. $-3(x + 1)^2 + 2 = -5$
38. $-7(x + 8)^2 + 4 = -1$

Find all x-intercepts.

39. $f(x) = x^2 - 17$ **40.** $f(x) = x^2 - 35$
41. $f(x) = 2(x - 3)^2 - 7$ **42.** $f(x) = 7(x - 6)^2 - 13$
43. $f(x) = -4(x - 2)^2 - 16$ **44.** $f(x) = -3(x - 5)^2 - 27$

Simplify.

45. $\sqrt{-36}$ **46.** $\sqrt{-25}$
47. $-\sqrt{-45}$ **48.** $-\sqrt{-40}$
49. $\sqrt{-\frac{5}{49}}$ **50.** $\sqrt{-\frac{7}{9}}$
51. $\sqrt{-\frac{13}{5}}$ **52.** $\sqrt{-\frac{2}{7}}$

Find all complex-number solutions.

53. $x^2 = -49$ **54.** $x^2 = -4$
55. $x^2 = -18$ **56.** $x^2 = -28$
57. $7x^2 + 26 = 5$ **58.** $4x^2 + 25 = 17$
59. $(m + 4)^2 = -8$ **60.** $(t + 8)^2 = -45$
61. $\left(x - \frac{5}{4}\right)^2 = -\frac{3}{16}$ **62.** $\left(x - \frac{1}{2}\right)^2 = -\frac{7}{4}$
63. $-2(y + 3)^2 + 1 = 9$ **64.** $-2(w + 1)^2 + 7 = 39$

65. The percentages of Americans who say they are "very happy" are shown in Table 28 for various years.

Table 28 Percentages of Americans Who Say They Are "Very Happy"

Year	Percent
1972	30
1980	35
1990	36
2000	34
2010	29

Source: *National Opinion Research Center*

Let $f(t)$ be the percentage of Americans who say they are "very happy" at t years since 1970.

a. Find a quadratic equation of f in vertex form.

b. Find $f(49)$. What does it mean in this situation?

c. Find t when $f(t) = 23$. What does it mean in this situation?

d. What is the vertex of the model? What does it mean in this situation?

66. The revenues from ringtones, ringbacks, and videos for cell phones are shown in Table 29 for various years.

Table 29 Revenues from Ringtones, Ringbacks, and Videos for Cell Phones

Year	Revenue (millions of dollars)
2005	422
2006	775
2007	879
2008	977
2009	729
2010	527

Source: *Recording Industry Association of America*

Let $f(t)$ be the annual revenue (in millions of dollars) from ringtones, ringbacks, and videos for cell phones at t years since 2000.

a. Find a quadratic equation of f in vertex form.

b. Use f to estimate the revenue in 2010. Find the error in your estimate.

c. Estimate when the annual revenue was \$70 million.

d. What is the vertex of the model? What does it mean in this situation?

67. The revenues from U.S. adult mattresses are shown in Table 30 for various years.

Table 30 Revenues from U.S. Adult Mattresses

Year	Revenue (billions of dollars)
2007	5.2
2008	4.7
2009	4.4
2010	4.6
2011	5.0

Source: *International Sleep Products Association*

Let $f(t)$ be the annual revenue (in billions of dollars) at t years since 2000. A model of the situation is the following: $f(t) = 0.16(t - 9.15)^2 + 4.45$.

a. Use a graphing calculator to draw the graph of the model f and, in the same viewing window, the scattergram of the data. Does the model fit the data well?

b. What is the vertex? What does it mean in this situation?

c. Predict the revenue in 2017.

d. Predict when the annual revenue will be \$10 billion.

68. The percentages of the U.S. federal debt owed to foreigners are shown in Table 31 for various years.

Table 31 Percentages of the U.S. Federal Debt Owed to Foreigners

Year	Percent
1985	15
1990	18
1995	22
2000	31
2005	45
2011	47

Source: *White House Office of Management and Budget*

Let $P(t)$ be the percentage of the U.S. federal debt owed to foreigners at t years since 1980. A model of the situation is $P(t) = 0.0157(t + 26.1)^2 - 1.7$

a. Use a graphing calculator to draw the graph of the model f and, in the same viewing window, the scattergram of the data. Does the model fit the data well?

b. Use P to predict the percentage of the federal debt owed to foreigners in 2018.

c. Assuming the federal debt will be \$25.2 trillion in 2018, use your result in part (b) to predict the number of trillions of dollars of that debt owed to foreigners in 2018.

d. Use P to predict when all of the debt will be owed to foreigners.

For Exercises 69–74, find approximate solutions of the given equation or system by referring to the graphs shown in Fig. 69. Round results or coordinates of results to the first decimal place.

69. $\frac{1}{2}(x - 2)^2 - 1 = -2(x - 3)^2 + 4$

70. $-2(x - 3)^2 + 4 = \frac{1}{2}x - 4$

71. $-2(x - 3)^2 + 4 = 2$

72. $\frac{1}{2}(x - 2)^2 - 1 = 0$

73. $y = -2(x - 3)^2 + 4$
$y = \frac{1}{2}x - 4$

74. $y = \frac{1}{2}(x - 2)^2 - 1$
$y = -2(x - 3)^2 + 4$

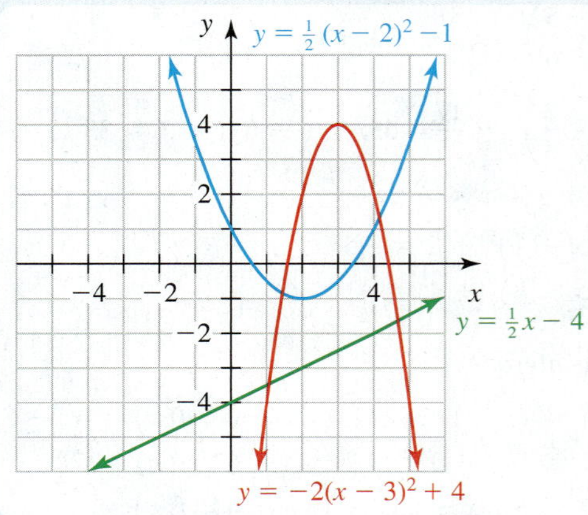

Figure 69 Exercises 69–74

For Exercises 75–82, solve for the specified variable. Assume the constants have values for which the equation has exactly two real-number solutions.

75. $a^2 + b^2 = c^2$, for a

76. $\dfrac{r^2}{b} = c$, for r

77. $\dfrac{w^2}{r} + p = b$, for w

78. $\dfrac{w^2 - k}{r} = b$, for w

79. $(x + b)^2 = k$, for x

80. $(x - b)^2 - q = k$, for x

81. $(py + a)^2 + b = c$, for y

82. $\dfrac{(py - a)^2}{b} = c$, for y

Concepts

83. A student tries to solve the equation $x^2 - 10x + 25 = 0$:

$$\begin{aligned} x^2 - 10x + 25 &= 0 \\ x^2 &= 10x - 25 \\ x &= \pm\sqrt{10x - 25} \end{aligned}$$

Describe any errors. Then solve the equation correctly.

84. A student tries to solve the equation $3(x - 5)^2 = 12$:

$$\begin{aligned} 3(x - 5)^2 &= 12 \\ (x - 5)^2 &= 4 \\ x - 5 &= 2 \\ x &= 7 \end{aligned}$$

Describe any errors. Then solve the equation correctly.

85. Let $f(x) = 2(x - 3)^2 + 5$.

a. What is the vertex of the graph of f?

b. Does the graph of f open upward or downward? Explain.

c. How many real-number solutions does each of the equations that follow have? Explain how you can use your results from parts (a) and (b).

i. $2(x - 3)^2 + 5 = 8$

ii. $2(x - 3)^2 + 5 = 5$

iii. $2(x - 3)^2 + 5 = 1$

86. Let $f(x) = -3(x + 2)^2 + 4$.

a. What is the vertex of the graph of f?

b. Does the graph of f open upward or downward? Explain.

c. How many real-number solutions does each of the equations that follow have? Explain how you can use your results from parts (a) and (b).

i. $-3(x + 2)^2 + 4 = 6$

ii. $-3(x + 2)^2 + 4 = 4$

iii. $-3(x + 2)^2 + 4 = 2$

87. a. Use factoring to solve the equation $25x^2 - 49 = 0$.

b. Use the square root property to solve the equation $25x^2 - 49 = 0$.

c. Compare your results for parts (a) and (b).

d. Is it easier to solve $25x^2 - 49 = 0$ by factoring or by using the square root property? Explain.

88. a. Use factoring to solve the equation $(2x - 3)^2 = 16$.

b. Use the square root property to solve the equation $(2x - 3)^2 = 16$.

c. Compare your results for parts (a) and (b).

d. Is it easier to solve $(2x - 3)^2 = 16$ by factoring or by using the square root property? Explain.

89. a. Can the equation $(x + 4)^2 = 5$ be solved with the square root property? If yes, do so.

b. Can the equation $(x + 4)^2 = 5$ be solved by factoring? If yes, do so.

c. Can all equations that can be solved with the square root property be solved by factoring? Explain.

90. a. Can the equation $3(x - 2)^2 + 1 = 22$ be solved by means of the square root property? If so, solve the equation by this method.

b. Can the equation $3(x - 2)^2 + 1 = 22$ be solved by factoring? If so, solve the equation by this method.

c. Can all equations that can be solved by means of the square root property be solved by factoring? Explain.

91. What forms of quadratic equations can you solve by using the square root property? By using factoring? For example, quadratic equations of the form $x^2 + 2bx + b^2 = 0$ can be solved by factoring. Include examples.

92. Give an example of a quadratic equation in one variable that has the given number and type of solutions. Then solve the equation.

a. two real-number solutions

b. one real-number solution

c. two imaginary-number solutions

Related Review

93. a. Solve $x^2 = 49$.

b. Find $\sqrt{49}$.

c. Explain why your results for parts (a) and (b) are different.

94. A student says 3 is a solution of the equation $-x^2 = 9$ because $-3^2 = 9$. Is the student correct? Explain.

Use factoring or the square root property to solve the equation. Use a graphing calculator to verify your work.

95. $x^2 = 4x + 12$

96. $x^2 = -12x - 20$

97. $y^2 - 81 = 0$

98. $m^2 - 49 = 0$

99. $(x - 2)^2 = 24$

100. $(x + 8)^2 = 32$

101. $3x^2 + 4 = 15$

102. $25x^2 - 1 = 16$

103. $2x^2 - 15 = -7x$

104. $3x^2 + 2 = 5x$

105. $5(t - 1)^2 - 3 = 4$

106. $3(y + 5)^2 + 2 = 13$

Expressions, Equations, Functions, and Graphs

Perform the indicated instruction. Then use words such as linear, quadratic, cubic, polynomial, degree, function, one variable, *and* two variables *to describe the expression, equation, or system.*

107. Factor $2w^3 + 3w^2 - 18w - 27$.

108. Solve $3(x - 4)^2 - 5 = -7$.

109. Solve $2w^3 + 3w^2 - 18w - 27 = 0$.

110. Simplify $3(x - 4)^2 - 5$.

111. Find the product $(5w^2 - 2)(4w + 3)$.

112. Graph $f(x) = 3(x - 4)^2 - 5$ by hand.

9.5 Solving Quadratic Equations by Completing the Square

Objectives

- Know the relationship between b and c in a perfect-square trinomial of the form $x^2 + bx + c$.
- Solve quadratic equations by *completing the square.*
- Know that any quadratic equation can be solved by completing the square.

So far, we have discussed how to solve quadratic equations by factoring and by using the square root property. In this section, we will discuss yet another way, called *completing the square.*

Perfect-Square Trinomials

To begin our study of completing the square, we simplify the square of a sum. For example,

$$(x + 3)^2 = x^2 + 6x + 9$$

So, $x^2 + 6x + 9$ is a perfect-square trinomial. Recall from Section 7.3 that a *perfect-square trinomial* is a trinomial that is equivalent to the square of a binomial.

For $x^2 + 6x + 9$, there is a special connection between the 6 and the 9. If we divide the 6 by 2 and then square the result, we get 9:

$$x^2 + 6x + 9$$

$$\left(\frac{6}{2}\right)^2 = 3^2 = 9$$

This is no coincidence. Consider simplifying $(x - 4)^2$: $(x - 4)^2 = x^2 - 8x + 16$. If we divide -8 by 2 and square the result, we get 16:

$$x^2 - 8x + 16$$

$$\left(\frac{-8}{2}\right)^2 = (-4)^2 = 16$$

For the general case, we simplify $(x + k)^2$, where k is a constant:

$$(x + k)^2 = x^2 + (2k)x + k^2$$

$$\left(\frac{2k}{2}\right)^2 = k^2$$

For $x^2 + (2k)x + k^2$, we see that if we divide the coefficient of x by 2 and square the result, we get the constant term k^2.

Perfect-Square Trinomial Property

For a perfect-square trinomial of the form $x^2 + bx + c$, dividing b by 2 and squaring the result gives c:

$$x^2 + bx + c$$

$$\left(\frac{b}{2}\right)^2 = c$$

Example 1 Factoring Perfect-Square Trinomials

Find the value of c such that the expression is a perfect-square trinomial. Then factor the perfect-square trinomial.

1. $x^2 + 10x + c$ **2.** $x^2 - 9x + c$ **3.** $x^2 + \frac{5}{3}x + c$

Solution

1. We divide 10 by 2 and square the result:

$$\left(\frac{10}{2}\right)^2 = 5^2 = 25 = c$$

The expression is $x^2 + 10x + 25$, with factored form $(x + k)^2$ for some positive integer k. So, we have

$$x^2 + 10x + 25 = (x + k)^2 = x^2 + 2kx + k^2$$

The constant terms 25 and k^2 are equal.

Here $k^2 = 25$, or $k = 5$ (k is positive). So, the factored form of $x^2 + 10x + 25$ is $(x + 5)^2$.

2. We divide -9 by 2 and square the result:

$$\left(\frac{-9}{2}\right)^2 = \frac{81}{4} = c$$

The expression is $x^2 - 9x + \frac{81}{4}$, with factored form $(x - k)^2$ for some positive integer k. So, we have

$$x^2 - 9x + \frac{81}{4} = (x - k)^2 = x^2 - 2kx + k^2$$

The constant terms $\frac{81}{4}$ and k^2 are equal.

Here $k^2 = \frac{81}{4}$, or $k = \frac{9}{2}$ (k is positive). So, the factored form of $x^2 - 9x + \frac{81}{4}$ is $\left(x - \frac{9}{2}\right)^2$.

3. Dividing by 2 is the same as multiplying by $\frac{1}{2}$. So, we multiply $\frac{5}{3}$ by $\frac{1}{2}$ and square the result:

$$\left(\frac{5}{3} \cdot \frac{1}{2}\right)^2 = \left(\frac{5}{6}\right)^2 = \frac{25}{36} = c$$

The expression is $x^2 + \frac{5}{3}x + \frac{25}{36}$, with factored form $\left(x + \frac{5}{6}\right)^2$.

Solving $x^2 + bx + c = 0$ by Completing the Square

In Example 2, we will begin to solve a quadratic equation by forming a perfect-square trinomial on one side of the equation.

Example 2 Solving by Completing the Square

Solve $x^2 + 8x = 3$.

Solution

Since $\left(\frac{8}{2}\right)^2 = 4^2 = 16$, we add 16 to both sides of $x^2 + 8x = 3$ so the left side will be a perfect-square trinomial:

$$
\begin{aligned}
x^2 + 8x &= 3 && \textit{Original equation} \\
x^2 + 8x + 16 &= 3 + 16 && \textit{Add 16 to both sides.} \\
(x + 4)^2 &= 19 && \textit{Factor the left side.} \\
x + 4 &= \pm\sqrt{19} && \textit{Square root property} \\
x &= -4 \pm \sqrt{19} && \textit{Subtract 4 from both sides.}
\end{aligned}
$$

We use graphing calculator tables to check that if $-4 - \sqrt{19}$ or $-4 + \sqrt{19}$ is substituted for x in the expression $x^2 + 8x$, the result is 3 (see Fig. 70).

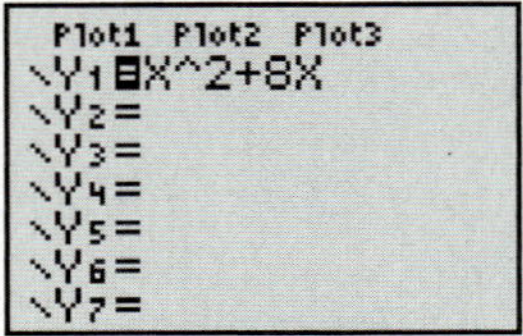

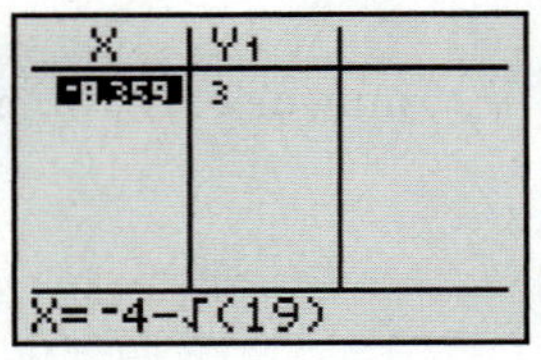

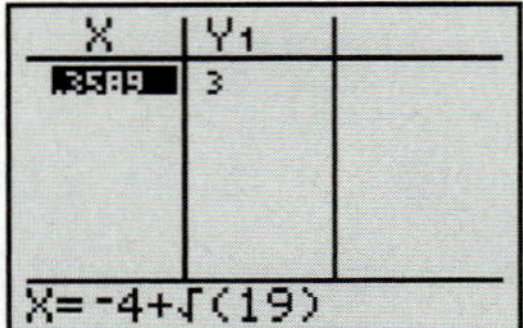

Figure 70 Verify the work

In Example 2, we added 16 to both sides of $x^2 + 8x = 3$ and then factored the left side to get $(x + 4)^2$. By adding 16 to both sides of the equation, we say we *completed the square* for $x^2 + 8x$.

WARNING In solving a quadratic equation such as $x^2 + 8x = 3$ by completing the square, it is a common error to add a number (in this case 16) to only the left side of the equation. Remember to add the number to *both* sides of the equation.

Example 3 Solving by Completing the Square

Solve $p^2 - 14p + 4 = 0$.

Solution

We're trying to get a perfect-square trinomial on the left side of the equation $p^2 - 14p + 4 = 0$, so we subtract 4 from both sides:

$$p^2 - 14p = -4$$

Since $\left(\frac{-14}{2}\right)^2 = (-7)^2 = 49$, we add 49 to both sides of $p^2 - 14p = -4$ to complete the square for $p^2 - 14p$:

$$\begin{aligned} p^2 - 14p + 49 &= -4 + 49 && \textit{Add 49 to both sides.} \\ (p - 7)^2 &= 45 && \textit{Factor left side.} \\ p - 7 &= \pm\sqrt{45} && \textit{Square root property} \\ p - 7 &= \pm 3\sqrt{5} && \textit{Simplify.} \\ p &= 7 \pm 3\sqrt{5} && \textit{Add 7 to both sides.} \end{aligned}$$

Example 4 Solving by Completing the Square

Solve $t^2 - 5t + 1 = 0$.

Solution

To solve $t^2 - 5t + 1 = 0$, we first put the equation into the form $t^2 + bt = k$ and then complete the square:

$$\begin{aligned} t^2 - 5t + 1 &= 0 && \textit{Original equation} \\ t^2 - 5t &= -1 && \textit{Subtract 1 from both sides.} \\ t^2 - 5t + \frac{25}{4} &= -1 + \frac{25}{4} && \textit{Add } \left(\frac{-5}{2}\right)^2 = \frac{25}{4} \textit{ to both sides.} \\ \left(t - \frac{5}{2}\right)^2 &= -\frac{4}{4} + \frac{25}{4} && \textit{Factor left side; find LCD.} \\ \left(t - \frac{5}{2}\right)^2 &= \frac{21}{4} && \textit{Add numerators and keep common denominator: } \frac{a}{b} + \frac{c}{b} = \frac{a + c}{b} \end{aligned}$$

Figure 71 Verify the work

$$t - \frac{5}{2} = \pm\sqrt{\frac{21}{4}} \qquad \text{Square root property}$$

$$t - \frac{5}{2} = \pm\frac{\sqrt{21}}{2} \qquad \text{Quotient property: } \sqrt{\frac{a}{b}} = \frac{\sqrt{a}}{\sqrt{b}}$$

$$t = \frac{5}{2} \pm \frac{\sqrt{21}}{2} \qquad \text{Add } \frac{5}{2} \text{ to both sides.}$$

$$t = \frac{5 \pm \sqrt{21}}{2} \qquad \text{Add/subtract numerators and keep common denominator.}$$

We use a graphing calculator to verify our work by storing an approximation of each solution as x and checking that $x^2 - 5x + 1$ is approximately 0 (see Fig. 71). For the calculator entry, the numerator of each fraction must be in parentheses. To store a value, see Appendix A.20.

Solving $ax^2 + bx + c = 0$ by Completing the Square

So far, we have worked with trinomials only of the form $ax^2 + bx + c$, where $a = 1$. What do we do when $a \neq 1$? Consider simplifying $(2x + 5)^2$:

$$(2x + 5)^2 = 4x^2 + 20x + 25$$

Dividing $b = 20$ by 2 and squaring the result does *not* give 25:

$$\left(\frac{20}{2}\right)^2 = 10^2 = 100 \neq 25$$

So, the perfect-square trinomial property given for $ax^2 + bx + c$, where $a = 1$, does not extend to trinomials with $a \neq 1$. However, **when solving a quadratic equation of the form $ax^2 + bx + c = 0$ with $a \neq 1$, we can first divide both sides by a to obtain an equation involving "$1x^2$"—an equation to which we *can* apply the property.**

Example 5 Solving by Completing the Square

Solve $2x^2 - 16x = -7$.

Solution

$$2x^2 - 16x = -7 \qquad \text{Original equation}$$

$$x^2 - 8x = -\frac{7}{2} \qquad \text{Divide both sides by 2.}$$

$$x^2 - 8x + 16 = -\frac{7}{2} + 16 \qquad \text{Add } \left(\frac{-8}{2}\right)^2 = (-4)^2 = 16 \text{ to both sides.}$$

$$(x - 4)^2 = -\frac{7}{2} + \frac{32}{2} \qquad \text{Factor left side; find LCD.}$$

$$(x - 4)^2 = \frac{25}{2} \qquad \text{Add numerators and keep common denominator: } \frac{a}{b} + \frac{c}{b} = \frac{a + c}{b}$$

$$x - 4 = \pm\sqrt{\frac{25}{2}} \qquad \text{Square root property}$$

$$x - 4 = \pm\frac{5\sqrt{2}}{2} \qquad \sqrt{\frac{25}{2}} = \frac{\sqrt{25}}{\sqrt{2}} = \frac{5}{\sqrt{2}} \cdot \frac{\sqrt{2}}{\sqrt{2}} = \frac{5\sqrt{2}}{2}$$

$$x = 4 \pm \frac{5\sqrt{2}}{2} \qquad \text{Add 4 to both sides.}$$

$$x = \frac{8}{2} \pm \frac{5\sqrt{2}}{2} \qquad \text{Find LCD.}$$

$$x = \frac{8 \pm 5\sqrt{2}}{2} \qquad \text{Add/subtract numerators and keep common denominator.}$$

Any quadratic equation can be solved by completing the square. Here is a summary of this method.

Solving a Quadratic Equation by Completing the Square

To solve a quadratic equation $ax^2 + bx + c = 0$ by **completing the square,**

1. If $a \neq 1$, divide both sides of the equation by a.
2. Write the equation in the form $x^2 + dx = k$, where d and k are constants.
3. Complete the square for the expression on the left side of the equation.
4. Solve the equation by using the square root property.

WARNING To solve an equation of the form $ax^2 + bx = k$, where $a \neq 1$, we must divide both sides of the equation by a before completing the square on the left side of the equation.

Example 6 Solving by Completing the Square

Solve $3x^2 + 7x - 5 = 0$.

Solution

$$3x^2 + 7x - 5 = 0 \quad \text{Original equation}$$

$$x^2 + \frac{7}{3}x - \frac{5}{3} = 0 \quad \text{Divide both sides by 3.}$$

$$x^2 + \frac{7}{3}x = \frac{5}{3} \quad \text{Add } \frac{5}{3} \text{ to both sides.}$$

$$x^2 + \frac{7}{3}x + \frac{49}{36} = \frac{5}{3} + \frac{49}{36} \quad \text{Add } \left(\frac{7}{3}\cdot\frac{1}{2}\right)^2 = \left(\frac{7}{6}\right)^2 = \frac{49}{36} \text{ to both sides.}$$

$$\left(x + \frac{7}{6}\right)^2 = \frac{60}{36} + \frac{49}{36} \quad \text{Factor left side; find LCD.}$$

$$\left(x + \frac{7}{6}\right)^2 = \frac{109}{36} \quad \text{Add numerators and keep common denominator: } \frac{a}{b} + \frac{c}{b} = \frac{a+c}{b}$$

$$x + \frac{7}{6} = \pm\sqrt{\frac{109}{36}} \quad \text{Square root property}$$

$$x + \frac{7}{6} = \pm\frac{\sqrt{109}}{6} \quad \text{Quotient property: } \sqrt{\frac{a}{b}} = \frac{\sqrt{a}}{\sqrt{b}}$$

$$x = -\frac{7}{6} \pm \frac{\sqrt{109}}{6} \quad \text{Subtract } \frac{7}{6} \text{ from both sides.}$$

$$x = \frac{-7 \pm \sqrt{109}}{6} \quad \text{Add/subtract numerators and keep common denominator.}$$

Solving a Quadratic Equation That Has Imaginary-Number Solutions

In Example 7, we will complete the square to solve a quadratic equation that has imaginary-number solutions.

Example 7 Solving by Completing the Square

Solve $5x^2 - 10x + 45 = 0$.

Solution

$$5x^2 - 10x + 45 = 0 \quad \text{Original equation}$$

$$x^2 - 2x + 9 = 0 \quad \text{Divide both sides by 5.}$$

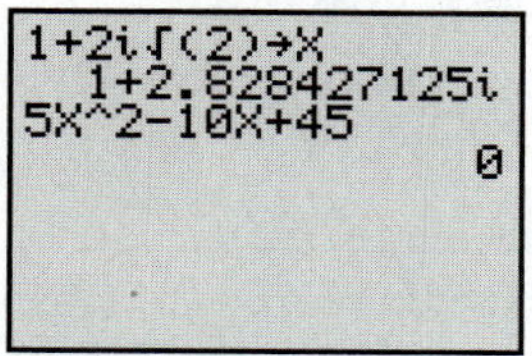

Figure 72 Verify the work

$$x^2 - 2x = -9 \quad \text{Subtract 9 from both sides.}$$

$$x^2 - 2x + 1 = -9 + 1 \quad \text{Add } \left(\frac{-2}{2}\right)^2 = (-1)^2 = 1 \text{ to both sides.}$$

$$(x-1)^2 = -8 \quad \text{Factor left side.}$$

$$x - 1 = \pm\sqrt{-8} \quad \text{Square root property}$$

$$x - 1 = \pm i\sqrt{8} \quad \sqrt{-p} = i\sqrt{p}, \text{ where } p > 0$$

$$x - 1 = \pm 2i\sqrt{2} \quad \sqrt{8} = \sqrt{4 \cdot 2} = 2\sqrt{2}$$

$$x = 1 \pm 2i\sqrt{2} \quad \text{Add 1 to both sides.}$$

To verify our work, we store each result as x and check that $5x^2 - 10x + 45$ is equal to 0 (see Fig. 72).

Group Exploration

Identifying errors in solving by completing the square

1. A student tries to solve the equation $x^2 + 6x - 5 = 0$:

$$x^2 + 6x - 5 = 0$$
$$x^2 + 6x = 5$$
$$x^2 + 6x + 9 = 5$$
$$(x+3)^2 = 5$$
$$x + 3 = \pm\sqrt{5}$$
$$x = -3 \pm \sqrt{5}$$

Describe any errors. Then solve the equation correctly.

2. A student tries to solve the equation $4x^2 - 8x = 12$:

$$4x^2 - 8x = 12$$
$$4x^2 - 8x + 16 = 12 + 16$$
$$(2x-4)^2 = 28$$
$$2x - 4 = \pm\sqrt{28}$$
$$2x - 4 = \pm 2\sqrt{7}$$
$$2x = 4 \pm 2\sqrt{7}$$
$$x = 2 \pm \sqrt{7}$$

Describe any errors. Then solve the equation correctly.

Group Exploration

Looking ahead: Deriving a formula for solving quadratic equations of the form $ax^2 + bx + c = 0$

1. Solve the equation $2x^2 + 9x + 3 = 0$.
2. Solve the quadratic equation $ax^2 + bx + c = 0$ for x. [**Hint:** Follow the same steps as in Problem 1.]
3. Use your result from Problem 2 to solve the equation $3x^2 + 11x + 5 = 0$. [**Hint:** Substitute the appropriate values of a, b, and c into your formula from Problem 2.]

Homework 9.5

For extra help ▶ MyMathLab® 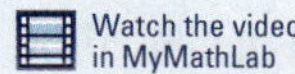 Watch the videos in MyMathLab  Download the MyDashboard App

Find the value of c for which the expression is a perfect-square trinomial. Then factor the perfect-square trinomial.

1. $x^2 + 12x + c$
2. $x^2 + 20x + c$
3. $x^2 - 14x + c$
4. $x^2 - 18x + c$
5. $x^2 + 7x + c$
6. $x^2 + 11x + c$
7. $x^2 - 3x + c$
8. $x^2 - 5x + c$
9. $x^2 + \frac{1}{2}x + c$
10. $x^2 + \frac{1}{7}x + c$
11. $x^2 - \frac{4}{5}x + c$
12. $x^2 - \frac{3}{4}x + c$

Solve by completing the square. All solutions are real numbers.

13. $x^2 + 6x = 1$

14. $x^2 + 8x = 3$

15. $p^2 - 2p = 19$

16. $r^2 - 10r = 2$

17. $x^2 + 12x = -4$

18. $x^2 + 10x = 3$

19. $x^2 - 10x - 7 = 0$

20. $x^2 - 14x - 11 = 0$

21. $x^2 + 4x - 24 = 0$

22. $x^2 + 6x - 9 = 0$

23. $x^2 - 7x = 3$

24. $x^2 - 3x = 12$

25. $t^2 + 5t - 4 = 0$

26. $w^2 + 9w - 9 = 0$

27. $x^2 - \frac{5}{2}x = \frac{1}{2}$

28. $x^2 - \frac{4}{3}x = \frac{5}{3}$

29. $5p^2 + 10p = 35$

30. $2k^2 + 8k = 16$

31. $4x^2 - 24x + 4 = 0$

32. $3x^2 - 6x - 33 = 0$

33. $2x^2 + 8x = 3$

34. $3x^2 + 12x = 1$

35. $2r^2 - r - 7 = 0$

36. $3m^2 - m - 5 = 0$

37. $3x^2 + 4x - 5 = 0$

38. $5x^2 + 2x - 2 = 0$

39. $6x^2 - 8x = -1$

40. $4x^2 - 6x = 3$

41. $8w^2 + 4w - 3 = 0$

42. $6p^2 + 9p + 2 = 0$

Find all complex-number solutions by completing the square.

43. $x^2 + 2x = -7$

44. $x^2 + 10x = -28$

45. $x^2 - 6x + 17 = 0$

46. $x^2 - 8x + 36 = 0$

47. $k^2 + 3k + 4 = 0$

48. $m^2 + 5m + 7 = 0$

49. $x^2 + \frac{2}{3}x + \frac{7}{3} = 0$

50. $x^2 + \frac{5}{2}x + \frac{7}{2} = 0$

51. $4r^2 - 3r = -5$

52. $3t^2 - 2t = -6$

53. $4p^2 + 6p + 3 = 0$

54. $6w^2 + 3w + 3 = 0$

Find all x-intercepts.

55. $f(x) = x^2 - 8x + 3$

56. $g(x) = x^2 - 12x + 1$

57. $h(x) = 2x^2 - 5x - 4$

58. $f(x) = 3x^2 + 2x - 7$

59. $g(x) = x^2 + 10x + 25$

60. $h(x) = x^2 - 8x + 16$

Solve by the method of your choice. Use a graphing calculator to verify your work.

61. $x^2 - 9 = 0$

62. $x^2 - 16 = 0$

63. $r^2 = 11r - 30$

64. $p^2 = 6p + 27$

65. $(x - 5)^2 = 32$

66. $(x + 3)^2 = 45$

67. $3x^2 + 5x = 12$

68. $2x^2 - 3x = 5$

69. $x^2 = 13$

70. $x^2 = 21$

71. $t^2 - 6t - 3 = 0$

72. $m^2 + 8m - 4 = 0$

73. Let $f(x) = x^2 + 6x + 13$.

a. Find x when $f(x) = 3$.

b. Find x when $f(x) = 4$.

c. Find x when $f(x) = 6$.

74. Let $f(x) = 2x^2 - 8x + 9$.

a. Find x when $f(x) = 2$.

b. Find x when $f(x) = 1$.

c. Find x when $f(x) = 0$.

For Exercises 75–80, solve the given equation or system by referring to the solutions of the functions shown in Table 32.

75. $x^2 - 5x + 4 = -\frac{1}{2}x^2 + x - \frac{1}{2}$

76. $-\frac{1}{2}x^2 + x - \frac{1}{2} = -\frac{1}{4}x^2 - \frac{1}{4}x - \frac{1}{2}$

77. $x^2 - 5x + 4 = -2$

78. $-\frac{1}{2}x^2 + x - \frac{1}{2} = -\frac{1}{2}$

79. $y = -\frac{1}{2}x^2 + x - \frac{1}{2}$
$y = -\frac{1}{4}x^2 - \frac{1}{4}x - \frac{1}{2}$

80. $y = x^2 - 5x + 4$
$y = -\frac{1}{2}x^2 + x - \frac{1}{2}$

Table 32 Some Solutions of Three Functions (Exercises 75–80)

x	0	1	2	3	4	5	6
$y = x^2 - 5x + 4$	4	0	−2	−2	0	4	10
$y = -\frac{1}{2}x^2 + x - \frac{1}{2}$	−0.5	0	−0.5	−2	−4.5	−8	−12.5
$y = -\frac{1}{4}x^2 - \frac{1}{4}x - \frac{1}{2}$	−0.5	−1	−2	−3.5	−5.5	−8	−11

Concepts

81. A student tries to solve the equation $4x^2 + 6x = 1$ by completing the square:

$$4x^2 + 6x = 1$$
$$4x^2 + 6x + 9 = 1 + 9$$
$$(2x + 3)^2 = 10$$
$$2x + 3 = \pm\sqrt{10}$$
$$2x = -3 \pm \sqrt{10}$$
$$x = \frac{-3 \pm \sqrt{10}}{2}$$

Describe any errors. Then solve the equation correctly.

82. A student tries to solve the equation $x^2 + 8x - 3 = 0$:

$$x^2 - 8x - 3 = 0$$
$$x^2 - 8x = 3$$
$$x^2 - 8x + 16 = 3$$
$$(x - 4)^2 = 3$$
$$x - 4 = \pm\sqrt{3}$$
$$x = 4 \pm \sqrt{3}$$

Describe any errors. Then solve the equation correctly.

83. a. Simplify $(x + k)^2$.

b. Consider the polynomial $x^2 + 2kx + k^2$, where k is a constant. Show that if you divide the coefficient of x by 2 and square the result, you get the constant term.

84. To solve $x^2 + 8x = 3$ by completing the square, we begin by computing $\left(\frac{8}{2}\right)^2 = 4^2 = 16$. Why must we add 16 to *both* sides of the equation $x^2 + 8x = 3$?

85. a. Solve the equation $x^2 - 6x + 8 = 0$ by factoring.

b. Solve the equation $x^2 - 6x + 8 = 0$ by completing the square.

c. In your opinion, is it easier to solve $x^2 - 6x + 8 = 0$ by factoring or by completing the square? Explain.

86. a. Solve the equation $x^2 + 4x = 12$ by factoring.

b. Solve the equation $x^2 + 4x = 12$ by completing the square.

c. In your opinion, is it easier to solve $x^2 + 4x = 12$ by factoring or by completing the square? Explain.

87. **a.** Can the equation $x^2 + 4x = 7$ be solved by completing the square? If so, solve the equation by using this method.
b. Can the equation $x^2 + 4x = 7$ be solved by factoring? If so, solve the equation by using this method.
c. Explain how to decide whether to solve a quadratic equation by completing the square or by factoring.

88. Find nonzero values of a, b, and c such that the equation $ax^2 + bx + c = 0$ has two imaginary-number solutions. Your equation should be different from those in the text. [**Hint:** Begin with an appropriate equation of the form $(x - h)^2 = k$, and simplify the left side of the equation.]

89. Compare the methods of solving a quadratic equation by factoring, by using the square root property, and by completing the square. Describe the methods, as well as their advantages and disadvantages. (See page 9 for guidelines on writing a good response.)

90. Describe how to solve a quadratic equation by completing the square. (See page 9 for guidelines on writing a good response.)

Related Review

91. Factor $w^2 - 10w + 25$.

92. Factor $t^2 + 14t + 49$.

93. Factor $x^2 + \frac{5}{3}x + \frac{25}{36}$.

94. Factor $x^2 - \frac{3}{2}x + \frac{9}{16}$.

Expressions, Equations, Functions, and Graphs

Perform the indicated instruction. Then use words such as linear, quadratic, cubic, power, radical, polynomial, degree, function, one variable, *and* two variables *to describe the expression, equation, or system.*

95. Simplify $-5x^2(x - 4)^2$.

96. Solve the following:

$$y = -3x + 1$$
$$5x + 2y = 1$$

97. Solve $2(x - 4)^2 + 5 = 11$.

98. Solve $4(2x - 5) = -3x + 1$.

99. Factor $p^2 - 8pq + 16q^2$.

100. Graph $y = -3x + 1$ by hand.

9.6 Using the Quadratic Formula to Solve Quadratic Equations

Objectives

» Solve quadratic equations by using the *quadratic formula.*

» Use the *discriminant* to determine the number and type of solutions of a quadratic equation.

» Decide which method to use to solve a quadratic equation.

» Use the quadratic formula to make predictions with a quadratic model.

Any quadratic equation can be solved by completing the square. However, this method may be difficult to use on most quadratic equations. An easier option is to use an important equation called the *quadratic formula,* which can also be used to solve *any* quadratic equation.

The Quadratic Formula

To find the quadratic formula, we solve the general quadratic equation

$$ax^2 + bx + c = 0$$

by completing the square. For now, we assume a is positive:

$$ax^2 + bx + c = 0 \qquad \text{General quadratic equation}$$

$$x^2 + \frac{b}{a}x + \frac{c}{a} = 0 \qquad \text{Divide both sides by } a.$$

$$x^2 + \frac{b}{a}x = -\frac{c}{a} \qquad \text{Subtract } \frac{c}{a} \text{ from both sides.}$$

$$x^2 + \frac{b}{a}x + \frac{b^2}{4a^2} = -\frac{c}{a} + \frac{b^2}{4a^2} \qquad \text{Add } \left(\frac{b}{a}\cdot\frac{1}{2}\right)^2 = \left(\frac{b}{2a}\right)^2 = \frac{b^2}{4a^2} \text{ to both sides.}$$

$$\left(x + \frac{b}{2a}\right)^2 = -\frac{c}{a}\cdot\frac{4a}{4a} + \frac{b^2}{4a^2} \qquad \text{Factor left side; find LCD.}$$

$$\left(x + \frac{b}{2a}\right)^2 = \frac{b^2 - 4ac}{4a^2} \qquad \text{Add numerators and keep common denominator: } \frac{A}{B} + \frac{C}{B} = \frac{A + C}{B}$$

$$x + \frac{b}{2a} = \pm\sqrt{\frac{b^2 - 4ac}{4a^2}} \qquad \text{Square root property}$$

$$x + \frac{b}{2a} = \pm\frac{\sqrt{b^2 - 4ac}}{2a} \qquad \text{Quotient property: } \sqrt{\frac{A}{B}} = \frac{\sqrt{A}}{\sqrt{B}}$$

$$x = -\frac{b}{2a} \pm \frac{\sqrt{b^2 - 4ac}}{2a}$$ Subtract $\frac{b}{2a}$ from both sides.

$$x = \frac{-b \pm \sqrt{b^2 - 4ac}}{2a}$$ Add/subtract numerators and keep common denominator.

We have found a formula (the last line) for the solutions of a quadratic equation $ax^2 + bx + c = 0$, where a is positive. In a similar way, we could derive the same formula for a quadratic equation where a is negative.

Quadratic Formula

The solutions of a quadratic equation $ax^2 + bx + c = 0$ are given by the **quadratic formula:**

$$x = \frac{-b \pm \sqrt{b^2 - 4ac}}{2a}$$

WARNING For the fraction in the quadratic formula, notice that the term $-b$ is part of the numerator:

$$\frac{-b \pm \sqrt{b^2 - 4ac}}{2a} \leftarrow \text{Correct}$$

$$-b \pm \frac{\sqrt{b^2 - 4ac}}{2a} \leftarrow \text{Incorrect}$$

Example 1 Solving by Using the Quadratic Formula

Solve $x^2 - 6x + 8 = 0$.

Solution

Comparing $x^2 - 6x + 8 = 0$ with $ax^2 + bx + c = 0$, we see $a = 1$, $b = -6$, and $c = 8$. We substitute these values for a, b, and c in the quadratic formula:

$$x = \frac{-(-6) \pm \sqrt{(-6)^2 - 4(1)(8)}}{2(1)}$$ Substitute 1 for a, -6 for b, and 8 for c.

$$x = \frac{6 \pm \sqrt{4}}{2}$$ Simplify.

$$x = \frac{6 \pm 2}{2}$$ $\sqrt{4} = 2$

$$x = \frac{6 - 2}{2} \quad \text{or} \quad x = \frac{6 + 2}{2}$$ Write as two equations.

$$x = 2 \quad \text{or} \quad x = 4$$

The solutions are 2 and 4.

Instead of using the quadratic formula, we could have solved $x^2 - 6x + 8 = 0$ by factoring:

$$x^2 - 6x + 8 = 0$$ Original equation

$$(x - 2)(x - 4) = 0$$ Factor left side.

$$x - 2 = 0 \quad \text{or} \quad x - 4 = 0$$ Zero factor property

$$x = 2 \quad \text{or} \quad x = 4$$

A benefit of the quadratic formula is that we can use it to solve equations that are difficult or even impossible to solve by factoring. In Example 2, we will solve an equation that is impossible to solve by factoring.

Example 2 Solving by Using the Quadratic Formula

Solve $5x^2 - x = 2$.

Solution

First, we write the equation in the form $ax^2 + bx + c = 0$:

$$5x^2 - x = 2 \quad \text{Original equation}$$
$$5x^2 - x - 2 = 0 \quad \text{Subtract 2 from both sides.}$$

Then we substitute $a = 5$, $b = -1$, and $c = -2$ into the quadratic formula:

$$x = \frac{-b \pm \sqrt{b^2 - 4ac}}{2a} \quad \text{Quadratic formula}$$
$$x = \frac{-(-1) \pm \sqrt{(-1)^2 - 4(5)(-2)}}{2(5)} \quad \text{Substitute into quadratic formula.}$$
$$x = \frac{1 \pm \sqrt{1 + 40}}{10} \quad \text{Simplify.}$$
$$x = \frac{1 \pm \sqrt{41}}{10} \quad \text{Add.}$$

We use graphing calculator tables to check that if $\frac{1 - \sqrt{41}}{10}$ or $\frac{1 + \sqrt{41}}{10}$ is substituted for x in the expression $5x^2 - x$, the result is 2 (see Fig. 73).

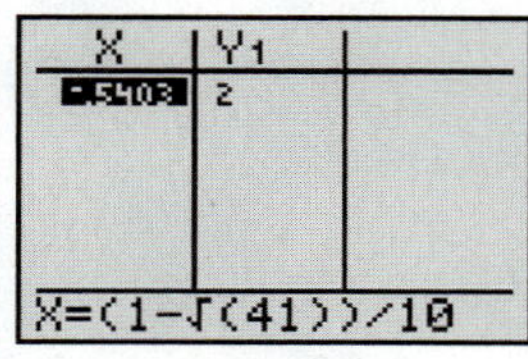

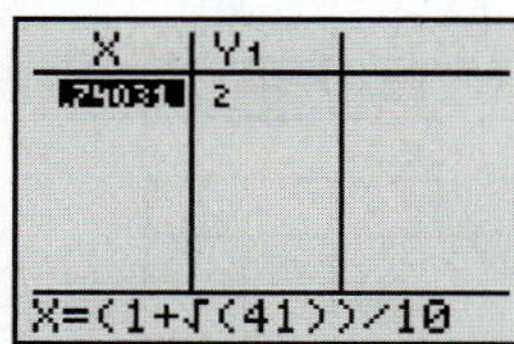

Figure 73 Verify the work

By the check in Fig. 73, we see that $\frac{1 - \sqrt{41}}{10}$ is about -0.54. So, $\frac{1 - \sqrt{41}}{10}$ is simply a number. Likewise, $\frac{1 + \sqrt{41}}{10}$ is a number (about 0.74).

Example 3 Solving by Using the Quadratic Formula

Solve $2x^2 = 10x - 3$.

Solution

First, we write $2x^2 = 10x - 3$ in the form $ax^2 + bx + c = 0$:

$$2x^2 - 10x + 3 = 0$$

So, $a = 2$, $b = -10$, and $c = 3$. By the quadratic formula,

$$x = \frac{-(-10) \pm \sqrt{(-10)^2 - 4(2)(3)}}{2(2)} \quad \text{Substitute 2 for } a, -10 \text{ for } b, \text{ and 3 for } c.$$
$$x = \frac{10 \pm \sqrt{76}}{4} \quad \text{Simplify.}$$
$$x = \frac{10 \pm 2\sqrt{19}}{4} \quad \sqrt{76} = \sqrt{4 \cdot 19} = \sqrt{4}\sqrt{19} = 2\sqrt{19}$$
$$x = \frac{10 - 2\sqrt{19}}{4} \quad \text{or} \quad x = \frac{10 + 2\sqrt{19}}{4} \quad \text{Write as two equations.}$$

$$x = \frac{2(5 - \sqrt{19})}{4} \quad \text{or} \quad x = \frac{2(5 + \sqrt{19})}{4} \qquad \textit{Factor out 2.}$$

$$x = \frac{5 - \sqrt{19}}{2} \quad \text{or} \quad x = \frac{5 + \sqrt{19}}{2} \qquad \textit{Simplify.}$$

The solutions are $\frac{5 \pm \sqrt{19}}{2}$.

Example 4 Solving by Using the Quadratic Formula

Solve $\frac{1}{2}x^2 - \frac{5}{4}x = \frac{3}{2}$.

Solution

To begin, we clear the equation of fractions by multiplying both sides by the LCD, 4:

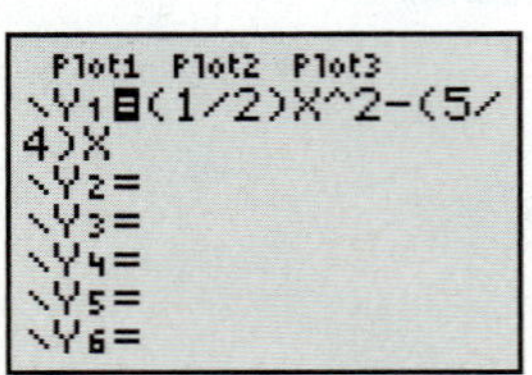

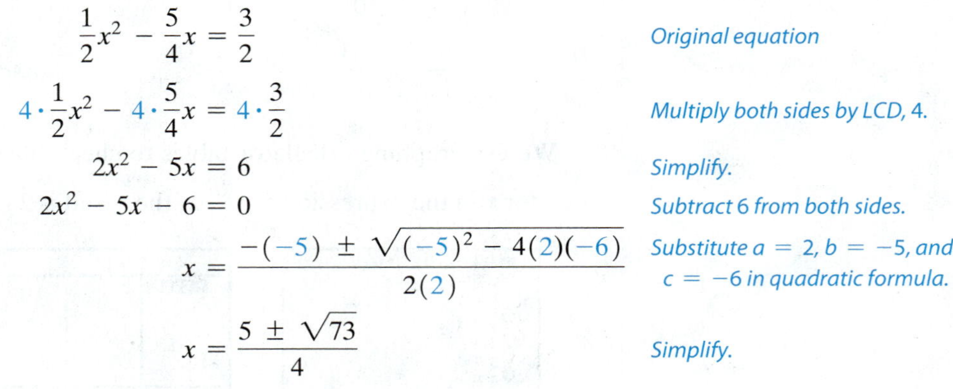

$$\frac{1}{2}x^2 - \frac{5}{4}x = \frac{3}{2} \qquad \textit{Original equation}$$

$$4 \cdot \frac{1}{2}x^2 - 4 \cdot \frac{5}{4}x = 4 \cdot \frac{3}{2} \qquad \textit{Multiply both sides by LCD, 4.}$$

$$2x^2 - 5x = 6 \qquad \textit{Simplify.}$$

$$2x^2 - 5x - 6 = 0 \qquad \textit{Subtract 6 from both sides.}$$

$$x = \frac{-(-5) \pm \sqrt{(-5)^2 - 4(2)(-6)}}{2(2)} \qquad \textit{Substitute } a = 2, b = -5, \textit{and } c = -6 \textit{ in quadratic formula.}$$

$$x = \frac{5 \pm \sqrt{73}}{4} \qquad \textit{Simplify.}$$

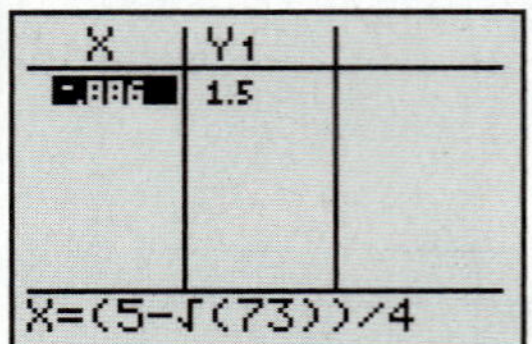

Figure 74 Verify the work

The solutions are $\frac{5 \pm \sqrt{73}}{4}$. We enter $y = \frac{1}{2}x^2 - \frac{5}{4}x$ in a graphing calculator and check that, for both inputs $\frac{5 \pm \sqrt{73}}{4}$, the output is $\frac{3}{2} = 1.5$ (see Fig. 74).

Example 5 Finding Approximate Solutions

Find approximate solutions of the equation $2.3x(1.4x - 5.3) = 6.9x - 7.2$.

Solution

First, we write the equation in the form $ax^2 + bx + c = 0$:

$$2.3x(1.4x - 5.3) = 6.9x - 7.2 \qquad \textit{Original equation}$$

$$3.22x^2 - 12.19x = 6.9x - 7.2 \qquad \textit{Distributive law}$$

$$3.22x^2 - 19.09x + 7.2 = 0 \qquad \textit{Subtract 6.9x from both sides; add 7.2 to both sides.}$$

Then we substitute 3.22 for a, -19.09 for b, and 7.2 for c in the quadratic formula:

$$x = \frac{-(-19.09) \pm \sqrt{(-19.09)^2 - 4(3.22)(7.2)}}{2(3.22)} \qquad \textit{Substitute into quadratic formula.}$$

$$x = \frac{19.09 \pm \sqrt{271.6921}}{6.44} \qquad \textit{Simplify.}$$

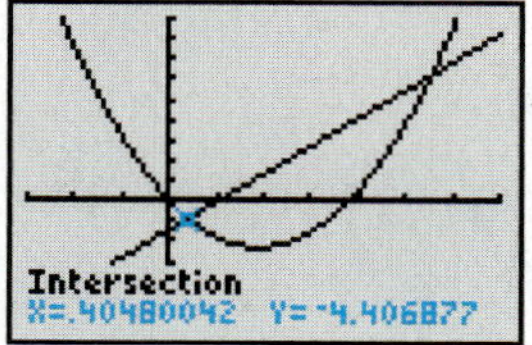

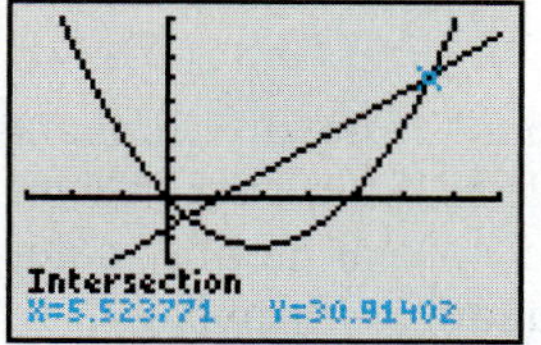

Figure 75 Verify the work

$$x \approx \frac{19.09 \pm 16.48}{6.44} \quad \textit{Approximate square root.}$$

$$x \approx 0.41 \quad \text{or} \quad x \approx 5.52 \quad \textit{Compute.}$$

We use graphing calculator graphs to verify our work (see Fig. 75).

Example 6 Finding x-intercepts

Find the x-intercepts of the parabola $f(x) = 6x^2 + 3x - 2$.

Solution

First, we substitute 0 for $f(x)$. Then, we use the quadratic formula to solve for x:

$$0 = 6x^2 + 3x - 2 \quad \textit{Substitute 0 for } f(x).$$

$$x = \frac{-3 \pm \sqrt{3^2 - 4(6)(-2)}}{2(6)} \quad \textit{Substitute } a = 6, b = 3, c = -2.$$

$$x = \frac{-3 \pm \sqrt{57}}{12} \quad \textit{Simplify.}$$

The x-intercepts are $\left(\frac{-3 - \sqrt{57}}{12}, 0\right)$ and $\left(\frac{-3 + \sqrt{57}}{12}, 0\right)$. We use "zero" on a graphing calculator to verify that the approximate x-intercepts are $(-0.88, 0)$ and $(0.38, 0)$. See Fig. 76.

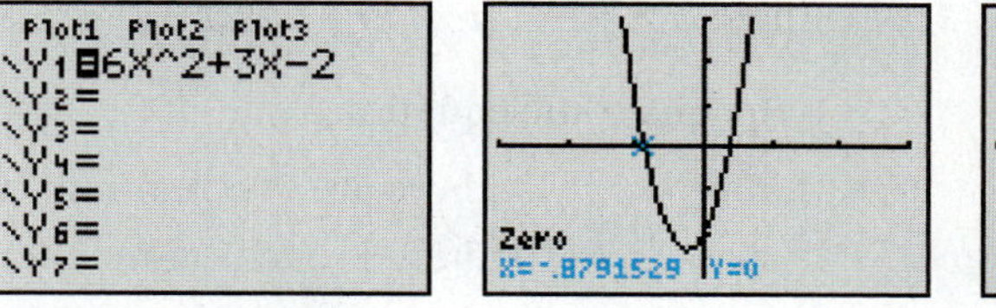

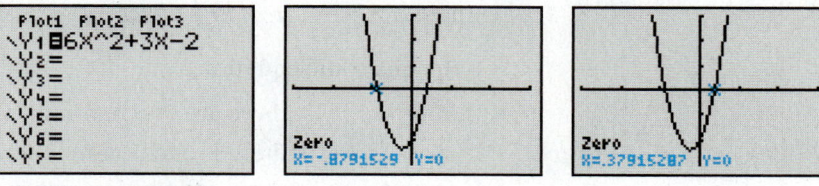

Figure 76 Verify the work

Solving a Quadratic Equation That Has Imaginary-Number Solutions

In Example 7, we will use the quadratic formula to solve a quadratic equation that has imaginary-number solutions.

Example 7 Solving by Using the Quadratic Formula

Solve $-3x^2 + 5x - 4 = 0$.

Solution

First, we multiply both sides of the equation by -1 so we can avoid having a negative denominator after we use the quadratic formula:

$$-3x^2 + 5x - 4 = 0 \quad \textit{Original equation}$$

$$3x^2 - 5x + 4 = 0 \quad \textit{Multiply both sides by } -1.$$

(Another benefit is that we have fewer negative numbers to substitute into the quadratic formula.) Then we substitute $a = 3$, $b = -5$, and $c = 4$ in the quadratic formula:

$$x = \frac{-(-5) \pm \sqrt{(-5)^2 - 4(3)(4)}}{2(3)} \quad \textit{Substitute into quadratic formula.}$$

$$x = \frac{5 \pm \sqrt{-23}}{6} \quad \textit{Simplify.}$$

$$x = \frac{5 \pm i\sqrt{23}}{6} \quad \sqrt{-p} = i\sqrt{p}, \textit{ where } p > 0$$

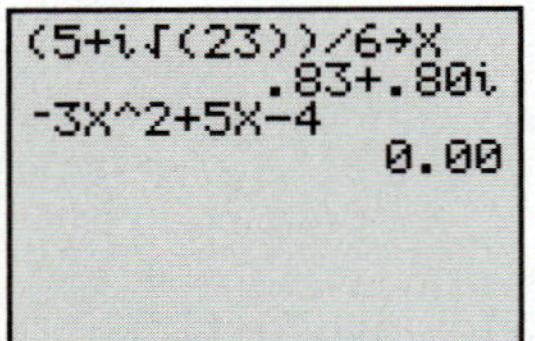

Figure 77 Verify the work

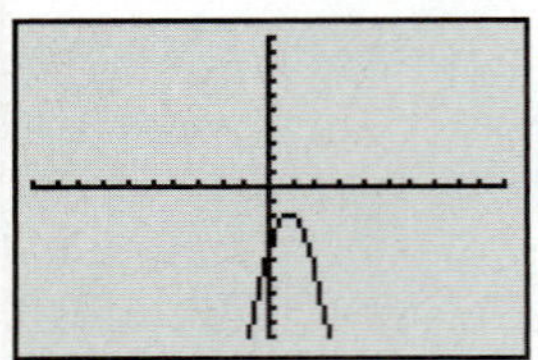

Figure 78 The graph of $y = -3x^2 + 5x - 4$ has no x-intercepts

To verify our work, we store each result as x and check that $-3x^2 + 5x - 4$ is equal to 0 (see Fig. 77). We press MODE and set FLOAT to 2 so the numbers in the result are rounded to the second decimal place.

The graphical check in Fig. 78 shows that the graph of $y = -3x^2 + 5x - 4$ does not have any x-intercepts. This means the equation $-3x^2 + 5x - 4 = 0$ has no real-number solutions, which checks.

Determining the Number of Real-Number Solutions

Recall from Section 8.5 that a quadratic equation can have two, one, or no real-number solutions. Let us see how this fact relates to the quadratic formula

$$x = \frac{-b \pm \sqrt{b^2 - 4ac}}{2a}$$

The answer lies with the number $b^2 - 4ac$, known as the *discriminant*. If the discriminant is positive, then there are two real-number solutions; see Example 5. If the discriminant is negative, then there are two imaginary-number solutions (and no real-number solutions); see Example 7. Finally, if the discriminant is 0, then the quadratic formula gives

$$x = \frac{-b \pm \sqrt{0}}{2a} = \frac{-b}{2a}$$

and therefore there is one real-number solution.

Determining the Number and Type of Solutions

For the quadratic equation $ax^2 + bx + c = 0$, the **discriminant** is $b^2 - 4ac$. Also,

- If $b^2 - 4ac > 0$, there are two real-number solutions.
- If $b^2 - 4ac = 0$, there is one real-number solution.
- If $b^2 - 4ac < 0$, there are two imaginary-number solutions (and no real-number solutions).

Example 8 Determining the Number and Type of Solutions

Determine the number and type of solutions of the equation $2x^2 - 3x + 5 = 0$.

Solution

Since $b^2 - 4ac = (-3)^2 - 4(2)(5) = -31 < 0$, we conclude that the quadratic equation $2x^2 - 3x + 5 = 0$ has two imaginary-number solutions (and no real-number solutions).

We can also use the discriminant to determine the number of points on a parabola at a given height.

Example 9 Finding the Number of Points at a Given Height

For $f(x) = x^2 - 6x + 12$, find the number of points that lie on the graph of f at the indicated height.

1. $y = 5$ **2.** $y = 3$ **3.** $y = 1$

Solution

1. We substitute 5 for $f(x)$ in the equation $f(x) = x^2 - 6x + 12$:

$$x^2 - 6x + 12 = 5 \quad \text{Substitute 5 for } f(x).$$
$$x^2 - 6x + 7 = 0 \quad \text{Subtract 5 from both sides.}$$

Since $b^2 - 4ac = (-6)^2 - 4(1)(7) = 8 > 0$, we conclude that there are two solutions of the equation $x^2 - 6x + 12 = 5$, which means two (symmetric) points have height $y = 5$.

2. We substitute 3 for $f(x)$:

$$x^2 - 6x + 12 = 3 \quad \text{Substitute 3 for } f(x).$$
$$x^2 - 6x + 9 = 0 \quad \text{Subtract 3 from both sides.}$$

Since $b^2 - 4ac = (-6)^2 - 4(1)(9) = 0$, we conclude that there is one solution of the equation $x^2 - 6x + 12 = 3$, which means one point has height $y = 3$. Since the point does not have a symmetric point, it must be the vertex of the parabola.

3. We substitute 1 for $f(x)$:

$$x^2 - 6x + 12 = 1 \quad \text{Substitute 1 for } f(x).$$
$$x^2 - 6x + 11 = 0 \quad \text{Subtract 1 from both sides.}$$

Since $b^2 - 4ac = (-6)^2 - 4(1)(11) = -8 < 0$, we conclude that there are no real-number solutions of the equation $x^2 - 6x + 12 = 1$, which means no points on the parabola have height $y = 1$.

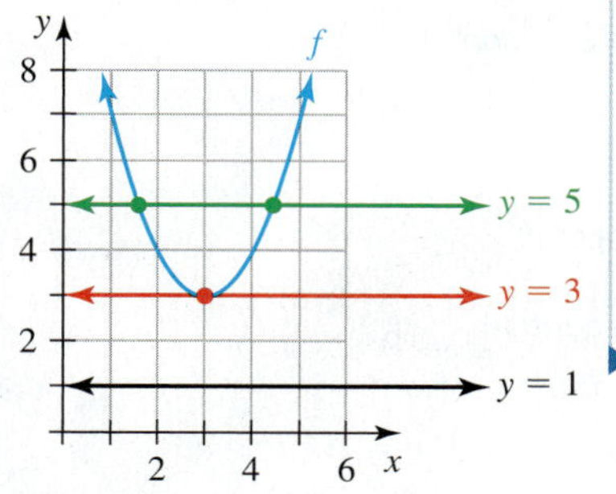

Figure 79 Graph of $f(x) = x^2 - 6x + 12$

In Example 9, we found that the parabola $f(x) = x^2 - 6x + 12$ has exactly two points at height $y = 5$, exactly one point at height $y = 3$, and no points at height $y = 1$. We indicate the three points on the graph of f in Fig. 79.

Deciding Which Method to Use to Solve a Quadratic Equation

In Chapter 8 and in this chapter, we have discussed four ways to solve quadratic equations: factoring, the square root property, completing the square, and the quadratic formula. How do we know which method to use?

Remember that **any quadratic equation can be solved by using the quadratic formula.** Although any equation can also be solved by completing the square, that method is much more difficult to use than the quadratic formula.

A low percentage of quadratic equations can be solved by factoring, because almost all polynomials are prime. However, if an equation *can* be solved by simple factoring techniques, then it is easier to solve by factoring than by any of the other three methods.

Here are some guidelines on deciding which method to use to solve a quadratic equation:

Method	When to Use
Factoring	For equations that can easily be put into the form $ax^2 + bx + c = 0$ and where $ax^2 + bx + c$ can easily be factored.
Square root property	For equations that can easily be put into the form $x^2 = k$ or $(x + p)^2 = k$.
Completing the square	When the directions require it.
Quadratic formula	For all equations except those that can easily be solved by factoring or by using the square root property.

Example 10 Deciding Which Method to Use

Solve.

1. $w^2 - 3w - 54 = 0$

2. $(x + 3)^2 = 5$

3. $(x + 3)(x - 2) = 2(x + 1)$

Solution

1. The polynomial $w^2 - 3w - 54$ is factorable, so we solve the quadratic equation $w^2 - 3w - 54 = 0$ by factoring:

$$w^2 - 3w - 54 = 0 \quad \text{Original equation}$$
$$(w + 6)(w - 9) = 0 \quad \text{Factor left side.}$$
$$w + 6 = 0 \quad \text{or} \quad w - 9 = 0 \quad \text{Zero factor property}$$
$$w = -6 \quad \text{or} \quad w = 9$$

2. The equation is of the form $(x + p)^2 = k$, so we solve it by using the square root property:

$$(x + 3)^2 = 5 \quad \text{Original equation}$$
$$x + 3 = \pm\sqrt{5} \quad \text{Square root property}$$
$$x = -3 \pm \sqrt{5}$$

3. First, we write the equation in $ax^2 + bx + c = 0$ form:

$$(x + 3)(x - 2) = 2(x + 1) \quad \text{Original equation}$$
$$x^2 + x - 6 = 2x + 2 \quad \text{Multiply; distributive law}$$
$$x^2 - x - 8 = 0 \quad \text{Write in } ax^2 + bx + c = 0 \text{ form.}$$

The polynomial $x^2 - x - 8$ is prime, so we can't solve the equation $x^2 - x - 8 = 0$ by factoring. The equation can't be put into the form $x^2 = k$, and it can't easily be put into the form $(x + p)^2 = k$, so we don't try to solve it by using the square root property. Instead, we substitute $a = 1$, $b = -1$, and $c = -8$ in the quadratic formula:

$$x = \frac{-(-1) \pm \sqrt{(-1)^2 - 4(1)(-8)}}{2(1)} \quad \text{Substitute into quadratic formula.}$$
$$x = \frac{1 \pm \sqrt{33}}{2} \quad \text{Simplify.}$$

Table 33 Bottled-Water Consumption

Year	Bottled-Water Consumption (billions of gallons)
1990	2.2
1995	3.1
2000	4.7
2005	7.5
2009	10.6

Source: *Beverage Marketing Corporation*

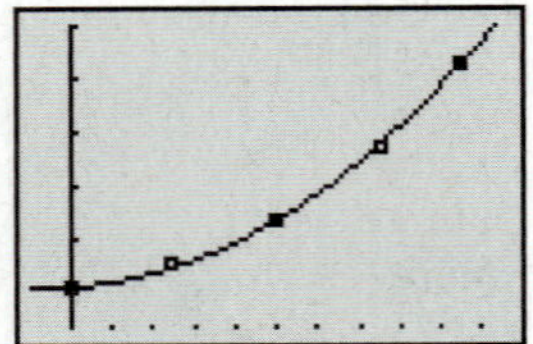

Figure 80 Check the fit

Using a Quadratic Function to Model a Situation

We can use the quadratic formula to make a prediction about the independent variable of a quadratic model.

Example 11 Modeling with a Quadratic Function

Annual consumption of bottled water in the United States has been increasing since 1990 (see Table 33). Let $f(t)$ be annual bottled-water consumption (in billions of gallons) at t years since 1990. A possible equation of f is

$$f(t) = 0.021t^2 + 0.04t + 2.25$$

1. Verify that f models the data well.
2. Predict when annual consumption will reach 20 billion gallons.

Solution

1. We draw the graph of f and the scattergram of the data in the same viewing window (see Fig. 80). It appears f is a reasonable model.
2. To predict when annual consumption will be 20 billion gallons, we substitute 20 for $f(t)$ and solve for t:

$$20 = 0.021t^2 + 0.04t + 2.25 \quad \text{Substitute 20 for } f(t).$$
$$0 = 0.021t^2 + 0.04t - 17.75 \quad \text{Subtract 20 from both sides.}$$

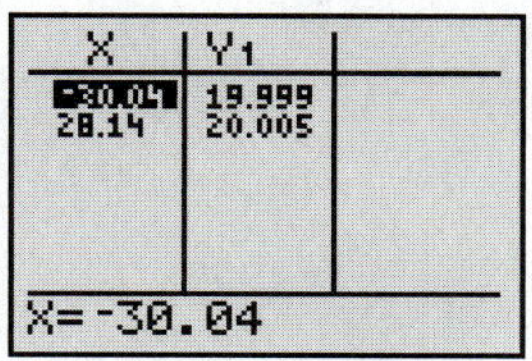

Figure 81 Verify the results

$$t = \frac{-0.04 \pm \sqrt{0.04^2 - 4(0.021)(-17.75)}}{2(0.021)}$$ *Substitute into quadratic formula.*

$$= \frac{-0.04 \pm \sqrt{1.4926}}{0.042}$$ *Simplify.*

$$\approx \frac{-0.04 \pm 1.2217}{0.042}$$ *Approximate square root.*

$$t \approx -30.04 \quad \text{or} \quad 28.14$$

We verify the results by entering $y = 0.021t^2 + 0.04t + 2.25$ in a graphing calculator and checking that the inputs -30.04 and 28.14 lead to outputs of about 20 (see Fig. 81).

The inputs -30.04 and 28.14 represent the years 1960 and 2018, respectively. The estimate of 1960 is a result of model breakdown, as a little research would show that bottled-water consumption in 1960 was much less than 20 billion gallons. Therefore, we predict it will be 2018 when bottled-water consumption reaches 20 billion gallons.

Recall from Section 8.6 that, to make a prediction about the dependent variable of a quadratic model, we substitute a value for the independent variable, then solve for the dependent variable. To make a prediction about the independent variable, we substitute a value for the dependent variable, then solve for the independent variable, usually by using the quadratic formula.

Group Exploration

Comparing methods of solving quadratic equations

1. Solve the equation $x^2 + 5x + 6 = 0$ by factoring.
2. Solve the equation $x^2 + 5x + 6 = 0$ by completing the square.
3. Solve the equation $x^2 + 5x + 6 = 0$ by using the quadratic formula.
4. Compare your results from Problems 1, 2, and 3. Which method was easiest?
5. Repeat Problems 1–4 for the equation $x^2 + 4x - 7 = 0$.
6. Compare the methods of solving quadratic equations by factoring, by completing the square, and by using the quadratic formula. What are the advantages and disadvantages of each?

Group Exploration

Looking ahead: Finding an equation of a parabola

In this exploration, you will find an equation of the parabola that passes through the points $(1, 8)$, $(2, 15)$, and $(3, 24)$.

1. The point $(1, 8)$ lies on the parabola, so the ordered pair $(1, 8)$ should satisfy the equation $y = ax^2 + bx + c$. Find the equation that results from substituting 1 for x and 8 for y. This equation will be in terms of a, b, and c. Find another equation by using the ordered pair $(2, 15)$. Finally, find a third equation by using the ordered pair $(3, 24)$.
2. You should now have three equations, each in terms of a, b, and c. Choose any two of these equations and eliminate c. Then choose another pair of equations and again eliminate c.
3. You should now have two equations, both in terms of a and b, forming a system of two equations in two variables. Solve this system by substitution or elimination.
4. You should now know the values of a and b. Find c by substituting the values of a and b in one of the three equations found in Problem 1.
5. You should now know the values of a, b, and c. Substitute these values into the equation $y = ax^2 + bx + c$ to obtain an equation of the parabola.
6. Verify that the graph of your equation passes through the points $(1, 8)$, $(2, 15)$, and $(3, 24)$ by using a graphing calculator table or graph.

Homework 9.6

For extra help ▶ MyMathLab® 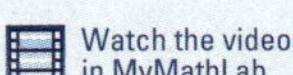 Watch the videos in MyMathLab 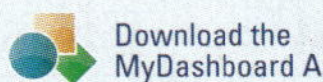 Download the MyDashboard App

Use the quadratic formula to solve the equation. All solutions are real numbers.

1. $2x^2 + 5x + 3 = 0$

2. $2x^2 + 3x + 1 = 0$

3. $x^2 + 3x - 5 = 0$

4. $x^2 + 7x - 2 = 0$

5. $3w^2 - 5w - 3 = 0$

6. $2p^2 - 7p - 2 = 0$

7. $3x^2 - 6x + 1 = 0$

8. $5x^2 - 4x - 3 = 0$

9. $t^2 = 4t + 3$

10. $w^2 = -2w + 5$

11. $-2x^2 + 5x = 3$

12. $-4x^2 + 7x = -2$

13. $3x^2 - 17 = 0$

14. $2x^2 - 15 = 0$

15. $2y^2 = -5y$

16. $5r^2 = -3r$

17. $\frac{2}{3}x^2 - \frac{5}{6}x = \frac{1}{3}$

18. $\frac{5}{8}x^2 + \frac{3}{4}x = \frac{1}{4}$

19. $(3x + 2)(x - 1) = 1$

20. $(4x - 3)(x - 1) = 4$

Find approximate solutions. Round any results to the second decimal place. All solutions are real numbers.

21. $2x^2 = 5x + 4$

22. $3x^2 = 9x + 2$

23. $2.85p^2 - 7.12p = 4.49$

24. $3.98r^2 - 2.17r = 3.68$

25. $-5.4x(x + 9.8) + 4.1 = 3.2 - 6.9x$

26. $7.1x(x - 4.9) - 7.1 = 2.5x + 6.3$

Find all x-intercepts.

27. $f(x) = 2x^2 - x - 7$

28. $f(x) = 3x^2 + 4x - 1$

29. $f(x) = 3x^2 + 2x + 5$

30. $f(x) = -5x^2 + 3x - 2$

31. $f(x) = x^2 + 2x - 5$

32. $f(x) = x^2 - 4x - 3$

Use the quadratic formula to find all complex-number solutions.

33. $x^2 - 3x + 8 = 0$

34. $x^2 - 3x + 15 = 0$

35. $-w^2 + 2w = 5$

36. $-r^2 + 4r = 7$

37. $\frac{1}{4}x^2 = 2x - \frac{9}{2}$

38. $\frac{5}{6}x^2 = 3x - \frac{7}{2}$

39. $3x(3x - 2) = -2$

40. $2x(5x - 2) = -1$

41. $3k^2 = 4k - 5$

42. $4y^2 = 2y - 1$

Solve by the method of your choice. All solutions are real numbers.

43. $4x^2 - 80 = 0$

44. $3x^2 - 36 = 0$

45. $5(w + 3)^2 + 2 = 8$

46. $3(w - 2)^2 - 1 = 6$

47. $m^2 = -12m - 36$

48. $t^2 = 14t - 49$

49. $-24x^2 + 18x = -60$

50. $-16x^2 + 20x = 4$

51. $\frac{1}{3}x^2 - \frac{3}{2}x = \frac{1}{6}$

52. $\frac{1}{8}x^2 - \frac{3}{4}x = \frac{1}{2}$

53. $(x - 5)(x + 2) = 3(x - 1) + 2$

54. $(x + 4)(x - 1) = 5(x + 2) - 1$

55. $25r^2 = 49$

56. $4p^2 = 81$

57. $(x - 1)^2 + (x + 2)^2 = 6$

58. $(x - 3)^2 + (x + 1)^2 = 17$

Use the method of your choice to find all complex-number solutions.

59. $4x^2 = -25$

60. $81x^2 = -49$

61. $-2t^2 + 5t = 6$

62. $-3w^2 + 4w = 2$

63. $(x - 6)^2 + 5 = -43$

64. $(x + 4)^2 - 3 = -66$

65. $(y - 2)(y - 5) = -4$

66. $(k + 3)(k - 2) = -25$

Determine the number and type of solutions.

67. $3x^2 + 4x - 5 = 0$

68. $x^2 - 5x - 8 = 0$

69. $2x^2 - 5x + 7 = 0$

70. $3x^2 - 2x + 5 = 0$

71. $4x^2 = 12x - 9$

72. $9x^2 = 6x - 1$

73. Let $f(x) = x^2 - 4x + 8$. Find the number of points that lie on the graph of f at the indicated height y.

a. $y = 3$ **b.** $y = 4$ **c.** $y = 5$

d. Use a graphing calculator to draw a graph of f, and sketch the graph on paper. Then explain why you found the number of points to be 0, 1, and 2 for parts (a), (b), and (c), respectively.

74. Let $g(x) = -x^2 + 6x - 2$. Find the number of points that lie on the graph of g at the indicated height y.

a. $y = 6$ **b.** $y = 7$ **c.** $y = 8$

d. Use a graphing calculator to draw a graph of g, and sketch the graph on paper. Then explain why you found the number of points to be 2, 1, and 0 for parts (a), (b), and (c), respectively.

75. Let $f(x) = x^2 - 6x + 7$. Find the coordinates of any points on the graph of f at height $y = 2$. Then find the vertex of the graph of f. Finally, sketch the graph of f.

76. Let $g(x) = x^2 + 8x + 6$. Find the approximate coordinates of any points on the graph of g at height $y = -5$. Then find the approximate vertex of the graph of g. Round all coordinates to the second decimal place. Finally, sketch the graph of g.

77. The *consumer confidence index* measures how optimistic consumers feel about the economy and their personal financial situation. The annual revenues from boats and accessories, new-home sales rates (in thousands of homes per year) in July, and the consumer confidence indexes in July are shown in Table 34 for various years.

Table 34 Annual Revenues from Boats and Accessories, July New-Home Sales Rates, and July Consumer Confidence Indexes

Year	Annual Revenue from Boats and Accessories (billions of dollars)	July New-Home Sales Rate (thousands of homes per year)	July Consumer Confidence Index
2007	37	800	106
2008	33	479	51
2009	31	402	48
2010	30	289	53
2011	32	300	59

Source: *National Marine Manufacturers Association*

Let $f(t)$ be the annual revenue from boats and accessories (in billions of dollars) at t years since 2005. A model of the situation is $f(t) = 0.929t^2 - 8.73t + 50.8$.

a. Use a graphing calculator to draw the graph of the model f and, in the same viewing window, the scattergram of the data. Does the model fit the data well?

b. Predict the revenue in 2014.

c. Predict when the annual revenue will be $55 billion.

d. For the years shown in Table 34, determine in which year the revenue from boats and accessories was the least. Do the same for the July new-home sales rate and for the July consumer confidence index. How much of a time lag does there seem to be between when consumers begin to feel better about the economy and when they begin to spend more?

78. The average state cigarette taxes per pack are listed in Table 35 for various years.

Table 35 Average State Cigarette Taxes per Pack

Year	Average State Cigarette Tax per pack (cents)
2001	43
2003	73
2005	93
2007	111
2009	134
2011	147
2012	149

Source: *Campaign for Tobacco-Free Kids*

Let $f(t)$ be the average state cigarette tax per pack (in cents) at t years since 2000. A model of the situation is $f(t) = -0.4t^2 + 14.97t + 29.22$.

a. Use a graphing calculator to draw the graph of the model and, in the same viewing window, the scattergram of the data. Does the model fit the data well?

b. Predict the average tax per pack in 2017.

c. Predict when the average tax per pack will be 169 cents.

d. About 14 billion packs of cigarettes were sold in 2010 (Source: *Federal Trade Commission*). Estimate *total* state cigarette taxes in 2010.

79. The percentages of police officers who are women are listed in Table 36 for various city populations.

Table 36 Percentages of Police Officers Who Are Women

City Population Group (in thousands)	Population Used to Represent City Population Group (in thousands)	Percent
0–9.999	5	8.3
10–24.999	17.5	7.5
25–49.999	37.5	8.5
50–99.999	75	9.4
100–249.999	175	11.7
250 or more	300	17.0

Source: *FBI Uniform Crime Report*

Let $f(n)$ be the percentage of police officers who are women in cities with populations of n thousand. A model of the situation is $f(n) = 0.00006n^2 + 0.012n + 7.88$.

a. Use a graphing calculator to draw the graph of the model and, in the same viewing window, the scattergram of the data. Does the model fit the data well?

b. Glen Ellyn, Illinois, has a population of 27.04 thousand. Estimate the percentage of police officers in Glen Ellyn who are women.

c. Find n when $f(n) = 10$. What does it mean in this situation?

80. The average weights of mako sharks are listed in Table 37 for various lengths.

Table 37 Average Weights of Mako Sharks

Length (inches)	Weight (pounds)
53.5	41.2
60.0	59.6
70.8	101.7
80.5	153.7
90.1	221.4
100.9	317.8

Source: *National Marine Fisheries Service*

Let $f(L)$ be the average weight (in pounds) of a mako shark with length L inches. A model of the situation is $f(L) = 0.0805L^2 - 6.64L + 167$.

a. Use a graphing calculator to draw the graph of the model f and, in the same viewing window, the scattergram of the data. Does the model fit the data well?

b. Find $f(75)$. What does it mean in this situation?
c. Find L when $f(L) = 75$. What does it mean in this situation?

81. A person throws a stone into the air. The height (in feet) $h(t)$ after t seconds is given by $h(t) = -16t^2 + 52t + 4$.
a. What is the height of the stone after 3 seconds?
b. When is the stone at a height of 30 feet?
c. When does the stone reach the ground?

82. A baseball is hit by a batter. The height (in feet) $h(t)$ of the ball after t seconds is given by $h(t) = -16t^2 + 125t + 4$.
a. What is the height of the ball after 2 seconds?
b. When is the ball at a height of 200 feet?
c. When does the ball reach the ground?

For Exercises 83–88, find approximate solutions of the given equation or system by referring to the graphs shown in Fig. 82. Round results or coordinates of results to the first decimal place.

83. $\frac{1}{2}x^2 + 1 = -x^2 - x + 5$

84. $-x^2 - x + 5 = \frac{2}{5}x - 2$

85. $\frac{1}{2}x^2 + 1 = 4$

86. $-x^2 - x + 5 = 1$

87. $y = \frac{2}{5}x - 2$
$y = -x^2 - x + 5$

88. $y = \frac{1}{2}x^2 + 1$
$y = -x^2 - x + 5$

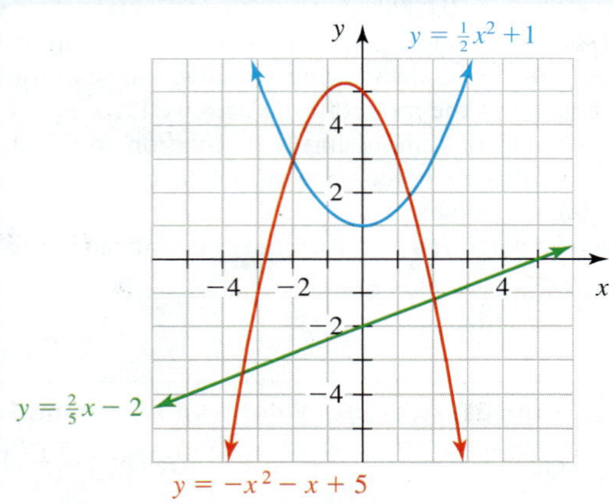

Figure 82 Exercises 83–88

Concepts

89. A student tries to solve $2x^2 + 5x = 1$:

$$x = \frac{-5 \pm \sqrt{5^2 - 4(2)(1)}}{2(2)}$$

$$x = \frac{-5 \pm \sqrt{17}}{4}$$

Describe any errors. Then solve the equation correctly.

90. A student tries to solve $3x^2 + 2x - 4 = 0$:

$$x = \frac{-2 \pm \sqrt{2^2 - 4(3)(-4)}}{2(3)}$$

$$x = \frac{-2 \pm \sqrt{52}}{6}$$

$$x = \frac{-2 \pm 2\sqrt{13}}{6}$$

$$x = \frac{-1 \pm 2\sqrt{13}}{3}$$

Describe any errors. Then solve the equation correctly.

91. A student tries to solve $5x^2 - 4x - 2 = 0$:

$$x = \frac{-(-4) \pm \sqrt{(-4)^2 - 4(5)(-2)}}{2(5)}$$

$$= \frac{4 \pm \sqrt{56}}{10}$$

Describe any errors. Then solve the equation correctly.

92. A student tries to solve $4x^2 + 3x - 2 = 0$:

$$x = -3 \pm \frac{\sqrt{3^2 - 4(4)(-2)}}{2(4)}$$

$$= -3 \pm \frac{\sqrt{41}}{8}$$

Describe any errors. Then solve the equation correctly.

93. The quadratic formula gives the solutions of any equation of the form $ax^2 + bx + c = 0$, where $a \neq 0$.
a. Find a "linear formula" that gives the solution of *any* equation of the form $mx + b = 0$, where $m \neq 0$.
b. Use your linear formula to solve $7x + 21 = 0$. Verify your result by solving $7x + 21 = 0$ in the usual way.

94. a. Use factoring to solve $3x^2 + 2x = 0$.
b. Use factoring to solve $ax^2 + bx = 0$ for x.
c. Use the formula you found in part (b) to solve the equation $3x^2 + 2x = 0$. Compare your results with your results from part (a).
d. Use the quadratic formula to solve $3x^2 + 2x = 0$. Compare your results with your results from part (a).
e. Use the quadratic formula to solve $ax^2 + bx = 0$ for x. Compare your results with your results from part (b).
f. Use the formula you found in part (e) to solve the equation $7x^2 - 3x = 0$.

95. Solve $x^2 - x - 20 = 0$ by factoring, by completing the square, and by using the quadratic formula. Compare your results.

96. Solve $3x^2 = 5x - 2$ by factoring, by completing the square, and by using the quadratic formula. Compare your results.

97. Explain how to determine whether to solve a quadratic equation by factoring, by using the square root property, by completing the square, or by using the quadratic formula. Give examples; for each example, describe the advantages of the method you chose and the disadvantages of the other methods.

98. Describe how to solve a quadratic equation by using the quadratic formula.

Related Review

Perform the indicated operations or solve, as appropriate.

99. $(x + 2)(x - 5)$

100. $2x(3x - 5) = 4$

101. $(x + 2)(x - 5) = 3$

102. $2x(3x - 5)$

103. $-4(x - 2)^2 + 3 = -1$

104. $(x + 3)^2$

105. $-4(x - 2)^2 + 3$

106. $(x + 3)^2 = 7$

Expressions, Equations, Functions, and Graphs

Perform the indicated instruction. Then use words such as linear, quadratic, cubic, polynomial, degree, function, one variable, *and* two variables *to describe the expression, equation, or system.*

107. Factor $8x^2 - 18x + 9$.

108. Simplify $-3(x + 1)^2 + 4$.

109. Find $f(-2)$, where $f(x) = 8x^2 - 18x + 9$.

110. Solve $-3(x + 1)^2 + 4 = -20$.

111. Solve $8x^2 - 18x + 9 = 0$.

112. Graph $f(x) = -3(x + 1)^2 + 4$ by hand.

9.7 Solving Systems of Linear Equations in Three Variables; Finding Quadratic Functions

Objectives

- Know the meaning of *linear equation in three variables.*
- Solve a *system of linear equations in three variables.*
- Find a quadratic equation, in standard form, of a parabola that contains three given points.

In this section, we will discuss how to solve a system of three equations. Then we will use this skill to help us find an equation of a parabola that contains three given points. In Section 9.8, we will use the skill to find an equation of a quadratic model.

Linear Equations in Three Variables

In Chapters 1–6, we worked with linear equations in *two* variables. Here we will work with linear equations in *three* variables.

Definition Linear equation in three variables

A **linear equation in three variables** is an equation that can be put into the form $Ax + By + Cz = D$, where A, B, C, and D are constants and A, B, and C are not all zero.

Here is an example of a linear equation in *three* variables:

$$3x - 5y + 2z = 8$$

An **ordered triple** (x, y, z) represents values of x, y, and z, just as an ordered pair (x, y) represents values of x and y. For example, the ordered triple $(2, -4, 7)$ represents the values $x = 2$, $y = -4$, and $z = 7$. An ordered triple (a, b, c) is a **solution** of an equation in terms of x, y, and z if the equation becomes a true statement when a, b, and c are substituted for x, y, and z, respectively. We say a solution **satisfies** the equation.

Example 1 Identifying Solutions of a Linear Equation in Three Variables

Decide whether the given ordered triple is a solution of the equation $5x - 2y + 4z = 11$.

1. $(-3, 1, 7)$ **2.** $(2, -9, 6)$

Solution

1. We substitute -3 for x, 1 for y, and 7 for z in $5x - 2y + 4z = 11$:

$$5(-3) - 2(1) + 4(7) \stackrel{?}{=} 11$$
$$11 \stackrel{?}{=} 11$$
$$\text{true}$$

The ordered triple $(-3, 1, 7)$ is a solution of the equation $5x - 2y + 4z = 11$.

2. We substitute 2 for x, -9 for y, and 6 for z in $5x - 2y + 4z = 11$:

$$5(2) - 2(-9) + 4(6) \stackrel{?}{=} 11$$
$$52 \stackrel{?}{=} 11$$
$$\text{false}$$

The ordered triple $(2, -9, 6)$ is not a solution of the equation $5x - 2y + 4z = 11$.

Solving a System of Linear Equations in Three Variables

A **system of linear equations in three variables** consists of two or more linear equations in three variables. Here is an example of a system of linear equations in three variables:

$$\begin{aligned} 2x - y + 3z &= 4 \\ x + 3y - 2z &= -1 \\ 3x - 5y + z &= 2 \end{aligned}$$

The **solution** of a system of linear equations in three variables is an ordered triple that satisfies *all* of the equations. We can use elimination to solve a system of equations in three variables.

Example 2 Solving a System of Three Equations

Solve the system

$$\begin{aligned} x + y - z &= -1 && \text{Equation (1)} \\ -4x - y + 2z &= -7 && \text{Equation (2)} \\ 2x - 2y - 5z &= 7 && \text{Equation (3)} \end{aligned}$$

Solution

By inspecting the coefficients of the nine variable terms, we see that it is easiest to eliminate y. We add the left sides and add the right sides of equations (1) and (2):

$$\begin{aligned} x + y - z &= -1 && \text{Equation (1)} \\ -4x - y + 2z &= -7 && \text{Equation (2)} \\ \hline -3x + z &= -8 && \text{Equation (4)} \end{aligned}$$

Next, we select equations (1) and (3) and eliminate y again. To do so, we multiply both sides of equation (1) by 2:

$$\begin{aligned} 2x + 2y - 2z &= -2 && \text{Multiply both sides of equation (1) by 2.} \\ 2x - 2y - 5z &= 7 && \text{Equation (3)} \\ \hline 4x - 7z &= 5 && \text{Equation (5)} \end{aligned}$$

Equations (4) and (5) form a system in two variables. To eliminate z, we multiply both sides of equation (4) by 7:

$$\begin{aligned} -21x + 7z &= -56 && \text{Multiply both sides of equation (4) by 7.} \\ 4x - 7z &= 5 && \text{Equation (5)} \\ \hline -17x &= -51 \\ x &= 3 \end{aligned}$$

Next, we substitute 3 for x in equation (4) and solve for z:

$$\begin{aligned} -3(3) + z &= -8 && \text{Substitute 3 for } x \text{ in equation (4).} \\ z &= 1 \end{aligned}$$

Then we substitute 3 for x and 1 for z in equation (1) and solve for y:

$$3 + y - 1 = -1 \quad \text{Substitute 3 for } x \text{ and 1 for } z \text{ in equation (1).}$$
$$y = -3$$

Therefore, $x = 3$, $y = -3$, and $z = 1$. So, the solution of the system is $(3, -3, 1)$. We check that the ordered triple satisfies *all three* original equations:

$$\begin{array}{ccc} x + y - z = -1 & -4x - y + 2z = -7 & 2x - 2y - 5z = 7 \\ 3 + (-3) - 1 \stackrel{?}{=} -1 & -4(3) - (-3) + 2(1) \stackrel{?}{=} -7 & 2(3) - 2(-3) - 5(1) \stackrel{?}{=} 7 \\ -1 \stackrel{?}{=} -1 & -7 \stackrel{?}{=} -7 & 7 \stackrel{?}{=} 7 \\ \text{true} & \text{true} & \text{true} \end{array}$$

Solving a System of Three Linear Equations in Three Variables

To solve a system of three linear equations in three variables,

1. Select a pair of equations and eliminate a variable.
2. Select any other pair of equations and eliminate the *same variable* as in step 1.
3. The equations you found in steps 1 and 2 form a system of linear equations in *two* variables. Use elimination or substitution to solve this system.
4. Substitute the values of the two variables you found in step 3 into one of the original equations that contains the third variable. Solve for the third variable.
5. Write your solution as an ordered triple.

WARNING Make sure you eliminate the *same* variable in steps 1 and 2. This way, your system in step 3 will be a system of equations in *two* variables (rather than three).

We can plot the point $(3, -3, 1)$ of Example 2 by using an x-axis, a y-axis, and a z-axis as shown in Fig. 83.

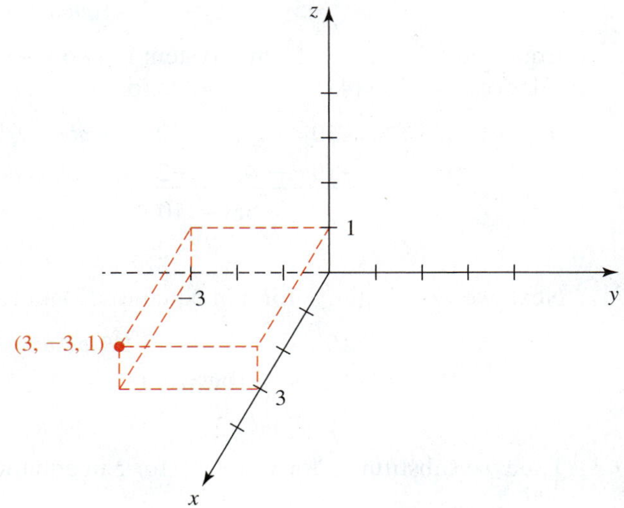

Figure 83 The point $(3, -3, 1)$

The graph of a linear equation in three variables is a plane. So, the graphs of equations (1), (2), and (3) of Example 2 are three planes that intersect at only the point $(3, -3, 1)$.

The intersection of three planes can be one point (see Fig. 84a), one line (see Fig. 84b), one plane (see Fig. 84c), or the empty set (see Fig. 84d). So, the solution set of a system of linear equations in three variables can contain exactly one ordered triple, an infinite number of ordered triples, or no ordered triples (the empty set). We will focus on systems whose solution set contains exactly one ordered triple.

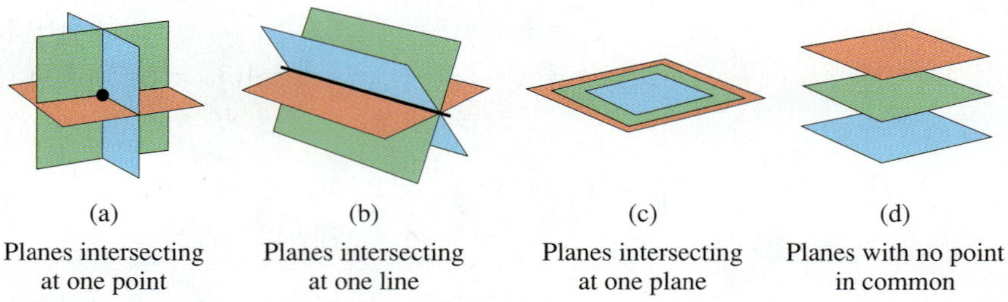

(a)	(b)	(c)	(d)
Planes intersecting at one point	Planes intersecting at one line	Planes intersecting at one plane	Planes with no point in common

Figure 84 Sketches of groups of three planes

It is difficult to graph linear equations in three variables, so we will not use graphing to solve systems of such equations.

Example 3 Solving a System of Three Equations

Solve the system

$$x + y + z = 2 \quad \text{Equation (1)}$$
$$10y - z = 12 \quad \text{Equation (2)}$$
$$2x - 3y = 3 \quad \text{Equation (3)}$$

Solution

Equation (2) does not contain the variable x. To obtain a second equation that does not contain x, we select equations (1) and (3) and eliminate x. We begin by multiplying equation (1) by -2:

$$-2x - 2y - 2z = -4 \quad \text{Multiply both sides of equation (1) by } -2.$$
$$2x - 3y = 3 \quad \text{Equation (3)}$$
$$-5y - 2z = -1 \quad \text{Equation (4)}$$

Equations (2) and (4) form a system in two variables. To eliminate y, we multiply both sides of equation (4) by 2 and solve for z:

$$10y - z = 12 \quad \text{Equation (2)}$$
$$-10y - 4z = -2 \quad \text{Multiply both sides of equation (4) by 2.}$$
$$-5z = 10$$
$$z = -2$$

Next, we substitute -2 for z in equation (2) and solve for y:

$$10y - (-2) = 12 \quad \text{Substitute } -2 \text{ for } z \text{ in equation (2).}$$
$$10y = 10$$
$$y = 1$$

Then we substitute 1 for y and -2 for z in equation (1) and solve for x:

$$x + 1 + (-2) = 2 \quad \text{Substitute 1 for } y \text{ and } -2 \text{ for } z \text{ in equation (1).}$$
$$x = 3$$

Therefore, $x = 3$, $y = 1$, and $z = -2$. So, the solution of the system is $(3, 1, -2)$. We can check that the ordered triple satisfies all three original equations.

Using Points That Are Not y-Intercepts to Find an Equation of a Parabola

Now that we know how to solve a system of linear equations in three variables, we can find an equation of a parabola that contains three given points. In Example 4, we will find an equation of a parabola for which none of the given points are y-intercepts.

Example 4 Finding an Equation of a Parabola

Find an equation of the parabola that contains the points $(1, 1)$, $(2, 3)$, and $(3, 9)$.

Solution

Our goal is to find values of the constants a, b, and c in the equation $y = ax^2 + bx + c$. Since the three given points lie on the parabola, each of the ordered pairs $(1, 1)$, $(2, 3)$, and $(3, 9)$ satisfies the equation $y = ax^2 + bx + c$:

$$1 = a(1)^2 + b(1) + c \quad \text{Substitute } (1, 1) \text{ into } y = ax^2 + bx + c.$$
$$3 = a(2)^2 + b(2) + c \quad \text{Substitute } (2, 3) \text{ into } y = ax^2 + bx + c.$$
$$9 = a(3)^2 + b(3) + c \quad \text{Substitute } (3, 9) \text{ into } y = ax^2 + bx + c.$$

We can simplify the right-hand sides of these equations:

$$a + b + c = 1 \quad \text{Equation (1)}$$
$$4a + 2b + c = 3 \quad \text{Equation (2)}$$
$$9a + 3b + c = 9 \quad \text{Equation (3)}$$

We select equations (1) and (2) and eliminate c by multiplying both sides of equation (1) by -1:

$$-a - b - c = -1 \quad \text{Multiply both sides of equation (1) by } -1.$$
$$4a + 2b + c = 3 \quad \text{Equation (2)}$$
$$3a + b = 2 \quad \text{Equation (4)}$$

Next, we select equations (1) and (3) and eliminate c again, once more multiplying both sides of equation (1) by -1:

$$-a - b - c = -1 \quad \text{Multiply both sides of equation (1) by } -1.$$
$$9a + 3b + c = 9 \quad \text{Equation (3)}$$
$$8a + 2b = 8 \quad \text{Equation (5)}$$

Then we divide both sides of equation (5) by 2:

$$4a + b = 4 \quad \text{Equation (6)}$$

Equations (4) and (6) form a system in two variables. To eliminate b, we multiply both sides of equation (4) by -1 and solve for a:

$$-3a - b = -2 \quad \text{Multiply both sides of equation (4) by } -1.$$
$$4a + b = 4 \quad \text{Equation (6)}$$
$$a = 2$$

Next, we substitute 2 for a in equation (4) and solve for b:

$$3(2) + b = 2$$
$$b = -4$$

Then we substitute 2 for a and -4 for b in equation (1) and solve for c:

$$2 + (-4) + c = 1 \quad \text{Substitute 2 for } a \text{ and } -4 \text{ for } b \text{ in equation (1).}$$
$$c = 3$$

Therefore, $a = 2$, $b = -4$, and $c = 3$, and the equation of the parabola is

$$y = 2x^2 - 4x + 3$$

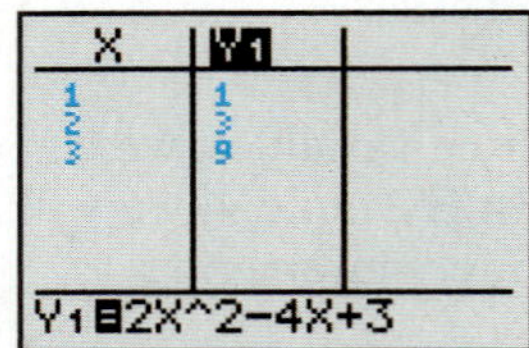

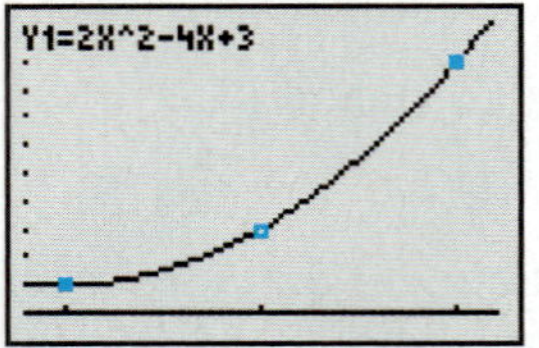

Figure 85 Verify the work

We use a graphing calculator table, as well as a scattergram and graph, to check that the parabola $y = 2x^2 - 4x + 3$ contains the points $(1, 1)$, $(2, 3)$, and $(3, 9)$. See Fig. 85.

To find a linear equation $y = mx + b$, we need *two* points to find the *two* constants m and b. To find a quadratic equation in standard form, $y = ax^2 + bx + c$, we need *three* points to find the *three* constants a, b, and c.

Finding an Equation of a Parabola

To find an equation of a parabola that contains three given points,

1. Obtain a system of three linear equations in three variables by substituting the coordinates of each of the three given points into the general equation $y = ax^2 + bx + c$.
2. Solve the system you found in step 1.
3. Substitute the values of a, b, and c you found in step 2 into the equation $y = ax^2 + bx + c$.

Using the y-Intercept and Two Other Points

The process of finding an equation of a parabola is considerably easier if one of the three given points is the y-intercept, as we shall see in Example 5.

Example 5 Finding an Equation when One of the Given Points Is the y-Intercept

Find an equation of the parabola that contains the points $(0, 1)$, $(3, 7)$, and $(4, 5)$.

Solution

We begin by substituting the ordered pairs $(0, 1)$, $(3, 7)$, and $(4, 5)$, into $y = ax^2 + bx + c$:

$$1 = a(0)^2 + b(0) + c \quad \text{Substitute } (0, 1) \text{ into } y = ax^2 + bx + c.$$
$$7 = a(3)^2 + b(3) + c \quad \text{Substitute } (3, 7) \text{ into } y = ax^2 + bx + c.$$
$$5 = a(4)^2 + b(4) + c \quad \text{Substitute } (4, 5) \text{ into } y = ax^2 + bx + c.$$

Next, we simplify the right-hand sides of these equations:

$$c = 1 \quad \text{Equation (1)}$$
$$9a + 3b + c = 7 \quad \text{Equation (2)}$$
$$16a + 4b + c = 5 \quad \text{Equation (3)}$$

Since $c = 1$, we substitute 1 for c in equations (2) and (3):

$$9a + 3b + 1 = 7 \quad \text{Substitute 1 for } c \text{ in equation (2).}$$
$$16a + 4b + 1 = 5 \quad \text{Substitute 1 for } c \text{ in equation (3).}$$

For each of these two equations, we subtract 1 on both sides:

$$9a + 3b = 6 \quad \text{Equation (4)}$$
$$16a + 4b = 4 \quad \text{Equation (5)}$$

To eliminate b, we multiply both sides of equation (4) by -4 and both sides of equation (5) by 3:

$$\begin{aligned} -36a - 12b &= -24 \quad \text{Multiply both sides of equation (4) by } -4. \\ 48a + 12b &= 12 \quad \text{Multiply both sides of equation (5) by 3.} \\ \hline 12a \quad &= -12 \\ a \quad &= -1 \end{aligned}$$

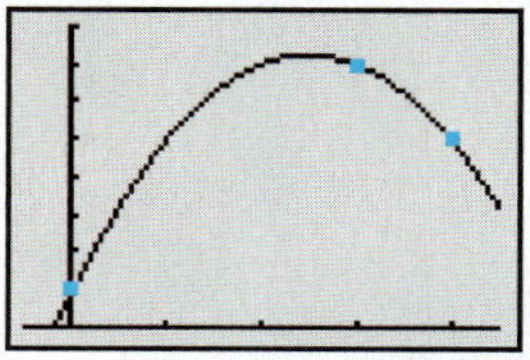

Figure 86 Verify the work

Next, we substitute -1 for a in equation (4) and solve for b:

$$9(-1) + 3b = 6 \quad \text{Substitute } -1 \text{ for } a \text{ in equation (4).}$$
$$3b = 15$$
$$b = 5$$

Therefore, $a = -1$, $b = 5$, and $c = 1$, and the equation of the parabola is

$$y = -x^2 + 5x + 1$$

We use a graphing calculator table, as well as a scattergram and graph, to verify that the graph of $y = -x^2 + 5x + 1$ contains the points $(0, 1)$, $(3, 7)$, and $(4, 5)$. See Fig. 86.

In Example 5, we were able to find an equation of the desired parabola by using elimination once, rather than three times. We need to use elimination only once when one of the three given points is the y-intercept.

Group Exploration

For any three points, is there a quadratic function that contains them?

1. Find the values of a, b, and c of the function $f(x) = ax^2 + bx + c$, where the graph of f contains the points $(0, 1)$, $(1, 4)$, and $(2, 7)$. What type of function is f? Why did this happen?
2. Do the same for the points $(0, 1)$, $(0, 8)$, and $(1, 4)$. What happens? Is there a function $f(x) = ax^2 + bx + c$ whose graph contains these points? Explain.
3. What must be true of three points so there is a quadratic function whose graph contains the points? Give an example of three such points, plot them, and sketch the graph of the quadratic function that contains them. Then find an equation of the function and use a graphing calculator to view the graph. Compare the two graphs.

Homework 9.7

For extra help ▶ MyMathLab® 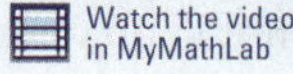 Watch the videos in MyMathLab  Download the MyDashboard App

Solve the system.

1. $x + y + z = 0$, $x - y + z = 6$, $x + 2y - z = -7$
2. $x + y + z = 4$, $-x + y + 2z = 1$, $-x + y - 3z = -4$
3. $x + y - z = -1$, $2x - 2y + 3z = 8$, $2x - y + 2z = 9$
4. $2x - 3y + z = -9$, $-2x + y - 3z = 7$, $x - y + 2z = -5$
5. $3x - y + 2z = 0$, $2x + 3y + 8z = 8$, $x + y + 6z = 0$
6. $-x + y + z = 6$, $x - 2y + 3z = 5$, $-2x + y - 2z = -1$
7. $2x + y + z = 3$, $2x - y - z = 9$, $x + y - z = 0$
8. $x + y + z = 6$, $2x - y + z = 3$, $x + 2y - 3z = -4$
9. $2x + 2y + z = 1$, $-x + y + 2z = 3$, $x + 2y + 4z = 0$
10. $2x - 3y + z = 2$, $x - 5y + 5z = 3$, $3x + y - 3z = 5$
11. $2x - y + 2z = 6$, $3x + y - z = 5$, $x + 2y + z = 3$
12. $2x + y + z = -2$, $2x - y + 3z = 6$, $3x - 5y + 4z = 7$
13. $x - 3z = 6$, $y + 2z = 2$, $7x - 3y - 5z = 14$
14. $2x + y = -2$, $3y - z = -14$, $x + 2z = 5$
15. $2x - y = -8$, $y + 3z = 22$, $x - z = -8$
16. $x + y = 1$, $y - 2z = 2$, $x - 3z = 14$

Find an equation of the parabola that contains the given points. Use a graphing calculator to verify that the graph of your equation contains the points.

17. $(1, 6)$, $(2, 11)$, $(3, 18)$
18. $(1, 1)$, $(2, 5)$, $(3, 15)$
19. $(1, 9)$, $(2, 7)$, $(4, -15)$
20. $(1, 4)$, $(2, 3)$, $(3, 0)$
21. $(2, 2)$, $(3, 11)$, $(4, 24)$
22. $(2, 3)$, $(3, -2)$, $(4, -11)$

23. $(1, -3), (3, 9), (5, 29)$

24. $(2, -1), (4, 19), (5, 38)$

25. $(3, 7), (4, 0), (5, -11)$

26. $(2, 4), (4, 30), (5, 49)$

27. $(2, -5), (4, 3), (5, 13)$

28. $(1, 3), (2, 3), (4, -9)$

29. $(0, 4), (2, 8), (3, 1)$

30. $(0, 0), (1, 4), (2, 14)$

31. $(0, -1), (1, 3), (2, 13)$

32. $(0, 5), (2, 13), (3, 26)$

33. $(1, 1), (2, 4), (3, 9)$

34. $(1, -1), (2, -4), (3, -9)$

35. The graph of a quadratic function has y-intercept $(0, 4)$ and x-intercepts $(1, 0)$ and $(2, 0)$. Find an equation of the function.

36. The graph of a quadratic function has y-intercept $(0, 8)$ and x-intercepts $(-4, 0)$ and $(2, 0)$. Find an equation of the function.

Concepts

37. Find an equation of the parabola sketched in Fig. 87. [**Hint:** Choose three points whose coordinates appear to be integers.]

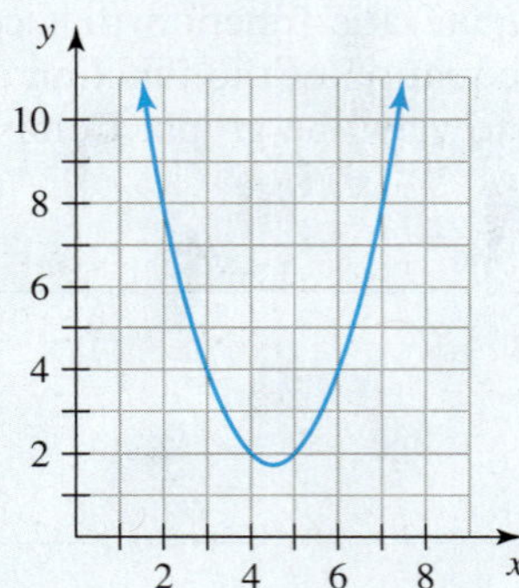

Figure 87 Exercise 37

38. Find an equation of the parabola sketched in Fig. 88. [**Hint:** Choose three points whose coordinates appear to be integers.]

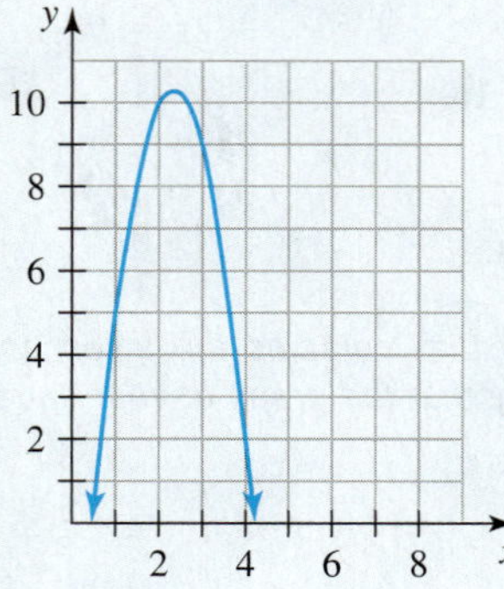

Figure 88 Exercise 38

39. Find an equation of the parabola sketched in Fig. 89.

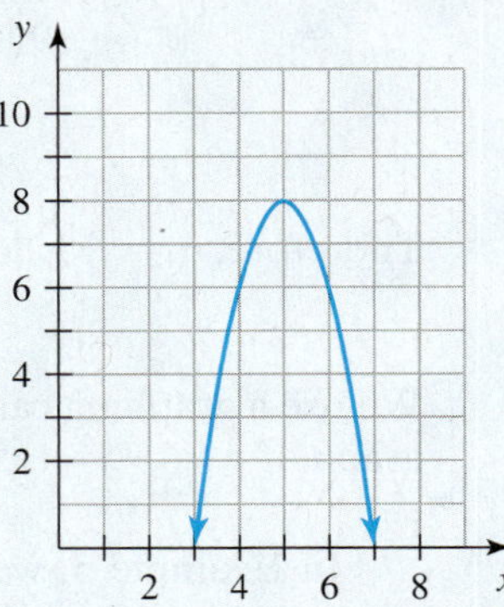

Figure 89 Exercise 39

40. Find an equation of the parabola sketched in Fig. 90.

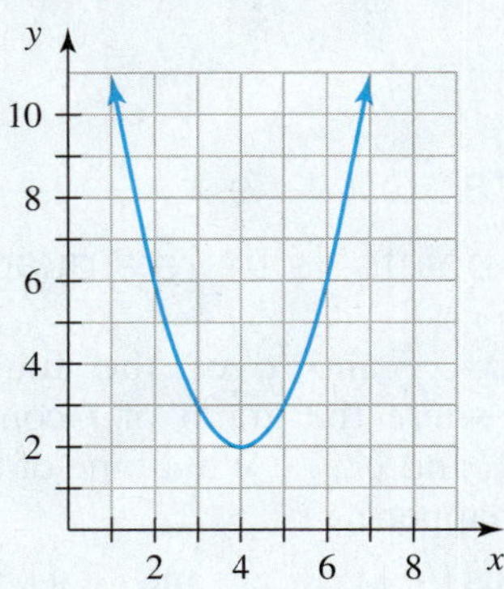

Figure 90 Exercise 40

41. In this exercise, you will show that two points do not determine a parabola.

a. Plot the points $(1, 4)$ and $(3, 4)$ on a coordinate system.

b. Sketch three parabolas, each of which contains both of the points $(1, 4)$ and $(3, 4)$. Find the vertex of each parabola.

c. Write an equation of each of your three sketched parabolas. [**Hint:** You can use the vertex form $y = a(x - h)^2 + k$.]

d. Explain why two points do not determine a parabola.

42. Find an equation of a parabola that is in quadrants I, III, and IV but not in quadrant II.

43. Choose three points that have different x-coordinates and do not all lie on a line. Find an equation of a parabola that contains the points. [**Hint:** To make the calculations easier, choose one of the points so it is the y-intercept.]

44. Create a system of three linear equations in three variables so $(2, 4, 3)$ is a solution.

45. Describe how to solve a system of three linear equations in three variables.

46. Describe how to find an equation of a parabola that contains three points that have different x-coordinates and do not all lie on a line.

Related Review

Find an equation of the line that contains the given points.

47. $(-2, 1), (1, -6)$

48. $(-5, -2), (3, 7)$

For Exercises 49 and 50, find equations of a linear function and a quadratic function such that the graph of each equation contains the given points.

49. $(0, 2), (1, 4)$

50. $(0, 4), (2, 36)$

51. Find the values of a, b, and c of $f(x) = ax^2 + bx + c$, where the graph of f contains the points $(1, 1)$, $(2, 2)$, and $(3, 3)$. What type of function is f?

52. Find the values of a, b, and c of $g(x) = ax^2 + bx + c$, where the graph of g contains the points $(1, 2)$, $(2, 5)$, and $(3, 8)$. What type of function is g?

53. Find an equation of a parabola that has vertex $(5, -7)$ and contains the point $(8, 11)$.

54. Find an equation of a parabola that has vertex $(-5, 8)$ and contains the point $(-7, -4)$.

Expressions, Equations, Functions, and Graphs

Perform the indicated instruction. Then use words such as linear, quadratic, cubic, polynomial, degree, function, one variable, *and* two variables *to describe the expression, equation, or system.*

55. Solve $2x^2 - 10x + 7 = 0$.

56. Simplify $-4x(3x - 2)^2$.

57. Find $f(2)$, where $f(x) = 2x^2 - 10x + 7$.

58. Factor $3x^3 - 21x^2y + 30xy^2$.

59. Graph $f(x) = 2x^2 - 10x + 7$ by hand.

60. Solve $t^2 - 3t - 13 = 0$.

9.8 Finding Quadratic Models

Objectives

» Find an equation of a quadratic model in standard form.

» Determine whether to model a situation by using a linear function, a quadratic function, or neither.

In Section 9.7, we used three given points to find an equation of a parabola. In this section, we will use this skill to find an equation of a quadratic model. We will also discuss how to determine whether a linear model, a quadratic model, or neither of these can be used to model an authentic situation.

Finding a Quadratic Model in Standard Form

In Example 1, we will find an equation of a quadratic model.

Example 1 Finding an Equation of a Quadratic Model

The percentages of adults ever diagnosed with high cholesterol are shown in Table 38 for various ages. Let $f(t)$ be the percentage of adults at age t years who have ever been diagnosed with high cholesterol. Find an equation of a model to describe the situation.

Table 38 Percentages of Adults Ever Diagnosed with High Cholesterol

Age (years)	Percent
50	29
60	43
70	50
80	47
90	39

Source: *The Gallup Organization*

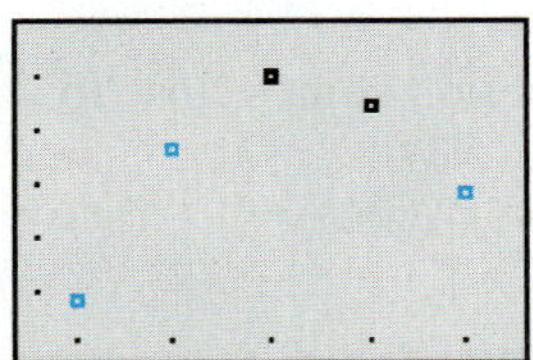

Figure 91 Cholesterol scattergram

Solution

We use a graphing calculator to draw a scattergram to describe the data (see Fig. 91). It looks like a parabola would fit the data well.

Through practice, we envision that the parabola containing the (blue) data points $(50, 29)$, $(60, 43)$, and $(90, 39)$ may be close to the other data points. To find an equation of this parabola, we substitute the three ordered pairs $(50, 29)$, $(60, 43)$, and $(90, 39)$ into the standard form $f(t) = at^2 + bt + c$:

$$29 = a(50)^2 + b(50) + c \quad \text{Substitute } (50, 29) \text{ into } f(t) = at^2 + bt + c.$$

$$43 = a(60)^2 + b(60) + c \quad \text{Substitute } (60, 43) \text{ into } f(t) = at^2 + bt + c.$$

$$39 = a(90)^2 + b(90) + c \quad \text{Substitute } (90, 39) \text{ into } f(t) = at^2 + bt + c.$$

We can simplify the right-hand sides of these equations:

$$2500a + 50b + c = 29 \quad \text{Equation (1)}$$

$$3600a + 60b + c = 43 \quad \text{Equation (2)}$$

$$8100a + 90b + c = 39 \quad \text{Equation (3)}$$

We select equations (1) and (2) and eliminate c by multiplying both sides of equation (1) by -1:

$$\begin{aligned} -2500a - 50b - c &= -29 && \text{Multiply both sides of equation (1) by } -1. \\ 3600a + 60b + c &= 43 && \text{Equation (2)} \\ \hline 1100a + 10b \quad &= 14 && \text{Equation (4)} \end{aligned}$$

Next, we select equations (2) and (3) and eliminate c again, this time multiplying both sides of equation (2) by -1:

$$\begin{aligned} -3600a - 60b - c &= -43 && \textit{Multiply both sides of equation (2) by } -1. \\ 8100a + 90b + c &= 39 && \textit{Equation (3)} \\ \hline 4500a + 30b \quad &= -4 && \textit{Equation (5)} \end{aligned}$$

Equations (4) and (5) form a system in two variables. To eliminate b, we multiply both sides of equation (4) by -3:

$$\begin{aligned} -3300a - 30b &= -42 && \textit{Multiply both sides of equation (4) by } -3. \\ 4500a + 30b &= -4 && \textit{Equation (5)} \\ \hline 1200a \quad &= -46 \\ a &\approx -0.03833 \end{aligned}$$

Next, we substitute -0.03833 for a in equation (4) and solve for b:

$$\begin{aligned} 1100(-0.03833) + 10b &= 14 && \textit{Substitute } -0.03833 \textit{ for a in equation (4).} \\ -42.163 + 10b &= 14 \\ 10b &= 56.163 \\ b &\approx 5.62 \end{aligned}$$

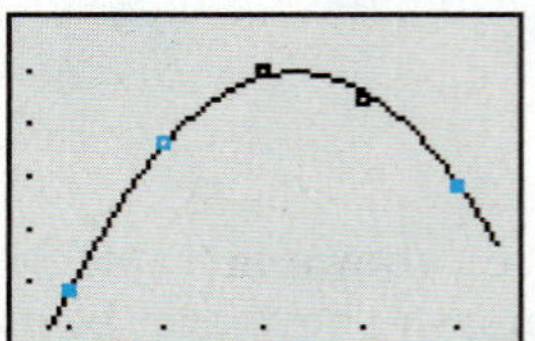

Figure 92 Check the fit

Then we substitute -0.03833 for a and 5.62 for b in equation (1) and solve for c:

$$\begin{aligned} 2500(-0.03833) + 50(5.62) + c &= 29 && \textit{Substitute } -0.03833 \textit{ for a and 5.62 for b in equation (1).} \\ 185.175 + c &= 29 \\ c &\approx -156.18 \end{aligned}$$

Finally, we substitute our approximate values of a, b, and c in the general equation $f(t) = at^2 + bt + c$ to obtain our quadratic model:

$$f(t) = -0.03833t^2 + 5.62t - 156.18$$

We verify the equation by observing that the graph of f appears to contain the points $(50, 29)$, $(60, 43)$, and $(90, 39)$. See Fig. 92. Since the graph appears to come close to the other data points, we conclude that f is a reasonable model of the situation.

Finding an Equation of a Quadratic Model

To find an equation of a quadratic model, given some data,

1. Create a scattergram of the data.
2. Imagine a parabola that comes close to the data points, and choose three points (not necessarily data points) that lie on or close to the parabola.
3. Use the three points to find an equation of the parabola.
4. Use a graphing calculator to verify that the graph of the equation comes close to the points of the scattergram.

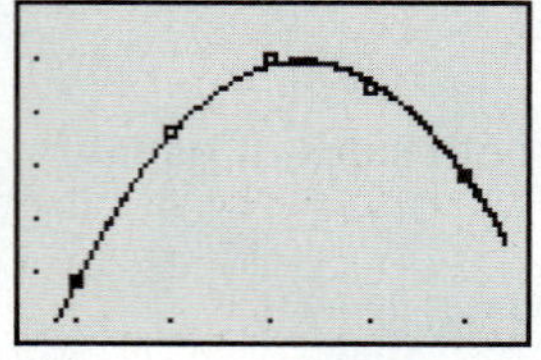

Figure 93 Compare the fit of models f and r

In Example 1, we could have used another procedure to find a quadratic model different from $f(t) = -0.03833t^2 + 5.62t - 156.18$. Instead of choosing three points and solving a system of equations, we could have used a graphing calculator's *quadratic regression*. Quadratic regression gives the model $r(t) = -0.03857t^2 + 5.64t - 156.49$. In Fig. 93, the graphs of the models f and r are so similar that ZoomStat appears to show just one curve. For graphing calculator instructions, see Appendix A.16.

We call the equation $r(t) = -0.03857t^2 + 5.64t - 156.49$ the **quadratic regression equation** for the given data. We refer to its graph as a **quadratic regression curve.**

Deciding Which Type of Model to Use

By graphing a scattergram of some data, we can determine whether to use a linear model, a quadratic model, or neither.

Example 2 Selecting a Model

Consider the scattergrams of data shown in Figs. 94, 95, and 96 for situations 1, 2, and 3, respectively. For each situation, determine whether a linear model, quadratic model, or neither would describe the situation well.

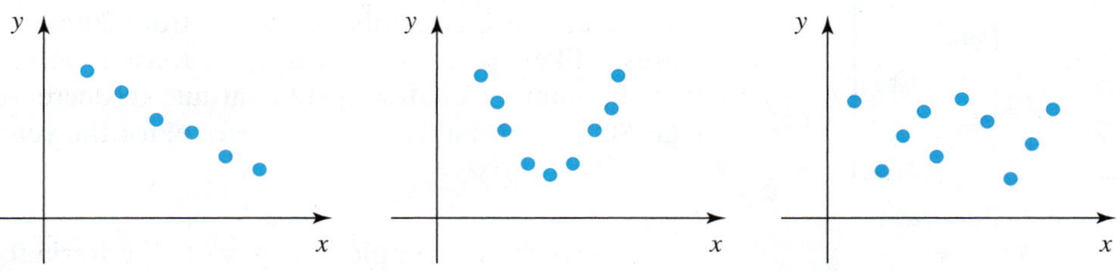

Figure 94 Use a linear model

Figure 95 Use a quadratic model

Figure 96 Do not use a linear or a quadratic model

Solution

A linear model would describe the data for situation 1 well. It appears that a quadratic model would fit the data for situation 2 well. Neither a linear model nor a quadratic model would fit the data well for situation 3.

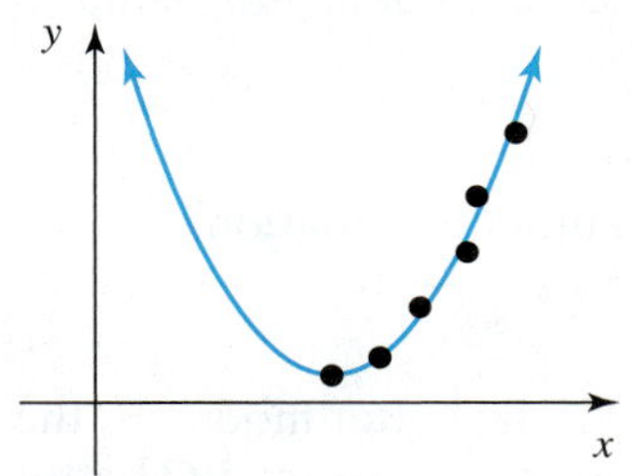

Figure 97 Scattergram and a quadratic model

Throughout this course, we have seen that usually only a part of a model describes a situation well. For the scattergram shown in Fig. 97, the sketched quadratic model fits the data well. At least part of the model that lies to the right of the vertex can be used to make reasonable estimates or predictions. The part of the model that lies to the left of the vertex may or may not describe the situation well. More data points could reveal whether this part of the parabola gives good estimates.

Example 3 Determining Which Model to Use

The numbers of specialty bicycle stores are shown in Table 39 for various years.

Table 39 Numbers of Specialty Bicycle Stores

Year	Number of Stores
2000	6195
2002	5505
2004	4982
2006	4600
2008	4349
2010	4256
2011	4178

Source: *The Bike Shop List*

Let n be the number of specialty bicycle stores at t years since 2000. Find an equation of a model to describe the situation.

Solution

First, we use a graphing calculator to draw a scattergram to describe the data (see Fig. 98). Since the points suggest a curve that "bends," the quadratic regression model will likely fit the data better than the linear regression model. Here are the two models:

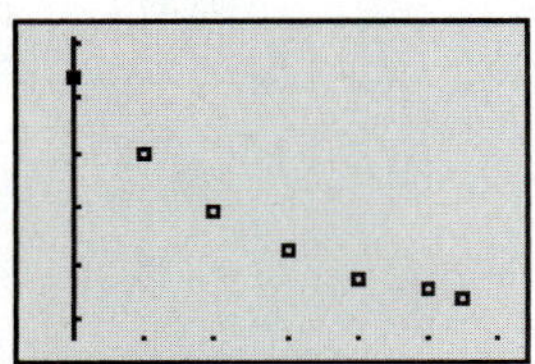

Figure 98 Bicycle store scattergram

$$L(t) = -175.25t + 5893 \quad \text{Linear regression equation}$$

$$Q(t) = 16.89t^2 - 365t + 6182 \quad \text{Quadratic regression equation}$$

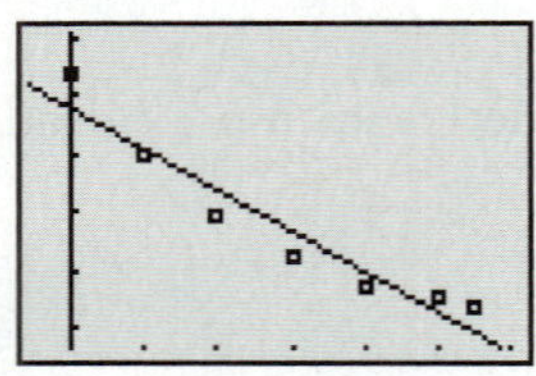

Figure 99 Linear regression model

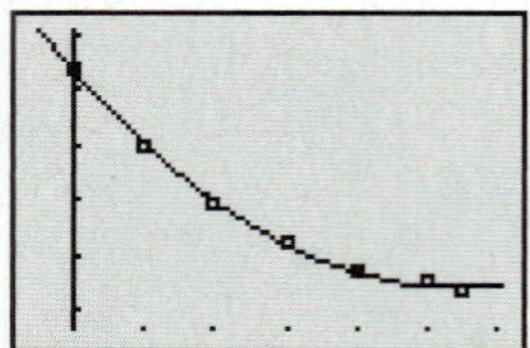

Figure 100 Quadratic regression model

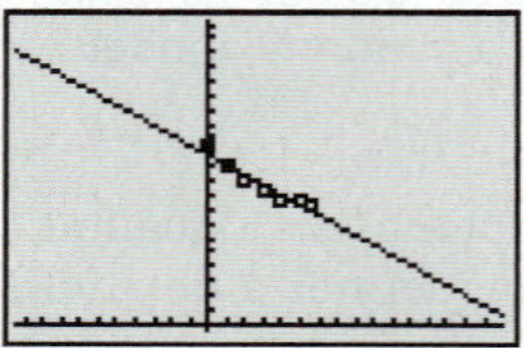

Figure 101 Linear regression model

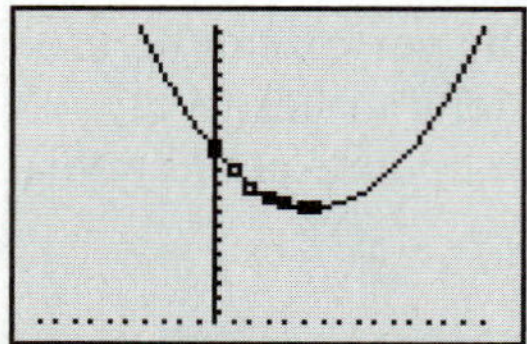

Figure 102 Quadratic regression model

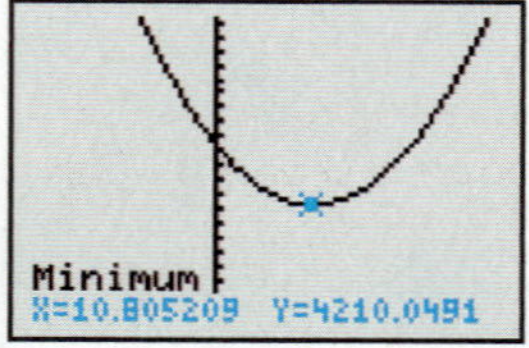

Figure 103 The vertex of the graph of Q

Next, we see how well both regression models fit the data (see Figs. 99 and 100).

It appears Q fits the data better than L does. So, for most years between 2000 and 2011, inclusive, Q estimates the number of specialty bicycle stores better than L does. This suggests Q estimates the number of stores better than L does for at least a few years before 2000.

Does Q predict the number of stores better than L does for years after 2011? To see, we use Zoom Out on a graphing calculator (see Figs. 101 and 102).

It will also help to use "minimum" on a graphing calculator to find the approximate vertex of the graph of Q, which is $(10.81, 4210)$. See Fig. 103. So, Q predicts that the number of specialty bicycle stores has increased since 2011, which is not consistent with the decrease in the number of stores from 2000 to 2011. The decrease in the number of stores is likely to continue due to increasing online bicycle sales. The model L predicts that the number of stores will continue to decrease, which will most likely prove to be true. So, L is probably the better model for the years after 2011.

Our work in Example 3 suggests two criteria for selecting a model.

Selecting a Model

When performing step 1 of the modeling process, we must decide whether a linear function, a quadratic function, or neither of these is suitable for modeling the situation. Here are the criteria for selecting a model:

- The graph of the model should fit the points well.
- The model should make sense within the context of the authentic situation.

In Example 3, we used the criterion of fit to determine that Q is the better model for the years between 2000 and 2011, inclusive. Using this same criterion, we assumed Q better estimates the number of specialty bicycle stores for at least a few years before 2000. We used the criterion of context (the number of stores has generally been decreasing over time) to determine L likely better predicts the number of stores for at least a few years after 2011.

Here, we outline the four-step modeling process again.

Four-Step Modeling Process

To find a model and make estimates and predictions,

1. Create a scattergram of the data. Decide whether a line, a parabola, or neither of these comes close to the points.
2. Find an equation of your model.
3. Verify that your equation has a graph that comes close to the points in the scattergram. If it doesn't, check for calculation errors or use different points to find the equation. An alternative is to reconsider your choice of model in step 1.
4. Use your equation of the model to draw conclusions, make estimates, and/or make predictions.

We will perform activities from step 4 in Section 9.9.

Group Exploration

Choosing three "good" points to find a quadratic model

Table 40 lists the average numbers of paid vacation days and holidays for full-time workers at medium-to-large companies for various years of experience. Let D be the average number of paid vacation days and holidays in one year for someone who has worked at a company for t years.

Table 40 Paid Vacation Days and Holidays

Years of Service	Days Off
1	9.4
3	11.2
5	13.6
10	16.6
15	18.8
20	20.4
25	21.6
30	21.9

Source: USA Today

1. Use a graphing calculator to create a scattergram of the vacation data. Does a linear function or a quadratic function seem to model the data better? Explain.
2. Use the three data points $(1, 9.4)$, $(15, 18.8)$, and $(30, 21.9)$ to find an equation $D = at^2 + bt + c$ of the parabola that comes close to the data points in your scattergram.
3. Draw the graph of your quadratic model and your scattergram in the same viewing window to verify that the parabola passes through the points $(1, 9.4)$, $(15, 18.8)$, and $(30, 21.9)$. Does your quadratic function seem to be a reasonable model of the vacation data?
4. The first three rows in Table 40 give the data points $(1, 9.4)$, $(3, 11.2)$, and $(5, 13.6)$. Had you used these three points, you would have found the equation $D = 0.075t^2 + 0.60t + 8.73$. Compare its graph with the graph you drew in Problem 3. Explain why the graphs look so different.
5. In the future, you will encounter other data sets that can be modeled well by using a quadratic function. Describe a general "game plan" for deciding which three points to use to find a quadratic model.

Homework 9.8

For extra help ▶ MyMathLab®  

1. Four scattergrams of data are sketched in Fig. 104. Decide whether a linear function, a quadratic function, or neither of these types of functions would be reasonable for modeling the data.

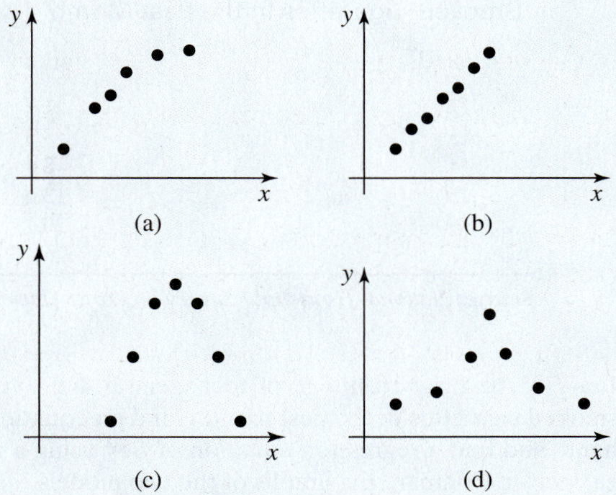

Figure 104 Scattergrams of data, Exercise 1

2. Make a sketch of each scattergram in Fig. 104. Then sketch the graph of the function you would use to model the data.
3. The percentages of employers giving workers a paid day off for Martin Luther King, Jr., Day are shown in Table 41 for various years.

Table 41 Percentages of Employers Giving Workers a Paid Day Off for Martin Luther King, Jr., Day

Years	Percent
1990	17
1995	22
2000	25
2005	29
2012	31

Source: *Bloomberg*

Let $f(t)$ be the percentage of employers who gave workers a paid day off for Martin Luther King, Jr., Day in the year that is t years since 1990. Find and verify an equation of f.

4. Internet gambling (e-gambling) is illegal in the United States. Estimates of e-gambling revenues are shown in Table 42 for various years.

Table 42 Global Internet Gambling Revenue Estimates

Year	Revenue (billions of U.S. dollars)
1998	2.4
2000	4.1
2002	6.3
2004	10.2
2006	18.0
2008	23.5
2010	30.3

Source: *HZ Gambling Capitol*

Let $f(t)$ be the e-gambling revenue (in billions of U.S. dollars) in the year that is t years since 1990. Find and verify an equation of f.

5. The percentages of law degrees earned by women are shown in Table 43 for various years.

Table 43 Percentages of Law Degrees Earned by Women

Year	Percent
2000	45.9
2002	48.0
2005	48.7
2007	47.6
2009	45.8

Source: *U.S. National Center for Education Statistics*

Let $f(t)$ be the percentage of law degrees earned by women in the year that is t years since 2000. Find and verify an equation of f.

6. The percentages of people working at age 65 or older are shown in Table 44 for various years.

Table 44 Percentages of People Working at Age 65 or Older

Year	Percent
1970	17
1975	14
1980	13
1985	11
1990	12
1995	12
2000	13
2005	15.1
2010	17.4
2011	16.7

Source: *Bureau of Labor Statistics*

Let $f(t)$ be the percentage of people working at age 65 or older at t years since 1970. Find and verify an equation of f.

7. The percentages of drivers who admit to running red lights are shown in Table 45 for various age groups. Let $f(t)$ be the percentage of drivers at age t years who admit to running red lights. Find an equation of f by hand, and find a regression equation of f by using a graphing calculator. Compare the graphs of the two models.

Table 45 Percentages of Drivers Who Admit to Running Red Lights

Age Group (years)	Age Used to Represent Age Group (years)	Percent
18–25	21.5	75
26–35	30.5	73
36–45	40.5	63
46–55	50.5	56
over 55	65.0	35

Source: *The Social Science Research Center*

8. Table 46 lists the percentages of U.S. households, by income groups, who have Internet access.

Table 46 Percentages of U.S. Households Who Have Internet Access

Household Income Group (in thousands of dollars)	Income Used to Represent Household Income Group (in thousands of dollars)	Percent
0–15	7.5	39.6
15–25	20	52.6
25–35	30	63.3
35–50	42.5	77.9
50–75	62.5	87.1
over 75	100	96.4

Source: *U.S. Census Bureau*

Let $f(d)$ be the percentage of U.S. households with income d thousand dollars who have Internet access. Find an equation of f by hand, and find a regression equation of f by using a graphing calculator. Compare the graphs of the two models.

9. The percentages of teenagers who smoked cigarettes in the past month are shown in Table 47 for various ages.

Table 47 Percentages of Teenagers Who Smoked Cigarettes in the Past Month

Age (years)	Percent
12	1.7
13	3.3
14	8.4
15	13.6
16	20.1
17	26.4

Source: *National Household Survey on Drug Abuse*

Let $f(t)$ be the percentage of teenagers at age t years who smoked cigarettes in the past month. Find an equation of f by hand, and find a regression equation of f by using a graphing calculator. Compare the graphs of the two models.

10. Domestic airline fuel prices are shown in Table 48 for various years. Let $f(t)$ be the domestic airline fuel price (in cents per gallon) at t years since 2000. Find an equation of f by hand, and find a regression equation of f by using a graphing calculator. Compare the graphs of the two models.

Table 48 Domestic Airline Fuel Prices

Year	Domestic Airline Fuel Price (cents per gallon)
2000	80
2001	78
2002	71
2003	84
2004	115
2005	165
2006	195
2007	209
2008	310

Source: *Bureau of Transportation Statistics*

11. The prices of an ounce of gold are listed in Table 49 for various years.

Table 49 Prices of an Ounce of Gold

Year	Price of an Ounce of Gold (dollars)
2001	271
2002	310
2003	363
2004	410
2005	445
2006	603
2007	695
2008	872
2009	972
2010	1116
2011	1572
2012	1669

Source: *Kitco, Inc.*

Let $f(t)$ be the price (in dollars) of an ounce of gold at t years since 2000.

a. Use the data for the years 2002, 2007, and 2012 to find a quadratic equation of f.

b. The data for the years 2005, 2006, and 2007 lead to the equation $f(t) = -33t^2 + 521t - 1335$. The data for the years 2010, 2011, and 2012 lead to the equation $f(t) = -179.5t^2 + 4226t - 23{,}189$. In part (a), you found yet another equation of f. Use a graphing calculator to determine which of these triples of years gives the best model of the data. Could you have guessed this before you found the equations? If so, explain how.

12. July is the most popular month for Americans to take vacations (see Table 50). Let $f(t)$ be the percentage of Americans who take a vacation in the month that is t months since July.

a. Use the data for August, October, and May to find a quadratic equation of f.

b. The data for July, August, and September lead to the equation $f(t) = -2t^2 - 7t + 43$. The data for November, December, and January lead to the equation $f(t) = -4t^2 + 36t - 69$. The data for December, March, and May lead to $f(t) = 2.07t^2 - 29.2t + 105.33$. In part (a), you found yet another equation of f. Use a graphing calculator to determine which of these four triples of months gives the best model of the data. Could you have guessed this before you found the equations? If so, explain how.

Table 50 Percentages of Americans Who Vacation, by Month

Month	Number of Months since July t	Percent
July	0	43
August	1	34
September	2	21
October	3	15
November	4	11
December	5	11
January	6	3
February	7	3
March	8	4
April	9	8
May	10	20
June	11	30

Source: *2001 American Express Leisure Travel Index*

13. The U.S. population is shown in Table 51 for various years.

Table 51 U.S. Population

Year	Population (millions)	Year	Population (millions)
1790	3.9	1910	92.2
1800	5.3	1920	106.0
1810	7.2	1930	123.2
1820	9.6	1940	132.2
1830	12.9	1950	151.3
1840	17.1	1960	179.3
1850	23.2	1970	203.3
1860	31.4	1980	226.5
1870	39.8	1990	248.7
1880	50.2	2000	281.4
1890	63.0	2010	308.7
1900	76.2		

Source: *U.S. Census Bureau*

a. Let $f(t)$ be the U.S. population (in millions) at t years since 1790. Find and verify an equation of f.

b. In Example 9 of Section 7.2, we used a linear function to model the U.S. population for the years shown in Table 52. An equation of such a linear model is $L(t) = 2.99t - 346.55$, where $L(t)$ is the U.S. population (in millions) at t years since 1790.

Table 52 U.S. Population

Year	Population (millions)
1990	248.7
1993	259.9
1996	269.4
1999	279.0
2002	287.8
2005	295.8
2008	304.4
2010	308.7

Source: *U.S. Census Bureau*

i. Use a graphing calculator to draw the graphs of f and L and, in the same viewing window, the scattergram of the data *for the years shown in Table 51*. Which function describes the population better for these years? Explain.

ii. Use a graphing calculator to draw the graphs of f and L and, in the same viewing window, the scattergram of the data *for the years shown in Table 52*. Which function describes the population better for these years? Explain.

iii. How can one function be a better model for the period 1790–2010 but the other function be a better model for part of that period?

iv. Does the portion of the graph of f you viewed in part (ii) for the years from 1990 to 2010 appear to be linear or quadratic? How is this possible?

14. The percentages of Americans who say they do volunteer work are listed in Table 53 for various age groups.

Table 53 Percentages of Americans Who Say They Volunteer

Age Group (years)	Age Used to Represent Age Group (years)	Percent
18–24	21.0	38
25–34	29.5	51
35–54	44.5	55
55–64	59.5	48
65–74	69.5	45
over 74	80.0	34

Source: *The Gallup Organization*

Let $f(t)$ be the percentage of Americans at age t years who say they volunteer. Find and verify an equation of f.

15. China has manufactured so many solar panels that it has created an enormous oversupply. The numbers of gigawatts of solar panels manufactured in China and the numbers of gigawatts of solar panels installed in the world are shown in Table 54 for various years.

Table 54 Solar Panels Manufactured in China and Worldwide Solar Panel Installations

Year	Solar Panels Manufactured in China (gigawatts)	Worldwide Solar Panel Installations (gigawatts)
2007	2	3
2008	3	6
2009	8	7
2010	16	18
2011	38	27
2012	50	31

Source: *GTM Research*

a. Let $f(t)$ be the number of gigawatts of solar panels manufactured in China and let $g(t)$ be the number of gigawatts of solar panels installed in the world, both in the year that is t years since 2000. Find a quadratic equation of f and a linear equation of g.

b. Use "intersect" on a graphing calculator to find the intersection points of the graphs of f and g. What do these points mean in this situation?

16. The percentages of California's population who are foreign born and the percentages who were born in other U.S. states are listed in Table 55 for various years.

Table 55 Californians Not Originally from California

	Percent	
Year	Foreign Born	Born in Other U.S. States
1970	8.8	47.9
1980	15.1	39.5
1990	21.7	31.8
2000	25.9	23.5
2010	27.2	18.0

Source: *William Frey analysis of U.S. Census Bureau sources*

a. Let $f(t)$ and $g(t)$ be the percentages of California's population that are foreign born and born in other U.S. states, respectively, at t years since 1900. Find and verify regression equations of f and g.

b. Use "intersect" on a graphing calculator to find the intersection points of the graphs of f and g. What do these points mean in this situation?

Concepts

17. A student believes the data listed in Table 56 suggest a quadratic relationship, because the values of y increase and then decrease. What would you tell the student?

Table 56 Is There a Quadratic Relationship? (Exercise 17)

x	y
0	3
1	4
2	7
3	12
4	20
5	35
6	20
7	12
8	7
9	4
10	3

18. A student uses the points $(2, 2.5)$, $(3, 4.1)$, and $(4, 6.4)$ to find an equation of a quadratic function to model the data in Table 57. Did the student make a good selection of points? If so, explain; then find the equation of those points. If not, explain; then make a better choice of points and find an equation.

Table 57 A Student Models Some Data (Exercise 18)

x	y
2	2.5
3	4.1
4	6.4
5	7.5
6	8.0
7	7.8
8	7.1
9	5.8
10	3.9
11	1.4
12	−1.7

19. Suppose that a situation can be modeled well by a parabola, where t is the independent variable and p is the dependent variable. If a data point (c, d) is below the parabola, does the model underestimate or overestimate the value of p when $t = c$? Explain.

20. Suppose that a situation can be modeled well by a parabola, where t is the independent variable and p is the dependent variable. If a data point (c, d) is above the parabola, does the model underestimate or overestimate the value of p when $t = c$? Explain.

21. Which is more desirable, finding a quadratic model whose graph contains several (but not all) data points or finding a quadratic model whose graph does not contain any data points but comes close to all data points? Include in your discussion some sketches of scattergrams and quadratic models.

22. If a quantity increases from one year to the next, explain how in some cases it might be better to model the situation by a quadratic equation rather than a linear equation. Include an example of a scattergram in your explanation.

23. Describe how to find an equation of a quadratic function that can be used to model data whose scattergram suggests a quadratic relationship.

24. Describe how to determine whether a linear function, a quadratic function, or neither of these can be used to model an authentic situation. Discuss at least two criteria you can use to help you select a type of model.

Related Review

25. The percentages of American college students who are minorities are listed in Table 58 for various years.

Table 58 Percentages of American College Students Who Are Minorities

Year	Percent
1976	15
1980	16
1990	20
2000	28
2010	39

Source: *Department of Education*

Let $f(t)$ be the percentage of American college students who are minorities at t years since 1970.

a. Find a linear equation and a quadratic equation of f. Compare how well the models fit the data.

b. For the period 1960–1972, which of the two models likely gives the better estimates of the percentages of American college students who are minorities? Explain. [**Hint:** Use a graphing calculator to sketch the graphs of the two equations. If you used ZoomStat to form your window, now use Zoom Out.]

26. The numbers of assaults in New York City are shown in Table 59 for various years.

Table 59 Assaults in New York City

Year	Number of Assaults (in thousands)
2000	40.9
2002	34.3
2004	29.3
2006	26.9
2008	24.8
2010	27.3

Source: *City-data.com*

Let $f(t)$ be the number (in thousands) of assaults in New York City in the year that is t years since 2000.

a. Find a linear equation and a quadratic equation of f. Compare how well the models fit the data.

b. For the period 2008–2026, which model would the mayor of New York City hope would describe the situation better? Explain. [**Hint:** Use a graphing calculator to sketch the graphs of the two equations. If you used ZoomStat to form your window, now use Zoom Out.]

Expressions, Equations, Functions, and Graphs

Perform the indicated instruction. Then use words such as linear, quadratic, cubic, polynomial, degree, function, one variable, *and* two variables *to describe the expression, equation, or system.*

27. Solve:

$$\frac{1}{2}x - \frac{2}{3}y = 2$$
$$\frac{4}{3}x + \frac{5}{2}y = 31$$

28. Solve $3x(x - 2) = 5 - 2x^2$.

29. Graph $\frac{1}{2}x - \frac{2}{3}y = 2$ by hand.

30. Find the product $(4x - 5)(2x^2 + x - 3)$.

31. Find an equation of a line that contains the points $(-5, -2)$ and $(-2, -7)$.

32. Graph $f(x) = 2(x - 3)^2 - 5$.

9.9 Modeling with Quadratic Functions

Objectives

- Use a quadratic model to make estimates and predictions.
- Estimate the maximum or minimum value of a quantity.
- Use a system of a quadratic model and another model to make estimates and predictions.
- Find the maximum revenue of a business venture.

In Section 9.8, we discussed how to find an equation of a quadratic model. In this section, we will use such an equation to make estimates and predictions about an authentic situation. We will also use a system of a quadratic model and another model to make predictions.

Using a Quadratic Function to Make Predictions

In Example 1, we use a quadratic model to make some predictions.

Example 1 Using a Quadratic Model to Make Predictions

The revenues of Microsoft are shown in Table 60 for various years. Let $f(t)$ be the annual revenue (in billions of dollars) at t years since 1990.

1. Find an equation of f.
2. In what years is there model breakdown for certain?
3. Predict the revenue in 2018.
4. Predict when the annual revenue will be \$100 billion.

Table 60 Revenues of Microsoft

Year	Revenue (billions of dollars)
1992	2.8
1995	5.9
2000	23.0
2005	39.8
2010	62.5
2012	73.7

Source: *Microsoft*

Solution

1. First, we use a graphing calculator to draw a scattergram to describe the data (see Fig. 105). Since the points suggest a curve that "bends," the scattergram suggests that we can model the data better with a quadratic function than with a linear function. We can use a graphing calculator to find the quadratic regression equation:

$$f(t) = 0.092t^2 + 1.36t - 1.25$$

To verify this result, we draw the scattergram and the graph of f in the same viewing window (see Fig. 106). It appears that f is a reasonable model of the data.

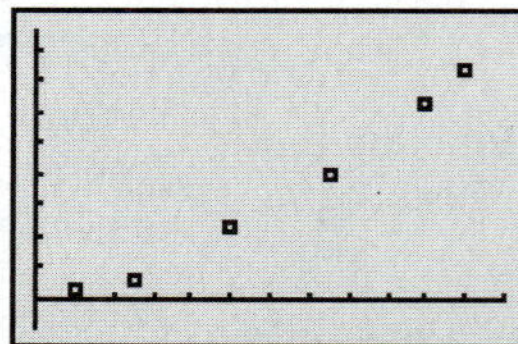

Figure 105 Microsoft scattergram

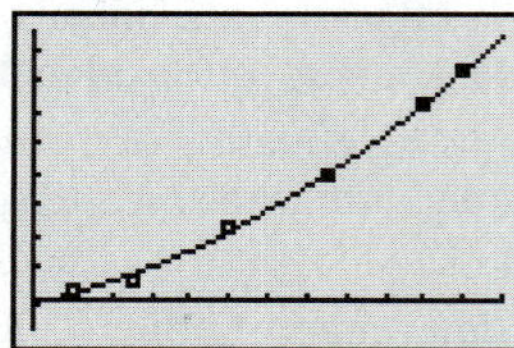

Figure 106 Verify the model

2. The part of the parabola in Fig. 107 that lies below the t-axis suggests the revenue was negative in those years, which is false—the revenue of a company is always nonnegative. So, model breakdown occurs for the part of the model between the t-intercepts.

```
WINDOW
 Xmin=-40
 Xmax=24
 Xscl=2
 Ymin=-20
 Ymax=85.753
 Yscl=10
 Xres=1
```

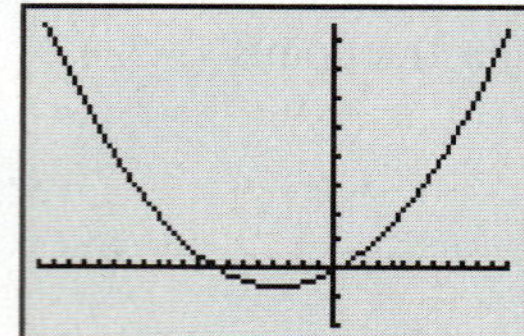

Figure 107 Part of the parabola lies below the t-axis

To find the t-intercepts, we substitute 0 for $f(t)$ in the equation $f(t) = 0.092t^2 + 1.36t - 1.25$ and solve for t:

$$0 = 0.092t^2 + 1.36t - 1.25 \qquad \text{Substitute 0 for } f(t).$$

$$t = \frac{-1.36 \pm \sqrt{1.36^2 - 4(0.092)(-1.25)}}{2(0.092)} \qquad \text{Substitute } a = 0.092,\ b = 1.36, \text{ and } c = -1.25 \text{ in quadratic formula.}$$

$$t \approx -15.65 \quad \text{or} \quad t \approx 0.87 \qquad \text{Compute.}$$

The t-intercepts are approximately $(-15.65, 0)$ and $(0.87, 0)$. To verify our work, we use "zero" on a graphing calculator (see Fig. 108 on the next page). See Appendix A.21 for graphing calculator instructions.

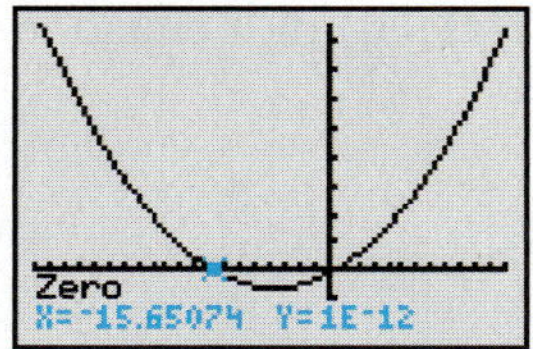

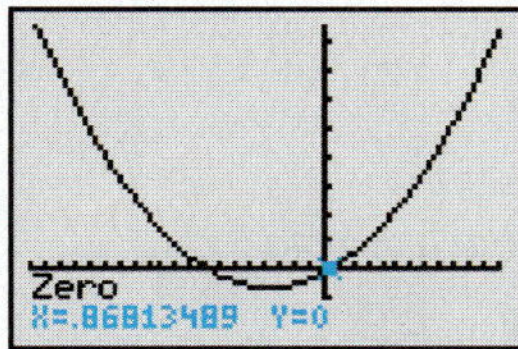

Figure 108 Verify the t-intercepts are approximately $(-15.65, 0)$ and $(0.87, 0)$

We conclude that model breakdown occurs for the years between 1974 and 1991. Model breakdown also occurs for the years up until 1974, because the company started in 1975. Therefore, model breakdown occurs for the years before 1991.

3. To find the revenue in 2018, we evaluate f at 28:

$$f(28) = 0.092(28)^2 + 1.36(28) - 1.25 \approx 108.96$$

So, the revenue will be about \$109 billion in 2018, according to the model.

4. To find when the revenue will be \$100 billion, we substitute 100 for $f(t)$ and solve for t:

$$100 = 0.092t^2 + 1.36t - 1.25 \qquad \textit{Substitute 100 for } Q(t).$$

$$0 = 0.092t^2 + 1.36t - 101.25 \qquad \textit{Subtract 100 from both sides.}$$

$$t = \frac{-1.36 \pm \sqrt{1.36^2 - 4(0.092)(-101.25)}}{2(0.092)} \qquad \textit{Substitute } a = 0.092, b = 1.36, \textit{ and } c = -101.25 \textit{ in quadratic formula.}$$

$$t \approx -41.38 \quad \text{or} \quad t \approx 26.60 \qquad \textit{Compute.}$$

We can use a graphing calculator to verify our work (see Fig. 109). The values of t we found represent the years 1949 and 2017. Model breakdown occurs for 1949, because the company began in 1975. In Problem 2, we decided model breakdown occurs for years before 1991, which checks. So, we predict the revenue will be \$100 billion in 2017.

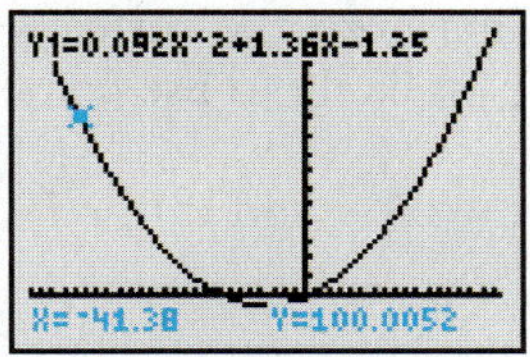

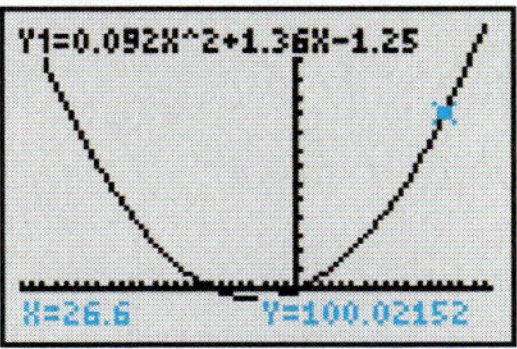

Figure 109 Verify that $t \approx -41.38$ and $t \approx 26.60$

In Example 1, we used a quadratic model to make predictions. Recall from Section 8.6 that, to make a prediction about the dependent variable, we substitute a value for the independent variable and solve for the dependent variable. To make a prediction about the independent variable, we substitute a value for the dependent variable and solve for the independent variable, usually by using the quadratic formula.

Finding the Maximum or Minimum Value of a Quantity

In Example 2, we will find the vertex of the graph of a quadratic function to help us determine the maximum value of the function.

Example 2 Making Estimates

Table 61 lists the percentages of workers in various age groups who use computers on the job. Let $p = f(t)$ be the percentage of workers who use computers at age t years.

Table 61 Workers Who Use Computers on the Job

Age Group (years)	Age Used to Represent Age Group (years)	Percent
18–25	21.5	34.4
25–29	27.0	48.3
30–39	34.5	50.7
40–49	44.5	51.3
50–59	54.5	43.9
over 59	62.5	27.2

Source: *National Center for Education Statistics*

1. Find an equation of f.
2. Use f to estimate the percentage of 22-year-old workers who use computers on the job.
3. Estimate the age(s) at which half of workers use computers on the job.
4. Estimate the age of workers who are *most likely* (the maximum percentage) to use computers on the job. What is that maximum percentage?
5. Find the t-intercepts. What do they mean in this situation?

Solution

1. We begin by drawing a scattergram of the data (see Fig. 110). It looks like a parabola would fit the data well. We can use a graphing calculator to find the quadratic regression model:

$$f(t) = -0.051t^2 + 4.09t - 28.14$$

To verify this result, we draw the scattergram and the graph of f in the same viewing window (see Fig. 111 on the next page). It appears f is a reasonable model of the data.

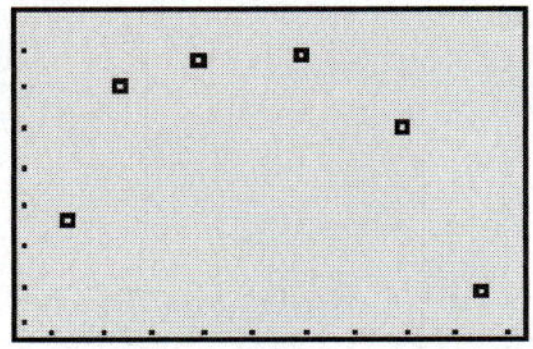

Figure 110 Computer worker scattergram

2. Since

$$f(22) = -0.051(22)^2 + 4.09(22) - 28.14 \approx 37.16$$

we estimate that about 37.2% of 22-year-old workers use computers at work. We can verify this computation by using a graphing calculator table or graph.

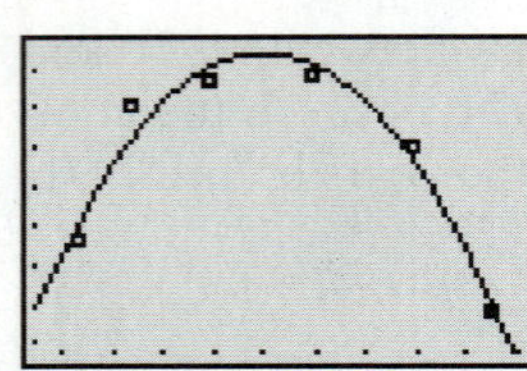

Figure 111 Verify computer worker model

3. Half of all workers is 50%, so we substitute 50 for $f(t)$ in the equation $f(t) = -0.051t^2 + 4.09t - 28.14$:

$$50 = -0.051t^2 + 4.09t - 28.14$$

Next, we write the equation in $at^2 + bt + c = 0$ form:

$$-0.051t^2 + 4.09t - 78.14 = 0$$

Then we apply the quadratic formula and solve for t:

$$t = \frac{-4.09 \pm \sqrt{4.09^2 - 4(-0.051)(-78.14)}}{2(-0.051)}$$ *Substitute $a = -0.051$, $b = 4.09$, and $c = -78.14$ in quadratic formula.*

$$t \approx 48.80 \quad \text{or} \quad t \approx 31.40$$ *Compute.*

So, according to our model, half of 31-year-old workers and half of 49-year-old workers use computers on the job. We can verify these results by using a graphing calculator table or graph.

4. The function f is of the form $f(t) = at^2 + bt + c$, with $a = -0.051 < 0$, so the graph of f is a parabola that opens downward (see Fig. 111). Therefore, the vertex is the maximum point. We can find the age when workers are most likely to use computers by finding the vertex.

To find the t-coordinate of the vertex, we substitute $a = -0.051$ and $b = 4.09$ in the vertex formula $t = -\frac{b}{2a}$:

$$t = -\frac{4.09}{2(-0.051)} \approx 40.10$$

The t-coordinate is about 40. To find the p-coordinate, we evaluate f at 40:

$$f(40) = -0.051(40)^2 + 4.09(40) - 28.14 = 53.86$$

The vertex is about $(40, 53.9)$. So, according to our model, about 53.9% of 40-year-old workers use computers on the job—the highest percentage for any age group. We can verify our computations by using "maximum" on a graphing calculator.

5. To find the t-intercepts, we substitute 0 for $f(t)$ in the quadratic equation $f(t) = -0.051t^2 + 4.09t - 28.14$ and solve for t:

$$0 = -0.051t^2 + 4.09t - 28.14$$ *Substitute 0 for $f(t)$.*

$$t = \frac{-4.09 \pm \sqrt{4.09^2 - 4(-0.051)(-28.14)}}{2(-0.051)}$$ *Substitute $a = -0.051$, $b = 4.09$, and $c = -28.14$ in quadratic formula.*

$$t \approx 72.60 \quad \text{or} \quad t \approx 7.60$$ *Compute.*

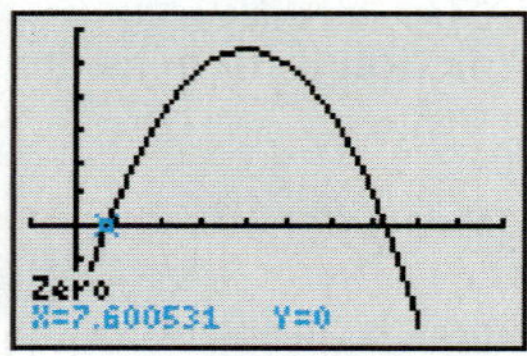

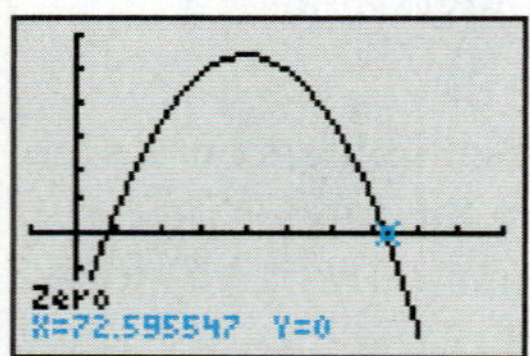

Figure 112 Verify the intercepts

The t-intercepts are approximately $(7.6, 0)$ and $(72.6, 0)$. We use "zero" on a graphing calculator to verify our work (see Fig. 112). See Appendix A.21 for graphing calculator instructions.

The model estimates that no 8-year-old workers use computers on the job; this is correct but only because, legally, 8-year-old children cannot work. The model also estimates that no 73-year-old workers use computers on the job. Model breakdown has occurred, because some 73-year-old workers do use computers.

Modeling with a System of a Quadratic Equation and Another Equation

In Example 3, we will make an estimate by working with a system of a quadratic model and a linear model.

Example 3 Modeling with a System of a Quadratic Equation and a Linear Equation

In Exercise 88 of Homework 7.1, the market shares of world manufacturing output $U(t)$ and $C(t)$ of the United States and China, respectively, are modeled by the system

$$p = U(t) = -0.47t + 24.68$$
$$p = C(t) = 0.13t^2 - 0.54t + 11.24$$

where t is the number of years since 2000 (see Table 62).

Table 62 United States' and China's Market Shares of Manufacturing Output

	Market Share (Percent)	
Year	United States	China
2004	22.8	11.1
2005	22.5	11.9
2006	21.7	12.8
2007	21.4	13.9
2008	20.8	15.0
2009	20.6	17.2
2010	20.0	18.9

Source: *United Nations*

Estimate in which year(s) the market share of world manufacturing output of the United States was equal to that of China.

Solution

We substitute $0.13t^2 - 0.54t + 11.24$ for p in the equation $p = -0.47t + 24.68$:

$$0.13t^2 - 0.54t + 11.24 = -0.47t + 24.68 \quad \text{Substitute } 0.13t^2 - 0.54t + 11.24 \text{ for } p.$$
$$0.13t^2 - 0.07t - 13.44 = 0 \quad \text{Write in } at^2 + bt + c = 0 \text{ form.}$$

The quadratic formula gives

$$t = \frac{-(-0.07) \pm \sqrt{(-0.07)^2 - 4(0.13)(-13.44)}}{2(0.13)}$$

$$t \approx -9.90 \quad \text{or} \quad t \approx 10.44$$

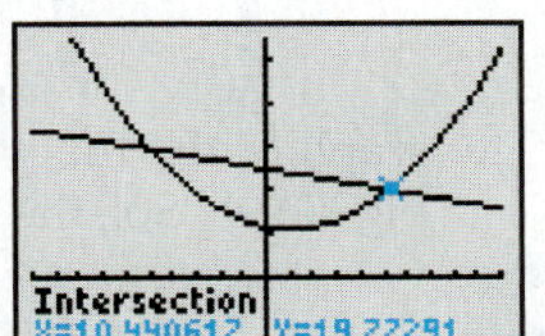

Figure 113 China's market share overtook the United States' in 2011

The values of t we found represent 1990 and 2010. However, a little research would show that China's market share of world manufacturing output was much less than the United States' in 1990. So, the market shares of the two countries were approximately equal in 2010, according to the models. Even though the 2010 market shares shown in Table 62 are not equal, the models suggest China's market share overtook the United States' by 2011 (see Fig. 113). We conclude the two countries' market shares were equal at some time in 2010.

In Example 3, we used substitution with a system of a quadratic equation and a linear equation, both in two variables. **We can solve such a system as well as a system of two quadratic equations if at least one quadratic equation is in standard form $y = ax^2 + bx + c$.**

Maximum Revenue

Recall from Section 4.6 that if each of n objects has value v, then their total value T is given by $T = vn$. In Example 4, we will find the maximum value of the revenue of a chartered flight.

Example 4 Finding the Maximum Value of a Quantity

A group charters a flight that normally costs \$800 per person. A group discount reduces the fare by \$10 for each ticket sold; the more tickets sold, the lower the per-person fare. There are 60 seats on the plane, including 4 seats for the crew. What size of group would maximize the airline's revenue?

Solution

We use the five-step problem-solving method of Section 6.5, but in step 2 we use a system of equations to build a function and in step 3 we find the maximum point of the graph of the function.

Step 1. Define each variable. We let n be the number of people in the group, p be the price (in dollars per person), and R be the revenue (in dollars).

Step 2. Use a system of two equations to build a function. A group of 3 people would be charged $800 - 3(10) = 770$ dollars per person. A group of 4 people would be charged $800 - 4(10) = 760$ dollars per person. So, a group of n people would be charged $800 - n(10)$ dollars per person. Our first equation is

$$p = -10n + 800$$

For our second equation, the revenue is equal to the price of one ticket times the number of tickets sold:

$$\overbrace{R}^{\text{revenue}} = \overbrace{p}^{\frac{\text{dollars}}{\text{ticket}}} \cdot \overbrace{n}^{\text{tickets}}$$

So, our system is

$$p = -10n + 800 \quad \textit{Equation (1)}$$
$$R = pn \quad \textit{Equation (2)}$$

Next, we will build a revenue function whose independent variable is n. To begin, we substitute $-10n + 800$ for p in equation (2):

$$R = pn \quad \textit{Equation (2)}$$
$$R = (-10n + 800)n \quad \textit{Substitute } -10n + 800 \textit{ for } p.$$
$$R = -10n^2 + 800n \quad \textit{Distributive law}$$

Step 3. Find the maximum point of the graph of the function. The graph of the function $R = -10n^2 + 800n$ is a parabola that opens downward (see Fig. 114). So, the parabola has a maximum point (at the vertex).

To find the n-coordinate of the vertex, we use the vertex formula $n = -\frac{b}{2a}$:

$$n = -\frac{800}{2(-10)} = 40$$

To find the R-coordinate of the vertex, we substitute 40 for n in the revenue equation $R = -10n^2 + 800n$:

$$R = -10(40)^2 + 800(40) = 16{,}000$$

To find the price, we substitute 40 for n in the equation $p = -10n + 800$:

$$p = -10(40) + 800 = 400$$

Step 4. Describe each result. The revenue is greatest if there are 40 people in a group. The price is then \$400 per person, and the revenue is \$16,000.

Step 5. Check. The revenue from 40 people, each paying \$400, is $40 \cdot 400 = 16{,}000$ dollars, which checks. To check that the largest revenue is \$16,000 dollars, we find the revenue for groups with 1 person fewer or 1 person more than 40 people.

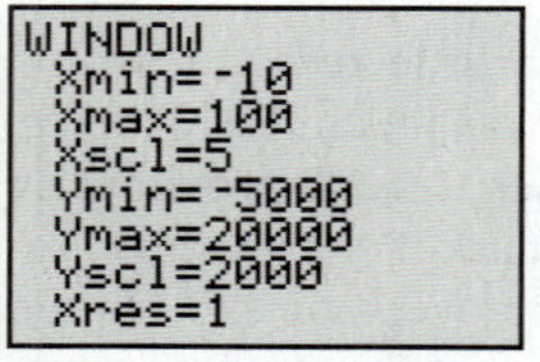

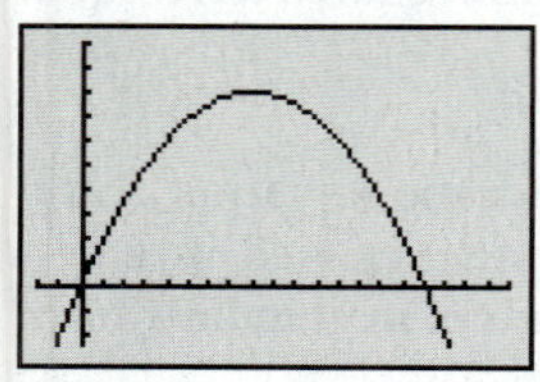

Figure 114 Graph of $R = -10n^2 + 800n$

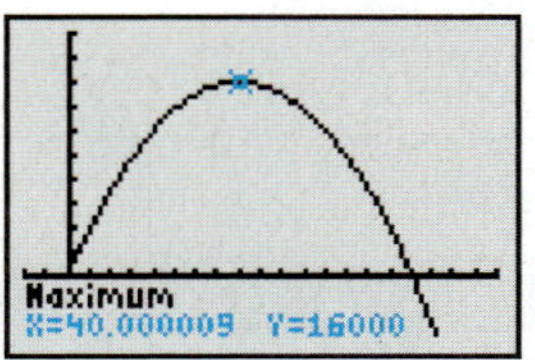

Figure 115 Verify that the maximum point is (40, 16,000)

39 people in group

price is $800 - 39(10) = 410$ dollars

revenue is $410 \cdot 39 = 15{,}990$ dollars

41 people in group

price is $800 - 41(10) = 390$ dollars

revenue is $390 \cdot 41 = 15{,}990$ dollars

Since a revenue of \$16,000 is more than either of these (equal) revenues of \$15,990, it seems reasonable that our work is correct. Finally, we use "maximum" on a graphing calculator to check that the maximum point of the parabola $R = -10n^2 + 800n$ is (40, 16,000). See Fig. 115.

Group Exploration

Modeling differences of quantities

Refer to Example 3 for background information on the market shares of world manufacturing output of the United States and China.

1. Complete Table 63.

Table 63 Differences in United States' and China's Market Shares of Manufacturing Output

	Market Share (Percent)		Difference in Market Shares of the
Year	United States	China	United States and China
2004	22.8	11.1	$22.8 - 11.1 = 11.7$
2005	22.5	11.9	
2006	21.7	12.8	
2007	21.4	13.9	
2008	20.8	15.0	
2009	20.6	17.2	
2010	20.0	18.9	

2. Let $D(t)$ be the difference in market shares of the manufacturing output of the United States and China at t years since 2000. Find a regression equation of D. Remember to use a graphing calculator to draw a scattergram of the data first.

3. Which of the statements that follow is correct? Explain.

$D(t) = (U + C)(t) \qquad D(t) = (U - C)(t)$

$D(t) = (C - U)(t)$

4. In Example 3, we used the following equations of U and C:

$$p = U(t) = -0.47t + 24.68$$

$$p = C(t) = 0.13t^2 - 0.54t + 11.24$$

Substitute $-0.47t + 24.68$ for $U(t)$ and substitute $0.13t^2 - 0.54t + 11.24$ for $C(t)$ in the equation $D(t) = U(t) - C(t)$ to find an equation of D. Simplify the right-hand side of your equation and compare the result with the equation you found in Problem 2.

5. Find t when $D(t) = 0$. What does your result mean in this situation? Compare your result with the result in Example 3.

Group Exploration

Looking ahead: Negative-integer exponents

In this exploration, you will explore the meaning of a negative-integer exponent.

1. Use your graphing calculator to compute 2^{-1}. Then press MATH 1 ENTER to convert your result into a fraction.

2. Use your graphing calculator to compute 3^{-1}. Then press MATH 1 ENTER to convert your result into a fraction.

3. For each part, use your graphing calculator to compute the expression and then convert the result into a fraction.

a. 4^{-1} **b.** 5^{-1} **c.** 6^{-1}

4. Review your work in Problems 1–3. What pattern do you notice? If you do not see a pattern, then keep working with more powers of the form x^{-1} until you do.

5. Without using any type of calculator, write 13^{-1} as a fraction.

6. Write x^{-1} as a fraction.

7. For each part, use your graphing calculator to compute the expression and then convert the result into a fraction.

a. 3^{-2} **b.** 4^{-2} **c.** 5^{-2}

8. Review your work in Problem 7. What pattern do you notice? If you do not see a pattern, then keep working with more powers of the form x^{-2} until you do.

9. Without using any type of calculator, write 9^{-2} as a fraction.

10. Write x^{-2} as a fraction.

11. Without using any type of calculator, write each expression as a fraction.

a. 8^{-1} **b.** 7^{-2} **c.** 4^{-3} **d.** 2^{-5}

12. Write each expression as a fraction.

a. x^{-3} **b.** x^{-4} **c.** x^{-5} **d.** x^{-6}

13. Write the expression x^{-n} as a fraction.

Homework 9.9

For extra help ▶ MyMathLab® 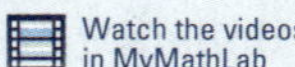 Watch the videos in MyMathLab  Download the MyDashboard App

1. In Exercise 49 of Homework 9.2, you worked with the model $f(t) = 0.058t^2 - 0.98t + 15.43$, where $p = f(t)$ is the percentage of households with outstanding student debt at t years since 1990 (see Table 64).

Table 64 Percentages of Households with Outstanding Student Debt

Year	Percent
1995	12
1998	11
2001	12
2004	13
2007	15
2010	19

Source: *Pew Research Center*

a. Find the p-intercept. What does it mean in this situation?
b. Estimate the percentage of households that had outstanding student debt in 2009. Did you perform interpolation or extrapolation? Explain.
c. Predict the percentage of households that will have outstanding student debt in 2018. Did you perform interpolation or extrapolation? Explain.
d. Predict when 36% of households will have outstanding student debt.

2. In Exercise 14 of Homework 9.8, you found an equation close to $f(t) = -0.0197t^2 + 1.86t + 9.90$, where $f(t)$ is the percentage of Americans at age t years who say they do volunteer work (see Table 65).

Table 65 Percentages of Americans Who Say They Volunteer

Age Group (years)	Age Used to Represent Age Group (years)	Percent
18–24	21.0	38
25–34	29.5	51
35–54	44.5	55
55–64	59.5	48
65–74	69.5	45
over 74	80.0	34

Source: *The Gallup Organization*

a. Estimate the percentage of 25-year-old Americans who say they volunteer. Did you perform interpolation or extrapolation?
b. Estimate the percentage of 16-year-old Americans who say they volunteer. Did you perform interpolation or extrapolation?
c. For what age do half of Americans say they volunteer?
d. For what age is the percentage who say they volunteer the greatest? What is that maximum percentage?

3. In Example 4 of Section 1.2, we used a scattergram to describe the percentages of householders who own a home for various age groups (see Table 66). Now we will use a function to describe the situation.

Table 66 Percentages of Householders Who Own a Home

Age Group (years)	Age Used to Represent Age Group (years)	Percent Who Own a Home
15–24	19.5	18
25–34	29.5	46
35–44	39.5	66
45–54	49.5	75
55–64	59.5	80
65–74	69.5	81
75–84	79.5	77

Source: *U.S. Census Bureau*

Let $f(t)$ be the percentage of householders who own a home at age t years.

a. Find an equation of f.
b. Find the t-intercepts. What do they mean in this situation?
c. For what values of t is there model breakdown for certain? Which ages are represented by these values?
d. Estimate the age of householders who are most likely to own a home. What percentage of householders at this age own a home?
e. Estimate all ages at which half of householders own homes.

4. A tennis ball is tossed upward, and its heights at various times are recorded (see Table 67).

Table 67 Heights of a Tennis Ball

Time (seconds)	Height (feet)
0.00	1.93
0.06	2.25
0.12	2.51
0.18	2.69
0.24	2.70
0.30	2.54
0.36	2.37

Source: *J. Lehmann*

Let $f(t)$ be the height (in feet) of the tennis ball at t seconds.

a. Find an equation of f.
b. Find the t-intercepts. What do they mean in this situation?
c. For what values of t is there model breakdown for certain? Explain.
d. Estimate the maximum height of the tennis ball. When did it reach this height?
e. Assume that, after the toss, the tennis ball bounces on the ground several times. Sketch a model that describes the relationship between the height of the tennis ball and the time until the tennis ball comes to rest. [**Hint:** The new model will be neither linear nor quadratic.]

5. In Exercise 13 of Homework 9.8, you found an equation close to $P = f(t) = 0.0068t^2 - 0.13t + 6.49$, where $f(t)$ is the U.S. population (in millions) at t years since 1790 (see Table 68).

Table 68 U.S. Population

Year	Population (millions)	Year	Population (millions)
1790	3.9	1910	92.2
1800	5.3	1920	106.0
1810	7.2	1930	123.2
1820	9.6	1940	132.2
1830	12.9	1950	151.3
1840	17.1	1960	179.3
1850	23.2	1970	203.3
1860	31.4	1980	226.5
1870	39.8	1990	248.7
1880	50.2	2000	281.4
1890	63.0	2010	308.7
1900	76.2		

Source: *U.S. Census Bureau*

a. Find $f(228)$. What does it mean in this situation?
b. Find t, where $f(t) = 335$. What does your result mean in this situation?
c. Graph f by hand.
d. For what values of t is there model breakdown? Which years are represented by these values?
e. Sketch a model that likely describes the relationship between t and P for the period 1700–2050 better than f does. [**Hint:** The new model will be neither linear nor quadratic.]

6. A *filibuster* is an action such as a prolonged speech or raising an objection that is used to delay the debate of a bill in the Senate. The numbers of filibusters are shown in Table 69 for various years.

Table 69 Numbers of Filibusters

Year	Number of Filibusters
1962	4
1972	24
1982	31
1992	60
2002	71
2012	110

Source: *U.S. Senate*

Let $f(t)$ be the number of filibusters in the year that is t years since 1950.
a. Find a quadratic equation of f.
b. Find $f(67)$. What does it mean in this situation?
c. Find t when $f(t) = 67$. What does it mean in this situation?
d. Find the t-intercepts of the model. What do they mean in this situation?
e. Assuming that the number of filibusters has increased throughout time, find the years in the past when model breakdown occurred for certain.

7. The numbers of career home runs hit by Barry Bonds are shown in Table 70 for various years. Let $f(t)$ be the number of career home runs hit by Barry Bonds at t years since 1980.
a. Find an equation of f.
b. Use f to estimate when Barry Bonds tied Hank Aaron's record of 755 lifetime home runs.
c. Due to injury, Bonds hit only 5 home runs in 2005. Find $f(25)$ and $f(26)$. Then find the difference $f(26) - f(25)$. Does your result suggest Bonds tied Aaron's record in 2006? Explain.

Table 70 Numbers of Career Home Runs Hit by Barry Bonds

Year	Number of Career Home Runs
1986	16
1990	117
1995	292
2000	494
2004	703

Source: *Baseball-reference.com*

d. Bonds turned 42 years old on July 24, 2006. Discuss any assumptions you made in part (c), and explain why Bonds may not have tied Aaron's record in 2006.

8. The percentages of Americans who say they've forgotten to do something special for their significant other on Valentine's Day are shown in Table 71 for various ages.

Table 71 Percentages of Americans Who Forgot to Do Something Special on Valentine's Day

Age Group (years)	Age Used to Represent Age Group (years)	Percent
25–34	29.5	11
35–44	39.5	17
45–54	49.5	24
55–64	59.5	27
over 64	70.0	29

Source: *CARAVAN for DHL*

Let $f(t)$ be the percentage of Americans at age t years who say they've forgotten to do something special for their significant other on Valentine's Day.
a. Find an equation of f.
b. Find the t-intercepts. What do they mean in this situation?
c. Find the vertex. What does it mean in this situation?
d. Estimate the percentage of 21-year-old Americans who say they've forgotten to do something special for their significant other on Valentine's Day. The actual percentage is 19%. Has model breakdown occurred? Explain.
e. For what values of t is there model breakdown for certain? Which years are represented by these values? Are there other values of t for which it is likely there is model breakdown? If so, find those values of t and the years represented by them.

9. Table 72 shows that people with more education tend to earn more money. Let $f(t)$ be the median annual income (in thousands of dollars) for people with a full-time equivalent of t years of education. A model for the situation is $f(t) = 0.187t^2 + 0.06t + 4.5$.
a. Use a graphing calculator to draw the graph of the model and, in the same viewing window, the scattergram of the data. Does the model fit the data well?
b. Estimate the full-time equivalent number of years of education for people whose median annual income is $47 thousand.

Table 72 Education versus Median Annual Income

Grade-Level Completion or Degree	Full-Time Equivalent Years in School	Median Annual Income (thousands of dollars)
9th to 12th Grade, no diploma	10	23.5
High school	12	33.2
Some college	13	37.4
Associate's	14	39.9
Bachelor's	16	54.8
Master's	18	65.7
Doctoral	20	80.7

Source: *U.S. Census Bureau*

c. In August 2009, a high school graduate is trying to decide whether to get a job or to go to a public college to get a bachelor's degree. For the 2009–2010 academic year, average tuition and fees for a public 4-year college are \$5964. In parts (i)–(iv), assume tuition and fees, as well as incomes, are constant.

i. If the student doesn't go to college, estimate the student's total earnings in four years.

ii. If the student goes to college, estimate the total cost of four years of college.

iii. If the student goes to college, estimate after how many years of work will the student be in the same financial position (total earnings minus total costs) as if the student hadn't gone to college and had gotten a job instead.

iv. Assume the student earns a bachelor's degree in 4 years and then works for 33 years until retirement. How much more will the student's lifetime earnings minus total college tuition and fees be than if the person had not gone to college (and had worked for 37 years)?

v. During the past 20 years, those with larger incomes have had larger growth in incomes. Is your estimate in part (iv) an underestimate or an overestimate? Explain.

10. Sales of downloaded albums are shown in Table 73 for various years.

Table 73 Downloaded Albums Annual Sales

Year	Annual Sales (millions of downloaded albums)
2004	4.6
2005	13.6
2006	27.6
2007	42.5
2008	63.6

Source: *Recording Industry Association of America*

Let $f(t)$ be the annual sales (in millions) of downloaded albums at t years since 2000.

a. Find an equation of f.

b. Predict when annual sales will reach 400 million downloaded albums.

c. Use f to estimate the sales in 2010. Sales in that year were only 83.1 million downloaded albums, most likely due to the poor economy during the period 2008–2010. Estimate lost sales of downloaded albums in 2010 due to the poor economy.

11. In Exercise 15 of Homework 9.8 you found a system close to

$$n = f(t) = 2.20t^2 - 31.65t + 115.51$$
$$n = g(t) = 6.11t - 42.75$$

where $f(t)$ is the number of gigawatts of solar panels manufactured in China and $g(t)$ is the number of gigawatts of solar panels installed in the world, both in the year that is t years since 2000 (see Table 74). Use substitution to estimate in which years the number of gigawatts of solar panels manufactured in China was equal to the number of gigawatts of solar panels installed in the world.

Table 74 Solar Panels Manufactured in China and Worldwide Solar Panel Installations

Year	Solar Panels Manufactured in China (gigawatts)	Worldwide Solar Panel Installations (gigawatts)
2007	2	3
2008	3	6
2009	8	7
2010	16	18
2011	38	27
2012	50	31

Source: *GTM Research*

12. Women tend to be younger than men in their first marriage. The median ages at which women and men were first married are shown in Table 75 for various years.

Table 75 Median Ages at First Marriages

	Median Age	
Year	Women (years)	Men (years)
1975	21.1	23.5
1980	22.0	24.7
1985	23.3	25.5
1990	23.9	26.1
1995	24.5	26.9
2000	25.1	26.8
2005	25.3	27.1
2010	26.1	28.2

Source: *U.S. Census Bureau*

Let $W(t)$ be the median age (in years) at which women were first married and $M(t)$ be the median age (in years) at which men were first married, both at t years since 1900.

a. Find regression equations of W and M.

b. Predict when the median age or ages at which women and men first marry will be equal, if ever.

13. In the 2000 presidential election, the race between George W. Bush and Al Gore was so close that there was concern about the accuracy of vote-counting systems, including punch cards, lever machines, and paper ballots. Punch cards and lever systems are gradually being replaced with optical scan or other modern electronic vote-counting systems (see Table 76).

a. Let $f(t)$ be the percentage of voters using punch cards or lever machines and $g(t)$ be the percentage of voters using optical scan or other modern electronic systems, both at t years since 1990. Find regression equations of f and g.

Table 76 Vote-Counting Systems

	Percent of Registered Voters	
Year	Punch Card or Lever Machine	Optical Scan or Other Modern Electronic System
1990	71.7	13.5
1992	69.4	16.5
1994	63.1	25.8
1996	58.0	32.3
1998	52.9	36.4
2002	35.7	54.3
2004	26.0	65.0
2008	11.0	80.0

Source: *Federal Election Commission*

b. Use substitution to estimate when the percentage of voters using optical scan or other modern electronic systems equaled the percentage of voters using punch cards or lever machines. Is your result a major (even-numbered) election year (congressional or presidential)? If not, in which even-numbered year were the percentages closest?

c. There are three other ways to vote: DataVote (punch holes in ballots that haven't been perforated), paper (count hand-marked ballots manually), and mixed systems. Let $h(t)$ be the percentage of voters using any of these three ways to vote at t years since 1990. Which of the statements that follow is correct? Explain.

$$h(t) = (f + g)(t) \quad h(t) = 100 - (f + g)(t)$$
$$h(t) = (f - g)(t) \quad h(t) = 100 - (f - g)(t)$$

d. Find an equation of h.

e. Find $h(17)$. What does it mean in this situation?

14. In Exercise 16 of Homework 9.8, the percentages of California's population $f(t)$ and $g(t)$ who are foreign born and born in other U.S. states, respectively, are modeled by the system

$$f(t) = -0.0089t^2 + 2.07t - 93.07$$
$$g(t) = 0.0037t^2 - 1.43t + 129.70$$

where t is the number of years since 1900 (see Table 77).

Table 77 Californians Not Originally from California

	Percent	
Year	Foreign Born	Born in Other U.S. States
1970	8.8	47.9
1980	15.1	39.5
1990	21.7	31.8
2000	25.9	23.5
2010	27.2	18.0

Source: *William Frey analysis of U.S. Census Bureau sources*

a. Use substitution to estimate when the percentage of Californians who were foreign born was equal to the percentage of Californians who were born in other U.S. states. What is that percentage? That year, what percentage of California's population was originally from California?

b. Let $h(t)$ be the percentage of California's population who were originally from California at t years since 1900. Which of the statements that follow is correct? Explain.

$$h(t) = 100 - (f - g)(t) \quad h(t) = (f - g)(t)$$
$$h(t) = 100 - (f + g)(t) \quad h(t) = (f + g)(t)$$

c. Find an equation of h.

d. Find $h(118)$. What does it mean in this situation?

e. Use a graphing calculator to sketch a graph of h. Is the curve increasing or decreasing for $t \geq 62$? What does it mean in this situation?

15. A ski club charters a bus that normally costs \$250 per person. A group discount reduces the fare by \$5 for each ticket sold; the more tickets sold, the lower is the per-person fare. There are 31 seats on the bus, including 1 seat for the driver. What size of group would maximize the bus company's revenue?

16. A company charters a party boat that normally costs \$60 per person. A group discount reduces the fare by \$0.50 for each ticket sold; the more tickets sold, the lower is the per-person fare. The maximum capacity of the boat is 80 people, including the crew of 10 people. What size of group would maximize the boat company's revenue?

17. Some students reserve the buffet room in a restaurant for a graduation party. The buffet dinner normally costs \$28 per person. A group discount reduces the price by \$0.20 for each student who attends; the more dinners sold, the lower is the per-person price. The room's maximum capacity is 80 people. What size of party would maximize the restaurant's revenue?

18. A group charters a flight that normally costs \$900 per person. A group discount reduces the fare by \$15 for each ticket sold; the more tickets sold, the lower is the per-person fare. There are 40 seats on the plane, including 4 seats for the crew. What size of group would maximize the airline's revenue?

19. A person plans to use 160 feet of fencing to enclose a rectangular play area. What dimensions of the rectangle would maximize the area? What is that area?

20. A rancher plans to use 1200 feet of fencing and the side of her barn to form a rectangular boundary for cattle (see Fig. 116). What dimensions of the rectangle would maximize the area? What is that area?

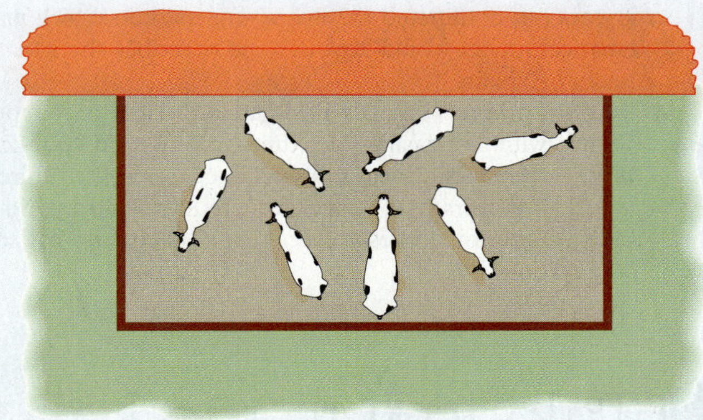

Figure 116 Boundary for cattle

Concepts

21. Assume the quadratic function $f(t) = at^2 + bt + c$ is a model of a situation where t is the number of years since 2000 and the vertex is (h, k). Explain why there is sometimes model breakdown for either $t < h$ or for $t > h$. Include a graph of f in your explanation.

22. Assume the quadratic function $f(t) = at^2 + bt + c$ is a model of a situation where t is the number of years since 2000 and $(2, 0)$ and $(18, 0)$ are t-intercepts. Explain why there is sometimes model breakdown for $t < 2$ and $t > 18$. Include a graph of f in your explanation.

23. For a quadratic model, discuss how to find intercepts, the vertex, and the maximum or minimum value; how to make predictions for the dependent or independent variable; and how to determine values of the independent variable where model breakdown occurs.

24. Describe how you can find a system of two quadratic equations in two variables to model an authentic situation. Explain how you can use the system to make an estimate or a prediction about the situation.

Related Review

25. Sales of gasoline–electric hybrid cars in the United States are shown in Table 78 for various years.

Table 78 Sales of Gasoline–Electric Hybrid Cars

Year	Sales (thousands of cars)
2007	352
2008	312
2009	290
2010	274
2011	269

Source: *HybridCars.com*

Let $f(t)$ be hybrid-car sales (in thousands of cars) in the year that is t years since 2000.

a. Find a linear regression equation and a quadratic regression equation of f. Compare how well the models fit the data.

b. Use first the linear model and then the quadratic model to estimate hybrid-car sales in 2009. Which is the better estimate? Explain.

c. Toyota has dominated the hybrid-car market. Which mathematical model would Toyota hope describes the situation better? Explain.

d. Because of a recession, hybrid cars aren't the only type of vehicle with decreasing sales during the period 2007–2011. Most experts believe car sales will increase in the future. In fact, J.D. Powers and Associates predicts 2016 hybrid-car sales will be four times greater than 2010 hybrid-car sales. Use the quadratic model to predict hybrid-car sales in 2016, and compare the result with J.D. Powers and Associates' prediction.

e. Use the quadratic model to estimate in which year hybrid-car sales were or will be 500 thousand cars.

26. The numbers of women who have run in the New York City (NYC) Marathon are shown in Table 79 for various years.

Table 79 Numbers of Women Who Have Run in the NYC Marathon

Year	Number of Women
1970	1
1980	1962
1990	5204
2000	8641
2010	16,253
2011	17,563

Source: *New York Road Runners Club*

Let $n = f(t)$ be the number of women who have run in the NYC Marathon at t years since 1970.

a. Find a linear regression equation and a quadratic regression equation of f. Compare how well the models fit the data.

b. Use first the linear model and then the quadratic model to estimate the number of women who ran in the marathon in 2000. Which is the better estimate? Explain.

c. Use first the linear model and then the quadratic model to predict in which year 24,000 women will run in the marathon—when about half of the runners will be women. Use the graphs of the models to explain why the quadratic model's result is earlier than the linear model's result.

Expressions, Equations, Functions, and Graphs

Give an example of each. Then solve, simplify, or graph, as appropriate.

27. quadratic equation in one variable that can be solved by factoring

28. linear equation in one variable

29. linear function

30. system of two linear equations in two variables

31. quadratic equation in one variable that can't be solved by factoring

32. cubic polynomial with five terms

33. quadratic function

Taking it to the Lab

Climate Change Lab (continued from Chapter 7)

Some developing countries' carbon dioxide emissions are increasing greatly as their economies continue to grow. In particular, China's future carbon dioxide emissions will affect the planet considerably, because of that country's large population, large population growth, and increasing per-person carbon dioxide emissions (see Table 80). Unless China takes measures to slow the growth of carbon dioxide emissions, its per-person emissions in 2050 could reach 41.5 metric tons—about 21 times the 1990 per-person emissions.

Table 80 China's Population and Carbon Dioxide Emissions

Year	Population (billions)	Per-Person Carbon Emissions (metric tons per person)
1990	1.15	1.98
1995	1.21	2.38
2000	1.27	2.86
2005	1.31	4.21
2010	1.34	6.15

Source: *U.S. Census Bureau; U.S. Energy Information Administration*

Although the United States has the largest gross domestic product (GDP) in the world, China's GDP is likely to surpass U.S. GDP as early as 2020.* So, it is not surprising China has large carbon dioxide emissions. In fact, China has already overtaken the United States as the top emitter of carbon dioxide in the world; in 2010, China's emissions were 8.2 billion metric tons and U.S. emissions were 5.6 billion metric tons.†

Even though China has larger carbon dioxide emissions than the United States, its per-person emissions are about one-third of the United States' (see Tables 80 and 81). If U.S. per-person emissions were to hold steady at 18.1 metric tons, China's per-person emissions could reach that level as early as 2029.

Table 81 U.S. Population and Carbon Dioxide Emissions

Year	Population (billions)	Per-Person Carbon Emissions (metric tons per person)
1950	0.15	16.0
1960	0.18	16.1
1970	0.21	20.5
1980	0.23	20.9
1990	0.25	20.0
2000	0.28	21.1
2010	0.31	18.1

Source: *U.S. Census Bureau; The Cambridge Factfinder*

Recall from the Climate Change Lab in Chapter 7 that if world population were to decline to 1.6 billion, then the IPCC's 2050 carbon dioxide emissions goal could be met if carbon dioxide emissions were limited to 5.4 metric tons per person each year. This per-person limit would create room for most developing countries to expand their economies.

Such a population–emissions strategy would require some developing countries to learn to expand their economies in new ways that would not increase per-person carbon dioxide emissions so much. Other developing countries such as China and oil-rich countries such as Qatar would actually have to reduce their per-person emissions. It would also require developed countries to learn to maintain their economies while reducing their per-person carbon dioxide emissions. Finally, it would require that world population be reduced by 77% from its 2010 level of 6.85 billion. All requirements would be great challenges.

*The New York Times.

†Carbon Dioxide Information Analysis Center

Analyzing the Situation

1. Assume that China's *per-person* annual carbon dioxide emissions are increasing linearly. Let $L(t)$ be China's per-person annual carbon emissions (in metric tons per person) at t years since 1950. Find the linear regression equation of L.
2. Now assume that China's *per-person* annual carbon emissions are increasing quadratically. Let $Q(t)$ be China's per-person annual carbon emissions (in metric tons per person) at t years since 1950. Find the quadratic regression equation of Q.
3. Find $L(70)$ and $Q(70)$. What do they mean in this situation? Compare your results with the claim that China's per-person emissions in 2050 could be as high as 41.5 metric tons of carbon dioxide per person.
4. Let $P(t)$ be China's population (in billions) at t years since 1950. Create a scattergram for China's population data. Find the linear regression equation of P.
5. Find an equation of $Q \cdot P$. Perform a unit analysis of the equation. Find $(Q \cdot P)(70)$. What does it mean in this situation?
6. In Problem 5 of the Climate Change Lab in Chapter 5, you found a model of U.S. annual carbon dioxide emissions (in billions of metric tons) $h(t)$ at t years since 1950. If, however, you didn't find such an equation, find it now. Then use h, $Q \cdot P$, and "intersect" on a graphing calculator to estimate when China and the United States had the same annual amount of carbon dioxide emissions. What is that annual amount of carbon dioxide emissions?
7. Use Q to verify the claim that China's *per-person* annual carbon dioxide emissions could reach the United States' 2010 level of 18.1 metric tons of carbon dioxide per person as early as 2029.
8. Explain why the year you found in Problem 6 is so much earlier than the year you found in Problem 7.

Projectile Lab (continued from Chapter 7)

In the Projectile Lab in Chapter 7, you found an equation of a quadratic model that describes the times and heights of a ball that you threw straight up. In this lab, you will analyze that situation further.

Analyzing the Situation

1. What model did you work with in the Projectile Lab in Chapter 7? Provide the equation and define the variables.
2. Use the model to estimate both times when the ball was at a height of 8 feet.
3. Find the average of the two times you found in Problem 2.
4. Use the model to estimate the height of the ball at the time you found in Problem 3.

5. Use a graphing calculator to draw a graph of the model. Copy the graph, and indicate on the graph the two points that represent when the ball was at a height of 8 feet and the point that represents when the ball was at the height you found in Problem 4.
6. By referring to your graph in Problem 5, explain why your result in Problem 4 is the maximum height of the ball.
7. Use "maximum" on a graphing calculator to estimate the maximum height of the ball. Compare this result with your result from Problem 4.
8. Find the average speed of the ball on its way up. That is, find the ball's average speed from the moment it was released to the moment it reached its maximum height.
9. Explain why it makes sense that your result in Problem 8 is less than the initial speed you found in the Projectile Lab in Chapter 7.

Projectile Lab (Using a CBR or CBL)

In this lab, you will toss a softball vertically into the air and record the softball's height at various times. If your instructor prefers, use the data listed in Table 82.

Table 82 Heights of a Softball

Time (seconds)	Height (feet)	Time (seconds)	Height (feet)
0.00	3.3994	0.32	4.4545
0.02	3.5650	0.34	4.4077
0.04	3.7271	0.36	4.3393
0.06	1.5000	0.38	4.2925
0.08	4.0620	0.40	4.2024
0.10	4.1340	0.42	4.0980
0.12	4.1916	0.44	3.9900
0.14	4.2709	0.46	3.8711
0.16	4.3429	0.48	3.7379
0.18	4.4005	0.50	3.5903
0.20	4.4473	0.52	3.4354
0.22	4.4797	0.54	3.2662
0.24	4.4977	0.56	3.0861
0.26	4.5049	0.58	2.9061
0.28	4.5013	0.60	2.7044
0.30	4.4797		

Source: *J. Lehmann*

Materials

If you are going to perform your own experiment, you will need the following materials:

1. a softball (or some other object)
2. a CBR data collection device, or a CBL data collection system with a Vernier motion detector probe

Analyzing the Data

1. If you collected your own softball data, display these data in a table. If not, use the data in Table 82.
2. Let $h = f(t)$ be the height (in feet) of the softball at t seconds. Use a graphing calculator to draw a scattergram of the softball data.
3. If you are using the data in Table 82, should the graph of your model come close to (0.06, 1.5), which is described in the table? If you performed your own experiment, should the graph of your model come close to all of your data points? Explain.
4. Find an equation of f.
5. Use a graphing calculator to draw a graph of your model and the scattergram in the same viewing window. Also, graph the model and scattergram by hand. How well does f model the data?
6. Use your equation of f to estimate when the softball reached the ground.
7. What is the h-intercept of your model? What does it mean in this situation?
8. Use your model to estimate the height of the softball at 0.7 second.
9. Use your model to estimate the height of the softball at 10 seconds.
10. For what values of t is there model breakdown? Explain.
11. Use your model to estimate when the softball reached its maximum height. What was the maximum height?

Water Flow Lab

In this experiment, you will fill a cylinder with water and allow the water to flow out of a small hole near the bottom. The point of the experiment is to explore the relationship between the volume of water that flows out of the hole and the time elapsed.

If your instructor prefers, use the data listed in Table 83.

Table 83 Heights of Water in a Cylinder with Radius 1.5 Inches

Time (seconds)	Height (inches)
0	30
9	25
18	20
29	15
42	10
58	5
79	1

Source: *J. Lehmann*

Materials

If you are going to perform your own experiment, you will need at least three people and the following items:

1. a timing device, a tape measure or ruler, and a marker
2. a transparent cylinder with a closed bottom. Near the bottom, there should be a hole that is large enough so that water flows out rather than drips out. The cylinder should be large enough and the hole small enough so that it takes at least a minute for the full cylinder to drain. For example, a cylinder that is 4 feet long with a diameter of 3 inches works well with a hole that has a diameter of about $\frac{1}{8}$ inch.
3. a bucket to catch the water that flows out of the cylinder

Preparation

Make about eight equally spaced marks along the cylinder. Fill the cylinder with water, keeping the hole near the bottom sealed until you are ready to begin timing. While two people are preparing the cylinder and water, the third person should prepare to record the times it takes for the water level to reach the various marks on the cylinder.

Recording of Data

Record the height of each mark, the amount of time (from the start of the experiment) it takes for the water level to reach each mark, and the radius of the cylinder.

Analyzing the Data

1. If you collect your own water-flow data, display these data in a table. If not, then use the data in Table 83.
2. Let $H = f(t)$ be the height (in inches) of the water at t seconds after the water begins to flow out of the cylinder. Use a graphing calculator to draw a scattergram of your data.
3. Find an equation of f. The water flow is likely to be a bit erratic at the end, so it's probably best to avoid using a data point that corresponds to a water level of zero.
4. Use a graphing calculator to graph your model and the scattergram in the same viewing window. Graph your model and the scattergram by hand. How well does f model the data?
5. Find the H-intercept of your model. What does it mean in this situation? Does model breakdown occur before the H-intercept, after the H-intercept, or neither? Explain.
6. Find the t-intercept(s) of your model. What does such a point mean in this situation? Does model breakdown occur before the t-intercept, after the t-intercept, both, or neither? Explain. If there are not any t-intercepts, what does your model imply?
7. Find the vertex of your model. Does model breakdown occur before the vertex, after the vertex, or neither? Explain.
8. Use your model to estimate the height of the water at 20 seconds.
9. Use your model to estimate how many seconds it took for the water level to reach a height of 7 inches.

Taking It One Step Further

10. For a cylinder of radius R and height H, the volume of the cylinder can be described by the equation $V = \pi R^2 H$. If R and H are in inches, then V is in cubic inches. Substitute your value of R into the equation $V = \pi R^2 H$.
11. You have $H = at^2 + bt + c$ and $V = \pi R^2 H$, where a, b, c, and R are known constants. Substitute $at^2 + bt + c$ for H into the equation $V = \pi R^2 H$ to find an equation that describes the volume of water in terms of the time elapsed since the start of the experiment.
12. Use your result from Part 11 to estimate the volume of water in the cylinder at $t = 10$ seconds.
13. How much water has flowed out of the cylinder during the time span from $t = 10$ to $t = 20$? From $t = 20$ to $t = 30$? Explain why the two amounts are not equal.

Quadratic Lab: Topic of Your Choice

Your objective in this lab is to find a quadratic function to model an authentic situation. Your function should model a situation that has not been discussed in this text. You must first find some data. Almanacs, newspapers, magazines, scientific journals, and the Internet are good resources for data. Or conduct an experiment to obtain your data.

Analyzing the Data

1. What two variables did you explore? [**Hint:** Describe the units of the variables.]
2. Does it makes sense that a quadratic function models your situation best? Explain.
3. Which variable is the dependent variable? Which variable is the independent variable?
4. State the source of your data. If you conducted an experiment, provide a careful description with specific details of how you ran your experiment.
5. Include a table of your data.
6. Use a graphing calculator to draw a scattergram of your data.
7. Find an equation of a quadratic model to describe the data.
8. Use a graphing calculator to graph your quadratic model and the scattergram in the same viewing window. Graph your model and the scattergram by hand. How well does the model fit the data?
9. Find all intercepts of your quadratic model. What do they mean in this situation? Has model breakdown occurred at the intercepts?
10. Find the vertex of your model. What does it mean in this situation? Has model breakdown occurred at the vertex?
11. Choose an input of your model. Find the output that comes from the chosen input. What does your result mean in this situation?
12. Choose an output of your model. Find the input that originates from the chosen output. What does your result mean in this situation?
13. Comment on your lab experience. For example,
    a. Address whether this lab was enjoyable, insightful, and so on.
    b. Were you surprised by any of your findings? If so, which ones?
    c. How would you improve your procedure for this lab if you did it again?
    d. How would you improve your procedure if you had more time and money?

Chapter Summary

Key Points of Chapter 9

Section 9.1 Graphing Quadratic Functions in Vertex Form

Vertex form of a quadratic function

Let $f(x) = a(x - h)^2 + k$, where $a \neq 0$. Then f is a quadratic function, and its graph is a parabola with vertex (h, k). We say the equation is in **vertex form.**

Graphs of quadratic functions of the form $f(x) = ax^2$

For a function of the form $f(x) = ax^2$,

- The graph is a parabola with vertex $(0, 0)$.
- If $|a|$ is a large number, then the parabola is steep. (It is narrow.)
- If a is near zero, then the parabola is not steep. (It is wide.)
- If $a > 0$, then the parabola opens upward.
- If $a < 0$, then the parabola opens downward.
- The graph of $y = -ax^2$ is the reflection of the graph of $f(x) = ax^2$ across the x-axis.

Three-step method of graphing a quadratic function in vertex form

To sketch the graph of $f(x) = a(x - h)^2 + k$, where $a \neq 0$,

1. Sketch the graph of $y = ax^2$.
2. Translate the graph from step 1 to the right by h units if $h > 0$ or to the left by $|h|$ units if $h < 0$.
3. Translate the graph from step 2 up by k units if $k > 0$ or down by $|k|$ units if $k < 0$.

Finding a quadratic model in vertex form

To find a quadratic model in vertex form, given some data,

1. Create a scattergram of the data.
2. Imagine a parabola that comes close to (or contains) the data points, and select a point (h, k) to be the vertex. Although it is not necessary to select a data point, it is often convenient and satisfactory to do so.
3. Select a nonvertex point (not necessarily a data point) of the parabola, substitute the point's coordinates into the equation $f(t) = a(t - h)^2 + k$, and solve for a.
4. Substitute the result you found for a in step 3 into $f(t) = a(t - h)^2 + k$.

Section 9.2 Graphing Quadratic Functions in Standard Form

Using two symmetric points to find the vertex

To find the vertex of a parabola in which (p, s) and (q, s) are symmetric points,

1. Find the x-coordinate by using the formula $x = \dfrac{p + q}{2}$. In words, the x-coordinate is the average of the x-coordinates of any two symmetric points.
2. Find the y-coordinate by evaluating f at the value found in step 1. That is, find $f\left(\dfrac{p + q}{2}\right)$.

Using symmetric points to graph a quadratic function in standard form (method 1)

To sketch a graph of a quadratic function $f(x) = ax^2 + bx + c$, where $b \neq 0$,

1. Find the y-intercept.
2. Find the y-intercept's symmetric point.
3. Average the x-coordinates of the two symmetric points to find the x-coordinate of the vertex.
4. Find the y-coordinate of the vertex.
5. Depending on how accurate your sketch is to be, find additional points on the parabola, as needed.
6. Sketch a parabola that contains the points found.

Section 9.2 Graphing Quadratic Functions in Standard Form (*Continued*)

Vertex formula	To find the vertex of the graph of a quadratic function $f(x) = ax^2 + bx + c$, **1.** Find the x-coordinate of the vertex by using the **vertex formula** $x = -\frac{b}{2a}$. **2.** Find the y-coordinate of the vertex by evaluating f at the value found in step 1. That is, find $f\left(-\frac{b}{2a}\right)$.
Maximum or minimum value of a function	For a quadratic function $f(x) = ax^2 + bx + c$ whose graph has vertex (h, k), • If $a < 0$, then the parabola opens downward and the maximum value of the function is k. • If $a > 0$, then the parabola opens upward and the minimum value of the function is k.

Section 9.3 Simplifying Radical Expressions

Principal square root	If a is a nonnegative number, then $\sqrt{a}$ is the nonnegative number we square to get a. We call $\sqrt{a}$ the **principal square root** of a.
Perfect square	A **perfect square** is a number whose principal square root is rational.
Product property for square roots	For nonnegative numbers a and b, $\sqrt{ab} = \sqrt{a}\sqrt{b}$
Simplified square root	A square root is **simplified** when the radicand does not have any perfect-square factors other than 1.
Simplifying a square root with a small radicand	To simplify a square root in which it is easy to determine the largest perfect-square factor of the radicand, **1.** Write the radicand as the product of the *largest* perfect-square factor and another number. **2.** Apply the product property for square roots.
Simplifying a square root with a large radicand	To simplify a square root in which it is difficult to determine the largest perfect-square factor of the radicand, **1.** Find the prime factorization of the radicand. **2.** Look for pairs of identical factors of the radicand and rearrange the factors to highlight them. **3.** Use the product property for square roots.
Quotient property for square roots	For a nonnegative number a and a positive number b, $\sqrt{\frac{a}{b}} = \frac{\sqrt{a}}{\sqrt{b}}$
Rationalize the denominator	To **rationalize the denominator** of a fraction of the form $\frac{p}{\sqrt{q}}$, where q is positive, we multiply the fraction by 1 in the form $\frac{\sqrt{q}}{\sqrt{q}}$.
Simplifying a radical expression	To simplify a radical expression, **1.** Use the quotient property for square roots so that no radicand is a fraction. **2.** Use the product property for square roots so that no radicands have perfect-square factors other than 1. **3.** Rationalize denominators so that no denominator is a radical expression. **4.** Use the property $\frac{a}{a} = 1$, where a is nonzero, to simplify. **5.** Continue applying steps 1–4 until the radical expression is simplified completely.

Section 9.4 Using the Square Root Property to Solve Quadratic Equations

Square root property	Let k be a real-number constant. Then $x^2 = k$ is equivalent to $x = \pm\sqrt{k}$.
Imaginary unit i	The **imaginary unit,** written i, is the number whose square is -1. That is, $i^2 = -1$ and $i = \sqrt{-1}$.
Square root of a negative number	If p is a positive real number, then $\sqrt{-p} = i\sqrt{p}$.
Complex number	A **complex number** is a number of the form $a + bi$, where a and b are real numbers.
Imaginary number	An **imaginary number** is a number $a + bi$, where a and b are real numbers and $b \neq 0$.

Section 9.5 Solving Quadratic Equations by Completing the Square

Perfect-square trinomial property	For a perfect-square trinomial of the form $x^2 + bx + c$, dividing b by 2 and squaring the result gives c: $$x^2 + bx + c$$ $$\left(\frac{b}{2}\right)^2 = c$$
When completing the square can be used	Any quadratic equation can be solved by completing the square.
Solving a quadratic equation by completing the square	To solve a quadratic equation $ax^2 + bx + c = 0$ by **completing the square,** **1.** If $a \neq 1$, divide both sides of the equation by a. **2.** Write the equation in the form $x^2 + dx = k$, where d and k are constants. **3.** Complete the square for the expression on the left side of the equation. **4.** Solve the equation by using the square root property.

Section 9.6 Using the Quadratic Formula to Solve Quadratic Equations

Quadratic formula	The solutions of a quadratic equation $ax^2 + bx + c = 0$ are given by the **quadratic formula:** $$x = \frac{-b \pm \sqrt{b^2 - 4ac}}{2a}$$
Determining the number and type of solutions	For the quadratic equation $ax^2 + bx + c = 0$, the **discriminant** is $b^2 - 4ac$. Also, • If $b^2 - 4ac > 0$, there are two real-number solutions. • If $b^2 - 4ac = 0$, there is one real-number solution. • If $b^2 - 4ac < 0$, there are two imaginary-number solutions (and no real-number solutions).
When the quadratic formula can be used	Any quadratic equation can be solved by using the quadratic formula.
Guidelines on solving quadratic equations	Here are some guidelines on deciding which method to use to solve a quadratic equation:

Method	When to Use
Factoring	For equations that can easily be put into the form $ax^2 + bx + c = 0$ and where $ax^2 + bx + c$ can easily be factored.
Square root property	For equations that can easily be put into the form $x^2 = k$ or $(x + p)^2 = k$.
Completing the square	When the directions require it.
Quadratic formula	For all equations except those that can easily be solved by factoring or by using the square root property.

Section 9.7 Solving Systems of Linear Equations in Three Variables; Finding Quadratic Functions

Linear equation in three variables

A **linear equation in three variables** is an equation that can be put into the form $Ax + By + Cz = D$, where A, B, C, and D are constants and A, B, and C are not all zero.

System of linear equations in three variables

A **system of linear equations in three variables** consists of two or more linear equations in three variables.

Solution

The **solution** of a system of linear equations in three variables is an ordered triple that satisfies *all* of the equations.

Solving a system of three linear equations in three variables

To solve a system of three linear equations in three variables,

1. Select a pair of equations and eliminate a variable.
2. Select any other pair of equations and eliminate the *same variable* as in step 1.
3. The equations you found in steps 1 and 2 form a system of linear equations in *two* variables. Use elimination or substitution to solve this system.
4. Substitute the values of the two variables you found in step 3 into one of the original equations that contains the third variable. Solve for the third variable.
5. Write your solution as an ordered triple.

Finding an equation of a parabola

To find an equation of a parabola that contains three given points,

1. Obtain a system of three linear equations in three variables by substituting the coordinates of each of the three given points into the general equation $y = ax^2 + bx + c$.
2. Solve the system you found in step 1.
3. Substitute the values of a, b, and c you found in step 2 into the equation $y = ax^2 + bx + c$.

Section 9.8 Finding Quadratic Models

Finding an equation of a quadratic model

To find an equation of a quadratic model, given some data,

1. Create a scattergram of the data.
2. Imagine a parabola that comes close to the data points, and choose three points (not necessarily data points) that lie on or close to the parabola.
3. Use the three points to find an equation of the parabola.
4. Use a graphing calculator to verify that the graph of the equation comes close to the points of the scattergram.

Selecting a model

When performing step 1 of the modeling process, we must decide whether a linear function, a quadratic function, or neither of these is suitable for modeling the situation. Here are the criteria for selecting a model:

- The graph of the model should fit the points well.
- The model should make sense within the context of the authentic situation.

Four-step modeling process

To find a model and make estimates and predictions,

1. Create a scattergram of the data. Decide whether a line, a parabola, or neither of these comes close to the points.
2. Find an equation of your model.
3. Verify that your equation has a graph that comes close to the points in the scattergram. If it doesn't, check for calculation errors or use different points to find the equation. An alternative is to reconsider your choice of model in step 1.
4. Use your equation of the model to draw conclusions, make estimates, and/or make predictions.

Section 9.9 Modeling with Quadratic Functions

Solving a system of quadratic equations

We can use substitution to solve any system of two quadratic equations if at least one equation is in standard form $y = ax^2 + bx + c$.

Chapter 9 Review Exercises

For Exercises 1–4, sketch the graph of the function by hand.

1. $y = -3x^2$

2. $y = 3(x + 1)^2 - 4$

3. $y = -2(x - 3)^2 + 5$

4. $y = \frac{1}{2}(x + 4)^2 - 2$

5. A function of the form $y = a(x - h)^2 + k$ has been sketched in Fig. 117. Describe the signs of the constants a, h, and k.

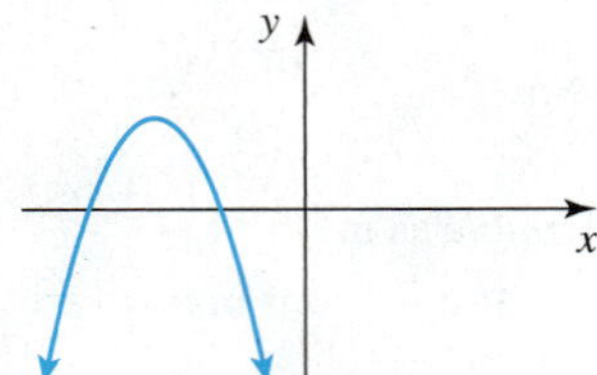

Figure 117 Exercise 5

Graph the function. Find the vertex; round coordinates to the second decimal place.

6. $y = 3x^2 - 12x + 7$

7. $y = -2x^2 + 8x + 5$

8. $y = x^2 - 5x + 3$

9. $1.7x + 2.6x^2 + y = 6.7x^2 - 10x + 2.1$

Simplify.

10. $\sqrt{196}$

11. $-\sqrt{64}$

Find an approximate value of the radical. Round your result to the second decimal place.

12. $\sqrt{95}$

13. $-7.29\sqrt{38.36}$

Simplify.

14. $\sqrt{18}$

15. $\sqrt{98}$

16. $-3\sqrt{50}$

For Exercises 17–20, simplify.

17. $\sqrt{\frac{5}{9}}$

18. $\frac{4}{\sqrt{7}}$

19. $\sqrt{\frac{7}{3}}$

20. $\sqrt{\frac{5}{32}}$

21. A student tries to simplify $\frac{3}{\sqrt{7}}$:

$$\frac{3}{\sqrt{7}} = \left(\frac{3}{\sqrt{7}}\right)^2 = \frac{3^2}{(\sqrt{7})^2} = \frac{9}{7}$$

Describe any errors. Then simplify the expression correctly.

Solve. Any solution is a real number.

22. $3x^2 - 2x - 2 = 0$

23. $5x^2 = 7$

24. $5(p - 3)^2 + 4 = 7$

25. $(t + 1)(t - 7) = 4$

26. $2x^2 = 4 - 5x$

27. $4x - x^2 = 1$

28. $2x^2 - x = \frac{3}{2}$

29. $2x(x - 1) = 5$

30. $5x^2 - 6x = 2$

31. $7r^2 - 20 = 0$

32. $(t + 2)^2 + (t - 3)^2 = 15$

33. $5(5x^2 - 8) = 9$

34. $\frac{3}{2}x^2 - \frac{3}{4}x = \frac{1}{2}$

Find approximate solutions. Round the results to the second decimal place.

35. $2.7x^2 - 5.1x = 9.8$

36. $1.7(x^2 - 2.3) = 3.4 - 2.8x$

Simplify.

37. $\sqrt{-45}$

38. $\sqrt{-\frac{7}{2}}$

Find all complex-number solutions.

39. $-2(x + 4)^2 = 9$

40. $2x^2 = 4x - 7$

Solve by completing the square. Any solution is a real number.

41. $x^2 + 6x - 4 = 0$

42. $x^2 - 5x = 2$

43. $3x^2 - 18x - 27 = 0$

44. $2t^2 = -3t + 6$

For Exercises 45 and 46, find all x-intercepts.

45. $h(x) = 3x^2 + 2x - 2$

46. $k(x) = -5x^2 + 3x - 1$

47. Solve $x^2 - 2x - 8 = 0$ by factoring, completing the square, and using the quadratic formula.

48. Find the number and type of solutions of the equation $3x^2 - 5x + 4 = 0$.

49. Let $f(x) = 3x^2 - 6x + 7$.

a. Find x when $f(x) = 3$.

b. Find x when $f(x) = 4$.

c. Find x when $f(x) = 5$.

d. Discuss, in terms of the graph of f, why you found zero, one, and two values of x for parts (a), (b), and (c), respectively.

For Exercises 50–52, find approximate solutions of the equation or system by referring to the graphs shown in Fig. 118. Round results or coordinates of results to the first decimal place.

50. $-\frac{1}{4}(x - 1)^2 + 1 = x^2$

51. $-\frac{1}{4}(x - 1)^2 + 1 = -3$

52. $y = -\frac{1}{4}(x - 1)^2 + 1$

$y = -\frac{1}{3}x - 2$

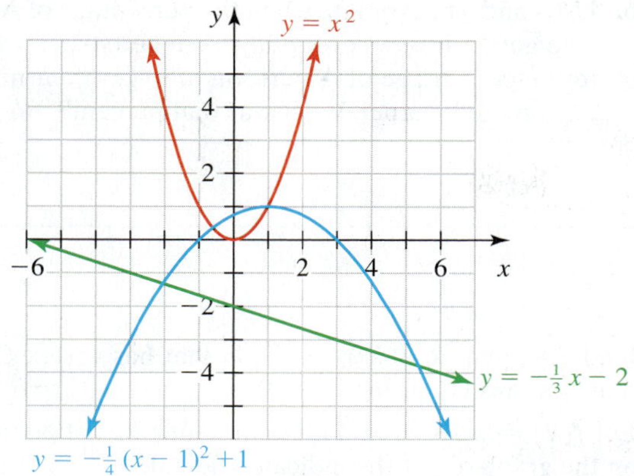

Figure 118 Exercises 50–52

53. Use "intersect" on a graphing calculator to solve the equation $7 - 2x^2 = x^2 - x - 8$. Round the solution(s) to the second decimal place.

Solve the system.

54. $\begin{aligned} x + 2y - 3z &= -4 \\ 2x - y + z &= 3 \\ 3x + 2y + z &= 10 \end{aligned}$

55. $\begin{aligned} 2x \quad\quad - 3z &= -4 \\ 3x + y \quad\quad &= 0 \\ x - 4y + 2z &= 17 \end{aligned}$

For Exercises 56–58, find an equation of the parabola that passes through the three given points.

56. $(1, 4), (2, 3), (3, -2)$

57. $(2, 9), (3, 18), (5, 48)$

58. $(0, 5), (2, 3), (4, -15)$

59. Find equations of a linear function and a quadratic function such that the graph of each function contains both of the points $(0, 4)$ and $(1, 2)$.

60. Find an equation of the parabola sketched in Fig. 119.

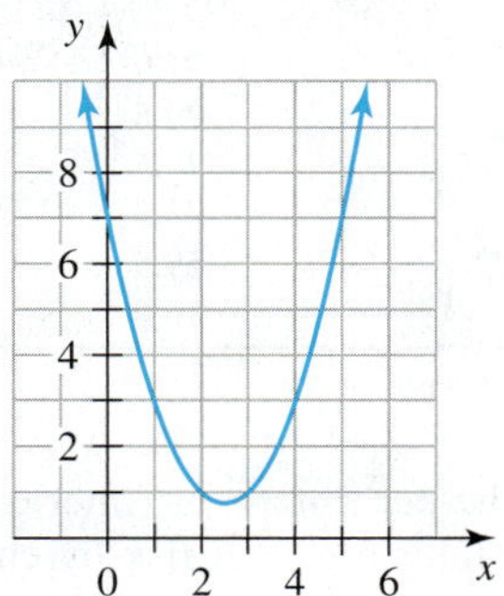

Figure 119 Exercise 60

61. A batter hits a baseball into the air. The height (in feet) $h(t)$ of the baseball after t seconds is given by

$$h(t) = -16t^2 + 100t + 3$$

a. Find the maximum height of the baseball. When does it reach that height?

b. A fielder catches the ball at a height of 3 feet. How many seconds after the batter hit the ball did the fielder have to get into position to catch the ball?

c. Graph h by hand.

62. A farmer plans to use 180 feet of fencing and a side of his barn to enclose a rectangular garden (see Fig. 120). What should the dimensions of the rectangle be so the area is as large as possible? What is that area?

Figure 120 Exercise 62

63. The percentages of military personnel in the Iraq and Afghanistan wars who have done more than one tour of duty are listed in Table 84 for various years.

Table 84 Percentages of Military Personnel Who Have Done More than One Tour of Duty

Year	Percent
2001	7
2003	18
2005	28
2007	33
2010	38

Source: *Department of Defense*

Let $p = f(t)$ be the percentage of military personnel in the Iraq and Afghanistan wars who have done more than one tour of duty at t years since 2000.

a. Find an equation of f.

b. Find the vertex. What does it mean in this situation?

c. Estimate when 37% of military personnel had done more than one tour of duty.

d. Estimate the percentage of military personnel in 2006 who had done more than one tour of duty.

64. The percentages of Americans who get their news every day from newspapers and the percentages of Americans who get their news every day on the Internet are shown in Table 85 for various years.

Table 85 Percentages of Americans Who Get Their News Every Day from Newspapers and from the Internet

Year	Percent from Newspapers	Year	Percent from Internet
1996	50	1998	6
1998	48	1999	8
2000	47	2002	15
2002	41	2004	20
2004	42	2006	23
2006	40	2008	29
2008	34	2010	34
2010	31	2012	39
2012	29		

Source: *Pew Research Center*

Let $f(t)$ be the percentage of Americans who get their news every day from newspapers, and let $g(t)$ be the percentage of Americans who get their news on the Internet, both at t years since 1990.

a. Find a quadratic regression equation of f and a linear regression equation of g.

b. Use f and g to estimate when the percentage of Americans who get their news every day from newspapers was equal to the percentage of Americans who get their news every day on the Internet. What was that percent?

Chapter 9 Test

1. Graph the function $f(x) = -2(x + 5)^2 - 1$ by hand.

2. What can you say about the values of a, h, and k if the equation of the parabola shown in Fig. 121 is $y = a(x - h)^2 + k$?

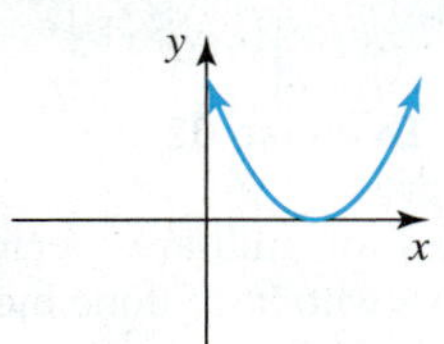

Figure 121 Exercise 2

3. Give an example of a quadratic function that has vertex $(2, -7)$ and no x-intercepts.

4. Graph $f(x) = -2x^2 - 4x + 3$ by hand. Find the vertex.

5. Let $f(x) = x^2 - 2x - 8$.

a. Find the x-intercepts.

b. Find the vertex.

c. Graph f by hand. Indicate the x-intercepts and vertex on the graph.

Simplify.

6. $\sqrt{32}$ **7.** $\sqrt{\dfrac{20}{75}}$

For Exercises 8–14, solve. All solutions are real numbers.

8. $x^2 - 3x - 10 = 0$ **9.** $6x^2 = 100$

10. $4(r - 3)^2 + 1 = 7$ **11.** $\frac{5}{6}x^2 - \frac{1}{2}x = \frac{2}{3}$

12. $(x - 3)(x + 5) = 6$ **13.** $2x(x + 5) = 4x - 3$

14. $3x^2 - 6x = 1$

15. Find approximate solutions of $3.7x^2 = 2.4 - 5.9x$. Round your results to the second decimal place.

Find all complex-number solutions.

16. $3x^2 - 6x = -5$ **17.** $-2(p + 4)^2 = 24$

For Exercises 18 and 19, solve by completing the square.

18. $x^2 - 8x - 2 = 0$ **19.** $2(x^2 - 4) = -3x$

20. Find the x-intercepts of the graph of $f(x) = 3x^2 - 8x + 1$.

21. Find the x-intercepts and vertex of the graph of $f(x) = -2(x - 3)^2 + 5$. Round the coordinates to the second decimal place. Then graph f by hand.

22. Find the nonzero value(s) of a such that the equation $ax^2 - 4x + 4a = 0$ has exactly one solution.

23. Find an equation of the parabola that passes through the points $(1, 4)$, $(2, 9)$, and $(3, 16)$.

24. Find an equation of the parabola that has vertex $(5, 3)$ and that contains the point $(3, 11)$.

25. Let $f(x) = x^2 - 6x + 11$. Find the number of points that lie on the graph of f at the indicated height.

a. $y = 1$ **b.** $y = 2$ **c.** $y = 3$

Solve the system.

26.
$$\begin{aligned} x + 4y + 3z &= 2 \\ 2x + y + z &= 10 \\ -x + y + 2z &= 8 \end{aligned}$$

27.
$$\begin{aligned} 2x - 3y \quad\quad &= 4 \\ 3y + 2z &= 2 \\ x \quad\quad - z &= -5 \end{aligned}$$

28. Let $f(t)$ be the height (in feet) of a baseball at t seconds after a batter has hit the ball. A reasonable equation of f is $f(t) = -16t^2 + 80t + 3$. At what time is the ball at its maximum height? What is that height?

29. The percentages of Americans who feel they are taking a great risk by entering personal information into a pop-up ad are shown in Table 86 for various age groups.

Table 86 Percentages of Americans Who Feel They Are Taking a Great Risk by Entering Personal Information into a Pop-Up Ad

Age Group (years)	Age Used to Represent Age Group (years)	Percent
18–24	21.0	27
25–34	29.5	33
35–44	39.5	37
45–54	49.5	44
55–64	59.5	39
over 64	70.0	23

Source: *Wells Fargo*

Let $f(t)$ be the percentage of Americans at age t years who feel they are taking a great risk by entering personal information into a pop-up ad.

a. Find an equation of f.

b. Find $f(31)$. What does it mean in this situation?

c. Find t when $f(t) = 31$. What does it mean in this situation?

d. Find the t-intercepts. What do they mean in this situation?

e. Find the vertex. What does it mean in this situation?

30. A company charters a party boat that normally costs \$40 per person. A group discount reduces the fare by \$0.25 for each ticket sold; the more tickets sold, the lower is the per-person fare. The maximum capacity of the boat is 90 people, including the crew of 5 people. What size of group would maximize the boat owner's revenue?

Cumulative Review of Chapters 1–9

1. Two hours ago the temperature was 6°F. Now the temperature is −3°F. What is the change in temperature over the past two hours?

Evaluate the expression for $a = -2$, $b = 8$, and $c = -4$.

2. $ac^2 - \frac{b}{c}$

3. $\frac{a + bc}{ab - c}$

4. Ski run A declines steadily for 130 yards over a horizontal distance of 610 yards. Ski run B declines steadily for 165 yards over a horizontal distance of 700 yards. Which run is steeper? Explain.

5. Find the slope of the line that contains the points $(-5, -4)$ and $(-1, -2)$. State whether the line is increasing, decreasing, horizontal, or vertical.

For Exercises 6–18, solve. Any solution is a real number.

6. $81x^2 - 49 = 0$

7. $\frac{5x}{6} - \frac{2}{3} = \frac{7}{4}$

8. $5x^2 - 2x = 4$

9. $2(3t^2 - 10) = -7t$

10. $(2p - 5)(3p + 4) = 5p - 2$

11. $2(5x + 2) - 1 = 9(x - 3) - (3x - 8)$

12. $2x(3x - 4) + 5 = 4 - x^2$

13. $w^2 = 5w + 24$

14. $(2r - 1)(3r - 2) = 1$

15. $3x^2 - 7 = 13$

16. $(x + 4)^2 = 60$

17. $3x^2 - 5x - 4 = 0$

18. $2x(x + 3) = -3$

19. Solve the formula $a = \frac{v - v_0}{t}$ for t.

20. Find all complex-number solutions of $2x^2 - 6x = -5$.

21. Solve $2x^2 + 3x - 6 = 0$ by completing the square.

Solve the system.

22.
$$\begin{aligned} y &= 3x - 1 \\ 2x - 3y &= -11 \end{aligned}$$

23.
$$\begin{aligned} 2x - 7y &= 3 \\ -5x + 3y &= 7 \end{aligned}$$

24.
$$\begin{aligned} \frac{1}{2}x - y &= \frac{5}{2} \\ \frac{2}{5}x - \frac{3}{5}y &= \frac{6}{5} \end{aligned}$$

25.
$$\begin{aligned} 2x - y + 3z &= 1 \\ 3x + 2y - z &= -6 \\ 4x - 3y + 2z &= -7 \end{aligned}$$

26. Solve the inequality $2(3x - 4) < 5 - 3(6x + 5)$. Describe the solution set as an inequality, in interval notation, and in a graph.

27. Graph the inequality $3x - 2y \geq 2$.

28. Graph the solution set of the system

$$\begin{aligned} y &> \frac{1}{4}x - 3 \\ y &\geq -2x \end{aligned}$$

For Exercises 29–32, perform the indicated operation.

29. $(3x - 4y)^2$

30. $(5p - 7q)(5p + 7q)$

31. $-3x(x^2 - 5)(x^3 + 8)$

32. $(x^2 - 3x - 4)(x^2 + 4x - 5)$

33. Write $f(x) = -2(x - 5)^2 + 3$ in standard form.

Factor.

34. $m^4 - 16n^4$

35. $x^3 - 13x^2 + 40x$

36. $8p^2 + 22pq - 21q^2$

37. $x^3 + 4x^2 - 9x - 36$

38. $128a^3 + 250$

39. $3p^3t - 6p^2t^2 - 45pt^3$

For Exercises 40–44, refer to Table 87.

40. Find a possible equation of f.

41. Find a possible equation of g.

42. Find the slope of the graph of k.

43. Find x when $g(x) = 4$.

44. Find $h(7)$.

Table 87 Values of Four Functions (Exercises 40–44)

x	$f(x)$	x	$g(x)$	x	$h(x)$	x	$k(x)$
0	20	1	4	4	7	0	−8
1	17	2	9	5	14	1	−4
2	14	3	14	6	28	2	0
3	11	4	19	7	56	3	4
4	8	5	24	8	112	4	8
5	5	6	29	9	224	5	12

Sketch the graph of the function.

45. $y = -2(x + 4)^2 + 3$

46. $y = -3x + 2$

47. $2x - 5y = 20$

48. $y = x^2 + 4x - 5$

49. Use the graph of the relation in Fig. 122 to determine the domain and range. Is the relation a function?

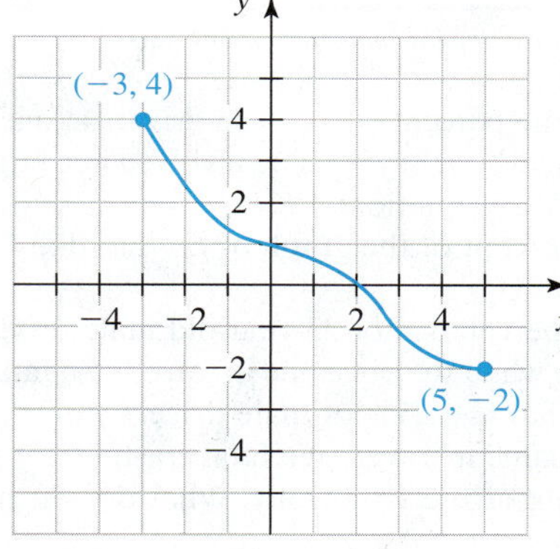

Figure 122 Exercise 49

50. Find the equation of the line that has slope $m = -\frac{2}{5}$ and contains the point $(-3, 4)$.

51. Find an equation of the line that passes through the points $(-2, 4)$ and $(6, -3)$.

52. Find the equation of the line that contains the point $(-2, 6)$ and is perpendicular to the line $3x - 4y = 5$.

53. Find an equation of a parabola that contains the points $(1, -1)$, $(2, 4)$, and $(4, 20)$.

54. Let f be the linear function and g be a quadratic function whose graphs contain the points $(0, 3)$ and $(1, 6)$.

a. Find possible equations of f and g.

b. Use a graphing calculator to draw the graphs of f and g in the same viewing window.

For Exercises 55–58, let $f(x) = -x^2 + 6x - 5$.

55. Find $f(-3)$.

56. Find x when $f(x) = 3$.

57. Find the x-intercepts.

58. Graph f by hand.

Simplify.

59. $5x^2y^6(2x^3y)^3$

60. $\left(\frac{3x^5y^7}{2w^3}\right)^4$

61. A person plans to invest a total of \$12,000. She will invest in both a First Funds TN Tax-Free I account at 6% annual interest and a W & R International Growth C account at 11% annual interest. How much should she invest in each account to earn \$845 total interest in one year?

62. The percentages of male drivers who were speeding when they became involved in fatal crashes are shown in Table 88 for various age groups.

Table 88 Speeding Male Drivers in Fatal Crashes

Age Group (years)	Age Used to Represent Age Group (years)	Percent
15–24	21.0	38
25–34	29.5	28
35–44	39.5	20
45–54	49.5	15
55–64	59.5	13
65–74	69.5	10
75+	80.0	7

Source: *National Center for Statistics & Analysis*

Let $p = f(t)$ be the percentage of male drivers at age t years who were speeding when they became involved in a fatal crash.

a. Find and verify an equation of f.

b. Find the p-intercept of the graph of f. What does it mean in this situation?

c. Estimate the percentage of 23-year-old male drivers who were speeding when they became involved in a fatal crash.

d. Estimate at what age 25% of male drivers were speeding when they became involved in a fatal crash.

e. Find the vertex of the graph of f. What does it mean in this situation?

63. Just before beginning its descent, an airplane is at an altitude of 27,500 feet. The airplane then descends steadily at a rate of 1350 feet per minute. Let $f(t)$ be the airplane's altitude (in feet) at t minutes after the plane has begun its descent.

a. Find an equation of f.

b. Estimate the airplane's altitude 7 minutes after the plane has begun its descent.

c. Estimate when the airplane reached an altitude of 1200 feet.

d. What is the slope? What does it mean in this situation?

e. What is the t-intercept? What does it mean in this situation?

64. As unions have attracted an increasing number of workers in the service industries, the percentage of union members who work in manufacturing has declined (see Table 89).

Table 89 Percentages of Union Members Who Work in Manufacturing

Year	Percent
1985	29
1990	25
1995	21
2000	17
2005	13
2010	13

Source: *Bureau of Labor Statistics*

Let $p = f(t)$ be the percentage of union members who work in manufacturing at t years since 1980.

a. Find an equation of f.

b. Find $f(37)$. What does it mean in this situation?

c. Find t when $f(t) = 5$. What does it mean in this situation?

d. Find an equation of f^{-1}.

e. Find $f^{-1}(100)$. What does it mean in this situation?

f. Find the t-intercept of the graph of f. What does it mean in this situation?

g. For what values of t is there model breakdown for certain?

65. The average ages of light trucks and passenger cars are shown in Table 90 for various years.

Table 90 Average Ages of Light Trucks and Passenger Cars

Year	Average Age (years)	
	Light Trucks	Passenger Cars
2007	9.0	10.4
2008	9.3	10.6
2009	9.8	10.8
2010	10.1	11.0
2011	10.4	11.1

Source: *R.L. Polk & Co.*

Let $f(t)$ and $g(t)$ be the average ages (in years) of light trucks and passenger cars, respectively, both at t years since 2000.

a. Find linear equations of f and g.

b. Find $f(17)$ and $g(17)$. What do they mean in this situation?

c. Predict when the average age of light trucks will be equal to the average age of passenger cars. What is that average age?

d. Predict when the average age of light trucks will be greater than the average age of passenger cars.

e. Find an equation of $f - g$. Then find $(f - g)(18)$. What does it mean in this situation?

f. Find $\frac{f(18) + g(18)}{2}$. Give at least two reasons why your result may *not* turn out to be the average age (in years) of all light trucks and passenger cars in 2018.

10 Exponential Functions

The most expensive slot for television advertising occurs during the Super Bowl. In 2012, a 30-second Super Bowl ad slot cost \$3.5 million—about \$117,000 per second! The costs of 30-second ad slots have increased greatly (see Table 1). In Exercise 50 of Homework 10.5, we will predict the cost of a 30-second ad slot during Super Bowl LII in 2018.

Table 1 Costs of Television Ad Slots during the Super Bowl

Super Bowl	Year	Cost for 30 Seconds (millions of dollars)
IX	1975	0.11
XIX	1985	0.5
XXIX	1995	1.0
XXXIX	2005	2.4
XLVI	2012	3.5

Source: *Ocean Media, Inc.*

In Section 7.5, we simplified expressions involving counting-number exponents. In this chapter, we will simplify expressions first involving integer exponents and then involving rational exponents. In addition to working with linear and quadratic functions, we will now work with *exponential functions*. We will graph and find equations of such functions. Finally, we will use exponential functions to make predictions about authentic situations, such as the number of subscribers to Showtime in 2017.

10.1 Integer Exponents

Objectives

- Know the meaning of a negative-integer exponent.
- Simplify expressions involving negative-integer exponents.
- Work with models whose equations involve negative-integer exponents.
- Use scientific notation.

In Section 7.4, we worked with power expressions with nonnegative-integer exponents. In this section, we will extend our work with power expressions to include negative-integer exponents.

Definition of a Negative-Integer Exponent

We can begin to see the meaning of a negative-integer exponent by considering the expression $\frac{b^3}{b^5}$. We write this expression in two different forms:

$$\frac{b^3}{b^5} = b^{3-5} = b^{-2}$$

$$\frac{b^3}{b^5} = \frac{b \cdot b \cdot b}{b \cdot b \cdot b \cdot b \cdot b} = \frac{1}{b^2}$$

Equating the two results, we have

$$b^{-2} = \frac{1}{b^2}$$

Likewise, we could write $\frac{b^2}{b^5}$ in two different forms to show that

$$b^{-3} = \frac{1}{b^3}$$

Next, we describe this pattern in general.

Definition Negative-integer exponent

If n is a counting number and $b \neq 0$, then

$$b^{-n} = \frac{1}{b^n}$$

In words, to find b^{-n}, take its reciprocal and change the sign of the exponent.

Simplifying Power Expressions Involving Negative-Integer Exponents

Simplifying a power expression includes writing it so that each exponent is positive.

Example 1 Simplifying Expressions Involving Exponents

Simplify.

1. 5^{-2} **2.** $9b^{-7}$ **3.** $3^{-1} + 4^{-1}$

Solution

1. $5^{-2} = \frac{1}{5^2} = \frac{1}{25}$

2. $9b^{-7} = 9 \cdot \frac{1}{b^7} = \frac{9}{b^7}$

3. $3^{-1} + 4^{-1} = \frac{1}{3} + \frac{1}{4} = \frac{4}{12} + \frac{3}{12} = \frac{7}{12}$

Next, we write $\frac{1}{b^{-n}}$ in another form, where b is nonzero and n is a counting number:

$$\frac{1}{b^{-n}} = 1 \div b^{-n} \qquad \frac{a}{b} = a \div b$$

$$= 1 \div \frac{1}{b^n} \qquad \text{Write power so that exponent is positive: } b^{-n} = \frac{1}{b^n}$$

$$= 1 \cdot \frac{b^n}{1} \qquad \text{Multiply by reciprocal of } \frac{1}{b^n}\text{, which is } \frac{b^n}{1}.$$

$$= b^n \qquad \text{Simplify.}$$

So, $\frac{1}{b^{-n}} = b^n$.

Negative-Integer Exponent in a Denominator

If n is a counting number and $b \neq 0$, then

$$\frac{1}{b^{-n}} = b^n$$

In words, to find $\frac{1}{b^{-n}}$, find its reciprocal and change the sign of the exponent.

Example 2 Simplifying Expressions Involving Exponents

Simplify.

1. $\dfrac{1}{2^{-3}}$

2. $\dfrac{5}{b^{-3}}$

Solution

1. $\dfrac{1}{2^{-3}} = 2^3 = 8$

2. $\dfrac{5}{b^{-3}} = 5 \cdot \dfrac{1}{b^{-3}} = 5b^3$

In Example 3, we will simplify some quotients of two powers.

Example 3 Simplifying Expressions Involving Exponents

Simplify.

1. $\dfrac{b^{-4}}{c^2}$

2. $\dfrac{b^3}{c^{-7}}$

Solution

1. $\dfrac{b^{-4}}{c^2} = b^{-4} \cdot \dfrac{1}{c^2}$ *Write quotient as a product:* $\frac{A}{B} = A \cdot \frac{1}{B}$

$= \dfrac{1}{b^4} \cdot \dfrac{1}{c^2}$ *Write powers so that exponents are positive:* $b^{-n} = \frac{1}{b^n}$

$= \dfrac{1}{b^4c^2}$ *Multiply numerators; multiply denominators:* $\frac{A}{B} \cdot \frac{C}{D} = \frac{AC}{BD}$

2. $\dfrac{b^3}{c^{-7}} = b^3 \cdot \dfrac{1}{c^{-7}}$ *Write quotient as a product:* $\frac{A}{B} = A \cdot \frac{1}{B}$

$= b^3c^7$ *Write powers so that exponents are positive:* $\frac{1}{b^{-n}} = b^n$

In Problem 1 of Example 3, we simplified $\dfrac{b^{-4}}{c^2}$. When simplifying such expressions, we usually skip the first two steps shown in the example and write

$$\frac{b^{-4}}{c^2} = \frac{1}{b^4c^2}$$

Likewise, we can skip the first step of our work in Problem 2 of Example 3 by simplifying $\dfrac{b^3}{c^{-7}}$ as follows:

$$\frac{b^3}{c^{-7}} = b^3c^7$$

Example 4 Simplifying Power Expressions

Simplify $\dfrac{4a^{-2}b^4}{7c^{-5}}$.

Solution

$\dfrac{4a^{-2}b^4}{7c^{-5}} = \dfrac{4c^5b^4}{7a^2}$ *Write powers so that exponents are positive:* $b^{-n} = \frac{1}{b^n}; \frac{1}{b^{-n}} = b^n$

In Section 7.4, we discussed various properties of counting-number exponents. It turns out that these properties are also true for all negative-integer exponents and the zero exponent.

Properties of Integer Exponents

If m and n are integers, $b \neq 0$, and $c \neq 0$, then

- $b^m b^n = b^{m+n}$ *Product property for exponents*
- $\dfrac{b^m}{b^n} = b^{m-n}$ *Quotient property for exponents*
- $(bc)^n = b^n c^n$ *Raising a product to a power*
- $\left(\dfrac{b}{c}\right)^n = \dfrac{b^n}{c^n}$ *Raising a quotient to a power*
- $(b^m)^n = b^{mn}$ *Raising a power to a power*

We can use these properties to expand our rules for simplifying expressions involving exponents to include those which involve negative-integer exponents.

Simplifying Expressions Involving Exponents

An expression involving exponents is simplified if

1. It includes no parentheses.
2. In any monomial, each variable or constant appears as a base at most once.
3. Each numerical expression (such as 7^2) has been calculated, and each numerical fraction has been simplified.
4. Each exponent is positive.

Example 5 Simplifying Expressions Involving Exponents

Simplify.

1. $2^{-1003}2^{1000}$ **2.** $(b^{-4})^3$ **3.** $(8b^6)(-3b^{-10})$

Solution

1.
$$2^{-1003}2^{1000} = 2^{-1003+1000}$$ *Add exponents: $b^m b^n = b^{m+n}$*
$$= 2^{-3}$$ *Simplify.*
$$= \frac{1}{2^3}$$ *Write powers so that exponents are positive: $b^{-n} = \frac{1}{b^n}$*
$$= \frac{1}{8}$$ *Simplify.*

2.
$$(b^{-4})^3 = b^{-12}$$ *Multiply exponents: $(b^m)^n = b^{mn}$*
$$= \frac{1}{b^{12}}$$ *Write power so that exponent is positive: $b^{-n} = \frac{1}{b^n}$*

3.
$$(8b^6)(-3b^{-10}) = 8(-3)b^6b^{-10}$$ *Rearrange factors.*
$$= -24b^{-4}$$ *Add exponents: $b^m b^n = b^{m+n}$*
$$= -\frac{24}{b^4}$$ *Write power so that exponent is positive: $b^{-n} = \frac{1}{b^n}$*

In Example 6, we will discuss how to simplify the quotient of two powers in which the bases are the same.

Example 6 Simplifying Expressions Involving Exponents

Simplify.

1. $\dfrac{b^4}{b^9}$
2. $\dfrac{4b^7}{3b^{-2}}$
3. $\dfrac{6b^{-3}c^{-2}}{9b^4c^{-5}}$

Solution

1. $\dfrac{b^4}{b^9} = b^{4-9}$ *Subtract exponents:* $\dfrac{b^m}{b^n} = b^{m-n}$

$= b^{-5}$ *Subtract.*

$= \dfrac{1}{b^5}$ *Write power so that exponent is positive:* $b^{-n} = \dfrac{1}{b^n}$

2. $\dfrac{4b^7}{3b^{-2}} = \dfrac{4b^{7-(-2)}}{3}$ *Subtract exponents:* $\dfrac{b^m}{b^n} = b^{m-n}$

$= \dfrac{4b^9}{3}$ *Simplify.*

3. $\dfrac{6b^{-3}c^{-2}}{9b^4c^{-5}} = \dfrac{2b^{-3-4}c^{-2-(-5)}}{3}$ *Simplify; subtract exponents:* $\dfrac{b^m}{b^n} = b^{m-n}$

$= \dfrac{2b^{-7}c^3}{3}$ *Simplify.*

$= \dfrac{2c^3}{3b^7}$ *Write powers so that exponents are positive:* $b^{-n} = \dfrac{1}{b^n}$

In the first step of Problem 2 of Example 6, we found that

$$\frac{4b^7}{3b^{-2}} = \frac{4b^{7-(-2)}}{3}$$

WARNING Note that we need a subtraction symbol *and* a negative symbol in the expression on the right-hand side. It is a common error to omit writing one of these two symbols in problems of this type.

Example 7 Simplifying Expressions Involving Exponents

Simplify.

1. $(2b^{-5})^{-3}$
2. $\left(\dfrac{2b^{-4}}{c^6}\right)^{-5}$

Solution

1. $(2b^{-5})^{-3} = 2^{-3}(b^{-5})^{-3}$ *Raise factors to power* -3*:* $(bc)^n = b^nc^n$

$= 2^{-3}b^{15}$ *Multiply exponents:* $(b^m)^n = b^{mn}$

$= \dfrac{b^{15}}{2^3}$ *Write powers so that exponents are positive:* $b^{-n} = \dfrac{1}{b^n}$

$= \dfrac{b^{15}}{8}$ *Simplify.*

2. $\left(\dfrac{2b^{-4}}{c^6}\right)^{-5} = \dfrac{(2b^{-4})^{-5}}{(c^6)^{-5}}$ *Raise numerator and denominator to power −5:* $\left(\dfrac{b}{c}\right)^n = \dfrac{b^n}{c^n}$

$= \dfrac{2^{-5}(b^{-4})^{-5}}{(c^6)^{-5}}$ *Raise each factor to power −5:* $(bc)^n = b^n c^n$

$= \dfrac{2^{-5}b^{20}}{c^{-30}}$ *Multiply exponents:* $(b^m)^n = b^{mn}$

$= \dfrac{b^{20}c^{30}}{2^5}$ *Write powers so that exponents are positive:* $b^{-n} = \dfrac{1}{b^n}; \dfrac{1}{b^{-n}} = b^n$

$= \dfrac{b^{20}c^{30}}{32}$ *Simplify.*

Example 8 Simplifying Expressions Involving Exponents

Simplify.

1. $\dfrac{(3bc^5)^2}{(2b^{-2}c^2)^3}$

2. $\left(\dfrac{18b^{-4}c^7}{6b^{-3}c^2}\right)^{-4}$

Solution

1. $\dfrac{(3bc^5)^2}{(2b^{-2}c^2)^3} = \dfrac{3^2b^2(c^5)^2}{2^3(b^{-2})^3(c^2)^3}$ *Raise factors to a power:* $(bc)^n = b^n c^n$

$= \dfrac{9b^2c^{10}}{8b^{-6}c^6}$ *Multiply exponents:* $(b^m)^n = b^{mn}$

$= \dfrac{9b^{2-(-6)}c^{10-6}}{8}$ *Subtract exponents:* $\dfrac{b^m}{b^n} = b^{m-n}$

$= \dfrac{9b^8c^4}{8}$ *Simplify.*

2. $\left(\dfrac{18b^{-4}c^7}{6b^{-3}c^2}\right)^{-4} = \left(3b^{-4-(-3)}c^{7-2}\right)^{-4}$ *Subtract exponents:* $\dfrac{b^m}{b^n} = b^{m-n}$

$= (3b^{-1}c^5)^{-4}$ *Simplify.*

$= 3^{-4}(b^{-1})^{-4}(c^5)^{-4}$ *Raise factors to nth power:* $(bc)^n = b^n c^n$

$= 3^{-4}b^4c^{-20}$ *Multiply exponents:* $(b^m)^n = b^{mn}$

$= \dfrac{b^4}{3^4c^{20}}$ *Write powers so that exponents are positive:* $b^{-n} = \dfrac{1}{b^n}$

$= \dfrac{b^4}{81c^{20}}$ *Simplify.*

Definition of an Exponential Function

In this chapter and Chapter 11, we will work with *exponential functions*. Here are some examples of such functions:

$$f(x) = 2(3)^x, \qquad g(x) = -7\left(\frac{1}{2}\right)^x, \qquad h(x) = 5^x$$

Notice that, in exponential functions, the variable appears as an exponent.

Definition Exponential function

An **exponential function** is a function whose equation can be put into the form

$$f(x) = ab^x$$

where $a \neq 0$, $b > 0$, and $b \neq 1$. The constant b is called the **base.**

Example 9 Evaluating Exponential Functions

For $f(x) = 3(2)^x$ and $g(x) = 5^x$, find the following.

1. $f(3)$ **2.** $f(-4)$ **3.** $g(a+3)$ **4.** $g(2a)$

Solution

1. $f(3) = 3(2)^3 = 3 \cdot 8 = 24$

2. $f(-4) = 3(2)^{-4} = \dfrac{3}{2^4} = \dfrac{3}{16}$

3.
$$\begin{aligned} g(a+3) &= 5^{a+3} && \text{Substitute } a + 3 \text{ for } x \text{ in } 5^x. \\ &= 5^a \cdot 5^3 && \text{Write as product: } b^{m+n} = b^m b^n \\ &= 125(5)^a && 5^3 = 125\text{; rearrange factors: } ab = ba \end{aligned}$$

4.
$$\begin{aligned} g(2a) &= 5^{2a} && \text{Substitute } 2a \text{ for } x \text{ in } 5^x. \\ &= (5^2)^a && b^{mn} = (b^m)^n \\ &= 25^a && 5^2 = 25 \end{aligned}$$

WARNING It is a common error to confuse exponential functions such as $E(x) = 2^x$ with linear functions such as $L(x) = 2x$. For the *exponential* function $E(x) = 2^x$, the variable x is an *exponent.* For the *linear* function $L(x) = 2x^1$, the variable x is a *base.*

Models Whose Equations Contain Negative-Integer Exponents

Many authentic situations can be modeled by equations that contain negative-integer exponents. Some examples of such situations are the sound level of a guitar, the intensity of illumination by a light bulb, and the gravitational force of the Sun acting on Earth.

Example 10 Using a Model Whose Equation Involves Negative-Integer Exponents

The intensity $f(d)$ (in watts per square meter) of a television signal at a distance d kilometers from the transmitter is described by the equation

$$f(d) = 250d^{-2}$$

1. Simplify the right-hand side of the equation.

2. Find $f(5)$. What does the result mean in this situation?

Solution

1. We use the definition of a negative-integer exponent to write the expression on the right-hand side without any negative exponents:

$$\begin{aligned} f(d) &= 250d^{-2} && \text{Original equation} \\ f(d) &= \frac{250}{d^2} && \text{Write power so exponent is positive: } b^{-n} = \frac{1}{b^n} \end{aligned}$$

2. We substitute 5 for d in the equation $f(d) = \dfrac{250}{d^2}$:

$$f(5) = \frac{250}{5^2} = \frac{250}{25} = 10$$

So, the intensity is 10 watts per square meter at a distance of 5 kilometers from the transmitter.

Scientific Notation

Now that we know how to work with negative-integer exponents, we can use exponents to describe numbers in *scientific notation*. This will enable us to describe compactly a number whose absolute value is very large or very small. For example, the distance to Proxima Centauri, the nearest star other than the Sun, is 40,100,000,000,000 kilometers. Here we write 40,100,000,000,000 in scientific notation:

$$4.01 \times 10^{13}$$

The symbol "×" stands for multiplication.

As another example, 1 square inch is approximately 0.00000016 acre. Here we write 0.00000016 in scientific notation:

$$1.6 \times 10^{-7}$$

Definition Scientific notation

A number is written in **scientific notation** if it has the form $N \times 10^k$, where k is an integer and the absolute value of N is between 1 and 10 or is equal to 1.

Here are more examples of numbers in scientific notation:

$$8.6 \times 10^{19} \qquad 2.159 \times 10^{8} \qquad -4.23 \times 10^{-14} \qquad 7.94 \times 10^{-97}$$

The problems in Example 11 suggest how to convert a number from scientific notation $N \times 10^k$ to standard decimal notation.

Example 11 Writing Numbers in Standard Decimal Notation

Simplify.

1. 7×10^3 **2.** 7×10^{-3} **3.** 9.48×10^{-4}

Solution

1. We simplify $7 \times 10^3 = 7.0 \times 10^3$ by *multiplying* 7.0 by 10 three times and hence moving the decimal point three places to the *right:*

$$7.0 \times 10^3 = 7000.0 = 7000$$

three places to the right

2. Since

$$7 \times 10^{-3} = 7 \times \frac{1}{10^3} = \frac{7}{1} \times \frac{1}{10^3} = \frac{7}{10^3}$$

we see that we can simplify 7.0×10^{-3} by *dividing* 7.0 by 10 three times and hence moving the decimal point three places to the *left:*

$$7.0 \times 10^{-3} = 0.007$$

three places to the left

3. We *divide* 9.48 by 10 four times and hence move the decimal point of 9.48 four places to the *left:*

$$9.48 \times 10^{-4} = 0.000948$$

four places to the left

Converting from Scientific Notation to Standard Decimal Notation

To write the scientific notation $N \times 10^k$ in standard decimal notation, we move the decimal point of the number N as follows:

- If k is *positive,* we multiply N by 10 k times and hence move the decimal point k places to the *right.*
- If k is *negative,* we divide N by 10 k times and hence move the decimal point k places to the *left.*

The problems in Example 12 suggest how to convert a number from standard decimal notation to scientific notation.

Example 12 Writing Numbers in Scientific Notation

Write the number in scientific notation.

1. 845,000,000
2. 0.0000382

Solution

1. In scientific notation, we would have 8.45×10^k, but what is k? If we move the decimal point of 8.45 eight places to the right, the result is 845,000,000. So, $k = 8$ and the scientific notation is 8.45×10^8.
2. In scientific notation, we would have 3.82×10^k, but what is k? If we move the decimal point of 3.82 five places to the left, the result is 0.0000382. So, $k = -5$ and the scientific notation is 3.82×10^{-5}.

Converting from Standard Decimal Notation to Scientific Notation

To write a number in scientific notation, count the number k of places that the decimal point must be moved so that the absolute value of the new number N is between 1 and 10 or is equal to 1:

- If the decimal point is moved to the left, then the scientific notation is written as $N \times 10^k$.
- If the decimal point is moved to the right, then the scientific notation is written as $N \times 10^{-k}$.

Example 13 Writing Numbers in Scientific Notation

Write the number in scientific notation.

1. 778,000,000 *(Jupiter's average distance, in kilometers, from the Sun)*
2. 0.000012 *(the diameter, in meters, of a white blood cell)*

Solution

1. For 778,000,000, the decimal point needs to be moved eight places to the left so the new number N is between 1 and 10. Therefore, the scientific notation is 7.78×10^8.

```
5.84*10^16
               5.84E16
7.25*10^-37
              7.25E-37
```

Figure 1 The numbers 5.84×10^{16} and 7.25×10^{-37}

2. For 0.000012, the decimal point needs to be moved five places to the right so the new number N is between 1 and 10. Therefore, the scientific notation is 1.2×10^{-5}.

Calculators express numbers in scientific notation so that the numbers "fit" on the screen. To represent 5.84×10^{16}, most calculators use the notation 5.84 E 16, where E stands for exponent (of 10). Calculators represent 7.25×10^{-37} as 7.25 E −37 (see Fig. 1).

Group Exploration

Properties of exponents

1. Since $b^{-5} = \dfrac{1}{b^5}$ for nonzero b, does it follow that $-5 = \dfrac{1}{5}$? Explain.

2. A student tries to simplify $7b^{-2}$:

$$7b^{-2} = \frac{1}{7b^2}$$

Describe any errors. Then simplify the expression correctly.

3. A student tries to simplify $\dfrac{b^8}{b^{-5}}$:

$$\frac{b^8}{b^{-5}} = b^{8-5} = b^3$$

Describe any errors. Then simplify the expression correctly.

4. A student tries to simplify $(5b^3)^{-2}$:

$$(5b^3)^{-2} = 5(b^3)^{-2} = 5b^{-6} = \frac{5}{b^6}$$

Describe any errors. Then simplify the expression correctly.

Group Exploration

Looking ahead: Definition of $b^{1/n}$

Throughout this exploration, assume that $(b^m)^n = b^{mn}$ for rational numbers m and n.

1. First, you will explore the meaning of $b^{1/2}$, where b is nonnegative.

a. For now, do not use a calculator. You will explore how you should define $9^{1/2}$. You can determine a reasonable value of $9^{1/2}$ by first finding its *square*:

$$(9^{1/2})^2 = 9^{\frac{1}{2}\cdot 2} = 9^1 = 9$$

What would be a good meaning of $9^{1/2}$? [**Hint:** Can you think of a positive number whose square equals 9?]

b. What would be a good meaning of $16^{1/2}$? Of $25^{1/2}$?

c. Now use a graphing calculator to find $9^{1/2}$, $16^{1/2}$, and $25^{1/2}$. For example, to find $9^{1/2}$, press **9** [^] [(] **1** [÷] **2** [)] [ENTER]. Is the calculator interpreting $b^{1/2}$ as you would expect?

d. What would be a good meaning of $b^{1/2}$, where b is nonnegative?

2. Now you will explore the meaning of $b^{1/3}$.

a. For now, do not use a calculator. You will explore how you should define $8^{1/3}$. You can first find the *cube* of $8^{1/3}$:

$$(8^{1/3})^3 = 8^{\frac{1}{3}\cdot 3} = 8^1 = 8$$

What would be a good meaning of $8^{1/3}$? Explain.

b. What would be a good meaning of $27^{1/3}$? Of $64^{1/3}$?

c. Use a graphing calculator to find $8^{1/3}$, $27^{1/3}$, and $64^{1/3}$. Is the calculator interpreting $b^{1/3}$ as you would expect?

d. What would be a good meaning of $b^{1/3}$?

3. What would be a good meaning of $b^{1/n}$, where n is a counting number and b is nonnegative?

Homework 10.1

For extra help ▶ MyMathLab®  Watch the videos in MyMathLab  Download the MyDashboard App

Simplify.

1. b^{-4} **2.** x^{-2} **3.** $\frac{1}{b^{-2}}$

4. $\frac{1}{b^{-5}}$ **5.** $\frac{b^{-3}}{c^5}$ **6.** $\frac{b^2}{c^{-7}}$

7. $\frac{c^{-4}}{b^{-2}}$ **8.** $\frac{b^{-4}}{b^{-8}}$ **9.** $\frac{4b^{-9}}{-6c^4d^{-1}}$

10. $\frac{-9b^{-1}}{6c^{-5}d^7}$

Simplify without using a calculator.

11. 2^{-1} **12.** 3^{-2}

13. $2^{-1} + 3^{-1}$ **14.** $\frac{1}{2^{-1}} + \frac{1}{3^{-1}}$

15. $7^{-902}7^{900}$ **16.** $4^{2003}4^{-2000}$

17. $13^{500}13^{-500}$ **18.** $(130^{-1})^{-1}$

Simplify.

19. $(b^{-2})^7$ **20.** $(b^5)^{-8}$

21. $(-4b^{-1})(3b^{-8})$ **22.** $(-5b^{-3})(-8b^{-5})$

23. $(-4b^3c^{-7})(-b^{-5}c^4)$ **24.** $(-3b^{-2}c^{-6})(-b^9c^{-1})$

25. $\frac{b^{-3}}{b^5}$ **26.** $\frac{b^{-4}}{b^7}$

27. $\frac{b^3}{b^{-2}}$ **28.** $\frac{b^6}{b^{-1}}$

29. $\frac{7b^{-3}}{4b^{-9}}$ **30.** $\frac{4b^{-2}}{9b^{-5}}$

31. $\frac{2^{-1}}{2^4}$ **32.** $\frac{3^{-2}}{3^2}$

33. $\frac{5^{-6}}{5^{-4}}$ **34.** $\frac{7^{-5}}{7^{-6}}$

35. $\frac{3^4b^{-8}}{3^2b^{-3}}$ **36.** $\frac{2^5b^{-7}}{2^2b^{-2}}$

37. $(2b^{-1})^{-5}$ **38.** $(2b^{-4})^{-4}$

39. $(b^{-2}c^5)^{-6}$ **40.** $(b^4c^{-7})^{-5}$

41. $3(b^5c)^{-2}$ **42.** $-6(bc^4)^{-3}$

43. $(2b^4c^{-2})^5(3b^{-3}c^{-4})^{-2}$ **44.** $(7b^{-4}c^{-1})^{-2}(2b^3c^{-2})^5$

45. $\frac{-12b^{-6}c^5}{14b^4c^5}$ **46.** $\frac{-28b^{-2}c^{-3}}{4b^{-3}c^{-1}}$

47. $\frac{15b^{-7}c^{-3}d^8}{-45c^2b^{-6}d^8}$ **48.** $\frac{18b^5c^3d^{-7}}{24b^{-6}c^3d^{-2}}$

49. $\frac{(-5b^{-3}c^4)(4b^{-5}c^{-1})}{80b^2c^{17}}$ **50.** $\frac{(3b^4c^{-1})(2b^{-7}c^{-8})}{42b^{-5}c^4}$

51. $\frac{(24b^3c^{-6})(49b^{-1}c^{-2})}{(28b^2c^4)(14b^{-5}c)}$ **52.** $\frac{(16b^{-2}c)(25b^4c^{-5})}{(15b^5c^{-1})(8b^{-7}c^{-2})}$

53. $\frac{(3b^5c^{-2})^3}{2^{-1}b^{-3}c}$ **54.** $\frac{(2b^{-7}c^4)^4}{5^{-1}b^2c^6}$

55. $\frac{(2b^{-4}c)^{-3}}{(2b^2c^{-5})^2}$ **56.** $\frac{(3bc^{-2})^{-2}}{(3b^{-3}c)^{-1}}$

57. $\left(\frac{6b^5c^{-2}}{7b^2c^4}\right)^2$ **58.** $\left(\frac{2bc^{-7}}{5b^{-1}c^{-2}}\right)^3$

59. $\left(\frac{5b^4c^{-3}}{15b^{-2}c^{-1}}\right)^{-4}$ **60.** $\left(\frac{8b^{-2}c^2}{12b^{-5}c^{-3}}\right)^{-3}$ **61.** $b^{-1}c^{-1}$

62. $\frac{1}{b^{-1}} \cdot \frac{1}{c^{-1}}$ **63.** $\frac{1}{b^{-1}} + \frac{1}{c^{-1}}$ **64.** $b^{-1} + c^{-1}$

Simplify. Assume n is a counting number.

65. $b^{4n}b^{3n}$ **66.** $b^{5n-1}b^{2n+4}$

67. $\frac{b^{7n-1}}{b^{2n+3}}$ **68.** $\frac{b^{3n+4}b^{n-5}}{b^{2n-3}}$

For Exercises 69–76, let $f(x) = 2(3)^x$ and $g(x) = 4^x$. Find the following.

69. $f(3)$ **70.** $f(2)$ **71.** $f(-4)$ **72.** $f(-1)$

73. $g(a + 2)$ **74.** $g(a + 3)$ **75.** $g(2a)$ **76.** $g(3a)$

77. a. Complete Table 2 with output values of the function $f(x) = 2^x$. Then use a graphing calculator to verify your results.

Table 2 Input–Output Pairs of $f(x) = 2^x$ (Exercise 77)

x	$f(x)$	x	$f(x)$
−3		1	
−2		2	
−1		3	
0		4	

b. Plot the ordered pairs you found in part (a). Then guess the graph of f and sketch it by hand. Use a graphing calculator to verify your graph.

c. Use your hand-drawn graph to estimate $2^{\frac{1}{2}}$.

78. a. Complete Table 3 with output values of the function $f(x) = \left(\frac{1}{2}\right)^x$. Then use a graphing calculator to verify your results.

Table 3 Input–Output Pairs of $f(x) = \left(\frac{1}{2}\right)^x$ (Exercise 78)

x	$f(x)$	x	$f(x)$
−4		0	
−3		1	
−2		2	
−1		3	

b. Plot the ordered pairs you found in part (a). Then guess the graph of f and sketch it by hand. Use a graphing calculator to verify your graph.

c. Use your hand-drawn graph to estimate $\left(\frac{1}{2}\right)^{\frac{1}{2}}$.

79. If an object is moving at a constant speed s (in miles per hour), then $s = dt^{-1}$, where d is the distance traveled (in miles) and t is the time (in hours).

a. Simplify the right-hand side of the equation.

b. Substitute 186 for d and 3 for t in the equation you found in part (a), and solve for s. What does your result mean in this situation?

80. The force $f(L)$ (in pounds) you must exert on a wrench handle of length L inches to loosen a bolt is described by the equation $f(L) = 720L^{-1}$.

a. Simplify the right-hand side of the equation.

b. Find $f(12)$. What does your result mean in this situation?

81. If a person plays an electric guitar outside, the sound level $f(d)$ (in decibels) at a distance d yards from the amplifier is described by the equation $f(d) = 5760d^{-2}$.

a. Simplify the right-hand side of the equation.

b. Find $f(8)$. What does your result mean in this situation?

82. The amount of light $f(d)$ (in milliwatts per square centimeter) of a 50-watt light bulb at a distance d centimeters from the bulb is described by the equation $f(d) = 8910d^{-2}$.

a. Simplify the right-hand side of the equation.

b. Find $f(80)$. What does your result mean in this situation?

Write the number in standard decimal form.

83. 4.9×10^4

84. 8.31×10^6

85. 8.59×10^{-3}

86. 6.488×10^{-5}

87. 2.95×10^{-4}

88. 8.7×10^{-2}

89. -4.512×10^8

90. -9.46×10^{10}

Write the number in scientific notation.

91. 45,700,000

92. 280,000

93. 0.0000659

94. 0.000023

95. −5,987,000,000,000

96. −308,000,000

97. 0.000001

98. 0.0004

For Exercises 99 and 100, numbers are displayed in a graphing calculator table's version of scientific notation. Write each number shown in the Y_1 column in standard decimal form.

99. See Fig. 2.

100. See Fig. 3.

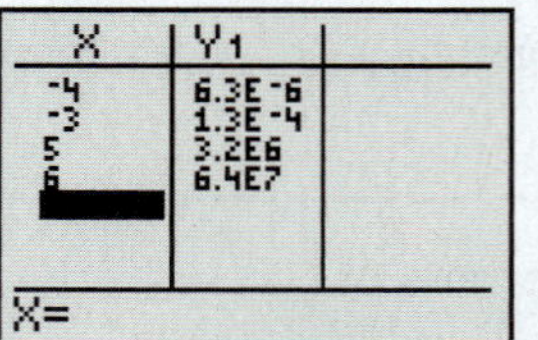

Figure 2 Exercise 99

Figure 3 Exercise 100

For Exercises 101–104, the given sentence contains a number written in scientific notation. Write the number in standard decimal form.

101. The first evidence of life on Earth dates back to 3.6×10^9 years ago.

102. The Moon has an average distance from Earth of approximately 2.389×10^5 miles.

103. The hydrogen ion concentration in human blood is about 6.3×10^{-8} mole per liter.

104. The faintest sound humans can hear has an intensity of about 10^{-12} watt per square meter.

For Exercises 105–108, the given sentence contains a number (other than a date) written in standard decimal form. Write the number in scientific notation.

105. The tanker *Exxon Valdez* spilled about 10,080,000 gallons of oil in Prince William Sound, Alaska, in 1989.

106. The average distance from Earth to Alpha Centauri is about 25,000,000,000,000 miles.

107. The wavelength of violet light is about 0.00000047 meter.

108. One second is about 0.0000000317 year.

109. The numbers of bald eagle pairs in the continental United States are shown in Table 4 for various years.

Table 4 Numbers of Bald Eagle Pairs

Year	Number of Bald Eagle Pairs (thousands)
1963	0.4
1974	0.8
1986	1.9
1995	4.7
2000	6.5
2007	9.8

Source: *U.S. Fish and Wildlife Service*

Let n be the number (in thousands) of bald eagle pairs in the continental United States at t years since 1960. The situation can be described by the linear function $n = 0.21t - 1.66$, the quadratic function $n = 0.0066t^2 - 0.12t + 0.85$, and the exponential function $n = 0.30(1.078)^t$.

a. Use a graphing calculator to draw the graphs of the three models and, in the same viewing window, the scattergram of the data. Compare how well the models fit the data.

b. Use the exponential model to predict the number of bald eagle pairs in 2017.

c. Use the linear model to predict the number of bald eagle pairs in 2017. Then use the quadratic model to do the same. Explain why your results are so much smaller than your result from part (b). [**Hint:** Zoom out.]

110. The revenues (in billions of dollars) of Amazon are shown in Table 5 for various years.

Table 5 Annual Revenues of Amazon

Year	Annual Revenue (billions of dollars)
2002	3.9
2004	6.9
2006	10.7
2008	19.2
2010	34.2
2011	48.1

Source: *Amazon*

Let r be the annual revenue (in billions of dollars) at t years since 2000. The situation can be described by the linear function $r = 4.67t - 11.38$, the quadratic function $r = 0.66t^2 - 4.06t + 10.59$, and the exponential function $r = 2.2(1.32)^t$.

a. Use a graphing calculator to draw the graphs of the three models and, in the same viewing window, the scattergram of the data. Compare how well the models fit the data.

b. Use the exponential model to predict the revenue in 2019.

c. Use the linear model to predict the revenue in 2019. Then use the quadratic model to do the same. Explain why your results are so much smaller than your result from part (b). [**Hint:** Zoom out.]

Concepts

111. Two students try to simplify $(5b^2)^{-1}$:

Student A

$$(5b^2)^{-1} = -5b^{-2} = \frac{-5}{b^2}$$

Student B

$$(5b^2)^{-1} = 5^{-1}(b^2)^{-1} = 5^{-1}b^{-2} = \frac{1}{5b^2}$$

Did either student simplify the expression correctly? Describe any errors.

112. Two students try to simplify an expression:

Student 1

$$\frac{7b^8}{b^{-3}} = 7b^{8-(-3)} = 7b^{11}$$

Student 2

$$\frac{7b^8}{b^{-3}} = 7b^{8-3} = 7b^5$$

Did either student simplify the expression correctly? Describe any errors.

113. A student tries to simplify $\dfrac{3b^{-2}c^4}{d^7}$:

$$\frac{3b^{-2}c^4}{d^7} = \frac{c^4}{3b^2d^7}$$

Describe any errors. Then simplify the expression correctly.

114. A student thinks that 4^{-3} is a negative number. Is the student correct? Explain.

115. It is common to confuse expressions such as 2^2, 2^{-1}, $2(-1)$, $\left(\frac{1}{2}\right)^2$, $\left(\frac{1}{2}\right)^{-1}$, -2^2, $(-2)^2$, and $\frac{1}{2}$. List these numbers from least to greatest. Are there any "ties"?

116. a. Simplify $\left(\dfrac{b}{c}\right)^{-2}$.

b. Simplify $\left(\dfrac{b}{c}\right)^{-n}$.

c. Use your result from part (b) to simplify $\left(\dfrac{b}{c}\right)^{-5}$ in one step.

117. Explore "0^0":

a. Simplify 5^0, 4^0, 3^0, 2^0, and 1^0. On the basis of these values, what would be a reasonable value of 0^0?

b. Simplify 0^5, 0^4, 0^3, 0^2, and 0^1. On the basis of these values, what would be a reasonable value of 0^0?

c. Why is it a good idea to leave 0^0 meaningless?

118. Simplify each expression.

a. b^{-1}
b. $(b^{-1})^{-1}$
c. $((b^{-1})^{-1})^{-1}$
d. $(((b^{-1})^{-1})^{-1})^{-1}$
e. $\underbrace{((((b^{-1})^{-1})^{-1})^{-1}\ldots)^{-1}}_{n \text{ exponents}}$

119. It is a common error to confuse the properties $b^mb^n = b^{m+n}$ and $(b^m)^n = b^{mn}$. Explain why each property makes sense, and compare the properties. Give examples to illustrate your comparison. (See page 9 for guidelines on writing a good response.)

120. Describe what it means to use exponential properties to simplify an expression. Include several examples in your description. (See page 9 for guidelines on writing a good response.)

Related Review

For $f(x) = 2x$, $g(x) = x^2$, and $h(x) = 2^x$, find the following.

121. $f(3)$ **122.** $f(-3)$ **123.** $g(3)$ **124.** $g(-3)$
125. $h(3)$ **126.** $h(-3)$

Expressions, Equations, Functions, and Graphs

Perform the indicated instruction. Then use words such as linear, quadratic, cubic, polynomial, degree, exponential, function, one variable, *and* two variables *to describe the expression, equation, or system.*

127. Solve:

$$y = 3x + 1$$
$$y = 2x - 4$$

128. Simplify $5(3x + 1) - 4(2x - 4)$.

129. Solve $3x + 1 = 2x - 4$.

130. Graph $f(x) = 3x + 1$ by hand.

131. Find the product $5(3x + 1)(2x - 4)$.

132. Graph $g(x) = 3x^2 - 6x - 2$ by hand.

10.2 Rational Exponents

Objectives

» Know definitions of *rational exponents.*

» Simplify expressions that have rational exponents.

In Section 10.1, we worked with integer exponents. In this section, we work with exponents that are rational numbers (Section 1.1).

Definitions of Rational Exponents

How should we define $b^{1/n}$, where n is a counting number? If the exponential property $(b^m)^n = b^{mn}$ is to be true for $m = \frac{1}{2}$ and $n = 2$, then

$$(9^{\frac{1}{2}})^2 = 9^{\frac{1}{2} \cdot 2} = 9^1 = 9$$

Since $(-3)^2 = 9$ and $3^2 = 9$, the statement suggests that a good meaning of $9^{1/2}$ is -3 or 3. We define $9^{1/2} = 3$. We call the nonnegative number 3 the *principal second root,* or **principal square root,** of 9, written $\sqrt{9}$.

Similarly, if the property $(b^m)^n = b^{mn}$ is to be true for $m = \frac{1}{3}$ and $n = 3$, then

$$(8^{\frac{1}{3}})^3 = 8^{\frac{1}{3} \cdot 3} = 8^1 = 8$$

Since $2^3 = 8$, the statement suggests that a good meaning of $8^{1/3}$ is 2. The number 2 is called the *third root,* or **cube root,** of 8, written $\sqrt[3]{8}$.

For $(-8)^{1/3}$, a good meaning is -2, since $(-2)^3 = -8$. We do not assign a real-number value to $(-9)^{1/2}$, since no real number squared is equal to -9.

> **Definition** $b^{1/n}$
>
> For the counting number n, where $n \neq 1$,
>
> - If n is odd, then $\boldsymbol{b^{1/n}}$ is the number whose nth power is b, and we call $b^{1/n}$ the ***n*th root of *b*.**
> - If n is even and $b \geq 0$, then $\boldsymbol{b^{1/n}}$ is the nonnegative number whose nth power is b, and we call $b^{1/n}$ the **principal *n*th root of *b*.**
> - If n is even and $b < 0$, then $b^{1/n}$ is not a real number.
>
> $\boldsymbol{b^{1/n}}$ may be represented by $\sqrt[n]{b}$.

Example 1 Simplifying Expressions Involving Rational Exponents

Simplify.

1. $25^{1/2}$
2. $64^{1/3}$
3. $(-64)^{1/3}$
4. $16^{1/4}$
5. $-16^{1/4}$
6. $(-16)^{1/4}$

Solution

1. $25^{1/2} = 5$, since $5^2 = 25$.
2. $64^{1/3} = 4$, since $4^3 = 64$.
3. $(-64)^{1/3} = -4$, since $(-4)^3 = -64$.
4. $16^{1/4} = 2$, since $2^4 = 16$.
5. $-16^{1/4} = -(16^{1/4}) = -2$.
6. $(-16)^{1/4}$ is not a real number, since the fourth power of any real number is nonnegative.

Graphing calculator checks for Problems 1, 2, and 3 are shown in Fig. 4. For example, to find $25^{1/2}$, press **25** [^] [(] **1** [÷] **2** [)] [ENTER]. If an exponent involves an operation, you must use parentheses.

```
25^(1/2)
                5
64^(1/3)
                4
(-64)^(1/3)
               -4
```

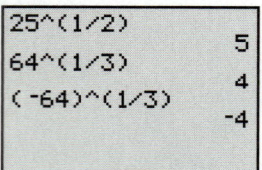

Figure 4 Checks for Problems 1, 2, and 3

What would be a reasonable definition of $b^{m/n}$? If the properties of exponents we discussed in Section 10.1 are to hold true for rational exponents, we have

$$8^{\frac{2}{3}} = 8^{\frac{1}{3}\cdot 2} = \left(8^{\frac{1}{3}}\right)^2 = 2^2 = 4 \quad \text{or} \quad 8^{\frac{2}{3}} = 8^{2\cdot\frac{1}{3}} = \left(8^2\right)^{\frac{1}{3}} = 64^{\frac{1}{3}} = 4$$

Likewise,

$$32^{\frac{3}{5}} = 32^{\frac{1}{5}\cdot 3} = \left(32^{\frac{1}{5}}\right)^3 = 2^3 = 8 \quad \text{or} \quad 32^{\frac{3}{5}} = 32^{3\cdot\frac{1}{5}} = \left(32^3\right)^{\frac{1}{5}} = 32{,}768^{\frac{1}{5}} = 8$$

Also,

$$32^{-\frac{3}{5}} = \frac{1}{32^{\frac{3}{5}}} = \frac{1}{8}$$

Definition Rational exponent

For the counting numbers m and n, where $n \neq 1$ and b is any real number for which $b^{1/n}$ is a real number,

- $\boldsymbol{b^{m/n}} = \left(b^{1/n}\right)^m = \left(b^m\right)^{1/n}$
- $\boldsymbol{b^{-m/n}} = \dfrac{1}{b^{m/n}}, \quad b \neq 0$

A power of the form $b^{m/n}$ or $b^{-m/n}$ is said to have a **rational exponent.**

Example 2 Simplifying Expressions Involving Rational Exponents

Simplify.

1. $25^{3/2}$ **2.** $(-27)^{2/3}$ **3.** $32^{-2/5}$ **4.** $(-8)^{-5/3}$

Solution

1. $25^{3/2} = (25^{1/2})^3 = 5^3 = 125$

2. $(-27)^{2/3} = ((-27)^{1/3})^2 = (-3)^2 = 9$

3. $32^{-2/5} = \dfrac{1}{32^{2/5}} = \dfrac{1}{\left(32^{1/5}\right)^2} = \dfrac{1}{2^2} = \dfrac{1}{4}$

4. $(-8)^{-5/3} = \dfrac{1}{(-8)^{5/3}} = \dfrac{1}{\left((-8)^{1/3}\right)^5} = \dfrac{1}{(-2)^5} = \dfrac{1}{-32} = -\dfrac{1}{32}$

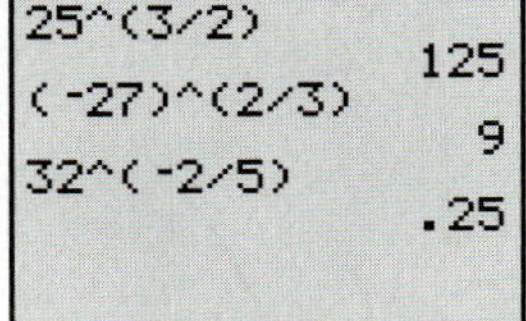

Figure 5 Checks for Problems 1, 2, and 3

Graphing calculator checks for Problems 1, 2, and 3 are shown in Fig. 5.

Example 3 Evaluating an Exponential Function

For $f(x) = 64^x$, $g(x) = 3(16)^x$, and $h(x) = -5(9)^x$, find the following.

1. $f\left(\dfrac{2}{3}\right)$ **2.** $g\left(\dfrac{3}{4}\right)$ **3.** $h\left(-\dfrac{1}{2}\right)$

Solution

1. $f\left(\dfrac{2}{3}\right) = 64^{2/3} = \left(64^{1/3}\right)^2 = 4^2 = 16$

2. $g\left(\dfrac{3}{4}\right) = 3(16)^{3/4} = 3\left(16^{1/4}\right)^3 = 3(2)^3 = 3\cdot 8 = 24$

3. $h\left(-\dfrac{1}{2}\right) = -5(9)^{-1/2} = \dfrac{-5}{9^{1/2}} = -\dfrac{5}{3}$

Properties of Rational Exponents

The properties of exponents we discussed in Section 10.1 are valid for *rational* exponents.

Properties of Rational Exponents

If m and n are rational numbers and b and c are any real numbers for which b^m, b^n, and c^n are real numbers, then

- $b^m b^n = b^{m+n}$ *Product property for exponents*
- $\dfrac{b^m}{b^n} = b^{m-n},\ b \neq 0$ *Quotient property for exponents*
- $(bc)^n = b^n c^n$ *Raising a product to a power*
- $\left(\dfrac{b}{c}\right)^n = \dfrac{b^n}{c^n},\ c \neq 0$ *Raising a quotient to a power*
- $(b^m)^n = b^{mn}$ *Raising a power to a power*

We can use properties of exponents to help us simplify expressions involving rational exponents.

Example 4 Simplifying Expressions Involving Rational Exponents

Simplify. Assume b is positive.

1. $(4b^6)^{3/2}$ **2.** $\dfrac{b^{2/7}}{b^{-3/7}}$

Solution

1.
$$\begin{aligned}(4b^6)^{3/2} &= 4^{3/2}(b^6)^{3/2} && \text{Raise factors to nth power: } (bc)^n = b^n c^n\\ &= (4^{1/2})^3 b^{\frac{6}{1}\cdot\frac{3}{2}} && b^{m/n} = (b^{1/n})^m\text{; multiply exponents: } (b^m)^n = b^{mn}\\ &= 2^3 b^9 && 4^{\frac{1}{2}} = 2\text{; multiply.}\\ &= 8b^9 && \text{Simplify.}\end{aligned}$$

2.
$$\begin{aligned}\frac{b^{2/7}}{b^{-3/7}} &= b^{\frac{2}{7}-\left(-\frac{3}{7}\right)} && \text{Subtract exponents: } \frac{b^m}{b^n} = b^{m-n}\\ &= b^{\frac{2}{7}+\frac{3}{7}} && \text{Simplify.}\\ &= b^{5/7} && \text{Add.}\end{aligned}$$

Example 5 Simplifying Expressions Involving Rational Exponents

Simplify. Assume b is positive.

1. $b^{2/3}b^{1/2}$ **2.** $\left(\dfrac{32b^2}{b^{12}}\right)^{2/5}$

Solution

1.
$$\begin{aligned}b^{2/3}b^{1/2} &= b^{\frac{2}{3}+\frac{1}{2}} && \text{Add exponents: } b^m b^n = b^{m+n}\\ &= b^{\frac{4}{6}+\frac{3}{6}} && \text{Find common denominator.}\\ &= b^{7/6} && \text{Add.}\end{aligned}$$

2.
$$\begin{aligned}\left(\frac{32b^2}{b^{12}}\right)^{2/5} &= (32b^{2-12})^{2/5} && \text{Subtract exponents: } \frac{b^m}{b^n} = b^{m-n}\\ &= (32b^{-10})^{2/5} && \text{Subtract.}\end{aligned}$$

$$= \left(\frac{32}{b^{10}}\right)^{2/5}$$ *Write powers so exponents are positive:* $b^{-n} = \frac{1}{b^n}$

$$= \frac{32^{2/5}}{(b^{10})^{2/5}}$$ *Raise numerator and denominator to nth power:* $\left(\frac{b}{c}\right)^n = \frac{b^n}{c^n}$

$$= \frac{(32^{1/5})^2}{b^{10\cdot\frac{2}{5}}}$$ $b^{m/n} = (b^{1/n})^m$*; multiply exponents:* $(b^m)^n = b^{mn}$

$$= \frac{2^2}{b^4}$$ $32^{1/5} = 2$*; multiply.*

$$= \frac{4}{b^4}$$ *Simplify.*

Example 6 Simplifying an Expression Involving Rational Exponents

Simplify $\frac{(81b^6c^{20})^{1/2}}{(27b^{12}c^9)^{2/3}}$. Assume b and c are positive.

Solution

$$\frac{(81b^6c^{20})^{1/2}}{(27b^{12}c^9)^{2/3}} = \frac{81^{1/2}(b^6)^{1/2}(c^{20})^{1/2}}{27^{2/3}(b^{12})^{2/3}(c^9)^{2/3}}$$ *Raise factors to a power:* $(bc)^n = b^nc^n$

$$= \frac{9b^{6\cdot\frac{1}{2}}c^{20\cdot\frac{1}{2}}}{(27^{1/3})^2 b^{12\cdot\frac{2}{3}}c^{9\cdot\frac{2}{3}}}$$ $81^{1/2} = 9$; $b^{m/n} = (b^{1/n})^m$*; multiply exponents:* $(b^m)^n = b^{mn}$

$$= \frac{9b^3c^{10}}{3^2b^8c^6}$$ $27^{1/3} = 3$*; multiply.*

$$= \frac{9b^{-5}c^4}{9}$$ *Subtract exponents:* $\frac{b^m}{b^n} = b^{m-n}$

$$= \frac{c^4}{b^5}$$ *Write powers so exponents are positive:* $b^{-n} = \frac{1}{b^n}$

Group Exploration

Looking ahead: Graphical significance of a and b for $y = ab^x$

1. Use ZDecimal to graph these equations of the form $y = b^x$ in order, and describe what you observe:

$$y = 1.2^x,\quad y = 1.5^x,\quad y = 2^x,\quad \text{and}\quad y = 5^x$$

If you want a better view, set Ymin = 0. To change window settings, see Appendix A.7.

Do the same with the equations

$$y = 0.3^x,\quad y = 0.5^x,\quad y = 0.7^x,\quad \text{and}\quad y = 0.9^x$$

If you want a better view, set Ymax = 0.

2. Use ZStandard to graph these equations of the form $y = a(1.1)^x$ in order, and describe what you observe:

$$y = 2(1.1)^x,\quad y = 3(1.1)^x,\quad y = 4(1.1)^x,\quad \text{and}\quad y = 5(1.1)^x$$

If you want a better view, set Ymin = 0.

Use ZStandard to do the same with the equations

$$y = -2(1.1)^x,\quad y = -3(1.1)^x,\quad y = -4(1.1)^x,\quad \text{and}\quad y = -5(1.1)^x$$

3. So far, you have sketched the graphs of equations of only the forms $y = b^x$ (where $a = 1$) and $y = a(1.1)^x$ (where $b = 1.1$). Graph more equations of the form $y = ab^x$, until you are confident you know the graphical significance of the constants a and b, for any possible combination of values of a and b. If you have any new insights into the graphical significance of a and b, describe those insights.

4. Describe the graph of $y = ab^x$ in the following situations.
   a. a is positive
   b. a is negative
   c. $b > 1$
   d. $0 < b < 1$
   e. $b = 1$
   f. b is negative

5. Describe the connection between the y-intercept of the graph of $y = ab^x$ and the values of a and b.

Group Exploration

Looking ahead: Numerical significance of a and b for $f(x) = ab^x$

In this exploration, you will investigate the nature of exponential functions of the form $f(x) = ab^x$.

1. Use a graphing calculator to create a table of ordered pairs for $f(x) = 2(3)^x$, $g(x) = 64\left(\frac{1}{2}\right)^x$, and a third exponential function of your choice. (See Figs. 6 and 7.) Use the following values for the x-coordinates: 0, 1, 2, . . . , 6.

Figure 6 Enter the three functions

Figure 7 Table setup

2. a. What connection do you notice between the y-coordinates of each function and the base b of the function $y = ab^x$?

b. Test the connection you described in part (a) by choosing yet another exponential function, and check whether it behaves as you think it should.

c. For $f(x) = ab^x$, we have $f(0) = a$, $f(1) = ab$, $f(2) = abb$, and $f(3) = abbb$. Explain why these results suggest that your response to part (a) is correct.

3. a. What connection do you notice between the y-coordinates of each function and the coefficient a of the function $y = ab^x$?

b. Test the connection you described in part (a) by choosing yet another exponential function, and check whether it behaves as you think it should.

c. Use pencil and paper to find $f(0)$, where $f(x) = ab^x$. Explain why your result shows that your response to part (a) is correct.

Homework 10.2

For extra help ▶ 

Download the MyDashboard App

*Simplify without using a calculator. Then use a graphing calculator to verify your result. [**Graphing Calculator:** Instructions for $x^{m/n}$: Press* X, T, Θ, n ^ (m ÷ n)*.]*

1. $16^{1/2}$ **2.** $27^{1/3}$ **3.** $1000^{1/3}$

4. $32^{1/5}$ **5.** $49^{1/2}$ **6.** $81^{1/4}$

7. $125^{1/3}$ **8.** $64^{1/6}$ **9.** $8^{4/3}$

10. $16^{3/4}$ **11.** $9^{3/2}$ **12.** $64^{2/3}$

13. $32^{2/5}$ **14.** $27^{4/3}$ **15.** $4^{5/2}$

16. $81^{3/4}$ **17.** $27^{-1/3}$ **18.** $16^{-1/4}$

19. $-36^{-1/2}$ **20.** $-32^{-1/5}$ **21.** $4^{-5/2}$

22. $9^{-3/2}$ **23.** $(-27)^{-4/3}$ **24.** $(-32)^{-3/5}$

Simplify without using a calculator. Then use a graphing calculator to verify your result.

25. $2^{1/4}2^{3/4}$ **26.** $3^{7/5}3^{3/5}$ **27.** $(3^{1/2}2^{3/2})^2$

28. $(2^{2/3}5^{1/3})^3$ **29.** $\dfrac{7^{1/3}}{7^{-5/3}}$ **30.** $\dfrac{5^{4/3}}{5^{1/3}}$

For $f(x) = 81^x$, $g(x) = 4(27)^x$, and $h(x) = -2(4)^x$, find the following.

31. $f\left(\frac{3}{4}\right)$ **32.** $f\left(\frac{1}{4}\right)$

33. $g\left(\frac{1}{3}\right)$ **34.** $g\left(\frac{2}{3}\right)$

35. $g\left(-\frac{1}{3}\right)$ **36.** $g\left(-\frac{2}{3}\right)$

37. $h\left(\frac{3}{2}\right)$ **38.** $h\left(\frac{5}{2}\right)$

39. Without using a calculator, complete Table 6 with values of the function $f(x) = 16^x$. Then use a graphing calculator to verify your results.

Table 6 Values of the Function $f(x) = 16^x$

x	$f(x)$	x	$f(x)$
$-\frac{3}{4}$		$\frac{1}{4}$	
$-\frac{1}{2}$		$\frac{1}{2}$	
$-\frac{1}{4}$		$\frac{3}{4}$	
0		1	

40. Without using a calculator, complete Table 7 with values of the function $f(x) = 64^x$. Then use a graphing calculator to verify your results.

Table 7 Values of the Function $f(x) = 64^x$

x	$f(x)$	x	$f(x)$
$-\frac{5}{6}$		$\frac{1}{6}$	
$-\frac{2}{3}$		$\frac{1}{3}$	
$-\frac{1}{2}$		$\frac{1}{2}$	
$-\frac{1}{3}$		$\frac{2}{3}$	
$-\frac{1}{6}$		$\frac{5}{6}$	
0		1	

Simplify. Assume b and c are positive.

41. $b^{7/6}b^{5/6}$

42. $b^{1/5}b^{3/5}$

43. $b^{3/5}b^{-13/5}$

44. $b^{2/7}b^{-6/7}$

45. $(16b^8)^{1/4}$

46. $(27b^{27})^{1/3}$

47. $4(25b^8c^{14})^{-1/2}$

48. $-(8b^{-6}c^{12})^{2/3}$

49. $(b^{3/5}c^{-1/4})(b^{2/5}c^{-7/4})$

50. $(b^{-4/3}c^{1/2})(b^{-2/3}c^{-3/2})$

51. $(5bcd)^{1/5}(5bcd)^{4/5}$

52. $(6bc^2)^{5/7}(6bc^2)^{2/7}$

53. $[(3b^5)^3(3b^9c^8)]^{1/4}$

54. $[(4b^3)^2(b^2c^{12})]^{1/4}$

55. $\dfrac{b^{-2/5}c^{11/8}}{b^{18/5}c^{-5/8}}$

56. $\dfrac{b^{3/4}c^{1/2}}{b^{-1/4}c^{-1/2}}$

57. $\left(\dfrac{9b^3c^{-2}}{25b^{-5}c^4}\right)^{-1/2}$

58. $\left(\dfrac{16b^{12}c^2}{2b^{-3}c^{-4}}\right)^{-1/3}$

59. $32^{1/5}b^{3/7}b^{2/5}$

60. $16^{1/4}b^{1/4}b^{1/3}$

61. $\dfrac{b^{5/6}}{b^{1/4}}$

62. $\dfrac{b^{-2/3}}{b^{1/7}}$

63. $\dfrac{(9b^5)^{3/2}}{(27b^4)^{2/3}}$

64. $\dfrac{(32b^3)^{3/5}}{(16b^3)^{3/2}}$

65. $\left(\dfrac{8b^{2/3}}{2b^{4/5}}\right)^{3/2}$

66. $\left(\dfrac{27b^{1/3}c^{3/4}}{8b^{-2/3}c^{1/2}}\right)^{4/3}$

67. $\dfrac{(8bc^3)^{1/3}}{(81b^{-5}c^3)^{3/4}}$

68. $\dfrac{(1000b^{-7}c^8)^{2/3}}{(32b^{15}c^4)^{3/5}}$

69. $b^{2/5}(b^{8/5} + b^{3/5})$

70. $c^{1/3}(c^{8/3} - c^{5/3})$

71. The amounts (in megawatts) of new solar power installed in the United States are shown in Table 8 for various years.

Table 8 Amounts of New Solar Power Installed in the United States

Year	Amount of New Solar Power Installed (megawatts)
2006	105
2007	160
2008	290
2009	435
2010	878
2011	1891
2012	3300

Source: *GTM Research*

Let A be the amount (in megawatts) of new solar power installed in the year that is t years since 2000. The situation can be described by the linear function $A = 486.96t - 3374$, the quadratic function $A = 140.25t^2 - 2038t + 7425$, and the exponential function $A = 2.73(1.8)^t$.

a. Use a graphing calculator to draw the graphs of the three models and, in the same viewing window, the scattergram of the data. Compare how well the models fit the data.

b. Which model likely describes the situation the best for the years before 2006? Explain.

c. Use the exponential model to predict the amount of solar power that will be installed in 2018.

d. Use the exponential model and "intersect" on a graphing calculator to predict in which year the amount of solar power installed will be 30,000 megawatts, the equivalent of more than 24 nuclear-power plants. [**Hint:** Graph the model and the horizontal line $A = 30{,}000$, and then use Zoom Out twice.]

72. The numbers of AP tests administered are shown in Table 9 for various years.

Table 9 Numbers of AP Tests Administered

Year	Number of AP Tests Administered (millions)
1980	0.2
1985	0.3
1990	0.5
1995	0.9
2000	1.3
2005	2.1
2010	3.3

Source: *The College Board*

Let n be the number (in millions) of AP tests administered in the year that is t years since 1980. The situation can be modeled by the linear function $n = 0.098t - 0.24$, the quadratic function $n = 0.004t^2 - 0.024t + 0.27$, and the exponential function $n = 0.2(1.099)^t$.

a. Use a graphing calculator to draw the graphs of the three models and, in the same viewing window, the scattergram of the data. Compare how well the models fit the data.

b. Which model likely describes the situation the best for the years before 1980? Explain.

c. Use the exponential model to predict the number of AP tests that will be administered in 2017.

d. Use the exponential model and "intersect" on a graphing calculator to predict in which year 8 million AP tests will be administered. [**Hint:** Graph the model and the horizontal line $n = 8$, and then use Zoom Out.]

Concepts

73. We can represent $\sqrt{5}$ by $5^{1/2}$. Explain.

74. To use a graphing calculator to find that $16^{1/2} = 4$, we press 16 [^] [(] 1 [÷] 2 [)]. If we omit the parentheses, we get the incorrect result 8. Explain.

75. A student tries to simplify $(36x^{36})^{1/2}$:

$$(36x^{36})^{1/2} = 36^{1/2}(x^{36})^{1/2} = 18x^{18}$$

Describe any errors. Then simplify the expression correctly.

76. A student tries to simplify $64^{2/3}$:

$$64^{2/3} = (64^{1/2})^3 = 8^3 = 512$$

Describe any errors. Then simplify the expression correctly.

77. Two students simplify $9^{3/2}$:

Student A	**Student B**
$9^{3/2} = (9^{1/2})^3 = 3^3 = 27$	$9^{3/2} = (9^3)^{1/2} = 729^{1/2} = 27$

Explain why it makes sense the students' results are equal. Which method is easier? Explain.

78. A student tries to simplify $25^{-1/2}$:

$$25^{-1/2} = -25^{1/2} = -5$$

Describe any errors. Then simplify the expression correctly.

79. a. Identify which of the following are *not* real numbers:

$$(-9)^{1/2}, (-27)^{1/3}, (-81)^{1/4}, (-32)^{1/5}, (-1)^{1/6}, (-1)^{1/7}$$

b. Describe all of the values of b and counting number n for which $b^{1/n}$ is not a real number. Explain.

80. Use a calculator to determine which is larger, $\left(\frac{1}{2}\right)^{1/3}$ or $\left(\frac{1}{3}\right)^{1/2}$. Explain why this makes sense. [**Hint:** Use the property $\left(\frac{b}{c}\right)^n = \frac{b^n}{c^n}$.]

81. Describe how to compute a numerical expression of the form $b^{m/n}$, assuming it is a real number.

82. List the exponent definitions and properties that are discussed in this section and Section 10.1. Explain how you can recognize which definition or property will help you simplify a given expression.

Related Review

For $f(x) = 8x$ *and* $g(x) = 8^x$, *find the following.*

83. $f\left(\frac{1}{3}\right)$

84. $f\left(\frac{4}{3}\right)$

85. $g\left(\frac{1}{3}\right)$

86. $g\left(\frac{4}{3}\right)$

87. $f\left(-\frac{5}{3}\right)$

88. $f\left(-\frac{2}{3}\right)$

89. $g\left(-\frac{5}{3}\right)$

90. $g\left(-\frac{2}{3}\right)$

Expressions, Equations, Functions, and Graphs

Perform the indicated instruction. Then use words such as linear, quadratic, cubic, polynomial, degree, exponential, function, one variable, *and* two variables *to describe the expression, equation, or system.*

91. Factor $2p^3 - 3p^2 - 18p + 27$.

92. Simplify $3(2x - 5) - (5x + 2) + 2(x - 3)$.

93. Graph $y = x^2 - 2x$ by hand.

94. Simplify $-2(4x - 6)^2$.

95. Solve $x(3x - 2) = 4$.

96. Graph $y = 3x^2$ by hand.

10.3 Graphing Exponential Functions

Objectives

- Sketch the graph of an exponential function.
- Know the graphical significance of a and b for a function of the form $f(x) = ab^x$.
- Know the *base multiplier property*, the *increasing or decreasing property*, and the *reflection property*.

Recall from Section 10.1 that an exponential function is a function whose equation can be put into the form $f(x) = ab^x$, where $a \neq 0$, $b > 0$, and $b \neq 1$. In this section, we discuss how to use the values of a and b to help us graph an exponential function.

Graphing Exponential Functions

When graphing a certain type of function for the first time, we often begin by finding outputs for integer inputs near zero.

Example 1 Graphing an Exponential Function with $b > 1$

Graph $f(x) = 2^x$ by hand.

Solution

First, we list input–output pairs of the function f in Table 10 on the next page. Note that as the value of x increases by 1, the value of y is multiplied by 2 (the base).

Next, we plot the solutions from Table 10 in Fig. 8 and sketch an increasing curve that contains the plotted points. The graph shows that as the value of x increases by 1, the value of y is doubled.

Table 10 Input–Output Pairs of $f(x) = 2^x$

x	$f(x)$
-3	$2^{-3} = \frac{1}{2^3} = \frac{1}{8}$
-2	$2^{-2} = \frac{1}{2^2} = \frac{1}{4}$
-1	$2^{-1} = \frac{1}{2^1} = \frac{1}{2}$
0	$2^0 = 1$
1	$2^1 = 2$
2	$2^2 = 4$
3	$2^3 = 8$

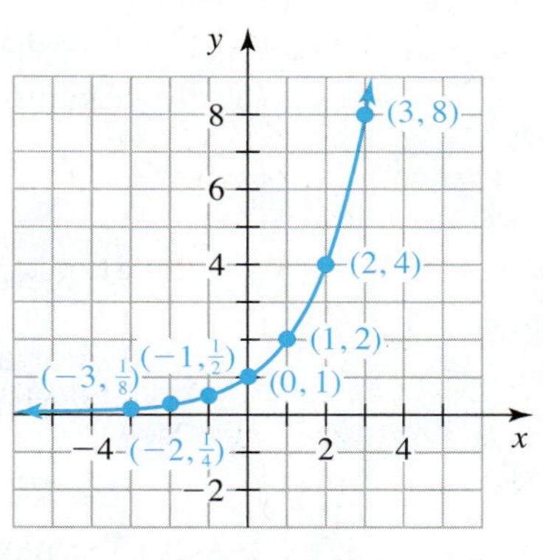

Figure 8 Graph of $f(x) = 2^x$

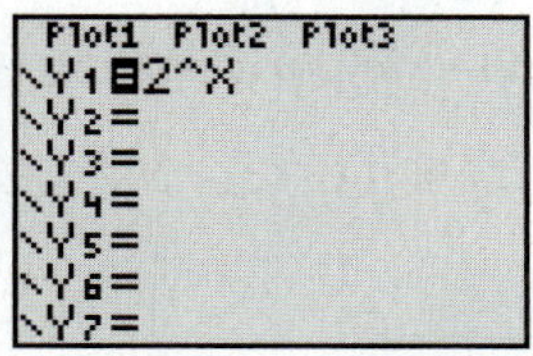

Figure 9 Enter the function

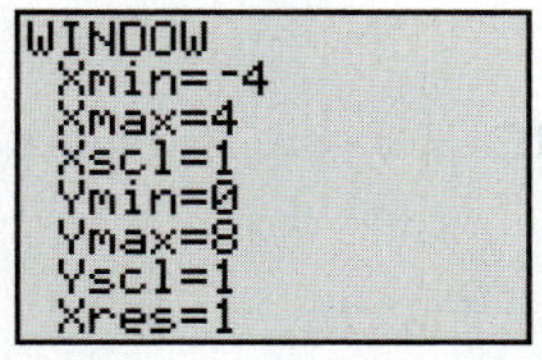

Figure 10 Set up the window

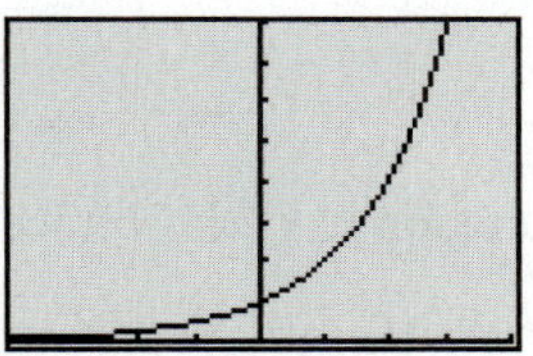
Figure 11 Graph the function

We can set up a window to verify our graph (see Figs. 9–11).

The smooth curve sketched in Fig. 8 implies that 2^x has meaning for *any* real-number exponent x. This is indeed true. In general, for $b > 0$, b^x has meaning for any real-number exponent x. The exponents are defined so the graph of any exponential function is a smooth graph. Also, the exponential properties we have discussed for rational exponents apply to real-number exponents as well. We can use a calculator to find *real-number powers* of numbers.

Recall from Section 3.1 that every point on the graph of an equation represents a solution of the equation. Every point *not* on the graph represents an ordered pair that is *not* a solution. The graph of an exponential function is called an **exponential curve.**

Example 2 Graphing an Exponential Function with $0 < b < 1$

Graph $g(x) = 4\left(\frac{1}{2}\right)^x$ by hand.

Solution

Input–output pairs of g are listed in Table 11. For example,

$$g(-1) = 4\left(\frac{1}{2}\right)^{-1} = 4\left(\frac{1}{2^{-1}}\right) = 4(2^1) = 8$$

Table 11 Input–Output Pairs of $g(x) = 4\left(\frac{1}{2}\right)^x$

x	$g(x)$
-1	8
0	4
1	2
2	1
3	$\frac{1}{2}$

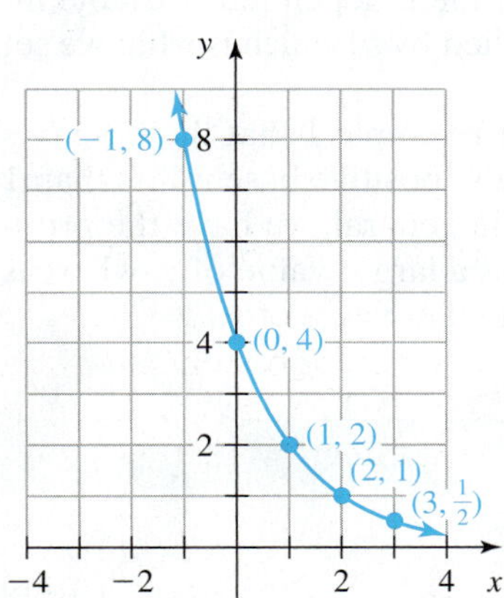

Figure 12 Graph of $y = 4\left(\frac{1}{2}\right)^x$

So, $(-1, 8)$ is an input–output pair. Note that as the value of x increases by 1, the value of y is multiplied by $\frac{1}{2}$.

We plot the found points in Fig. 12 and sketch a decreasing exponential curve that contains the plotted points. The graph shows that as the value of x increases by 1, the value of y is halved.

We can use a graphing calculator to verify our graph (see Fig. 13).

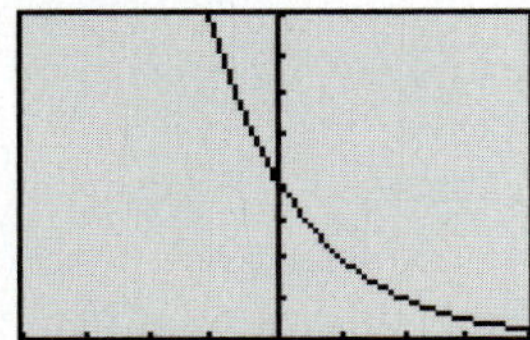
Figure 13 Graph of $y = 4\left(\frac{1}{2}\right)^x$

Base Multiplier Property; Increasing or Decreasing Property

Examples 1 and 2 suggest the *base multiplier property* and the *increasing or decreasing property*.

> **Base Multiplier Property**
>
> For an exponential function of the form $y = ab^x$, if the value of the independent variable increases by 1, the value of the dependent variable is multiplied by b.

We have seen two examples of this property in Examples 1 and 2. Here are two more examples of the base multiplier property:

1. For the function $f(x) = 2(3)^x$, if the value of x increases by 1, the value of y is multiplied by 3.
2. For the function $f(x) = 5\left(\frac{3}{4}\right)^x$, if the value of x increases by 1, the value of y is multiplied by $\frac{3}{4}$.

To prove the base multiplier property for the exponential function $f(x) = ab^x$, we compare outputs for the inputs k and $k + 1$, which differ by 1:

$$f(k) = ab^k \qquad \begin{aligned} f(k+1) &= ab^{k+1} \\ &= ab^k b^1 \\ &= f(k)b \end{aligned}$$

Since $f(k + 1) = f(k)b$, we conclude that if the value of the independent variable increases by 1, the value of the dependent variable is multiplied by b, which is what we set out to show.

For the increasing or decreasing property, we note in Example 1 that the base b is greater than 1 and the graph is increasing. In Example 2, the positive base is less than 1 and the graph is decreasing. For $f(x) = ab^x$ with $a > 0$, in general, we have the property that each multiplication by a base greater than 1 gives a larger value of y, whereas each multiplication by a positive base less than 1 gives a smaller value of y.

> **Increasing or Decreasing Property**
>
> Let $f(x) = ab^x$, where $a > 0$. Then
>
> - If $b > 1$, then the function f is increasing. We say the function **grows exponentially** (see Fig. 14).
> - If $0 < b < 1$, then the function f is decreasing. We say the function **decays exponentially** (see Fig. 15).

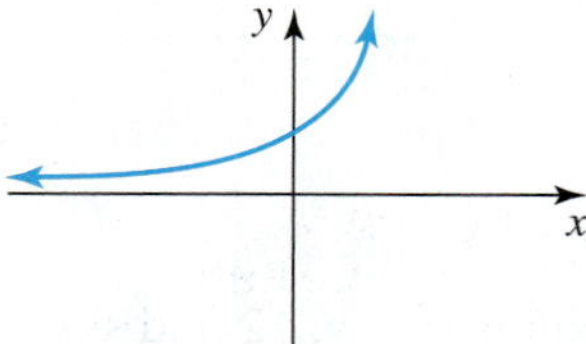

Figure 14 Typical graph of $f(x) = ab^x$, where $a > 0$ and $b > 1$

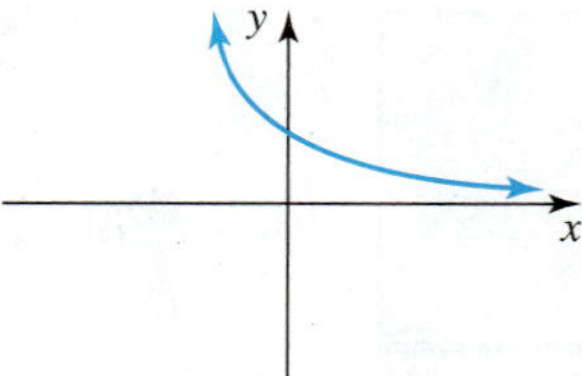

Figure 15 Typical graph of $f(x) = ab^x$, where $a > 0$ and $0 < b < 1$

Intercepts

When we sketch the graph of an exponential function, it is helpful to plot the y-intercept first. Substituting 0 for x in the general equation $y = ab^x$ gives

$$y = ab^0 = a(1) = a$$

So, the y-intercept is $(0, a)$.

y-Intercept of an Exponential Function

For an exponential function of the form

$$y = ab^x,$$

the y-intercept is $(0, a)$.

For the function $y = 5(8)^x$, the y-intercept is $(0, 5)$. For the function $y = 4\left(\frac{1}{7}\right)^x$, the y-intercept is $(0, 4)$.

WARNING

For an exponential function of the form $y = b^x$ (rather than $y = ab^x$), the y-intercept is *not* $(0, b)$. By writing $y = b^x = 1b^x$, we see the y-intercept is $(0, 1)$. For example, for $y = 2^x$, the y-intercept is $(0, 1)$. See Example 1.

Example 3 Intercepts and Graph of an Exponential Function

Let $f(x) = 6\left(\frac{1}{2}\right)^x$.

1. Find the y-intercept of the graph of f.
2. Find the x-intercept of the graph of f.
3. Graph f by hand.

Solution

1. Since $f(x) = 6\left(\frac{1}{2}\right)^x$ is of the form $f(x) = ab^x$, the y-intercept is $(0, a)$, or $(0, 6)$.
2. By the base multiplier property, as the value of x increases by 1, the value of y is multiplied by $\frac{1}{2}$ (see Table 12).

   When we halve a number, it becomes smaller. But no number of halvings will give a result that is zero. So, as x grows large, y will become extremely close to, but never equal, 0. Likewise, the graph of f gets arbitrarily close to, but never reaches, the x-axis (see Fig. 16). In this case, we call the x-axis a **horizontal asymptote.** We conclude that the function f has no x-intercepts.
3. We plot five solutions from Table 12 and sketch a decreasing exponential curve that contains the five points (see Fig. 16). If we had not already found a table of solutions, we could have plotted the y-intercept and plotted additional solutions by increasing the value of x by 1 and going half as high for the value of y each time.

Table 12 Input–Output Pairs of $f(x) = 6\left(\frac{1}{2}\right)^x$

x	$f(x)$
0	6
1	3
2	$\frac{3}{2}$
3	$\frac{3}{4}$
4	$\frac{3}{8}$

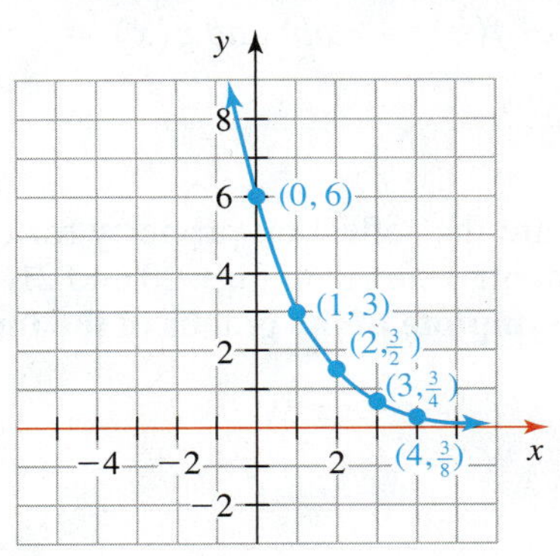

Figure 16 Graph of $f(x) = 6\left(\frac{1}{2}\right)^x$

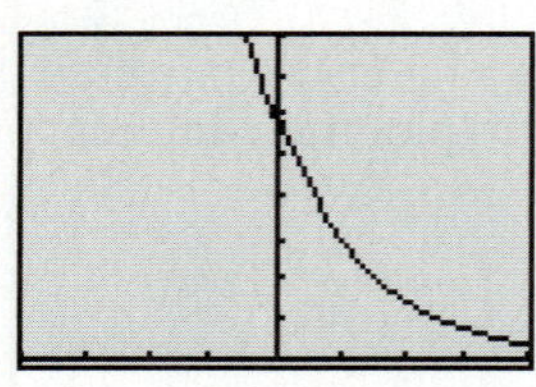

Figure 17 Graph of $y = 6\left(\frac{1}{2}\right)^x$

As a check, we note that according to the increasing or decreasing property, the function f is decreasing, since $a > 0$ and the base, $\frac{1}{2}$, is between 0 and 1. For a more thorough check, we can use a graphing calculator to verify our graph (see Fig. 17).

In Fig. 18, the graphs of both exponential functions get closer and closer to, but never reach, the x-axis. For both functions, the x-axis is a horizontal asymptote.

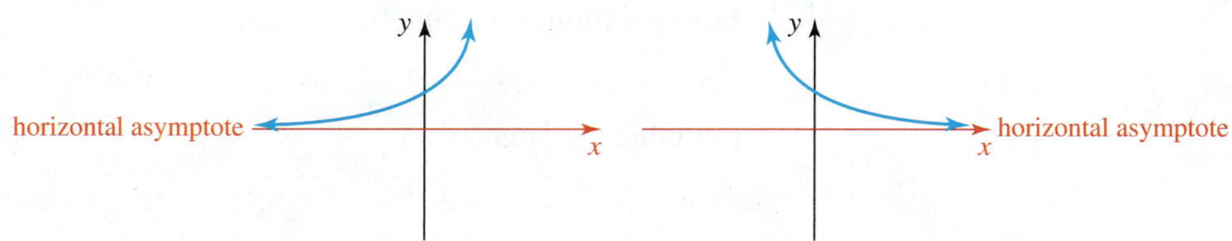

Figure 18 For both exponential functions, the x-axis is a horizontal asymptote

Reflection Property

In Example 4, we will graph two related exponential functions that will help us understand the *reflection property for exponential functions.*

Example 4 Graphs of Functions of the Form $y = ab^x$ and $y = -ab^x$

1. Sketch and compare the graphs of $f(x) = 5(3)^x$ and $g(x) = -5(3)^x$.
2. Find the domain and range of f.
3. Find the domain and range of g.

Table 13 Input–Output Pairs of $f(x) = 5(3)^x$ and $g(x) = -5(3)^x$

x	$f(x)$	$g(x)$
0	5	−5
1	15	−15
2	45	−45
3	135	−135
4	405	−405

Solution

1. Input–output pairs of f and g are listed in Table 13 and plotted in Fig. 19. In Table 13, we see that for each value of x, the outputs of g are the opposites of the outputs of f. Because of this, the graph of g is the reflection of the graph of f, with the mirror along the x-axis. We can find the graph of g by reflecting the graph of f across the x-axis.
2. The expression $5(3)^x$ is defined for any real number x. So, the domain of f is the set of all real numbers. From Fig. 19, we see that the range of f (the set of all outputs of f) is the set of all positive real numbers.
3. The expression $-5(3)^x$ is defined for any real number x. So, the domain of g is the set of all real numbers. From Fig. 19, we see that the range is the set of all negative real numbers.

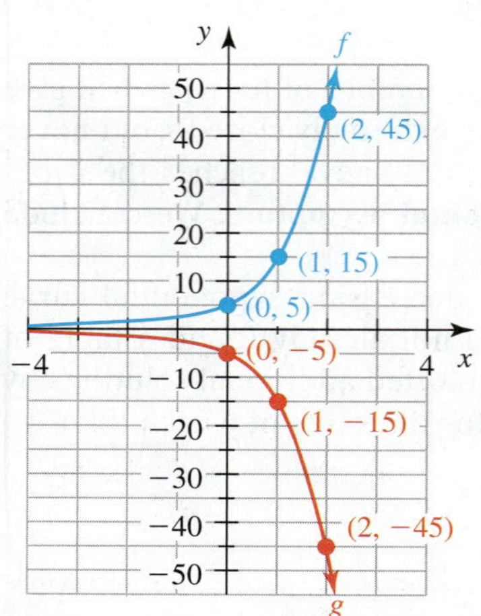

Figure 19 Graphs of $f(x) = 5(3)^x$ and $g(x) = -5(3)^x$

Reflection Property for Exponential Functions

The graphs of $f(x) = -ab^x$ and $g(x) = ab^x$ are reflections of each other across the x-axis.

We illustrate the reflection property for exponential functions and summarize four types of exponential curves in Figs. 20 and 21. **For all exponential functions, the x-axis is a horizontal asymptote of the graphs of the functions.**

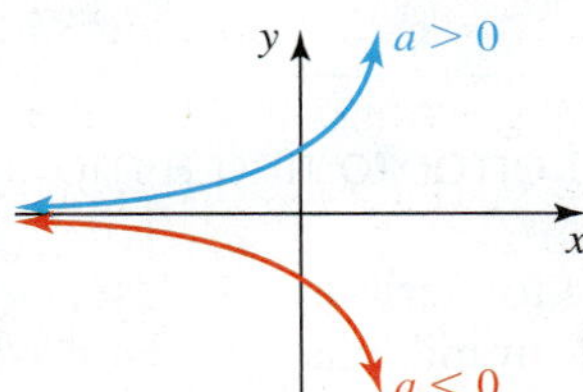

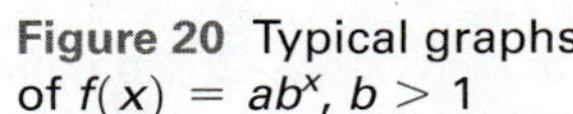

Figure 20 Typical graphs of $f(x) = ab^x$, $b > 1$

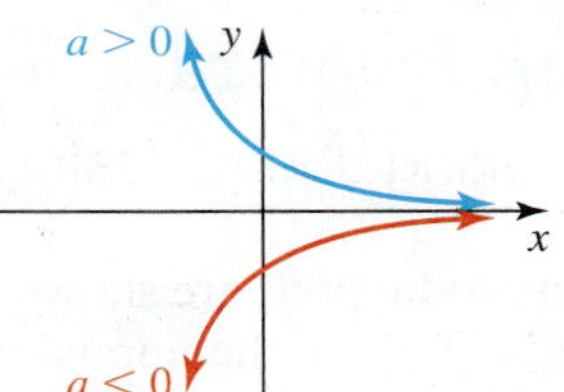

Figure 21 Typical graphs of $f(x) = ab^x$, $0 < b < 1$

Recall that for $b > 0$, b^x has meaning for any real-number exponent x. So, **the domain of any exponential function $f(x) = ab^x$ is the set of real numbers.**

Further, Figs. 20 and 21 show that an exponential function $f(x) = ab^x$ has positive outputs if $a > 0$ and negative outputs if $a < 0$. Therefore, **the range of an exponential function $f(x) = ab^x$ is the set of all positive real numbers if $a > 0$, and the range is the set of all negative real numbers if $a < 0$.**

In Example 5, we will use the graph of an exponential function f to find input or output values of f.

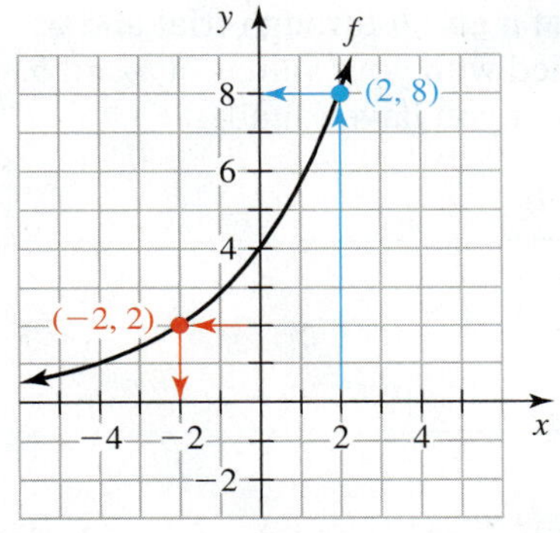

Figure 22 Graph of f

Example 5 Finding Values of a Function from Its Graph

The graph of an exponential function f is shown in Fig. 22.

1. Find $f(2)$.
2. Find x when $f(x) = 2$.
3. Find x when $f(x) = 0$.

Solution

1. The blue arrows in Fig. 22 show that the input $x = 2$ leads to the output $y = 8$. We conclude that $f(2) = 8$.
2. The red arrows in Fig. 22 show that the output $y = 2$ originates from the input $x = -2$. We conclude that $x = -2$ when $f(x) = 2$.
3. Recall that the graph of an exponential function gets close to, but never reaches, the x-axis. So, there is no value of x where $f(x) = 0$.

Group Exploration

Drawing families of exponential curves

For each problem, use a graphing calculator to graph a family of curves.

1. List the equations of a family of exponential curves like the ones shown in Fig. 23.

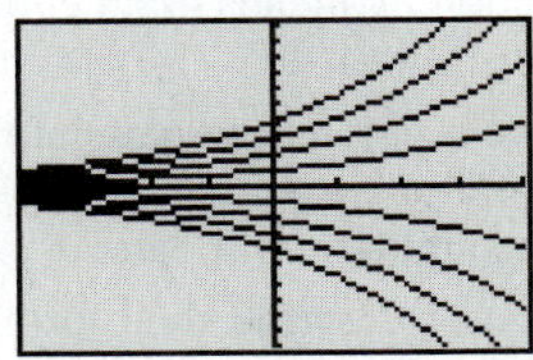

Figure 23 A family of exponential curves

2. List the equations of a family of exponential curves like the ones shown in Fig. 24. All of these curves pass through the point $(0, 2)$.

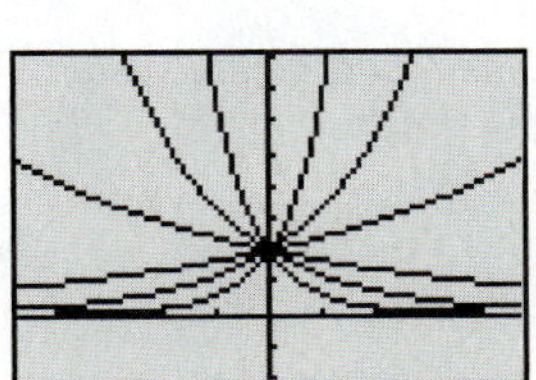

Figure 24 A family of exponential curves passing through $(0, 2)$

3. Summarize what you have learned from this exploration and this section about the coefficient a and the base b in functions of the form $f(x) = ab^x$.

Group Exploration

Looking ahead: Using trial and error to find a model

Ethanol fuel productions are shown in Table 14 for various years. Let $p = f(t)$ be the annual production (in millions of gallons) of ethanol fuel at t years since 2000.

Table 14 Annual Ethanol Fuel Productions

Year	Annual Production (millions of gallons)
2000	1.6
2001	1.8
2002	2.1
2003	2.8
2004	3.4
2005	3.9
2006	4.9
2007	6.5
2008	9.0
2009	10.6
2010	13.2
2011	13.9

Source: *Renewable Fuels Association*

1. Use a graphing calculator to draw a scattergram of the data. Would it be better to model the data with a linear or an exponential function? Explain.
2. Imagine an exponential function $f(t) = ab^t$ whose graph comes close to the data points in your scattergram. What is the p-intercept? What does this tell you about the value of a or b? Explain.
3. Guess a reasonable value of b for your function $f(t) = ab^t$. [**Hint:** The base multiplier property may help.]
4. Substitute your values of a and b from Problems 2 and 3 into the equation $f(t) = ab^t$.
5. Graph f and the scattergram in the same viewing window to see how well your model fits the data.
6. Now find better values of a and b through trial and error. When you are satisfied with your values of a and b, write the equation of f that you have found.

Homework 10.3

For extra help ▶ MyMathLab® 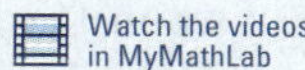 Watch the videos in MyMathLab  Download the MyDashboard App

Graph the given function by hand. Then use a graphing calculator to verify your graph.

1. $y = 3^x$ **2.** $y = 4^x$ **3.** $y = 10^x$

4. $y = 5^x$ **5.** $y = 3(2)^x$ **6.** $y = 2(3)^x$

7. $y = 6(3)^x$ **8.** $y = 3(5)^x$ **9.** $y = 15\left(\frac{1}{3}\right)^x$

10. $y = 20\left(\frac{1}{4}\right)^x$ **11.** $y = 12\left(\frac{1}{2}\right)^x$ **12.** $y = 18\left(\frac{1}{3}\right)^x$

Graph both functions by hand on the same coordinate system. Then use a graphing calculator to verify your graphs.

13. $f(x) = 2^x, g(x) = -2^x$

14. $f(x) = 3^x, g(x) = -3^x$

15. $f(x) = 4(3)^x, g(x) = -4(3)^x$

16. $f(x) = 2(10)^x, g(x) = -2(10)^x$

17. $f(x) = 8\left(\frac{1}{2}\right)^x, g(x) = -8\left(\frac{1}{2}\right)^x$

18. $f(x) = 6\left(\frac{1}{3}\right)^x, g(x) = -6\left(\frac{1}{3}\right)^x$

For Exercises 19–22, graph the function by hand. Then use a graphing calculator to verify your graph. Find the domain and range of the function.

19. $f(x) = 5(2)^x$

20. $f(x) = -3(3)^x$

21. $f(x) = -8\left(\frac{1}{4}\right)^x$

22. $f(x) = 9\left(\frac{1}{3}\right)^x$

23. Recall that we can describe some or all of the input–output pairs of a function by means of an equation, a graph, a table, or words. Let $f(x) = 4(2)^x$.

a. Describe five input–output pairs of f by using a table.
b. Describe the input–output pairs of f by using a graph.
c. Describe the input–output pairs of f by using words.

24. Recall that we can describe some or all of the input–output pairs of a function by means of an equation, a graph, a table, or words. Let $g(x) = 16\left(\frac{1}{2}\right)^x$.

a. Describe five input–output pairs of g by using a table.
b. Describe the input–output pairs of g by using a graph.
c. Describe the input–output pairs of g by using words.

25. Input–output pairs of four exponential functions are listed in Table 15. Complete the table.

Table 15 Complete the Table (Exercise 25)

x	$f(x)$	$g(x)$	$h(x)$	$k(x)$
0	162	3	2	800
1	54	12	10	400
2	18	48		
3	6			
4				

26. Input–output pairs of four exponential functions are listed in Table 16. Complete the table.

Table 16 Complete the Table (Exercise 26)

x	$f(x)$	$g(x)$	$h(x)$	$k(x)$
0	3	64	2	100
1	6	32	6	10
2	12	16		
3	24			
4				

27. Input–output pairs of four exponential functions are listed in Table 17. Complete the table.

Table 17 Complete the Table (Exercise 27)

x	$f(x)$	$g(x)$	$h(x)$	$k(x)$
0	5			
1		80	54	
2	20			
3		20		192
4			2	768

28. Input–output pairs of four exponential functions are listed in Table 18. Complete the table.

Table 18 Complete the Table (Exercise 28)

x	$f(x)$	$g(x)$	$h(x)$	$k(x)$
0			3	400
1		3		
2	25			
3		147		
4	1		30,000	25

For Exercises 29–36, refer to Fig. 25.

29. Find $f(-3)$.

30. Find $f(-1)$.

31. Find $f(0)$.

32. Find $f(1)$.

33. Find x when $f(x) = 4$.

34. Find x when $f(x) = 2$.

35. Find x when $f(x) = 1$.

36. Find x when $f(x) = -2$.

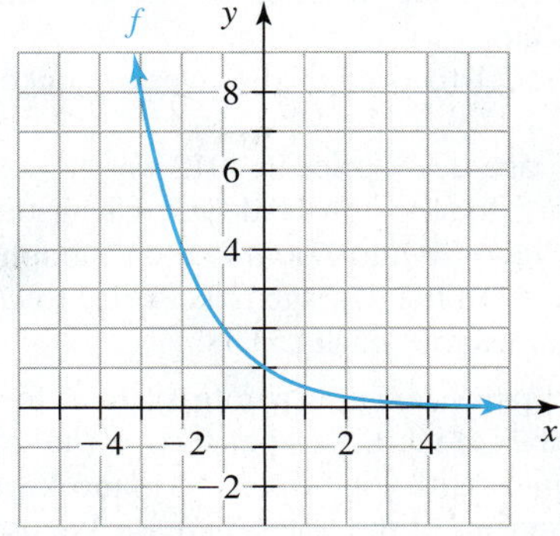

Figure 25 Graph of f—Exercises 29–36

For Exercises 37–44, refer to Table 19.

37. Find $f(3)$.

38. Find $f(6)$.

39. Find $f(5)$.

40. Find $f(0)$.

41. Find x when $f(x) = 3$.

42. Find x when $f(x) = 6$.

43. Find x when $f(x) = 24$.

44. Find x when $f(x) = 96$.

Table 19 Some Values of an Exponential Function f (Exercises 37–44)

x	$f(x)$
0	3
1	6
2	12
3	24
4	48
5	96
6	192

45. The average ticket prices to major league baseball games are shown in Table 20 for various years.

Table 20 Average Ticket Prices to Major League Baseball Games

Year	Average Ticket Price (dollars)
1950	1.54
1960	1.96
1970	2.72
1980	4.45
1991	8.84
2000	16.22
2011	26.91

Sources: *The Sporting News and the Sporting News Baseball Dope Book, 1950–85; Team Marketing Report, 1991–2004; ESPN*

Let $f(t)$ be the average ticket price (in dollars) to major league baseball games for the year that is t years since 1950. The situation can be described by the linear function $p = 0.39t - 2.88$, the quadratic function $p = 0.01t^2 - 0.24t + 2.4$, and the exponential function $p = 1.22(1.051)^t$.

a. Use a graphing calculator to draw the graphs of the three models and, in the same viewing window, a scattergram of the data. Compare how well the models fit the data.

b. Which model likely describes the situation the best for the years before 1950? Explain.

c. Use the exponential model to predict the average ticket price in 2018.

d. The most expensive average ticket price in 2012 was $53.38, at Fenway Park, home of the Boston Red Sox. Using the exponential model, use TRACE and Zoom Out on a graphing calculator to predict when the average ticket price to *all* major league baseball games will reach $53.38.

46. If you place your hand on a piano and play a note, you will feel the piano vibrate. The number of vibrations per second (hertz) of a note is called its *frequency.* If you strike the piano keys from left to right, the frequencies of the notes increase. We use some of the letters of the alphabet, sometimes in conjunction with the "sharp" symbol #, to refer to these notes (see Fig. 26). The frequencies of 13 notes in a row are listed in Table 21.

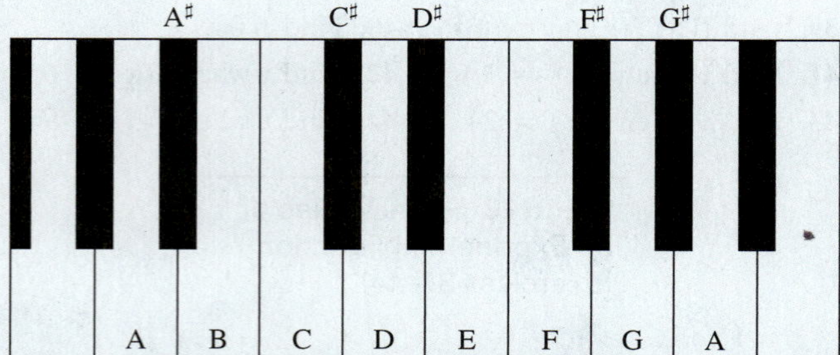

Figure 26 Notes on a piano

Table 21 Frequencies of Notes on a Piano

Note	Number of Notes above A	Frequency (in hertz)
A	0	220.0
A#	1	233.1
B	2	246.9
C	3	261.6
C#	4	277.2
D	5	293.7
D#	6	311.1
E	7	329.6
F	8	349.2
F#	9	370.0
G	10	392.0
G#	11	415.3
A	12	440.0

Source: Math and Music *by Garland and Kahn*

a. Let $f(n)$ be the frequency (in hertz) of the note that is n notes above the note A (the one with frequency 220.0 hertz). Draw the graph of the function $f(n) = 220(2)^{n/12}$ and a scattergram of the data listed in Table 21 in the same viewing window. Does the graph of f come close to the data points? [***Graphing Calculator:*** For the exponential expression $220(2)^{n/12}$, press **220** [(] **2** [)] [^] [(] [X, T, Θ, n] [÷] **12** [)].]

b. Estimate the frequency of the note D that is 17 notes above the note A (the one with frequency 220.0 hertz).

c. Use TRACE to find which note has a frequency of 523.25 hertz.

d. Use a graphing calculator table to find $f(0)$, $f(12)$, $f(24)$, $f(36)$, and $f(48)$. What pattern do you notice? Describe this pattern in terms of the situation.

Find the x- and y-intercepts of the graph of the function.

47. $y = 7^x$

48. $y = 2(5)^x$

49. $y = 3\left(\frac{1}{5}\right)^x$

50. $y = -9\left(\frac{2}{3}\right)^x$

For Exercises 51–54, let $f(x) = 2^x + 3^x$.

51. Find $f(2)$.

52. Find $f(0)$.

53. Find $f(-2)$.

54. Find $f(-1)$.

For Exercises 55–58, let $f(x) = 3^x$.

55. Find x when $f(x) = 3$.

56. Find x when $f(x) = 9$.

57. Find x when $f(x) = 1$.

58. Find x when $f(x) = \frac{1}{3}$.

*For Exercises 59–70, use graphing calculator tables to compare each pair of functions f and g. What do you observe? Use exponential properties to show why this is so. [**Graphing Calculator:** For 2^{3x}, press 2 [^] [(] 3 [X, T, Θ, n] [)]. Recall that if an exponent involves an operation, you must use parentheses.]*

59. $f(x) = 3^x 3^x, g(x) = 3^{2x}$

60. $f(x) = \frac{3^{2x}}{3^x}, g(x) = 3^x$

61. $f(x) = 2^{3x}, g(x) = 8^x$

62. $f(x) = 2^{-x}, g(x) = \left(\frac{1}{2}\right)^x$

63. $f(x) = 2^{x+3}, g(x) = 8(2)^x$

64. $f(x) = \frac{6^x}{3^x}, g(x) = 2^x$

65. $f(x) = 2^0,\ g(x) = 3^0$

66. $f(x) = 2^x 3^x, g(x) = 6^x$

67. $f(x) = 5^{x/3}, g(x) = (5^{1/3})^x$

68. $f(x) = 2^x, g(x) = 8^{x/3}$

69. $f(x) = x^{1/2}, g(x) = \sqrt{x}$ [***Graphing Calculator:*** For $\sqrt{x}$, press [2nd] [x^2] [X, T, Θ, n] [)].]

70. $f(x) = 25^{x/2} \cdot 5^x, g(x) = 25^x$

Concepts

71. Graphs of four functions of the form $y = ab^x$ are shown in Fig. 27. Describe the constants a and b of each function.

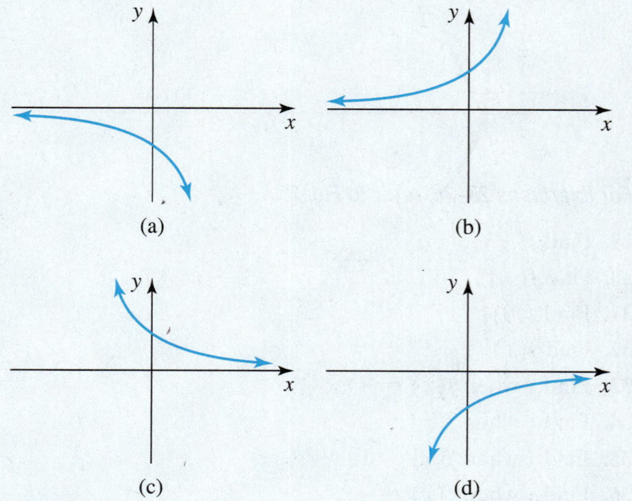

Figure 27 Exercise 71

72. The graphs of functions $f(x) = ab^x$ and $g(x) = cd^x$ are shown in Fig. 28.

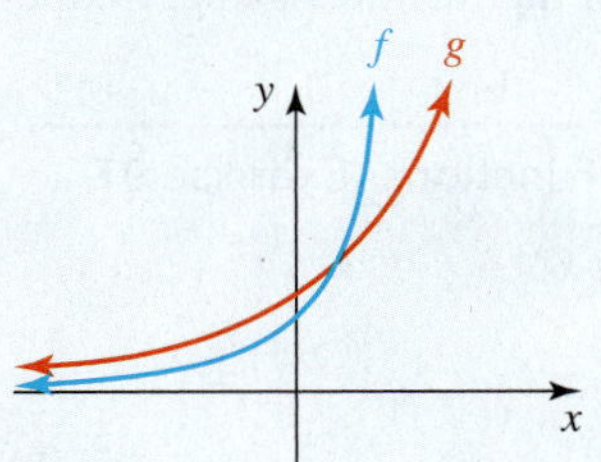

Figure 28 Graphs of $f(x) = ab^x$ and $g(x) = cd^x$

a. Which coefficient is greater, a or c? Explain.
b. Which base is greater, b or d? Explain.

73. Use a graphing calculator to graph a family of exponential curves similar to the family graphed in Fig. 29. List the equations of that family.

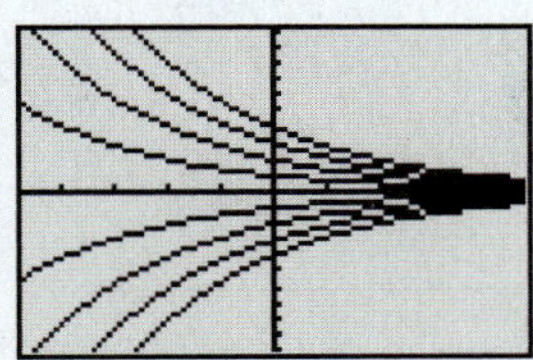

Figure 29 A family of exponential curves

74. Use a graphing calculator to graph a family of exponential curves similar to the family graphed in Fig. 30. All of these curves pass through the point $(0, -2)$. List the equations of that family.

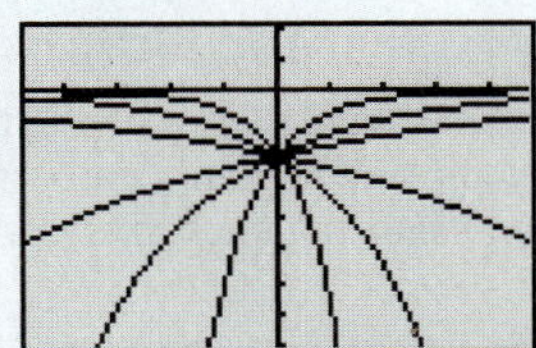

Figure 30 A family of exponential curves passing through $(0, -2)$

75. Use a graphing calculator to draw a graph similar to the one in Fig. 31. Use an equation of the form $f(x) = ab^x$, where a and b are constants you specify. What equation works?

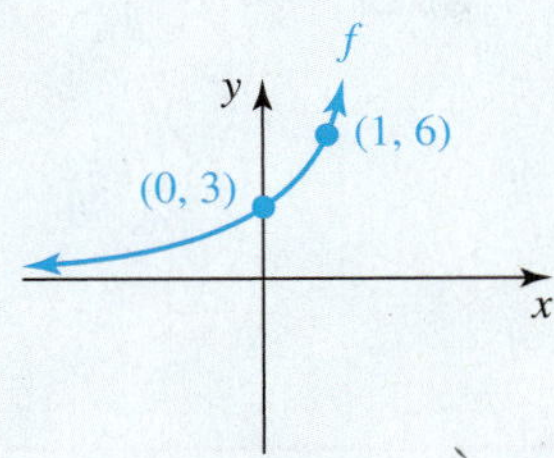

Figure 31 Exercise 75

76. Use a graphing calculator to draw a graph similar to the one in Fig. 32. Use an equation of the form $g(x) = ab^x$, where a and b are constants you specify. What equation works? Use trial and error.

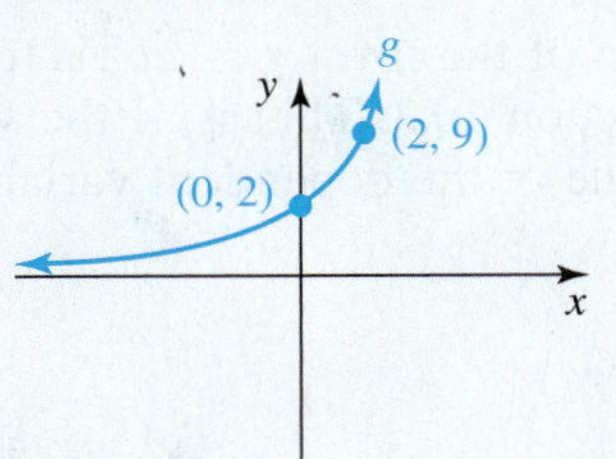

Figure 32 Exercise 76

77. Find equations of exponential functions that could correspond to the graphs shown in Fig. 33.

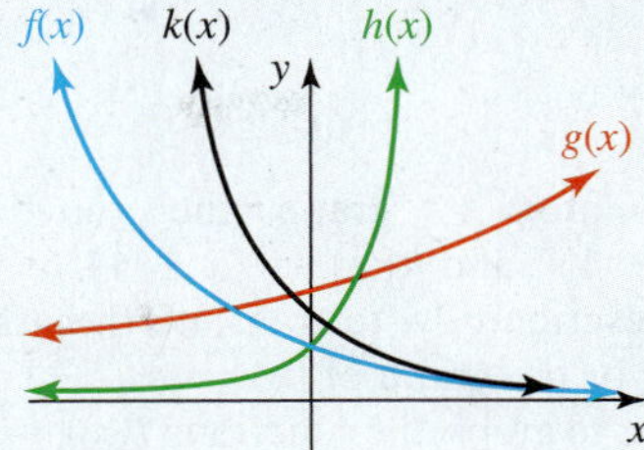

Figure 33 Exercise 77

78. a. For each part that follows, simplify and compare both expressions.
i. $4(3)^2$ and 12^2 **ii.** $4^2 \cdot 3^2$ and 12^2

b. For each part that follows, build a graphing calculator table that shows the same input values for both functions. Explain in terms of order of operations or exponential properties why the tables are the same or different.
i. $f(x) = 4(3)^x$ and $g(x) = 12^x$
ii. $f(x) = 4^x \cdot 3^x$ and $g(x) = 12^x$

79. In this exercise, you will compare the function $f(x) = 100(2)^x$ with the function $g(x) = 5(3)^x$.
a. Find the y-intercept of the graph of each function.
b. What does the base multiplier property tell you about each function?
c. On the basis of your comments in parts (a) and (b), which function's outputs will eventually be much greater than the other's outputs? Explain.
d. Use a graphing calculator table to verify your comments to parts (a)–(c). To do this, enter the functions and set up a table as indicated in Figs. 34 and 35, respectively.

Figure 34 Enter the functions

Figure 35 Set up the table

80. What are the x-intercepts and y-intercepts of the graph of a function of the form $y = ab^x$, where $b > 0$?

81. Is the statement true for $f(x) = 2^x$?
a. $f(3 + 4) = f(3) + f(4)$ **b.** $f(x + y) = f(x) + f(y)$

82. Is the statement true for $g(x) = 3^x$?
a. $g(2 + 5) = g(2) \cdot g(5)$ **b.** $g(4 + 6) = g(4) \cdot g(6)$
c. $g(2 + 4) = g(2) \cdot g(4)$ **d.** $g(x + y) = g(x) \cdot g(y)$

83. Let $f(x) = ab^x$, where $a > 0$. Explain why f is increasing if $b > 1$ and f is decreasing if $0 < b < 1$.

84. In an exponential function $f(x) = b^x$, the base b is a positive number not equal to 1. In this exercise, you will explore what happens if we try to define a function whose base is negative. Consider $f(x) = (-4)^x$.
a. Explain why $f\left(\frac{1}{2}\right)$ is undefined.
b. Explain why $f\left(\frac{1}{4}\right)$ is undefined.
c. List three more values of x that result in undefined outputs.

85. The graphs of the exponential functions $f(x) = -ab^x$ and $g(x) = ab^x$ are reflections of each other across the x-axis. Explain why this makes sense.

86. Explain how to sketch the graph of a function of the form $f(x) = ab^x$, where $b > 0$. Include the effect of a value of a or b on the graph.

Related Review

87. a. Use a graphing calculator to graph the functions $f(x) = x^3$, $g(x) = (x - 4)^3$, and $h(x) = (x + 4)^3$ in the same viewing screen. Describe how to translate the graph of f to get the graph of g or the graph of h.

b. Use a graphing calculator to graph the functions $f(x) = x^3$, $g(x) = x^3 - 4$, and $h(x) = x^3 + 4$ in the same viewing screen. Describe how to translate the graph of f to get the graph of g or the graph of h.

c. Do the types of translation of $f(x) = x^3$ you performed in parts (a) and (b) match the types of translation we performed with $k(x) = x^2$ in Section 9.1? Explain.

d. Graph $f(x) = 2^x$ *by hand*. Then translate this graph to graph the following functions.

i. $g(x) = 2^{x+3}$ **ii.** $h(x) = 2^{x-4}$

iii. $k(x) = 2^x + 2$ **iv.** $p(x) = 2^x - 1$

88. a. Use a graphing calculator to graph the functions $f(x) = 2^x$ and $g(x) = -2^x$ in the same viewing screen. Describe how to reflect the graph of f to get the graph of g.

b. Use a graphing calculator to graph the functions $f(x) = x^4$ and $g(x) = -x^4$ in the same viewing screen. Describe how to reflect the graph of f to get the graph of g.

c. Do the types of reflections you performed in parts (a) and (b) match the type of reflection we performed with $k(x) = x^2$ in Section 9.1? Explain.

d. Graph $f(x) = x^3$ *by hand*. Then graph $g(x) = -x^3$ by hand.

Graph $y = 3^x$ by hand. Then use translations and reflections, as appropriate, to graph the given equation. Use a graphing calculator to verify your graph.

89. $f(x) = 3^{x+2} + 1$ **90.** $f(x) = 3^{x-3} - 2$

91. $f(x) = -3^x + 4$ **92.** $f(x) = -3^{x+3}$

For Exercises 93–96, find all x-intercepts and y-intercepts.

93. $y = 8 + 4x$ **94.** $y = 8(4)^x$

95. $y = 8\left(\frac{1}{4}\right)^x$ **96.** $y = 8 - 4x^2$

97. Some input–output pairs of the functions f, g, h, and k are provided in Table 22. For each function, determine whether the given values suggest that the function is linear, exponential, or neither.

Table 22 Identifying Functions (Exercise 97)

x	$f(x)$	$g(x)$	$h(x)$	$k(x)$
0	13	4	48	5
1	9	12	24	55
2	5	36	12	555
3	1	108	6	5555
4	−3	324	3	55555

98. The graphs of the equation $y = \frac{1}{3}x + 2$ and an equation of the form $y = ab^x$ are shown in Fig. 36. Find the values of a and b.

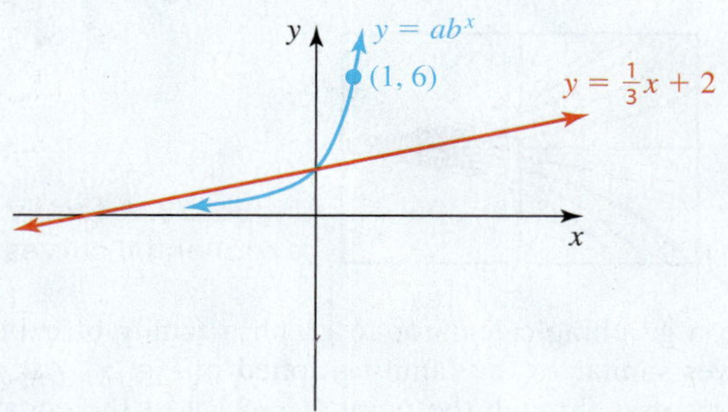

Figure 36 Exercise 98

Expressions, Equations, Functions, and Graphs

Perform the indicated instruction. Then use words such as linear, quadratic, cubic, polynomial, degree, exponential, function, one variable, *and* two variables *to describe the expression, equation, or system.*

99. Graph $f(x) = 6\left(\frac{1}{2}\right)^x$ by hand.

100. Graph $y = -3x + 4$ by hand.

101. Find $f(-2)$, where $f(x) = 6\left(\frac{1}{2}\right)^x$.

102. Find the product $(3x^2 - x + 2)(2x^2 - 4x - 1)$.

103. Simplify $\frac{(8b^{-6}c)^{1/3}}{(16b^{12}c^{-2})^{-1/4}}$.

104. Factor $12x^2 - 23x + 10$.

10.4 Finding Equations of Exponential Functions

Objectives

» Use the base multiplier property to find an exponential equation.

» Solve an equation of the form $ab^n = k$ for the base b.

» Use two points to find an exponential equation.

In this section, we will discuss two ways to find an equation of an exponential function.

Using the Base Multiplier Property to Find Exponential Functions

One way to find an equation of an exponential function of the form $y = ab^x$ is to use the base multiplier property, which we discussed in Section 10.3. That is, if the value of the independent variable increases by 1, then the value of the dependent variable is multiplied by the base b.

Example 1 Finding an Equation of an Exponential Curve

An exponential curve contains the points listed in Table 23. Find an equation of the curve.

Table 23 Solutions of an Exponential Equation

x	$f(x)$
0	3
1	6
2	12
3	24
4	48

Solution

For $f(x) = ab^x$, recall from Section 10.3 that the y-intercept is $(0, a)$. From Table 23, we see that the y-intercept is $(0, 3)$, so $a = 3$. As the value of x increases by 1, the value of y is multiplied by 2. By the base multiplier property, $b = 2$. Therefore, an equation of the curve is

$$f(x) = 3(2)^x$$

We check our result with a graphing calculator table (see Fig. 37).

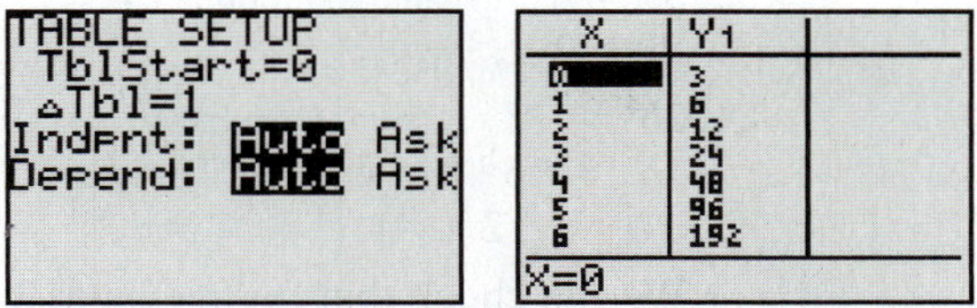

Figure 37 Verify the exponential equation $f(x) = 3(2)^x$

For Example 2, it will be helpful to review the slope addition property from Section 3.5: For a linear function of the form $y = mx + b$, if the value of the independent variable increases by 1, then the value of the dependent variable changes by the slope m.

Example 2 Linear versus Exponential Functions

Table 24 Input–Output Pairs for f

x	$f(x)$
0	162
1	54
2	18
3	6
4	2

Table 25 Input–Output Pairs for g

x	$g(x)$
0	50
1	46
2	42
3	38
4	34

1. Find a possible equation of a function whose input–output pairs are listed in Table 24.
2. Find a possible equation of a function whose input–output pairs are listed in Table 25.

Solution

1. As the value of x increases by 1 throughout Table 24, the value of y is multiplied by $\frac{1}{3}$. This relationship suggests there is an exponential function $f(x) = a\left(\frac{1}{3}\right)^x$ whose graph contains the points in Table 24. Since the y-intercept is $(0, 162)$, we have $f(x) = 162\left(\frac{1}{3}\right)^x$.
2. As the value of x increases by 1 throughout Table 25, the value of y changes by adding -4. This relationship suggests there is a linear function $g(x) = -4x + b$ whose graph contains the points in Table 25. Since the y-intercept is (0, 50), we have $g(x) = -4x + 50$.

Solving Equations of the Form $ab^n = k$ for b

So far, we have discussed how to find an equation of an exponential curve that contains points whose x-coordinates are *consecutive* integers. Later in this section, we will discuss how to find an equation of an exponential curve that contains two given points, such as (2, 5) and (5, 63), whose x-coordinates are *not* consecutive integers. To use this method, we first need to discuss how to solve equations of the form $ab^n = k$ for the base b.

Example 3 One-Variable Equations Involving Exponents

Find all real-number solutions.

1. $b^2 = 25$
2. $b^3 = 8$
3. $2b^4 = 32$
4. $10b^5 = 90$
5. $b^6 = -28$

Solution

1. $b^2 = 25$ *Original equation*

$b = -5$ or $b = 5$ $(-5)^2 = 25$ *and* $5^2 = 25$

So, the solutions are -5 and 5. We can use the notation ± 5 to stand for the numbers -5 and 5.

2. $b^3 = 8$ *Original equation*

$b = 2$ $2^3 = 8$

3. $2b^4 = 32$ *Original equation*

$b^4 = 16$ *Divide both sides by 2.*

$b = \pm 2$ $(-2)^4 = 16$ *and* $2^4 = 16$

We can check that both -2 and 2 satisfy the equation $2b^4 = 32$.

4. $10b^5 = 90$ *Original equation*

$b^5 = 9$ *Divide both sides by 10.*

$b = 9^{1/5}$ $9^{1/5}$ *is the number whose 5th power is 9.*

$b \approx 1.55$ $1.55^5 \approx 9$

```
10(1.55)^5
         89.46609688
```

Figure 38 Checking that 1.55 approximately satisfies $10b^5 = 90$

We can check that 1.55 approximately satisfies the equation $10b^5 = 90$ by using a graphing calculator to verify that $10(1.55)^5 \approx 90$ (see Fig. 38).

5. The equation $b^6 = -28$ has no real-number solutions, since an even-numbered exponent gives a positive number.

The problems in Example 3 suggest how to solve equations of the form $b^n = k$ for b.

Solving Equations of the Form $b^n = k$ for b

To solve an equation of the form $b^n = k$ for b,

1. If n is odd, the real-number solution is $k^{1/n}$.
2. If n is even and $k \geq 0$, the real-number solutions are $\pm k^{1/n}$.
3. If n is even and $k < 0$, there is no real-number solution.

Example 4 One-Variable Equations Involving Exponents

Find all real-number solutions. Round any results to the second decimal place.

1. $5.42b^6 - 3.19 = 43.74$
2. $\dfrac{b^9}{b^4} = \dfrac{70}{3}$

Solution

1. $5.42b^6 - 3.19 = 43.74$ *Original equation*

$5.42b^6 = 43.74 + 3.19$ *Add 3.19 to both sides.*

$5.42b^6 = 46.93$ *Add.*

$b^6 = \dfrac{46.93}{5.42}$ *Divide both sides by 5.42.*

$b = \pm\left(\dfrac{46.93}{5.42}\right)^{1/6}$ *The solutions of* $b^6 = k$ *are* $\pm k^{1/6}$ *if* $k \geq 0$.

$b \approx \pm 1.43$ *Compute.*

2. $\dfrac{b^9}{b^4} = \dfrac{70}{3}$ *Original equation*

$b^5 = \dfrac{70}{3}$ *Subtract exponents:* $\dfrac{b^m}{b^n} = b^{m-n}$

$b = \left(\dfrac{70}{3}\right)^{1/5}$ *The solution of* $b^5 = k$ *is* $k^{1/5}$.

$b \approx 1.88$ *Compute.*

Example 5 Solving General Equations Involving Exponents

Solve the equations for b, where n and $m - n$ are odd and $a, d + c$, and $d - c$ are nonzero.

1. $ab^n + c = d$ **2.** $\dfrac{b^m}{b^n} - c = d$

Solution

1. $ab^n + c = d$ *Original equation*

$ab^n + c - c = d - c$ *Subtract c from both sides.*

$ab^n = d - c$ *Combine like terms.*

$\dfrac{ab^n}{a} = \dfrac{d - c}{a}$ *Divide both sides by a.*

$b^n = \dfrac{d - c}{a}$ *Simplify.*

$b = \left(\dfrac{d - c}{a}\right)^{1/n}$ *The solution of* $b^n = p$ *is* $p^{1/n}$ *when n is odd.*

2. $\dfrac{b^m}{b^n} - c = d$ *Original equation*

$\dfrac{b^m}{b^n} - c + c = d + c$ *Add c to both sides.*

$\dfrac{b^m}{b^n} = d + c$ *Combine like terms.*

$b^{m-n} = d + c$ *Subtract exponents:* $\dfrac{b^m}{b^n} = b^{m-n}$.

$b = (d + c)^{\frac{1}{m-n}}$ *The solution of* $b^k = p$ *is* $p^{1/k}$ *when k is odd.*

Using Two Points to Find Equations of Exponential Functions

Now that we have discussed how to solve equations of the form $ab^n = k$ for b, we can discuss a second way to find an equation of an exponential function.

Example 6 Finding an Equation of an Exponential Curve

Find an approximate equation $y = ab^x$ of the exponential curve that contains the points $(0, 3)$ and $(4, 70)$. Round the value of b to two decimal places.

Solution

Since the y-intercept is $(0, 3)$, the equation has the form $y = 3b^x$. Next, we substitute $(4, 70)$ in the equation $y = 3b^x$ and solve for b:

$$70 = 3b^4 \quad \text{Substitute 4 for } x \text{ and 70 for } y.$$
$$3b^4 = 70 \quad \text{If } c = d, \text{ then } d = c.$$
$$b^4 = \frac{70}{3} \quad \text{Divide both sides by 3.}$$
$$b = \pm\left(\frac{70}{3}\right)^{1/4} \quad \text{The solutions of } b^4 = k \text{ are } \pm k^{1/4} \text{ if } k \geq 0.$$
$$b \approx 2.20 \quad \text{Compute; base of an exponential function is positive.}$$

So, our equation is $y = 3(2.20)^x$; its graph contains the given point (0, 3). Since we rounded the value b, the graph of the equation comes close to, but does not pass through, the given point (4, 70).

We use a graphing calculator to verify our work (see Fig. 39).

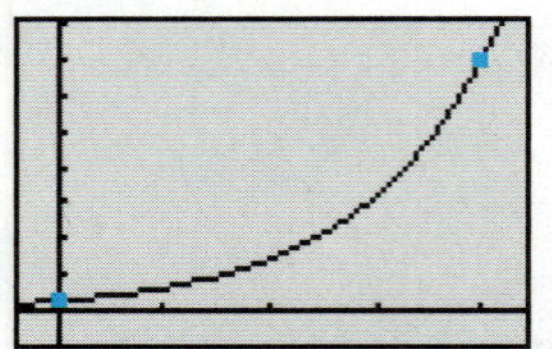

Figure 39 Verify the work

In Example 7, we will find an equation of a curve that approximates the exponential curve containing two given points. Neither point will be the y-intercept. To do this, we will use the following property.

Dividing Left Sides and Right Sides of Two Equations

If $a = b$, $c = d$, $c \neq 0$, and $d \neq 0$, then

$$\frac{a}{c} = \frac{b}{d}$$

In words, the quotient of the left sides of two equations is equal to the quotient of the right sides.

For example, if we divide the left sides and divide the right sides of the equations $2 = 2$ and $3 = 3$, we obtain the true statement $\frac{2}{3} = \frac{2}{3}$.

Example 7 Finding an Equation of an Exponential Curve

Find an approximate equation $y = ab^x$ of the exponential curve that contains (2, 5) and (5, 63). Round the values of a and b to two decimal places.

Solution

Since both of the ordered pairs (2, 5) and (5, 63) must satisfy the equation $y = ab^x$, we have the following system of equations:

$$5 = ab^2 \quad \text{Substitute 2 for } x \text{ and 5 for } y.$$
$$63 = ab^5 \quad \text{Substitute 5 for } x \text{ and 63 for } y.$$

It will be slightly easier to solve this system if we switch the equations to list the equation with the greater exponent of b first:

$$63 = ab^5$$
$$5 = ab^2$$

We divide the left sides and divide the right sides of the two equations to get the following result for nonzero a and b:

$$\frac{63}{5} = \frac{ab^5}{ab^2}$$

By then applying the properties $\frac{b^m}{b^n} = b^{m-n}$ and $\frac{a}{a} = 1$, where a and b are nonzero, to the right-hand side of the equation, we have an equation in terms of b (and not a):

$$\frac{63}{5} = b^3$$

We can now solve for b by finding the cube root of $\frac{63}{5}$:

$$\begin{aligned} b^3 &= \frac{63}{5} && \textit{If } c = d, \textit{ then } d = c. \\ b &= \left(\frac{63}{5}\right)^{1/3} && \textit{The solution of } b^3 = k \textit{ is } k^{1/3}. \\ &\approx 2.33 && \textit{Compute.} \end{aligned}$$

So, we can substitute 2.33 for the constant b in the equation $y = ab^x$:

$$y \approx a(2.33)^x$$

To find a, we substitute the coordinates of the given point (2, 5) into $y = a(2.33)^x$:

$$\begin{aligned} 5 &= a(2.33)^2 && \textit{Substitute 2 for x and 5 for y.} \\ \frac{5}{2.33^2} &= a && \textit{Divide both sides by } 2.33^2. \\ a &\approx 0.92 && \textit{Compute.} \end{aligned}$$

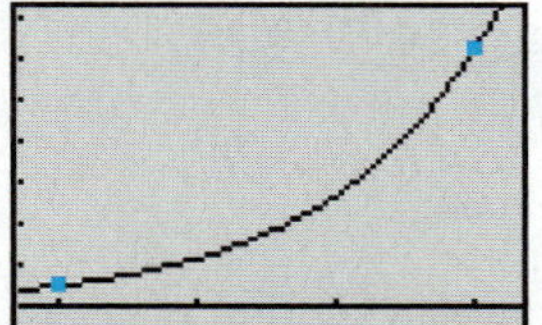

Figure 40 Check that the curve approximately contains (2, 5) and (5, 63)

So, an equation that approximates the exponential curve that passes through (2, 5) and (5, 63) is $y = 0.92(2.33)^x$.

We use a graphing calculator to verify our work (see Fig. 40).

In summary, **we can find an equation of an exponential function by using the base multiplier property or by using two points. Both methods give the same result.**

Group Exploration

Comparing three ways to find exponential equations

An exponential curve contains the points listed in Table 26.

Table 26 Solutions of an Exponential Equation

x	$f(x)$
0	5
2	20
4	80
6	320
8	1280

1. Use the point (0, 5) and one other point in Table 26 to find an equation of the curve (see Example 6).
2. Use two points in Table 26 other than (0, 5) to find an equation of the curve (see Example 7).
3. Use the base multiplier property to find an equation of the curve. [**Hint:** First find $f(1)$ by recognizing a pattern.]
4. Compare your equations from Problems 1, 2, and 3.
5. An exponential curve contains the points listed in Table 27. Which method would you use to find an equation $y = ab^x$ that approximates the exponential curve? Explain. Also, find the equation. Round the value of b to two decimal places.

Table 27 Solutions of an Exponential Equation

x	$f(x)$
0	400
3	200
6	100
9	50
12	25

Homework 10.4

For extra help ▶ MyMathLab® 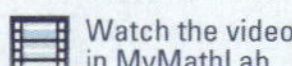 Watch the videos in MyMathLab  Download the MyDashboard App

1. Some values of functions f, g, h, and k are provided in Table 28. Find a possible equation of each function. Verify your results with a graphing calculator table.

Table 28 Values of Four Functions (Exercise 1)

x	$f(x)$	$g(x)$	$h(x)$	$k(x)$
0	4	36	5	250
1	8	12	50	50
2	16	4	500	10
3	32	$\frac{4}{3}$	5000	2
4	64	$\frac{4}{9}$	50,000	$\frac{2}{5}$

2. Some values of functions f, g, h, and k are provided in Table 29. Find a possible equation of each function. Verify your results with a graphing calculator table.

Table 29 Values of Four Equations (Exercise 2)

x	$f(x)$	$g(x)$	$h(x)$	$k(x)$
0	80	4	3	3700
1	40	12	15	370
2	20	36	75	37
3	10	108	375	3.7
4	5	324	1875	0.37

3. Some values of functions f, g, h, and k are provided in Table 30. Find a possible equation of each function. Verify your results with a graphing calculator table. [**Hint:** Use linear or exponential equations.]

Table 30 Values of Four Functions (Exercise 3)

x	$f(x)$	$g(x)$	$h(x)$	$k(x)$
0	100	100	2	2
1	50	50	6	6
2	25	0	10	18
3	12.5	−50	14	54
4	6.25	−100	18	162

4. Some values of functions f, g, h, and k are provided in Table 31. Find a possible equation of each function. Verify your results with a graphing calculator table. [**Hint:** Use linear or exponential equations.]

Table 31 Values of Four Functions (Exercise 4)

x	$f(x)$	$g(x)$	$h(x)$	$k(x)$
0	3	19	2	512
1	12	13	9	128
2	48	7	16	32
3	192	1	23	8
4	768	−5	30	2

Find all real-number solutions. Round your result(s) to the second decimal place. Verify that your results satisfy the equation.

5. $b^2 = 16$

6. $b^4 = 81$

7. $b^3 = 27$

8. $b^5 = 100{,}000$

9. $3b^5 = 96$

10. $5b^2 = 45$

11. $35b^4 = 15$

12. $44b^3 = 12$

13. $3.6b^3 = 42.5$

14. $1.7b^4 = 86.4$

15. $32.7b^6 + 8.1 = 392.8$

16. $2.1b^5 - 8.2 = 237.5$

17. $\frac{1}{4}b^3 - \frac{1}{2} = \frac{9}{4}$

18. $\frac{1}{6}b^4 + \frac{5}{3} = \frac{11}{2}$

19. $\frac{b^6}{b^2} = 81$

20. $\frac{b^{10}}{b^3} = 2187$

21. $\frac{b^8}{b^3} = \frac{79}{5}$

22. $\frac{b^9}{b^6} = \frac{2}{9}$

Solve the equation for b, where n and m − n are odd and a, d, d + c, and d − c are nonzero.

23. $b^n + c = d$

24. $b^n - c = d$

25. $ab^n - c = d$

26. $\frac{b^n}{a} + c = d$

27. $\frac{b^m}{b^n} = d$

28. $\frac{ab^m}{b^n} = d$

29. $\frac{ab^m}{b^n} + c = d$

30. $\frac{b^m}{ab^n} - c = d$

Find an approximate equation $y = ab^x$ of the exponential curve that contains the given pair of points. Round the value of b to two decimal places. Verify your result with a graphing calculator.

31. $(0, 4)$ and $(1, 8)$

32. $(0, 5)$ and $(1, 15)$

33. $(0, 3)$ and $(5, 100)$

34. $(0, 8)$ and $(4, 79)$

35. $(0, 87)$ and $(6, 14)$

36. $(0, 256)$ and $(7, 23)$

37. $(0, 5.5)$ and $(2, 73.9)$

38. $(0, 2.1)$ and $(5, 9.7)$

39. $(0, 7.4)$ and $(3, 1.3)$

40. $(0, 97.2)$ and $(4, 17.1)$

41. $(0, 39.18)$ and $(15, 3.66)$

42. $(0, 12.94)$ and $(20, 357.03)$

Find an approximate equation $y = ab^x$ of the exponential curve that contains the given pair of points. Round the values of a and b to two decimal places. Verify your result with a graphing calculator.

43. $(1, 4)$ and $(2, 12)$

44. $(2, 5)$ and $(3, 10)$

45. $(3, 4)$ and $(5, 9)$

46. $(2, 1)$ and $(5, 7)$

47. $(10, 329)$ and $(30, 26)$

48. $(11, 492)$ and $(17, 8)$

49. $(5, 8.1)$ and $(9, 2.4)$

50. $(1, 3.5)$ and $(5, 1.3)$

51. $(13, 24.71)$ and $(21, 897.35)$

52. $(4, 6.3)$ and $(10, 250.8)$

53. $(2, 73.8)$ and $(7, 13.2)$

54. $(8, 39.43)$ and $(12, 6.52)$

Concepts

55. Find an equation of the exponential curve sketched in Fig. 41. [**Hint:** Choose two points whose coordinates appear to be integers.]

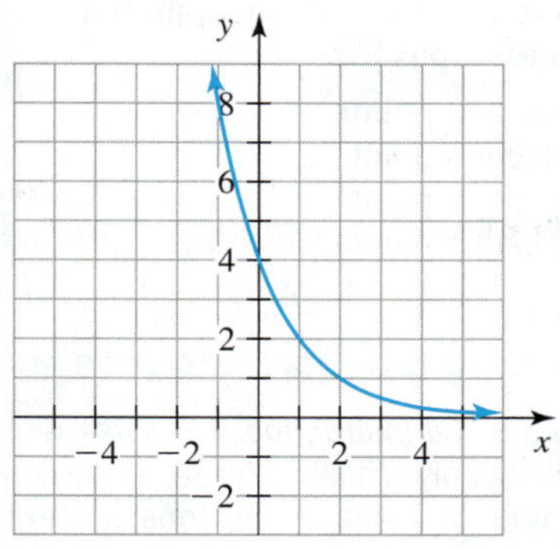

Figure 41 Graph of *f*—Exercise 55

56. Find an equation of the exponential curve sketched in Fig. 42.

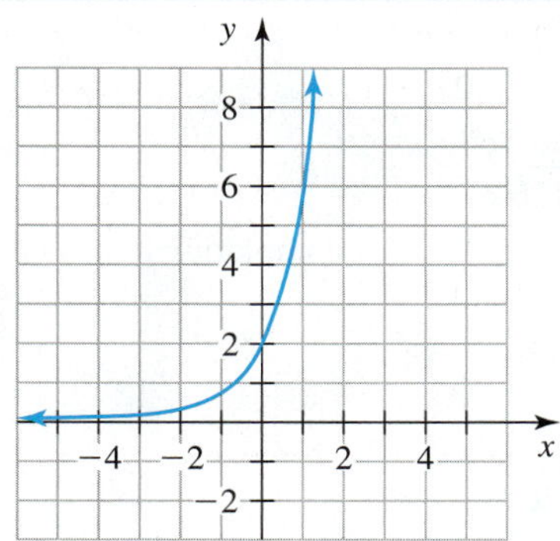

Figure 42 Graph of *f*—Exercise 56

57. Find an equation of the exponential curve sketched in Fig. 43.

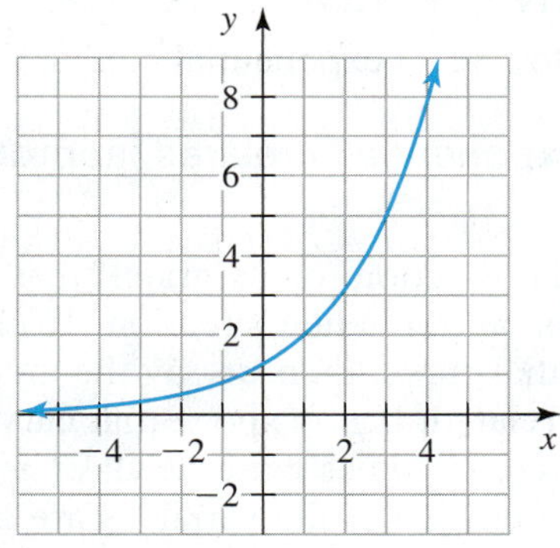

Figure 43 Graph of *f*—Exercise 57

58. Find an equation of the exponential curve sketched in Fig. 44.

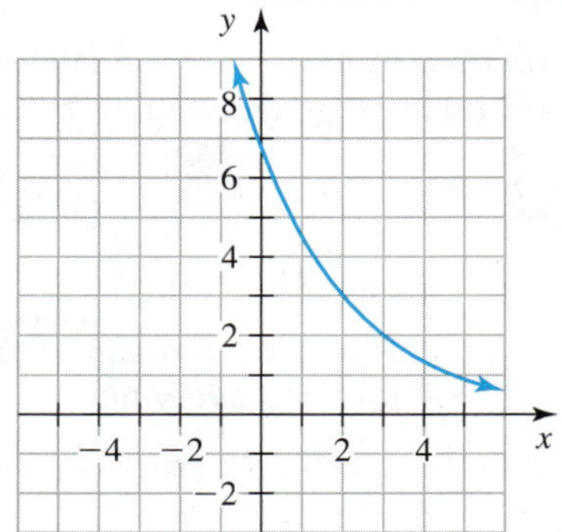

Figure 44 Graph of *f*—Exercise 58

59. Solve the system. [**Hint:** Think graphically.]

$$y = 6(4)^x$$
$$y = 6\left(\frac{1}{3}\right)^x$$

60. Solve the system. [**Hint:** Think graphically.]

$$y = 7(3)^x$$
$$y = 4(3)^x$$

61. **a.** Is there an exponential curve that contains the given point? If so, find an equation of the curve. If not, explain.

i. $(0, 2)$ **ii.** $(2, 0)$

b. Is there an exponential curve that contains the points $(2, -1)$ and $(3, 1)$? If so, find an equation of the curve. If not, explain.

62. **a.** Is there an exponential curve that passes through the given points? If so, find an equation of the curve. If not, explain.

i. $(0, 5)$ and $(1, 5)$ **ii.** $(0, 3)$ and $(7, 3)$

b. Is there an exponential curve that passes through two given points that have the same *y*-coordinate? Explain.

63. Describe the base multiplier property and explain why it makes sense. Include an example.

64. Describe how to find an equation of an exponential curve that contains two given points. Include both the case in which one of the points is the *y*-intercept and the case in which neither of the points is the *y*-intercept.

Related Review

Find all real-number solutions or simplify, whichever is appropriate. Round your solution(s) to the second decimal place.

65. $\dfrac{b^7}{b^2}$

66. $\dfrac{b^8}{b^4}$

67. $\dfrac{b^7}{b^2} = 76$

68. $\dfrac{b^8}{b^4} = \dfrac{65}{3}$

69. $\dfrac{8b^3}{6b^{-1}}$

70. $\dfrac{10b^{-7}}{15b^{-2}}$

71. $\dfrac{8b^3}{6b^{-1}} = \dfrac{3}{7}$

72. $\dfrac{10b^{-7}}{15b^{-2}} = \dfrac{4}{7}$

Find equations of a linear function L, an exponential function E, and a quadratic function Q such that the graph of each equation contains the given points. Use a graphing calculator to verify your work.

73. $(0, 2)$ and $(1, 6)$

74. $(2, 8)$ and $(5, 2)$

For Exercises 75 and 76, the graph of a function contains the given pair of points. Could the function be linear, exponential, either linear or exponential, or neither? Explain.

75. $(5, 3)$ and $(7, 6)$

76. $(2, 6)$ and $(4, 6)$

77. In this exercise, you will compare the linear function $L(x) = 2x + 100$ with the exponential function $E(x) = 3(2)^x$.

a. Find the *y*-intercept of the graph of both functions.

b. For functions *L* and *E*, describe what happens to the value of *y* as the value of *x* increases by 1.

c. On the basis of your responses to parts (a) and (b), which function's outputs will eventually dominate the other's outputs? Explain.

d. Use a graphing calculator table to verify your responses to parts (a)–(c). To do this, enter the functions and set up a table as indicated in Figs. 45 and 46, respectively.

Figure 45 Enter the functions

Figure 46 Set up a table

78. Is it possible for a linear function and an exponential function to have the indicated number of intersection points? If so, give equations of the two functions. If not, explain. [**Hint:** First sketch some graphs.]

a. 3 intersection points

b. 2 intersection points

c. 1 intersection point

d. 0 intersection points

Expressions, Equations, Functions, and Graphs

Perform the indicated instruction. Then use words such as linear, quadratic, cubic, polynomial, degree, exponential, function, one variable, *and* two variables *to describe the expression, equation, or system.*

79. Graph $f(x) = 3(2)^x$ by hand.

80. Factor $16p^4 - q^4$.

81. Find $f(-3)$, where $f(x) = 3(2)^x$.

82. Solve $3x^2 = 2(2x + 1)$.

83. Simplify $\dfrac{8b^{-3}c^6}{12b^2c^3}$.

84. Solve $2x^2 - 5x - 4$ by completing the square.

10.5 Using Exponential Functions to Model Data

Objectives

- Find an equation of an *exponential model* by using the base multiplier property.
- Model a *half-life* situation.
- Find an equation of an exponential model by using data described in words.
- Find an equation of an exponential model by using data displayed in a table.
- For a model $f(t) = ab^t$, know the meaning of the coefficient a and the base b in terms of the situation being modeled.
- Make estimates and predictions by using an exponential model.
- Determine whether to model a situation by using a linear function, a quadratic function, an exponential function, or none of these.

In Section 10.4, we found equations of exponential functions. In this section, we will use that skill to model authentic situations.

Using the Base Multiplier Property to Find a Model

We can use the base multiplier property to find an exponential model.

Definition Exponential model, exponentially related, approximately exponentially related

An **exponential model** is an exponential function, or its graph, that describes the relationship between two quantities for an authentic situation. If all of the data points for a situation lie on an exponential curve, then we say the independent and dependent variables are **exponentially related.** If no exponential curve contains all of the data points, but an exponential curve comes close to all of the data points (and perhaps contains some of them), then we say the variables are **approximately exponentially related.**

Example 1 Modeling with an Exponential Function

Suppose that a peach has 3 million bacteria on it at noon on Monday and that one bacterium divides into two bacteria every hour, on average (see Fig. 47).

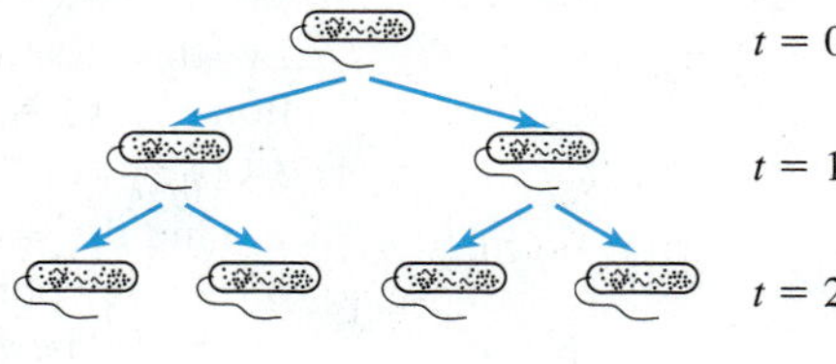

Figure 47 A result of bacteria dividing every hour

Table 32 Values of a Bacteria Model

t (hours)	$B = f(t)$ (millions)
0	3
1	6
2	12
3	24
4	48

X	Y1
0	3
1	6
2	12
3	24
4	48
5	96
6	192

X=0

Figure 48 Table for bacteria model

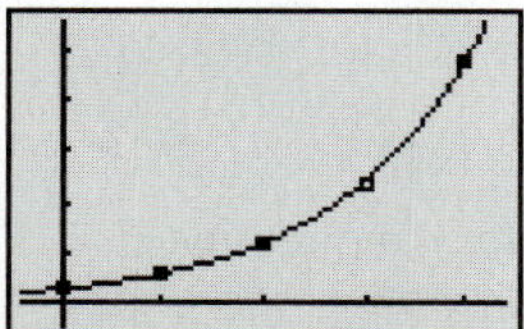

Figure 49 Graph of bacteria model

Let $B = f(t)$ be the number of bacteria (in millions) on the peach at t hours after noon on Monday.

1. Find an equation of f.
2. Predict the number of bacteria on the peach at noon on Tuesday.

Solution

1. We complete a table of values of f based on the assumption that one bacterium divides into two bacteria every hour (see Table 32). As the value of t increases by 1, the value of B changes by greater and greater amounts, so it would *not* be appropriate to model the data by using a linear function. Note, though, that as the value of t increases by 1, the value of B is multiplied by 2, so we *can* model the situation by using an *exponential* model of the form $f(t) = a(2)^t$. The B-intercept is $(0, 3)$, so $f(t) = 3(2)^t$.

   We use a graphing calculator table and graph to verify our work (see Figs. 48 and 49).
2. We use $t = 24$ to represent noon on Tuesday. We substitute 24 for t in our equation of f:

$$f(24) = 3(2)^{24} = 50{,}331{,}648$$

According to the model, there would be 50,331,648 million bacteria. To omit writing "million," we must add six zeros to 50,331,648—that is, 50,331,648,000,000. There would be about 50 trillion bacteria at noon on Tuesday.

For an exponential function $y = ab^t$, the y-intercept of the graph of the function is $(0, a)$. So, **if $y = ab^t$ is an exponential model where y is a quantity at time t, then the coefficient a is the value of that quantity present at time $t = 0$.** For example, the bacteria model $f(t) = 3(2)^t$ has coefficient 3, which represents the 3 million bacteria that were present at time $t = 0$ (noon on Monday).

We can use an exponential function to model the value of an investment that earns r percent interest compounded annually. The term ***r* percent interest compounded annually** means the interest earned each year equals r percent of the principal plus *any interest earned in previous years* (all of which becomes part of the investment).

Example 2 Modeling with an Exponential Function

A person invests \$5000 in an account that earns 6% interest compounded annually.

1. Let $V = f(t)$ be the value (in dollars) of the account at t years after the money is invested. Find an equation of f.
2. What will be the value after 10 years?

Solution

Table 33 Values of a Compounded-Interest (6%) Account

t	$V = f(t)$
0	5000.00
1	$5000.00(1.06) = 5300.00$
2	$5300.00(1.06) = 5618.00$
3	$5618.00(1.06) = 5955.08$
4	$5955.08(1.06) \approx 6312.38$

1. Each year, the investment value is equal to the previous year's value (100% of it) plus 6% of the previous year's value. So, the value is equal to 106% of the previous year's value. For example, after one year, the value will be 106% of \$5000, or $1.06(5000) = 5300$ dollars. After two years, the value will be $1.06(5300) = 5618$ dollars. See Table 33.

   As the value of t increases by 1, the value of V is multiplied by 1.06. So, f is the exponential function $f(t) = a(1.06)^t$. Since the value of the account at the start is \$5000, we have $a = 5000$. So, $f(t) = 5000(1.06)^t$.
2. To find the value in 10 years, we substitute 10 for t:

$$f(10) = 5000(1.06)^{10} \approx 8954.24$$

The value will be \$8954.24 in 10 years.

In Example 2, we used the function $f(t) = 5000(1.06)^t$ to model the value of the 6% compounded-interest account. Note that subtracting 1 from the base 1.06 gives the interest rate in decimal form:

$$b - 1 = 1.06 - 1 = 0.06 = \text{interest rate (in decimal form)}$$

Half-Life Applications

If a quantity decays exponentially, we can describe how quickly it decays by its *half-life*.

Definition Half-life

If a quantity decays exponentially, the **half-life** is the amount of time it takes for that quantity to be reduced to half (see Fig. 50).

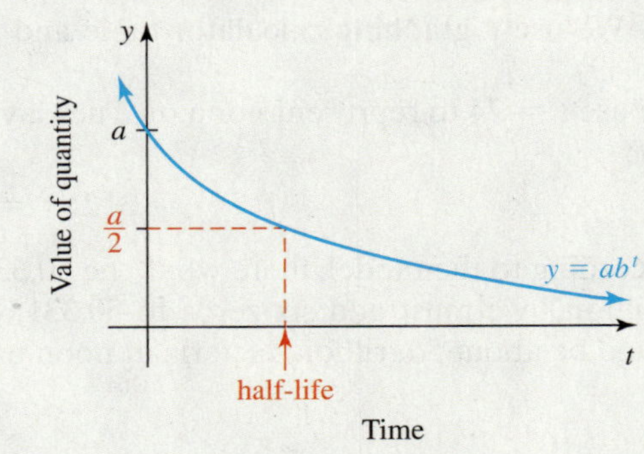

Figure 50 Half-life of a quantity

For example, the half-life of the radioactive element hydrogen-3 is 12.3 years, which means every 12.3 years the number of hydrogen-3 atoms is reduced to half. The half-lives of some radioactive elements are much different from that of hydrogen-3. For instance, polonium-214 has a half-life of 0.164 millisecond, and uranium-238 has a half-life of 4.5 billion years!

Example 3 Modeling with an Exponential Function

The world's worst nuclear accident occurred in Chernobyl, Ukraine, on April 26, 1986. Immediately afterward, 28 people died from acute radiation sickness. So far, about 25,000 people have died from the accident, mostly due to the release of the radioactive element cesium-137 (Source: *Medicine Worldwide*).

Cesium-137 has a half-life of 30 years. Let $P = f(t)$ be the percent of the cesium-137 that remains at t years since 1986.

1. Find an equation of f.
2. Describe the meaning of the base of f.
3. What percent of the cesium-137 will remain in 2014?

Solution

1. We discuss two methods of finding an equation of f.

Method 1 At time $t = 0$, 100% of the cesium-137 is present. At time $t = 30$, there will be $\frac{1}{2}(100) = 50$ percent. At time $t = 60$, there will be $\frac{1}{2} \cdot \frac{1}{2}(100) = 25$ percent. We organize these results, and one more calculation, in Table 34.

Table 34 Percentages of Cesium-137 That Remain

Year t	Percent P
0	$100 = 100\left(\frac{1}{2}\right)^0$
30	$100 \cdot \frac{1}{2} = 100 \cdot \left(\frac{1}{2}\right)^1$
60	$100 \cdot \frac{1}{2} \cdot \frac{1}{2} = 100\left(\frac{1}{2}\right)^2$
90	$100 \cdot \frac{1}{2} \cdot \frac{1}{2} \cdot \frac{1}{2} = 100\left(\frac{1}{2}\right)^3$
t	$100\left(\frac{1}{2}\right)^{t/30}$

From Table 34, we see that the situation can be modeled well with an exponential function. Each exponent in the second column of the table is equal to the value of t in the first column, divided by 30. Thus, the equation of f is

$$f(t) = 100\left(\frac{1}{2}\right)^{t/30}$$

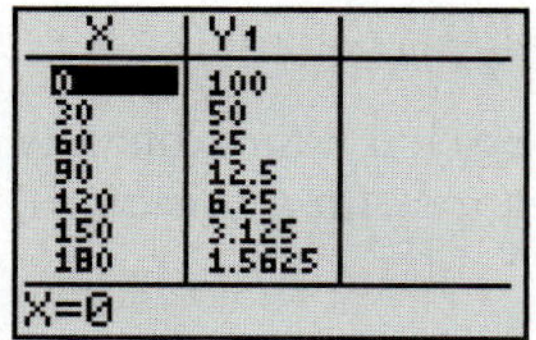

Figure 51 Table for f

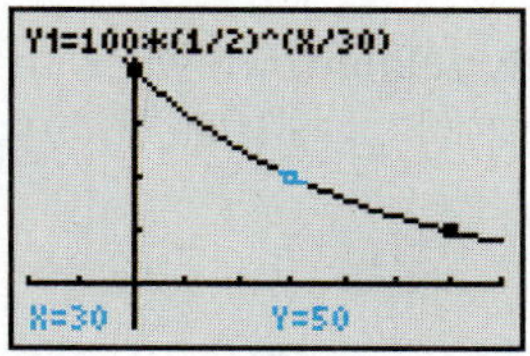

Figure 52 Graph of f and the points $(0, 100)$, $(30, 50)$, and $(60, 25)$

We can use a graphing calculator table and graph to verify our equation (see Figs. 51 and 52). We can write this equation in the form $f(t) = ab^t$:

$$f(t) = 100\left(\frac{1}{2}\right)^{t/30} = 100\left(\left(\frac{1}{2}\right)^{\frac{1}{30}}\right)^t$$

Since $\left(\frac{1}{2}\right)^{1/30} \approx 0.977$, we can write

$$f(t) = 100(0.977)^t$$

Method 2 Instead of recognizing a pattern from a table, we can find an equation of f by using the points $(0, 100)$ and $(30, 50)$. Since the P-intercept is $(0, 100)$, we have

$$P = f(t) = 100b^t$$

To find b, we substitute the coordinates of $(30, 50)$ into the equation $f(t) = 100b^t$:

$$
\begin{aligned}
50 &= 100b^{30} && \text{Substitute 30 for } t \text{ and 50 for } f(t).\\
100b^{30} &= 50 && \text{If } c = d, \text{ then } d = c.\\
b^{30} &= \frac{50}{100} && \text{Divide both sides by 100.}\\
b^{30} &= \frac{1}{2} && \text{Simplify.}\\
b &= \pm\left(\frac{1}{2}\right)^{1/30} && \text{The solution of } b^{30} = k \text{ is } \pm k^{1/30}.\\
b &\approx 0.977 && \text{Compute; base of an exponential function is positive.}
\end{aligned}
$$

So, an equation of f is $f(t) = 100(0.977)^t$, the same equation we found earlier.

2. The base of f is 0.977. Each year, 97.7% of the previous year's cesium-137 is present. In other words, the cesium-137 decays by 2.3% each year.
3. Since $2014 - 1986 = 28$, we substitute 28 for t in the equation $f(t) = 100(0.977)^t$:

$$f(28) = 100(0.977)^{28} \approx 52.13$$

In 2014, about 52.1% of the cesium-137 will remain.

Meaning of the Base of an Exponential Model

If $f(t) = ab^t$, where $a > 0$, models a quantity at time t, then the percent rate of change is constant. In particular,

- If $b > 1$, then the quantity grows exponentially at a rate of $b - 1$ percent (in decimal form) per unit of time.
- If $0 < b < 1$, then the quantity decays exponentially at a rate of $1 - b$ percent (in decimal form) per unit of time.

In Example 2, we found the compounded-interest model $f(t) = 5000(1.06)^t$. Since $1.06 - 1 = 0.06$, the base 1.06 indicates that the value grows exponentially by 6% per year.

In Example 3, we found the cesium-137 model $f(t) = 100(0.977)^t$. Since $1 - 0.977 = 0.023$, the base 0.977 indicates that the cesium-137 decays exponentially by 2.3% per year.

Finding a Model by Using Data Described in Words

Instead of using the base multiplier property, we can sometimes find an equation of a model by using two data points described in words.

Table 35 Known Values of t and n

Years since 2000	Number of Severe Near Collisions
0	67
10	16

Example 4 Modeling with an Exponential Function

The number of severe near collisions on airplane runways has decayed approximately exponentially from 67 in 2000 to 16 in 2010 (Source: *Federal Aviation Administration*). Predict the number of severe near collisions in 2018.

Solution

Let n be the number of severe near collisions on airplane runways in the year that is t years since 2000. Known values of t and n are shown in Table 35.

Because the variables t and n are approximately exponentially related, we want an equation of the form $n = ab^t$. From Table 35, we see that the n-intercept is $(0, 67)$. So, the equation is of the form

$$n = 67b^t$$

To find b, we substitute the coordinates of $(10, 16)$ into the equation $n = 67b^t$ and then solve for b:

$$\begin{aligned} 16 &= 67b^{10} && \text{Substitute 10 for } t \text{ and 16 for } n. \\ 67b^{10} &= 16 && \text{If } c = d, \text{ then } d = c. \\ b^{10} &= \frac{16}{67} && \text{Divide both sides by 67.} \\ b &= \left(\frac{16}{67}\right)^{1/10} && \text{The solution of } b^{10} = k \text{ is } k^{1/10}. \\ b &\approx 0.867 \end{aligned}$$

Then we substitute 0.867 for b in the equation $n = 67b^t$:

$$n = 67(0.867)^t$$

Finally, to predict the number of severe near collisions in 2018, we substitute $2018 - 2000 = 18$ for t in the equation $n = 67(0.867)^t$ and solve for n:

$$n = 67(0.867)^{18} \approx 5.13$$

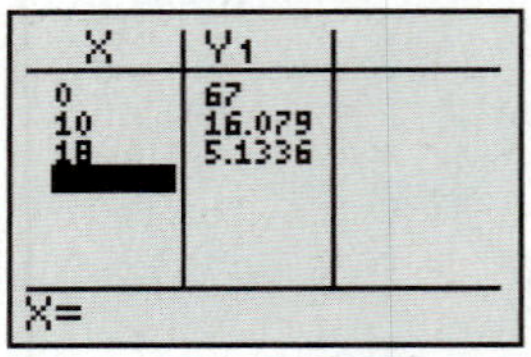

The model predicts that there will be about 5 severe near collisions in 2018. We use a graphing calculator table (see Fig. 53) to check that each of the three ordered pairs $(0, 67)$, $(10, 16)$, and $(18, 5.13)$ approximately satisfies the equation $n = 67(0.867)^t$.

X	Y1
0	67
10	16.079
18	5.1336

X=

Figure 53 Verify the work

Finding an Exponential Model by Using Data Displayed in a Table

In Example 5, we will find an equation of an exponential model by using data shown in a table.

Example 5 Modeling with an Exponential Function

Table 36 Numbers of Viewers of MLB All-Star Game

Year	Number of Viewers (millions)
1982	34
1985	28
1990	24
1995	20
2000	15
2005	12
2010	12
2011	11

Source: *Nielsen Media Research, Inc.*

The numbers of viewers of the Major League Baseball (MLB) All-Star Game are shown in Table 36 for various years. Let $n = f(t)$ be the number of viewers (in millions) of the All-Star Game at t years since 1980.

1. Find an exponential equation of f.
2. What is the coefficient a of the model $f(t) = ab^t$? What does it mean in this situation?
3. What is the base b of the model $f(t) = ab^t$? What does it mean in this situation?
4. Predict the number of viewers of the All-Star Game in 2018.

Solution

1. We use a graphing calculator to view a scattergram of the data (see Fig. 54). If we imagine an exponential curve that contains the points $(15, 20)$ and $(31, 11)$, it appears the curve might come close to the other data points. To find an equation of

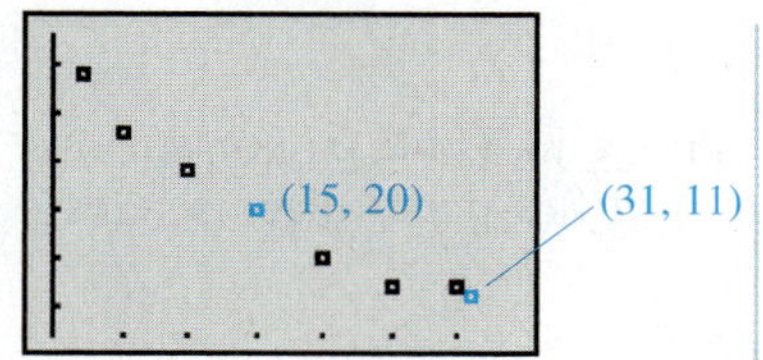

Figure 54 All-Star Game scattergram

this curve, we substitute the coordinates of the points $(15, 20)$ and $(31, 11)$ into the equation $f(t) = ab^t$:

$$11 = ab^{31}$$
$$20 = ab^{15}$$

Next, we divide the two left sides and divide the two right sides and solve for b:

$$\frac{11}{20} = \frac{ab^{31}}{ab^{15}} \qquad \text{Divide left sides and divide right sides.}$$

$$\frac{11}{20} = b^{16} \qquad \text{Simplify; subtract exponents: } \frac{b^m}{b^n} = b^{m-n}$$

$$b = \pm\left(\frac{11}{20}\right)^{1/16} \qquad \text{The solutions of } b^{16} = k \text{ are } \pm k^{1/16}.$$

$$\approx 0.963 \qquad \text{Compute; base of an exponential function is positive.}$$

So, an equation is $f(t) = a(0.963)^t$. To find a, we substitute the coordinates of $(15, 20)$ into the equation:

$$20 = a(0.963)^{15} \qquad \text{Substitute 15 for } t \text{ and 20 for } f(t).$$

$$a = \frac{20}{0.963^{15}} \qquad \text{Divide both sides by } 0.963^{15}.$$

$$\approx 35.21 \qquad \text{Compute.}$$

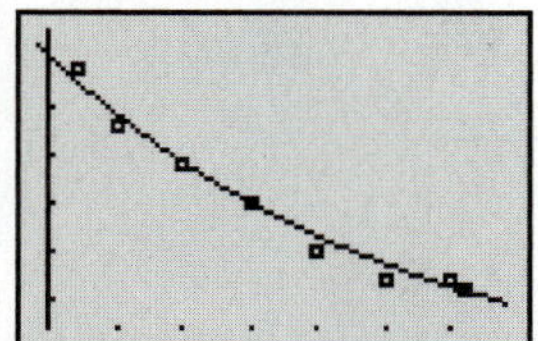

Figure 55 Check how well the model fits the data

The equation is $f(t) = 35.21(0.963)^t$. The graph in Fig. 55 shows that our exponential model fits the data well.

2. The coefficient of $f(t) = 35.21(0.963)^t$ is 35.21. So, the n-intercept is $(0, 35.21)$. This means there were about 35 million viewers of the All-Star Game in 1980. A little research would show that the actual number of viewers was 36 million, so the model's estimate is pretty accurate.
3. The base of $f(t) = 35.21(0.963)^t$ is $b = 0.963$. Because $1 - b = 1 - 0.963 = 0.037$, we conclude the number of viewers is decaying exponentially by 3.7% per year.
4. To predict the number of viewers in 2018, we evaluate f at $2018 - 1980 = 38$:

$$f(38) = 35.21(0.963)^{38} \approx 8.40$$

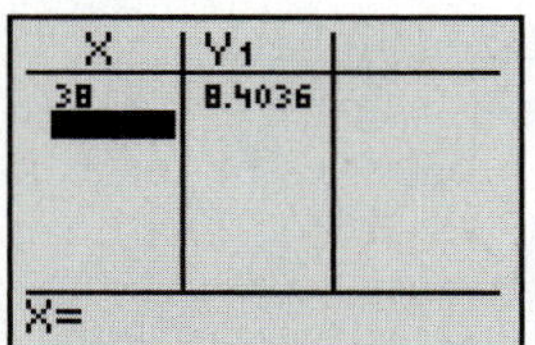

Figure 56 Verify the work

There will be about 8 million viewers in 2018. We use a graphing calculator table to verify our work (see Fig. 56).

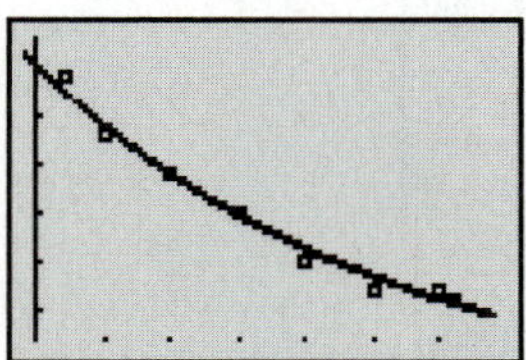

Figure 57 Compare the fit of models f and r

In Example 5, we used two points to find the exponential model $f(t) = 35.21(0.963)^t$. We can also find an exponential model by using a graphing calculator's *exponential regression*. Exponential regression gives the model $r(t) = 34.86(0.962)^t$. In Fig. 57, we see that both of the models fit the data well. For graphing calculator instructions, see Appendix A.16.

We call the equation $r(t) = 34.86(0.962)^t$ the **exponential regression equation** for the given data. We refer to its graph as an **exponential regression curve.**

The bases of both models are approximately equal. The base 0.963 of f estimates that the number of viewers of the MLB All-Star Game is decaying exponentially by 3.7% each year. The base 0.962 of r estimates that the number of viewers is decaying exponentially by 3.8% each year.

Deciding Which Type of Model to Use

Now that we know how to work with exponential models, we can choose among linear functions, quadratic functions, and exponential functions to model authentic situations. Recall from Section 9.8 that we use the following criteria for selecting a model:

- The graph of the model should fit the points well.
- The model should make sense within the context of the authentic situation.

Example 6 Determining Which Type of Model to Use

Table 37 Average Prices of Flat-Panel Plasma Televisions

Year	Average Price (thousands of dollars)
2000	9.8
2001	6.8
2003	4.6
2005	2.5
2006	1.7

Source: *DisplaySearch*

The average prices of flat-panel plasma televisions are shown in Table 37 for various years. Let p be the average price (in thousands of dollars) of flat-panel plasma televisions at t years since 2000.

1. Find an equation of a model to describe the situation.
2. Predict the average price of a flat-panel plasma television in 2014.

Solution

1. First, we use a graphing calculator to draw a scattergram to describe the data (see Fig. 58). Since the points suggest a curve that "bends," we will not use a linear function to model the data.

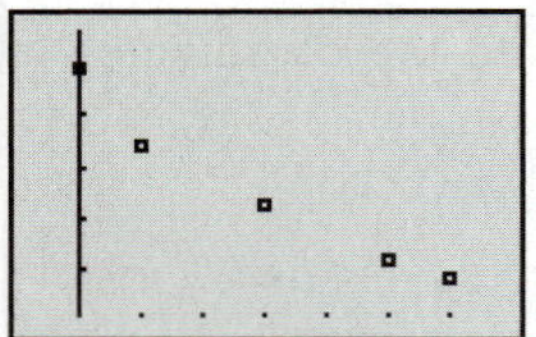

Figure 58 Television scattergram

To decide between a quadratic model and an exponential model, we use a graphing calculator to find the quadratic regression equation and the exponential regression equation:

$$Q(t) = 0.147t^2 - 2.14t + 9.43 \quad \textit{Quadratic regression model}$$

$$E(t) = 9.70(0.756)^t \quad \textit{Exponential regression model}$$

Next, we see how well both regression models fit the data (see Figs. 59 and 60).

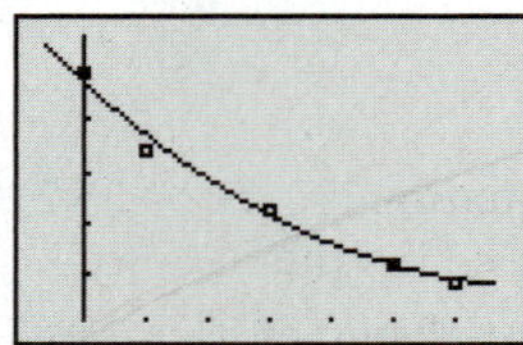

Figure 59 Quadratic regression model

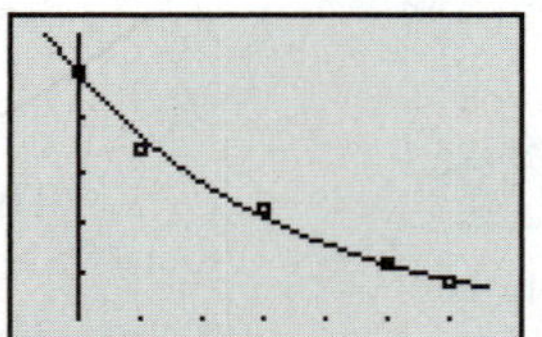

Figure 60 Exponential regression model

Both models fit the data well, but it appears that E fits the data a bit better than Q does. In fact, evaluating the functions at $t = 0, t = 1, t = 3, t = 5$, and $t = 6$ would show that E and Q give about the same estimate for 2005 and E gives better estimates for three of the other four years shown in Table 37 (2000, 2001, 2006). We conclude that E is the best model for the period 2000–2006.

Does E likely predict the average prices better than Q does for years after 2006? To see, we Zoom Out on a graphing calculator (see Figs. 61 and 62).

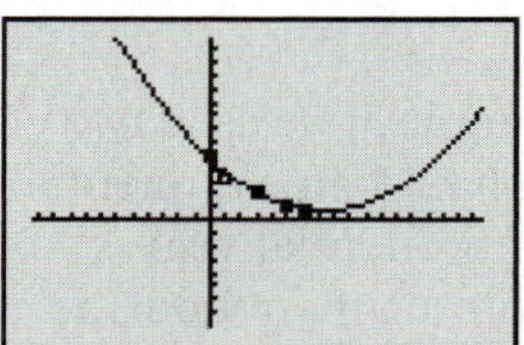

Figure 61 Quadratic regression model

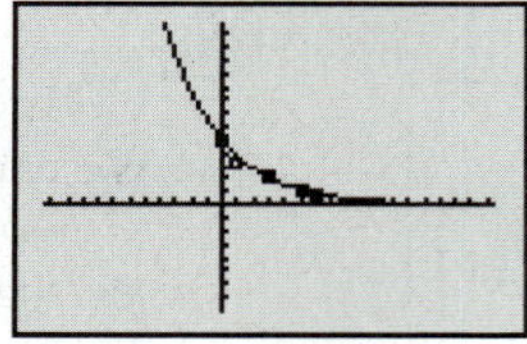

Figure 62 Exponential regression model

It will also help to use "minimum" on a graphing calculator to find the approximate vertex of the graph of Q, which is (7.28, 1.64). See Fig. 63. So, Q predicts that the average price will be increasing for the years after 2007. Model breakdown has likely occurred, because usually the average price of a technological product decreases over time. Since the model E predicts that the average price will decrease over time, we conclude that E is likely the best model for the years after 2006.

Minimum X=7.27891 Y=1.6415646

Figure 63 The vertex of the graph of Q

Finally, since E fits the data best for the period 2000–2006 and makes the most sense for the years after 2006, we assume that it will give the best estimates for the period 1997–1999. (Flat-panel plasma televisions were first sold to the public in 1997 by Pioneer®.)

2. We substitute 14 for t into the equation $E(t) = 9.70(0.756)^t$:

$$E(14) = 9.70(0.756)^{14} \approx 0.193$$

The model predicts that the average price of a flat-panel plasma television will be about \$0.193 thousand, or \$193, in 2014.

Here we outline the four-step modeling process again.

Four-Step Modeling Process

To find a model and make estimates and predictions,

1. Create a scattergram of the data. Decide whether a line, a parabola, an exponential curve, or none of these comes close to the points.
2. Find an equation of your model.
3. Verify that your equation has a graph that comes close to the points in the scattergram. If it doesn't, check for calculation errors or use different points to find the equation. An alternative is to reconsider your choice of model in step 1.
4. Use your equation of the model to draw conclusions, make estimates, and/or make predictions.

Group Exploration

Comparing a linear model with an exponential model

In 1950, world population was 2.5 billion. In 1987, it was 5.0 billion (Source: *U.S. Census Bureau*).

1. First, assume world population is growing exponentially. Let $E(t)$ be the world's population (in billions) at t years since 1950. Find an equation of E.
2. Now assume world population is growing linearly. Let $L(t)$ be the world's population (in billions) at t years since 1950. Find an equation of L.
3. Use your equations of E and L to make two predictions of the world's population for each of the years that follow.
   a. 2020 **b.** 2050 **c.** 2150
4. Use the window settings shown in Fig. 64 to compare the graphs of E and L.

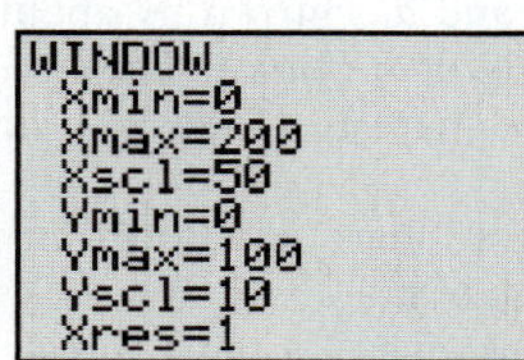

Figure 64 Compare the graphs of the exponential and linear models

5. Will there be much difference in the world's population if it grows exponentially or linearly in the short run? in the long run? Explain.

Homework 10.5

For extra help ▶ MyMathLab®  Watch the videos in MyMathLab  Download the MyDashboard App

1. Suppose a rumor is spreading that eating pickles will raise your IQ. Assume 40 Americans have heard the rumor as of today and each day the number of Americans (past and present) who have heard it triples. Let $f(t)$ be the number of Americans (past and present) who have heard the rumor at t days since today.
   a. Find an equation of f.
   b. How many Americans (past and present) will have heard the rumor 10 days from now?
   c. How many Americans (past and present) will have heard the rumor 15 days from now? Has model breakdown occurred? [**Hint:** Assume the U.S. population is 306 million.]
2. Suppose a flu epidemic has broken out at your school. Assume that on February 10, 20 people have the flu and that each day the number of people (past and present) who have gotten the flu triples. Let $f(t)$ be the number of people (past

and present) who have gotten the flu by the day that is t days since February 10.

a. Find an equation of f.

b. Find $f(4)$. What does it mean in this situation?

c. Find $f(10)$. What does it mean in this situation?

3. The name of the search engine Google™ is a play on the word "googol," which refers to the number 1 followed by one hundred zeros.

a. In 2012, Google's index contained 30 trillion web pages. According to founders Larry Page and Sergey Brin, the size of the index doubles every year. Let $f(t)$ be the number of web pages (in trillions) in Google's index at t years since 2012. Find an equation of f.

b. Predict the number of web pages Google's index will contain in 2018.

c. If the web pages in Google's index in 2018 were printed and stacked in one pile, what would be the height (in miles) of that pile? [**Hint:** Take a stack of 500 pages of computer paper to be 2 inches tall.]

4. The total number of hours people spent listening to Pandora Radio was 1 billion hours in 2009 and has approximately doubled each year (Source: *Pandora Media, Inc.*).

a. Let $H = f(t)$ be the total number (in billions) of listener hours in the year that is t years since 2009. Find an equation of f.

b. What is the H-intercept of the model? What does it mean in this situation?

c. Predict the total number of listener hours in 2017. Assuming the world population will be 7.3 billion in that year, predict the average annual number of listener hours per person.

5. The market share of eBooks (the percentage of revenue from all books) was 8.3% in 2010 and has grown by about 108% per year since then (Source: *German Book Office New York*). That is, each year the market share is about 2.08 times the previous year's market share.

a. Let $M = f(t)$ be the market share of eBooks at t years since 2010. Find an equation of f.

b. What is the M-intercept? What does it mean in this situation?

c. What is the base b of $f(t) = ab^t$? What does it mean in this situation?

d. Predict the market share in 2013. Assuming the revenue of all books will be \$5 billion in that year, predict the revenue from eBooks.

6. The number of U.S. natural catastrophes (events that cost at least \$1 million) was 250 catastrophes in 2010 and has grown by about 5% per year (Source: *Munich Reinsurance Company*). That is, each year there are about 1.05 times the number of natural catastrophes as in the previous year.

a. Let $n = f(t)$ be the number of U.S. natural catastrophes in the year that is t years since 2010. Find an equation of f.

b. What is the n-intercept? What does it mean in this situation?

c. What is the base b of $f(t) = ab^t$? What does it mean in this situation?

d. Predict the number of U.S. natural catastrophes in 2018. Use your result to predict the average number of U.S. natural catastrophes per day in 2018.

7. a. About 2.5 million TiVo® subscribers got TiVo through DIRECTV® in 2007, and that number has grown by about 50% per year (Source: *TiVo*). That is, each year there are about 1.5 times as many subscribers as in the preceding year. Let $D(t)$ be the number (in millions) of such subscribers at t years since 2007. Find an equation of D.

b. About 1.71 million people were stand-alone TiVo subscribers in 2007, and that number has grown by about 120% per year (Source: *TiVo*). That is, each year there are about 2.2 times as many subscribers as in the preceding year. Let $S(t)$ be the number (in millions) of such subscribers at t years since 2007. Find an equation of S.

c. Use "intersect" on a graphing calculator to find the intersection point of the graphs of D and S. What does it mean in this situation? [**Hint:** Use the window settings shown in Fig. 65.]

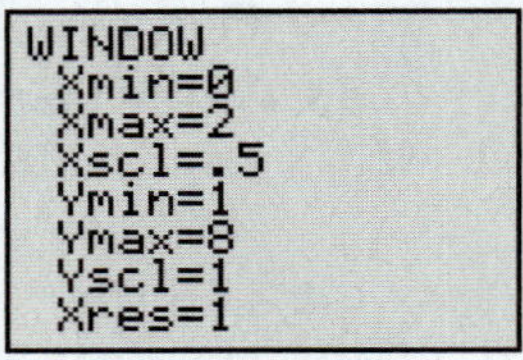

Figure 65 Exercise 7c

8. The average number of cable channels per household was 135 channels in 2010 and has grown by about 8% per year (Source: *Nielsen Media*). That is, each year, there have been about 1.08 times the number of channels as in the previous year.

a. Let $f(t)$ be the average number of cable channels at t years since 2010. Find an equation of f.

b. Predict the average number of cable channels per household in 2017.

c. Currently, people must subscribe to bundles of channels rather than individual channels. On a weekly basis, subscribers watch only 12% of the channels to which they subscribe, on average. The Federal Communications Commission has urged Congress to eliminate bundling. Predict the average number of channels per household in 2017 if that happens.

9. Someone invests \$3000 in an account at 8% interest compounded annually. Let $f(t)$ be the value (in dollars) of the account at t years after she has invested the \$3000.

a. Find an equation of f.

b. What is the base b of your model $f(t) = ab^t$? What does it mean in this situation?

c. What is the coefficient a of your model $f(t) = ab^t$? What does it mean in this situation?

d. What will be the account's value in 15 years?

10. A person invests \$7000 at 10% interest compounded annually. Let $f(t)$ be the value (in dollars) of the account at t years after he has deposited the \$7000.

a. Find an equation of f.

b. What is the base b of your model $f(t) = ab^t$? What does it mean in this situation?

c. What is the coefficient a of $f(t) = ab^t$? What does it mean in this situation?

d. What will be the account's value in 10 years? Explain why the value has more than doubled, even though the investment earned 10% for 10 years.

11. A person invests \$4000 in stocks today, and their value doubles every 6 years. Let $f(t)$ be the value (in dollars) of the investment at t years from now.
a. Find an equation of f.
b. Find the value of the investment 20 years from now.

12. Someone invests \$2800 in stocks today, and their value doubles every 11 years. Let $g(t)$ be the value (in dollars) of the investment at t years from now.
a. Find an equation of g.
b. Find the value of the investment 25 years from now.

13. A person invests \$5000 at 6% interest compounded annually for 3 years and then invests the balance (the \$5000 plus the interest earned) in an account at 8% interest for 5 years. Find the value of the investment after the 8 years.

14. A person invests \$3500 at 3% interest compounded annually for 4 years and then invests the balance (the \$3500 plus the interest earned) in an account at 7% interest for 9 years. Find the value of the investment after the 13 years have elapsed.

15. Suppose a college bookstore sells 984 copies of a new (not used) textbook in 2011. Each year from 2011 to 2014, the new-textbook sales are half of the previous year's sales (due to used-textbook sales). Let $s = g(t)$ be the annual sales of copies of the new textbook at t years since 2011.
a. Find an equation of g.
b. What is the s-intercept of the model? What does it mean in this situation?
c. Find $g(3)$. What does it mean in this situation?
d. What is the half-life of new-textbook sales?

16. A storage tank contains niobium-97m, a radioactive element. The percentage of the element that remains is halved each hour. Let $p = f(t)$ be the percentage of niobium-97m that remains at t hours since the element was placed in the tank.
a. Find an equation of f.
b. What is the p-intercept of the model? What does it mean in this situation?
c. Find $f(9)$. What does it mean in this situation?
d. What is the half-life of niobium-97m?

17. A storage tank contains radium-226, a radioactive element with a half-life of 1600 years. Let $f(t)$ be the percentage of radium-226 that remains at t years since the element was placed in the tank.
a. Find an equation of f.
b. What percentage of the radium-226 will remain after 100 years?
c. What percentage of the radium-226 will remain after 3200 years? Explain how you can find this result without using the equation of f.

18. A storage tank contains californium-251, a radioactive element with a half-life of about 900 years. Let $g(t)$ be the percentage of californium-251 that remains at t years since the element was placed in the tank.
a. Find an equation of g.
b. What percentage of the californium-251 will remain after 600 years?
c. What percentage of the californium-251 will remain after 3600 years? Explain how you can find this result without using the equation of g.

19. A thyroid cancer patient ingests a single dose of radioactive iodine-131 to kill the cancer cells. Iodine-131 has an *effective half-life* of 7.56 days—some is lost to radioactive decay, and some is removed through urination. Let $f(t)$ be the percentage of the iodine-131 that remains in the patient's body at t days since he ingested the iodine-131.
a. Find an equation of f.
b. For 3 days after ingesting the iodine-131, the patient must stay at least 1 meter away from other people, because the radiation he emits could be harmful. What percentage of the iodine-131 will remain in his body after 3 days?
c. The patient can safely spend a lot of time near a child when at most 5% of the iodine-131 remains. Use "intersect" on a graphing calculator to estimate when that time will be. [**Hint:** Use ZStandard followed by Zoom Out.]

20. A physician injects a patient with thallium-201 to determine how well blood flows to the patient's heart muscle. Thallium-201 has an *effective half-life* of 2.3 days—some is lost to exponential decay, and some is removed through the digestive and urinary tracts. Let $f(t)$ be the percentage of the thallium-201 that remains in the patient's body at t days since she was injected.
a. Find an equation of f.
b. What percentage of thallium-201 will remain after 5 days?
c. Use "intersect" on a graphing calculator to estimate when only 10% of the thallium-201 will remain. [**Hint:** Use ZStandard followed by Zoom Out.]

21. The half-life of caffeine in a person's bloodstream is about 6 hours. If a person's bloodstream contains 80 milligrams of caffeine, how much of that caffeine will remain after 14 hours?

22. The half-life of aspirin in a person's bloodstream is about 15 minutes. If a person's bloodstream contains 200 milligrams of aspirin, how much of that aspirin will remain after 40 minutes?

23. The number of wireless Internet users in the United States grew approximately exponentially from 12 million users in 2003 to 105 million users in 2010 (Source: *Computer Industry Almanac*). Estimate the number of wireless Internet users in 2012.

24. The enrollment in private, for-profit colleges grew approximately exponentially from 0.6 million students in 2000 to 2.3 million students in 2011 (Source: *National Center for Education Statistics*). Predict the enrollment in 2017.

25. For those who have private health insurance, per-person monthly spending on healthcare grows approximately exponentially with age. The average monthly spending of 21-year-old adults is \$198 and the average monthly spending of 60-year-old adults is \$715 (Source: *Thomson Reuters*). Estimate the average monthly spending on healthcare by 30-year-old adults who have private health insurance.

26. Hard-copy sales of *Encyclopedia Britannica* decayed approximately exponentially from 108 thousand encyclopedias in 1994 to 8 thousand encyclopedias in 2011 (Source: *Kellogg School of Management*). Estimate the hard-copy sales in 2012. Explain why it is not surprising that *Encyclopedia Britannica* went out of print in 2012.

27. The numbers of subscribers to Showtime are shown in Table 38 for various years.

Table 38 Numbers of Showtime Subscribers

Year	Number of Showtime Subscribers (in millions)
2004	13
2006	14
2008	16
2010	19
2012	22

Source: *SNL Kagan*

Let $f(t)$ be the number (in millions) of Showtime subscribers at t years since 2000.

a. Find an exponential equation of f.
b. Predict the number of Showtime subscribers in 2017.
c. AT&T charged \$14 per month for Showtime in 2013. Predict the total amount of money subscribers paid for Showtime in the *entire* year 2013, assuming all Showtime subscribers paid \$14 per month and they subscribed for the entire year.

28. The numbers of North American cruise ship passengers are shown in Table 39 for various years.

Table 39 Numbers of North American Cruise Ship Passengers

Year	Number of Passengers (in millions)
1980	1.4
1985	2.2
1990	3.8
1995	4.7
2000	7.2
2005	11.2
2010	14.8

Source: *Cruise Lines International Association*

Let $f(t)$ be the number (in millions) of North American cruise ship passengers in the year that is t years since 1980.

a. Find an exponential equation of f.
b. Predict the number of North American cruise ship passengers in 2017.
c. The average amount of money spent per passenger is \$1770 per week (Source: *American Association of Port Authorities*). Predict the total amount of money spent by all passengers on North American cruise ships in 2019, assuming the weekly expenditure of \$1770 stays the same and the average time passengers will be on cruises is one week.

29. World population is provided in Table 40 for various years.

Table 40 World Population

Year	Population (billions)
1930	2.070
1940	2.295
1950	2.500
1960	3.050
1970	3.700
1980	4.454
1990	5.279
2000	6.080
2012	7.044

Source: *U.S. Census Bureau*

Let $p = f(t)$ be the world's population (in billions) at t years since 1900.

a. Find a linear equation, a quadratic equation, and an exponential equation of f. Compare how well the models fit the data.
b. Which function best models the situation for years before 1930? Explain.
c. Use the exponential model to find $f(110)$. What does it mean in this situation? Have you performed interpolation or extrapolation? Explain.
d. Use the exponential model to find $f(118)$. What does it mean in this situation? Have you performed interpolation or extrapolation? Explain.

30. The numbers of men's colleges, not including seminaries, are shown in Table 41 for various years.

Table 41 Numbers of Men's Colleges

Year	Number of Men's Colleges
1967	145
1975	80
1985	27
1995	11
2012	4

Source: *National Association of Independent Colleges and Universities*

Let $n = f(t)$ be the number of men's colleges at t years since 1960.

a. Find a linear equation, a quadratic equation, and an exponential equation of f. Compare how well the models fit the data.
b. Assuming the number of men's colleges continues to decrease, which function best models the situation for years after 2012? Explain.
c. Find $f(45)$. What does it mean in this situation? Have you performed interpolation or extrapolation? Explain.
d. Find $f(58)$. What does it mean in this situation? Have you performed interpolation or extrapolation? Explain.

31. The number of Starbucks stores worldwide has increased substantially since 1991 (see Table 42).

Table 42 Numbers of Starbucks Stores Worldwide

Year	Number of Stores
1991	116
1993	272
1995	676
1997	1412
1999	2135
2001	4709
2003	7225

Source: *Starbucks Corporation*

Let $n = f(t)$ be the number of Starbucks stores worldwide at t years since 1990.

a. Find an exponential equation of f.
b. What is the percentage rate of growth of Starbucks stores?
c. Find $f(4)$. What does it mean in this situation?
d. Use f to estimate the number of stores in 2008. At the end of 2008, there were 15,256 stores (after 500 store

I'll have a double decaf blended vanilla mocha frappuccino, please.
I hate tourists.
Donuts $1.50
Coffee $.75
NEW! Decaf Coffee $.25
Dan's Donuts

closures during 2008). Is the percentage rate of growth of stores for the period 1991–2003 the same as for the period 2003–2008? Explain.

e. Use "intersect" on a graphing calculator to estimate when there were 3,000 stores. [**Hint:** Graph $n = 3{,}000$.]

32. Percentages of adults surveyed who plan to attend a Halloween party this year are shown in Table 43 for various age groups.

Table 43 Percentages of Adults Who Plan to Attend a Halloween Party

Age Group (years)	Age Used to Represent Age Group (years)	Percent
18–24	21.0	44
25–34	29.5	34
35–44	39.5	25
45–54	49.5	14
55–64	59.5	10
65 or over	70.0	6

Source: *International Mass Retail Association*

Let $p = f(t)$ be the percentage of adults at age t years who plan to attend a Halloween party this year.

a. Find a linear equation, a quadratic equation, and an exponential equation of f. Compare how well the models fit the data.

b. For ages of adults over 80 years, which of the three models likely gives the best estimates of the percentage of adults who plan to attend a Halloween party this year? Explain.

c. What is the base b of the exponential model $f(t) = ab^t$? What does it mean in this situation?

d. Use the exponential model to find $f(26)$. What does it mean in this situation?

e. Use the exponential model to estimate the *number* of 42-year-old adults who plan to attend a Halloween party. Assume that there are about 4.2 million 42-year-old adults.

f. Use "intersect" on a graphing calculator to estimate at what age 12% of adults plan to attend a Halloween party. [**Hint:** Graph $p = 12$.]

33. The University of Michigan offers a $500,000 life insurance policy. Monthly rates for nonsmoking faculty are shown in Table 44 for various ages.

Table 44 University of Michigan Life Insurance Monthly Rates for Nonsmoking Faculty

Age Group (years)	Age Used to Represent Age Group (years)	Monthly Rate (dollars)
30–34	32	15.00
35–39	37	18.50
40–44	42	26.00
45–49	47	46.00
50–54	52	75.50
55–59	57	118.00
60–64	62	195.50
65–69	67	326.60

Source: *University of Michigan*

Let $g(t)$ be the monthly rate (in dollars) for a nonsmoking faculty member at t years of age.

a. Find an equation of g.

b. What is the coefficient a of your model $g(t) = ab^t$? What does it mean in this situation?

c. What is the base b of your model $g(t) = ab^t$? What does it mean in this situation?

d. Find $g(35)$. What does it mean in this situation?

e. For many life insurance policies, monthly rates for women are different from monthly rates for men. Assume these rates depend on life expectancy only. Given that the life expectancy of women is higher than that of men, would women or men pay higher monthly rates? Explain.

34. A person's heart attack risk can be estimated by using *Framingham point scores,* which are based on such factors as age, cholesterol level, blood pressure, and smoking habits. Men's risks of having a heart attack in the next 10 years are shown in Table 45 for various scores.

Table 45 Risks of Having a Heart Attack

Framingham Point Scores	Risk (percent)
0	1
5	2
10	6
15	20
17	30

Sources: *The Journal of the American Medical Association; Framingham Heart Study*

Let $f(s)$ be a man's risk of having a heart attack in the next 10 years if his score is s points.

a. Find an exponential equation of f.

b. A 47-year-old man with high cholesterol has high blood pressure but does not smoke. His score is 11 points. What is the risk he will have a heart attack in the next 10 years?

c. Another 47-year-old man has the same cholesterol level and blood pressure as the man described in part (b). However, this man's score is 5 points higher, because he smokes. What is the risk that he will have a heart attack in the next 10 years?

d. What is the coefficient a of your model $f(s) = ab^s$? What does it mean in this situation?

e. What is the base b of your model $f(s) = ab^s$? What does it mean in this situation?

35. From 1790 to 1860, U.S. population grew rapidly (see Table 46).

Table 46 U.S. Population

Year	Population (millions)	Population Ratio (current to previous)
1790	3.9	—
1800	5.3	1.36
1810	7.2	
1820	9.6	
1830	12.9	
1840	17.1	
1850	23.2	
1860	31.4	

Source: *U.S. Census Bureau*

a. Complete the third column of Table 46. The first entry is 1.36, since $\frac{\text{1800 population}}{\text{1790 population}} = \frac{5.3}{3.9} \approx 1.36$.

b. What do you observe about the ratios in the third column?

c. On the basis of your observation in part (b), is it best to use a linear function, a quadratic function, or an exponential function to model the data? Explain.

d. Let $f(t)$ be the U.S. population (in millions) at t years since 1790. Find an equation of an exponential function that models the data from 1790 to 1860.

e. Complete the third column of Table 47.

Table 47 U.S. Population

Year	Population (millions)	Population Ratio (current to previous)
1860	31.4	—
1870	39.8	
1880	50.2	
1890	62.9	
1900	76.0	

Source: *U.S. Census Bureau*

f. Is it likely f gives reasonable population estimates after 1860? Explain.

g. Use f to estimate the population in 2012. The actual population was 313.9 million. What is the error in your estimate?

36. The amounts of the federal debt are listed in Table 48 for various years.

Table 48 Federal Debt Amounts

Year	Federal Debt (billions of dollars)
1960	291
1970	381
1980	909
1990	3206
2000	5629
2010	13,529

Source: *U.S. Office of Management and Budget*

a. Let $D = f(t)$ be the federal debt (in billions of dollars) at t years since 1960. Find an equation of f.

b. Predict the federal debt in 2018.

c. If the federal debt were paid off in 2018 by each U.S. citizen contributing an equal amount of money, how much would each person have to pay? Assume the population will be about 335 million in 2018.

d. Predict the federal debt in 2050.

e. If the federal debt were paid off in 2050 by each U.S. citizen contributing an equal amount of money, how much would each person have to pay? Assume the population will be about 439 million in 2050. Explain why some people want to reduce or eliminate the debt now rather than later.

37. The average numbers of lightning deaths per million people per year are shown in Table 49 for various decades, and the percentages of Americans who live in rural areas are shown in the table for various years.

Table 49 Average Numbers of Lightning Deaths per Million People per Year; Percentages of Americans Who Live in Rural Areas

Decade	Year Used to Represent Decade	Lightning Fatality Rate	Year	Percentage of Americans Who Live in Rural Areas
1940–1949	1945	2.4	1950	36.0
1950–1959	1955	1.1	1960	30.1
1960–1969	1965	0.7	1970	26.3
1970–1979	1975	0.5	1980	26.3
1980–1989	1985	0.4	1990	24.8
1990–1999	1995	0.2	2000	22.0
2000–2010	2005	0.1	2010	21.0

Source: *Lopez and Holle 1998; Storm Data; U.S. Census Bureau*

a. Let $f(t)$ be the lightning fatality rate (average number of lightning deaths per million people per year) at t years since 1900. Use your graphing calculator to find the regression equation of f.

b. Predict the lightning fatality rate in 2017. Round your result to the second decimal place.

c. Use your result in part (b) to predict the *number* of lightning deaths in 2017. Assume the U.S. population will be 332 million in that year.

d. The ratio of injuries to deaths, both from lightning, appears to be 10 to 1 (Source: *Cherington et al. 1999*). Use your result in part (c) to predict the number of injuries from lightning in 2017.

e. Let $g(t)$ be the percentage of Americans who live in rural areas at t years since 1900. Use your graphing calculator to find the exponential regression equation of g.

f. It has been hypothesized that the lightning fatality rate has decreased due to the migration of Americans from rural areas to urban ones (Source: *Annual Rates of Lightning Fatalities by Country, Holle 2008*). Explain why this is probably not the only reason by comparing the meaning of the bases of the functions f and g.

38. Table 50 compares the economic strength of a country with the percentage of the population that was involved in producing agriculture. *Gross national product* (*GNP*) is a measure of the amount of goods and services a country produces.

Table 50 Percentage of Population in Agriculture versus GNP

Country	Percent of Population in Agriculture	GNP per Person (dollars)
United States	1	43,743
Great Britain	1	37,632
France	2	35,854
Canada	2	32,546
Australia	3	32,170
Italy	3	29,999
Japan	5	38,984
New Zealand	9	25,942
Slovenia	9	17,352
Korea (North and South)	12	10,975
Latvia	16	6757
Chile	17	5865
Panama	23	4626
Brazil	25	3455
Columbia	32	2292
Bolivia	36	1009
Bangladesh	50	467
Vietnam	60	623

Source: *United Nations*

a. Let $f(p)$ be the GNP per person (in dollars) for a country in which p percent of the population was involved in agriculture. Use a graphing calculator to find the regression equation of f.

b. Use a graphing calculator to sketch a scattergram of the data and your model on the same coordinate system. Does the model fit the data well?

c. What is the base b of your function $f(p) = ab^p$? What does it mean in this situation? Explain why this makes sense in terms of productivity.

d. Which country's data point is farthest from the regression curve? What does the position of the point in relation to the other data points and the regression curve suggest about this situation? [**Hint:** Zoom in.]

Concepts

39. A storage tank contains a radioactive element. Let $p = f(t)$ be the percentage of the element that remains at t years since the element was placed in the tank. A graph of f is shown in Fig. 66.

a. What is the half-life of the element?

b. What percentage of the element will remain in the tank after 40 years?

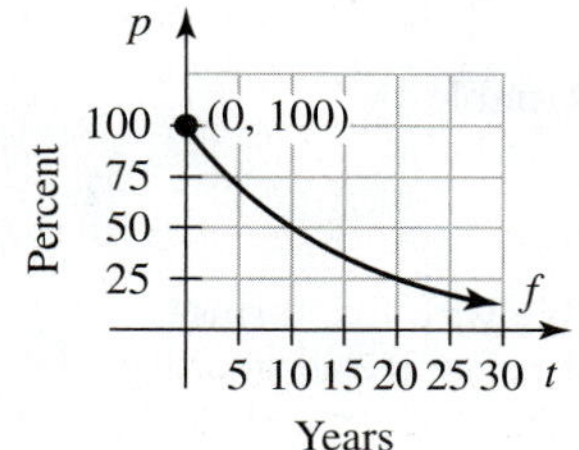

Figure 66 The graph of the model f

40. A storage tank contains a radioactive element. Let $p = g(t)$ be the percentage of the element that remains at t years since the element was placed in the tank. Some values of g are shown in Table 51.

Table 51 Percentages of a Radioactive Element

Year t	Percent $g(t)$
0	100.0
1	79.4
2	63.0
3	50.0
4	39.7
5	31.5
6	25.0

a. What is the half-life of the element?

b. What percentage of the element will remain in the tank after 12 years?

Consider the scattergram of data and the graph of the model $f(t) = ab^t$ in the figure. Sketch the graph of an exponential model that describes the data better; then explain how you would adjust the values of a and b of the original model to describe the data better.

41. See Fig. 67.

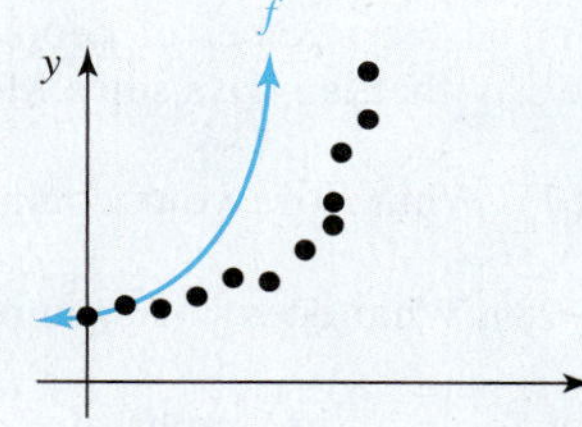

Figure 67 Exercise 41

42. See Fig. 68.

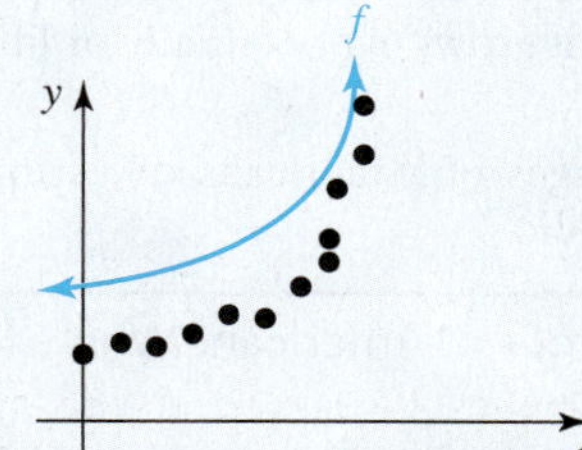

Figure 68 Exercise 42

43. Explain how you can tell whether to model a situation with a linear model, a quadratic model, an exponential model, or none of these. Discuss at least two criteria you can use to help you select a type of model.

44. Assume $f(t) = ab^t$, where $a > 0$, models a quantity at time t.

a. Explain why the quantity grows exponentially at a rate of $b - 1$ percent (in decimal form) if $b > 1$.

b. Explain why the quantity decays exponentially at a rate of $1 - b$ percent (in decimal form) if $0 < b < 1$.

45. Describe how to use the base multiplier property to find an exponential model for a given situation.

46. Explain how to find an exponential model for a situation described by a table of data. Also, explain how to use the model to make an estimate of, or prediction about, the situation.

Related Review

47. In this exercise, you will explore two types of interest-bearing accounts.

a. Suppose \$800 is deposited into a savings account that earns 3% interest compounded annually. Let $C(t)$ be the value (in dollars) of the account at t years after the \$800 has been deposited. Find an equation of C.

b. Now suppose \$800 is deposited into a savings account that earns 3% simple interest. Recall from Section 6.5 that the term *simple interest* means the interest earned each year is 3% of the \$800 only. Let $S(t)$ be the value (in dollars) of the account at t years after the \$800 has been deposited. Find an equation of S. [**Hint:** Each year, the balance increases by $800(0.03) = 24$ dollars.]

c. Find $C(1)$, $C(2)$, $S(1)$, and $S(2)$. Explain in terms of the situation why it makes sense that $C(1)$ is equal to $S(1)$ but that $C(2)$ is not equal to $S(2)$.

d. Compare $C(20)$ with $S(20)$. What does your comparison mean in this situation?

48. On Monday, 20 people receive a prank e-mail warning them zombies are on the loose. On Tuesday, 40 more people receive the e-mail.

a. Assume the number of people receiving the e-mail is growing exponentially. Let $E(t)$ be the number of people who receive the e-mail on the day that is t days since Monday. Find an equation of E.

b. Now assume the number of people receiving the e-mail is growing linearly. Let $L(t)$ be the number of people who receive the e-mail on the day that is t days since Monday. Find an equation of L.

c. Compare $E(7)$ with $L(7)$. What does your comparison mean in this situation?

d. Compare $E(28)$ with $L(28)$. What does your comparison mean in this situation?

e. Use a graphing calculator to draw the graphs of E and L on the same coordinate system. Compare the graphs.

f. Is there much difference in the numbers of people who will receive the e-mail if it grows exponentially or linearly? Explain.

49. Table 52 lists the percentages of Americans of various age groups who listen to talk radio.

Table 52 Percentages of Americans Who Listen to Talk Radio

Age Group (years)	Age Used to Represent Age Group (years)	Percent
12–17	14.5	6
18–34	26.0	16
35–44	39.5	25
45–54	49.5	26
55–64	59.5	21
over 64	75.0	6

Source: *Talkers Magazine: Mediamark*

Let $f(a)$ be the percentage of Americans of age a years who listen to talk radio.

a. Find and verify an equation of f.

b. Estimate the percentage of 20-year-old Americans who listen to talk radio.

c. Estimate the age(s) at which 20% of Americans listen to talk radio.

d. What is the vertex of the model? What does it mean in this situation?

e. What are the a-intercepts? What do they mean in this situation?

50. The costs of 30-second ad slots during the Super Bowl have increased greatly (see Table 53).

Table 53 Costs of Television Ad Slots during the Super Bowl

Super Bowl	Year	Cost for 30 Seconds (millions of dollars)
IX	1975	0.11
XIX	1985	0.5
XXIX	1995	1.0
XXXIX	2005	2.4
XLVI	2012	3.5

Source: *Ocean Media, Inc.*

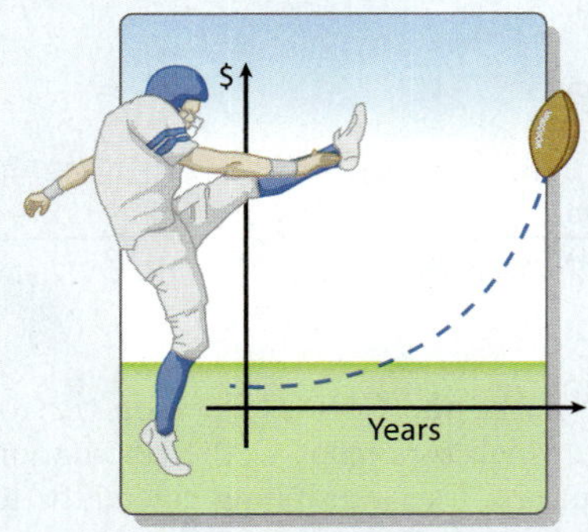

a. First, assume the 30-second ad cost is growing linearly. Let $L(t)$ be the 30-second ad cost (in millions of dollars) at t years since 1950. Find an equation of L.

b. Now assume the 30-second ad cost is growing exponentially. Let $E(t)$ be the 30-second ad cost (in millions of dollars) at t years since 1950. Find an equation of E.

c. Describe how close each model comes to the points in the scattergram.

d. Use L and E to estimate the cost of a 30-second ad slot during Super Bowl I in 1967. The actual cost was \$40,000. Which function estimates that cost better?

e. Find the slope of the graph of L and the base b of $E(t) = ab^t$, and describe what they mean in this situation.

f. In Super Bowl XLIV in 2010, a 30-second ad slot cost \$2.65 million, the first decline in cost since 2003. Use E to help estimate the total loss in revenue from ad slots during Super Bowl XLIV, likely due to a poor economy. [**Hint:** First find $E(60)$.]

g. Use E to predict the cost of a 30-second ad slot during Super Bowl LII in 2018.

Expressions, Equations, Functions, and Graphs

Give an example of the following. Then solve, simplify, or graph, as appropriate.

51. exponential equation in one variable

52. quadratic equation in one variable

53. linear function

54. cubic polynomial in one variable with five terms

55. exponential function

56. system of two linear equations in two variables

57. expression involving exponents

58. quadratic function

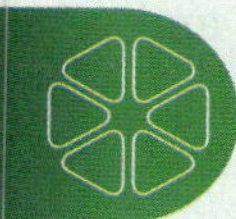

Taking it to the Lab

Stringed Instrument Lab

Many stringed instruments (such as guitars, banjos, and basses) have thin metal strips called *frets* across the neck and underneath the strings. Frets are precisely placed so that the instruments produce the 12 chromatic notes of our Western musical scale. In this lab, you will discover where the frets of a stringed instrument must be placed to produce the 12 chromatic notes. You can apply what you learn to determine where violinists and cellists must put their fingers to produce these 12 notes.

Materials

You will need the following materials:

1. a stringed instrument with frets
2. a meterstick or tape measure

Recording of Data

Measure the length (in centimeters) of one of the strings of the instrument from the *nut* to the *bridge* (see Fig. 69). This is the length of an *open string*. Then measure the length (in centimeters) of the same string from the 12th fret to the bridge.

Figure 69 A bass guitar

Analyzing the Data

1. Compare the sound produced by plucking the open string with that produced by plucking the string when it is pressed just before the 12th fret (toward the nut). The higher pitched note is called the *octave* of the lower note. How do distances between the nut and the bridge and between the 12th fret and the bridge compare?
2. You probably found that when you halve the length of an open string, you can produce the octave of the open string. Since the Western chromatic musical scale has 12 notes from each note to its octave, this means that the 12th fret should be placed in the middle of the nut and the bridge. Where should the 24th fret be placed to achieve the next octave?
3. Complete Table 54.

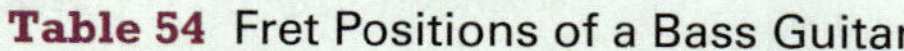

Table 54 Fret Positions of a Bass Guitar

Number of Frets	Length of String from Fret to Bridge (centimeters)
0	
12	
24	
36	
48	
n	

4. Let $f(n)$ be the distance (in centimeters) from the nth fret to the bridge. On the basis of the entries in Table 54, is f linear, quadratic, exponential, or none of these? Explain.
5. Find an equation of f.
6. Use your equation of f to find the distance between the fifth fret and the bridge.
7. Use a graphing calculator table to find the appropriate distances for all the frets of the instrument. Compare these values with the actual distances.
8. On a 1974 Fender® Jazz bass, one of the strings has an open-string length of 86.5 cm. Explain how the Fender music technician knew where to put the frets.
9. In this lab, you observed that an octave is achieved by halving a string. Explain why it follows that the frets of an instrument are closer together when they are closer to the bridge.

Cooling Water Lab

In this lab, you will relate the temperature of some heated water to the amount of time that the water has been cooling.

In case the necessary measuring devices are not available for this experiment, some sample data are listed in Table 55.

Table 55 Temperatures of Water

Time (minutes)	Water Temperature (°C)	Difference Between Water Temperature and Room Temperature (°C)
0	83.49	
5	71.22	
10	63.09	
15	57.23	
20	52.65	
25	48.83	
30	45.63	

Source: *J. Lehmann*
*Room temperature = 21.7°C

Materials

To do this lab, you will need the following materials:

1. hot water
2. a coffee cup
3. a temperature probe that measures temperatures in degrees Celsius. Or measure temperatures in degrees Fahrenheit and use the formula $C = \frac{F - 32}{1.8}$ to convert from degrees Fahrenheit to degrees Celsius. *If you are using a thermometer as the temperature probe, make sure that it can handle hot water; otherwise it might break or even explode.*
4. a timing device

Preparation

Heat some water and pour it into a coffee cup. Set the cup on a counter or table to cool. Choose an environment where the temperature of the air will not change much during the 30-minute experiment.

Recording of Data

Record the temperature of the air at the start of the experiment and then again at the end. If the room temperatures at the beginning and end of the experiment are significantly different, redo the experiment in an environment where the room temperature will not change much. Also, record the water temperature every five minutes for 30 minutes.

Analyzing the Data

1. Find the average of the room temperature in degrees Celsius at the start and end of the experiment. This average will be referred to as *the* room temperature. If you are using the sample data in Table 55, the room temperature when it was collected was 21.7°C.
2. Display the water temperature data in the first two columns of a table like Table 55.
3. For each water temperature reading, compute the difference between the water temperature and the room temperature. Enter these differences in the third column of your table.
4. Let $D = f(t)$ be the difference between the water and room temperatures (in degrees Celsius) at t minutes after the water is allowed to cool. Create a scattergram by hand, comparing D with t for your cooling-water data.
5. Use a graphing calculator to draw a scattergram of the data.
6. Find an equation of f.
7. Use a graphing calculator to draw a graph of your model and the scattergram in the same viewing window. Also, graph the model and scattergram by hand. How well does f model the data?
8. If your model is linear, what is the meaning of the slope? If your model is exponential, what is the meaning of the base?
9. Use your model to estimate the water temperature at 22 minutes.
10. Use your model to estimate the water temperature at 1 hour.
11. What will be the temperature of the water when it stops cooling? Does f predict when this temperature will be reached? According to f, how much time will it take for this to happen? Is this a reasonable prediction? Explain.

Exponential Lab: Topic of Your Choice

Your objective in this lab is to find some authentic data that appear to be exponential and then model them with an exponential function. Your function should model a situation that has not been discussed in this text. Your first task will be to find the data. Almanacs, newspapers, magazines, scientific journals, and the Internet are good resources. Or you can conduct an experiment to obtain your data.

Analyzing the Data

1. What two variables did you explore? [**Hint:** Describe the units of the variables.]
2. Which variable is the dependent variable? Which variable is the independent variable? Explain.
3. Does it make sense to you why an exponential function would model your situation best? Explain.
4. State the source of your data. If you conducted an experiment, provide a careful description with specific details of how you ran your experiment.
5. Include a table of your data.
6. Use a graphing calculator to draw a scattergram of your data.
7. Find an equation of an exponential model to describe the data.
8. Use a graphing calculator to graph your exponential model and the scattergram in the same viewing window. Graph your model and the scattergram by hand. How well does your model fit the data?

9. What is the base b of your exponential model? What does it mean in this situation?

10. What is the coefficient a of your exponential model? What does it mean in this situation?

11. Find any intercepts of your exponential model. What do they mean in this situation?

12. Choose an input of your model. Find the output that comes from the input. What does your result mean in this situation?

13. Comment on your lab experience.

a. For example, you might address whether this lab was enjoyable, insightful, and so on.

b. Were you surprised by any of your findings? If so, which ones?

c. How would you improve your lab procedure if you did this lab again?

d. How would you improve your procedure if you had more time and money?

Chapter Summary

Key Points of Chapter 10

Section 10.1 Integer Exponents

Negative-integer exponent	If n is a counting number and $b \neq 0$, then $b^{-n} = \dfrac{1}{b^n}$.
Negative-integer exponent in a denominator	If n is a counting number and $b \neq 0$, then $\dfrac{1}{b^{-n}} = b^n$.
Properties of integer exponents	If m and n are integers, $b \neq 0$, and $c \neq 0$, then • $b^m b^n = b^{m+n}$ **Product property for exponents** • $\dfrac{b^m}{b^n} = b^{m-n}$ **Quotient property for exponents** • $(bc)^n = b^n c^n$ **Raising a product to a power** • $\left(\dfrac{b}{c}\right)^n = \dfrac{b^n}{c^n}$ **Raising a quotient to a power** • $(b^m)^n = b^{mn}$ **Raising a power to a power**
Simplifying expressions involving exponents	An expression involving exponents is simplified if **1.** It includes no parentheses. **2.** Each variable or constant appears as a base at most once. **3.** Each numerical expression has been calculated, and each numerical fraction has been simplified. **4.** Each exponent is positive.
Exponential function	An **exponential function** is a function whose equation can be put into the form $f(x) = ab^x$, where $a \neq 0$, $b > 0$, and $b \neq 1$. The constant b is called the **base.**
Scientific notation	A number is written in **scientific notation** if it has the form $N \times 10^k$, where k is an integer and the absolute value of N is between 1 and 10 or is equal to 1.
Converting from scientific notation to standard decimal notation	To write the scientific notation $N \times 10^k$ in standard decimal notation, we move the decimal point of the number N as follows: • If k is *positive,* we multiply N by 10 k times; hence, we move the decimal point k places to the *right.* • If k is *negative,* we divide N by 10 k times; hence, we move the decimal point k places to the *left.*
Converting from standard decimal notation to scientific notation	To write a number in scientific notation, count the number k of places that the decimal point must be moved so that the absolute value of the new number N is between 1 and 10 or is equal to 1: • If the decimal point is moved to the left, then the scientific notation is written as $N \times 10^k$. • If the decimal point is moved to the right, then the scientific notation is written as $N \times 10^{-k}$.

Section 10.2 Rational Exponents

Definition of $b^{1/n}$	For the counting number n, where $n \neq 1$, • If n is odd, then $b^{1/n}$ is the number whose nth power is b, and we call $b^{1/n}$ the ***n*th root of *b*.** • If n is even and $b \geq 0$, then $b^{1/n}$ is the nonnegative number whose nth power is b, and we call $b^{1/n}$ the **principal *n*th root of *b*.** • If n is even and $b < 0$, then $b^{1/n}$ is not a real number. $b^{1/n}$ may be represented by $\sqrt[n]{b}$.
Rational exponent	For the counting numbers m and n, where $n \neq 1$ and b is any real number for which $b^{1/n}$ is a real number, • $b^{m/n} = (b^{1/n})^m = (b^m)^{1/n}$ • $b^{-m/n} = \dfrac{1}{b^{m/n}}, b \neq 0$
Properties of rational exponents	If m and n are rational numbers and b and c are any real numbers for which b^m, b^n, and c^n are real numbers, then • $b^m b^n = b^{m+n}$ **Product property for exponents** • $\dfrac{b^m}{b^n} = b^{m-n}, b \neq 0$ **Quotient property for exponents** • $(bc)^n = b^n c^n$ **Raising a product to a power** • $\left(\dfrac{b}{c}\right)^n = \dfrac{b^n}{c^n}, c \neq 0$ **Raising a quotient to a power** • $(b^m)^n = b^{mn}$ **Raising a power to a power**

Section 10.3 Graphing Exponential Functions

Base multiplier property	For an exponential function of the form $y = ab^x$, if the value of the independent variable increases by 1, the value of the dependent variable is multiplied by b.
Increasing or decreasing property	Let $f(x) = ab^x$, where $a > 0$. Then • If $b > 1$, then the function f is increasing. We say the function **grows exponentially.** • If $0 < b < 1$, then the function f is decreasing. We say the function **decays exponentially.**
***y*-intercept of an exponential function**	For an exponential function of the form $y = ab^x$, the y-intercept is $(0, a)$.
Reflection property	The graphs of $f(x) = -ab^x$ and $g(x) = ab^x$ are **reflections** of each other across the x-axis.
Horizontal asymptote	For all exponential functions, the x-axis is a horizontal asymptote.
Domain	The domain of any exponential function $f(x) = ab^x$ is the set of real numbers.
Range	The range of an exponential function $f(x) = ab^x$ is the set of all positive real numbers if $a > 0$, and the range is the set of all negative real numbers if $a < 0$.

Section 10.4 Finding Equations of Exponential Functions

Solving $b^n = k$ for b	To solve an equation of the form $b^n = k$ for b, • If n is odd, the real-number solution is $k^{1/n}$. • If n is even and $k \geq 0$, the real-number solutions are $\pm k^{1/n}$. • If n is even and $k < 0$, there is no real-number solution.
Dividing left sides and right sides of two equations	If $a = b$, $c = d$, $c \neq 0$, and $d \neq 0$, then $\dfrac{a}{c} = \dfrac{b}{d}$.
Finding an equation	We can find an equation of an exponential function by using the base multiplier property or by using two points. Both methods give the same result.

Section 10.5 Using Exponential Functions to Model Data

Exponential model, exponentially related, approximately exponentially related	An **exponential model** is an exponential function, or its graph, that describes the relationship between two quantities for an authentic situation. If all of the data points for a situation lie on an exponential curve, then we say the independent and dependent variables are **exponentially related.** If no exponential curve contains all of the data points, but an exponential curve comes close to all of the data points (and perhaps contains some of them), then we say the variables are **approximately exponentially related.**
Quantity present at time $t = 0$	If $y = ab^t$ is an exponential model where y is a quantity at time t, then the coefficient a is the value of that quantity present at time $t = 0$.
***r* percent interest compounded annually**	The term ***r* percent interest compounded annually** means the interest earned each year equals r percent of the principal plus any interest earned in previous years (all of which becomes part of the investment).
Half-life	If a quantity decays exponentially, the **half-life** is the amount of time it takes for that quantity to be reduced to half.
Meaning of the base *b*	If $f(t) = ab^t$, where $a > 0$, models a quantity at time t, then the percent rate of change is constant. In particular, • If $b > 1$, then the quantity grows exponentially at a rate of $b - 1$ percent (in decimal form) per unit of time. • If $0 < b < 1$, then the quantity decays exponentially at a rate of $1 - b$ percent (in decimal form) per unit of time.
Four-step modeling process	To find a model and then make estimates and predictions, **1.** Create a scattergram of the data. Decide whether a line, a parabola, an exponential curve, or none of these comes close to the points. **2.** Find an equation of your model. **3.** Verify your equation by checking that the graph comes close to all of the data points. **4.** Use your equation of the model to draw conclusions, make estimates, and/or make predictions.

Chapter 10 Review Exercises

Simplify. Assume b and c are positive.

1. $\dfrac{2^{-400}}{2^{-405}}$

2. $(8b^{-3}c^5)(6b^{-9}c^{-2})$

3. $\dfrac{4b^{-3}c^{12}}{16b^{-4}c^3}$

4. $(2b^{-5}c^{-2})^3(3b^4c^{-6})^{-2}$

5. $(37b^{-3}c^4)^{-97}(37b^{-3}c^4)^{97}$

6. $\dfrac{(20b^{-2}c^{-9})(27b^5c^3)}{(18b^3c^{-1})(30b^{-1}c^{-4})}$

7. $32^{4/5}$

8. $16^{-3/4}$

9. $b^{-4/5}b^{2/3}$

10. $\dfrac{b^{-1/3}}{b^{4/3}}$

11. $\dfrac{(16b^8c^{-4})^{1/4}}{(25b^{-6}c^4)^{3/2}}$

12. $\left(\dfrac{32b^2c^5}{2b^{-6}c^1}\right)^{1/4}$

13. $(8^{2/3}b^{-1/3}c^{3/4})(64^{-1/3}b^{1/2}c^{-5/2})$

For Exercises 14 and 15, simplify. Assume n is a counting number.

14. $b^{2n-1}b^{4n+3}$

15. $\dfrac{b^{n/2}}{b^{n/3}}$

16. Use properties of exponents to show why $3^{2x} = 9^x$.

For $f(x) = 3(5)^x$ and $g(x) = 6^x$, find the following.

17. $f(-2)$

18. $g(a + 2)$

For $f(x) = 49^x$ and $g(x) = 2(81)^x$, find the following.

19. $f\left(\dfrac{1}{2}\right)$

20. $g\left(-\dfrac{3}{4}\right)$

Write the number in standard decimal form.

21. 4.4487×10^7

22. 3.85×10^{-5}

Write the number in scientific notation.

23. 54,000,000

24. -0.00897

Graph the equation by hand.

25. $f(x) = 2(3)^x$

26. $k(x) = -18\left(\dfrac{1}{3}\right)^x$

Graph the function by hand. Find its domain and range.

27. $h(x) = -3(2)^x$

28. $g(x) = 12\left(\dfrac{1}{2}\right)^x$

For Exercises 29–34, find all real-number solutions of the equation. Round any result(s) to the second decimal place.

29. $b^3 = 8$

30. $2b^5 = 60$

31. $3.9b^7 = 283.5$

32. $5b^4 - 13 = 67$

33. $\dfrac{1}{3}b^2 - \dfrac{1}{5} = \dfrac{2}{3}$

34. $\dfrac{b^7}{b^2} = \dfrac{83}{6}$

35. Some values of the functions f, g, h, and k are provided in Table 56. For each function, determine whether the given values suggest that the function is linear, exponential, or neither. If the function could be linear or exponential, find a possible equation for it.

Table 56 Values of Four Functions (Exercises 35–39)

x	$f(x)$	$g(x)$	$h(x)$	$k(x)$
1	30	5	2	96
2	26	15	3	48
3	22	45	6	24
4	18	135	11	12
5	14	405	18	6

For Exercises 36–39, refer to Table 56.

36. Find $f(4)$.

37. Find $h(3)$

38. Find x when $g(x) = 5$.

39. Find x when $k(x) = 6$.

For Exercises 40–43, find an approximate equation $y = ab^x$ of the exponential curve that contains the given pair of points. Round the values of a and b to two decimal places.

40. $(0, 2)$ and $(5, 3)$

41. $(0, 3.8)$ and $(4, 113.2)$

42. $(3, 30)$ and $(9, 7)$

43. $(5, 6.9)$ and $(20, 78.6)$

44. Consider the scattergram of data and the graph of the model $f(x) = ab^t$ in Fig. 70. Sketch the graph of an exponential model that describes the data better; then explain how you would adjust the values of a and b of the original model to describe the data better.

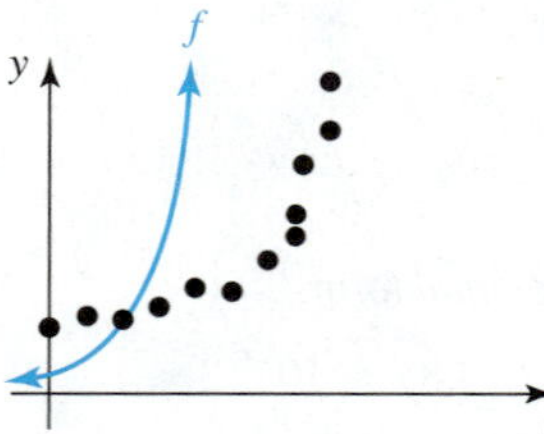

Figure 70 Exercise 44

45. Suppose $2000 is deposited into an account that earns 7% interest compounded annually. Let $f(t)$ be the value (in dollars) of the account at t years after the $2000 is deposited.

a. Find an equation of f.

b. Find the value of the account after 5 years.

46. A corporation's annual revenue has doubled every year. The revenue in 2010 was $17 thousand.

a. Let $g(t)$ be the annual revenue (in thousands of dollars) at t years since 2010. Find an equation of g.

b. Predict the revenue in 2018.

47. A storage tank contains carbon-14, a radioactive element with a half-life of 5730 years. Let $f(t)$ be the percentage of carbon-14 that remains at t years since the element was placed in the tank.

a. Find an equation of f.

b. Predict the percentage of carbon-14 remaining after 100 years.

48. The number of homes that are unoccupied has grown approximately exponentially from 8.5 million homes in 1980 to 14.3 million homes in 2010 (Source: *U.S. Census Bureau*). Predict the number of homes that will be unoccupied in 2017.

49. The prices of one ounce of gold are shown in Table 57 on January 1 of various years.

Table 57 Prices of Gold

Year	Price (dollars per ounce)
2000	282
2002	278
2004	416
2006	530
2008	846
2010	1121
2012	1599

Source: *Kitco Metals, Inc.*

a. Let $f(t)$ be the price (in dollars per ounce) of gold on January 1 at t years since 2000. Find an exponential equation of f.

b. What is the coefficient a of your model $f(t) = ab^t$? What does it mean in this situation?

c. What is the base b of your model $f(t) = ab^t$? What does it mean in this situation?

d. Find $f(18)$. What does it mean in this situation?

50. Zimride is a website where people at least 18 years old can offer and get paid for shared car rides. The numbers of users of the site are shown in Table 58 for various years.

Table 58 Numbers of Users of Zimride

Year	Number of Users (thousands)
2007	6
2008	10
2009	35
2010	105
2011	200
2012	400

Source: *Zimride*

a. Let $n = f(t)$ be the number (in thousands) of users of Zimride at t years since 2000. Find an exponential equation of f.

b. Find the percentage rate of growth of the number of users of Zimride.

c. Predict the number of users of Zimride in 2017.

d. Use "intersect" on a graphing calculator to predict when all people at least 18 years old will be users of Zimride. Assume there will be 260 million people at least 18 years old in that year. [**Hint:** Enter the model and the horizontal line $n = 260{,}000$, use ZoomFit, and then Zoom Out. See Appendix A.6 for graphing calculator instructions.]

Chapter 10 Test

Simplify.

1. $32^{2/5}$

2. $-8^{-4/3}$

For Exercises 3–8, simplify. Assume b and c are positive.

3. $(2b^3c^8)^3$

4. $\left(\dfrac{4b^{-3}c}{25b^5c^{-9}}\right)^0$

5. $\dfrac{b^{1/2}}{b^{1/3}}$

6. $\dfrac{25b^{-9}c^{-8}}{35b^{-10}c^{-3}}$

7. $\left(\dfrac{6b(b^3c^{-2})}{3b^2c^5}\right)^2$

8. $\dfrac{(25b^8c^{-6})^{3/2}}{(7b^{-2})(2c^3)^{-1}}$

9. Use properties of exponents to show that $8^{x/3}2^{x+3} = 8(4)^x$.

For $f(x) = 4^x$, find the following.

10. $f(-2)$

11. $f\left(-\dfrac{3}{2}\right)$

For Exercises 12 and 13, graph the function by hand. Find its domain and range.

12. $f(x) = -5(2)^x$

13. $f(x) = 18\left(\dfrac{1}{3}\right)^x$

14. For each graph in Fig. 71, find an equation of an exponential function that could fit the graph.

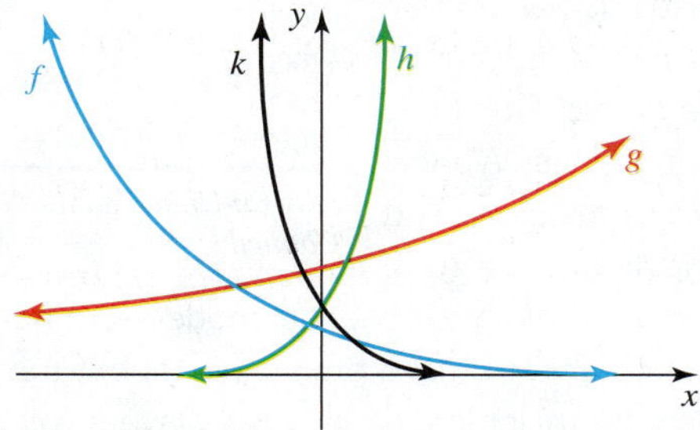

Figure 71 Exercise 14

15. Some values of a function f are provided in Table 59. Find an equation of f in terms of t.

Table 59 Values of a Function f (Exercise 15)

t	$f(t)$
0	160
1	80
2	40
3	20
4	10
5	5

16. Find all real-number solutions of $3b^6 + 5 = 84$. Round any result(s) to the second decimal place.

Find an approximate equation $y = ab^x$ of an exponential curve that contains the given pair of points. Round the values of a and b to two decimal places.

17. $(0, 70)$ and $(6, 20)$

18. $(4, 9)$ and $(7, 50)$

For Exercises 19–21, refer to Fig. 72.

19. Find $f(0)$.

20. Find x when $f(x) = 3$.

21. Find an equation of f.

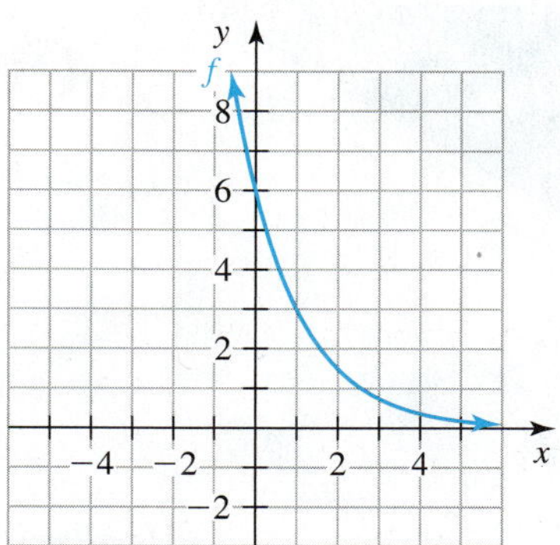

Figure 72 Exercises 19–21

22. On March 1, there are 400 leaves on a tree. For a while, the total number of leaves triples each week. Let $f(t)$ be the total number of leaves on the tree at t weeks since March 1.

a. Find an equation of f.

b. Find $f(6)$. What does it mean in this situation?

c. Find $f(52)$. What does it mean in this situation?

23. The numbers of fraud complaints by consumers are shown in Table 60 for various years.

Table 60 Numbers of Fraud Complaints by Consumers

Year	Number of Complaints (millions)
2005	0.6
2006	0.7
2007	0.7
2008	0.9
2009	1.0
2010	1.2
2011	1.5

Source: *Federal Trade Commission*

a. Let $f(t)$ be the number (in millions) of fraud complaints by consumers in the year that is t years since 2000. Find an exponential equation of f.

b. What is the base b of your model $f(t) = ab^t$? What does it mean in this situation?

c. What is the coefficient a of your model $f(t) = ab^t$? What does it mean in this situation?

d. Predict the number of complaints in 2018.

e. Predict what *percentage* of the U.S. population will file fraud complaints in 2020. Assume the U.S. population will be 341 million in that year and that no one will file more than one complaint.

11 Logarithmic Functions

Have you heard of "green" buildings? They are buildings designed to meet certain environmental standards. The numbers of buildings with Leadership in Energy and Environmental Design (LEED) certification for being green are shown in Table 1 for various years. In Exercise 22 of Homework 11.5, you will predict when there will be an average of 2000 LEED-certified green buildings per state.

Table 1 Numbers of LEED-Certified Green Buildings

Year	Number of Buildings with LEED Certification
2002	28
2004	183
2006	660
2008	2113
2010	7267

Source: *United States Green Building Council*

In Chapter 10, we worked with exponential functions. In this chapter, we will work with *logarithmic functions,* which are closely related to exponential functions. We will also discuss how to simplify *logarithmic expressions* and solve *logarithmic equations in one variable.* In Section 10.5, we made predictions for the dependent variable of an exponential model—but not for the independent variable. In Section 11.5, we will be able to make predictions about the independent variable, such as when the minimum salary for major league baseball players will be $1 million.

11.1 Composite Functions

Objectives

- » Know the meaning of *composite function.*
- » Use tables to evaluate a composite function.
- » Use equations to evaluate a composite function.
- » Find an equation of a composite function.
- » Use graphs to evaluate a composite function.
- » Use a composite function to model an authentic situation.

In this section, we will discuss how to combine two functions to make a new function called the *composite function.*

Definition of a Composite Function

Years ago, vending machines did not accept dollar bills. Suppose a student puts a dollar bill into a change machine and gets four quarters. Then the student puts the quarters into an old-fashioned vending machine and gets a bag of chips (see Fig. 1). So the output (quarters) of the change maker is used as the input of the old-fashioned vending machine.

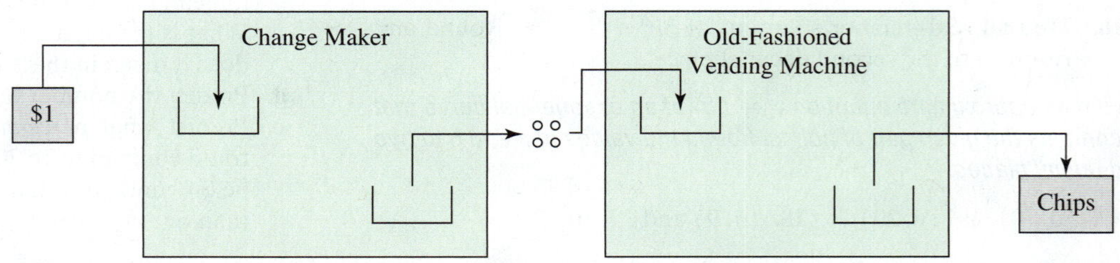

Figure 1 Action of change maker followed by action of old-fashioned vending machine

Now suppose the student puts the dollar bill directly into a modern vending machine and gets the bag of chips (see Fig. 2).

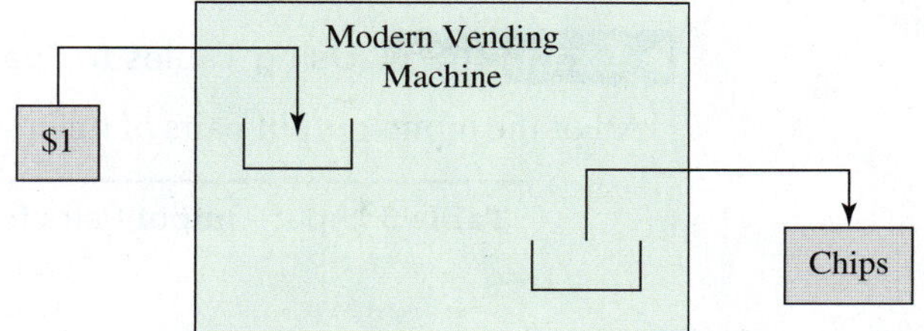

Figure 2 Action of modern vending machine

Note that the modern vending machine combines the actions of making change and then selling chips.

Similarly, we can create a function that combines the "actions" of two functions, where the output of the first function is used as the input of the second function. For example, we will next combine the tasks of finding the total cost of some tickets and then computing the sales tax. Let $C = g(n)$ be the total cost of n tickets and let $f(C)$ be the sales tax for C dollars worth of tickets (see Tables 2 and 3).

Table 2 Input–Output Pairs of g

Number of Tickets n	Total Cost (dollars) $g(n)$
2	40
3	60
4	80
5	100

Table 3 Input–Output Pairs of f

Total Cost of Tickets (dollars) C	Sales Tax (dollars) $f(C)$
40	3.20
60	4.80
80	6.40
100	8.00

Table 4 Input–Output Pairs of the Composition of f and g

Number of Tickets n	Sales Tax (dollars) $(f \circ g)(n)$
2	3.20
3	4.80
4	6.40
5	8.00

To find the sales tax on 2 tickets, we first refer to the first row of Table 2 and see that the total cost of 2 tickets is \$40. Then we refer to the first row of Table 3 and see that the sales tax on \$40 worth of tickets is \$3.20. In conclusion, the sales tax on 2 tickets is \$3.20.

We can repeat the process for 3, 4, and 5 tickets and summarize the information in Table 4, leaving out the middle step of showing the cost of the tickets. The pairs of numbers in Table 4 are input–output values of a function called the *composition of f and g*, written $f \circ g$.

In Fig. 3, we show the input–output action of g followed by the input–output action of f. Since the output of g, $g(n)$, is used as the input of f, the final output is $f(g(n))$.

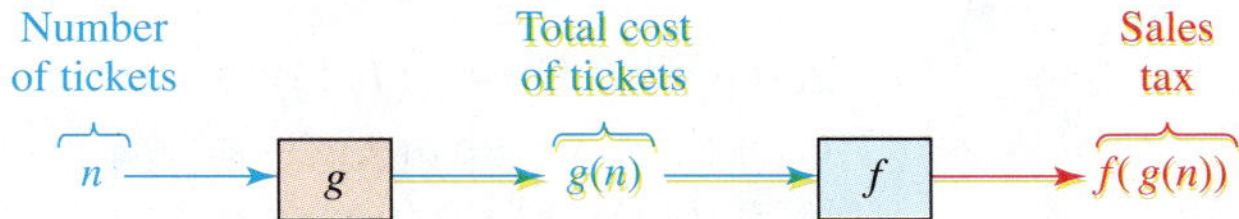

Figure 3 Action of g followed by action of f

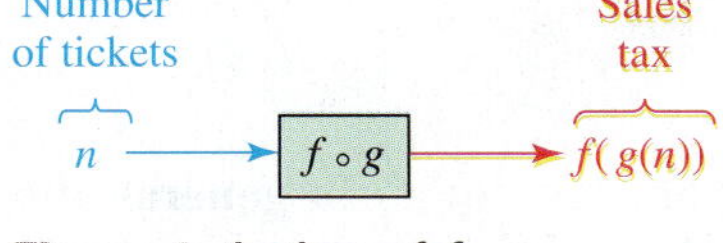

Figure 4 Action of $f \circ g$

In Fig. 4, we show the input–output action of $f \circ g$. Note that $f \circ g$ combines the actions of g and f, much like the modern vending machine combines the actions of the change maker and the old-fashioned vending machine.

Definition Composite function

If f and g are functions, x is in the domain of g, and $g(x)$ is in the domain of f, then we can form the **composite function $f \circ g$:**

$$(f \circ g)(x) = f(g(x))$$

We say $f \circ g$ is the **composition of f and g.**

Using Tables to Evaluate a Composite Function

In Example 1, we will use tables to evaluate the composition of two functions.

Example 1 Using Tables to Evaluate a Composite Function

All of the input–output pairs of functions g and f are shown in Tables 5 and 6, respectively.

Table 5 Input–Output Pairs for g

x	$g(x)$
0	8
1	5
2	9
3	6
4	7

Table 6 Input–Output Pairs for f

x	$f(x)$
5	30
6	27
7	24
8	21
9	18

1. Find $(f \circ g)(0)$.
2. Use a table to describe five input–output pairs of $f \circ g$.

Solution

1. $(f \circ g)(0) = f(g(0))$ *Definition of $f \circ g$*
$= f(8)$ *From the first row of Table 5, we see that $g(0) = 8$.*
$= 21$ *From the fourth row of Table 6, we see that $f(8) = 21$.*

2. We repeat the process used in Problem 1 for the inputs 1, 2, 3, and 4 and show the work in Table 7, including the work from Problem 1.

Table 7 Input–Output Pairs for $f \circ g$

x	$h(x)$
0	$f(g(0)) = f(8) = 21$
1	$f(g(1)) = f(5) = 30$
2	$f(g(2)) = f(9) = 18$
3	$f(g(3)) = f(6) = 27$
4	$f(g(4)) = f(7) = 24$

Evaluating a Composite Function

Instead of using tables, we can use equations of functions to evaluate their composition.

Example 2 Evaluating a Composite Function

Let $f(x) = 2x - 5$ and $g(x) = 3x + 6$.

1. Find $(f \circ g)(4)$. **2.** Find $(g \circ f)(4)$.

Solution

1. $(f \circ g)(4) = f(g(4))$ *Definition of composite function*
$= f(18)$ *$g(4) = 3(4) + 6 = 18$*
$= 2(18) - 5$ *Substitute 18 for x in $f(x) = 2x - 5$.*
$= 31$ *Compute.*

2. $(g \circ f)(4) = g(f(4))$ *Definition of composite function*
$= g(3)$ *$f(4) = 2(4) - 5 = 3$*
$= 3(3) + 6$ *Substitute 3 for x in $g(x) = 3x + 6$.*
$= 15$ *Compute.*

Note that the results of Problems 1 and 2 of Example 2 are not equal. **In general, the outputs $(f \circ g)(a)$ and $(g \circ f)(a)$ may or may not be equal.**

Finding an Equation of a Composite Function

To prepare for Example 3, consider the function $f(x) = 5x$. To find $f(4)$, we substitute 4 for x in $f(x) = 5x$:

$$f(4) = 5(4)$$

Similarly, to find $f(x - 3)$, we substitute $x - 3$ for x in $f(x) = 5x$:

$$f(x - 3) = 5(x - 3)$$

Example 3 Finding Equations of Composite Functions

Let $f(x) = 5x$ and $g(x) = x - 3$.

1. Find an equation of $f \circ g$. **2.** Find an equation of $g \circ f$.

Solution

1. Since $g(x) = x - 3$, we can substitute $x - 3$ for $g(x)$ in the second step:

$$\begin{aligned}(f \circ g)(x) &= f(g(x)) && \text{Definition of composite function}\\ &= f(x - 3) && \text{Substitute } x - 3 \text{ for } g(x).\\ &= 5(x - 3) && \text{Substitute } x - 3 \text{ for } x \text{ in } f(x) = 5x.\\ &= 5x - 15 && \text{Distributive law}\end{aligned}$$

2. Since $f(x) = 5x$, we can substitute $5x$ for $f(x)$ in the second step:

$$\begin{aligned}(g \circ f)(x) &= g(f(x)) && \text{Definition of composite function}\\ &= g(5x) && \text{Substitute } 5x \text{ for } f(x).\\ &= 5x - 3 && \text{Substitute } 5x \text{ for } x \text{ in } g(x) = x - 3.\end{aligned}$$

Similar to what we noticed in Example 2, the results of Problems 1 and 2 in Example 3 are different. **In general, $f \circ g$ and $g \circ f$ may or may not be the same function.**

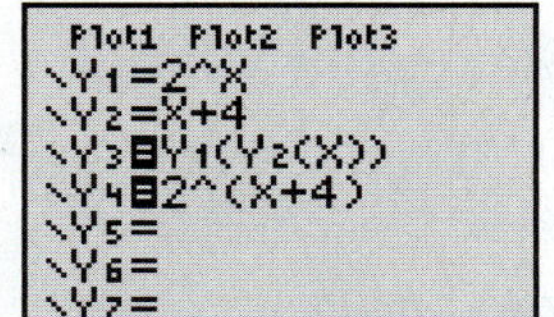

X	Y3	Y4
0	16	16
1	32	32
2	64	64
3	128	128
4	256	256
5	512	512
6	1024	1024

X=0

Figure 5 Verify the work

Example 4 Finding Equations of Composition Functions

Let $f(x) = 2^x$ and $g(x) = x + 4$.

1. Find an equation of $f \circ g$. **2.** Find an equation of $g \circ f$.

Solution

1. Since $g(x) = x + 4$, we can substitute $x + 4$ for $g(x)$ in the second step:

$$\begin{aligned}(f \circ g)(x) &= f(g(x)) && \text{Definition of composite function}\\ &= f(x + 4) && \text{Substitute } x + 4 \text{ for } g(x).\\ &= 2^{x+4} && \text{Substitute } x + 4 \text{ for } x \text{ in } f(x) = 2^x.\end{aligned}$$

We verify our work by creating a graphing calculator table for $y = f(g(x))$ and $y = 2^{x+4}$. See Fig. 5. (The command $Y_1(Y_2(X))$ is the calculator's notation for composing functions. To enter an equation by using Y_n references, see Appendix A.25.)

2. Since $f(x) = 2^x$, we can substitute 2^x for $f(x)$ in the second step:

$$\begin{aligned}(g \circ f)(x) &= g(f(x)) && \text{Definition of composite function}\\ &= g(2^x) && \text{Substitute } 2^x \text{ for } f(x).\\ &= 2^x + 4 && \text{Substitute } 2^x \text{ for } x \text{ in } g(x) = x + 4.\end{aligned}$$

Example 5 Evaluating and Finding an Equation of a Composite Function

Let $f(x) = x^2$ and $g(x) = 4x - 5$.

1. Find $(f \circ g)(2)$.
2. Find an equation of $f \circ g$.
3. Use the equation of $f \circ g$ to find $(f \circ g)(2)$.

Solution

1.
$$\begin{aligned}(f \circ g)(2) &= f(g(2)) && \text{Definition of composite function}\\ &= f(3) && g(2) = 4(2) - 5 = 3.\\ &= 3^2 && \text{Substitute 3 for } x \text{ in } f(x) = x^2.\\ &= 9 && \text{Compute.}\end{aligned}$$

2. Since $g(x) = 4x - 5$, we can substitute $4x - 5$ for $g(x)$ in the second step:

$(f \circ g)(x) = f(g(x))$	*Definition of composite function*
$= f(4x - 5)$	*Substitute $4x - 5$ for $g(x)$.*
$= (4x - 5)^2$	*Substitute $4x - 5$ for x in $f(x) = x^2$.*
$= 16x^2 - 40x + 25$	$(A - B)^2 = A^2 - 2AB + B^2$

3.

$(f \circ g)(2) = 16(2)^2 - 40(2) + 25$	*Substitute 2 for x in $(f \circ g)(x) = 16x^2 - 40x + 25$.*
$= 9$	*Compute.*

So, $(f \circ g)(2)$ is equal to 9, which is the same result we found in Problem 1.

Example 6 Comparing a Composite Function and a Product Function

Let $f(x) = 2x^2 - 4x$ and $g(x) = 3x - 2$.

1. Find an equation of the composite function $f \circ g$.
2. Find an equation of the product function $f \cdot g$.

Solution

1.

$(f \circ g)(x) = f(g(x))$	*Definition of composition*
$= f(3x - 2)$	*Substitute $3x - 2$ for $g(x)$.*
$= 2(3x - 2)^2 - 4(3x - 2)$	*Substitute $3x - 2$ for x in $f(x) = 2x^2 - 4x$.*
$= 2(9x^2 - 12x + 4) - 4(3x - 2)$	$(A - B)^2 = A^2 - 2AB + B^2$
$= 18x^2 - 24x + 8 - 12x + 8$	*Distributive law*
$= 18x^2 - 36x + 16$	*Combine like terms.*

2.

$(f \cdot g)(x) = f(x) \cdot g(x)$	*Definition of product function*
$= (2x^2 - 4x)(3x - 2)$	*Substitute $2x^2 - 4x$ for $f(x)$ and $3x - 2$ for $g(x)$.*
$= 6x^3 - 4x^2 - 12x^2 + 8x$	*Multiply pairs of terms.*
$= 6x^3 - 16x^2 + 8x$	*Combine like terms.*

WARNING It is a common error to confuse the composition function $f \circ g$ with the product function $f \cdot g$. Our work in Example 6 shows that these two functions are not the same in general.

Example 7 Expressing a Function as a Composition of Two Functions

If $h(x) = (7x + 3)^5$, find equations of the functions f and g such that $h(x) = (f \circ g)(x)$.

Solution

To form $f(g(x))$, we substitute $g(x)$ for x in $f(x)$. Similarly, to form $(7x + 3)^5$, we can substitute $7x + 3$ for x in x^5. This suggests that $g(x) = 7x + 3$ and $f(x) = x^5$. We check by performing the composition:

$$(f \circ g)(x) = f(g(x)) = f(7x + 3) = (7x + 3)^5$$

There are other possible answers. For example, $g(x) = 7x$ and $f(x) = (x + 3)^5$ also work:

$$(f \circ g)(x) = f(g(x)) = f(7x) = (7x + 3)^5$$

Using Graphs to Evaluate a Composite Function

Instead of using tables or equations to evaluate a composition, we can use graphs of functions.

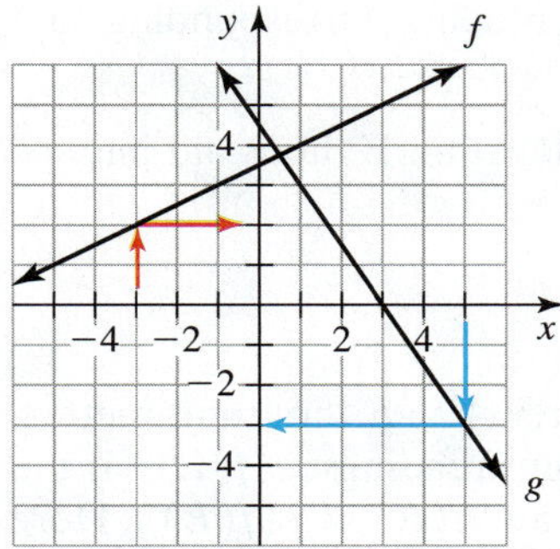

Figure 6 Graphs of f and g

Example 8 Using Graphs to Evaluate a Composite Function

Refer to the graphs in Fig. 6 to find $(f \circ g)(5)$.

Solution

The blue arrows in Fig. 6 show that $g(5) = -3$. So, we can substitute -3 for $g(5)$ in the second step:

$$\begin{aligned}(f \circ g)(5) &= f(g(5)) && \text{Definition of composite function}\\ &= f(-3) && \text{Substitute } -3 \text{ for } g(5).\\ &= 2 && \text{The red arrows in Fig. 6 show that } f(-3) = 2.\end{aligned}$$

Using a Composite Function to Model an Authentic Situation

Near the start of this section, we used tables to evaluate the composition of two models. In Example 9, we will use an equation to evaluate the composition of two models.

Example 9 Using a Composite Function to Model a Situation

Let $f(Q)$ be the number of cups in Q quarts, and let $g(x)$ be the number of ounces in x cups.

1. Find an equation of f.
2. Find an equation of g.
3. Find an equation of $g \circ f$.
4. Find $(g \circ f)(5)$. What does it mean in this situation?

Solution

1. Since there are 4 cups in one quart, $f(Q) = 4Q$.
2. Since there are 8 ounces in one cup, $g(x) = 8x$.
3. $(g \circ f)(Q) = g(f(Q)) = g(4Q) = 8(4Q) = 32Q$
4. $(g \circ f)(5) = 32(5) = 160$

   The function f converts units of quarts to units of cups, and the function g converts units of cups to units of ounces, so $g \circ f$ converts units of quarts to units of ounces (see Fig. 7).

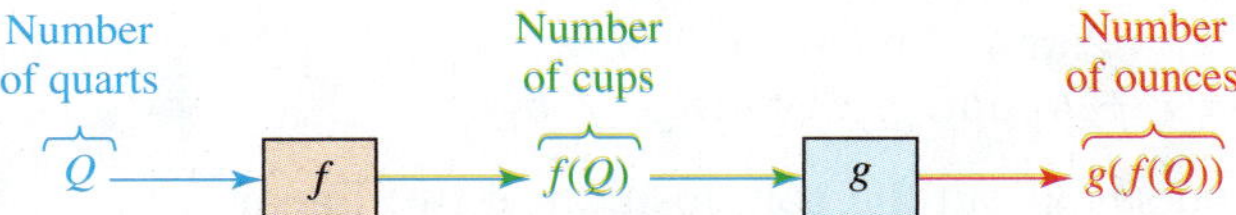

Figure 7 Action of f followed by action of g

So, $(g \circ f)(5) = 160$ means there are 160 ounces in 5 quarts.

Example 10 Using a Composition Function to Model a Situation

In Exercise 8 of Homework 1.4, you used a line to describe the relationship between elevation and boiling point (see Table 8). A reasonable model is $f(E) = -5.9E + 211.3$, where $F = f(E)$ is the boiling point (in Fahrenheit degrees) at an elevation of E thousand meters. Table 9 shows equivalent Fahrenheit and Celsius temperatures. A reasonable

Table 8 Boiling Points of Water

Elevation (thousands of meters)	Boiling Point (°F)
0	212
1	205
2	200
5	181
10	151
15	123

Table 9 Equivalent Temperature Readings

Fahrenheit Reading (°F)	Celsius Reading (°C)
32	0
68	20
104	40
140	60
176	80
212	100

model is $c(F) = 0.56F - 17.78$, where $c(F)$ is the Celsius reading corresponding to a Fahrenheit reading of F degrees.

1. Let $h(E)$ be the boiling point (in Celsius degrees) at elevation E thousand meters. Find an equation of h.
2. Find $h(7)$. What does it mean in this situation?

Solution

1. Note that $f(E)$ is the boiling point (in Fahrenheit degrees) at E thousand meters. Since c converts Fahrenheit readings to equivalent Celsius readings, $c(f(E))$ is the boiling point (in *Celsius* degrees) at E thousand meters. So, $h(E) = c(f(E))$. Here we find the equation of h:

$$
\begin{aligned}
h(E) &= c(f(E)) && \text{h is the composition of c and f.} \\
&= c(-5.9E + 211.3) && \text{Substitute } -5.9E + 211.3 \text{ for } f(E). \\
&= 0.56(-5.9E + 211.3) - 17.78 && \text{Substitute } -5.9E + 211.3 \text{ for } F \text{ in } c(F) = 0.56F - 17.78. \\
&= -3.304E + 118.328 - 17.78 && \text{Distributive law} \\
&= -3.304E + 100.548 && \text{Combine like terms.}
\end{aligned}
$$

2. $h(7) = -3.304(7) + 100.548 = 77.42$

The model estimates that at an elevation of 7 thousand meters, the boiling point is about 77 degrees Celsius. We use a graphing calculator to verify our work in Problems 1 and 2 (see Fig. 8).

Figure 8 Verify the work

Group Exploration

Composition of two linear functions

1. For each pair of functions given in parts (a), (b), and (c), find an equation of $f \circ g$. Is $f \circ g$ a linear function? If it is, compare the slope of the graph of $f \circ g$ to the product of the slopes of the graphs of f and g.
   a. $f(x) = 2x + 6$ and $g(x) = 4x + 3$
   b. $f(x) = 3x - 2$ and $g(x) = 5x + 4$
   c. $f(x) = m_1x + b_1$ and $g(x) = m_2x + b_2$
2. What can you say about the composition of two linear functions? What can you say about the slope of the graph of the composition? [**Hint:** See part (c).]
3. The graphs of two functions are sketched in Fig. 9. Sketch the graph of $f \circ g$. [**Hint:** Find $(f \circ g)(0)$. Then refer to Problem 2.]

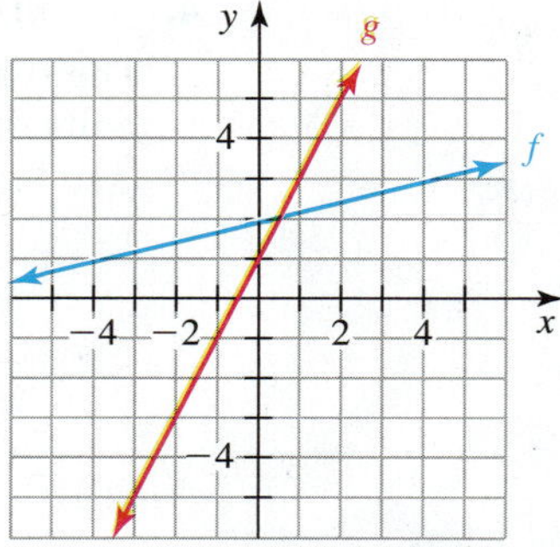

Figure 9 Graphs of two linear functions

Group Exploration

Looking ahead: Inverse function

A company's profit in 2010 was \$5 million. Each year, the profit increases by \$2 million. Let $p = f(t)$ be the profit (in millions of dollars) for the year that is t years after 2010.

1. Find an equation of f. Write the equation in function notation. Also, write the equation with the variables p and t.
2. Use an equation of f to predict when the company will have a profit of \$21 million.
3. In Problem 1, you found an equation with p and t. Solve this equation for t.
4. Substitute 21 for p in the equation you found in Problem 3, and solve for t.
5. Compare your results for Problems 2 and 4.

6. Enter the equation you found in Problem 3 into a graphing calculator to help you complete Table 10.

Table 10 From Profit to Year

Profit (millions of dollars) p	Years since 2010 t
21	
22	
23	
24	
25	

7. In Problems 2 and 4, you used two different methods to find when the company will have a profit of \$21 million. If the company wants to know when it might attain 15 different profit levels, which method would be better to use? Explain.

Homework 11.1

For extra help ▶ MyMathLab®  Watch the videos in MyMathLab 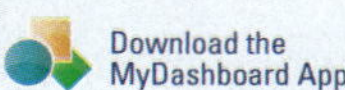 Download the MyDashboard App

All of the values of functions g and f are shown in Table 11. For Exercises 1–8, refer to this table.

1. Find $(f \circ g)(0)$.
2. Find $(f \circ g)(3)$.
3. Find $(g \circ f)(0)$.
4. Find $(g \circ f)(3)$.
5. Find $(g \circ g)(1)$.
6. Find $(f \circ f)(4)$.
7. Use a table to describe five input–output pairs of $f \circ g$.
8. Use a table to describe five input–output pairs of $g \circ f$.

Table 11 Input–Output Pairs for g and f (Exercises 1–8)

x	$g(x)$	x	$f(x)$
0	4	0	1
1	0	1	3
2	3	2	0
3	2	3	4
4	1	4	2

All of the values of functions g and f are shown in Table 12. For Exercises 9–16, refer to this table.

9. Find $(f \circ g)(8)$.
10. Find $(f \circ g)(7)$.
11. Find $(g \circ f)(8)$.
12. Find $(g \circ f)(7)$.
13. Find $(f \circ f)(6)$.
14. Find $(g \circ g)(6)$.
15. Use a table to describe five input–output pairs of $g \circ f$.
16. Use a table to describe five input–output pairs of $f \circ g$.

Table 12 Input–Output Pairs for g and f (Exercises 9–16)

x	$g(x)$	x	$f(x)$
5	7	5	8
6	5	6	5
7	6	7	9
8	9	8	6
9	8	9	7

*For each pair of functions, find **(a)** $(f \circ g)(2)$; **(b)** $(g \circ f)(2)$.*

17. $f(x) = x^2$ and $g(x) = 4x - 3$
18. $f(x) = x^2$ and $g(x) = 2x + 1$
19. $f(x) = 3x^2 - 2x$ and $g(x) = 2x + 1$
20. $f(x) = 2x^2 + 3x$ and $g(x) = 3x - 8$
21. $f(x) = 2(3)^x$ and $g(x) = 4x - 5$
22. $f(x) = 3(2)^x$ and $g(x) = 3x - 2$
23. $f(x) = \dfrac{x + 1}{5x - 7}$ and $g(x) = 16 - 3x^2$
24. $f(x) = \dfrac{3x + 6}{x + 2}$ and $g(x) = 2x^2 - 5$

*For each pair of functions, find **(a)** an equation of $f \circ g$; **(b)** an equation of $g \circ f$; **(c)** $(f \circ g)(3)$; **(d)** $(g \circ f)(3)$.*

25. $f(x) = 3x - 4$ and $g(x) = 2x - 1$
26. $f(x) = 5x + 2$ and $g(x) = 4x - 3$
27. $f(x) = 2x + 4$ and $g(x) = x^2$
28. $f(x) = 3x - 5$ and $g(x) = x^2$
29. $f(x) = x^2 - 2x$ and $g(x) = x - 4$
30. $f(x) = x^2 - 4x$ and $g(x) = x + 2$
31. $f(x) = 2^x$ and $g(x) = x + 2$
32. $f(x) = 3^x$ and $g(x) = 5 - x$

For Exercises 33–40, refer to Fig. 10. Find the following.

33. $(f \circ g)(-2)$
34. $(f \circ g)(-6)$
35. $(g \circ f)(5)$
36. $(g \circ f)(-3)$
37. $(f \circ f)(-5)$
38. $(g \circ g)(4)$
39. The y-intercept of the graph of $f \circ g$
40. The y-intercept of the graph of $g \circ g$

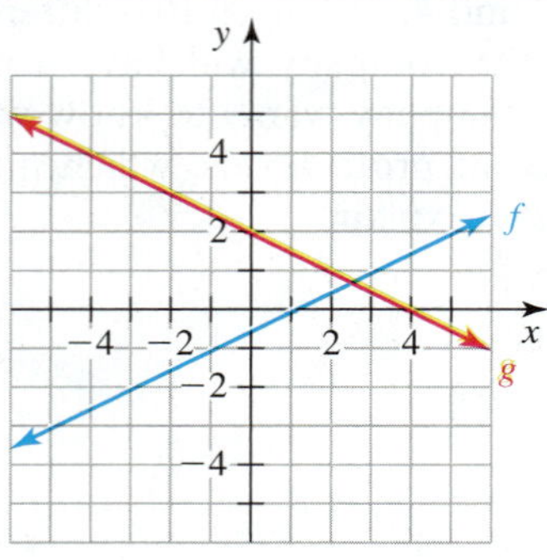

Figure 10 Exercises 33–40

For Exercises 41–46, refer to Fig. 11. Find the following.

41. $(f \circ g)(2)$ **42.** $(f \circ g)(6)$ **43.** $(g \circ f)(2)$

44. $(g \circ f)(1)$ **45.** $(f \circ f)(0)$ **46.** $(g \circ g)(0)$

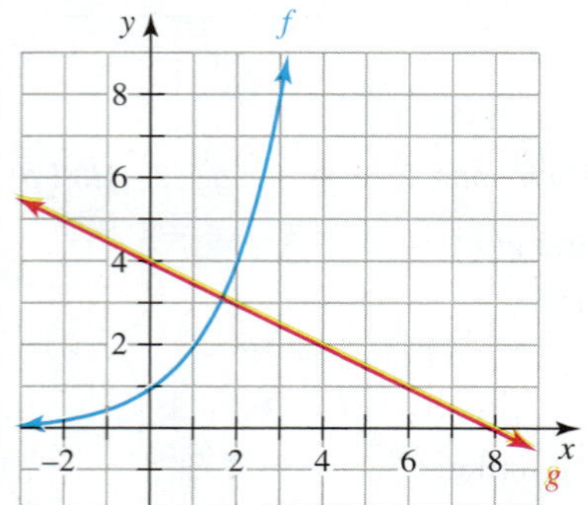

Figure 11 Exercises 41–46

Find equations of f and g such that $h(x) = f(g(x))$.

47. $h(x) = 5^{x-9}$ **48.** $h(x) = 8^{x+3}$

49. $h(x) = 2^x + 6$ **50.** $h(x) = 7^x - 5$

51. $h(x) = (5x - 2)^2$ **52.** $h(x) = (4x + 8)^3$

53. $h(x) = \dfrac{1}{3x - 7}$ **54.** $h(x) = \dfrac{1}{4x + 5}$

55. Let $f(t)$ be the U.S. population (in millions) at t years since 2000. Let $g(p)$ be the weight (in billions of pounds) of french fries consumed annually by p million Americans. What does $(g \circ f)(19) = 9.8$ mean in this situation?

56. Let $f(I)$ be the amount (in thousands of dollars) of federal income tax an American has to pay if their personal income is I thousand dollars. Let $g(t)$ be the median personal income (in thousands of dollars) of an American at t years since 2005. What does $(f \circ g)(12) = 14$ mean in this situation?

57. Let $f(p)$ be the number (in thousands) of Americans who will purchase The Kills album *No Wow* at price p dollars. Let $g(n)$ be the revenue (in millions of dollars) from selling n thousand albums. What does $(g \circ f)(8.99) = 2.3$ mean in this situation?

58. Let $g(t)$ be the number (in thousands) of couches sold by a company in the year that is t years since 2010. Let $f(n)$ be the revenue (in millions of dollars) from selling n thousand couches. What does $(f \circ g)(8) = 5$ mean in this situation?

59. Let $f(y)$ be the number of feet in y yards, and let $g(x)$ be the number of inches in x feet.

a. Find equations of f and g.

b. Find an equation of $g \circ f$.

c. Find $(g \circ f)(5)$. What does it mean in this situation?

60. Let $f(Q)$ be the value (in cents) of Q quarters, and let $g(D)$ be the number of quarters worth D dollars.

a. Find equations of f and g.

b. Find an equation of $f \circ g$.

c. Find $(f \circ g)(3)$. What does it mean in this situation?

61. Let $f(M)$ be the number of seconds in M minutes, and let $g(H)$ be the number of minutes in H hours.

a. Find equations of f and g.

b. Find an equation of $f \circ g$.

c. Find $(f \circ g)(3)$. What does it mean in this situation?

62. There are 2.54 centimeters in 1 inch. Let $f(I)$ be the number of centimeters in I inches, and let $g(x)$ be the number of inches in x feet.

a. Find equations of f and g.

b. Find an equation of $f \circ g$.

c. Find $(f \circ g)(4)$. What does it mean in this situation?

63. The gas mileage of a car is 25 miles per gallon. Assume gasoline costs \$4.25 per gallon. Let $f(d)$ be the amount (in gallons) of gasoline used in driving d miles. Let $g(x)$ be the cost (in dollars) of x gallons of gasoline.

a. Find equations of f and g.

b. Find an equation of $g \circ f$.

c. Find $(g \circ f)(300)$. What does it mean in this situation?

64. A U.S. dollar is worth 0.63 British pound. A British pound is worth 132 Japanese yen. Let $f(d)$ be the number of British pounds worth d U.S. dollars. Let $g(p)$ be the number of Japanese yen worth p British pounds.

a. Find equations of f and g.

b. Find an equation of $g \circ f$.

c. Find $(g \circ f)(10)$. What does it mean in this situation?

65. In Exercise 17 of Homework 5.5, you found an equation close to $f(t) = 0.76t - 42.04$, where $f(t)$ is the percentage of births outside marriage at t years since 1900 (see Table 13).

Table 13 Births Outside Marriage

Year	Percent of Births outside Marriage
1970	10.7
1975	14.3
1980	18.4
1985	22.0
1990	28.0
1995	32.2
2000	33.2
2005	36.8
2010	40.8

Source: *National Center for Health Statistics*

a. The percentage $g(p)$ of births from *married* couples is given by $g(p) = 100 - p$, where p is the percentage of births outside of marriage. Explain why this makes sense.

b. Let $h(t)$ be the percentage of births from *married* couples at t years since 1900. Determine which of the following is true: $h(t) = (f \circ g)(t)$ or $h(t) = (g \circ f)(t)$. Explain.

c. Find an equation of h.

d. Find $h(117)$. What does it mean in this situation?

e. Find t when $h(t) = 52$. What does it mean in this situation?

66. In Exercise 22 of Homework 5.5, you found an equation close to $f(a) = 0.42a - 13.91$, where $f(a)$ is the percentage of Americans at age a years who have been diagnosed with diabetes at some point in their lives (see Table 14).

Table 14 Percentages of Americans Diagnosed with Diabetes, by Age Group

Age Group (years)	Age Used to Represent Age Group (years)	Percent
35–39	37	2
40–44	42	4
45–49	47	5
50–54	52	8
55–59	57	10
60–64	62	13
65–69	67	14

Source: *National Health Interview Survey*

a. The percentage $g(p)$ of Americans who have *never* been diagnosed with diabetes is given by $g(p) = 100 - p$, where p is the percentage of Americans who have been diagnosed with diabetes. Explain why this makes sense.
b. Let $h(a)$ be the percentage of Americans at age a years who have *never* been diagnosed with diabetes. Determine which of the following is true: $h(a) = (f \circ g)(a)$ or $h(a) = (g \circ f)(a)$. Explain.
c. Find an equation of h.
d. Find $h(50)$. What does it mean in this situation?
e. What is the slope of the graph of h? What does it mean in this situation? Explain why it makes sense that the slope is the opposite of the slope of the graph of f.

67. The revenues from downloaded music are shown in Table 15 for various years. Let $D(t)$ be the annual revenue (in billions of dollars) from downloaded music at t years since 2005. A reasonable model is $D(t) = 0.34t + 0.56$.

Table 15 Annual Revenues from Downloaded Music

Year	Annual Revenue (billions of dollars)
2006	0.86
2007	1.23
2008	1.67
2009	1.98
2010	2.20

Source: *Recording Industry Association of America*

a. The value (in *millions* of dollars) $M(b)$ of b billion dollars is given by $M(b) = 1000b$. Explain why this makes sense.
b. Let $h(t)$ be the annual revenue (in *millions* of dollars) from downloaded music at t years since 2005. Determine which of the following is true: $h(t) = (D \circ M)(t)$ or $h(t) = (M \circ D)(t)$. Explain.
c. Find an equation of h.
d. Find $h(-1)$. What does it mean in this situation?
e. What is the t-intercept of the graph of h? What does it mean in this situation? Compare the t-intercept of the graph of h with the t-intercept of the graph of D. Explain why your comparison makes sense.

68. In Exercise 24 of Chapter 3 Test, you found the equation $C(t) = 37t + 870$, where $C(t)$ is the median compensation (in thousands of dollars) of college presidents in the year that is t years since 2010.

a. The value (in *millions* of dollars) $M(d)$ of d thousand dollars is given by $M(d) = \frac{d}{1000}$. Explain why this makes sense.
b. Let $h(t)$ be the median compensation (in *millions* of dollars) of college presidents in the year that is t years since 2010. Determine which of the following is true: $h(t) = (C \circ M)(t)$ or $h(t) = (M \circ C)(t)$. Explain.
c. Find an equation of h.
d. Find $h(7)$. What does it mean in this situation?
e. What is the t-intercept of the graph of h? Compare it with the t-intercept of the graph of C. Explain why your comparison makes sense.

69. In Exercise 11 of Homework 5.6, you found the equation $f(C) = 1.8C + 32$, where $f(C)$ is the Fahrenheit reading that corresponds to the Celsius reading of C degrees (see Table 16).

Table 16 Equivalent Temperature Readings

Celsius Reading (°C)	Fahrenheit Reading (°F)
0	32
20	68
40	104
60	140
80	176
100	212

In Exercise 12 of Homework 5.6, you found the equation $g(F) = 4.3F - 172$, where $g(F)$ is the number of chirps per minute a cricket makes when the temperature is F degrees Fahrenheit (see Table 17).

Table 17 Rates of Cricket Chirping

Temperature (°F)	Rate (number of chirps per minute)
50	43
60	86
70	129
80	172
90	215

a. Let $h(C)$ be the number of chirps per minute a cricket makes when the temperature is C degrees Celsius. Determine which of the following is true: $h(C) = (f \circ g)(C)$ or $h(C) = (g \circ f)(C)$. Explain.
b. Find an equation of h.
c. Estimate the rate crickets chirp when the temperature is 23°C.
d. Estimate the temperature in Celsius degrees at which crickets chirp 200 times per minute.

70. In Exercise 10 of Homework 5.6, you found an equation close to $f(t) = -0.64t + 31.48$, where $f(t)$ is the average gasoline tax (in 2010 dollars) per 1000 miles driven at t years since 1990 (see Table 18).

a. If the average gasoline tax were \$15 per 1000 miles driven, how much would a person who drives 12 thousand miles per year pay in gasoline taxes *per year*?
b. Let $g(R)$ be the average gasoline tax (in 2010 dollars) paid per person *per year*, where R is the average gasoline tax (in 2010 dollars) per 1000 miles driven. The average number

of miles Americans drive per year is 13.5 thousand miles (Source: *Federal Highway Administration*). Find an equation of g. [**Hint:** Part (a) might suggest what to do.]

c. Let $h(t)$ be the average gasoline tax (in 2010 dollars) paid per person *per year* at t years since 1990. Determine which of the following is true: $h(t) = (f \circ g)(t)$ or $h(t) = (g \circ f)(t)$. Explain.

d. Find an equation of h.

e. Find $h(27)$. What does it mean in this situation?

f. Find t when $h(t) = 175$. What does it mean in this situation?

Table 18 Average Gasoline Taxes per 1000 Miles Driven

Year	Average Gasoline Tax per 1000 Miles Driven (2010 dollars)
1995	28
1998	27
2001	24
2004	23
2007	20
2010	19

Source: *Bureau of Economic Analysis, Bureau of Transportation Statistics, Bureau of Labor Statistics*

Concepts

71. Let $f(x) = x + 8$ and $g(x) = x + 5$. A student tries to find an equation of $f \circ g$:

$$\begin{aligned}(f \circ g)(x) &= (x + 8)(x + 5)\\ &= x^2 + 13x + 40\end{aligned}$$

Describe any errors. Then find an equation of $f \circ g$ correctly.

72. Let $f(x) = 2^x$ and $g(x) = x + 3$. A student tries to find an equation of $f \circ g$:

$$\begin{aligned}(f \circ g)(x) &= f(x + 3)\\ &= 2^x + 3\end{aligned}$$

Describe any errors. Then find an equation of $f \circ g$ correctly.

73. Let $f(x) = 4x - 2$ and $g(x) = -7x + 3$. A student tries to find an equation of $f \circ g$:

$$\begin{aligned}(f \circ g)(x) &= -7(4x - 2) + 3\\ &= -28x + 14 + 3\\ &= -28x + 17\end{aligned}$$

Describe any errors. Then find an equation of $f \circ g$ correctly.

74. Let $f(x) = 2x^2 - 9$ and $g(x) = x + 4$. A student tries to find an equation of $f \circ g$:

$$\begin{aligned}(f \circ g)(x) &= 2(x + 4)^2\\ &= 2(x^2 + 8x + 16)\\ &= 2x^2 + 16x + 32\end{aligned}$$

Describe any errors. Then find an equation of $f \circ g$ correctly.

75. Let $f(x) = 5x - 6$ and $g(x) = 4x - 10$.

a. Find $(f \circ g)(3)$.

b. Find $(g \circ f)(3)$.

c. Are your results from parts (a) and (b) equal?

76. Let $f(x) = x^2$ and $g(x) = 5x + 2$.

a. Find an equation of $f \circ g$.

b. Find an equation of $g \circ f$.

c. Are the functions $f \circ g$ and $g \circ f$ the same? Explain.

77. Let $f(x) = 4x - 7$ and $g(x) = 3x - 2$.

a. Find $g(2)$. Then use the result to find $(f \circ g)(2)$.

b. Find an equation of $f \circ g$.

c. Use your result from part (b) to find $(f \circ g)(2)$. Compare the result with your result from part (a).

78. a. For each part, use the given definitions of f and g to find an equation of $f \circ g$.

i. $f(x) = x - 3$ and $g(x) = x + 3$

ii. $f(x) = \dfrac{x}{4}$ and $g(x) = 4x$

iii. $f(x) = 5x$ and $g(x) = \dfrac{x}{5}$

b. What do you notice about your results of parts (a.i), (a.ii), and (a.iii)? Why does this make sense?

c. Let $g(x) = x - 2$. Find an equation of a function f such that $(f \circ g)(x) = x$.

79. Let $f(x) = 2x$. Find an equation for the function.

a. $f \circ f$

b. $f \circ (f \circ f)$

c. $f \circ (f \circ (f \circ f))$

d. $\underbrace{f \circ (f \circ (f \circ \ldots \circ f) \ldots)}_{n \text{ functions}}$

80. Let $f(x) = x + 2$. Find an equation for the function.

a. $f \circ f$

b. $f \circ (f \circ f)$

c. $f \circ (f \circ (f \circ f))$

d. $\underbrace{f \circ (f \circ (f \circ \ldots \circ f) \ldots)}_{n \text{ functions}}$

81. Give an example of functions f and g such that $f \circ g$ and $g \circ f$ are different functions. Your example should be different than those in the textbook.

82. What is the meaning of a composite function? Give an example.

Related Review

For $f(x) = x^2$, $g(x) = 4x - 5$, $h(x) = 2x^2 - 3x + 1$, and $k(x) = x + 3$, find an equation of the given function; then evaluate the function at the indicated value.

83. $g - h$, $(g - h)(-2)$

84. $g + h$, $(g + h)(-2)$

85. $h \cdot k$, $(h \cdot k)(-1)$

86. $g \cdot h$, $(g \cdot h)(-1)$

87. $f \circ g$, $(f \circ g)(3)$

88. $g \circ f$, $(g \circ f)(3)$

89. Suppose that the function f has domain set A and range set B and that the function g has domain set B and range set C.

a. Explain why $g \circ f$ is a function. [**Hint:** Recall that a function is a relation in which each input leads to exactly one output.]

b. What are the domain and range of $g \circ f$?

90. Table 19 lists all of the input–output pairs of functions f and g. Find the domain and range of $f \circ g$. [**Hint:** Recall that for $(f \circ g)(x)$ to be defined, x must be in the domain of g and $g(x)$ must be in the domain of f.]

Table 19 Input–Output Pairs for g and f (Exercise 90)

x	$g(x)$	x	$f(x)$
0	10	3	9
1	9	4	10
2	8	5	11
3	7	6	12
4	6	7	13

Expressions, Equations, Functions, and Graphs

Perform the indicated instruction. Then use words such as linear, quadratic, cubic, polynomial, degree, exponential, function, one variable, *and* two variables *to describe the expression, equation, or system.*

91. Graph $f(x) = 2(3)^x$ by hand.

92. Find an approximate equation $y = ab^x$ of an exponential curve that contains the points $(4, 15)$ and $(9, 12)$. Round a and b to the second decimal place.

93. Graph $f(x) = 2 + 3x$ by hand.

94. Find an equation of a line that contains the points $(4, 15)$ and $(9, 12)$.

95. Find $f(-4)$, where $f(x) = 2(3)^x$.

96. Find an equation of the line that contains the point $(4, 15)$ and is perpendicular to the line $3x - 2y = 12$.

11.2 Inverse Functions

Objectives

- Know the meaning of *inverse of a function, invertible function,* and *one-to-one function.*
- Know the *reflection property of inverse functions.*
- Graph the inverse of a function.
- Find an equation of the inverse of a model.
- Find an equation of the inverse of a function that is not a model.

Table 20 Celsius and Fahrenheit Equivalent Readings

Celsius (°C)	Fahrenheit (°F)
0	32
20	68
40	104
60	140
80	176
100	212

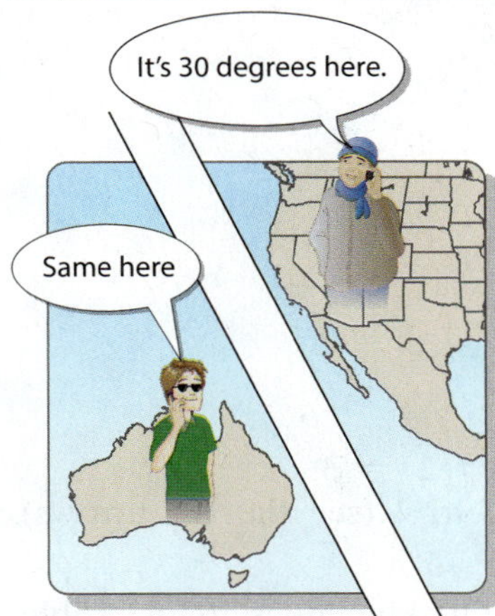

In this section, we will study a type of function that has a special relationship to a given function. It is called the *inverse* of a function. We will begin the section with an example that will allow us to develop a definition of the inverse of a function. Then we will describe inverses of functions by using first tables, then graphs, and finally equations.

Definition of an Inverse of a Function

Although the United States and several other countries use the Fahrenheit scale (in °F) to measure temperature, most countries use the Celsius scale (in °C). A comparison of the two scales is shown in Table 20.

If an American visiting Europe hears the local temperature will reach 20°C, it would be helpful to be able to convert 20°C to 68°F. There is a function g that converts Celsius inputs to Fahrenheit outputs (see Fig. 12). We let C be the Celsius temperature and F be the Fahrenheit temperature.

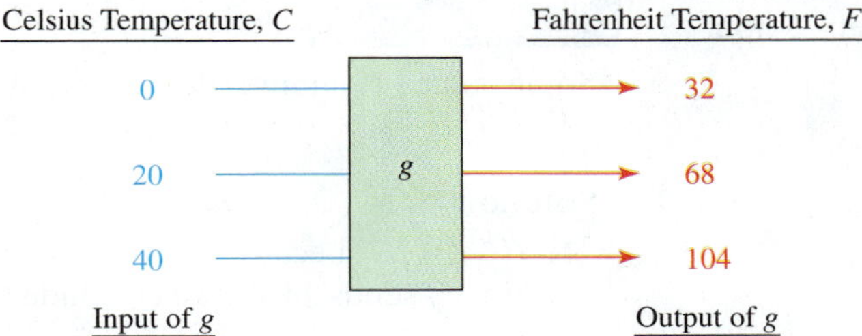

Figure 12 The function g converts Celsius temperatures to Fahrenheit temperatures

If a European visiting the United States hears the local temperature will reach 68°F, it would be helpful to be able to convert 68°F to 20°C. If we reverse the arrows of Fig. 12, we have an input–output diagram of a new relation (see Fig. 13). For each Fahrenheit temperature, there is exactly one Celsius temperature, so the new relation is also a function. We call this function the *inverse* of g and show it as "g^{-1}" (read "g-inverse").

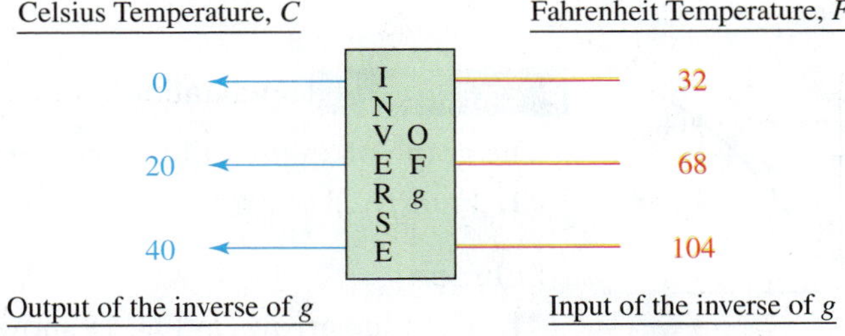

Figure 13 The inverse of g converts Fahrenheit temperatures back to Celsius temperatures

There are two key observations we can make about g^{-1}:

1. g^{-1} sends outputs of g to inputs of g. For example, g sends the input 0 to the output 32 and g^{-1} sends 32 to 0 (see Figs. 12 and 13). Using symbols, we write

$$g(0) = 32 \quad \text{and} \quad g^{-1}(32) = 0$$

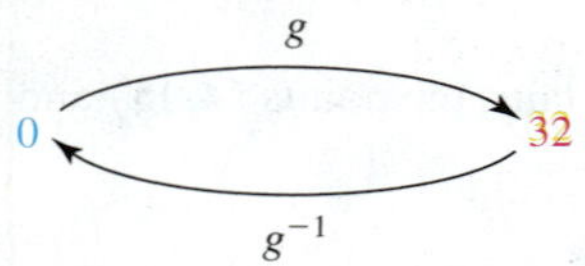

Figure 14 g^{-1} undoes g

We say these two statements are **equivalent,** which means one statement implies the other and vice versa.

2. g^{-1} undoes g. For example, g sends 0 to 32 and g^{-1} undoes this action by sending 32 *back* to 0 (see Fig. 14).

The **inverse of a function** f is a relation that sends b to a if $f(a) = b$. The inverse of a function is not necessarily a function. If the inverse of a function f is also a function, we say f is **invertible** and use "f^{-1}" as a name for the inverse of f. We say f^{-1} is the **inverse function of** f.

Property of an Inverse Function

For an invertible function f, the following statements are equivalent:

$$f(a) = b \quad \text{and} \quad f^{-1}(b) = a$$

In words: If f sends a to b, then f^{-1} sends b to a. If f^{-1} sends b to a, then f sends a to b.

Example 1 Evaluating an Inverse Function

Let f be an invertible function where $f(2) = 5$. Find $f^{-1}(5)$.

Solution

Since f sends 2 to 5, it follows that f^{-1} sends 5 back to 2. So, $f^{-1}(5) = 2$.

Table 21 Input–Output Values of f

x	$f(x)$
0	1
1	3
2	9
3	27
4	81

Example 2 Evaluating f and f^{-1}

Some values of an invertible function f are shown in Table 21. Find the following.

1. $f(3)$
2. $f^{-1}(9)$

Solution

1. $f(3) = 27$, because f sends 3 to 27.
2. Since f sends 2 to 9, we conclude that f^{-1} sends 9 back to 2. Therefore, $f^{-1}(9) = 2$.

WARNING

The -1 in "$f^{-1}(x)$" is *not* an exponent. It is part of the function notation "f^{-1}"—which stands for the inverse of the function f. Here, we simplify 3^{-1} and use the values of f shown in Table 21 to find $f^{-1}(3)$:

$$3^{-1} = \frac{1}{3} \quad \text{"}-1\text{" is an exponent: Use } b^{-n} = \frac{1}{b^n}$$

$$f^{-1}(3) = 1 \quad \text{"}f^{-1}\text{" stands for the inverse of } f\text{; } f \text{ sends 1 to 3, so } f^{-1} \text{ sends 3 back to 1.}$$

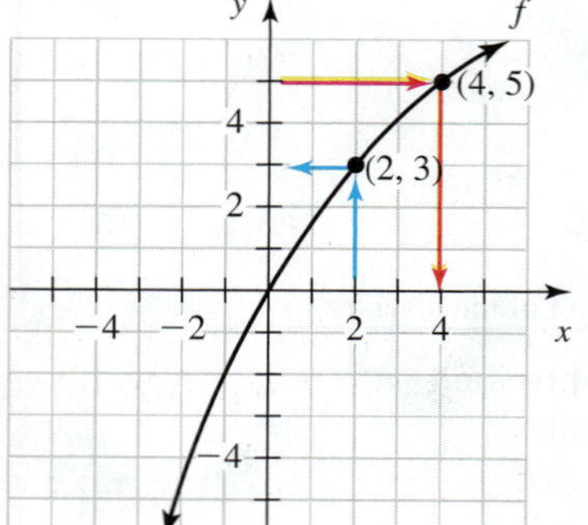

Figure 15 Graph of an invertible function f

Example 3 Evaluating f and f^{-1}

The graph of an invertible function f is shown in Fig. 15.

1. Find $f(2)$.
2. Find $f^{-1}(5)$.

Solution

1. The blue arrows in Fig. 15 show that f sends 2 to 3. So, $f(2) = 3$.
2. The function f sends 4 to 5. So, f^{-1} sends 5 back to 4 (see the red arrows). Therefore, $f^{-1}(5) = 4$.

All linear functions with nonzero slope and all exponential functions are invertible. (We will see why at the end of this section.) Therefore, we can use the notation f^{-1} whenever we describe the inverse of either of these two types of functions.

Table 22 Input–Output Values of f

x	$f(x)$
0	16
1	8
2	4
3	2
4	1

Table 23 Input–Output Values of f^{-1}

x	$f^{-1}(x)$
1	4
2	3
4	2
8	1
16	0

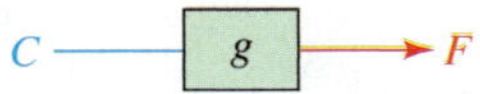

Figure 16 g sends values of C to values of F

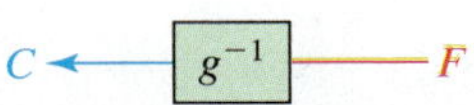

Figure 17 g^{-1} sends values of F to values of C

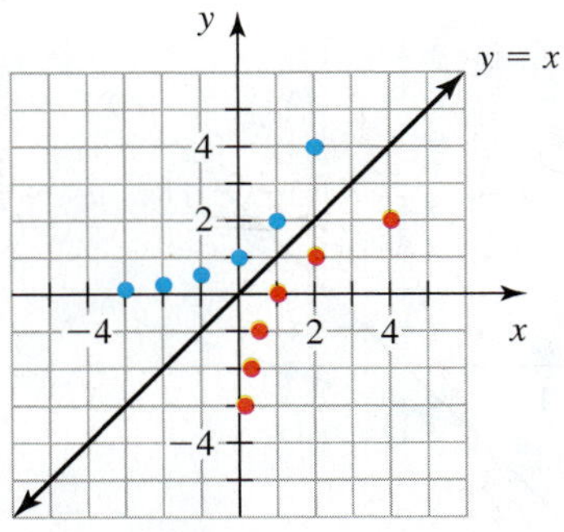

Figure 18 Some points of the graph of f (in blue), some points of the graph of f^{-1} (in red), and the line $y = x$

Example 4 Finding Input–Output Values of an Inverse Function

Let $f(x) = 16\left(\frac{1}{2}\right)^x$.

1. Find five input–output values of f^{-1}.
2. Find $f^{-1}(8)$.

Solution

1. We begin by finding input–output values of f (see Table 22). Since f^{-1} sends outputs of f to inputs of f, we conclude that f^{-1} sends 16 to 0, 8 to 1, 4 to 2, 2 to 3, and 1 to 4. We list these results, from the smallest to the largest input, in Table 23.
2. From Table 23, we see that f^{-1} sends the input 8 to the output 1, so $f^{-1}(8) = 1$.

Let's return to the function g that converts Celsius temperatures to Fahrenheit temperatures. Consider the input–output diagrams of g and g^{-1} in Figs. 16 and 17. We've noted that to find the inverse of g, we reverse the arrow for g. But also notice that if we reverse the arrow for g^{-1}, we get the arrow for g. This suggests that the inverse of g^{-1} is g, which is true. So, g and g^{-1} are inverses of each other.

f and f^{-1} Are Inverses of Each Other

If f is an invertible function, then

- f^{-1} is invertible and
- f and f^{-1} are inverses of each other.

Graphing Inverse Functions

Next, we discuss how to graph the inverse of an invertible function.

Example 5 Comparing the Graphs of a Function and Its Inverse

Sketch the graphs of $f(x) = 2^x$, f^{-1}, and $y = x$ on the same set of axes.

Solution

We list some input–output values of f and f^{-1} in Tables 24 and 25.

Table 24 Input–Output Pairs for f

x	$f(x)$
−3	$\frac{1}{8}$
−2	$\frac{1}{4}$
−1	$\frac{1}{2}$
0	1
1	2
2	4

Table 25 Input–Output Pairs for f^{-1}

x	$f^{-1}(x)$
$\frac{1}{8}$	−3
$\frac{1}{4}$	−2
$\frac{1}{2}$	−1
1	0
2	1
4	2

From Tables 24 and 25, we see that if a point (a, b) is on the graph of f, then the point (b, a) is on the graph of f^{-1}. This observation will lead us to a key step in graphing inverses in future problems.

Next, we plot the input–output pairs of f with blue dots, plot the input–output pairs of f^{-1} with red dots, and sketch the line $y = x$ (see Fig. 18).

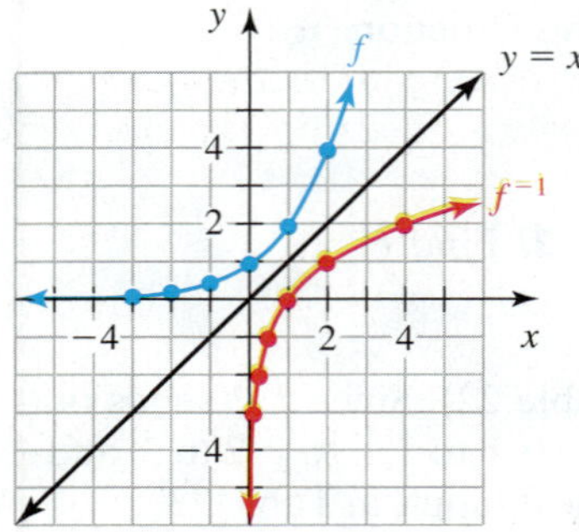

Figure 19 The graphs of f, f^{-1}, and $y = x$

If we were to draw the blue dots in wet ink and fold the paper along the line $y = x$, the ink would make dots where the red dots are. We say the red dots are the *reflection* of the blue dots across the line $y = x$.

In Fig. 19, we sketch a graph of $f(x) = 2^x$ with a blue exponential curve. By reflecting all the blue points on the graph of f across the line $y = x$, we obtain the red graph of f^{-1}. The graph of f^{-1} is the reflection of the graph of $f(x) = 2^x$ across the line $y = x$.

Reflection Property of Inverse Functions

For an invertible function f, the graph of f^{-1} is the reflection of the graph of f across the line $y = x$.

In Example 5, we saw that if a point (a, b) is on the graph of f, then the point (b, a) is on the graph of the inverse of f.

Graphing an Inverse Function

For an invertible function f, we sketch the graph of f^{-1} by the following steps:

1. Sketch the graph of f.
2. Choose several points that lie on the graph of f.
3. For each point (a, b) chosen in step 2, plot the point (b, a).
4. Sketch the curve that contains the points plotted in step 3.

Example 6 Graphing an Inverse Function

Let $f(x) = \frac{1}{3}x - 1$. Sketch the graph of f, f^{-1}, and $y = x$ on the same set of axes.

Solution

We apply the four steps to graph the inverse function:

Step 1: Sketch the graph of f: See Fig. 20.

Step 2: Choose several points that lie on the graph of f: $(-6, -3)$, $(-3, -2)$, $(0, -1)$, $(3, 0)$, and $(6, 1)$.

Step 3: For each point (a, b) chosen in step 2, plot the point (b, a): We plot $(-3, -6)$, $(-2, -3)$, $(-1, 0)$, $(0, 3)$, and $(1, 6)$ in Fig. 21.

Step 4: Sketch the curve that contains the points plotted in step 3: The points from step 3 lie on a line. See Fig. 21.

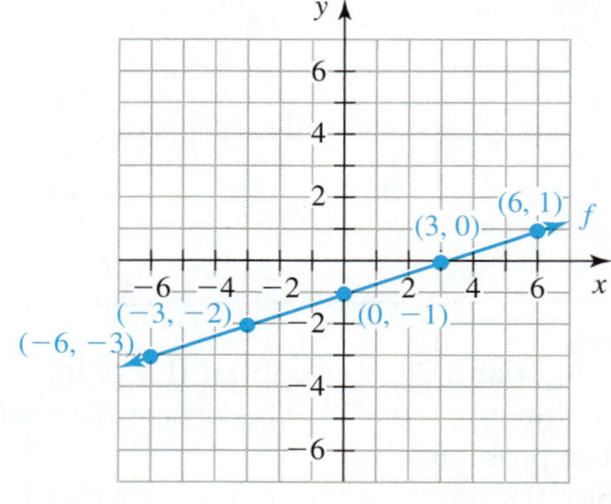

Figure 20 Graph of $f(x) = \frac{1}{3}x - 1$

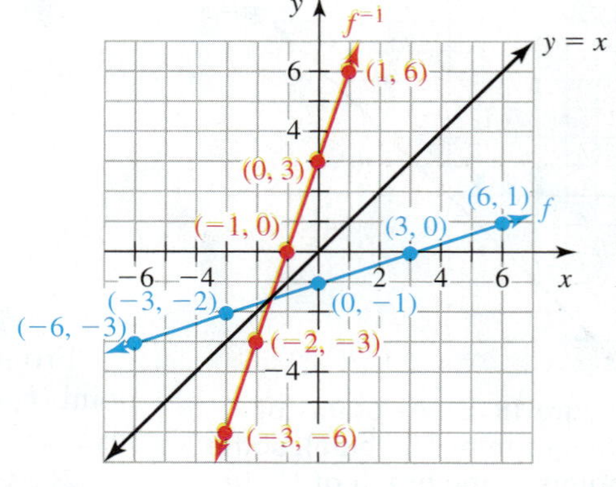

Figure 21 Graphs of $f(x) = \frac{1}{3}x - 1$, $y = x$, and f^{-1}

Finding an Equation of the Inverse of a Model

So far, we have described inverse functions by using tables and graphs. We now describe them by using equations. To begin, we will use a three-step process to find an equation of the inverse of a model.

Table 26 Numbers of Worldwide E-Mails Sent Daily

Year	Number of E-Mails Sent Daily (billions)
2000	15
2002	35
2004	65
2006	88
2007	97

Source: *IDC*

Example 7 Finding an Equation of the Inverse of a Model

The numbers (in billions) of worldwide e-mail messages sent daily are shown in Table 26 for various years. Let $n = f(t)$ be the number (in billions) of e-mails sent daily at t years since 2000. A reasonable model is

$$f(t) = 12.1t + 14$$

1. Find an equation of f^{-1}.
2. Find $f(15)$. What does it mean in this situation?
3. Find $f^{-1}(210)$. What does it mean in this situation?
4. What is the slope of f? What does it mean in this situation?
5. What is the slope of f^{-1}? What does it mean in this situation?

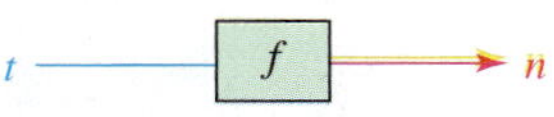

Figure 22 f sends values of t to values of n

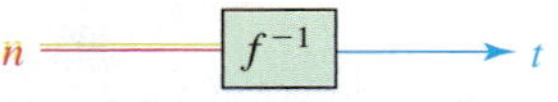

Figure 23 f^{-1} sends values of n to values of t

Solution

1. Since f sends values of t to values of n, f^{-1} sends values of n to t (see Figs. 22 and 23). To find an equation of f^{-1}, we want to write t in terms of n. Here are three steps to follow to find an equation of f^{-1}:

   Step 1: We replace $f(t)$ with n: $n = 12.1t + 14$.

   Step 2: We solve the equation for t:

   $$n = 12.1t + 14 \quad \text{Equation from step 1}$$
   $$n - 14 = 12.1t \quad \text{Subtract 14 from both sides.}$$
   $$\frac{n}{12.1} - \frac{14}{12.1} = t \quad \text{Divide both sides by 12.1.}$$

   An approximate equation is $t = 0.083n - 1.16$.

   Step 3: Since f^{-1} sends values of n to values of t, we have $f^{-1}(n) = t$. So, we can substitute $f^{-1}(n)$ for t in the equation $t = 0.083n - 1.16$:

   $$f^{-1}(n) = 0.083n - 1.16$$

   To verify our work, we use ZStandard followed by ZSquare to check that the graph of f^{-1} is the reflection of the graph of f across the line $y = x$ (see Fig. 24).

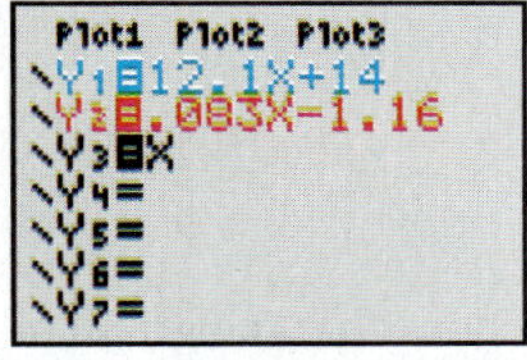

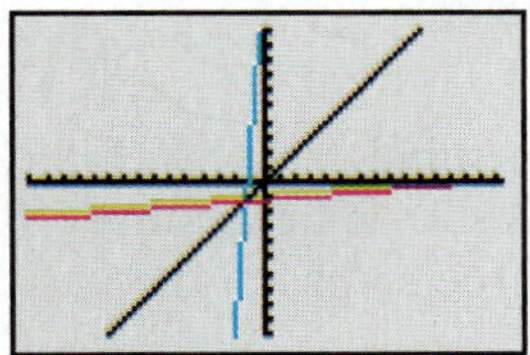

Figure 24 Check that the graph of f^{-1} is the reflection of f across the line $y = x$

2. $f(15) = 12.1(15) + 14 = 195.5$. Since f sends values of t to values of n, it follows that $n = 195.5$ when $t = 15$. According to the model f, 195.5 billion e-mails will be sent daily in 2015.
3. $f^{-1}(210) = 0.083(210) - 1.16 = 16.27$. Since f^{-1} sends values of n to values of t, it follows that $t = 16.27$ when $n = 210$. According to the model f^{-1}, 210 billion e-mails will be sent daily in 2016.
4. The slope of the graph of $f(t) = 12.1t + 14$ is 12.1. This means the rate of change of n with respect to t is 12.1. According to the model f, the number of e-mails sent daily increases by 12.1 billion e-mails each year.
5. The slope of the graph of $f^{-1}(n) = 0.083n - 1.16$ is 0.083. This means the rate of change of t with respect to n is 0.083. According to the model f^{-1}, 0.083 year passes each time the number of e-mails sent daily increases by 1 billion e-mails.

In Example 7, we performed three steps to find an equation of the inverse of a model.

Three-Step Process for Finding the Inverse of a Model

To find the inverse of an invertible *model* f, where $p = f(t)$,

1. Replace $f(t)$ with p.
2. Solve for t.
3. Replace t with $f^{-1}(p)$.

Finding an Equation of the Inverse of a Function That Is Not a Model

We can take three similar steps, followed by one more step, to find an equation of the inverse of a function that is *not* a model.

Example 8 Finding the Inverse of a Function That Is Not a Model

Find the inverse of $f(x) = 2x - 3$.

Solution

Step 1: Substitute y for $f(x)$: $y = 2x - 3$.

Step 2: Solve for x:

$$y = 2x - 3 \quad \text{Equation from step 1}$$

$$y + 3 = 2x \quad \text{Add 3 to both sides.}$$

$$2x = y + 3 \quad \text{If } a = b \text{, then } b = a.$$

$$\frac{2x}{2} = \frac{y}{2} + \frac{3}{2} \quad \text{Divide both sides by 2.}$$

$$x = \frac{1}{2}y + \frac{3}{2} \quad \text{Simplify.}$$

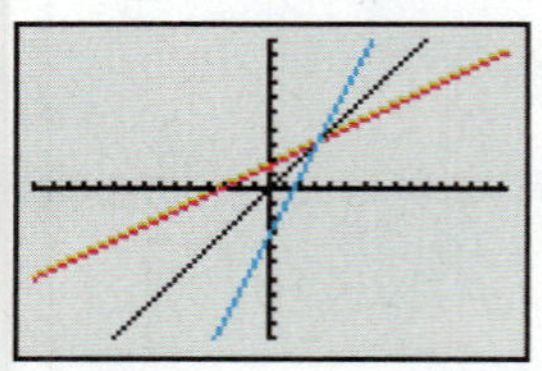

Figure 25 Check that the graph of f^{-1} is the reflection of f across the line $y = x$

Step 3: Replace x with $f^{-1}(y)$: $f^{-1}(y) = \frac{1}{2}y + \frac{3}{2}$.

Step 4: When a function is not a model, we usually want the input variable to be x. So, we rewrite the equation $f^{-1}(y) = \frac{1}{2}y + \frac{3}{2}$ in terms of x:

$$f^{-1}(x) = \frac{1}{2}x + \frac{3}{2}$$

To verify our work, we use ZStandard followed by ZSquare to check that the graph of f^{-1} is the reflection of the graph of f across the line $y = x$ (see Fig. 25).

When finding the inverse of a function f, we must keep in mind whether f is a model. If the function is not a model, we perform all four steps as shown in Example 8. However, if the function is a model, we perform only the first three steps, so the variables retain their original meaning.

Four-Step Process for Finding the Inverse of a Function That Is Not a Model

Let f be an invertible function that is *not* a model. To find the inverse of f, where $y = f(x)$,

1. Replace $f(x)$ with y.
2. Solve for x.
3. Replace x with $f^{-1}(y)$.
4. Write the equation of f^{-1} in terms of x.

So far, we have used tables and graphs to describe the inverse of an exponential function, such as $f(x) = 2^x$. Can we find an equation of the inverse of $f(x) = 2^x$? To find such an equation, we would have to solve the equation $y = 2^x$ for x. We cannot do this by using familiar operations such as adding or subtracting. Nonetheless, the inverse of an exponential function is a powerful tool that we will continue to explore throughout the rest of this chapter.

Composing an Invertible Function with Its Inverse

Let's return one last time to the function g that converts Celsius temperatures to Fahrenheit temperatures. In Fig. 26, we show the input–output action of g followed by the input–output action of g^{-1}.

Figure 26 Action of g followed by action of g^{-1}

Note that the final output, C, is the same as the original input. This means $(g^{-1} \circ g)(C) = C$, which is a symbolic way of saying g^{-1} undoes g.

In Fig. 27, we show the input–output action of g^{-1} followed by the input–output action of g.

Figure 27 Action of g^{-1} followed by action of g

Note that the final output, F, is the same as the original input. This means $(g \circ g^{-1})(F) = F$, which is a symbolic way of saying g undoes g^{-1}.

Composing an Invertible Function with Its Inverse

If f is an invertible function, then

- $(f^{-1} \circ f)(x) = x$, where x is in the domain of f, and
- $(f \circ f^{-1})(x) = x$, where x is in the domain of f^{-1}.

Example 9 Composing an Invertible Function with Its Inverse

In Example 8, we found that the inverse of $f(x) = 2x - 3$ is the function $f^{-1}(x) = \frac{1}{2}x + \frac{3}{2}$. Show that the following are true.

1. $(f^{-1} \circ f)(x) = x$
2. $(f \circ f^{-1})(x) = x$

Solution

1.

$$\begin{aligned}
(f^{-1} \circ f)(x) &= f^{-1}(f(x)) && \textit{Definition of composite function} \\
&= f^{-1}(2x - 3) && \textit{Substitute } 2x - 3 \textit{ for } f(x). \\
&= \frac{1}{2}(2x - 3) + \frac{3}{2} && \textit{Substitute } 2x - 3 \textit{ for } x \textit{ in } f^{-1}(x) = \frac{1}{2}x + \frac{3}{2}. \\
&= \frac{1}{2} \cdot 2x - \frac{1}{2} \cdot 3 + \frac{3}{2} && \textit{Distributive law} \\
&= x - \frac{3}{2} + \frac{3}{2} && \textit{Multiply; simplify.} \\
&= x && \textit{Combine like terms.}
\end{aligned}$$

2. $(f \circ f^{-1})(x) = f(f^{-1}(x))$ *Definition of composite function*

$= f\left(\frac{1}{2}x + \frac{3}{2}\right)$ *Substitute $\frac{1}{2}x + \frac{3}{2}$ for $f^{-1}(x)$.*

$= 2\left(\frac{1}{2}x + \frac{3}{2}\right) - 3$ *Substitute $\frac{1}{2}x + \frac{3}{2}$ for x in $f(x) = 2x - 3$.*

$= 2 \cdot \frac{1}{2}x + 2 \cdot \frac{3}{2} - 3$ *Distributive law*

$= x + 3 - 3$ *Multiply; simplify.*

$= x$ *Combine like terms.*

After finding the inverse of an invertible function f, we can verify our work by one of two ways. We can check that the graph of f^{-1} is the reflection of the graph of f across the line $y = x$, or we can check that $(f^{-1} \circ f)(x) = x$ and that $(f \circ f^{-1})(x) = x$.

One-to-One Functions

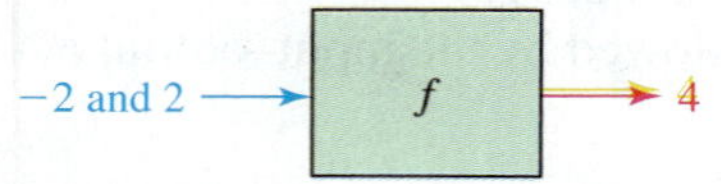

Figure 28 f sends both −2 and 2 to 4

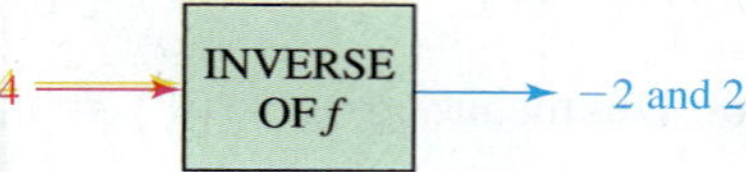

Figure 29 The inverse of f sends 4 to both −2 and 2

We said earlier in this section that the inverse of a function is not necessarily a function. Consider $f(x) = x^2$. Note that $f(-2) = 4$ and $f(2) = 4$. So, f sends both of the inputs -2 and 2 to the output 4 (see Fig. 28). Therefore, the inverse of f sends the input 4 to the *two* outputs -2 and 2 (see Fig. 29). Thus, the inverse of f is not a function. Note that the inverse of f is not a function because an output of f originates from more than one input of f.

If each output of a function originates from exactly one input, we say the function is **one-to-one. A one-to-one function is invertible.** All exponential functions are one-to-one and, hence, invertible. To see why, recall that all exponential functions are either increasing functions or decreasing functions. If f is an increasing exponential function, then $a < b$ implies that $f(a) < f(b)$. See Fig. 30. If f is a decreasing exponential function, then $a < b$ implies that $f(a) > f(b)$. See Fig. 31. Both implications show that if two inputs a and b are different, then their outputs $f(a)$ and $f(b)$ are different, too. It follows that each output originates from exactly one input.

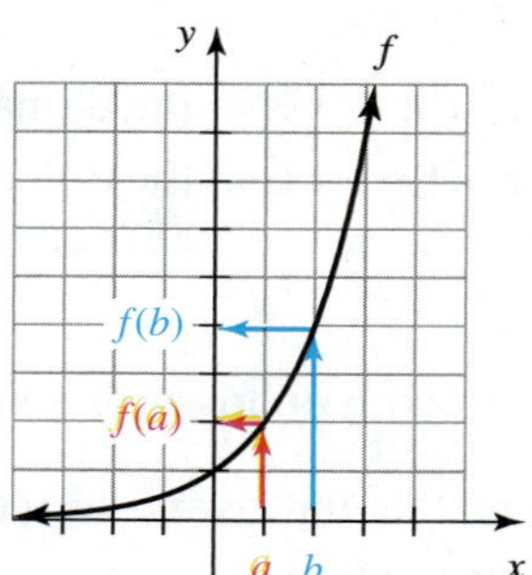

Figure 30 Graph of an increasing exponential function

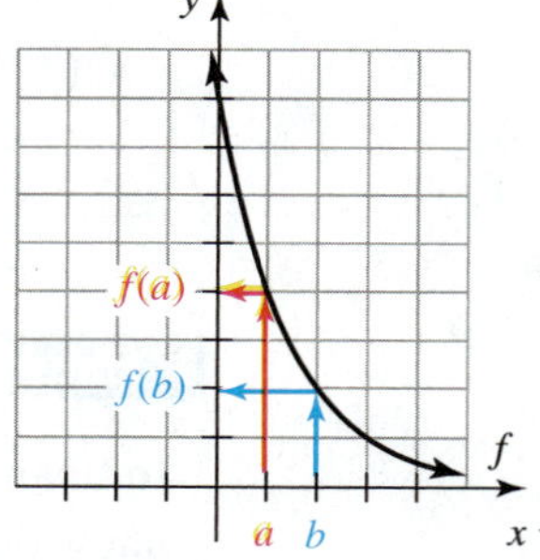

Figure 31 Graph of a decreasing exponential function

Using similar approaches, we could show that all linear functions with nonzero slope are one-to-one and, hence, invertible. Or we could show that such a function is invertible by finding the equation of the inverse of the function; you are asked to do this in Exercise 99.

Group Exploration

Looking ahead: A logarithm is an exponent

1. **a.** Solve $2^x = 16$. **b.** Solve $2^x = 32$.
   c. Approximate the solution of $2^x = 24$ by using trial and error. Your result should be correct up to four decimal places. **[Hint:** Parts (a) and (b) should suggest a reasonable first guess. Then you can use calculator tables to speed up the trial-and-error process.]

2. Approximate the solution of $3^x = 15$ by using trial and error. We call the solution $\log_3(15)$, where *log* is shorthand for *logarithm.*

3. Approximate the solution of $10^x = 500$. We call the solution $\log_{10}(500)$.

Tips for Success Review Material

At various times throughout this course, you can improve your understanding of algebra by reviewing material you have learned so far. Re-solve problems, redo explorations, and consider again key points from previous chapters.

Homework 11.2

For extra help ▶ MyMathLab® 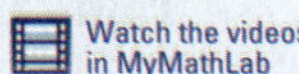 Watch the videos in MyMathLab 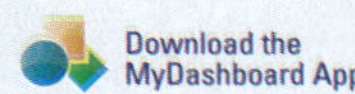 Download the MyDashboard App

1. Let f be an invertible function where $f(4) = 7$. Find $f^{-1}(7)$.

2. Let g be an invertible function where $g^{-1}(3) = 2$. Find $g(2)$.

Some values of an invertible function f are given in Table 27. For Exercises 3–10, refer to this table.

3. Find $f(4)$.

4. Find $f(2)$.

5. Find $f^{-1}(4)$.

6. Find $f^{-1}(2)$.

7. Use a table to describe five input–output values of f^{-1}.

8. Find x when $f(x) = 2$.

9. Find $f^{-1}(f(6))$.

10. Find $f(f^{-1}(6))$.

Table 27 Values of f (Exercises 3–10)

x	$f(x)$
2	10
3	8
4	6
5	4
6	2

Some values of an invertible function g are given in Table 28. For Exercises 11–18, refer to this table.

11. Find $g(2)$.

12. Find $g(6)$.

13. Find $g^{-1}(2)$.

14. Find $g^{-1}(6)$.

15. Use a table to describe six input–output values of g^{-1}.

16. Find x when $g(x) = 6$.

17. Find $g^{-1}(g(4))$.

18. Find $g(g^{-1}(486))$.

Table 28 Values of g (Exercises 11–18)

x	$g(x)$
1	2
2	6
3	18
4	54
5	162
6	486

19. Complete Table 29 by using the table of values of f to complete the table of values of f^{-1}.

Table 29 Finding Values of f^{-1} (Exercise 19)

x	$f(x)$	x	$f^{-1}(x)$
1	34	4	
2	28	10	
3	22	16	
4	16	22	
5	10	28	
6	4	34	

20. Complete Table 30 by using the table of values of f to complete the table of values of f^{-1}.

Table 30 Finding Values of f^{-1} (Exercise 20)

x	$f(x)$	x	$f^{-1}(x)$
2	5	2	
3	2	3	
4	3	4	
5	6	5	
6	4	6	

Let $f(x) = 3(2)^x$.

21. Use a table to describe five input–output values of f^{-1}.

22. Find $f(2)$.

23. Find $f(3)$.

24. Find $f^{-1}(24)$.

25. Find $f^{-1}(3)$.

26. Find x when $f(x) = 6$.

Graph the given function, its inverse, and $y = x$ *by hand on the same set of axes. Label each graph as* f, f^{-1}, *or* $y = x$.

27. $f(x) = 3(2)^x$

28. $f(x) = 2(3)^x$

29. $f(x) = 3^x$

30. $f(x) = 4^x$

31. $f(x) = 2x$

32. $f(x) = 3x$

33. $f(x) = 3x - 2$

34. $f(x) = 2x - 5$

35. $f(x) = \frac{1}{2}x + 1$

36. $f(x) = \frac{2}{3}x - 1$

37. $f(x) = 4\left(\frac{1}{2}\right)^x$

38. $f(x) = 6\left(\frac{1}{3}\right)^x$

39. $f(x) = \left(\frac{1}{3}\right)^x$

40. $f(x) = \left(\frac{1}{2}\right)^x$

For Exercises 41–46, refer to Fig. 32, which shows the graph of an invertible function g.

41. Find $g(2)$.
42. Find $g(-1)$.
43. Find $g^{-1}(2)$.
44. Find $g^{-1}(-3)$.
45. Find $g^{-1}(0)$.
46. Graph g^{-1} by hand.

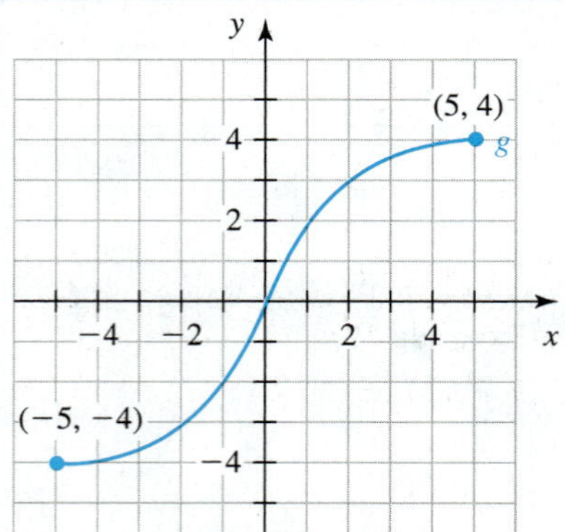

Figure 32 Exercises 41–46

For Exercises 47–52, refer to Fig. 33, which shows the graph of an invertible function f.

47. Find $f(2)$.
48. Find $f(-1)$.
49. Find $f^{-1}(4)$.
50. Find $f^{-1}(0)$.
51. Graph f^{-1} by hand.
52. Find $f^{-1}(1)$.

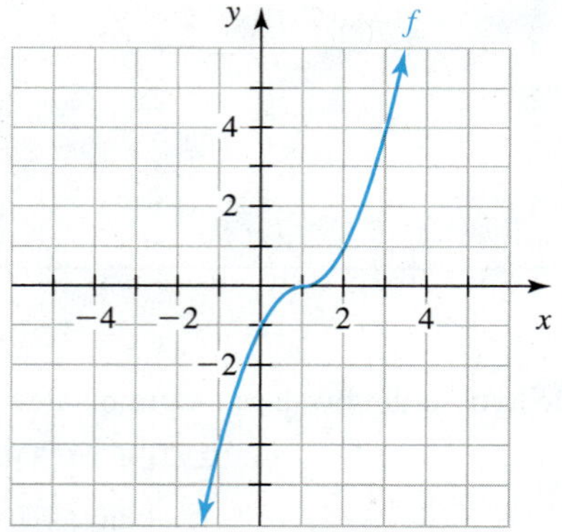

Figure 33 Exercises 47–52

53. In Exercise 17 of Homework 5.5, you found an equation close to $p = f(t) = 0.76t - 42.04$ that models the percentage p of births outside marriage at t years since 1900 (see Table 31).

Table 31 Births outside Marriage

Year	Percent
1970	10.7
1975	14.3
1980	18.4
1985	22.0
1990	28.0
1995	32.2
2000	33.2
2005	36.8
2010	40.8

Source: *National Center for Health Statistics*

a. Find an equation of f^{-1}.
b. Find $f(100)$. What does it mean in this situation?
c. Find $f^{-1}(100)$. What does it mean in this situation?
d. What is the slope of f^{-1}? What does it mean in this situation?

54. In Exercise 80 of Homework 5.7, you found an equation close to $p = f(t) = 1.17t + 12.07$ that models the percentage p of obstetricians and gynecologists who are women at t years since 1980 (see Table 32).

Table 32 Percentages of Obstetricians and Gynecologists Who Are Women

Year	Percent
1980	12
1985	19
1990	23
1995	30
2000	34
2002	38
2009	47

Source: *American Medical Association*

a. Find an equation of f^{-1}.
b. Find $f(36)$. What does it mean in this situation?
c. Find $f^{-1}(36)$. What does it mean in this situation?
d. What is the slope of f^{-1}? What does it mean in this situation?

55. The percentages of households with two or more working computers are shown in Table 33 for various years.

Table 33 Percentages of Households with Two or More Working Computers

Year	Percent
2004	26
2005	30
2006	32
2007	37
2008	37
2009	42
2010	45

Source: *Edison Research and Arbitron*

Let $p = f(t)$ be the percentage of households with two or more working computers at t years since 2000.

a. Use a graphing calculator to draw a scattergram of the data. Can the data be modeled best by using a linear function, a quadratic function, or an exponential function? Explain.
b. Find an equation of f.
c. Find an equation of f^{-1}.
d. Use f to predict when 69% of households will have two or more working computers.
e. Now use f^{-1} to predict when 69% of households will have two or more working computers.
f. Compare your results in parts (d) and (e).

56. The numbers of cremations in the United States are shown in Table 34 for various years. Let $p = f(t)$ be the percentage of bodies that are cremated in the year that is t years since 1990.

a. Use a graphing calculator to draw a scattergram of the data. Can the data be modeled best by using a linear function, a quadratic function, or an exponential function? Explain.

Table 34 Percentages of Bodies That Are Cremated

Year	Percent
1995	19.2
1998	24.1
2000	26.2
2003	29.5
2005	32.3
2007	34.3
2010	40.6

Source: *Cremation Association of North America*

b. Find an equation of f.
c. Find an equation of f^{-1}.
d. Use f to predict in which year half of people who die will be cremated.
e. Now use f^{-1} to predict in which year half of people who die will be cremated.
f. Compare your results from parts (d) and (e).

57. Credit scores measure financial responsibility. Credit scores of Americans are shown in Table 35 for various age groups.

Table 35 Credit Scores of Americans

Age Group (years)	Age Used to Represent Age Group (years)	Average Credit Score (points)
18–29	23.5	637
30–39	34.5	654
40–49	44.5	675
50–59	54.5	697
60–69	64.5	722
70 or more	75	747

Source: *Experian*

Let $c = f(a)$ be the average credit score (in points) of adults at a years of age.
a. Find a linear equation of f.
b. Find an equation of f^{-1}.
c. The average credit score of all adults is 677 points. Use f^{-1} to estimate the (single) age of adults whose average credit score is 677 points.
d. The highest possible score is 830 points. Use f^{-1} to estimate the age of adults whose average credit score is 830 points. (The oldest American ever was Sarah Knauss, who died at age 119 years.)
e. What is the slope of the graph of f^{-1}? What does it mean in this situation?

58. The percentages of Americans who went to the movies at least once in the past year are shown in Table 36 for various age groups. Let $p = f(a)$ be the percentage of Americans at age a years who go to the movies.
a. Use a graphing calculator to draw a scattergram of the data. Can the data be modeled best by using a linear function, a quadratic function, or an exponential function? Explain.

Table 36 Percentages of Americans Who Go to the Movies

Age Group (years)	Age Used to Represent Age Group (years)	Percent
18–24	21.0	88
25–34	29.5	79
35–44	39.5	73
45–54	49.5	65
55–64	59.5	46
65–74	69.5	38
over 74	80	28

Source: *U.S. National Endowment for the Arts*

b. Find an equation of f.
c. Find an equation of f^{-1}.
d. Use f^{-1} to estimate at what age half of Americans go to the movies.
e. What is the slope of the graph of f^{-1}? What does it mean in this situation?

Find the inverse of the given function. Use a graphing calculator to verify your work by graphing f, f^{-1}, and y = x on the same set of axes.

59. $f(x) = x + 8$
60. $f(x) = x - 6$
61. $f(x) = -4x$
62. $f(x) = 5x$
63. $f(x) = \frac{x}{7}$
64. $f(x) = -\frac{x}{2}$
65. $f(x) = -6x - 2$
66. $f(x) = 3x - 8$
67. $f(x) = 0.4x - 7.9$
68. $f(x) = -6.25x + 12.5$
69. $f(x) = \frac{7}{3}x + 1$
70. $f(x) = \frac{8}{5}x + 4$
71. $f(x) = -\frac{5}{6}x - 3$
72. $f(x) = -\frac{2}{5}x - 8$
73. $f(x) = \frac{6x - 2}{5}$
74. $f(x) = \frac{2x - 7}{4}$
75. $f(x) = 7 - 8(x + 1)$
76. $f(x) = 2(x - 1) + 5$
77. $f(x) = x$
78. $f(x) = -x$
79. $f(x) = x^3$
80. $f(x) = x^5$

For Exercises 81–86, show that **(a)** $(f^{-1} \circ f)(x) = x$; **(b)** $(f \circ f^{-1})(x) = x$.

81. $f(x) = x + 7$ and $f^{-1}(x) = x - 7$
82. $f(x) = 4x$ and $f^{-1}(x) = \frac{1}{4}x$
83. $f(x) = 2x - 5$ and $f^{-1}(x) = \frac{1}{2}x + \frac{5}{2}$
84. $f(x) = 4x + 1$ and $f^{-1}(x) = \frac{1}{4}x - \frac{1}{4}$
85. $f(x) = \frac{3}{4}x - 2$ and $f^{-1}(x) = \frac{4}{3}x + \frac{8}{3}$

86. $f(x) = \frac{7}{6}x + 3$ and $f^{-1}(x) = \frac{6}{7}x - \frac{18}{7}$

87. Let $f(x) = 5x - 9$.
a. Find an equation of f^{-1}.
b. Find $f(4)$.
c. Find $f^{-1}(4)$.

88. Let $g(x) = \frac{3}{5}x - 1$.
a. Find an equation of g^{-1}.
b. Find $g(20)$.
c. Find $g^{-1}(20)$.

89. Recall that we can describe some or all of the input–output pairs of a function by means of an equation, a graph, a table, or words. Let $f(x) = 3x - 5$.
a. Describe the input–output pairs of f^{-1} by using an equation.
b. Describe five input–output pairs of f^{-1} by using a table.
c. Describe the input–output pairs of f^{-1} by using a graph.
d. Describe the input–output pairs of f^{-1} by using words.

90. Recall that we can describe some or all of the input–output pairs of a function by means of an equation, a graph, a table, or words. Let $g(x) = \frac{4}{5}x + 2$.
a. Describe the input–output pairs of g^{-1} by using an equation.
b. Describe five input–output pairs of g^{-1} by using a table.
c. Describe the input–output pairs of g^{-1} by using a graph.
d. Describe the input–output pairs of g^{-1} by using words.

Concepts

91. A student says that $g(x) = \frac{x}{2}$ is the inverse of the function $f(x) = 2^x$. Is the student correct? Explain.

92. A student tries to show $f^{-1}(f(x)) = x$ for $f(x) = x + 3$:

$$f^{-1}(f(x)) = \frac{1}{f}(f(x)) = x$$

Describe any errors. Then do the work correctly.

93. Explain why it makes sense that the function $g(x) = x - 5$ is the inverse of the function $f(x) = x + 5$.

94. Explain why it makes sense that the function $g(x) = \frac{x}{3}$ is the inverse of the function $f(x) = 3x$.

95. Explain why it makes sense that if a function g is the inverse of an invertible function f, then f is the inverse function of g.

96. If a function f is invertible, then $(f^{-1} \circ f)(x) = x$ and $(f \circ f^{-1})(x) = x$. Explain why this makes sense.

97. Is the function $f(x) = 3$ one-to-one? Is it invertible? Explain.

98. Is the function $f(x) = x^4$ one-to-one? Is it invertible? Explain.

99. In this exercise, you will show that all linear functions with nonzero slope are invertible.
a. Let $f(x) = mx + b$, where $m \neq 0$. Find an equation of the inverse of f.
b. Explain why your work in part (a) shows that the inverse of f is a function.

100. Explain how to find the inverse of an invertible linear function. Also, explain the meaning of an inverse function. (See page 9 for guidelines on writing a good response.)

Related Review

101. Let $f(x) = 2x - 3$ and $g(x) = \frac{1}{2}x + 3$.
a. Find any points of intersection of the graphs of f and g.
b. Find an equation of f^{-1}.
c. Find an equation of g^{-1}.
d. Find any points of intersection of the graphs of f^{-1} and g^{-1}.
e. What do you observe about your results from parts (a) and (d)? Explain why this makes sense in terms of a property of an inverse function.

102. Let $f(x) = 3x + 5$.
a. Solve the equation $3x + 5 = 11$.
b. Find an equation of f^{-1}.
c. Find $f^{-1}(11)$.
d. What do you observe about your results from parts (a) and (c)? Explain why this makes sense in terms of a property of an inverse function.
e. Use f^{-1} to solve each of the following equations.
  i. $3x + 5 = 17$
  ii. $3x + 5 = 2$
  iii. $3x + 5 = -6$

Expressions, Equations, Functions, and Graphs

Perform the indicated instruction. Then use words such as linear, quadratic, cubic, polynomial, degree, exponential, function, one variable, *and* two variables *to describe the expression, equation, or system.*

103. Solve:
$$3x - 5y = 10$$
$$7x + 4y = 39$$

104. Simplify $\left(\frac{25b^7c^{-3}}{49b^3c^{-9}}\right)^{1/2}$.

105. Graph $3x - 5y = 10$ by hand.

106. Find all real-number solutions of $6b^5 = 349$. Round any results to the second decimal place.

107. Factor $12x^3 - 20x^2 - 27x + 45$.

108. Graph $f(x) = -3(2)^x$ by hand.

11.3 Logarithmic Functions

Objectives

- » Know the meaning of *logarithm* and *logarithmic function.*
- » Find logarithms and evaluate logarithmic functions.
- » Know properties of logarithmic functions.
- » Graph logarithmic functions.
- » Use logarithms to model authentic situations.

In Section 11.2, we discussed the inverse of a function. In this section, we will focus on the inverse of an exponential function, which can be used to measure the energy released by earthquakes and the noise level of sounds.

Definition of a Logarithm

Consider the function $f(x) = 2^x$. Input–output values of f and f^{-1} are shown in Tables 37 and 38.

From Table 38, we see that $f^{-1}(2^3) = 3$. The output 3 is the *exponent* of the input 2^3. In fact, all of the outputs in Table 38 are *exponents* of the corresponding inputs. We know $f^{-1}(2^6) = 6$, since 6 is the exponent of 2^6. Likewise, $f^{-1}(2^7) = 7$.

For $f(x) = 2^x$, the function f^{-1} is so useful we give it a specific name and call it $\log_2$ (read "logarithm, base 2"). So, we write $\log_2(2^3) = 3$, $\log_2(2^6) = 6$, and so on.

To find $\log_2(32)$, we first write 32 as a power of 2:

$$\log_2(32) = \log_2(2^5) = 5$$

So, $\log_2(32) = 5$ because $2^5 = 32$. Likewise, $\log_2(16) = 4$ because $2^4 = 16$. In general,

$$\log_2(a) = k \quad \text{if} \quad 2^k = a$$

We define $\log_3$ in a similar way:

$$\log_3(a) = k \quad \text{if} \quad 3^k = a$$

For example, $\log_3(9) = 2$, since $3^2 = 9$. The logarithm 2 is the *exponent* on the base 3 that gives 9.

Table 37 Input–Output Pairs of $f(x) = 2^x$

x	$f(x)$
0	2^0
1	2^1
2	2^2
3	2^3
4	2^4

Table 38 Input–Output Pairs of f^{-1}

x	$f^{-1}(x)$
2^0	0
2^1	1
2^2	2
2^3	3
2^4	4

Definition Logarithm

For $b > 0$, $b \neq 1$, and $a > 0$,

the **logarithm** $\boldsymbol{\log_b(a)}$ is the number k such that $b^k = a$.

In words, $\log_b(a)$ is the exponent on the base b that gives a. We call b the **base** of the logarithm.

In short, a logarithm is an exponent.

Example 1 Finding Logarithms

Find the logarithm.

1. $\log_7(49)$
2. $\log_8(64)$
3. $\log_3(81)$
4. $\log_5(125)$
5. $\log_4(64)$
6. $\log_{10}(100{,}000)$

Solution

1. $\log_7(49) = 2$, since $7^2 = 49$.
2. $\log_8(64) = 2$, since $8^2 = 64$.
3. $\log_3(81) = 4$, since $3^4 = 81$.
4. $\log_5(125) = 3$, since $5^3 = 125$.
5. $\log_4(64) = 3$, since $4^3 = 64$.
6. $\log_{10}(100{,}000) = 5$, since $10^5 = 100{,}000$.

Definition Common logarithm

A **common logarithm** is a logarithm with base 10. We write $\log(a)$ to represent $\log_{10}(a)$.

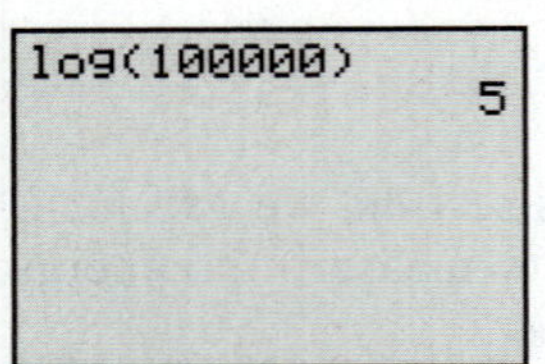

Figure 34 Use a graphing calculator to find that $\log(100{,}000) = 5$

For example, $\log(100{,}000)$ represents $\log_{10}(100{,}000)$. We can use the "log" key on a calculator to find base-10 logarithms. In Fig. 34, we find that $\log(100{,}000) = 5$.

It is important to keep in mind that a logarithm is an exponent. In Example 2, we will use properties of exponents to find more logarithms.

Example 2 Finding Logarithms

Find the logarithm.

1. $\log_7(7)$
2. $\log_4(1)$
3. $\log_3\left(\frac{1}{9}\right)$
4. $\log_6(\sqrt{6})$
5. $\log_b(b^5)$
6. $\log(0.001)$

Solution

1. $\log_7(7) = 1$, since $7^1 = 7$.
2. $\log_4(1) = 0$, since $4^0 = 1$.
3. $\log_3\left(\frac{1}{9}\right) = -2$, since $3^{-2} = \frac{1}{3^2} = \frac{1}{9}$.
4. $\log_6(\sqrt{6}) = \frac{1}{2}$, since $6^{\frac{1}{2}} = \sqrt{6}$.
5. $\log_b(b^5) = 5$, since $b^5 = b^5$.
6. Remember that "$\log(a)$" is shorthand for $\log_{10}(a)$, so $\log(0.001) = -3$, because $10^{-3} = \frac{1}{10^3} = \frac{1}{1000} = 0.001$.

Properties of Logarithms

What are the general properties of logarithms? In the discussion that follows, we assume that $b > 0$ and $b \neq 1$.

In Example 2, we found that $\log_7(7) = 1$. In general, $\log_b(b) = 1$, since $b^1 = b$. We also found that $\log_4(1) = 0$. In general, $\log_b(1) = 0$, since $b^0 = 1$.

Properties of Logarithms

For $b > 0$ and $b \neq 1$,

- $\log_b(b) = 1$
- $\log_b(1) = 0$

Input | Output

$8 \longrightarrow \boxed{g(x) = \log_2(x)} \longrightarrow 3$

Figure 35 Illustration of $g(8) = \log_2(8) = 3$

Recall that $\log_2$ is the name of a function. When we write $\log_2(8) = 3$, we mean the function $\log_2$ sends the input 8 to the output 3 (see Fig. 35). What follows is the general definition of a logarithmic function.

Definition Logarithmic function

A **logarithmic function, base b,** is a function that can be put into the form

$$f(x) = \log_b(x)$$

where $b > 0$ and $b \neq 1$.

The nonpositive numbers are not in the domain of $\log_2$. Consider $\log_2(0)$. No exponent on the base 2 gives 0. Now consider $\log_2(-8)$. No exponent on the base 2 gives a negative number. In general, **the domain of a logarithmic function $\log_b$ is the set of all positive real numbers.**

The function $\log_2$ is the inverse of $f(x) = 2^x$. From our work in Section 11.2, we can conclude that $f(x) = 2^x$ is also the inverse of $\log_2$.

Logarithmic and Exponential Functions Are Inverses of Each Other

- For an exponential function $f(x) = b^x$, $f^{-1}(x) = \log_b(x)$.
- For a logarithmic function $g(x) = \log_b(x)$, $g^{-1}(x) = b^x$.

In words, $g(x) = \log_b(x)$ and $f(x) = b^x$ are inverse functions of each other.

Example 3 Finding an Inverse Function

Find the inverse of the function.

1. $f(x) = 4^x$ **2.** $h(x) = \log_9(x)$

Solution

1. $f^{-1}(x) = \log_4(x)$
2. $h^{-1}(x) = 9^x$

Example 4 Evaluating f and f^{-1}

Let $f(x) = 3^x$.

1. Find $f(4)$. **2.** Find $f^{-1}(9)$.

Solution

1. $f(4) = 3^4 = 81$
2. $f^{-1}(9) = \log_3(9) = 2$

Recall from Section 11.2 that if a function f is invertible, then $f^{-1}(f(x)) = x$, where x is in the domain of f. For the exponential function $f(x) = b^x$ and its inverse $f^{-1}(x) = \log_b(x)$, this implies that

$$\log_b(b^x) = x$$

Also recall the property $f(f^{-1}(x)) = x$, where x is in the domain of f^{-1}. This implies that

$$b^{\log_b(x)} = x, \text{ where } x > 0$$

Composing Logarithmic and Exponential Functions with the Same Base

For $b > 0$ and $b \neq 1$,

- $\log_b(b^x) = x$
- $b^{\log_b(x)} = x$, where $x > 0$

Example 5 Composing Logarithmic and Exponential Functions with the Same Base

Simplify.

1. $\log_5(5^9)$ **2.** $6^{\log_6(3)}$

Solution

1. $\log_5(5^9) = 9$, because $\log_b(b^x) = x$.
2. $6^{\log_6(3)} = 3$, because $b^{\log_b(x)} = x$.

Graphing a Logarithmic Function

Because a logarithmic function is the inverse of an exponential function, we can use the four-step graphing method of Section 11.2 to help us graph a logarithmic function.

Example 6 Graphing a Logarithmic Function

Sketch the graph of $y = \log_3(x)$.

Solution

The inverse of $f(x) = 3^x$ is $f^{-1}(x) = \log_3(x)$. So, we can apply the four-step method to graph the inverse of f:

Step 1: Sketch the graph of f: See Fig. 36.

Step 2: Choose several points on the graph of f: $\left(-1, \frac{1}{3}\right)$, $(0, 1)$, $(1, 3)$, and $(2, 9)$.

Step 3: For each point (a, b) chosen in step 2, plot point (b, a): We plot $\left(\frac{1}{3}, -1\right)$, $(1, 0)$, $(3, 1)$, and $(9, 2)$ in Fig. 37.

Step 4: Sketch the curve that contains the points plotted in step 3: See Fig. 37. The red curve is the graph of $y = \log_3(x)$.

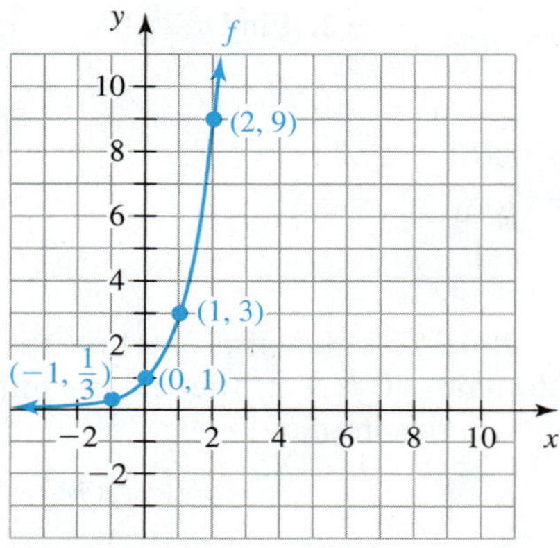

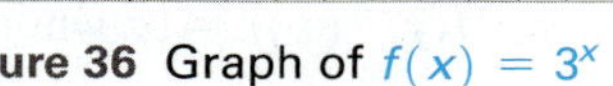
Figure 36 Graph of $f(x) = 3^x$

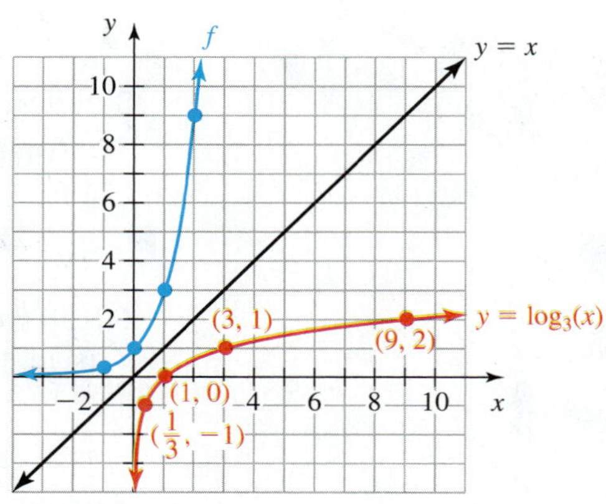

Figure 37 Graphs of $f(x) = 3^x$, $y = \log_3(x)$, and $y = x$

To graph a logarithmic function $y = \log_b(x)$, we use the four-step graphing method from Section 11.2 to sketch the inverse of $f(x) = b^x$.

Using Logarithms to Model Authentic Situations

Scientists often use logarithms to rescale measurements of objects or phenomena when the measurements tend to be very small $\left(\text{e.g., } 3.2 \times 10^{-8}\right)$ or very large $\left(\text{e.g., } 7.9 \times 10^{13}\right)$. For example, scientists use logarithms for measurements of amplitudes of earthquakes, noise levels of sounds, and pH values of solutions.

The energy released by an earthquake is sometimes measured on the *Richter scale.* The *Richter number, R,* of an earthquake is given by

$$R = \log\left(\frac{A}{A_0}\right)$$

where A is the amplitude (maximum value) of a seismic wave and A_0, called the *reference amplitude,* is the amplitude of the smallest seismic wave that a seismograph can detect.

Oh, NO!!
A mean trick to play on a seismographer
Hee hee!

Example 7 Richter Numbers

In 1906, an earthquake in San Francisco had an amplitude 2×10^8 times the reference amplitude A_0. In 1989, an earthquake in San Francisco had an amplitude 8×10^6 times A_0.

1. Find the Richter number of both earthquakes.
2. Find the ratio of the amplitudes of the 1906 and 1989 earthquakes.

Solution

1. The Richter number of the 1906 earthquake is

$$\begin{aligned} R &= \log\left(\frac{2 \times 10^8 A_0}{A_0}\right) && \text{Substitute } 2 \times 10^8 A_0 \text{ for } A. \\ &= \log(2 \times 10^8) && \text{Simplify.} \\ &\approx 8.3 && \text{Compute.} \end{aligned}$$

The Richter number of the 1989 earthquake is

$$\begin{aligned} R &= \log\left(\frac{8 \times 10^6 A_0}{A_0}\right) && \text{Substitute } 8 \times 10^6 A_0 \text{ for } A. \\ &= \log(8 \times 10^6) && \text{Simplify.} \\ &\approx 6.9 && \text{Compute.} \end{aligned}$$

2. The ratio of the amplitudes of the 1906 earthquake and the 1989 earthquake is

$$\frac{2 \times 10^8 A_0}{8 \times 10^6 A_0} = 25$$

So, the 1906 earthquake had an amplitude 25 times greater than that of the 1989 earthquake.

Group Exploration

Looking ahead: Power property for logarithms

1. Use a calculator to compare $\log(3^2)$ with $2\log(3)$.
2. Use a calculator to compare $\log(7^4)$ with $4\log(7)$.
3. Use a graphing calculator table to compare values of $f(x) = \log(x^3)$ and $g(x) = 3\log(x)$. Also, compare the graphs of f and g in the same viewing window.
4. Use a graphing calculator table to compare values of $f(x) = \log(x^5)$ and $g(x) = 5\log(x)$. Also, compare the graphs of f and g in the same viewing window.
5. What do Problems 1–4 suggest about $\log(x^p)$? Test your observation.

Homework 11.3

For extra help ▶ MyMathLab®

Watch the videos in MyMathLab

Download the MyDashboard App

Find the logarithm.

1. $\log_9(81)$
2. $\log_6(36)$
3. $\log_3(27)$
4. $\log_5(625)$
5. $\log_4(256)$
6. $\log_3(243)$
7. $\log_6(216)$
8. $\log_2(64)$
9. $\log(100)$
10. $\log(1000)$
11. $\log_4\left(\frac{1}{4}\right)$
12. $\log_3\left(\frac{1}{3}\right)$
13. $\log_2\left(\frac{1}{8}\right)$
14. $\log_3\left(\frac{1}{81}\right)$
15. $\log\left(\frac{1}{10{,}000}\right)$
16. $\log\left(\frac{1}{100}\right)$
17. $\log_5(1)$
18. $\log_8(1)$

19. $\log_9(9)$

20. $\log_4(4)$

21. $\log_9(3)$

22. $\log_{36}(6)$

23. $\log_8(2)$

24. $\log_{32}(2)$

25. $\log_7(\sqrt{7})$

26. $\log_2(\sqrt{2})$

27. $\log_5(\sqrt[4]{5})$

28. $\log_7(\sqrt[3]{7})$

29. $\log_2(\log_2(16))$

30. $\log_2(\log_3(81))$

31. $\log(\log(10))$

32. $\log_3(\log_3(27))$

33. $\log_b(b)$

34. $\log_b(1)$

35. $\log_b(b^4)$

36. $\log_b(b^6)$

37. $\log_b\left(\frac{1}{b^5}\right)$

38. $\log_b\left(\frac{1}{b^3}\right)$

39. $\log_b(\sqrt{b})$

40. $\log_b(\sqrt[5]{b})$

41. $\log_b(\log_b(b))$

42. $\log_2(\log_b(\sqrt{b}))$

Find the inverse of the given function.

43. $f(x) = 3^x$

44. $g(x) = 8^x$

45. $h(x) = 10^x$

46. $g(x) = \left(\frac{1}{3}\right)^x$

47. $f(x) = \log_5(x)$

48. $g(x) = \log_4(x)$

49. $h(x) = \log(x)$

50. $f(x) = \log_{\frac{1}{2}}(x)$

Let $f(x) = 2^x$.

51. Find $f(2)$.

52. Find $f(4)$.

53. Find $f^{-1}(2)$.

54. Find $f^{-1}(4)$.

Let $g(x) = \log_3(x)$.

55. Find $g(3)$.

56. Find $g(81)$.

57. Find $g^{-1}(3)$.

58. Find $g^{-1}(2)$.

For Exercises 59–62, refer to the values of the function $f(x) = 3^x$ listed in Table 39.

59. Find $f(1)$.

60. Find $f(3)$.

61. Find $f^{-1}(1)$. Also, write your result as a logarithm.

62. Find $f^{-1}(3)$. Also, write your result as a logarithm.

Table 39 Values of $f(x) = 3^x$ (Exercises 59–62)

x	$f(x)$
0	1
1	3
2	9
3	27
4	81

Simplify.

63. $\log_2(2^6)$

64. $\log_4(4^3)$

65. $5^{\log_5(8)}$

66. $3^{\log_3(9)}$

67. $\log(10^7)$

68. $\log(10^2)$

69. $10^{\log(3)}$

70. $10^{\log(4)}$

Graph the function by hand.

71. $y = \log_2(x)$

72. $y = \log_4(x)$

73. $y = \log(x)$

74. $y = \log_6(x)$

75. $y = \log_{\frac{1}{2}}(x)$

76. $y = \log_{\frac{1}{3}}(x)$

77. Recall that we can describe some or all of the input–output pairs of a function by means of an equation, a graph, a table, or words. Let $f(x) = \log_5(x)$.
   a. Describe five input–output pairs of f by using a table.
   b. Describe the input–output pairs of f by using a graph.
   c. Describe the input–output pairs of f by using words.

78. Recall that we can describe some or all of the input–output pairs of a function by means of an equation, a graph, a table, or words. Let $g(x) = \log_{\frac{1}{4}}(x)$.
   a. Describe five input–output pairs of g by using a table.
   b. Describe the input–output pairs of g by using a graph.
   c. Describe the input–output pairs of g by using words.

79. **a.** Solve the equation $5^x = 25$.
   b. Find $\log_5(25)$.
   c. Explain why the results of parts (a) and (b) are the same.

80. **a.** Find $\log(100)$.
   b. Complete Table 40.

Table 40 Values of $\log(x)$ (Exercise 80)

x	$\log(x)$
0.001	
0.01	
0.1	
1	
10	
100	
1000	

   c. Examine the entries in Table 40, and describe all patterns that you observe.

81. In 2004, an earthquake in the Indian Ocean had an amplitude 1.6×10^9 times the reference amplitude A_0. In 1985, an earthquake in Mexico City had an amplitude 6.3×10^7 times A_0.
   a. Find the Richter number of the Indian Ocean earthquake.
   b. Find the Richter number of the Mexico City earthquake.
   c. Find the ratio of the amplitudes of the Indian Ocean and Mexico City earthquakes.

82. In 1920, an earthquake in Gansu, China, had an amplitude 4.0×10^8 times the reference amplitude A_0. In 1980, an earthquake in Naples, Italy, had an amplitude 1.6×10^7 times A_0.
   a. Find the Richter number of the Gansu earthquake.
   b. Find the Richter number of the Naples earthquake.
   c. Find the ratio of the amplitudes of the Gansu and Naples earthquakes.

83. Recall from Exercise 11 of Homework 1.4 that the loudness of sound can be measured on a *decibel scale*. The sound level L (in decibels) of a sound is given by

$$L = 10\log\left(\frac{I}{I_0}\right)$$

where I is the intensity of the sound (in watts per square meter, W/m^2) and $I_0 = 10^{-12}$ W/m^2. The constant I_0 is the approximate intensity of the softest sound a human can hear. Find the decibel values of the sounds listed in Table 41.

Table 41 Examples of Sound Intensities

Sound	Intensity of Sound (W/m^2)
Faintest sound heard by humans	10^{-12}
Whisper	10^{-10}
Inside a running car	10^{-8}
Conversation	10^{-6}
Noisy street corner	10^{-4}
Soft-rock concert	10^{-2}
Threshold of pain	1

84. The acidity or alkalinity of a solution is measured on a *pH scale*. The pH of a solution is given by

$$\text{pH} = -\log(H^+)$$

where H^+ is the hydrogen ion concentration (in moles per liter) of the solution. Distilled water has a pH of 7. Acidic solutions have a pH less than 7, and basic (alkaline) solutions have a pH greater than 7. Most solutions have a pH between 1 and 14. Find the pH of each solution listed in Table 42, and determine whether the solution is acidic or basic. Round your results to the first decimal place.

Table 42 Hydrogen Ion Concentrations of Some Solutions

Solution	Hydrogen Ion Concentration (moles per liter)
Vinegar	1.6×10^{-3}
Human blood	6.3×10^{-8}
Shampoo	7.4×10^{-10}
Orange juice	6.3×10^{-4}
Hydrochloric acid	2.5×10^{-2}

Concepts

For Exercises 85–88, match the function with its graph in Fig. 38.

85. $f(x) = b^x, 0 < b < 1$

86. $g(x) = b^x, b > 1$

87. $h(x) = \log_b(x), 0 < b < 1$

88. $k(x) = \log_b(x), b > 1$

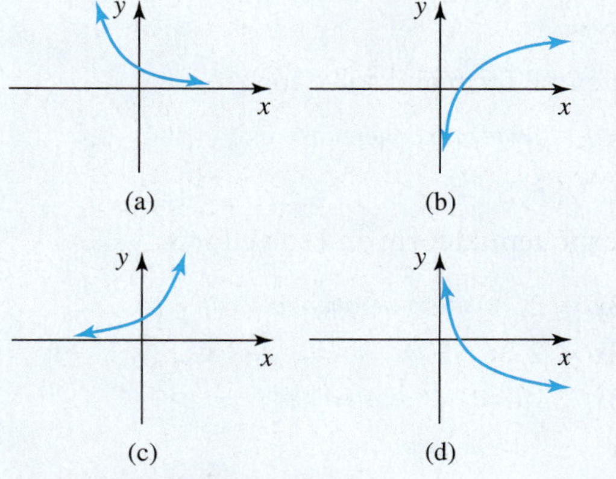

Figure 38 Exercises 85–88

89. Without computing the values of $\log_2(7)$ and $\log_3(7)$, determine which is larger. Explain.

90. **a.** Is $\log_3(0)$ defined? Explain.
b. Is $\log_3(-27)$ defined? Explain.
c. What is the domain of $f(x) = \log_3(x)$? Explain.

91. Give an example of an exponential function f of the form $f(x) = b^x$. List five input–output pairs of f. Then list five input–output pairs of f^{-1}. What is another name for f^{-1}?

92. Explain how to find a logarithm. Also, explain why a logarithmic function is the inverse of a function.

Related Review

93. In this exercise, you will compare the logarithmic function $f(x) = \log_2(x)$, the linear function $g(x) = 2x$, the quadratic function $h(x) = x^2$, and the exponential function $k(x) = 2^x$.

a. Complete Table 43.

Table 43 Complete the Table (Exercise 93)

x	$f(x)$	$g(x)$	$h(x)$	$k(x)$
1				
2				
4				
8				
16				

b. As the inputs in Table 43 increase, for which function are the outputs growing at the fastest rate? the next-fastest rate? the slowest rate?

94. **a.** Sketch the graphs of the functions $f(x) = 3(2)^x$ and $g(x) = -3(2)^x$ on the same coordinate system. Are the graphs reflections of each other across the x-axis?
b. Sketch the graphs of the functions $f(x) = 2x^2$ and $g(x) = -2x^2$ on the same coordinate system. Are the graphs reflections of each other across the x-axis?
c. Considering your observations in parts (a) and (b), graph the functions $f(x) = \log_2(x)$ and $g(x) = -\log_2(x)$ by hand on the same coordinate system.

Show that **(a)** $(f^{-1} \circ f)(x) = x$; **(b)** $(f \circ f^{-1})(x) = x$.

95. $f(x) = 3^x + 5$ and $f^{-1}(x) = \log_3(x - 5)$

96. $f(x) = 4^x - 7$ and $f^{-1}(x) = \log_4(x + 7)$

97. $f(x) = 5(2)^x - 6$ and $f^{-1}(x) = \log_2\left(\dfrac{x + 6}{5}\right)$

98. $f(x) = 8^{x-1} + 2$ and $f^{-1}(x) = \log_8(x - 2) + 1$

Expressions, Equations, Functions, and Graphs

Perform the indicated instruction. Then use words such as linear, quadratic, cubic, polynomial, degree, exponential, logarithmic, function, one variable, *and* two variables *to describe the expression, equation, or system.*

99. Solve $x^2 = 5 - 3x$.

100. Graph $y = -6\left(\dfrac{1}{3}\right)^x$ by hand.

101. Factor $6x^2 - x - 1$.

102. Solve $8x^2 - 18x + 9 = 0$.

103. Simplify $(2x - 3)^2 - (3x - 2)^2$.

104. Find the product $(x^2 - 3)(x^3 + 2)$.

11.4 Properties of Logarithms

Objectives

- Convert equations in *logarithmic form* to *exponential form* and vice versa.
- Know the *power property for logarithms.*
- Use properties of logarithms to solve exponential and logarithmic equations.
- Solve equations in one variable by graphing.

In this section, we will discuss some properties of logarithms and how to use these properties to solve exponential and logarithmic equations in one variable.

Exponential/Logarithmic Forms Property

In Section 11.3, we discussed how to find logarithms. For example,

$$\log_2(8) = 3 \quad \text{since} \quad 2^3 = 8$$

We say that the equation $\log_2(8) = 3$ is in *logarithmic form* and that the equation $2^3 = 8$ is in *exponential form.* The forms $\log_2(8) = 3$ and $2^3 = 8$ are equivalent.

Here are more examples of equations in equivalent logarithmic and exponential forms:

Logarithmic form	*Exponential form*
$\log_2(16) = 4$	$2^4 = 16$
$\log_3(9) = 2$	$3^2 = 9$
$\log_5(125) = 3$	$5^3 = 125$
$\log(100{,}000) = 5$	$10^5 = 100{,}000$

Exponential/Logarithmic Forms Property

For $a > 0$, $b > 0$, and $b \neq 1$, the equations

$$\log_b(a) = c \quad \text{and} \quad b^c = a$$

are equivalent.

The equation $\log_b(a) = c$ is in **logarithmic form** and the equation $b^c = a$ is in **exponential form.** Either form can replace the other when you solve a problem.

Example 1 Solving Equations in Logarithmic Form

Solve for x.

1. $\log_4(x) = 3$ **2.** $\log(3x - 2) = 2$

Solution

1. We write $\log_4(x) = 3$ in exponential form and solve for x:

$$4^3 = x \quad \text{Write in exponential form.}$$
$$64 = x \quad \text{Simplify.}$$

2. We write $\log(3x - 2) = 2$ in exponential form and solve for x:

$$10^2 = 3x - 2 \quad \text{Write in exponential form.}$$
$$100 = 3x - 2 \quad \text{Simplify.}$$
$$102 = 3x \quad \text{Add 2 to both sides.}$$
$$34 = x \quad \text{Divide both sides by 3.}$$

A logarithmic equation in one variable is an equation in one variable that contains one or more logarithms. Here are some examples of logarithmic equations in one variable:

$$\log_3(x) = 4 \qquad \log_b(87) = 6 \qquad 3\log_2(t) - 7 = 8 \qquad \log_5(x^4) + \log_5(3x) = 3$$

Example 2 Solving Logarithmic Equations in One Variable

Solve for x.

1. $6\log_9(t) + 1 = 4$ 2. $\log_3(x^4) = 2$

Solution

1. We get $\log_9(t)$ alone on the left side of the equation and solve for t:

$$\begin{aligned} 6\log_9(t) + 1 &= 4 && \textit{Original equation} \\ 6\log_9(t) &= 3 && \textit{Subtract 1 from both sides.} \\ \log_9(t) &= \frac{1}{2} && \textit{Divide both sides by 6.} \\ 9^{1/2} &= t && \textit{Write in exponential form.} \\ 3 &= t && \textit{Simplify.} \end{aligned}$$

2. We write $\log_3(x^4) = 2$ in exponential form and solve for x:

$$\begin{aligned} 3^2 &= x^4 && \textit{Write in exponential form.} \\ x^4 &= 9 && \textit{If } c = d, \textit{ then } d = c; \textit{ simplify.} \\ x &= \pm 9^{1/4} && \textit{The solution of } x^4 = k \textit{ is } \pm k^{1/4} \textit{ if } k \ge 0. \\ x &\approx \pm 1.7321 && \textit{Compute.} \end{aligned}$$

In most cases throughout this chapter, we will round approximate solutions to the fourth decimal place.

In Example 3, we will solve some more logarithmic equations—this time for the *base* of a logarithm. A key step will still be to write an equation in logarithmic form in exponential form instead.

Example 3 Solving for the Base of a Logarithm

Solve for b.

1. $\log_b(81) = 4$ 2. $\log_b(67) = 5$

Solution

1. We write $\log_b(81) = 4$ in exponential form and solve for b:

$$\begin{aligned} b^4 &= 81 && \textit{Write in exponential form.} \\ b &= \pm 81^{1/4} && \textit{The solution of } b^4 = k \textit{ is } \pm k^{1/4}. \\ b &= 3 && \textit{Simplify; the base of a logarithm is positive.} \end{aligned}$$

2. We write $\log_b(67) = 5$ in exponential form and solve for b:

$$\begin{aligned} b^5 &= 67 && \textit{Write in exponential form.} \\ b &= 67^{1/5} && \textit{The solution of } b^5 = k \textit{ is } k^{1/5}. \\ b &\approx 2.3185 && \textit{Compute.} \end{aligned}$$

In summary, **for an equation of the form $\log_b(x) = k$, we can solve for b or x by writing the equation in exponential form.**

Power Property for Logarithms

An exponential equation in one variable is an equation in one variable in which an exponent contains a variable. Here are some examples of exponential equations in one variable:

$$3^x = 50 \qquad 5(2)^x = 71 \qquad 4(7)^x + 5 = 785 \qquad 4^{3x-2} = 391$$

An important property called the **power property for logarithms** will help us solve exponential equations.

Power Property for Logarithms

For $x > 0, b > 0$, and $b \neq 1$,

$$\log_b(x^p) = p\log_b(x)$$

In words, a logarithm of a power of x is the exponent times the logarithm of x.

For example, $\log_3(x^5) = 5\log_3(x)$. Also, $\log_2(x^7) = 7\log_2(x)$.

A proof of the power property for logarithms follows: Let $k = \log_b(x)$. The exponential form of this equation is $b^k = x$. Taking the exponent p on both sides and simplifying gives

$$(b^k)^p = x^p \quad \textit{Raise both sides to the power p.}$$
$$b^{kp} = x^p \quad \textit{Multiply exponents: } (b^m)^n = b^{mn}$$

A logarithmic form of this equation is

$$\begin{aligned} \log_b(x^p) &= kp \\ &= pk && \textit{Commutative law} \\ &= p\log_b(x) && \textit{Substitute } \log_b(x) \textit{ for } k. \end{aligned}$$

Therefore

$$\log_b(x^p) = p\log_b(x)$$

This statement is what we set out to prove.

Next, we describe the **logarithm property of equality,** which also is helpful in solving exponential equations.

Logarithm Property of Equality

For positive real numbers a, b, and c, where $b \neq 1$, the equations

$$a = c \quad \text{and} \quad \log_b(a) = \log_b(c)$$

are equivalent.

If we have the equation $a = c$ and then write $\log(a) = \log(c)$, we say we "take the log of both sides" of the equation $a = c$.

To solve an equation in exponential form, such as $3^x = 17$, we first take the log of both sides and then apply the power property for logarithms.

Example 4 Solving an Exponential Equation

Solve the equation $2^x = 12$.

Solution

$$\begin{aligned} 2^x &= 12 && \textit{Original equation} \\ \log(2^x) &= \log(12) && \textit{Take the log of both sides.} \\ x\log(2) &= \log(12) && \textit{Power property: } \log_b(x^p) = p\log_b(x) \\ x &= \frac{\log(12)}{\log(2)} && \textit{Divide both sides by } \log(2). \\ x &\approx 3.5850 && \textit{Compute.} \end{aligned}$$

We check that 3.5850 approximately satisfies the equation $2^x = 12$:

$$2^{3.5850} \approx 12.0003 \approx 12$$

WARNING When we compute a quotient of logarithms on a graphing calculator, it pays to watch the use of parentheses. Here we show a correct and an incorrect way to compute $\frac{\log(12)}{\log(2)}$:

Figure 39 Compute $\frac{\log(12)}{\log(2)}$

Calculator Entry	Calculator's Interpretation	
log(12) [÷] log(2) [ENTER]:	$\frac{\log(12)}{\log(2)}$	Correct
log(12 [÷] log(2 [ENTER]:	$\log\left(\frac{12}{\log(2)}\right)$	Incorrect

The correct computation is shown in Fig. 39.

Example 5 Solving an Exponential Equation

Solve $3(4)^x = 71$.

Solution

$$3(4)^x = 71 \quad \text{Original equation}$$

$$4^x = \frac{71}{3} \quad \text{Divide both sides by 3.}$$

$$\log(4^x) = \log\left(\frac{71}{3}\right) \quad \text{Take the log of both sides.}$$

$$x\log(4) = \log\left(\frac{71}{3}\right) \quad \text{Power property: } \log_b(x^p) = p\log_b(x)$$

$$x = \frac{\log\left(\frac{71}{3}\right)}{\log(4)} \quad \text{Divide both sides by log(4).}$$

$$x \approx 2.2824 \quad \text{Compute.}$$

We check that 2.2824 approximately satisfies the equation $3(4)^x = 71$:

$$3(4)^{2.2824} \approx 71.0008 \approx 71$$

WARNING Since $3(4)^x \neq (3 \cdot 4)^x$, we *cannot* begin to solve $3(4)^x = 71$ in Example 5 by saying

$$3(4)^x = 71$$
$$\log[3(4)^x] = \log(71)$$
$$x\log(3 \cdot 4) = \log(71) \quad \text{Cannot do this, since } 3(4)^x \neq (3 \cdot 4)^x.$$

That is why we began by dividing both sides of $3(4)^x = 71$ by 3.

In general,

$$\log_b(ax^p) \neq p\log_b(ax)$$

To solve some equations of the form $ab^x = c$ for x, we divide both sides of the equation by a, and then take the log of both sides. Next, we use the power property for logarithms.

Example 6 Solving an Exponential Equation

Solve $5^{3w-1} = 17$.

Figure 40 Verify the work

Solution

$$
\begin{aligned}
5^{3w-1} &= 17 && \text{Original equation} \\
\log(5^{3w-1}) &= \log(17) && \text{Take the log of both sides.} \\
(3w-1)\log(5) &= \log(17) && \text{Power property: } \log_b(x^p) = p\log_b(x) \\
3w - 1 &= \frac{\log(17)}{\log(5)} && \text{Divide both sides by } \log(5). \\
3w &= \frac{\log(17)}{\log(5)} + 1 && \text{Add 1 to both sides.} \\
w &= \frac{\frac{\log(17)}{\log(5)} + 1}{3} && \text{Divide both sides by 3.} \\
w &\approx 0.9201 && \text{Compute.}
\end{aligned}
$$

We use a graphing calculator table to check that, for the function $y = 5^{3x-1}$, the input 0.9201 leads approximately to the output 17 (see Fig. 40).

Example 7 Solving an Exponential Equation

Solve $7(2)^x - 4 = 20 + 3(2)^x$.

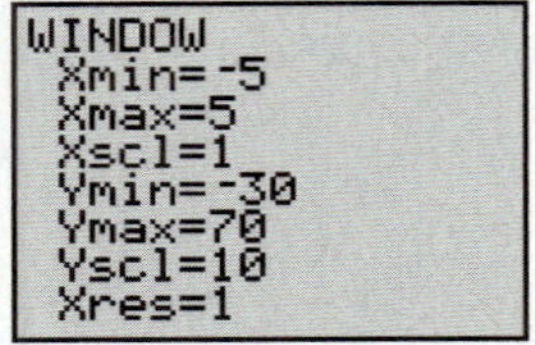

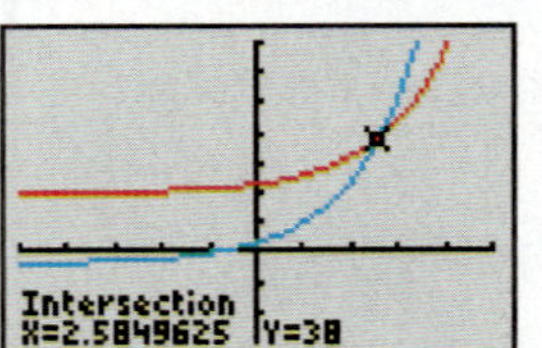

Intersection X=2.5849625 Y=38

Figure 41 Verify the work

Solution

$$
\begin{aligned}
7(2)^x - 4 &= 20 + 3(2)^x && \text{Original equation} \\
7(2)^x - 3(2)^x &= 24 && \text{Subtract } 3(2)^x \text{ from both sides; add 4 to both sides.} \\
4(2)^x &= 24 && 7(2)^x - 3(2)^x = (7-3)(2)^x = 4(2)^x \\
2^x &= 6 && \text{Divide both sides by 4.} \\
\log(2^x) &= \log(6) && \text{Take the log of both sides.} \\
x\log(2) &= \log(6) && \text{Power property: } \log_b(x^p) = p\log_b(x) \\
x &= \frac{\log(6)}{\log(2)} && \text{Divide both sides by } \log(2). \\
x &\approx 2.5850 && \text{Compute.}
\end{aligned}
$$

An approximate solution is 2.5850. Recall from Section 4.3 that we can use "intersect" on a graphing calculator to solve an equation in one variable. In Fig. 41, we graph the equations $y = 7(2)^x - 4$ and $y = 20 + 3(2)^x$ and find an approximate intersection point (2.5850, 38), which has x-coordinate 2.5850. So, an approximate solution of the original equation is 2.5850, which checks.

Example 8 Solving a General Exponential Equation

Solve the equation $ab^x + c = d$ for x, where $b > 0$, $b \neq 1$, and the constants have values for which the equation has exactly one real-number solution.

Solution

$$
\begin{aligned}
ab^x + c &= d && \text{Original equation} \\
ab^x + c - c &= d - c && \text{Subtract } c \text{ from both sides.} \\
ab^x &= d - c && \text{Combine like terms.} \\
\frac{ab^x}{a} &= \frac{d-c}{a} && \text{Divide both sides by } a. \\
b^x &= \frac{d-c}{a} && \text{Simplify.}
\end{aligned}
$$

$$\log(b^x) = \log\left(\frac{d-c}{a}\right) \quad \textit{Take the log of both sides.}$$

$$x\log(b) = \log\left(\frac{d-c}{a}\right) \quad \textit{Power property: } \log_b(x^p) = p\log_b(x)$$

$$x = \frac{\log\left(\frac{d-c}{a}\right)}{\log(b)} \quad \textit{Divide both sides by } \log(b).$$

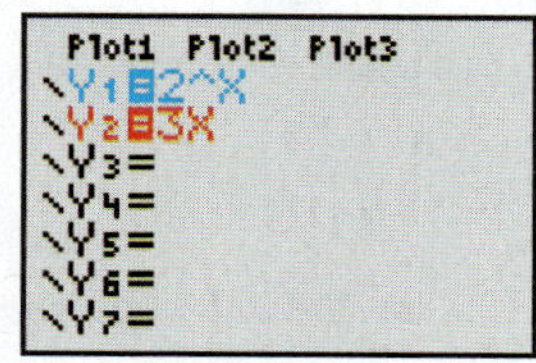

Solving Equations in One Variable by Using Graphs

In Example 9, we will use graphing to solve an equation that is impossible to solve by any combination of performing operations on both sides and taking the log of both sides.

Example 9 Using Graphing to Solve an Equation in One Variable

Use graphing to solve $2^x = 3x$.

Solution

We use "intersect" on a graphing calculator to find the solutions of the system

$$y = 2^x$$
$$y = 3x$$

See Fig. 42.

The approximate solutions of the system are (0.4578, 1.3735) and (3.3132, 9.9395). The x-coordinates of these ordered pairs, 0.4578 and 3.3132, are the approximate solutions of the equation $2^x = 3x$.

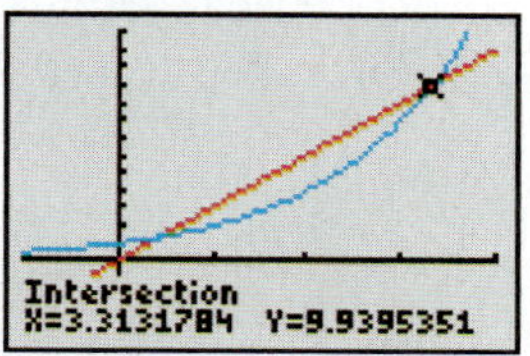

Figure 42 Solve the system

Group Exploration

Comparing the power property with other statements

Consider the following equations, where $x > 0$, $a > 0$, $b > 0$, and $b \neq 1$:

$$\log_b(x^p) = p\log_b(x)$$
$$\log_b[a(x^p)] = p\log_b(ax)$$
$$\log_b[(ax)^p] = p\log_b(ax)$$

1. Which, if any, of these equations are true in general? Explain why in terms of the power property for logarithms.
2. Show that the other equation or equations are false by using the substitutions $b = 10$, $a = 10$, $x = 10$, and $p = 2$.
3. Three students tried to solve $5(4)^x = 30$. Which students, if any, solved the equation correctly? Describe any errors and where they occurred.

Student 1's work

$$5(4)^x = 30$$
$$4^x = 6$$
$$\log(4^x) = \log(6)$$
$$x\log(4) = \log(6)$$
$$x = \frac{\log(6)}{\log(4)}$$
$$x \approx 1.2925$$

Student 2's work

$$5(4)^x = 30$$
$$\log[5(4)^x] = \log(30)$$
$$x\log[5(4)] = \log(30)$$
$$x\log(20) = \log(30)$$
$$x = \frac{\log(30)}{\log(20)}$$
$$x \approx 1.1353$$

Student 3's work

$$5(4)^x = 30$$
$$20^x = 30$$
$$\log(20^x) = \log(30)$$
$$x\log(20) = \log(30)$$
$$x = \frac{\log(30)}{\log(20)}$$
$$x \approx 1.1353$$

Homework 11.4

For extra help ▶ MyMathLab® 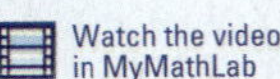 Watch the videos in MyMathLab 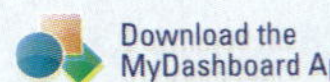 Download the MyDashboard App

Write the equation in exponential form. Assume all constants are positive and not equal to 1.

1. $\log_3(243) = 5$
2. $\log_2(32) = 5$
3. $\log(100) = 2$
4. $\log(10{,}000) = 4$
5. $\log_b(a) = c$
6. $\log_r(s) = t$
7. $\log(m) = n$
8. $\log(y) = z$

Write the equation in logarithmic form. Assume all constants are positive and not equal to 1.

9. $5^3 = 125$
10. $2^5 = 32$
11. $10^3 = 1000$
12. $10^5 = 100{,}000$
13. $y^w = x$
14. $r^s = t$
15. $10^p = q$
16. $10^x = y$

Solve.

17. $\log_4(x) = 2$
18. $\log_2(x) = 3$
19. $\log(x) = -2$
20. $\log(x) = -3$
21. $\log_4(x) = 0$
22. $\log_9(x) = 0$
23. $\log_{27}(t) = \frac{4}{3}$
24. $\log_{16}(p) = \frac{3}{4}$
25. $2\log_8(2x - 5) = 4$
26. $3\log_5(4x + 1) = 9$
27. $4\log_{81}(x) - 3 = -2$
28. $3\log_8(x) + 5 = 6$
29. $\log_2(\log_3(y)) = 3$
30. $\log_3(\log_2(p)) = -1$

Solve. Round any solutions to the fourth decimal place.

31. $\log_6(x^3) = 2$
32. $\log_4(x^5) = 3$

Solve for b. Round any approximate solutions to the fourth decimal place.

33. $\log_b(49) = 2$
34. $\log_b(16) = 2$
35. $\log_b(8) = 3$
36. $\log_b(125) = 3$
37. $\log_b(16) = 5$
38. $\log_b(95) = 9$

Solve. Round any approximate solutions to the fourth decimal place. Verify your result by checking that it satisfies or approximately satisfies the original equation.

39. $4^x = 9$
40. $10^x = 50$
41. $5(4)^x = 80$
42. $3(2)^x = 17$
43. $3.83(2.18)^t = 170.91$
44. $1.73(4.09)^w = 526.44$
45. $8 + 5(2)^x = 79$
46. $20 = -3 + 4(2)^x$
47. $2^{4x+5} = 17$
48. $8 = 3^{7x-1}$
49. $6(3)^x - 7 = 85 + 4(3)^x$
50. $5^x + 7 = 50 - 3(5)^x$
51. $4^{3p} \cdot 4^{2p-1} = 100$
52. $3^{2r} \cdot 3^{r-4} = 97$
53. $3^x = -8$
54. $1^x = 13$

Solve. Round any approximate solutions to the fourth decimal place.

55. $\log_4(x) = 3$
56. $\log_3(x) = 4$
57. $3(4)^t + 15 = 406$
58. $2(6)^w - 17 = 3$
59. $\log_b(73) = 5$
60. $\log_b(19) = 4$
61. $3\log_{27}(y - 1) = 2$
62. $2\log_4(3r + 2) = 5$
63. $3(2)^{4x-2} = 83$
64. $5(4)^{2x+1} = 974$

For Exercises 65–70, estimate any solutions of the equation or system by referring to the graphs shown in Fig. 43.

65. $2^x = 4\left(\frac{1}{2}\right)^x$
66. $2^x = 4 - x$
67. $4\left(\frac{1}{2}\right)^x = 4 - x$
68. $2^x = 5$
69. $4\left(\frac{1}{2}\right)^x = 1$
70. $y = 2^x$, $y = 4\left(\frac{1}{2}\right)^x$

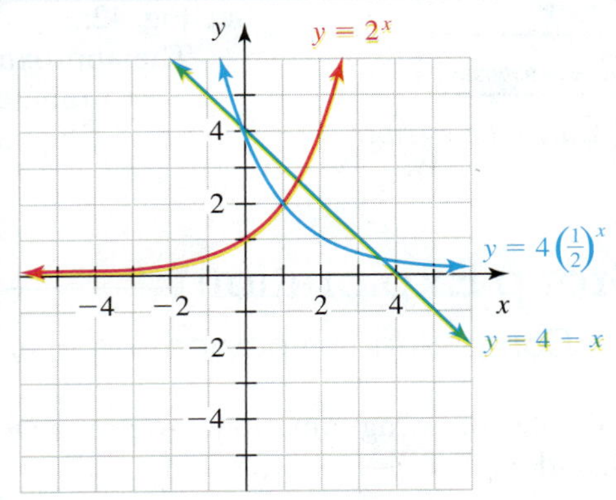

Figure 43 Exercises 65–70

Use "intersect" on a graphing calculator to solve the equation. Round any solutions to the fourth decimal place.

71. $3^x = 5 - x$
72. $2^x = 5 - 2x$
73. $7\left(\frac{1}{2}\right)^x = 2x$
74. $8\left(\frac{1}{3}\right)^x = x$
75. $\log(x + 1) = 3 - \frac{2}{5}x$
76. $6 - x = 3\log(x + 5)$

For Exercises 77–82, solve the given equation or system by referring to the values of $y = 3^{x-1}$, $y = 12\left(\frac{1}{2}\right)^x$, and $y = x - \frac{3}{2}$ shown in Table 44.

77. $3^{x-1} = 12\left(\frac{1}{2}\right)^x$
78. $12\left(\frac{1}{2}\right)^x = x - \frac{3}{2}$
79. $3^{x-1} = 81$
80. $12\left(\frac{1}{2}\right)^x = 6$

81. $y = 12\left(\frac{1}{2}\right)^x$

$y = x - \frac{3}{2}$

82. $y = 3^{x-1}$

$y = 12\left(\frac{1}{2}\right)^x$

Table 44 Some Values of Three Functions

x	1	2	3	4	5	6
$y = 3^{x-1}$	1	3	9	27	81	243
$y = 12\left(\frac{1}{2}\right)^x$	6	3	1.5	0.75	0.375	0.1875
$y = x - \frac{3}{2}$	−0.5	0.5	1.5	2.5	3.5	4.5

In Exercises 83–90, solve for x. Assume $b > 0$, $b \neq 1$, and the constants have values for which the equation has exactly one real-number solution.

83. $ab^x = c$

84. $\frac{b^x}{a} = c$

85. $b^x + c = d$

86. $b^x - c = d$

87. $ab^x - c = d$

88. $\frac{b^x}{a} + c = d$

89. $ab^{x+p} - c = d$

90. $ab^{kx} + c = d$

Concepts

91. A student tries to solve $3(8^x) = 7$:

$$\begin{aligned} 3(8)^x &= 7 \\ \log[3(8)^x] &= \log(7) \\ x\log[3(8)] &= \log(7) \\ x\log(24) &= \log(7) \\ x &= \frac{\log(7)}{\log(24)} \\ x &\approx 0.6123 \end{aligned}$$

Describe any errors. Then solve the equation correctly.

92. A student tries to solve $2(3)^x = 10$:

$$\begin{aligned} 2(3)^x &= 10 \\ 6^x &= 10 \\ \log(6^x) &= \log(10) \\ x\log(6) &= 1 \\ x &= \frac{1}{\log(6)} \\ x &\approx 1.2851 \end{aligned}$$

Describe any errors. Then solve the equation correctly.

Let $f(x) = 4^x$. Round any approximate results to the fourth decimal place.

93. Find $f(4)$.

94. Find $f^{-1}(16)$.

95. Find x when $f(x) = 3$.

96. Find x when $f^{-1}(x) = 3$.

For Exercises 97–100, let $g(x) = \log_2(x)$. Round any approximate results to the fourth decimal place.

97. Find $g(8)$.

98. Find $g^{-1}(4)$.

99. Find a when $g(a) = 5$.

100. Find a when $g^{-1}(a) = 5$.

101. Determine whether the statement is true. Explain.

a. $\frac{\log_2(4)}{\log_2(16)} = \frac{4}{16}$ **b.** $\frac{\log_3(1)}{\log_3(27)} = \frac{1}{27}$

c. $\frac{\log(1000)}{\log(10{,}000)} = \frac{1000}{10{,}000}$

d. $\frac{\log_b(c)}{\log_b(d)} = \frac{c}{d}$, where b, c, and d are positive, $b \neq 1$, and $d \neq 1$.

102. Determine whether the statement is true. Explain. [**Hint:** For parts (a), (b), and (c), use the "log" key on a calculator.]

a. $\log(4\cdot 3^2) = 2\log(4\cdot 3)$

b. $\log(5\cdot 2^3) = 3\log(5\cdot 2)$

c. $\log(6\cdot 10^4) = 4\log(6\cdot 10)$

d. $\log(ax^p) = p\log(ax)$, where a and x are positive.

103. a. Use ZDecimal to graph $f(x) = \log(x^3) - 3\log(x) + 1$. Describe the graph in words.

b. Explain why the graph of f is in neither quadrant II nor quadrant III.

c. Use properties of logarithms to write the right-hand side of the equation of f as a constant. Use your result to explain why the graph that you found in part (a) makes sense.

104. The incorrect work that follows shows that the logarithm of any positive number is 0. Describe any errors. Assume $b > 0$, $b \neq 1$, and $k > 0$:

$$\log_b(k) = \log_b(k\cdot 1) = \log_b(k\cdot 5^0) = 0\cdot\log_b(5k) = 0$$

105. Describe the power property for logarithms. Give an example. Does the power property imply that $x^p = px$? Explain.

106. Describe how to use the power property for logarithms to solve an exponential equation.

Related Review

Solve. Find the exact solution if the equation is linear. For other types of equations, round the solution(s) to the fourth decimal place.

107. $5(3p - 7) - 9p = -4p + 23$

108. $99 - 2(3)^x = 12$

109. $5b^6 - 88 = 56$

110. $\log_b(75) = 4$

111. $6x^2 - 3x = 5 - x$

112. $\log_2(3x + 7) = 4$

Expressions, Equations, Functions, and Graphs

Perform the indicated instruction. Then use words such as linear, quadratic, cubic, polynomial, degree, exponential, logarithmic, function, one variable, *and* two variables *to describe the expression, equation, or system.*

113. Solve $\log_2(x) = -5$.

114. Find an equation of a line that contains the points $(-2, -4)$ and $(3, 5)$.

115. Graph $y = \log_2(x)$ by hand.

116. Find the inverse of the function $f(x) = 3x - 7$.

117. Find $f(16)$, where $f(x) = \log_2(x)$.

118. Graph $3x + 2(y - 2x) = 3 - y$.

11.5 Using the Power Property with Exponential Models to Make Predictions

Objectives

- Use the power property for logarithms with exponential models to make predictions.
- Use the half-life of carbon-14 to date archeological artifacts and fossils.

In Section 10.5, we used exponential functions to model data. In this section, we will use exponential functions with the power property for logarithms to make predictions about the independent variable of the function.

Example 1 Using the Power Property to Make a Prediction

A person invests \$7000 in a bank account with a yearly interest rate of 6% compounded annually. When will the balance be \$10,000?

Solution

Let $B = f(t)$ be the balance (in thousands of dollars) after t years or any fraction thereof. From our work with compounded-interest accounts in Section 10.5, we know we can model the situation well by using an exponential model of the form $f(t) = ab^t$ with y-intercept $(0, a)$. The B-intercept is $(0, 7)$, for \$7000 when $t = 0$, so $a = 7$ and $f(t) = 7b^t$. By the end of each year, the account has increased by 6% of the previous year's balance, so $b = 1.06$. Thus,

$$f(t) = 7(1.06)^t$$

To find when the balance is \$10,000 ($B = 10$), we substitute 10 for $f(t)$ and solve for t:

$$10 = 7(1.06)^t \quad \text{Substitute 10 for } f(x).$$

$$\frac{10}{7} = 1.06^t \quad \text{Divide both sides by 7.}$$

$$\log\left(\frac{10}{7}\right) = \log\left(1.06^t\right) \quad \text{Take the log of both sides.}$$

$$\log\left(\frac{10}{7}\right) = t\log(1.06) \quad \text{Power property: } \log_b(x^p) = p\log_b(x)$$

$$t = \frac{\log\left(\frac{10}{7}\right)}{\log(1.06)} \quad \text{If } c = d\text{, then } d = c\text{; divide both sides by } \log(1.06).$$

$$t \approx 6.1212 \quad \text{Compute.}$$

Figure 44 Verify the work

So, it will take about 6 years and 45 days for the balance to reach \$10,000. We use a graphing calculator table to check that, for the function $y = 7(1.06)^x$, the input 6.1212 leads approximately to the output 10 (see Fig. 44).

To make a prediction about the independent variable t of an exponential model of the form $f(t) = ab^t$, we substitute a value for $f(t)$ and divide both sides of the equation by the coefficient a. Next, we take the log of both sides of the equation and use the power property to help solve for t.

Table 45 Infant Mortality Rates

Year	Rate (number of deaths per 1000 infants)
1915	99.9
1920	85.8
1930	64.6
1940	47.0
1950	29.2
1960	26.0
1970	20.0
1980	12.6
1990	9.2
2000	6.9
2010	6.2
2012	6.0

Source: *National Center for Health Statistics*

Example 2 Using the Power Property to Make a Prediction

The infant mortality rate is the number of deaths of infants under one year old per 1000 births.* In 1915, the rate was almost 100 deaths per 1000 infants, or 1 death per 10 infants. The infant mortality rate has decreased substantially since then (see Table 45).

1. Let $I = f(t)$ be the infant mortality rate (number of deaths per 1000 infants) at t years since 1900. Find an equation of f.
2. What is the percent rate of decay for infant mortality rates?
3. Find $f^{-1}(4.7)$. What does the result mean in this situation?

*The rate does not include fetal deaths.

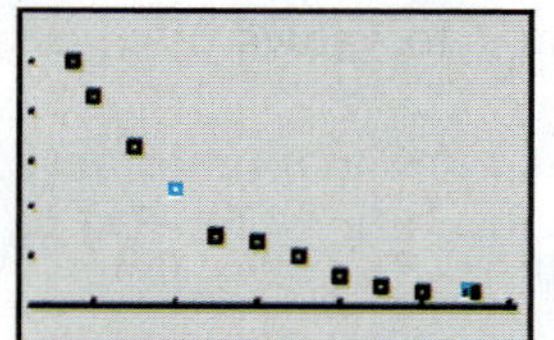
Figure 45 Scattergram of the data

Solution

1. We use a graphing calculator to draw a scattergram to describe the data (see Fig. 45). Since the points suggest a curve that "bends," we will not use a linear function to model the data. So, we will explore whether an exponential function or a quadratic function models the situation well.

   First, we use the points $(40, 47)$ and $(110, 6.2)$ to find an exponential equation of the form $f(t) = ab^t$. We substitute the coordinates of the two chosen points into $f(t) = ab^t$:

$$6.2 = ab^{110}$$
$$47 = ab^{40}$$

We divide the left sides and divide the right sides of the equations and solve for b:

$$\frac{6.2}{47} = \frac{ab^{110}}{ab^{40}}, \text{ where } a \neq 0 \text{ and } b \neq 0 \quad \textit{Divide left sides and divide right sides.}$$

$$\frac{6.2}{47} = b^{70} \quad \textit{Simplify; subtract exponents: } \frac{b^m}{b^n} = b^{m-n}$$

$$b = \left(\frac{6.2}{47}\right)^{1/70} \quad \textit{The solution of } b^{70} = k \textit{ is } k^{1/70}.$$

$$b \approx 0.971 \quad \textit{Compute.}$$

So, $f(t) = a(0.971)^t$. We substitute the coordinates of $(110, 6.2)$ into the equation $f(t) = a(0.971)^t$ and solve for a:

$$6.2 = a(0.971)^{110} \quad \textit{Substitute 110 for t and 6.2 for } f(t).$$

$$a = \frac{6.2}{0.971^{110}} \quad \textit{Divide both sides by } 0.971^{110}.$$

$$a \approx 157.86 \quad \textit{Compute.}$$

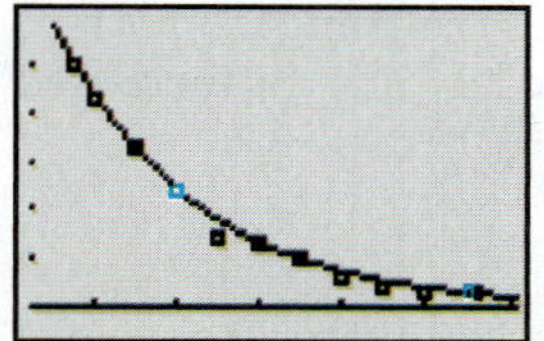
Figure 46 Exponential model

Thus, an exponential equation is $f(t) = 157.86(0.971)^t$.

Next, we use a graphing calculator to find the quadratic regression equation:

$$f(t) = 0.0131t^2 - 2.56t + 131.56$$

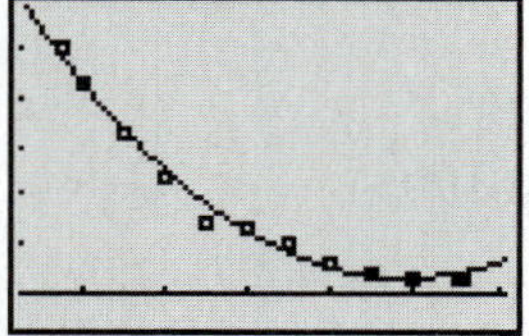
Figure 47 Quadratic model

Then, we see how well the exponential function and the quadratic function fit the data (see Figs. 46 and 47). Both models fit the data well, but it appears that the exponential model fits the data a bit better than the quadratic model does.

Finally, we consider the context of the situation. We use Zoom Out on a graphing calculator and then use "minimum" to find the approximate vertex (97.71, 6.49) of the graph of the quadratic model (see Fig. 48). So, the quadratic model estimates that the infant mortality rate has been increasing since 1998, which is false. To make matters worse, the quadratic model predicts that the mortality rate will continue to increase, which is hard to believe since medical advances are likely to continue to occur.

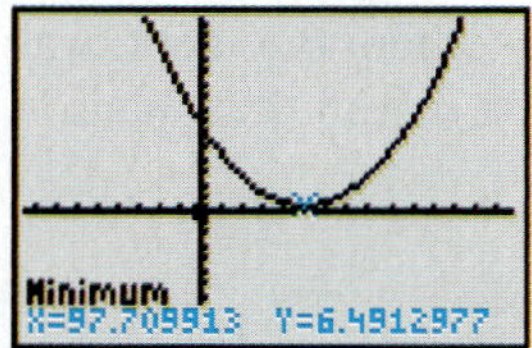

Figure 48 The vertex of the graph of the quadratic model

The exponential model correctly estimates that the infant mortality rate decreased over the period 1915–2012. Further, the exponential model predicts that the infant mortality rate will continue to decrease, which matches our expectation.

Since the exponential model both fits the data better and makes more sense within the context of the situation than the quadratic model does, we conclude that the exponential model is the best model. So, our equation of f is

$$f(t) = 157.86(0.971)^t$$

2. The base b is 0.971. Since $1 - 0.971 = 0.029$, we conclude that the model estimates the infant mortality rate has decayed by 2.9% per year.

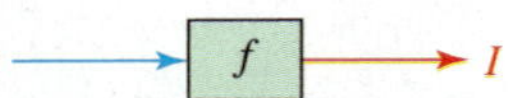

Figure 49 f sends values of t to values of I

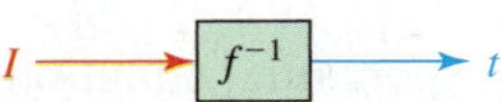

Figure 50 f^{-1} sends values of I to values of t

3. Since f sends values of t to values of I, f^{-1} sends values of I to values of t (see Figs. 49 and 50).

Therefore, $f^{-1}(4.7)$ represents the year (since 1900) when the infant mortality rate will be 4.7 deaths per 1000 infants. To find the year, we substitute 4.7 for $f(t)$ in the equation $f(t) = 157.86(0.971)^t$ and solve for t:

$$4.7 = 157.86(0.971)^t \quad \text{Substitute 4.7 for } f(t).$$

$$\frac{4.7}{157.86} = 0.971^t \quad \text{Divide both sides by 157.86.}$$

$$\log\left(\frac{4.7}{157.86}\right) = \log(0.971^t) \quad \text{Take the log of both sides.}$$

$$\log\left(\frac{4.7}{157.86}\right) = t\log(0.971) \quad \text{Power property: } \log_b(x^p) = p\log_b(x)$$

$$t = \frac{\log\left(\frac{4.7}{157.86}\right)}{\log(0.971)} \quad \text{Divide both sides by log (0.971).}$$

$$t \approx 119.41 \quad \text{Compute.}$$

The model predicts the infant mortality rate will be 4.7 deaths per 1000 infants in 2019.

Before we perform more modeling, it will be helpful to know a fact related to the base multiplier property (Section 10.3). For an exponential function $y = f(t) = ab^t$, we know by the base multiplier property that, as the value of t increases by 1, the value of y is multiplied by the base b. Also, if the value of y is multiplied by M in going from $t = 0$ to $t = k$, then the value of y will continue to be multiplied by M each time t is increased by k. To see this, recall that $f(0) = a$ and note that

$$f(k) = ab^k = aM, \quad \text{so} \quad b^k = M$$

We can use the fact that $b^k = M$ to show that if $t = 0$ is increased by k a total of n times, then y is multiplied by M a total of n times:

$$f(nk) = ab^{nk} = a(b^k)^n = a(M)^n$$

We use this idea in Example 3.

Table 46 Federal Debt Amounts

Year	Federal Debt (billions of dollars)
1960	291
1970	381
1980	909
1990	3206
2000	5629
2010	13,529

Source: *U.S. Office of Management and Budget*

Example 3 Using the Power Property to Estimate Doubling Time

In Exercise 36 of Homework 10.5, you may have found the equation $f(t) = 291(1.08)^t$, where $f(t)$ is the federal debt (in billions of dollars) at t years since 1960 (see Table 46). Estimate how often the federal debt doubles.

Solution

In 1960, the federal debt was \$291 billion. We find the year when the debt was $2(291) = 582$ billion dollars (twice as large) by substituting 582 for $f(t)$ in the equation $f(t) = 291(1.08)^t$ and solving for t:

$$582 = 291(1.08)^t \quad \text{Substitute 582 for } f(t).$$

$$2 = 1.08^t \quad \text{Divide both sides by 291.}$$

$$\log(2) = \log(1.08^t) \quad \text{Take the log of both sides.}$$

$$\log(2) = t\log(1.08) \quad \text{Power property: } \log_b(x^p) = p\log_b(x)$$

$$t = \frac{\log(2)}{\log(1.08)} \quad \text{Divide both sides by log (1.08).}$$

$$t \approx 9.01 \quad \text{Compute.}$$

According to the exponential function f, it took about 9 years to double the 1960 debt. By the discussion preceding this example, we can conclude that f predicts the debt will double *every* 9 years.

Recall from Section 10.5 that the half-life of an element is the amount of time it takes for the number of atoms to be reduced to half. All organisms are, in part, composed of the elements carbon-12 and carbon-14. Carbon-14 is radioactive. After an animal or plant dies, its carbon-14 decays exponentially with a half-life of 5730 years. However, the amount of carbon-12 remains constant. Scientists know the ratio of carbon-14 to carbon-12 in *living* organisms. Hence, scientists can determine how long ago an organism lived by measuring the decreased ratio of carbon-14 to carbon-12 in a bone, a piece of wood, or another artifact that was once living and was owned or used by the organism.

Example 4 Using the Power Property to Make an Estimate

A violent volcanic eruption and subsequent collapse of the former Mount Mazama created Crater Lake, the deepest lake in the United States. Scientists found a charcoal sample from a tree that burned in the eruption. If only 39.40% of the carbon-14 remains in the sample, when did Crater Lake form?

Solution

Let $P = f(t)$ be the percentage of carbon-14 that remains at t years after the sample formed. Since the percentage is halved every 5730 years, we will find an exponential equation of the form

$$f(t) = ab^t$$

At time $t = 0$, 100% (all) of the carbon-14 remained, so the P-intercept is (0, 100). Therefore, $a = 100$ and $f(t) = 100b^t$. At time $t = 5730$, $\frac{1}{2}(100) = 50\%$ of the carbon-14 remained. So, we substitute the coordinates of the point (5730, 50) into the equation $f(t) = 100b^t$ and solve for b:

$$\begin{aligned} 50 &= 100b^{5730} && \text{Substitute 5730 for } t \text{ and 50 for } f(t). \\ 0.5 &= b^{5730} && \text{Divide both sides by 100.} \\ b &= \pm 0.5^{1/5730} && \text{The solutions of } b^{5730} = k \text{ are } \pm k^{1/5730}. \\ b &\approx 0.999879 && \text{Compute; the base of an exponential function is positive.} \end{aligned}$$

The equation is $f(t) = 100(0.999879)^t$. We use more digits than usual for the base, as even a small change in it would greatly affect estimates well into the past (or future).

To estimate the age of the sample, we substitute 39.40 for $f(t)$ and solve for t:

$$\begin{aligned} 39.40 &= 100(0.999879)^t && \text{Substitute 39.40 for } f(t). \\ 0.3940 &= 0.999879^t && \text{Divide both sides by 100.} \\ \log(0.3940) &= \log(0.999879^t) && \text{Take the log of both sides.} \\ \log(0.3940) &= t\log(0.999879) && \text{Power property: } \log_b(x^p) = p\log_b(x) \\ t &= \frac{\log(0.3940)}{\log(0.999879)} && \text{Divide both sides by log (0.999879).} \\ t &\approx 7697 && \text{Compute.} \end{aligned}$$

So, the age of Crater Lake (and the sample) is approximately 7697 years.

Group Exploration

Finding an equation of the inverse of an exponential model

In Example 1, we found the model $B = f(t) = 7(1.06)^t$, where B is the balance (in thousands of dollars) of an account after t years. The starting balance is \$7000, and the interest rate is 6% compounded annually. In this exploration, you will find the equation of the inverse function f^{-1}.

1. Substitute B for $f(t)$ in the equation $f(t) = 7(1.06)^t$.
2. Solve your equation for t.
3. Replace t with $f^{-1}(B)$ in your equation. You now have an equation of f^{-1}.
4. Use your equation of f^{-1} to find $f^{-1}(12)$. What does it mean in this situation?
5. Use a graphing calculator to help you complete Table 47.

Table 47 Values of f^{-1}

B	$f^{-1}(B)$
7	
8	
9	
10	
11	
12	
13	
14	

6. How could the information in your completed table help an investor?

Group Exploration

Looking ahead: Product and quotient properties for logarithms

1. **a.** Use a calculator to compare $\log(2 \cdot 3)$ with $\log(2) + \log(3)$.
   b. Use a calculator to compare $\log(7 \cdot 2)$ with $\log(7) + \log(2)$.
   c. Use a calculator to compare $\log(4 \cdot 6)$ with $\log(4) + \log(6)$.
   d. What do parts (a)–(c) of this exploration suggest about $\log(xy)$? Check whether your observation is true for other values of x and y. (Your observation is referred to as the *product property for logarithms*.)

2. Determine how $\log\left(\frac{x}{y}\right)$ can be expressed in terms of two logarithms. [**Hint:** Choose specific values for x and y. Then compute $\log\left(\frac{x}{y}\right)$, $\log(x)$, and $\log(y)$, and compare the values.] Check whether your observation is true for other values of x and y. (Your observation is referred to as the *quotient property for logarithms*.)

Homework 11.5

For extra help ▶ MyMathLab®

1. A person invests \$2000 in an account at 5% interest compounded annually. Let $V = f(t)$ be the value (in dollars) of the account after t years or any fraction thereof.
   a. Find an equation of f.
   b. What is the V-intercept? What does it mean in this situation?
   c. What will be the value of the investment in five years?
   d. When will the value of the investment be \$3000?

2. A person invests \$12,500 in an account at 8% interest compounded annually. Let $V = f(t)$ be the value (in dollars) of the account after t years or any fraction thereof.
   a. Find an equation of f.
   b. What is the V-intercept? What does it mean in this situation?
   c. What will be the value of the investment in four years?
   d. When will the value of the investment be \$20,000?

3. A person invests \$9300 in an account at 6% interest compounded annually. When will the value of the investment be \$13,700?

4. A person invests \$4500 in an account at 3% interest compounded annually. When will the value of the investment be \$5900?

5. A person invests \$6000 in an account at 10% interest compounded annually. When will the value of the investment be doubled? Explain why it will take less than 10 years, even though the rate is 10%.

6. A person invests \$4000 in an account at 5% interest compounded annually. When will the value of the investment be doubled? Explain why it will take less than 20 years, even though the rate is 5%.

7. The U.S. annual production of ethanol, used as fuel for automobiles, was 13.9 billion gallons in 2011 and has grown by about 19% per year since then (Source: *Department of Energy*). Predict when annual ethanol production will reach 40 billion gallons.

8. The number of Iraqi students enrolled at U.S. colleges and universities was 616 students in 2011 and has grown by about 27% per year since then (Source: *Institute of International Education*). Predict when 3000 Iraqi students will enroll in U.S. colleges and universities.

9. Suppose a rumor is spreading in the United States that tomato juice causes hair loss. Assume 30 people have heard the rumor as of today and each day the number of people (both past and present) who will have heard the rumor triples. Let $f(t)$ be the number of people (both past and present) who will have heard the rumor at t days since today.

a. Find an equation of f.

b. Find $f(8)$. What does it mean in this situation?

c. Find $f^{-1}(21{,}870)$. What does it mean in this situation?

d. Predict when all Americans (both past and present) will have heard the rumor. Assume the U.S. population is 315 million.

10. There are 4 million bacteria on a peach at noon on Tuesday. Assume a bacterium divides into 2 bacteria every hour, on average. Let $f(t)$ be the number of bacteria (in millions) on the peach at t hours after Tuesday noon.

a. Find an equation of f.

b. Find $f(24)$. What does it mean in this situation?

c. Find $f^{-1}(8000)$. What does it mean in this situation?

11. According to the U.S. Occupational Safety and Health Administration standard, an average person can listen to 8 hours of sound per day at a sound level of 90 decibels without experiencing hearing loss. Recall from Exercise 83 of Homework 11.3 that a decibel is a unit for measurement of sound intensity. For each increase of 5 decibels, the exposure time must be cut in half. For example, an average person can listen to 4 hours of sound per day at a sound level of 95 decibels without experiencing hearing loss. (One overexposure may result in temporary, but probably not permanent, hearing loss.)

Some examples of sound being made at various decibel levels are listed in Table 48.

Table 48 Examples of Decibel Levels

Sound Level (decibels)	Example
0	Faintest sound heard by humans
20	Whisper
40	Inside a running car
60	Conversation
80	Noisy street corner
100	Soft-rock concert
120	Threshold of pain

Source: *Math and Music, by Garland and Kahn. Dale Seymour Publications, 1985, Pearson Education, Inc.*

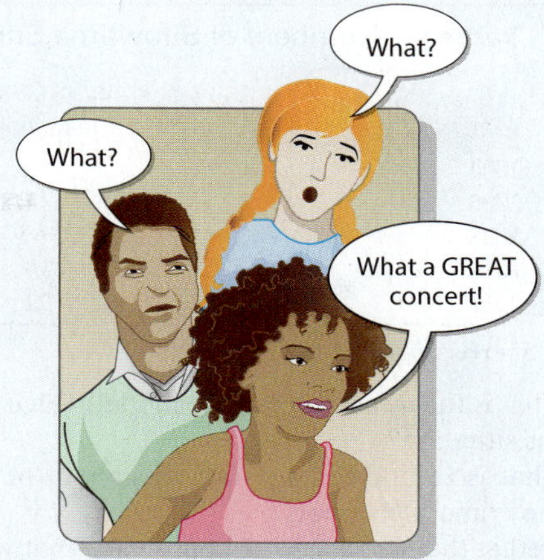

a. Let $T = f(d)$ be the number of hours of safe exposure time in one day to a sound at a level d decibels *above 90 decibels*. Find an equation of f.

b. Many rock bands play at about 114 decibels. Use your equation of f to predict how long they can play at concerts so their fans do not experience hearing loss. On the basis of your result, do you think these types of fans experience hearing loss? Assume most people are not wearing earplugs.

c. Many rock concerts last about 3 hours. At what sound level should the bands play so fans who attend a lot of concerts will not experience hearing loss?

12. In a weightless environment, astronauts lose about 1.5% of the calcium in their bodies per month (Source: *NASA*). Suppose some astronauts flew to Mars. Assuming the astronauts did not lose or gain calcium while on Mars, they would lose 17% of the calcium in their bodies due to the round trip. Estimate the number of months the round trip would take (not counting the time spent on Mars).

13. First prize for the World Series of Poker's main event has grown approximately exponentially from \$0.21 million in 1975 to \$8.53 million in 2012 (Source: *Harrah's Entertainment*). Predict when the first prize will be \$19 million.

14. Electronic cigarette sales have grown approximately exponentially from 0.05 million units in 2008 to 2.5 million units in 2011 (Source: *Tobacco Vapor Electronic Cigarette Association*). Each unit has three parts: battery, charger, and nicotine-containing cartridge. Predict in which year sales will reach 1.676 *billion* units.

15. The timber harvests in the Tongass National Forest in Alaska have decayed approximately exponentially from 471 million board feet in 1990 to 36 million board feet in 2010 (Source: *U.S. Forest Service*). Predict when the timber harvest will be 13 million board feet.

16. The revenue from personal breathalyzers to test for alcohol impairment has grown approximately exponentially from \$27.9 million in 2005 to \$215.2 million in 2011 (Source: *WinterGreen Research*). Predict in which year the revenue will be \$641.9 million. WinterGreen Research predicts this will happen in 2016. Does this market-analysis firm believe the percentage rate of change of revenue per year will be greater than, less than, or equal to that during the period 2005–2011? Explain.

17. In Exercise 27 of Homework 10.5, you found the model $f(t) = 9.58(1.07)^t$, when $n = f(t)$ is the number (in millions) of Showtime subscribers at t years since 2000 (see Table 49).

Table 49 Numbers of Showtime Subscribers

Year	Number of Subscribers (in millions)
2004	13
2006	14
2008	16
2010	19
2012	22

Source: *SNL Kagan*

a. What is the n-intercept of the model? What does it mean in this situation?
b. What is the percentage rate of growth of the number of Showtime subscribers?
c. Predict the *percentage* of households that will subscribe to Showtime in 2019. Assume there will be 120 million households in that year.
d. There were 28 million subscribers to HBO in 2012. Predict when there will be that many subscribers to Showtime.

18. In Exercise 28 of Homework 10.5 you found the model $f(t) = 1.51(1.08)^t$, where $n = f(t)$ is the number of North American cruise ship passengers in the year that is t years since 1980 (see Table 50).

Table 50 Numbers of North American Cruise Ship Passengers

Year	Number of Passengers (in millions)
1980	1.4
1985	2.2
1990	3.8
1995	4.7
2000	7.2
2005	11.2
2010	14.8

Source: *Cruise Lines International Association*

a. What is the n-intercept of the model? What does it mean in this situation?
b. What is the percentage rate of growth of the number of passengers?
c. Assuming no one goes on more than one cruise in a single year, predict the *percentage* of Americans who will go on a cruise in 2018. Assume the U.S. population will be 338 million in that year.
d. Predict in which year the number of passengers will equal Texas's population, which is 25 million.

19. The numbers of polio cases in the world are shown in Table 51 for various years.

Table 51 Numbers of Polio Cases Worldwide

Year	Number of Polio Cases (thousands)
1988	350
1992	138
1996	33
2000	4
2004	1.3
2008	1.7
2011	0.7

Source: *World Health Organization*

Let $f(t)$ be the number of polio cases (in thousands) in the year that is t years since 1980.

a. Find an exponential equation of f.
b. Predict the number of polio cases in 2018.
c. Predict in which year there will be 1 case of polio.
d. Find the approximate half-life of the number of polio cases. [**Hint:** Use f to estimate the number of cases in 1980. Then use f to estimate when there was half that number of cases.]

20. The numbers of Russian children adopted by American parents are shown in Table 52 for various years.

Table 52 Numbers of Russian Children Adopted by American Parents

Year	Number of Children
2004	5862
2005	4631
2006	3702
2007	2303
2008	1857
2009	1586
2010	1079
2011	962

Source: *U.S. Department of State*

Let $f(t)$ be the number of Russian children adopted by American parents in the year that is t years since 2000.

a. Find an exponential equation of f.
b. Predict the number of Russian children that will be adopted by American parents in 2018. How many adoptions will that be per state, on average?
c. Predict in which year only one Russian child will be adopted by American parents.
d. Find the approximate half-life of the number of Russian children adopted by American parents. [**Hint:** Use f to estimate the number of adoptions in 2000. Then use f to estimate in which year the number of adoptions was half that amount.]

21. The numbers of Twitter employees are shown in Table 53 for various years.

Table 53 Numbers of Twitter Employees

Year	Number of Employees
2008	8
2009	29
2010	130
2011	350
2012	900

Source: *AllThingsD.com*

Let $f(t)$ be the number of Twitter employees at t years since 2005.

a. Find a linear equation, a quadratic equation, and an exponential equation of f. Compare how well the models fit the data.
b. Twitter began in 2006 with 4 employees. For years before 2008, which function best models the situation? Explain.
c. What is the base b of your exponential model $f(t) = ab^t$? What does it mean in this situation?
d. Use the exponential model to predict the number of Twitter employees in 2014.
e. Use the exponential model to predict when the number of Twitter employees will equal 394 thousand.

22. The numbers of LEED-certified green buildings in the United States are shown in Table 54 for various years.

Table 54 Numbers of LEED-Certified Green Buildings

Year	Number of Buildings with LEED Certification
2002	28
2004	183
2006	660
2008	2113
2010	7267

Source: *United States Green Building Council*

Let $f(t)$ be the number of LEED-certified green buildings in the United States at t years since 2000.

a. Find a linear equation, a quadratic equation, and an exponential equation of f. Compare how well the models fit the data.

b. What is the base b of your exponential model $f(t) = ab^t$? What does it mean in this situation?

c. Predict the number of LEED-certified green buildings in 2015.

d. Predict when there will be an average of 2000 LEED-certified green buildings *per state*.

23. In Exercise 29 of Homework 10.5, you may have found an equation close to $f(t) = 1.21(1.0161)^t$, where $p = f(t)$ models world population (in billions) at t years since 1900 (see Table 55).

Table 55 World Population

Year	World Population (billions)
1930	2.070
1940	2.295
1950	2.500
1960	3.050
1970	3.700
1980	4.454
1990	5.279
2000	6.080
2012	7.044

Source: *U.S. Census Bureau*

a. The United Nations predicts world population will reach 9.3 billion in 2050. Use f to predict when world population will reach 9.3 billion.

b. Use a graphing calculator to draw a scattergram to describe the part of the data in Table 55 *from 1970 to 2012*. Is it better to use a linear function, a quadratic function, or an exponential function to model the data? Explain. Find an equation for such a function. Use the function notation "g" for this function.

c. Use g to predict when world population will reach 9.3 billion. Is that year before or after the predicted year you found in part (a)? Why does this make sense?

d. The United Nations describes a possible post-2050 scenario in which world population will reach 9.3 billion in 2050 and increase to 10.1 billion in 2100. Draw by hand a scattergram of all of the data shown in Table 55, and plot points for the scenario's predictions for 2050 and 2100. Graph the functions f and g by hand. Then sketch a model that describes the scenario better than the functions f and g do.

24. The minimum salaries for major league baseball players are shown in Table 56 for various years.

Table 56 Salaries for Major League Baseball Players

Year	Minimum Salary (thousands of dollars)
1970	12
1975	16
1980	30
1985	60
1990	100
1995	109
2000	200
2005	316
2010	400
2012	480

Source: *baseball-reference.com*

Let $M(t)$ be the minimum salary (in thousands of dollars) at t years since 1970.

a. Find an exponential equation of M.

b. What is the percentage rate of growth of minimum salaries?

c. Find $f(48)$. What does it mean in this situation?

d. Find $f^{-1}(1000)$. What does it mean in this situation?

e. In Exercise 45 of Homework 10.3, you may have worked with the model $f(t) = 1.22(1.051)^t$, where $f(t)$ is the average ticket price (in dollars) to major league baseball games at t years since 1950. What is the percentage rate of growth of average ticket prices? Can the growth in minimum salaries be accounted for by the growth in average ticket prices alone? Explain.

25. The percentages of seniors with severe memory impairment (based on memory tests) are shown in Table 57 for various age groups.

Table 57 Percentages of Seniors with Severe Memory Impairment

Age Group (years)	Age Used to Represent Age Group (years)	Percent
65–69	67	1.1
70–74	72	2.5
75–79	77	4.5
80–84	82	6.4
over 84	88	12.9

Source: *Federal Interagency Forum on Aging-Related Statistics*

Let $p = f(t)$ be the percentage of seniors at age t years with severe memory impairment.

a. Find an exponential equation of f.

b. What is the base b of your model $f(t) = ab^t$? What does it mean in this situation?

c. Find $f(70)$. What does it mean in this situation?

d. Find $f^{-1}(10)$. What does it mean in this situation?

e. In Exercise 7 of Homework 6.4, you may have found that there is a linear relationship between an adult's age and an adult's score on a test measuring memory and information-processing speed. Would that *linear* relationship necessarily conflict with an *exponential* relationship between a senior's age and the percentage of seniors with severe memory impairment? Explain.

26. Saks Fifth Avenue® offered a promotional sale in which customers could receive a gift card. The values of the gift cards are shown in Table 58 for various expenditures.

Table 58 Saks Fifth Avenue Gift Card Values

Expenditure Group (dollars)	Expenditure Used to Represent Expenditure Group (dollars)	Gift Card Value (dollars)
250–499	375	25
500–999	750	50
1000–1999	1500	100
2000–2999	2500	200
3000 or more	3500	450

Source: *Saks Fifth Avenue*

Let $v = f(s)$ be the value (in dollars) of a gift card that a customer who spends s dollars will receive.

a. Find an exponential equation of f.
b. What is the coefficient a of your model $f(s) = ab^s$? What does it mean in this situation?
c. What is the base b of your model $f(s) = ab^s$? What does it mean in this situation?
d. Customer A spends \$2000, customer B spends \$2500, and customer C spends \$2999. According to your model f, what are the values of the gift cards that these customers will receive? Compare these values with the actual values of the gift cards they will receive.
e. Use your model to estimate for what expenditure a customer would receive a \$700 gift card.
f. If a new promotion is to include a gift card for \$700, as well as gift cards for the values shown in Table 58, determine reasonable expenditure groups for \$450 and \$700 gift cards.

27. In a study of 10 of the most selective U.S. colleges and universities, researchers found that a student applicant has a better chance of being accepted to a college through early decision (students apply early and colleges decide early) than by regular decision. Table 59 shows a comparison of SAT scores (out of 1600) and acceptance rates by both systems.

Table 59 Percentages of Applicants Accepted by Early Decision and Regular Decision

SAT Score Group	Score Used to Represent SAT Score Group	Percent	
		Early Decision	Regular Decision
1100–1190	1145	21	10
1200–1290	1245	35	17
1300–1390	1345	52	31
1400–1490	1445	70	48
1500–1600	1550	93	72

Source: *Professor Christopher Avery, Kennedy School of Government, Harvard University*

For students who score s points, let $E(s)$ and $R(s)$ be the percentages of early-decision and regular-decision applicants, respectively, who are accepted.

a. Find exponential regression equations of E and R.
b. What percentage of early-decision applicants who score 1425 points get accepted? What about regular-decision applicants who score 1425 points?
c. For what score do half of early-decision applicants get accepted? What about regular-decision applicants?
d. The study concluded that students who apply for early decision have the equivalent of 100 points added to their SAT score, compared with students applying for regular decision. What do your results from part (c) suggest the equivalent number of added points to be?
e. Use "intersect" on a graphing calculator to find the intersection point of the graphs of E and R. What does it mean in this situation? [Challenge: Use substitution and the property $\frac{b^t}{c^t} = \left(\frac{b}{c}\right)^t$, where b and c are positive, to find the intersection point.]

You're planning pretty far ahead, but by applying to our ultra-early decision system, your newborn need score only 1 point on the SAT to be accepted!

28. New York Life offers a \$250,000 life insurance policy. Quarterly rates for women and men are shown in Table 60 for various ages.

Table 60 New York Life Quarterly Rates for a \$250,000 Policy

Age (years)	Quarterly Rate (dollars)	
	Women	Men
35	25.00	28.75
40	33.75	35.75
45	51.25	57.50
50	70.00	87.50
55	104.50	145.00
60	145.75	230.75
64	220.00	355.00

Source: *New York Life*

Let $W(t)$ and $M(t)$ be the quarterly rates (in dollars) for women and men, respectively, both at t years of age.

a. Find exponential regression equations of W and M.
b. For a \$250,000 policy, how much would a 52-year-old woman pay per quarter? How much would a 52-year-old man pay per quarter?
c. At what age would a woman pay \$120 per quarter for a \$250,000 policy? At what age would a man pay that much for a \$250,000 policy?
d. Due to Montana insurance regulations, both sexes must pay the same quarterly rates. So, New York Life uses the male rates for all residents of Montana. Estimate how much more a 62-year-old woman would pay per quarter for a \$250,000 policy if she lived in Montana rather than in some other state.

e. Use "intersect" on a graphing calculator to find the intersection point of the graphs of W and M. What does it mean in this situation? [Challenge: Use substitution and the property $\frac{b^t}{c^t} = \left(\frac{b}{c}\right)^t$, where b and c are positive, to find the intersection point.]

29. Physicians use gallium citrate-67 to detect certain types of cancer, including lymphoma. Gallium citrate-67 has an effective half-life of 3.25 days—some is lost to radioactive decay, and some is removed through the digestive and urinary tracts. A patient who is breast-feeding is injected with the radioactive element.

a. Let $f(t)$ be the percentage of the gallium citrate-67 that remains in the patient's body at t days since she was injected. Find an equation of f.

b. A scan of the gallium citrate-67 is performed 2 days after the injection. What percentage of the element remains?

c. The patient can resume breast-feeding when only 0.39% of the gallium citrate-67 remains. When can she resume breast-feeding?

30. Physicians use technetium-99m to locate stress fractures in bones. Technetium-99m has an effective half-life of 5.3 hours—some is lost to radioactive decay, and some is removed through urination. A patient with a possible stress fracture in his foot is injected with the radioactive element.

a. Let $f(t)$ be the percentage of the technetium-99m that remains in the patient's body at t hours since he was injected. Find an equation of f.

b. What percentage of the technetium-99m will remain after 1 day?

c. When will only 1% of the technetium-99m remain?

31. Scientists used a sample of spruce wood from the Two Creeks Forest Bed in Wisconsin to date an advance of the continental ice sheet into the United States during the last Ice Age. If 24.46% of the carbon-14 remains in the sample, when did the ice sheet advance? (Assume this advance killed the tree.) The half-life of carbon-14 is 5730 years.

32. A mummy was on display at a museum in Niagara Falls until it was sold in 1999. A few years later, researchers identified the mummy as the ancient Egyptian pharaoh Rameses I. The mummy was eventually returned to Egypt. If 69.57% of the carbon-14 in the mummy remains, estimate how long ago Rameses I lived. The half-life of carbon-14 is 5730 years.

33. An archeologist discovers a tool made of wood.

a. If 50% of the wood's carbon-14 remains, how old is the wood? Explain how you can find this result without using an equation. The half-life of carbon-14 is 5730 years.

b. If 25% of the wood's carbon-14 remains, how old is the wood? Explain how you can find this result without using an equation.

c. If 10% of the wood's carbon-14 remains, how old is the wood? First, guess an approximate age without solving an equation. Explain how you decided on your estimate. Then, use an equation to find the age.

34. A person drinks a cup of coffee. Assume the caffeine enters his bloodstream immediately and there was no caffeine in his bloodstream before he drank the coffee. The half-life of caffeine in a person's bloodstream is about 6 hours. A cup of coffee contains about 240 milligrams of caffeine.

a. Let $f(t)$ be the number of milligrams of caffeine in the person's bloodstream at t hours after he drank the coffee. Find an equation of f.

b. The person drinks the coffee at 8 A.M. and goes to bed at 11 P.M. Use f to predict the amount of caffeine in his bloodstream when he goes to bed.

c. The person drinks another cup of coffee 24 hours after the first cup. How much caffeine will be in his bloodstream from these 2 cups of coffee just after he drank the second cup? Explain how you can find this result without using an equation.

d. Now assume that the person drinks the cup of coffee at 8 A.M. on March 1 and then drinks a cup of coffee every morning at 8 A.M. from then on. Sketch a model that describes the relationship between caffeine in his bloodstream and time from 8 A.M. on March 1 to 8 A.M. on March 3. Describe any assumptions that you make.

35. A storage tank contains a liquid radioactive element with a half-life of 100 years. It will be relatively safe for the contents to leak from the tank when 0.01% of the radioactive element remains. How long must the tank remain intact for this storage procedure to be safe?

36. A storage tank contains a liquid radioactive element with a half-life of 500 years. It will be relatively safe for the contents to leak from the tank when 0.02% of the radioactive element remains. How long must the tank remain intact for this storage procedure to be safe?

Concepts

37. Suppose the same amount of principal is deposited in an account at 3% interest compounded annually as in an account at 6% interest compounded annually. After how many years will there be twice as much money in the 6% account as in the 3% account? [**Hint:** For each account, find an expression that describes the value of the account, where t is the number of years the principal P (in dollars) has been invested. Set the ratio of the expression for the 6% account and the expression for the 3% account equal to 2 and solve for t.]

38. Describe how you can use the power property for logarithms to make estimates and predictions.

Related Review

39. A teacher ran an experiment to compare the weight of a bar of soap with the number of days he had used it in the shower (see Table 61).

Table 61 Weight of a Bar of Soap versus Number of Days of Use

Number of Days	Weight (grams)
0	124
4	103
7	84
11	58
17	27
20	12
22	6

Source: *Rex Boggs, Glenmore State High School, Rockhampton, Queensland, Australia*

Let $w = f(t)$ be the weight (in grams) of the bar of soap after t days of use.

a. Find an equation of f.

b. Estimate when the bar of soap weighed 45 grams.

c. If your model is linear, find the slope. If your model is quadratic, find the vertex. If your model is exponential, find the base b of $f(t) = ab^t$. What does your result mean in this situation?

d. On day 23, the bar of soap broke into two pieces and went down the drain. Use the model to estimate when there would have been no soap left. Has model breakdown occurred? Explain.

40. Due to inflation, an item that cost \$10,000 in 1980 cost \$26,720 in 2012. Costs comparable to \$10,000 in 1980 are shown in Table 62 for various years.

Table 62 Costs Comparable to \$10,000 in 1980

Year	Comparable Cost (dollars)
1980	10,000
1985	12,893
1990	15,559
1995	18,059
2000	20,367
2005	23,030
2010	25,279
2012	26,720

Source: *Bureau of Labor Statistics*

Let $c = f(t)$ be the cost (in dollars) at t years since 1980 that is comparable to \$10,000 in 1980.

a. Find an equation of f.

b. If your model is linear, find the slope. If your model is quadratic, find the vertex. If your model is exponential, find the base b of $f(t) = ab^t$. What does your result mean in this situation?

c. What is the c-intercept? What does it mean in this situation?

d. Use f to predict when the cost of \$29,000 would be comparable to the cost of \$10,000 in 1980.

Expressions, Equations, Functions, and Graphs

Perform the indicated instruction. Then use words such as linear, quadratic, cubic, polynomial, degree, exponential, logarithmic, function, one variable, *and* two variables *to describe the expression, equation, or system.*

41. Solve $(x - 5)(x + 2) = 3x$.

42. Graph $f(x) = x^2 - 6x + 5$.

43. Solve $7x^2 + 3 = 8$.

44. For $f(x) = x^2 - 6x + 5$, find x when $f(x) = 21$.

45. Solve $\frac{2}{3}x^2 = \frac{1}{2}x + \frac{5}{6}$.

46. Find an equation of the parabola that contains the points $(2, 1)$, $(3, 6)$, and $(4, 15)$.

11.6 More Properties of Logarithms

Objectives

- Know the *product, quotient,* and *change-of-base properties* for logarithms.
- Use properties of logarithms to simplify expressions and solve equations.
- Use a calculator to evaluate a logarithm with a base other than 10.

In Section 11.4, we studied some properties of logarithms. In this section, we will discuss three more.

Product Property for Logarithms

We can use the **product property for logarithms** to add two logarithms that have the same base.

Product Property for Logarithms

For $x > 0$, $y > 0$, $b > 0$, and $b \neq 1$,

$$\log_b(x) + \log_b(y) = \log_b(xy)$$

In words, the sum of logarithms is the logarithm of the product of their inputs.

For example, for $x > 0$, $\log_3(5) + \log_3(x) = \log_3(5x)$. A proof of the product property for logarithms follows.

Let $m = \log_b(x)$ and $n = \log_b(y)$. Writing both equations in exponential form, we have

$$x = b^m$$
$$y = b^n$$

Multiplying the left sides and multiplying the right sides yields

$$xy = (b^m)(b^n) \quad \text{Multiply left sides and multiply right sides.}$$
$$= b^{m+n} \quad \text{Add exponents: } b^m b^n = b^{m+n}$$

Writing $xy = b^{m+n}$ in logarithmic form gives

$$m + n = \log_b(xy)$$

Substituting $\log_b(x)$ for m and $\log_b(y)$ for n yields

$$\log_b(x) + \log_b(y) = \log_b(xy)$$

This statement is what we set out to prove.

Example 1 Using the Product Property for Logarithms

Simplify. Write the sum of logarithms as a single logarithm.

1. $\log_b(2x) + \log_b(x)$ **2.** $3\log_b(x^2) + 2\log_b(6x)$

Solution

1. $\log_b(2x) + \log_b(x) = \log_b(2x \cdot x)$ *Product property:* $\log_b(x) + \log_b(y) = \log_b(xy)$

$= \log_b(2x^2)$ *Add exponents:* $b^m b^n = b^{m+n}$

2. $3\log_b(x^2) + 2\log_b(6x) = \log_b(x^2)^3 + \log_b(6x)^2$ *Power property:* $p\log_b(x) = \log_b(x^p)$

$= \log_b\left[(x^2)^3 \cdot (6x)^2\right]$ *Product property:* $\log_b(x) + \log_b(y) = \log_b(xy)$

$= \log_b\left[x^6 \cdot 36x^2\right]$ *Multiply exponents; raise factors to 2nd power.*

$= \log_b(36x^8)$ *Add exponents:* $b^m b^n = b^{m+n}$

WARNING For us to apply the product property for logarithms, the coefficient of each logarithm must be 1. So, in Problem 2 of Example 1, we first applied the power property to get coefficients of 1:

$$3\log_b(x^2) + 2\log_b(6x) = \log_b(x^2)^3 + \log_b(6x)^2$$

Then we applied the product property.

Quotient Property

We use the product property to simplify the sum of two logarithms with the same base. We use the **quotient property for logarithms** to simplify the *difference* of two logarithms with the same base.

Quotient Property for Logarithms

For $x > 0$, $y > 0$, $b > 0$, and $b \neq 1$,

$$\log_b(x) - \log_b(y) = \log_b\left(\frac{x}{y}\right)$$

In words, the difference of two logarithms is the logarithm of the quotient of their inputs.

For example, for $x > 0$, $\log_4(x) - \log_4(7) = \log_4\left(\frac{x}{7}\right)$. You are asked to prove the quotient property in Exercise 50.

Example 2 Product and Quotient Properties

Simplify. Write the result as a single logarithm with a coefficient of 1.

1. $\log_b(6w^7) - \log_b(w^2)$ **2.** $2\log_b(3p) + 3\log_b(p^2) - 4\log_b(2p)$

Solution

1. $\log_b(6w^7) - \log_b(w^2) = \log_b\left(\frac{6w^7}{w^2}\right)$ *Quotient property:* $\log_b(x) - \log_b(y) = \log_b\left(\frac{x}{y}\right)$

$= \log_b(6w^5)$ *Subtract exponents:* $\frac{b^m}{b^n} = b^{m-n}$

2. $2\log_b(3p) + 3\log_b(p^2) - 4\log_b(2p)$

$= \log_b(3p)^2 + \log_b(p^2)^3 - \log_b(2p)^4$ *Power property:* $p\log_b(x) = \log_b(x^p)$

$= \log_b\left[(3p)^2(p^2)^3\right] - \log_b(2p)^4$ *Product property:* $\log_b(x) + \log_b(y) = \log_b(xy)$

$= \log_b\frac{(3p)^2(p^2)^3}{(2p)^4}$ *Quotient property:* $\log_b(x) - \log_b(y) = \log_b\left(\frac{x}{y}\right)$

$= \log_b\frac{9p^2 \cdot p^6}{16p^4}$ *Raise factors to a power; multiply exponents.*

$= \log_b\frac{9p^8}{16p^4}$ *Add exponents:* $b^m b^n = b^{m+n}$

$= \log_b\frac{9p^4}{16}$ *Subtract exponents:* $\frac{b^m}{b^n} = b^{m-n}$

Solving Logarithmic Equations

We can use the power, product, and quotient properties to solve logarithmic equations.

Example 3 Solving a Logarithmic Equation

Solve $2\log_5(3x) + 4\log_5(2x) = 3$.

Solution

$2\log_5(3x) + 4\log_5(2x) = 3$ *Original equation*

$\log_5(3x)^2 + \log_5(2x)^4 = 3$ *Power property:* $p\log_b(x) = \log_b(x^p)$

$\log_5\left[(3x)^2(2x)^4\right] = 3$ *Product property:* $\log_b(x) + \log_b(y) = \log_b(xy)$

$\log_5\left[9x^2(16x^4)\right] = 3$ *Raise factors to a power:* $(bc)^n = b^n c^n$

$\log_5(144x^6) = 3$ *Add exponents:* $b^m b^n = b^{m+n}$

$5^3 = 144x^6$ *Write in exponential form.*

$x^6 = \frac{125}{144}$ *Divide both sides by 144.*

Although there is a negative sixth root of $\frac{125}{144}$, the original equation contains $4\log_5(2x)$, and the domain of a logarithmic function is the set of *positive* numbers. So, $2x$ must be positive; hence, x must be positive:

$$x = \left(\frac{125}{144}\right)^{1/6}$$

$$x \approx 0.9767$$

Example 4 Solving a Logarithmic Equation

Solve $5\log_7(t^3) - 2\log_7(3t) = 2$

Solution

$$
\begin{aligned}
5\log_7(t^3) - 2\log_7(3t) &= 2 && \text{Original equation} \\
\log_7(t^3)^5 - \log_7(3t)^2 &= 2 && \text{Power property: } p\log_b(x) = \log_b(x^p) \\
\log_7\frac{(t^3)^5}{(3t)^2} &= 2 && \text{Quotient property: } \log_b(x) - \log_b(y) = \log_b\left(\frac{x}{y}\right) \\
\log_7\frac{t^{15}}{9t^2} &= 2 && \text{Multiply exponents; raise factors to 2nd power.} \\
\log_7\frac{t^{13}}{9} &= 2 && \text{Subtract exponents: } \frac{b^m}{b^n} = b^{m-n} \\
7^2 &= \frac{t^{13}}{9} && \text{Write in exponential form.} \\
t^{13} &= 441 && \text{Multiply both sides by 9.} \\
t &= 441^{1/13} && \text{The solution of } b^{13} = k \text{ is } k^{1/13}. \\
t &\approx 1.5974 && \text{Compute.}
\end{aligned}
$$

We solved the equations in Examples 3 and 4 by first applying the power property so the coefficient of each logarithm was 1. Next, we combined the logarithms on one side of the equation by using the product property or the quotient property. Then we solved the equation by using techniques discussed in Section 11.4.

Change-of-Base Property

The "log" key on a calculator finds logarithms, base 10. We use the **change-of-base property** to find logarithms for bases other than 10.

Change-of-Base Property

For $a > 0$, $b > 0$, $a \neq 1$, $b \neq 1$, and $x > 0$,

$$\log_b(x) = \frac{\log_a(x)}{\log_a(b)}$$

For example, $\log_3(5) = \frac{\log_2(5)}{\log_2(3)}$. Also, $\log_3(5) = \frac{\log_4(5)}{\log_4(3)}$ and $\log_3(5) = \frac{\log(5)}{\log(3)}$. We are free to write a logarithm in terms of any new base, including base 10.

To prove the change-of-base property, we let $k = \log_b(x)$. In exponential form, we have

$$b^k = x$$

Next, we take $\log_a$ of both sides and solve for k:

$$
\begin{aligned}
\log_a(b^k) &= \log_a(x) && \text{Take the } \log_a \text{ of both sides.} \\
k\log_a(b) &= \log_a(x) && \text{Power property: } \log_a(b^k) = k\log_a(b) \\
k &= \frac{\log_a(x)}{\log_a(b)} && \text{Divide both sides by } \log_a(b).
\end{aligned}
$$

But $k = \log_b(x)$, so, by substitution, we have

$$\log_b(x) = \frac{\log_a(x)}{\log_a(b)}$$

which is what we set out to prove.

To find a logarithm to a base other than 10, we use the change-of-base property to convert to $\log_{10}$; then we can use the log key on a calculator.

Example 5 Converting to $\log_{10}$

Find $\log_2(12)$.

Solution

We can use the change-of-base property to write $\log_2(12)$ in terms of base 10:

$$\log_2(12) = \frac{\log(12)}{\log(2)}$$

log(12)/log(2)
3.584962501

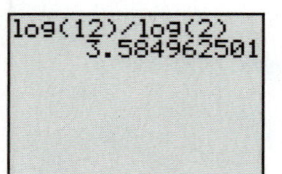

Figure 51 Compute $\dfrac{\log(12)}{\log(2)}$

Using the log key on a calculator, we compute $\dfrac{\log(12)}{\log(2)} \approx 3.5850$ (see Fig. 51). So, $\log_2(12) \approx 3.5850$.

Example 6 Using the Change-of-Base Property

Write $\dfrac{\log_7(x)}{\log_7(4)}$ as a single logarithm.

Solution

By the change-of-base property, we have $\dfrac{\log_7(x)}{\log_7(4)} = \log_4(x)$.

In Section 11.3, we sketched the graph of a logarithmic function. From now on, we can use a graphing calculator to verify such a graph by converting the logarithmic function to $\log_{10}$.

Example 7 Using a Graphing Calculator to Graph a Logarithmic Function

Use a graphing calculator to draw the graph of $y = \log_3(x)$.

Solution

By the change-of-base property, we have $y = \log_3(x) = \dfrac{\log(x)}{\log(3)}$. Using the log key on a graphing calculator, we enter the function and draw the graph (see Fig. 52). This graph verifies the graph that we sketched by hand in Example 6 of Section 11.3.

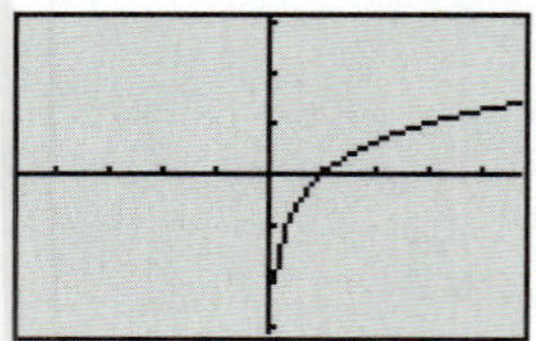

Figure 52 Graph of $y = \dfrac{\log(x)}{\log(3)}$

Recall from Section 11.4 that some equations in one variable that are impossible to solve by performing operations on both sides can be solved by graphing. We will work with one such equation in Example 8.

Example 8 Using Graphing to Solve an Equation in One Variable

Use graphing to solve $\log_2(x + 1) + \log_3(x + 2) = 6 - x$.

Solution

We use the change-of-base property on the left side of the equation to write

$$\frac{\log(x+1)}{\log(2)} + \frac{\log(x+2)}{\log(3)} = 6 - x$$

Then we use "intersect" on a graphing calculator to solve the system

$$y = \frac{\log(x+1)}{\log(2)} + \frac{\log(x+2)}{\log(3)}$$
$$y = 6 - x$$

See Fig. 53.

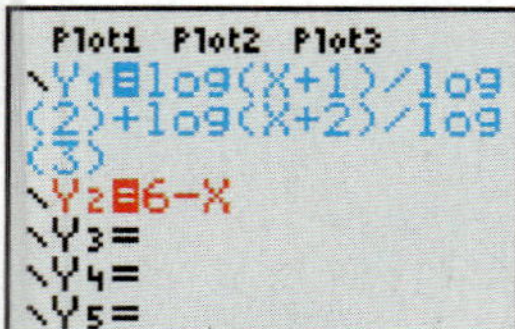

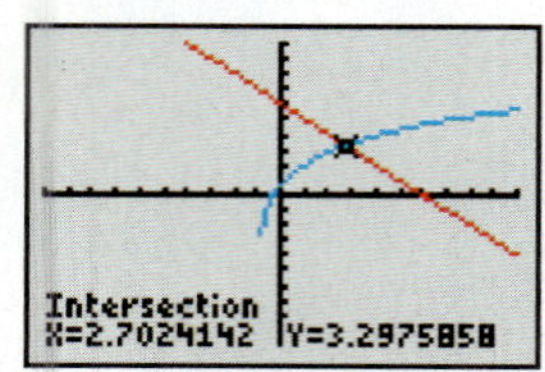

Figure 53 Solve the system

The approximate solution of the system is (2.7024, 3.2976). The x-coordinate, 2.7024, is the approximate solution of the equation $\log_2(x+1) + \log_3(x+2) = 6 - x$.

Comparing Properties of Logarithms

How do the properties of logarithms compare? Throughout this discussion, we assume that x, y, a, and b are positive and that a and b are not equal to 1.

The quotient property for logarithms tells us that a difference of logarithms is equal to a logarithm of a quotient:

$$\log_b(x) - \log_b(y) = \log_b\left(\frac{x}{y}\right)$$

The change-of-base property tells us that a logarithm is equal to a quotient of logarithms (with a "new" base):

$$\log_b(x) = \frac{\log_a(x)}{\log_a(b)}$$

WARNING It is a common error to confuse the quotient property and the change-of-base property for logarithms. In general,

$$\log_b(x) - \log_b(y) \neq \frac{\log_b(x)}{\log_b(y)}$$

and

$$\log_b\left(\frac{x}{y}\right) \neq \frac{\log_b(x)}{\log_b(y)}$$

Group Exploration

Function of a sum

1. Substitute 2 for x and 3 for y and use a calculator to help you decide whether the resulting statement is true or false.
   a. $\log(x+y) = \log(x) + \log(y)$
   b. $2^{x+y} = 2^x + 2^y$
   c. $(x+y)^2 = x^2 + y^2$
   d. $\sqrt{x+y} = \sqrt{x} + \sqrt{y}$
2. All of the statements in Problem 1 are of the form $f(x+y) = f(x) + f(y)$. Is the statement $f(x+y) = f(x) + f(y)$ true for every function f? Explain.
3. According to the distributive law, $a(x+y) = ax + ay$. Explain why this statement is true for all values of a but the statement $f(x+y) = f(x) + f(y)$ is not true for all functions f.
4. Give an example of a function f such that the statement $f(x+y) = f(x) + f(y)$ *is* true.

Homework 11.6

For extra help ▶ MyMathLab®

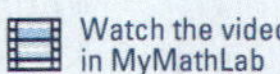
Watch the videos in MyMathLab

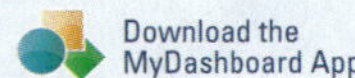
Download the MyDashboard App

Simplify. Write your result as a single logarithm with a coefficient of 1.

1. $\log_b(x) + \log_b(3x)$
2. $\log_b(5x) + \log_b(x)$
3. $\log_b(8x) - \log_b(2)$
4. $\log_b(24x) - \log_b(6)$
5. $4\log_b(t) + \log_b(5t)$
6. $\log_b(7w) + 3\log_b(w)$
7. $\log_b(3x^2) - 5\log_b(x)$
8. $\log_b(6x^4) - 7\log_b(x)$
9. $2\log_b(3x) + 3\log_b(x^3)$
10. $4\log_b(x^2) + 5\log_b(x)$
11. $3\log_b(2m) + 5\log_b(m^2) - \log_b(3m)$
12. $2\log_b(3k) + 4\log_b(k^3) - \log_b(5k)$

Solve. Round any solutions to the fourth decimal place.

13. $\log_5(6x) + \log_5(x) = 2$
14. $\log_3(x) + \log_3(6x) = 3$
15. $\log_2(9x) - \log_2(3) = 5$
16. $\log_4(12x) - \log_4(6) = 3$
17. $\log_7(w^2) + 2\log_7(3w) = 2$
18. $4\log_3(2r) + \log_3(r^3) = 4$
19. $\log(x^{13}) - 2\log(x^4) = 1$

20. $\log(x^9) - 3\log(x^2) = 3$

21. $3\log(x^2) + 4\log(2x) = 2$

22. $2\log(2x) + 3\log(x^4) = 4$

23. $3\log_5(p^4) - 5\log_5(2p) = 3$

24. $4\log_2(k^2) - 3\log_2(2k) = 4$

Evaluate. Round your result to the fourth decimal place.

25. $\log_3(7)$ **26.** $\log_2(11)$

27. $\log_9(3.58)$ **28.** $\log_{12}(2.88)$

29. $\log_8\left(\frac{1}{70}\right)$ **30.** $\log_{\frac{1}{2}}(7)$

Solve by using "intersect" on a graphing calculator. Round any solutions to the fourth decimal place.

31. $\log(x + 5) + \log(x + 2) = 3 - x$

32. $\log(x + 1) + \log(x + 4) = 8 - 2x$

33. $\log_5(x + 3) + \log_2(x + 4) = -2x + 9$

34. $\log_3(x + 2) + \log_4(x + 1) = -x + 7$

35. $\log_2(x + 4) + \log_3(x + 5) = 2^x + 1$

36. $\log_3(x + 6) + \log_2(x + 8) = 3^x + 2$

For Exercises 37–40, write the expression as a single logarithm.

37. $\dfrac{\log_2(x)}{\log_2(7)}$ **38.** $\dfrac{\log_4(x)}{\log_4(5)}$

39. $\dfrac{\log_b(r)}{\log_b(s)}$ **40.** $\dfrac{\log_b(p)}{\log_b(q)}$

For Exercises 41–44, let $g(x) = \log_{12}(x)$. Find each output. Round your result to the fourth decimal place.

41. $g(17)$ **42.** $g(50)$ **43.** $g(8)$ **44.** $g(5)$

Concepts

45. Three students try to solve the equation $3(2^x) = 7$:

Student 1's work

$$3(2)^x = 7$$
$$2^x = \frac{7}{3}$$
$$\log(2^x) = \log\left(\frac{7}{3}\right)$$
$$x\log(2) = \log\left(\frac{7}{3}\right)$$
$$x = \frac{\log\left(\frac{7}{3}\right)}{\log(2)}$$

Student 2's work

$$3(2)^x = 7$$
$$2^x = \frac{7}{3}$$
$$x = \log_2\left(\frac{7}{3}\right)$$

Student 3's work

$$3(2)^x = 7$$
$$\log[3(2)^x] = \log(7)$$
$$\log(3) + \log(2^x) = \log(7)$$
$$\log(2^x) = \log(7) - \log(3)$$
$$x\log(2) = \log(7) - \log(3)$$
$$x = \frac{\log(7) - \log(3)}{\log(2)}$$

Which student(s) solved the equation correctly? Explain.

46. A student tries to write the expression $3\log_2(x) + \log_2(x^2)$ as a single logarithm:

$$3\log_2(x) + \log_2(x^2) = 3\log_2(x \cdot x^2)$$
$$= 3\log_2(x^3)$$

Describe any errors. Then write the expression correctly as a single logarithm.

47. Which of the following expressions are equal?

$$\log_b(b^2) \qquad \log_b(b^6) - \log_b(b^4) \qquad \log_b(b^6) \qquad 2$$
$$\log_b\left(\frac{b^6}{b^4}\right) \qquad \frac{\log_b(b^6)}{\log_b(b^4)}$$

48. a. Use ZDecimal to graph $f(x) = \log(100x) - \log(x)$. Describe the graph in words.

b. Explain why the graph of f is in neither quadrant II nor quadrant III.

c. Use properties of logarithms to write the right-hand side of the equation of f as a constant. Use your result to explain why the graph that you found in part (a) makes sense.

49. Clearly, $\log_b(x) - \log_b(x) = 0$. Use a property of logarithms to write $\log_b(x) - \log_b(x)$ in another form to show that $\log_b(1) = 0$. Assume that $b > 0, x > 0$, and $b \neq 1$.

50. Prove the quotient property for logarithms. [**Hint:** Try to find a creative way to use the product property, followed by the power property to write the expression $\log_b\left(\frac{x}{y}\right)$ as a difference of logarithms.]

51. a. Simplify $\log_2(x^3) + \log_2(x^5)$.

b. Solve $\log_2(x^3) + \log_2(x^5) = 7$. Round any solutions to the fourth decimal place.

c. Compare the process of simplifying an expression with solving an equation.

d. Explain how simplifying an expression can help when you are solving an equation.

52. List the properties for logarithms discussed in this section and in Section 11.4. Explain how each property can be used. Give examples to illustrate your points.

Related Review

Simplify by writing your result as a single logarithm with a coefficient of 1, or solve, as appropriate. Round any solutions to the fourth decimal place.

53. $\log_2(x^4) + \log_2(x^3)$ **54.** $\log_2(t^9) - \log_2(t^5) = 5$

55. $\log_2(x^4) + \log_2(x^3) = 4$ **56.** $\log_2(t^9) - \log_2(t^5)$

57. $2\log_9(x^3) - 3\log_9(2x) = 2$

58. $3\log_5(3x) + 2\log_5(x^2)$

59. $2\log_9(x^3) - 3\log_9(2x)$

60. $3\log_5(3x) + 2\log_5(x^2) = 3$

Simplify. If an expression contains two logarithms, write your result as a single logarithm with a coefficient of 1.

61. $(16b^{16}c^{-7})^{1/4}(27b^{27}c^5)^{1/3}$ **62.** $\dfrac{(25b^9c^{-6})^{1/2}}{(81b^3c^{-8})^{1/4}}$

63. $3\log_b(2x^5) + 2\log_b(3x^4)$ **64.** $4\log_b(3x^2) - 5\log_b(2x^7)$

Solve the system.

65. $y = 3x - 7$
$y = -2x + 3$

66. $y = \frac{1}{2}x - 4$
$y = -\frac{2}{3}x + 3$

67. $y = \log_2(4x^2) - 3$
$y = \log_2(x) + 2$

68. $y = 2 + \log_3(3x^2)$
$y = 8 - \log_3(9x)$

69. Solve:

$$2x - 3y = 6$$
$$y = \frac{2}{3}x - 2$$

70. Solve $\log_3(2m - 1) = 5$.

71. Graph $2x - 3y = 6$ by hand.

72. Find $f(5)$, where $f(x) = \log_3(2x - 1)$.

73. Find $f^{-1}(2)$, where $f(x) = -\frac{4}{3}x - 1$.

74. Find an approximate equation $y = ab^x$ of an exponential curve that contains the points (3, 5) and (7, 89). Round a and b to the second decimal place.

Expressions, Equations, Functions, and Graphs

Perform the indicated instruction. Then use words such as linear, quadratic, cubic, polynomial, degree, exponential, logarithmic, function, one variable, *and* two variables *to describe the expression, equation, or system.*

11.7 Natural Logarithms

Objectives

- Know the meaning of a *natural logarithm.*
- Evaluate natural logarithms.
- Use properties of natural logarithms to simplify expressions and solve equations.
- Use exponential models with base e to make estimates and predictions.

In Chapter 10 and in this chapter, we have worked with exponential and logarithmic functions with various values of the base. In statistics and calculus, it is helpful to use a special constant called e as the base for these two types of functions. In this section, we will describe this constant and use it as the base for logarithmic and exponential functions.

Definition of Natural Logarithm

In this section, we discuss a logarithm with a special base called e, where e is an irrational number:

$$e \approx 2.718281828459045\ldots$$

To the nearest ten-thousandth, e = 2.7183.

Many equations for useful models contain e. For example, the equation for one type of "bell curve" is

$$f(x) = \frac{e^{-0.5x^2}}{\sqrt{2\pi}}$$

The graph of f has the shape of a bell (see Fig. 54). Bell curves can be used to model heights of women (or men), IQs, widths of trunks of redwood trees, and many other quantities.

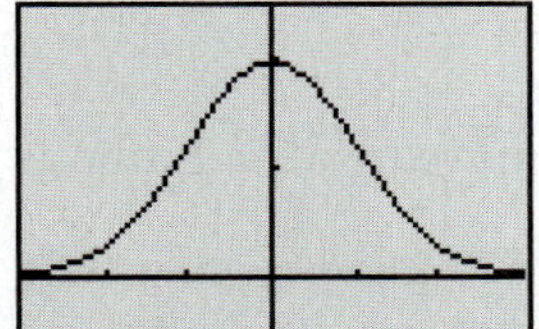

Figure 54 Graph of $f(x) = \frac{e^{-0.5x^2}}{\sqrt{2\pi}}$

> **Definition** Natural logarithm
>
> A **natural logarithm** is a logarithm with base e. We write $\ln(a)$ to represent $\log_e(a)$.

Throughout the discussion that follows, assume that $a > 0, b > 0$, and $b \neq 1$.

Recall from Section 11.3 that $\log_b(a)$ is the exponent on the base b that gives a. So, **for $a > 0$, $\ln(a)$ is the exponent on the base e that gives a.**

Recall from Section 11.4 that $\log_b(a) = c$ and $b^c = a$ are equivalent forms. In terms of base e, this means $\ln(a) = c$ and $e^c = a$ are equivalent forms.

> **Exponential/Natural Logarithmic Forms Property**
>
> For $a > 0$, the equations
>
> $$\ln(a) = c \quad \text{and} \quad e^c = a$$
>
> are equivalent.

The equation $\ln(a) = c$ is in logarithmic form, and the equation $e^c = a$ is in exponential form. Either form can replace the other when you solve a problem.

The key on most graphing calculators that is labeled "ln" or "LN" will give you the natural logarithm of a number.

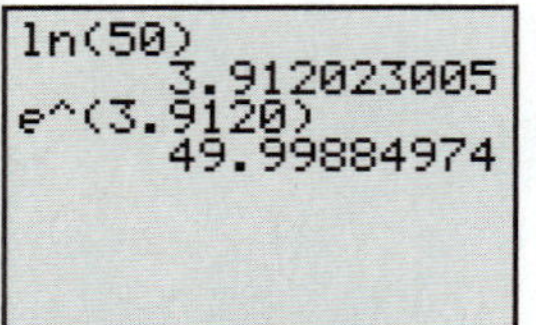

Figure 55 Computing $\ln(50)$

Example 1 Finding a Natural Logarithm

Use a calculator to find $\ln(50)$.

Solution

By pressing [LN] 50 [)] [ENTER], we get $\ln(50) \approx 3.9120$ (see Fig. 55). This means that $e^{3.9120} \approx 50$. We check that $e^{3.9120} \approx 50$ by pressing [2nd] [LN] 3.9120 [)] [ENTER].

We can find the natural logarithm of powers of e without using a calculator.

Example 2 Finding a Natural Logarithm

Find $\ln(e^5)$.

Solution

$$\begin{aligned} \ln(e^5) &= \log_e(e^5) && \text{Definition of natural logarithm} \\ &= 5 && \text{Simplify.} \end{aligned}$$

From Example 2, we see that for the function ln, the input e^5 leads to the output 5. So, the *positive* real number e^5 is in the domain of ln. Recall from Section 11.3 that for *any* logarithmic function, the domain is the set of all positive real numbers.

Solving Logarithmic and Exponential Equations

In Example 3, we will solve two logarithmic equations.

Example 3 Solving Logarithmic Equations

Solve the equation.

1. $\ln(x) = 4$
2. $3\ln(4x) - 2 = 5$

Figure 56 Computing e^4

Solution

1. We write $\ln(x) = 4$ in the exponential form $e^4 = x$. The approximate solution is 54.5982 (see Fig. 56).
2. $$\begin{aligned} 3\ln(4x) - 2 &= 5 && \text{Original equation} \\ 3\ln(4x) &= 7 && \text{Add 2 to both sides.} \\ \ln(4x) &= \frac{7}{3} && \text{Divide both sides by 3.} \\ e^{7/3} &= 4x && \text{Write in exponential form.} \\ \frac{e^{7/3}}{4} &= x && \text{Divide both sides by 4.} \\ x &\approx 2.5781 && \text{Compute.} \end{aligned}$$

We check that 2.5781 approximately satisfies the equation $3\ln(4x) - 2 = 5$:

$$3\ln[4(2.5781)] - 2 \approx 5.00004 \approx 5$$

Example 4 Solving an Exponential Equation

Solve $5e^{x-1} = 100$.

Solution

$$
\begin{aligned}
5e^{x-1} &= 100 && \text{Original equation} \\
e^{x-1} &= 20 && \text{Divide both sides by 5.} \\
\ln(20) &= x - 1 && \text{Write in logarithmic form.} \\
\ln(20) + 1 &= x && \text{Add 1 to both sides.} \\
x &\approx 3.9957 && \text{Compute.}
\end{aligned}
$$

We check that 3.9957 approximately satisfies the equation $5e^{x-1} = 100$:

$$5e^{3.9957-1} \approx 99.9968 \approx 100$$

How do the properties for $\log_b(x)$ correspond to the properties for $\ln(x)$? Assume that $x > 0$, $y > 0$, $b > 0$, and $b \neq 1$ unless stated otherwise. The following properties apply:

Properties of Logarithms	**Properties of Natural Logarithms**	
$\log_b(1) = 0$	$\ln(1) = 0$	*The natural logarithm of 1 is 0.*
$\log_b(b) = 1$	$\ln(e) = 1$	*The natural logarithm of e is 1.*
$\log_b(b^x) = x$, for real number x	$\ln(e^x) = x$, for real number x	*Composing $y = \ln(x)$ with $y = e^x$*
$b^{\log_b(x)} = x$	$e^{\ln(x)} = x$	*Composing $y = e^x$ with $y = \ln(x)$*
$\log_b(x^p) = p\log_b(x)$	$\ln(x^p) = p\ln(x)$	*Power property*
$\log_b(x) + \log_b(y) = \log_b(xy)$	$\ln(x) + \ln(y) = \ln(xy)$	*Product property*
$\log_b(x) - \log_b(y) = \log_b\left(\frac{x}{y}\right)$	$\ln(x) - \ln(y) = \ln\left(\frac{x}{y}\right)$	*Quotient property*

We can use the power property for natural logarithms to solve exponential equations.

Example 5 Solving an Equation

Solve $2(5)^t + 3 = 63$.

Solution

$$
\begin{aligned}
2(5)^t + 3 &= 63 && \text{Original equation} \\
2(5)^t &= 60 && \text{Subtract 3 from both sides.} \\
5^t &= 30 && \text{Divide both sides by 2.} \\
\ln(5^t) &= \ln(30) && \text{Take the natural logarithm of both sides.} \\
t\ln(5) &= \ln(30) && \text{Power property: } \ln(x^p) = p\ln(x) \\
t &= \frac{\ln(30)}{\ln(5)} && \text{Divide both sides by } \ln(5). \\
t &\approx 2.1133 && \text{Compute.}
\end{aligned}
$$

We check that 2.1133 approximately satisfies the equation $2(5)^t + 3 = 63$:

$$2(5)^{2.1133} + 3 \approx 63.0017 \approx 63$$

In Example 5, we used ln to solve an exponential equation. In Section 11.4, we used log to solve exponential equations. It does not matter whether we use ln or log to solve an exponential equation such as $2(5)^t + 3 = 63$. Both methods require about the same amount of work and give the same result.

In Example 6, we will use the power and quotient properties for logarithms to simplify a logarithmic expression.

Example 6 Power and Quotient Properties

Write $5\ln(x^3) - 3\ln(2x)$ as a single logarithm with a coefficient of 1. Simplify the result.

Solution

$$5\ln(x^3) - 3\ln(2x) = \ln(x^3)^5 - \ln(2x)^3 \quad \text{Power property: } p\ln(x) = \ln(x^p)$$

$$= \ln\frac{(x^3)^5}{(2x)^3} \quad \text{Quotient property: } \ln(x) - \ln(y) = \ln\left(\frac{x}{y}\right)$$

$$= \ln\frac{x^{15}}{8x^3} \quad \text{Multiply exponents; raise factors to 3rd power.}$$

$$= \ln\frac{x^{12}}{8} \quad \text{Subtract exponents: } \frac{b^m}{b^n} = b^{m-n}$$

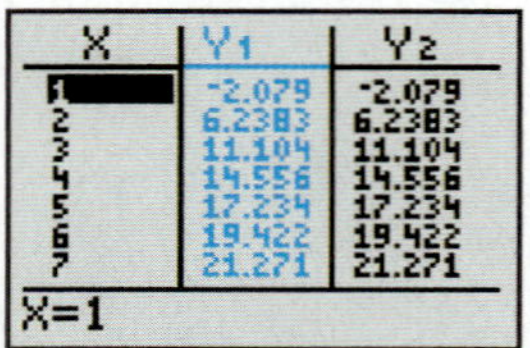

X	Y1	Y2
1	-2.079	-2.079
2	6.2383	6.2383
3	11.104	11.104
4	14.556	14.556
5	17.234	17.234
6	19.422	19.422
7	21.271	21.271

X=1

Figure 57 Verify the work

We verify our work by comparing tables for the functions $y = 5\ln(x^3) - 3\ln(2x)$ and $y = \ln\frac{x^{12}}{8}$ (see Fig. 57).

Example 7 Solving an Equation

Solve $3\ln(4x) + \ln(5x) = 7$.

Solution

$$3\ln(4x) + \ln(5x) = 7 \quad \text{Original equation}$$

$$\ln(4x)^3 + \ln(5x) = 7 \quad \text{Power property: } p\ln(x) = \ln(x^p)$$

$$\ln[(4x)^3(5x)] = 7 \quad \text{Product property: } \ln(x) + \ln(y) = \ln(xy)$$

$$\ln[64x^3(5x)] = 7 \quad \text{Raise factors to nth power: } (bc)^n = b^nc^n$$

$$\ln(320x^4) = 7 \quad \text{Add exponents: } b^mb^n = b^{m+n}$$

$$e^7 = 320x^4 \quad \text{Write in exponential form.}$$

$$x^4 = \frac{e^7}{320} \quad \text{Divide both sides by 320.}$$

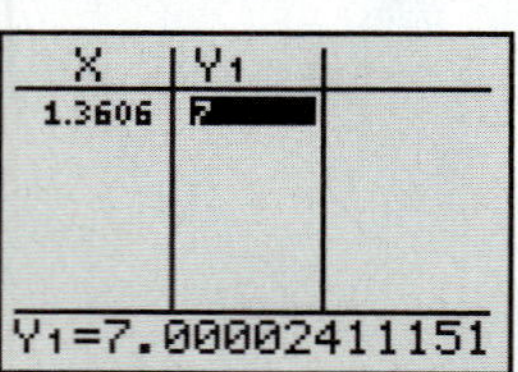

Figure 58 Verify the work

Although there is a negative fourth root of $\frac{e^7}{320}$, the original equation contains $3\ln(4x)$, and the domain of a (natural) logarithm function is the set of *positive* numbers. So, $4x$ must be positive. Hence, x must be positive:

$$x = \left(\frac{e^7}{320}\right)^{1/4}$$

$$x \approx 1.3606$$

We use a graphing calculator table to check that, for the function $y = 3\ln(4x) + \ln(5x)$, the input 1.3606 leads approximately to the output 7 (see Fig. 58).

Table 63 Numbers of BlackBerry Subscribers

Year	Number of Subscribers (millions)
2000	0.03
2002	0.32
2004	1.07
2006	5.0
2008	16
2010	50

Source: *Research in Motion*

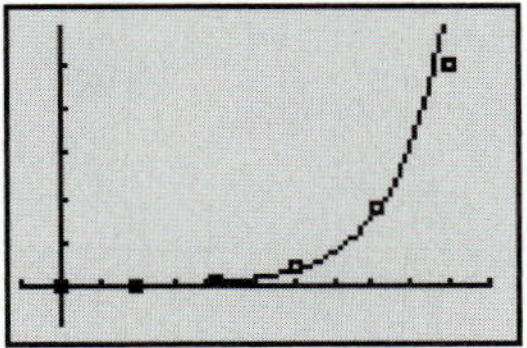

Figure 59 Verify the fit

Exponential Models with Base *e*

So far, we have worked with exponential models of the form $f(t) = ab^t$. In calculus, it is often helpful to write such models in the form $f(t) = ae^{kt}$, where a and k are constants.

Example 8 Using a Model with Base *e* to Make a Prediction

The BlackBerry® is a wireless handheld device. The numbers of BlackBerry subscribers are shown in Table 63 for various years. Let $f(t)$ be the number of BlackBerry subscribers (in millions) at t years since 2000. A possible equation of f is

$$f(t) = 0.05e^{0.72t}$$

1. Verify that f models the situation well.
2. Predict when there will be 500 million subscribers.

Solution

1. We draw the graph of f and the scattergram of the data in the same viewing window (see Fig. 59). It appears that f is a reasonable model.
2. To predict when there will be 500 million subscribers, we substitute 500 for $f(t)$ and solve for t:

$$500 = 0.05e^{0.72t} \quad \text{Substitute 500 for } f(t).$$

$$\frac{500}{0.05} = e^{0.72t} \quad \text{Divide both sides by 0.05.}$$

$$\ln\left(\frac{500}{0.05}\right) = \ln\left(e^{0.72t}\right) \quad \text{Take the natural logarithm of both sides.}$$

$$\ln\left(\frac{500}{0.05}\right) = 0.72t \quad \ln(e^a) = a$$

$$t = \frac{\ln\left(\frac{500}{0.05}\right)}{0.72} \quad \text{Divide both sides by 0.72.}$$

$$t \approx 12.79 \quad \text{Compute.}$$

The model predicts there will be 500 million subscribers in 2013.

Group Exploration

Newton's law of cooling

A hot potato is taken out of an oven and allowed to cool to room temperature. Let p be the temperature (in degrees Fahrenheit) of the potato at t minutes after it is removed from the oven.

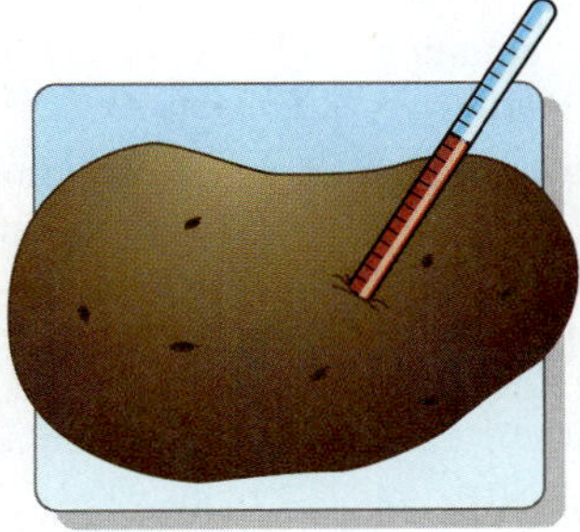

1. Newton's law of cooling states that
$$p - r = ae^{-kt}$$
where r is room temperature (in degrees Fahrenheit) and a and k are constants. The room temperature is 70°F. Substitute $r = 70$ into the equation.
2. The temperature of the potato was 350°F when it was removed from the oven. Find the value of the constant a, and substitute it into your equation.
3. The temperature of the potato was 200°F 5 minutes later. Find the value of k, and substitute it into your equation.
4. Isolate p on one side of your equation. Then use a graphing calculator to draw the graph of your equation.
5. What will be the temperature of the potato after a long time? Explain.
6. Estimate at what temperature a potato can be comfortably eaten. How long will it take for the potato to reach that temperature?

Group Exploration

Looking ahead: Simplifying rational expressions

In Sections 9.3 and 9.6, we simplified radical quotients by factoring the numerator and the denominator and then using the property $\frac{a}{a} = 1$, where $a \neq 0$. Here we use similar steps to simplify $\frac{14x}{21x}$:

$$\frac{14x}{21x} = \frac{2(7x)}{3(7x)} = \frac{2}{3} \cdot \frac{7x}{7x} = \frac{2}{3} \cdot 1 = \frac{2}{3}$$

Simplify the following expressions:

1. $\frac{3x}{5x}$
2. $\frac{15x}{25x}$
3. $\frac{(x+2)(x-5)}{(x-3)(x-5)}$
4. $\frac{x^2 - 9}{x^2 + 7x + 12}$ **[Hint:** First, factor the numerator and the denominator.]
5. $\frac{3x^2 - 12x}{x^2 - 8x + 16}$

Homework 11.7

For extra help ▶ MyMathLab®   Watch the videos in MyMathLab Download the MyDashboard App

Use a calculator to find the natural logarithm. Round your result to the fourth decimal place.

1. $\ln(54.8)$ **2.** $\ln(37.28)$ **3.** $\ln\left(\frac{1}{2}\right)$ **4.** $\ln\left(\frac{5}{8}\right)$

Simplify. Verify your result by using a graphing calculator.

5. $\ln(e^4)$ **6.** $\ln(e^6)$ **7.** $\ln(e)$ **8.** $\ln(1)$

9. $\ln\left(\frac{1}{e}\right)$ **10.** $\ln\left(\frac{1}{e^2}\right)$ **11.** $\frac{1}{2}\ln(e^6)$ **12.** $\ln(\sqrt{e})$

13. $e^{\ln 7}$ **14.** $e^{\ln 4}$

Solve the equation. Round any solutions to the fourth decimal place. Use graphing calculator tables or graphs to verify your result.

15. $\ln(x) = 2$ **16.** $\ln(x) = 5$

17. $\ln(p + 5) = 3$ **18.** $\ln(t - 4) = 5$

19. $7e^x = 44$ **20.** $3e^x = 85$

21. $5\ln(3x) + 2 = 7$ **22.** $2\ln(5x) - 3 = 1$

23. $4e^{3m-1} = 68$ **24.** $7e^{2p+10} = 100$

25. $e^{3x-5} \cdot e^{2x} = 135$ **26.** $e^{2x-3} \cdot e^x = 83$

27. $3.1^x = 49.8$ **28.** $2.4^x = 63.5$

29. $3(6)^x - 1 = 97$ **30.** $5(2)^x + 3 = 264$

31. $5e^x - 20 = 2e^x + 67$ **32.** $7e^x - 12 = 3e^x + 44$

Simplify. Write the expression as a single logarithm with a coefficient of 1. Use graphing calculator tables or graphs to verify your result.

33. $\ln(4x) + \ln(3x^4)$ **34.** $\ln(8x^2) + \ln(4x)$

35. $\ln(25x^4) - \ln(5x^3)$ **36.** $\ln(6x^3) - \ln(2x^4)$

37. $2\ln(w^4) + 3\ln(2w)$ **38.** $4\ln(3r) + 5\ln(r^3)$

39. $3\ln(3x) - 2\ln(x^2)$ **40.** $2\ln(x^3) - 3\ln(2x)$

41. $3\ln(2k) + 4\ln(k^2) - \ln(k^7)$

42. $5\ln(p^2) + 2\ln(3p) - \ln(p^9)$

Solve the equation. Round any solutions to the fourth decimal place. Check your result.

43. $\ln(3x) + \ln(x) = 4$ **44.** $\ln(2x) + \ln(5x) = 6$

45. $\ln(4w^5) - 2\ln(w^2) = 5$ **46.** $\ln(7r^{15}) - 3\ln(r^4) = 1$

47. $2\ln(3x) + 2\ln(x^3) = 8$ **48.** $4\ln(2x) + 3\ln(x^5) = 9$

49. $5\ln(2m) - 3\ln(m^4) = 7$ **50.** $3\ln(4y) - 4\ln(y^3) = 5$

Solve by using "intersect" on a graphing calculator. Round any solutions to the fourth decimal place.

51. $e^x = 5 - x$ **52.** $2e^x = 9 - 2x$

53. $3\ln(x + 2) = -2x + 6$ **54.** $2\ln(x + 1) = -x + 7$

55. $3\ln(x + 3) = 0.7x + 2$ **56.** $4\ln(x + 5) = x + 5$

For Exercises 57–60, let $f(x) = 4\ln(x)$.

57. Find $f(e^5)$. **58.** Find $f\left(\frac{1}{e^3}\right)$.

59. Find x when $f(x) = -8$. **60.** Find x when $f(x) = 2$.

61. Assume the equation $ae^{bx} = c$ has a solution for x, where $a \neq 0$ and $b \neq 0$. Solve for x.

62. Assume the equation $ae^{bx+d} + k = c$ has a solution for x, where $a \neq 0$ and $b \neq 0$. Solve for x.

63. Pointing a laser at aircraft, which can temporarily blind pilots, is a serious offense with a maximum punishment of 20 years in prison and a $250,000 fine. The numbers of laser incidents involving aircraft are shown in Table 64 for various years. Let $n = f(t)$ be the number of laser incidents involving aircraft in the year that is t years since 2000. An equation of f is

$$f(t) = 29.89e^{0.45t}$$

a. Use a graphing calculator to draw the graph of the model and, in the same viewing window, the scattergram of the data. Does the model fit the data well?

b. Find the n-intercept. What does it mean in this situation?

c. Predict when there will be 60,000 laser incidents involving aircraft.

Table 64 Numbers of Laser Incidents Involving Aircraft

Year	Number
2005	283
2006	446
2007	675
2008	988
2009	1527
2010	2836

Source: *Federal Aviation Administration*

d. Cockpits have been illuminated in 67% of laser incidents involving aircraft. Predict the number of times cockpits will be illuminated in 2018.

64. Americans for Prosperity is a conservative interest group that promotes less regulation and lower government spending. The annual budgets of the group are shown in Table 65 for various years.

Table 65 Annual Budgets of Americans for Prosperity

Year	Annual Budget (millions of dollars)
2007	7
2008	12
2009	28
2010	40
2011	50
2012	129

Source: *Americans for Prosperity*

Let $B = f(t)$ be the annual budget (in millions of dollars) of Americans for Prosperity at t years since 2000. An equation of f is

$$f(t) = 0.16e^{0.55t}$$

a. Use a graphing calculator to draw the graph of the model and, in the same viewing window, the scattergram of the data. Does the model fit the data well?
b. Find the B-intercept. What does it mean in this situation?
c. Predict the annual budget in 2017.
d. Predict when the annual budget will reach \$1.8 *billion*.

65. A person buys a cup of coffee and drinks it in the store. The coffee's temperature y (in degrees Fahrenheit) is given by

$$y = 70 + 137e^{-0.06t}$$

where t is the number of minutes since he bought the coffee.
a. What was the temperature of the coffee when the person bought it?
b. If the person begins drinking the coffee when it reaches 180°F, how much time must he wait after buying it?
c. Use a graphing calculator table or graph to estimate the room temperature of the *store*.

66. A person makes a cup of tea. The tea's temperature y (in degrees Fahrenheit) is given by

$$y = 68 + 132e^{-0.05t}$$

where t is the number of minutes since the person made the tea.
a. What was the temperature of the tea when the person made it?
b. If the person waits 5 minutes to begin drinking the tea, what is the temperature of the tea then?
c. The tea is *lukewarm* at a temperature of about 98.6°F. If the person lets the tea sit until it is lukewarm, how much time has gone by since she made it?

67. A cable hangs between two poles that are 20 feet apart (see Fig. 60). The height of the cable (in feet) is given by

$$h(x) = 10\left(e^{0.03x} + e^{-0.03x}\right), \quad -10 \le x \le 10$$

where x is the horizontal position (in feet) as indicated in Fig. 60.

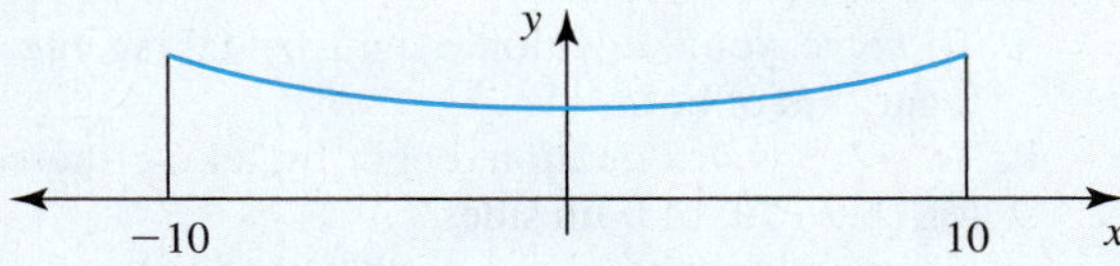

Figure 60 Exercise 67

a. Find the height of the cable at either pole.
b. Find $h(6)$. What does it mean in this situation?
c. How high is the cable where it is closest to the ground?

68. A cable hangs between two poles that are 30 feet apart (see Fig. 61). The height of the cable (in feet) is given by

$$h(x) = 20\left(e^{0.05x} + e^{-0.05x}\right), \quad -15 \le x \le 15$$

where x is the horizontal position (in feet) as indicated in Fig. 61.

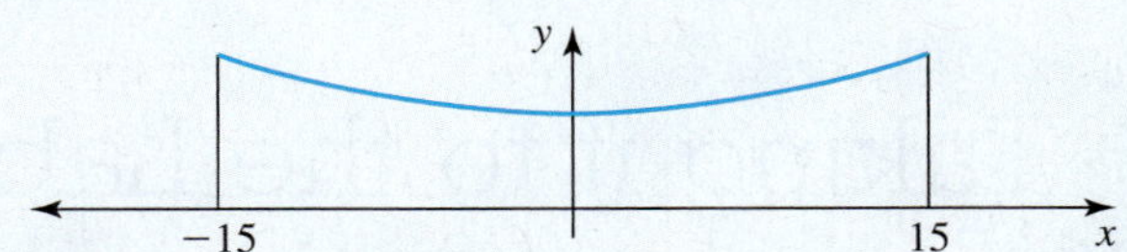

Figure 61 Exercise 68

a. Find $h(-8)$. What does it mean in this situation?
b. How high is the cable where it is closest to the ground?
c. Explain why there is model breakdown for $x < -15$ and for $x > 15$.

Concepts

69. Two students try to solve the equation $2^x = 7$:

Student 1's work	Student 2's work
$2^x = 7$	$2^x = 7$
$\ln(2^x) = \ln(7)$	$\log(2^x) = \log(7)$
$x\ln(2) = \ln(7)$	$x\log(2) = \log(7)$
$x = \dfrac{\ln(7)}{\ln(2)}$	$x = \dfrac{\log(7)}{\log(2)}$

Which student(s) solved the equation correctly? Explain.

70. A student tries to simplify $\log\left(e^8\right)$:

$$\log\left(e^8\right) = 8$$

Is the work correct? Explain.

71. Which expressions are equal? Assume that $x > 0$ and $x \ne 1$.

$3\ln(x)$ $\quad \ln\left(x^7\right) - \ln\left(x^4\right)$ $\quad 2\ln(x)\ln(x)$ $\quad \dfrac{\ln\left(x^7\right)}{\ln\left(x^4\right)}$

$\ln\left(x^3\right)$ $\quad \ln(3x)$

72. a. Solve $e^x = 30$ by writing the equation in logarithmic form.

b. Solve $e^x = 30$ by taking the natural logarithm of both sides of the equation.

c. Compare your results in parts (a) and (b).

73. Explain in your own words why $\ln(e) = 1$.

74. Explain in your own words why $\ln(1) = 0$.

75. a. **i.** To solve $3^x = 58$, begin by taking the natural logarithm of both sides.

ii. To solve $3^x = 58$, begin by taking the common log (base 10) of both sides.

iii. Compare your results in parts (i) and (ii).

b. Create an equation of the form $b^x = c$, where b and c are positive and $b \neq 1$.

i. To solve your equation, begin by taking the natural logarithm of both sides.

ii. To solve your equation, begin by taking the common log (base 10) of both sides.

iii. Compare your results in parts (i) and (ii).

c. Summarize the main point of this exercise.

76. Explain how to use the power property for natural logarithms to solve an exponential equation in one variable.

Related Review

Simplify by writing your result as a single logarithm with a coefficient of 1, or solve, as appropriate. Round any solutions to the fourth decimal place.

77. $\ln(x^8) - \ln(x^3)$

78. $2\ln(4x) + 3\ln(x^2)$

79. $\ln(x^8) - \ln(x^3) = 4$

80. $2\ln(4x) + 3\ln(x^2) = 7$

Solve. Find the exact solution if the equation is linear. For any other type of equation, round the solution(s) to the fourth decimal place.

81. $3e^x - 5 = 7$

82. $4\log_2(x^2) - 2\log_2(3x) = 5$

83. $7 - 3(2t - 4) = 5t + 6$

84. $\log(w - 17) = 2$

85. $25x^3 - 4x = 12 - 75x^2$

86. $2x(2x - 1) = 5$

87. $\dfrac{b^7}{b^3} = 16$

88. $3\ln(2x) + 4\ln(x^2) = 7$

Expressions, Equations, Functions, and Graphs

Give an example of the following. Then solve, simplify, or graph, as appropriate.

89. expression involving exponents

90. system of two linear equations in two variables

91. logarithmic equation in one variable

92. quadratic function

93. sum of two natural logarithmic expressions

94. linear function

95. trinomial in one variable with five terms

96. exponential function

Taking it to the Lab

China and India Populations Lab

In this lab, you will compare the populations of China and India.*

Collecting the Data

For both countries, you will find populations for every five years since 1950, plus make a projection for the current year.

1. Go to www.census.gov/population/international/data/idb/informationGateway.php. Under "Select Country(ies)," hold the CTRL key and select China and India. Under "Select Year(s)," hold the CTRL key and select the years that are multiples of 5 between 1950 and the present (1950, 1955, 1960, etc.). The current year should already be highlighted. (Mac users: If the control key doesn't work, try using the command key.) Then click "Submit."
2. Record the data or print the screen.

Analyzing the Data

1. Include tables of data for China's population and India's population.
2. Define the variables for your models. [**Hint:** Describe the units of the variables.]
3. Which variables are the independent variables? Which variables are the dependent variables? If you round the data, describe how.
4. Use a graphing calculator to draw scattergrams of the data. Copy the scattergrams on paper. For both countries, discuss whether a linear or an exponential function is the better choice of model. Explain.
5. For both countries, find an equation of a model.
6. For both countries, use a graphing calculator to graph your model and your scattergram in the same viewing window. Also, graph your model and scattergram by hand. How well does your model fit the data?
7. Which country has the larger current population? Use your models to estimate the difference in the current populations.
8. Use "intersect" on a graphing calculator to predict when the populations will be equal.
9. Go to www.census.gov to find an estimate of the current world population. Use a model to predict when China's population will reach that size. Use a model to predict when India's population will reach that size. In terms of the types of functions used to model the populations, explain why the two results are so different.

*"Taking it to the Lab: China and India Populations Lab" adapted from a lab written by Cheryl Gregory, College of San Marco, CA. Used by permission of Cheryl Gregory.

10. Use your models to predict when the sum of the populations of China and India will reach the current world population. Explain how you found your result.

Folding Paper Lab

In this lab, you will investigate the thickness of a sheet of paper by folding it many times.

Materials

To do this lab, you will need the following materials:

1. an $8\frac{1}{2}$-inch-by-11-inch piece of paper
2. a ruler

Preparation

Fold the piece of paper in half six times very carefully, each time without unfolding.

Recording of Data

Use the ruler to measure the thickness of the folded paper.

Analyzing the Data

1. What is the thickness of the folded paper (after six folds)? Include units.
2. Use your answer to Part 1 to estimate the thickness of the paper when it is unfolded. Include units.
3. Let $f(n)$ be the thickness of the paper if it has been folded n times. Find an equation of f. What are the units of $f(n)$?
4. Can you fold the paper a seventh time? If not, use f to predict the thickness if it could be folded seven times. If you can, keep folding the paper until you cannot fold it any more and predict the thickness if it could be folded one more time.
5. How thick would the folded paper be if you could fold it 15 times? Would the folded paper be taller or shorter than you?
6. After how many folds would the folded paper be at least as tall as a football field is long (that is, 120 yards long if you include the end zones)?
7. After how many folds would the thickness of the folded paper match the distance to the Moon? (The average distance to the Moon is approximately 238,857 miles. There are 5280 feet in 1 mile.)
8. The situation in this lab is limited by your inability to fold the paper many times. Instead of folding a piece of paper, you could cut a piece of paper in half, then stack the two halves. Next, you could cut the stack of papers and restack the two piles of papers. Cutting the stack in two each time and then restacking achieves the same thickness as folding the paper in two. By cutting and stacking, can the result described in Part 7 be achieved? Explain.

Exponential/Logarithmic Lab: Topic of Your Choice

Repeat the Exponential Lab: Topic of Your Choice, but choose a different situation. Also, choose an output of your model. Find the input that originates from that output. What does the result mean in this situation?

Chapter Summary

Key Points of Chapter 11

Section 11.1 Composite Functions

Composite function	If f and g are functions, x is in the domain of g, and $g(x)$ is in the domain of f, then we can form the **composite function $f \circ g$:** $(f \circ g)(x) = f(g(x))$. We say $f \circ g$ is the **composition of f and g.**

Section 11.2 Inverse Functions

Inverse of a function f	The **inverse of a function** f is a relation that sends b to a if $f(a) = b$.
Invertible function	If the inverse of a function f is also a function, we say f is **invertible** and use "f^{-1}" as a name for the inverse of f.
Property of an inverse function	For an invertible function f, the following statements are equivalent: $f(a) = b$ and $f^{-1}(b) = a$.

Section 11.2 Inverse Functions (*Continued*)

f and f^{-1} are inverses of each other	If f is an invertible function, then • f^{-1} is invertible, • f and f^{-1} are inverses of each other, • $(f^{-1} \circ f)(x) = x$, where x is in the domain of f, and • $(f \circ f^{-1})(x) = x$, where x is in the domain of f^{-1}.
Reflection property of inverse functions	For an invertible function f, the graph of f^{-1} is the reflection of the graph of f across the line $y = x$.
Graphing an inverse function	For an invertible function f, we sketch the graph of f^{-1} by the following steps: **1.** Sketch the graph of f. **2.** Choose several points that lie on the graph of f. **3.** For each point (a, b) chosen in step 2, plot the point (b, a). **4.** Sketch the curve that contains the points plotted in step 3.
Three-step process for finding the inverse of a model	To find the inverse of an invertible *model* f, where $p = f(t)$, **1.** Replace $f(t)$ with p. **2.** Solve for t. **3.** Replace t with $f^{-1}(p)$.
Four-step process for finding the inverse of a function that is not a model	Let f be an invertible function that is *not* a model. To find the inverse of f, where $y = f(x)$, **1.** Replace $f(x)$ with y. **2.** Solve for x. **3.** Replace x with $f^{-1}(y)$. **4.** Write the equation of f^{-1} in terms of x.
One-to-one function	If each output of a function originates from exactly one input, we say that the function is **one-to-one.**

Section 11.3 Logarithmic Functions

Definition of a logarithm	For $b > 0, b \neq 1$, and $a > 0$, the **logarithm $\log_b(a)$** is the number k such that $b^k = a$. In words, $\log_b(a)$ is the exponent on the base b that gives a. We call b the **base** of the logarithm.
Common logarithm	A **common logarithm** is a logarithm with base 10. We write $\log(a)$ to represent $\log_{10}(a)$.
Properties of logarithms	For $b > 0$ and $b \neq 1$, • $\log_b(b) = 1$ • $\log_b(1) = 0$ • $\log_b(b^x) = x$ • $b^{\log_b(x)} = x$, where $x > 0$
Logarithmic function	A **logarithmic function, base b,** is a function that can be put into the form $f(x) = \log_b(x)$, where $b > 0$ and $b \neq 1$.
Domain of a logarithmic function	The domain of a logarithmic function $\log_b$ is the set of all positive real numbers.
Logarithmic and exponential functions are inverses of each other	For an exponential function $f(x) = b^x$, $f^{-1}(x) = \log_b(x)$. For a logarithmic function $g(x) = \log_b(x)$, $g^{-1}(x) = b^x$.
Graphing $y = \log_b(x)$	To graph a logarithmic function $y = \log_b(x)$, use the four-step graphing method from Section 11.2 to sketch the inverse of $f(x) = b^x$.

Section 11.3 Logarithmic Functions (*Continued*)

Richter number	The *Richter number, R,* of an earthquake is given by $$R = \log\left(\frac{A}{A_0}\right)$$ where A is the amplitude (maximum value) of a seismic wave and A_0, called the *reference amplitude,* is the amplitude of the smallest seismic wave that a seismograph can detect.

Section 11.4 Properties of Logarithms

Exponential/logarithmic forms property	For $a > 0, b > 0$, and $b \neq 1, \log_b(a) = c$ and $b^c = a$ are equivalent.
Logarithmic equation in one variable	A **logarithmic equation in one variable** is an equation in one variable that contains one or more logarithms.
Solving $\log_b(x) = k$	For an equation of the form $\log_b(x) = k$, we can solve for b or x by writing the equation in exponential form.
Exponential equation in one variable	An **exponential equation in one variable** is an equation in one variable in which an exponent contains a variable.
Power property for logarithms	For $x > 0, b > 0$, and $b \neq 1, \log_b(x^p) = p\log_b(x)$.
Logarithm property of equality	For positive real numbers a, b, and c, where $b \neq 1$, the equations $a = c$ and $\log_b(a) = \log_b(c)$ are equivalent.
Solving $ab^x = c$ for x	To solve some equations of the form $ab^x = c$ for x, we divide both sides of the equation by a and then take the log of both sides. Next, we use the power property for logarithms.

Section 11.5 Using the Power Property with Exponential Models to Make Predictions

Making a prediction	To make a prediction about the independent variable t of an exponential model of the form $f(t) = ab^t$, we substitute a value for $f(t)$ and divide both sides of the equation by the coefficient a. Next, we take the log of both sides of the equation and use the power property to help solve for t.

Section 11.6 More Properties of Logarithms

Properties of logarithms	For $x > 0, y > 0, a > 0, b > 0, a \neq 1$, and $b \neq 1$, • $\log_b(x) + \log_b(y) = \log_b(xy)$ Product property for logarithms • $\log_b(x) - \log_b(y) = \log_b\left(\frac{x}{y}\right)$ Quotient property for logarithms • $\log_b(x) = \frac{\log_a(x)}{\log_a(b)}$ Change-of-base property
Finding $\log_b(a)$ where $b \neq 10$	To find a logarithm to a base other than 10, we use the change-of-base property for logarithms to convert to $\log_{10}$; then we can use the log key on a calculator.

Section 11.7 Natural Logarithms

Approximation of e	To the nearest ten-thousandth, $e = 2.7183$.
Definition of a natural logarithm	A **natural logarithm** is a logarithm with base e. We write $\ln(a)$ to represent $\log_e(a)$.
Meaning of $\ln(a)$	For $a > 0, \ln(a)$ is the exponent on the base e that gives a.
Exponential/natural logarithmic forms property	For $a > 0$, the equations $\ln(a) = c$ and $e^c = a$ are equivalent.

Section 11.7 Natural Logarithms (*Continued*)

Properties of natural logarithms	$\ln(1) = 0$ $\ln(e) = 1$ $\ln(e^x) = x$ For $x > 0$ and $y > 0$, • $e^{\ln(x)} = x$	
	• $\ln(x^p) = p\ln(x)$	Power property for natural logarithms
	• $\ln(x) + \ln(y) = \ln(xy)$	Product property for natural logarithms
	• $\ln(x) - \ln(y) = \ln\left(\frac{x}{y}\right)$	Quotient property for natural logarithms

Chapter 11 Review Exercises

All of the values of functions g and f are shown in Table 66. For Exercises 1–7, refer to this table.

1. Find $f(2)$.
2. Find $f^{-1}(2)$.
3. Find $(f \circ g)(1)$.
4. Find $(g \circ f)(1)$.
5. Find $(g^{-1} \circ g)(3)$.
6. Find $(f \circ g^{-1})(0)$.
7. Use a table to describe five input–output pairs of $f \circ g$.

Table 66 Input–Output Pairs for f and g (Exercises 1–7)

x	$f(x)$	x	$g(x)$
0	2	0	3
1	4	1	2
2	3	2	1
3	0	3	4
4	1	4	0

For each pair of functions in Exercises 8–10, find **(a)** *an equation of* $f \circ g$; **(b)** *an equation of* $g \circ f$; **(c)** $(f \circ g)(3)$; **(d)** $(g \circ f)(3)$.

8. $f(x) = x^2 + 4$ and $g(x) = 3x - 5$
9. $f(x) = 4(2)^x$ and $g(x) = 2x - 4$
10. $f(x) = \log_3(x)$ and $g(x) = x + 6$
11. Let $h(x) = e^{x-5}$. Find equations of f and g so that $h(x) = f(g(x))$.
12. A book sells for \$8. The sales tax in Michigan is 6%. Let $f(n)$ be the total cost (in dollars) of n of the books. Let $g(d)$ be the sales tax (in dollars) on a purchase in Michigan worth d dollars.
    a. Find equations of f and g.
    b. Find an equation of $g \circ f$.
    c. Find $g(f(7))$. What does it mean in this situation?

Graph f, f^{-1}, and y = x by hand on the same set of axes.

13. $f(x) = 2x - 3$
14. $f(x) = 3^x$
15. The numbers of FBI background checks for firearm purchases are shown in Table 67 for various years.

Table 67 Numbers of FBI Background Checks for Firearm Purchases

Year	Number of Background Checks (millions)
2005	9
2006	10
2007	11
2008	13
2009	14
2010	15
2011	16

Source: *FBI*

Let $n = f(t)$ be the number (in millions) of FBI background checks for firearm purchases in the year that is t years since 2000.

a. Find a linear equation of f.
b. Find $f(18)$. What does it mean in this situation?
c. Find $f^{-1}(18)$. What does it mean in this situation?
d. Although states charge various amounts for background checks, assume the average charge is \$15 per check. Let $C(n)$ be the total cost (in millions of dollars) of n million background checks. Find an equation of C.
e. Let $h(t)$ be the total cost (in millions of dollars) of FBI background checks for all firearm purchases in the year that is t years since 2000. Determine which of the following is true: $h(t) = (f \circ C)(t)$ or $h(t) = (C \circ f)(t)$. Explain.
f. Find an equation of h.
g. Find $h(17)$. What does it mean in this situation?

For Exercises 16 and 17, **(a)** *find the inverse of the given function;* **(b)** *show that* $(f^{-1} \circ f)(x) = x$; **(c)** *show that* $(f \circ f^{-1})(x) = x$.

16. $f(x) = 3x$
17. $f(x) = \frac{5}{6}x - 2$

Evaluate the logarithmic function at the given value. Round approximate results to the fourth decimal place.

18. $\log_5(25)$

19. $\log(100{,}000)$

20. $\log_3\left(\frac{1}{9}\right)$

21. $\ln\left(\frac{1}{e^3}\right)$

22. $\log_4\left(\sqrt[3]{4}\right)$

23. $\log_3(7)$

24. $\ln(5)$

25. $\log_b\left(b^7\right)$

For Exercises 26 and 27, find the inverse of the function.

26. $h(x) = 3^x$

27. $h(x) = \log(x)$

28. Sketch the graph of $y = \log_4(x)$ by hand.

29. Write the equation $d^x = k$ in logarithmic form.

30. Write the equation $\log_y(w) = r$ in exponential form.

Solve. Round any approximate solutions to the fourth decimal place.

31. $6(2)^x = 30$

32. $\log_3(x) = -4$

33. $4.3(9.8)^x - 3.3 = 8.2$

34. $\log_b(83) = 6$

35. $5\log_{32}(m) - 3 = -1$

36. $5(4)^{3r-7} = 40$

37. $2^{4t} \cdot 2^{3t-5} = 94$

For Exercises 38–40, find any solutions of the equation or system by referring to the graphs shown in Fig. 62.

38. $\log_2(x) = -\frac{3}{4}x + 5$

39. $2^x - 3 = -2$

40. $y = \log_2(x)$

$y = -\frac{3}{4}x + 5$

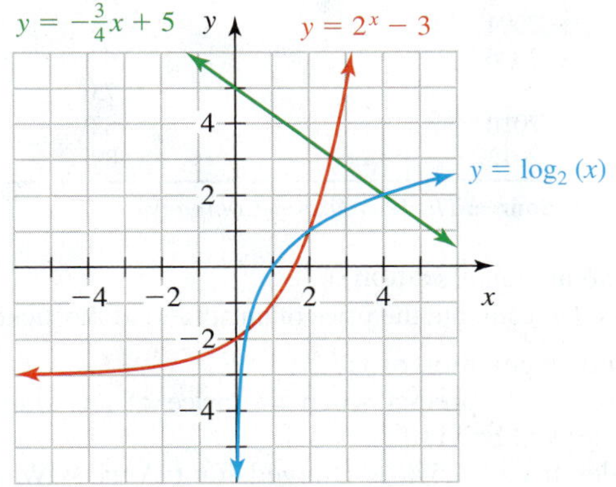

Figure 62 Exercises 38–40

For Exercises 41–44, let $f(x) = 3^x$.

41. Find $f(4)$.

42. Find $f^{-1}(25)$.

43. Find x when $f(x) = 6$.

44. Find x when $f^{-1}(x) = 6$.

45. Suppose \$8000 is deposited into an account that earns 5% interest compounded annually. Let $f(t)$ be the value (in dollars) of the account after t years or any fraction thereof.

a. Find a formula of f.

b. Find the value of the account after 9 years.

c. After how many years will the value of the account double?

46. On April 1, a tree has 30 leaves. Each week, the number of leaves quadruples (increases by four times). Let $f(t)$ be the number of leaves on the tree at t weeks since April 1.

a. Find an equation of f.

b. Predict the number of leaves at 5 weeks after April 1.

c. Predict when there will be 100,000 leaves.

47. National health spendings from all sources, public and private, are shown in Table 68 for various years.

Table 68 Annual National Health Spendings

Year	Annual National Health Spending (trillions of dollars)
1970	0.1
1975	0.2
1980	0.3
1985	0.4
1990	0.7
1995	1.0
2000	1.3
2005	2.0
2010	2.6

Source: *Centers for Medicare and Medicaid Services*

Let $s = f(t)$ be the annual national health spending (in trillions of dollars) at t years since 1970.

a. Find an exponential equation of f.

b. What is the s-intercept of the model? What does it mean in this situation?

c. What is the percentage rate of growth of annual national health spending?

d. Predict the national health spending in 2019.

e. Predict when the annual national health spending will be \$5 trillion.

48. A student ran an experiment to investigate the relationship between the length of a rubber band and the weight applied to one end of it. He hooked one end of a rubber band to a horizontal pole supported between two chairs. He attached a bag to the other end of the rubber band. Then, he recorded the lengths of the rubber band with various numbers of tape cassettes in the bag (see Table 69).

Table 69 Lengths of a Rubber Band Stretched by Tape Cassettes

Number of Cassettes	Length of Rubber Band (inches)
0	10.00
1	12.00
2	15.38
3	19.50
4	28.31
5	33.50
6	45.45
7	64.15

Source: *Michael S.*

a. Let $f(n)$ be the length (in inches) of the rubber band stretched by n cassettes. Find an exponential equation of f.

b. What is the base b of your function $f(n) = ab^n$? What does it mean in this situation?

c. What is the coefficient a of your function $f(n) = ab^n$? What does it mean in this situation?
d. Estimate the length of the rubber band if it were stretched by 8 cassettes. Describe two scenarios in which model breakdown might occur for your estimate.
e. Estimate the number of cassettes needed to stretch the rubber band to 139 inches. If model breakdown occurs for your estimate from part (d), does that imply model breakdown occurs for the estimate you made in this part? Explain.

49. A storage tank contains cobalt-60, which has a half-life of 5.3 years. Predict when 15% of the cobalt-60 will remain.

For Exercises 50–53, simplify. Write your result as a single logarithm with a coefficient of 1.

50. $\log_b(p) + \log_b(6p) - \log_b(2p)$
51. $3\log_b(2x) + 2\log_b(3x)$
52. $4\log_b(x^2) - 2\log_b(x^5)$
53. $\dfrac{\log_b(w)}{\log_b(y)}$

54. Which of the following expressions are equal?

$$\log_b(b^5) - \log_b(b^2) \qquad 3 \qquad \log_b(b^5)$$
$$\log_b(b^3) \qquad \frac{\log_b(b^5)}{\log_b(b^2)} \qquad \log_b\left(\frac{b^5}{b^2}\right)$$

Solve. Round any solutions to the fourth decimal place.

55. $2\log_9(3w) + 3\log_9(w^2) = 5$
56. $5\log_6(2x) - 3\log_6(4x) = 2$

Simplify. Write your result as a single logarithm with a coefficient of 1.

57. $3\ln(4x) + 2\ln(2x)$
58. $\ln(2m^7) - 4\ln(m^3) + 3\ln(m^2)$

Solve. Round any solutions to the fourth decimal place.

59. $4e^x = 75$
60. $-3\ln(p) + 7 = 1$
61. $3\ln(t^5) - 5\ln(2t) = 7$
62. $e^{3x-8} = 12$
63. $\ln(3x + 1) = 2$

Chapter 11 Test

For each pair of functions in Exercises 1 and 2, find **(a)** *an equation of $f \circ g$;* **(b)** *an equation of $g \circ f$;* **(c)** $(f \circ g)(2)$; **(d)** $(g \circ f)(2)$.

1. $f(x) = 3x^2$ and $g(x) = 2x - 5$
2. $f(x) = 3^x$ and $g(x) = x - 4$

For Exercises 3–6, refer to Fig. 63, which shows the graphs of invertible functions f and g.

3. Find $(f \circ g)(1)$.
4. Find $(g \circ f)(1)$.
5. Find $g^{-1}(3)$.
6. Graph f^{-1} by hand.

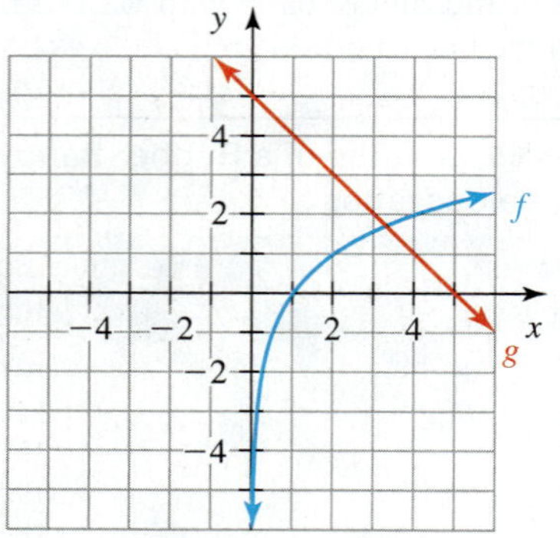

Figure 63 Exercises 3–6

7. Graph $f(x) = 3x - 6$, f^{-1}, and $y = x$ by hand on the same set of axes.

8. The prices of an adult one-day ticket to Walt Disney World® are shown in Table 70 for various years. Let $p = f(t)$ be the price (in dollars) of an adult one-day ticket at t years since 2000.

Table 70 Prices of an Adult One-Day Ticket to Walt Disney World

Year	Ticket Price (dollars)
2000	46
2002	50
2004	58
2006	67
2008	75
2010	82
2012	89

Source: *The Walt Disney Company*

a. Find an equation of f.
b. Use f to estimate the price of an adult one-day ticket in 2011.
c. Find an equation of f^{-1}.
d. Use f^{-1} to predict when the price of an adult one-day ticket will be \$110.
e. Sales tax of 6.5% is charged for tickets at Walt Disney World. Let $S(d)$ be the sales tax (in dollars) on a ticket worth d dollars. Find an equation of S.
f. Let $h(t)$ be the sales tax (in dollars) on an adult one-day ticket at t years since 2000. Determine which of the following is true: $h(t) = (f \circ S)(t)$ or $h(t) = (S \circ f)(t)$. Explain.
g. Find an equation of h.
h. Find $h(17)$. What does it mean in this situation?

9. Find the inverse of the function $g(x) = \dfrac{2x - 9}{5}$.

Evaluate. Round approximate results to the fourth decimal place.

10. $\log_2(16)$
11. $\log_4\left(\dfrac{1}{64}\right)$
12. $\log_7(10)$
13. $\log(0.1)$

14. $\log_b(\sqrt{b})$ **15.** $\ln\left(\frac{1}{e^2}\right)$

For Exercises 16 and 17, find the inverse of the function.

16. $h(x) = 4^x$

17. $f(x) = \log_5(x)$

18. Write $s^k = w$ in logarithmic form.

19. Write $\log_c(a) = d$ in exponential form.

For Exercises 20–22, solve. Round any solutions to the fourth decimal place.

20. $\log_b(50) = 4$ **21.** $6(2)^x - 9 = 23$

22. $\log_4(7p + 5) = -\frac{3}{2}$

23. Use "intersect" on a graphing calculator to solve the equation $4^x - 8 = -\frac{1}{2}x + 3$. Round any solutions to the second decimal place.

24. The numbers of patients in public psychiatric hospitals per 100,000 people are shown in Table 71 for various years.

Table 71 Numbers of Patients in Public Psychiatric Hospitals per 100,000 People

Year	Number of Patients per 100,000 People
1960	298
1970	166
1980	58
1990	37
2000	19
2010	14

Source: *Treatment Advocacy Center*

Let $f(t)$ be the number of patients in psychiatric hospitals per 100,000 people at t years since 1960.

a. Find a linear equation, a quadratic equation, and an exponential equation of f. Compare how well the models fit the data.

b. Assuming the number of patients in public psychiatric hospitals per 100,000 people continues to decrease, which function best models the data?

c. What is the percentage rate of decay of the number of patients in psychiatric hospitals per 100,000 people?

d. Predict the number of patients in psychiatric hospitals in 2019. Assume the U.S. population will be 338 million in that year.

e. Predict when there will be 8 patients in public psychiatric hospitals per 100,000 people.

25. Scientists wanted to date a sample of cloth wrappings of a mummified bull from a pyramid in Dashur, Egypt. If 78.04% of the carbon-14 remains in the sample, estimate the age of the mummy. The half-life of carbon-14 is 5730 years.

Simplify. Write your result as a single logarithm with a coefficient of 1.

26. $\log_b(x^3) + \log_b(5x)$

27. $3\log_b(4p^2) - 2\log_b(8p^5) + \log_b(2p^4)$

For Exercises 28 and 29, solve. Round any solutions to the fourth decimal place.

28. $\log_3(x) + \log_3(2x) = 5$

29. $2\log_4(x^4) - 3\log_4(3x) = 3$

30. Simplify $2\ln(5w) + 3\ln(w^6)$. Write your result as a single logarithm with a coefficient of 1.

Solve. Round any solutions to the fourth decimal place.

31. $2e^{3x-1} = 54$ **32.** $7\ln(t - 2) - 1 = 4$

Cumulative Review of Chapters 1–11

For Exercises 1–11, solve. Any solution is a real number. Round approximate results to the fourth decimal place.

1. $2(4)^{5x-1} = 17$ **2.** $(x - 2)(x + 4) = 7$

3. $\log_3(x - 5) = 4$ **4.** $3b^7 - 18 = 7$

5. $3x^2 = 7 - 2x$

6. $8 + 2e^x = 15$

7. $2(x - 5)^2 - 1 = 6$

8. $4\log_5(3x^2) + 3\log_5(6x^4) = 3$

9. $7 - 3(4w - 2) = 2(3w + 5) - 4(2w + 1)$

10. $3x(2x - 1) + 1 = 2 - 4x^2$

11. $2\ln(12x^9) - \ln(3x^3) = 5$

12. Find all complex-number solutions of $5x^2 - 4x = -3$.

13. Solve $2x^2 + 5x - 2 = 0$ by completing the square.

14. Solve $(at - b)^2 = c$ for t. Assume that the constants have values for which the equation has exactly two real-number solutions.

For Exercises 15 and 16, estimate any solutions of the equation by referring to the graphs of $f(x) = 3^x$, $g(x) = 9\left(\frac{1}{3}\right)^x$, and $h(x) = x - 1$ shown in Fig. 64.

15. $3^x = 9\left(\frac{1}{3}\right)^x$ **16.** $9\left(\frac{1}{3}\right)^x = x - 1$

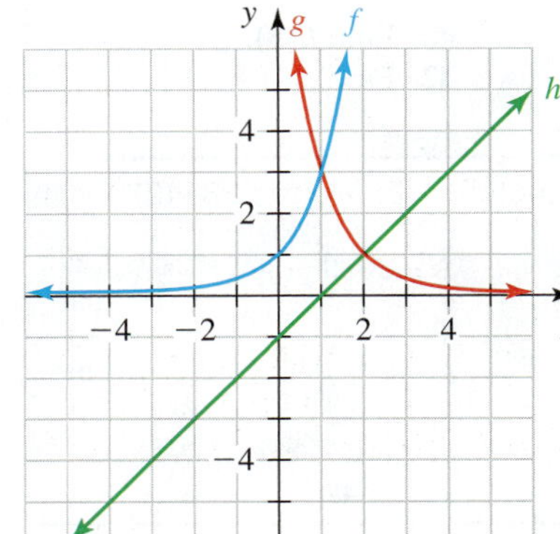

Figure 64 Exercises 15 and 16

For Exercises 17–19, solve the system.

17. $x = 2y - 5$
$4x - 5y = -14$

18. $3(2 - 4x) = -10 - 2y$
$2x - 3y = -8$

19. $2x + 3y - 2z = 3$
$3x - y + 4z = 2$
$5x - 2y + 3z = -5$

20. Find the x-intercepts of the graph of $y = 2x^2 - 5x - 12$.

Perform the indicated operations.

21. $2x^2(3x - 5)(2x - 4)$

22. $(4mt + 7r)(4mt - 7r)$

23. $(5m - 3n)^2$

For Exercises 24 and 25, let $f(x) = 2x^2 - x + 5$ and $g(x) = 3x^2 + 2x - 4$. Find an equation of the given function. Then evaluate the result at the indicated value.

24. $f - g$, $(f - g)(-3)$

25. $f \cdot g$, $(f \cdot g)(2)$

26. For $f(x) = x^2 - 3x$ and $g(x) = x + 5$, find an equation of the function $f \circ g$. Then find $(f \circ g)(2)$.

27. Perform long division: $\dfrac{12x^3 - 11x^2 + 8x - 4}{3x - 2}$

For Exercises 28–30, factor when possible.

28. $12x^3 - 20x^2 - 27x + 45$

29. $3x^3y^2 + 6x^2y^3 - 24xy^4$

30. $27x^3 + 64$

31. Solve the inequality $8x - 3 \geq -3(4x - 5)$. Describe the solution set as an inequality, in interval notation, and in a graph.

32. Graph $5x - 3y > 6$ by hand.

33. Graph the solution set of the system

$$y \leq -2x + 5$$
$$y \geq \frac{3}{4}x - 4$$

Simplify.

34. $(4b^{-3}c^2)^3(5b^{-7}c^{-1})^2$

35. $\dfrac{8b^{1/3}c^{-1/2}}{6b^{-1/2}c^{3/4}}$

Simplify. Write your result as a single logarithm with a coefficient of 1.

36. $4\log_b(x^7) - 2\log_b(7x)$

37. $3\ln(p^6) + 4\ln(p^2)$

For Exercises 38–42, refer to Table 72.

38. Find an equation of f.

39. Find an equation of g.

40. Find $h(5)$.

41. Find $k^{-1}(5)$.

42. Find $(g \circ k)(8)$.

Table 72 Values of Four Functions (Exercises 38–42)

x	$f(x)$	x	$g(x)$	x	$h(x)$	x	$k(x)$
0	5	0	25	4	83	3	160
1	15	1	28	5	76	4	80
2	45	2	31	6	69	5	40
3	135	3	34	7	62	6	20
4	405	4	37	8	55	7	10
5	1215	5	40	9	48	8	5

For Exercises 43–48, graph the function by hand.

43. $y = 8\left(\frac{1}{2}\right)^x$

44. $y = 2(x - 3)^2 - 5$

45. $y = -\frac{2}{5}x + 3$

46. $y = -2x^2 + 4x + 3$

47. $2(2x - y) + 2y = 3(4 + y)$

48. $y = \log_2(x)$

49. Find an equation of a line that contains the points $(-4, 7)$ and $(5, -3)$.

50. Find an equation of the parabola that contains the points $(2, 4)$, $(3, 14)$, and $(4, 30)$.

51. Find an equation of an exponential curve that contains the points $(3, 85)$ and $(7, 13)$.

Let $f(x) = 2(3)^x$. Round approximate results to the fourth decimal place.

52. Find $f(-4)$.

53. Graph f by hand.

54. Graph f^{-1} by hand.

55. Find $f^{-1}(35)$.

Find the logarithm.

56. $\log_3\left(\frac{1}{81}\right)$

57. $\log_b(\sqrt[7]{b})$

For Exercises 58–62, refer to Fig. 65.

58. Find $g(-1)$.

59. Find $(f \circ g)(3)$.

60. Find an equation of f.

61. Graph f^{-1} by hand.

62. Find $f^{-1}(2)$.

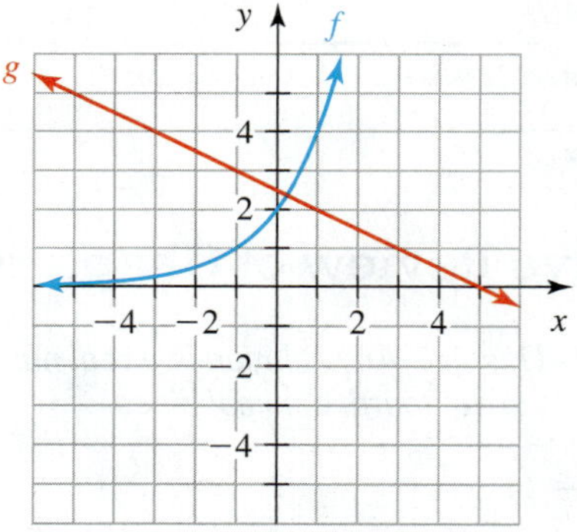

Figure 65 Exercises 58–62

For Exercises 63 and 64, find the inverse of the function.

63. $f(x) = \frac{2}{7}x - 3$

64. $g(x) = 8^x$

65. Compare the function $f(x) = 3x + 2$ with the function $g(x) = 2(3)^x$.

a. Find the y-intercept of the graphs of both functions.

b. For both f and g, describe what happens to the value of y as the value of x increases by 1.

c. For a large input value of x, which function will have a greater output value of y? Explain.

d. Graph f and g by hand on the same coordinate system.

66. Let $f(x) = 3x$ and $g(x) = 3^x$.
a. Find $f(2)$ and $g(2)$.
b. Find an equation of f^{-1} and an equation of g^{-1}.
c. Find $f^{-1}(81)$ and $g^{-1}(81)$.

67. One U-Haul office rents pickup trucks for a one-day fee of \$19.95 plus \$0.69 per mile. One Budget office charges a one-day fee of \$29.95 plus \$0.45 per mile (Sources: *U-Haul; Budget*).
a. Let $U(x)$ be U-Haul's charge and $B(x)$ be Budget's charge, both in dollars, for driving x miles in one day. Find equations of U and B.
b. Find the slopes of the graphs of U and B. What do they mean in this situation?
c. For how many miles driven is the one-day charge at U-Haul equal to the one-day charge at Budget?

68. A 15,000-seat amphitheater has tickets for sale at \$43 and \$60. How many tickets should be sold at each price for a sellout performance to generate a total revenue of \$721,500?

69. The annual revenue of Whole Foods Market was \$9.0 billion in 2010 and has grown by about 14% per year since then (Source: *Whole Foods Market*).
a. Let $r = f(t)$ be the annual revenue (in billions of dollars) of Whole Foods Market at t years since 2010. Find an equation of f.
b. What is the r-intercept of the model? What does it mean in this situation?
c. What is the base b of your function $f(t) = ab^t$? What does it mean in this situation?
d. Predict when the annual revenue will be \$25 billion.

70. The numbers of tuberculosis cases in the United States are shown in Table 73 for various years.

Table 73 Tuberculosis Cases in the United States

Year	Number of Cases (thousands)
1955	69.9
1960	55.5
1970	37.1
1980	27.7
1990	25.7
2000	16.3
2010	11.2

Source: *Centers for Disease Control and Prevention*

Let $n = f(t)$ be the number (in thousands) of tuberculosis cases in the year that is t years since 1950.
a. Find a linear equation, a quadratic equation, and an exponential equation of f. Compare how well the models fit the data.
b. Assuming the number of tuberculosis cases continues to decrease, which function best models the situation?
c. What is the n-intercept of the graph of the exponential model? What does it mean in this situation?
d. What is the base b of the exponential model $f(t) = ab^t$? What does it mean in this situation?
e. Use the exponential model to find $f(9)$. What does it mean in this situation?
f. Use the exponential model to find $f^{-1}(9)$. What does it mean in this situation?
g. Find the approximate half-life of the number of cases. [**Hint:** Use f to estimate the number of cases in 1950. Then use f to estimate when the number of cases was half that amount.]

71. The number of women who placed in the top 100 of the New York City Marathon are shown in Table 74 for various years.

Table 74 Numbers of Women Who Placed in the Top 100 of the NYC Marathon

Year	Number
1981	0
1986	3
1991	5
1996	8
2001	14
2006	13
2011	15

Source: *nycmarathon.org*

Let $n = f(t)$ be the number of women who placed in the top 100 of the New York City Marathon at t years since 1980.
a. Find an equation of f.
b. If your model is linear, find the slope. If your model is quadratic, find the vertex. If your model is exponential, find the base b of $f(t) = ab^t$. What does your result mean in this situation?
c. What is the t-intercept? What does it mean in this situation?
d. The marathon was canceled in 2012 due to Hurricane Sandy. Estimate the number of women who would have placed in the top 100 if the marathon had not been canceled.
e. Find an equation of f^{-1}.
f. Find $f^{-1}(20)$. What does it mean in this situation?
g. The number of men who placed in the top 100 is given by $g(n) = 100 - n$, where n is the number of women who placed in the top 100. Let $h(t) = (g \circ f)(t)$. Find an equation of h.
h. Find $h(37)$. What does it mean in this situation?

72. Table 75 lists the percentages of Americans in various age groups who visit online trading sites.

Table 75 Americans Who Visit Online Trading Sites

Age Group (years)	Age Used to Represent Age Group (years)	Percent
2–17	9.5	4.7
18–24	21.0	9.7
25–34	29.5	22.9
35–44	39.5	22.2
45–54	49.5	20.6
55–64	59.5	14.6
65+	70.0	5.3

Source: *comScore Media Matrix*

Let $f(t)$ be the percentage of Americans of age t years who visit online trading sites.
a. Find an equation of f.
b. What are the t-intercepts? What do they mean in this situation?
c. What is the vertex of the model? What does it mean in this situation?
d. Estimate the percentage of 19-year-old Americans who visit online trading sites.
e. Estimate the age(s) at which 18% of Americans visit online trading sites.

Rational Functions

If you drive a car, how much did you pay for gasoline last year? In Exercise 8 of Homework 12.6, you will find a model that you will use to predict how much an American will pay, on average, for gasoline in 2018 (see Table 1).

In Chapters 7–9, we worked with polynomial functions. In this chapter, we will discuss how to divide these functions to form new ones called *rational functions*. We will also work with *rational expressions* and discuss how to simplify them and perform operations with them. We will solve *rational equations*, which will help us use rational functions to make predictions about authentic situations. For example, we will use a rational function to estimate the percentage of revenue that was from downloaded music in 2011.

Table 1 Average Prices of Gasoline; Vehicle Fuel Consumption

Year	Price of Gasoline (dollars per gallon)	Year	Amount of Fuel Consumed (millions of gallons)
1995	1.21	1975	108,900
1997	1.29	1980	115,000
1999	1.22	1985	121,300
2001	1.53	1990	130,800
2003	1.64	1995	143,800
2005	2.34	2000	162,600
2007	2.85	2005	174,800
2009	2.40	2010	169,700
2010	2.84		

Source: *Federal Highway Administration*

12.1 Finding the Domains of Rational Functions and Simplifying Rational Expressions

Objectives

- Know the meaning of *rational function* and *vertical asymptote.*
- Find the domain of a rational function.
- Simplify a *rational expression.*
- Know the connection between *vertical asymptotes* of graphs of rational functions and domains of rational functions.
- Form *quotient functions.*
- Use a *rational model* to describe the percentage of a quantity.

In this section, we will discuss the meaning of a *rational function* and how to find its domain. We will simplify *rational expressions.* We will also discuss how to use a *rational model* to make estimates about authentic situations.

Meaning of a Rational Function

In Chapters 7–9, we worked with polynomials. If P and Q are polynomials, where Q is nonzero, we call the ratio $\frac{P}{Q}$ a **rational expression.** The name "rational" refers to "ratio." Here are some examples:

$$\frac{x^3 - 3x + 6}{x^7 - 8} \qquad \frac{-2x^2 + 17}{5x - 1} \qquad -\frac{2}{7x^3} \qquad \frac{3x^2 + 5x - 4}{2x^2 - x + 9}$$

A rational expression is part of an equation of a rational function.

Definition Rational function

A **rational function** is a function whose equation can be put into the form

$$f(x) = \frac{P(x)}{Q(x)}$$

where $P(x)$ and $Q(x)$ are polynomials and $Q(x)$ is nonzero.

In Example 1, we will evaluate a rational function at some input values.

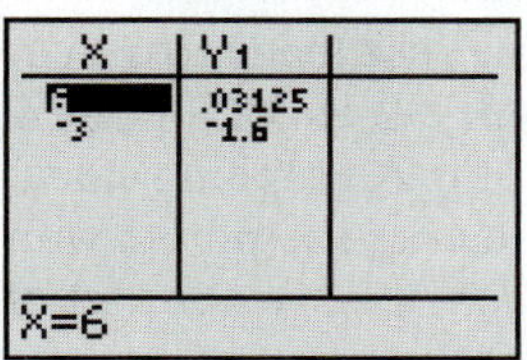

Figure 1 Verify the work

Example 1 Evaluating a Rational Function

Evaluate the function $f(x) = \dfrac{x-5}{x^2-4}$ at the indicated values.

1. $f(6)$ **2.** $f(-3)$

Solution

1. $f(6) = \dfrac{6-5}{6^2-4} = \dfrac{1}{32}$

2. $f(-3) = \dfrac{-3-5}{(-3)^2-4} = \dfrac{-8}{5} = -\dfrac{8}{5}$

We use a graphing calculator table to check that $f(6) = \dfrac{1}{32} = 0.03125$ and $f(-3) = -\dfrac{8}{5} = -1.6$ (see Fig. 1).

WARNING To enter a rational function such as $\dfrac{x-5}{x^2-4}$ into a graphing calculator, we enclose both the numerator and the denominator in parentheses (see Fig. 1).

Domain of a Rational Function

Our work in Example 1 shows that the numbers 6 and -3 are in the domain of the function $f(x) = \dfrac{x-5}{x^2-4}$. Are all real numbers in the domain of f? We will explore this issue in Example 2.

To learn about the domain of a rational function, it will help to consider the simpler rational function $f(x) = \dfrac{5}{x}$ first. Note that $f(0)$ is undefined, since $\dfrac{5}{0}$ is undefined. (We cannot divide by 0.) So, 0 is *not* in the domain of f. Since division by nonzero numbers *is* defined, the domain of f is the set of (all) real numbers except 0.

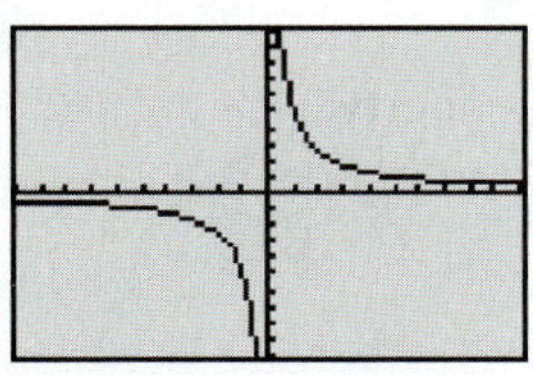

Figure 2 Graph of $f(x) = \dfrac{5}{x}$

Figure 2 displays a graphing calculator graph of $f(x) = \dfrac{5}{x}$. The x-axis appears to be a horizontal asymptote of the graph of f; that is, in fact, true. The graph of f also appears to get arbitrarily close to, but never intersect, the y-axis; that is also true. We say the y-axis is a **vertical asymptote** of the graph of f. Just as with a horizontal asymptote, a vertical asymptote is *not* part of the graph of a function.

Example 2 Finding the Domain of a Rational Function

Find the domain of the function.

1. $f(x) = \dfrac{2}{x-3}$ **2.** $g(x) = \dfrac{x-5}{x^2-4}$

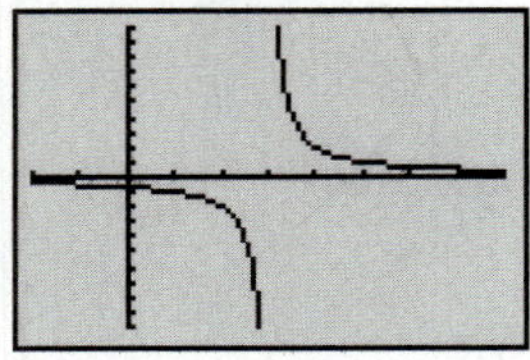

X	Y4
-3	-.3333
-2	-.4
-1	-.5
0	-.6667
1	-1
2	-2
3	ERROR

X=-3

Figure 3 The graph of f has a vertical asymptote at $x = 3$, and 3 is not in the domain of f

Solution

1. For $f(x) = \frac{2}{x - 3}$, the number 3 is not in the domain, because $\frac{2}{3 - 3}$ involves a division by zero. No other value of x leads to a division by 0, so the domain is the set of real numbers except 3. In Fig. 3, we draw a graph of f and build a table of input–output pairs of f.

   From Fig. 3, it appears the graph of f has a vertical asymptote at $x = 3$; that is true. The "ERROR" message across from $x = 3$ in the table supports the idea that the value 3 for x leads to division by 0. (The TI-83 and TI-84 display either "ERROR" or "ERR:.")

2. To find which values of x lead to a division by 0, we set the denominator of $\frac{x - 5}{x^2 - 4}$ equal to 0 and solve for x:

$$x^2 - 4 = 0 \quad \text{Set denominator equal to 0.}$$
$$(x + 2)(x - 2) = 0 \quad \text{Factor left side.}$$
$$x + 2 = 0 \quad \text{or} \quad x - 2 = 0 \quad \text{Zero factor property}$$
$$x = -2 \quad \text{or} \quad x = 2$$

   The numbers -2 and 2 are *not* in the domain, since each value of x leads to a division by zero. The domain of g is the set of real numbers except -2 and 2. In Fig. 4, we draw a graph of g and build a table of input–output pairs of g. A graphing calculator approximates graphs by plotting many points and connecting the points with curves. The steep lines that are "almost" the lines $x = -2$ and $x = 2$ are *not* part of the graph. Some TI graphing calculators do not show these lines.

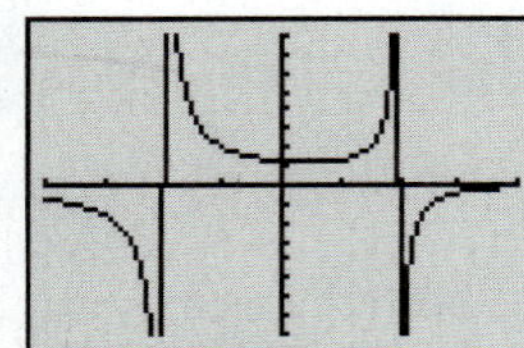

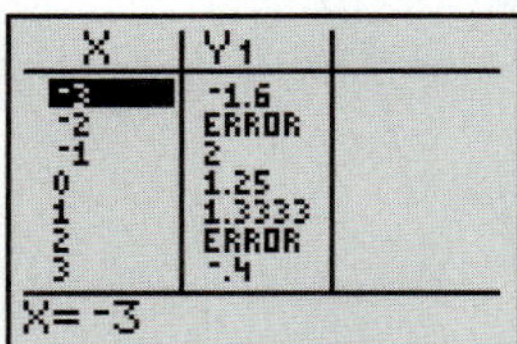

X	Y1
-3	-1.6
-2	ERROR
-1	2
0	1.25
1	1.3333
2	ERROR
3	-.4

X=-3

Figure 4 The graph of g has vertical asymptotes at $x = -2$ and $x = 2$, and -2 and 2 are not in the domain of g

   From Fig. 4, it appears the graph of g has vertical asymptotes at $x = -2$ and $x = 2$; that is true. The "ERROR" messages across from $x = -2$ and $x = 2$ in the table support the idea that each of the values -2 and 2 for x lead to a division by zero.

Our work in Problem 2 of Example 2 suggests the following property.

Domain of a Rational Function

The domain of a rational function $f(x) = \frac{P(x)}{Q(x)}$ is the set of real numbers except for those numbers that, when substituted for x, give $Q(x) = 0$.

Example 3 Finding the Domain of a Rational Function

Find the domain of the function.

1. $f(x) = \frac{x - 4}{x^2 + 5x + 2}$
2. $g(x) = \frac{x^2 - 5x + 1}{x^2 + 10}$

Solution

1. We set the denominator of the right-hand side of $f(x) = \dfrac{x-4}{x^2+5x+2}$ equal to 0:

$$x^2 + 5x + 2 = 0$$

Then we substitute $a = 1$, $b = 5$, and $c = 2$ in the quadratic formula:

$$x = \frac{-5 \pm \sqrt{5^2 - 4(1)(2)}}{2(1)} \quad \text{Substitute into quadratic formula.}$$

$$= \frac{-5 \pm \sqrt{17}}{2} \quad \text{Simplify.}$$

The domain of f is the set of real numbers except $\dfrac{-5-\sqrt{17}}{2}$ and $\dfrac{-5+\sqrt{17}}{2}$.

2. We set the denominator of the right-hand side of $g(x) = \dfrac{x^2-5x+1}{x^2+10}$ equal to 0 and solve:

$$x^2 + 10 = 0$$
$$x^2 = -10$$

Since a real number squared is nonnegative, we conclude there are no real-number solutions of $x^2 + 10 = 0$. Therefore, for the function g, no values of x lead to division by 0. So, the domain of g is the set of (all) real numbers.

In Problem 1 of Example 2, we found that substituting 3 in the expression $\dfrac{2}{x-3}$ leads to a division by 0. We call the number 3 an excluded value of the expression.

Definition Excluded value

A number is an **excluded value** of a rational expression if substituting the number into the expression leads to a division by 0.

For instance, 5 is an excluded value of the expression $\dfrac{7}{x-5}$. This means 5 is *not* in the domain of the function $f(x) = \dfrac{7}{x-5}$.

WARNING

For the expression $\dfrac{7}{x-5}$, 0 itself is *not* an excluded value, because $\dfrac{7}{0-5} = \dfrac{7}{-5}$ does not contain a division by 0.

For the expression $\dfrac{x}{3}$, again 0 is not an excluded value, because $\dfrac{0}{3}$ does not have a division by 0. In fact, $\dfrac{0}{3}$ is defined and $\dfrac{0}{3} = 0$.

Simplifying a Rational Expression

In Sections 9.3 and 9.6, we simplified radical quotients by factoring the numerator and the denominator and then using the property $\dfrac{a}{a} = 1$, where $a \neq 0$. Here we use similar steps to simplify $\dfrac{15}{35}$:

$$\frac{15}{35} = \frac{5\cdot 3}{5\cdot 7} = \frac{5}{5}\cdot\frac{3}{7} = 1\cdot\frac{3}{7} = \frac{3}{7}$$

We simplify the rational expression $\frac{3x + 6}{x^2 - 4}$ in a similar manner:

$$\begin{aligned}
\frac{3x+6}{x^2-4} &= \frac{3(x+2)}{(x-2)(x+2)} && \text{Factor numerator and denominator.} \\
&= \frac{3}{x-2} \cdot \frac{x+2}{x+2} && \frac{AC}{BD} = \frac{A}{B} \cdot \frac{C}{D} \\
&= \frac{3}{x-2} \cdot 1 && \text{Simplify: } \frac{x+2}{x+2} = 1 \\
&= \frac{3}{x-2} && a \cdot 1 = a
\end{aligned}$$

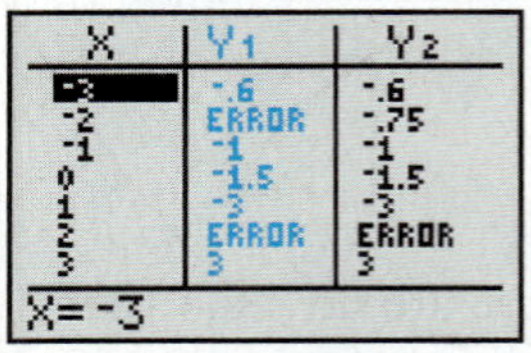

Figure 5 Verify the work

The number 2 is the only excluded value of our result, $\frac{3}{x-2}$. The numbers -2 and 2 are the only excluded values of the original expression, $\frac{3x+6}{x^2-4}$. (Try it.) When we write

$$\frac{3x+6}{x^2-4} = \frac{3}{x-2}$$

we mean that the two expressions yield the same number for each real number substituted for x, except for any excluded values of either expression (-2 and 2). In Fig. 5, we use a graphing calculator table to verify our work.

A rational expression is in **lowest terms** if the numerator and denominator have no common factors other than 1 or -1. We **simplify a rational expression** by writing it in lowest terms. We also write the numerator and the denominator in factored form.

Simplifying a Rational Expression

To simplify a rational expression,

1. Factor the numerator and the denominator.
2. Use the property

$$\frac{AB}{AC} = \frac{A}{A} \cdot \frac{B}{C} = 1 \cdot \frac{B}{C} = \frac{B}{C}$$

where A and C are nonzero, so that the expression is in lowest terms.

Throughout the rest of this chapter, you may assume that the form $\frac{A}{B}$ represents a rational expression.

For the expression $\frac{AB}{AC}$, note that the polynomial A is a factor of both the numerator and the denominator. The expression $\frac{2x}{5x}$ can be simplified to $\frac{2}{5}$ when $x \neq 0$, because x is a factor of both the numerator and the denominator.

WARNING The expression

$$\frac{2x+7}{5x}$$

is in lowest terms already. Although x is a factor of the term $2x$, it is not a factor of $2x + 7$, the (entire) *numerator*.

Likewise, the expression

$$\frac{(x+3)(x+5)+4}{7(x+3)}$$

is in lowest terms already. The expression $x + 3$ is not a factor of $(x + 3)(x + 5) + 4$, the (entire) numerator. To see that the rational expression is in lowest terms, we simplify the numerator to get $\frac{x^2 + 8x + 19}{7(x+3)}$. (Try it.)

Example 4 Simplifying Rational Expressions

Simplify.

1. $\dfrac{20x^2}{12x^4}$

2. $\dfrac{5t + 15}{t^2 + 5t + 6}$

Plot1 Plot2 Plot3
\Y1■(5X+15)/(X^2+5X+6)
\Y2■5/(X+2)
\Y3=
\Y4=
\Y5=
\Y6=

X	Y1	Y2
-3	ERROR	-5
-2	ERROR	ERROR
-1	5	5
0	2.5	2.5
1	1.6667	1.6667
2	1.25	1.25
3	1	1

X=-3

Figure 6 Verify the work

Solution

1. $$\frac{20x^2}{12x^4} = \frac{2\cdot 2\cdot 5\cdot x\cdot x}{2\cdot 2\cdot 3\cdot x\cdot x\cdot x\cdot x}$$ *Factor numerator and denominator.*

$$= \frac{5}{3x^2}$$ *Simplify:* $\frac{2\cdot 2\cdot x\cdot x}{2\cdot 2\cdot x\cdot x} = 1$

2. $$\frac{5t + 15}{t^2 + 5t + 6} = \frac{5(t + 3)}{(t + 2)(t + 3)}$$ *Factor numerator and denominator.*

$$= \frac{5}{t + 2}$$ *Simplify:* $\frac{t + 3}{t + 3} = 1$

We use a graphing calculator table to verify our work (see Fig. 6). There are "ERROR" messages across from $x = -3$ and $x = -2$ in the table, because both -3 and -2 are excluded values of the original expression and -2 is an excluded value of our result.

Example 5 Simplify the Right-Hand Side of an Equation

Simplify the right-hand side of the equation $f(x) = \dfrac{2x^2 - 6x - 20}{2x^2 - 50}$.

Solution

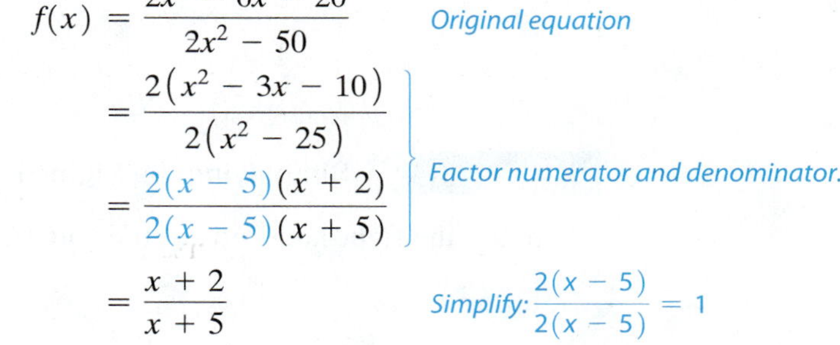

$$f(x) = \frac{2x^2 - 6x - 20}{2x^2 - 50}$$ *Original equation*

$$= \frac{2(x^2 - 3x - 10)}{2(x^2 - 25)}$$

$$= \frac{2(x - 5)(x + 2)}{2(x - 5)(x + 5)}$$ *Factor numerator and denominator.*

$$= \frac{x + 2}{x + 5}$$ *Simplify:* $\frac{2(x - 5)}{2(x - 5)} = 1$

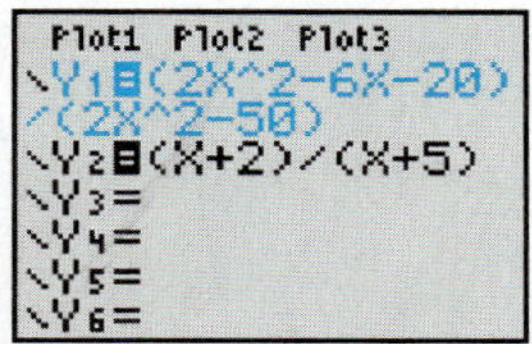

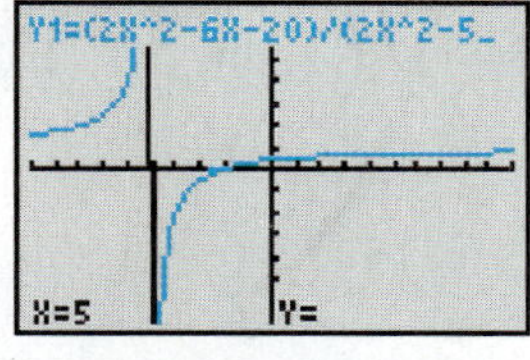

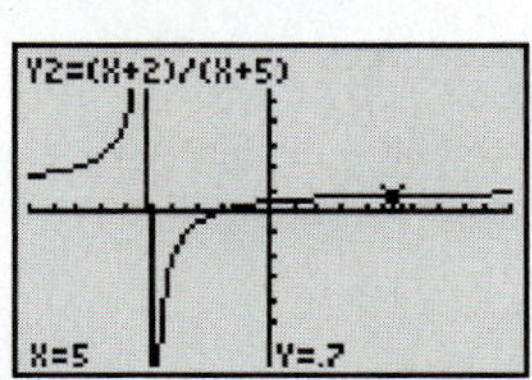

Figure 7 Verify the work

So, we can describe f by the equation $f(x) = \dfrac{x + 2}{x + 5}$, where the domain is all real numbers except -5 and 5.

To check, we compare graphing calculator graphs for the functions $y = \dfrac{2x^2 - 6x - 20}{2x^2 - 50}$ and $y = \dfrac{x + 2}{x + 5}$. See Fig. 7. To hand-sketch a graph of $f(x) = \dfrac{2x^2 - 6x - 20}{2x^2 - 50}$, we use an open circle at the point $(5, 0.7)$ to indicate that the point is not part of the graph.

The graphs of the functions are the same for $x \neq -5$ and $x \neq 5$.

Connection Between Vertical Asymptotes and Domains

Example 2 suggests that if the graph of a rational function has a vertical asymptote at $x = k$, then k is not in the domain of the function, which is true.

In Example 5, we wrote f in the form

$$f(x) = \frac{2(x - 5)(x + 2)}{2(x - 5)(x + 5)}$$

by factoring the numerator and the denominator of the right-hand side of the original equation of f. From this form, we see that the domain of f is the set of real numbers except

-5 and 5. Yet, $x = -5$ is the only vertical asymptote of the graph of f (see Fig. 7). The reason $x = 5$ is not a vertical asymptote has to do with the fact that both the numerator and the denominator of f contain the factor $x - 5$.

Vertical Asymptotes and Domains of Rational Functions

If the graph of a rational function f has a vertical asymptote $x = k$, then k is not in the domain of f. If k is not in the domain of a rational function g, then $x = k$ may or may not be a vertical asymptote of the graph of g.

Example 6 Simplifying the Right-Hand Side of an Equation

Simplify the right-hand side of the equation $f(x) = \dfrac{x^3 - 5x^2 + 6x}{x^3 - 2x^2 - 9x + 18}$.

Solution

$$\begin{aligned}
f(x) &= \frac{x^3 - 5x^2 + 6x}{x^3 - 2x^2 - 9x + 18} && \textit{Original equation} \\
&= \frac{x(x^2 - 5x + 6)}{x^2(x-2) - 9(x-2)} && \\
&= \frac{x(x-3)(x-2)}{(x^2-9)(x-2)} && \textit{Factor numerator and denominator.} \\
&= \frac{x(x-3)(x-2)}{(x+3)(x-3)(x-2)} && \\
&= \frac{x}{x+3} && \textit{Simplify: } \frac{(x-3)(x-2)}{(x-3)(x-2)} = 1
\end{aligned}$$

Example 7 Simplifying the Right-Hand Side of an Equation

Simplify the right-hand side of the equation $f(x) = \dfrac{x^2 - x - 12}{4 - x}$.

Solution

$$\begin{aligned}
f(x) &= \frac{x^2 - x - 12}{4 - x} && \textit{Original equation} \\
&= \frac{x^2 - x - 12}{-x + 4} && \textit{Write denominator in descending order.} \\
&= \frac{(x-4)(x+3)}{-1(x-4)} && \textit{Factor numerator and denominator.} \\
&= \frac{x+3}{-1} && \textit{Simplify: } \frac{x-4}{x-4} = 1 \\
&= -x - 3 && \frac{x+3}{-1} = -\frac{x+3}{1} = -(x+3) = -x-3
\end{aligned}$$

In Example 8, we will simplify a rational expression in two variables.

Example 8 Simplifying a Rational Expression in Two Variables

Simplify $\dfrac{x^2 - 8xy + 16y^2}{x^2 - 16y^2}$.

Solution

$$\frac{x^2 - 8xy + 16y^2}{x^2 - 16y^2} = \frac{(x-4y)(x-4y)}{(x+4y)(x-4y)} \quad \text{Factor numerator and denominator.}$$

$$= \frac{x-4y}{x+4y} \quad \text{Simplify: } \frac{x-4y}{x-4y} = 1$$

Quotient Function

In Sections 7.1 and 7.2, we worked with sum, difference, and product functions. Now we will work with *quotient functions.*

Definition Quotient function

If f and g are functions, x is in the domain of both functions, and $g(x)$ is nonzero, then we can form the **quotient function** $\frac{f}{g}$:

$$\left(\frac{f}{g}\right)(x) = \frac{f(x)}{g(x)}$$

For example, if $f(x) = 3^x$ and $g(x) = x + 5$, then $\left(\frac{f}{g}\right)(x) = \frac{f(x)}{g(x)} = \frac{3^x}{x+5}$.

Example 9 Quotient Functions

Let $f(x) = 8x^3 - 125$ and $g(x) = 4x^2 - 25$.

1. Find an equation of $\frac{f}{g}$. Simplify the right-hand side of the equation.
2. Find $\left(\frac{f}{g}\right)(3)$.

Solution

1. $$\left(\frac{f}{g}\right)(x) = \frac{f(x)}{g(x)} \quad \text{Definition of quotient function}$$

$$= \frac{8x^3 - 125}{4x^2 - 25} \quad \text{Substitute } 8x^3 - 125 \text{ for } f(x) \text{ and } 4x^2 - 25 \text{ for } g(x).$$

$$= \frac{(2x-5)(4x^2+10x+25)}{(2x-5)(2x+5)} \quad \text{Factor numerator and denominator.}$$

$$= \frac{4x^2+10x+25}{2x+5} \quad \text{Simplify: } \frac{2x-5}{2x-5} = 1$$

2. $$\left(\frac{f}{g}\right)(3) = \frac{4(3)^2 + 10(3) + 25}{2(3) + 5} = \frac{91}{11}$$

By the definition of a rational function, we see that if P and Q are polynomial functions, then the quotient function $\frac{P}{Q}$ is rational.

Use a Rational Model to Describe the Percentage of a Quantity

A **rational model** is a rational function, or its graph, that describes an authentic situation. In this section, we will focus on using a rational model to describe the percentage of a quantity.

Suppose a student takes a quiz and answers 6 out of 8 questions correctly. To compute the percentage of questions answered correctly, we divide the number of questions answered correctly by the total number of questions and multiply the result by 100:

$$\text{Percentage of correct answers} = \frac{6}{8} \cdot 100\% = 75\%$$

We take similar steps to calculate a percentage in general.

Percentage Formula

If m items out of n items have a certain attribute, then the percentage p (written $p\%$) of the n items that have the attribute is

$$p = \frac{m}{n} \cdot 100$$

We call this equation the **percentage formula.**

We will use the percentage formula to form a rational model in Example 10.

Table 2 Numbers of Internet Users in the United States

Year	Number of Internet Users (millions)
2006	204
2007	212
2008	220
2009	228
2010	240

Source: *The Nielsen Company*

Isn't the Internet great? You can shop without leaving home. I used it to send all of my gifts this year.
That's nothing–last week I went on a virtual vacation and never had to board a plane!

Example 10 Using a Rational Model to Make a Prediction

In Exercise 7 of Homework 1.4, you modeled the number of Internet users in the United States (see Table 2). A reasonable model is $I(t) = 8.8t + 150.4$, where $I(t)$ is the number of Internet users (in millions) at t years since 2000. In Exercise 13 of Homework 9.8, you modeled the U.S. population. A reasonable model is $U(t) = 0.0068t^2 + 2.71t + 277.89$, where $U(t)$ is the U.S. population (in millions) at t years since 2000.

1. Let $P(t)$ be the percentage of Americans who are Internet users at t years since 2000. Find an equation of P.
2. Use P to estimate the percentage of Americans who were Internet users in 2010. Then compute the actual percentage by referring to Table 2 and using the 2010 U.S. population as 308.7 million. Is your result from using the model an underestimate or an overestimate?
3. Find $P(18)$. What does it mean in this situation?

Solution

1. To find the percentage of Americans who are Internet users, we divide the number of Internet users by the number of Americans and multiply the result by 100:

$$P(t) = \frac{I(t)}{U(t)} \cdot 100 \qquad \textit{Percentage formula: } p = \frac{m}{n} \cdot 100$$

$$= \frac{8.8t + 150.4}{0.0068t^2 + 2.71t + 277.89} \cdot \frac{100}{1} \qquad \textit{Substitute } 8.8t + 150.4 \textit{ for } I(t) \textit{ and } 0.0068t^2 + 2.71t + 277.89 \textit{ for } U(t).$$

$$= \frac{880t + 15{,}040}{0.0068t^2 + 2.71t + 277.89} \qquad \textit{Multiply numerators; multiply denominators.}$$

2. $P(10) = \dfrac{880(10) + 15{,}040}{0.0068(10)^2 + 2.71(10) + 277.89} \approx 77.99$

The model estimates that about 78.0% of Americans were Internet users in 2010. To find the actual percentage, we divide the number of Internet users in 2010 (from Table 2) by the U.S. population (given) and multiply the result by 100:

$$\frac{240{,}000{,}000}{308{,}700{,}000} \cdot 100\% \approx 77.75\%$$

Our result from using the model is a slight overestimate.

3. $P(18) = \dfrac{880(18) + 15{,}040}{0.0068(18)^2 + 2.71(18) + 277.89} \approx 93.90$

The model estimates that about 93.9% of Americans will be Internet users in 2018.

Group Exploration

Connection between the domain and vertical asymptotes

1. Find the domain of $f(x) = \dfrac{6}{x}$. Use ZStandard to graph f. Explain why it makes sense that the graph does not have a y-intercept. Find any vertical asymptotes.

2. Use ZStandard to graph $f(x) = \dfrac{6}{x}$ and $g(x) = \dfrac{6}{x-4}$ on the same viewing screen. Describe how you can translate the graph of f to get the graph of g.

3. What is the vertical asymptote of the graph of $g(x) = \dfrac{6}{x-4}$? What do you observe about the connection between the vertical asymptote and the domain of g?

4. Use ZStandard to graph $h(x) = \dfrac{7x-7}{x^2+2x-8}$. What is (are) the vertical asymptote(s)? What is the connection between the vertical asymptote(s) and the domain of h?

5. Find the domain of the function $f(x) = \dfrac{x-2}{2-x}$. Use ZDecimal to graph f. Explain why the graph is so different from the graphs you drew in Problems 1, 2, and 4. [**Hint:** Simplify the right-hand side of the equation of f.] Describe what happens when you use TRACE to try to find the value of y when $x = 2$. Explain.

6. Describe the connection between the vertical asymptote(s) of the graph of the function $g(x) = \dfrac{5x-10}{x^2-7x+10}$ and the domain of g. [**Hint:** Simplify the right-hand side of the equation of g.]

7. True or False? Explain.

a. If the graph of a rational function f has a vertical asymptote $x = k$, then k is not in the domain of f.

b. If k is not in the domain of a rational function g, then $x = k$ is a vertical asymptote of the graph of g.

Group Exploration

Simplifying rational expressions

We can write $\dfrac{2x}{5x} = \dfrac{2}{5}$ because x is a common factor of the numerator and the denominator. For each problem, use this concept to help you determine whether the work is correct. If you are unsure, it may help to decide whether you can show work similar to the following:

$$\frac{2x}{5x} = \frac{2}{5} \cdot \frac{x}{x} = \frac{2}{5} \cdot 1 = \frac{2}{5}$$

1. a. $\dfrac{4x}{9x} = \dfrac{4}{9}$ **b.** $\dfrac{4+x}{9x} = \dfrac{4+1}{9} = \dfrac{5}{9}$

2. a. $\dfrac{3x \cdot 5}{7x} = \dfrac{3 \cdot 5}{7} = \dfrac{15}{7}$ **b.** $\dfrac{3x+5}{7x} = \dfrac{3+5}{7} = \dfrac{8}{7}$

3. a. $\dfrac{8+x^5}{6+x^2} = \dfrac{4+x^3}{3}$ **b.** $\dfrac{8x^5}{6x^2} = \dfrac{4x^3}{3}$

4. a. $\dfrac{(x+2)(x-4)}{5(x-4)} = \dfrac{x+2}{5}$

b. $\dfrac{(x+2)(x-4)+1}{5(x-4)} = \dfrac{(x+2)+1}{5} = \dfrac{x+3}{5}$

Homework 12.1

For extra help ▶ MyMathLab®  Watch the videos in MyMathLab  Download the MyDashboard App

Evaluate the function at the indicated values if possible. If an indicated value is not in the domain, say so.

1. $f(x) = \dfrac{x+1}{x^2-9}; f(-1), f(2), f(3)$

2. $f(x) = \dfrac{x-2}{x^2-1}; f(-3), f(1), f(2)$

3. $f(x) = \dfrac{x^3-8}{2x^2+3x-1}; f(-1), f(0), f(3)$

4. $f(x) = \dfrac{x^3+5}{3x^2-x-6}; f(-2), f(0), f(1)$

Find the domain of the function. Verify your result with a graphing calculator table or graph.

5. $f(x) = \dfrac{8}{x}$

6. $f(x) = \dfrac{9}{x}$

7. $f(x) = \dfrac{x}{2}$

8. $f(x) = \dfrac{x}{5}$

9. $f(x) = \dfrac{x-5}{x+3}$

10. $f(x) = \dfrac{x+4}{x-9}$

11. $f(x) = \dfrac{x-3}{2x+1}$

12. $f(x) = \dfrac{x+4}{5x-7}$

13. $f(x) = \dfrac{x-9}{x^2-3x-10}$

14. $f(x) = \dfrac{x-1}{x^2+5x-24}$

15. $f(x) = \dfrac{3x-1}{x^2-16}$

16. $f(x) = \dfrac{x+8}{x^2-25}$

17. $f(x) = -\dfrac{2}{4x^2-25}$

18. $f(x) = -\dfrac{7}{81x^2-49}$

19. $f(x) = \dfrac{x+3}{x^2+1}$

20. $f(x) = \dfrac{x+8}{x^2+4}$

21. $f(x) = \dfrac{x-10}{2x^2-7x-15}$

22. $f(x) = \dfrac{x-3}{6x^2+7x-20}$

23. $f(x) = -\dfrac{x+3}{x^2-3x+6}$

24. $f(x) = -\dfrac{x+6}{3x^2-x+4}$

25. $f(x) = \dfrac{x^2+5x-1}{3x^2-2x-7}$

26. $f(x) = \dfrac{x^2-x+3}{6x^2+4x-1}$

27. $f(x) = \dfrac{2x-14}{4x^3-8x^2-9x+18}$

28. $f(x) = \dfrac{5x+10}{x^3+5x^2-4x-20}$

Simplify the right-hand side of the equation of f. Use a graphing calculator table to verify your work.

29. $f(x) = \dfrac{4x}{6}$

30. $f(x) = \dfrac{12}{15x}$

31. $f(x) = \dfrac{20x^7}{15x^4}$

32. $f(x) = \dfrac{12x^3}{16x^{10}}$

33. $f(x) = \dfrac{4x-28}{5x-35}$

34. $f(x) = \dfrac{-4x-20}{6x+30}$

35. $f(x) = \dfrac{x^2+7x+10}{x^2-7x-18}$

36. $f(x) = \dfrac{x^2+8x+12}{x^2+9x+18}$

37. $f(x) = \dfrac{x^2-49}{x^2-14x+49}$

38. $f(x) = \dfrac{x^2-25}{x^2+10x+25}$

39. $f(x) = \dfrac{16x^2-25}{8x^2-22x+15}$

40. $f(x) = \dfrac{9x^2-4}{3x^2-17x+10}$

41. $f(x) = \dfrac{x-5}{5-x}$

42. $f(x) = \dfrac{2-x}{x-2}$

43. $f(x) = \dfrac{4x-12}{18-6x}$

44. $f(x) = \dfrac{-2x+10}{4x-20}$

45. $f(x) = \dfrac{6x-18}{9-x^2}$

46. $f(x) = \dfrac{9-x}{x^2-81}$

47. $f(x) = \dfrac{x^3+4x^2}{7x^2+28x}$

48. $f(x) = \dfrac{4x^2-24x}{x^3-6x^2}$

49. $f(x) = \dfrac{x^2+2x-35}{-x^2+3x+10}$

50. $f(x) = \dfrac{x^2+x-20}{-x^2+11x-28}$

51. $f(x) = \dfrac{3x^3+21x^2+36x}{x^2-9}$

52. $f(x) = \dfrac{x^2-49}{2x^3-8x^2-42x}$

53. $f(x) = \dfrac{3x^2+9x+6}{6x^2+5x-1}$

54. $f(x) = \dfrac{2x^2+6x-20}{4x^2-5x-6}$

55. $f(x) = \dfrac{x^2-2x-8}{4x^3+8x^2-9x-18}$

56. $f(x) = \dfrac{25x^3-75x^2-4x+12}{x^2-4x+3}$

57. $f(x) = \dfrac{x^3+8}{x^2-4}$

58. $f(x) = \dfrac{x^3+27}{x^2-9}$

59. $f(x) = \dfrac{3x^2+7x-6}{27x^3-8}$

60. $f(x) = \dfrac{4x^2-11x+6}{64x^3-27}$

Simplify.

61. $\dfrac{18x^3y}{27x^2y^4}$

62. $\dfrac{35x^2y^3}{25x^5y}$

63. $\dfrac{x^2-6xy+9y^2}{x^2-3xy}$

64. $\dfrac{x^2+10xy+25y^2}{xy+5y^2}$

65. $\dfrac{6a^2+ab-2b^2}{3a^2-7ab-6b^2}$

66. $\dfrac{2p^2-9pt+10t^2}{8p^2-18pt-5t^2}$

67. $\dfrac{p^3-q^3}{p^2-q^2}$

68. $\dfrac{mn+n^2}{m^3+n^3}$

For the functions $f(x) = x^2+2x-8$, $g(x) = x^2-8x+12$, $h(x) = 3x^2+17x+20$, and $k(x) = 3x^2-13x-30$, find an equation of the given quotient function. Evaluate the quotient function at the indicated value.

69. $\dfrac{f}{g}, \left(\dfrac{f}{g}\right)(3)$

70. $\dfrac{g}{k}, \left(\dfrac{g}{k}\right)(2)$

71. $\dfrac{h}{f}, \left(\dfrac{h}{f}\right)(4)$

72. $\dfrac{k}{h}, \left(\dfrac{k}{h}\right)(3)$

For $f(x) = 3x^3 - x^2$, $g(x) = 18x^3 + 12x^2 + 2x$, $h(x) = 9x^2 - 1$, and $k(x) = 27x^3 + 1$, find an equation of the given quotient function. Evaluate the quotient function at the indicated value.

73. $\frac{f}{h}, \left(\frac{f}{h}\right)(-2)$

74. $\frac{g}{f}, \left(\frac{g}{f}\right)(-1)$

75. $\frac{k}{g}, \left(\frac{k}{g}\right)(-1)$

76. $\frac{h}{k}, \left(\frac{h}{k}\right)(0)$

77. Levels of participation in the Supplemental Nutrition Assistance Program (SNAP) in August are shown in Table 3 for various years.

Table 3 Levels of Participation in August in SNAP

Year	Participation (millions)
2008	29.2
2009	36.2
2010	42.4
2011	45.8
2012	47.0

Source: *U.S. Department of Agriculture*

Let $F(t)$ be the participation (in millions) in August of the year that is t years since 2000. A reasonable model is $F(t) = -1.03t^2 + 25.09t - 105.88$. In Exercise 13 of Homework 9.8, you modeled the U.S. population. A reasonable model is $U(t) = 0.0068t^2 + 2.71t + 277.9$, where $U(t)$ is the U.S. population (in millions) at t years since 2000.

a. Let $P(t)$ be the percentage of Americans who participated in SNAP in August of the year that is t years since 2000. Find an equation of P.

b. Use P to estimate the percentage of Americans who participated in SNAP in August of 2012. Then compute the actual percentage by referring to Table 3 and using the 2012 U.S. population as 314.5 million. Is your result from using the model an underestimate or an overestimate?

c. Find $P(16)$. What does it mean in this situation?

78. In Exercise 67 of Homework 11.1, you worked with the model $D(t) = 0.34t + 0.56$, where $D(t)$ is the annual revenue (in billions of dollars) from downloaded music at t years since 2005 (see Table 4).

Table 4 Annual Revenues from Downloaded Music and All Music

	Annual Revenue (billions of dollars)	
Year	Downloads	All Music
2006	0.86	10.25
2007	1.23	8.70
2008	1.67	7.15
2009	1.98	6.26
2010	2.20	5.57

Source: *Recording Industry Association of America*

The annual revenue (in billions of dollars) from all music (downloaded and CDs) $A(t)$ is given by $A(t) = -1.18t + 11.13$, where t is the number of years since 2005.

a. Let $P(t)$ be the percentage of the annual revenue from all music that is from downloaded music at t years since 2005. Find an equation of P.

b. Use P to estimate the percentage of the annual revenue from all music that was from downloaded music in 2010. Then compute the actual percentage. Is your result an underestimate or an overestimate?

c. Find $P(6)$. What does it mean in this situation?

79. The numbers of Latinos who are registered to vote and the numbers of them who are eligible to vote are shown in Table 5.

Table 5 Numbers of Latinos Who Are Registered to Vote and Eligible to Vote

	Millions of Latinos	
Year	Registered to Vote	Eligible to Vote
1988	4.6	7.7
1992	5.1	8.3
1996	6.6	11.2
2000	7.5	13.2
2004	9.3	16.1
2008	11.6	19.5
2010	11.0	21.3

Source: *Pew Hispanic Center*

a. Let $R(t)$ be the number (in millions) of Latinos who are registered to vote at t years since 1980. Find a quadratic equation of R.

b. Let $E(t)$ be the number (in millions) of Latinos who are eligible to vote at t years since 1980. Find a quadratic equation of E.

c. Let $P(t)$ be the percentage of voter-eligible Latinos who are registered to vote at t years since 1980. Find an equation of P.

d. Find the percentage of voter-eligible Latinos who will be registered to vote in 2017.

e. Use the window settings in Fig. 8 to draw a graph of P. Then use "maximum" to find the maximum point of the graph of P for values of t between 0 and 40. What does it mean in this situation?

Figure 8 Window settings for Exercise 79e

f. Has the percentage of voter-eligible Latinos who are registered to vote been increasing, decreasing, or neither since 1996? How is this possible, given that the number of Latinos who were registered to vote grew by over 75% from 1996 to 2008?

80. The enrollments of men and of all students at U.S. colleges are shown in Table 6 for various years. In Example 11 of Section 7.1, we found the equation $M(t) = 0.012t^2 - 0.12t + 6.65$, where $M(t)$ is the enrollment (in millions) of men at t years since 1990.

Table 6 College Enrollments

Year	Enrollment (millions) Men	All Students
1995	6.3	14.2
2000	6.7	15.3
2005	7.5	17.5
2008	8.2	19.1
2010	9.1	21.2

Source: *National Center for Education Statistics*

a. Let $E(t)$ be the enrollment (in millions) of all college students at t years since 1990. Find an equation of E.

b. Let $P(t)$ be the percentage of college students who are men at t years since 1990. Find an equation of P.

c. Find the percentage of college students who will be men in 2017.

d. Use the window settings in Fig. 9 to draw a graph of P. Is P increasing, decreasing, or neither for the values of t between 5 and 20? What does that mean in this situation? How is this possible, given that the college enrollment of men has been increasing since 1995?

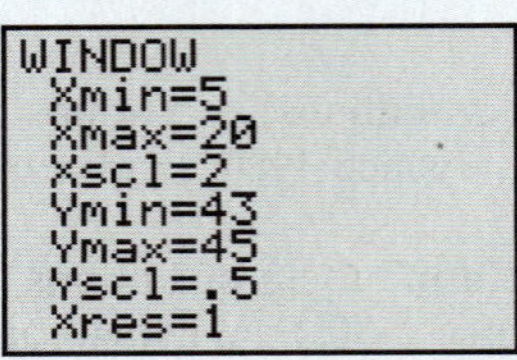

Figure 9 Window settings for Exercise 80d

Concepts

81. Give examples of three rational functions, each of whose domains are the set of real numbers except -3 and 3.

82. Give examples of three rational functions, each of whose domains are the set of real numbers except $\frac{1}{2}$ and $\frac{1}{3}$.

83. A graphing calculator table for a rational function is shown in Fig. 10. List seven members of the domain. List seven members of the range.

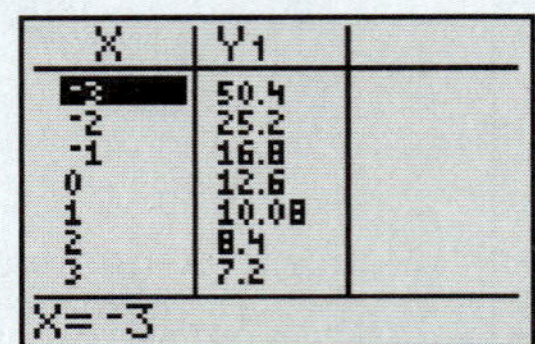

Figure 10 Exercise 83

84. A student decides that the domain of the function

$$f(x) = \frac{6}{x(x-3) + 2(x-3)}$$

is the set of real numbers except 3. After viewing the table shown in Fig. 11, the student believes his work is correct. What would you tell this student?

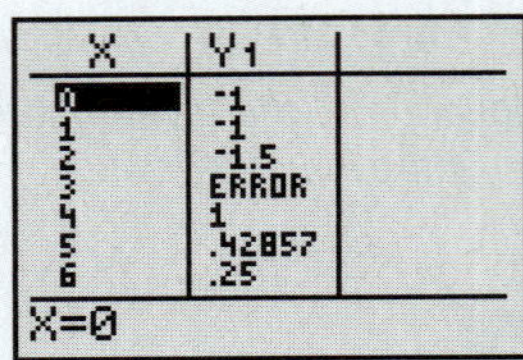

Figure 11 Exercise 84

85. A student states that the domain of the function

$$f(x) = \frac{(x-2)(x-4)}{(x-5)(x-1)}$$

is the set of real numbers except 1, 2, 4, and 5. Is the student correct? Explain.

86. It is a common error to think 0 is not in the domain of the function $f(x) = \frac{7}{x+4}$. Evaluate f at 0 to show that 0 is in the domain of f.

87. A student tries to simplify the expression $\frac{2(x+4)+3}{(x+4)(x-1)}$:

$$\frac{2(x+4)+3}{(x+4)(x-1)} = \frac{2+3}{x-1} = \frac{5}{x-1}$$

Substitute 3 for x in the original expression and in the student's result. What can you conclude about the student's work, based on the results of your substitutions? Explain.

88. A student tries to simplify the expression $\frac{(x-2)(x-1)}{4(x-2)+5}$:

$$\frac{(x-2)(x-1)}{4(x-2)+5} = \frac{x-1}{4+5} = \frac{x-1}{9}$$

Next, she substitutes 3 for x in the original expression and in her result:

$$\frac{(3-2)(3-1)}{4(3-2)+5} = \frac{2}{9}; \quad \frac{3-1}{9} = \frac{2}{9}$$

The student concludes that the result $\frac{x-1}{9}$ is correct. What would you tell her?

89. A person plans to drive at a constant speed from San Francisco to Los Angeles. The driving time (in hours) $T(s)$ is given by the equation

$$T(s) = \frac{420}{s}$$

where s is the constant speed (in miles per hour).

a. Find $T(50)$. What does it mean in this situation?

b. Find $T(55)$, $T(60)$, $T(65)$, and $T(70)$.

c. Is T an increasing function or a decreasing function for $s > 0$? Explain why that makes sense in this situation. Use ZStandard and then Zoom Out once or twice to verify your answer.

90. Some students agree to share equally in the expense of renting a beach house for \$1200 during spring break.

a. What is the per-student expense if 10 students rent it?

b. What is the per-student expense if 12 students rent it?

c. Let $p(n)$ be the per-student expense (in dollars) for n students to rent the house. Find an equation of p. [**Hint:** Reflect on your work in parts (a) and (b).]

d. Use ZStandard and then Zoom Out twice to draw a graph of p. Is p a decreasing function or an increasing function for $n > 0$? Explain why that makes sense in this situation.

e. Find $p(15)$. What does it mean in this situation?

91. Describe how to find the domain of a rational function. If you believe a number is not in the domain, describe how you can verify that belief. (See page 9 for guidelines on writing a good response.)

92. Describe how to simplify a rational expression. Explain how you can check that the result and the original expression are equivalent. (See page 9 for guidelines on writing a good response.)

Related Review

For Exercises 93–96, find the domain.

93. $f(x) = 4x - 8$

94. $f(x) = 3(2)^x$

95. $f(x) = 2(x - 4)^2 - 5$

96. $f(x) = \dfrac{x - 3}{x^2 - 49}$

97. a. Use factoring to simplify $\dfrac{x^2 + 5x + 6}{x + 3}$.

b. Use long division or synthetic division to simplify $\dfrac{x^2 + 5x + 6}{x + 3}$.

c. Which method from parts (a) and (b) is easier to use to simplify $\dfrac{x^2 + 5x + 6}{x + 3}$? Explain.

98. a. Use factoring to simplify $\dfrac{4x^3 + 8x^2 - 9x - 18}{2x^2 + x - 6}$.

b. Use long division to simplify $\dfrac{4x^3 + 8x^2 - 9x - 18}{2x^2 + x - 6}$.

c. Which method from parts (a) and (b) is easier to use to simplify $\dfrac{4x^3 + 8x^2 - 9x - 18}{2x^2 + x - 6}$? Explain.

Expressions, Equations, Functions, and Graphs

Perform the indicated instruction. Then use words such as linear, quadratic, cubic, exponential, logarithmic, rational, polynomial, degree, function, one variable, *and* two variables *to describe the expression, equation, or system.*

99. Find the domain of $f(x) = \dfrac{x - 5}{x^2 - 4x - 21}$.

100. Solve $-2x^2 - 4x + 3 = 0$.

101. Solve $3x^3 + 5x^2 = 12x + 20$.

102. Graph $f(x) = -2x^2 - 4x + 3$ by hand.

103. Factor $8x^3 - 125$.

104. Find an equation of a parabola that contains the points $(2, 6), (3, 1)$, and $(4, -8)$.

12.2 Multiplying and Dividing Rational Expressions; Converting Units

Objectives

- » Multiply and divide rational expressions.
- » Convert units of quantities.

In Section 12.1, we simplified rational expressions. In this section, we will multiply and divide rational expressions.

Multiplication of Rational Expressions

Recall from Section 2.2 that we multiply fractions by using the property $\dfrac{a}{b} \cdot \dfrac{c}{d} = \dfrac{ac}{bd}$, where b and d are nonzero. For example,

$$\frac{2}{5} \cdot \frac{3}{7} = \frac{2 \cdot 3}{5 \cdot 7} = \frac{6}{35}$$

We multiply more complicated rational expressions in a similar way.

Multiplying Rational Expressions

If $\dfrac{A}{B}$ and $\dfrac{C}{D}$ are rational expressions and B and D are nonzero, then

$$\frac{A}{B} \cdot \frac{C}{D} = \frac{AC}{BD}$$

In words: To multiply two rational expressions, write the numerators as a product and write the denominators as a product.

Example 1 Finding the Product of Two Rational Expressions

Find the product $\dfrac{4x}{5} \cdot \dfrac{7}{6x^4}$. Simplify the result.

Solution

$$\frac{4x}{5} \cdot \frac{7}{6x^4} = \frac{4x \cdot 7}{5 \cdot 6x^4}$$ *Write numerators as a product and denominators as a product:* $\frac{A}{B} \cdot \frac{C}{D} = \frac{AC}{BD}$

$$= \frac{2 \cdot 2 \cdot x \cdot 7}{5 \cdot 2 \cdot 3 \cdot x \cdot x \cdot x \cdot x}$$ *Factor numerator and denominator.*

$$= \frac{14}{15x^3}$$ *Simplify:* $\frac{2 \cdot x}{2 \cdot x} = 1$

To find the product of rational expressions, we usually begin by factoring the numerators and the denominators if possible. That will put us in a good position to simplify after we have found the product.

Example 2 Finding the Product of Two Rational Expressions

Find the product $\frac{2x + 6}{3} \cdot \frac{6x}{5x + 15}$. Simplify the result.

Solution

$$\frac{2x + 6}{3} \cdot \frac{6x}{5x + 15} = \frac{2(x + 3)}{3} \cdot \frac{2 \cdot 3 \cdot x}{5(x + 3)}$$ *Factor numerators and denominators.*

$$= \frac{2(x + 3) \cdot 2 \cdot 3 \cdot x}{3 \cdot 5(x + 3)}$$ *Write both numerators and denominators as a product:* $\frac{A}{B} \cdot \frac{C}{D} = \frac{AC}{BD}$

$$= \frac{4x}{5}$$ *Simplify:* $\frac{(x + 3) \cdot 3}{(x + 3) \cdot 3} = 1$

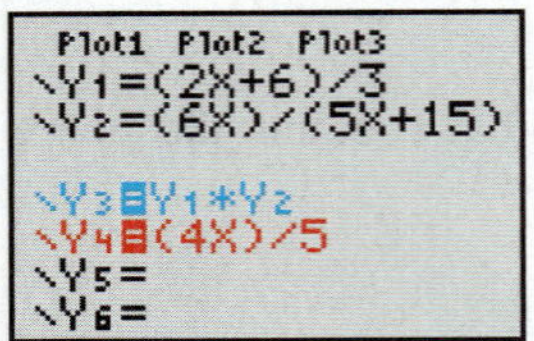

X	Y3	Y4
-3	ERROR	-2.4
-2	-1.6	-1.6
-1	-.8	-.8
0	0	0
1	.8	.8
2	1.6	1.6
3	2.4	2.4

X=-3

Figure 12 Verify the work

We verify our work by creating a graphing calculator table for $y = \frac{2x + 6}{3} \cdot \frac{6x}{5x + 15}$ and $y = \frac{4x}{5}$. See Fig. 12. To enter an equation by using Y_n references, see Appendix A.25. There is an "ERROR" message across from $x = -3$ in the table, because -3 is an excluded value of the original expression.

Example 3 Finding the Product of Two Rational Expressions

Find the product $\frac{x^2 - 49}{5x - 10} \cdot \frac{x^2 - 2x}{x^2 - 9x + 14}$. Simplify the result.

Solution

$$\frac{x^2 - 49}{5x - 10} \cdot \frac{x^2 - 2x}{x^2 - 9x + 14} = \frac{(x - 7)(x + 7)}{5(x - 2)} \cdot \frac{x(x - 2)}{(x - 7)(x - 2)}$$ *Factor numerators and denominators.*

$$= \frac{(x - 7)(x + 7) \cdot x(x - 2)}{5(x - 2)(x - 7)(x - 2)}$$ *Multiply numerators; multiply denominators.*

$$= \frac{x(x + 7)}{5(x - 2)}$$ *Simplify:* $\frac{(x - 7)(x - 2)}{(x - 7)(x - 2)} = 1$

How to Multiply Rational Expressions

To multiply two rational expressions,

1. Factor the numerators and the denominators.
2. Multiply by using the property $\frac{A}{B} \cdot \frac{C}{D} = \frac{AC}{BD}$, where B and D are nonzero.
3. Simplify the result.

Example 4 Finding the Product of Two Rational Expressions

Find the product $\dfrac{4x^2-9}{2x-10}\cdot\dfrac{x^2-10x+25}{2x^2-7x+6}$. Simplify the result.

Solution

$$\begin{aligned}\frac{4x^2-9}{2x-10}\cdot\frac{x^2-10x+25}{2x^2-7x+6} &= \frac{(2x-3)(2x+3)}{2(x-5)}\cdot\frac{(x-5)(x-5)}{(2x-3)(x-2)} && \text{Factor numerators and denominators.}\\ &= \frac{(2x-3)(2x+3)(x-5)(x-5)}{2(x-5)(2x-3)(x-2)} && \text{Multiply numerators; multiply denominators.}\\ &= \frac{(x-5)(2x+3)}{2(x-2)} && \text{Simplify: } \frac{(2x-3)(x-5)}{(2x-3)(x-5)}=1\end{aligned}$$

Example 5 Finding a Product Function

Let $f(x)=\dfrac{35x^2-25x}{x^2-36}$ and $g(x)=\dfrac{6-x}{15x^4}$.

1. Find an equation of the product function $f\cdot g$.
2. Find $(f\cdot g)(2)$.

Solution

1.
$$\begin{aligned}(f\cdot g)(x) &= f(x)\cdot g(x) && \text{Definition of } f\cdot g\\ &= \frac{35x^2-25x}{x^2-36}\cdot\frac{6-x}{15x^4} && \text{Substitute } \frac{35x^2-25x}{x^2-36} \text{ for } f(x) \text{ and } \frac{6-x}{15x^4} \text{ for } g(x).\\ &= \frac{5x(7x-5)}{(x-6)(x+6)}\cdot\frac{-(x-6)}{3\cdot 5x^4} && \text{Factor numerators and denominators; } 6-x=-x+6=-(x-6)\\ &= -\frac{5x(7x-5)(x-6)}{(x-6)(x+6)\cdot 3\cdot 5\cdot x^4} && \text{Multiply numerators; multiply denominators; } \frac{-a}{b}=-\frac{a}{b}\\ &= -\frac{7x-5}{3x^3(x+6)} && \text{Simplify: } \frac{5x(x-6)}{5x(x-6)}=1\end{aligned}$$

2. $(f\cdot g)(2) = -\dfrac{7(2)-5}{3(2)^3(2+6)} = -\dfrac{9}{192} = -\dfrac{3}{64}$

Division of Rational Expressions

Recall from Section 2.2 that, to divide two fractions, we use the property $\dfrac{a}{b}\div\dfrac{c}{d}=\dfrac{a}{b}\cdot\dfrac{d}{c}$, where b, c, and d are nonzero. For example,

$$\frac{2}{5}\div\frac{3}{7}=\frac{2}{5}\cdot\frac{7}{3}=\frac{14}{15}$$

Note that dividing by $\dfrac{3}{7}$ is the same as multiplying by the reciprocal of $\dfrac{3}{7}$, which is $\dfrac{7}{3}$. We divide more complicated rational expressions in a similar way.

Dividing Rational Expressions

If $\frac{A}{B}$ and $\frac{C}{D}$ are rational expressions and B, C, and D are nonzero, then

$$\frac{A}{B} \div \frac{C}{D} = \frac{A}{B} \cdot \frac{D}{C}$$

In words: To divide by a rational expression, multiply by its reciprocal.

Example 6 Finding the Quotient of Two Rational Expressions

Find the quotient $\frac{15x^3}{2} \div \frac{35x^7}{4}$.

Solution

$$\frac{15x^3}{2} \div \frac{35x^7}{4} = \frac{15x^3}{2} \cdot \frac{4}{35x^7}$$ *Multiply by reciprocal of $\frac{35x^7}{4}$, which is $\frac{4}{35x^7}$.*

$$= \frac{3 \cdot 5x^3}{2} \cdot \frac{2 \cdot 2}{5 \cdot 7x^7}$$ *Factor numerators and denominators.*

$$= \frac{3 \cdot 5 \cdot x^3 \cdot 2 \cdot 2}{2 \cdot 5 \cdot 7 \cdot x^7}$$ *Multiply numerators; multiply denominators.*

$$= \frac{6}{7x^4}$$ *Simplify: $\frac{2 \cdot 5 \cdot x^3}{2 \cdot 5 \cdot x^3} = 1$*

How to Divide Rational Expressions

To divide two rational expressions,

1. Write the quotient as a product by using the property $\frac{A}{B} \div \frac{C}{D} = \frac{A}{B} \cdot \frac{D}{C}$, where B, C, and D are nonzero.
2. Find the product.
3. Simplify.

Example 7 Finding the Quotient of Two Rational Expressions

Find the quotient.

1. $\frac{x+1}{x^2-x-6} \div \frac{3x+3}{x^2-9}$

2. $\frac{x^3-y^3}{9x^2-y^2} \div \frac{x^3+x^2y+xy^2}{6x^2-xy-y^2}$

Solution

1. $\frac{x+1}{x^2-x-6} \div \frac{3x+3}{x^2-9}$

$$= \frac{x+1}{x^2-x-6} \cdot \frac{x^2-9}{3x+3}$$ *Multiply by reciprocal of $\frac{3x+3}{x^2-9}$.*

$$= \frac{x+1}{(x-3)(x+2)} \cdot \frac{(x-3)(x+3)}{3(x+1)}$$ *Factor numerators and denominators.*

$$= \frac{(x+1)(x-3)(x+3)}{(x-3)(x+2) \cdot 3(x+1)}$$ *Multiply numerators; multiply denominators.*

$$= \frac{x+3}{3(x+2)}$$ *Simplify: $\frac{(x+1)(x-3)}{(x+1)(x-3)} = 1$*

We verify our work by creating a graphing calculator table for the equations $y = \dfrac{x+1}{x^2 - x - 6} \div \dfrac{3x+3}{x^2 - 9}$ and $y = \dfrac{x+3}{3(x+2)}$. See Fig. 13. There are "ERROR" messages across from the values $x = -3$, $x = -2$, $x = -1$, and $x = 3$, because each of these values is an excluded value of either the original expression or our result (why?).

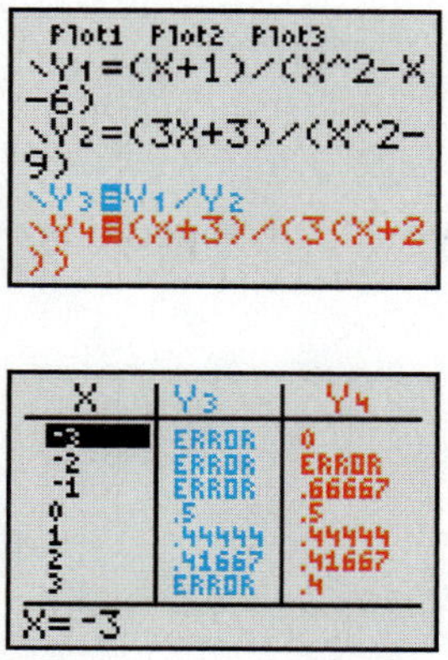

2. $$\frac{x^3 - y^3}{9x^2 - y^2} \div \frac{x^3 + x^2y + xy^2}{6x^2 - xy - y^2} = \frac{x^3 - y^3}{9x^2 - y^2} \cdot \frac{6x^2 - xy - y^2}{x^3 + x^2y + xy^2}$$ *Multiply by reciprocal of $\dfrac{x^3 + x^2y + xy^2}{6x^2 - xy - y^2}$.*

$$= \frac{(x-y)(x^2 + xy + y^2)}{(3x+y)(3x-y)} \cdot \frac{(2x-y)(3x+y)}{x(x^2 + xy + y^2)}$$ *Factor numerators and denominators.*

$$= \frac{(x-y)(x^2 + xy + y^2)(2x-y)(3x+y)}{(3x+y)(3x-y) \cdot x(x^2 + xy + y^2)}$$ *Multiply numerators; multiply denominators.*

$$= \frac{(x-y)(2x-y)}{x(3x-y)}$$ *Simplify: $\dfrac{(x^2 + xy + y^2)(3x+y)}{(x^2 + xy + y^2)(3x+y)} = 1$*

Since our result is in lowest terms, we are done.

```
Plot1 Plot2 Plot3
\Y1=(X+1)/(X^2-X-6)
\Y2=(3X+3)/(X^2-9)
\Y3=Y1/Y2
\Y4=(X+3)/(3(X+2))
```

X	Y3	Y4
-3	ERROR	0
-2	ERROR	ERROR
-1	ERROR	.66667
0	.5	.5
1	.44444	.44444
2	.41667	.41667
3	ERROR	.4

X=-3

Figure 13 Verify the work

Example 8 Finding the Quotient of Two Rational Expressions

Find the quotient $\dfrac{x^2 + 6x + 9}{x - 4} \div (x^2 + x - 6)$.

Solution

$$\frac{x^2 + 6x + 9}{x - 4} \div (x^2 + x - 6) = \frac{x^2 + 6x + 9}{x - 4} \cdot \frac{1}{x^2 + x - 6}$$ *Multiply by reciprocal of $\dfrac{x^2 + x - 6}{1}$.*

$$= \frac{(x+3)(x+3)}{x-4} \cdot \frac{1}{(x+3)(x-2)}$$ *Factor numerators and denominators.*

$$= \frac{(x+3)(x+3)}{(x-4)(x+3)(x-2)}$$ *Multiply numerators; multiply denominators.*

$$= \frac{x+3}{(x-4)(x-2)}$$ *Simplify: $\dfrac{x+3}{x+3} = 1$*

Example 9 Finding a Quotient Function

Let $f(x) = \dfrac{81x^2 - 49}{3x^2 + 16x + 5}$ and $g(x) = \dfrac{7 - 9x}{18x + 6}$.

1. Find an equation of the quotient function $\dfrac{f}{g}$.

2. Find $\left(\dfrac{f}{g}\right)(3)$.

Solution

1. $$\frac{f}{g}(x) = \frac{f(x)}{g(x)}$$ *Definition of $\dfrac{f}{g}$*

$$= f(x) \div g(x)$$ *$\dfrac{R}{S} = R \div S$*

$$= \frac{81x^2 - 49}{3x^2 + 16x + 5} \div \frac{7 - 9x}{18x + 6}$$ *Substitute $\dfrac{81x^2 - 49}{3x^2 + 16x + 5}$ for $f(x)$ and $\dfrac{7 - 9x}{18x + 6}$ for $g(x)$.*

$$= \frac{81x^2 - 49}{3x^2 + 16x + 5} \cdot \frac{18x + 6}{7 - 9x}$$ *Multiply by reciprocal of $\dfrac{7 - 9x}{18x + 6}$.*

$$= \frac{(9x-7)(9x+7)}{(3x+1)(x+5)} \cdot \frac{6(3x+1)}{-(9x-7)}$$ *Factor numerators and denominators;* $7-9x=-9x+7=-(9x-7)$

$$= -\frac{(9x-7)(9x+7)\cdot 6(3x+1)}{(3x+1)(x+5)(9x-7)}$$ *Multiply numerators; multiply denominators;* $\frac{a}{-b} = -\frac{a}{b}$

$$= -\frac{6(9x+7)}{x+5}$$ *Simplify:* $\frac{(9x-7)(3x+1)}{(9x-7)(3x+1)} = 1$

2. $\left(\frac{f}{g}\right)(3) = -\frac{6[9(3)+7]}{3+5} = -\frac{204}{8} = -\frac{51}{2}$

In Example 10, we will combine three rational expressions by using two operations.

Example 10 Combining Three Rational Expressions

Perform the indicated operations:

$$\left(\frac{x^2-2x-48}{x^2+8x+16} \div \frac{3x^2-9x}{x^2-16}\right)\cdot\frac{6x+24}{5x+30}$$

Solution

$$\left(\frac{x^2-2x-48}{x^2+8x+16} \div \frac{3x^2-9x}{x^2-16}\right)\cdot\frac{6x+24}{5x+30} = \left(\frac{x^2-2x-48}{x^2+8x+16} \cdot \frac{x^2-16}{3x^2-9x}\right)\cdot\frac{6x+24}{5x+30}$$

$$= \left(\frac{(x-8)(x+6)}{(x+4)^2}\cdot\frac{(x-4)(x+4)}{3x(x-3)}\right)\cdot\frac{2\cdot 3\cdot(x+4)}{5(x+6)}$$

$$= \frac{2\cdot 3\cdot(x-8)(x+6)(x-4)(x+4)^2}{3\cdot 5\cdot x\cdot(x+4)^2(x-3)(x+6)}$$

$$= \frac{2(x-8)(x-4)}{5x(x-3)}$$

Converting Units of Quantities

Suppose we're ordering a rug measured in feet, but we measured the width and the length of the floor in inches. There is no need to remeasure: We can convert the units of a quantity to equivalent units by multiplying by ratios of units, where each ratio is equal to 1. For example, since there are 12 inches in 1 foot, the following ratio is equal to 1:

$$\frac{12 \text{ inches}}{1 \text{ foot}} = 1$$

The reciprocal is also equal to 1:

$$\frac{1 \text{ foot}}{12 \text{ inches}} = 1$$

Suppose the width of a floor is 110 inches. Here we convert from units of inches to feet:

$$\frac{110 \text{ inches}}{1}\cdot\frac{1 \text{ foot}}{12 \text{ inches}} \approx 9.2 \text{ feet}$$

When converting, we can eliminate "inches," because $\frac{\text{inches}}{\text{inches}} = 1$.

We can eliminate a pair of the same units even if one is in singular form and the other is in plural form. For example, $\frac{\text{feet}}{\text{foot}} = 1$. However, $\frac{\text{inch}^2}{\text{inch}} = \text{inch}$, not 1.

Some equivalent units are shown on the next page.

Equivalent Units

Length

1 inch = 2.54 centimeters
1 foot = 12 inches
1 yard = 3 feet
1 mile = 5280 feet
1 mile ≈ 1.61 kilometers

Volume

1 cup = 8 ounces
1 quart = 4 cups
1 quart ≈ 0.946 liter
1 gallon = 4 quarts

Weight

1 gram = 1000 milligrams
1 pound = 16 ounces

Time

1 year ≈ 365 days

Example 11 Converting Units

Make the indicated unit conversions. Round the results to two decimal places for Problems 2–4.

1. The official height of a basketball hoop is 10 feet. What is its height in yards?
2. An electric 1974 Fender® Jazz Bass® is 46.25 inches long. What is the length in centimeters?
3. In 2011, Americans consumed an average of 171 pounds of meat (Source: *U.S. Department of Agriculture*). What is this average in ounces per day?
4. A person walks at a speed of 4 miles per hour. What is the person's speed in feet per second?

Solution

1. Since there are 3 feet in 1 yard, we can multiply 10 feet by $\frac{1\text{ yard}}{3\text{ feet}} = 1$. By doing so, we can eliminate "feet," because $\frac{\text{feet}}{\text{feet}} = 1$:

$$\frac{10\text{ feet}}{1} \cdot \frac{1\text{ yard}}{3\text{ feet}} = \frac{10}{3}\text{ yards}$$

The official height of the hoop is $3\frac{1}{3}$ yards.

2. Since there are 2.54 centimeters in 1 inch, we multiply 46.25 inches by $\frac{2.54\text{ centimeters}}{1\text{ inch}} = 1$ so that the inches are eliminated:

$$\frac{46.25\text{ inches}}{1} \cdot \frac{2.54\text{ centimeters}}{1\text{ inch}} \approx 117.48\text{ centimeters}$$

The bass is approximately 117.48 centimeters long.

3. There are 16 ounces in 1 pound and approximately 365 days in 1 year. To convert, we multiply by ratios equal to 1 or approximately equal to 1. We arrange the ratios so that the units we want to eliminate appear in one numerator and one denominator:

$$\frac{171\text{ pounds}}{1\text{ year}} \cdot \frac{16\text{ ounces}}{1\text{ pound}} \cdot \frac{1\text{ year}}{365\text{ days}} \approx 7.50\,\frac{\text{ounces}}{\text{day}}$$

On average, Americans consume about 7.50 ounces of meat per day.

4. There are 5280 feet in 1 mile, 60 minutes in 1 hour, and 60 seconds in 1 minute. To convert, we multiply by ratios equal to 1. Again, we arrange the ratios so that the units we want to eliminate appear in one numerator and one denominator:

$$\frac{4\text{ miles}}{1\text{ hour}} \cdot \frac{5280\text{ feet}}{1\text{ mile}} \cdot \frac{1\text{ hour}}{60\text{ minutes}} \cdot \frac{1\text{ minute}}{60\text{ seconds}} \approx 5.87\,\frac{\text{feet}}{\text{second}}$$

So, the person is walking at a speed of about 5.87 feet per second.

Converting Units

To convert the units of a quantity,

1. Write the quantity in the original units.
2. Multiply by fractions equal to 1 so that the units you want to eliminate appear in one numerator and one denominator.

Looking ahead: Adding and subtracting rational expressions

1. Compare graphing calculator tables for $y = \frac{2}{x} + \frac{3}{x}$ and $y = \frac{5}{x}$. What does this comparison suggest about finding the sum $\frac{A}{B} + \frac{C}{B}$?
2. Compare graphing calculator tables for $y = \frac{5}{x} - \frac{2}{x}$ and $y = \frac{3}{x}$. What does this comparison suggest about finding the difference $\frac{A}{B} - \frac{C}{B}$?
3. Use a graphing calculator table to find outputs of the function $y = \frac{x}{5} - \frac{x+10}{5}$. What do you notice about the outputs? Explain why it makes sense that the outputs are what they are.
4. Compare graphing calculator tables for $y = \frac{2}{x} + \frac{x}{3}$ and $y = \frac{2+x}{x+3}$. Is the statement

$$\frac{A}{B} + \frac{C}{D} = \frac{A+C}{B+D}, \quad \text{where } B, D, \text{ and } B+D \text{ are nonzero}$$

true or false, in general? Explain. [**Hint:** In what way is this sum different from the sum and differences in Problems 1–3?]
5. State any concepts suggested by this exploration.

Homework 12.2

For extra help ▶ MyMathLab®  Watch the videos in MyMathLab  Download the MyDashboard App

Perform the indicated operation. Simplify the result.

1. $\frac{3}{x} \cdot \frac{5}{x}$
2. $\frac{x}{2} \cdot \frac{x}{7}$
3. $\frac{x}{6} \div \frac{3}{2x}$
4. $\frac{8}{x} \div \frac{10x}{7}$
5. $\frac{6a^2}{7} \cdot \frac{21}{5a^8}$
6. $\frac{15}{3t^4} \cdot \frac{2t^9}{40}$
7. $\frac{2}{x-3} \cdot \frac{x-4}{x+5}$
8. $\frac{x-1}{x+7} \cdot \frac{x-6}{5}$
9. $\frac{k-2}{k-6} \div \frac{k+6}{k+4}$
10. $\frac{y-4}{y+8} \div \frac{y+3}{y+4}$
11. $\frac{6}{7x-14} \cdot \frac{5x-10}{9}$
12. $\frac{3x+15}{8} \cdot \frac{3}{4x+20}$
13. $\frac{3x+18}{x-6} \div \frac{x+6}{2x-12}$
14. $\frac{x+4}{7x-7} \div \frac{3x+12}{x-1}$
15. $\frac{4w^6}{w+3} \cdot \frac{w+5}{2w^2}$
16. $\frac{b-1}{6b^3} \cdot \frac{9b^7}{b+4}$
17. $\frac{(x-4)(x+1)}{(x-7)(x+2)} \cdot \frac{5(x+2)}{3(x-4)}$
18. $\frac{(x+3)(x+9)}{(x-6)(x-1)} \cdot \frac{x(x-6)}{6(x+3)}$
19. $\frac{4(x-4)^2}{(x+5)^2} \div \frac{14(x-4)}{15(x+5)}$
20. $\frac{9(x+2)^2}{8(x-1)^2} \div \frac{3(x+2)}{10(x-1)}$
21. $\frac{4t^7}{3t-9} \cdot \frac{5t-15}{8t^3}$
22. $\frac{10w+10}{18w^5} \cdot \frac{27w^3}{5w+5}$
23. $\frac{8x^2}{x^2-49} \div \frac{4x^5}{3x+21}$
24. $\frac{33x^6}{5x-30} \div \frac{11x^{10}}{x^2-36}$
25. $\frac{15a^4b}{8ab^5} \div \frac{25ab^3}{4a}$
26. $\frac{4xy^3}{9x^3y} \div \frac{10y}{3xy^5}$
27. $\frac{2x-12}{x+1} \cdot \frac{4x+4}{18-3x}$
28. $\frac{2x+4}{4x-20} \cdot \frac{12-3x}{x+2}$
29. $\frac{2k^2-32}{k^2-2k-24} \div \frac{k+6}{k^2-7k+6}$
30. $\frac{3b^2-27}{b^2-12b+35} \div \frac{b+3}{b^2-b-20}$
31. $\frac{2a^2+3ab}{3a-6b} \cdot \frac{a^2-4b^2}{2ab+3b^2}$
32. $\frac{2a+10b}{3a^2+ab} \cdot \frac{9a^2-b^2}{ab+5b^2}$
33. $\frac{4-x}{x^2+10x+25} \div \frac{3x^2-9x-12}{25-x^2}$
34. $\frac{6-x}{x^2+8x+16} \div \frac{4x^2-16x-48}{16-x^2}$
35. $\frac{t^2-8t+16}{t^2-2t-3} \cdot \frac{3-t}{t^2-16}$
36. $\frac{p^2-3p-10}{p^2-4p-12} \cdot \frac{6-p}{p^2-25}$
37. $\frac{-x^2+7x-10}{2x^2+5x-12} \div \frac{-x^2+4}{8x^2-18}$
38. $\frac{-x^2-5x+14}{3x^2-5x-12} \div \frac{-x^2+4}{5x^2-45}$

39. $\dfrac{-4x-6}{36-x^2}\cdot\dfrac{4x+24}{6x^2+x-12}$

40. $\dfrac{-6x-15}{9-x^2}\cdot\dfrac{6x+18}{4x^2+4x-15}$

41. $\dfrac{9x^2-16}{x+2}\div\left(3x^2+5x-12\right)$

42. $\dfrac{16x^2-25}{x+3}\div\left(4x^2-13x+10\right)$

43. $\dfrac{6m^2-17m-14}{m^2+6m+9}\cdot\dfrac{9-m^2}{4m^2-49}$

44. $\dfrac{3r^2+7r-20}{r^2+4r+4}\cdot\dfrac{4-r^2}{9r^2-25}$

45. $\dfrac{x^2-4x-32}{x^2+7x+12}\div\dfrac{x^2-2x-48}{x^2+3x-4}$

46. $\dfrac{x^2-6x+8}{x^2-9x+14}\div\dfrac{x^2-x-12}{x^2+x-56}$

47. $\dfrac{p^2+4pt-12t^2}{p^2+pt-12t^2}\cdot\dfrac{p^2+7pt+12t^2}{p^2-7pt+10t^2}$

48. $\dfrac{m^2-6mn+9n^2}{m^2+3mn-10n^2}\cdot\dfrac{m^2+10mn+25n^2}{m^2-4mn+3n^2}$

49. $\dfrac{2x^2-xy-3y^2}{3xy-5y^2}\div\dfrac{4x^2-9y^2}{3x^2-14xy+15y^2}$

50. $\dfrac{4t^2-9tw+2w^2}{9t^2-4w^2}\div\dfrac{4t^2-5tw+w^2}{6tw-4w^2}$

51. $\dfrac{3x^3-15x^2+18x}{x^2+16x+64}\cdot\dfrac{x^2-64}{4x^4-28x^3+40x^2}$

52. $\dfrac{6x^3+6x^2-12x}{x^2-9}\cdot\dfrac{x^2-6x+9}{2x^4+10x^3-12x^2}$

53. $\dfrac{w^2-2w-8}{12w^4+32w^3-12w^2}\div\dfrac{w^2-9w+20}{12w^3+54w^2+54w}$

54. $\dfrac{t^2+9t+18}{6t^4-8t^3-8t^2}\div\dfrac{t^2-5t-24}{12t^3+20t^2+8t}$

55. $\dfrac{x^2+4x-5}{x^3+6x^2-4x-24}\cdot\dfrac{x^2+8x+12}{x^2+10x+25}$

56. $\dfrac{x^2-3x-10}{x^2-12x+36}\cdot\dfrac{x^2-3x-18}{x^3+2x^2-25x-50}$

57. $\dfrac{18x^3+27x^2-8x-12}{3x^2-x-2}\div\left(6x^2+5x-6\right)$

58. $\dfrac{4x^3-8x^2-x+2}{2x^2-11x+5}\div\left(2x^2-3x-2\right)$

59. $\dfrac{k^3-8}{k^3+27}\cdot\dfrac{k^2-9}{k^2-4}$

60. $\dfrac{y^3-1}{y^2-4}\cdot\dfrac{y^3-8}{y^2-1}$

61. $\dfrac{8x^3-27}{3x^2-6x+12}\div\dfrac{8x^2+12x+18}{6x^3+48}$

62. $\dfrac{64x^3+125}{3x^2-3x+3}\div\dfrac{32x^2-40x+50}{6x^3+6}$

63. $\dfrac{a^2+ab-2b^2}{a^3+b^3}\cdot\dfrac{a^2+2ab+b^2}{a^2-b^2}$

64. $\dfrac{6p+4q}{4p^2+2pq+q^2}\cdot\dfrac{8p^3-q^3}{3p^2+2pq}$

For $f(x)=\dfrac{x^2-6x-16}{x^2+3x-40}$ and $g(x)=\dfrac{x^2-64}{x^2-3x-10}$, find an equation of the given function. Evaluate the function at the indicated value.

65. $f\cdot g,\ (f\cdot g)(6)$

66. $\dfrac{f}{g},\ \left(\dfrac{f}{g}\right)(9)$

67. $\dfrac{g}{f},\ \left(\dfrac{g}{f}\right)(7)$

For $f(x)=\dfrac{1-x^2}{x^2-3x-28}$ and $g(x)=\dfrac{x^2-8x+7}{x^2+5x+4}$, find an equation of the given function. Evaluate the function at the indicated value.

68. $f\cdot g,\ (f\cdot g)(2)$

69. $\dfrac{f}{g},\ \left(\dfrac{f}{g}\right)(4)$

70. $\dfrac{g}{f},\ \left(\dfrac{g}{f}\right)(3)$

Perform the indicated operations. Simplify the result.

71. $\left(\dfrac{20x^7}{x^2-9}\div\dfrac{x^2-14x+24}{5x-15}\right)\cdot\dfrac{x^2+x-6}{8x^{13}}$

72. $\left(\dfrac{8x^2+10x-3}{-2x^2-8x}\cdot\dfrac{x^2+x}{16x^2-1}\right)\div\dfrac{-10x-15}{8x^5}$

73. $\dfrac{12k^3}{k^2-4}\div\left(\dfrac{22k^6}{-6k+12}\cdot\dfrac{k}{11k+22}\right)$

74. $\dfrac{3-p}{20p^5}\div\left(\dfrac{p^2-9}{30p^2}\div\dfrac{8p^2+6p}{3p}\right)$

75. $\left(\left(\dfrac{x-4}{x+5}\right)^2\cdot\left(\dfrac{x+5}{x-1}\right)^2\right)\div\left(\dfrac{x-4}{x-1}\right)^2$

76. $\dfrac{3x+6}{4x+20}\div\left(\dfrac{x^2-4}{x^2-25}\div\dfrac{x-2}{x-5}\right)^2$

For Exercises 77–86, round approximate results to the second decimal place. Refer to the list of equivalent units in the margin of page 747 as needed.

77. The average height of a woman in the United States is 63.8 inches. What is that height in feet?

78. A German stein is 23 centimeters tall. What is the mug's height in inches?

79. A person buys 15 gallons of gasoline. How many liters of gasoline is that?

80. The speed limit on motorways in England is 113 kilometers per hour. What is the speed limit in miles per hour?

81. In Trader Giotto's Roasted Garlic Spaghetti Sauce, there are approximately 1.63 grams of salt in 1 pound of sauce. How many milligrams of salt are there in 1 ounce of sauce?

82. In Barbara's Puffins Cinnamon Cereal, there are 42.5 milligrams of potassium in 1 ounce of cereal. How many grams of potassium are there in 1 pound of the cereal?

83. DairyCo, a UK dairy company, reports that the 2011 average yield of one of their cows is 7533 liters of milk per year. What is that yield in gallons per day?

84. A Volkswagen Golf BlueMotion has gas mileage equal to 100 kilometers per 3.2 liters. What is the car's gas mileage in miles per gallon?

85. A person drives at a speed of 67 miles per hour. How fast is the person traveling in feet per second?

86. A person drives at a speed of 25 meters per second. How fast is the person traveling in miles per hour?

87. Find the product of $\frac{x^3}{12}$ and $\frac{3}{x}$.

88. Find the product of $\frac{15}{x^2}$ and $\frac{x^6}{20}$.

89. Find the quotient of $\frac{x-2}{x^2-3x-18}$ and $\frac{x+4}{x^2+4x+3}$.

90. Find the quotient of $\frac{x^2-4}{6x-42}$ and $\frac{5x-10}{3x-21}$.

Concepts

91. A student tries to find the product $\frac{x+2}{x+4}\cdot\frac{x+4}{x+6}$:

$$\frac{x+2}{x+4}\cdot\frac{x+4}{x+6}=\frac{(x+2)(x+4)}{(x+4)(x+6)}=\frac{x^2+6x+8}{x^2+10x+24}$$

Describe any errors. Then find the product correctly.

92. A student tries to find the quotient $\frac{x+4}{x-5}\div\frac{x-5}{x+8}$:

$$\frac{x+4}{x-5}\div\frac{x-5}{x+8}=\frac{x+4}{x+8}$$

Describe any errors. Then find the quotient correctly.

93. A student tries to find the product $\frac{x}{3}\cdot\frac{x+4}{x-7}$:

$$\frac{x}{3}\cdot\frac{x+4}{x-7}=\frac{x^2+4}{3x-7}$$

Find a value to substitute for x to show that the student's work is incorrect. Then perform the multiplication correctly. Use a graphing calculator table to verify your result.

94. A student tries to find the product $\frac{x+3}{x+7}\cdot\frac{x+3}{x-7}$:

$$\frac{x+3}{x+7}\cdot\frac{x+3}{x-7}=\frac{x^2+9}{x^2-49}$$

Find a value to substitute for x to show that the student's work is incorrect. Then perform the multiplication correctly. Use a graphing calculator table to verify your result.

95. Perform the indicated operations.

a. $\frac{1}{x}\div\frac{1}{x}$ **b.** $\frac{1}{x}\div\left(\frac{1}{x}\div\frac{1}{x}\right)$

c. $\frac{1}{x}\div\left(\frac{1}{x}\div\left(\frac{1}{x}\div\frac{1}{x}\right)\right)$

d. $\frac{1}{x}\div\left(\frac{1}{x}\div\left(\frac{1}{x}\div\left(\frac{1}{x}\div\frac{1}{x}\right)\right)\right)$

e. $\underbrace{\frac{1}{x}\div\left(\frac{1}{x}\div\left(\frac{1}{x}\div\cdots\div\left(\frac{1}{x}\div\left(\frac{1}{x}\div\frac{1}{x}\right)\right)\cdots\right)\right)}_{n \text{ division symbols}}$

96. Explain how dividing rational expressions is similar to dividing rational numbers.

97. When converting units, why can we only multiply by ratios equal to 1?

98. Describe how to multiply two rational expressions, and describe how to divide two rational expressions.

Related Review

Find the quotient twice, first by using factoring and then by using long division. Compare your results.

99. $\frac{x^2-7x+12}{x-3}$ **100.** $\frac{x^2-49}{x+7}$

Find equations of $f\cdot g$ and $\frac{f}{g}$. $\left[\textbf{Hint: } b^nc^n=(bc)^n, \frac{b^n}{c^n}=\left(\frac{b}{c}\right)^n\right]$

101. $f(x)=8^x, g(x)=2^x$ **102.** $f(x)=6^x, g(x)=3^x$

103. $f(x)=12(6)^x, g(x)=3(2)^x$

104. $f(x)=10(8)^x, g(x)=2(4)^x$

For Exercises 105 and 106, refer to the list of equivalent units in the margin of page 747 as needed.

105. The width of a picture is 2.5 yards.
a. Use ratios equal to 1 to find the width in inches.
b. Let $f(w)$ be the number of inches in w feet, and let $g(x)$ be the number of feet in x yards.
i. Find equations of f and g.
ii. Find an equation of $f\circ g$.
iii. Find $(f\circ g)(2.5)$. What does it mean in this situation? Compare the result with your result in part (a).

106. A bowl contains 3.5 quarts of punch.
a. Use ratios equal to 1 to describe the volume of punch in ounces.
b. Let $f(C)$ be the number of ounces in C cups, and let $g(Q)$ be the number of cups in Q quarts.
i. Find equations of f and g.
ii. Find an equation of $f\circ g$.
iii. Find $(f\circ g)(3.5)$. What does it mean in this situation? Compare the result with your result in part (a).

Expressions, Equations, Functions, and Graphs

Perform the indicated instruction. Then use words such as linear, quadratic, cubic, exponential, logarithmic, rational, polynomial, degree, function, one variable, *and* two variables *to describe the expression, equation, or system.*

107. Write $4\log_b(2x^2)-2\log_b(3x^3)$ as a single logarithm.

108. Find the product $(5t-3)(4t-6)$.

109. Solve $\log_2(x-3)+\log_2(x-2)=3$. Round any solutions to the fourth decimal place.

110. Factor $6t^2-19t+10$.

111. Solve $5(4)^x-23=81$. Round any solutions to the fourth decimal place.

112. Solve $6t^2-19t+10=0$.

12.3 Adding and Subtracting Rational Expressions

Objectives

- » Add rational expressions with a common denominator.
- » Add rational expressions with different denominators.
- » Subtract rational expressions with a common denominator.
- » Subtract rational expressions with different denominators.

In Section 12.2, we multiplied and divided rational expressions. How do we add rational expressions?

Addition of Rational Expressions with a Common Denominator

Recall from Section 2.2 that we add fractions with a common denominator by using the property $\frac{a}{b} + \frac{c}{b} = \frac{a + c}{b}$, where b is nonzero. For example,

$$\frac{2}{7} + \frac{3}{7} = \frac{5}{7}$$

We add more complicated rational expressions with a common denominator in a similar way.

Adding Rational Expressions with a Common Denominator

If $\frac{A}{B}$ and $\frac{C}{B}$ are rational expressions, where B is nonzero, then

$$\frac{A}{B} + \frac{C}{B} = \frac{A + C}{B}$$

In words: To add two rational expressions with a common denominator, add the numerators and keep the common denominator.

Example 1 Adding Two Rational Expressions with a Common Denominator

Perform the addition $\frac{4x + 3}{x} + \frac{2}{x}$.

Solution

$$\frac{4x + 3}{x} + \frac{2}{x} = \frac{4x + 3 + 2}{x}$$ *Add numerators and keep common denominator:* $\frac{A}{B} + \frac{C}{B} = \frac{A + C}{B}$

$$= \frac{4x + 5}{x}$$ *Combine like terms.*

After we add two rational expressions, we may be able to simplify the result.

Example 2 Adding Two Rational Expressions with a Common Denominator

Perform the addition $\frac{x^2 - 10x}{x - 5} + \frac{2x + 15}{x - 5}$.

Solution

$$\frac{x^2 - 10x}{x - 5} + \frac{2x + 15}{x - 5} = \frac{x^2 - 10x + 2x + 15}{x - 5}$$ *Add numerators and keep common denominator:* $\frac{A}{B} + \frac{C}{B} = \frac{A + C}{B}$

$$= \frac{x^2 - 8x + 15}{x - 5}$$ *Combine like terms.*

$$= \frac{(x - 5)(x - 3)}{x - 5}$$ *Factor numerator.*

$$= x - 3$$ *Simplify:* $\frac{x - 5}{x - 5} = 1$

We use a graphing calculator table to verify our work (see Fig. 14).

X	Y1	Y2
-3	-6	-6
-2	-5	-5
-1	-4	-4
0	-3	-3
1	-2	-2
2	-1	-1
3	0	0

X=-3

Figure 14 Verify the work

Addition of Rational Expressions with Different Denominators

When adding two fractions with different denominators in Section 2.2, we listed multiples of the denominators to help us find the least common denominator (LCD) of the fractions. In this section, we discuss another way to find the LCD of two fractions.

Suppose that two brothers, John and Paul, own the following numbers and types of musical instruments:

John	Paul
3 guitars	1 guitar
1 bass guitar	2 bass guitars
1 sitar	1 sitar

The brothers will not give their instruments to each other, but each brother wants to own the same numbers and types of musical instruments that the other brother has.

Since Paul has 1 more bass than John, John wants 1 more bass guitar. Since John has 2 more guitars than Paul, Paul wants 2 more guitars. John and Paul have the same number of sitars, so neither brother wants another sitar.

Let's compare this situation with finding the LCD of the sum $\frac{7}{24} + \frac{5}{18}$. First, we find the prime factorization of each denominator:

$$24 = 2 \cdot 2 \cdot 2 \cdot 3$$
$$18 = 2 \cdot 3 \cdot 3$$

What factors does each of the denominators need so that the denominators can become the same? Since 18 has one more 3 factor than 24, the denominator 24 needs one 3 factor. Since 24 has two more 2 factors than 18, the denominator 18 needs two 2 factors.

We use the fact that $\frac{A}{A} = 1$ when $A \neq 0$ to introduce the one 3 factor for 24 and the two 2 factors for 18:

$$\begin{aligned}
\frac{7}{24} + \frac{5}{18} &= \frac{7}{2 \cdot 2 \cdot 2 \cdot 3} + \frac{5}{2 \cdot 3 \cdot 3} && \text{Find prime factorization of each denominator.} \\
&= \frac{7}{2 \cdot 2 \cdot 2 \cdot 3} \cdot \frac{3}{3} + \frac{5}{2 \cdot 3 \cdot 3} \cdot \frac{2 \cdot 2}{2 \cdot 2} && \text{Introduce missing factors.} \\
&= \frac{7 \cdot 3}{2 \cdot 2 \cdot 2 \cdot 3 \cdot 3} + \frac{5 \cdot 2 \cdot 2}{2 \cdot 3 \cdot 3 \cdot 2 \cdot 2} && \text{Find products.} \\
&= \frac{21}{72} + \frac{20}{72} && \text{Simplify.} \\
&= \frac{41}{72} && \text{Add numerators and keep common denominator: } \frac{A}{B} + \frac{C}{B} = \frac{A + C}{B}
\end{aligned}$$

Note that this method not only helps us find the LCD, but also suggests what forms of $\frac{A}{A}$ to use to introduce missing factors.

Example 3 Adding Two Rational Expressions with Different Denominators

Find the sum $\frac{1}{6} + \frac{9}{2w}$.

Solution

We begin by factoring the denominators:

$$6 = 2 \cdot 3$$
$$2w = 2 \cdot w$$

Since $2w$ has a w factor and 6 has no w factors, the denominator 6 needs a w factor. Since 6 has a 3 factor and $2w$ has no 3 factors, the denominator $2w$ needs a 3 factor.

We use the fact that $\frac{A}{A} = 1$, where A is nonzero, to introduce the missing factors:

$$\frac{1}{6} + \frac{9}{2w} = \frac{1}{2 \cdot 3} + \frac{9}{2 \cdot w} \quad \text{Factor denominators.}$$

$$= \frac{1}{2 \cdot 3} \cdot \frac{w}{w} + \frac{9}{2 \cdot w} \cdot \frac{3}{3} \quad \text{Introduce missing factors.}$$

$$= \frac{w}{6w} + \frac{27}{6w} \quad \text{Find products.}$$

$$= \frac{w + 27}{6w} \quad \text{Add numerators and keep common denominator: } \frac{A}{B} + \frac{C}{B} = \frac{A + C}{B}$$

Notice that the result is in lowest terms.

Example 4 Adding Two Rational Expressions with Different Denominators

Find the sum $\frac{7}{12x} + \frac{5}{8x^3}$.

Solution

We begin by factoring the denominators:

$$12x = 2 \cdot 2 \cdot 3 \cdot x$$
$$8x^3 = 2 \cdot 2 \cdot 2 \cdot x \cdot x \cdot x$$

Since $8x^3$ has one more 2 factor and two more x factors than $12x$, the denominator $12x$ needs one more 2 factor and two more x factors. Since $12x$ has a 3 factor and $8x^3$ does not have any 3 factors, the denominator $8x^3$ needs a 3 factor.

We use the fact that $\frac{A}{A} = 1$, where A is nonzero, to introduce the missing factors:

$$\frac{7}{12x} + \frac{5}{8x^3} = \frac{7}{2 \cdot 2 \cdot 3 \cdot x} + \frac{5}{2 \cdot 2 \cdot 2 \cdot x \cdot x \cdot x} \quad \text{Factor denominators.}$$

$$= \frac{7}{2 \cdot 2 \cdot 3 \cdot x} \cdot \frac{2 \cdot x \cdot x}{2 \cdot x \cdot x} + \frac{5}{2 \cdot 2 \cdot 2 \cdot x \cdot x \cdot x} \cdot \frac{3}{3} \quad \text{Introduce missing factors.}$$

$$= \frac{14x^2}{24x^3} + \frac{15}{24x^3} \quad \text{Find products.}$$

$$= \frac{14x^2 + 15}{24x^3} \quad \text{Add numerators and keep common denominator.}$$

Notice that the result is in lowest terms.

Example 5 Adding Two Rational Expressions with Different Denominators

Find the sum $\frac{2}{x - 4} + \frac{6}{x + 7}$.

Solution

The denominator $x - 4$ needs an $x + 7$ factor, and the denominator $x + 7$ needs an $x - 4$ factor. We use the fact that $\frac{A}{A} = 1$, where A is nonzero, to introduce the missing factors:

$$\frac{2}{x-4} + \frac{6}{x+7} = \frac{2}{x-4} \cdot \frac{x+7}{x+7} + \frac{6}{x+7} \cdot \frac{x-4}{x-4} \quad \text{Introduce missing factors.}$$

$$= \frac{2(x+7)}{(x-4)(x+7)} + \frac{6(x-4)}{(x+7)(x-4)} \quad \text{Find products.}$$

$$= \frac{2(x+7) + 6(x-4)}{(x-4)(x+7)} \quad \text{Add numerators and keep common denominator: } \frac{A}{B} + \frac{C}{B} = \frac{A+C}{B}$$

$$= \frac{2x + 14 + 6x - 24}{(x-4)(x+7)} \quad \text{Distributive law}$$

$$= \frac{8x - 10}{(x-4)(x+7)} \quad \text{Combine like terms.}$$

$$= \frac{2(4x - 5)}{(x-4)(x+7)} \quad \text{Factor numerator.}$$

Notice that the result is in lowest terms.

For the sum of two expressions with different denominators, we first factor the denominators, if possible, to help us find the LCD.

Example 6 Adding Two Rational Expressions with Different Denominators

Find the sum $\frac{5}{6} + \frac{7}{2x - 8}$.

Solution

First, we factor the denominators:

$$6 = 2 \cdot 3$$
$$2x - 8 = 2(x - 4)$$

Since $2x - 8$ has an $x - 4$ factor and 6 does not, the denominator 6 needs an $x - 4$ factor. Since 6 has a 3 factor and $2x - 8$ does not, the denominator $2x - 8$ needs a 3 factor. Thus,

$$\frac{5}{6} + \frac{7}{2x-8} = \frac{5}{2 \cdot 3} + \frac{7}{2(x-4)} \quad \text{Factor denominators.}$$

$$= \frac{5}{2 \cdot 3} \cdot \frac{x-4}{x-4} + \frac{7}{2(x-4)} \cdot \frac{3}{3} \quad \text{Introduce missing factors.}$$

$$= \frac{5(x-4)}{6(x-4)} + \frac{21}{6(x-4)} \quad \text{Find products.}$$

$$= \frac{5(x-4) + 21}{6(x-4)} \quad \text{Add numerators and keep common denominator: } \frac{A}{B} + \frac{C}{B} = \frac{A+C}{B}$$

$$= \frac{5x - 20 + 21}{6(x-4)} \quad \text{Distributive law}$$

$$= \frac{5x + 1}{6(x-4)} \quad \text{Combine like terms.}$$

Example 7 Adding Two Rational Expressions with Different Denominators

Find the sum $\frac{x}{2x - 4} + \frac{-4}{x^2 - 4}$.

Solution

First, we factor the denominators:

$$2x - 4 = 2(x - 2)$$
$$x^2 - 4 = (x - 2)(x + 2)$$

Since $x^2 - 4$ has an $x + 2$ factor and $2x - 4$ does not, the denominator $2x - 4$ needs an $x + 2$ factor. Since $2x - 4$ has a 2 factor and $x^2 - 4$ does not, the denominator $x^2 - 4$ needs a 2 factor. Thus,

$$\frac{x}{2x-4} + \frac{-4}{x^2-4} = \frac{x}{2(x-2)} + \frac{-4}{(x-2)(x+2)} \qquad \textit{Factor denominators.}$$

$$= \frac{x}{2(x-2)} \cdot \frac{x+2}{x+2} + \frac{-4}{(x-2)(x+2)} \cdot \frac{2}{2} \qquad \textit{Introduce missing factors.}$$

$$= \frac{x(x+2)}{2(x-2)(x+2)} + \frac{-8}{2(x-2)(x+2)} \qquad \textit{Find products.}$$

$$= \frac{x(x+2) - 8}{2(x-2)(x+2)} \qquad \textit{Add numerators and keep common denominator.}$$

$$= \frac{x^2 + 2x - 8}{2(x-2)(x+2)} \qquad \textit{Distributive law}$$

$$= \frac{(x-2)(x+4)}{2(x-2)(x+2)} \qquad \textit{Factor numerator.}$$

$$= \frac{x+4}{2(x+2)} \qquad \textit{Simplify: } \frac{x-2}{x-2} = 1$$

X	Y1	Y2
-3	-.5	-.5
-2	ERROR	ERROR
-1	1.5	1.5
0	1	1
1	.83333	.83333
2	ERROR	.75
3	.7	.7

X=-3

Figure 15 Verify the work

We use a graphing calculator table to verify our work (see Fig. 15). There are "ERROR" messages across from $x = -2$ and $x = 2$ in the table, because -2 and 2 are excluded values of either the original expression or our result (why?).

How to Add Two Rational Expressions with Different Denominators

To add two rational expressions with different denominators,

1. Factor the denominators of the expressions if possible. Determine which factors are missing.
2. Use the property $\frac{A}{A} = 1$, where A is nonzero, to introduce missing factors.
3. Add the expressions by using the property $\frac{A}{B} + \frac{C}{B} = \frac{A + C}{B}$, where B is nonzero.
4. Simplify.

Example 8 Adding Two Rational Expressions That Have Different Denominators

Find the sum $\frac{3x}{x^2 + 2xy + y^2} + \frac{2y}{x^2 - y^2}$.

Solution

First, we factor the denominators:

$$x^2 + 2xy + y^2 = (x + y)(x + y)$$
$$x^2 - y^2 = (x + y)(x - y)$$

Since $x^2 - y^2$ has an $x - y$ factor but $x^2 + 2xy + y^2$ does not, the denominator $x^2 + 2xy + y^2$ needs an $x - y$ factor. Since $x^2 + 2xy + y^2$ has one more $x + y$ factor than $x^2 - y^2$ has, the denominator $x^2 - y^2$ needs an $x + y$ factor. Thus,

$$\frac{3x}{x^2 + 2xy + y^2} + \frac{2y}{x^2 - y^2} = \frac{3x}{(x + y)(x + y)} + \frac{2y}{(x + y)(x - y)} \quad \text{Factor denominators.}$$

$$= \frac{3x}{(x + y)(x + y)} \cdot \frac{x - y}{x - y} + \frac{2y}{(x + y)(x - y)} \cdot \frac{x + y}{x + y} \quad \text{Introduce missing factors.}$$

$$= \frac{3x(x - y)}{(x + y)(x + y)(x - y)} + \frac{2y(x + y)}{(x + y)(x + y)(x - y)} \quad \text{Find products.}$$

$$= \frac{3x(x - y) + 2y(x + y)}{(x + y)(x + y)(x - y)} \quad \text{Add numerators and keep common denominator.}$$

$$= \frac{3x^2 - 3xy + 2xy + 2y^2}{(x + y)^2(x - y)} \quad \text{Distributive law}$$

$$= \frac{3x^2 - xy + 2y^2}{(x + y)^2(x - y)} \quad \text{Combine like terms.}$$

The result is in lowest terms.

Example 9 Finding a Sum Function

Let $f(x) = \dfrac{3}{12x^3 - 22x^2 + 6x}$ and $g(x) = \dfrac{x + 1}{30x^2 - 10x}$.

1. Find an equation of $f + g$.
2. Find $(f + g)(2)$.

Solution

1. To start, we use the definition of $f + g$:

$$(f + g)(x) = f(x) + g(x) \quad \text{Definition of } f + g$$

$$= \frac{3}{12x^3 - 22x^2 + 6x} + \frac{x + 1}{30x^2 - 10x} \quad \text{Substitute } \frac{3}{12x^3 - 22x^2 + 6x} \text{ for } f(x) \text{ and } \frac{x + 1}{30x^2 - 10x} \text{ for } g(x).$$

Next, we factor the denominators:

$$12x^3 - 22x^2 + 6x = 2x(6x^2 - 11x + 3) = 2 \cdot x \cdot (3x - 1)(2x - 3)$$
$$30x^2 - 10x = 10x(3x - 1) = 2 \cdot 5 \cdot x \cdot (3x - 1)$$

Since $30x^2 - 10x$ has a 5 factor but $12x^3 - 22x^2 + 6x$ does not, the denominator $12x^3 - 22x^2 + 6x$ needs a 5 factor. Since $12x^3 - 22x^2 + 6x$ has a $2x - 3$ factor but $30x^2 - 10x$ does not, the denominator $30x^2 - 10x$ needs a $2x - 3$ factor. Thus,

$$(f + g)(x) = \frac{3}{2 \cdot x \cdot (3x - 1)(2x - 3)} + \frac{x + 1}{2 \cdot 5 \cdot x \cdot (3x - 1)} \quad \text{Factor denominators.}$$

$$= \frac{3}{2 \cdot x \cdot (3x - 1)(2x - 3)} \cdot \frac{5}{5} + \frac{x + 1}{2 \cdot 5 \cdot x \cdot (3x - 1)} \cdot \frac{2x - 3}{2x - 3} \quad \text{Introduce missing factors.}$$

$$= \frac{15}{10x(3x - 1)(2x - 3)} + \frac{(x + 1)(2x - 3)}{10x(3x - 1)(2x - 3)} \quad \text{Find products.}$$

$$= \frac{15 + (x + 1)(2x - 3)}{10x(3x - 1)(2x - 3)} \quad \text{Add numerators and keep common denominator.}$$

$$= \frac{15 + 2x^2 - x - 3}{10x(3x - 1)(2x - 3)} \quad \text{Find product.}$$

$$= \frac{2x^2 - x + 12}{10x(3x - 1)(2x - 3)} \quad \text{Combine like terms.}$$

The result is in lowest terms.

2. $(f + g)(2) = \dfrac{2(2)^2 - (2) + 12}{10(2)[3(2) - 1][2(2) - 3]} = \dfrac{18}{100} = \dfrac{9}{50}$

Subtraction of Rational Expressions with a Common Denominator

Recall from Section 2.2 that we subtract fractions with a common denominator by using the property $\frac{a}{b} - \frac{c}{b} = \frac{a - c}{b}$, where b is nonzero. For example,

$$\frac{5}{7} - \frac{2}{7} = \frac{3}{7}$$

We subtract more complicated rational expressions with a common denominator in a similar way.

Subtracting Rational Expressions with a Common Denominator

If $\frac{A}{B}$ and $\frac{C}{B}$ are rational expressions, where B is nonzero, then

$$\frac{A}{B} - \frac{C}{B} = \frac{A - C}{B}$$

In words: To subtract two rational expressions with a common denominator, subtract the numerators and keep the common denominator.

After subtracting two rational expressions, it may be possible to simplify the result.

Example 10 Subtracting Two Rational Expressions with a Common Denominator

Find the difference $\dfrac{x}{x^2 - 4} - \dfrac{2}{x^2 - 4}$.

Solution

$$\frac{x}{x^2 - 4} - \frac{2}{x^2 - 4} = \frac{x - 2}{x^2 - 4} \quad \text{Subtract numerators and keep common denominator: } \frac{A}{B} - \frac{C}{B} = \frac{A - C}{B}$$

$$= \frac{x - 2}{(x - 2)(x + 2)} \quad \text{Factor denominator.}$$

$$= \frac{1}{x + 2} \quad \text{Simplify: } \frac{x - 2}{x - 2} = 1$$

Example 11 Subtracting Two Rational Expressions with a Common Denominator

Find the difference $\dfrac{x^2}{x + 4} - \dfrac{3x + 28}{x + 4}$.

Solution

$$\frac{x^2}{x+4} - \frac{3x+28}{x+4} = \frac{x^2-(3x+28)}{x+4} \quad \text{Subtract numerators and keep common denominator: } \frac{A}{B}-\frac{C}{B}=\frac{A-C}{B}$$

$$= \frac{x^2-3x-28}{x+4} \quad \text{Simplify.}$$

$$= \frac{(x-7)(x+4)}{x+4} \quad \text{Factor numerator.}$$

$$= x-7 \quad \text{Simplify: } \frac{x+4}{x+4}=1$$

WARNING It is a common error to write

$$\frac{x^2}{x+4} - \frac{3x+28}{x+4} = \frac{x^2-3x+28}{x+4} \quad \text{Incorrect}$$

This work is incorrect. **When subtracting rational expressions, be sure to subtract the *entire* numerator:**

$$\frac{x^2}{x+4} - \frac{3x+28}{x+4} = \frac{x^2-(3x+28)}{x+4} = \frac{x^2-3x-28}{x+4}$$

See Example 11 for the rest of the work.

Subtraction of Rational Expressions with Different Denominators

When subtracting rational expressions with different denominators, we use the method discussed earlier in this section to find the LCD.

Example 12 Subtracting Two Rational Expressions That Have Different Denominators

Find the difference $\frac{5}{4ab^2} - \frac{3}{2a^3b}$.

Solution

We begin by factoring the denominators:

$$4ab^2 = 2 \cdot 2 \cdot a \cdot b \cdot b$$
$$2a^3b = 2 \cdot a \cdot a \cdot a \cdot b$$

Since $2a^3b$ has two more a factors than $4ab^2$ has, the denominator $4ab^2$ needs two more a factors. Since $4ab^2$ has one more 2 factor and one more b factor than $2a^3b$ has, the denominator $2a^3b$ needs one 2 factor and one b factor. Thus,

$$\frac{5}{4ab^2} - \frac{3}{2a^3b} = \frac{5}{2\cdot 2\cdot a\cdot b\cdot b} - \frac{3}{2\cdot a\cdot a\cdot a\cdot b} \quad \text{Factor denominators.}$$

$$= \frac{5}{2\cdot 2\cdot a\cdot b\cdot b}\cdot\frac{a\cdot a}{a\cdot a} - \frac{3}{2\cdot a\cdot a\cdot a\cdot b}\cdot\frac{2\cdot b}{2\cdot b} \quad \text{Introduce missing factors.}$$

$$= \frac{5a^2}{4a^3b^2} - \frac{6b}{4a^3b^2} \quad \text{Find products.}$$

$$= \frac{5a^2-6b}{4a^3b^2} \quad \text{Subtract numerators and keep common denominator.}$$

The result is in lowest terms.

The steps we take to subtract two rational expressions are similar to the steps we take to add two rational expressions.

How to Subtract Two Rational Expressions That Have Different Denominators

To subtract two rational expressions that have different denominators,

1. Factor the denominators of the expressions if possible. Determine which factors are missing.
2. Use the property $\frac{A}{A} = 1$, where A is nonzero, to introduce missing factors.
3. Subtract the expressions by using the property $\frac{A}{B} - \frac{C}{B} = \frac{A - C}{B}$, where B is nonzero.
4. Simplify.

Example 13 Subtracting Two Rational Expressions That Have Different Denominators

Find the difference $\frac{3x - 1}{2x^2 - 7x - 4} - \frac{5}{x^2 - 8x + 16}$.

Solution

$$
\begin{aligned}
\frac{3x - 1}{2x^2 - 7x - 4} - \frac{5}{x^2 - 8x + 16} &= \frac{3x - 1}{(2x + 1)(x - 4)} - \frac{5}{(x - 4)(x - 4)} && \text{Factor denominators.} \\
&= \frac{3x - 1}{(2x + 1)(x - 4)} \cdot \frac{x - 4}{x - 4} - \frac{5}{(x - 4)(x - 4)} \cdot \frac{2x + 1}{2x + 1} && \text{Introduce missing factors.} \\
&= \frac{(3x - 1)(x - 4) - 5(2x + 1)}{(2x + 1)(x - 4)(x - 4)} && \text{Find products; subtract numerators and keep common denominator.} \\
&= \frac{3x^2 - 13x + 4 - 10x - 5}{(2x + 1)(x - 4)^2} && \text{Find products.} \\
&= \frac{3x^2 - 23x - 1}{(2x + 1)(x - 4)^2} && \text{Combine like terms.}
\end{aligned}
$$

The result is in lowest terms.

Example 14 Subtracting Two Rational Expressions That Have Different Denominators

Find the difference $\frac{y}{y - 2} - \frac{3}{2 - y}$.

Solution

$$
\begin{aligned}
\frac{y}{y - 2} - \frac{3}{2 - y} &= \frac{y}{y - 2} - \frac{3}{-(y - 2)} && 2 - y = -y + 2 = -(y - 2) \\
&= \frac{y}{y - 2} - \left(-\frac{3}{y - 2}\right) && \frac{A}{-B} = -\frac{A}{B} \\
&= \frac{y}{y - 2} + \frac{3}{y - 2} && A - (-B) = A + B \\
&= \frac{y + 3}{y - 2} && \text{Add numerators and keep common denominator: } \frac{A}{B} + \frac{C}{B} = \frac{A + C}{B}
\end{aligned}
$$

WARNING The work in Example 14 shows that, to find the difference $\frac{y}{y-2} - \frac{3}{2-y}$, it is *not* necessary to introduce the factor $2 - y$ into the denominator $y - 2$ or to introduce the factor $y - 2$ into the denominator $2 - y$. This method applies to any sum or difference of two rational expressions in which the denominators are of the form $A - B$ and $B - A$.

In Example 15, we will combine three rational expressions by performing two operations.

Example 15 Performing Operations with Three Rational Expressions

Perform the indicated operations: $\left(\frac{x+2}{x^2-x} - \frac{6}{x^2-1}\right) + \frac{3}{x^2+x}$.

Solution

$$\left(\frac{x+2}{x^2-x} - \frac{6}{x^2-1}\right) + \frac{3}{x^2+x}$$

$$= \left(\frac{x+2}{x(x-1)} - \frac{6}{(x-1)(x+1)}\right) + \frac{3}{x(x+1)} \qquad \text{Factor denominators.}$$

$$= \left(\frac{x+2}{x(x-1)} \cdot \frac{x+1}{x+1} - \frac{6}{(x-1)(x+1)} \cdot \frac{x}{x}\right) + \frac{3}{x(x+1)} \cdot \frac{x-1}{x-1} \qquad \text{Introduce missing factors.}$$

$$= \frac{(x+2)(x+1) - 6x}{x(x-1)(x+1)} + \frac{3(x-1)}{x(x-1)(x+1)} \qquad \text{Find products; subtract numerators.}$$

$$= \frac{(x+2)(x+1) - 6x + 3(x-1)}{x(x-1)(x+1)} \qquad \text{Add numerators.}$$

$$= \frac{x^2 + 3x + 2 - 6x + 3x - 3}{x(x-1)(x+1)} \qquad \text{Find products.}$$

$$= \frac{x^2 - 1}{x(x-1)(x+1)} \qquad \text{Combine like terms.}$$

$$= \frac{(x-1)(x+1)}{x(x-1)(x+1)} \qquad \text{Factor numerator.}$$

$$= \frac{1}{x} \qquad \text{Simplify.}$$

Example 16 Finding a Difference Function

Let $f(x) = \frac{x-1}{x+1}$ and $g(x) = \frac{x+1}{x-1}$.

1. Find an equation of $f - g$.
2. Find $(f - g)(5)$.

Solution

1. $(f-g)(x) = f(x) - g(x)$ — Definition of $f - g$

$$= \frac{x-1}{x+1} - \frac{x+1}{x-1} \qquad \text{Substitute } \frac{x-1}{x+1} \text{ for } f(x) \text{ and } \frac{x+1}{x-1} \text{ for } g(x).$$

$$= \frac{x-1}{x+1} \cdot \frac{x-1}{x-1} - \frac{x+1}{x-1} \cdot \frac{x+1}{x+1} \qquad \text{Introduce missing factors.}$$

X	Y1	Y2
-3	1.5	1.5
-2	2.6667	2.6667
-1	ERROR	ERROR
0	0	0
1	ERROR	ERROR
2	-2.667	-2.667
3	-1.5	-1.5

X=-3

Figure 16 Verify the work

$$= \frac{(x-1)(x-1) - (x+1)(x+1)}{(x-1)(x+1)}$$ *Find products; subtract numerators and keep common denominator:* $\frac{A}{B} - \frac{C}{B} = \frac{A-C}{B}$

$$= \frac{x^2 - 2x + 1 - (x^2 + 2x + 1)}{(x-1)(x+1)}$$ *Find products.*

$$= \frac{x^2 - 2x + 1 - x^2 - 2x - 1}{(x-1)(x+1)}$$ *Subtract trinomial.*

$$= \frac{-4x}{(x-1)(x+1)}$$ *Combine like terms.*

The result is in lowest terms. We use a graphing calculator table to verify our work (see Fig. 16). The values $x = -1$ and $x = 1$ give "ERROR" messages, because -1 and 1 are not in the domain of either the function $y = \frac{x-1}{x+1} - \frac{x+1}{x-1}$ or the function $y = \frac{-4x}{(x-1)(x+1)}$ (why?).

2. $(f - g)(5) = \frac{-4(5)}{(5-1)(5+1)} = \frac{-20}{24} = -\frac{5}{6}$

Group Exploration

Adding and subtracting rational expressions

In Problems 1–3, a student tries to perform an operation and simplify the result. If the work is correct, decide whether there is a more efficient way to do the problem. If the work is incorrect, describe any errors and do the problem correctly.

1. Find the sum $\frac{5}{x+1} + \frac{2}{x+3}$. Simplify the result.

$$\frac{5}{x+1} + \frac{2}{x+3} = \left(\frac{5}{x+1} + \frac{2}{x+3}\right) \cdot (x+1)(x+3)$$
$$= \frac{5}{x+1} \cdot (x+1)(x+3) + \frac{2}{x+3} \cdot (x+1)(x+3)$$
$$= 5(x+3) + 2(x+1)$$
$$= 7x + 17$$

2. Find the difference $\frac{5x}{x-7} - \frac{3x+4}{x-7}$. Simplify the result.

$$\frac{5x}{x-7} - \frac{3x+4}{x-7} = \frac{5x - 3x + 4}{x-7}$$
$$= \frac{2x+4}{x-7}$$

3. Find the sum $\frac{4}{(x-2)(x+3)} + \frac{1}{x-2}$. Simplify the result.

$$\frac{4}{(x-2)(x+3)} + \frac{1}{x-2}$$
$$= \frac{4}{(x-2)(x+3)} \cdot \frac{x-2}{x-2} + \frac{1}{x-2} \cdot \frac{(x-2)(x+3)}{(x-2)(x+3)}$$
$$= \frac{4(x-2) + (x-2)(x+3)}{(x-2)(x-2)(x+3)}$$
$$= \frac{4x - 8 + x^2 + x - 6}{(x-2)^2(x+3)}$$
$$= \frac{x^2 + 5x - 14}{(x-2)^2(x+3)}$$
$$= \frac{(x-2)(x+7)}{(x-2)^2(x+3)}$$
$$= \frac{x+7}{(x-2)(x+3)}$$

Homework 12.3

For extra help ▶ MyMathLab®  Watch the videos in MyMathLab  Download the MyDashboard App

Perform the indicated operation. Simplify the result.

1. $\dfrac{7}{x}+\dfrac{2}{x}$
2. $\dfrac{3}{x^2}+\dfrac{5}{x^2}$
3. $\dfrac{9x}{x-2}-\dfrac{2x}{x-2}$
4. $\dfrac{4x}{x+7}-\dfrac{6x}{x+7}$
5. $\dfrac{t^2}{t+5}+\dfrac{7t+10}{t+5}$
6. $\dfrac{a^2+9a}{a+3}+\dfrac{18}{a+3}$
7. $\dfrac{x}{x^2-9}-\dfrac{3}{x^2-9}$
8. $\dfrac{x}{x^2-25}-\dfrac{5}{x^2-25}$
9. $\dfrac{x^2}{x+1}-\dfrac{2x+3}{x+1}$
10. $\dfrac{x^2}{x-7}-\dfrac{3x+28}{x-7}$
11. $\dfrac{c^2-5c}{c^2+5c+6}+\dfrac{4c-12}{c^2+5c+6}$
12. $\dfrac{m^2-4m}{m^2+3m-4}+\dfrac{6m-8}{m^2+3m-4}$
13. $\dfrac{3x^2+9x}{x^2+10x+21}-\dfrac{2x^2+x-15}{x^2+10x+21}$
14. $\dfrac{2x^2-4x}{3x^2-6x}-\dfrac{x^2+x-6}{3x^2-6x}$
15. $\dfrac{3}{x}+\dfrac{5}{2x}$
16. $\dfrac{8}{3x}+\dfrac{2}{x}$
17. $\dfrac{5}{2b}-\dfrac{3}{8}$
18. $\dfrac{2}{3t}-\dfrac{5}{6}$
19. $\dfrac{5x}{6}+\dfrac{3}{4x}$
20. $\dfrac{4}{9x}+\dfrac{7x}{6}$
21. $\dfrac{2}{x^6}-\dfrac{4}{x^2}$
22. $\dfrac{4}{x^5}-\dfrac{7}{x^3}$
23. $\dfrac{3}{10x^6}+\dfrac{5}{12x^4}$
24. $\dfrac{3}{14x^2}+\dfrac{4}{21x^9}$
25. $\dfrac{7}{4a^2b}-\dfrac{5}{6ab^3}$
26. $\dfrac{5}{6ab^3}-\dfrac{2}{9a^4b^2}$
27. $\dfrac{5}{4x}+\dfrac{2}{x+3}$
28. $\dfrac{7}{x-5}+\dfrac{4}{3x}$
29. $\dfrac{3}{x+1}+\dfrac{4}{x-2}$
30. $\dfrac{3}{x-2}+\dfrac{2}{x+3}$
31. $\dfrac{3x}{x-2}-\dfrac{5}{x+3}$
32. $\dfrac{4}{x-4}-\dfrac{3x}{x+2}$
33. $\dfrac{1}{a+b}+\dfrac{1}{a-b}$
34. $\dfrac{1}{x+y}+\dfrac{1}{x+2y}$
35. $\dfrac{c}{2c-8}-\dfrac{3}{c-4}$
36. $\dfrac{y}{4y-20}-\dfrac{6}{y-5}$
37. $\dfrac{2x}{5x-25}+\dfrac{4}{3x-15}$
38. $\dfrac{3x}{4x+12}+\dfrac{2}{6x+18}$
39. $\dfrac{6}{k^2-1}+\dfrac{3}{k+1}$
40. $\dfrac{1}{p+2}+\dfrac{4}{p^2-4}$
41. $\dfrac{3x-1}{x^2+2x-15}-\dfrac{2}{x+5}$
42. $\dfrac{6x}{x^2-8x+15}-\dfrac{9}{x-3}$
43. $\dfrac{3}{x^2-25}+\dfrac{5}{x^2-5x}$
44. $\dfrac{-2}{x^2-2x-3}+\dfrac{3}{x^2-9}$
45. $\dfrac{2}{x^2-9}+\dfrac{3}{x^2-7x+12}$
46. $\dfrac{4}{x^2-2x}+\dfrac{1}{x^2-5x+6}$
47. $\dfrac{5}{3t-6}-\dfrac{2}{5t+15}$
48. $\dfrac{b}{8b-4}-\dfrac{2}{3b-6}$
49. $2+\dfrac{k-3}{k+1}$
50. $\dfrac{w+2}{w-5}+3$
51. $2-\dfrac{2x+4}{x^2+3x+2}$
52. $\dfrac{6x^2+2x-4}{x^2-1}-5$
53. $\dfrac{8}{x-6}-\dfrac{4}{6-x}$
54. $\dfrac{x}{x-3}-\dfrac{2}{3-x}$
55. $\dfrac{2x+1}{x^2-4x-21}+\dfrac{3}{14-2x}$
56. $\dfrac{4x+5}{x^2+3x-40}+\dfrac{2}{15-3x}$
57. $\dfrac{-2c}{7-2c}-\dfrac{c+1}{4c^2-49}$
58. $\dfrac{-5m}{5-3m}-\dfrac{m+2}{9m^2-25}$
59. $\dfrac{2b}{a^2-b^2}+\dfrac{a}{ab-b^2}$
60. $\dfrac{n}{m^2+3mn}+\dfrac{3m}{2mn+6n^2}$
61. $\dfrac{x}{x^2+5x+6}-\dfrac{3}{x^2+7x+12}$
62. $\dfrac{x}{x^2+11x+30}-\dfrac{5}{x^2+9x+20}$
63. $\dfrac{x-1}{x+2}+\dfrac{x+2}{x-1}$
64. $\dfrac{x+4}{x-6}+\dfrac{x-6}{x+4}$
65. $\dfrac{y-5}{y-3}-\dfrac{y+3}{y+5}$
66. $\dfrac{k+2}{k-1}-\dfrac{k+1}{k-2}$
67. $\dfrac{x+4}{x^2-7x+10}-\dfrac{5}{x^2-25}$
68. $\dfrac{x-2}{x^2-x-2}-\dfrac{4}{x^2-1}$
69. $\dfrac{x+2}{(x-4)(x+3)^2}+\dfrac{x-1}{(x-4)(x+1)(x+3)}$
70. $\dfrac{x+1}{(x+2)^2(x+3)}+\dfrac{x+5}{(x+2)(x+3)(x+4)}$
71. $\dfrac{c+2}{c^2-4}+\dfrac{3c}{c^2-2c}$
72. $\dfrac{w+5}{w^2-25}+\dfrac{4w}{w^2-5w}$

73. $\dfrac{x-1}{4x^2+20x+25} - \dfrac{x+4}{6x^2+17x+5}$

74. $\dfrac{x-3}{2x^2+11x-6} - \dfrac{x+2}{2x^2+7x-4}$

75. $\dfrac{3x-1}{x^2+4x+4} + \dfrac{2x+1}{3x^2+5x-2}$

76. $\dfrac{x-3}{3x^2-x-4} + \dfrac{2x+5}{6x^2+x-12}$

77. $\dfrac{3p}{p^2-2pq-24q^2} - \dfrac{2q}{p^2-3pq-18q^2}$

78. $\dfrac{2x}{x^2-6xy+8y^2} - \dfrac{5y}{x^2+3xy-10y^2}$

79. $\dfrac{x-1}{6x^2-24x} + \dfrac{5}{3x^3-6x^2-24x}$

80. $\dfrac{3}{2x^3+14x^2+20x} + \dfrac{x+2}{14x^2+28x}$

Perform the indicated operations. Simplify your result.

81. $\left(\dfrac{2}{x^2-4} + \dfrac{3}{x+2}\right) - \dfrac{1}{2x-4}$

82. $\left(\dfrac{5}{3x-9} - \dfrac{2x-1}{x^2-9}\right) + \dfrac{4}{x+3}$

83. $\dfrac{3}{t+1} - \left(\dfrac{2t-3}{t^2+6t+5} + \dfrac{2}{t+5}\right)$

84. $\dfrac{2k+11}{k^2+k-6} - \left(\dfrac{2}{k+3} + \dfrac{3}{2-k}\right)$

Let $f(x) = \dfrac{x+3}{x-4}$ *and* $g(x) = \dfrac{x+4}{x-3}$*. Find an equation of the given function.*

85. $f+g$ **86.** $f-g$ **87.** $g-f$

Let $f(x) = \dfrac{x-2}{x^2-2x-8}$ *and* $g(x) = \dfrac{x+1}{3x+6}$*. Find an equation of the given function.*

88. $f+g$ **89.** $f-g$ **90.** $g-f$

91. There are two 100-watt lights in a room. A person is twice as far from one light as from the other. The illumination (the brightness of the light) can be measured in watts per square meter (W/m^2). The illumination (in W/m^2) from the closer light source is

$$\frac{18}{d^2}$$

where d is the distance (in meters) to that light. The illumination (in W/m^2) from the other light source is

$$\frac{18}{(2d)^2}$$

where $2d$ is the distance (in meters) to that light.

a. Find an expression for the total illumination (in W/m^2) from the two light sources.

b. Write your result from part (a) as a single fraction.

c. Evaluate the expression you found in part (b) for $d = 1.2$. What does your result mean in this situation?

92. A student plans to drive from El Camino College in Torrance, California, to Chandler-Gilbert Community College in Sun Lakes, Arizona. The trip involves 231 miles in California, followed by 174 miles in Arizona. The speed limit is 70 mph in California and 75 mph in Arizona. The student plans to drive at a mph above the speed limits. The driving time (in hours) in California is

$$\frac{231}{a+70}$$

The driving time (in hours) in Arizona is

$$\frac{174}{a+75}$$

a. Find an expression for the total driving time (in hours).

b. Write your result from part (a) as a single fraction.

c. Evaluate the expression you found in part (b) for $a = 5$. What does your result mean in this situation?

Concepts

93. A student tries to find the sum $\dfrac{2}{x+1} + \dfrac{3}{x+2}$:

$$\begin{aligned}\frac{2}{x+1} + \frac{3}{x+2} &= \frac{2}{x+1}\cdot\frac{1}{x+2} + \frac{3}{x+2}\cdot\frac{1}{x+1}\\ &= \frac{2}{(x+1)(x+2)} + \frac{3}{(x+2)(x+1)}\\ &= \frac{5}{(x+1)(x+2)}\end{aligned}$$

Describe any errors. Then find the sum correctly.

94. A student tries to find the difference $\dfrac{6x}{x+4} - \dfrac{3x+2}{x+4}$:

$$\begin{aligned}\frac{6x}{x+4} - \frac{3x+2}{x+4} &= \frac{6x-(3x+2)}{x+4}\\ &= \frac{6x-3x+2}{x+4}\\ &= \frac{3x+2}{x+4}\end{aligned}$$

Describe any errors. Then find the difference correctly.

95. A student tries to find the difference $\dfrac{9x}{x-3} - \dfrac{5x+1}{x-3}$:

$$\begin{aligned}\frac{9x}{x-3} - \frac{5x+1}{x-3} &= \frac{9x-5x+1}{x-3}\\ &= \frac{4x+1}{x-3}\end{aligned}$$

Find any errors. Then find the difference correctly.

96. Two students try to find the sum $\dfrac{3}{x-4} + \dfrac{2}{4-x}$:

Student A's Work

$$\begin{aligned}\frac{3}{x-4} + \frac{2}{4-x} &= \frac{3}{x-4} + \frac{2}{-(x-4)}\\ &= \frac{3}{x-4} - \frac{2}{x-4}\\ &= \frac{1}{x-4}\end{aligned}$$

Student B's Work

$$\frac{3}{x-4}+\frac{2}{4-x}=\frac{3}{x-4}\cdot\frac{4-x}{4-x}+\frac{2}{4-x}\cdot\frac{x-4}{x-4}$$
$$=\frac{3(4-x)+2(x-4)}{(x-4)(4-x)}$$
$$=\frac{12-3x+2x-8}{(x-4)(4-x)}$$
$$=\frac{4-x}{(x-4)(4-x)}$$
$$=\frac{1}{x-4}$$

Compare the two methods. Are both correct? Explain. Discuss why student A's method is shorter.

97. Explain how adding rational expressions is similar to adding rational numbers.

98. Why do we factor the denominators of rational expressions before finding the LCD?

99. Describe how to add two rational expressions that have different denominators. Then describe how to subtract two such expressions.

100. When subtracting rational expressions, we subtract the *entire* numerator. Give an example to show how to do this and what can go wrong if we subtract only part of the numerator.

Related Review

Find equations of $f+g$ and $f-g$.

101. $f(x)=6x^2-4x+3, g(x)=2x^2-7x-5$

102. $f(x)=3x^2+8x-2, g(x)=-5x^2-3x+4$

103. $f(x)=2(5)^x, g(x)=-3(5)^x$

104. $f(x)=4(3)^x, g(x)=7(3)^x$

Perform the indicated operations. Simplify your result.

105. $\dfrac{4x+5}{x+2}+\left(\dfrac{3x+15}{x^2-4}\cdot\dfrac{x^2-2x}{x^2+7x+10}\right)$

106. $\dfrac{2x-7}{x-5}-\left(\dfrac{2x+10}{x^2+9x+20}\div\dfrac{x^2-25}{3x+12}\right)$

107. $\dfrac{5x+5}{3x+6}\cdot\left(\dfrac{x^2+4x}{x^2+2x+1}+\dfrac{4}{x^2+2x+1}\right)$

108. $\dfrac{x^2-16}{2x-12}\div\left(\dfrac{x^2}{x^2-36}-\dfrac{2x+8}{x^2-36}\right)$

Expressions, Equations, Functions, and Graphs

Perform the indicated instruction. Then use words such as linear, quadratic, cubic, exponential, logarithmic, rational, polynomial, degree, function, one variable, *and* two variables *to describe the expression, equation, or system.*

109. Graph $f(x)=3\left(\frac{1}{3}\right)^x$ by hand.

110. Find the product $(3p-2)(9p^2+6p+4)$.

111. Find all real-number solutions of $5b^4=66$. Round any results to the fourth decimal place.

112. Factor $64p^3-27$.

113. Find an approximate equation $y=ab^x$ of an exponential curve that contains the points $(3, 95)$ and $(7, 2)$. Round the values of a and b to the second decimal place.

114. Solve $6p^3+21p^2=12p$.

12.4 Simplifying Complex Rational Expressions

Objective

» Simplify *complex rational expressions.*

In this section, we will work with complex rational expressions. A **complex rational expression** is a rational expression whose numerator or denominator (or both) is a rational expression. Here are some examples of such expressions:

$$\frac{\frac{x^2}{2}-\frac{3}{x^3}}{\frac{x}{6}+\frac{7}{x^2}}\qquad\frac{\frac{3x}{x-1}}{\frac{x^3}{x-2}}\qquad\frac{5x}{\frac{x^2-2x+1}{x+4}}$$

Here we find the values of two numerical complex rational expressions:

$$\frac{2}{\frac{2}{2}}=\frac{2}{1}=2\qquad\frac{\frac{2}{2}}{2}=\frac{1}{2}$$

From these two examples, we see that it is important to keep track of the main fraction bar (the longest one, in bold) of the complex fraction.

We will discuss two methods for simplifying complex rational expressions. Ask your instructor whether you are required to know method 1, method 2, or both methods. If the choice is yours, compare the use of method 1 in Examples 1 and 2 with the use of method 2 in Examples 3 and 4. The complex rational expressions in these examples are simplified by both methods so you can get a sense of the advantages and disadvantages of each.

Method 1: Writing a Complex Rational Expression as a Quotient of Two Rational Expressions

An expression in the form $\frac{R}{S}$, where R and S are themselves expressions, can be written in the form $R \div S$. We use this fact to help simplify a complex rational expression:

$$\frac{\frac{5}{3}}{\frac{7}{2}} = \frac{5}{3} \div \frac{7}{2} \qquad \frac{R}{S} = R \div S$$

$$= \frac{5}{3} \cdot \frac{2}{7} \qquad \text{Multiply by reciprocal of } \frac{7}{2}.$$

$$= \frac{10}{21} \qquad \text{Multiply numerators; multiply denominators.}$$

We **simplify a complex rational expression** by writing it as a rational expression $\frac{P}{Q}$, with $\frac{P}{Q}$ in lowest terms.

Example 1 Simplifying a Complex Rational Expression by Method 1

Simplify by method 1.

1. $\dfrac{\frac{12}{x}}{\frac{8}{x^3}}$

2. $\dfrac{\frac{x^2-9}{x^2+2x+1}}{\frac{2x-6}{4x+4}}$

Solution

1. $$\frac{\frac{12}{x}}{\frac{8}{x^3}} = \frac{12}{x} \div \frac{8}{x^3} \qquad \frac{R}{S} = R \div S$$

$$= \frac{12}{x} \cdot \frac{x^3}{8} \qquad \text{Multiply by reciprocal of } \frac{8}{x^3}\text{, which is } \frac{x^3}{8}.$$

$$= \frac{2 \cdot 2 \cdot 3}{x} \cdot \frac{x^3}{2 \cdot 2 \cdot 2} \qquad \text{Factor numerator and denominator.}$$

$$= \frac{2 \cdot 2 \cdot 3 \cdot x^3}{x \cdot 2 \cdot 2 \cdot 2} \qquad \text{Multiply numerators; multiply denominators.}$$

$$= \frac{3x^2}{2} \qquad \text{Simplify: } \frac{2 \cdot 2 \cdot x}{2 \cdot 2 \cdot x} = 1$$

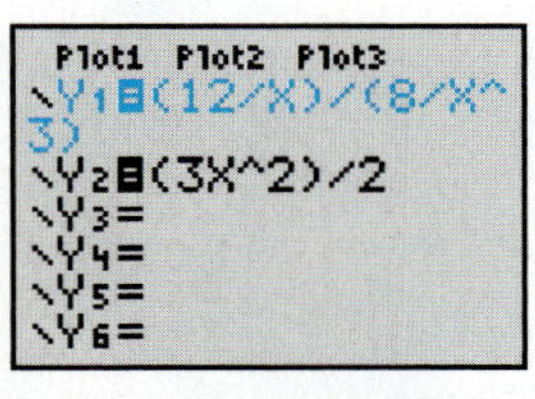

X	Y1	Y2
-3	13.5	13.5
-2	6	6
-1	1.5	1.5
0	ERROR	0
1	1.5	1.5
2	6	6
3	13.5	13.5

X=-3

Figure 17 Verify the work

We use a graphing calculator table to verify our work (see Fig. 17). The value $x = 0$ gives an "ERROR" message, because 0 is an excluded value of the numerator (and the denominator) of the original expression.

2. $$\frac{\frac{x^2-9}{x^2+2x+1}}{\frac{2x-6}{4x+4}} = \frac{x^2-9}{x^2+2x+1} \div \frac{2x-6}{4x+4} \qquad \frac{R}{S} = R \div S$$

$$= \frac{x^2-9}{x^2+2x+1} \cdot \frac{4x+4}{2x-6} \qquad \text{Multiply by reciprocal of } \frac{2x-6}{4x+4}.$$

$$= \frac{(x-3)(x+3)}{(x+1)(x+1)} \cdot \frac{2 \cdot 2(x+1)}{2(x-3)}$$ Factor numerators and denominators.

$$= \frac{(x-3)(x+3) \cdot 2 \cdot 2 \cdot (x+1)}{(x+1)(x+1) \cdot 2(x-3)}$$ Multiply numerators; multiply denominators.

$$= \frac{2(x+3)}{x+1}$$ Simplify: $\frac{2(x-3)(x+1)}{2(x-3)(x+1)} = 1$

Next, we use method 1 to simplify a complex rational expression that has two rational expressions in the numerator and two rational expressions in the denominator.

Example 2 Simplifying a Complex Rational Expression by Method 1

Simplify $\dfrac{\frac{1}{y^2} + \frac{3}{2x}}{\frac{2}{y} - \frac{1}{3x}}$.

Solution

We write both the numerator and the denominator as fractions and simplify as before:

$$\frac{\frac{1}{y^2} + \frac{3}{2x}}{\frac{2}{y} - \frac{1}{3x}} = \frac{\frac{1}{y^2} \cdot \frac{2x}{2x} + \frac{3}{2x} \cdot \frac{y^2}{y^2}}{\frac{2}{y} \cdot \frac{3x}{3x} - \frac{1}{3x} \cdot \frac{y}{y}}$$ Introduce missing factors to get a common denominator, $2xy^2$. Introduce missing factors to get a common denominator, $3xy$.

$$= \frac{\frac{2x}{2xy^2} + \frac{3y^2}{2xy^2}}{\frac{6x}{3xy} - \frac{y}{3xy}}$$ Find products.

$$= \frac{\frac{2x + 3y^2}{2xy^2}}{\frac{6x - y}{3xy}}$$ Add numerators and keep common denominator. Subtract numerators and keep common denominator.

$$= \frac{2x + 3y^2}{2xy^2} \div \frac{6x - y}{3xy}$$ $\frac{R}{S} = R \div S$

$$= \frac{2x + 3y^2}{2xy^2} \cdot \frac{3xy}{6x - y}$$ Multiply by reciprocal of $\frac{6x - y}{3xy}$.

$$= \frac{(2x + 3y^2) \cdot 3xy}{2xy^2(6x - y)}$$ Multiply numerators; multiply denominators.

$$= \frac{3(2x + 3y^2)}{2y(6x - y)}$$ Simplify: $\frac{xy}{xy} = 1$

Since our result is in lowest terms, we are done.

Using Method 1 to Simplify a Complex Rational Expression

To simplify a complex rational expression by method 1,

1. Write both the numerator and the denominator as fractions.
2. Write the complex rational expression as the quotient of two rational expressions:

$$\frac{\frac{A}{B}}{\frac{C}{D}} = \frac{A}{B} \div \frac{C}{D}, \quad \text{where } B, C, \text{ and } D \text{ are nonzero.}$$

3. Divide the rational expressions.

Method 2: Multiplying by $\frac{\text{LCD}}{\text{LCD}}$

Alternatively, we can simplify a complex rational expression by first finding the LCD of all of the fractions in the numerator and denominator. Then, we multiply by 1 in the form $\frac{\text{LCD}}{\text{LCD}}$.

In Example 3, we will simplify the same complex rational expressions that we simplified in Example 1, but now we will use method 2.

Example 3 Simplifying a Complex Rational Expression by Method 2

Simplify by method 2.

1. $\dfrac{\frac{12}{x}}{\frac{8}{x^3}}$

2. $\dfrac{\frac{x^2 - 9}{x^2 + 2x + 1}}{\frac{2x - 6}{4x + 4}}$

Solution

1. The LCD of $\frac{12}{x}$ and $\frac{8}{x^3}$ is x^3. So, we multiply the complex rational expression by 1 in the form $\frac{x^3}{x^3}$:

$$\frac{\frac{12}{x}}{\frac{8}{x^3}} = \frac{\frac{12}{x}}{\frac{8}{x^3}} \cdot \frac{\frac{x^3}{1}}{\frac{x^3}{1}} \qquad \textit{Multiply by } \frac{\text{LCD}}{\text{LCD}}, \frac{x^3}{x^3} = 1.$$

$$= \frac{\frac{12x^3}{x}}{\frac{8x^3}{x^3}} \qquad \textit{Simplify.}$$

$$= \frac{12x^2}{8} \qquad \textit{Simplify fractions in numerator and in denominator.}$$

$$= \frac{2 \cdot 2 \cdot 3 \cdot x^2}{2 \cdot 2 \cdot 2} \qquad \textit{Factor numerator and denominator.}$$

$$= \frac{3x^2}{2} \qquad \textit{Simplify: } \frac{2 \cdot 2}{2 \cdot 2} = 1$$

The result is the same as our result in Problem 1 of Example 1.

2. $$\frac{\dfrac{x^2-9}{x^2+2x+1}}{\dfrac{2x-6}{4x+4}} = \frac{\dfrac{(x-3)(x+3)}{(x+1)(x+1)}}{\dfrac{2(x-3)}{4(x+1)}}$$ *Factor numerators and denominators of fractions.*

$$= \frac{\dfrac{(x-3)(x+3)}{(x+1)(x+1)}}{\dfrac{2(x-3)}{4(x+1)}} \cdot \frac{\dfrac{4(x+1)(x+1)}{1}}{\dfrac{4(x+1)(x+1)}{1}}$$ *Multiply by* $\frac{\text{LCD}}{\text{LCD}}$, $\frac{4(x+1)(x+1)}{4(x+1)(x+1)} = 1$.

$$= \frac{\dfrac{(x-3)(x+3)\cdot 4(x+1)(x+1)}{(x+1)(x+1)}}{\dfrac{2(x-3)\cdot 4(x+1)(x+1)}{4(x+1)}}$$ *Multiply numerators; multiply denominators.*

$$= \frac{2\cdot 2(x-3)(x+3)}{2(x-3)(x+1)}$$ *Simplify numerator and denominator.*

$$= \frac{2(x+3)}{x+1}$$ *Simplify:* $\frac{2(x-3)}{2(x-3)} = 1$

The result is the same as our result in Problem 2 of Example 1.

In comparing our work in Examples 1 and 3, we see that methods 1 and 2 required about the same number of steps. One advantage that method 1 has over method 2 for *these* complex rational expressions is that method 1 does not require us to find an LCD.

In Example 4, we will simplify the same expression as in Example 2, but now we will use method 2.

Example 4 Simplifying a Complex Rational Expression by Method 2

Simplify $\dfrac{\dfrac{1}{y^2}+\dfrac{3}{2x}}{\dfrac{2}{y}-\dfrac{1}{3x}}$ by method 2.

Solution

The LCD of the rational expressions in the numerator and the denominator is $6xy^2$. To simplify, we multiply by $\dfrac{6xy^2}{6xy^2}$:

$$\frac{\dfrac{1}{y^2}+\dfrac{3}{2x}}{\dfrac{2}{y}-\dfrac{1}{3x}} = \frac{\dfrac{1}{y^2}+\dfrac{3}{2x}}{\dfrac{2}{y}-\dfrac{1}{3x}} \cdot \frac{6xy^2}{6xy^2}$$ *Multiply by* $\frac{\text{LCD}}{\text{LCD}}$, $\frac{6xy^2}{6xy^2}$.

$$= \frac{\dfrac{1}{y^2}\cdot\dfrac{6xy^2}{1}+\dfrac{3}{2x}\cdot\dfrac{6xy^2}{1}}{\dfrac{2}{y}\cdot\dfrac{6xy^2}{1}-\dfrac{1}{3x}\cdot\dfrac{6xy^2}{1}}$$ *Distributive law*

$$= \frac{6x+9y^2}{12xy-2y^2}$$ *Simplify.*

$$= \frac{3(2x+3y^2)}{2y(6x-y)}$$ *Factor numerator and denominator.*

The result is the same as our result in Example 2.

WARNING For the first step in Example 4, it would be incorrect to multiply by the fraction $\frac{\text{LCD of the numerator}}{\text{LCD of the denominator}} = \frac{2xy^2}{3xy}$:

$$\frac{\frac{1}{y^2} + \frac{3}{2x}}{\frac{2}{y} - \frac{1}{3x}} = \frac{\frac{1}{y^2} + \frac{3}{2x}}{\frac{2}{y} - \frac{1}{3x}} \cdot \frac{2xy^2}{3xy} \quad \text{Incorrect}$$

This is incorrect because the expression $\frac{2xy^2}{3xy}$ is not equivalent to 1. It *is* correct to multiply by $\frac{6xy^2}{6xy^2} = 1$.

In comparing our work in Examples 2 and 4, we see that method 2 required fewer steps than method 1 did. In general, when the numerator, denominator, or both contain two rational expressions, method 2 is usually easier to use.

Using Method 2 to Simplify a Complex Rational Expression

To simplify a rational expression by method 2,

1. Find the LCD of all of the fractions in the numerator and denominator.
2. Multiply by 1 in the form $\frac{\text{LCD}}{\text{LCD}}$.
3. Simplify the numerator and the denominator to polynomials.
4. Simplify the rational expression.

Example 5 Simplifying a Complex Rational Expression by Method 2

Simplify $\frac{3 - \frac{12}{x^2}}{2 - \frac{4}{x}}$ by method 2.

Solution

The LCD of the rational expressions in the numerator and the denominator is x^2. To simplify, we multiply by $\frac{x^2}{x^2}$:

$$\begin{aligned}
\frac{3 - \frac{12}{x^2}}{2 - \frac{4}{x}} &= \frac{3 - \frac{12}{x^2}}{2 - \frac{4}{x}} \cdot \frac{x^2}{x^2} && \text{Multiply by } \frac{\text{LCD}}{\text{LCD}}, \frac{x^2}{x^2}. \\
&= \frac{3x^2 - \frac{12}{x^2} \cdot \frac{x^2}{1}}{2x^2 - \frac{4}{x} \cdot \frac{x^2}{1}} && \text{Distributive law} \\
&= \frac{3x^2 - 12}{2x^2 - 4x} && \text{Simplify.} \\
&= \frac{3(x^2 - 4)}{2x(x - 2)} && \text{Factor numerator and denominator.} \\
&= \frac{3(x - 2)(x + 2)}{2x(x - 2)} && \text{Factor numerator.} \\
&= \frac{3(x + 2)}{2x} && \text{Simplify: } \frac{x - 2}{x - 2} = 1
\end{aligned}$$

In Example 6, we will form the quotient function of two rational functions.

Example 6 Finding a Quotient Function by Method 2

Let $f(x) = 2 - \frac{5}{x+2}$ and $g(x) = \frac{x}{x+2} + \frac{x+1}{x^2 - 4x - 12}$.

1. Find an equation of $\frac{f}{g}$.

2. Find $\left(\frac{f}{g}\right)(4)$.

Solution

1. $\left(\frac{f}{g}\right)(x) = \frac{f(x)}{g(x)}$ *Definition of $\frac{f}{g}$*

$$= \frac{2 - \frac{5}{x+2}}{\frac{x}{x+2} + \frac{x+1}{x^2 - 4x - 12}}$$ *Substitute for $f(x)$ and $g(x)$.*

Since the denominator (and the numerator) contains two rational expressions, method 2 should be easier to use than method 1:

$$\left(\frac{f}{g}\right)(x) = \frac{2 - \frac{5}{x+2}}{\frac{x}{x+2} + \frac{x+1}{(x-6)(x+2)}}$$ *Factor $x^2 - 4x - 12$.*

$$= \frac{2 - \frac{5}{x+2}}{\frac{x}{x+2} + \frac{x+1}{(x-6)(x+2)}} \cdot \frac{(x-6)(x+2)}{(x-6)(x+2)}$$ *Multiply by $\frac{\text{LCD}}{\text{LCD}}$.*

$$= \frac{2(x-6)(x+2) - \frac{5}{x+2} \cdot \frac{(x-6)(x+2)}{1}}{\frac{x}{x+2} \cdot \frac{(x-6)(x+2)}{1} + \frac{x+1}{(x-6)(x+2)} \cdot \frac{(x-6)(x+2)}{1}}$$ *Distributive law*

$$= \frac{2(x-6)(x+2) - 5(x-6)}{x(x-6) + (x+1)}$$ *Simplify.*

$$= \frac{2x^2 - 8x - 24 - 5x + 30}{x^2 - 6x + x + 1}$$ *Find products.*

$$= \frac{2x^2 - 13x + 6}{x^2 - 5x + 1}$$ *Combine like terms.*

$$= \frac{(2x-1)(x-6)}{x^2 - 5x + 1}$$ *Factor numerator.*

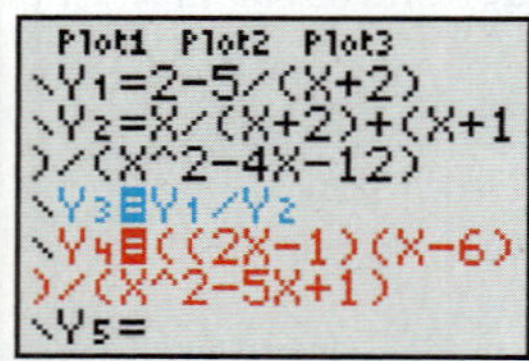

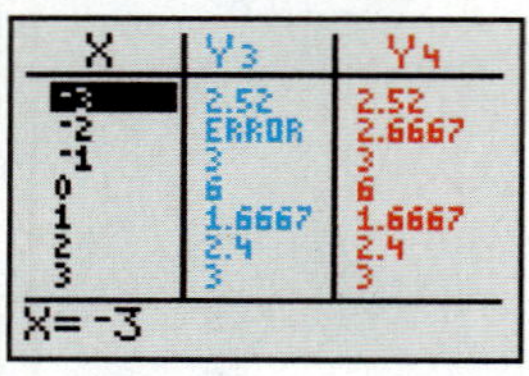

Figure 18 Verify the work

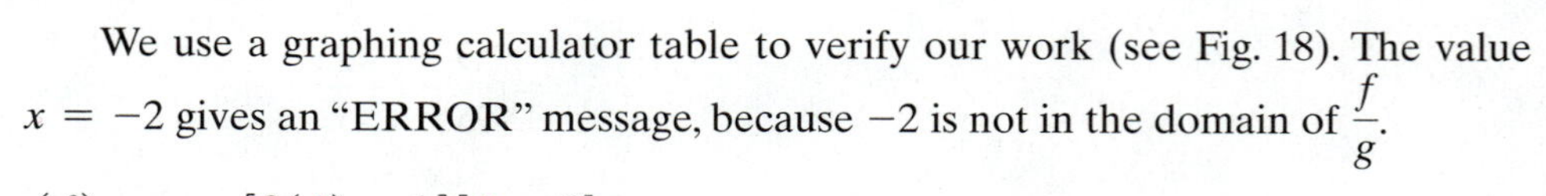

We use a graphing calculator table to verify our work (see Fig. 18). The value $x = -2$ gives an "ERROR" message, because -2 is not in the domain of $\frac{f}{g}$.

2. $\left(\frac{f}{g}\right)(4) = \frac{[2(4) - 1][4 - 6]}{(4)^2 - 5(4) + 1} = \frac{-14}{-3} = \frac{14}{3}$

Our work in Example 6 suggests that if two expressions R and S are rational, then the quotient $\frac{R}{S}$ is rational. This is, indeed, true if the numerator of S is nonzero.

Group Exploration

Looking ahead: Solving rational equations

1. Solve $\frac{1}{2} + \frac{x}{3} = \frac{5}{6}$.

2. Solve $\frac{1}{x-2} + \frac{2}{3} = \frac{5}{x-2}$. **[Hint:** Multiply both sides by the LCD of all three fractions.]

3. Solve $\frac{x}{x-2} - 5 = \frac{2}{x-2}$. Check whether your result satisfies the equation. What is the solution set? Explain.

4. The equations in Problems 1–3 are called *rational equations in one variable.* What does your work in Problem 3 suggest you should always do when solving a rational equation in one variable?

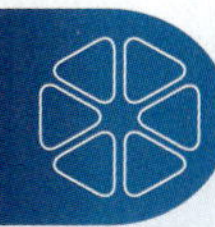

Homework 12.4

For extra help ▶   Watch the videos in MyMathLab   Download the MyDashboard App

Simplify.

1. $\dfrac{\frac{4}{5}}{\frac{8}{3}}$

2. $\dfrac{\frac{3}{4}}{\frac{5}{6}}$

3. $\dfrac{\frac{x}{4}}{\frac{x}{7}}$

4. $\dfrac{\frac{3}{x}}{\frac{5}{x}}$

5. $\dfrac{\frac{6}{5x}}{\frac{3}{7x}}$

6. $\dfrac{\frac{4}{3x}}{\frac{8}{5x}}$

7. $\dfrac{\frac{w}{6}}{\frac{w^2}{9}}$

8. $\dfrac{\frac{10}{y}}{\frac{12}{y^2}}$

9. $\dfrac{\frac{15x^3}{8}}{\frac{25x^5}{12}}$

10. $\dfrac{\frac{7}{6x^8}}{\frac{21}{4x^7}}$

11. $\dfrac{\frac{4a^2}{5b}}{\frac{6a}{15b^3}}$

12. $\dfrac{\frac{9b^3}{8a}}{\frac{3b}{20a^4}}$

13. $\dfrac{\frac{14}{x^2-9}}{\frac{21}{x-3}}$

14. $\dfrac{\frac{8}{x^2-16}}{\frac{20}{x+4}}$

15. $\dfrac{\frac{5a-10}{4}}{\frac{3a-6}{2}}$

16. $\dfrac{\frac{3t-12}{14}}{\frac{4t-16}{21}}$

17. $\dfrac{\frac{3x-3}{2x+10}}{\frac{6x^2-6}{4x+20}}$

18. $\dfrac{\frac{6x-12}{5x-5}}{\frac{3x^2-12}{2x-6}}$

19. $\dfrac{\frac{x^2-49}{3x^2-9x}}{\frac{x^2-5x-14}{7x-21}}$

20. $\dfrac{\frac{x^2-25}{5x-30}}{\frac{x^2+x-30}{2x^2-12x}}$

21. $\dfrac{\frac{25x^2-4}{9x^2-16}}{\frac{25x^2-20x+4}{9x^2-24x+16}}$

22. $\dfrac{\frac{4x^2+28x+49}{25x^2-30x+9}}{\frac{4x^2-49}{25x^2-9}}$

23. $\dfrac{\frac{4}{x}+\frac{2}{x}}{\frac{9}{x}-\frac{7}{x}}$

24. $\dfrac{\frac{5}{x}-\frac{2}{x}}{\frac{3}{x}+\frac{4}{x}}$

25. $\dfrac{\frac{3}{8}+\frac{1}{4}}{\frac{1}{2}+\frac{5}{8}}$

26. $\dfrac{\frac{2}{3}+\frac{5}{6}}{\frac{7}{12}-\frac{1}{6}}$

27. $\dfrac{\frac{7}{4x}+\frac{1}{x}}{\frac{3}{2x}+\frac{5}{x}}$

28. $\dfrac{\frac{2}{x}+\frac{5}{3x}}{\frac{8}{5x}-\frac{1}{x}}$

29. $\dfrac{\frac{5}{3r}-\frac{3}{2r}}{\frac{1}{2r}-\frac{4}{3r}}$

30. $\dfrac{\frac{7}{2t}+\frac{3}{5t}}{\frac{3}{5t}+\frac{5}{2t}}$

31. $\dfrac{\frac{2}{3}-\frac{4}{3x}}{\frac{5}{6x}}$

32. $\dfrac{\frac{5}{4}-\frac{7}{4x}}{\frac{3}{8x}}$

33. $\dfrac{\frac{2}{x^3}-\frac{3}{x}}{\frac{5}{x^3}+\frac{4}{x^2}}$

34. $\dfrac{\frac{3}{x}-\frac{1}{x^2}}{\frac{3}{x^2}-\frac{1}{x^3}}$

35. $\dfrac{4+\frac{3}{x}}{\frac{2}{x}-3}$

36. $\dfrac{\frac{3}{x}+5}{\frac{2}{x}+4}$

37. $\dfrac{\frac{5}{2x^3}-4}{\frac{1}{6x^3}-3}$

38. $\dfrac{\frac{3}{4x^2}-2}{4-\frac{5}{8x^2}}$

39. $\dfrac{\frac{a^2}{b}-b}{\frac{1}{b}-\frac{1}{a}}$

40. $\dfrac{\frac{m^2}{9n}-n}{\frac{1}{n}+\frac{3}{m}}$

41. $\dfrac{\frac{1}{x}-\frac{8}{x^2}+\frac{15}{x^3}}{\frac{1}{x}-\frac{5}{x^2}}$

42. $\dfrac{\frac{4}{x}-\frac{1}{x^2}+\frac{3}{x^3}}{\frac{2}{x}-\frac{5}{x^2}}$

43. $\dfrac{\frac{x}{x-4}-\frac{2x}{x+1}}{\frac{x}{x+1}-\frac{2x}{x-4}}$

44. $\dfrac{\frac{2}{x-3}+\frac{5}{x-2}}{\frac{3}{x-2}-\frac{4}{x-3}}$

45. $\dfrac{p + \dfrac{2}{p-4}}{p - \dfrac{3}{p-4}}$

46. $\dfrac{\dfrac{3}{w+1} + 2}{4 - \dfrac{5}{w-1}}$

47. $\dfrac{\dfrac{1}{x+3} - \dfrac{1}{x}}{3}$

48. $\dfrac{\dfrac{2}{x+4} - \dfrac{2}{x}}{4}$

49. $\dfrac{\dfrac{3}{a+b} - \dfrac{3}{a-b}}{2ab}$

50. $\dfrac{\dfrac{4}{2a-b} - \dfrac{2}{a-3b}}{5ab^2}$

51. $\dfrac{\dfrac{1}{(x+2)^2} - \dfrac{1}{x^2}}{2}$

52. $\dfrac{\dfrac{5}{(x+3)^2} - \dfrac{5}{x^2}}{3}$

53. $\dfrac{\dfrac{6}{2x-8} + \dfrac{10}{x^2-4x}}{\dfrac{1}{x^2-x-12} - \dfrac{2}{x^2-16}}$

54. $\dfrac{\dfrac{4}{3x-9} - \dfrac{1}{x^2-3x}}{\dfrac{2}{x^2-2x-15} + \dfrac{5}{x^2-9}}$

55. $\dfrac{\dfrac{x+7}{x^2+7x+10} - \dfrac{6}{x^2+2x}}{\dfrac{x+1}{x^2+7x+10} + \dfrac{6}{x^2+5x}}$

56. $\dfrac{\dfrac{x-1}{x^2+x-12} + \dfrac{2}{x^2-3x}}{\dfrac{x+2}{x^2+x-12} - \dfrac{3}{x^2+4x}}$

Use the given functions to find an equation of $\dfrac{f}{g}$.

57. $f(x) = \dfrac{5x+10}{x^2-6x+9}$ and $g(x) = \dfrac{4x+8}{x^2-4x+3}$

58. $f(x) = \dfrac{5x-10}{2x^2+x-15}$ and $g(x) = \dfrac{7x-14}{x^2+5x+6}$

59. $f(x) = \dfrac{x}{2} - \dfrac{2}{x}$ and $g(x) = \dfrac{3}{2} + \dfrac{3}{x}$

60. $f(x) = \dfrac{x}{5} - \dfrac{5}{x}$ and $g(x) = \dfrac{8}{5} - \dfrac{8}{x}$

61. $f(x) = \dfrac{2x}{x^2-25} + \dfrac{x+5}{x-5}$ and $g(x) = \dfrac{x-5}{x+5} + \dfrac{3x}{x^2-25}$

62. $f(x) = \dfrac{6x}{x^2-9} + \dfrac{x-3}{x+3}$ and $g(x) = \dfrac{x+3}{x-3} - \dfrac{6x}{x^2-9}$

Concepts

63. A student tries to simplify the complex rational expression $\dfrac{x}{\dfrac{1}{x}+\dfrac{1}{2}}$:

$$\frac{x}{\frac{1}{x}+\frac{1}{2}} = x \div \left(\frac{1}{x}+\frac{1}{2}\right) = x\left(\frac{x}{1}+\frac{2}{1}\right) = x \cdot (x+2) = x^2 + 2x$$

Describe any errors. Then simplify the expression correctly.

64. A student tries to simplify the complex rational expression $\dfrac{\dfrac{5}{2}+\dfrac{3}{x}}{\dfrac{7}{x}+\dfrac{4}{x^2}}$:

$$\frac{\frac{5}{2}+\frac{3}{x}}{\frac{7}{x}+\frac{4}{x^2}} = \frac{\frac{5}{2}+\frac{3}{x}}{\frac{7}{x}+\frac{4}{x^2}} \cdot \frac{2x}{x^2} = \frac{\frac{5}{2}\cdot\frac{2x}{1}+\frac{3}{x}\cdot\frac{2x}{1}}{\frac{7}{x}\cdot\frac{x^2}{1}+\frac{4}{x^2}\cdot\frac{x^2}{1}} = \frac{5x+6}{7x+4}$$

Describe any errors. Then simplify the expression correctly.

65. Simplify the given expression by method 1; then simplify it again by method 2. Decide which method you prefer for this expression. Explain.

$$\frac{\frac{6}{x^2}-\frac{5}{x}}{\frac{2}{x^2}+\frac{3}{2x}}$$

66. Simplify the given expression by method 1; then simplify it again by method 2. Decide which method you prefer for this expression. Explain.

$$\frac{\frac{x^2-3x-28}{3x-6}}{\frac{x^2-14x+49}{6x-12}}$$

67. The equation $\dfrac{8}{4} = 2$ implies that $2 \cdot 4 = 8$. What does the equation $\dfrac{2}{\frac{1}{3}} = 6$ imply? Is your result true?

68. Let $f(x) = \dfrac{x}{x+2}$. Find an equation of a function g so $\left(\dfrac{f}{g}\right)(x) = x + 3$.

69. Describe a complex rational expression. Give an example. Then simplify it.

70. Describe how to simplify a complex rational expression.

Related Review

Simplify.

71. $\dfrac{8x^{-2}y^5}{6x^{-7}y^8}$

72. $\dfrac{6x^4y^{-5}}{15x^{-3}y^{-2}}$

73. $\dfrac{x^{-1}+x^{-2}}{x^{-2}-x^{-1}}$ **[Hint:** First write as a complex rational expression.]

74. $\dfrac{x^{-3}-x^{-1}}{x^{-1}+x^{-2}}$ **[Hint:** First write as a complex rational expression.]

75. $\dfrac{2b^{-2} - 4b}{8b^{-1} - 6b}$

76. $\dfrac{3y + 2y^{-1}}{6y - 4y^{-3}}$

For $f(x) = \dfrac{x+2}{3-x}$ *and* $g(x) = \dfrac{x}{x-1}$, *find an equation of the given composite function.*

77. $f \circ g$

78. $g \circ f$

79. When a patient is treated with a radioactive element, the element decreases in amount exponentially from both radioactive (physical) decay and biological means, such as urination. The effective half-life of an element is given by

$$H_e = \frac{1}{\dfrac{1}{H_p} + \dfrac{1}{H_b}}$$

where H_p and H_b are the physical and biological half-lives, respectively.

a. Simplify the right-hand side of the formula.

b. Use the result you found in part (a) to find the effective half-life of sulfur-35, which has a physical half-life of 87.4 days and a biological half-life of 623 days.

80. a. Solve the equation $b^{mx+c} = k$ for x.

b. Students A and B try to solve the equation in part (a). Student A's result is $\dfrac{\log(k) - c\log(b)}{m\log(b)}$. Student B's result is $\dfrac{\dfrac{\log(k)}{\log(b)} - c}{m}$. Simplify student B's result. Are the students' results equivalent?

Expressions, Equations, Functions, and Graphs

Perform the indicated instruction. Then use words such as linear, quadratic, cubic, exponential, logarithmic, rational, polynomial, degree, function, one variable, *and* two variables *to describe the expression, equation, or system.*

81. Find the sum $\dfrac{x-2}{x^2 - 2x - 24} + \dfrac{x+4}{x^2 - 8x + 12}$.

82. Solve:
$$y = -\frac{3}{2}x + 4$$
$$\frac{3}{4}x - \frac{1}{2}y = 4$$

83. Find the product $\dfrac{x-2}{x^2 - 2x - 24} \cdot \dfrac{x+4}{x^2 - 8x + 12}$.

84. Graph $f(x) = -\dfrac{3}{2}x + 4$ by hand.

85. Find the domain of $f(x) = \dfrac{2x-1}{x^2 - 2x - 24}$.

86. Find the inverse of $f(x) = -\dfrac{3}{2}x + 4$.

12.5 Solving Rational Equations

Objectives

- Solve *rational equations in one variable.*
- Solve formulas involving rational expressions.
- Compare solving rational equations with simplifying rational expressions.
- Use a rational model to make estimates and predictions about the independent variable.

In Sections 12.1–12.4, we worked mostly with rational expressions. In this section, we will solve rational *equations* in one variable.

Solving Rational Equations in One Variable

A **rational equation in one variable** is an equation in one variable in which both sides can be written as rational expressions. Here are some examples of rational equations in one variable:

$$\frac{3x}{x-4} + \frac{x-2}{5x+3} = 2 \qquad \frac{5}{x} = 9 \qquad \frac{x}{x-3} + \frac{3}{x^2 - 6x + 9} = \frac{x-5}{x-3}$$

As we have done in earlier sections, we will clear an equation of fractions by multiplying both sides of the equation by the LCD.

With rational equations, it is possible to take the usual steps for solving equations, yet arrive at x values that are excluded values for one or more of the fractions in the equation. These values of x are *not* solutions. We call such values **extraneous solutions.**

Example 1 Solving a Rational Equation

Solve $7 - \dfrac{4}{x} = 2 + \dfrac{6}{x}$.

Solution

We note that 0 is an excluded value. We clear the equation of fractions by multiplying both sides of the equation by x, which is the LCD of both of the fractions $\frac{4}{x}$ and $\frac{6}{x}$:

$$7 - \frac{4}{x} = 2 + \frac{6}{x} \qquad \textit{Original equation}$$

$$x \cdot \left(7 - \frac{4}{x}\right) = x \cdot \left(2 + \frac{6}{x}\right) \qquad \textit{Multiply both sides by LCD, x.}$$

$$x \cdot 7 - x \cdot \frac{4}{x} = x \cdot 2 + x \cdot \frac{6}{x} \qquad \textit{Distributive law}$$

$$7x - 4 = 2x + 6 \qquad \textit{Simplify.}$$

$$5x = 10 \qquad \textit{Subtract 2x on both sides; add 4 to both sides.}$$

$$x = 2 \qquad \textit{Divide both sides by 5.}$$

Since 2 is not an excluded value, we conclude that 2 is the solution of the equation. We check that 2 satisfies the original equation:

$$7 - \frac{4}{x} = 2 + \frac{6}{x}$$

$$7 - \frac{4}{2} \stackrel{?}{=} 2 + \frac{6}{2}$$

$$7 - 2 \stackrel{?}{=} 2 + 3$$

$$5 \stackrel{?}{=} 5$$

true

Example 2 Solving a Rational Equation

Solve $\frac{x}{4} - \frac{3}{2x} = \frac{5}{4}$.

Solution

We note that 0 is an excluded value. We clear the equation of fractions by multiplying both sides of the equation by $4x$, which is the LCD of all of the fractions:

$$\frac{x}{4} - \frac{3}{2x} = \frac{5}{4} \qquad \textit{Original equation}$$

$$4x \cdot \left(\frac{x}{4} - \frac{3}{2x}\right) = 4x \cdot \frac{5}{4} \qquad \textit{Multiply both sides by the LCD, 4x.}$$

$$4x \cdot \frac{x}{4} - 4x \cdot \frac{3}{2x} = 4x \cdot \frac{5}{4} \qquad \textit{Distributive law}$$

$$x^2 - 6 = 5x \qquad \textit{Simplify.}$$

$$x^2 - 5x - 6 = 0 \qquad \textit{Write in } ax^2 + bx + c = 0 \textit{ form.}$$

$$(x + 1)(x - 6) = 0 \qquad \textit{Factor left side.}$$

$$x + 1 = 0 \quad \text{or} \quad x - 6 = 0 \qquad \textit{Zero factor property}$$

$$x = -1 \quad \text{or} \quad x = 6$$

Since neither of our results, -1 and 6, is an excluded value, we conclude that the solutions are -1 and 6. We use "intersect" on a graphing calculator to check our work (see Fig. 19).

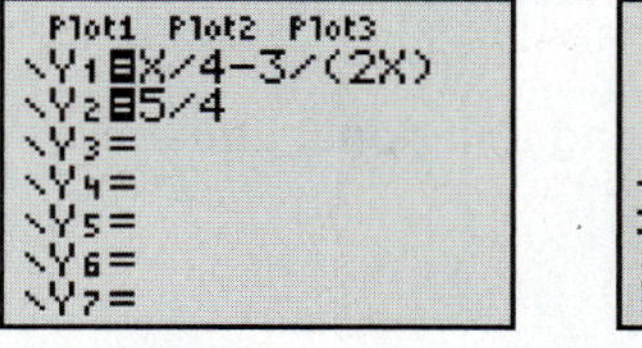

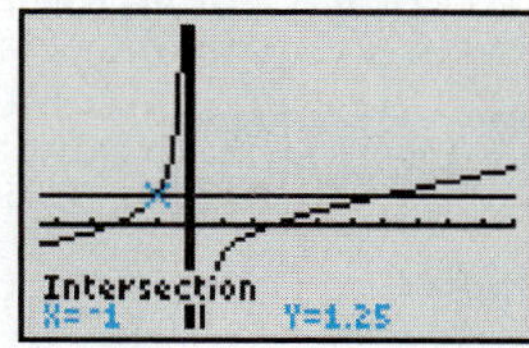

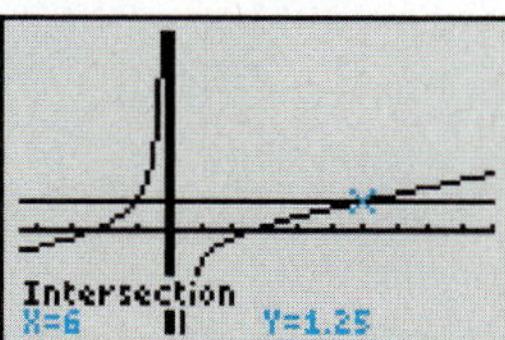

Figure 19 Verify the work

In Example 3, we will see the importance of keeping track of excluded values.

Example 3 Solving a Rational Equation

Solve $\dfrac{4}{w-3} + 2 = \dfrac{w+1}{w-3}$.

Solution

We note that 3 is an excluded value. We clear the equation of fractions by multiplying both sides of the equation by the LCD, $w-3$:

$$\frac{4}{w-3} + 2 = \frac{w+1}{w-3} \quad \textit{Original equation}$$

$$(w-3)\cdot\left(\frac{4}{w-3} + 2\right) = (w-3)\cdot\left(\frac{w+1}{w-3}\right) \quad \textit{Multiply both sides by LCD, } w-3.$$

$$(w-3)\cdot\frac{4}{w-3} + (w-3)\cdot 2 = (w-3)\cdot\frac{w+1}{w-3} \quad \textit{Distributive law}$$

$$4 + (w-3)\cdot 2 = w+1 \quad \textit{Simplify.}$$

$$4 + 2w - 6 = w+1 \quad \textit{Distributive law}$$

$$2w - 2 = w+1 \quad \textit{Combine like terms.}$$

$$w = 3 \quad \textit{Subtract w from both sides; add 2 to both sides.}$$

Our result, 3, is *not* a solution, because 3 is an excluded value. Since the only possibility is not a solution of the original equation, we conclude that no number is a solution. We say that the solution set is the *empty set*.

In Example 3, we multiplied both sides of the rational equation $\dfrac{4}{w-3} + 2 = \dfrac{w+1}{w-3}$ by the LCD and simplified both sides. We got the equation $4 + (w-3)\cdot 2 = w+1$, which has the solution 3. However, 3 is *not* a solution of the original rational equation.

To see where we introduced the extraneous solution 3, notice that 3 does not satisfy the equation

$$(w-3)\cdot\frac{4}{w-3} + (w-3)\cdot 2 = (w-3)\cdot\frac{w+1}{w-3}$$

because 3 is an excluded value of the expressions $\dfrac{4}{w-3}$ and $\dfrac{w+1}{w-3}$. However, 3 does satisfy the next equation, $4 + (w-3)\cdot 2 = w+1$:

$$4 + (w-3)\cdot 2 = w+1$$

$$4 + (3-3)\cdot 2 \stackrel{?}{=} 3+1$$

$$4 \stackrel{?}{=} 4$$

$$\text{true}$$

Since multiplying both sides of a rational equation by the LCD and then simplifying both sides may introduce extraneous solutions, we must always check that any proposed solution is not an excluded value.

Example 4 Solving a Rational Equation

Solve $\dfrac{x+4}{x-3} = \dfrac{x-6}{x+1}$.

Solution

We note that 3 and -1 are excluded values. We clear the equation of fractions by multiplying both sides of the equation by the LCD, $(x-3)(x+1)$:

$$\frac{x+4}{x-3} = \frac{x-6}{x+1} \qquad \textit{Original equation}$$

$$(x-3)(x+1)\cdot\left(\frac{x+4}{x-3}\right) = (x-3)(x+1)\cdot\left(\frac{x-6}{x+1}\right) \qquad \textit{Multiply both sides by LCD, } (x-3)(x+1).$$

$$(x+1)(x+4) = (x-3)(x-6) \qquad \textit{Simplify.}$$

$$x^2+5x+4 = x^2-9x+18 \qquad \textit{Find the products.}$$

$$5x+4 = -9x+18 \qquad \textit{Subtract } x^2 \textit{ from both sides.}$$

$$14x = 14 \qquad \textit{Add 9x to both sides; subtract 4 from both sides.}$$

$$x = 1 \qquad \textit{Divide both sides by 14.}$$

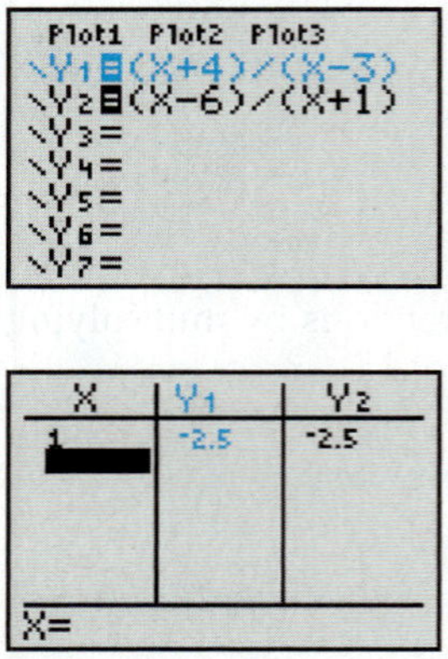

Since 1 is not an excluded value, the solution is 1. We use a graphing calculator table to check our work (see Fig. 20).

Figure 20 Verify the work

To solve a rational equation, we factor the denominators of fractions to help us determine any excluded values, to find the LCD, and, later, to help us simplify rational expressions.

Example 5 Solving a Rational Equation

Solve $\dfrac{x}{x-5} + \dfrac{2}{x-6} = \dfrac{2}{x^2-11x+30}$.

Solution

We begin by factoring the denominator of the expression on the right-hand side of the equation:

$$\frac{x}{x-5} + \frac{2}{x-6} = \frac{2}{x^2-11x+30} \qquad \textit{Original equation}$$

$$\frac{x}{x-5} + \frac{2}{x-6} = \frac{2}{(x-6)(x-5)} \qquad \textit{Factor denominator.}$$

The excluded values are 5 and 6. Next, we clear the equation of fractions by multiplying both sides by the LCD, $(x-6)(x-5)$:

$$(x-6)(x-5)\cdot\left(\frac{x}{x-5} + \frac{2}{x-6}\right) = (x-6)(x-5)\cdot\frac{2}{(x-6)(x-5)}$$

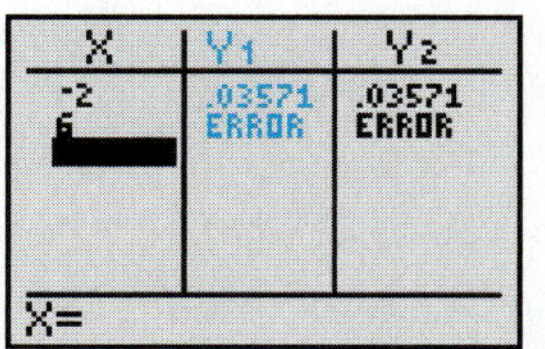

Figure 21 Verify the work

On the left-hand side, we use the distributive law. On the right-hand side, we simplify:

$$(x-6)(x-5)\cdot\frac{x}{x-5}+(x-6)(x-5)\cdot\frac{2}{x-6}=2$$

$$(x-6)\cdot x+(x-5)\cdot 2=2 \quad \textit{Simplify.}$$

$$x^2-6x+2x-10=2 \quad \textit{Distributive law}$$

$$x^2-4x-10=2 \quad \textit{Combine like terms.}$$

$$x^2-4x-12=0 \quad \textit{Write in } ax^2+bx+c=0 \textit{ form.}$$

$$(x-6)(x+2)=0 \quad \textit{Factor left-hand side.}$$

$$x-6=0 \quad \text{or} \quad x+2=0 \quad \textit{Zero factor property}$$

$$x=6 \quad \text{or} \quad x=-2$$

Since 6 is an excluded value, it is *not* a solution. The only solution is -2. We use a graphing calculator table to check our work (see Fig. 21).

Solving a Rational Equation in One Variable

To solve a rational equation in one variable,

1. Factor the denominator(s) if possible.
2. Identify any excluded values.
3. Find the LCD of all of the fractions.
4. Multiply both sides of the equation by the LCD, which gives a simpler equation to solve.
5. Solve the simpler equation.
6. Discard any proposed solutions that are excluded values.

Example 6 Solving a Rational Equation

Solve $\dfrac{x}{x+2}-\dfrac{7}{5-x}=\dfrac{14}{x^2-3x-10}$.

Solution

We begin by factoring the denominators:

$$\frac{x}{x+2}-\frac{7}{5-x}=\frac{14}{x^2-3x-10} \quad \textit{Original equation}$$

$$\frac{x}{x+2}-\frac{7}{-(x-5)}=\frac{14}{(x-5)(x+2)} \quad \textit{Factor denominators; } 5-x=-x+5=-(x-5)$$

$$\frac{x}{x+2}+\frac{7}{(x-5)}=\frac{14}{(x-5)(x+2)} \quad \frac{a}{-b}=-\frac{a}{b}$$

The excluded values are -2 and 5. Next, we clear the equation of fractions by multiplying both sides by the LCD, $(x-5)(x+2)$:

$$(x-5)(x+2)\left(\frac{x}{x+2}+\frac{7}{x-5}\right)=(x-5)(x+2)\cdot\frac{14}{(x-5)(x+2)} \quad \textit{Multiply both sides by LCD.}$$

On the left-hand side, we use the distributive law. On the right-hand side, we simplify:

$$(x-5)(x+2)\cdot\frac{x}{x+2}+(x-5)(x+2)\cdot\frac{7}{x-5}=14 \quad \textit{Distributive law; simplify.}$$

$$(x-5)\cdot x+(x+2)\cdot 7=14 \quad \textit{Simplify.}$$

$$x^2-5x+7x+14=14 \quad \textit{Distributive law}$$

$$x^2+2x+14=14 \quad \textit{Combine like terms.}$$

$$x^2+2x=0 \quad \textit{Subtract 14 from both sides.}$$

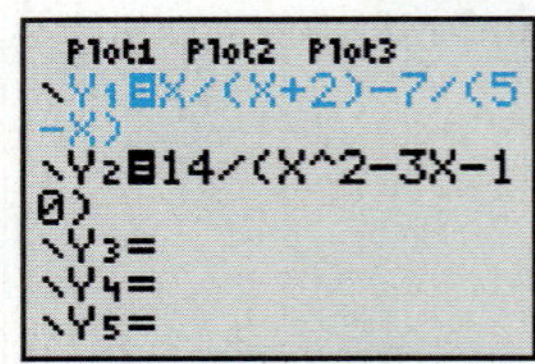

Figure 22 Verify the work

$$x(x+2) = 0 \quad \textit{Factor left-hand side.}$$
$$x = 0 \quad \text{or} \quad x + 2 = 0 \quad \textit{Zero factor property}$$
$$x = 0 \quad \text{or} \quad x = -2$$

Since -2 is an excluded value, it is *not* a solution. The only solution is 0. We use a graphing calculator table to check our work (see Fig. 22).

Example 7 Finding an Input Value of a Rational Function

Let $f(x) = \dfrac{x+1}{x-3} - \dfrac{x-2}{x+3}$. Find x when $f(x) = 1$.

Solution

We note that the domain of f is the set of real numbers except -3 and 3. We substitute 1 for $f(x)$ in the equation $f(x) = \dfrac{x+1}{x-3} - \dfrac{x-2}{x+3}$ and solve for x:

$$1 = \frac{x+1}{x-3} - \frac{x-2}{x+3}$$
$$(x-3)(x+3)\cdot 1 = (x-3)(x+3)\left(\frac{x+1}{x-3} - \frac{x-2}{x+3}\right)$$
$$(x-3)(x+3) = (x-3)(x+3)\cdot\frac{x+1}{x-3} - (x-3)(x+3)\cdot\frac{x-2}{x+3}$$
$$(x-3)(x+3) = (x+3)(x+1) - (x-3)(x-2) \quad \textit{Simplify.}$$
$$x^2 - 9 = x^2 + 4x + 3 - (x^2 - 5x + 6) \quad \textit{Find products.}$$
$$x^2 - 9 = x^2 + 4x + 3 - x^2 + 5x - 6 \quad \textit{Subtract trinomial.}$$
$$x^2 - 9 = 9x - 3 \quad \textit{Combine like terms.}$$
$$x^2 - 9x - 6 = 0 \quad \textit{Write in } ax^2 + bx + c = 0 \textit{ form.}$$
$$x = \frac{-(-9) \pm \sqrt{(-9)^2 - 4(1)(-6)}}{2(1)} \quad \textit{Substitute into quadratic formula.}$$
$$x = \frac{9 \pm \sqrt{105}}{2} \quad \textit{Simplify.}$$

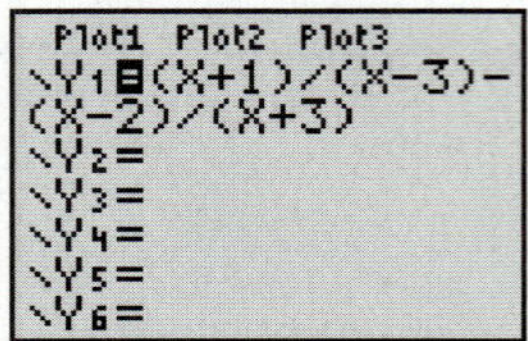

Figure 23 Verify the work

Since both of our results are in the domain of f, we conclude that if $f(x) = 1$, then $x = \dfrac{9 - \sqrt{105}}{2}$ or $x = \dfrac{9 + \sqrt{105}}{2}$.

We enter $y = \dfrac{x+1}{x-3} - \dfrac{x-2}{x+3}$ in a graphing calculator and check that, for both inputs $\dfrac{9 - \sqrt{105}}{2}$ and $\dfrac{9 + \sqrt{105}}{2}$, the output is 1 (see Fig. 23).

Solving Formulas Involving Rational Expressions

Formulas that contain rational expressions are useful in many fields, such as finance, physics, meteorology, mathematics, electronics, and chemistry. It can be helpful to solve such a formula for one of its variables.

Example 8 Solving a Formula Involving a Rational Expression

The following formula is useful in electronics:

$$I = \frac{\mathcal{E}}{R + r}$$

Here I is the current in an electrical circuit, $\mathcal{E}$ is the electromotive force, R is the circuit's resistance, and r is the battery's resistance. Solve the formula for the variable R.

Solution

To begin, we multiply both sides of the equation by the LCD, $R + r$:

$$I = \frac{\mathcal{E}}{R + r} \qquad \text{Original formula}$$

$$(R + r) \cdot I = (R + r) \cdot \frac{\mathcal{E}}{R + r} \qquad \text{Multiply both sides by LCD, } R + r.$$

$$RI + rI = \mathcal{E} \qquad \text{Distributive law; simplify.}$$

$$RI = \mathcal{E} - rI \qquad \text{Subtract } rI \text{ from both sides.}$$

$$\frac{RI}{I} = \frac{\mathcal{E} - rI}{I} \qquad \text{Divide both sides by } I.$$

$$R = \frac{\mathcal{E} - rI}{I} \qquad \text{Simplify.}$$

Solving Rational Equations versus Simplifying Rational Expressions

Throughout this course, we have solved equations and simplified expressions. In solving an equation, our objective is to find any *numbers* that satisfy the equation. In simplifying an expression, our objective is to find a simpler, yet equivalent, *expression*.

Solving a Rational Equation versus Simplifying a Rational Expression

To solve a rational equation, clear the fractions in it by multiplying both sides of the equation by the LCD. To simplify a rational expression, do *not* multiply it by the LCD—the only multiplication permissible is multiplication by 1, usually in the form $\frac{A}{A}$, where A is a nonzero polynomial.

Here we compare solving a rational equation with simplifying a rational expression:

Solving the Equation $\frac{2}{3} = \frac{4}{x}$

The number 0 is an excluded value.

$$\frac{2}{3} = \frac{4}{x} \qquad \text{Original equation}$$

$$3x \cdot \frac{2}{3} = 3x \cdot \frac{4}{x} \qquad \text{Multiply both sides by LCD, } 3x.$$

$$2x = 12 \qquad \text{Simplify.}$$

$$x = 6$$

The result is a number.

Simplifying the Expression $\frac{2}{3} + \frac{4}{x}$

$$\frac{2}{3} + \frac{4}{x} = \frac{2}{3} \cdot \frac{x}{x} + \frac{4}{x} \cdot \frac{3}{3} \qquad \text{Introduce missing factors.}$$

$$= \frac{2x}{3x} + \frac{12}{3x} \qquad \text{Find products.}$$

$$= \frac{2x + 12}{3x} \qquad \frac{A}{B} + \frac{C}{B} = \frac{A + C}{B}$$

The result is an expression.

For the equation $\frac{2}{3} = \frac{4}{x}$, the solution is the *number* 6. We simplify the expression $\frac{2}{3} + \frac{4}{x}$ by writing it as the *expression* $\frac{2x + 12}{3x}$. In general, **the result of solving a rational equation is the empty set or a set of one or more numbers. The result of simplifying a rational expression is an expression.**

Using a Rational Model to Make Predictions about the Independent Variable

In Section 12.1, we modeled the percentage of a quantity by means of a rational model, which we used to make a prediction about the dependent variable. Now that we know how to solve a rational equation in one variable, we can use such a model to make a prediction about the independent variable.

Table 7 Numbers of Internet Users

Year	Number of Internet Users (millions)
2006	204
2007	212
2008	220
2009	228
2010	240

Source: *The Nielsen Company*

Example 9 Using a Rational Model to Make a Prediction

In Example 10 of Section 12.1, we found the model $P(t) = \dfrac{880t + 15{,}040}{0.0068t^2 + 2.71t + 277.89}$, where $P(t)$ is the percentage of Americans who are Internet users at t years since 2000 (see Table 7). Predict when 92% of Americans will be Internet users.

Solution

We substitute 92 for $P(t)$ in the equation of P and solve for t:

$$92 = \frac{880t + 15{,}040}{0.0068t^2 + 2.71t + 277.89}$$

$$(0.0068t^2 + 2.71t + 277.89) \cdot 92 = (0.0068t^2 + 2.71t + 277.89) \cdot \frac{880t + 15{,}040}{0.0068t^2 + 2.71t + 277.89}$$

$$0.6256t^2 + 249.32t + 25{,}565.88 = 880t + 15{,}040$$

$$0.6256t^2 - 630.68t + 10{,}525.88 = 0$$

Now we substitute $a = 0.6256$, $b = -630.68$, and $c = 10{,}525.88$ in the quadratic formula:

$$t = \frac{-(-630.68) \pm \sqrt{(-630.68)^2 - 4(0.6256)(10{,}525.88)}}{2(0.6256)} \qquad \text{Substitute into quadratic formula.}$$

$$t \approx 16.98 \quad \text{or} \quad t \approx 991.14 \qquad \text{Compute.}$$

We verify the results by entering $y = \dfrac{880t + 15{,}040}{0.0068t^2 + 2.71t + 277.89}$ in a graphing calculator and checking that the inputs 16.98 and 991.14 lead to outputs of about 92 (see Fig. 24).

The inputs 16.98 and 991.14 represent the years 2017 and 2991, respectively. Model breakdown has occurred for the prediction 2991: The year is too far into the future for us to have any faith in this prediction! Therefore, we predict 92% of Americans will be Internet users in 2017.

Figure 24 Verify the work

Group Exploration

Simplifying versus solving

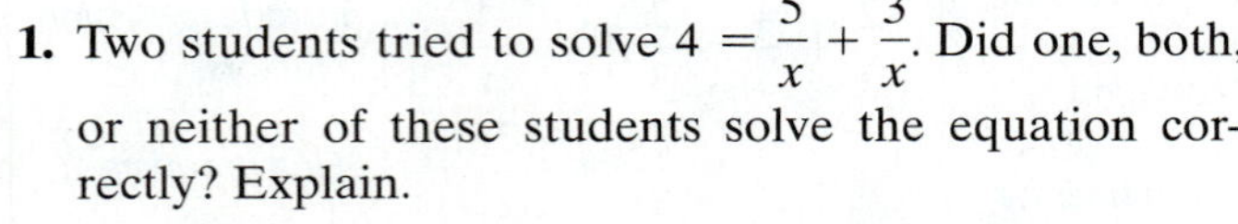

1. Two students tried to solve $4 = \dfrac{5}{x} + \dfrac{3}{x}$. Did one, both, or neither of these students solve the equation correctly? Explain.

Student A

$$4 = \frac{5}{x} + \frac{3}{x}$$
$$4x = x\left(\frac{5}{x} + \frac{3}{x}\right)$$
$$4x = x \cdot \frac{5}{x} + x \cdot \frac{3}{x}$$
$$4x = 5 + 3$$
$$4x = 8$$
$$x = 2$$

Student B

$$\begin{aligned} 4 &= \frac{5}{x} + \frac{3}{x} \\ &= \frac{5}{x} + \frac{3}{x} - 4 \\ &= \frac{8}{x} - 4 \\ &= \frac{8}{x} - 4 \cdot \frac{x}{x} \\ &= \frac{8}{x} - \frac{4x}{x} \\ &= \frac{-4x + 8}{x} \end{aligned}$$

2. Three students tried to simplify $4 + \dfrac{5}{x} + \dfrac{3}{x}$. Which students, if any, simplified the expression correctly? Explain.

Student C

$$\begin{aligned} 4 + \frac{5}{x} + \frac{3}{x} &= x\left(4 + \frac{5}{x} + \frac{3}{x}\right) \\ &= 4x + x \cdot \frac{5}{x} + x \cdot \frac{3}{x} \\ &= 4x + 5 + 3 \\ &= 4x + 8 \end{aligned}$$

Student D

$$\begin{aligned} 4 + \frac{5}{x} + \frac{3}{x} &= 4 \cdot \frac{x}{x} + \frac{5}{x} + \frac{3}{x} \\ &= \frac{4x}{x} + \frac{5}{x} + \frac{3}{x} \\ &= \frac{4x + 8}{x} \end{aligned}$$

Student E

$$4 + \frac{5}{x} + \frac{3}{x} = 0$$
$$x\left(4 + \frac{5}{x} + \frac{3}{x}\right) = x \cdot 0$$
$$4x + 5 + 3 = 0$$
$$4x = -8$$
$$x = -2$$

3. a. What is the difference in your goals in solving a rational equation versus simplifying a rational expression?
   b. Explain how that difference in goals relates to the techniques you use to solve an equation versus simplify an expression.

Group Exploration

Looking ahead: Modeling the mean of a quantity

Recall from Section 1.1 that to find the mean (or average) of a group of numbers, we divide the sum of the numbers by the number of numbers in the group.

1. Suppose that five friends go out to eat and the total bill is \$100. Consider the following possibilities.
   a. Each person pays \$20. Find the mean of the amounts \$20, \$20, \$20, \$20, and \$20.
   b. The amounts contributed for dinner are \$19, \$22, \$20, \$18, and \$21. Compute the mean amount contributed.
   c. One person pays \$100 and the other four friends eat for free. Compute the mean amount contributed.
   d. In which scenario (part (a), (b), or (c)) does the mean give the exact per-person amount? In which scenario does the mean give a reasonable estimate of the per-person amount? In which scenario does the mean give a poor estimate of the per-person amount?
2. For a ten-year high school reunion, graduates rent out a dance hall that charges a flat fee of \$800 for a band, plus \$40 per person for drinks and appetizers.
   a. Let $T(n)$ be the total cost (in dollars) for n people to attend the reunion. Find an equation of T.
   b. Let $M(n)$ be the mean cost per person (in dollars per person) if n people attend the reunion. Find an equation of M.
   c. Find $M(200)$. What does it mean in this situation?
   d. Find n when $M(n) = 43$. What does it mean in this situation?

Homework 12.5

For extra help ▶ MyMathLab®

Watch the videos in MyMathLab

Download the MyDashboard App

Solve.

1. $\frac{3}{x} - 2 = \frac{7}{x}$
2. $\frac{9}{x} - 3 = \frac{3}{x}$
3. $5 - \frac{4}{x} = 3 + \frac{2}{x}$
4. $6 - \frac{8}{x} = 3 + \frac{4}{x}$
5. $\frac{5}{p - 1} = \frac{2p + 1}{p - 1}$
6. $\frac{7}{y + 3} = \frac{3y + 4}{y + 3}$
7. $\frac{8x + 4}{x + 2} = \frac{5x - 2}{x + 2}$
8. $\frac{3x - 2}{x - 5} = \frac{x + 8}{x - 5}$
9. $\frac{w + 2}{w - 4} + 3 = \frac{2}{w - 4}$
10. $\frac{t - 5}{t + 1} + 2 = \frac{3}{t + 1}$
11. $\frac{2}{x} + \frac{5}{4} = \frac{3}{x}$
12. $\frac{3}{x} - \frac{7}{2} = \frac{6}{x}$
13. $\frac{5}{6x} - \frac{1}{2} = \frac{3}{4x}$
14. $\frac{3}{8x} - \frac{3}{4} = \frac{1}{6}$
15. $\frac{4}{x - 2} = \frac{2}{x + 3}$
16. $\frac{3}{x + 4} = \frac{4}{x - 1}$
17. $\frac{2r + 7}{4r} = \frac{5}{3}$
18. $\frac{5p - 2}{3p} = \frac{6}{7}$
19. $\frac{5}{x + 3} + \frac{3}{4} = 2$
20. $\frac{4}{x - 3} + \frac{5}{2} = 3$
21. $\frac{2}{x - 3} + \frac{1}{x + 3} = \frac{5}{x^2 - 9}$
22. $\frac{3}{x + 5} + \frac{2}{x - 5} = \frac{1}{x^2 - 25}$
23. $\frac{4}{x + 2} + \frac{3}{x + 1} = \frac{3}{x^2 + 3x + 2}$
24. $\frac{2}{x - 3} + \frac{4}{x + 5} = \frac{16}{x^2 + 2x - 15}$
25. $\frac{5}{x^2 - 4} + \frac{2}{x + 2} = \frac{4}{x - 2}$
26. $\frac{3}{x^2 - 1} + \frac{7}{x - 1} = \frac{4}{x + 1}$
27. $2 + \frac{4}{k - 2} = \frac{8}{k^2 - 2k}$
28. $1 + \frac{4}{m - 5} = \frac{2}{m^2 - 5m}$
29. $\frac{-48}{x^2 - 2x - 15} - \frac{6}{x + 3} = \frac{7}{x - 5}$
30. $\frac{-36}{x^2 + x - 20} - \frac{2}{x - 4} = \frac{4}{x + 5}$
31. $\frac{x^2 - 23}{2x^2 - 5x - 3} + \frac{2}{x - 3} = \frac{-1}{2x + 1}$
32. $\frac{4x^2 - 24x}{3x^2 - x - 2} + \frac{3}{3x + 2} = \frac{-4}{x - 1}$
33. $\frac{w + 7}{w^2 - 9} = \frac{-w + 2}{w - 3}$
34. $\frac{t - 6}{t^2 - 4} = \frac{-t + 1}{t - 2}$
35. $3 + \frac{2}{x} = \frac{4}{x^2}$
36. $\frac{5}{x} = \frac{3}{x^2} - 4$
37. $\frac{5}{r^2 - 3r + 2} - \frac{1}{r - 2} = \frac{r + 6}{3r - 3}$
38. $\frac{3}{p^2 - 6p + 9} + \frac{p - 2}{3p - 9} = \frac{p}{2p - 6}$

39. $\dfrac{2x}{x+1} - \dfrac{3}{2} = \dfrac{-2}{x+2}$

40. $-\dfrac{1}{2} + \dfrac{x}{x-1} = \dfrac{-1}{x+3}$

41. $\dfrac{x-4}{x^2-7x+12} - \dfrac{x+2}{x-3} = 0$

42. $\dfrac{x+2}{x^2+x-30} - \dfrac{x+3}{x-5} = 0$

43. $\dfrac{t}{t-3} = 2 - \dfrac{5}{3-t}$

44. $4 - \dfrac{k}{k-5} = \dfrac{1}{5-k}$

45. $\dfrac{12}{9-x^2} + \dfrac{3}{x+3} = \dfrac{-2}{x-3}$

46. $\dfrac{-2}{x+5} + \dfrac{3}{x-5} = \dfrac{-20}{25-x^2}$

47. $\dfrac{x+2}{x-3} - \dfrac{x-3}{x+2} = \dfrac{5x}{x^2-x-6}$

48. $\dfrac{x-4}{x+1} + \dfrac{x+1}{x-4} = \dfrac{13x}{x^2-3x-4}$

49. $\dfrac{2y}{y-2} - \dfrac{2y-5}{y^2-7y+10} = \dfrac{-4}{y-5}$

50. $\dfrac{3p}{p+1} - \dfrac{4p-1}{p^2-2p-3} = \dfrac{-2}{p-3}$

51. $\dfrac{x-2}{x^2-2x-3} + \dfrac{x+5}{x^2-1} = \dfrac{x+3}{x^2-4x+3}$

52. $\dfrac{x+4}{x^2+x-2} + \dfrac{x+1}{x^2-4} = \dfrac{x-3}{x^2-3x+2}$

Find all complex-number solutions.

53. $\dfrac{5}{x} - \dfrac{2}{x^2} = 4$

54. $\dfrac{7}{x} + 5 = -\dfrac{6}{x^2}$

55. $\dfrac{2}{t-5} - \dfrac{3t}{t+5} = \dfrac{35}{t^2-25}$

56. $\dfrac{3}{k+4} - \dfrac{2k}{k-4} = \dfrac{2}{k^2-16}$

57. $\dfrac{x-1}{3x-12} + \dfrac{-x+1}{x-5} = \dfrac{4x}{x^2-9x+20}$

58. $\dfrac{-x+6}{2x-6} + \dfrac{x-1}{x+4} = \dfrac{2x}{x^2+x-12}$

Find x when y is equal to the indicated value.

59. $f(x) = \dfrac{3}{x-5}, y = 4$

60. $g(x) = \dfrac{7}{x+4}, y = 3$

61. $f(x) = \dfrac{5}{x-1} + \dfrac{3}{x+1}, y = -1$

62. $g(x) = \dfrac{2}{x+1} - \dfrac{4}{x-2}, y = -1$

Find all x-intercepts.

63. $f(x) = \dfrac{x-1}{x-5} - \dfrac{x+2}{x+3}$

64. $g(x) = \dfrac{x+4}{x-2} - \dfrac{x-3}{x+6}$

For Exercises 65–70, solve for the specified variable.

65. $F = \dfrac{mv^2}{r}$, for r

66. $P = \dfrac{nrT}{V}$, for V

67. $F = -\dfrac{GMm}{r^2}$, for M

68. $\dfrac{P_1V_1}{T_1} = \dfrac{P_2V_2}{T_2}$, for P_1

69. $P = \dfrac{A}{1+rt}$, for t

70. $P = \dfrac{A}{1+rt}$, for r

71. In Exercise 79 of Homework 12.1, you found the rational model $P(t) = \dfrac{0.48t^2 + 14t + 295}{0.015t^2 + 0.055t + 6.05}$, where $P(t)$ is the percentage of voter-eligible Latinos who are registered to vote at t years since 1980 (see Table 8).

Table 8 Numbers of Latinos Who Are Registered to Vote and Eligible to Vote

	Millions of Latinos	
Year	Registered to Vote	Eligible to Vote
1988	4.6	7.7
1992	5.1	8.3
1996	6.6	11.2
2000	7.5	13.2
2004	9.3	16.1
2008	11.6	19.5
2010	11.0	21.3

Source: *Pew Hispanic Center*

Predict when exactly half of voter-eligible Latinos will be registered to vote.

72. In Exercise 80 of Homework 12.1, you found the rational model $P(t) = \dfrac{1.2t^2 - 12t + 665}{0.026t^2 - 0.2t + 14.6}$, where $P(t)$ is the percentage of college students who are men at t years since 1990 (see Table 9).

Table 9 College Enrollments

	Enrollment (millions)	
Year	Men	All Students
1995	6.3	14.2
2000	6.7	15.3
2005	7.5	17.5
2008	8.2	19.1
2010	9.1	21.2

Source: *National Center for Education Statistics*

Predict when 43.2% of college students will be men.

73. The amounts employees pay for health insurance and the total costs (paid by employees and employers) are shown in Table 10 for various years.

a. Let $E(t)$ be the amount (in thousands of dollars) paid by employees for health insurance in the year that is t years since 2000. Find a linear equation of E.

b. Let $C(t)$ be the total cost (in thousands of dollars) of health insurance in the year that is t years since 2000. Find an equation of C.

Table 10 Health Insurance Costs

	Health Insurance Cost (thousands of dollars)	
Year	Employee Cost	Total Cost
2007	1.98	8.6
2008	2.22	9.23
2009	2.27	9.78
2010	2.38	10.39
2011	2.53	10.98

Source: *TW/NBGH Value Purchasing Survey*

c. Let $P(t)$ be the percentage of the total cost of health insurance paid by employees in the year that is t years since 2000. Find an equation of P.
d. Find $P(17)$. What does it mean in this situation?
e. Find t when $P(t) = 23$. What does it mean in this situation?
f. Use the window settings in Fig. 25 to graph P. Is P increasing, decreasing, or neither for the values of t between 7 and 20? What does that mean in this situation? How is this possible, given that the amount employees pay for health insurance has been increasing since 2007?

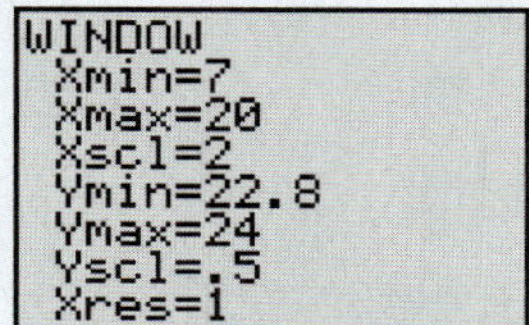

Figure 25 Window settings for Exercise 73f

74. The numbers of prisoners and the numbers of releases from prison are shown in Table 11 for various years.

Table 11 Prisoners and Releases from Prisons

Year	Numbers of Prisoners Released from Prison (thousands)	Total Number of Prisoners (thousands)
2000	605	1391
2002	630	1440
2004	672	1497
2006	713	1570
2008	735	1610
2009	729	1614

Source: *Bureau of Justice Statistics*

a. Let $r(t)$ be the number (in thousands) of releases from prison in the year that is t years since 2000. Find a linear equation of r.
b. Let $n(t)$ be the total number (in thousands) of prisoners at t years since 2000. Find a linear equation of n.
c. Let $P(t)$ be the percentage of prisoners who are released in the year that is t years since 2000. Find an equation of P.
d. Find $P(11)$. What does it mean in this situation?
e. Find t when $P(t) = 47$. What does it mean in this situation?
f. Use the window settings in Fig. 26 to graph P. Is P increasing, decreasing, or neither for the values of t between 0 and 20? What does that mean in this situation?

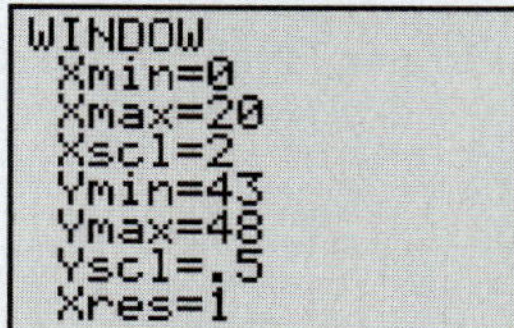

Figure 26 Window settings for Exercise 74f

For Exercises 75–80, find approximate solutions of the given equation or system by referring to the graphs shown in Fig. 27. Round results or coordinates of results to the first decimal place.

75. $\dfrac{5}{x-2} = x^2 - 6x + 10$

76. $\dfrac{5}{x-2} = -x^2 + x - 1$

77. $\dfrac{5}{x-2} = 4$

78. $\dfrac{5}{x-2} = -1$

79. $y = \dfrac{5}{x-2}$
$y = -x^2 + x - 1$

80. $y = \dfrac{5}{x-2}$
$y = x^2 - 6x + 10$

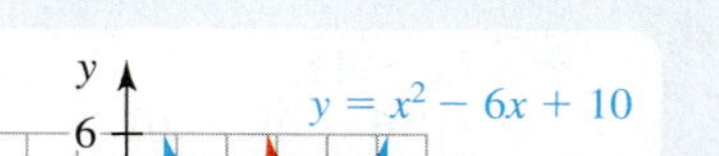

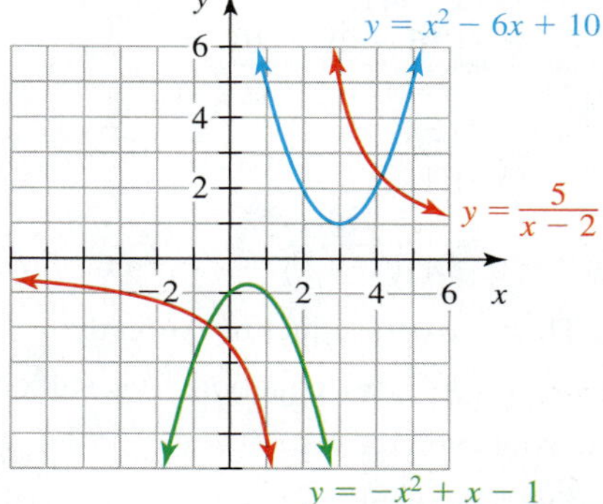

Figure 27 Exercises 75–80

81. Let $f(x) = \dfrac{x+a}{x+b}$. Find values of a and b such that $f(0) = 2$ and $f(1) = \dfrac{5}{2}$.

82. Let $f(x) = \dfrac{x+a}{x+b}$. Find values of a and b such that $f(1) = -2$ and $f(3) = -8$.

Concepts

83. A student tries to solve the equation $\dfrac{4}{x+2} - \dfrac{2}{x} = \dfrac{1}{x+2}$:

$$\frac{4}{x+2} - \frac{2}{x} = \frac{1}{x+2}$$
$$\frac{4}{x+2}\cdot\frac{x}{x} - \frac{2}{x}\cdot\frac{x+2}{x+2} = \frac{1}{x+2}\cdot\frac{x}{x}$$
$$\frac{4x - 2(x+2)}{x(x+2)} = \frac{x}{x(x+2)}$$
$$\frac{4x - 2x - 4}{x(x+2)} = \frac{x}{x(x+2)}$$
$$\frac{2x - 4}{x(x+2)} = \frac{x}{x(x+2)}$$
$$2x - 4 = x$$
$$x = 4$$

Describe a more efficient way to solve the equation.

84. A student tries to simplify the expression $\frac{3}{x+1} + \frac{3}{x-1}$:

$$\begin{aligned}\frac{3}{x+1} + \frac{3}{x-1} &= (x-1)(x+1)\left(\frac{3}{x+1} + \frac{3}{x-1}\right)\\ &= (x-1)(x+1)\cdot\frac{3}{x+1} + (x-1)(x+1)\cdot\frac{3}{x-1}\\ &= 3(x-1) + 3(x+1)\\ &= 3x - 3 + 3x + 3\\ &= 6x\end{aligned}$$

Describe any errors. Then simplify the expression correctly.

85. A student tries to solve $\frac{7}{x^2+x-20} = \frac{4}{x+5} - \frac{2}{x-4}$:

$$\begin{aligned}\frac{7}{x^2+x-20} &= \frac{4}{x+5} - \frac{2}{x-4}\\ &= \frac{4}{x+5}\cdot\frac{x-4}{x-4} - \frac{2}{x-4}\cdot\frac{x+5}{x+5}\\ &= \frac{4(x-4) - 2(x+5)}{(x-4)(x+5)}\\ &= \frac{4x - 16 - 2x - 10}{(x-4)(x+5)}\\ &= \frac{2x - 26}{(x-4)(x+5)}\\ &= \frac{2(x-13)}{(x-4)(x+5)}\end{aligned}$$

Describe any errors. Then solve the equation correctly.

86. A student tries to solve a rational equation. The student's result is $\frac{x+1}{x-4}$. What would you tell the student?

87. A student tries to solve a rational equation in one variable. The result is $\frac{x-2}{x^2-4x+1}$. What would you tell the student?

88. Why must we check that any proposed solution of a rational equation is an excluded value?

89. When simplifying a rational expression, can we multiply it by the LCD? Explain. When solving a rational equation, can we multiply both sides of the equation by the LCD? Explain. (See page 9 for guidelines on writing a good response.)

90. Describe how to solve a rational equation in one variable. (See page 9 for guidelines on writing a good response.)

Related Review

Solve or simplify, as appropriate. For equations, all solutions are real numbers.

91. $\frac{5}{x} + \frac{4}{x+1} - \frac{3}{x}$

92. $\frac{2}{k+3} + \frac{k}{k-3} + \frac{-10}{k^2-9}$

93. $\frac{5}{x} + \frac{4}{x+1} = \frac{3}{x}$

94. $\frac{2}{k+3} + \frac{k}{k-3} = \frac{-10}{k^2-9}$

95. $\frac{x+2}{x^2-5x+6} - \frac{x+1}{x^2-4} = \frac{4}{x^2-x-6}$

96. $\frac{3}{x^2-4x-12} - \frac{x-3}{x^2-7x+6}$

97. $\frac{x+2}{x^2-5x+6} - \frac{x+1}{x^2-4} + \frac{4}{x^2-x-6}$

98. $\frac{3}{x^2-4x-12} - \frac{x-3}{x^2-7x+6} = \frac{x}{x^2+x-2}$

Solve. All solutions are real numbers. Round approximate solutions to the fourth decimal place.

99. $2p^3 - p^2 = 8p - 4$

100. $2(2m^2 - m) = 1 - 3m$

101. $2(4)^x + 3 = 106$

102. $\frac{3}{x^2-9} - \frac{x-2}{x^2-x-12} = \frac{-1}{x-4}$

103. $\log_3(5x-4) - \log_3(2x-3) = 2$

104. $2\log_2(x+2) - \log_2(x+3) = 1$

Expressions, Equations, Functions, and Graphs

Perform the indicated instruction. Then use words such as linear, quadratic, cubic, exponential, logarithmic, rational, polynomial, degree, function, one variable, *and* two variables *to describe the expression, equation, or system.*

105. Graph $f(x) = -2(x+2)^2 + 3$ by hand.

106. Solve $\log_5(2x^4) = 3$. Round any solutions to the fourth decimal place.

107. Solve $-2(x+2)^2 + 3 = -15$.

108. Graph $f(x) = \log_2(x)$ by hand.

109. Simplify $-2(x+2)^2 + 3$.

110. Simplify $\log_b(b^3)$.

12.6 Modeling with Rational Functions

Objectives

- Use a rational function to model the *mean* of a quantity.
- Model the percentage of a quantity.
- Use a rational function to model the time it takes to travel a given distance at a constant speed.

In this section, we will use rational functions to model authentic situations.

Modeling the Mean of a Quantity

To begin, suppose four students go on a road trip during spring break. The total cost for gas is \$20. Consider the following possibilities:

A. Each person pays \$5 for gas.

B. The amounts contributed for gas are \$4, \$4, \$6, and \$6.

C. One student pays \$20, and the other three students ride for free.

We compute the mean amount of money each student spent for gas by dividing the total spent by the number of students (Section 1.1):

$$\begin{aligned}\text{mean amount spent per student} &= \frac{\text{total amount spent for gas}}{\text{number of students}}\\ &= \frac{20 \text{ dollars}}{4 \text{ students}} = 5 \text{ dollars per student}\end{aligned}$$

So, the mean amount of money spent per student is \$5.

For scenarios A, B, and C, we make the following observations:

A. The mean gives the (exact) per-person amount if all students pay an equal amount.

B. The mean gives a reasonable estimate of the per-person amount if the students pay nearly the same amount.

C. The mean gives a poor estimate of the per-person amount if the students pay very different amounts.

Computing the Mean

If a quantity Q is divided into n parts, the **mean** amount M of the quantity per part is given by

$$M = \frac{Q}{n}$$

As another example, if a student makes 21 phone calls in 7 days, the mean number of calls he makes per day is $\frac{21}{7} = 3$ calls. The mean is a fairly good estimate of the number of calls on a given day, provided the student made about the same number of calls each day.

Example 1 Modeling the Mean of a Quantity

The underground band Melted Zipper wants to make and sell a CD of its original songs. It costs about \$1000 to record the music onto a digital audiotape (DAT), \$100 to rearrange the music and improve the sound quality, \$350 for artwork for the cover and inside leaflet, and \$350 to set up production. In addition, it will cost \$2.50 for each CD manufactured.

1. What is the total cost of making 300 CDs?
2. Let $C(n)$ be the total cost (in dollars) of making n CDs. Find an equation of C.
3. Let $P(n)$ be the price (in dollars) the band should set for each CD so it breaks even by making and selling n CDs. Find an equation of P.
4. Find $P(300)$. What does it mean in this situation?
5. Find n when $P(n) = 10$. What does it mean in this situation?
6. Describe the values of $P(n)$ for large values of n.

Solution

1. First, we compute the total *fixed costs*—the costs that do not depend on how many CDs are manufactured:

$$1000 + 100 + 350 + 350 = 1800 \text{ dollars}$$

The band must also pay \$2.50 per CD manufactured. If the band manufactures 300 CDs, this cost, called the *variable cost*, is $2.50(300) = 750$ dollars.

To find the total cost, we add the variable cost and the fixed costs:

$$2.50(300) + 1800 = 2550 \text{ dollars}$$

2. The total cost is equal to \$2.50 times the number of CDs, plus the fixed cost of \$1800:

$$C(n) = 2.50n + 1800$$

3. If the band makes and sells n CDs, it can break even by selling the CD for the amount found by dividing the total cost into n parts. This amount is the mean cost per CD:

$$\begin{aligned} P(n) &= \text{mean cost per CD} \\ &= \frac{\text{total cost}}{\text{number of CDs manufactured}} \\ &= \frac{2.50n + 1800}{n} \end{aligned}$$

4. $P(300) = \dfrac{2.50(300) + 1800}{300} = 8.50$

If the band makes and sells 300 CDs, it must sell each CD for \$8.50 to break even.

5. We substitute 10 for $P(n)$ in the equation of P and solve for n:

$$\begin{aligned} 10 &= \frac{2.50n + 1800}{n} && \textit{Substitute 10 for P(n).} \\ n \cdot 10 &= n \cdot \frac{2.50n + 1800}{n} && \textit{Multiply both sides by LCD, n.} \\ 10n &= 2.50n + 1800 && \textit{Simplify.} \\ 7.5n &= 1800 \\ n &= 240 \end{aligned}$$

If the band can sell each CD for \$10, then it must make 240 CDs to break even.

6. First, we use graphing calculator tables to display input–output pairs of P (see Figs. 28, 29, and 30).

Figure 28 Enter the function

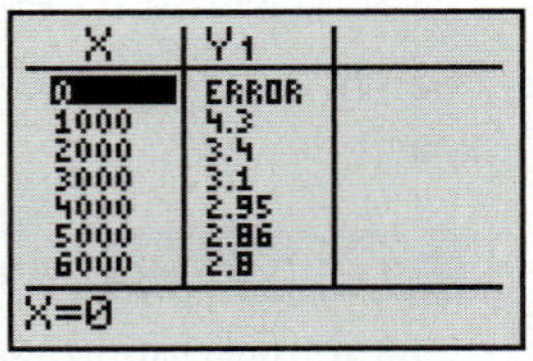

X	Y1
0	ERROR
1000	4.3
2000	3.4
3000	3.1
4000	2.95
5000	2.86
6000	2.8

X=0

Figure 29 Inputs increasing by 1000

X	Y1
0	ERROR
10000	2.68
20000	2.59
30000	2.56
40000	2.545
50000	2.536
60000	2.53

X=0

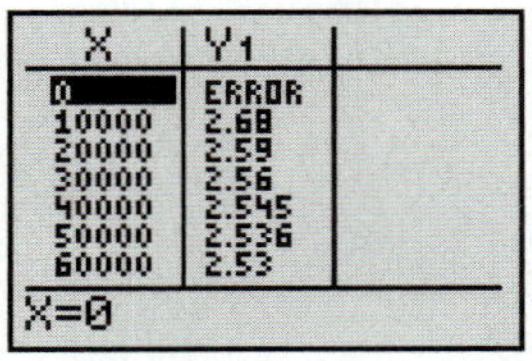

Figure 30 Inputs increasing by 10,000

From the tables, we see that as n increases, the mean price required to break even decreases. This happens because, as the number of CDs manufactured increases, the more the fixed cost is spread out, so the smaller the fixed cost that each sale has to cover. In fact, if we continue to scroll down a table for larger and larger inputs n, the outputs $P(n)$ approach 2.50, the variable cost in dollars per CD.

We can also study a graphing calculator graph to observe that the break-even CD price approaches \$2.50 for large manufacturing runs (see Figs. 31, 32, and 33). It appears that, for large inputs, the height of the graph of P gets close to 2.50.

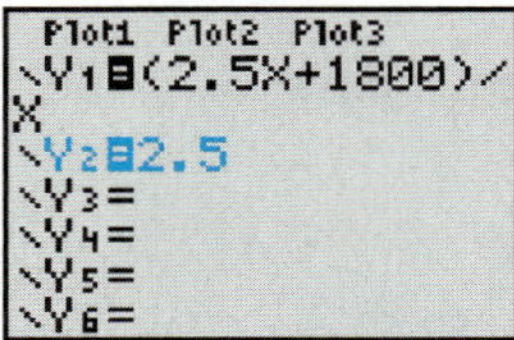

Figure 31 Enter the functions; Y_2 is the variable cost per disc

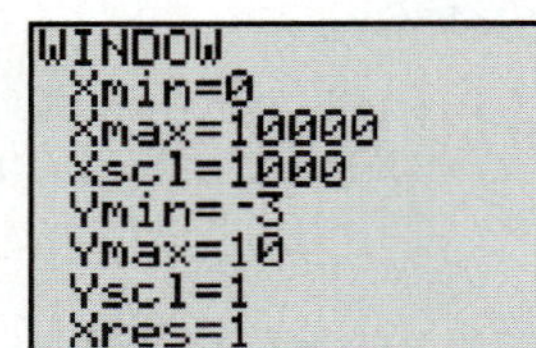

Figure 32 Set up the window to allow for large n

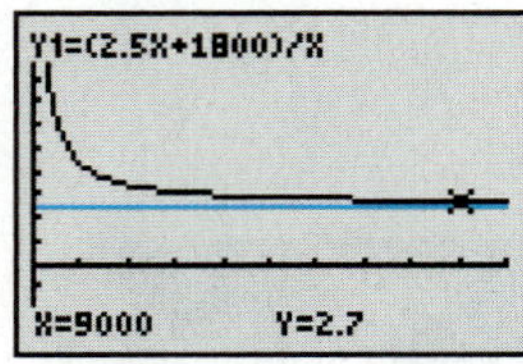

Figure 33 Graph the functions

This means that if the band could sell tens of thousands of CDs, it could price them at a few cents above the \$2.50 cost and still break even.

Table 12 Bottled-Water Consumption

Year	Bottled-Water Consumption (millions of gallons)
1990	2200
1995	3100
2000	4700
2005	7500
2009	10,600

Source: *Beverage Marketing Corporation*

Example 2 Modeling the Mean of a Quantity

In Example 11 of Section 9.6, we modeled the annual U.S. consumption of bottled water. A reasonable model is $B(t) = 21t^2 + 42t + 2250$, where $B(t)$ is the annual U.S. bottled-water consumption (in millions of gallons) at t years since 1990 (see Table 12).

In Exercise 13 of Homework 9.8, you modeled the U.S. population. A reasonable model is $U(t) = 0.0068t^2 + 2.58t + 251.4$, where $U(t)$ is the U.S. population (in millions) at t years since 1990.

1. Let $M(t)$ be the annual mean consumption of bottled water per person (in gallons per person) at t years since 1990. Find an equation of M.
2. Perform a unit analysis of the equation of M.
3. Find $M(27)$. What does it mean in this situation?
4. Find t when $M(t) = 60$. What does it mean in this situation?

Solution

1. The annual mean consumption of bottled water per person is equal to the total annual consumption divided by the U.S. population:

$$M(t) = \frac{B(t)}{U(t)} = \frac{21t^2 + 42t + 2250}{0.0068t^2 + 2.58t + 251.4}$$

2. Here is a unit analysis of the equation of M:

$$\text{gallons per person} \rightarrow M(t) = \frac{B(t)}{U(t)} = \frac{21t^2 + 42t + 2250}{0.0068t^2 + 2.58t + 251.4} \begin{matrix} \leftarrow \text{millions of gallons} \\ \leftarrow \text{millions of people} \end{matrix}$$

The units on both sides of the equation are gallons per person, which suggests the equation is correct.

3. $$M(27) = \frac{21(27)^2 + 42(27) + 2250}{0.0068(27)^2 + 2.58(27) + 251.4} \approx 57.34$$

The annual mean consumption of bottled water will be about 57.3 gallons per person in 2017, according to the model.

4. To begin, we substitute 60 for $M(t)$ in the equation of M:

$$60 = \frac{21t^2 + 42t + 2250}{0.0068t^2 + 2.58t + 251.4}$$

$$(0.0068t^2 + 2.58t + 251.4) \cdot 60 = (0.0068t^2 + 2.58t + 251.4) \cdot \frac{21t^2 + 42t + 2250}{0.0068t^2 + 2.58t + 251.4}$$

$$0.408t^2 + 154.8t + 15{,}084 = 21t^2 + 42t + 2250$$

$$0 = 20.592t^2 - 112.8t - 12{,}834$$

Now we substitute $a = 20.592$, $b = -112.8$, and $c = -12{,}834$ in the quadratic formula and solve for t:

$$t = \frac{-(-112.8) \pm \sqrt{(-112.8)^2 - 4(20.592)(-12{,}834)}}{2(20.592)} \quad \textit{Substitute into quadratic formula.}$$

$$t \approx -22.38 \quad \text{or} \quad t \approx 27.85 \quad \textit{Compute.}$$

We verify the results by entering $y = \dfrac{21t^2 + 42t + 2250}{0.0068t^2 + 2.58t + 251.4}$ in a graphing calculator and checking that the inputs -22.38 and 27.85 lead to outputs of about 60 (see Fig. 34).

The inputs -22.38 and 27.85 represent the years 1968 and 2018, respectively. The 1968 estimate indicates model breakdown: Research would show that the annual mean consumption of bottled water in 1968 was much less than 60 gallons per person. Therefore, we predict it will be 2018 when the annual mean consumption of bottled water reaches 60 gallons per person.

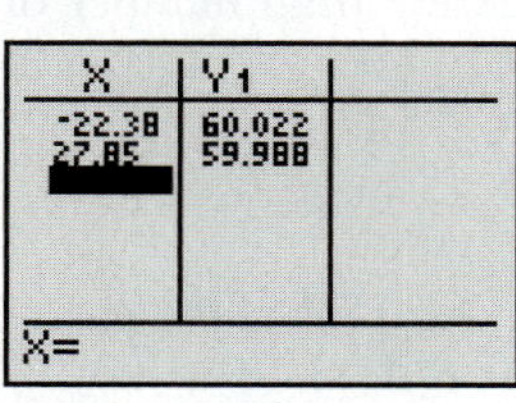

Figure 34 Verify the work

Modeling the Percentage of a Quantity

In Sections 12.1 and 12.5, we modeled the percentage of a quantity. In Example 3, we will first find a sum function that will help us find another function that describes the percentage of a quantity.

Table 13 Numbers of Broadband Cable and DSL and Fiber Optic Subscribers

	Number of Subscribers (millions)	
Year	Broadband Cable	DSL and Fiber Optic
2003	13	6
2005	22	13
2007	32	22
2009	40	27
2011	44	31

Source: *Kagan*

Example 3 Modeling the Percentage of a Quantity

The numbers of broadband cable subscribers and the numbers of DSL and fiber optic subscribers in the United States are shown in Table 13 for various years. Let $B(t)$ be the number of broadband cable subscribers and $D(t)$ be the number of DSL and fiber optic subscribers, both in millions, at t years since 1990.

1. Find equations of B and D.
2. Find an equation of the sum function $B + D$. What do the inputs and outputs of $B + D$ mean in this situation?
3. The numbers of U.S. households are shown in Table 14 for various years. A reasonable model is $H(t) = 1.3t + 92.7$, where $H(t)$ is the total number (in millions) of U.S. households at t years since 1990. Let $P(t)$ be the percentage of U.S. households that are broadband cable, DSL, or fiber optic subscribers. Find an equation of P. Assume broadband cable subscribers do not subscribe to DSL or fiber optic services.
4. Find t when $P(t) = 68$. What does it mean in this situation?

Table 14 Numbers of Households

Year	Number of Households (millions)
1995	99.0
1997	101.0
1999	103.6
2001	108.2
2003	111.3
2005	111.3
2007	116.0
2009	117.2
2010	117.5

Source: *U.S. Census Bureau*

Solution

1. Scattergrams of the data suggest we use a quadratic function to model both the number of broadband cable subscribers and the number of DSL and fiber optic subscribers (see Figs. 35 and 36). The quadratic regression equation of B and the quadratic regression equation of D are, respectively,

$$B(t) = -0.214t^2 + 11.29t - 98$$
$$D(t) = -0.179t^2 + 9.27t - 85$$

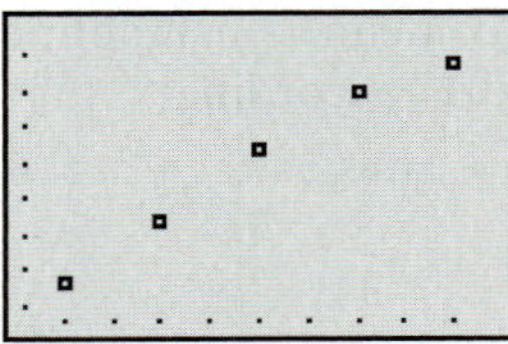

Figure 35 Broadband cable scattergram

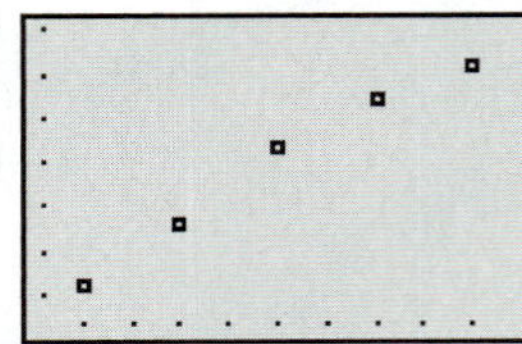

Figure 36 DSL and fiber optic scattergram

The models fit the points in the scattergrams of the two data sets quite well (see Figs. 37 and 38).

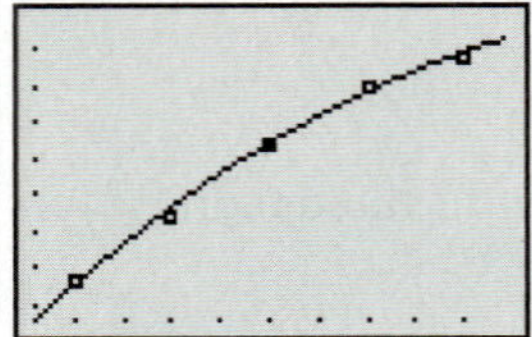

Figure 37 Broadband cable model

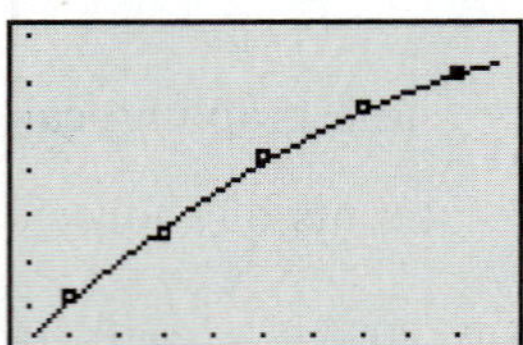

Figure 38 DSL and fiber optic model

2. $$\begin{aligned}(B + D)(t) &= B(t) + D(t) && \textit{Definition of sum function}\\ &= (-0.214t^2 + 11.29t - 98)\\ &\quad + (-0.179t^2 + 9.27t - 85) && \textit{Substitute.}\\ &= -0.393t^2 + 20.56t - 183 && \textit{Combine like terms.}\end{aligned}$$

The inputs of $B + D$ are the numbers of years since 1990, and the outputs are the total numbers of subscribers (in millions).

3. To find the percentage of households that subscribe, we divide the total number of subscribers by the number of households and multiply the result by 100:

$$\begin{aligned}P(t) &= \frac{(B + D)(t)}{H(t)} \cdot 100 && \textit{Percent formula: } p = \frac{m}{n} \cdot 100\\ &= \frac{-0.393t^2 + 20.56t - 183}{1.3t + 92.7} \cdot \frac{100}{1} && \textit{Substitute } -0.393t^2 + 20.56t - 183 \textit{ for } (B + D)(t) \textit{ and } 1.3t + 92.7 \textit{ for } H(t).\\ &= \frac{-39.3t^2 + 2056t - 18{,}300}{1.3t + 92.7} && \textit{Multiply numerators; multiply denominators.}\end{aligned}$$

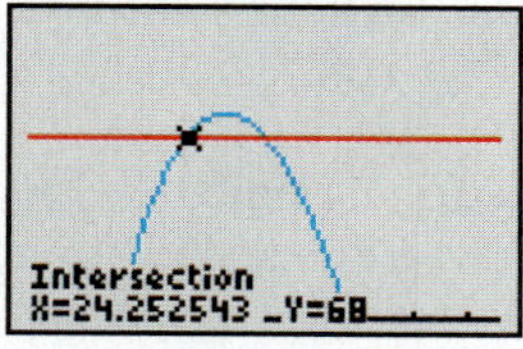

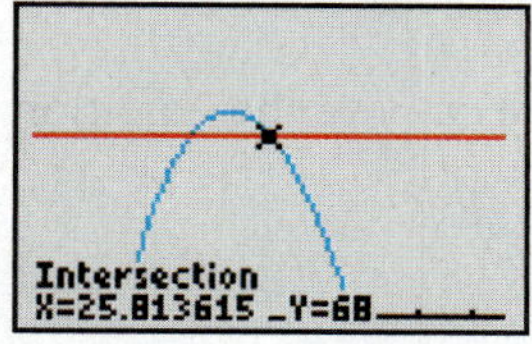

Figure 39 Verify the work

4. We substitute 68 for $P(t)$ in the equation of P:

$$68 = \frac{-39.3t^2 + 2056t - 18{,}300}{1.3t + 92.7}$$

$$(1.3t + 92.7) \cdot 68 = (1.3t + 92.7) \cdot \frac{-39.3t^2 + 2056t - 18{,}300}{1.3t + 92.7}$$

$$88.4t + 6303.6 = -39.3t^2 + 2056t - 18{,}300$$

$$39.3t^2 - 1967.6t + 24{,}603.6 = 0$$

Now we substitute $a = 39.3$, $b = -1967.6$, and $c = 24{,}603.6$ in the quadratic formula:

$$t = \frac{-(-1967.6) \pm \sqrt{(-1967.6)^2 - 4(39.3)(24{,}603.6)}}{2(39.3)} \quad \text{Substitute into quadratic formula.}$$

$$t \approx 24.25 \quad \text{or} \quad t \approx 25.81 \quad \text{Compute.}$$

To verify the results, we can enter the equations $y = \dfrac{-39.3t^2 + 2056t - 18{,}300}{1.3t + 92.7}$ and $y = 68$ and use "intersect" to find the intersection points, which are approximately $(24.25, 68)$ and $(25.81, 68)$. See Fig. 39.

The t-coordinates 24.25 and 25.81 represent the years 2014 and 2016, respectively. The 2016 prediction most likely indicates model breakdown: It is unlikely the percentage will decline and return to the 2014 level (see Fig. 39). Therefore, we predict 68% of households will be broadband cable, DSL, or fiber optic subscribers in 2014.

Distance-Speed-Time Applications

How do we model the distance traveled by an object moving at a constant speed? For instance, if a car is driven at 50 mph for 2 hours, it will travel 50 miles in the first hour and 50 miles in the second hour, for a total distance of $50 \cdot 2 = 100$ miles. If a car is driven at 60 miles per hour for 3 hours, it will travel $60 \cdot 3 = 180$ miles.

In general, the (constant) speed of a car multiplied by the amount of time the car is in motion gives the distance traveled.

Distance-Speed-Time Relationship

If an object is moving at a constant speed s for an amount of time t, then the distance d traveled is given by

$$d = st$$

and the time t is given by

$$t = \frac{d}{s}$$

Example 4 Distance-Speed-Time Relationship

A person plans to drive a steady 55 mph on an 80-mile trip. Compute the driving time.

Solution

Since the person is traveling at a constant speed, we use the equation

$$t = \frac{d}{s}$$

We substitute 80 for d and 55 for s in the equation:

$$t = \frac{80}{55} \approx 1.45$$

So, the driving time will be about 1.5 hours.

Example 5 Modeling Driving Time

A student at Seattle Central Community College plans to drive from Seattle, Washington, to Eugene, Oregon. The speed limit is 70 mph in Washington and 65 mph in Oregon. She will drive 164 miles in Washington, then 121 miles in Oregon.

1. If the student drives steadily at the speed limits, compute the driving time.
2. If the student exceeds the speed limits, let $T(a)$ be the driving time (in hours) at a mph above the speed limits. Find an equation of T.
3. Find $T(0)$. Compare this result with the result in Problem 1.
4. If the student drives 5 mph over the speed limits, compute the driving time.
5. If the student wants the driving time to be 4 hours, how much over the speed limits would she have to drive?

Solution

1. Since the student drives at a constant speed in Washington, we can use the equation $t = \frac{d}{s}$ to compute the time (in hours) spent driving in Washington:

$$t = \frac{\text{distance in Washington}}{\text{speed in Washington}} = \frac{164}{70}$$

We can also compute the time (in hours) spent driving in Oregon:

$$t = \frac{\text{distance in Oregon}}{\text{speed in Oregon}} = \frac{121}{65}$$

The total driving time is the sum of our two computed times:

$$\frac{164}{70} + \frac{121}{65} \approx 4.2 \text{ hours}$$

2. If the student drives, say, 5 mph over the speed limits, then she will drive $5 + 70 = 75$ mph in Washington and $5 + 65 = 70$ mph in Oregon. If she drives a miles per hour over the speed limits, she will drive $(a + 70)$ mph in Washington and $(a + 65)$ mph in Oregon. We use these expressions for speeds to find an equation of T:

$$T(a) = \frac{\text{distance in Washington}}{\text{speed in Washington}} + \frac{\text{distance in Oregon}}{\text{speed in Oregon}} = \frac{164}{a + 70} + \frac{121}{a + 65}$$

3. $T(0) = \frac{164}{0 + 70} + \frac{121}{0 + 65} \approx 4.2$

The driving time will be about 4.2 hours if the student drives at the speed limits. We found the same result in Problem 1.

4. If the student drives 5 mph over the speed limits, then $a = 5$:

$$T(5) = \frac{164}{5 + 70} + \frac{121}{5 + 65} \approx 3.9$$

The driving time will be about 3.9 hours.

5. If the trip is to take 4 hours, then $T(a) = 4$. Thus,

$$\begin{aligned} 4 &= \frac{164}{a+70} + \frac{121}{a+65} \\ (a+65)(a+70)\cdot 4 &= (a+65)(a+70)\cdot\left(\frac{164}{a+70} + \frac{121}{a+65}\right) \\ (a^2+135a+4550)\cdot 4 &= (a+65)(a+70)\cdot\frac{164}{a+70} + (a+65)(a+70)\cdot\frac{121}{a+65} \\ 4a^2+540a+18{,}200 &= (a+65)\cdot 164 + (a+70)\cdot 121 \\ 4a^2+540a+18{,}200 &= 164a + 10{,}660 + 121a + 8470 \\ 4a^2+540a+18{,}200 &= 285a + 19{,}130 \\ 4a^2+255a-930 &= 0 \\ a &= \frac{-255 \pm \sqrt{255^2 - 4(4)(-930)}}{2(4)} \\ a \approx -67.2 \quad &\text{or} \quad a \approx 3.5 \end{aligned}$$

The value $a = -67.2$ represents driving under the speed limits by 67.2 mph—model breakdown has occurred. So, the student would have to drive about 3.5 mph over the speed limits for the driving time to be 4 hours.

Group Exploration

Looking ahead: Proportions

Recall from Section 2.5 that the ratio of a to b is the quotient $\frac{a}{b}$.

1. a. The price of a notebook is \$3. For the given number of notebooks, find the ratio of the total cost to the number of notebooks. *Do not simplify the ratio.*
   i. 2 notebooks [**Hint:** First find the total cost of 2 notebooks. Then divide your result by 2. Include units in your result.]
   ii. 5 notebooks

   b. Compare the results that you found in parts (i) and (ii). In particular, describe how the following equation is related to your results:
   $$\frac{6}{2} = \frac{15}{5}$$
   When an equation says that two ratios are equal, we call the equation a *proportion*.

   c. For any number of notebooks, what is the ratio of the total cost to the number of notebooks? Explain.

2. a. A student earns \$8 per hour. For the given number of hours worked, find the ratio of the total income to the number of hours worked. *Do not simplify the ratio.*
   i. 3 hours
   ii. 5 hours

   b. Write a proportion that is related to the results you found in parts (i) and (ii).

   c. For any number of hours worked, what is the ratio of the total income to the number of hours worked? Explain how this ratio is related to the rate of change of income.

Homework 12.6

For extra help ▶ MyMathLab®

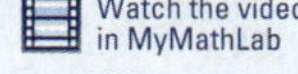

1. The ski club at a community college plans to spend \$1250 to charter a bus for a ski trip. The cost will be split evenly among the students who sign up for the trip. Each student will also pay \$350 for food, lodging, and ski lift tickets.
   a. Let $C(n)$ be the total cost (in dollars) for n students going on the trip. Find an equation of C.
   b. Let $M(n)$ be the mean cost per student (in dollars per student) if n students go on the trip. Find an equation of M.
   c. What is the mean cost per student if 30 students go?
   d. The ski trip will be cancelled unless the mean cost per student is \$400 or less. What is the minimum number of students needed to go on the trip?

2. Manhattan is one of the most densely populated regions in the United States. This borough consists of 660,700,000 square feet of land. Let $f(P)$ be the number of square feet

of Manhattan each resident would own if Manhattan were divided equally among its P residents.

a. Find an equation of f.
b. Use f to estimate the amount of land per Manhattanite if 1.6 million people live in Manhattan.
c. If the land in the United States were divided equally among U.S. residents, each resident would own about 437,000 square feet of land. How many people could live in Manhattan if each resident owned 437,000 square feet of Manhattan? Compare your answer with the number of people given in part (b).

3. For a five-year high school reunion, graduates rent out a restaurant that charges a flat fee of $500 for a band, plus $50 per person for food and two drinks.
a. Let $T(n)$ be the total cost (in dollars) for n people to attend the reunion. Find an equation of T.
b. Let $M(n)$ be the mean cost per person (in dollars per person) if n people attend the reunion. Find an equation of M.
c. Find $M(270)$. What does it mean in this situation?
d. Find n when $M(n) = 60$. What does it mean in this situation?
e. Complete Table 15 by using a graphing calculator table.

Table 15 Mean Cost per Person for the Reunion

Number of People n	Mean Cost per Person $M(n)$
100	
200	
300	
400	
500	

f. What do the values of $M(n)$ get close to as the values of n get very large? What does the result mean in this situation?

4. A student agrees to throw a party, provided that each guest shares the expenses equally with him. A four-person local band will play for $200 and free drinks and snacks. The student estimates the mean cost of drinks and snacks will be $3 per person.
a. Suppose n people (including the host and band members) attend the party. Let $C(n)$ be the party's total cost (in dollars). Find an equation of C.
b. Let $P(n)$ be the equal share of expenses (in dollars) each guest and host contributes. Find an equation of P. [**Hint:** Recall that band members get free drinks and snacks.]
c. If the host and guests are willing to pay at most $5 each, how many guests must attend to cover expenses?
d. What do the values of $P(n)$ get close to as the values of n get very large? What does the result mean in this situation?
e. If the host and guests are willing to pay only $2 each, how many guests must attend to cover expenses? Explain why this makes sense.

5. Leasing expenses, equipment maintenance, salaries, electricity, and marketing cost a car manufacturer an average of $90,000 per day to produce a certain type of car. Materials, invoices, and shipping cost the manufacturer an average of $7000 per car.
a. Suppose the car manufacturer produces and sells n cars of that type per day. Let $C(n)$ be the total daily cost (in dollars). Find an equation of C.
b. Suppose the car manufacturer produces and sells n cars of that type per day. Let $B(n)$ be the amount (in dollars) the manufacturer should charge per car to break even by selling n cars. Find an equation of B.
c. Suppose the car manufacturer produces and sells n cars of that type per day. Let $P(n)$ be the amount (in dollars) the manufacturer should charge per car to make a profit of $2000 per car. Find an equation of P. [**Hint:** Build on your equation from part (b) to find an equation of P.]
d. Find $P(40)$. What does it mean in this situation?
e. What do the values of $P(n)$ get close to as the values of n get very large? What does that mean in this situation?

6. In Example 1 of this section, we found the rational function

$$P = f(n) = \frac{2.50n + 1800}{n}$$

where $f(n)$ models the price (in dollars) the band Melted Zipper should set for its CD to break even by making and selling n CDs. [**Note:** We use different notation than in Example 1. Here we use f for the name of the function and P for the name of the dependent variable.]
a. If Melted Zipper sets the CD's price at $7, how many CDs must be made and sold for the band to break even?
b. Find an equation of f^{-1}. [**Hint:** Perform steps similar to those in part (a), but do not substitute a value for P. After a couple of steps, factor out n on one side of the equation.]
c. Find $f^{-1}(7)$. Compare this result with that in part (a).

7. The total income of all U.S. households and the numbers of U.S. households are shown in Table 16 for various years.

Table 16 Total Annual Income of All Households; Numbers of Households

Year	Total Annual Income of All Households (billions of dollars)	Number of Households (billions)
1995	6201	0.0990
1997	6937	0.1010
1999	7787	0.1036
2001	8685	0.1082
2003	9164	0.1113
2005	10,486	0.1133
2007	11,912	0.1160
2009	12,175	0.1172
2010	12,547	0.1175

Source: *U.S. Census Bureau*

a. Let $I(t)$ be the total annual income (in billions of dollars) of all households at t years since 1990. Find an equation of I.

b. In Example 3 of this section, we modeled the number of U.S. households. A reasonable model is $H(t) = 0.0013t + 0.093$, where $H(t)$ is the number (in billions) of U.S. households at t years since 1990. Let $M(t)$ be the mean annual income per household (in dollars per household) at t years since 1990. Find an equation of M.

c. Perform a unit analysis of your equation of M.

d. Predict when the mean annual income per household will be \$125,000.

e. Use the window settings in Fig. 40 to graph M. Is M increasing, decreasing, or neither for values of t between 5 and 30? What does that mean in this situation?

Figure 40 Window settings for Exercise 7e

8. Levels of fuel consumption (in millions of gallons) by vehicles in the United States are listed in Table 17 for various years.

Table 17 Vehicle Fuel Consumption

Year	Amount of Fuel Consumed (millions of gallons)
1975	108,900
1980	115,000
1985	121,300
1990	130,800
1995	143,800
2000	162,600
2005	174,800
2010	169,700

Source: *Federal Highway Administration*

a. Let $F(t)$ be the total amount of fuel (in millions of gallons) used during the year that is t years since 1970. Find a regression equation of F.

b. In Exercise 13 of Homework 9.8, you modeled the U.S. population. A good model is $U(t) = 0.0068t^2 + 2.31t + 202.61$, where $U(t)$ is the U.S. population (in millions) at t years since 1970. Let $M(t)$ be the mean amount of fuel used per person (in gallons per person) during the year that is t years since 1970. Find an equation of M.

c. Find t when $M(t) = 591$. What does it mean in this situation?

d. Use the window settings in Fig. 41 to graph M. Is M increasing, decreasing, or neither for values of t between -5 and 50? What does that mean in this situation?

e. Let $P(t)$ be the average price (in dollars per gallon) of gasoline at t years since 1970 (see Table 18). The quadratic regression model is $P(t) = 0.0036t^2 - 0.12t + 1.79$. Find $(P \cdot M)(48)$. What does it mean in this situation?

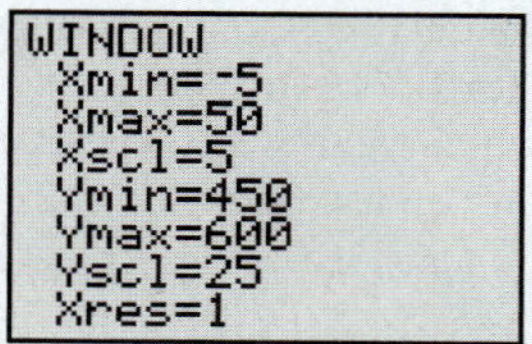

Figure 41 Window settings for Exercise 8d

Table 18 Average Prices of Gasoline

Year	Price (dollars per gallon)
1995	1.21
1997	1.29
1999	1.22
2001	1.53
2003	1.64
2005	2.34
2007	2.85
2009	2.40
2010	2.84

Source: *Department of Energy*

f. Is $M(t)$ an underestimate or an overestimate of the mean amount of fuel used per *driver*? Explain.

9. Textbook revenues are shown in Table 19 for various years.

Table 19 Textbook Revenues

Year	Revenue (millions of dollars)
2004	9198
2005	9977
2006	10,126
2007	10,697
2008	11,162

Source: *U.S. Census Bureau*

a. Let $R(t)$ be the annual revenue (in millions of dollars) from textbooks at t years since 1990. Find a linear equation of R.

b. In Exercise 80 of Homework 12.1, you found an equation close to $E(t) = 0.026t^2 - 0.2t + 14.61$, where $E(t)$ is the college enrollment (in millions) at t years since 1990 (see Table 20).

Table 20 College Enrollments

Year	Enrollment (millions)
1995	14.2
2000	15.3
2005	17.5
2008	19.1
2010	21.2

Source: *National Center for Education Statistics*

Let $A(t)$ be the mean amount of money spent on textbooks per student (in dollars per student) during the year that is t years since 1990. Find an equation of A.

c. Predict the mean amount of money per student that will be spent on textbooks in 2018.

d. During which year will the mean amount of money students spend on textbooks be $525 per student?

10. The numbers of colleges are shown in Table 21 for various years.

Table 21 Numbers of Colleges

Year	Number of Colleges (thousands)
1980	3.231
1985	3.340
1990	3.559
1995	3.706
2000	4.182
2005	4.276
2010	4.495

Source: *National Center for Education Statistics*

a. Let $N(t)$ be the total number (in thousands) of colleges at t years since 1980. Find a linear equation of N.

b. College enrollment (in *thousands*), $E(t)$, can be modeled by the equation $E(t) = 26t^2 - 726t + 19{,}257$, where t is the number of years since 1980 (see Table 20). Let $M(t)$ be the mean enrollment per college at t years since 1980. Find an equation of M.

c. Predict when the mean enrollment will be 5800 students per college.

d. Use the window settings shown in Fig. 42 to graph M. Then use "minimum" on a graphing calculator to find when the mean enrollment was the lowest. What was that enrollment?

```
WINDOW
 Xmin=0
 Xmax=40
 Xscl=5
 Ymin=3000
 Ymax=7000
 Yscl=500
 Xres=1
```

Figure 42 Window settings for Exercise 10d

11. The numbers of morning daily newspapers and evening daily newspapers are shown in Table 22 for various years.

Table 22 Numbers of Morning Dailies and Evening Dailies

Year	Number of Daily Newspapers: Morning	Evening
1980	387	1388
1985	482	1220
1990	559	1084
1995	656	891
2000	766	727
2005	817	645
2009	869	528

Source: *Editor & Publisher Co.*

a. Let $M(t)$ be the number of morning daily newspapers and $E(t)$ be the number of evening daily newspapers, both at t years since 1980. Find linear equations of M and E.

b. Find an equation of the sum function $M + E$. What do the inputs and outputs of $M + E$ mean in this situation?

c. Let $P(t)$ be the percentage of newspapers that are morning dailies at t years since 1980. Find an equation of P.

d. Find $P(36)$. What does it mean in this situation?

e. Find t when $P(t) = 80$. What does it mean in this situation?

12. The numbers of women and men who live alone are shown in Table 23 for various years.

Table 23 Numbers of Women and Men Who Live Alone

Year	Number Living Alone (millions): Women	Men
1980	11.3	7.0
1985	12.7	7.9
1990	14.0	9.0
1995	14.6	10.1
2000	15.6	11.2
2005	17.3	12.8
2010	17.4	14.0

Source: *U.S. Census Bureau*

a. Let $W(t)$ be the number of women who live alone and $M(t)$ be the number of men who live alone, both in millions, at t years since 1980. Find linear equations of W and M.

b. Find an equation of the sum function $W + M$. What do the inputs and outputs of $W + M$ mean in this situation?

c. Let $P(t)$ be the percentage of people living alone who are women at t years since 1980. Find an equation of P.

d. Predict in which year 55% of people who live alone will be women.

e. Use the window settings shown in Fig. 43 to graph P. Is P increasing, decreasing, or neither for values of t between 0 and 40? How is this possible, given that the number of women who live alone has been increasing since 1980?

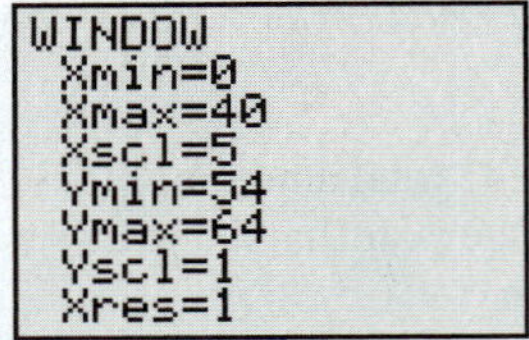

Figure 43 Window settings for Exercise 12e

13. The numbers of U.S. women and men who earned a bachelor's degree are listed in Table 24 for various years.

a. Let $W(t)$ be the number of women and $M(t)$ be the number of men, both in thousands, who earned a bachelor's degree in the year that is t years since 1980. Find quadratic equations of W and M.

b. Find an equation of the sum function $W + M$. What do the inputs and outputs of $W + M$ mean in this situation?

c. Among people who earned a bachelor's degree at t years since 1980, let $P(t)$ be the percentage who are men. Find an equation of P.

Table 24 Women and Men Who Earned a Bachelor's Degree

Year	Number of People Who Earned a Bachelor's Degree (thousands)	
	Women	Men
1980	456	474
1985	497	483
1990	560	492
1995	634	526
2000	708	530
2005	826	613
2009	916	685

Source: *National Center for Education Statistics*

d. Predict in which year 42.1% of people who earn a bachelor's degree will be men.

e. Use the window settings shown in Fig. 44 to graph P. Is P increasing, decreasing, or neither for values of t between 0 and 35? How is this possible, given that the number of bachelor's degrees earned by men has been increasing since 1980?

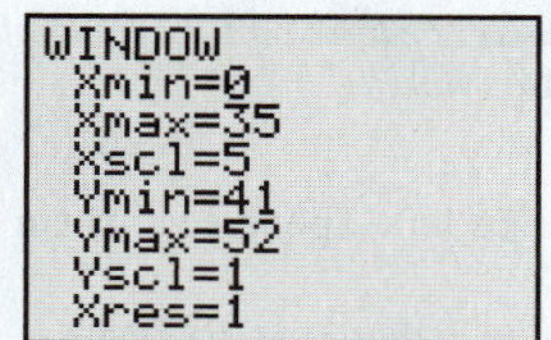

Figure 44 Window settings for Exercise 13e

14. The annual U.S. per-person consumptions of whole milk and the annual U.S. per-person consumptions of lower fat and skim milk are shown in Table 25 for various years.

Table 25 Annual U.S. Per-Person Consumptions of Whole Milk, Lower Fat and Skim Milk

Year	Annual Consumption (gallons per person)	
	Whole Milk	Lower Fat and Skim Milk
1955	33.5	2.9
1965	28.8	3.7
1975	21.7	8.1
1985	14.3	12.3
1995	8.6	15.3
2005	7.0	14.1
2010	5.6	14.8

Source: *USDA/Economic Research Service*

a. Let $W(t)$ be the annual consumption of whole milk and $L(t)$ be the annual consumption of lower fat and skim milk, both in gallons per person, at t years since 1900. Find equations of W and L.

b. Find an equation of the sum function $W + L$. What do the inputs and outputs of $W + L$ mean in this situation?

c. Let $P(t)$ be the percentage of milk consumed that is whole milk in the year that is t years since 1900. Find an equation of P.

d. Predict the percentage of milk consumed in 2016 that will be whole milk.

e. Predict in which year 6% of the milk consumed will be whole milk.

15. A person drives 60 mph for 85 miles. Compute the driving time.

16. A person drives at a constant speed for 100 miles in 1.7 hours. What is that speed?

17. A student plans to drive from Cuyahoga Community College in Cleveland, Ohio, to Pittsburgh, Pennsylvania. He will drive 85 miles in Ohio, then 53 miles in Pennsylvania. The speed limit is 70 mph in Ohio and 65 mph in Pennsylvania.

a. Let $T(a)$ be the driving time (in hours) if the student drives a mph above the speed limits. Find an equation of T.

b. If he drives 5 mph over the speed limits, compute the driving time.

c. By how much would he have to exceed the speed limits for the driving time to be 1.8 hours? Verify your answer by using a graphing calculator table.

18. A student plans to drive from Oklahoma City, Oklahoma, to Little Rock, Arkansas. She will drive 183 miles in Oklahoma, then 161 miles in Arkansas. The speed limit is 75 mph in Oklahoma and 70 mph in Arkansas.

a. Let $T(a)$ be the driving time (in hours) if the student drives a mph above the speed limits. Find an equation of T.

b. If she drives 5 mph over the speed limits, compute the driving time.

c. By how much would she have to exceed the speed limits for the driving time to be 4 hours? Verify your answer by using a graphing calculator table.

19. A student plans to drive from Ivy Tech Community College in Indianapolis, Indiana, to the University of Illinois in Champaign–Urbana, Illinois. She will drive 83 miles in Indiana, then 37 miles in Illinois. The speed limit is 70 mph in Indiana and 65 mph in Illinois.

a. Let $T(a)$ be the driving time (in hours) if the student drives a mph above the speed limits. Find an equation of T.

b. Find $T(0)$ and $T(10)$. What do they mean in this situation?

c. Find $T(0) - T(10)$. What does it mean in this situation?

d. Find a when $T(a) = 1.6$. What does it mean in this situation?

20. A student plans to drive from Madison Area Technical College in Madison, Wisconsin, to Fergus Falls, Minnesota. He will drive 204 miles in Wisconsin, then 240 miles in Minnesota. The speed limit is 65 mph in Wisconsin and 70 mph in Minnesota.

a. Let $T(a)$ be the driving time (in hours) if the student drives a mph above the speed limits. Find an equation of T.

b. Find $T(0)$ and $T(5)$. What do they mean in this situation?

c. Find $T(0) - T(5)$. What does it mean in this situation?

d. Find a when $T(a) = 6$. What does it mean in this situation?

21. In Example 5 of this section, we found the equation

$$T(a) = \frac{164}{a + 70} + \frac{121}{a + 65}$$

where $T(a)$ is the driving time (in hours) for a student to drive from Seattle, Washington, to Eugene, Oregon, at a mph above the speed limits.

a. Perform the addition on the right side of the equation of T.

b. Use your result in part (a) to find the driving time if the student drives 10 mph over the speed limits.

Concepts

22. If a person is traveling at 0 miles per hour, how long will it take the person to travel 5 miles? Explain why this suggests we cannot divide by 0 by referring to the formula $t = \frac{d}{s}$, where t is the number of hours it takes to travel d miles at s miles per hour.

23. a. Describe a situation where the mean salary of a group of people estimates the salaries of the people well.

b. Describe a situation where the mean salary of a group of people does not estimate the salaries of the people well.

24. Describe how to use two or more models to form a function that models the mean of a quantity. Compare this process with using two or more models to form a function that models the percentage of a quantity. For both types of modeling, give an example that is different from those in this textbook.

Related Review

25. The risk of having shingles (a painful rash related to chicken pox) increases with age (see Table 26). Let $f(t)$ be the percentage of Americans who have shingles at age t years.

Table 26 Percentages of Americans Who Have Shingles

Age (years)	Percent
10	0.5
20	1.3
30	2.7
40	4.8
50	7.5
60	11.9
70	19.7
80	31.8
90	46.1

Source: *J. C. Donahue et al.,* Archives of Internal Medicine, *1995*

a. Find a linear equation, an exponential equation, and a quadratic equation of f. Compare how well the models fit the data.

b. For ages less than 23 years, explain why there is model breakdown for both the linear model and the quadratic model. How well does the exponential model fit the data for these ages?

c. Use the exponential model to estimate at what age 25% of Americans have shingles.

d. What is the base b of the exponential model $f(t) = ab^t$? What does it mean in this situation?

e. A study showed that a new vaccine can reduce the number of shingles cases by 51%. If the vaccine is approved and all 85-year-old Americans receive it, use the exponential model to predict the percentage of 85-year-old Americans who will have shingles.

26. World land speed records are shown in Table 27 for various years.

Table 27 Land Speed Records

Year	Record (mph)
1904	91
1914	124
1927	175
1935	301
1947	394
1960	407
1970	622
1983	633
1997	763

Source: *Fédération Internationale de l'Automobile*

Let $r = f(t)$ be the land speed record (in miles per hour) at t years since 1900.

a. Use a graphing calculator to draw a scattergram of the data. Which type of function would best model the data—linear, exponential, or quadratic? Explain.

b. Find an equation of a linear model to describe the data.

c. What is the slope? What does it mean in this situation?

d. The speed of sound is about 748 mph. Use the model you found in part (b) to estimate when a car first broke the sound barrier (i.e., exceeded 748 mph). For which year(s) shown in Table 27 did a car actually break the barrier?

e. Find an equation of f^{-1}.

f. Find $f^{-1}(1000)$. What does it mean in this situation?

Expressions, Equations, Functions, and Graphs

Perform the indicated instruction. Then use words such as linear, quadratic, cubic, exponential, logarithmic, rational, polynomial, degree, function, one variable, *and* two variables *to describe the expression, equation, or system.*

27. Factor $75x^3 - 50x^2 - 12x + 8$.

28. Solve $2x^2 + 12x + 13 = 0$.

29. Solve $75x^3 - 50x^2 = 12x - 8$.

30. Find the vertex of the graph of $f(x) = 2x^2 + 12x + 13$.

31. Find the product $(x^2 + 2x - 3)(3x^2 - x - 4)$.

32. Graph $f(x) = 2x^2 + 12x + 13$ by hand.

12.7 Proportions; Similar Triangles

Objectives

» Use *proportions* to make estimates.

» Find the length of a side of a *similar triangle*.

In Section 2.5, we worked with ratios. In this section, we will discuss an equation called a *proportion,* which contains ratios. We will use proportions to make estimates of quantities and to find the lengths of the sides of special pairs of triangles called *similar triangles*.

Throughout this chapter, we have been working with rational expressions and rational equations. Recall from Section 12.1 that the word "rational" refers to a ratio. Recall from Section 2.5 that the ratio of a to b can be written as the fraction $\frac{a}{b}$ or as $a:b$. For example, if there are 9 women and 5 men on a coed softball team, then the ratio of the number of women to the number of men is

$$\frac{9 \text{ women}}{5 \text{ men}}$$

Proportions

A **proportion** is a statement of the equality of two ratios. For example,

$$\frac{4}{6} = \frac{2}{3}$$

is a proportion that says the ratios $\frac{4}{6}$ and $\frac{2}{3}$ are equal.

There are many situations in which proportions can describe equal ratios. For example, suppose that a certain type of pen costs $2. Then the ratio of the total cost of some pens to the number of pens is constant for any group of pens. Here is the ratio for a group of 3 pens:

$$\begin{array}{r} \text{Total cost} \longrightarrow \\ \text{Number of pens} \longrightarrow \end{array} \frac{\$6}{3 \text{ pens}} = \$2 \text{ per pen}$$

Here is the ratio for a group of 5 pens:

$$\begin{array}{r} \text{Total cost} \longrightarrow \\ \text{Number of pens} \longrightarrow \end{array} \frac{\$10}{5 \text{ pens}} = \$2 \text{ per pen}$$

The proportion

$$\frac{\$6}{3 \text{ pens}} = \frac{\$10}{5 \text{ pens}}$$

correctly states that the ratio of the total cost to the number of pens is the same for a group of 3 pens and a group of 5 pens. We say the total cost and the number of pens are *proportional*.

In general, if the ratio of two related quantities is constant, we say the two quantities are **proportional.**

When we work with a proportion of the form

$$\frac{a}{b} = \frac{c}{d}$$

if we know the values of three of the four variables, we can solve for the fourth variable.

Example 1 Using a Proportion to Make an Estimate

While commuting to work and running errands, a person travels 576 miles in a 3-week period.

1. Estimate over what period the person will travel 1000 miles while commuting to work and running errands.
2. Discuss any assumptions that you made in Problem 1. Describe a scenario in which the result in Problem 1 is an overestimate.

Solution

1. We let t be the period (in weeks) required for the person to travel 1000 miles while commuting to work and running errands. Assuming that the ratio of the total miles traveled to the number of weeks is constant, we set up a proportion:

$$\begin{matrix}\text{Total miles traveled} \longrightarrow \\ \text{Number of weeks} \longrightarrow\end{matrix} \quad \frac{576}{3} = \frac{1000}{t} \quad \begin{matrix}\longleftarrow \text{Total miles traveled} \\ \longleftarrow \text{Number of weeks}\end{matrix}$$

Next, we multiply both sides of the equation by the LCD, $3t$:

$$\frac{576}{3} \cdot \frac{3t}{1} = \frac{1000}{t} \cdot \frac{3t}{1} \qquad \textit{Multiply both sides by 3t.}$$

$$576t = 3000 \qquad \textit{Simplify.}$$

$$t \approx 5.21 \qquad \textit{Divide both sides by 576.}$$

So, the person will drive 1000 miles over about a 5.2-week period.

2. We assumed that the ratio of the total miles traveled to the number of weeks is constant. This will not be the case if the person's driving patterns over the course of the 1000 miles are significantly different from those for the 576 miles. For example, if the person is transferred to a work location that is farther from home and the person's driving patterns while running errands remained the same, our estimate in Problem 1 would be an overestimate.

In Example 1, the person traveled 576 miles in a 3-week period. For this period, the unit ratio of total miles traveled to number of weeks is

$$\frac{576 \text{ miles}}{3 \text{ weeks}} = \frac{192 \text{ miles}}{1 \text{ week}}$$

To estimate over what period the person would travel 1000 miles, we assumed the ratio of total miles to number of weeks would remain $\frac{192 \text{ miles}}{1 \text{ week}}$. If that ratio were not close to $\frac{192 \text{ miles}}{1 \text{ week}}$, then our result would not be accurate.

If we use a proportion to estimate the value of one of two quantities, the accuracy of our result depends on how much the ratio of the two quantities varies. If the ratio is constant, the estimate will be accurate. If the ratio is approximately constant, the estimate will likely be reasonable. If the ratio varies a great deal, the estimate will likely be inaccurate.

Example 2 Using a Proportion to Make an Estimate

In a poll of 2209 adults, 1679 adults said professional baseball players who use steroids should be banned from the National Baseball Hall of Fame (Source: *Harris Interactive, Inc.*).

1. Estimate how many of the 20,000 students at Cincinnati State Technical and Community College would say that professional baseball players who use steroids should be banned from the National Baseball Hall of Fame.
2. Discuss any assumptions you made in Problem 1. Describe a scenario in which the result in Problem 1 is an underestimate.

Solution

1. We let n be the number of students at Cincinnati State Technical and Community College who would say professional baseball players who use steroids should be banned from the National Baseball Hall of Fame. Assuming the ratio of people

who support the ban to the total number of people is the same at the university as in the poll, we set up a proportion:

Number of students in favor of ban $\longrightarrow$ / Total number of students $\longrightarrow$

$$\frac{n}{20{,}000} = \frac{1679}{2209}$$

$\longleftarrow$ Number of polled adults in favor of ban / $\longleftarrow$ Total number of polled adults

Next, we multiply both sides of the equation by 20,000:

$$\frac{n}{20{,}000} \cdot 20{,}000 = \frac{1679}{2209} \cdot 20{,}000 \quad \textit{Multiply both sides by 20,000.}$$

$$n \approx 15{,}201 \quad \textit{Simplify.}$$

So, about 15,201 of the 20,000 students would say that professional baseball players who use steroids should be banned from the National Baseball Hall of Fame.

2. We assumed the ratio of people who support the ban to the total number of people is the same at the college as in the poll. This may not be the case. For example, if students at the college are more disapproving of steroid use than the adults in the survey, our estimate would be an underestimate.

Similar Triangles

Two triangles are called **similar triangles** if their corresponding angles are equal in measure. For example, the two triangles in Fig. 45 are similar triangles because the measures of angles A and X are equal, the measures of angles B and Y are equal, and the measures of angles C and Z are equal.

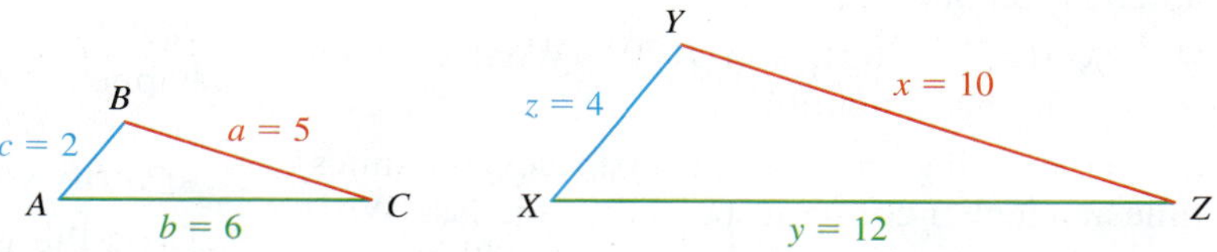

Figure 45 Two similar triangles

Here we find the ratio of the lengths of corresponding sides of the two triangles in Fig. 45:

$$\frac{c}{z} = \frac{2}{4} = \frac{1}{2}, \quad \frac{a}{x} = \frac{5}{10} = \frac{1}{2}, \quad \frac{b}{y} = \frac{6}{12} = \frac{1}{2}$$

Notice that all three ratios are equal.

Side Lengths of Similar Triangles

The lengths of the corresponding sides of two similar triangles are proportional.

Similar triangles have the same shape, but not necessarily the same size.

Example 3 Finding the Length of the Side of a Triangle

The two triangles shown in Fig. 46 are similar. Find the length of the side labeled "x feet" on the larger triangle.

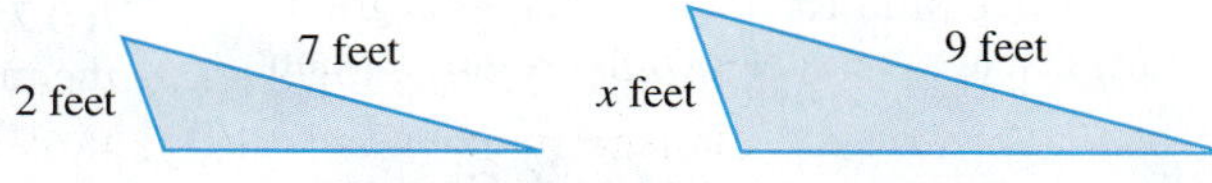

Figure 46 Two similar triangles

Solution

Since the triangles are similar, the lengths of corresponding sides are proportional. Thus,

$$\frac{2}{x} = \frac{7}{9} \quad \text{Use a proportion.}$$

$$\frac{2}{x} \cdot 9x = \frac{7}{9} \cdot 9x \quad \text{Multiply both sides by LCD, } 9x.$$

$$18 = 7x \quad \text{Simplify.}$$

$$\frac{18}{7} = x \quad \text{Divide both sides by 7.}$$

$$2.57 \approx x \quad \text{Round to second decimal place.}$$

So, the length of the side labeled "x feet" is approximately 2.57 feet.

Group Exploration

Looking ahead: Inverse variation

Suppose you intend to drive 100 miles. Let $f(s)$ be the time (in hours) it will take you to drive 100 miles if you drive at a constant speed of s miles per hour.

1. Find an equation of f.
2. Find $f(50)$, $f(55)$, $f(60)$, and $f(70)$. What do your results mean in this situation?
3. Consider completing the 100-mile trip several times, each time at a higher constant speed than the last. What happens to the travel time as the speed gets extremely high? Use a graphing calculator table and graph to verify your answer. (Recall from Section 10.3 that when a function behaves like this, we say the horizontal axis is a *horizontal asymptote* of the graph of the function.)
4. Consider completing the 100-mile trip several times, each time at a *lower* constant speed than the last. What happens to the travel time as the speed gets extremely close to 0? Use a graphing calculator table and graph to verify your answer.
5. Does the graph of f have a vertical asymptote? If so, what is it? How does your answer relate to Problem 4?

Homework 12.7

For extra help ▶ MyMathLab®

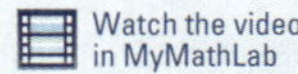

Watch the videos in MyMathLab

Download the MyDashboard App

1. A band pays \$720 for 6 hours of recording time in a studio. How much will the band pay for 8 hours of recording time?
2. A household pays \$146.25 for 3 months' use of cable TV. How much will the household pay for 5 months' use of cable TV?
3. If 0.75 cup of Post Grape-Nuts Flakes® cereal contains 4 grams of sugar, what amount, in cups, of that cereal contains 7 grams of sugar?
4. If 0.5 cup of Chef Boyardee Mini Ravioli® contains 4.5 grams of fat, what amount, in cups, of that ravioli would contain 11.25 grams of fat?
5. A 10-ounce solution contains 4 ounces of acid. How many ounces of acid are in 6 ounces of that solution?
6. A 7-ounce solution contains 3 ounces of lemon juice. How many ounces of lemon juice are in 5 ounces of the solution?
7. If 3 out of 5 students at a college are full-time students, what is the enrollment at the college if it has 21,720 full-time students?
8. If 2 out of 7 students at a college are business majors, what is the enrollment at the college if it has 4172 business majors?
9. If 1.25 inches on a map represent 50 miles, how many inches on the map represent 270 miles?
10. If 0.75 inch on a map represents 3 miles, how many inches on the map represent 10 miles?
11. If 5 U.S. dollars can be exchanged for 4.05 European euros, how many U.S. dollars can be exchanged for 260 euros?

12. If 15 U.S. dollars can be exchanged for 158.68 Mexican pesos, how many U.S. dollars can be exchanged for 3000 pesos?

13. The weight of an object on Earth and the weight of the same object on the Moon are proportional. An astronaut who weighs 180 pounds on Earth weighs 29.8 pounds on the Moon. What is the weight of a person on Earth if she weighs 28.5 pounds on the Moon?

14. The weight of an object on Earth and the weight of the same object on Jupiter are proportional. An astronaut who weighs 150 pounds on Earth would weigh 379.9 pounds on Jupiter. What is the weight of a person on Earth if he would weigh 450 pounds on Jupiter?

15. An American pays $175.50 for 3 months of phone service.
a. Estimate how much the person will pay for 7 months of phone service.
b. Discuss any assumptions you made in part (a). If the person begins to make a lot of international calls during the 7-month period and the person's U.S.-call patterns remain the same, is the result you found in part (a) an underestimate or an overestimate? Explain.

16. A family spends $475 on groceries in a 5-month period.
a. Estimate the family's total money for groceries during a 3-month period.
b. Discuss any assumptions you made in part (a). If hungry children were away during part of the 5-month period but were home during the 3-month period, is the result you found in part (a) an underestimate or an overestimate? Explain.

17. In a poll of 1017 adults in the United States, 478 adults believed the U.S. government should redistribute wealth by heavy taxes on the rich (Source: *Gallup Organization*).
a. Estimate how many of the 12,758 adults in Deerfield, Illinois, believe the U.S. government should redistribute wealth by heavy taxes on the rich.
b. Discuss any assumptions you made in part (a). The 2009 median household income was $128,149 in Deerfield and $49,777 in the United States overall (Source: *U.S. Census Bureau*). Do you think the result you found in part (a) is an underestimate or an overestimate? Explain.

18. In a poll of 2242 adults in the United States, 1143 of the adults believed the Olympics should be restricted to amateur athletes (Source: *Harris Interactive*).
a. Estimate how many of the 14,953 students at the Rockville Campus of Montgomery College believe the Olympics should be restricted to amateur athletes.
b. A much higher percentage of adults over age 40 believe in the restriction than do adults under age 40. Do you think that the result you found in part (a) is an underestimate or an overestimate? Explain.

19. When a 4000-pound car travels at 40 miles per hour and then hits a stationary object, such as a concrete wall, the force of impact is 26 tons. When the speed is 60 miles per hour instead, the force of impact is 55 tons. Are the speed and the force of impact proportional for a 4000-pound car? Explain.

20. If you park your car with the windows closed in a sunny location, the temperature inside the car will increase by 29°F in 20 minutes and by 34°F in 30 minutes (Source: *Jan Null*). Are the time elapsed and the increase in temperature proportional? Explain.

21. In fall 2003, the full-time equivalent (FTE) enrollment at Louisiana State University (LSU) was 27,764 students and the number of FTE faculty was 1294. The ratio of the FTE enrollment to the number of FTE faculty at 23 peer institutions is 17:1 (Source: *LSU*; U.S. News & World Report).
a. Find the ratio of the FTE enrollment to the number of FTE faculty at LSU in fall 2003. Is the ratio greater than, equal to, or less than 17:1?
b. The LSU deans and department chairs met to determine the cost of reducing their ratio of FTE enrollment to the number of FTE faculty to 17:1. How many FTE faculty would need to be hired to do that? The average cost (salary plus benefits) of adding an FTE faculty member at LSU in 2003 was $73,125. What would be the total cost of hiring enough faculty to reduce the ratio to 17:1?
c. How much would the FTE enrollment at LSU have to be reduced to reduce the ratio of the FTE enrollment to the number of FTE faculty to 17:1? The revenue (tuition plus an academic excellence fee) from an FTE student was $3345 at LSU in 2003. What would be the total cost of reducing the FTE enrollment to reduce the ratio to 17:1?
d. Which is the cheaper way to reduce the ratio, hiring FTE faculty or reducing the FTE enrollment?

22. The ratio of 2 and a number is equal to the ratio of the number and 8. Find the number. (The result is called the *geometric mean* of 2 and 8.)

For Exercises 23–28, the given figure shows two similar triangles. Find the length of the side labeled "x inches," "x feet," or "x meters." Round your result to the second decimal place.

23. See Fig. 47.

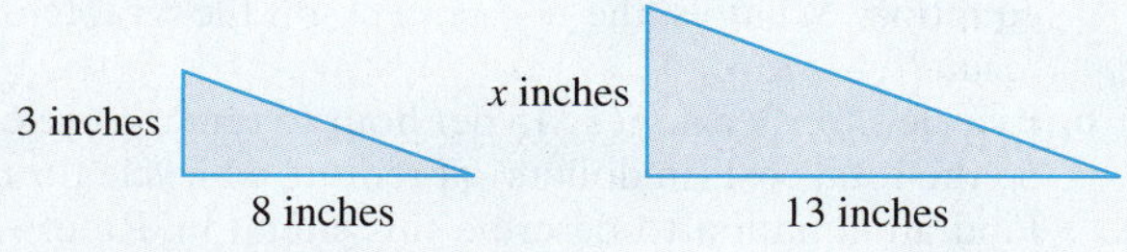

Figure 47 Exercise 23

24. See Fig. 48.

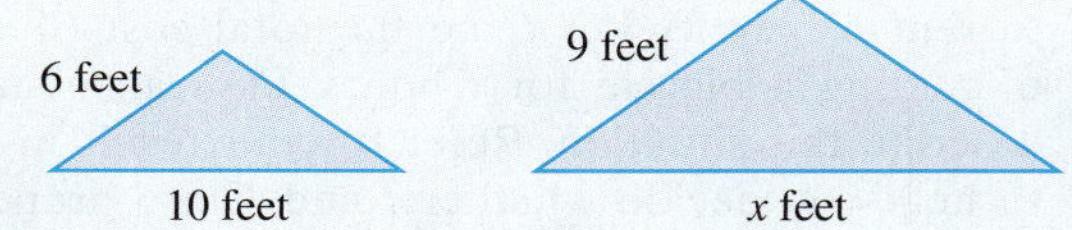

Figure 48 Exercise 24

25. See Fig. 49.

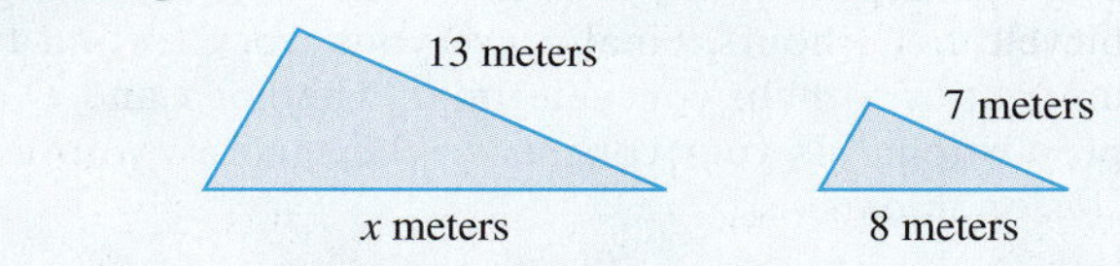

Figure 49 Exercise 25

26. See Fig. 50.

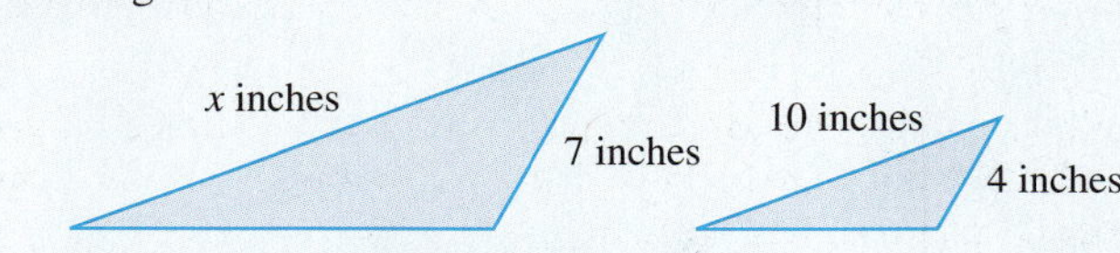

Figure 50 Exercise 26

27. See Fig. 51.

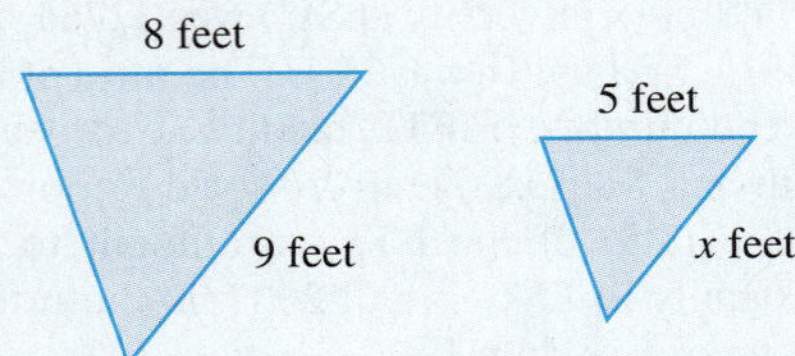

Figure 51 Exercise 27

28. See Fig. 52.

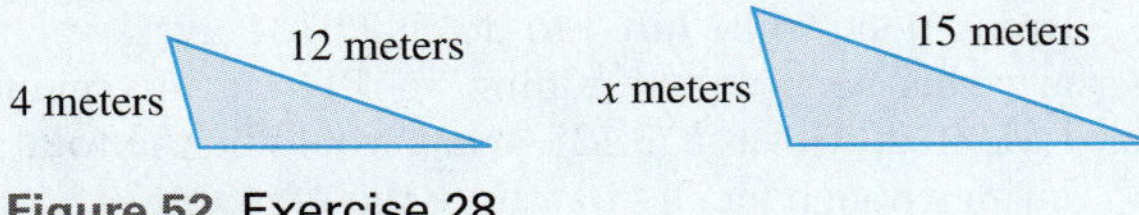

Figure 52 Exercise 28

Concepts

29. Compare the meaning of *ratio* with the meaning of *proportion*.

30. Give an example of an authentic situation in which two quantities are proportional. Why are the quantities proportional?

Related Review

31. If x and y are proportional, then the ratio of x and y is constant. That is,

$$\frac{y}{x} = k$$

where k is a constant.

a. For the equation $\frac{y}{x} = k$, get y alone on one side of the equation. What is the y-intercept of the graph of your result?

b. Bicycle shop A charges \$15 per hour to rent a bicycle. Let C be the total cost (in dollars) of renting a bicycle for t hours. Find an equation to describe this situation. Refer to your result in part (a) to help you decide whether t and C are proportional. Explain.

c. Bicycle shop B charges a flat fee of \$25, plus \$10 per hour, to rent a bicycle. Let C be the total cost (in dollars) of renting a bicycle for t hours. Find an equation to describe this situation. Refer to your result in part (a) to help you decide whether t and C are proportional. Explain.

d. Find the total cost of renting a bicycle for 2 hours at bicycle shop B. Next, find the total cost of renting a bicycle for 3 hours. Finally, use your work to find two unit ratios to help you determine whether t and C are proportional. Is your conclusion the same as your conclusion in part (c)?

32. A student earns \$72 for working 8 hours at a clothing store.

a. Use a proportion to find the student's total earnings for working 40 hours.

b. What is the rate of change of the student's total earnings?

c. Let $E(t)$ be the student's total earnings (in dollars) for working t hours. Find an equation of E.

d. Use E to find the student's total earnings for working 40 hours. Compare the result with the result that you found in part (a).

33. Consider the line shown in Fig. 53.

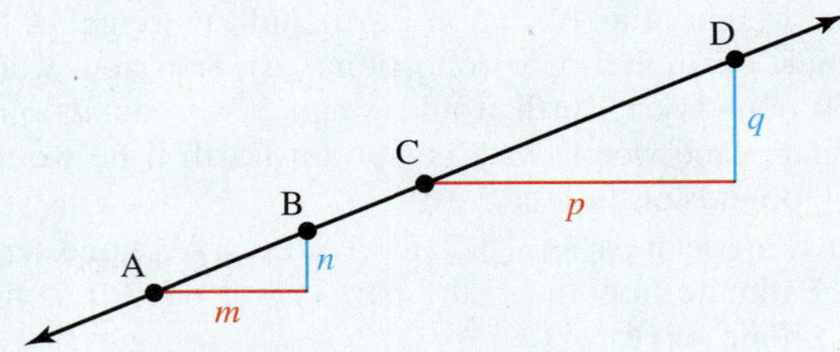

Figure 53 Exercise 33

a. Use the points A and B to find the slope of the line in terms of m and n.

b. Use the points C and D to find the slope of the line in terms of p and q.

c. Use the concept of similar triangles to help you explain why your results in parts (a) and (b) are equal.

d. Explain why your work in parts (a), (b), and (c) shows that no matter which two distinct points on a line are used to find the slope of the line, the result will be the same.

34. Two similar right triangles are shown in Fig. 54. Find the length of the side labeled "x feet." [**Hint**: First use the Pythagorean theorem, and then set up a proportion.]

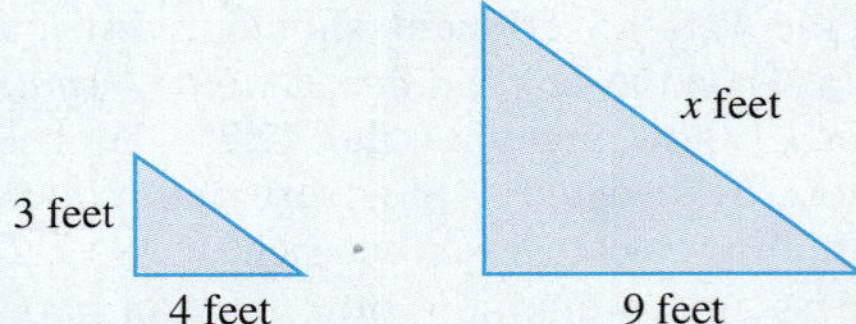

Figure 54 Exercise 34

Expressions, Equations, Functions, and Graphs

Perform the indicated instruction. Then use words such as linear, quadratic, cubic, power, radical, rational, polynomial, degree, function, one variable, *and* two variables *to describe the expression, equation, or system.*

35. Solve $2x^2 + 7x - 15 = 0$.

36. Find the sum $\frac{2x}{x-4} + \frac{5}{x+3}$.

37. Factor $2x^2 + 7x - 15$.

38. Find the difference $\frac{2x}{x-4} - \frac{5}{x+3}$.

39. For $f(x) = 2x^2 + 7x - 15$, find $f(-2)$.

40. Find the quotient $\frac{2x}{x-4} \div \frac{5}{x+3}$.

12.8 Variation

Objectives

- » Know the meaning of *direct variation* and *inverse variation.*
- » In direct variation and inverse variation, know how a change in the independent variable affects the value of the dependent variable.
- » Use a single point to find a *direct variation equation* or an *inverse variation equation.*
- » Use direct variation models and inverse variation models to make estimates.
- » Use ratios to find a direct variation constant, and use products to find an inverse variation constant.

In this section, we will discuss two simple, yet important, types of equations in two variables: *direct variation equations* and *inverse variation equations.*

Direct Variation

Recall from Section 12.7 that two related quantities are proportional if the ratio of the two quantities is constant. For example, x and y are proportional if

$$\frac{y}{x} = k$$

where k is a constant. If we multiply both sides of this equation by x, we have

$$y = kx$$

Definition Direct variation

If $y = kx$ for some constant k, we say that **y varies directly as x** or that **y is proportional to x.** We call k the **variation constant** or the **constant of proportionality.** The equation $y = kx$ is called a **direct variation equation.**

For $y = 3x$, we say y varies directly as x with variation constant 3. For $w = 8t$, we say w varies directly as t with variation constant 8.

In Example 1, we will model a situation by using a direct variation equation.

Table 28 Drop and Bounce Heights of a Golf Ball

Drop Height (inches)	Bounce Height (inches)
6	4.8
12	10.0
18	15.0
24	20.3
30	26.4
36	31.0
42	37.5
48	44.5
54	47.3
60	52.0

Source: *J. Lehmann*

Example 1 Using a Direct Variation Equation to Model Data

A golf ball is dropped from various heights, and the bounce height is recorded each time (see Table 28). Let B be the bounce height (in inches) of the golf ball after it was dropped from an initial height of d inches. Find an equation of a function that models the situation well.

Solution

We begin by drawing a scattergram of the data (see Fig. 55). It appears the variables d and B are approximately linearly related. The linear regression equation is

$$B = 0.90d - 0.77$$

Although this model fits the data very well (see Fig. 56), it estimates that very small drop heights will have *negative* bounce heights (see Fig. 57). Model breakdown has occurred. Also, as the drop heights get closer to 0 inches, the bounce heights should get close to 0 inches. However, the model estimates that as the drop heights get close to 0, the bounce heights will get close to -0.77 inch.

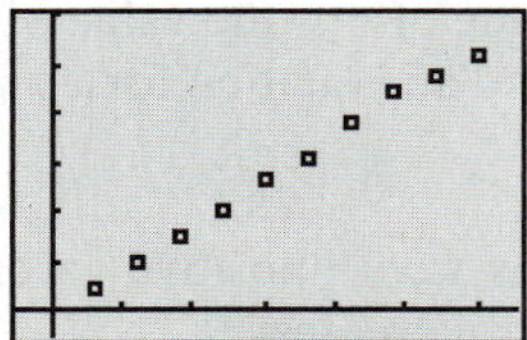

Figure 55 A scattergram of the data

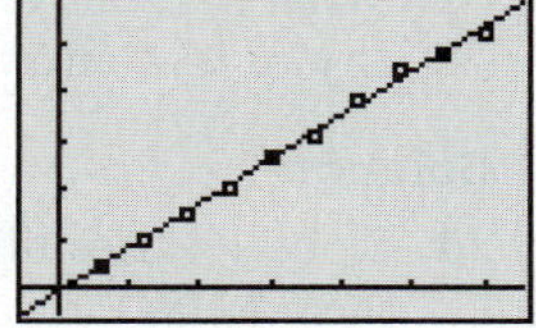

Figure 56 Regression line fits the data well

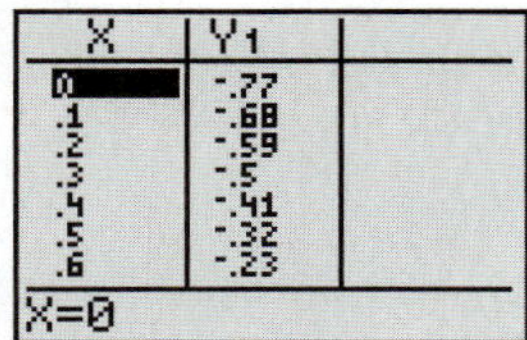

Figure 57 Regression line estimates negative bounce heights

It would be better to find a model of the form $B = md$, where $m > 0$ (see Fig. 58). Such a model estimates a *positive* bounce height for any drop height, including small ones. It also estimates that as drop heights get close to 0 inches, the bounce heights will get close to 0 inches.

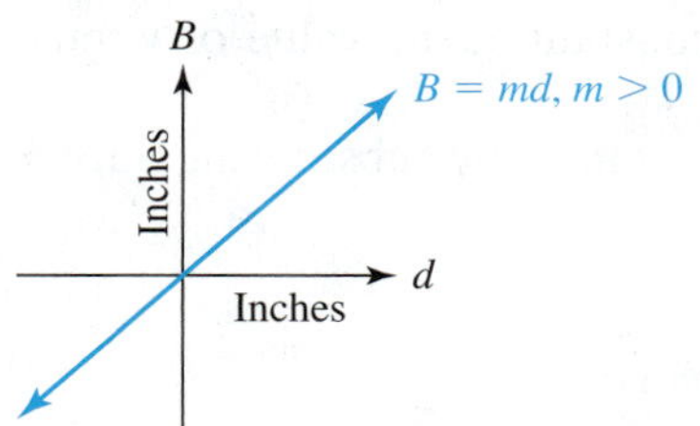

Figure 58 The graph of $B = md$ with $m > 0$

Figure 59 A line that contains the origin and the data point (30, 26.4)

If we imagine the line that contains the origin $(0, 0)$ and the data point $(30, 26.4)$, it appears the line might come close to the other data points (see Fig. 59). To find a slope m for the model $B = md$, we substitute the coordinates of the data point $(30, 26.4)$ into the equation $B = md$:

$$26.4 = m(30) \qquad \textit{Substitute 30 for d and 26.4 for B.}$$

$$\frac{26.4}{30} = m \qquad \textit{Divide both sides by 30.}$$

$$m = 0.88$$

The equation is $B = 0.88d$. The model fits the data quite well (see Fig. 59).

In Example 1, we found the model $B = 0.88d$. This equation is an example of a direct variation equation. We conclude that the bounce height varies directly as the drop height.

Changes in Values for Direct Variation

The graph of a direct variation equation $y = kx$ is a line with slope k and y-intercept $(0, 0)$. In Fig. 60, we sketch three such lines, where $k = \frac{1}{2}$, $k = 1$, and $k = 2$.

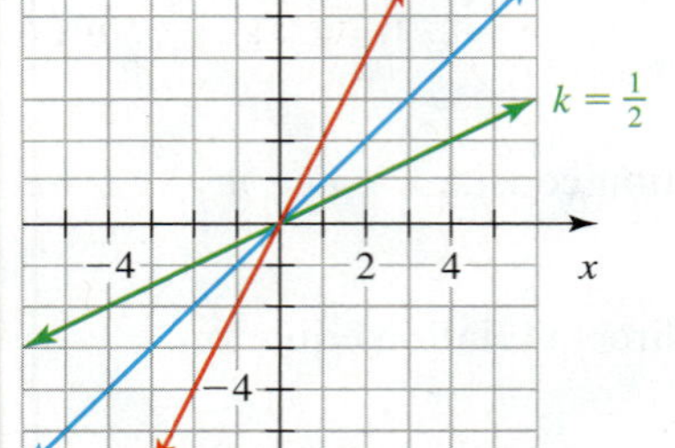

Figure 60 Graphs of $y = kx$, where $k = \frac{1}{2}$, $k = 1$, and $k = 2$

If k is positive (positive slope), then $y = kx$ is an increasing function. So, if the value of x increases, the value of y increases.

Changes in Values of Variables for Direct Variation

Assume y varies directly as x with some positive variation constant k:

- If the value of x increases, then the value of y increases.
- If the value of x decreases, then the value of y decreases.

For example, consider the golf ball model $B = 0.88d$ in Example 1. As drop heights increase, so do the bounce heights. As drop heights decrease, so do the bounce heights.

Using One Point to Find a Direct Variation Equation

If one variable varies directly as another variable and we know one point that lies on the graph, we can find the variation constant k as well as the direct variation equation.

Example 2 Finding a Direct Variation Equation

The variable y varies directly as x with positive variation constant k.

1. What happens to the value of y as the value of x increases?
2. If $y = 5$ when $x = 3$, find an equation for x and y.

Solution

1. Since y varies directly as x with positive variation constant k, the value of y must increase as the value of x increases.
2. The equation is of the form $y = kx$. We find the constant k by substituting 3 for x and 5 for y:

$$y = kx \qquad \textit{y varies directly as x.}$$

$$5 = k(3) \qquad \textit{Substitute 3 for x and 5 for y.}$$

$$\frac{5}{3} = k \qquad \textit{Divide both sides by 3.}$$

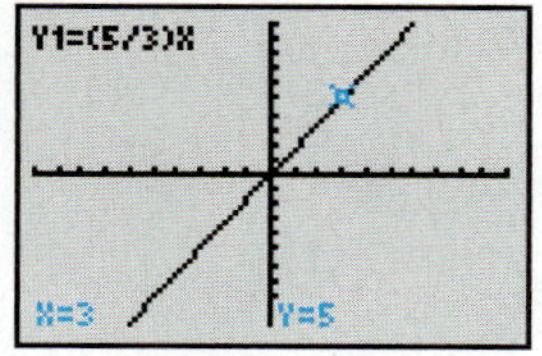

Figure 61 Verify the work

The equation is $y = \frac{5}{3}x$. We use a graphing calculator graph to check that the curve $y = \frac{5}{3}x$ contains the point $(3, 5)$. See Fig. 61.

Example 3 Using a Direct Variation Equation

The variable w varies directly as t. If $w = 7$ when $t = 2$, find the value of t when $w = 9$.

Solution

Since w varies directly as t, we can describe the relationship between t and w by the equation $w = kt$. We find the constant k by substituting 2 for t and 7 for w:

$$w = kt \quad \textit{w varies directly as t.}$$

$$7 = k(2) \quad \textit{Substitute 2 for t and 7 for w.}$$

$$\frac{7}{2} = k \quad \textit{Divide both sides by 2.}$$

The equation is $w = \frac{7}{2}t$. We find the value of t when $w = 9$ by substituting 9 for w in the equation $w = \frac{7}{2}t$ and solving for t:

$$9 = \frac{7}{2}t \quad \textit{Substitute 9 for w.}$$

$$18 = 7t \quad \textit{Multiply both sides by LCD, 2.}$$

$$\frac{18}{7} = t \quad \textit{Divide both sides by 7.}$$

So, $t = \frac{18}{7}$ when $w = 9$.

In Example 1, we used the function $B = 0.88d$ to model bounce heights of a golf ball. In Exercise 46, you will show that the function $B = 0.80d$ reasonably models bounce heights of a racquetball. Since B varies directly as d for a golf ball and a racquetball, it is a reasonable conjecture that B varies directly as d for another type of ball similar in construction, such as another golf ball, another racquetball, or perhaps even a tennis ball.

Example 4 Finding a Direct Variation Model

Assume the bounce height B (in inches) of a tennis ball varies directly as the drop height d (in inches). The bounce height of the tennis ball is 20 inches when the ball is dropped from an initial height of 30 inches.

1. Find an equation of B and d.
2. Estimate the bounce height if the drop height is 50 inches.

Solution

1. The equation is of the form $B = kd$. We substitute 30 for d and 20 for B to find the constant k:

$$20 = k(30) \quad \textit{Substitute 30 for d and 20 for B.}$$

$$\frac{20}{30} = k \quad \textit{Divide both sides by 30.}$$

$$k \approx 0.67$$

The equation is $B = 0.67d$.

2. We substitute 50 for d in the equation $B = 0.67d$:

$$B = 0.67(50) \quad \text{Substitute 50 for } d.$$
$$B = 33.5$$

The bounce height will be 33.5 inches, according to the model.

Finding and Using a Direct Variation Model

Assume a quantity p varies directly as a quantity t. To make estimates about an authentic situation,

1. Substitute the values of a data point into the equation $p = kt$; then solve for k.
2. Substitute the value of k into the equation $p = kt$.
3. Use the equation from step 2 to make estimates of quantity t or p.

In Example 4, we did not run an experiment and perform the usual modeling steps to find the tennis ball model $B = 0.67d$. Instead, we assumed the model was of the form $B = kd$ because that is the form of our models of the golf ball and the racquetball. The tennis ball model may or may not be accurate. The only way to know for sure is to run an experiment.

Next, we take a closer look at the significance of $k = 0.88$ for the golf ball model $B = 0.88d$. This equation $B = 0.88d$ tells us the bounce height is equal to 88% of the drop height. We list the values of k and their meanings for the three balls we have investigated:

Type of Ball	*Model*	*Value of k*	*Height of Bounce*
Golf ball	$B = 0.88d$	0.88	88% of drop height
Racquetball	$B = 0.80d$	0.80	80% of drop height
Tennis ball	$B = 0.67d$	0.67	67% of drop height

The value of k takes into account how "bouncy" the ball is.

A Variable Varying Directly as an Expression

So far we have described functions in which the dependent variable varies directly as the independent *variable*. Now, for nonzero constant k, we list some examples of functions in which the dependent variable varies directly as an *expression*:

$$y = kx^2 \quad y \text{ varies directly as } x^2.$$
$$p = k\sqrt{t} \quad p \text{ varies directly as } \sqrt{t}.$$
$$F = k\log(r) \quad F \text{ varies directly as } \log(r).$$

So, we use "varies directly" to mean the dependent variable is equal to a constant times an expression containing the independent variable.

Inverse Variation

In Sections 12.1, 12.5, and 12.6, we worked with rational equations in two variables. Now we will focus on a very simple type of rational equation in two variables: an inverse variation equation.

Definition Inverse variation

If $y = \dfrac{k}{x}$ for some nonzero constant k, we say **y varies inversely as x** or that **y is inversely proportional to x.** We call k the **variation constant** or the **constant of proportionality.** The equation $y = \dfrac{k}{x}$ is called an **inverse variation equation.**

For $y = \frac{5}{x}$, we say y varies inversely as x with variation constant 5. For $B = \frac{8}{v}$, we say B varies inversely as v with variation constant 8.

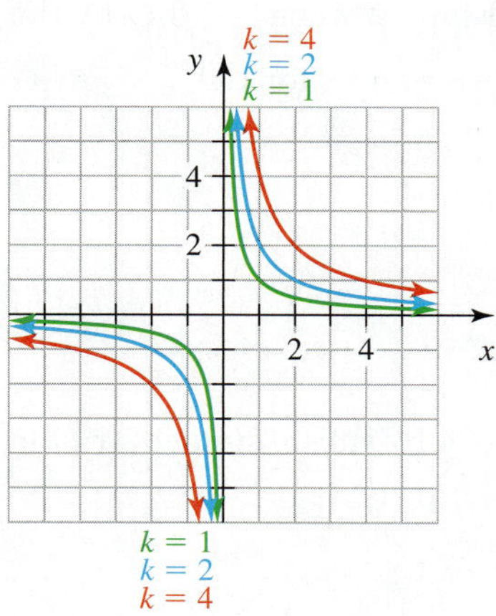

Figure 62 Graphs of $y = \frac{k}{x}$, where $k = 1$, $k = 2$, and $k = 4$

Changes in Values for Inverse Variation

In Fig. 62, we graph three equations of the form $y = \frac{k}{x}$, where $k = 1$, $k = 2$, and $k = 4$.

If k is positive, then the function $y = \frac{k}{x}$ is decreasing for positive values of x. So, if the value of x increases, the value of y decreases.

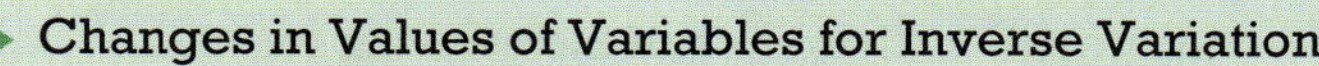

Changes in Values of Variables for Inverse Variation

Assume y varies inversely as x with some positive variation constant k. For positive values of x,

- If the value of x increases, then the value of y decreases.
- If the value of x decreases, then the value of y increases.

Use One Point to Find an Inverse Variation Equation

If one variable varies inversely as another variable and we know one point that lies on the graph, we can find the variation constant k as well as the inverse variation equation.

Example 5 Finding an Inverse Variation Equation

The variable y varies inversely as x with the positive variation constant k.

1. For positive values of x, what happens to the value of y as the value of x increases?
2. If $y = 4$ when $x = 2$, find an equation for x and y.

Solution

1. Since y varies inversely as x with positive variation constant k, and since the values of x are positive, the value of y must decrease as the value of x increases.
2. The equation is of the form $y = \frac{k}{x}$. We find the constant k by substituting 2 for x and 4 for y in the equation $y = \frac{k}{x}$:

$$y = \frac{k}{x} \quad \textit{y varies inversely as x.}$$

$$4 = \frac{k}{2} \quad \textit{Substitute 2 for x and 4 for y.}$$

$$8 = k \quad \textit{Multiply both sides by LCD, 2.}$$

The equation is $y = \frac{8}{x}$. We use a graphing calculator graph to check that the curve $y = \frac{8}{x}$ contains the point $(2, 4)$. See Fig. 63.

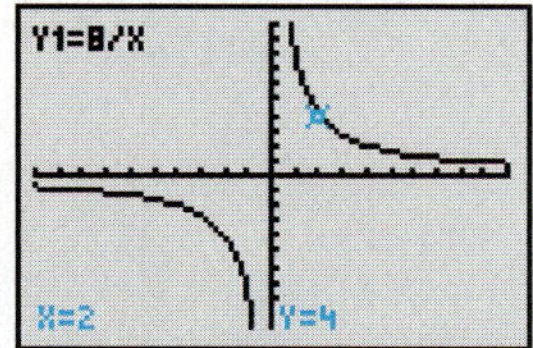

Figure 63 Verify the work

Example 6 Using an Inverse Variation Equation

The variable G varies inversely as r. If $G = 8$ when $r = 3$, find the value of r when $G = 4$.

Solution

Since G varies inversely as r, we can describe the relationship between r and G by the equation $G = \frac{k}{r}$. We find the constant k by substituting 3 for r and 8 for G:

$$G = \frac{k}{r} \quad \textit{G varies inversely as r.}$$

$$8 = \frac{k}{3} \quad \textit{Substitute 3 for r and 8 for G.}$$

$$24 = k \quad \textit{Multiply both sides by LCD, 3.}$$

The equation is $G = \frac{24}{r}$. We find the value of r when $G = 4$ by substituting 4 for G in the equation $G = \frac{24}{r}$ and then solving for r:

$$4 = \frac{24}{r} \quad \textit{Substitute 4 for G.}$$

$$4r = 24 \quad \textit{Multiply both sides by LCD, r.}$$

$$r = 6 \quad \textit{Divide both sides by 4.}$$

So, $r = 6$ when $G = 4$.

Table 29 Volumes and Pressures in a Syringe

Volume (cm^3)	Pressure (atm)
3	2.23
4	1.76
5	1.46
6	1.23
7	1.05
8	0.93
9	0.83
10	0.74
11	0.67
12	0.60
13	0.56
14	0.52
15	0.48
16	0.44
17	0.42
18	0.39
19	0.37
20	0.35

Source: *J. Lehmann*

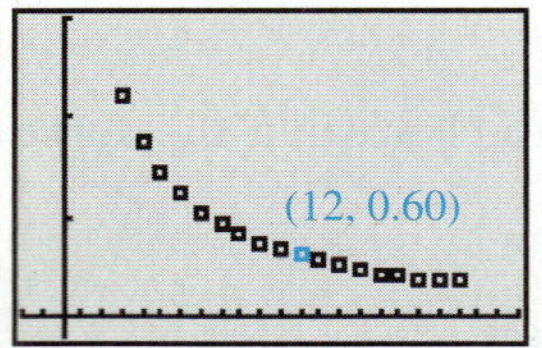

Figure 64 Scattergram of the data

Example 7 Finding an Inverse Variation Model

The more you squeeze a sealed syringe filled with air, the harder it gets to squeeze it farther. Some air volumes in cubic centimeters (cm^3) and corresponding pressures in atmospheres (atm) in a sealed syringe are given in Table 29.

Let P be the pressure (in atm) in the syringe at air volume V (in cm^3).

1. Find an equation for V and P.
2. As the value of V increases, what happens to the value of P, according to the model? What does that pattern mean in this situation?
3. Estimate at what volume the pressure will be 5 atm.

Solution

1. We draw a scattergram of the data in Fig. 64. Since the points suggest a curve that "bends," we will not use a linear function to model the data. To decide among an exponential function, a quadratic function, and an inverse variation function, we find equations of each.

   First, we find an inverse variation equation of the form $P = \frac{k}{V}$. If we imagine an inverse variation curve that contains the data point (12, 0.60), it appears the curve might come close to the other data points (see Fig. 64). To find k, we substitute the data point (12, 0.60) in the equation $P = \frac{k}{V}$:

$$0.60 = \frac{k}{12} \quad \textit{Substitute 12 for V and 0.60 for P.}$$

$$7.2 = k \quad \textit{Multiply both sides by LCD, 12.}$$

   The variation equation is $P = \frac{7.2}{V}$. Next, we use a graphing calculator to find the exponential regression equation and the quadratic regression equation. We list all three equations:

$$P = 2.32(0.90)^V \quad \textit{Exponential regression equation}$$

$$P = 0.0086V^2 - 0.29V + 2.77 \quad \textit{Quadratic regression equation}$$

$$P = \frac{7.2}{V} \quad \textit{Inverse variation equation}$$

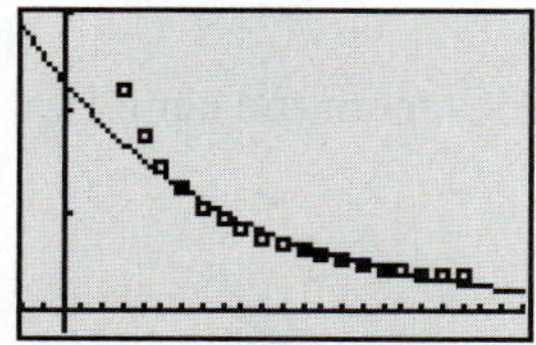
Figure 65 Exponential regression model

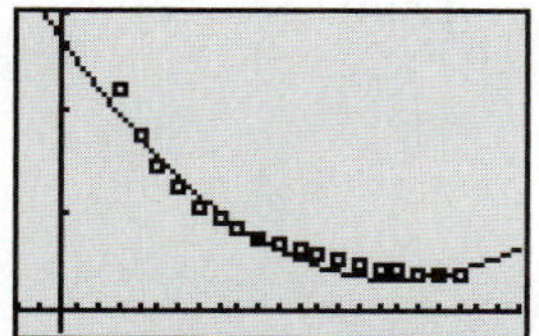
Figure 66 Quadratic regression model

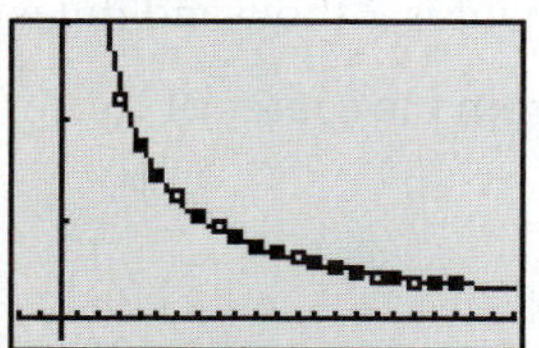
Figure 67 Inverse variation model

Now we see how well each model fits the data (see Figs. 65, 66, and 67). It appears the inverse variation model $P = \frac{7.2}{V}$ fits the data better than the other two models do.

2. The variable P varies inversely as V, according to our model. So, as the value of V increases, the value of P decreases. The larger the volume of air, the less the pressure will be.
3. We substitute 5 for P in the equation $P = \frac{7.2}{V}$ and solve for V:

$$5 = \frac{7.2}{V} \quad \textit{Substitute 5 for P.}$$

$$5V = 7.2 \quad \textit{Multiply both sides by LCD, V.}$$

$$V = 1.44 \quad \textit{Divide both sides by 5.}$$

The pressure is 5 atm when the air volume is 1.44 cubic centimeters.

The equation $P = \frac{k}{V}$ is an excellent model for many situations, as long as the temperature and number of molecules in the container are constant. This equation is a form of **Boyle's law.** The constant k takes into account the temperature and the number of molecules.

Finding and Using an Inverse Variation Model

Assume a quantity p varies inversely as a quantity t. To make estimates about an authentic situation,

1. Substitute the values of a data point into the equation $p = \frac{k}{t}$; then solve for k.
2. Substitute the value of k into the equation $p = \frac{k}{t}$.
3. Use the equation from step 2 to make estimates of quantity t or p.

A Variable Varying Inversely as an Expression

For nonzero constant k, we list some examples of functions in which the dependent variable varies inversely as an *expression*:

$$y = \frac{k}{x^3} \quad \textit{y varies inversely as } x^3.$$

$$A = \frac{k}{r - 5} \quad \textit{A varies inversely as } r - 5.$$

$$w = \frac{k}{2^t} \quad \textit{w varies inversely as } 2^t.$$

So, we use "varies inversely" to mean that the dependent variable is equal to a constant divided by an expression containing the independent variable.

Example 8 Finding an Inverse Variation Model

The weight of an object varies inversely as the square of its distance from the center of Earth. An astronaut weighs 180 pounds at sea level (about 4 thousand miles from Earth's center). How much does the astronaut weigh 2 thousand miles above Earth's surface?

Solution

We let w be the weight (in pounds) of the astronaut d thousand miles from the center of Earth. Our desired model has the form

$$w = \frac{k}{d^2}$$

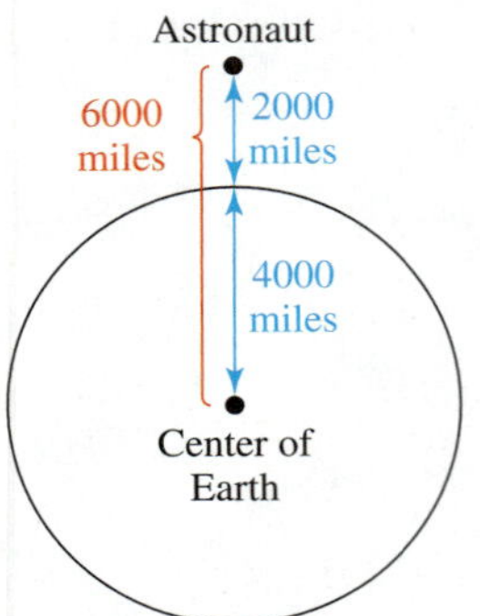

Figure 68 The astronaut is 6000 miles from Earth's center

The denominator of the fraction on the right-hand side is d^2, because weight varies inversely as the *square* of distance d.

Next, we substitute 4 for d and 180 for w in the equation $w = \frac{k}{d^2}$ and solve for k:

$$180 = \frac{k}{4^2} \quad \text{Substitute 4 for } d \text{ and 180 for } w.$$

$$180 = \frac{k}{16} \quad 4^2 = 16$$

$$2880 = k \quad \text{Multiply both sides by LCD, 16.}$$

The model is $w = \frac{2880}{d^2}$. We see from Fig. 68 that when the astronaut is 2 thousand miles above Earth's surface, the astronaut is $4 + 2 = 6$ thousand miles from the center of Earth. To find w, we substitute 6 for d in the equation $w = \frac{2880}{d^2}$:

$$w = \frac{2880}{6^2} = 80$$

The astronaut weighs 80 pounds 2 thousand miles above Earth's surface.

Using Ratios and Products to Find Variation Constants

First, we will discuss another way to find a direct variation constant. Then we will discuss another way to find an inverse variation constant.

Using Ratios to Find a Direct Variation Constant

In Example 9, we will use ratios to find a direct variation constant.

Example 9 Finding a Direct Variation Constant by Using Ratios

Use ratios of bounce heights and drop heights to find a model of the golf ball data shown in Example 1.

Solution

From our work in Example 1, we know we want an equation of the form $B = kd$. To find k, we begin by solving the equation $B = kd$ for k:

$$k = \frac{B}{d}$$

Next, we use the first two columns of Table 30 (on the next page) to calculate the approximate ratios $\frac{B}{d}$ for the third column.

According to our model $B = kd$, the ratios $\frac{B}{d}$ should be equal to a constant k. The reason the ratios vary a bit is most likely due to errors in measuring the drop and bounce heights. Or there might be imperfections in the golf ball or the surface of the floor.

Next, we find the mean of the ratios $\frac{B}{d}$ in the third column:

$$\frac{0.80 + 0.83 + 0.83 + 0.85 + 0.88 + 0.86 + 0.89 + 0.93 + 0.88 + 0.87}{10} \approx 0.86$$

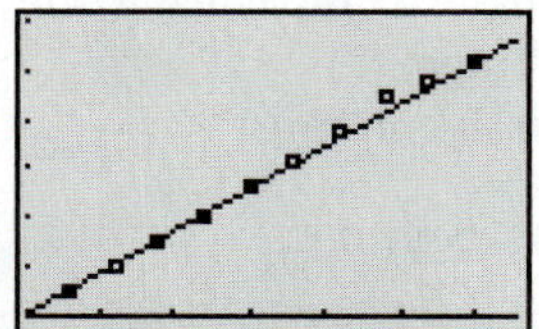

Figure 69 The model $B = 0.86d$

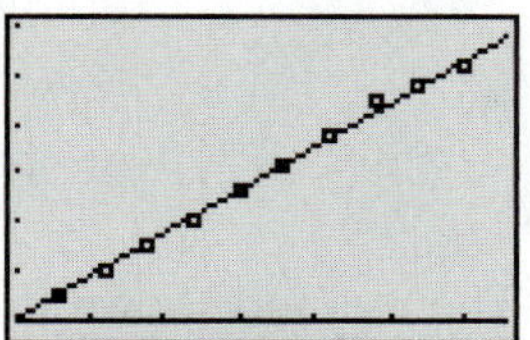

Figure 70 The model $B = 0.88d$

Table 30 Ratios of Drop and Bounce Heights of a Golf Ball

Drop Height d (inches)	Bounce Height B (inches)	$\frac{B}{d}$
6	4.8	0.80
12	10.0	0.83
18	15.0	0.83
24	20.3	0.85
30	26.4	0.88
36	31.0	0.86
42	37.5	0.89
48	44.5	0.93
54	47.3	0.88
60	52.0	0.87

Finally, we substitute 0.86 for k in the equation $B = kd$:

$$B = 0.86d$$

The model fits the data reasonably well (see Fig. 69). However, it appears the model we found in Example 1 fits the data slightly better (see Fig. 70).

We now have two ways to find a direct variation model. The method shown in Example 1 requires few calculations, but finding a good value of k depends on a good selection of an ordered pair. The method shown in Example 9 requires more calculations, but it is not necessary to select an ordered pair to find k.

Using Products to Find an Inverse Variation Constant

In Example 10, we will use products to find an inverse variation constant.

Example 10 Finding an Inverse Variation Constant by Using Products

Use products of volumes and pressures to find a model of the syringe data shown in Example 7.

Solution

From our work in Example 7, we know we want an equation of the form $P = \frac{k}{V}$. To find k, we begin by solving the equation $P = \frac{k}{V}$ for k:

$$k = VP$$

Next, we use the values of V and P in the first two columns and fourth and fifth columns of Table 31 to calculate the products VP for the third and sixth columns, respectively.

Table 31 Products of Volumes and Pressures in a Syringe

Volume V (cm^3)	Pressure P (atm)	VP	Volume V (cm^3)	Pressure P (atm)	VP
3	2.23	6.69	12	0.60	7.20
4	1.76	7.04	13	0.56	7.28
5	1.46	7.30	14	0.52	7.28
6	1.23	7.38	15	0.48	7.20
7	1.05	7.35	16	0.44	7.04
8	0.93	7.44	17	0.42	7.14
9	0.83	7.47	18	0.39	7.02
10	0.74	7.40	19	0.37	7.03
11	0.67	7.37	20	0.35	7.00

According to our model $P = \frac{k}{V}$, the products VP should be equal to a constant k. The reason the products vary a bit is most likely that errors arise in measuring the volumes and pressures. Or there might be imperfections in the syringe.

Next, we find the mean of the products PV in the third and sixth columns:

$$\frac{6.69 + 7.04 + 7.30 + 7.38 + 7.35 + \cdots + 7.00}{18} \approx 7.20$$

Finally, we substitute 7.20 for k in the equation $P = \frac{k}{V}$:

$$P = \frac{7.20}{V}$$

This is the same equation we found in Example 7.

We now have two ways to find an inverse variation model. The method shown in Example 7 requires few calculations, but finding a good value of k depends on a good selection of an ordered pair. The method shown in Example 10 requires more calculations, but it is not necessary to select an ordered pair to find k.

Group Exploration

Comparing methods of finding a direct variation equation

In this exploration, you will use three different methods to find an equation for a direct variation model.

1. Suppose that an employee's total earnings E (in dollars) varies directly as the number t of hours worked. The employee earns \$72 for an 8-hour workday.

a. Use the method discussed in this section to find an equation of a direct variation model to describe the data.

b. What does the ordered pair $(0, 0)$ mean in this situation? Does that make sense? Use the slope formula $m = \frac{E_2 - E_1}{t_2 - t_1}$ with the ordered pairs $(0, 0)$ and $(8, 72)$ to help you find an equation of a model to describe the data.

c. Use the information that the employee earns \$72 for an 8-hour workday to compute the rate of change of earnings per hour. Then use the fact that the slope is a rate of change to help you find an equation of a model to describe the data.

2. Compare the three equations you found in Problem 1. Which method was easiest for you to use?

3. If a variable y varies directly as a variable x, does it follow that the rate of change of y with respect to x is constant? Explain.

Group Exploration

Looking ahead: Simplifying radical expressions

Recall from Section 10.2 that $\sqrt[n]{a} = a^{1/n}$. So,

$$\sqrt[5]{a^3} = \left(a^3\right)^{1/5} = a^{3 \cdot \frac{1}{5}} = a^{3/5}$$

We say $\sqrt[5]{a^3}$ is in *radical form* and the expression $a^{3/5}$ is in *exponential form.*

1. Write $\sqrt[7]{a^4}$ in exponential form.

2. Write $\sqrt[9]{a^2}$ in exponential form.

3. Write $\sqrt{a^7}$ in exponential form. [**Hint:** $\sqrt{a}$ is shorthand for $\sqrt[2]{a}$.]

4. Write $\sqrt[n]{a^m}$ in exponential form.

5. Write each of the following in exponential form and simplify.

a. $\sqrt{a^2}$, $\sqrt{a^4}$, $\sqrt{a^6}$, $\sqrt{a^8}$

b. $\sqrt[3]{a^3}$, $\sqrt[3]{a^6}$, $\sqrt[3]{a^9}$, $\sqrt[3]{a^{12}}$

6. Simplify $\sqrt{x^7}$. [**Hint:** $x^7 = x^6 \cdot x$]

7. Simplify $\sqrt{x^{13}}$.

Tips for Success Plan for the Final Exam

Don't wait until the last minute to begin studying for your final exam. Look at your finals schedule and decide how you will allocate your time to prepare for each final.

It is important that you are well rested so you can fully concentrate during your final exam. Plan to do some fun activities that involve exercise—a great way to neutralize stress.

Homework 12.8

For extra help ▶ MyMathLab®  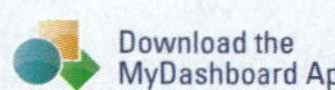

Translate the sentence into an equation.

1. I varies directly as t.

2. z varies inversely as d.

3. w varies inversely as $x + 4$.

4. V varies directly as r cubed.

Translate the equation into a sentence by using the phrase "varies directly" or "varies inversely."

5. $w = \frac{k}{r}$

6. $d = kt$

7. $T = k\sqrt{w}$

8. $y = \frac{k}{2^x}$

Find an equation that meets the given conditions.

9. c varies directly as u, and $c = 12$ when $u = 3$.

10. p varies inversely as d, and $p = 3$ when $d = 5$.

11. w varies inversely as $\sqrt{t}$, and $w = 3$ when $t = 16$.

12. A varies directly as r^2, and $A = 4\pi$ when $r = 2$.

For Exercises 13–20, find the requested value of the variable.

13. If y varies directly as x, and $y = 12$ when $x = 4$, find y when $x = 9$.

14. If p varies directly as t, and $p = 18$ when $t = 3$, find p when $t = 5$.

15. If G varies inversely as r, and $G = 8$ when $r = 3$, find G when $r = 4$.

16. If W varies inversely as u, and $W = 15$ when $u = 2$, find W when $u = 6$.

17. If p varies directly as x^2, and $p = 6$ when $x = 2$, find x when $p = 24$.

18. If y varies directly as w^2, and $y = 6$ when $w = 3$, find w when $y = 54$.

19. If I varies inversely as $r + 2$, and $I = 9$ when $r = 3$, find r when $I = 7$.

20. If D varies inversely as $t + 3$, and $D = 5$ when $t = 4$, find t when $D = 9$.

21. The variable B varies directly as w with positive variation constant k. Describe what happens to the value of B as the value of w increases.

22. The variable y varies directly as t with positive variation constant k. Describe what happens to the value of y as the value of t decreases.

23. The variable w varies inversely as p with positive variation constant k. For positive values of p, what happens to the value of w as the value of p increases?

24. The variable F varies inversely as r with positive variation constant k. For positive values of r, what happens to the value of F as the value of r decreases?

25. The transaction demand is the amount of money demanded in an immediate exchange for goods and services. The transaction demand varies directly as the gross domestic product (GDP). If the GDP increases, what happens to the transaction demand?

26. A training apparatus for improving the performance of swimmers produces a pulsating signal whose frequency varies directly as the swimmer's speed. If a swimmer's speed decreases, what happens to the frequency of the signal?

27. Nerve conduction in muscles varies inversely as a person's height. Will a tall person have more or less nerve conduction than a short person?

28. The frequency of a vibrating guitar string varies inversely with the diameter of the string. Will a large-diameter string have a higher or lower frequency than a small-diameter string?

29. Bernice Fitz-Gibbon, U.S. advertising executive, said, "Creativity varies inversely with the number of cooks involved in the broth." What did she mean?

30. Craig Bruce said, "Time is a resource whose supply is inversely proportional to its demand." What did he mean?

31. For a Montgomery County (Maryland) resident, the cost of tuition at Montgomery College varies directly as the number of credit hours taken. For spring 2013, the cost of 15 credit hours was \$1680. What did 12 credit hours cost?

32. For a Jackson County (Michigan) resident, the cost of tuition at Jackson Community College varies directly as the number of billing contact hours taken. For academic year 2012–2013, the cost of 13 billing contact hours was \$1378. What did 15 billing contact hours cost?

33. When a stone is tied to a string and whirled in a circle at constant speed, the tension in the string varies inversely as the radius of the circle. If the radius is 60 centimeters, the tension is 80 newtons. Find the tension if the radius is 50 centimeters.

34. The current flowing in an electrical circuit at a constant potential varies inversely as the resistance of the circuit. If the current is 25 amperes when the resistance is 4 ohms, what is the current when the resistance is 5 ohms?

35. The distance that an object falls varies directly as the square of the time the object is in motion. If an object falls for 3 seconds, it will fall 144.9 feet. To estimate the height of a cliff, a person drops a stone at the edge of the cliff and measures how long it takes the stone to reach the base. If it takes 3.4 seconds, what is the height of the cliff?

36. A car is traveling at speed s (in mph) on a dry asphalt road, and the brakes are suddenly applied. The braking distance d (in feet) varies directly as the square of the speed s. If a car traveling at 60 mph has a braking distance of 226.8 feet, what is the braking distance of a car traveling at 70 mph?

37. The intensity of the radiation used to treat a tumor varies inversely as the square of the distance from the machine that emits the radiation. If the intensity is 90 milliroentgens per hour (mr/hr) at a distance of 2.5 meters, at what distance is the intensity 45 mr/hr?

38. The volume of a sphere varies directly as the cube of its radius. A sphere with a 3-foot radius has a volume of 36π cubic feet. How much air is required to fill a beach ball with radius 1.6 feet?

39. The force F (in pounds) required to push a sofa across a floor varies directly as the weight w (in pounds) of the sofa.

a. A person can push a 120-pound sofa across a wood floor by pushing with a force of 50 pounds. Find an equation that describes the relationship between w and F.

b. How much force is required to push a 150-pound sofa across a wood floor?

c. How would an equation of a model for a *carpeted* floor compare with the model you found in part (a)? In particular, how would the variation constants compare?

40. The distance d (in miles) that a student travels by car varies directly as the travel time t (in hours). The student travels 93 miles in 1.5 hours.

a. Is the student traveling at a constant speed? Explain.

b. Find an equation that describes the relationship between time and distance.

c. How far will the student travel in 2 hours?

41. The time T (in seconds) it takes to hear thunder after you see lightning varies directly as the distance d (in feet) from the lightning if the temperature does not vary much during the storm. In a certain storm, it takes 3 seconds to hear thunder when lightning is seen 3313 feet away.

a. Find an equation that describes the relationship between d and T for the storm.

b. If it takes 4 seconds for you to hear thunder after you see lightning, how far away is the lightning?

c. What does the constant k represent in this situation? Explain. [**Hint:** Recall that slope is the rate of change of the *dependent* variable with respect to the *independent* variable.]

d. There is a rule of thumb that the number of seconds it takes you to hear thunder after you see lightning is equal to the number of miles that you are from the lightning. Is this rule of thumb a good approximation? If yes, explain. If no, find a better rule of thumb. [**Hint:** There are 5280 feet in 1 mile.]

42. The force F (in pounds) you must exert on a wrench handle to loosen a bolt on a bike varies inversely as the length L (in inches) of the handle. A force of 40 pounds is needed when the handle is 6 inches long.

a. Find an equation that describes the relationship between handle length and force.

b. If you use an 8-inch-long wrench to loosen the bolt, how much force do you need to apply to the handle?

c. Is it easier to use a wrench with a short or long handle? Explain in terms of the equation you found in part (a).

43. The weight (in pounds) $w = f(d)$ of an object varies inversely as the square of its distance (in thousands of miles) d from the center of Earth.

a. An astronaut weighs 200 pounds at sea level (about 4 thousand miles from Earth's center). Find an equation of f.

b. How much would the astronaut weigh 1 thousand miles above Earth's surface?

c. At what distance from the center of Earth would the astronaut weigh 1 pound?

d. Use f to estimate how much the astronaut would weigh on the surface of the Moon. The Moon is a mean distance of about 239 thousand miles from Earth. Has model breakdown occurred? Explain.

e. Without finding an equation, discuss how an equation of a model for a 190-pound astronaut would compare with the equation you found in part (a). Discuss how the variation constants would compare.

44. The intensity (in watts per square meter, W/m^2) $I = f(d)$ of a television signal varies inversely as the square of the distance d (in kilometers) from the transmitter. The intensity of a television signal is 30 W/m^2 at a distance of 2.6 km.

a. Find an equation of f.

b. Find $f(1)$, $f(2)$, $f(3)$, and $f(4)$. What do they mean in this situation?

c. Use a graphing calculator to draw a graph of f. What happens to the value of I as the value of d increases, for $d > 0$? What does that mean in this situation?

d. What can you say about the value of $f(d)$ for an extremely large value of d? Explain. What does that mean in this situation?

45. When a guitarist picks the second-thickest string on a six-string guitar (the "open 'A' string"), the string vibrates at a frequency of 110 hertz—that is, 110 times a second. By firmly pressing the "A" string against the fret board, the guitarist shortens the effective length of the string. As a result, the frequency increases (and so does the pitch of the note). See Fig. 71.

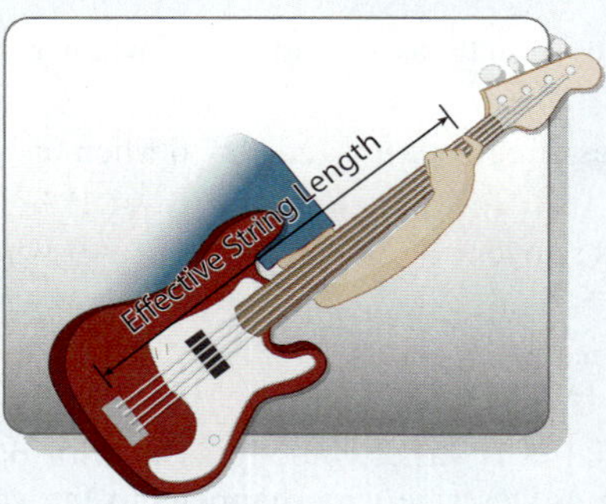

Figure 71 Exercise 45

We use some of the letters of the alphabet, sometimes in conjunction with the "sharp" symbol, ♯, to refer to these notes. The frequencies for the open "A" string and the next 12 notes are listed in Table 32.

Table 32 Effective Lengths and Frequencies of 13 Notes on the "A" String

Note	Effective Length of "A" String (inches)	Frequency (hertz)
A	25.50	110.0
A♯	24.07	116.6
B	22.72	123.5
C	21.44	130.8
C♯	20.24	138.6
D	19.10	146.9
D♯	18.03	155.6
E	17.02	164.8
F	16.06	174.6
F♯	15.16	185.0
G	14.31	196.0
G♯	13.51	207.7
A	12.75	220.0

Sources: Math and Music, *Garland and Kahn; J. Lehmann*

a. Let F be the frequency (in hertz) of the "A" string when the string's effective length is L inches. Use a graphing calculator to draw a scattergram of the data.

b. Find an equation of a reasonable model to describe the data. Does your model fit the data well?

c. In a sentence that uses the phrase "varies directly" or "varies inversely," describe how the effective length and frequency of the "A" string are related.

d. When the "A" string is vibrating, what is its frequency if the effective length is 7.58 inches?

e. Use your equation from part (b) to show that if you halve any effective length of the "A" string, the frequency will double. [**Hint:** First, substitute a for L in your model's equation. Then substitute $\frac{1}{2}a$ for L.]

46. A racquetball is dropped from various heights, and the bounce height is recorded each time (see Table 33).

Table 33 Drop and Bounce Heights of a Racquetball

Drop Height (inches)	Bounce Height (inches)
6	5.0
12	9.3
18	15.0
24	19.6
30	24.0
36	27.6
42	32.8
48	38.0

Source: *J. Lehmann*

Let $B = f(d)$ be the bounce height (in inches) of the racquetball after it is dropped from an initial height of d inches.

a. Use a graphing calculator to draw a scattergram of the data.

b. Find an equation of f.

c. In a sentence using the phrase "varies directly" or "varies inversely," describe how the initial and bounce heights of the racquetball are related.

d. What is the slope of the graph of f? What does it mean in this situation?

e. What is the B-intercept? What does it mean in this situation?

47. As you move away from an object, it appears to decrease in height. To describe this relationship, a math professor stood 10 feet from his garage, held a yardstick 1 foot away, and measured the image of his garage. The image of the garage had an *apparent height* of 16.0 inches. He collected apparent heights of the garage at various distances from it (see Table 34).

Table 34 Apparent Heights of a Car Garage

Distance from Garage (feet)	Apparent Height of Garage (inches)
10	16.0
20	7.3
30	4.8
40	3.8
50	3.0
60	2.5
70	2.0

Source: *J. Lehmann*

Let $a = f(d)$ be the apparent height (in inches) of the garage when the professor was d feet from the garage.

a. Use a graphing calculator to draw a scattergram of the data.

b. Find an equation of f.

c. In a sentence using the phrase "varies directly" or "varies inversely," describe how the distance from the garage and the apparent height of the garage are related.

d. Your model f is a decreasing function for $d > 0$. Explain why that makes sense in this situation.

e. Find the apparent height from a distance of 100 feet.

f. Estimate the actual height of the garage. [**Hint:** Think about how the apparent heights were recorded.]

48. A pizza with diameter 12 inches (a "12-inch pizza") and three toppings weighs 36 ounces at Toto's Restaurant and Pizzeria in San Bruno, California. The weight of a three-topping pizza varies directly as the square of the pizza's diameter.

a. Let $W(d)$ be the weight (in ounces) of a three-topping pizza with diameter d inches. Find an equation of W.

b. Table 35 lists the prices for various sizes of three-topping pizzas at Toto's Restaurant and Pizzeria. Let $P(d)$ be the price (in dollars) of a three-topping pizza with diameter d inches. Find an equation of P.

Table 35 Prices of Pizzas

Diameter (inches)	Price (dollars)
12	18.75
14	20.75
16	24.25
18	26.25

Source: *Toto's Restaurant and Pizzeria*

c. In a sentence using the phrase "varies directly," "varies linearly," or "varies inversely," describe how the diameter and price of the pizza are related.

d. Let $C(d)$ be the cost per ounce (in dollars per ounce) of a three-topping pizza with diameter d inches. Find an equation of C.

e. Find the value of d where $C(d) = 0.372$. What does it mean in this situation?

f. Use a graphing calculator to draw a graph of C. Is C increasing or decreasing for $d > 0$? What does that mean in this situation?

49. A *pendulum* is an object hanging from a lightweight cord. A pendulum can be made by tying one end of some thread to a weight—say, a washer—and attaching the other end of the thread to a surface so that the weight is suspended and can swing freely. The *period* of the pendulum is the amount of time it takes for the weight to swing forward and backward once. Let T be the period (in seconds) of a pendulum and L be the length (in centimeters) of the thread. Some values of L and T generated from an experiment are shown in Table 36.

Table 36 Periods of a Pendulum

Length L (cm)	Period T (seconds)	k
5.0	0.50	
10.0	0.63	
15.0	0.88	
20.0	1.00	
25.0	1.13	
32.5	1.25	
45.0	1.50	
60.0	1.75	
85.0	2.00	
110.0	2.25	

Source: *J. Lehmann*

a. The period of a pendulum varies directly as the square root of the length of the thread. Write an equation involving L, T, and the variation constant k.

b. Solve the equation you found in part (a) for k.

c. Use the first two columns of Table 36 to complete the third column. What is a reasonable value of k? How did you find this value?

d. Substitute your value of k into your equation from part (a).

e. Draw the graph of your model and a scattergram of the data in the same viewing window. Does your model fit the data well?

f. Estimate the period if the thread is 130 inches long.

50. The illumination from a light bulb decreases as the distance from the bulb increases. Let I be the illumination in milliwatts per square centimeter (mW/cm^2) at d centimeters from a 25-watt light bulb. Some values of d and I generated by an experiment are shown in Table 37.

a. The illumination from a light source varies inversely as the square of the distance from the source. Write an equation involving d, I, and the variation constant k.

b. Solve the equation you found in part (a) for k.

c. Use the first two columns of Table 37 to complete the third column. What is a reasonable value of k? How did you find this value?

Table 37 Illumination from a Light Bulb

Distance d (cm)	Illumination I (mW/cm^2)	k
70	0.845	
80	0.677	
90	0.546	
100	0.435	
110	0.349	
120	0.293	
130	0.260	
140	0.214	

Source: *J. Lehmann*

d. Substitute your value of k into your equation from part (a) to find an equation of a model.

e. Draw the graph of your model and a scattergram of the data in the same viewing window. Does your model fit the data well?

f. Estimate the illumination at 160 centimeters from the bulb.

g. How would an equation of a model for a 50-watt bulb compare with the model you found in part (d)? How would the variation constants compare?

Concepts

For Exercises 51–56, write a variation equation in the given variables. State the value of the variation constant k.

51. Let $f(L)$ be the area (in square inches) of a rectangle with width 5 inches and length L (in inches).

52. Let $f(s)$ be the area (in square meters) of a square with sides of length s (in meters).

53. A group of n people wins a total of \$2 million from a state lottery drawing. Let $f(n)$ be each person's share (in millions of dollars).

54. Let $f(s)$ be the time (in hours) it takes a person to drive 100 miles at a constant speed s (in mph).

55. Let $f(r)$ be the circumference (in inches) of a circle with radius r (in inches).

56. Let $f(r)$ be the area (in square feet) of a circle with radius r (in feet).

57. Suppose the latest Foster the People album is about to be released. Let $n = f(a)$ be the number of albums (in millions) that will be sold, where a is the advertising budget (in thousands of dollars). A graph of f is sketched in Fig. 72.

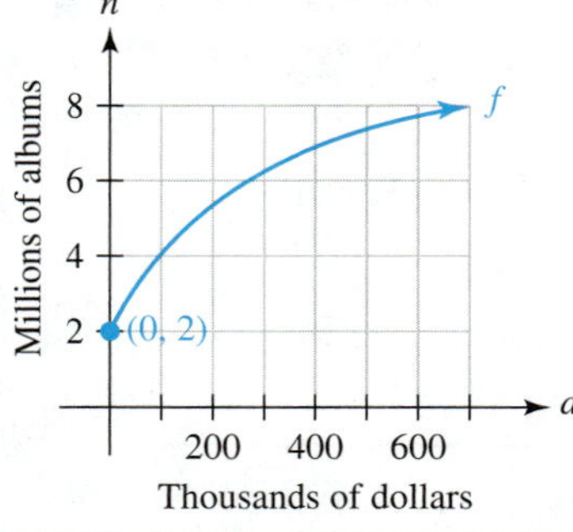

Figure 72 Advertising budget and album sales

a. Is f an increasing function? Explain.

b. Does the number of albums sold vary directly as the amount of money spent on advertising? Explain.

c. Compare the meanings of these two statements: "One quantity varies directly as another quantity." "One quantity will increase if the other quantity increases."

58. For a direct variation equation $y = kx$, where k is positive, we know that as the value of x increases, the value of y increases. If k is negative for $y = kx$, what happens to the value of y as the value of x increases? Explain. [**Hint:** Sketch a graph.]

For Exercises 59–62, decide whether the statement is true or false. Explain.

59. The height of a person varies directly as the person's age.

60. The area of a circle varies directly as the radius of the circle.

61. The temperature of hot coffee varies inversely as the time since it was poured into a cup.

62. The height of a baseball hit straight up varies inversely as the time since it was hit.

63. Assume y varies directly as x and that the variation constant is k. Does it follow that x varies directly as y? If yes, what is the variation constant? If no, explain. [**Hint:** Solve the equation $y = kx$ for x.]

64. Assume y varies inversely as x and that the variation constant is k. Does it follow that x varies inversely as y? If yes, what is the variation constant? If no, explain. [**Hint:** Solve the equation $y = \frac{k}{x}$ for x.]

65. Five miles is approximately the same distance as 8.05 kilometers. One student finds the model $y = 1.61x$, and another student finds the model $y = 0.62x$. Even though their results are different, both students are correct. How is this possible? What does 1.61 of the model $y = 1.61x$ represent? What does 0.62 of the equation $y = 0.62x$ represent?

66. Describe the meanings of *direct variation equation* and *inverse variation equation*. Compare these two types of equations. Compare some properties of these types of equations.

Related Review

67. a. If y varies directly as x, are x and y linearly related? Explain.

b. If w and t are linearly related, does w vary directly as t? Explain.

68. a. If a quantity y varies directly as a quantity x, how many data points do you need to find an equation to describe the situation?

b. If the quantities w and t are linearly related, how many data points do you need to find an equation to describe the situation?

69. The number n of words a typist can type varies directly as the amount of time t (in minutes) he types. He can type 310 words in 5 minutes. What is the slope of the model that describes this situation? What does it mean in this situation?

70. The total cost c (in dollars) of some 5-inch × 7-inch school photos shot by Lifetouch® varies directly as the number n of photos that are purchased. It costs \$27.96 for 4 such photos. What is the slope of the model that describes this situation? What does it mean in this situation?

Expressions, Equations, Functions, and Graphs

Give an example of the following. Then solve, simplify, or graph, as appropriate.

71. system of two linear equations in two variables

72. linear equation in one variable

73. difference of two rational expressions

74. quadratic equation in one variable

75. quadratic function

76. rational equation in one variable

77. exponential function

78. the sum of two logarithmic expressions with the same base

Taking it to the Lab

Climate Change Lab (continued from Chapter 9)

In previous Climate Change Labs, we found various models related to carbon dioxide emissions and human populations. We can use these models to find meaningful ratios and percentages of quantities.

Analyzing the Situation

1. Let $u(t)$ be the U.S. population and $w(t)$ be the world population (both in billions), both at t years since 1950. Here are some reasonable models:

$$u(t) = 0.0025t + 0.155$$
$$w(t) = -0.00046t^2 + 0.132t + 0.63$$

Recall from the Climate Change Lab in Chapter 7 that the world population model works well only for the years 2000–2100. Find a model of the percentage $p(t)$ of the world population that is in the United States at t years since 1950.

2. Use a graphing calculator table to find $p(50)$, $p(60)$, $p(70)$, $p(80)$, $p(90)$, and $p(100)$. What do your results tell you about this situation?

3. Let $f(t)$ be U.S. carbon dioxide emissions and $g(t)$ be world carbon dioxide emissions (both in billions of metric tons), both in the year that is t years since 1950. Here are some reasonable models:

$$f(t) = 0.06t + 2.61$$
$$g(t) = 0.4t + 5.75$$

Find a model of the percentage $h(t)$ of worldwide carbon dioxide emissions emitted in the United States in the year that is t years since 1950.

4. Use a graphing calculator table to find $h(50)$, $h(60)$, $h(70)$, $h(80)$, $h(90)$, and $h(100)$. What do your results tell you about this situation?
5. A model of U.S. carbon dioxide emissions $c(t)$ (in billions of metric tons) in the year that is t years since 1950 is $f(t) = 0.06t + 2.61$. A model of the U.S. population $u(t)$ (in billions) at t years since 1950 is $u(t) = 0.0025t + 0.155$. Find a model of the U.S. per-person carbon dioxide emissions $r(t)$ (in metric tons per person) in the year that is t years since 1950.
6. Use a graphing calculator table to find $r(50)$, $r(60)$, $r(70)$, $r(80)$, $r(90)$, and $r(100)$. What do your results tell you about this situation?
7. Explain how it is possible for the U.S. share of annual worldwide carbon dioxide emissions to decline even if U.S. per-person annual carbon dioxide emissions increase. [**Hint:** There may be two reasons. Consider your result in Problem 1. Consider also what is happening in developing countries.]

Illumination Lab

The illumination from a light bulb decreases as the distance from the bulb increases. Illumination can be measured in milliwatts per square centimeter (mW/cm^2). In this experiment, you will study the relationship between the illumination I (in mW/cm^2) of a 25-watt light bulb and the distance d (in cm) from the bulb.

Materials

You will need the following materials:

1. a 25-watt light bulb
2. a metric tape measure or a meterstick
3. a light sensor probe. (A Texas Instruments CBL unit and light sensor probe used in conjunction with a TI-83, TI-84, TI-86, or TI-Nspire graphing calculator works well.)

Preparation

It is best to perform this experiment at night in a room with all lights other than the 25-watt bulb turned off, the shades or curtains closed, and all doors shut. Place the meterstick so that you can measure horizontal distances from the center of the light bulb.

Recording of Data

Record the intensity of the light at distances of 70 cm, 80 cm, 90 cm, and every 10 centimeters thereafter until you reach 140 cm.

Analyzing the Data

1. Adjust your illumination readings for any small amount of light in the room due to sources other than the 25-watt light bulb. If you are using the CBL unit, an initial step will allow you to input the illumination of the room with the bulb off. The unit will subtract this value from your illumination readings. Otherwise, you will have to subtract this small amount of "background" illumination from your illumination readings.
2. Complete the second column of Table 38.

Table 38 Illumination by a Light Bulb

Distance d (cm)	Illumination I (mW/cm^2)	d^2I
70		
80		
90		
100		
110		
120		
130		
140		

Source: *J. Lehmann*

3. Use a graphing calculator to draw a scattergram of your light data.
4. The illumination from a light bulb varies inversely as the distance squared. Write an equation involving d, I, and k.
5. Solve your equation from Problem 4 for k. Use the first two columns of Table 38 to complete the third column. What is a reasonable value of k? How did you find this value?
6. Substitute your value of k into your equation from Problem 4. Let $f(d) = I$. Write your equation with the function name f.
7. Use a graphing calculator to graph your model and the scattergram in the same viewing window. Graph your model and the scattergram by hand. How well does f model the data?
8. Use your model to estimate the illumination from the bulb at a distance of 150 centimeters.
9. Use your model to estimate the distance at which the illumination from the bulb is 0.1 mW/cm^2.
10. For what values of d is there model breakdown?
11. Explain why it makes sense in this situation that f is a decreasing function for positive values of d.

Boyle's Law Lab

The more you squeeze a sealed syringe filled with air, the harder it gets to squeeze it farther. That's because, as the volume of air in the syringe decreases, the pressure of the air inside the syringe increases. In this experiment, you will study the relationship between the pressure P in atmospheres (atm) of the air inside a syringe and the volume V in cubic centimeters (cm^3) of the air in the syringe.

Materials

To do this lab, you will need the following materials:

1. a Texas Instruments CBL unit
2. a TI-83, TI-84, TI-86, or TI-Nspire graphing calculator with graph link cable
3. a syringe apparatus (made by the Vernier Company)
4. a CBL-DIN adapter (made by the Vernier Company; Vernier offers the syringe apparatus and CBL-DIN adapter as a package.)

Preparation

Follow the instructions that come with the syringe apparatus to set up the equipment. An initial "at-rest" volume setting of around 8 cm^3 works well. You will have to enter a pressure-gathering program into your CBL unit. (The code for the program is included in the syringe/CBL-DIN package.)

Recording of Data

Record the pressures of the air at volumes 3 cm^3, 4 cm^3, 5 cm^3, and so on until you reach 20 cm^3. Take one pressure reading with each run of the pressure-gathering program.

Analyzing the Data

1. Complete the second and fifth columns of Table 39.

Table 39 Pressures and Volumes in a Syringe

Volume V (cm^3)	Pressure P (atm)	PV	V (cm^3)	P (atm)	PV
3			12		
4			13		
5			14		
6			15		
7			16		
8			17		
9			18		
10			19		
11			20		

Source: *J. Lehmann*

2. Use a graphing calculator to draw a scattergram of the data.
3. The pressure of air inside a syringe varies inversely as the volume of that air. Write an equation involving V, P, and k.
4. Solve your equation from Problem 3 for k. Use the first two columns and the fourth and fifth columns of Table 39 to complete the third and sixth columns, respectively. What is a reasonable value of k? How did you find this value?
5. Substitute your value of k into your equation from Problem 3. Let $f(V) = P$. Write your equation with the function name f.
6. Use a graphing calculator to graph your model and the scattergram in the same viewing window. Graph your model and the scattergram by hand. How well does f model the data?
7. Use your model to estimate the air pressure in the syringe if the volume is 2 cm^3.
8. Use your model to estimate the volume in the syringe at which the air pressure would be 4 atm.
9. Use your model to predict what the air pressure would be if all the air were squeezed out of the syringe. Do you think this could be done in reality? Explain.

Estimating π Lab

One of the great moments in early mathematics was the discovery of π and its usefulness. Long before there were calculators, many great mathematicians tried to estimate π. In this lab, you will build on your work from the Volume Lab in Chapter 1 to find an estimate of π. If you haven't done that lab yet, do it now. Then respond to the questions that follow.

Analyzing the Situation

1. Recall from the Volume Lab in Chapter 1 that V is the volume of water (in ounces) in a cylinder when the height of the water in the cylinder is h centimeters (cm). In the Volume Lab, you likely sketched a linear model that contains the origin $(0, 0)$. If not, sketch a new linear model that does.
2. Find an equation of a model to describe the data.
3. What does the slope mean in this situation? Make sure that you include units in your description.
4. There are 29.57353 cm^3 in one ounce. Use this fact to help you find the slope of the model, in units of cm^2.
5. The formula for the volume of a cylinder is $V = \pi r^2 h$, where r is the cylinder's radius and h is the cylinder's height. In Problem 2, you found an equation of the form $V = mh$, where m is the slope. Explain why we can conclude that $m = \pi r^2$.
6. Substitute the slope of your model for m and the radius of the cylinder for r in the equation $m = \pi r^2$. Then treat π as if it is an unknown constant and solve for it. If you are using the data provided in the Volume Lab, the base radius of the cylinder is 4.45 cm. What is your estimate of π?
7. Use a calculator to help you find the error in your estimate of π.

Chapter Summary

Key Points of Chapter 12

Section 12.1 Finding the Domains of Rational Functions and Simplifying Rational Expressions

	Throughout these key points, assume A, B, C, and D are polynomials.
Rational expression	If P and Q are polynomials with Q nonzero, we call the ratio $\frac{P}{Q}$ a **rational expression.**
Rational function	A **rational function** is a function whose equation can be put into the form $f(x) = \frac{P(x)}{Q(x)}$, where $P(x)$ and $Q(x)$ are polynomials and $Q(x)$ is nonzero.
Domain of a rational function	The domain of a rational function $f(x) = \frac{P(x)}{Q(x)}$ is the set of all real numbers except for those numbers that, when substituted for x, give $Q(x) = 0$.
Excluded value	A number is an **excluded value** of a rational expression if substituting the number into the expression leads to a division by 0.
Simplify a rational expression	To simplify a rational expression, **1.** Factor the numerator and the denominator. **2.** Use the property $\frac{AB}{AC} = \frac{A}{A} \cdot \frac{B}{C} = 1 \cdot \frac{B}{C} = \frac{B}{C}$, where A and C are nonzero, so the expression is in lowest terms.
Vertical asymptotes of the graphs of rational functions and domains of rational functions	If the graph of a rational function f has a vertical asymptote $x = k$, then k is not in the domain of f. If k is not in the domain of a rational function g, then $x = k$ may or may not be a vertical asymptote of the graph of g.
Quotient function	If f and g are functions, x is in the domain of both functions, and $g(x)$ is nonzero, then we can form the **quotient function** $\frac{f}{g}$: $\left(\frac{f}{g}\right)(x) = \frac{f(x)}{g(x)}$.
Rational model	A **rational model** is a rational function, or its graph, that describes an authentic situation.
Percentage formula	If m items out of n items have a certain attribute, then the percentage p (written $p\%$) of the n items that have the attribute is $p = \frac{m}{n} \cdot 100$. We call this equation the **percentage formula.**

Section 12.2 Multiplying and Dividing Rational Expressions; Converting Units

Multiplying rational expressions	If $\frac{A}{B}$ and $\frac{C}{D}$ are rational expressions and B and D are nonzero, then $\frac{A}{B} \cdot \frac{C}{D} = \frac{AC}{BD}$.
How to multiply rational expressions	To multiply two rational expressions, **1.** Factor the numerators and the denominators. **2.** Multiply by using the property $\frac{A}{B} \cdot \frac{C}{D} = \frac{AC}{BD}$, where B and D are nonzero. **3.** Simplify the result.
Dividing rational expressions	If $\frac{A}{B}$ and $\frac{C}{D}$ are rational expressions and B, C, and D are nonzero, then $\frac{A}{B} \div \frac{C}{D} = \frac{A}{B} \cdot \frac{D}{C}$.

Section 12.2 Multiplying and Dividing Rational Expressions; Converting Units (*Continued*)

How to divide rational expressions	To divide two rational expressions, **1.** Write the quotient as a product by using the property $\frac{A}{B} \div \frac{C}{D} = \frac{A}{B} \cdot \frac{D}{C}$, where B, C, and D are nonzero. **2.** Find the product. **3.** Simplify.
Converting units	To convert the units of a quantity, **1.** Write the quantity in the original units. **2.** Multiply by fractions equal to 1 so the units you want to eliminate appear in one numerator and one denominator.

Section 12.3 Adding and Subtracting Rational Expressions

Adding rational expressions that have a common denominator	If $\frac{A}{B}$ and $\frac{C}{B}$ are rational expressions and B is nonzero, then $\frac{A}{B} + \frac{C}{B} = \frac{A + C}{B}$.
Subtracting rational expressions that have a common denominator	If $\frac{A}{B}$ and $\frac{C}{B}$ are rational expressions and B is nonzero, then $\frac{A}{B} - \frac{C}{B} = \frac{A - C}{B}$.
Subtract entire numerator	When subtracting rational expressions, be sure to subtract the *entire* numerator.
How to add or subtract two rational expressions that have different denominators	To add or subtract two rational expressions that have different denominators, **1.** Factor the denominators of the expressions if possible. Determine which factors are missing. **2.** Use the property $\frac{A}{A} = 1$, where A is nonzero, to introduce missing factors. **3.** Add the expressions by using the property $\frac{A}{B} + \frac{C}{B} = \frac{A + C}{B}$, where B is nonzero; or subtract the expressions by using the property $\frac{A}{B} - \frac{C}{B} = \frac{A - C}{B}$, where B is nonzero. **4.** Simplify.

Section 12.4 Simplifying Complex Rational Expressions

Complex rational expression	A **complex rational expression** is a rational expression whose numerator or denominator (or both) is a rational expression.
Using method 1 to simplify a complex rational expression	To simplify a complex rational expression by method 1, **1.** Write both the numerator and the denominator as fractions. **2.** Write the complex rational expression as the quotient of two rational expressions: $\frac{\frac{A}{B}}{\frac{C}{D}} = \frac{A}{B} \div \frac{C}{D}$, where B, C, and D are nonzero. **3.** Divide the rational expressions.
Using method 2 to simplify a complex rational expression	To simplify a complex rational expression by method 2, **1.** Find the LCD of all of the fractions in the numerator and denominator. **2.** Multiply by 1 in the form $\frac{\text{LCD}}{\text{LCD}}$. **3.** Simplify the numerator and the denominator to polynomials. **4.** Simplify the rational expression.

Section 12.5 Solving Rational Equations

Rational equation in one variable	A **rational equation in one variable** is an equation in one variable in which both sides can be written as rational expressions.
Check proposed solutions	Since multiplying both sides of a rational equation by the LCD and then simplifying both sides may introduce extraneous solutions, we must always check that any proposed solution is not an excluded value.
Solving a rational equation in one variable	To solve a rational equation in one variable, **1.** Factor the denominator(s) if possible. **2.** Identify any excluded values. **3.** Find the LCD of all of the fractions. **4.** Multiply both sides of the equation by the LCD, which gives a simpler equation to solve. **5.** Solve the simpler equation. **6.** Discard any proposed solutions that are excluded values.
Solving a rational equation versus simplifying a rational expression	To solve a rational equation, clear the fractions in it by multiplying both sides of the equation by the LCD. To simplify a rational expression, do *not* multiply it by the LCD—the only multiplication permissible is multiplication by 1, usually in the form $\frac{A}{A}$, where A is a nonzero polynomial.
Results of solving rational equations and simplifying rational expressions	The result of solving a rational equation is the empty set or a set of one or more numbers. The result of simplifying a rational expression is an expression.

Section 12.6 Modeling with Rational Functions

Mean	If a quantity Q is divided into n parts, the **mean** amount M of the quantity per part is given by $M = \frac{Q}{n}$.
Distance-speed-time relationship	If an object is moving at a constant speed s for an amount of time t, then the distance d traveled is given by $d = st$ and the time t is given by $t = \frac{d}{s}$.

Section 12.7 Proportions; Similar Triangles

Proportion	A **proportion** is a statement of the equality of two ratios.
Proportional	If the ratio of two related quantities is constant, we say that the two quantities are **proportional.**
Similar triangles	Two triangles are called **similar triangles** if their corresponding angles are equal in measure.
Side lengths of similar triangles	The lengths of the corresponding sides of two similar triangles are proportional.

Section 12.8 Variation

Direct variation	If $y = kx$ for some nonzero constant k, we say that ***y* varies directly as *x*** or that ***y* is proportional to *x*.** We call k the **variation constant** or the **constant of proportionality.** The equation $y = kx$ is called a **direct variation equation.**
Changes in values of variables for direct variation	Assume y varies directly as x with some positive variation constant k: • If the value of x increases, then the value of y increases. • If the value of x decreases, then the value of y decreases.
Finding and using a direct variation model	Assume a quantity p varies directly as a quantity t. To make estimates about an authentic situation, **1.** Substitute the values of a data point into the equation $p = kt$; then solve for k. **2.** Substitute the value of k into the equation $p = kt$. **3.** Use the equation from step 2 to make estimates of quantity t or p.

Section 12.8 Variation (*Continued*)

Inverse variation	If $y = \frac{k}{x}$ for some nonzero constant k, we say that ***y* varies inversely as *x*** or that ***y* is inversely proportional to *x*.** We call k the **variation constant** or the **constant of proportionality.** The equation $y = \frac{k}{x}$ is called an **inverse variation equation.**
Changes in values of variables for inverse variation	Assume y varies inversely as x with some positive variation constant k. For positive values of x, • If the value of x increases, then the value of y decreases. • If the value of x decreases, then the value of y increases.
Finding and using an inverse variation model	Assume a quantity p varies inversely as a quantity t. To make estimates about an authentic situation, **1.** Substitute the values of a data point into the equation $p = \frac{k}{t}$; then solve for k. **2.** Substitute the value of k into the equation $p = \frac{k}{t}$. **3.** Use the equation from step 2 to make estimates of quantity t or p.

Chapter 12 Review Exercises

1. For $f(x) = \frac{5x - 3}{2x^2 - 3x + 1}$, find $f(0)$ and $f(2)$.

Find the domain of the rational function.

2. $f(x) = \frac{9}{3x - 5}$

3. $f(x) = \frac{5}{4x^2 - 49}$

4. $f(x) = \frac{x - 5}{x^2 - 6x + 8}$

5. $f(x) = \frac{x^2 - 4}{12x^2 + 13x - 35}$

6. $f(x) = \frac{3x + 7}{9x^3 + 18x^2 - x - 2}$

For Exercises 7–11, Simplify the right-hand side of the equation.

7. $f(x) = \frac{3x - 12}{x^2 - 6x + 8}$

8. $f(x) = \frac{16 - x^2}{2x^3 - 16x^2 + 32x}$

9. $f(x) = \frac{x^2 - 8x + 12}{3x^2 - 16x - 12}$

10. $f(x) = \frac{x^2 + 10x + 25}{2x^3 - 3x^2 - 50x + 75}$

11. $f(x) = \frac{x + 2}{x^3 + 8}$

12. Simplify $\frac{6a^2 - 17ab + 5b^2}{3a^2 - 4ab + b^2}$.

For Exercises 13–30 perform the indicated operation(s).

13. $\frac{3x + 6}{2x - 4} \cdot \frac{5x - 10}{6x + 12}$

14. $\frac{25b^3}{b^2 - b} \cdot \frac{b^2 - 1}{35b}$

15. $\frac{x^2 - 49}{9 - x^2} \cdot \frac{2x^3 + 8x^2 - 42x}{5x - 35}$

16. $\frac{p^3 - t^3}{p^2 - t^2} \cdot \frac{p^2 + 6pt + 5t^2}{p^2t + pt^2 + t^3}$

17. $\frac{x^2 - 4}{x^2 + 3x + 2} \div \frac{4x^2 - 24x + 32}{x^2 - 5x + 4}$

18. $\frac{7t + 14}{t - 7} \div (3t^2 + 2t - 8)$

19. $\frac{4 - x}{4x} \div \frac{16 - x^2}{16x^2}$

20. $\frac{8x^3 + 4x^2 - 18x - 9}{x^2 - 6x + 9} \div \frac{4x^2 + 8x + 3}{x^2 - 9}$

21. $\frac{5}{x^2 + 7x + 6} + \frac{2x}{x^2 - 3x - 4}$

22. $\frac{x}{x^2 - 5x + 6} + \frac{3}{3 - x}$

23. $\frac{x}{2x^3 - 3x^2 - 5x} + \frac{2}{x^3 - x}$

24. $\frac{x - 1}{x^2 - 4} + \frac{x + 3}{x^2 - 4x + 4}$

25. $\frac{3}{4x - 12} - \frac{x}{x^2 - 2x - 3}$

26. $\frac{x - 4}{x^2 + 2x - 3} - \frac{x + 2}{x^2 - 6x + 5}$

27. $\frac{x + 1}{25 - x^2} - \frac{x - 4}{2x^2 - 14x + 20}$

28. $\frac{2m}{m^2 - 3mn - 10n^2} - \frac{4n}{m^2 + 8mn + 12n^2}$

29. $\frac{2}{x - 5} - \left(\frac{x^2 + 5x + 6}{3x^2 - 75} \div \frac{x^2 + 2x}{3x + 15}\right)$

30. $\frac{2x - 8}{3x + 4} \cdot \left(\frac{x - 2}{x + 1} - \frac{x + 3}{x - 4}\right)$

For Exercises 31–34, let $f(x) = \dfrac{x^2 - x - 2}{x^2 + 5x + 6}$ *and* $g(x) = \dfrac{x + 3}{x + 2}$. *Find an equation of the given function.*

31. $f \cdot g$ **32.** $\dfrac{f}{g}$ **33.** $f + g$ **34.** $f - g$

35. A student tries to find the difference $\dfrac{2x}{x+4} - \dfrac{7x-3}{x+4}$:

$$\frac{2x}{x+4} - \frac{7x-3}{x+4} = \frac{2x - 7x - 3}{x+4} = \frac{-5x-3}{x+4}$$

Describe any errors. Then find the difference correctly.

For Exercises 36 and 37, round approximate results to two decimal places. Refer to the list of equivalent units in the margin of page 747 as needed.

36. A TI-84 graphing calculator weighs 0.62 pound. What is the calculator's weight in ounces?

37. Americans consume an average of 121 gallons of water per year (Source: *Wirthlin Worldwide*). What is this average in cups per day?

Simplify.

38. $\dfrac{\frac{12}{x^2}}{\frac{9}{x^3}}$ **39.** $\dfrac{\frac{x-2}{x^2-9}}{\frac{x^2-4}{x+3}}$

40. $\dfrac{5 - \frac{2}{w}}{1 - \frac{3}{w}}$ **41.** $\dfrac{\frac{4}{3x^4} - \frac{2}{6x^2}}{\frac{1}{2x} + \frac{1}{4}}$

For Exercises 42–46, solve. Any solution is a real number.

42. $\dfrac{1}{x+5} - \dfrac{2}{x-2} = \dfrac{-14}{x^2 + 3x - 10}$

43. $\dfrac{x}{x+2} + \dfrac{3}{x+4} = \dfrac{14}{x^2 + 6x + 8}$

44. $\dfrac{5}{w} + 3 = \dfrac{4}{w^2}$ **45.** $\dfrac{-3}{x+6} + \dfrac{2}{x+1} = \dfrac{x-2}{x+6}$

46. $\dfrac{x-3}{2x^2 - 7x - 4} - \dfrac{5}{2x^2 + 3x + 1} = \dfrac{x-1}{x^2 - 3x - 4}$

47. Find all complex-number solutions of the equation $\dfrac{2x}{x+6} - \dfrac{4}{x-3} = \dfrac{-37}{x^2 + 3x - 18}$.

For Exercises 48–51, solve or simplify, whichever is appropriate.

48. $\dfrac{3}{2-x} - \dfrac{7}{x-2} - 4 = 0$

49. $\dfrac{3}{x^2 - 25} + \dfrac{1}{x^2 - x - 30}$ **50.** $\dfrac{3}{2-x} - \dfrac{7}{x-2} - 4$

51. $\dfrac{3}{x^2 - 25} + \dfrac{1}{x^2 - x - 30} = \dfrac{2}{x^2 - 11x + 30}$

52. Find the x-intercept(s) of the graph of $f(x) = \dfrac{x-7}{x+1} - \dfrac{x+3}{x-4}$.

53. Solve the formula $S = \dfrac{a}{1-r}$ for r.

54. If a recipe for 4 servings of chicken cacciatore calls for 14 ounces of diced tomatoes, how many ounces of diced tomatoes should be used to make 7 servings?

55. Two similar triangles are shown in Fig. 73. Find the length of the side labeled "x yards." Round your result to the second decimal place.

Figure 73 Exercise 55

Translate the equation with positive constant k by using the phrase "varies directly" or "varies inversely" in a sentence.

56. $H = ku^2$ **57.** $w = \dfrac{k}{\log(t)}$

For Exercises 58 and 59, find an equation that meets the given conditions.

58. y varies directly as $\sqrt{x}$, and $y = 2$ when $x = 49$.

59. B varies inversely as r^3, and $B = 9$ when $r = 2$.

60. The number of inches of water varies directly as the number of inches of snow. If 20 inches of snow will melt to 2.24 inches of water, to how many inches of water will 37 inches of snow melt?

61. Let m be the mass (in grams) of a ball bearing with radius r (in cm). Table 40 shows values of r and m for ball bearings of various sizes.

Table 40 Radii and Masses of Ball Bearings

Radius r (cm)	Mass m (grams)	k
1.0	17.1	
1.2	29.4	
1.4	46.7	
1.6	69.6	
1.8	99.1	
2.0	135.9	

a. The mass of a ball bearing is directly proportional to the radius cubed. Write an equation involving r, m, and k.

b. Solve the equation that you found in part (a) for k.

c. Use the first two columns of Table 40 to complete the third column. What is a reasonable value of k? How did you find this value?

d. Substitute your value of k into your equation from part (a) to find an equation of a model.

e. Draw the graph of your model and a scattergram of the data in the same viewing window. Does your model fit the data well?

f. What is the mass of a ball bearing with radius 2.3 cm?

62. A hotel offers a one-day rental of a conference room (capacity 300) for a flat fee of \$600, plus a per-person charge of \$40 for lunch.

a. Let $C(n)$ be the total cost (in dollars) of renting the room if n people use the room for one day. Find an equation of C.

b. Let $M(n)$ be the mean cost per person (in dollars per person) if n people use the room for one day. Find an equation of M.
c. Find $M(270)$. What does it mean in this situation?
d. Find n when $M(n) = 50$. What does it mean in this situation?

63. In Exercise 38 of Homework 6.1, the average annual U.S. per-person consumption (in pounds per person) $C(t)$ and $R(t)$ of chicken and red meat, respectively, are modeled by the system

$$C(t) = 1.13t + 39.29$$
$$R(t) = -0.94t + 144.88$$

where t is the number of years since 1970 (see Table 41).
a. Find an equation of the sum function $C + R$. What do the inputs and outputs of $C + R$ mean in this situation?
b. Let $P(t)$ be the percentage of chicken and red meat consumed that is chicken at t years since 1970. Find an equation of P.
c. Use P to estimate the percentage of chicken and red meat consumed that was chicken in 2011. Then compute the actual percentage. Is your result from using the model an underestimate or an overestimate?
d. Find $P(47)$. What does it mean in this situation?
e. Predict the year when half of the chicken and red meat consumed will be chicken.

Table 41 Average Annual U.S. Per-Person Consumption of Chicken and Red Meat

	Average Annual Consumption (pounds per person)	
Year	Chicken	Red Meat
1970	40.3	145.8
1980	48.0	136.8
1990	61.5	120.0
2000	78.0	120.7
2010	83.7	108.7
2011	84.2	104.3

Source: *U.S. Department of Agriculture*

64. A student plans to drive 75 miles on an undivided highway and another 40 miles on a divided highway. The speed limits are 50 mph for the undivided highway and 65 mph for the divided highway.
a. Let $T(a)$ be the driving time (in hours) if the student drives a mph above the speed limits. Find an equation of T.
b. Find $T(5)$. What does it mean in this situation?
c. By how much over the speed limits would the student need to drive for the driving time to be 2 hours? Use a graphing calculator table to verify your result.

Chapter 12 Test

For Exercises 1–3, find the domain of the function.

1. $f(x) = \dfrac{5}{6x^2 + 11x - 10}$

2. $g(x) = \dfrac{2}{72 - 2x^2}$ **3.** $h(x) = \dfrac{x}{3}$

4. Give examples of three functions, each with a domain that is the set of real numbers except -3 and 7.

Simplify the right-hand side of the equation.

5. $f(x) = \dfrac{6 - 3x}{x^2 - 5x + 6}$ **6.** $f(x) = \dfrac{9x^2 - 1}{18x^3 - 12x^2 + 2x}$

For Exercises 7–10, perform the indicated operation.

7. $\dfrac{5x^4}{3x^2 + 6x + 12} \cdot \dfrac{x^3 - 8}{15x^7}$

8. $\dfrac{p^2 - 4t^2}{p^2 + 6pt + 9t^2} \div \dfrac{p^2 - 3pt + 2t^2}{p^2 + 3pt}$

9. $\dfrac{5x + 12}{-2x^2 - 8x} - \dfrac{2x + 1}{x^2 + 2x - 8}$

10. $\dfrac{x + 2}{x^2 - 9} + \dfrac{3}{x^2 + 11x + 24}$

11. Perform the indicated operations:

$$\frac{3}{x^2 - 2x} \div \left(\frac{x}{5x - 10} - \frac{x - 1}{x^2 - 4}\right)$$

12. Let $f(x) = \dfrac{x + 1}{x - 5}$ and $g(x) = \dfrac{x - 2}{x + 4}$. Find an equation of $f - g$. Then find $(f - g)(0)$.

13. One ounce of Hormel Turkey Chili with Beans, contains 83.33 milligrams of sodium. How many grams of sodium are there in 1 pound of the chili? Round you answer to the second decimal place. Refer to the list of equivalent units in the margin of page 747 as needed.

14. Simplify $\dfrac{5 + \dfrac{2}{x}}{3 - \dfrac{4}{x - 1}}$.

For Exercises 15 and 16, solve.

15. $\dfrac{2}{x - 1} - \dfrac{5}{x + 1} = \dfrac{4x}{x^2 - 1}$

16. $\dfrac{5}{w - 3} = \dfrac{w}{w - 2} + \dfrac{w}{w^2 - 5w + 6}$

17. Let $f(x) = \dfrac{2}{x - 4} + \dfrac{3}{x + 1}$. Find x when $f(x) = 5$.

For Exercises 18–20, let $f(x) = \dfrac{(x - 5)(x + 2)}{(x - 1)(x + 3)}$.

18. Find $f(-2)$. **19.** Find $f(1)$.

20. Find x when $f(x) = 0$.

21. A 2009 Toyota Prius® uses 3 gallons of gasoline to travel 153 miles on highways. Estimate how many gallons of gasoline are required to travel 400 miles on highways.

22. Two similar triangles are shown in Fig. 74. Find the length of the side labeled "x meters." Round your result to the second decimal place.

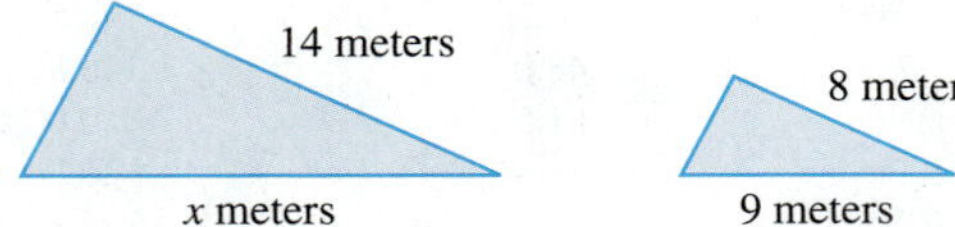

Figure 74 Exercise 22

For Exercises 23 and 24, find an equation that meets the given conditions.

23. W varies directly as t^2, and $W = 3$ when $t = 7$.

24. y varies inversely as $\sqrt{x}$, and $y = 8$ when $x = 25$.

25. It costs a bike manufacturer about \$200 per bike for the materials to manufacture a line of mountain bikes. It also costs \$10,000 each month for the manufacturing plant's lease, electricity, salaries, and so on.
   a. Let $C(n)$ be the total monthly cost (in dollars) if n bikes are manufactured in a certain month. Find an equation of C.
   b. Let $B(n)$ be the price (in dollars) the manufacturer should charge for each bike to break even by making and selling n bikes in a month. Find an equation of B.
   c. Let $P(n)$ be the price (in dollars) the manufacturer should charge for each bike to earn a profit of \$150 per bike by making and selling n bikes in a month. Find an equation of P.
   d. Find $P(100)$. What does it mean in this situation?

26. A student plans to drive from San Francisco, California, to Salt Lake City, Utah. She will drive 400 miles in California, then 920 miles in Nevada and Utah. The speed limit is 70 mph in California and 75 mph in Nevada and Utah.
   a. Let $T(a)$ be the driving time (in hours) if the student drives a mph above the speed limits. Find an equation of T.
   b. Find $T(5)$. What does it mean in this situation?
   c. Find a when $T(a) = 17$. What does your result mean in this situation?

27. The frequency $g(L) = F$ (in hertz) of a tuning fork varies inversely as the square of the length L (in cm) of the prongs. The frequency is 50 hertz when the prong length is 8 cm.
   a. Find an equation of g.
   b. Find the frequency if the prong length is 6 cm.
   c. What is the prong length if the frequency is 200 hertz?
   d. Is g increasing or decreasing for $L > 0$? What does that mean in this situation?

28. Community centers provide health care to millions of mostly poor people. The numbers of such centers and the numbers of patients served are shown in Table 42 for various years.

Table 42 Numbers of Community Health Centers and Patients

Year	Number of Centers (thousands)	Number of Patients (thousands)
2001	0.76	10,000
2003	0.88	12,000
2005	0.96	14,000
2007	1.06	16,000
2009	1.15	18,000
2011	1.14	20,000

Source: *National Association of Community Health Centers*

   a. Let $C(t)$ be the number (in thousands) of community centers at t years since 2000. Find an equation of C.
   b. Let $P(t)$ be the number (in thousands) of patients at community centers at t years since 2000. Find an equation of P.
   c. Let $M(t)$ be the mean number of patients per community center at t years since 2000. Find an equation of M.
   d. Predict the mean number of patients per community center in 2017.
   e. Predict when the mean number of patients per community center will be 21,000 patients.

Radical Functions

How much spam do you receive? The percentage of e-mail that is spam has more than quadrupled since 1999 (see Table 1). In Exercise 75 of Homework 13.5, you will predict when 95% of e-mail will be spam.

Table 1 Percentages of E-Mail That Is Spam

Year	Percent
1999	21
2002	56
2004	68
2006	80
2008	81
2010	85

Source: *IronPort*

In previous chapters, we worked with linear, exponential, logarithmic, polynomial, and rational functions. In Section 9.3, we worked with radical expressions. In this chapter, we will discuss how to simplify more types of radical expressions, as well as how to perform operations with them. We will also solve *radical equations in one variable*. Finally, we will use *square root functions* to make predictions, such as when the average credit card debt per household will be $13 thousand.

13.1 Simplifying Radical Expressions

Objectives

- Convert expressions in radical form to exponential form and vice versa.
- Know the *product property for radicals* and the *power property for radicals.*
- Simplify radical expressions.
- Know the meaning of *radical function, square root function, radical model,* and *square root model.*
- Use a radical model to make an estimate or prediction.

In this section, we will discuss the meaning of a *radical expression* and how to use properties to simplify such an expression.

Radical Expressions

Recall from Section 10.2 that we can represent a principal nth root such as $x^{1/n}$ as $\sqrt[n]{x}$, where n is a counting number greater than 1. We say n is the **index** and the symbol $\sqrt{}$ is the **radical sign.** The notation for the principal square root of x, $\sqrt{x}$, is shorthand for $\sqrt[2]{x}$. An expression under the radical sign is called a **radicand.** In $\sqrt[5]{3x+6}$, the radicand is $3x + 6$. A radical sign together with an index and a radicand is called a **radical.** Here we label the radical sign, index, and radicand for the radical $\sqrt[4]{2x-7}$:

$$\underbrace{\sqrt[4]{2x-7}}_{\text{Radical}}$$

Index → 4; Radical sign → $\sqrt{}$; Radicand → $2x - 7$

Some more radicals are $\sqrt[3]{21}$, $\sqrt{x}$, $\sqrt[7]{x^4}$, $\sqrt{5x-8}$, and $\sqrt[4]{(x+1)^3}$.

An expression that contains a radical is called a **radical expression.** Here are some radical expressions:

$$\sqrt[4]{85} \quad \sqrt{x} \quad \sqrt[3]{4x+9} \quad 2\sqrt[3]{2x}+5\sqrt[4]{7} \quad (8\sqrt{x}-5)(3\sqrt[3]{x}+4) \quad \frac{2\sqrt{x}-7}{6\sqrt{x}-4}$$

Recall from Section 10.2 that, for nonnegative a, $\sqrt[n]{a}$ is the nonnegative number whose nth power is a. If a is negative and n is odd, then $\sqrt[n]{a}$ is the (negative) number whose nth power is a. If a is negative and n is even, then $\sqrt[n]{a}$ is not a real number.

Example 1 Evaluating Radicals

Evaluate the radical.

1. $\sqrt[3]{-8}$
2. $\sqrt[4]{81}$

Solution

1. $\sqrt[3]{-8} = -2$, since $(-2)^3 = -8$.
2. $\sqrt[4]{81} = 3$, since $3^4 = 81$.

We use a graphing calculator to verify our work (see Fig. 1). To enter $\sqrt[3]{-8}$, press MATH **4** (−) **8**) ENTER. To enter $\sqrt[4]{81}$, press **4** MATH **5** (**81**) ENTER.

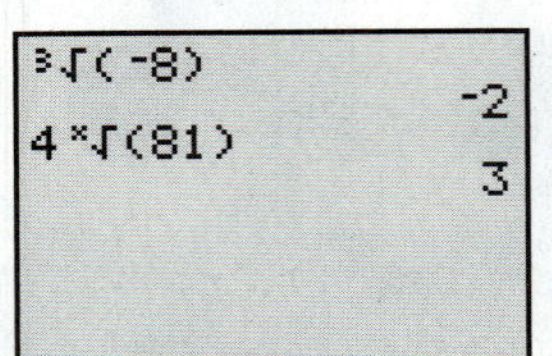

Figure 1 Verify the work

Recall from Section 9.3 that a number that has a square root that is rational is called a *perfect square*. More generally, a number that has an nth root that is rational is called a **perfect nth power.** For example, -8 is a perfect 3rd power, or **perfect cube,** because $\sqrt[3]{-8} = -2 = \frac{-2}{1}$ is rational. The number 81 is a perfect 4th power, because $\sqrt[4]{81} = 3 = \frac{3}{1}$ is rational.

Recall from Section 10.2 that if $a^{1/n}$ is defined, then

$$a^{m/n} = \left(a^{1/n}\right)^m = \left(a^m\right)^{1/n}$$

or we can write

$$a^{m/n} = \left(\sqrt[n]{a}\right)^m = \sqrt[n]{a^m}$$

For example, $x^{3/5} = \left(\sqrt[5]{x}\right)^3 = \sqrt[5]{x^3}$. We say the expressions $\left(\sqrt[n]{x}\right)^m$ and $\sqrt[n]{x^m}$ are in *radical form* and the expression $x^{m/n}$ is in *exponential form.*

Example 2 Writing Exponential and Radical Forms

If the expression is in exponential form, write it in radical form. If it is in radical form, write it in exponential form.

1. $x^{3/7}$
2. $\sqrt{x^5}$
3. $(3w+1)^{4/5}$
4. $\sqrt[8]{(2x+5)^3}$

Solution

1. $x^{3/7} = \sqrt[7]{x^3}$
2. $\sqrt{x^5}$ is shorthand for $\sqrt[2]{x^5}$. So, $\sqrt{x^5} = x^{5/2}$.
3. $(3w+1)^{4/5} = \sqrt[5]{(3w+1)^4}$
4. $\sqrt[8]{(2x+5)^3} = (2x+5)^{3/8}$

Simplifying Radical Expressions

It will be helpful, at times, to write a radical expression in exponential form and use exponential properties to simplify the result.

Example 3 Simplifying Radical Expressions

Write the expression in exponential form; then simplify the result.

1. $\sqrt[3]{x^3}$
2. $\sqrt[3]{x^{12}}$
3. $\sqrt[3]{x^{21}}$

Solution

1. $\sqrt[3]{x^3} = x^{3/3} = x^1 = x$
2. $\sqrt[3]{x^{12}} = x^{12/3} = x^4$
3. $\sqrt[3]{x^{21}} = x^{21/3} = x^7$

In addition to the numbers that are perfect nth powers, some variable expressions are perfect nth powers. As in Example 3, we can eliminate the radical sign from expressions of the form $\sqrt[3]{x^k}$ when k is a multiple of 3. In general, if $\sqrt[n]{x}$ is defined, we can eliminate the radical sign from $\sqrt[n]{x^k}$ when k is a multiple of the index n. If n is a counting number greater than 1 and k is a multiple of n, we say x^k is a **perfect nth power.**

Here are perfect nth powers for $n = 2$, $n = 3$, and $n = 4$:

Value of n	Name	Examples	Description of Exponents
2	perfect square	$x^2, x^4, x^6, x^8, \dots$	multiples of 2
3	perfect cube	$x^3, x^6, x^9, x^{12}, \dots$	multiples of 3
4	perfect 4th power	$x^4, x^8, x^{12}, x^{16}, \dots$	multiples of 4

The base of a perfect nth power can be any variable expression. For example, $(7x + 2)^{15}$ is a perfect 5th power; furthermore, we can find the 5th root of it:

$$\sqrt[5]{(7x+2)^{15}} = (7x+2)^{15/5} = (7x+2)^3$$

In Section 9.3, we studied the product property for square roots,

$$\sqrt{ab} = \sqrt{a}\sqrt{b}$$

where $a \geq 0$ and $b \geq 0$. The more general product property for radicals,

$$\sqrt[n]{ab} = \sqrt[n]{a}\sqrt[n]{b}$$

is true for any counting-number index n greater than 1 where $\sqrt[n]{a}$ and $\sqrt[n]{b}$ are defined. Here is the proof:

$$\begin{aligned} \sqrt[n]{ab} &= (ab)^{1/n} && \text{Write in exponential form.} \\ &= a^{1/n}b^{1/n} && (ab)^m = a^m b^m \\ &= \sqrt[n]{a}\sqrt[n]{b} && \text{Write in radical form.} \end{aligned}$$

Product Property for Radicals

If $\sqrt[n]{a}$ and $\sqrt[n]{b}$ are defined, then

$$\sqrt[n]{ab} = \sqrt[n]{a}\sqrt[n]{b}$$

In words, the nth root of a product is the product of the nth roots.

A radical with index n is **simplified** when the radicand does not have any factors that are perfect nth powers (other than -1 or 1) and the index is as small as possible. We will discuss how to find the smallest possible index later in this section.

Here we use the product property for radicals to simplify $\sqrt{25x^6}$:

$$\begin{aligned} \sqrt{25x^6} &= \sqrt{25}\sqrt{x^6} && \text{Product property: } \sqrt[n]{ab} = \sqrt[n]{a}\sqrt[n]{b} \\ &= 5x^3 && \sqrt{25} = 5;\ \sqrt{x^6} = x^3 \text{ for } x \geq 0 \end{aligned}$$

Notice that the radicand $25x^6$ is the product of the perfect squares 25 and x^6. We can use the product property to simplify any radical with index n whose radicand is a product of nth powers.

We can also use the product property to simplify some radicals with index n whose radicand is *not* a product of perfect nth powers only. For those radicals, we write the radicand as a product of one or more perfect nth powers and another expression with no factors that are perfect nth powers. Then we apply the product property.

Example 4 Simplifying Radical Expressions

Simplify. Assume x is nonnegative.

1. $\sqrt{200}$ **2.** $\sqrt{x^7}$ **3.** $\sqrt{12w^{11}}$

Solution

1. The numbers 4, 25, and 100 are all perfect-square factors of 200. Recall from Section 9.3 that it is most efficient to work with the largest one, which is 100. So, we write 200 as $100 \cdot 2$ and apply the product property for radicals:

$$\begin{aligned} \sqrt{200} &= \sqrt{100 \cdot 2} && \text{100 is largest perfect-square factor.} \\ &= \sqrt{100}\sqrt{2} && \text{Product property: } \sqrt[n]{ab} = \sqrt[n]{a}\sqrt[n]{b} \\ &= 10\sqrt{2} && \sqrt{100} = 10 \end{aligned}$$

2. The powers x^2, x^4, and x^6 are all perfect-square factors of x^7. Since x^6 is the perfect-square factor with the largest exponent, we write x^7 as $x^6 \cdot x$ and apply the product property for radicals:

$$\begin{aligned} \sqrt{x^7} &= \sqrt{x^6 \cdot x} && x^6 \text{ is perfect-square factor with largest exponent.} \\ &= \sqrt{x^6}\sqrt{x} && \text{Product property: } \sqrt[n]{ab} = \sqrt[n]{a}\sqrt[n]{b} \\ &= x^3\sqrt{x} && \sqrt{x^6} = x^3 \text{ for } x \ge 0 \end{aligned}$$

3. The number 4 is the largest perfect-square factor of 12, and the power w^{10} is the perfect-square factor of w^{11} with the largest exponent:

$$\begin{aligned} \sqrt{12w^{11}} &= \sqrt{4 \cdot 3 \cdot w^{10} \cdot w} && \text{4 and } w^{10} \text{ are perfect squares.} \\ &= \sqrt{4 \cdot w^{10} \cdot 3w} && \text{Rearrange factors.} \\ &= \sqrt{4}\sqrt{w^{10}}\sqrt{3w} && \text{Product property: } \sqrt[n]{ab} = \sqrt[n]{a}\sqrt[n]{b} \\ &= 2w^5\sqrt{3w} && \sqrt{4} = 2;\ \sqrt{w^{10}} = w^5 \text{ for } w \ge 0 \end{aligned}$$

In Example 4, we had to decide which perfect nth powers to use to factor the radicands. Our work in Example 4 suggests two guidelines:

- **If the radicand has more than one numerical factor that is a perfect nth power, select the largest one.**
- **If the radicand has more than one factor that is a perfect nth power with the same variable base, select the one with the largest exponent.**

Example 5 Simplifying Radical Expressions

Simplify. Assume all variables are nonnegative.

1. $\sqrt{48x^4y^{13}}$

2. $\sqrt{(5x+3)^9}$

Solution

1.
$$\begin{aligned} \sqrt{48x^4y^{13}} &= \sqrt{16 \cdot 3 \cdot x^4 \cdot y^{12} \cdot y} && 16,\ x^4, \text{ and } y^{12} \text{ are perfect squares.} \\ &= \sqrt{16 \cdot x^4 \cdot y^{12} \cdot 3y} && \text{Rearrange factors.} \\ &= \sqrt{16}\sqrt{x^4}\sqrt{y^{12}}\sqrt{3y} && \text{Product property: } \sqrt[n]{ab} = \sqrt[n]{a}\sqrt[n]{b} \\ &= 4x^2y^6\sqrt{3y} && \sqrt{16} = 4;\ \sqrt{x^4} = x^2 \text{ for } x \ge 0;\ \sqrt{y^{12}} = y^6 \text{ for } y \ge 0 \end{aligned}$$

2.
$$\begin{aligned} \sqrt{(5x+3)^9} &= \sqrt{(5x+3)^8(5x+3)} && (5x+3)^8 \text{ is a perfect-square factor.} \\ &= \sqrt{(5x+3)^8}\sqrt{5x+3} && \text{Product property: } \sqrt[n]{ab} = \sqrt[n]{a}\sqrt[n]{b} \\ &= (5x+3)^4\sqrt{5x+3} && \sqrt{(5x+3)^8} = (5x+3)^4 \end{aligned}$$

We use a graphing calculator table to verify our work (see Fig. 2).

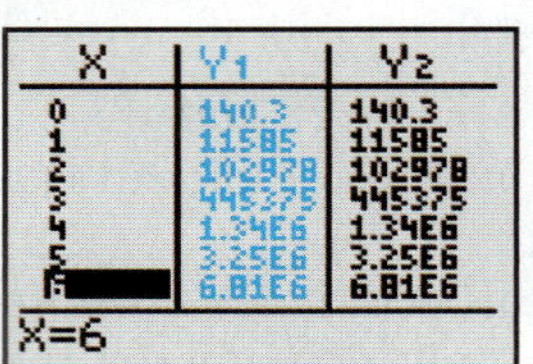

Figure 2 Verify the work

In Example 5, we simplified square root expressions. In Example 6, we will simplify radical expressions with index n greater than 2.

Example 6 Simplifying Radical Expressions

Simplify. Assume all variables are nonnegative.

1. $\sqrt[3]{40x^{17}}$ **2.** $\sqrt[4]{80x^{20}y^{15}}$ **3.** $\sqrt[5]{(2x+7)^{34}}$

Solution

1.
$$\begin{aligned}
\sqrt[3]{40x^{17}} &= \sqrt[3]{8 \cdot 5 \cdot x^{15} \cdot x^2} && \textit{8 and } x^{15} \textit{ are perfect 3rd powers (perfect cubes).}\\
&= \sqrt[3]{8 \cdot x^{15} \cdot 5x^2} && \textit{Rearrange factors.}\\
&= \sqrt[3]{8}\sqrt[3]{x^{15}}\sqrt[3]{5x^2} && \sqrt[n]{ab} = \sqrt[n]{a}\sqrt[n]{b}\\
&= 2x^5\sqrt[3]{5x^2} && \sqrt[3]{8} = 2;\ \sqrt[3]{x^{15}} = x^5
\end{aligned}$$

2.
$$\begin{aligned}
\sqrt[4]{80x^{20}y^{15}} &= \sqrt[4]{16 \cdot 5 \cdot x^{20} \cdot y^{12} \cdot y^3} && 16,\ x^{20}, \textit{ and } y^{12} \textit{ are perfect 4th powers.}\\
&= \sqrt[4]{16 \cdot x^{20} \cdot y^{12} \cdot 5y^3} && \textit{Rearrange factors.}\\
&= \sqrt[4]{16}\sqrt[4]{x^{20}}\sqrt[4]{y^{12}}\sqrt[4]{5y^3} && \sqrt[n]{ab} = \sqrt[n]{a}\sqrt[n]{b}\\
&= 2x^5y^3\sqrt[4]{5y^3} && \sqrt[4]{16} = 2;\ \sqrt[4]{x^{20}} = x^5 \textit{ for } x \ge 0;\ \sqrt[4]{y^{12}} = y^3 \textit{ for } y \ge 0
\end{aligned}$$

3.
$$\begin{aligned}
\sqrt[5]{(2x+7)^{34}} &= \sqrt[5]{(2x+7)^{30}(2x+7)^4} && (2x+7)^{30} \textit{ is a perfect 5th power.}\\
&= \sqrt[5]{(2x+7)^{30}}\sqrt[5]{(2x+7)^4} && \sqrt[n]{ab} = \sqrt[n]{a}\sqrt[n]{b}\\
&= (2x+7)^6\sqrt[5]{(2x+7)^4} && \sqrt[5]{(2x+7)^{30}} = (2x+7)^6
\end{aligned}$$

Consider the radical expression $\sqrt[8]{x^6}$, $x \ge 0$. Although the radicand x^6 does not have factors that are perfect 8th powers, we can write the radical with a smaller index:

$$\sqrt[8]{x^6} = x^{6/8} = x^{3/4} = \sqrt[4]{x^3}$$

If $\sqrt[n]{x}$ is defined and the fraction $\frac{m}{n}$ can be simplified, then we can decrease the index of $\sqrt[n]{x^m}$ by writing $\sqrt[n]{x^m} = x^{m/n}$ and simplifying the exponent $\frac{m}{n}$.

Simplifying a radical expression includes writing the result with as small an index as possible.

Example 7 Simplifying Radical Expressions

Simplify. Assume $x \ge 0$.

1. $\sqrt[12]{(3x+7)^8}$ **2.** $\sqrt[10]{4}$ **3.** $\sqrt[4]{81x^6}$ **4.** $\sqrt[3]{\sqrt{t}}$

Solution

1.
$$\begin{aligned}
\sqrt[12]{(3x+7)^8} &= (3x+7)^{8/12} && \textit{Write in exponential form.}\\
&= (3x+7)^{2/3} && \textit{Simplify exponent.}\\
&= \sqrt[3]{(3x+7)^2} && \textit{Write in radical form.}
\end{aligned}$$

2.
$$\begin{aligned}
\sqrt[10]{4} &= \sqrt[10]{2^2} && 4 = 2^2\\
&= 2^{2/10} && \textit{Write in exponential form.}\\
&= 2^{1/5} && \textit{Simplify exponent.}\\
&= \sqrt[5]{2} && \textit{Write in radical form.}
\end{aligned}$$

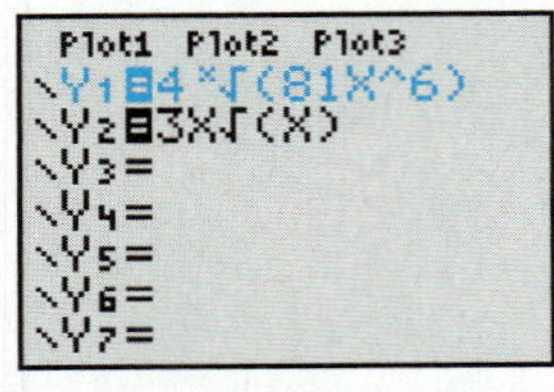

X	Y1	Y2
0	0	0
1	3	3
2	8.4853	8.4853
3	15.588	15.588
4	24	24
5	33.541	33.541
6	44.091	44.091

X=0

Figure 3 Verify the work

3. $\sqrt[4]{81x^6} = \sqrt[4]{81 \cdot x^4 \cdot x^2}$ *81 and x^4 are perfect 4th powers.*

$= \sqrt[4]{81}\sqrt[4]{x^4}\sqrt[4]{x^2}$ *$\sqrt[n]{ab} = \sqrt[n]{a}\sqrt[n]{b}$*

$= 3x\sqrt[4]{x^2}$ *$\sqrt[4]{81} = 3$; $\sqrt[4]{x^4} = x$ for $x \geq 0$*

$= 3x \cdot x^{2/4}$ *Write in exponential form.*

$= 3x \cdot x^{1/2}$ *Simplify exponent.*

$= 3x\sqrt{x}$ *Write in radical form.*

We verify our work by comparing graphing calculator tables for $y = \sqrt[4]{81x^6}$ and $y = 3x\sqrt{x}$ for $x \geq 0$ (see Fig. 3).

4. $\sqrt[3]{\sqrt{t}} = \left(t^{1/2}\right)^{1/3}$ *Write in exponential form.*

$= t^{\frac{1}{2}\cdot\frac{1}{3}}$ *Multiply exponents: $(b^m)^n = b^{mn}$*

$= t^{1/6}$ *Multiply numerators; multiply denominators.*

$= \sqrt[6]{t}$ *Write in radical form.*

Simplifying a Radical Expression

To simplify a radical expression with index n,

1. Find perfect nth-power factors of the radicand.
2. Apply the product property for radicals.
3. Find the nth root of each perfect nth power.
4. Write the radical with as small an index as possible.

Here we consider the expression $\sqrt[n]{x^n}$, where x is negative. First, we compare $\sqrt{x^2}$ for $x = -4$ and for $x = 4$:

$$\sqrt{(-4)^2} = \sqrt{16} = 4 \qquad \sqrt{4^2} = \sqrt{16} = 4$$

So, when -4 or 4 is substituted for x in the expression $\sqrt{x^2}$, we get the same result: 4. This is precisely what happens if we substitute -4 or 4 in the absolute value expression $|x|$:

$$|-4| = 4 \qquad |4| = 4$$

These examples suggest that $\sqrt{x^2} = |x|$. It turns out that $\sqrt[n]{x^n} = |x|$ for any even index n.

Here are some examples where n is odd:

$$\sqrt[3]{4^3} = \sqrt[3]{64} = 4 \qquad \sqrt[5]{2^5} = \sqrt[5]{32} = 2$$

$$\sqrt[3]{(-4)^3} = \sqrt[3]{-64} = -4 \qquad \sqrt[5]{(-2)^5} = \sqrt[5]{-32} = -2$$

These examples suggest that $\sqrt[n]{x^n} = x$ if n is odd.

Power Property for Radicals

Let n be a counting number greater than 1:

- If n is even, then $\sqrt[n]{x^n} = |x|$.
- If n is odd, then $\sqrt[n]{x^n} = x$.

For example, $\sqrt[8]{(-5)^8} = |-5| = 5$ and $\sqrt[7]{(-3)^7} = -3$.

Radical Functions

A **radical function** is a function whose equation contains a radical with a variable in the radicand. Here are some examples of radical functions:

$$f(x) = \sqrt[4]{x} \qquad g(x) = 8\sqrt{x} - 3 \qquad h(x) = 6\sqrt[3]{(x+7)^2} \qquad k(x) = \frac{\sqrt[3]{x} - 5}{\sqrt{x} + 1}$$

Example 8 Evaluating Radical Functions

For $f(x) = \sqrt[3]{2x - 1}$ and $g(x) = -3\sqrt{x} + 7$, find the following.

1. $f(14)$ **2.** $f(6)$ **3.** $g(16)$

Solution

1. $f(14) = \sqrt[3]{2(14) - 1} = \sqrt[3]{27} = 3$

2. $f(6) = \sqrt[3]{2(6) - 1} = \sqrt[3]{11}$

3. $g(16) = -3\sqrt{16} + 7 = -3(4) + 7 = -5$

A **square root function** is a radical function in which any radicals are square root radicals. For example, the function $h(x) = 9\sqrt{x - 3} + 4$ is a square root function. In Example 9, we will graph the square root function $f(x) = \sqrt{x}$.

Example 9 Graphing a Square Root Function

Sketch the graph of $f(x) = \sqrt{x}$.

Solution

We list some input–output pairs in Table 2. We choose perfect-square inputs because we can find their outputs mentally. Since the radicand of $\sqrt{x}$ must be nonnegative, we cannot choose any negative numbers as inputs. Then we sketch the graph of f (see Fig. 4).

Table 2 Input–Output Pairs of $f(x) = \sqrt{x}$

x	y
0	0
1	1
4	2
9	3
16	4

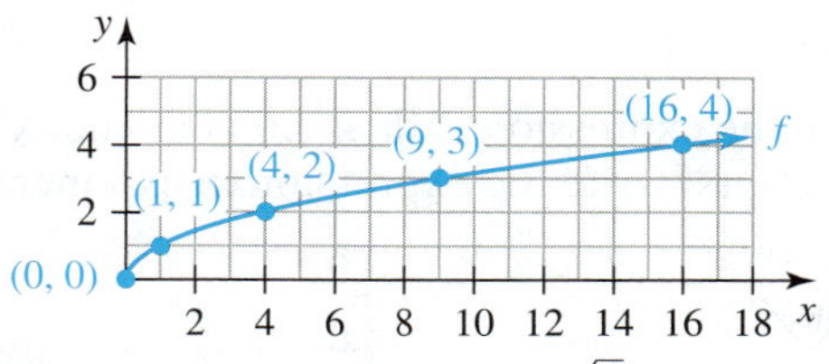

Figure 4 Graph of $f(x) = \sqrt{x}$

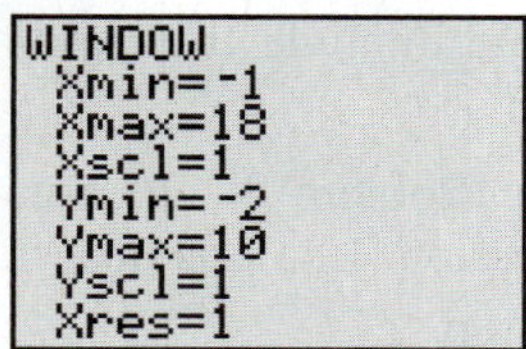

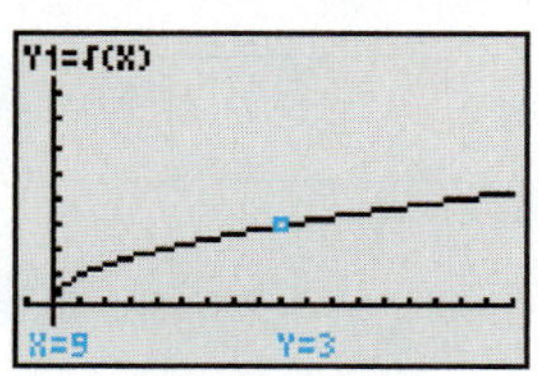

Figure 5 Verify the graph of $f(x) = \sqrt{x}$

We use a graphing calculator to verify our graph (see Fig. 5).

We use a graphing calculator to draw the graphs of $y = \sqrt{x}$, $y = \sqrt[3]{x}$, and $y = \sqrt[4]{x}$ (see Fig. 6).

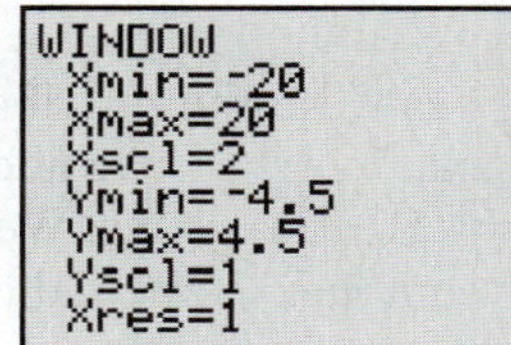

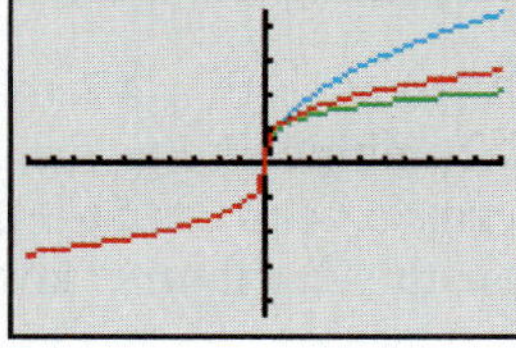
Figure 6 Graphs of $y = \sqrt{x}$, $y = \sqrt[3]{x}$, and $y = \sqrt[4]{x}$

Radical Model

A **radical model** is a radical function, or its graph, that describes an authentic situation. Radical functions can model a variety of situations, including the rise in temperature of an enclosed car, the period of a planet, the length of a braking car's skid marks, and the percentage of e-mail that is spam.

Table 3 Percentages of Foundations That Compensate All of Their Board Members

Asset Group (millions of dollars)	Asset Used to Represent Asset Group (millions of dollars)	Percent
0–5	2.5	4
5–10	7.5	8
10–25	17.5	10
25–50	37.5	15
50–100	75	20
100–250	175	21
250–500	375	32

Source: *New York Times*

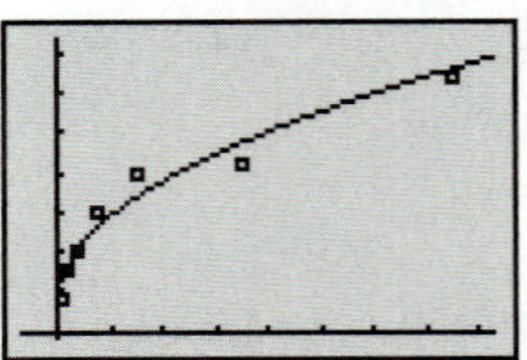

Figure 7 Check the fit

A **square root model** is a square root function, or its graph, that describes an authentic situation. We will work with such a model in Example 10.

Example 10 Using a Square Root Model to Make a Prediction

The percentages of foundations that compensate all of their board members are shown in Table 3 for various assets. Let $f(a)$ be the percentage of foundations with assets of a million dollars that compensate all of their board members. A model of the situation is $f(a) = 1.5\sqrt{a} + 4$.

1. Use a graphing calculator to draw the graph of f and, in the same viewing window, the scattergram of the data. Does the graph of f fit the data well?
2. Estimate the percentage of foundations with assets of \$30 million that compensate all of their board members.

Solution

1. The graph of f and the scattergram of the data are shown in Fig. 7. The model appears to fit the data well.
2. We evaluate f at 30:

$$f(30) = 1.5\sqrt{30} + 4 \approx 12.22$$

About 12.2% of foundations with assets of \$30 million compensate all of their board members, according to the model.

Group Exploration

Index property for radicals

Assume $x \geq 0$.

1. Write the expression with as small an index as possible. [**Hint:** First write the expression in exponential form.]
   a. $\sqrt[8]{x^6}$ b. $\sqrt[6]{x^4}$
   c. $\sqrt[16]{x^{10}}$ d. $\sqrt[9]{x^3}$
   e. $\sqrt[22]{x^4}$ f. $\sqrt[30]{x^{25}}$
2. Let k, m, and n be counting numbers, where m and n have no common factors. Write the expression $\sqrt[kn]{x^{km}}$ with as small an index as possible. Include the exponential form of the expression $\sqrt[kn]{x^{km}}$.
3. Describe what your result from Problem 2 tells you about simplifying a radical expression. Use this observation to simplify $\sqrt[20]{x^{16}}$ in one step.

Group Exploration

Looking ahead: Combining like radicals

Recall from Section 4.2 that we can use the distributive law to combine like terms. For example, here we add the like terms $5x$ and $3x$: $5x + 3x = (5 + 3)x = 8x$. We can also use the distributive law to "combine" some radicals.

1. Use the distributive law to perform the operations.
   a. $5\sqrt{x} + 3\sqrt{x}$ b. $9\sqrt{2} - 6\sqrt{2}$
   c. $4\sqrt{7} + \sqrt{7} - 2\sqrt{7}$
2. Can you use the distributive law to "combine" $4\sqrt{3}$ and $6\sqrt{7}$ in the sum $4\sqrt{3} + 6\sqrt{7}$? If yes, show how. If no, explain why not.
3. Can you use the distributive law to "combine" $8\sqrt{2}$ and $3\sqrt{5}$ in the difference $8\sqrt{2} - 3\sqrt{5}$? If yes, show how. If no, explain why not.
4. Describe in general when you can use the distributive law to "combine" radicals in a sum or difference of two radicals.
5. Use the distributive law to perform the operations. [**Hint:** To begin, simplify the radicals.]
   a. $\sqrt{8} + \sqrt{18}$
   b. $\sqrt{27} - \sqrt{12}$

Tips for Success Use 3-by-5 Cards

Do you have trouble memorizing definitions and properties? If so, try writing a word or phrase on one side of a 3-by-5 card. On the other side, put its definition or state a property and how it can be applied. For example, you could write "product property for radicals" on one side of a card and "$\sqrt[n]{ab} = \sqrt[n]{a}\sqrt[n]{b}$, where n is a counting number and $\sqrt[n]{a}$ and $\sqrt[n]{b}$ are defined" on the other side. You could also describe, in your own words, the meaning of the property and how you can apply it. Once you have completed a card for each definition and property, shuffle the cards and quiz yourself until you are confident you know the definitions and properties and how to apply them. Quiz yourself again later to make sure you have retained the information.

In addition to memorizing definitions and properties, it is important you continue to strive to understand their meanings and how to apply them.

Homework 13.1

For extra help ▶ MyMathLab®

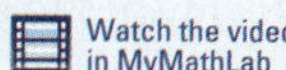

Watch the videos in MyMathLab

Download the MyDashboard App

If the expression is in exponential form, write it in radical form. If it is in radical form, write it in exponential form.

1. $x^{2/5}$ **2.** $x^{3/8}$ **3.** $\sqrt[4]{x^3}$

4. $\sqrt[9]{x^5}$ **5.** $\sqrt{w}$ **6.** $\sqrt[3]{t}$

7. $(2x + 9)^{3/7}$ **8.** $(5x + 1)^{5/6}$

9. $\sqrt[7]{(3k + 2)^4}$ **10.** $\sqrt[6]{8m + 3}$

Simplify. Assume each variable is nonnegative.

11. $\sqrt{50}$ **12.** $\sqrt{20}$ **13.** $\sqrt{x^8}$ **14.** $\sqrt{x^{18}}$

15. $\sqrt{36x^6}$ **16.** $\sqrt{4x^4}$ **17.** $\sqrt{5a^2b^{12}}$ **18.** $\sqrt{7a^{10}b^{14}}$

19. $\sqrt{x^9}$ **20.** $\sqrt{x^{15}}$ **21.** $\sqrt{24x^5}$ **22.** $\sqrt{12x^{13}}$

23. $\sqrt{80x^3y^8}$ **24.** $\sqrt{27x^{10}y^7}$ **25.** $\sqrt{200a^3b^5}$ **26.** $\sqrt{75a^{15}b^9}$

27. $\sqrt{(2x + 5)^8}$ **28.** $\sqrt{(3x + 4)^2}$ **29.** $\sqrt{(6t + 3)^5}$

30. $\sqrt{(7w + 1)^{13}}$ **31.** $\sqrt[3]{27}$ **32.** $\sqrt[5]{32}$

33. $\sqrt[6]{x^6}$ **34.** $\sqrt[9]{x^9}$ **35.** $\sqrt[3]{8x^3}$

36. $\sqrt[4]{16x^4}$ **37.** $\sqrt[5]{-32x^{20}}$ **38.** $\sqrt[3]{-27x^{18}}$

39. $\sqrt[4]{81a^{12}b^{28}}$ **40.** $\sqrt[5]{32a^{15}b^{30}}$ **41.** $\sqrt[5]{x^{17}}$

42. $\sqrt[8]{x^{25}}$ **43.** $\sqrt[3]{-125a^{17}b^{12}}$ **44.** $\sqrt[3]{-8a^{21}b^{29}}$

45. $\sqrt[5]{64x^{39}y^7}$ **46.** $\sqrt[4]{32x^{19}y^{13}}$ **47.** $\sqrt[5]{(6xy)^5}$

48. $\sqrt[7]{(4x^2y)^7}$ **49.** $\sqrt[4]{(3x + 6)^4}$ **50.** $\sqrt[8]{(5x + 2)^8}$

51. $\sqrt[5]{(4p + 7)^{20}}$ **52.** $\sqrt[3]{(3k + 5)^{12}}$

53. $\sqrt[6]{(2x + 9)^{31}}$ **54.** $\sqrt[5]{(4x + 5)^{43}}$

Simplify. (Write your result with as small an index n as possible.) Assume each variable is nonnegative.

55. $\sqrt[8]{x^6}$ **56.** $\sqrt[6]{x^3}$ **57.** $\sqrt[6]{x^4}$ **58.** $\sqrt[9]{x^6}$

59. $\sqrt[12]{(2m + 7)^{10}}$ **60.** $\sqrt[21]{(3r + 5)^{14}}$

61. $\sqrt[6]{x^{14}}$ **62.** $\sqrt[8]{x^{22}}$ **63.** $\sqrt[6]{27}$ **64.** $\sqrt[8]{25}$

65. $\sqrt[4]{\sqrt[3]{p}}$ **66.** $\sqrt[5]{\sqrt{t}}$ **67.** $\sqrt[10]{16x^8}$ **68.** $\sqrt[12]{125x^9}$

69. $\sqrt[4]{\sqrt{ab}}$ **70.** $\sqrt[3]{\sqrt[5]{3w}}$

For $f(x) = \sqrt[5]{x}$, $g(x) = \sqrt[3]{3x + 2}$, and $h(x) = 2\sqrt{x} - 5$, find the following.

71. $f(-32)$ **72.** $f(32)$ **73.** $g(2)$ **74.** $g(-22)$

75. $g(-7)$ **76.** $g(5)$ **77.** $h(49)$ **78.** $h(25)$

79. Graph $f(x) = 2\sqrt[3]{x}$ by hand. [**Hint:** Evaluate f at some perfect cubes.]

80. Graph $g(x) = 3\sqrt[4]{x}$ by hand. [**Hint:** Evaluate g at some perfect 4th powers.]

81. The average temperature rises above the *ambient temperature* (outside temperature) in an enclosed vehicle are shown in Table 4 for various elapsed times.

Table 4 Average Temperature Rises in an Enclosed Vehicle (for Ambient Temperatures between 72°F and 96°F)

Elapsed Time (minutes)	Average Temperature Rise (°F)
0	0
10	19
20	29
30	34
40	38
50	41
60	43

Source: *Jan Null, Department of Geosciences, San Francisco State University*

Let $f(t)$ be the average temperature rise (°F) in a vehicle at t minutes after the vehicle is enclosed. ("Cracking" the windows had little effect on the data.) A model of the situation is $f(t) = 8.5\sqrt[5]{t^2}$.

a. Use a graphing calculator to draw the graph of the model and, in the same viewing window, the scattergram of the data. Does the model fit the data well?

b. Estimate the average temperature rise 24 minutes after the vehicle is enclosed if the ambient temperature is 85°F.

c. Estimate the *temperature* inside the car 45 minutes after the vehicle is enclosed if the ambient temperature is 90°F.

d. Since 1998, more than 300 children have died of hyperthermia after being left inside a hot vehicle. A body core temperature of 107°F is usually fatal. Use TRACE to estimate how long it would take for the temperature in an enclosed vehicle to reach 107°F if the ambient temperature is 80°F.

82. The *Beaufort wind scale,* which ranges from 0 to 12, describes wind intensities. Some of the even-numbered *Beaufort numbers* and the corresponding wind speeds and weather conditions are shown in Table 5.

Table 5 Beaufort Numbers, Wind Speeds, and Weather Conditions

Beaufort Number	Wind Speed Group (miles per hour)	Wind Speed Used to Represent Wind Speed Group (miles per hour)	Weather Conditions
0	0	0	calm
2	4–7	5.5	light breeze
4	13–18	15.5	moderate breeze
6	25–31	28.0	strong breeze
8	39–46	42.5	gale
10	55–63	59.0	storm

Source: Stormfax® Weather Almanac

Let $V = f(B)$ be the wind speed (in miles per hour) that corresponds to the Beaufort number B. A model of the situation is $f(B) = 1.87\sqrt{B^3}$.

a. Use a graphing calculator to draw the graph of the model and, in the same viewing window, the scattergram of the data. Does the model fit the data well?

b. The U.S. Coast Guard issues a small-craft advisory when the Beaufort number is 6 or 7. Estimate the wind speed of a near gale, which has Beaufort number 7.

c. Estimate the wind speed in a violent storm, which has Beaufort number 11.

d. The quadratic function $Q(B) = 0.33B^2 + 2.72B - 0.46$ is a model of the situation. Does Q fit the data well? Which model, f or Q, describes the wind speed better when the weather is calm? Explain.

83. A *tsunami* is a fast-moving sea wave typically caused by an underwater earthquake. An earthquake in the Indian Ocean in 2004 caused tsunamis that killed more than 265,000 people. The speed (in meters per second) $f(d)$ of a tsunami can be modeled by the function $f(d) = \sqrt{gd}$, where g is a constant approximately equal to 9.8 and d is the average depth (in meters) of the water.

a. The Indian Ocean has an average depth of 3890 meters. Estimate the speed in meters per second of a tsunami in the Indian Ocean.

b. Find $f(1000)$, $f(2000)$, and $f(3000)$. Is f an increasing or a decreasing function? What does that mean in this situation?

c. As a tsunami slows down, it gets taller. As a tsunami approaches the shore, what happens to the speed and height of the tsunami?

d. Convert your result in part (a) to units of miles per hour. (One mile is approximately 1609 meters.)

84. Suppose a driver traveling on a dry road slams on the brakes. Let D be the distance (in feet) that the car will skid and S be the speed (in miles per hour) before braking. The relationship between D and S is described by the model $S = \sqrt{30FD}$, where F is a drag factor—a measure of the roughness of the road surface. The drag factor on new concrete is 0.95; on polished concrete or asphalt, it is 0.75.

a. A motorist involved in an accident claims that he was driving at the speed limit, 50 miles per hour. The car's skid marks on the new concrete road are 120 feet long. Assuming the motorist applied the brakes suddenly, estimate his speed before braking.

b. If the motorist first applied the brakes lightly, then forcefully applied them after a few seconds, explain why your result in part (a) is an underestimate of his speed before braking.

Concepts

85. A student says $\sqrt{x^{16}}$ is equal to x^4, since $\sqrt{16} = 4$. What would you tell this student?

86. Is the statement $\sqrt[n]{a+b} = \sqrt[n]{a} + \sqrt[n]{b}$ true or false for all values of a and b where $\sqrt[n]{a}$ and $\sqrt[n]{b}$ are defined? [**Hint:** Substitute values of a, b, and n in the two expressions and compare the results.]

87. A student tries to solve $\sqrt[6]{x^3}$:

$$\sqrt[6]{x^3} = x^{6/3}$$
$$= x^2$$

Describe any errors. Then simplify the expression correctly.

88. A student tries to simplify $\sqrt[3]{(a+b)^3}$:

$$\sqrt[3]{(a+b)^3} = \sqrt[3]{a^3 + b^3}$$
$$= \sqrt[3]{a^3} + \sqrt[3]{b^3}$$
$$= a + b$$

Describe any errors. Then simplify the expression correctly.

89. Write the expression $\sqrt[n]{\sqrt[n]{x}}$ with one radical sign. [**Hint:** Write the expression in exponential form.]

90. a. Use a graphing calculator to draw a graph of the given function. What do you notice about the graph?

i. $y = \sqrt[3]{x^3}$

ii. $y = \sqrt[5]{x^5}$

iii. $y = \sqrt[7]{x^7}$

b. Compare your graphs in part (a) with the graph of $y = x$. Explain how your observation relates to the power property for radicals.

c. Use a graphing calculator to draw a graph of the given function. What do you notice about the graph?

i. $y = \sqrt{x^2}$ **ii.** $y = \sqrt[4]{x^4}$ **iii.** $y = \sqrt[6]{x^6}$

d. Compare your graphs in part (c) with the graph of $y = |x|$. Explain how your observation relates to the power property for radicals. [***Graphing Calculator:*** The absolute value choice "abs(" is located in the "NUM" menu. ("NUM" is in the "MATH" menu.)]

91. In this exercise, you will explore another version of the power property. Simplify.

a. $(\sqrt{x})^2$ **b.** $(\sqrt[3]{x})^3$ **c.** $(\sqrt[n]{x})^n$

92. Describe how to simplify a radical expression. (See page 9 for guidelines on writing a good response.)

Related Review

Assume each variable is nonnegative.

93. a. Simplify $\sqrt{16x^4y^6}$ by using the product property for radicals.

b. Simplify $\sqrt{16x^4y^6}$ first by writing the expression in exponential form, then by using properties of exponents.

c. Compare the results that you found in parts (a) and (b).

94. a. Simplify $\sqrt[3]{8x^6y^9}$ by using the product property for radicals.

b. Simplify $\sqrt[3]{8x^6y^9}$ first by writing the expression in exponential form, then by using properties of exponents.

c. Compare the results you found in parts (a) and (b).

Expressions, Equations, Functions, and Graphs

Perform the indicated instruction. Then use words such as linear, quadratic, cubic, exponential, logarithmic, rational, radical, polynomial, degree, function, one variable, *and* two variables *to describe the expression, equation, or system.*

95. Solve $\dfrac{2x}{x^2 + x - 6} - \dfrac{3x - 1}{x^2 + 6x + 9} = \dfrac{-3}{x + 3}$.

96. Solve $2x^2 - 4x - 3 = 0$ by completing the square.

97. Find the difference $\dfrac{2x}{x^2 + x - 6} - \dfrac{3x - 1}{x^2 + 6x + 9}$.

98. Graph $f(x) = 2x^2 - 4x - 3$ by hand.

99. Find the domain of $f(x) = \dfrac{2x}{x^2 + x - 6}$.

100. Solve $4(2x - 5)^2 = 48$.

13.2 Adding, Subtracting, and Multiplying Radical Expressions

Objectives

- Add, subtract, and multiply radical expressions.
- Know another version of the power property for radicals.
- Simplify the square of a radical expression with two terms.

In this section, we add, subtract, and multiply radical expressions. We also simplify the square of a radical with two terms.

Adding and Subtracting Radical Expressions

Recall from Section 4.2 that we can use the distributive law to add like terms, such as $2x$ and $7x$:

$$2x + 7x = (2 + 7)x = 9x$$

How do we add (or subtract) radical expressions? We can again use the distributive law if the radicals are like radicals. We say $2\sqrt[3]{x}$ and $7\sqrt[3]{x}$ are like radicals, because they have the same index *and* the same radicand. In general, radicals that have the same index and the same radicand are called **like radicals.**

We add the like radicals $2\sqrt[3]{x}$ and $7\sqrt[3]{x}$ as follows:

$$2\sqrt[3]{x} + 7\sqrt[3]{x} = (2 + 7)\sqrt[3]{x} = 9\sqrt[3]{x}$$

To add or subtract like radicals, we use the distributive law. When we add or subtract like radicals, we say we *combine like radicals.*

Example 1 Combining Like Radicals

Combine like radicals.

1. $3\sqrt{x} + 6\sqrt{x}$ **2.** $8\sqrt[4]{x} - 5\sqrt[4]{x}$ **3.** $4\sqrt[5]{3xy^2} - 2\sqrt[5]{3xy^2}$

4. $4\sqrt[3]{x} + 5\sqrt[6]{x}$ **5.** $3\sqrt[4]{x} - 2\sqrt[4]{x+1}$

Solution

1. $3\sqrt{x} + 6\sqrt{x} = (3 + 6)\sqrt{x}$ *Distributive law*

$= 9\sqrt{x}$ *Add.*

2. $8\sqrt[4]{x} - 5\sqrt[4]{x} = (8 - 5)\sqrt[4]{x}$ *Distributive law*

$= 3\sqrt[4]{x}$ *Subtract.*

3. $4\sqrt[5]{3xy^2} - 2\sqrt[5]{3xy^2} = (4 - 2)\sqrt[5]{3xy^2}$ *Distributive law*

$= 2\sqrt[5]{3xy^2}$ *Subtract.*

4. Since the radicals $4\sqrt[3]{x}$ and $5\sqrt[6]{x}$ have different indexes, we cannot use the distributive law. The expression $4\sqrt[3]{x} + 5\sqrt[6]{x}$ is already in simplified form.

5. Since the radicals $3\sqrt[4]{x}$ and $2\sqrt[4]{x+1}$ have different radicands, we cannot use the distributive law. The expression $3\sqrt[4]{x} - 2\sqrt[4]{x+1}$ is already in simplified form.

Example 2 Performing Operations with Radical Expressions

Perform the indicated operations.

1. $3\sqrt[4]{x} + 4\sqrt{x} + 2\sqrt[4]{x} + 7\sqrt{x}$ **2.** $3(5\sqrt[3]{x+1} - 2) - 4\sqrt[3]{x+1}$

Solution

1.
$$\begin{aligned} 3\sqrt[4]{x} + 4\sqrt{x} + 2\sqrt[4]{x} + 7\sqrt{x} &= (4\sqrt{x} + 7\sqrt{x}) + (3\sqrt[4]{x} + 2\sqrt[4]{x}) && \text{Group like radicals.} \\ &= (4+7)\sqrt{x} + (3+2)\sqrt[4]{x} && \text{Distributive law} \\ &= 11\sqrt{x} + 5\sqrt[4]{x} && \text{Add.} \end{aligned}$$

2.
$$\begin{aligned} 3(5\sqrt[3]{x+1} - 2) - 4\sqrt[3]{x+1} &= 3 \cdot 5\sqrt[3]{x+1} - 3 \cdot 2 - 4\sqrt[3]{x+1} && \text{Distributive law} \\ &= 15\sqrt[3]{x+1} - 4\sqrt[3]{x+1} - 6 && \text{Group like radicals.} \\ &= (15-4)\sqrt[3]{x+1} - 6 && \text{Distributive law} \\ &= 11\sqrt[3]{x+1} - 6 && \text{Subtract.} \end{aligned}$$

Sometimes, simplifying radicals will allow us to combine like radicals.

Example 3 Adding or Subtracting Radical Expressions

Perform the indicated operation.

1. $\sqrt{8} + 5\sqrt{2}$ **2.** $\sqrt{45w} + \sqrt{20w}$ **3.** $5b\sqrt{3a^3} - a\sqrt{12ab^2}$

Solution

1.
$$\begin{aligned} \sqrt{8} + 5\sqrt{2} &= \sqrt{4 \cdot 2} + 5\sqrt{2} && \text{4 is a perfect square.} \\ &= \sqrt{4}\sqrt{2} + 5\sqrt{2} && \sqrt{ab} = \sqrt{a}\sqrt{b} \\ &= 2\sqrt{2} + 5\sqrt{2} && \sqrt{4} = 2 \\ &= (2+5)\sqrt{2} && \text{Combine like radicals.} \\ &= 7\sqrt{2} && \text{Add.} \end{aligned}$$

2.
$$\begin{aligned} \sqrt{45w} + \sqrt{20w} &= \sqrt{9 \cdot 5w} + \sqrt{4 \cdot 5w} && \text{9 and 4 are perfect squares.} \\ &= \sqrt{9}\sqrt{5w} + \sqrt{4}\sqrt{5w} && \sqrt[n]{ab} = \sqrt[n]{a}\sqrt[n]{b} \\ &= 3\sqrt{5w} + 2\sqrt{5w} && \sqrt{9} = 3; \sqrt{4} = 2 \\ &= (3+2)\sqrt{5w} && \text{Distributive law} \\ &= 5\sqrt{5w} && \text{Add.} \end{aligned}$$

3.
$$\begin{aligned} 5b\sqrt{3a^3} - a\sqrt{12ab^2} &= 5b\sqrt{a^2 \cdot 3a} - a\sqrt{4 \cdot b^2 \cdot 3a} && a^2\text{, 4, and } b^2 \text{ are perfect squares.} \\ &= 5b\sqrt{a^2}\sqrt{3a} - a\sqrt{4}\sqrt{b^2}\sqrt{3a} && \sqrt[n]{ab} = \sqrt[n]{a}\sqrt[n]{b} \\ &= 5b \cdot a \cdot \sqrt{3a} - a \cdot 2 \cdot b \cdot \sqrt{3a} && \sqrt{a^2} = a \text{ for } a \ge 0; \sqrt{4} = 2; \sqrt{b^2} = b \text{ for } b \ge 0 \\ &= 5ab\sqrt{3a} - 2ab\sqrt{3a} && \text{Rearrange factors.} \\ &= (5ab - 2ab)\sqrt{3a} && \text{Distributive law} \\ &= 3ab\sqrt{3a} && \text{Combine like terms.} \end{aligned}$$

Example 4 Subtracting Radical Expressions

Find the difference $2\sqrt[3]{16x^4} - 4x\sqrt[3]{54x}$.

Solution

$$\begin{aligned} 2\sqrt[3]{16x^4} - 4x\sqrt[3]{54x} &= 2\sqrt[3]{8 \cdot x^3 \cdot 2x} - 4x\sqrt[3]{27 \cdot 2x} && 8, x^3\text{, and 27 are perfect cubes.} \\ &= 2\sqrt[3]{8}\sqrt[3]{x^3}\sqrt[3]{2x} - 4x\sqrt[3]{27}\sqrt[3]{2x} && \sqrt[n]{ab} = \sqrt[n]{a}\sqrt[n]{b} \end{aligned}$$

$$
\begin{aligned}
&= 2 \cdot 2 \cdot x \cdot \sqrt[3]{2x} - 4x \cdot 3 \cdot \sqrt[3]{2x} && \sqrt[3]{8} = 2, \sqrt[3]{x^3} = x, \sqrt[3]{27} = 3 \\
&= 4x\sqrt[3]{2x} - 12x\sqrt[3]{2x} && \text{Multiply.} \\
&= (4x - 12x)\sqrt[3]{2x} && \text{Distributive law} \\
&= -8x\sqrt[3]{2x} && \text{Combine like terms.}
\end{aligned}
$$

Multiplying Radical Expressions

Next, we multiply radical expressions. We will use the product property

$$\sqrt[n]{ab} = \sqrt[n]{a}\sqrt[n]{b}, \qquad \text{where } \sqrt[n]{a} \text{ and } \sqrt[n]{b} \text{ are defined}$$

Here we multiply $\sqrt{2}$ and $\sqrt{3}$:

$$\sqrt{2} \cdot \sqrt{3} = \sqrt{2 \cdot 3} = \sqrt{6}$$

It is good practice to check whether the product of radical expressions can be simplified.

Example 5 Finding Products of Radical Expressions

Find the product.

1. $5\sqrt{3} \cdot \sqrt{6}$ **2.** $2\sqrt{6x} \cdot 5\sqrt{2x}$

Solution

1.
$$
\begin{aligned}
5\sqrt{3} \cdot \sqrt{6} &= 5\sqrt{18} && \sqrt{a}\sqrt{b} = \sqrt{ab} \\
&= 5\sqrt{9 \cdot 2} && \text{9 is a perfect square.} \\
&= 5\sqrt{9}\sqrt{2} && \sqrt{ab} = \sqrt{a}\sqrt{b} \\
&= 5 \cdot 3\sqrt{2} && \sqrt{9} = 3 \\
&= 15\sqrt{2} && \text{Multiply.}
\end{aligned}
$$

2.
$$
\begin{aligned}
2\sqrt{6x} \cdot 5\sqrt{2x} &= 2 \cdot 5\sqrt{6x} \cdot \sqrt{2x} && \text{Rearrange factors.} \\
&= 2 \cdot 5 \cdot \sqrt{6x \cdot 2x} && \text{Product property} \\
&= 10 \cdot \sqrt{12x^2} && \text{Multiply.} \\
&= 10 \cdot \sqrt{4 \cdot x^2 \cdot 3} && \text{4 and } x^2 \text{ are perfect squares.} \\
&= 10 \cdot 2x\sqrt{3} && \sqrt{4} = 2, \sqrt{x^2} = x \text{ for } x \ge 0 \\
&= 20x\sqrt{3} && \text{Multiply.}
\end{aligned}
$$

X	Y1	Y2
0	0	0
1	34.641	34.641
2	69.282	69.282
3	103.92	103.92
4	138.56	138.56
5	173.21	173.21
6	207.85	207.85

X=0

Figure 8 Verify the work

We verify our work by comparing graphing calculator tables for $y = 2\sqrt{6x} \cdot 5\sqrt{2x}$ and $y = 20x\sqrt{3}$, for $x \ge 0$ (see Fig. 8).

Example 6 Finding the Product of Radical Expressions

Find the product.

1. $\sqrt{5}(3 - \sqrt{2})$ **2.** $3\sqrt{5x}(4\sqrt{x} - \sqrt{5})$

Solution

1.
$$
\begin{aligned}
\sqrt{5}(3 - \sqrt{2}) &= \sqrt{5} \cdot 3 - \sqrt{5} \cdot \sqrt{2} && \text{Distributive law} \\
&= 3\sqrt{5} - \sqrt{10} && \text{Rearrange factors; } \sqrt{a}\sqrt{b} = \sqrt{ab}
\end{aligned}
$$

2.
$$
\begin{aligned}
3\sqrt{5x}(4\sqrt{x} - \sqrt{5}) &= 3\sqrt{5x} \cdot 4\sqrt{x} - 3\sqrt{5x} \cdot \sqrt{5} && \text{Distributive law} \\
&= 3 \cdot 4 \cdot \sqrt{5x}\sqrt{x} - 3\sqrt{5x}\sqrt{5} && \text{Rearrange factors.} \\
&= 12\sqrt{5x \cdot x} - 3\sqrt{5x \cdot 5} && \text{Product property} \\
&= 12\sqrt{x^2 \cdot 5} - 3\sqrt{25x} && x^2 \text{ and 25 are perfect squares.} \\
&= 12 \cdot x \cdot \sqrt{5} - 3 \cdot 5 \cdot \sqrt{x} && \sqrt{x^2} = x \text{ for } x \ge 0, \sqrt{25} = 5 \\
&= 12x\sqrt{5} - 15\sqrt{x} && \text{Multiply.}
\end{aligned}
$$

Next, we find the product of two radical expressions for which each factor has two terms.

Example 7 Finding the Product of Radical Expressions

Find the product.

1. $(5 - \sqrt{3})(4 + \sqrt{3})$ **2.** $(2\sqrt{x} - 7)(3\sqrt{x} + 4)$

Solution

1. We begin by multiplying each term in the first radical expression by each term in the second radical expression:

$$\begin{aligned}(5 - \sqrt{3})(4 + \sqrt{3}) &= 5 \cdot 4 + 5 \cdot \sqrt{3} - \sqrt{3} \cdot 4 - \sqrt{3} \cdot \sqrt{3} && \text{Multiply pairs of terms.}\\ &= 20 + 5\sqrt{3} - 4\sqrt{3} - 3 && \sqrt{x}\sqrt{x} = x \text{ for } x \ge 0\\ &= 17 + \sqrt{3} && \text{Combine like radicals.}\end{aligned}$$

2. Again, we begin by multiplying each term in the first radical expression by each term in the second radical expression:

$$\begin{aligned}(2\sqrt{x} - 7)(3\sqrt{x} + 4) &= 2\sqrt{x} \cdot 3\sqrt{x} + 2\sqrt{x} \cdot 4 - 7 \cdot 3\sqrt{x} - 7 \cdot 4 && \text{Multiply pairs of terms.}\\ &= 6\sqrt{x^2} + 8\sqrt{x} - 21\sqrt{x} - 28 && \text{Simplify.}\\ &= 6x - 13\sqrt{x} - 28 && \sqrt{x^2} = x \text{ for } x \ge 0; \text{ combine like radicals.}\end{aligned}$$

Product of Two Radical Expressions, Each with Two Terms

To find the product of two radical expressions in which each factor has two terms,

1. Multiply each term in the first radical expression by each term in the second radical expression.
2. Combine like radicals.

Note that if $\sqrt[n]{x}$ is defined, then

$$\left(\sqrt[n]{x}\right)^n = x^{n/n} = x^1 = x$$

This property is helpful in simplifying powers or products of radical expressions.

Another Version of the Power Property for Radicals

If $\sqrt[n]{x}$ is defined, then

$$\left(\sqrt[n]{x}\right)^n = x$$

In words, the nth power of the nth root of a number is that number.

In particular, we have $\left(\sqrt{x}\right)^2 = x$ if $x \ge 0$.

Example 8 Finding the Product of Two Radical Expressions

Find the product $(2\sqrt{3x} + 5)(2\sqrt{3x} - 5)$.

Solution

We use the property $(A + B)(A - B) = A^2 - B^2$:

$$\begin{aligned}(2\sqrt{3x} + 5)(2\sqrt{3x} - 5) &= \left(2\sqrt{3x}\right)^2 - 5^2 && (A + B)(A - B) = A^2 - B^2\\ &= 2^2\left(\sqrt{3x}\right)^2 - 5^2 && (AB)^2 = A^2B^2\end{aligned}$$

$$= 4(3x) - 25 \quad (\sqrt{x})^2 = x \text{ for } x \geq 0$$
$$= 12x - 25 \quad \text{Multiply.}$$

We can verify our result by comparing tables for $y = (2\sqrt{3x} + 5)(2\sqrt{3x} - 5)$ and $y = 12x - 25$ for $x \geq 0$.

Example 9 Simplifying the Square of a Radical Expression with Two Terms

Simplify $(x - \sqrt{3})^2$.

Solution

To begin, we substitute x for A and $\sqrt{3}$ for B in the property for the square of a difference:

$$(A - B)^2 = A^2 - 2\,A\,B + B^2$$
$$\downarrow \quad \downarrow \qquad \downarrow \quad \downarrow\ \downarrow\ \downarrow \qquad \downarrow$$
$$(x - \sqrt{3})^2 = x^2 - 2(x)\sqrt{3} + (\sqrt{3})^2 \quad \text{Substitute } x \text{ for } A \text{ and } \sqrt{3} \text{ for } B.$$
$$= x^2 - 2x\sqrt{3} + 3 \quad (\sqrt{x})^2 = x \text{ for } x \geq 0$$

Another way to simplify $(x - \sqrt{3})^2$ is to use the fact that $C^2 = CC$, multiply pairs of terms, and combine like radicals:

$$(x - \sqrt{3})^2 = (x - \sqrt{3})(x - \sqrt{3}) \quad C^2 = CC$$
$$= x^2 - x\sqrt{3} - x\sqrt{3} + \sqrt{3}\sqrt{3} \quad \text{Multiply pairs of terms.}$$
$$= x^2 - 2x\sqrt{3} + 3 \quad \text{Combine like radicals; } \sqrt{x}\sqrt{x} = x \text{ for } x \geq 0$$

Simplifying the Square of a Radical Expression

To simplify the square of a radical expression with two terms,

- Use the square-of-a-sum property $(A + B)^2 = A^2 + 2AB + B^2$ or the square-of-a-difference property $(A - B)^2 = A^2 - 2AB + B^2$,

or

- Use $C^2 = CC$ to write the square as a product of two identical expressions, and then multiply each term in the first radical expression by each term in the second radical expression.

Example 10 Simplifying the Square of a Radical Expression with Two Terms

Simplify $(\sqrt{a} + 3\sqrt{b})^2$.

Solution

We use the property for the square of a sum:

$$(\sqrt{a} + 3\sqrt{b})^2 = (\sqrt{a})^2 + 2 \cdot \sqrt{a} \cdot 3\sqrt{b} + (3\sqrt{b})^2 \quad (A + B)^2 = A^2 + 2AB + B^2$$
$$= (\sqrt{a})^2 + 6\sqrt{ab} + 3^2(\sqrt{b})^2 \quad \sqrt[n]{a}\sqrt[n]{b} = \sqrt[n]{ab};\ (xy)^n = x^n y^n$$
$$= a + 6\sqrt{ab} + 9b \quad (\sqrt{x})^2 = x \text{ for } x \geq 0;\ 3^2 = 9$$

WARNING When we simplify $(x + k)^2$, it is important to remember the middle term of $x^2 + 2kx + k^2$. Likewise, when we simplify $(x - k)^2$, it is important to remember the middle term of $x^2 - 2kx + k^2$. Do not make the following typical error in simplifying $(\sqrt{x} + \sqrt{5})^2$:

$$(\sqrt{x} + \sqrt{5})^2 = (\sqrt{x})^2 + (\sqrt{5})^2 = x + 5 \quad \text{Incorrect}$$
$$(\sqrt{x} + \sqrt{5})^2 = (\sqrt{x})^2 + 2\sqrt{5}\sqrt{x} + (\sqrt{5})^2 = x + 2\sqrt{5x} + 5 \quad \text{Correct}$$

In Example 11, we will find products of radical expressions with indexes other than $n = 2$.

Example 11 Multiplying Radical Expressions

Find the product.

1. $(2\sqrt[5]{x^2})(7\sqrt[5]{x^4})$ **2.** $(\sqrt[4]{x^3} + 5)(\sqrt[4]{x^3} - 6)$

Solution

1.
$$\begin{aligned}(2\sqrt[5]{x^2})(7\sqrt[5]{x^4}) &= 2 \cdot 7 \cdot \sqrt[5]{x^2}\sqrt[5]{x^4} && \text{Rearrange factors.}\\ &= 2 \cdot 7\sqrt[5]{x^2 \cdot x^4} && \sqrt[n]{a}\sqrt[n]{b} = \sqrt[n]{ab}\\ &= 14\sqrt[5]{x^6} && b^m b^n = b^{m+n}\\ &= 14x\sqrt[5]{x} && \sqrt[5]{x^6} = \sqrt[5]{x^5 x^1} = x\sqrt[5]{x}\end{aligned}$$

2. $(\sqrt[4]{x^3} + 5)(\sqrt[4]{x^3} - 6)$

$$\begin{aligned}&= \sqrt[4]{x^3}\sqrt[4]{x^3} - \sqrt[4]{x^3} \cdot 6 + 5 \cdot \sqrt[4]{x^3} - 5 \cdot 6 && \text{Multiply pairs of terms.}\\ &= \sqrt[4]{x^3 \cdot x^3} - 6\sqrt[4]{x^3} + 5\sqrt[4]{x^3} - 30 && \text{Product property}\\ &= \sqrt[4]{x^6} - \sqrt[4]{x^3} - 30 && \text{Multiply; combine like radicals.}\\ &= x\sqrt[4]{x^2} - \sqrt[4]{x^3} - 30 && \sqrt[4]{x^6} = \sqrt[4]{x^4 \cdot x^2} = x\sqrt[4]{x^2} \text{ for } x \ge 0\\ &= x\sqrt{x} - \sqrt[4]{x^3} - 30 && \sqrt[4]{x^2} = x^{2/4} = x^{1/2} = \sqrt{x} \text{ for } x \ge 0\end{aligned}$$

We cannot combine the radicals $x\sqrt{x}$ and $-\sqrt[4]{x^3}$, since the indexes (and the radicands) are different. So, we are done.

To multiply two radicals that have the same index, we use the product property. How do we multiply two radicals with *different* indexes? Here we find the product $\sqrt[3]{x} \cdot \sqrt[4]{x}$:

$$\begin{aligned}\sqrt[3]{x} \cdot \sqrt[4]{x} &= x^{\frac{1}{3}} \cdot x^{\frac{1}{4}} && \text{Write in exponential form.}\\ &= x^{\frac{1}{3} + \frac{1}{4}} && a^m a^n = a^{m+n}\\ &= x^{\frac{4}{12} + \frac{3}{12}} && \text{Get a common denominator.}\\ &= x^{\frac{7}{12}} && \text{Add numerators; keep common denominator.}\\ &= \sqrt[12]{x^7} && \text{Write in radical form.}\end{aligned}$$

Multiplying Two Radicals That Have Different Indexes but the Same Radicand

To multiply two radicals that have different indexes but the same radicand,

1. Write the radicals in exponential form.

2. Use exponential properties to simplify the expression involving exponents.

3. Write the simplified expression in radical form.

Example 12 Simplifying Radical Expressions

Perform the operations. Assume $x \ge 0$.

1. $2\sqrt{x}(\sqrt[3]{x} - 5)$ **2.** $(\sqrt[3]{x} + 3\sqrt[5]{x^2})^2$

X	Y1	Y2
0	0	0
1	-8	-8
2	-10.58	-10.58
3	-12.32	-12.32
4	-13.65	-13.65
5	-14.71	-14.71
6	-15.59	-15.59
X=0		

Figure 9 Verify the work

Solution

1.
$$
\begin{aligned}
2\sqrt{x}\left(\sqrt[3]{x}-5\right) &= 2\sqrt{x}\sqrt[3]{x} - 2\sqrt{x}\cdot 5 && \text{Distributive law}\\
&= 2x^{\frac{1}{2}}x^{\frac{1}{3}} - 10\sqrt{x} && \text{Write in exponential form.}\\
&= 2x^{\frac{1}{2}+\frac{1}{3}} - 10\sqrt{x} && a^m a^n = a^{m+n}\\
&= 2x^{\frac{3}{6}+\frac{2}{6}} - 10\sqrt{x} && \text{Get a common denominator.}\\
&= 2x^{\frac{5}{6}} - 10\sqrt{x} && \text{Add numerators; keep common denominator.}\\
&= 2\sqrt[6]{x^5} - 10\sqrt{x} && \text{Write in radical form.}
\end{aligned}
$$

We can verify our result by comparing tables for $y = 2\sqrt{x}\left(\sqrt[3]{x}-5\right)$ and $y = 2\sqrt[6]{x^5} - 10\sqrt{x}$, for $x \ge 0$ (see Fig. 9).

2. $\left(\sqrt[3]{x} + 3\sqrt[5]{x^2}\right)^2$

$$
\begin{aligned}
&= \left(\sqrt[3]{x}\right)^2 + 2\left(\sqrt[3]{x}\right)\left(3\sqrt[5]{x^2}\right) + \left(3\sqrt[5]{x^2}\right)^2 && (A+B)^2 = A^2 + 2AB + B^2\\
&= \left(\sqrt[3]{x}\right)^2 + 6x^{\frac{1}{3}}x^{\frac{2}{5}} + 3^2\left(\sqrt[5]{x^2}\right)^2 && \text{Write in exponential form.}\\
&= \sqrt[3]{x^2} + 6x^{\frac{1}{3}+\frac{2}{5}} + 9\sqrt[5]{\left(x^2\right)^2} && \left(\sqrt[n]{x}\right)^m = \sqrt[n]{x^m}\\
&= \sqrt[3]{x^2} + 6x^{\frac{5}{15}+\frac{6}{15}} + 9\sqrt[5]{x^4} && \text{Get a common denominator.}\\
&= \sqrt[3]{x^2} + 6x^{\frac{11}{15}} + 9\sqrt[5]{x^4} && \text{Add numerators; keep common denominator.}\\
&= \sqrt[3]{x^2} + 6\sqrt[15]{x^{11}} + 9\sqrt[5]{x^4} && \text{Write in radical form.}
\end{aligned}
$$

Another way to simplify $\left(\sqrt[3]{x} + 3\sqrt[5]{x^2}\right)^2$ is to use the fact that $C^2 = CC$ and multiply pairs of terms. (Try it.)

Simplifying a Radical Expression

To simplify a radical expression,

1. Perform any indicated multiplications.
2. Combine like radicals.
3. For any radical with index n, write the radicand as a product of one or more perfect nth powers and another expression that has no factors that are perfect nth powers. Then apply the product property for radicals.
4. Write any radicals with as small an index as possible.

Depending on the radical expression, we may need to perform these steps in a different order or return to a step at a later stage in the process of simplifying the expression. We will discuss more ways to simplify radical expressions in Section 13.3.

Group Exploration

Looking ahead: Rationalizing the denominator

In Section 9.3, you "rationalized the denominator" of fractions of the form $\frac{1}{\sqrt{a}}$ by finding an equivalent expression that does not have a radical in any denominator. Here you will explore how to rationalize the denominator of a fraction with a denominator that is a sum or a difference involving radicals.

1. Perform the indicated multiplication.
   a. $\left(x-\sqrt{2}\right)\left(x+\sqrt{2}\right)$ **b.** $\left(x+\sqrt{5}\right)\left(x-\sqrt{5}\right)$
   c. $\left(\sqrt{x}-4\right)\left(\sqrt{x}+4\right)$ **d.** $\left(\sqrt{x}+3\right)\left(\sqrt{x}-3\right)$
2. What patterns do you notice from your work in Problem 1?

3. Rationalize the denominator of $\dfrac{1}{\sqrt{x} - 7}$ by performing the multiplication

$$\frac{1}{\sqrt{x} - 7} \cdot \frac{\sqrt{x} + 7}{\sqrt{x} + 7}$$

Use graphing calculator tables to verify your work.

4. Rationalize the denominator of the expression $\dfrac{1}{\sqrt{x} + 5}$.

5. Describe how to rationalize the denominator of a radical expression.

Homework 13.2

For extra help ▶ MyMathLab®
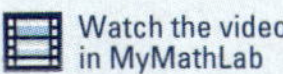
Watch the videos in MyMathLab
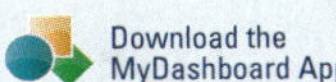
Download the MyDashboard App

Simplify. Use a graphing calculator table to compare your result with the original expression when possible. Assume each variable is nonnegative.

1. $4\sqrt{3} + 5\sqrt{3}$

2. $2\sqrt{7} + 6\sqrt{7}$

3. $-3\sqrt{5} + 9\sqrt{5}$

4. $-5\sqrt{6} + 3\sqrt{6}$

5. $2\sqrt[3]{5x^2y} - 6\sqrt[3]{5x^2y}$

6. $5\sqrt[4]{2xy^3} - 7\sqrt[4]{2xy^3}$

7. $3\sqrt{5a} + 2\sqrt{3b} - 6\sqrt{3b} + 7\sqrt{5a}$

8. $4\sqrt{7b} - \sqrt{2a} - 4\sqrt{2a} + 9\sqrt{7b}$

9. $2\sqrt{x} + 5 - 7\sqrt[3]{x} - 9 + 5\sqrt[3]{x}$

10. $4 - 6\sqrt[4]{x} + 3\sqrt[3]{x} - 1 - 8\sqrt[4]{x}$

11. $6\sqrt[3]{x - 1} - 3\sqrt[3]{x - 1} - 2\sqrt{x - 1}$

12. $4\sqrt[7]{3x + 1} + 3\sqrt[7]{3x + 1} - 5\sqrt[3]{3x + 1}$

13. $3.7\sqrt[4]{x} - 1.1\sqrt[4]{x} - 4.2\sqrt[6]{x} + 4.2\sqrt[6]{x}$

14. $4.1\sqrt{x} - 2.9\sqrt[3]{x} - 5.8\sqrt[3]{x} + 2.3\sqrt{x}$

15. $3(7 - \sqrt{x} + 2) - (\sqrt{x} + 2)$

16. $4(1 - 3\sqrt{x} - 8) - (5\sqrt{x} - 3)$

17. $7(\sqrt[3]{x} + 1) - 7(\sqrt[3]{x} - 1)$

18. $5(\sqrt[4]{a} - 2) + 5(2 - \sqrt[4]{a})$

19. $5\sqrt{6} - \sqrt{24}$

20. $7\sqrt{5} - \sqrt{45}$

21. $\sqrt{12b} + \sqrt{75b}$

22. $\sqrt{8x} - \sqrt{18x}$

23. $5\sqrt{12} + 4\sqrt{75} - 2\sqrt{3}$

24. $3\sqrt{8} + 2\sqrt{18} - 8\sqrt{2}$

25. $\sqrt{18x^5} + 2x\sqrt{50x^3}$

26. $\sqrt{27x^7} + 2x^2\sqrt{12x^3}$

27. $5\sqrt{4x^3} - x\sqrt{36x}$

28. $2\sqrt{9x^3} - x\sqrt{49x}$

29. $3\sqrt{81x^2} - 2\sqrt{100x^2}$

30. $2\sqrt{36x^2} + 5\sqrt{16x^2}$

31. $a\sqrt{12b^3} + b\sqrt{75a^2b}$

32. $b\sqrt{20a^5} - a\sqrt{45a^3b^2}$

33. $\sqrt[3]{27x^5} - x\sqrt[3]{8x^2}$

34. $\sqrt[3]{54x^7} - x\sqrt[3]{16x^4}$

35. $y\sqrt[4]{16x^{11}y^4} - 3x\sqrt[4]{x^7y^8}$

36. $7x^2\sqrt[5]{x^3y^{10}} + 3y\sqrt[5]{32x^{13}y^5}$

Simplify. Use a graphing calculator table to verify your result when possible. Assume each variable is nonnegative.

37. $\sqrt{2} \cdot \sqrt{5}$

38. $\sqrt{7} \cdot \sqrt{3}$

39. $3\sqrt{x} \cdot 2\sqrt{x}$

40. $-5\sqrt{x} \cdot 4\sqrt{x}$

41. $2\sqrt{5} \cdot \sqrt{10}$

42. $2\sqrt{15} \cdot \sqrt{3}$

43. $-2\sqrt{5x} \cdot 4\sqrt{3x}$

44. $-3\sqrt{10x} \cdot 2\sqrt{5x}$

45. $\sqrt{5}(1 + \sqrt{7})$

46. $\sqrt{2}(4 + \sqrt{3})$

47. $2\sqrt{7t}(\sqrt{7t} - \sqrt{2t})$

48. $4\sqrt{2k}(\sqrt{8} - 3\sqrt{k})$

49. $(4 + \sqrt{5})(2 - \sqrt{5})$

50. $(6 + \sqrt{2})(3 - \sqrt{2})$

51. $(2\sqrt{x} + 6)(5\sqrt{x} + 4)$

52. $(3\sqrt{x} + 7)(2\sqrt{x} + 5)$

53. $(4\sqrt{x} + \sqrt{3})(2\sqrt{x} - \sqrt{5})$

54. $(5\sqrt{x} - \sqrt{2})(3\sqrt{x} - \sqrt{3})$

55. $(5\sqrt{a} + \sqrt{b})(\sqrt{a} - 2\sqrt{b})$

56. $(2\sqrt{a} + \sqrt{b})(\sqrt{a} - \sqrt{b})$

57. $(1 - \sqrt{w})(1 + \sqrt{w})$

58. $(2 + 3\sqrt{p})(2 - 3\sqrt{p})$

59. $(7x + \sqrt{5})(7x - \sqrt{5})$

60. $(4\sqrt{x} + \sqrt{3})(4\sqrt{x} - \sqrt{3})$

61. $(2\sqrt{a} - \sqrt{b})(2\sqrt{a} + \sqrt{b})$

62. $(3\sqrt{a} - \sqrt{b})(3\sqrt{a} + \sqrt{b})$

63. $(4 + \sqrt{7})^2$

64. $(6 + \sqrt{2})^2$

65. $(5 + 6\sqrt{x})^2$

66. $(3\sqrt{x} + 2)^2$

67. $(4\sqrt{x} - \sqrt{5})^2$

68. $(2\sqrt{x} - \sqrt{7})^2$

69. $(\sqrt{a} + 2\sqrt{b})^2$

70. $(3\sqrt{a} - \sqrt{b})^2$

71. $\left(\sqrt{2x-5}+3\right)^2$

72. $\left(\sqrt{3x+4}-5\right)^2$

73. $\sqrt{x}\sqrt[5]{x}$

74. $\sqrt[4]{x}\sqrt[6]{x}$

75. $\sqrt[5]{x^4}\sqrt[5]{x^3}$

76. $\sqrt[4]{3x^2}\sqrt[4]{3x^2}$

77. $-5\sqrt{m}\left(\sqrt[4]{2m}-4\right)$

78. $-4\sqrt[3]{t}\left(\sqrt{t}+3\right)$

79. $\left(\sqrt[3]{x}+1\right)^2$

80. $\left(\sqrt[4]{x}-5\right)^2$

81. $\left(\sqrt[4]{k}-\sqrt[3]{k}\right)^2$

82. $\left(2\sqrt[5]{r}+\sqrt{r}\right)^2$

83. $\left(2\sqrt{x}-6\right)\left(3\sqrt[3]{x}+1\right)$

84. $\left(4\sqrt[3]{x^2}+1\right)\left(5\sqrt[4]{x^2}+2\right)$

85. $\left(3\sqrt[4]{x}+5\right)\left(3\sqrt[4]{x}-5\right)$

86. $\left(2\sqrt[5]{x}+1\right)\left(3\sqrt[5]{x}-2\right)$

87. Because of DVRs and other technologies, many people are postponing viewing their television programs (see Table 6).

Table 6 Percentages of Viewers Who Watch Their Programs within the Specified Numbers of Days

Number of Days	Percent
0	46
1	62
2	76
3	84
5	95
7	100

Source: *Nielsen Media Research*

Let $f(d)$ be the percentage of viewers who watch their programs within d days. A model of the situation is $f(d) = 21\sqrt{d} + 46$.

a. Use a graphing calculator to draw the graph of the model and, in the same viewing window, the scattergram of the data. Does the model fit the data well?

b. Estimate the percentage of viewers who watch their programs within 6 days.

c. Use TRACE on a graphing calculator to estimate after how many days have 88% of viewers watched their programs?

88. The percentages of adults who got a flu shot are shown in Table 7 for various years.

Table 7 Percentages of Adults Who Got a Flu Shot

Year	Percent
2005	27
2007	35
2008	36
2009	40
2010	42

Source: *Harris Interactive*

Let $f(t)$ be the percentage of adults who got a flu shot in the year that is t years since 2005. A model of the situation is $f(t) = 6.2\sqrt{t} + 27$.

a. Use a graphing calculator to draw the graph of the model and, in the same viewing window, the scattergram of the data. Does the model fit the data well?

b. Data are not available for 2006. Estimate the percentage of adults who got flu shots in 2006.

c. Use TRACE on a graphing calculator to predict in which year exactly half of adults will get a flu shot.

89. The flow rate r (in gallons per minute) of water from the nozzle of a firefighter's hose can be modeled by the formula $r = 30d^2\sqrt{P}$, where d is the nozzle diameter (in inches) and P is the nozzle pressure (in pounds per square inch). The flow rates of solid bore nozzles are shown in Table 8 for various nozzle pressures and diameters.

Table 8 Flow Rates of Solid Bore Nozzles (gallons per minute)

Nozzle Pressure (pounds per square inch)	Nozzle Diameter (inches)				
	0.5	1.0	1.5	2.0	2.5
40	47	188	423	752	1174
60	58	230	518	921	1438
80	66	266	598	1063	1661
100	74	297	668	1188	1857
120	81	325	732	1302	2034
140	88	352	791	1406	2197
175	98	393	884	1572	2456
200	105	420	945	1681	2626

Source: *Firetactics.com*

a. If the value of P is constant and the value of d is increased, what happens to the value of r? Explain how you can tell this from Table 8, the model's equation, and thinking about the situation.

b. **i.** If the nozzle pressure is 100 pounds per square inch, use the model to estimate the flow rates of water for nozzle diameters of 0.5 inch, 1 inch, 1.5 inches, 2 inches, and 2.5 inches.

ii. Which of your results in part (i) has the largest error? What is that error?

iii. Which of your results in part (i) has the largest percentage error? What is that percentage error? [**Hint:** To find each percentage error, divide the error by the actual amount and multiply the result by 100.]

c. In Virginia, the flow rate of water for firefighting must be at least 500 gallons per minute for one- and two-family dwellings that do not exceed 3600 square feet in area. Use the model to determine whether the requirement will be met if a 1.75-inch-diameter nozzle has 45 pounds per square inch of pressure. What is the estimated flow rate?

90. The time it takes for a planet to make one revolution around the Sun is the planet's *period*. The period $f(d)$ (in years) of a planet whose average distance from the Sun is d million kilometers is modeled by the equation

$$f(d) = 0.0005443\sqrt{d^3}$$

a. What is the period of Neptune, whose average distance from the Sun is 4498 million kilometers?
b. Use a graphing calculator table to find Earth's average distance from the Sun.
c. Suppose in the future we colonize Mars, whose average distance from the Sun is 228 million kilometers. What is the period of Mars? If a person is 20 years old in "Earth years," how old is the person in "Mars years"?

Concepts

91. A student tries to simplify $(x + \sqrt{7})^2$:

$$(x + \sqrt{7})^2 = x^2 + (\sqrt{7})^2 = x^2 + 7$$

Describe any errors. Then simplify the expression correctly.

92. A student tries to simplify $(x - \sqrt{3})^2$:

$$(x - \sqrt{3})^2 = x^2 - (\sqrt{3})^2 = x^2 - 3$$

Describe any errors. Then simplify the expression correctly.

93. A student tries to find the product $7(2\sqrt{3})$:

$$7(2\sqrt{3}) = 14\sqrt{21}$$

Describe any errors. Then find the product correctly.

94. A student tries to find the product $(3\sqrt{5})(4\sqrt{5})$:

$$(3\sqrt{5})(4\sqrt{5}) = 12\sqrt{5}$$

Describe any errors. Then find the product correctly.

*For Exercises 95 and 96, write the expression as a single radical. [**Hint:** Write the expression in exponential form.]*

95. $\dfrac{\sqrt{x}}{\sqrt[3]{x}}$

96. $\sqrt[3]{\sqrt{x}}$

97. **a.** Write the expression $\sqrt[4]{x}\sqrt[5]{x}$ as a single radical.
b. Write the expression $\sqrt[k]{x}\sqrt[n]{x}$ as a single radical. [**Hint:** Perform steps similar to your work in part (a).]
c. Use your result from part (b) to find the product $\sqrt[4]{x}\sqrt[5]{x}$. Compare your result with your result from part (a).
d. Use your result from part (b) to find the product $\sqrt[3]{x}\sqrt{x}$.

98. We cannot factor $x^2 - 3$ over the integers. We *can* factor $x^2 - 3$ over the real numbers:

$$x^2 - 3 = (x - \sqrt{3})(x + \sqrt{3})$$

a. Factor $x^2 - 2$ over the real numbers.
b. Factor $x^2 - 5$ over the real numbers.
c. Simplify $\dfrac{x^2 - 2}{x - \sqrt{2}}$.
d. Simplify $\dfrac{x^2 - 7}{x + \sqrt{7}}$.

99. **a.** Let n be a counting number. Decide whether each of the following is true or false:
i. $\sqrt[n]{ab} = \sqrt[n]{a}\sqrt[n]{b}$
ii. $\sqrt[n]{a + b} = \sqrt[n]{a} + \sqrt[n]{b}$
iii. $(ab)^n = a^n b^n$
iv. $(a + b)^n = a^n + b^n$
v. $\dfrac{1}{ab} = a^{-1}b^{-1}$
vi. $\dfrac{1}{a + b} = a^{-1} + b^{-1}$

b. Compare the types of equations that are true and the types of equations that are false in part (a). What patterns do you notice?

100. Why can we write the product of $\sqrt[5]{2}$ and $\sqrt[5]{3}$ as one radical but we cannot write the sum of $\sqrt[5]{2}$ and $\sqrt[5]{3}$ as one radical?

101. Describe how to multiply two radical expressions. Include a discussion of various types of formulas, laws, and techniques you can use to find such products.

102. Find two radical expressions whose sum is $9\sqrt{x} + 7$ and whose difference is $\sqrt{x} + 3$.

Related Review

Simplify. Assume $x \ge 0$.

103. $3\sqrt{x} - 5\sqrt{x}$

104. $(3\sqrt{x})^2$

105. $(3\sqrt{x})(-5\sqrt{x})$

106. $(3 + \sqrt{x})^2$

Expressions, Equations, Functions, and Graphs

Perform the indicated instruction. Then use words such as linear, quadratic, cubic, exponential, logarithmic, rational, radical, polynomial, degree, function, one variable, *and* two variables *to describe the expression, equation, or system.*

107. Write $\log_b(x^2 + 3x - 40) - \log_b(x^2 - 64)$ as a single logarithm.

108. Factor $2x^2 + 5x - 12$.

109. Solve $\log_2(3x - 4) - \log_2(2x - 3) = 3$.

110. Solve $2x(2x - 3) = 15 - 2x$.

111. Solve $2(3)^{5x-1} = 35$. Round any solutions to the fourth decimal place.

112. Let $f(x) = 2x^2 + 5x - 12$. Find x when $f(x) = -5$.

13.3 Rationalizing Denominators and Simplifying Quotients of Radical Expressions

Objectives

- » Rationalize the denominator of a radical expression.
- » Know the *quotient property for radicals.*
- » Use the quotient property to simplify radical expressions.
- » Use a *radical conjugate* to rationalize the denominator of a radical expression.

In Section 13.2, we discussed how to add, subtract, and multiply radical expressions. How do we simplify quotients of radical expressions?

Rationalizing Denominators of Radical Expressions

Recall from Section 9.3 that we simplify an expression of the form $\frac{p}{\sqrt{q}}$ by leaving no denominator as a radical expression and that we call this process *rationalizing the denominator.* For example, to rationalize the denominator of $\frac{2}{\sqrt{5}}$, we multiply by $1 = \frac{\sqrt{5}}{\sqrt{5}}$:

$$\frac{2}{\sqrt{5}} = \frac{2}{\sqrt{5}} \cdot \frac{\sqrt{5}}{\sqrt{5}}$$
$$= \frac{2\sqrt{5}}{5}$$

In this section, we will rationalize the denominator of each of the following radical expressions:

$$\frac{4}{5\sqrt{3x}} \qquad \frac{3y}{\sqrt[5]{8x^2}} \qquad \frac{5}{3+\sqrt{x}} \qquad \frac{\sqrt{x}+4}{3\sqrt{x}-\sqrt{2}}$$

That is, we will write each expression so that no denominator is a radical expression.

Example 1 Rationalizing a Denominator

Simplify $\frac{4}{5\sqrt{3x}}$.

Solution

Since $\sqrt{3x} \cdot \sqrt{3x} = 3x$ where $x \ge 0$, we rationalize the denominator of $\frac{4}{5\sqrt{3x}}$ by multiplying by $\frac{\sqrt{3x}}{\sqrt{3x}}$:

$$\frac{4}{5\sqrt{3x}} = \frac{4}{5\sqrt{3x}} \cdot \frac{\sqrt{3x}}{\sqrt{3x}} \qquad \text{Rationalize denominator.}$$
$$= \frac{4\sqrt{3x}}{5\left(\sqrt{3x}\right)^2} \qquad \text{Multiply numerators; multiply denominators.}$$
$$= \frac{4\sqrt{3x}}{5(3x)} \qquad \left(\sqrt{x}\right)^2 = x \text{ for } x \ge 0$$
$$= \frac{4\sqrt{3x}}{15x} \qquad \text{Multiply.}$$

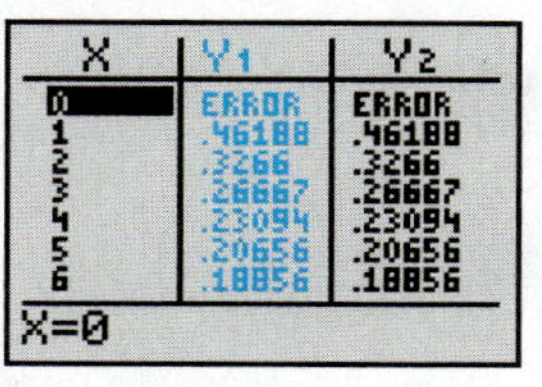

Figure 10 Verify the work

We use a graphing calculator table to verify our work (see Fig. 10). The table has "ERROR" messages across from $x = 0$, because the original expression and our result are not defined at 0 (why?).

In Example 2, we will rationalize the denominator for indexes other than 2. With any index n, our intermediate goal is the same: to write the denominator so that the radicand of its radical is a perfect nth power.

Example 2 Rationalizing Denominators

Simplify.

1. $\dfrac{1}{\sqrt[3]{x}}$ **2.** $\dfrac{3y}{\sqrt[5]{8x^2}}$

Solution

1. For the radicand to be a perfect cube, x must be multiplied by $x \cdot x = x^2$. So, we multiply $\dfrac{1}{\sqrt[3]{x}}$ by $\dfrac{\sqrt[3]{x^2}}{\sqrt[3]{x^2}}$:

$$\frac{1}{\sqrt[3]{x}} = \frac{1}{\sqrt[3]{x}} \cdot \frac{\sqrt[3]{x^2}}{\sqrt[3]{x^2}} \quad \text{Rationalize denominator.}$$

$$= \frac{\sqrt[3]{x^2}}{\sqrt[3]{x^3}} \quad \text{Multiply numerators; multiply denominators.}$$

$$= \frac{\sqrt[3]{x^2}}{x} \quad \sqrt[3]{x^3} = x$$

2. To become a perfect 5th power, the radicand $8x^2 = 2 \cdot 2 \cdot 2 \cdot x \cdot x$ must be multiplied by $2 \cdot 2 \cdot x \cdot x \cdot x = 4x^3$:

$$\frac{3y}{\sqrt[5]{8x^2}} = \frac{3y}{\sqrt[5]{8x^2}} \cdot \frac{\sqrt[5]{4x^3}}{\sqrt[5]{4x^3}} \quad \text{Rationalize denominator.}$$

$$= \frac{3y\sqrt[5]{4x^3}}{\sqrt[5]{8x^2 \cdot 4x^3}} \quad \text{Multiply numerators; multiply denominators.}$$

$$= \frac{3y\sqrt[5]{4x^3}}{\sqrt[5]{32x^5}} \quad \text{Multiply.}$$

$$= \frac{3y\sqrt[5]{4x^3}}{\sqrt[5]{32}\sqrt[5]{x^5}} \quad \sqrt[n]{ab} = \sqrt[n]{a}\sqrt[n]{b}$$

$$= \frac{3y\sqrt[5]{4x^3}}{2x} \quad \sqrt[5]{32} = 2, \sqrt[5]{x^5} = x$$

As shown in Example 2, **to rationalize the denominator of a radical expression of the form $\dfrac{A}{\sqrt[n]{x^m}}$, we multiply the expression by a fraction of the form $\dfrac{\sqrt[n]{x^k}}{\sqrt[n]{x^k}}$ so the radical in the denominator has a perfect nth-power radicand.**

Quotient Property

In Section 9.3, we worked with the quotient property for square roots:

$$\sqrt{\frac{a}{b}} = \frac{\sqrt{a}}{\sqrt{b}}, \quad \text{where } a \geq 0 \text{ and } b > 0$$

Next, we describe the quotient property for any index n.

Quotient Property for Radicals

If $\sqrt[n]{a}$ and $\sqrt[n]{b}$ are defined and b is nonzero, then

$$\sqrt[n]{\frac{a}{b}} = \frac{\sqrt[n]{a}}{\sqrt[n]{b}}$$

In words, the nth root of a quotient is the quotient of the nth roots.

For example, $\sqrt[3]{\frac{8}{27}} = \frac{\sqrt[3]{8}}{\sqrt[3]{27}} = \frac{2}{3}$.

You will prove the quotient property for radicals in Exercise 74.

Using the Quotient Property to Simplify Radical Expressions

If a radical expression has a fractional radicand, we simplify the expression by writing it as an expression in which no radicand is a fraction. We can use the quotient property to help us do this.

Example 3 Simplifying Radical Expressions

Simplify.

1. $\sqrt{\frac{5}{k}}$
2. $\sqrt[3]{\frac{7y}{2x^2}}$

Solution

1. $\sqrt{\frac{5}{k}} = \frac{\sqrt{5}}{\sqrt{k}}$ *Quotient property*

$= \frac{\sqrt{5}}{\sqrt{k}} \cdot \frac{\sqrt{k}}{\sqrt{k}}$ *Rationalize denominator.*

$= \frac{\sqrt{5k}}{k}$ *Multiply numerators; multiply denominators.*

2. $\sqrt[3]{\frac{7y}{2x^2}} = \frac{\sqrt[3]{7y}}{\sqrt[3]{2x^2}}$ *Quotient property*

$= \frac{\sqrt[3]{7y}}{\sqrt[3]{2x^2}} \cdot \frac{\sqrt[3]{4x}}{\sqrt[3]{4x}}$ *To become a perfect cube, $2x^2 = 2 \cdot x \cdot x$ must be multiplied by $2 \cdot 2 \cdot x = 4x$.*

$= \frac{\sqrt[3]{28xy}}{\sqrt[3]{8x^3}}$ *Multiply numerators; multiply denominators.*

$= \frac{\sqrt[3]{28xy}}{2x}$ *$\sqrt[3]{8} = 2, \sqrt[3]{x^3} = x$*

Using Radical Conjugates to Rationalize Denominators

Recall from Section 7.3 that we call binomials such as $5x + 2$ and $5x - 2$ binomial conjugates of each other. Similarly, we call the radical expressions $\sqrt{5} + \sqrt{2}$ and $\sqrt{5} - \sqrt{2}$ radical conjugates of each other. We say the sum of two radicals and the difference of the same two radicals are **radical conjugates** of each other.

What happens when we find the product of two radical conjugates? Here we use the property $(A + B)(A - B) = A^2 - B^2$ to find the product $(\sqrt{5} + \sqrt{2})(\sqrt{5} - \sqrt{2})$:

$$(\sqrt{5} + \sqrt{2})(\sqrt{5} - \sqrt{2}) = (\sqrt{5})^2 - (\sqrt{2})^2 \quad (A + B)(A - B) = A^2 - B^2$$
$$= 5 - 2 \quad (\sqrt{x})^2 = x \text{ for } x \geq 0$$
$$= 3 \quad \text{Subtract.}$$

The result contains no radicals. We next list a few expressions, their conjugates, and the products of the expressions and their conjugates:

Expression	Conjugate	Product
$4 - \sqrt{x}$	$4 + \sqrt{x}$	$16 - x$
$7 + 3\sqrt{x}$	$7 - 3\sqrt{x}$	$49 - 9x$
$2\sqrt{x} - 3\sqrt{5}$	$2\sqrt{x} + 3\sqrt{5}$	$4x - 45$

Just as before, we notice that the products of the conjugates contain no radicals. We will use this observation to help us rationalize a denominator in Example 4.

Example 4 Rationalizing a Denominator

Simplify $\dfrac{5}{3 + \sqrt{x}}$.

Solution

The conjugate of the denominator is $3 - \sqrt{x}$. We can rationalize the denominator of $\dfrac{5}{3 + \sqrt{x}}$ by multiplying by $\dfrac{\text{conjugate}}{\text{conjugate}} = \dfrac{3 - \sqrt{x}}{3 - \sqrt{x}}$:

$$\frac{5}{3+\sqrt{x}} = \frac{5}{3+\sqrt{x}} \cdot \frac{3-\sqrt{x}}{3-\sqrt{x}} \qquad \textit{Multiply by } \frac{3-\sqrt{x}}{3-\sqrt{x}}.$$

$$= \frac{5(3-\sqrt{x})}{(3+\sqrt{x})(3-\sqrt{x})} \qquad \textit{Multiply numerators; multiply denominators.}$$

$$= \frac{5(3-\sqrt{x})}{3^2-(\sqrt{x})^2} \qquad (A+B)(A-B) = A^2 - B^2$$

$$= \frac{15-5\sqrt{x}}{9-x} \qquad \textit{Distributive law; } (\sqrt{x})^2 = x \textit{ for } x \geq 0$$

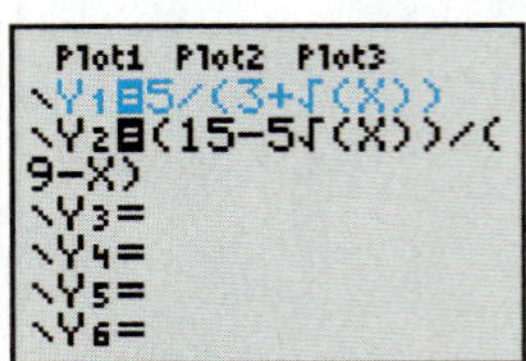

X	Y1	Y2
0	1.6667	1.6667
1	1.25	1.25
2	1.1327	1.1327
3	1.0566	1.0566
4	1	1
5	.95492	.95492
6	.91752	.91752

X=0

Figure 11 Verify the work

We use a graphing calculator table to verify our work (see Fig. 11).

WARNING When we try to rationalize the denominator of $\dfrac{5}{3+\sqrt{x}}$, it is *not* helpful to multiply the fraction by $\dfrac{\sqrt{x}}{\sqrt{x}}$:

$$\frac{5}{3+\sqrt{x}} = \frac{5}{3+\sqrt{x}} \cdot \frac{\sqrt{x}}{\sqrt{x}} = \frac{5\sqrt{x}}{3\sqrt{x}+x}$$

Rather, the conjugate of the denominator is $3 - \sqrt{x}$, so we multiply the fraction by $\dfrac{\text{conjugate}}{\text{conjugate}} = \dfrac{3-\sqrt{x}}{3-\sqrt{x}}$, as shown in Example 4.

Rationalizing a Denominator by Using a Radical Conjugate

To rationalize the denominator of a square root expression if the denominator is a sum or difference involving radicals,

1. Determine the radical conjugate of the denominator.
2. Multiply the original fraction by the fraction $\dfrac{\text{conjugate}}{\text{conjugate}}$.
3. Find the product of the denominators by using $(A + B)(A - B) = A^2 - B^2$.

Example 5 Rationalizing a Denominator

Simplify $\dfrac{\sqrt{x}+4}{3\sqrt{x}-\sqrt{2}}$.

Solution

$$\frac{\sqrt{x}+4}{3\sqrt{x}-\sqrt{2}} = \frac{\sqrt{x}+4}{3\sqrt{x}-\sqrt{2}} \cdot \frac{3\sqrt{x}+\sqrt{2}}{3\sqrt{x}+\sqrt{2}}$$ *Conjugate of $3\sqrt{x}-\sqrt{2}$ is $3\sqrt{x}+\sqrt{2}$.*

$$= \frac{(\sqrt{x}+4)(3\sqrt{x}+\sqrt{2})}{(3\sqrt{x}-\sqrt{2})(3\sqrt{x}+\sqrt{2})}$$ *Multiply numerators; multiply denominators.*

$$= \frac{3\sqrt{x}\sqrt{x}+\sqrt{x}\sqrt{2}+4\cdot 3\sqrt{x}+4\sqrt{2}}{(3\sqrt{x})^2-(\sqrt{2})^2}$$ *Multiply pairs of terms; $(A-B)(A+B)=A^2-B^2$*

$$= \frac{3x+\sqrt{2x}+12\sqrt{x}+4\sqrt{2}}{9x-2}$$ *Simplify.*

Example 6 Rationalizing a Denominator

Simplify $\dfrac{\sqrt{a}+\sqrt{b}}{\sqrt{a}-\sqrt{b}}$.

Solution

$$\frac{\sqrt{a}+\sqrt{b}}{\sqrt{a}-\sqrt{b}} = \frac{\sqrt{a}+\sqrt{b}}{\sqrt{a}-\sqrt{b}} \cdot \frac{\sqrt{a}+\sqrt{b}}{\sqrt{a}+\sqrt{b}}$$ *Conjugate of $\sqrt{a}-\sqrt{b}$ is $\sqrt{a}+\sqrt{b}$.*

$$= \frac{(\sqrt{a}+\sqrt{b})(\sqrt{a}+\sqrt{b})}{(\sqrt{a}-\sqrt{b})(\sqrt{a}+\sqrt{b})}$$ *Multiply numerators; multiply denominators.*

$$= \frac{(\sqrt{a}+\sqrt{b})^2}{(\sqrt{a})^2-(\sqrt{b})^2}$$ *$CC=C^2$; $(A-B)(A+B)=A^2-B^2$*

$$= \frac{(\sqrt{a})^2+2\sqrt{a}\sqrt{b}+(\sqrt{b})^2}{(\sqrt{a})^2-(\sqrt{b})^2}$$ *$(A+B)^2=A^2+2AB+B^2$*

$$= \frac{a+2\sqrt{ab}+b}{a-b}$$ *$(\sqrt{x})^2=x$ for $x\ge 0$; $\sqrt[n]{a}\sqrt[n]{b}=\sqrt[n]{ab}$*

Group Exploration

Looking ahead: Sketching graphs of square root functions

1. Use a graphing calculator to draw a graph of $y=\sqrt{x}$.
2. Use a graphing calculator to compare graphs of $y=0.5\sqrt{x}$, $y=\sqrt{x}$, $y=2\sqrt{x}$, and $y=-2\sqrt{x}$. Describe the effect a has on the graph of $y=a\sqrt{x}$, where $a\ne 0$.
3. Use a graphing calculator to compare graphs of $y=\sqrt{x-2}$, $y=\sqrt{x}$, and $y=\sqrt{x+4}$. Describe the effect h has on the graph of $y=\sqrt{x-h}$.
4. Use a graphing calculator to compare graphs of $y=\sqrt{x}-2$, $y=\sqrt{x}$, and $y=\sqrt{x}+4$. Describe the effect k has on the graph of $y=\sqrt{x}+k$.
5. Use a graphing calculator to graph
   $y=\sqrt{x}$, $y=0.5\sqrt{x}$, $y=0.5\sqrt{x+3}$, and $y=0.5\sqrt{x+3}-2$
   in order, and explain how these graphs relate to your observations in Problems 2, 3, and 4.
6. Sketch the graph of $y=2\sqrt{x-3}+1$. Use a graphing calculator to verify your sketch.
7. Describe how a, h, and k affect the graph of $f(x)=a\sqrt{x-h}+k$, where $a\ne 0$. Compare their effects for this function with their effects for the quadratic function $g(x)=a(x-h)^2+k$.

Homework 13.3

For extra help ▶ MyMathLab® Watch the videos in MyMathLab  Download the MyDashboard App

Simplify. Use a graphing calculator table to verify your result when possible. Assume each variable is nonnegative.

1. $\frac{8}{\sqrt{x}}$
2. $\frac{2}{\sqrt{x}}$
3. $\frac{3}{\sqrt{5p}}$
4. $\frac{2}{\sqrt{7r}}$
5. $\frac{4}{3\sqrt{2x}}$
6. $\frac{7}{6\sqrt{3x}}$
7. $\frac{10}{\sqrt{8k}}$
8. $\frac{6}{\sqrt{27t}}$
9. $\sqrt{\frac{4}{x}}$
10. $\sqrt{\frac{25}{x}}$
11. $\sqrt{\frac{7}{2}}$
12. $\sqrt{\frac{5}{3}}$
13. $\sqrt{\frac{2y}{x}}$
14. $\sqrt{\frac{11y}{x}}$
15. $\sqrt{\frac{x}{12y}}$
16. $\sqrt{\frac{x}{18y}}$
17. $\frac{3}{\sqrt{x-4}}$
18. $\frac{5}{\sqrt{2x+1}}$
19. $\frac{\sqrt{2a^3}}{\sqrt{3b}}$
20. $\frac{\sqrt{5b^5}}{\sqrt{7a}}$
21. $\frac{2}{\sqrt[3]{5}}$
22. $\frac{5}{\sqrt[3]{2}}$
23. $\frac{5}{\sqrt[3]{4}}$
24. $\frac{1}{\sqrt[3]{25}}$
25. $\frac{4}{5\sqrt[3]{x}}$
26. $\frac{7}{4\sqrt[3]{x^2}}$
27. $\frac{6}{\sqrt[3]{2x^2}}$
28. $\frac{1}{\sqrt[3]{9x}}$
29. $\frac{7t}{\sqrt[4]{4t^3}}$
30. $\frac{2w}{\sqrt[5]{16w^2}}$
31. $\frac{\sqrt[3]{x}}{\sqrt{x}}$
32. $\frac{\sqrt[5]{x}}{\sqrt[4]{2x}}$
33. $\sqrt[5]{\frac{2}{x^3}}$
34. $\sqrt[3]{\frac{4}{x^2}}$
35. $\sqrt[4]{\frac{4}{9x^2}}$
36. $\sqrt[3]{\frac{7}{25x}}$
37. $\sqrt[5]{\frac{3w}{4x^4y^2}}$
38. $\sqrt[6]{\frac{5w}{8x^2y^3}}$

Simplify. Use a graphing calculator to verify your result when possible.

39. $\frac{1}{5+\sqrt{3}}$
40. $\frac{1}{1+\sqrt{5}}$
41. $\frac{2}{\sqrt{3}+\sqrt{7}}$
42. $\frac{4}{\sqrt{2}+\sqrt{5}}$
43. $\frac{1}{3\sqrt{r}-7}$
44. $\frac{6}{5\sqrt{t}-2}$
45. $\frac{\sqrt{x}}{\sqrt{x}-1}$
46. $\frac{\sqrt{x}}{\sqrt{x}+1}$
47. $\frac{3\sqrt{x}}{4\sqrt{x}-\sqrt{5}}$
48. $\frac{4\sqrt{x}}{2\sqrt{x}+\sqrt{6}}$
49. $\frac{\sqrt{x}}{\sqrt{x}-y}$
50. $\frac{\sqrt{y}}{x-2\sqrt{y}}$
51. $\frac{\sqrt{x}-5}{\sqrt{x}+5}$
52. $\frac{\sqrt{x}+9}{\sqrt{x}+9}$
53. $\frac{2\sqrt{x}+5}{3\sqrt{x}+1}$
54. $\frac{4\sqrt{x}-3}{2\sqrt{x}+5}$
55. $\frac{6\sqrt{x}+\sqrt{5}}{3\sqrt{x}-\sqrt{7}}$
56. $\frac{8\sqrt{x}-\sqrt{3}}{4\sqrt{x}-\sqrt{2}}$
57. $\frac{\sqrt{x}-\sqrt{y}}{\sqrt{x}+\sqrt{y}}$
58. $\frac{2\sqrt{x}-\sqrt{y}}{3\sqrt{x}-\sqrt{y}}$
59. $\frac{1}{\sqrt{x+1}-\sqrt{x}}$
60. $\frac{2}{\sqrt{x+3}+\sqrt{x}}$

61. The distance (in miles) $f(h)$ to the horizon at an altitude h feet above sea level is given by the equation

$$f(h) = \sqrt{\frac{3h}{2}}$$

   a. Simplify the right-hand side of the horizon–distance equation.
   b. The Willis Tower in Chicago is 1450 feet tall. What would be the distance to the horizon from the top of this skyscraper? Assume the base of the building is at sea level.
   c. If an airplane flies at an altitude of 30,000 feet above sea level, what is the distance to the horizon from the airplane?

62. The time (in seconds) $f(d)$ that it takes for an object to fall d feet can be modeled by the equation

$$f(d) = \sqrt{\frac{2d}{g}}$$

where g is the constant 32.2 feet per second squared.
   a. Simplify the right-hand side of the model's equation.
   b. Find $f(100)$. What does it mean in this situation?
   c. In 2002, sky divers jumped off the 1483-foot-tall Petronas Towers (in Malaysia), the tallest habitable buildings in the world. Estimate how long a sky diver was in free fall by finding the time it would take to fall 1483 feet with a closed parachute. Ideally, the parachute opens. If so, is your estimate an underestimate or an overestimate? Explain.
   d. Is f an increasing or a decreasing function? Explain why that makes sense in this situation.

63. In the ISO paper-size system, the length-to-width ratio of all pages is $\frac{\sqrt{2}}{1}$ (see Fig. 12).

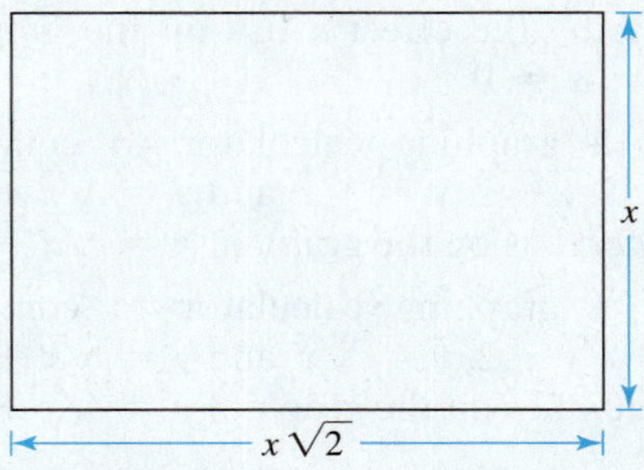

Figure 12 ISO-sized paper

a. Show that if you cut a piece of ISO-sized paper parallel to its shorter side to form two pieces with equal area, each piece will also have a length-to-width ratio of $\frac{\sqrt{2}}{1}$ (see Fig. 13). (This property allows two ISO pages of equal size to be photocopied onto one page by setting the magnification factor on a copying machine to $\sqrt{\frac{1}{2}} \approx 0.71 = 71\%$.)

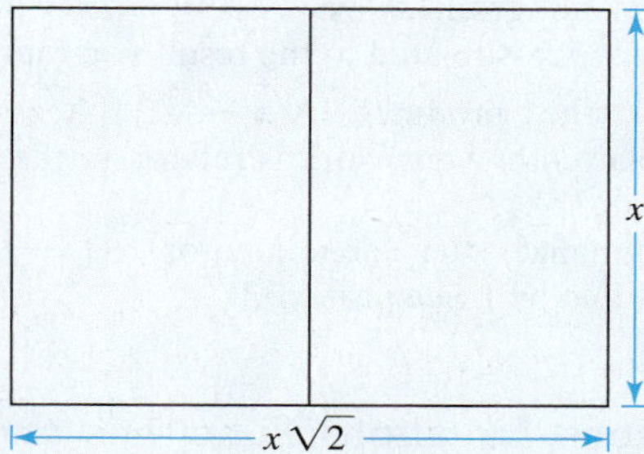

Figure 13 ISO-sized paper cut in two

b. The largest ISO paper size has an area of 1 square meter. Find the (exact) width of a page of this size. Then round your result to the third decimal place.

64. *Escape velocity* is the initial velocity that an object needs in order to break free of a planet's or moon's gravitational pull. If we ignore the effects of air resistance and the rotation of the planet or moon and assume there is no continued propulsion, as by a rocket, the escape velocity v (in meters per second) is given by $v = \sqrt{\frac{2GM}{r}}$, where G is a constant equal to 6.67×10^{-11}, M is the mass (in kilograms) of the planet or moon, and r is the object's distance (in meters) from the center of the planet or moon.

a. Simplify the right-hand side of the model's equation.

b. Use Table 9 to find the escape velocity on the planet's or moon's surface at the equator. Round your result to the first decimal place.

i. Earth **ii.** the Moon **iii.** Jupiter

Table 9 Masses and Equatorial Radii of Earth, the Moon, and Jupiter

Planet/Moon	Mass of Planet/Moon (kilograms)	Radius of Planet/Moon (meters)
Earth	5.976×10^{24}	6.378×10^{6}
The Moon	7.349×10^{22}	1.737×10^{6}
Jupiter	1.899×10^{27}	7.149×10^{7}

c. In terms of blastoffs, would a round trip to Jupiter or the Moon require more fuel?

d. Convert your result of part (b, i) to units of miles per hour. (There are approximately 1609.3 meters in 1 mile.)

Concepts

65. Two students try to rationalize the denominator of the expression $\frac{2}{\sqrt{x}}$:

Student 1's work

$$\frac{2}{\sqrt{x}} = \frac{2}{\sqrt{x}} \cdot \frac{\sqrt{x}}{\sqrt{x}} = \frac{2\sqrt{x}}{\sqrt{x}\sqrt{x}} = \frac{2\sqrt{x}}{x}$$

Student 2's work

$$\frac{2}{\sqrt{x}} = \left(\frac{2}{\sqrt{x}}\right)^2 = \frac{2^2}{\left(\sqrt{x}\right)^2} = \frac{4}{x}$$

Did one, both, or neither of these students rationalize the denominator correctly? Describe any errors and where they occurred.

66. Two students try to rationalize the denominator of $\frac{3}{\sqrt{x^3}}$:

Student 1's work

$$\frac{3}{\sqrt{x^3}} = \frac{3}{\sqrt{x^3}} \cdot \frac{\sqrt{x^3}}{\sqrt{x^3}} = \frac{3\sqrt{x^3}}{\sqrt{x^3}\sqrt{x^3}} = \frac{3\left(x\sqrt{x}\right)}{x^3} = \frac{3\sqrt{x}}{x^2}$$

Student 2's work

$$\frac{3}{\sqrt{x^3}} = \frac{3}{x\sqrt{x}} = \frac{3}{x\sqrt{x}} \cdot \frac{\sqrt{x}}{\sqrt{x}} = \frac{3\sqrt{x}}{x \cdot x} = \frac{3\sqrt{x}}{x^2}$$

Did one, both, or neither of the students rationalize the denominator correctly? Explain.

67. A student tries to rationalize the denominator of $\frac{5}{\sqrt[3]{x}}$:

$$\frac{5}{\sqrt[3]{x}} = \frac{5}{\sqrt[3]{x}} \cdot \frac{\sqrt[3]{x}}{\sqrt[3]{x}} = \frac{5\sqrt[3]{x}}{x}$$

Describe any errors. Then rationalize the denominator correctly.

68. Two students try to rationalize the denominator of the expression $\frac{4}{2 + \sqrt{x}}$:

Student 1's work

$$\frac{4}{2 + \sqrt{x}} = \frac{4}{2 + \sqrt{x}} \cdot \frac{\sqrt{x}}{\sqrt{x}} = \frac{4\sqrt{x}}{2 + \sqrt{x}\sqrt{x}} = \frac{4\sqrt{x}}{2 + x}$$

Student 2's work

$$\frac{4}{2 + \sqrt{x}} = \frac{4}{2 + \sqrt{x}} \cdot \frac{2 - \sqrt{x}}{2 - \sqrt{x}} = \frac{8 - 4\sqrt{x}}{2^2 - \left(\sqrt{x}\right)^2} = \frac{8 - 4\sqrt{x}}{4 - x}$$

Did one, both, or neither of these students rationalize the denominator correctly? Describe any errors and where they occurred.

We rationalize the numerator of a radical expression by finding an equivalent expression whose numerator contains no radicals. For Exercises 69–72, rationalize the numerator of the given expression.

69. $\dfrac{\sqrt{x}}{3}$

70. $\dfrac{\sqrt{x}}{\sqrt{2}}$

71. $\dfrac{\sqrt{x+2}-\sqrt{x}}{2}$

72. $\dfrac{\sqrt{x+3}+\sqrt{x}}{3}$

73. Simplify the expression.

$$\frac{\dfrac{1}{\sqrt{x}}-\dfrac{3}{x}}{\dfrac{2}{\sqrt{x}}+\dfrac{1}{x}}$$

74. Prove the quotient property for radicals—that is,

$$\sqrt[n]{\frac{a}{b}}=\frac{\sqrt[n]{a}}{\sqrt[n]{b}}$$

where $\sqrt[n]{a}$ and $\sqrt[n]{b}$ are defined and b is nonzero.

75. Find and simplify the exact solution of $x\sqrt{2}+3\sqrt{5}=9\sqrt{5}$.

76. Find the exact solution of $x^2\sqrt{2}+x\sqrt{17}+\sqrt{2}=0$. Simplify your result. [**Hint:** Use the quadratic formula.]

77. Describe how to rationalize the denominator of a radical expression in which the denominator is a radical.

78. Describe how to rationalize the denominator of a radical expression in which the denominator is a sum or difference involving radicals.

Related Review

79. a. Factor A^3+B^3.

b. Find the product $(A+B)(A^2-AB+B^2)$. Explain how your work is related to your work in part (a).

c. Find the product $(x+2)(x^2-2x+4)$. Explain how your work is related to the result you found in part (b).

d. Find the product $(\sqrt[3]{x}+\sqrt[3]{2})(\sqrt[3]{x^2}-\sqrt[3]{2x}+\sqrt[3]{4})$. Explain how your work is related to the result you found in part (b).

e. Rationalize the denominator of $\dfrac{1}{\sqrt[3]{x}+\sqrt[3]{2}}$. [**Hint:** Multiply by 1. See part (d).]

80. a. Factor A^3-B^3.

b. Find the product $(A-B)(A^2+AB+B^2)$. Explain how your work is related to your work in part (a).

c. Find the product $(x-2)(x^2+2x+4)$. Explain how your work is related to the result you found in part (b).

d. Find the product $(\sqrt[3]{x}-\sqrt[3]{2})(\sqrt[3]{x^2}+\sqrt[3]{2x}+\sqrt[3]{4})$. Explain how your work is related to the result you found in part (b).

e. Rationalize the denominator of $\dfrac{1}{\sqrt[3]{x}-\sqrt[3]{2}}$. [**Hint:** Multiply by 1. See part (d).]

Expressions, Equations, Functions, and Graphs

Perform the indicated instruction. Then use words such as linear, quadratic, cubic, exponential, logarithmic, rational, radical, polynomial, degree, function, one variable, *and* two variables *to describe the expression, equation, or system.*

81. Find the product $(5x-4)(3x^2-2x-1)$.

82. Graph $f(x)=-4(3)^x$ by hand.

83. Factor $24x^3-3000$.

84. Simplify $\dfrac{(16b^{2/3}c^3)^{1/4}}{(27b^{3/4}c^{-5})^{2/3}}$.

85. Solve $5x^2-3=4x-1$.

86. Find any real-number solutions of $5b^4-43=76$. Round any results to the fourth decimal place.

13.4 Graphing and Combining Square Root Functions

Objectives

- Know the graphical significance of a, h, and k for a square root function of the form $y=a\sqrt{x-h}+k$.
- Sketch graphs of square root functions.
- Find the domain and range of a square root function.
- Find the sum function, difference function, product function, and quotient function of two square root functions.

In this section, we will graph square root functions and perform operations with square root functions.

Graphing Square Root Functions

Recall from Section 9.1 that, to graph a function such as $h(x)=(x-3)^2$, we translate the graph of $k(x)=x^2$ to the right by 3 units. Can we graph the square root function $g(x)=\sqrt{x-3}$ by translating the function $f(x)=\sqrt{x}$ in some way? We will explore this question in Example 1.

Example 1 Translating a Graph Horizontally

Compare the graph of $g(x)=\sqrt{x-3}$ with the graph of $f(x)=\sqrt{x}$.

Solution

We list input–output pairs of g in Table 10. We choose inputs that lead to easily found outputs. Then we sketch graphs of g and f (see Fig. 14).

Table 10 Input–Output Pairs of $g(x) = \sqrt{x - 3}$

x	$g(x)$
3	0
4	1
7	2
12	3
19	4

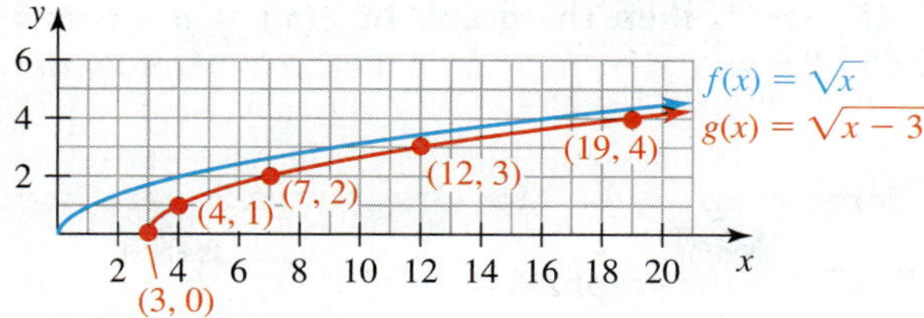

Figure 14 Graphs of $g(x) = \sqrt{x - 3}$ and $f(x) = \sqrt{x}$

The graph of $g(x) = \sqrt{x - 3}$ is the translation of the graph of $f(x) = \sqrt{x}$ to the right by 3 units. To see why this makes sense, we solve both equations for x:

$$\begin{aligned} f(x) &= \sqrt{x} \\ y &= \sqrt{x} \\ \sqrt{x} &= y \\ (\sqrt{x})^2 &= y^2 \\ x &= y^2 \end{aligned} \qquad\qquad \begin{aligned} g(x) &= \sqrt{x - 3} \\ y &= \sqrt{x - 3} \\ \sqrt{x - 3} &= y \\ (\sqrt{x - 3})^2 &= y^2 \\ x - 3 &= y^2 \\ x &= y^2 + 3 \end{aligned}$$

For each positive value of y, the input value of x for g is 3 more than the input value of x for f. Therefore, the graph of $g(x) = \sqrt{x - 3}$ lies 3 units to the right of the graph of $f(x) = \sqrt{x}$.

In Example 1, we found that, to graph $g(x) = \sqrt{x - 3}$, we translate the graph of $f(x) = \sqrt{x}$ to the right by 3 units. This is the same pattern we observed with quadratic functions: To graph the quadratic function $g(x) = (x - 3)^2$, we translate the graph of $f(x) = x^2$ to the right by 3 units.

In fact, the values of a, h, and k have similar effects on a square root function $g(x) = a\sqrt{x - h} + k$, where $a \neq 0$, as on a quadratic function $Q(x) = a(x - h)^2 + k$. Here are some examples of how we can translate the graph of $y = \sqrt{x}$ to obtain the graph of an equation of the form $y = \sqrt{x - h} + k$:

Function	To graph the function $g(x) = \sqrt{x - h} + k$,
$g(x) = \sqrt{x - 2}$	translate the graph of $y = \sqrt{x}$ to the right by 2 units.
$g(x) = \sqrt{x + 2}$	translate the graph of $y = \sqrt{x}$ to the left by 2 units.
$g(x) = \sqrt{x} - 2$	translate the graph of $y = \sqrt{x}$ down by 2 units.
$g(x) = \sqrt{x} + 2$	translate the graph of $y = \sqrt{x}$ up by 2 units.

The graphs of $f(x) = -a\sqrt{x}$ and $g(x) = a\sqrt{x}$ are reflections of each other across the x-axis (see Fig. 15).

If $a > 0$, then $g(x) = a\sqrt{x - h} + k$ is an increasing function and (h, k) is the minimum point (see Fig. 16). **If $a < 0$, then g is a decreasing function and (h, k) is the maximum point** (see Fig. 17).

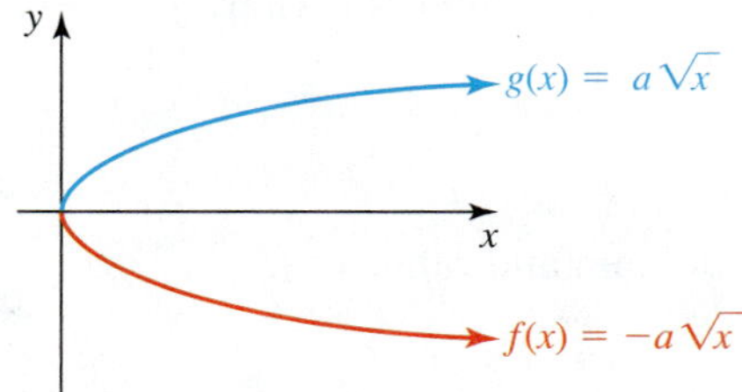

Figure 15 Typical graphs of $f(x) = -a\sqrt{x}$ and $g(x) = a\sqrt{x}$, where $a > 0$

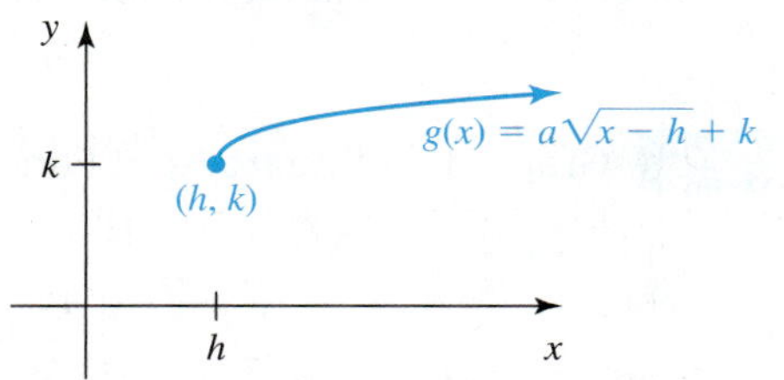

Figure 16 If $a > 0$, then g is increasing with minimum point (h, k)

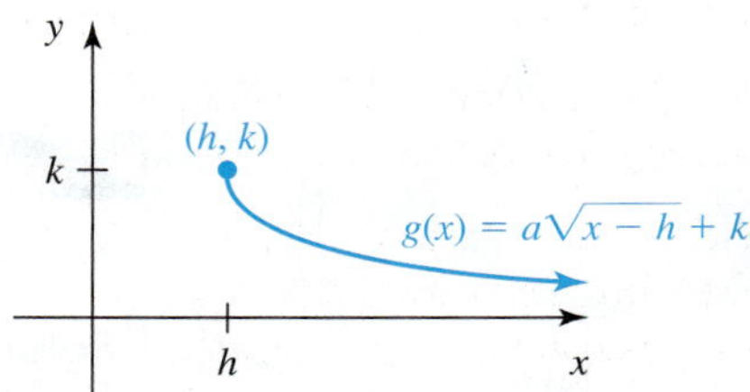

Figure 17 If $a < 0$, then g is decreasing with maximum point (h, k)

If $a > 1$, then the graph of $g(x) = a\sqrt{x - h} + k$ rises more quickly than the graph of $y = \sqrt{x - h} + k$. If $0 < a < 1$, then the graph rises more slowly.

Three-Step Method for Graphing a Square Root Function

To sketch the graph of $f(x) = a\sqrt{x - h} + k$, where $a \neq 0$,

1. Sketch the graph of $y = a\sqrt{x}$.
2. Translate the graph sketched in step 1 to the right by h units if $h > 0$ and to the left by $|h|$ units if $h < 0$.
3. Translate the graph sketched in step 2 up by k units if $k > 0$ and down by $|k|$ units if $k < 0$.

The graph of a square root function is called a **square root curve.**

Example 2 Graphing a Square Root Function

Sketch the graph of $g(x) = -2\sqrt{x + 4} - 1$. Find the domain and range of g.

Solution

First, we sketch the graph of $y = -2\sqrt{x}$ in Fig. 18. Next, we translate the graph to the left by 4 units and down by 1 unit.

We check that the solutions of g listed in Table 11 lie on our sketch of the graph of g. Also, we use a graphing calculator to verify our sketch (see Fig. 19).

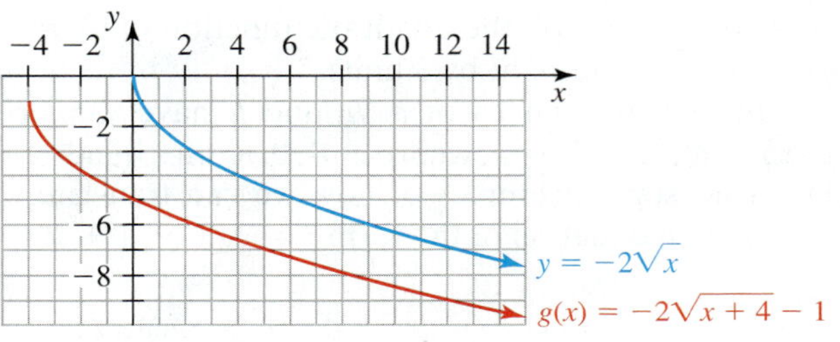

Figure 18 Graphs of $y = -2\sqrt{x}$ and $g(x) = -2\sqrt{x + 4} - 1$

Table 11 Input–Output Pairs of $g(x) = -2\sqrt{x + 4} - 1$

x	$g(x)$
−4	−1
−3	−3
0	−5
5	−7
12	−9

```
WINDOW
 Xmin=-5
 Xmax=15
 Xscl=1
 Ymin=-10
 Ymax=1
 Yscl=1
 Xres=1
```

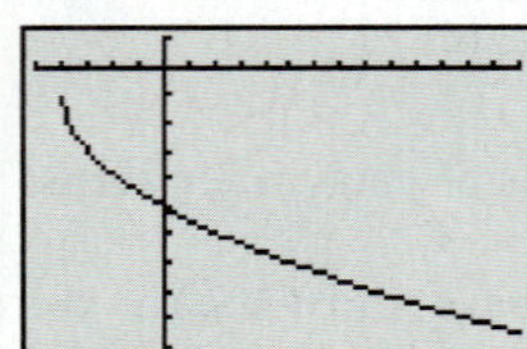

Figure 19 Verify the graph of $g(x) = -2\sqrt{x + 4} - 1$

From the graph of g, we see that the values of x are greater than or equal to -4. So, the domain is the set of numbers x where $x \geq -4$. We can also find the domain by using the fact that the radicand $x + 4$ must be nonnegative:

$$
\begin{aligned}
x + 4 &\geq 0 && \text{Radicand must be nonnegative.} \\
x + 4 - 4 &\geq 0 - 4 && \text{Subtract 4 from both sides.} \\
x &\geq -4
\end{aligned}
$$

From the graph of g, we see that the maximum point is $(-4, -1)$, so the values of y are less than or equal to -1. Therefore, the range is the set of numbers y where $y \leq -1$.

Example 3 Graphing a Square Root Function

Sketch the graph of $f(x) = 2\sqrt{x - 3} + 5$. Find the domain and range of f.

Solution

First, we sketch the graph of $y = 2\sqrt{x}$ in Fig. 20. Next, we translate the graph to the right by 3 units and up by 5 units.

We check that the solutions of f listed in Table 12 lie on our sketch of the graph of f. Also, we use a graphing calculator to verify our sketch (see Fig. 21).

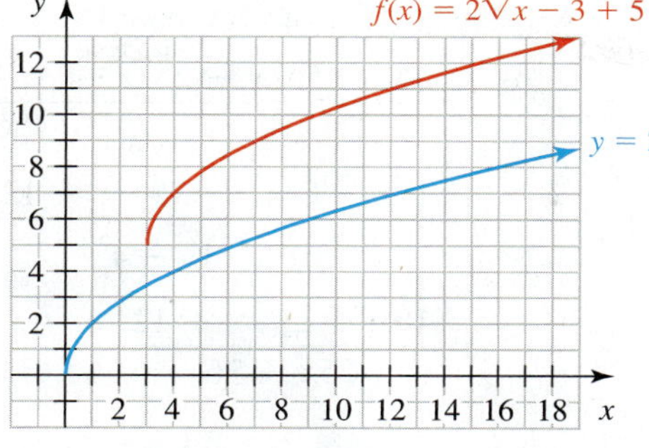

Figure 20 Graphs of $y = 2\sqrt{x}$ and $f(x) = 2\sqrt{x-3} + 5$

Table 12 Input–Output Pairs of $f(x) = 2\sqrt{x-3} + 5$

x	$f(x)$
3	5
4	7
7	9
12	11
19	13

```
WINDOW
 Xmin=-2
 Xmax=19
 Xscl=1
 Ymin=-2
 Ymax=13
 Yscl=1
 Xres=1
```

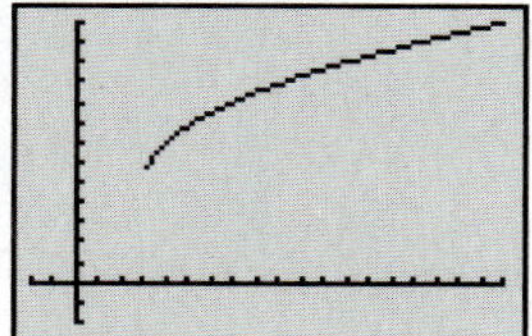

Figure 21 Verify the graph of $f(x) = 2\sqrt{x-3} + 5$

From the graph of f, we see that the values of x are greater than or equal to 3. So, the domain is the set of numbers x where $x \geq 3$. We can also find the domain by using the fact that the radicand $x - 3$ must be nonnegative:

$$x - 3 \geq 0 \quad \text{Radicand must be nonnegative.}$$
$$x - 3 + 3 \geq 0 + 3 \quad \text{Add 3 to both sides.}$$
$$x \geq 3$$

From the graph of f, we see that the minimum point is $(3, 5)$, so the values of y are greater than or equal to 5. Therefore, the range is the set of numbers y where $y \geq 5$.

Performing Operations with Square Root Functions

In Example 4, we will use two square root functions to form a sum function, a difference function, a product function, and a quotient function.

Example 4 Performing Operations with Square Root Functions

Let $f(x) = 2\sqrt{x} - 3$ and $g(x) = 5\sqrt{x} + 4$. For each of the following functions, find an equation of the function.

1. $f + g$ **2.** $f - g$ **3.** $f \cdot g$ **4.** $\dfrac{f}{g}$

Solution

1.
$$(f + g)(x) = f(x) + g(x) = (2\sqrt{x} - 3) + (5\sqrt{x} + 4) \quad \text{Substitute for } f(x) \text{ and } g(x).$$
$$= 7\sqrt{x} + 1 \quad \text{Combine like radicals.}$$

2.
$$(f - g)(x) = f(x) - g(x) = (2\sqrt{x} - 3) - (5\sqrt{x} + 4) \quad \text{Substitute for } f(x) \text{ and } g(x).$$
$$= 2\sqrt{x} - 3 - 5\sqrt{x} - 4 \quad \text{Subtract.}$$
$$= -3\sqrt{x} - 7 \quad \text{Combine like radicals.}$$

3.
$$(f \cdot g)(x) = f(x)g(x) = (2\sqrt{x} - 3)(5\sqrt{x} + 4) \quad \text{Substitute for } f(x) \text{ and } g(x).$$
$$= 2\sqrt{x} \cdot 5\sqrt{x} + 2\sqrt{x} \cdot 4 - 3 \cdot 5\sqrt{x} - 12 \quad \text{Multiply pairs of terms.}$$
$$= 10x + 8\sqrt{x} - 15\sqrt{x} - 12 \quad \text{Simplify.}$$
$$= 10x - 7\sqrt{x} - 12 \quad \text{Combine like radicals.}$$

X	Y1	Y2
0	-.75	-.75
1	-.1111	-.1111
2	-.0155	-.0155
3	.03666	.03666
4	.07143	.07143
5	.09698	.09698
6	.11688	.11688

X=0

Figure 22 Verify the work

4. $\left(\frac{f}{g}\right)(x) = \frac{f(x)}{g(x)} = \frac{2\sqrt{x} - 3}{5\sqrt{x} + 4}$ *Substitute for $f(x)$ and $g(x)$.*

$= \frac{2\sqrt{x} - 3}{5\sqrt{x} + 4} \cdot \frac{5\sqrt{x} - 4}{5\sqrt{x} - 4}$ *Rationalize denominator.*

$= \frac{2\sqrt{x} \cdot 5\sqrt{x} - 2\sqrt{x} \cdot 4 - 3 \cdot 5\sqrt{x} + 12}{\left(5\sqrt{x}\right)^2 - 4^2}$ *Multiply numerators; multiply denominators.*

$= \frac{10x - 8\sqrt{x} - 15\sqrt{x} + 12}{5^2\left(\sqrt{x}\right)^2 - 4^2}$ *Simplify.*

$= \frac{10x - 23\sqrt{x} + 12}{25x - 16}$ *Simplify.*

We verify our work by comparing graphing calculator tables for $y = \frac{2\sqrt{x} - 3}{5\sqrt{x} + 4}$ and $y = \frac{10x - 23\sqrt{x} + 12}{25x - 16}$ for $x > 0$ and $x \neq \frac{16}{25}$ (see Fig. 22).

Group Exploration

Translating and reflecting the absolute value function

1. Complete Table 13 for the absolute value function $y = |x|$.

Table 13 Values of the Function $y = |x|$

x	y
−3	
−2	
−1	
1	
2	
3	

2. Sketch a graph of $y = |x|$. Use a graphing calculator to verify your graph. To enter $|x|$, press MATH ▷ 1 X,T,Θ,n).

3. Translate and/or reflect the graph of $y = |x|$ to sketch the graph of the given function. Use a graphing calculator to verify your sketch.
   a. $y = |x| - 2$ **b.** $y = |x - 4|$
   c. $y = -|x + 3|$ **d.** $y = -|x - 2| + 5$

4. Sketch the graph of the given function.
   a. $y = 2|x|$ **b.** $y = -3|x|$

5. Describe the graphical significance of a, h, and k for a function of the form $y = a|x - h| + k$, where $a \neq 0$.

Homework 13.4

For extra help ▶ MyMathLab®  Watch the videos in MyMathLab  Download the MyDashboard App

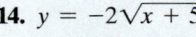

Graph the function by hand. Use a graphing calculator to verify your sketch.

1. $y = 2\sqrt{x}$
2. $y = -3\sqrt{x}$
3. $y = -\sqrt{x}$
4. $y = -\frac{1}{2}\sqrt{x}$
5. $y = \sqrt{x} + 3$
6. $y = \sqrt{x} - 5$
7. $y = 2\sqrt{x} - 5$
8. $y = 3\sqrt{x} - 2$
9. $y = -3\sqrt{x} + 4$
10. $y = -2\sqrt{x} + 1$
11. $y = \sqrt{x - 2}$
12. $y = \sqrt{x - 5}$
13. $y = -\sqrt{x + 2}$
14. $y = -2\sqrt{x + 5}$
15. $y = \frac{1}{2}\sqrt{x - 4}$
16. $y = \frac{1}{4}\sqrt{x - 1}$
17. $y = \sqrt{x + 3} + 2$
18. $y = \sqrt{x + 1} + 3$
19. $y = -2\sqrt{x + 3} - 4$
20. $y = -3\sqrt{x + 2} - 1$
21. $y = 4\sqrt{x - 1} - 3$
22. $y = 2\sqrt{x - 6} - 2$
23. $\sqrt{x} + y = 4$
24. $2\sqrt{x} - y = 3$
25. $2y - 6\sqrt{x} = 8$
26. $3y - 6\sqrt{x} = 9$

Graph the function by hand. Use a graphing calculator to verify your sketch. Also, find the domain and range of the function.

27. $y = -2\sqrt{x}$

28. $y = 3\sqrt{x}$

29. $y = \sqrt{x + 2}$

30. $y = \sqrt{x - 6}$

31. $y = \sqrt{x} + 2$

32. $y = \sqrt{x} - 1$

33. $y = \sqrt{x - 5} - 3$

34. $y = \sqrt{x + 1} + 2$

35. $y = 2\sqrt{x + 5} + 1$

36. $y = 3\sqrt{x - 3} - 6$

37. $y = -\sqrt{x - 2} + 4$

38. $y = -2\sqrt{x + 4} - 2$

39. Recall that we can describe some or all of the input–output pairs of a function by means of an equation, a graph, a table, or words. Let $f(x) = 2\sqrt{x - 3}$.
a. Describe five input–output pairs of f by using a table.
b. Describe the input–output pairs of f by using a graph.
c. Describe the input–output pairs of f by using words.

40. Recall that we can describe some or all of the input–output pairs of a function by means of an equation, a graph, a table, or words. Let $g(x) = -\sqrt{x + 4}$.
a. Describe five input–output pairs of g by using a table.
b. Describe the input–output pairs of g by using a graph.
c. Describe the input–output pairs of g by using words.

Evaluate the function $f(x) = 7\sqrt{x} - 3$ at the indicated value of x. Assume $c \geq 0$.

41. $f(4)$

42. $f(0)$

43. $f(9c)$

44. $f(4c)$

For $f(x) = 5\sqrt{x} - 9$ and $g(x) = 4\sqrt{x} + 1$, find an equation of the given function.

45. $f + g$

46. $f - g$

47. $f \cdot g$

48. $\dfrac{f}{g}$

For $f(x) = 2\sqrt{x} - 3\sqrt{5}$ and $g(x) = 2\sqrt{x} + 3\sqrt{5}$, find an equation of the given function.

49. $f - g$

50. $f + g$

51. $\dfrac{f}{g}$

52. $f \cdot g$

For $f(x) = \sqrt{x + 1} - 2$ and $g(x) = \sqrt{x + 1} + 2$, find an equation of the given function.

53. $f + g$

54. $f - g$

55. $f \cdot g$

56. $\dfrac{f}{g}$

57. The percentages of e-mail that is spam are shown in Table 14 for various years.

Table 14 Percentages of E-Mail That Is Spam

Year	Percent
1999	21
2002	56
2004	68
2006	80
2008	81
2010	85

Source: *IronPort*

Let $f(t)$ be the percentage of e-mail that is spam at t years since 1999. A reasonable model is $f(t) = 20.4\sqrt{t} + 21$.
a. Use a graphing calculator to draw the graph of the model and, in the same viewing window, the scattergram of the data. Does the model fit the data well?
b. Estimate the percentage of e-mail that was spam in 2005. Did you perform interpolation or extrapolation? Explain.
c. Predict the percentage of e-mail that will be spam in 2013. Did you perform interpolation or extrapolation? Explain.

58. The global airline industry's revenues from fees not included in fares are shown in Table 15 for various years.

Table 15 Global Airline Industry's Revenues from Fees Not Included in Fares

Year	Revenue (billions of dollars)
2007	2.5
2008	10.3
2009	13.5
2010	21.5
2011	22.6

Source: *IdeaWorks and Amadeus*

Let $f(t)$ be the global airline industry's annual revenue (in billions of dollars) from fees not included in fares at t years since 2007. A reasonable model is $f(t) = 10\sqrt{t} + 2.5$.
a. Use a graphing calculator to draw the graph of the model and, in the same viewing window, the scattergram of the data. Does the model fit the data well?
b. Estimate the global airline industry's revenue from fees not included in fares in 2010. Is your result an underestimate or an overestimate? Did you perform interpolation or extrapolation? Explain.
c. Predict the global airline industry's revenue from fees not included in fares in 2018. Did you perform interpolation or extrapolation? Explain.

For Exercises 59–66, refer to Fig. 23.

59. Estimate $f(-6)$.

60. Estimate $f(-2)$.

61. Estimate $f(0)$.

62. Estimate $f(5)$.

63. Estimate x when $f(x) = 0$.

64. Estimate x when $f(x) = 2$.

65. Estimate x when $f(x) = 3$.

66. Estimate x when $f(x) = 4$.

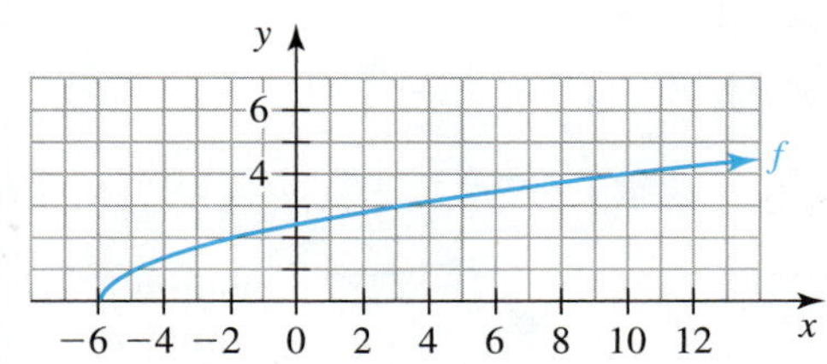

Figure 23 Exercises 59–66

Concepts

67. Figure 24 shows four functions of the form $y = a\sqrt{x - h} + k$. For each, describe the signs of the constants a, h, and k.

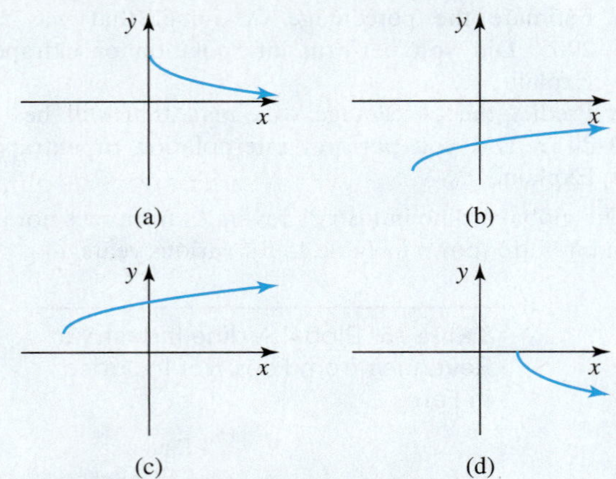

Figure 24 Exercise 67

68. For each part, copy the graph of $f(x) = a\sqrt{x - h} + k$ as shown in Fig. 25. On the same coordinate system, use the graph of f to sketch the graph of the given function. Label each graph.

a. $g(x) = a\sqrt{x - h} + 2k$
b. $h(x) = a\sqrt{x - 2h} + k$
c. $k(x) = -a\sqrt{x - h} + k$
d. $r(x) = 2a\sqrt{x + h} - k$

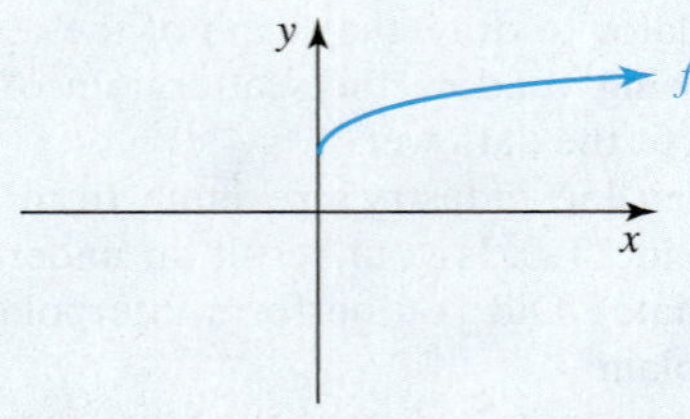

Figure 25 Exercise 68

69. Use a graphing calculator to graph a family of square root curves similar to the one in Fig. 26. List the equations of your square root curves.

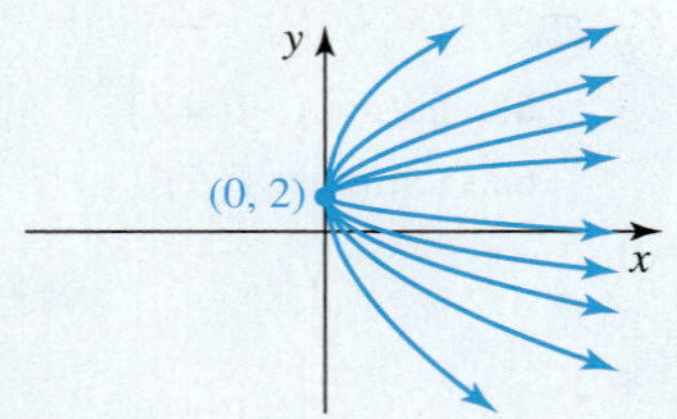

Figure 26 A family of square root curves—Exercise 69

70. Use a graphing calculator to graph a family of square root curves similar to the one in Fig. 27. List the equations of your square root curves.

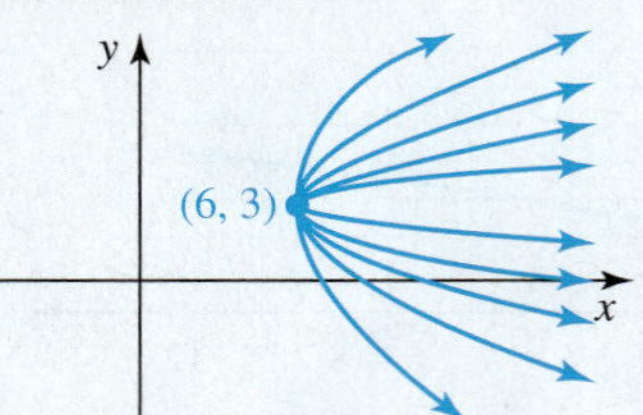

Figure 27 Another family of square root curves—Exercise 70

71. For what values of a, h, and k for the square root function $f(x) = a\sqrt{x - h} + k$, where $a \neq 0$, is there a point on the graph of f that is higher than all other points on the graph? (Recall that this point is the maximum point.) For what values of a, h, and k does f have a minimum point? Describe the maximum or minimum point in terms of a, h, and k.

72. Solve the system

$$y = 2\sqrt{x - 3} + 1$$
$$y = -2\sqrt{x - 3} + 1$$

73. Use ZStandard followed by ZSquare to draw the graphs of $f(x) = 2\sqrt{x + 3} + 2$ and $g(x) = -2\sqrt{x + 3} + 2$ on the same coordinate system. Consider the combined graph as the graph of a single relation. What do you notice about the graph? Is the relation described by this graph a function? Explain.

74. Find two functions that have the same domain and range as the function $f(x) = \sqrt{x + 7} - 2$.

75. Explain why the graphs of $f(x) = -a\sqrt{x}$ and $g(x) = a\sqrt{x}$ are reflections of each other across the x-axis.

76. Let k be a positive constant. To graph the function $g(x) = \sqrt{x} + k$, we translate the graph of $f(x) = \sqrt{x}$ up by k units. Explain why this makes sense.

77. Describe how to graph the function $g(x) = a\sqrt{x - h} + k$, where $a \neq 0$, given the graph of $f(x) = \sqrt{x}$.

78. Compare the process of graphing a square root function of the form $f(x) = a\sqrt{x - h} + k$, where $a \neq 0$, with that of graphing a quadratic function of the form $g(x) = a(x - h)^2 + k$. How are the processes similar? different?

Related Review

Graph the function by hand.

79. $2x - 5y = 20$

80. $y = -2(x - 4)^2 + 3$

81. $y = 2\sqrt{x + 3} - 4$

82. $y = -3\sqrt{x - 1} + 3$

83. $y = 8\left(\frac{1}{2}\right)^x$

84. $y = 3x^2 - 12x + 9$

For $f(x) = \sqrt{3x + 2} - 7$ and $g(x) = 4x + 5$, find an equation of the given composite function.

85. $f \circ g$

86. $g \circ f$

For $f(x) = \sqrt{x + 3}$ and $g(x) = x^2 + 2$, find an equation of the given composite function.

87. $f \circ g$

88. $g \circ f$

Expressions, Equations, Functions, and Graphs

Perform the indicated instruction. Then use words such as linear, quadratic, cubic, exponential, logarithmic, rational, radical, polynomial, degree, function, one variable, *and* two variables *to describe the expression, equation, or system.*

89. Factor $6x^2 - 5x - 6$.

90. Find the product $(3\sqrt{x} - 5)(2\sqrt{x} + 4)$.

91. Let $f(x) = 3x^2 - 2x + 4$. Find x when $f(x) = 6$.

92. Graph $f(x) = -2\sqrt{x - 3} + 1$ by hand.

93. Find an equation of a parabola that contains the points $(1, 4)$, $(3, 14)$, and $(4, 25)$.

94. Graph $f(x) = -2(x - 3)^2 + 1$.

13.5 Solving Radical Equations

Objectives

- » Solve radical equations.
- » Find the x-intercept of the graph of a square root function.
- » Use a radical model to make predictions about the independent variable.

In Sections 13.1–13.3, we worked with radical *expressions*. In this section, we solve radical *equations*. A **radical equation in one variable** is an equation in one variable that contains a radical with the variable in the radicand. Here are some examples of a radical equation in one variable:

$$\sqrt{x} = 5 \qquad \sqrt[3]{7x - 3} = 2 \qquad \sqrt{x - 2} = x - 4 \qquad \sqrt{x + 3} - \sqrt{2x - 4} = 6$$

We begin by solving square root equations in one variable.

Solving Square Root Equations in One Variable

Consider the square root equation

$$\sqrt{x} = 3$$

If x is nonnegative, then $\left(\sqrt{x}\right)^2 = x$. To get the left side of the equation $\sqrt{x} = 3$ to be x, we square both sides:

$$\left(\sqrt{x}\right)^2 = 3^2$$
$$x = 9$$

So, the solution is 9. This checks out, because $\sqrt{9} = 3$.

Squaring Property of Equality

If A and B are expressions, then all solutions of the equation $A = B$ are *among* the solutions of the equation $A^2 = B^2$. That is, the solutions of an equation are among the solutions of the equation obtained by squaring both sides.

Recall from Section 12.5 that if we clear a rational equation of fractions and arrive at a value of x that is an excluded value, then we call that result an extraneous solution. In general, if a proposed solution of any type of equation is *not* a solution, we call it an **extraneous solution.**

Squaring both sides of an equation can introduce extraneous solutions. Consider the simple equation $x = 5$, whose only solution is 5. Here we square both sides of $x = 5$ and solve by using the square root property (Section 9.4):

$x = 5$	*The only solution is 5.*
$x^2 = 5^2$	*Square both sides.*
$x^2 = 25$	$5^2 = 25$
$x = \pm\sqrt{25}$	*Square root property*
$x = \pm 5$	$\sqrt{25} = 5$

Squaring both sides of the equation $x = 5$ introduced the extraneous solution -5 (which is *not* a solution of the original equation).

Checking Proposed Solutions

Because squaring both sides of a square root equation may introduce extraneous solutions, it is essential to check that each proposed solution satisfies the original equation.

Example 1 Solving a Square Root Equation

Solve $\sqrt{6x - 2} = 4$.

Solution

$$\begin{aligned}
\sqrt{6x-2} &= 4 && \textit{Original equation}\\
\left(\sqrt{6x-2}\right)^2 &= 4^2 && \textit{Square both sides.}\\
6x-2 &= 16 && (\sqrt{x})^2 = x \text{ for } x \ge 0\\
6x &= 18 && \textit{Add 2 to both sides.}\\
x &= 3 && \textit{Divide both sides by 6.}
\end{aligned}$$

We check that 3 satisfies the original equation:

$$\begin{aligned}
\sqrt{6x-2} &= 4 && \textit{Original equation}\\
\sqrt{6(3)-2} &\stackrel{?}{=} 4 && \textit{Substitute 3 for x.}\\
\sqrt{16} &\stackrel{?}{=} 4 && \textit{Simplify.}\\
4 &\stackrel{?}{=} 4 && \sqrt{16} = 4\\
&\text{true}
\end{aligned}$$

So, the solution is 3.

Example 2 Solving a Square Root Equation

Solve $\sqrt{x+5} = -3$.

Solution

$$\begin{aligned}
\sqrt{x+5} &= -3 && \textit{Original equation}\\
\left(\sqrt{x+5}\right)^2 &= (-3)^2 && \textit{Square both sides.}\\
x+5 &= 9 && (\sqrt{x})^2 = x \text{ for } x \ge 0\\
x &= 4 && \textit{Subtract 5 from both sides.}
\end{aligned}$$

We check that 4 satisfies the original equation:

$$\begin{aligned}
\sqrt{x+5} &= -3 && \textit{Original equation}\\
\sqrt{4+5} &\stackrel{?}{=} -3 && \textit{Substitute 4 for x.}\\
\sqrt{9} &\stackrel{?}{=} -3 && \textit{Add.}\\
3 &\stackrel{?}{=} -3 && \sqrt{9} = 3\\
&\text{false}
\end{aligned}$$

The result, 4, is an extraneous solution—it is *not* a solution. The solution set is the empty set.

In Example 2, we introduced an extraneous solution by squaring both sides of an equation. Remember that if you square both sides of an equation, you must check that each proposed solution satisfies the original equation.

Example 3 Solving a Square Root Equation

Solve the equation $2\sqrt{x} + 5 = 13$.

Solution

First, we isolate $\sqrt{x}$ (get $\sqrt{x}$ alone) on one side of the equation:

$$\begin{aligned}
2\sqrt{x} + 5 &= 13 && \textit{Original equation}\\
2\sqrt{x} &= 8 && \textit{Subtract 5 from both sides.}\\
\sqrt{x} &= 4 && \textit{Divide both sides by 2.}
\end{aligned}$$

$$(\sqrt{x})^2 = 4^2 \quad \textit{Square both sides.}$$
$$x = 16 \quad (\sqrt{x})^2 = x \text{ for } x \geq 0; 4^2 = 16$$

We check that 16 satisfies the original equation:

$$2\sqrt{x} + 5 = 13 \quad \textit{Original equation}$$
$$2\sqrt{16} + 5 \stackrel{?}{=} 13 \quad \textit{Substitute 16 for x.}$$
$$2 \cdot 4 + 5 \stackrel{?}{=} 13 \quad \sqrt{16} = 4$$
$$13 \stackrel{?}{=} 13$$
true

So, the solution is 16.

We see from Example 3 that, to solve a square root equation, we isolate a square root term on one side of the equation before we square both sides.

Example 4 Solving a Square Root Equation

Solve $x = \sqrt{x-1} + 3$.

Solution

$$x = \sqrt{x-1} + 3 \quad \textit{Original equation}$$
$$x - 3 = \sqrt{x-1} \quad \textit{Isolate radical.}$$
$$(x-3)^2 = (\sqrt{x-1})^2 \quad \textit{Square both sides.}$$
$$x^2 - 6x + 9 = x - 1 \quad (A-B)^2 = A^2 - 2AB + B^2; (\sqrt{x})^2 = x \text{ for } x \geq 0$$
$$x^2 - 7x + 10 = 0 \quad \textit{Write in } ax^2 + bx + c = 0 \textit{ form.}$$
$$(x-2)(x-5) = 0 \quad \textit{Factor left side.}$$
$$x - 2 = 0 \quad \text{or} \quad x - 5 = 0 \quad \textit{Zero factor property}$$
$$x = 2 \quad \text{or} \quad x = 5$$

We check that 2 and 5 satisfy the original equation:

Check $x = 2$	***Check*** $x = 5$
$x = \sqrt{x-1} + 3$	$x = \sqrt{x-1} + 3$
$2 \stackrel{?}{=} \sqrt{2-1} + 3$	$5 \stackrel{?}{=} \sqrt{5-1} + 3$
$2 \stackrel{?}{=} 4$	$5 \stackrel{?}{=} 5$
false	true

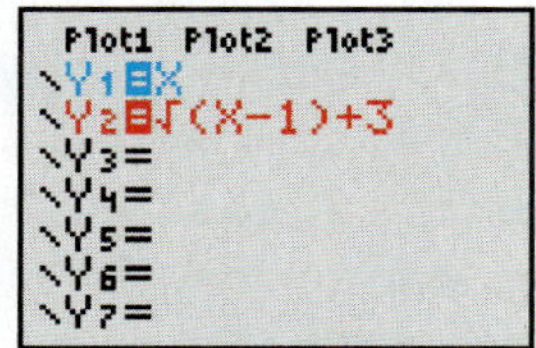

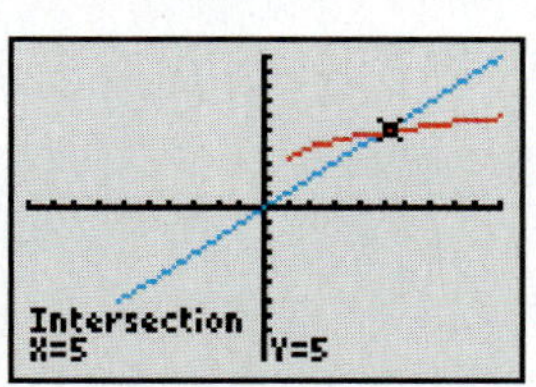

Figure 28 Verify the work

Since 2 does not satisfy the original equation, it is an extraneous solution. Therefore, the only solution is 5.

We use "intersect" on a graphing calculator to verify our work (see Fig. 28). The point $(5, 5)$ is the only point of intersection. This supports our conclusion that the only solution of the original equation is 5.

Example 5 Solving a Square Root Equation

Solve $2\sqrt{x-3} - \sqrt{x} = 0$.

Solution

We isolate $2\sqrt{x-3}$, square both sides of the equation, and solve for x:

$$2\sqrt{x-3} - \sqrt{x} = 0 \quad \textit{Original equation}$$
$$2\sqrt{x-3} = \sqrt{x} \quad \textit{Isolate } 2\sqrt{x-3}.$$

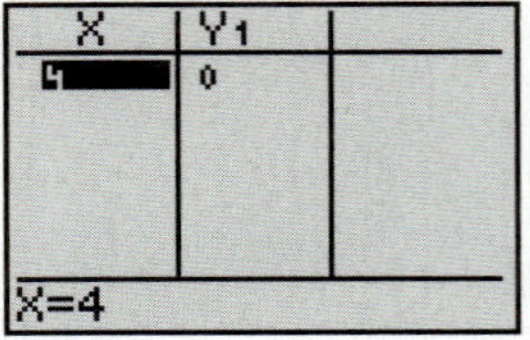

Figure 29 Verify that 4 is a solution

$$
\begin{aligned}
\left(2\sqrt{x-3}\right)^2 &= \left(\sqrt{x}\right)^2 && \text{Square both sides.}\\
2^2\left(\sqrt{x-3}\right)^2 &= \left(\sqrt{x}\right)^2 && (AB)^2 = A^2B^2\\
4(x-3) &= x && \left(\sqrt{x}\right)^2 = x \text{ for } x \geq 0\\
4x - 12 &= x && \text{Distributive law}\\
3x &= 12\\
x &= 4
\end{aligned}
$$

To verify that 4 is a solution, we enter the function $y = 2\sqrt{x-3} - \sqrt{x}$. Then we check that, for the input 4, the output is 0 (see Fig. 29).

In Example 5, after squaring both sides of $2\sqrt{x-3} = \sqrt{x}$, we simplified $\left(2\sqrt{x-3}\right)^2$:

$$\left(2\sqrt{x-3}\right)^2 = 2^2\left(\sqrt{x-3}\right)^2 = 4(x-3) \quad \text{Correct}$$

WARNING It is a common error to forget to square the coefficient 2 and write

$$\left(2\sqrt{x-3}\right)^2 = 2(x-3) \quad \text{Incorrect}$$

Recall from Section 7.3 that when we simplify $(AB)^2$, we square both A and B:

$$(AB)^2 = A^2B^2$$

If an equation contains two or more square root terms, we may need to use the squaring property of equality twice.

Example 6 Solving a Square Root Equation

Solve $\sqrt{t-5} - \sqrt{t} = -1$.

Solution

First, we isolate the radical $\sqrt{t-5}$; then we square both sides of the equation:

$$
\begin{aligned}
\sqrt{t-5} - \sqrt{t} &= -1 && \text{Original equation}\\
\sqrt{t-5} &= \sqrt{t} - 1 && \text{Isolate radical } \sqrt{t-5}.\\
\left(\sqrt{t-5}\right)^2 &= \left(\sqrt{t} - 1\right)^2 && \text{Square both sides.}\\
t - 5 &= \left(\sqrt{t}\right)^2 - 2\sqrt{t}\cdot 1 + 1^2 && \left(\sqrt{x}\right)^2 = x \text{ for } x \geq 0;\ (A-B)^2 = A^2 - 2AB + B^2\\
t - 5 &= t - 2\sqrt{t} + 1 && \left(\sqrt{x}\right)^2 = x \text{ for } x \geq 0
\end{aligned}
$$

Next, we isolate the radical $\sqrt{t}$ and square both sides:

$$
\begin{aligned}
2\sqrt{t} &= t + 1 - t + 5 && \text{Isolate } 2\sqrt{t}.\\
2\sqrt{t} &= 6 && \text{Combine like terms.}\\
\sqrt{t} &= 3 && \text{Divide both sides by 2.}\\
\left(\sqrt{t}\right)^2 &= 3^2 && \text{Square both sides.}\\
t &= 9 && \left(\sqrt{x}\right)^2 = x \text{ for } x \geq 0
\end{aligned}
$$

Now we check that 9 satisfies the original equation:

$$
\begin{aligned}
\sqrt{t-5} - \sqrt{t} &= -1 && \text{Original equation}\\
\sqrt{9-5} - \sqrt{9} &\stackrel{?}{=} -1 && \text{Substitute 9 for } t.\\
2 - 3 &\stackrel{?}{=} -1\\
-1 &\stackrel{?}{=} -1\\
&\text{true}
\end{aligned}
$$

So, the solution is 9.

Solving a Square Root Equation in One Variable

To solve a square root equation in one variable,

1. Isolate a square root term on one side of the equation.
2. Square both sides.
3. Repeat steps 1 and 2 until no square root terms remain.
4. Solve the new equation.
5. Check that each proposed solution satisfies the original equation.

Example 7 Solving a Square Root Equation

Solve $\sqrt{2x+1} - \sqrt{x+2} = 1$.

Solution

$$\sqrt{2x+1} - \sqrt{x+2} = 1 \quad \text{Original equation}$$
$$\sqrt{2x+1} = \sqrt{x+2} + 1 \quad \text{Isolate radical } \sqrt{2x+1}.$$
$$\left(\sqrt{2x+1}\right)^2 = \left(\sqrt{x+2}+1\right)^2 \quad \text{Square both sides.}$$
$$2x+1 = \left(\sqrt{x+2}\right)^2 + 2\sqrt{x+2}\cdot 1 + 1^2 \quad \left(\sqrt{x}\right)^2 = x \text{ for } x \ge 0, \ (A+B)^2 = A^2 + 2AB + B^2$$
$$2x+1 = x+2+2\sqrt{x+2}+1 \quad \left(\sqrt{x}\right)^2 = x \text{ for } x \ge 0$$
$$x-2 = 2\sqrt{x+2} \quad \text{Isolate radical } 2\sqrt{x+2}.$$
$$(x-2)^2 = \left(2\sqrt{x+2}\right)^2 \quad \text{Square both sides.}$$
$$x^2 - 4x + 4 = 4(x+2) \quad (A-B)^2 = A^2 - 2AB + B^2; \ (bc)^n = b^n c^n$$
$$x^2 - 4x + 4 = 4x + 8 \quad \text{Distributive law}$$
$$x^2 - 8x - 4 = 0 \quad \text{Write in } ax^2 + bx + c = 0 \text{ form.}$$
$$x = \frac{-(-8) \pm \sqrt{(-8)^2 - 4(1)(-4)}}{2(1)} \quad \text{Substitute } a = 1, b = -8, c = -4 \text{ into quadratic formula.}$$
$$= \frac{8 \pm \sqrt{80}}{2} \quad \text{Simplify.}$$
$$= \frac{8 \pm 4\sqrt{5}}{2} \quad \sqrt{80} = \sqrt{16\cdot 5} = 4\sqrt{5}$$
$$= \frac{2\left(4 \pm 2\sqrt{5}\right)}{2} \quad \text{Factor out 2.}$$
$$= 4 \pm 2\sqrt{5} \quad \text{Simplify.}$$

So, $x = 4 - 2\sqrt{5} \approx -0.47$ or $x = 4 + 2\sqrt{5} \approx 8.47$. We check that the approximations of the solutions approximately satisfy the original equation $\sqrt{2x+1} - \sqrt{x+2} = 1$:

Check x ≈ −0.47	***Check x ≈ 8.47***
$\sqrt{2(-0.47)+1} - \sqrt{-0.47+2} \approx -0.9920$	$\sqrt{2(8.47)+1} - \sqrt{8.47+2} \approx 0.9998$
not close to 1	close to 1

So, the only solution is $4 + 2\sqrt{5}$.

Now we store $4 - 2\sqrt{5}$ for the variable "X" in a graphing calculator and perform a similar, but more precise, check. We do the same for $4 + 2\sqrt{5}$ (see Fig. 30).

```
4-2√(5)→X
      -.472135955
√(2X+1)-√(X+2)
               -1
```

```
4+2√(5)→X
      8.472135955
√(2X+1)-√(X+2)
                1
```

Figure 30 Verify that $4 - 2\sqrt{5}$ is an extraneous solution and $4 + 2\sqrt{5}$ is a solution

In Example 7, we simplified the right-hand side of the equation

$$\left(\sqrt{2x+1}\right)^2 = \left(\sqrt{x+2}+1\right)^2$$

by using the property for the square of a sum, $(A+B)^2 = A^2 + 2AB + B^2$:

$$\left(\sqrt{x+2}+1\right)^2 = \left(\sqrt{x+2}\right)^2 + 2\sqrt{x+2}\cdot 1 + 1^2 \quad \text{Correct}$$

WARNING Remember that, in general, $(A + B)^2$ is *not* equal to $A^2 + B^2$, so it is incorrect to say that

$$\left(\sqrt{x+2}+1\right)^2 = \left(\sqrt{x+2}\right)^2 + 1^2 \quad \textit{Incorrect}$$

Example 8 Solving a General Square Root Equation

Solve the equation $F = \frac{1}{2L}\sqrt{\frac{T}{p}}$ for p. Assume the constants have values for which the equation has exactly one real-number solution.

Solution

First, we square both sides of the equation:

$$F = \frac{1}{2L}\sqrt{\frac{T}{p}} \qquad \textit{Original equation}$$

$$F^2 = \left(\frac{1}{2L}\sqrt{\frac{T}{p}}\right)^2 \qquad \textit{Square both sides.}$$

$$F^2 = \left(\frac{1}{2L}\right)^2\left(\sqrt{\frac{T}{p}}\right)^2 \qquad (AB)^2 = A^2B^2$$

$$F^2 = \frac{1}{4L^2}\cdot\frac{T}{p} \qquad \left(\frac{1}{2L}\right)^2 = \frac{1^2}{(2L)^2} = \frac{1}{2^2L^2} = \frac{1}{4L^2}; \left(\sqrt{x}\right)^2 = x \textit{ for } x \ge 0$$

$$F^2 = \frac{T}{4L^2p} \qquad \textit{Multiply numerators; multiply denominators.}$$

$$F^2 \cdot p = \frac{T}{4L^2p}\cdot p \qquad \textit{Multiply both sides by } p.$$

$$F^2p = \frac{T}{4L^2} \qquad \textit{Simplify.}$$

$$\frac{1}{F^2}\cdot F^2p = \frac{1}{F^2}\cdot\frac{T}{4L^2} \qquad \textit{Multiply both sides by } \frac{1}{F^2}.$$

$$p = \frac{T}{4F^2L^2} \qquad \textit{Simplify; multiply numerators; multiply denominators.}$$

The equation $\sqrt{x+8} = x^2 - 3x - 4$ would be very difficult to solve by using properties such as the squaring property of equality. In Example 9, we will use graphing to find approximate solutions of the equation.

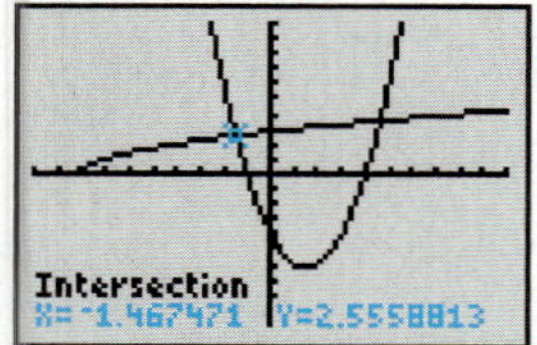

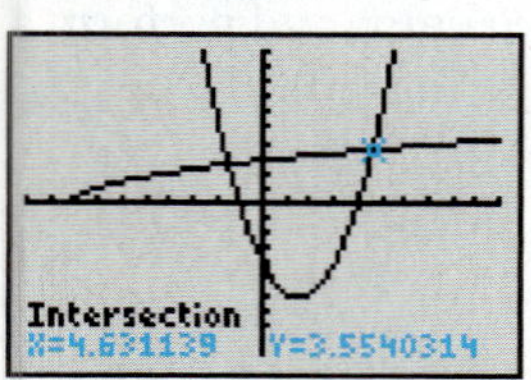

Figure 31 Solve the system

Example 9 Using Graphing to Solve a Square Root Equation in One Variable

Use graphing to solve $\sqrt{x+8} = x^2 - 3x - 4$.

Solution

We use "intersect" on a graphing calculator to find the solutions of the system

$$y = \sqrt{x+8}$$
$$y = x^2 - 3x - 4$$

See Fig. 31.

The approximate solutions of the system are $(-1.47, 2.56)$ and $(4.63, 3.55)$. The x-coordinates of these ordered pairs, -1.47 and 4.63, are the approximate solutions of the equation $\sqrt{x+8} = x^2 - 3x - 4$.

Finding x-Intercepts of the Graph of a Square Root Function

To find all x-intercepts of the graph of a square root equation in x and y, we substitute 0 for y and solve for x.

Example 10 Finding the x-Intercept of a Square Root Function

Find the x-intercepts of the graph of $f(x) = \sqrt{3x - 4} - \sqrt{x + 2}$.

Solution

We substitute 0 for $f(x)$ and solve for x:

$$\begin{aligned} \sqrt{3x-4} - \sqrt{x+2} &= 0 && \textit{Substitute 0 for } f(x). \\ \sqrt{3x-4} &= \sqrt{x+2} && \textit{Isolate at least one radical.} \\ \left(\sqrt{3x-4}\right)^2 &= \left(\sqrt{x+2}\right)^2 && \textit{Square both sides.} \\ 3x - 4 &= x + 2 && \left(\sqrt{x}\right)^2 = x \text{ for } x \geq 0 \\ 2x &= 6 \\ x &= 3 \end{aligned}$$

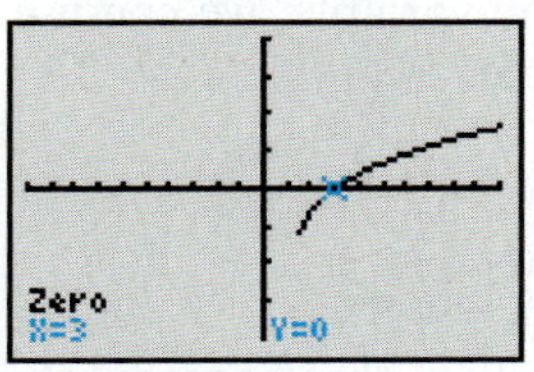

Figure 32 Verify the work

We check that 3 satisfies the equation $\sqrt{3x - 4} - \sqrt{x + 2} = 0$:

$$\begin{aligned} \sqrt{3x-4} - \sqrt{x+2} &= 0 && \textit{Original equation} \\ \sqrt{3(3)-4} - \sqrt{3+2} &\stackrel{?}{=} 0 && \textit{Substitute 3 for } x. \\ \sqrt{5} - \sqrt{5} &\stackrel{?}{=} 0 \\ 0 &\stackrel{?}{=} 0 \\ &\text{true} \end{aligned}$$

The x-intercept is (3, 0). We use "zero" on a graphing calculator to verify our work (see Fig. 32).

Solving Radical Equations

So far, we have solved square root equations in one variable. We can solve other types of radical equations in one variable in a similar way.

Power Property of Equality

If A and B are expressions and n is a counting number greater than 1, then all solutions of the equation $A = B$ are *among* the solutions of the equation $A^n = B^n$. That is, the solutions of an equation are among the solutions of the equation obtained by raising both sides to the nth power.

Example 11 Solving a Radical Equation

Solve $\sqrt[3]{2w - 4} + 7 = 9$.

Solution

We isolate $\sqrt[3]{2w - 4}$ on one side of the equation:

$$\begin{aligned} \sqrt[3]{2w-4} + 7 &= 9 && \textit{Original equation} \\ \sqrt[3]{2w-4} &= 2 && \textit{Subtract 7 from both sides.} \end{aligned}$$

From the power property for radicals (Section 13.2), $\left(\sqrt[3]{2w - 4}\right)^3 = 2w - 4$. To get the left side of the equation $\sqrt[3]{2w - 4} = 2$ to be $2w - 4$, we cube both sides:

$$\begin{aligned} \left(\sqrt[3]{2w-4}\right)^3 &= 2^3 && \textit{Cube both sides.} \\ 2w - 4 &= 8 && \left(\sqrt[3]{x}\right)^3 = x \\ 2w &= 12 \\ w &= 6 \end{aligned}$$

We check that 6 satisfies the original equation:

$$\sqrt[3]{2w-4}+7=9 \quad \text{Original equation}$$
$$\sqrt[3]{2(6)-4}+7 \stackrel{?}{=} 9 \quad \text{Substitute 6 for } w.$$
$$\sqrt[3]{8}+7 \stackrel{?}{=} 9$$
$$2+7 \stackrel{?}{=} 9 \quad \sqrt[3]{8}=2$$
$$9 \stackrel{?}{=} 9$$
$$\text{true}$$

So, the solution is 6.

It is only when we raise both sides of an equation to an *even* power that we might introduce extraneous solutions. However, no matter what type of equation we are solving, it is a good idea to check that any results satisfy the original equation to make sure the work is correct.

Using a Radical Model to Make Predictions about the Independent Variable

Now that we have discussed how to solve radical equations in one variable, we can use radical models to make predictions about the independent variable.

Table 16 Percentages of Foundations That Compensate All of Their Board Members

Asset Group (millions of dollars)	Asset Used to Represent Asset Group (millions of dollars)	Percent
0–5	2.5	4
5–10	7.5	8
10–25	17.5	10
25–50	37.5	15
50–100	75	20
100–250	175	21
250–500	375	32

Source: *The New York Times*

Example 12 Using a Radical Model to Make Predictions

In Example 10 of Section 13.1, we worked with the model $f(a)=1.5\sqrt{a}+4$, where $f(a)$ is the percentage of foundations with assets of a million dollars that compensate all of their board members (see Table 16). Estimate the assets of foundations of which 25% of the foundations compensate all of their board members.

Solution

We substitute 25 for $f(a)$ and solve for a:

$$25=1.5\sqrt{a}+4 \quad \text{Substitute 25 for } f(a).$$
$$21=1.5\sqrt{a} \quad \text{Subtract 4 from both sides.}$$
$$14=\sqrt{a} \quad \text{Divide both sides by 1.5.}$$
$$14^2=\left(\sqrt{a}\right)^2 \quad \text{Square both sides.}$$
$$a=196$$

The model estimates that 25% of foundations with assets of $196 million compensate all of their board members.

Group Exploration

Solving square root equations

1. A student tries to solve $\sqrt{x}+3=5$:

$$\sqrt{x}+3=5$$
$$(\sqrt{x})^2+3^2=5^2$$
$$x+9=25$$
$$x=16$$

Describe any errors. Then solve the equation correctly.

2. A student tries to solve $\sqrt{2x-5}=-3$:

$$\sqrt{2x-5}=-3$$
$$(\sqrt{2x-5})^2=(-3)^2$$
$$2x-5=9$$
$$2x=14$$
$$x=7$$

Describe any errors. Then solve the equation correctly.

3. A student tries to solve $3\sqrt{x} - 2 = 5$:

$$3\sqrt{x} - 2 = 5$$
$$3\sqrt{x} = 7$$
$$(3\sqrt{x})^2 = 7^2$$
$$3x = 49$$
$$x = \frac{49}{3}$$

Describe any errors. Then solve the equation correctly.

Group Exploration

Extraneous solutions

1. Solve the equation $\sqrt{3x-2} + 2 = x$. Record each step of your work carefully.

2. In Problem 1, you found that 1 is an extraneous solution and that 6 is the only solution. Now substitute 1 for x in each step you recorded in Problem 1. Which of the equations are satisfied by 1?

3. What does it mean to say that 1 is an extraneous solution? Why do we sometimes get extraneous solutions when we solve square root equations but not when we solve linear, exponential, or quadratic equations?

Homework 13.5

For extra help ▶ MyMathLab®  Watch the videos in MyMathLab  Download the MyDashboard App

Solve.

1. $\sqrt{x} = 5$

2. $\sqrt{x} = 8$

3. $\sqrt{x} = -2$

4. $\sqrt{x} = -7$

5. $\sqrt[3]{t} = -2$

6. $\sqrt[4]{w} = -3$

7. $\sqrt{x-1} = 2$

8. $\sqrt{x-5} = 3$

9. $\sqrt[4]{r+2} = 2$

10. $\sqrt[3]{b+7} = 3$

11. $3\sqrt{x} - 1 = 5$

12. $5\sqrt{x} + 2 = 37$

13. $\sqrt{5x-7} + 7 = 3$

14. $\sqrt{15+x} + 8 = 2$

15. $\sqrt[3]{2x-5} + 3 = 7$

16. $\sqrt[4]{3x-1} + 5 = 8$

17. $2 - 10\sqrt{6x+3} = -98$

18. $7 - 8\sqrt{2x-1} = -17$

19. $\sqrt{3k+1} = \sqrt{2k+6}$

20. $\sqrt{10m-3} = \sqrt{6m+2}$

21. $\sqrt[4]{6x-3} = \sqrt[4]{2x+17}$

22. $\sqrt[3]{5x-8} = \sqrt[3]{3x+6}$

23. $2\sqrt{1-x} - \sqrt{2x+5} = 0$

24. $2\sqrt{x-1} - \sqrt{3x-1} = 0$

25. $\sqrt{3w+3} = w - 5$

26. $\sqrt{t+10} = t - 2$

27. $\sqrt{12x+13} + 2 = 3x$

28. $\sqrt{x+2} - x = 2$

29. $\sqrt{3x-4} - x = 3$

30. $1 + \sqrt{2x+5} = 2x$

31. $\sqrt{r^2 - 5r + 1} = r - 3$

32. $\sqrt{k^2 + 2k} = k + 5$

33. $\sqrt{x} - 1 = \sqrt{5-x}$

34. $3 = \sqrt{6+x} + \sqrt{x}$

35. $\sqrt{x} - \sqrt{2x} = -1$

36. $\sqrt{x} = \sqrt{3x} - 2$

37. $\sqrt{x-3} + \sqrt{x+5} = 4$

38. $\sqrt{x+6} - \sqrt{x-2} = 2$

39. $\sqrt{2p-1} + \sqrt{3p-2} = 2$

40. $\sqrt{3t-1} - \sqrt{4t+1} = -1$

41. $\sqrt{\sqrt{x} - 2} = 3$

42. $\sqrt{\sqrt{x} + 1} = 1$

43. $\dfrac{1}{\sqrt{x+2}} = 3 - \sqrt{x+2}$

44. $\dfrac{1}{\sqrt{x-5}} = 2 - \sqrt{x-5}$

Find an approximate solution of the equation. Round your result to the second decimal place.

45. $5.2\sqrt{x} - 2.8 = 13.9$

46. $4.7\sqrt{x} + 3.1 = 46.9$

47. $1.52 - 4.91\sqrt{3.18x - 7.14} = -0.69$

48. $-7.93 = 5.61 - 3.79\sqrt{4.42 - 9.87x}$

Solve for the specified variable. Assume the constants have values for which the equation has exactly one real-number solution.

49. $S = \sqrt{gd}$, for d

50. $r = 30d^2\sqrt{P}$, for P

51. $d = \sqrt{\dfrac{3h}{2}}$, for h

52. $T = \sqrt{\dfrac{2d}{g}}$, for d

53. $v = \sqrt{\dfrac{2GM}{R}}$, for R

54. $S = 2\pi\sqrt{\dfrac{L}{32}}$, for L

Use "intersect" on a graphing calculator to solve the equation. Round any solutions to the second decimal place.

55. $\sqrt{x+1} = 4 - \sqrt{x+3}$

56. $\sqrt{x+6} = 6 - \sqrt{x+4}$

57. $\sqrt{x+3} = x^2 - 4x - 2$

58. $\sqrt{x+7} = x^2 - 2x - 7$

For Exercises 59–64, estimate all solutions of the equation or system by referring to the graphs shown in Fig. 33. Round results or coordinates of results to the first decimal place.

59. $\sqrt{x+5} = x^2 - 2x - 4$

60. $\sqrt{x+5} = -2x - 5$

61. $\sqrt{x+5} = 1$

62. $\sqrt{x+5} = 2$

63. $y = \sqrt{x + 5}$
$y = -2x - 5$

64. $y = \sqrt{x + 5}$
$y = x^2 - 2x - 4$

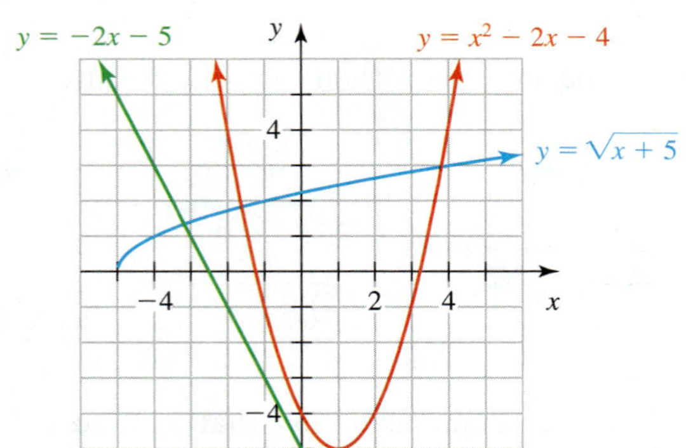

Figure 33 Exercises 59–64

Find all x-intercepts. Use a graphing calculator to verify your work.

65. $h(x) = 3\sqrt{-3x + 4} - 15$

66. $k(x) = 2\sqrt{4x + 1} - 22$

67. $f(x) = \sqrt{3x - 2} - \sqrt{x + 8}$

68. $g(x) = \sqrt{2x + 5} - \sqrt{4x + 1}$

69. $h(x) = 2\sqrt{x + 4} + 3\sqrt{x - 5}$

70. $k(x) = 3\sqrt{x - 1} + 5\sqrt{x - 4}$

Let $f(x) = 3\sqrt{x} - 7$.

71. Find x when $f(x) = -1$.

72. Find x when $f(x) = -8$.

For Exercises 73 and 74, let $f(x) = -2\sqrt{x - 4} + 5$.

73. Find x when $f(x) = -3$.

74. Find x when $f(x) = 1$.

75. In Exercise 57 of Homework 13.4, you worked with the model $f(t) = 20.4\sqrt{t} + 21$, where $f(t)$ is the percentage of e-mail that is spam at t years since 1999 (see Table 17).

Table 17 Percentages of E-Mail That Is Spam

Year	Percent
1999	21
2002	56
2004	68
2006	80
2008	81
2010	85

Source: *IronPort*

a. Find $f(4)$. What does it mean in this situation?
b. Find t when $f(t) = 95$. What does it mean in this situation?
c. Predict when all e-mail will be spam.
d. In 2011, corporate employees received an average of 72 e-mails per day (Source: *Radicati Group*). Estimate the average daily *number* of spam e-mails received by corporate employees in 2011.
e. A corporation loses about \$1250 in productivity per employee per year due to employees reading and deleting spam (Source: *Alinean*). Assuming employees work 50 weeks per year, use your result in part (d) to estimate the number of cents lost in productivity *per e-mail*.

76. In Exercise 58 of Homework 13.4, you worked with the model $f(t) = 10\sqrt{t} + 2.5$, where $f(t)$ is the global airline industry's revenue (in billions of dollars) from fees not included in fares at t years since 2007 (see Table 18).

Table 18 Global Airline Industry's Revenues from Fees Not Included in Fares

Year	Revenue (billions of dollars)
2007	2.5
2008	10.3
2009	13.5
2010	21.5
2011	22.6

Source: *IdeaWorks and Amadeus*

a. Find $f(10)$. What does it mean in this situation?
b. Find t when $f(t) = 37$. What does it mean in this situation?
c. The global airline industry's total revenue from fares and fees was \$598 billion in 2011. Find the ratio of \$598 billion to \$22.6 billion, which is the revenue from fees not included in fares in 2011 (see Table 18). Then use that ratio to help you estimate the total revenue from fares and fees in 2012. Describe any assumptions you have made.

77. For the academic year 2005–2006, tuition at Princeton Day School ranged from \$19,200 to \$23,600 by grade (kindergarten through 8th grade). In addition, the school charges for supplies and field-trip expenses (see Table 19).

Table 19 Charges for Supplies and Field Trips at Princeton Day School

Grade	Per-Student Charge for Supplies and Field Trips (dollars)
kindergarten	260
1	310
3	365
5	410
6	425
7	425

Source: *Princeton Day School*

Let $f(n)$ be the per-student charge (in dollars) for supplies and field trips for nth grade; kindergarten is represented by $n = 0$. A model of the situation is $f(n) = 257\sqrt[4]{n + 1}$.

a. Use a graphing calculator to draw the graph of the model and, in the same viewing window, the scattergram of the data. Does the model fit the data well?
b. Estimate the per-student charge for supplies and field trips for 2nd grade. The actual charge is \$365. Is your result an underestimate or an overestimate?
c. Estimate for what grade the per-student charge for supplies and field trips is \$385.
d. The actual per-student charge for 4th grade is \$450. By referring to the data shown in Table 19, explain why this charge is surprising.

78. The percentages of registered voters who voted in 2008 are shown in Table 20 for various household income groups. Let $f(d)$ be the percentage of registered voters with a household income of d thousand dollars who voted in 2008. A model of the situation is $f(d) = 4.2\sqrt{d} + 30$.

Table 20 Percentages of Registered Voters Who Voted in 2008, by Household Income Groups

Income Group ($1000s)	Income Used to Represent Income Group ($1000s)	Percent
0–9.999	5.0	41.3
10–14.999	12.5	41.2
15–19.999	17.5	44.3
20–29.999	25.0	48.0
30–39.999	35.0	54.4
40–49.999	45.0	58.2
50–74.999	62.5	65.9
75–99.999	87.5	72.6
100–149.999	125.0	74.9

Source: *U.S. Census Bureau*

a. Use a graphing calculator to draw the graph of the model and, in the same viewing window, the scattergram of the data. Does the model fit the data well?

b. Estimate the percentage of registered voters with a household income of $38 thousand who voted in 2008.

c. Estimate at which household income 60% of registered voters voted in 2008.

d. Is f an increasing function, decreasing function, or neither? What does that mean in this situation?

Concepts

79. A student tries to solve $\sqrt{x^2 + 4x + 5} = x + 3$:

$$\begin{aligned} \sqrt{x^2 + 4x + 5} &= x + 3 \\ \left(\sqrt{x^2 + 4x + 5}\right)^2 &= (x + 3)^2 \\ x^2 + 4x + 5 &= x^2 + 9 \\ 4x &= 4 \\ x &= 1 \end{aligned}$$

Describe any errors. Then solve the equation correctly.

80. A student tries to solve $\sqrt{x - 2} = x - 4$:

$$\begin{aligned} \sqrt{x - 2} &= x - 4 \\ \left(\sqrt{x - 2}\right)^2 &= (x - 4)^2 \\ x - 2 &= x^2 - 8x + 16 \\ 0 &= x^2 - 9x + 18 \\ 0 &= (x - 3)(x - 6) \end{aligned}$$

$$x - 3 = 0 \quad \text{or} \quad x - 6 = 0$$

$$x = 3 \quad \text{or} \quad x = 6$$

He says the solutions are 3 and 6. Is he correct? Explain.

81. Solve the system; then use "intersect" on a graphing calculator to verify your work:

$$\begin{aligned} y &= 3\sqrt{x} - 4 \\ y &= -2\sqrt{x} + 6 \end{aligned}$$

82. Create a system of two square root equations that has (4, 5) as its only solution. Verify your system graphically.

83. Find nonzero values of a, h, and k so the equation $\sqrt{x} = a\sqrt{x - h} + k$ has no real-number solutions. [**Hint:** Think about the graphs of $y = \sqrt{x}$ and $y = a\sqrt{x - h} + k$.]

84. For the equation $\sqrt{x} + 2 = 5$, why do we first subtract 2 from both sides, rather than first square both sides?

85. The first of the following statements is true, yet the last statement is false:

$$\begin{aligned} 2x - x &= x \\ (2x)^2 - x^2 &= x^2 \\ 4x^2 - x^2 &= x^2 \\ 3x^2 &= x^2 \\ 3 &= 1 \end{aligned}$$

Describe any errors.

86. Describe how to solve square root equations that contain one square root. Also, describe how to solve square root equations that contain two or more square roots.

Related Review

Solve or simplify, as appropriate.

87. $3\sqrt{x} + 4 - 7\sqrt{x} + 1$

88. $2\sqrt{x + 1} - 2 + 5\sqrt{x + 1} - 9 = 3$

89. $3\sqrt{x} + 4 - 7\sqrt{x} + 1 = -7$

90. $2\sqrt{x + 1} - 2 + 5\sqrt{x + 1} - 9$

91. $(\sqrt{p} + 3)(\sqrt{p} + 1) = 3$

92. $(\sqrt{m} - 2)(\sqrt{m} - 3)$

93. $(\sqrt{p} + 3)(\sqrt{p} + 1)$

94. $(\sqrt{m} - 2)(\sqrt{m} - 3) = 2$

Solve. Round any approximate solutions to the fourth decimal place.

95. $50 - 4(2)^x = -83$

96. $\dfrac{1}{x - 2} - \dfrac{2}{x + 3} = \dfrac{11}{x^2 + x - 6}$

97. $\sqrt{x + 3} - \sqrt{x - 2} = 1$

98. $3x^2 + 2x - 4 = 0$

99. $-3(2k - 5) + 1 = 2(4x + 3)$

100. $3b^4 - 29 = 83$

101. $\log_2(5t - 1) = 5$

102. $\log_2(2y + 1) - 2\log_2(y - 1) = 1$

Expressions, Equations, Functions, and Graphs

Perform the indicated instruction. Then use words such as linear, quadratic, cubic, exponential, logarithmic, rational, radical, polynomial, degree, function, one variable, *and* two variables *to describe the expression, equation, or system.*

103. Find the quotient $\dfrac{3x^2 - x - 10}{x^3 - x^2 - x + 1} \div \dfrac{3x^2 - 12}{2x^2 + x - 3}$.

104. Solve $\log_4(7x - 2) = 3$.

105. Find the sum $\dfrac{6}{b - 2} + \dfrac{3b}{b^2 - 7b + 10}$.

106. Solve $2(5)^t + 14 = 249$. Round any solutions to the fourth decimal place.

107. Solve $\dfrac{6}{x - 2} + \dfrac{3x}{x^2 - 7x + 10} = \dfrac{x}{x - 5}$.

108. Write $3\log_b(2x^2) + 2\log_b(3x^7)$ as a single logarithm.

13.6 Modeling with Square Root Functions

Objectives

- Find an equation of a square root curve that contains two given points.
- Find an equation of a square root model.
- Use a square root model to make estimates and predictions.

In this section, we will discuss how to find the equation of a square root function of the form $f(x) = a\sqrt{x} + b$ whose graph contains two given points. Then we will find a model of this form that describes an authentic situation and use the model to make estimates and predictions.

Finding a Square Root Function

Recall that the first step in finding a quadratic function (Section 9.7) or an exponential function (Section 10.4) is to substitute any given ordered pairs into a general equation of the function. We will do this same first step to find a square root function.

Example 1 Finding an Equation of a Square Root Function

Find an equation of a square root curve that contains the points $(2, 3)$ and $(5, 7)$.

Solution

We substitute the ordered pairs $(2, 3)$ and $(5, 7)$ into the equation $f(x) = a\sqrt{x} + b$:

$$3 = a\sqrt{2} + b \quad \text{Substitute 2 for } x \text{ and 3 for } y.$$

$$7 = a\sqrt{5} + b \quad \text{Substitute 5 for } x \text{ and 7 for } y.$$

We then calculate approximate values for $\sqrt{2}$ and $\sqrt{5}$:

$$1.41a + b = 3 \quad \text{Equation (1)}$$

$$2.24a + b = 7 \quad \text{Equation (2)}$$

Next, we multiply both sides of equation (1) by -1:

$$-1.41a - b = -3 \quad \text{Equation (3)}$$

To eliminate b, we add the left sides and add the right sides of equations (2) and (3) and solve for a:

$$\begin{aligned} 2.24a + b &= 7 \\ -1.41a - b &= -3 \\ \hline 0.83a + 0 &= 4 \\ 0.83a &= 4 \quad C + 0 = C \\ a &= \frac{4}{0.83} \\ a &\approx 4.82 \end{aligned}$$

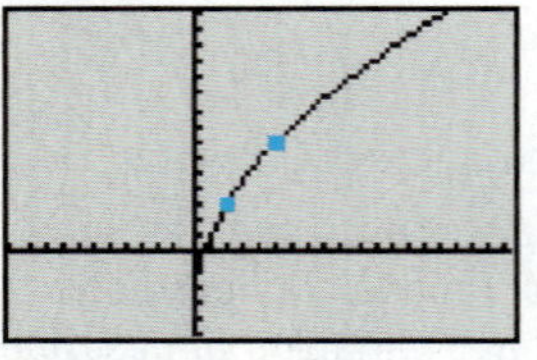

Figure 34 Verify that the graph of $f(x) = 4.82\sqrt{x} - 3.82$ approximately contains the points $(2, 3)$ and $(5, 7)$

So, our equation has the form $f(x) = 4.82\sqrt{x} + b$. To find b, we substitute the ordered pair $(2, 3)$ into the equation $f(x) = 4.82\sqrt{x} + b$:

$$\begin{aligned} 3 &= 4.82\sqrt{2} + b && \text{Substitute 2 for } x \text{ and 3 for } y. \\ 3 - 4.82\sqrt{2} &= b && \text{Isolate } b. \\ b &\approx -3.82 && \text{Compute.} \end{aligned}$$

So, the equation is $f(x) = 4.82\sqrt{x} - 3.82$.

We use a graphing calculator to verify that the graph of f approximately contains the two given points (see Fig. 34).

To find an equation of a square root curve that contains two given points, we substitute the points' ordered pairs into the equation $f(x) = a\sqrt{x} + b$ and get a system of two equations. We find values of a and b by solving the system by elimination or substitution.

Finding an Equation of a Square Root Curve

To find an equation of a square root curve that contains two given points,

1. Obtain a system of two linear equations in two variables by substituting the coordinates of both points into the general equation $y = a\sqrt{x} + b$.
2. Solve the system you found in step 1.
3. Substitute the values of a and b you found in step 2 into the equation $y = a\sqrt{x} + b$.

Using the y-Intercept and One Other Point to Find an Equation

For an equation of the form $y = a\sqrt{x} + b$, we can find the y-intercept by substituting 0 for x:

$$y = a\sqrt{0} + b = b$$

So, the y-intercept is $(0, b)$.

y-Intercept of the Graph of $y = a\sqrt{x} + b$

The graph of an equation of the form $y = a\sqrt{x} + b$ has y-intercept $(0, b)$.

For example, the graph of the equation $y = 3\sqrt{x} + 7$ has y-intercept $(0, 7)$.

The process of finding an equation of a square root function is easier if one of the two given points is the y-intercept, as we shall see in Example 2.

Example 2 Finding an Equation when One Given Point Is the y-Intercept

Find an equation of a square root curve that contains the points $(0, 5)$ and $(6, 19)$.

Solution

Since the y-intercept of the curve is $(0, 5)$, we can find an equation of the form

$$y = a\sqrt{x} + 5$$

To find a, we substitute the ordered pair $(6, 19)$ in the equation $y = a\sqrt{x} + 5$:

$$19 = a\sqrt{6} + 5 \quad \text{Substitute 6 for x and 19 for y.}$$
$$14 = a\sqrt{6} \quad \text{Subtract 5 from both sides.}$$
$$\frac{14}{\sqrt{6}} = a \quad \text{Divide both sides by } \sqrt{6}.$$
$$a \approx 5.72 \quad \text{Compute.}$$

So, the equation is $f(x) = 5.72\sqrt{x} + 5$.

We use a graphing calculator table, as well as TRACE on a graphing calculator, to check that the graph of $f(x) = 5.72\sqrt{x} + 5$ approximately contains the points $(0, 5)$ and $(6, 19)$. See Fig. 35.

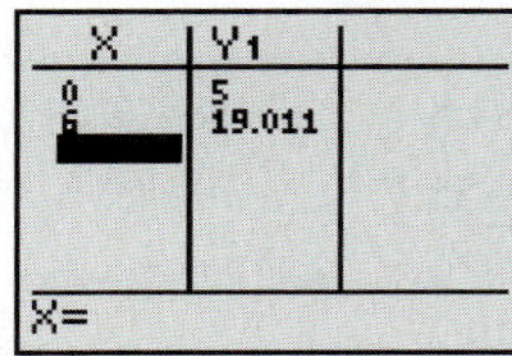

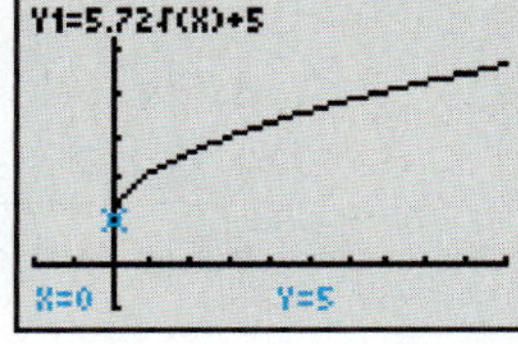

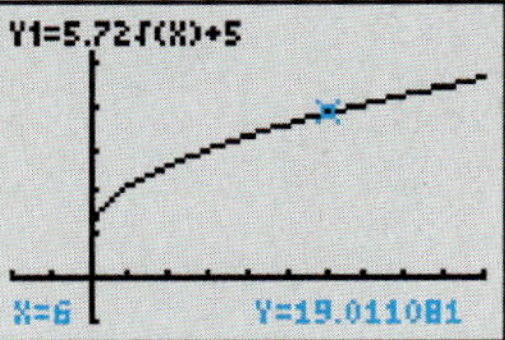

Figure 35 Verify the work

Finding a Square Root Model

Now that we have discussed how to find an equation of a square root function, we can find an equation of a square root model.

Example 3 Finding an Equation of a Square Root Model

The percentages of American adults who watch cable television are shown in Table 21 for various income groups. Let $f(I)$ be the percentage of American adults with an annual income of I thousand dollars who watch cable television. Find an equation of f.

Table 21 Percentages of American Adults Who Watch Cable Television

Thousands of Dollars		
Annual Income Group	Income Used to Represent Income Group	Percent
0–9.999	5.0	55.9
10–19.999	15.0	62.3
20–29.999	25.0	67.8
30–34.999	32.5	72.2
35–39.999	37.5	74.2
40–49.999	45.0	77.2
50 or over	70.0	84.8

Source: *Mediamark Research, Inc.*

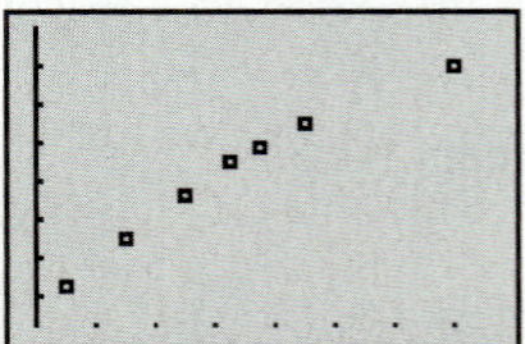

Figure 36 Cable television scattergram

Solution

First, we use a graphing calculator to draw a scattergram of the data (see Fig. 36). It appears the data might be modeled well by a square root function or a quadratic function. To decide which model to use, we find an equation of both types of functions and compare the fit of each model.

For the square root model, we use a function of the form

$$f(I) = a\sqrt{I} + b$$

We use the data points $(5, 55.9)$ and $(70, 84.8)$ to find the values of a and b. To begin, we substitute the ordered pairs $(5, 55.9)$ and $(70, 84.8)$ in the equation $f(I) = a\sqrt{I} + b$:

$$55.9 = a\sqrt{5} + b \quad \text{Substitute } (5, 55.9) \text{ in } f(I) = a\sqrt{I} + b.$$

$$84.8 = a\sqrt{70} + b \quad \text{Substitute } (70, 84.8) \text{ in } f(I) = a\sqrt{I} + b.$$

We then calculate $\sqrt{5} \approx 2.24$ and $\sqrt{70} \approx 8.37$:

$$2.24a + b = 55.9 \quad \text{Equation (1)}$$

$$8.37a + b = 84.8 \quad \text{Equation (2)}$$

Next, we multiply both sides of equation (1) by -1:

$$-2.24a - b = -55.9 \quad \text{Equation (3)}$$

To eliminate b, we add the left sides and add the right sides of equations (2) and (3) and solve for a:

$$\begin{aligned} 8.37a + b &= 84.8 && \text{Equation (2)} \\ -2.24a - b &= -55.9 && \text{Equation (3)} \\ \hline 6.13a + 0 &= 28.9 && \text{Add left sides and add right sides; combine like terms.} \\ a &= \frac{28.9}{6.13} \\ a &\approx 4.71 \end{aligned}$$

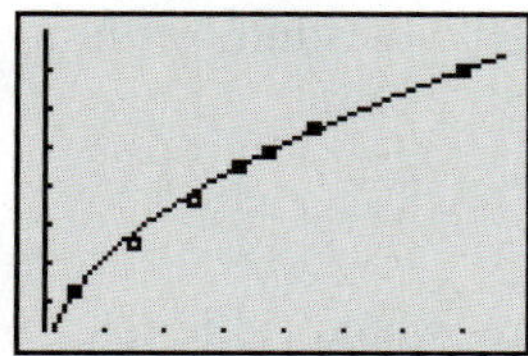
Figure 37 Square root model

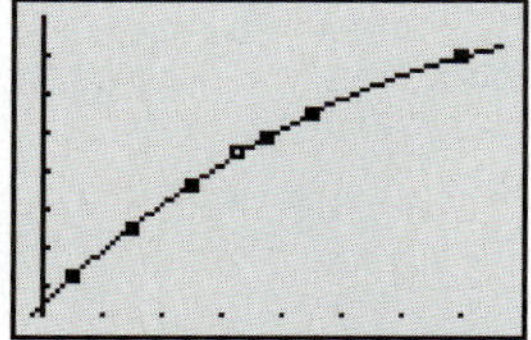
Figure 38 Quadratic model

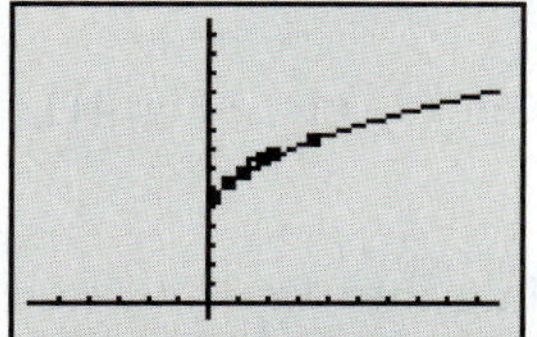
Figure 39 Square root model

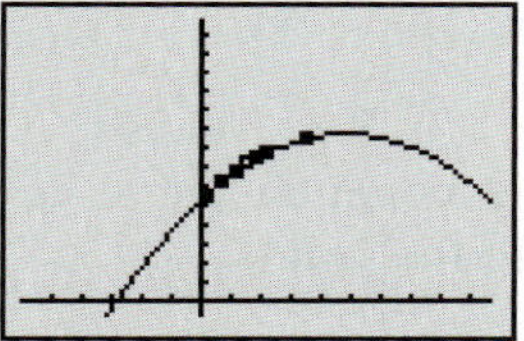
Figure 40 Quadratic model

So, our equation has the form $f(I) = 4.71\sqrt{I} + b$. To find b, we substitute the ordered pair $(5, 55.9)$ in the equation $f(I) = 4.71\sqrt{I} + b$:

$$55.9 = 4.71\sqrt{5} + b \quad \text{Substitute } (5, 55.9) \text{ in } f(I) = 4.71\sqrt{I} + b.$$
$$55.9 - 4.71\sqrt{5} = b \quad \text{Subtract } 4.71\sqrt{5} \text{ from both sides.}$$
$$45.37 \approx b \quad \text{Compute.}$$

So, the square root model is $f(I) = 4.71\sqrt{I} + 45.37$.

We use a graphing calculator to find the quadratic regression equation:

$$f(I) = -0.0037I^2 + 0.72I + 52.36$$

Next, we use a graphing calculator to compare the fits of the two models (see Figs. 37 and 38).

The quadratic model appears to fit the data slightly better than the square root model. However, recall from Section 9.8 that we should also consider whether either model makes sense within the context of the situation. In this situation, it will help to consider what each model estimates for Americans with incomes well over $50 thousand. So, we Zoom Out and compare the graphs again (see Figs. 39 and 40).

The square root model estimates that the percentage of adults who watch cable television continues to increase as income increases, whereas the quadratic model estimates that the percentage of adults who watch cable television will decrease significantly as income increases, which is model breakdown, as a little research would show. Model breakdown also occurs for the square root model, because it estimates the percentage of adults who watch cable television is over 100% for very large incomes (over $135 thousand—as we can show, with work). But for incomes up to $135 thousand, the square root model is the better of the two models.

Finding an Equation of a Square Root Model

To find an equation of a square root model, given some data,

1. Create a scattergram of the data.
2. Imagine a square root curve that comes close to the data points, and choose two points (not necessarily data points) that lie on or close to the square root curve.
3. Use the two points to find an equation of the square root curve.
4. Use a graphing calculator to verify that the graph of the equation comes close to the points of the scattergram.

Using a Square Root Model to Make Estimates and Predictions

Once we have found a square root model, we can use it to make estimates and predictions.

Example 4 Using a Square Root Model to Make Predictions

In Example 3, we found the equation $f(I) = 4.71\sqrt{I} + 45.37$, where $f(I)$ is the percentage of American adults with an annual income of I thousand dollars who watch cable television. Let $p = f(I)$.

1. Find the p-intercept. What does it mean in this situation?
2. Estimate the percentage of adults with an annual income of $30 thousand who watch cable television.
3. At what level of income do 80% of Americans watch cable television?

Solution

1. To find the p-intercept, we find $f(0)$:

$$f(0) = 4.71\sqrt{0} + 45.37 = 45.37$$

The p-intercept is $(0, 45.37)$. The model estimates about 45.4% of adults with no income watch cable television.

2. We evaluate f at 30:

$$f(30) = 4.71\sqrt{30} + 45.37 \approx 71.17$$

About 71.2% of adults with an annual income of \$30 thousand watch cable television, according to the model.

3. We substitute 80 for $f(I)$ in the equation $f(I) = 4.71\sqrt{I} + 45.37$ and solve for I:

$$80 = 4.71\sqrt{I} + 45.37 \quad \text{Substitute 80 for } f(I).$$

$$34.63 = 4.71\sqrt{I} \quad \text{Subtract 45.37 from both sides.}$$

$$\frac{34.63}{4.71} = \sqrt{I} \quad \text{Divide both sides by 4.71.}$$

$$\left(\frac{34.63}{4.71}\right)^2 = (\sqrt{I})^2 \quad \text{Square both sides.}$$

$$I \approx 54.06 \quad \text{Compute.}$$

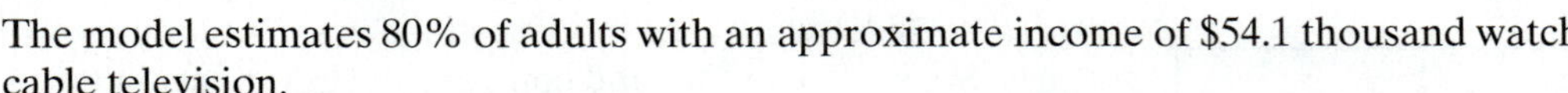

The model estimates 80% of adults with an approximate income of \$54.1 thousand watch cable television.

We use a graphing calculator table to verify our work in Problems 1–3 (see Fig. 41).

Figure 41 Verify the work

We close this section by incorporating square root modeling into the four-step modeling process.

Four-Step Modeling Process

To find a model and make estimates and predictions,

1. Create a scattergram of the data. Decide whether a line, an exponential curve, a parabola, a square root curve, or none of these comes close to the points.
2. Find an equation of your model.
3. Verify that your equation has a graph that comes close to the points in the scattergram. If it doesn't, check for calculation errors or use different points to find the equation. An alternative is to reconsider your choice of model in step 1.
4. Use your equation of the model to draw conclusions, make estimates, and/or make predictions.

Group Exploration

Looking ahead: Arithmetic sequences

A math tutor charges \$25 per hour to tutor one student, plus \$8 per hour for each additional student.

1. Let $f(n)$ be the amount of money (in dollars) the tutor will charge per hour to work with n students. Find an equation of f.
2. Evaluate f at each given value of n. Explain what each result means in this situation.
   a. $f(3.8)$ **b.** $f(-2)$ **c.** $f(0)$
3. On the basis of your results in Problem 2, determine a domain of the *model* f. [**Hint:** Which inputs make sense in this situation?]
4. On the basis of your domain of f, find the range of f. List the values of the range of f in this order: $f(1)$, $f(2)$, $f(3)$, $f(4)$, ... What do you notice about these numbers?
5. Sketch a graph of f, but plot only points whose n-coordinates are in the domain. [**Hint:** Your graph will look like a scattergram.]

Tips for Success Create a Mind Map for the Final Exam

In preparing for a final exam, consider how all the concepts you have learned are interconnected. One way to help yourself do this is to make a *mind map*. Put the main topic in the middle. Around it, attach concepts that relate to it, then concepts that relate to those concepts, and so on.

A portion of a mind map that describes this course is illustrated in Fig. 42. Many more "concept rectangles" could be added to it. You could make one mind map showing an overview of the course and several mind maps for the components of the overview mind map.

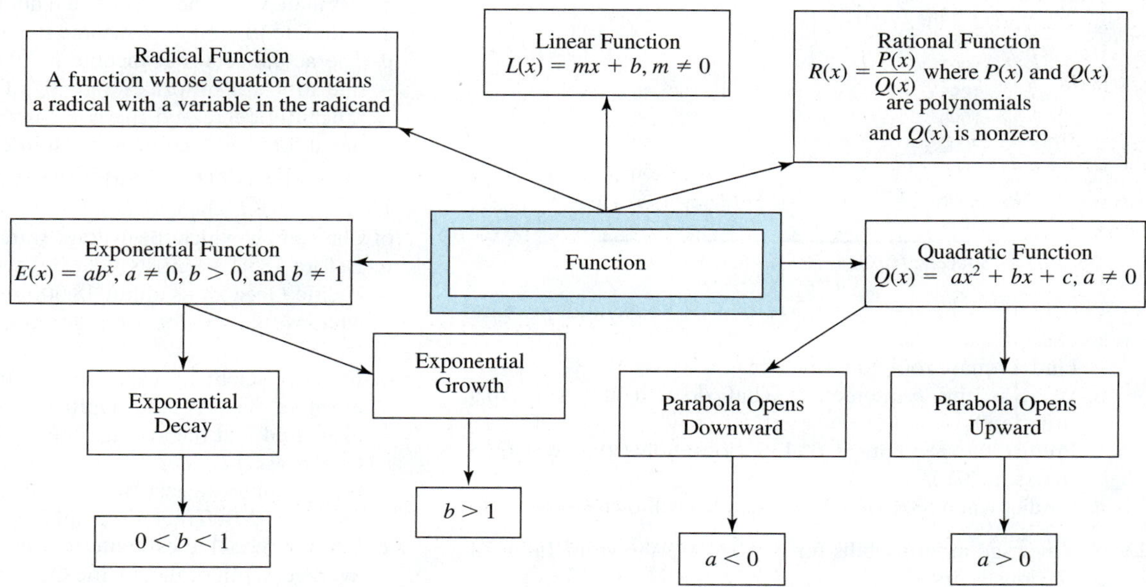

Figure 42 A portion of a mind map describing this course

Homework 13.6

For extra help ▶ MyMathLab® Watch the videos in MyMathLab Download the MyDashboard App

Find an equation of a square root curve of the form $y = a\sqrt{x} + b$ that approximately contains the given points. Round the values of a and b to the second decimal place.

1. $(0, 3)$ and $(4, 5)$
2. $(0, 5)$ and $(4, 3)$
3. $(0, 2)$ and $(9, 6)$
4. $(0, 3)$ and $(1, 2)$
5. $(0, 4)$ and $(5, 7)$
6. $(0, 1)$ and $(2, 3)$
7. $(0, 9)$ and $(3, 2)$
8. $(0, 8)$ and $(6, 3)$
9. $(1, 2)$ and $(4, 3)$
10. $(4, 5)$ and $(9, 8)$
11. $(2, 4)$ and $(3, 5)$
12. $(5, 2)$ and $(7, 4)$
13. $(2, 6)$ and $(5, 4)$
14. $(3, 8)$ and $(6, 5)$
15. $(5, 7)$ and $(13, 21)$
16. $(3, 9)$ and $(15, 29)$
17. $(7, 31)$ and $(10, 6)$
18. $(6, 43)$ and $(14, 5)$
19. $(15, 3)$ and $(35, 18)$
20. $(17, 6)$ and $(26, 19)$

21. The numbers of American female troops in Iraq and Afghanistan are shown in Table 22 for various years. Let $n = f(t)$ be the number (in thousands) of American female troops in Iraq and Afghanistan at t years since 2004.

Table 22 Numbers of American Female Troops in Iraq and Afghanistan

Year	Number of Female Troops (thousands)
2004	10
2005	15
2006	20
2007	20
2008	25
2009	25
2010	26

Source: *U.S. Army*

a. Find a square root equation of f.
b. What is the n-intercept? What does it mean in this situation?
c. Estimate when there were 28 thousand female troops in Iraq and Afghanistan.

d. By December 18, 2011, the United States had withdrawn all its troops from Iraq. Use the fact that there were 15 thousand American female troops in Afghanistan in 2012 to estimate the number of American female troops that would have been in Iraq in 2012 if the withdrawal had not occurred.

22. The percentages of U.S. households owning stocks are shown in Table 23 for various years.

Table 23 Percentages of U.S. Households Owning Stocks

Year	Percent
1983	19
1989	32
1992	37
1995	40
1999	48
2002	50
2005	50
2008	52
2011	52

Sources: *ICI/SIA; Federal Reserve Board*

Let $p = f(t)$ be the percentage of U.S. households owning stocks at t years since 1983.

a. Find a square root equation of f.

b. What is the p-intercept? What does it mean in this situation?

c. Predict the percentage of U.S. households that will own stocks in 2017.

d. Predict when 58% of U.S. households will own stocks.

23. The average monthly bills for pay-TV are shown in Table 24 for various years.

Table 24 Average Monthly Bills for Pay-TV

Year	Average Monthly Bill (dollars)
2006	56.0
2007	63.8
2008	65.6
2009	68.0
2010	70.8
2011	72.8

Source: *IHS Screen Digest*

Let $M = f(t)$ be the average monthly bill (in dollars) for pay-TV in the year that is t years since 2006.

a. Find an equation of f.

b. What is the M-intercept? What does it mean in this situation?

c. Find $f(13)$. What does it mean in this situation?

d. Find t when $f(t) = 81$. What does it mean in this situation?

e. In 2011, there were 100.9 million subscribers to pay-TV (Source: *The NPD Group*). Assuming the number of subscribers does not change, predict the total *annual* revenue from *all* subscribers of pay-TV in 2017.

24. The total annual net incomes (in billions of dollars) of the top five oil companies are shown in Table 25 for various years. Let $I = f(t)$ be the total annual net income (in billions of dollars) of the top five oil companies at t years since 2003.

a. Find a square root equation of f.

b. Find the I-intercept. What does it mean in this situation?

Table 25 Total Annual Net Incomes of the Top Five Oil Companies

Year	Total Annual Net Income (billions of dollars)
2003	58.2
2004	82.4
2005	111.4
2006	119.6
2007	123.4

Source: *Evaluate Energy*

c. Estimate when the total annual net income was $135 billion.

d. Estimate the total net income in 2010.

e. The actual total net income in 2010 was only $78 billion, due to lower oil prices and BP's Gulf oil spill, which significantly decreased the company's profits. Estimate how much total net income was lost in 2010 due to these events.

25. In Exercise 83 of Homework 13.1, you worked with the model $f(d) = \sqrt{9.8d}$, where $f(d)$ is the speed (in meters per second) of a tsunami in which the average water depth is d meters.

a. Before 1856, scientists believed the average depth of the Pacific Ocean was about 18,000 meters. If that were true, what would be the average speed of a tsunami in the Pacific Ocean?

b. In 1856, scientists used their knowledge of tsunamis to estimate the average depth of the Pacific Ocean. They estimated that the average speed of tsunamis in the Pacific Ocean was between 203 and 210 meters per second. Does this estimate suggest that the average depth of the Pacific Ocean is 18,000 meters? Explain.

c. Use the model to estimate between what two depths is the average depth of the Pacific Ocean.

d. Most current estimates of the average depth of the Pacific Ocean are close to 4280 meters. Is this estimate between the two depths you found in part (c)?

26. A study of children who were adopted after being in temporary foster care explored the relationship between the children's age when separated from their foster parents and the percentage of children showing problems immediately after the separation (see Table 26).

Table 26 Adopted Children Separated from Foster Parents

Age when Separated from Foster Parents (years)	Middle of Age Group (years)	Percent Showing Problems
<3	1.5	4
3–4	3.5	40
4–5	4.5	70
6–8	7.0	90
9	9.0	100

Source: The Immediate Impact of Separation: Reactions of Infants to a Change in Mother Figures, *by Yarrow and Goodwin*

Let $p = f(t)$ be the percentage of adopted children showing problems immediately after separation if they are separated from foster parents at age t years.

a. Find a square root equation of f.

b. Is f an increasing function, a decreasing function, or neither? What does that mean in this situation?

c. Estimate the percentage of adopted children who show problems if separated at age 10 years. Is this a reasonable estimate? If so, explain why. If not, explain why not and suggest a better value.

d. Sketch a model that likely describes the relationship between t and p for *all* ages of adopted children better than f does. [**Hint:** Your model will be different from those discussed in this textbook.]

27. Some students ran an experiment to explore the relationship between the time it takes a baseball to fall to the ground and the various heights from which it was dropped (see Table 27). Let $T = S(h)$ be the time (in seconds) it takes a baseball to fall to the ground when dropped from h feet above the ground.

Table 27 Drop Heights and Falling Times of a Baseball*

Drop Height (feet)	Falling Time (seconds)
0.00	0.00
3.28	0.53
13.10	0.96
26.30	1.40
39.40	1.70
52.50	1.94

Source: *J. Lehmann*

* Although a ball cannot be dropped from height zero feet, the data point (0, 0) is included because drop heights near zero will correspond to falling times near zero.

a. Find a square root equation of S.

b. A linear model of the baseball drop data is given by $L(h) = 0.034h + 0.327$. A quadratic model is given by $Q(h) = -0.00058h^2 + 0.064h + 0.165$.

i. Describe how well each of the functions S, L, and Q fits the data points in the scattergram.

ii. Find $S(0)$, $L(0)$, and $Q(0)$. Which function models the baseball situation the best near $h = 0$? Explain.

iii. Use Zoom Out and decide which function could not be a good model for the baseball situation when the drop heights are very large. Explain.

iv. On the basis of your responses to parts (i)–(iii), determine which of S, L, or Q best models the baseball drop situation. Explain.

v. An equation that is often used to model drop times is $T = \sqrt{\frac{2h}{g}}$, where g is the constant 32.2 feet per second squared. Find an approximate form $T = a\sqrt{h}$ of this equation so you can compare the model with S.

c. Suppose you wish to estimate the height of a sheer cliff. You drop a stone from the top of the cliff, and it takes 3 seconds to reach the foot of the cliff. Use S to estimate the height of the cliff.

d. Use S to estimate how long it would take for a baseball to reach the ground if it were to be dropped from the top of the Empire State Building, which is 1250 feet tall.

28. The percentages of ex-convicts who have been arrested for a new crime after being out of a state prison for various numbers of years are shown in Table 28. Let $p = S(t)$ be the percentage of convicts released from a state prison who are arrested for a new crime after being out of prison for t years.

a. Find a square root equation of S.

Table 28 Ex-Convicts Who Have Been Arrested for a New Crime

Number of Years since Release	Percent
0	0
0.5	30
1	44
2	59
3	68

Source: *U.S. Department of Justice*

b. A quadratic model of the data is given by

$$Q(t) = -9.05t^2 + 48.09t + 3.47$$

i. Describe how well both of the functions S and Q fit the data points in the scattergram.

ii. Explain in terms of the situation why the model should be increasing for $t \ge 0$. Is function S or Q increasing for $t \ge 0$?

c. Use S to estimate the percentage of ex-convicts who have been arrested for a new crime after being out of prison for 4 years.

d. Use S to estimate after how many years since release all ex-convicts will have been arrested for a new crime.

29. A study of births among 1444 fertile couples recorded the birth order and claims by the parents that conception occurred despite the use of contraception (see Table 29). The percentage, for example, of first births that occurred despite contraception, according to the parents, was 37.6%.

Table 29 Percentages of Births Despite Contraception

Birth Order	Percentage of Conceptions Despite Contraception
1	37.6
2	54.3
3	66.5
4	73.0
5	84.1
6	81.2

Source: Social and Psychological Factors Affecting Fertility, *by Whelpton and Kiser*

Let $p = f(n)$ be the percentage of births of the nth-born child that the parents claimed occurred despite the use of contraception at the time of conception.

a. Find a square root equation of f. [**Hint:** Use the points $(2, 54.3)$ and $(4, 73.0)$.]

b. Find $f(7)$. What does it mean in this situation?

c. Find n when $f(n) = 100$. What does your result mean in this situation?

d. Note that f is an increasing function. What does that mean in this situation? Form a theory to explain why this happens.

30. The numbers of McDonald's restaurants around the world are shown in Table 30 for various years. Let $n = f(t)$ be the number (in thousands) of McDonald's restaurants at t years since 1998.

a. Find an equation of f.

b. Predict the number of McDonald's restaurants in 2019.

c. There are 192 countries in the world. What will be the average number of McDonald's restaurants per country in 2019?

d. Predict when there will be 35 thousand McDonald's restaurants.

Table 30 Numbers of McDonald's Restaurants

Year	Number of Restaurants (thousands)
1998	24.5
2000	27.9
2002	30.2
2004	30.5
2006	31.0
2008	32.0
2010	32.7
2011	33.5

Source: *McDonald's*

31. The number of acres devoted to genetically modified crops has increased greatly since 1997 (see Table 31).

Table 31 Genetically Modified Crops

Year	Genetically Modified Crops (millions of acres)
1997	8
2000	80
2003	106
2006	135
2008	154
2011	165
2012	170

Sources: *Clive James; Department of Agriculture*

Let $c = f(t)$ be the amount of land used for genetically modified crops (in millions of acres) at t years since 1997.

a. Find an equation of f.

b. What is the c-intercept? What does it mean in this situation?

c. Predict the number of acres of genetically modified crops in 2016.

d. Predict when there will be 200 million acres of genetically modified crops.

32. Tire dumps are unsightly and sometimes catch fire, releasing hazardous chemicals. In 2003 alone, Americans threw out 300 million tires, one per person in the United States. Table 32 shows the percentages of tires that were reused or recycled for various years.

Table 32 Reused or Recycled Tires

Year	Percent
1990	11
1994	55
1998	67
2001	78
2003	83
2007	89
2008	90

Source: *Rubber Manufacturers Association*

Let $f(t)$ be the percentage of tires that have been reused or recycled at t years since 1990.

a. Find an equation of f.

b. Estimate when all tires were reused or recycled.

c. In 2006, there were two U.S. plants that burned tires to create energy. A third plant had been proposed that would have burned 10 million tires per year and produced enough energy for up to 20 thousand homes. However, there were concerns about the release of harmful chemicals from the plant. Should the plant have been built? To decide, make the following assumptions:

- Construction would have begun in 2006, and it would have taken two years for the plant to be built.
- It is better to reuse or to recycle a tire than to burn it for energy.
- If a tire can't be reused or recycled, it is better to burn it for energy than to dump it in a landfill or tire pile.

[**Hint:** Refer to your result in part (b).]

Concepts

Consider the scattergram of data and the graph of the model $f(t) = a\sqrt{t} + b$ in the indicated figure. For Exercises 33 and 34, sketch the graph of a square root model that describes the data better. Then explain how you would adjust the values of a and b of the original model to describe the data better.

33. See Fig. 43.

Figure 43 Exercise 33

34. See Fig. 44.

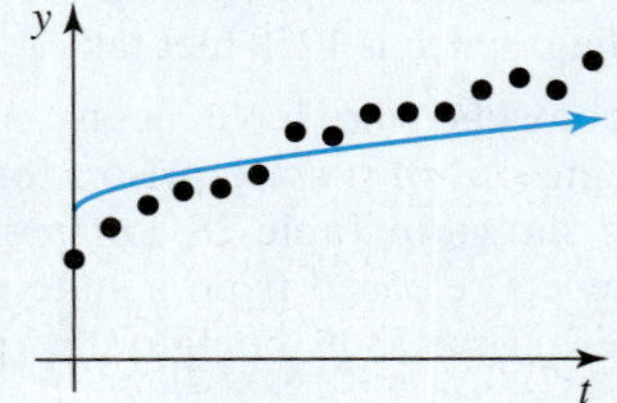

Figure 44 Exercise 34

35. Explain why the graph of an equation of the form $y = a\sqrt{x} + b$ has y-intercept $(0, b)$.

36. Describe how to find an equation of a square root function that contains two given points. How is this process different if one point is the y-intercept?

Related Review

37. The numbers of U.S. communities with cameras that photograph drivers who run red lights are shown in Table 33 for various years.

Table 33 Numbers of U.S. Communities with Red-Light Cameras

Year	Number of Communities with Red-Light Cameras
1999	19
2001	28
2003	74
2005	120
2007	243
2009	445
2010	501

Source: *Insurance Institute for Highway Safety*

Let $f(t)$ be the number of U.S. communities with red-light cameras at t years since 1990.

a. Use a graphing calculator to draw a scattergram to describe the data. By inspecting the scattergram alone, determine which two of the following functions *might* fit the data well: linear, exponential, quadratic, and square root.

b. Find equations of the two types of functions that you selected in part (a).

c. Compare how well each of your two models fits the data.

d. Which of your two models describes the situation better for years before 1999?

e. Use both of your models to predict when there will be 6000 U.S. communities with red-light cameras. Refer to the graphs of the two models to explain why your two results are so different.

f. **i.** Use the exponential model to find $f(15) - f(14)$. What does it mean in this situation? (Assume $f(t)$ is the number of communities with red-light cameras by the *end* of the year that is t years since 1990.)

ii. Use the exponential model to find $f(27) - f(26)$. What does it mean in this situation?

iii. In 2005, Redflex Traffic Systems sold \$26 million worth of red-light cameras. Assuming Redflex continues to sell about 40% of the red-light cameras used in the United States, predict Redflex's revenue from sales of red-light cameras in 2017. Describe all the assumptions you have made. [**Hint:** Use the ratio of your result in part (ii) to your result in part (i).]

38. Annual revenues from portable media and MP3 players in the United States are shown in Table 34 for various years. Let $f(t)$ be the annual revenue (in billions of dollars) from portable media and MP3 players at t years since 2000.

Table 34 Revenues from Portable Media and MP3 Players

Year	Revenue (billions of dollars)
2005	4.23
2006	5.56
2007	5.97
2008	5.84
2009	5.21
2010	4.73

Source: *Consumer Electronics Association*

a. Find an equation of f.

b. Estimate the revenue in 2011.

c. Estimate when the annual revenue was \$1.4 billion.

d. What are the t-intercepts? What do they mean in this situation?

e. For what values of t is there model breakdown for certain?

Expressions, Equations, Functions, and Graphs

Give an example of the following. Then solve, simplify, or graph, as appropriate.

39. rational equation in one variable

40. system of two linear equations in two variables

41. difference of two rational expressions

42. quadratic equation in one variable

43. square root equation in one variable

44. expression involving exponents

45. exponential function

46. difference of two logarithmic expressions with the same base

Taking it to the Lab

Pendulum Lab

Do you know what a *pendulum* is? You can construct a pendulum by tying one end of some thread to a washer and attaching the other end to a surface so that the washer is suspended and swings forward and backward freely.

The *period* of the pendulum is the amount of time it takes for the washer to swing forward *and* backward once. The period of a pendulum depends on the length of the thread. In this lab, you will discover the relationship between the period of a pendulum and the length of its thread.

Materials

You will need the following materials:

1. thread
2. scissors
3. a timing device
4. tape
5. a washer (or some other small, dense object that can be tied to the thread)
6. a meterstick

Preparation

Knot one end of the thread to the washer. Tape the other end of the thread to a surface well so that the washer is suspended and swings freely. The distance from the middle of the washer to the tape should be at least 100 centimeters.

Recording of Data

Record the distance from the middle of the washer to the tape. Then time how long it takes for the washer to swing back and forth four times. Divide this time by 4 to find the period of the pendulum. Repeat the procedure several times. Discard the times that are very different from most of the times, and average the remaining times.

Repeat the procedure for various lengths of thread. When the thread is quite short, time how long it takes for the washer to swing back and forth eight times and divide this time by 8 to find the period.

Analyzing the Data

1. Display your data in a table.
2. Let $f(L)$ be the period (in seconds) of the pendulum, where L is the length (in centimeters) of the thread. Use a graphing calculator to draw a scattergram of your data.
3. Find an equation of f.
4. Use a graphing calculator to graph your model and the scattergram in the same viewing window. Graph your model and the scattergram by hand. How well does f model the data?
5. Find all intercepts of your model, and describe what they mean in this situation.
6. Is f an increasing function, a decreasing function, or neither? What does that mean in this situation?
7. Use your model to estimate the period of the pendulum if the thread's length is 150 centimeters.
8. Use your model to estimate the length of the thread if the pendulum's period is 0.1 second.
9. Suppose that a big chunk of concrete attached to some thin, strong cable is suspended from the skydeck of the Willis Tower in Chicago. The skydeck is 1353 feet above Wacker Drive. Assume that the concrete, when at rest, almost touches the street. Estimate the period of this pendulum. (*Note:* You can use f to model the Willis Tower pendulum even though a concrete chunk weighs a lot more than a washer.)
10. What is the length of a pendulum that has a period of 1 minute?

Chapter Summary

Key Points of Chapter 13

Section 13.1 Simplifying Radical Expressions

Perfect *n*th power	If n is a counting number greater than 1, then each of the following is a **perfect *n*th power:** • A number that has an nth root that is rational. • A power x^k, where k is a multiple of n.
Product property for radicals	If $\sqrt[n]{a}$ and $\sqrt[n]{b}$ are defined, then $\sqrt[n]{ab} = \sqrt[n]{a}\sqrt[n]{b}$.
Simplified radical	A radical with index n is **simplified** when the radicand does not have any factors that are perfect nth powers (other than -1 or 1) and the index is as small as possible.
Selecting perfect *n*th-power factors	To select perfect nth-power factors of radicands, • If the radicand has more than one numerical factor that is a perfect nth power, select the largest one. • If the radicand has more than one factor that is a perfect nth power with the same variable base, select the one with the largest exponent.
Decreasing the index	If $\sqrt[n]{x}$ is defined and the fraction $\frac{m}{n}$ can be simplified, then we can decrease the index of $\sqrt[n]{x^m}$ by writing $\sqrt[n]{x^m} = x^{m/n}$ and simplifying the exponent $\frac{m}{n}$.

Section 13.1 Simplifying Radical Expressions (*Continued*)

Simplifying a radical expression	To simplify a radical expression with index n, **1.** Find perfect nth-power factors of the radicand. **2.** Apply the product property for radicals. **3.** Find the nth root of each perfect nth power. **4.** Write the radical with as small an index as possible.
Power property for radicals	Let n be a counting number greater than 1: • If n is even, then $\sqrt[n]{x^n} = \lvert x \rvert$. • If n is odd, then $\sqrt[n]{x^n} = x$.
Radical function	A **radical function** is a function whose equation contains a radical with a variable in the radicand.
Square root function	A **square root function** is a radical function in which any radicals are square root radicals.
Radical model	A **radical model** is a radical function, or its graph, that describes an authentic situation.
Square root model	A **square root model** is a square root function, or its graph, that describes an authentic situation.

Section 13.2 Adding, Subtracting, and Multiplying Radical Expressions

Like radicals	Radicals that have the same index and the same radicand are called **like radicals.**
Adding or subtracting like radicals	To add or subtract like radicals, we use the distributive law.
Product of two radical expressions, each with two terms	To find the product of two radical expressions in which each factor has two terms, **1.** Multiply each term in the first radical expression by each term in the second radical expression. **2.** Combine like radicals.
Power property for radicals	If $\sqrt[n]{x}$ is defined, then $\left(\sqrt[n]{x}\right)^n = x$.
Simplifying the square of a radical expression	To simplify the square of a radical expression with two terms, • Use the square-of-a-sum property $(A + B)^2 = A^2 + 2AB + B^2$ or the square-of-a-difference property $(A - B)^2 = A^2 - 2AB + B^2$, or • Use $C^2 = CC$ to write the square as a product of two identical expressions, and then multiply each term in the first radical expression by each term in the second radical expression.
Multiplying radicals that have the same index	To multiply two radicals that have the same index, we use the product property.
Multiplying two radicals that have different indexes but the same radicand	To multiply two radicals that have different indexes but the same radicand, **1.** Write the radicals in exponential form. **2.** Use exponential properties to simplify the expression involving exponents. **3.** Write the simplified expression in radical form.
Simplifying a radical expression	To simplify a radical expression, **1.** Perform any indicated multiplications. **2.** Combine like radicals. **3.** For any radical with index n, write the radicand as a product of one or more perfect nth powers and another expression that has no factors that are perfect nth powers. Then apply the product property for radicals. **4.** Write any radicals with as small an index as possible.

Section 13.3 Rationalizing Denominators and Simplifying Quotients of Radical Expressions

Rationalizing the denominator of $\dfrac{A}{\sqrt[n]{x^m}}$	To rationalize the denominator of a radical expression of the form $\dfrac{A}{\sqrt[n]{x^m}}$, we multiply the expression by a fraction of the form $\dfrac{\sqrt[n]{x^k}}{\sqrt[n]{x^k}}$ so the radical in the denominator has a perfect nth-power radicand.
Quotient property for radicals	If $\sqrt[n]{a}$ and $\sqrt[n]{b}$ are defined and b is nonzero, then $\sqrt[n]{\dfrac{a}{b}} = \dfrac{\sqrt[n]{a}}{\sqrt[n]{b}}$.
Radical conjugates	We say the sum of two radicals and the difference of the same two radicals are **radical conjugates** of each other.

Section 13.3 Rationalizing Denominators and Simplifying Quotients of Radical Expressions (*Continued*)

Rationalizing a denominator by using a radical conjugate	To rationalize the denominator of a square root expression if the denominator is a sum or difference involving radicals, **1.** Determine the radical conjugate of the denominator. **2.** Multiply the original fraction by the fraction $\frac{\text{conjugate}}{\text{conjugate}}$. **3.** Find the product of the denominators by using $(A + B)(A - B) = A^2 - B^2$.

Section 13.4 Graphing and Combining Square Root Functions

Reflections across x-axis	The graphs of $f(x) = -a\sqrt{x}$ and $g(x) = a\sqrt{x}$ are reflections of each other across the x-axis.
Increasing or decreasing	If $a > 0$, then $g(x) = a\sqrt{x - h} + k$ is an increasing function and (h, k) is the minimum point. If $a < 0$, then g is a decreasing function and (h, k) is the maximum point.
Graph rises more quickly or more slowly	If $a > 1$, then the graph of $g(x) = a\sqrt{x - h} + k$ rises more quickly than the graph of $y = \sqrt{x - h} + k$. If $0 < a < 1$, then the graph rises more slowly.
Three-step method for graphing a square root function	To sketch the graph of $f(x) = a\sqrt{x - h} + k$, where $a \neq 0$, **1.** Sketch the graph of $y = a\sqrt{x}$. **2.** Translate the graph sketched in step 1 to the right by h units if $h > 0$ and to the left by $\lvert h \rvert$ units if $h < 0$. **3.** Translate the graph sketched in step 2 up by k units if $k > 0$ and down by $\lvert k \rvert$ units if $k < 0$.

Section 13.5 Solving Radical Equations

Squaring property of equality	If A and B are expressions, then all solutions of the equation $A = B$ are *among* the solutions of the equation $A^2 = B^2$. That is, the solutions of an equation are among the solutions of the equation obtained by squaring both sides.
Checking proposed solutions	Because squaring both sides of a square root equation may introduce extraneous solutions, it is essential to check that each proposed solution satisfies the original equation.
Solving a square root equation in one variable	To solve a square root equation in one variable, **1.** Isolate a square root term on one side of the equation. **2.** Square both sides. **3.** Repeat steps 1 and 2 until no square root terms remain. **4.** Solve the new equation. **5.** Check that each proposed solution satisfies the original equation.
Power property of equality	If A and B are expressions and n is a counting number greater than 1, then all solutions of the equation $A = B$ are *among* the solutions of the equation $A^n = B^n$. That is, the solutions of an equation are among the solutions of the equation obtained by raising both sides to the nth power.

Section 13.6 Modeling with Square Root Functions

Finding an equation of a square root curve	To find an equation of a square root curve that contains two given points, **1.** Obtain a system of two linear equations in two variables by substituting the coordinates of both points into the general equation $y = a\sqrt{x} + b$. **2.** Solve the system you found in step 1. **3.** Substitute the values of a and b you found in step 2 into the equation $y = a\sqrt{x} + b$.
y-Intercept of the graph of $y = a\sqrt{x} + b$	The graph of an equation of the form $y = a\sqrt{x} + b$ has y-intercept $(0, b)$.
Finding an equation of a square root model	To find an equation of a square root model, given some data, **1.** Create a scattergram of the data. **2.** Imagine a square root curve that comes close to the data points, and choose two points (not necessarily data points) that lie on or close to the square root curve. **3.** Use the two points to find an equation of the square root curve. **4.** Use a graphing calculator to verify that the graph of the equation comes close to the points of the scattergram.

Section 13.6 Modeling with Square Root Functions (*Continued*)

Four-step modeling process	To find a model and make estimates and predictions, **1.** Create a scattergram of the data. Decide whether a line, an exponential curve, a parabola, a square root curve, or none of these comes close to the points. **2.** Find an equation of your model. **3.** Verify that your equation has a graph that comes close to the points in the scattergram. If it doesn't, check for calculation errors or use different points to find the equation. An alternative is to reconsider your choice of model in step 1. **4.** Use your equation of the model to draw conclusions, make estimates, and/or make predictions.

Chapter 13 Review Exercises

If the expression is in exponential form, then write it in radical form. If it is in radical form, then write it in exponential form.

1. $x^{3/7}$

2. $\sqrt[5]{(3k+4)^7}$

Simplify. Assume each variable is nonnegative.

3. $\sqrt{8x^6}$

4. $\sqrt{18x^7y^{10}}$

5. $\sqrt[8]{x^6}$

6. $\sqrt[3]{24x^{10}y^{24}}$

7. $\sqrt[5]{(6x+11)^{27}}$

8. $2\sqrt{5} - 3\sqrt{7} - 8\sqrt{5}$

9. $5\sqrt{20x} - 2\sqrt{45x} + 7\sqrt{5x}$

10. $5\sqrt{3x^2} - 3x\sqrt{48}$

11. $b\sqrt[3]{16a^5b} + a\sqrt[3]{2a^2b^4}$

12. $5(4\sqrt{x} - \sqrt[3]{x}) - 2\sqrt[3]{x} + 8\sqrt{x}$

13. $3\sqrt{x}(\sqrt{x} - 7)$

14. $2\sqrt{7}(5\sqrt{3} + \sqrt{7})$

15. $(t + \sqrt{3})(t + \sqrt{5})$

16. $(4\sqrt{x} - 3)(2\sqrt{x} + 1)$

17. $(2\sqrt{a} - \sqrt{b})(5\sqrt{a} + \sqrt{b})$

18. $(b - \sqrt{3})(b + \sqrt{3})$

19. $(5\sqrt{a} - 7\sqrt{b})(5\sqrt{a} + 7\sqrt{b})$

20. $(4\sqrt{x} + 3)^2$

21. $(3\sqrt{5} - 4\sqrt{2})^2$

22. $(2\sqrt[3]{x} - 5)^2$

23. $\sqrt[4]{x}\sqrt[5]{x}$

24. $\sqrt[3]{\sqrt[6]{x}}$

25. $\dfrac{\sqrt[4]{x}}{\sqrt[6]{x}}$

26. $\sqrt{\dfrac{3}{x}}$

27. $\dfrac{5t}{\sqrt[3]{t}}$

28. $\sqrt[5]{\dfrac{7y}{27x^2}}$

29. $\dfrac{5}{3+\sqrt{x}}$

30. $\dfrac{\sqrt{a}}{\sqrt{a} - 2\sqrt{b}}$

31. $\dfrac{5\sqrt{x} - 4}{2\sqrt{x} + 3}$

Graph the function by hand.

32. $y = -2\sqrt{x}$

33. $y = 3\sqrt{x} + 1$

34. $y = -\sqrt{x-5} + 3$

35. $y = 2\sqrt{x+4} - 1$

For $f(x) = 3\sqrt{x} + 5$ and $g(x) = 2 - 4\sqrt{x}$, find an equation of the given function.

36. $f + g$

37. $f - g$

38. $f \cdot g$

39. $\dfrac{f}{g}$

For Exercises 42–47, solve.

40. $\sqrt{3r+8} = -7$

41. $2\sqrt{x} - 5 = 11$

42. $\sqrt{2x+1} + 4 = 7$

43. $\sqrt{2x-4} - x = -2$

44. $\sqrt{x} + 6 = x$

45. $\sqrt{13t+4} = \sqrt{5t+20}$

46. $\sqrt{2x-1} = 1 + \sqrt{x+3}$

47. $\sqrt{x+2} + \sqrt{x+9} = 7$

48. Solve $3.57 + 2.99\sqrt{8.06x - 6.83} = 14.55$. Round your result to the second decimal place.

49. Solve $\sqrt{x+5} = x^2 - 3x - 4$ by using "intersect" on a graphing calculator. Round any solutions to the second decimal place.

For Exercises 50 and 51, find all x-intercepts.

50. $f(x) = \sqrt{4x-7} - \sqrt{2x+1}$

51. $g(x) = -3\sqrt{x+2} + 9$

52. Consider the scattergram of data and the graph of the model $f(x) = a\sqrt{x} + b$ in Fig. 45. Draw a square root model that fits the data better. Explain how you would adjust the values of a and b of the original model to describe the data better.

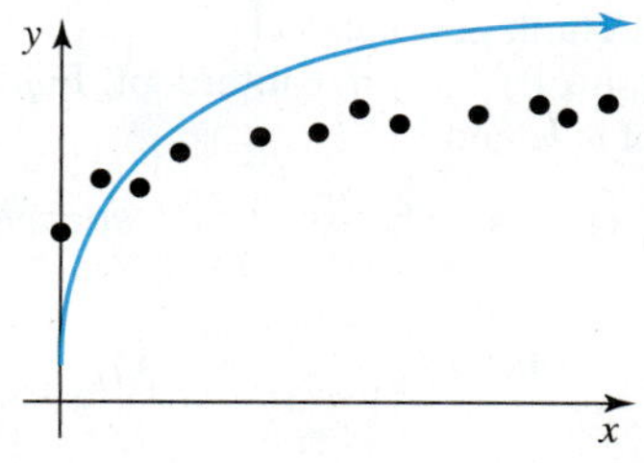

Figure 45 Exercise 52

For Exercises 53–56, find an equation of a square root curve of the form $y = a\sqrt{x} + b$ that approximately contains the given points. Round the values of a and b to the second decimal place.

53. (0, 3) and (4, 8)

54. (0, 7) and (9, 3)

55. (3, 7) and (5, 4)

56. (2, 5) and (3, 6)

57. Average credit card debts per household are shown in Table 35 for various years.

Table 35 Average Credit Card Debts per Household

Year	Average Credit Card Debt per Household (thousands of dollars)
1992	3.8
1996	6.9
2000	8.3
2004	9.6
2008	10.7

Source: *CardTrak.com*

Let $D = f(t)$ be the average credit card debt (in thousands of dollars) per household at t years since 1992.

a. Find a square root equation of f.

b. Find the D-intercept. What does it mean in this situation?

c. Find $f(26)$. What does it mean in this situation?

d. Find t when $f(t) = 13$. What does it mean in this situation?

Chapter 13 Test

For Exercises 1–10, simplify. Assume $x \geq 0$.

1. $\sqrt{32x^9y^{12}}$
2. $\sqrt[3]{64x^{22}y^{14}}$
3. $\sqrt[4]{(2x+8)^{27}}$
4. $\dfrac{4\sqrt[3]{x}}{6\sqrt[5]{x}}$
5. $\dfrac{\sqrt{x}+1}{2\sqrt{x}-3}$
6. $4\sqrt{12x^3} - 2x\sqrt{75x} + \sqrt{3x^3}$
7. $3\sqrt{x}(6\sqrt{x}-5)$
8. $(2+4\sqrt{x})(3-5\sqrt{x})$
9. $(3\sqrt{a}-5\sqrt{b})(3\sqrt{a}+5\sqrt{b})$
10. $(4\sqrt[5]{x}-3)^2$
11. Show that if n and k are counting numbers greater than 1 and $x > 0$, then the statement
$$\frac{\sqrt[n]{x}}{\sqrt[k]{x}} = \sqrt[kn]{x^{k-n}}$$
is true.
12. Graph the function $y = -2\sqrt{x+3} + 1$ by hand.
13. Let $f(x) = a\sqrt{x-h} + k$, where $a \neq 0$.
    a. What must be true of a, h, and k for the graph of f to have an x-intercept? [**Hint:** Think graphically.]
    b. Now assume the graph of f has an x-intercept. Find the x-intercept in terms of a, h, and k.

For $f(x) = 7 - 3\sqrt{x}$ and $g(x) = 4 + 5\sqrt{x}$, find an equation of the given function.

14. $f + g$
15. $f - g$
16. $f \cdot g$
17. $\dfrac{f}{g}$

Solve.

18. $2\sqrt{x} + 3 = 13$
19. $3\sqrt{5x-4} = 27$
20. $3 - 2\sqrt{x} + \sqrt{9-x} = 0$

For Exercises 21 and 22, let $f(x) = 6 - 4\sqrt{x+1}$.

21. Find $f(8)$.
22. Find a value of x such that $f(x) = -2$.

For Exercises 23 and 24, estimate all solutions of the equation by referring to the graphs shown in Fig. 46. Round results to the first decimal place.

23. $\sqrt{x+4} = x^2 - 4x + 5$
24. $\sqrt{x+4} = 1$

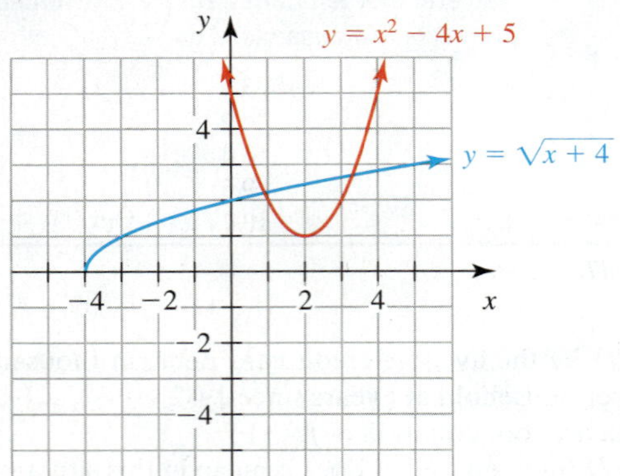

Figure 46 Exercises 23 and 24

25. Find all x-intercepts of the graph of the square root function $f(x) = 3\sqrt{2x-4} - 2\sqrt{2x+1}$.
26. Consider the scattergram of the data and the graph of the model $y = a\sqrt{x} + b$ sketched in Fig. 47. Draw a square root model that fits the data better; then explain how you would adjust the values of a and b of the original model to describe the data better.

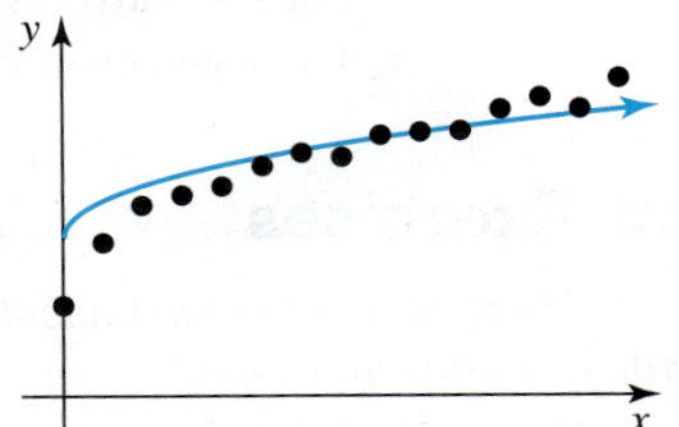

Figure 47 Exercise 26

27. Find an equation of a square root curve of the form $y = a\sqrt{x} + b$ that approximately contains the points $(2, 4)$ and $(5, 6)$.
28. The median heights of boys in the United States are listed in Table 36 for various ages, up to 5 years.

Table 36 Boys' Median Heights

Age (months)	Height (inches)
0	20.5
6	27.0
12	30.8
18	32.9
24	35.0
36	37.5
48	40.8
60	43.4

Source: The Portable Pediatrician for Parents, *by Laura Walther Nathanson*

Let $h = f(t)$ be the median height (in inches) of boys who are t months of age.

a. Find an equation of f.
b. Estimate the median height of boys who are 6 years old.
c. Estimate the age at which the median height of boys is 3 feet.
d. Find the h-intercept. What does it mean in this situation?

14 Sequences and Series

Did you know many organizations spend millions of dollars to try to influence decisions in government and politics? Members of Congress said the pharmaceutical industry was the fifth most effective special-interest group (Source: *National Journal*). See Table 1. In Exercise 35 of Homework 14.3, you will predict the total spending by the pharmaceutical industry on government and politics from 2000 through 2017.

In this chapter, we reexamine linear and exponential functions from a different perspective—that of lists of numbers called *sequences* and sums of numbers called *series*. We will use these ideas to help us model new situations. We will also discuss how to find sums of quantities efficiently, such as the total spending on pets from 2000 to 2018.

Table 1 Spending by the Pharmaceutical Industry on Government and Politics

Year	Spending (millions of dollars)
2002	14.3
2004	15.5
2006	18.1
2008	20.2
2010	21.7

Source: *Political Moneyline*

14.1 Arithmetic Sequences

Objectives

- Know the meaning of *sequence, term, term number,* and *arithmetic sequence.*
- Find a formula, term, or term number of an arithmetic sequence.
- Use an arithmetic sequence to make estimates and predictions.

In this section, we use lists of numbers to reexamine linear functions from a different perspective. We will also use these ideas to make estimates and predictions.

Definition of Arithmetic Sequence

Suppose a math tutor charges $23 per hour for one student, plus $6 per hour for each additional student. The tutor will take up to 10 students. We list the total charges (in dollars per hour) for 1 through 10 students in order:

$$23, 29, 35, 41, 47, 53, 59, 65, 71, 77$$

We call this list of numbers a sequence.

> **Definition** Sequence
>
> Any ordered list of numbers is called a **sequence.** Each number is a **term** of the sequence.

A sequence that has a last term, such as 77 in the math tutor sequence, is called a **finite sequence.** A sequence that does not have a last term is called an **infinite sequence.** For example, the sequence of odd numbers

$$1, 3, 5, 7, 9, \ldots$$

is an infinite sequence. The three dots mean the pattern of numbers continues without ending.

For the math tutor sequence, notice that the difference between any term and the preceding term is 6:

$$29 - 23 = 6, 35 - 29 = 6, 41 - 35 = 6, \dots, 77 - 71 = 6$$

We call 6 the common difference of this sequence and say the sequence is arithmetic.

Definition Arithmetic Sequence

If the difference between any term of a sequence and the preceding term is a constant d for every such pair of terms, then the sequence is an **arithmetic sequence.** We call the constant d the **common difference.**

Example 1 Identifying Arithmetic Sequences

Determine whether the sequence is arithmetic. If it is, find the common difference d.

1. 2, 6, 10, 14, 18, ...
2. 80, 77, 74, 71, 68, ...
3. 3, 6, 12, 24, 48, ...

Solution

1. The sequence is arithmetic, because it has a common difference of 4:

$$6 - 2 = 4, 10 - 6 = 4, 14 - 10 = 4, 18 - 14 = 4, \dots$$

2. The sequence is arithmetic, because it has a common difference of -3:

$$77 - 80 = -3, 74 - 77 = -3, 71 - 74 = -3, 68 - 71 = -3, \dots$$

3. The sequence is not arithmetic, because we can see from the first two differences that the sequence does not have a common difference:

$$6 - 3 = 3, 12 - 6 = 6$$

We use the notation $a_1, a_2, a_3, \dots$ to denote the terms of a sequence. We say that a_n is the ***n*th term** of the sequence and that its **term number** is n. For the math tutor sequence, we write

$$a_1 = 23, a_2 = 29, a_3 = 35, \dots, a_{10} = 77$$

where a_n is the charge (in dollars per hour) for n students. For instance, the term 35 is the 3rd term, and its term number is 3.

Finding a Formula of an Arithmetic Sequence

Can we find a formula that describes the terms of an arithmetic sequence? Since the math tutor sequence has a common difference of 6, we add 6 to the first term, 23, to find the second term; we add 6 two times to 23 to find the third term; we add 6 three times to 23 to find the fourth term; and so on:

$$23 + 6 = 29, 23 + 6 + 6 = 35, 23 + 6 + 6 + 6 = 41, \dots$$

In general, for an arithmetic sequence with common difference d, we have the terms

$$a_1, a_1 + d, a_1 + d + d, a_1 + d + d + d, \dots$$

Simplifying, we have

$$a_1, a_1 + d, a_1 + 2d, a_1 + 3d, \dots$$

We can use a pattern to find a formula that describes any term a_n of an arithmetic sequence:

$a_1 = a_1$

$a_2 = a_1 + d$ *Add d once to the first term to get the second term.*

$a_3 = a_1 + 2d$ *Add d twice to the first term to get the third term.*

$a_4 = a_1 + 3d$ *Add d three times to the first term to get the fourth term.*

$\vdots$

$a_n = a_1 + (n-1)d$ *Add d a total of (n − 1) times to the first term to get the nth term.*

Formula That Describes the *n*th Term of an Arithmetic Sequence

If an arithmetic sequence $a_1, a_2, a_3, \ldots, a_n, \ldots$ has the common difference d, then

$$a_n = a_1 + (n-1)d$$

In words, the *n*th term of an arithmetic sequence is equal to the first term plus $n-1$ times the common difference.

Example 2 Finding a Formula

Find a formula that describes the terms of the math tutor sequence.

Solution

To find a formula that describes the math tutor sequence 23, 29, 35, 41, 47,..., 77, we substitute $a_1 = 23$ and $d = 6$ in the formula $a_n = a_1 + (n-1)d$:

$$\begin{aligned} a_n &= 23 + (n-1)(6) && \text{Substitute 23 for } a_1 \text{ and 6 for } d. \\ &= 23 + 6n - 6 && \text{Distributive law} \\ &= 6n + 17 \end{aligned}$$

The tutor equation $a_n = 6n + 17$ describes a linear function whose inputs are only the numbers of students 1, 2, 3,..., 10 and whose outputs are only the dollars-per-hour charges 23, 29, 35,..., 77. For instance,

$$\begin{aligned} a_1 &= 6(1) + 17 = 23 \\ a_2 &= 6(2) + 17 = 29 \\ a_3 &= 6(3) + 17 = 35 \\ a_{10} &= 6(10) + 17 = 77 \end{aligned}$$

We can verify our formula by entering $y = 6x + 17$ in a graphing calculator and checking that the inputs 1, 2, 3,..., 10 give the outputs 23, 29, 35,..., 77 (see Fig. 1).

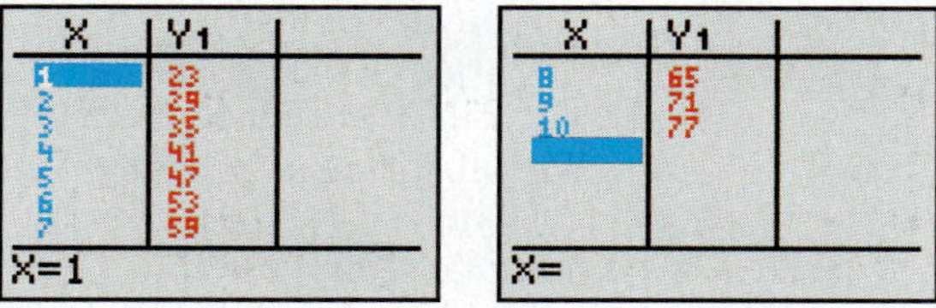

Figure 1 Verify the formula $a_n = 6n + 17$

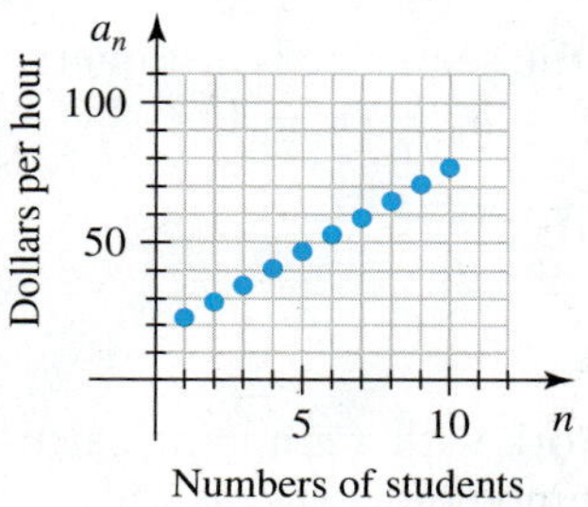

Figure 2 Math tutor sequence

We can also sketch a graph of the math tutor sequence. The graph consists of the 10 input–output pairs $(1, 23), (2, 29), (3, 35), \ldots, (10, 77)$. For $a_n = 6n + 17$, the coefficient of the independent variable n is 6, so the points lie on an increasing line with slope 6 (see Fig. 2).

Finding a Term or a Term Number of an Arithmetic Sequence

For an arithmetic sequence, if we know the values of three of the four variables a_1, a_n, n, and d, we can find the value of the fourth variable from the formula $a_n = a_1 + (n-1)d$.

Because the formula $a_n = a_1 + (n - 1)d$ is valid for arithmetic sequences only, we must first check that a sequence is arithmetic before we use the formula.

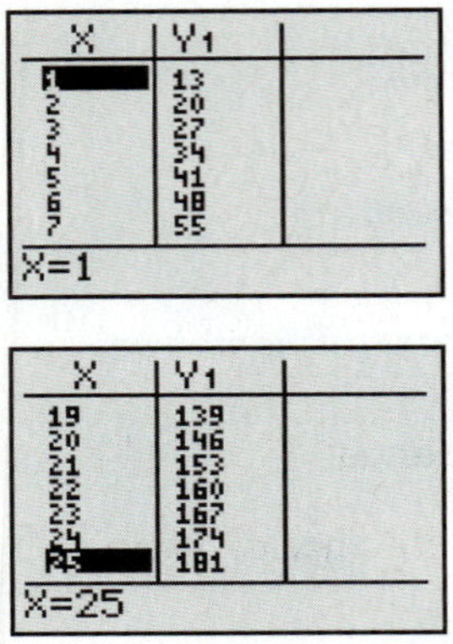

X	Y1
1	13
2	20
3	27
4	34
5	41
6	48
7	55

X=1

X	Y1
19	139
20	146
21	153
22	160
23	167
24	174
25	181

X=25

Figure 3 Verify the work

Example 3 Finding a Term

Find the 25th term of the sequence 13, 20, 27, 34, 41, ...

Solution

The sequence has a common difference $(d = 7)$, so the sequence is arithmetic. We substitute $a_1 = 13$, $n = 25$, and $d = 7$ in the formula $a_n = a_1 + (n - 1)d$:

$$\begin{aligned} a_{25} &= 13 + (25 - 1)(7) && \text{Substitute 13 for } a_1\text{, 25 for } n\text{, and 7 for } d. \\ &= 13 + 24(7) \\ &= 181 \end{aligned}$$

So, the 25th term is $a_{25} = 181$. To verify our work, we enter $y = 13 + (x - 1)(7)$ in a graphing calculator; we check that the first five terms of the sequence are 13, 20, 27, 34, and 41 and that the 25th term is 181 (see Fig. 3).

Example 4 Finding a Term Number

The number 23 is a term in the sequence 155, 151, 147, 143, 139, ... What is its term number?

Solution

The sequence has a common difference $(d = -4)$, so the sequence is arithmetic. We substitute $a_1 = 155$, $d = -4$, and $a_n = 23$ in the formula $a_n = a_1 + (n - 1)d$ and solve for n:

$$\begin{aligned} 23 &= 155 + (n - 1)(-4) && \text{Substitute 155 for } a_1\text{, } -4 \text{ for } d\text{, and 23 for } a_n. \\ 23 &= 155 - 4n + 4 && \text{Distributive law} \\ 4n &= 136 && \text{Isolate } 4n. \\ n &= 34 \end{aligned}$$

So, 23 is the 34th term. In symbols, $a_{34} = 23$. We can use a graphing calculator table to verify our work.

Example 5 Modeling with an Arithmetic Sequence

A person's salary is \$25,000 in the first year. It will increase by \$750 each year. Let a_n be the salary (in dollars) in the nth year.

1. Find a formula that describes a_n.
2. What will be the salary in the 32nd year?
3. In what year will the salary be \$40,000?

Solution

1. The salary sequence has a common difference of 750, so the sequence is arithmetic. We substitute $a_1 = 25{,}000$ and $d = 750$ in the formula $a_n = a_1 + (n - 1)d$:

$$\begin{aligned} a_n &= 25{,}000 + (n - 1)(750) && \text{Substitute 25,000 for } a_1 \text{ and 750 for } d. \\ &= 25{,}000 + 750n - 750 && \text{Distributive law} \\ &= 750n + 24{,}250 \end{aligned}$$

The formula is $a_n = 750n + 24{,}250$. We can verify our work with a graphing calculator table by checking that $a_1 = 25{,}000$ and that the common difference is 750.

2. To find the salary in the 32nd year, we substitute $n = 32$ in the formula $a_n = 750n + 24{,}250$:

$$a_{32} = 750(32) + 24{,}250 = 48{,}250$$

The salary will be \$48,250 in the 32nd year.

3. To determine when the salary will be $40,000, we substitute $a_n = 40{,}000$ in the formula $a_n = 750n + 24{,}250$ and solve for n:

$$\begin{aligned} 40{,}000 &= 750n + 24{,}250 && \text{Substitute 40,000 for } a_n. \\ 15{,}750 &= 750n \\ 21 &= n && \text{Divide.} \end{aligned}$$

The salary will be $40,000 in the 21st year.

Connection Between a Linear Function and an Arithmetic Sequence

We close this section by noticing a connection between a linear function and an arithmetic sequence. Note that the math tutor formula $a_n = 6n + 17$ describes a linear function whose graph has a slope of 6, which is the common difference of the sequence. If we let $f(n)$ be the tutor's charge (in dollars per hour) for n students, then

$$f(n) = 6n + 17$$

and the sequence 23, 29, 35, 41, 47, …, 77 is given by

$$f(1), f(2), f(3), \ldots, f(10)$$

> **Connection Between a Linear Function and an Arithmetic Sequence**
>
> If f is a linear function of the form $f(x) = mx + b$, then
>
> $$f(1), f(2), f(3), \ldots$$
>
> is an arithmetic sequence with common difference equal to the slope m.

Example 6 Identifying an Arithmetic Sequence

Let $f(x) = -5x + 38$. Is the sequence $f(1), f(2), f(3), \ldots$ arithmetic? Explain.

Solution

Since f is a linear function whose graph has a slope of -5, the sequence $f(1)$, $f(2)$, $f(3), \ldots$ must be an arithmetic sequence with common difference -5. By computing the outputs, we see that the sequence is 33, 28, 23, 18, 13, ….

We know from the slope addition property (Section 3.5) that if the inputs of a linear function increase by 1, the outputs change by the value of the slope. In terms of the slope addition property, it makes sense that the slope of the graph of a linear function and the common difference of the corresponding arithmetic sequence are equal.

Group Exploration

Connection between combining linear functions and forming new sequences

1. Let $f(x) = 3x + 5$. Use a graphing calculator table to find the values of the sequence $f(1), f(2), f(3), f(4), f(5)$. Determine whether the sequence is arithmetic. Explain. If the sequence is arithmetic, find the common difference and compare it with the slope of the graph of f.

2. For $f(x) = 3x + 5$, find the difference

$f(x + 1) - f(x)$. Compare your result with your result in Problem 1.

3. Let $f(x) = 2x + 1$ and $g(x) = 3x - 4$. For each of the following definitions of h, decide whether the sequence $h(1), h(2), h(3), h(4), h(5)$ is arithmetic. Explain.

a. $h(x) = f(x) + g(x)$

b. $h(x) = f(x) - g(x)$

c. $h(x) = f(x)g(x)$

4. Assume that the sequence $a_1, a_2, a_3, \ldots$ is arithmetic with common difference d_a and the sequence $b_1, b_2, b_3, \ldots$ is arithmetic with common difference d_b. Decide whether the sequence $a_1 + b_1, a_2 + b_2, a_3 + b_3, \ldots$ is arithmetic. If so, what is the common difference?

Homework 14.1

For extra help ▶ MyMathLab® Watch the videos in MyMathLab  Download the MyDashboard App

Check whether the sequence is arithmetic. If so, find the common difference d.

1. 3, 11, 19, 27, 35, ...
2. 40, 38, 36, 34, 32, ...
3. 1, 5, 7, 11, 13, ...
4. 7, 2, −5, −8, −11, ...
5. −20, −13, −6, 1, 8, ...
6. −2, −5, −8, −11, −14, ...
7. 4, 44, 444, 4444, 44,444, ...
8. 1, 1, 2, 2, 3, 3, ...

Using a_n notation, find a formula of the sequence. Use a graphing calculator table to verify your result.

9. 5, 11, 17, 23, 29, ...
10. 7, 11, 15, 19, 23, ...
11. −4, −15, −26, −37, −48, ...
12. −3, −7, −11, −15, −19, ...
13. 100, 94, 88, 82, 76, ...
14. 72, 69, 66, 63, 60, ...
15. 1, 3, 5, 7, 9, ...
16. 1, 2, 3, 4, 5, ...

Find the indicated term of the sequence. Verify your result with a graphing calculator table.

17. 37th term of 5, 8, 11, 14, 17, ...
18. 52nd term of 4, 19, 34, 49, 64, ...
19. 45th term of 200, 191, 182, 173, 164, ...
20. 21st term of 83, 79, 75, 71, 67, ...
21. a_{96} of 4.1, 5.7, 7.3, 8.9, 10.5, ...
22. a_{31} of 23.8, 21.5, 19.2, 16.9, 14.6, ...
23. a_{400} of 1, 2, 3, 4, 5, ...
24. a_{235} of −2, −4, −6, −8, −10, ...

For Exercises 25–34, find the term number n of the last term of the finite sequence. Verify your result with a graphing calculator table.

25. 3, 8, 13, 18, 23, ..., 533
26. 4, 10, 16, 22, 28, ..., 1426
27. 7, 15, 23, 31, 39, ..., 695
28. 10, 19, 28, 37, 46, ..., 415
29. −27, −19, −11, −3, 5, ..., 2469
30. −11, −5, 1, 7, 13, ..., 409
31. 29, 25, 21, 17, 13, ..., −14,251
32. 35, 32, 29, 26, 23, ..., −703
33. −8, −13, −18, −23, −28, ..., −493
34. −27, −39, −51, −63, −75, ..., −999

35. Is 2537 a term in the sequence 8, 15, 22, 29, 36, ...? Explain.

36. Is 3901 a term in the sequence 5, 14, 23, 32, 41, ...? Explain.

37. Find an equation of a function f such that $f(1), f(2), f(3), f(4), f(5), \ldots$ is the sequence 8, 17, 26, 35, 44, ...

38. Find an equation of a function g such that $g(1), g(2), g(3), g(4), g(5), \ldots$ is the sequence 75, 65, 55, 45, 35, ...

39. A person's starting salary is $27,500. Each year, he receives an $800 raise.
a. Let a_n be the person's salary (in dollars) for the nth year. Find a formula that describes a_n.
b. What will be his salary for the 22nd year?
c. In what year will his salary first exceed $50,000?

40. A person's starting salary is $30,700. At the end of each of the first 9 years, she will get a $950 raise. After that, she will get an $1150 raise at the end of each year. What will be her salary for the 17th year?

41. A math instructor estimates that it takes an average of 10 minutes per student to grade students' quizzes and tests each week. She also spends a total of 35 hours per week in classroom activities, holding office hours, planning for classes, and attending committee meetings. Let a_n be the number of hours that the instructor works per week when n students are enrolled in her courses.
a. Find a formula that describes a_n.
b. Find the values of a_1, a_2, a_3, and a_4. What do these four terms mean in this situation?
c. If the instructor has 130 students, how many hours does she work per week?
d. What is the greatest number of students the instructor can have without having to work over 60 hours per week?

42. A full bottle of household glass cleaner holds 22 fluid ounces. It takes about 500 squeezes of the bottle's trigger to use all the cleaner.
a. Let a_n be the number of ounces of cleaner remaining in the bottle after the trigger has been squeezed n times. Find a formula that describes a_n.
b. Find a_1, a_2, a_3, and a_4. What do these four terms mean in this situation?
c. Assume it takes about 7 squeezes of the trigger to clean one side of a 4×3-foot window. If the bottle starts out full, how much liquid would remain in it after both sides of the 4×3-foot windows have been cleaned in a building that has 32 such windows?

43. The underground rock band Little Muddy spends $50 to send postcards announcing its latest gig at a club. Three bands are playing that night. Each band gets 30% of the money collected from a cover charge of $6 per person. Let a_n be Little Muddy's profit (in dollars) if n people pay the cover charge.
a. Find a formula that describes a_n.
b. If the band's profit is $256, how many people paid the cover charge?
c. The club's maximum capacity is 200 people. Assume 18 people are on the guest list, 11 people are in the three bands, 6 people work for the club, and all of these people get in free. What is the greatest profit Little Muddy can earn, assuming no one leaves the club until closing time?
d. For what values of n will Little Muddy lose money?

44. The main library in San Francisco has a five-story-high glass sculpture created in 1996 by Nayland Blake. It looks like a star constellation made up of white lights. Each light is actually an illuminated disk displaying the name of an author etched into the glass. The sculpture originally had 160 names, but Blake left room for 200 more names, 5 to be added each year.

Let a_n be the number of names in the sculpture in the nth year, where 1996 is the first year. So, $a_1 = 160$, $a_2 = 165$, $a_3 = 170$, and so on.
a. Find a formula that describes a_n.

b. Predict the number of names there will be in the 33rd year (2028).

c. Graph by hand the library sculpture sequence.

45. The pharmaceutical industry's spending on government and politics is shown in Table 2 for various years.

Table 2 Spending by the Pharmaceutical Industry on Government and Politics

Year	Spending (millions of dollars)
2002	14.3
2004	15.5
2006	18.1
2008	20.2
2010	21.7

Source: *Political Moneyline*

Let $f(t)$ be the pharmaceutical industry's spending (in millions of dollars) on government and politics in the year that is t years since 2000.

a. Find an equation of f.

b. Use a graphing calculator table to find the values of the sequence $f(13), f(14), f(15), f(16), f(17)$. What do they mean in this situation?

c. Predict in which year the pharmaceutical industry's spending on government and politics will be \$30 million.

46. The number of Chihuahuas in Los Angeles city and county shelters has more than doubled since 2006 (see Table 3).

Table 3 Numbers of Chihuahuas in Los Angeles City and County Shelters

Year	Number of Chihuahuas (thousands)
2006	6
2007	7
2008	10
2009	12
2010	14
2011	16

Source: *Los Angeles Animal Services*

Let $f(t)$ be the number (in thousands) of Chihuahuas in Los Angeles city and county shelters at t years since 2000.

a. Find an equation of f.

b. Use a graphing calculator table to find the values of the sequence $f(14), f(15), f(16), f(17), f(18)$. What do they mean in this situation?

c. Despite large demand for Chihuahuas in the rest of the country, about a third of Chihuahuas in Los Angeles city and county shelters are euthanized (Source: *Los Angeles Animal Services*). Estimate the number of Chihuahuas euthanized in 2012.

47. Postage for a large envelope depends on its weight. In 2012, the cost to mail a large envelope weighing 1 ounce or less was \$0.90. The postage increased by \$0.20 for each additional ounce through 13 total ounces.

a. Let a_n be the postage (in dollars) for a large envelope that weighs n ounces. Find a formula that describes a_n.

b. Find the postage for a large envelope that weighs 13 ounces.

c. Postage for a 1-pound large envelope sent by priority mail is \$6.30 (there is no first-class service for this weight). Is this a better deal than the postage that would be required by your formula? (There are 16 ounces in 1 pound.)

d. Postage from Boston to Chicago for a 5-pound package is \$14.55. According to your formula, how much would the postage be?

Concepts

48. Describe an arithmetic sequence. Also, given the first few terms of an arithmetic sequence, explain how to find

- A term with a known term number.
- The term number of a known term.

(See page 9 for guidelines on writing a good response.)

49. If $a_{41} = 500$ and $a_{81} = 500$ are terms of an arithmetic sequence, find a_{990}.

50. If $a_{12} = 12$ and $a_{78} = 78$ are terms of an arithmetic sequence, find a_{103}.

51. Let $f(x) = 4x - 2$. Is the sequence $f(1), f(2), f(3), \ldots$ arithmetic? Explain.

52. Let $f(x) = -5x + 1$. Is the sequence $f(1), f(2), f(3), \ldots$ arithmetic? Explain.

53. Let $f(x) = x^2$. Is the sequence $f(1), f(2), f(3), \ldots$ arithmetic? Explain.

54. Let $f(x) = \sqrt{x}$. Is the sequence $f(1), f(2), f(3), \ldots$ arithmetic? Explain.

55. A student tries to find a_{54} of the sequence 2, 7, 11, 16, 20, 25, 29, 34, 38, ...:

$$\begin{aligned} a_{54} &= 2 + (54 - 1)(5) \\ &= 2 + 53(5) \\ &= 267 \end{aligned}$$

Is the student's work correct? Explain.

56. An arithmetic sequence is described by $a_n = 3n + 7$. A student concludes that the first term is 7 and that the common difference is 3. What would you tell the student?

Related Review

57. a. Find the common difference of the arithmetic sequence 5, 7, 9, 11, 13, ...

b. Find the slope of the line that contains the points $(1, 5)$, $(2, 7)$, $(3, 9)$, $(4, 11)$, and $(5, 13)$.

c. Compare your results in parts (a) and (b). Explain.

58. a. Using a_n notation, find a formula of the sequence 20, 17, 14, 11, 8, ...

b. Find an equation of the line that contains the points $(2, 17)$ and $(5, 8)$.

c. Compare your results in parts (a) and (b). Explain.

Expressions, Equations, Functions, and Graphs

Perform the indicated instruction. Then use words such as linear, quadratic, cubic, exponential, logarithmic, rational, radical, polynomial, degree, function, one variable, *and* two variables *to describe the expression, equation, or system.*

59. Solve $2\sqrt{x+3} - 1 = 5$.

60. Find the product $(4x - 5)(4x + 5)$.

61. Graph $f(x) = 2\sqrt{x+3} - 1$ by hand.

62. Factor $9x^2 - 49$.

63. Find the product $(4\sqrt{x} - 5)(3\sqrt{x} - 2)$.

64. Solve $100x^2 + 81 = 0$.

14.2 Geometric Sequences

Objectives

- Know the meaning of *geometric sequence.*
- Find a formula, a term, or the term number of a geometric sequence.
- Use a geometric sequence to make estimates and predictions.

In Section 14.1, we worked with arithmetic sequences. In this section, we will study another type of sequence: the geometric sequence.

Definition of Geometric Sequence

Consider the sequence

$$3, 6, 12, 24, 48, \ldots$$

Notice that the *ratio* of any term to its preceding term is 2:

$$\frac{6}{3} = 2, \frac{12}{6} = 2, \frac{24}{12} = 2, \frac{48}{24} = 2, \ldots$$

We call 2 the common ratio, and we call the sequence a geometric sequence.

Definition Geometric sequence

If the ratio of any term of a sequence to the preceding term is a constant r for every such pair of terms, then the sequence is a **geometric sequence.** We call the constant r the **common ratio.**

Example 1 Identifying Geometric Sequences

Determine whether the sequence is geometric. If so, find the common ratio r.

1. 2, 6, 18, 54, 162, ...
2. 4, 8, 24, 48, 240, ...
3. 32, 16, 8, 4, 2, ...

Solution

1. The sequence is geometric, because it has a common ratio of 3:

$$\frac{6}{2} = 3, \frac{18}{6} = 3, \frac{54}{18} = 3, \frac{162}{54} = 3, \ldots$$

2. The sequence is not geometric, because we can see from the first two ratios that it does not have a common ratio:

$$\frac{8}{4} = 2, \frac{24}{8} = 3$$

3. The sequence is geometric, because it has a common ratio of $\frac{1}{2}$:

$$\frac{16}{32} = \frac{1}{2}, \frac{8}{16} = \frac{1}{2}, \frac{4}{8} = \frac{1}{2}, \frac{2}{4} = \frac{1}{2}, \ldots$$

Finding a Formula of a Geometric Sequence

Can we find a general formula that describes the terms of a geometric sequence? Earlier, we determined that the geometric sequence 3, 6, 12, 24, 48, ... has the common ratio 2. This means we multiply the first term, 3, by 2 to find the second term; we multiply 3 by 2 two times to find the third term; we multiply 3 by 2 three times to find the fourth term; and so on:

$$3 \cdot 2 = 6, 3 \cdot 2 \cdot 2 = 12, 3 \cdot 2 \cdot 2 \cdot 2 = 24, \ldots$$

In general, for a geometric sequence with common ratio r, we have the terms

$$a_1, a_1 \cdot r, a_1 \cdot r \cdot r, a_1 \cdot r \cdot r \cdot r, \ldots$$

Using exponents, we have

$$a_1, a_1r, a_1r^2, a_1r^3, \ldots$$

We can use a pattern to find a formula that describes any term a_n of a geometric sequence:

$$\begin{aligned} a_1 &= a_1 \\ a_2 &= a_1r && \textit{Multiply } a_1 \textit{ by } r \textit{ once to get the second term.} \\ a_3 &= a_1r^2 && \textit{Multiply } a_1 \textit{ by } r \textit{ twice to get the third term.} \\ a_4 &= a_1r^3 && \textit{Multiply } a_1 \textit{ by } r \textit{ three times to get the fourth term.} \\ &\vdots \\ a_n &= a_1r^{n-1} && \textit{Multiply } a_1 \textit{ by } r \textit{ a total of } (n-1) \textit{ times to get the nth term.} \end{aligned}$$

Formula That Describes the *n*th Term of a Geometric Sequence

If a geometric sequence $a_1, a_2, a_3, \ldots, a_n, \ldots$ has the common ratio r, then

$$a_n = a_1r^{n-1}$$

In words, the nth term of a geometric sequence is equal to the first term times the $(n-1)$th power of r.

Example 2 Finding a Formula

Find a formula that describes the terms of the sequence 12, 36, 108, 324, 972, ...

Solution

The sequence has a common ratio ($r = 3$), so the sequence is geometric. We substitute $a_1 = 12$ and $r = 3$ in the formula $a_n = a_1r^{n-1}$:

$$a_n = 12(3)^{n-1}$$

We verify our formula by entering $y = 12(3)^{x-1}$ in a graphing calculator and checking that the first five terms are 12, 36, 108, 324, and 972 (see Fig. 4).

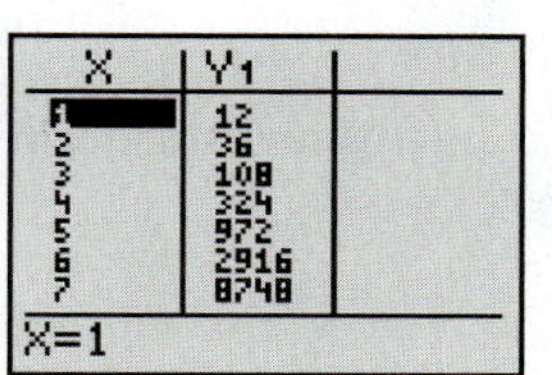

Figure 4 Verify the formula $a_n = 12(3)^{n-1}$

Here we write the formula $a_n = 12(3)^{n-1}$ in the form $a_n = ab^n$:

$$\begin{aligned} a_n &= 12(3)^{n-1} && \textit{Original formula} \\ &= 12(3)^n(3)^{-1} && b^{m+n} = b^mb^n \\ &= \frac{12(3)^n}{3} && b^{-n} = \frac{1}{b^n} \\ &= 4(3)^n && \textit{Simplify.} \end{aligned}$$

The formula $a_n = 4(3)^n$ describes an exponential function whose only inputs are the counting numbers 1, 2, 3, 4, 5, ... and whose only outputs are the terms of the sequence 12, 36, 108, 324, 972, ...

In Fig. 5, we sketch a graph of the first five terms of the sequence 12, 36, 108, 324, 972, ... For $a_n = 4(3)^n$, the base 3 is greater than 1, so the points lie on an *increasing* exponential curve.

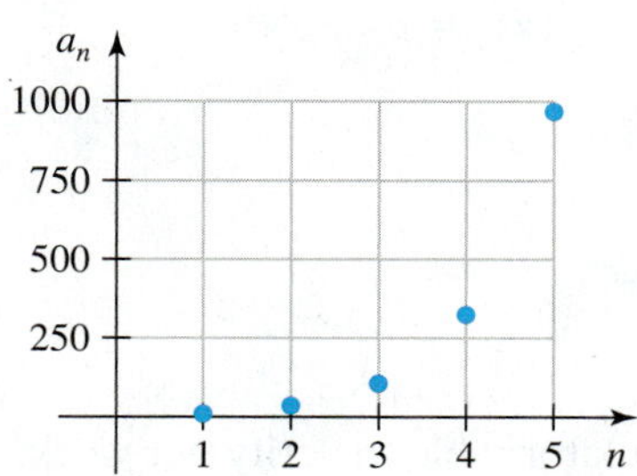

Figure 5 The first five terms of the sequence 12, 36, 108, 324, 972, ...

Finding a Term or a Term Number of a Geometric Sequence

For a geometric sequence, if we know values of three of the four variables a_1, a_n, n, and r, we can find the value of the fourth variable by using the formula $a_n = a_1r^{n-1}$.

Example 3 Finding a Term

Find the 12th term of the sequence 160, 80, 40, 20, 10, ...

Solution

The sequence has a common ratio $\left(r = \frac{1}{2}\right)$, so the sequence is geometric. We substitute $a_1 = 160$, $n = 12$, and $r = \frac{1}{2}$ in the formula $a_n = a_1 r^{n-1}$:

$$a_{12} = 160\left(\frac{1}{2}\right)^{12-1} \quad \text{Substitute 160 for } a_1\text{, 12 for } n\text{, and } \frac{1}{2} \text{ for } r.$$

$$= 160\left(\frac{1}{2}\right)^{11} \quad \text{Subtract.}$$

$$= 160\left(\frac{1}{2^{11}}\right) \quad \left(\frac{a}{b}\right)^n = \frac{a^n}{b^n}$$

$$= \frac{160}{2^{11}} \quad \text{Multiply numerators; multiply denominators.}$$

$$= 0.078125 \quad \text{Divide.}$$

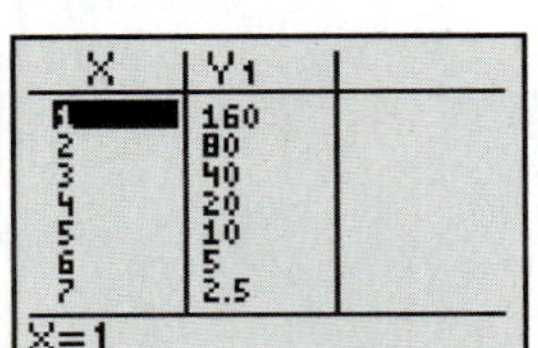

Figure 6 Verify the work

We enter $y = 160\left(\frac{1}{2}\right)^{x-1}$ in a graphing calculator and check that the first five terms are 160, 80, 40, 20, and 10 and that the 12th term is 0.078125 (see Fig. 6). The entry across from $x = 12$ has been rounded. However, the bottom entry of the table displays the exact value $a_{12} = 0.078125$.

In working with a sequence, we must first determine whether it is arithmetic, geometric, or neither. An arithmetic sequence has a common difference, whereas a geometric sequence has a common ratio.

Example 4 Finding a Term Number

The number 1,572,864 is a term in the sequence 6, 24, 96, 384, 1536, ... What is its term number?

Solution

The sequence has a common ratio $(r = 4)$, so the sequence is geometric. We substitute $a_1 = 6$, $a_n = 1{,}572{,}864$, and $r = 4$ in the formula $a_n = a_1 r^{n-1}$ and solve for n:

$$1{,}572{,}864 = 6(4)^{n-1} \quad \text{Substitute 6 for } a_1\text{, 1,572,864 for } a_n\text{, and 4 for } r.$$

$$262{,}144 = 4^{n-1} \quad \text{Divide both sides by 6.}$$

$$\log(262{,}144) = \log(4^{n-1}) \quad \text{Take the logarithm of both sides.}$$

$$\log(262{,}144) = (n-1)\log(4) \quad \text{Power property: } \log_b(x)^p = p\log_b(x)$$

$$\frac{\log(262{,}144)}{\log(4)} = n - 1 \quad \text{Divide both sides by } \log(4).$$

$$\frac{\log(262{,}144)}{\log(4)} + 1 = n \quad \text{Add 1 to both sides.}$$

$$10 = n \quad \text{Compute.}$$

So, 1,572,864 is the 10th term. We can use a graphing calculator table to verify our work.

Example 5 Modeling with a Geometric Sequence

A person's salary is \$25,000 in the first year. It will increase by 3% each year. Let a_n be the salary (in dollars) in the nth year.

1. Find a formula that describes a_n.
2. Predict the salary in the 32nd year.
3. Compare the result from Problem 2 with the result from Section 14.1, Example 5, Problem 2, where we assumed the salary increases by a constant \$750 each year.

Figure 7 Verify the comparison of the two scenarios

Solution

1. The salary in the second year is 103% of \$25,000, or $25{,}000(1.03) = 25{,}750$ dollars. Each year, the salary is equal to 1.03 times the salary in the preceding year. So, a_n is a geometric sequence with the common ratio 1.03. We substitute $a_1 = 25{,}000$ and $r = 1.03$ in the formula $a_n = a_1 r^{n-1}$:

$$a_n = 25{,}000(1.03)^{n-1}$$

2. To find the salary in the 32nd year, we substitute $n = 32$ in the formula $a_n = 25{,}000(1.03)^{n-1}$:

$$a_{32} = 25{,}000(1.03)^{32-1} \approx 62{,}502.01$$

The salary will be \$62,502.01 in the 32nd year.

3. First, we note that 3% of \$25,000 is \$750, so the first raise is the same in both scenarios. In Example 5 of Section 14.1, we found that if the salary increases by \$750 each year, the salary in the 32nd year will be \$48,250, considerably less than the salary of \$62,502.01 from receiving 3% raises each year.

We can verify our salary comparison by entering the constant-raise formula $a_n = 750n + 24{,}250$ and the percentage-raise formula $a_n = 25{,}000(1.03)^{n-1}$ in a graphing calculator and comparing tables (see Fig. 7).

Connection Between an Exponential Function and a Geometric Sequence

Now we write the geometric sequence $a_n = 25{,}000(1.03)^{n-1}$ of Example 5 in the form $a_n = ab^n$:

$$\begin{aligned} a_n &= 25{,}000(1.03)^{n-1} && \textit{Original formula} \\ &= 25{,}000(1.03)^n(1.03)^{-1} && b^{m+n} = b^m b^n \\ &= \frac{25{,}000(1.03)^n}{1.03} && b^{-n} = \frac{1}{b^n} \\ &= \frac{25{,}000}{1.03}(1.03)^n && \textit{Write right-hand side in } ab^n \textit{ form.} \end{aligned}$$

Notice that the base of the exponential function $a_n = \frac{25{,}000}{1.03}(1.03)^n$ is 1.03, which is also the common ratio of the geometric sequence. If we let $f(n)$ be the person's salary (in dollars) in the nth year, then

$$f(n) = \frac{25{,}000}{1.03}(1.03)^n$$

and the geometric sequence of the salary is

$$f(1), f(2), f(3), \ldots$$

Connection Between an Exponential Function and a Geometric Sequence

If f is an exponential function of the form $f(x) = ab^x$, then

$$f(1), f(2), f(3), \ldots$$

is a geometric sequence with common ratio equal to the base b of f.

Example 6 Identifying a Geometric Sequence

Let $f(x) = 4(3)^x$. Is the sequence $f(1), f(2), f(3), \ldots$ geometric? Explain.

Solution

Since f is an exponential function with base 3, the sequence $f(1), f(2), f(3), \ldots$ must be a geometric sequence with common ratio 3. By computing the outputs, we see that the sequence is 12, 36, 108, 324, 972, ...

Recall from Section 10.3 that the base multiplier property for exponential functions states that if the value of the independent variable increases by 1, then the value of the dependent variable is multiplied by the base of the exponential function. If we think in terms of the base multiplier property, it makes sense that the base of an exponential function is equal to the common ratio of the corresponding geometric sequence.

Group Exploration

Connection between combining exponential functions and forming new sequences

1. Let $f(x) = 3(2)^x$. Use a graphing calculator table to find the values of the sequence $f(1), f(2), f(3), f(4), f(5)$. Determine whether the sequence is geometric. Explain. If the sequence is geometric, find the common ratio and compare it with the base of f.

2. For $f(x) = 3(2)^x$, find the ratio $\dfrac{f(x+1)}{f(x)}$. Compare your result with your result in Problem 1.

3. Let $f(x) = 4(2)^x$ and $g(x) = 2(3)^x$. For each of the following definitions of h, decide whether the sequence $h(1), h(2), h(3), h(4), h(5)$ is geometric. Explain.
   a. $h(x) = f(x) + g(x)$
   b. $h(x) = f(x)g(x)$
   c. $h(x) = \dfrac{f(x)}{g(x)}$

4. Assume that the sequence $a_1, a_2, a_3, \ldots$ is geometric with common ratio r_a and the sequence $b_1, b_2, b_3, \ldots$ is geometric with common ratio r_b. Is the sequence $a_1b_1, a_2b_2, a_3b_3, \ldots$ geometric? If so, what is the common ratio?

Homework 14.2

For extra help ▶ MyMathLab®

Watch the videos in MyMathLab

Download the MyDashboard App

Check whether the sequence is arithmetic, geometric, or neither. If the sequence is geometric, find the common ratio r. If the sequence is arithmetic, find the common difference d.

1. 4, 28, 196, 1372, 9604, ...
2. 0.08, 0.8, 8, 80, 800, ...
3. 13, 6, −1, −8, −15, ...
4. 3, 7, 11, 15, 19, ...
5. 3, 4, 6, 9, 13, ...
6. 62, 57, 54, 42, 39, ...
7. $200, 40, 8, \frac{8}{5}, \frac{8}{25}, \ldots$
8. 96, 48, 24, 12, 6, ...

Find a formula of the sequence. Use a_n notation. Use a graphing calculator table to verify your formula.

9. 3, 6, 12, 24, 48, ...
10. 4, 20, 100, 500, 2500, ...
11. 800, 200, 50, 12.5, 3.125, ...
12. 162, 54, 18, 6, 2, ...
13. 100, 50, 25, 12.5, 6.25, ...
14. 1250, 250, 50, 10, 2, ...
15. 1, 4, 16, 64, 256, ...
16. 5, 15, 45, 135, 405, ...

Find the indicated term of the sequence. Write the result in scientific notation $N \times 10^k$, with N rounded to the fourth decimal place. Use a graphing calculator table to verify your result.

17. 34th term of 4, 20, 100, 500, 2500, ...
18. 103rd term of 2, 8, 32, 128, 512, ...
19. 27th term of 80, 40, 20, 10, 5, ...
20. 25th term of $36, 12, 4, \frac{4}{3}, \frac{4}{9}, \ldots$
21. a_{23} of 8, 16, 32, 64, 128, ...
22. a_{60} of $1, \frac{3}{2}, \frac{9}{4}, \frac{27}{8}, \frac{81}{16}, \ldots$

Find the term number n of the last term of the finite sequence. Verify your result with a graphing calculator.

23. 240, 120, 60, 30, 15, ..., 0.46875
24. 80, 20, 5, 1.25, 0.3125, ..., 0.01953125

25. 0.00224, 0.0112, 0.056, 0.28, 1.4, …, 109,375

26. 0.046875, 0.09375, 0.1875, 0.375, 0.75, …, 192

For Exercises 27–30, the given number is a term in the sequence that follows. Find the term number of that term. Use a graphing calculator table to verify your result.

27. 3,407,872; 13, 26, 52, 104, 208, …

28. 2,470,629; 3, 21, 147, 1029, 7203, …

29. 28,697,814; 2, 6, 18, 54, 162, …

30. 25,165,824; 6, 24, 96, 384, 1536, …

31. Find an equation of a function f such that $f(1)$, $f(2)$, $f(3)$, $f(4)$, $f(5)$, … is the sequence 8, 24, 72, 216, 648, ….

32. Find an equation of a function g such that $g(1)$, $g(2)$, $g(3)$, $g(4)$, $g(5)$, … is the sequence 48, 24, 12, 6, 3, ….

33. Is 9,238,946 a term in the sequence 13, 26, 52, 104, 208, …? Explain.

34. Is 1,240,029 a term in the sequence 7, 21, 63, 189, 567, …? Explain.

35. A person's salary is $27,000 in the first year and the salary increases by 4% each year.

a. Let a_n be the salary (in dollars) in the nth year. Find a formula that describes a_n.
b. What will be the person's salary in the 10th year?
c. In what year will the salary first exceed $50,000?

36. A person's salary is $24,000 in the first year.

a. If the salary increases by $960 each year, calculate the salary in the 2nd year and in the 30th year.
b. If the salary increases by 4% each year, calculate the salary in the 2nd year and in the 30th year.
c. Compare your results for parts (a) and (b). Explain why the salaries are the same in the 2nd year but different in the 30th year.

37. Your ancestors one generation back are your natural parents. Your ancestors two generations back are your natural grandparents. Let a_n be the number of ancestors that you had in the nth generation back.

a. List the first five terms of the sequence a_n.
b. Find a formula that describes a_n.
c. Use your formula to find the number of ancestors in the 8th generation back.
d. Use your formula to find the number of ancestors in the 35th generation back. Explain why model breakdown has occurred. Describe any assumptions you made.

38. A rubber ball is dropped. The height of the ball is measured from the floor to the bottom of the ball. The ball's maximum height after one bounce is 4 feet. The ball's maximum height after the second bounce is 70% of 4 feet, or 2.8 feet. This pattern continues; that is, the maximum height after each bounce is 70% of the maximum height of the preceding bounce. Let a_n be the maximum height (in feet) of the ball after the nth bounce.

a. Find a formula that describes a_n.
b. Predict the ball's maximum height after the 5th bounce.
c. For which bounce does the ball reach at least half a foot for the last time?
d. Graph by hand the first five terms of the bouncing-ball sequence.
e. Does model breakdown occur? Explain.

39. The *origination volume* of some loans is the total amount of money originally borrowed. The origination volumes of federal student loans are shown in Table 4 for various years.

Table 4 Federal Student Loan Origination Volume

Year	Origination Volume (billions of dollars)
2005	55.8
2006	58.9
2007	64.4
2008	75.8
2009	94.5
2010	104.3

Source: *President's 2012 Budget*

Let $f(t)$ be the origination volume (in billions of dollars) of federal student loans in the year that is t years since 2000.

a. Find an exponential equation of f.
b. Use a graphing calculator table to find the values of the sequence $f(11)$, $f(12)$, $f(13)$, $f(14)$, $f(15)$. What do they mean in this situation?
c. Predict in which year the origination volume of federal student loans will be $250 billion.

40. The numbers of Iraq and Afghanistan War veterans who are homeless or at risk of becoming homeless are shown in Table 5 for various years.

Table 5 Numbers of Iraq and Afghanistan War Veterans Who Are Homeless or at Risk of Becoming Homeless

Year	Number (thousands)
2006	1
2007	2
2008	4
2009	7
2010	13
2011	20

Source: *Department of Veterans Affairs*

Let $f(t)$ be the number (in thousands) of Iraq and Afghanistan War veterans who are homeless or at risk of becoming homeless at t years since 2000.

a. Find an exponential equation of f.
b. Use a graphing calculator table to find the values of the sequence $f(12)$, $f(13)$, $f(14)$, $f(15)$, $f(16)$. What do they mean in this situation?
c. Predict when there will be 840 thousand Iraq and Afghanistan War veterans who are homeless or at risk of becoming homeless.

41. Suppose a rumor is spreading on campus that students will not have to take any final exams this semester. On the first day, 5 students hear the rumor. Each person who hears the rumor tells it, approximately 24 hours later, to exactly 3 students who have not yet heard it. Let a_n be the number of students who hear the rumor on the nth day.

a. Find a formula that describes a_n.
b. How many students will hear the rumor on the 5th day?

c. Use your formula to predict the number of students who will hear the rumor on the 11th day. Has model breakdown occurred? Explain.

d. To find your formula, you made some assumptions about the way the rumor would spread. Describe each assumption, and discuss whether you think it is reasonable.

Concepts

42. Describe a geometric sequence. Also, given the first few terms of a geometric sequence, explain how to find

- A term with a known term number.
- The term number of a known term.

For f, is the sequence $f(1), f(2), f(3), \ldots$ arithmetic, geometric, or neither? Explain.

43. $f(x) = 2(5)^x$

44. $f(x) = 8\left(\frac{1}{2}\right)^x$

45. $f(x) = 7x - 3$

46. $f(x) = 3x^2$

47. A student tries to find a_{17} for the sequence 2, 6, 10, 14, 18, ...:

$$\begin{aligned} a_n &= a_1 r^{n-1} \\ a_{17} &= 2(3)^{17-1} \\ &= 2(3)^{16} \\ &= 86{,}093{,}442 \end{aligned}$$

Describe any errors. Then find the term correctly.

48. A geometric sequence is described by the formula $a_n = 4(6)^n$. A student concludes that the first term is 4 and the common ratio is 6. What would you tell the student?

Related Review

49. **a.** Find the common ratio of the geometric sequence 7, 14, 28, 56, 112, ...

b. Find the base b of the exponential function $y = ab^x$ that contains the points $(1, 7)$, $(2, 14)$, $(3, 28)$, $(4, 56)$, and $(5, 112)$.

c. Compare your results in parts (a) and (b). Explain why this happened.

50. **a.** Find a formula of the sequence 486, 162, 54, 18, 6, ... Write your result in the form $a_n = cb^n$. [**Hint:** $b^{n-1} = b^n b^{-1}$]

b. Find an equation of the exponential curve that contains the points $(3, 54)$ and $(5, 6)$.

c. Compare your results in parts (a) and (b). Explain why this happened.

Find a formula of the sequence. Use a_n notation.

51. 14, 19, 24, 29, 34, ...

52. 57, 49, 41, 33, 25, ...

53. 448, 224, 112, 56, 28, ...

54. 4, 12, 36, 108, 324, ...

Find the indicated term of the sequence.

55. a_9 of 2, 10, 50, 250, 1250, ...

56. a_{10} of 3200, 640, 128, 25.6, 5.12, ...

57. a_{99} of 17, 12, 7, 2, −3, ...

58. a_{96} of 9.5, 12.9, 16.3, 19.7, 23.1, ...

Find the term number n of the last term of the finite sequence.

59. 4, 7, 10, 13, 16, ..., 367

60. 88, 81, 74, 67, 60, ..., −801

61. 8192, 2048, 512, 128, 32, ..., 0.0078125

62. 5, 15, 45, 135, 405, ..., 2,657,205

Expressions, Equations, Functions, and Graphs

Perform the indicated instruction. Then use words such as linear, quadratic, cubic, exponential, logarithmic, rational, radical, polynomial, degree, function, one variable, *and* two variables *to describe the expression, equation, or system.*

63. Solve $-3(4)^x = -44$. Round any solutions to the fourth decimal place.

64. Solve: $2x - 7y = 14$

$$y = \frac{2}{7}x - 2$$

65. Graph $f(x) = -3(4)^x$ by hand.

66. Graph $2x - 7y = 14$ by hand.

67. Write $2\log_b(5x^3) - 3\log_b(2x^7)$ as a single logarithm.

68. Find the inverse of $f(x) = \frac{2}{7}x - 2$.

14.3 Arithmetic Series

Objectives

- Know the meaning of *arithmetic series*.
- Evaluate the sum of an arithmetic series.
- Use arithmetic series to make estimates and predictions.

So far in this chapter, we have worked with arithmetic sequences and geometric sequences. In the next two sections, we will discuss sums that are related to these sequences.

Definition of Arithmetic Series

Suppose a person's salary is \$23,000 in the first year and increases by \$2000 each year. Here we use an arithmetic sequence to describe the salaries (in thousands of dollars) for the first 32 years:

$$23, 25, 27, \ldots, 81, 83, 85$$

To find the *total* earnings (in thousands of dollars) during the first 32 years, we find the sum

$$23 + 25 + 27 + \cdots + 81 + 83 + 85$$

We call this sum an arithmetic series.

Definition Arithmetic series

If the sequence $a_1, a_2, a_3, \ldots, a_n$ is an arithmetic sequence, then the sum $a_1 + a_2 + a_3 + \cdots + a_n$ is an **arithmetic series.** We say a_i is the ***i*th term** of the series and its **term number** is i.

For example, for the series $23 + 25 + 27 + \cdots + 81 + 83 + 85$, the number 27 is the third term and its term number is 3. We use the notation S_n to represent the sum of the first n terms of an arithmetic sequence:

$$S_n = a_1 + a_2 + a_3 + \cdots + a_n$$

Finding the Sum of an Arithmetic Series

Next, we find the total earnings, S_{32} (in thousands of dollars), for 32 years, where

$$S_{32} = 23 + 25 + 27 + \cdots + 81 + 83 + 85$$

We find the sum in a way that suggests a general formula of S_n for any arithmetic series.

To find the sum, we write the equation for S_{32} twice, the second time with the terms in reverse order. Then we add the left-hand sides and add the right-hand sides of the two equations:

$$\begin{aligned} S_{32} &= 23 + 25 + 27 + \cdots + 81 + 83 + 85 \\ S_{32} &= 85 + 83 + 81 + \cdots + 27 + 25 + 23 \\ \hline 2S_{32} &= 108 + 108 + 108 + \cdots + 108 + 108 + 108 \end{aligned}$$

On the right-hand side of the last equation, the number 108 appears 32 times, so the sum equals 32(108):

$$\begin{aligned} 2S_{32} &= 32(108) && \text{Equation (1)} \\ S_{32} &= \frac{32(108)}{2} && \text{Equation (2)} \\ S_{32} &= 1728 && \text{Equation (3)} \end{aligned}$$

The total earnings for 32 years will be $1,728,000.

Notice that 108 is the sum of the first and last terms of the series:

$$108 = 23 + 85$$

So, we can write equation (2) as

$$S_{32} = \frac{32(23 + 85)}{2}$$

Our process and result suggest we can find S_n for any arithmetic series by first multiplying the sum of the first and last terms, $a_1 + a_n$, by the number of terms n and dividing the product by 2.

Formula for Sum of an Arithmetic Series

If $S_n = a_1 + a_2 + a_3 + \cdots + a_n$ is an arithmetic series, then

$$S_n = \frac{n(a_1 + a_n)}{2}$$

We derive the formula $S_n = \frac{n(a_1 + a_n)}{2}$ at the end of this section.

To evaluate the sum $S_n = a_1 + a_2 + a_3 + \cdots + a_n$, we must first check that the sequence $a_1, a_2, a_3, \ldots, a_n$ is arithmetic before we use the formula $S_n = \frac{n(a_1 + a_n)}{2}$.

Example 1 Evaluating Sums

1. Evaluate S_{50}, where $S_{50} = 3 + 7 + 11 + 15 + 19 + \cdots + 199$.
2. Evaluate S_{80}, where $S_{80} = 60 + 53 + 46 + 39 + 32 + \cdots + (-493)$.

Solution

1. The sequence 3, 7, 11, 15, 19, ..., 199 is arithmetic with common difference $d = 4$. We substitute $n = 50$, $a_1 = 3$, and $a_n = 199$ in the equation $S_n = \frac{n(a_1 + a_n)}{2}$:

$$S_{50} = \frac{50(3 + 199)}{2} = 5050$$

2. The sequence 60, 53, 46, 39, 32, ..., −493 is arithmetic with common difference $d = -7$. We substitute $n = 80$, $a_1 = 60$, and $a_n = -493$ in the equation $S_n = \frac{n(a_1 + a_n)}{2}$:

$$S_{80} = \frac{80(60 + (-493))}{2} = -17{,}320$$

In evaluating S_n for an arithmetic series, we sometimes must use the formula $a_n = a_1 + (n - 1)d$ to find n, a_1, or a_n before we can use the formula $S_n = \frac{n(a_1 + a_n)}{2}$.

Example 2 Evaluating a Sum

Evaluate S_{43}, where $S_{43} = 150 + 147 + 144 + 141 + 138 + \cdots + a_{43}$.

Solution

The sequence 150, 147, 144, 141, 138, ..., a_{43} is arithmetic with common difference $d = -3$. Although we know that $a_1 = 150$ and $n = 43$, we must first find a_{43} before we can use the formula $S_n = \frac{n(a_1 + a_n)}{2}$. We find a_{43} by substituting $a_1 = 150$, $n = 43$, and $d = -3$ in the equation $a_n = a_1 + (n - 1)d$:

$$a_{43} = 150 + (43 - 1)(-3) = 24$$

Next, we substitute $n = 43$, $a_1 = 150$, and $a_n = 24$ in the equation $S_n = \frac{n(a_1 + a_n)}{2}$:

$$S_{43} = \frac{43(150 + 24)}{2} = 3741$$

Example 3 Evaluating a Sum

Evaluate the sum $2 + 8 + 14 + 20 + 26 + \cdots + 338$.

Solution

The sequence 2, 8, 14, 20, 26, ..., 338 is arithmetic with common difference $d = 6$. Although we know that $a_1 = 2$ and $a_n = 338$, we must first find n before we can use the formula $S_n = \frac{n(a_1 + a_n)}{2}$. We find n by substituting $a_1 = 2$, $a_n = 338$, and $d = 6$ in the equation $a_n = a_1 + (n - 1)d$:

$$338 = 2 + (n - 1)6 \quad \text{Substitute 2 for } a_1\text{, 338 for } a_n\text{, and 6 for } d.$$

$$336 = (n - 1)6 \quad \text{Subtract 2 from both sides.}$$

$$56 = n - 1 \quad \text{Divide both sides by 6.}$$

$$57 = n$$

Now we substitute $n = 57$, $a_1 = 2$, and $a_n = 338$ in the equation $S_n = \frac{n(a_1 + a_n)}{2}$:

$$S_{57} = \frac{57(2 + 338)}{2} = 9690$$

Example 4 Modeling with an Arithmetic Series

A person's salary is \$30,000 in the first year and increases by \$1200 each year. Find the person's total earnings for the first 25 years.

Solution

Let a_n be the person's salary (in dollars) in the nth year. The salary sequence a_1, a_2, $a_3, \ldots,$ a_n is arithmetic with common difference 1200. First, we find the salary in the 25th year by substituting $a_1 = 30{,}000$, $n = 25$, and $d = 1200$ in the equation $a_n = a_1 + (n - 1)d$:

$$a_{25} = 30{,}000 + (25 - 1)(1200) = 58{,}800$$

Next, we find S_{25} by substituting $n = 25$, $a_1 = 30{,}000$, and $a_n = 58{,}800$ in the equation $S_n = \frac{n(a_1 + a_n)}{2}$:

$$S_{25} = \frac{25(30{,}000 + 58{,}800)}{2} = 1{,}110{,}000$$

The total earnings for 25 years will be \$1,110,000.

How do we derive the formula $S_n = \frac{n(a_1 + a_n)}{2}$? To begin, consider an arithmetic series:

$$S_n = a_1 + a_2 + a_3 + \cdots + a_n$$

Since the series is arithmetic, each term of the series is found by adding d (the common difference) to the preceding term. This means a_2 is found by adding d to a_1, so $a_2 = a_1 + d$. Also, a_3 is found by adding d twice to a_1, so $a_3 = a_1 + 2d$. The pattern continues, so the series can be expressed as

$$S_n = a_1 + (a_1 + d) + (a_1 + 2d) + \cdots + a_n$$

If we list the terms backward from a_n, we find the term before a_n by subtracting d, so the term before a_n is $a_n - d$. Likewise, the term before that is $a_n - 2d$. This means the series can be expressed as

$$S_n = a_1 + (a_1 + d) + (a_1 + 2d) + \cdots + (a_n - 2d) + (a_n - d) + a_n$$

Now, just as we did at the start of the section, we write the equation that describes S_n twice, the second time with the terms in reverse order. Then we add the left-hand sides and add the right-hand sides of the two equations:

$$\begin{aligned} S_n &= a_1 + (a_1 + d) + (a_1 + 2d) + \cdots + (a_n - 2d) + (a_n - d) + a_n \\ S_n &= a_n + (a_n - d) + (a_n - 2d) + \cdots + (a_1 + 2d) + (a_1 + d) + a_1 \\ \hline 2S_n &= (a_1 + a_n) + (a_1 + a_n) + (a_1 + a_n) + \cdots + (a_1 + a_n) + (a_1 + a_n) + (a_1 + a_n) \end{aligned}$$

Notice that the expression $a_1 + a_n$ appears n times on the right-hand side of the last equation. Thus, the sum on the right-hand side is $n(a_1 + a_n)$. We have

$$2S_n = n(a_1 + a_n)$$

$$S_n = \frac{n(a_1 + a_n)}{2} \quad \text{Divide both sides by 2.}$$

This is the formula we set out to derive, that of the sum of an arithmetic series.

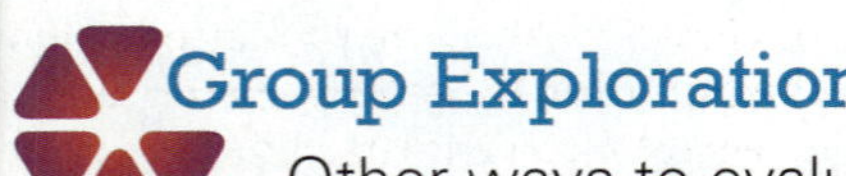

Group Exploration

Other ways to evaluate the sum of an arithmetic series

A person's salary is \$26,000 for the first year and it increases by \$1500 each year.

1. Find the person's total earnings for the first 19 years.
2. Find the person's earnings for the 19th year. Find the average of the salaries for the 1st year and the 19th year. Multiply the average by 19.
3. Find the person's earnings for the 10th year (the "middle" year). Multiply the result by 19.
4. Explain why it makes sense that your results in Problems 1, 2, and 3 are equal.
5. Your work in Problems 1–3 suggests three ways to compute the person's total earnings for the first 19 years, an odd number of years. Which of these methods will give correct results for calculating the total earnings for an *even* number of years (such as 40 years)? Explain.

Homework 14.3

For extra help ▶ MyMathLab® 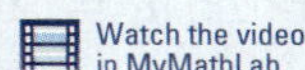 Watch the videos in MyMathLab  Download the MyDashboard App

Evaluate the sum of the arithmetic series with the given values of a_1, a_n, and n.

1. $a_1 = 2, a_n = 447$, and $n = 90$
2. $a_1 = 7, a_n = 187$, and $n = 61$
3. $a_1 = 13, a_n = 548$, and $n = 108$
4. $a_1 = 38, a_n = 605$, and $n = 82$
5. $a_1 = 37, a_n = -1099$, and $n = 72$
6. $a_1 = 208, a_n = -386$, and $n = 67$

For Exercises 7–28, evaluate the sum of the series.

7. $S_{74} = 5 + 13 + 21 + 29 + 37 + \cdots + 589$
8. $S_{59} = 14 + 17 + 20 + 23 + 26 + \cdots + 188$
9. $S_{101} = 93 + 89 + 85 + 81 + 77 + \cdots + (-307)$
10. $S_{45} = 131 + 129 + 127 + 125 + 123 + \cdots + 43$
11. $S_{117} = 4 + 4 + 4 + 4 + 4 + \cdots + 4$
12. $S_{46} = -6 + (-6) + (-6) + (-6) + (-6) + \cdots + (-6)$
13. $S_{125} = 3 + 13 + 23 + 33 + 43 + \cdots + a_{125}$
14. $S_{125} = 4 + 14 + 24 + 34 + 44 + \cdots + a_{125}$
15. $S_{81} = 8 + 19 + 30 + 41 + 52 + \cdots + a_{81}$
16. $S_{87} = 11 + 17 + 23 + 29 + 35 + \cdots + a_{87}$
17. $(-15) + (-28) + (-41) + (-54) + (-67) + \cdots + a_{152}$
18. $(-23) + (-26) + (-29) + (-32) + (-35) + \cdots + a_{85}$
19. $(-40) + (-37) + (-34) + (-31) + (-28) + \cdots + a_{137}$
20. $(-29) + (-24) + (-19) + (-14) + (-9) + \cdots + a_{214}$
21. $19 + 25 + 31 + 37 + 43 + \cdots + 247$
22. $14 + 26 + 38 + 50 + 62 + \cdots + 794$
23. $900 + 892 + 884 + 876 + 868 + \cdots + (-900)$
24. $207 + 203 + 199 + 195 + 191 + \cdots + 3$
25. $4 + 7 + 10 + 13 + 16 + \cdots + 340$
26. $1 + 3 + 5 + 7 + 9 + \cdots + 10{,}001$
27. $1 + 2 + 3 + 4 + 5 + \cdots + 10{,}000$
28. $2 + 3 + 4 + 5 + 6 + \cdots + 10{,}001$
29. A first-year salary is \$28,500. Each year there is a raise of \$1100.
    a. Find the salary in the 28th year of work.
    b. Find the total earnings for 28 years of work.
30. A first-year salary is \$35,100. Each year there is a raise of \$1400.
    a. Find the salary in the 30th year of work.
    b. Find the total earnings for 30 years of work.
31. Two companies have made you job offers. Company A offers a first-year salary of \$35,000 with a \$700 raise at the end of each year. Company B offers a first-year salary of \$27,000 with a \$1500 raise at the end of each year. At which company would your total earnings over 20 years be greater? By how much?
32. Two companies have made you job offers. Company A offers a first-year salary of \$24,500 with a \$1700 raise at the end of each year. Company B offers a first-year salary of \$33,200 with a \$600 raise at the end of each year. At which company would your total earnings over 20 years be greater? By how much?
33. An auditorium has 30 rows of seats. There are 20 seats in the front row, 24 in the second row, 28 in the third row, and so on. In other words, each row has four more seats than the row in front of it.
    a. How many seats are in the back row?
    b. How many seats are in the auditorium?
34. An auditorium has 50 rows of seats. There are 16 seats in the front row, 18 in the second row, 20 in the third row, and so on. In other words, each row has two more seats than the row in front of it.
    a. How many seats are in the auditorium?
    b. If a ticket costs \$20 for a seat in the first 10 rows and \$15 for a seat in the remaining rows, what is the revenue for a sellout performance?
    c. If 2900 people buy tickets for one performance, describe all possibilities for the revenue from the performance.
35. In Exercise 45 of Homework 14.1, you found the model $f(t) = 0.98t + 12.11$, where $f(t)$ is the pharmaceutical

industry's spending (in millions of dollars) on government and politics in the year that is t years since 2000 (see Table 6).

Table 6 Spending by the Pharmaceutical Industry on Government and Politics

Year	Spending (millions of dollars)
2002	14.3
2004	15.5
2006	18.1
2008	20.2
2010	21.7

Source: *Political Moneyline*

a. Use f to estimate the pharmaceutical industry's spending on government and politics in 2000.

b. Predict the pharmaceutical industry's spending on government and politics in 2017.

c. Predict the pharmaceutical industry's total spending on government and politics from 2000 through 2017. [**Hint:** When finding the sum, think carefully about the value of n.]

36. The median revenues from Division I-A athletic departments are shown in Table 7 for various years.

Table 7 Median Annual Revenues from Division I-A Athletic Departments

Year	Median Annual Revenue (millions of dollars)
2005	32
2006	36
2007	38
2008	41
2009	46
2010	49
2011	52

Source: *NCAA Division I Revenues and Expense Report*

Let $f(t)$ be the median annual revenue (in millions of dollars) at t years since 2000.

a. Find an equation of f.

b. Find $f(0)$. What does it mean in this situation?

c. Find $f(18)$. What does it mean in this situation?

d. Find $f(0) + f(1) + f(2) + \cdots + f(18)$. What does it mean in this situation? [**Hint:** When finding the sum, think carefully about the value of n.]

37. A first-year salary is \$24,800. Each year there is a raise of \$1200.

a. What will be the total amount of money earned in 26 years?

b. What will be the total amount of money earned from raises alone in 26 years?

c. What is the mean amount of money earned per year for the 26 years? For which of the 26 years will this mean be greater than the actual amount of money earned? For which years will this mean be less than the actual amount of money earned?

d. Assume, for each of the 26 years, *taxable income* is equal to salary minus \$4250. Assume also the federal income tax rate is 15.016% on the first \$25,000 of taxable income and 17.04% on the remaining taxable income. Estimate the total amount paid in federal income tax for the 26 years.

Concepts

38. Describe an arithmetic series. Also, explain how to evaluate the sum of an arithmetic series $S_n = a_1 + a_2 + a_3 + \cdots + a_n$ if you know a_1, a_n, and the common difference d of the arithmetic sequence $a_1, a_2, a_3, \ldots, a_n$.

*For Exercises 39–42, let S_n be the sum of an arithmetic series. For the given conditions, determine whether S_n is positive or negative. Explain. [**Hint:** Try experimenting with specific values of a_1, d, and n that meet the stated conditions. Then explain why your answer makes sense for any values that meet those conditions.]*

39. $a_1 > 0$, $d > 0$, and n is any counting number

40. $a_1 < 0$, $d < 0$, and n is any counting number

41. $a_1 = -20$, $d = 8$, and n is a very large counting number

42. $a_1 = 10$, $d = -4$, and n is a very large counting number

43. If $f(x) = 7x - 1$, is the series

$$f(1) + f(2) + f(3) + \cdots + f(100)$$

arithmetic? Explain.

44. If $g(x) = 4(3)^x$, is the series

$$g(1) + g(2) + g(3) + \cdots + g(50)$$

arithmetic? Explain.

Related Review

For each of the following, if it is a sequence, find the 15th term; if it is a series, find the sum.

45. $8, 24, 40, 56, 72, \ldots$

46. $8, 24, 72, 216, 648, \ldots$

47. $8 + 24 + 40 + 56 + 72 + \cdots + a_{15}$

48. $8 + (-8) + (-24) + (-40) + (-56) + \cdots + a_{15}$

Expressions, Equations, Functions, and Graphs

Perform the indicated instruction. Then use words such as linear, quadratic, cubic, exponential, logarithmic, rational, radical, polynomial, degree, function, one variable, *and* two variables *to describe the expression, equation, or system.*

49. Find the sum $\dfrac{x-5}{x^2-9} + \dfrac{x+3}{x^2-8x+15}$.

50. Solve $3x(x-2) = 5(x-1)$.

51. Find the product $\dfrac{x-5}{x^2-9} \cdot \dfrac{x+3}{x^2-8x+15}$.

52. Factor $8x^3 + 12x^2 - 2x - 3$.

53. Solve $\dfrac{x-5}{x^2-9} + \dfrac{x+3}{x^2-8x+15} = \dfrac{2}{x-5}$.

54. Find an equation of a parabola that contains the points $(2, 7)$, $(4, 15)$, and $(5, 22)$.

14.4 Geometric Series

Objectives

- Know the meaning of *geometric series.*
- Evaluate the sum of a geometric series.
- Use geometric series to make estimates and predictions.

In Section 14.3, we worked with arithmetic series. In this section, we will discuss another type of series: the geometric series.

Definition of Geometric Series

Consider the geometric sequence

$$3, 6, 12, 24, 48, \ldots, 1536$$

We call the sum

$$3 + 6 + 12 + 24 + 48 + \cdots + 1536$$

a geometric series.

Definition Geometric series

If the sequence $a_1, a_2, a_3, \ldots, a_n$ is a geometric sequence, then the sum $a_1 + a_2 + a_3 + \cdots + a_n$ is a **geometric series.** We say a_i is the ***i*th term** of the series and its **term number** is i.

Finding the Sum of a Geometric Series

Can we derive a general formula that describes the sum of a geometric series $S_n = a_1 + a_2 + a_3 + \cdots + a_n$? In Section 14.2, we described the terms of a geometric sequence $a_1, a_2, a_3, \ldots, a_n$ in terms of a_1 and r:

$$\begin{aligned} a_1 &= a_1 \\ a_2 &= a_1 r \\ a_3 &= a_1 r^2 \\ a_4 &= a_1 r^3 \\ &\vdots \\ a_n &= a_1 r^{n-1} \end{aligned}$$

In each case, the exponent of r is one less than the term number. So, we can express the series $S_n = a_1 + a_2 + a_3 + \cdots + a_n$ as

$$S_n = a_1 + a_1 r + a_1 r^2 + \cdots + a_1 r^{n-1}$$

If we list the terms backward from $a_1 r^{n-1}$, we find that the term before $a_1 r^{n-1}$ will have an exponent of r that is one less than $n - 1$. So, the term before $a_1 r^{n-1}$ is $a_1 r^{n-2}$. Similarly, the term before that is $a_1 r^{n-3}$. This means the series can be expressed as

$$S_n = a_1 + a_1 r + a_1 r^2 + \cdots + a_1 r^{n-3} + a_1 r^{n-2} + a_1 r^{n-1} \quad \textit{Equation (1)}$$

We multiply both sides of this equation by r to obtain

$$rS_n = a_1 r + a_1 r^2 + a_1 r^3 + \cdots + a_1 r^{n-2} + a_1 r^{n-1} + a_1 r^n \quad \textit{Equation (2)}$$

Next, we multiply both sides of equation (2) by -1, add the left-hand sides, and add the right-hand sides of the resulting equation and equation (1):

$$\begin{aligned} S_n &= a_1 + a_1 r + a_1 r^2 + \cdots + a_1 r^{n-2} + a_1 r^{n-1} \\ -rS_n &= \quad\;\; - a_1 r - a_1 r^2 - \cdots - a_1 r^{n-2} - a_1 r^{n-1} - a_1 r^n \\ \hline S_n - rS_n &= a_1 + 0 + 0 + \cdots + 0 + 0 - a_1 r^n \end{aligned}$$

We can simplify both sides of the last equation:

$$S_n - rS_n = a_1 - a_1 r^n$$

We now factor out S_n from the left-hand side and a_1 from the right-hand side:

$$\begin{aligned} S_n(1 - r) &= a_1(1 - r^n) && \textit{Factor both sides.} \\ S_n &= \frac{a_1(1 - r^n)}{1 - r}, \quad r \neq 1 && \textit{Divide both sides by } 1 - r. \end{aligned}$$

Formula for Sum of a Geometric Series

If $S_n = a_1 + a_2 + a_3 + \cdots + a_n$ is a geometric series with common ratio $r \neq 1$, then

$$S_n = \frac{a_1(1 - r^n)}{1 - r}.$$

To find the sum $S_n = a_1 + a_2 + a_3 + \cdots + a_n$, we must first determine whether the series is arithmetic, geometric, or neither. If the series is arithmetic, we use $S_n = \frac{n(a_1 + a_n)}{2}$. If the series is geometric with common ratio $r \neq 1$, we use $S_n = \frac{a_1(1 - r^n)}{1 - r}$.

Example 1 Evaluating Sums

Evaluate the sum of the series. Round the result to the fourth decimal place.

1. $S_{15} = 4 + 12 + 36 + 108 + 324 + \cdots + a_{15}$
2. $S_{13} = 486 + 162 + 54 + 18 + 6 + \cdots + a_{13}$

Solution

1. The sequence 4, 12, 36, 108, 324, ..., a_{15} is geometric with common ratio $r = 3$. We substitute $a_1 = 4$, $r = 3$, and $n = 15$ in the equation $S_n = \frac{a_1(1 - r^n)}{1 - r}$:

$$S_{15} = \frac{4(1 - 3^{15})}{1 - 3} = 28{,}697{,}812$$

2. The sequence 486, 162, 54, 18, 6, ..., a_{13} is geometric with common ratio $r = \frac{1}{3}$. We substitute $a_1 = 486$, $r = \frac{1}{3}$, and $n = 13$ in the equation $S_n = \frac{a_1(1 - r^n)}{1 - r}$:

$$S_{13} = \frac{486\left(1 - \left(\frac{1}{3}\right)^{13}\right)}{1 - \frac{1}{3}} \approx 728.9995$$

In evaluating S_n for a geometric series with common ratio $r \neq 1$, we sometimes must use the formula $a_n = a_1 r^{n-1}$ to find a_1, r, or n before we can use the formula $S_n = \frac{a_1(1 - r^n)}{1 - r}$.

Example 2 Evaluating a Sum

Evaluate the sum of the series $24{,}576 + 12{,}288 + 6144 + 3072 + 1536 + \cdots + 3$.

Solution

The sequence 24,576, 12,288, 6144, 3072, 1536, ..., 3 is geometric with common ratio $r = \frac{1}{2}$. First, we find the term number n of the last term, 3; then we find S_n. To find n, we substitute $a_1 = 24{,}576$, $a_n = 3$, and $r = \frac{1}{2}$ in the equation $a_n = a_1 r^{n-1}$ and solve for n:

$$3 = 24{,}576\left(\frac{1}{2}\right)^{n-1} \quad \text{Substitute 24,576 for } a_1\text{, 3 for } a_n\text{, and } \frac{1}{2} \text{ for } r.$$

$$\frac{3}{24{,}576} = \left(\frac{1}{2}\right)^{n-1} \quad \text{Divide both sides by 24,576.}$$

$$\log\left(\frac{3}{24{,}576}\right) = \log\left(\frac{1}{2}\right)^{n-1}$$ *Take the logarithm of both sides.*

$$\log\left(\frac{3}{24{,}576}\right) = (n-1)\log\left(\frac{1}{2}\right)$$ *Power property: $\log_b(x)^p = p\log_b(x)$*

$$\frac{\log\left(\frac{3}{24{,}576}\right)}{\log\left(\frac{1}{2}\right)} = n - 1$$ *Divide both sides by $\log\left(\frac{1}{2}\right)$.*

$$\frac{\log\left(\frac{3}{24{,}576}\right)}{\log\left(\frac{1}{2}\right)} + 1 = n$$ *Add 1 to both sides.*

$$14 = n$$ *Compute.*

Next, we substitute $a_1 = 24{,}576$, $r = \frac{1}{2}$, and $n = 14$ in the equation $S_n = \frac{a_1(1 - r^n)}{1 - r}$:

$$S_{14} = \frac{24{,}576\left(1 - \left(\frac{1}{2}\right)^{14}\right)}{1 - \frac{1}{2}} = 49{,}149$$

Example 3 Modeling with a Geometric Series

A person's salary is \$30,000 in the first year and increases by 4% at the end of each year.

1. Calculate the person's total earnings for the first 25 years.
2. Compare the result from Problem 1 with the result from Example 4 of Section 14.3, where we assumed the person's salary increases by a constant \$1200 each year.

Solution

1. Let a_n be the person's salary (in dollars) in the nth year. Since the salary in each year is 104% of the salary in the previous year, the sequence $a_1, a_2, a_3, \ldots, a_n$ is geometric with common ratio 1.04. To find the total earnings, we substitute $a_1 = 30{,}000$, $r = 1.04$, and $n = 25$ in the equation $S_n = \frac{a_1(1 - r^n)}{1 - r}$:

$$S_{25} = \frac{30{,}000(1 - 1.04^{25})}{1 - 1.04} \approx 1{,}249{,}377.25$$

So, the total earnings will be about \$1,249,377.25.

2. First, note that 4% of \$30,000 is \$1200, so the first raise is the same in both scenarios. In Example 4 of Section 14.3, we found that if the person receives constant raises of \$1200, the total earnings will be \$1,110,000 in 25 years, which is \$139,377.25 less than the total earnings of \$1,249,377.25 from earning 4% raises each year.

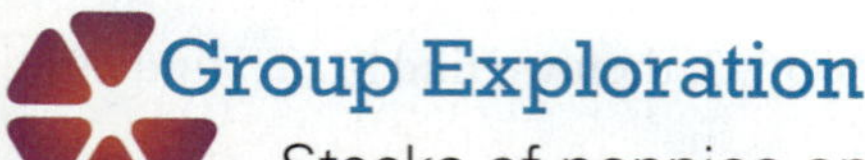

Group Exploration

Stacks of pennies on a chessboard

A chessboard (or checkerboard) has 64 squares. Suppose you have won the lottery and may choose between payment plans A and B. By plan A, you will receive \$50 million. By plan B, you will receive a chessboard with 1 penny on the first square, 2 pennies stacked on the second square, 4 pennies stacked on the third square,

8 pennies stacked on the fourth square, 16 pennies stacked on the fifth square, and so on, where each square has twice as many pennies as the previous square.

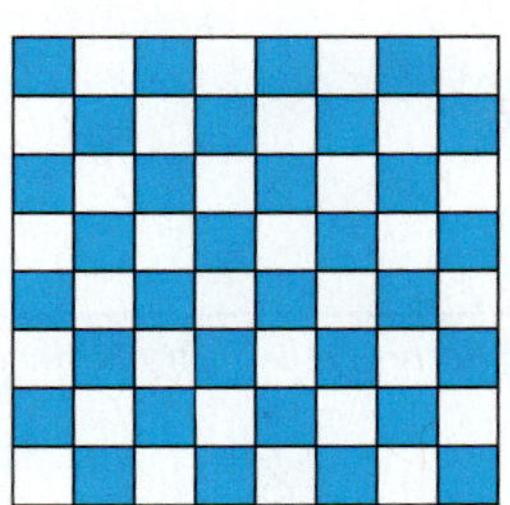

1. What is the total number of pennies paid under plan B? How much are they worth in dollars?
2. By which plan would you receive more money?

Homework 14.4

For extra help ▶ MyMathLab®  Watch the videos in MyMathLab  Download the MyDashboard App

Evaluate the sum of the geometric series with the given values of a_1, r, and n. Round any approximate results to the fourth decimal place.

1. $a_1 = 5, r = 2$, and $n = 13$

2. $a_1 = 6, r = 3$, and $n = 9$

3. $a_1 = 6, r = 1.3$, and $n = 12$

4. $a_1 = 10, r = 1.5$, and $n = 15$

5. $a_1 = 13, r = 0.8$, and $n = 13$

6. $a_1 = 9, r = 0.7$, and $n = 12$

7. $a_1 = 2.3, r = 0.9$, and $n = 10$

8. $a_1 = 4, r = 0.6$, and $n = 11$

For Exercises 9–24, find the sum of the series. Round any approximate results to the fourth decimal place.

9. $2 + 10 + 50 + 250 + 1250 + \cdots + a_{13}$

10. $1 + 2 + 4 + 8 + 16 + \cdots + a_{18}$

11. $600 + 180 + 54 + 16.2 + 4.86 + \cdots + a_{11}$

12. $625 + 500 + 400 + 320 + 256 + \cdots + a_{12}$

13. $3 + 2 + \frac{4}{3} + \frac{8}{9} + \frac{16}{27} + \cdots + a_{10}$

14. $10 + 6 + \frac{18}{5} + \frac{54}{25} + \frac{162}{125} + \cdots + a_{12}$

15. $1 + 4 + 16 + 64 + 256 + \cdots + 67{,}108{,}864$

16. $7 + 21 + 63 + 189 + 567 + \cdots + 33{,}480{,}783$

17. $5 + 6 + 7.2 + 8.64 + 10.368 + \cdots + 21.4990848$

18. $800 + 1120 + 1568 + 2195.2 + 3073.28 + \cdots + 11{,}806.312448$

19. $10{,}000 + 5000 + 2500 + 1250 + 625 + \cdots + 4.8828125$

20. $2500 + 2000 + 1600 + 1280 + 1024 + \cdots + 335.54432$

21. $S_{100} = 1 + 1 + 1 + 1 + 1 + \cdots + 1$

22. $3 + 30 + 300 + 3000 + 30{,}000 + \cdots + 3{,}000{,}000{,}000{,}000$

23. $324 + 108 + 36 + 12 + 4 + \cdots + \frac{4}{729}$

24. $80 + 40 + 20 + 10 + 5 + \cdots + \frac{5}{1024}$

25. A person's starting salary is \$23,500. Each year, the salary increases by 4%. What will be the person's total earnings after 20 years of work?

26. A person's first-year salary is \$32,000. The salary increases by 3% each year. What will be the person's total earnings after 30 years of work?

27. Two companies make you job offers. Company A offers a first-year salary of \$26,000 and a 5% raise at the end of each year. Company B offers a first-year salary of \$31,000 and a 3% raise at the end of each year. At which company would your total earnings for 30 years be greater? By how much?

28. Two companies are bidding against each other to hire you. Company A offers a first-year salary of \$25,000, a 4% raise at the end of each year, and a \$500 bonus at the end of each year. (The 4% raise is based on the salary, not on the bonus.) Company B offers a first-year salary of \$30,000 and a 3% raise at the end of each year. At which company would your total earnings for 26 years be greater? By how much?

29. In Exercise 37 of Homework 14.2, you found the (greatest possible) number of ancestors a person has in the nth generation back. Find the total number of ancestors a person has through 10 generations back.

30. Suppose a rumor is spreading in the United States that chlorine in swimming pools causes skin cancer. On the first day, 4 people hear the rumor. Approximately 24 hours after hearing the rumor, each person who hears it tells it to exactly 5 people who have not yet heard it.

a. How many people will have heard the rumor after 10 days?

b. After how many days will everyone in the United States have heard the rumor? Assume the U.S. population is 315 million.

c. To model the spread of the chlorine-causes-cancer rumor, you made some assumptions about the way the rumor would spread. Describe each assumption, and discuss whether you think it is reasonable.

31. An entrepreneur writes letters to 8 people (the first round of letters), explaining she has found a way for herself and many other people to get rich. On each letter, she has written her name and address. She asks each of the 8 people to send her \$5 and to add their name and address below hers, so each letter will now have two names on it. The entrepreneur also instructs each of the 8 people to send the list of two names, together with the instructions, to 8 more people (the second round). Then all these people should send her \$5, add their names and addresses to the list, send the list of three names to 8 more people (the third round), and so on. Each person who receives a letter is instructed to send \$5 to the name at the top of the list. When there are 10 people on the list, the next person should send \$5 to the name at the top of the list, scratch that name off the list, and add his or her name to the bottom of the list. The instructions include a warning that something terrible will happen to those who do not send the money as well as the eight letters. (These letters are called *chain letters* and are illegal.)

Assume the letters of any one round are received at approximately the same time and no one receives more than one letter.

a. In which round would the entrepreneur's name be taken off the list? How much money could she receive?

b. By which round would everyone in the world (about 7.0 billion people) have received a letter?

c. How many people will receive money from the chain letters? How much will each of them receive? [**Hint:** With 7.0 billion people, there would be only \$35 billion to go around.]

32. Suppose you win a contest and choose between two award plans. If you choose plan A, you will receive \$100,000 per day for 30 days. If you choose plan B, you will receive 1 cent the first day, 2 cents the second day, 4 cents the third day, and so on (each day you receive twice as much as you did on the preceding day) for 30 days. Which plan would you choose? Explain your reasoning.

33. The name Nevaeh has made the fastest climb in U.S. girls' names in more than a century (see Table 8). "Nevaeh" is "Heaven" spelled backward.

Table 8 Numbers of Nevaehs Born

Year	Number of Nevaehs Born
2001	1191
2002	1692
2003	2287
2004	3156
2005	4457
2006	5922
2007	6784

Source: *Social Security Administration*

a. Let $f(t)$ be the number of Nevaehs born in the year that is t years since 2000. Find an exponential equation of f.

b. Find $f(1)$. What does it mean in this situation?

c. Find $f(17)$. What does it mean in this situation?

d. Find $f(1) + f(2) + f(3) + \cdots + f(17)$. What does it mean in this situation?

34. Prior to CDs and cassettes, popular recordings were sold on *eight-track* tape cartridges. In 1980, 89.5 million eight-track cartridges were sold. After that year, sales dropped off sharply due to consumers' preference for cassettes over eight-tracks. (Sound recordings are no longer made on eight-tracks.) The numbers of eight-track cartridges (in millions) sold are listed in Table 9 for various years.

Table 9 Eight-Track Cartridge Sales

Year	Sales (millions)
1980	89.5
1981	32.0
1982	20.0
1983	10.0
1984	5.0
1985	1.5

Source: *Recording Industry Association of America*

a. Let $f(t)$ be the number of eight-track cartridges (in millions) sold in the year that is t years since 1980. Find an exponential equation of f.

b. Find $f(0)$. What does it mean in this situation?

c. Use the formula for the sum of a geometric series to find

$$f(0) + f(1) + f(2) + f(3) + f(4) + f(5)$$

[**Hint:** Think carefully about the value of n.]

d. Compare your result from part (c) with the actual total.

e. Predict the total number of eight-track cartridges that would have been sold from 1980 through 2018. Explain why this total is not much more than your total found in part (c).

Concepts

*For Exercises 35 and 36, let S_n be the sum of a geometric series. For the given conditions, determine whether S_n is positive or negative. Explain. [**Hint:** Try experimenting with specific values of a_1, r, and n that meet the conditions stated. Then explain why your response makes sense for any values that meet those conditions.]*

35. $a_1 > 0$, $r > 0$, and n is a counting number

36. $a_1 < 0$, $r > 0$, and n is a counting number

37. If $f(x) = 7 - x$, is the series

$$f(1) + f(2) + f(3) + \cdots + f(30)$$

arithmetic, geometric, or neither? Explain.

38. If $f(x) = 2(4)^x$, is the series

$$f(1) + f(2) + f(3) + \cdots + f(70)$$

arithmetic, geometric, or neither? Explain.

39. a. Find the sum

$$5 + 10 + 20 + 40 + 80 + \cdots + 2560$$

b. Solve the equation

$$a_n = a_1 r^{n-1}$$

for n.

c. Use the equation you found in part (b) to help you make a substitution for n in the equation

$$S_n = \frac{a_1(1 - r^n)}{1 - r}$$

d. Use the equation you found in part (c) to find the sum

$$5 + 10 + 20 + 40 + 80 + \cdots + 2560$$

e. Compare the methods you used in parts (a) and (d) to find the sum

$$5 + 10 + 20 + 40 + 80 + \cdots + 2560$$

Which method do you prefer? Explain.

40. Describe a geometric series. Also, explain how to evaluate the sum of a geometric series

$$S_n = a_1 + a_2 + a_3 + \cdots + a_n$$

if you know a_1, a_n, and the common ratio r of the geometric sequence

$$a_1, a_2, a_3, \ldots, a_n$$

Related Review

Find the sum of the series. Round any approximate results to the fourth decimal place.

41. $3 + 9 + 15 + 21 + 27 + \cdots + 351$

42. $351 + 347 + 343 + 339 + 335 + \cdots + 103$

43. $10 + 9 + 8.1 + 7.29 + 6.561 + \cdots + 3.486784401$

44. $7 + 28 + 112 + 448 + 1792 + \cdots + 469{,}762{,}048$

Expressions, Equations, Functions, and Graphs

For each of the following, give an example and then solve, simplify, or graph, as appropriate.

45. quadratic function

46. square root function

47. quotient of two rational expressions

48. linear function

49. quadratic equation in one variable

50. product of two radical expressions

51. exponential function

52. rational equation in one variable

53. system of two linear equations in two variables

54. exponential equation in one variable

Taking it to the Lab

Bouncing Ball Lab

In this lab, you will analyze the heights reached by a bouncing ball. The heights represent the distance from the floor to the bottom of the ball.

Check with your instructor whether you should collect your own data or use the data listed in Table 10. In this table, $f(n)$ is the maximum height (in centimeters) reached by a ball after n bounces.

Table 10 Maximum Heights of a Bouncing Ball

Number of Bounces	Maximum Height (centimeters)
1	72.930
2	48.673
3	33.745
5	18.379
6	11.301
7	10.805

Source: *J. Lehmann*

Materials

If you are going to perform your own experiment, you will need the following materials:

1. a rubber ball
2. a Texas Instruments CBL unit
3. a TI-83, TI-84, TI-86, or TI-Nspire graphing calculator
4. a Vernier motion detector
5. (optional) a Texas Instruments CBR unit (to be used in place of the CBL unit and motion detector)

If you don't have items 2, 4, and/or 5, another option is to use a video camera to tape the bouncing ball. You can make a background indicating heights in large print so you can estimate the heights by watching the video recording and using "pause" on your video player.

Preparation

First, find a level surface for bouncing the ball. For the experimenting, it is ideal for the ball to bounce almost straight up and down at least seven times. Attach the motion detector to a fixed object so the motion detector is above where the ball will bounce, facing directly downward (see Fig. 8). If you have a low ceiling, you can tape the motion detector to it.

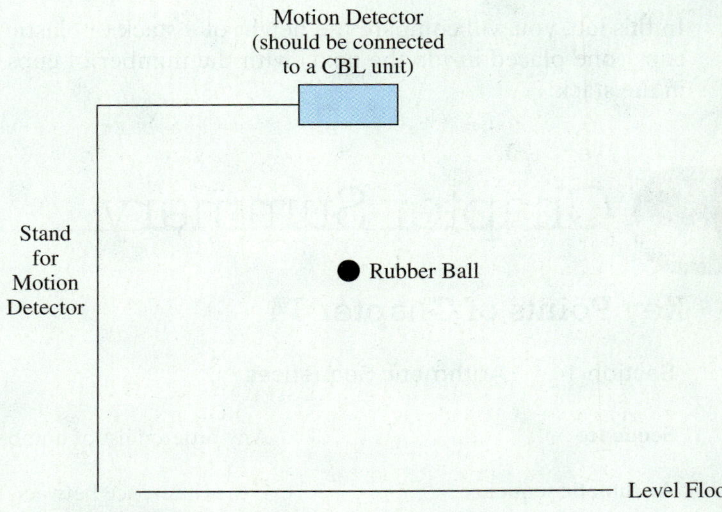

Figure 8 Equipment setup

Recording of Data

Use the CBL unit and motion detector to measure the maximum height (in centimeters) of the ball after each bounce. It is ideal to measure the heights reached by the ball for at least seven bounces.

Analyzing the Data

1. Let $f(n)$ be the maximum height (in centimeters) of the ball after n bounces. Display your data in a table or use the data in Table 10.
2. Recall that we have defined the ball's "height" to be the distance from the floor to the bottom of the ball. If you measured the distance from the floor to the top of the ball, adjust your data accordingly. (The data in Table 10 have already been adjusted.)
3. Find a quadratic and an exponential equation of f. Which model fits the data better? Which model is likely to make better predictions? Decide which equation of f you will use for the rest of this lab.
4. Is the sequence $f(1), f(2), f(3), f(4), \ldots, f(7)$ arithmetic, geometric, or neither? Explain.
5. If you performed your own experiment, you may skip this question. If you are using the data from Table 10, notice that no information is given for the maximum height between the 4th and 5th bounces. That's because the motion detector malfunctioned. Use your equation of f to estimate the maximum height reached after the 4th bounce.
6. Use your equation of f to estimate the height of the ball after the 8th bounce.
7. After which bounces will the ball reach a height of at least 1 foot?
8. Use your equation of f to estimate the height reached by the ball after the 30th bounce.
9. If your function f is an exponential function, then describe what the base means in this situation.

Stacked Cups Lab

In this lab, you will compare the height of a stack of plastic cups (one placed inside the next) with the number of cups in the stack.

Check with your instructor whether you should collect your own data or use the data listed in Table 11.

Table 11 Heights of Stacks of Cups

Number of Cups in Stack	Height of Stack (centimeters)
1	12.00
2	14.75
3	17.50
4	20.25
5	23.00

Source: *J. Lehmann*

Materials

If you are going to perform your own experiment, you will need the following materials:

1. Some plastic cups that can be stacked
2. A ruler

Recording of Data

Measure (in centimeters) the height of one cup. Then measure the heights of stacks of two, three, four, and five cups.

Analyzing the Data

1. Display your data in a table, or use the data in Table 11.
2. Let $h = f(n)$ be the height (in centimeters) of n cups. Find an equation of f.
3. Use f to estimate the height of a stack of 70 cups.
4. Find the h-intercept. What does it mean in this situation?
5. Find the slope. What does it mean in this situation?
6. Sketch a graph of your model. Sketch only the portion for which your model makes reasonably good estimates.
7. Suppose that $f(n) = 0.3n + 20$ for a different set of cups.
   a. What does this equation tell you about the cups?
   b. What is the arithmetic sequence that corresponds to f?

Chapter Summary

Key Points of Chapter 14

Section 14.1 Arithmetic Sequences

Sequence	Any ordered list of numbers is called a **sequence.** Each number is a **term** of the sequence.
Arithmetic sequence	If the difference between any term of a sequence and the preceding term is a constant d for every such pair of terms, then the sequence is an **arithmetic sequence.** We call the constant d the **common difference.**

Section 14.1 Arithmetic Sequences (*Continued*)

Formula that describes the *n*th term of an arithmetic sequence	If an arithmetic sequence $a_1, a_2, a_3, \ldots, a_n, \ldots$ has the common difference d, then $a_n = a_1 + (n - 1)d$.
Check that a sequence is arithmetic	Because the formula $a_n = a_1 + (n - 1)d$ is valid for arithmetic sequences only, we must first check that a sequence is arithmetic before we use the formula.
Connection between a linear function and an arithmetic sequence	If f is a linear function of the form $f(x) = mx + b$, then $f(1), f(2), f(3), \ldots$ is an arithmetic sequence with common difference equal to the slope m.

Section 14.2 Geometric Sequences

Geometric sequence	If the ratio of any term of a sequence to the preceding term is a constant r for every such pair of terms, then the sequence is a **geometric sequence.** We call the constant r the **common ratio.**
Formula that describes the *n*th term of a geometric sequence	If a geometric sequence $a_1, a_2, a_3, \ldots, a_n, \ldots$ has the common ratio r, then $a_n = a_1r^{n-1}$.
Check that a sequence is geometric	In working with a sequence, we must first determine whether it is arithmetic, geometric, or neither. An arithmetic sequence has a common difference, whereas a geometric sequence has a common ratio.
Connection between an exponential function and a geometric sequence	If f is an exponential function of the form $f(x) = ab^x$, then $f(1), f(2), f(3), \ldots$ is a geometric sequence with common ratio equal to the base b of f.

Section 14.3 Arithmetic Series

Arithmetic series	If the sequence $a_1, a_2, a_3, \ldots, a_n$ is an arithmetic sequence, then the sum $a_1 + a_2 + a_3 + \cdots + a_n$ is an **arithmetic series.** We say a_i is the ***i*th term** of the series and its **term number** is i.
Formula for sum of an arithmetic series	If $S_n = a_1 + a_2 + a_3 + \cdots + a_n$ is an arithmetic series, then $S_n = \dfrac{n(a_1 + a_n)}{2}$.
Check that a series is arithmetic	To evaluate the sum $S_n = a_1 + a_2 + a_3 + \cdots + a_n$, we must first check that the sequence $a_1, a_2, a_3, \ldots, a_n$ is arithmetic before we use the formula $S_n = \dfrac{n(a_1 + a_n)}{2}$.
Using a combination of formulas	In evaluating S_n for an arithmetic series, we sometimes must use the formula $a_n = a_1 + (n - 1)d$ to find n, a_1, or a_n before we can use the formula $S_n = \dfrac{n(a_1 + a_n)}{2}$.

Section 14.4 Geometric Series

Geometric series	If the sequence $a_1, a_2, a_3, \ldots, a_n$ is a geometric sequence, then the sum $a_1 + a_2 + a_3 + \cdots + a_n$ is a **geometric series.** We say a_i is the ***i*th term** of the series and its **term number** is i.
Formula for sum of a geometric series	If $S_n = a_1 + a_2 + a_3 + \cdots + a_n$ is a geometric series with common ratio $r \neq 1$, then $S_n = \dfrac{a_1(1 - r^n)}{1 - r}$.
Check that a series is geometric	To find the sum $S_n = a_1 + a_2 + a_3 + \cdots + a_n$, we must first determine whether the series is arithmetic, geometric, or neither. If the series is arithmetic, we use $S_n = \dfrac{n(a_1 + a_n)}{2}$. If the series is geometric with common ratio $r \neq 1$, we use $S_n = \dfrac{a_1(1 - r^n)}{1 - r}$.
Using a combination of formulas	In evaluating S_n for a geometric series with common ratio $r \neq 1$, we sometimes must use the formula $a_n = a_1r^{n-1}$ to find a_1, r, or n before we can use the formula $S_n = \dfrac{a_1(1 - r^n)}{1 - r}$.

Chapter 14 Review Exercises

Determine whether the following is an arithmetic sequence, an arithmetic series, a geometric sequence, a geometric series, or none of these types of sequences or series.

1. 160, 40, 10, 2.5, 0.625, ...

2. $13 + 24 + 35 + 46 + 57 + \cdots + 299$

3. 101, 95, 89, 83, 77, ...

4. $9 + 18 + 36 + 72 + 144 + \cdots$

5. $7 + \frac{7}{5} + \frac{7}{25} + \frac{7}{125} + \frac{7}{625} + \cdots + \frac{7}{390{,}625}$

6. 3, 4, 6, 9, 13, ...

Using a_n notation, find a formula of the sequence.

7. 2, 6, 18, 54, 162, ...

8. 25, 28, 31, 34, 37, ...

9. 9, 4, −1, −6, −11, ...

10. 200, 100, 50, 25, 12.5, ...

11. 3.2, 5.9, 8.6, 11.3, 14, ...

12. 800, 560, 392, 274.4, 192.08, ...

Find the indicated term of the sequence. Find the exact value, or write the result in scientific notation $N \times 10^k$ with N rounded to the fourth decimal place.

13. 47th term of 6, 12, 24, 48, 96, ...

14. 9th term of 768, 192, 48, 12, 3, ...

15. 98th term of 87, 84, 81, 78, 75, ...

16. 87th term of 2.3, 4.9, 7.5, 10.1, 12.7, ...

For Exercises 17 and 18, find the term number of the last term in the finite sequence.

17. 7, 11, 15, 19, 23, ..., 2023

18. 501, 493, 485, 477, 469, ..., −107

19. The number 470,715,894,135 is a term in the sequence 5, 15, 45, 135, 405, ... What is its term number?

20. If $a_5 = 52$ and $a_9 = 36$ are terms of an arithmetic sequence, find a_{69}.

21. Find the sum of the first 43 terms of an arithmetic series with $a_1 = 52$ and $a_{43} = -200$.

22. Find the sum of the first 22 terms of a geometric series with $a_1 = 4$, $r = 1.7$, and $n = 22$. Round your result to the fourth decimal place.

For Exercises 23–26, evaluate the sum of the series.

23. $3 + 6 + 12 + 24 + 48 + \cdots + 1{,}610{,}612{,}736$

24. $30 + 36 + 42 + 48 + 54 + \cdots + 1200$

25. $11 + 7 + 3 + (-1) + (-5) + \cdots + a_{33}$

26. $531{,}441 + 177{,}147 + 59{,}049 + 19{,}683 + 6561 + \cdots + a_{13}$

27. If $f(x) = 4(5)^x$, is $f(1) + f(2) + f(3) + \cdots + f(80)$ an arithmetic sequence, an arithmetic series, a geometric sequence, or a geometric series? Explain.

28. If $f(x) = -9x + 40$, is $f(1), f(2), f(3), \ldots, f(80)$ an arithmetic sequence, an arithmetic series, a geometric sequence, or a geometric series? Explain.

29. Two companies have made you job offers. Company A offers a first-year salary of \$28,000 with a 4% raise at the end of each year. Company B offers a first-year salary of \$34,000 with a constant raise of \$1500 each year.

a. What would be the salary in the 25th year at company A? at company B?

b. What would be the total earnings for 25 years of work at company A? at company B?

c. Explain how it is possible for the salary in the 25th year to be greater at company A than at company B, yet the total earnings for 25 years to be greater at company B than at company A.

30. Levels of spending (in billions of dollars) on pets in the United States are shown in Table 12 for various years.

Table 12 Levels of Spending on Pets in the United States

Year	Spending on Pets (billions of dollars)
2002	29.5
2004	34.4
2006	38.5
2008	43.2
2010	48.4
2011	51.0

Source: *American Pet Products Manufacturers Association*

Let $f(t)$ be the spending (in billions of dollars) on pets in the United States in the year that is t years since 2000.

a. Find a linear equation of f.

b. Find the slope of the graph of f. What does it mean in this situation?

c. Use f to predict the spending on pets in 2018.

d. Use f to estimate the total spending on pets from 2000 through 2018.

Chapter 14 Test

Determine whether the following is an arithmetic sequence, an arithmetic series, a geometric sequence, a geometric series, or none of these types of sequences or series.

1. 3, 6, 12, 24, 48, ...

2. 20, 19, 17, 14, 10, ...

3. $7 + 35 + 175 + 875 + 4375 + \cdots + 546{,}875$

4. $69 + 61 + 53 + 45 + 37 + \cdots + 5$

Using a_n notation, find a formula of the sequence.

5. 31, 25, 19, 13, 7, ...

6. 6, 24, 96, 384, 1536, ...

7. Find the 87th term of the sequence 4, 7, 10, 13, 16, ...

8. Find the 16th term of the sequence 6144, 3072, 1536, 768, 384, ...

Find the term number of the last term of the finite sequence.

9. $-27, -23, -19, -15, -11, \ldots, 1789$
10. $200, 220, 242, 266.2, 292.82, \ldots, 428.717762$

For Exercises 11–14, evaluate the sum of the series. Round any approximate results to the fourth decimal place, or write such results in scientific notation $N \times 10^k$ with N rounded to the fourth decimal place.

11. $27 + 9 + 3 + 1 + \frac{1}{3} + \cdots + a_{20}$
12. $4 + 8 + 16 + 32 + 64 + \cdots + 2{,}147{,}483{,}648$
13. $50 + 46 + 42 + 38 + 34 + \cdots + (-78)$
14. $19 + 33 + 47 + 61 + 75 + \cdots + a_{400}$
15. Evaluate the sum of the series [**Hint:** Begin by writing the series as a sum of two series]:
$$(7 + 2) + (7 \cdot 2 + 2^2) + (7 \cdot 3 + 2^3) + (7 \cdot 4 + 2^4) + (7 \cdot 5 + 2^5) + \cdots + (7 \cdot 20 + 2^{20})$$
16. Let $f(x) = 3x^2 + 1$. Is $f(1) + f(2) + f(3) + \cdots + f(100)$ an arithmetic sequence, an arithmetic series, a geometric sequence, a geometric series, or none of these types of sequences or series? Explain.
17. Let S_n be the sum of an arithmetic series. Determine whether S_n is positive or negative if $a_1 = 10$, $d = -3$, and n is a very large counting number. Explain.
18. Online retail sales are shown in Table 13 for various years. Let $f(t)$ be the online retail sales (in billions of dollars) in the year that is t years since 2000.

Table 13 Online Retail Sales

Year	Online Retail Sales (billions of dollars)
2002	42
2004	67
2006	102
2008	130
2010	142
2011	162

Source: *Forrester*

a. Find a linear equation of f.
b. Find $f(1)$. What does it mean in this situation?
c. Find $f(18)$. What does it mean in this situation?
d. Find $f(1) + f(2) + f(3) + \cdots + f(18)$. What does it mean in this situation?

19. Assume a person's salary is \$32 thousand for the first year and increases by 3% each year.
a. Let a_n be the person's salary (in thousands of dollars) in the nth year. Find a formula that describes a_n.
b. When will the salary first exceed \$40 thousand?
c. What will the salary be in the 25th year?
d. What will the total earnings be for the first 25 years?

Cumulative Review of Chapters 1–14

Solve. Any solution is a real number.

1. $4x^2 - 49 = 0$
2. $6x^2 + 13x = 5$
3. $\log_3(4x - 7) = 4$
4. $(t + 3)(t - 4) = 5$
5. $\frac{1}{w^2 - w - 6} - \frac{w}{w + 2} = \frac{w - 2}{w - 3}$
6. $5(3x - 2)^2 + 7 = 17$
7. $\log_6(3x) + \log_6(x - 1) = 1$
8. $20 - 4x = 7(2x + 9)$
9. $\sqrt{x + 1} - \sqrt{2x - 5} = 1$
10. $3(2x - 5) + 4 = (x - 3)^2$
11. $\log_b(81) = 4$

For Exercises 12–15, solve. Any solution is a real number. Round any results to the fourth decimal place.

12. $2b^7 - 3 = 51$
13. $6(3)^x - 5 = 52$
14. $5e^x = 98$
15. $4\ln(2x^3) - \ln(8x^6) = 9$
16. Solve $3x^2 - 5x + 1 = 0$ by completing the square.
17. Find all complex-number solutions of $2x^2 = 4x - 3$.
18. Solve the formula $\frac{x}{a} + \frac{y}{b} = 1$ for y.

For Exercises 19–21, solve the system.

19. $2x + 4y = 0$
$5x + 3y = 7$
20. $y = 3x + 9$
$4x + 2y = -2$
21. $2x - 3y + 4z = 19$
$5x + y - 5z = -6$
$3x - y + 2z = 13$
22. Solve the inequality $5 - 2(3x - 5) + 1 \geq 2 - 4x$. Describe the solution set as an inequality, in interval notation, and in a graph.
23. Graph the inequality $4x - 6y - 6 \geq 0$ by hand.
24. Graph by hand the solution set of the system
$$2x - 4y \leq 8$$
$$3x + 5y \leq 10$$

Simplify. Assume that any variable is positive.

25. $(3b^{-2}c^{-3})^4(6b^{-5}c^2)^2$
26. $\left(\frac{14b^3c^5}{21b^{-9}c^{-4}}\right)^2$
27. $\frac{8b^{1/2}c^{-4/3}}{10b^{3/4}c^{-7/3}}$
28. $3y\sqrt{8x^3} - 2x\sqrt{18xy^2}$
29. $\sqrt[4]{(5x - 7)^{21}}$
30. $\sqrt{12x^7y^{14}}$
31. $\sqrt[3]{\frac{4}{x}}$
32. $\frac{3\sqrt{x} - \sqrt{y}}{2\sqrt{x} + \sqrt{y}}$

Simplify. Write your result as a single logarithm with a coefficient of 1.

33. $2\ln(x^4) + 3\ln(x^9)$
34. $4\log_b(x^5) - 5\log_b(2x)$

Perform the indicated operations.

35. $(3a - 5b)^2$
36. $(3\sqrt{k} - 4)(2\sqrt{k} + 7)$
37. $(2x^2 - x + 3)(x^2 + 2x - 1)$
38. $-2x(x^2 + 1)(x^3 + 5)$

For Exercises 39–42, perform the indicated operations.

39. $\dfrac{x^3 - 27}{2x^2 - 3x + 1} \div \dfrac{2x^3 + 6x^2 + 18x}{4x^2 - 1}$

40. $\dfrac{3x}{x^2 - 10x + 25} - \dfrac{x + 2}{x^2 - 7x + 10}$

41. $\dfrac{4x - x^2}{6x^2 + 10x - 4} \cdot \dfrac{7 - 21x}{x^2 - 8x + 16}$

42. $\dfrac{1}{x^2 + 12x + 27} + \dfrac{x + 2}{x^3 + x^2 - 9x - 9}$

43. Simplify $\dfrac{\dfrac{x + 2}{x^2 - 64}}{\dfrac{x^2 + 4x + 4}{3x + 24}}$.

44. Write $f(x) = -3(x + 3)^2 - 7$ in standard form.

Factor:

45. $4x^3 - 8x^2 - 25x + 50$ **46.** $2x^3 - 4x^2 - 30x$

47. $6w^2 + 2wy - 20y^2$ **48.** $100p^2 - 1$

For Exercises 49–53, refer to Fig. 9.

49. Find $f(2)$. **50.** Find x when $f(x) = 3$.

51. Find an equation of f. **52.** Find the domain of f.

53. Find the range of f.

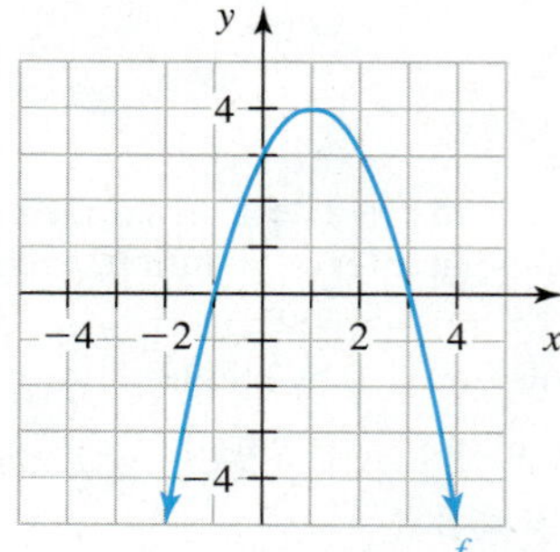

Figure 9 Exercises 49–53

For Exercises 54–60, graph the equation by hand.

54. $y = 5(2)^x$ **55.** $y = -3(x - 4)^2 + 3$

56. $y = 2\sqrt{x + 5} - 4$ **57.** $y = 15\left(\dfrac{1}{3}\right)^x$

58. $y = 2x^2 + 5x - 1$ **59.** $y = -\dfrac{3}{5}x + 4$

60. $2x(x - 3) + y = 5(x + 1)$

61. Find an equation of the line that contains the points $(-3, 2)$ and $(2, -5)$.

62. Find an approximate equation $y = ab^x$ of an exponential curve that contains the points (3, 95) and (6, 12). Round the values of a and b to the second decimal place.

63. Find an equation of a parabola that contains the points $(2, 1)$, $(3, 6)$, and $(4, 15)$.

64. Find an approximate equation of a square root curve $y = a\sqrt{x} + b$ that contains the points $(2, 5)$ and $(6, 17)$. Round the values of a and b to the second decimal place.

65. Let f be the linear function, g be the exponential function, and h be a quadratic function whose graphs contain the points $(0, 2)$ and $(1, 4)$.
  a. Find possible equations of f, g, and h.
  b. Use a graphing calculator to draw the graphs of f, g, and h in the same viewing window.

Find the logarithm.

66. $\log_3(81)$ **67.** $\log_b(\sqrt{b})$ **68.** $\log(0.001)$

For Exercises 69 and 70, find the inverse of the function.

69. $g(x) = \log_2(x)$ **70.** $f(x) = -4x - 7$

71. Find the domain of $f(x) = \dfrac{x - 3}{x^2 - 2x - 35}$.

72. Find the 10th term of the sequence 2, 8, 32, 128, 512, ...

73. Find the term number of the last term of the finite sequence $-86, -82, -78, -74, -70, \ldots, 170$.

For Exercises 74 and 75, find the sum of the series.

74. $98{,}304 + 49{,}152 + 24{,}576 + 12{,}288 + 6144 + \cdots + 3$

75. $11 + 14 + 17 + 20 + 23 + \cdots + 182$

76. A chemist wants to mix a 15% acid solution and a 30% acid solution to make a 25% acid solution. How many liters of each solution must be mixed to make 6 liters of the 25% solution?

77. The numbers of slot machines and video poker machines are shown in Table 14 for various years.

Table 14 Numbers of Slot Machines and Video Poker Machines

Year	Number of Slot Machines and Video Poker Machines (in thousands)
2001	561
2003	618
2005	689
2007	730
2008	770

Source: *Datamonitor's Productscan Online*

Let $n = f(t)$ be the number (in thousands) of slot machines and video poker machines at t years since 2000.
  a. Find an equation of f.
  b. Find an equation of f^{-1}.
  c. Find $f(18)$. What does it mean in this situation?
  d. Find $f^{-1}(1100)$. What does it mean in this situation?
  e. Find the slope of the graph of f. What does it mean in this situation?

78. In 2011, the two most populous nations were China and India, with populations of 1.347 billion and 1.210 billion, respectively. India is expected to surpass China as the most populous nation within the next 50 years. The population of India in 1980 was 0.687 billion.
  a. First, assume India's population is growing linearly. Let $L(t)$ be India's population (in billions) at t years since 1980. Find an equation of L.
  b. Now assume India's population is growing exponentially. Let $E(t)$ be India's population (in billions) at t years since 1980. Find an equation of E.

c. Find $L(70)$ and $E(70)$. What do they mean in terms of India's population?

d. Find $(E - L)(70)$. What does this result mean in terms of India's population? To get an idea of the size of your result, compare it with 0.439, a prediction of the U.S. population (in billions) in 2050 (Source: *U.S. Census Bureau*).

e. The U.S. Census Bureau predicts China's population will reach 1.424 billion in 2050. Use first L and then E to predict when India's population will reach that level. Compare your results.

79. The sales of U.S. Irish whiskey are shown in Table 15 for various years. Let $f(t)$ be annual U.S. Irish whiskey sales (in millions of 9-liter cases) at t years since 2000.

Table 15 U.S. Irish Whiskey Sales

Year	Sales (millions of 9-liter cases)
2005	0.6
2006	0.7
2007	0.8
2008	1.0
2009	1.1
2010	1.4

Source: *Distilled Spirits Council of the United States*

a. Find an exponential equation of f and a quadratic equation of f. Compare how well the two models fit the data.

b. If sales have generally increased, which of the two models describes the situation better for years before 2005?

c. Use the exponential model to predict when the annual sales will be 5 million 9-liter cases.

d. Use the quadratic model to predict when the annual sales will be 5 million 9-liter cases.

e. Explain why the year predicted in part (d) is later than the year you predicted in part (c).

80. Average annual per-person expenditures on books and all forms of recreation in the United States are shown in Table 16 for various years.

Table 16 Average per-Person Expenditures on Books and All Forms of Recreation

	Average per-Person Expenditure (dollars)	
Year	Books	All Forms of Recreation
2000	146	1863
2002	139	2079
2004	130	2218
2006	117	2376
2008	116	2835
2009	110	2693

Source: *U.S. Bureau of Labor Statistics*

a. Let $B(t)$ be the average annual per-person expenditure (in dollars) on books at t years since 2000. Find a linear equation of B.

b. Let $R(t)$ be the average annual per-person expenditure (in dollars) on all forms of recreation at t years since 2000. Find a linear equation of R.

c. Let $P(t)$ be the percentage of total recreational expenditures that consist of book purchases in the year that is t years since 2000. Find an equation of P.

d. Use the window settings in Fig. 10 to graph P. Is P increasing, decreasing, or neither for values of t between 0 and 20? What does that mean in this situation?

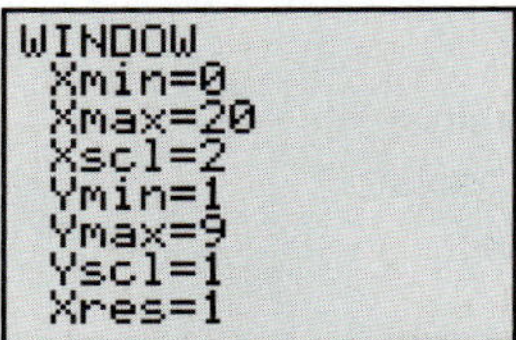

Figure 10 Window settings for Exercise 80d

e. Predict when 2% of recreational expenditures will consist of book purchases.

15 Additional Topics

15.1 Absolute Value: Equations and Inequalities

Objectives

- » Know the meaning of *absolute value function, absolute value equation in one variable,* and *absolute value inequality in one variable.*
- » Solve absolute value equations and absolute value inequalities.

In this section, we will work with *absolute value functions*. We will also solve equations and inequalities that have absolute values.

Absolute Value Function

Recall from Section 2.3 that the absolute value of a number a, written $|a|$, is the distance from a to 0 on the number line. For example, $|-6| = 6$, because -6 is a distance of 6 units from 0 (see Fig. 1). Also, $|6| = 6$, because 6 is a distance of 6 units from 0 (see Fig. 1).

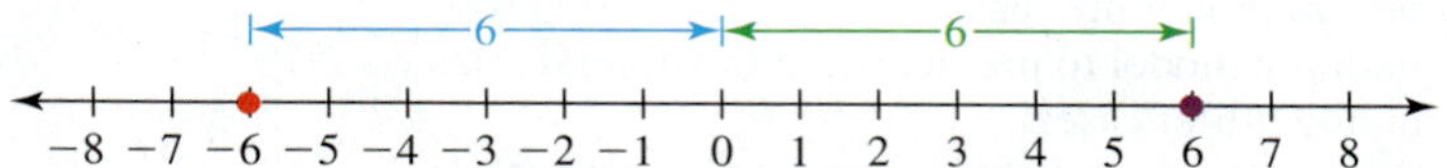

Figure 1 Both the numbers −6 and 6 are a distance of 6 units from 0

Recall from Section 2.3 that the absolute value of any number is nonnegative.

An **absolute value function** is a function whose equation contains the absolute value of a variable expression. Here are some examples of absolute value functions:

$$f(x) = |x| \qquad g(x) = |5x - 3| + 4 \qquad h(x) = \left|\frac{3x - 7}{2}\right|$$

Table 1 Input–Output Pairs of $f(x) = |x|$

x	$f(x)$
−3	3
−2	2
−1	1
0	0
1	1
2	2
3	3

Example 1 Graphing an Absolute Value Function

Graph $f(x) = |x|$.

Solution

First, we list some input–output pairs of f in Table 1. Then we plot the corresponding points and sketch a "V"-shaped curve through them (see Fig. 2).

We use ZStandard followed by ZSquare to verify our graph (see Fig. 3). To enter $|x|$, press MATH ▷ 1 X,T,Θ,n).

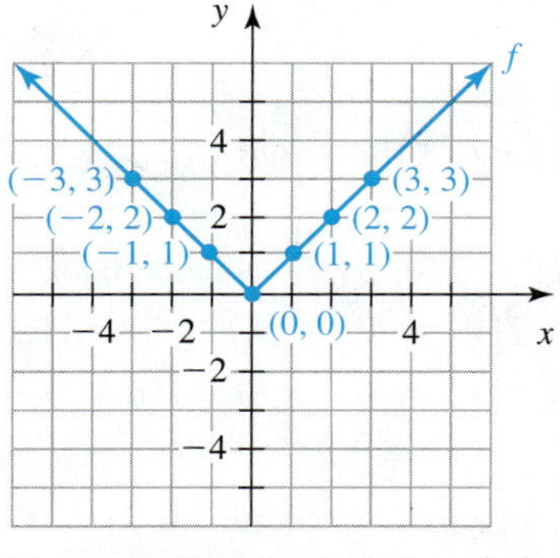

Figure 2 Graph of $f(x) = |x|$

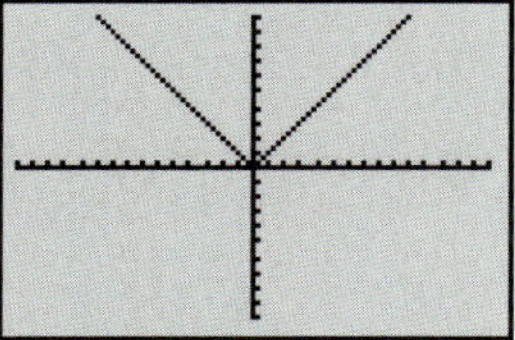

Figure 3 Verify the work

Solving Absolute Value Equations in One Variable

We will now solve some equations in *one* variable that contain absolute values. An **absolute value equation in one variable** is an equation in one variable that contains the absolute value of a variable expression. Here are some examples:

$$|x| = 3 \qquad |3x - 7| + 8 = 2 \qquad |4x - 5| = |3 - x| \qquad \left|\frac{2x}{3} - \frac{1}{2}\right| = \frac{5}{6}$$

To solve the absolute value equation $|x| = 4$, we must determine all numbers that are a distance of 4 units from 0. There are two such numbers: -4 and 4. So, the statements $|x| = 4$ and $x = \pm 4$ are equivalent statements.

Absolute Value Property for Equations

For an expression A and a positive constant k, the equation $|A| = k$ is equivalent to

$$A = -k \text{ or } A = k$$

Example 2 Solving an Absolute Value Equation

Solve $|2x + 1| = 11$.

Solution

In $|2x + 1| = 11$, the expression $2x + 1$ represents numbers that are a distance of 11 from 0. These are the numbers -11 and 11:

$$\begin{aligned} 2x + 1 &= -11 \quad \text{or} \quad 2x + 1 = 11 \qquad &&\textit{Absolute value property for equations} \\ 2x &= -12 \quad \text{or} \quad 2x = 10 \\ x &= -6 \quad \text{or} \quad x = 5 \end{aligned}$$

We check that both -6 and 5 satisfy the original equation:

Check $x = -6$	***Check*** $x = 5$
$\|2x + 1\| = 11$	$\|2x + 1\| = 11$
$\|2(-6) + 1\| \stackrel{?}{=} 11$	$\|2(5) + 1\| \stackrel{?}{=} 11$
$\|-11\| \stackrel{?}{=} 11$	$\|11\| \stackrel{?}{=} 11$
$11 \stackrel{?}{=} 11$	$11 \stackrel{?}{=} 11$
true	true

Example 3 Solving an Absolute Value Equation

Solve $2|x| - 3 = 5$.

Solution

We isolate $|x|$ on one side of the equation; then we use the absolute value property for equations:

$$\begin{aligned} 2|x| - 3 &= 5 && \text{Original equation} \\ 2|x| &= 8 && \text{Add 3 to both sides.} \\ |x| &= 4 && \text{Divide both sides by 2.} \\ x &= \pm 4 && \text{Absolute value property for equations} \end{aligned}$$

We use "intersect" on a graphing calculator to verify our work (see Fig. 4).

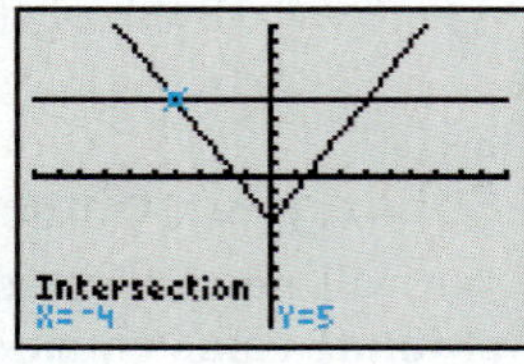

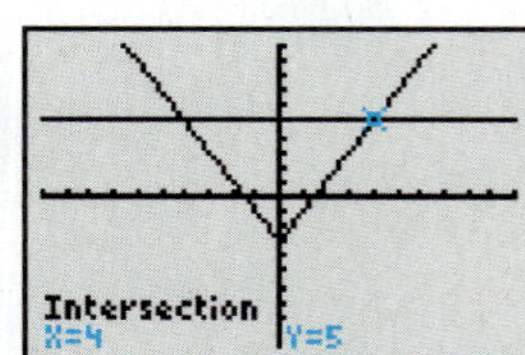

Figure 4 Verify the work

Example 4 Solving Absolute Value Equations

Solve.

1. $|4p + 12| = 0$ **2.** $|2x - 8| = -3$

Solution

1. In $|4p + 12| = 0$, the expression $4p + 12$ represents the one number that is a distance of 0 units from 0. This is the number 0:

$$\begin{aligned} 4p + 12 &= 0 \\ 4p &= -12 \\ p &= -3 \end{aligned}$$

2. Since $|2x - 8|$ is nonnegative, the solution set of $|2x - 8| = -3$ is the empty set.

The graphical check in Fig. 5 shows that the graphs of $y = |2x - 8|$ and $y = -3$ do not intersect. This means the solution set of the equation $|2x - 8| = -3$ is the empty set, which checks.

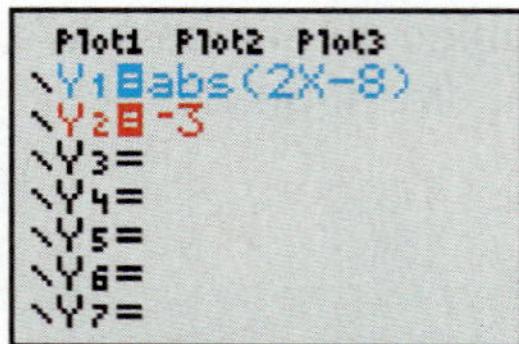

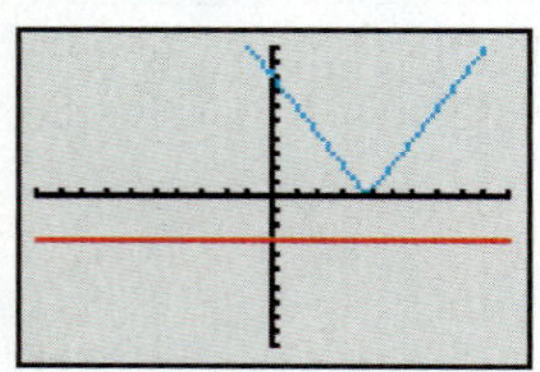

Figure 5 Verify the work

If $|a| = |b|$, what can we say about the numbers a and b? Since the absolute value of a number is its distance from 0, this means that a and b are the same distance from 0 on the number line. So, the numbers must be opposites of each other $(a = -b)$ or equal to each other $(a = b)$.

Solving an Equation of the Form $|A| = |B|$

For expressions A and B, the equation $|A| = |B|$ is equivalent to

$$A = -B \text{ or } A = B$$

For example, $|x| = |3|$ is equivalent to $x = -3$ or $x = 3$. This makes sense, because the only solutions of $|x| = 3$ are ± 3.

Example 5 Solving an Absolute Value Equation

Solve $|6x - 2| = |4x + 5|$.

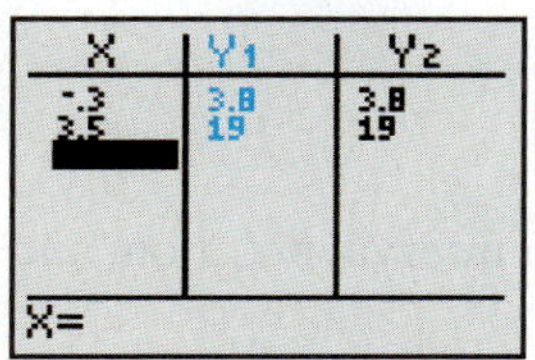

Figure 6 Verify the work

Solution

$$
\begin{aligned}
|6x - 2| &= |4x + 5| && \text{Original equation}\\
6x - 2 = -(4x + 5) \quad &\text{or} \quad 6x - 2 = 4x + 5 && \text{If } |A| = |B|, \text{ then } A = -B \text{ or } A = B.\\
6x - 2 = -4x - 5 \quad &\text{or} \quad 6x - 2 = 4x + 5 && \text{Simplify right-hand side of equation./}\\
10x - 2 = -5 \quad &\text{or} \quad 2x - 2 = 5 && \text{Add 4x to both sides./Subtract 4x}\\
10x = -3 \quad &\text{or} \quad 2x = 7 && \text{from both sides.}\\
x = -\frac{3}{10} \quad &\text{or} \quad x = \frac{7}{2}
\end{aligned}
$$

We use a graphing calculator table to verify that $-\frac{3}{10} = -0.3$ and $\frac{7}{2} = 3.5$ are solutions (see Fig. 6).

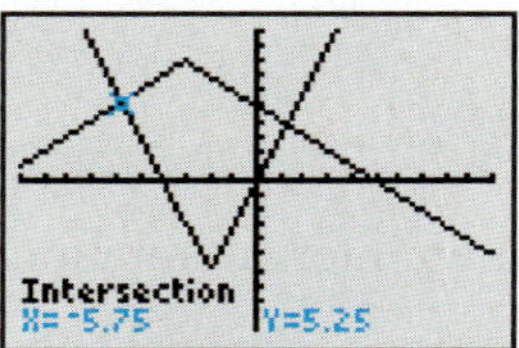

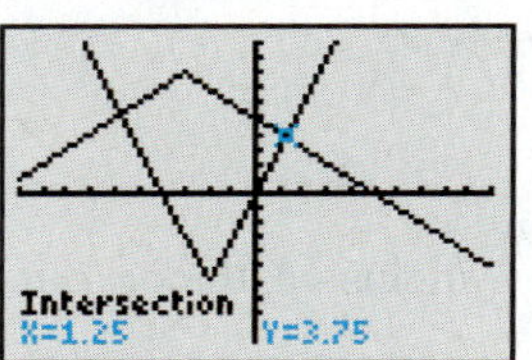

Figure 7 Solve the system

Example 6 Using Graphing to Solve an Equation in One Variable

Use graphing to solve $8 - |x + 3| = 3|x + 2| - 6$.

Solution

We use "intersect" on a graphing calculator to find the solutions of the system

$$
\begin{aligned}
y &= 8 - |x + 3|\\
y &= 3|x + 2| - 6
\end{aligned}
$$

See Fig. 7.

The solutions of the system are $(-5.75, 5.25)$ and $(1.25, 3.75)$. The x-coordinates of these ordered pairs, -5.75 and 1.25, are the solutions of $8 - |x + 3| = 3|x + 2| - 6$.

Solving Absolute Value Inequalities in One Variable

We now turn our attention from solving *equations* to solving *inequalities.* An **absolute value inequality in one variable** is an inequality in one variable that contains the absolute value of a variable expression. Here are some absolute value inequalities in one variable:

$$
|x| < 5 \qquad 3|x| - 7 \le 2 \qquad |4x + 1| > 6 \qquad \left|\frac{8x - 5}{3}\right| \ge 2
$$

To solve the absolute value inequality $|x| < 3$, we find all numbers whose distance from 0 is less than 3 units. So, the solutions of $|x| < 3$ are all the numbers between -3 and 3 (see Fig. 8). The solution set is the set of numbers x where $-3 < x < 3$. In interval notation (Section 5.7), the solution set is $(-3, 3)$.

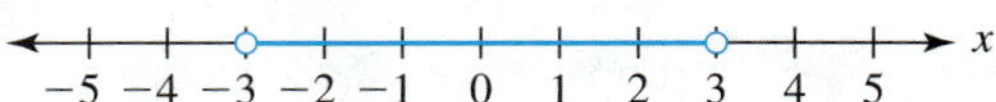

Figure 8 Graph of numbers whose distance from 0 is less than 3 units

The solution set of an absolute value inequality is sometimes best described as the union of two sets. If A and B are sets, then the **union of A and B**, denoted $A \cup B$, is the set of all members of A together with all members of B. So, $(-\infty, 1) \cup (3, \infty)$ is the set of numbers less than 1 together with numbers greater than 3.

To solve $|x| > 3$, we find all numbers whose distance from 0 is more than 3 units. So, the solutions of $|x| > 3$ are all numbers that are either less than -3 *or* greater than 3 (see Fig. 9). The solution set is the set of numbers x where $x < -3$ or $x > 3$ or, in interval notation, $(-\infty, -3) \cup (3, \infty)$.

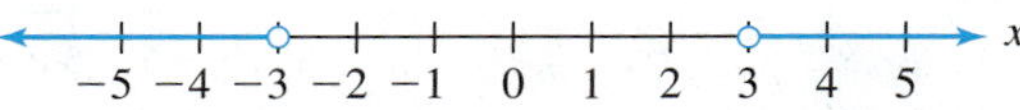

Figure 9 Graph of numbers whose distance from 0 is more than 3 units

Absolute Value Property for Inequalities

For an expression A and a positive constant k,

- The inequality $|A| < k$ is equivalent to $-k < A < k$.
- The inequality $|A| > k$ is equivalent to $A < -k$ or $A > k$.

Example 7 Solving an Absolute Value Inequality

Solve $|2x - 3| \leq 9$. Describe the solution set as an inequality, in interval notation, and in a graph.

Solution

In $|2x - 3| \leq 9$, the expression $2x - 3$ represents numbers whose distance from 0 is less than or equal to 9. Such a number is between -9 and 9, inclusive:

$$-9 \leq 2x - 3 \leq 9$$

Next, we solve the inequality $-9 \leq 2x - 3 \leq 9$:

$$-9 \leq 2x - 3 \leq 9 \quad \textit{Original inequality}$$
$$-9 + 3 \leq 2x - 3 + 3 \leq 9 + 3 \quad \textit{Add 3 to all parts.}$$
$$-6 \leq 2x \leq 12 \quad \textit{Add.}$$
$$-3 \leq x \leq 6 \quad \textit{Divide all parts by 2.}$$

So, the solution set is the set of numbers x where $-3 \leq x \leq 6$ or, in interval notation, $[-3, 6]$. We graph the solution set in Fig. 10.

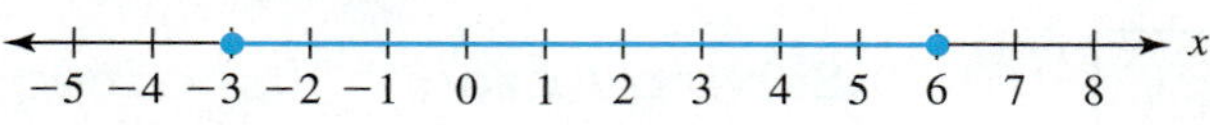

Figure 10 Graph of $-3 \leq x \leq 6$

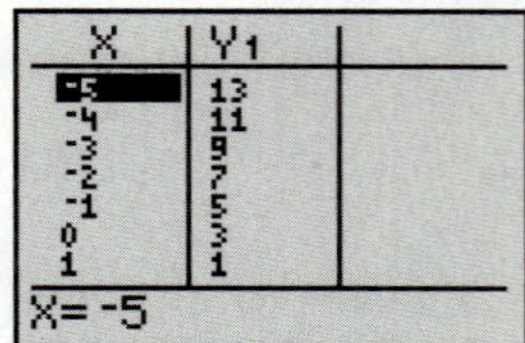

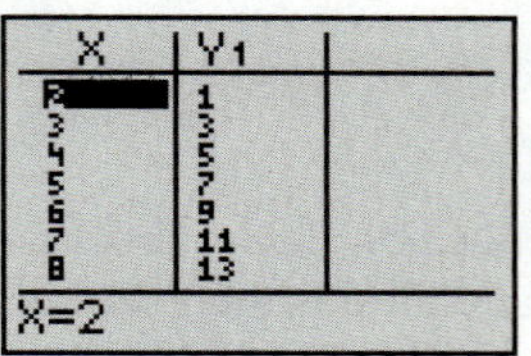

Figure 11 Verify the work

To verify our result, we check that, for inputs between -3 and 6, inclusive, the outputs of $y = |2x - 3|$ are less than or equal to 9 (see Fig. 11).

Example 8 Solving an Absolute Value Inequality

Solve $|3t + 4| > 12$. Describe the solution set as an inequality, in a graph, and in interval notation.

Solution

For $|3t + 4| > 12$, the expression $3t + 4$ represents numbers whose distance is more than 12 units from 0. These numbers are less than -12 or greater than 12:

$$3t + 4 < -12 \quad \text{or} \quad 3t + 4 > 12$$
$$3t < -16 \quad \text{or} \quad 3t > 8 \quad \textit{Subtract 4 from both sides.}$$
$$t < -\frac{16}{3} \quad \text{or} \quad t > \frac{8}{3} \quad \textit{Divide both sides by 3.}$$

We can graph the solution set on a number line (see Fig. 12), or we can describe the solution set in interval notation as $\left(-\infty, -\frac{16}{3}\right) \cup \left(\frac{8}{3}, \infty\right)$.

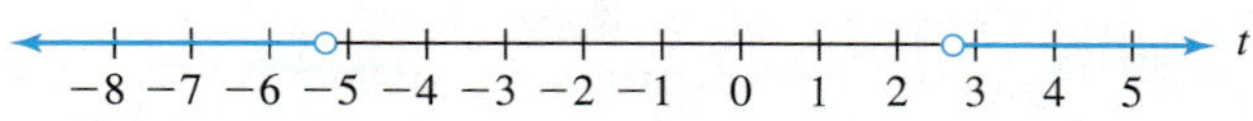

Figure 12 Graph of $t < -\frac{16}{3}$ or $t > \frac{8}{3}$

Recall from Section 5.7 that when we multiply or divide both sides of an inequality by a negative number, we reverse the inequality symbol.

Example 9 Solving an Absolute Value Inequality

Solve $7 - |x + 2| > 3$. Describe the solution set as an inequality, in a graph, and in interval notation.

Solution

To begin, we isolate $|x + 2|$ on the left-hand side of the inequality:

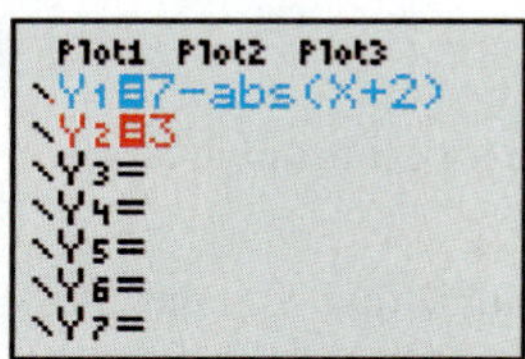

$$7 - |x + 2| > 3 \quad \textit{Original inequality}$$
$$-|x + 2| > -4 \quad \textit{Subtract 7 from both sides.}$$
$$|x + 2| < 4 \quad \textit{Multiply both sides by } -1\textit{; reverse inequality symbol.}$$

So, $x + 2$ represents numbers whose distance is less than 4 units from 0. These numbers are between -4 and 4:

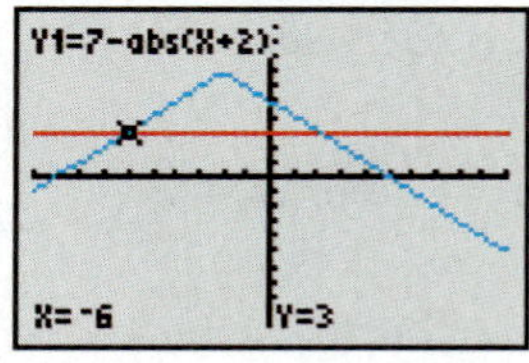

$$-4 < x + 2 < 4$$
$$-4 - 2 < x + 2 - 2 < 4 - 2 \quad \textit{Subtract 2 from all three parts.}$$
$$-6 < x < 2$$

We can graph the solution set on a number line (see Fig. 13), or we can describe the solution set in interval notation as $(-6, 2)$.

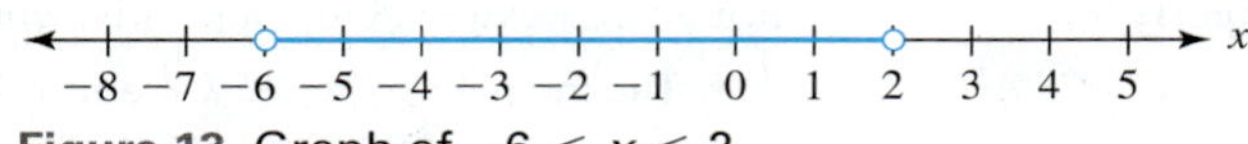

Figure 13 Graph of $-6 < x < 2$

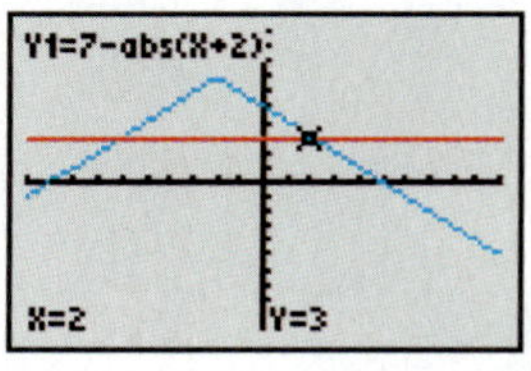

Figure 14 Verify the work

To verify our work, we check that the graph of $y = 7 - |x + 2|$ is above the horizontal line $y = 3$ for values of x between -6 and 2 (see Fig. 14).

Example 10 Solving Absolute Value Inequalities

Solve.

1. $|7x - 10| \le -2$
2. $|5x - 8| > -1$

Solution

1. Since $|7x - 10|$ is nonnegative, the inequality $|7x - 10| \le -2$ has an empty-set solution.
2. Since $|5x - 8|$ is nonnegative for *any* real number x, the solution set of $|5x - 8| > -1$ is the set of all real numbers or, in interval notation, $(-\infty, \infty)$.

Group Exploration

Graphical meaning of $|a - b|$

In this exploration, you will explore the graphical meaning of $|a - b|$.

1. Plot the points 1 and 6 on a number line. What is the distance between 1 and 6? Compare your result with $|1 - 6|$ and with $|6 - 1|$.
2. Plot the points -2 and 3 on a number line. What is the distance between -2 and 3? Compare your result with $|(-2) - 3|$ and with $|3 - (-2)|$.
3. Find the distance between -7 and -3, and compare your result with $|(-7) - (-3)|$ and with $|(-3) - (-7)|$.

4. Describe the graphical meaning of $|a - b|$.

5. Solve the equation. Then find the distance between 5 and each solution. Explain why your result makes sense in terms of the graphical meaning of $|x - 5|$.
   a. $|x - 5| = 1$
   b. $|x - 5| = 2$
   c. $|x - 5| = 3$

6. Solve the equation or inequality and graph the solutions. Explain why your result makes sense in terms of the graphical meaning of $|x - 4|$.
   a. $|x - 4| = 3$ [**Hint:** For the graph, plot the two solutions.]
   b. $|x - 4| < 3$
   c. $|x - 4| > 3$

Key Points of Section 15.1

Absolute value function	An **absolute value function** is a function whose equation contains the absolute value of a variable expression.
Absolute value equation in one variable	An **absolute value equation in one variable** is an equation in one variable that contains the absolute value of a variable expression.
Absolute value property for equations	For an expression A and a positive constant k, the equation $\lvert A\rvert = k$ is equivalent to $A = -k$ or $A = k$.
Solving an equation of the form $\lvert A\rvert = \lvert B\rvert$	For expressions A and B, the equation $\lvert A\rvert = \lvert B\rvert$ is equivalent to $A = -B$ or $A = B$.
Absolute value inequality in one variable	An **absolute value inequality in one variable** is an inequality in one variable that contains the absolute value of a variable expression.
Absolute value property for inequalities	For an expression A and a positive constant k, • The inequality $\lvert A\rvert < k$ is equivalent to $-k < A < k$. • The inequality $\lvert A\rvert > k$ is equivalent to $A < -k$ or $A > k$.

Homework 15.1

For extra help ▶ MyMathLab®

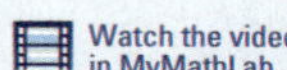

Watch the videos in MyMathLab

Download the MyDashboard App

Solve. Use a graphing calculator table or graph to verify your work.

1. $|x| = 7$ **2.** $|x| = 4$ **3.** $|x| = -3$
4. $|x| = -1$ **5.** $5|p| - 3 = 15$ **6.** $-7|w| + 6 = 4$
7. $|x + 2| = 5$ **8.** $|x - 3| = 8$
9. $|x - 5| = 0$ **10.** $|x + 1| = 0$
11. $|3t - 1| = 11$ **12.** $|6k + 4| = 7$
13. $|2x + 9| = -6$ **14.** $|5x - 1| = -3$
15. $|4x| + 1 = 9$ **16.** $|6x| - 5 = 7$
17. $2|a + 5| = 8$ **18.** $-3|m - 4| = -15$
19. $|2x - 5| - 4 = -3$ **20.** $|5x + 3| - 2 = 5$
21. $|4x - 5| = |3x + 2|$ **22.** $|3x + 7| = |2x - 1|$
23. $|5w + 1| = |3 - w|$ **24.** $|2p - 4| = |5 - p|$
25. $\left|\frac{4x + 3}{2}\right| = 5$ **26.** $\left|\frac{3x - 5}{6}\right| = 2$
27. $\left|\frac{1}{2}x - \frac{5}{3}\right| = \frac{7}{6}$ **28.** $\left|\frac{3}{4}x + \frac{7}{2}\right| = \frac{1}{3}$
29. $\left|\frac{2}{3}k + \frac{4}{9}\right| = \left|\frac{5}{6}k - \frac{1}{3}\right|$ **30.** $\left|\frac{5}{6}t - \frac{3}{4}\right| = \left|\frac{1}{4}t + \frac{3}{2}\right|$

Solve. Round any solutions to the second decimal place.

31. $4.7|x| - 3.9 = 8.8$ **32.** $1.9|x| + 4.1 = 12.8$
33. $|2.1x + 5.8| - 9.7 = 10.2$ **34.** $|3.6x - 2.1| + 2.8 = 9.4$

Solve by using "intersect" on a graphing calculator. Round any solutions to the second decimal place.

35. $|x| - 3 = 7 - |x + 1|$
36. $|x| - 1 = 8 - |x - 3|$
37. $|x + 4| + 3 = 9 - 2|x + 5|$
38. $2|x - 2| + 1 = 6 - |x - 3|$

For Exercises 39–42, use the graphs shown in Fig. 15 to solve the given equation or system.

39. $1 - |x| = -3$

40. $|x + 1| - 4 = 0$

41. $|x + 1| - 4 = 1 - |x|$

42. $y = |x + 1| - 4$
$y = 1 - |x|$

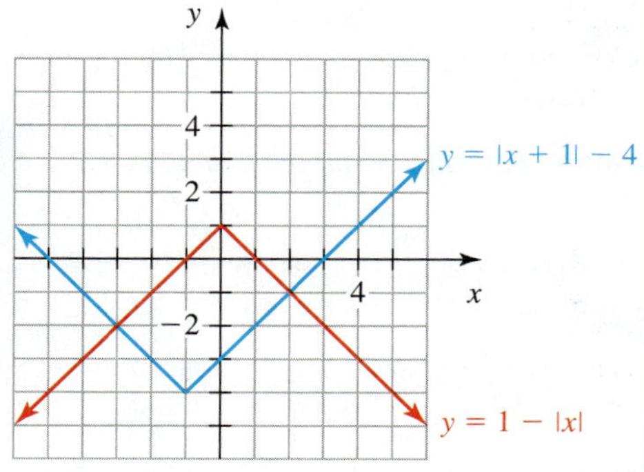

Figure 15 Exercises 39–42

Let $f(x) = 2|x| - 11$.

43. Find $f(-5)$.

44. Find $f(-3)$.

45. Find x, where $f(x) = -5$.

46. Find x, where $f(x) = -3$.

Let $f(x) = |4x + 7| - 9$.

47. Find $f(-3)$.

48. Find $f(-5)$.

49. Find x, where $f(x) = -3$.

50. Find x, where $f(x) = -5$.

Solve. Describe the solution set as an inequality, in a graph, and in interval notation.

51. $|x| < 4$

52. $|x| \leq 1$

53. $|x| \geq 3$

54. $|x| > 2$

55. $|r| < -3$

56. $|m| > -5$

57. $|x| > 0$

58. $|x| \leq 0$

59. $|7x + 15| > -4$

60. $|2x - 1| \geq -2$

61. $|0.25t - 1.3| \geq 1.1$

62. $|0.8w - 3.1| < 2.9$

63. $2|x| - 5 > 3$

64. $3|x| - 1 \geq 11$

65. $2 - 5|p| \leq -8$

66. $10 - 4|b| < -2$

67. $|x - 6| \geq 7$

68. $|x + 1| > 3$

69. $|2x + 5| < 15$

70. $|3x - 4| \leq 25$

71. $7 - |x + 3| \leq 2$

72. $5 - |x - 2| < 4$

73. $\left|\frac{x + 4}{3}\right| \geq 2$

74. $\left|\frac{x - 1}{2}\right| > 4$

75. $\left|\frac{2x}{5} + \frac{3}{2}\right| \leq \frac{9}{20}$

76. $\left|\frac{5x}{3} - \frac{1}{4}\right| > \frac{7}{12}$

Concepts

77. Assume $m \neq 0$ and the equation $|mx + b| + c = k$ has at least one solution for x. Solve the equation for x.

78. What must be true of the constants m, b, c, and k if the equation $|mx + b| + c = k$ has exactly one solution?

79. A student tries to solve the equation $|x - 5| = 7$:

$$|x - 5| = 7$$
$$x + 5 = 7$$
$$x = 2$$

Describe any errors. Then solve the equation correctly.

80. A student tries to solve the equation $|x + 6| = -3$:

$$|x + 6| = -3$$
$$x + 6 = 3 \quad \text{or} \quad x + 6 = -3$$
$$x = -3 \quad \text{or} \quad x = -9$$

Describe any errors. Then solve the equation correctly.

81. A student tries to solve the inequality $|x + 3| < 10$:

$$|x + 3| < 10$$
$$x + 3 < -10 \quad \text{or} \quad x + 3 < 10$$
$$x < -13 \quad \text{or} \quad x < 7$$

Describe any errors. Then solve the inequality correctly.

82. A student tries to solve the inequality $|x + 2| < 7$:

$$|x + 2| < 7$$
$$x + 2 < 7$$
$$x < 5$$

Describe any errors. Then solve the inequality correctly.

83. **a.** Solve $|2x + 3| = 13$.
b. Solve $|2x + 3| < 13$.
c. Solve $|2x + 3| > 13$.
d. Graph the solutions in parts (a), (b), and (c) on the same number line. Use three colors to identify the different solutions. Make some observations about the solutions. Explain these observations.

84. List three numbers that satisfy the inequality $-2|3x - 4| < -4$.

85. The inequality $|x| < 3$ is equivalent to $-3 < x < 3$. Graph the equations $y = |x|$ and $y = 3$ on the same coordinate system to help explain.

86. The inequality $|x| > 2$ is equivalent to $x < -2$ or $x > 2$. Graph the equations $y = |x|$ and $y = 2$ to help explain.

87. Is the statement "$|a + b| = |a| + |b|$ for all real numbers a and b" true or false? Explain. [**Hint:** Try substituting a positive number for a and a negative number for b.]

88. Explain how to solve an inequality of the form $|mx + b| + c < k$, where $m \neq 0$. (See page 9 for guidelines on writing a good response.)

Related Review

Solve. Round any approximate solutions to the fourth decimal place.

89. $|x - 5| = 4$

90. $|x^2 - 5| = 4$

91. $|2^y - 5| = 4$

92. $|\sqrt{w} - 5| = 4$

93. $\left|\frac{2x + 3}{x - 2} - 5\right| = 4$

94. $|\log_4(x) - 5| = 4$

Solve. Describe the solution set as an inequality, in a graph, and in interval notation.

95. $3(2x) - 5 \leq 7$

96. $2(2x - 3) > 6$

97. $3|2x| - 5 \leq 7$

98. $2|2x - 3| > 6$

Expressions, Equations, Functions, and Graphs

Perform the indicated instruction. Then use words such as absolute value, linear, quadratic, cubic, exponential, logarithmic, rational, radical, polynomial, degree, function, one variable, *and* two variables *to describe the expression, equation, or system.*

99. Graph $y = 3(x - 2) + 1$ by hand.

100. Factor $4x^3 - 20x^2 + 24x$.

101. Solve $3|x - 2| + 1 = 7$.

102. Solve $4x^3 - 20x^2 + 24x = 0$.

103. Find an equation of a line that contains the points $(-4, 2)$ and $(5, -3)$.

104. Find the product $(2x - 6)(3x^2 + 4x - 2)$.

Section 15.1 Quiz

For Exercises 1–5, solve.

1. $3|t| - 4 = 11$

2. $5|6r - 5| = 15$

3. $|7x + 1| = -3$

4. $|5x - 2| = |3x + 6|$

5. $\left|\frac{3}{4}x - \frac{1}{2}\right| = \frac{7}{8}$

6. Is the statement "$|a - b| = |a| - |b|$ for all real numbers a and b" true or false? Explain.

Solve. Describe the solution set as an inequality, in a graph, and in interval notation.

7. $3|k| - 4 \geq 2$

8. $|4c - 8| > 12$

9. $7|3x - 2| \leq 42$

10. $|x - 5| < -7$

15.2 Performing Operations with Complex Numbers

Objective

» Perform operations with complex numbers.

Recall from Section 9.4 that the *imaginary unit i* is the number whose square is -1. That is,

$$i^2 = -1 \quad \text{and} \quad i = \sqrt{-1}$$

If b is a nonzero real number, then an expression of the form bi is a *pure imaginary number.* A *complex number* is a number of the form $a + bi$, where a and b are real numbers. An *imaginary number* is a number $a + bi$, where a and b are real numbers and $b \neq 0$.

Finally, recall from Section 9.4 that if p is a positive real number, then $\sqrt{-p} = i\sqrt{p}$. For example, $\sqrt{-25} = i\sqrt{25} = 5i$.

In this section, we will perform operations with complex numbers.

Adding, Subtracting, and Multiplying Complex Numbers

Since $i = \sqrt{-1}$, **we perform operations with complex numbers in much the same way as we do with radical expressions.** For example,

$$(2 + 3\sqrt{7}) + (1 + 5\sqrt{7}) = 3 + 8\sqrt{7} \quad \textit{Add two radical expressions.}$$
$$(2 + 3i) + (1 + 5i) = 3 + 8i \quad \textit{Add two complex numbers.}$$

If a radicand is a negative real number, we first write the radical in terms of *i* before performing any operations:

$$\sqrt{-1}\sqrt{-1} = i \cdot i = i^2 = -1 \quad \textit{Correct}$$
$$\sqrt{-1}\sqrt{-1} = \sqrt{-1 \cdot -1} = \sqrt{1} = 1 \quad \textit{Incorrect}$$

So, there is no product property $\sqrt{a}\sqrt{b} = \sqrt{ab}$ when a and b are both negative. To find $\sqrt{-2}\sqrt{-3}$, we first write each radical in terms of i:

$$\sqrt{-2}\sqrt{-3} = i\sqrt{2} \cdot i\sqrt{3} = i^2\sqrt{6} = -\sqrt{6}$$

When we use an operation to combine two complex numbers, we write the result in the form $a + bi$, where a and b are in lowest terms.

Example 1 Performing Operations with Complex Numbers

Perform the indicated operation. Simplify the result.

1. $(5 + 9i) + (3 - 2i)$

2. $(3 - \sqrt{-36}) - (2 - \sqrt{-16})$

3. $4i \cdot 6i$

4. $\sqrt{-4}\sqrt{-9}$

Solution

1. $(5 + 9i) + (3 - 2i) = 5 + 3 + 9i - 2i$ *Rearrange terms.*
$= 8 + 7i$ *Write in $a + bi$ form.*

2. $(3 - \sqrt{-36}) - (2 - \sqrt{-16}) = (3 - i\sqrt{36}) - (2 - i\sqrt{16})$ *Write in terms of i.*
$= (3 - 6i) - (2 - 4i)$ *$\sqrt{36} = 6, \sqrt{16} = 4$*
$= 3 - 6i - 2 + 4i$ *Distributive law*
$= 3 - 2 - 6i + 4i$ *Rearrange terms.*
$= 1 - 2i$ *Write in $a + bi$ form.*

3. $4i \cdot 6i = 24i^2$ *Simplify.*
$= 24(-1)$ *$i^2 = -1$*
$= -24$

4. $\sqrt{-4}\sqrt{-9} = i\sqrt{4} \cdot i\sqrt{9}$ *Write in terms of i.*
$= 2i \cdot 3i$ *$\sqrt{4} = 2, \sqrt{9} = 3$*
$= 6i^2$ *Simplify.*
$= 6(-1)$ *$i^2 = -1$*
$= -6$

Example 2 Performing Operations with Complex Numbers

Perform the indicated operations. Simplify the result.

1. $9 - 4i(2 - 7i)$ **2.** $(2 + 5i)(3 - 7i)$ **3.** $(3 - 5i)^2$

Solution

1. $9 - 4i(2 - 7i) = 9 - 8i + 28i^2$ *Distributive law*
$= 9 - 8i + 28(-1)$ *$i^2 = -1$*
$= 9 - 8i - 28$ *Multiply.*
$= -19 - 8i$

2. $(2 + 5i)(3 - 7i) = 2 \cdot 3 - 2 \cdot 7i + 5i \cdot 3 - 5i \cdot 7i$ *Multiply pairs of terms.*
$= 6 - 14i + 15i - 35i^2$ *Simplify.*
$= 6 + i - 35(-1)$ *Simplify; $i^2 = -1$*
$= 6 + i + 35$ *Multiply.*
$= 41 + i$

(2+5i)(3-7i)
41+i

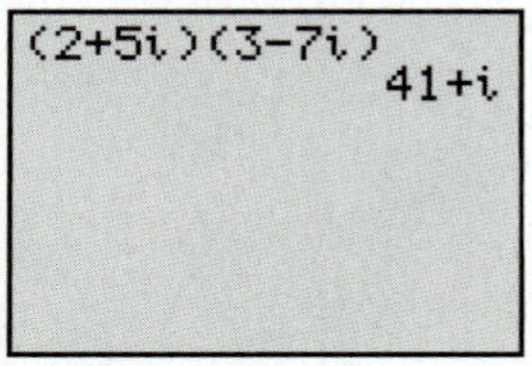

Figure 16 Verify that $(2 + 5i)(3 - 7i) = 41 + i$

We use a graphing calculator to verify our work (see Fig. 16). Press [2nd] [.] to enter the imaginary unit i.

3. $(3 - 5i)^2 = 3^2 - 2(3)(5i) + (5i)^2$ *$(A - B)^2 = A^2 - 2AB + B^2$*
$= 9 - 30i + 25i^2$ *$(bc)^n = b^n c^n$*
$= 9 - 30i + 25(-1)$ *$i^2 = -1$*
$= 9 - 30i - 25$ *Multiply.*
$= -16 - 30i$

Recall from Section 13.3 that we call radical expressions such as $7 + \sqrt{3}$ and $7 - \sqrt{3}$ radical conjugates of each other. Similarly, we call the complex numbers $7 + 3i$ and $7 - 3i$ complex conjugates of each other.

Definition Complex conjugate

The complex numbers $a + bi$ and $a - bi$ are called **complex conjugates** of each other.

For example, the complex conjugate of $2 - 8i$ is $2 + 8i$. The complex conjugate of $6i$ is $-6i$.

Example 3 Finding the Product of Two Complex Conjugates

Find the product $(4 + 7i)(4 - 7i)$.

Solution

$$\begin{aligned}(4 + 7i)(4 - 7i) &= 4^2 - (7i)^2 && (A + B)(A - B) = A^2 - B^2 \\ &= 16 - 7^2 i^2 && (bc)^n = b^n c^n \\ &= 16 - 49(-1) && i^2 = -1 \\ &= 65\end{aligned}$$

Dividing Complex Numbers

We **simplify** a quotient of two complex numbers by removing i from the denominator of the quotient. In Example 3, we found that the product of two particular complex conjugates is a real number. In general, **the product of *any* two complex conjugates is a real number.** We can use this generalization to remove i from the denominator of a quotient of two complex numbers.

Example 4 Simplifying the Quotient of Two Complex Numbers

Simplify $\frac{5}{2 + 3i}$.

Solution

Since the product of the complex conjugates $2 + 3i$ and $2 - 3i$ is a real number, we can remove i from the denominator of $\frac{5}{2 + 3i}$ by multiplying the quotient by $\frac{2 - 3i}{2 - 3i}$:

$$\begin{aligned}\frac{5}{2 + 3i} &= \frac{5}{2 + 3i} \cdot \frac{2 - 3i}{2 - 3i} && \text{The conjugate of } 2 + 3i \text{ is } 2 - 3i. \\ &= \frac{10 - 15i}{4 - 9i^2} && \text{Distributive law; } (A + B)(A - B) = (A^2 - B^2) \\ &= \frac{10 - 15i}{4 - 9(-1)} && i^2 = -1 \\ &= \frac{10 - 15i}{13} && \text{Simplify.} \\ &= \frac{10}{13} - \frac{15}{13}i && \text{Write in } a + bi \text{ form.}\end{aligned}$$

To simplify the quotient of two complex numbers, we multiply the quotient by $\dfrac{\textbf{complex conjugate of the denominator}}{\textbf{complex conjugate of the denominator}}$. Notice that this process is similar to the way we rationalize the denominator of a radical expression such as $\frac{5}{2 + 3\sqrt{7}}$.

Example 5 Simplifying the Quotient of Two Complex Numbers

Simplify $\frac{2 + 4i}{3 - 5i}$.

Solution

$$\frac{2+4i}{3-5i} = \frac{2+4i}{3-5i} \cdot \frac{3+5i}{3+5i}$$ *The conjugate of $3 - 5i$ is $3 + 5i$.*

$$= \frac{(2+4i)(3+5i)}{(3-5i)(3+5i)}$$ *Multiply numerators; multiply denominators.*

$$= \frac{6 + 10i + 12i + 20i^2}{9 - 25i^2}$$ *Multiply pairs of terms; $(A - B)(A + B) = A^2 - B^2$*

$$= \frac{6 + 22i + 20(-1)}{9 - 25(-1)}$$ *Simplify; $i^2 = -1$*

$$= \frac{-14 + 22i}{34}$$ *Simplify.*

$$= -\frac{14}{34} + \frac{22}{34}i$$ *Write in $a + bi$ form.*

$$= -\frac{7}{17} + \frac{11}{17}i$$ *Simplify.*

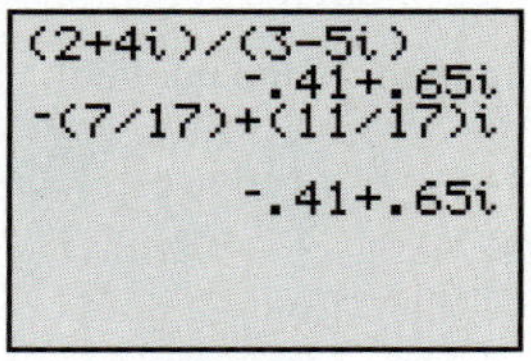

```
(2+4i)/(3-5i)
          -.41+.65i
-(7/17)+(11/17)i
          -.41+.65i
```

Figure 17 Verify the work

We use a graphing calculator to verify our work (see Fig. 17). Here we set the float to 2, so the numbers in the result are rounded to the second decimal place.

When the denominator of a fraction is a pure imaginary number, the easiest way to remove i from the denominator is to multiply the fraction by $\frac{i}{i}$.

Example 6 Simplifying the Quotient of Two Complex Numbers

Simplify $\frac{2 - 5i}{3i}$.

Solution

Since the denominator $3i$ is a pure imaginary number, we can remove i from the denominator by multiplying by $\frac{i}{i}$:

$$\frac{2-5i}{3i} = \frac{2-5i}{3i} \cdot \frac{i}{i}$$ *Multiply by $\frac{i}{i}$.*

$$= \frac{2i - 5i^2}{3i^2}$$ *Multiply numerators; multiply denominators.*

$$= \frac{2i - 5(-1)}{3(-1)}$$ *$i^2 = -1$*

$$= \frac{5 + 2i}{-3}$$ *Multiply.*

$$= -\frac{5}{3} - \frac{2}{3}i$$ *Write in $a + bi$ form.*

Group Exploration

Finding powers of i

1. Find the indicated power of i. Verify your work with a graphing calculator.
   a. i^2
   b. i^3 [**Hint:** $i^3 = i^2 \cdot i$]
   c. i^4 [**Hint:** $i^4 = i^3 \cdot i$]
   d. i^5
2. Continue finding powers of i, such as $i^6, i^7, i^8, \ldots$, until you see a pattern in your results. Describe the pattern.
3. Find the indicated power of i. Verify your result with a graphing calculator.
   a. i^{23}
   b. i^{41}
   c. i^{102}
   d. i^{400}

Key Points of Section 15.2

Performing operations with complex numbers	We perform operations with complex numbers in much the same way as we do with radical expressions.
First write radicals in terms of *i*	If a radicand is a negative real number, we first write the radical in terms of i before performing any operations.
Complex conjugate	The complex numbers $a + bi$ and $a - bi$ are called **complex conjugates** of each other.
Simplify a quotient	We **simplify** a quotient of two complex numbers by removing i from the denominator of the quotient.
Product of complex conjugates	The product of *any* two complex conjugates is a real number.
Simplifying the quotient of two complex numbers	To simplify the quotient of two complex numbers, we multiply the quotient by $\frac{\text{complex conjugate of the denominator}}{\text{complex conjugate of the denominator}}$.
Multiplying by $\frac{i}{i}$	When the denominator of a fraction is a pure imaginary number, the easiest way to remove i from the denominator is to multiply the fraction by $\frac{i}{i}$.

Homework 15.2

Perform the indicated operation. If your result is an imaginary number, write it in a + bi form. Use a graphing calculator to verify your work when possible.

1. $(4 - 7i) + (3 + 10i)$
2. $(15 - 2i) + (6 + 17i)$
3. $(5 - \sqrt{-9}) + (2 - \sqrt{-25})$
4. $(7 - \sqrt{-4}) + (9 - \sqrt{-36})$
5. $(6 - 5i) - (2 - 13i)$
6. $(9 - 4i) - (8 - 2i)$
7. $(6 - \sqrt{-49}) - (1 + \sqrt{-81})$
8. $(3 - \sqrt{-1}) - (3 + \sqrt{-64})$
9. $2i \cdot 9i$
10. $4i \cdot 6i$
11. $-10i(-5i)$
12. $-7i(4i)$
13. $\sqrt{-4}\sqrt{-25}$
14. $\sqrt{-49}\sqrt{-16}$
15. $\sqrt{-3}\sqrt{-5}$
16. $\sqrt{-2}\sqrt{-11}$
17. $(8i)^2$
18. $(-4i)^2$
19. $5i(3 - 2i)$
20. $6i(1 + 3i)$
21. $20 - 3i(2 - 7i)$
22. $2 + 4i(3 - 8i)$
23. $(2 + 5i)(3 + 4i)$
24. $(7 + 3i)(10 + 2i)$
25. $(3 - 6i)(5 + 2i)$
26. $(4 - 3i)(2 + 7i)$
27. $(-6 + 4i)(-2 + 7i)$
28. $(-5 + 7i)(-3 + 2i)$
29. $(5 + 4i)(5 - 4i)$
30. $(8 + 3i)(8 - 3i)$
31. $(2 - 9i)(2 + 9i)$
32. $(3 - 7i)(3 + 7i)$
33. $(1 + i)(1 - i)$
34. $(3 + i)(3 - i)$
35. $(2 + 7i)^2$
36. $(6 + 3i)^2$
37. $(4 - 5i)^2$
38. $(7 - 4i)^2$
39. $(-4 + 3i)^2$
40. $(-5 + 2i)^2$
41. $\frac{3}{2 + 5i}$
42. $\frac{4}{4 + 3i}$
43. $\frac{3i}{7 - 2i}$
44. $\frac{2i}{3 - 8i}$
45. $\frac{2 + 3i}{7 + i}$
46. $\frac{5 + 8i}{3 + i}$
47. $\frac{3 + 4i}{3 - 4i}$
48. $\frac{4 - 7i}{4 + 7i}$
49. $\frac{3 - 5i}{2 - 9i}$
50. $\frac{4 - 3i}{1 - 5i}$
51. $\frac{5 + 7i}{4i}$
52. $\frac{2 + 8i}{3i}$
53. $\frac{7}{5i}$
54. $\frac{3}{2i}$

Concepts

55. Two students try to find the product $\sqrt{-2}\sqrt{-8}$:

Student 1's work

$$\sqrt{-2}\sqrt{-8} = \sqrt{-2(-8)} = \sqrt{16} = 4$$

Student 2's work

$$\sqrt{-2}\sqrt{-8} = i\sqrt{2} \cdot i\sqrt{8} = i^2\sqrt{16} = -4$$

Did one, both, or neither student find the product correctly? Explain.

56. A student tries to find the product $(3 + i)(3 - i)$:

$$\begin{aligned}(3 + i)(3 - i) &= 3^2 - i^2\\ &= 9 - 1\\ &= 8\end{aligned}$$

Describe any errors. Then find the product correctly.

57. Find nonzero real-number values of a, b, c, and d such that the sum $(a + bi) + (c + di)$
a. is an imaginary number (not pure).
b. is a real number.
c. is a pure imaginary number.

58. Find nonzero real-number values of a, b, c, and d such that the product $(a + bi)(c + di)$
a. is an imaginary number (not pure).
b. is a real number.
c. is a pure imaginary number.

59. If a radicand is a negative number, explain why it is important we first write the radical in terms of i before performing any operations.

60. Give an example of a quadratic equation in which $3i$ and $-3i$ are solutions.

61. The square of a real number is a nonnegative real number. What can you say about the square of a pure imaginary number? Explain.

62. Describe how to multiply two complex numbers. Describe how to simplify the quotient of two complex numbers.

Related Review

Simplify. If your result is an imaginary number, write it in a + bi form.

63. $\dfrac{4}{3 + 2\sqrt{x}}$ **64.** $\dfrac{7 + 4\sqrt{x}}{2 - 5\sqrt{x}}$

65. $\dfrac{4}{3 + 2i}$ **66.** $\dfrac{7 + 4i}{2 - 5i}$

Find all complex-number solutions of the equation.

67. $3x^2 - 2x + 3 = 0$ **68.** $x^2 - 2x + 5 = 0$

69. $5x^2 - 4x = -1$ **70.** $4x^2 - x = -2$

71. $(p - 3)(2p + 1) = -10$ **72.** $(k + 1)(k + 2) = k$

73. $x(3x - 2) = 2 + 2(x - 3)$

74. $5 - 2(x - 4) = -2x(x - 2)$

75. $(5w + 3)^2 = -20$ **76.** $(4m - 7)^2 = -18$

Expressions, Equations, Functions, and Graphs

Perform the indicated instruction. Then use words such as linear, quadratic, cubic, exponential, logarithmic, rational, radical, polynomial, degree, function, imaginary number, one variable, *and* two variables *to describe the expression, equation, or system.*

77. Solve $4x^2 - 2x + 3 = 0$.

78. Simplify $2\sqrt{20x^3} - 3x\sqrt{45x} + 4x\sqrt{24}$.

79. Factor $10x^2 - 19x + 6$.

80. Simplify $\dfrac{2\sqrt{x} - 5}{3\sqrt{x} + 4}$.

81. Find the product $(3i - 7)(4i + 6)$.

82. Solve $4\sqrt{3x - 1} - 3 = 5$.

Section 15.2 Quiz

For Exercises 1–9, perform the indicated operations. If your result is an imaginary number, write it in a + bi form.

1. $(6 - 2i) + (3 - 4i)$ **2.** $(3 + 7i) - (8 - 2i)$

3. $-4i \cdot 3i$ **4.** $\sqrt{-2}\sqrt{-7}$

5. $(5 - 3i)(7 + i)$ **6.** $(4 - 3i)^2$

7. $(8 + 5i)(8 - 5i)$ **8.** $\dfrac{3 + 2i}{5 - 4i}$

9. $\dfrac{5 - 7i}{6i}$

10. True or false? A complex number times a pure imaginary number must be an imaginary number. Explain.

15.3 Pythagorean Theorem, Distance Formula, and Circles

Objectives

» Use the Pythagorean theorem to find the length of a side of a right triangle.

» Use the Pythagorean theorem to make estimates about authentic situations.

» Use the distance formula to find the distance between two points.

» Find or graph the equation of a circle.

In this section, we will discuss a special type of triangle called a *right triangle* and a useful formula relating to right triangles: the Pythagorean theorem. We will also find the distance between two points. Finally, we will find and graph equations of circles.

Pythagorean Theorem

We begin by working with an important type of triangle called a *right triangle*. An angle of 90° is called a *right angle*. If one angle of a triangle measures 90°, the triangle is a **right triangle** (see Fig. 18). The side opposite the right angle is the triangle's longest side. We call that side the **hypotenuse,** and we call the two shorter sides the **legs.**

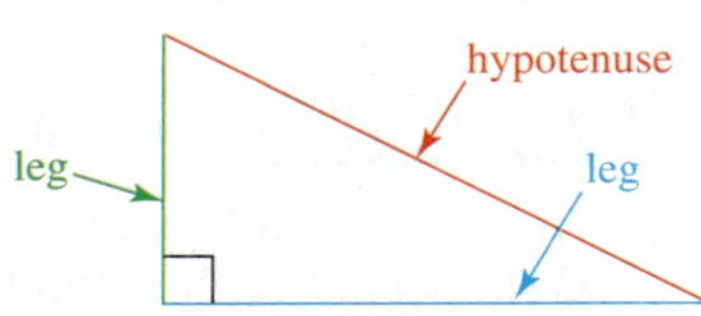

Figure 18 A right triangle

The **Pythagorean theorem** describes the relationship between the lengths of the legs and the length of the hypotenuse of a right triangle.

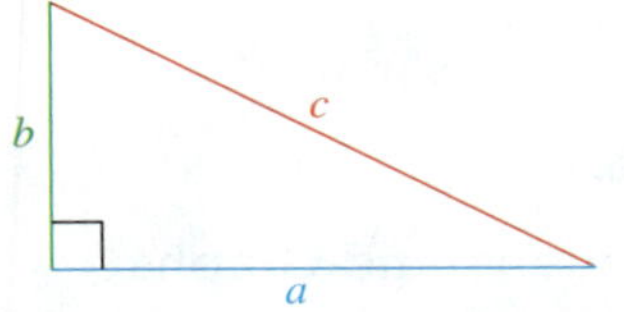

Figure 19 The Pythagorean theorem: $a^2 + b^2 = c^2$

> **Pythagorean Theorem**
>
> If a and b are the lengths of the legs of a right triangle and c is the length of the hypotenuse, then
>
> $$a^2 + b^2 = c^2$$
>
> In words, the sum of the squares of the lengths of the legs is equal to the square of the length of the hypotenuse (see Fig. 19).

If we know the lengths of two of the three sides of a right triangle, how can we use the Pythagorean theorem to find the length of the third side?

Example 1 Finding the Length of a Side of a Right Triangle

The lengths of two sides of a right triangle are given. Find the length of the third side.

1.

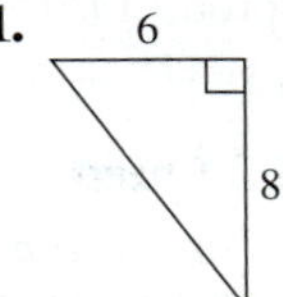

2.

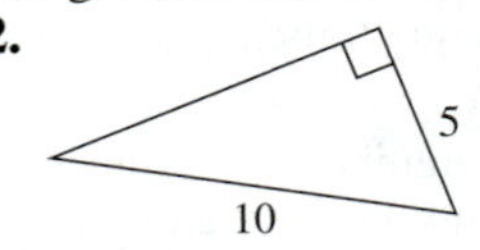

Solution

1. Since the lengths of the legs are given, we are to find the length of the hypotenuse. We substitute $a = 6$ and $b = 8$ into the equation $c^2 = a^2 + b^2$ and solve for c:

$$c^2 = 6^2 + 8^2 \quad \text{Substitute 6 for } a \text{ and 8 for } b.$$
$$c^2 = 36 + 64 \quad \text{Simplify.}$$
$$c^2 = 100$$
$$c = 10 \quad c \text{ is nonnegative.}$$

The length of the hypotenuse is 10 units.

2. The length of the hypotenuse is 10 units, and the length of one of the legs is 5 units. We substitute $a = 5$ and $c = 10$ into the equation $a^2 + b^2 = c^2$ and solve for b:

$$5^2 + b^2 = 10^2 \quad \text{Substitute 5 for } a \text{ and 10 for } c.$$
$$25 + b^2 = 100 \quad \text{Simplify.}$$
$$b^2 = 75 \quad \text{Subtract 25 from both sides.}$$
$$b = \sqrt{75} \quad b \text{ is nonnegative.}$$
$$b = 5\sqrt{3} \quad \sqrt{75} = \sqrt{25 \cdot 3} = 5\sqrt{3}$$

The length of the other leg is $5\sqrt{3}$ units (about 8.66 units).

Using the Pythagorean Theorem to Make Estimates about Authentic Situations

The Pythagorean theorem is used in numerous fields, including architecture, physics, engineering, graphic design, mathematics, surveying, chemistry, and aeronautics.

A
5.3 miles
L
4.4 miles
B

Figure 20 Surveying a lake

Example 2 Using the Pythagorean Theorem to Find a Distance

A surveyor wants to estimate the distance (in miles) across a lake from point A to point B as shown in Fig. 20. Find that distance.

Solution

We define L to be the distance (in miles) between points A and B (see Fig. 20). The triangle is a right triangle in which the hypotenuse has length 5.3 miles and one of its legs has length 4.4 miles. We use the Pythagorean theorem to find L:

$$\begin{aligned} 4.4^2 + L^2 &= 5.3^2 && \textit{Pythagorean theorem} \\ 19.36 + L^2 &= 28.09 && \textit{Find the squares.} \\ L^2 &= 8.73 && \textit{Subtract 19.36 from both sides.} \\ L &= \sqrt{8.73} && \textit{L is nonnegative.} \\ L &\approx 3.0 && \textit{Compute.} \end{aligned}$$

So, the distance across the lake between points A and B is approximately 3.0 miles.

Distance Between Two Points

We can use the Pythagorean theorem to find a formula for the distance between two points in a coordinate system. Let (x_1, y_1) and (x_2, y_2) represent two points, where $x_2 > x_1$ and $y_2 > y_1$ (see Fig. 21). We let d be the distance between the two points.

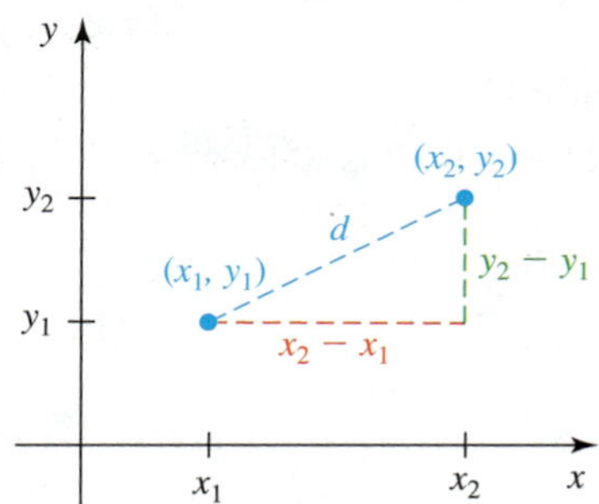

Figure 21 Find the distance between points (x_1, y_1) and (x_2, y_2)

Notice that the triangle shown in Fig. 21 is a right triangle with hypotenuse of length d and legs of lengths $x_2 - x_1$ and $y_2 - y_1$. We apply the Pythagorean theorem:

$$\begin{aligned} d^2 &= (x_2 - x_1)^2 + (y_2 - y_1)^2 \\ d &= \sqrt{(x_2 - x_1)^2 + (y_2 - y_1)^2} && \textit{d is nonnegative.} \end{aligned}$$

Although we assumed $x_2 > x_1$ and $y_2 > y_1$ to find the *distance formula,* it can be shown that the formula gives the correct distance between *any* two points.

Distance Formula

The distance d between points (x_1, y_1) and (x_2, y_2) is given by the **distance formula:**

$$d = \sqrt{(x_2 - x_1)^2 + (y_2 - y_1)^2}$$

Example 3 Finding the Distance Between Two Points

Find the distance between $(-2, 5)$ and $(3, -1)$.

Solution

We substitute $x_1 = -2$, $y_1 = 5$, $x_2 = 3$, and $y_2 = -1$ into the distance formula:

$$\begin{aligned} d &= \sqrt{(3 - (-2))^2 + (-1 - 5)^2} && \textit{Substitute into distance formula.} \\ &= \sqrt{5^2 + (-6)^2} && \textit{Subtract.} \\ &= \sqrt{61} && \textit{Simplify.} \end{aligned}$$

The distance between $(-2, 5)$ and $(3, -1)$ is $\sqrt{61}$ units (about 7.81 units).

Equation of a Circle

We can use the distance formula to find an equation whose graph is a circle. To see how, we first state the definition of a circle in terms of its *center* and *radius*.

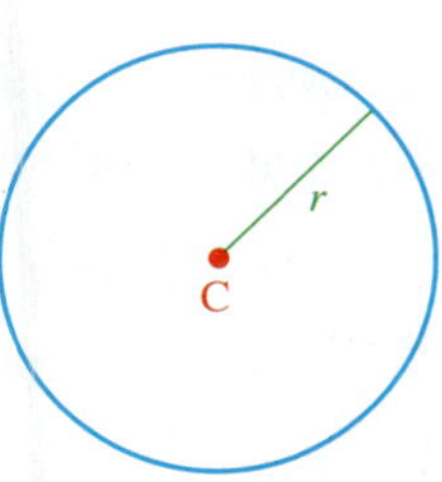

Figure 22 Circle with center C and radius r

Definition Circle

A **circle** with **center** point C and **radius** r, where $r > 0$, is the set of all points in a plane that are r units from point C in that plane (see Fig. 22).

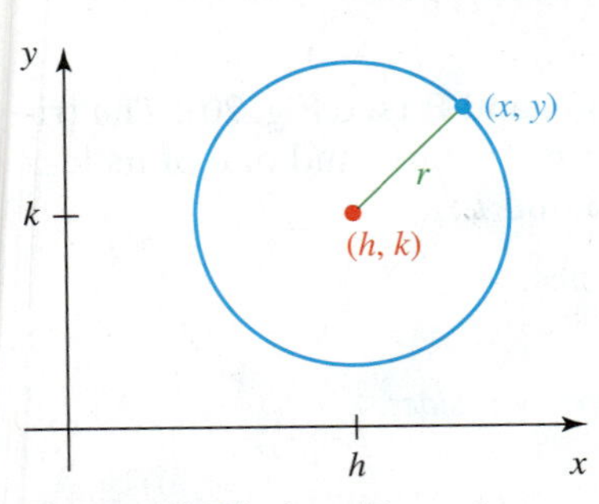

Figure 23 Circle with center (h, k) and radius r

Now we find an equation of a circle with center (h, k) and radius r (see Fig. 23). If (x, y) is a point on the circle, then the distance between (x, y) and (h, k) is the radius r:

$$\sqrt{(x-h)^2 + (y-k)^2} = r \quad \textit{Distance formula}$$

Squaring both sides of the equation gives

$$(x-h)^2 + (y-k)^2 = r^2$$

Equation of a Circle

If a circle has center (h, k) and radius r, then an equation of the circle is

$$(x-h)^2 + (y-k)^2 = r^2$$

The graph of an equation of this form with $r > 0$ is a circle with center (h, k) and radius r.

Example 4 Finding an Equation of a Circle

Find an equation of the circle with the given center and radius.

1. Center $(0, 0)$, radius 3
2. Center $(2, -5)$, radius $\sqrt{13}$

Solution

1. We substitute $h = 0$, $k = 0$, and $r = 3$ in $(x-h)^2 + (y-k)^2 = r^2$:

$$(x-0)^2 + (y-0)^2 = 3^2 \quad \textit{Substitute 0 for h, 0 for k, and 3 for r.}$$
$$x^2 + y^2 = 9 \quad \textit{Simplify.}$$

2. We substitute $h = 2$, $k = -5$, and $r = \sqrt{13}$ in $(x-h)^2 + (y-k)^2 = r^2$:

$$(x-2)^2 + (y-(-5))^2 = \left(\sqrt{13}\right)^2 \quad \textit{Substitute 2 for h, −5 for k, and } \sqrt{13} \textit{ for r.}$$
$$(x-2)^2 + (y+5)^2 = 13 \quad \textit{Simplify.}$$

The equation of a circle centered at the origin $(0, 0)$ with radius r is

$$(x-0)^2 + (y-0)^2 = r^2$$
$$x^2 + y^2 = r^2$$

Equation of a Circle Centered at the Origin

If a circle has center $(0, 0)$ and radius r, then an equation of the circle is

$$x^2 + y^2 = r^2$$

Example 5 Finding the Center and Radius of a Circle

Determine the center and radius of the circle. Also, sketch the circle.

1. $x^2 + y^2 = 19$
2. $(x+5)^2 + (y-3)^2 = 4$

Solution

1. The equation has the form $x^2 + y^2 = r^2$. Thus, the circle is centered at the origin $(0, 0)$, and we have

$$r^2 = 19$$
$$r = \sqrt{19} \quad \textit{r is positive.}$$

So, the radius is $\sqrt{19} \approx 4.36$. We sketch the circle in Fig. 24.

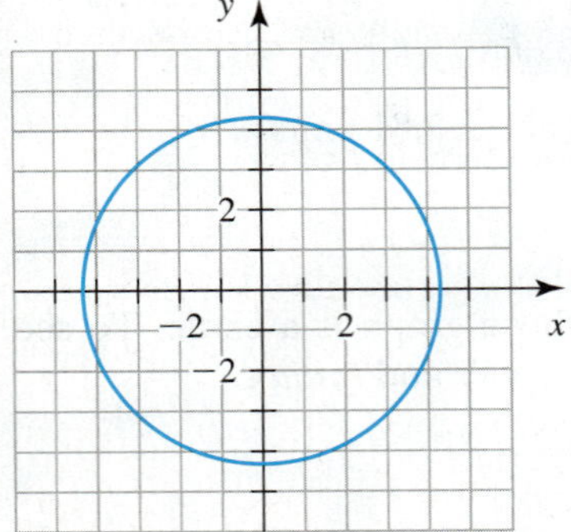

Figure 24 Graph of $x^2 + y^2 = 19$

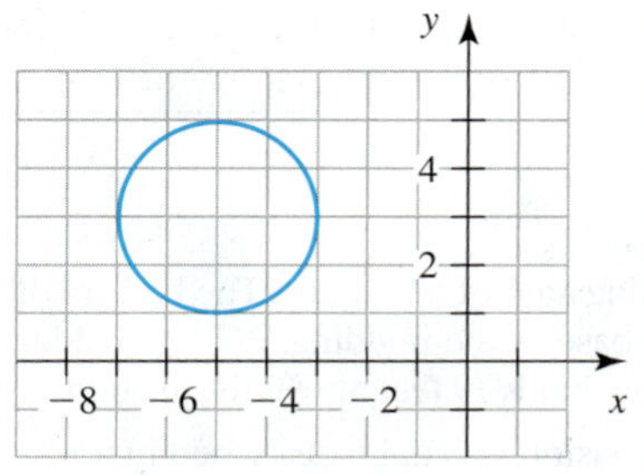

Figure 25 Graph of $(x + 5)^2 + (y - 3)^2 = 4$

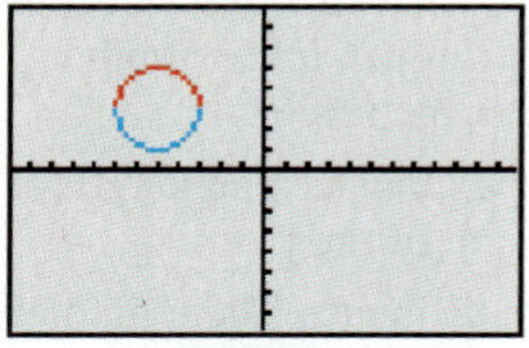

Figure 26 Use ZStandard, then ZInteger, and then Zoom In

2. We write $(x + 5)^2 + (y - 3)^2 = 4$ in the form $(x - h)^2 + (y - k)^2 = r^2$:

$$(x - (-5))^2 + (y - 3)^2 = 2^2$$

Here $h = -5$, $k = 3$, and $r = 2$. So, the circle has center $(-5, 3)$ and radius 2. We sketch the circle in Fig. 25.

To use a graphing calculator to draw the circle $(x + 5)^2 + (y - 3)^2 = 4$, we first isolate y:

$$\begin{aligned} (x + 5)^2 + (y - 3)^2 &= 4 && \textit{Original equation} \\ (y - 3)^2 &= 4 - (x + 5)^2 && \textit{Subtract } (x + 5)^2 \textit{ from both sides.} \\ y - 3 &= \pm\sqrt{4 - (x + 5)^2} && \textit{Square root property} \\ y &= 3 \pm \sqrt{4 - (x + 5)^2} && \textit{Add 3 to both sides.} \end{aligned}$$

Then we enter $y = 3 - \sqrt{4 - (x + 5)^2}$ and $y = 3 + \sqrt{4 - (x + 5)^2}$ and draw the graphs of both functions on the same coordinate system (see Fig. 26).

Group Exploration

Pythagorean theorem and its converse

For this exploration, you will need a ruler, scissors, and paper. It will help to have a tool for drawing right angles, such as a protractor or graph paper, although a corner of a piece of paper will suffice. For each triangle, assume c is the length of (one of) the *longest* side(s) and a and b are the lengths of the other sides.

1. Sketch three right triangles of different sizes. Measure the sides and show that, for each right triangle, $a^2 + b^2 = c^2$.
2. Now sketch three triangles of different sizes that are *not* right triangles. For these triangles, check whether $a^2 + b^2 = c^2$.
3. Sketch a triangle that has an angle close, but not equal, to 90°. Check whether $a^2 + b^2 \approx c^2$. If you cannot show this for your triangle, repeat the problem with a triangle that has an angle even closer to 90°.
4. If $a = 3$, $b = 5$, and $c = \sqrt{34}$, then $a^2 + b^2 = c^2$. Cut three thin strips of paper that are about 3, 5, and $\sqrt{34}$ inches in length. Form a triangle with the three strips of paper. Is the triangle a right triangle?
5. Find values of a, b, and c, other than the ones in Problem 4, such that $a^2 + b^2 = c^2$. Then repeat Problem 4 with your values.
6. Find three more values of a, b, and c such that $a^2 + b^2 = c^2$. Then repeat Problem 4 with your values.
7. Summarize at least three concepts addressed in this exploration.

Key Points of Section 15.3

Pythagorean theorem	If a and b are the lengths of the legs of a right triangle and c is the length of the hypotenuse, then $a^2 + b^2 = c^2$.
Distance formula	The distance d between points (x_1, y_1) and (x_2, y_2) is given by the **distance formula:** $d = \sqrt{(x_2 - x_1)^2 + (y_2 - y_1)^2}$.
Definition of a circle	A **circle** with **center** point C and **radius** r, where $r > 0$, is the set of all points in a plane that are r units from point C in that plane.
Equation of a circle	If a circle has center (h, k) and radius r, then an equation of the circle is $$(x - h)^2 + (y - k)^2 = r^2$$ The graph of an equation of this form with $r > 0$ is a circle with center (h, k) and radius r.
Equation of a circle centered at the origin	If a circle has center $(0, 0)$ and radius r, then an equation of the circle is $x^2 + y^2 = r^2$.

Homework 15.3

For extra help ▶ MyMathLab® 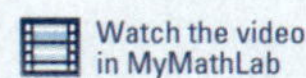 Watch the videos in MyMathLab  Download the MyDashboard App

The lengths of two sides of a right triangle are given. Find the length of the third side. (The triangle is not drawn to scale.)

1. See Fig. 27.

2. See Fig. 28.

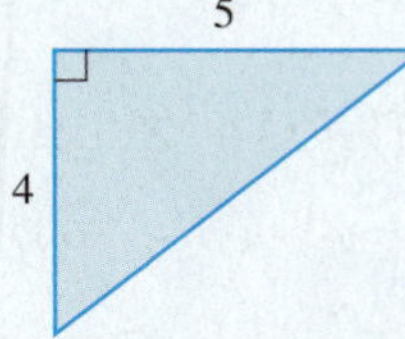

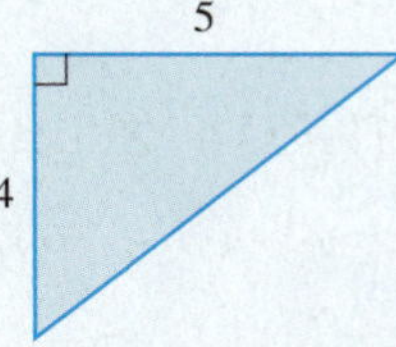

Figure 27 Exercise 1

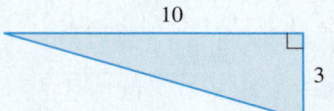

Figure 28 Exercise 2

3. See Fig. 29.

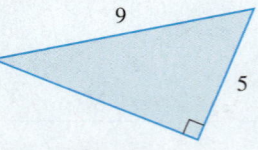

4. See Fig. 30.

Figure 29 Exercise 3

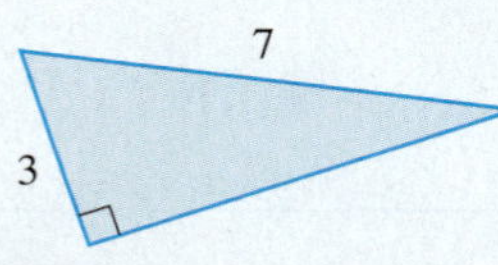

Figure 30 Exercise 4

5. See Fig. 31.

6. See Fig. 32.

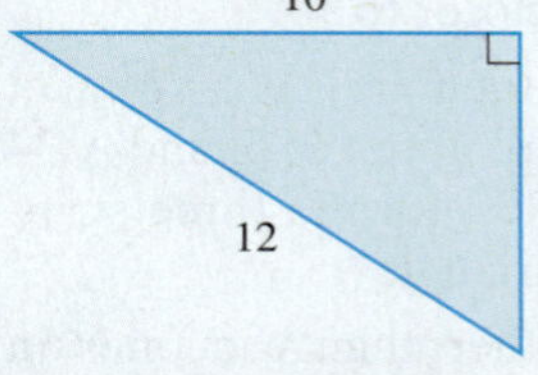

Figure 31 Exercise 5

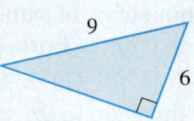

Figure 32 Exercise 6

Let a and b be the lengths of the legs of a right triangle, and let c be the length of the hypotenuse. Values of two of the three lengths are given. Find the third length.

7. $a = 5$ and $b = 12$

8. $a = 9$ and $b = 12$

9. $a = 6$ and $b = 7$

10. $a = 2$ and $b = 7$

11. $a = 3$ and $c = 8$

12. $a = 2$ and $c = 9$

13. $b = 5$ and $c = 7$

14. $b = 4$ and $c = 10$

15. $a = \sqrt{2}$ and $b = \sqrt{5}$

16. $a = \sqrt{3}$ and $b = \sqrt{11}$

For Exercises 17–30, round approximate results to the first decimal place.

17. A student drives 8 miles west from home and then 17 miles north to get to school. What would be the length of the trip if it were possible to drive along a straight line from home to school?

18. A commuter drives 5 miles south from home and then 9 miles east to get to work. What would be the length of the trip if it were possible to drive along a straight line from home to work?

19. A 12-foot-long ladder is leaning against a building. The bottom of the ladder is 4 feet from the base of the building. Will the ladder reach the bottom of a window that is 11 feet above the ground?

20. A 20-foot-long ladder is leaning against a building. The bottom of the ladder is 5 feet from the base of the building. Will the ladder reach the bottom of a window that is 19 feet above the ground?

21. The size of a rectangular television is usually described in terms of the length of its diagonal. If a 21-inch television screen has a height of 13 inches, what is the width of the screen?

22. The size of a rectangular television is usually described in terms of the length of its diagonal. If a 26-inch television has a height of 15 inches, what is the width of the screen?

23. A rectangular painting has a width of 24 inches. The length of the diagonal is 37 inches. Find the length of the painting.

24. A rectangular painting has a length of 42 inches. The length of the diagonal is 53 inches. Find the width of the painting.

25. A surveyor wants to estimate the distance (in miles) across a lake from point A to point B as shown in Fig. 33. Find that distance.

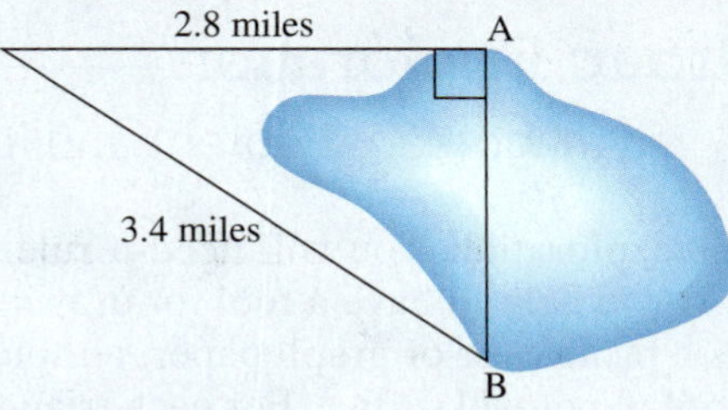

Figure 33 Exercise 25

26. A surveyor wants to estimate the distance (in miles) across a lake from point A to point B as shown in Fig. 34. Find that distance.

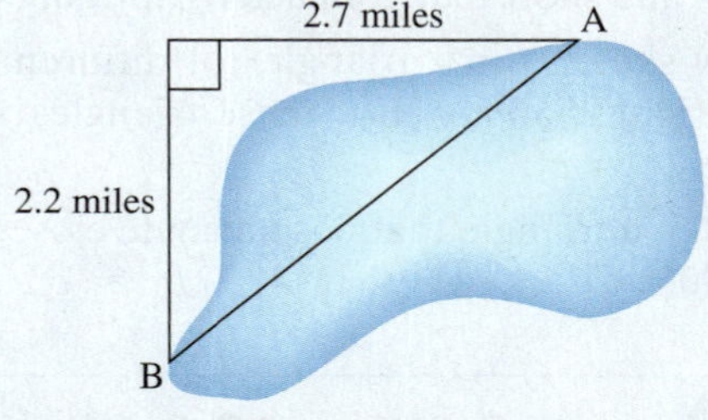

Figure 34 Exercise 26

27. Not including the end zones, an official football field is rectangular with a length of 100 yards and a width of 53 yards, 1 foot. If a player runs diagonally across this rectangle, how far will he run? [**Hint:** Use units of feet to perform your calculations.]

28. A soccer field is rectangular. For international matches, the length must be between 110 and 120 yards, inclusive, and the width must be between 70 and 80 yards, inclusive.

a. Find the length of the diagonal for the largest permitted field for international matches.

b. Find the length of the diagonal for the smallest permitted field for international matches.

29. Los Angeles is about 465 miles almost directly south of Reno, Nevada, and is almost directly west of Albuquerque, New Mexico. The distance between Albuquerque and Reno is about 964 miles. What would be the total distance of a trip from Los Angeles to Reno to Albuquerque to Los Angeles?

30. Salt Lake City, Utah, is about 498 miles almost directly south of Helena, Montana, and is almost directly west of Omaha, Nebraska. The distance between Omaha and Helena is about 1144 miles. What would be the total distance of a trip from Salt Lake City to Omaha to Helena to Salt Lake City?

Find the distance between the two given points.

31. $(2, 9)$ and $(8, 1)$ **32.** $(3, 2)$ and $(7, 5)$

33. $(-3, 5)$ and $(4, 2)$ **34.** $(-3, 4)$ and $(2, 7)$

35. $(-6, -3)$ and $(-4, 1)$ **36.** $(-7, 2)$ and $(-5, -6)$

37. $(-4, -5)$ and $(-8, -9)$ **38.** $(-5, -4)$ and $(-2, -7)$

Find the distance between the two given points. Round your result to the second decimal place.

39. $(2.1, 8.9)$ and $(5.6, 1.7)$ **40.** $(3.2, 7.1)$ and $(6.6, 8.4)$

41. $(-2.18, -5.74)$ and $(3.44, 6.29)$

42. $(-6.41, 1.12)$ and $(2.89, -3.55)$

Find an equation of the circle with the given center and radius.

43. Center $(0, 0)$, radius 7 **44.** Center $(0, 0)$, radius 10

45. Center $(0, 0)$, radius 6.7 **46.** Center $(0, 0)$, radius 2.3

47. Center $(5, 3)$, radius 2 **48.** Center $(4, 7)$, radius 5

49. Center $(-2, 1)$, radius 4 **50.** Center $(3, -4)$, radius 6

51. Center $(-7, -3)$, radius $\sqrt{3}$

52. Center $(-6, -1)$, radius $\sqrt{2}$

For Exercises 53–62, find the center and radius of the circle. Graph the equation by hand.

53. $x^2 + y^2 = 25$ **54.** $x^2 + y^2 = 9$

55. $x^2 + y^2 = 8$ **56.** $x^2 + y^2 = 17$

57. $(x - 3)^2 + (y - 5)^2 = 16$ **58.** $(x - 2)^2 + (y - 4)^2 = 4$

59. $(x + 6)^2 + (y - 1)^2 = 7$ **60.** $(x - 5)^2 + (y + 2)^2 = 3$

61. $(x + 3)^2 + (y + 2)^2 = 1$ **62.** $(x + 1)^2 + (y + 1)^2 = 1$

63. Find an equation of the circle shown in Fig. 35.

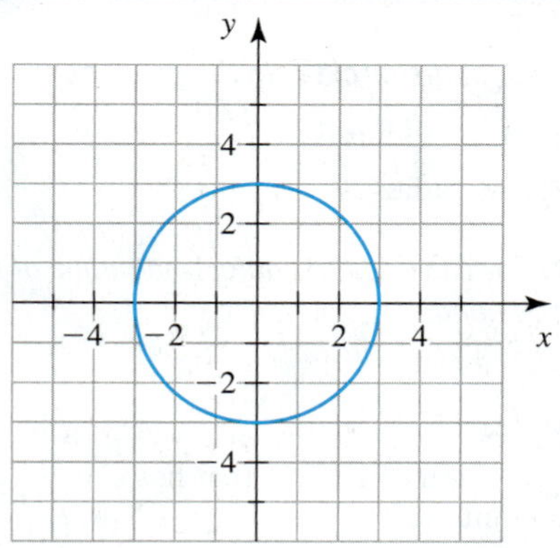

Figure 35 Exercise 63

64. Find an equation of the circle shown in Fig. 36.

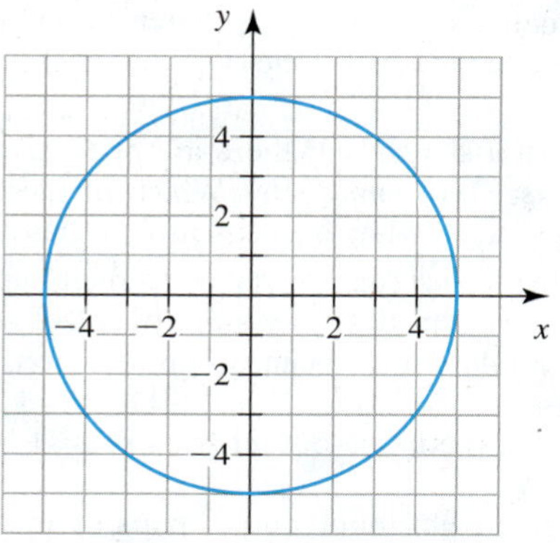

Figure 36 Exercise 64

65. Find an equation of the circle shown in Fig. 37.

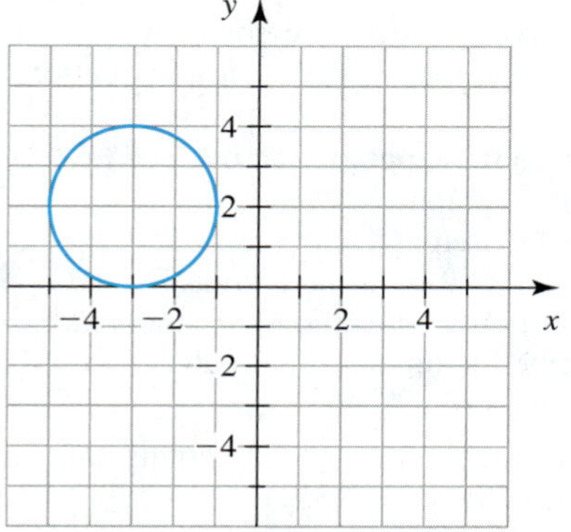

Figure 37 Exercise 65

66. Find an equation of the circle shown in Fig. 38.

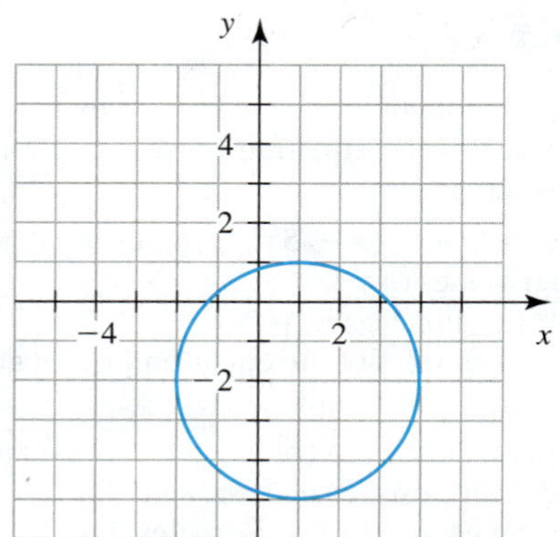

Figure 38 Exercise 66

Concepts

67. A circle with center $(3, 2)$ contains the point $(5, 6)$. Find an equation of the circle.

68. A circle with center $(-4, 3)$ contains the point $(2, -1)$. Find an equation of the circle.

69. Find equations of two distinct circles that contain the point $(5, 3)$. Sketch the two circles by hand in the same coordinate system.

70. Show with a sketch that there are many circles that contain the points $(2, 1)$ and $(4, 6)$. Which of these circles has the smallest radius? What is an equation of this circle?

71. The Rule of Four can be applied to relations as well as functions. So, we can describe some or all of the input–output pairs of a relation by means of an equation, a graph, a table, or words.

a. Describe the input–output pairs of $x^2 + y^2 = 16$ by using a graph.

b. Describe eight input–output pairs of $x^2 + y^2 = 16$ by using a table. Round approximate inputs and outputs to the second decimal place.

c. Describe the input–output pairs of $x^2 + y^2 = 16$ by using words.

72. The Rule of Four can be applied to relations as well as functions. So, we can describe some or all of the input–output pairs of a relation by means of an equation, a graph, a table, or words.

a. Describe the input–output pairs of $(x + 1)^2 + (y - 3)^2 = 9$ by using a graph.

b. Describe eight input–output pairs of $(x + 1)^2 + (y - 3)^2 = 9$ by using a table. Round approximate inputs and outputs to the second decimal place.

c. Describe the input–output pairs of $(x + 1)^2 + (y - 3)^2 = 9$ by using words.

73. Give the coordinates of five points that are a distance of 4 units from the point $(3, 2)$.

74. Give the coordinates of five points that are a distance of 1 unit from the point $(-4, 1)$.

75. Is the relation $x^2 + y^2 = 49$ a function? Explain.

76. Is the relation $(x - 4)^2 + (y + 2)^2 = 16$ a function? Explain.

77. **a.** For $y = \sqrt{25 - x^2}$, explain why $y \geq 0$ for real-number values of y.

b. Graph the function $y = \sqrt{25 - x^2}$ by hand. [**Hint:** Square both sides of the equation.] Use a graphing calculator to verify your graph.

78. **a.** For $y = \sqrt{9 - (x - 5)^2}$, explain why $y \geq 0$ for real-number values of y.

b. Graph the function $y = \sqrt{9 - (x - 5)^2}$ by hand. [**Hint:** Square both sides of the equation.] Use a graphing calculator to verify your graph.

79. If the lengths of the legs of a right triangle are equal, we call the triangle an *isosceles right triangle.*

a. Sketch an example of an isosceles right triangle.

b. Show that the length of the hypotenuse of an isosceles right triangle is $\sqrt{2}$ times the length of either leg of the triangle. [**Hint:** Let $a = k$ and $b = k$, and apply the Pythagorean theorem.]

c. If the length of a leg of an isosceles right triangle is 3 units, what is the length of the hypotenuse?

d. If the length of the hypotenuse of an isosceles right triangle is 5 units, what is the length of each leg?

80. Explain how to graph by hand an equation of the form $(x - h)^2 + (y - k)^2 = r$, where $r > 0$.

Related Review

Sketch the graph.

81. $x + y = 4$

82. $x^2 + y = 4$

83. $x^2 + y^2 = 4$

84. $x^{1/2} + y = 4$

85. $2^x + y = 0$

86. $\left(\frac{1}{2}\right)^x + y = 0$

Expressions, Equations, Functions, and Graphs

Perform the indicated instruction. Then use words such as linear, quadratic, cubic, exponential, logarithmic, rational, radical, polynomial, degree, function, one variable, *and* two variables *to describe the expression, equation, or system.*

87. Graph $f(x) = 2(x - 4)^2 - 3$ by hand.

88. Find the domain of $f(x) = \dfrac{x + 7}{27x^3 + 18x^2 - 12x - 8}$.

89. Factor $6x^2 - 16x + 8$.

90. Find the difference $\dfrac{3x + 5}{x^2 + 5x - 14} - \dfrac{2x}{x^2 - 4x - 21}$.

91. Solve $x(5x - 3) = 3(x + 1)$.

92. Solve $\dfrac{x + 5}{x} - \dfrac{2}{3x^2} = 7$.

Section 15.3 Quiz

1. The length of a leg of a right triangle is 4 inches, and the length of the hypotenuse is 8 inches. Find the length of the other leg.

2. The size of a rectangular television screen is usually described as the length of a diagonal. If a 19-inch screen has a width of 16 inches, what is the screen's height?

Find the distance between the two given points.

3. $(-2, -5)$ and $(3, -1)$

4. $(-3, 2)$ and $(-7, -2)$

Find an equation of the circle with the given center and radius.

5. Center $(-3, 2)$, radius 6

6. Center $(0, 0)$, radius 2.8

For Exercises 7 and 8, find the center and radius of the circle. Graph the equation by hand.

7. $x^2 + y^2 = 12$

8. $(x + 4)^2 + (y - 3)^2 = 25$

9. Find an equation of a circle that has center $(2, -1)$ and contains the point $(4, 7)$.

10. Find equations of two circles that contain the point $(0, 0)$. Sketch the two circles in the same coordinate system.

15.4 Ellipses and Hyperbolas

Objective

» Graph equations of ellipses and hyperbolas.

In this text, we work with four types of curves that are cross sections of cones: circles, *ellipses,* parabolas, and *hyperbolas* (see Fig. 39).

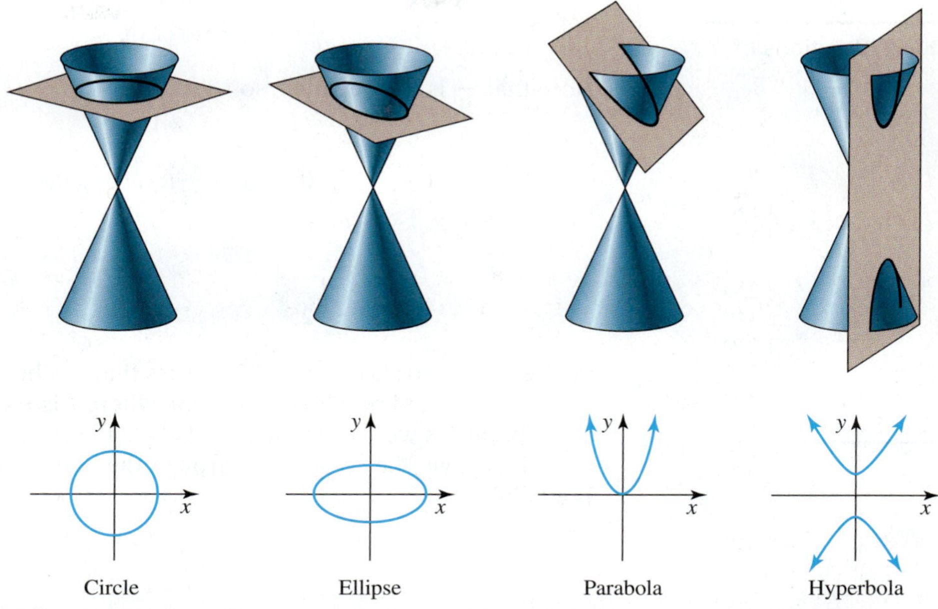

Figure 39 Circles, ellipses, parabolas, and hyperbolas are cross sections of cones

In Chapters 7–9, we worked with parabolas. In Section 15.3, we studied circles. In this section, we will first discuss ellipses; then we will discuss hyperbolas.

Graphing Ellipses

To begin our study of ellipses, we sketch a graph of

$$\frac{x^2}{25} + \frac{y^2}{9} = 1$$

First, we find the y-intercepts by substituting 0 for x and solving for y:

$$\begin{aligned} \frac{0^2}{25} + \frac{y^2}{9} &= 1 && \text{Substitute 0 for } x. \\ \frac{y^2}{9} &= 1 && \text{Simplify.} \\ y^2 &= 9 && \text{Multiply both sides by 9.} \\ y &= \pm 3 && \text{Square root property} \end{aligned}$$

So, the y-intercepts are $(0, -3)$ and $(0, 3)$.

Then we find the x-intercepts by substituting 0 for y and solving for x:

$$\begin{aligned} \frac{x^2}{25} + \frac{0^2}{9} &= 1 && \text{Substitute 0 for } y. \\ \frac{x^2}{25} &= 1 && \text{Simplify.} \\ x^2 &= 25 && \text{Multiply both sides by 25.} \\ x &= \pm 5 && \text{Square root property} \end{aligned}$$

So, the x-intercepts are $(-5, 0)$ and $(5, 0)$.

All of the points on the graph of the relation have x-coordinates between -5 and 5, inclusive. To see why, we isolate the term $\frac{y^2}{9}$ in the equation $\frac{x^2}{25} + \frac{y^2}{9} = 1$:

$$\frac{y^2}{9} = 1 - \frac{x^2}{25}$$

Table 2 Solutions of $\frac{x^2}{25} + \frac{y^2}{9} = 1$

x	y
-4	± 1.8
-3	± 2.4
-2	± 2.75
-1	± 2.94
0	± 3
1	± 2.94
2	± 2.75
3	± 2.4
4	± 1.8

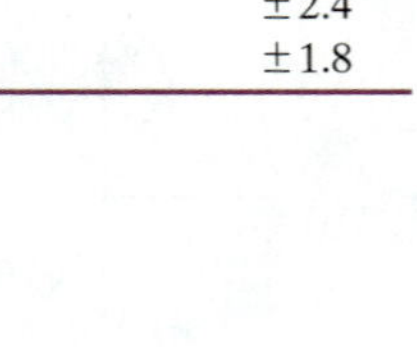

Note that $\frac{y^2}{9}$ is nonnegative, so

$$1 - \frac{x^2}{25} \geq 0 \qquad 1 - \frac{x^2}{25} \text{ is nonnegative.}$$

$$-\frac{x^2}{25} \geq -1 \qquad \text{Subtract 1 from both sides.}$$

$$x^2 \leq 25 \qquad \text{Multiply both sides by } -25\text{; reverse inequality symbol.}$$

As we set out to show, $x^2 \leq 25$ implies that x is between -5 and 5, inclusive.

Next, we find points on the graph where x is $-4, -3, -2, \ldots, 4$ (see Table 2) and plot these points as well as the intercepts (see Fig. 40).

Finally, we sketch a curve through the points we've plotted (see Fig. 41). The graph is an ellipse.

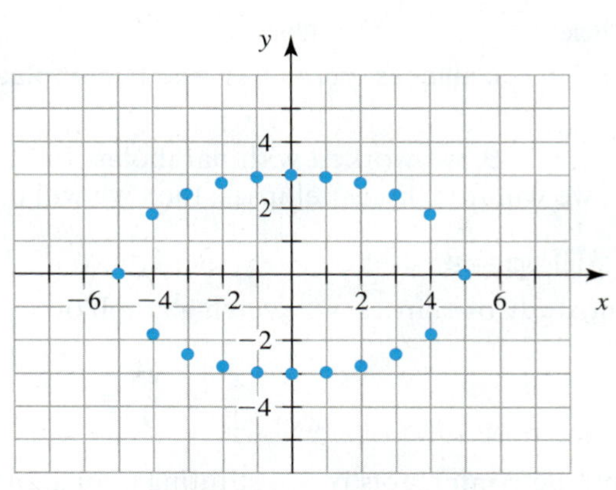

Figure 40 Plot the points

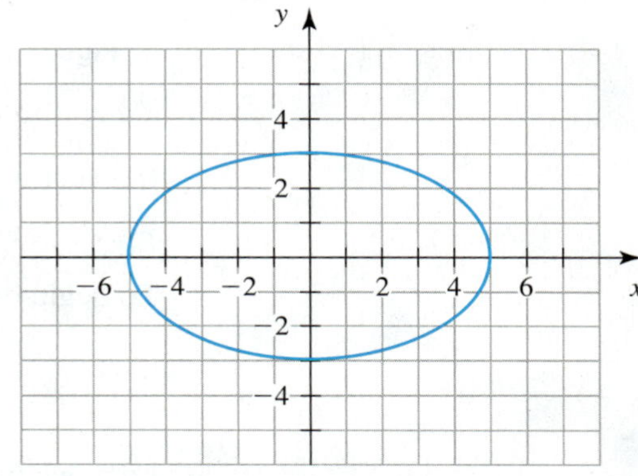

Figure 41 Graph of $\frac{x^2}{25} + \frac{y^2}{9} = 1$

Equation of an Ellipse

An equation that can be put into the form

$$\frac{x^2}{a^2} + \frac{y^2}{b^2} = 1, \qquad \text{where } a > 0 \text{ and } b > 0$$

has an **ellipse** as its graph.

Intercepts of an Ellipse

There is an easier way to sketch the graph of an equation in the form $\frac{x^2}{a^2} + \frac{y^2}{b^2} = 1$, where $a > 0$ and $b > 0$. We begin by finding the x-intercepts:

$$\frac{x^2}{a^2} + \frac{0^2}{b^2} = 1 \quad \text{Substitute 0 for } y.$$

$$\frac{x^2}{a^2} = 1 \quad \text{Simplify.}$$

$$x^2 = a^2 \quad \text{Multiply both sides by } a^2.$$

$$x = \pm\sqrt{a^2} \quad \text{Square root property}$$

$$x = \pm a \quad \sqrt{a^2} = a, \text{ where } a \geq 0$$

So, the x-intercepts are $(-a, 0)$ and $(a, 0)$.

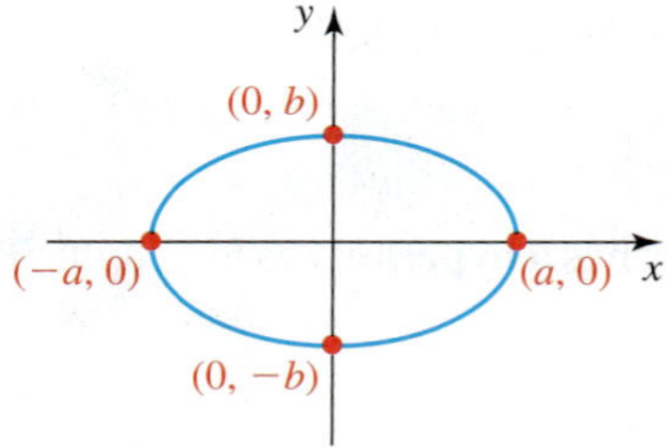

Figure 42 Intercepts of the ellipse $\frac{x^2}{a^2} + \frac{y^2}{b^2} = 1$

The y-intercepts are $(0, -b)$ and $(0, b)$. You will show this in Exercise 59.

> **Intercepts of an Ellipse**
>
> The ellipse described by
>
> $$\frac{x^2}{a^2} + \frac{y^2}{b^2} = 1$$
>
> has x-intercepts $(-a, 0)$ and $(a, 0)$ and y-intercepts $(0, -b)$ and $(0, b)$. See Fig. 42.

Example 1 Sketching the Graph of an Ellipse

Sketch the graph of $9x^2 + 4y^2 = 36$.

Solution

First, we divide both sides of the equation $9x^2 + 4y^2 = 36$ by 36 so that the right-hand side of the equation is 1:

$$\frac{9x^2}{36} + \frac{4y^2}{36} = \frac{36}{36} \quad \text{Divide both sides by 36.}$$

$$\frac{x^2}{4} + \frac{y^2}{9} = 1 \quad \text{Simplify.}$$

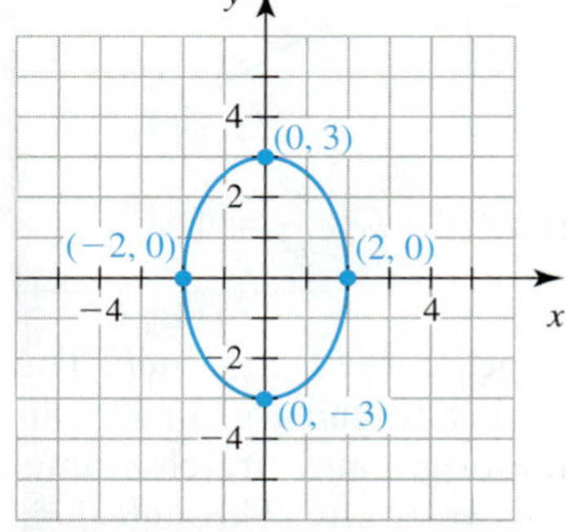

Figure 43 Graph of $9x^2 + 4y^2 = 36$

The equation is of the form $\frac{x^2}{a^2} + \frac{y^2}{b^2} = 1$, where $a > 0$ and $b > 0$, with $a^2 = 4$ and $b^2 = 9$. Since $a^2 = 4$, we have $a = 2$. So, the x-intercepts are $(-2, 0)$ and $(2, 0)$. Because $b^2 = 9$, we have $b = 3$. So, the y-intercepts are $(0, -3)$ and $(0, 3)$. We plot the intercepts and sketch an ellipse that contains them (see Fig. 43).

To use a graphing calculator to draw the ellipse $9x^2 + 4y^2 = 36$, we begin by isolating y:

$$9x^2 + 4y^2 = 36 \quad \text{Original equation}$$

$$4y^2 = 36 - 9x^2 \quad \text{Subtract } 9x^2 \text{ from both sides.}$$

$$y^2 = \frac{36 - 9x^2}{4} \quad \text{Divide both sides by 4.}$$

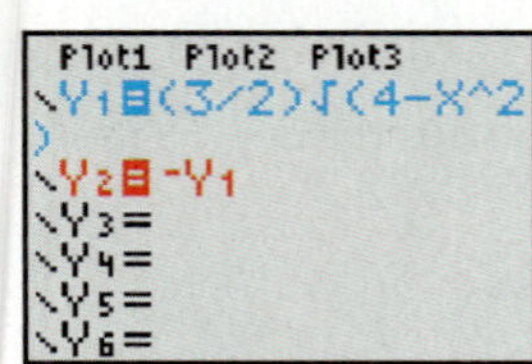

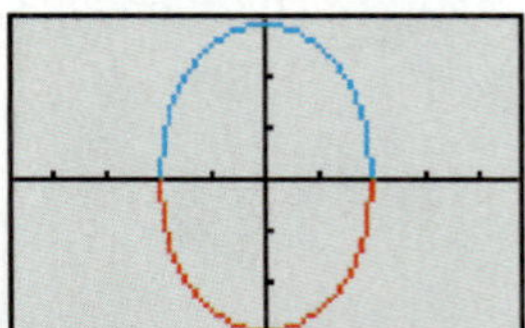

Figure 44 Graph of $9x^2 + 4y^2 = 36$, using ZDecimal

$$y = \pm\sqrt{\frac{36 - 9x^2}{4}} \quad \text{Square root property}$$

$$y = \pm\sqrt{\frac{9(4 - x^2)}{4}} \quad \text{Factor.}$$

$$y = \pm\frac{3}{2}\sqrt{4 - x^2} \quad \text{Simplify.}$$

Then we enter the functions $y = \frac{3}{2}\sqrt{4 - x^2}$ and $y = -\frac{3}{2}\sqrt{4 - x^2}$ (or $Y_2 = -Y_1$) and graph both functions in the same coordinate system (see Fig. 44).

Graphing Hyperbolas

We now turn our attention to sketching graphs of hyperbolas. The general equation of a hyperbola is similar to the general equation of an ellipse, except that the left side of the equation is a difference rather than a sum.

Equations of a Hyperbola

An equation that can be put into one of the following forms has a **hyperbola** as its graph:

- An equation that can be put into the form

$$\frac{x^2}{a^2} - \frac{y^2}{b^2} = 1, \qquad \text{where } a > 0 \text{ and } b > 0$$

is a hyperbola with x-intercepts $(-a, 0)$ and $(a, 0)$. See Fig. 45. There are no y-intercepts.

- An equation that can be put into the form

$$\frac{y^2}{b^2} - \frac{x^2}{a^2} = 1, \qquad \text{where } a > 0 \text{ and } b > 0$$

is a hyperbola with y-intercepts $(0, -b)$ and $(0, b)$. See Fig. 46. There are no x-intercepts.

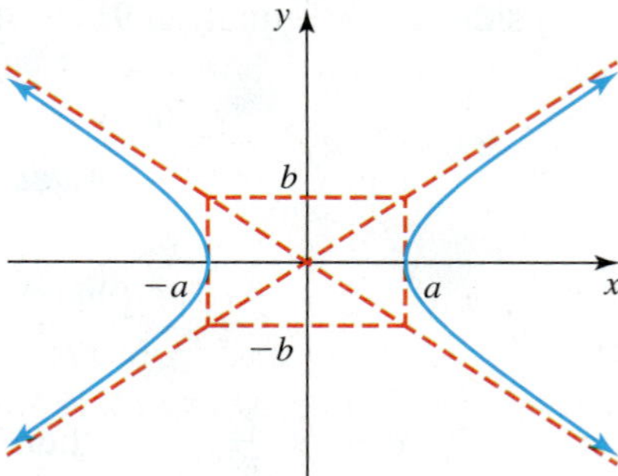

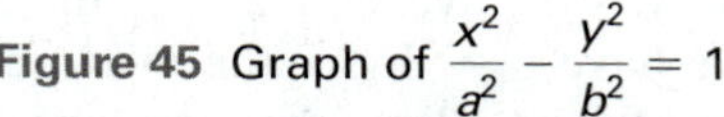

Figure 45 Graph of $\frac{x^2}{a^2} - \frac{y^2}{b^2} = 1$

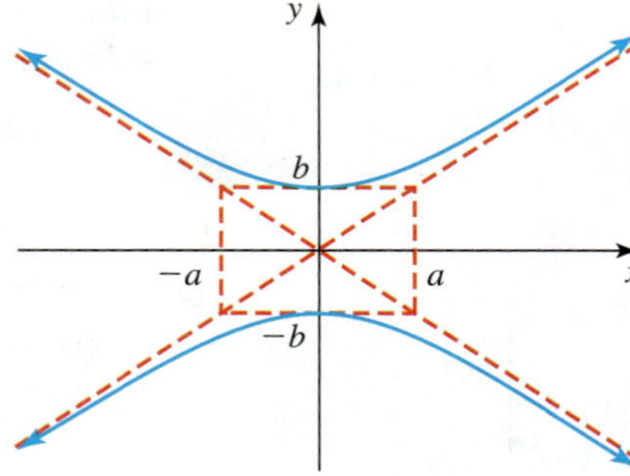

Figure 46 Graph of $\frac{y^2}{b^2} - \frac{x^2}{a^2} = 1$

Each *pair* of curves in Figs. 45 and 46 is a hyperbola. Each curve is a *branch*. The red dashes are *not* parts of the hyperbola; they are simply tools to guide us in sketching the branches. The dashed rectangles are centered at the origin and stretch a units in both directions horizontally and b units in both directions vertically. Through their opposite corners, we draw the dashed lines that are *inclined asymptotes*. The branches of the hyperbolas approach the inclined asymptotes as $|x|$ gets large.

Graphing Hyperbolas

To sketch a hyperbola on the basis of its equation,

1. Sketch a dashed rectangle whose sides are parallel to the axes and contain the points $(-a, 0)$, $(a, 0)$, $(0, -b)$, and $(0, b)$.
2. Sketch two dashed lines (the inclined asymptotes) that contain the diagonals of the rectangle.
3. Plot the intercepts of the hyperbola.
4. Sketch the branches to contain the intercepts and get closer to the asymptotes as $|x|$ gets large.

Example 2 Sketching the Graph of a Hyperbola

Sketch the graph of $\frac{y^2}{4} - \frac{x^2}{9} = 1$.

Solution

Since the equation is of the form $\frac{y^2}{b^2} - \frac{x^2}{a^2} = 1$, where $a > 0$ and $b > 0$, we have $a^2 = 9$ and $b^2 = 4$. So, $a = 3$ and $b = 2$. We sketch a dashed rectangle that contains the points $(-3, 0)$, $(3, 0)$, $(0, -2)$, and $(0, 2)$; then we sketch the inclined asymptotes (see Fig. 47).

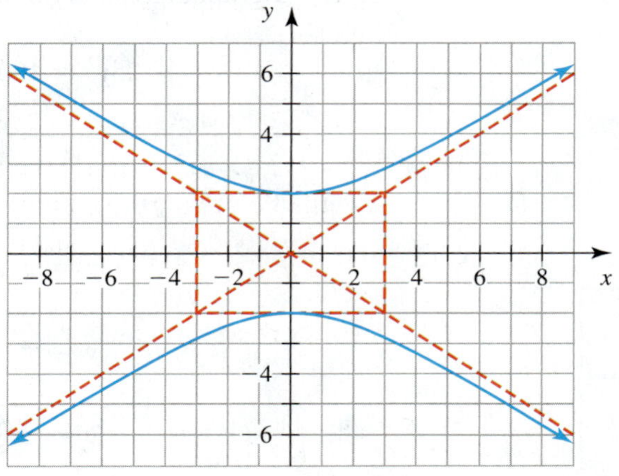

Figure 47 Graph of $\frac{y^2}{4} - \frac{x^2}{9} = 1$

Since the equation is of the form $\frac{y^2}{b^2} - \frac{x^2}{a^2} = 1$ (with $\frac{y^2}{b^2}$ first), the graph has y-intercepts at $(0, -2)$ and $(0, 2)$ and there are no x-intercepts. The branches contain the y-intercepts and approach the inclined asymptotes for large $|x|$.

Example 3 Sketching the Graph of a Hyperbola

Sketch the graph of $4x^2 - 25y^2 = 100$.

Solution

We divide both sides of the equation by 100 so the right-hand side of the equation is equal to 1:

$$4x^2 - 25y^2 = 100 \quad \textit{Original equation}$$

$$\frac{4x^2}{100} - \frac{25y^2}{100} = \frac{100}{100} \quad \textit{Divide both sides by 100.}$$

$$\frac{x^2}{25} - \frac{y^2}{4} = 1 \quad \textit{Simplify.}$$

The equation is of the form $\frac{x^2}{a^2} - \frac{y^2}{b^2} = 1$, where $a > 0, b > 0, a^2 = 25$, and $b^2 = 4$. So, $a = 5$ and $b = 2$. We sketch the dashed rectangle that contains $(-5, 0)$, $(5, 0)$, $(0, -2)$, and $(0, 2)$ and then sketch the inclined asymptotes (see Fig. 48).

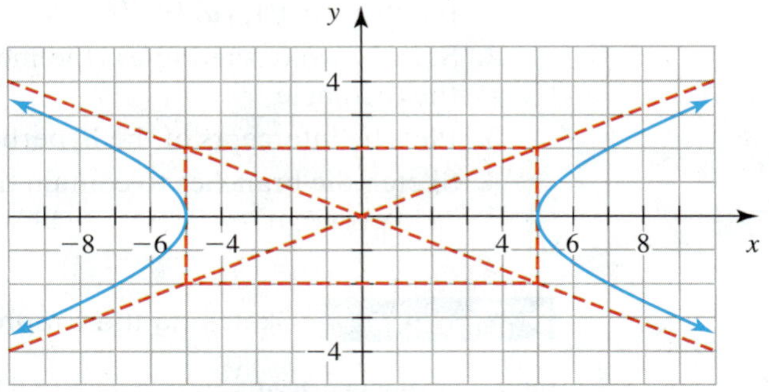

Figure 48 Graph of $4x^2 - 25y^2 = 100$

Since the equation is in $\frac{x^2}{a^2} - \frac{y^2}{b^2} = 1$ form (with $\frac{x^2}{a^2}$ first), the graph has x-intercepts at $(-5, 0)$ and $(5, 0)$ and there are no y-intercepts. The branches contain the intercepts and approach the inclined asymptotes for large $|x|$.

To use a graphing calculator to draw a graph of $4x^2 - 25y^2 = 100$, we begin by isolating y:

$$4x^2 - 25y^2 = 100 \qquad \textit{Original equation}$$

$$4x^2 - 100 = 25y^2 \qquad \textit{Isolate } 25y^2.$$

$$\frac{4x^2 - 100}{25} = y^2 \qquad \textit{Divide both sides by 25.}$$

$$y = \pm\sqrt{\frac{4x^2 - 100}{25}} \qquad \textit{Square root property}$$

$$y = \pm\sqrt{\frac{4(x^2 - 25)}{25}} \qquad \textit{Factor.}$$

$$y = \pm\frac{2}{5}\sqrt{x^2 - 25} \qquad \textit{Simplify.}$$

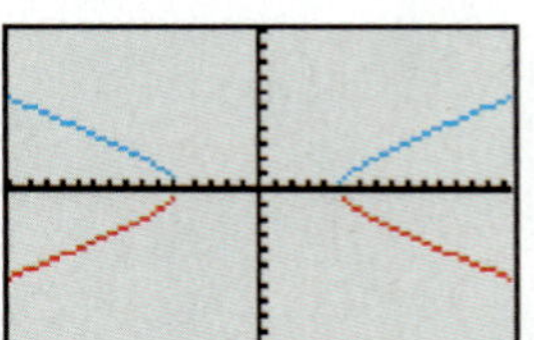

Figure 49 Use ZStandard followed by ZSquare to graph $4x^2 - 25y^2 = 100$

Then we enter the functions $y = \frac{2}{5}\sqrt{x^2 - 25}$ and $y = -\frac{2}{5}\sqrt{x^2 - 25}$ (or $Y_2 = -Y_1$) and draw the graphs in the same coordinate system (see Fig. 49).

Group Exploration

Graphical significance of *a* and *b* for ellipses and hyperbolas

1. Sketch two ellipses that
   a. intersect in four points.
   b. intersect in two points.
   c. intersect in no points.
2. Write and graph equations to correspond to each of your sketches in Problem 1.
3. Sketch an ellipse and a hyperbola that
   a. intersect in four points.
   b. intersect in two points.
   c. intersect in no points.
4. Write and graph equations to correspond to each of your sketches in Problem 3.
5. Sketch two hyperbolas that
   a. intersect in four points.
   b. intersect in two points.
   c. have no intersection points.
6. Write and graph equations to correspond to each of your sketches in Problem 5.

Key Points of Section 15.4

Throughout these Key Points, assume a and b are positive.

Equation of an ellipse An equation that can be put into the form $\frac{x^2}{a^2} + \frac{y^2}{b^2} = 1$ has an **ellipse** as its graph.

Intercepts of an ellipse The ellipse described by $\frac{x^2}{a^2} + \frac{y^2}{b^2} = 1$ has x-intercepts $(-a, 0)$ and $(a, 0)$ and y-intercepts $(0, -b)$ and $(0, b)$.

Equations of a hyperbola An equation that can be put into one of the following forms has a **hyperbola** as its graph:

- An equation that can be put into the form $\frac{x^2}{a^2} - \frac{y^2}{b^2} = 1$ is a hyperbola with x-intercepts $(-a, 0)$ and $(a, 0)$. There are no y-intercepts.
- An equation that can be put into the form $\frac{y^2}{b^2} - \frac{x^2}{a^2} = 1$ is a hyperbola with y-intercepts $(0, -b)$ and $(0, b)$. There are no x-intercepts.

Graphing hyperbolas To sketch a hyperbola on the basis of its equation,

1. Sketch a dashed rectangle whose sides are parallel to the axes and contain the points $(-a, 0)$, $(a, 0)$, $(0, -b)$, and $(0, b)$.
2. Sketch two dashed lines (the inclined asymptotes) that contain the diagonals of the rectangle.
3. Plot the intercepts of the hyperbola.
4. Sketch the branches to contain the intercepts and get closer to the asymptotes as $|x|$ gets large.

Homework 15.4

For extra help ▶ MyMathLab®  Watch the videos in MyMathLab  Download the MyDashboard App

For Exercises 1–16, graph the equation by hand.

1. $\frac{x^2}{36} + \frac{y^2}{9} = 1$ **2.** $\frac{x^2}{49} + \frac{y^2}{16} = 1$ **3.** $\frac{x^2}{4} + \frac{y^2}{36} = 1$

4. $\frac{x^2}{9} + \frac{y^2}{64} = 1$ **5.** $\frac{x^2}{100} + \frac{y^2}{16} = 1$ **6.** $\frac{x^2}{81} + \frac{y^2}{25} = 1$

7. $25x^2 + 4y^2 = 100$ **8.** $4x^2 + 16y^2 = 64$

9. $9x^2 + 100y^2 = 900$ **10.** $16x^2 + 25y^2 = 400$

11. $x^2 + y^2 = 36$ **12.** $2x^2 + 2y^2 = 50$

13. $x^2 + 25y^2 = 25$ **14.** $64x^2 + y^2 = 64$

15. $5x^2 + 16y^2 = 80$ **16.** $22x^2 + 4y^2 = 88$

17. Find an equation of the ellipse shown in Fig. 50.

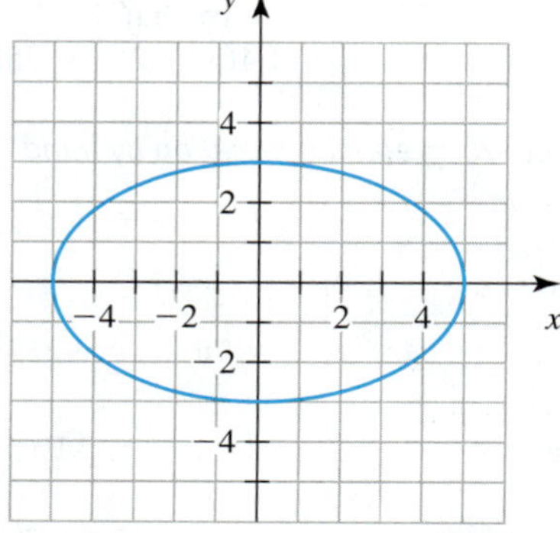

Figure 50 Exercise 17

18. Find an equation of the ellipse shown in Fig. 51.

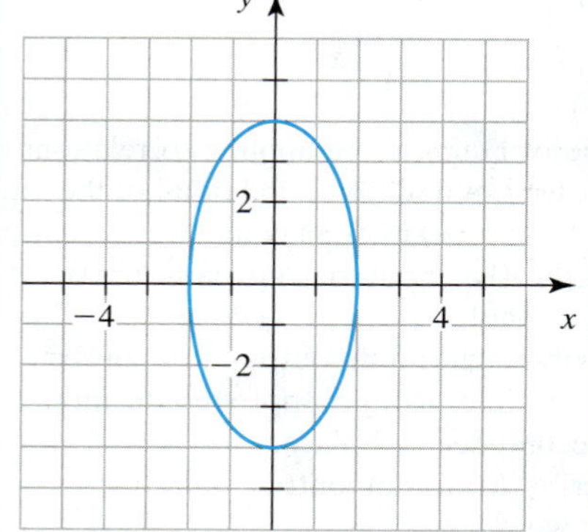

Figure 51 Exercise 18

Graph the equation by hand.

19. $\frac{x^2}{16} - \frac{y^2}{4} = 1$ **20.** $\frac{y^2}{25} - \frac{x^2}{9} = 1$

21. $\frac{y^2}{16} - \frac{x^2}{25} = 1$ **22.** $\frac{x^2}{49} - \frac{y^2}{16} = 1$

23. $\frac{x^2}{25} - \frac{y^2}{81} = 1$ **24.** $\frac{x^2}{64} - \frac{y^2}{9} = 1$

25. $16x^2 - 4y^2 = 64$ **26.** $25x^2 - 16y^2 = 400$

27. $x^2 - 9y^2 = 9$ **28.** $y^2 - 4x^2 = 4$

29. $y^2 - x^2 = 4$

30. $4x^2 - 4y^2 = 36$

31. $16y^2 - x^2 = 16$

32. $25x^2 - y^2 = 25$

33. $25x^2 - 7y^2 = 175$

34. $30x^2 - 9y^2 = 270$

Graph the equation by hand.

35. $\frac{x^2}{64} + \frac{y^2}{4} = 1$

36. $\frac{x^2}{49} + \frac{y^2}{100} = 1$

37. $x^2 - y^2 = 1$

38. $x^2 - y^2 = 9$

39. $81x^2 + 49y^2 = 3969$

40. $4x^2 + 36y^2 = 144$

41. $x^2 + y^2 = 1$

42. $x^2 + y^2 = 49$

43. $9y^2 - 4x^2 = 144$

44. $4x^2 - 25y^2 = 100$

45. $\frac{x^2}{25} - \frac{y^2}{25} = 1$

46. $\frac{x^2}{9} - \frac{y^2}{25} = 1$

47. $x^2 + y^2 = 16$

48. $5x^2 + 5y^2 = 45$

49. $9x^2 + 16y^2 = 144$

50. $4x^2 + 9y^2 = 36$

51. $\frac{x^2}{16} + \frac{y^2}{16} = 1$

52. $\frac{x^2}{36} + \frac{y^2}{36} = 1$

Concepts

53. a. Graph the equation $\frac{x^2}{c} + \frac{y^2}{d} = 1$ by hand for the given values of the constants c and d.

i. $c = 4$ and $d = 16$.

ii. $c = 4$ and $d = -16$.

iii. $c = -4$ and $d = 16$.

iv. $c = 4$ and $d = 4$.

b. In terms of the values of c and d, discuss whether the graph of the equation $\frac{x^2}{c} + \frac{y^2}{d} = 1$ is a circle, an ellipse, a hyperbola with x-intercepts, or a hyperbola with y-intercepts.

54. Graph the following equations by hand in the same coordinate system.

a. $\frac{x^2}{36} + \frac{y^2}{9} = 1$ **b.** $\frac{x^2}{36} - \frac{y^2}{9} = 1$ **c.** $\frac{y^2}{9} - \frac{x^2}{36} = 1$

55. The Rule of Four can be applied to relations as well as functions. So, we can describe some or all of the input–output pairs of a relation by means of an equation, a graph, a table, or words.

a. Describe the input–output pairs of $4x^2 + 25y^2 = 100$ by using a graph.

b. Describe eight input–output pairs of $4x^2 + 25y^2 = 100$ by using a table. Round approximate inputs and outputs to the second decimal place.

c. Describe the input–output pairs of $4x^2 + 25y^2 = 100$ by using words.

56. The Rule of Four can be applied to relations as well as functions. So, we can describe some or all of the input–output pairs of a relation by means of an equation, a graph, a table, or words.

a. Describe the input–output pairs of $x^2 - 4y^2 = 4$ by using a graph.

b. Describe six input–output pairs of $x^2 - 4y^2 = 4$ by using a table. Round approximate inputs and outputs to the second decimal place.

c. Describe the input–output pairs of $x^2 - 4y^2 = 4$ by using words.

57. a. For $y = \frac{5}{2}\sqrt{4 - x^2}$, explain why $y \geq 0$ for real-number values of y.

b. Graph the function $y = \frac{5}{2}\sqrt{4 - x^2}$ by hand. [**Hint:** Square both sides of the equation.] Use a graphing calculator to verify your graph.

58. a. For $y = \frac{2}{3}\sqrt{x^2 - 9}$, explain why $y \geq 0$ for real-number values of y.

b. Graph the function $y = \frac{2}{3}\sqrt{x^2 - 9}$ by hand. Use a graphing calculator to verify your graph. [**Hint:** Square both sides of the equation.]

59. Show that the y-intercepts of the graph of $\frac{x^2}{a^2} + \frac{y^2}{b^2} = 1$ are the points $(0, -b)$ and $(0, b)$.

60. Assume $a > 0$ and $b > 0$. Show that the graph of $\frac{x^2}{a^2} - \frac{y^2}{b^2} = 1$ has no y-intercepts.

61. Is the graph of the equation $x^2 + y^2 = r^2$ with $r > 0$ a circle, an ellipse, both, or neither? Explain. [**Hint:** Is it possible to write the equation in the form $\frac{x^2}{a^2} + \frac{y^2}{b^2} = 1$?]

62. Find equations of five ellipses that do not intersect each other. Sketch the five ellipses in the same coordinate system.

63. Assume $a > 0$ and $b > 0$. Describe how to graph by hand an equation of the form $\frac{x^2}{a^2} + \frac{y^2}{b^2} = 1$.

64. Describe how to graph by hand an equation of the form $\frac{x^2}{a^2} - \frac{y^2}{b^2} = 1$.

Related Review

65. a. Graph the equations $x^2 + y^2 = 1$ and $(x - 3)^2 + (y - 2)^2 = 1$ by hand in the same coordinate system. How can the graph of $x^2 + y^2 = 1$ be translated to get the graph of $(x - 3)^2 + (y - 2)^2 = 1$?

b. Graph the equations $x^2 + y^2 = 1$ and $(x + 3)^2 + (y + 2)^2 = 1$ by hand in the same coordinate system. How can the graph of $x^2 + y^2 = 1$ be translated to get the graph of $(x + 3)^2 + (y + 2)^2 = 1$?

c. Explain how you can translate the graph of an equation of the form $x^2 + y^2 = r^2$ to get the graph of $(x - h)^2 + (y - k)^2 = r^2$, where h, k, and r are constants and $r > 0$.

d. Graph $4x^2 + 25y^2 = 100$ by hand. Then translate your graph to get the graph of $4(x + 2)^2 + 25(y - 5)^2 = 100$.

e. Graph $4x^2 - 25y^2 = 100$ by hand. Then translate your graph to get the graph of $4(x + 2)^2 - 25(y - 5)^2 = 100$.

For Exercises 66–70, graph the equation by hand.

66. $3x - 2y = 8$

67. $y = \log_2(x)$

68. $y = -2(x - 4)^2 + 6$

69. $y = 3\sqrt{x + 5} - 4$

70. $y = -2(3)^x$

71. Is the relation $\frac{x^2}{4} + \frac{y^2}{81} = 1$ a function? Explain.

72. Is the relation $\frac{x^2}{9} - \frac{y^2}{81} = 1$ a function? Explain.

Expressions, Equations, Functions, and Graphs

Perform the indicated instruction. Then use words such as linear, quadratic, cubic, exponential, logarithmic, rational, radical, polynomial, degree, function, one variable, *and* two variables *to describe the expression, equation, or system.*

73. Graph $y = 2x^2 - 8x + 3$ by hand.
74. Solve $\log_3(7x^3) = 5$. Round any solutions to the fourth decimal place.
75. Solve $2x^2 - 8x + 3 = 0$.
76. Graph $f(x) = \log_3(x)$ by hand.
77. Find the product $-5x(2x - 1)(3x - 1)$.
78. Write $4\log_b(2x^2) - 3\log_b(4x^5)$ as a single logarithm.

Section 15.4 Quiz

For Exercises 1–8, graph the equation by hand.

1. $\frac{x^2}{9} + \frac{y^2}{25} = 1$
2. $\frac{y^2}{49} - \frac{x^2}{9} = 1$
3. $4x^2 - y^2 = 16$
4. $16x^2 + 3y^2 = 48$
5. $x^2 - 9y^2 = 81$
6. $4y^2 - 4x^2 = 16$
7. $\frac{x^2}{5} + \frac{y^2}{14} = 1$
8. $\frac{x^2}{8} + \frac{y^2}{3} = 1$
9. Is the relation $\frac{x^2}{9} - \frac{y^2}{4} = 1$ a function? Explain.
10. Find equations of three distinct ellipses that all contain the points $(0, 3)$ and $(0, -3)$.

15.5 Solving Nonlinear Systems of Equations

Objectives

- Know the meaning of a *nonlinear system*.
- Solve nonlinear systems by graphing, substitution, or elimination.

In this section, we will solve nonlinear systems. A **nonlinear system of equations** is a system of equations in which *at least* one of the equations is not linear. Here is an example of a nonlinear system:

$$x^2 + y^2 = 4$$
$$y = x^2$$

The graph of $x^2 + y^2 = 4$ is a circle, and the graph of $y = x^2$ is a parabola. Just as with solutions of linear systems, a *solution* of a nonlinear system is an ordered pair that satisfies *all* of the equations in the system. The *solution set* of a nonlinear system is the set of all solutions of the system.

When solving nonlinear systems in this text, we find only solutions that have real-number coordinates.

Solving a Nonlinear System by Graphing and by Substitution

Just as we solve linear systems, we can solve a nonlinear system by graphing. **The solution set of a system of nonlinear equations can be found by locating the intersection of the graphs of *all* of the equations.**

We can also solve some nonlinear systems by substitution.

Example 1 Solving a Nonlinear System

Solve the system

$$y = x^2 - 5$$
$$y = -x + 1$$

by graphing and by substitution.

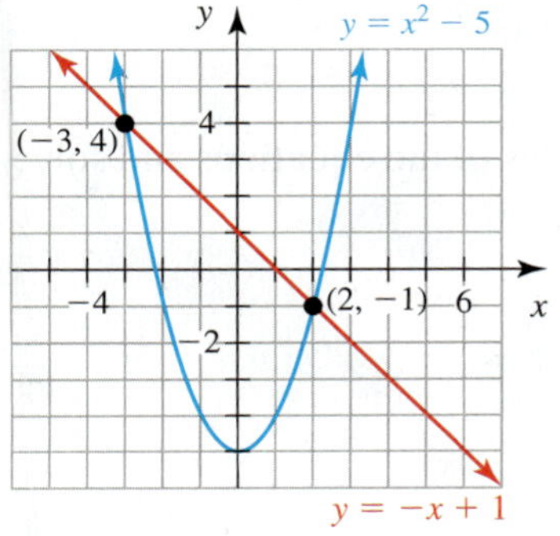

Figure 52 Graphs of $y = x^2 - 5$ and $y = -x + 1$

Solution

The graph of $y = x^2 - 5$ is a parabola, and the graph of $y = -x + 1$ is a line. We sketch graphs of both equations in the same coordinate system (see Fig. 52). The graphs appear to intersect at $(-3, 4)$ and $(2, -1)$. The two intersection points are the solutions of the system.

Next, we solve the system

$$y = x^2 - 5$$
$$y = -x + 1$$

by substitution. To begin, we substitute $x^2 - 5$ for y in the equation $y = -x + 1$ and solve for x:

$$
\begin{aligned}
x^2 - 5 &= -x + 1 && \text{Substitute } x^2 - 5 \text{ for } y. \\
x^2 + x - 6 &= 0 && \text{Write in } ax^2 + bx + c = 0 \text{ form.} \\
(x + 3)(x - 2) &= 0 && \text{Factor left side.} \\
x + 3 = 0 \quad \text{or} \quad x - 2 &= 0 && \text{Zero factor property} \\
x = -3 \quad \text{or} \quad x &= 2
\end{aligned}
$$

We substitute $x = -3$ and $x = 2$ into the equation $y = -x + 1$ to find the corresponding values of y:

$$
\begin{aligned}
y &= -(-3) + 1 & \qquad y &= -2 + 1 \\
&= 4 & &= -1
\end{aligned}
$$

So, $(-3, 4)$ and $(2, -1)$ are the solutions of the system.

We can check that $(-3, 4)$ and $(2, -1)$ satisfy both equations in the original system. However, the fact that we have solved the system in two different ways and gotten the same result is itself a check of our work.

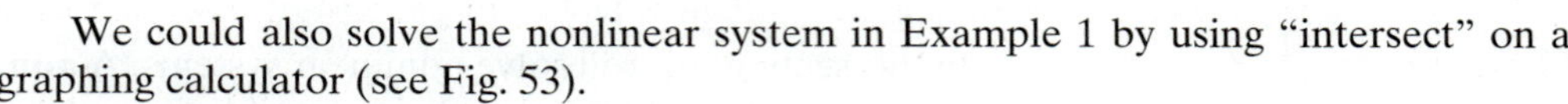

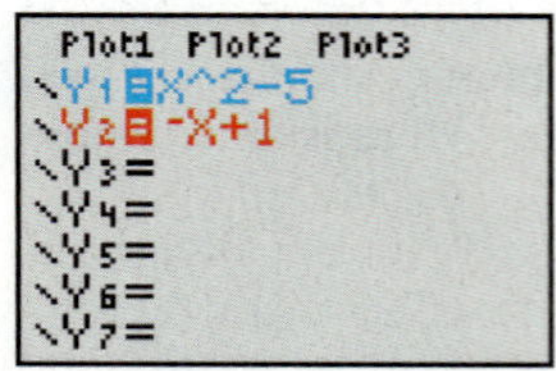

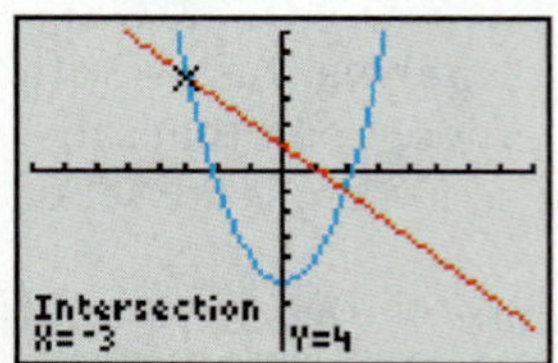

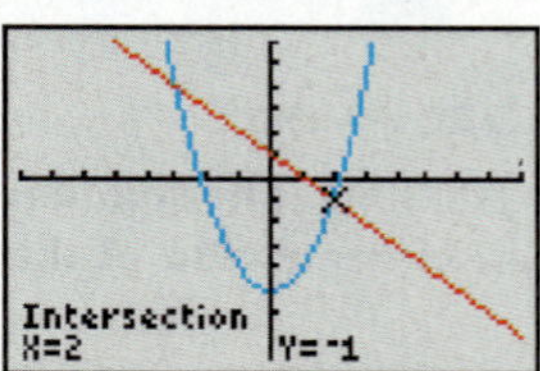

Figure 53 Verify that $(-3, 4)$ and $(2, -1)$ are solutions

We could also solve the nonlinear system in Example 1 by using "intersect" on a graphing calculator (see Fig. 53).

Solving a Nonlinear System by Elimination

We can solve some nonlinear systems by elimination.

Example 2 Solving a Nonlinear System

Solve the system

$$
\begin{aligned}
x^2 + y^2 &= 9 \\
9x^2 + 4y^2 &= 36
\end{aligned}
$$

by graphing and by elimination.

Solution

The graph of $x^2 + y^2 = 9$ is a circle, and the graph of $9x^2 + 4y^2 = 36$ is an ellipse. We sketch the graphs in the same coordinate system (see Fig. 54). (We sketched the ellipse $9x^2 + 4y^2 = 36$ in Example 1 of Section 15.4.) The intersection points appear to be $(0, -3)$ and $(0, 3)$. These points are the solutions of the system.

Now we solve the system

$$
\begin{aligned}
x^2 + y^2 &= 9 && \text{Equation (1)} \\
9x^2 + 4y^2 &= 36 && \text{Equation (2)}
\end{aligned}
$$

by elimination. First, we multiply both sides of equation (1) by -4, yielding the system

$$
\begin{aligned}
-4x^2 - 4y^2 &= -36 && \text{Multiply both sides of equation (1) by } -4. \\
9x^2 + 4y^2 &= 36 && \text{Equation (2)}
\end{aligned}
$$

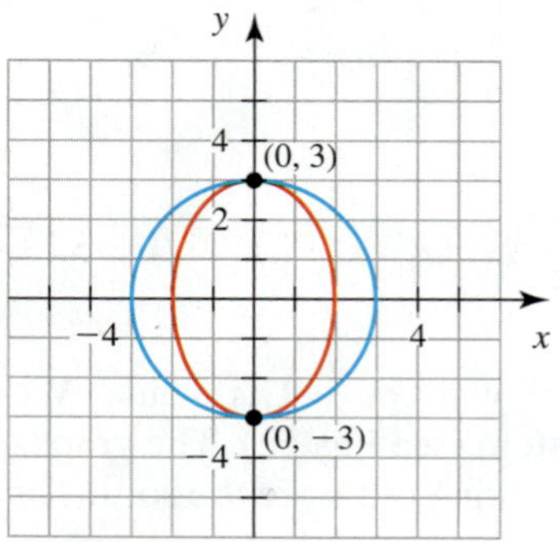

Figure 54 Graphs of $x^2 + y^2 = 9$ and $9x^2 + 4y^2 = 36$

Next, we add the left-hand sides and add the right-hand sides of the equations and solve for x:

$$
\begin{aligned}
-4x^2 - 4y^2 &= -36 \\
9x^2 + 4y^2 &= 36 \\
\hline
5x^2 + 0 &= 0 \\
5x^2 &= 0 && a + 0 = a \\
x^2 &= 0 && \text{Divide both sides by 5.} \\
x &= 0 && \text{Square root property}
\end{aligned}
$$

Then we substitute 0 for x in equation (1) and solve for y:

$$0^2 + y^2 = 9 \quad \text{Substitute 0 for } x.$$
$$y^2 = 9 \quad \text{Simplify.}$$
$$y = \pm 3 \quad \text{Square root property}$$

The solutions are $(0, -3)$ and $(0, 3)$. We got the same result when we solved the system by graphing.

If it is reasonable to do so, we first graph the equations of a nonlinear system to determine the number of solutions and to find approximate coordinates of the solutions. Then we solve the system by substitution or elimination.

Example 3 Solving a Nonlinear System

Solve the system

$$x^2 + y^2 = 25$$
$$-4x^2 + 9y^2 = 36$$

by graphing and by elimination. Round coordinates of any solutions to the second decimal place.

Solution

The graph of $x^2 + y^2 = 25$ is a circle, and the graph of $-4x^2 + 9y^2 = 36$ is a hyperbola. We sketch both graphs in the same coordinate system (see Fig. 55). (If we divide both sides of $-4x^2 + 9y^2 = 36$ by 36, we have $-\frac{x^2}{9} + \frac{y^2}{4} = 1$. We sketched the graph of this equation in Example 2 of Section 15.4.)

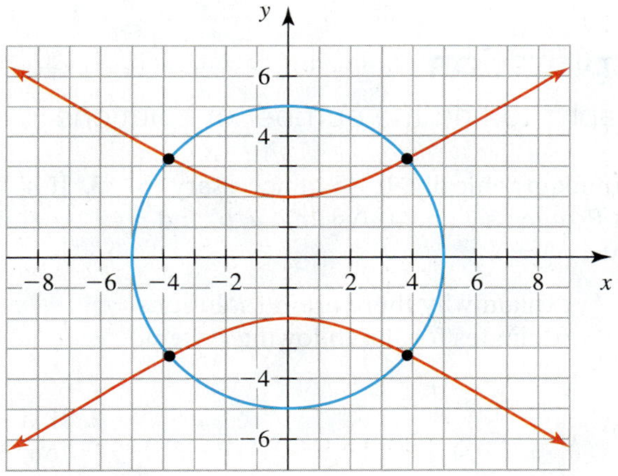

Figure 55 Graphs of $x^2 + y^2 = 25$ and $-4x^2 + 9y^2 = 36$

The graphs suggest there are four solutions: $(-3.8, -3.2)$, $(-3.8, 3.2)$, $(3.8, -3.2)$, and $(3.8, 3.2)$.

Now we solve the system

$$x^2 + y^2 = 25 \quad \text{Equation (1)}$$
$$-4x^2 + 9y^2 = 36 \quad \text{Equation (2)}$$

by elimination. First, we multiply both sides of equation (1) by 4, yielding the system

$$4x^2 + 4y^2 = 100 \quad \text{Multiply both sides by 4.}$$
$$-4x^2 + 9y^2 = 36 \quad \text{Equation (2)}$$

Next, we add the left-hand sides and add the right-hand sides of the equations and solve for y:

$$\begin{aligned} 4x^2 + 4y^2 &= 100 \\ -4x^2 + 9y^2 &= 36 \\ \hline 0 + 13y^2 &= 136 \\ 13y^2 &= 136 && 0 + a = a \\ y^2 &= \frac{136}{13} && \text{Divide both sides by 13.} \\ y &= \pm\sqrt{\frac{136}{13}} && \text{Square root property} \\ y &\approx \pm 3.234 && \text{Compute.} \end{aligned}$$

Then we substitute -3.234 and 3.234 for y in equation (1) and solve for x:

$$\begin{aligned} x^2 + (-3.234)^2 &= 25 \\ x^2 + 3.234^2 &= 25 \\ x^2 &= 25 - 3.234^2 \\ x &= \pm\sqrt{25 - 3.234^2} \\ x &\approx \pm 3.81 \end{aligned}$$

$$\begin{aligned} x^2 + 3.234^2 &= 25 \\ x^2 &= 25 - 3.234^2 \\ x &= \pm\sqrt{25 - 3.234^2} \\ x &\approx \pm 3.81 \end{aligned}$$

So, the four approximate solutions are $(-3.81, -3.23)$, $(-3.81, 3.23)$, $(3.81, -3.23)$, and $(3.81, 3.23)$, which agrees with what we found graphically.

Group Exploration

Using graphs to find the number of solutions

For each problem, think graphically. It is not necessary to solve the systems in Problems 2 and 4. Assume a, b, and r are positive constants.

1. If $r < a$ and $r < b$, explain why there are no solutions with real-number coordinates for the following system:

$$x^2 + y^2 = r^2$$
$$\frac{x^2}{a^2} + \frac{y^2}{b^2} = 1$$

2. If $a < r < b$, explain why the following system has four solutions:

$$x^2 + y^2 = r^2$$
$$\frac{x^2}{a^2} + \frac{y^2}{b^2} = 1$$

3. If $a < r < b$, explain why there are no solutions with real-number coordinates for the following system:

$$x^2 + y^2 = r^2$$
$$\frac{y^2}{b^2} - \frac{x^2}{a^2} = 1$$

4. If $a < r < b$, explain why the following system has four solutions:

$$x^2 + y^2 = r^2$$
$$\frac{x^2}{a^2} - \frac{y^2}{b^2} = 1$$

Key Points of Section 15.5

Nonlinear system of equations	A **nonlinear system of equations** is a system of equations in which *at least* one of the equations is not linear.
Solution set of a nonlinear system	The solution set of a system of nonlinear equations can be found by locating the intersection of the graphs of *all* of the equations.
First graph a nonlinear system	If it is reasonable to do so, we first graph the equations of a nonlinear system to determine the number of solutions and to find approximate coordinates of the solutions. Then we solve the system by substitution or elimination.

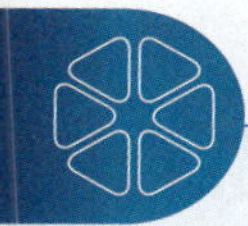

Homework 15.5

For extra help ▶ MyMathLab®  Watch the videos in MyMathLab  Download the MyDashboard App

Solve the system by graphing the equations by hand. Also, solve the system by substitution or elimination.

1. $x^2 + y^2 = 25$
$4x^2 + 25y^2 = 100$

2. $x^2 + y^2 = 4$
$4x^2 + 16y^2 = 64$

3. $y = x^2 + 1$
$y = -x + 3$

4. $y = x^2 - 3$
$y = 2x$

5. $y = x^2 - 2$
$y = -x^2 + 6$

6. $y = 2x^2 - 8$
$y = x^2 + 1$

7. $x^2 + y^2 = 49$
$x^2 + y^2 = 16$

8. $4x^2 + 9y^2 = 36$
$9x^2 + 25y^2 = 225$

9. $x^2 + y^2 = 25$
$y = -x - 1$

10. $x^2 + y^2 = 36$
$y = x + 6$

11. $y^2 - x^2 = 16$
$y + x^2 = 4$

12. $y^2 - 4x^2 = 4$
$4x^2 + y^2 = 4$

13. $25x^2 - 9y^2 = 225$
$4x^2 + 9y^2 = 36$

14. $y^2 - x^2 = 9$
$4x^2 + 100y^2 = 400$

15. $9x^2 + y^2 = 9$
$y = 3x + 3$

16. $x^2 - y^2 = 16$
$y = -x + 1$

17. $4x^2 + 9y^2 = 36$
$16x^2 + 25y^2 = 225$

18. $16x^2 + 9y^2 = 144$
$x^2 + 4y^2 = 4$

19. $y = \sqrt{x} - 3$
$y = -x - 1$

20. $y = \sqrt{x} + 1$
$y = -\sqrt{x} + 5$

21. $y = 2x^2 - 5$
$y = x^2 - 2$

22. $y = -3x^2 + 7$
$y = -x^2 + 3$

Solve the system by graphing the equations by hand. Also, solve the system by substitution or elimination. Round the coordinates of your solution(s) to the second decimal place.

23. $25y^2 - 4x^2 = 100$
$9x^2 + y^2 = 9$

24. $16x^2 - 4y^2 = 64$
$x^2 + 16y^2 = 16$

25. $25x^2 + 9y^2 = 225$
$x^2 + y^2 = 16$

26. $36x^2 + 4y^2 = 144$
$x^2 + y^2 = 9$

Solve the system by substitution or elimination. Check that any results satisfy both equations.

27. $9x^2 + y^2 = 85$
$2x^2 - 3y^2 = 6$

28. $x^2 - 6y = 34$
$x^2 + y^2 = 25$

29. $x^2 + 4y^2 = 25$
$y = -x + 5$

30. $x^2 + 9y^2 = 13$
$y = x - 1$

31. $y = x^2 - 3x + 2$
$y = 2x - 4$

32. $y = 2x^2 - 5x - 11$
$y = x^2 - 3x + 4$

Solve the system of three equations by graphing the equations by hand. Check that any results satisfy each equation.

33. $x^2 + y^2 = 25$
$4x^2 - 25y^2 = 100$
$4x^2 + 25y^2 = 100$

34. $x^2 + y^2 = 1$
$9x^2 + y^2 = 9$
$y = x + 1$

Concepts

35. Create a nonlinear system of two equations in two variables whose solutions are $(-4, 0)$ and $(4, 0)$.

36. Create a nonlinear system of two equations in two variables whose solutions are $(0, -3)$ and $(0, 3)$.

37. Consider the system

$$y = x^2$$
$$y = -x^2 + c$$

For what values of c does the system have

a. two solutions?
b. one solution?
c. no solutions?

38. Consider the system

$$y = ax^2$$
$$y = x^2 + 3$$

For what values of a does the system have

a. two solutions?
b. one solution?
c. no solutions?

39. Find values of c and d such that $(1, 4)$ is a solution of the system

$$\begin{aligned} 2x^2 + cy^2 &= 82 \\ y &= x^2 + dx + 5 \end{aligned}$$

40. Find values of c and d such that $(2, -5)$ is a solution of the system

$$\begin{aligned} y &= cx^2 - 4x^2 - 5 \\ y &= dx - 13 \end{aligned}$$

41. Explain how to solve a nonlinear system.

42. In your own words, describe a linear system and a nonlinear system. Also, compare the numbers of possible solutions for both types of systems.

Related Review

Solve the system by substitution or elimination.

43. $y = 2^x$

$y = 4\left(\frac{1}{2}\right)^x$

[**Hint:** $4 = 2^2$ and $\frac{1}{2} = 2^{-1}$]

44. $y = \log_2(x + 1) + 2$

$y = \log_2(3x + 13) - 1$

Expressions, Equations, Functions, and Graphs

For each description that follows, give an example. Then solve, simplify, or graph, as appropriate.

45. linear function

46. quotient of two radical expressions

47. rational equation in one variable

48. system of two linear equations in two variables

49. exponential function

50. square root function

51. quadratic function

52. difference of two rational expressions

53. quadratic equation in one variable

54. logarithmic equation in one variable

Section 15.5 Quiz

For Exercises 1–4, solve the system.

1. $9x^2 + y^2 = 81$

$x^2 + y^2 = 9$

2. $y = x^2 - 2$

$y = -2x + 1$

3. $y = x^2 + 3$

$y = x^2 - 6x + 9$

4. $25x^2 - 4y^2 = 100$

$9x^2 + y^2 = 9$

5. Solve the system of three equations by graphing the equations by hand:

$$\begin{aligned} x^2 - y^2 &= 16 \\ x^2 + y^2 &= 16 \\ y &= (x + 4)^2 \end{aligned}$$

6. Create a nonlinear system of two equations whose solution is $(0, 5)$.

Answers to Odd-Numbered Exercises

Answers to most discussion exercises and to exercises in which answers may vary have been omitted.

Chapter 1

Homework 1.1 **1.** 25 thousand fans attended the concert. **3.** In 2011, 274 million Americans had cell phones. **5.** In 2010, 14.8 million iPads were sold. **7.** The company lost \$45 thousand that year. **9.** The statement $t = 9$ represents the year 2019. **11.** The statement $t = -3$ represents the year 2002. **13.** h; 67, 72; -5, 0; answers may vary. **15.** p; 50, 60; -2, -8; answers may vary. **17.** T; 15, 40; 240, -10; answers may vary. **19.** s; 25, 32; -15, -9; answers may vary.

21. a. The rectangles are not drawn to scale. Answers may vary. 4 inches by 6 inches; 3 inches by 8 inches; 1 inch by 24 inches **b.** W, L **c.** A

23. a. The rectangles are not drawn to scale. Answers may vary. 5 feet by 5 feet; 3 feet by 7 feet; 1 foot by 9 feet **b.** W, L **c.** P

25. a. The rectangles are not drawn to scale. Answers may vary. 1 inch by 4 inches; 2 inches by 5 inches; 3 inches by 6 inches **b.** W, L, A **c.** None

27. a. The rectangles are not drawn to scale. Answers may vary. 2 yards by 2 yards; 2 yards by 3 yards; 2 yards by 4 yards **b.** L, P **c.** W

29. **31.** **33.**

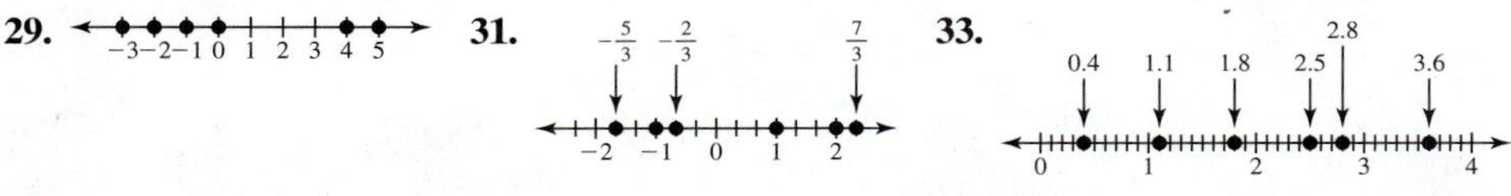

35. **37.** **39.**

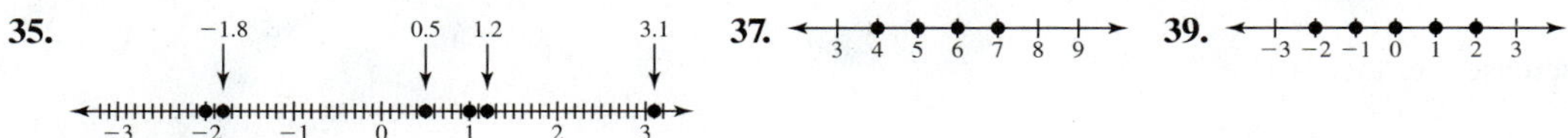

41. **43.** **45.** 3, 356 **47.** -4 **49.** $\sqrt{7}, \pi$ **51.** $-2, -5, -7$; answers may vary. **53.** $-8, -9, -27$; answers may vary. **55.** $-2, -5, -40$; answers may vary. **57.** $\frac{5}{4}, \frac{3}{2}, \frac{7}{4}$; answers may vary. **59.** $-2.1, -2.3, -2.8$; answers may vary. **61.** **63.** **65.**

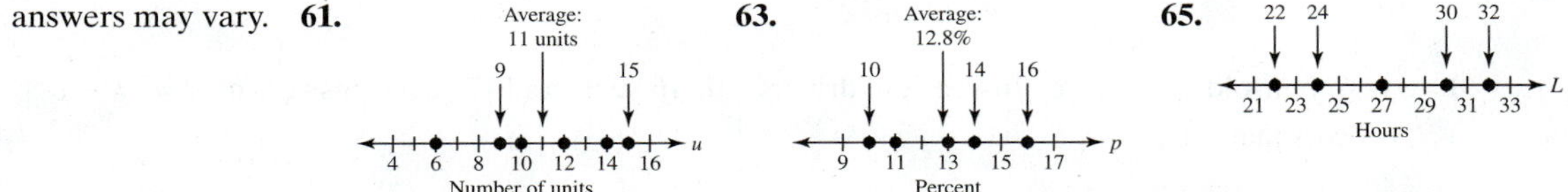

67. **69. a.**

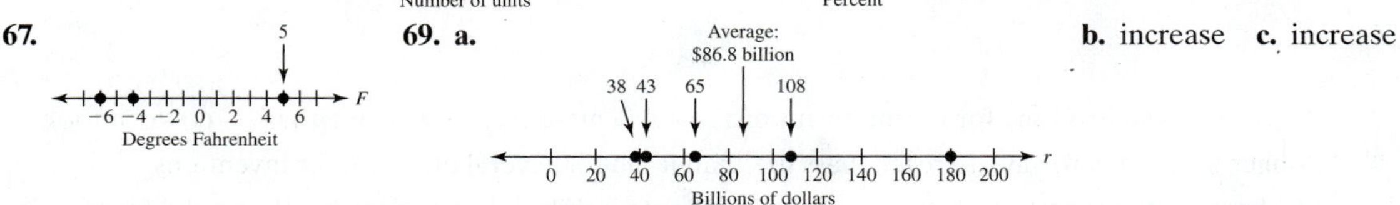

b. increase **c.** increase

71. a.

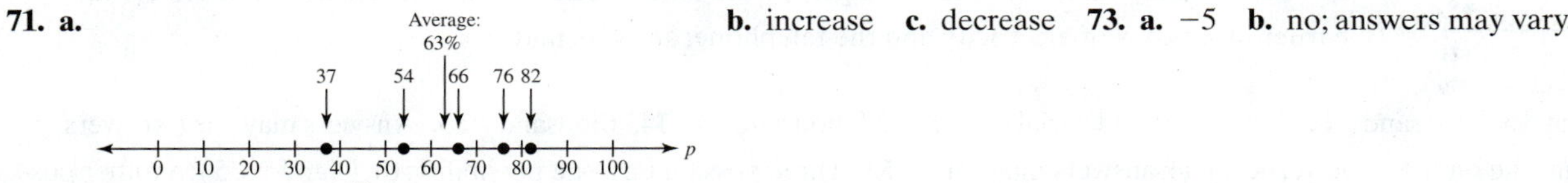

b. increase **c.** decrease **73. a.** -5 **b.** no; answers may vary.

75. a. i. 8 **ii.** 3 **iii.** 5 **b.** Answers may vary. **c.** infinitely many **77.** Answers may vary. **79.** Answers may vary.

Homework 1.2 **1–15 odd.**

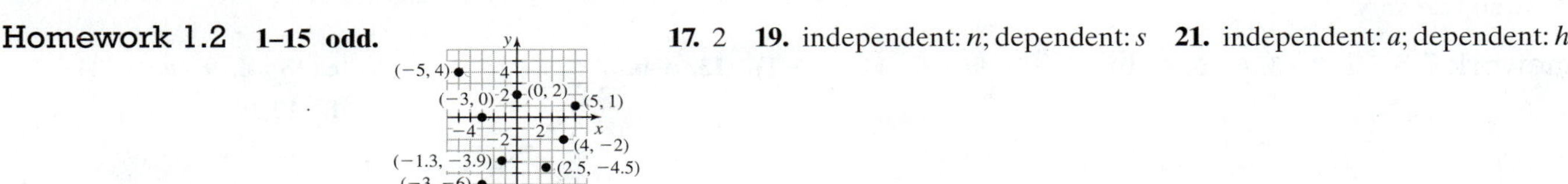

17. 2 **19.** independent: n; dependent: s **21.** independent: a; dependent: h

23. independent: c; dependent: T **25.** independent: A; dependent: n **27.** independent: t; dependent: h **29.** A telemarketer who works 32 hours per week will sell an average of 43 magazine subscriptions per week. **31.** 38% of Americans at age 21 years say that they volunteer. **33.** \$698 billion was spent on defense in 2010. **35.** In 2009, 33% of Americans believed travel websites did a good job of presenting travel choices.

37.

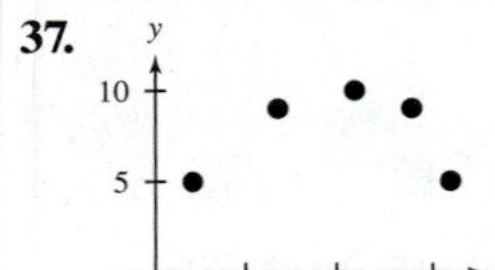

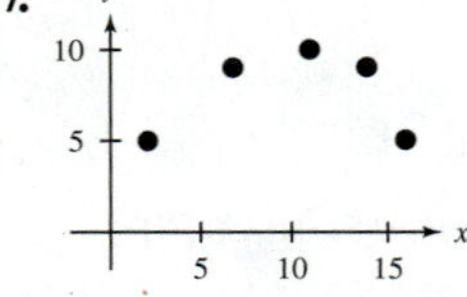

39. $A(-4,-3), B(-5,0), C(-2,4), D(1,3), E(0,-2), F(5,-4)$

41. a.

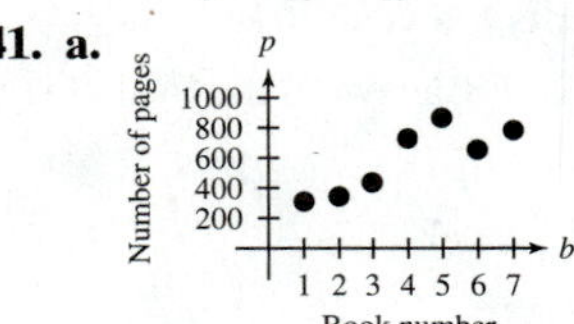

b. The fifth book **c.** The third book to the fourth book; answers may vary.

43. a.

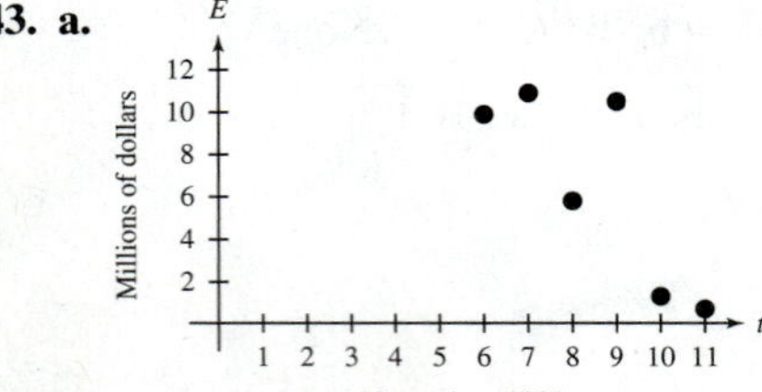

b. 2011; \$0.7 million **c.** 2007; \$10.9 million **d.** no; answers may vary.

45. a.

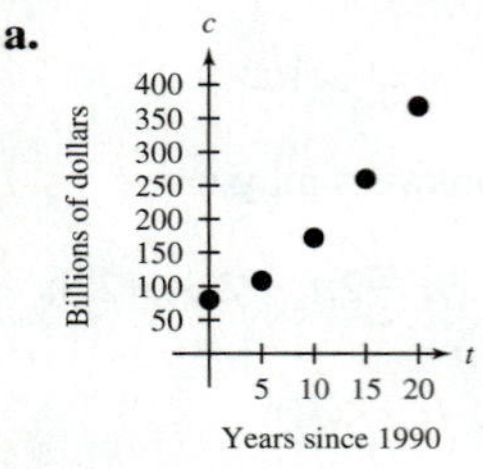

b. increase **c.** increase

47. a.

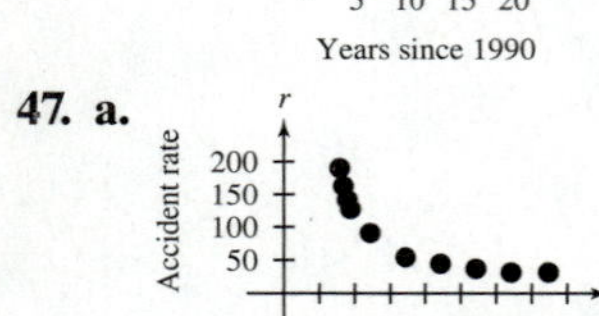

b. 60–69-year-old drivers **c.** 16-year-old drivers **d.** 16 years and 17 years; answers may vary. **e.** Answers may vary.

49. a.

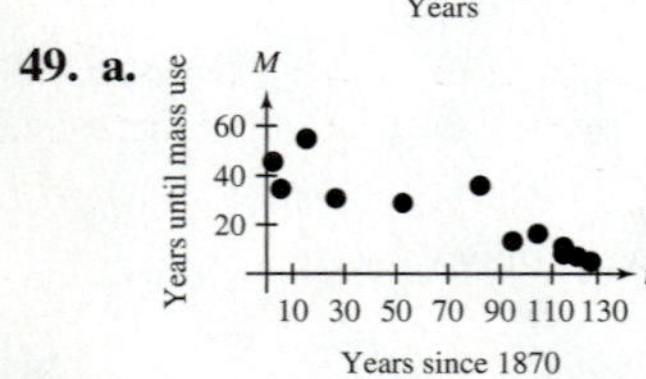

b. It has taken less time for recent inventions to reach mass use; answers may vary. **c.** No; it took longer for the microwave to reach mass use than it did for several other earlier inventions. **d.** Answers may vary. **e.** It took longer for the automobile to reach mass use than it did for the earlier inventions of electricity and the telephone; answers may vary.

51. a. engineering; \$62 thousand **b.** humanities and social science; \$37 thousand **c.** \$43 thousand **53.** Answers may vary; answers may vary; the points lie on the same vertical line; answers may vary. **55.** Three possibilities; one possibility: $(6,1)$ and $(6,5)$; another possibility: $(-2,1)$ and $(-2,5)$ **57. a.** x-coordinate: positive; y-coordinate: positive **b.** x-coordinate: negative; y-coordinate: positive **c.** x-coordinate: negative; y-coordinate: negative **d.** x-coordinate: positive; y-coordinate: negative **59.** The points on the x-axis **61.** Answers may vary.

Homework 1.3

1. 2 **3.** 6 **5.** $(2,0)$ **7.** -2 **9.** -6 **11.** $(0,-1)$ **13. a–b.**

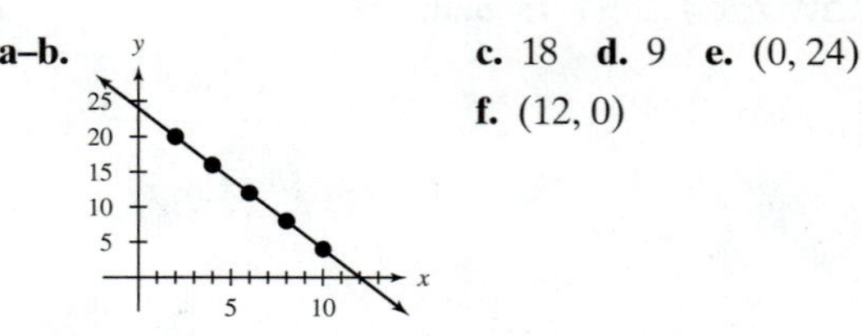

c. 18 **d.** 9 **e.** $(0,24)$ **f.** $(12,0)$

15. a. 18 thousand gallons **b.** 4.2 hours **c.** 30 thousand gallons **d.** 5 hours **17.** No; answers may vary.

19. a. 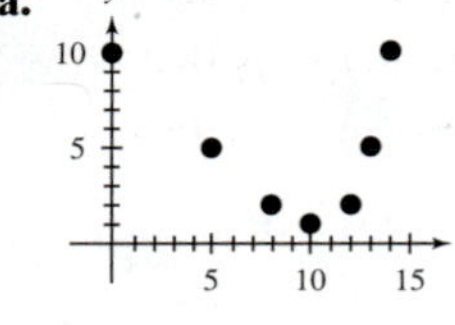 **b.** No; answers may vary. **21. a.** 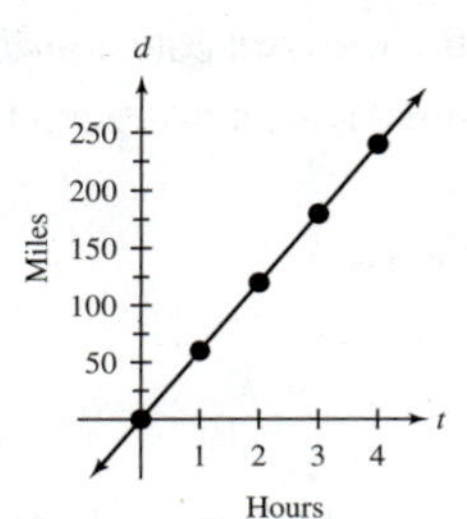 **b.** 150 miles **c.** 3.5 hours

23. a. 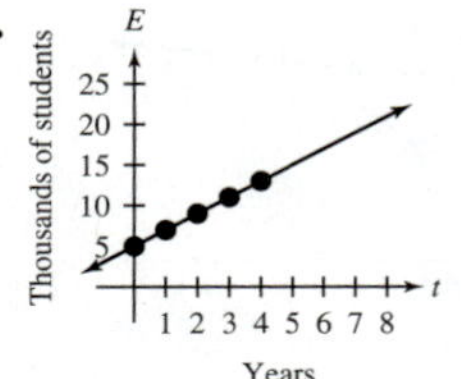 **b.** 17 thousand students **c.** 7 years **25. a.** 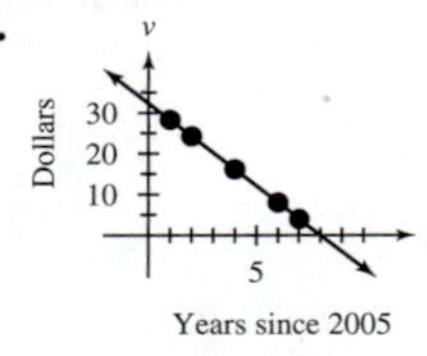 **b.** 2010 **c.** $(8, 0)$; the stock will have no value in 2013. **d.** $(0, 32)$; the value of the stock was \$32 in 2005.

27. a. 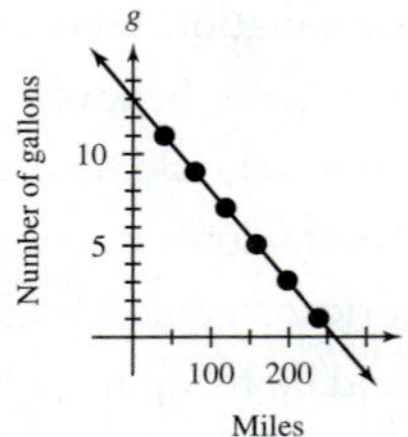 **b.** 6 gallons **c.** 220 miles **d.** $(260, 0)$; the gasoline tank will be empty after 260 miles of driving (if no refueling takes place). **e.** $(0, 13)$; the car has a 13-gallon gasoline tank.

29. a. **b.** \$38 million **c.** 2007 **d.** $(0, 8)$; the revenue was \$8 million in 2005. **31. a.** 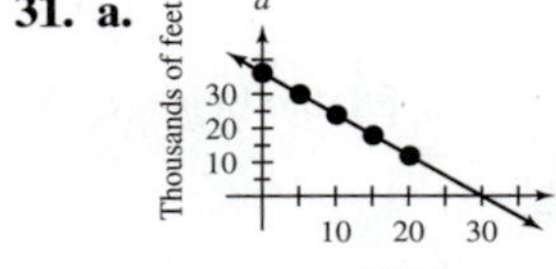 **b.** 22 thousand feet **c.** 30 minutes **d.** underestimate; answers may vary.

33. a. 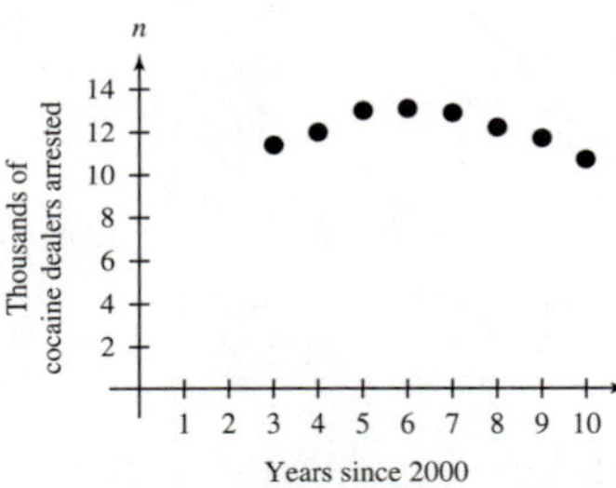 **b.** No; answers may vary. **35.** No; the y-coordinate (not the y-intercept) of $(2, 5)$ is 5.

37. No; the y-coordinate of an x-intercept must be 0. **39.** No; an x-intercept is a point that corresponds to an ordered pair with two coordinates (not a single number). **41.** Answers may vary. **43. a. i.** Answers may vary. **ii.** Answers may vary. **iii.** Answers may vary. **b.** One; answers may vary. **c.** One; answers may vary. **45.** Answers may vary. **47.** Answers may vary.

Homework 1.4

Throughout this section, answers may vary.

1. a. and c. 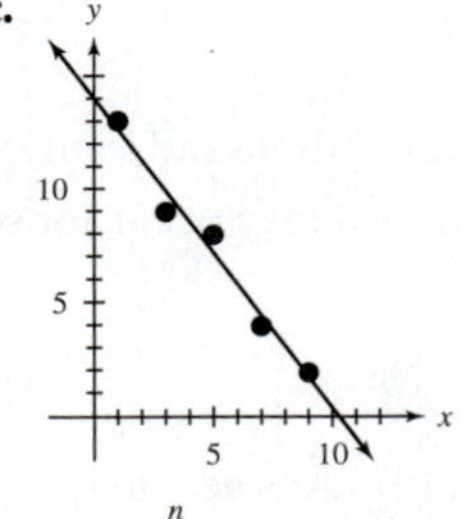 **b.** approximately linearly related **d.** $(8, 3.2)$ **e.** $(5.9, 6)$ **f.** $(0, 14)$ **g.** $(10.3, 0)$

3. a. and c. 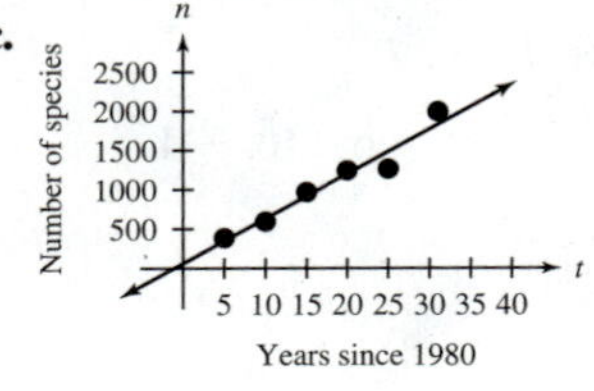 **b.** approximately linearly related **d.** 1993 **e.** 2228 species

5. a–b.

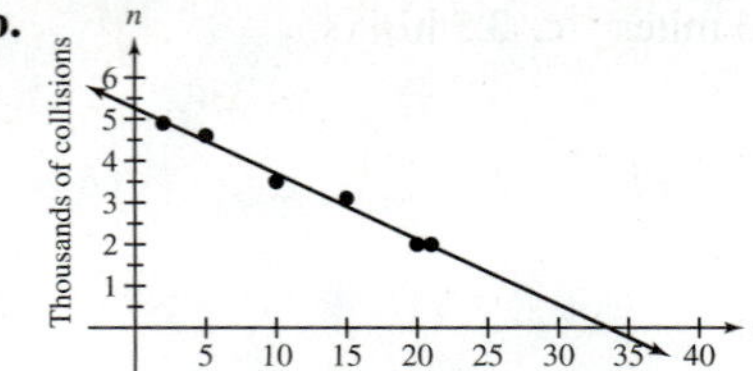

c. 2.6 thousand collisions; interpolation **d.** 2017; extrapolation **e.** (0, 5.3); there were 5.3 thousand collisions in 1990.

f. (34, 0); there will be no collisions in 2024; model breakdown has likely occurred. **7. a–b.**

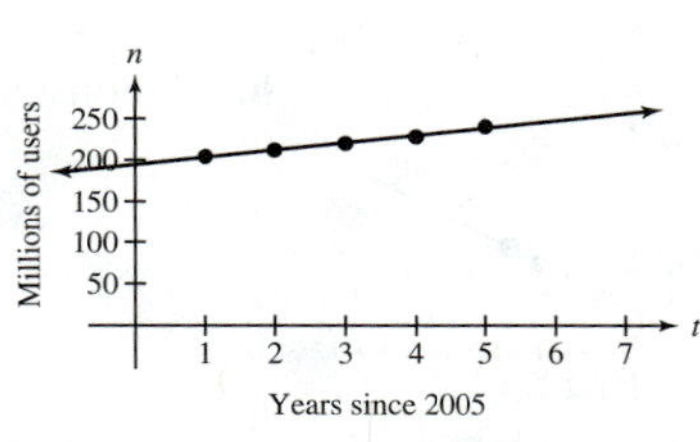

c. (0, 194); there were 194 million Internet users in 2005; extrapolation. **d.** 9 million users per year **e.** 2019; extrapolation; model breakdown has likely occurred. **9. a–b.**

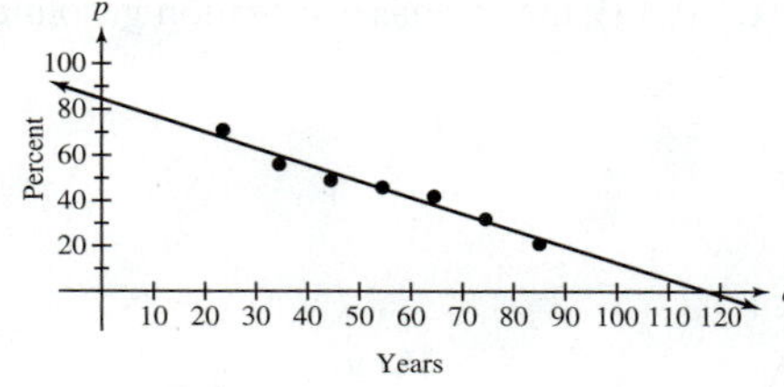

c. 27 years **d.** (0, 85); 85% of newborns believe marriages between same-sex couples should be recognized by the law as valid; model breakdown has occurred. **e.** (117, 0); no 117-year-old Americans believe marriages between same-sex couples should be recognized by law as valid; model breakdown has occurred.

11. a. and b.

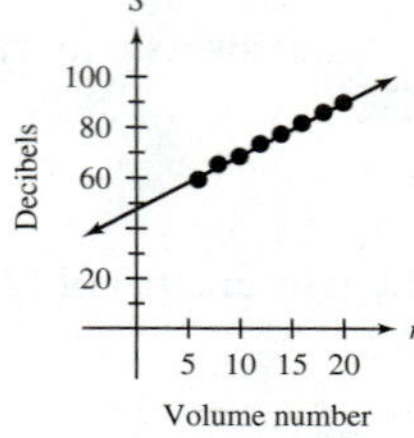

c. 88 decibels **d.** Volume number 15 **13. a–b.**

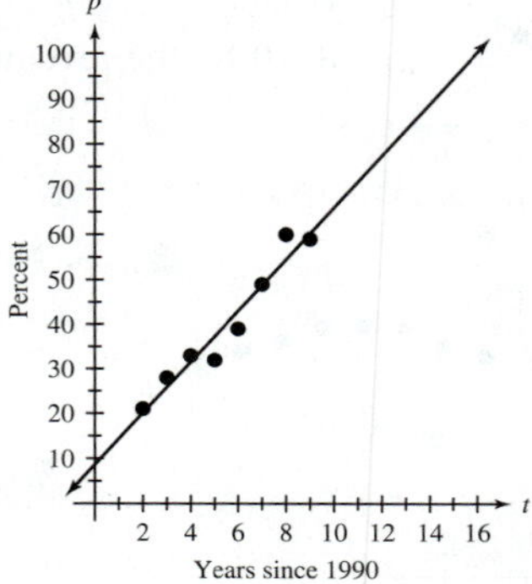

c. 100% **d.**

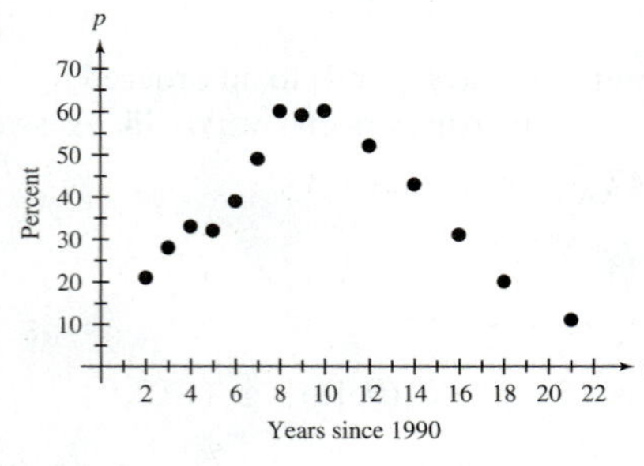

e. 69 percentage points; answers may vary.

15. a. (2, 0); the profit was 0 dollars in 2004.
b. (0, −68); the profit was −68 million dollars in 2002.

17. a. 2.0 thousand injuries **b.** 1.6 thousand injuries **c.** overestimate; the line is above the data point; 0.4 thousand injuries **19.** overestimate **21.** Answers may vary. **23.** Answers may vary. **25.** Not necessarily; model breakdown may occur for some points on the line. **27.** interpolation; answers may vary.

Chapter 1 Review Exercises

1. The total box office gross was $10.58 billion in 2010. **2.** 2016 **3.** *p*; 60, 70 (Answers may vary.); −12, 107 (Answers may vary.) **4. a.** The rectangles are not drawn to scale. Answers may vary. 10 inches by 10 inches; 7 inches by 13 inches; 2 inches by 18 inches **b.** *W*, *L* **c.** *P*

5. $-\frac{3}{2}$, $\frac{5}{2}$; −3 −2 −1 0 1 2 3 **6.** −6 −5 −4 −3 −2 −1 0 **7.** −5 −4 −3 −2 −1 0 1 2 3 4 5 *p*; Millions of dollars **8.** *y*; 4 2 −2 −4; 2 4 *x* **9.** −6 **10.** −4

11. independent: *a*; dependent: *p* **12.** independent: *t*; dependent: *a* **13.** There were 413 U.S. billionaires in 2011.
14. In 2010, the revenue from ADHD drugs was $7 billion. **15.**

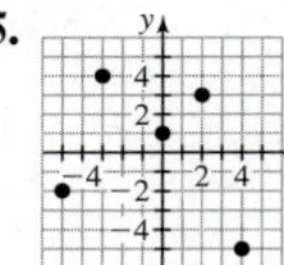

16. a. **b.** 2009 **c.** 1970

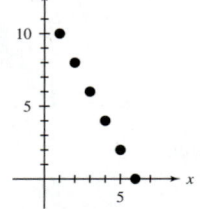

17. a. France; 78% **b.** India; 4% **c.** 40% **18.** -1 **19.** -5 **20.** 4 **21.** -6 **22.** $(0, -2)$ **23.** $(-4, 0)$
24. a. and b.

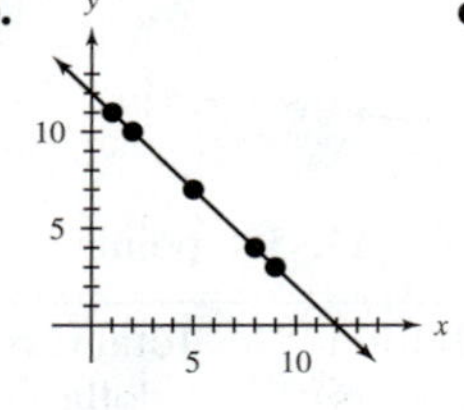

c. 1 **d.** 7 **e.** $(12, 0)$ **f.** $(0, 12)$ **25. a.** **b.** linearly related

26. a. **b.** \$2 million **c.** 2007 **d.** $(0, 22)$; in 2005, the profit was \$22 million. **e.** $(11, 0)$; in 2016, the profit will be 0 dollars.

27. 0 **28. a. and c.** **b.** approximately linearly related **d.** $(5, 13.5)$ **e.** $(2, 20)$ **f.** $(0, 24.3)$ **g.** $(11.2, 0)$

29. a. and b. **c.** 39% **d.** 2017 **30. a. and b.**

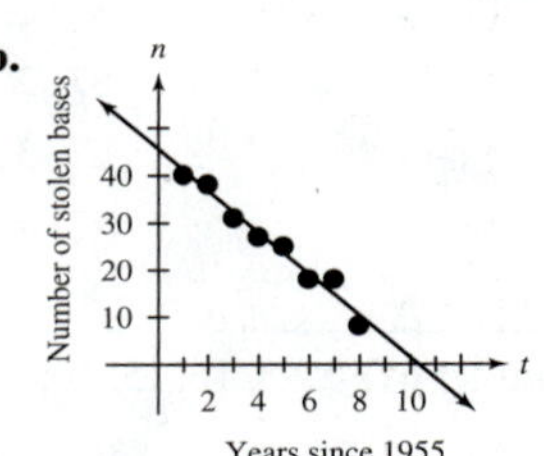

c. $(0, 45.2)$; Mays stole 45 bases in 1955, according to the model. **d.** $(10.4, 0)$; Mays did not steal any bases in 1965, according to the model. **e.** overestimate; yes; answers may vary. **f.** underestimate; yes; answers may vary.

Chapter 1 Test

1. a. The rectangles are not drawn to scale. Answers may vary. 6 feet 6 feet 4 feet 9 feet 2 feet 18 feet **b.** W, L **c.** A

2.

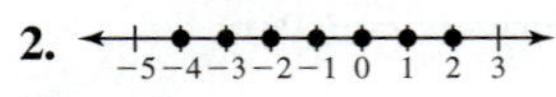

3.

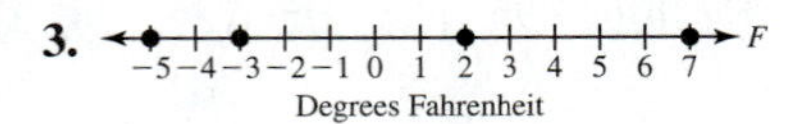

4.

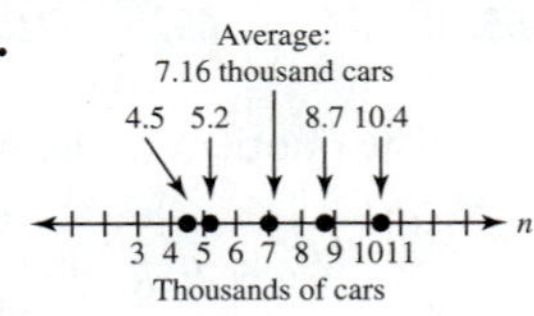

5. *c*; answers may vary. **6.** In 2012, Alex Rodriguez's salary was \$30 million.
7. a. **b.** $(21, 50)$; Americans in the age group 18–24 are the most likely to be without health insurance. **c.** $(59.5, 17)$; Americans in the age group 55–64 are the least likely to be without health insurance.

8. -3 **9.** 4 **10.** $(0, -1)$ **11.** $(2, 0)$ **12. a.**

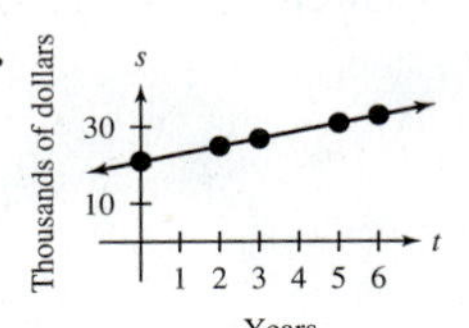

b. \$29 thousand **c.** 7 years **d.** $(0, 21)$; when the person was hired, her salary was \$21 thousand.

13. Answers may vary. **14. a. and b.**

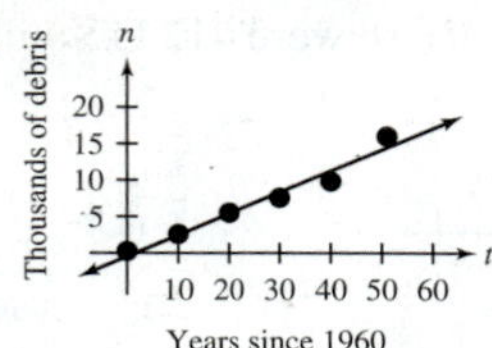

c. 1993 **d.** 17.1 thousand debris

15. overestimate; answers may vary.

Chapter 2

Homework 2.1 **1.** 8 **3.** 3 **5.** 42 **7.** 2 **9.** 12 **11.** 36 **13.** 36; the total cost of 4 albums is \$36. **15.** 500 points

17. a.

Number of Shares	Total Value (dollars)
1	$5 \cdot 1$
2	$5 \cdot 2$
3	$5 \cdot 3$
4	$5 \cdot 4$
n	$5n$

$5n$ **b.** 35; the total value of 7 shares is \$35. **19. a.**

Tuition (dollars)	Total Cost (dollars)
400	400 + 12
401	401 + 12
402	402 + 12
403	403 + 12
t	$t + 12$

$t + 12$

b. 429; if the tuition is \$417, then the total cost is \$429.

21. a.

Number of Hours of Courses	Total Cost (dollars)
1	$87 \cdot 1$
2	$87 \cdot 2$
3	$87 \cdot 3$
4	$87 \cdot 4$
n	$87n$

$87n$ **b.** 1305; the total cost for 15 hours of classes is \$1305. **23.** $x + 4$; 12 **25.** $x \div 2$; 4 **27.** $x - 5$; 3 **29.** $7x$; 56 **31.** $16 \div x$; 2 **33.** The number divided by 2 **35.** 7 minus the number **37.** The number plus 5 **39.** The product of 9 and the number **41.** The difference of the number and 7 **43.** The number times 2 **45.** 9 **47.** 3 **49.** 18 **51.** xy; 27 **53.** $x - y$; 6 **55.** 186; the car traveled 186 miles when driven for 3 hours at 62 mph. **57.** 20; if a car can travel 240 miles on 12 gallons of gasoline, the car's gas mileage is 20 miles per gallon.

59. \$170 thousand **61. a.** 5, 10, 15, 20; the person earns \$5, \$10, \$15, and \$20 for working 1, 2, 3, and 4 hours, respectively. **b.** \$5 per hour **c.** Answers may vary. **63. a.** 50, 100, 150, 200; the person drives 50, 100, 150, and 200 miles in 1, 2, 3, and 4 hours, respectively. **b.** 50 mph **c.** Answers may vary. **65.** Answers may vary. **67.** Answers may vary. **69.** Answers may vary.

Homework 2.2 **1.** 7 **3.** $2 \cdot 2 \cdot 5$ **5.** $2 \cdot 2 \cdot 3 \cdot 3$ **7.** $3 \cdot 3 \cdot 5$ **9.** $2 \cdot 3 \cdot 13$ **11.** $\frac{3}{4}$ **13.** $\frac{1}{4}$ **15.** $\frac{3}{5}$ **17.** $\frac{2}{5}$ **19.** $\frac{1}{5}$ **21.** $\frac{5}{6}$ **23.** $\frac{2}{15}$ **25.** $\frac{3}{10}$ **27.** $\frac{5}{3}$ **29.** $\frac{5}{6}$ **31.** $\frac{2}{3}$ **33.** $\frac{2}{15}$ **35.** $\frac{5}{7}$ **37.** $\frac{3}{4}$ **39.** $\frac{1}{5}$ **41.** $\frac{1}{3}$ **43.** $\frac{3}{4}$ **45.** $\frac{19}{12}$ **47.** $\frac{14}{3}$ **49.** $\frac{1}{9}$ **51.** $\frac{17}{63}$ **53.** $\frac{11}{5}$ **55.** 1 **57.** 599 **59.** Undefined **61.** 0 **63.** 1 **65.** 0 **67.** $\frac{1}{3}$ **69.** $\frac{9}{5}$ **71.** $\frac{1}{3}$ **73.** 0.17 **75.** 1.33 **77.** 0.43 **79.** Answers may vary. **81.** $\frac{1}{10}$ square mile **83.** $\frac{1}{4}$ of the course points **85.** The quotient of the number and 3

87.

Number of People	Cost per Person (dollars)
2	$\frac{19}{2}$
3	$\frac{19}{3}$
4	$\frac{19}{4}$
5	$\frac{19}{5}$
n	$\frac{19}{n}$

$\frac{19}{n}$ **89. a. i.** $\frac{5}{9}$ **ii.** $\frac{5}{4}$ **iii.** $\frac{3}{2}$ **iv.** $\frac{1}{6}$ **b.** Answers may vary. **91.** Answers may vary. **93.** Answers may vary. **95. a. and b.**

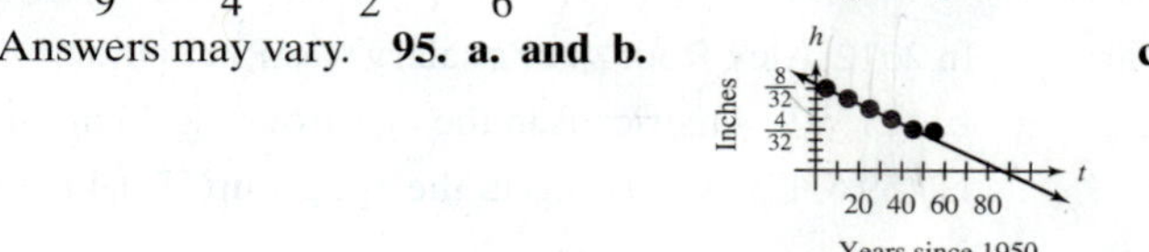

c. 0.04 inch

d. 2016 **e.** (87, 0); there will be no grass on putting surfaces in 2037. This prediction is highly unlikely. **97. a. i.** −6 **ii.** −8 **iii.** −7 **b.** The results are all negative. **c.** −9 **d.** Answers may vary. **99.** Answers may vary.

Homework 2.3 **1.** 4 **3.** −7 **5.** 3 **7.** 8 **9.** −4 **11.** −7 **13.** −5 **15.** −5 **17.** 2 **19.** −3 **21.** −10 **23.** −3 **25.** 0 **27.** 0 **29.** −13 **31.** −22 **33.** −1145 **35.** 0 **37.** −6.7 **39.** −4.8 **41.** −97.3 **43.** $\frac{2}{7}$ **45.** $-\frac{1}{4}$ **47.** $-\frac{3}{4}$ **49.** $\frac{7}{12}$ **51.** 6221.4 **53.** −97,571.14 **55.** −0.11 **57.** −1 **59.** −6 **61.** $x + 2$; −4 **63.** $-5 + x$; −11 **65.** $175

67.

Check No.	Date	Description of Transaction	Payment	Deposit	Balance
					−89.00
	7/18	Transfer		300.00	211.00
3021	7/22	State Farm	91.22		119.78
3022	7/22	MCI	44.26		75.52
	7/31	Paycheck		870.00	945.52

69. −2871 dollars **71.** −1633 dollars **73.** 4°F

75. a.

Weight before Diet (pounds)	Weight after Diet (pounds)
160	160 + (−20)
165	165 + (−20)
170	170 + (−20)
175	175 + (−20)
B	$B + (-20)$

$B + (-20)$ **b.** 149; the person's current weight is 149 pounds.

77. a.

Deposit (dollars)	New Balance (dollars)
50	−80 + 50
100	−80 + 100
150	−80 + 150
200	−80 + 200
d	$-80 + d$

$-80 + d$ **b.** 45; the new balance is $45. **79.** negative **81.** The numbers are equal in absolute value and opposite in sign, or the numbers are both 0. **83. a.** 3 **b.** 4 **c.** 6 **d.** no; answers may vary.

Homework 2.4 **1.** −2 **3.** −6 **5.** 9 **7.** −1 **9.** −3 **11.** 11 **13.** −6 **15.** −79 **17.** 420 **19.** −5.4 **21.** −11.3 **23.** 5.7 **25.** 15.98 **27.** −1 **29.** $\frac{1}{2}$ **31.** $\frac{3}{4}$ **33.** $-\frac{13}{24}$ **35.** 2 **37.** −2 **39.** $-\frac{1}{4}$ **41.** −2.7 **43.** −7 **45.** −2 **47.** −3128.17 **49.** 112,927.91 **51.** −0.95 **53.** −12°F **55.** 11°F **57. a.** −12°F **b.** −6°F **c.** Answers may vary. **59.** 20,602 feet **61. a.** 0.7 percentage point, −2.5 percentage points, 4.5 percentage points, −7.7 percentage points, 0.5 percentage point, 9.1 percentage points, −0.2 percentage point, −6.1 percentage points **b.** 9.1 percentage points **c.** no **63. a.** 140 thousand cars **b.** From 2003 to 2005, from 2006 to 2007, from 2009 to 2010 **c.** From 2005 to 2006, from 2007 to 2009

65. a.

Score on the Second Exam (points)	Change in Score (points)
80	80 − 87
85	85 − 87
90	90 − 87
95	95 − 87
p	$p - 87$

$p - 87$ **b.** −6; the score decreased by 6 points.

67. a.

Change in Enrollment	Current Enrollment
100	100 + 24,500
200	200 + 24,500
300	300 + 24,500
400	400 + 24,500
c	$c + 24{,}500$

$c + 24{,}500$ **b.** 23,800; the current enrollment is 23,800 students due to a decrease in enrollment of 700 students in the past year. **69.** −3 **71.** −7 **73.** 9 **75.** $-3 - x$; 2 **77.** $x - 8$; −13 **79.** $x - (-2)$; −3 **81.** Answers may vary. **83. a. i.** 2 **ii.** 8 **iii.** 5 **b.** Answers may vary. **85. a. i.** −10 **ii.** −24 **iii.** −63 **b.** Answers may vary. **c.** −21 **d.** Answers may vary.

87. a. 3 **b.** −3 **c.** They are equal in absolute value and opposite in sign. **d.** −6, 6; they are equal in absolute value and opposite in sign. **e.** Answers may vary. **f.** They are equal in absolute value and opposite in sign. **89.** Positive

Homework 2.5 **1.** 0.63 **3.** 0.09 **5.** 8% **7.** 0.073 **9.** 5.2% **11.** \$2.80 **13.** 125 students **15.** 175 cars **17.** −12 **19.** 18 **21.** −1 **23.** −8 **25.** −5 **27.** 8 **29.** 555 **31.** −39 **33.** 0.08 **35.** −0.975 **37.** −0.3 **39.** −9 **41.** 4 **43.** $-\frac{1}{10}$ **45.** $\frac{1}{15}$ **47.** $-\frac{9}{14}$ **49.** $\frac{15}{14}$ **51.** −3 **53.** 13 **55.** 6 **57.** −100 **59.** $-\frac{1}{4}$ **61.** $\frac{4}{3}$ **63.** $-\frac{11}{12}$ **65.** $-\frac{9}{20}$ **67.** $-\frac{4}{5}$ **69.** $\frac{3}{4}$ **71.** $-\frac{1}{2}$ **73.** 1 **75.** $-\frac{1}{24}$ **77.** 10,252.84 **79.** −6.78 **81.** 0.48 **83.** −8.07 **85.** −24 **87.** $-\frac{3}{2}$ **89.** −48 **91.** $\frac{1}{2}$ **93.** $\frac{w}{2}$; −4 **95.** $w(-5)$; 40 **97.** $\frac{3}{4}$ **99.** $\frac{2.25}{1}$; the Freedom Tower is 2.25 times taller than the John Hancock Tower. **101.** $\frac{1.58}{1}$; the number of U.S. billionaires in 2011 is 1.58 times the number of U.S. billionaires in 2001. **103. a.** $\frac{0.8 \text{ red bell pepper}}{1 \text{ black olive}}$; for each black olive used, 0.8 bell pepper is needed. **b.** $\frac{1.25 \text{ black olives}}{1 \text{ red bell pepper}}$; for each red bell pepper used, 1.25 olives are required. **105. a.** $\frac{12.77}{1}$; the FTE enrollment at Texas A&M University is 12.77 times larger than that at St. Olaf College. **b.** $\frac{2.93}{1}$; the number of FTE faculty at University of Massachusetts Amherst is 2.93 times greater than that at Butler University. **c.** Butler University: $\frac{12.44}{1}$; St. Olaf College: $\frac{11.87}{1}$; Stonehill College: $\frac{13.21}{1}$; University of Massachusetts Amherst: $\frac{17.32}{1}$; Texas A&M University: $\frac{21.1}{1}$ **d.** Texas A&M University; St. Olaf College **e.** Answers may vary. **107. a.** $\frac{2.39}{1}$ **b.** For each \$1 the person pays to her MasterCard account, she should pay about \$2.39 to her Discover account. **109.** −3162 dollars **111.** −29.52 dollars **113. a.** −6 **b.** 8 **c.** A negative number times a negative number is equal to a positive number. **d.** Answers may vary. **115.** $\frac{a}{b} = \frac{-a}{-b}, \frac{-a}{b} = \frac{a}{-b} = -\frac{a}{b} = -\frac{-a}{-b}$ **117.** Answers may vary. **119.** One number is positive and one number is negative. **121.** a or b is zero. **123. a.** 16 **b.** 1 **c.** yes; answers may vary. **125. a.** Answers may vary. **b.** Answers may vary. **c.** 0

Homework 2.6 **1.** 64 **3.** 32 **5.** −64 **7.** 64 **9.** $\frac{36}{49}$ **11.** 12 **13.** −18 **15.** 10 **17.** −3 **19.** $-\frac{5}{2}$ **21.** $-\frac{2}{3}$ **23.** 10 **25.** −17 **27.** −50 **29.** −10 **31.** −14 **33.** 20 **35.** 15 **37.** −27 **39.** −57 **41.** $\frac{1}{2}$ **43.** 27 **45.** −48 **47.** 1 **49.** −17 **51.** 5 **53.** 2 **55.** −41 **57.** 3 **59.** $-\frac{9}{7}$ **61.** −9 **63.** 48 **65.** −613.37 **67.** −1.54 **69.** −1.33 **71.** −14 **73.** −8 **75.** −5 **77.** 40 **79.** $\frac{5}{4}$ **81.** $-\frac{13}{7}$ **83.** $-\frac{5}{2}$ **85.** $\frac{3}{5}$ **87.** −27 **89.** −12 **91.** 32 **93.** $5 + (-6)x$; 29 **95.** $\frac{x}{-2} - 3$; −1

97. a.

Years since 1975	Congressional Pay (thousands of dollars)
0	$3.8 \cdot 0 + 44.6$
1	$3.8 \cdot 1 + 44.6$
2	$3.8 \cdot 2 + 44.6$
3	$3.8 \cdot 3 + 44.6$
4	$3.8 \cdot 4 + 44.6$
t	$3.8t + 44.6$

$3.8t + 44.6$ **b.** 158.6; congressional pay was about \$158.6 thousand in 2005. **c.** \$175.5 thousand

99. a.

Years since 1980	Population (thousands)
0	$-2.2 \cdot 0 + 145$
1	$-2.2 \cdot 1 + 145$
2	$-2.2 \cdot 2 + 145$
3	$-2.2 \cdot 3 + 145$
4	$-2.2 \cdot 4 + 145$
t	$-2.2 \cdot t + 145$

$-2.2t + 145$ **b.** 63.6; Gary's population will be 63.6 thousand in 2017. **101.** 188 billion messages **103.** \$200 million **105.** 4254 shops **107.** \$4.3 billion **109.** 8 cubic feet **111.** Answers may vary; 25. **113.** no; answers may vary. **115. a.** −2 **b.** 3 **c.** Answers may vary; −2. **117. a.** 24 **b.** 24 **c.** The results are equal. **d.** −40; −40; the results are equal. **e.** Answers may vary. **f.** yes **g.** Answers may vary. **119. a.** 4; 1; 0; 1; 4 **b.** nonnegative **c.** nonnegative **121. a.** Answers may vary. **b.** Answers may vary. **c.** 0, 1

Chapter 2 Review Exercises

1. 6 **2.** −12 **3.** −3 **4.** 10 **5.** −16 **6.** −4 **7.** −3 **8.** 12 **9.** −1 **10.** $-\frac{3}{2}$ **11.** −11 **12.** −13 **13.** −16 **14.** −4 **15.** −20 **16.** −26 **17.** −12 **18.** 4 **19.** 14 **20.** 4 **21.** 10.9 **22.** $-\frac{2}{15}$ **23.** $\frac{5}{6}$ **24.** $\frac{7}{9}$ **25.** $\frac{1}{24}$ **26.** $-\frac{1}{6}$ **27.** 64 **28.** −64 **29.** 16 **30.** $\frac{27}{64}$ **31.** −54 **32.** 3 **33.** 0 **34.** $\frac{2}{3}$ **35.** $-\frac{8}{11}$ **36.** −19 **37.** −3 **38.** 58 **39.** $\frac{3}{4}$ **40.** $-\frac{4}{5}$ **41.** −8.68 **42.** 0.62 **43.** $2\frac{1}{6}$ yards **44.** \$4095.49 **45.** −4700 feet **46. a.** −12°F **b.** −4°F **c.** Answers may vary. **47. a.** \$5.6 million **b.** −39 million dollars **c.** from 2004 to 2008; \$239 million **d.** from 2000 to 2004; \$167.7 million **48.** 1.40; the number of messages sent or received per day in 2011 is 1.40 times larger than the number of messages sent or received per day in 2009. **49.** 0.75 **50.** 0.029 **51.** \$37.41 **52.** 74 students **53.** −4394.4 dollars **54.** −10 **55.** 57 **56.** −2 **57.** $-\frac{11}{4}$ **58.** 55 **59.** $-\frac{1}{2}$ **60.** $x + 5$; 2 **61.** $-7 - x$; −4 **62.** $2 - x(4)$; 14 **63.** $1 + \frac{-24}{x}$; 9 **64.** 50; if the total cost is \$650 and there are 13 players on the team, the cost is \$50 per player.

65. a.

Time (hours)	Volume of Water (cubic feet)
0	$-50 \cdot 0 + 400$
1	$-50 \cdot 1 + 400$
2	$-50 \cdot 2 + 400$
3	$-50 \cdot 3 + 400$
4	$-50 \cdot 4 + 400$
t	$-50t + 400$

$-50t + 400$ **b.** 50; there will be 50 cubic feet of water in the basement after 7 hours of pumping. **66.** 10.8 million people

Chapter 2 Test

1. −13 **2.** 63 **3.** −6 **4.** −8 **5.** $\frac{1}{2}$ **6.** 3 **7.** −25 **8.** −0.08 **9.** $-\frac{45}{4}$ **10.** $\frac{13}{40}$ **11.** 81 **12.** −16 **13.** 6 **14.** −17 **15.** $-\frac{21}{4}$ **16.** −4°F **17. a.** 0.7 audit per 1000 tax returns **b.** −3.2 audits per 1000 returns **c.** from 2003 to 2005; 3.2 audits per 1000 returns **18.** 2.95; the average ticket price in 2012 was 2.95 times larger than the average ticket price in 1991. **19.** −33 **20.** $-\frac{2}{3}$ **21.** 11 **22.** 124 **23.** $2x - 3x$; 5 **24.** $\frac{-10}{x} - 6$; −4

25. a.

Years since 2010	First-Class Mail Volume (billions of pieces)
0	$-4.9(0) + 78.2$
1	$-4.9(1) + 78.2$
2	$-4.9(2) + 78.2$
3	$-4.9(3) + 78.2$
4	$-4.9(4) + 78.2$
t	$-4.9t + 78.2$

$-4.9t + 78.2$ **b.** 39; the first-class mail volume will be 39 billion pieces in 2018. **26.** 118 labs

Cumulative Review of Chapters 1 and 2

1. a. The rectangles are not drawn to scale. Answers may vary. 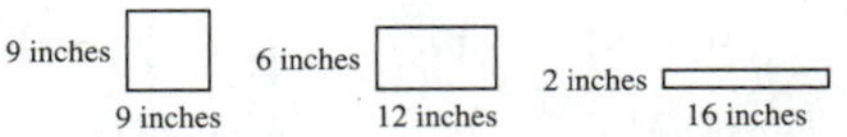

b. W, L **c.** P

2. 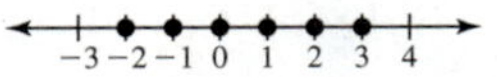

3. 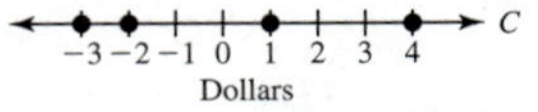

4. −5 **5.** independent variable: t; dependent variable: V

6. a. 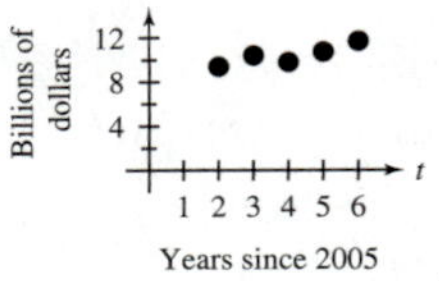

b. 2011 **c.** 2007 **d.** from 2008 to 2009; −0.6 billion dollars **e.** from 2007 to 2008 and from 2010 to 2011; 1.0 billion dollars **7.** −3 **8.** 4 **9.** (0, −1) **10.** (2, 0)

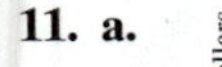

11. a.

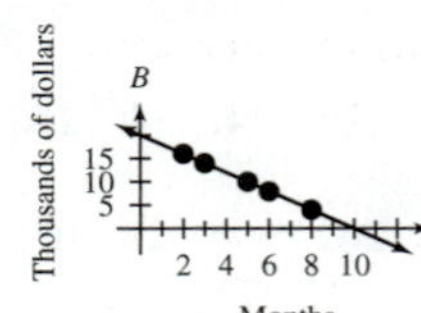

b. \$12 thousand **c.** 7 months after the person was laid off **d.** (0, 20); when the person was laid off, the balance was \$20 thousand. **e.** (10, 0); there was no money in the account 10 months after the person was laid off.

12. a. and b.

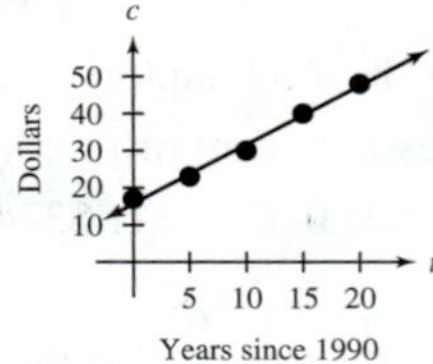

c. (0, 16); the average monthly basic rate for cable TV in 1990 was \$16. **d.** 2020 **e.** \$60 **13.** $\frac{3}{4}$ **14.** -4 **15.** $\frac{18}{25}$ **16.** $-\frac{11}{24}$ **17.** -17 **18.** $-\frac{2}{9}$ **19.** -8°F **20.** \$1865 **21.** $-\frac{1}{2}$ **22.** 49 **23.** $x - \frac{-12}{x}$; -7 **24.** $7 + (-2)x$; 15 **25.** 7.14; the percent growth of the investment was 7.14%.

26. a.

Years since 2005	Sales (thousands motorcycles)
0	$4 \cdot 0 + 15$
1	$4 \cdot 1 + 15$
2	$4 \cdot 2 + 15$
3	$4 \cdot 3 + 15$
4	$4 \cdot 4 + 15$
t	$4t + 15$

$4t + 15$ **b.** 63; the sales will be 63 thousand motorcycles in 2017.

Chapter 3

Homework 3.1 **1.** $(-3, -10), (2, 0)$ **3.** $(0, 7), (4, -5)$

5. (0, 2) **7.** (0, −4) **9.** (0, 0) **11.** (0, 0) **13.** (0, 0) **15.** (0, 0) **17.** (0, 0)

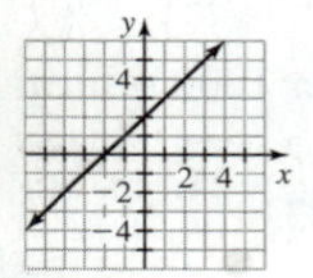

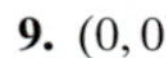

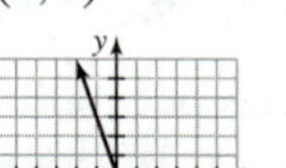

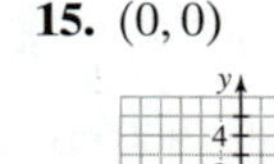

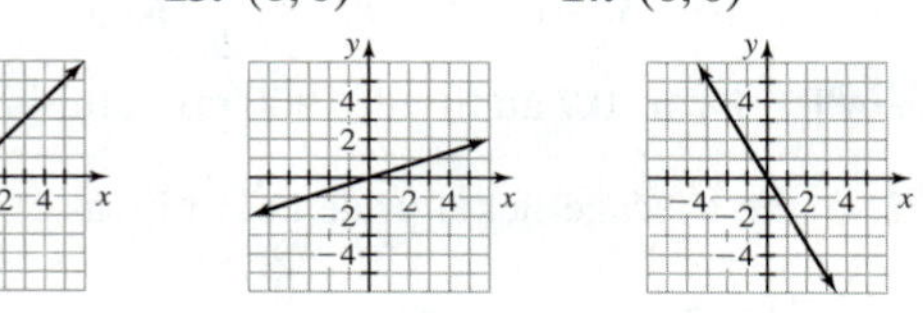

19. (0, 1) **21.** (0, −3) **23.** (0, 5) **25.** (0, −3) **27.** (0, −3) **29.** (0, 1)

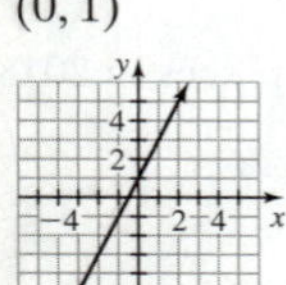

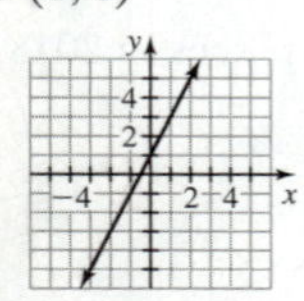

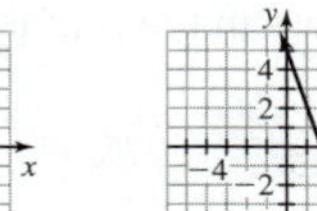

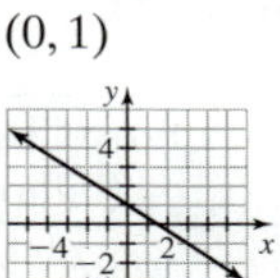

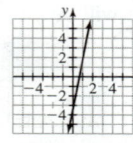

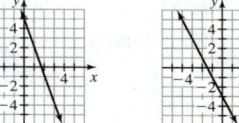

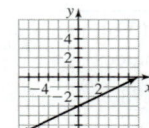

31. a.

x	y
0	-3
1	-1
2	1

Answers may vary. **b.**

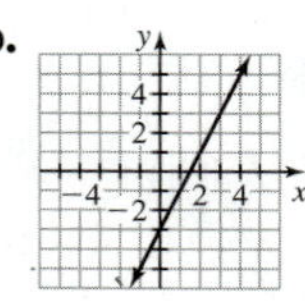

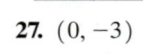

c. For each solution, the y-coordinate is 3 less than twice the x-coordinate.

33. a. i. 7; one **ii.** 13; one **iii.** -5; one **b.** one; answers may vary. **c. i.** Answers may vary; one **ii.** Answers may vary; one **iii.** Answers may vary; one **d.** one; answers may vary. **e.** one; answers may vary.

35. a. i.

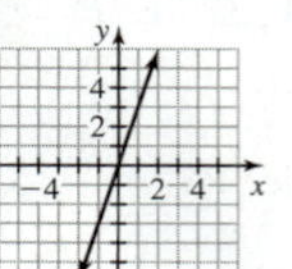

x-intercept: (0, 0); y-intercept: (0, 0)

ii.

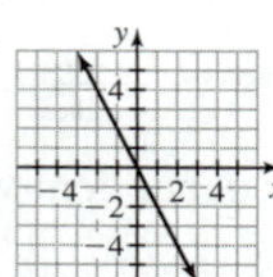

x-intercept: (0, 0); y-intercept: (0, 0)

iii.

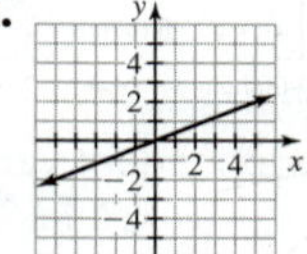

x-intercept: (0, 0); y-intercept: (0, 0)

b. x-intercept: (0, 0); y-intercept: (0, 0)

37. Answers may vary. **39.** 3 **41.** 0 **43.** 4 **45.** -2 **47.** C, D, E **49.** Answers may vary; infinitely many **51.** $y = x + 3$ **53.** $y = x$ **55. a.** Answers may vary. **b.** $y = 3x$ **57.** [graph] **59.** Answers may vary.

61. a. i. 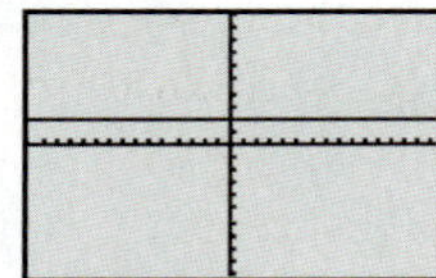 **ii.** 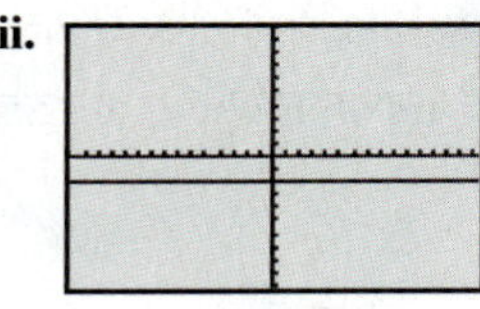 **iii.** 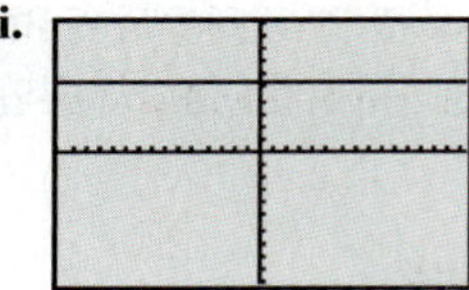 **b.** The graph is a horizontal line with y-intercept $(0, b)$.

63. Answers may vary. **65.** Answers may vary.

Homework 3.2

1. a.

Drink Cost (dollars) d	Total Cost (dollars) T
2	$2 + 3$
3	$3 + 3$
4	$4 + 3$
5	$5 + 3$
d	$d + 3$

$T = d + 3$ **b.** The units for both of the expressions T and $d + 3$ are dollars. **c.**

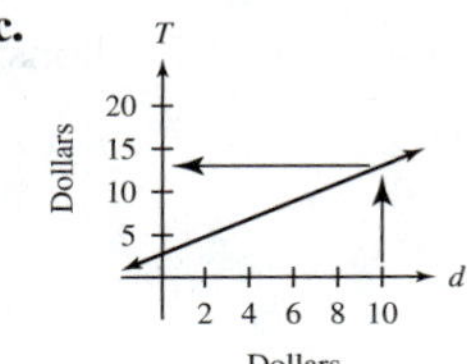

d. $(0, 3)$; if the person does not buy any drinks, then the total cost is \$3. **e.** \$13

3. a.

Number of Credits c	Total Cost (dollars) T
3	$87 \cdot 3$
6	$87 \cdot 6$
9	$87 \cdot 9$
12	$87 \cdot 12$
c	$87 \cdot c$

$T = 87c$ **b.** The units for both of the expressions T and $87c$ are dollars. **c.**

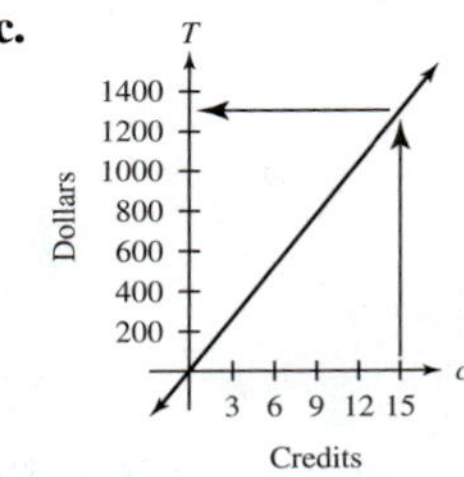

d. \$1305

5. a.

Time at Company (years) t	Salary (thousands of dollars) s
0	$3 \cdot 0 + 24$
1	$3 \cdot 1 + 24$
2	$3 \cdot 2 + 24$
3	$3 \cdot 3 + 24$
4	$3 \cdot 4 + 24$
t	$3 \cdot t + 24$

$s = 3t + 24$ **b.** The units for both of the expressions $3t + 24$ and s are thousands of dollars. **c.**

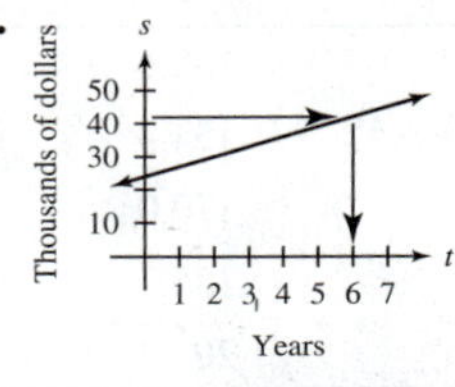

d. $(0, 24)$; the starting salary is \$24 thousand. **e.** 6 years

7. a.

Years Since 2010 t	Number of Traffic Deaths (thousands) n
0	$33 - 2 \cdot 0$
1	$33 - 2 \cdot 1$
2	$33 - 2 \cdot 2$
3	$33 - 2 \cdot 3$
4	$33 - 2 \cdot 4$
t	$33 - 2 \cdot t$

$n = 33 - 2t$ **b.** The units for both of the expressions $33 - 2t$ and n are thousands of deaths. **c.**

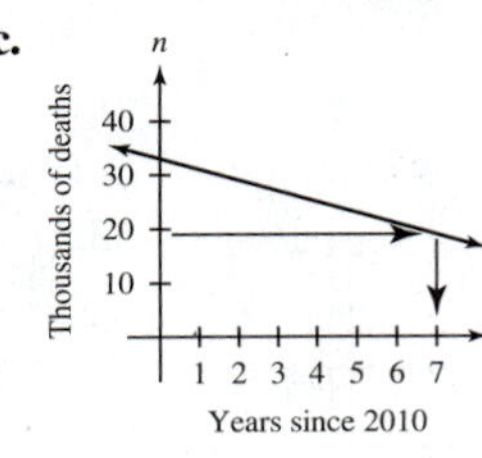
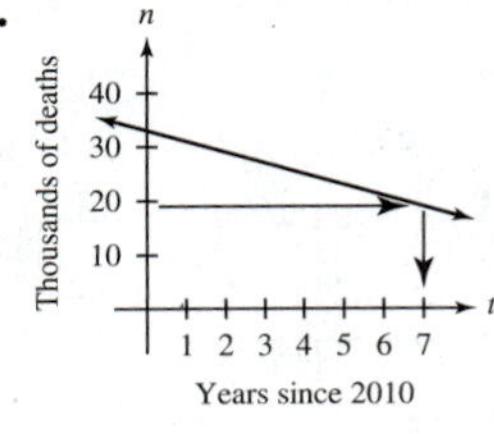

d. $(0, 33)$, in 2010, there were 33 thousand traffic deaths. **e.** 2017

9. a. $a = r - 5$ **b.** The units for both of the expressions a and $r - 5$ are minutes. **c.**

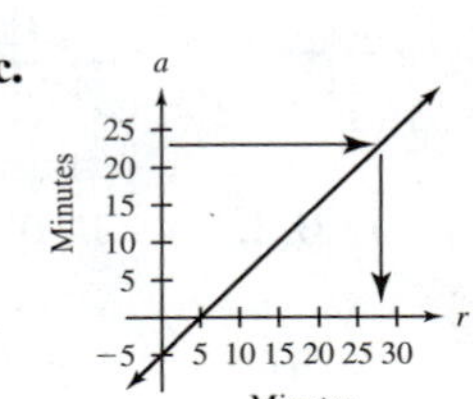

d. 28 minutes

11. a. $d = 60t$ **b.** The units for both of the expressions d and $60t$ are miles.

c.

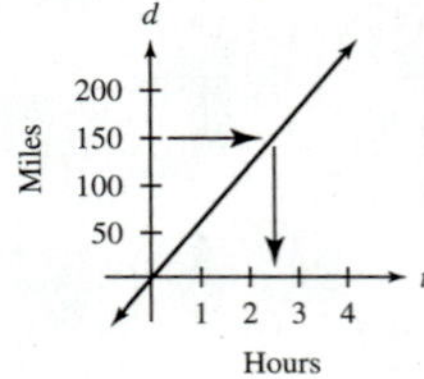

d. $(0, 0)$; the person will not travel any distance in 0 hours of driving. **e.** 2.5 hours

13. a. $T = 36u + 31$ **b.** The units for both of the expressions T and $36u + 31$ are dollars. **c.** \$571 **15. a.** $p = 3n + 62$, where p is the pressure (in psi) after n pumps; answers may vary. **b.** The units for both of the expressions p and $3n + 62$ are psi; answers may vary. **c.** 113 psi **17. a.** $g = 11 - 2t$ **b.** The units for both of the expressions g and $11 - 2t$ are gallons.

c.

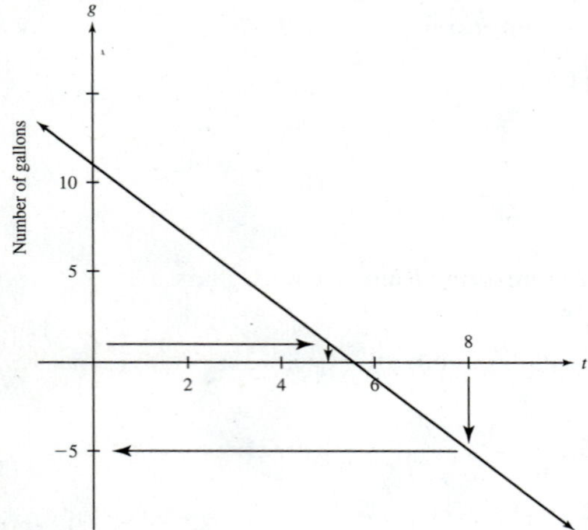

d. 5 hours **e.** -5 gallons; model breakdown has occurred.

19. a.

Years Since 2011 t	Sales (millions of pounds) s
0	138.6
1	154.3
2	170.0
3	185.7
4	201.4

Answers may vary. **b.** $s = 15.7t + 138.6$ **c.**

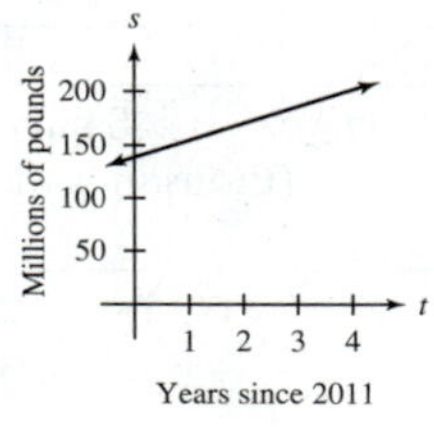

21. $n = 45$;

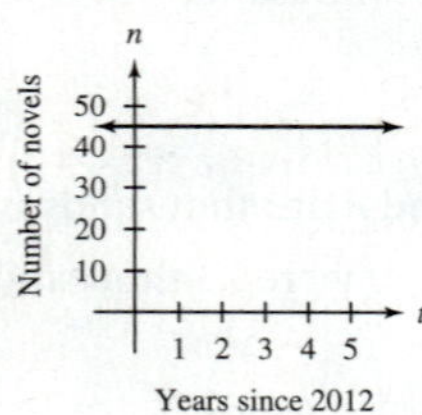

23. a. 2007 **b.** 0.8 thousand independent CD and record stores **c.** $(0, 3.1)$; in 2005, there were 3.1 thousand independent CD and record stores. **d.** $(13.3, 0)$; there will be neither independent CD stores nor independent record stores in 2018; model breakdown has likely occurred.

25.

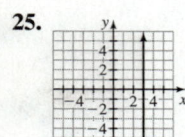

27.

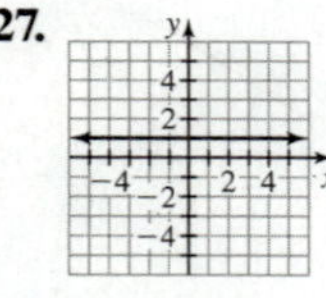

29.

31.

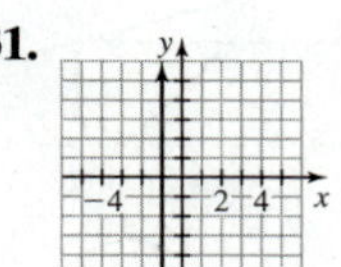

33. $x = 0$

35.

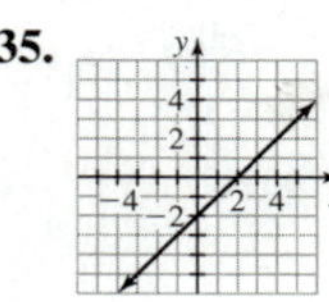

37.

39.

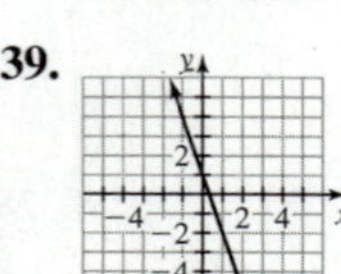

41.

43.

45.

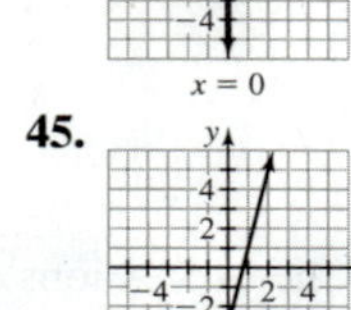

47.

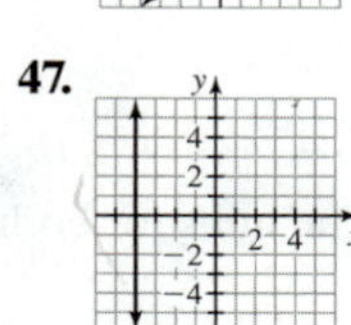

49. $x = -3$ **51. a. i.** 4 **ii.** 0 **b. i.** -4 **ii.** 0 **53. a. i.** 5 **ii.** 3 **b. i.** -5 **ii.** -3

55. a. 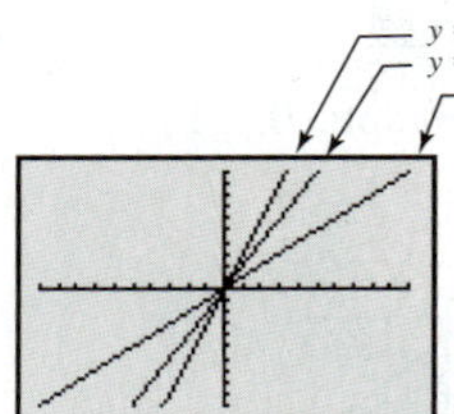

b. $y = x, y = 2x, y = 3x$ **c.** Answers may vary. **d.** Answers may vary. **57.** Vertical line; answers may vary. **59.** no; answers may vary. **61.** Answers may vary.

Homework 3.3

1. road A **3.** ski run A **5.** 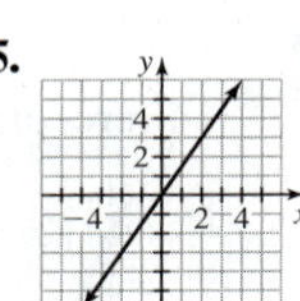 run: 2; rise: 3; slope: $\frac{3}{2}$ **7.** 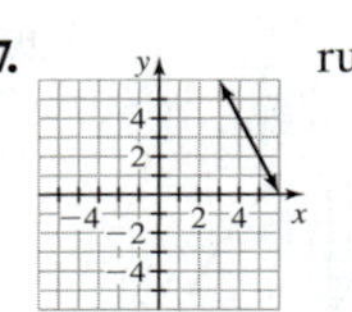 run: 2; rise: -4; slope: -2

9. 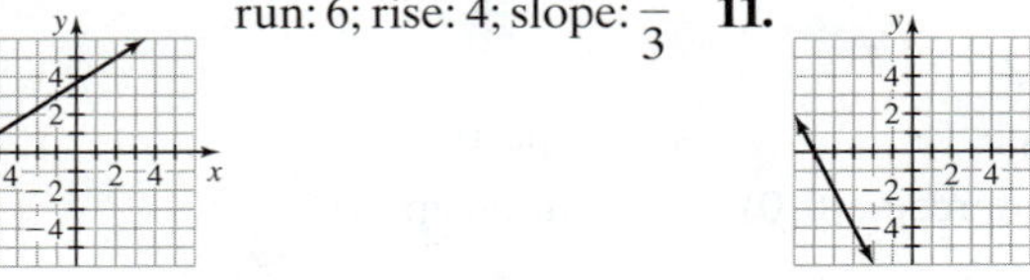 run: 6; rise: 4; slope: $\frac{2}{3}$ **11.** run: 2; rise: -4; slope: -2 **13.** 2; increasing **15.** -4; decreasing

17. $\frac{1}{2}$; increasing **19.** $-\frac{1}{3}$; decreasing **21.** -1; decreasing **23.** $-\frac{1}{2}$; decreasing **25.** 2; increasing **27.** $\frac{3}{2}$; increasing

29. $-\frac{4}{5}$; decreasing **31.** $\frac{2}{3}$; increasing **33.** $-\frac{1}{2}$; decreasing **35.** 0; horizontal **37.** undefined slope; vertical **39.** -9.25; decreasing

41. 1.14; increasing **43.** -0.21; decreasing **45.** 0.71; increasing **47.** $\frac{2}{3}$ **49.** -3 **51. a.** negative **b.** positive **c.** undefined **d.** zero **53.** Answers may vary. **55.** Answers may vary. **57.** Answers may vary. **59.** Answers may vary. **61.** Answers may vary; $\frac{4}{3}$

63. Answers may vary; $\frac{13}{4}$ **65.** Answers may vary; yes **67. a.** Answers may vary. **b.** no **c.** yes **d.** Answers may vary.

69. Answers may vary. **71.** Answers may vary. **73.** Answers may vary.

75. a. i. 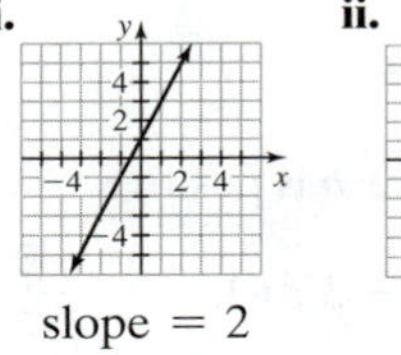
slope = 2

ii. 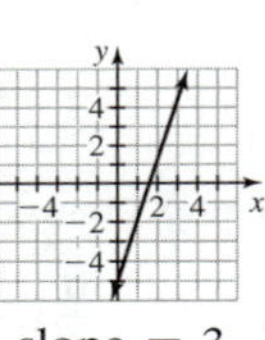
slope = 3

iii. 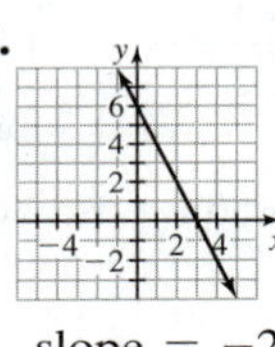
slope = -2

b. The slope is equal to the coefficient of x.

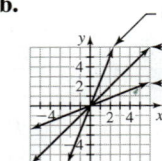

77. a. 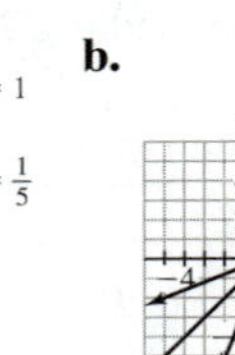

b. 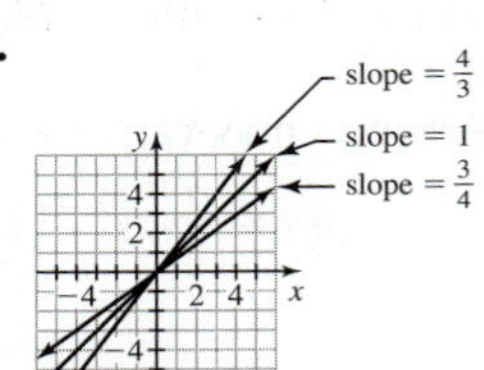

c. slope = $\frac{4}{3}$, slope = 1, slope = $\frac{3}{4}$

d. For $m \neq 0$, the lines $y = mx$ and $y = \frac{1}{m}x$ are mirror reflections of each other across the line $y = x$.

e.

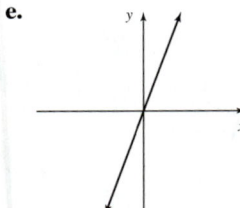

79. yes; yes; answers may vary. **81.** Answers may vary.

Homework 3.4

1. **3.** **5.** **7.** **9.** **11.** 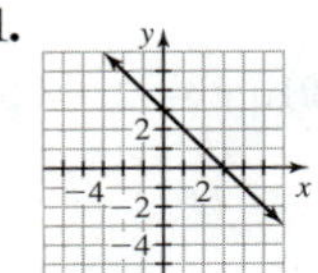 **13.** 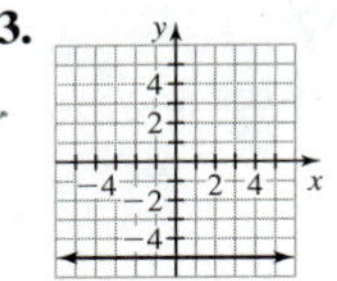

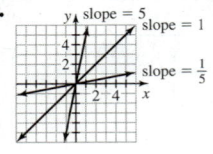

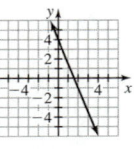

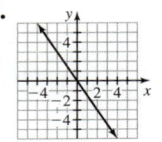

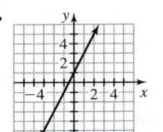

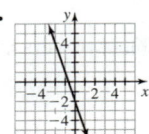

15.

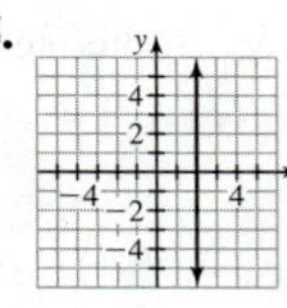

17. slope: $\frac{2}{3}$; y-intercept: $(0, -1)$

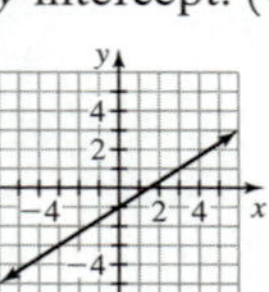

19. slope: $-\frac{1}{3}$; y-intercept: $(0, 4)$

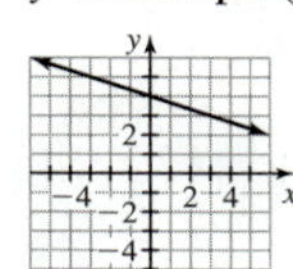

21. slope: $\frac{4}{3}$; y-intercept: $(0, 2)$

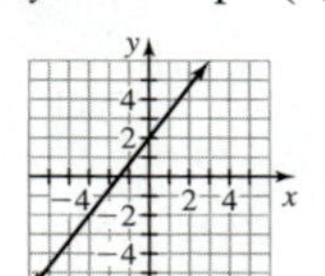

23. slope: $-\frac{4}{5}$; y-intercept: $(0, -1)$

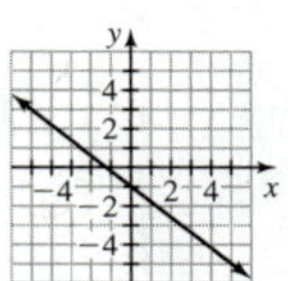

25. slope: $\frac{1}{2}$; y-intercept: $(0, 0)$

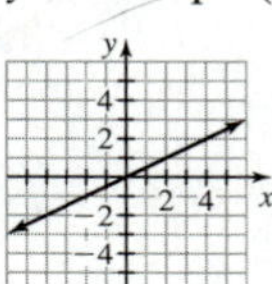

27. slope: $-\frac{5}{3}$; y-intercept: $(0, 0)$

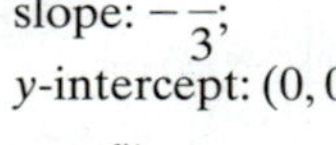

29. slope: 4; y-intercept: $(0, -2)$

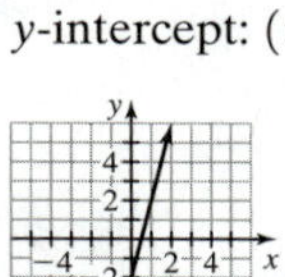

31. slope: -2; y-intercept: $(0, 4)$

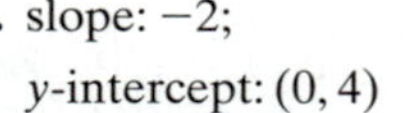

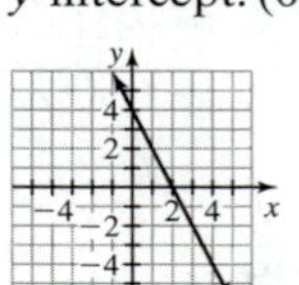

33. slope: -4; y-intercept: $(0, -1)$

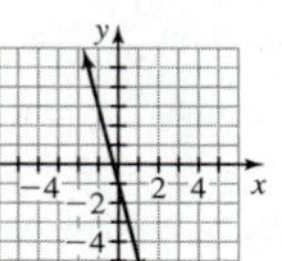

35. slope: 1; y-intercept: $(0, 1)$

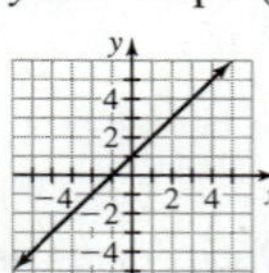

37. slope: -1; y-intercept: $(0, 3)$

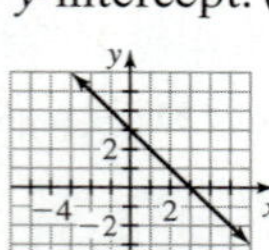

39. slope: -3; y-intercept: $(0, 0)$

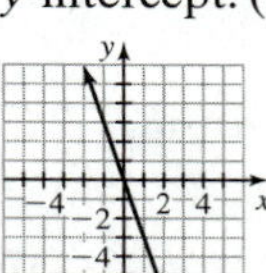

41. slope: 1; y-intercept: $(0, 0)$

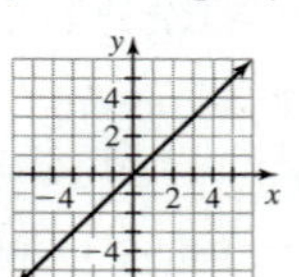

43. slope: 0; y-intercept: $(0, -3)$

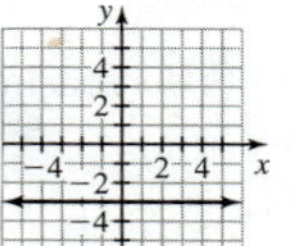

45. slope: 0; y-intercept: $(0, 0)$

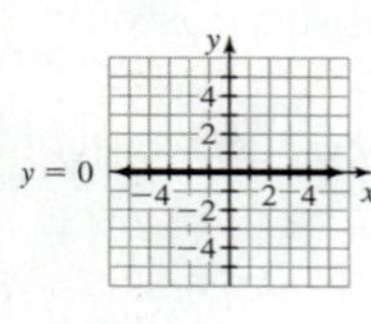

47. a.

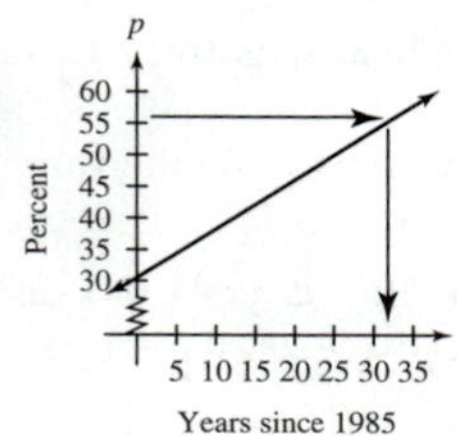

b. 2017

49. a.

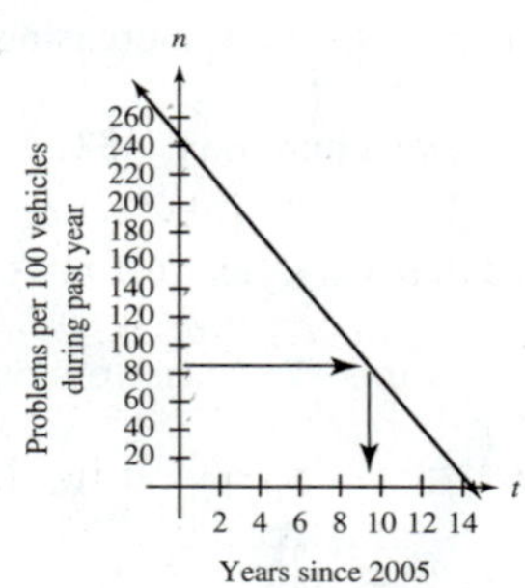

b. 2014

51. a. m is positive; b is negative **b.** m is zero; b is negative **c.** m is negative; b is positive **d.** m is positive; b is positive

53. Answers may vary. **55.** Answers may vary. **57.** Answers may vary. **59.** $y = 3x - 4$ **61.** $y = -\frac{6}{5}x + 3$ **63.** $y = -\frac{2}{7}x$

65. $y = 2x + 3$ **67.** perpendicular **69.** neither **71.** parallel **73.** neither **75.** parallel **77.** perpendicular **79.** no; answers may vary.

81. -2 **83.** 3 **85.** $\frac{1}{3}$ **87.** no; answers may vary. **89.** Answers may vary. **91. a.**

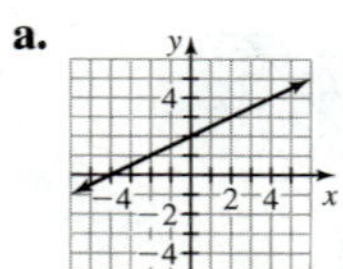

b.

x	y
-2	1
0	2
2	3

Answers may vary.

c. For each solution, the y-coordinate is two more than half the x-coordinate. **93.** Answers may vary;

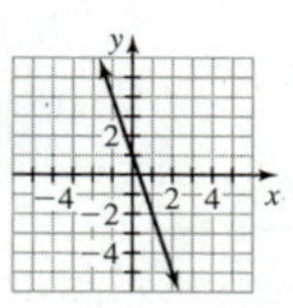

95. a.

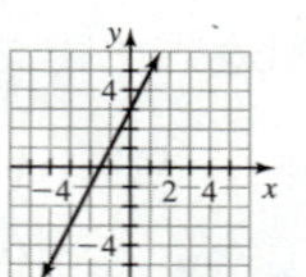

b. $y = 2x + 3$ **97. a.** The slope of each line is zero. **b.** 0 **99. a.** k; answers may vary.

b. b; answers may vary. **101.** $y = 2x - 3$ **103.** Answers may vary.

Homework 3.5 **1.** \$1550 per year **3.** −1650 feet per minute **5.** 10 percentage points per year **7.** −8.13 thousand Steller sea lions per year **9.** −10.6 meals per year **11.** \$98 per credit hour **13.** \$5940 per person **15. a.** yes; 4; the revenue increases by \$4 million per year. **b. i.** $r = 4t + 3$

ii.

Years since 2010 t	Revenue (millions of dollars) r
0	3
1	7
2	11
3	15
4	19

Answers may vary. **iii.**

17. a. yes; −4.9; the number of sites is decreasing by about 4.9 sites per year. **b.** (0, 381); in 2009, there were 381 sites. **c.** $n = -4.9t + 381$ **d.** The units for both of the expressions n and $-4.9t + 381$ are sites. **e.** 337 sites **19. a.** −650; the balance declines by \$650 per month. **b.** (0, 4700); the balance is \$4700 on September 1. **c.** $B = -650t + 4700$ **d.** The units for both of the expressions B and $-650t + 4700$ are dollars. **e.** \$800 **21. a.** $T = 550c + 15$, where T is the total one-semester cost (in dollars) of c credits of classes plus the part-time student fee; answers may vary. **b.** The units for both of the expressions T and $550c + 15$ are dollars; answers may vary. **c.** 550; the tuition increases by \$550 per credit; answers may vary. **d.** \$4965 **23. a.** −0.02; the car uses 0.02 gallon of gasoline per mile. **b.** (0, 11.9); there were 11.9 gallons of gasoline in the tank at the start of the trip. **c.** $G = -0.02d + 11.9$ **d.** The units for both of the expressions G and $-0.02d + 11.9$ are gallons. **e.** 10.5 gallons **25. a.** −18.59; the number of new U.S. offshore wells is decreasing by about 18.59 wells per year. **b.** (0, 146.86); there were about 147 new U.S. offshore wells drilled in 2005. **c.** 17 wells

27. a. 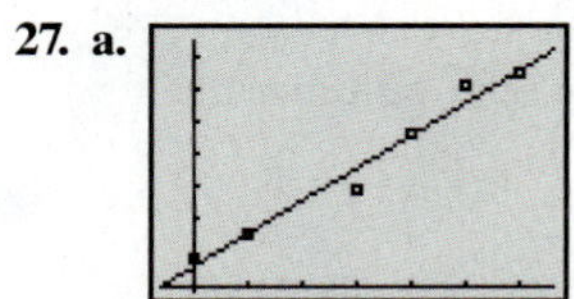 yes **b.** 10; the percentage of adult Internet users who use social networking sites is increasing by 10 percentage points per year. **c.** 8, 10, 9.5 (all in percentage points per year); answers may vary.

d. (0, 5.7); in 2005, 5.7% of adult Internet users were using social networking sites. **e.** 105.7%; model breakdown has occurred.

29. a. yes **b.** 3.56; the result overestimates the qualifying core GPA by 0.01. **c.** −0.00254; the qualifying core GPA decreases by 0.00254 for an increase of 1 point on the SAT. **d.** (0, 4.58); the qualifying core GPA is 4.58 for an SAT score of 0; model breakdown has occurred, because the highest possible GPA is 4.0 and the lowest possible SAT score is 400 points.

31. a. yes; answers may vary. **b.** 60 miles per hour **33. a.** yes; answers may vary. **b.** −1.5 gallons per hour **35.** −10; the value of the stock decreases by \$10 per week. **37.** 100; the median sales price increased by \$100 thousand per bedroom. **39.** yes; 30; the total charge increases by \$30 per hour. **41.** equations 1 and 3 **43.**

Equation 1		Equation 2		Equation 3		Equation 4	
x	y	x	y	x	y	x	y
0	3	0	99	21	16	43	17
1	8	1	92	22	14	44	20
2	13	2	85	23	12	45	23
3	18	3	78	24	10	46	26
4	23	4	71	25	8	47	29

45. Set 1: $y = 2x + 5$; Set 2: $y = -3x + 20$; Set 3: $y = 8x + 21$; Set 4: $y = -5x + 9$ **47.** y increases by 7. **49.** Answers may vary. **51. a.** The slope 3 is the number multiplied by x. **b.** If the run is 1, the rise is 3. **c.** As the value of x increases by 1, the value of y increases by 3. **d.** Answers may vary. **53.** Answers may vary. **55.** $2m$; answers may vary. **57.** Answers may vary.

Chapter 3 Review Exercises

1. (−3, 9), (4, −5) **2.** −1 **3.** −3 **4.** −2 **5.** −2 **6.** −4 **7.** 4 **8.** airplane A **9.** 2; increasing **10.** −1; decreasing **11.** $\frac{1}{2}$; increasing **12.** $-\frac{1}{2}$; decreasing **13.** −1; decreasing **14.** $-\frac{1}{3}$; decreasing **15.** $-\frac{9}{8}$; decreasing **16.** $-\frac{2}{3}$; decreasing **17.** undefined slope; vertical **18.** 0; horizontal **19.** −3.62; decreasing **20.** 0.94; increasing **21.** Answers may vary.

22. 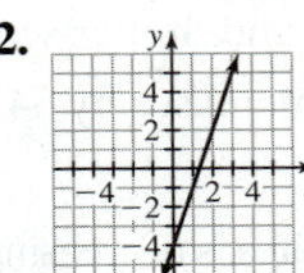 **23.** 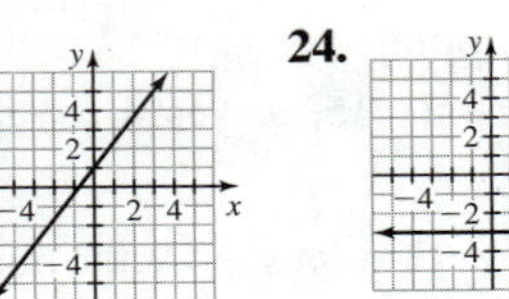 **24.** 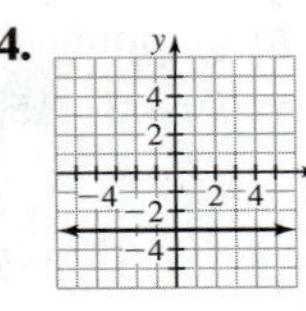 **25.** slope: $\frac{3}{4}$; y-intercept: $(0, -1)$ 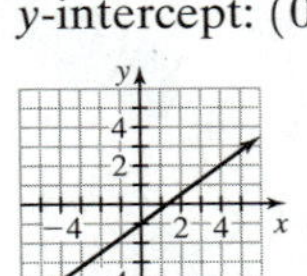 **26.** slope: $-\frac{1}{2}$; y-intercept: $(0, 3)$ 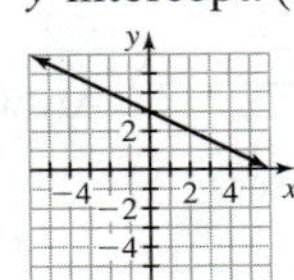

27. slope: $-\frac{2}{5}$; y-intercept: $(0, -1)$ 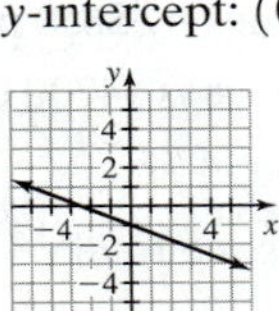 **28.** slope: $\frac{2}{3}$; y-intercept: $(0, 0)$ **29.** slope: -4; y-intercept: $(0, 0)$ **30.** slope: 2; y-intercept: $(0, -4)$ **31.** slope: -3; y-intercept: $(0, 1)$ 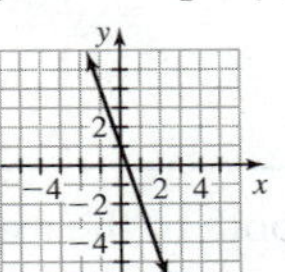

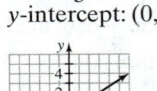

32. slope: 1; y-intercept: $(0, 2)$ 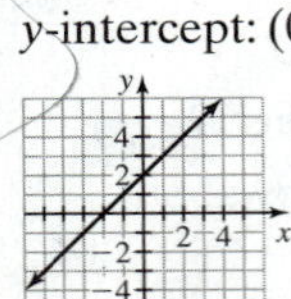 **33.** slope: 0; y-intercept: $(0, -5)$ 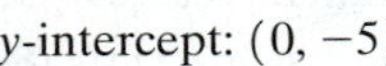 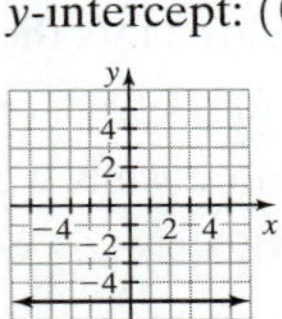 **34.** 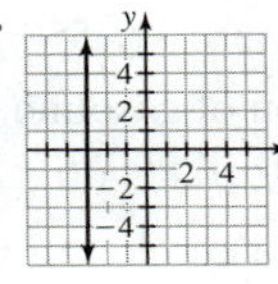 **35.** 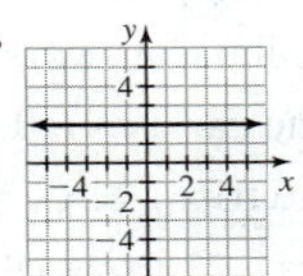

36. a.

x	y
-1	3
0	1
1	-1

Answers may vary. **b.** 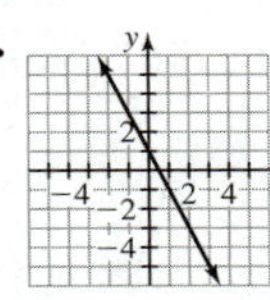 **c.** For each solution, the y-coordinate is 1 more than -2 times the x-coordinate.

37. a. $B = -3t + 19$ **b.** 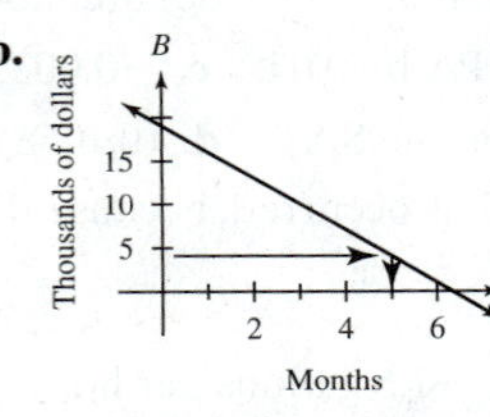

c. $(0, 19)$; when the person first lost his job, the balance was \$19 thousand. **d.** 5 months **38.** $x = 5$ **39.** $y = -\frac{2}{3}x + 4$ **40.** neither **41.** perpendicular **42.** perpendicular **43.** parallel **44.** -1.5°F per hour **45.** -117 million dollars per year

46. a. $c = 2d + 2.5$ **b.** The units for both of the expressions c and $2d + 2.5$ are dollars. **c.** \$36.50 **47. a.** -4; the person loses 4 pounds per month. **b.** $w = -4t + 195$ **c.** 171 pounds **48. a.** 45.60; the average monthly cost increases by \$45.60 per year. **b.** $c = 45.60t + 972$ **c.** \$1291.20 **49.** yes; 139; the cost is \$139 per calculator. **50. a.** 4.6 thousand deaths **b.** 2012 **c.** $(0, 5.0)$; there were 5.0 thousand pedestrian deaths in 2005. **d.** -0.2; the number of pedestrian deaths is decreasing by 0.2 thousand deaths per year. **51.** equations 1, 3, and 4

52.

Equation 1		Equation 2		Equation 3		Equation 4	
x	y	x	y	x	y	x	y
0	50	0	12	61	25	26	-4
1	41	1	16	62	23	27	-1
2	32	2	20	63	21	28	2
3	23	3	24	64	19	29	5
4	14	4	28	65	17	30	8

53. The value of y decreases by 6 when the value of x is increased by 1.

Chapter 3 Test

1. 3 **2.** 3 **3.** $(0, 1)$ **4.** $(1.5, 0)$ **5.** ski run B **6.** 3; increasing **7.** $-\frac{1}{2}$; decreasing **8.** 0; horizontal **9.** undefined slope; vertical

10. 1.29; increasing **11.** 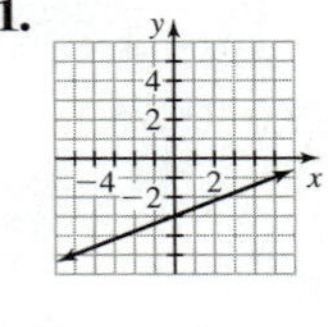 **12.** slope: $-\frac{3}{2}$; y-intercept: $(0, 2)$ 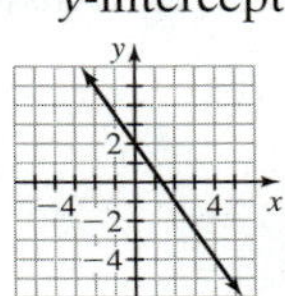 **13.** slope: $\frac{5}{6}$; y-intercept: $(0, 0)$ 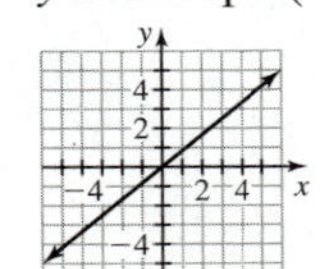 **14.** slope: 3; y-intercept: $(0, -4)$ 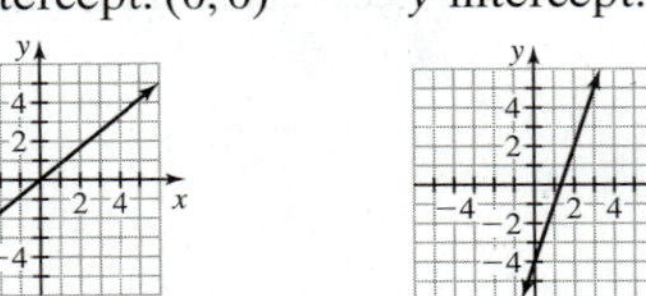

15. slope: 0; y-intercept: $(0, 2)$ **16.** slope: -2; y-intercept: $(0, 3)$ 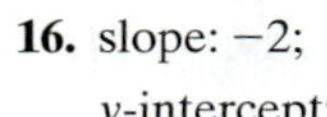 **17.** $y = \frac{1}{2}x + 1$ **18. a.** $v = -2t + 17$ **b.**

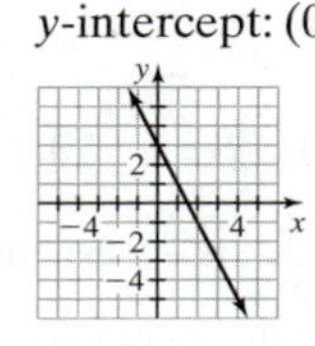

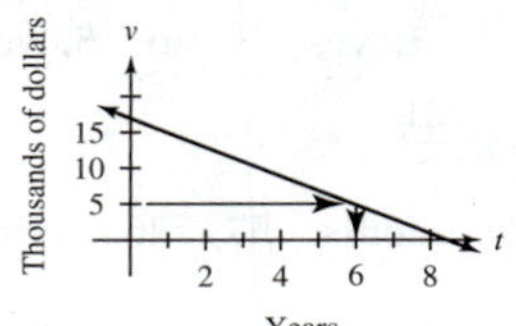

c. $(0, 17)$; the used car is currently worth \$17 thousand. **d.** 6 years from now **19. a.** m is positive; b is positive **b.** m is zero; b is negative **c.** m is negative; b is positive **d.** m is negative; b is negative **20.** neither **21.** parallel **22.** \$832.5 million **23.** -2.1 presents per year **24. a.** yes; 37; the median compensation is increasing by about \$37 thousand per year. **b.** $(0, 870)$; the median compensation was \$870 thousand in 2010. **c.** $C = 37t + 870$ **d.** \$1166 thousand (\$1.166 million) **25. a.** yes

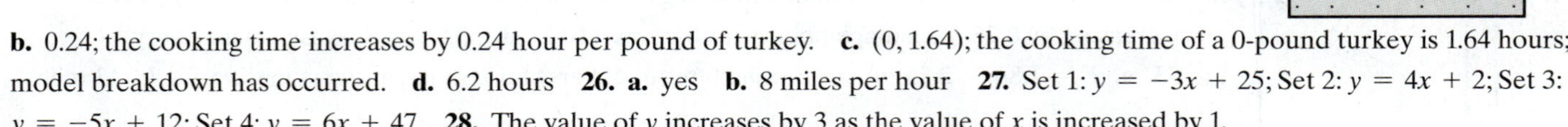

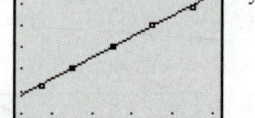

b. 0.24; the cooking time increases by 0.24 hour per pound of turkey. **c.** $(0, 1.64)$; the cooking time of a 0-pound turkey is 1.64 hours; model breakdown has occurred. **d.** 6.2 hours **26. a.** yes **b.** 8 miles per hour **27.** Set 1: $y = -3x + 25$; Set 2: $y = 4x + 2$; Set 3: $y = -5x + 12$; Set 4: $y = 6x + 47$ **28.** The value of y increases by 3 as the value of x is increased by 1.

Chapter 4

Homework 4.1 **1.** $x + 5$ **3.** $7 + 2p$ **5.** yx **7.** $x(-2)$ **9.** $15 + m \cdot 4$ **11.** $x + (4 + y)$ **13.** $4(bc)$ **15.** $(x + y) + 3$ **17.** $(ab)c$ **19.** $10x$ **21.** $p + 7$ **23.** $36x$ **25.** $3b + 11$ **27.** $-4x$ **29.** $\frac{7x}{4}$ **31.** $-3x + 2$ **33.** $2k + 8p - 7$ **35.** $x + y - 2$ **37.** $3x + 27$ **39.** $2x - 10$ **41.** $-2t - 10$ **43.** $10x - 30$ **45.** $-24x - 42$ **47.** $6x - 10y$ **49.** $-25x - 15y + 40$ **51.** $-0.3x - 0.06$ **53.** $2x + 4$ **55.** $21x - 27$ **57.** $2x + 5$ **59.** $4a + 19$ **61.** $-12x + 9$ **63.** $-9a + 19$ **65.** $10.5x - 38.56$ **67.** $-t - 2$ **69.** $-8x + 9y$ **71.** $-5x - 8y + 1$ **73.** $-3x + 3$ **75.** $-x + 5$ **77.** $-2t + 3$ **79.** $-8x + 14y - 3$ **81.** $8x$; 16; answers may vary. **83.** $5x - 35$; -25; answers may vary. **85.** $-3x - 7$; -13; answers may vary. **87.** $3(x + 2)$; $3x + 6$ **89.** $-4(2x - 5)$; $-8x + 20$ **91.** $5 - 2(x - 4)$; $-2x + 13$ **93.** $-2 + 7(2x + 1)$; $14x + 5$ **95. a.** $2x - 2$ **b.** 8 **c.** 8 **d.** The results are equal; the result in part (a) might be correct. **97.** 18; 10; the work is incorrect. **99. a.** 24; 48 **b.** Answers may vary. **c.** $(ab)c$ **101.** $y = 2x - 6$ **103.** $y = -3x - 3$ **105.** $y = -6x$

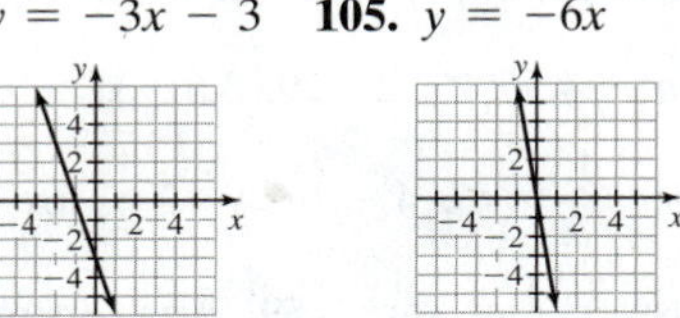

107. $(7 + 3 + 6)10 = 160$ dollars; $7(10) + 3(10) + 6(10) = 160$ dollars; answers may vary. **109.** Answers may vary. **111.** $3 = 1$; answers may vary. **113.** commutative law for addition; associative law for addition; commutative law for addition; associative law for addition **115.** commutative law for multiplication; distributive law; commutative law for multiplication **117.** $-a = -1 \cdot a$; associative law for multiplication; the product of two real numbers with different signs is negative; $-1 \cdot a = -a$; $-(-a) = a$ **119.** Answers may vary.

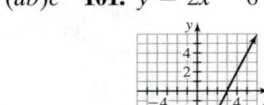

Homework 4.2 **1.** $7x$ **3.** $5x$ **5.** $-13w$ **7.** $4t$ **9.** $-0.5x$ **11.** $\frac{7}{3}x$ **13.** $-3x - 3$ **15.** $-2p - 7$ **17.** $3x + y + 1$ **19.** $-9.9x + 1.1y + 2.1$ **21.** $-a + 15$ **23.** $43.16x + 23.08$ **25.** $7a - 8b$ **27.** $-x + 2$ **29.** $2x - 2y$ **31.** $-13t - 5$ **33.** $-2x - 4y + 6$ **35.** $-8x - 14$ **37.** $-24x - 38y$ **39.** 0 **41.** $-4x + 7y - 21$ **43.** $7x - 13y - 2$ **45.** $-\frac{2}{7}a + \frac{6}{7}$ **47.** $3x - 3$ **49.** $x + 5x$; $6x$ **51.** $4(x - 2)$; $4x - 8$ **53.** $x + 3(x - 7)$; $4x - 21$ **55.** $2x - 4(x + 6)$; $-2x - 24$ **57.** Twice the number plus 6 times the number (answers may vary); $8x$ **59.** 7 times the difference of the number and 5 (answers may vary); $7x - 35$ **61.** The number, plus 5 times the sum of the number and 1 (answers may vary); $6x + 5$ **63.** Twice the number, minus 3 times the difference of the number and 9 (answers may vary); $-x + 27$ **65.** $8x - 5$ **67.** $-3x + 9$ **69.** $5x + 2$; 22; answers may vary. **71.** $3x + 11$; 23; answers may vary. **73. a.** $8x + 12$ **b.** 28 **c.** 28 **d.** The results are equal; the result in part (a) might be correct. **75.** $-2(x - 3)$, $2(3 - x)$, $-3(x - 2) + x$, $-2x + 6$ **77.** 5, 14, 14; 14; $9x - 13$ **79.** Answers may vary. **81.** $y = -2x$ **83.** $y = 2x - 4$ **85.** $y = 3x - 4$

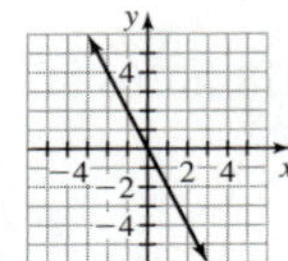

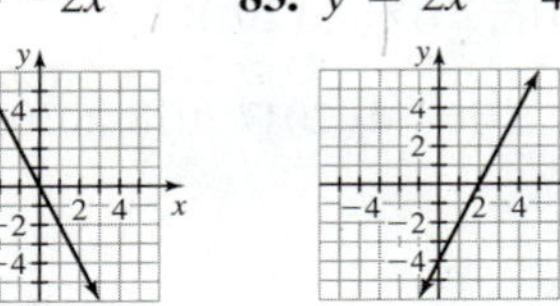

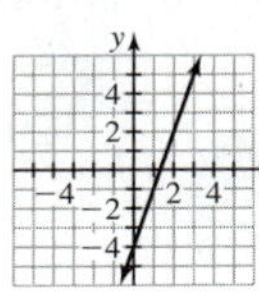

87. a. Answers may vary. **b.** Answers may vary. **c.** $2(x - 3) - x + 6$; x **89.** Answers may vary.

Homework 4.3 **1.** yes **3.** no **5.** yes **7.** 5 **9.** −14 **11.** 24 **13.** −28 **15.** −3 **17.** 3 **19.** 3 **21.** −4 **23.** 5 **25.** $\frac{4}{3}$ **27.** $\frac{6}{5}$ **29.** 0 **31.** 15 **33.** $\frac{21}{2}$ **35.** −12 **37.** $-\frac{10}{3}$ **39.** 6 **41.** −3 **43.** $\frac{1}{2}$ **45.** −11.1 **47.** 41.76 **49.** 2.3 **51.** \$820 **53.** 40 points **55.** 930 thousand students **57.** 726 servings **59.** −4 **61.** 4 **63.** 5 **65.** 4 **67.** 3 **69.** −3 **71.** 3 **73.** −2 **75.** 2 **77.** 2 **79.** Answers may vary; 5 **81.** 4(3) is equal to 12; $\frac{4(3)}{4} = 3$ is equal to $\frac{12}{4} = 3$; 3 is equal to 3. **83.** yes; answers may vary. **85.** Answers may vary. **87.** Answers may vary. **89.** no **91.** Answers may vary. **93.** Answers may vary. **95.** yes **97. a.** 5 **b.** 4 **c.** $k - b$ **99.** Answers may vary. **101.** Answers may vary.

Homework 4.4 **1.** 5 **3.** −5 **5.** $-\frac{4}{3}$ **7.** 12 **9.** 2 **11.** 6 **13.** −5 **15.** 2 **17.** 3 **19.** $\frac{3}{4}$ **21.** −1 **23.** $\frac{5}{4}$ **25.** $\frac{5}{2}$ **27.** $\frac{8}{5}$ **29.** 6 **31.** $-\frac{41}{3}$ **33.** $-\frac{27}{10}$ **35.** $\frac{12}{13}$ **37.** −7 **39.** $\frac{11}{3}$ **41.** 1.67 **43.** 6.34 **45.** 21.85 **47. a.** $n = 1.04t + 46.7$ **b.** 2018 **c.** 50.9 million **49. a.** $v = 3.36t + 8.16$ **b.** \$18.24 **c.** 6 months after Stewart was sentenced.

51. a. 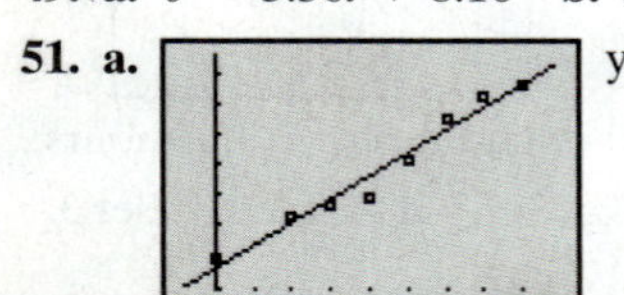 yes **b.** 0.76; the percentage of freshmen in college whose average grade in high school was an A is increasing by 0.76 percentage point per year. **c.** 2012 **d.** 53.8% **e.** Answers may vary. **53. a.** $F = 1.8d + 2.25$, where F is the fare (in dollars) for d miles; answers may vary. **b.** 13 miles

55. a. $V = -1976t + 25{,}772$, where V is the value (in dollars) of the car at t years since 2012; answers may vary. **b.** 2022 **57.** 11 weeks **59.** 2018 **61.** \$1.2 trillion **63.** 29.8 pounds **65.** 6.6 million pints **67.** 225 square feet **69.** 3 **71.** −4 **73.** −4 **75.** −7 **77.** Twice the number, minus 3, is 7 (answers may vary); 5 **79.** Six times the number, minus 3, is equal to 8 times the number, minus 4 (answers may vary); $\frac{1}{2}$ **81.** Twice the difference of the number and 4 is 10 (answers may vary); 9 **83.** Four, minus 7 times the sum of the number and 1 is 2 (answers may vary); $-\frac{5}{7}$ **85.** −1 **87.** −2 **89.** 2.83 **91.** 3.45 **93.** −5.43 **95.** 2 **97.** −2 **99.** 4 **101.** −1 **103.** 2 **105.** −3 **107.** set of all real numbers; identity **109.** empty set; inconsistent equation **111.** 2; conditional equation **113.** set of all real numbers; identity **115.** empty set; inconsistent equation **117.** Answers may vary; answers may vary; 7 **119.** Answers may vary; 7 **121.** Answers may vary. **123.** 3; answers may vary. **125.** 3; 3; answers may vary. **127. a.** m can be any real number except 7; b can be any real number. **b.** $m = 7; b = 4$ **c.** $m = 7$; b can be any real number except 4.

Homework 4.5 **1.** linear equation **3.** linear expression **5.** linear expression **7.** linear equation **9.** 2 **11.** $7x$ **13.** $-4b + 5$ **15.** $\frac{5}{4}$ **17.** $\frac{7}{19}$ **19.** $19x - 7$ **21.** $-\frac{3}{4}$ **23.** $-16x - 12$ **25.** $5.9p - 4.5$ **27.** 2 **29.** 3.5 **31.** $-6.1x + 35.28$ **33.** $-\frac{3w}{4}$ **35.** −2 **37.** $\frac{x}{12} + \frac{1}{2}$ **39.** −6 **41.** $\frac{14}{33}$ **43.** $\frac{17}{4}x - \frac{7}{6}$ **45.** $3 + 2x = -10; -\frac{13}{2}$ **47.** $4 - 6(x - 2); -6x + 16$ **49.** $-9x = x - 5; \frac{1}{2}$ **51.** $\frac{x}{2} = 3(x - 5); 6$ **53.** $x + x(6); 7x$ **55.** $x + \frac{x}{2}; \frac{3x}{2}$ **57.** Answers may vary. **59.** Answers may vary. **61.** no; answers may vary. **63.** Answers may vary; $\frac{7}{12}x$ **65.** Answers may vary. **67.** Answers may vary. **69.** Answers may vary. **71.** 1; answers may vary. **73.** Answers may vary.

Homework 4.6 **1.** $P = 4S$ **3.** $P = 2H + 2S + B$ **5.** $P = 2A + 2B + 2C + 2D$ **7.** $V = 5$ **9.** $B = 4$ **11.** $t \approx 1.86$ **13.** $L = 3$ **15.** $b = 7$ **17.** 14.5 feet **19.** 16 inches **21.** 59 feet **23. a.** $A = 3x$ **b.** $P = 2x + 6$ **c.** The units for both of the expressions P and $2x + 6$ are inches. **25. a.** Answers may vary. **b.** Answers may vary. **c.** Answers may vary. **27. a.** 30 cents **b.** 40 cents **c.** $T = 10d$ **d.** The units for both of the expressions T and $10d$ are cents. **29.** $T = 725n$ **31. a.** $T = 20x$ **b.** 1500; if 1500 people buy tickets in advance, the total sales will be \$30,000. **33. a.** $C = 15k$ **b.** $E = 575n$ **c.** $T = 15k + 575n$ **d.** 100; for 8000 \$15 tickets and 100 \$575 tickets, the total cost is \$177,500. **35. a.** $V = LWH$ **b.** 8 feet **37. a.** $I = Prt$ **b.** \$600 **c.** $B = P + Prt$ **d.** \$2400 **39.** 82 **41.** $W = \frac{A}{L}$ **43.** $T = \frac{PV}{nR}$ **45.** $M = -\frac{Ur}{Gm}$ **47.** $B = \frac{2A}{H}$ **49.** $t = \frac{v - v_0}{g}$ **51.** $r = \frac{A - P}{Pt}$ **53.** $b = 3A - a - c$ **55.** $x = \frac{y - k + mh}{m}$ **57.** $y = a - x$ **59.** $y = -\frac{3}{4}x + 4$ **61.** $y = -\frac{1}{2}x + 2$ **63.** $y = \frac{5}{2}x - 3$ **65.** $y = -\frac{3}{7}x - \frac{5}{7}$ **67. a.** 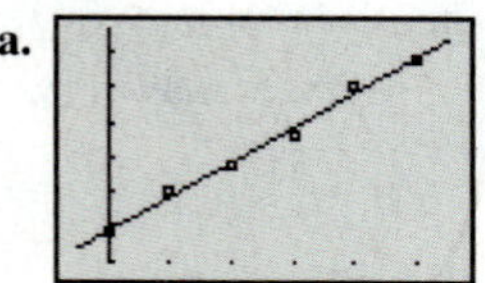 yes **b.** $t = \frac{p - 54.2}{2.5}$ **c.** 2009 **d.** 2014, 2015, 2016, 2017, and 2018 **69. a.** $s = 20.1t + 356$ **b.** $t = \frac{s - 356}{20.1}$ **c.** 2016 **d.** 2017, 2018, 2020, 2021, and 2022

71. a. $p = -0.8t + 71$ **b.** 2016 **c.** $t = \frac{p - 71}{-0.8}$ **d.** 2016 **e.** The results are the same; $t = \frac{p - 71}{-0.8}$; answers may vary. **f.** $p = -0.8t + 71$; answers may vary; 65% **73. a.** The entries in the third column are $50 \cdot 4$, $70 \cdot 3$, $65 \cdot 2$, $55 \cdot 5$, st, all in miles; $d = st$ **b.** The units for both of the expressions d and st are miles. **c.** $t = \frac{d}{s}$ **d.** 4.5; it takes 4.5 hours to travel 315 miles at 70 miles per hour. **e.** 6.44 hours **75.** The units of s are miles per hour and the units of dt are miles-hours. Since the units are different, the formula is incorrect. **77. a.** It doubles the perimeter. **b.** The area is multiplied by 4. **79.** Answers may vary. **81.** Answers may vary.

Chapter 4 Review Exercises

1. $9 + 5w$ **2.** $8 + pw$ **3.** $(2 + k) + y$ **4.** $b(xw)$ **5.** $-20x$ **6.** $-24x - 12$ **7.** $12y - 28$ **8.** $-3x + 6y + 8$ **9.** $\frac{7}{9}x$ **10.** $4a - 9b - 7$ **11.** $-18x - 8y$ **12.** $-8.06x - 20.2$ **13.** $-5m - 4$ **14.** $-3a - 40b$ **15.** $-4(x - 7)$; $-4x + 28$ **16.** $-7 + 2(x + 8)$; $2x + 9$ **17.** Answers may vary; $a(b + c) = ab + ac$ **18.** Answers may vary. **19.** $-5(x - 4)$, $5(4 - x)$, $-2(x - 10) - 3x$, $-5x + 20$ **20.**

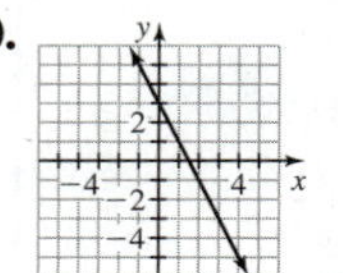

21. **22.** no **23.** 7 **24.** -5 **25.** 3 **26.** -6 **27.** 2.7 **28.** $-\frac{7}{2}$ **29.** $\frac{5}{11}$ **30.** 1 **31.** $\frac{15}{8}$ **32.** $\frac{38}{3}$ **33.** $\frac{44}{3}$ **34.** Answers may vary; 7 **35.** Answers may vary. **36.** -8.31

37. a. $n = -38t + 3189$ **b.** 2017 **c.** 2923 prisoners **38.** $\frac{39}{4}$ **39.** -39 **40.** -1.64 **41.** 3 **42.** 1 **43.** 2 **44.** 0 **45.** the set of all real numbers; identity **46.** $\frac{22}{15}$; conditional equation **47.** empty set; inconsistent equation **48.** Answers may vary; 12 **49.** linear expression **50.** linear equation **51.** $-2t$ **52.** 3 **53.** -5 **54.** $0.2a + 0.1$ **55.** $-3p - 54$ **56.** $-\frac{11}{32}$ **57.** -18 **58.** $\frac{13}{3}r - \frac{11}{12}$ **59.** no; answers may vary. **60.** Answers may vary; $\frac{2}{3}x + \frac{7}{5}$ **61.** $4(6 - x) = 17$; $\frac{7}{4}$ **62.** $x - \frac{x}{2}$; $\frac{x}{2}$ **63.** $P = 2A + 2B + 2C + 2D + 2E$ **64. a.** $T = 15n + 25w$ **b.** 220; if the total cost is \$11,050 and 370 \$15 tickets were sold, 220 \$25 tickets were sold. **65.** $r = \frac{C}{2\pi}$ **66.** $c = P - a - b$ **67.** $y = \frac{1}{2}x - 3$ **68.** $T = \frac{2A - HB}{H}$ **69. a.** $v = 0.05t + 1.59$ **b.** $t = \frac{v - 1.59}{0.05}$ **c.** 2017 **d.** 2017 **e.** The results are the same; $t = \frac{v - 1.59}{0.05}$; answers may vary. **f.** $v = 0.05t + 1.59$; answers may vary; 2.09 billion visits **70.** 2017 **71.** 2.3 million participants

Chapter 4 Test

1. $3p + 4$ **2.** $(3x)y$ **3.** $-4x + 6$ **4.** $-4.08x + 0.96$ **5.** $-22w + 53$ **6.** $-11a - 3b - 2$ **7.** Answers may vary; $-2x - 10$ **8.** **9.** $\frac{11}{3}$ **10.** 10 **11.** 7 **12.** $\frac{10}{13}$ **13.** $-\frac{41}{4}$ **14.** $-\frac{32}{25}$ **15.** 1.18 **16.** -32 **17.** $23x + 24$ **18.** $-\frac{12}{11}$ **19.** no; answers may vary. **20.** Answers may vary. **21.** $5(x - 2) = 29$; $\frac{39}{5}$ **22.** $2 + 4(3 + x)$; $4x + 14$ **23.** 2 **24.** 4 **25.** -2 **26.** 4 **27. a.** $n = 21.7t + 520$ **b.** 693.6 thousand **c.** 2016 **28.** 11 months **29.** 304 thousand complaints **30.** 18 feet **31.** $a = 2A - b$

Cumulative Review of Chapters 1–4

1. n; 275, 300; 0, -150; answers may vary. **2.**

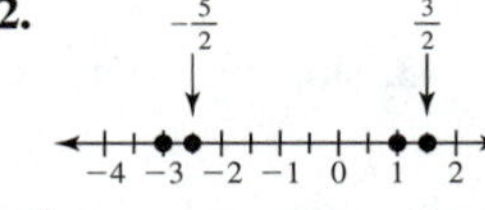

3. In 2011, the average number of unique monthly visitors was 790 million visitors. **4.** -2 **5.** 4 **6.** $-\frac{3}{10}$ **7.** $-\frac{47}{40}$ **8.** $-\frac{3}{5}$ **9.** $\frac{19}{6}$ feet **10.** -7 points **11.** -27 **12.** $-\frac{1}{2}$; decreasing **13.** undefined slope; vertical **14.** road A

15.

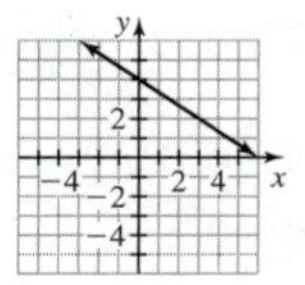

16.

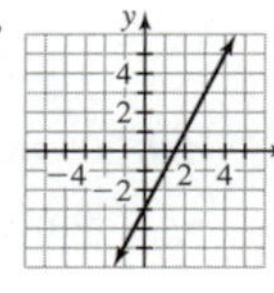

17.

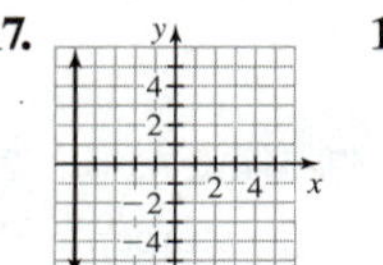

18.

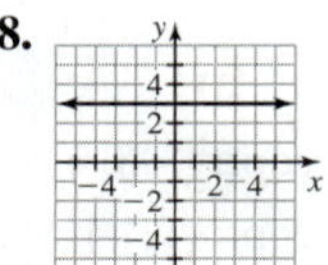

19. $y = 3x - 2$ **20.** -932 dollars per month **21.** 56.2 thousand reports per year **22. a.** 0.14; sales increase by \$0.14 billion per year. **b.** $s = 0.14t + 1.06$ **c.** \$1.9 billion **d.** 2018

23. equations 2, 3, and 4 **24. a.**

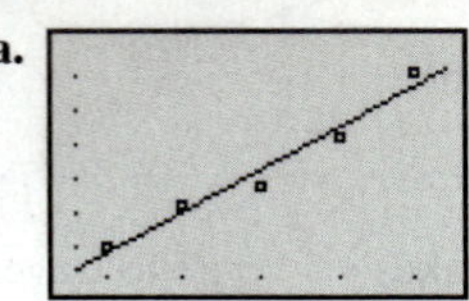

yes **b.** 120.9; the number of firearms discovered at checkpoints is increasing by 120.9 firearms per year. **c.** (0, 544.4); in 2005, 544.4 firearms were discovered at checkpoints. **d.** 2017 **e.** 2116 firearms; 5.8 firearms **25.** 3 **26.** -8 **27.** $-3a - 7b + 6$ **28.** $14p + 7$ **29.** 2 **30.** $-6a - 4$

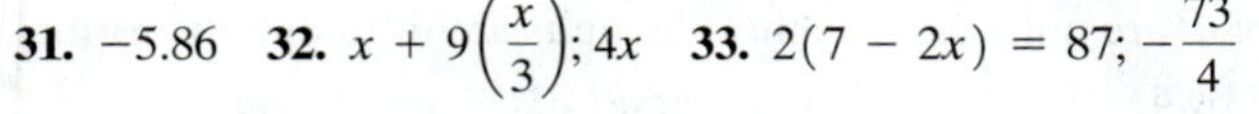

31. -5.86 **32.** $x + 9\left(\frac{x}{3}\right)$; $4x$ **33.** $2(7 - 2x) = 87$; $-\frac{73}{4}$ **34.** $h = \frac{A}{2\pi r}$ **35.** $y = \frac{2}{3}x - 2$

Chapter 5

Homework 5.1 **1.** slope: 2; y-intercept: $(0, -3)$

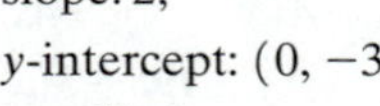

3. slope: $-\frac{3}{5}$; y-intercept: $(0, -2)$

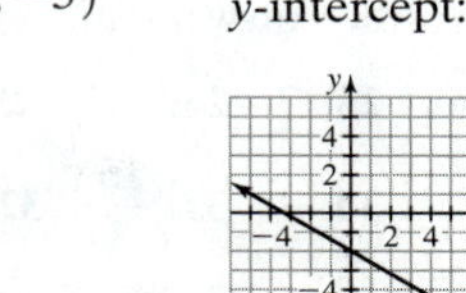

5. slope: 0; y-intercept: $(0, -4)$

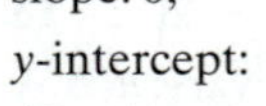

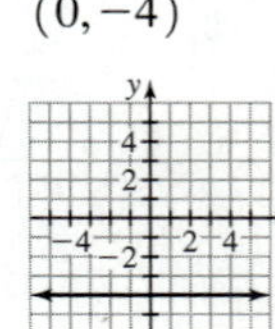

7. slope: -1; y-intercept: $(0, 3)$

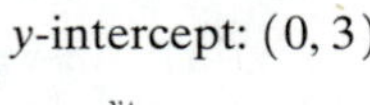

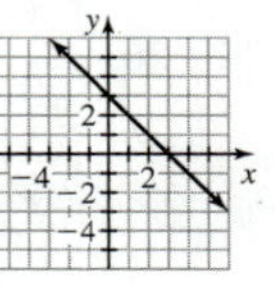

9. slope: -2; y-intercept: $(0, 4)$

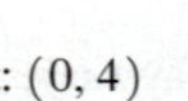

11. slope: 2; y-intercept: $(0, -1)$

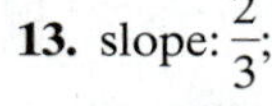

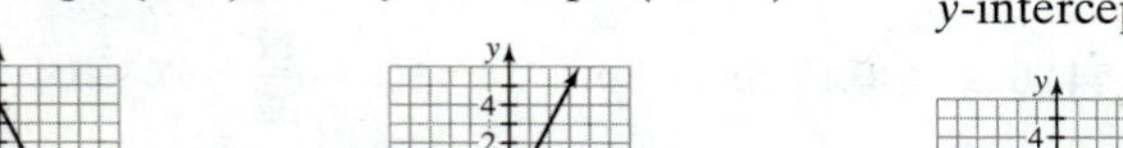

13. slope: $\frac{2}{3}$; y-intercept: $(0, 0)$

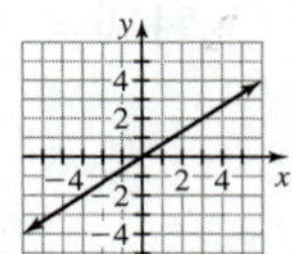

15. slope: $-\frac{3}{2}$; y-intercept: $(0, -2)$

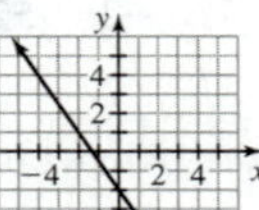

17. slope: $\frac{4}{5}$; y-intercept: $(0, -3)$

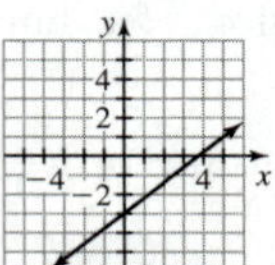

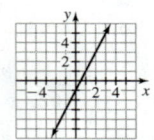

19. slope: $\frac{3}{4}$; y-intercept: $(0, -2)$

21. slope: $\frac{2}{5}$; y-intercept: $(0, -2)$

23. slope: $-\frac{1}{4}$; y-intercept: $(0, 1)$

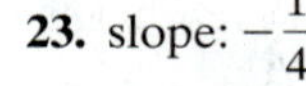

25. slope: $-\frac{1}{2}$; y-intercept: $(0, -2)$

27. slope: -4; y-intercept: $(0, -2)$

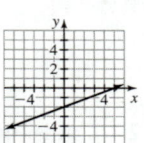

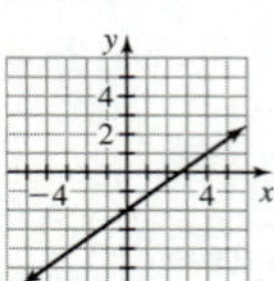

29. slope: $\frac{3}{2}$; y-intercept: $(0, 2)$

31. slope: $-\frac{5}{3}$; y-intercept: $(0, 0)$

33. slope: 0; y-intercept: $(0, 3)$

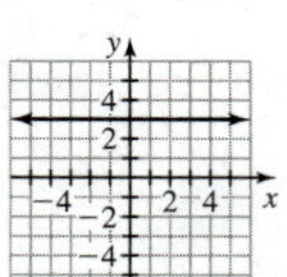

35. slope: $\frac{1}{2}$; y-intercept: $\left(0, -\frac{3}{2}\right)$

37. slope: $\frac{2}{3}$; y-intercept: $(0, 2)$

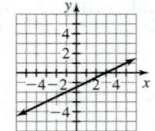

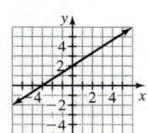

39. slope: $-\frac{1}{2}$; y-intercept: $(0, 1)$ **41.** slope: $\frac{a}{b}$; y-intercept: $\left(0, -\frac{c}{b}\right)$ **43.** slope: $\frac{b}{a}$; y-intercept: $\left(0, -\frac{bd}{a}\right)$ **45.** slope: $\frac{1}{a}$; y-intercept: $(0, -b)$

47. slope: 1; y-intercept: $\left(0, -\frac{d}{a}\right)$ **49.** slope: -1; y-intercept: $(0, a)$

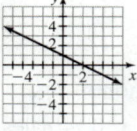

51. x-intercept: $(6, 0)$; y-intercept: $(0, -2)$

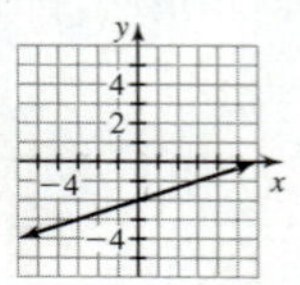

53. x-intercept: $(5, 0)$; y-intercept: $(0, 3)$

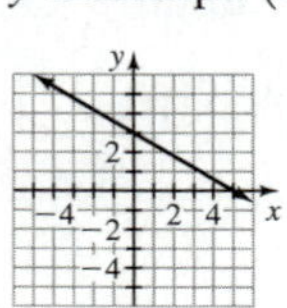

55. x-intercept: $(-6, 0)$; y-intercept: $(0, 4)$

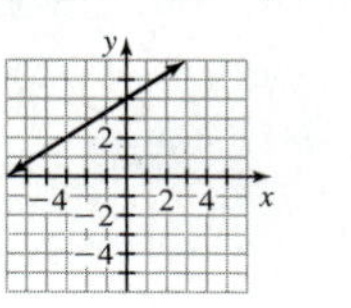

57. x-intercept: $(2, 0)$; y-intercept: $(0, 6)$

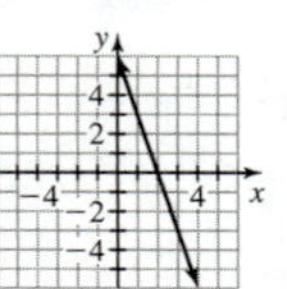

59. x-intercept: $(4, 0)$; y-intercept: $(0, -6)$

61. x-intercept: $(3, 0)$; y-intercept: $(0, 5)$

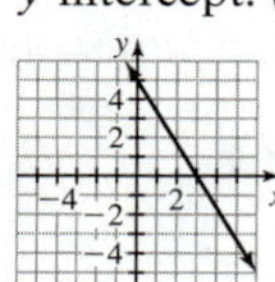

63. x-intercept: $(1.21, 0)$; y-intercept: $(0, 2.68)$ **65.** x-intercept: $(-2.05, 0)$; y-intercept: $(0, 3.47)$ **67.** x-intercept: $(2.07, 0)$; y-intercept: $(0, 9.32)$ **69.** x-intercept: $(-14.94, 0)$; y-intercept: $(0, -37.21)$

71. x-intercept: $\left(-\frac{b}{m}, 0\right)$; y-intercept: $(0, b)$ **73.** x-intercept: $\left(\frac{c}{ab}, 0\right)$; y-intercept: $\left(0, \frac{c}{a}\right)$ **75.** x-intercept: $\left(-\frac{bd}{a}, 0\right)$; y-intercept: $\left(0, \frac{d}{c}\right)$ **77.** x-intercept: $(a, 0)$; y-intercept: $(0, b)$ **79.**

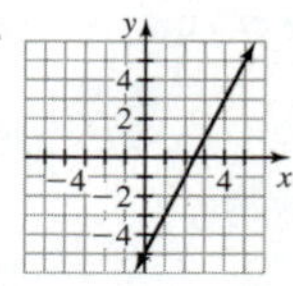

81.

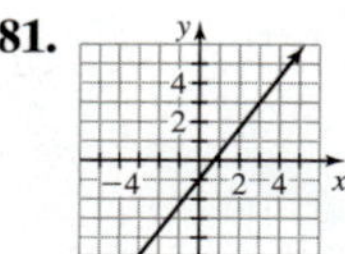

83.

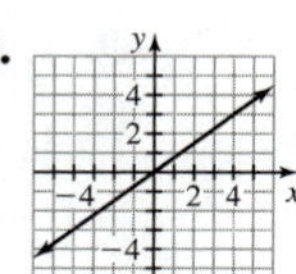

85.

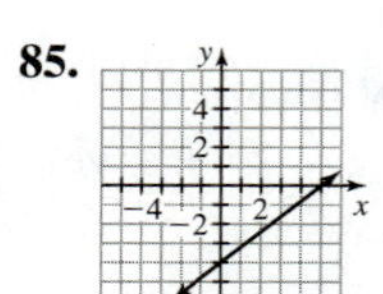

87. a. 1 **b.** 2 **c.**

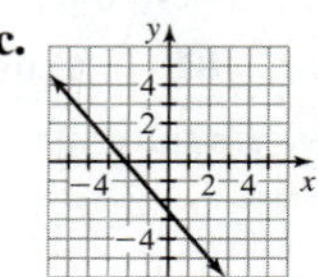

89. no; $-\frac{2}{3}$ **91. a.**

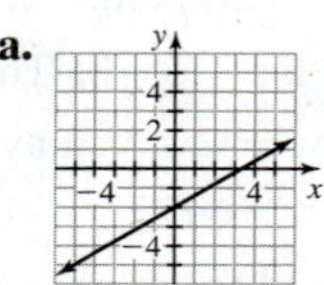

b. Answers may vary. **c.** For each solution, the difference of three times the x-coordinate and five times the y-coordinate is equal to 10.

93. a. x-intercept: $(5, 0)$; y-intercept: $(0, 7)$ **b.** x-intercept: $(4, 0)$; y-intercept: $(0, 6)$ **c.** x-intercept: $(a, 0)$; y-intercept: $(0, b)$ **d.** $\frac{x}{2} + \frac{y}{5} = 1$ **95.** Answers may vary. **97.** Answers may vary. **99.** 1 **101.** 2 **103.** $(4, 0)$ **105.** $\frac{1}{2}$ **107.** $-6x + 26$; linear expression in one variable **109.** 3; linear equation in one variable

Homework 5.2 **1.** relations 2 and 3 **3.** no **5.** yes **7.** yes **9.** no **11.** no **13.** yes **15.** yes **17.** yes **19.** yes **21.** no **23.** yes **25.** yes; the graph will pass the vertical line test. **27.** no; the graph will not pass the vertical line test. **29. a.** Answers may vary. **b.**

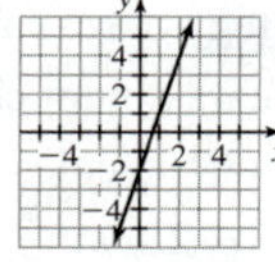

c. For each input–output pair, the output is 2 less than 3 times the input.

31. domain: $-4 \le x \le 5$; range: $-2 \le y \le 3$ **33.** domain: $-5 \le x \le 4$; range: $-2 \le y \le 3$ **35.** domain: $-4 \le x \le 4$; range: $-2 \le y \le 2$ **37.** domain: $0 \le x \le 4$; range: $0 \le y \le 2$ **39.** domain: all real numbers; range: $y \le 4$ **41.** domain: $x \ge 0$; range: $y \ge 0$ **43.** yes **45.** no **47.** Answers may vary. **49.** Answers may vary. No; an input of $x = 2$ gives two different outputs. **51.** Answers may vary. **53.** Answers may vary. **55.** nonvertical lines; answers may vary. **57.** no; no input corresponds to two different outputs, so the definition of a function is not violated. **59.** yes **61.** no **63.** $-4x - 18$; linear expression in one variable **65.** $-\frac{9}{2}$; linear equation in one variable

Homework 5.3 **1.** 26 **3.** 0 **5.** $6a + 8$ **7.** -2 **9.** 33 **11.** $\frac{1}{6}$ **13.** $\frac{3a - 13}{5a - 13}$ **15.** -16 **17.** -3 **19.** -4 **21.** $-20a - 7$ **23.** $-2a - 7$ **25.** $-4a - 23$ **27.** $-4a + 4h - 7$ **29.** $\frac{1}{3}$ **31.** $\frac{3}{2}$ **33.** $\frac{7 - a}{3}$ **35.** 118.31 **37.** 14.10 **39.** 4 **41.** 1, 3 **43.** 4 **45.** 1.2 **47.** 6 **49.** -3 **51.** 4.5 **53.** all real numbers **55.** 1 **57.** $-4 \le x \le 5$ **59.** -3 **61.** $-5 \le x \le 4$ **63.** x-intercept: $\left(\frac{8}{5}, 0\right)$; y-intercept: $(0, -8)$ **65.** x-intercept: $(0, 0)$; y-intercept: $(0, 0)$ **67.** no x-intercept; y-intercept: $(0, 5)$ **69.** x-intercept: $(6, 0)$; y-intercept: $(0, -3)$ **71.** x-intercept: $(17.52, 0)$; y-intercept: $(0, -45.21)$ **73.** $f(t) = 1.04t + 46.7$ **75.** $h(d) = -0.02d + 11.9$ **77. a.** Answers may vary. **b.** **c.** The output is 4 more than -3 times the input. **79.** The student should substitute 5 for $f(x)$, not x; 3 **81. a.** 12; 20; 32; yes **b.** 4; 9; 25; no **c.** 3; 4; 5; no **d.** no **83.** no; answers may vary. **85.** $-\frac{3}{4}$ **87.** x-intercept: $(4, 0)$; y-intercept: $(0, -3)$ **89.** 3; linear function **91.** 5; linear equation in one variable

Homework 5.4 **1.** $y = 2x - 1$ **3.** $y = -3x + 1$ **5.** $y = -6x - 15$ **7.** $y = \frac{2}{5}x - \frac{1}{5}$ **9.** $y = -\frac{3}{4}x - \frac{13}{2}$ **11.** $y = 3$ **13.** $x = -2$ **15.** $y = 2.1x - 13.67$ **17.** $y = -6.59x - 17.95$ **19.** $y = 2x - 4$ **21.** $y = 5x - 2$ **23.** $y = -2x - 14$ **25.** $y = -4x + 9$ **27.** $y = 2$ **29.** $x = -4$ **31.** $y = \frac{1}{2}x + 1$ **33.** $y = -\frac{1}{6}x + \frac{3}{2}$ **35.** $y = -\frac{2}{7}x + \frac{3}{7}$ **37.** $y = \frac{3}{5}x + \frac{2}{5}$ **39.** $y = \frac{3}{2}x - 2$ **41.** $y = -1.70x - 5.46$ **43.** $y = 0.48x - 6.11$ **45.** $y = 3x - 7$ **47.** $y = -2x + 2$ **49.** $y = \frac{3}{4}x + \frac{7}{4}$ **51.** $y = \frac{1}{6}x - \frac{3}{2}$ **53.** $y = 3$ **55.** $x = -5$ **57.** $y = -\frac{1}{2}x + \frac{19}{2}$ **59.** $y = \frac{1}{3}x + \frac{22}{3}$ **61.** $y = -\frac{5}{4}x + \frac{31}{2}$ **63.** $y = -\frac{3}{2}x - \frac{11}{2}$ **65.** $y = 3$ **67.** $x = 2$ **69.** $y = -\frac{4}{3}x + \frac{23}{3}$ **71.** Answers may vary. **73. a.** $y = \frac{3}{2}x - 3$ **b.** **c.** Answers may vary.

75. a. possible; answers may vary. **b.** possible; answers may vary. **c.** not possible; answers may vary. **d.** possible; $y = 0$ **77.** $E = 2t + 5$ **79. a. i.** $y = 3x - 5$ **ii.** $y = 3x - 5$ **b.** The results are the same. **81. a.** Answers may vary. **b.** Answers may vary. **c.** Answers may vary. **d.** no such line; answers may vary. **83. a–c.** Answers may vary. **85.** $y = 2x + 1$; $y = 2x + 1$; yes; answers may vary. **87.** Set 1: $y = -2x + 25$; Set 2: $y = 4x + 12$; Set 3: $y = -5x + 77$; Set 4: $y = 3$ **89.** Answers may vary.

91. a.

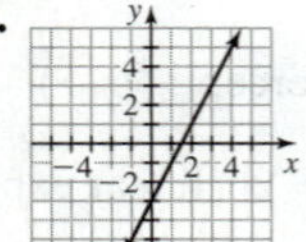

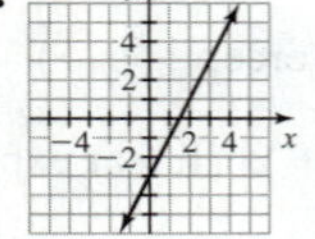

b. $y = 2x - 3$; it is the same equation. **93.** linear equation in two variables

95. 2; linear expression in two variables

Homework 5.5

For many exercises in this section, answers may vary.

1. $n = 15t + 1121$ **3.** $n = -85.2t + 3232.2$ **5. a.** $p = 0.41t + 8.29$ **b.** 0.41; the percentage of sexual harassment charges filed by men is increasing by 0.41 percentage point per year. **c.** $(0, 8.29)$; in 1990, 8.29% of sexual harassment charges were filed by men. **7. a.** $n = 14.67t + 8.33$ **b.** 14.67; the number of suicides among soldiers is increasing by 14.67 suicides per year. **c.** $(0, 8.33)$; there were about 8 suicides among soldiers in 2000. **9.** $L = 1.96a + 15.21$ **11.** $y = 2.5x - 2.2$; answers may vary. **13.** $y = -1.11x + 20.83$; answers may vary. **15. a.**

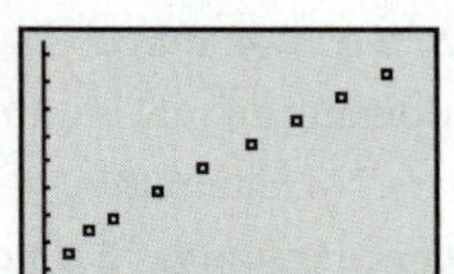

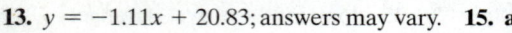

b. $H = 4.5d + 15.3$; answers may vary. **c.**

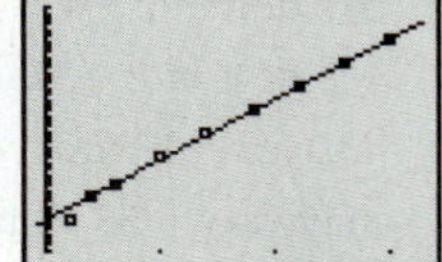

17. a.

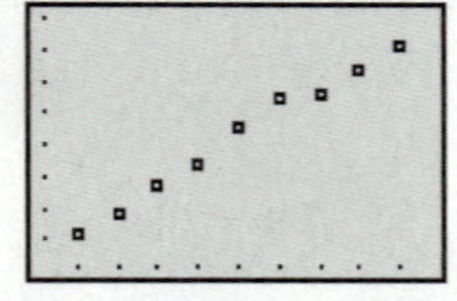

b. $p = 0.76t - 42.04$; answers may vary. **c.**

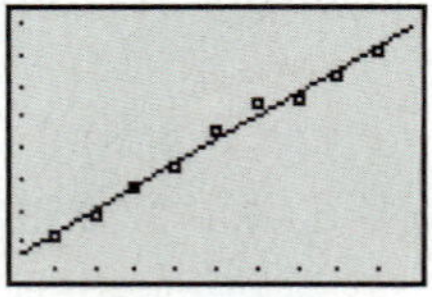

19. a.

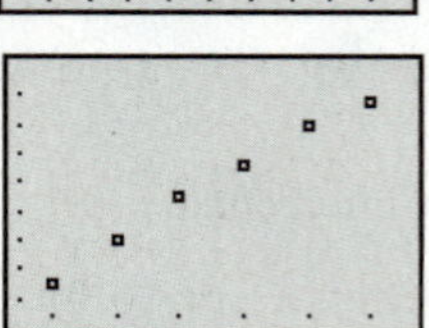

b. $p = 12.74n + 4.40$; answers may vary. **c.**

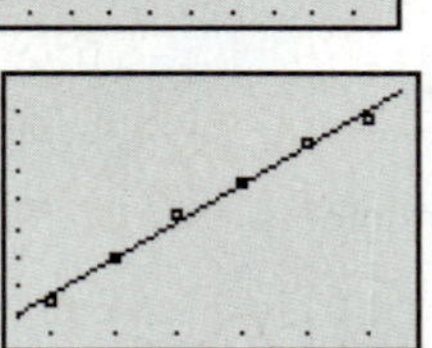

21. a.

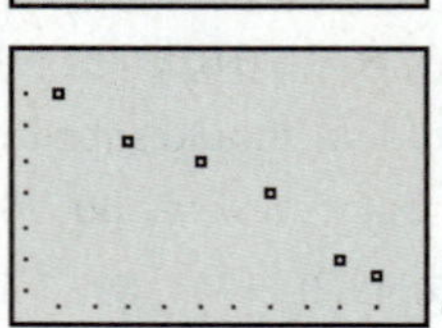

b. $p = -1.19t + 64.86$; answers may vary. **c.**

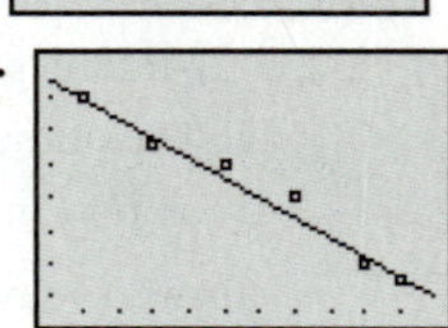

23. a.

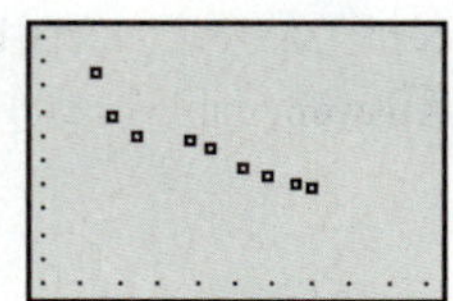

b. $r = -0.27t + 70.45$; answers may vary. **c.**

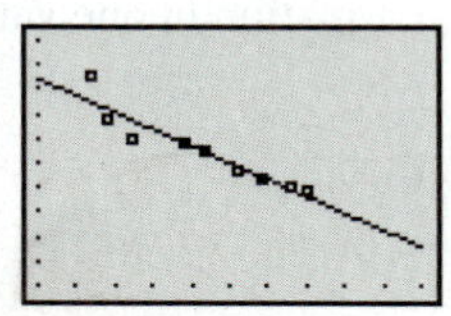

25. a.

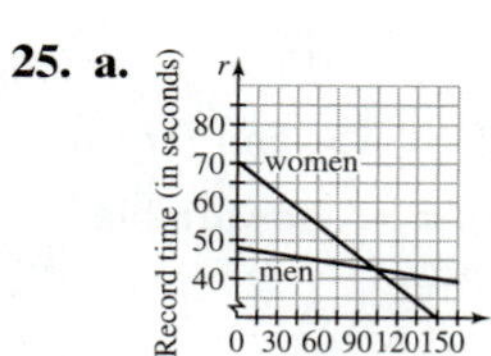

b. yes; 42.62 seconds; 2003 **c.** yes; after 2003 **27. a.**

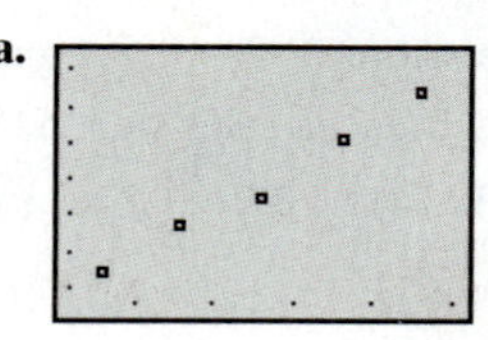

b. $p = 2.48x - 23.64$; answers may vary.

c.

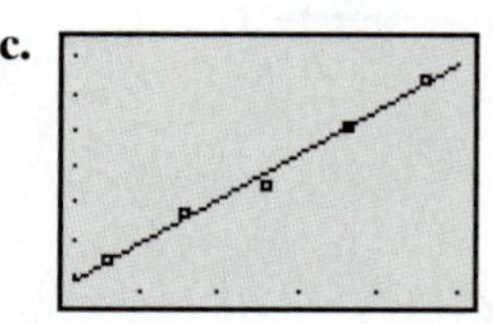

29. student B **31.**

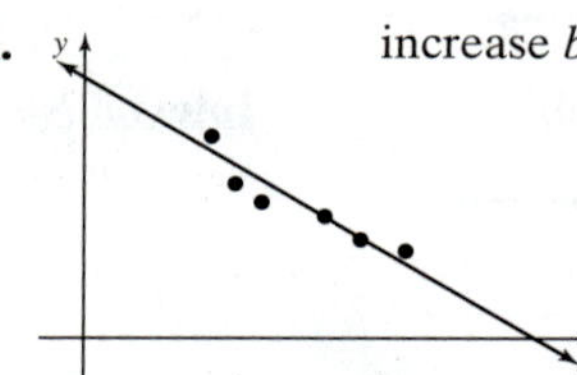

increase b **33.** Answers may vary. **35. a.** $V = 2t + 10$; the units for both of the expressions $2t + 10$ and V are dollars.

b.

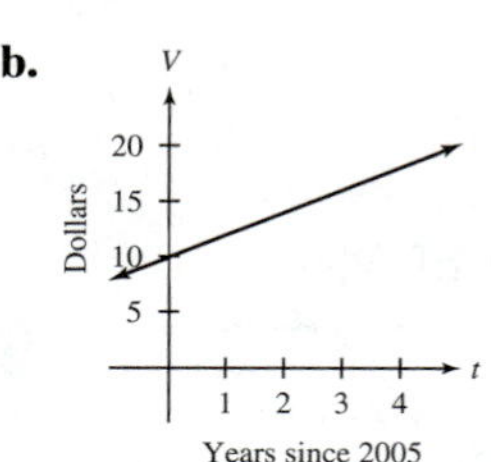
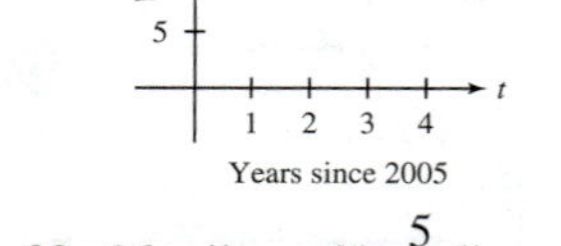

c. Answers may vary. **37. a.** $n = -731t + 50{,}435$ **b.** -731; the average attendance is decreasing by 731 spectators per year. **c.** The units for both of the expressions $-731t + 50{,}435$ and n are number of spectators.

39. 86 miles **41.** $\frac{5}{2}$; linear equation in one variable **43.**

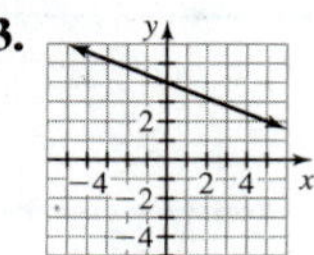

linear equation in two variables

Homework 5.6

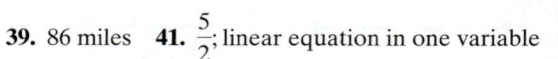

1. a. $f(t) = 0.76t - 42.04$ **b.** 47.64; in 2018, about 47.6% of births will be outside marriage. **c.** 117.16; in 2017, 47% of births will be outside marriage. **d.** 2087; model breakdown has likely occurred. **e.** 31.7%; -0.7 percentage point **3. a.** $f(t) = -1.19t + 64.86$ **b.** 31.54; in 2018, baseball will be the favorite sport of about 32% of Americans. **c.** 30.97; in 2021, baseball will be the favorite sport of 28% of Americans. **d.** $(0, 64.86)$; in 1990, baseball was the favorite sport of about 65% of Americans. **e.** $(54.50, 0)$; in 2045, baseball will not be the favorite sport of any Americans; model breakdown has likely occurred. **5. a.** $f(n) = 12.74n + 4.40$ **b.** $(0, 4.40)$; the price of renting skis for 0 days is \$4.40; model breakdown has occurred. **c.** \$93.58 **d.**

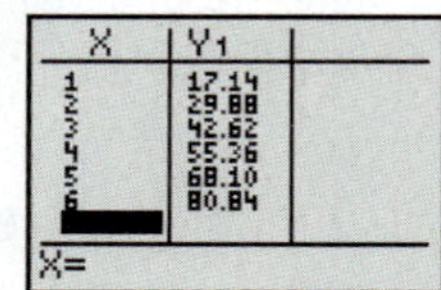

X	Y1
1	17.14
2	29.88
3	42.62
4	55.36
5	68.10
6	80.84

X=

12.74; equal to slope of the graph of f; the price increases by \$12.74 for each additional day. **e.** 17.14, 14.94, 14.21, 13.84, 13.62, 13.47; 13.47; the cost per day is lowest for a 6-day package (\$13.47 per day). **f.** The cost increase for each additional day is less than the \$13 charge per day; the cost per day for the 6-day package is more than the \$13 charge per day.

7. a.

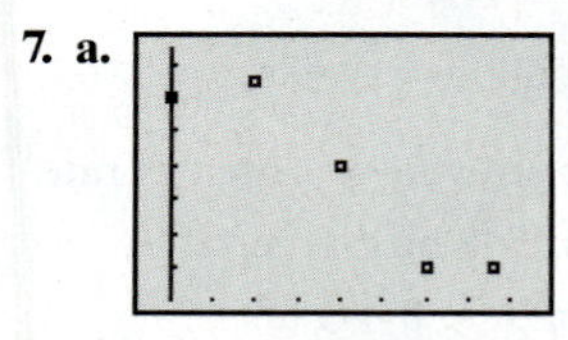

b. $f(t) = -0.16t + 67.88$ **c.** 2019 **d.** $(0, 67.88)$; in 1970, about 67.9% of married persons said they were "very happy" with their marriages. **e.** $(424.25, 0)$; in 2394, no married persons will say they are "very happy" with their marriages; model breakdown has likely occurred.

9. a.

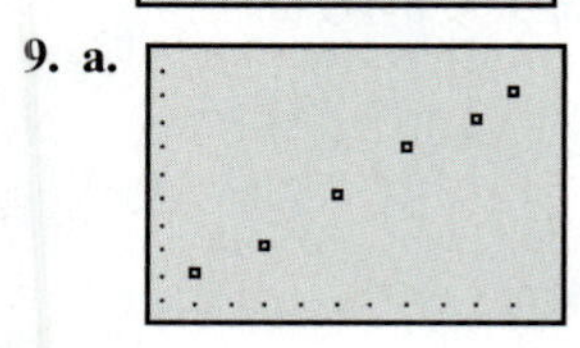

b. $f(t) = 24.07t + 423.92$; yes **c.** 689 million; 67 million boardings **d.** \$5.7 billion **11. a.** $f(C) = 1.8C + 32$ **b.** 75.2°F **c.** 17.78°C **d.** -459.67°F **13. a.** $f(d) = 4.5d + 15.3$ **b.** 14 dog years **c.** 146 human years **d.** 23 dog years **e.** 4.5; a dog aging 1 year is equivalent to a human aging 4.5 years. **f.** $H = 7d$ **g.** 7; a dog aging 1 year is equivalent to a human aging 7 years; the slope of the graph of the equation in part (f) is greater than the slope found in part (e).

15. a. $f(x) = 2.48x - 23.64$ **b.** 50 points **c.** less than 10 points **d.** 30 students; no **e.** 83 students **17.** \$13.7 thousand **19.** 2018 **21. a.** \$4.7 billion **b.** 2021 **23.** 218 points **25. a.** $f(t) = -160t + 640$ **b.**

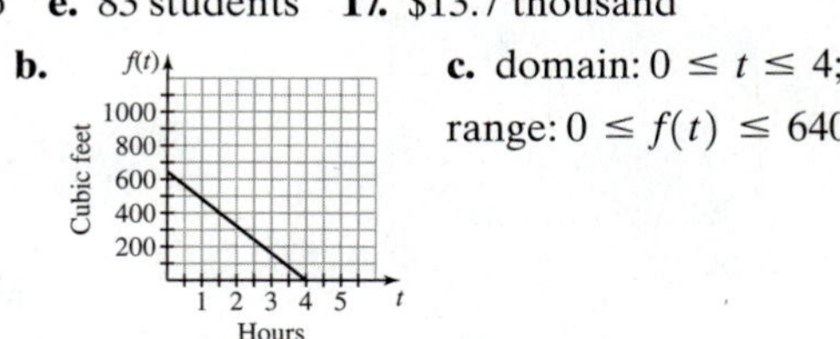

c. domain: $0 \le t \le 4$; range: $0 \le f(t) \le 640$

27. Answers may vary. **29. a.** $f(t) = 0.42t + 31.4$ **b.** $(0, 31.4)$; in 2010, there were 31.4 million Americans who lived alone.

c. The units for both of the expressions $0.42t + 31.4$ and $f(t)$ are millions of people. **d.** 2024 **31. a.** $f(t) = -6.9t + 132$ **b.** $(0, 132)$; in 2009, the annual echinacea sales were \$132 million. **c.** The units for both of the expressions $-6.9t + 132$ and $f(t)$ are millions of dollars. **d.** 2017 **33.** 2019 **35.** $\frac{17}{18}$; linear equation in one variable **37.** $-18x + 17$; linear expression in one variable

Homework 5.7

1. true **3.** true **5.** [graph] **7.** [graph] **9.** [graph] **11.** [graph]

13.

In Words	Inequality	Graph	Interval Notation
numbers greater than or equal to 4	$x \geq 4$	[graph]	$[4, \infty)$
numbers less than or equal to -2	$x \leq -2$	[graph]	$(-\infty, -2]$
numbers less than 1	$x < 1$	[graph]	$(-\infty, 1)$
numbers greater than -5	$x > -5$	[graph]	$(-5, \infty)$

15. 3, 6 **17.** -4 **19.** $x > 1; (1, \infty)$; **21.** $x < -3; (-\infty, -3)$; **23.** $x \leq 3; (-\infty, 3]$; **25.** $x \geq -2; [-2, \infty)$; **27.** $t \leq -2; (-\infty, -2]$; **29.** $x < -\frac{1}{2}; \left(-\infty, -\frac{1}{2}\right)$; **31.** $x \leq 0; (-\infty, 0]$; **33.** $x > -2; (-2, \infty)$; **35.** $x \leq -3; (-\infty, -3]$; **37.** $x \geq 1; [1, \infty)$; **39.** $x > 4; (4, \infty)$; **41.** $c \geq -3; [-3, \infty)$; **43.** $x \geq -3; [-3, \infty)$; **45.** $x < 2.5; (-\infty, 2.5)$; **47.** $b < 2; (-\infty, 2)$; **49.** $x > -5; (-5, \infty)$; **51.** $x \leq 1; (-\infty, 1]$; **53.** $a < -1; (-\infty, -1)$; **55.** $x \leq \frac{5}{2}; \left(-\infty, \frac{5}{2}\right]$; **57.** $x \leq \frac{4}{3}; \left(\infty, \frac{4}{3}\right]$ **59.** $x \leq -1.7; (-\infty, -1.7]$; **61.** $y \geq \frac{5}{3}; \left[\frac{5}{3}, \infty\right)$; **63.** $x > 7; (7, \infty)$; **65.** $x \leq -\frac{7}{12}; \left(-\infty, -\frac{7}{12}\right]$; **67.** $c \geq -31; [-31, \infty)$; **69.** $x < -5; (-\infty, -5)$; **71.** $1 < x < 5; (1, 5)$; **73.** $-5 \leq x \leq 6; [-5, 6]$; **75.** $-3 \leq x < 5; [-3, 5)$; **77.** $3 < x \leq \frac{11}{2}; \left(3, \frac{11}{2}\right]$; **79. a.** 5.9; total student loan amount is increasing by \$5.9 billion per year. **b.** after 2018 **81. a.** $f(t) = -0.31t + 52.66$ **b.** before 2012 **c.** The number of nonmarried households is growing at a greater rate than the number of married households. **83. a.** $f(t) = -1.33t + 61.02$ **b.** 21.1 births per 1000 women; 232,248 births **c.** after 2028 **85.** Answers may vary; $x > -5$ **87. a.** Answers may vary. **b.** Answers may vary. **89. a.** $x = 3$ **b.** $x < 3$ **c.** $x > 3$ **d.** **91.** Answers may vary. **93.** Answers may vary. **95.** 4 **97.** $x < 4; (-\infty, 4)$;

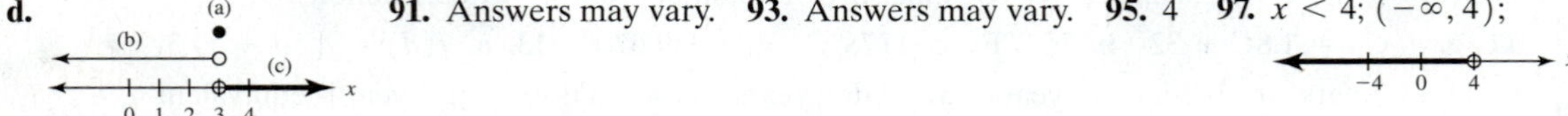

99. $x + 5 > 2; x > -3; (-3, \infty)$; **101.** $2x \leq 5x - 6; x \geq 2; [2, \infty)$; **103.** Answers may vary. **105.** Answers may vary.

Chapter 5 Review Exercises

1. slope: $\frac{5}{3}$; y-intercept: $(0, 0)$

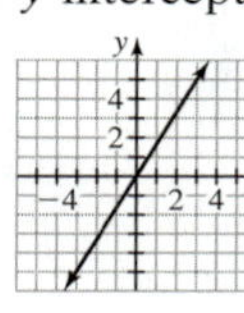

2. slope: $\frac{3}{2}$; y-intercept: $(0, 3)$

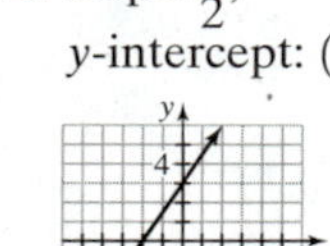

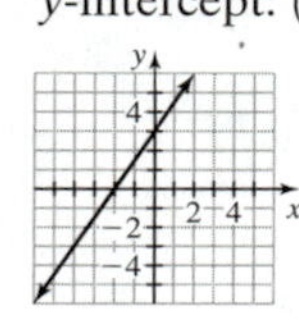

3. slope: $-\frac{1}{3}$; y-intercept: $(0, 2)$

4. slope: $-\frac{2}{5}$; y-intercept: $(0, 4)$

5. slope: 0; y-intercept: $(0, 4)$

6. slope: $-\frac{2}{3}$; y-intercept: $(0, -5)$

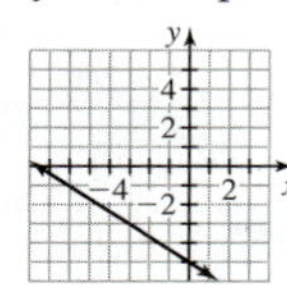

7. slope: $-\frac{1}{2}$; y-intercept: $(0, 2)$

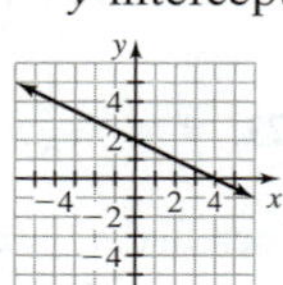

8. slope: 1; y-intercept: $\left(0, -\frac{c}{a}\right)$

9. x-intercept: $(5, 0)$; y-intercept: $(0, -4)$

10. x-intercept: $(-4, 0)$; y-intercept: $(0, -3)$

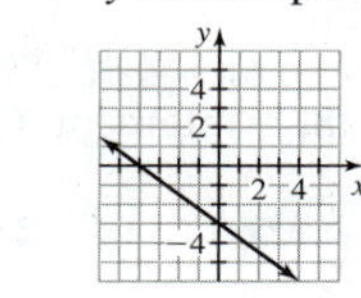

11. x-intercept: $(2, 0)$; y-intercept: $(0, -4)$

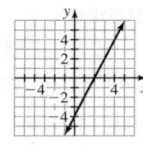

12. x-intercept: $(3, 0)$; y-intercept: $(0, -2)$

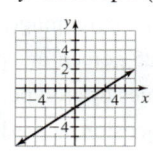

13. x-intercept: $\left(-\frac{b}{a}, 0\right)$; y-intercept: $\left(0, \frac{b}{c}\right)$ **14.** x-intercept: $(9.48, 0)$; y-intercept: $(0, -22.95)$ **15.** x-intercept: $(-37.99, 0)$; y-intercept: $(0, 97.25)$

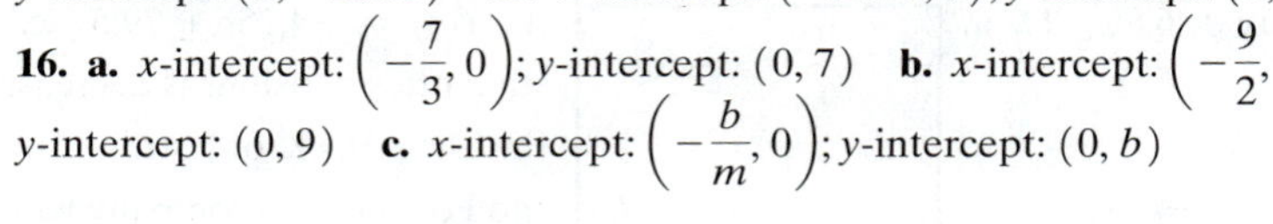

16. a. x-intercept: $\left(-\frac{7}{3}, 0\right)$; y-intercept: $(0, 7)$ **b.** x-intercept: $\left(-\frac{9}{2}, 0\right)$; y-intercept: $(0, 9)$ **c.** x-intercept: $\left(-\frac{b}{m}, 0\right)$; y-intercept: $(0, b)$

17. a. slope: $\frac{1}{2}$; y-intercept: $(0, -2)$

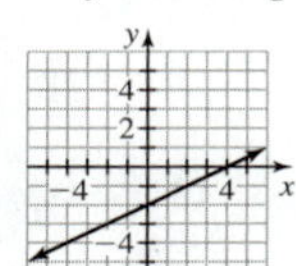

b. x-intercept: $(4, 0)$; y-intercept: $(0, -2)$

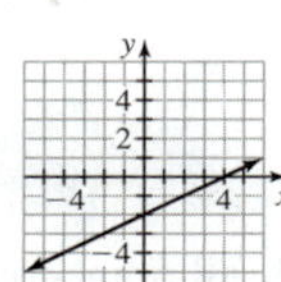

c. The graphs are the same; answers may vary.

18. relations 1 and 3 **19.** no **20.** yes **21.** no **22.** no **23.** domain: all real numbers; range: $y \le 4$ **24.** 20 **25.** $\frac{3}{4}$ **26.** $-10a - 33$ **27.** $-\frac{7}{6}$ **28.** $\frac{a+4}{2}$ **29.** 1 **30.** 3.6 **31.** 2 **32.** 4 **33.** $-5 \le x \le 6$ **34.** $-2 \le y \le 4$ **35.** 4 **36.** 1 **37.** x-intercept: $\left(\frac{3}{7}, 0\right)$; y-intercept: $(0, 3)$; **38.** x-intercept: $\left(\frac{7}{2}, 0\right)$; y-intercept: $(0, 2)$ **39.** $y = -4x + 7$ **40.** $y = -\frac{2}{3}x - 8$ **41.** $x = 2$ **42.** $y = -4$ **43.** $y = -5.29x - 17.26$ **44.** $y = 1.45x - 3.08$ **45.** $y = 3x - 1$ **46.** $y = 5x - 15$ **47.** $y = -\frac{5}{3}x + 4$ **48.** $y = \frac{3}{2}x - 4$ **49.** $x = 5$ **50.** $y = -3$ **51.** $y = -0.85x + 12.16$ **52.** $y = -0.92x - 2.31$ **53.** $y = 3x + 11$ **54.** $y = -\frac{1}{5}x - \frac{17}{5}$ **55.** $y = -2.13x + 30.38$ **56. a.** $f(t) = 2.02t + 32.32$ **b.** 2.02; each year, the IRS standard mileage rate for businesses has increased by about 2.0 cents per mile. **c.** $(0, 32.32)$; in 2000, the standard mileage rate was about 32.3 cents per mile. **d.** 2018 **e.** \$8332.50 **f.** 54.5 is the average of 50.5 and 58.5. **57. a.** $f(t) = -2.11t + 39.02$ **b.** -2.11; the percentage of Americans who think the First Amendment goes too far in the rights it guarantees decreases by 2.11 percentage points per year. **c.** 2017 **d.** 22.14; in 2008, 22% of Americans thought the First Amendment went too far in the rights it guarantees. **e.** 14.70; in 2015, 8% of Americans will think the First Amendment goes too far in the rights it guarantees. **f.** $(18.49, 0)$; in 2018, no Americans will think the First Amendment goes too far in the rights it guarantees; model breakdown has occurred. **58.** \$29,052; overestimate; answers may vary. **59.** 2018

60. $x \ge -1$; $[-1, \infty)$; **61.** $x > -2$; $(-2, \infty)$; **62.** $w > -3$; $(-3, \infty)$; **63.** $a \le -3$; $(-\infty, -3]$; **64.** $b \ge -4$; $[-4, \infty)$; **65.** $-2 \le x < 3$; $[-2, 3)$; **66. a.** $f(t) = -17.14t + 731.93$ **b.** -17.14; the violent-crime rate is decreasing by about 17 violent crimes per 100,000 people per year. **c.** after 2017

Chapter 5 Test

1. slope: $\frac{2}{5}$; y-intercept: $(0, -2)$

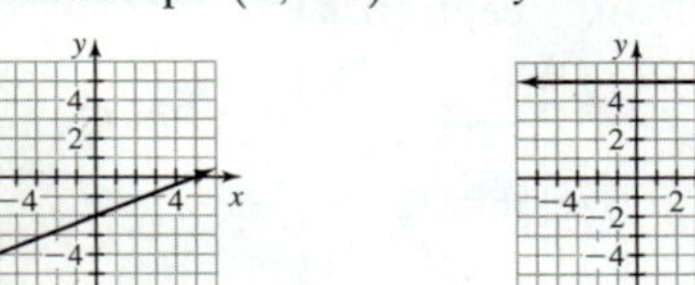

2. slope: 0; y-intercept: $(0, 5)$

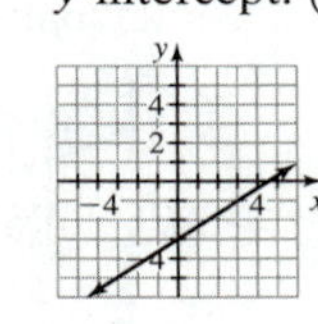

3. slope: $\frac{2}{3}$; y-intercept: $(0, -3)$

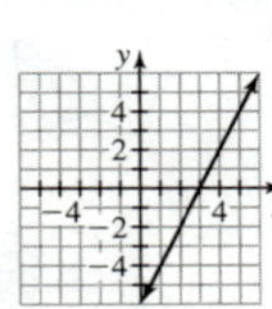

4. x-intercept: $(3, 0)$; y-intercept: $(0, -6)$

5. x-intercept: $(2, 0)$; y-intercept: $(0, 7)$ **6.** no **7.** yes **8.** domain: $-3 \le x \le 5$; range: $-3 \le y \le 4$; yes **9.** -2 **10.** 3 **11.** $-6 \le x \le 6$ **12.** $-3 \le y \le 1$ **13.** 19 **14.** $-4a + 27$ **15.** $\frac{5}{4}$ **16.** $-\frac{a-7}{4}$ or $\frac{7-a}{4}$ **17.** x-intercept: $\left(\frac{7}{3}, 0\right)$; y-intercept: $(0, -7)$ **18.** x-intercept: $(24, 0)$; y-intercept: $(0, -8)$ **19.** $y = 7x + 10$ **20.** $y = -\frac{2}{3}x + 3$ **21.** $y = -\frac{1}{2}x + 4$ **22.** $y = -1.92x - 3.64$ **23.** $y = -\frac{5}{3}x + \frac{17}{3}$ **24.** $y = -\frac{1}{3}x - \frac{7}{3}$ **25.** decrease m and increase b

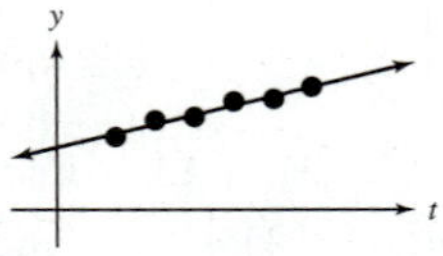

26. $y = 1.15x + 6.88$ **27. a.**

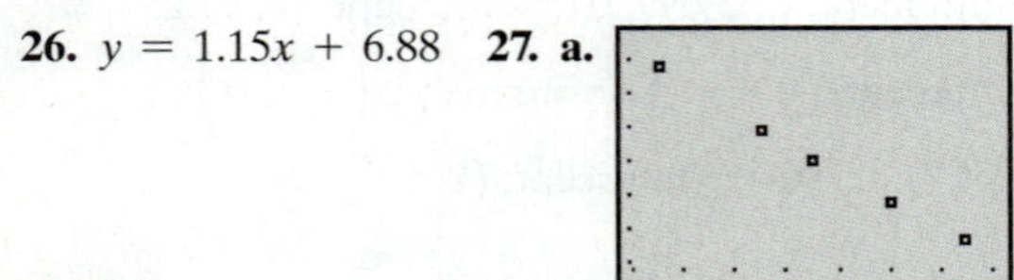

b. $f(t) = -4.25t + 79.5$ **c.** -4.25; the percentage of Fortune 100 companies that offer pensions is decreasing by 4.25 percentage points per year. **d.** $(0, 79.5)$; in 1995, 79.5% of Fortune 100 companies offered pensions. **e.** $(18.71, 0)$; in 2014, no Fortune 100 company will offer a pension, according to the model. **f.** 1990; a little research would show that this is false. **g.** before 2002

28. $x \le 2$; $(-\infty, 2]$; **29.** $-4 \le x < 1$; $[-4, 1)$;

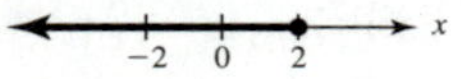

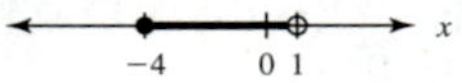

Chapter 6

Homework 6.1 **1.** $(1, -1)$ **3.** $(3, -2)$ **5.** $(1, 4)$ **7.** $(-6, 6)$ **9.** $(0, 0)$ **11.** $(2, 3)$ **13.** all points on the line $y = -2x + 3$; dependent system **15.** empty set; inconsistent system **17.** $(4, 4)$ **19.** $(3, -2)$ **21.** all points on the line $y = 2x - 1$; dependent system **23.** $(0, 5)$ **25.** $(4, 2)$ **27.** $(1.16, -2.81)$ **29.** $(-4.67, -3.83)$ **31.** $(4.14, -4.90)$ **33.** $(-4.83, 1.07)$ **35.** $(3.33, 1.33)$ **37. a.** 37.07 seconds; 34.24 seconds; -0.98 second; -0.67 second **b.** The absolute value of the slope of W is greater than the absolute value of the slope of M; the women's winning times are decreasing at a faster rate than the men's winning times. **c.** Answers may vary. **d.** 2167; 13.02 seconds; model breakdown has likely occurred. **39. a.** $C(t) = -2.7t + 53.1$; $W(t) = 2.5t + 21.5$ **b.** 2006; 37% **c.** 71% **41. a.** $I(t) = 2.3t + 47.6$; $B(t) = 6.83t + 1.4$ **b.** 2010 **c.** Answers may vary. **43.** $(-1.9, -2.8)$ **45.** $(10, -1)$ **47.** $(3, -4)$ **49.** $(3.5, 19.5)$ **51.** 0 **53.** 5 **55.** -1 **57. a.** B and E **b.** E and F **c.** E **d.** A, C, and D **59. a–c.** Answers may vary. **61.** $(2, 3)$ **63.** Answers may vary. **65.** The ordered pair $(1, 2)$ does not satisfy the equation $y = -2x + 9$, so $(1, 2)$ is not a solution of the system; $(2, 5)$ **67.** Answers may vary. **69. a.** The lines are parallel. **b.** no **c.** empty set solution **71.** 4 **73.** 1 **75.** $(-2, 1)$ **77.** $(-2, 4)$; system of two linear equations in two variables **79.** -2; linear equation in one variable

Homework 6.2 **1.** $(2, 4)$ **3.** $(1, 1)$ **5.** $(-2, 3)$ **7.** $(-3, -1)$ **9.** $(4, 1)$ **11.** $(-2, -4)$ **13.** $(4, -3)$ **15.** $(-2, 3)$ **17.** $(0, 0)$ **19.** $(1, -6)$ **21.** $(-2.07, 1.76)$ **23.** $(3.35, -1.71)$ **25.** $(-2, -5)$ **27.** $(2, -1)$ **29.** $(-1, 3)$ **31.** $(-2, -5)$ **33.** $(1, -3)$ **35.** infinite number of solutions of the equation $x = 4 - 3y$; dependent system **37.** $(2, -4)$ **39.** empty set solution; inconsistent system **41.** infinite number of solutions of the equation $x = 3y - 1$; dependent system **43.** empty set solution; inconsistent system **45.** $\left(\frac{47}{10}, \frac{109}{5}\right)$ **47.** A: $(0, 0)$; B: $(0, 8)$; C: $(5, 3)$; D: $\left(\frac{7}{2}, 0\right)$ **49. a.** $(1, 2)$ **b.** $(1, 2)$ **c.** The results are the same. **51.** Answers may vary; $(1, 3)$ **53.** Answers may vary. **55. a.** $(2, 1)$ **b.** $(2, 1)$ **c.** 2 **d.** They are the same. **e.** 3; 3; they are the same. **f.** Answers may vary. **57.** 1.57 **59.** -2.42 **61.** -2.33 **63.** linear equation in two variables **65.** $(-3, 14)$; system of two linear equations in two variables

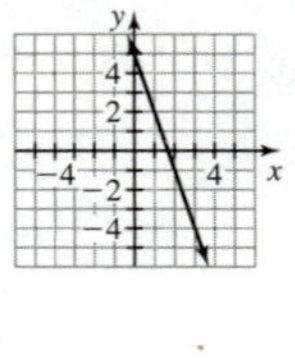

Homework 6.3 **1.** $(2, 1)$ **3.** $(2, 4)$ **5.** $(2, -3)$ **7.** $(1, -2)$ **9.** $(-1, -2)$ **11.** $(1, -3)$ **13.** $(-2, -5)$ **15.** $(1, -3)$ **17.** $(-4, 5)$ **19.** $(3, -2)$ **21.** $(2, 1)$ **23.** $(-2, 3)$ **25.** $(3, 10)$ **27.** the infinite number of solutions of the equation $4x - 7y = 3$; dependent system **29.** empty set solution; inconsistent system **31.** $(-4, 1)$ **33.** empty set solution; inconsistent system **35.** the infinite number of solutions of the equation $3x - 9y = 12$; dependent system **37.** $(3.36, 5.51)$ **39.** $(1.52, 5.11)$

41. $(-2, 4)$ **43.** $(5, 1)$ **45.** $(-4, 3)$ **47.** $(-4, -1)$ **49.** $(-2, 1)$ **51.** $(-1, 3)$ **53.** $(3, 2)$ **55.** all the points on the line $y = \frac{1}{2}x + 3$; dependent system **57.** $(3, 2)$ **59.** $(1, 2)$ **61.** $(2, 1)$; answers may vary. **63. a.** $(1, 2)$ **b.** $(1, 2)$ **c.** The results are the same. **65.** $(23, 24)$ **67.** A: $(0, 0)$; B: $(0, 3)$; C: $(3, 9)$; D: $(6, 8)$; E: $\left(\frac{36}{5}, \frac{22}{5}\right)$; F: $(5, 0)$ **69. a.** $\left(\frac{cp - bd}{ap - bk}, \frac{ad - ck}{ap - bk}\right)$, assuming that $ap - bk \neq 0$ **b.** $\left(\frac{14}{11}, -\frac{4}{11}\right)$ **71.** Answers may vary. **73. a.** $5 = 2m + b$ **b.** $9 = 4m + b$ **c.** $m = 2; b = 1$ **d.** $y = 2x + 1$ **e.** Answers may vary. **75.** 5 **77.** 3 **79.** $(1, 2.8)$ **81.** $-8x - 4$; linear expression in one variable **83.** $-\frac{1}{2}$; linear equation in one variable

Homework 6.4 **1.** 2167; 13.02 seconds; model breakdown has likely occurred. **3. a.** 2.30; 6.83; the percentage of households that have Internet access is increasing by 2.30 percentage points per year; the percentage of households that have broadband Internet access is increasing by 6.83 percentage points per year. **b.** Answers may vary. **c.** 2010 **5. a.** $M(t) = -0.28t + 50.87$; $S(t) = 0.82t - 32.00$ **b.** 1975; 29.8 gallons per person **c.**

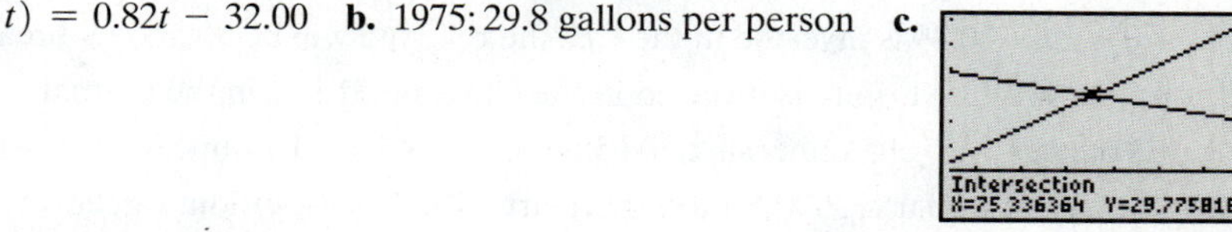

d. Answers may vary. **e.** Answers may vary. **7. a.** $K(a) = 0.014a - 0.75$; $M(a) = -0.028a + 1.66$ **b.** 57 years; 0.1 point **c.**

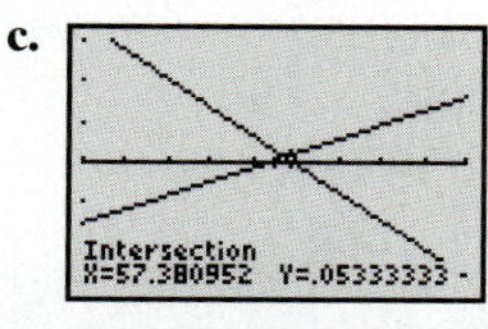

9. a. 1997 **b.** The competition heated up because the two newspapers had approximately equal circulations. **c.** 147 thousand bonus issues **d.** 688 thousand newspapers **e.** overestimate; answers may vary. **11. a.** $F(t) = -1414t + 14{,}290$; $D(t) = -3740t + 30{,}450$ **b.** 2019; \$4466 **c.**

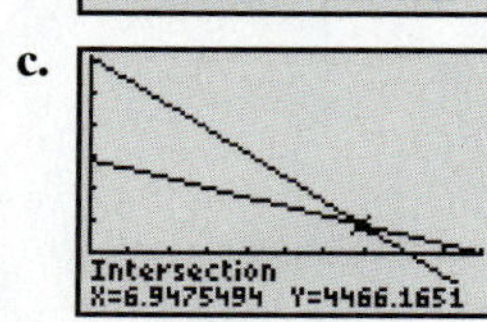

13. a. $N(t) = 91t$; $W(t) = 87t + 20$ **b.** For each equation, the units of the expressions on both sides of the equation are dollars. **c.** 5 weeks; \$455 **d.**

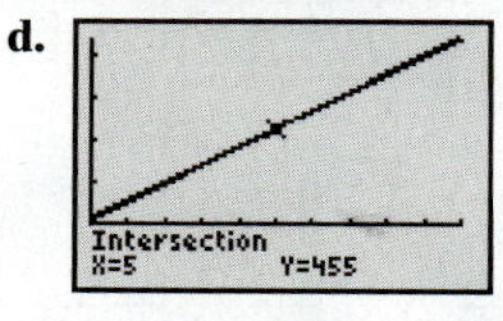

e. Weight Watchers **15. a.** $C(t) = 2.0t + 83.1$; $E(t) = -0.7t + 101.7$ **b.** 2016; 96.9 thousand degrees **c.**

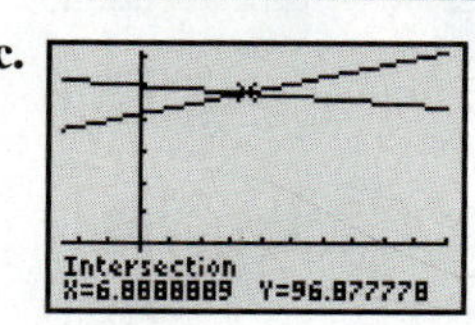

17. 2023; 35.3% **19.** 2009; 72 million visitors **21. a–b.**

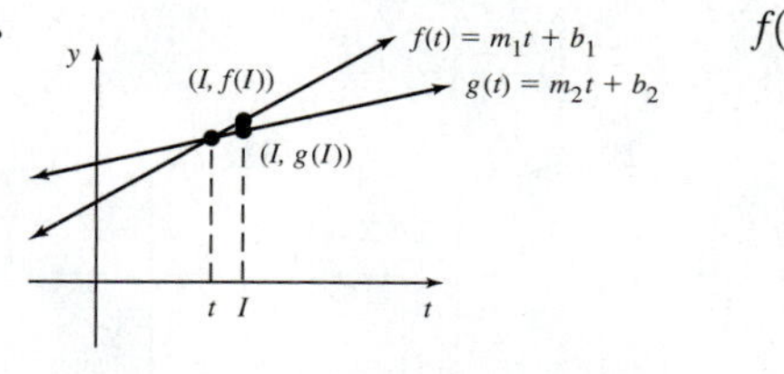

$f(I)$

23. 2007; 10% **25.** $(1, 4)$; system of two linear equations in two variables **27.** 1; linear equation in one variable

Homework 6.5 **1.** width: 114.50 feet, length: 185.50 feet **3.** width: 8 feet, length: 13 feet **5.** width: 36 feet, length: 78 feet **7.** width: 19 inches, length: 35 inches **9.** 1500 \$15 tickets, 500 \$22 tickets **11.** 39 *The Best of Radiohead* albums, 214 *The King of Limbs* albums **13.** main level: \$26, balcony: \$14 **15.** general: \$37, reserved: \$62 **17. a.** $f(x) = -25x + 1{,}500{,}000$ **b.**

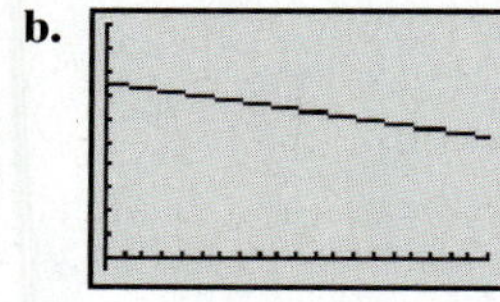

-25; the more \$50 tickets are sold (so, the fewer \$75 tickets are sold), the lower the total revenue will be (by \$25 for each additional \$50 ticket sold). **c.** 1,100,000; if 16,000 \$50 tickets are sold (and 4000 \$75 tickets are sold), the total revenue will be \$1,100,000. **d.** 17,000 \$50 tickets, 3000 \$75 tickets

19. a. $f(x) = -25x + 840{,}000$ **b.**

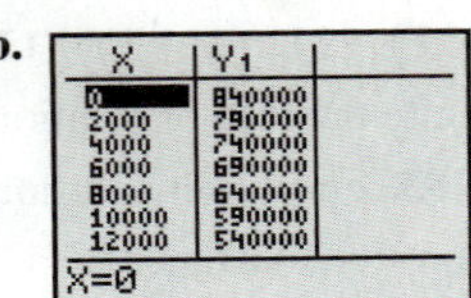

X	Y1
0	840000
2000	790000
4000	740000
6000	690000
8000	640000
10000	590000
12000	540000

X=0

See second column of table. If the number of \$45 tickets sold is 0, 2000, 4000, 6000, 8000, 10,000, or 12,000, the total revenue will be \$840,000, \$790,000, \$740,000, \$690,000, \$640,000, \$590,000, or \$540,000, respectively. **c.** The total revenue will be between \$540,000 and \$840,000, inclusive.

d. 9500 \$45 tickets, 2500 \$70 tickets **21. a.** $f(x) = 134x + 1936$ **b.** 134; if the prices of all tickets are increased by \$1, the total revenue increases by \$134. **c.** coach: \$91; first class: \$333 **23. a.** \$200 **b.** \$280 **c.** $0.08d$ **25. a.** \$480 **b.** \$420 **c.** $0.03x + 0.06y$ **27.** First Funds TN Tax-Free I account: \$14,000, W & R International Growth C account: \$6000 **29.** Middlesex Savings Bank CD account: \$2500, First Funds Growth & Income I account: \$6000 **31.** Limited Term NY Municipal X account: \$7500, Calvert Income A account: \$2500 **33.** \$5000 should be invested in each account. **35. a.** $f(x) = -0.0523x + 810$

b.

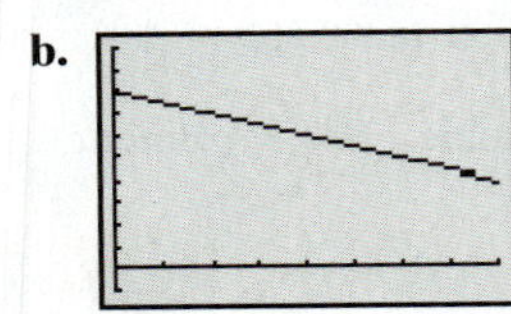

-0.0523; for each additional \$1 invested in the CD (so, \$1 less invested in the mutual fund), the total interest decreases by 5.23 cents. **c.** CD: \$7839.39; mutual fund: \$2160.61 **37. a.** $f(x) = -0.0695x + 850.5$ **b.** 815.75; if \$500 is invested in the CD (and \$8500 in the mutual fund), the total interest will be \$815.75. **c.** 5043.17; if \$5043.17 is invested in the CD (and \$3956.83 in the mutual fund), the total interest will be \$500.

d. 155.5; if \$10,000 is invested in the CD, the total interest will be \$155.50; model breakdown has occurred, because only \$9000 is being invested. **39. a.** $f(x) = -0.101x + 928$ **b.** The total interest will be between \$120 and \$675.50, inclusive. **c.** \$5227.72 should be invested in the CD and \$2772.28 in the mutual fund. **41. a.** $f(x) = -0.0615x + 540$ **b.** $(0, 540)$; if all of the \$6000 is invested in the mutual fund, the total interest will be \$540. **c.** $(8780.49, 0)$; if \$8780.49 is invested in the CD, the revenue will be \$0; model breakdown has occurred, because only \$6000 is being invested. **d.** -0.0615; if \$1 more is invested in the CD (and \$1 less in the mutual fund), the total interest will decrease by 6.15 cents. **43. a.** 1.3 ounces **b.** 1.95 ounces **c.** $0.65x$ ounces **45. a.** 1.7 ounces **b.** 35% solution: 6 ounces, 10% solution: 9 ounces **47.** 20% solution: 4 quarts, 30% solution: 1 quart **49.** 10% solution: 1 gallon, 25% solution: 2 gallons **51.** 15% solution: 1 quart, 35% solution: 3 quarts **53.** 20% solution: 3 ounces, water: 2 ounces **55. a.** \$3000 **b.** \$4500 **c. i.** no; answers may vary. **ii.** no; answers may vary. **iii.** yes; 250 \$10 tickets, 50 \$15 tickets **57.** 15%; answers may vary. **59.** Answers may vary. **61.** Answers may vary. **63.** $L = \dfrac{P - 2W}{2}$ **65.** $r = \dfrac{A - P}{Pt}$ **67.** $-28x + 12$; linear expression in one variable **69.** $\dfrac{10}{33}$; linear equation in one variable

Homework 6.6 **1.** $(4, 1)$ **3.** $(-1, -4), (5, 0)$

5.
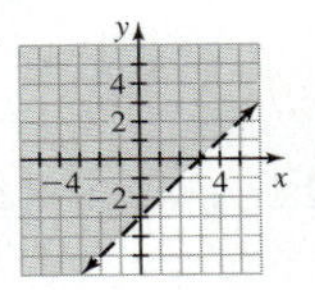

7.
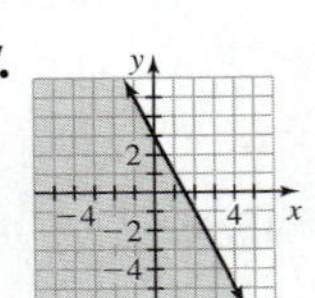

9.
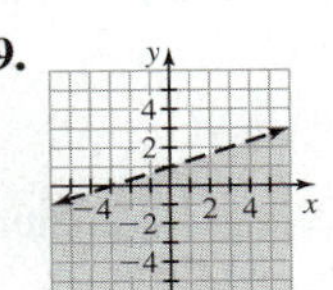

11.
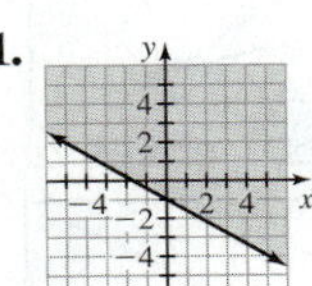

13.
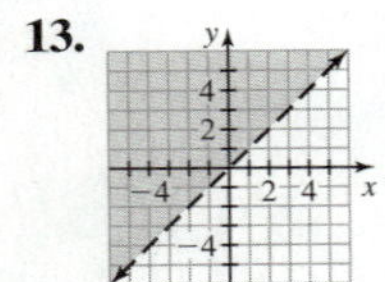

15.
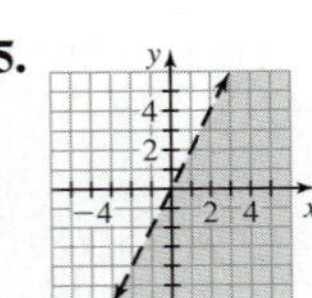

17.
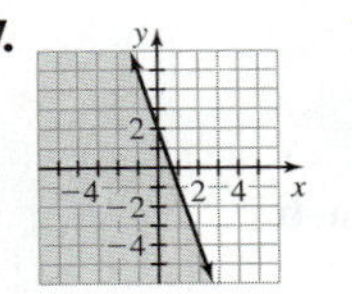

19.
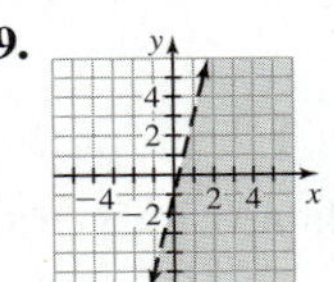

21.
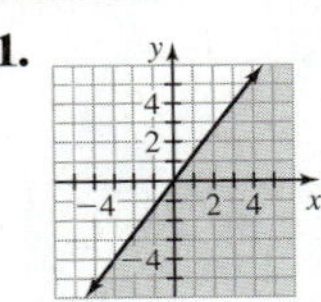

23.
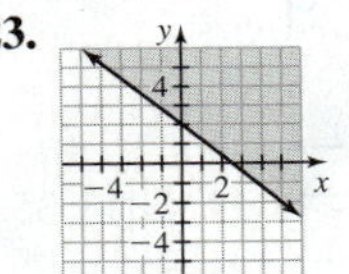

25.
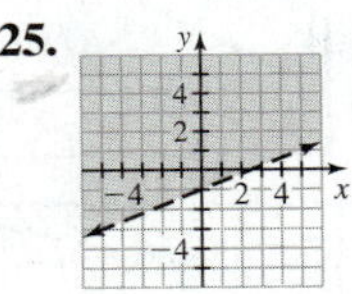

27.
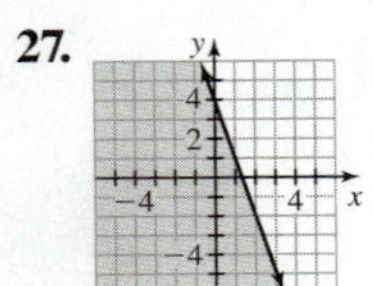

29.
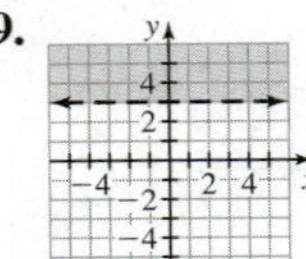

31.
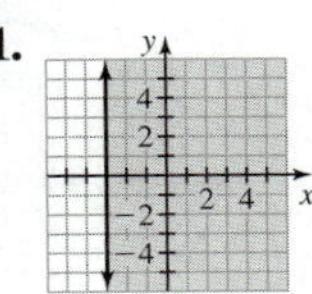

33.
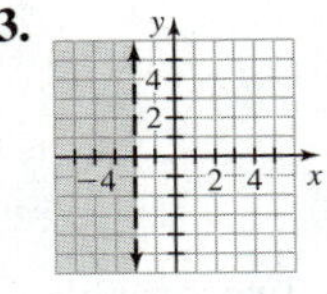

35.
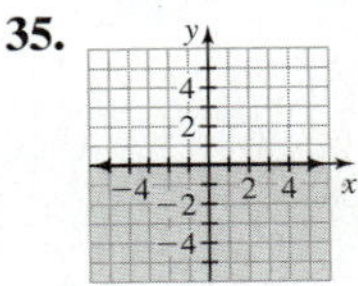

37.
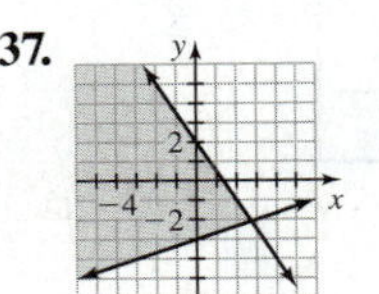

39.
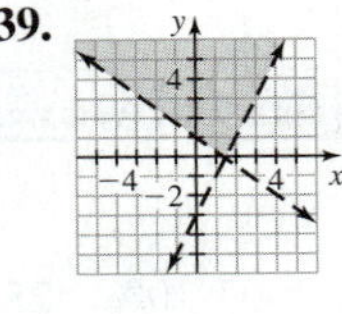

41.
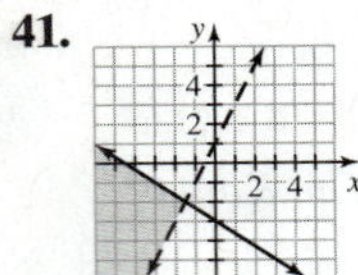

43.
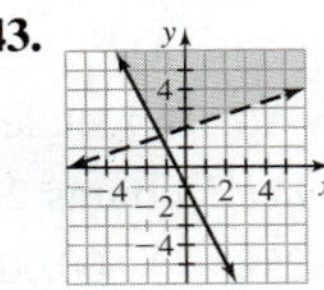

45.
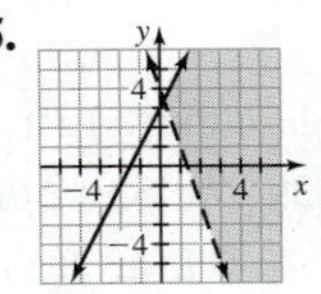

47.
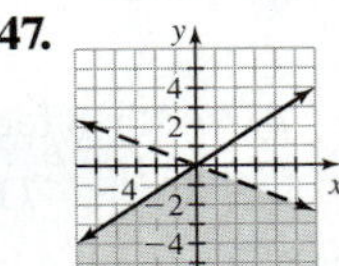

49.
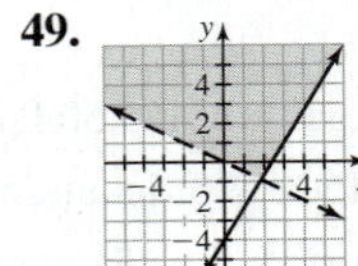

51.
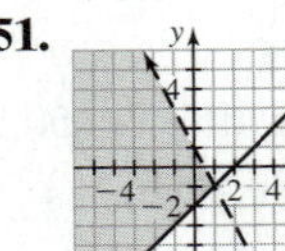

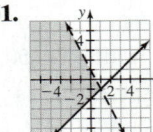

53.
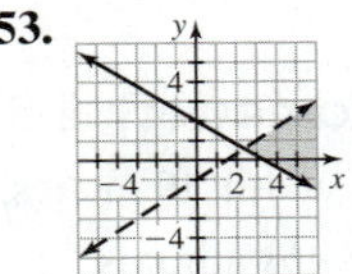

55.
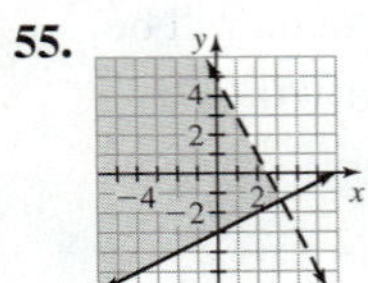

57.
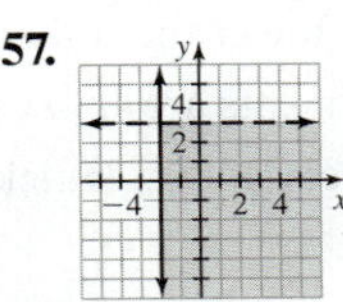

59.
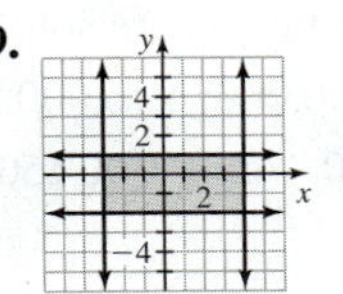

61.
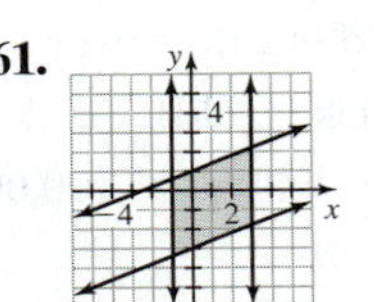

63.
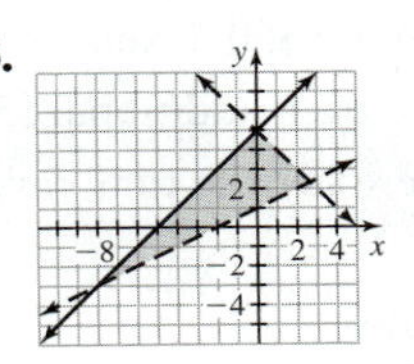

65. a. $B(t) = 0.196t + 70.00$; $T(t) = 0.167t + 51.73$ **b.** The system consists of the inequalities $L \le 0.196t + 70.00$, $L \ge 0.167t + 51.73$, $t \ge 0$, and $t \le 40$.

c.

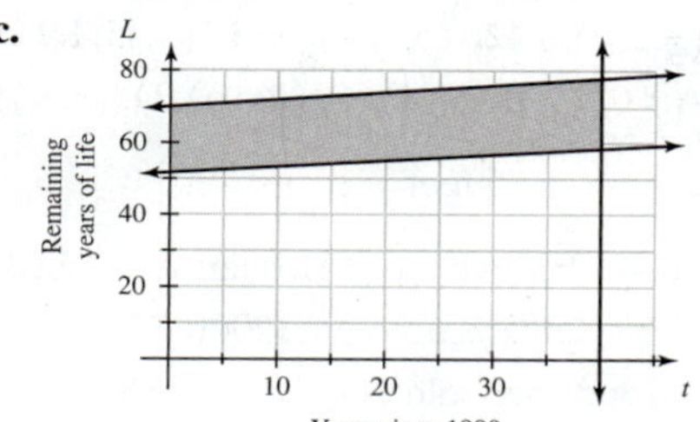

d. between 57.6 years and 76.9 years **67. a.** $B(w) = 0.44w + 84.21$; $I(w) = 0.44w + 89.21$ **b.** The system consists of the inequalities $L \ge 0.44w + 84.21$,

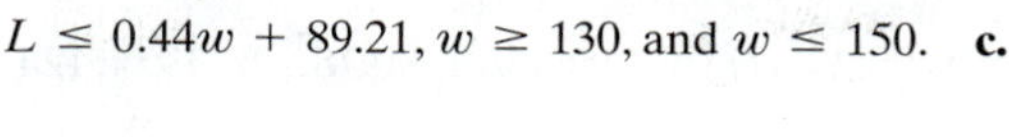

$L \le 0.44w + 89.21$, $w \ge 130$, and $w \le 150$. **c.**

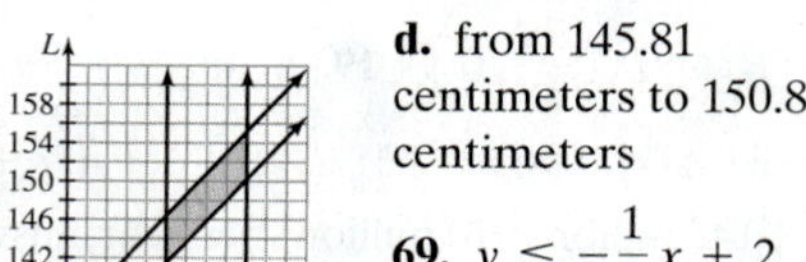

d. from 145.81 centimeters to 150.81 centimeters **69.** $y \le -\frac{1}{2}x + 2$

71. no; answers may vary. **73. a.** A, B, C, D **b.** A, B, C, F, G, H **c.** A, B, C **d.** E **75.** Answers may vary. **77.** Answers may vary.

79.

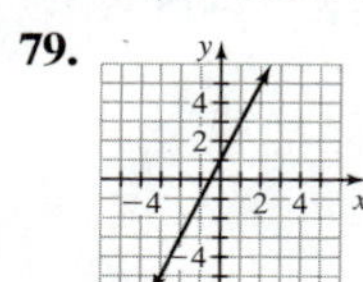

81. Answers may vary.

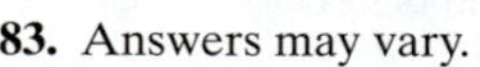

83. Answers may vary. **85.**

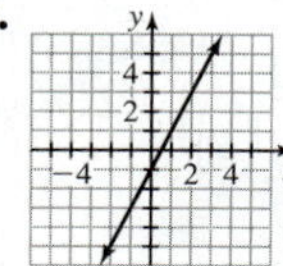

87.

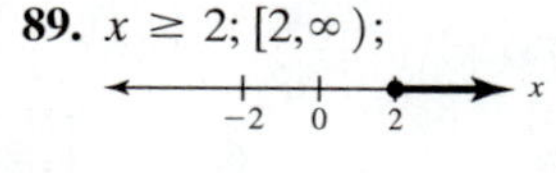

89. $x \ge 2$; $[2, \infty)$;

91.

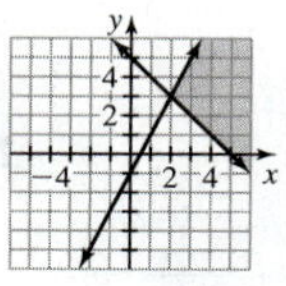

93. $(2, 3)$ **95.** Answers may vary. **97.** Answers may vary. **99.** Answers may vary.

Chapter 6 Review Exercises

1. $(2, 1)$ **2.** $(-4, -2)$ **3.** $(0, 0)$ **4.** empty set; inconsistent system **5.** all points on the line $y = -2x + 8$; dependent system **6.** $(-3, -2)$ **7.** $(3, -1)$ **8.** $(2, 1)$ **9.** $(0, 0)$ **10.** $(2, -12)$ **11.** $(1, -2)$ **12.** $(-3, 4)$ **13.** $(2.61, -2.20)$ **14.** $(-1.19, 4.76)$ **15.** $(3, 2)$ **16.** $(-4, -1)$ **17.** $(-2, 1)$ **18.** $(-3, -2)$ **19.** $(4, -2)$ **20.** $(2, 1)$ **21.** $(-1, 2)$ **22.** $(-2, 1)$ **23.** $(0.77, 4.79)$ **24.** $(2.14, -1.62)$ **25.** $(-3, 1)$ **26.** $(-5, 4)$ **27.** $(-2, 0)$ **28.** $(4, 3)$ **29.** $(-4, 2)$ **30.** $(5, 3)$ **31.** $(-3, -2)$ **32.** all solutions of the equation $y = -4x + 3$; dependent system **33.** empty set solution; inconsistent system **34.** $(1, -2)$ **35.** $(2, -1)$ **36.** $(10, 3)$ **37. a.** Answers may vary. **b.** Answers may vary. **c.** Answers may vary. **d.** $(2, 3)$ **38.** $(-1.4, 2.8)$ **39.** $(0.4, 15.1)$, $(2.6, 21.9)$ **40.** $a = 19$, $b = 18$ **41.** $(1, 3)$; answers may vary. **42. a.** $L(t) = -4.23t + 113.20$; $W(t) = 4.28t - 15.94$ **b.** -4.23; 4.28; the percentage of households with phone landlines is decreasing by 4.23 percentage points per year; the percentage of households with only wireless phones is increasing by 4.28 percentage points per year. **c.** 2015; 49.0% **d.** Some households do not have phones. **e.** $t > 15.18$; the percentage of households with phone landlines will be less than the percentage of households with only wireless phones after 2015. **43. a.** $w(t) = -0.58t + 31.57$; $p(t) = -0.37t + 20.76$ **b.** 2021; 1.7 **44. a.** $R(d) = 0.22d + 75$; $U(d) = 0.69d + 29.95$ **b.** 95.85 miles; $96.09 **c.** for miles driven over 95.85 miles **45.** 2019 **46.** width: 5 feet, length: 17 feet **47.** 6500 $22 tickets, 1500 $39 tickets **48.** 4 gallons of 10% solution, 6 gallons of 20% solution **49. a.** $f(x) = -0.062x + 1040$ **b.** 1004.35; if $575 is invested in Hartford Global Leaders Y (and $7425 in Mutual Discovery Z), the total interest will be $1004.35. **c.** 7500; if $7500 is invested in Hartford Global Leaders Y (and $500 in Mutual Discovery Z), the total interest will be $575.

50. **51.**

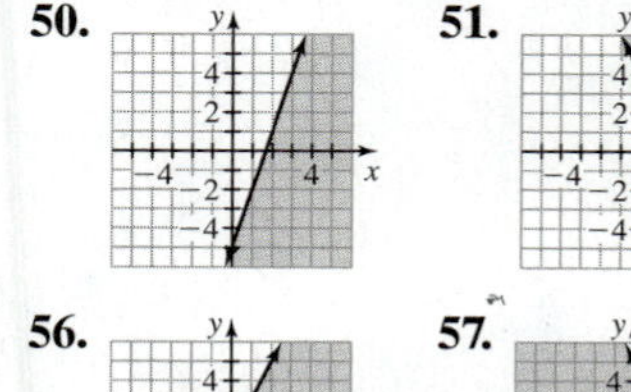

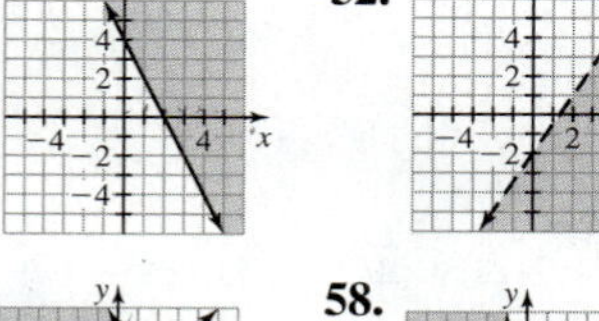

52.

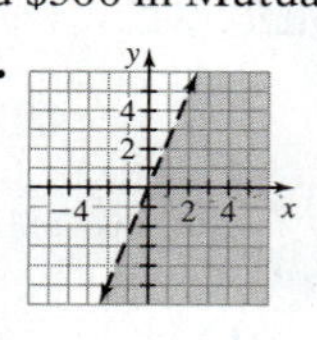

53.

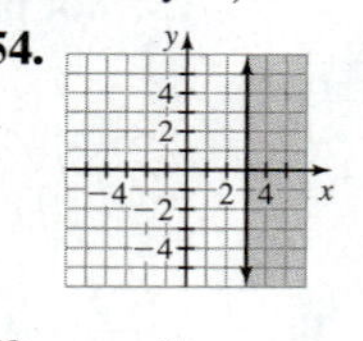

54.

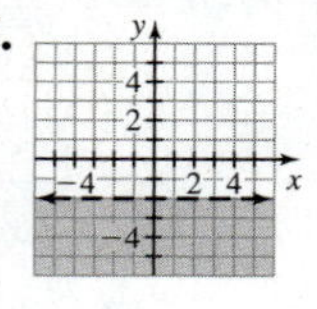

55.

56.

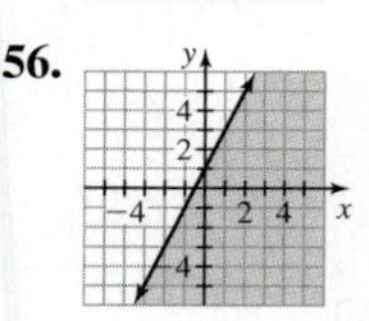

57.

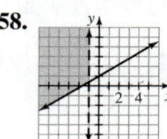

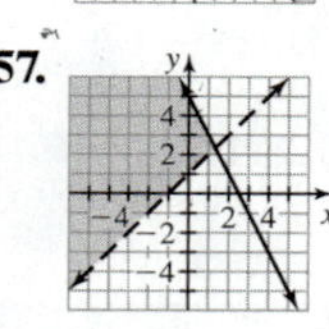

58. **59.**

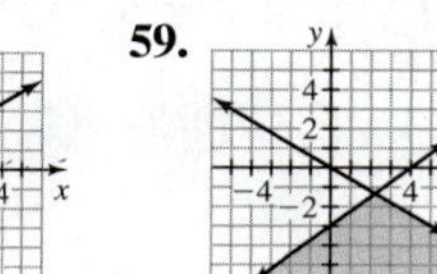

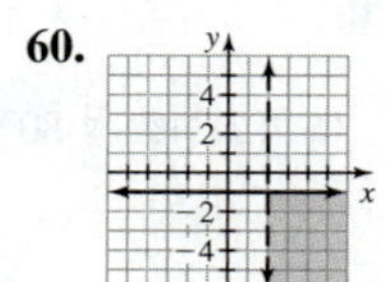

60.

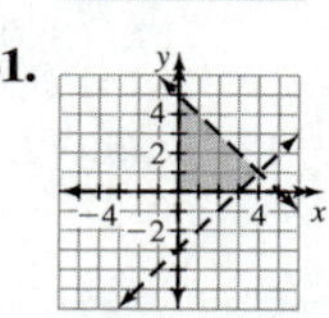

61.

62. a. $L(h) = 3.08h - 64.14$; $U(h) = 3.50h - 80.97$ **b.** $w \ge 3.08h - 64.14$, $w \le 3.50h - 80.97$, $h \ge 63$, $h \le 78$

c.

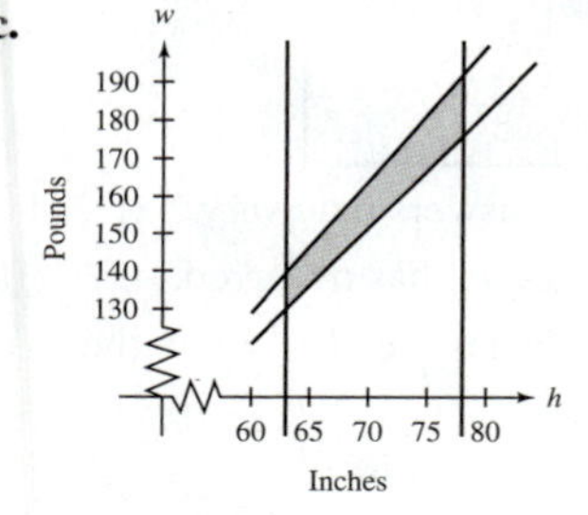

d. The ideal weights are between 145 pounds and 157 pounds, inclusive.

63. Answers may vary.

Chapter 6 Test

1. $(5, -3)$ **2.** $(-3.18, 1.88)$ **3.** $(-2, -7)$ **4.** $(-1, 3)$ **5.** $(2, -3)$ **6.** $(-4, 2)$ **7.** $(-2, 3)$ **8.** all solutions of the equation $2x - 3y = 4$; dependent system **9.** empty set solution; inconsistent system **10.** $(5, 2)$ **11.** $(-2, 4)$ **12.** $(5, -8)$ **13.** $(3.45, 1.93)$ **14.** $(3, 1)$ **15.** $m = 5, b \neq -13$ **16. a.** C, D **b.** D, F **c.** D **d.** A, B, E **17.** $(24, 5)$ **18.** A: $(0, 0)$; B: $(0, 4)$; C: $(2, 10)$; D: $(5, 8)$; E: $(6, 4)$; F: $\left(\frac{14}{3}, 0\right)$ **19. a.** $w(t) = -0.73t + 98.28$; $s(t) = -0.94t + 90.78$ **b.** 1934; 124.4 feet **c.** 54.5 feet **20. a.** $A(t) = 0.8t - 1$; $B(t) = -0.47t + 6.97$ **b.** 2006; 4.0 billion **c.** years before 2006 **d.** -0.47; the revenue of Borders decreased by \$0.47 billion (470 million) per year; answers may vary. **21.** 13,500 tickets at \$55, 6500 tickets at \$70 **22.** 3% account: \$2000, 7% account: \$5000 **23. a.** $f(x) = -15x + 500{,}000$ **b.**

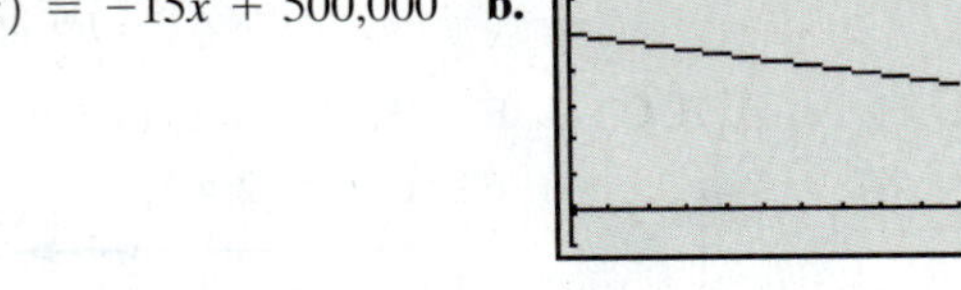

-15; the more \$35 tickets are sold (so, the fewer \$50 tickets are sold), the lower the total revenue will be (by \$15 for each additional \$35 ticket sold). **c.** 7300 \$35 tickets, 2700 \$50 tickets

24. 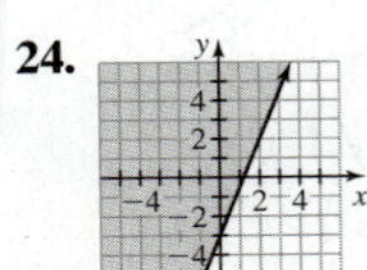 **25.** 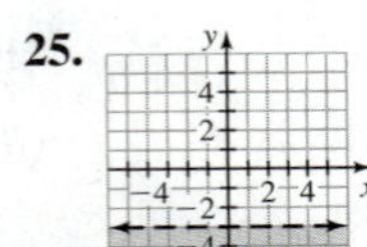 **26.** 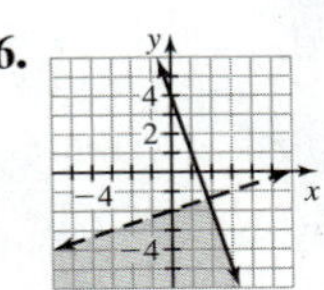 **27.** 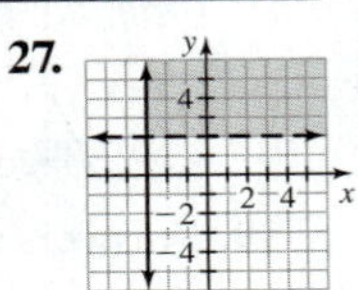

Cumulative Review of Chapters 1–6

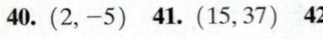

1. Fahrenheit degrees **2.** A cricket chirps 129 times per minute when the temperature is 70°F. **3.** $-\frac{8}{9}$ **4.** $\frac{46}{35}$ **5.** 21 **6.** $-\frac{3}{4}$; decreasing **7.** equation 1: $y = -8x + 49$; equation 2: $y = 4x + 11$; equation 3: $y = -2x + 45$; equation 4: $y = 3x + 8$ **8. a.** m is negative; b is positive. **b.** m is positive; b is negative. **c.** m is zero; b is positive. **d.** m is negative; b is negative. **9.** $-\frac{10}{13}$ **10.** $-13p + w + 5$ **11.** $4w + 6y - 10$ **12.** $\frac{22}{9}$ **13.** $y = \frac{c - ax}{b}$ **14.** $6 + 3(4 + x)$; $3x + 18$ **15.** $4 - \frac{x}{3} = 2$; 6

16. slope: 2; y-intercept: $(0, -4)$

17. slope: $\frac{1}{2}$; y-intercept: $(0, -3)$

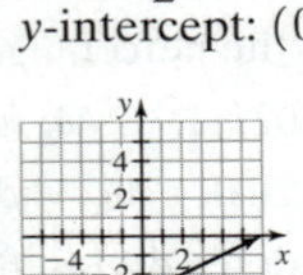

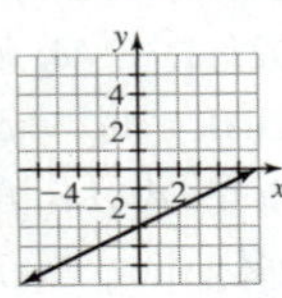

18. slope: $-\frac{5}{2}$; y-intercept: $(0, 6)$

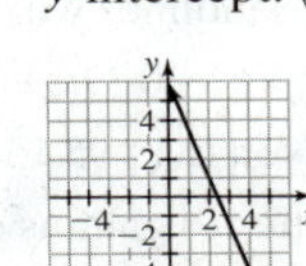

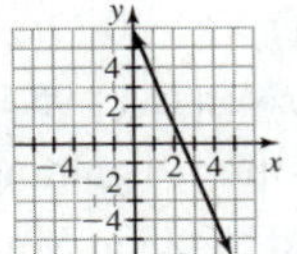

19. slope: 0; y-intercept: $(0, -3)$

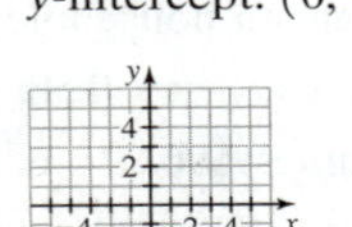

20. x-intercept: $(5, 0)$; y-intercept: $(0, -2)$

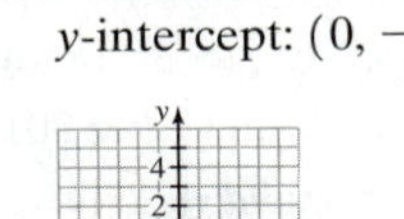

21. x-intercept: $(2, 0)$; y-intercept: $(0, 4)$ **22.**

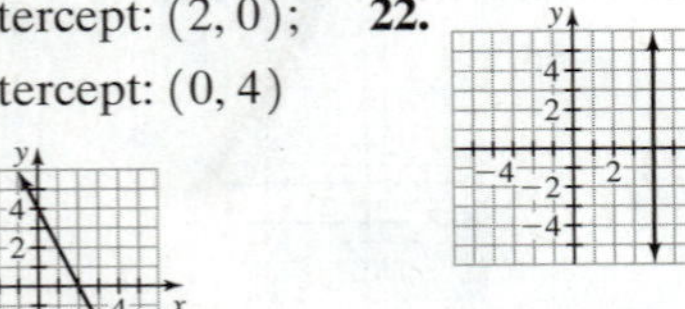

23. not parallel **24.** $y = -\frac{2}{5}x - \frac{4}{5}$ **25.** $y = -\frac{4}{3}x - \frac{17}{3}$ **26.** $x = -2$ **27.** 2 **28.** 6

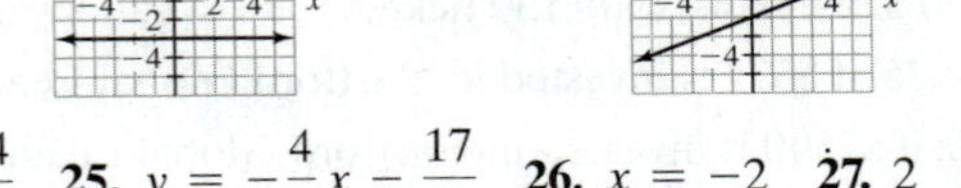

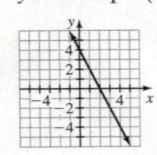

29. $(3, 0)$ **30.** $y = -\frac{1}{3}x + 1$ **31.** $-5 \le x \le 5$ **32.** $-2 \le y \le 3$ **33.** yes

34. $x < -1$; $(-\infty, -1)$ **35.** $(2, 1)$ **36.** $(3.31, -1.80)$ **37.** $(3, -2)$ **38.** $(2, -1)$ **39.** $(23.90, 28.49)$

40. $(2, -5)$ **41.** $(15, 37)$ **42.** 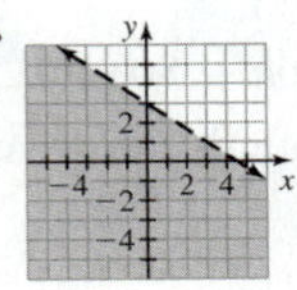 **43.** **44. a.** Answers may vary. **b.** 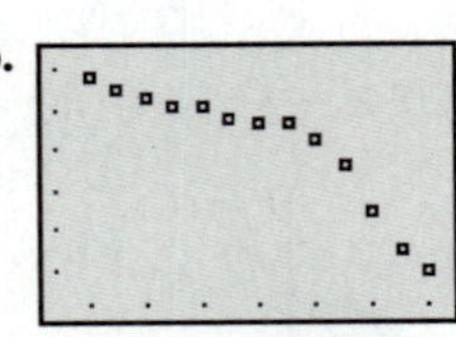

c. $f(t) = -0.085t + 2.79$ **d.** $(0, 2.79)$; there were 2.79 deaths per 100 million miles traveled in 1990; no; answers may vary. **e.** 2011 **f.** 0.50 death per 100 million miles traveled **g.** $(32.82, 0)$; there will be no fatalities in 2023; model breakdown has occurred. **45. a.** $f(t) = -9t + 82$ **b.** -9; each year 9 fewer bicyclists younger than 16 are hit and killed by motor vehicles. **c.** $(0, 82)$; in the year 2009, 82 bicyclists younger than 16 were hit and killed by motor vehicles. **d.** $(9.11, 0)$; in 2018, no bicyclists younger than 16 will be hit and killed by motor vehicles; model breakdown has likely occurred. **e.** years after 2018 **46. a.** $H(t) = 2.2t + 86.8$; $L(t) = 3.2t + 95.9$ **b.** 2.2; 3.2; the DSI of Home Depot is increasing by 2.2 days per year; the DSI of Lowes is increasing by 3.2 days per year; negative slope; answers may vary. **c.** 2001; 66.8 days **d.** $t > -9.1$; the DSI of Home Depot has been less than the DSI of Lowes after 2001. **47. a.** $W(t) = -0.064t + 27.00$; $M(t) = -0.026t + 21.86$ **b.** $W(118) = 19.45$; $M(118) = 18.79$; in 2018, the women's record time will be 19.45 seconds and the men's record time will be 18.79 seconds. **c.** The absolute value of W's slope is greater than the absolute value of M's slope. Women's record times are decreasing at a greater rate than men's record times. **d.** Answers may vary. **e.** 2035; 18.34 seconds **f.** $t < 135$; women's record times are greater than men's record times for years before 2035. **g.** W: $(421.88, 0)$; M: $(840.77, 0)$; the women's record time will be 0 seconds in 2322, and the men's record time will be 0 seconds in 2741; model breakdown has occurred. **48.** 16% solution: 8 quarts, 28% solution: 4 quarts

Chapter 7

Homework 7.1 **1.** quadratic (or second-degree) polynomial in one variable **3.** cubic (or third-degree) polynomial in one variable **5.** seventh-degree polynomial in two variables **7.** $8t^2$ **9.** $-11a^4b^3$ **11.** $5x^2$ **13.** The terms $7x^2$ and $-3x$ cannot be combined. **15.** $-3b^3$ **17.** $6x^6$ **19.** The terms $2t^3w^5$ and $4t^5w^3$ cannot be combined. **21.** $4x^2 + x$ **23.** $-12x^3 + 2x^2 - 5x + 5$ **25.** $-5a^4b^2 - 5ab^3$ **27.** $2x^4 - 3x^3y + xy^3$ **29.** $9x^2 - 3x - 9$ **31.** $3x^3 - 6x^2 + 4x - 1$ **33.** $11a^2 - 3ab - 5b^2$ **35.** $2m^4p + 2m^3p^2 - 8mp^3$ **37.** $-7x^2 + 9x - 11$ **39.** $13x^3 - 3x^2 - x + 5$ **41.** $10m^2 + 10mp - p^2$ **43.** $a^3b - 10a^2b^2 + 8ab^3 - b^3$ **45.** -30 **47.** 79 **49.** 3 **51.** 42 **53.** -8 **55.** 3 **57.** $-1, 3$ **59.** 1 **61.** 3 **63.** 0, 6 **65.** 3 **67. a.** 1, 5 **b.** Answers may vary.

69. 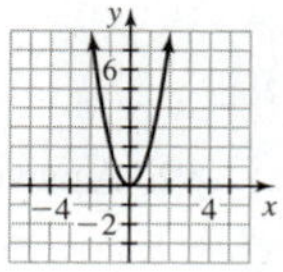 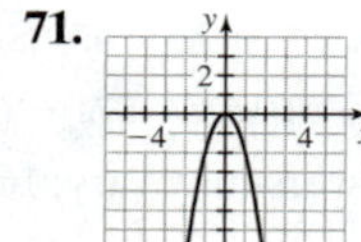
71. 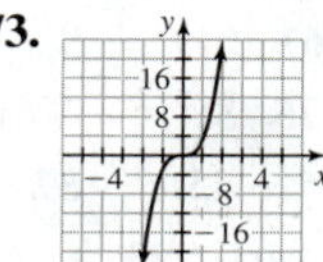
73. 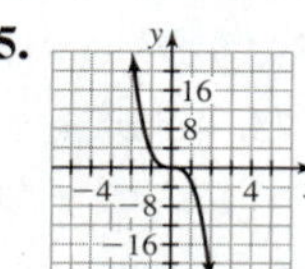
75.
77. $(f + g)(x) = 11x^2 + 3x + 7$; 115

79. $(f - h)(x) = 7x^2 + 2x + 17$; 137 **81.** $(f + g)(x) = 2x^3 - 3x^2 + x - 2$; 4 **83.** $(f - h)(x) = x^3 + 3x^2 - 6x + 1$; 9 **85. a.** $(M + S)(t) = 0.54t + 45.63$ **b.** gallons per person **c.** 74.3; in 2003, 74.3 gallons of milk and soft drinks (combined) were consumed per person. **d.** $(M - S)(t) = -1.1t + 27.71$ **e.** -30.6; in 2003, 30.6 fewer gallons of milk than soft drinks were consumed per person. **87. a.** $R(s) = 2.2s$ **b.** $(R + B)(s) = 0.063s^2 + 2.2s$ **c.** feet **d.** 99.79; if you are driving at 26 miles per hour, it will take about 100 feet to stop. **e.** yes; she would need about 175 feet to stop. **89.** Answers may vary; $4x^2 + 4x + 2$ **91. a.** $(f - g)(x) = -2x + 5$; $(g - f)(x) = 2x - 5$ **b.** $1, -1$; the answers are opposites. **c.** $-3, 3$; the answers are opposites. **d.** $-9, 9$; the answers are opposites. **e.** In general, $(f - g)(x) = -(g - f)(x)$. **93.** Answers may vary. **95.** Answers may vary. **97.** (c) **99.** (b) **101.** $x^2 - 3x + 5x^2$; $6x^2 - 3x$ **103.** $4x^2 - (x^2 + x)$; $3x^2 - x$ **105.** $(2, 5)$; system of two linear equations in two variables **107.** linear equation in two variables

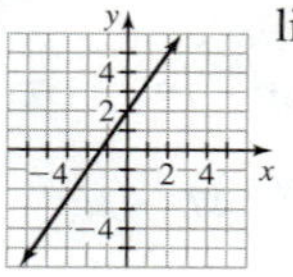

Homework 7.2 **1.** x^7 **3.** w^9 **5.** $36x^{12}$ **7.** $30x^7$ **9.** $x^{15}y^7$ **11.** $24c^{14}d^8$ **13.** $-36p^5t^3$ **15.** $-\frac{14}{5}x^5$ **17.** $3w^2 - 6w$ **19.** $-8x^3 - 12x$ **21.** $6m^3n^2 + 10mn^3$ **23.** $6x^3 - 4x^2 + 14x$ **25.** $-6t^4 - 12t^3 + 6t^2$ **27.** $6x^3y^2 - 8x^2y^3 + 10xy^4$ **29.** $x^2 + 6x + 8$ **31.** $x^2 + 3x - 10$ **33.** $a^2 - 5a + 6$ **35.** $x^2 - 36$ **37.** $x^2 - 14.5x + 48.76$ **39.** $15y^2 + 14y - 8$ **41.** $4x^2 + 16x + 16$ **43.** $9x^2 - 6x + 1$ **45.** $12x^2 - 17xy - 5y^2$ **47.** $6a^2 - 32ab + 32b^2$ **49.** $9x^2 - 16$ **51.** $81x^2 - 16y^2$ **53.** $11.5x^2 + 22.61x - 70.07$ **55.** $x^3 + 6x^2 - 3x - 18$ **57.** $6t^3 - 4t^2 - 15t + 10$ **59.** $6a^4 + a^2b^2 - 15b^4$ **61.** $24x^4 - 54x^3 - 15x^2$ **63.** $5x^4 - 20x^3 + 15x^2 - 60x$ **65.** $x^3 + 5x^2 + 11x + 10$ **67.** $x^3 + 8$ **69.** $2b^3 - 11b^2 + 14b - 8$ **71.** $a^3 + b^3$ **73.** $8x^3 - 10x^2y + 23xy^2 - 15y^3$ **75.** $6x^4 - 2x^3 + 17x^2 - 3x + 12$ **77.** $6x^4 + 10x^3 - 3x^2 + 9x - 2$ **79.** $2x^4 - 3x^3y - 3x^2y^2 + 7xy^3 - 3y^4$ **81.** $(f \cdot g)(x) = 6x^2 - 5x - 6$; 33 **83.** $(f \cdot h)(x) = 4x^3 - 14x^2 + 18x - 9$; 3 **85.** $(f \cdot f)(x) = 4x^2 - 12x + 9$; 25 **87.** $(f \cdot g)(x) = 20x^2 + 17x + 3$; 6 **89.** $(f \cdot h)(x) = 12x^3 - x^2 - 9x - 2$; -84 **91.** $(h \cdot h)(x) = 9x^4 - 6x^3 - 11x^2 + 4x + 4$; 0

93. a. 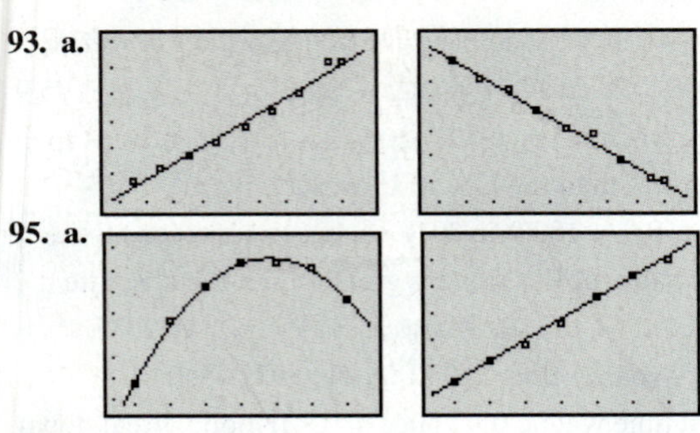 **b.** $(V \cdot A)(t) = -204.6t^2 + 58{,}959.6t + 496{,}824$ **c.** millions of dollars **d.** 1,939,579.8; in 2017, the total value of U.S. farmland will be \$1,939,579.8 million, or about \$1.94 trillion. **e.** increasing; between 1990 and 2017, the total value of U.S. farmland was increasing and will continue to increase every year; answers may vary.

95. a. **b.** $(B \cdot N)(t) = -3.94t^3 + 141.124t^2 - 505.8t - 276$ **c.** millions of dollars per month **d.** 14,406; in 2014, the total monthly revenue from cell phones will be \$14,406 million, or about \$14.4 billion. **e.** increasing; between 2004 and 2010, the total monthly revenue from cell phones increased every year; answers may vary.

97. Answers may vary; $-24x^2$ **99. a. i.** $8x^2 + 22x + 15$; quadratic **ii.** $15x^2 - 29x - 14$; quadratic **b.** Answers may vary; quadratic polynomial **c.** quadratic polynomial; answers may vary. **101. a.** $x^2 + 11x + 28$ **b.** $x^2 + 11x + 28$ **c.** Answers may vary. **103.** $(2x - 5)(3x + 4) = 3x(2x - 2) - x - 20 = 6x^2 - 7x - 20 = (3x + 4)(2x - 5)$; $6x^2 + 7x - 20 = (3x - 4)(2x + 5)$ **105.** $6x^3 - 22x^2 + 26x - 10$ **107.** $-2x^2 + 7x - 7$ **109.** $y = 3x^2 - 6x$; quadratic; parabola **111.** $y = -2x - 3$; linear; line **113.** $y = 10x^2 + x - 2$; quadratic; parabola **115.** -2; linear equation in one variable **117.** $14x^2 - 25x - 25$; quadratic (or second-degree) polynomial in one variable

Homework 7.3

1. x^8y^8 **3.** $36x^2$ **5.** $64p^3$ **7.** $64x^2$ **9.** $-27x^3$ **11.** $-a^5$ **13.** $16m^4p^4$ **15.** $81x^4y^4$ **17.** $x^2 + 10x + 25$ **19.** $x^2 - 8x + 16$ **21.** $4x^2 + 12x + 9$ **23.** $25y^2 - 20y + 4$ **25.** $4a^2 + 20ab + 25b^2$ **27.** $64x^2 - 48xy + 9y^2$ **29.** $4x^4 - 24x^2y^2 + 36y^4$ **31.** $-8x^3 - 40x^2 - 50x$ **33.** $25b^2 - 15b$ **35.** $c^2 + 5c + 4$ **37.** $b^2 - 9b + 18$ **39.** $4a - 2$ **41.** $h^2 + 2ah - 3h$ **43.** $f(x) = x^2 + 12x + 36$ **45.** $f(x) = x^2 - 6x + 10$ **47.** $f(x) = 2x^2 + 16x + 29$ **49.** $f(x) = -3x^2 + 6x - 5$ **51.** $x^2 - 16$ **53.** $t^2 - 49$ **55.** $49a^2 - 81$ **57.** $4x^2 - 9y^2$ **59.** $9r^2t^2 - 81w^2$ **61.** $64a^4 - 9b^4$ **63.** $x^4 - 16$ **65.** $81a^4 - 16b^4$ **67.** $-2x^3 + 4x^2 + 3x$ **69.** $-10t^3$ **71.** $3x^3 + x^2 + 2x + 8$ **73.** $4t^2w^2 - 9p^2$ **75.** $8x^3y^2 - 16x^2y^2 - 10xy^2$ **77.** $-8x^2 - x - 3$ **79.** $16w^2 - 64w + 64$ **81.** $6x^2 - 5xy - 21y^2$ **83.** $36x^2 - 49$ **85.** $4t^4 + 20t^2w^2 + 25w^4$ **87.** Answers may vary; $16x^2$ **89.** Answers may vary; $x^2 + 14x + 49$ **91.** Answers may vary; $x^2 - 10x + 25$; answers may vary **93. a.** Answers may vary. **b.** $x^2 + 8x + 16$ **c.** Answers may vary. **95.** $25 = 13$; answers may vary; $(A + B)^2 = A^2 + 2AB + B^2$ **97.** $(x - 2)^2, x(x - 4) + 4, x^2 - 4x + 4$ **99.** Answers may vary. **101. a.** $A^2 + B^2 + C^2 + 2AB + 2AC + 2BC$ **b.** Answers may vary. **103. a. i.** 32, 16, 8, 4, 2 **ii.** Answers may vary. **iii.** 1 **b.** 81, 27, 9, 3; 1 **c.** 1 **105.** $f(x) = 10x^2 - 6x$; quadratic; parabola **107.** $f(x) = 24x^2 - 38x + 15$; quadratic; parabola **109.** $f(x) = 6x - 9$; linear; line **111.** $(5, -1)$; system of two linear equations in two variables **113.** linear equation in two variables

Homework 7.4

1. x^8 **3.** r^6 **5.** $15x^9$ **7.** $32b^8$ **9.** $54a^6b^8$ **11.** r^7t^7 **13.** $64x^2$ **15.** $32x^5y^5$ **17.** $16a^4$ **19.** 1 **21.** a^3 **23.** $2x^4$ **25.** $\frac{5x^3y^7}{4}$ **27.** $\frac{t^7}{w^7}$ **29.** $\frac{27}{t^3}$ **31.** 1 **33.** r^8 **35.** x^{36} **37.** $36x^6$ **39.** t^{12} **41.** $8a^{27}$ **43.** $x^{13}y^{20}$ **45.** $45x^{16}$ **47.** $-3c^{26}$ **49.** x^9y^{19} **51.** $\frac{5t^8}{4}$ **53.** $\frac{3}{4}$ **55.** $\frac{y^3}{8x^3}$ **57.** $\frac{x^8}{y^{20}}$ **59.** $\frac{r^{12}}{36}$ **61.** $\frac{8a^{12}}{27b^6}$ **63.** 1 **65.** $\frac{8a^{18}b^3}{27c^{15}}$ **67.** $x^{11}y^4$ **69.** $\frac{w^7}{32}$ **71.** $2x^2y^7$ **73.** \$6719.58 **75.** 400 feet **77.** 507.66 watts **79.** 423.33 cubic centimeters **81.** Answers may vary; x^8 **83.** Answers may vary; $25x^6$ **85.** $\left(\frac{x}{y}\right)^3 = \frac{x}{y} \cdot \frac{x}{y} \cdot \frac{x}{y} = \frac{x \cdot x \cdot x}{y \cdot y \cdot y} = \frac{x^3}{y^3}$ **87.** Answers may vary; $16x^8$ **89.** x^5 **91.** $x^3 + x^2$ **93.** $5x^4$ **95.** $6x^8$ **97.** $9x^2$ **99.** $x^2 + 6x + 9$ **101.** 5; linear equation in one variable **103.** $\frac{1}{6}x - \frac{5}{6}$; linear (or first-degree) polynomial in one variable

Homework 7.5

1. $3x^4 + 7x^2$ **3.** $2x^2 + 4x$ **5.** $4p^6 - p^3 + 3$ **7.** $-2x^3 + \frac{3}{8}x^2 + 3x$ **9.** $2x - 4 + \frac{7}{2x^4}$ **11.** $2k^2 + \frac{3}{2}k - \frac{3}{k^2}$ **13.** $x^3 + x^2y - y^3$ **15.** $5m^3 + \frac{3}{2}mr + 2r^2$ **17.** $-3x^3y^3 - \frac{9}{2}x^2y^2 + 2x$ **19.** $3x + 2$ **21.** $4x - 2$ **23.** $2p + 5 + \frac{2}{3p + 2}$ **25.** $2x - 4 - \frac{3}{2x + 5}$ **27.** $2x - 1 + \frac{2}{5x - 1}$ **29.** $5m + 3 - \frac{3}{4m - 2}$ **31.** $2x^2 + 3x + 3 + \frac{4}{2x + 3}$ **33.** $4p^2 - 2p + 3 - \frac{2}{3p - 4}$

35. $x - 3 + \frac{16}{x+3}$ **37.** $4x^2 + 10x + 25$ **39.** $x^2 - 3x - 1 + \frac{6}{3x-1}$ **41.** $3x - 4 - \frac{4}{x^2+3}$ **43.** $2y + 3 - \frac{1}{3y^2-2}$

45. $3x^2 - 4x + 2$ **47.** $-x^2 + 5 + \frac{3}{x+4}$ **49.** $3k - 4 - \frac{3}{k-1}$ **51.** $2x^2 - x + 3 - \frac{2}{x-5}$ **53.** $3x^2 - x + 4 - \frac{2}{x+4}$

55. The student did not divide $7x^3$ by $2x^2$; $3x^2 + \frac{7x}{2}$ **57.** The student did not change the signs before adding; $2x + 4$

59. Answers may vary; $4x^2 + 2x - 3 + \frac{6}{x-3}$ **61.** Answers may vary. **63.** $2x^2 - 11x + 19$ **65.** Answers may vary.

67. Answers may vary. **69.** Answers may vary. **71.** $18x^3 - 15x^2 - 4x + 4$ **73.** $6x^2 - 4x$ **75.** $2x + 1$ **77.** $9x^2 - 12x + 4$ **79.** Answers may vary. **81.** Answers may vary. **83.** Answers may vary.

Chapter 7 Review Exercises

1. $-3x^3 - 2x^2 - 3x$ **2.** $3a^4b - 9a^3b^2 - 4a^2b^3$ **3.** $-5x^3 - 3x^2 + 3x - 9$ **4.** $-3a^3b - 6a^2b^2 + 10ab^3$ **5.** 24 **6.** -3 **7.** no such value **8.** 1 **9.** $-3, 5$ **10.** 9 **11.** 2, 4 **12.** 3 **13.** no such value **14.** $(f + g)(x) = x^3 - 2x^2 - 7x + 3; -11$ **15.** $(f - g)(x) = 5x^3 - 12x^2 - x + 1; -239$ **16. a.** $(B + A)(t) = -4.71t^2 + 157.6t - 303$ **b.** thousands of bank tellers and ATMs **c.** 884.56; in 2012, there were a total of about 885 thousand bank tellers and ATMs. **d.** $(B - A)(t) = -0.13t^2 + 4t + 151$ **e.** 176.08; in 2012, there were about 176 thousand more bank tellers than ATMs. **17.** $x^2 - 49$ **18.** $-40a^5b^6$ **19.** $8p^2 - 2pt - 45t^2$ **20.** $20x^3 - 23x^2 + 22x - 12$ **21.** $9x^2 + 42xy + 49y^2$ **22.** $36p^4 - 81t^6$ **23.** $-6r^3t^3 + 15r^2t^4 - 9rt^5$ **24.** $-36x^3 + 48x^2 - 16x$ **25.** $6m^4 + 7m^3p - 11m^2p^2 + 10mp^3 - 8p^4$ **26.** $10a^4 + 11a^2b^2 - 6b^4$ **27.** $81p^4 - 256t^4$ **28.** $x^2 - 8x + 16$ **29.** $8b^3 - 30b^2 + 13b - 21$ **30.** $w^2 - 12w + 27$ **31.** $-5c^2 - 20c - 20$ **32.** $16m^2 - 49n^2$ **33.** $a^2 - 10a + 24$ **34.** $6a + 3$ **35.** $f(x) = -2x^2 + 16x - 29$ **36.** $(f \cdot g)(x) = 6x^3 - 26x^2 + 37x - 21; 18$ **37. a.** $(G \cdot P)(t) = 1.316t^2 + 110.98t - 762$ **b.** millions of gallons of sports drinks consumed by Americans per year **c.** 3193.8; in 2017, Americans will consume a total of 3193.8 million (or about 3.19 billion) gallons of sports drinks. **d.** increasing; the total U.S. consumption of sports drinks will increase from 2000 to 2020. **38.** cubic polynomial; answers may vary. **39.** x^{10} **40.** $12x^7$ **41.** $-40a^6b^{12}$ **42.** $\frac{xy^3}{2}$ **43.** $\frac{x^3}{8}$ **44.** $32x^{45}y^{15}$ **45.** $75x^{14}$ **46.** $\frac{3c^5}{2}$ **47.** $\frac{a^8}{81}$ **48.** 1 **49.** $\frac{3x^3y^5}{2}$ **50.** $\frac{27x^9}{64}$ **51.** $-2x^3 + \frac{2}{3}x - \frac{3}{x^2}$ **52.** $4w^2 + 3wp - 7p^2$ **53.** $4x - 5 - \frac{3}{2x+1}$ **54.** $2x^2 - 4x + 3 + \frac{4}{3x-2}$ **55.** $16b^2 + 12b + 9$ **56.** $2x^2 - 4x + 3 + \frac{5}{x-3}$ **57.** $3y^2 - y - 4$ **58.** $x^2 - 2x + 6 - \frac{2}{x+3}$

Chapter 7 Test

1. $-a^3b - 5a^2b^2 + ab^3$ **2.** $-4x^3 - x^2 - 2x$ **3.** 5 **4.** $-3, -1$ **5.** -2 **6.** no such value **7.** $(f - g)(x) = -2x^2 + 8x - 16; -40$ **8. a.** $(W + M)(t) = 0.45t + 18.33$ **b.** millions of people who live alone **c.** 35.43; in 2018, about 35.4 million people will live alone. **d.** $(W - M)(t) = -0.03t + 4.83$ **e.** 3.69; in 2018, there will be about 3.7 million more women than men living alone. **9.** $-14x^3y^2 + 6x^2y^3 - 12xy^4$ **10.** $12x^2 - xy - 35y^2$ **11.** $6w^3 - 17w^2t + 13wt^2 - 20t^3$ **12.** $x^2 - 14x + 49$ **13.** $12x^3 + 36x^2 + 27x$ **14.** $6x^4 + 14x^3 - 9x^2 - 21x + 5$ **15.** $16x^4 - 81y^4$ **16.** $a^2 - 13a + 40$ **17.** $(f \cdot g)(x) = 6x^3 - 19x^2 + 22x - 8; 49$

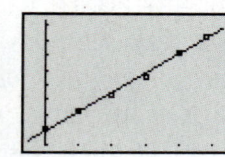

18. a. **b.** $(R \cdot P)(t) = -202.8t^2 + 748.8t + 1{,}006{,}176$ **c.** number of deaths **d.** 367,387.2; in 2018, there will be about 367,387 deaths from heart disease. **e.** decreasing; from 1965 to 2020, the number of deaths from heart disease decreased and will continue decreasing; answers may vary.

19. $f(x) = -3x^2 + 6x + 2$ **20.** $\frac{3x^4}{4y^5}$ **21.** $64a^{15}b^{16}$ **22.** $\frac{x^{18}}{y^{24}}$ **23.** $49x^6$ **24.** $\frac{x^{12}y^{24}}{16w^{12}}$ **25.** $2p^{13}t^3$ **26.** 64.44 decibels

27. $3w^3 - 4w^2y^2 + 2wy^3$ **28.** $3x^2 - x + 2 + \frac{4}{4x-1}$ **29.** $x^2 - 5x + 3 - \frac{2}{x-2}$

Making Sure You're Ready for Intermediate Algebra: A Review of Chapters 1–7

1. a. independent: a, dependent: p **b.** 85% of 45-year-old Americans work. **2.** -6°F **3.** -42 **4.** 15 **5.** 84 **6.** $-\frac{21}{8}$

7.

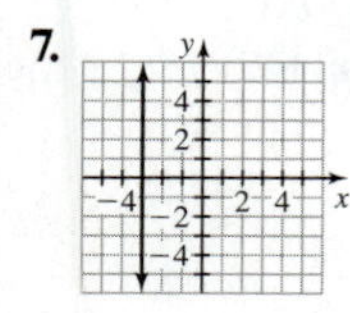

8. $\frac{1}{2}$; increasing **9.**

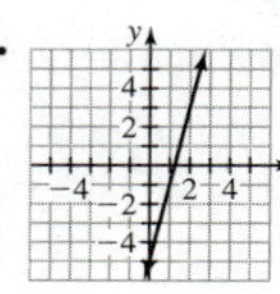

10.

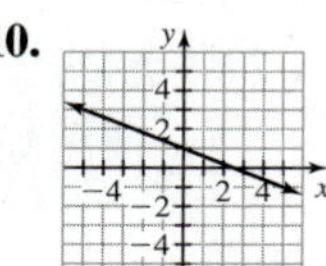

11. $y = \frac{1}{2}x + 1$ **12.** about -4.58 executions per year

13. a. yes; 5.3; the number of households that pay bills online has increased by 5.3 million per year. **b.** (0, 66); in 2010, 66 million households paid bills online. **c.** $n = 5.3t + 66$ **d.** The units of the expressions on both sides of the equation are millions of households. **e.** 97.8 million households **14.** $-10x - 47y$ **15.** $7 - 4\left(\frac{x}{2}\right); 7 - 2x$ **16.** $\frac{7}{6}$ **17.** $\frac{19}{3}$ **18.** $\frac{28}{3}$ **19.** -2 **20.** 6 **21.** -2

22. $h = \frac{S - 2\pi r^2}{r}$ **23.** 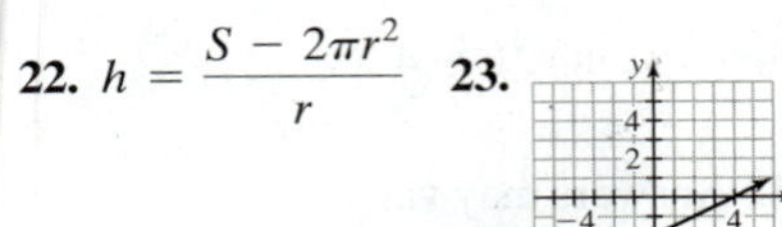 **24.** x-intercept: $(10, 0)$; y-intercept: $(0, -4)$ **25.** $-4 \le x \le 5$ **26.** $-3 \le x \le 3$

27. yes; answers may vary. **28.** -26 **29.** $-\frac{1}{2}$ **30.** $y = \frac{2}{3}x - \frac{14}{3}$ **31.** $y = -\frac{5}{3}x + \frac{1}{3}$ **32. a.**

b. $f(t) = -0.58t + 14.32$ **c.** 2018 **d.** $(0, 14.32)$; in 2000, about 14.3% of high school students dropped out of school. **e.** $(24.69, 0)$; in 2025, no high school students will drop out of school; model breakdown has likely occurred. **33.** 2018 **34.** $x > -4$; $(-4, \infty)$ **35.** $-1 \le x \le 3$; $[-1, 3]$ **36.** $(9, 0)$ **37.** $(2, -1)$ **38.** $(4, -2)$ **39. a.** $D(t) = 184t + 1247$ **b.** $P(t) = -91t + 231$ **c.** 2006; 567 million units **40.** 4200 \$20 tickets, 1800 \$35 tickets **41.** **42.** **43.** $-4x^2 - 7x + 9$ **44.** $5a^2 + 4ab - 7b^2$ **45.** $(f - g)(x) = 6x^2 - 7x - 8; 67$ **46.** -5 **47.** 1

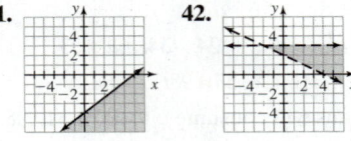

48. 0, 2 **49.** no such value **50.** $16p^2 + 34pt - 15t^2$ **51.** $2x^4 - 3x^3 + 3x^2 + 10x - 12$ **52.** $(f \cdot g)(x) = 6x^3 - 22x^2 + 29x - 15; 3$ **53. a.** $(S \cdot N)(t) = 0.06t^2 + 4.82t + 87.73$ **b.** billions of dollars **c.** 269.73; the total money paid for teacher salaries will be \$269.73 billion in 2018. **54.** $x^2 + 12x + 36$ **55.** $9p^2 - 42pq + 49q^2$ **56.** $f(x) = -3x^2 + 12x - 7$ **57.** $64b^2 - 25c^2$ **58.** $24x^{11}$ **59.** $\frac{x^{20}w^5}{32y^{40}}$ **60.** $81xy^{10}$ **61.** $-\frac{4}{3}x^3 + \frac{5}{3}x - \frac{7}{6x^2}$ **62.** $5x^2 + x - 2 - \frac{3}{3x - 2}$

Chapter 8

Homework 8.1 **1.** $(x + 2)(x + 3)$ **3.** $(t + 4)(t + 5)$ **5.** $(x + 4)^2$ **7.** $(x - 4)(x + 2)$ **9.** $(a - 8)(a + 2)$ **11.** $(x + 8)(x - 3)$ **13.** prime **15.** $(t - 4)(t + 7)$ **17.** $(x - 8)(x - 2)$ **19.** $(x - 8)(x - 3)$ **21.** prime **23.** $(r - 5)^2$ **25.** $(x + y)(x + 9y)$ **27.** $(m - 3n)(m + 2n)$ **29.** $(a - 6b)(a - b)$ **31.** $(p - 4q)(p + 7q)$ **33.** $(p - 8q)(p + 2q)$ **35.** $(x - 5)(x + 5)$ **37.** $(x - 9)(x + 9)$ **39.** prime **41.** $(2x - 5)(2x + 5)$ **43.** $(9r - 1)(9r + 1)$ **45.** prime **47.** $(7p - 10q)(7p + 10q)$ **49.** $(8m - 3n)(8m + 3n)$ **51.** $(4x^2 + 9)(2x + 3)(2x - 3)$ **53.** $(t^2 + w^2)(t + w)(t - w)$ **55.** $(x - 6)(x + 3)$ **57.** $(x + 7)^2$ **59.** $(a - 2)(a + 2)$ **61.** prime **63.** $(x - 6)(x - 2)$ **65.** $(w - 8)(w + 6)$ **67.** $(t - 3k)(t + 3k)(t^2 + 9k^2)$ **69.** prime **71.** $(m - 9n)(m + 3n)$ **73.** $(x - 16)(x - 2)$ **75.** $(10p - 3t)(10p + 3t)$ **77.** $(a + 6b)^2$ **79.** Answers may vary; the polynomial is prime. **81.** $(x - 3)(x + 7), x^2 + 4x - 21, (x + 7)(x - 3)$ **83.** $(x - 8)(x + 3); x^2 - 5x - 24$; answers may vary. **85. a.** $(x - 4)(x - 1)$ **b.** $(1, 0), (4, 0)$ **c.** Answers may vary. **87.** Answers may vary. **89.** $-13, -8, -7, 7, 8, 13$ **91.** 0; 0; answers may vary. **93.** Answers may vary. **95.** $x^2 - 7x - 18$ **97.** $(x - 5)(x - 10)$ **99.** $9x^2 - 49$ **101.** $(5x - 6)(5x + 6)$ **103.** $3p + 11w$; linear (or first-degree) polynomial in two variables **105.** $10p^2 - 6pw - 28w^2$; quadratic (or second-degree) polynomial in two variables **107.** $(p - 9w)(p - 2w)$; quadratic (or second-degree) polynomial in two variables

Homework 8.2 **1.** $2(3x + 4)$ **3.** $5w(4w + 7)$ **5.** $6x^2(2x - 5)$ **7.** $3ab(2a - 3)$ **9.** $4x^2y^2(2x + 3y)$ **11.** $5(3x^3 - 2x - 6)$ **13.** $4t(3t^3 + 2t^2 - 4)$ **15.** $5ab(2a^3 - 3a^2 + 5)$ **17.** $2(x - 3)(x + 3)$ **19.** $3(m + 2)(m + 5)$ **21.** $2(x - 6)(x - 3)$ **23.** $4r(r - 5)(r + 1)$ **25.** $6x^2(x - 2)(x + 2)$ **27.** $2m^2n(2m - 3)(2m + 3)$ **29.** $5x^2(x - 4)(x + 6)$ **31.** $4t(t + 1)(t + 8)$ **33.** $-3x(2x - 3)(2x + 3)$ **35.** $-3x(x - 2)(x + 8)$ **37.** $-(x - 1)(x - 10)$ **39.** $6a^2b(a + 3)^2$ **41.** $4x^2y(x - 5y)(x + 2y)$ **43.** $-2xy^2(x - 4y)^2$ **45.** $(x - 3)(5x^2 + 2)$ **47.** $(2x + 5)(6x^2 - 7)$ **49.** $(p + 3)(2p^2 + 5)$ **51.** $(3x - 1)(2x^2 + 7)$

53. $(3w+1)(5w^2-2)$ **55.** $(2x-3)(2x+3)(4x-3)$ **57.** $(b-3)(b+3)(2b-5)$ **59.** $(x-1)^2(x+1)$ **61.** $(x-3y)(a-2b)$ **63.** $(5x+2y)(a^2-b)$ **65.** $(9x-5)(9x+5)$ **67.** $(w-8)(w-2)$ **69.** $(x-6)(x-4)$ **71.** $5ab(4a-3b^2)$ **73.** $(x-5y)(x+6y)$ **75.** $-6r(r-2)^2$ **77.** $x(8x-7)(8x+7)$ **79.** $-(m-3)^2$ **81.** $(x-2)(x+2)(x+9)$ **83.** $2mn(m-2n)(m-3n)$ **85.** Answers may vary; $(2x^2+5)(3x+4)$ **87.** Answers may vary; $4x(x+2)(x+5)$ **89.** Answers may vary. **91.** Answers may vary. **93.** Answers may vary. **95.** $2x^3+2x^2-24x$ **97.** $5x(x-4)^2$ **99.** $(2x-3)(3x^2-2)$ **101** $x^3-3x^2+5x-15$ **103.**

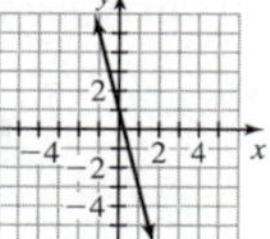

linear equation in two variables

105. 1; linear equation in one variable **107.** $(1,-3)$; system of two linear equations in two variables

Homework 8.3

1. $(x+3)(2x+1)$ **3.** $(x+2)(5x+1)$ **5.** $(x+2)(3x+2)$ **7.** $(t+2)(2t-3)$ **9.** $(2x-3)(3x-2)$ **11.** $(2x+5)^2$ **13.** prime **15.** $(3x+4)(6x-1)$ **17.** $(m-6)(3m-4)$ **19.** $(x-8)(2x-5)$ **21.** $(3w-1)^2$ **23.** $(2a+3b)(a+b)$ **25.** $(5x-2y)(x+4y)$ **27.** $3(2b-c)(b-2c)$ **29.** $(2r-5y)^2$ **31.** $2(x+5)(2x+3)$ **33.** $5(2a-3)(2a-1)$ **35.** $3(x+1)(8x-3)$ **37.** $-2(2x-3)(5x+2)$ **39.** $-3(4x+3)(x-1)$ **41.** $4x(2x-3)(2x-1)$ **43.** $2w^2(2w^2-3w-6)$ **45.** $5x^2(x+2)(2x-5)$ **47.** $2(a-6b)(3a+b)$ **49.** $4r(r+2w)(3r+4w)$ **51.** $10ab^2(2a+7b)(a-2b)$ **53.** $(x-9)(x+3)$ **55.** $-8x(6x-5)$ **57.** prime **59.** $(2x-3)^2$ **61.** $-17(p+1)(p-1)$ **63.** $(x+4)(x+6)$ **65.** $(b-7c)(b+4c)$ **67.** $(4t-3)(2t-1)$ **69.** $7x^2(x-2)(x+2)$ **71.** $3x^2(x-9y)(x+2y)$ **73.** $(2p+3)(2p-3)(3p-1)$ **75.** prime **77.** $3x^2(x-5)(x-2)$ **79.** $2x^2(8x-5)(x-2)$ **81.** $(6a+7b)(6a-7b)$ **83.** $-2y(x-6)(x+2)$ **85.** $2pt^2(5p-4t)(p+3t)$ **87.** no; answers may vary. **89.** Answers may vary. **91.** Answers may vary. **93.** Answers may vary. **95.** $2(x-2)(x-6)=2(x^2-8x+12)=2x^2-16x+24=(x-2)(2x-12)=2(x-4)^2-8=(2x-4)(x-6)$ **97.** $(x+6)(3x-2)$ **99.** $12x^2-25x+7$ **101.** $2x^3-3x^2-14x+15$ **103.** $2x(3x-1)(x+2)$ **105.**

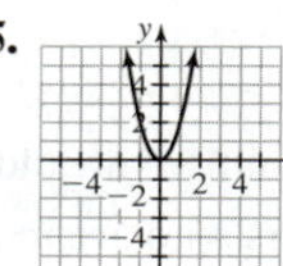

quadratic equation in two variables **107.** $(x-3)(x+1)$; quadratic (or second-degree) polynomial in one variable **109.** 32; quadratic (or second-degree) polynomial in one variable

Homework 8.4

1. $(x+3)(x^2-3x+9)$ **3.** $(x+5)(x^2-5x+25)$ **5.** $(x-2)(x^2+2x+4)$ **7.** $(x-1)(x^2+x+1)$ **9.** $(2t+3)(4t^2-6t+9)$ **11.** $(3x-2)(9x^2+6x+4)$ **13.** $5(x+2)(x^2-2x+4)$ **15.** $2(x-3)(x^2+3x+9)$ **17.** $(2x+3y)(4x^2-6xy+9y^2)$ **19.** $(4a-3b)(16a^2+12ab+9b^2)$ **21.** $2x(x-3y)(x^2+3xy+9y^2)$ **23.** $(k+1)(k^2-k+1)(k-1)(k^2+k+1)$ **25.** $(2x+y)(4x^2-2xy+y^2)(2x-y)(4x^2+2xy+y^2)$ **27.** $(x-8)(x+8)$ **29.** $2(t-3)(t+4)$ **31.** prime **33.** $-3(x-5)(x-3)$ **35.** $(3p-1)(5p-1)$ **37.** $(x-1)^2$ **39.** $4(2r+1)(3r-1)$ **41.** $-2ab^2(2b-3a)$ **43.** $(a-5b)(a+4b)$ **45.** $(4a+3)(16a^2-12a+9)$ **47.** $(2x-5)(2x-1)(2x+1)$ **49.** $-4x^3(3x+1)$ **51.** $5a^2(a+1)(3a+2)$ **53.** $(x-2)(x-12)$ **55.** $2w^2(w^2+2w-4)$ **57.** $3x^2(2x-3)(2x+3)$ **59.** $(x+5)^2$ **61.** $(x-3)(2x+7)$ **63.** $(9p^2+4q^2)(3p+2q)(3p-2q)$ **65.** $3(x+2)(x^2-2x+4)$ **67.** $m(m-9n)(m-4n)$ **69.** prime **71.** $(10x-3y)(10x+3y)$ **73.** $(2x+3)^2$ **75.** $(2x+3)(3x-2)(3x+2)$ **77.** $a(a-2b)(3a-4b)$ **79.** prime **81.** $(x-10)(x^2+10x+100)$ **83.** $2y(2x-5y)(3x+y)$ **85.** Answers may vary. **87.** Answers may vary. **89.** Answers may vary; $x^3-27=(x-3)(x^2+3x+9)$ **91.** no; answers may vary. **93.** $(A-B)(A^2+AB+B^2)=A^3+A^2B+AB^2-A^2B-AB^2-B^3=A^3-B^3$. **95.** Answers may vary. **97.** $25x^2-49$ **99.** $(9x-4)(9x+4)$ **101.** $3x(x-1)(x+4)$ **103.** $-14x^3+10x^2-2x$ **105.** linear equation in two variables

107. 7; linear equation in one variable **109.** $12x^2-7x-10$; quadratic (or second-degree) polynomial in one variable

Homework 8.5

1. $-4, 7$ **3.** $-5, -2$ **5.** $-4, 3$ **7.** $3, 5$ **9.** -7 **11.** $-4, 6$ **13.** $\pm\frac{7}{5}$ **15.** $\frac{1}{3}, \frac{3}{2}$ **17.** $-6, 5$ **19.** $-1, 0, \frac{5}{2}$ **21.** $-2, 7$ **23.** $-2, 4$ **25.** 6 **27.** $-7, 7$ **29.** $\pm\frac{5}{4}$ **31.** $\pm2, 0$ **33.** $0, 2$ **35.** $-5, \frac{1}{2}$ **37.** $-3, 0, 6$ **39.** $-\frac{2}{3}, 0, \frac{1}{2}$ **41.** $-\frac{2}{3}, 1$

43. $-3, 4$ **45.** $\pm\frac{1}{5}$ **47.** $-10, 3$ **49.** $\pm 3, \frac{1}{2}$ **51.** $-3, \pm\frac{2}{3}$ **53.** $0, \frac{37}{3}$ **55.** $-6, 4$ **57.** $-3, -2, 8$ **59.** $(4, 0), (5, 0)$

61. $\left(-\frac{5}{6}, 0\right), \left(\frac{5}{6}, 0\right)$ **63.** $\left(-\frac{2}{3}, 0\right), (0, 0), \left(\frac{5}{4}, 0\right)$ **65.** $(-2, 0), (-1, 0), (1, 0)$ **67.** 0 **69.** $-4, 5$ **71.** $-5, 3$ **73.** -1 **75.** $-1, 2$

77. $\pm 1, 3$ **79.** $-1.24, 3.24$ **81.** $-0.81, 1.47, 3.34$ **83.** $-2, 4$ **85.** no real solution **87.** Answers may vary; 0, 1 **89.** Answers may vary; 5, 8 **91.** Answers may vary. **93.** Answers may vary. **95.** Answers may vary. **97. a.** 1, 5 **b.** 3 **c.**

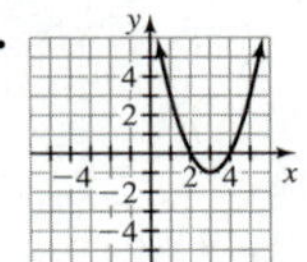

99. Answers may vary. **101.** $(x + 2)(x + 3)$ **103.** $-2, -3$ **105.** $-2, -\frac{2}{3}, 0$ **107.** $p(3p + 2)(p + 2)$ **109.** $P = \frac{A}{1 + RT}$

111. $-18x^3 + 60x^2 - 50x$; a cubic (or third-degree) polynomial in one variable **113.** $2x(2x - 5)^2$; a cubic (or third-degree) polynomial in one variable **115.** -324; cubic function

Homework 8.6

1. a. 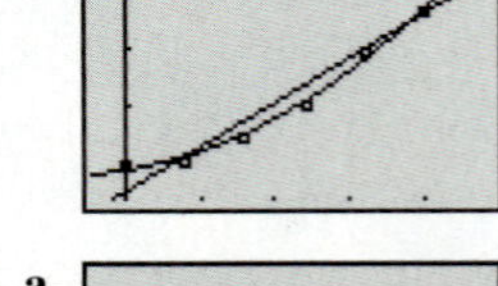 Q **b.** Q **c.** L: (0,1301); Q: (0,1376); that of the quadratic model **d.** 1980, 1992

3. a. 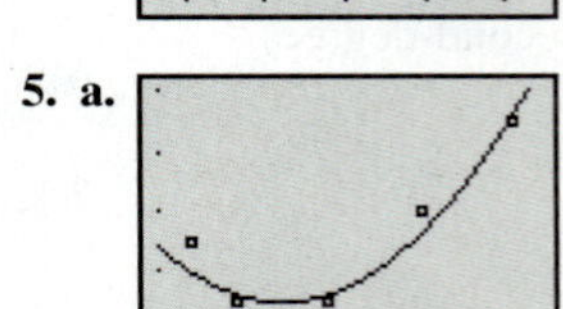 yes **b.** 35; 7 years after being rated B2, 35% of companies default on their bonds. **c.** 1, 21; 1 year and 21 years after being rated B2, 7% of companies default on their bonds; model breakdown has occurred for the estimate of 21 years. **d.** (0, 0), (22, 0); no companies default 0 years and 22 years after being rated B2; model breakdown has occurred for the estimate of 22 years.

5. a. 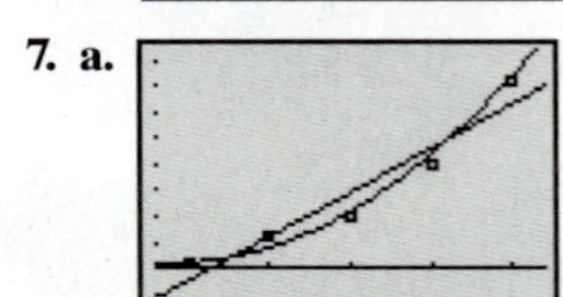 yes **b.** (0, 28); in 2000, 28% of Americans thought that labor unions would becomes stronger. **c.** 49% **d.** 2016 **e.** The percentage of private-sector workers who are in a union decreased from 2007 to 2011.

7. a. 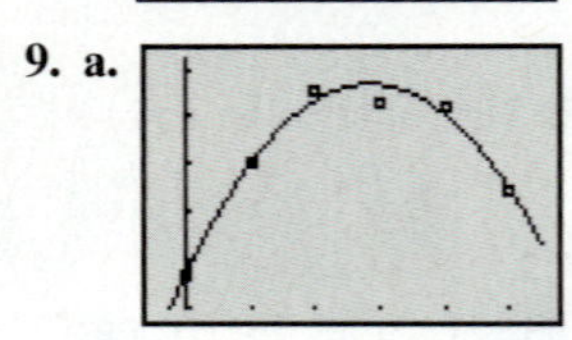 quadratic equation **b.** 29.2 million iPhones; 22.0 million iPhones; 22.0 million iPhones is the better estimate; answers may vary. **c.** 422 million iPhones **d.** 2014

9. a. 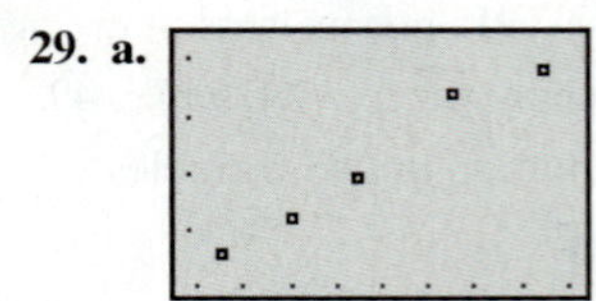 yes **b.** 10,661 thousand (10.661 million) people **c.** 2005, 2011 **d.** no; the model predicts attendance will continue to decrease. **11.** 2007, 2016 **13.** 2012, 2015 **15. a.** 0 seconds, 4 seconds; answers may vary. **b.** 1 second, 3 seconds; answers may vary. **c.** 2 seconds; answers may vary. **17.** width: 5 feet, length: 12 feet **19.** width: 5 feet, length: 12 feet **21.** width: 2 centimeters, length: 6 centimeters **23.** 2 feet **25.** 1 inch **27.** Answers may vary.

29. a. linear equation **b.** $f(t) = 123.02t + 427.78$ **c.** 2017 **31. a.** $g(t) = 0.27t + 3.34$ **b.** 2018 **33.** \$11,247 **35.** Answers may vary. **37.** Answers may vary. **39.** Answers may vary. **41.** Answers may vary.

Chapter 8 Review Exercises

1. $(x + 4)(x + 5)$ **2.** $2(3x - 4)(x + 1)$ **3.** $(x + 7)^2$ **4.** $-3t^2(2t + 5)(3t - 2)$ **5.** $(p - 9q)(p + 6q)$ **6.** $(x - 8)(x - 4)$ **7.** $-(3x - 2)(3x + 2)$ **8.** prime **9.** $5mn(4m - 9n^2)$ **10.** $2x(x + 7)(x + 1)$ **11.** $16x(x - 1)^2$ **12.** $8x^2(3x - 4)$ **13.** $5x^2y(x - 3)(x - 4)$ **14.** $-(m - 5)(m + 7)$ **15.** $(2r^2 + 3)(2r - 5)$ **16.** $(9t^2 + 4w^2)(3t + 2w)(3t - 2w)$ **17.** $2(y - 3)(y^2 + 3y + 9)$ **18.** $(x - 5)(x - 4)$ **19.** $(3t - 2y)(2t + 5y)$ **20.** $2x(x - 5)(x + 5)$ **21.** $(x - 5)^2$ **22.** $(p - 9)(p + 9)$ **23.** prime **24.** $(2x + 5)^2$ **25.** $2w(6w - 1)(w - 4)$ **26.** $(7a - 3b)(7a + 3b)$ **27.** prime

28. $(x-2)(x+2)(x+3)$ **29.** $(r+2)(r^2-2r+4)$ **30.** $(x-5y)(2a-3b)$ **31.** Answers may vary; the polynomial is prime. **32.** Answers may vary; $5x(x+3)(x+4)$ **33.** $-25, -14, -11, -10, 10, 11, 14, 25$ **34.** $-6, -2, 0$ **35.** $-1, 2$ **36.** 3 **37.** 3, 5 **38.** $-\frac{9}{5}, \frac{9}{5}$ **39.** $-1, 1, \frac{7}{2}$ **40.** $-\frac{2}{3}, \frac{1}{2}$ **41.** 0, 5 **42.** $\frac{3}{2}, \frac{3}{4}$ **43.** $-5, 7$ **44.** $-5, 6$ **45.** $-3, 3, \frac{2}{3}$ **46.** $-2, 2$ **47.** $-6, 2$ **48.** $\frac{1}{3}, 2$ **49.** $\frac{3}{4}$ **50.** $(-3,0), (0,0), (2,0)$ **51.** $-2.56, 1.56$ **52.** 1 **53.** $-2, 4$ **54.** no real-number solutions **55.** 0, 2 **56.** $(-7,0), (7,0)$ **57.** $\left(-\frac{3}{4}, 0\right), \left(\frac{5}{2}, 0\right)$ **58.** Answers may vary. **59.** Answers may vary.

60. a. quadratic equation **b.** 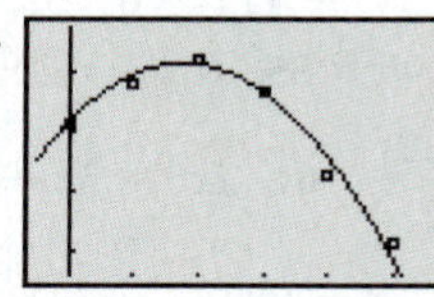 yes **c.** 78 thousand deaths **d.** 2002, 2005 **e.** (3.50, 807); in 2004, the number of malaria deaths was 807 thousand deaths—the largest in any year—according to the model. **f.** no; answers may vary.

61. 1 second, 4 seconds **62.** width: 3 feet, length: 10 feet

Chapter 8 Test

1. $(x-8)(x+5)$ **2.** $(x-6)(x-4)$ **3.** $2m^2n(4n^2-5m)$ **4.** $(p-10q)(p-4q)$ **5.** $(5p-6y)(5p+6y)$ **6.** $3x^2y(x-4)(x-3)$ **7.** $(2x-3)(2x+3)(2x+5)$ **8.** $(4x-3)(2x-5)$ **9.** $-2(8x-3)(x+2)$ **10.** $2a^2b(2a-3b)(4a-3b)$ **11.** $2(3m+4p)(9m^2-12mp+16p^2)$ **12.** $(x-5)(x+2), x^2-3x-10, (x+2)(x-5)$ **13.** Answers may vary; $(x-2)(x+2)(5x+3)$ **14.** 4, 9 **15.** $-\frac{3}{7}, \frac{3}{7}$ **16.** $1, \frac{11}{2}$ **17.** $-4, 6$ **18.** $-2, -\frac{2}{3}, 2$ **19.** $-1, 0, 5$ **20.** $\frac{2}{3}, \frac{3}{2}$ **21.** $\left(\frac{2}{5}, 0\right), \left(\frac{3}{2}, 0\right)$ **22.** $-5, -1$ **23.** $-4, -2$ **24.** -3 **25.** no real-number solution **26.** $\left(-\frac{2}{5}, 0\right), \left(\frac{3}{2}, 0\right)$ **27.** Answers may vary. **28. a.** 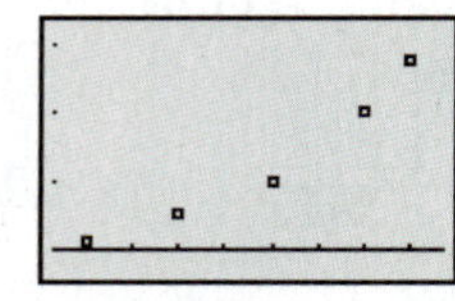 quadratic equation; answers may vary. **b.** 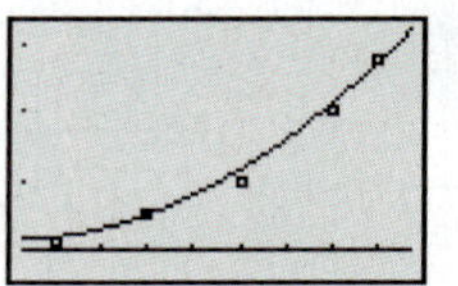 yes **c.** 2008 **d.** 208 thousand troops **29.** 2 inches

Making Sure You're Ready for Intermediate Algebra: A Review of Chapters 1–8

1. a. independent: a; dependent: p **b.** 75% of 50-year-old Americans own a home. **2.** \$210 **3.** $-\frac{7}{5}$ **4.** 40 **5.** **6.** $-\frac{4}{3}$; decreasing **7.** 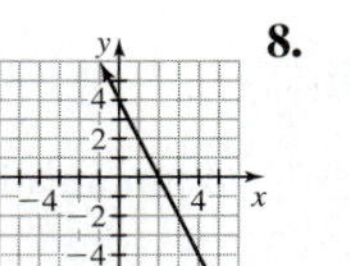 **8.** **9.** $y=-\frac{2}{3}x+2$ **10.** about 1.77 shredder models per year **11. a.** -0.05; the car uses 0.05 gallon of gas per mile. **b.** $g(x)=-0.05x+15.3$ **c.** (306, 0); after she has driven 306 miles, the tank will be empty. **d.** domain: $0 \le x \le 306$; range: $0 \le g(x) \le 15.3$ **e.** 286 miles **12.** $8x+11y-9$ **13.** $5-3(x-4)$; $-3x+17$ **14.** $\frac{20}{11}$ **15.** $\frac{3}{7}$ **16.** -2 **17.** $\frac{53}{4}$ **18.** -5 **19.** 1 **20.** $x=\frac{y-y_1+mx_1}{m}$ **21.** 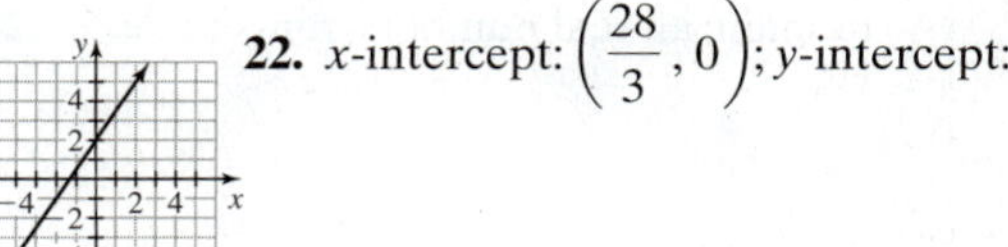 **22.** x-intercept: $\left(\frac{28}{3}, 0\right)$; y-intercept: $(0,-4)$ **23.** $-5 \le x \le 4$ **24.** $-3 \le y \le 3$ **25.** yes **26.** 33 **27.** $-\frac{27}{2}$ **28.** $y=-\frac{3}{5}x+\frac{2}{5}$ **29.** $y=-\frac{1}{2}x-\frac{11}{2}$

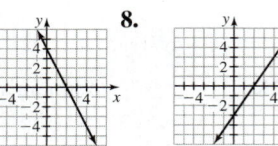

30. a. **b.** $f(t)=-1.41t+21.65$ **c.** -1.41; the percentage of Americans who say they have "a great deal of confidence" in Congress decreases by 1.41 percentage points per year. **d.** 2% **e.** 2006 **f.** (0, 21.65); in 2000, 22% of Americans said they have "a great deal of confidence" in Congress. **g.** (15.35, 0); in 2015, no one will say they have "a great deal of confidence" in Congress; model breakdown has occurred.

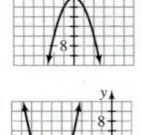

31. 2017 **32.** $x \geq 3$; $[3, \infty)$ **33.** $-4 < x < 2$; $(-4, 2)$ **34.** $(-2, 3)$ **35.** $(-1, 2)$
36. $(3, -2)$ **37. a.** $W(t) = 0.45t + 12.35$; $M(t) = -0.23t + 97.14$ **b.** 2025; 68.5% **c.** Answers may vary.
38. $f(t) = -1903t + 18{,}249$; $g(t) = -1225t + 14{,}564$ **b.** 2017; $7906 **39.** $5000 in UBS Global Equity Y, $2500 in Fidelity Worldwide
40. **41.** **42.** $-6x^3 + 5x^2 - 6x$ **43.** $-3a^2 + 9ab - 10b^2$ **44.** $(f - g)(x) = -2x^2 + 7x + 1; -21$
45. -2 **46.** $-3, -1$ **47.** -2 **48.** no such value
49. **50.** $15r^2 - 28rt + 12t^2$ **51.** $8x^4 + 2x^3 + 5x^2 + 11x - 6$ **52.** $(f \cdot g)(x) = 12x^3 - 14x^2 - 14x - 2; -14$
53. $4x^2 - 20x + 25$ **54.** $16a^2 + 24ab + 9b^2$ **55.** $f(x) = 2x^2 - 12x + 16$ **56.** $9x^4 - 64$ **57.** $162x^{16}$
58. $\dfrac{27x^9}{y^{21}w^{12}}$ **59.** $16x^7y^3$ **60.** $\dfrac{4}{3}p^2 - \dfrac{1}{9} - \dfrac{2}{3p^3}$ **61.** $2x^2 + 3x + 2 - \dfrac{2}{4x + 3}$ **62.** $(w + 7)(w - 2)$
63. $(2m - 7n)(2m + 7n)$ **64.** $(a - 8b)(a + 5b)$ **65.** $3x^2(x - 9y)(x - 2y)$ **66.** $(2x - 3)(4x^2 + 6x + 9)$ **67.** $-5, 7$
68. $-4, \dfrac{2}{5}$ **69.** $-2, -\dfrac{3}{2}, \dfrac{3}{2}$ **70.** $(-7, 0), (3, 0)$ **71.** 2006, 2016 **72.** width: 6 feet, length: 14 feet **73. a.**

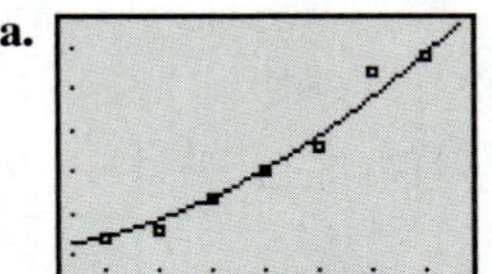

yes **b.** 85% **c.** 2012

Chapter 9

Homework 9.1

1. $(0, 0)$ **3.** $(0, 0)$ **5.** $(0, 5)$ **7.**

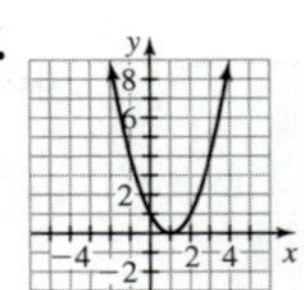

$(1, 0)$ **9.**

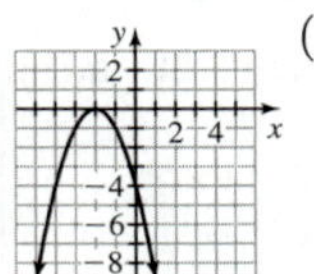

$(-2, 0)$

11.

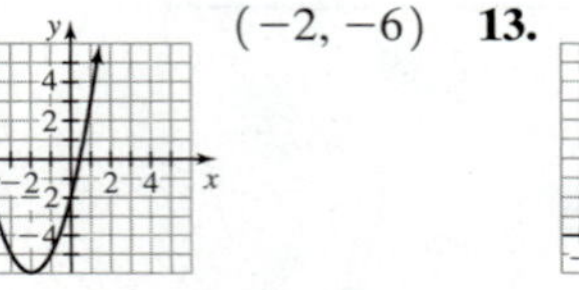

$(-2, -6)$ **13.** $(1, 3)$ **15.**

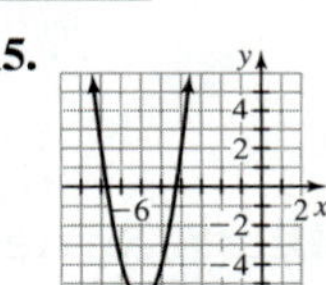

$(-6, -6)$ **17.**

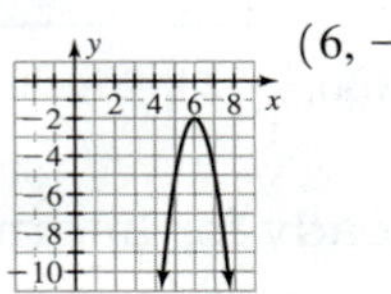

$(6, -2)$

19.

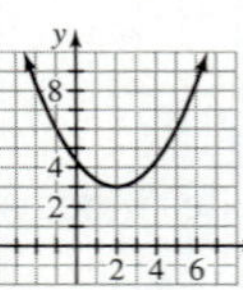

$(2, 3)$ **21.**

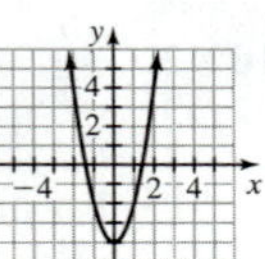

domain: all real numbers; range: $y \geq -4$

23. domain: all real numbers; range: $y \leq -3$ **25.**

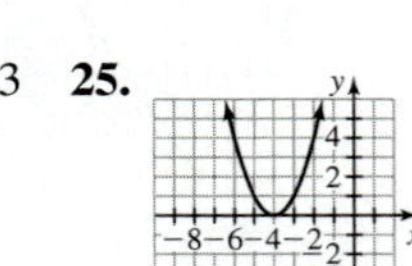

domain: all real numbers; range: $y \geq 0$

27. domain: all real numbers; range: $y \geq 2$ **29.** domain: all real numbers; range: $y \geq -4$

31. domain: all real numbers; range: $y \leq 2$ **33. a.** $f(t) = 2.25(t - 5.7)^2 + 265$ **b.** $(5.7, 265)$; in 1996, the U.S. Department of Defense spent the least, $265 billion. **c.** $605 billion **d.** 11.6
35. a. $f(t) = -0.016(t - 52.3)^2 + 31$ **b.** $(52.3, 31)$; the largest percentage of Americans who are obese, 31%, occurs at age 52 years. **c.** 24% **d.** $(8.28, 0)$, $(96.32, 0)$; no 8-year-old Americans and no 96-year-old Americans are obese; model breakdown has occurred.

37. a. 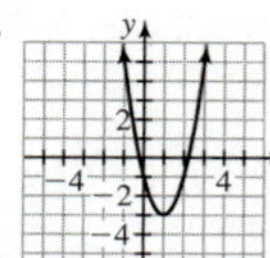 **b.** Answers may vary. **c.** For each input–output pair, the output variable is 3 less than twice the square of the difference of the input variable and 1. **39. a.** **b.** 2, 4 **c.** 3 **d.** no such value **41.** Answers may vary. **43. a.** $a > 0, h < 0, k < 0$ **b.** $a < 0, h < 0, k > 0$ **c.** $a > 0, h > 0, k = 0$ **d.** $a < 0, h = 0, k < 0$ **45.** Answers may vary; functions are of the form $f(x) = a(x + 5)^2 + 3$, where $a \neq 0$. **47.** $f(x) = \frac{5}{8}(x - 5)^2 - 6$ **49.** $f(x) = -2.1(x + 7)^2 + 3.71$ **51.** yes; answers may vary. **53.** yes; answers may vary. **55.** $(2, 5)$ **57. a.** 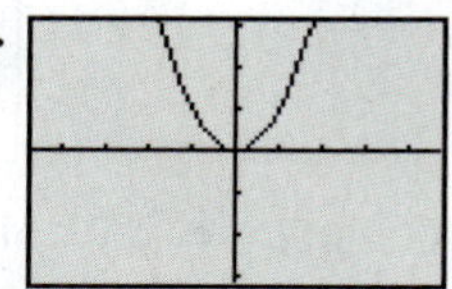 **b.** 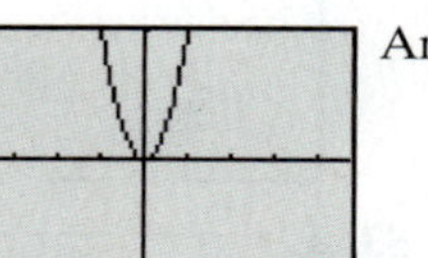 Answers may vary.

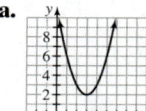

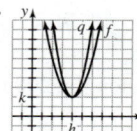

c. 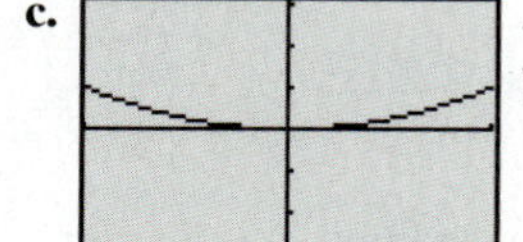 Answers may vary. **d.** Answers may vary. **59.** No, the graph of $y = x^2$ should be translated to the right by 4 units. **61.** Answers may vary. **63. a.** 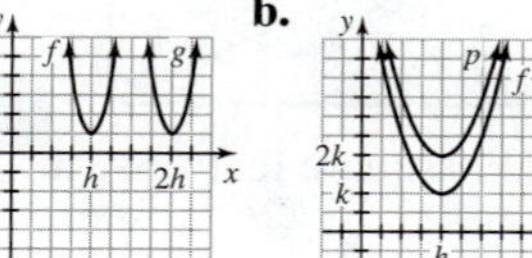 **b.** **c.**

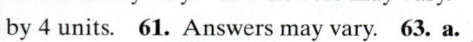

d. **65.** Answers may vary. **67.** 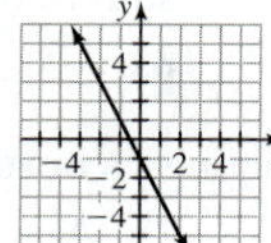 **69.** **71. a.** 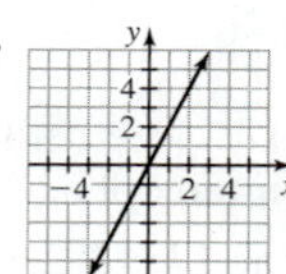 **b.**

c. The y-intercept is $(0, 0)$ for both graphs. **d.** $y = x^2$; answers may vary. **73.** $-\frac{1}{5}$; linear equation in one variable **75.** $20x + 4$; linear (or first-degree) polynomial in one variable **77.** $24x^2 - 136x + 80$; quadratic (or second-degree) polynomial in one variable

Homework 9.2

1. 5 **3.** 3 **5.** -3.5 **7.** 3.65 **9.** $(4, 9)$

11. 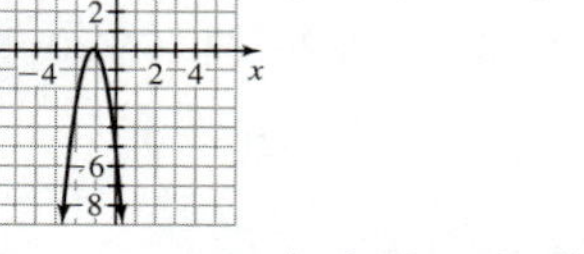 $(3, -2)$ **13.** 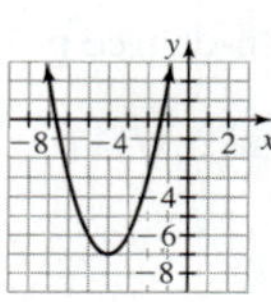 $(-4, -7)$ **15.** 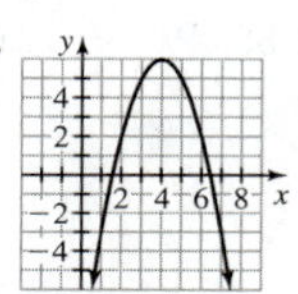 $(4, 6)$ **17.** 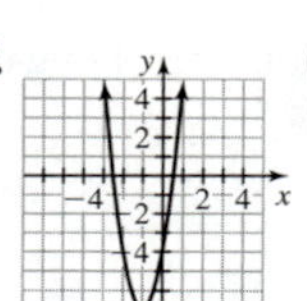 $(-1, -7)$ **19.** 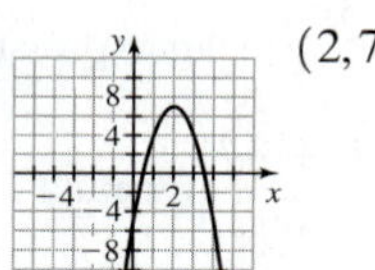 $(2, 7)$

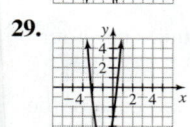

21. 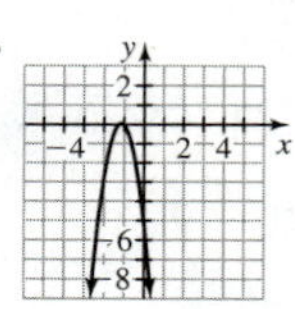 $(-1.13, 0.06)$ **23.** 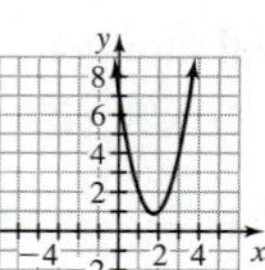 $(1.75, 0.88)$ **25.** 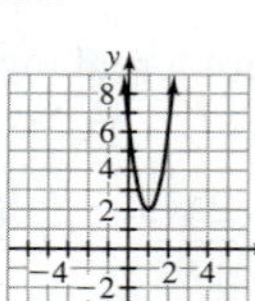 $(1, 2)$ **27.** 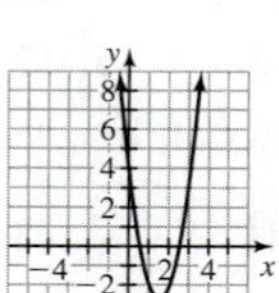 $(1.55, -2.76)$

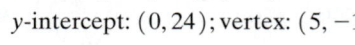

29. 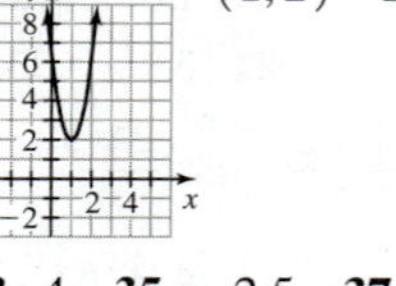 $(-0.88, -6.45)$ **31.** 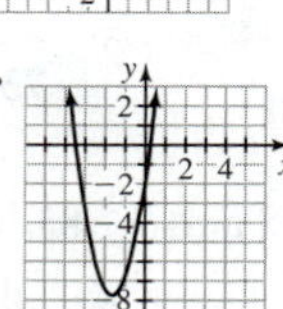 $(-1.58, -7.76)$ **33.** 4 **35.** -2.5 **37.** x-intercepts: $(0, 0)$, $(2, 0)$; y-intercept: $(0, 0)$; vertex: $(1, -5)$ 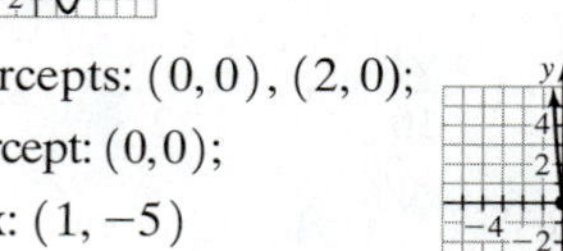

39. x-intercepts: $(0, 0)$, $(3, 0)$; y-intercept: $(0, 0)$; vertex: $(1.5, 4.5)$ 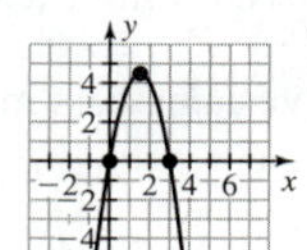

41. x-intercepts: $(4, 0)$, $(6, 0)$; y-intercept: $(0, 24)$; vertex: $(5, -1)$ 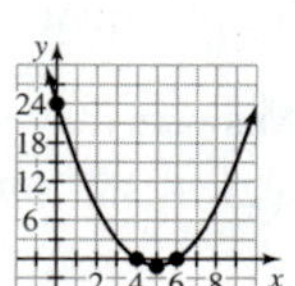

43. x-intercepts: $(1, 0)$, $(7, 0)$; y-intercept: $(0, 7)$; vertex: $(4, -9)$ 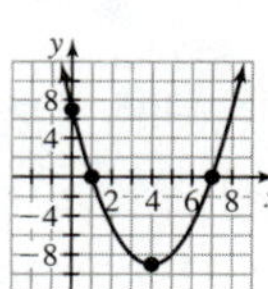

45. x-intercepts: $(-3, 0)$, $(3, 0)$; y-intercept: $(0, -9)$; vertex: $(0, -9)$ 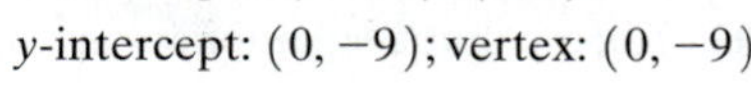 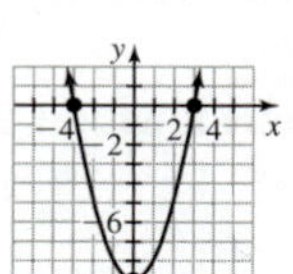

47. a. 3 feet **b.** 309.25 feet; 4.375 seconds

c. 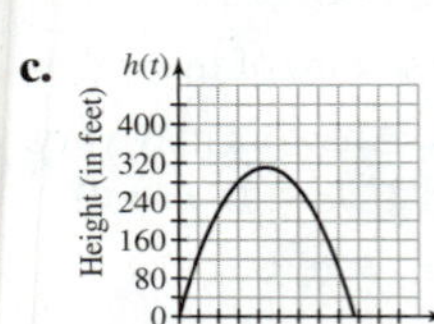

49. a. 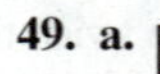 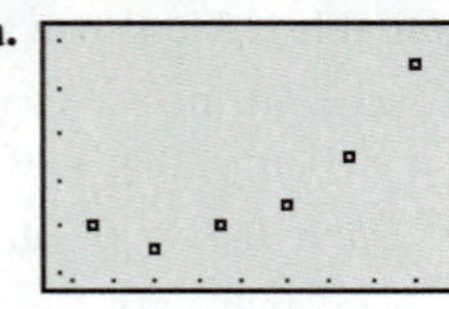 quadratic function **b.** 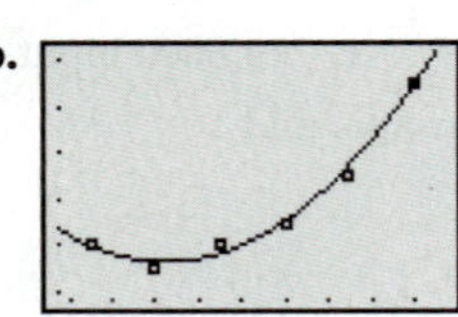 yes **c.** 31% **d.** 1998; 11% **e.** 21 million households **51. a.** quadratic function **b.** 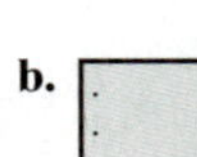 yes **c.** \$20.8 thousand **d.** 46 years; \$49.1 thousand **53. a.** 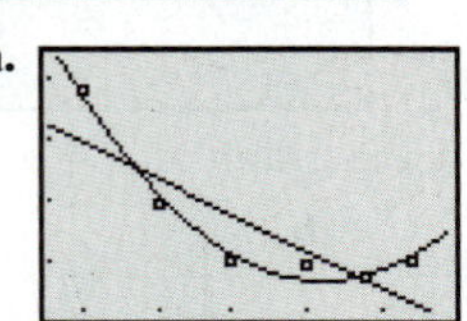 Q **b.** L **c.** 2022 **d.** 2001; 10.6 **e.** 150 professors **55.** width: 20 feet, length: 20 feet; 400 square feet

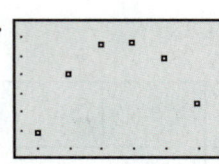

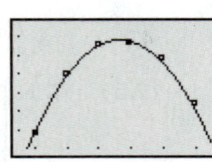

57. width: 100 feet, length: 200 feet; 20,000 square feet **59. a.** 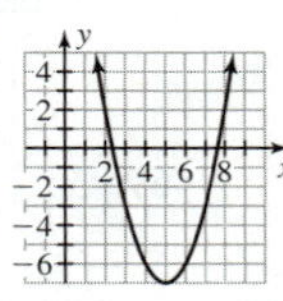

b. Answers may vary. **c.** For each input–output pair, the output is 18 more than the difference between the square of the input and 10 times the input. **61.** −1 **63.** −3 **65.** −2, −4 **67.** 3 **69. a.** −2 **b.** −2 **c.** yes **d.** averaging x-coordinates of y-intercept and its symmetric point **e.** averaging x-coordinates of y-intercept and its symmetric point **f.** Answers may vary. **71.** for f and k: $(3, 2)$; for g: $(2.7, 1.8)$; for h: $(3.3, 1.7)$ **73.** Answers may vary. **75.** Answers may vary.

77. 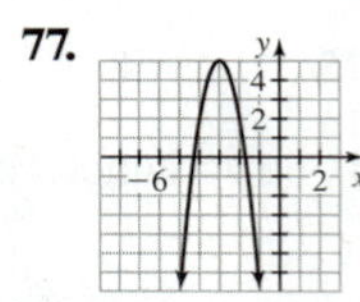

79. 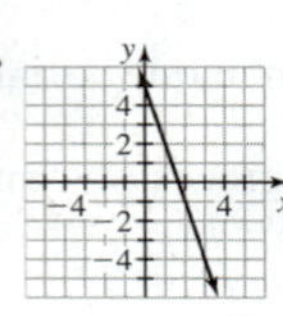

81. a. $a(x-h)^2+k = a(x^2-2xh+h^2)+k = ax^2-2ahx+ah^2+k = ax^2-2ahx+c$ **b.** x-coordinate of vertex: $-\dfrac{b}{2a} = -\dfrac{-2ah}{2a} = h$; y-coordinate: $f(h) = ah^2-2ah^2+ah^2+k = k$

83. $\dfrac{3x^5y^{13}}{2}$; an expression in two variables involving exponents **85.** $9x^4-12x^2y^2+4y^4$; fourth-degree polynomial in two variables

87. $\dfrac{b+cd}{a}$; formula

Homework 9.3 **1.** 9 **3.** 11 **5.** 12 **7.** −9 **9.** not a real number **11.** not a real number **13.** irrational; 5.48 **15.** irrational; 8.83 **17.** rational; 14 **19.** $2\sqrt{5}$ **21.** $3\sqrt{5}$ **23.** $3\sqrt{3}$ **25.** $5\sqrt{2}$ **27.** $10\sqrt{3}$ **29.** $-7\sqrt{2}$ **31.** $24\sqrt{2}$ **33.** $6\sqrt{30}$ **35.** $\dfrac{5}{6}$ **37.** $\dfrac{11}{2}$ **39.** $\dfrac{\sqrt{7}}{5}$ **41.** $\dfrac{\sqrt{19}}{8}$ **43.** $-\dfrac{2\sqrt{2}}{7}$ **45.** $\dfrac{2\sqrt{5}}{9}$ **47.** $\dfrac{5\sqrt{3}}{6}$ **49.** $\dfrac{4\sqrt{5}}{7}$ **51.** $\dfrac{2\sqrt{3}}{3}$ **53.** $\dfrac{\sqrt{6}}{4}$ **55.** $\dfrac{\sqrt{14}}{7}$ **57.** $\dfrac{\sqrt{22}}{2}$ **59.** $\dfrac{\sqrt{6}}{4}$ **61.** $\dfrac{\sqrt{6}}{10}$ **63.** $\dfrac{3+\sqrt{2}}{2}$ **65.** $2-\sqrt{7}$ **67.** $\dfrac{4+6\sqrt{13}}{3}$ **69.** $\dfrac{2+\sqrt{3}}{4}$ **71.** $\dfrac{2-\sqrt{2}}{4}$ **73.** $\dfrac{3-\sqrt{5}}{2}$ **75.** 68.7 miles per hour **77. a.** $f(h) = \dfrac{\sqrt{h}}{4}$ **b.** 9.5 seconds **79.** no; answers may vary. **81.** 4 and 5 **83.** 8 and 9 **85. a. i.** 2 **ii.** 5 **iii.** 8 **b.** Each of the numbers 2, 5, and 8 is larger than its principal square root. **c. i.** $\sqrt{0.2}$ **ii.** $\sqrt{0.5}$ **iii.** $\sqrt{0.8}$ **d.** Each of the numbers 0.2, 0.5, and 0.8 is smaller than its principal square root. **e.** a number greater than 1; a positive number less than 1

87. **89.** Answers may vary. **91.** Answers may vary; $\dfrac{5\sqrt{3}}{3}$ **93.** yes; answers may vary. **95.** Answers may vary. **97.** $4(2x+3)$ **99.** $4(3+2\sqrt{3})$

101. 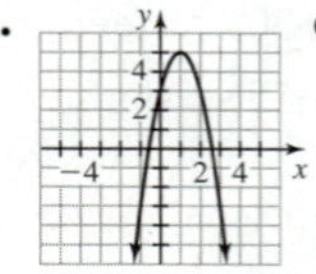

quadratic function **103.** 2, 4; quadratic equation in one variable **105.** $2\sqrt{17}$; radical expression

Homework 9.4 **1.** ± 2 **3.** 0 **5.** $\pm\sqrt{15}$ **7.** $\pm 2\sqrt{5}$ **9.** $\pm 2\sqrt{7}$ **11.** $\pm\frac{\sqrt{5}}{2}$ **13.** $\pm\frac{\sqrt{35}}{5}$ **15.** $\pm\frac{\sqrt{10}}{4}$ **17.** $\pm\frac{\sqrt{6}}{2}$ **19.** $\pm\frac{\sqrt{70}}{5}$ **21.** $-6, 2$ **23.** $7 \pm \sqrt{13}$ **25.** $-2 \pm 3\sqrt{2}$ **27.** $-\frac{9}{8}, \frac{3}{8}$ **29.** $\frac{5}{9}$ **31.** $\frac{-3 \pm \sqrt{41}}{4}$ **33.** $\frac{7 \pm \sqrt{5}}{3}$ **35.** $6 \pm \sqrt{6}$ **37.** $\frac{-3 \pm \sqrt{21}}{3}$ **39.** $(-\sqrt{17}, 0), (\sqrt{17}, 0)$ **41.** $\left(\frac{6 - \sqrt{14}}{2}, 0\right), \left(\frac{6 + \sqrt{14}}{2}, 0\right)$ **43.** no x-intercepts **45.** $6i$ **47.** $-3i\sqrt{5}$ **49.** $\frac{i\sqrt{5}}{7}$ **51.** $\frac{i\sqrt{65}}{5}$ **53.** $\pm 7i$ **55.** $\pm 3i\sqrt{2}$ **57.** $\pm i\sqrt{3}$ **59.** $-4 \pm 2i\sqrt{2}$ **61.** $\frac{5 \pm i\sqrt{3}}{4}$ **63.** $-3 \pm 2i$ **65. a.** $f(t) = -0.018(t - 19.9)^2 + 36.2$ **b.** 20.96; in 2019, about 21% of Americans will say they are "very happy." **c.** $-7.18, 46.98$; in 1963, about 23% of Americans said they were "very happy," and in 2017, about 23% of Americans will say they are "very happy." **d.** $(19.9, 36.2)$; the largest percentage of Americans who said they were "very happy," about 36%, occurred in 1990.

67. a. 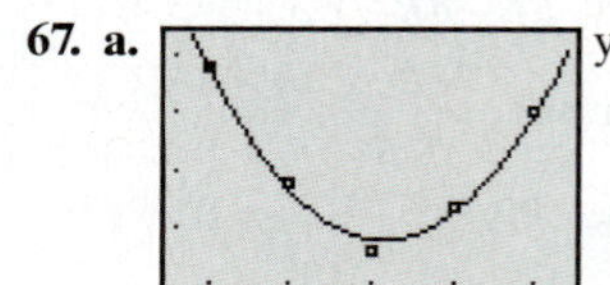 yes **b.** $(9.15, 4.45)$; the smallest annual revenue from U.S. adult mattresses, about \$4.5 billion, occurred in 2009. **c.** \$14.3 billion **d.** 2015 **69.** 1.4, 4.2 **71.** 2, 4 **73.** $(1.1, -3.5), (4.7, -1.7)$ **75.** $\pm\sqrt{c^2 - b^2}$ **77.** $\pm\sqrt{rb - rp}$ **79.** $-b \pm \sqrt{k}$ **81.** $\frac{-a \pm \sqrt{c - b}}{p}$ **83.** There is still an x on the right-hand side; 5

85. a. $(3, 5)$ **b.** upward **c. i.** 2 **ii.** 1 **iii.** 0 **87. a.** $\pm\frac{7}{5}$ **b.** $\pm\frac{7}{5}$ **c.** They are the same. **d.** Answers may vary. **89. a.** yes; $-4 \pm \sqrt{5}$ **b.** no **c.** no; answers may vary. **91.** Answers may vary. **93. a.** ± 7 **b.** 7 **c.** Answers may vary. **95.** $-2, 6$ **97.** ± 9 **99.** $2 \pm 2\sqrt{6}$ **101.** $\pm\frac{\sqrt{33}}{3}$ **103.** $-5, \frac{3}{2}$ **105.** $\frac{5 \pm \sqrt{35}}{5}$ **107.** $(2w + 3)(w + 3)(w - 3)$; a cubic (or third-degree) polynomial in one variable **109.** $\pm 3, -\frac{3}{2}$; a cubic equation in one variable **111.** $20w^3 + 15w^2 - 8w - 6$; a cubic (or third-degree) polynomial in one variable

Homework 9.5 **1.** 36; $(x + 6)^2$ **3.** 49; $(x - 7)^2$ **5.** $\frac{49}{4}; \left(x + \frac{7}{2}\right)^2$ **7.** $\frac{9}{4}; \left(x - \frac{3}{2}\right)^2$ **9.** $\frac{1}{16}; \left(x + \frac{1}{4}\right)^2$ **11.** $\frac{4}{25}; \left(x - \frac{2}{5}\right)^2$ **13.** $-3 \pm \sqrt{10}$ **15.** $1 \pm 2\sqrt{5}$ **17.** $-6 \pm 4\sqrt{2}$ **19.** $5 \pm 4\sqrt{2}$ **21.** $-2 \pm 2\sqrt{7}$ **23.** $\frac{7 \pm \sqrt{61}}{2}$ **25.** $\frac{-5 \pm \sqrt{41}}{2}$ **27.** $\frac{5 \pm \sqrt{33}}{4}$ **29.** $-1 \pm 2\sqrt{2}$ **31.** $3 \pm 2\sqrt{2}$ **33.** $\frac{-4 \pm \sqrt{22}}{2}$ **35.** $\frac{1 \pm \sqrt{57}}{4}$ **37.** $\frac{-2 \pm \sqrt{19}}{3}$ **39.** $\frac{4 \pm \sqrt{10}}{6}$ **41.** $\frac{-1 \pm \sqrt{7}}{4}$ **43.** $-1 \pm i\sqrt{6}$ **45.** $3 \pm 2i\sqrt{2}$ **47.** $\frac{-3 \pm i\sqrt{7}}{2}$ **49.** $\frac{-1 \pm 2i\sqrt{5}}{3}$ **51.** $\frac{3 \pm i\sqrt{71}}{8}$ **53.** $\frac{-3 \pm i\sqrt{3}}{4}$ **55.** $(4 - \sqrt{13}, 0), (4 + \sqrt{13}, 0)$ **57.** $\left(\frac{5 - \sqrt{57}}{4}, 0\right), \left(\frac{5 + \sqrt{57}}{4}, 0\right)$ **59.** $(-5, 0)$ **61.** ± 3 **63.** 5, 6 **65.** $5 \pm 4\sqrt{2}$ **67.** $-3, \frac{4}{3}$ **69.** $\pm\sqrt{13}$ **71.** $3 \pm 2\sqrt{3}$ **73. a.** no such real-number value **b.** -3 **c.** $-3 \pm \sqrt{2}$ **75.** 1, 3 **77.** 2, 3 **79.** $(0, -0.5), (5, -8)$ **81.** To complete the square, the leading coefficient must be 1; $\frac{-3 \pm \sqrt{13}}{4}$ **83. a.** $x^2 + 2kx + k^2$ **b.** $\left(\frac{2k}{2}\right)^2 = k^2$ **85. a.** 2, 4 **b.** 2, 4 **c.** Answers may vary. **87. a.** yes; $-2 \pm \sqrt{11}$ **b.** no **c.** Answers may vary. **89.** Answers may vary. **91.** $(w - 5)^2$ **93.** $\left(x + \frac{5}{6}\right)^2$ **95.** $-5x^4 + 40x^3 - 80x^2$; fourth-degree polynomial in one variable **97.** $4 \pm \sqrt{3}$; quadratic equation in one variable **99.** $(p - 4q)^2$; quadratic (or second-degree) polynomial in two variables

Homework 9.6 **1.** $-1, -\frac{3}{2}$ **3.** $\frac{-3 \pm \sqrt{29}}{2}$ **5.** $\frac{5 \pm \sqrt{61}}{6}$ **7.** $\frac{3 \pm \sqrt{6}}{3}$ **9.** $2 \pm \sqrt{7}$ **11.** $1, \frac{3}{2}$ **13.** $\pm\frac{\sqrt{51}}{3}$ **15.** $-\frac{5}{2}, 0$ **17.** $\frac{5 \pm \sqrt{57}}{8}$ **19.** $\frac{1 \pm \sqrt{37}}{6}$ **21.** $-0.64, 3.14$ **23.** $-0.52, 3.02$ **25.** $-8.54, 0.02$ **27.** $\left(\frac{1 - \sqrt{57}}{4}, 0\right), \left(\frac{1 + \sqrt{57}}{4}, 0\right)$ **29.** no x-intercepts **31.** $(-1 - \sqrt{6}, 0), (-1 + \sqrt{6}, 0)$ **33.** $\frac{3 \pm i\sqrt{23}}{2}$ **35.** $1 \pm 2i$ **37.** $4 \pm i\sqrt{2}$ **39.** $\frac{1 \pm i}{3}$ **41.** $\frac{2 \pm i\sqrt{11}}{3}$ **43.** $\pm 2\sqrt{5}$ **45.** $\frac{-15 \pm \sqrt{30}}{5}$ **47.** -6 **49.** $-\frac{5}{4}, 2$ **51.** $\frac{9 \pm \sqrt{89}}{4}$ **53.** $3 \pm 3\sqrt{2}$ **55.** $\pm\frac{7}{5}$ **57.** $\frac{-1 \pm \sqrt{3}}{2}$ **59.** $\pm\frac{5}{2}i$ **61.** $\frac{5 \pm i\sqrt{23}}{4}$ **63.** $6 \pm 4i\sqrt{3}$ **65.** $\frac{7 \pm i\sqrt{7}}{2}$ **67.** 2 real solutions **69.** 2 imaginary solutions **71.** 1 real solution

73. a. 0 **b.** 1 **c.** 2 **d.** 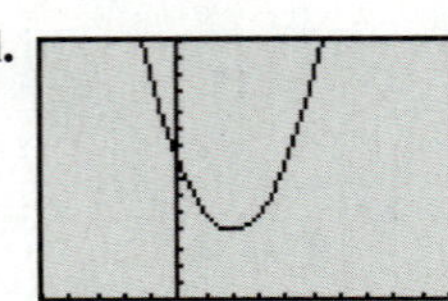 Answers may vary. **75.** $(1, 2), (5, 2); (3, -2);$ 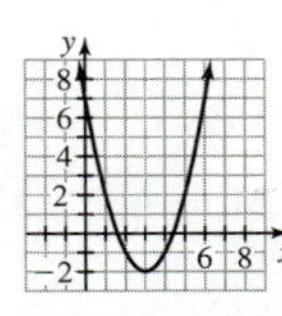

77. a. 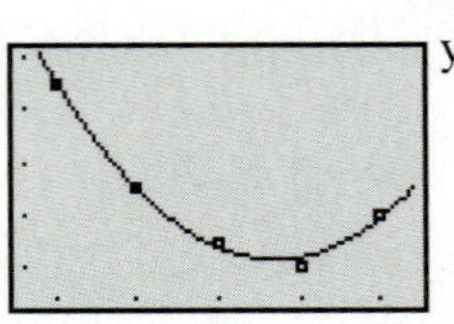 yes **b.** \$47 billion **c.** 2015 **d.** 2010; 2010; 2009; 1 year **79. a.** 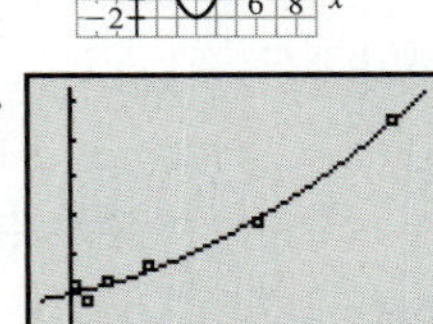 yes

b. 8.2% **c.** $-312.92, 112.92$; 10% of police officers are women in cities with populations of about -312.9 thousand (model breakdown) and about 112.9 thousand. **81. a.** 16 feet **b.** 0.62 second, 2.63 seconds **c.** 3.33 seconds **83.** $-2, 1.3$ **85.** ± 2.4 **87.** $(-3.4, -3.4)$, 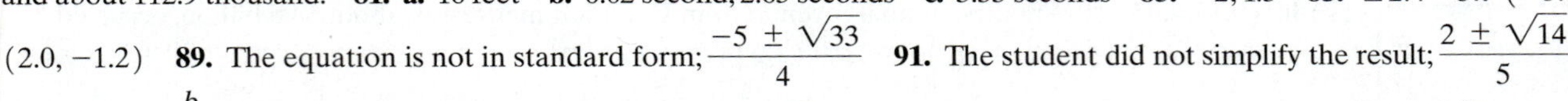 $(2.0, -1.2)$ **89.** The equation is not in standard form; $\frac{-5 \pm \sqrt{33}}{4}$ **91.** The student did not simplify the result; $\frac{2 \pm \sqrt{14}}{5}$

93. a. $x = -\frac{b}{m}$ **b.** -3 **95.** The results are the same by all three methods: $-4, 5$ **97.** Answers may vary. **99.** $x^2 - 3x - 10$

101. $\frac{3 \pm \sqrt{61}}{2}$ **103.** 1, 3 **105.** $-4x^2 + 16x - 13$ **107.** $(4x - 3)(2x - 3)$; a quadratic (or second-degree) polynomial in one variable

109. 77; a quadratic function **111.** $\frac{3}{4}, \frac{3}{2}$; a quadratic equation in one variable

Homework 9.7

1. $(1, -3, 2)$ **3.** $(-2, 15, 14)$ **5.** $(2, 4, -1)$ **7.** $(3, -3, 0)$ **9.** $(-2, 3, -1)$ **11.** $(2, 0, 1)$ **13.** $(3, 4, -1)$ **15.** $(-2, 4, 6)$ **17.** $y = x^2 + 2x + 3$ **19.** $y = -3x^2 + 7x + 5$ **21.** $y = 2x^2 - x - 4$ **23.** $y = x^2 + 2x - 6$ **25.** $y = -2x^2 + 7x + 4$ **27.** $y = 2x^2 - 8x + 3$ **29.** $y = -3x^2 + 8x + 4$ **31.** $y = 3x^2 + x - 1$ **33.** $y = x^2$ **35.** $y = 2x^2 - 6x + 4$ **37.** $y = x^2 - 9x + 22$ **39.** $y = -2x^2 + 20x - 42$ **41. a.** 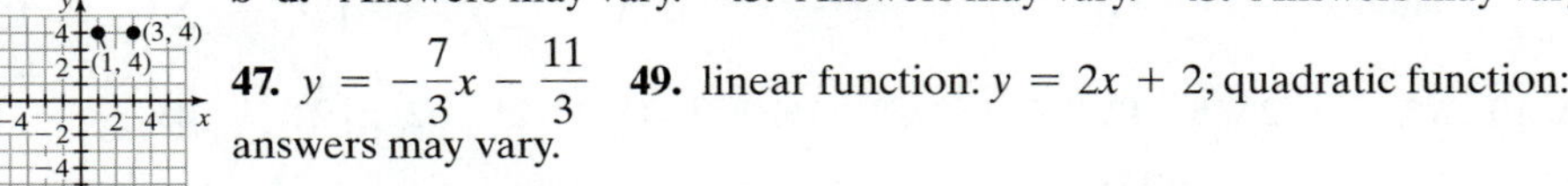

b–d. Answers may vary. **43.** Answers may vary. **45.** Answers may vary. **47.** $y = -\frac{7}{3}x - \frac{11}{3}$ **49.** linear function: $y = 2x + 2$; quadratic function: answers may vary.

51. $f(x) = x$; linear **53.** $y = 2(x - 5)^2 - 7$, or $y = 2x^2 - 20x + 43$ **55.** $x = \frac{5 \pm \sqrt{11}}{2}$; a quadratic equation in one variable

57. -5; a quadratic function **59.**  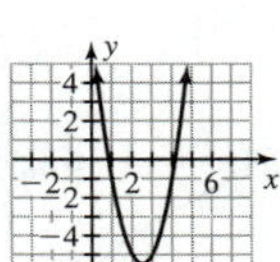

a quadratic function

Homework 9.8

1. a. quadratic **b.** linear **c.** quadratic **d.** none **3.** $f(t) = -0.018t^2 + 1.04t + 17.02$ **5.** $f(t) = -0.142t^2 + 1.25t + 45.96$ **7.** by hand, using points $(21.5, 75)$, $(50.5, 56)$, and $(65, 35)$: $f(t) = -0.018t^2 + 0.66t + 69.29$; by regression: $f(t) = -0.015t^2 + 0.38t + 74.10$; 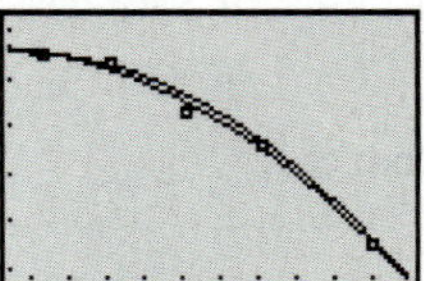 ; answers may vary.

9. by hand, using points $(14, 8.4)$, $(15, 13.6)$, and $(16, 20.1)$: $f(t) = 0.65t^2 - 13.65t + 72.10$; by regression: $f(t) = 0.52t^2 - 9.95t + 45.79$; ; answers may vary. **11. a.** $f(t) = 11.78t^2 - 29.02t + 320.92$ **b.** 2002, 2007, and 2012; yes; answers may vary. 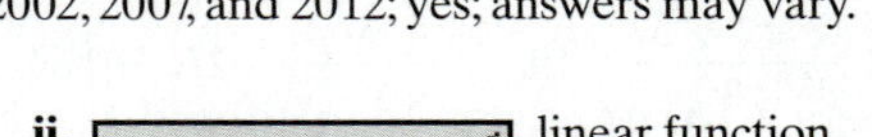
13. a. $f(t) = 0.0068t^2 - 0.13t + 6.49$
b. i. 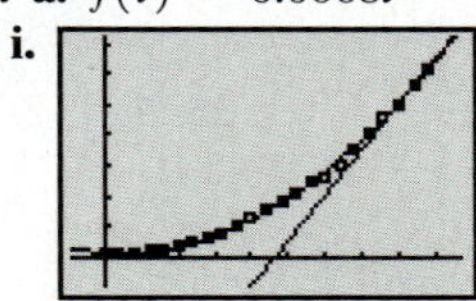 quadratic function **ii.** 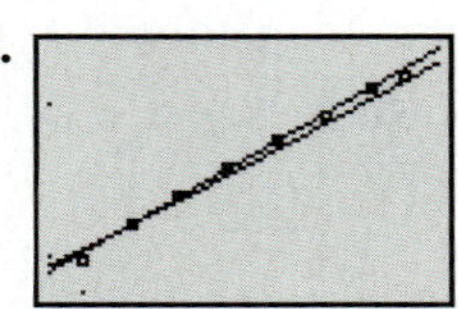 linear function
iii. Answers may vary. **iv.** linear; answers may vary.

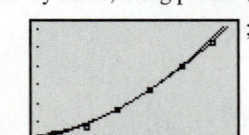

15. a. $f(t) = 2.20t^2 - 31.65t + 115.51$; $g(t) = 6.11t - 42.75$ **b.** $(7.27, 1.69)$, $(9.89, 17.68)$; in 2007, the numbers of gigawatts of solar panels manufactured in China and installed in the world were both about 2 gigawatts; in 2010, the numbers of gigawatts of solar panels manufactured in China and installed in the world were both about 18 gigawatts. **17.** A scattergram of the data does not suggest a quadratic relationship. **19.** overestimate; answers may vary. **21.** It is more desirable to find a quadratic model whose graph does not contain any data points but comes close to all data points; answers may vary. **23.** Answers may vary. **25. a.** linear: $f(t) = 0.70t + 8.86$; quadratic: $f(t) = 0.017t^2 - 0.08t + 14.94$; the quadratic model fits the data well. The linear model does not. **b.** linear model **27.** $(12, 6)$; a linear system in two variables **29.**

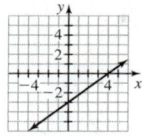

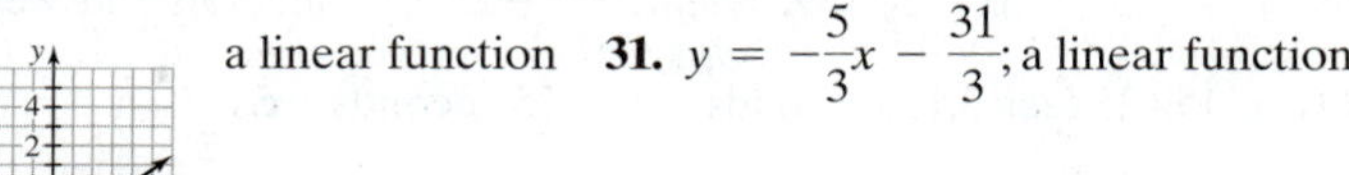

a linear function **31.** $y = -\frac{5}{3}x - \frac{31}{3}$; a linear function

Homework 9.9

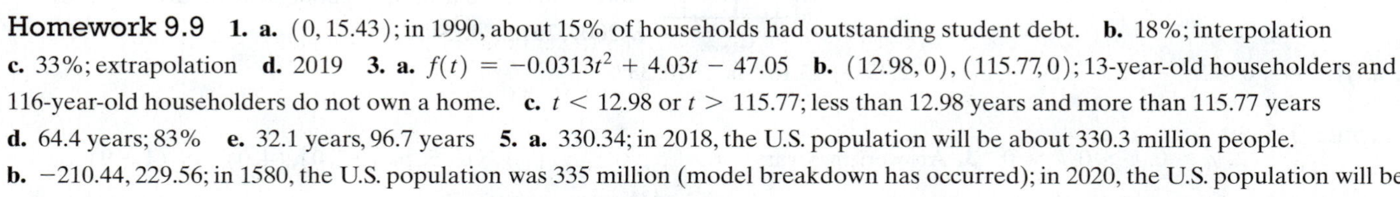

1. a. $(0, 15.43)$; in 1990, about 15% of households had outstanding student debt. **b.** 18%; interpolation **c.** 33%; extrapolation **d.** 2019 **3. a.** $f(t) = -0.0313t^2 + 4.03t - 47.05$ **b.** $(12.98, 0)$, $(115.77, 0)$; 13-year-old householders and 116-year-old householders do not own a home. **c.** $t < 12.98$ or $t > 115.77$; less than 12.98 years and more than 115.77 years **d.** 64.4 years; 83% **e.** 32.1 years, 96.7 years **5. a.** 330.34; in 2018, the U.S. population will be about 330.3 million people. **b.** -210.44, 229.56; in 1580, the U.S. population was 335 million (model breakdown has occurred); in 2020, the U.S. population will be 335 million people. **c.**

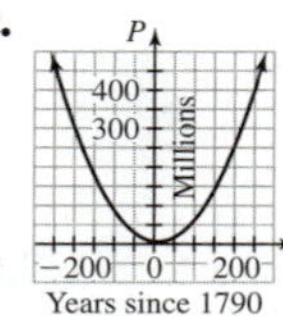

d. $t < 9.56$; years before 1800 **e.**

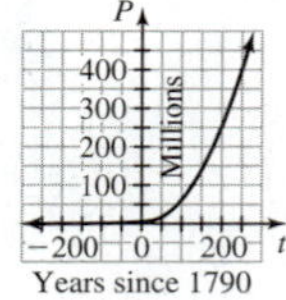

7. a. $f(t) = 0.88t^2 + 11.63t - 85.63$ **b.** 2005 **c.** 755.12, 811.63; 56.51; yes **d.** Answers may vary. **9. a.**

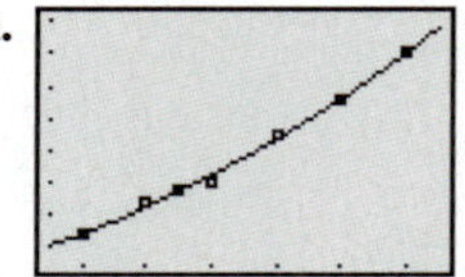

yes **b.** 15 years **c. i.** \$132,800 **ii.** \$23,856 **iii.** 7.25 years **iv.** \$556,144 **v.** underestimate; answers may vary.

11. 2007, 2010 **13. a.** $f(t) = -0.081t^2 - 2.06t + 72.85$; $g(t) = 3.80t + 10.11$ **b.** 1999; no; 2000 **c.** $h(t) = 100 - (f + g)(t)$ **d.** $h(t) = 0.081t^2 - 1.74t + 17.04$ **e.** 10.87; in 2007, about 10.9% of registered voters used voting methods other than punch cards, lever machines, or optical scan or other modern electronic systems. **15.** 25 people **17.** 70 people **19.** width: 40 feet, length: 40 feet; 1600 square feet **21.** Answers may vary. **23.** Answers may vary. **25. a.** linear model: $f(t) = -20.4t + 483$; quadratic model: $f(t) = 5.43t^2 - 118.11t + 911.86$; the quadratic model describes the situation better. **b.** 299 thousand hybrid cars; 289 thousand hybrid cars; 289 thousand hybrid cars; answers may vary. **c.** quadratic model; answers may vary. **d.** 412 thousand hybrid cars; the result is much less than J. D. Powers and Associates' prediction of 1.096 million hybrid cars. **e.** 2004; 2017; model breakdown has occurred for 2004. **27.** Answers may vary. **29.** Answers may vary. **31.** Answers may vary. **33.** Answers may vary.

Chapter 9 Review Exercises

1.

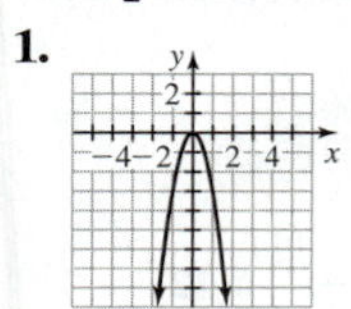

2.

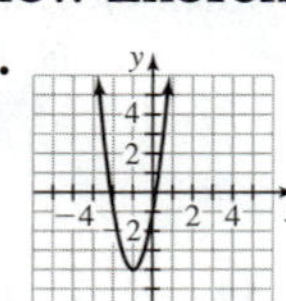

3.

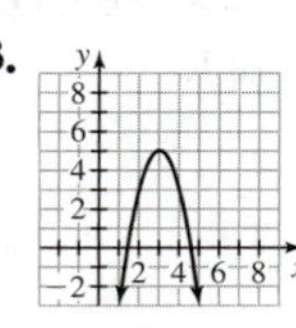

4.

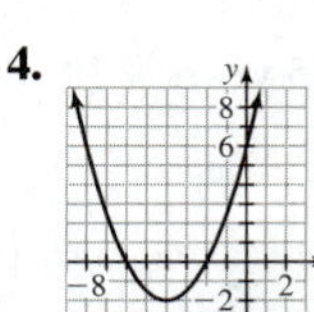

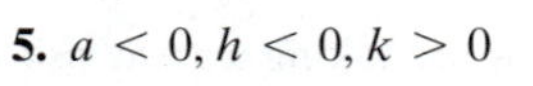

5. $a < 0, h < 0, k > 0$ **6.**

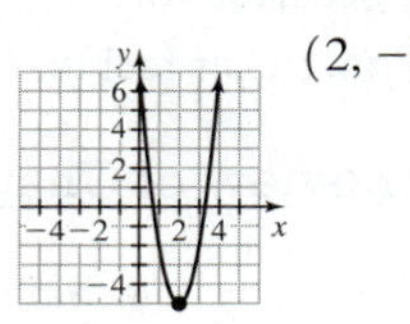

$(2, -5)$

7. $(2, 13)$ **8.**

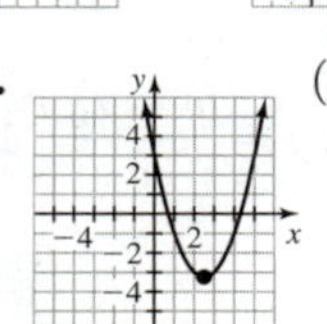

$(2.5, -3.25)$ **9.**

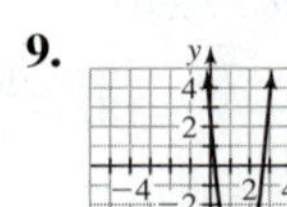

$(1.43, -6.25)$ **10.** 14 **11.** -8 **12.** 9.75 **13.** -45.15 **14.** $3\sqrt{2}$ **15.** $7\sqrt{2}$ **16.** $-15\sqrt{2}$ **17.** $\frac{\sqrt{5}}{3}$

18. $\frac{4\sqrt{7}}{7}$ **19.** $\frac{\sqrt{21}}{3}$ **20.** $\frac{\sqrt{10}}{8}$ **21.** Answers may vary; $\frac{3\sqrt{7}}{7}$ **22.** $\frac{1 \pm \sqrt{7}}{3}$ **23.** $\pm\frac{\sqrt{35}}{5}$ **24.** $\frac{15 \pm \sqrt{15}}{5}$

25. $3 \pm 2\sqrt{5}$ **26.** $\frac{-5 \pm \sqrt{57}}{4}$ **27.** $2 \pm \sqrt{3}$ **28.** $\frac{1 \pm \sqrt{13}}{4}$ **29.** $\frac{1 \pm \sqrt{11}}{2}$ **30.** $\frac{3 \pm \sqrt{19}}{5}$ **31.** $\pm\frac{2\sqrt{35}}{7}$ **32.** $\frac{1 \pm \sqrt{5}}{2}$

33. $\pm\frac{7}{5}$ **34.** $\frac{3 \pm \sqrt{57}}{12}$ **35.** $-1.18, 3.07$ **36.** $-3.05, 1.41$ **37.** $3i\sqrt{5}$ **38.** $\frac{i\sqrt{14}}{2}$ **39.** $\frac{-8 \pm 3i\sqrt{2}}{2}$ **40.** $\frac{2 \pm i\sqrt{10}}{2}$

41. $-3 \pm \sqrt{13}$ **42.** $\frac{5 \pm \sqrt{33}}{2}$ **43.** $3 \pm 3\sqrt{2}$ **44.** $\frac{-3 \pm \sqrt{57}}{4}$ **45.** $\left(\frac{-1-\sqrt{7}}{3}, 0\right), \left(\frac{-1+\sqrt{7}}{3}, 0\right)$ **46.** no x-intercepts

47. $-2, 4$ **48.** 2 imaginary solutions **49. a.** no such value **b.** 1 **c.** $\frac{3 \pm \sqrt{3}}{3}$ **d.** Answers may vary. **50.** $-0.6, 1$ **51.** $-3, 5$

52. $(-2.0, -1.3), (5.4, -3.8)$ **53.** $-2.08, 2.41$ **54.** $(1, 2, 3)$ **55.** $(1, -3, 2)$ **56.** $y = -2x^2 + 5x + 1$ **57.** $y = 2x^2 - x + 3$

58. $y = -2x^2 + 3x + 5$ **59.** linear: $y = -2x + 4$; quadratic: answers may vary. **60.** $y = x^2 - 5x + 7$

61. a. 159.25 feet; 3.125 seconds **b.** 6.25 seconds **c.**

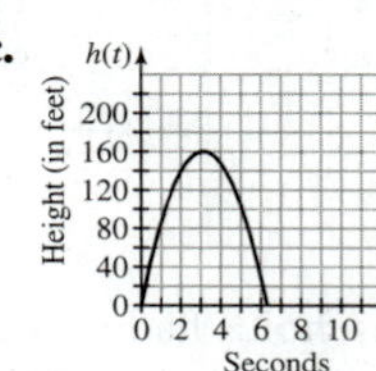

62. width: 45 feet, length: 90 feet; 4050 square feet **63. a.** $f(t) = -0.33t^2 + 7.06t + 0.19$ **b.** $(10.70, 37.95)$; the largest percentage of military personnel who had done more than one tour of duty, about 38%, occurred in 2011. **c.** 2009, 2012 **d.** 31%

64. a. $f(t) = -0.027t^2 - 0.59t + 54.54$; $g(t) = 2.35t - 13.16$ **b.** 2010; 33%

Chapter 9 Test

1.

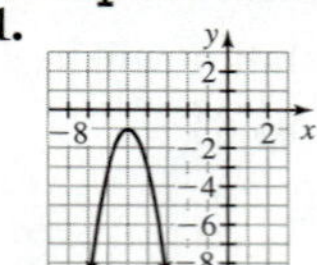

2. $a > 0, h > 0, k = 0$ **3.** Answers may vary. **4.**

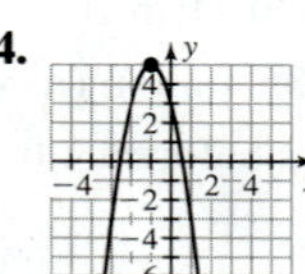

$(-1, 5)$ **5. a.** $(-2, 0), (4, 0)$ **b.** $(1, -9)$

c.

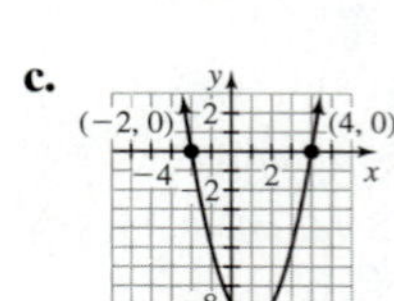

6. $4\sqrt{2}$ **7.** $\frac{2\sqrt{15}}{15}$ **8.** $-2, 5$ **9.** $\pm\frac{5\sqrt{6}}{3}$ **10.** $\frac{6 \pm \sqrt{6}}{2}$ **11.** $\frac{3 \pm \sqrt{89}}{10}$ **12.** $-1 \pm \sqrt{22}$ **13.** $\frac{-3 \pm \sqrt{3}}{2}$

14. $\frac{3 \pm 2\sqrt{3}}{3}$ **15.** $-1.93, 0.34$ **16.** $\frac{3 \pm i\sqrt{6}}{3}$ **17.** $-4 \pm 2i\sqrt{3}$ **18.** $4 \pm 3\sqrt{2}$ **19.** $\frac{-3 \pm \sqrt{73}}{4}$ **20.** $\left(\frac{4-\sqrt{13}}{3}, 0\right)$, $\left(\frac{4+\sqrt{13}}{3}, 0\right)$ **21.** x-intercepts: $(1.42, 0), (4.58, 0)$; vertex: $(3, 5)$;

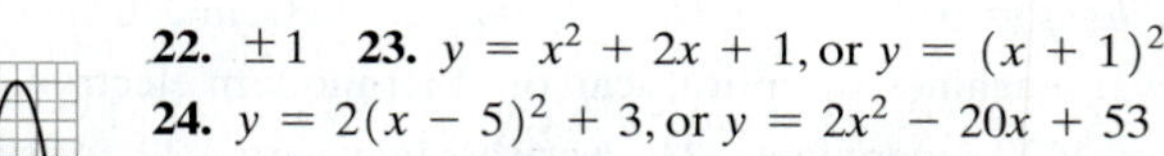

22. ± 1 **23.** $y = x^2 + 2x + 1$, or $y = (x+1)^2$ **24.** $y = 2(x-5)^2 + 3$, or $y = 2x^2 - 20x + 53$ **25. a.** 0 **b.** 1 **c.** 2 **26.** $(4, -8, 10)$ **27.** $(-1, -2, 4)$ **28.** 2.5 seconds; 103 feet

29. a. $f(t) = -0.028t^2 + 2.54t - 15.92$ **b.** 35.91; about 36% of 31-year-old Americans feel that they are taking a great risk by entering personal information in a pop-up ad. **c.** 25.82, 64.89; 31% of 26-year-old and 31% of 65-year-old Americans feel that they are taking a great risk. **d.** $(6.77, 0), (83.94, 0)$; no 7-year-old or 84-year-old Americans feel that they are taking a great risk; model breakdown has likely occurred for the estimate about 84-year-old Americans. **e.** $(45.36, 41.68)$; the age at which the maximum percentage, about 42%, of Americans feel that they are taking a great risk is 45 years. **30.** 80 people

Cumulative Review of Chapters 1–9

1. −9°F **2.** −30 **3.** $\frac{17}{6}$ **4.** Ski run B **5.** $\frac{1}{2}$; increasing **6.** $\pm\frac{7}{9}$ **7.** $\frac{29}{10}$ **8.** $\frac{1 \pm \sqrt{21}}{5}$ **9.** $-\frac{5}{2}, \frac{4}{3}$ **10.** $-1, 3$ **11.** $-\frac{11}{2}$ **12.** $\frac{1}{7}, 1$

13. $-3, 8$ **14.** $\frac{1}{6}, 1$ **15.** $\pm\frac{2\sqrt{15}}{3}$ **16.** $-4 \pm 2\sqrt{15}$ **17.** $\frac{5 \pm \sqrt{73}}{6}$ **18.** $\frac{-3 \pm \sqrt{3}}{2}$ **19.** $t = \frac{v - v_0}{a}$ **20.** $\frac{3 \pm i}{2}$

21. $\frac{-3 \pm \sqrt{57}}{4}$ **22.** $(2, 5)$ **23.** $(-2, -1)$ **24.** $(-3, -4)$ **25.** $(-2, 1, 2)$ **26.** $x < -\frac{1}{12}$; $\left(-\infty, -\frac{1}{12}\right)$;

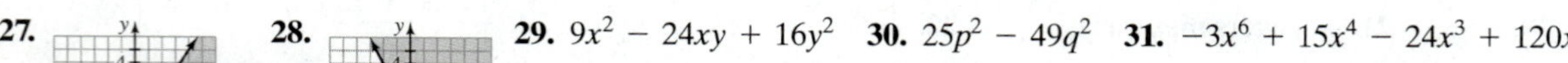

27.

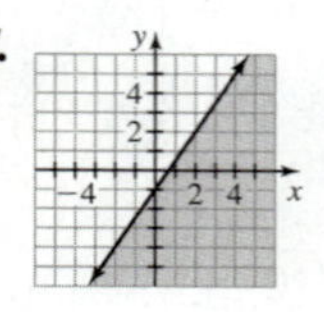

28.

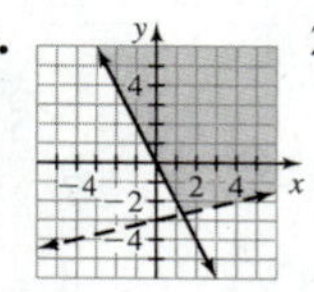

29. $9x^2 - 24xy + 16y^2$ **30.** $25p^2 - 49q^2$ **31.** $-3x^6 + 15x^4 - 24x^3 + 120x$

32. $x^4 + x^3 - 21x^2 - x + 20$ **33.** $f(x) = -2x^2 + 20x - 47$ **34.** $(m^2 + 4n^2)(m + 2n)(m - 2n)$ **35.** $x(x - 5)(x - 8)$ **36.** $(4p - 3q)(2p + 7q)$ **37.** $(x + 4)(x + 3)(x - 3)$ **38.** $2(4a + 5)(16a^2 - 20a + 25)$ **39.** $3pt(p - 5t)(p + 3t)$ **40.** $f(x) = -3x + 20$ **41.** $g(x) = 5x - 1$ **42.** 4 **43.** 1 **44.** 56 **45.**

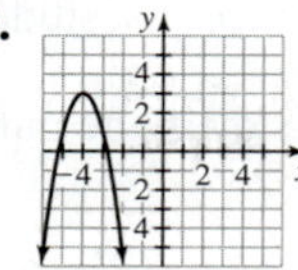

46.

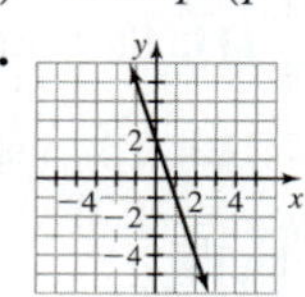

47.

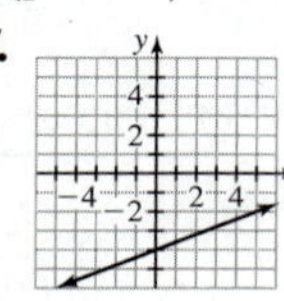

48. **49.** domain: $-3 \le x \le 5$; range: $-2 \le y \le 4$; yes **50.** $y = -\frac{2}{5}x + \frac{14}{5}$ **51.** $y = -\frac{7}{8}x + \frac{9}{4}$ **52.** $y = -\frac{4}{3}x + \frac{10}{3}$

53. $y = x^2 + 2x - 4$ **54. a.** $f(x) = 3x + 3$; $g(x) = 3x^2 + 3$ (answers may vary) **b.**

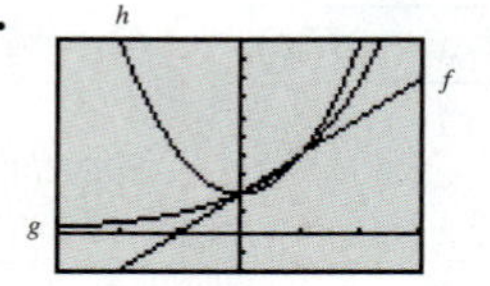

55. -32

56. 2, 4 **57.** (1, 0), (5, 0) **58.**

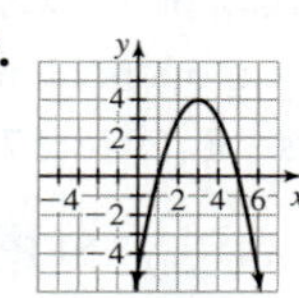

59. $40x^{11}y^9$ **60.** $\frac{81x^{20}y^{28}}{16w^{12}}$ **61.** \$9500 at 6%, \$2500 at 11% **62. a.** $f(t) = 0.0086t^2 - 1.35t + 61.33$ **b.** (0, 61.33); approximately 61% of male drivers at age 0 years were speeding when they became involved in a fatal crash; model breakdown has occurred. **c.** 35% **d.** 34 years, 122 years (model breakdown) **e.** (78.49, 8.35); approximately 8% of 78-year-old male drivers were speeding when they became involved in a fatal crash—the lowest percentage for any age; model breakdown has occurred. **63. a.** $f(t) = -1350t + 27{,}500$ **b.** 18,050 feet **c.** 19.48 minutes **d.** -1350; the airplane descends by 1350 feet per minute. **e.** (20.37, 0); the airplane will land 20.4 minutes after beginning its descent. **64. a.** $f(t) = -0.69t + 31.67$ **b.** 6.14; in 2017, about 6% of union members will work in manufacturing. **c.** 38.65; in 2019, 5% of union members will work in manufacturing. **d.** $f^{-1}(p) = -1.45p + 45.90$ **e.** -99.1; in 1881, 100% of union members worked in manufacturing; model breakdown has occurred. **f.** (45.90, 0); in 2026, no union members will be working in manufacturing; model breakdown has likely occurred. **g.** $t \le -99.1, t > 45.90$ **65. a.** $f(t) = 0.36t + 6.48$; $g(t) = 0.18t + 9.16$ **b.** 12.6; 12.22; in 2017, the average age of light trucks will be 12.6 years; in 2017, the average age of passenger cars will be about 12.2 years. **c.** 2015; 11.8 years **d.** after 2015 **e.** $(f - g)(t) = 0.18t - 2.68$; 0.56; in 2018, the average age of light trucks will be 0.56 year more than the average age of passenger cars. **f.** 12.68; answers may vary.

Chapter 10

Homework 10.1 **1.** $\frac{1}{b^4}$ **3.** b^2 **5.** $\frac{1}{b^3c^5}$ **7.** $\frac{b^2}{c^4}$ **9.** $-\frac{2d}{3b^9c^4}$ **11.** $\frac{1}{2}$ **13.** $\frac{5}{6}$ **15.** $\frac{1}{49}$ **17.** 1 **19.** $\frac{1}{b^{14}}$ **21.** $-\frac{12}{b^9}$ **23.** $\frac{4}{b^2c^3}$ **25.** $\frac{1}{b^8}$ **27.** b^5 **29.** $\frac{7b^6}{4}$ **31.** $\frac{1}{32}$ **33.** $\frac{1}{25}$ **35.** $\frac{9}{b^5}$ **37.** $\frac{b^5}{32}$ **39.** $\frac{b^{12}}{c^{30}}$ **41.** $\frac{3}{b^{10}c^2}$ **43.** $\frac{32b^{26}}{9c^2}$ **45.** $-\frac{6}{7b^{10}}$ **47.** $-\frac{1}{3bc^5}$ **49.** $-\frac{1}{4b^{10}c^{14}}$ **51.** $\frac{3b^5}{c^{13}}$ **53.** $\frac{54b^{18}}{c^7}$ **55.** $\frac{b^8c^7}{32}$ **57.** $\frac{36b^6}{49c^{12}}$ **59.** $\frac{81c^8}{b^{24}}$ **61.** $\frac{1}{bc}$ **63.** $b + c$ **65.** b^{7n} **67.** b^{5n-4} **69.** 54 **71.** $\frac{2}{81}$ **73.** $16(4^a)$ **75.** 16^a

77. a.

x	$f(x)$	x	$f(x)$
-3	$\frac{1}{8}$	1	2
-2	$\frac{1}{4}$	2	4
-1	$\frac{1}{2}$	3	8
0	1	4	16

b.

c. 1.4 **79. a.** $s = \frac{d}{t}$ **b.** 62; an object that travels 186 miles in 3 hours at a constant speed is traveling at a speed of 62 miles per hour. **81. a.** $f(d) = \frac{5760}{d^2}$ **b.** 90; the sound level is 90 decibels at a distance of 8 yards from the amplifier. **83.** 49,000 **85.** 0.00859 **87.** 0.000295 **89.** $-451{,}200{,}000$ **91.** 4.57×10^7 **93.** 6.59×10^{-5} **95.** -5.987×10^{12} **97.** 1×10^{-6} **99.** 0.0000063; 0.00013; 3,200,000; 64,000,000 **101.** 3,600,000,000 years **103.** 0.000000063 mole per liter **105.** 1.008×10^7 gallons **107.** 4.7×10^{-7} meter

109. a. 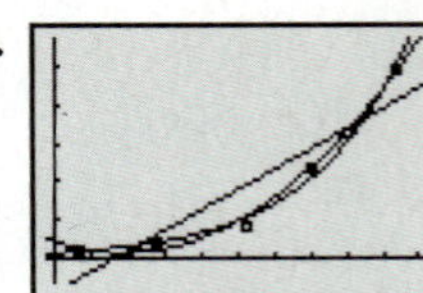
The exponential and quadratic models both fit the data well. The linear model does not fit the data as well. **b.** 21.7 thousand pairs **c.** 10.3 thousand pairs; 15.5 thousand pairs; answers may vary. **111.** Student B was correct; student A should have 5 in the denominator, not -5 in the numerator. **113.** The 3 should stay in the numerator; $\frac{3c^4}{b^2d^7}$ **115.** -2^2, which is -4; $2(-1)$, which is -2; $\left(\frac{1}{2}\right)^2$, which is $\frac{1}{4}$; $2^{-1} = \frac{1}{2}$ (tie); $\left(\frac{1}{2}\right)^{-1}$, which is 2; $(-2)^2 = (2)^2$, which are 4 (tie) **117. a.** 1; 1; 1; 1; 1; 1 **b.** 0; 0; 0; 0; 0; 0 **c.** Answers may vary. **119.** Answers may vary. **121.** 6 **123.** 9 **125.** 8 **127.** $(-5, -14)$; a linear system in two variables **129.** -5; a linear equation in one variable **131.** $30x^2 - 50x - 20$; quadratic (or second-degree) polynomial in one variable

Homework 10.2

1. 4 **3.** 10 **5.** 7 **7.** 5 **9.** 16 **11.** 27 **13.** 4 **15.** 32 **17.** $\frac{1}{3}$ **19.** $-\frac{1}{6}$ **21.** $\frac{1}{32}$ **23.** $\frac{1}{81}$ **25.** 2 **27.** 24 **29.** 49 **31.** 27 **33.** 12 **35.** $\frac{4}{3}$ **37.** -16

39.

x	$f(x)$	x	$f(x)$
$-\frac{3}{4}$	$\frac{1}{8}$	$\frac{1}{4}$	2
$-\frac{1}{2}$	$\frac{1}{4}$	$\frac{1}{2}$	4
$-\frac{1}{4}$	$\frac{1}{2}$	$\frac{3}{4}$	8
0	1	1	16

41. b^2 **43.** $\frac{1}{b^2}$ **45.** $2b^2$ **47.** $\frac{4}{5b^4c^7}$ **49.** $\frac{b}{c^2}$ **51.** $5bcd$ **53.** $3b^6c^2$ **55.** $\frac{c^2}{b^4}$ **57.** $\frac{5c^3}{3b^4}$ **59.** $2b^{29/35}$ **61.** $b^{7/12}$ **63.** $3b^{29/6}$ **65.** $\frac{8}{b^{1/5}}$ **67.** $\frac{2b^{49/12}}{27c^{5/4}}$ **69.** $b^2 + b$

71. a. 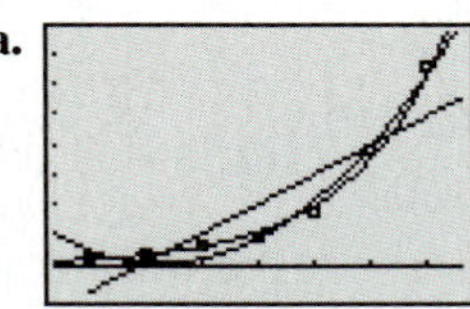
The exponential model fits the data extremely well. The quadratic model fits the data pretty well. The linear model does not fit the data well. **b.** exponential model; answers may vary. **c.** 107,416 megawatts **d.** 2016 **73.** Answers may vary.

75. The student did not compute $36^{1/2}$ correctly; $6x^{18}$ **77.** Answers may vary. **79. a.** $(-9)^{1/2}, (-81)^{1/4}, (-1)^{1/6}$ **b.** b is negative and n is even. **81.** Answers may vary. **83.** $\frac{8}{3}$ **85.** 2 **87.** $-\frac{40}{3}$ **89.** $\frac{1}{32}$ **91.** $(p-3)(p+3)(2p-3)$; cubic (or third-degree) polynomial in one variable **93.** 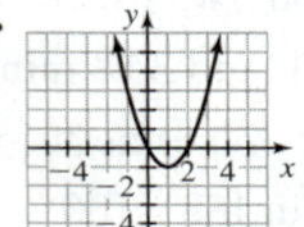
quadratic function **95.** $\frac{1 \pm \sqrt{13}}{3}$; quadratic equation in one variable

Homework 10.3

1. 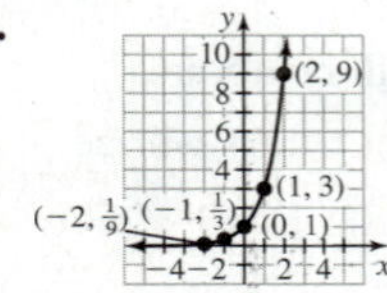

3. 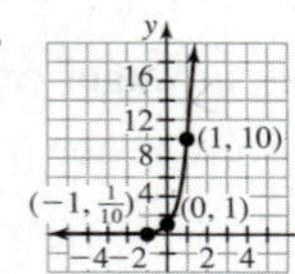

5. 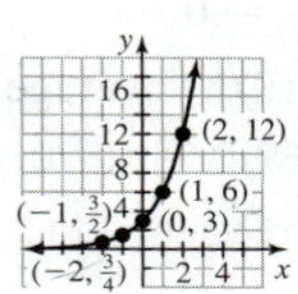

7. 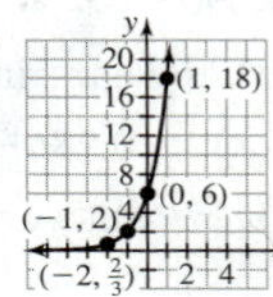

9. 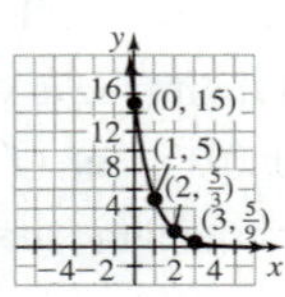

11. 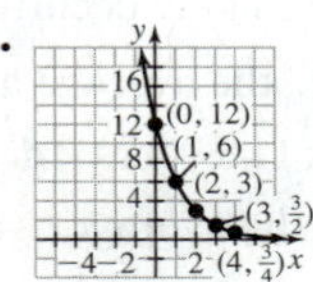

13. 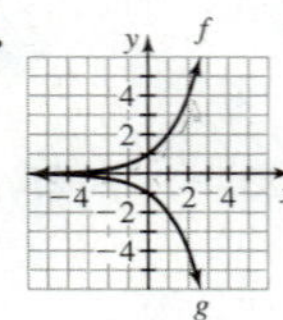

15. 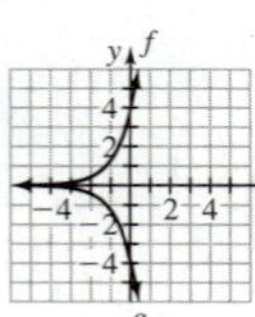

17. 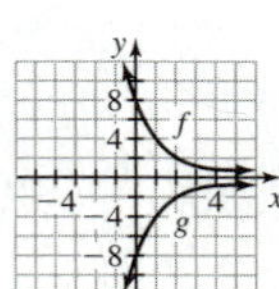

19. 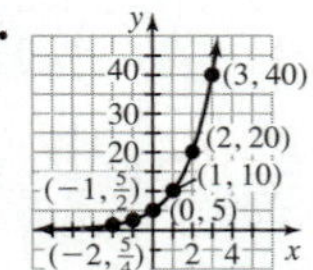

domain: all real numbers; range: $y > 0$

21. 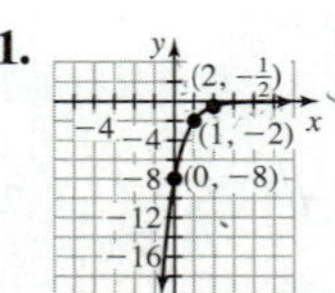

domain: all real numbers; range: $y < 0$

23. a. Answers may vary. **b.** 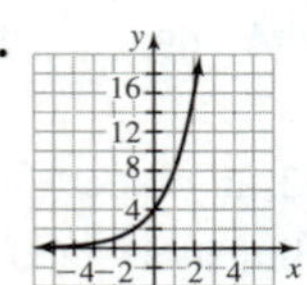
c. For each input–output pair, the output is 4 times 2 raised to the power equal to the input.

25.

x	$f(x)$	$g(x)$	$h(x)$	$k(x)$
0	162	3	2	800
1	54	12	10	400
2	18	48	50	200
3	6	192	250	100
4	2	768	1250	50

27.

x	$f(x)$	$g(x)$	$h(x)$	$k(x)$
0	5	160	162	3
1	10	80	54	12
2	20	40	18	48
3	40	20	6	192
4	80	10	2	768

29. 8 **31.** 1 **33.** -2 **35.** 0 **37.** 24 **39.** 96 **41.** 0 **43.** 3

45. a. 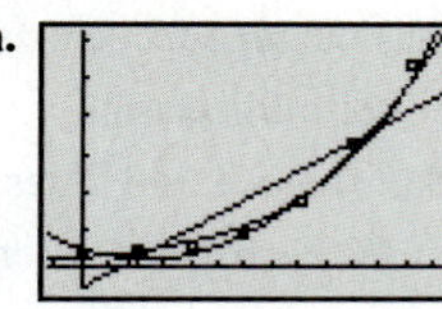 The quadratic and exponential models fit the data fairly well. The linear model does not fit the data well. **b.** exponential model; answers may vary. **c.** \$35.92 **d.** 2026 **47.** no x-intercept; y-intercept: $(0, 1)$ **49.** no x-intercept; y-intercept: $(0, 3)$ **51.** 13 **53.** $\frac{13}{36}$ **55.** 1 **57.** 0 **59.** $f(x) = g(x)$ **61.** $f(x) = g(x)$ **63.** $f(x) = g(x)$ **65.** $f(x) = g(x)$ **67.** $f(x) = g(x)$ **69.** $f(x) = g(x)$ **71. a.** $a < 0, b > 1$ **b.** $a > 0, b > 1$ **c.** $a > 0, 0 < b < 1$ **d.** $a < 0, 0 < b < 1$ **73.** Answers may vary. **75.** $f(x) = 3(2)^x$ **77.** Answers may vary. **79. a.** for f: $(0, 100)$; for g: $(0, 5)$ **b.** For f, as the value of x increases by 1, the value of $f(x)$ is multiplied by 2. For g, as the value of x increases by 1, the value of $g(x)$ is multiplied by 3. **c.** The outputs of g will eventually be much greater. **d.**

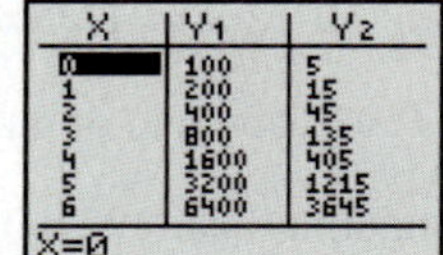

X	Y1	Y2
0	100	5
1	200	15
2	400	45
3	800	135
4	1600	405
5	3200	1215
6	6400	3645

X=0

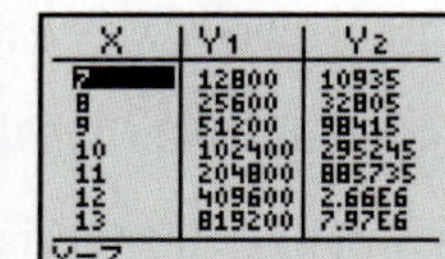

X	Y1	Y2
7	12800	10935
8	25600	32805
9	51200	98415
10	102400	295245
11	204800	885735
12	409600	2.66E6
13	819200	7.97E6

X=7

81. a. no **b.** no **83.** Answers may vary. **85.** Answers may vary.

87. a. 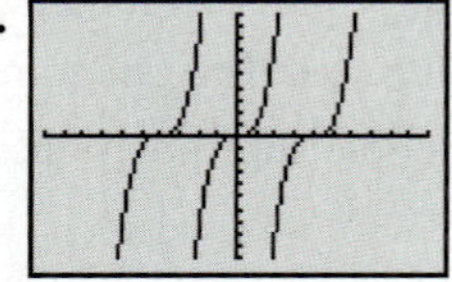 g: Translate the graph of f by 4 units to the right; h: Translate the graph of f by 4 units to the left. **b.** 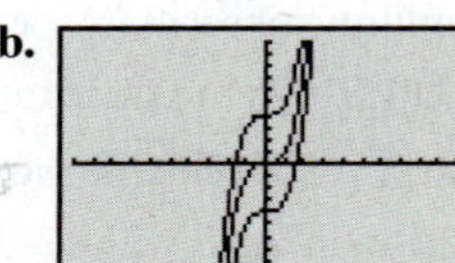 g: Translate the graph of f by 4 units down; h: Translate the graph of f by 4 units up.

c. yes; answer may vary. **d.** 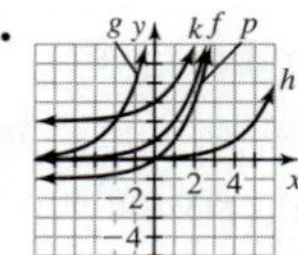

89. 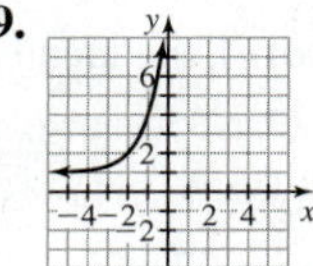 **91.** 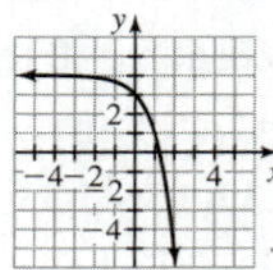 **93.** x-intercept: $(-2, 0)$; y-intercept: $(0, 8)$ **95.** no x-intercept; y-intercept: $(0, 8)$ **97.** f might be linear; g might be exponential; h might be exponential; k is neither linear nor exponential.

99. 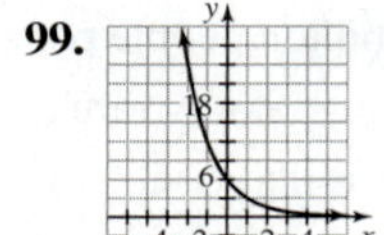 an exponential function **101.** 24; an exponential function **103.** $\frac{4b}{c^{1/6}}$; an expression in two variables involving exponents

Homework 10.4 **1.** $f(x) = 4(2)^x, g(x) = 36\left(\frac{1}{3}\right)^x, h(x) = 5(10)^x, k(x) = 250\left(\frac{1}{5}\right)^x$ **3.** $f(x) = 100\left(\frac{1}{2}\right)^x$, $g(x) = -50x + 100, h(x) = 4x + 2, k(x) = 2(3)^x$ **5.** ± 4 **7.** 3 **9.** 2 **11.** ± 0.81 **13.** 2.28 **15.** ± 1.51 **17.** 2.22 **19.** ± 3 **21.** 1.74 **23.** $(d - c)^{1/n}$ **25.** $\left(\frac{c + d}{a}\right)^{1/n}$ **27.** $d^{1/(m-n)}$ **29.** $\left(\frac{d - c}{a}\right)^{1/(m-n)}$ **31.** $y = 4(2)^x$ **33.** $y = 3(2.02)^x$ **35.** $y = 87(0.74)^x$ **37.** $y = 5.5(3.67)^x$ **39.** $y = 7.4(0.56)^x$ **41.** $y = 39.18(0.85)^x$ **43.** $y = 1.33(3)^x$ **45.** $y = 1.19(1.50)^x$ **47.** $y = 1170.33(0.88)^x$ **49.** $y = 37.05(0.74)^x$ **51.** $y = 0.072(1.57)^x$ **53.** $y = 146.91(0.71)^x$ **55.** $y = 4\left(\frac{1}{2}\right)^x$ **57.** $y = 1.26(1.58)^x$ **59.** $(0, 6)$ **61. a. i.** yes; answers may vary. **ii.** no; answers may vary. **b.** no; answers may vary. **63.** Answers may vary. **65.** b^5 **67.** 2.38 **69.** $\frac{4b^4}{3}$ **71.** ± 0.75 **73.** $L(x) = 4x + 2$; $E(x) = 2(3)^x$; $Q(x) = 2x^2 + 2x + 2$ (answers may vary for the equation of Q) **75.** could be linear or exponential **77. a.** L: $(0, 100)$; E: $(0, 3)$ **b.** $L(x)$ increases by 2; $E(x)$ is multiplied by 2. **c.** $E(x)$ will eventually dominate over $L(x)$. **d.**

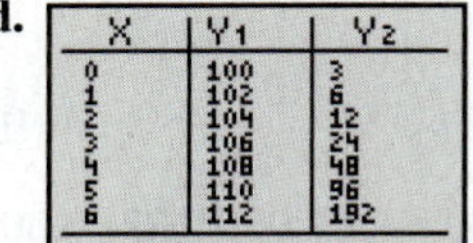

X	Y1	Y2
0	100	3
1	102	6
2	104	12
3	106	24
4	108	48
5	110	96
6	112	192

79. 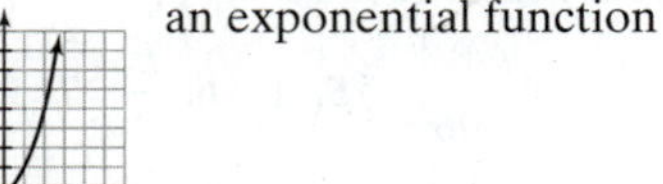 an exponential function

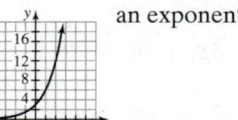

81. $\frac{3}{8}$; an exponential function **83.** $\frac{2c^3}{3b^5}$; an expression in two variables involving exponents

Homework 10.5 **1. a.** $f(t) = 40(3)^t$ **b.** 2,361,960 people **c.** 573,956,280 people; model breakdown has occurred, because this number exceeds the U.S. population. **3. a.** $f(t) = 30(2)^t$ **b.** 1920 trillion (1.92 quadrillion) web pages **c.** 121 million miles **5. a.** $f(t) = 8.3(2.08)^t$ **b.** $(0, 8.3)$; in 2010, the market share of eBooks was 8.3%. **c.** 2.08; each year, the market share is 2.08 times that of the previous year. **d.** 74.7%; \$3.74 billion **7. a.** $D(t) = 2.5(1.5)^t$ **b.** $S(t) = 1.71(2.2)^t$ **c.** $(0.99, 3.74)$; in 2008, the subscribers who got TiVo through DIRECTV equaled the number of stand-alone TiVo subscribers, 3.74 million subscribers. **9. a.** $f(t) = 3000(1.08)^t$ **b.** 1.08; the account balance increases by 8% per year. **c.** 3000; the initial amount invested was \$3000. **d.** \$9516.51 **11. a.** $f(t) = 4000(2)^{t/6}$ or $f(t) = 4000(1.1225)^t$ **b.** \$40,317.47 **13.** \$8749.97 **15. a.** $g(t) = 984\left(\frac{1}{2}\right)^t$ **b.** $(0, 984)$; in 2011, 984 new copies of the textbook were sold. **c.** 123; in 2014, 123 new copies of the textbook will be sold. **d.** 1 year **17. a.** $f(t) = 100\left(\frac{1}{2}\right)^{t/1600}$ or $f(t) = 100(0.999567)^t$ **b.** 95.76% **c.** 25%; 3200 years is two half-lives; so, half of 100% is 50%, and half of that is 25%. **19. a.** $f(t) = 100\left(\frac{1}{2}\right)^{t/7.56}$ or $f(t) = 100(0.9124)^t$ **b.** 75.95% **c.** 32.7 days **21.** 15.87 milligrams **23.** 190 million users **25.** \$263 **27. a.** $f(t) = 9.58(1.07)^t$ **b.** 30 million subscribers **c.** \$3.9 billion **29. a.** linear: $f(t) = 0.063t - 0.37$; quadratic: $f(t) = 0.00047t^2 - 0.004t + 1.67$; $f(t) = 1.21(1.016)^t$; the quadratic and exponential models fit the data quite well. The linear model fits the data fairly well. **b.** exponential function **c.** 6.936; the world population in 2010 was 6.936 billion people; interpolation; answers may vary. **d.** 7.875; the world population in 2018 will be 7.875 billion people; extrapolation; answers may vary. **31. a.** $f(t) = 100.84(1.41)^t$ **b.** 41% growth per year **c.** 398.57; in 1994, there were 399 stores. **d.** 48,930 stores; no; answers may vary. **e.** 2000 **33. a.** $g(t) = 0.66(1.096)^t$ **b.** 0.66; at $t = 0$ (a newborn), the faculty member pays \$0.66; model breakdown has occurred. **c.** 1.096; the rate increases by 9.6% each year of age. **d.** 16.33; a 35-year-old faculty member must pay \$16.33 per month. **e.** men **35. a.** right-hand column: 1.36, 1.36, 1.33, 1.34, 1.33, 1.36, 1.35 **b.** They are approximately equal. **c.** exponential function; answers may vary. **d.** $f(t) = 3.94(1.03)^t$ **e.** right-hand column: 1.27, 1.26, 1.25, 1.21 **f.** no; answers may vary. **g.** 2788.4 million (2.7884 billion) people; 2474.5 million (2.4745 billion) people **37. a.** $f(t) = 18.47(0.95)^t$ **b.** 0.05 death per million people per year **c.** 17 lightning deaths **d.** 170 injuries **e.** $g(t) = 50.68(0.992)^t$ **f.** The base of f is 0.95, which means the lightning fatality rate is decreasing by 5% per year, but the base of g is 0.992, which means the percentage of Americans who live in rural areas is decreasing by only 0.8%, so migration of Americans from rural areas to urban ones can't be the only reason the lightning fatality rate is decreasing. **39. a.** 10 years **b.** 6.25% **41.** Decrease b. **43.** Answers may vary. **45.** Answers may vary. **47. a.** $C(t) = 800(1.03)^t$ **b.** $S(t) = 24t + 800$ **c.** $C(1) = 824$; $C(2) = 848.72$; $S(1) = 824$; $S(2) = 848$; answers may vary **d.** 1444.89, 1280; compound interest gives a larger balance than simple interest does. **49. a.** $f(a) = -0.0217t^2 + 1.96t - 18.63$ **b.** 12% **c.** 29 years, 61 years **d.** $(45.16, 25.63)$; 26% of 45-year-old Americans listen to talk radio—the largest percentage of any age. **e.** $(10.80, 0)$, $(79.53, 0)$; no 11-year-old Americans and no 80-year-old Americans listen to talk radio; model breakdown has occurred. **51.** Answers may vary. **53.** Answers may vary. **55.** Answers may vary. **57.** Answers may vary.

Chapter 10 Review Exercises

1. 32 **2.** $\frac{48c^3}{b^{12}}$ **3.** $\frac{bc^9}{4}$ **4.** $\frac{8c^6}{9b^{23}}$ **5.** 1 **6.** $\frac{b}{c}$ **7.** 16 **8.** $\frac{1}{8}$ **9.** $\frac{1}{b^{2/15}}$ **10.** $\frac{1}{b^{5/3}}$ **11.** $\frac{2b^{11}}{125c^7}$ **12.** $2b^2c$ **13.** $\frac{b^{1/6}}{c^{7/4}}$ **14.** b^{6n+2} **15.** $b^{n/6}$ **16.** $3^{2x} = (3^2)^x = 9^x$ **17.** $\frac{3}{25}$ **18.** $36(6^a)$ **19.** 7 **20.** $\frac{2}{27}$ **21.** 44,487,000 **22.** 0.0000385 **23.** 5.4×10^7 **24.** -8.97×10^{-3}

25. **26.** **27.** domain: all real numbers; range: $y < 0$ **28.** domain: all real numbers; range: $y > 0$

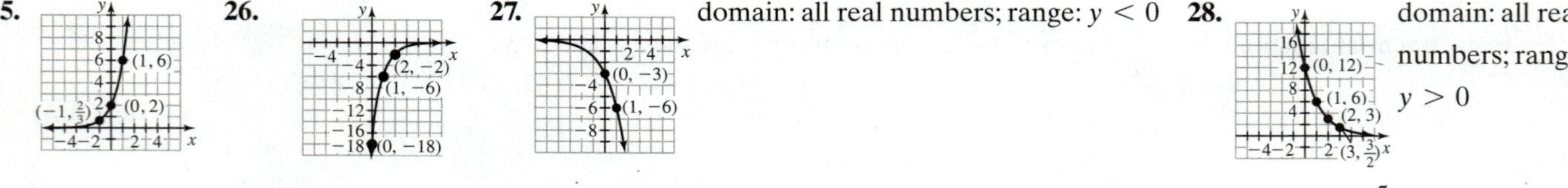

29. 2 **30.** 1.97 **31.** 1.84 **32.** ± 2 **33.** ± 1.61 **34.** 1.69 **35.** f is linear, $f(x) = -4x + 34$; g is exponential, $g(x) = \frac{5}{3}(3)^x$; h is neither; k is exponential, $k(x) = 192\left(\frac{1}{2}\right)^x$. **36.** 18 **37.** 6 **38.** 1 **39.** 5 **40.** $y = 2(1.08)^x$ **41.** $y = 3.8(2.34)^x$ **42.** $y = 62.11(0.78)^x$

43. $y = 3.07(1.18)^x$ **44.**

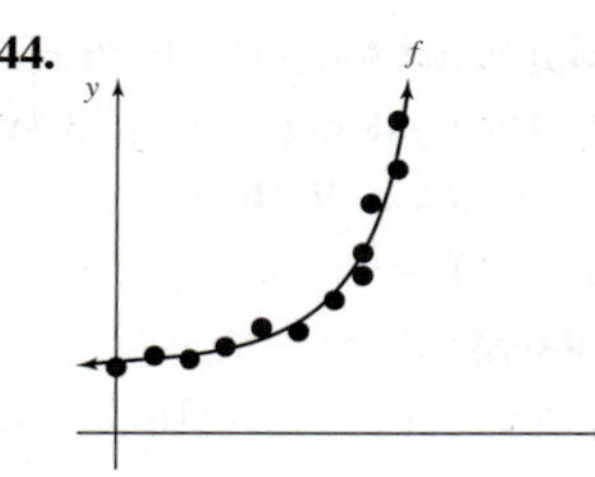

Increase a and decrease b. **45. a.** $f(t) = 2000(1.07)^t$ **b.** \$2805.10 **46. a.** $g(t) = 17(2)^t$ **b.** \$4,352,000 **47. a.** $f(t) = 100\left(\frac{1}{2}\right)^{t/5730}$ or $f(t) = 100(0.999879)^t$ **b.** 98.8% **48.** 15.9 million homes

49. a. $f(t) = 233.91(1.17)^t$ **b.** 233.91; the price of one ounce of gold was about \$234 in 2000. **c.** 1.17; the price of one ounce of gold is growing exponentially by 17% per year. **d.** 3948.16; the price of one ounce of gold will be about \$3948 in 2018. **50. a.** $f(t) = 0.011(2.43)^t$ **b.** 143% **c.** 39,509 thousand (39.509 million) users **d.** 2019; model breakdown has occurred.

Chapter 10 Test

1. 4 **2.** $-\frac{1}{16}$ **3.** $8b^9c^{24}$ **4.** 1 **5.** $b^{1/6}$ **6.** $\frac{5b}{7c^5}$ **7.** $\frac{4b^4}{c^{14}}$ **8.** $\frac{250b^{14}}{7c^6}$ **9.** $8^{x/3}2^{x+3} = (2^3)^{x/3}2^{x+3} = 2^x2^{x+3} = 2^{2x+3} = 2^{2x}2^3 = 8(2^2)^x = 8(4)^x$ **10.** $\frac{1}{16}$ **11.** $\frac{1}{8}$ **12.**

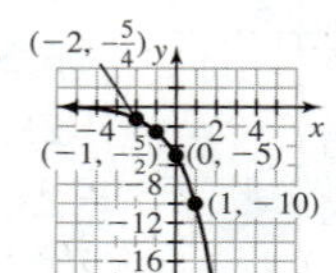

domain: all real numbers; range: $y < 0$ **13.**

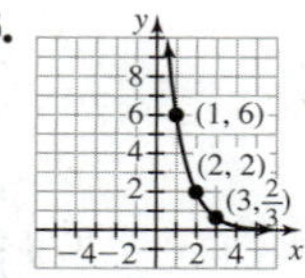

domain: all real numbers; range $y > 0$

14. Answers may vary. **15.** $f(t) = 160\left(\frac{1}{2}\right)^t$ **16.** ± 1.72 **17.** $y = 70(0.81)^x$ **18.** $y = 0.91(1.77)^x$ **19.** 6 **20.** 1

21. $f(x) = 6\left(\frac{1}{2}\right)^x$ **22. a.** $f(t) = 400(3)^t$ **b.** 291,600; there will be 291,600 leaves on the tree six weeks after March 1. **c.** approximately 2.58×10^{27}; one year after March 1, there will be about 2.58×10^{27} leaves on the tree; model breakdown has occurred. **23. a.** $f(t) = 0.27(1.16)^t$ **b.** 1.16; the number of fraud complaints by consumers is increasing by 16% per year. **c.** 0.27; there were 0.27 million (270 thousand) fraud complaints by consumers in 2000. **d.** 3.9 million complaints **e.** 1.5%

Chapter 11

Homework 11.1 **1.** 2 **3.** 0 **5.** 4 **7.**

x	$(f \circ g)(x)$
0	2
1	1
2	4
3	0
4	3

9. 7 **11.** 5 **13.** 8 **15.**

x	$(g \circ f)(x)$
5	9
6	7
7	8
8	5
9	6

17. a. 25 **b.** 13

19. a. 65 **b.** 17 **21. a.** 54 **b.** 67 **23. a.** $\frac{5}{13}$ **b.** 13 **25. a.** $(f \circ g)(x) = 6x - 7$ **b.** $(g \circ f)(x) = 6x - 9$ **c.** 11 **d.** 9 **27. a.** $(f \circ g)(x) = 2x^2 + 4$ **b.** $(g \circ f)(x) = 4x^2 + 16x + 16$ **c.** 22 **d.** 100 **29. a.** $(f \circ g)(x) = x^2 - 10x + 24$ **b.** $(g \circ f)(x) = x^2 - 2x - 4$ **c.** 3 **d.** −1 **31. a.** $(f \circ g)(x) = 2^{x+2}$ **b.** $(g \circ f)(x) = 2^x + 2$ **c.** 32 **d.** 10 **33.** 1 **35.** 1 **37.** −2 **39.** (0, 0.5) **41.** 8 **43.** 2 **45.** 2 **47.** $f(x) = 5^x$; $g(x) = x - 9$; answers may vary. **49.** $f(x) = x + 6$; $g(x) = 2^x$; answers may vary. **51.** $f(x) = x^2$; $g(x) = 5x - 2$; answers may vary. **53.** $f(x) = \frac{1}{x}$; $g(x) = 3x - 7$; answers may vary.

55. Americans will eat 9.8 billion pounds of french fries in 2019. **57.** The revenue will be \$2.3 million if the price of the album is \$8.99. **59. a.** $f(y) = 3y$; $g(x) = 12x$ **b.** $(g \circ f)(y) = 36y$ **c.** 180; there are 180 inches in 5 yards. **61. a.** $f(M) = 60M$; $g(H) = 60H$ **b.** $(f \circ g)(H) = 3600H$ **c.** 10,800; there are 10,800 seconds in 3 hours. **63. a.** $f(d) = \frac{d}{25}$; $g(x) = 4.25x$ **b.** $(g \circ f)(d) = 0.17d$ **c.** 51; for a 300-mile trip, the cost of gasoline is \$51. **65. a.** Answers may vary. **b.** $h(t) = (g \circ f)(t)$; answers may vary. **c.** $h(t) = -0.76t + 142.04$ **d.** 53.12; about 53.1% of births will be from married couples in 2017. **e.** 118.47; 52% of births will be from married couples in 2018. **67. a.** Answers may vary. **b.** $h(t) = (M \circ D)(t)$; answers may vary. **c.** $h(t) = 340t + 560$ **d.** 220; in 2004, the revenue from downloaded music was \$220 million. **e.** (−1.65, 0); there was no revenue from downloaded music in 2003; the t-intercept of the graph of h is the same as the t-intercept of the graph of D; answers may vary.

69. a. $h(C) = (g \circ f)(C)$; answers may vary. **b.** $h(C) = 7.74C - 34.4$ **c.** 144 chirps per minute **d.** 30°C **71.** The student found the product of $x + 8$ and $x + 5$, which is incorrect; $(f \circ g)(x) = x + 13$ **73.** The student substituted $4x - 2$ for x in $g(x) = -7x + 3$, which is incorrect; $(f \circ g)(x) = -28x + 10$ **75. a.** 4 **b.** 26 **c.** no **77. a.** 9 **b.** $(f \circ g)(x) = 12x - 15$ **c.** 9; the result is equal to the one in part (a). **79. a.** $(f \circ f)(x) = 4x$ **b.** $(f \circ (f \circ f))(x) = 8x$ **c.** $(f \circ (f \circ (f \circ f)))(x) = 16x$ **d.** $\underbrace{(f \circ (f \circ (f \circ \cdots \circ f) \cdots))}_{n \text{ functions}}(x) = 2^n x$ **81.** Answers may vary. **83.** $(g - h)(x) = -2x^2 + 7x - 6; -28$ **85.** $(h \cdot k)(x) = 2x^3 + 3x^2 - 8x + 3; 12$ **87.** $(f \circ g)(x) = 16x^2 - 40x + 25; 49$

89. a. Each input x of f has exactly one output $f(x)$, because f is a function, and each input $f(x)$ of g has exactly one output $g(f(x))$, because g is a function. **b.** domain: set A; range: set C **91.**

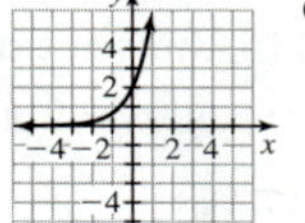

exponential function **93.** linear function

95. $\frac{2}{81}$; exponential function

Homework 11.2

1. 4 **3.** 6 **5.** 5 **7.**

x	$f^{-1}(x)$
2	6
4	5
6	4
8	3
10	2

9. 6 **11.** 6 **13.** 1 **15.**

x	$g^{-1}(x)$
2	1
6	2
18	3
54	4
162	5
486	6

17. 4 **19.** fourth column: 6, 5, 4, 3, 2, and 1 **21.** Answers may vary. **23.** 24 **25.** 0

27.

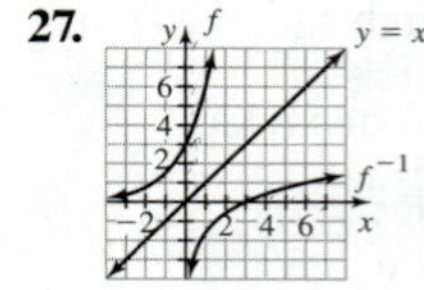

29.

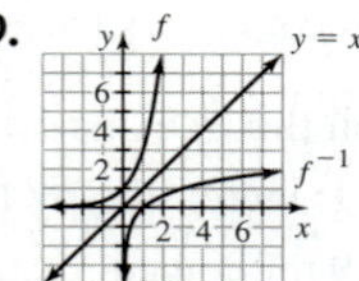

31.

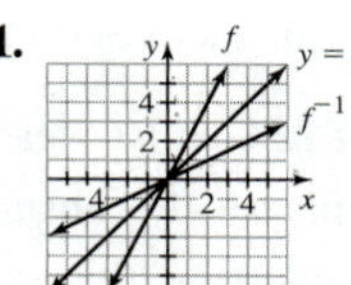

33.

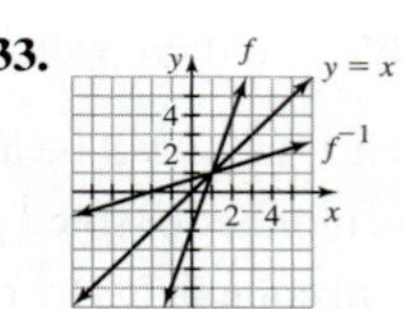

35.

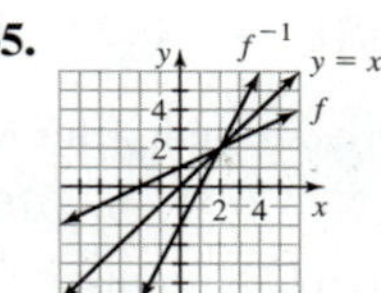

37.

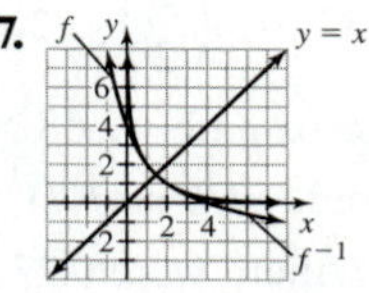

39.

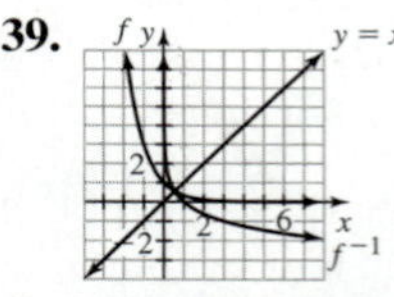

41. 3 **43.** 1 **45.** 0 **47.** 1 **49.** 3 **51.**

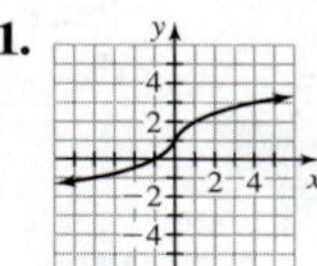

53. a. $f^{-1}(p) = 1.32p + 55.32$ **b.** 33.96; in 2000, about 34.0% of births were outside marriage. **c.** 187.32; in 2087, 100% of births will be outside marriage; model breakdown has likely occurred. **d.** 1.32; the percentage of births outside marriage increases by 1 percentage point every 1.32 years.

55. a. linear function **b.** $f(t) = 3.07t + 14.07$ **c.** $f^{-1}(p) = 0.33p - 4.58$ **d.** 2018 **e.** 2018 **f.** They are the same. **57. a.** $f(a) = 2.17a + 581.49$ **b.** $f^{-1}(c) = 0.46c - 267.97$ **c.** 43 years **d.** 114 years **e.** 0.46; the credit score increases by 1 unit for each age increase of 0.46 year.

59. $f^{-1}(x) = x - 8$ **61.** $f^{-1}(x) = -\frac{1}{4}x$ **63.** $f^{-1}(x) = 7x$ **65.** $f^{-1}(x) = -\frac{1}{6}x - \frac{1}{3}$ **67.** $f^{-1}(x) = 2.5x + 19.75$

69. $f^{-1}(x) = \frac{3}{7}x - \frac{3}{7}$ **71.** $f^{-1}(x) = -\frac{6}{5}x - \frac{18}{5}$ **73.** $f^{-1}(x) = \frac{5}{6}x + \frac{1}{3}$ **75.** $f^{-1}(x) = -\frac{1}{8}x - \frac{1}{8}$ **77.** $f^{-1}(x) = x$ **79.** $f^{-1}(x) = x^{\frac{1}{3}}$

81. a. $(f^{-1} \circ f)(x) = f^{-1}(f(x)) = f^{-1}(x + 7) = x + 7 - 7 = x$ **b.** $(f \circ f^{-1})(x) = f(f^{-1}(x)) = f(x - 7) = x - 7 + 7 = x$

83. a. $(f^{-1} \circ f)(x) = f^{-1}(f(x)) = f^{-1}(2x - 5) = \frac{1}{2}(2x - 5) + \frac{5}{2} = \frac{1}{2} \cdot 2x - \frac{1}{2} \cdot 5 + \frac{5}{2} = x - \frac{5}{2} + \frac{5}{2} = x$

b. $(f \circ f^{-1})(x) = f(f^{-1}(x)) = f\left(\frac{1}{2}x + \frac{5}{2}\right) = 2\left(\frac{1}{2}x + \frac{5}{2}\right) - 5 = 2 \cdot \frac{1}{2}x + 2 \cdot \frac{5}{2} - 5 = x + 5 - 5 = x$

85. a. $(f^{-1} \circ f)(x) = f^{-1}(f(x)) = f^{-1}\left(\frac{3}{4}x - 2\right) = \frac{4}{3}\left(\frac{3}{4}x - 2\right) + \frac{8}{3} = \frac{4}{3} \cdot \frac{3}{4}x - \frac{4}{3} \cdot 2 + \frac{8}{3} = x - \frac{8}{3} + \frac{8}{3} = x$

b. $(f \circ f^{-1})(x) = f(f^{-1}(x)) = f\left(\frac{4}{3}x + \frac{8}{3}\right) = \frac{3}{4}\left(\frac{4}{3}x + \frac{8}{3}\right) - 2 = \frac{3}{4} \cdot \frac{4}{3}x + \frac{3}{4} \cdot \frac{8}{3} - 2 = x + 2 - 2 = x$

87. a. $f^{-1}(x) = \frac{1}{5}x + \frac{9}{5}$ **b.** 11 **c.** $\frac{13}{5}$ **89. a.** $f^{-1}(x) = \frac{1}{3}x + \frac{5}{3}$ **b.** Answers may vary. **c.**

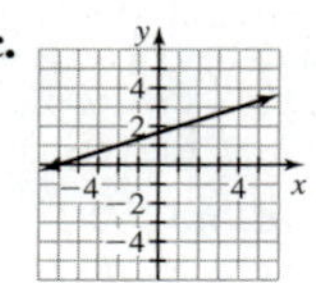

d. For each input–output pair, the output variable is $\frac{5}{3}$ more than $\frac{1}{3}$ the input variable. **91.** no; answers may vary. **93.** Answers may vary. **95.** Answers may vary. **97.** no; no; answers may vary. **99. a.** $f^{-1}(x) = \frac{1}{m}x - \frac{b}{m}$ **b.** Answers may vary. **101. a.** $(4, 5)$ **b.** $f^{-1}(x) = \frac{1}{2}x + \frac{3}{2}$ **c.** $g^{-1}(x) = 2x - 6$ **d.** $(5, 4)$ **e.** The coordinates are interchanged; answers may vary.

103. $(5, 1)$; a linear system in two variables **105.**

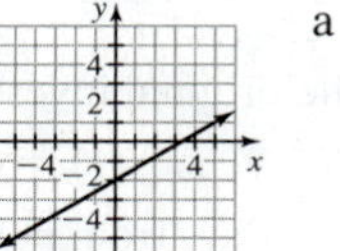

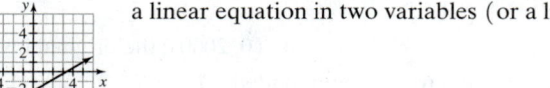

a linear equation in two variables (or a linear function)

107. $(2x - 3)(2x + 3)(3x - 5)$; cubic (or third-degree) polynomial in one variable

Homework 11.3

1. 2 **3.** 3 **5.** 4 **7.** 3 **9.** 2 **11.** −1 **13.** −3 **15.** −4 **17.** 0 **19.** 1 **21.** $\frac{1}{2}$ **23.** $\frac{1}{3}$ **25.** $\frac{1}{2}$ **27.** $\frac{1}{4}$ **29.** 2 **31.** 0 **33.** 1 **35.** 4 **37.** −5 **39.** $\frac{1}{2}$ **41.** 0 **43.** $f^{-1}(x) = \log_3(x)$ **45.** $h^{-1}(x) = \log(x)$ **47.** $f^{-1}(x) = 5^x$ **49.** $h^{-1}(x) = 10^x$ **51.** 4 **53.** 1 **55.** 1 **57.** 27 **59.** 3 **61.** 0; $\log_3(1)$ **63.** 6 **65.** 8 **67.** 7 **69.** 3 **71.** **73.**

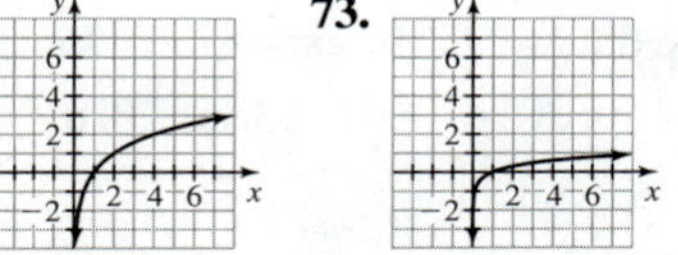

75.

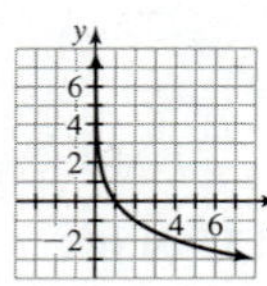

77. a. Answers may vary. **b.**

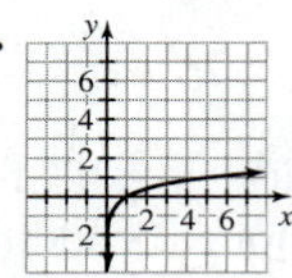

c. For each input–output pair, the output variable is the logarithm, base 5, of the input variable. **79. a.** 2 **b.** 2 **c.** Answers may vary. **81. a.** 9.2 **b.** 7.8 **c.** 25.4 **83.** 0, 20, 40, 60, 80, 100, 120

85. (a) **87.** (d) **89.** $\log_2(7)$; because 2 is less than 3, 2 requires a larger exponent to get 7 than 3 does. **91.** Answers may vary.

93. a.

x	$f(x)$	$g(x)$	$h(x)$	$k(x)$
1	0	2	1	2
2	1	4	4	4
4	2	8	16	16
8	3	16	64	256
16	4	32	256	65,536

b. $k; h; f$

95. a. $(f^{-1} \circ f)(x) = f^{-1}(f(x)) = f^{-1}(3^x + 5) = \log_3(3^x + 5 - 5) = \log_3(3^x) = x$
b. $(f \circ f^{-1})(x) = f(f^{-1}(x)) = f(\log_3(x - 5)) = 3^{\log_3(x-5)} + 5 = x - 5 + 5 = x$

97. a. $(f^{-1} \circ f)(x) = f^{-1}(f(x)) = f^{-1}(5 \cdot 2^x - 6) = \log_2\left(\frac{5 \cdot 2^x - 6 + 6}{5}\right) = \log_2\left(\frac{5 \cdot 2^x}{5}\right) = \log_2(2^x) = x$

b. $(f \circ f^{-1})(x) = f(f^{-1}(x)) = f\left(\log_2\left(\frac{x+6}{5}\right)\right) = 5(2)^{\log_2\left(\frac{x+6}{5}\right)} - 6 = 5 \cdot \frac{x+6}{5} - 6 = x + 6 - 6 = x$

99. $\frac{-3 \pm \sqrt{29}}{2}$; quadratic equation in one variable **101.** $(2x - 1)(3x + 1)$; quadratic (or second-degree) polynomial in one variable **103.** $-5x^2 + 5$; quadratic (or second-degree) polynomial in one variable

Homework 11.4

1. $3^5 = 243$ **3.** $10^2 = 100$ **5.** $b^c = a$ **7.** $10^n = m$ **9.** $\log_5(125) = 3$ **11.** $\log(1000) = 3$ **13.** $\log_y(x) = w$ **15.** $\log(q) = p$ **17.** 16 **19.** $\frac{1}{100}$ **21.** 1 **23.** 81 **25.** $\frac{69}{2}$ **27.** 3 **29.** 6561 **31.** 3.3019 **33.** 7 **35.** 2 **37.** 1.7411 **39.** 1.5850 **41.** 2 **43.** 4.8738 **45.** 3.8278 **47.** −0.2281 **49.** 3.4850 **51.** 0.8644 **53.** no real-number solution **55.** 64 **57.** 3.5130 **59.** 2.3587 **61.** 10 **63.** 1.6975 **65.** 1 **67.** 0, 3.7 **69.** 2 **71.** 1.2122 **73.** 1.3618 **75.** 5.4723 **77.** 2 **79.** 5 **81.** $(3, 1.5)$ **83.** $\frac{\log\left(\frac{c}{a}\right)}{\log(b)}$ **85.** $\frac{\log(d - c)}{\log(b)}$

87. $\dfrac{\log\left(\dfrac{d+c}{a}\right)}{\log(b)}$ **89.** $\dfrac{\log\left(\dfrac{c+d}{a}\right) - p\log(b)}{\log(b)}$ **91.** $\log[3(8)^x] \neq x\log[3(8)]$; 0.4075 **93.** 256 **95.** 0.7925 **97.** 3 **99.** 32

101. a. no **b.** no **c.** no **d.** no **103. a.** the part of the line $y = 1$ where $x > 0$ **b.** We cannot take the logarithm of a negative number. **c.** $f(x) = \log(x^3) - 3\log(x) + 1 = 3\log(x) - 3\log(x) + 1 = 1$; answers may vary. **105.** Answers may vary; no. **107.** $\frac{29}{5}$ **109.** ± 1.7508

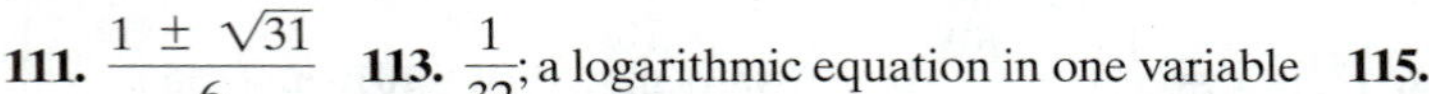

111. $\frac{1 \pm \sqrt{31}}{6}$ **113.** $\frac{1}{32}$; a logarithmic equation in one variable **115.**

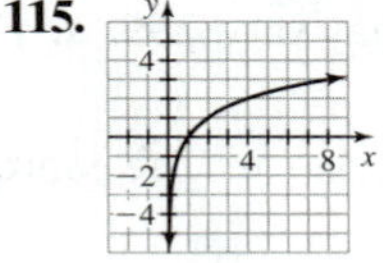

logarithmic function **117.** 4; logarithmic function

Homework 11.5 **1. a.** $f(t) = 2000(1.05)^t$ **b.** $(0, 2000)$; the original investment was \$2000 **c.** \$2552.56 **d.** 8.31 years **3.** 6.65 years **5.** 7.27 years; the interest is compounded **7.** 2017 **9. a.** $f(t) = 30(3)^t$ **b.** 196,830; in 8 days from now, 196,830 people will have heard the rumor (both past and present). **c.** 6; in 6 days from now, 21,870 people will have heard the rumor (both past and present). **d.** 15 days **11. a.** $f(d) = 8\left(\frac{1}{2}\right)^{d/5}$ or $f(d) = 8(0.8706)^d$ **b.** 0.29 hour; yes **c.** 97 decibels **13.** 2018 **15.** 2018 **17. a.** $(0, 9.58)$; there were about 10 million subscribers in 2000. **b.** 7% per year **c.** 29% **d.** 2016 **19. a.** $f(t) = 2726(0.75)^t$ **b.** 0.049 thousand cases, or 49 cases **c.** 2032 **d.** 2.41 years **21. a.** linear: $f(t) = 210.5t - 769.1$; quadratic: $f(t) = 84.07t^2 - 630.21t + 1164.54$; $f(t) = 0.25(3.30)^t$; the exponential and quadratic models fit the data quite well. The linear model does not fit the data well. **b.** exponential function **c.** 3.30; the number of Twitter employees is growing exponentially by 230% per year. **d.** 11,603 employees **e.** 2017 **23. a.** 2028 **b.** linear function; $g(t) = 0.080t - 1.92$ **c.** 2040; after; answers may vary.

d.

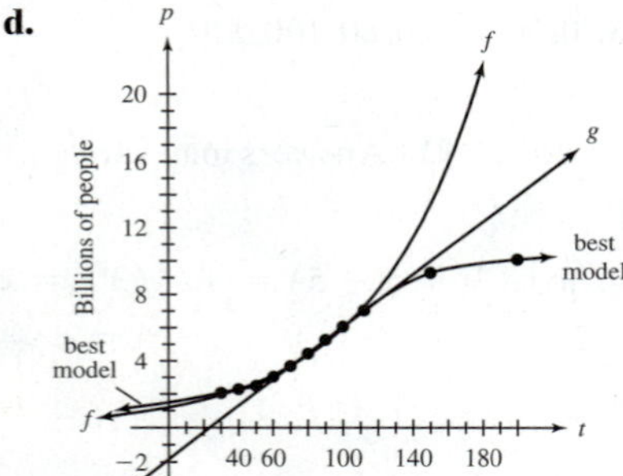

25. a. $f(t) = 0.00067(1.12)^t$ **b.** 1.12; the percentage of seniors with severe memory impairment increases by 12% for each additional year of age. **c.** 1.87; about 1.9% of 70-year-old seniors have severe memory impairment. **d.** 84.80; 10% of 85-year-old seniors have severe memory impairment. **e.** no; answers may vary. **27. a.** $E(s) = 0.36(1.0036)^s$; $R(s) = 0.037(1.0049)^s$ **b.** 60%; 39% **c.** 1373 points; 1475 points **d.** 102 points **e.** $(1757.58, 199.19)$; students who score 1758 points have the same chance (199%) of being selected by early decision as by regular decision; model breakdown has occurred.

29. a. $f(t) = 100\left(\frac{1}{2}\right)^{t/3.25}$ or $f(t) = 100(0.8079)^t$ **b.** 65.28% **c.** 26.0 days **31.** 11,641 years ago **33. a.** 5730 years old **b.** 11,460 years old **c.** 19,035 years old **35.** 1329 years **37.** 24.14 years **39. a.** $f(t) = -5.51t + 122.84$ **b.** 14 days **c.** -5.51; the weight of the soap decreases by 5.51 grams per day **d.** 22 days; yes; answers may vary. **41.** $3 \pm \sqrt{19}$; quadratic equation in one variable **43.** $\pm\frac{\sqrt{35}}{7}$; quadratic equation in one variable **45.** $\frac{3 \pm \sqrt{89}}{8}$; quadratic equation in one variable

Homework 11.6 **1.** $\log_b(3x^2)$ **3.** $\log_b(4x)$ **5.** $\log_b(5t^5)$ **7.** $\log_b\left(\frac{3}{x^3}\right)$ **9.** $\log_b(9x^{11})$ **11.** $\log_b\left(\frac{8m^{12}}{3}\right)$ **13.** 2.0412 **15.** 10.6667 **17.** 1.5275 **19.** 1.5849 **21.** 1.2011 **23.** 3.2702 **25.** 1.7712 **27.** 0.5804 **29.** -2.0431 **31.** 1.6204 **33.** 2.6031 **35.** -2.6876, 1.6964 **37.** $\log_7(x)$ **39.** $\log_s(r)$ **41.** 1.1402 **43.** 0.8368 **45.** all three students **47.** $\log_b(b^2) = \log_b(b^6) - \log_b(b^4) = 2 = \log_b\left(\frac{b^6}{b^4}\right)$ **49.** $\log_b(x) - \log_b(x) = \log_b\left(\frac{x}{x}\right) = \log_b(1)$; hence, $\log_b(1) = 0$ **51. a.** $\log_2(x^8)$ **b.** 1.8340 **c, d.** Answers may vary. **53.** $\log_2(x^7)$ **55.** 1.4860 **57.** 8.6535 **59.** $\log_9\left(\frac{x^3}{8}\right)$ **61.** $\frac{6b^{13}}{c^{1/12}}$ **63.** $\log_b(72x^{23})$ **65.** $(2, -1)$ **67.** $(8, 5)$ **69.** all points on the line

$y = \frac{2}{3}x - 2$; dependent linear system of equations in two variables **71.** 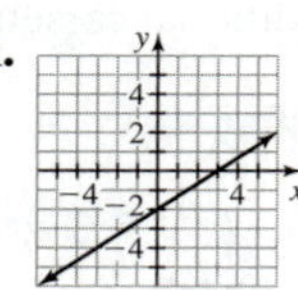 a linear equation in two variables (or a linear function)

73. $-\frac{9}{4}$; linear function (or a linear equation in two variables)

Homework 11.7

1. 4.0037 **3.** −0.6931 **5.** 4 **7.** 1 **9.** −1 **11.** 3 **13.** 7 **15.** 7.3891 **17.** 15.0855 **19.** 1.8383 **21.** 0.9061 **23.** 1.2777 **25.** 1.9811 **27.** 3.4541 **29.** 1.9458 **31.** 3.3673 **33.** $\ln(12x^5)$ **35.** $\ln(5x)$ **37.** $\ln(8w^{11})$ **39.** $\ln\left(\frac{27}{x}\right)$ **41.** $\ln(8k^4)$ **43.** 4.2661 **45.** 37.1033 **47.** 2.0654 **49.** 0.6036 **51.** 1.3066 **53.** 1.2377 **55.** −1.6856, 7.0194 **57.** 20 **59.** 0.1353 **61.** $\frac{\ln\left(\frac{c}{a}\right)}{b}$, or $\frac{\ln(c) - \ln(a)}{b}$ **63. a.** 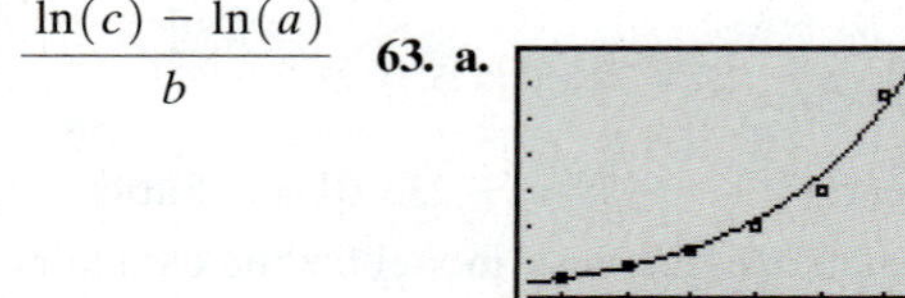 yes **b.** (0, 29.89); there were about 30 laser incidents involving aircraft in 2000. **c.** 2017 **d.** 65,976 **65. a.** 207°F **b.** 3.66 minutes **c.** 70°F **67. a.** 20.91 feet **b.** 20.32; the cable's height 4 feet to the left of the right pole is 20.32 feet. **c.** 20 feet **69.** both students; answers may vary.

71. $3\ln(x) = \ln(x^7) - \ln(x^4) = \ln(x^3)$ **73.** Answers may vary. **75. a. i.** 3.6960 **ii.** 3.6960 **iii.** They are the same. **b. i.** Answers may vary. **ii.** Answers may vary. **iii.** They are the same. **c.** Answers may vary. **77.** $\ln(x^5)$ **79.** 2.2255 **81.** 1.3863 **83.** $\frac{13}{11}$ **85.** $-3, \pm\frac{2}{5}$ **87.** ± 2 **89.** Answers may vary. **91.** Answers may vary. **93.** Answers may vary. **95.** Answers may vary.

Chapter 11 Review Exercises

1. 3 **2.** 0 **3.** 3 **4.** 0 **5.** 3 **6.** 1 **7.**

x	$(f \circ g)(x)$
0	0
1	3
2	4
3	1
4	2

8. a. $(f \circ g)(x) = 9x^2 - 30x + 29$ **b.** $(g \circ f)(x) = 3x^2 + 7$ **c.** 20 **d.** 34 **9. a.** $(f \circ g)(x) = 4(2)^{2x-4}$ **b.** $(g \circ f)(x) = 8(2)^x - 4$ **c.** 16 **d.** 60 **10. a.** $(f \circ g)(x) = \log_3(x + 6)$ **b.** $(g \circ f)(x) = \log_3(x) + 6$ **c.** 2 **d.** 7 **11.** $f(x) = e^x$; $g(x) = x - 5$; answers may vary. **12. a.** $f(n) = 8n$; $g(d) = 0.06d$ **b.** $g(f(n)) = 0.48n$ **c.** 3.36; the sales tax on the purchase of 7 books is \$3.36.

13. 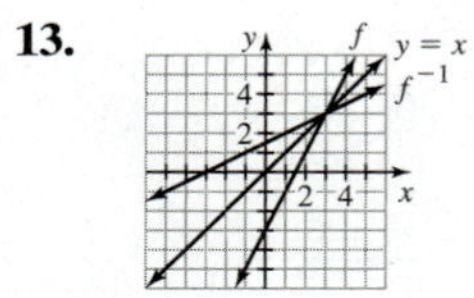

14. 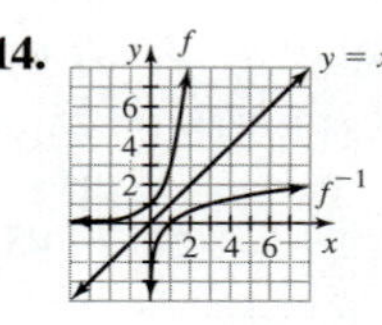

15. a. $f(t) = 1.21t + 2.86$ **b.** 24.64; there will be about 25 million background checks in 2018. **c.** 12.51; in 2013, there were about 18 million background checks. **d.** $C(n) = 15n$ **e.** $h(t) = (C \circ f)(t)$; answers may vary. **f.** $h(t) = 18.15t + 42.9$ **g.** 351.45; in 2017, the total cost of background checks will be about \$351 million.

16. a. $f^{-1}(x) = \frac{1}{3}x$ **b.** $(f^{-1} \circ f)(x) = f^{-1}(f(x)) = f^{-1}(3x) = \frac{1}{3} \cdot 3x = x$ **c.** $(f \circ f^{-1})(x) = f(f^{-1}(x)) = f\left(\frac{1}{3}x\right) = 3\left(\frac{1}{3}x\right) = x$

17. a. $f^{-1}(x) = \frac{6}{5}x + \frac{12}{5}$ **b.** $(f^{-1} \circ f)(x) = f^{-1}(f(x)) = f^{-1}\left(\frac{5}{6}x - 2\right) = \frac{6}{5}\left(\frac{5}{6}x - 2\right) + \frac{12}{5} = \frac{6}{5} \cdot \frac{5}{6}x - \frac{6}{5} \cdot 2 + \frac{12}{5} = x - \frac{12}{5} + \frac{12}{5} = x$ **c.** $(f \circ f^{-1})(x) = f(f^{-1}(x)) = f\left(\frac{6}{5}x + \frac{12}{5}\right) = \frac{5}{6}\left(\frac{6}{5}x + \frac{12}{5}\right) - 2 = \frac{5}{6} \cdot \frac{6}{5}x + \frac{5}{6} \cdot \frac{12}{5} - 2 = x + 2 - 2 = x$ **18.** 2 **19.** 5 **20.** −2 **21.** −3 **22.** $\frac{1}{3}$ **23.** 1.7712 **24.** 1.6094 **25.** 7

26. $h^{-1}(x) = \log_3(x)$ **27.** $h^{-1}(x) = 10^x$ **28.** 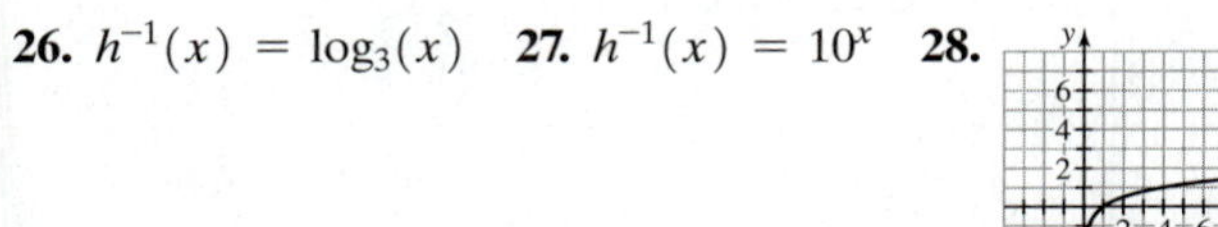 **29.** $\log_d(k) = x$ **30.** $y^r = w$ **31.** 2.3219 **32.** $\frac{1}{81}$ **33.** 0.4310 **34.** 2.0886 **35.** 4 **36.** 2.8333 **37.** 1.6507

38. 4 **39.** 0 **40.** (4, 2) **41.** 81 **42.** 2.9299 **43.** 1.6309 **44.** 729 **45. a.** $f(t) = 8000(1.05)^t$ **b.** \$12,410.63 **c.** 14.2 years **46. a.** $f(t) = 30(4)^t$ **b.** 30,720 leaves **c.** 5.9 weeks after April 1 **47. a.** $f(t) = 0.12(1.083)^t$ **b.** (0, 0.12); the national health spending in 1970 was 0.12 trillion (120 billion) dollars. **c.** 8.3% per year **d.** 6.0 trillion dollars **e.** 2017

48. a. $f(n) = 9.33(1.31)^n$ **b.** 1.31; for each additional cassette, the length increases by 31%. **c.** 9.33; the initial length of the rubber band was 9.33 inches. **d.** 80.92 inches; answers may vary. **e.** 10 cassettes; yes **49.** 14.5 years **50.** $\log_b(3p)$ **51.** $\log_b(72x^5)$ **52.** $\log_b\left(\frac{1}{x^2}\right)$ **53.** $\log_y(w)$ **54.** $\log_b(b^5) - \log_b(b^2) = 3 = \log_b(b^3) = \log_b\left(\frac{b^5}{b^2}\right)$ **55.** 3 **56.** 8.4853 **57.** $\ln(256x^5)$ **58.** $\ln(2m)$ **59.** 2.9312 **60.** 7.3891 **61.** 2.8479 **62.** 3.4950 **63.** 2.1297

Chapter 11 Test

1. a. $(f \circ g)(x) = 12x^2 - 60x + 75$ **b.** $(g \circ f)(x) = 6x^2 - 5$ **c.** 3 **d.** 19 **2. a.** $(f \circ g)(x) = 3^{x-4}$ **b.** $(g \circ f)(x) = 3^x - 4$ **c.** $\frac{1}{9}$ **d.** 5 **3.** 2 **4.** 5 **5.** 2 **6.** **7.**

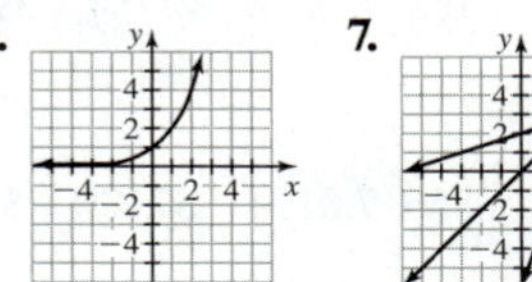

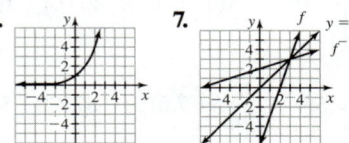

8. a. $f(t) = 3.75t + 44.21$ **b.** \$85.46 **c.** $f^{-1}(p) = 0.27p - 11.79$ **d.** 2018 **e.** $S(d) = 0.065d$ **f.** $h(t) = (S \circ f)(t)$; answers may vary. **g.** $h(t) = 0.24t + 2.87$ **h.** 6.95; in 2017, the sales tax on an adult one-day ticket will be \$6.95.

9. $g^{-1}(x) = \frac{5}{2}x + \frac{9}{2}$ **10.** 4 **11.** -3 **12.** 1.1833 **13.** -1 **14.** $\frac{1}{2}$ **15.** -2 **16.** $h^{-1}(x) = \log_4(x)$ **17.** $f^{-1}(x) = 5^x$ **18.** $\log_s(w) = k$ **19.** $c^d = a$ **20.** 2.6591 **21.** 2.4150 **22.** -0.6964 **23.** 1.67 **24. a.** linear: $f(t) = -5.38t + 233.10$; quadratic: $f(t) = 0.18t^2 - 14.26t + 292.32$; $f(t) = 270.26(0.94)^t$; the exponential model fits the data best. The quadratic model fits the data fairly well. The linear model does not fit the data well. **b.** exponential function **c.** 6% per year **d.** 23.7 thousand patients **e.** 2017 **25.** 2050 years old **26.** $\log_b(5x^4)$ **27.** $\log_b(2)$ **28.** 11.0227 **29.** 4.4413 **30.** $\ln(25w^{20})$ **31.** 1.4319 **32.** 4.0427

Chapter 11 Cumulative Review Chapters 1–11

1. 0.5087 **2.** $-5, 3$ **3.** 86 **4.** 1.3538 **5.** $\frac{-1 \pm \sqrt{22}}{3}$ **6.** 1.2528 **7.** $\frac{10 \pm \sqrt{14}}{2}$ **8.** ± 0.7811 **9.** $\frac{7}{10}$ **10.** $-\frac{1}{5}, \frac{1}{2}$ **11.** 1.0782 **12.** $\frac{2}{5} \pm i\frac{\sqrt{11}}{5}$ **13.** $\frac{-5 \pm \sqrt{41}}{4}$ **14.** $\frac{b \pm \sqrt{c}}{a}$ **15.** 1 **16.** 2 **17.** $(-1, 2)$ **18.** $(2, 4)$ **19.** $(-1, 3, 2)$ **20.** $\left(-\frac{3}{2}, 0\right), (4, 0)$ **21.** $12x^4 - 44x^3 + 40x^2$ **22.** $16m^2t^2 - 49r^2$ **23.** $25m^2 - 30mn + 9n^2$ **24.** $(f - g)(x) = -x^2 - 3x + 9; 9$ **25.** $(f \cdot g)(x) = 6x^4 + x^3 + 5x^2 + 14x - 20; 132$ **26.** $(f \circ g)(x) = x^2 + 7x + 10; 28$ **27.** $4x^2 - x + 2$ **28.** $(2x - 3)(2x + 3)(3x - 5)$ **29.** $3xy^2(x + 4y)(x - 2y)$ **30.** $(3x + 4)(9x^2 - 12x + 16)$

31. $x \geq \frac{9}{10}$; $\left[\frac{9}{10}, \infty\right)$; **32.**

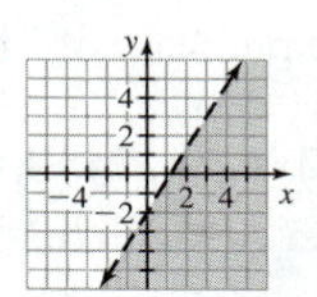

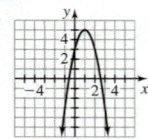

33. **34.** $\frac{1600c^4}{b^{23}}$ **35.** $\frac{4b^{5/6}}{3c^{5/4}}$ **36.** $\log_b\left(\frac{x^{26}}{49}\right)$ **37.** $\ln(p^{26})$ **38.** $f(x) = 5(3)^x$ **39.** $g(x) = 3x + 25$ **40.** 76 **41.** 8

42. 40 **43.**

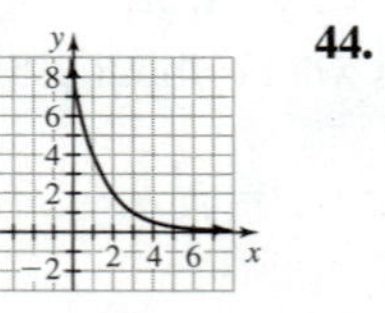

44. **45.**

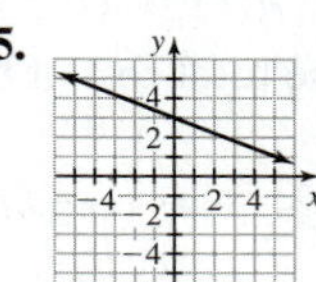

46. **47.**

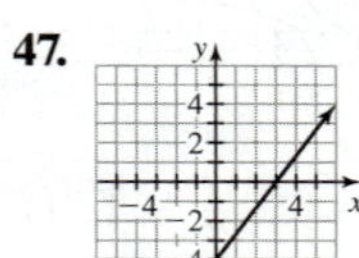

48.

49. $y = -\frac{10}{9}x + \frac{23}{9}$ **50.** $y = 3x^2 - 5x + 2$ **51.** $y = 347.56(0.63)^x$ **52.** $\frac{2}{81}$ **53.**

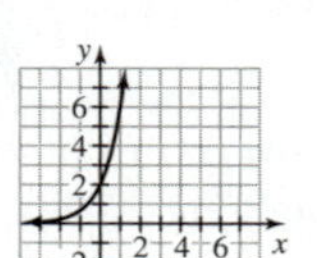

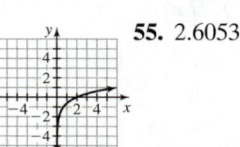

54. **55.** 2.6053

56. -4 **57.** $\frac{1}{7}$ **58.** 3 **59.** 4 **60.** $f(x) = 2(2)^x$, or $f(x) = (2)^{x+1}$ **61.**

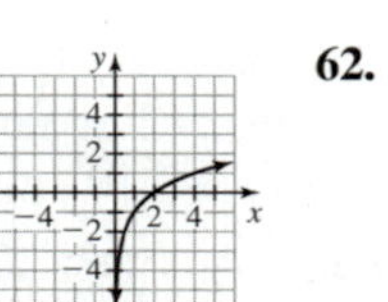

62. 0 **63.** $f^{-1}(x) = \frac{7}{2}x + \frac{21}{2}$

64. $g^{-1}(x) = \log_8(x)$ **65. a.** $f{:}(0, 2)$; $g{:}(0, 2)$ **b.** As x increases by 1, $f(x)$ increases by 3 and $g(x)$ is multiplied by 3. **c.** g

d. 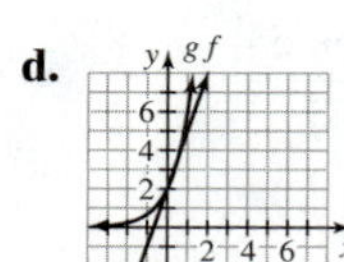

66. a. $f(2) = 6; g(2) = 9$ **b.** $f^{-1}(x) = \frac{1}{3}x; g^{-1}(x) = \log_3(x)$ **c.** $f^{-1}(81) = 27; g^{-1}(81) = 4$

67. a. $U(x) = 0.69x + 19.95; B(x) = 0.45x + 29.95$ **b.** 0.69, 0.45; U-Haul charges $0.69 per additional mile, and Budget charges $0.45 per additional mile. **c.** 41.67 miles **68.** 10,500 tickets at $43, 4500 tickets at $60

69. a. $f(t) = 9.0(1.14)^t$ **b.** (0, 9.0); in 2010, the revenue was $9.0 billion. **c.** 1.14; the revenue is growing exponentially by 14% per year. **d.** 2018 **70. a.** linear: $f(t) = -0.98t + 65.02$; quadratic: $f(t) = 0.018t^2 - 2.10t + 76.51$; exponential: $f(t) = 76.58(0.97)^t$; the quadratic and exponential models fit the data quite well. The linear model does not fit the data as well. **b.** exponential function **c.** (0, 76.58); there were 76,580 tuberculosis cases in 1950. **d.** 0.97; the number of tuberculosis cases decreases by 3% per year. **e.** 58.22; there were 58,220 tuberculosis cases in 1959. **f.** 70.29; there will be 9 thousand tuberculosis cases in 2020. **g.** 22.76 years **71. a.** $f(t) = 0.53t - 0.17$ **b.** 0.53; the number of women who place in the top 100 is increasing by 0.53 woman per year, on average. **c.** (0.32, 0); no women placed in the top 100 in 1980. **d.** 17 women **e.** $f^{-1}(n) = 1.89n + 0.32$ **f.** 38.12; in 2018, 20 women will place in the top 100. **g.** $h(t) = -0.53t + 100.17$ **h.** 80.56; in 2017, 81 men will place in the top 100. **72. a.** $f(t) = -0.0195t^2 + 1.58t - 10.20$ **b.** (7.07, 0), (73.95, 0); no 7-year-old children visit online trading sites. Also, no 74-year-old adults visit online trading sites; model breakdown has occurred. **c.** (40.51, 21.81); this means that about 21.8% of 41-year-old adults visit online trading sites, the highest percentage for any age group, according to the model. In reality, 22.9% of Americans between the ages of 25 and 34 years, inclusive, visit online trading sites. **d.** 12.8% **e.** 27-year-old and 54-year-old Americans

Chapter 12

Homework 12.1

1. $0, -\frac{3}{5}$, undefined **3.** $\frac{9}{2}, 8, \frac{19}{26}$ **5.** all real numbers except 0 **7.** all real numbers **9.** all real numbers except -3 **11.** all real numbers except $-\frac{1}{2}$ **13.** all real numbers except -2 and 5 **15.** all real numbers except ± 4 **17.** all real numbers except $\pm\frac{5}{2}$ **19.** all real numbers **21.** all real numbers except $-\frac{3}{2}$ and 5 **23.** all real numbers **25.** all real numbers except $\frac{1 \pm \sqrt{22}}{3}$ **27.** all real numbers except $\pm\frac{3}{2}$ and 2 **29.** $f(x) = \frac{2x}{3}$ **31.** $f(x) = \frac{4x^3}{3}$ **33.** $f(x) = \frac{4}{5}$ **35.** $f(x) = \frac{x+5}{x-9}$ **37.** $f(x) = \frac{x+7}{x-7}$ **39.** $f(x) = \frac{4x+5}{2x-3}$ **41.** $f(x) = -1$ **43.** $f(x) = -\frac{2}{3}$ **45.** $f(x) = -\frac{6}{x+3}$ **47.** $f(x) = \frac{x}{7}$ **49.** $f(x) = -\frac{x+7}{x+2}$ **51.** $f(x) = \frac{3x(x+4)}{x-3}$ **53.** $f(x) = \frac{3(x+2)}{6x-1}$ **55.** $f(x) = \frac{x-4}{(2x+3)(2x-3)}$ **57.** $f(x) = \frac{x^2-2x+4}{x-2}$ **59.** $f(x) = \frac{x+3}{9x^2+6x+4}$ **61.** $\frac{2x}{3y^3}$ **63.** $\frac{x-3y}{x}$ **65.** $\frac{2a-b}{a-3b}$ **67.** $\frac{p^2+pq+q^2}{p+q}$ **69.** $\frac{f}{g}(x) = \frac{x+4}{x-6}; -\frac{7}{3}$ **71.** $\frac{h}{f}(x) = \frac{3x+5}{x-2}; \frac{17}{2}$ **73.** $\frac{f}{h}(x) = \frac{x^2}{3x+1}; -\frac{4}{5}$ **75.** $\frac{k}{g}(x) = \frac{9x^2-3x+1}{2x(3x+1)}; \frac{13}{4}$ **77. a.** $P(t) = \frac{-103t^2 + 2509t - 10{,}588}{0.0068t^2 + 2.71t + 277.9}$; answers may vary. **b.** 15.1%; 14.9%; overestimate **c.** 9.9; in August of 2016, 9.9% of Americans will participate in SNAP. **79. a.** $R(t) = 0.0048t^2 + 0.14t + 2.95$ **b.** $E(t) = 0.015t^2 + 0.055t + 6.05$ **c.** $P(t) = \frac{0.48t^2 + 14t + 295}{0.015t^2 + 0.055t + 6.05}$ **d.** 51.4% **e.** (12.73, 60.02); the largest percentage of voter-eligible Latinos who were registered, about 60.0%, occurred in 1993. **f.** decreasing; the number of Latinos who are eligible to vote has been growing at a faster rate than the number of Latinos who are registered to vote. **81.** Answers may vary. **83.** domain: $-3, -2, -1, 0, 1, 2, 3$; range: 50.4, 25.2, 16.8, 12.6, 10.08, 8.4, 7.2 **85.** no; answers may vary. **87.** $\frac{17}{14}, \frac{5}{2}$; the work is incorrect; answers may vary. **89. a.** 8.4; the drive takes 8.4 hours at a constant speed of 50 mph. **b.** 7.64, 7, 6.46, 6 **c.** decreasing; answers may vary. **91.** Answers may vary. **93.** all real numbers **95.** all real numbers **97. a.** $x + 2$ **b.** $x + 2$ **c.** Answers may vary. **99.** all real numbers except -3 and 7; a rational function **101.** $-\frac{5}{3}, -2, 2$; a cubic equation in one variable **103.** $(2x - 5)(4x^2 + 10x + 25)$; a cubic (or third-degree) polynomial in one variable

Homework 12.2 **1.** $\frac{15}{x^2}$ **3.** $\frac{x^2}{9}$ **5.** $\frac{18}{5a^6}$ **7.** $\frac{2(x-4)}{(x-3)(x+5)}$ **9.** $\frac{(k-2)(k+4)}{(k-6)(k+6)}$ **11.** $\frac{10}{21}$ **13.** 6 **15.** $\frac{2w^4(w+5)}{w+3}$ **17.** $\frac{5(x+1)}{3(x-7)}$ **19.** $\frac{30(x-4)}{7(x+5)}$ **21.** $\frac{5t^4}{6}$ **23.** $\frac{6}{x^3(x-7)}$ **25.** $\frac{3a^3}{10b^7}$ **27.** $-\frac{8}{3}$ **29.** $\frac{2(k-4)(k-1)}{k+6}$ **31.** $\frac{a(a+2b)}{3b}$ **33.** $\frac{x-5}{3(x+5)(x+1)}$ **35.** $-\frac{t-4}{(t+1)(t+4)}$ **37.** $\frac{2(x-5)(2x+3)}{(x+4)(x+2)}$ **39.** $\frac{8}{(x-6)(3x-4)}$ **41.** $\frac{3x+4}{(x+2)(x+3)}$ **43.** $-\frac{(3m+2)(m-3)}{(m+3)(2m+7)}$ **45.** $\frac{(x+4)(x-1)}{(x+3)(x+6)}$ **47.** $\frac{(p+6t)(p+3t)}{(p-3t)(p-5t)}$ **49.** $\frac{(x+y)(x-3y)}{y(2x+3y)}$ **51.** $\frac{3(x-3)(x-8)}{4x(x-5)(x+8)}$ **53.** $\frac{3(w+2)(2w+3)}{2w(3w-1)(w-5)}$ **55.** $\frac{x-1}{(x-2)(x+5)}$ **57.** $\frac{1}{x-1}$ **59.** $\frac{(k-3)(k^2+2k+4)}{(k+2)(k^2-3k+9)}$ **61.** $(2x-3)(x+2)$ **63.** $\frac{a+2b}{(a^2-ab+b^2)}$ **65.** $(f\cdot g)(x)=\frac{(x-8)^2}{(x-5)^2}$; 4 **67.** $\left(\frac{g}{f}\right)(x)=\frac{(x+8)^2}{(x+2)^2}$; $\frac{25}{9}$ **69.** $\left(\frac{f}{g}\right)(x)=-\frac{(x+1)^2}{(x-7)^2}$; $-\frac{25}{9}$ **71.** $\frac{25}{2x^6(x-12)}$ **73.** $-\frac{36}{k^4}$ **75.** 1 **77.** 5.32 feet **79.** 56.76 liters **81.** 101.88 milligrams **83.** 5.45 gallons per day **85.** 98.27 feet per second **87.** $\frac{x^2}{4}$ **89.** $\frac{(x-2)(x+1)}{(x-6)(x+4)}$ **91.** Answers may vary; $\frac{x+2}{x+6}$ **93.** Answers may vary; $\frac{x(x+4)}{3(x-7)}$ **95. a.** 1 **b.** $\frac{1}{x}$ **c.** 1 **d.** $\frac{1}{x}$ **e.** 1 if n is odd, $\frac{1}{x}$ if n is even **97.** Answers may vary. **99.** $x-4$; $x-4$; the results are the same. **101.** $(f\cdot g)(x)=16^x$; $\left(\frac{f}{g}\right)(x)=4^x$ **103.** $(f\cdot g)(x)=36(12)^x$; $\left(\frac{f}{g}\right)(x)=4(3)^x$ **105 a.** 90 inches **b. i.** $f(w)=12w$; $g(x)=3x$ **ii.** $(f\circ g)(x)=36x$ **iii.** 90; the width is 90 inches; the two results are equal. **107.** $\log_b\left(\frac{16x^2}{9}\right)$; a logarithmic expression in one variable **109.** 5.3723; a logarithmic equation in one variable **111.** 2.1893; an exponential equation in one variable

Homework 12.3 **1.** $\frac{9}{x}$ **3.** $\frac{7x}{x-2}$ **5.** $t+2$ **7.** $\frac{1}{x+3}$ **9.** $x-3$ **11.** $\frac{c-4}{c+2}$ **13.** $\frac{x+5}{x+7}$ **15.** $\frac{11}{2x}$ **17.** $\frac{-3b+20}{8b}$ **19.** $\frac{10x^2+9}{12x}$ **21.** $\frac{2(1-2x^4)}{x^6}$ **23.** $\frac{25x^2+18}{60x^6}$ **25.** $\frac{21b^2-10a}{12a^2b^3}$ **27.** $\frac{13x+15}{4x(x+3)}$ **29.** $\frac{7x-2}{(x-2)(x+1)}$ **31.** $\frac{3x^2+4x+10}{(x-2)(x+3)}$ **33.** $\frac{2a}{(a+b)(a-b)}$ **35.** $\frac{c-6}{2(c-4)}$ **37.** $\frac{2(3x+10)}{15(x-5)}$ **39.** $\frac{3}{k-1}$ **41.** $\frac{1}{x-3}$ **43.** $\frac{8x+25}{x(x-5)(x+5)}$ **45.** $\frac{5x+1}{(x-4)(x-3)(x+3)}$ **47.** $\frac{19t+87}{15(t-2)(t+3)}$ **49.** $\frac{3k-1}{k+1}$ **51.** $\frac{2x}{x+1}$ **53.** $\frac{12}{x-6}$ **55.** $\frac{1}{2(x+3)}$ **57.** $\frac{4c^2+13c-1}{(2c-7)(2c+7)}$ **59.** $\frac{a^2+ab+2b^2}{b(a+b)(a-b)}$ **61.** $\frac{x-2}{(x+2)(x+4)}$ **63.** $\frac{2x^2+2x+5}{(x-1)(x+2)}$ **65.** $-\frac{16}{(y-3)(y+5)}$ **67.** $\frac{x^2+4x+30}{(x-2)(x-5)(x+5)}$ **69.** $\frac{2x^2+5x-1}{(x-4)(x+1)(x+3)^2}$ **71.** $\frac{4}{c-2}$ **73.** $\frac{x^2-15x-21}{(2x+5)^2(3x+1)}$ **75.** $\frac{11x^2-x+3}{(x+2)^2(3x-1)}$ **77.** $\frac{3p^2+7pq-8q^2}{(p-6q)(p+3q)(p+4q)}$ **79.** $\frac{x^2+x+8}{6x(x-4)(x+2)}$ **81.** $\frac{5}{2(x+2)}$ **83.** $\frac{-t+16}{(t+1)(t+5)}$ **85.** $(f+g)(x)=\frac{2x^2-25}{(x-4)(x-3)}$ **87.** $(g-f)(x)=-\frac{7}{(x-4)(x-3)}$ **89.** $(f-g)(x)=\frac{-x^2+6x-2}{3(x-4)(x+2)}$ **91. a.** $\frac{18}{d^2}+\frac{18}{(2d)^2}$ **b.** $\frac{45}{2d^2}$ **c.** 15.63; the total illumination is 15.63 W/m^2 when the person is 1.2 meters away from the closer light and 2.4 meters away from the other light. **93.** First fraction should be multiplied by $\frac{x+2}{x+2}$, second fraction by $\frac{x+1}{x+1}$; $\frac{5x+7}{(x+1)(x+2)}$ **95.** Numerator should be $9x-(5x+1)=9x-5x-1=4x-1$; $\frac{4x-1}{x-3}$ **97.** Answers may vary. **99.** Answers may vary. **101.** $(f+g)(x)=8x^2-11x-2$; $(f-g)(x)=4x^2+3x+8$

103. $(f+g)(x) = -5^x$; $(f-g)(x) = (5)^{x+1}$ **105.** $\frac{2(2x^2+8x+5)}{(x+2)^2}$ **107.** $\frac{5(x+2)}{3(x+1)}$ **109.** an exponential function

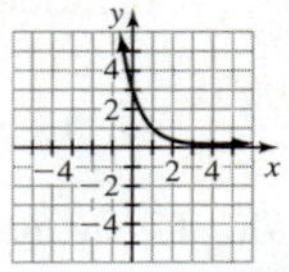

111. ± 1.9061; a fourth-degree polynomial equation in one variable **113.** $y = 1718.87(0.38)^x$; an exponential function

Homework 12.4 **1.** $\frac{3}{10}$ **3.** $\frac{7}{4}$ **5.** $\frac{14}{5}$ **7.** $\frac{3}{2w}$ **9.** $\frac{9}{10x^2}$ **11.** $2ab^2$ **13.** $\frac{2}{3(x+3)}$ **15.** $\frac{5}{6}$ **17.** $\frac{1}{x+1}$ **19.** $\frac{7(x+7)}{3x(x+2)}$ **21.** $\frac{(5x+2)(3x-4)}{(3x+4)(5x-2)}$ **23.** 3 **25.** $\frac{5}{9}$ **27.** $\frac{11}{26}$ **29.** $-\frac{1}{5}$ **31.** $\frac{4(x-2)}{5}$ **33.** $\frac{-3x^2+2}{4x+5}$ **35.** $-\frac{4x+3}{3x-2}$ **37.** $\frac{3(8x^3-5)}{18x^3-1}$ **39.** $a(a+b)$ **41.** $\frac{x-3}{x}$ **43.** $\frac{x-9}{x+6}$ **45.** $\frac{p^2-4p+2}{p^2-4p-3}$ **47.** $-\frac{1}{x(x+3)}$ **49.** $-\frac{3}{a(a+b)(a-b)}$ **51.** $-\frac{2(x+1)}{x^2(x+2)^2}$ **53.** $-\frac{(x+3)(x+4)(3x+10)}{x(x+2)}$ **55.** $\frac{(x+6)(x-5)}{(x+4)(x+3)}$ **57.** $\left(\frac{f}{g}\right)(x) = \frac{5(x-1)}{4(x-3)}$ **59.** $\left(\frac{f}{g}\right)(x) = \frac{x-2}{3}$ **61.** $\left(\frac{f}{g}\right)(x) = \frac{x^2+12x+25}{x^2-7x+25}$ **63.** The reciprocal of $\left(\frac{1}{x}+\frac{1}{2}\right)$ is not $\left(\frac{x}{1}+\frac{2}{1}\right)$; $\frac{2x^2}{x+2}$ **65.** $-\frac{2(5x-6)}{3x+4}$; answers may vary. **67.** $6 \cdot \frac{1}{3} = 2$; yes **69.** Answers may vary. **71.** $\frac{4x^5}{3y^3}$ **73.** $-\frac{x+1}{x-1}$ **75.** $\frac{2b^3-1}{b(3b^2-4)}$ **77.** $(f \circ g)(x) = \frac{3x-2}{2x-3}$ **79. a.** $H_e = \frac{H_pH_b}{H_p+H_b}$ **b.** 76.6 days **81.** $\frac{2(x^2+2x+10)}{(x-6)(x+4)(x-2)}$; a rational expression in one variable **83.** $\frac{1}{(x-6)^2}$; a rational expression in one variable **85.** all real numbers except -4 and 6; a rational function

Homework 12.5 **1.** -2 **3.** 3 **5.** 2 **7.** empty-set solution **9.** 3 **11.** $\frac{4}{5}$ **13.** $\frac{1}{6}$ **15.** -8 **17.** $\frac{3}{2}$ **19.** 1 **21.** $\frac{2}{3}$ **23.** empty-set solution **25.** $-\frac{7}{2}$ **27.** -2 **29.** empty set **31.** -8 **33.** -1 **35.** $\frac{-1 \pm \sqrt{13}}{3}$ **37.** $-10, 3$ **39.** $\frac{-3 \pm \sqrt{17}}{2}$ **41.** -1 **43.** 1 **45.** empty set **47.** 1 **49.** $\frac{4 \pm \sqrt{22}}{2}$ **51.** $\frac{5 \pm \sqrt{89}}{2}$ **53.** $\frac{5 \pm i\sqrt{7}}{8}$ **55.** $\frac{17 \pm i\sqrt{11}}{6}$ **57.** $\frac{-3 \pm i\sqrt{47}}{4}$ **59.** $\frac{23}{4}$ **61.** $-4 \pm \sqrt{15}$ **63.** $\left(-\frac{7}{5}, 0\right)$ **65.** $r = \frac{mv^2}{F}$ **67.** $M = -\frac{r^2F}{mG}$ **69.** $t = \frac{A-P}{rP}$ **71.** 2021 **73. a.** $E(t) = 0.13t + 1.14$ **b.** $C(t) = 0.59t + 4.47$ **c.** $P(t) = \frac{13t+114}{0.59t+4.47}$ **d.** 23.10; in 2017, 23.1% of the total cost of health insurance will be paid by employees. **e.** 19.63; employees will pay 23% of the total cost of health insurance in 2020. **f.** decreasing; the percentage of the total cost of health insurance that employees pay is decreasing; the rate of change of the amount of money employers pay for employees' health insurance is greater than the rate of change of the amount of money employees pay for health insurance. **75.** 4.2 **77.** 3.3 **79.** $(-0.6, -1.9)$ **81.** $a = -6, b = -3$ **83.** Answers may vary. **85.** Answers may vary; $\frac{33}{2}$ **87.** Answers may vary. **89.** Answers may vary. **91.** $\frac{2(3x+1)}{x(x+1)}$ **93.** $-\frac{1}{3}$ **95.** $-\frac{15}{2}$ **97.** $\frac{10x-1}{(x-3)(x+2)(x-2)}$ **99.** $\pm 2, \frac{1}{2}$ **101.** 2.8433 **103.** $\frac{23}{13}$

105. a quadratic function (or a quadratic equation in two variables) **107.** $-5, 1$; a quadratic equation in one variable

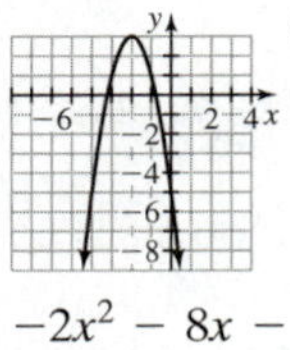

109. $-2x^2 - 8x - 5$; a quadratic (or second-degree) polynomial in one variable

Homework 12.6 **1. a.** $C(n) = 350n + 1250$ **b.** $M(n) = \frac{350n+1250}{n}$ **c.** \$391.67 **d.** 25 students **3. a.** $T(n) = 50n + 500$ **b.** $M(n) = \frac{50n+500}{n}$ **c.** 51.85; if 270 people attend, the mean cost per person is \$51.85. **d.** 50; if 50 people attend, the mean cost per person is \$60. **e.** second column: 55, 52.5, 51.67, 51.25, 51 **f.** 50; If tens of thousands of people attend the reunion, the mean cost per person would be a few cents above \$50; model breakdown has occurred. **5. a.** $C(n) = 7000n + 90{,}000$ **b.** $B(n) = \dfrac{7000n + 90{,}000}{n}$ **c.** $P(n) = \dfrac{9000n + 90{,}000}{n}$ **d.** 11,250; if 40 cars are produced and sold each day, the price should be \$11,250 per car for the profit to be \$2000 per car. **e.** 9000; if very many cars are produced and sold, a price set a few cents more than \$9000 ensures a profit of \$2000 per car. **7. a.** $I(t) = 441.53t + 3852.97$ **b.** $M(t) = \dfrac{441.53t + 3852.97}{0.0013t + 0.093}$ **c.** The units of the expressions on both sides of the equation are dollars per household. **d.** 2018 **e.** increasing; between 1995 and 2020, the mean annual income per household is increasing. **9. a.** $R(t) = 464.8t + 2795.2$ **b.** $A(t) = \dfrac{464.8t + 2795.2}{0.026t^2 - 0.2t + 14.61}$ **c.** \$537.85 **d.** 2020 **11. a.** $M(t) = 16.98t + 395.66$; $E(t) = -29.87t + 1369.89$ **b.** $(M + E)(t) = -12.89t + 1765.55$; the inputs of $M + E$ are the number of years since 1980, and the outputs are the total number of daily newspapers. **c.** $P(t) = \dfrac{1698t + 39{,}566}{-12.89t + 1765.55}$ **d.** 77.37; in 2016, about 77.4% of dailies will be morning newspapers. **e.** 37.26; in 2017, 80% of dailies will be morning newspapers. **13. a.** $W(t) = 0.31t^2 + 6.93t + 456.43$; $M(t) = 0.33t^2 - 2.96t + 482.26$ **b.** $(W + M)(t) = 0.64t^2 + 3.97t + 938.69$; the inputs of $W + M$ are the number of years since 1980, and the outputs are the total number (in thousands) of people who have earned a bachelor's degree. **c.** $P(t) = \dfrac{33t^2 - 296t + 48{,}226}{0.64t^2 + 3.97t + 938.69}$ **d.** 2013, 2023 **e.** decreasing; the number of bachelor's degrees earned has been increasing at a faster rate for women than for men. **15.** 1.42 hours **17. a.** $T(a) = \dfrac{85}{a + 70} + \dfrac{53}{a + 65}$ or $T(a) = \dfrac{138a + 9235}{(a + 65)(a + 70)}$ **b.** 1.89 hours **c.** 8.67 mph **19. a.** $T(a) = \dfrac{83}{a + 70} + \dfrac{37}{a + 65}$ or $T(a) = \dfrac{120a + 7985}{(a + 65)(a + 70)}$ **b.** 1.75, 1.53; the trip will take 1.75 hours at the speed limits and 1.53 hours at 10 mph over the speed limits. **c.** 0.22; the trip will take 0.22 hour less at 10 mph over the speed limits than it would at the speed limits. **d.** −66.61, 6.61; the trip will take 1.6 hours at 66.61 mph below the speed limits (model breakdown) and 1.6 hours at 6.61 mph over the speed limits. **21. a.** $T(a) = \dfrac{285a + 19{,}130}{(a + 70)(a + 65)}$ **b.** 3.66 hours **23. a.** Answers may vary. **b.** Answers may vary. **25. a.** $L(t) = 0.53t - 12.22$; $E(t) = 0.43(1.056)^t$; $Q(t) = 0.01t^2 - 0.47t + 6.10$; both the exponential and quadratic models fit the data well. The linear model does not. **b.** Answers may vary; well. **c.** 75 years **d.** 1.056; the percentage of Americans who have shingles increases by 5.6% with each year a person ages. **e.** 21.6% **27.** $(3x - 2)(5x + 2)(5x - 2)$; a cubic (or third-degree) polynomial in one variable **29.** $\pm\frac{2}{5}, \frac{2}{3}$; a cubic equation in one variable **31.** $3x^4 + 5x^3 - 15x^2 - 5x + 12$; a fourth-degree polynomial in one variable

Homework 12.7 **1.** \$960 **3.** 1.31 cups **5.** 2.4 ounces **7.** 36,200 students **9.** 6.75 inches **11.** 320.99 U.S. dollars **13.** 172 pounds **15. a.** \$409.50 **b.** Answers may vary; underestimate **17. a.** 5996 adults **b.** Answers may vary; overestimate **19.** no; answers may vary. **21. a.** 21.5:1; greater **b.** 339 FTE faculty; \$24.8 million **c.** 5766 students; \$19.3 million **d.** reduce FTE enrollment **23.** 4.88 inches **25.** 14.86 meters **27.** 5.63 feet **29.** Answers may vary. **31. a.** $y = kx$; $(0, 0)$ **b.** $C = 15t$; the variables t and C are proportional; answers may vary. **c.** $C = 10t + 25$; the variables t and C are not proportional; answers may vary. **d.** \$45; \$55; the variables t and C are not proportional; yes **33. a.** $\dfrac{n}{m}$ **b.** $\dfrac{q}{p}$ **c.** Answers may vary. **d.** Answers may vary. **35.** $-5, \frac{3}{2}$; quadratic equation in one variable **37.** $(2x - 3)(x + 5)$; quadratic (or second-degree) polynomial in one variable **39.** −21; quadratic function

Homework 12.8 **1.** $I = kt$ **3.** $w = \dfrac{k}{x + 4}$ **5.** w varies inversely as r. **7.** T varies directly as the square root of w. **9.** $c = 4u$ **11.** $w = \dfrac{12}{\sqrt{t}}$ **13.** 27 **15.** 6 **17.** ±4 **19.** $\dfrac{31}{7}$ **21.** It increases. **23.** It decreases. **25.** It increases. **27.** less **29.** Too many cooks spoil the broth. **31.** \$1344 **33.** 96 newtons **35.** 186.1 feet **37.** 3.5 meters **39. a.** $F = \frac{5}{12}w$ **b.** 62.5 pounds **c.** The variation constant should increase, because more force would be needed to move the sofa. **41. a.** $T = 0.000906d$ **b.** 4415 feet **c.** The time it takes to hear thunder increases by 0.000906 second for each 1 foot from the lightning strike. **d.** no; the number of seconds divided by 5 is approximately the number of miles to the strike. **43. a.** $f(d) = \dfrac{3200}{d^2}$ **b.** 128 pounds **c.** 56,569 miles **d.** 0.056 pound; yes; the astronaut would weigh much more than 0.056 pound due to the Moon's gravitational field. **e.** Answers may vary.

45. a. 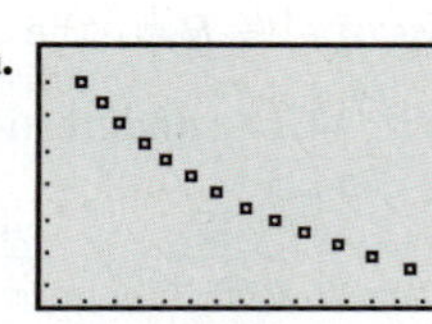 **b.** $F = \frac{2805.21}{L}$; yes **c.** The frequency varies inversely as the effective length. **d.** 370.1 hertz **e.** Answers may vary.

47. a. **b.** $f(d) = \frac{148.86}{d}$ **c.** The apparent height varies inversely as the distance. **d.** As the distance increases, the apparent height decreases. **e.** 1.5 inches **f.** 148.9 inches

49. a. $T = k\sqrt{L}$ **b.** $k = \frac{T}{\sqrt{L}}$ **c.** third column: 0.2236, 0.1992, 0.2272, 0.2236, 0.2260, 0.2193, 0.2236, 0.2259, 0.2169, 0.2145; 0.22; average all values in third column. **d.** $T = 0.22\sqrt{L}$ **e.** 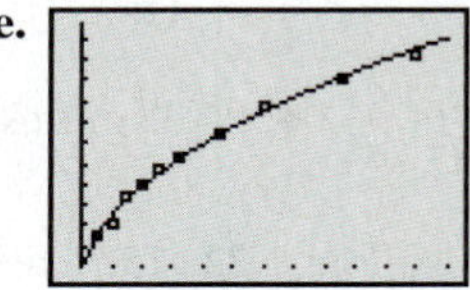 yes **f.** 2.51 seconds **51.** $f(L) = 5L; 5$

53. $f(n) = \frac{2}{n}; 2$ **55.** $f(r) = 2\pi r; 2\pi$ **57. a.** yes **b.** no **c.** Answers may vary. **59.** false **61.** false **63.** yes; $\frac{1}{k}$ **65.** The students defined the variables x and y differently; 1 mile is approximately equal to 1.61 kilometers; 1 kilometer is approximately equal to 0.62 mile. **67. a.** yes **b.** not necessarily; answers may vary. **69.** 62; the typist can type 62 words per minute. **71.** Answers may vary. **73.** Answers may vary. **75.** Answers may vary. **77.** Answers may vary.

Chapter 12 Review Exercises

1. $-3; \frac{7}{3}$ **2.** all real numbers except $\frac{5}{3}$ **3.** all real numbers except $\pm\frac{7}{2}$ **4.** all real numbers except 2 and 4 **5.** all real numbers except $-\frac{7}{3}$ and $\frac{5}{4}$ **6.** all real numbers except -2 and $\pm\frac{1}{3}$ **7.** $f(x) = \frac{3}{x-2}$ **8.** $f(x) = -\frac{x+4}{2x(x-4)}$ **9.** $f(x) = \frac{x-2}{3x+2}$ **10.** $f(x) = \frac{x+5}{(x-5)(2x-3)}$ **11.** $f(x) = \frac{1}{x^2-2x+4}$ **12.** $\frac{2a-5b}{a-b}$ **13.** $\frac{5}{4}$ **14.** $\frac{5b(b+1)}{7}$ **15.** $-\frac{2x(x+7)^2}{5(x+3)}$ **16.** $\frac{p+5t}{t}$ **17.** $\frac{x-1}{4(x+1)}$ **18.** $\frac{7}{(t-7)(3t-4)}$ **19.** $\frac{4x}{x+4}$ **20.** $\frac{(x+3)(2x-3)}{x-3}$ **21.** $\frac{2x^2+17x-20}{(x-4)(x+1)(x+6)}$ **22.** $-\frac{2}{x-2}$ **23.** $\frac{(x+5)(x-2)}{x(x+1)(x-1)(2x-5)}$ **24.** $\frac{2(x^2+x+4)}{(x-2)^2(x+2)}$ **25.** $-\frac{1}{4(x+1)}$ **26.** $-\frac{14}{(x-5)(x+3)}$ **27.** $-\frac{(x-3)(3x+8)}{2(x+5)(x-5)(x-2)}$ **28.** $\frac{2(m^2+4mn+10n^2)}{(m+6n)(m+2n)(m-5n)}$ **29.** $\frac{x-3}{x(x-5)}$ **30.** $-\frac{10(2x-1)}{(x+1)(3x+4)}$ **31.** $(f \cdot g)(x) = \frac{(x-2)(x+1)}{(x+2)^2}$ **32.** $\left(\frac{f}{g}\right)(x) = \frac{(x-2)(x+1)}{(x+3)^2}$ **33.** $(f+g)(x) = \frac{2x^2+5x+7}{(x+2)(x+3)}$ **34.** $(f-g)(x) = -\frac{7x+11}{(x+2)(x+3)}$ **35.** Answers may vary; $\frac{-5x+3}{x+4}$ **36.** 9.92 ounces **37.** 5.30 cups per day **38.** $\frac{4x}{3}$ **39.** $\frac{1}{(x-3)(x+2)}$ **40.** $\frac{5w-2}{w-3}$ **41.** $-\frac{4(x-2)}{3x^3}$ **42.** empty set **43.** $-8, 1$ **44.** $\frac{-5 \pm \sqrt{73}}{6}$ **45.** $\pm\sqrt{11}$ **46.** $-3 \pm 3\sqrt{3}$ **47.** $\frac{5 \pm i}{2}$ **48.** $-\frac{1}{2}$ **49.** $\frac{4x-23}{(x-6)(x-5)(x+5)}$ **50.** $-\frac{2(2x+1)}{x-2}$ **51.** $\frac{33}{2}$ **52.** $\left(\frac{5}{3}, 0\right)$ **53.** $r = \frac{S-a}{S}$ **54.** 24.5 ounces **55.** 27.5 yards **56.** H varies directly as the square of u. **57.** w varies inversely as the common logarithm of t. **58.** $y = \frac{2}{7}\sqrt{x}$ **59.** $B = \frac{72}{r^3}$ **60.** 4.14 inches **61. a.** $m = kr^3$ **b.** $k = \frac{m}{r^3}$ **c.** third column: 17.1, 17.01, 17.02, 16.99, 16.99, 16.99; 17.02; average all of the values in the third column. **d.** $m = 17.02r^3$ **e.** yes **f.** 207.1 grams **62. a.** $C(n) = 40n + 600$ **b.** $M(n) = \frac{40n + 600}{n}$ **c.** 42.22; when 270 people use the room, the mean cost per person is \$42.22. **d.** 60; when the mean cost per person is \$50, 60 people are using the room.

63. a. $(C + R)(t) = 0.19t + 184.17$; the inputs of $C + R$ are the number of years since 1970, and the outputs of $C + R$ are the total average annual per-person consumptions of chicken and red meat. **b.** $P(t) = \dfrac{113t + 3929}{0.19t + 184.17}$ **c.** 44.6%; 44.7%; underestimate **d.** 47.85; in 2017, about 47.9% of chicken and red meat consumed will be chicken. **e.** 2021 **64. a.** $T(a) = \dfrac{75}{a + 50} + \dfrac{40}{a + 65}$ **b.** 1.94; when the student drives 5 mph above the speed limits, the driving time is about 1.9 hours. **c.** 3.1 mph over the speed limits

Chapter 12 Test

1. all real numbers except $-\dfrac{5}{2}$ and $\dfrac{2}{3}$ **2.** all real numbers except ± 6 **3.** all real numbers **4.** Answers may vary. **5.** $f(x) = -\dfrac{3}{x - 3}$ **6.** $f(x) = \dfrac{3x + 1}{2x(3x - 1)}$ **7.** $\dfrac{x - 2}{9x^3}$ **8.** $\dfrac{p(p + 2t)}{(p + 3t)(p - t)}$ **9.** $\dfrac{-9x^2 - 4x + 24}{2x(x - 2)(x + 4)}$ **10.** $\dfrac{x^2 + 13x + 7}{(x + 3)(x - 3)(x + 8)}$ **11.** $\dfrac{15(x + 2)}{x(x^2 - 3x + 5)}$ **12.** $(f - g)(x) = \dfrac{6(2x - 1)}{(x - 5)(x + 4)}; \dfrac{3}{10}$ **13.** 1.33 grams **14.** $\dfrac{(5x + 2)(x - 1)}{x(3x - 7)}$ **15.** empty set **16.** 5 **17.** $2 \pm \sqrt{6}$ **18.** 0 **19.** undefined **20.** $-2, 5$ **21.** 7.84 gallons **22.** 15.75 meters **23.** $W = \dfrac{3}{49}t^2$ **24.** $y = \dfrac{40}{\sqrt{x}}$ **25. a.** $C(n) = 200n + 10{,}000$ **b.** $B(n) = \dfrac{200n + 10{,}000}{n}$ **c.** $P(n) = \dfrac{350n + 10{,}000}{n}$ **d.** 450; if the manufacturer makes and sells 100 bikes in a month, the price should be \$450 per bike to make a profit of \$150 per bike. **26. a.** $T(a) = \dfrac{400}{a + 70} + \dfrac{920}{a + 75}$ **b.** 16.83; when she drives 5 mph above the speed limits, the trip takes about 16.8 hours. **c.** $-71.58, 4.23$; when the trip takes 17 hours, she is driving 71.58 mph below the speed limits (model breakdown) or 4.23 mph above the speed limits. **27. a.** $g(L) = \dfrac{3200}{L^2}$ **b.** 88.89 hertz **c.** 4 cm **d.** decreasing; the longer the prongs, the lower is the frequency. **28. a.** $C(t) = -0.0027t^2 + 0.073t + 0.68$ **b.** $P(t) = 1000t + 9000$ **c.** $M(t) = \dfrac{1000t + 9000}{-0.0027t^2 + 0.073t + 0.68}$ **d.** 22,793 patients per center **e.** 2015

Chapter 13

Homework 13.1 **1.** $\sqrt[5]{x^2}$ **3.** $x^{3/4}$ **5.** $w^{1/2}$ **7.** $\sqrt[7]{(2x + 9)^3}$ **9.** $(3k + 2)^{4/7}$ **11.** $5\sqrt{2}$ **13.** x^4 **15.** $6x^3$ **17.** $ab^6\sqrt{5}$ **19.** $x^4\sqrt{x}$ **21.** $2x^2\sqrt{6x}$ **23.** $4xy^4\sqrt{5x}$ **25.** $10ab^2\sqrt{2ab}$ **27.** $(2x + 5)^4$ **29.** $(6t + 3)^2\sqrt{6t + 3}$ **31.** 3 **33.** x **35.** $2x$ **37.** $-2x^4$ **39.** $3a^3b^7$ **41.** $x^2\sqrt[6]{x^5}$ **43.** $-5a^5b^4\sqrt[3]{a^2}$ **45.** $2x^7y\sqrt[5]{2x^4y^2}$ **47.** $6xy$ **49.** $3x + 6$ **51.** $(4p + 7)^4$ **53.** $(2x + 9)^5\sqrt[6]{2x + 9}$ **55.** $\sqrt[4]{x^3}$ **57.** $\sqrt[3]{x^2}$ **59.** $\sqrt[6]{(2m + 7)^5}$ **61.** $x^2\sqrt[3]{x}$ **63.** $\sqrt{3}$ **65.** $\sqrt[12]{p}$ **67.** $\sqrt[5]{4x^4}$ **69.** $\sqrt[8]{ab}$ **71.** -2 **73.** 2 **75.** $-\sqrt[3]{19}$ **77.** 9 **79.** **81. a.** 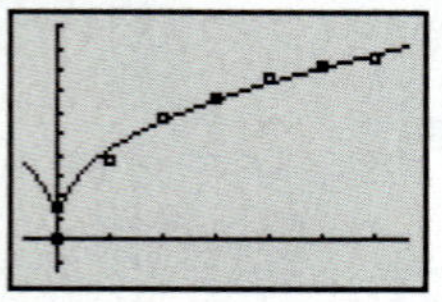 yes **b.** 30°F **c.** 129°F **d.** 18 minutes **83. a.** 195 meters per second **b.** 99, 140, 171; increasing; the greater the depth, the greater is the tsunami's speed. **c.** The speed decreases, and the height increases. **d.** 436 miles per hour **85.** Answers may vary. **87.** The exponent of x should be $\dfrac{3}{6}$, not $\dfrac{6}{3}$; $\sqrt{x}$ **89.** $\sqrt[n^2]{x}$ **91. a.** x **b.** x **c.** x **93. a.** $4x^2y^3$ **b.** $4x^2y^3$ **c.** They are the same. **95.** $-4 \pm \sqrt{26}$; a rational equation in one variable **97.** $-\dfrac{x^2 - 13x + 2}{(x + 3)^2(x - 2)}$; a rational expression in one variable **99.** all real numbers except -3 and 2; a rational function

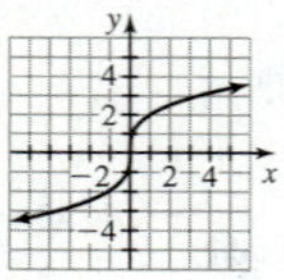

Homework 13.2 **1.** $9\sqrt{3}$ **3.** $6\sqrt{5}$ **5.** $-4\sqrt[3]{5x^2y}$ **7.** $10\sqrt{5a} - 4\sqrt{3b}$ **9.** $-4 - 2\sqrt[3]{x} + 2\sqrt{x}$ **11.** $3\sqrt[3]{x - 1} - 2\sqrt{x - 1}$ **13.** $2.6\sqrt[4]{x}$ **15.** $25 - 4\sqrt{x}$ **17.** 14 **19.** $3\sqrt{6}$ **21.** $7\sqrt{3b}$ **23.** $28\sqrt{3}$ **25.** $13x^2\sqrt{2x}$ **27.** $4x\sqrt{x}$ **29.** $7x$ **31.** $7ab\sqrt{3b}$ **33.** $x\sqrt[3]{x^2}$ **35.** $-x^2y^2\sqrt[4]{x^3}$ **37.** $\sqrt{10}$ **39.** $6x$ **41.** $10\sqrt{2}$ **43.** $-8x\sqrt{15}$ **45.** $\sqrt{5} + \sqrt{35}$ **47.** $14t - 2t\sqrt{14}$ **49.** $3 - 2\sqrt{5}$ **51.** $10x + 38\sqrt{x} + 24$ **53.** $8x - 4\sqrt{5x} + 2\sqrt{3x} - \sqrt{15}$ **55.** $5a - 9\sqrt{ab} - 2b$ **57.** $1 - w$ **59.** $49x^2 - 5$ **61.** $4a - b$ **63.** $23 + 8\sqrt{7}$ **65.** $36x + 60\sqrt{x} + 25$ **67.** $16x - 8\sqrt{5x} + 5$ **69.** $a + 4\sqrt{ab} + 4b$ **71.** $2x + 6\sqrt{2x - 5} + 4$ **73.** $\sqrt[10]{x^7}$

75. $x\sqrt[5]{x^2}$ **77.** $-5\sqrt[4]{2m^3} + 20\sqrt{m}$ **79.** $\sqrt[3]{x^2} + 2\sqrt[3]{x} + 1$ **81.** $\sqrt{k} - 2\sqrt[12]{k^7} + \sqrt[3]{k^2}$ **83.** $6\sqrt[6]{x^5} + 2\sqrt{x} - 18\sqrt[3]{x} - 6$

85. $9\sqrt{x} - 25$ **87. a.** 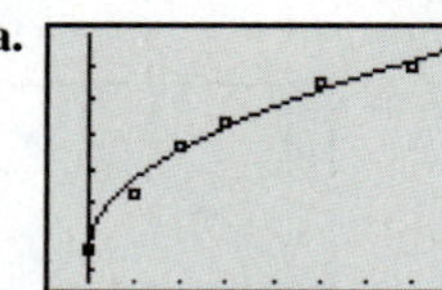 yes **b.** 97% **c.** 4 days **89. a.** increases; answers may vary. **b. i.** 75, 300, 675, 1200, 1875, all in gallons per minute **ii.** 1875 gallons per minute; 18 gallons per minute **iii.** 75 gallons per minute; 1.35% **c.** yes; 616 gallons per minute **91.** The middle term is missing; $x^2 + 2x\sqrt{7} + 7$ **93.** The radicand should not be multiplied by 7; $14\sqrt{3}$

95. $\sqrt[6]{x}$ **97. a.** $\sqrt[20]{x^9}$ **b.** $\sqrt[kn]{x^{k+n}}$ **c.** $\sqrt[20]{x^9}$; they are the same. **d.** $\sqrt[21]{x^{10}}$ **99. a. i.** true **ii.** false **iii.** true **iv.** false **v.** true **vi.** false **b.** Answers may vary. **101.** Answers may vary. **103.** $-2\sqrt{x}$ **105.** $-15x$ **107.** $\log_b\left(\frac{x-5}{x-8}\right)$; a logarithmic expression in one variable **109.** $\frac{20}{13}$; a logarithmic equation in one variable **111.** 0.7211; an exponential equation in one variable

Homework 13.3 **1.** $\frac{8\sqrt{x}}{x}$ **3.** $\frac{3\sqrt{5p}}{5p}$ **5.** $\frac{2\sqrt{2x}}{3x}$ **7.** $\frac{5\sqrt{2k}}{2k}$ **9.** $\frac{2\sqrt{x}}{x}$ **11.** $\frac{\sqrt{14}}{2}$ **13.** $\frac{\sqrt{2xy}}{x}$ **15.** $\frac{\sqrt{3xy}}{6y}$ **17.** $\frac{3\sqrt{x-4}}{x-4}$ **19.** $\frac{a\sqrt{6ab}}{3b}$ **21.** $\frac{2\sqrt[3]{25}}{5}$ **23.** $\frac{5\sqrt[3]{2}}{2}$ **25.** $\frac{4\sqrt[3]{x^2}}{5x}$ **27.** $\frac{3\sqrt[3]{4x}}{x}$ **29.** $\frac{7\sqrt[4]{4t}}{2}$ **31.** $\frac{\sqrt[6]{x^5}}{x}$ **33.** $\frac{\sqrt[5]{2x^2}}{x}$ **35.** $\frac{\sqrt{6x}}{3x}$ **37.** $\frac{\sqrt[5]{24wxy^3}}{2xy}$ **39.** $\frac{5-\sqrt{3}}{22}$ **41.** $\frac{\sqrt{7}-\sqrt{3}}{2}$ **43.** $\frac{3\sqrt{r}+7}{9r-49}$ **45.** $\frac{x+\sqrt{x}}{x-1}$ **47.** $\frac{12x+3\sqrt{5x}}{16x-5}$ **49.** $\frac{x+y\sqrt{x}}{x-y^2}$ **51.** $\frac{x-10\sqrt{x}+25}{x-25}$ **53.** $\frac{6x+13\sqrt{x}-5}{9x-1}$ **55.** $\frac{18x+6\sqrt{7x}+3\sqrt{5x}+\sqrt{35}}{9x-7}$ **57.** $\frac{x-2\sqrt{xy}+y}{x-y}$ **59.** $\sqrt{x+1}+\sqrt{x}$ **61. a.** $f(h) = \frac{\sqrt{6h}}{2}$ **b.** 47 miles **c.** 212 miles **63. a.** Answers may vary. **b.** $\frac{\sqrt[4]{8}}{2}$ meter; 0.841 meter **65.** student 1; answers may vary. **67.** Answers may vary; $\frac{5\sqrt[3]{x^2}}{x}$ **69.** $\frac{x}{3\sqrt{x}}$ **71.** $\frac{1}{\sqrt{x+2}+\sqrt{x}}$ **73.** $\frac{2x-7\sqrt{x}+3}{4x-1}$ **75.** $3\sqrt{10}$ **77.** Answers may vary. **79. a.** $(A+B)(A^2-AB+B^2)$ **b.** A^3+B^3; answers may vary. **c.** x^3+8; answers may vary. **d.** $x+2$; answers may vary. **e.** $\frac{\sqrt[3]{x^2}-\sqrt[3]{2x}+\sqrt[3]{4}}{x+2}$ **81.** $15x^3-22x^2+3x+4$; a cubic (or third-degree) polynomial in one variable **83.** $24(x-5)(x^2+5x+25)$; a cubic (or third-degree) polynomial in one variable **85.** $\frac{2\pm\sqrt{14}}{5}$; a quadratic equation in one variable

Homework 13.4 **1.** 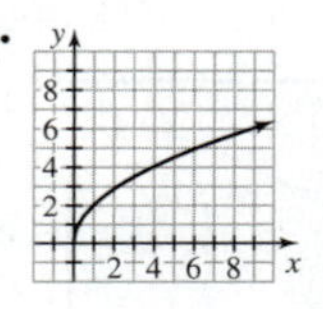 **3.** **5.** 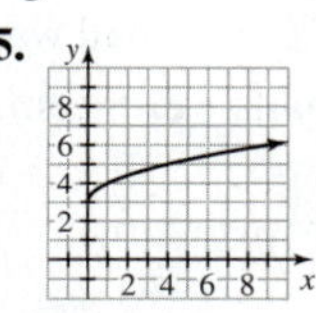 **7.** **9.** 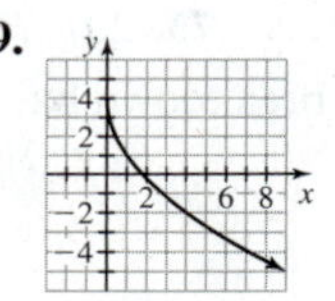

11. 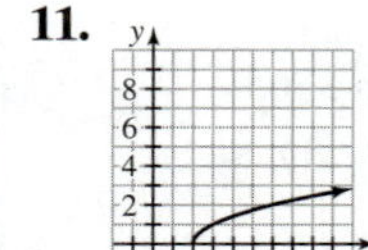 **13.** 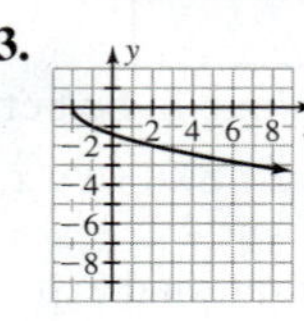 **15.** **17.** 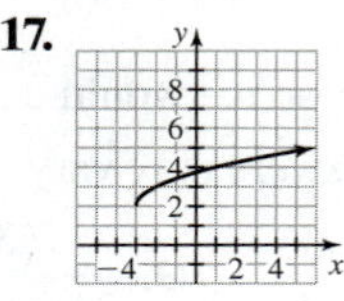 **19.** 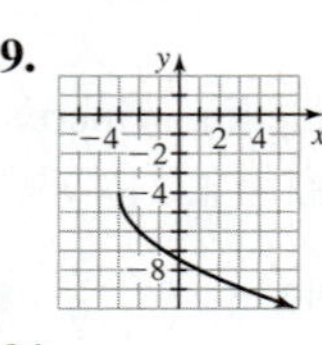 **21.** 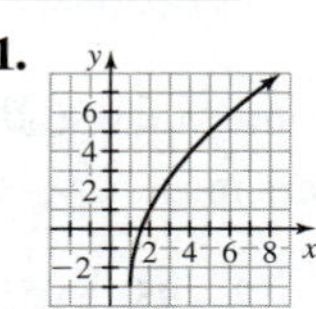

23. 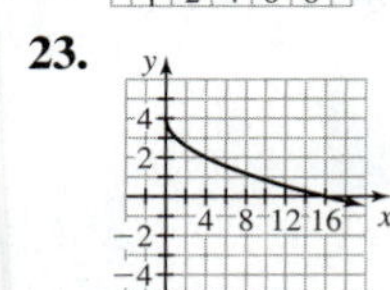 **25.** 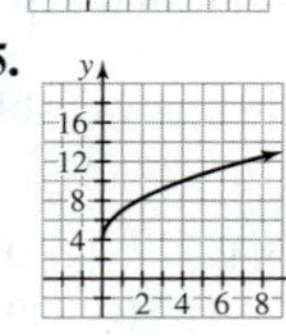 **27.** domain: $x \geq 0$; range: $y \leq 0$ **29.** 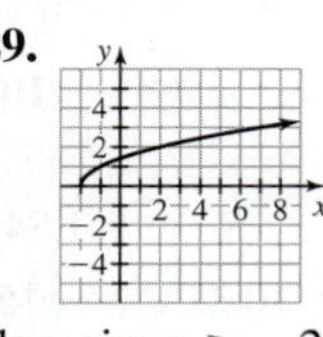 domain: $x \geq -2$; range: $y \geq 0$ **31.** 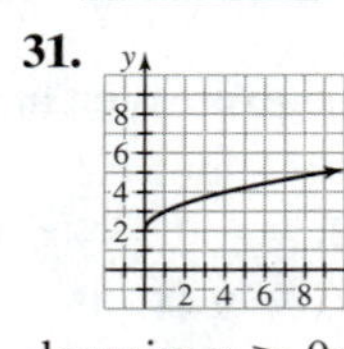 domain: $x \geq 0$; range: $y \geq 2$ **33.** 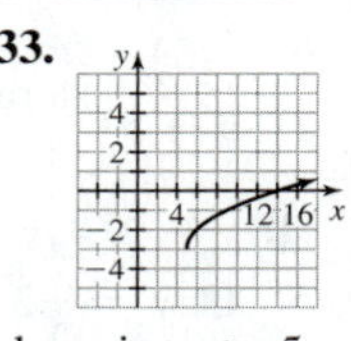 domain: $x \geq 5$; range: $y \geq -3$

35. domain: $x \geq -5$; range: $y \geq 1$ **37.** 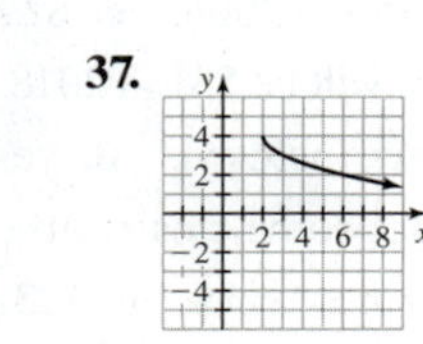 domain: $x \geq 2$; range: $y \leq 4$ **39. a.** Answers may vary. **b.** 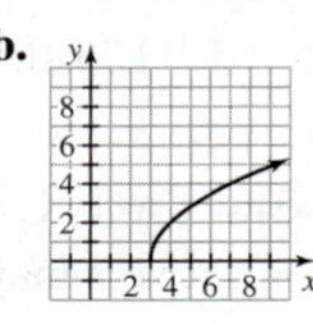 **c.** For each input–output pair, the output variable is equal to 2 times the square root of 3 less than the input variable.

41. 11 **43.** $21\sqrt{c} - 3$ **45.** $(f + g)(x) = 9\sqrt{x} - 8$ **47.** $(f \cdot g)(x) = 20x - 31\sqrt{x} - 9$ **49.** $(f - g)(x) = -6\sqrt{5}$

51. $\dfrac{f}{g}(x) = \dfrac{4x - 12\sqrt{5x} + 45}{4x - 45}$ **53.** $(f + g)(x) = 2\sqrt{x + 1}$ **55.** $(f \cdot g)(x) = x - 3$ **57. a.** 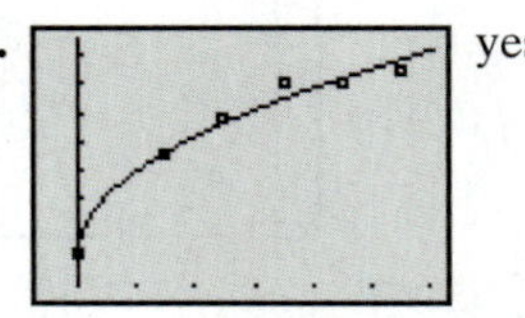 yes

b. 71%; interpolation **c.** 97%; extrapolation **59.** 0 **61.** 2.4 **63.** −6 **65.** 3 **67. a.** $a < 0, h = 0, k > 0$ **b.** $a > 0, h < 0, k < 0$ **c.** $a > 0, h < 0, k > 0$ **d.** $a < 0, h > 0, k = 0$ **69.** Answers may vary. **71.** If $a < 0$, f has a maximum point, (h, k). If $a > 0$, f has a minimum point, (h, k).

73. 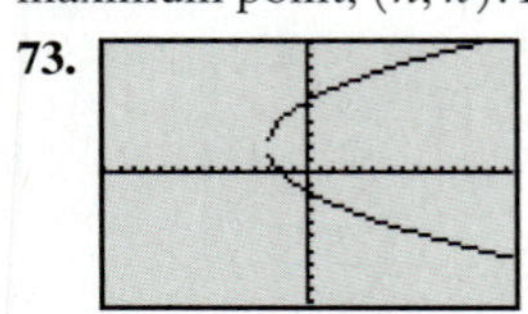 It is a parabola; no **75.** Answers may vary. **77.** Answers may vary. **79.** 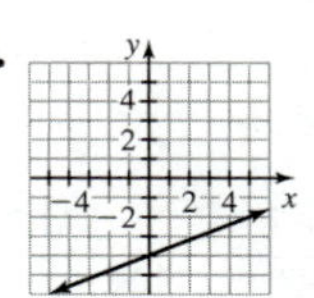

81. 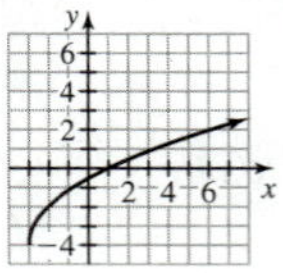 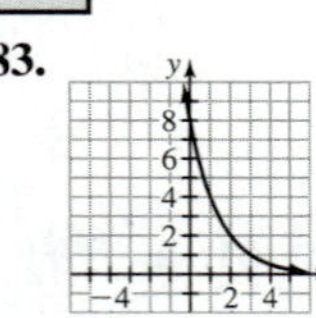 **83.** **85.** $(f \circ g)(x) = \sqrt{12x + 17} - 7$ **87.** $(f \circ g)(x) = \sqrt{x^2 + 5}$ **89.** $(3x + 2)(2x - 3)$; a quadratic (or second-degree) polynomial in one variable

91. $\dfrac{1 \pm \sqrt{7}}{3}$; a quadratic function **93.** $y = 2x^2 - 3x + 5$; a quadratic function

Homework 13.5 **1.** 25 **3.** empty set **5.** −8 **7.** 5 **9.** 14 **11.** 4 **13.** empty set **15.** $\dfrac{69}{2}$ **17.** $\dfrac{97}{6}$ **19.** 5 **21.** 5 **23.** $-\dfrac{1}{6}$

25. 11 **27.** 3 **29.** empty set **31.** 8 **33.** 4 **35.** $3 + 2\sqrt{2}$ **37.** 4 **39.** 1 **41.** 121 **43.** $\dfrac{3 \pm 3\sqrt{5}}{2}$ **45.** 10.31 **47.** 2.31

49. $d = \dfrac{S^2}{g}$ **51.** $h = \dfrac{2d^2}{3}$ **53.** $R = \dfrac{2GM}{v^2}$ **55.** 2.06 **57.** −0.74, 4.97 **59.** −1.6, 3.8 **61.** −4 **63.** (−3.2, 1.4) **65.** (−7, 0) **67.** (5, 0)

69. no x-intercepts **71.** 4 **73.** 20 **75. a.** 61.8; in 2003, 62% of e-mail was spam. **b.** 13.16; in 2012, 95% of e-mail was spam. **c.** 2014; model breakdown has occurred. **d.** 66 e-mails **e.** 8 cents **77. a.** 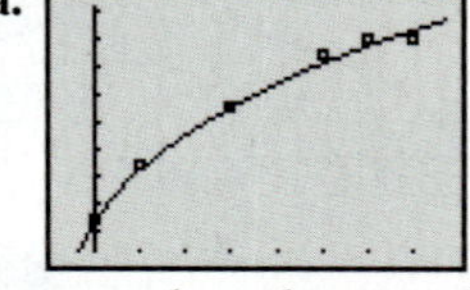 yes **b.** \$338; underestimate **c.** 4th grade

d. The table seems to indicate that the function is increasing, so we would expect the per-student charge to be between \$365 and \$410. **79.** Answers may vary; −2 **81.** (4, 2) **83.** Answers may vary. **85.** Answers may vary. **87.** $-4\sqrt{x} + 5$ **89.** 9

91. 0 **93.** $p + 4\sqrt{p} + 3$ **95.** 5.0553 **97.** 6 **99.** $\dfrac{5}{7}$ **101.** $\dfrac{33}{5}$ **103.** $\dfrac{(3x + 5)(2x + 3)}{3(x + 1)(x - 1)(x + 2)}$; a rational expression in one variable **105.** $\dfrac{3(3b - 10)}{(b - 2)(b - 5)}$; a rational expression in one variable **107.** 6; a rational equation in one variable

Homework 13.6 **1.** $y = \sqrt{x} + 3$ **3.** $y = 1.33\sqrt{x} + 2$ **5.** $y = 1.34\sqrt{x} + 4$ **7.** $y = -4.04\sqrt{x} + 9$ **9.** $y = \sqrt{x} + 1$ **11.** $y = 3.15\sqrt{x} - 0.45$ **13.** $y = -2.43\sqrt{x} + 9.44$ **15.** $y = 10.22\sqrt{x} - 15.86$ **17.** $y = -48.40\sqrt{x} + 159.06$ **19.** $y = 7.34\sqrt{x} - 25.43$ **21. a.** $f(t) = 6.6\sqrt{t} + 10$ **b.** (0, 10); there were 10 thousand American female troops in 2004. **c.** 2011 **d.** 14 thousand American female troops **23. a.** $f(t) = 7.3\sqrt{t} + 56$ **b.** (0, 56); the average monthly bill for pay-TV was \$56 in 2006. **c.** 82.32; the average monthly bill for pay-TV will be about \$82.32 in 2019. **d.** 11.73; the average monthly bill for pay-TV will be \$81 in 2018. **e.** \$97.1 billion **25. a.** 420 meters per second **b.** no; answers may vary. **c.** between 4205 meters and 4500 meters, inclusive **d.** yes **27. a.** $S(h) = 0.27\sqrt{h}$ **b. i.** S fits the data points quite well; Q fits the data fairly well; L does not come close to the data points. **ii.** 0, 0.327, 0.165; S **iii.** Q; answers may vary. **iv.** S; answers may vary. **v.** $T = 0.25\sqrt{h}$; answers may vary. **c.** 123.46 feet **d.** 9.55 seconds **29. a.** $f(n) = 31.92\sqrt{n} + 9.15$ **b.** 93.60; 93.6% of 7th births occurred despite contraception. **c.** 8.10; all 8th births occurred despite contraception; model breakdown has likely occurred. **d.** The higher the birth order, the higher is the percentage of births that occurred despite contraception; answers may vary.

31. a. $f(t) = 42.0\sqrt{t} + 8$ **b.** $(0, 8)$; there were 8 million acres of genetically modified crops in 1997. **c.** 191 million acres **d.** 2018

33.

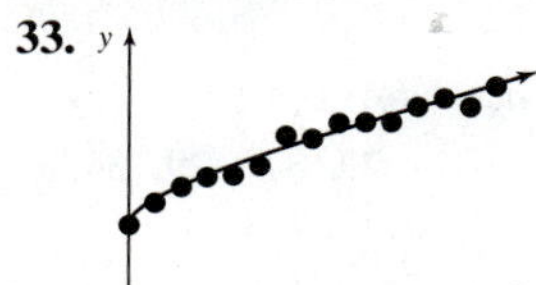

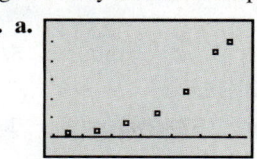

Increase b. **35.** Answers may vary. **37. a.** exponential and quadratic **b.** exponential: $f(t) = 1.09(1.37)^t$; quadratic: $f(t) = 5.17t^2 - 105.43t + 556$ **c.** Both fit the data well. **d.** exponential **e.** exponential: 2017; quadratic: 2034; answers may vary. **f. i.** 33.09; About 33 communities installed red-light cameras in 2005. **ii.** 1446.65; About 1447 communities will install red-light cameras in 2017. **iii.** \$1137 million (\$1.137 billion); we have assumed that the price of a red-light camera in 2017 will be equal to the price of one in 2005 and that the average number of red-light cameras sold per community in 2005 is equal to the average number of red-light cameras sold per community in 2017. **39.** Answers may vary. **41.** Answers may vary. **43.** Answers may vary. **45.** Answers may vary.

Chapter 13 Review Exercises

1. $\sqrt[7]{x^3}$ **2.** $(3k + 4)^{7/5}$ **3.** $2x^3\sqrt{2}$ **4.** $3x^3y^5\sqrt{2x}$ **5.** $\sqrt[4]{x^3}$ **6.** $2x^3y^8\sqrt[3]{3x}$ **7.** $(6x + 11)^5\sqrt[5]{(6x + 11)^2}$ **8.** $-6\sqrt{5} - 3\sqrt{7}$ **9.** $11\sqrt{5x}$ **10.** $-7x\sqrt{3}$ **11.** $3ab\sqrt[3]{2a^2b}$ **12.** $28\sqrt{x} - 7\sqrt[3]{x}$ **13.** $3x - 21\sqrt{x}$ **14.** $10\sqrt{21} + 14$ **15.** $t^2 + t\sqrt{5} + t\sqrt{3} + \sqrt{15}$ **16.** $8x - 2\sqrt{x} - 3$ **17.** $10a - 3\sqrt{ab} - b$ **18.** $b^2 - 3$ **19.** $25a - 49b$ **20.** $16x + 24\sqrt{x} + 9$ **21.** $-24\sqrt{10} + 77$ **22.** $4\sqrt[3]{x^2} - 20\sqrt[3]{x} + 25$ **23.** $\sqrt[20]{x^9}$ **24.** $\sqrt[18]{x}$ **25.** $\sqrt[12]{x}$ **26.** $\dfrac{\sqrt{3x}}{x}$ **27.** $5\sqrt[3]{t^2}$ **28.** $\dfrac{\sqrt[5]{63x^3y}}{3x}$ **29.** $\dfrac{15 - 5\sqrt{x}}{9 - x}$ **30.** $\dfrac{a + 2\sqrt{ab}}{a - 4b}$

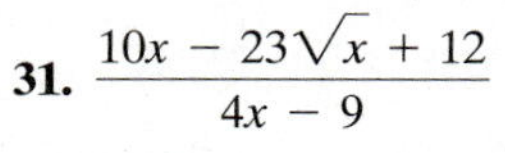

31. $\dfrac{10x - 23\sqrt{x} + 12}{4x - 9}$ **32.**

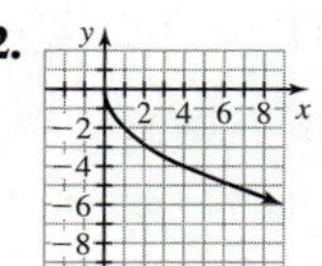

33.

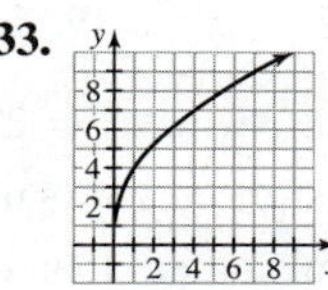

34.

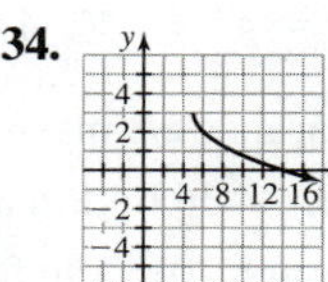

35.

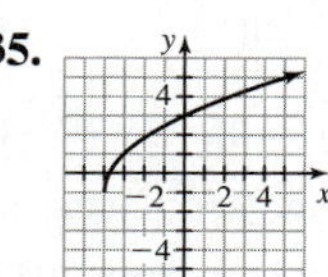

36. $(f + g)(x) = -\sqrt{x} + 7$ **37.** $(f - g)(x) = 7\sqrt{x} + 3$ **38.** $(f \cdot g)(x) = -12x - 14\sqrt{x} + 10$ **39.** $\dfrac{f}{g}(x) = \dfrac{6x + 13\sqrt{x} + 5}{2 - 8x}$ **40.** empty set solution **41.** 64 **42.** 4 **43.** 2, 4 **44.** 9 **45.** 2 **46.** 13 **47.** 7 **48.** 2.52 **49.** −1.36, 4.56 **50.** $(4, 0)$ **51.** $(7, 0)$ **52.**

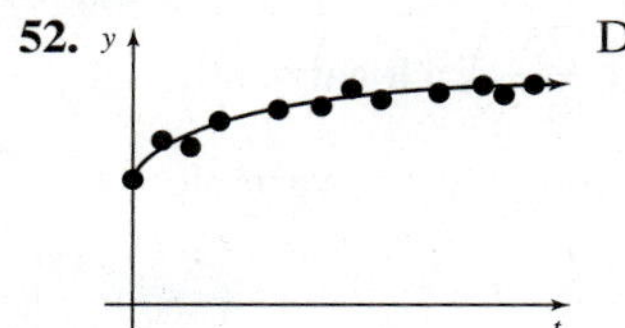

Decrease a, increase b. **53.** $y = 2.5\sqrt{x} + 3$ **54.** $y = -1.33\sqrt{x} + 7$ **55.** $y = -5.95\sqrt{x} + 17.31$ **56.** $y = 3.15\sqrt{x} + 0.55$ **57. a.** $f(t) = 1.67\sqrt{t} + 3.8$ **b.** $(0, 3.8)$; in 1992, the average credit card debt per household was \$3.8 thousand. **c.** 12.32; in 2018, the average credit card debt per household will be \$12.3 thousand. **d.** 30.35; the average credit card debt per household will be \$13 thousand in 2022.

Chapter 13 Test

1. $4x^4y^6\sqrt{2x}$ **2.** $4x^7y^4\sqrt[3]{xy^2}$ **3.** $(2x + 8)^6\sqrt[4]{(2x + 8)^3}$ **4.** $\dfrac{2\sqrt[15]{x^2}}{3}$ **5.** $\dfrac{2x + 5\sqrt{x} + 3}{4x - 9}$ **6.** $-x\sqrt{3x}$ **7.** $18x - 15\sqrt{x}$ **8.** $-20x + 2\sqrt{x} + 6$ **9.** $9a - 25b$ **10.** $16\sqrt[5]{x^2} - 24\sqrt[5]{x} + 9$ **11.** Answers may vary. **12.**

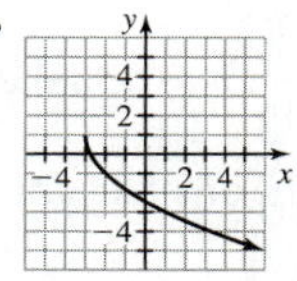

13. a. $a < 0$ and $k \geq 0$, or $a > 0$ and $k \leq 0$ **b.** $\left(\dfrac{k^2 + a^2h}{a^2}, 0\right)$ **14.** $(f + g)(x) = 2\sqrt{x} + 11$ **15.** $(f - g)(x) = -8\sqrt{x} + 3$ **16.** $(f \cdot g)(x) = -15x + 23\sqrt{x} + 28$ **17.** $\left(\dfrac{f}{g}\right)(x) = \dfrac{15x - 47\sqrt{x} + 28}{-25x + 16}$ **18.** 25 **19.** 17 **20.** $\dfrac{144}{25}$ **21.** −6 **22.** 3 **23.** 0.9, 3.3 **24.** −3 **25.** $(4, 0)$ **26.**

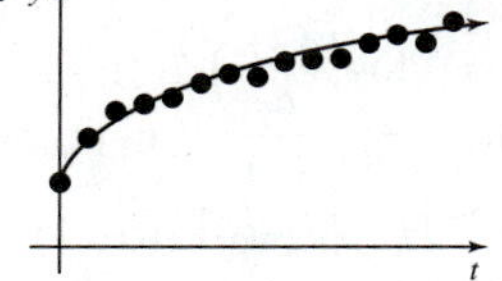

Increase a, decrease b. **27.** $y = 2.43\sqrt{x} + 0.56$

28. a. $f(t) = 2.90\sqrt{t} + 20.5$ **b.** 45.1 inches **c.** 29 months **d.** $(0, 20.5)$; the median height of boys at birth is 20.5 inches.

Chapter 14

Homework 14.1 **1.** arithmetic; $d = 8$ **3.** not arithmetic **5.** arithmetic; $d = 7$ **7.** not arithmetic **9.** $a_n = 6n - 1$ **11.** $a_n = -11n + 7$ **13.** $a_n = -6n + 106$ **15.** $a_n = 2n - 1$ **17.** 113 **19.** −196 **21.** 156.1 **23.** 400 **25.** 107 **27.** 87 **29.** 313 **31.** 3571 **33.** 98 **35.** no; answers may vary. **37.** $f(n) = 9n - 1$ **39. a.** $a_n = 800n + 26{,}700$ **b.** \$44,300 **c.** 30th year **41. a.** $a_n = \frac{1}{6}n + 35$ **b.** 35.17, 35.33, 35.50, 35.67; they represent the number of hours the instructor would work if she had 1, 2, 3, or 4 students, respectively. **c.** 56.67 hours per week **d.** 150 students **43. a.** $a_n = 1.8n - 50$ **b.** 170 people **c.** \$247 **d.** integers between 0 and 27, inclusive **45. a.** $f(t) = 0.98t + 12.11$ **b.** 24.85, 25.83, 26.81, 27.79, 28.77; from 2013 through 2017, the pharmaceutical industry spent (in millions of dollars) about 24.9, 25.8, 26.8, 27.8, 28.8, respectively, on government and politics. **c.** 2018 **47. a.** $a_n = 0.2n + 0.7$ **b.** \$3.30 **c.** no **d.** \$16.70 **49.** 500 **51.** yes; answers may vary. **53.** no; answers may vary. **55.** no; answers may vary. **57. a.** 2 **b.** 2 **c.** They are the same; answers may vary. **59.** 6; a radical equation in one variable **61.** [graph] a radical function **63.** $12x - 23\sqrt{x} + 10$; a radical expression in one variable

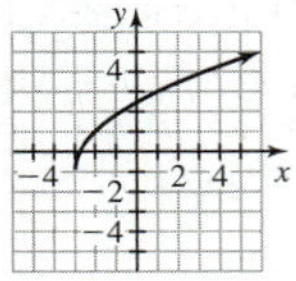

Homework 14.2 **1.** geometric; $r = 7$ **3.** arithmetic; $d = -7$ **5.** neither **7.** geometric; $r = \frac{1}{5}$ **9.** $a_n = 3(2)^{n-1}$ **11.** $a_n = 800\left(\frac{1}{4}\right)^{n-1}$ **13.** $a_n = 100\left(\frac{1}{2}\right)^{n-1}$ **15.** $a_n = 4^{n-1}$ **17.** 4.6566×10^{23} **19.** 1.1921×10^{-6} **21.** 3.3554×10^{7} **23.** 10 **25.** 12 **27.** 19 **29.** 16 **31.** $f(n) = 8(3)^{n-1}$ **33.** no; answers may vary. **35. a.** $a_n = 27{,}000(1.04)^{n-1}$ **b.** \$38,429.42 **c.** 17th year **37. a.** 2, 4, 8, 16, 32 **b.** $a_n = 2^n$ **c.** 256 ancestors **d.** 34.36 billion ancestors; answers may vary. **39. a.** $f(t) = 26.81(1.144)^t$ **b.** 117.76, 134.71, 154.11, 176.30, 201.69; from 2011 through 2015, the federal student loan origination volumes (in billions of dollars) will be about 117.8, 134.7, 154.1, 176.3, 201.7, respectively. **c.** 2017 **41. a.** $a_n = 5(3)^{n-1}$ **b.** 405 students **c.** 295,245 students; yes; answers may vary. **d.** Answers may vary. **43.** geometric **45.** arithmetic **47.** The sequence is arithmetic, not geometric; 66 **49. a.** 2 **b.** 2 **c.** They are the same; answers may vary. **51.** $a_n = 5n + 9$ **53.** $a_n = 448\left(\frac{1}{2}\right)^{n-1}$ **55.** 781,250 **57.** −473 **59.** 122 **61.** 11 **63.** 1.9372; an exponential equation in one variable **65.** [graph] an exponential function **67.** $\log_b\left(\frac{25}{8x^{15}}\right)$; a logarithmic expression in one variable

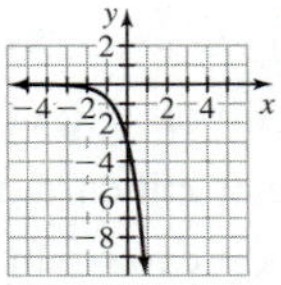

Homework 14.3 **1.** 20,205 **3.** 30,294 **5.** −38,232 **7.** 21,978 **9.** −10,807 **11.** 468 **13.** 77,875 **15.** 36,288 **17.** −151,468 **19.** 22,468 **21.** 5187 **23.** 0 **25.** 19,436 **27.** 50,005,000 **29. a.** \$58,200 **b.** \$1,213,800 **31.** A; \$8000 **33. a.** 136 seats **b.** 2340 seats **35. a.** \$12.1 million **b.** \$28.8 million **c.** \$368.1 million **37. a.** \$1,034,800 **b.** \$390,000 **c.** \$39,800; years 1 through 13; years 14 through 26 **d.** \$144,559 **39.** positive **41.** positive **43.** yes; answers may vary. **45.** 232 **47.** 1800 **49.** $\frac{2(x^2 - 2x + 17)}{(x+3)(x-3)(x-5)}$; a rational expression in one variable **51.** $\frac{1}{(x-3)^2}$; a rational expression in one variable **53.** 13; a rational equation in one variable

Homework 14.4 **1.** 40,955 **3.** 445.9617 **5.** 61.4266 **7.** 14.9804 **9.** 610,351,562 **11.** 857.1413 **13.** 8.8439 **15.** 89,478,485 **17.** 103.9945 **19.** 19,995.1172 **21.** 100 **23.** 485.9973 **25.** \$699,784.85 **27.** A; \$252,572.15 **29.** 2046 ancestors **31. a.** 11th round; approximately \$6.14 billion **b.** There will be 10 full rounds and part of an 11th round. **c.** nine people; the entrepreneur will get approximately \$6.14 billion; the other eight people will get an average of \$3.61 billion. **33. a.** $f(t) = 927.46(1.35)^t$ **b.** 1252.07; 1252 Nevaehs were born in 2001. **c.** 152,394.49; 152,394 Nevaehs will be born in 2017. **d.** 584,229.49; 584,229 Nevaehs will be born from 2001 to 2017, inclusive. **35.** positive **37.** arithmetic **39. a.** 5115 **b.** $n = \frac{\log\left(\frac{a_n r}{a_1}\right)}{\log(r)}$ **c.** $S_n = \frac{a_1\left(1 - r^{\log(a_n r/a_1)/\log(r)}\right)}{1 - r}$ **d.** 5115 **e.** Answers may vary. **41.** 10,443 **43.** 68.6189 **45.** Answers may vary. **47.** Answers may vary. **49.** Answers may vary. **51.** Answers may vary. **53.** Answers may vary.

Chapter 14 Review Exercises

1. geometric sequence **2.** arithmetic series **3.** arithmetic sequence **4.** geometric series **5.** geometric series **6.** none of these **7.** $a_n = 2(3)^{n-1}$ **8.** $a_n = 3n + 22$ **9.** $a_n = -5n + 14$ **10.** $a_n = 200\left(\frac{1}{2}\right)^{n-1}$ **11.** $a_n = 2.7n + 0.5$ **12.** $a_n = 800(0.7)^{n-1}$ **13.** 4.2221×10^{14} **14.** 1.1719×10^{-2} **15.** −204 **16.** 225.9 **17.** 505 **18.** 77 **19.** 24 **20.** −204 **21.** −3182 **22.** 671,173.0723 **23.** 3,221,225,469 **24.** 120,540 **25.** −1749 **26.** 797,161 **27.** geometric series **28.** arithmetic sequence **29. a.** \$71,772.52; \$70,000 **b.** \$1,166,085.43; \$1,300,000 **c.** Answers may vary. **30. a.** $f(t) = 2.37t + 24.65$ **b.** 2.37; the spending on pets in the United States increased by \$2.37 billion per year. **c.** \$67.3 billion **d.** \$873.6 billion

Chapter 14 Test

1. geometric sequence **2.** none **3.** geometric series **4.** arithmetic series **5.** $a_n = -6n + 37$ **6.** $a_n = 6(4)^{n-1}$ **7.** 262 **8.** 0.1875 **9.** 455 **10.** 9 **11.** 40.5000 **12.** 4.2950×10^9 **13.** −462 **14.** 1.1248×10^6 **15.** 2,098,620 **16.** none **17.** negative **18. a.** $f(t) = 13.13t + 17.81$ **b.** 30.94; in 2001, online retail sales were about \$31 billion. **c.** 254.15; in 2018, online retail sales will be about \$254 billion. **d.** 2565.81; total online retail sales from 2001 through 2018 will be about \$2566 billion, or about \$2.6 trillion. **19. a.** $a_n = 32(1.03)^{n-1}$ **b.** 9th year **c.** \$65,049.41 **d.** \$1,166,696.46

Cumulative Review of Chapters 1–14

1. $\pm\frac{7}{2}$ **2.** $-\frac{5}{2}, \frac{1}{3}$ **3.** 22 **4.** $\frac{1 \pm \sqrt{69}}{2}$ **5.** $-1, \frac{5}{2}$ **6.** $\frac{2 \pm \sqrt{2}}{3}$ **7.** 2 **8.** $-\frac{43}{18}$ **9.** 3 **10.** 2, 10 **11.** 3 **12.** 1.6013 **13.** 2.0492 **14.** 2.9755 **15.** 3.9927 **16.** $\frac{5 \pm \sqrt{13}}{6}$ **17.** $\frac{2 \pm i\sqrt{2}}{2}$ **18.** $y = \frac{ab - bx}{a}$ **19.** $(2, -1)$ **20.** $(-2, 3)$ **21.** $(2, -1, 3)$ **22.** $x \le 7; (-\infty, 7]$;

23.

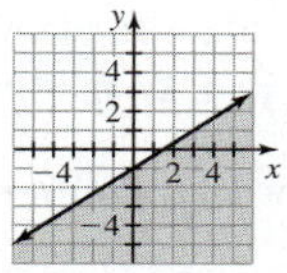

24.

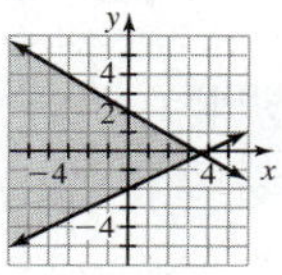

25. $\frac{2916}{b^{18}c^8}$ **26.** $\frac{4b^{24}c^{18}}{9}$ **27.** $\frac{4c}{5b^{1/4}}$ **28.** 0 **29.** $(5x - 7)^5\sqrt[4]{5x - 7}$ **30.** $2x^3y^7\sqrt{3x}$ **31.** $\frac{\sqrt[3]{4x^2}}{x}$ **32.** $\frac{6x - 5\sqrt{xy} + y}{4x - y}$ **33.** $\ln(x^{35})$ **34.** $\log_b\left(\frac{x^{15}}{32}\right)$ **35.** $9a^2 - 30ab + 25b^2$ **36.** $6k + 13\sqrt{k} - 28$ **37.** $2x^4 + 3x^3 - x^2 + 7x - 3$ **38.** $-2x^6 - 2x^4 - 10x^3 - 10x$ **39.** $\frac{(2x + 1)(x - 3)}{2x(x - 1)}$ **40.** $\frac{2x^2 - 3x + 10}{(x - 5)^2(x - 2)}$ **41.** $\frac{7x}{2(x - 4)(x + 2)}$ **42.** $\frac{2x^2 + 9x + 15}{(x - 3)(x + 1)(x + 3)(x + 9)}$ **43.** $\frac{3}{(x - 8)(x + 2)}$ **44.** $f(x) = -3x^2 - 18x - 34$ **45.** $(x - 2)(2x - 5)(2x + 5)$ **46.** $2x(x - 5)(x + 3)$ **47.** $2(3w - 5y)(w + 2y)$ **48.** $(10p + 1)(10p - 1)$ **49.** 3 **50.** 0, 2 **51.** $f(x) = -x^2 + 2x + 3$ **52.** all real numbers **53.** $y \le 4$

54.

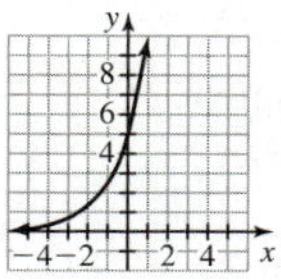

55.

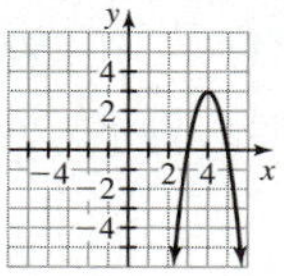

56.

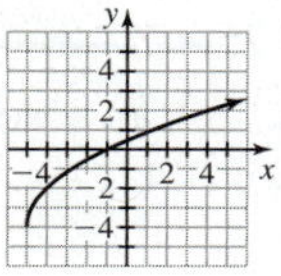

57.

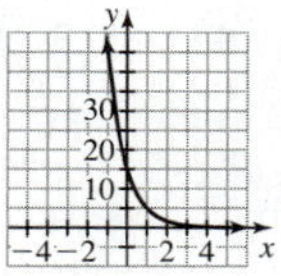

58.

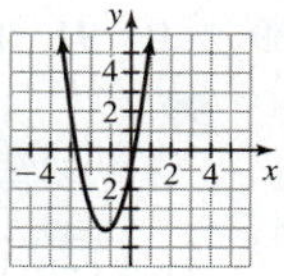

59.

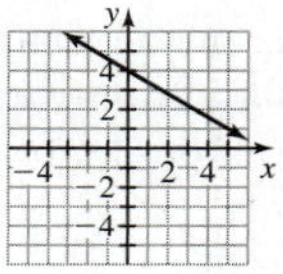

60.

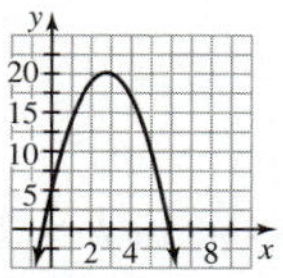

61. $y = -\frac{7}{5}x - \frac{11}{5}$ **62.** $y = 752.08(0.50)^x$ **63.** $y = 2x^2 - 5x + 3$ **64.** $y = 11.59\sqrt{x} - 11.39$

65. a. $f(x) = 2x + 2; g(x) = 2(2)^x; h(x) = 2x^2 + 2$; answers may vary for h. **b.**

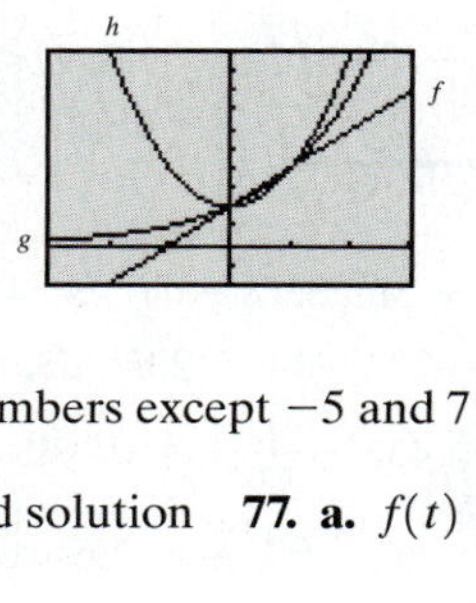

66. 4 **67.** $\frac{1}{2}$ **68.** −3 **69.** $g^{-1}(x) = 2^x$ **70.** $f^{-1}(x) = -\frac{1}{4}x - \frac{7}{4}$ **71.** all real numbers except −5 and 7 **72.** 524,288 **73.** 65 **74.** 196,605 **75.** 5597 **76.** 2 liters of the 15% acid solution, 4 liters of the 30% acid solution **77. a.** $f(t) = 29.38t + 532.59$

b. $f^{-1}(n) = 0.034n - 18.13$ **c.** 1061.4; there will be about 1061 thousand slot machines and video poker machines in 2018. **d.** 19.27; there will be 1100 thousand (1.1 million) slot machines and video poker machines in 2019. **e.** 29.38; the number of slot machines and video poker machines is increasing by 29.38 thousand machines per year. **78. a.** $L(t) = 0.0169t + 0.687$ **b.** $E(t) = 0.687(1.0184)^t$ **c.** 1.870; 2.462; in 2050, India's population will be 1.870 billion according to the linear model and 2.462 billion according to the exponential model. **d.** 0.592; the difference in India's population in 2050 between the models is 592 million, which is greater than the predicted U.S. population of 439 million for 2050. **e.** linear: 2024; exponential: 2020; answers may vary. **79. a.** exponential: $f(t) = 0.26(1.18)^t$; quadratic: $f(t) = 0.0179t^2 - 0.114t + 0.73$; both models fit the data well. **b.** exponential **c.** 2018 **d.** 2019 **e.** Answers may vary. **80. a.** $B(t) = -4.04t + 145.85$ **b.** $R(t) = 102.36t + 1849.25$ **c.** $P(t) = \dfrac{-404t + 14{,}585}{102.36t + 1849.25}$ **d.**

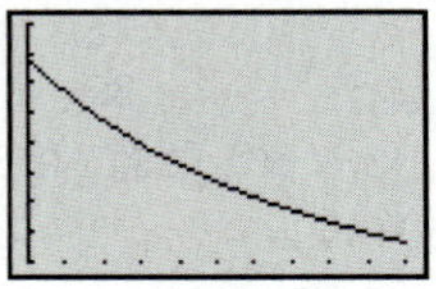

decreasing; between 2000 and 2020, the percentage of total recreational expenditures that consists of book purchases is decreasing. **e.** 2018

Chapter 15

Homework 15.1

1. ± 7 **3.** empty set **5.** $\pm\frac{18}{5}$ **7.** $-7, 3$ **9.** 5 **11.** $-\frac{10}{3}, 4$ **13.** empty set **15.** ± 2 **17.** $-1, -9$ **19.** 2, 3 **21.** $\frac{3}{7}, 7$ **23.** $-1, \frac{1}{3}$ **25.** $-\frac{13}{4}, \frac{7}{4}$ **27.** $1, \frac{17}{3}$ **29.** $-\frac{2}{27}, \frac{14}{3}$ **31.** ± 2.70 **33.** $-12.24, 6.71$ **35.** $-5.5, 4.5$ **37.** $-6.67, -2.67$ **39.** ± 4 **41.** $-3, 2$ **43.** -1 **45.** ± 3 **47.** -4 **49.** $-\frac{13}{4}, -\frac{1}{4}$ **51.** $-4 < x < 4$; (number line: open circles at −4 and 4, 0 marked); $(-4, 4)$ **53.** $x \le -3$ or $x \ge 3$; (number line: −3, 0, 3); $(-\infty, -3] \cup [3, \infty)$ **55.** empty set **57.** $x < 0$ or $x > 0$; (number line: open circle at 0); $(-\infty, 0) \cup (0, \infty)$ **59.** The set of real numbers; (number line: 0); $(-\infty, \infty)$ **61.** $t \le 0.8$ or $t \ge 9.6$; (number line: 0.8, 9.6); $(-\infty, 0.8] \cup [9.6, \infty)$ **63.** $x < -4$ or $x > 4$; (number line: −4, 0, 4); $(-\infty, -4) \cup (4, \infty)$ **65.** $p \le -2$ or $p \ge 2$; (number line: −2, 0, 2); $(-\infty, -2] \cup [2, \infty)$ **67.** $x \le -1$ or $x \ge 13$; (number line: −1, 13); $(-\infty, -1] \cup [13, \infty)$ **69.** $-10 < x < 5$; (number line: −10, 0, 5); $(-10, 5)$ **71.** $x \le -8$ or $x \ge 2$; (number line: −8, 0, 2); $(-\infty, -8] \cup [2, \infty)$ **73.** $x \le -10$ or $x \ge 2$; (number line: −10, 0, 2); $(-\infty, -10] \cup [2, \infty)$ **75.** $-\frac{39}{8} \le x \le -\frac{21}{8}$; (number line: $-\frac{39}{8}$, $-\frac{21}{8}$, 0); $\left[-\frac{39}{8}, -\frac{21}{8}\right]$ **77.** $x = \dfrac{-b \pm (k - c)}{m}$ **79.** Answers may vary; $-2, 12$ **81.** Answers may vary; $-13 < x < 7$ **83. a.** $-8, 5$ **b.** $-8 < x < 5$ **c.** $x < -8$ or $x > 5$ **d.**

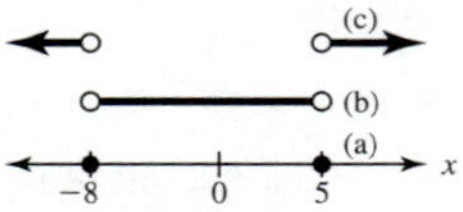

85. Answers may vary. **87.** false; answers may vary. **89.** 1, 9 **91.** 0, 3.1699 **93.** $-5, 3$ **95.** $x \le 2$; (number line: 0, 2); $(-\infty, 2]$ **97.** $-2 \le x \le 2$; (number line: −2, 0, 2); $[-2, 2]$ **99.**

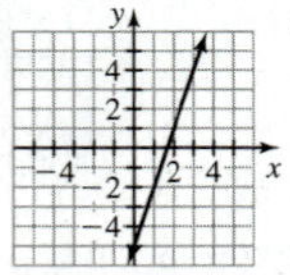

a linear function **101.** 0, 4; an absolute value equation in one variable **103.** $y = -\frac{5}{9}x - \frac{2}{9}$; a linear equation in two variables

Section 15.1 Quiz

1. ± 5 **2.** $\frac{1}{3}, \frac{4}{3}$ **3.** empty set **4.** $-\frac{1}{2}, 4$ **5.** $-\frac{1}{2}, \frac{11}{6}$ **6.** false; answers may vary. **7.** $k \le -2$ or $k \ge 2$; (number line: −2, 0, 2); $(-\infty, -2] \cup [2, \infty)$ **8.** $c < -1$ or $c > 5$; (number line: −1, 0, 5); $(-\infty, -1) \cup (5, \infty)$ **9.** $-\frac{4}{3} \le x \le \frac{8}{3}$; (number line: $-\frac{4}{3}$, 0, $\frac{8}{3}$); $\left[-\frac{4}{3}, \frac{8}{3}\right]$ **10.** empty set

Homework 15.2

1. $7 + 3i$ **3.** $7 - 8i$ **5.** $4 + 8i$ **7.** $5 - 16i$ **9.** -18 **11.** -50 **13.** -10 **15.** $-\sqrt{15}$ **17.** -64 **19.** $10 + 15i$ **21.** $-1 - 6i$ **23.** $-14 + 23i$ **25.** $27 - 24i$ **27.** $-16 - 50i$ **29.** 41 **31.** 85 **33.** 2 **35.** $-45 + 28i$ **37.** $-9 - 40i$ **39.** $7 - 24i$ **41.** $\frac{6}{29} - \frac{15}{29}i$ **43.** $-\frac{6}{53} + \frac{21}{53}i$ **45.** $\frac{17}{50} + \frac{19}{50}i$ **47.** $-\frac{7}{25} + \frac{24}{25}i$ **49.** $\frac{3}{5} + \frac{1}{5}i$ **51.** $\frac{7}{4} - \frac{5}{4}i$ **53.** $-\frac{7}{5}i$ **55.** Student 2's work is correct; answers may vary. **57. a–c.** Answers may vary. **59.** Answers may vary. **61.** The result will be a negative real number;

answers may vary. **63.** $\dfrac{12 - 8\sqrt{x}}{9 - 4x}$ **65.** $\dfrac{12}{13} - \dfrac{8}{13}i$ **67.** $\dfrac{1 \pm 2i\sqrt{2}}{3}$ **69.** $\dfrac{2 \pm i}{5}$ **71.** $\dfrac{5 \pm i\sqrt{31}}{4}$ **73.** $\dfrac{2 \pm 2i\sqrt{2}}{3}$ **75.** $\dfrac{-3 \pm 2i\sqrt{5}}{5}$ **77.** $\dfrac{1 \pm i\sqrt{11}}{4}$; a quadratic equation in one variable **79.** $(5x - 2)(2x - 3)$; a quadratic (or second-degree) polynomial in one variable **81.** $-54 - 10i$; an imaginary number

Section 15.2 Quiz

1. $9 - 6i$ **2.** $-5 + 9i$ **3.** 12 **4.** $-\sqrt{14}$ **5.** $38 - 16i$ **6.** $7 - 24i$ **7.** 89 **8.** $\dfrac{7}{41} + \dfrac{22}{41}i$ **9.** $-\dfrac{7}{6} - \dfrac{5}{6}i$ **10.** false; answers may vary.

Homework 15.3

1. $\sqrt{41}$ **3.** $2\sqrt{14}$ **5.** $2\sqrt{11}$ **7.** $c = 13$ **9.** $c = \sqrt{85}$ **11.** $b = \sqrt{55}$ **13.** $a = 2\sqrt{6}$ **15.** $c = \sqrt{7}$ **17.** 18.8 miles **19.** yes **21.** 16.5 inches **23.** 28.2 inches **25.** 1.9 miles **27.** 340 feet (113 yards and 1 foot) **29.** 2273.4 miles **31.** 10 **33.** $\sqrt{58}$ **35.** $2\sqrt{5}$ **37.** $4\sqrt{2}$ **39.** 8.01 **41.** 13.28 **43.** $x^2 + y^2 = 49$ **45.** $x^2 + y^2 = 44.89$ **47.** $(x - 5)^2 + (y - 3)^2 = 4$ **49.** $(x + 2)^2 + (y - 1)^2 = 16$ **51.** $(x + 7)^2 + (y + 3)^2 = 3$ **53.** $(0, 0), 5$

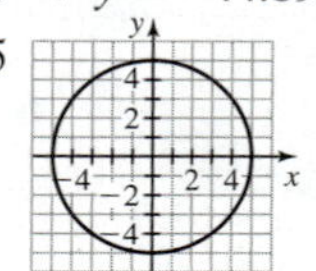

55. $(0, 0), 2\sqrt{2}$

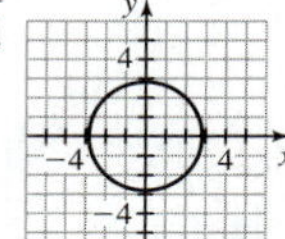

57. $(3, 5), 4$ **59.** $(-6, 1), \sqrt{7}$

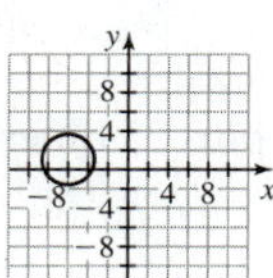

61. $(-3, -2), 1$

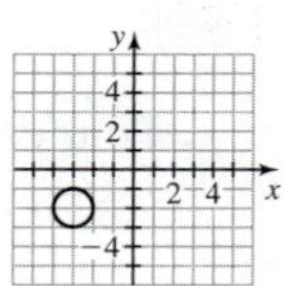

63. $x^2 + y^2 = 9$

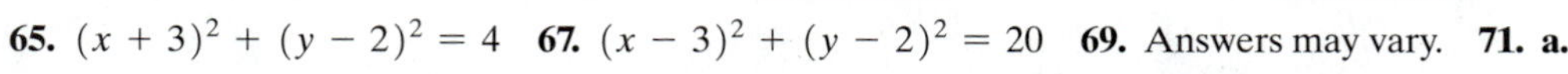

65. $(x + 3)^2 + (y - 2)^2 = 4$ **67.** $(x - 3)^2 + (y - 2)^2 = 20$ **69.** Answers may vary. **71. a.**

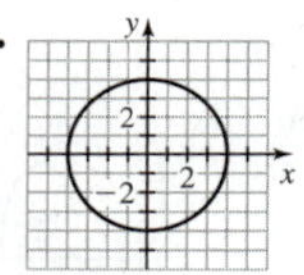

b. Answers may vary. **c.** The sum of the squares of both x and y equals 16. **73.** Answers may vary. **75.** no; answers may vary. **77. a.** Answers may vary. **b.** **79. a.** Answers may vary. **b.** Answers may vary. **c.** $3\sqrt{2}$ **d.** $\dfrac{5\sqrt{2}}{2}$ **81.**

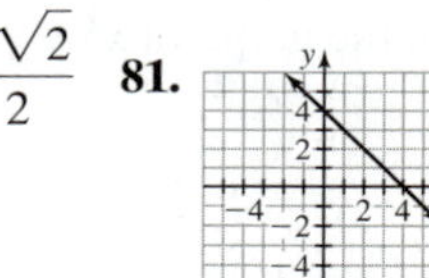

83.

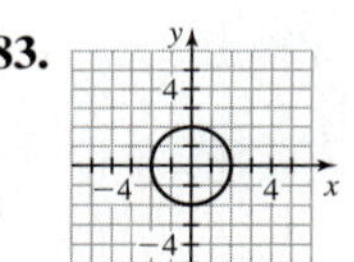

85. **87.** a quadratic function **89.** $2(3x - 2)(x - 2)$; a quadratic (or second-degree) polynomial in one variable **91.** $x = \dfrac{3 \pm 2\sqrt{6}}{5}$; a quadratic equation in one variable

Section 15.3 Quiz

1. $4\sqrt{3}$ inches ≈ 6.9 inches **2.** $\sqrt{105}$ inches ≈ 10.2 inches **3.** $\sqrt{41}$ **4.** $4\sqrt{2}$ **5.** $(x + 3)^2 + (y - 2)^2 = 36$ **6.** $x^2 + y^2 = 7.84$ **7.** $(0, 0), 2\sqrt{3}$

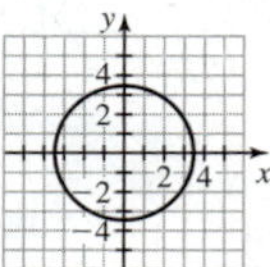

8. $(-4, 3), 5$

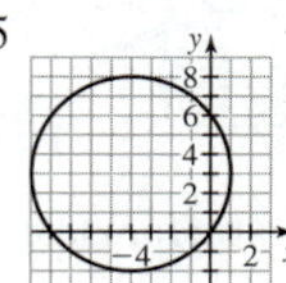

9. $(x - 2)^2 + (y + 1)^2 = 68$ **10.** Answers may vary.

Homework 15.4

1. **3.**

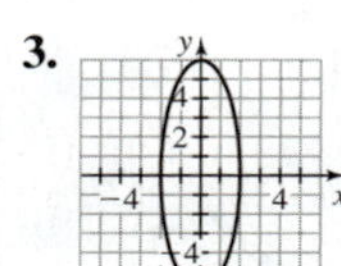

5. **7.**

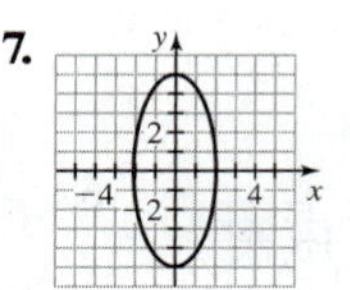

9.

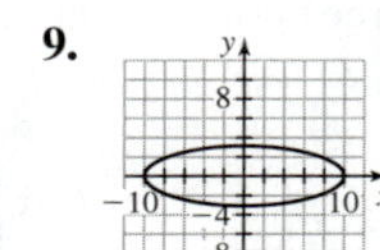

11.

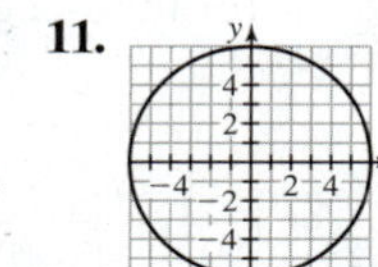

13.

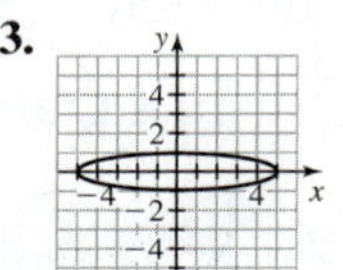

15.  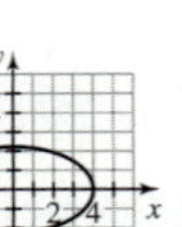 **17.** $\frac{x^2}{25} + \frac{y^2}{9} = 1$ **19.** 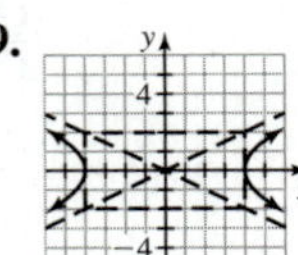 **21.** 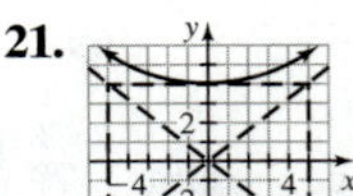 **23.** 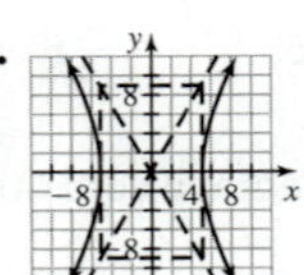 **25.** **27.** 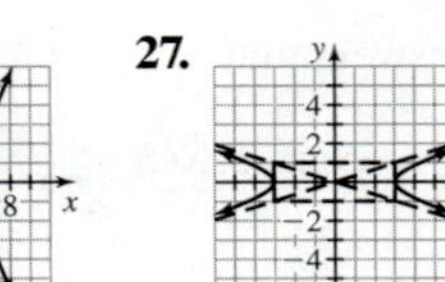

29. 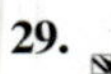 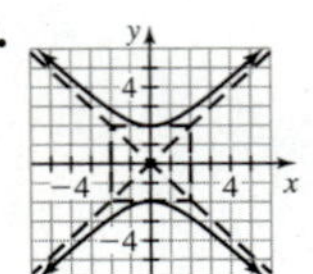 **31.** 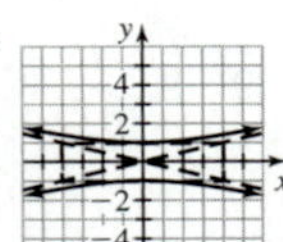 **33.** 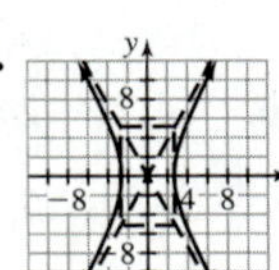 **35.** 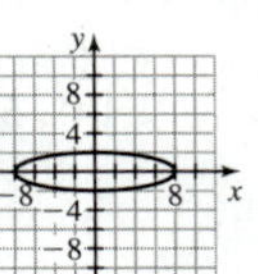 **37.** 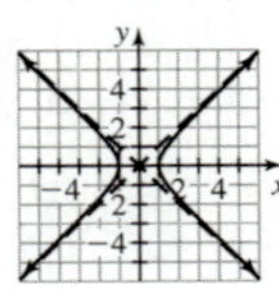 **39.** **41.** 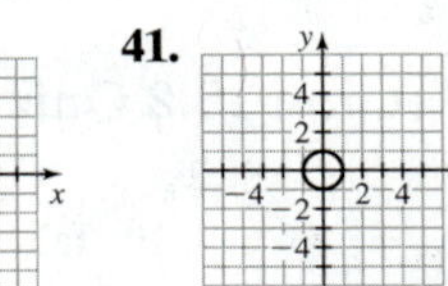

43.  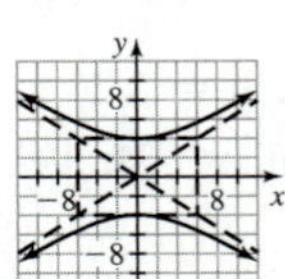 **45.** 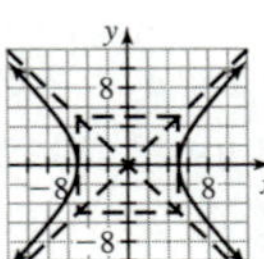 **47.** 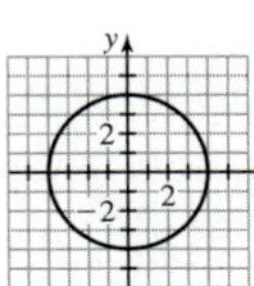 **49.** **51.** 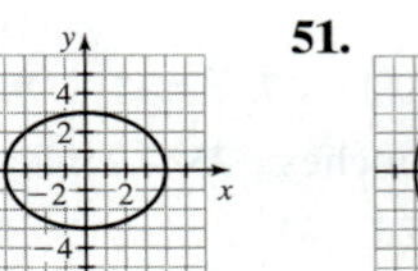 **53. a. i.** 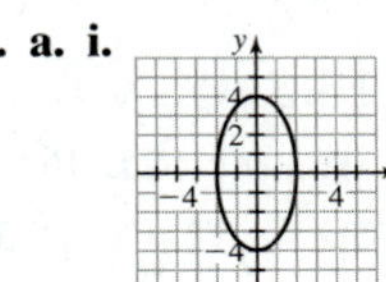

ii. 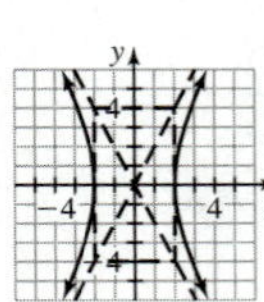 **iii.** **iv.** 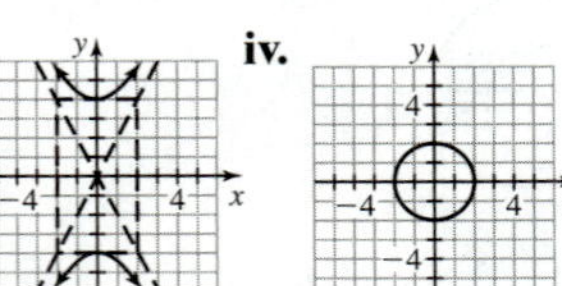 **b.** circle if $c = d$, both positive; ellipse if $c \neq d$, both positive; hyperbola with x-intercepts if $c > 0$ and $d < 0$; hyperbola with y-intercepts if $c < 0$ and $d > 0$

55. a. 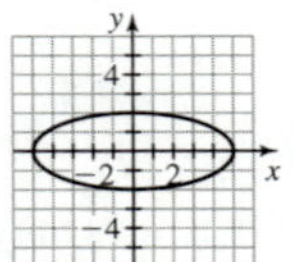 **b.** Answers may vary. **c.** Four times the square of x plus 25 times the square of y equals 100.

57. a. Answers may vary. **b.** 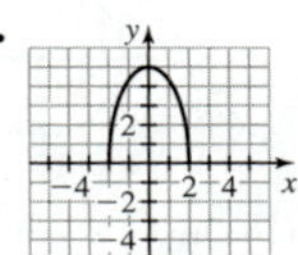 **59.** Answers may vary. **61.** both a circle and an ellipse **63.** Answers may vary.

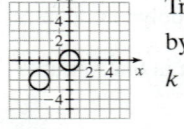

65. a. 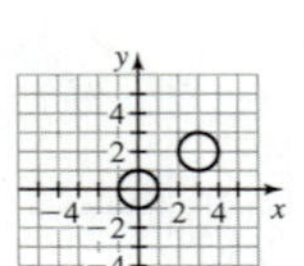 Translate the graph of $x^2 + y^2 = 1$ by 3 units to the right and 2 units up.

b. Translate the graph of $x^2 + y^2 = 1$ by 3 units to the left and 2 units down. **c.** Translate the graph of $x^2 + y^2 = r^2$ by h units to the right if $h > 0$ or by $|h|$ units to the left if $h < 0$, then by k units up if $k > 0$ or by $|k|$ units down if $k < 0$.

d. 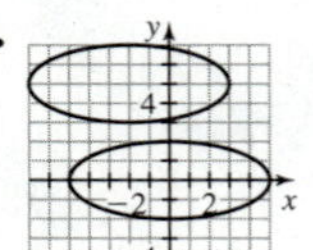 **e.** 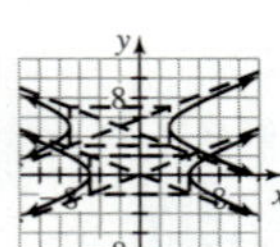 **67.** 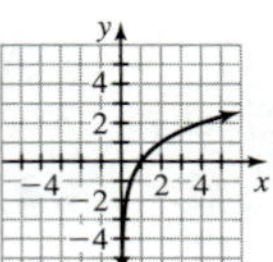 **69.** 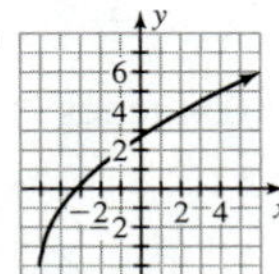 **71.** no; answers may vary.

73. 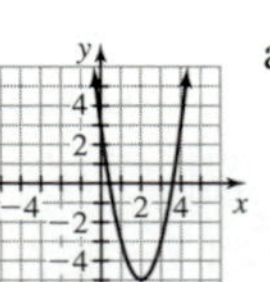 a quadratic function **75.** $x = \frac{4 \pm \sqrt{10}}{2}$; a quadratic equation in one variable

77. $-30x^3 + 25x^2 - 5x$; a cubic (or third-degree) polynomial in one variable

Section 15.4 Quiz

1. 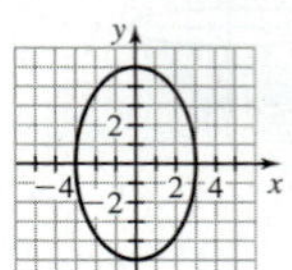 **2.** 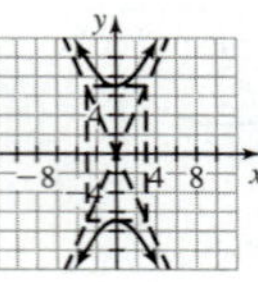 **3.** 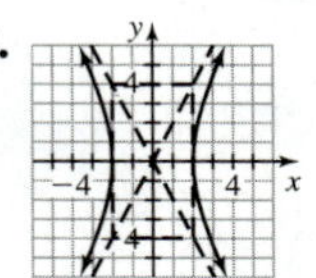 **4.** 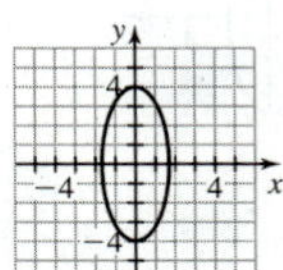 **5.** 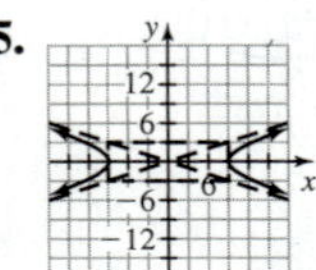 **6.** 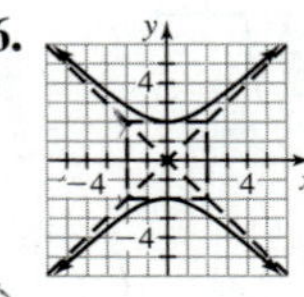 **7.** 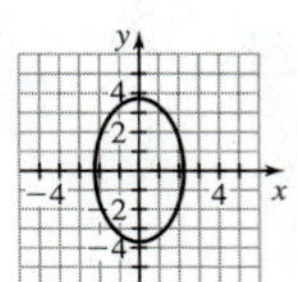

8. 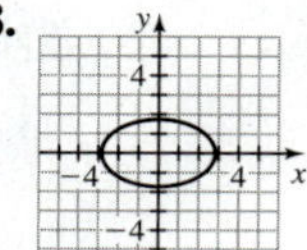

9. no; answers may vary. **10.** Answers may vary.

Homework 15.5 **1.** $(-5, 0), (5, 0)$ **3.** $(-2, 5), (1, 2)$ **5.** $(-2, 2), (2, 2)$ **7.** empty set **9.** $(-4, 3), (3, -4)$ **11.** $(-3, -5), (0, 4), (3, -5)$ **13.** $(-3, 0), (3, 0)$ **15.** $(-1, 0), (0, 3)$ **17.** empty set **19.** $(1, -2)$ **21.** $(-\sqrt{3}, 1), (\sqrt{3}, 1)$ **23.** $(-0.74, -2.02), (-0.74, 2.02), (0.74, -2.02), (0.74, 2.02)$ **25.** $(-2.25, -3.31), (-2.25, 3.31), (2.25, -3.31), (2.25, 3.31)$ **27.** $(-3, -2), (-3, 2), (3, -2), (3, 2)$ **29.** $(3, 2), (5, 0)$ **31.** $(3, 2), (2, 0)$ **33.** $(-5, 0), (5, 0)$ **35.** Answers may vary. **37. a.** $c > 0$ **b.** $c = 0$ **c.** $c < 0$ **39.** $c = 5, d = -2$ **41.** Answers may vary. **43.** $(1, 2)$ **45.** Answers may vary. **47.** Answers may vary. **49.** Answers may vary. **51.** Answers may vary. **53.** Answers may vary.

Section 15.5 Quiz **1.** $(-3, 0), (3, 0)$ **2.** $(-3, 7), (1, -1)$ **3.** $(1, 4)$ **4.** empty set **5.** $(-4, 0)$ **6.** Answers may vary.

Index

F

N

O

P

Q

R

T